SEAT BELTS:
THE DEVELOPMENT OF AN ESSENTIAL SAFETY FEATURE

PT-92 (SEAT BELTS: THE DEVELOPMENT OF AN ESSENTIAL SAFETY FEATURE) – RELATED TITLES

2002 SAE Active Safety Technology Collection on CD-ROM
(Order No. ACTSAFE2002)

Air Bag Development and Performance – New Perspectives from Industry, Government and Academia
by Richard W. Kent
(Order No. PT-88)

Automotive Vehicle Safety
by George A. Peters and Barbara J. Peters
(Order No. R-341)

Neck Injury: The Use of X-Rays, CTs and MRIs to Study Crash-Related Injury Mechanisms
by Jeffrey A. Pike
(Order No. R-268)

Occupant Protection and Crashworthiness Technology Collection on CD-ROM
(updated annually)

Role of the Seat in Rear Crash Safety
by David C. Viano
(Order No. R-317)

For information on these or other related books, contact SAE by phone at (724) 776-4970, fax (724) 776-0790, e-mail: publications@sae.org, or the SAE website at www.sae.org.

SEAT BELTS:
THE DEVELOPMENT OF AN ESSENTIAL SAFETY FEATURE

PT-92

Edited by

David C. Viano

Published by
Society of Automotive Engineers, Inc.
400 Commonwealth Drive
Warrendale, PA 15096-0001
U.S.A.
Phone (724) 776-4841
Fax: (724) 776-5760
www.sae.org

For permission and licensing requests contact:

SAE Permissions
400 Commonwealth Drive
Warrendale, PA 15096-0001-USA
Email: permissions@sae.org
Fax: 724-772-4891
Tel: 724-772-4028

For multiple print copies contact:

SAE Customer Service
Tel: 877-606-7323 (inside USA and Canada)
Tel: 724-776-4970 (outside USA)
Fax: 724-776-1615
Email: CustomerService@sae.org

ISBN 0-7680-1122-1
Library of Congress Catalog Card Number: 2003104662
SAE/PT-92

SAE Order No. PT-92

Printed in USA

Table of Contents

Please note:
Some papers may not be included in their entirety. Appendices and other portions may have been deleted due to space limitations. Copies of entire papers may be purchased from SAE, if desired.

Please note:
Some papers may not be included in their entirety. Appendices and other portions may have been deleted due to space limitations. Copies of entire papers may be purchased from SAE, if desired.

Introduction

Lap-Shoulder Belts: Some Historical Aspects

David C. Viano
ProBiomechanics LLC

ABSTRACT

The origin of modern lap-shoulder belts is often cited to the 1955 patent by R. Griswold and H. DeHaven derived from their extensive military and aviation crash safety research. This system included an integrated lap and diagonal shoulder belt configuration and convenient latching buckle. They declined to include an inertial locking reel because of cost and complexity, but noted the existence of such an added feature. Their configuration is typical of today's belt systems, but the history of safety belts has much earlier roots. Belt restraints were considered for aviation and passenger vehicle safety from 1910s-30s, but safety belt harnesses had their beginnings in designs from the 1880s to secure individuals in wagons, climbers to poles and trees, and restrain children. By the 1930s, lap belt designs were patented for motor vehicles with flexible connections to the seat, easy latching and inertia locking or take-up reels. By the 1940s, 4-pt safety belts emerged for the protection of pilots in aviation crashes. These also included inertia locking reels for the shoulder belts either fashioned vertically or diagonally across the torso.

One of the earliest inertia locking belt reels was patented by Shively in 1933 for vehicle lap belts and there were many others in the 1940s. By the mid-1950s refinements in the belt anchorages to the vehicle, latching buckles and inertia locking reels had progressed sufficiently that the elements of the modern lap-shoulder belt were known. However, the motor vehicle industry varied in its effort to adopt the new belt systems that are now commonplace. An effort in Sweden built upon the US and European innovations in belt systems, and cooperation between Vattenfalls and Saab led to the first field trial of lap-shoulder belts in a fleet of 100 vehicles. The results were so promising that Bohlin at Volvo converted their restraints to lap-shoulder belts in 1959 to provided enough vehicles for a full field study of safety effectiveness. He published a comprehensive study on belt crash testing in 1964, and another on 28,000 field crashes including lap-shoulder belts in 1967, both of which changed opinions in favor of lap-shoulder belts, which were found to be convenient to put on, comfortable to wear with the inertia locking retractor and gave a high degree of crash safety by virtually eliminating ejection and reducing impacts. These results were the turning point that started the phase-in of lap-shoulder belt system in modern vehicles.

Over the years, refinements have been introduced to lap-shoulder belts to further improve comfort, convenience and safety function. These include pretensioners, adjustable guide loops, load limiters, web clamps and energy absorbing anchors. The background on these enhancements is summarized. In addition, the interaction between the occupant and belt system was refined by including an anti-submarining ramp in the seat to support the pelvis during lap-belt loading.

Today, lap-shoulder belts are commonplace and have a proven safety record. They are a fundamental part of safety systems that include vehicle crashworthiness and airbags. They were spawned by early use in aviation safety and later adapted in the form of lap-shoulder belts with inertial locking reel for passenger vehicles. Possible future developments are also considered.

ROOTS OF MODERN LAP-SHOULDER BELTS

Much of the early research on belt restraint systems surrounded aviation crash investigations and safety research. Dr. Hugh DeHaven and Col. John Paul Stapp were pioneers in the field among many others around the world. DeHaven's interest in crash injury prevention began in 1917, when his Curtiss Model JN4 airplane was rammed from behind by another airplane during gunnery practice and he suffered two broken legs and a ruptured liver, pancreas and gallbladder in the crash. At the time, he was a young American volunteer cadet in the Canadian Royal Flying Corps.

DeHaven concluded that his internal injuries were caused by a poorly designed "safety belt" and observed that solid structures in front of the pilot's head were often the cause of serious injury. This experience led to his lifetime of research in the area of crashworthiness. In April 1942, he was appointed a research associate at Cornell University Medical College, which led to the formation of the Cornell Crash Injury Research (CIR) Program, which led to the basic understanding of the causes of occupant injury in car crashes.

The concept that vehicle and roadway design could increase safety was promoted by Dr. Hugh DeHaven, an ex-Royal Air Force medical officer conducting research at Cornell University. In 1942, DeHaven published a report on seven people who had survived falls from as high as 150 feet. He concluded his report by observing that "structural provisions to reduce impact and distribute pressure could enhance survival and modify injury within wide limits in aircraft and automobile accidents."

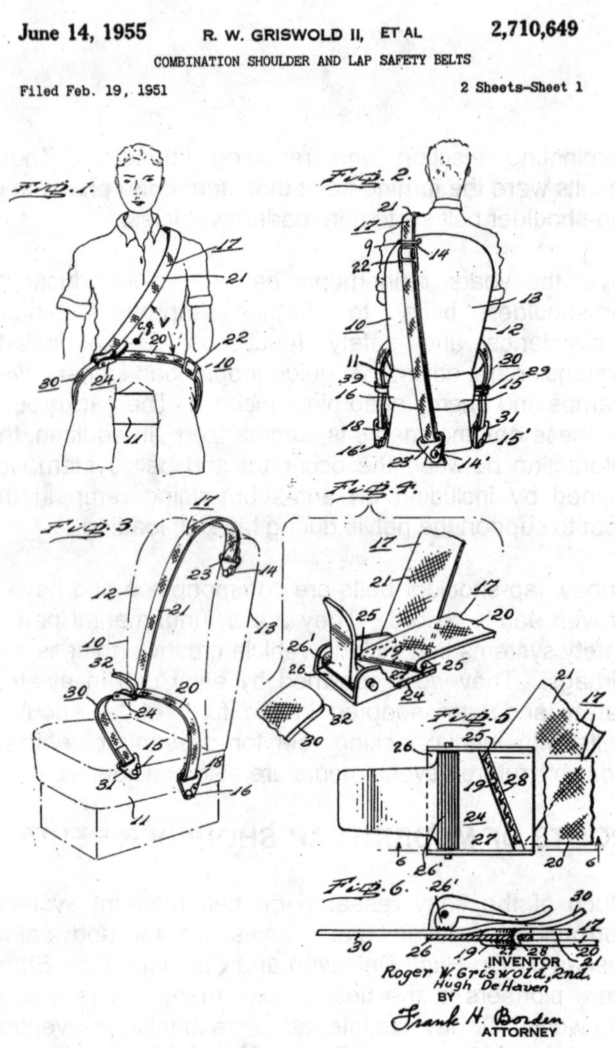

Figure 1: R. Griswold and H. DeHaven (1955) lap-shoulder belt system considered to be the basis for modern belt restraints in widespread use today.

In the early 1950s, a series of rocket sled tests were conducted by Col. John Paul Stapp, a medical officer in the U.S. Army. These studies established the foundation for vehicle-safety design. Using himself as a test subject, Stapp ran a series of tests to determine the human body tolerance to forward-facing decelerations (sudden stops from high speeds).

Col. Stapp is well known for his rocket sled experiments to establish human tolerance in rapid deceleration. His most severe exposure involved a 1000 km/h deceleration in 1.25 s leading to approximately 50 g deceleration for the well restrained military volunteer. While the most severe exposure resulted in a detached retina and need for emergency assistance, the testing established limits for human tolerance that have lasted 50 years.

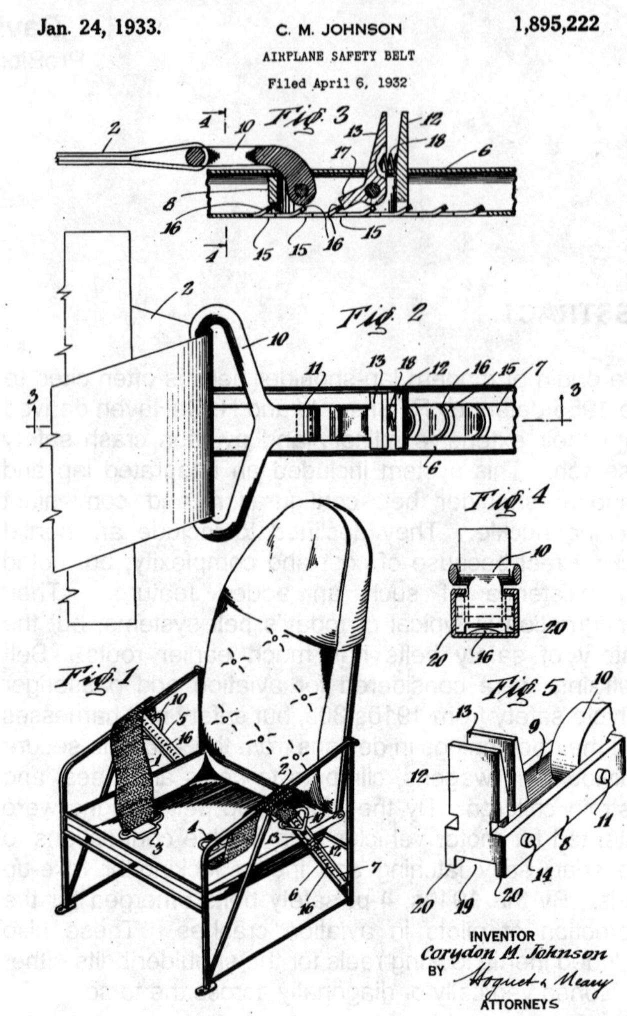

Figure 2: Johnson (1933) lap belt with adjustable attachment to the vehicle seat.

The research led to the understanding of safety benefits of belt restraint systems offering occupants ride-down of a crash and prevention of secondary impacts in the vehicle. While most of the research aimed to improve protection of military pilots in aviation crashes, the application to passenger cars was obvious. Based on years of experience in aviation and military safety, R. Griswold and H. DeHaven (1955) patented a lap-shoulder belt system for passenger safety in car crashes. This design included the essential embodiments of modern safety belt systems in today's vehicles (Figure 1). It offered the convenience and comfort of wearing and effectiveness in crashes with a single belt loop and lockable sliding latch plate at the buckle.

Table 1: Examples of Safety Belt Related Patents Through the 1950s.

Patent #	Date	Inventor	Title	Innovation
199,545	22-Jan-1878	Hickey	Vehicle Seat	Safety guard for fire engine seat.
312,085	10-Feb-1885	Claghorn	Safety Belt	Safety harness for climbers
534,674	26-Feb-1895	Schwarzmiller	Child Guard for Vehicle Seat	Restraining bar to hold child
1,369,456	22-Feb-1921	Meredith	Childs' Safety Harness	Shoulder-torso harness for standing child
1,895,222	24-Jan-1933	Johnson	Airplane Safety Belt	Adjustable lap belt
1,971,264	22-Aug-1934	Irwin	Safety Belt	Torso seat belt
1,973,011	11-Sep-1934	Morrison	Stabilizing Device for Passengers in Automobiles	Torso stabilizing bar/belt
1,991,633	19-Feb-1935	Serpico	Restraining Device	Adjustable torso belt for children
2,071,903	23-Feb-1937	Shively	Inertia Operated Device	Inertia locking belt restraint system
2,195,334	26-Mar-1940	Lethern	Safety Apparatus for Retaining Persons in the Seats of Aircraft and Other Vehicles	4-pt safety belt with inertia shoulder belt locking
2,212,746	27-Aug-1940	Nunn	Safety Harness for Children	4 pt Diagonal shoulder harness system for children
2,263,348	18-Nov-1941	Barros	Safety Belt Attachment for Vehicle	Load yielding attachment of belts to vehicle
2,275,450	10-Mar-1942	Manson	Safety Belt	4-pt safety belt with single point attachment to vehicle
2,280,694	21-Apr-1942	Embree	Vehicle Passenger Safety Device	Adjustable lap belts with anchor loops to vehicle
2,312,946	2-Mar-1943	Watter	Seat Harness	4-pt safety belt system
2,365,626	19-Dec-1944	Carlisle	Safety Holding Device for Vehicle	Crossing diagonal belts
2,365,625	19-Dec-1944	Carlisle	Safety Device for Vehicle Passengers	4-pt safety belt
2,372,558	27-Mar-1945	Dowd	Safety Belt	Belt latching mechanism and adjustable guide
2,394,523	5-Feb-1946	Pancoe	Safety Harness and Control Apparatus Therefor	4-pt belts with spring take-up reel and manual locking
2,403,653	9-Jul-1946	Geohegan	Shoulder Harness Take-up and Inertia Lock	4-pt belts with inertia locking and take-up reel for shoulder straps
2,433,950	6-Jan-1948	Henderson	Crash Seat for High Speed Conveyances	Swingable seat to decelerate occupant in crash
2,434,119	6-Jan-1948	Nordmark	Automatic Reel	Inertia actuated should belt reel
2,475,588	12-Jul-1949	Bierman	Deceleration Harness Comprising Undrawn Synthetic Fibers or the like	Woven belt material
2,480,335	30-Aug-1949	Nordmark	Automatic Reel	Inertia actuated belt reel
2,480,915	6-Sep-1949	George	Retractable Safety Belt	Stowable lap belt concept
2,499,993	7-Mar-1950	Gregg	Inertia Actuated Vehicle Safety Device	Steering system inertia locking device for airplane controls
2,557,313	19-Jun-1951	Quilter	Safety Device for Chairs	Inertia locking should belt retractor
2,650,655	1-Sep-1953	Neahr	Multidirectional Inertia Operated Safety Device for Vehicle Chairs	4-pt airplane inertia harness and chair
2,670,967	2-Mar-1954	Kean	Vehicle Passenger Safety Strap	Large width torso strap for vehicle seat
2,576,867	27-Nov-1951	Wilson	Safety Harness	4-pt safety belt system
2,665,143	5-Jan-1954	Rasmussen	Passenger Safety Belt Device for Automobiles	Lap belt with sliding latching buckle.
2,710,649	14-Jun-1955	Grisswold-DeHaven	Combination Shoulder and Lap Safety Belt	3-pt safety belt with diagonal should belt
2,705,044	29-Mar-1955	Nolen	Safety Belts for Seats of Vehicles	Diagonal shoulder belt harness
2,794,654	4-Jun-1957	Sullivan	Safety Belt for Vehicle Seats	Lap-shoulder belts
2,804,313	27-Aug-1957	Gilles	Safety Harness for Use in Vehicles	Lap-shoulder belt and 4-pt belts
2,880,788	7-Apr-1959	Phillips	Safety Belt for Automotive Vehicles	Diagonal shoulder belt harness with door latching/release
2,898,976	11-Apr-1959	Barecki	Safety Equipment for Vehicle Occupants	4-pt belt with roof attachment of the shoulder belts

EARLY HISTORY OF SAFETY BELTS

From the 1880s, there have been numerous patents on belt harnesses and restraints. Table 1 summarizes some of the belt related patents through the 1950s. The patent by Claghorn (1885) showed a safety belt harness for climbers that could secure tools and the individual. Other harnesses were considered for child restraints; the earliest considered for securing the child to a vehicle was by Meredith (1921). Other harness designs were patented to shackle prisoners, control aggressive outbursts or provide gyves.

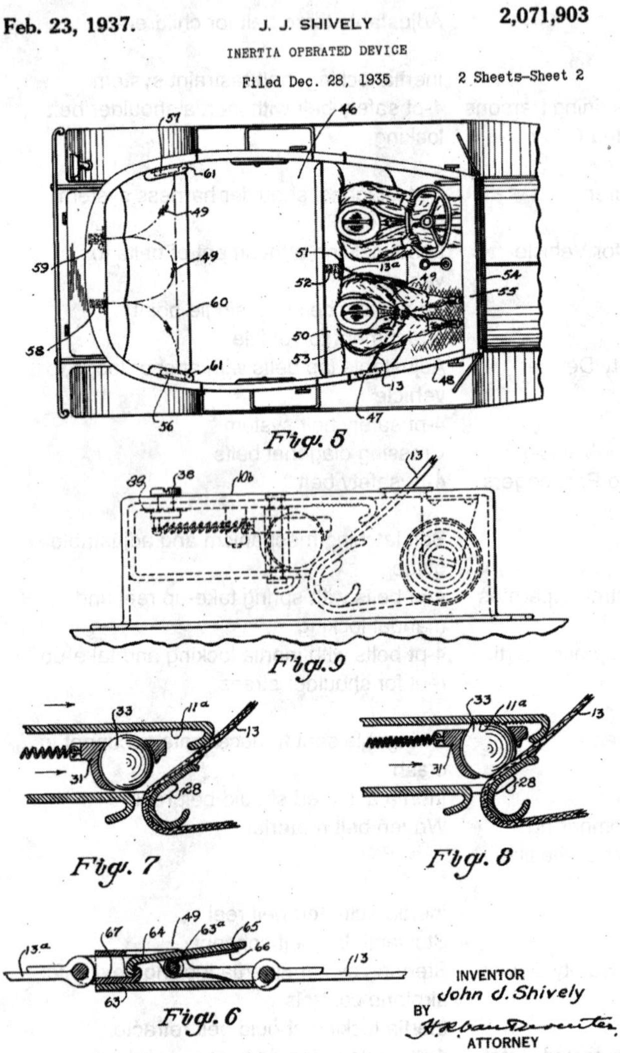

Figure 3: Shively (1937) inertial locking safety belt.

In 1933, Johnson demonstrated a lap belt system for airplane or vehicle seats with a simple latching arrangement. It is one of the early US safety belt patents to include a conveniently adjustable attachment to the chair (Figure 2). Andreasson (2000) notes the earliest safety belt patent by Claghorn (1885) in the US, and ones by Leveau in France with a diagonal shoulder belt (1903), Pounce in the UK with an adjustable diagonal shoulder and lap belt (1903). However, the Claghorn patent was found to be only a tool harness arrangement with a security loop to fasten around poles or tress while climbing. The European patents have not been reviewed. In 1935, Serpico demonstrated an adjustable safety belt for standing children that stabilized the child in moving carts.

Shively (1937) demonstrated the essential principles for an inertia locking lap belt that provided ease of movement until activation of a lock in emergency braking or a crash, and release after deceleration (Figure 3). This seems to be an early application of a take-up reel and inertia lock for seat belts. The basis for this design was a rich history of reel and wind-up devices for fishing, and winding suspended cords from the 1870s through the 1920s. Andreaason (2000) sites the later patent by Geohegan (1946) with an inertia locking shoulder harness, but there were earlier patents.

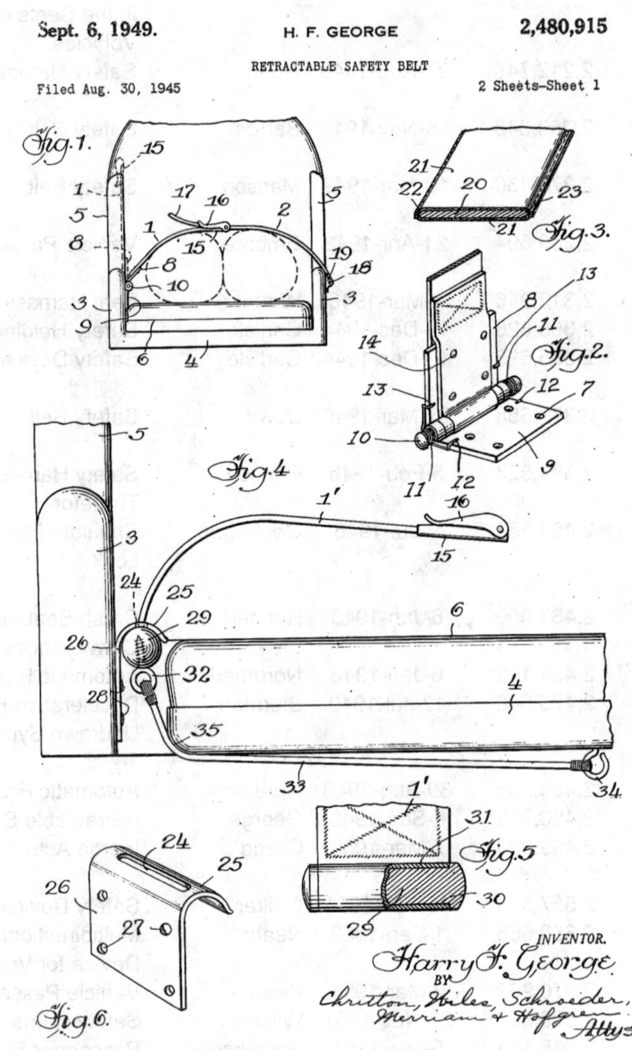

Figure 4: George (1949) retractable lap belt.

Many safety belt patents were filed in the 1930s-40s, including one by George (1949) demonstrating a retractable lap belt using an elastic cord to stow the restraint when not in use (Figure 4). There was a conventional buckle for latching the restraint.

4-POINT AVIATION PILOT RESTRAINTS

In the 1940s, various 4-point and 6-point lap-shoulder belt designs were patented for airplane seats, although most claims included other motor vehicles. Lethern (1940) seems to be one of the earliest with an inertial locking shoulder belt arrangement, although the double shoulder belt design appears complex today (Figure 5). Over a half dozen other 4-pt safety belt designs were patented in the 1940s-50s with various arrangements for take-up reels, latches, inertia locks and belt configurations.

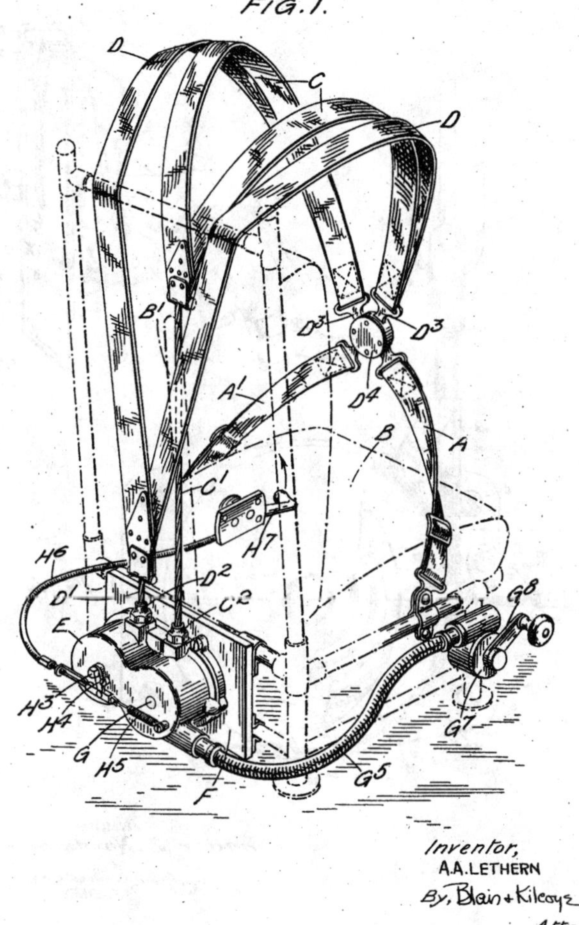

Figure 5: 6-pt lap-shoulder belt system by Lethern (1940) for airplane seat with inertia locking shoulder belts.

Lethern (1940) included a quick release fastening and straps that had slack for comfort allowing the occupant to lean forward under slight tension. Inertia locking of the take-up reel provided restraint during deceleration. These features became common in modern lap-shoulder belt system in motor vehicles.

Manson (1942) showed a 4-point aviation restraint with a single latching and release point in front of the pilot and single tether point behind the spine (Figure 6).

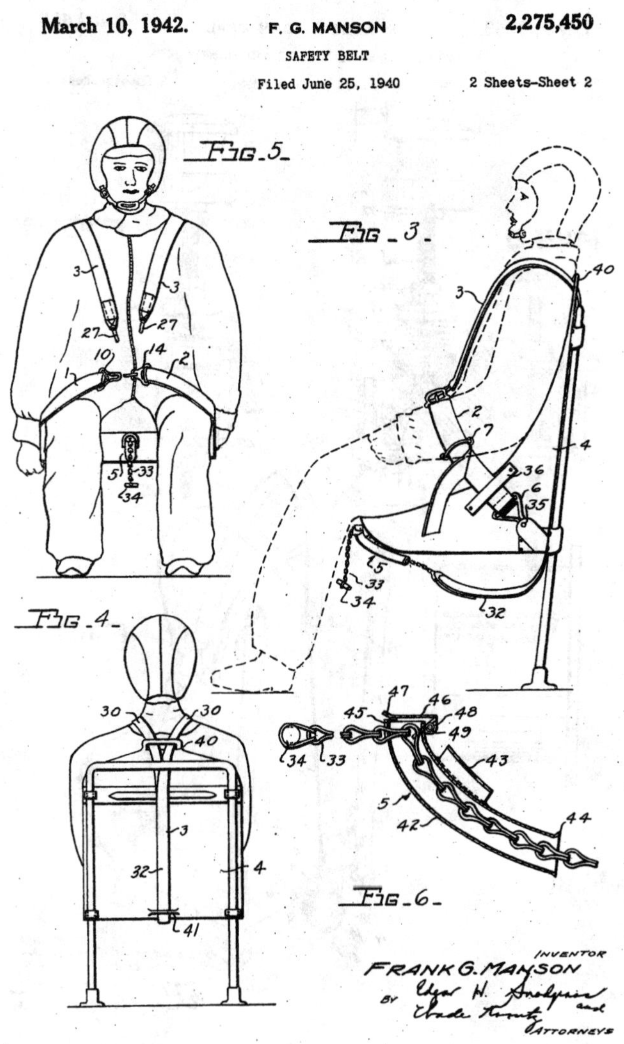

Figure 6: Manson (1942) 4-pt safety belt arrangement with simple attachment to an aviation seat.

Geohegan (1946) demonstrated a shoulder harness take-up reel and inertia lock for an aviation or motor vehicle seat (Figure 7). In a comprehensive description of the design, the inertia operated control prevented yielding of the pilot seat harness during excessive deceleration. Otherwise, the straps were under tension but allowed free movement of the pilot and provide slack under spring tension. These early pilot restraint systems demonstrated the essential features of the eventual lap-shoulder belt system. They were more complex for the aviation application, and needed more development for use by the motoring public.

July 9, 1946. W. A. GEOHEGAN ET AL 2,403,653

SHOULDER HARNESS TAKE-UP AND INERTIA LOCK

Filed April 5, 1944 2 Sheets-Sheet 1

FIG.5 FIG.6 FIG.1

SPRING TENSIONED SHOULDER HARNESS RETRIEVING DEVICE

INVENTORS WILLIAM A. GEOHEGAN RAYMOND E. PHIPPS BY ATTORNEYS

Figure 7: Geohegan et al. (1946) patent for a 4-pt lap-shoulder belt system with inertia locking reel for the shoulder belts of an aviation seat.

Nordmark (1949) demonstrated an automatic take-up reel for shoulder belts in an aviation seat. It offered the convenience of movement until a sudden deceleration when the reel locked the restraints through a cable and wind-up reel (Figure 8). This followed earlier 4-point designs for pilots.

These designs offered reeling in of the belt harness whenever tension was relaxed on the cable and permitted free movement of the pilot and an inertia locking of the cable to secure the restraint system in a crash.

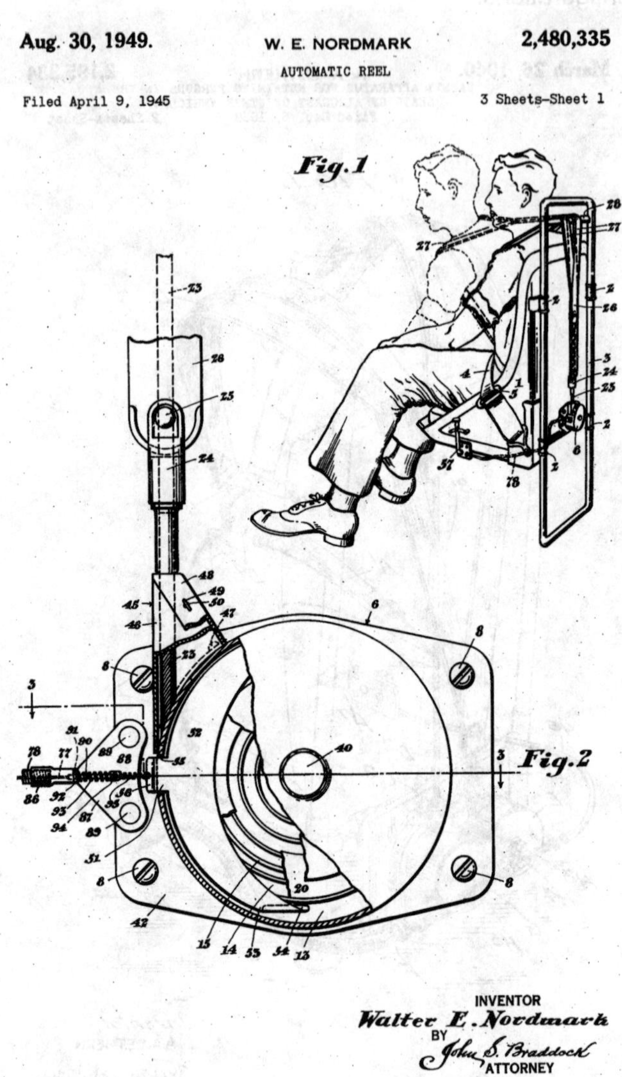

Aug. 30, 1949.
W. E. NORDMARK
2,480,335
AUTOMATIC REEL
Filed April 9, 1945
3 Sheets-Sheet 1
Fig.1
Fig.2
INVENTOR
Walter E. Nordmark
BY
John S. Braddock
ATTORNEY

Figure 8: 4-pt lap-shoulder belt system by Nordmark (1949) for an airplane seat with inertia locking shoulder belts.

The development of 4-point restraint continued into the 1950s. Neahr et al. (1953) refined pilot restraints with the inclusion of a multidirectional inertial reel for safety in plane crashes (Figure 9). This inertia reel locked under any direction of crash acceleration rather than the typical frontal acceleration activation of earlier designs.

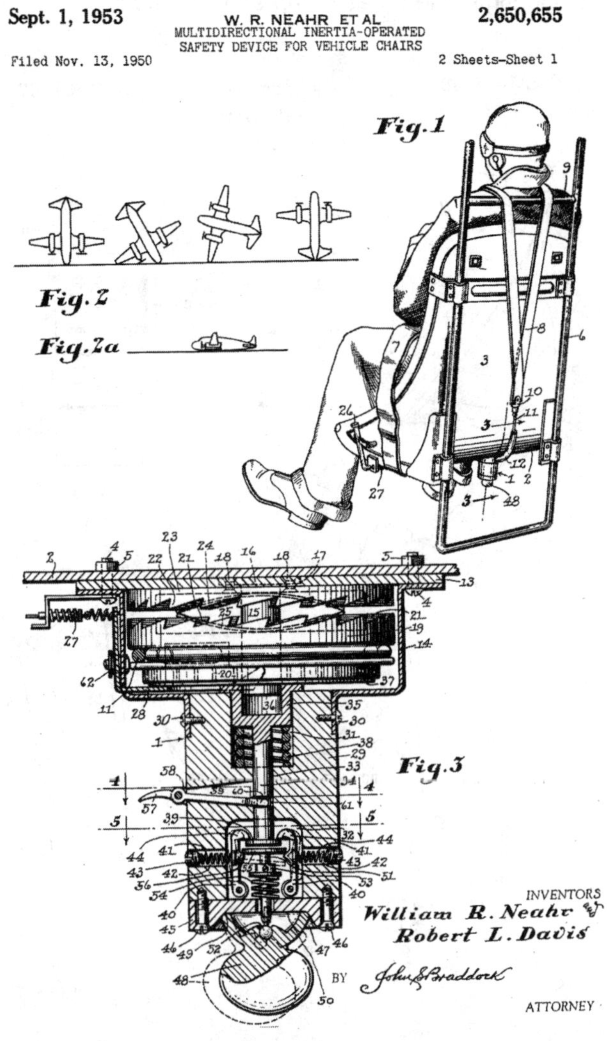

Figure 9: Aviation seat belts system by Neahr et al. (1953) showing a 4-pt restraint for pilot seat.

MODERN LAP-SHOULDER BELT SYSTEMS

Belt restraint systems appeared in many patents of the 1930s-40s and many of the characteristics of the designs in use today were demonstrated during those years. Embree et al. (1942) designed secure belt anchor attachments to the vehicle so they would be firmly held and not yield under occupant loads in a crash.

Carlisle (1944) was granted two patents on various belt restraint configurations including crossing shoulder belts (Figure 10) and vertical belts with attachments secured to the vehicle. He clearly understood the dynamics of an occupant in a crash, the need for adjustments for comfort and the need to stow the belts when not in use.

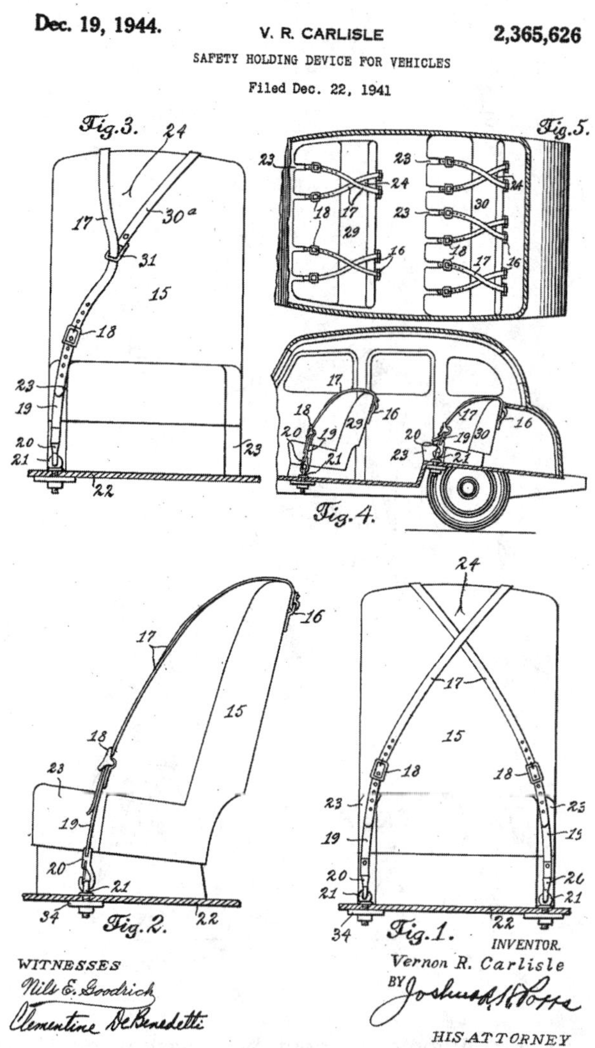

Figure 10: Automotive seat belt system by Carlisle (1944) showing crossing or vertical shoulder belts.

Further developments of belt concepts occurred and Wilson (1951) showed refinements in aviation and motor vehicle belt systems. Figure 11 shows vertical or crossing shoulder belts, but the patent emphasized features for the convenience of putting the restraints on and off so they would be more acceptable by the passengers.

R. Griswold worked on belt restraints for the American Air Force in 1945. H. DeHaven was a combat pilot during World War I who later conducted research on deceleration. Further research on the ability of humans to tolerate rapid and sudden deceleration was conducted by another American, Air Force Colonel John Paul Stapp. Roger W. Griswold and Hugh DeHaven (1955) designed and patented the lap-shoulder belt system shown in Figure 1.

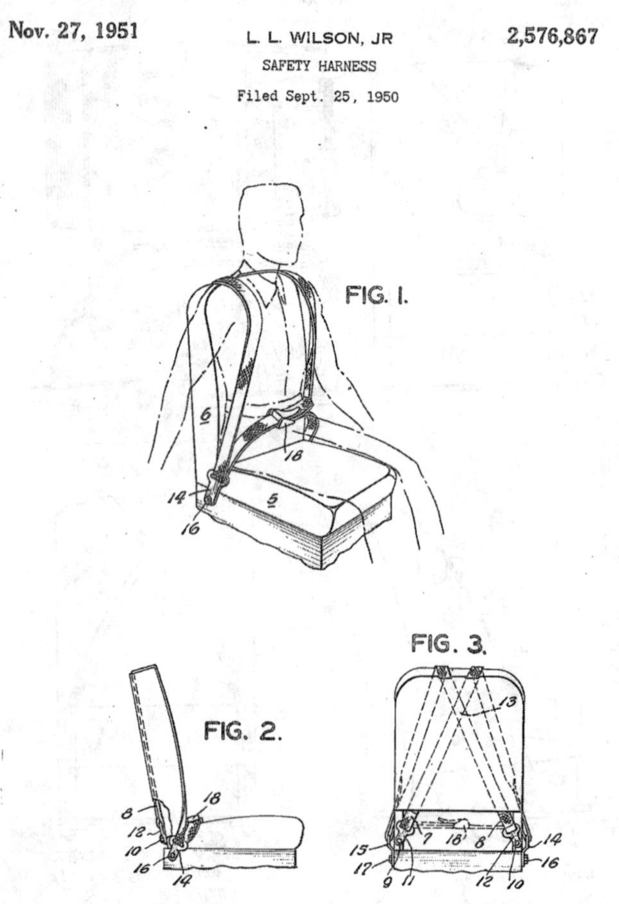

Figure 11: Automotive seat belt system by Wilson (1951) showing a 4-pt occupant restraint.

After considering the need for restraint systems in military and aviation crashes from studies in the 1930s-40s, the Griswold-DeHaven patent includes detailed descriptions of the form and function of a convenient restraint system for motor vehicle passengers. Their concept shows the essential features of belt system in use today. It included a combination lap belt and diagonal shoulder portion designed as a continuous strap. This design was the basis for the single-band lap-shoulder belt offered as standard equipment by Volvo in 1959.

There were other patents in the late 1950s showing the characteristics of modern lap-shoulder belts. Springs (1958) showed a lap-shoulder belt system, which typifies the appearance of today's belt systems (Figure 12) and a reading of the patent showed a refinement in the use of a continuous loop of belt webbing for the 3-point belt with light tension to reduce slack and allow movement until an inertia locking of the retractor. This patent references others by Gilles (1957) and Sullivan (1957) showing similar features of lap-shoulder belts.

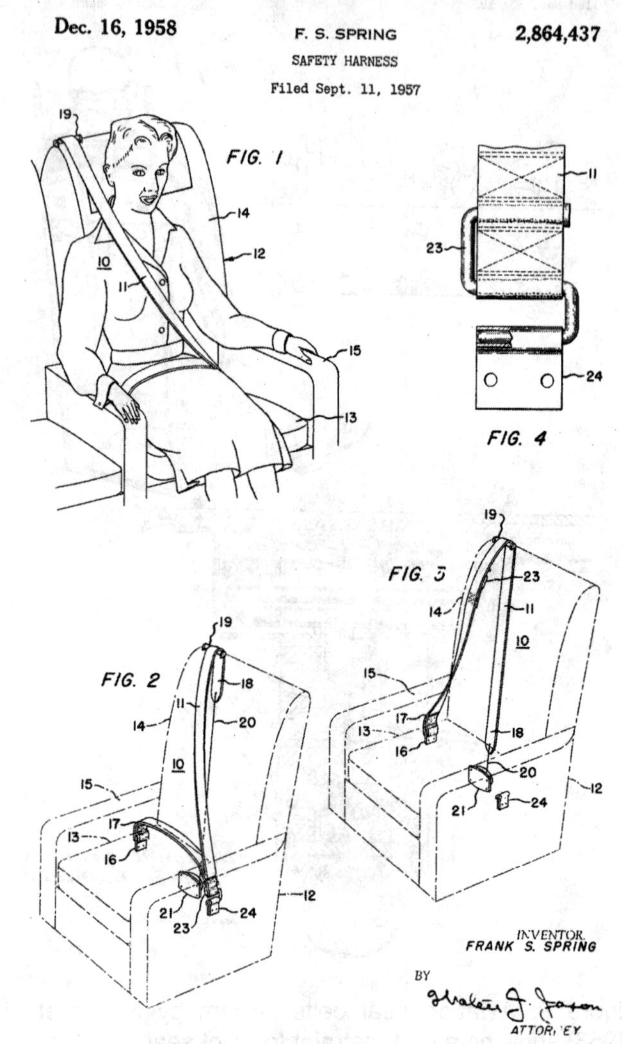

Figure 12: Automotive seat belt system by Spring (1958) showing a 3-pt occupant restraint.

There is no doubt that the technology for modern lap-shoulder belts was well defined by the late 1950s. By 1955-58, a number of patents were filed on lap-shoulder belts building on the earlier aviation and military concepts. These patents covered virtually every aspect of the systems that eventually went into production. However, it took more than a decade for these restraint systems to become commonplace in motor vehicles, and developments in Sweden were pivotal to the acceptance of lap-shoulder belts in motor vehicles.

SEAT BELT DEVELOPMENT IN SWEDEN

From the work by DeHaven and others, safety belt development switched to Sweden in the early 1950s as part of an industrial safety plan of the State Power Board in Sweden, now known as Vattenfall. Andreasson and Backstrom (2000) explain that Vattenfall was interested in protecting workers in company vehicles as part of its occupational safety program. Using the research of DeHaven, Stapp and others, two engineers constructed a 2-point diagonal safety belt. At the same time, they indicated that a three-point belt in a combination of the two-point belt plus a lap belt provided more effective protection.

Also by the mid-1950s, Saab had undertaken the development of seat belts in Sweden (Andreasson 2000). Saab introduced a continuous diagonal belt in test vehicles in 1956. However, crash test studies were also pursued at Saab on 3-point designs. By 1957, they tested a lap and diagonal shoulder belt attached at the B-pillar, as an alternative to the diagonal only belt. These systems were evaluated in a test fleet.

The use of a diagonal belt over the shoulder and collarbone was eventually found to be the preferred orientation, as earlier studies found less favorable results with lateral chest belts and other configurations that loaded the more compliant regions of the body. At the same time, the researchers at Vattenfalls had filed a patent on a 3-point belt system (Swedish patent 176,102, 1957). By 1959, Vattenfalls had tested lap-shoulder belts made of the Polyester Fiber Terylene, which had minimal resilience reducing recoil by loads in the belt webbing.

The safety research at Vattenfalls emphasized the need for belt restraint systems for car crash safety. Diagonal belts were finding their way into vehicles sold in Sweden and Europe. According to Andreasson and Backstrom (2000), a medical advisor to Vattenfall contacted the head of Volvo in 1956 and presented the idea of equipping cars with safety belts. As a result, Volvo installed two-point belts in 1958, as did Saab.

In 1958, Nils Bohlin assumed responsibility for safety belt development at Volvo and moved the company from a 2-point system, which had shown some field injury problems, to the lap-shoulder belts with an inertia locking reel. Lap-shoulder belts went into production in 1959. Volvo made three-point belts standard after extensive development work and demonstration of the increased safety effectiveness of the lap-shoulder belt system. This led to a comprehensive field evaluation of the restraint effectiveness in preventing injury. Tourin, Aldman (1962) later provided an overview of belt restraint characteristics for seat belts. This covered the preferred attachment of the belts to the vehicle, orientation of the diagonal shoulder belt, and testing requirements to ensure control of forward movement in crashes of 18, 21 and 25 mph.

Bohlin (1964) later published a comprehensive, 62 page study on the laboratory evaluation and crash testing of lap-shoulder belt systems by Volvo. This work included evaluation of different slip and non-slip joints, seat cushion hardness requirements, slack in the harness loop, restraint performance in front through ± 30^0 crash direction, and comparative tests up to 15 g with dummies and volunteers.

Three years later, Bohlin (1967) published the seminal study of restraint effectiveness of lap-shoulder belts based on 28,000 real world crashes. The Volvo system was a lap-shoulder belt with inertia reel and slip joint, which was found to be 40%-90% effective in reducing injury depending on the type of crash and severity of injury. The system essentially eliminated ejection and was not found to cause serious injury to wearers. However, it was worn by only 26% of the occupants. At the 11th Stapp Conference, Haddon (1967), Director of the National Highway Safety Bureau, commented that "Bohlin presents definitive evidence of the great reductions in death and injury that can be achieved through the use of a simple lap and upper torso safety belt."

SEAT BELTS IN THE UNITED STATES

Belt installation in production vehicles lagged in the US, however lap belts were required in 1964 and lap-shoulder belts in 1968 in the front seats of cars. But no specific design was mandated, and most domestic models were equipped with separate lap and shoulder belts. Occupants had to buckle each separately, and the failure to use one or the other portion compromised protection. Three-point safety belts weren't widely available in the US until a federal law required them. Now such belts are acknowledged as the single most effective safety device in passenger vehicles, when used.

Interestingly lap belts were common for US pilots in the 1910s-20s, and as reported by Johannessen (1984), Barney Oldfield installed lap belts in 1922 for motor sport racing after years of race car crashes and fatalities. However, it wasn't until 30 years later that belts became required by the Sport Car Club of America in motor car racing. In 1958, lap belts were required in sanctioned races, but it wasn't until 1966 that lap-shoulder belts became required equipment. This was well after cars had been fit with lap-shoulder belts in 1962-63 races (States, Benedict 1963).

In 1949, Nash was the first car company to install factory equipped lap belts. More than a decade later Ford and Chrysler provided optional seat belt installations. By the mid-1960s, five US states passed laws requiring standard installation of front seat belts and the transition to standard belt systems started. States, Benedict (1963) demonstrated the essential characteristics of lap-shoulder belts for occupant restraint in car crashes. This included the preferred orientation and configuration of the lap and shoulder belt, and they summarized the earlier configurations.

Neff (1962) provided a comprehensive review of seat belt webbing for automotive restraints that included strength, elongation, abrasion, weather resistance and weave patterns for polyester belts. Neff (1964) later summarized work by SAE in cooperation with the National Bureau of Standards aimed at setting up minimum standards for seat belts for use in motor vehicles. The work addressed the lap and shoulder harnesses, as well as belts for child restraints. SAE's belt committee was formed in 1954 and issued initial recommendations (J4a 1955) that were dramatically upgraded in a later revision (J4b 1964). The standard referenced three basic types of restraining devices:

Type 1	the lap belt
Type 2	a combination lap and upper torso restraint
Type 2a	upper torso restraint to be used with a type 1 belt
Type 3	combination pelvic and upper torso belt for children weighing <50 lbs

The standard included webbing strength requirements, hardware, child restraints and assembly testing. By 1968, the US Department of Transportation required lap-shoulder belts in front outboards seats as standard equipment after the clear demonstration of field effectiveness presented by Bohlin (1967).

In the late 1950s-60s, American carmakers showed interest in safety belts and conducted research on occupant restraints; but, they did not quickly become a standard feature in passenger vehicles. At the time this research was going on, US manufacturers were offering lap belts only as optional equipment. Lap belts became a standard feature in the United States in 1964 largely in response to state laws requiring them and later lap-shoulder belts in 1968.

As belts became more available and states considered mandatory use laws, a series of arguments were raised for not wearing safety belts. A common concern was for entrapment in the restraint system after a crash, and the inability to release the belts and extricate oneself from the vehicle. The common reply was that belts gave occupants the best chance to remain conscious increasing their ability to unbuckle the belts, open the door and get out of the vehicle after a crash. There were medical concerns as well. Individuals with implanted cardiac pacemakers and others with chest or abdominal surgery were concerned that belt loads may aggravate old incisions and implants and be a source of injury with their condition. Some people sought physicians to write letters so they would not have to use safety belts. In the end, it was determined that there were no reasons not to wear lap-shoulder belts, although special exemptions remain for cab drivers and other special circumstances.

LAP-SHOULDER BELT ENHANCEMENTS

Lap-shoulder belt systems continued to evolve after the introduction of mandatory installation in vehicles in the 1960s and mandatory use laws passed by countries around the world starting in the 1970s. There are many features that have been introduced to enhance the comfort and safety performance of lap-shoulder belt restraints. The most notable will be summarized, including pretensioners, belt web clamps, shoulder belt height adjusters and load limiters developed to enhance the convenience of putting the restraint on, the comfort while wearing it and performance in real-world crashes.

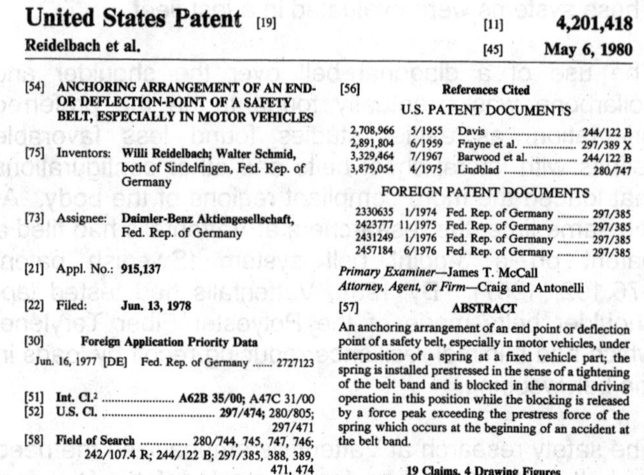

United States Patent [19] — Reidelbach et al.

[11] **4,201,418**

[45] **May 6, 1980**

[54] **ANCHORING ARRANGEMENT OF AN END- OR DEFLECTION-POINT OF A SAFETY BELT, ESPECIALLY IN MOTOR VEHICLES**

[75] Inventors: **Willi Reidelbach; Walter Schmid,** both of Sindelfingen, Fed. Rep. of Germany

[73] Assignee: **Daimler-Benz Aktiengesellschaft,** Fed. Rep. of Germany

[21] Appl. No.: **915,137**

[22] Filed: **Jun. 13, 1978**

[30] **Foreign Application Priority Data**

Jun. 16, 1977 [DE] Fed. Rep. of Germany 2727123

[51] **Int. Cl.**[2] **A62B 35/00;** A47C 31/00

[52] **U.S. Cl.** **297/474;** 280/805; 297/471

[58] **Field of Search** .... 280/744, 745, 747, 746; 242/107.4 R; 244/122 B; 297/385, 388, 389, 471, 474

[56] **References Cited**

U.S. PATENT DOCUMENTS

2,708,966	5/1955	Davis	244/122 B
2,891,804	6/1959	Frayne et al.	297/389 X
3,329,464	7/1967	Barwood et al.	244/122 B
3,879,054	4/1975	Lindblad	280/747

FOREIGN PATENT DOCUMENTS

2330635	1/1974	Fed. Rep. of Germany	297/385
2423777	11/1975	Fed. Rep. of Germany	297/385
2431249	1/1976	Fed. Rep. of Germany	297/385
2457184	6/1976	Fed. Rep. of Germany	297/385

Primary Examiner—James T. McCall
Attorney, Agent, or Firm—Craig and Antonelli

[57] **ABSTRACT**

An anchoring arrangement of an end point or deflection point of a safety belt, especially in motor vehicles, under interposition of a spring at a fixed vehicle part; the spring is installed prestressed in the sense of a tightening of the belt band and is blocked in the normal driving operation in this position while the blocking is released by a force peak exceeding the prestress force of the spring which occurs at the beginning of an accident at the belt band.

19 Claims, 4 Drawing Figures

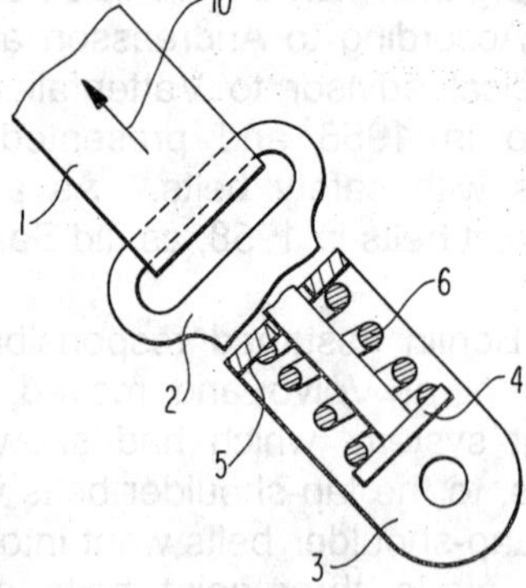

Figure 13: Automotive seat belt pretensioner attachment by Reidelbach et al. (1980).

Belt Pretensioners: From the earliest testing of safety belts, there has been recognition of the value in taking out slack in the restraint. This tightens the belts around the occupant, compresses clothing and adjusts the belt loading on the shoulder and pelvis. This has led to a range of concepts for belt tensioning and pretensioning early in a crash (Table 2). Belt pretensioners require a trigger signal indicating a rapid velocity change of the vehicle. The trigger activates a stored energy source to tighten the belt.

The torsion bar concept of Andersson et al. (1987) had sufficient stored energy to pull down the buckle early in a crash (Figure 14). This had the advantage of pulling the shoulder and lap belt on the inboard side of the occupant. This concept saw limited production until the introduction of pyrotechnic systems that had higher energy capacity and increased the velocity of belt tightening.

United States Patent [19] — Andersson et al.

[11] **Patent Number:** **4,705,296**
[45] **Date of Patent:** **Nov. 10, 1987**

[54] **SEAT BELT PRE-TENSIONING DEVICE**

[75] Inventors: **Sture R. Andersson**, Nacka; **Lars Y. Haland**, Falsterbo; **Bengt O. J. S. Mörner**, Hovas, all of Sweden

[73] Assignee: **Autoliv Development AB**, Vargarda, Sweden

[21] Appl. No.: **831,523**

[22] Filed: **Feb. 20, 1986**

[30] **Foreign Application Priority Data**

Feb. 21, 1985 [GB] United Kingdom 8504452
Jan. 9, 1986 [GB] United Kingdom 8600469

[51] **Int. Cl.**[4] **B60R 21/10**
[52] **U.S. Cl.** **280/806**; 297/480
[58] **Field of Search** 280/801, 806; 297/480

[56] **References Cited**

U.S. PATENT DOCUMENTS

4,257,626 3/1980 Addmeit 280/806
4,435,000 3/1984 Chiba et al. 280/806
4,458,921 7/1984 Chiba et al. 280/806

Primary Examiner—David M. Mitchell
Assistant Examiner—Kenneth R. Rice
Attorney, Agent, or Firm—Spencer & Frank

[57] **ABSTRACT**

A seat belt pre-tensioning device is adapted to be mounted on a vehicle and includes a buckle to receive a seat belt buckle. The device incorporates components which are responsive to a predetermined deceleration of the vehicle and which move to permit spring-loaded components to apply a force to move the buckle to apply a tension to the seat belt.

13 Claims, 9 Drawing Figures

United States Patent [19] — Schwanz et al.

[11] **3,871,470**
[45] **Mar. 18, 1975**

[54] **SAFETY BELT TENSIONING DEVICE**

[75] Inventors: **Wilfried Schwanz; Ernst Fiala,** both of Braunschweig; **Rolf Warnecke,** Ribbesbuttel; **Peter Buchholz,** Braunschweig, all of Germany

[73] Assignee: **Volkswagenwerk Aktiengesellschaft,** Wolfsburg, Germany

[22] Filed: **Oct. 24, 1972**

[21] Appl. No.: **299,783**

[30] **Foreign Application Priority Data**

May 12, 1972 Germany........................... 2223061

[52] U.S. Cl.. **180/82 C,** 280/150 SB, 242/107.4 SB, 297/388, 60/26.1
[51] **Int. Cl.** **B60r 21/10**
[58] **Field of Search**..... 280/150 SB; 180/82 C, 103; 188/67; 60/26.1; 242/107.4 SB

[56] **References Cited**

UNITED STATES PATENTS

2,825,581 3/1958 Knight......................... 280/150 SB
2,845,902 8/1958 Anderson............................. 188/67
2,883,123 4/1959 Finnigan.................... 280/250 SB X
2,994,563 8/1961 Ruggiero.......................... 60/26.1 X
3,077,324 2/1963 Strickland........................ 60/26.1 X
3,178,136 4/1965 Bayer .... 280/150 SB
3,219,361 11/1965 Brown .... 280/150 SB
3,533,659 10/1970 Major.......................... 280/150 SB
3,554,577 1/1971 Hane.............................. 280/150 SB
3,559,400 2/1971 Kleiner et al. 60/26.1
3,583,726 6/1971 Lindblad .... 280/150 SB
3,713,506 1/1973 Lipschutz .... 280/150 SB
3,715,130 2/1973 Harada et al. .... 60/26.1 X

Primary Examiner—Kenneth H. Betts
Attorney, Agent, or Firm—Brumbaugh, Graves, Donohue & Raymond

[57] **ABSTRACT**

The invention concerns an apparatus for tensioning restraining devices, such as safety belts. The tensioning is effected by a piston arranged to move within a cylinder when actuated by a pressure medium. A return stopping device is provided to maintain tension after actuation.

3 Claims, 5 Drawing Figures

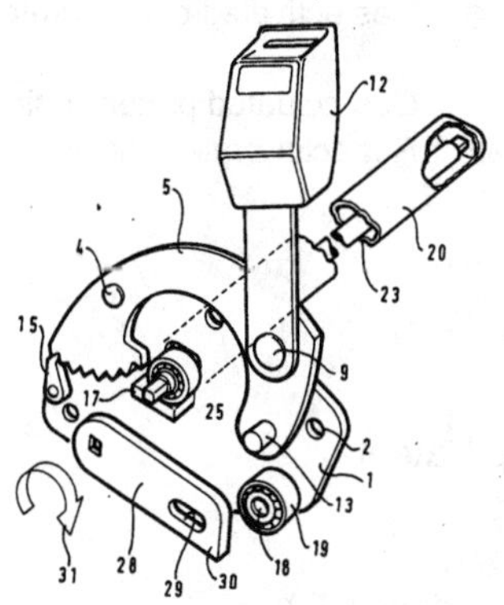

Figure 14: Spring actuated seat belt buckle pretensioner by Andersson et al. (1987).

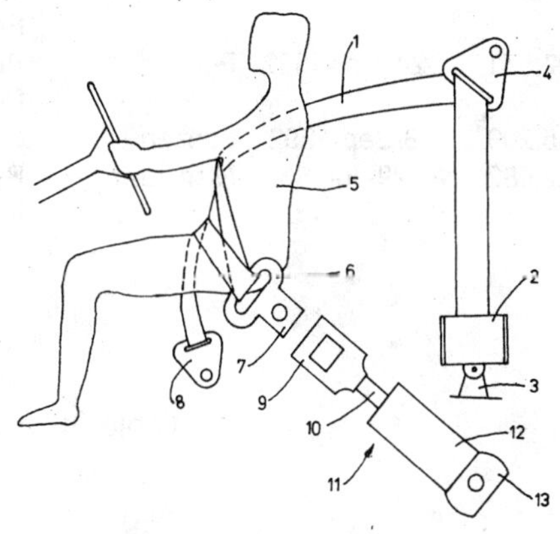

Figure 15: Automotive seat belt pretensioner by Schwanz et al. (1975).

The first belt tensioning concepts were stretched springs that could be actuated and pull the belt toward the anchorage on the vehicle. The concepts of Brown (1965), Lipschultz (1973), Reidelbach (1980) involved spring actuation of belt tension. An example of this technology is shown in Figure 13. The belt anchor included a tensioned spring that was activated in a crash. However, because of the high forces in automotive crashes, these systems had limited energy to pull sufficient belt once the occupant accelerated into the restraint system.

Schwanz (1975) and Lenzen (1992) demonstrated pyrotechnic belt pretensioners either acting on the buckle or on the retractor. The Schwanz (1975) approach was to pull on the belt anchors (Figure 15) and the concept of Lenzen (1992) was to pull on the spindle of the belt retractor winding in belt webbing (Figure 16). This approach had the capacity to pull in slack and overcome some the play-out of webbing as the belt tightens around the spindle in the retractor.

Table 2: Examples of Belt Pretensioning Patents.

Patent #	Date	Inventor	Title	Innovation
2,883,123	21-Apr-1959	Finnigan	Safety Apparatus of Use on a Motor Vehicle	Automatic electronic take-up reel.
3,077,324	12-Feb-1963	Strickland	Harness Inertia Reel	
3,219,361	23-Nov-1965	Brown	Safety Belt for Vehicle	Spring retractor at door attachment
3,559,400	2-Feb-1971	Kleiner	Actuator	Dynamic pull device
3,713,506	30-Jan-1973	Lipschultz	Safety Belt with Deceleration Release Control	Spring release of belt tension
3,871,470	18-Mar-1975	Schwanz	Safety Belt Tensioning Device	Piston actuated belt tensioning
4,008,780	22-Feb-1977	Bender	Device for Tensioning Safety Belts in Vehicles	Piston tensioning device for belts
4,201,418	6-May-1980	Reidelbach	Anchoring Arrangement of an End- or Deflection-Point of a Safety Belt, Especially in Motor Vehicles	Spring tensioning in buckle attachment
4,257,626	24-Mar-1981	Adomeit	Device for Tightening a Safety Belt for a Motor Vehicle	Spring actuated belt tightening
4,435,000	6-Mar-1984	Chiba	Seatbelt Tightening Device	Spring actuated belt tensioning
4,705,296	10-Nov-1987	Andersson	Seat Belt Pre-Tensioning Device	Spring actuated buckle tensioner
4,750,686	14-Jun-1988	Fohl	Belt Tightener on a Safety Belt Retractor	Retractor belt tensioner
4,932,603	12-Jun-1990	Yamanoi	Safety Belt Arrangement with Pretensioning Mechanism	Retractor belt pretensioner
5,097,571	24-Mar-1992	Fohl	Buckle for a Safety Belt System Provided with a Belt Pretensioner	Gas activated buckle pretensioner
5,145,209	8-Sep-1992	Lenzen	Seat Belt Pretensioner	Gas actuated pretensioning at retractor
5,222,682	29-Jun-93	Nishizawa	Belt Pretensioner for Vehicle Seat Belt Retractor	Retractor pretensioner

Table 3: Examples of Belt Webbing Clamp Patents.

Patent #	Date	Inventor	Title	Innovation
3,817,473	18-Jun-1974	Board	Retractable Safety Belts	Automatic locking belt webbing
4,120,466	17-Oct-1978	Adomeit	Safety Belt Clamping Device	Web clamping mechanism in retractor
4,168,810	25-Sep-1979	Sack	Novel Locking Mechaniam and Reel for Retractor	Belt locking retractor
4,241,886	30-Dec-1980	Maekawa	Retractor Device for Seat Belt Mechanism	Belt clamping in retractor
4,328,933	11-May-1982	Loose	Belt Clamping Retractor	Web clamp in retractor
4,381,085	26-Apr-1983	Stephenson	SeatBelt Retractor with Reduced Spooling	Web clamping of belt in retractor

Belt pretensioners have found widespread application in restraint systems and are activated as early as practical in a crash to give tensioning benefits before the occupant experiences deceleration into the restraint. At that point, the forces developed are quite high and it is difficult to pull with sufficient force and not add risks of injury to frailer occupants in other situations.

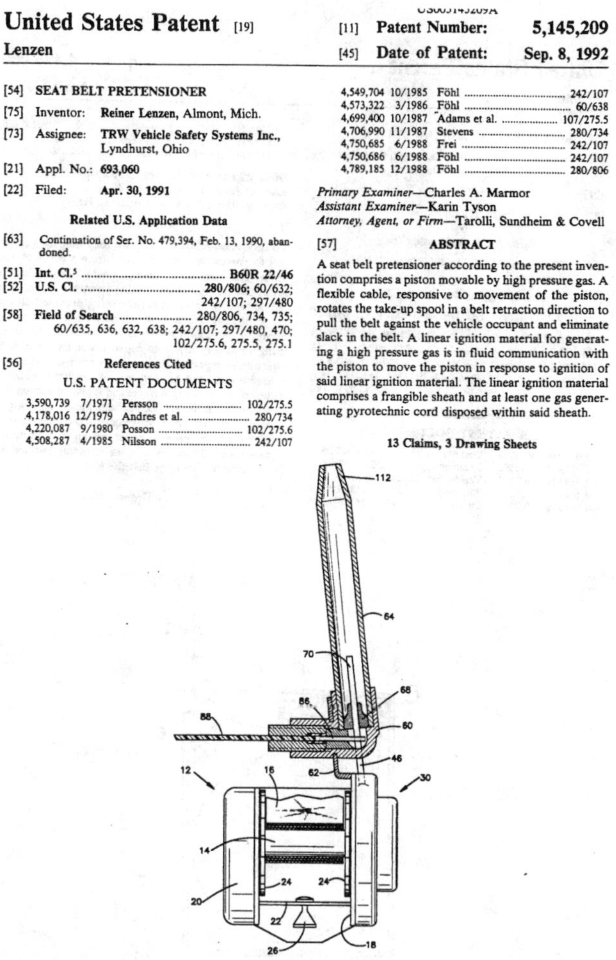

United States Patent [19]
Lenzen

[11] **Patent Number: 5,145,209**
[45] **Date of Patent: Sep. 8, 1992**

[54] **SEAT BELT PRETENSIONER**
[75] Inventor: **Reiner Lenzen,** Almont, Mich.
[73] Assignee: **TRW Vehicle Safety Systems Inc.,** Lyndhurst, Ohio
[21] Appl. No.: **693,060**
[22] Filed: **Apr. 30, 1991**

Related U.S. Application Data

[63] Continuation of Ser. No. 479,394, Feb. 13, 1990, abandoned.
[51] **Int. Cl.**[5] **B60R 22/46**
[52] **U.S. Cl.** **280/806;** 60/632; 242/107; 297/480
[58] **Field of Search** 280/806, 734, 735; 60/635, 636, 632, 638; 242/107; 297/480, 470; 102/275.6, 275.5, 275.1

[56] **References Cited**

U.S. PATENT DOCUMENTS

3,590,739 7/1971 Persson 102/275.5
4,178,016 12/1979 Andres et al. 280/734
4,220,087 9/1980 Posson 102/275.6
4,508,287 4/1985 Nilsson 242/107
4,549,704 10/1985 Föhl 242/107
4,573,322 3/1986 Föhl 60/638
4,699,400 10/1987 Adams et al. 107/275.5
4,706,990 11/1987 Stevens 280/734
4,750,685 6/1988 Frei 242/107
4,750,686 6/1988 Föhl 242/107
4,789,185 12/1988 Föhl 280/806

Primary Examiner—Charles A. Marmor
Assistant Examiner—Karin Tyson
Attorney, Agent, or Firm—Tarolli, Sundheim & Covell

[57] **ABSTRACT**

A seat belt pretensioner according to the present invention comprises a piston movable by high pressure gas. A flexible cable, responsive to movement of the piston, rotates the take-up spool in a belt retraction direction to pull the belt against the vehicle occupant and eliminate slack in the belt. A linear ignition material for generating a high pressure gas is in fluid communication with the piston to move the piston in response to ignition of said linear ignition material. The linear ignition material comprises a frangible sheath and at least one gas generating pyrotechnic cord disposed within said sheath.

13 Claims, 3 Drawing Sheets

Figure 16: Pyrotechnic seat belt pretensioner by Lenzen (1992) at retractor spindle.

Today, pretensioners can be found pulling at the buckle on the inboard side of the restraint system or at the spindle of the retractor on the shoulder belt. Because of the use of sliding latch plates, the retractor pretensioners have enough force to tighten both the shoulder and lap belt when activated early in a crash, and they generally have a greater amount of belt that can be pulled in than on the buckle side, which is limited by packaging constraints.

A current aim of pretensioner is quicker activation and greater capacity of pulling belt webbing. Both these aims are challenges due to the very high accelerations of components in the actuator and pretensioner hardware that limit the speed that belt can be drawn against the occupant. There is also the inevitable tradeoff between safe performance in low-speed crashes and adequate capacity for pulling belt webbing in the most severe crashes.

Pretensioners are typically single-action safety devices that need to be replaced after the crash by dealer service. Other concepts have been pursued including automatically resettable belt tensioning devices. These systems use an electric motor with gears to rapidly tension the belts. While these concepts do not pull as fast as pyrotechnic pretensioners, they can be activated earlier in crashes and even in emergencies since the systems are resettable without dealer repair.

In the long term, forward collision warning radar and advance vehicle dynamics systems may provide signals to activate resettable belt tensioning devices. This would tighten the belts in high risks driving situations, emergencies and potential crashes. In the short term, other vehicle based signals can be used to activate slack take-up systems. About half of frontal crashes involve panic braking by the driver, so consideration has been given to using brake activation as a trigger. The speed of brake pedal motion is a good indicator of panic braking. However, an even earlier signal is available.

Brake pedal acceleration occurs earlier than pedal velocity and appears to be the earliest reliable signal to differentiate panic from normal or hard braking based on brake pedal responses. Studies have shown that a pedal acceleration threshold is a more reliable differentiator of panic braking than a pedal velocity threshold. The accelerometer signal from the pedal is relatively noise free and a 10 g, 1 ms duration criterion may account for any noise that might occur in the system. However, a rigorous evaluation of pedal acceleration needs to be carried out on rough and off road driving, or during abuse testing of the vehicle.

An even earlier indication of panic braking would be beneficial. There is 0.2 s between the driver removing their foot from the throttle and contacting the brake pedal. Consideration has been given to using a minimum time-window between throttle and brake contact to indicate panic braking. While such an approach may be possible for drivers who use only the right foot for throttle and brake pedal action, it would not be practical for drivers who brake with the left foot. Yet, there may be some other indications from the speed with which the driver moves their foot from the throttle to the brake.

Shinar (1995) developed an interesting approach using the throttle release speed to trigger an Automatic Brake Warning (ABW) light for following vehicles. ABW anticipates a possible panic or emergency braking situation and gives following drivers a 0.22 s earlier warning than the normal brake light. This is similar to the time window for movement of the foot from the throttle to the brake pedal, so it would substantially reduce the trigger time for resettable safety systems. A vehicle was equipped with ABW and driven 61,668 km where all braking actions were recorded. Normal braking occurred 95,394 times and ABW was activated 820 times, or 13.3 times per 1,000 km. With this system, there was a 23% false alarm rate, which was not considered excessive in this application. Additional study was conducted on the approach showing further positive results (Shinar et al. 1997).

A motorized seatbelt retractor can rapidly take-up slack in the restraint system in emergency situations. In most cases of hard or panic braking, the system is well tolerated by the occupants and is viewed as an approach increasing the security of the belt restraints. It is an approach to improve the restraint of occupants wearing their belts, and improves the performance of belt pretensioners that would activate in a crash to increase ride-down benefits and lower deceleration forces on the restrained occupants.

Slack take-up systems must function before the occupant decelerates due to the braking action of the vehicle. Once the occupant decelerates in the seatbelt, the force necessary to overcome the occupant motion exceeds the levels acceptable for some occupants. For example, the retraction force necessary to remove slack for a large occupant leaning forward in the seat during rapid vehicle deceleration may be >500 N. However, if this force is applied to the restraint system of a small adult or child, it may exceed the comfort level of the individuals, even in an emergency.

If motorized retractors are activated in an emergency or crash, they should remain on until there is a tension release signal from removal of the foot from the brake, changing the gear shift or another indicating a driver intention. The benefit of leaving the tensioning on is to pull the occupant back into normal seating position in case of a subsequent rear crash. This would align the occupant with the seat and head restraint, and reduce the risk of whiplash or more serious injury.

Web Clamping Devices: One of the obvious consequences of using a belt retractor is the loose winding around the reel that occurs with normal use. Once a crash occurs, force on the belt rises rapidly causing the webbing to tighten on the spindle and belt webbing to displace out of the retractor. This is called belt payout and adds slack to the restraint system delaying restraining loads on the occupant.

United States Patent [19] — [11] **4,120,466**

Adomeit — [45] **Oct. 17, 1978**

[54] **SAFETY BELT CLAMPING DEVICE**

[76] Inventor: **Heinz Dieter Adomeit,** Knesebeckstr. 6/7, Berlin, Germany

[21] Appl. No.: **789,607**

[22] Filed: **Apr. 21, 1977**

Related U.S. Application Data

[63] Continuation of Ser. No. 593,710, Jul. 7, 1975, abandoned.

[30] **Foreign Application Priority Data**

Jul. 9, 1974 [DE] Fed. Rep. of Germany 2432956

[51] **Int. Cl.**[2] **A62B 35/00; B65H 75/48**
[52] **U.S. Cl.** **242/107.2; 280/744**
[58] **Field of Search** .... 242/107.2, 107.4 R–107.4 E, 242/75.2; 280/744–747; 297/388, 385–387, 389; 188/65.1, 72.9, 188

[56] **References Cited**

U.S. PATENT DOCUMENTS

3,003,589	10/1961	Desbrow	188/72.9 X
3,323,831	6/1967	Buechler	280/150 SB X
3,389,874	6/1968	Van Noord	242/107.2
3,467,337	9/1969	Putman	242/107.2 X
3,692,253	9/1972	Curran	242/107.4 R
3,722,824	3/1973	Hayashi	242/107.4 A
3,817,473	6/1974	Board et al.	242/107.2
3,847,434	11/1974	Weman	242/107.2 X

Primary Examiner—Harvey C. Hornsby
Assistant Examiner—John M. Jillions
Attorney, Agent, or Firm—Patrick L. Henry

[57] **ABSTRACT**

The invention relates to a device for clamping a safety belt to prevent it being unwound from a belt reel, the device consisting of a housing in which is provided a lever, one end of the lever being pivotably mounted in the housing, a clamping surface being provided on the lever intermediate its ends which co-operates with a clamping surface provided on the housing, the lever being resiliently urged to move about its pivot in a direction to separate the clamping surfaces. The belt is guided around the pivoted end of the lever, extends between the clamping surfaces and is guided around the free end of the lever, the resilient means applying a tension force to the belt. Movement of the lever against the force of the resilient means moving the clamping surface on the lever towards the clamping surface on the housing to clamp the belt therebetween.

4 Claims, 6 Drawing Figures

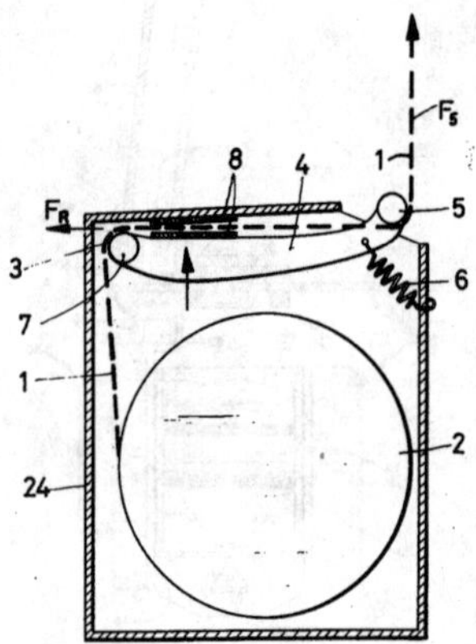

Figure 17: Web clamping of seat belt in the retractor by Adomeit (1978).

Web clamps have been considered to counter the action of webbing payout under occupant loading (Table 3). These devices have the aim of locking the belt between the reel and occupant thereby avoiding loads on the belt wrapped around the spindle in the retractor. Adomeit (1978) presented a concept as part of the retractor that would clamp belt webbing at the exit from the retractor housing in a crash. Figure 17 shows the approach of clamping the belt webbing.

With frequent activation of web clamps over the course of belt use and activation of the clamp, there can be fraying of the webbing with unwanted degradation in the belt system. Other clamping concepts have been presented, such as the cam actuated clamp by Loose (1982) in Figure 18; but, virtually all designs grab and compress the webbing with a sufficiently rough interface to damage the belts.

United States Patent [19] — Loose

[11] **4,328,933**
[45] **May 11, 1982**

[54] **BELT CLAMPING RETRACTOR**

[75] Inventor: **Richard D. Loose,** Birmingham, Mich.

[73] Assignee: **General Motors Corporation,** Detroit, Mich.

[21] Appl. No.: **191,284**

[22] Filed: **Sep. 26, 1980**

[51] **Int. Cl.**[3] **A62B 35/02;** B65H 75/48
[52] **U.S. Cl.** .... **242/107.2;** 242/107.4 A
[58] **Field of Search** .... 242/107.2, 107.3, 107.4 R, 242/107.4 E, 156.1; 280/801–808; 297/474–480

[56] **References Cited**

U.S. PATENT DOCUMENTS

384,187	6/1888	Blackman	242/107.3
3,384,415	5/1968	Monroe	242/107.4 B X
3,471,100	10/1969	Arcari	242/107.4 B
3,480,228	11/1969	Ulert	242/107.4
3,482,798	12/1969	Kawaguchi	242/107.3
3,942,740	3/1976	Torphammar et al.	242/107.4 B
4,120,466	10/1978	Adomeit	242/107.2
4,211,377	7/1980	Yasumatsu	242/107.2
4,273,301	6/1981	Frankila	242/107.2

FOREIGN PATENT DOCUMENTS

1531356 11/1978 United Kingdom .

Primary Examiner—John M. Jillions
Attorney, Agent, or Firm—Charles E. Leahy

[57] **ABSTRACT**

A seat belt retractor includes a belt reel rotated in the belt winding direction by a winding spring. An inertia actuated reel locking device acts to selectively lock the reel against belt unwinding rotation so that a load imposed on the belt during restraint of an occupant tends to tighten or spool down the wound-up belt stored on the reel. A belt clamp mechanism for limiting the extension of belt from the reel subsequent to reel lockup includes a lever pivotally mounted on the housing and having a belt clamping surface at one end thereof overlying the belt wound-up on the reel. A spring urges the lever to a normal position in which the belt clamping surface is spaced from the wound-up belt. The belt passes around a belt roller at the other end of the lever so that the imposition of occupant restraint load on the belt simultaneous with lockup of the reel by the inertia locking means forcibly pivots the lever to carry the belt clamping surface into clamping engagement of the belt wound upon the reel and thereby prevent belt from spooling off the reel.

1 Claim, 2 Drawing Figures

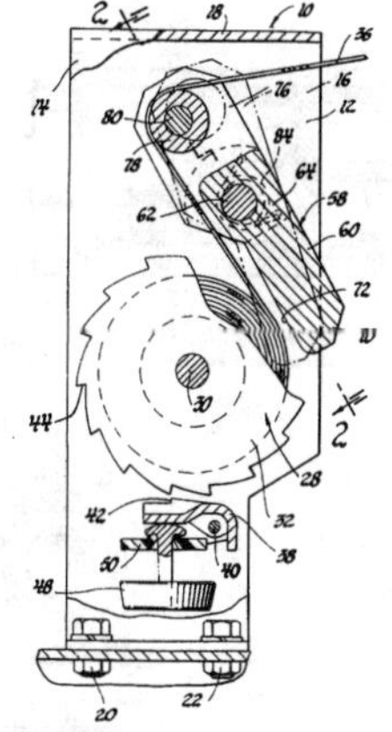

Figure 18: Web clamping mechanism in retractor by Loose (1982).

In spite of these shortcomings, web clamps reduce the effective length of belt webbing involved in crash restraint. This also reduces the webbing elongation that occurs under occupant loading. If a successful web clamp is developed, it may be combined with other belt features like load-limiting.

Height Adjustment of the Shoulder Belt: With vehicle mounted seat belt attachments and guides at the D-ring, the belt geometry is fixed to the vehicle. This changes the orientation of the belt with respect to the occupant depending on the size and seating position of the occupant and position of the seat in the vehicle. There are also differences between 2 and 4 door vehicles because of the available placement of the D-ring on the B-pillar. For these reasons, some adjustment of the belt anchors is desired.

United States Patent [19] — Burleigh

[11] **3,964,798**
[45] **June 22, 1976**

[54] **SAFETY BELT ANCHORAGES**

[75] Inventor: **David William Burleigh,** Yateley, England

[73] Assignee: **Britax (London) Limited,** Byfleet, England

[22] Filed: **Feb. 7, 1975**

[21] Appl. No.: **548,031**

[30] **Foreign Application Priority Data**

Feb. 14, 1974 United Kingdom................. 6695/74

[52] **U.S. Cl.** .... **303/1 A;** 280/744
[51] **Int. Cl.**[2] **F16C 29/02**
[58] **Field of Search** .... 308/1 R, 1 A, 3 R; 280/150 SB; 105/464, 475, 478, 482, 502; 248/40 B, 242, 243, 423

[56] **References Cited**

UNITED STATES PATENTS

2,736,272	2/1956	Elsner	105/482
2,880,788	4/1959	Phillips et al.	280/150 SB
3,366,996	2/1968	Springer	105/475 X
3,618,975	11/1971	Bombach	280/150 SB
3,746,393	7/1973	Andres et al.	280/150 SB X

Primary Examiner—Kenneth H. Betts
Assistant Examiner—Gene A. Church
Attorney, Agent, or Firm—Baldwin, Wight & Brown

[57] **ABSTRACT**

An anchorage member for adjusting the position of a safety belt with regard to body shape and seat position. The anchorage member is attached to a door pillar or alongside a seat and comprises a slide member for selectively engaging a series of abutments. The slide member is biased into engagement with the abutments and may be locked in position.

15 Claims, 4 Drawing Figures

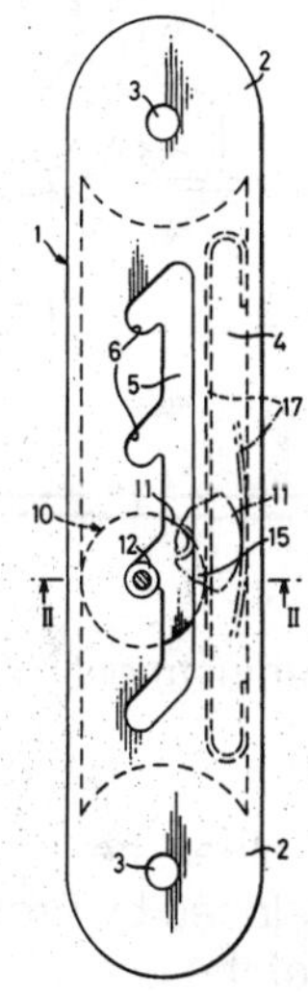

Figure 19: Height adjustment of the shoulder belt by Burleigh (1976).

Burleigh (1976) showed a mechanism connected to the D-ring and integrated in the B-pillar and trim of the vehicle (Figure 19). This approach involved a ladder with several heights that the D-ring could be placed to change the orientation of the shoulder belt as it routes over the chest. In the lowest position, it favors the shortest occupants, whereas the taller occupant wants the height adjustment at its highest level to give a better placement over the shoulder. This type of height adjuster is common today.

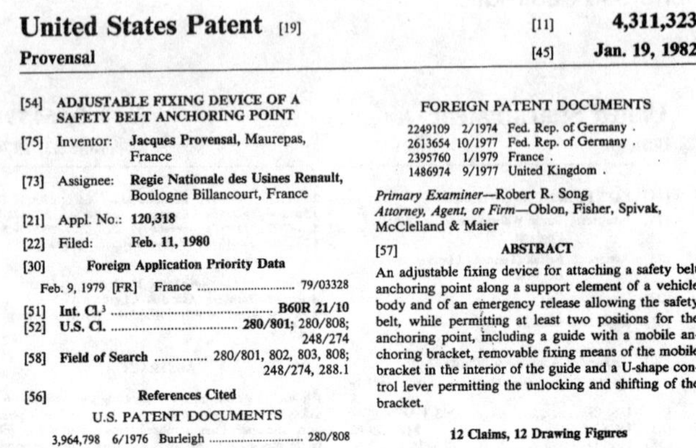

United States Patent [19]

Provensal

[11] **4,311,323**

[45] **Jan. 19, 1982**

[54] **ADJUSTABLE FIXING DEVICE OF A SAFETY BELT ANCHORING POINT**

[75] Inventor: **Jacques Provensal,** Maurepas, France

[73] Assignee: **Regie Nationale des Usines Renault,** Boulogne Billancourt, France

[21] Appl. No.: **120,318**

[22] Filed: **Feb. 11, 1980**

[30] **Foreign Application Priority Data**

Feb. 9, 1979 [FR] France 79/03328

[51] **Int. Cl.**[3] **B60R 21/10**

[52] **U.S. Cl.** **280/801;** 280/808; 248/274

[58] **Field of Search** 280/801, 802, 803, 808; 248/274, 288.1

[56] **References Cited**

U.S. PATENT DOCUMENTS

3,964,798 6/1976 Burleigh 280/808

FOREIGN PATENT DOCUMENTS

2249109 2/1974 Fed. Rep. of Germany .
2613654 10/1977 Fed. Rep. of Germany .
2395760 1/1979 France .
1486974 9/1977 United Kingdom .

Primary Examiner—Robert R. Song
Attorney, Agent, or Firm—Oblon, Fisher, Spivak, McClelland & Maier

[57] **ABSTRACT**

An adjustable fixing device for attaching a safety belt anchoring point along a support element of a vehicle body and of an emergency release allowing the safety belt, while permitting at least two positions for the anchoring point, including a guide with a mobile anchoring bracket, removable fixing means of the mobile bracket in the interior of the guide and a U-shape control lever permitting the unlocking and shifting of the bracket.

12 Claims, 12 Drawing Figures

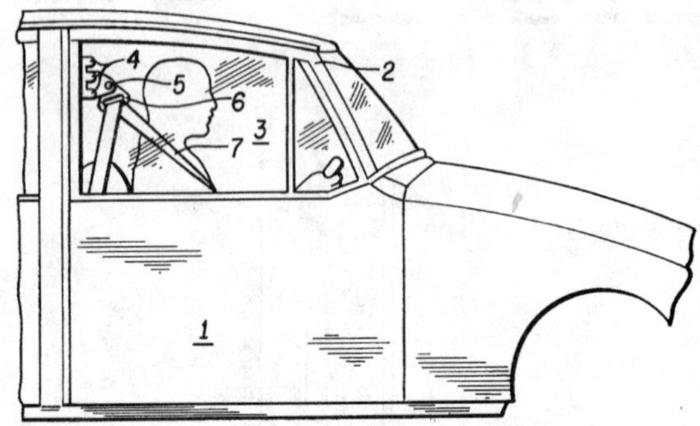

Figure 20: Height adjustment of the shoulder belt by Provensal (1982).

There have been other height adjustment concepts (Table 4). One developed by Provensal (1982) is shown in Figure 20. Most of the concepts involve adjusting the vertical height of the shoulder belt guide, but others have included horizontal adjustment of the shoulder belt attachment at the roof rail (Manz 1979) and adjustability of the other anchor points of the belt to the seat and vehicle.

Another design for height adjustment used a vertical spindle for the retractor placed at the typical D-ring location on the B-pillar. The addition of a belt routing trim cover automatically adjusted the anchor height by the orientation of the belt routing over the shoulder and chest. This approach also reduced the effective length of belt webbing, since the retractor was now at shoulder height.

Load Limiting Belt Tension: Once safety belts became common and worn in crashes, there was data on the field performance of the restraints. It became clear that in severe crashes, high loads could develop in the belts and injure the occupant. This was particularly the case with older occupants and those with frailer skeletal structures in the ribcage. The approach of limiting the shoulder belt load was considered early in the use of safety belts. Renault pioneered the use of tear stitching in the shoulder belt that progressively broke loose limiting shoulder belt loads at incrementally higher levels (Foret-Bruno et al. 1998). This also allowed crash investigators to compare occupant injuries with the level of belt tension, adding insight on human tolerance to belt loading.

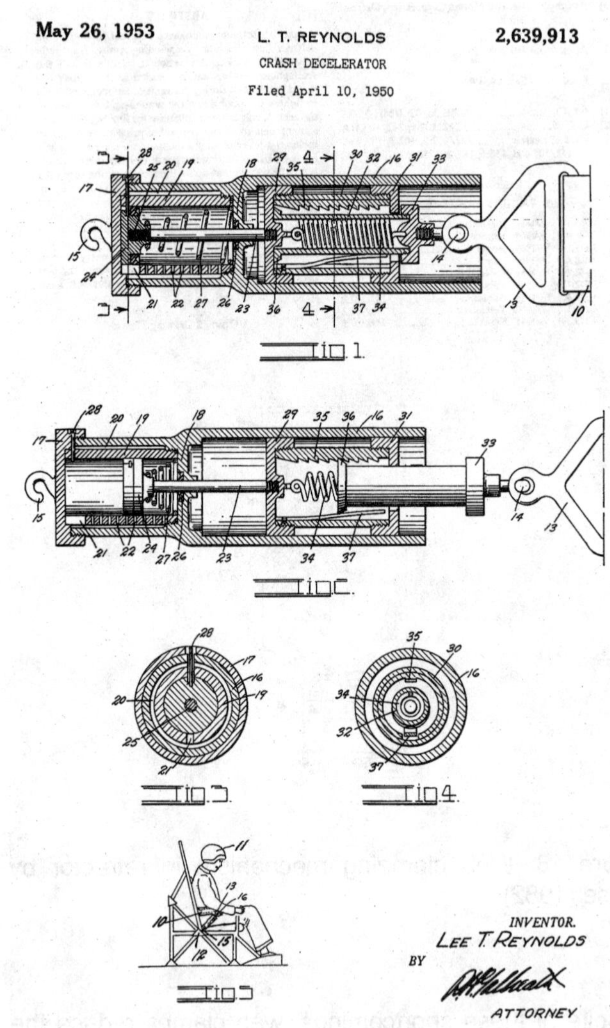

Figure 21: Load limiting belt anchor by Reynolds (1953).

Table 4: Examples of Belt Height or Position Adjusting Patents.

Patent #	Date	Inventor	Title	Innovation
3,964,798	22-Jun-1976	Burleigh	Seat Belt Anchorages	Adjustable shoulder belt anchor
4,159,848	3-Jul-1979	Manz	Linear Locking Seat Belt Retractor	Horizontal adjustment of the shoulder belt anchor
4,225,185	30-Sep-1980	Krzok	Vertically Adjustable Mounting Arrangement for Safety Belts at a Side Rail of an Automotive Vehicle	Height adjusting shoulder belt anchor
4,311,323	19-Jan-1982	Provensal	Adjustable Fixing Device of a Safety Belt Anchoring Point	Adjustable upper anchor
4,398,749	16-Aug-1983	Hipp	Device for Adjustably Fastening a Belt Fitting to an Automobile Body	Adjustable upper anchor concept
4,469,352	4-Sep-1984	Korner	Motor Vehicle with a Vertically Adjustable Safety Belt Guide Fitting	Adjustable shoulder belt anchor
4,500,115	19-Feb-1985	Ono	Mounting Arrangement for Through Anchor	Adjustbale shoulder belt guide
4,538,832	3-Sep-1985	Anderson	Adjustable Seat Belt Anchorage	Shoulder belt guide adjsutment device
4,652,012	24-Mar-1987	Biller	Fitting for a Safety Belt for Motor Vehicles	Adjustable upper guide anchor
4,989,900	5-Feb-1991	Steinhuser	Apparatus for Adjusting the Level of the Deflector for a Shoulder Belt in the Occupant Restraint System of an Automotive Vehicle	Upper adjustable guide
5,482,325	9-Jan-1996	Moller	Height-Adjustable Guide Fitting for a Sae Belt of a Motor Vehicle	Upper adjustable guide

There have been numerous papers on the optimum limit load for the shoulder belts in crashes even before the introduction of supplemental airbag systems. However, once the combination belt and airbag acted to restrain the occupant, there was interest to provide the earliest restraint from the belt system and then later reduce the shoulder belt loads as the airbag provides the primary restraint of the upper body and head. This gave further motivation for belt load limiting in a crash.

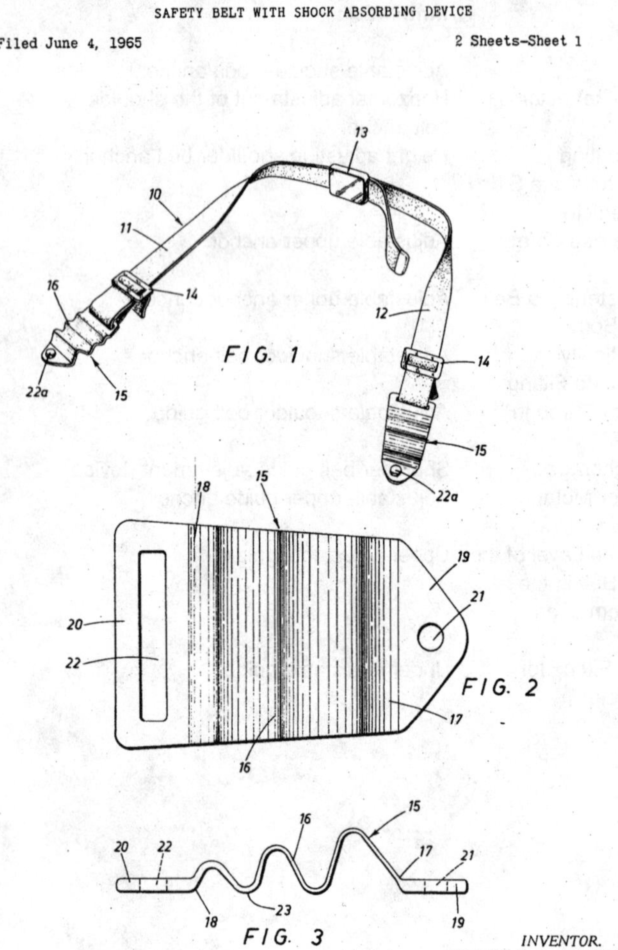

Figure 22: Load limiting belt anchor by Villiers (1968).

There is a rich patent history on belt load limiting concepts (Table 5). Some of the earliest concepts were from Reynolds (1953) and Hendry (1962), which involved deforming metal or stroking a piston in a cylinder under occupant load (Figure 21). Most of the systems act like a spring or an elastic-plastic device that yields at a prescribed load. There have been other concepts involving a rate dependent load limit that increased the limit force with the severity of the crash (Adomeit 1975).

Oftentimes simple but elegant concepts were proposed to allow yielding at the anchorage attachments to the vehicle or seat. Villiers (1968) showed a simple corrugated belt attachment plate that is an example of the various concepts patented (Figure 22). Other approaches have been used including a tearing ladder attachment of the retractor to the vehicle and other yielding components in the restraint system, seat or vehicle body. Doty (1996) demonstrated an approach of load limiting at the retractor attachment to the vehicle using deformable metal tabs in a slot (Figure 23).

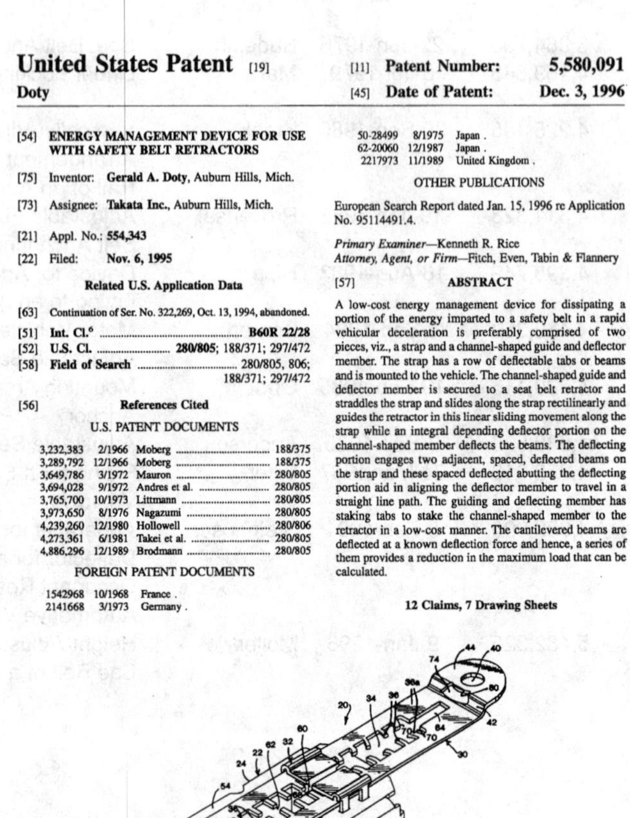

United States Patent [19] — **Doty**

[11] **Patent Number: 5,580,091**

[45] **Date of Patent: Dec. 3, 1996**

[54] **ENERGY MANAGEMENT DEVICE FOR USE WITH SAFETY BELT RETRACTORS**

[75] Inventor: **Gerald A. Doty**, Auburn Hills, Mich.

[73] Assignee: **Takata Inc.**, Auburn Hills, Mich.

[21] Appl. No.: **554,343**

[22] Filed: **Nov. 6, 1995**

Related U.S. Application Data

[63] Continuation of Ser. No. 322,269, Oct. 13, 1994, abandoned.

[51] **Int. Cl.**^{6} **B60R 22/28**

[52] **U.S. Cl.** **280/805**; 188/371; 297/472

[58] **Field of Search** 280/805, 806; 188/371; 297/472

[56] **References Cited**

U.S. PATENT DOCUMENTS

3,232,383	2/1966	Moberg	188/375
3,289,792	12/1966	Moberg	188/375
3,649,786	3/1972	Mauron	280/805
3,694,028	9/1972	Andres et al.	280/805
3,765,700	10/1973	Littmann	280/805
3,973,650	8/1976	Nagazumi	280/805
4,239,260	12/1980	Hollowell	280/806
4,273,361	6/1981	Takei et al.	280/805
4,886,296	12/1989	Brodmann	280/805

FOREIGN PATENT DOCUMENTS

1542968	10/1968	France .
2141668	3/1973	Germany .
50-28499	8/1975	Japan .
62-20060	12/1987	Japan .
2217973	11/1989	United Kingdom .

OTHER PUBLICATIONS

European Search Report dated Jan. 15, 1996 re Application No. 95114491.4.

Primary Examiner—Kenneth R. Rice
Attorney, Agent, or Firm—Fitch, Even, Tabin & Flannery

[57] **ABSTRACT**

A low-cost energy management device for dissipating a portion of the energy imparted to a safety belt in a rapid vehicular deceleration is preferably comprised of two pieces, viz., a strap and a channel-shaped guide and deflector member. The strap has a row of deflectable tabs or beams and is mounted to the vehicle. The channel-shaped guide and deflector member is secured to a seat belt retractor and straddles the strap and slides along the strap rectilinearly and guides the retractor in this linear sliding movement along the strap while an integral depending deflector portion on the channel-shaped member deflects the beams. The deflecting portion engages two adjacent, spaced, deflected beams on the strap and these spaced deflected abutting the deflecting portion aid in aligning the deflector member to travel in a straight line path. The guiding and deflecting member has staking tabs to stake the channel-shaped member to the retractor in a low-cost manner. The cantilevered beams are deflected at a known deflection force and hence, a series of them provides a reduction in the maximum load that can be calculated.

12 Claims, 7 Drawing Sheets

Figure 23: Load limiting retractor attachment by Doty (1996).

An adjustable, load limiting belt restraint is a key feature of smart restraint systems. This would involve adjustments of the restraining load by crash or occupant characteristics, and would vary the load limit during the crash. This is an area of future safety research.

Table 5: Examples of Load Limiting Belt Patents.

Patent #	Date	Inventor	Title	Innovation
2,401,748	11-Jun-1946	Dillon	Aircraft Seat Supporting Structure	Load limiting seat strut
2,682,931	6-Jul-1954	Young	Means for Absorbing Energy due to Sudden Impact	Seat strut load limiter
2,639,913	26-May-1953	Reynolds	Crash Decelerator	Seat belt load limiter
2,880,815	7-Apr-1959	Apfelbaum	Vehicle Brake Operation Responsive Safety Seat Device	Load limiting restraint belt
3,026,972	27-Mar-1962	Hendry	Energy Absorbing Seat Belt Attachment	Seat belt load limiter
3,087,584	30-Apr-1963	Jackson	Load Limiting Shock Strut	Load limiting strut attachment
3,106,989	15-Oct-1963	Fuchs	Energy Absorbing Device	Load limiting belt attachment
3,232,383	1-Feb-1966	Moberg	Energy Absorbing Means	Shoulder belt load limiter
3,280,942	25-Oct-1966	Millington	Energy Absorber-Metal Bender Type	Load limiter
3,361,475	2-Jan-1968	Villiers	Safety Belt with Shock Absorbing Device	Deformable belt anchor
3,438,674	15-Apr-1969	Radke	Safety Seat Belt Device with Shear Strip Energy Absorbing Means	Deformable belt anchor
3,547,468	15-Dec-1970	Giuffrida	Energy Absorbing Restraint Apparatus	Shoulder belt load limiter
3,680,913	1-Aug-1972	Seybold	Attachment Device for a Retaining Belt	Deformable belt anchor
3,765,700	16-Oct-1973	Littmann	Reusable and Automatically Resettable Energy Dissipating Device	Load limiting at the retractor
3,927,846	23-Dec-1975	Meissner	Safety Belt Unwinding Device with Energy Dissipator	Retractor spindle load limiting
3,973,650	10-Aug-1976	Nagazumi	Mechanical Energy Absorbing Device and Safety Harness Using the Same	Deformable belt anchor
4,027,905	7-Jun-1977	Shimogawa	Seat Belt System for Vehicle	Deformable belt anchor
4,886,296	12-Dec-1989	Brodmann	Belt Force Limiting Device	Deformable belt anchor
4,978,139	18-Dec-1990	Andres	Energy Absorbing Device for Safety Belts of Motor Vehicle	Deformable belt anchor
5,005,909	9-Apr-1991	Fohl	Force-Limiting Energy Absorber for Safety Belt Systems	Load limiting link
5,580,091	3-Dec-1996	Doty	Energy Management Device for Use with Safety Belt Retractor	Load limiting retractor attachment
5,611,498	18-Mar-1997	Miller	Seat Belt Retractor with Auxiliary Shaft Load Limiting	Load limiting retractor spindle
5,618,006	8-Apr-1997	Sayles	Seat belt Retractor with Energy Management	Retractor load limiter
5,785,269	28-Jul-1998	Miller	Dual Level Retractor for Oblique or Offset Impacts	Dual level load limiting retractor

AUTOMATIC CRASH PROTECTION

In August 1973, the Department of Transportation (DOT) issued a notice that cars not equipped with "passive" restraints would be required to install an ignition interlock tied to the driver's use of the lap-shoulder belt. This started less than an 18 month experience with belt interlocks that ended 1974 as Congress voided the rule.

Interlock Systems: For a brief period in the 1970s, the US implemented a belt interlock requirement to increase safety belt use. The starter interlock was introduced for the period August 15, 1973 until August 15, 1975 for vehicles without automatic protection produced during that period. The systems were connected to both front seats so that if any front seat belt in an occupied seat was not locked, the starter was disabled. If a buckle was opened later, a buzzer-light system was activated. All 1974 model year cars sold in the United States came with this ignition interlock except a few thousand GM models that came with airbags that met the automatic protection requirement.

In March 1974, NHTSA described the public reaction to the ignition interlock as "Public resistance to the belt-starter interlock system currently required has been substantial with current tallies of proper lap-shoulder belt usage at or below the 60% level. Even that figure is probably optimistic as a measure of results to be achieved, in light of the likelihood that as time passes the awareness that the forcing systems can be disabled, and the means for doing so will become more widely disseminated, ..." There were also speeches on the floor of both houses of Congress expressing the public's anger at the interlock system. On October 27, 1974 President Ford signed into law a bill that prohibited any Federal Motor Vehicle Safety Standard from requiring or permitting the use of any seatbelt interlock system. NHTSA then deleted the interlock option from October 31, 1974. Thus, interlock systems were required in the USA for 14.5 months instead of the 24 months that was originally intended (Kratzke 1995).

There have been many studies on the failure of the safety belt interlock legislation. The analysis by the Swedish Road Authority (Turbell, Larsson 1998) is the most informative and comprehensive, and shows many positive aspects about the interlock experience and the public's attitude about legislative means to increase belt use. The failure of interlock systems in the US may be explained by the following factors. Many people felt that it was an infringement of personal freedom. This was an expected reaction that may not be valid today, particularly for countries outside the US. Voluntary seatbelt use was very low at the time and the belt systems at that time were difficult to use, and had a bad fit and were inconvenient. The interlock system itself was too unsophisticated. It did not allow low speed maneuvers or sitting in the car with the engine idling. Furthermore, most of the complaint came from only a few vehicles with particularly bad systems.

Some basic principles for possible new approaches to interlock systems have been proposed by the Swedish Road Authority (Turbell, Larsson 1998). New systems have been established so the normal seatbelt user does not notice the interlock system. It is also more difficult and cumbersome to cheat on the system than using the belt. There must be a means to permanently disconnection the system but it should be hard to activate. The system must be very reliable and have a long lifetime. All seating positions in the car may be covered by the system. The accident and injury risk must not increase and retrofit systems should also be available for old cars.

What also remains a perplexing issue is that observational studies of safety belt use in normal driving regularly show very high belt use compliance of 80%-95% in Sweden, Germany, Canada and some states in the US; however, crash studies in each of these countries show less than 50% belt use in fatal crashes. Obviously, those most in need of the belt restraint system are the most unlikely to use the benefits of safety belts in serious crashes.

For more than 25 years, the approach of belt interlocks has been avoided. However, recently discussions have emerged in various countries about the value of such systems that require belt use before benefiting from other vehicle functions. At the lowest level, the lack of belt use could disable the use of the radio or other conveniences, and at a higher level, it could prevent more important vehicle functions including blocking vehicle starting except in emergency situation. The use of interlocks is still a consideration.

Automatic (Passive) Belts: While automatic protection regulation started in 1970 and led to the interim rule for the ignition interlock, the approach returned later in a series of NHTSA/DOT decisions (Kratzke 1995). Eventually, the Brock Adams' decision called for the installation of front seat passive restraints for big cars in 1982 and all cars by 1984. This decision was later revised by Ray Peck, who reversed the order of phase-in to small cars first. All of these decisions were challenged in court and the last by the insurance industry rose to the Supreme Court, which remanded the issue back to DOT and then Secretary Dole. NHTSA then worked to setup performance criteria for new cars so that front seat occupants did not absorb excessive crash forces. The targets were set so that there were only two ways to meet the test, either airbags or passive belts.

The issue continued to fluctuate, until July 1984 when Secretary Dole made a thoughtful decision to reinstate the Carter rule for automatic crash protection with a phase-in of 10% in 1987, 25% in 1988, 40% in 1989 and 100% in 1990. Two incentives were added to encourage changes. If the automatic crash protection was either an airbag or friendly interior, each vehicle was given 1.5 credits toward the phase-in requirement. This encouraged the use of airbags. Also, if states covering two-thirds of the US population adopted mandatory seatbelt use laws with certain penalties by April 1, 1989, the passive restraint rule would be rescinded.

This was an artful decision that promoted safety at various levels, since after its issuance there was a rapid adoption of state safety belt use laws. Secretary Dole's decision for automatic restraints activated requirements for passive systems unless the US states enacted sufficient safety belt use laws to cover 67% of the driving population by 1989. The initial reaction of the automotive and restraint supplier industries was the development of passive or automatic safety belts.

One of the earliest passive belt systems was the 2-pt. passive shoulder belt. It was developed and introduced in 1975 in the VW Rabbit and then GM Chevette. It used an automatic diagonal shoulder belt with an active lap belt and an energy absorbing knee bolster for those that only used the passive belt. The system had initial acceptance, but crash problems were noted including posterior cruciate ligament injury of the knee and ejection without lap belt use that made the overall performance less than expected. Although there was a reported 30%-50% detachment rate of the passive shoulder belt, the system encouraged greater belt use.

The most popular passive belt designs involved a motorized shoulder belt that acted on a track running along the roof rail. In the stowed position, the upper anchor of the shoulder belt was forward in the vehicle close to the steering wheel. In the restraint position the track moved rearward on the roof rail bringing the upper anchor of the shoulder belt rearward in position to restrain the occupant in crashes. This approach facilitated ingress and egress and passive shoulder belt restraint. However, motorized system proved complex for reliability and function through the normal use of vehicles and many years of vehicle life.

GM introduced a 3-point passive belt system that was non-motorized to comply with the passive restraint requirement. When the door was open the shoulder and lap belt were attached to the rear edge of the door and swung outward with the door. This allowed ingress when the belts were pushed forward to give access to the seat. When the door was closed, the belts moved closer to the occupant and formed the restraint system. The shoulder belt anchor was on the upper rear edge of the door and the lap belt on the lower rear edge of the door frame at the sill. This formed a passive 3-point belt system.

There was an emergency release buckle at the inboard buckle that could release the shoulder and lap belts against the door. The system was often used as an active lap-shoulder belt configuration. Despite this, the field studies of belt use showed an increase in wearing rates with the door mounted system showing a positive safety benefit. The attachment of belts to the door was considered, and it was felt that the added load from occupant restraint would hold the door closed in crashes. However, if the door came opening a crash, there was no longer lap belt restraint and ejection could occur. The response was to install automatic door locks to further secure door latching.

Each of the passive belt systems introduced in production had an emergency release buckle that could allow the system to operate as a manual belt system. Nonetheless, there are many passive belt systems still in use today with the belts engaged in the passive mode indicating an acceptance of the restraint configuration if comfortable and convenient. There were several years of passive belt production until airbags emerged to fulfill the passive restraint requirement in FMVSS 208.

The regulation was formulated that compliance with the passive requirement was adequate, so once airbags were introduced, efforts to advance safety belt use were diminished with the lack of regulatory pressure. The US and international governments kept up the pressure to increase belt wearing rates and refinements in the performance of belt systems in combination with airbags.

PRINCIPLES OF BELT RESTRAINTS

The primary role of a lap belt is to restrain an occupant in the vehicle during rollover, side and other crashes where ejection may be a factor. Although much of the development work on restraints has involved frontal barrier and sled testing, the importance of the lap belt was recognized by crash investigations by DeHaven and others as preventing ejection, which was the leading cause of fatality in the 1950s. In rollover crashes, safety belt use virtually eliminates the risk of paralyzing cervical injury and ejection, and provides a 77% reduction in fatality risk with 63% of it due to the lap belt. Another early use of the lap belt was to complement crash protection of the driver's chest by a load-limiting steering system or airbag. In the absence of pelvic restraint, loads would be applied through the knees and seat pan to restrain the occupant and control the upright posture of the driver during restraint. The lap belt complemented early airbag designs for passenger protection by a padded dashboard and also found their way into rear seating positions to supplement padded seatbacks.

In the late 1960s, the combination of lap-shoulder belts gained rapid acceptance, because it provided occupant restraint by routing safety belt loads over the boney structures of the pelvis and shoulder. This took advantage of the relatively high tolerance to impact force for these regions of the skeleton and avoided concentrating load on the more complaint abdominal and thoracic regions. By adding the convenience of an inertial locking retractor to the lap-shoulder belt, Bohlin (1967) proved the system would be worn in the field and would be protective of injury in crashes. This set in motion a series of international efforts to require safety belt use by the motoring public, thus defining the lap-shoulder belt system as the preferred restraint configuration.

Among the first countries to adopt safety belt wearing requirements were Sweden and Australia, and this necessitated the development of a better scientific understanding of the principals of belt restraints. The fundamentals of a high quality belt restraint system involve kinematic controls on occupant movement in a crash. Adomeit (1975, 1977 and 1979) showed one goal was to maintain the lap belt low on the pelvis through adequate seat cushion support and use of an anti-submarining ramp. This minimized pelvic rotation and reduced the tendency for the lap-belt to slide off the ileum and directly load the abdomen (Figure 24a). A second goal was forward rotation of the upper torso to slightly greater than 90° upright posture. This directed a major portion of the upper torso restraint into the shoulder, which can tolerate higher restraining loads than the chest or abdomen.

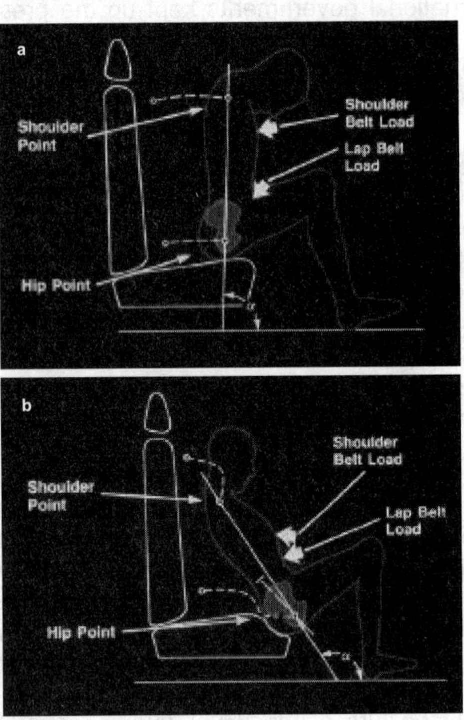

Figure 24: Showing Adomeit's biomechanical quality of belt restraint with a) high quality restraint and b) poor quality restraint kinematics.

In contrast, Figure 24b shows a poor biomechanical quality belt restraint. The pelvis has more forward motion because of its rearward rotation as the lap belt migrates up to the abdomen and a weak seat cushion allows the pelvis to drop. Both pelvic rotation and drop contribute to submarining, which involves lap belt loading of the abdomen and risk of hollow organ and spinal injury in the most severe crashes. The kinematics with pelvic lead also drop the shoulders and cause belt loading on the chest and abdomen, which increases loads on the ribcage and risks of rib fracture.

In a frontal impact, a snug-fitting lap-shoulder belt ties the occupant directly to the passenger compartment and allows the occupant to "ride-down" the crash as the vehicle front-end crushes. Coupling of the occupant to the vehicle and ride-down decelerates the occupant more gradually than possible with energy-absorbing interiors. Even with shoulder belt restraint, however, there is forward excursion of the torso and, in particular, of the head and neck (Figure 25). While this kinematic is more favorable than the consequence of an unrestrained occupant striking interior surfaces at high speed, there is still the potential for head/face contact with the vehicle interior and for inertial injuries of the neck. An increase in whiplash injuries was observed after the widespread use of lap-shoulder belts in the UK (Rutherford 1985). This pointed to the concentration of loads on the neck with belt restraint in the absence of head contact.

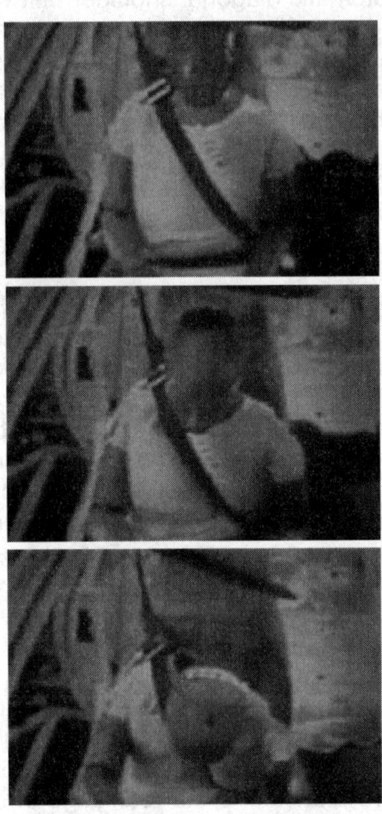

Figure 25: Kinematics of belt restraint for a passenger dummy in a frontal crash with forward lean of the upper torso, neck flexion and head excursion.

BELTS WITH AIRBAGS

In the late 1980s, airbags emerged in vehicles as an alternative to passive belts to gain the 1.5 credit incentive toward the passive restraint requirements and because they were the best solution for automatic protection. This started the inexorable introduction of inflatable restraints as part of the safety package of vehicles. It also altered the role of safety belts in the occupant restraint system.

Safety belt systems have been refined over the years of laboratory testing and field evaluations of belts and airbags. The belt systems of today are seen as providing the initial restraint in a severe frontal crash. With rapid pretensioning, the goal is to get restraining loads very early for occupant ridedown, as the crash progresses and the airbag inflates. Then, there is a handover of restraining function from the belts to the airbag. This requires a degressive shoulder belt tension so that high belt loads on the occupant do not occur with high airbag loads. This combination would increase loading on the occupant.

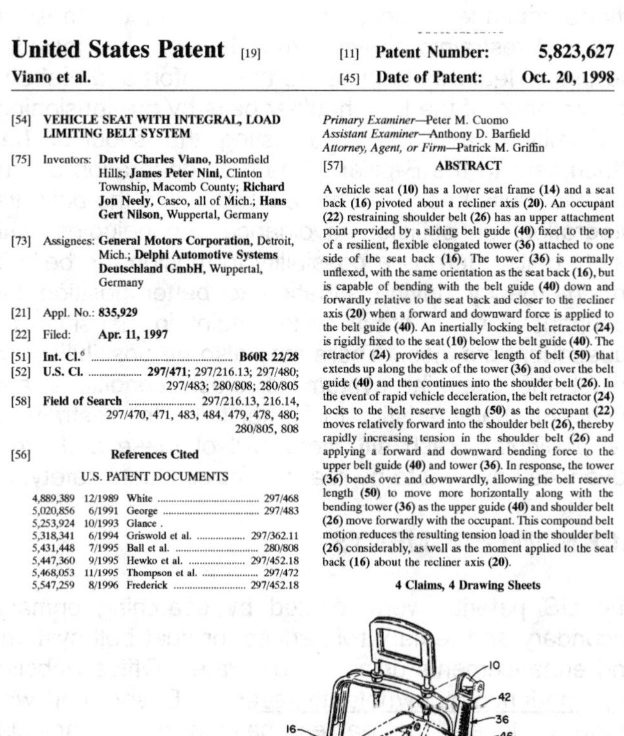

United States Patent [19]
Viano et al.

[11] **Patent Number: 5,823,627**
[45] **Date of Patent: Oct. 20, 1998**

[54] **VEHICLE SEAT WITH INTEGRAL, LOAD LIMITING BELT SYSTEM**

[75] Inventors: **David Charles Viano**, Bloomfield Hills; **James Peter Nini**, Clinton Township, Macomb County; **Richard Jon Neely**, Casco, all of Mich.; **Hans Gert Nilson**, Wuppertal, Germany

[73] Assignees: **General Motors Corporation**, Detroit, Mich.; **Delphi Automotive Systems Deutschland GmbH**, Wuppertal, Germany

[21] Appl. No.: **835,929**

[22] Filed: **Apr. 11, 1997**

[51] **Int. Cl.**[6] **B60R 22/28**
[52] **U.S. Cl.** **297/471**; 297/216.13; 297/480; 297/483; 280/808; 280/805
[58] **Field of Search** 297/216.13, 216.14, 297/470, 471, 483, 484, 479, 478, 480; 280/805, 808

[56] **References Cited**

U.S. PATENT DOCUMENTS

4,889,389	12/1989	White	297/468
5,020,856	6/1991	George	297/483
5,253,924	10/1993	Glance .	
5,318,341	6/1994	Griswold et al.	297/362.11
5,431,448	7/1995	Ball et al.	280/808
5,447,360	9/1995	Hewko et al.	297/452.18
5,468,053	11/1995	Thompson et al.	297/472
5,547,259	8/1996	Frederick	297/452.18

Primary Examiner—Peter M. Cuomo
Assistant Examiner—Anthony D. Barfield
Attorney, Agent, or Firm—Patrick M. Griffin

[57] **ABSTRACT**

A vehicle seat (**10**) has a lower seat frame (**14**) and a seat back (**16**) pivoted about a recliner axis (**20**). An occupant (**22**) restraining shoulder belt (**26**) has an upper attachment point provided by a sliding belt guide (**40**) fixed to the top of a resilient, flexible elongated tower (**36**) attached to one side of the seat back (**16**). The tower (**36**) is normally unflexed, with the same orientation as the seat back (**16**), but is capable of bending with the belt guide (**40**) down and forwardly relative to the seat back and closer to the recliner axis (**20**) when a forward and downward force is applied to the belt guide (**40**). An inertially locking belt retractor (**24**) is rigidly fixed to the seat (**10**) below the belt guide (**40**). The retractor (**24**) provides a reserve length of belt (**50**) that extends up along the back of the tower (**36**) and over the belt guide (**40**) and then continues into the shoulder belt (**26**). In the event of rapid vehicle deceleration, the belt retractor (**24**) locks to the belt reserve length (**50**) as the occupant (**22**) moves relatively forward into the shoulder belt (**26**), thereby rapidly increasing tension in the shoulder belt (**26**) and applying a forward and downward bending force to the upper belt guide (**40**) and tower (**36**). In response, the tower (**36**) bends over and downwardly, allowing the belt reserve length (**50**) to move more horizontally along with the bending tower (**36**) as the upper guide (**40**) and shoulder belt (**26**) move forwardly with the occupant. This compound belt motion reduces the resulting tension load in the shoulder belt (**26**) considerably, as well as the moment applied to the seat back (**16**) about the recliner axis (**20**).

4 Claims, 4 Drawing Sheets

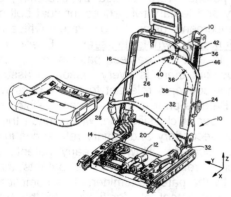

Figure 26: ABTS with load limiting tower and standard lap-shoulder belt and retractor (Viano et al. 1998).

The range of potential future belt restraint systems is varied and yet undefined in combination with airbags. In one approach, the lap-shoulder belt system is integrated in the seat, called ABTS (all belts to seat). This reflects the early concepts of the belt system envisioned by R. Griswold and H DeHaven in the 1950s. However, there are very high loads on the seat with the integrated shoulder belt, which has led to heavier and more costly ABTS seats than conventional designs. Even though, there remain reasons to consider ABTS. An example of an integration of restraint concepts is shown in Figure 26 where a flexible belt tower is used with a conventional belt retractor in the seat to give a load-limiting belt. There are obviously many other approaches for the future.

EFFECTIVENESS OF RESTRAINTS

From the early field effectiveness study of lap-shoulder belts by Bohlin (1967) until today, the performance of belt restraints has been widely explored. This has included the performance by direction and severity of crash, age and gender of occupants and other factors and circumstances.

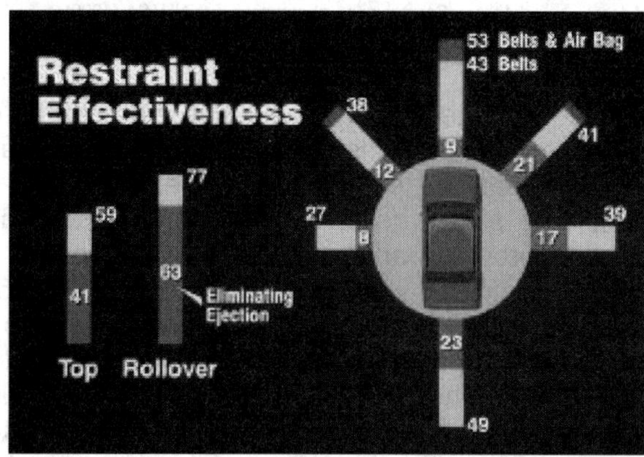

Figure 27: Effectiveness of occupant restraints in preventing fatalities.

The 1980s saw the introduction of statistical methods for epidemiological analyses of field accident data and the use of the double-pair comparison method to isolate the effectiveness of belt use from other confounding factors in automotive crashes. The work has been extended to investigate how rear lap belt use, airbags, alcohol use, occupant age, seating position, and direction of the crash affect fatality risks (Evans 1988, 1991). The fatality prevention effectiveness of front seat lap-shoulder belts was found to be (41 ± 4)% in preventing fatality for front-seat occupants, (42 ± 4)% for drivers and (39 ± 4)% for right-front passengers.

Safety belts provide at least two essential components of occupant protection. One is protection against ejection, which is primarily due to the lap belt, and the other is mitigation of interior impact, which is largely due to upper body restraint by the shoulder belt. Figure 27 shows the effectiveness of driver lap-shoulder belt restraint for different crash types and directions. In frontal crashes, belts are 43% effective in preventing fatality, 9% of the effectiveness comes by prevention of ejection from the lap belt. The remainder comes from the shoulder belt reducing upper body impacts.

The relative safety contribution with the lap belt in eliminating ejection depends on the type of crash. It is highest at 63% in rollovers where ejection risks are the greatest. Overall, safety belts are 77% in rollovers. Lap-shoulder belts are least effective in preventing fatalities in near-side impacts for the driver. Nonetheless, they are at 27%. Importantly, lap-shoulder belts are effective in all crash modes ranging from 27%-77% in preventing fatality.

Table 6: Belt Restraint Performance in the Field

Lap-Shoulder Belts	Belts + Airbag	Effectiveness Measure	Reference
31%		Fatality	Wilson, Savage (1973)
37%	57%	Driver fatality, frontals	NHTSA (1974)
32%		Fatality	Huelke et al. (1979)
40%-50%	45%-55%	Fatality	NHTSA (1984)
42% ± 4%	46%±4%	Fatality	Evans (1991)
45%	50%	Fatalities in all crashes	NHTSA (1999)
64%	64%	Serious injury, all crashes	NHTSA (1999)
39% ± 2%		Fatality	Cummings et al. (2003)

Zador and Ciccone (1991) evaluated crash data on airbag equipped vehicles and provided the early estimates of fatality and injury prevention. As a supplement to safety belts, the driver airbag and belt combination provides an overall 47% effectiveness in preventing fatalities. While this represents a 5% incremental improvement over the 42% effectiveness of belts alone, the supplemental airbag reduces the risk of fatality by 9%. The 47% level coincides with earlier projections of safety benefit. Fatalities of unbelted drivers are 18% with airbag restraint. This level is higher than earlier estimates, but within the expected confidence interval. In addition, injury claims data indicate substantially lower hospital admissions and seriousness of injuries. The airbag is providing a 24% lower hospitalization rate and a 25%-29% lower incidence of serious injuries for the current mix of safety belt use in field accidents.

In summary, the overall effectiveness of occupant restraints in preventing fatalities is 42% with lap-shoulder belts, 46% with the combination lap-shoulder belts and airbag. For the rear seat, the lap only belt is 18% effective and the combination lap-shoulder belt is 27% effective. Lap-shoulder belts are the most effective restraint in protecting occupants in severe crashes and the reductions in risk are even higher for serious injury than fatalities (Table 6).

Cummings et al. (2003) used 1986-98 FARS and a matched-pair analysis method to estimate the relative risk of death with safety belts. Belted occupants had a risk of 0.39 (95% CI 0.37–0.41) compared to 1.00 for unbelted occupants. The estimate is lower than previous, but it is within earlier confidence intervals. Even in states without seat belt laws from 1986-98, the relative risk was 0.45 (95% CI 0.39–0.52). The difference between this and earlier estimates is explained by misclassification of seatbelt use. Relative risk moves lower if misclassification of both the belted and unbelted decrease over time, or if the degree of misclassification remains constant as the prevalence of belt use increased.

The fact that a lap-shoulder belt and airbag combination is only 47% effective in preventing fatalities underscores that absolute protection is not achievable by occupant restraints. Injury and fatality will continue to occur to belt wearers even with an overall net safety gain as the use of lap-shoulder belt restraints increases and airbags are universally available. The limit reflects the severity of many fatal crashes which may involve extreme vehicle damage and forces on the passenger compartment, unusual crash configurations and causes of death, and unique situations associated with particular seating positions, crash dynamics and human tolerance limits.

Efforts continue to increase the overall effectiveness of occupant restraints above current levels. This includes the use of features to increase the comfort and restraint performance of the lap-shoulder belts by pre-tensioning, load-limiting, and height adjusting the shoulder belt attachment to the B-pillar. The next generation of lap-shoulder belts has yet to be realized. With the potential integration of collision avoidance technologies with safety belts, there is a possibility to tension the belts in an emergency driving situation to better position the occupant for a crash and maintain tension for subsequent crashes. There may also be possibilities to adjust the restraint system to crash conditions and occupant characteristics to further tailor the restraint to the particular crash situation. All of these and other advances have yet to be realized for occupant safety.

ACKNOWLEDGMENTS

The US patents were located by searching primary, secondary and tertiary references for seat belt systems and enhancements using the US Patent Office website: http://patft.uspto.gov/netahtml/search. Every effort was made to locate the earliest patents on a particular technology, but there is every reason to assume that many other concepts relevant to safety belts were not located. Patents prior to 1975 are accessible only by a patent number or classification, which made the search depend on secondary and tertiary references in patents that were located. However, early patents did not include cited references to earlier patents that were reviewed by the patent examiner. This complicated the search. Any patent of significance to the history of safety belts that is not included here was an inadvertent oversight, and the items located and reported here should be considered examples of the history of technology and patents on safety belts.

REFERENCES

Adomeit D, Heger Al. Motion Sequence Criteria and Design Proposals for Restraint Devices in Order to Avoid Unfavorable Biomechanic Conditions and Submarining. 19th Stapp Car Crash Conference, 139-166, SAE 751146, Society of Automotive Engineers, Warrendale, PA. 1975.

Adomeit D. Seat Design: A Significant Factor for Safety Belt Effectiveness. 23rd Stapp Car Crash Conference, 39-68, SAE 791004, Society of Automotive Engineers, Warrendale, PA, 1979.

Adomeit D. Evaluation Methods for the Biomechanical Quality of Restraint Systems During Frontal Impact. 21st Stapp Car Crash Conference, 911-932 SAE 770936, Society of Automotive Engineers, Warrendale, PA, 1977.

Andreasson R, Backstrom CG. The Seat Belt: Swedish Research and Development for Global Automotive Safety. Vattenfall Support AB, B. Lagerstrom, SE-16287 Stockholm, Sweden (fax 46-8-739-5627 or email: info@vattenfall.se).

Andreasson, Rune. Arguments against the seat belt and the facts. Paper No. 1992-15-0011, presented at the Proceedings of International Symposium on Road Traffic Accidents, Saudi Arabia, February 1992.

Bohlin NI. Studies of Three-Point Restraint Harness Systems in Full-Scale Barrier Crashes and Sled Runs. 8th Stapp Car Crash Conference, p:258-319, 1963.

Bohlin NI. Studies of Three-Point Restraint Harness Systems in Full-Scale Barrier Crashes and Sled Runs. 11th Stapp Car Crash Conference, p:455-478, 1967.

Crandall J, Kent R, Viano DC. The Biomechanics of Air Bags - Occupant Protection and Induced Injury. SAE Special Publication, Society of Automotive Engineers, Warrendale, PA, 2003.

Cummings P, Wells JD, Rivara FP. Estimating seat belt effectiveness using matched-pair cohort methods. Accident Analysis and Prevention 35 (2003) 143–149.

DeHaven H. War Medicine 2:586, 1942.

Evans L. The Effectiveness of Safety Belts in Preventing Fatalities. Accident Analysis and Prevention 18:229:241, 1986.

Evans L. Traffic Safety and the Driver. Van Nostrad Reinhold, January, 1991.

Foret-Bruno, J.Y., et.al. Thoracic Injury Risk in Frontal Car Crashes with Occupant Restrained with Belt Load Limiter. SAE 983166, 42nd Stapp Car Crash Conference, 1998.

Haddon W. Federal Safety Standards and Highway Safety Program. 11th Stapp Car Crash Conference, p:2-4, 1967.

Huelke D, Sherman H, Murphy M. (1979) Effectiveness of current and future restraint systems in fatal and serious injury automobile crashes, SAE 790323, Society of Automotive Engineers, Warrendale, PA.

Johannessen, George H. "Historical review of automatic seat belt restraint systems." Paper No. 870221, presented at the SAE International Congress and Exposition, Detroit, MI, February 1987.

Johannessen, H. George and Pulley, Charles H. Seat belt use laws: a solution to occupant fatality and injury reduction. SAE 750189, presented at the Automotive Engineering Congress and Exposition, Detroit, MI, February 1975.

Johannessen, H. George and Vos, Thomas H. The changing shape of seat belt systems. SAE 820796, presented at the SAE Passenger Car Meeting and Exposition, Troy, MI, June 1982.

Johannessen, H. George. Historical perspective on seat belt restraint systems. Paper No. 840392, presented at the SAE International Congress and Exposition, Detroit, MI, February 1984.

Kent R, Crandall J, Viano DC. Field Performance of Airbags. SAE Special Publication, Society of Automotive Engineers, Warrendale, PA, 2003.

Neff RJ. Federal Seat-Belt Regulation: A Progress Report. 8th Stapp Car Crash Conference, p:72-78, 1964.

Neff RJ. Webbing for Use in Automotive Seat Belts. 6th Stapp Car Crash Conference p:134-137, 1962.

NHTSA (1974) Analysis of effects of proposed changes to passenger car requirements of MVSS 208. National Highway Traffic Safety Administration, U.S. Department of Transportation, Washington DC.

NHTSA (1984) FMVSS 208 regulatory impact analysis. National Highway Traffic Safety Administration, U.S. Department of Transportation, Washington DC.

NHTSA (1999) Fourth report to Congress – effectiveness of occupant protection systems and their use. National Highway Traffic Safety Administration, U.S. Department of Transportation, Washington DC.

Ruff S, Strughold B. German Aviation Medicine. World War II. US Government Printing Office 1950.

SAE J4a. Motor Vehicle Seat Belt Assemblies. Society of Automotive Engineers, Warrendale, PA, 1955.

SAE J4b. Motor Vehicle Seat Belt Assemblies. Society of Automotive Engineers, Warrendale, PA, 1964.

Shinar D, Rotenberg E, Cohen T. Crash Reduction with an Advanced Brake Warning System: A Digital Simulation. Human Factors 39(2): 296-302, June 1997.

Shinar D. Field Evaluation of an Advanced Brake Warning System. Human Factors 37(4): 746-751, December 1995.

States JD, Benedict JF. Safe and Unsafe Torso Restraints for Occupant Protection in Motor Vehicles, 7th Stapp Car Crash Conference, p:312-323, 1963.

Struble D. Airbag Technology: What It Is and How It Came to Be. SAE 980648, Society of Automotive Engineers, Warrendale, PA, 1998.

Tourin B, Aldman B. Restraining Characteristics of Harnesses. 6th Stapp Car Crash Conference, p:138-148, 1962.

Turbell T, Larsson R. Optimizing seat belt usage by interlock systems. SAE 976007, presented at the Fifteenth International Technical Conference on the Enhanced Safety of Vehicles, Melbourne, Australia, May 1997.

Viano DC. Cause and Control of Automotive Trauma. Bulletin of the New York Academy of Medicine, Second Series 64(5):376-421, 1988.

Viano DC. Limits and Challenges of Crash Protection. Accident Analysis and Prevention 20(6):421-429, 1988.

Viano, D.C. Effectiveness of Safety Belts and Airbags in Preventing Fatal Injury. SAE 910901, Society of Automotive Engineers, 1991.

Wilson R, Savage C. Restraint system effectiveness – a study of fatal accidents. Proceedings of Automotive Safety Engineering Seminar, General Motors, Warren, MI, 1973.

Zador P, Ciccone M, Driver Fatalities in Frontal Impacts: Comparisons Between Cars with Airbags and Manual Belts. Insurance Institute for Highway Safety, Arlington VA, October 1991.

BIBLIOGRAPHY

SAFETY BELT SYSTEMS AND PERFORMANCE

BACKGROUND STUDIES

Horsch J.D., "Evaluation of Occupant Protection from Responses Measured in Laboratory Tests." 870222, SAE Congress, 1987.

Johannessen, H. George and Pulley, Charles H. "Seat belt use laws: a solution to occupant fatality and injury reduction." Paper No. 750189, presented at the Automotive Engineering Congress and Exposition, Detroit, MI, February 1975.

Johannessen, H. George and Vos, Thomas H. "The changing shape of seat belt systems." Paper No. 820796, presented at the SAE Passenger Car Meeting and Exposition, Troy, MI, June 1982.

Johannessen, HG. "Historical perspective on seat belt restraint systems." Paper No. 840392, presented at the SAE International Congress and Exposition, Detroit, MI, February 1984.

Kratzke, SR. "Regulatory History of Automatic Crash Protection in FMVSS 208." Paper No. 950865, presented at the SAE Congress and Exposition, Detroit, MI 1995.

Mackay M, Hassan AM. Adaptive restraints - Their characteristics and benefits. SAE 974147, presented at the Automotive Environmental Impact and Safety Autotech Congress '97, Birmingham, UK, November 1997.

Norin, Hans, Carlsson, Gerd, and Korner, Johnny. "Seat belt usage in Sweden and its injury reducing effect." Paper No. 840194, presented at the SAE International Congress and Exposition, Detroit, MI, February 1984.

Searle, John. "Optimum occupant restraints." Paper No. 700422, presented at the International Automobile Safety Conference, Detroit, MI, May 1970.

Takada, Juichiro. "Development of energy-absorbing safety belt webbing." Paper No. 740581, presented at the 3rd International Conference on Occupant Protection, Troy, MI, July 1974.

Williams, Allan F. and O'Neill Brian. " Seat Belt Laws: Implications for Occupant Protection." Paper No. 790683, presented at the Passenger Car Meeting, Dearborn, MI 1979.

LAP-SHOULDER BELTS AND EVALUATION

Adomeit, D. and Appel, H. "Influence of seat design on the kinematics and loadings of the belted dummy." Paper No. 1979-13-0025, presented at the International IRCOBI Conference on the Biomechanics of Impacts, Göteborg, Sweden, September 1979.

Adomeit, D., Goegler, H., and Vu Han, V. "Expected belt-specific injury patterns dependent on the angle of impact." Paper No. 1977-13-0018, presented at the 3rd International Conference on Impact Trauma, Berlin, Germany, September 1977.

Adomeit, Dieter. "Seat design~a significant factor for safety belt effectiveness." Paper No. 791004, presented at the 23rd Stapp Car Crash Conference, San Diego, CA , October 1979.

Adomeit, Dieter. "Evaluation Methods for the Biomechanical Quality of Restraint Systems During Frontal Impact. 770936, 21st Stapp Car Crash Conference, 1977.

Andreasson, Rune. "Arguments against the seat belt and the facts." Paper No. 1992-15-0011, presented at the Proceedings of International Symposium on Road Traffic Accidents, Saudi Arabia, February 1992.

Appleby, M.R. and Bintz, L.J. "Increased seat belt use as a result of improved seat belt systems." Paper No. 740048, presented at the Automotive Engineering Congress and Exposition, Detroit, MI, February 1974.

Appleby, M.R. and Bintz, L.J. "Techniques to determine occupant restraint usage and the effect of improved restraint systems on usage." Paper No. 1972-12-0027, presented at the annual meeting of the American Association for Automotive Medicine, Chapel Hill, NC, October 1972.

Araszewski, Michael, Roenitz, Eric, and Toor, Amrit. "Maximum head displacement of vehicle occupants restrained by lap and torso seat belts in frontal impacts." Paper No. 1999-01-0443, presented at the SAE International Congress and Exposition, Detroit, MI, March 1999.

Araszewski, Michael, Toor, Amrit, and Happer, Andrew. "Knee and hip displacements of vehicle occupants restrained by seat belts in frontal impacts." Paper No. 2001-01-0180, presented at the 2001 World Congress, Detroit, MI, March 2001.

Ariyoshi, Tomohiko and Sogo, Toshiharu. "Development of a seat belt anchorage deformation analysis model~Application to a rear floor anchorage." Paper No. 918027, presented at the JSAE Spring Conference, Urayasu, Japan, May 1991.

Ariyoshi, Tomohiko, Sogo, Toshiharu, and Fukusima, Naomi. "Development of a seat belt anchorage deformation analysis model~Large plastic deformation analysis with a dynamic analysis method." Paper No. 918176, presented at the JSAE Autumn Conference, Hokkaido, Japan, October 1991.

Arndt, Mark W. "Testing of seats and seat belts for rollover protection systems in motor vehicles." Paper No. 982295, presented at the International Body Engineering Conference and Exposition, Detroit, MI, September 1998.

Arndt, Mark W., et.al. "The development of a method for determining effective slack in motor vehicle restraint systems for rollover protection." Paper No. 970781, presented at the SAE International Congress and Exposition, Detroit, MI, February 1997.

Backaitis S.H., Hicks, M.E., et al. "Variability of Hybrid III Clearance Dimensions within the FMVSS 208 and NCAP Vehicle Test Fleets and the Effects of Clearance Dimensions on Dummy Impact Responses. 952710, Society of Automotive Engineers, 1995.

Backaitis, Stanley H. and St. Laurent, A. "Chest deflection characteristics of volunteers and Hybrid III dummies." Paper No. 861884, presented at the 30th Stapp Car Crash Conference, San Diego, CA, October 1986.

Bacon, D.G.C. "The effect of restraint design and seat position on the crash trajectory of the Hybrid III dummy." Paper No. 896052, presented at the Twelfth International Technical Conference on Experimental Safety Vehicles, Göteborg, Sweden, 1989.

Baudrit, Pascal, et.al. "Comparative studies of dummy and human body models behavior in frontal and lateral impact conditions. " Paper No. 99SC05, presented at the 43rd Stapp Car Crash Conference, San Diego, CA, October 1999.

Begeman, P.C., et.al. "Biodynamic response of the musculoskeletal system to impact acceleration." Paper No. 801312, presented at the 24th Stapp Car Crash Conference, Troy, MI, October 1980.

Bendjella, F., et.al. "Measurement of seat-belt loads sustained by the dummy in frontal impact with critical pelvic motion~Development of a 2-D transducer." Paper No. 896080, presented at the Twelfth International Technical Conference on Experimental Safety Vehicles, Göteborg, Sweden, 1989.

Biss, D.J. and Pompa, J.A. "A kinematic and dynamic analysis of occupant responses to lap belt only restraint forces." Paper No. 1987-12-0021, presented at the annual meeting of the American Assocation for Automotive Medicine, New Orleans, LA, September 1987.

Cesari, Dominique and Bouquet, Robert. "Behaviour of human surrogates thorax under belt loading." Paper No. 902310, presented at the 34th Stapp Car Crash Conference, Orlando, FL, November 1990.

Chabert, Laurent, et.al. "Anatomical study and three-dimensional reconstruction of the belted human body in seated position." Paper No. 976155, presented at the Fifteenth International Technical Conference on the Enhanced Safety of Vehicles, Melbourne, Australia, May 1996.

Chillon, Claude, et.al. "Comparison between the three-point belt and the air cushion evaluation and discussion of their cost-efficiency ratio." Paper No. 876072, presented at the Eleventh International Technical Conference on Experimental Safety Vehicles, Washington, DC, 1987.

Chuo, Young-Woo, et.al. "Strength of a passenger car body with relocated seat belt anchorage points." Paper No. 958478, presented at The Eighth International Pacific Conference on Automotive Engineering, Harmony and Efficiency~The Technical Challenge, Yokohama, Japan, November 1995.

Cichowski, William G. and Silver, Jeffrey N. "Effective use of restraint systems in passenger cars." Paper No. 680032, presented at the Automotive Engineering Congress and Exposition, Detroit, MI, January 1968.

Cooper E.D., Croteau J.L. Parenteau C., et al. "Head Excursion of Seat Belted Cadavers, Volunteers and Hybrid III ATD in a Dynamic/Static Rollover Fixture." 973347 Society of Automotive Engineers, 1997.

Cotte, J.P. "Semi static loading of baboon torsos." Paper No. 1977-13-0013, presented at the 3rd International Conference on Impact Trauma, Berlin, Germany, September 1977.

Crandall, Jeff R., et.al. "A comparison of two and three point belt restraint systems." Paper No. 1994-12-0049, presented at the Advances in Occupant Restraint Technologies AAAM/IRCOBI, Lyon, France, September 1994.

Cromack, J.R. and Mason, R.L. "Restraint system use and misuse." Paper No. 1976-12-0031, presented at the annual meeting of the American Association for Automotive Medicine, Atlanta, GA, November 1976.

Culver C.C., Viano D.C. "Influence of Crush Orientation on the knee Bolster Function in Barrier Crash Simulations. 800852, Passenger Car Meeting, Dearborn, 1980.

Dalmotas, D. and Krzyzewski, J. "Restraint system effectiveness as a function of seating position." Paper No. 870489, presented at the SAE International Congress and Exposition, Detroit, MI, February 1987.

Dalmotas, D.J. and Welbourne, E.R. "Improving the protection of restrained front seat occupants in frontal crashes." Paper No. 916133, presented at the 13th International Technical Conference on Experimental Safety Vehicles, Paris, France, November 1991.

Dalmotas, D.J., Hurley, R.M., and German, A. "Supplemental restraint systems: friend or foe to belted occupants?" Paper No. 1996-12-0005, presented at the 40th Annual Proceedings, British Columbia, Canada, October 1996.

Dance, Murray and Enserink, Bert. "Safety performance evaluation of seat belt retractors." Paper No. 790680, presented at the SAE Passenger Car Meeting and Exposition, Dearborn, MI, June 1979.

Dejeammes, M., et.al. "Restraint system study with a living subject: methodology and results on three point belt restraints." Paper No. 1978-13-0021, paper presented at the 3rd International Meeting on the Simulation and Reconstruction of Impacts in Collosion, Lyon, France, September 1978.

DeJeammes, M., et.al. "Restraint systems comparison in frontal crashes using a living animal." Paper No. 800297, presented at the Automotive Engineering Congress and Exposition, Detroit, MI, February 1980.

Deng, Yih Charng. "Analytical study of the interaction between the seat belt and a Hybrid III dummy in sled tests." Paper No. 880648, presented at the SAE International Congress and Exposition, Detroit, MI, February 1988.

Deng, Yih-Charng. "How airbags and seat belts work together in frontal crashes." Paper No. 952702, presented at the 39th Stapp Car Crash Conference, San Diego, CA, November 1995.

DeRosa, D. and Larsonneur, J.F. "Seat belt improvements." Paper No. 840400, presented at the SAE International Congress and Exposition, Detroit, MI, February 1984.

DeSantis Klinich, Kathleen, et.al. "Challenges in frontal crash protection of pregnant drivers based on anthropometric considerations." Paper No. 1999-01-0711, presented at the SAE International Congress and Exposition, Detroit, MI, March 1999.

Deutsch, D., Sameth, S., and Akinyemi, J. "Seat belt usage and risk-taking behavior at two major traffic intersections." Paper No. 1980-12-0036, presented at the annual meeting of the American Association for Automotive Medicine, Rochester, NY, October 1980.

Dischinger, P., Cushing, B., and Kerns, T. "Lower extremity fractures in motor vehicle collisions~Influence of direction of impact and seatbelt use." Paper No. 1992-12-0020, presented at the annual meeting of the Association for the Advancement of Automotive Medicine, Portland, OR, October 1992.

Ensslen, Arnold, Schwant, Wilfrid, and Zieglschmidt, Reiner. "Can we improve the crash performance of seat belts?" Paper No. 851202, presented at the SAE Government Industry Meeting and Exposition, Washington, DC, May 1985.

Estep, Christina R. and Lund, Adrian K. "Dummy kinematics in offset-frontal crash tests." Paper No. 976042, presented at the Fifteenth International Technical Conference on the Enhanced Safety of Vehicles, Melbourne, Australia, May 1996.

Faerber, E. and Glaeser, K.P. "Results of crash tests according to the ECE/GRCS regulation proposal concerning 50km/h frontal impact against the rigid 30°barrier." Paper No. 856059, presented at the Tenth International Technical Conference on Experimental Safety Vehicles, Oxford, England, 1985.

Fayon, A. "Thorax of 3-point belt wearers during a crash (experiments with cadavers)." Paper No. 751148, presented at the 19th Stapp Car Crash Conference, San Diego, CA, November 1975.

Fulton, J.L. "Performance of seat belts and shoulder harnesses in high performance vehicles used in hazardous police traffic assignments." Paper No. 660536, presented at the National West Coast Meeting, Los Angeles, CA, August 1966.

Galer, I.A.R. "The acceptability of car seat belts." Paper No. 770186, presented at the International Automotive Engineering Congress and Exposition, Detroit, MI, February 1977.

Gallup, B.M., et.al. "The development of two prototype seat belt systems for improved lap belt fit." Paper No. 1984-12-0031, presented at the annual meeting of the American Assocation for Automotive Medicine, Denver, CO, October 1984.

Green, R.N. and Sharpe, G.S. "Seat belt legislation." Paper No. 1979-12-0002, presented at the annual meeting of the American Association for Automotive Medicine, Louisville, KY, October 1979.

Grosch, Lothar, Katz, Egon, and Kassing, Lothar. "Chest compression response of Hybrid III with combined restraint." Paper No. 876043, presented at the Eleventh International Technical Conference on Experimental Safety Vehicles, Washington, DC, 1987.

Haley, J.L., Jr. "Effect of rapid loading rates on the stress-strain properties of restraint webbing." Paper No. 660798, presented at the 10th Stapp Car Crash Conference, Holloman Air Force Base, New Mexico, November 1966.

Hartemann, F., et.al. "Ten years of safety due to the three-point seat belt." Paper No. 840193, presented at the SAE International Congress and Exposition, Detroit, MI, February 1984.

Haworth, Narelle L., et.al. "Truck seat belts." Paper No. 976186, presented at the Fifteenth International Technical Conference on the Enhanced Safety of Vehicles, Melbourne, Australia, May 1996.

Herbert, David C., Stott, John D., and Corben, Christopher W. "Head space requirements for seat belt wearers." Paper No. 751164, presented at the 19th Stapp Car Crash Conference, San Diego, CA, November 1975.

Herbst, Brian, et.al. "The ability of 3-point safety belts to restrain occupants in rollover crashes." Paper No. 976081, presented at the Fifteenth International Technical Conference on the Enhanced Safety of Vehicles, Melbourne, Australia, May 1996.

Hill, J., Mackay, G.M., and Henderson, S. "Seat belt limitations in collisions that involve no compromise of the passenger compartment." Paper No. 970118, presented at the SAE International Congress and Exposition, Detroit, MI, February 1997.

Hilton, B.C. "Development of safety seating for injury prevention." Paper No. 660797, presented at the 10th Stapp Car Crash Conference, Holloman Air Force Base, New Mexico, November 1966.

Hontschik, H. and Rüter, G. "Investigations into the efficacy of three-point seat-belts in oblique impact experiments." Paper No. 1980-13-0011, presented at the International IRCOBI Conference on the Biomechanics of Impacts, Birmingham, England, September 1980.

Horsch J.D., Melvin J.W., Viano D.C., Mertz, H.J., "Thoracic Injury Assessment of Belt Restraint Systems Based on Hybrid III Chest Compression." 912895, Society f Automotive Engineers, 1991.

Horsch, John D. "Occupant dynamics as a function of impact angle and belt restraint." Paper No. 801310, presented at the 24th Stapp Car Crash Conference, Troy, MI, October 1980.

Horsch, John D., et.al. "Response of belt restrained subjects in simulated lateral impact." Paper No. 791005, presented at the 23rd Stapp Car Crash Conference, San Diego, CA , October 1979.

Horsch, John H., et.al. "Thoracic injury assessment of belt restraint systems based on hybrid III chest compression." Paper No. 912895, presented at the 35th Stapp Car Crash Conference, San Diego, CA, November 1991.

Johannessen, H. George. "Historical perspective on seat belt restraint systems." Paper No. 840392, presented at the SAE International Congress and Exposition, Detroit, MI, February 1984.

Jonah, Brain A., et.al. "Promoting seat belt use~A comparison of three approaches." Paper No. 1982-12-0010, presented at the annual meeting of the American Assocation for Automotive Medicine, Ottawa, Ontario, October 1982.

Joubert, P.N. "Compulsory wearing of seat belts." Paper No. 725034, presented at the 14th FISITA Congress, London, England, 1972.

Kallieris D., Rizzetti A., Mattern R., et al. "On the Synergism of the Driver Air Bag and the 3-Point Belt in Frontal Collisions." 952700, Society of Automotive Engineers, 1995,

Kallieris, Dimitrios, et.al. "Comparison between frontal impact tests with cadavers and dummies in a simulated true car restrained environment." Paper No. 821170, presented at the 26th Stapp Car Crash Conference, Ann Arbor, MI, October 1982.

Katoh, Hiroshi and Nakahama, Ryoji. "A study on the ride-down evaluation." Paper No. 826026, presented at the Ninth International Technical Conference on Experimental Safety Vehicles, Kyoto, Japan, 1982.

Katz E., Grosch L., Kassing, L. "Chest Compression Response of a Modified Hybrid III with Different Restraint Systems." 8722215, Society of Automotive Engineers, 1987.

Keenan, Lori, et.al. "On the synergism of the driver air bag and the 3-point belt in frontal collisions." Paper No. 952700, presented at the 39th Stapp Car Crash Conference, San Diego, CA, November 1995.

Kent, Richard, et.al. "Restrained Hybrid III dummy-based criteria for thoracic hard-tissue injury prediction." Paper No. 2001-13-0017, presented at the International IRCOBI Conference on the Biomechanics of Impact, Isle of Man, United Kingdom, October 2001.

Kidd, Edwin A. and Walsh, Michael J. "Effectiveness of ATDs and cadavers in evaluation of restraint system." Paper No. 766037, presented at the Sixth International Technical Conference on Experimental Safety Vehicles, Washington, DC, 1976.

Knapp, R.A. "Data acquisition results from Los Angeles traffic tests." Paper No. 850056, presented at the SAE International Congress and Exposition, Detroit, MI, February 1985.

Kuechenmeister , Timothy J., Morrison, Andrew J., and Cohen, Mitchell E. "A comparative analysis of factors impacting on seat belt use." Paper No. 790687, presented at the SAE Passenger Car Meeting and Exposition, Dearborn, MI, June 1979.

L'Abbé, R.J., Dainty, D.A., and Newman, J.A.. "An experimental analysis of thoracic deflection response to belt loading." Paper No. 1982-13-0015, presented at the International IRCOBI Conference on the Biomechanics of Impacts, Cologne, West Germany, September 1982.

Langwieder, Klaus, et.al. "Comparison of passenger injuries in frontal car collisions with Dummy Loadings in equivalent simulations." Paper No. 791009, presented at the 23rd Stapp Car Crash Conference, San Diego, CA, October 1979.

Lee, H.G. "Study of the characteristics of automobile collision and seat belts." Paper No. 943845, presented at the KSAE Spring Conference, Korea, June 1994.

Lowne, R.W. and Wall, J.G. "A procedure for estimating injury tolerance levels for car occupants." Paper No. 760820, presented at the 20th Stapp Car Crash Conference, Dearborn, MI, October 1976.

Mackay, G.M., et.al. "Restrained occupants on the non-struck side in lateral collisions." Paper No. 1991-12-0009, presented at the annual meeting of the Association for the Advancement of Automotive Medicine, Ontario, Canada, October 1991.

Mackay, G.M., et.al. "Restrained occupants on the non-struck side in lateral collisions." Paper No. 916092, presented at the 13th International Technical Conference on Experimental Safety Vehicles, Paris, France, November 1991.

Mackay, M., Hassan, A.M., and Hill, J.R. "Observational studies of car occupants' positions." Paper No. 986141, presented at the 16th International Technical Conference on the Enhanced Safety of Vehicles, Ontario, Canada, May 1998.

Mackay, S.M., Hill, J., and Parkin, S. "Restrained occupants on the non-struck side in lateral collisions." Paper No. 925218, presented at the 24th FISITA Congress, London, England, June 1992.

McElhaney, James H., et.al. "Biomechanics of seat belt design." Paper No. 720972, presented at the 16th Stapp Car Crash Conference, Detroit, MI, November 1972.

McHenry, Ray. "Analysis of the dynamics of automobile passenger-restraint systems." Paper No. 1963-12-0017, presented at the Seventh Stapp Car Crash Conference, Los Angeles, CA, November 1963.

Moffatt, Edward A., et.al. "Head excursion of seat-belted cadaver, volunteers and Hybrid III ATD in a dynamic/static rollover fixture." Paper No. 973347, presented at the 41st Annual Stapp Car Crash Conference, Lake Buena Vista, FL, November 1997.

Morgan, L.E. "Current developments in restraint systems." Paper No. 730291, presented at the National Business Aircraft Meeting and Exposition, Wichita, KS, April 1973.

Morita, Satoshi, Fukuda, Shigeo, and Iwata, Toru. "A study of the body strength on seat belt systems." Paper No. 796028, presented at the Seventh International Technical Conference on Experimental Safety Vehicles, Paris, France, 1979.

Mortimer, R.G. "Use of seat belts during and after heightened enforcement of the seat belt law." Paper No. 1991-12-0006, presented at the annual meeting of the Association for the Advancement of Automotive Medicine, Ontario, Canada, October 1991.

Mutoh, Miki and Nishida, Yasushi. "Study on relationship between driver and front seat passenger seat belt usage." Paper No. 988124, presented at the JSAE Spring Conference, Tokyo, Japan, May 1998.

Neff, Russell J. "Webbings for use in automobile seat belts." Paper No. 1962-12-0016, presented at the Sixth Stapp Car Crash Conference, New Mexico, November 1962.

Nichol, E. "Seat belt systems for the future." Paper No. 726026, presented at the Third International Technical Conference on Experimental Safety Vehicles, Washington, DC, 1972.

Niederer, P, Walz, F., and Zollinger, U. "Adverse effects of seat belts and causes of belt failures in severe car accidents in Switzerland during 1976." Paper No. 770916, presented at the 21st Stapp Car Crash Conference, New Orleans, LA, October 1977.

Nielsen, H.V., Nordentoft, E., and Weeth, R. "Different behavioral patterns of seat belt users and non-users." Paper No. 1978-13-0001, presented at the 3rd International Meeting on the Simulation and Reconstruction of Impacts in Collosion, Lyon, France, September 1978.

Nilson, Gert, et.al. "Rear-end collisions~The effect of the seat belt and the crash pulse on occupant motion." Paper No. 946175, presented at the 14th International Technical Conference on the Enhanced Safety of Vehicles, Munich, Germany, May 1994.

O'Riordan, Shan, et.al. "A comparison of hybrid III and cadaver thorax response under diagonal belt loading." Paper No. 916112, presented at the 13th International Technical Conference on Experimental Safety Vehicles, Paris, France, November 1991.

Otte D., "Comparison and Realism of Crash Simulation Tests and Real Accident Situations for the Biomechanical Movements in Car Collisions." 902329. Society of Automotive Engineers, 1990.

Park B.T., Morgan R.M., Hackney, J.R., et al. Upper Neck Response of the Belt and Air Bag Restrained 50th Percentile Hybrid III Dummy in the USA's New Car Assessment Program.." 983184, 42nd Stapp Car Crash Conference. 1998.

Preston, F.L. "A comparison of contacts for unrestrained and lap belted occupants in automobile accidents." Paper No. 1973-12-0008, presented at the annual meeting of the American Association for Automotive Medicine, Oklahoma City, OK, November 1973.

Preston, Mark and Sparke, Laurie. "Race car safety development." Paper No. 976067, presented at the the Fifteenth International Technical Conference on the Enhanced Safety of Vehicles, Melbourne, Australia, May 1996.

Pywell, James F., et.al. "Characterization of belt restraint systems in quasistatic vehicle rollover tests." Paper No. 973334, presented at the 41st Annual Stapp Car Crash Conference, Lake Buena Vista, FL, November 1997.

Rangarajan, N., et.al. "Response of THOR in frontal sled testing in different restraint conditions." Paper No. 1998-13-0036, presented at the International IRCOBI Conference on the Biomechanics of Impact, Göteborg, Sweden, September 1998.

Rattenbury, S.J., et.al. "The biomechanical limits of seat belt protection." Paper No. 1979-12-0014, presented at the American Association for Automotive Medicine, Louisville, KY, October 1979.

Rouhana, Stephen W., Horsch, John D., and Kroell, Charles K. "Assessment of lap-shoulder belt restraint performance in laboratory testing." Paper No. 892439, presented at the 33rd Stapp Car Crash Conference, Washington, DC, October 1989.

Ryan, James J. and Tobias, James R. "Reduction in crash forces." Paper No. 650974, presented at the Ninth Stapp Car Crash Conference, Minneapolis, MN, October 1965.

Saki, Hideo, et.al. "A study of car and driver in the very small commuter car made of FRP in collision~The influence of characteristic of seat belts." Paper No. 938108, presented at the JSAE Spring Conference, Kanagawa, Japan, May 1993.

Sano, S. "Safety belt development." Paper No. 856020, presented at the Tenth International Technical Conference on Experimental Safety Vehicles, Oxford, England, 1985.

Schmidt, Georg, et.al. "Results of 49 cadaver tests simulating frontal collision of front seat passengers." Paper No. 741182, presented at the 18th Stapp Car Crash Conference, Ann Arbor, MI, November 1974.

Schubert, Hiltmar and Ziegahn, Karl-Friedrich. "Technological trends in occupant protection systems~Recent research challenges from the German point of view." Paper No. 960663, presented at the SAE International Congress and Exposition, Detroit, MI, February 1996.

Seiffert, Ulrich W., Carbon , Erich, and Ristau, Helmut. "Three-point energy-absorbing seat belt system with combined vehicle- and webbing-sensitive emergency retractor." Paper No. 720434, presented at the National Automobile Engineering Meeting, Detroit, MI, May 1972.

Serizawa, Yoshio. "Evaluation of seat belt system and dummy characteristics." Paper No. 746042, presented at the Fifth International Technical Conference on Experimental Safety Vehicles, London, England, 1974.

Sharp, Jonathan E. and Stapp, John Paul. "Automobile seats and their role as an adjunct to restraint systems." Paper No. 920129, presented at the SAE International Congress and Exposition, Detroit, MI, February 1992.

Shaw, G., Crandall, J., and Butcher, J. "Biofidelity evaluation of the THOR advanced frontal crash test dummy." Paper No. 2000-13-0002, presented at the 2000 International IRCOBI Conference on The Biomechanics of Impact, Montpellier, France, September 2000.

Shi Qinzhong. "Optimization for the behavior of vehicle cabin and safety belt using most probable optimal design method." Paper No. 1999-08-0047, presented at the JSAE Spring Conference, Yokohama, Japan, May 1999.

Shimoda, Nobuyoshi, Nishida, Yousuke, and Akiyama, Akihiko. "Effects of belt restraint systems on occupant protection performance in side impact crashes." Paper No. 896116, presented at the Twelfth International Technical Conference on Experimental Safety Vehicles, Göteborg, Sweden, 1989.

Sparke, L.J. "Restraint system optimization for minimum societal harm." Paper No. 986012, presented at the 16th International Technical Conference on the Enhanced Safety of Vehicles, Ontario, Canada, May 1998.

Stapp, John P. "Medical aspects of safety seat belt development." Paper No. 1962-12-0018, presented at the Sixth Stapp Car Crash Conference, New Mexico, November 1962.

States, J.D. "Safety belts and seat design - an insight from racing." Paper No. 1980-12-0038, presented at the annual meeting of the American Association for Automotive Medicine, Rochester, NY, October 1980.

Stolinski, Richard, et.al. "Response of far-side occupants in car-to-car impacts with standard and modified restraint systems using Hybrid III and US- SID." Paper No. 1999-01-1321, presented at the SAE International Congress and Exposition, Detroit, MI, March 1999.

Strien, W. "A new driver-seat system for motor vehicles." Paper No. 805105, presented at the 18th FISITA Congress, Hamburg, West Germany, 1980.

Strother, Charles, Fitzpatrick, Michael U., and Egbert, Timothy P. "Development of Advanced Restraint Systems for Minicars RSV." Paper No. 766043, presented at the Sixth International Technical Conference on Experimental Safety Vehicles, Washington, DC, 1976.

Svensson, Lars G. "Means for effective improvement of the three-point seat belt in frontal crashes." Paper No. 780898, presented at the 22nd Stapp Car Crash Conference, Ann Arbor, MI, October 1978.

Tarriere, C. "Analysis of interrelation of vehicle to seat belt as a function of rigidity of the vehicle." Paper No. 680777, presented at the 12th Stapp Car Crash Conference, October 1968.

Tarriere, C. "Priorities for a better quality of the occupants protection in real accidents." Paper No. 92A113, presented at the 3rd International Conference Innovation and Reliability in Automotive Design and Testing, Firenze, Italy, April 1992.

Tarriere, C., et.al. "Optimum utilization of the vehicle available occupant space to ensure passenger protection." Paper No. 700360, presented at the International Automobile Safety Conference, Detroit, MI, May 1970.

Thomas, C., et.al. "The main causes of risk differences for restrained drivers and their restrained front passenger in frontal collisions." Paper No. 826078, presented at the Ninth International Technical Conference on Experimental Safety Vehicles, Kyoto, Japan, 1982.

Trinca, Gordon W. "Thirteen years of seat belt usage-how great the benefits." Paper No. 840192, presented at the SAE International Congress and Exposition, Detroit, MI, February 1984.

Verriest, J.P. and Chapon, A. "Multiple cinephotogrammetric determination of thorax deformation under belt load." Paper No. 1979-13-0017, presented at the International IRCOBI Conference on the Biomechanics of Impacts, Göteborg, Sweden, September 1979.

Verriest, J.P., Chapon, A., and Trauchessec, R. "Cinephotogrammetrical study of porcine thoracic response to belt applied load in frontal impact~comparison between living and dead subjects." Paper No. 811015, presented at the 25th Stapp Car Crash Conference, San Francisco, CA, September 1981.

Viano, David C. "Restraint of a belted or unbelted occupant by the seat in rear-end impacts." Paper No. 922522, presented at the 36th Stapp Car Crash Conference, Seattle, WA, November 1992.

Viano, David C. and Arepally, Sudhakar. "Assessing the safety performance of occupant restraint systems." Paper No. 902328, presented at the 34th Stapp Car Crash Conference, Orlando, FL, November 1990.

Viano, David C. and Culver, Clyde C. "Performance of a shoulder belt and knee restraint in barrier crash simulations." Paper No. 791006, presented at the 23rd Stapp Car Crash Conference, San Diego, CA , October 1979.

Viano, David C. and Culver, Clyde C. "Test dummy interaction with a shoulder or lap belt." Paper No. 811017, presented at the 25th Stapp Car Crash Conference, San Francisco, CA, September 1981.

Viano, David, et.al. "Belt and airbag testing with a pregnant hybrid III female dummy." Paper No. 976154, presented at the Fifteenth International Technical Conference on the Enhanced Safety of Vehicles, Melbourne, Australia, May 1996.

von Buseck, Calvin R., et.al. "Seat belt usage and risk taking in driving behavior." Paper No. 800388, presented at the Automotive Engineering Congress and Exposition, Detroit, MI, February 1980.

W.W. Hunter, et.al. "Non-sanction seat belt law enforcement: a modern day tale of two cities." Paper No. 1991-12-0005, presented at the annual meeting of the Association for the Advancement of Automotive Medicine, Ontario, Canada, October 1991.

Wagenaar, Alexander, C., Maybee, Richard G., and Sullivan, Kathleen P. "Effects of mandatory seatbelt laws on traffic fatalities in the United States." Paper No. 876028, presented at the Eleventh International Technical Conference on Experimental Safety Vehicles, Washington, DC, 1987.

Walfisch, G., Chamouard, F., et al. "Predictive Functions for Thoracic Injuries to Belt Waerers in Frontal Collisions and Their Conversion into Protection Criteria." 851722 Society of Automotive Engineers, 1985.

Wall, J., Lowne, R.W., and Harris, J. "The determination of tolerable loadings for car occupants in impacts." Paper No. 766071, presented at the Sixth International Technical Conference on Experimental Safety Vehicles, Washington, DC, 1976.

Wall, J.G. and Lowne, R.W. "Human injury tolerance level determination from accident data using the OPAT dummy." Paper No. 746034, presented at the Fifth International Technical Conference on Experimental Safety Vehicles, London, England, 1974.

Walsh, Michael J. and Romeo, David J. "Results of cadaver and anthropomorphic dummy tests in identical crash situations." Paper No. 760803, presented at the 20th Stapp Car Crash Conference, Dearborn, MI, October 1976.

Ward, Carley C., et.al. "Investigation of restraint function on male and female occupants in rollover events." Paper No. 2001-01-0177, presented at the 2001 World Congress, Detroit, MI, March 2001.

Warner, Charles Y., et.al. "Occupant protection in rear-end collisions: I. safety priorities and seat belt effectiveness." Paper No. 912913, presented at the 35th Stapp Car Crash Conference, San Diego, CA, November 1991.

Waters, Wendy, Macnabb, Michael J., and Brown, Betty. "A half century of attempts to resolve vehicle occupant safety: Understanding seatbelt and airbag technology." Paper No. 986129, presented at the 16th International Technical Conference on the Enhanced Safety of Vehicles, Ontario, Canada, May 1998.

Wee, Sang Wan. "Seat belt wearing in drivers and front seat passengers." Paper No. 1992-15-0054, presented at the 12th World Congress of the International Association for Accident and Traffic Medicine and 7th Nordic Congress on Traffic Medicine, Helsinki, Finland, June 1992.

Weissner, Rüdiger. "A comparison of advanced belt systems regarding their effectiveness." Paper No. 780414, presented at the Automotive Engineering Congress and Exposition, Detroit, MI, February 1978.

Werber, Stephen J. "Legal effects of seat belt non-use." Paper No. 840330, presented at the SAE International Congress and Exposition, Detroit, MI, February 1984.

Wiechel, John F., Sens, Michael J., and Guenther, Dennis A. "Critical review of the use of seat belts by pregnant women." Paper No. 890752, presented at the SAE International Congress and Exposition, Detroit, MI, February 1989.

Yang, Jikuang, et.al. "A methodology using a combined injury criteria index to study the performance of various driver restraint system configurations." Paper No. 2001-06-0166, presented at the International Technical Conference on the Enhanced Safety of Vehicles, Amsterdam, The Netherlands, June 2001.

Zhou Q., Rouhana S.W., Melvin, J.W. "Age Effects on Thoracic Injury Tolerance." 962421, Society of Automotive Engineers, 1996.

FIELD EFFECTIVENESS

Bohlin N.I. "A Statistical Analysis of 28,000 Accident Cases with Emphasis on Occupant Restraint Value." 670925, Society of Automotive Engineers, 1967.

Campbell, B.J. and Kihlberg, Jaakko, K. "Seat belt effectiveness in the non-ejection situation." Paper No. 1963-12-0015, presented at the Seventh Stapp Car Crash Conference, Los Angeles, CA, November 1963.

Evans, Leonard. "Occupant protection device effectiveness in preventing fatalities." Paper No. 876032, presented at the Eleventh International Technical Conference on Experimental Safety Vehicles, Washington, DC, January 1987.

Hoxie, Paul and Skinner, David. "Fatality reductions from mandatory seatbelt usage laws." Paper No. 870219, presented at the SAE International Congress and Exposition, Detroit, MI, February 1987.

Huelke, D.F. et.al. "Effectiveness of current and future restraint systems in fatal and serious injury automobile crashes. Data from on-scene field accident investigations." Paper No. 70323, presented at the Automotive Engineering Congress and Exposition, Detroit, MI February 1979.

Huelke, D.F., Lawson, T. E., et al. "The Effectiveness of Belt Systems in Frontal and Rollover Crashes." 770148, Society of Automotive Engineers, 1977.

Huelke, Donald F. and Gikas, Paul W. "Determination of seat belt effectiveness for survival in fatal highway collisions," Paper No. 1963-12-0028, presented at the Seventh Stapp Car Crash Conference, Los Angeles, CA, November 1963.

Janssen, E.G., Huijskens, C.G., and Beusenberg, M.C. "Reduction in seat belt effectiveness due to misuse." Paper No. 1992-13-0027, presented at the International IRCOBI Conference on the Biomechanics of Impacts, Verona, Italy, September 1992.

Jones I.S., Whitfield R.A., Carroll, D.M., "New Car Assessment Program Results and the Risk of Injury in Actual Accidents." 856056, Society of Automotive Engineers, 1985.

Kahane C.J., Hackney, J.R., Berkowitz, A.M., "Correlation of Vehicle Performance in the New Car Assessment Program with Fatality Risk in Actual Head-On Collisions." 946150, Society of Automotive Engineers, 1994.

Kallieris, D., Stein, K. and Mattern, E. "The effectiveness of available restraint systems in frontal collisions." Paper No. 1992-15-0066, presented at the 12th World Congress of the International Association for Accident and Traffic Medicine and 7th Nordic Congress on Traffic Medicine, Helsinki, Finland, June 1992.

MacKay, Murray. "The effectiveness and limitations of seat belts in collisions." Paper No. 1992-15-0016, presented at the Proceedings of International Symposium on Road Traffic Accidents, Saudi Arabia, February 1992.

McCarthy, Roger L. et. al. "Seat belts: effectiveness of mandatory use requirements." Paper No. 840329, presented at the SAE International Congress and Exposition, Detroit, MI, February 1984.

Mohan, D. et.al. "Air bag and lap/shoulder belts - a comparison of their effectiveness in real world, frontal crashes." Paper No. 1976-12-0028, presented at the American Association for Automotive Medicine, Annual Meeting, Atlanta, GA, November 1976.

Thomas, C. et.al. "Comparative study of 1624 belted and 3242 non-belted occupants: results on the effectiveness of seat belts." Paper No. 1980-12-0037, presented at the American Association for Automotive Medicine, Annual Meeting, Rochester, NY, October 1980.

Vallet, G. et. al. "Seat belt efficiency~Paired case study with unbelted and belted occupants." Paper No. 1986-13-0005, presented at the International IRCOBI Conference on the Biomechanics of Impacts, Zurich, Switzerland, September 1986.

Viano, D.C. "Effectiveness of Safety Belts and Airbags in Preventing Fatal Injury." 910901, Society of Automotive Engineers, 1991.

Viano, David C. "Crash injury prevention: a case study of fatal crashes of lap- shoulder belted occupants." Paper No. 922523, presented at the 36th Stapp Car Crash Conference, Seattle, WA, November 1992.

Wegman, F., Bos, J. and Bijleveld, F. "Estimating effectiveness of increased seat belt usage on the number of fatalities." Paper No. 896135, presented at the Twelfth International Technical Conference on Experimental Safety Vehicles, Goteborg, Sweden, January 1989.

INJURIES OF BELTED OCCUPANTS

Aibe, Tsuyoshi. "Influence of occupant seating posture and size on head and chest injuries in frontal collision." Paper No. 826032, presented at the Ninth International Technical Conference on Experimental Safety Vehicles, Kyoto, Japan, 1982.

Augenstein, J., Perdeck, E., et al. "Injury Patters Among Belted Drivers Protected by Air Bags in 30 to 35 mph Crashes." 1999-01-1062, SAE Congress, 1999.

Behrens, S., et.al. "Injury patterns caused by seat belts." Paper No. 1977-13-0006, presented at the 3rd International Conference on Impact Trauma, Berlin, Germany, September 1977.

Bohman, Katarina. "A study of AIS1 neck injury parameters in 168 frontal collisions using a restrained hybrid III dummy." Paper No. 2000-01-SC08, presented at the 44th Stapp Car Crash Conference, Atlanta, GA, November 2000.

Boström, Ola, et.al. "New AIS1 long-term neck injury criteria candidates based on real frontal crash analysis." Paper No. 2000-13-0019, presented at the 2000 International IRCOBI Conference on The Biomechanics of Impact, Montpellier, France, September 2000.

Campbell, Horace E. "Twenty-three fatal crashes with seat belts." Paper No. 1963-12-0040, presented at

Cooper, Kirby D., et.al. "Seat belt syndrome: An examination of patterns of injury associated with seat belt use." Paper No. 1994-12-0002, presented at the 38th Annual Proceedings Association for the Advancement of Automotive Medicine, Lyon, France, September 1994.

Crandall, Jeff R. "Differing patterns of head and facial injury with air bag and/or belt restrained drivers in frontal collisions." Paper No. 1994-12-0052, presented at the Advances in Occupant Restraint Technologies AAAM/IRCOBI, September 1994.

Dalmotas D.J./ "Injury Mechanisms to Occupants Restrained by Three-Point Seat Belts in Side Impact." 830462, SAE Congress, Detroit, MI, 1983.

Dalmotas, Dainius, J. "Mechanisms of injury to vehicle occupants restrained by three-point seat belts." Paper No. 801311, presented at the 24th Stapp Car Crash Conference, Troy, MI, October 1980.

Damholt, W., Nielsen, H.V., and Nordentoft, E.L. "Correlations between accident circumstances and the type and grade of injuries in traffic accidents." Paper No. 1975-13-0009, presented at the International Conference on the Biomechanics of Serious Trauma, Birmingham, England, September 1975.

Danner, M., et.al. "Experience from the analysis of accidents with a high belt usage rate and aspects of continued increase in passenger safety." Paper No. 876030, presented at Eleventh International Technical Conference on Experimental Safety Vehicles, Washington, DC, 1987.

Digges, Kennerly and Dalmotas, Dainius. "Injuries to restrained occupants in far-side crashes." Paper No. 2001-06-0149, presented at the International Technical Conference on the Enhanced Safety of Vehicles, Amsterdam, The Netherlands, June 2001.

Fisher, P. "Critical injuries produced by seat belts. Case report." Paper No. 1964-12-0004, presented at the American Association of Automotive Medicine, Louisville, KY, October 1964.

Foret-Bruno, J.Y., Faverjomn, G., et al. "Females More Vulnerable Than Males in Road Accidents. 905122, Society of Automotive Engineers, 1990.

Frampton, R.J. and Mackay, G.M. "The characteristics of fatal collisions for belted occupants." Paper No. 945167, presented at the 25th FISITA Congress-Automobile in Harmony with Human Society, Beijing, China, October 1994.

Grattan, E. and Hobbs, J.A. "Some patterns and causes of injury in car occupants." Paper No. 746028, presented at the Fifth International Technical Conference on Experimental Safety Vehicles, London, England, 1974.

Green, R.N., et.al. "Case studies of severe frontal collisions involving fully-restrained occupants." Paper No. 1987-12-0020, presented at the annual meeting of the American Association for Automotive Medicine, New Orleans, LA, September 1987.

Green, R.N., et.al. "Misuse of three-point occupant restraints in real-world collisions." Paper No. 1987-13-0008, presented at the International IRCOBI Conference on the Biomechanics of Impacts, Birmingham, England, September 1987.

Harms, P.L. "Injuries to restrained car occupants~What are the outstanding problems." Paper No. 876029, presented at the Eleventh International Technical Conference on Experimental Safety Vehicles, Washington, DC, 1987.

Hartemann, H., et.al. "Belted or not belted: the only difference between two matched samples of 200 car occupants." Paper No. 770917, presented at the 21st Stapp Car Crash Conference, New Orleans, LA, October 1977.

Hassan, A.M. and Mackay, M. "Injuries of moderate severity to restrained drivers in frontal crashes." Paper No. 2000-04-0337, presented at the International Conference on Vehicle Safety 2000, London, England, June 2000.

Henderson, Michael J. and Wyllie, James M. "Seat belts~limits of protection: a study of fatal injuries among belt wearers." Paper No. 730964, presented at the 17th Stapp Car Crash Conference, Oklahoma City, OK, November 1973.

Hill, J.R. "Car occupant injury patterns with special reference to chest and abdominal injuries caused by seat belt loading." Paper No. 1992-13-0028, presented at the International IRCOBI Conference on the Biomechanics of Impacts, Verona, Italy, September 1992.

Hill, J.R., Mackay, G.M., and Morris, A.P. "Chest and abdominal injuries caused by seat belt loading." Paper No. 1992-12-0002, presented at the annual meeting of the Association for the Advancement of Automotive Medicine, Portland, OR, October 1992.

Hill, Julian, Parkin, Steven, and Mackay, Murray. "Injuries caused by seat belt loads to drivers and front seat passengers." Paper No. 1996-12-0004, presented at the 40th Annual Proceedings, Vancouver, BC, October 1996.

Huelke, Donald F. and Chewning, William A. "Comparison of occupant injuries with and without seat belts." Paper No. 690244, presented at the International Automotive Engineering Congress and Exposition, Detroit, MI, January 1969.

James, Michael B., et.al. "Injury mechanisms and field accident data analysis in rollover accidents." Paper No. 970396, presented at the SAE International Congress and Exposition, Detroit, MI, February 1997.

Kim, Carl and Li, Leu. "Seat belt usage and injuries sustained in motor vehicle collisions." Paper No. 1994-12-0004, presented at the 38th Annual Proceedings Association for the Advancement of Automotive Medicine, Lyon, France, September 1994.

Kramer, F. "Abdominal and pelvic injuries of vehicle occupants wearing safety belts incurred in frontal collisions~Mechanism and protection." Paper No. 1991-13-0022, presented at the International IRCOBI Conference on the Biomechanics of Impacts, Berlin, Germany, September 1991.

Krantz, P. and Löwenhielm, P. "Injury response in belted and unbelted car occupants related to the car crash energy in 458 accidents~A study of all fatal automobile accidents in Sweden, 1975." Paper No. 1980-13-0025, presented at the International IRCOBI Conference on the Biomechanics of Impacts, Birmingham, England, September 1980.

Langwieder, K., et.al. "Comparative studies of neck injuries of car occupants in frontal collisions in the United States and in the Federal Republic of Germany." Paper No. 811030, presented at the 25th Stapp Car Crash Conference, San Francisco, CA, September 1981.

Larder, D.R., et.al. "Neck injury to car occupants using seat belts." Paper No. 1985-12-0011, presented at the annual meeting of the American Association for Automotive Medicine, Washington, DC, October 1985.

Maag, U., et.al. "Seat belts and neck injuries." Paper No. 1990-13-0001, presented at the International IRCOBI Conference on the Biomechanics of Impacts, Bron-Lyon, France, September 1990.

MacKay, G.M. "Airbag effectiveness~a case for the compulsory use of seat belts." Paper No. 725042, presented at the 14th FISITA Congress, London, England, 1972.

Mackay, G.M., et.al. "Restrained front seat car occupant fatalities - the nature and circumstances of their injuries." Paper No. 1990-12-0010, presented at the annual meeting of the Association for the Advancement of Automotive Medicine, Scottsdale, AZ, October 1990.

Mackay, G.M., et.al. "Serious trauma to car occupants wearing seat belts." Paper No. 1975-13-0001, presented at the International Conference on the Biomechanics of Serious Trauma, Birmingham, England, September 1975.

Mackay, Murray and Ashton, Steve. "Injuries in collisions involving small cars in Europe." Paper No. 730284, presented at the International Automotive Engineering Congress and Exposition, Detroit, MI, January 1973.

Maeda, Kuozo. "Scientific analysis of the circumstances and causes of traffic accidents~Analysis concerning the injuries and fatalities involving occupants of motor vehicles." Paper No. 958354, presented at the JSAE Autumn Conference, Tokyo, Japan, September 1995.

Nakahama, Ryoji and Katoh, Hiroshi. "Study on the relationship between seat belt anchorage location and occupant injury." Paper No. 876073, presented at the Eleventh International Technical Conference on Experimental Safety Vehicles, Washington, DC, 1987.

Otte, D., "Assessment of Measures Reducing Residual Severe and Fatal Injuries MAIS 3+ of Car Occupants." 976064, Society of Automotive Engineers, 1997.

Otte, D., et.al. "Variations of injury patterns of seat-belt users." Paper No. 870226, presented at the SAE International Congress and Exposition, Detroit, MI, February 1987.

Rubinstein, E. "An analysis of fatal car crashes in which the victim was wearing a seat belt." Paper No. 1973-13-0005, presented at the International Conference on the Biokinetics of Impacts, June 1973.

Ryan, G. Anthony. "A study of seat belts and injuries." Paper No. 730965, presented at the 17th Stapp Car Crash Conference, Oklahoma City, OK, November 1973.

Schuller, Erich, Beier, Gundolf, and Steiger, Thomas. "Injury patterns of restrained car occupants in near-side impacts." Paper No. 890376, presented at the SAE International Congress and Exposition, Detroit, MI, February 1989.

Shortridge, R.M. "A comparison of injury severity patterns for unrestrained, lap belted, and torso restrained occupants in automobile accidents." Paper No. 1973-12-0009, presented at the annual meeting of the American Association for Automotive Medicine, Oklahoma City, OK, November 1973.

Thomas, C., et.al. "Influence of age and restraint force value on the seriousness of thoracic injuries sustained by belted occupants in real accidents." Paper No. 1979-13-0005, presented at the International IRCOBI Conference on the Biomechanics of Impacts, Göteborg, Sweden, September 1979.

Walfisch, G. "Synthesis of abdominal injuries in frontal collisions with belt-wearing cadavers compared with injuries sustained by real-life accident victims. Problems of simulation with dummies and protection criteria." Paper No. 1979-13-0013, presented at the International IRCOBI Conference on the Biomechanics of Impacts, Göteborg, Sweden, September 1979.

Walz, Felix, H., et.al. "Frequency and significance of seat belt induced neck injuries in lateral collisions." Paper No. 811031, presented at the 25th Stapp Car Crash Conference, San Francisco, CA, September 1981.

Yoganandan, N, Pintar, F.A., Gennarelli T.A., "Mechanisms and Factors Involved in Hip Injuries During Frontal Crashes." Stapp Car Crash Journal, 45:437-448, 2001.

SUBMARINING

Adomeit, Dieter and Heger, Alfred. "Motion sequence criteria and design proposals for restraint devices in order to avoid unfavorable biomechanic conditions and submarining." Paper No. 751146, presented at the 19th Stapp Car Crash Conference, San Diego, CA, November 1975.

Begeman, Paul C., Levine, Robert S. and King, Albert I. "Belt slip measurements on human volunteers and the Part 572 dummy in low-Gximpact acceleration." Paper No. 831635, presented at the 27th Stapp Car Crash Conference with IRCOBI and Child Injury and Restraint Conference with IRCOBI, San Diego, CA, October 1983.

Bendjella, F. et. al. "Measurement of seat-belt loads sustained by the dummy in frontal impact with critical pelvic motion~Development of a 2-D transducer." Paper No. 896080, presented at the Twelfth International Technical Conference on Experimental Safety Vehicles, Goteborg, Sweden, January 1989.

Biard, Roger, Cesari, Dominique and Derrien, Yves. "Advisability and reliability of submarining detection." Paper No. 870484, presented at the SAE International Congress and Exposition, Detroit, MI, February 1987.

Chamouard, F., Tarriere, C. and Baudrit, P. "Protection of children on board vehicles~Influence of pelvis design and thigh and abdomen stiffness on the submarining risk for dummies installed on a booster." Paper No. 976100, presented at the Fifteenth International Technical Conference on the Enhanced Safety of Vehicles, Melbourne, Australia, May 1996.

DeJeammes, M., Biard, R., and Derrien, Y. "Factors influencing the estimation of submarining on the dummy." Paper No. 811021, presented at the 25th Stapp Car Crash Conference, San Francisco, CA, September 1981.

Gallup, B.M., St-Laurent, A.M. and Newman, J.A. "Abdominal injuries to restrained front seat occupants in frontal collisions." Paper No. 1982-12-0008, presented at the American Association for Automotive Medicine, Annual Meeting, Ottawa, Ontario, Canada, October 1982.

Horsch, John D. and Hering, William E. "A kinematic analysis of lap-belt submarining for test dummies." Paper No. 892441, presented at the 33rd Stapp Car Crash Conference, Washington, DC, October 1989.

Huelke, Donald F., MacKay, G. Murray and Morris, Andrew. "Intraabdominal injuries associated with lap-shoulder belt usage." Paper No. 930639, presented at the SAE International Congress and Exposition, Detroit, MI, March 1993.

Kramer, F. "Experimental determination of the pelvis rotational angle and its relation to a protection criterion for pelvic and abdominal injuries." Paper No. 1994-13-0015, presented at the International Conference on the Biomechanics of Impacts, Lyon, France, September 1994.

Leung, Y.C. et. al. "A comparison between part 572 dummy and human subject in the problem of submarining." Paper No. 791026, presented at the 23rd Stapp Car Crash Conference, San Diego, CA, October 1979.

Leung, Y.C. et. al. "Submarining injuries of 3 pt. belted occupants in frontal collisions~description, mechanisms and protection." Paper No. 821158, presented at the 26th Stapp Car Crash Conference, Ann Arbor, MI, October 1982.

Leung, Y.C. et.al. "An anti-submarining scale determined from theoretical and experimental studies using three-dimensional geometrical definition of the lap-belt." Paper No. 811020, presented at the 25th Stapp Car Crash Conference, San Francisco, CA, September 1981.

Levine, R.S. et. al. "Effect of quadriceps function on submarining." Paper No. 1978-12-0026 presented at the American Association for Automotive Medicine, Annual Meeting, Ann Arbor, MI, July 1978.

MacLaughlin, Thomas F., Sullivan, Lisa K. and O'Connor, Christopher S. "Experimental investigation of rear seat submarining." Paper No. 896032, presented at the Twelfth International Technical Conference on Experimental Safety Vehicles, Goteborg, Sweden, January 1989.

Rouhana, Stephen W. et. al. "Assessing submarining and abdominal injury risk in the Hybrid III family of dummies." Paper No. 892440, presented at the 33rd Stapp Car Crash Conference, Washington, DC, October 1989.

Rouhana, Stephen W., Jedrzejczak, Edward A. and McCleary, Joseph D. "Assessing submarining and abdominal injury risk in the hybrid III family of dummies part II~Development of the small female frangible abdomen." Paper No. 902317, presented at the 34th Stapp Car Crash Conference, Orlando, FL, November 1990.

Tarriere, C.H. "Proposal for a protection criterion as regards abdominal internal organs." Paper No. 1973-12-0023, presented at the American Association for Automotive Medicine, Annual Meeting, Oklahoma City, OK, November 1973.

Uriot, Jerome et. al. "Measurment of submarining on Hybrid III 50°& 5°percentile crash test dummies." Paper No. 946040, presented at the 14th International Technical Conference on the Enhanced Safety of Vehicles, Munich, Germany, May 1994.

Walfisch, G. et. al. "Synthesis of abdominal injuries in frontal collisions with belt-wearing cadavers compared with injuries sustained by real-life accident victims. Problems of simulation with dummies and protection criteria." Paper No. 1979-13-0013, presented at the International IRCOBI Conference on the Biomechanics of Impacts, Goteborg, Sweden, September 1979.

SYSTEM ENHANCEMENTS/FEATURES

PRETENSIONER

Häland, Yngve and Skänberg, Torbjörn. "A mechanical buckle pretensioner to improve a three point seat belt." Paper No. 896134, presented at the Twelfth International Technical Conference on Experimental Safety Vehicles, Göteborg, Sweden, 1989.

Haland, Yngye and Nilson, Gert. "Seat belt pretensioners to avoid the risk of submarining~A study of lap- belt slippage factors." Paper No. 916137, presented at the 13th International Technical Conference on Experimental Safety Vehicles, Paris, France, November 1991.

Kullgren, Anders, et.al. "Influence of airbags and seatbelt pretensioners on AIS1 neck injuries for belted occupants in frontal impacts." Paper No. 2000-01-SC09, presented at the 44th Stapp Car Crash Conference, November 2000.

Lane, Wendell C., Jr. "The interaction of pretensioner with restraints and supplemental restraints." Paper No. 1994-20-0042, presented at the International Body Engineering Conference and Exposition, Detroit, MI, September 1994.

Lorenz, Berend, et.al. "Volunteer tests on human tolerance levels of pretension for reversible seatbelt tensioners in the pre-crash phase~Phase 1 results: Tests using a stationary vehicle." Paper No. 2001-13-0024, presented at the International IRCOBI Conference on the Biomechanics of Impact, Isle of Man, UK, October 2001.

Mitzkus, Jüergen E. and Eyrainer, Heinz. "Three-point belt improvements for increased occupant protection." Paper No. 840395, presented at the SAE International Congress and Exposition, Detroit, MI, February 1984.

Müller, Helmut E. and "Seat belt pretensioners." Paper No. 980557, presented at the SAE International Congress and Exposition, Detroit, MI, February 1998.

Quincy, R. and Dejeammes, M. "Analysis of the preloaded safety belt restraint with an animal." Paper No. 1976-12-0007, presented at the annual meeting of the American Association for Automotive Medicine, Atlanta, GA, November 1976.

Reidelbach, W. and Scholz, H. "Advanced restraint system concepts." Paper No. 790321, presented at the Automotive Engineering Congress and Exposition, Detroit, MI, March 1979.

Tabata, Hiroshi and Kinoshita, Yoshihiko. "The development of a new-type seat belt pretensioner." Paper No. 918258, presented at the JSAE Autumn Conference, Sapporo, Hokkaido, Japan, October 1991.

Walsh, Michael J. and Kelleher, Barbara J. "Development of a preloaded, force-limited passive belt system for small cars." Paper No. 800300, presented at the Automotive Engineering Congress and Exposition, Detroit, MI, February 1980.

Willumeit, H.P. "Passive preloaded energy-absorbing seat belt system." Paper No. 720433, presented at the National Automobile Engineering Meeting, Detroit, MI, May 1972.

Xunnan He, Simon and Wilkins, Michael D. "A method to evaluate the energy capability of seat belt pretensioners." Paper No. 1999-01-0080, presented at the SAE International Congress and Exposition, Detroit, MI, March 1999.

Zuppichini, F. "Effectiveness of a mechanical pretensioner on the performance of seat belts." Paper No. 1989-13-0008, presented at the International IRCOBI Conference on the Biomechanics of Impacts, Stockholm, Sweden, September 1989.

Zuppichini, F. and Häland, Y. "Effectiveness of pretensioners on the performance of seat belts." Paper No. 905139, presented at the 23rd FISITA Congress, Torino, Italy, 1990.

LOAD LIMITERS, WEB CLAMPS AND LOCKS

Adhikari, Prasenjit and Renfroe, David A. "An overview of emergency locking retractor performances of shoulder harness safety belt systems in American made vehicles." Paper No. 940528, presented at the SAE International Congress and Exposition, Detroit, MI, February 1994.

Adomeit, Dieter and Balser, Werner. "Items of an engineering program on an advanced web-clamp device." Paper No. 870328, presented at the SAE International Congress and Exposition, Detroit, MI, February 1987.

Adomeit, Dieter. "A force limiting system on a three-point-belt system depending on crash velocity." Paper No. 740582, presented at the 3rd International Conference on Occupant Protection, Troy, MI, July 1974.

Bendjellal, Farid et. al. "The combination of a new air bag technology with a belt load limiter." Paper No. 986101, presented at the 16th International Technical Conference on the Enhanced Safety of Vehicles meeting, Windsor, Ontario, Canada, May 1998.

Bendjellal, Farid et. al. "The programmed restraint system~A lesson from accidentology." Paper No. 973333, presented at the 41st Annual Stapp Car Crash Conference, Lake Buena Vista, FL, November 1997.

Compton, Charles P. and Baker, Leonard L. "The influence of the seat belt tension reliever feature on seat belt usage and injury severity." Paper No. 901752, presented at the SAE Passenger Car Meeting and Exposition, Dearborn, MI, September 1990.

Foret-Bruno, J.Y. et. al. "Correlation between thoracic lesions and force values measured at the shoulder of 92 belted occupants involved in real accidents." Paper No. 780892, presented at the 22nd Stapp Car Crash Conference, Ann Arbor, MI, October 1978.

Foret-Bruno, J.Y., et.al. "Thoracic deflection of Hybrid III~Dummy response for simulations of real accidents." Paper No. 896077, presented at the Twelfth International Technical Conference on Experimental Safety Vehicles, Göteborg, Sweden, 1989.

Foret-Bruno, J.Y., et.al. "Thoracic Injury Risk in Frontal Car Crashes with Occupant Restrained with Belt Load Limiter." 983166, 42nd Stapp Car Crash Conference, 1998.

Gamble, James F. "What's next in energy absorption of restraint systems." Paper No. 740372, presented at the National Business Aircraft Meeting and Exposition, Wichita, KS, April 1974.

Got, C. et. al. "Thoracic injury risk in frontal car crashes with occupant restrained with belt load limiter." Paper No. 983166, presented at the 42nd Annual Stapp Car Crash Conference, Tempe, AZ, November 1998.

Hontschik, Heinrich, Müller, Egbert, and Rüter, Gert. "Necessities and possibilities of improving the protective effect of three-point seat belts." Paper No. 770933, presented at the 21st Stapp Car Crash Conference, New Orleans, LA, October 1977.

Kallieris, Dimitrios et. al. "The influence of force limiter to the injury severity by using a 3-point belt and driver airbag in frontal collisions." Paper No. 976038, presented at the Fifteenth International Technical Conference on the Enhanced Safety of Vehicles, Melbourne, Australia, May 1996.

Koon Soon Kang, et. al. "Tension control device of the lap-shoulder belt." Paper No. 1992-15-0067, presented at the 12th World Congress of the International Association for Accident and Traffic Medicine and 7th Nordic Congress on Traffic Medicine, Helsinki, Finland, June 1992.

Mertz, Harold J. et. al. "Hybrid III sternal deflection associated with thoracic injury severities of occupants restrained with force-limiting shoulder belts." Paper No. 910812, presented at the SAE International Congress and Exposition, Detroit, MI, February 1991.

Mertz, Harold J., Williamsom, James E. and Vander Lugt, Donald A. "The effect of limiting shoulder belt load with airbag restraint." Paper No. 950886, presented at the SAE International Congress and Exposition, Detroit, MI, February 1995.

Meyer, Steven E. et.al. "Dynamic analysis of ELR retractor spoolout." Paper No. 2001-01-3312, presented at the Automotive and Transportation Technology Congress and Exhibition, Barcelona, Spain, October 2001.

Miller, John H. "Occupant performance with constant force restraint systems." Paper No. 960502, presented at the SAE International Congress and Exposition, Detroit, MI, February 1996.

Nilson, G. and Haland, Y. "A force-limiting device to reduce the seat-belt loading to the chest and the abdomen in frontal impacts."

Paper No. 1993-13-0027 presented at the International Conference on the Biomechanics of Impacts meeting, Eindhoven, Netherlands, September 1993.

Zuppichini, F. "Seat belt performance and after-market web-locking devices: an experimental study." Paper No. 1990-13-0002, presented at the International IRCOBI Conference on the Biomechanics of Impacts meeting, Bron-Lyon, France, September 1990.

SMART RESTRAINTS

Adomeit, H.-D., Wils, O. and Heym, A. "Adaptive airbag-belt-restraints~An analysis of biomechanical benefits." Paper No. 970776, presented at the SAE International Congress and Exposition, Detroit, MI, February 1997.

Bendjellal, Farid et. al. "The programmed restraint system~A lesson from accidentology." Paper No. 973333, presented at the 41st Annual Stapp Car Crash Conference, Lake Buena Vista, FL, November 1997.

Bigi, Dante. "Smart restraint systems: the next step in occupant protection." Paper No. 963335, presented at the International Conference on Car Interior Components, Paris, France, March 1996.

Clute, Gunter. "Potentials of adaptive load limitation presentation and system validation of the adaptive load limiter." Paper No. 2001-06-0194, presented at the International Technical Conference on the Enhanced Safety of Vehicles meeting, Amsterdam, The Netherlands, June 2001.

Cuerden, Richard. "The potential effectiveness of adaptive restraints." Paper No. 2001-13-0025, presented at the International IRCOBI Conference on the Biomechanics of Impact meeting, Isle of Man, UK, October 2001.

Cullen, E. et. al. "How people sit in cars: implications for driver and passenger safety in frontal collisions~The case for smart restraints." Paper No. 1996-12-0006, presented at the Association for the Advancement of Automotive Medicine 40th Annual Proceedings, Vancouver, British Columbia, Canada, October 1996.

Habib, M.S. "A smart restraint-seat system for occupant's protection in front- and rear- end collisions." Paper No. 2000-25-0072, presented at the ISATA 2000 – Safety, Crashworthiness, Mobility, Occupant Safety/Automotive Interiors – Materials, Ergonomics, Manufacturing, ICE and Comfort meeting, Dublin, Ireland, September 2000.

Heym, A. and Adomeit, H.-D. "A smart-restraint-system concept to limited complexity." Paper No. 977084, presented at the Innovative Passenger Protection in Private Automobiles meeting, Berlin, Germany, October 1997.

Johannessen, George H. and Mackay, Murray. "Why "intelligent" automotive occupant restraint systems?" Paper No. 1995-12-0033, presented at the 39th Annual Proceedings of the Association for the Advancement of Automotive Medicine, Chicago, IL, October 1995.

Kozyreff, Michel. "Adaptive restraint system for an intelligent occupant protection." Paper No. 953346, presented at the European Automotive Industry Meets the Challenges of the Year 2000-5th International Congress of the European Automobile Engineers Cooperation, Strasbourgh, France, June 1995.

Mackay, M. "Smart seatbelts - what they offer." Paper No. 964148, presented at the Automotive Passenger Safety meeting, London, England, November 1995.

Mackay, M. and Hassan, A.M. "Adaptive restraints~Their characteristics and benefits." Paper No. 974147, presented at the Automotive Environmental Impact and Safety Autotech Congress '97, Birmingham, UK, November 1997.

Mackay, Murray, Parkin, Stephen and Scott, Andrew. "Intelligent restraint systems--What characteristics should they have." Paper No. 1994-12-0053, presented at the Advances in Occupant Restraint Technologies AAAM/IRCOBI meeting, Lyon, France, September 1994.

Mackay, Murray. "Smart seat belts~Some population considerations applied to intelligent restraint systems." Paper No. 940531, presented at the SAE International Congress and Exposition, Detroit, MI, February 1994.

Mackay, Murray. "What variables should intelligent restraint systems address." Paper No. 953447, presented at the European Automotive Industry Meets the Challenges of the Year 2000-5th International Congress of the European Automobile Engineers Cooperation, Strasbourgh, France, June 1995.

Miller, John H. "Injury reduction with smart restraint systems." Paper No. 1995-12-0034, presented at the 39th Annual Proceedings of the Association for the Advancement of Automotive Medicine, Chicago, IL, October 1995.

Miller, John H. and Maripudi, Vivek. "Restraint force optimization for a smart restraint system." Paper No. 960662, presented at the SAE International Congress and Exposition, Detroit, MI, February 1996.

Monroe, S. and Obesser, F. "Innovations in technology leading to an intelligent safety system." Paper No. 2000-25-0099, presented at the ISATA 2000 – Safety, Crashworthiness, Mobility, Occupant Safety/Automotive Interiors – Materials, Ergonomics, Manufacturing, ICE and Comfort meeting, Dublin, Ireland, September 2000.

Schaub, S. and Bosio, A.C. "Smart restraint systems for the European market." Paper No. 977083, presented at the Innovative Passenger Protection in Private Automobiles meeting, Berlin, Germany, October 1997.

Steurer, H., Steiner, P. and Zoister, P. ""Delta V"~Sensor used for optimized control of restraint systems." Paper No. 977096, presented at the Innovative Passenger Protection in Private Automobiles meeting, Berlin, Germany, October 1997.

Trosseille, Xavier, et. al. "Evaluation of secondary risk with a new programmed restraining system (PRS2)." Paper No. 986107, presented at the 16th International Technical Conference on the Enhanced Safety of Vehicles, Windsor, Ontario, Canada, May 1998.

Van Voorhies, Kurt L. and Narwani, Gopal. "Optimization of an intelligent total restraint system." Paper No. 976039, presented at the Fifteenth International Technical Conference on the Enhanced Safety of Vehicles, Melbourne, Australia, May 1996.

STARTER INTERLOCK

Appleby, Michael R., Bintz, Louis J. and Wolfe, John C. "Seat belt use inducing system effectiveness in fleet automobiles," Paper No. 751006, presented at the SAE Automobile Engineering and Manufacturing meeting, Detroit, MI, October 1975.

Cutshaw, E. "Design considerations in developing a seat belt interlock module." Paper No. 741012, presented at the International Automobile Engineering and Manufacturing Meeting, Toronto, Ontario, Canada, October 1974.

Houser, David E. "Design of a seat belt interlock circuit." Paper No. 741013, presented at the International Automobile Engineering and Manufacturing Meeting, Toronto, Ontario, Canada, October 1974.

Pulley, C.H. "Increased seat belt-shoulder harness usage by a starter interlock system." Paper No. 1971-12-0014, presented at the American Association for Automotive Medicine, Colorado Springs, CO, October 1971.

Rukavina, D.M. and Wilke, R.A. "Ford ignition interlock design considerations." Paper No. 741099, presented at the International Automobile Engineering and Manufacturing Meeting, Toronto, Ontario, Canada, October 1974.

States, J.D. "Restraint system usage~education, electronic inducement systems or mandatory usage legislation." Paper No. 1973-12-0027, presented at the American Association for Automotive Medicine, Oklahoma City, OK, November 1973.

Turbell, Thomas et. al. "Optimizing seat belt usage by interlock systems." Paper No. 976007, presented at the Fifteenth International Technical Conference on the Enhanced Safety of Vehicles, Melbourne, Australia, May 1996.

Wada, Akihiro, Sugiura, Fumio, and Okamoto, Kazuo. "The 1974 Toyota belt interlock system." Paper No. 740047, presented at the Automotive Engineering Congress and Exposition, Detroit, MI, February 1974.

ISSUES OF BELT RESTRAINTS

COMFORT AND FIT

Balci, Rana, Shen, Wengi and Vertiz, Alicia. "Comfort and usability of the seat belts." Paper No. 2001-01-0051, presented at the SAE 2001 World Congress, Detroit, MI, March 2001.

Belli Jr., Lino. "Safety and comfort in a retention system~Is there a commitment." Paper No. 952294, presented at the SAE Brazil 95 meeting, Sao Paulo, Brazil, October 1995.

Brown, Christina M., Nov Ian Y. and Pruett, Casey J. "The electronic belt-fit test device (eBTD): A method for certifying safe seat belt fit." Paper No. 2001-06-0081, presented at the International Technical Conference on the Enhanced Safety of Vehicles, Amsterdam, The Netherlands, June 2001.

Chabert, Laurent, Ghannouchi, Slah and Cavallero, Claude. "Geometrical characterization of a seated occupant." Paper No. 986197, presented at the 16th International Technical Conference on the Enhanced Safety of Vehicles, Windsor, Ontario, Canada, May 1998.

Gardner, W.T., Pedder, J.D., and Gallup, B.M. "Development and further refinement of the belt deployment test device." Paper No. 870327, presented at the SAE International Congress and Exposition, Detroit, MI, February 1987.

Johannessen, H.G. and Pulley, C.H. "Safety belt restraint systems: comfort, convenience, use laws as factors in increased usage." Paper No. 1975-12-0002, presented at the American Association for Automotive Medicine Annual Meeting, San Diego, CA, November 1975.

Kendall, Doug. "The development of a computer program to enhance the fit of seat belts." Paper No. 916150, presented at the 13th International Technical Conference on Experimental Safety Vehicles, Paris, France, November 1991.

McCarthy, R.L., Padmanaban, J.A. and Ray, R.M. "An analysis of the safety related impact of "comfort feature" introduction in GM vehicles." Paper No. 1989-12-0021, presented at the Association for the Advancement of Automotive Medicine Annual Meeting, Baltimore, MD, October 1989.

Nov, Ian Y. and Batista, Vittoria. "Prospects for electronic compliance with belt fit requirements." Paper No. 986193, presented at the 16th International Technical Conference on the Enhanced Safety of Vehicles, Windsor, Ontario, Canada, May 1998.

Nov, Y.I., Battista, V. and Carrier, R. "Development of an electronic belt fit test device." Paper No. 971137, presented at the SAE International Congress and Exposition, Detroit, MI, February 1997.

Off-Highway Engineering. "Improving seat-belt comfort." Society of Automotive Engineers, Inc., Warrendale, PA, April 1999.

Pritz, Howard B. and Ulman, Marian S. "FMVSS 208 belt fit evaluation~Possible modification to accommodate larger people." Paper No. 890883, presented at the SAE International Congress and Exposition, Detroit, MI, February 1989.

Pruett, Casey J., Balzulat, Jochen and Brown, Christina M. "Development of an electronic belt fit test device (eBTD) for digitally certifying seat belt fit compliance." Paper No. 2001-01-2087, presented at the SAE Digital Human Modeling Conference and Engineering, Arlington, VA, June 2001.

Sheppard, David. "The BELTFIT program for making seat belts safer and more comfortable." Paper No. 820795, presented at the SAE Passenger Car Meeting and Exposition, Troy, MI, June 1982.

Wilcoxon, Kerry T. "Improving seat belt comfort on off-highway vehicles." Paper No. 982058, presented at the SAE International Off-Highway and Powerplant Congress and Exposition, Milwaukee, WI, September 1998.

Zimmerman, Karen A. "Laboratory test device for the optimization of seat belt system component design and installation geometry." Paper No. 860056, presented at the SAE International Congress and Exposition, Detroit, MI, February 1986.

PERFORMANCE ISSUES: RETRACTORS, AGING BELTS, ETC.

Andrews, Stanley B., et.al. "Dynamic characteristics of end release seatbelt buckles." Paper No. 2001-06-0071, presented at the International Technical Conference on the Enhanced Safety of Vehicles, Amsterdam, The Netherlands, June 2001.

Arndt, S.M. et. al. "Characterization of automotive seat belt buckle inertial release." Paper No. 1993-12-0030, presented at the Association for the Advancement of Automotive Medicine Annual Meeting, San Antonio, TX, November 1993.

Bready, Jon E. et al. "Characteristics of seat belt restraint system markings." Paper No. 2000-01-1317, presented at the SAE 2000 World Congress, Detroit, MI, March 2000.

Bready, Jon E., Nordhagen, Ronald P. and Kent, Richard W. "Seat belt survey: Identification and assessment of noncollision markings." Paper No. 1999-01-0441, presented at the SAE International Congress and Exposition, Detroit, MI, March 1999.

Duignan, Paul and Griffiths, Michael. "Aged, used, crashed seat belts." Paper No. 916161, presented at the 13th International Technical Conference on Experimental Safety Vehicles, Paris, France, November 1991.

Eisentraut, Donald K. et. al. "Assessment of timely retractor lockup in automotive seat belt systems." Paper No. 971515, presented at the X International Conference on Vehicle Structural Mechanics and CAE, Troy, MI, April 1997.

Griffiths, M. and Skidmore, A. "The performance of aged used seatbelts." Paper No. 1987-13-0010, presented at the International IRCOBI Conference on the Biomechanics of Impacts meeting, Birmingham, England, September 1987.

James, Michael B. et. al. "Inertial seatbelt release." Paper No. 930641, presented at the SAE International Congress and Exposition, Detroit, MI, March 1993.

Lee, Kwangho. "The influence of belt tension on buckle response during car crash." Paper No. 932914, presented at the SAE Worldwide Passenger Car Conference and Exposition, Dearborn, MI, October 1993.

Malinow, Ivanna and Perkins, N.C. "Seat belt retractor rattle: Understanding root sources and testing methods." Paper No. 1999-01-1729, presented at the SAE Noise and Vibration Conference and Exposition, Traverse City, MI, May 1999.

Moffatt C.A., Moffatt, E.A., Weiman, T.R. "Diagnosis of Seat Belt Usage in Accidents." 840396, Society of Automotive Engineers, 1984.

Moffatt, Edward A., Thomas, Terry M. and Cooper, Eddie R. "Safety belt buckle inertial responses in laboratory and crash tests." Paper No. 950887, presented at the SAE International Congress and Exposition, Detroit, MI, February 1995.

Nov, Ian Y. "Seat belt buckle release force: cause for entrapment." Paper No. 930342, presented at the SAE International Congress and Exposition, Detroit, MI, March 1993.

Renfroe, David A. "Rollover ejection while wearing lap and shoulder harness: the role of the retractor." Paper No. 960096, presented at the SAE International Congress and Exposition, Detroit, MI, February 1996.

SIMULATION

Andreatta, Dale, et.al. "An analytical model of the inertial opening of seatbelt latches." Paper No. 960436, presented at the SAE International Congress and Exposition, Detroit, MI, February 1996.

Bapu, Girish. "CAE simulation of FMVSS 207/210 test for seat belt anchorages." Paper No. 1995-20-0003, presented at the International Body Engineering Conference and Exposition, Detroit, MI, October 1995.

Deng Y.C., "An Improved Belt Model in CAL3D and its Applications." 900549, Society of Automotive Engineers, 1990.

Deng, Yih-Charng. "Development of a submarining model in the CAL3D program." Paper No. 922530, presented at the 36th Stapp Car Crash Conference, Seattle, WA, November 1992.

Fraterman, E. and Lupker, H.A. "Evaluation of belt modelling techniques." Paper No. 930635, presented at the SAE International Congress and Exposition, Detroit, MI, March 1993.

Freeman, C.M. and Bacon, D.G.C. "The 3-dimensional trajectories of dummy car occupants restrained by seat belts in crash simulations." Paper No. 881727, presented at the 32nd Stapp Car Crash Conference, Atlanta, GA, October 1988.

Hayashi, Seiji and Kumagai, Kuoshi. "Development of a set belt anchorage strength analysis method using dynamic explicit FEM code." Paper No. 2000-08-0137, presented at the JSAE Autumn Conference, Tokyo, Japan, October 2000.

Hiller, M., et.al. "The use of simulation for development of intelligent restraint systems." Paper No. 977097, presented at the Innovative Passenger Protection in Private Automobiles, Berlin, Germany, October 1997.

Holding, P.N., Chinn, B.P., and Happian-Smith, J. "An evaluation of the benefits of active restraint systems in frontal impacts through computer modelling and dynamic testing." Paper No. 2001-06-0094, presented at the International Technical Conference on the Enhanced Safety of Vehicles, Amsterdam, The Netherlands, June 2001.

Igarashi, M. and Atsumi, M. "An analysis of 3 pt. belted occupant impact dynamics in frontal collision and its application." Paper No. 850436, presented at the SAE International Congress and Exposition, Detroit, MI, February 1985.

Kurimoto, Koji, Hirai, Hayao, and Sugawara, Tsutsomu. "On the optimum parameters of a seat belt pre-loading device." Paper No. 826028, presented at the Ninth International Technical Conference on Experimental Safety Vehicles, Kyoto, Japan, 1982.

Mechling, F. "Mathematical modelling of the automobile occupant~Use and limitation of the model." Paper No. 1980-13-0006, presented at the International IRCOBI Conference on the Biomechanics of Impacts, Birmingham, England, September 1980.

Mori, Yoshiyuki, Nakaho, Toshihiro, Yoshimura, Tatsuhiko. "Evaluation of vibration isolating rubber crack growth properties." Paper No. 1999-08-0137, presented at the JSAE Spring Conference, Yokohama, Japan, May 1999.

Nilson, Gert and Håland, Yngve. "An analytical method to assess the risk of the lap-belt slipping off the pelvis in frontal impacts." Paper No. 952708, presented at the 39th Stapp Car Crash Conference, San Diego, CA, November 1995.

O'Connor, C.S. and Rao, M.K. "Dynamic simulations of belted occupants with submarinin." Paper No. 901749, presented at the SAE Passenger Car Meeting and Exposition, Dearborn, MI, September 1990.

Shaw, Greg, et.al. "Spinal kinematics of restrained occupants in frontal impacts." Paper No. 2001-13-0019, presented at the International IRCOBI Conference on the Biomechanics of Impact, Isle of Man, United Kingdom, October 2001.

Shimamura, Munemasa, Omura, Hideo, and Isobe, Hisaaki. "An occupant movement analysis using improved input data for MVMA-2D simulation." Paper No. 870332, presented at the SAE International Congress and Exposition, Detroit, MI, February 1987.

Song, D. "Finite element simulation of the occupant/belt interaction: chest and pelvis deformation, belt sliding and submarining." Paper No. 933108, presented at the 37th Stapp Car Crash Conference, San Antonio, TX, November 1993.

Stephens, V.M., Piper, A.J., and Mascia, A.M. "Modelling of the safety and comfort of rear seat car occupants." Paper No. 925219, presented at the 24th FISITA Congress, London, England, June 1992.

Viano, David C., Culver, Clyde C., and Prisk, Bert C. "Influence of initial length of lap-shoulder belt on occupant dynamics~a comparison of sled testing and MVMA-2D modeling." Paper No. 801309, presented at the 24th Stapp Car Crash Conference, Troy, MI, October 1980.

TESTING

Seiffert, U.W., "Description of a Universal Pulling Machine." 720223, SAE Congress, 1972.

ALTERNATIVES TO MANUAL, BODY MOUNTED 3-POINT BELTS

AUTOMATIC BELTS / PASSIVE BELTS

Appleby, Michael R., Pratt, Joseph J. and Hodge, Bruce E. "Insurance losses in relation to the safety characteristics of automobiles demonstrated by an automatic vs manual seat belt study." Paper No. 810216, presented at the SAE International Congress and Exposition, Detroit, MI, February 1981.

Ashworth, R. A. et. al. "A review of development of passive restraint systems." Paper No. 746040, presented at the Fifth International Technical Conference on Experimental Safety Vehicles, London, England, January 1974.

Ashworth, Roger A. "The development of a passive seatbelt restraint system." Paper No. 766034, presented at the Sixth International Technical Conference on Experimental Safety Vehicles, Washington, DC, January 1976.

Bohlin N., Pilhall, S., "Consumer Acceptance of the Volvo Passive Belt System." 720428, 2nd International Conference on Passive Restraints, Detroit, MI 1972.

Bradford, G.M., Broughton, J. and Jakob, H. "Design approach and crash performance of automatic seat belt systems." Paper No. 720436, presented at the National Automobile Engineering Meeting, Detroit, MI, May 1972.

Broadhead, William G. and Weiss, Kurt D. "Driver usage patterns of non-motorized, door-mounted, passive 3-point seat belts." Paper No. 970119, presented at the SAE International Congress and Exposition, Detroit, MI, February 1997.

Coenen, Michael J.W. "How to turn a 3-point inertia belt into a passive seat-belt system." Paper No. 720523, presented at the International Automotive Engineering Congress and Exposition, Detroit, MI, January 1973.

Evans, L. "Motorized two-point safety belt effectiveness in preventing fatalities." Paper No. 1990-12-0012, presented at the Association for the Advancement of Automotive Medicine, Annual Meeting, Scottsdale, AZ, October 1990.

Finch, Peter and Giffen, William. "The United Kingdom technical presentation ~ a review of seat belt effectiveness and investigation of potential improvements, including passive systems." Paper No. 736035, presented at the Fourth International Technical Conference on Experimental Safety Vehicles, Kyoto, Japan, January 1973.

Hellriegel, Edmund and Rauthmann, Axel. "The Ford automatic safety belt system." Paper No. 746052, presented at the Fifth International Technical Conference on Experimental Safety Vehicles, London, England, January 1974

Johannessen, George H. "Historical review of automatic seat belt restraint systems." Paper No. 870221, presented at the SAE International Congress and Exposition, Detroit, MI, February 1987.

Johannessen, George H. "Passive and semi-passive seat belts for increased occupant safety." Paper No. 720438, presented at the National Automobile Engineering Meeting, Detroit, MI, May 1972.

Johannessen, George, H. "Automatic belt restraint systems for motor vehicle occupants." Paper No. 856019, presented at the Tenth International Technical Conference on Experimental Safety Vehicles, Oxford, England, January 1985.

Kallieris, Dimitrios et al. "The performance of active and passive driver restraint systems in simulated frontal collisions." Paper No. 942216, presented at the 38th Stapp Car Crash Conference, Ft. Lauderdale, FL, October 1994.

Laughery., Kenneth R., Laugherty, Keith A. and Lowoll, David R. "Automatic shoulder belt manual lap belt restraint systems: human factors analyses of case studies data." Paper No. 1996-14-0030, presented at the 40th Annual Meeting of the Human Factors Society, Philadelphia, PA, August 1996.

Nash, Carl E. and Eisemann, Barry. "Fatality rates in Toyota Cressidas with automatic belts." Paper No. 856079, presented at theTenth International Technical Conference on Experimental Safety Vehicles, Oxford, England, January 1985.

Padmanaban, Jeya and Ray, Rose M. "Comparison of automatic front-seat-outboard occupant restraint system performance. " Paper No. 1994-12-0048, presented at the Advances in Occupant Restraint Technologies AAAM/IRCOBI meeting, Lyon, France, September 1994.

Peck, Derek P. "Semipassive seat belt system." Paper No. 720435, presented at the National Automobile Engineering Meeting, Detroit, MI, May 1972.

Phillips, N.S. "Fully passive restraint systems: alternatives to inflatable systems." Paper No. 1973-12-0004, presented at the American Association for Automotive Medicine Annual Meeting, Oklahoma City, OK, November 1973.

Pilhall, Stig and Bohlin, Nils. "Passive safety belt system." Paper No. 720440, presented at the National Automobile Engineering Meeting, Detroit, MI, May 1972.

Reinfurt, D.W., Cyr, C.L. and Hunter, W.W. "Usage patterns and misuse rates of automatic seat belts by system type." Paper No. 1990-12-0011, presented at the Association for the Advancement of Automotive Medicine, Annual Meeting, Scottsdale, AZ, October 1990.

Rosenau, Wolfgang and Welkey, George M. "Field performance of Volkswagen automatic restraint system." Paper No. 806033, presented at the Eighth International Technical Conference on Experimental Safety Vehicles, Wolfsburg, Germany, January 1980.

Shingleton, J.A. "Passive seat belt systems." Paper No. 716018, presented at the Second International Technical Conference on Experimental Safety Vehicles, Sindelfingen, Germany, January 1971.

Takada, Juichiro. "Seat belt type passive restraint system." Paper No. 720680, presented at the National Automobile Engineering Meeting, Detroit, MI, May 1972.

Ventre, P, Rullier, J.C., et al. "An Objective Analysis of the Protection Offered by Active and Passive Restraint Systems." 750393, SAE Congress, 1975.

ABTS (ALL BELTS TO SEAT)

Automotive Engineering. "Audi's cabriolet will feature integral seat." Society of Automotive Engineers, Inc., Warrendale, PA, July 1993.

Automotive Engineering. "Belt-integrated safety seat." Society of Automotive Engineers, Inc., Warrendale, PA, September 2000.

Cole, Joanne H. "Developing a cost effective integrated structural seat." Paper No. 930109, presented at the SAE International Congress and Exposition, Detroit, MI, March 1993.

Cremer, Heinz P. "Seat-integrated safety belt." Paper No. 860053, presented at the SAE International Congress and Exposition, Detroit, MI, February 1986.

Harberl, J., Ritzl, F. and Eichinger, S. "The effect of fully seat-integrated front seat belt systems on vehicle occupants in frontal crashes." Paper No. 896136, presented at the Twelfth International Technical Conference on Experimental Safety Vehicles, Goteborg, Sweden, January 1989.

Journal of Automotive Engineering. "Crash analyses and design of a belt-integrated seat for occupant safety." Institution of Mechanical Engineers, London, England, January 2001.

Moscarini, F. and Biagi, A. "Study of a safety seat with incorporated belts." Paper No. 796019, presented at the Seventh International Technical Conference on Experimental Safety Vehicles, Paris, France, January 1979.

Rashidy, Mostafa et. al. "Analytical evaluation of an advanced integrated safety seat design in frontal, rear, side, and rollover crashes." Paper No. 2001-06-0017, presented at the International Technical Conference on Enhanced Safety of Vehicles, Amsterdam, The Netherlands, June 2001.

Ruter, Gert and Hontschik, Heinrich. "Protection of occupants of commercial vehicles by integrated seat/belt systems." Paper No. 791002, presented at the 23rd Stapp Car Crash Conference, San Diego, CA, October 1979.

Saczalski, Kenneth J. et. al. "Belt-integrated vehicular seat rear impact studies." Paper No. 2000-05-0189, presented at the 2000 FISITA World Automotive Congress-Automotive Innovation for the New Millennium, Seoul, Korea, June 2000.

Walter, R., Bartha, E. and Ruter, G. "Requirements to the deformation behavior of passenger seats for coaches with integrated seat belts." Paper No. 977136, presented at the Commercial Vehicles-Safety, Environment and Efficiency Innovations, Mannheim, Germany, June 1997.

Zhou, Rongrong, Hong, Wei and Lakshminaravan, Venkat. "Design targets of seat integrated restraint system for optimal occupant protection." Paper No. 2001-01-0158, presented at the SAE 2001 World Congress, March 2001.

BELT INNOVATIONS: INFLATABELTS, 4 POINTS, ETC.

Automotive Engineering. "New seatbelt fiber increases safety." Society of Automotive Engineers, Inc,. Warrendale, PA, October 1999.

Automotive Engineering. "The state of inflatable belt restraints." Society of Automotive Engineers, Inc., Warrendale, PA, February 1992.

Billault, P. et. al. "The inflatable diagonal belt (improvement of protection in the case of frontal impact)." Paper No. 796029, presented at the Seventh International Technical Conference on Experimental Safety Vehicles, Paris, France, January 1979.

Danese, Gaetano. "A research study for an energy absorbing sliding seat." Paper No. 746055, presented at the Fifth International Technical Conference on Experimental Safety Vehicles, London, England, January 1974.

Hammer, David R., McClenathan, V. Robert and Karigiri, Shyam S. "Performance requirements for an inflatable seatbelt assembly." Paper No. 1999-01-3233, presented at the International Body Engineering Conference and Exposition, Detroit, MI, September 1999.

Horsch, John H., Horn, Gerald and McCleary, Joseph D. "Investigation of inflatable restraints." Paper No. 912905, presented at the 35th Stapp Car Crash Conference, San Diego, CA, November 1991.

Karigiri, Shyam S. et.al. "Injury mitigating benefits of an inflatable shoulder belt for seat integrated application." Paper No. 1999-01-0085, presented at the SAE International Congress and Exposition, Detroit, MI, March 1999.

Natalini, Tiffani et. al. "Variables influencing shoulder belt positioning of four-point safety belts." Paper No. 2001-01-0382, presented at the SAE 2001 World Congress, Detroit, MI, March 2001.

Rodewald, Hanns-Ludecke, Kuhnel IA V, Arne and Franzmann Keiper Recaro, Kaiserslautern, Gunter. "Investigations made on riding-up of the lap belt of a safety harness system." Paper No. 860052, presented at the SAE International Congress and Exposition, Detroit, MI, February 1986.

Sadeghi, M.M. and Mellander, H. "Alternative passenger car seat-seat belt design for crash safety." Paper No. 880900, presented at the International Conference on Vehicle Structural Mechanics, Detroit, MI, April 1988.

Seiffert, Ulrich and Schwanz, Wilfried. "Performance matrices of four restraint systems." Paper No. 740583, presented at the 3rd International Conference on Occupant Protection, Troy, MI, July 1974.

Snyder, Richard G. "Survey of automotive occupant restraint systems: we've been, where we are and our current problems." Paper No. 690243, presented at the International Automotive Engineering Congress and Exposition, Detroit, MI, January 1969.

REAR SEAT OCCUPANTS AND CHILDREN

REAR SEAT OCCUPANTS

Campbell, B.J. "The effectiveness of rear-seat lap-belts in crash injury reduction." Paper No. 870480, presented at the SAE International Congress and Exposition, Detroit, MI, February 1987.

Cuerden, Richard W. et. al. "The injury experience of adult rear-seat car passengers." Paper No. 1997-13-0018, presented at the International IRCOBI Conference on the Biomechanics of Impacts, Hannover, Germany, September 1997.

Evans, Leonard. "Rear compared to front seat restraint system effectiveness in preventing fatalities." Paper No. 870485, presented at the SAE International Congress and Exposition, Detroit, MI, February 1987.

Foret-Bruno, J.Y. et. al. "Could a lap belt in the rear center position save human lives?" Paper No. 916159, presented at the 13th International Technical Conference on Experimental Safety Vehicles, Paris, France, November 1991.

Gikas, Paul W. and Huelke, Donald F. "Pathogenesis of fatal injuries to rear seat occupants of automobiles." Paper No. 650970, presented at the Ninth Stapp Car Crash Conference, Minneapolis, MN, October 1965.

Green, R.N. et. al. "Abdominal injuries associated with the use of rear-seat lap belts in real-world collisions." Paper No. 1987-13-0009, presented at the International IRCOBI Conference on the Biomechanics of Impacts, Birmingham, England, September 1987.

Haberl, Josef, Eichinger, Siegfried and Wintershoff, Werner. "New rear safety belt geometry ~ a contribution to increase belt usage and restraint effectiveness." Paper No. 870488, presented at the SAE International Congress and Exposition, Detroit, MI, February 1987.

Heulke, Donald F. and Compton, Charles P. "The effects of seat belts on injury severity of front and rear seat occupants in the same frontal crash." Paper No. 1994-12-0001, presented at the 38th Annual Association for the Advancement of Automotive Medicine, Lyon, France, September 1994.

Huelke, D.F. and Compton, P. "Rear seat occupants in frontal crashes - adults and children - the effects of restraint systems." Paper No. 1995-13-0030, presented at the International IRCOBI Conference on the Biomechanics of Impacts, Brunnen, Switzerland, September 1995.

Huelke, D.F. and Lawson, T.E. "The rear seat automobile passenger in frontal crashes." Paper No. 1978-12-0013, presented at the American Association for Automotive Medicine, Annual Meeting, Ann Arbor, MI, July 1978.

Huelke, Donald F., Sherman, Harold W. and Elliott, Audrey F. "The rear seat occupant from data analysis of selected clinical case studies." Paper No. 870487, presented at the SAE International Congress and Exposition, Detroit, MI, February 1987.

Johannessen, George H. and Pilarski, Regis, V. "Rear seat occupant protection." Paper No. 810797, presented at the SAE Passenger Car Meeting and Exposition, Dearborn, MI, June 1981.

Kahane, Charles J. "Fatality and injury reducing effectiveness of lap belts for back seat occupants." Paper No. 870486, presented at the SAE International Congress and Exposition, Detroit, MI, February 1987.

Kallieris, Dimitrios and Schmidt, Georg. "Neck response and injury assessment using cadavers and the US-SID for far-side lateral impacts of rear seat occupants with inboard-anchored shoulder belts." Paper No. 902313, presented at the 34th Stapp Car Crash Conference, Orlando, FL, November 1990.

Karlbring, Leif and Mellander, Hugo. "A three-point belt in the rear center seating position as accessories." Paper No. 870483, presented at the SAE International Congress and Exposition, Detroit, MI, February 1987.

Lundell, B., Mellander, H. and Carlsson, I. "Safety performance of a rear seat belt system with optimized seat cushion design." Paper No. 810796, presented at the SAE Passenger Car Meeting and Exposition, Dearborn, MI, June 1981.

Norin, H. et. al. "Injury-reducing effect of seat belts on rear seat passengers." Paper No. 806063, presented at the Eighth International Technical Conference on Experimental Safety Vehicles, Wolfsburg, Germany, January 1980.

Otte, Dietmar, Sudkamp, Norbert and Appel, Hermann. "Residual injuries to restrained car occupants in front- and rear- seat positions." Paper No. 876031, presented at the Eleventh International Technical Conference on Experimental Safety Vehicles, Washington, DC, January 1987.

Padmanaban, Jeya and Ray, Rose M. "Rear seat belt effectiveness for passenger cars." Paper No. 1992-15-0057, presented at the 12th World Congress of the International Association for Accident and Traffic Medicine and 7th Nordic Congress on Traffic Medicine, Helsinki, Finland, June 1992.

Padmanaban, Jeya and Ray, Rose M. "Safety performance of rear seat occupant resistraint systems." Paper No. 922524, presented at the 36th Stapp Car Crash Conference, Seattle, WA, November 1992.

Smith, Lynne. "Crash protection for rear seat occupants." Paper No. 870479, presented at the SAE International Congress and Exposition, Detroit, MI, February 1987.

Stephens, V.M. and Piper, A.J. "An integrated methodology for improving the safety of rear seat passengers." Paper No. 934215, presented at the Autotech '93 Safety and the Automobile, Birmingham, England, November 1993.

Stephens, V.M., Piper, A.J. and Mascia, A.M. "Modelling of the safety and comfort of rear seat car occupants." Paper No. 925219, presented at the 24th FISITA Congress, London, England, June 1992.

Valle, Harold. "Accidents analysis of rear seat-belt protection." Paper No. 893102, SIA, France, October 1989.

CHILDREN, BELT RESTRAINTS AND SEATS

Agran, P.F. and Winn, D. "Traumatic injuries among children using lap belts and lap/shoulder belts in motor vehicle collisions." Paper No. 1987-12-0018, presented at the American Association for Automotive Medicine, Annual Meeting, New Orleans, LA, September 1987.

Agran, P.F. and Winn, D.G. "Injuries among 4 to 9 year old restrained motor vehicle occupants by seat location and crash impact site." Paper No. 1988-12-0004, presented at the Association for the Advancement of Automotive Medicine, Annual Meeting, Seattle, WA, September 1988.

Agran, P.F., Dunkle, D.E. and Winn, D.G. "Restraint usage patterns of children less than four years of age evaluated in a medical setting after a motor vehicle accident." Paper No. 1984-12-0030, presented at the American Association for Automotive Medicine, Annual Meeting, Denver, CO, October 1984.

Bacon, D.G.C. "Crash restraint of children by adult seat belts and booster cushions." Paper No. 854997, Institution of Mechanical Engineers, London, England, January 1985.

Bastiaanse, J.C., Malatha, J. and Tak, A.G.M. "Evaluation of the dynamic test requirements for child restraints according to the ECE 44 regulation." Paper No. 1982-13-0020, presented at the International IRCOBI Conference on the Biomechanics of Impacts, Cologne, West Germany, September 1982.

Chamouard, F., Tarriere, C. and Baudrit, P. "Protection of children on board vehicles~Influence of pelvis design and thigh and abdomen stiffness on the submarining risk for dummies installed on a booster." Paper No. 976100, presented at the Fifteenth International Technical Conference on the Enhanced Safety of Vehicles, Melbourne, Australia, May 1996.

Chipman, Mary L., Li, Jehui and Hu, Xiaohan. "The effectiveness of safety belts in preventing fatalities and major injuries among school-aged children." Paper No. 1995-12-0010, presented at the 39th Annual Association for the Advancement of Automotive Medicine, Chicago, IL, October 1995.

Clark, Carl C. "Learning from child protection devices and concepts from outside of the United States." Paper No. 831666, presented at the 27th Stapp Car Crash Conference with IRCOBI and Child Injury and Restraint Conference with IRCOBI, San Diego, CA, October 1983.

Czernakowski Romer-Britax, Waldemar. "Usage of adult belts in conjunction with child safety systems as a means to optimize convenience." Paper No. 840527, presented at the SAE International Congress and Exposition, Detroit, MI, February 1984.

Czernakowski, Waldemar and Bell, Robert. "The effects of belt pretensioners on various child restraint designs in frontal impacts." Paper No. 973314, presented at the Second Child Occupant Protection Symposium, Lake Buena Vista, FL, November 1997.

Czernakowski, Waldemar and Otte, Dietmar. "The effect of pre-impact braking on the performance of child restraint systems in real life accidents and under varying test conditions." Paper No. 933097, presented at the Child Occupant Protection Symposium, San Antonio, TX, November 1993.

Dalmotas, D.J. et. al. "Current activities in Canada relating to the protection of children in automobile accidents." Paper No. 840529, presented at the SAE International Congress and Exposition, Detroit, MI, February 1984.

DeSantis-Klinich, Kathleen, et. al. "Survey of older children in automotive restraints." Paper No. 942222, presented at the 38th Stapp Car Crash Conference, Ft. Lauderdale, FL, October 1994.

Edwards, Jack and Sullivan, Kaye. "Where are all the children seated and when are they restrained?" Paper No. 971550, presented at the SAE Government Industry Meeting and Exposition, Washington, DC, May 1997.

Gotschall, Catherine S. et. al. "Injuries to children restrained in 2- and 3-point belts." Paper No. 1998-12-0003, presented at the 42nd Annual Association for the Advancement of Automotive Medicine, Charlottesville, VA, October 1998.

Hall, W.L. and Stewart, J.R. "Effects of child passenger safety and seat belt legislation on fatalities and serious injuries to children involved in North Carolina crashes." Paper No. 1991-12-0002, presented at the Association for the Advancement of Automotive Medicine, Annual Meeting, Toronto, Ontario, Canada, October 1991.

Heap, Samuel A. and Grenier, Emile P. "Restraint system for small children riding in passenger cars." Paper No. 680002, presented at the Automotive Engineering Congress and Exposition, Detroit, MI, January 1968.

Henderson, Michael, Brown, Julie and Griffiths, Michael. "Adult seat belts: how safe are they for children?" Paper No. 976101, presented at the Fifteenth International Technical Conference on the Enhanced Safety of Vehicles, Melbourne, Australia, May 1996.

Henderson, Michael, Brown, Julie and Griffiths, Michael. "Children in adult seat belts and child harnesses: crash sled comparisons of dummy responses." Paper No. 973308, presented at the Second Child Occupant Protection Symposium, Lake Buena Vista, FL, November 1997.

Huijskens, C.G., Janssen, E.G. and Verschut, R. "The influence of asymmetrical lower belt anchorage locations on the crash performance of child restraint systems." Paper No. 1993-13-0029, presented at the International Conference on the Biomechanics of Impacts, Eindhoven, Netherlands, September 1993.

Janssen, E.G. et. al. "Cervical spine loads induced in restrained child dummies." Paper No. 912919, presented at the 35th Stapp Car Crash Conference, San Diego, CA, November 1991.

Janssen, E.G. et. al. "Cervical spine loads induced in restrained child dummies II." Paper No. 933102, presented at the Child Occupant Protection Symposium, San Antonio, TX, November 1993.

Khaewpong, Nopporn, et. al. "Injury severity in restrained children in motor vehicle crashes." Paper No. 952711, presented at the 39th Stapp Car Crash Conference, San Diego, CA, November 1995.

Krafft, M., Nygren, C. and Tingyall, C. "Rear seat occupant protection - A study of children and adults in the rear seat of cars in relation to restraint use and car characteristics." Paper No. 896137, presented at the Twelfth International Technical Conference on Experimental Safety Vehicles, Goteborg, Sweden, January 1989.

Lane, J.C. "The seat belt syndrome in children." Paper No. 933098, presented at the Child Occupant Protection Symposium, San Antonio, TX, November 1993.

Lowne, Richard et. al. "The effect of the UK seat belt legislation on restraint usage by children." Paper No. 840526, presented at the SAE International Congress and Exposition, Detroit, MI, February 1984.

Newman, James A. and Dalmotas, Daninius. "Atlanto-occipital fracture dislocation in lap-belt restrained children." Paper No. 933099, presented at the Child Occupant Protection Symposium, San Antonio, TX, November 1993.

Pincemaille, Y. et. al. "Booster cushions: from experimentation to usage in France." Paper No. 933096, presented at the Child Occupant Protection Symposium, San Antonio, TX, November 1993.

Robbins, D.H., Henke, A.W. and Roberts, V.L. "Study of concepts in child seating and restraint systems." Paper No. 700041, presented at the Automotive Engineering Congress and Exposition, Detroit, MI, January 1970.

Schneider, Lawrence W., Melvin, J.W. and Cooney, C. Ernest. "Impact sled test evaluation of restraint systems used in transportation of handicapped children." Paper No. 790074, presented at the Automotive Engineering Congress and Exposition, Detroit, MI, February 1979.

Thompson, A.L. and Verreault, R. "Who are using child restraint systems and why - data from four major Canadian cities." Paper No. 1979-12-0023, presented at the American Association for Automotive Medicine, Annual Meeting, Louisville, KY, October 1979.

Warren Bidez, Martha and Syson, Stephen R. "Kinematics, injury mechanisms and design considerations for older children in adult torso belts." Paper No. 2001-01-0173, presented at the SAE 2001 World Congress, Detroit, MI, March 2001.

Waters, P.E. "Development of a seat belt booster cushion standard." Paper No. 831653, presented at the 27th Stapp Car Crash Conference with IRCOBI and Child Injury and Restraint Conference with IRCOBI, San Diego, CA, October 1983.

Weber, Kathleen and Radovich, Vladislav G. "Performance evaluation of child restraints relative to vehicle lap- belt anchorage location." Paper No. 870324, presented at the SAE International Congress and Exposition, Detroit, MI, February 1987.

Weinstein, Elaine B., et. al. "The effect of size appropriate and proper restraint use on injury severity of children." Paper No. 973310, presented at the Second Child Occupant Protection Symposium, Lake Buena Vista, FL, November 1997.

SECTION 1:

SAFETY BELT SYSTEMS AND PERFORMANCE

CHAPTER 1:

BACKGROUND STUDIES

840392

Historical Perspective on Seat Belt Restraint Systems

H. George Johannessen
OmniSafe Associates, Inc.

ABSTRACT

Landmarks in the chronology of the development of seat belt restraint systems are highlighted and examined on the bases of the influence of requirements of legislation, product performance, and marketing considerations. Recent developments in component hardware are identified.

SEAT BELT HISTORY will soon be in its hundredth year. As early as 1885 belts were used on horse-drawn vehicles to prevent passengers from being ejected from the vehicles on rough roads. The "incubation" period before seat belts were installed in all seating positions in all new passenger cars was over 80 years - until 1968 - and nearly 100 years will have been passed before seat belts are present in virtually all seating positions in all passenger cars in the total car population in the United States. This long lag is difficult to understand in light of the early use of seat belts in airplanes and racing cars. U.S. Army Plane No. 1 was equipped with a seat belt in 1910. Open cockpit commercial aircraft were required to be equipped with seat belts as of 31 December 1926, and all commercial aircraft had this requirement a few years later. Barney Oldfield equipped his racer with a seat belt in 1922. All race drivers in recognized races in the United States are now required to wear seatbelts. Many passenger car occupants in the United States continue to disregard the seat belts installed in their cars and fail to use them.

Seat belt history may be considered from the standpoint of seat belt geometry and components, seat belt performance requirements, comfort and convenience, usage rates, legislation and regulations affecting seat belts, and litigation relating to the use or non-use of seat belts.

SEAT BELT HISTORY - SYSTEM GEOMETRY AND COMPONENTS

The very early seat belts were nothing more than conventional leather belts secured by primitive means to the vehicle seat or structure. No outstanding improvements were seen until the 1950s and 1960s when a significant amount of development effort began to be directed to improved designs. The webbing-to-metal buckle design (Fig. 1), characteristic of earlier airplane applications, was supplanted quite quickly by the metal-to-metal buckle designs. The non-locking

Fig. 1 - Webbing-to-metal buckle

retractor (Fig. 2) was added for improved stowage of the webbing when the belt was not in use. Soon afterward, the automatic-locking retractor (Fig. 3) became available. At about this time, in the mid-1960s, upper torso restraint was receiving increased attention. A limited production application of a Y-yoke shoulder harness with a webbing-sensitive emergency-locking retractor (Fig. 4) appeared in the Shelby-American GT 350 and GT 500 cars. As of 1 January 1968 all seating positions in all passenger cars manufac-

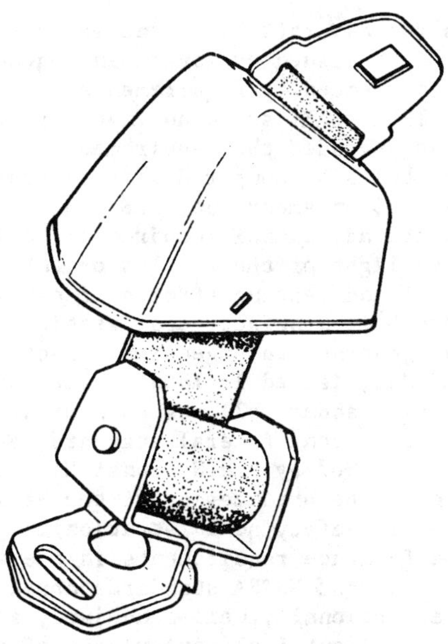

Fig. 2 - Non-locking retractor

Fig. 3 - Automatic-locking retractor

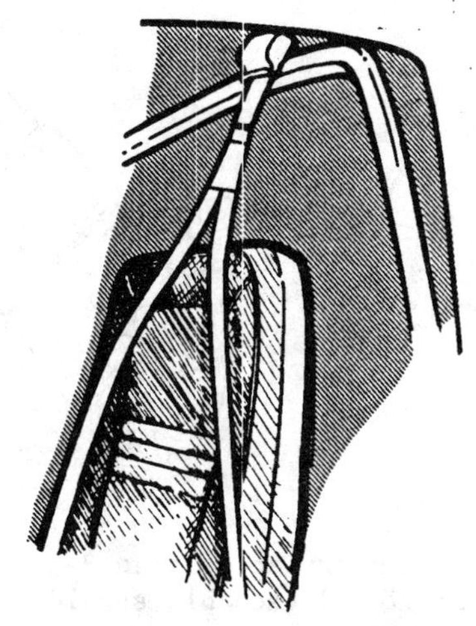

Fig. 4 - Shoulder harness with emergency-locking retractor

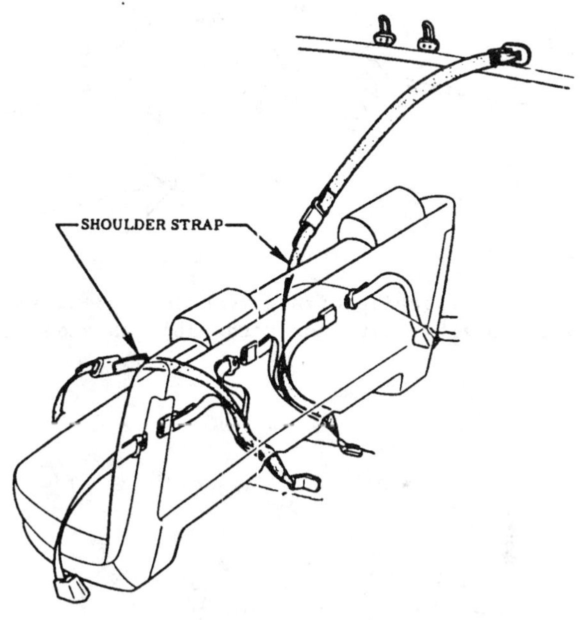

Fig. 5 - 4-Point front outboard seat belt system

tured for sale in the United States were required to have seat belts and front seat outboard positions were required to have shoulder straps as well. The first installations in front outboard positions were 4-point systems (Fig. 5). These were superseded by 3-point systems (Fig. 6) in which the detachable shoulder strap was connected to the lap belt by means of a terminal fitting on the shoulder strap that could be attached or detached from the tongue of the lap belt. As of the start of production of the 1974 Model Year cars - at the same time that the starter-interlock was introduced - both shoulder strap and lap belt were required to terminate in a common tongue that engaged the seat belt buckle (Fig. 7). Shortly thereafter, a modified 3-point system (the "continuous-loop", or "unibelt" system) was introduced in which a continuous length of webbing passing through a slot in the tongue comprised the webbing for both lap and upper-torso portions of the belt system (Fig. 8). The introduction of a new, thinner polyester webbing made it possible to store the additional webbing required by the continuous-loop system on the retractor without unduly increasing the package size of the retractor. The polyester webbing provided strength equal to that of the Nylon webbing. In addition, it provided lower elongation of the webbing under load, which would result in decreased excursion of the restrained occupant in the event of an accident.

SEAT BELT HISTORY - PERFORMANCE REQUIREMENTS

Early model automotive seat belts were based

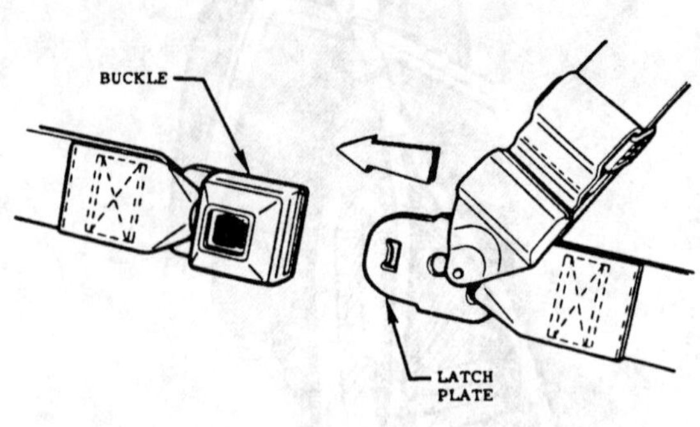

Fig. 6 - Latching arrangement in 3-point seat belt assembly with detachable shoulder strap

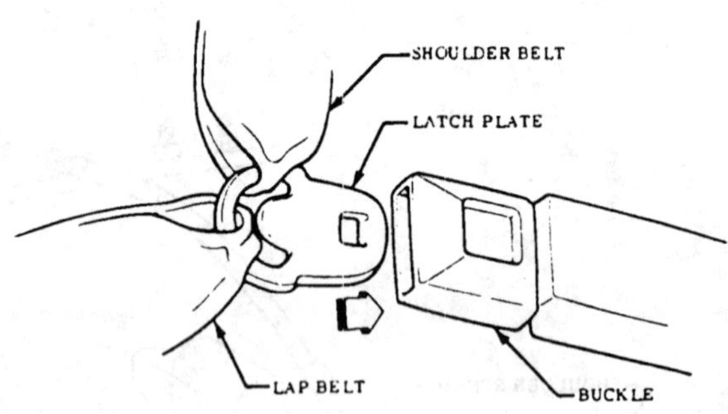

Fig. 7 - Latching arrangement in 3-point seat belt assembly with non-detachable shoulder strap

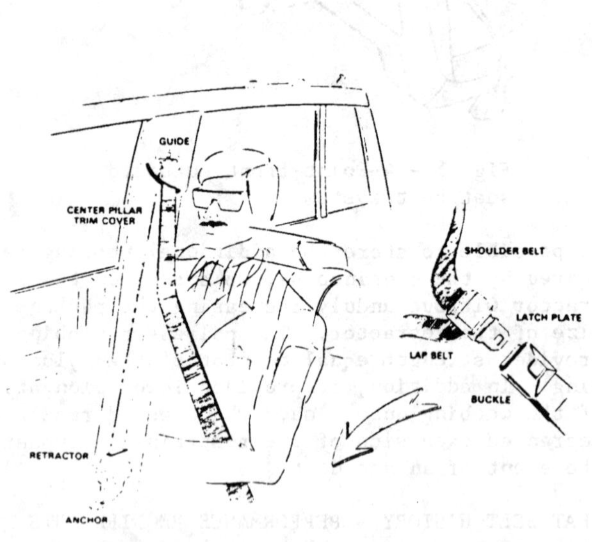

Fig. 8 - "Continuous-loop" seat belt system

on designs for aircraft belts and early standards for these belts issued by government agencies (1)* were influenced by requirements for aircraft belts. In 1961, SAE issued an industry standard, SAE J4, that upgraded the requirements in automotive seat belts as compared with aircraft belts. In 1963 an amendment, SAE J4a, was issued to incorporate additional requirements considered necessary in light of the results of intensive developmental and testing efforts of car manufacturers, seat belt manufacturers, webbing manufacturers, and governmental agencies. Another revision, SAE J4c, issued in 1965 became the basis for a federal standard (2) issued shortly thereafter and the current federal standard, FMVSS 209, issued in 1967 by the National Highway Safety Bureau (the predecessor of the National Highway Traffic Safety Administration).

The performance requirements in the early SAE and the current NHTSA standards appear to have been exceptionally well-conceived, as borne out by the excellent field experience of seat belts in the past twenty five years. Seat belts are not a panacea to prevent all crash fatalities and injuries, but they appear to be able to prevent about half the crash fatalities and two thirds of the serious and severe injuries (3). With further analysis of the substantial number of definitive crash statistics now being acquired in the FARS (4) and NASS (5) data files at NHTSA, it may be found that improved seat belt performance at higher crash velocities maybe not only be desirable but also cost-effective. VW demonstrated in barrier crash tests that an improved belt restraint system incorporating a belt pretensioner would provide occupant protection at 40 mph Barrier Equivalent Velocity (6). The VW system was displayed in 1972 at "Transpo 72" in Washington, D.C. Daimler-Benz has offered an optional advanced supplemental restraint system in Germany since late 1980 having a pretensioner on a 3-point belt restraint in the front passenger seating position (7). Similar systems are now being offered on an optional basis in the United States on Mercedes cars. Field data for cars having these improved systems should soon provide a basis for determining the cost-effectiveness of these systems.

SEAT BELT HISTORY - COMFORT AND CONVENIENCE

Since Model Year 1964, when seat belts began to be factory-installed in all passenger cars, changes in system geometry and components ahve occurred quite rapidly to improve the comfort and convenience of the systems. In 1964 non-locking retractors (Fig. 2) became available to roll up and stow the webbing of the outboard portion of the lapbelt when it was not in use. Non-locking retractors, however, did not provide for automatic adjustment of the lap belt length. Automatic adjustment was first provided in 1965

*Numbers in parentheses designate References at end of paper.

by automatic-locking retractors (ALR) which provided for automatic take-up of paid-out webbing until it was snug on the occupant and automatic lock-up of the webbing at this length, in addition to retracting and stowing the webbing on a covered roller when the seat belt was doffed. With this device on the outboard portion of the seat belt assembly, the inboard portion containing the buckle could be shortened to project only a short distance above the seat. The buckle end could then also be enclosed and supported by a stiffening sleeve which would position the buckle for ready accessibility. Automatic-locking retractors were quite ideally suited for use in lap belts, but they were subject to one criticism. They could "cinch up" - that is, tighten up more than a desirable amount - when the occupant moved downward and compressed the seat when travelling over excessively bumpy roads or terrain. Also, automatic-locking retractors were not useful in upper torso restraints because they did not permit occupant movement.

Emergency-locking retractors (ELR) were developed to overcome the deficiencies of the automatic-locking retractors. They were well-suited for application in shoulder straps since they did not "cinch up" and did provide for free movement of the webbing unless lock-up occurred because of quick evasive movement of the car or an emergency event such as an accident or a panic stop. Emergency-locking retractors may lock up as a result of webbing acceleration or vehicle acceleration. Emergency-locking retractors sensitive to vehicle acceleration (VSR) are basically simple in concept and design (Fig. 9), reliable, and well-suited for applications in passenger cars. (They are not as well-suited for certain specialized applications, such as on the air-suspension seat on a heavy truck where the retractor could be subject to frequent unwanted lock-ups as a result of the bouncing and "chucking" movements of the seat to which it is mounted. A webbing-sensitive ELR is better suited for this application.) Lock-up in webbing-sensitive emergency-locking retractors is initiated by an inertial device which responds to webbing acceleration (Fig. 9). Emergency-locking retractors were applied on a low-production basis in 1967 in the Shelby-American GT 350 and GT 500, and were used in shoulder straps of all new U.S. passenger cars in Model Year 1974 and thereafter.

Most foreign cars sold currently in the United States have emergency-locking retractors that have two inertial sensing modes. One is sensitive to webbing acceleration, and the other to vehicle acceleration. This dual-sensing redundancy has been retained to secure the earlier response and lock-up of the vehicle-acceleration sensing mode, and at the same time permit the user to test the operability of the device by jerking on the webbing to effect lock-up by means of the webbing-acceleration sensing mode.

In 1975, a lap-shoulder belt having only a single emergency-locking retractor appeared in some U.S. passenger cars (Fig. 8). This design used a continuous length of webbing for both the lap portion and the upper torso portion of the seat belt assembly. A tongue sliding on the webbing was positioned manually by the user to permit insertion into the buckle. In some versions of this design the tongue incorporated an adjustment feature to permit snug adjustment of the lap-belt portion (Fig. 8). The upper-torso portion was subject to the action of the emergency-locking retractor. This system design provided adequate performance, good stowage when not in use, and lower cost. Surveys conducted for NHTSA have shown that the subjects queried considered this system less convenient to access and don than a two-retractor system. (8)

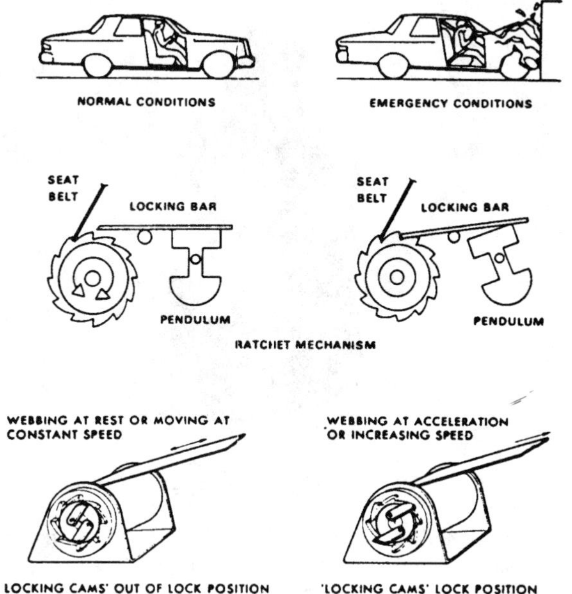

Upper: Vehicle acceleration sensitive mode

Lower: Webbing acceleration sensitive mode

Fig. 9 - Operation of emergency-locking retractors

The "dual-spool" two-retractor system (Fig. 10) was developed to retain the comfort and convenience advantages of the two-retractor system and be more cost-competitive with the "continuous-loop" single-retractor system. This system first appeared in Model Year 1979 in vans and in light trucks in Model Year 1980.

SEAT BELT HISTORY - LEGISLATION AND REGULATIONS

Legislative and regulatory actions have had significant effects on seat belt development and installation. The enactment in 23 states of legislation effective 1 January 1964 requiring installation of seatbelts in front outboard seating positions in all new cars sold in those states prompted all car manufacturers to begin installation of these seat belts in all passenger cars regardless of destination at the start of produc-

Fig. 10 - Dual-spool retractor

tion for Model Year 1964. This action wrought a fundamental change in the seat belt supply scene. Before 1963, most seat selts were produced in relatively small production quantities by a large number of small manufacturers for "after-market" sales. After 1963, most seat belts were produced by "OEM" (original equipment manufacturers) who were generally well-established suppliers of the car manufacturers. Thereafter, seatbelt development benefitted from the broader base of these manufacturers' facilities and funding. Also in 1963, the Congress passed Public Law 88-201 to enable the issuance of standards for seat belts sold or shipped in interstate commerce. The resulting standard (9) was issued by the Department of Commerce in 1965 and a revision (10), incorporating the new requirements of J4c, was issued in 1966.

In 1966, the Congress enacted Public Law 89-593 (11), the "National Highway Traffic Safety Act of 1966", to "reduce traffic accidents and deaths and injuries. . . resulting from traffic accidents". This enabling legislation was implemented by the National Highway Safety Bureau (NHSB), which issued "Initial Federal Motor Vehicle Safety Standards" in 1967. These included MVSS 208, "Seat Belt Installations - Passenger Cars" (12), which designated the number and kind of seat belts to be installed, and MVSS 209, "Seat Belt Assemblies - Passenger Cars, Multipurpose Vehicles, Trucks, and Buses" (13), which specified the requirements for the seat belts. MVSS 209 became effective in March of 1967 and has continued with minimal changes and interpretations since then. MVSS 208, which included a requirement for seat belt installations in all forward-facing seating positions and shoulder straps in front outboard seating positions, became effective on 1 January 1968 and has had a complicated and colorful history ever since.

Dr. William Haddon, M.D., the original director of the National Highway Safety Bureau, viewed the high incidence of highway traffic deaths and injuries as an "epidemic" and held that the most effective way to bring the "epidemic" under control would be by "passive" means of car occupant protection, that is, occupant crash protection requiring no action on the part of the car occupants. The available means for such "passive" or "automatic" protection were air bags and automatic seat belt systems. The direction set by Dr. Haddon was essentially continued by his successors, the Administrators of the National Highway Traffic Safety Administration (the successor agency to the NHSB), until October 1981 when rulemaking was issued during the tenure of Mr. Raymond Peck that amended MVSS 208 to rescind the requirement for automatic restraints for front seat occupants. This rulemaking action was subsequently challenged in the United States Supreme Court and remanded to NHTSA by action of the Court. It is currently undergoing reconsideration. A parallel action by NHTSA is a comprehensive, nationwide effort to increase seat belt usage.

SEAT BELT HISTORY - USAGE RATES

Seat Belt usage rates have been carefully monitored for the past several years in the NHTSA "19-City" program (14). In earlier years, usage rates were determined by numerous local studies which varied considerably in quality and coverage. Many surveys were also conducted to determine the reasons for the low usage rates. Before Model Year 1964, usage of installed lap belts was undoubtedly high because all belt installations resulted from a conscious decision by the car owner to pay extra for the belts as either a retrofit or dealer-installed option. Early statistics from various sources tended to show usage rates after 1964 in the order of 30-35% in the period from 1964 to 1976 (Fig. 11).

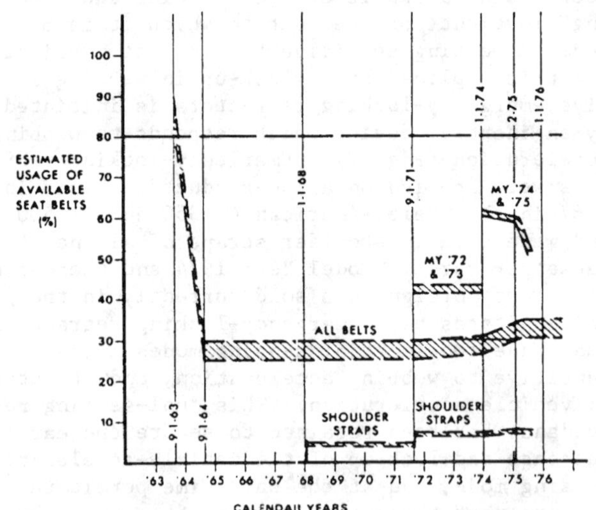

Fig. 11 - Seat belt usage rates from early surveys

Statistics from the "19-City" program indicate a small but significant average increase in usage, from 10.9% to 14.6%, in the period from 1979 to 1983. The range in usage rates from one locality to another is very large, however, and is explained on the basis of the demographics of the human and car populations. In general, for example, usage rates tend to be higher among females, higher socio-economic and better educated groups, users of foreign and small cars, and freeway travellers. Drivers of VW Rabbit cars with automatic restraints are a unique group, with usage rates of about 80% persisting through the years (Fig. 12).

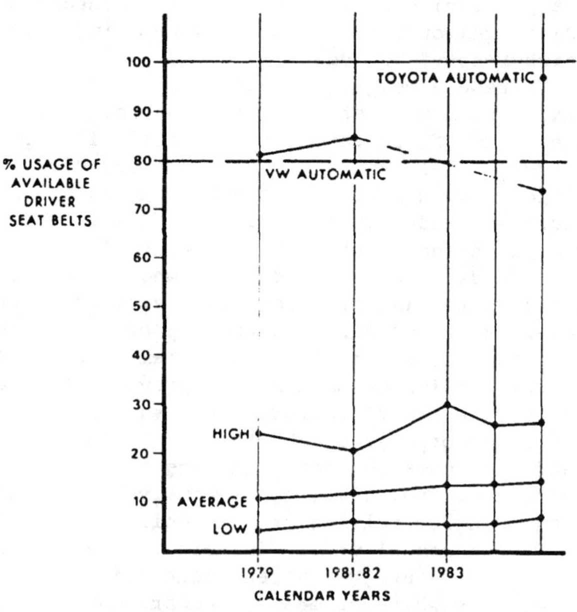

Fig. 12 - Seat belt usage rates from "19-City" program

The possibility for increasing usage rates in defined groups through local programs has been demonstrated by General Motors and others (15). In the General Motors program at the GM Technical Center, the usage rate increased from about 36% to over 70% during the course of the program and maintained a residual rate of about 60% after the termination of the program.

Surveys to analyze seat belt usage rates as affected by public perception of seat belts consistently show that people generally feel that seat belts provide effective protection, but that present seat belt systems are considered uncomfortable and inconvenient. Some investigators have concluded from sophisticated studies (16) that comfort and convenience are not really significant factors in determining an individual's decision to use or not use a seat belt. To a degree this is probably true. A committed user of seat belts (a "hard-core user") will not be deterred by inconvenience or discomfort of the available system. A determined non-user (a "hard-core non-user") will not don the most comfortable and convenient system conceivable.

These two special groups taken together, however, constitute only a third of the total vehicle-borne population (Fig. 13). The remaining two-thirds are the target of programs aimed to increase seat belt usage. The intuitively obvious truism that discretionary actions will be done more often when they can be done more easily would support the contention that the perception of an individual in this group as to the comfort and convenience of the seat belt system will significantly influence his dicision to use or not use the seat belt.

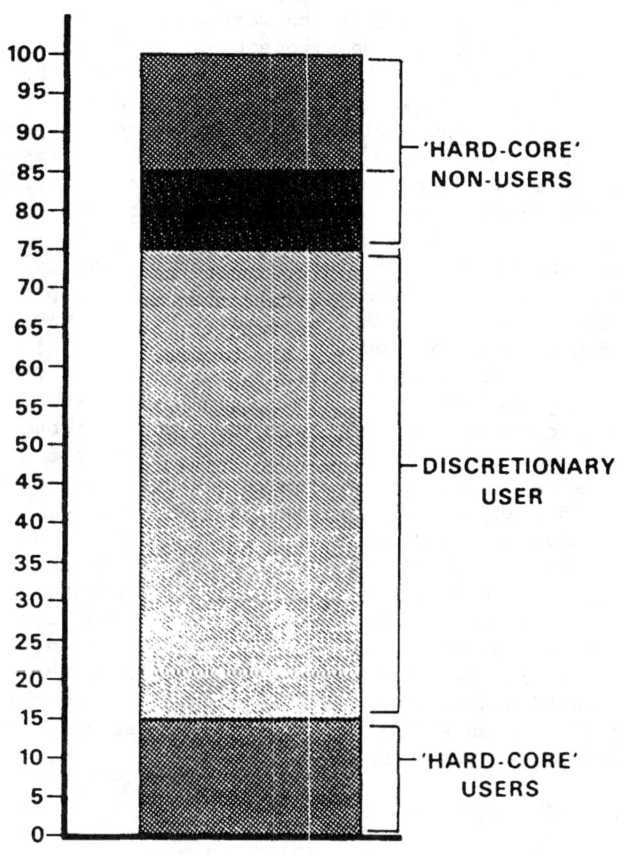

Fig. 13 - Percentage distribution of seat belt user population

Existing statistics provide some measure of usage rates that may be anticipated with certain populations under specified conditions. The "19-City" study, for example, shows that the average usage rate for those nineteen cities is increasing and is now approaching 15% (Fig 12). The range for the individual cities in the study extends from a low of 5-6% for Fargo, North Dakota to a high of about 25% for Seattle. The GM study showed that a well-conceived, well-executed program for a specific, well-defined group could achieve a usage rate of about 70%. A continuing usage rate of about 80% of the VW automatic restraints demonstrates that a satisfactorily high usage rate can be achieved with automatic belt restraints with car owners who have made a conscious decision to order and pay extra for these systems. Experience in foreign countries having mandatory belt usage laws shows that usage rates of 80-85% are achieved with reasonable enforcement of the laws and nominal fines.

SEAT BELT HISTORY - LITIGATION

Litigation involving seat belts is increasing and can be expected to continue to increase. The seat belt involvement in lawsuits may arise from either the use or non-use of the available seat belts.

Injured plaintiffs who were wearing seat belts may have had the erroneous perception that the seat belt would be a panacea to protect them from all harm in all events. They may have had unrealistic expectations as to the degree of protection at excessive speeds, or in certain crash modes (e.g., certain side or rear impacts), or in the event of excessive intrusion into the occupant space. Realistic appraisals of protective capabilities of seat belts in many typical crash modes have been documented in reports presented before the SAE, AAAM, and other technical and professional groups (17) (18).

Injured plaintiffs may allege that they were wearing seat belts that were defective in some way and did not afford the protection that could reasonably be expected. Such allegations are addressed to determine whether the seat belt was indeed used, was operational, and did afford reasonable protection.

Non-use of seat belts is becoming increasingly important as an issue in civil litigation (19). Recent studies undertaken for NHTSA (20) (21) have held that the prudent person can increasingly be expected to avail himself of the protection afforded by the seat belt available in his car and that he exhibits some degree of negligence in not doing so.

SEAT BELT HISTORY - SUMMARY

The chronology of some of the landmark events in the history of seat belts is summarized graphically in Figure 14. It is apparent from the graph that most of the significant developments in seat belts have occurred in the last twenty years in spite of the recognition of the utility and effectiveness of seat belts nearly one hundred years ago.

RECENT DEVELOPMENTS IN SEAT BELT HARDWARE

Developmental efforts with seat belt hardware are proceeding continuously to improve performance, reliability, comfort and convenience, universality in application, and amenability to packaging in the vehicle; and to decrease complexity and cost.

With the continued downsizing of cars and increasing numbers of cars with front wheel drives, seat belt buckles are being designed to sustain the higher loads that may be experienced by the restraint systems in accidents involving these cars. In addition, with the increase in bucket seats in downsized cars, buckle designs and buckle installation arrangements are undergoing intensive review to permit design modifications that will improve convenience in donning and doffing the seat belt in these installations.

The belt pretensioners introduced in production cars by Mercedes (22) may very well prove to provide a new dimension in occupant protection against fatalities and severe injuries in higher-speed accidents.

To reduce weight, complexity, and cost of retractors in two-retractor systems and to increase comfort, convenience, and reliability, the "dual-spool" retractor has been broadly applied in vans and light trucks. An additional feature was added to a "dual-spool" retractor for application on the 1983 Corvette (Fig. 15). To accommodate dedicated drivers who appreciate the secure feeling of a snug lap belt, the emergency-locking retractor on the lap belt portion of the two-retractor, 3-point system incorporated a manual locking mechanism that provides the continuously locked retractor. This same feature could be incorporated into the front passenger location of more conventional cars to provide a convenient, positive, continuously-locked condition in a lap belt securing a child restraint.

Developmental effort in belt restraint systems in the future, when seat belt usage increases - by whatever means - toward more acceptable rates, will continue to be driven by the same objectives as those currently operative: increased performance and reliability, comfort and convenience, and reduced weight, size, complexity, and cost.

Fig. 15 - Corvette "dual-spool" retractor assembly - Emergency-locking lap belt retractor has manual optional locking provision

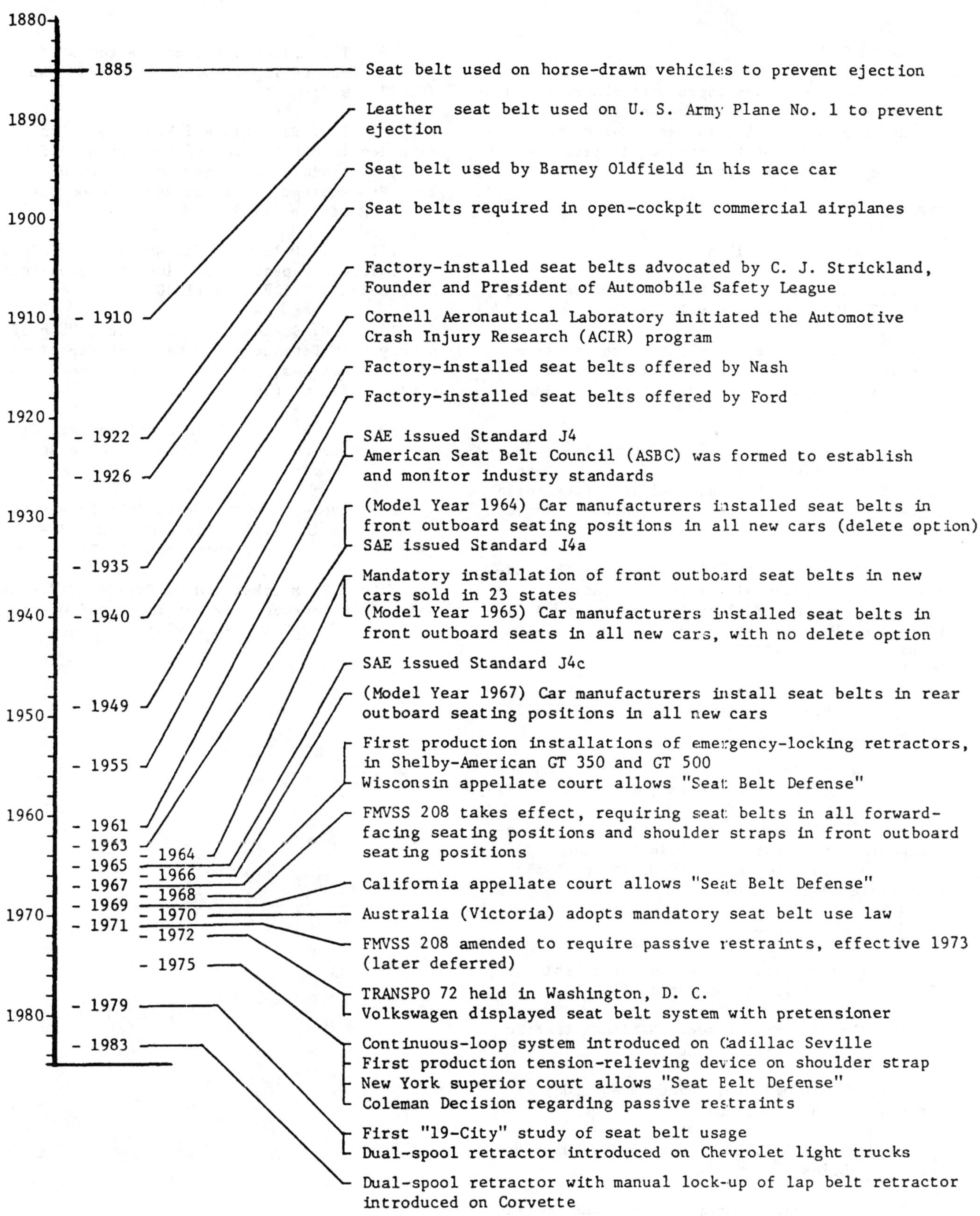

Fig. 14 - Seat belt history

ACKNOWLEDGEMENT

The author acknowledges with thanks the support provided by the American Seat Belt Council in the preparation of this report, and the cooperation of ASBC member companies in providing hardware for the illustrations.

REFERENCES

1. GSA - JJB - 185a

2. "Seat Belts for Use in Motor Vehicles", 15 CFR 9, 30 FR 5132 (July 1965)

3. D.A. Huelke, et al, "Effectiveness of Current and Future Restraint Systems in Fatal and Serious Injury Automobile Crashes" - SAE 790323 (1979)

4. FATAL ACCIDENT REPORTING SYSTEM, administered by National Center for Statistics and Analysis, National Highway Traffic Safety Administration, U.S. Department of Transportation, Washington, D.C. 20590

5. NATIONAL ACCIDENT SAMPLING SYSTEM, administered by National Center for Statistics and Analysis, National Highway Traffic Safety Administration, U.S. Department of Transportation, Washington, D.C. 20590

6. H.P. Willumeit, "Passive Preloaded Energy-Absorbing Seat Belt System", SAE 720433 (1972)

7. W. Reidelbach and H.J. Scholz, "Advanced Restraint System Concepts", SAE 790321 (1979)

8. "Evaluation of the Comfort and Convenience of Safety Belt Systems in 1980 and 1981 Model Vehicles", DOT HS-805 860, (March 1981) (available through NTIS)

9. See 2 above

10. "Seat Belts for Use in Motor Vehicles", 15 CFR 9, 31 FR 11528 (August 1966)

11. 15 USC 1391 et seq., National Traffic and Motor Vehicle Safety Act

12. 49 CFR 571.208, Standard No. 208; Occupant Crash Protection

13. 49 CFR 571.209, Standard No. 209; Seat Belt Assemblies

14. "Restraint System Usage in the Traffic Population", NHTSA Contract DT-NH 22-82-C-07126

15. T.D. Horne and C.T. Terry, "Seat Belt Sweepstakes - An Incentive Program", SAE 830474 (1983)

16. T.J. Kuechenmeister, "A Comparative Analysis of Factors Impacting on Seat Belt Use", SAE 790687 (1979)

17. D.J. Dalmotas and P.M. Keyl, "An Investigation into the Level of Protection Afforded to Fully Restrained Passenger Vehicle Occupants", 22nd Proceedings, American Association for Automotive Medicine (1978)

18. D.J. Dalmotas, "Mechanisms of Injuries to Vehicle Occupants Restrained by Three-Point Seat Belts", SAE 801311 (1980)

19. J.D. States and T.G. Smith, "Use of the Safety Belt Defense - The New York Experience", 27th Proceedings, American Association for Automotive Medicine (1983)

20. Unpublished NHTSA Staff Report, "The Seat Belt Defense - The Education of a Reasonable Person", Summer 1981

21. "Non-Use of Motor Vehicle Safety Belts as an Issue in Civil Litigation", DOT HS-806 443, (August 1983), (available through NTIS)

22. J.E. Mitzkus and H. Eyrainer, "Three-point Belt Improvements for Increased Occupant Protection", SAE 840395 (1984)

950865

Regulatory History of Automatic Crash Protection in FMVSS 208

Stephen R. Kratzke
National Highway Traffic Safety Administration

ABSTRACT

This paper summarizes the regulatory history of the automatic crash protection requirements in Federal Motor Vehicle Safety Standard 208. It is intended to give the reader an overview of the regulatory history involved in Standard 208, from its beginning in 1968 as a requirement that passenger cars be equipped with seat belts to its present requirement that, as of 1998, all passenger cars and light trucks must be equipped with air bags. It also discusses and summarizes the various court cases that have challenged different aspects of the automatic crash protection requirements.

INTRODUCTION

This paper traces the evolution of the occupant protection requirements in Federal Motor Vehicle Safety Standard 208 from its beginnings as a requirement for seat belts to be installed in passenger cars to its current requirements that each passenger car provide an air bag and a manual lap/shoulder belt for the driver and right front passenger position beginning in the 1998 model year (September 1, 1997) and that light trucks and vans provide an air bag and a manual lap/shoulder belt for the driver and right front passenger beginning in the 1999 model year (September 1, 1998). The purpose of this paper is to give the reader an understanding of how these requirements evolved into their present form and what purpose they are intended to serve.

THE FIRST OCCUPANT PROTECTION REQUIREMENTS IN STANDARD 208

ORIGINAL OCCUPANT PROTECTION REQUIREMENTS - Standard 208 was one of the 19 original Federal motor vehicle safety standards. It required that passenger cars provide a seat belt at every forward-facing designated seating position. This requirement took effect on January 1, 1968. There were no crash testing requirements to evaluate the protection afforded to vehicle occupants.

INITIAL AUTOMATIC PROTECTION REQUIREMENTS - It was not long, however, before the National Highway Traffic Safety Administration (NHTSA) began to explore the possibility of requiring automatic crash protection in motor vehicles. Vehicles that provide automatic crash protection protect their occupants by means that require no action by the vehicle occupants. The two types of automatic crash protection that have been offered for sale on production vehicles are automatic seat belts and air bags. The effectiveness of a vehicle's automatic crash protection is assessed through crash testing. A vehicle must comply with specified injury criteria, as measured on a test dummy, when tested in a 30 mph crash test.

On July 2, 1969, NHTSA published an advance notice of proposed rulemaking (ANPRM) requesting comments on the merits of crash protection systems that protect vehicle occupants by means that require no action on the part of the occupants.[1] This notice specifically mentions the possibility of meeting such a requirement by means of air bags. After evaluating these comments, NHTSA published a notice proposing to require automatic crash protection for all passenger cars beginning January 1, 1972 and for all light trucks and vans beginning January 1, 1974.[2]

On November 3, 1970, NHTSA published a final rule that required automatic crash protection for all passenger cars as of July 1, 1973, and for most light trucks and vans as of July 1, 1974. Compliance would have been determined by a crash test with test dummies in the front outboard seats.[3] In the preamble to this rule, the agency made clear its position that automatic protection would supplant the need for vehicle occupants to use manual seat belts. NHTSA said: "Several comments recommended that the requirement for seat belts be retained, citing the benefits of keeping the driver in his seat during violent maneuvers and the possibility of failure of a passive system. It is the [NHTSA's] position that the possible benefits of required seat belts would not justify the costs to the manufacturers and to the public. Only a small percentage of the public uses the upper torso restraints that are presently furnished with passenger

---

[1] 34 FR 11148

[2] 35 FR 7187; May 7, 1970

[3] 35 FR 16927

cars."[4] In other words, passenger cars, light trucks, and vans would no longer be required to provide manual seat belts at each designated seating position once the automatic protection requirements took effect.

NHTSA received many petitions for reconsideration of this rule. In response to those petitions, NHTSA published a notice that postponed the effective date of the automatic protection requirements for passenger cars from July 1, 1973 until August 15, 1973, and granted a similar 45 day postponement for light trucks and vans, to correspond more precisely to the manufacturers' changeovers for a new model year's production.[5] This notice also repeated NHTSA's previous conclusion that the existing test dummy specifications published by the Society of Automotive Engineers were "the best available." The notice went on to state that, "NHTSA is sponsoring further research and examining all available data, however, with a view to issuance of further specifications for these devices."[6]

As indicated above, the automatic protection requirements did not require the use of any particular technology to achieve the desired result. Instead, manufacturers were free to use any means they chose so long as their vehicles provided the specified level of protection with no action by vehicle occupants. Up to this point, however, the regulatory notices dealing with automatic protection and the public comments responding to those notices had focused almost exclusively on air bag systems. In a July 8, 1971 final rule,[7] NHTSA added explicit language to Standard 208 to acknowledge that automatic belts could be used to meet the automatic protection requirements and to state the applicability of various requirements in the Standard to automatic belts (called "passive belts" at that time). On that same day, the agency published a proposal that automatic protection systems have a means of emergency release.[8] NHTSA suggested that automatic belts could use a spool-out mechanism and air bags would meet this requirement by deflating.

In response to petitions for reconsideration of the July 8, 1971 final rule on automatic belts, NHTSA excluded automatic belts from the assembly performance and webbing requirements. The agency explained that this change was made to allow manufacturers as much freedom in the design of automatic belt systems to fit the particular crash pulse of each car as they have in the design of other types of automatic protection systems.[9]

In addition to broadening the focus of its automatic protection rulemaking to recognize means other than air bags, the agency also introduced into its automatic protection rulemaking the concept of an ignition interlock system. Several vehicle manufacturers asked NHTSA to delay the date by which automatic protection had to be installed in passenger cars because of unresolved technical problems with automatic protection systems. On October 1, 1971,[10] NHTSA proposed to postpone the effective date for mandatory automatic protection from August 15, 1973 until August 15, 1975. However, if a car produced between those dates did not provide automatic protection, it had to be equipped with an interlock system that would prevent the engine from starting if any front seat occupants did not have their manual seat belts fastened. Front outboard seating positions would also be subject to a crash test with a test dummy in each such seating position and the manual belt system fastened around the test dummy. The agency explained its proposal as follows: "It is intended by this option to provide a high level of seat belt usage, and to increase the life- and injury-saving effectiveness of installed belt systems, in the interim period before [automatic] systems are required." The ignition interlock option was adopted as proposed in a rule published February 24, 1972.[11]

THE COURT DECISION IN CHRYSLER V. DOT

Shortly after the March 10, 1971 rule was published requiring automatic protection in new vehicles, Chrysler, Jeep, American Motors, Ford, and the Automobile Importers of America filed lawsuits in the U.S. Court of Appeals for the Sixth Circuit challenging the automatic protection requirements. These lawsuits raised three primary arguments:

1. The automatic protection requirements were not "practicable," as required by NHTSA's authorizing legislation, because the technology needed to comply with automatic protection was not sufficiently developed as of that time. The manufacturers argued that NHTSA had no authority to establish a safety standard that required the industry to improve upon the existing technology. Under this view, automatic protection should not be required until devices to meet the requirement were sufficiently developed as of the date of the rulemaking so as to permit ready installation.

2. The automatic protection requirements did not "meet the need for motor vehicle safety," as required by NHTSA's authorizing legislation, because seat belts offered better occupant protection than automatic protection.

3. The automatic protection requirements were not "objective," as required by NHTSA's authorizing legislation, because the existing SAE Recommended Practice did not adequately specify sufficient details for the construction of the test dummy.

The Sixth Circuit announced its decision on these lawsuits on Dec 5, 1972 in an opinion titled Chrysler v. DOT, 472 F.2d 659 (6th Cir. 1972). The court ruled in favor of NHTSA on the first argument, stating that "the agency is empowered to issue safety standards which require improvements in existing technology or which require the development of new technology, and it is not limited to issuing standards based solely

[4] 35 FR 16928

[5] 36 FR 4600; March 10, 1971

[6] 36 FR 4602

[7] 36 FR 12858

[8] 36 FR 12866; July 8, 1971

[9] 36 FR 23725; December 14, 1971

[10] 36 FR 19266

[11] 37 FR 3911

on devices already fully developed."[12] The court also upheld the agency on the second point raised by the manufacturers, ruling that the question of whether to require automatic protection was delegated to NHTSA, that there was substantial support in the record for the decision to mandate automatic protection, and so the court had no basis for substituting its judgment for that of the agency.[13] However, the court found in favor of the manufacturers on the third argument. The court concluded that the SAE J963 test dummy incorporated in Standard 208 was not defined with sufficient specificity to meet the statutory requirement for "objectivity." Because of this shortcoming, this issue was remanded to the agency with instructions to delay the automatic crash protection requirement until a reasonable time after objective dummy specifications had been issued.[14]

The Sixth Circuit was petitioned by Ford to clarify the effect of its Chrysler decision on the requirement in Standard 208 for crash testing of the manual belts at front outboard seating positions in cars to be equipped with an ignition interlock. The Sixth Circuit announced its decision on this petition on February 2, 1973 in an opinion titled Ford v. NHTSA, 473 F.2d 1241 (6th Cir. 1973). The court ruled that its conclusion that the test dummy was not objective was equally applicable to this crash testing, and ordered the agency to delay the crash testing requirements for manual belt systems at front outboard seating positions in cars equipped with an ignition interlock.

The Ford v. NHTSA opinion was particularly problematic, because it overturned on February 2 an option that was scheduled to take effect on August 15 of that same year. All car manufacturers except General Motors had intended to choose the ignition interlock option, but it had now been declared invalid. On April 20, 1973, NHTSA addressed this problem by proposing to delete the requirement for crash testing of manual belts at the front outboard seats in cars equipped with an ignition interlock system.[15] NHTSA adopted this proposal in a June 20, 1973 final rule.[16] That June 20 final rule also announced NHTSA's position that the decision in the Chrysler case invalidated the automatic protection requirements, regardless of whether the language mandating automatic protection remained in the text of the standard. The agency also announced in the June 20 rule that additional rulemaking would be needed to reestablish an effective date for automatic protection requirements.

NHTSA proposed to adopt much more detailed test dummy specifications in a notice published April 2, 1973. [17] In fact, these proposed specifications were the existing specifications General Motors used for its Hybrid II test dummy. The Hybrid II test dummy was adopted as the test dummy to be used in Standard 208 compliance testing in a final rule published on August 1, 1973.[18] That August 1 rule also repeated the agency's previous announcement that it would give the public further notice and opportunity for comment before making a final decision on whether to reinstate the automatic protection requirements.

THE IGNITION INTERLOCK EXPERIENCE

As of August 15, 1973, all new cars had to be equipped either with automatic protection or an ignition interlock for both front outboard seating positions. General Motors sold a few thousand of its 1974 model year cars equipped with air bags that met the automatic protection requirement. Every other 1974 model year car sold in the United States came with an ignition interlock, which prevented the engine from operating if either the driver or front seat outboard passenger failed to fasten their manual seat belt.

In a March 19, 1974 notice, NHTSA described the public reaction to the ignition interlock as follows: "Public resistance to the belt-starter interlock system currently required (except on vehicles providing [automatic] protection) has been substantial, with current tallies of proper lap-shoulder belt usage on 1974 models running at or below the 60% level. Even that figure is probably optimistic as a measure of results to be achieved, in light of the likelihood that as time passes the awareness that the forcing systems can be disabled, and the means for doing so will become more widely disseminated, ...[19] There were also speeches on the floor of both houses of Congress expressing the public's anger at the interlock requirement. On October 27, 1974, President Ford signed into law a bill that prohibited any Federal motor vehicle safety standard from requiring or permitting the use of any seat belt interlock system. In response to this change in the law, NHTSA published a final rule on October 31, 1974 that deleted the interlock option from Standard 208 effective immediately.[20]

While the interlock option was still in place, NHTSA also addressed the subject of automatic belts in response to a petition from Volkswagen. That company was going to introduce in its 1975 model year Rabbit models an automatic belt system that consisted of an upper torso restraint and knee bolsters, and asked for some changes and clarifications to the requirements of Standard 208 to ensure that this new belt system would comply with the standard. NHTSA proposed to require a manual single point emergency release mechanism on all automatic belts to allow vehicle occupants postcrash egress from the vehicle.[21] In the final rule adopting this proposal[22], NHTSA noted that one commenter had objected to the emergency release mechanism for automatic belts because the mechanism could be used to

---

[12] 472 F.2d 673

[13] 472 F.2d 674-675

[14] 472 F.2d 681

[15] 38 FR 9830

[16] 38 FR 16072

[17] 38 FR 8455

[18] 38 FR 20449

[19] 39 FR 10272

[20] 39 FR 38380

[21] 39 FR 3834; January 30, 1974

[22] 39 FR 14593; April 25, 1974

disconnect the belts in non-emergency situations. NHTSA responded that the advantages of having an emergency release mechanism outweighed the disadvantages of possible abuse.[23] Another commenter suggested that a lever or pushbutton that allowed the belt to spool out from the retractor instead of separating would be a more appropriate emergency release mechanism. NHTSA responded by stating that it believed the uniformity of having push-button action to release all motor vehicle seat belt assemblies was more compelling than the advantages suggested by the commenter of permitting a spool-out release.[24] The agency also provided more information about how it would determine whether an automatic belt system provided the necessary protection with "no action by vehicle occupants," as specified in Standard 208. Volkswagen did in fact introduce automatic belts as an optional piece of equipment on its 1975 Rabbit models.

THE COLEMAN DECISION

Standard 208 had never allowed the ignition interlock option to be anything more than an interim measure on the way to full automatic protection. The interlock was scheduled to expire as a permissible option on August 15, 1975 in any event. Thus, the change in the law eliminated the ignition interlock 10 months before it would have expired anyway. The more significant question at this point was whether rulemaking would be initiated to reinstate the automatic protection requirements and when any such requirements would take effect.

In a March 19, 1974 notice,[25] when the interlock option was still in place, NHTSA proposed to reinstate the automatic protection requirements for front outboard seating positions in passenger cars as of September 1, 1976. A little more than one year later, NHTSA again proposed to extend until September 1, 1976 the period during which manufacturers could comply with Standard 208 simply by installing manual belts at front seating positions.[26] However, this later notice did not propose any specific date for reinstating automatic protection requirements. In fact, this April 1975 notice announced that "A decision has not yet been made on the long-term requirements for occupant crash protection." The proposed extension of existing requirements until September 1, 1976 was made final in an August 13, 1975 rule.[27] The August 1975 rule stated: "While the NHTSA recognizes that the present crash protection options will in all likelihood be in effect for some period after August 31, 1976, the agency has not proposed more than the 1-year extension. ... The NHTSA intends to propose the long-term requirements for occupant crash protection, both for passenger cars and for light trucks and MPV's, as soon as possible."

The Secretary of Transportation, William T. Coleman, Jr., published a proposal on June 14, 1976 that had several noteworthy aspects.[28] This notice announced that Secretary Coleman was personally taking responsibility for the decision on automatic crash protection requirements in Standard 208. All previous decisions had been left to the NHTSA Administrator, subject to the review and approval of the Secretary. This notice proposed five options for dealing with occupant protection in frontal crashes. These were:

1. Continue existing manual belt requirements and continue further research to identify effective means of automatic protection.
2. Continue existing manual belt requirements and try to encourage States to pass mandatory belt use laws, which would substantially increase use of manual belts.
3. Continue existing manual belt requirements and conduct a Federally sponsored field test of automatic protection in vehicles used on the public roads.
4. Reinstate automatic protection requirements.
5. Require manufacturers to provide consumers with the option of ordering automatic protection in some or all of their models.

In accordance with this decision, NHTSA published a proposal[29] and final rule[30] extending the period during which manufacturers had the option of complying with Standard 208's occupant protection requirements by simply installing manual seat belts at all designated seating positions. That option was now extended until August 31, 1977, to allow time for the Secretary to announce his final decision on the automatic protection requirements.

On December 6, 1976, Secretary Coleman announced his decision on automatic protection. This decision was not published in the Federal Register, but written copies were placed in the public docket at the Department of Transportation. Secretary Coleman called on auto manufacturers to join with the Federal government in a demonstration program so that approximately 500,000 cars with passive restraint systems would be offered for sale at a reasonable cost to consumers in the 1979 and 1980 model years (i.e., the period beginning Sept. 1, 1978 and ending August 31, 1980). Secretary Coleman also announced that the Department of Transportation would make additional efforts to promote seat belt use during this period. At the end of this demonstration project, the Secretary concluded that the Department could make a more informed choice about the need for automatic protection requirements, based upon the real world experience gained from the 500,000 production vehicles that would be equipped with automatic protection and updated estimates of likely manual belt use in the future.

A notice appeared in the January 27, 1977 Federal Register[31] announcing Secretary Coleman's December 6, 1976 decision and "incorporating it by reference" in the Federal Register

---

[23] 39 FR 14593-14594

[24] 39 FR 14594

[25] 39 FR 10271

[26] 40 FR 16217; April 10, 1975

[27] 40 FR 33977

[28] 41 FR 24070

[29] 41 FR 29715; July 19, 1976

[30] 41 FR 36494; August 30, 1976

[31] 42 FR 5071; January 27, 1977

notice. The January 27 notice also extended indefinitely the option for manufacturers to comply with Standard 208 by installing belts at all designated seating positions in the vehicle.

THE ADAMS DECISION

Less than two months after the publication of the Federal Register notice announcing Secretary Coleman's final decision on the automatic protection requirements in Standard 208, a notice was published announcing that the new Secretary of Transportation was reexamining that decision. Thus, on March 24, 1977, a notice was published in the Federal Register[32] announcing that the new Secretary of Transportation, Brock Adams, was conducting a reexamination of Secretary Coleman's decision on automatic protection. This notice asked for public comments on the following three alternatives:

1. Leave the Coleman decision in place;
2. Reinstate the automatic protection requirements; or
3. Seek to raise the usage level for manual belts by encouraging the States to pass belt use laws.

Before a final decision was announced on the automatic protection requirements for cars, NHTSA published a notice extending indefinitely the existing occupant protection requirements for light trucks and vans. This notice was published on June 2, 1977[33], and allowed manufacturers to either install automatic protection for the front outboard seating positions or to install manual seat belts at those seating positions. NHTSA explained in this notice that the agency had originally intended that manufacturers would have had the benefit of the experience with installing automatic protection in passenger cars before automatic protection systems were required in light trucks and vans. Since the Secretary of Transportation was in the process of deciding whether automatic protection systems should be required in passenger cars, it seemed premature to require those systems in light trucks and vans. Hence, NHTSA announced an indefinite extension of the option for light trucks and vans to meet the occupant protection requirements simply by installing manual seat belts at the front outboard seating positions. The agency noted that this indefinite extension did not preclude future rulemaking to modify the occupant protection requirements for light trucks and vans, but promised that notice and opportunity for comment would be provided prior to any modifications.[34]

On July 5, 1977, a final rule reinstating automatic protection requirements for passenger cars was published in the Federal Register.[35] The Department of Transportation concluded that automatic protection was necessary even though manual seat belts were "highly effective" at preventing injury and ejection, because so few vehicle occupants used their manual belts.[36] The Department rejected the option of seeking mandatory seat belt use laws in the various States, because the prospects for success looked so poor.[37] The Department concluded that the demonstration programs called for in the Coleman decision were unnecessary, because they would have further delayed the mandatory introduction of occupant protection systems the Department had already found to be technologically feasible, practicable, and capable of offering substantial life-saving potential at reasonable costs.[38]

Accordingly, the July 5, 1977 rule required that all 1982 model year cars with a wheelbase over 114 inches be equipped with automatic protection. All 1983 model year cars with a wheelbase over 100 inches would be required to offer automatic protection, and all 1984 model year cars would be required to offer automatic protection. This gradual phased-in approach was intended to give vehicle manufacturers additional leadtime to overcome the greater difficulties of installing air bags in smaller cars and to give those efforts the benefit of the manufacturers' experience with installing air bags in larger cars.

Six petitions for reconsideration of this rule were filed. In addition, a group called the Pacific Legal Foundation filed a petition for a review of the rule in the U.S. Court of Appeals for the District of Columbia Circuit and asked the Department to stay the effectiveness of the automatic protection requirements for a period of time equal to the length of the judicial review. The vehicle manufacturers that filed petitions for reconsideration questioned the Department's analyses of the effectiveness of automatic protection systems. The Pacific Legal Foundation charged that the Department had failed to consider the public reaction to automatic protection and had ignored potential hazards posed by automatic protection systems. Ralph Nader and the Center for Auto Safety charged that the Department had improperly delayed implementation of the automatic protection requirements and that the Department had no authority to phase-in the automatic protection requirements gradually rather than requiring full compliance by the effective date. These petitions were denied in all significant respects by a December 5, 1977 notice in the Federal Register.[39]

THE COURT DECISION IN <u>PACIFIC LEGAL FOUNDATION v. DOT</u>

As indicated above, a group called Pacific Legal Foundation filed a lawsuit challenging the Adams decision to reinstate automatic protection requirements even before the Department had responded to the petitions for reconsideration of the Adams decision. After the Department denied their petitions, Ralph Nader and the Center for Auto Safety filed their own lawsuit challenging the Adams decision in the same court.

---

[32] 42 FR 15935

[33] 42 FR 28135

[34] 42 FR 28136; June 2, 1977

[35] 42 FR 34289

[36] 42 FR 34290; July 5, 1977

[37] 42 FR 34291-34292; July 5, 1977

[38] 42 FR 34291

[39] 42 FR 61466

The Pacific Legal Foundation argued that the Adams decision should be overturned because of three major shortcomings. First, the group argued that the data did not support the Secretary's findings on the effectiveness of air bags. Second, Pacific Legal Foundation argued that the Adams decision was unlawful because it failed to consider public reaction to the automatic protection requirements. Third, the group charged that the rule ignored collateral dangers to public safety posed by air bags. Ralph Nader and the Center for Auto Safety also challenged the rule, arguing that the Department had no authority to delay the effective date for the automatic protection requirements until the 1982 model year or to phase-in the requirements over three successive model years.

The D. C. Circuit announced its decision on this matter on Feb 1, 1979 in an opinion titled Pacific Legal Foundation v. DOT, 593 F.2d 1338 (D.C. Cir. 1979). The court upheld the Adams decision on all grounds. This court decision contains two especially significant findings. First, the court agreed with Pacific Legal Foundation that the Department must consider the likely public reaction to mandates in the safety standards in order to fulfill its responsibility to ensure that the standard is "practicable" and "meets the need for safety."[40] In this case, the court found that the Department had adequately considered the anticipated public reaction to the automatic protection requirements, notwithstanding the Department's claim that it was not required to consider the public reaction when promulgating safety standards. Second, the court expressly found that the Department could use a phase-in schedule when needed "to tailor safety standards to engineering reality."[41]

THE PECK DECISION

There had been only minor rulemaking notices from NHTSA dealing with the subject of automatic protection after the December 5, 1977 response to petitions for reconsideration of the Adams decision. Perhaps the most noteworthy rulemaking on automatic protection between 1978-1980 had to do with the emergency release mechanism on automatic belts. On May 22, 1978,[42] NHTSA published a notice proposing to allow automatic belts to use an emergency release mechanism other than the push-button, single-point release that had been required since the April 1974 final rule. This proposal was in response to a GM petition to allow a "spool release" design as the emergency release mechanism for automatic belts. The agency followed up this proposal with a November 13, 1978 final rule[43] that allowed automatic belts to use any emergency release mechanism that is a single-point release and that is accessible to a seated occupant. The agency explained its action thus: "This amendment will allow manufacturers to experiment with various [automatic] belt designs before the effective date of the [automatic protection] requirements and determine which designs are the most effective and at the same time acceptable to the public."[44]

However, NHTSA's actions in 1981 ended this period of relative stability regarding the regulatory requirements for automatic protection. On February 12, 1981,[45] a notice signed by Drew Lewis, the new Secretary of Transportation, was published in the Federal Register. This notice proposed to delay the start of the phase-in for cars to be equipped with automatic protection for one year (from model year 1982 to model year 1983). The proposal to delay the phase-in was based upon the economic difficulties then confronting the automobile industry.

On April 9, 1981, two notices signed by Secretary Lewis were published in the Federal Register. The first was a final rule delaying for one additional year the start of the phase-in of automatic protection.[46] The second was a notice proposing three alternative courses of action regarding the future status of automatic protection requirements.[47] The three alternatives on which the Department sought comment were:

1. Retain the new phase-in schedule, but reverse the sequence of vehicles. In other words, small cars would be required to provide automatic protection first (in model year 1983), then mid-size cars (in model year 1984), and finally large cars (in model year 1985).
2. Allow one additional year of leadtime, but eliminate the phase-in. In other words, all cars would have to provide automatic protection beginning in model year 1984.
3. Rescind the requirements for automatic protection.

A final rule signed by NHTSA Administrator Raymond Peck that rescinded the automatic protection requirements for cars was published on October 29, 1981.[48] This rule indicated that there was significant uncertainty about the public acceptability of automatic protection. Vehicle manufacturers had overwhelmingly indicated that they would comply with the automatic protection requirements by installing detachable automatic belts in their new cars. The agency found substantial uncertainty about the likely use rates of these detachable automatic belts and announced that it could not reliably predict that detachable automatic belts would produce even a 5 percentage point increase in belt use over the existing belt use rates for manual belts.[49] The uncertain benefits combined with the relatively substantial costs for automatic protection led NHTSA to conclude that the automatic restraint requirement was no longer reasonable or practicable.

LEGAL BATTLE CULMINATING WITH SUPREME COURT DECISION ON AUTOMATIC PROTECTION

40 593 F.2d 1345-1346

41 593 F.2d 1348

42 43 FR 21912

43 43 FR 52493

44 43 FR 52494

45 46 FR 12033

46 46 FR 21172

47 46 FR 21205

48 46 FR 53419

49 46 FR 53423

Throughout the course of this standard, groups have looked to the Federal courts to overturn decisions on the automatic protection requirements with which the group is dissatisfied. This rescission of the automatic protection requirements proved no different. On November 23, 1981, State Farm Mutual Automobile Insurance Company, joined by several other petitioners, filed a lawsuit in the U.S. Court of Appeals for the District of Columbia Circuit challenging the rescission of the automatic protection requirements.

That court announced its decision on June 1, 1982 in an opinion titled State Farm v. DOT, 680 F.2d 206 (D.C. Cir. 1982). The court unanimously reversed NHTSA's rescission of the automatic protection requirements. The court found that NHTSA had failed to consider or analyze obvious alternatives to rescission and had offered no evidence to show that the 1977 conclusion that seat belt usage would increase with automatic belts was no longer true.[50]

On September 8, 1982, NHTSA filed a petition with the U.S. Supreme Court asking the Supreme Court to review the D.C. Circuit Court's decision. The Supreme Court granted that petition on November 8, 1982.[51] The Supreme Court announced its decision on this matter on June 24, 1983 in an opinion titled Motor Vehicle Manufacturers Association v. State Farm, 463 U.S. 29 (1983). The Supreme Court unanimously ruled that the rescission of the automatic protection requirements was unlawful, or "arbitrary and capricious," because the agency had failed to consider obvious alternatives to rescission and explain why alternatives short of rescission were not chosen.[52] The Court noted that air bags, nondetachable automatic belts, and detachable automatic belts were three existing technologies that could be used to comply with the automatic protection requirement. The Supreme Court said that, even if NHTSA were correct that detachable automatic belts would yield few benefits, that fact alone would not justify rescission. Instead, the Court said that fact would justify only the modification of the automatic protection requirements to prohibit compliance by means of detachable automatic belts.[53] The Court reasoned that the necessary next step to justify rescission was for NHTSA to adequately explain why it no longer believed that compliance with the automatic protection requirements by means of either air bags or nondetachable automatic belts would be an effective and cost-beneficial way of saving lives and preventing injuries.[54] Since NHTSA had not offered such an explanation, its rescission of the automatic protection requirements was unlawful. Accordingly, the Department was ordered to conduct further rulemaking on the status of the automatic protection requirements.

THE DOLE RULE AND THE RESULTING LEGAL CHALLENGE

Rulemaking on the automatic protection requirements was begun again on October 19, 1983, when the Department of Transportation published a notice seeking comments on a wide range of alternative actions the Department might take with respect to automatic protection in response to the Supreme Court's decision.[55] This notice sought comments on three broad possibilities - retain the automatic protection requirements in their existing form and establish a new compliance date, amend the automatic protection requirements, e.g., to preclude detachable automatic belts, or rescind the automatic protection requirements. In addition, this notice asked for comments on three other supplementary actions that could be taken in conjunction with any of these three broad possibilities. The supplementary actions were: (1) conduct a demonstration program for automatic protection, along the lines of the Coleman decision, (2) seek legislation to encourage States to pass mandatory seat belt use laws, and (3) seek legislation to require vehicle manufacturers to provide consumers with an option to select air bags or automatic belts, instead of manual belts, in their new cars.[56]

The Department received more than 6,000 comments on this notice. After reviewing the comments, the Department published a supplemental notice on May 14, 1984.[57] This notice asked for further comment on alternatives being considered regarding State laws mandating seat belt use, mandatory demonstration programs for automatic crash protection, and the possibility of requiring driver's side air bags on all small cars.

A final rule reinstituting automatic crash protection for cars was signed by Elizabeth Dole, the Secretary of Transportation and published on July 17, 1984.[58] This rule provides for a phase-in of automatic protection in cars beginning September 1, 1986 (the 1987 model year). All cars manufactured after September 1, 1989 (the 1990 model year) are required to provide automatic protection. During the phase-in of automatic protection, the rule encourages manufacturers to choose to install air bags instead of automatic belts by providing a 1.5 car credit for cars with driver air bags and any type of automatic protection

---

[50] 680 F.2d 230

[51] 459 U.S. 987 (1982)

[52] All nine Justices joined in the opinion finding the rescission of the automatic protection requirements was arbitrary and capricious because it failed to consider alternatives to rescission, such as permitting compliance only by means of air bags or nondetachable automatic belts, and explain why those alternatives were not adopted. However, in a separate opinion for four of the nine Justices, Justice Rehnquist expressed the view that the rescission of requirements permitting compliance with automatic protection by means of detachable automatic belts was satisfactorily justified and would not be considered "arbitrary and capricious." 463 U.S. 57-59 (1983).

[53] 463 U.S. 47 (1983)

[54] 463 U.S. 49-51 (1983)

[55] 48 FR 48622

[56] 48 FR 48632

[57] 49 FR 20460

[58] 49 FR 28962

for the passenger. This rule also encouraged States to pass mandatory seat belt use laws. It did so by including a provision that the automatic protection requirements might be eliminated if the Secretary of Transportation made a determination by April 15, 1989 that enough States had enacted mandatory seat belt use laws that met criteria specified in Standard 208.

There were 16 petitions for reconsideration of this rule. NHTSA published a notice responding to those petitions on August 30, 1985.[59] This notice denied requests to change the phase-in schedule and to eliminate the possibility that automatic protection requirements might be rescinded if enough States enacted mandatory belt use laws. This notice also expanded NHTSA's efforts to encourage manufacturers to install air bags instead of automatic belts by adding a new one car credit. Under this new one car credit provision, manufacturers could comply with the automatic protection requirement by installing an air bag for the driver and manual lap/shoulder belt for the passenger. The one car credit provision was scheduled to remain in effect until the phase-in for automatic protection ended on August 31, 1989.

Shortly after the publication of this response to the petitions for reconsideration, a lawsuit challenging the July 1984 final rule was filed in the U.S. Court of Appeals for D.C. Circuit by State Farm Insurance Company and the State of New York. State Farm's primary argument was that the rule was unlawful because it would revoke automatic crash protection if enough States enacted mandatory belt use laws. New York argued that the rule was unlawful because it failed to require air bags or nondetachable automatic belts as the only permissible means of automatic protection.

The court announced its decision on September 18, 1986 in an opinion titled State Farm v. Dole, 802 F.2d 474 (D.C. Cir. 1986). The court upheld the Secretary's 1984 rule in all respects. The court indicated that State Farm's objection was based upon that insurer's speculation about what the Secretary might do if States passed laws and if those law were determined to comply with certain criteria. The court said that State Farm was not entitled to any legal relief until these potential events and determinations had actually occurred. In response to New York's argument that NHTSA should not have allowed detachable belts to be permitted as automatic protection, the court said: "The Safety Act does not require the Secretary to adopt the technological alternative providing the greatest degree of safety. The Act expressly permits the Secretary to consider such factors as reasonableness and practicality in addition to safety features. Both the Supreme Court and this court, moreover, have recognized the Secretary's authority to consider such factors as cost and public acceptance."[60]

THREE SIGNIFICANT RULES THAT GREW OUT OF THE DOLE RULE ON AUTOMATIC PROTECTION

USE OF A MORE ADVANCED TEST DUMMY - While the rulemaking that led to Secretary Dole's decision was underway, General Motors filed a petition for rulemaking asking that a new test dummy it had developed, called the Hybrid III dummy, be permitted to be used in testing compliance with the provisions of Standard 208, including the automatic protection requirements. NHTSA granted this petition on July 20, 1984. On April 12, 1985,[61] NHTSA published a notice proposing to incorporate the Hybrid III test dummy as a permissible alternative for Standard 208 compliance testing. NHTSA explained its proposal as follows: "Based on its review of the available test data, the agency recognizes that the Hybrid III test dummy represents an appreciable advancement in the state-of-the-art of human simulation and that its use would be beneficial for continued improvement in vehicle safety."[62] This notice proposed to take advantage of the enhanced capabilities of the Hybrid III by adding new injury criteria for the neck, lower leg, facial laceration, and chest deflection. This notice also proposed to require the use of the Hybrid III test dummy for all Standard 208 compliance testing for cars manufactured on or after September 1, 1991.

NHTSA published a final rule in this area on July 25, 1986.[63] This rule adopted the Hybrid III test dummy for Standard 208 compliance testing and adopted the proposal that the Hybrid III be the exclusive dummy for Standard 208 compliance testing as of September 1, 1991. However, the July 1986 rule adopted only one additional injury criterion for use with the Hybrid III test dummy, the proposed one for chest deflection. In response to petitions for reconsideration of this July 1986 rule, NHTSA postponed the date for mandatory use of the Hybrid III test dummy to allow more time to examine technical issues that might arise from the use of this new test dummy.[64]

After completing its technical examination of these issues, NHTSA published a notice on December 10, 1992[65] proposing that the Hybrid III test dummy would be the only dummy used in Standard 208 compliance testing beginning September 1, 1996. After reviewing the comments on this proposal, NHTSA published a final rule giving one additional year of leadtime before requiring exclusive use of the Hybrid III test dummy.[66] The Hybrid III test dummy is now the only dummy that will be used for Standard 208 compliance testing of vehicles manufactured on or after September 1, 1997 (the 1998 model year).

CRASH TESTING OF MANUAL BELTS IN LIGHT TRUCKS AND VANS - On April 12, 1985, NHTSA published a notice that proposed, among other things, that the occupant protection afforded by manual seat belts installed in front outboard seating positions of light trucks and vans be evaluated according to the same crash test used to evaluate automatic

---

[59] 50 FR 35233

[60] 802 F.2d 474, at fn.23, 486-87 (1986)

[61] 50 FR 14602

[62] 50 FR 14603

[63] 51 FR 26688

[64] 53 FR 8755; March 17, 1988

[65] 57 FR 58437

[66] 58 FR 59189; November 8, 1993

protection.[67] NHTSA adopted this proposal in a final rule published November 23, 1987.[68] The November 1987 rule required light trucks and vans manufactured on or after September 1, 1991 (the 1992 model year) to be certified as complying with this crash testing requirement.

"ONE CAR CREDIT" FOR DRIVER AIR BAGS EXTENDED UNTIL AUGUST 31, 1993 - On June 11, 1986, Ford Motor Company filed a petition asking NHTSA to permit the production of cars with driver air bags and no automatic protection for the front seat passenger after September 1, 1989, the date on which this "one car credit" provision was scheduled to expire. In its petition, Ford said that, if this request were granted, Ford would "in all likelihood" install driver air bags in the majority of its North American-designed cars. NHTSA published a notice proposing to grant Ford's petition on November 25, 1986.[69] In that notice, NHTSA indicated that the proposed extension of the one car credit "would encourage the orderly development and production of passenger cars with full-front air bag systems."[70]

Comments supporting the proposed extension of the one car credit were submitted by air bag suppliers, insurance companies and their trade associations, vehicle manufacturers and their trade associations, and researchers and other organizations involved in highway safety issues. The only comments opposing this extension were submitted by the Center for Auto Safety and Robert Phelps, a private citizen. After considering these comments, NHTSA published a final rule extending the one car credit until August 31, 1993.[71] The agency explained that the extension of the one car credit would promote the widespread introduction of air bags. In addition, the agency concluded that there were a number of technical issues that still needed to be resolved before widespread introduction of passenger air bags would occur.

One petition for reconsideration of this rule was filed by Public Citizen, a group that had not previously participated in this rulemaking. NHTSA denied this petition in a notice published November 5, 1987.[72] In the denial, NHTSA again explained that it was seeking to encourage manufacturers to install air bags, instead of automatic belts, and that all available evidence indicated that the extension of the one car credit had increased the likelihood of widespread use of air bags.

Public Citizen filed a lawsuit in the U.S. Court of Appeals for the D.C. Circuit challenging NHTSA's extension of the one car credit until 1993. The court announced its decision on July 15, 1988 in an opinion titled Public Citizen v. Steed, 851 F.2d 444 (D.C. Cir. 1988). The court unanimously upheld the extension of the one car credit provision. The court specifically found that NHTSA had extended the one car credit to encourage greater installation of air bags, because the agency believed that air bags would offer long-term overall safety gains for vehicle occupants. The court said that, even if it accepted Public Citizen's assertion that the one car credit extension would permit a near-term reduction in front seat occupant protection, "[i]t is within NHTSA's province to balance estimated long-term safety benefits against the possibility of a marginal short-term reduction in safety."[73]

AUTOMATIC PROTECTION IN LIGHT TRUCKS AND VANS

NHTSA had not taken up the question of automatic protection in light trucks and vans again since its June 1977 notice indefinitely suspending the requirements for automatic protection in those vehicles. However, the agency's rulemaking on crash testing of manual belt systems in light trucks and vans had raised the issue of occupant protection in those vehicles. On January 9, 1990, NHTSA published a notice proposing to require automatic protection to be phased in for light trucks and vans in a manner that closely paralleled the recently completed phase-in of automatic protection for passenger cars.[74] A final rule requiring automatic protection in light trucks and vans was published on March 26, 1991.[75] This rule requires that 20 percent of each manufacturer's model year 1995 production of light trucks and vans provide automatic protection, 50 percent of model year 1996 production provide automatic protection, 90 percent of model year 1997 production provide automatic protection, and all 1998 light trucks and vans must provide automatic protection.

AIR BAGS REQUIRED AS THE MEANS OF AUTOMATIC PROTECTION

On December 18, 1991, then President Bush signed into law the Intermodal Surface Transportation Efficiency Act (ISTEA). Among other things, ISTEA requires that all passenger cars manufactured on or after September 1, 1997 (the 1998 model year) and light trucks and vans manufactured on or after September 1, 1998 (the 1999 model year) provide air bags at the driver and right front passenger positions. In response to this mandate, NHTSA published a notice on December 14, 1992.[76] This notice proposed to require that passenger cars and light trucks and vans comply with the automatic protection requirements by installing air bags and manual lap/shoulder seat belts at the driver and right front passenger positions. This NPRM also proposed to require that these vehicles have a label on the sun visor providing occupants with important safety information about the air bags and advising occupants that they must always wear their safety belts for maximum safety

[67] 50 FR 14589; April 12, 1985

[68] 52 FR 44898

[69] 51 FR 42598

[70] 51 FR 42599

[71] 52 FR 10096; March 30, 1987

[72] 52 FR 42440

[73] 851 F.2d 444, at 449 (1988).

[74] 55 FR 747

[75] 56 FR 12472

[76] 57 FR 59043

protection in all types of crashes. This proposal was adopted as a final rule in a notice published September 2, 1993.[77]

SUMMARY

The regulatory history of the automatic protection requirements in Standard 208 is obviously a long and complicated one that begins in the late 1960's and will continue into the late 1990's. However, there are eight major events that have been especially significant in that history. These are:

1. The 1972 court decision in Chrysler v. DOT. In this case, the court overturned the automatic protection requirements, although the court found that NHTSA had authority to promulgate such requirements and that the automatic protection requirements that were then established met the need for motor vehicle safety. However, the requirements were invalid because the specifications for a test dummy were inadequate.

2. The 1973 ignition interlock option and the 1974 Congressional disapproval of such an option. NHTSA thought it had an alternative to automatic protection that would allow it to achieve roughly the same benefits as automatic protection without getting into the technical and cost issues associated with automatic protection. However, the ignition interlock was so unpopular that Congress amended Federal law to provide that NHTSA could not require or even permit manufacturers to comply with a safety standard by means of an interlock.

3. The 1976 decision by Secretary Coleman to implement a demonstration program. This program was intended to resolve any lingering concerns about the public acceptability and real world effectiveness of automatic protection.

4. The 1977 decision by Secretary Adams to reimpose the automatic protection requirements on a phased-in schedule. This would have required passenger cars to provide automatic protection beginning in the 1982 model year.

5. The 1981 decisions by Secretary Lewis and Administrator Peck to rescind the automatic protection requirements. This rescission was based on the changed economic circumstances and the likely insignificance of the safety benefits that would result if vehicles provided automatic protection by means of detachable automatic belts.

6. The 1983 decision by the Supreme Court declaring the 1981 rescission of the automatic protection requirements unlawful. This decision guided the Department's subsequent consideration of these requirements.

7. The 1984 decision by Secretary Dole to reinstate the automatic protection requirements on a phased in schedule. This decision became the first requirement for automatic protection that actually went into effect.

8. The 1991 Federal law requiring that air bags, supplemented by manual lap/shoulder seat belts, be the means of automatic protection offered in all new cars by the 1998 model year and in all new light trucks and vans by the 1999 model year.

Most of the comments on the NHTSA proposal to implement the 1991 Federal law mandating air bags were directed toward the agency's proposed language for labels to be required on sun visors and on the proposed exemption procedures, with almost nothing said about air bags. This probably reflects the fact that all the commenters knew that Federal law mandated an air bag requirement, regardless of the comments. It is nevertheless ironic that 24 years after an automatic protection requirement was first proposed, the requirement to provide air bags in all passenger cars and light trucks and vans was adopted with so little comment. One would not have predicted this after all the regulatory notices, lawsuits, and other high profile actions that have been associated with the automatic protection requirements.

---

[77] 58 FR 46551

CHAPTER 2:

LAP-SHOULDER BELTS AND EVALUATIONS

770936

Evaluation Methods for the Biomechanical Quality of Restraint Systems During Frontal Impact

Dieter Adomeit
Institute of Automotive Engineering
Technische Universitat Berlin (Germany)

Abstract

From a biomechanical point of view, test criteria in current safety standards for passenger protection do not insure a sufficient over-all protection of quality.

First, the deficiencies of data and criteria, responsible for biomechanical problems, are being analysed. Secondly, additional criteria are being defined, which we think are significant for a better over-all evaluation of restraint devices considering biomechanical facts. As a result of a dummy-crash-series, an analysis is presented, demonstrating correlations between the new defined and former criteria.

The final aim is to develop a complete system of criteria. By the use of simple evaluation methods it would guarantee clear results concerning biomechanical properties of passenger protection systems. With this technique one can have correct biomechanical evaluations of restraint systems, gained from dynamic tests even with anthropometric dummies.

PROBLEMS IN BIOMECHANICAL RESEARCH

In biomechanical research of the automobile occcupant, tests of single human body parts are not adequate any more. Today, cadaver tests give much better information about injuries and test data.

Cadaver tests are complicated because of the preparation of the cadaver. Measurements, evaluation and calculation of these tests is very complicated. The results and the repeatability of the tests are markedly influenced by age, sex and constitution of the cadaver. Furthermore, in cadaver tests the control of geometry of the restraint system during the test is more difficult than in dummy tests.

The main problem is caused by the evaluation of cadaver tests: the problem of correlating injury patterns to measured data. The final aim is to get exact information to judge special problems.

Current methods to compare results of autopsy findings with the measured decelerations are not adequate. Often, the correlating factor between autopsy findings and measured decelerations is unknown. Characteristic data (1)* from high speed films enable us to analyse the way the load is applied to the cadaver during the test.

The importance of this question can be clearly explained by the example of "submarining" of belted passengers. With submarining severe abdominal or other typical injuries can occur during deceleration. But the same hip deceleration without submarining would be very favourable (3) (4).

This paper tries to clarify the importance and feasibility of measuring data from motion sequences during crash tests. It shows that only by use of additional motion sequence criteria (1) can an over-all evaluation of the biomechanical quality of restraint systems be possible in dummy tests and not only in cadaver tests.

AIMS OF BIOMECHANICAL RESEARCH - Two aims of biomechanical research are:

First, to get tolerance limits of the human body under phemenological and statistical aspects. Second, to evaluate the efficiency of the restraint system taking into account current biomachanical criteria (e.g. FMVSS 208). The first aim is basic; it helps to develop criteria for safety standards which guarantee a minimum of protection for the passenger. The second aim - the rating of the biomechanical quality of restraint systems - is of increasing importance,

---

* Numbers in parentheses designate references at end of paper

caused by the advance of safety-technology which reduces the importance of criteria of current standards. If new definitions of protection criteria are discussed, the actual state of safety technology must always be considered, especially if values of adverse passenger loading can occur far below current biomechanical tolerance levels.

From this it can be concluded that the importance of biomechanical tolerance limits is decreasing, and decisions concerning the absolute protection level of a standard are more and more dependent upon cost/benefit considerations and political aspects.

BIOMECHANICAL QUALITY OF RESTRAINT SYSTEMS

PROBLEMS OF CURRENT SAFETY STANDARDS - Properties of restraint systems are currently rated - in Europe as well as in the U.S. - following an incomplete criteria system (7).

The FMVSS 208 e.g. uses evaluated and calculated dummy-decelerations of head, chest and hip. Except for a longitudinal loading criterion for the upper legs, these criteria are calculated secondary values, specially SI and HIC. A clarification of origin of these values is impossible. Specially the way the restraint load is applied to the body e.g. directions and locations of loading, is not specified.

Characteristic data from the motion sequence make an evaluation of restraint load application possible (1).

Fig. 1 is used for further explanations; it demonstrates the human skeleton with high load resistant areas. We call these areas "high load resistant" because basic biomechanical results show that other areas of the body like facial, cervical, and abdominal areas have much lower load resistance (5), as much as 5 to 20 times less. The tolerable ranges in the direction of load application lead to a further very important and often neglected aspect, the dependence of load resistance on the mechanical structure of the skeleton.

The defined direction ranges in Fig. 1 mainly result from mechanical-structural considerations and have been confirmed by in-depth analyses of typical real-world accidents with three-point-belt usage.

Our findings proved a loading of thorax according to case a) of Fig. 2 to be very unfavourable and to cause a significant accumulation of serial rib-fractures of ribs 6 - 12, often combined with severe injuries of inner organs (6). The reason, a submarining-similar motion sequence has been explained (1). As far as the spinal column is concerned, Fig. 1 clarifies the philosophy of achieving longitudinal loading of the column as low as possible, resulting from components of restraint loads (2).

MOTION-SEQUENCE-CRITERIA - Following this anatomical-physiological basis, we defined a concept of describing characteristic biomechanical properties of passengers' motion sequence (1) (Fig. 3). This data should give a rating of the biomechanical quality of restraint systems using original values of passengers' loading. As a result of crash-test-series we defined the following motion-sequence-criteria:

Torso angle $\alpha \leq 90° \pm [10°]$

(at the moment of maximum forward displacement = maximum restraint load).

The tolerance-range of the angle in square-brackets should be investigated further. The general aim is to keep the resulting restraint load on the thorax at the sternum area within the above defined angles and to keep longitudinal loading of the spine as low as possible.

It is obvious that a torso-angle-criterion alone cannot be sufficient because it does not include information about vertical load components caused by vertical components of motion. Therefore, a criterion for the vertical component of motion must be defined.

With respect to the H-point (Hip-point) we defined

vertical displacement $h = 0 \pm [30]$ mm

The tolerance-range in square-brackets also has to be further investigated.

Generally, this criterion seems to be adequate from two biomechanical aspects:
First: it helps to keep vertical components of loading low as explained above;
second: it reduces the danger of submarining , sliding of the lap belt over the iliac crests,

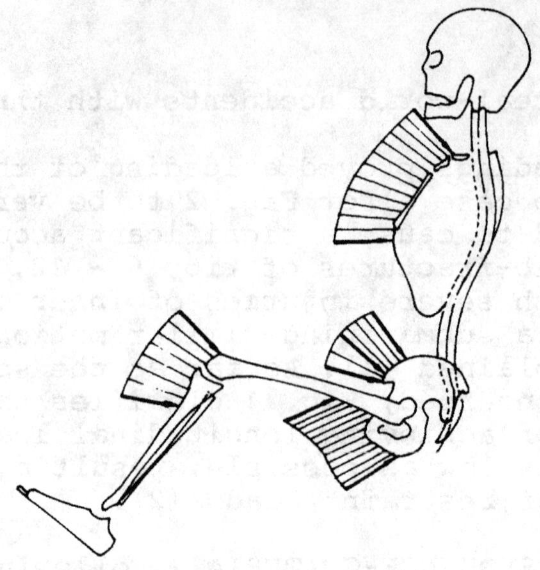

Fig. 1 - Favorable ranges of deceleration load directions on high load resistant body areas

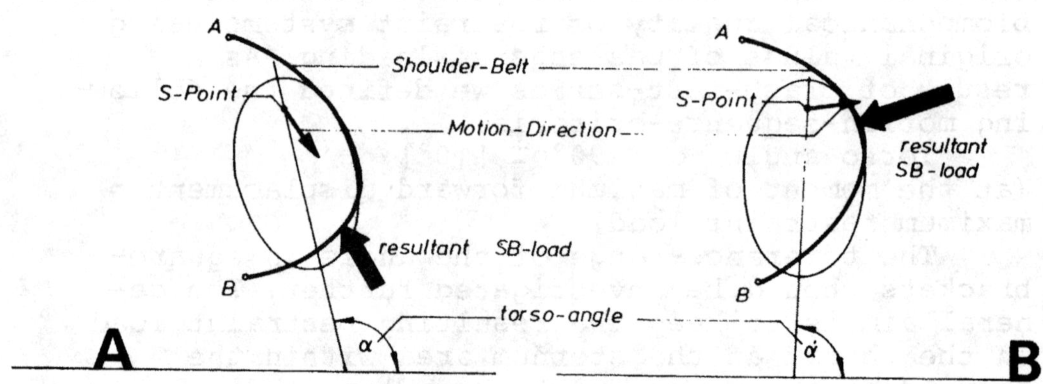

Fig. 2A&B - Sketch of thorax with shoulder-belt (SB) A) unfavorable compression of the lower thorax, B) favorable loading of sternum

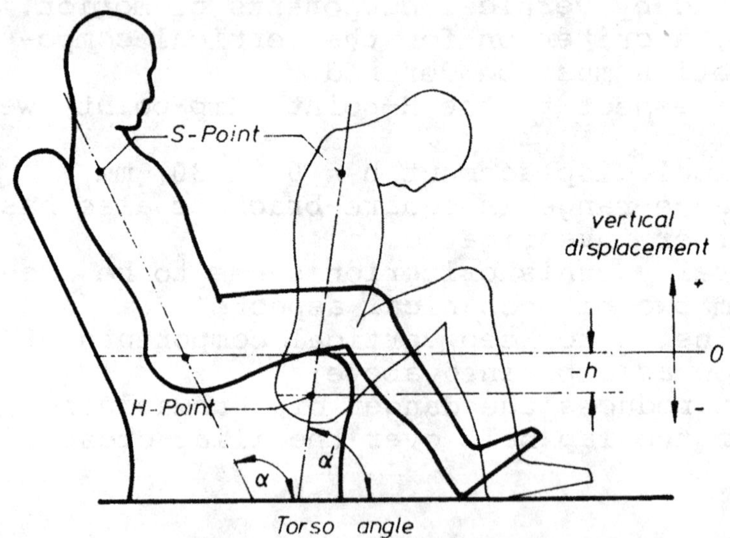

Fig. 3 - Characteristic data of motion-sequence

due to favourable lap belt/iliac crest geometry during forward motion of the hip.

HIGH BIOMECHANICAL QUALITY - A restraint system of high biomechanical quality protects according to the criteria of FMVSS 208 as well as according to the additional motion sequence criteria. Only such a restraint system guarantees that the 208-criteria have been fulfilled with restraint load application on the high load resistant body areas under favourable biomechanical directions and not sacrifying other injury protection criteria. By this, a minimization of restraint specific injuries should be achieved.

Following this philosophy, only a restraint system of high biomechanical quality offers good prerequisites to optimize the energy-absorbing components with the intention to come as far as possible below the maximum tolerable loads on the passenger.

DUMMY-CRASHTEST-RESULTS

TYPES OF EVALUATED MOTION SEQUENCES - In a dummy-crash-test series two characteristic types of motion sequences have been carried out, evaluated and compared. The final aim of this evaluation consisted of a correlation between resulting decelerations (according to FMVSS 208) and the new defined motion sequence criteria. To keep tests comparable, any test has been dropped in which classical submarining occured.

Fig. 4a shows the final position and the trajectories of the shoulder-point (S) and hip-point (H) of the motion sequence type A. Fig. 4b explains that this type A sequence is in line with the motion sequence criteria; at final position

$$\alpha \leq 90°$$
$$h = 0 \pm [30] \text{ mm}$$

The reason for this type A sequence was a stiff energy absorbing cushion of the seat in the area of upper leg contact.

The restraint system, a three-point-belt, was kept constant during all tests as far as belt geometry and webbing are concerned.

Fig. 5a explains motion sequence type B. Such sequences could be observed on current soft spring cushion seats. Specially Fig. 5 b clarifies that this type B sequence does not follow the motion sequence criteria. The unfavourable loading of the passenger's body, caused by the vertical component of the torso-motion and the positioning of the body is obvious. Nevertheless, it is necessary to analyse and to compare measured and calculated resultant dummy-decelerations. But even before that we can state that motion sequence type A is more favourable than type B with respect to the "biomechanical quality" even in the case of equal deceleration.

CORRELATIONS BETWEEN DUMMY-DECELERATIONS AND MOTION SEQUENCE CRITERIA - Table 2 demonstrates the conclusion of results for head (Fig. 6), HIC (Table 1), chest (Fig. 7) and hip (Fig. 8). The more favourable result of the head-neck- and chest-loading during type A-sequence (Fig. 6 and Table 1) is significant. However, type A leads to a certain increase of loading of the hip (Fig. 8). But taking into account the differences in forward displacement the type A-sequence is also more favourable and offers possibilities for further optimization.

The analysis of the time-histories of horizontal and vertical deceleration components proved that the vertical component of motion and of deceleration was responsible for higher values of the resultant dummy type B-sequences.

Fig. 6, 7 and 8 show two characteristic peaks in the time-histories of type B-sequences. The second one was caused by a bottoming out of the hip after vertical downward motion.

During classical submarining an excessive forward displacement of the hip occurs resulting in low hip decelerations in the vertical direction. A torso rotation opposite to the head rotation submarining often leads to low head-neck loadings and low HIC-values. To the contrary, during type B-sequence without classical submarining, the above mentioned high vertical deceleration of the hip at the end of vertical displacement can be observed, which influences the types of loadings of chest and head (see Fig. 6, 7, 8). Furthermore, an additional longitudinal loading of the lower spine of about 30% could be calculated for type B-sequences. This calculation was managed through a simplified rigid torso model using the differences of vertical decelerations of chest and hip.

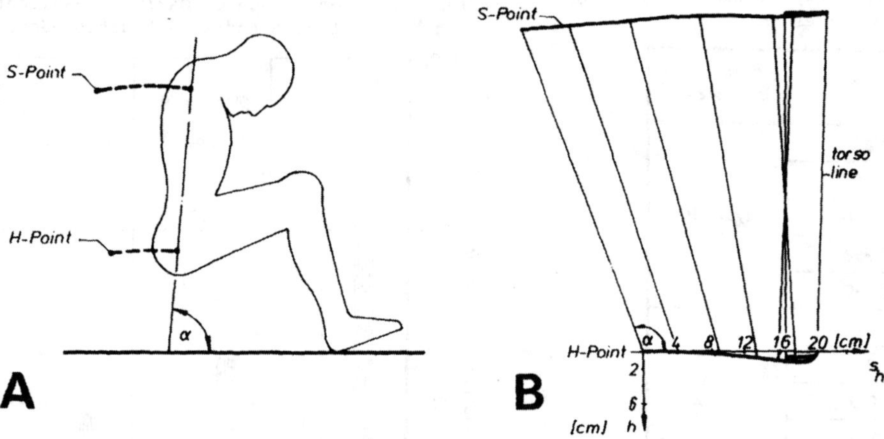

Fig. 4A&B - A) Trajectories of S-point and H-point during type A motion-sequence (stiff seating), B) motion-sequence type A in terms of motion-sequence data recorded by high-speed-film

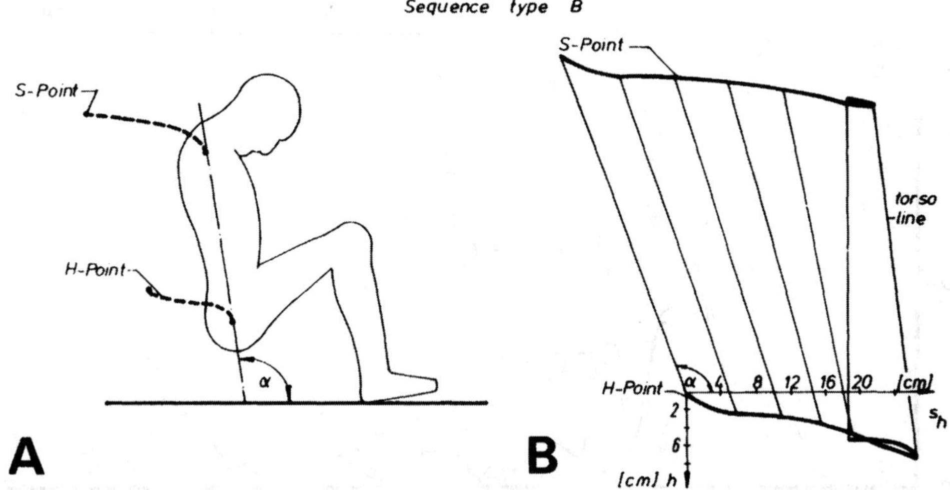

Fig. 5A&B - A) Trajectories of S-point and H-point during type B motion-sequence (soft seating), B) motion-sequence type B in terms of motion-sequence data recorded by high-speed-film

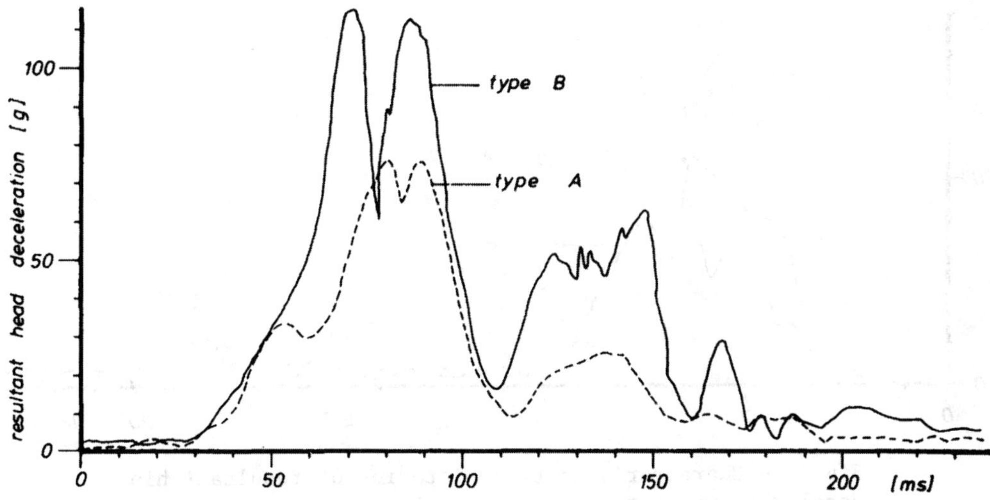

Fig. 6 - Characteristic time-histories of resultant head decelerations of type A and type B motion-sequence

Table 1 - Maximum HIC-Values of Type A and Type B Motion-Sequences Calculated from Dummy-Test-Runs

test-No.	HIC	motion sequence
9	2 813	type B
10	2 885	
11	2 726	
12	2 952	
13	3 085	
22	678	type A
23	879	
25	978	
26	395	
27	723	

Table 2 - Comparison of Maximum Resultant Decelerations and HIC of Type A and Type B Motion-Sequence

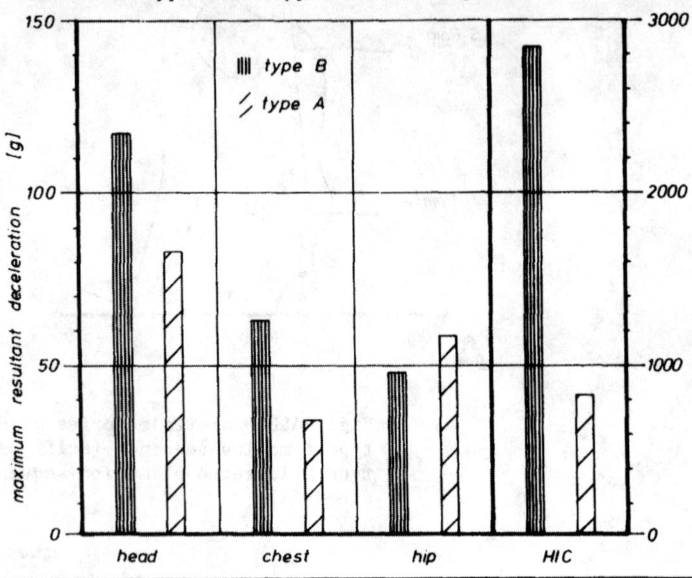

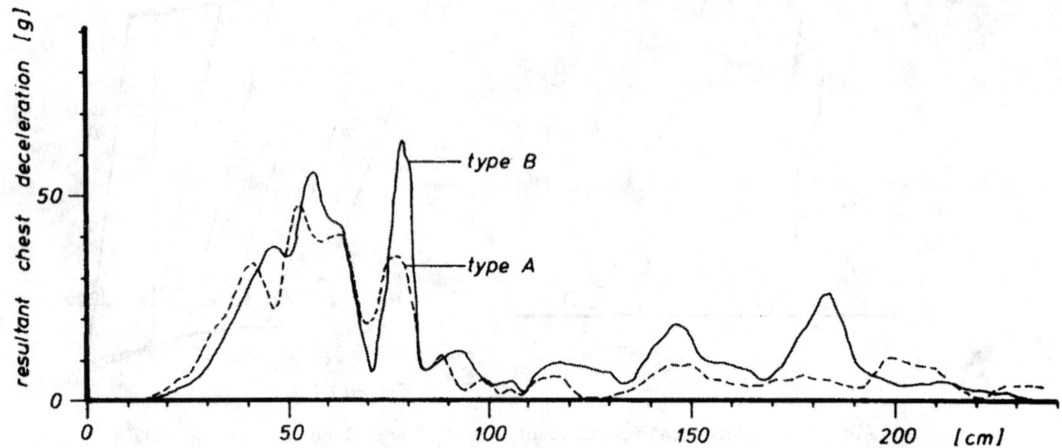

Fig. 7 - Characteristic time-histories of resultant chest decelerations of type A and type B motion-sequence

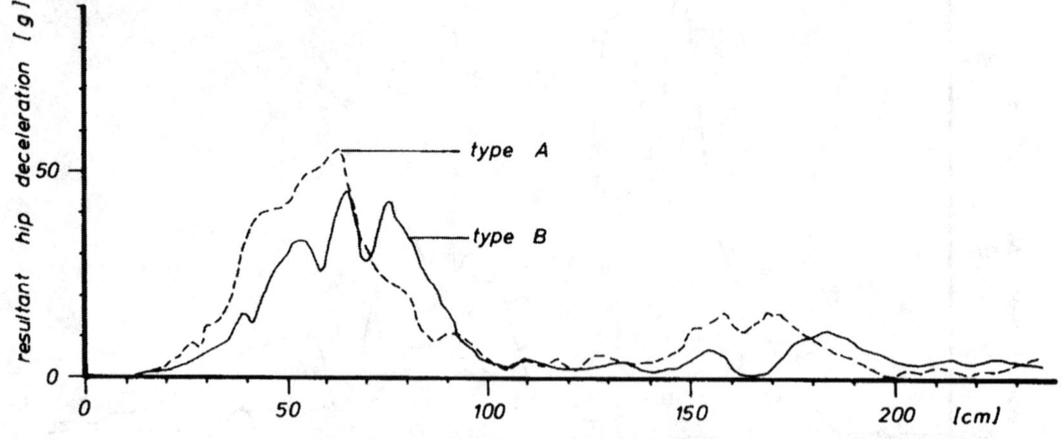

Fig. 8 - Characteristic time-histories of resultant hip decelerations of type A and type B motion-sequence

In a summary we found that motion sequence type A complies with the motion sequence criteria. In this way, favourable biomechanical basic conditions have been reached with respect to the load application on the body. The resultant decelerations have also clearly been more favourable during type A sequences. Type A offers further possibilities of optimization.

CONSEQUENCES FOR STANDARDISATION OF PASSENGER PROTECTION

GENERAL - It is wellknown that criteria of different current standards are not fully sufficient to judge biomechanical quality of passenger protection during frontal impact. There is a need of criteria to ensure a correct restraint load application on the passenger's body from a biomechanical point of view (1) (2). The additional data and criteria of motion sequence seem to have good properties for that purpose. On the other hand, additional data evaluation generally leads to more complicated method of analysis. Therefore, it is necessary to ask for further simplification of criteria with respect to an improved reproducibility of the test and data evaluation procedure.

This question has a special importance with respect to legal restraint approval: Too often, unfavourable compromises of protection criteria for practical test and evaluation procedures have been accepted, from a biomechanical point of view. In future discussions, this should be avoidable by means of a system of primary and descriptive criteria. In this way, the state of safety-technology as well as the state of biomechanical research can more directly be introduced into standards.

The motion sequence criteria should be a primary biomechanical prerequisite for further steps of development and optimization of restraint components by design-measures. A definition of the priority of criteria can be helpful. An analysis of relations between criteria can furthermore cause that single criteria can be completely dropped.

RELATION BETWEEN CRITERIA OF PASSENGER PROTECTION - In stating that motion sequence criteria are valid, some consequences for existing criteria can be observed; the head-neck motion (head decelerations, loading of the neck) will in this case only be dependent on the motion of the chest (chest deceleration) and on design properties of the dummy's head-neck or the head-neck properties of the passenger involved. So the head deceleration criterion can be dropped because this would only define design of test-dummies. Nevertheless, the HIC should be preserved for the case of head contact.

Therefore, the definition of a chest-loading criterion is of basic importance for the thorax and head-neck-area. The horizontal component of chest deceleration which is responsible for rotational head accelerations and thorax deflection should be defined as low as possible under cost benefit considerations as a result of the state of safety technology.

Due to a limited maximum forward chest displacement, a low maximum deceleration value automatically causes an advantageous deceleration-displacement-history with an early onset (low relative velocity) and continuous characteristic. By this, extreme chest-deceleration-peaks - influencing the characteristic of rotational head acceleration - should be prohibited. Thereby, loading of the head-neck region as well as of the thorax could be minimized.

The maximum allowable vertical component of deceleration has to be dependent on the special biomechanical properties of thorax and especially the spinal column, for which evidently lower values have to be provided.

Because of the vertical displacement criterion, deceleration loading of the hip has to be defined in the horizontal direction only.

The criterion of maximum tolerable longitudinal loading of the upper legs has to remain valid for cases of direct knee contact:

These should be the criteria:

First: Motion sequence criteria:

maximum torso angle $\alpha = 90° \pm [10]°$

maximum vertical hip displacement $h = 0 \pm [30]$ mm

(at moment of maximum restraint-loads = maximum displacements)

Second: Quantitative loading criteria:
maximum HIC (in case of head contact)
maximum horizontal chest deceleration
maximum vertical chest deceleration
maximum horizontal hip deceleration
maximum longitudinal loading of upper legs

CONCLUSION

This system of criteria represents a possible contribution towards a pure functional requirement. Only measurements and evaluations of simple, primary and reproducible values from a high-speed-film or electronic measurements are necessary.

This system seems to be descriptive and enables approval authorities to come to clear judgements of biomechanical quality of restraint systems.

The deficiency of criteria which currently allow insufficient or even dangerous passenger protection systems on the market could be eliminated by motion sequence criteria.

REFERENCES

1. D. Adomeit, A. Heger, "Motion Sequence Criteria and Design Proposals for Restraint Devices in Order to Avoid Unfavourable Biomechanic Conditions and Submarining". Paper 751146, 19th Stapp Car Crash Conference, San Diego, 1975

2. J.D. States, "Trauma Evaluation Needs". G.M. Symposium, Edit. Plenum Press, N.Y., 1973

3. C.H. Tarriere, "Proposal for a Protection Criterion as Regards Abdominal Internal Organs". Proceedings of 17th AAAM

4. G. Schmidt, D. Kallieris, J. Barz, B. Meister, "Injuries Despite Restraints". 5th International Conference of the International Association for Accident and Traffic Medicine, 1975

5. E. Faerber, H.A. Gülich, A. Heger, G. Rüter, "Biomechanische Grenzwerte". Unfall- und Sicherheitsforschung Straßenverkehr, Heft 3/76

6. L.M. Patrick, A. Andersson, "Three Point Harness Accident and Laboratory Data Comparison". Paper 741181

7. D.F. Huelke, J. O'Day, "The Federal Motor Vehicle Safety Standards: Recommendations for Increased Occupant Safety". Proceedings of Fourth International Congress on Automotive Safety, 1975

791004

Seat Design— A Significant Factor for Safety Belt Effectiveness

Dieter Adomeit
Institute for Automotive Engineering
Technical University Berlin

Abstract

Production seats and specially designed research seats were analyzed with respect to seat-safety belt interaction under frontal crash conditions. The objective was to evaluate seat influences on effectiveness of safety belts, or any other restraint system. We determined that classical measurement techniques alone are insufficient to completely cover problems of seat-safety belt interaction. Supplementary evaluation parameters of dummy kinematics were therefore defined to clarify

- the poor safety design of current production seats
- necessary seat design development for increased safety.

Proposals for possible effective seat design were then derived to satisfy necessary new safety requirements for seating.

BACKGROUND: THE CURRENT LEGAL SITUATION AND STAGE OF RESEARCH WITH RESPECT TO THE INTERACTION OF SEATING AND THE SAFETY BELT

THE AUTOMOBILE SEAT has, as an important component of the overall restraint system for vehicle occupants, a decisive role in the protective effectiveness of the safety belt. This is well known and has been proved in numerous dummy crash test series. It has also been evidenced, with the increasing amount of data and experience gained, that the seat also plays an important role in the occurrence of belt-related injury phenomena (1, 3, 8).*

Investigation of the characteristics and of the influence of vehicle seats in their interworking with the three-point belt is therefore especially justified in view of the fact that no national or international safety regulations or standards exist on the structure or the operation of vehicle seats under these conditions.

There exist only strength requirements for the seat mountings and fittings as prescribed in ECE R 17/FMVSS 207—requirements which, however, are not directly related to the deformation characteristics of the seat and the seat frame which are of interest in study of the head-on crash.

Up to now, legal safety regulations concerned with head-on crashes have concentrated essentially on the safety belt in the over-simplifying assumption that the quality of this protective system alone dictates the degree of auto occupant protection.

The seat, however, plays a decisive role—precisely for the belt-protected occupant. The seat determines the nature and the extent of the vertical collapse of the occupant's body caused by the lap belt, anchored as it is in the floor of the vehicle. The seat structure furthermore can under certain conditions determine the point in time and the force of the vertical impact of the hips onto the lower frame parts of the bottom of the seat structure. This fact can have critical significance for the resulting head and cervical spine forces.

In this manner, the lower seat framework controls the entire sequence of movement and forces and is critical with respect to the "biomechanical quality" of the motion-loading sequence of the strapped-in occupant; i.e., the manner of introduction of restraining forces into the occupant's body. The seat characteristics furthermore

---

*Numbers in parentheses designate References at end of paper.

determine varying values for the absolute degree of occupant loading (accelerations, forces, etc.) with seat-belt systems which are in all other features identical. The series of head-on crash tests presented here and utilizing production and research seats were performed in order to evaluate the effect of the seating structure on the effectiveness of the three-point automatic belt.

Evaluation criteria utilized were partly the protective criteria according to US FMVSS 208. Utilization was furthermore made of kinematic variables with the aid of which the manner of introduction of the safety system restraining forces into the body of the occupant—i.e., the "biomechanical quality" of the system—could be evaluated. Work preparatory to this methodology has already been presented (1, 2).

TEST APPARATUS; MEASURING AND EVALUATION TECHNIQUE

The Δv values chosen for the crash series were approximately 40 km/h. A deceleration curve lying within the range of tolerance is shown in Fig. 1 and indicates an approximate deceleration peak of 28 *g* as well as a deceleration period of approx. 65—70 milliseconds. Distribution in the impact velocity and deceleration pulse resulted from slightly differing values for the mass of the seat and mounting, as well as from the differences in the interaction among the three-point automatic belt, the seat, and the dummy. These factors, however, do not affect the basic kinematic test relationships.

Since the impact velocities were lower than under test conditions specified in the US regulation no. 208, direct comparison of the dummy acceleration values with FMVSS 208 empirical values is not possible. Despite this, however, the velocity of impact chosen corresponds more closely to accident reality (3).

The basis for the series of experiements was a 50% Hybrid II dummy and a crash sled system at the Institute of Automotive Engineering at the Technical University of Berlin. Both dummy and sled were fully equipped with respect to measurement technology.

The technical evaluation of data gathered was performed in the conventional manner. Mention should be made here, however, that the high-speed film (500 frames per second) was subjected to highly sophisticated evaluation techniques for determination of the trajectories and rotation of the dummy head, chest, pelvis, and knee over the duration of the crash. The dummy was equipped for marking in the positions shown in Fig. 2: the point S' is fixed to the upper thorax segment, point B is fixed to

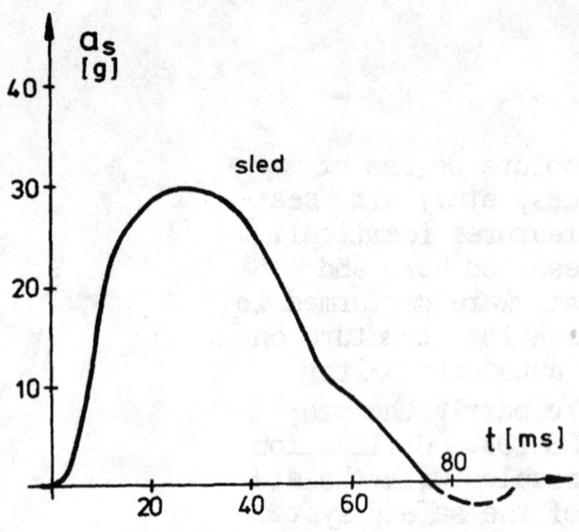

Fig. 1 - Sled deceleration curve

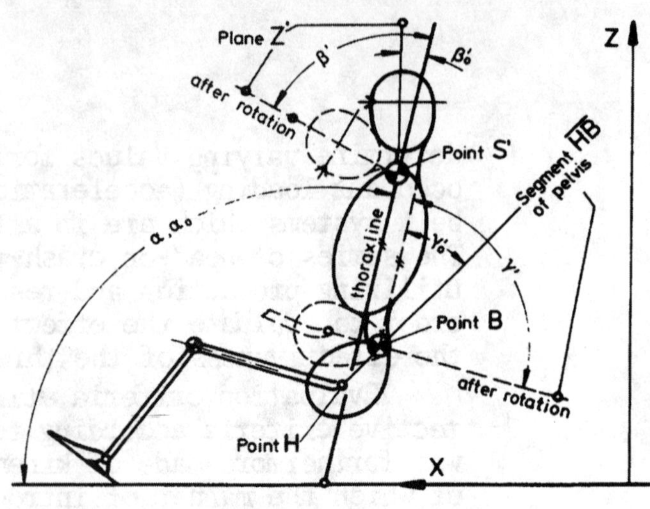

Fig. 2 - Marking scheme and angle definitions

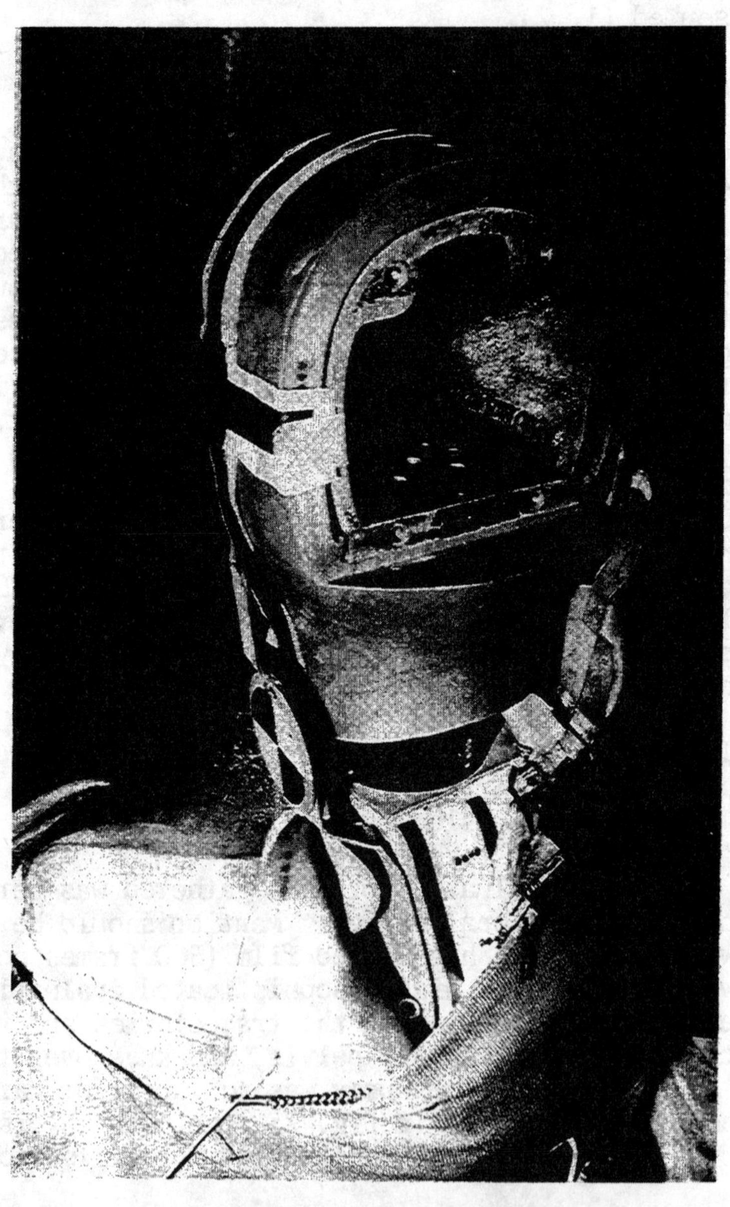

Fig. 2a - Marking and direction arrow at dummy's upper thorax

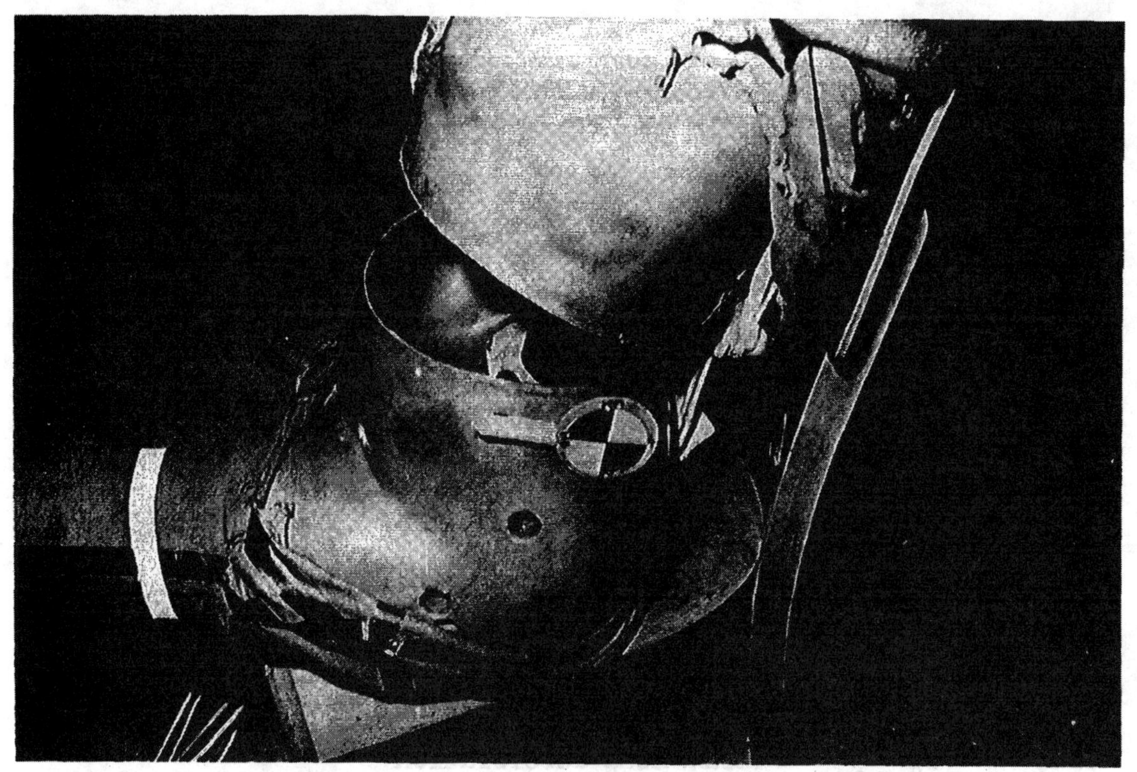

Fig. 2b - Marking and direction arrow at dummy's upper pelvis

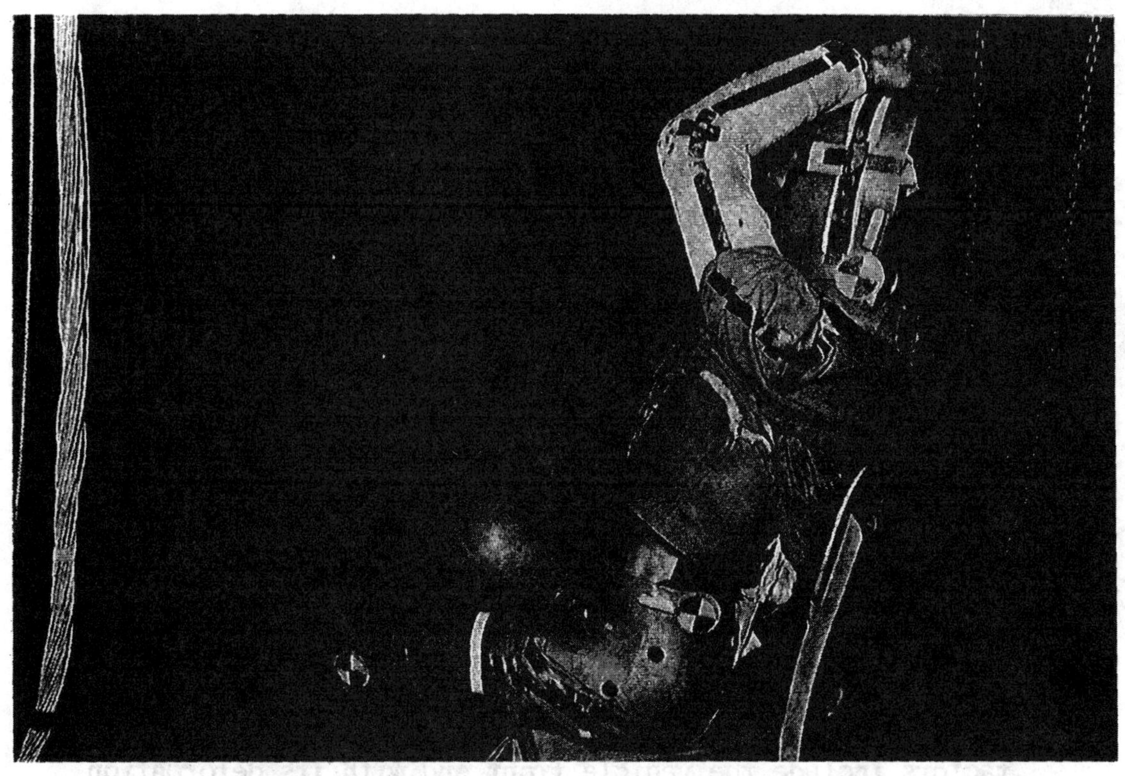

Fig. 2c - Marking and direction arrow configuration on the dummy

the top rear of the iliac crest, and point K is fastened to the knee joint. These special positions of the reference points on the dummy skeleton enable exact measurement of the angular relation of the individual body segments.

Parallax distortions, always present owing to the relative motion of the test object to the camera lens, were eliminated during the evaluation by means of geometric plan view reconstruction.

The following values were either measured, reconstructed, or calculated (see also Tables 1 and 2):

Forces and accelerations:

a_s = sled acceleration

v_{os} = sled impact velocity

$a_{H\ x,y,z}$ = dummy head acceleration

$a_{H\ rot}$ = dummy head rotational acceleration

$a_{Th\ x,z}$ = dummy thorax acceleration

$a_{P\ x,z}$ = dummy pelvic acceleration

HIC = Head Injury Criterion

SI = Severity Index (chest)

Kinematic variables (dummy):

x_H, z_H = displacement of the head

x_{Th}, z_{Th} = displacement of the thorax

x_P, z_P = displacement of the pelvis

β' = bending angle of the neck

α = angular position of the thorax in stationary coordinate system

γ' = bending angle of the lumbar spine

TEST CONDITIONS

It is well known that, in addition to the influencing variables to be discussed in this paper, there are further structural parameters which have graduated influence on the interaction of the seat and the safety belt. Such factors include the vehicle front end with its deformation characteristics, the rigidity of the interior space structure, the safety belt anchor points, and the interior floor complex. For this reason, care must be taken in formulation of the sled test conditions that excessive simplification does not change, or even reverse, the basic tendencies obtained in the results. In the following, presentation and justification of the sled test conditions are made.

SEAT AND VEHICLE INTERIOR - The production seats were tested in their original mode of slide positioning and mounting, and in their original installation configuration (distance above the vehicle floor, angular position, etc.). The slide mountings were fixed on the sled frame in such a way that any vehicle floor deformations would not be taken into consideration. The test loading values were therefore more severe for the seating console area and the mountings than under identical impact conditions with an actual vehicle.

The research seats were rigidly fixed onto the sled frame so that comparisons with the results with the production seats were possible.

The angle of the seat back for all tests was a constant of 110° to the horizontal, measured as a tangent or as a secant over the support area of the lumbar spine of the seat back curvature in the seat back center line, and referenced from the initial dummy position. A foot support enabled positioning of the legs of the dummy in a realistic manner.

The sled tests were conducted in all cases without an instrument panel. This is of considerable significance, since the lower part of the instrument panel is very often struck by the knees of the strapped-in occupant as he moves forward, and is thereby deformed, as safety-belt accident studies show. Only by disregard of the panel, however, can the interaction of the belt (and therefore the characteristics of the seat) be measured and determined as an isolated, undisturbed phenomenon. The characteristics of the trajectories are, however, not basically modified by the instrument panel, since a knee impact will take place only in late phases of the movements. This is as a result of the free space measured to the panel contour in various European vehicles. Evidence can therefore be presented that such knee impact (uncontrolled in any individual case) can have no tendency-changing effect on the results.

In individual cases, indeed, the knee impact even has a mitigating effect. Recent accident evidence shows namely that this impact has the fortunate result of preventing expected abdominal injuries due to submarining. The frequency of knee impact in reality furthermore shows that even if classical submarining does not occur, a type of trajectory is observed which can be characterized as indeed similar to the submarining phenomenon, as evidenced by the patterns of injuries (1, 3).

The currently valid FMVSS 208 limit for axial forces in the femur is, in this connection, moreover, merely an auxiliary criterion and cannot serve by itself as a basis for optimization, since kinematic requirements there are not sufficiently taken into account.

SEATBELT GEOMETRY AND THE SEAT - Tests for all seats were conducted with the same safety belt geometry, with fixed anchoring points on the sled. The influence of the belt system on the trajectory can therefore be considered constant, with the result than an analysis of the seats for their effectiveness, and in their comparison with each other, was possible.

The safety-belt geometry on the test sleds, when judged in relation to the actual installation conditions on the respective motor vehicles, was, in each of the cases identical, or, in isolated cases, more effective, i.e., steeper lap belt angle, shortened buckle part.

The following represents the belt geometry in figures, in relation to point H, with a constant 50% dummy seating position:

Lap belt angle:

Buckle side $= \beta_B = 55 - 60^o$

External side $= \beta_E = 60 - 65^o$

Length of the buckle part, with belt force transducer:

$$1 = 220 \text{ mm}$$

The buckle was therefore situated considerably lower than the seat sitting surface level in the conduct of the sled tests. This thereby eliminated the possibility that the shoulder belt could pull the lap belt up at the buckle side over the iliac crest—an occurrence surely leading to submarining.

The upper shoulder belt anchor point was approx. 900 mm above the plane of the sled frame—a position similar to the level as prescribed in ECE R 16. The transversal plane of the upper belt anchor lay approx. 200 mm behind the common transversal plane of the lower anchor points.

SUBJECTED TO TESTING: PRODUCTION SEATS AND RESEARCH SEATS

For the case of the head-on crash, only the structure of the seat cushion and of the lower cushion framework is of significance for our study. Testing revealed that, in the case of production seats, the seat back fittings and seat back structure remained as a rule without significant deformation and/or without effect on the trajectory of the dummy.

The types of production seat structures can be broken down into three structural groups, in accordance with Fig. 3, as follows:

Type SC	Sheet metal frame with spring center in cushion form, and cover: the classical spring—cushion seat
Type SC/SF	Sheet metal frame with steel coil netting or elastic belting in support of a soft foam cushion in seat form
Type SF	Sheet metal pan (self-supporting or mounted on tubular frame) with soft foam cushion in seat form.

Basic differences in the design of the seat cushion structure can be immediately recognized. This is evident both as far as comfort and subjective sitting evaluation are concerned, as well as for the compression characteristics observed for strapped-in passengers in the case of head-on crashes. These differences are of fundamental nature and cannot be explained away by alleged "adaptation" to the frequency response of the vehicle system alone. This fact assumes greater significance once one has become involved with the design of the following research seats and has established a reference to the production seats.

Fig. 4 shows the basic designs of the research seats, described as follows:

Type S	A simple, thickly upholstered soft foam structure (*"soft"*)
Type EA M	A soft foam upholstery, into which a medium-hard energy-absorbing wedge of styrofoam material has been integrated into the forward seating surface region (*"energy-absorbing, medium"*)
Type EA St	A soft foam upholstery, into which a stiff energy-absorbing wedge-formed piece of styrofoam material has been integrated into the forward seating region (*"energy-absorbing, stiff"*).

The basic design and the contour of the frame, of the metal pan, and of the seat upholstery are the same for all three graduated types above. The soft foam layer is installed for comfort and is made of polyurethane with a unit weight of approx. 18 kg/m³. The energy-absorbing styrofoam supporting wedge has the same cross-section in both of the last two EA types. Type EA M has a density of 34 kg/m³,

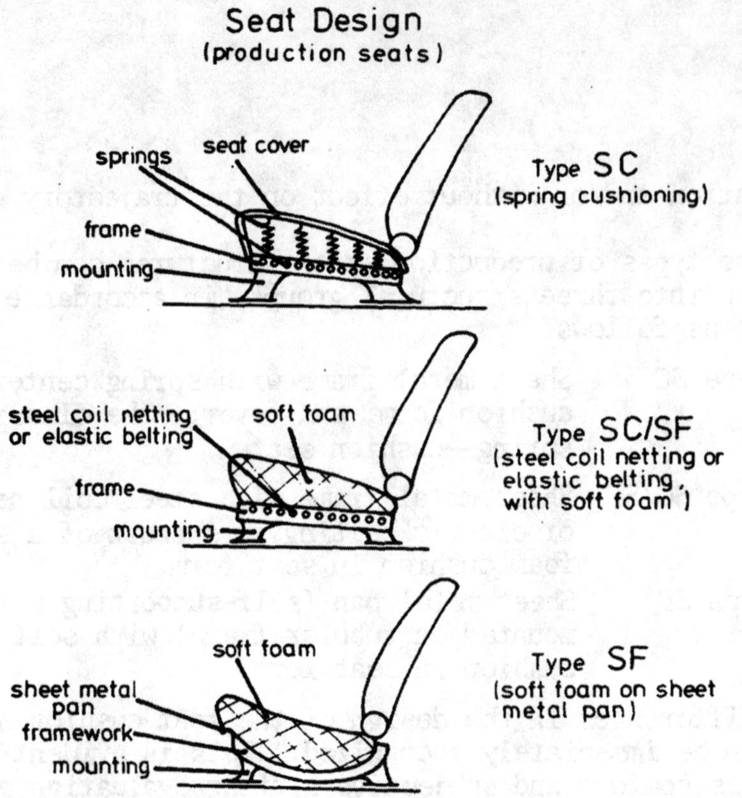

Seat Design
(production seats)
springs
seat cover
frame
mounting
Type SC
(spring cushioning)
steel coil netting
or elastic belting
soft foam
frame
mounting
Type SC/SF
(steel coil netting or
elastic belting,
with soft foam)
soft foam
sheet metal
pan
framework
mounting
Type SF
(soft foam on sheet
metal pan)

Fig. 3 - Three current production seat types used for testing

Seat Design
(research seats)
soft foam
sheet metal pan
frame
mounting
Type S
(soft)
energy
absorbing
support
(medium-hard)
soft foam
Type EA M
(energy-absorbing:
medium-hard)
energy
absorbing
support
(stiff)
soft foam
Type EA St
(energy-absorbing:
stiff)

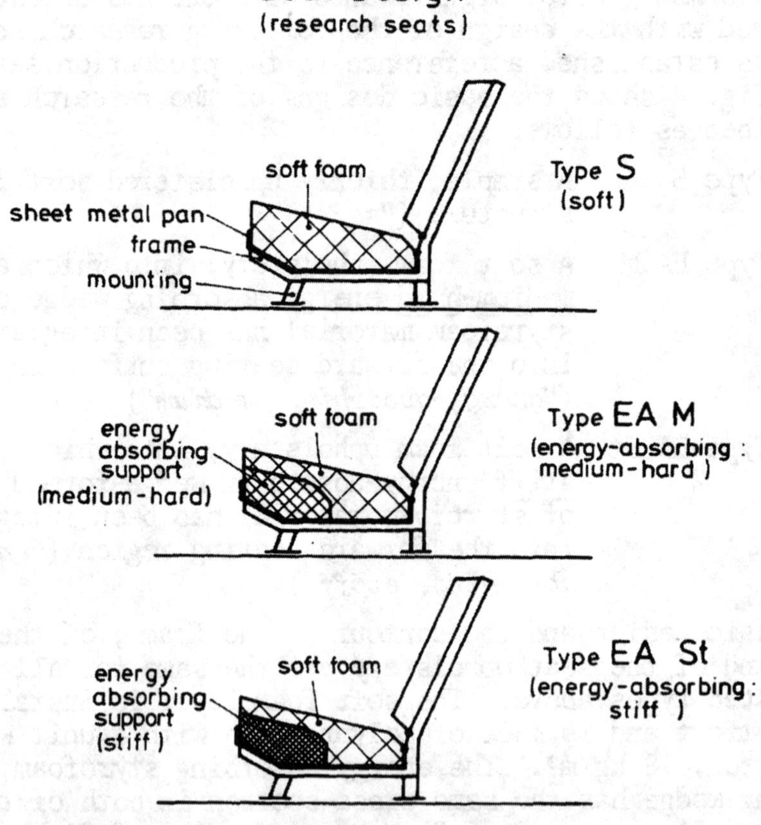

Fig. 4 - Research seat design

whereas type EA St is considerably heavier, with 60 kg/m³. The wedge form of the EA elements is designed in such a way that, with the exception of an overall more rigid seat feeling, there are no basic effects evident on the vertical cushioning characteristics in the more important rear part of the seating region. Subjective seating comfort comparisons do not show marked differences in the three research types from corresponding type groups of production seats.

ANALYTICAL STUDY AND DEFINITION OF THE EVALUATION PARAMETERS

Evaluation parameters were defined for valuation of the loads as well as of the kinematics of the loading process.

LOADING CRITERIA - The FMVSS 208 loading criteria alone are not sufficient for valuation of the influence of the seat on the effectiveness of the safety belt (2). These well-known criteria, nevertheless, were measured, calculated, and listed.

KINEMATIC EVALUATION PARAMETERS - The kinematic evaluation parameters serve the purpose of classifying the manner of occurrence of the loads on the passenger, i.e., to judge the more or less biomechanically favorable points and directions of applications for the restraint forces on the occupant's body.

Fig. 2 represents the dummy skeleton in its initial position, in the stationary x-z coordinate system. The measuring points, rigidly fixed to the skeleton, can be seen at thc thorax, pelvis, and knee. The positions of installation for the points S' and B have been chosen such that these points form the instantaneous center in first approximation for the crash-caused relative movements (bending between head and thorax, and between thorax and pelvis).

[Explanation may be inserted here that the measuring points S' and B are positioned approximately at the level of the lower rubber element flange of the cervical rubber cylinder and the lumbar cylinder of the dummy. See Fig. 2a, 2b, and 2c with S' and B supplemented by rigidly fixed white auxiliary arrows for indication of relative rotation of the body segments.]

On the basis of the crash tests, the measured angles further defined in Fig. 2 proved to be especially valuable for the following conclusions:

- The cervical spine bending angle (β') is the difference between the angle of initial head position (α_0) and the angle of final head position with respect to the thorax (α). As can be seen in the above-mentioned figure, it is defined as the angle between the thorax line and the z' plane of the head.

For the evaluation of the results for β', it must be remarked that, although a human has an anteflexion limit of 75 to 80° (threshold of pain) in a static case, the rubber neck of the dummy easily reaches 90° since there is no energy absorption through the contact of cervical vertebrae. Nevertheless, the measured values in relation to each other can be utilized for the drawing of evaluative conclusions.

- The thorax angle α is defined as the angle between the vertical axis of the thorax and the stationary x axis. It is measured in the forward-facing direction and indicates to which angular position the thorax ellipse has turned in the shoulder belt with respect to the horizontal x axis. Final positions with the angles of $\alpha \leq 90°$ are advantageous because, under the assumption of purely horizontal movement, provision is made in this way that the resulting shoulder belt force is applied at the significantly more resistant upper thorax segment in the sternum area (1, 2).

- The angle γ' deserves special attention. It is measured as the angle between the thorax line when displaced parallel up to the point B, and the segment $\overline{HB}$ of the pelvis. The pelvic rotation, determined primarily by the seat characteristics, influences the entire dummy kinematics. The mechanical pelvic model in accordance with Fig. 5 highlights the following fundamentally critical relationships under the influence of the lap belt:

Because the lap belt force F_{LB} always acts above the common center of gravity of pelvis and femur, the force of reaction of the seat $F_{Seating}$ provided via the lower seat framework has the only possible (and therefore decisive) influence on the degree of loading of the lumbar spine, i.e., the internal forces M, N, and Q of the lumbar spine. The crux of the problem is the imbalance of the moments produced, which lead to excessive bending of the lumbar spine. Finally, bending of the lumbar spine has as a result other biomechanically unfavorable and dangerous movement and loading configurations in the abdominal area and in the lower thorax, as well as in the areas of head and neck (5, 6, 8).

On the left side of Fig. 7, these interrelations are demonstrated using the example of a test performed on a soft seat. The dummy is shown in the initial position and at the end of the forward displacement. This is not an especially extreme case: it is merely the tendency generally observed with production seats and soft research seats.

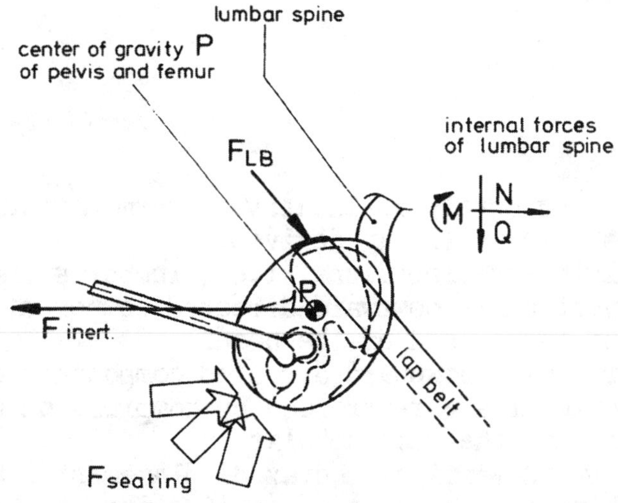

Fig. 5 - External forces on a lap-belted model of pelvis and lumbar spine

Fig. 6 - Motion-sequence comparison for seat—safety belt interaction on production seat, and on soft and on stiff research seats

The following purely qualitative determinations and valuations were initially derived:

• The pelvic rotation alone, i.e., lumbar spine bending, causes a vertically downward-directed component of movement of the thorax and of the head.

• The vertically downward-directed components of the upper torso movement are reinforced by compression of the pelvic contour into the seat upholstery.

• The downward vertical thorax displacement leads also to a sinking of the point of application of the shoulderbelt loop force into the lower thorax segment, despite an otherwise favorable thorax angle of $\alpha \leq 90^\circ$. The resulting shoulderbelt loop forces are thereby applied not horizontally via the sternum, but from below via the 5th to 12th ribs.

• Large pelvic rotation angles of $\gamma' \geq 50$ to 60° endanger the lumbar spine through excessive bending (fracture of vertebrae, lumbar in juries, etc.) (6, 8).

• Large values of γ', furthermore, lead to averting of the iliac crests, with the danger of submarining or other biomechanically dangerous loading configurations.

• A pelvic rotation angle of $\gamma' \cong 80^\circ$ leads to extension of the femur segment $\overline{HK}$ by the distance $\overline{HB}$ (see Fig. 2). Pelvic rotation alone can therefore be a parameter for knee contact with the lower instrument panel.

Table 1 shows an outline of the kinematic evaluation parameters and their biomechanical significance.

ANALYSIS OF THE RESULTS

Table 2 shows a summary of the numerical results of the individual vehicle seat tests on the basis of the above-explained, constant, and easily reproducible test conditions.

Two representative tests for each type of seat were chosen for this listing. It must be conceded once again at this point that the actual conditions in the interior of any particular vehicle involved could possibly shift the numerical values slightly; after careful analysis, however, there are in our opinion no indications whatsoever of a reversal in the tendencies discussed below on the basis of such possible conditions.

DUMMY LOADING - The numerical results with respect to the loading on the dummy are shown in columns 3 to 7 of Table 2; nothing extraordinary is to be observed here for the group of production seats. The loading values measured are noncritical as judged by FMVSS 208, if consideration is taken of the normal distribution to be found in crash test measurements.

Table 1 - Evaluation Parameters for Kinematics

Evaluation parameter	Description	Responsible for influence on:
	HEAD / NECK	
$x_{H\ max}$	Maximum horizontal head displacement	Head to steering system and dashboard impact
$z_{H\ max}$	Maximum vertical head displacement	Head to steering system and dashboard impact
β'_{max}	Maximum angle of neck anteflexion relative to the thorax line	Neck injuries caused by hyperanteflexion
	THORAX	
$z_{Th\ max} = f(x_{Th\ max})$	Maximum vertical thorax displacement at moment of maximum horizontal thorax displacement $x_{Th\ max}$	Loading of lower thorax area (ribs 5 - 12) Increase of vertical upward component of loading direction on lower thorax area
α_{max}	Thorax angle to the horizontal at the moment of maximum upper belt loading	Loading of lower thorax area (ribs 5 - 12) Increase of vertical upward component of loading direction on lower thorax area
	PELVIS	
$x_{P\ max}$	Maximum horizontal displacement of pelvis (point H)	Thorax positioning at the moment of the maximum shoulderbelt force
$z_{P\ max} = f(x_{P\ max})$	Maximum vertical displacement of the pelvis (point H) at the moment of maximum horizontal pelvis displacement	Compression and bending of the lumbar spine Vertical impact severity
γ'_{max}	Maximum angle of pelvis rotation relative to the thorax line	Basic guidance of motion; lumbar spine loading; possibility of knee impact

Table 2 - Results of Seat Tests: Comparison of Production Seats and Research Seats Under 40 km/h and 30 g Impact Conditions

	1	2	3	4	5	6	7	8	9	10	11	12	13	14	15	16
			LOADINGS OF:							KINEMATICS OF:						
			Head	Thorax		Pelvis		Head		Neck	Thorax			Pelvis/lumbar spine		
				Maximum decelerations		Max. decelerations		Maximum displacements		Max. bending	Maximum displacement		Max. angle	Maximum displacement		Max. bending
	Type of seat	Test no.	HIC	$a_{Th\ res}$ [g]	SI	$a_{P\ x}$ [g]	$a_{P\ z}$ [g]	$x_{H\ max}$ [mm]	$z_{H\ max}$ [mm]	β'_{max} [°]	$x_{Th\ max}$ [mm]	$z_{Th\ max}$ [mm]	α_{max} [°]	$x_{P\ max}$ [mm]	$z_{P\ max}$ [mm]	γ'_{max} [°]
PRODUCTION SEATS	SC	5	813	47	330	25	33	510	480	104	370	110	80	260	70	100
	SC	23	536	49	290	37	37	550	440	96	410	160	90	350	190	82
	SC/SF	3	677	51	360	30	41	540	460	97	340	120	72	270	80	99
	SC/SF	4	875	50	390	20	38	480	500	110	290	90	74	260	60	87
	SF	11	1036	47	270	26	28	550	500	109	390	100	72	290	160	93
	SF	20	688	55	320	32	35	580	510	108	420	130	74	310	110	96
				$a_{Th\ x}$	$a_{Th\ z}$											
RESEARCH SEATS	S	7'	1114	52	25	28	40	500	440	113	320	140	94	335	50	87
	S	8'	1208	52	28	22	48	500	450	114	340	190	98	376	25	91
	EA M	13'	517	52	8	38	40	550	410	89	350	60	71	215	0	58
	EA M	14'	542	50	7	34	48	500	400	90	340	10	68	220	0	58
	EA St	23'	178	50	14	52	36	480	360	72	330	10	68	175	-10	50
	EA St	24'	272	48	14	42	38	490	410	75	340	20	77	163	-28	56

Legend: SC = spring cushion; SC/SF = spring cushion/soft foam; SF = soft foam; S = soft; EA M = energ absorbing, medium-hard; EA St = energy-absorbing, stiff

Attention must be called, however, to type SC of the production seats, in test no. 23, to the extraordinarily low HIC value and the small pelvic decelerations. This can be explained as part of an extreme, classical submarining sequence and therefore stands out in comparison to the other production seats. This example is renewed evidence of the well-known and significant phenomenon that, precisely under classic submarining conditions, the FMVSS 208 limits can be especially easily observed—with margin to spare—*with the unfortunate result, however, that the occupant will have it particularly rough.*

Furthermore, it can be easily seen that our research seat type S (see Fig. 4) easily fits into the range of production seats. This seat, moreover, is identical to the future seat provided for the child restraint approval test in the context of an expected ECE regulation. A significant characteristic here—a desirable one within this context as well—is the tendency toward a certain increase of loading, especially in the head and neck region (HIC values). This is caused by the rigid, nondeformable lower seat framework, which leads to a "hard" vertical final impact of the pelvis.

Moreover, determination can on the other hand be made of the good coincidence of the dummy loading values for the type S research seat and for the group of production seats—a phenomenon to be desired within this context. This similarity permits formulation of a fundamental and conclusive correlation between the production seats and the two further-developed stages of the research seat—types EA M and EA St.

With regard to the loading values, types EA M and EA St (the energy-absorbing research seats) show more marked trends than had been expected:

- The HIC decreases markedly with increasing hardness of the energy-absorbing element.
- The vertical acceleration component of the thorax, a_{Thz} , is clearly reduced.
- The vertical acceleration component of the pelvis, a_{Pz} , decreases slightly.
- The horizontal acceleration component of the pelvis, a_{Px} , increases, as expected, but remains noncritical.

This overall positive influence on the scheme of loading has already been discussed in other studies (2).

An analysis of the measured parameters of the associated kinematic processes, however, reveals significant consequences from the standpoint of the biomechanics involved.

It is already clear at this point that the loading

criteria alone do not allow the safety engineer a real picture of the kinematics involved in the loading process. They do not reveal the manner in which the forces measured at the occupant have arisen as a result of the protective restraint system. The following analysis is an attempt to clarify the significance of the evaluation parameters for the kinematics in accordance with Table 1.

KINEMATICS OF THE DUMMY - Columns 8 to 16 of Table 2 allow an overall picture .

In accordance with our experience, one can obtain the first typical impression from the various sequences of movement of the dummy with the aid of the values for the maximum vertical thorax displacement, $z_{Th\ max}$, as given in column 12. These values show just how markedly the thorax has vertically displaced as a result of the components of acceleration and force.

The reasons for this are the following, as also explained by Fig. 6 and Table 1:

• The compression characteristics of the seat upholstery in the forward seat area

• The rotation of the pelvis as a result of the bending of the lumbar spine.

The results associated with the bending of the lumbar spine are listed in column 16 under γ'_{max}. It hardly requires the evaluation of orthopedic specialists to conclude that the relative bending angle of $\gamma'_{max} = 90 - 100°$ for the production seat group and for type S of the research seats is dangerous and unacceptable. Accident statistics (3, 4, and 5) as well as human cadaver experiments (6, 7, and 8) vertify the above.

An overall evaluation of the kinematic parameters also shows the clear similarity of the research seat, type S, to the production seats. The orders of magnitude of the displacements of the individual body segments of the cervical spine (β'_{max} in column 10) and of the lumbar spine (γ'_{max} in column 16) show no significant deviations with respect to each other.

The data measured for this group of soft seats (lumbar spine flexures of $\gamma'_{max} = 90 - 100°$; cervical spine bending of $\beta'_{max} = 96\text{-}115°$; and x and z displacements of the head, thorax, and pelvis) require that the associated loading values (columns 3 to 7) be subjected to more differentiated valuation. Clearly, a resulting restraint force introduced into the lower thorax half (ribs 5 to 12) should especially be derived from these kinematic measured values, in addition to the absolute bending loads on the spine. Under consideration of the differentiated strengths of various parts of the thorax, the loading values as presented may therefore by no means be designated as noncritical.

In contrast to these results, a comparison with the research seats type EA M and EA St shows the following favorable biomechanical-anatomic trends:

- Lumbar spine bending is greatly reduced (γ'_{max} in column 16).
- Vertical thorax displacements decrease significantly ($z_{Th\ max}$ in column 12).
- Vertical head displacements decrease ($z_{H\ max}$ in column 9).
- Cervical spine bending is reduced (β'_{max} in column 10).
- The thorax shows more favorable positioning under maximum shoulder belt force (α_{max} in column 13).
- Reduction or complete avoidance of vertical components of movement leads to a fundamental lower vertical loading on the spine and the thorax.

A summary of the trends mentioned above is reflected in the following figures, 6 to 8.

Fig. 6 shows the motion sequence in test no. 20 (production seat type SF), test no. 07' (research seat, type S), and test no. 24' (research seat, type EA St) in adjacent, synchronized representation. The close similarity between test no. 20 and no. 07' can clearly be seen. The skeleton configurations during the phase of maximum loading are, on the other hand, not so obvious, because the dummy clothing prevents a better view. In contrast, however, Fig. 7 gives an idea of the difference in the loading caused by kinematic conditions in the comparison of the initial and final dummy skeleton positions for the type S research seat (examples: test no. 20 and no. 07') and for the type EA St research seat (example: test no. 24'). With the aid of evaluation of the above-explained dummy markings by high-speed film, Fig. 9 schematically shows typical trajectories of the head, the thorax, the pelvis, and the knee on the type S research seat (similar to the production seats), and on the type EA St research seat. The vertical axis of the head, thorax, and pelvis planes has been entered onto the respective trajectories of these body parts in the time-dependent angular position in order to represent the relative rotation of the individual body segments.

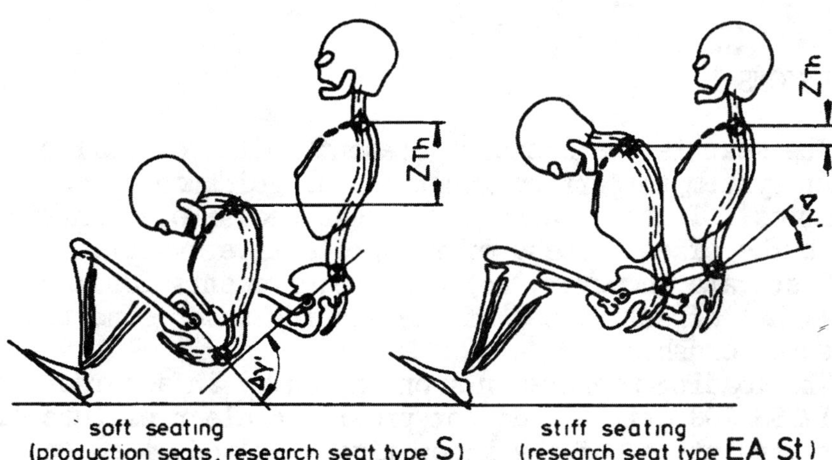

Fig. 7 - Seat—safety belt interaction: final position demonstrating lumbar spine bending and vertical thorax travel on soft and on stiff seating

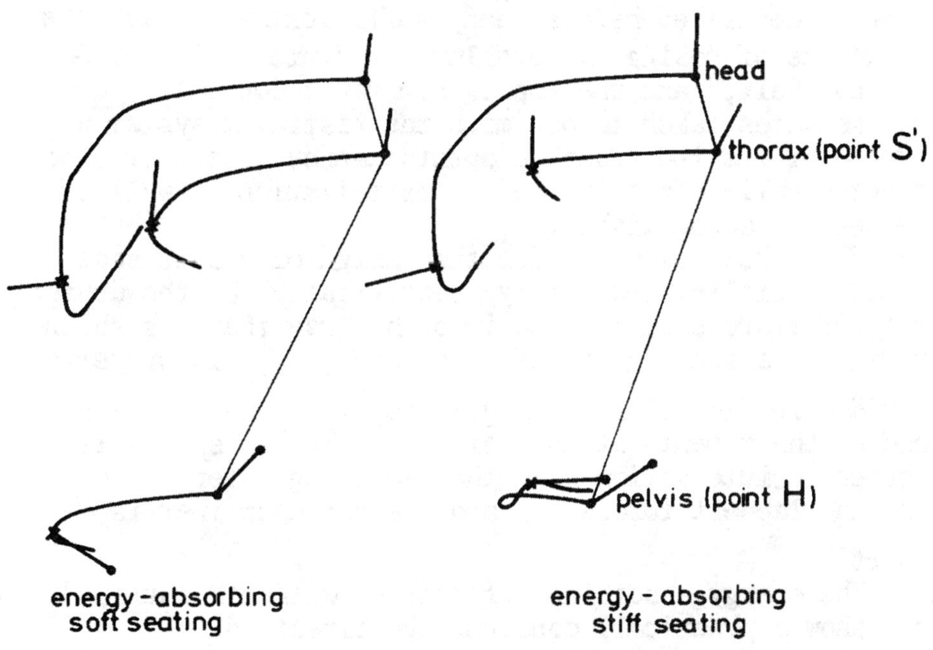

Fig. 8 - Typical trajectories and rotations of head, thorax (point S), and pelvis (point H) on soft and stiff seating

CONCLUSIONS

The seat is the decisive element in the overall protection system to perform functions of guidance of occupant kinematics. In this manner, the seat decisively contributes to the effectiveness of the protective system.

There are, however, no legal requirements for the operational effectiveness of the seating under conditions of head-on crash.

The loading measurements on the dummy in accordance with FMVSS 208 criteria do not provide a clear picture of the interrelations of the kinematics involved, i.e., on the manner in which the restraint system introduces its forces into the occupant's body, and on the manner in which seat and seat belt interact.

The kinematic evaluation parameters defined and introduced here as supplementary aids represent a motivation for further consideration. No effort was made in this stage of the work toward the subsequent and necessary definition of criteria for these kinematic evaluation parameters, as would be required for safety regulations. The trends alone as described here already sufficiently clarify the importance of posing the problems of "Interaction of Seat and Seat Belt," and the topic "Kinematic Control."

Measures taken to optimize the restraint system at any one of its individual elements cannot, therefore, be rendered fully effective before optimization as well of the seat is accomplished.

The first objective for the design of a safe seat should initially, and purely theoretically, be the design of the forward seating area in such a way that—as shown in Fig. 5—a seating reaction force $F_{Seating}$ is properly provided in direction and magnitude, i.e., so that a balance of the moments at the pelvis and lumbar spines is created in interaction with the remaining external forces—the lapbelt forces F_{LB} and the force of inertia F_{inert}.

The energy-absorbing deformation elements outlined here show a practical, constructive direction.

The advantages to be gained would show up in the form of further reduction in the frequency and severity of restraint-specific injuries.

REFERENCES

1. D. Adomeit and A. Heger, "Motion Sequence Criteria and Design Proposals for Restraint Devices to Avoid Unfavourable Biomechanic Conditions and Submarin-

ing." SAE Paper No. 751146.

2. D. Adomeit, "Evaluation Methods for the Biomechanical Quality of Restraint Systems During Frontal Impact." SAE Paper No. 770936.

3. D. Adomeit, Behrens, and H. Appel, "Bewertung von typischen Verletzungsmustern gurtgesicherter Insassen im realen Frontalunfall." Published in DER VERKEHRSUNFALL, vol. 9, 1978.

4. Goegler, "Sicherheitsgurt und Mitverschulden." Presentation at the 16th Convention of the German Traffic Courts, 1978, held in Goslar, West Germany.

5. D. Havemann and L. Schröder, "Verletzungen unter Dreipunkt-Sicherheitsgurten: Ergebnisse einer Studie an Fahrzeuginsassen." Printed in the Report of the 1979 Annual Convention of the Deutsche Gesellschaft für Verkehrsmedizin e.V., held in Cologne, West Germany.

6. G. Heess, "Wirbelsäulenverletzungen menschlicher Leichen bei simulierten Frontalaufprallen." Doctoral dissertation, University of Heidelberg, West Germany, 1977.

7. L.M. Patrick and A. Andersson, "Three Point Harness Accident and Laboratory Data Comparison." SAE Paper No. 741181.

8. Schmidt, Kallieris, Barz, Mattern, and Schulz, "Untersuchungen zur Ermittlung von Belastbarkeitsgrenze und Verletzungsmechanik des angegurteten Fahrzeuginsassen." FAT Research Project No. 3102, Frankfurt, West Germany.

912895

Thoracic Injury Assessment of Belt Restraint Systems Based on Hybrid III Chest Compression

John D. Horsch, John W. Melvin, and David C. Viano
General Motors Research Labs.
Biomedical Science Dept.
Warren, MI

Harold J. Mertz
General Motors Corp.
Current Product Engrg.
Warren, MI

ABSTRACT

Measurement of chest compression is vital to properly assessing injury risk for restraint systems. It directly relates chest loading to the risk of serious or fatal compression injury for the vital organs protected by the rib cage. Other measures of loading such as spinal acceleration or total restraint load do not separate how much of the force is applied to the rib cage, shoulders, or lumbar and cervical spines. Hybrid III chest compression is biofidelic for blunt impact of the sternum, but is "stiff" for belt loading. In this study, an analysis was conducted of two published crash reconstruction studies involving belted occupants. This provides a basis for comparing occupant injury risks with Hybrid III chest compression in similar exposures. Results from both data sources were similar and indicate that belt loading resulting in 40 mm Hybrid III chest compression represents a 20-25% risk of an AIS≥3 thoracic injury.

AS EARLY AS 1968, Gadd and Patrick had discussed an important principle of occupant protection. The upper body can sustain more force than the rib cage alone if force is applied to the shoulders and other upper body regions. They recognized that injury risk assessment for the rib cage and the underlying vital organs, such as the heart and lungs, must be independent of measures of the total restraining force applied to the upper body. Thus, thoracic spine acceleration, which sums the effects of the force inputs from the rib cage, shoulders and arms, abdomen, neck, and lumbar spine, cannot provide direct injury assessment of loading to the rib cage and the resulting compression injury risk to vital underlying organs. Patrick's and many following studies have developed a basis for thoracic injury assessment which clearly demonstrates that the risk of rib cage and underlying organ injury can be directly and objectively based on chest compression.

Vehicle restraint systems such as air bags, shoulder belts, and steering wheels involve frontal loading of the rib cage, shoulders, and other torso regions. For these restraints, it is important to separate rib cage loading from loads on the other torso regions to properly assess injury risk to this vital body region. Knowledge of chest acceleration or total restraint load is not sufficient to determine this important load distribution issue.

The higher tolerance to total impact force when upper body loads are shared between the chest and shoulders was a critical factor in the development of the energy absorbing (EA) steering system. Much of the safety literature has focused on the force limiting element, such as an axial compression column, and FMVSS 203 evaluates "impact-force". However, the load distributing properties of the steering wheel were also fundamental to the success of the system in saving lives and preventing injuries, Horsch et al. (1991). The role of load distribution was twofold: 1) to have sufficient area of the hub

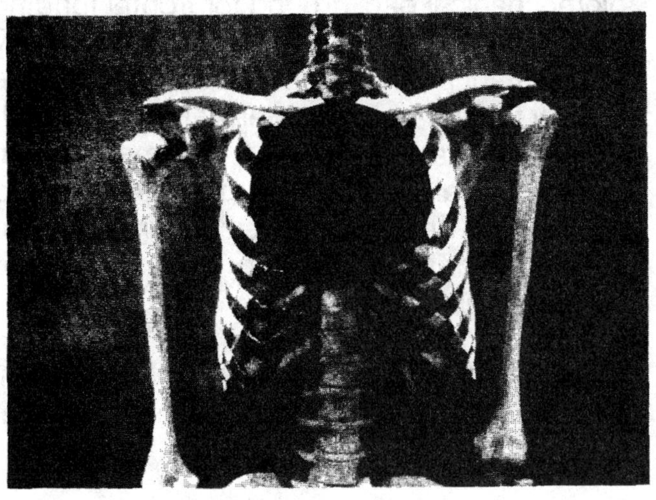

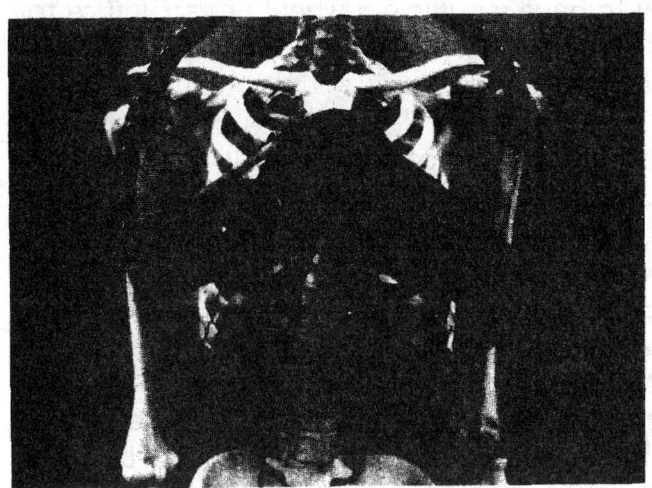

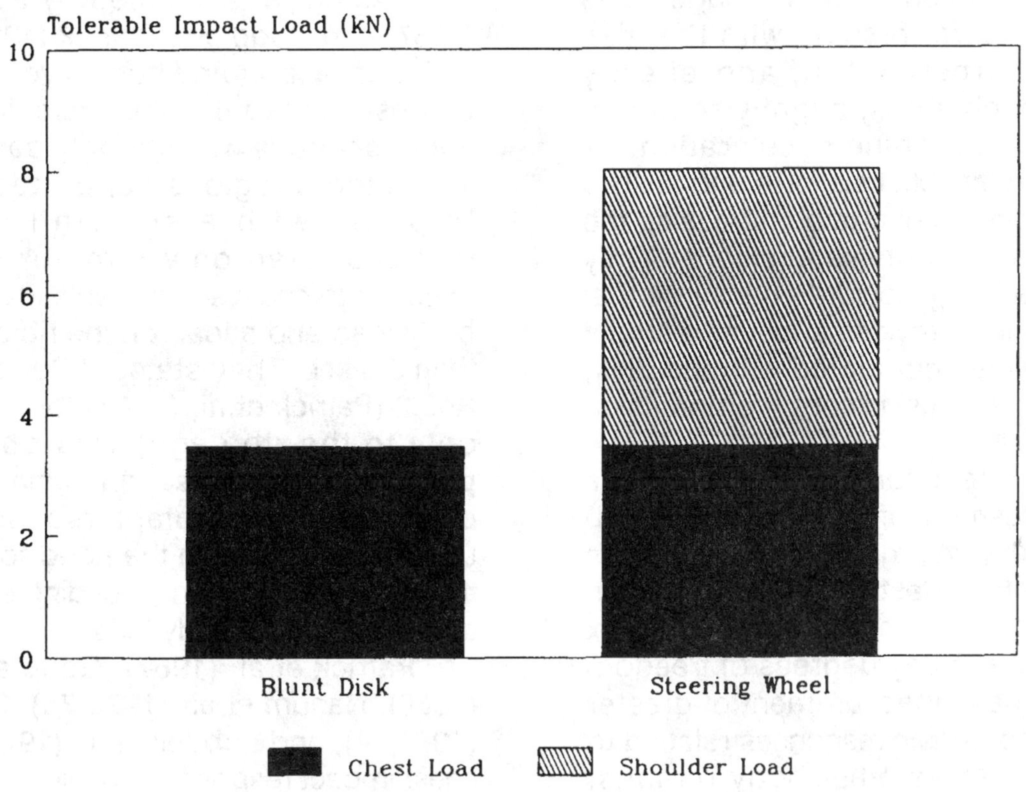

FIGURE 1. Demonstration of the importance of load sharing between the rib cage and shoulders for upper body restraint based on Gadd and Patrick (1968). They concluded that the upper body could sustain more than twice the load than that by the rib cage alone. Chest compression can be used to objectively assess the risk of thoracic injury related to the chest load. In these two conditions, the chest load and compression are similar even though the overall tolerable load is substantially greater with combined loading.

to distribute impact force on the rib cage and 2) to have a sufficient wheel rim structure to simultaneously apply load on the shoulder region, Figure 1.

Horsch and Viano (1984) demonstrated the importance of the Hybrid III dummy and chest compression response to assess the load distributing performance of steering wheels. Sled tests were conducted with an EA steering column and two types of steering wheels: a steering wheel which loaded both the rib cage and the shoulders; and a flat, 152 mm diameter rigid disk which loaded only the rib cage. Although the impact force and chest acceleration were slightly less for the disk, all of the load with the disk was applied to the rib cage with a corresponding 80% greater compression of the rib cage. The greater chest compression with the disk provided a force limiting and energy absorbing mechanism, slightly reducing impact force and spine acceleration. A restraint system should limit force and absorb energy. It should minimize compressing the occupant's chest for these functions. Only the Hybrid III dummy could demonstrate that the steering wheel provided superior restraint for the torso when compared with the disk, and then only when using chest compression response.

Although chest compression provides objective assessment of injury risk for the rib cage and underlying organs, other responses are also essential. Restraint of the human in car crashes involves loading of a complex structure having many degrees of freedom and injury modes. Measurement of greater numbers of appropriate responses relating to the torso, as well as other body regions, provide increased information for assessment of restraint performance.

A dummy with a biofidelic chest compression response provides a measure of the risk of injury to the rib cage independent of the total restraint load and can thus discriminate among restraint designs as to the portion of the restraining load applied to the rib cage. Because the shoulder belt passes over the shoulder, the load sharing between the shoulder and the rib cage is an important issue for shoulder belt restraint protection. This paper reviews the biomechanical background for frontal loading of the chest and associated injury assessment methods and then examines the special features of shoulder belt loading with respect to that background. Biomechanical tests and real-world crash data are used to evaluate the appropriateness of dummy chest deflection criteria for injury assessment of shoulder belt performance.

REVIEW OF THE BASIS FOR CHEST COMPRESSION RESPONSE AND INJURY ASSESSMENT

An important understanding of chest impact response, injury, and safety assessment was initiated by Patrick et al. (1967) and Gadd and Patrick (1968). After a series of cadaver and human volunteer tests, they established that the tolerable loading of the upper body was strongly dependent on which torso regions supported the load. Impacts with a steering assembly demonstrated only a minor risk of rib fractures to cadavers with an 8 kN total upper body load and a load on the rib cage of less than 3.3 kN. They stated, "Correlation with Ref. 2 (Patrick et al. (1967) for load applied only to the rib cage) was obtained by providing means to separate the load on the chest alone from total torso load." Their understanding led to the need to determine the injury risk of rib cage loading independent of the total upper body load.

Patrick et al. (1967), Gadd and Patrick (1968), Nahum et al. (1970,75), Kroell et al. (1971,74), and Lobdell et al. (1973) pursued chest impact response and injury assessment from static and dynamic loadings of volunteer and human cadaver chests. The most comprehensive of these studies was conducted at the University of California, San Diego (UCSD) and is summarized by Kroell et al. (1971,74) and Nahum et al. (1970,75). Frontal impacts were made at mid-sternum with a rigid ballistic mass having a 152 mm diameter, flat-faced disk, resulting in all of the impact load on the rib cage. These tests employed a range of impact mass, velocity, and subject size, age and condition. The

UCSD data demonstrated that blunt injury assessment should be based on compression of the chest, and that spinal acceleration or impact force are not the most accurate predictors of chest injury, Kroell et al. (1971,74) and Neathery (1974).

INJURY ASSESSMENT FOR BLUNT LOADING - Injury assessment recommendations for the rib cage and underlying organs were developed from the UCSD human cadaver data by Neathery et al. (1975) for blunt frontal loading. The experiments demonstrated that when a distributed impact load is entirely directed on the rib cage, severe to critical injury occurred in the cadavers with thoracic spine accelerations of less than half of the FMVSS 208 limit of the acceleration level of 60 G for cumulative durations greater than 3 ms. However, they demonstrated that injury risk for this loading could be objectively based on maximum chest compression, scaled for subject size and age.

Neathery et al. (1975) recommended a chest injury assessment value of three inches (76mm) maximum skeletal compression for a 50 percentile male in blunt frontal impact (assuming an age of 45 years old). This recommendation was based on an injury severity of AIS=3, thus representing a 50% risk of an AIS≥3 for a 45 year old human cadaver, or greater than 50% risk for the UCSD subjects whose mean age was 62 years. It should be noted that AIS scaling for thoracic injury has been modified since Neathery's analysis.

Subsequently, Viano (1978) showed a discontinuity in injury risk as a function of chest compression. For low levels of compression, the rib cage deformed but supported the applied load. However, with multiple rib fractures, high compression occurred which resulted in direct loading of thoracic organs and critical soft tissue injuries. To avoid crush injury to internal organs, an external compression limit of 32% was recommended to avoid rib cage collapse due to multiple fractures. There was a high risk of an AIS≥4 injury associated with 40% external compression.

More recently Viano and Lau (1988) analyzed the UCSD chest impact data and recommended a 35% external compression based on a logist analysis of injury risk and a 25% probability of an AIS≥4 in the cadaver. Assuming an external dimension of 229 mm for the 50th percentile male, this percentage represents 80 mm of external compression. They used a factor of 1.21 to adjust external compression to internal (skeletal) deflection. On this basis, a 35% compression represents 66 mm internal skeletal deflection. This value is likely conservative in representing the tolerance of typical car occupants because it is based on human cadavers, which lack muscle tone and reflect an older population than car occupants. However, it might well be representative of the important elderly segment of the driving population and thus representative of a reasonable tolerance threshold.

Lau and Viano (1986) also identified that the velocity of compression is an important factor in assessing soft tissue injury risks from blunt impact when the velocity of chest compression exceeds 3 m/s. The Viscous criterion (VC) is an injury assessment parameter which considers the product of compression amplitude and velocity at each instant of time. They recommended a VC of 1.0 m/s as a threshold assessment level which represents a 25% risk of an AIS≥4 injury. For compression velocities below 3 m/s, they recommended 35% peak external compression discussed above for injury assessment of crushing injury risks. Both human cadaver and animal subjects are included in the basis of their recommendations.

CHEST COMPRESSION RESPONSE - Analysis of the UCSD cadaver impact data by Neathery (1974) resulted in an empirical model and recommendations for human-like frontal impact biofidelity in terms of impact force vs chest compression at 4.3 and 6.7 m/s impact velocities with a 23.4 kg impact mass for the 50th percentile occupant. A "muscle- tensing" force of 0.67 kN was added to the force determined from the human cadaver data. This "muscle-tensing" force was estimated from measurements of human volunteers reported by Lobdell et al. (1973). The compression response was an external measurement made by film analysis. It was adjusted to approximate skeletal compression by subtracting 12.7 mm from the compression to account for "skin" compression.

Foster et al. (1977) showed that the Hybrid II dummy is very stiff compared to Neathery's recommendation, Figure 2a. They designed the Hybrid III thorax to meet the proposed response requirements. Horsch and Schneider (1988) demonstrated that the Hybrid III thorax had reasonable biofidelity beyond the calibration velocities of 4.3 and 6.7 m/s by comparing responses with human cadavers in 13.4 m/s blunt frontal thoracic impacts, Figure 2b. The Hybrid III dummy appeared stiffer than the cadavers, approximately by the muscle tensing that Neathery had added to cadaver responses.

Clearly, the Hybrid III dummy represents a major improvement in thoracic impact response and injury assessment in comparison to the Hybrid II dummy, and can be considered to have biofidelity for blunt frontal impact.

THORACIC INJURY ASSESSMENT FOR SHOULDER BELT LOADING

Estimates of an injury assessment relationship for Hybrid III chest compression response with shoulder belt loading was made by comparing the chest compression response of the Hybrid III dummy with injuries to car occupants in "similar" exposures. Importantly, there are limitations due to the lack of exposure details for car occupants involved in crashes. However, two well documented studies of belted occupants exist in the literature which provide a basis for estimating the severity of loading to determine "equivalent" Hybrid III responses.

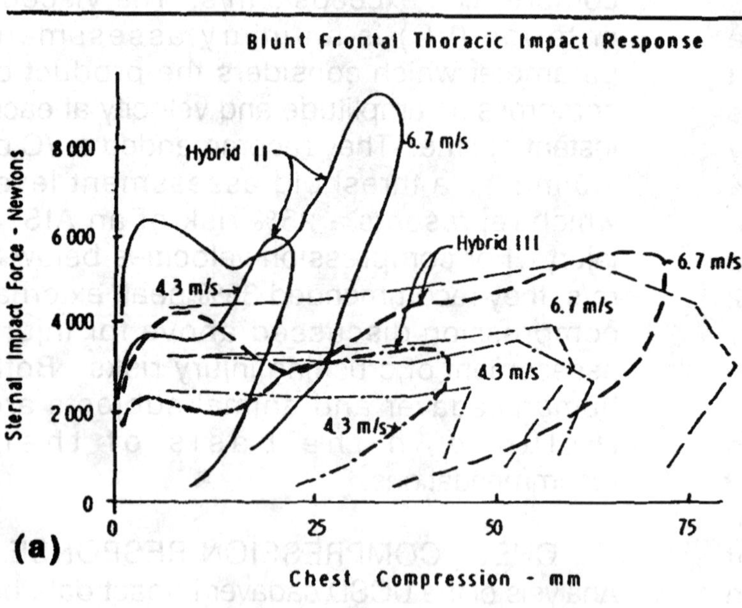

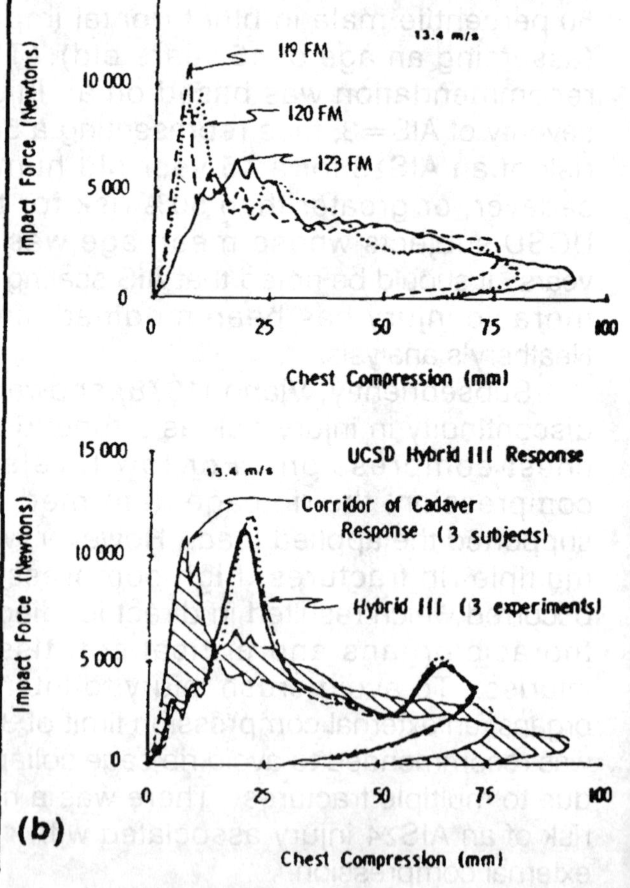

FIGURE 2. **(a)** Comparison of Hybrid III and Part 572 dummy blunt frontal thoracic response with responses recommended by Neathery (1974). **(b)** Hybrid III and human cadaver thoracic response at 13.4 m/s blunt frontal impact, Horsch (1988).

APR CAR CRASH DATA - An important set of car crash data has been provided by the Association Peugeot/Renault (APR) in which the actual loading of the occupant by the shoulder belt can be estimated by the amount of deformation in a shoulder belt force limiting element, Foret-Bruno et al. (1978) and Mertz et al. (1991). The force limiting or energy absorbing (EA) element consisted of a series of stitched webs that tear at a predetermined force and thus act to limit the load on the occupant by allowing a controlled elongation of the shoulder belt. Car crashes involving the use of these EA belt systems were investigated in depth by APR. Examination of the EA portion of the belt gave information on the severity of loads sustained by the occupants in those crashes. Combining this information with the occupant injury data provided an objective basis for comparing Hybrid III dummy responses for similar belt loads when tested with the APR restraint configuration.

Sled tests reported by Mertz et al. (1991) were conducted with the Hybrid III dummy restrained by a lap-shoulder belt containing a type B Renault shoulder belt force-limiting element. A range of sled severities was selected which resulted in tearing 2, 4, and 5 stitch levels. Chest compression was compared with belt tension, indicating a nominally linear relation between chest compression and belt tension. For these tests, an approximate fit was found with 5 mm of compression for each kN of shoulder belt tension. On this basis, 8 kN of belt tension would result in 40 mm of Hybrid III chest compression.

AIS≥3 injury risk as a function of Hybrid III chest compression from the APR car crash data analyzed by Mertz et al. (1991) is summarized in Figure 3. Probit analysis indicates that 40 mm of Hybrid III chest compression due to belt loading corresponds to a 25% risk of an AIS≥3 chest injury and a 20% risk of an AIS≥4 thoracic injury. Most of the APR real world exposures were at a lower severity than that represented by 40 mm of Hybrid III chest compression. Extrapolation to higher levels of compression should consider the reported confidence levels.

VOLVO CAR CRASH DATA - A comprehensive study of Volvos with lap-shoulder belts involved in frontal crashes was reported by Patrick et al. (1974). The study is unique in that the vehicles involved were similar in structure and had similar belt systems, the crash investigators were specially trained by Volvo, and a standard fixture was used to measure vehicle deformation. These features served to minimize the effect of variability in such factors on the results. Patrick et al. (1974), Nyquist et al. (1980), Kallieris et al. (1982) and Saul et al. (1983) modeled the Volvo exposures in sled experiments to determine dummy responses and cadaver injuries relative to this well documented field study. However, test responses are not in complete agreement among the various sled based studies.

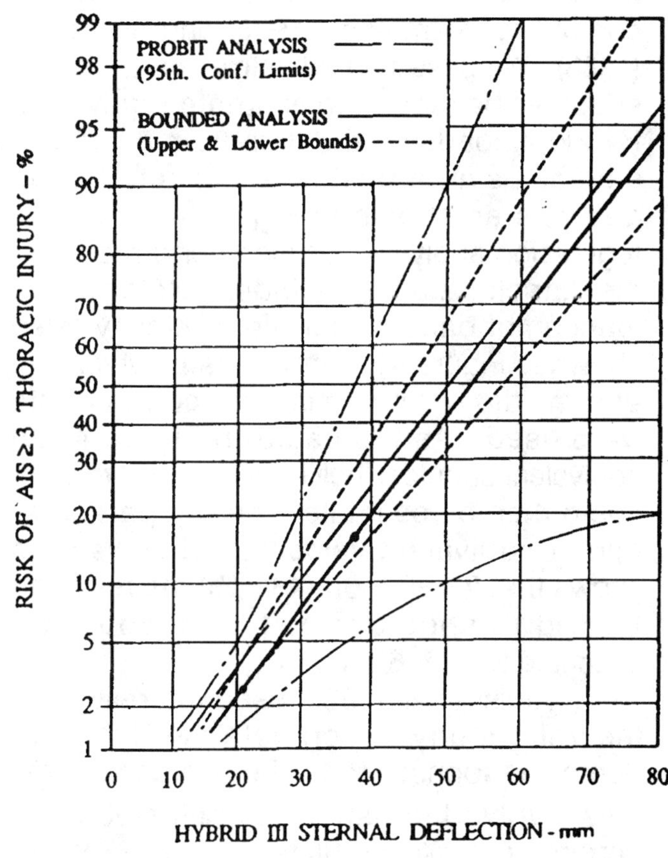

FIGURE 3. Injury risk relationship of APR occupant injury to Hybrid III chest compression in equivalent exposures, Mertz et al. (1991).

Patrick et al. (1974) reported similar dummy responses and belt loads between staged car crashes at Volvo's crash facilities and sled tests at WSU over a range of severities. The sled tests used a Volvo vehicle for the fixture. On this basis, we have judged that the WSU sled tests reported by Patrick, with Bohlin and Andersson from Volvo as a co-authors, represent the most reliable sled simulation for relating dummy responses to Volvo belted occupant injury risk. In general, their study produced much lower responses than those reported in the studies by Nyquist, Kallieris, and Saul. Since Patrick's study used a Sierra 1050 dummy, we have used the results of the other, more severe exposure studies to estimate the Hybrid III responses for the Volvo environment.

Shoulder belt loads produced by the older design 50th percentile dummies are similar but slightly lower than those produced by the Hybrid III dummy under the same test conditions, Alem et al. (1978) and Saul et al. (1983). We based an adjustment of Hybrid II shoulder belt loads to estimate equivalent Hybrid III loads on the data from these two studies using a best fit linear regression analysis as shown in Figure 4. A linear regression analysis was also performed with the Patrick data for shoulder belt load as a function of barrier equivalent velocity (also shown in Figure 4). Only tests in which the dummy did not strike interior components were used. As indicated in Figure 4, an equivalent shoulder belt load for the Hybrid III could then be estimated for any particular barrier equivalent velocity. The example shows that for 30 mph (48 kph) the resulting Hybrid III shoulder belt load would be expected to be 6.6 kN.

Nyquist et al. (1980) used the results of the Volvo study to correlate the Hybrid III dummy responses to the human occupants. This involved lap-shoulder belt sled tests performed at Wayne State University (WSU). Other sled tests at the Vehicle Research and Test Center (VRTC) were reported by Saul et al. (1983). The Nyquist study reported a mean chest deflection of 29 mm at a 48 kph barrier velocity while the Saul study reported 40 mm at nominally the same test severity. The peak shoulder belt tensions in both studies were similar at 8.5 kN for the WSU tests and 8.2 kN for the VRTC tests. These values are much higher than the expected value of 6.6 kN based on the adjusted Patrick data. The much lower chest compression (29 vs 40 mm) found in the WSU study at a slightly higher belt load (8.5 vs 8.2 kN)

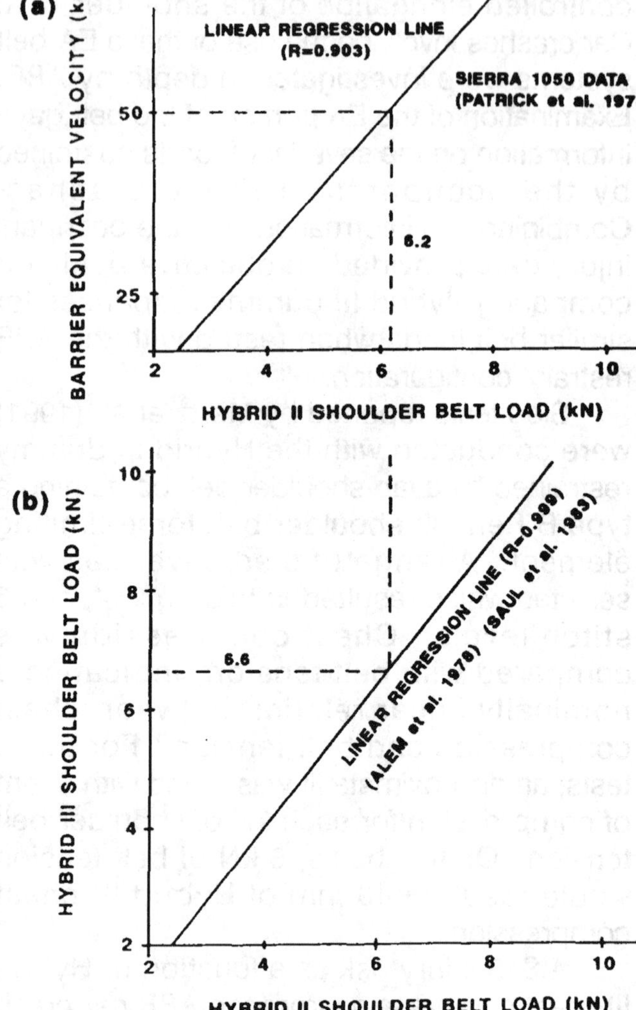

FIGURE 4. **(a)** Method for estimating Hybrid III dummy shoulder belt loads for Volvo barrier equivalent velocities using shoulder belt loads from the Patrick et al. 1974 study and **(b)** then converting them to equivalent Hybrid III dummy loads with Hybrid II/Hybrid III data from Alem et al. (1978) and Saul et al. (1983).

appears to be resolved when we studied the films of the WSU tests. In these tests the belt was positioned against the neck, a non-humanlike position as discussed later in this paper. Our tests have shown a 34% reduction of chest compression for the belt positioned against the Hybrid III neck compared with the belt outside the neck "belt-guide".

Accordingly, we have chosen to use the ratio between Hybrid III chest compression and shoulder belt load determined from the VRTC study as being the most representative of the two simulations of the Hybrid III dummy with the Volvo belt system. This ratio is 40 mm/8.2 kN = 4.89 mm/kN and was applied to adjusted belt loads reported by Patrick. Thus, at a barrier equivalent velocity of 30 mph (48kph) the expected Hybrid III chest deflection would be 6.6 kN x 4.89 mm/kN = 32 mm.

There are data on 98 restrained occupants in the Volvo study which we re-analyzed to estimate the probability of chest injury associated with a particular crash severity. The chest injuries were recoded using AIS (1990) injury scaling. The injury ratings were analyzed using a Logistic regression to produce probabilities of an AIS≥1, an AIS≥2, or an AIS≥3 chest injury as a function of crash severity. The relationship between Hybrid III chest deflection and crash severity, as estimated using the method described above for the Volvo restraint system, allowed the probability curves to be interpreted in terms of Hybrid III chest compression. The results are shown in Figure 5. A compression of 40 mm would have probabilities of an AIS≥2 injury of 38% and an AIS≥3 injury of 24% for actual car occupants by this analysis. The probability of AIS≥3 injury estimates associated with Hybrid III chest compressions in the 30 to 50 mm range agree quite well with the estimates from the APR data. As with the APR field data, most of the Volvo real world exposures were at a lower severity than that represented by 40 mm of Hybrid III chest compression and thus has reduced confidence in extrapolation of injury risk to higher levels of compression.

COMPARISON OF THORACIC INJURY ASSESSMENT FOR BELT AND BLUNT LOADING

The objective of this study was to evaluate biomechanical considerations for thoracic injury assessment based on Hybrid III chest compression with belt loading. The APR and Volvo field data indicated injury risk relationships with Hybrid III chest compression that are mutually supportive. A 25% risk of an AIS≥3 injury for car occupants is associated with a belt loading exposure resulting in about 40 mm of Hybrid III chest compression. This is more conservative than the 66 mm recommended by Viano and Lau (1988) for a 25% risk of an AIS≥4 injury to cadavers for blunt loading. This is consistent with observations that the Hybrid III exhibits less deflection of the chest than human cadavers or volunteers with belt loading, Backaitis and St.-Laurent (1986), and Cesari and Bouquet (1990).

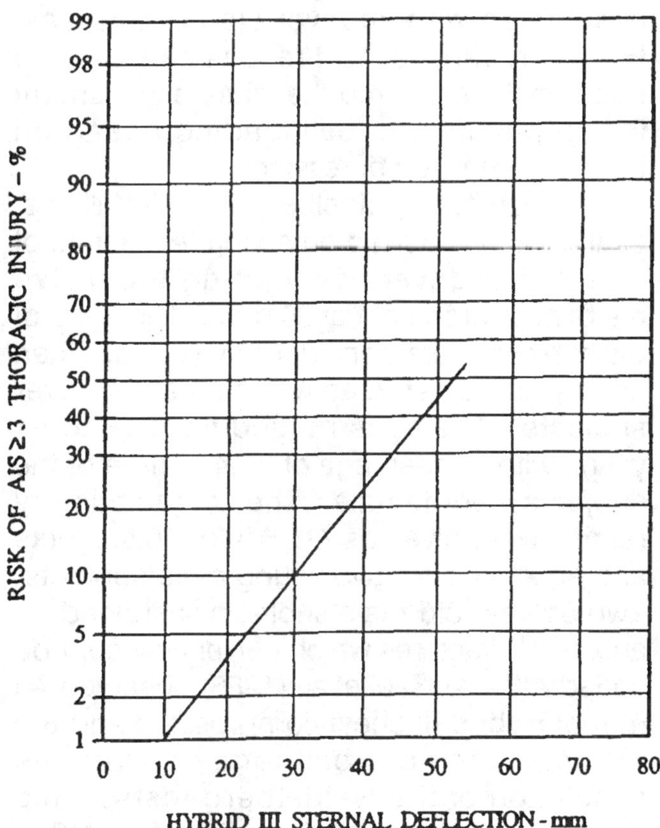

FIGURE 5. Injury risk relationship for Volvo occupant injury as a function of Hybrid III chest compression in equivalent exposures.

To additionally test this hypothesis, we compared injury to human cadaver subjects for both blunt and belt loading as a function of Hybrid III chest compression in similar exposures. Also, we critically examined the Hybrid III chest compression response for belt loading.

ANALYSIS OF BELT RESTRAINED HUMAN CADAVER TESTS - Exposure conditions for sled tests with human cadaver subjects can be objectively defined and thus test conditions for determining "equivalent" Hybrid III responses can also be objectively defined. However, there are limitations in extrapolating human cadaver injury response to represent car occupant injury risk due to deficiencies of the human cadaver to represent car occupants in terms of impact response and injury response. Human cadaver tests also differ from field data due to the range of exposure severities in car crashes contrasted with specific and generally severe exposure conditions associated with cadaver sled exposures. However, cadaver test results can provide additional insight into the injury assessment for both blunt and belt loading based on Hybrid III chest compression.

University of Heidelberg Tests - Kallieris, et al. (1982) reported on a sled test series of 12 fresh cadavers simulating the Volvo restraint environment at 30 mph. The ages of the subjects were considerably younger than other reported studies with human cadaver subjects. The ages ranged from 18 to 51 years, with a mean age of 31.4. Three of the subjects were female. The thoracic injury ratings were given as AIS=3 for 10 subjects and an AIS=2 for two young male subjects. Two 51 year old male subjects sustained 12 and 14 rib fractures which currently would be coded AIS=4. Saul et al. (1983) observed 40 mm of Hybrid III chest compression and 8.2 kN peak shoulder belt tension from his simulation of the Heidelberg tests. This comparison indicates a high risk of an AIS≥3 thoracic injury to "younger" cadavers exposed to a test condition resulting in 40 mm chest compression for the Hybrid III.

HSRI Whole Body Response Study - A sled based study was conducted at the Highway Safety Research Institute (HSRI), now UMTRI (University of Michigan Transportation Research Institute), with belt restrained human cadavers, a Hybrid II dummy, and a Hybrid III dummy (Alem et al. (1978)). The data provide a comparison of Hybrid III chest compression and injuries to human cadavers under similar belt restraint and sled exposure conditions. Three exposure severities of 15, 20, and 30 mph were used. Subjects tested at the lowest severity had multiple exposures and thus cannot be used in the following analysis of injury risk. Only cadavers exposed to a single test are considered.

Chest compression for the Hybrid III dummy was reported as 12 mm and 25 mm and peak shoulder belt tension as 5.3 kN and 8.3 kN respectively in the intermediate and high severity tests. Chest compression is low compared to other test data with the dummy for these shoulder belt tensions. The HSRI test films show that the shoulder belt was against the dummy neck. The dummy neck is narrower than the human neck and thus this belt routing cannot be obtained for a human. Test results discussed later indicates a 34% reduction of chest compression for the belt against the Hybrid III neck compared with the belt outside the margin of the human neck. Chest compression for the highest severity HSRI test condition on this basis would adjust to 38 mm from 25 mm (25/.66) and to 23 from 15 mm in the intermediate severity test and would reflect a more appropriate chest compression response for the Hybrid III dummy in those test exposures.

Three human cadaver subjects were exposed to the high test severity. Their ages were 57, 72, and 73 years. These subjects had 12, 40 and 18 rib fractures respectively, all having at least an AIS=4 injury. This indicates an almost certain risk of an AIS≥4 for the subjects at an exposure severity resulting in 38 mm Hybrid III chest compression.

Four cadavers were exposed to the intermediate test severity, ages 47, 50, 57, and 88 years. These subjects had 6, 7, 10, and 13 rib fractures respectively. The ratio to adjust chest compression from 15 to 23 mm for belt routing was determined at much higher exposure severities. Extrapolating by the ratio of Hybrid III chest compression to shoulder belt tension found in the APR analysis indicates 26 mm for 5.3 kN tension or 25 mm using the ratio found by Saul. Although these extrapolations are large, they appear mutually supportive. Thus, rib fractures to human cadavers appear to be associated with shoulder belt loading situations resulting in under 30 mm of Hybrid III chest compression.

Injury Risk for Cadavers - The results suggest that human cadavers are more easily injured than car occupants for similar exposures. Analysis for young cadaver subjects suggests nearly a 100% risk of AIS≥3 injuries for an exposure severity resulting in 40 mm of Hybrid III chest compression and AIS≥4 for older subjects. This constrasts markedly with the estimated 20-25% risk of an AIS≥3 injury for similarly exposed car occupants. The apparent difference in tolerance between car occupant and human cadaver data bases is consistent with a study by Foret-Bruno et al. (1978). Their discussion suggests that age is only part of the difference between human cadaver test subjects and car occupants, and that bone condition is important and is partly dependent on age.

Cadaver injuries are more severe for belt loading than for blunt loading based on Hybrid III chest compression observed in similar test exposures. Injury risk for the cadavers as a function of Hybrid III chest compression in a similar exposure is provided in Figure 6 for both belt restraint and blunt loading. This comparison of injury risk supports a more conservative injury assessment relationship for belt loading based on Hybrid III chest compression.

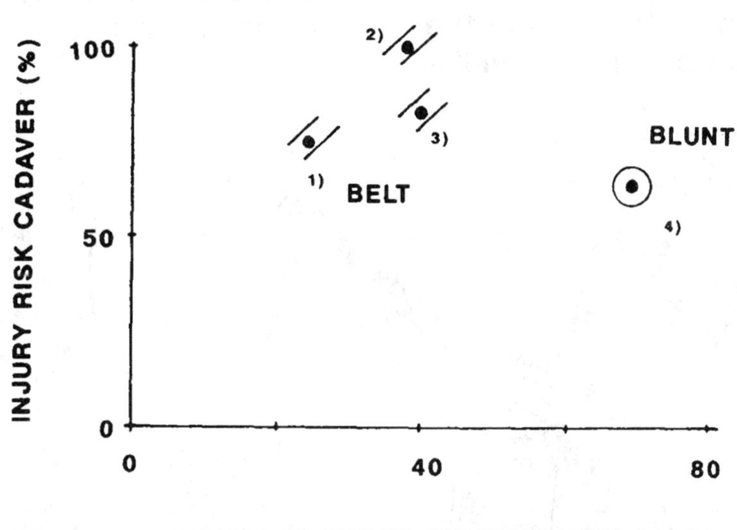

1) BELTED - AIS≥3, 4 subjects, 60.3 years mean age, HSRI mid severity;

2) BELTED - AIS≥4, 3 subjects, 67.3 years mean age, HSRI high severity;

3) BELTED - AIS≥3, 12 subjects, 31.4 years mean age, Heidelberg;

4) BLUNT - AIS≥4, 11 subjects, 61.1 years mean age, UCSD.

FIGURE 6. Comparison of cadaver injury risk as a function of Hybrid III chest compression for blunt and belt loading. Belt loading indicated by the diagonal symbol, blunt loading by a circle.

DUMMY BIOFIDELITY - The following discussion addresses reasons for a more conservative Hybrid III chest compression injury risk function for belt loading than for blunt loading. The Hybrid II is also discussed because chest compression can be measured in the Hybrid II, Horsch (1984).

Comparison of Skeletal Structures - The upper torso structure of the Hybrid II and III dummies have differences from that of a human skeleton. The thoracic spine is rigid and the human thoracic spine is not. The sternal structure differs between these dummies, neither of which model the human. The Hybrid II and III rib cages lack ribs in the upper and lower thorax.

In the human, the shoulder is supported by the rib cage. The clavicle attaches to the top of the sternum in the front and the scapula in the rear. Loads on the shoulder are supported by the rib cage, not directly by the thoracic spine. The Hybrid II and III shoulders are directly attached to the spine. A shoulder pivot protrudes forward from the spine in the approximate location of the proximal end of the clavicle. Loads on the shoulder and clavicle region are supported by the spine in both dummies. Importantly, these structures provide a direct non human-like load path to the thoracic spine of the dummies and in some situations could result in unrealistic thoracic spine accelerations.

Restraint systems such as air bags, belts, and energy absorbing steering systems apply loads not only to the rib cage, but also to other regions such as the clavicle and shoulders. Thus, the shoulder acts to share an important portion of the upper body restraining load. Because the Hybrid II and III clavicle and shoulder structures are not anatomically and structurally similar to the human, they might not distribute load in a manner similar to humans in some loading situations. The blunt frontal calibration test is aligned at mid-sternum and all of the load is below the shoulders, Figure 7. Accordingly, the load path for the calibration impact does not directly interact with the shoulders as the chest compresses and, thus, does not address biofidelity of the shoulder and clavicle regions.

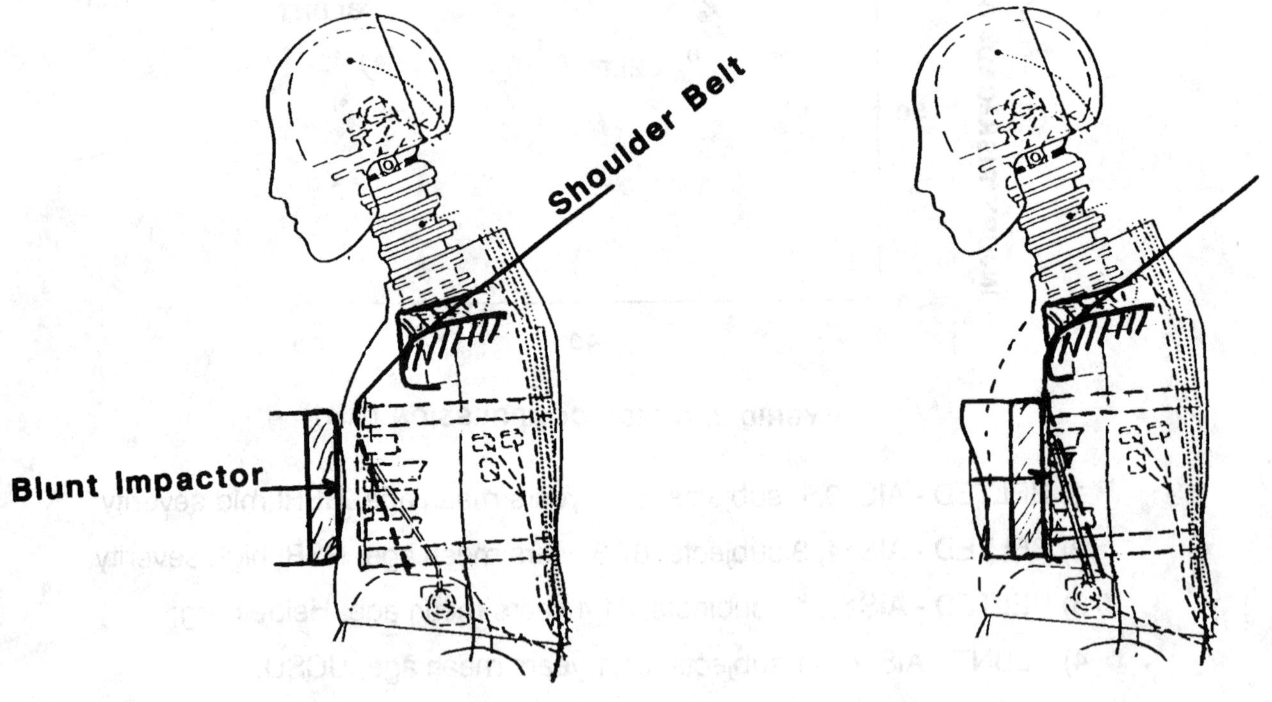

FIGURE 7. Schematic of the influence of the shoulder for blunt mid-sternal loading with a 152 mm diameter disk and for shoulder belt restraint. Blunt impactor loads only the rib cage, the belt loads both the shoulder and rib cage.

Thoracic loading experience with the Hybrid III dummy includes energy absorbing steering systems. Horsch and Viano (1984) judged the Hybrid III as the best of frontal dummies to evaluate steering system performance. Morgan et al. (1987) have demonstrated similar responses between the Hybrid III dummy and human cadavers for impacts with steering assemblies, further confirming the suitability of Hybrid III to assess thoracic injury risk for complex loading situations. Begeman et al. (1990) has shown that the Hybrid III more closely represents human cadaver response and interaction with steering assemblies than does the Hybrid II dummy.

Belt Routing - The Hybrid III neck region, unlike the Hybrid II, does not have a vinyl flesh covering. This was done to ensure repeatable neck bending response. As a result, the neck of the Hybrid III is narrower than the human neck, Figure 8. Thus, a shoulder belt can be initially placed or dynamically moved to a position "inside" the anatomic margin of the human neck. This can result in a belt routing that cannot be attained for a car occupant. The Hybrid III incorporates a "belt-guide" at the margin of a 50th percentile male neck. However, due to padding and the flesh covering on the shoulder, this does not always assure that the belt remains outside of the guide. The belt placed against the neck results in the belt passing over the rigid sterno-clavicular junction, reducing chest compression.

Sled tests were conducted to assess the significance of belt placement at the neck on thoracic responses and belt tensions. A lap-shoulder belt restrained Hybrid III dummy was tested with the belt placed against the neck and again with everything similar except the belt placed 50 mm laterally away from the neck. Chest compression was reduced 34% by locating the belt against the neck, Figure 9. Increasing the effective width of the Hybrid III neck to that of a 50th percentile male would prevent the possibility for this nonhuman-like belt routing and the resulting underestimation of chest compression amplitude. This would be among the advantages of developing a neck skin for the Hybrid III (Horsch et al. (1990)).

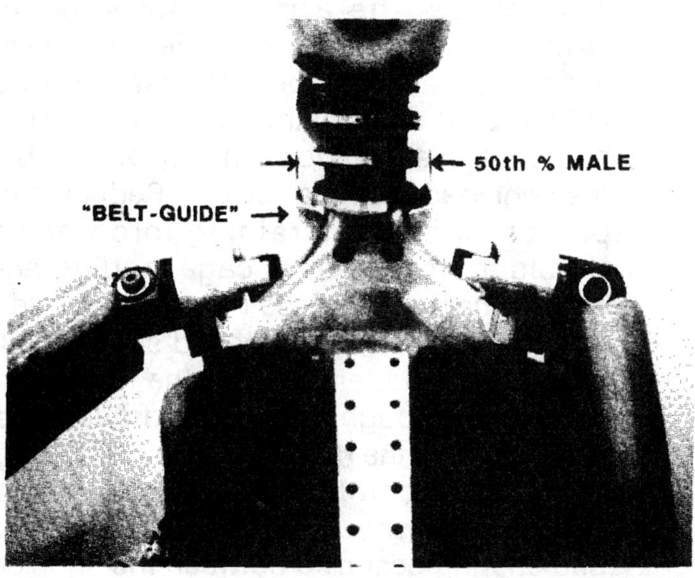

FIGURE 8. Comparison of neck widths for a 50th percentile male and the Hybrid III dummy indicating a narrower neck for the Hybrid III which influences the shoulder belt path if it moves inside the "belt-guide".

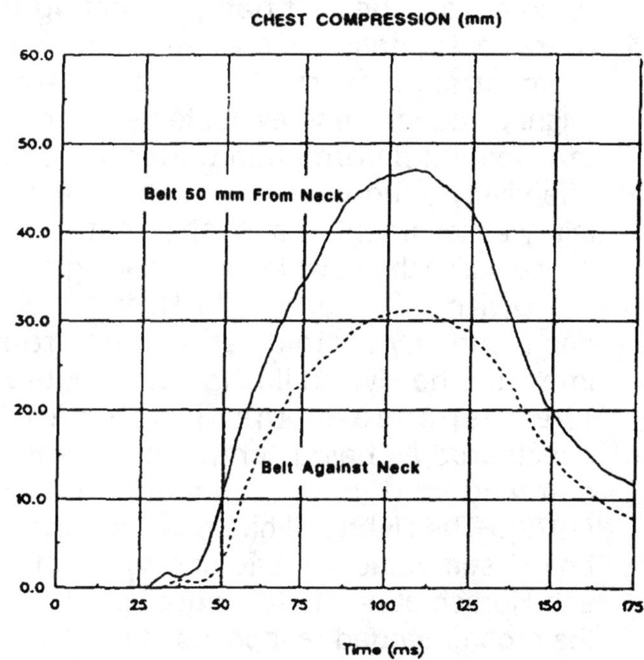

FIGURE 9. Influence of belt routing on chest compression for belt paths against and away from the neck of Hybrid III.

Measurement of Compression - The Hybrid II does not typically contain instrumentation to measure chest compression. The Hybrid III thorax contains a transducer to measure the axial deflection of the sternum. The response of the chest and the method of measuring sternal deflection were developed for blunt frontal impact of the rib cage, a situation having midsagittal symmetry with a flat-faced impactor which tends to control sternal alignment. However, loading with a shoulder belt is nonsymmetrical and can contain a lateral component. The rib cage deformation under the belt likely has significant local and overall deformations which are not fully reflected by the "X" direction sternal deflection measurement. Nonuniform deformation of the rib cage with belt loading was reported by Backaitis (1986) and Cesari (1990).

Because rib fractures tend to follow the belt path across the chest in human cadaver tests, Alem et al. (1978), and belt restrained occupants in severe crashes, Patrick et al. (1974), it is likely that "local" stresses and, thus, deformations are associated with belt loading. However, some rib fractures occur remote from the belt path, indicating that overall deformation is also an issue. Thus, the central measure of sternal deflection might be adequate to evaluate belt loading if the local rib deformation under the belt is effectively accounted for by the use of an injury assessment relationship that is more conservative than that for blunt loading.

Dynamic Response - The Hybrid III chest has biofidelity for midsternal blunt frontal impact. The Hybrid II is too stiff, Figure 2. The response of the thorax can be considered as being composed of spring, damping, and inertial elements, Lobdell (1973). The Hybrid III has been reported to be too stiff statically, Lau and Viano (1988) and Horsch et al. (1990), but approximates the recommended responses at the 4.3 and 6.7 m/s impact velocities used to calibrate the dynamic compression response. The Hybrid III chest appears to have too great of a spring force and, on this basis, too little damping. Thus, the Hybrid III might be too stiff at compression rates observed in belt tests, which are typically less than 50% of the compression rate in the 4.3 and 6.9 m/s chest calibration tests. On this basis the Hybrid III would underestimate the amount of compression for a human. This is consistent with belt loading experiments reported by Cesari and Bouquet (1990) and Backaitis and St-Laurent (1986) in which they observed less rib deflection for the Hybrid III than for human cadavers or volunteers.

DISTRIBUTION OF LOAD ON THE SHOULDER AND RIB CAGE

Distribution of restraint loads between the rib cage and other upper body regions is important for steering wheel, air bag, or shoulder belt loading. Chest compression provides a measure of the specific loading to the rib cage in terms of an injury risk. The shoulder acts to share an important portion of the upper body load for a belt restraint, providing upper body restraint.

The influence of the shoulder on rib cage loading by a belt was demonstrated by conducting tests with and without the shoulder for a lap- shoulder belt restraint. In both cases, the arm on that side was removed to minimize mass differences in the two test situations. The responses provided in Figure 10 demonstrate that the shoulder and clavicle provide an important belt restraint load-carrying region. Reducing the portion of belt restraining force on the shoulder increases rib cage loading and, thus, compression of the rib cage. Hybrid III chest compression increased substantially without a shoulder to share the restraining force, even though the belt load decreased slightly. Thus, the ratio of chest compression to shoulder belt load was increased by removing the shoulder, demonstrating that load sharing occurred between the shoulder and rib cage and that chest compression is influenced by this load sharing.

INFLUENCE OF PELVIC DISPLACEMENT

INFLUENCE OF PELVIC DISPLACEMENT - Although removing the shoulder demonstrated the influence of belt restraint load sharing by the shoulder on chest compression, it was not a practical demonstration. In restraint system testing, the proportion of belt load distributed between the shoulder and the rib cage is influenced by the torso angle in the x-z plane. Adomeit and Heger (1975) recommended that the upper body be tipped forward at the top to "avoid unfavorable biomechanic conditions". The belt load on the torso shifts higher on the thorax as the torso is tipped forward, thereby increasing the load on the shoulder and reducing the relative loading of the rib cage.

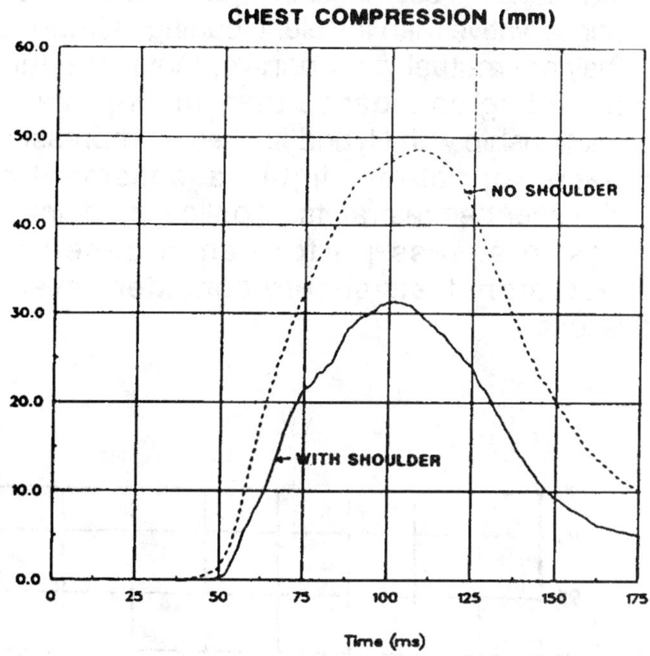

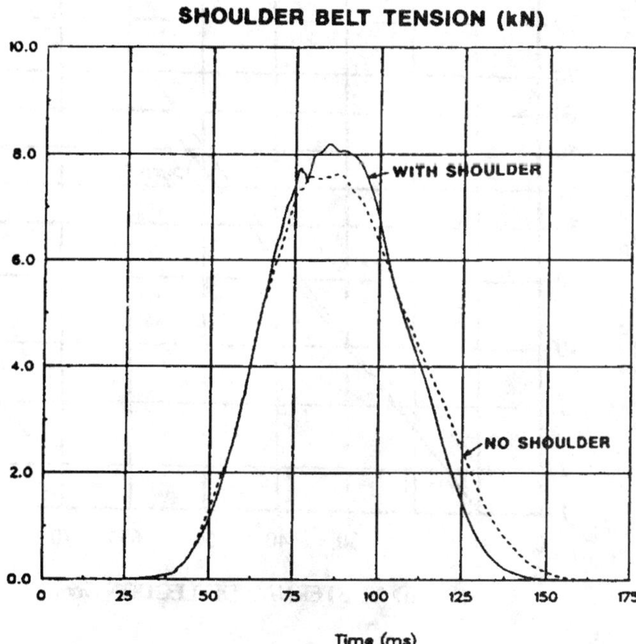

FIGURE 10. Test responses with belt restraint with and without shoulder (arm removed for both conditions to minimize mass differences between test situations).

Greater displacement of the pelvis than the chest tends to reduce forward angulation of the torso and, thus, reduce the portion of belt load on the shoulder. Submarining (the lap belt slipping over the pelvic spines) or lap belt slack can increase pelvic forward displacement for a lap-shoulder belt restraint. Rouhana et al. (1989) found greater chest compression and slightly lower shoulder belt loads associated with adding 76 mm slack to the lap belt to cause submarining. Increased spacing between the knees and knee-bolster can increase pelvic forward displacement for upper torso belts (without a lap belt). On this basis, it might be expected that Hybrid III would exhibit greater chest compression for greater pelvic forward displacement.

To explore this hypothesis, 53 km/h sled tests were conducted with a lap-shoulder belt restraint or with an upper-torso belt and knee bolster restraint. For both restraints, the pelvic forward displacement was varied by lap- belt slack (snug or 100 mm) or knee-bolster spacing (140, 190, and 240 mm). Test films indicated that the torso was angulated more rearward at the top for tests having greater forward pelvic displacement. For both restraint configurations, chest compression was greater for tests with the greatest forward displacement of the pelvis and thus with the torso angulated more rearward at the top, Figure 11. Importantly, the increase of Hybrid III chest compression as the pelvis was allowed to displace further forward was primarly due to more of the belt load acting on the rib cage, increasing the ratio of chest compression to belt tension, and minimally due to increased belt load, Figure 11. These factors demonstrate the importance of the chest compression response in improving the evaluation of restraint load distribution on the upper torso.

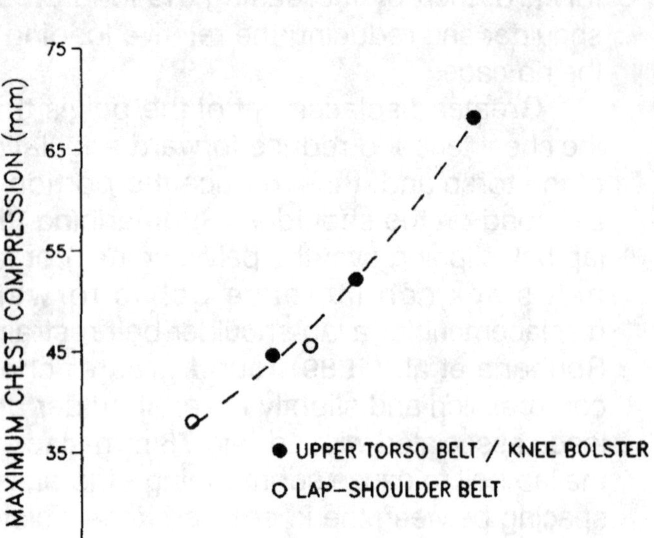

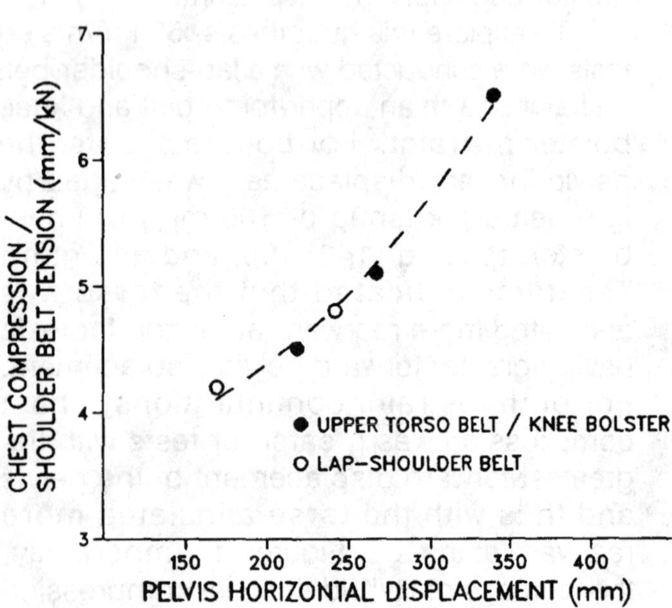

FIGURE 11. Influence of maximum pelvic horizontal displacement on peak chest compression and ratio of chest compression to belt tension for a lap- shoulder belt restraint and an upper-torso belt with knee bolster restraint.

DISCUSSION

The APR and Volvo field data bases provided independent estimates of the relationship between thoracic injury risk for car occupants and Hybrid III chest compression in similar exposures. These independent estimates of injury risk are more conservative than for blunt loading. However, they are mutually supportive, Figure 12, thus providing confidence that this injury risk relationship with Hybrid III chest compression is appropriate for injury assessment of shoulder belt restraints. Application of injury risk to assess performance based on laboratory tests should consider several issues.

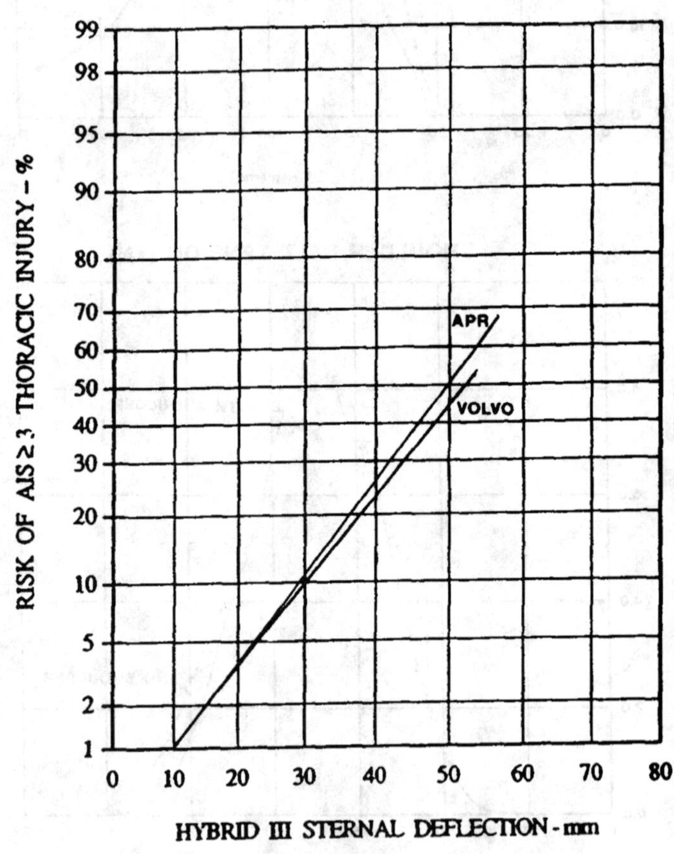

FIGURE 12. Injury risk summary for APR and Volvo car occupants as a function of Hybrid III chest compression in similar exposures.

A BASIS FOR INJURY ASSESSMENT

Crash tests using instrumented dummies provide a primary means of evaluating occupant protection systems in the laboratory. Dummies are not actually "injured" in such tests, instead, a potential for injury is judged from the amplitude of injury assessment parameters which are transformed from physical measurements such as force, acceleration, velocity or displacement. The relationship between injury risk and the assessment parameter must be known if an objective inference of injury is to be associated with a safety test. The biofidelity of the dummy and the associated injury assessment measurements for important body regions are critical issues in the evaluation of occupant protection systems.

Injury data indicates that people have a distribution in tolerance to impact loading, age and sex being among the important factors, Patrick et al. (1974), Foret-Bruno et al. (1978), Eppinger (1976), and Evans (1988a and 1988b). Thus, there is not a unique "tolerance", but an increasing risk of injury at greater amplitudes of the injury assessment parameter. Figure 12 provides examples. The relationship of an assessment parameter amplitude to injury outcome can be stated as a risk or probability of a defined injury as a function of loading severity.

The objectivity of a given injury risk relationship depends on many factors. The quantity and quality of biomechanical data are obvious factors. A rigorous injury risk relationship cannot be developed with a "few" exposures or with marginal experiments with poorly chosen parameters. The loading environment used for the injury studies might not fully relate to the conditions for which the injury assessment parameter is being used to judge performance. The population at risk is important and depends on crash type and severity, Viano (1987), Horsch (1987), Evans (1988a and 1991), and Viano et al. (1990).

The selection of acceptable response amplitudes to be associated with a particular issue or test situation is in part based on biomechanics but also includes other considerations. The selection of an injury risk and severity might depend on body region. For example, an occupant having an AIS = 4 brain injury and an occupant having an AIS = 4 liver injury might have similar chances of survival, but with very different long-term consequences to these injured survivors.

Other considerations for the selection of injury risk and injury severity for an assessment value depend to some extent on the particular test situation and evaluation issues, test severity being an important issue. The arbitrary use of a single "tolerance" response amplitude, independent of the test situation or severity, is easy for the user and he might perceive that no one is being injured by his strategy if he considers his choice as a tolerable loading limit for all occupants. However, this approach does not account for the spectrum of real world car crashes and occupant tolerances, and results in incorrect perceptions of performance, (Horsch, 1987).

Both the relevance of the injury assessment parameter and the biofidelity of the test dummy used to measure the injury assessment parameter are important for objective injury assessment. The availability of injury assessment instrumentation for discrete body regions is also an important issue. The Hybrid III dummy has more features that have biofidelity and has a much larger range of injury assessment instrumentation than does the Hybrid II dummy. Thus the Hybrid III is a more effective tool to assess injury risk in a crash test. However, an injury risk assessment should not automatically be applied without consideration of the biofidelity of the dummy's response in the body region of interest.

INJURY RISK RELATIONSHIP TO HYBRID III CHEST COMPRESSION

The APR and Volvo studies of real world crashes provide an estimate for injury risk relationship based on Hybrid III chest compression for belt loading, Figure 11. We believe that the APR data provides a more direct measure of the severity of occupant loading than the Volvo data because the APR study bases severity on shoulder belt loading of the occupant while the Volvo study bases severity on the loading of the vehicle. The analysis of both the APR and Volvo data indicate that a belt loading of the Hybrid III thorax resulting in 40 mm of chest compression would represent about a 25% risk of an AIS≥3 thoracic injury.

Assessment for belt loading appears to be driven by peak chest compression, not the viscous response (VC) because the rate of chest compression with belt restraints is typically well below 3 m/s compression velocity. VC becomes predominate above 3 m/s rates. However, VC should also be considered, since combined interactions with the belt and air bag or steering wheel might result in the rate of compression introducing an injury risk by a viscous mechanism.

POPULATION AT RISK - The population that is represented in an injury risk function is important. These populations should not be based on the "average" person or occupant and probably differ from the distribution of car occupants. The population of the APR and Volvo field studies of car crashes represent the APR and Volvo occupants actually exposed in the investigated crashes. Injury outcome is a function of not only an occupant's tolerance, but also the severity of loading. Low tolerance occupants might experience different exposure severities than high tolerance occupants.

Some information which describes the APR population (Foret-Bruno et al. (1989)) and the equivalent information for the Volvo population are provided in Table 1. The populations are divided into three age groups with the total number of occupants in each age group and the number of occupants that incurred rib fractures in each age group. The APR data contains more than twice as many occupants as the Volvo data, but the overall percentage of occupants sustaining rib fractures is similar (7.7% for APR vs 7.8% for

Table 1

AGE DISTRIBUTIONS FOR THE APR AND VOLVO ACCIDENT SAMPLES

	Age Groups			Total
APR	<30 yrs.	30 to 49	≥ 50 yrs.	
No. Occupants/% Total	160/41%	169/44%	57/15%	386/100%
No. Occ. with Rib Fx/% Group	3/1.9%	10/5.9%	17/29.8%	30/7.7%
Volvo				
No. Occupants/% Total	41/25%	64/38%	61/37%	166/100%
No. Occ. with Rib Fx/% Group	1/2.4%	6/9.4%	6/9.8%	13/7.8%
APR (at least one band broken)				
No. Occupants/% Total	70/41.2%	72/42.4%	28/16.4%	170/100%
No. Occ. with Rib Fx/% Group	3/4.3%	6/8.3%	10/35.7%	19/11.2%

Volvo). There are differences between the data sets when the individual age groups are examined, however. APR occupant age distribution is skewed toward younger age than that of the Volvo data with 37% of the Volvo occupants being in the oldest category compared with 15% in the APR data.

In both data sets the youngest group had the lowest frequency of rib fractures, on the order of 2%. The greatest difference was in the oldest category where the APR data indicated a 29.8% frequency of rib fracture and the Volvo data had a frequency of only 9.8%. The reasons for such a large difference could be due to different accident sampling procedures, vehicle and restraint characteristics, and population differences related to various crash severities.

Table 1 also contains a subset of 170 APR occupants which broke at least one of the shoulder belt force-limiting bands. The occupants in this subset can be considered to have definitely loaded the shoulder belt while there could always be some indeterminacy in the loading in the other APR cases and in the Volvo cases. The distribution of occupants between age groups is very similar to the overall APR sample. The frequency of occurrence of rib fractures is slightly higher, which is to be expected because this subset contains only those occupants who sustained shoulder belt loads greater than 2.1 kN (the minimum load to break one band). This subset shows a clear trend of the risk of rib fracture increasing with the age range of the group.

Injury risks based on the APR and Volvo populations are relevant to the real world since they are the populations at risk in real world exposures. Because each vehicle and region in which it is operated might have unique demographics, there might not be a unique "exposed population" to apply to all situations. However, the APR and Volvo data sets provide better estimates of injury risk than basing risk on human cadavers, the average male or female, or car occupants in general.

INJURY RISK FOR LARGE COMPRESSIONS - Probit Analysis of the APR data reported by Mertz et al. (1991) indicates a risk of an AIS≥3 injury is 50% for 50 mm of Hybrid III chest compression and greater than 90% for 75 mm of compression, Figure 3. There is greatly reduced confidence at these larger amplitudes of Hybrid III chest compression because the APR and Volvo field exposures were infrequent at exposure severities sufficient to produce these levels of Hybrid III chest compression. Mertz et al. (1991) concluded that "only 5% of the field accidents investigated by APR would produce a Hybrid III sternal deflection of 45 mm or greater." Our analysis of the Volvo data indicates that only 5 of 98 restrained occupants were exposed sufficiently to have resulted in greater than 40 mm of Hybrid III chest compression.

The APR and Volvo "windows" to real world distributions of exposure severity and injury have most of the exposures and many of the injuries in crashes consistent with a "low" injury risk. This is not atypical, since, for example, the velocity change (ΔV) associated with a 49 kph barrier crash test is above the 98th percentile velocity change in the National Accident Sampling System (NASS), Malliaris et al. (1985), and that more than 75% of NASS occupant HARM is associated with a lower velocity change, Malliaris (1988).

A Simplistic Performance Model - Projection of injuries over the distribution of real world crash severities provides a different perception of performance than that based on injury risk in a "severe" crash test, Horsch (1987). Using a simplistic analysis, the number of injured occupants at each exposure severity can be estimated by multiplying the risk of injury per exposed occupant by the number of occupants at that exposure severity.

We have used a similar analysis to indicate performance differences between two hypothetical belt restraint systems. One of the belt systems would result in 33 mm of Hybrid III chest compression in a 49 km/h ΔV frontal barrier crash, the other, 50 mm of compression.

Since our analysis is for a hypothetical vehicle model, NASS was used to provide an exposure distribution. We used NASS frontal exposure frequency as a function of reported velocity change for the estimate of exposed occupants at each crash severity. However, each vehicle model will have a unique real world exposure distribution, depending on factors such as mass, size, driver demographics, and geographic area, Evans (1991).

Hybrid III chest compression as a function of exposure severity is based on the Patrick et al. (1974) sled based Volvo study and our analysis of the various Volvo simulations. The hypothetical belt restraint system that results in 33 mm of Hybrid III chest compression in a 49 km/h ΔV barrier crash approximates our analysis of the Volvo data.

The hypothetical belt restraint that results in 50 mm of Hybrid III chest compression in a 49 km/h ΔV barrier crash was assumed to exhibit 50% more Hybrid III chest compression at all test severities, than the belt restraint that resulted in 33 mm. In this hypothetical situation, the increased chest compression could be due, for instance, to upper body kinematics caused by greater pelvic forward displacement as demonstrated in Figure 11.

The risk of injury at each exposure severity uses the injury risk function for Hybrid III chest compression from Mertz et al. (1991) based on the APR belted occupant field study, Figure 3. Injury risk below 1% was zeroed, with a "smoothed" transition because of insufficient data to objectively define this region.

The projected thoracic injury at each exposure severity is the product of NASS exposure and injury risk as plotted in Figure 13. Injury distributions by this analysis are similar to real world injury, in that injury is broadly distributed as a function of exposure severity and much of the injury is associated with low to moderate risk exposure conditions, Horsch (1987). Although the higher injury risks are strongly biased toward severe exposures, the highest incidence of crashes is strongly biased to lower severities.

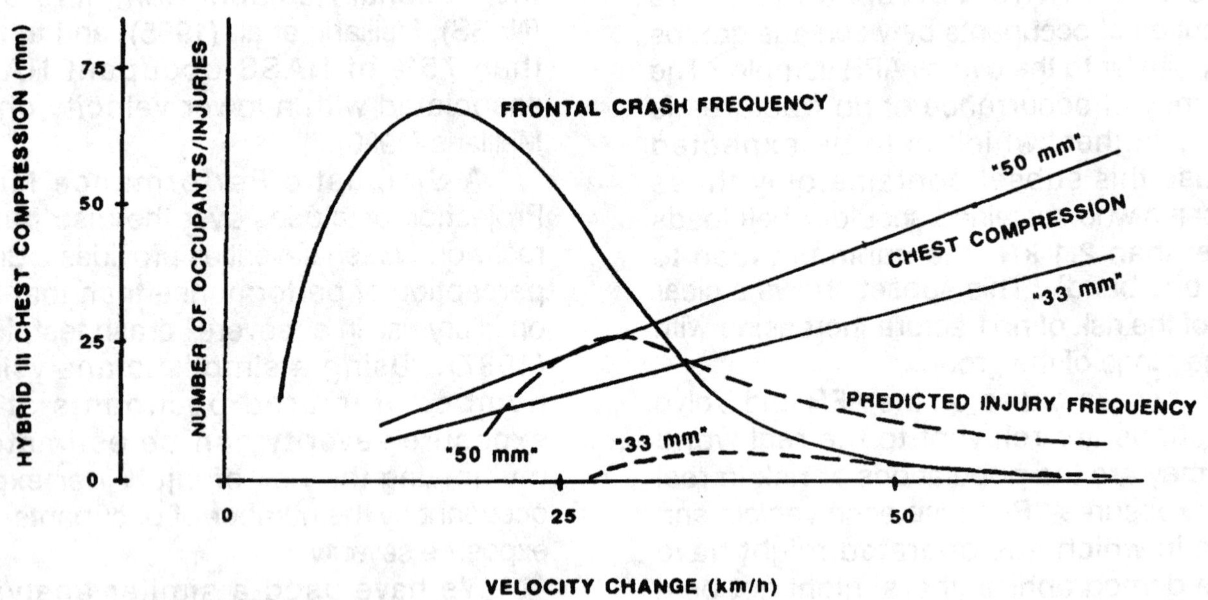

FIGURE 13. Simplistic performance model based on: 1) NASS frontal exposure; 2) two hypothetical belt restraint systems that result in either 33 or 50 mm of Hybrid III chest compression in a 49 km/h velocity change barrier crash test (chest compression based on our discussion of laboratory simulations of the Volvo field study for the "33 mm" restraint); and 3) injury risk of belt induced AIS≥3 thoracic injuries as a function of Hybrid III chest compression based on analysis of APR field data by Mertz et al. (1991), (see Figure 3). Predicted injury frequency was determined as the risk of injury per exposed occupant multiplied by the number of exposed occupants at each exposure severity.

This results in the broad distribution of injury across the range of exposure.

Comparison of injuries for the two hypothetical belt restraints indicate a difference over a broad range of exposure severities. Integration of the injury distribution provides the number of occupants having thoracic injury due to belt loading across the real world exposure range. The hypothetical system that results in 50 mm of Hybrid III chest compression in a 49 km/h ΔV barrier crash has a much greater number of injured occupants and has injured occupants at lower exposure severities than the system that results in 33 mm of Hybrid III chest compression.

In the APR and Volvo data, most of the injury is associated with exposure severities that would result in less than 50 mm of Hybrid III chest compression. Our model demonstrates a similar trend. The injury risk associated with 50, 60 or 75 mm of Hybrid III chest compression is not the most relevant performance issue for belt systems with this level of performance. However, with other belt restraint configurations, compressions of 50 or even 75 mm might be relevant if the belt system can produce this level of chest compression in the more frequent lower severity crashes.

COMBINED THORACIC LOADING - Laboratory tests with belted dummies and supplemental air bags will involve combined thoracic restraint loads. Air bags provide a more distributed thoracic load than a shoulder belt. A dummy with human-like response for combined air bag and shoulder belt loading on the upper body and measures of relevant injury modes could directly assess the injury risk for the test situation. The Hybrid III dummy represents an important improvement over the Hybrid II dummy towards achieving these objectives in both response and injury assessment measurements.

Additional state-of-the-art improvements in a more human-like upper torso response for chest compression and load sharing by the shoulder and clavicle regions could refine injury assessment in laboratory tests. However, the restraint systems for many millions of vehicles will be designed with current dummies before further advanced dummies are sufficiently mature for routine assessment of automotive restraints. Using the best of mature injury assessment technology should always provide the best approach to develop restraint systems.

Hybrid III chest compression provides an important insight to load sharing between the rib cage and other upper torso regions such as the shoulder. On this basis, the single compression measurement and design of the rib cage represents a significant improvement over chest acceleration, but does not provide a complete assessment of load distribution on the rib cage. In this study, we have recommended a more conservative injury risk function for belt loading than for blunt loading of the rib cage for the Hybrid III. Importantly, the Hybrid II does not provide biofidelity in response or indication of load distribution.

For assessment of combined belt and air bag loading with the Hybrid III dummy, the risk function for either belt or blunt loading could apply. If 99% of the load was from the belt, then the more conservative injury risk function for belt loading would apply. If 99% of the load was from the air bag, the less conservative injury risk function for blunt loading would apply.

The distributed load of the air bag compared with that of a shoulder belt suggests a lower injury risk for a given load if more of the restraining load is by the air bag than by the shoulder belt, particularly in severe crashes and for lower tolerance occupants. The ratio of restraining forces will depend not only on the vehicle design, but also on such factors as occupant size and seat adjustment. Limiting shoulder belt tension might be a method to influence the relative loading.

We have not directly addressed thoracic injury assessment for combined loading for the Hybrid III. Both blunt and belt loads cause rib cage deformation which can result in fractures away from the loaded region. Injury risk assessment can probably be based on blunt loading for this injury mode. Additionally, rib fractures tend to follow the belt path due to a more concentrated belt load compared with blunt loading. Injury risk assessment should probably be based on belt loading for this injury mode.

The relative amount of restraint by the shoulder belt and air bag cannot be directly determined from Hybrid III responses. However, shoulder belt tension does provide a measure of restraint by the shoulder belt. Our analysis of various studies has demonstrated a relationship of shoulder belt tension with Hybrid III chest compression. Thus, if the relationship between shoulder belt tension and Hybrid III chest compression is known for a particular belt restraint, then the relative loading of the rib cage by each restraint can be estimated. One assessment strategy might be based on the blunt injury risk assessment for Hybrid III chest compression while independently considering the influence of belt loading from shoulder belt tension and its relationship with Hybrid III chest compression for that belt restraint. This would allow the application of the injury risk function for belt loading in combination with the blunt injury risk assessment.

SUMMARY

Chest compression is one of the vital responses for development of restraint systems. This response directly relates to loading of the rib cage and the risk of serious or fatal injury to the heart and lungs protected by the rib cage. The "tolerable" loading of the upper body is more than twice that of the rib cage alone.

The chest compression response of the Hybrid III provides an objective measure of this important aspect of torso load distribution for blunt loading. Our analysis indicates that the chest compression of the Hybrid III dummy is important for belt restraint if an appropriate injury risk relationship is used.

We found similar estimates of thoracic injury risk as a function of chest compression based on two in depth studies of crashes involving belted car occupants. An exposure severity that results in 40 mm of Hybrid III chest compression would have a 20-25% risk of an AIS≥3 thoracic injury for car occupants. Importantly, the field experience of APR and Volvo indicate that crashes severe enough to result in greater than 40 mm of Hybrid III chest compression are infrequent and that injuries in these classic field studies are mostly associated with crash severities which would result in less than 40 mm of Hybrid III chest compression.

ACKNOWLEDGEMENTS

The authors thank Larry Schneider of UMTRI and Paul Begeman of Wayne State University for providing Hybrid III test films, Kazunari Ueno for the statistical analysis of the Volvo field data, Stephen Rouhana for technical review of the paper, Gerald Horn and Joseph McCleary for technical support, and Mary Rouhana for formatting this paper.

REFERENCES

Adomeit, D. and Heger, A. (1975) "Motion Sequence Criteria and Design Proposals for Restraint Devices to Avoid Unfavorable Biomechanical Conditions and Submarining". Proceedings of the 19th Stapp Car Crash Conference, November 1975, SAE No. 751146, Society of Automotive Engineers, Warrendale, PA.

Alem, N.M., Melvin, J.W., Bowman, B.M. and Benson J.B. (1978) "Whole-Body Human Surrogate Response to Three-Point Harness Restraint". Proceedings of the 22nd Stapp Car Crash Conference, pp 359-399. Society of Automotive Engineers, Warrendale, PA.

Backaitis, S.H., and St-Laurent, A. (1986) "Chest Deflection Characteristics of Volunteers and Hybrid III Dummies". Proceedings of the 30th Stapp Car Crash Conference, pp 157-166. Society of Automotive Engineers, Warrendale, PA.

Begeman, P.C., Kopacz, J.M., and King A.I. (1990) "Steering Assembly Impacts Using Cadavers and Dummies". Proceedings of the 34th Stapp Car crash Conference, pp 123-144. Society of Automotive Engineers, Warrendale, PA.

Cesari, D. and Bouquet, R. (1990) "Behavior of Human Surrogates under Belt Loading". Proceedings of the 34th Stapp Car Crash Conference, pp 73-82. Society of Automotive Engineers, Warrendale, PA.

Eppinger, R.H. (1976) "Prediction of Thoracic Injury Using Measurable Experimental Parameters". Report 6th International Technical Conference on Experimental Safety Vehicles, pp 770-779. National Highway Traffic Safety Administration, Washington, D.C.

Evans, L. (1988a) "Older Driver Involvement in Fatal and Severe Traffic Crashes". The Journal of Gerontology: Social Sciences, 43 , S186-S193, 1988.

Evans, L. (1988b) "Risk of Fatality from Physical Trauma Versus Sex and Age". The Journal of Trauma, 28, 368-378, 1988.

Evans, L. (1991) "Traffic Safety and the Driver". Van Nostrand Reinhold, January, 1991.

Fayon, A., Tarriere, C., Walfisch, G., Got, C. and Patel, A. (1975) "Thorax of 3-Point Belt Wearers During a Crash". Proceedings of the 19th Stapp Car Crash Conference, Society of Automotive Engineers, Warrendale, PA.

Foret-Bruno, J.Y., Hartemann, F., Thomas, C., Fayon, A., Tarriere, C., Got, C. and Patel, A. (1978) "Correlation Between Thoracic Lesions and Force Values Measured at the Shoulder of 92 Belted Occupants Involved in Real Accidents". Proceedings of the 22nd Stapp Car Crash Conference, pp 271- 292. Society of Automotive Engineers, Warrendale, PA.

Foret-Bruno, J.Y., Brun-Cassan, F., Brigout, C. and Tarriere, C. (1989) "Thoracic Deflection of Hybrid III Dummy Response for Simulations of Real Accidents". 12th Experimental Safety Vehicle Conference, Gothenburg, Sweden.

Foster, J.K., Kortge, J.O., and Wolanin, M.J. (1977) "Hybrid III: A Biomechanically-Based Crash Test Dummy". Proceedings of the 21st Stapp Car Crash Conference, pp. 975-1014. Society of Automotive Engineers, Warrendale, PA.

Gadd, C.W. and Patrick, L.M. (1968) "System Versus Laboratory Impact Tests for Estimating Injury Hazard". SAE No. 680053, Society of Automotive Engineers Automotive Engineering Congress, January, 1968.

Horsch, J.D. and Viano, D.C. (1984) "Influence of the Surrogate in Laboratory Evaluation of Energy-Absorbing Steering System". Proceedings of the 28th Stapp Car Crash Conference, SAE 841660, Society of Automotive Engineers, Warrendale, PA.

Horsch, J.D. (1987) "Evaluation of Occupant Protection from Responses Measured in Laboratory Tests". SAE No. 870222, Society of Automotive Engineers International Congress and Exposition, February, 1987.

Horsch, J.D. and Schneider, D. (1988) "Biofidelity of the Hybrid III Thorax in High-Velocity Frontal Impact". SAE No. 880718, Society of Automotive Engineers International Congress and Exposition, February, 1988.

Horsch, J.D., Lau, I.V., Andrzejak, D.V., Viano, D.C., Melvin, J.W., Pearson, J., Cok, D. and Miller, G. (1990) "Assessment of Air Bag Deployment Loads". Proceedings of the 34th Stapp Car Crash Conference, pp 267-288. Society of Automotive Engineers, Warrendale, PA.

Horsch, J.D., Viano, D.C. and DeCou, J.M. (1991) "The History of the General Motors Energy Absorbing Steering System". Proceedings of the 35th Stapp Car Crash Conference, pp.xx-yy, SAE Technical Paper No. 91xxxx, Society of Automotive Engineers, Warrendale, PA.

Kallieris, D., Mellander, H., Schmidt, G., Barz, J. and Mattern, R. (1982) "Comparison Between Frontal Impact Tests with Cadavers and Dummies in a Simulated True Car Restrained Environment". Proceedings of the 26th Stapp Car Crash Conference, pp 353-367. Society of Automotive Engineers, Warrendale, PA.

Kroell, C.K., Schneider, D.C. and Nahum, A.M. (1971) "Impact Tolerance and Response of the Human Thorax". Proceedings of the 15th Stapp Car Crash Conference, pp 84-134. Society of Automotive Engineers, Warrendale, PA.

Kroell, C.K., Schneider, D.C. and Nahum, A.M. (1974) "Impact Tolerance Of the Human Thorax II". Proceedings of the 18th Stapp Car Crash Conference, pp 383- 457. Society of Automotive Engineers, Warrendale, PA.

Lau, I.V. and Viano D.C. (1986) "The Viscous Criterion -- Bases and Applications of an Injury Severity Index for Soft Tissue". Proceedings of the 30th Stapp Car Crash Conference, pp 123-142. Society of Automotive Engineers, Warrendale, PA.

Lau, I.V. and Viano D.C. (1988) "How and When Blunt Injury Occurs: Implications to Frontal and Side Impact Protection". Proceedings of the 32nd Stapp Car Crash Conference, SAE No. 881714, pp. 81-100, October, 1988. Society of Automotive Engineers, Warrendale, PA.

Lobdell, T.E., Kroell, C.K., Schneider, D.C., Hering, W.E. and Nahum, A.M. (1973) "Impact Response of the Human Thorax". Human Impact Response: Measurement and Simulation, pp. 201- 245. Plenum Press, New York.

Malliaris, A.C., Hitchcock, R. and Hansen, M. (1985) "Harm Causation and Ranking in Car Crashes". SAE Technical Paper No.850090, Society of Automotive Engineers, Warrendale, PA.

Malliaris, A.C. (1988) "National Estimates of Car Occupant Injury Rates and Injury Counts". Contract Report CR-88/13/BI, October 1988.

Mertz, H.J., Horsch, J.D. and Horn, G. (1991) "Hybrid III Sternal Deflection Associated with Thoracic Injury Severities of Occupants Restrained with Force-Limiting Shoulder Belts". SAE No. 910812, Society of Automotive Engineers International Congress and Exposition, February, 1991.

Morgan, R.M., Schneider, D.C., Eppinger, R.H., Nahum, A.M., Marcus, J.H., Awad, J., Dainty, D. and Forrest, S. (1987) "Interaction of Human Cadaver and Hybrid III Subjects with a Steering Assembly". Proceedings of the 31st Stapp Car Crash Conference, pp.79-94. Society of Automotive Engineers, Warrendale, PA.

Nahum,A.M., Gadd, C.W., Schneider, D.C. and Kroell, C.K. (1970) "Deflection of the Human Thorax Under Sternal Impact". 1970 International Automotive Safety Conference Compendium, pp 797- 807. Society of Automotive Engineers, New York.

Nahum, A.M., Schneider, D.C. and Kroell, C.K. (1975) "Cadaver Skeletal Response to Blunt Thoracic Impact". Proceedings of the 19th Stapp Car Crash Conference, pp. 259-293. Society of Automotive Engineers, Warrendale, PA.

Neathery, R.F. (1974) "An Analysis of Chest Impact Response Data and Scaled Performance Recommendations". Proceedings fo the 18th Stapp Car Crash Conference, pp. 459-493. Society of Automotive Engineers, Warrendale, PA.

Neathery, R.F. and Lobdell, T.E. (1973) "Mechanical Simulation of Human Thorax Under Impact". Proceedings of the 17th Stapp Car Crash Conference, pp. 451-466. Society of Automotive Engineers, New York.

Neathery, R.F., Kroell, C.K. and Mertz, H.J. (1975) "Prediction of Thoracic Injury from Dummy Responses". Proceedings of the 19th Stapp Car Crash Conference, pp. 295-316. Society of Automotive Engineers, Warrendale, PA.

Nyguist, G.W., Begeman, P.C., King, A.I. and Mertz, H.J. (1980) "Correlation of Field Injuries and GM Hybrid III Dummy Responses for Lap-Shoulder Belt Restraint". Journal of Biomechanical Engineering, May 1980, Vol. 102, pp. 103-109.

Patrick, L.M., Kroell, C.K. and Mertz, H.J. (1965) "Forces on the Human Body in Simulated Crashes". Proceedings of the 9th Stapp Car Crash Conference, pp 237-260, University of Minnesota.

Patrick, L.M., Mertz, H.J. and Kroell, C.K. (1967) "Cadaver Knee, Chest and Head Impact Loads". Proceedings of the 11th Stapp Car Crash Conference, pp. 106-117. Society of Automotive Engineers, New York.

Patrick , L.M., Bohlin, N.I. and Andersson, A. (1974) "Three-Point Harness Accident and Laboratory Data Comparison". Proceedings of the 18th Stapp Car Crash Conference, pp. 201-282. Society of Automotive Engineers, Warrendale, PA.

Rouhana, S.W., Viano, D.C., Jedrzejczak, E.A. and McCleary, J.D. (1989) "Assessing Submarining and Abdominal Injury Risk in the Hybrid III Family of Dummies". Proceedings of the 33rd Stapp Car Crash Conference, pp. 257-279. SAE No. 892440, Society of Automotive Engineers, Warrendale, PA.

Saul, R.A., Sullivan, L.K., Marcus, J.H. and Morgan, R.M. (1983) "Comparison of Current Anthropomorphic Test Devices in a Three-Point Belt Restraint System". Proceedings of the 27th Stapp Car Crash Conference, pp 445-462. Society of Automotive Engineers, Warrendale, PA.

Viano, D.C. (1978) "Thoracic Injury Potential". Proceedings of the 3rd International Meeting on the Simulation and Reconstruction of Impacts in Collisions, pp. 142-156. IRCOBI, Bron, France.

Viano, D.C. (1987) "Evaluation of the Benefit of Energy-Absorbing Material for Side Impact Protection: Part II". Proceedings of the 31st Stapp Car Crash Conference. November, 1987. pp. 205-224, SAE No. 872213, Society of Automotive Engineers, Warrendale, PA.

Viano, D.C. and Lau, I.V. (1988) "A Viscous Tolerance Criterion for Soft Tissue Injury Assessment". Journal of Biomechanics, 21(5):387-399, 1988.

Viano, D.C. and Arepally, S. (1990) "Assessing the Safety Performance of Occupant Restraint Systems". Proceedings of the 34th Stapp Car Crash Conference, 1990, pp. 301-328, SAE No. 902328, Society of Automotive Engineers, Warrendale, PA.

Viano, D.C., Culver, C.C., Evans, L., Frick, M. and Scott, R. (1990) "Involvement of Older Drivers in Multi-Vehicle Side Impact Crashes". Accident Analysis and Prevention, 22(2):177-188, 1990.

910812

Hybrid III Sternal Deflection Associated with Thoracic Injury Severities of Occupants Restrained with Force-Limiting Shoulder Belts

Harold J. Mertz, John D. Horsch, and Gerald Horn
General Motors Corp.

Richard W. Lowne
Transport and Road Research Laboratory

ABSTRACT

A relationship between the risk of significant thoracic injury (AIS ≥ 3) and Hybrid III dummy sternal deflection for shoulder belt loading is developed. This relationship is based on an analysis of the Association Peugeot-Renault accident data of 386 occupants who were restrained by three-point belt systems that used a shoulder belt with a force-limiting element. For 342 of these occupants, the magnitude of the shoulder belt force could be estimated with various degrees of certainty from the amount of force-limiting band ripping. Hyge sled tests were conducted with a Hybrid III dummy to reproduce the various degrees of band tearing. The resulting Hybrid III sternal deflections were correlated to the frequencies of AIS ≥ 3 thoracic injury observed for similar band tearing in the field accident data. This analysis indicates that for shoulder belt loading a Hybrid III sternal deflection of 50 mm corresponds to a 40 to 50% risk of an AIS ≥ 3 thoracic injury.

THE PRIMARY DESIGN OBJECTIVE for the Hybrid III thorax was to develop a structure that could be used to assess the restraint efficacy of energy absorbing steering assemblies. Gadd and Patrick (1)* conducted tests of energy absorbing steering columns using cadavers. They noted that the total tolerable restraint load applied to the torso could be more than doubled if the wheel was designed so that part of the restraint load was applied to the shoulder structure by the steering wheel rim. This led to the understanding of how the total upper torso restraint load could be increased without increasing the risk of thoracic injury by increasing the load transmitted to the shoulder structure. Nahum, Kroell et al (2, 3, and 4) conducted cadaver tests to determine thoracic tolerance to sternal loading. They chose to impact the cadavers at mid sternum using a rigid face impactor. The impactor face was 152 mm in diameter which represented the area of a hub of a steering wheel used by Gadd and Patrick (1). The mass of the impactor was 23.4 kg, the average mass of the thorax. They found that thoracic compression was the primary thoracic injury parameter. The Hybrid III thorax was developed to mimic these blunt, sternal impact responses of the cadavers adjusted by 667 N to account for the lack of muscle tone (5, 6 and 7). The chest was instrumented to measure gross sternal to spine deflection. This measurement was to be used to assess the potential for thoracic injury due to distributed sternal impacts produced by steering wheel hub loading or air cushion loading. The Hybrid III thorax was not specifically designed to mimic human thoracic response for more concentrated, asymmetric loading of a shoulder belt. Thus the relationship between sternal compression and thoracic injury potential for shoulder belt loading may indeed be different than the relationship noted by Neathery et al (8) for blunt, frontal sternal impacts.

One method of estimating the relationship between thoracic injury risk and Hybrid III sternal deflection for shoulder belt loadings is to conduct a series of simulated car crashes for which the thoracic injuries to shoulder belted occupants are known. A major difficulty with this method has been the lack of exposure details for the car crashes. An important set of car crash data has been published by Foret-Bruno et al (9). They investigated accidents of Renault and Peugeot vehicles that were equipped with force limiters in the shoulder belt webbings. The force limiting elements consisted of a series of stitched webbing bands of different lengths. The bands were designed to tear at specified force levels. When the shortest band tore, the belt load was transmitted through the next shortest band. Published accident data are available on the level of band tearing and the corresponding thoracic

*Numbers in parentheses designate references listed at end of paper.

injury for 386 occupants (9, 10 and 11). This is the set of data that will be used to develop a relationship between thoracic injury due to shoulder belt loading and Hybrid III sternal deflection.

Association Peugeot-Renault (APR) provided force-limiting elements for simulating various belt loadings with a Hybrid III dummy. This report provides a summary of Hyge sled tests of these force-limiting elements conducted using a Hybrid III dummy, an analysis of the field accident data for belt systems using force limiters, and the development of a relationship between Hybrid III sternal deflection and the risk of thoracic injury for shoulder belt loading.

HYBRID III SLED TESTS

TEST CONDITIONS - The sled fixture consisted of a "hard" seat, toe pan, and belt anchor supports. The "hard" seat contained a rigid metal plate under the cushion, which provides a reusable and repeatable seat resulting in a nominal horizontal pelvic motion. The selection of anchor locations was based on a tabulation of anchor locations for selected crashes investigated by APR, Figure 1. Dimensions of our test configuration with the toe pan and "hard" seat location are also provided. The belt restraint used current automotive webbing, fixed at all three anchors (no

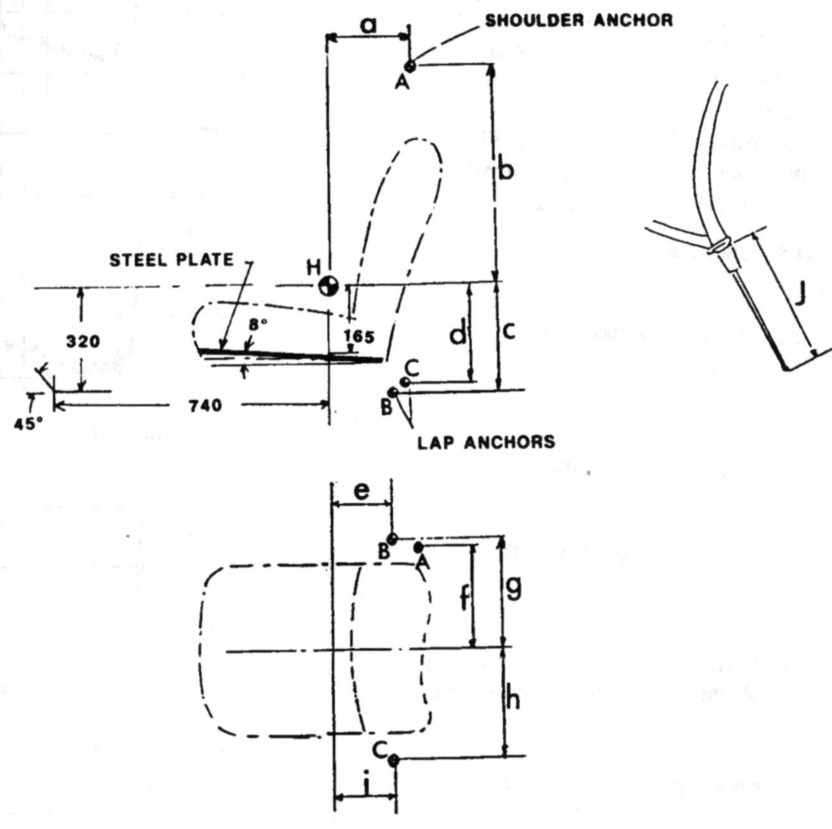

	CASES	2091	2165	APR 3168	DATA 3805	4061	4549	SLED TESTS
D	a	220	430	220	125	500	205	270
I	b	600	600	600	600	600	600	600
M	c	220	220	220	270	220	270	230
E	d	270	300	270	310	300	310	280
N	e	170	150	170	100	220	180	170
S	f	270	250	270	270	250	270	260
I	g	350	250	350	280	250	280	280
O	h	300	300	300	300	300	300	300
N	i	170	150	170	100	220	180	170
S	j	310	360	310	340	360	340	380

Dimensions in mm; A, B, C are belt anchor locations.

Figure 1 - Dimensions Used in the Sled Tests Compared with Vehicle Dimensions Provided by APR.

retractors), and had a locking latch-plate. The force-limiting element was located at the shoulder belt anchor, Figure 2. Adjustment of the belt restraint was "snug" for the lap belt and 25 mm slack in the shoulder belt. Placement of the belt on the shoulder was distal to the neck belt-guard. Dummy positioning included a pelvic angle of 22 degrees and head X-axis horizontal.

The force-limiting elements (Type B) provided by APR have 5 individual bands which are loaded in sequence such that when the first loaded band is torn, the load shifts to the next band. Figure 3 provides published APR data for the Type B element tested, indicating the force limit associated with tearing of each band (9).

A range of test severities was used which resulted in the tearing of 2, 4, and 5 bands. Tests were also conducted without the force-limiting element for comparison of responses.

Generic sled acceleration pulses were used, Figure 4. Since shoulder belt load was to be the controlling parameter of severity, there was no need to try to

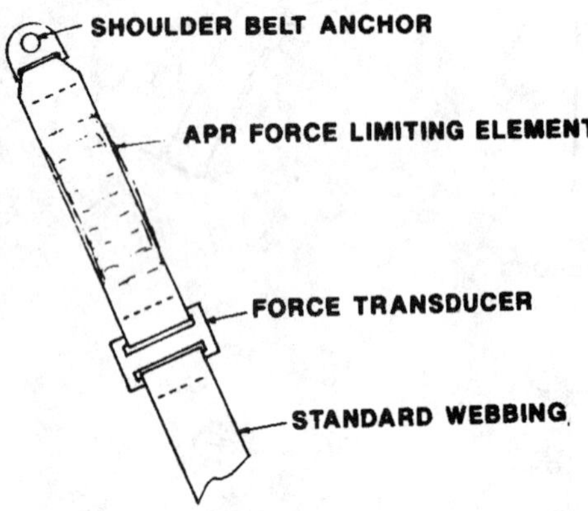

Figure 2 - Method of Attaching the Force Limiting Element to Standard Shoulder Belt Webbing.

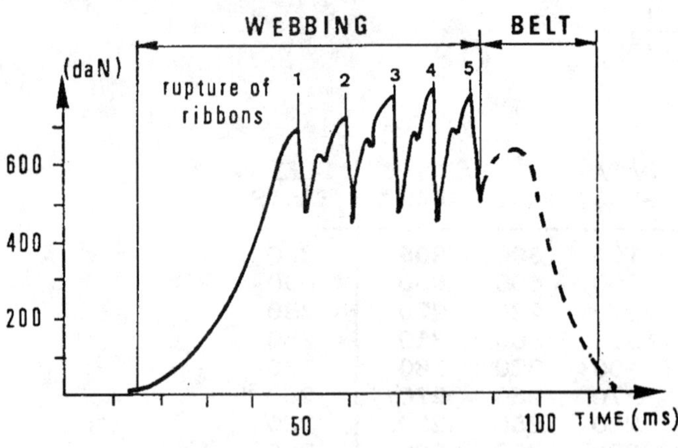

Figure 3 - APR Data Characterizing the Dynamic Load Response of a Type B Force Limiting Element (9).

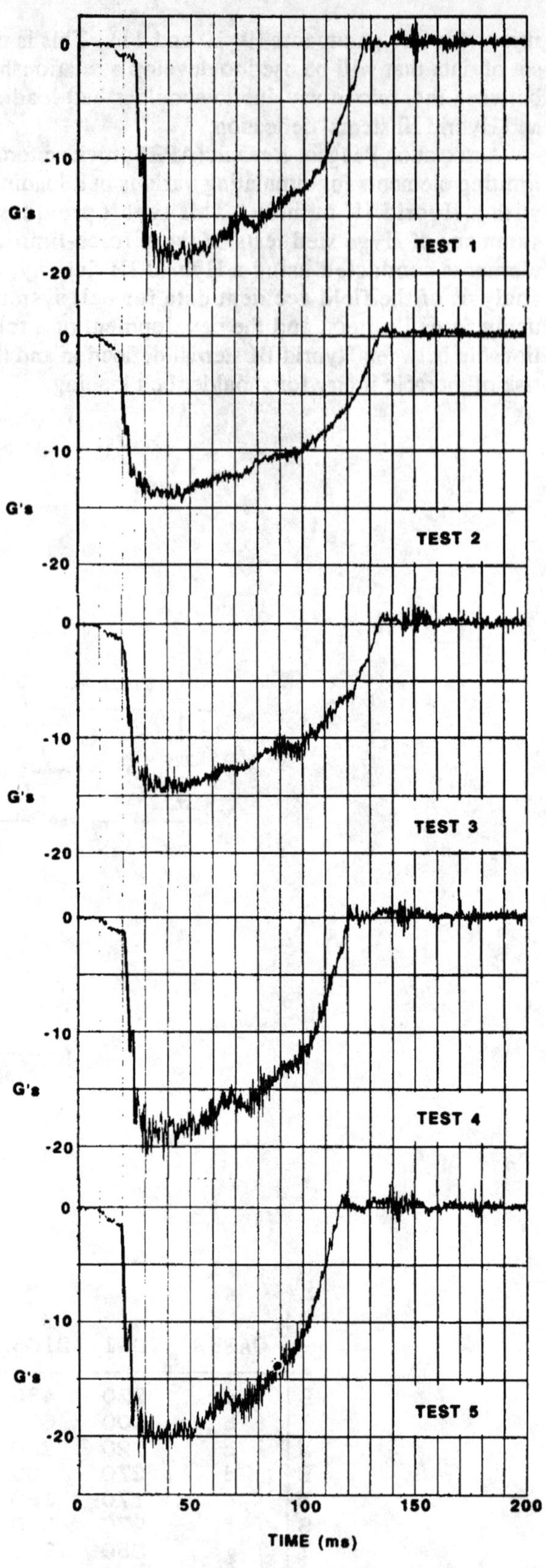

Figure 4 - Acceleration Profiles of Sled Tests. Profiles Chosen to Produce Prescribed Amount of Band Tearing. Filter Frequency of 2000 Hz.

duplicate any specific collision pulse. The pulse duration was constrained to 100-120 ms. Increase in test severity was obtained by increasing the level of sled acceleration. The test conditions are provided in Table 1.

Instrumentation included Hybrid III sternal deflection, thoracic acceleration, and shoulder belt tension. The shoulder belt tension transducer location is shown in Figure 2. Data plots are not filtered and have a nominal frequency response associated with the 2000 Hz anti- aliasing filter to better illustrate responses related to rapid release of load due to stitching tearing.

TEST RESULTS - Plots of shoulder belt force and sternal deflection as functions of time are provided in Figures 5 and 6, respectively. The tearing of each band is clearly evident by a sudden drop in shoulder belt load with a lesser influence on the sternal deflection.

Belt tensions for tearing of each band and the corresponding sternal deflections are provided in Table 1. Belt force required to tear the bands ranged from 5.5 to 6.9 kN with a trend that "early" tearings were at a lower force level. Sternal deflection associated with band tearing ranged from 26 mm to 35 mm. Most of the variation of sternal deflection is due to the variation in belt loads that caused band tearing.

Figure 7 contains cross-plots of belt force with sternal deflection. Note the linear relationship between shoulder belt loading force and Hybrid III sternal deflection. Figure 8 is a plot of Hybrid III sternal deflection corresponding to the belt force that produced band tearing; i.e., the data from Table 1. Forty millimeters of sternal deflection correspond to 8 kN of belt tension, a slope of 5 mm/kN. For this belt geometry, this slope is essentially constant over the range of sled test severities used and for shoulder belts with or without the load limiting element. This observation implies that if the shoulder belt load is known for a given accident then the Hybrid III sternal deflection can be estimated. This is the relationship that is needed in order to correlate Hybrid III sternal deflection with thoracic injuries experienced by car occupants.

In all five tests, the rate of sternal deflection was less than 3 m/s resulting in viscous criterion levels of less than 0.25. Thoracic injury due to rate of loading associated with the belt loading or band tearing is not a primary concern with this belt system.

FIELD ACCIDENT DATA

The field accident data used in the analysis were compiled from the studies of Foret-Bruno et al (9 and 10), case histories of seven accidents that were reviewed by Working Group 6 of ISO/TC22/SC12 (11) and personal communications with Mr. Foret-Bruno. The cars were 1970-1977 Peugeots or Renaults that were equipped with front seat three-point belts with a force-limiting shoulder belt. Three types of force limiters were used and are denoted as A, B and C. The multidisciplinary team from the Laboratory of Physiology and Biomechanics of the Association Peugeot/Renault has investigated accidents where 427 occupants were wearing a three-point belt with a force limiter in the shoulder belt. Foret-Bruno et al (10) have provided data for the accidents of 386 occupants for which the age of the occupant is known. Table 2 provides a summary of accident data for these 386 occupants. Note that 6 occupants (1.6%) suffered AIS = 3 thoracic injuries and 4 occupants (1%) suffered AIS ≥ 4 thoracic injuries. Table 3 is a summary of data for 15 occupants involved in accidents wearing a Type B load limiter. This Table contains the data for Type B accidents where part or

TABLE 1 - Summary of Sled Test Conditions and Results

Run No.	Belt Type	Sled Kinematics Acc. (G)	Sled Kinematics Vel. (km/h)	Stitch Bands Torn	Tear Force (kN) / Sternal Def. (mm)				
1	STD	18.0	51	-	-				
2	STD	14.0	44	-	-				
3	B	14.5	45	2	6.8 / 32	6.9 / 33			
4	B	18.5	52	4	5.6 / 26	5.9 / 27	7.2 / 35	6.9 / 34	
5	B	19.0	53	5	5.5 / 28	5.7 / 29	5.8 / 29	6.3 / 29	6.3 / 29

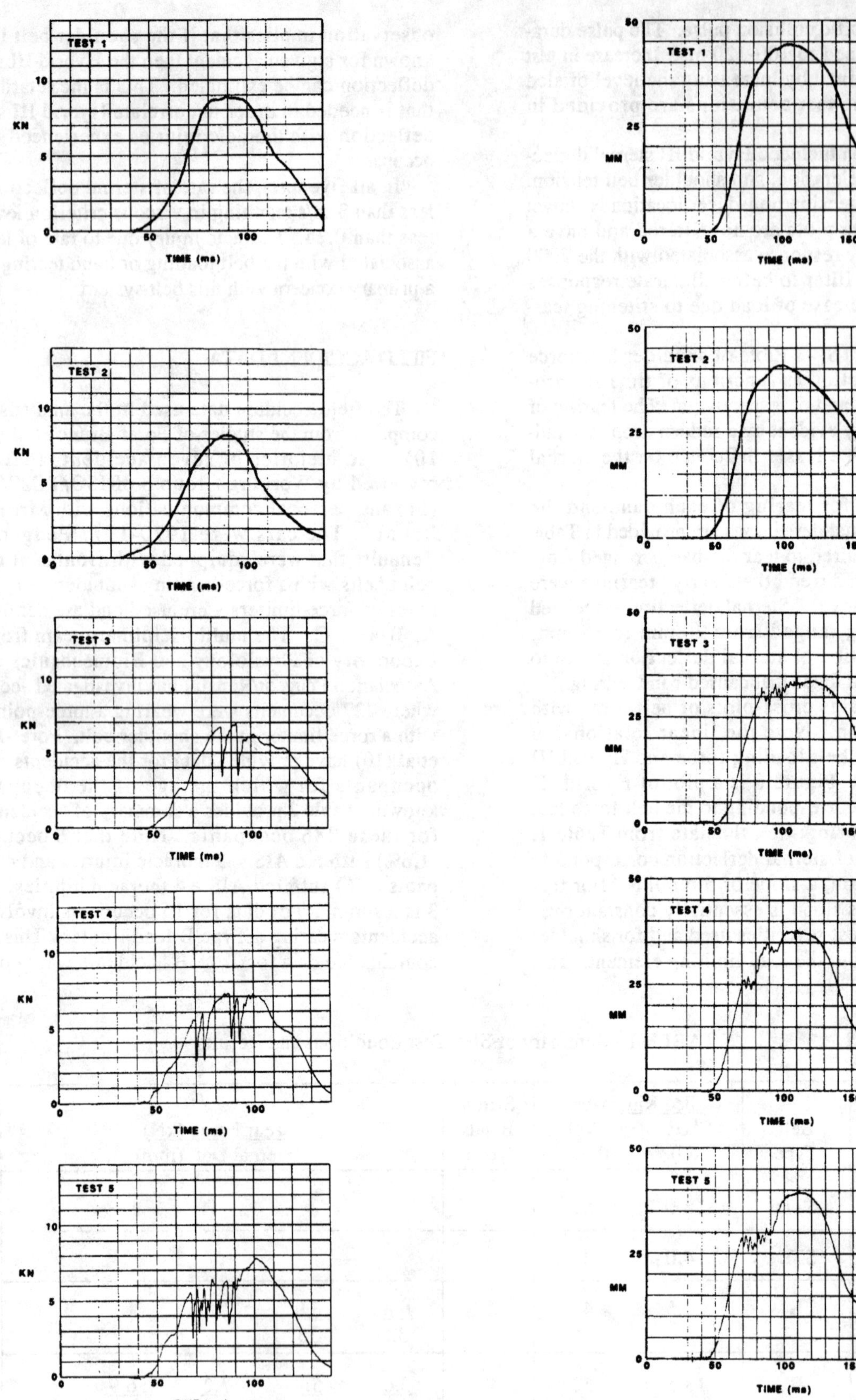

Figure 5 - Shoulder Belt Loadings from Sled Tests. Filter Frequency of 2000 Hz.

Figure 6 - Hybrid III Sternal Deflections from Sled Tests. Filter Frequency of 2000 Hz.

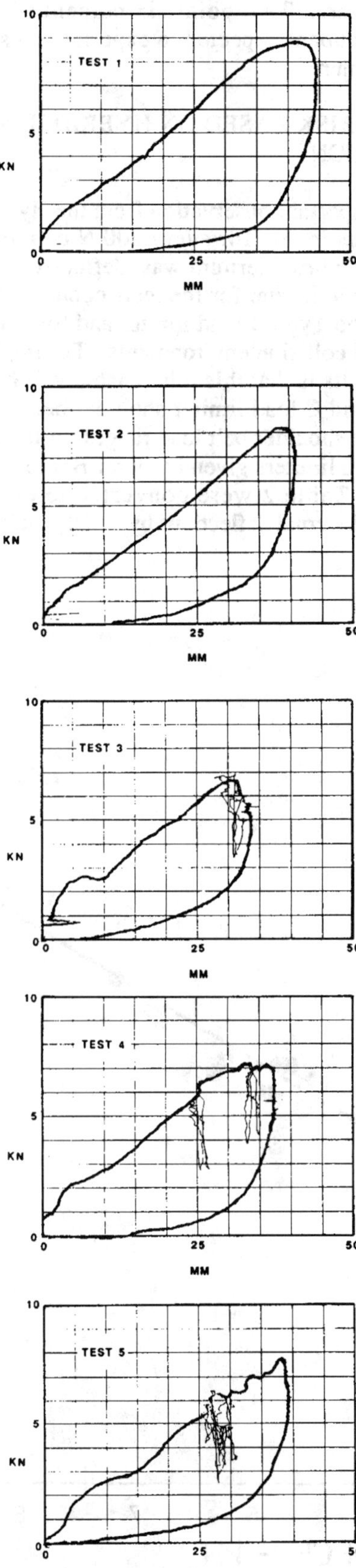

Figure 7 - Crossplots of Shoulder Belt Force vs. Hybrid III Sternal Deflection from Sled Tests.

all of the bands tore and accidents where the bands did not tear, but an occupant suffered rib fracture.

TYPE "A" LOAD LIMITER - The Type A load limiter had five discrete bands. Each band tore at a prescribed load level. The possible shoulder belt load range for tearing of each band is given in Table 2 and are based on the data given by Foret-Bruno (9). For example, if none of the bands tore, then the shoulder belt load could be any value from zero to 2.1 kN which is the average load required to tear the first band. If the first band tore, then the shoulder belt load could be any value between 2.1 kN which is the load required to tear the first band and 3.85 kN which is the load required to tear the second band. If the second band tore, then the shoulder belt load must have been 3.85 kN which is the load required to tear the second band. A higher load is not possible since the load required to tear the third band is only 3.25 kN. The shoulder belt load ranges for the tearing of bands 3 and 4 were deduced in a similar fashion. If all 5 bands tore, then the shoulder belt load can be any value greater than 4.4 kN.

There were 273 occupants involved in accidents with Type A force limiters. One hundred forty-nine occupants (54.6%) experienced shoulder belt loads between zero and 3.85 kN without any thoracic injury greater than AIS = 2. Ninety-one occupants (33.3%) experienced shoulder belt loads between 3.85 kN and 4.40 kN. Two of these occupants (2.2%) experienced thoracic injuries greater than AIS = 2. Thirty-three occupants (12.1%) tore all five bands. The shoulder belt loads for these occupants are not known; however, in all cases the load exceeded 4.4 kN. Of these occupants, two occupants (6.1%) experienced thoracic injuries greater than AIS = 2.

TYPE "B" LOAD LIMITER - The Type B load limiter also had five discrete bands of stitching, but the tearing loads for each band were approximately twice the tearing loads of the Type A. The load ranges for each band are based on the data given by Foret-Bruno et al (9) and were calculated in the same manner as was indicated for the Type A load limiter. Although not specified in References 9 or 10, Foret-Bruno has noted that there were 79 occupants involved in accidents with Type B load limiters. Seventy-two occupants (91.1%) experienced shoulder belt loads between zero and 7.4 kN. Three of these occupants (4.5%) suffered thoracic injuries which were classified as AIS greater than 2. Two of these occupants, a 39 year old male and a 67 year old male (Table 3), suffered flail chests (AIS = 5). Six occupants (7.6%) experienced shoulder belt loads ranging from 7.4 kN to 8.0 kN. One of these occupants (16.7%), a 43 year old female (Table 3), suffered a flail chest (AIS = 5). One occupant (1.3%), a 23 year old male (Table 3), experienced a shoulder belt load greater than 8.0 kN. He suffered a chest contusion classified as AIS = 1.

TYPE "C" LOAD LIMITER - The Type C load limiter had a single band of stitching that began to tear at a shoulder belt load of 5.5 kN (9). Although not noted in Reference 9 or 10, Foret-Bruno has indicated that thirty-four occupants were involved in accidents with this type of load limiter. Twenty-one occupants (61.8%) had no tearing of the stitching. Three occupants (8.8%) had partial tearing of the stitching so their shoulder belt load was limited to 5.5 kN. Ten occupants (29.4%) had complete tearing of the stitching. Their shoulder belt loads exceeded 5.5 kN, but the exact values are not known. None of the occupants involved in accidents with the Type C load limiter experienced thoracic injuries classified as AIS 3 or greater. This lack of an occupant with a significant thoracic injury is due to two facts. Most of the 386 accidents were not severe since only 10 occupants (2.6%) suffered AIS ≥ 3 thoracic injuries. There were only 34 occupants involved in accidents with Type C load limiters. Two point six percent of 34 is less than 1 occupant expected to experience a significant thoracic injury.

INJURY RISK BASED ON HYBRID III STERNAL DEFLECTION

An important observation from the Hybrid III sled test results is that for every 200 N of shoulder belt load the Hybrid sternum was deflected 1 mm. This relationship is true for the tests conducted with and without the Type B load limiter and for two levels of simulated collision environments. The implication of these results is that this relationship will hold for the Type A and C load limiter shoulder belt load data as well. The shoulder belt load ranges for the Type A, B and C load limiters given by Foret-Bruno et al (9) and listed in Table 2 were converted to estimates of Hybrid III sternal deflections by simply multiplying the

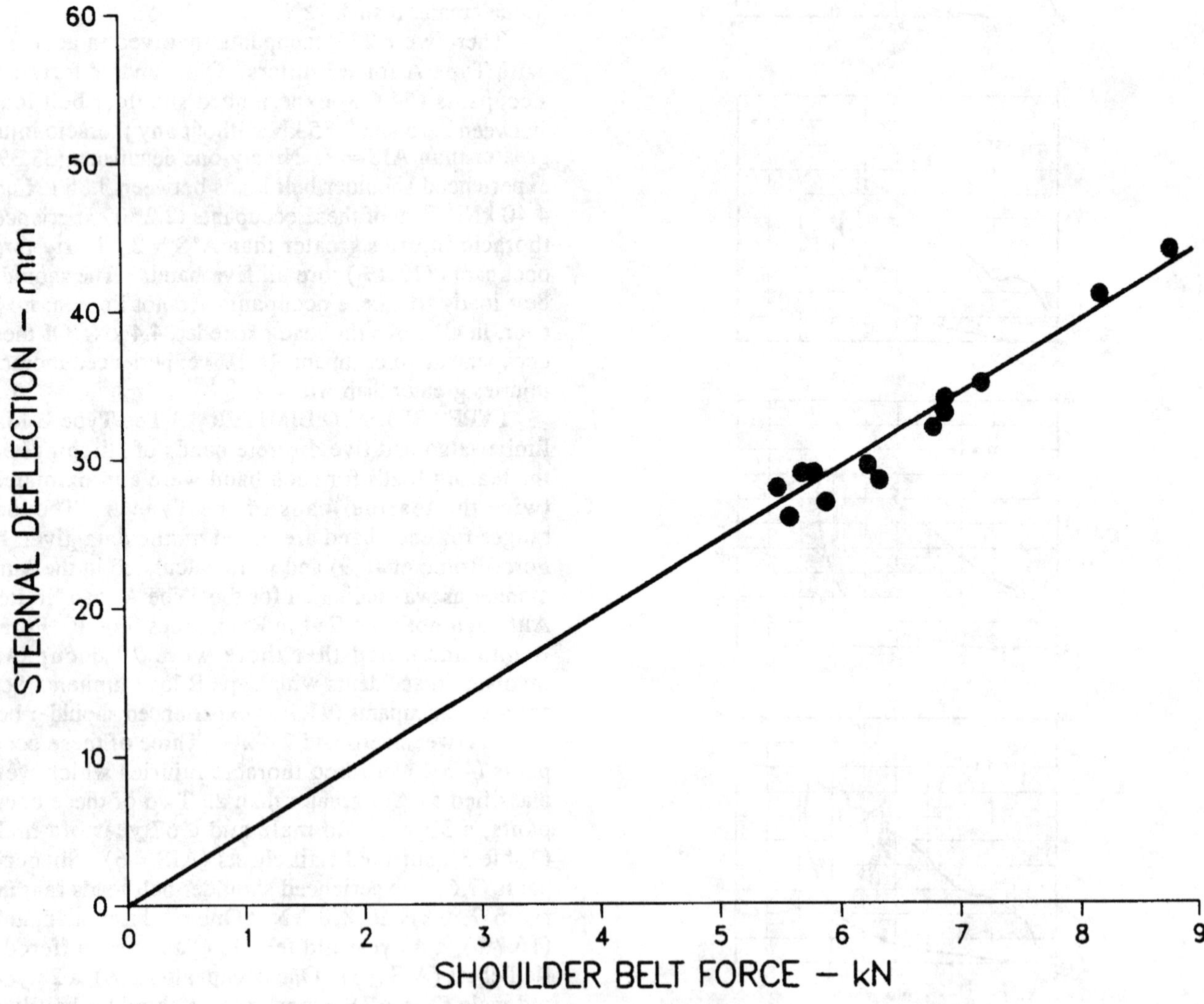

Figure 8 - Correlation Between Hybrid III Sternal Deflection and Shoulder Belt Force Corresponding to Band Tearing.

TABLE 2 - Data Summary of Accidents Involving Load Limiting Shoulder Belts. Shoulder Belt Load Ranges Taken from Foret-Bruno et al (9).

Type A Load Limiter

Stitch Bands Torn	Load Range (kN)	No. Occupants with Thoracic Injury			No. Cases
		AIS = 0 to 2	AIS = 3	AIS = 4 to 6	
0	0 - 2.10	123	0	0	123
1	2.10 - 3.85	26	0	0	26
2	3.85				
3	3.85 - 4.00	89	1	1	91
4	4.00 - 4.40				
5	> 4.40	31	2	0	33
				Total	273

Type B Load Limiter

Stitch Bands Torn	Load Range (kN)	No. Occupants with Thoracic Injury			No. Cases
		AIS = 0 to 2	AIS = 3	AIS = 4 to 6	
0	0 - 7.40	67	3	2	72
1	7.40	2	0	0	2
2	7.40 - 7.90	1	0	0	1
3	7.90 - 8.00	1	0	0	1
4	8.00	1	0	1	2
5	> 8.00	1	0	0	1
				Total	79

Type C Load Limiter

Torn Stitching	Load Range (kN)	No. Occupants with Thoracic Injury			No. Cases
		AIS = 0 to 2	AIS = 3	AIS = 4 to 6	
None	0 - 5.50	21	0	0	21
Partial	5.50	3	0	0	3
Complete	> 5.50	10	0	0	10
				Total	34

load values by 5 mm/kN. These estimated Hybrid III sternal deflections and the corresponding thoracic injury data are given in Table 4. The data for tearing 1 to 4 bands of the Type B load limiter have been grouped together for a Hybrid III sternal deflection range of 37 to 40 mm. The accident data observations for those occupants where all the bands or stitching tore are not given in Table 4. These data cannot be used in the injury risk analysis, since the shoulder load was not bounded and consequently the Hybrid III sternal deflection is not bounded.

SENSITIVITY ANALYSIS - The sensitivities of the sample sizes of the field accident data that were used to calculate the various injury risk values given in Table 4 can be determined as follows. For example, there were 123 occupants restrained by a Type A force limiter where no bands were torn. Since none of these occupants experienced an AIS ≥ 3 thoracic injury, a definitive injury risk value can not be calculated. However, an upper bound of the risk can be calculated by assuming that the next occupant involved in an accident where there is no tearing of a band of a Type A load limiter would experience an AIS ≥ 3 thoracic injury. If such were the case, then the risk of AIS ≥ 3 thoracic injury would be 1 in 124 occupants or 0.8% which is an upper bound. A lower bound of

TABLE 3 - Summary of Accident Data for Type B Load Limiter. All Cases Where Bands Broke. All Cases Where Bands Did Not Break, But There Was Rib Fracture.

Stitch Bands Torn	Occupant Information						Loading By Rear Occupant
	Age	Sex	Ht. (mm)	Mass (kg)	No. Rib Fracture	Thorax AIS	
0	31	F	-	-	1	1	Yes
0	68	M	-	-	1	1	No
0	72	F	-	-	2	1	No
0	52	F	-	-	4	3	No
0	33	F	-	-	4	3	Yes
0	56	F	-	-	4	3	Yes
0	67	M	-	-	14	5	Yes
0	39	M	-	-	15	5	Yes
1	28	M	180	76	0	1	No
1	34	F	157	55	0	1	Yes
2	33	M	185	92	0	1	No
3	59	M	175	80	2	1	No
4	24	M	176	63	0	0	No
4	43	F	160	70	14	5	-
5	23	M	188	75	0	1	Yes

risk can be calculated by assuming that this next occupant does not experience an AIS ≥ 3 thoracic injury. For this case, the injury risk would be zero in 124 occupants or 0.0% which is a lower bound. This procedure for determining upper and lower bounds was applied to all the accident data sets given in Table 4. The results of these calculations are summarized in Table 5.

INJURY RISK CURVES - Two methods were used to obtain curves that can be used to estimate the risk of significant thoracic injury due to shoulder belt loading as a function of Hybrid III sternal deflection. In the first method, the mean Hybrid III sternal deflections and associated risks for AIS ≥ 3 and AIS ≥ 4 thoracic injury given in Table 4 were subjected to Probit analyses (13, 14). The resulting risk curves for AIS ≥ 3 and AIS ≥ 4 thoracic injury are shown in Figure 9. Figure 10 shows the AIS ≥ 3 risk curve and its associated 95% confidence limits. These confidence limits reflect the spread in the upper and lower bounds of the risk values given in Table 5; that is, the curve is well determined for Hybrid III sternal deflections of 20 mm or less with greater uncertainty for larger chest deflections.

A second approach for obtaining an injury risk curve is to use only those data points from Table 4 for which the Hybrid III sternal deflection is closely bounded and the injury risk is also closely bounded. Only two Hybrid III sternal deflection ranges (19.3 to 22.0 mm and 37.0 to 40.0 mm) meet these requirements. The average sternal deflections for these ranges (20.7 mm and 38.5 mm) and their corresponding risks of AIS ≥ 3 thoracic injury (2.2% and 16.7%) were plotted on normal distribution graph paper, Figure 11. A straight line drawn through these two points provide a risk curve of AIS ≥ 3 thoracic injury as a function of Hybrid III sternal deflection for shoulder belt loading. Also shown on Figure 11 are dashed lines representing estimates of the upper and lower bounds of the injury risk curve. These boundaries were determined by plotting the upper and lower bounds of the injury risk values for mean Hybrid III sternal deflections of 20.7 mm and 38.5 mm given in Table 5. The dashed lines were drawn through the upper bounds and lower bounds of these mean sternal deflection levels. Note that the risk curve is well bounded from below and is quite well determined for sternal deflections of less than 20 mm.

TABLE 4 - Estimated Hybrid III Sternal Deflections Produced by Shoulder Belt Loading Associated with Risks of Various Thoracic Injury Severity Levels. Shoulder Belt Load Ranges and Occupant Injury Data Taken from Foret-Bruno et al (9).

Force Limiter Type	Stitch Bands Torn	Load Range (kN)	Hybrid III Sternal Defl Range (mm)	Hybrid III Sternal Defl Mean (mm)	Total Occupants	Thoracic Injury AIS ≥ 3 No.	Thoracic Injury AIS ≥ 3 %	Thoracic Injury AIS ≥ 4 No.	Thoracic Injury AIS ≥ 4 %
A	None	0 - 2.10	0 - 10.5	5.3	123	0	0.0	0	0.0
A	1	2.10 - 3.85	10.5 - 19.3	14.9	26	0	0.0	0	0.0
A	2 to 4	3.85 - 4.40	19.3 - 22.0	20.7	91	2	2.2	1	1.1
C	None	0 - 5.50	0 - 27.5	13.8	21	0	0.0	0	0.0
C	Partial	5.50	27.5	27.5	3	0	0.0	0	0.0
B	None	0 - 7.40	0 - 37.0	18.5	72	5	6.9	2	2.8
B	1 to 4	7.40 - 8.00	37.0 - 40.0	38.5	6	1	16.7	1	16.7

TABLE 5 - Bounds of Risk of Significant Thoracic Injury Due to Shoulder Belt Loading for Various Hybrid III Sternal Deflection Levels.

Hybrid III Sternal Defl.		Bounds of Risk of Thoracic Injury (%)			
Range (mm)	Mean (mm)	AIS ≥ 3		AIS ≥ 4	
		Lower	Upper	Lower	Upper
0 - 10.5	5.3	0.0	0.8	0.0	0.8
10.5 - 19.3	14.9	0.0	3.7	0.0	3.7
19.3 - 22.0	20.7	2.2	3.3	1.1	2.2
0 - 27.5	13.8	0.0	4.5	0.0	4.5
27.5	27.5	0.0	25.0	0.0	25.0
0 - 37.0	18.5	6.8	8.2	2.7	4.1
37.0 - 40.0	38.5	14.3	28.6	14.3	28.6

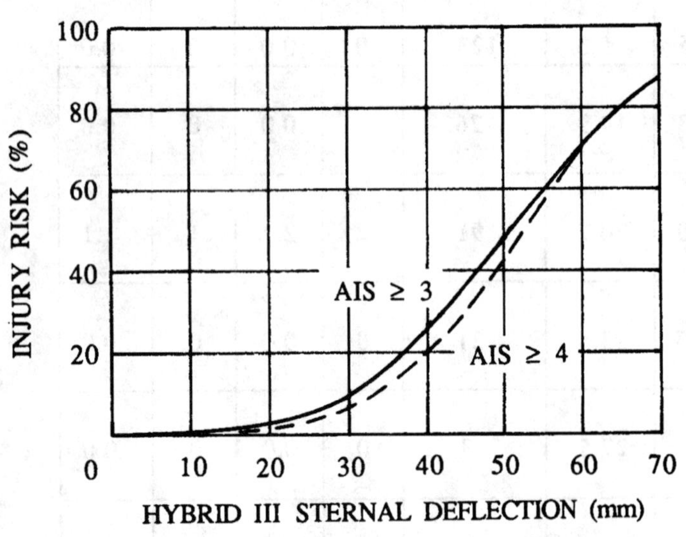

Figure 9 - Risks of AIS ≥ 3 and AIS ≥ 4 Thoracic Injury as a Function of Hybrid III Sternal Deflection for Shoulder Belt Loading. Risk Curves Determined by Probit Analysis.

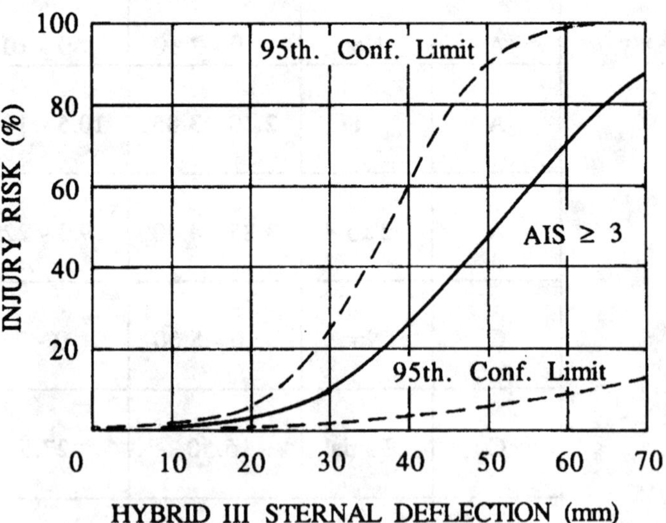

Figure 10 - Injury Risk Curve and Associated 95% Confidence Bands for AIS ≥ 3 Thoracic Injury Based on Probit Analysis.

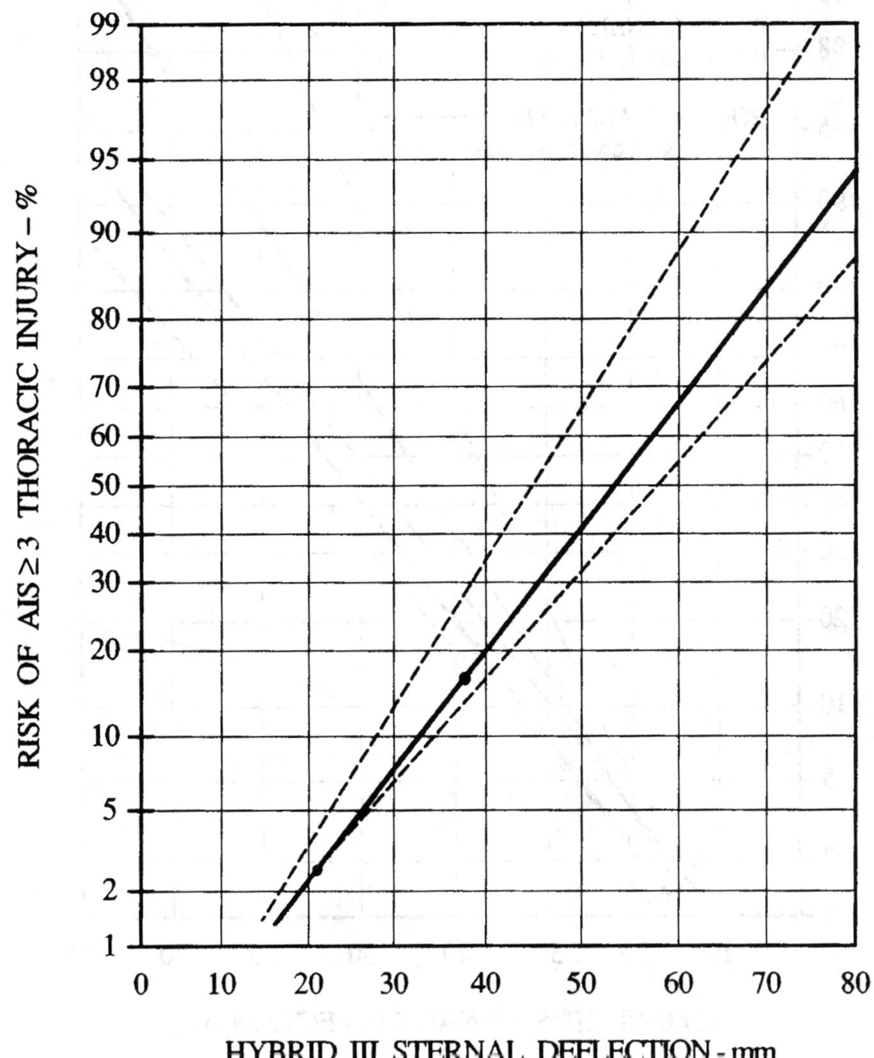

Figure 11 - Risk of AIS ≥ 3 Thoracic Injury Due to Shoulder Belt Loading as a Function of Hybrid III Sternal Deflection. Risk Curve Determined Using Closely Bounded Data. Dash Lines are Bounds Due to Sensitivity of Accident Sample Size.

Additional field data are needed for the 38.5 mm deflection level to decrease the sample size sensitivity of the upper bound. A risk curve for AIS ≥ 4 thoracic injury was not plotted using this method because there were only four such injury observations in the accident sample given in Table 4.

Figure 12 compares the injury risk curves for AIS ≥ 3 thoracic injury calculated by the two methods. Note the close agreement of the two curves, especially at low and high injury risk levels where the differences between the curves are less than one percent. Both curves lie within the 95% confidence bands determined by the Probit analysis. They also lie within the sample size sensitivity bounds for Hybrid III sternal deflections greater than 20 mm. For Hybrid III sternal deflections of less than 20 mm, the Probit curve lies slightly outside of sample size sensitivity bounds probably because of the greater uncertainty of the data used in the Probit analysis. Both curves are in excellent agreement and well determined for Hybrid III sternal deflections corresponding to low injury risks.

DISCUSSION

Foret-Bruno et al (10) conducted a series of Hybrid III sled tests at UTAC using the Type B load limiter. In the UTAC tests, the average Hybrid III sternal deflection during tearing of the bands was 60 mm (Test 1943) compared to 29 mm (Run 5) in our tests. Upon review of the UTAC data, an error was found in the sternal deflection measurements. To confirm this finding two additional sled tests were conducted. The corrected results along with the results of the two additional tests have been summa-

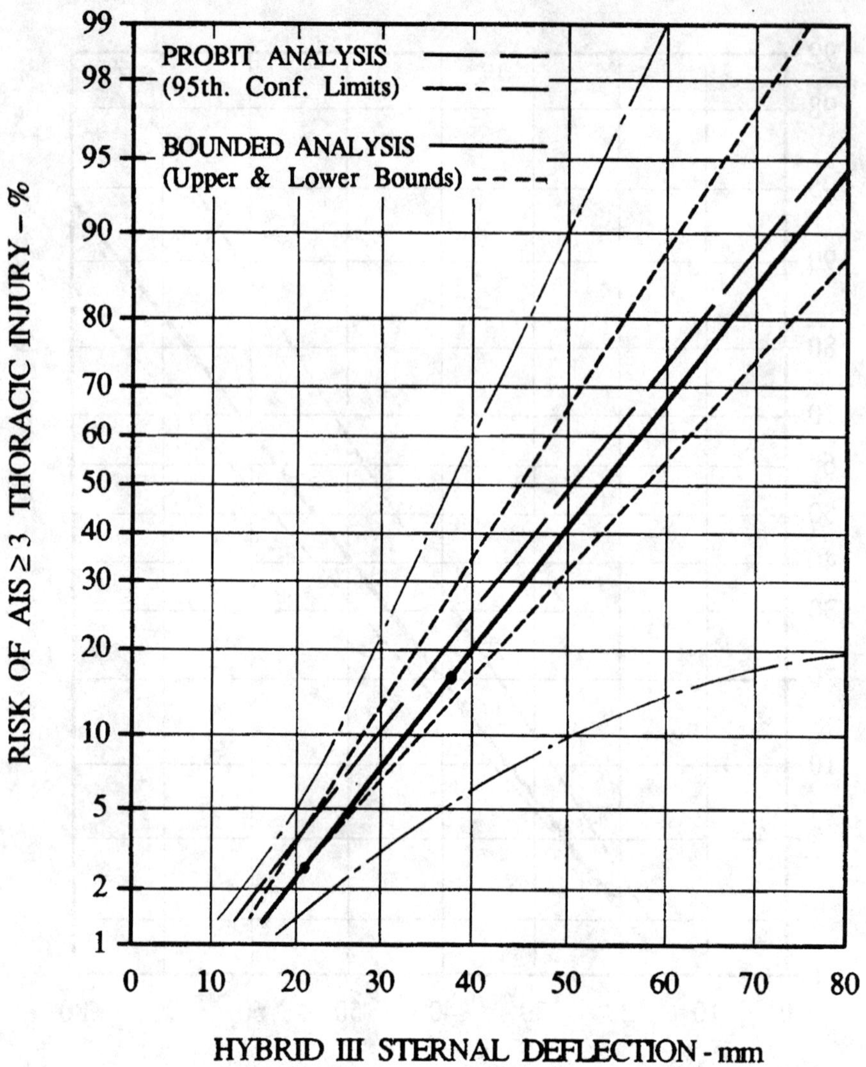

Figure 12 - Comparison of Injury Risk Curves Determined by Probit Analysis and Bounded Sample Analysis.

rized by Foret-Bruno and Bendjellal (12). Hybrid III sternal deflections for their corrected results ranged from 30 to 43 mm during band tearing which compare favorably with our sternal deflections which ranged from 26 to 35 mm. Based on these corrected results the Foret-Bruno et al (10) conclusion relative to tolerable sternal deflection for young people should be restated as, "a load of 8.0 kN corresponding to a sternal deflection of 40 to 50 mm for the Hybrid III can be withstood without thoracic injury by young adults in the population". This conclusion is now in agreement with the injury risk curves shown in Figure 12 which indicate that sternal deflections of 40 to 50 mm of the Hybrid III dummy produced by shoulder belt loading pose a risk of AIS ≥ 3 thoracic injury to 20 to 50 percent of the population.

Foret-Bruno et al (10) have eliminated from their analysis the accident cases where a rear seat occupant loaded the front seat occupant who was wearing the three-point belt equipped with the shoulder belt load limiter. We chose to include these data since the shoulder harness limits the loading. The loading of the rear passenger is simulated by simply increasing the level of the sled pulse to produce the observed level of band tearing. Foret-Bruno et al (10) also chose to segregate the field accident by age groups. We chose to do our risk analysis for the population as a whole. Obviously, people with lower bone strength, either due to the effects of aging, health or physical geometric size, will have a greater risk of experiencing a significant thoracic injury for a given shoulder belt load. It is because of these types of factors that we have the risk curves shown in Figure 12.

The field accident data for the Type A, B and C load limiters given in Table 2 can be used to estimate the collision severity distribution of real world accidents as measured by Hybrid II sternal deflection. Note that for the Type A load limiter, 123 of the 273 occupants (45.1%) had shoulder belt loadings that would produce Hybrid III sternal deflections of less than 10.5 mm. For Type B load limiter, 72 of 79 occupants (91.1%) experienced shoulder belt loadings that would have produced Hybrid III sternal deflections of less than 37 mm. For Type C limiter, 21 of 34 occupants (61.8%) had shoulder belt loadings corresponding to Hybrid III sternal deflections of less than 27.5 mm. Using these values, a cumulative frequency distribution curve of collision severity of the accidents involving occupants wearing three-point harnesses that were investigated by APR can be plotted, Figure 13. FMVSS 208 barrier tests of cars equipped with three-point belt systems can produce shoulder belt loads ranging from 9 to 10 kN and corresponding Hybrid III sternal deflections ranging from 45 to 50 mm. For example, Run 1 with standard webbing (Table 1) was slightly less severe than a 30 mph FMVSS 208 barrier test. It produced a peak belt load of 8.9 kN and Hybrid III sternal deflection of 44 mm. From Figure 12, Hybrid III sternal deflections of 45 to 50 mm correspond to a 30 to 50% risk of AIS ≥ 3 injury. Note from Figure 13 that less than 5% of the field accidents investigated by APR would produce a Hybrid III sternal deflection of 45 mm or greater. This implies that very few of the accidents investigated are as severe as a FMVSS 30 mph barrier test. Also note that the average risk of an AIS ≥ 3 thoracic injury for the APR accident sample is 10 in 386 or 2.6%. From Figure 12, this risk level corresponds to a Hybrid III sternal deflection of 17 to 21 mm which is much less than the deflection for a FMVSS 208 barrier test. Clearly, it would be inappropriate to associate a Hybrid III sternal deflection that is measured in a FMVSS 208 frontal barrier test with the average risk of AIS ≥ 3 thoracic injury for the APR accident sample.

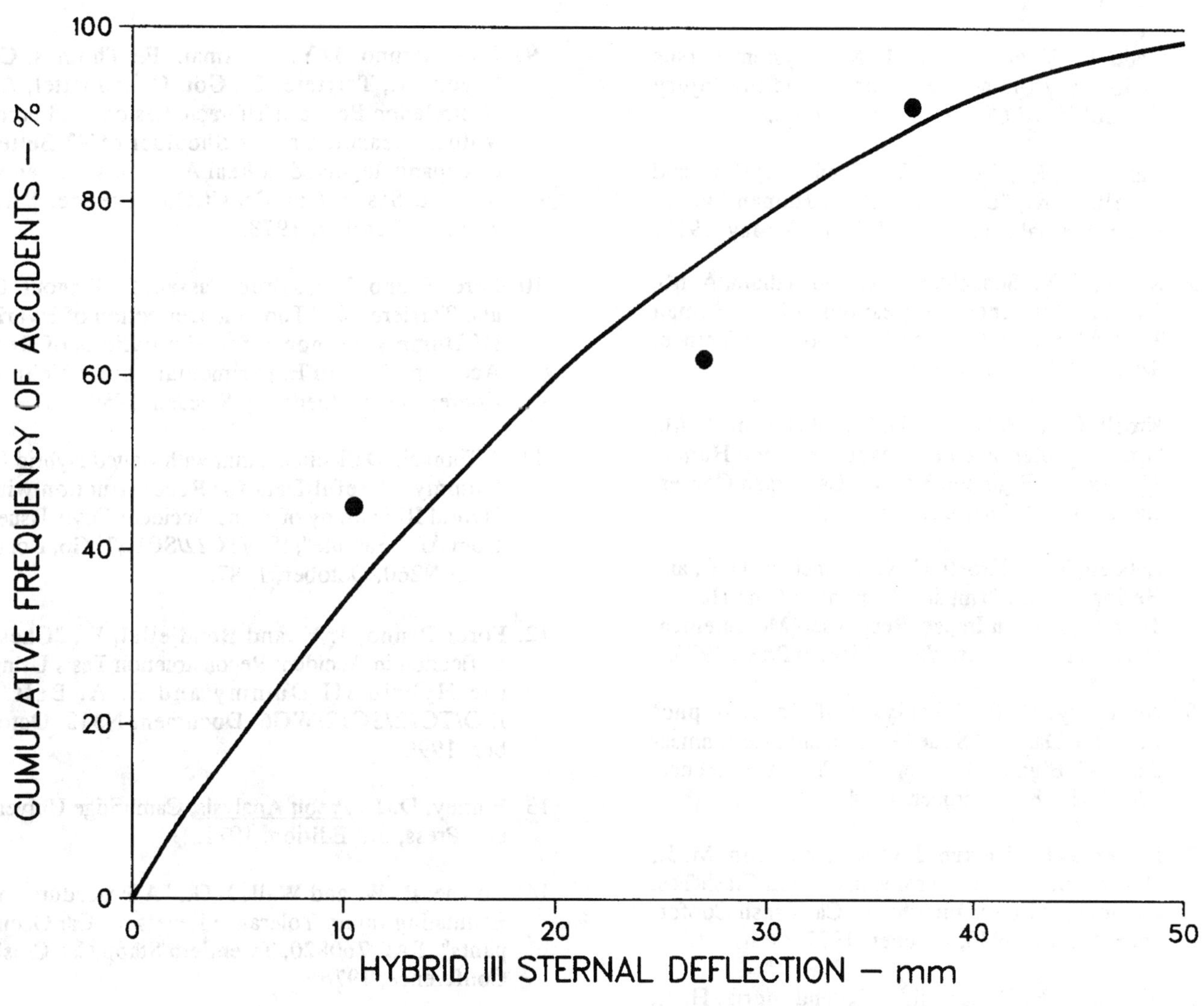

Figure 13 - Cumulative Frequency of Accidents Involving 3-Point Restrained Occupants Investigated by APR as a Function of Equivalent Hybrid III Sternal Deflection.

SUMMARY

Field accidents data obtained by the Association Peugeot-Renault of 386 occupants who were involved in frontal collisions while being restrained by three-point belt systems that used a shoulder belt with a force-limiting element were analyzed. For 342 of these occupants, the magnitude of the shoulder belt force could be estimated with various degrees of certainty from the amount of force-limiting band ripping. These data were used to calculate the risks of AIS ≥ 3 and AIS ≥ 4 thoracic injury for the various peak shoulder belt loads. Hyge sled tests were conducted with a Hybrid III dummy to reproduce the various degrees of band tearing. The results of these tests indicated that Hybrid III sternal deflection was well correlated with shoulder belt load. Every 5 mm of Hybrid III sternal deflection corresponded to 1 kN of shoulder belt load. This relationship was used to convert the estimated shoulder belt loads of the field accident data to equivalent Hybrid III sternal deflections. Statistical methods were used to develop relationships between Hybrid III sternal deflections and the risks of AIS ≥ 3 or AIS ≥ 4 thoracic injury produced by shoulder belt loading. This analysis indicates that for shoulder belt loading of the Hybrid III thorax a sternal deflection of 50 mm corresponds to a 40 to 50% risk of an AIS ≥ 3 thoracic injury.

ACKNOWLEDGMENTS

The authors wish to acknowledge the cooperation of the following personnel of the Association of Peugeot-Renault: Dr. Tarriere for providing the force limiters used in our test program, Mr. Bendjellal who participated in a joint review of the APR and our sled test data, and Mr. Foret-Bruno for providing additional details of the APR field accident data. The contribution from Mr. Lowne is published with permission of the Director of TRRL.

REFERENCES

1. Gadd, C. W. and Patrick, L. M., "System Versus Laboratory Impact Tests for Estimating Injury Hazard", SAE 680053, January, 1968.

2. Nahum, A. M., Gadd, C. W., Schneider, D. C. and Kroell, C. K., "Deflection of the Human Thorax Under Sternal Impact", SAE 700400, May, 1970.

3. Kroell, C. K., Schneider, D. C. and Nahum, A. M., "Impact Tolerance and Response of the Human Thorax", Fifteenth Stapp Car Crash Conference, SAE 710851, November, 1971.

4. Kroell, C. K., Schneider, D. C. and Nahum, A. M., "Impact Tolerance and Response of the Human Thorax-II", Eighteenth Stapp Car Crash Conference, SAE 741187, December, 1974.

5. Lobdell, T. E., Kroell, C. K., Schneider, D. C. and Hering, W. E., "Impact Response of the Human Thorax", Human Impact Response - Measurement and Simulation, New York: Plenum Press, 1973.

6. Neathery, R. F., "Analysis of Chest Impact Response Data and Scaled Performance Recommendations", Eighteenth Stapp Car Crash Conference, SAE 741188, December, 1974.

7. Foster, J. K., Kortge, J. O. and Wolanin, M. J., "Hybrid III - A Biomechanically-Based Crash Test Dummy", Twenty-First Stapp Car Crash Conference, SAE 770938, October, 1977.

8. Neathery, R. F., Kroell, C. K. and Mertz, H. J., "Prediction of Thoracic Injury from Dummy Responses", Nineteenth Stapp Car Crash Conference, SAE 751151, November, 1975.

9. Foret-Bruno, J. Y., Hartman, F., Thomas, C., Fayon, A., Tarriere, C., Got, C. and Patel, A., "Correlation Between Thoracic Lesions and Force Values Measured at the Shoulder of 92 Belted Occupants Involved in Real Accidents", Twenty-Second Stapp Car Crash Conference, SAE 780892, October, 1978.

10. Foret-Bruno, J. Y., Brun-Cassan, F., Brigout, C. and Tarriere, C., "Thoracic Deflection of Hybrid III Dummy Response for Simulations of Real Accidents", 12th Experimental Safety Vehicle Conference, Gothenburg, Sweden, 1989.

11. " Thoracic Deflection Limit with Belted Hybrid III Dummy: Useful Data for Reconstruction with Hybrid III Dummy of Some Accident Cases Issued from APR Sample", ISO/TC22/SC12/WG6, Document N260, October, 1987.

12. Foret-Bruno, J. Y. and Bendjellal, F., "Chest Deflection in Accident Reconstruction Tests Using the Hybrid III Dummy and E. A. Belt", ISO/TC22/SC12/WG6, Document N313, October, 1990.

13. Finney, D. J., Probit Analysis, Cambridge University Press, 3rd Edition, 1971.

14. Lowne, R. W. and Wall, J. G., "A Procedure for Estimating Injury Tolerance Levels for Car Occupants", SAE 760820, Twentieth Stapp Car Crash Conference, 1976.

892439

Assessment of Lap-Shoulder Belt Restraint Performance in Laboratory Testing

Stephen W. Rouhana and John D. Horsch
Biomedical Science Dept.
General Motors Research Labs.

Charles K. Kroell
General Motors Research Labs. (Retired)

ABSTRACT

Hyge sled tests were conducted using a rear-seat sled fixture to evaluate submarining responses (the lap belt of a lap-shoulder belt restraint loads the abdominal region instead of the pelvis). Objectives of these tests included: an evaluation of methods to determine the occurrence of submarining; an investigation into the influence of restraint system parameters, test severity, and type of anthropomorphic test device on submarining response; and an exploration of the mechanics of submarining.

This investigation determined that:

1. Slippage of the lap belt off the pelvis due to dynamic loading of the dummy and the resulting kinematics can cause abdominal loading to the dummy in laboratory crash testing.
2. The 5th female dummy submarined more easily than did the Hybrid III in the test environment.
3. Motion of the pelvis was controlled using a "pelvic stop", which reduced the submarining tendency for both the 5th female and Hybrid III dummies.
4. Shortening the buckle strap length (in conjunction with a pelvic stop) increased the submarining "threshold".
5. Modifying the belt restraint by removing the retractor from the lap-belt and adding a retractor to the shoulder belt (with the standard seat) increased the "threshold" severity for submarining.
6. The occurrence of the belt slipping off the pelvis and loading the abdomen was well correlated to rearward rotation of the pelvis.
7. In addition to standard film analysis, detection of the belt slipping off the pelvis is strongly enhanced by use of a submarining indicating pelvis, pelvic in-line accelerometry, and belt force transducers.
8. The development of a method to unequivocally detect the occurrence of submarining, <u>and</u> to predict the severity of any resulting injury would greatly strengthen the analysis of restraint performance.

LABORATORY TESTING OF BELT-RESTRAINT systems may result in submarining. This involves the belt slipping off the dummy pelvis and directly loading the abdominal region. There are many issues related to assessing belt restraint performance associated with direct abdominal loading. One group of these issues is directed toward field relevance, another is directed toward laboratory assessment of submarining, and still another is directed toward development of counter-measures to prevent abdominal loading for a range of occupant sizes, seating positions, and crash severities.

Field relevance includes such considerations as: the frequency and severity of abdominal injuries due to belt loading observed in the field; and differences between laboratory and field conditions that might influence interpretation of the test results or the setting of performance goals (e.g., clothing, impact configuration and severity, occupant characteristics, etc.) [1-8].

Considerations in laboratory assessment include: the biofidelity of dummy response; the risk of injury if direct abdominal loading by the belt restraint does occur; as well as the differences between laboratory and field conditions (e.g., adjustment and placement of the belt-restraint, occupant initial posture, etc.) [1-8].

Various methods have been developed to detect motion of a lap belt past the anterior superior iliac spines of the pelvis in laboratory tests. The traditional method has been to review high speed movies. But, such reviews are subjective in nature and might not have sufficient visibility on which to base judgments (due to obscuration by other dummies, intervening doors or panels, or by loose clothing on the dummy, etc.). Several approaches have been proposed which utilize instrumentation in the dummy to measure belt loading, either as the belt slides over the iliac spines or after it slips off the iliac spines and loads the abdomen.

Studies of the tendency of the belt to slip off the dummy pelvis in laboratory tests have demonstrated that at least five groups of parameters can influence submarining: 1) the belt restraints; 2) the vehicle environment; 3) the exposure severity; 4) the characteristics of the dummy; and 5) the test setup parameters such as dummy positioning or belt adjustments. A variety of recommendations exist in the literature which propose system configurations that may reduce the tendency for the belt to slip off the pelvis, and which demonstrate that in addition to the various belt restraint characteristics, other vehicle structures can also be a primary factor in providing restraint [9-14]. Various studies [10-14] have also recommended a kinematic analysis of the dummy motions to better assess the restraint performance.

METHODOLOGY

Hyge sled tests were conducted with a rear-seat buck having reinforced belt anchors. The "baseline" configuration consisted of a bench seat and an add-on shoulder belt to a rear-seat lap-belt restraint. The upper anchor of the shoulder belt was shelf-mounted directly behind the shoulder and did not have a retractor. Thus the shoulder belt was shorter and had less potential elongation than typical for front seat applications. The lap-belt had a retractor at the out-board anchor location. The buckle strap extended from the junction of the seat cushions. All anchors were reinforced and rigid.

Various modifications of the restraint system were made. A Styrofoam pelvic stop was used to reduce the tendency of the dummy to submarine. The stop had to be tightly coupled to the floor-pan to prevent large forward displacements of the pelvis. An alternative pelvic stop was constructed of wood, which was also tightly coupled to the floor pan.

Modifications of the belt restraint were also tested. The buckle strap was shortened so that the buckle was at the junction of the seat pan and seat back and not near the pelvic anterior iliac spine. A retractor was added at the shoulder-belt anchor and removed from the lap-belt to reduce the excursion of the pelvis and increase the forward bending of the upper torso as per Adomeit's kinematic criteria [10-12]. While the aim of these modifications was to improve restraint system performance, they did not comprehend other issues such as comfort and convenience.

A generic sled acceleration characteristic was used for these tests and did not model any particular vehicle nor some aspects of car crashes such as pitch. Tests were conducted at several severities to determine the submarining threshold for this test environment. A typical pulse is shown in Figure 1.

High speed movies were taken of all but the last four tests and included: an offboard wide-angle view, an onboard view of the inboard side of the dummy, and an offboard view of the outboard side of the dummy. In the last four tests the wide-angle view was replaced with an onboard view of the outboard side.

A Hybrid III dummy and a 5th percentile female dummy were used in these tests. The Hybrid III dummy had more instrumentation than the 5th female dummy (Table 1). The pelvic rotation instrumentation consisted of 5 in-line accelerometers, Figure 2, and used the analysis from Ref [15].

The Hybrid III was equipped with a load-bolt pelvis [16] which provided an indication of belt slip over the anterior superior iliac spines of the pelvis. Each iliac spine of the load-bolt pelvis has three load transducers, situated vertically to indicate the superior/inferior location of the belt load, Figure 3. During the test, the pelvis rotates rearward and the belt load shifts from the lower load bolts to the higher load bolts. At the time of submarining, as the lap-belt slips over the anterior-superior iliac spine, the upper bolts suddenly unload. Note that the load bolts would not detect a belt placed directly on the abdomen (i.e., a misplaced belt).

The forces in the shoulder and lap belts were recorded during these tests using GSE Model 2500 Seat Belt Force Sensors. The shoulder belt load cell was positioned on the outboard side of the dummy between the shoulder and the anchor or retractor. The lap belt load cell was placed on the inboard side of the dummy between the buckle and the seat back.

Previous studies [13,14] have demonstrated that submarining is associated with rearward rotation of the pelvis (the iliac crests move in a posterior direction relative to the pubic rami). If sufficient, this rearward rotation allows "unhooking" of the lap-belt from the pelvic iliac spines. (Also involved, but to a lesser extent, is the decreasing lap-belt angle caused by the the forward and downward translation of the pelvis.) Because of the rear-seat environment, the test movies could not be used to determine changes in lap-belt angle for many of the tests. However, rotation of the pelvis could be determined using an array of accelerometers known as an in-line array. The in-line array takes advantage of the measurable difference in the accelerations experienced by accelerometers mounted along a line, and constrained to rotate as a unit. The information in these accelerations can be used to obtain rotational: acceleration, velocity, and displacement. A least squares fit of these parameters through the five accelerometers gives accurate results. [15]

RESULTS

The test matrix is provided in Table 2 and includes occurrence of submarining. In tests where the belt slipped off the pelvis on one side only, it was always on the buckle strap (inboard) side.

SUBMARINING THRESHOLD/BASELINE SYSTEMS - The Hybrid III dummy had a submarining threshold of about 42 km/h in the sled tests using the conventional seat and "add-on" shoulder belt. For test severities slightly greater than the submarining threshold severity, the belt slipped over the pelvic iliac spine only on the buckle-strap side --- likely due to lifting of the lap-belt by the shoulder-belt.

The 5th female dummy submarined on both sides of the pelvis at all severities tested, indicating that the submarining threshold for the baseline system is less than 35 km/h. Thus the 5th female dummy tendency to submarine is greater in this environment than that of the Hybrid III dummy.

SUBMARINING THRESHOLD/MODIFIED SYSTEMS - Hardware modifications influence the submarining threshold. Use of the pelvic stop, or the modified belt restraint raised the submarining threshold from about 42 km/h to greater than 52 km/h for the Hybrid III dummy.

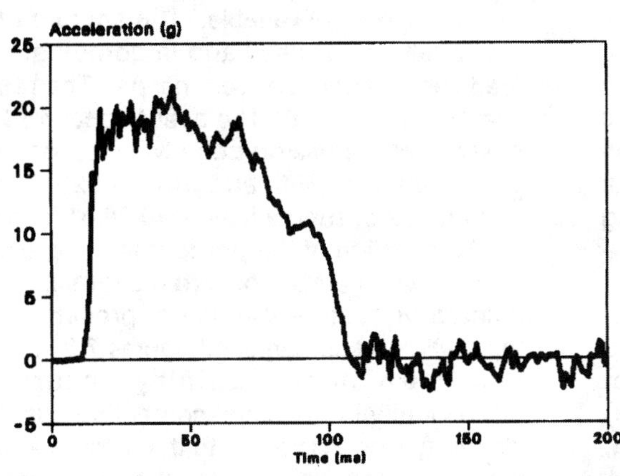

Figure 1. Typical sled acceleration-time history.

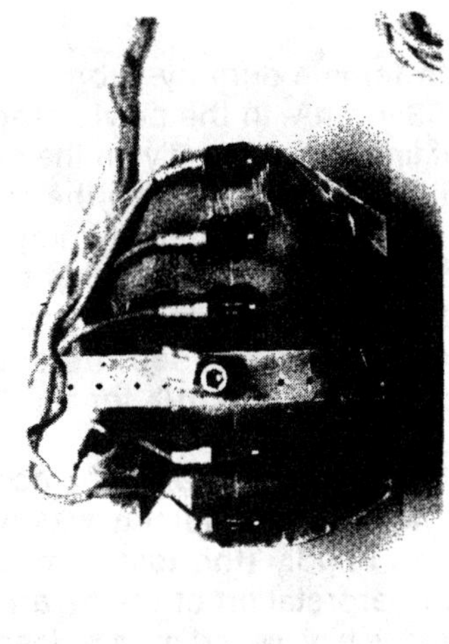

Figure 2. In-line accelerometer installation.

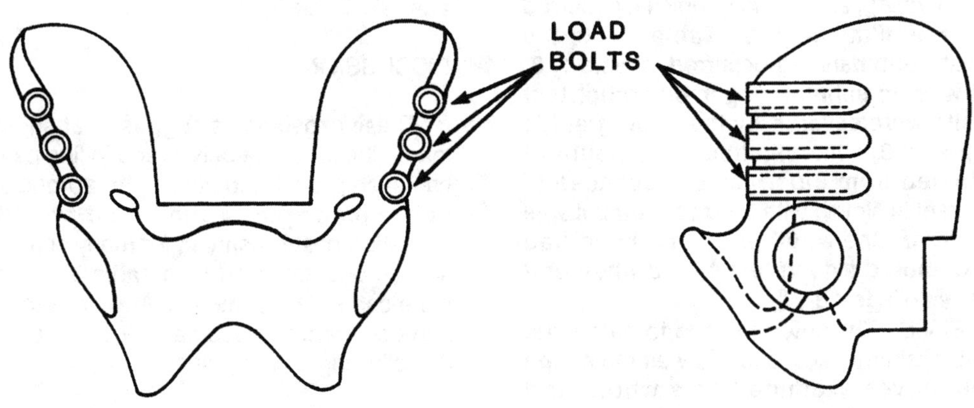

Figure 3. Load-bolt pelvis: a) front view, b) side view

TABLE 1 - DUMMY INSTRUMENTATION

Dummy	Instrumentation
5th Female	Pelvic Rotational Acceleration (some tests) Belt Tensions (except short buckle-strap)
Hybrid III	"208" Instrumentation Pelvic Rotational Acceleration (in-line) Pelvic Translational Acceleration Belt Tensions (except short buckle-strap) Load-Bolt Pelvis

TABLE 2 - SLED TEST MATRIX

Dummy	Restraint System	Test Velocity	Submarining?
Hybrid III	"Baseline"	55 km/h	Both Sides
		52 km/h	Both Sides
		44 km/h	Inboard Side
		40 km/h	None
	Styrofoam Pelvic Stop	52 km/h	None
	Wood Pelvic Stop	52 km/h	None
	Modified Belt	52 km/h	None
5th Female	"Baseline"	52 km/h	Both Sides
		44 km/h	Both Sides
		40 km/h	Both Sides
		35 km/h	Both Sides
	Styrofoam Pelvic Stop	55 km/h	Inboard Side
		53 km/h	Inboard Side
	Styrofoam Pelvic Stop & Short Buckle Strap	56 km/h	None

However, the 5th female dummy submarined on the inboard side at 53 km/h with the pelvic stop in place. Because submarining occurred only on the buckle-strap side, 53 km/h must be slightly above the submarining threshold. Thus, for the 5th female dummy, the pelvic stop increased the submarining threshold from below 35 km/h to almost 53 km/h. Tests with the pelvic stop and a shortened buckle-strap increased the submarining threshold to greater than 56 km/h for the 5th female dummy.

SUBMARINING DETERMINATION - The occurrence of submarining for the Hybrid III dummy was determined independently, and blindly (the test number was not known when the interpretation of the data was made) through analysis of the high speed movies, load bolt data, belt load data, and the pelvic in-line acceleration data. A summary of the results of the individual determinations follows, with a comparison of the various methods in the discussion section.

FILM ANALYSIS - The pelvic angular displacement, the occurrence of submarining, and frame in which it occurred were recorded during film analysis. Table 3 lists the results for each test. Submarining occurred in tests 1, 3, 4, and 6. There was no submarining in any other test (n.b., the lap belts were severed by the photographic targets in tests 2 and 9, which limited the amount of information obtained from those tests). Because of inadequacies inherent in filming with fixed cameras, it was very difficult to determine whether submarining had occurred on the outboard side in test 4, and whether it occurred on either side in test 8.

LOAD-BOLT PELVIS - The criterion used to determine whether or not submarining had occurred was the shape of the load time curves examined as a whole and individually, Figure 4. The combination of factors assumed to indicate submarining was asymmetry in the curves which normally approximate bell shapes, rapid unloading of the belt, and shifting of the load from the inferior to the superior load bolts (Because the belt is wider than the inter-bolt spaces, there were usually forces on two bolts recorded simultaneously).

Table 4 displays the determination of the occurrence of submarining using the load-bolt data. In some tests, absolute determination of the occurrence or non-occurrence of submarining was not possible. These included test 6 where a determination could not be made on the outboard side, and test 8 where the clothing of the dummy made it very difficult to be sure.

BELT LOADS - The criterion used to judge the occurrence of submarining for the belt load data was similar to that for the load bolt data, viz. asymmetry of the force-time curves. The belt load curves are very similar to bell shaped curves when the dummy kinematics yield no submarining. Therefore, rapid unloading of the belts, and/or multiple peaks or deviations from the bell shape were assumed to indicate the occurrence of submarining, Figure 5.

The results of the belt load data analysis are presented in Table 5.

PELVIS ANGULAR RESPONSES - Table 6 indicates the maximum rearward pelvis rotation associated with the Hybrid III tests. A logistic regression analysis was performed on the pelvic rotation data using submarining as the dependent variable. The amount of pelvic rotation was analyzed by itself and in combination with the belt loads as independent parameters. The results have been graphed in Figure 6. The best fit occurred with the single parameter regression on pelvic rotation, which yielded a $\chi^2 = 8.88$, $r = .655$, and $p = 0.0029$. The ED_{50} for the occurrence of submarining is at 28.0° for these tests.

Examination of the pelvic rotation graphs for the tests in which submarining occurred shows that all curves have a common feature that is not present in the curves for tests with no submarining, Figures 7,8. In each graph of a test in which submarining occurred, there is a discontinuity in the displacement versus time curve. For simplicity if one fits a line to the rising portion of the curve, the slope changes abruptly at some point in the curve of all the tests in which submarining occurred. Further examination shows that in 3 out of the 4 tests the change in slope took place at or near the angle of the pelvis at which submarining occurred as determined in the film analysis, Table 7).

DISCUSSION

These experiments suggest that shoulder and lap belt loads, the load-bolt pelvis, and in-line pelvic accelerometry can each provide indications of lap-belt slippage over the pelvic iliac spines and could provide an objective indication of submarining for many situations. Even when all methods are used (see Table 8), in any particular test there can still be some uncertainty as to the occurrence or non-occurrence of submarining. The errors made fall into the following 3 categories:

(I) False Positives, which occur when interpretation of the transducer output indicates that submarining has occurred but, in actuality, it has not.

- The load bolts and belt loads gave false positives when the pelvic angle photographic target cut through the belt.

(II) Unknowns, which occur when interpretation of the transducer output does not unequivocally indicate whether submarining has occurred.

- The film analysis resulted in unknown determination of outboard submarining when the view was obscured in test 4.
- The belt load data yielded an unknown determination in tests 1 and 8.

(III) False Negatives, which occur when interpretation of the transducer output indicates that submarining has not occurred when, in actuality, it has.

- The film analysis in test 8 showed no submarining while the load-bolts and in-line accelerometers indicated submarining, and the belt loads indicated a high possibility of submarining.
- The in-line data in test 3 indicated no submarining had occurred when in actuality it had. This was probably because the mechanism of submarining was shoulder belt lifting and not pelvic rotation. The belt loads show a high shoulder belt load which peaks before the lap belt load.

TABLE 3 - SUMMARY OF FILM ANALYSIS RESULTS

Test Number	Submarined ? (Yes/No)	Pelvic Displacement*	Time (ms)	Remarks
1	Yes	25°	58	Inboard; Outboard + (See below)
2	unk			
3	Yes	22°	52	Inboard side only
4	Yes	25°	62	Inboard; Outboard?
5	No	10°		
6	Yes	29°	92	Inboard side only
7	No	26°		
8	No?	45°		Difficult to determine + (See below)
9	unk			
10	No	27°		
11	No	7°		
12	No	10°		
13	No	10°		
14	No	17°		
15	No	13°		

* Pelvic angular displacement (Degrees) is given at the time of submarining for tests in which submarining occurred, or, for tests with no submarining, the maximum observed pelvic angular displacement is given.

+ Lap belts were severed by the pelvic angle photographic target in tests 2 and 9.

TABLE 4 - SUMMARY OF LOAD-BOLT RESULTS

Test	Submarining ?	Remarks
1	Yes	Inboard side only
2	Yes	Inboard side; Outboard side
3	Yes	Inboard side; Outboard side
4	Yes	Inboard side; Outboard side
5	No	
6	Yes	Inboard side; Outboard possible
7	No	
8	Yes ?	Inboard side possible
9	Yes ?	Inboard possible; Outboard possible
10	No	
11	No	
12	No	
13	No	
14	No	
15	No	

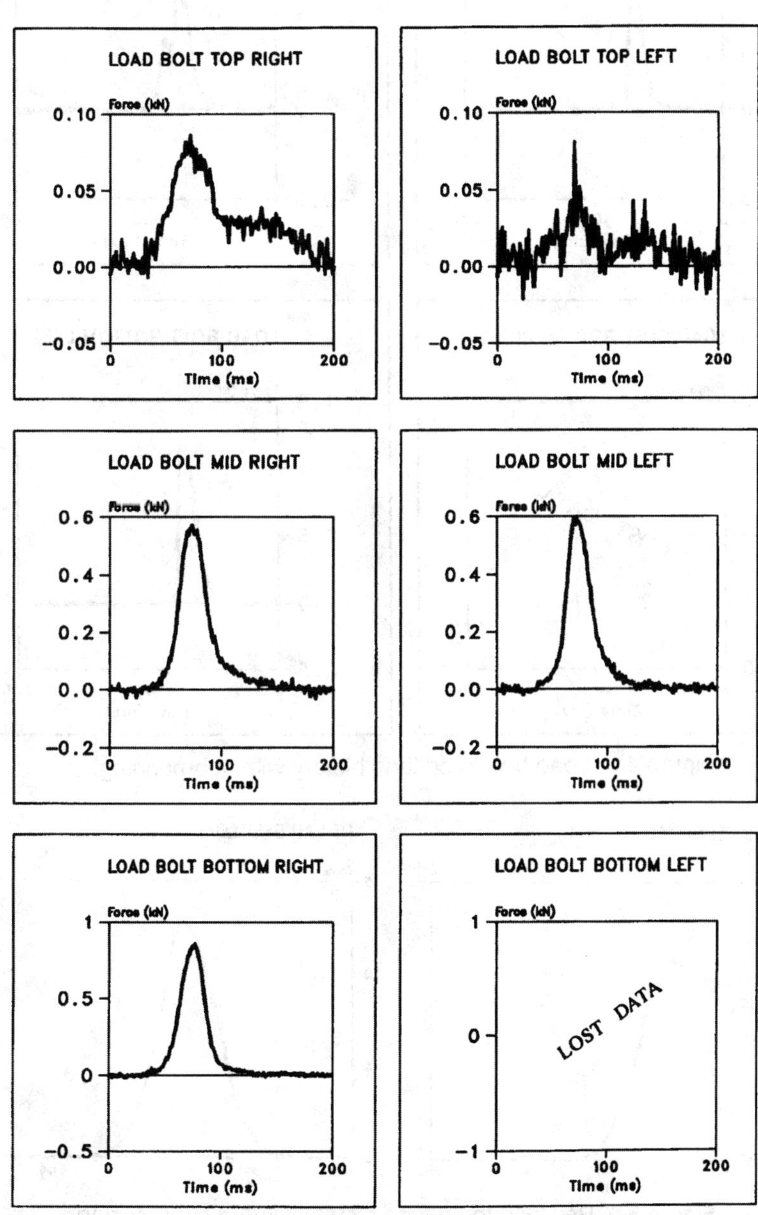

Figure 4a) Load-bolt force-time history without submarining.

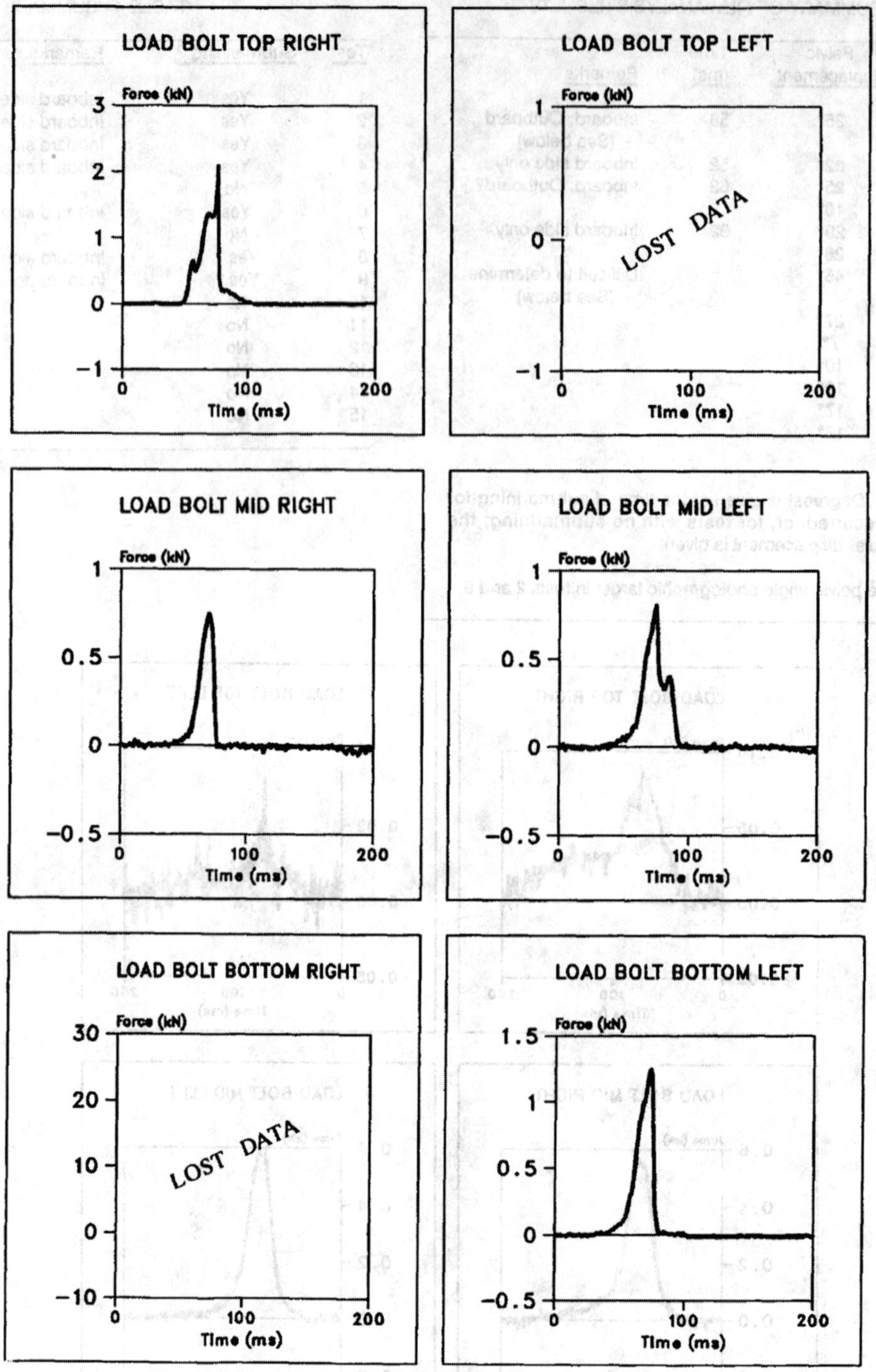

LOAD BOLT TOP RIGHT

Force (kN)

3, 2, 1, 0, −1

0, 100, 200

Time (ms)

LOAD BOLT TOP LEFT

Force (kN)

1, 0, −1

LOST DATA

0, 100, 200

Time (ms)

LOAD BOLT MID RIGHT

Force (kN)

1, 0.5, 0, −0.5

0, 100, 200

Time (ms)

LOAD BOLT MID LEFT

Force (kN)

1, 0.5, 0, −0.5

0, 100, 200

Time (ms)

LOAD BOLT BOTTOM RIGHT

Force (kN)

30, 20, 10, 0, −10

LOST DATA

0, 100, 200

Time (ms)

LOAD BOLT BOTTOM LEFT

Force (kN)

1.5, 1, 0.5, 0, −0.5

0, 100, 200

Time (ms)

Figure 4b) Load-bolt force-time history with submarining.

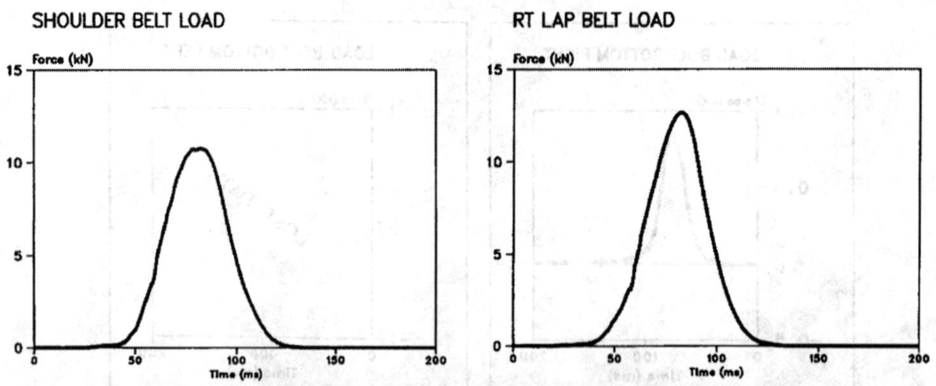

Figure 5a) Belt load force-time history without submarining.

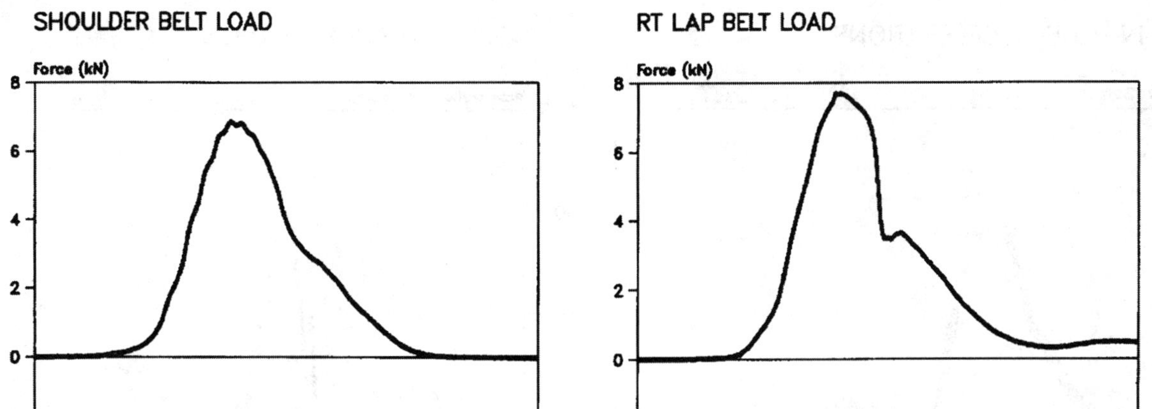

Figure 5b) Belt load force-time history with submarining.

TABLE 5 - SUMMARY OF BELT LOAD DATA

Test	Submarining ?	Remarks
1	Yes	Rapid unloading of SB; No LB
2	Yes	Severe; Very rapid unloading of SB; No LB
3	Yes	Possible; Rapid unloading of LB
4	Yes	SB asymmetry; Rapid unloading of LB
5	No	Symmetric curves
6	Yes	Rapid unloading of LB
7	No	Symmetric curves
8	Yes	Asymmetric LB & SB; Rapid unloading of LB
9	Yes	Very rapid unloading of LB & SB
10	No	Symmetric curves
11	No	Probably not; Anomalous loading curve
12	No	Symmetric SB; No LB
13	No	Symmetric SB; No LB
14	No	Symmetric SB; No LB
15	No	Symmetric curves

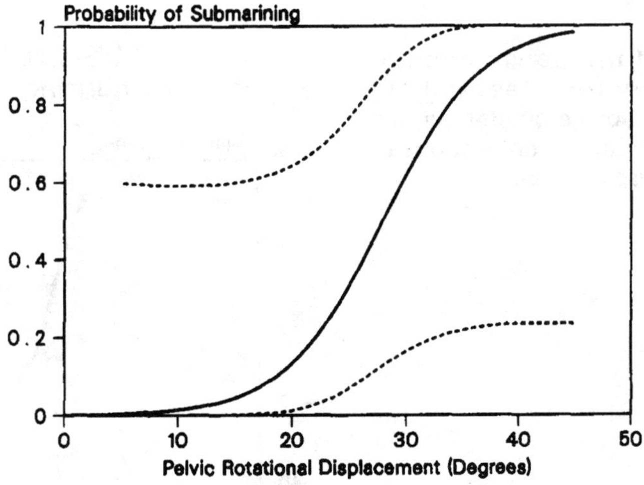

Figure 6. Logistic regression of pelvic rotational displacement versus probability of submarining.

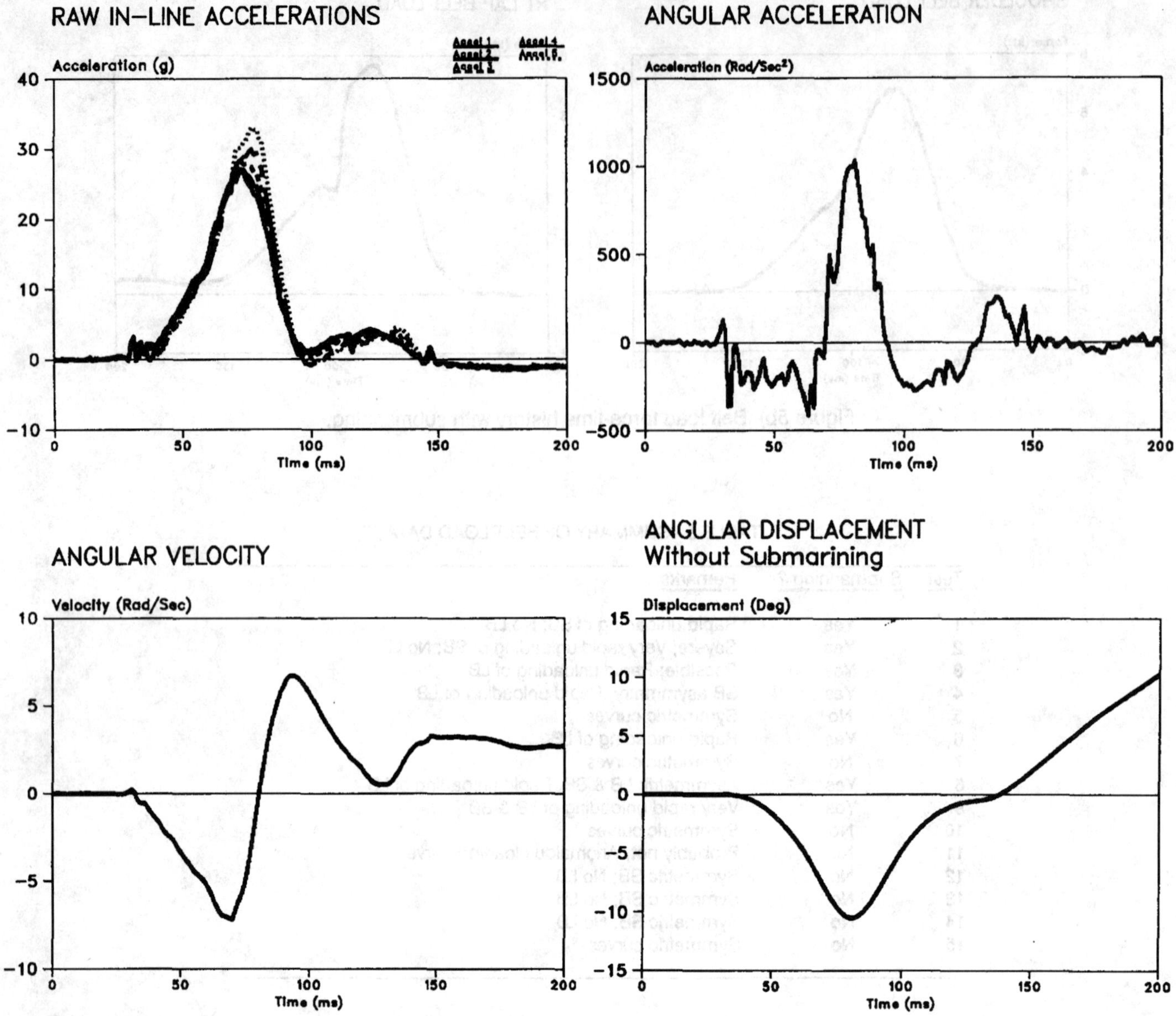

RAW IN-LINE ACCELERATIONS

Accel 1 Accel 2 Accel 3 Accel 4 Accel 5

Acceleration (g)

Time (ms)

ANGULAR ACCELERATION

Acceleration (Rad/Sec²)

Time (ms)

ANGULAR VELOCITY

Velocity (Rad/Sec)

Time (ms)

ANGULAR DISPLACEMENT
Without Submarining

Displacement (Deg)

Time (ms)

Figure 7. Derivation of the pelvic angular displacement-time history for a test without submarining: a) Five in-line accelerometer outputs overlayed, b) Resultant angular acceleration, c) Angular velocity, d) Angular displacement.

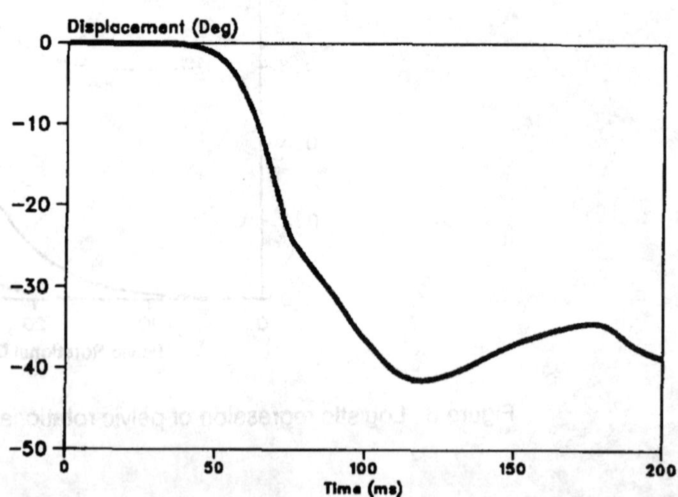

Figure 8. Pelvic angular displacement time-history with submarining.

TABLE 6 - SUMMARY OF PELVIC IN-LINE ACCELEROMETER DATA

Test Number	Maximum Pelvic Rotation	Time (ms)	Results from film Submarined ?
1	41.4°	119	Yes
2	11.7°	74	unk
3	22.8°	102	Yes
4	42.6°	119	Yes
5	6.9°	81	No
6	31.0°	116	Yes
7	25.0°	109	No
8	30.5°	122	No (Possibly Yes)
9	15.0°	77	unk
10	14.8°	77	No
11	5.2°	68	No
12	10.4°	81	No
13	14.0°	71	No
14	17.0°	74	No
15	11.2°	82	No

TABLE 7 - ROTATION SLOPE CHANGE & SUBMARINING

Test Number	In-line - Maximum Pelvic Rotation	In-line - Angle of Slope Change	Film - Angle of Submarining
1	41°	25°	25°
3	23°	19°	22°
4	43°	32°	25°
6	31°	29°	29°
8	**	**	**

** Test 8 also shows a change in slope, at approximately 15°. The Hybrid III probably submarined in this test, but it could not be determined from the film analysis (this will be discussed in more detail later.)

TABLE 8 - COMPARISON OF METHODS TO DETERMINE THE OCCURRENCE OF SUBMARINING

Test Number	Occurrence of Submarining as Determined By: Film	Belt Load	Load Bolt	In-line
1	Yes	unk	Yes	Yes
2	*	Yes*	Yes*	No*
3	Yes	Yes	Yes	No
4	Yes?	Yes	Yes	Yes
5	No	No	No	No
6	Yes	Yes	Yes	Yes
7	No	No	No	No
8	No?	unk	Yes	Yes
9	*	Yes*	Yes*	No*
10	No	No	No	No
11	No	No	No	No
12	No	No	No	No
13	No	No	No	No
14	No	No	No	No
15	No	No	No	No

* Lap belt severed by pelvic angle photographic target - unknown if submarining would have occurred.

This could have caused the shoulder belt to "lift" the lap belt off the pelvis before the lap belt could become "hooked" beneath the anterior superior iliac spines.

LOAD-BOLT PELVIS - The load-bolt installation and the load transducers change the shape of the iliac spines (Figure 3) and thus could change the "critical" angle, between the lap-belt and pelvis, at which the belt slips off the pelvis. This may introduce a change in the dummy's tendency to submarine. This pelvis also appears to be more sensitive to tears in the simulated flesh covering the iliac spines. However, the load bolt pelvis was as reliable an indicator of submarining as the other instrumentation investigated in this study.

IN-LINE ACCELEROMETRY - These results indicate that the belt will slip from the pelvis after approximately 28° of rearward rotation of the pelvis from the 22°-23° initial position (total angle of 50°-51° relative to vertical). Of course this does not allow for changes in the lap-belt angle which increases the tendency to submarine as the pelvis translates forward and downward. On this basis, the 40 km/h test with the standard seat and belt restraint was close to submarining (25° rearward pelvic rotation) as was concluded by the submarining threshold severity analysis. In contrast, in the tests with the pelvic stops, rearward pelvic rotation was reduced to 10°-11° by providing a force which tended to rotate the pelvis in a direction opposite that of the belt. This aspect of belt restraint systems has been discussed by others [9-14] and is corroborated in this study.

THRESHOLD SEVERITY - A "threshold" severity concept for submarining provides a method to compare experimental situations on the basis of submarining tendency. This method has the disadvantage of requiring multiple tests to establish the submarining threshold severity, but could be a powerful tool for some situations. The submarining tendency among types of dummies could be compared by determining the submarining threshold severity for each of the dummies in one or more restraint situations (i.e., the submarining tendency of the Standard vs the Hybrid III version of the 5th percentile female dummy could be effectively determined by this technique). This technique would be effective in comparing the submarining tendency of design modifications of a particular dummy, and could have been used in our study to determine if the load bolt pelvis influenced submarining tendency for restraint configurations similar to that tested.

ANALYSIS OF SUBMARINING PERFORMANCE - The analysis of submarining performance can be expanded beyond a yes/no evaluation of submarining. The kinematics of the dummy's pelvis interacting with the belt restraint can be a powerful tool for comparison of submarining performance of prototype systems. The degree of pelvic rotation can be used to estimate how close the exposure came to actual submarining, or how much a system's performance is above the submarining threshold. Other measures of submarining performance can likewise be effective for comparison of submarining tendency among restraint designs or among test dummies. The uncertainty in the determination of the occurrence of submarining shown by the methods evaluated in this study, and the absence of any information on the significance of the abdominal loading by the belt, underscore the need for an improved and objective method of dealing with lap belt submarining. The development of a method to detect abdominal loading <u>and</u> to predict the severity of any resulting injury would greatly strengthen analysis of restraint performance.

ACKNOWLEDGEMENT

The authors are grateful to Gerald Horn, Edward Jedrzejczak, Joseph McCleary, and Brian Stouffer for their technical assistance in these studies.

REFERENCES

1. Dalmotas, D., "Mechanisms of Injury to Vehicle Occupants Restrained by Three-Point Seat Belts", Proceedings of the 24th Stapp Car Crash Conference, October 1980, SAE Paper No 801311.

2. Leung, Y.C., Tarriere, C., Fayon, A., Mairesse, P., Delmas, A., and Banzet, P., "A Comparison Between Part 572 Dummy and Human Subject in the Problem of Submarining", Proceedings of the 23rd Stapp Car Crash Conference, October 1979, SAE Paper No 791026.

3. Cromack, J. and Ziperman, H., "Three-Point Belt Induced Injuries: A Comparison Between Laboratory Surrogates and Real World Accident Victims", Proceedings of the 19th Stapp Car Crash Conference, November 1975, SAE Paper No 751141.

4. Marsh, J., Scott, R.E. and Melvin, J.W., "Injury Patterns by Restraint Usage in 1973 and 1974 Passenger Cars", Proceedings of the 19th Stapp Car Crash Conference, November 1975, Sae Paper No 751143.

5. Patrick, L. and Andersson, A., "Three-Point Harness Accident and Laboratory Data Comparison", Proceedings of the 18th Stapp Car Crash Conference, December 1974, SAE Paper No 741181.

6. Patrick, L., and Levine, R., "Injury to Unembalmed Belted Cadavers in Simulated Collisions", Proceedings of the 19th Stapp Car Crash Conference, November 1975, SAE Paper No 751144.

7. Schmidt, G., Kallieris, D., Barz, J. and Mattern, R., "Results of 49 Cadaver Tests Simulating Frontal Collision of Front Seat Passengers", Proceedings of the 18th Stapp Car Crash Conference, December 1974, SAE Paper No 741182.

8. Miller, M., "The Biomechanical Response of the Lower Abdomen to Belt-Restraint Loading," J. Trauma, In Press.

9. Svensson, L., "Means for Effective Improvement of the Three-Point Seat Belt in Frontal Crashes", Proceedings of the 22nd Stapp Car Crash Conference, October 1978, SAE Paper No 780898.

10. Adomeit, D., "Seat Design - A Significant Factor for Safety Belt Effectiveness", Proceedings of the 23rd Stapp Car Crash Conference, October 1979, SAE Paper No 791004.

11. Adomeit, D. and Heger, A., "Motion Sequence Criteria and Design Proposals for Restraint Devices to Avoid Unfavorable Biomechanical Conditions and Submarining", Proceedings of the 19th Stapp Car Crash Conference, November 1975, SAE Paper No 751146.

12. Adomeit, D., "Evaluation Methods for the Biomechanical Quality of Restraint Systems During Frontal Impact", Proceedings of the 21st Stapp Car Crash Conference, October 1979, SAE Paper No 770936.

13. Freeman, C. and Bacon, D., "The 3-Dimensional Trajectories of Dummy Car Occupants Restrained by Seat Belts in Crash Simulations", Proceedings of the 32nd Stapp Car Crash Conference, October 1988, SAE Paper No 881727.

14. Horsch J. and Hering, W., "A Kinematic Analysis of Lap-Belt Submarining in Test Dummies", Proceedings of the 33rd Stapp Car Crash Conference, October 1989.

15. Viano, D.C., Melvin, J.W., McCleary, J.D., Madeira, R.G., Shee, T.R. and Horsch, J.D., "Measurement of Head Dynamics and Facial Contact Forces in the Hybrid III Dummy", Proceedings of the 30th Stapp Car Crash Conference, October 1986, SAE Paper No 861891.

16. Daniel, R., "Test Dummy Submarining Indicator System", United States Patent 3,841,163, Oct 1974.

861884

Chest Deflection Characteristics of Volunteers and Hybrid III Dummies

Stanley H. Backaitis
National Highway Traffic Safety Administration

Andre St-Laurent
Biokinetics and Associates Ltd.

ABSTRACT

This two part study investigates differences in thorax deflections between volunteers and dummies when they are dynamically loaded by diagonal shoulder belts and it shows how internal measurements at the Hybrid III dummy's sternum relate to external compressions at various points of the rib cage. Test results reveal that the thorax of the Hybrid III dummy, when loaded by a diagonal belt, is somewhat stiffer than that of volunteers for both tensed and relaxed conditions. The thorax of volunteers under dynamic belt loading deflects underneath the belt in an action similar to that of a flat rigid plate being pushed into the thorax with increasing deflection towards the lower part of the ribcage, while the dummy's deflected profile assumes a parabolic curvature.

The Hybrid III dummy's thorax deflection gage underestimates the compressions which are administered externally to the thorax by small area loading probes except in those cases in which the loading device is aligned with the sensitive axis of the deflection gage. The internal deflection gage, however, overestimates externally produced rib cage compressions when loading occurs by large area test probes which are concentric with the sensitive axis of the deflection gage. Angled compressions by the large area test probe are underestimated by the dummy's internal deflection gage measurements.

PROBLEM STATEMENT

Research conducted by Kroell [1],* Lobdell [2], Stalnaker [3], Fayon [4], Walfisch [5] and Viano [6] has shown that chest deflection is an important cause of occupant injury. To evaluate properly the effectiveness of occupant protection systems that limit injuries caused by chest deflection, it is imperative that whatever surrogates are used in crash tests that they are capable of responding to thoracic impact with either correct chest deflection or a correctable chest deflection measurement.

The projected infusion of the Hybrid III dummy into vehicle crash testing and the intention to measure chest deflection for injury assessment has brought about the need to understand how well the dummy fulfills this requirement. Specifically, it is essential to know how the Hybrid III dummy's chest deflection gage senses chest deformation under various loading conditions and how the deflections relate to real world vehicle occupants who are subjected in accidents to similar loading conditions.

SCOPE OF THIS STUDY

This paper reports and discusses test results of a series of experiments performed in Canada and in the United States in which chest deflections under controlled loading conditions were measured on volunteers and Hybrid III test dummies. The first study, performed in Canada by Biokinetics and Associates, investigated chest deflection characteristics of volunteers and test dummies who were subjected to shoulder belt loading. The second study performed in the United States by the Vehicle Research Test Center (VRTC) investigated the ability of the thorax internal deflection measurement mechanism to assess deflections on various portions of the chest produced by several types of loading devices.

*/ Numbers in brackets denote references at the end of paper

TEST METHOD

1. Description of the Canadian experiment.

The Canadian test procedure was described by R. J. L'Abbe in the paper titled, "An Experimental Analysis of Thoracic Deflection Response to Belt Loading." [7] In this experiment ten volunteer test subjects were asked to lay supine on a test table. Shoulder belts were located on the test subject's torso in a geometric layout similar to that of a passenger car occupant wearing shoulder belts. Chest loading through the belt was imparted via impact mechanism as shown in Figure 1. The tests were carried out with tensed and relaxed test subjects.

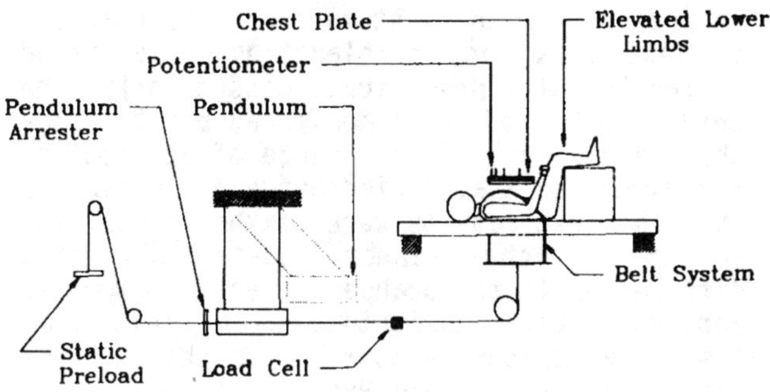

Figure 1 - Test set-up

In this experiment, L'Abbe measured chest deflections at 11 predetermined sights on the thorax as shown in Figure 2. However, the points of specific interest for this study are the mid-clavicle, the mid-sternum and the 7th rib. The mid-clavicle and the 7th rib deflections, based on external measurements, are diagonally opposite from each other and lie under the belt.

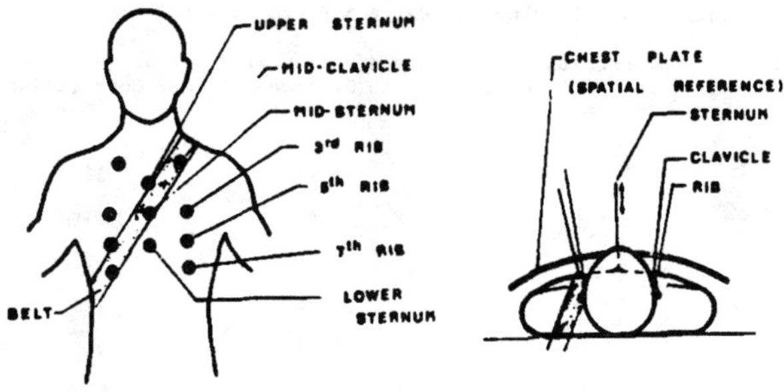

Figure 2 - Layout of test points and of the upper torso belt

Deflection measurements on the Hybrid III dummy were taken at the clavicle, at mid-sternum and at the 5th dummy rib (the equivalent of the 7th human rib). All measured values represent externally sensed deflections rather than internal measurements. The deflection measurements were conducted with and without the soft tissue covering the thorax. However, for the purposes of this study, only the results from the thorax with soft tissue coverings were considered since comparable deflections on the volunteers included the compression of the flesh and of the ribs.

Because these particular data sets were not reported in the L'Abbe paper [7], the data bank from this test series was re-interrogated and specific data points extracted to serve the needs of this study. The data are presented in Appendix 1.

2. Description of the SRL Thorax Deflection Test Experiment

The VRTC test series was designed to examine the Hybrid III ribcage deflection magnitudes measured by the centrally mounted internal deflection gage within the dummy's thorax as a result of: 1. loadings by a one inch dia. area compression device at various portions of the ribcage and 2. loadings by a compression device with a 6 inch dia. surface area.

The ribcage was mounted for small area load applications in an Instron Universal Test Machine as shown in Figure 3 and for large area applications as shown in Figure 4. Because the primary interest of this portion of the study was to determine how realistically the centrally located interior deflection sensor estimates compressions at remote portions of the chest, it was decided to use for this analysis the data from tests in which the ribcage was stripped from the flesh covering. The compressions were produced at quasistatic rates to eliminate any inertial effects of the structure. The above method does not in any way imply that mass effects are not important under impact conditions and that dynamic load deflection rates need not be determined. It does provide however, a convenient way to compare the characteristics of rib cage structures. The measurements consisted of recorded external penetrations by the loading device, and the deflections at the sternum were sensed by the internally located deflection gage.

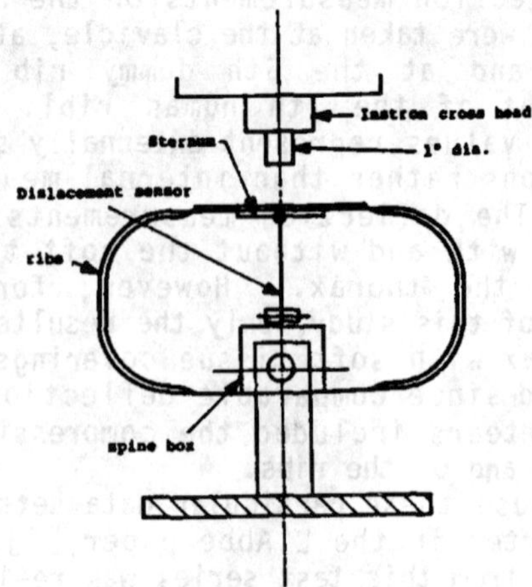

Figure 3 - Hybrid III Dummy thorax mounted for small area compression tests

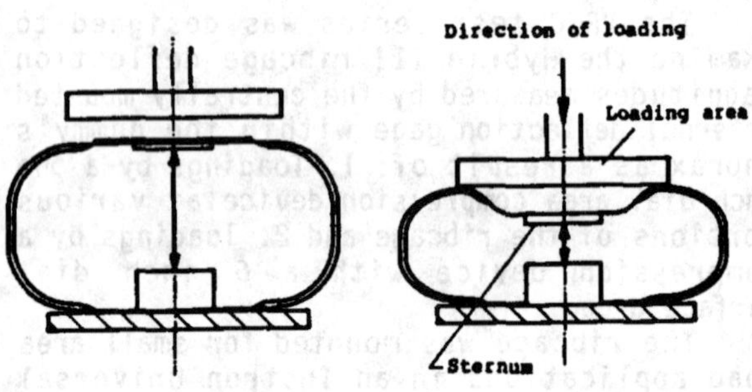

Figure 4 - Hybrid III Dummy thorax mounted for large area compression tests

TEST RESULTS

The results from the Canadian test series are summarized in Table 1. For the purposes of this study chest deflections are compared to a point when the combined loads at each end of the belt reached a force level of 3 kN. Although some volunteers were exposed to force levels as high as 3.6 kN, the 3.0 kN level was selected for this study because the data set at this loading was most voluminous and because the volunteers could maintain at that level a sharper differentiation between relaxed and tensed states.

TABLE 1

Deflection of Thorax due to Diagonal Shoulder Belt Loading at 2.8-3.2 kN Level

Measured at	Volunteer* Relaxed	Volunteer* Tensed	Dummy
Left Clavicle (in.)	0.39-0.72 (0.58)	0.22-0.67 (0.47)	0.74
Mid Sternum (in.)	0.61-1.22 (0.88)	0.42-1.22 (0.73)	0.66
Right 7th Rib (in.)	0.94-1.42 (1.16)	0.73-1.18 (0.94)	0.82

* Individual measurements are shown in Appendix

It should also be noted that the volunteers varied in stature and their levels of tenseness and or relaxation could not be controlled with precision. Consequently, the test results of each monitored point on the chest are presented as a range of deflections for loadings that varied between 2.8 and 3.2 kN. The test results were neither normalized nor in any other manner adjusted to reflect differences in the occupants' anthropometry, bony condition, soft tissue content, etc. Also, the measurements reflect the results from tests of a single Hybrid III dummy.

The results from the VRTC test series are given in Tables 2 and 3 for tests with the one inch diameter loading probe and in Table 4 for tests with the six inch test probe. The loads on the dummy were applied parallel to the midsagittal plane in all tests except for angled compressions. The one inch probe was pushed into the chest at the designated points one at a time (Figure 5). The internal deflection at the mid-sternum was recorded when the probe produced one and two inches of compression.

Table 2

Deflection of Hybrid III Dummy's Thorax at Mid Sternum due to Single Point Loading on Various Parts of the Ribcage

Point* Loaded	Compression at Point of Loading (in.)	Deflection at Mid Sternum (in.)	Load at Point of Compression (lbs)
1	1.0	0.97	180
2	1.0	1.00	170
3	1.0	0.91	183
4	1.0	0.81	220
		>(0.84)	
8	1.0	0.87	235
6	1.0	0.64	180
		>(0.61)	
7	1.0	0.58	167
5	1.0	0.78	195
		>(0.77)	
9	1.0	0.76	194

* Ref. Figure 5

TABLE 3

Deflection of Hybrid III Dummy's Thorax at Mid Sternum due to Single Point Loading on Various Parts of the Ribcage

Point* Loaded	Compression at Point of Loading (in.)	Deflection at Mid Sternum (in.)	Load at Point of Compression (lbs)
1	2.0	2.09	325
2	2.0	1.97	291
3	2.0	1.64	300
4	2.0	1.47 >(1.54)	352
8	2.0	1.61	395
6	2.0	1.21 >(1.19)	325
7	2.0	1.17	319
5	2.0	1.53 >(1.54)	327
9	2.0	1.54	345

* Ref. Figure 5

Table 4

Deflection of Hybrid III Dummy Thorax due to Large Area Rigid Flat Face Loading

Point Loaded	Compression at Point of Loading (in.)	Deflection at Mid Sternum (in.)
Mid-sternum*	2.0	2.03 >(2.04)
	2.0	2.06
Mid-sternum*	3.0	3.83
	3.0	3.44>(3.55)
	3.0	3.38
45 deg.**	2.0	1.21 >(1.30)
	2.0	1.38

* Ref. Figure 4

** Loading surface located at mid-sternum level and aimed at center of spinal box as shown in Fig. 9

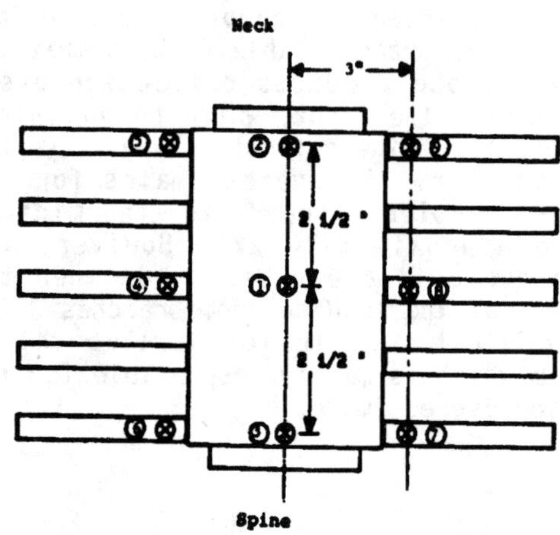

Figure 5 - Loading points on the Hybrid III Dummy's ribcage

In the first test series with the large area loading device, the test probe was centered on the mid-sternum and pushed in the anterior-posterior (A-P) direction into the thorax until two and then three inch compressions were obtained. In the second test series, the results represent the compression of the rib cage with the loading probe applied at 45° with respect to the midsagital plane of the dummy while the probe's motion, depicted by the centerline of the probe, was aligned to coincide with the center of the spinal box. (Fig. 9). Measurements include concurrent recordings of deflections at midsternum registered by the chest deflection gage and by external penetrations by the test probe at selected test points.

DISCUSSION

Test results from Table 1 are graphically shown in Figures 6 and 7. The graphs reveal an interesting deflection pattern of the ribcage of shoulder belt restrained volunteers under dynamic loading. As expected, the relaxed volunteers experienced larger chest deflections for similar loadings than those that were tensed. The deflections at the clavicle were the smallest. They increased as the belt descended toward the bottom end of the ribcage. What is perhaps more surprising is that the relaxed volunteers exhibited almost a constant 20% to 25% larger deflection than tensed volunteers at all three monitored locations. Furthermore, it was observed that deflections underneath the belt increased between measured points by approximately 45% for both tensed and relaxed volunteers as the belt descended down the ribcage.

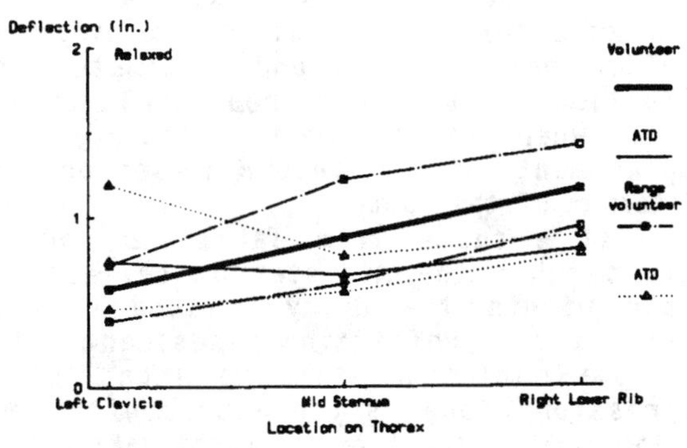

Figure 6 - Thorax deflection due to diagonal shoulder belt at 3 kN loading

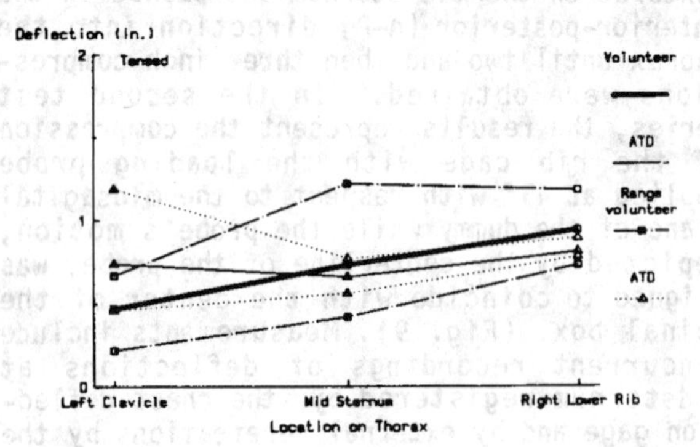

Figure 7 - Thorax deflection due to diagonal shoulder belt at 3 kN loading

However, under similar belt loading conditions the deflection pattern on the dummy is somewhat different. Test data indicate that the dummy's thorax, when loaded by a diagonal shoulder belt, assumes a somewhat parabolically shaped deflection pattern with maximum deflections occurring at the clavicle and at the 7th rib levels and minimum deflections at mid-sternum. The dummy thorax, as a whole, appears to be stiffer than that of the average of the tensed volunteers both at mid-sternum and at the 7th rib and softer at the clavicle. Test results indicate that the dummy at mid-sternum is approximately 10% stiffer than the average of the tensed volunteers and at least 25% stiffer than the average of the relaxed test subjects. These stiffness relationships are, of course, valid only for voluntary loading levels and there is no way of knowing from this data set whether similar relationships would exist at loading levels four to five times as high, such as those experienced in 30-35 mph vehicle collisions. Nevertheless, these results suggest that dummy based deflection measurements due to diagonal shoulder belt loading underestimate the deflections occurring to real world crash victims when the assessment is based on the displacement of an internal sensor at midsternum of the dummy.

Test results from Tables 2 and 3 indicate that the mid-sternum deflection sensor within the dummy's ribcage will provide a true deflection assessment only when the motion of the one inch dia. compression probe is concentric with the sensing axis of the internal chest deflection gage. Externally induced deflections at remote portions of the thorax are underestimated by the mid-sternum deflection sensor from 3% up to 40% depending on the location of the load application. The test data show that the underestimates are smallest when the input compressions occur on the sternum in the midsagittal plane. Of these three measurements, the largest underestimate is produced by compression at the bottom of the ribcage (point 3 in Figure 5). The underestimates become considerably larger when the loads are applied on either side of the midsagittal plane. For all of the points measured on either side of the ribcage the underestimates are maximum for deflections originating at the bottom of the ribcage (points 6 and 7 in Figure 5), and are least for compression produced at the midpoint height level of the sternum (points 4 and 8). The above deflection patterns are summarized in Figure 8. Comparison of deflections at different loading levels show that the centermounted internal deflection gage senses changes by nearly identical proportions as are the variations in externally induced compression magnitudes at the nine test points.

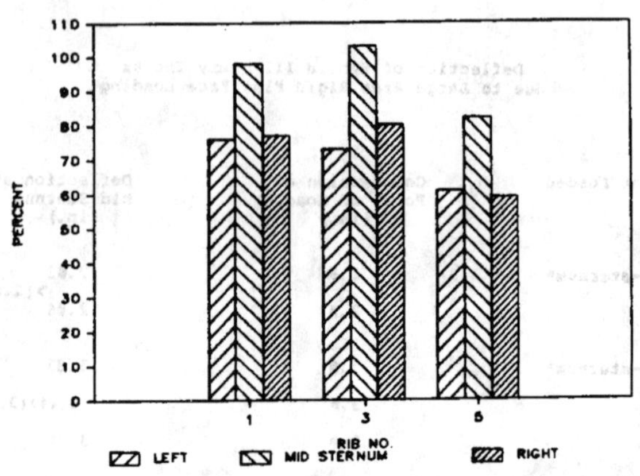

Figure 8 - Ratio of deflection at sternum vs. deflections at other points of the Hybrid III dummy's thorax

Test results from chest compression in the A-P plane with the 6 in. dia. test probe yield deflection patterns opposite from those which were observed for compressions by the one inch dia. probe. Table 4 shows that the large area probe produces deflection overestimates when the input, which is concentric with the axis of the deflection gage, begins to exceed 2 in. The overestimates for two inch penetration appear to be minimal (average approximately 2%). However, the average overestimate exceeds 18% when the deflection of the loading probe reaches 3 in. Figure 4 illustrates the structural buckling mechanism which is mainly responsible for the deflection overestimate.

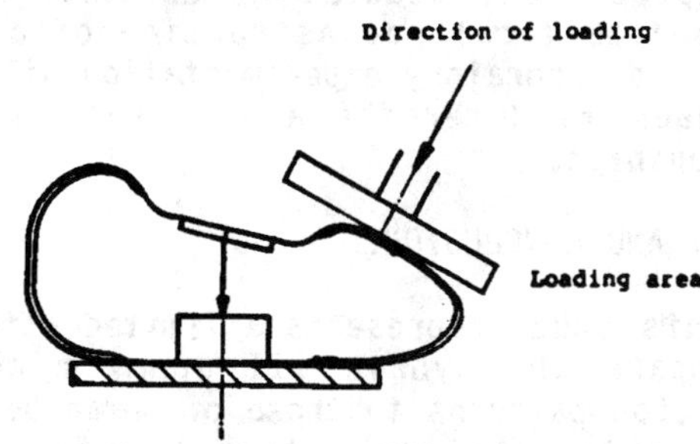

Figure 9 - Hybrid III dummy thorax mounted for large area compression tests at 45 deg. loading angle

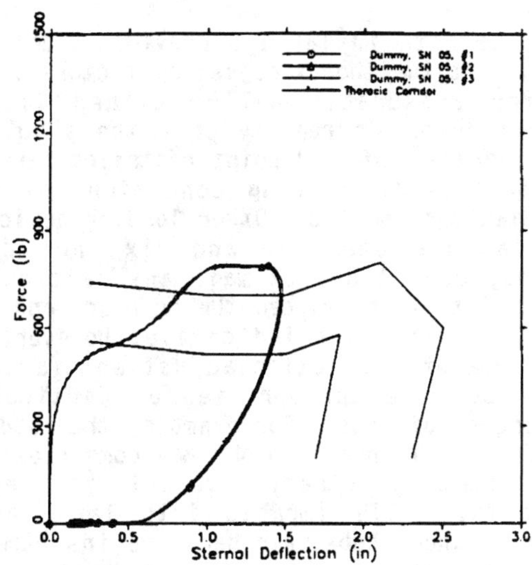

Figure 10 - Hybrid III dummy thorax biomechanical response in 14 fps pendulum impact tests

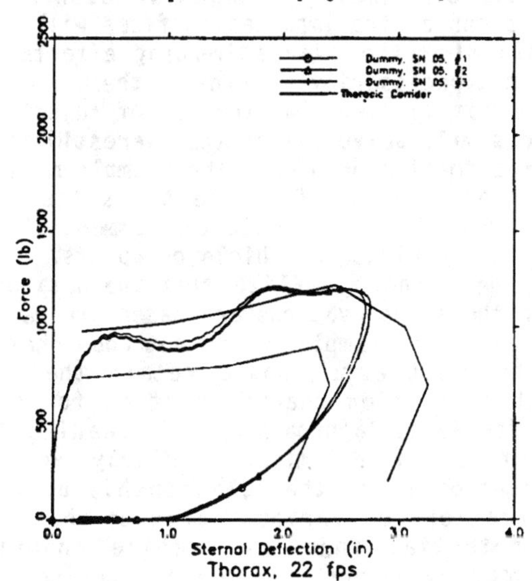

Figure 11 - Hybrid III dummy thorax biomechanical response in 22 fps pendulum impact tests

Figure 9 shows the thorax being compressed by the 6 inch probe with a 45 degree input orientation. Table 4 indicates that the angled external compression is being sensed by the internally mounted deflection sensor as a much lower value. As would be expected, the underestimate occurs due to the fact that deflection induced by the angled loading device is diminished by the effects of the cosine function of the angle between the axis of the loading probe and the A-P plane of the ribcage and the bending of the sternum plate.

The fact that the Hybrid III dummy's thorax design was based on cadaver response characteristics is documented in reference 8. These characteristics were derived from results of pendulum impact tests of cadavers during which the forces on the impacting pendulum and its penetration into the thorax cavity were measured. The dummy appears to reflect the cadaver impact responses reasonably well as shown in Figures 10 and 11. However, the vehicle crash environment in which the dummy is used and the occupant protection systems with which the dummy interacts seldom offer impact exposures that were used to develop the impact response of the thorax. The impact exposures in the Canadian test series and the static deflection tests at VRTC clearly show that dummy based deflection measurements will have considerably different meanings for the same amount of measured deflection depending on the type of restraint or impact surface intercepted by the occupant. Table 5 summarizes these effects. The summary suggests that interpretation of the thorax deflection data for injury assessment purposes must reflect the type of restraint system used to contain the occupant.

Table 5

Summary of Hybrid III Dummy Thorax Deflection Patterns by Type of Loading

Type of Loading	Loading Location or Orientation	Assesment by Thorax Deflection Sensor	Amount
Shoulder Belt	Diagonal	Underestimates	12% to 24%
Small Area*	On Center	No Effect	0%
	Off Center	Underestimates	3% to 40%
Large Area*	On Center	Overestimates	2% to 18%
	Off Center	Underestimates	up to 35%

* Rigid Flat Surface

The data in Table 5 provide guidelines as to what the thorax adjusting factors might be for reasonably well- defined impact environments. In reality only the shoulder belt portion of a 3 point restraint harness approaches the loading conditions of the Canadian test series. Other loading devices, such as the one inch and six inch dia. probes, do not approximate any particular impact interface between the occupant and the vehicle. They are indicative, however, of what type of deflection adjustment factors might be more appropriate for particular loading conditions. For example, the loading by the one inch dia. probe may come close to approximating steering wheel rim edge contacts, while impacts into large area steering wheel hubs, air bags and instrument panel surfaces may be presentedbest by the six inch dia. loading surface. Of course, the use of padding or special cushioning provisions on the impacted surfaces will tend to minimize the overestimating effects but will not necessarily eliminate them. While it is not claimed that the use of adjustment factors will solve all of the overestimation/underestimation problems, their employment in the interpretation of the test results will produce a more realistic assessment of the true injury risks to vehicle occupants.

The authors realize that the data upon which these observations are based are rather limited. For example, we do not know whether the test subjects would retain the same chest deflection characteristics for compression levels approaching or exceeding the impact tolerance limits. Similarly, we lack information about the relationship between deformations on various parts of the chest and potential injuries. While injuries produced by large area loading devices are somewhat better understood due to extensive experiments with cadavers, very little is known about injury mechanisms associated with small loading surfaces. Questions arise as to whether the observed buckling of the Hybrid III dummy's ribcage, when loaded by the six inch diameter probe, is a realistic representation of what happens to the real accident victim under similar loading conditions. While it may not be important to know the specific mechanisms by which ribcage deflection occurs, it is necessary to assure that the measured deflections either reflect the correct magnitudes or that they are correctable in case of over-or underestimation.

The authors hope that this study and analysis will trigger further investigation into the deflection characteristics of the Hybrid III dummy's ribcage and will bring about a better understanding of how dummy based ribcage deflections correlate with those occurring to injured car occupants. Even more important is the need to establish the relationship between these measurements and specific injury severities. It is hoped, that appropriate studies of accident injury data, mathematical reconstruction of accidents, and laboratory experimentation will be continued to determine and clarify these relationships.

SUMMARY AND CONCLUSION:

This study represents a limited attempt to compare the Hybrid III dummy's chest deflection patterns to those of human beings and to determine how internal deflection measurements at mid-sternum reflect compressions at other locations on the chest. Various limitations precluded the expansion of this study at this time to investigate a broader spectrum of possible impact configurations on the chest. Nevertheless, even these limited series of tests showed that single point deflection measurements require careful interpretation. The study yielded the following conclusions:

1. Diagonal shoulder belts produce on volunteers a linearly increasing chest surface deflection underneath the belt as the belt descends towards the bottom of the ribcage. The deflection pattern of the Hybrid III dummy's thorax has a parabolic shape, with the least amount of deflection occurring at mid-sternum. In this test environment the Hybrid III based measurements by the thorax deflection gage underestimate the depth of compressions occurring to the human test subjects.
2. For shoulder belt loading conditions, the Hybrid III dummy appears to have a stiffer thorax than human occupants. The dummy's thorax stiffness exceeds that of tensed volunteers by approximately 10% and that of relaxed test subjects by up to 25%.
3. Dummy thorax deflection readings are highly affected by the surface area of the loading probe and the location and direction of the applied load. The following was observed: (1) for small area load applicators, the deflection gage underestimates the input compressions produced at any point of the ribcage except for mid-sternum; (2) compressions by small area loading devices which are colinear with the sensitive axis of the deflection gage are measured correctly; (3) for large area loading devices, the existing Hybrid III thorax deflection gage overestimates human occupant thorax compressions due to the buckling effects of the sternum; (4) for angled compressions by large area loading devices, the recorded deflections by the dummy's internal gage underestimate the amount of compression occurring on the ribcage.
4. This study suggests that dummy based thorax deflection measurements should be

interpreted within the framework of loading conditions. This work, however, neither substantiates nor refutes the thorax deflection as a relevant and effective injury predictor for the various loading conditions.

ACKNOWLEDGEMENT

The results presented in this paper are based on cooperative contributions by NHTSA's Vehicle Research Test Center in East Liberty, Ohio and Biokinetics in Ottawa, Canada. The authors are greatly indebted to Messrs. James Hofferberth and Howard Pritz of VRTC and Mr. L'Abbe of Biokinetics and Associates, Ltd.for directing the test work and for compiling the test results in a usable format.

The views, data analysis and conclusions expressed in this presentation are those of the authors and not necessarily those of the National Highway Traffic Safety Administration and Biokinetics & Associates, Ltd.

BIOLIOGRAPHY

1. Kroell, C.K., Thoracic Responses to Blunt Frontal Loading, SAE Special Publication P-67, Society of Automotive Engineers, Inc. Warrendale, PA., 1976.

2. Lobdell, T.E. et al., Impact Response of the Human Thorax; Human Impact Response, Measurement and Simulation; Plenum Press, New York - London, 1973.

3. Stalnaker, R. et al., Human Torso Response to Blunt Trauma, Human Impact Response, Measurement and Simulation. Plenum Press, New York - London, 1973.

4. Fayon, A., Tarriere C., Walfisch G., Got C. and Patel A. Thorax of 3-Point Belt Wearers During a Crash; SAE 751148. Proceedings, Nineteenth Stapp Car Crash Conference, San Diego, CA., November 1975.

5. G. Walfisch et al. Tolerance Limits and Mechanical Characteristic of the Human Thorax in Front and Side Impact and Transposition of These Characteristics into Protection Criteria; Proceedings, 7th IRCOBI Conference on the Biomechanics of of Impacts, Cologne, West Germany, 1982.

6. Viano, D.C. "Chest Anatomy, Types and Mechanics of Injury, Tolerance Criteria and Units, and Injury Factors," Symposium on Biomechanics of Impact Trauma, San Diego, CA. October 1983.

7. L'Abbe, R.J., Dainty, D.A., Newman, J.A., An Experimental Analysis of Thoracic Deflection Response to Belt Loading, Proceedings, 7th IRCOBI Conference on the Biomechanics of Impacts, Cologne, West Germany, 1982.

8. Foster, K.J., Kortge, J.O., Wolamin, M.J. Hybrid III - A Biomechanically Based Crash Test Dummy; SAE 770938, Proceedings, 21st Stapp Car Crash Conference, New Orleans, LA., 1977.;

851722

Predictive Functions for Thoracic Injuries to Belt Wearers in Frontal Collisions and Their Conversion into Protection Criteria

G. Walfisch, F. Chamouard, D. Lestrelin, C. Tarriere, and F. Brun Cassan
Laboratory of Physiology and Biomechanics
Peugeot S.A.—
Renault, France
P. Mack, C. Got, F. Guillon, and A. Patel
I.R.B.A., Raymond Poincare Hospital
Garches, France
J. Hureau
University Rene Descartes
Paris, France

ABSTRACT

The data presented in this paper were yielded by tests performed on unembalmed human cadavers fitted with three-point seat belts and subjected to frontal collisions. The purpose is to define one or more functions predictive of thoracic injuries to cadavers whose rib "resistance" is known (i.e. BCF parameter (1)*).

These functions predict the number of rib fractures and the thoracic AIS in terms of :

- anthropometrical data on the cadavers,
- data representative of the thoracic resistance of the cadavers and physical parameters arising from the deceleration pulses measured on the cadaver vertebrae during the occurrence of impact.

By integrating the BCF data which characterize the ribs of the population exposed to the risk of thoracic injury, it is possible satisfactorily to define the tolerance of living road users, in terms of their age.

Provided that maximum admissible injury level, and the age for which this limit is required are set, a tolerance criterion can then be defined.

DEFINING A TOLERANCE CRITERION is especially important because each condition imposed upon the thorax can have consequences upon the characteristics to be required of seat belts and of vehicles, and also upon the mechanical loadings imposed on the other body areas.

Several parameters and methods have been used in order to characterize thoracic tolerance. In some cases, published data have been obtained from the impact of a disk against the sternum, but thoracic behaviour differs depending on whether the forces are applied to the thorax with a disk or with a seat belt.

The authors set forth below the results of a statistical analysis of data obtained on 29 fresh human cadavers fitted with measuring instruments, subjected to frontal collisions and restrained by a three-point seat belt. These data are restated as protection criteria measured on an experimental Part 572 dummy.

TEST METHODOLOGY

The test subjects were recently deceased, unembalmed cadavers of individuals whose death had occurred four days prior to the test. Pending their use in the laboratory, they had been stored in a cold room (temperature approximately 2° Celsius), from which they were removed several hours before the test.

TEST CONDITIONS - The principal data pertaining to the test conditions are shown in Table 1.

Depending on the tests, the cadavers were placed either in a vehicle or in a passenger compartment attached to a sled. The sled deceleration in terms of time was obtained by deceleration tubes secured to the front part of the sled. The impact velocities ranged from 44 to 67 km/h, and the sled's stopping distances from 500 to 900 mm.

The restraint systems used were three-point seat belts of several types. The cadavers were in either the front-seat passenger position or in the driver's position.

PREPARATION OF THE HUMAN CADAVERS - The human cadavers were prepared as already described in the literature (2).

ANTHROPOMETRIC MEASUREMENTS - Precise anthropometric measurements preceded each test. They are shown in Table 1. In addition to heights (symbol TAI in Table 1) and weights

*Numbers in parentheses designate references at end of paper.

Table 1 : Cadaver Test Conditions - Cadaver Anthropometrical Data

Cadaver Ref.	Vehicle	Δv	Age (Years)	Sex	Poi (Kg)	CTH (cm)	LTH (cm)	ETH (cm)	HEP (cm)	APBO
33	R15	49.000	51	1	50.000	80.000	27.500	18.000	58.500	0.858
41	R15	50.000	60	1	50.000	*	*	*	62.000	*
44	R15	49.800	53	1	65.500	88.000	*	21.000	60.000	*
47	R15	48.000	64	1	63.000	90.000	28.000	22.000	63.000	-0.100
53	R15	44.800	46	1	63.000	94.000	28.000	22.000	69.000	-0.334
54	R15	48.600	34	1	60.000	88.000	29.000	22.000	74.000	-0.807
185	R9	64.550	56	1	53.000	82.900	27.200	20.500	71.000	0.292
190	R9	49.900	39	2	51.000	78.500	26.400	16.200	56.000	1.839
223	R18	50.600	52	1	44.000	88.400	29.500	18.500	60.500	0.821
224	R18	49.800	34	2	40.000	75.000	22.000	15.000	64.100	1.380
246	R20	50.100	62	1	52.000	88.500	29.500	22.500	71.000	0.135
247	R20	50.500	42	1	58.000	93.000	32.500	28.000	72.000	-0.970
248	*	*	66	1	63.000	104.000	32.000	23.000	70.000	-0.849
254	R18	49.500	63	1	52.000	85.200	25.900	18.900	62.000	0.725
255	R18	50.900	68	1	56.000	86.500	27.800	23.000	69.000	-0.178
257	R18	67.100	42	2	53.000	88.200	29.400	16.500	62.500	1.060
258	R18	65.500	42	1	68.000	93.500	30.600	17.500	69.000	0.129
267	R18	60.000	68	1	71.000	92.000	31.500	20.900	66.000	0.024
268	R18	66.800	62	1	66.000	94.000	34.500	19.200	67.000	-0.190
276	R18	61.100	55	1	85.000	101.000	31.000	23.000	72.000	-1.365
277	R18	67.750	52	1	50.000	81.500	27.000	18.000	63.000	0.718
286	R9	56.500	47	1	74.000	97.000	32.500	18.500	73.000	-0.452
1152	504	50.200	52	1	63.000	92.500	30.000	22.000	92.000	-1.652
1174	R30	50.200	60	1	53.000	87.700	27.000	21.000	90.000	-1.010
1242	R30	66.360	61	1	52.000	89.000	27.000	22.500	86.500	-0.970
1261	R30	65.100	51	1	67.000	95.000	29.000	18.500	93.000	-1.345
1352	R14	48.500	54	2	50.000	77.500	25.000	16.000	88.500	-0.047
1362	R14	64.100	55	1		92.000	28.000	26.000	83.000	-1.457
1482	R30	50.100	65	2	59.000	80.000	27.000	22.000	81.500	-0.332
1484	R30	50.100	62	1	67.000	87.500	26.000	24.000	84.000	-1.296
1841	R14	64.100	66	1	42.000	78.000	26.300	19.200	62.000	*
3521	FUE	56.700	49	1	83.000	104.000	31.000	26.700	66.500	-1.085
3532	FUE	56.700	49	1	75.000	97.500	32.000	25.700	64.000	-0.992

Symbol meanings :

Δ v : Vehicle variation speed
POI : Weight
CTH : Thoracic circumference
LTH : Thoracic width
Sex : 1 Male - 2 Female

ETH : Thoracic thickness
HEP : Height of the bust
APBO : Anthropometrical parameter for belted occupant

(symbol POI in Table 1), the other data used in the analysis that follows were the following :

- the depth, width and circumference of the thorax measured at the mesosternal level (in Table 1, ETH is the symbol used for the thoracic depth, LTH for the thoracic width and CTH for the thoracic circumference) ;
- the height of the torso, i.e. the distance between the ischium and the acromion when the subject is lying horizontally on its back with its arms along the body and its thigh forming a 90° angle with the horizontal (symbol HEP in Table 1).

This last datum was used because it partly determines the position of the seat belt with respect to the thorax.

After each test, the cadavers were X-rayed and autopsied. Rib fragments were collected for the purpose of characterizing cadaver bone conditions.

PHYSICAL MEASUREMENTS - Depending on the test, one to four triaxial acceleration transducers were attached to the first, fourth, seventh and twelfth dorsal vertebrae. The procedure used for the acceleration measurements conformed to standard SAE J 211b pertaining to measurements of thoracic accelerations.

DESCRIPTION OF PARAMETERS USED IN THE STATISTICAL ANALYSIS

The predictive functions established are multiple linear correlations. They are intended to correlate the severity of thoracic lesions, i.e. AIS (3) and the number of rib fractures, with three groups of parameters which would appear to have an influence on injury data :

- cadaver anthropometry,
- thorax impact violence, and
- bone condition.

The BCF (1) is used to characterize the bone condition of the test cadavers. Cadaver BCF are indicated in Table 2. Its equation is as follows :

$$BCF = -\Big[0.117\,\frac{MMY-31}{6.7} + 0.128\,\frac{LMY-0.24}{0.07} + 0.141\,\frac{MFS-199}{118} + 0.100\,\frac{EFS-614}{426} + 0.126\,\frac{CFS-8.6}{5.5} + 0.121\,\frac{MCS-709}{443} + 0.134\,\frac{WCS-2.2}{1.33} + 0.131\,\frac{PFS-158}{93}\Big]$$

(MMY) : rate of mineralization
(LMY) : ash mass/unit of length (mean of the four fragments)
(MFS) : maximum bending force
(MCS) : maximum shearing force
(WCS) : shearing energy
(PFS) : slope of force/deformation curve in bending test
(CFS) : maximum bending stress
(EFS) : young's module

Anthropometry and impact violence to the thorax are represented by parameters described below.

DESCRIPTION OF THORACIC LESIONS - It should here be noted that this paper is exclusively concerned with 3-point belt wearing occupants involved in frontal collisions. They are indicated in Table 2. In this sample, the test cadavers sustained only rib fractures which were sometimes associated with a fracture of the sternum and/or the clavicle. Injuries of the same kind have also been observed by other teams specialized in human cadaver experiments (4 to 10).

Moreover, in our sample, in the absence of flail chest (i.e. AIS $\leqslant$ 4), we have not observed any intra-thoracic visceral lesions (for example, haemothorax - AIS = 3). These results are confirmed by data from real world accidents which show that for 3-point seat belt wearers, the probability of occurrence of such internal injuries, except when associated with flail chest, is very low.

Table 3 shows an overview of a sample of 1057 thoracic lesions sustained by 3-point seat belt wearers (11). These data are broken down by ranges of velocity and by age group. It can be seen that for velocity $\leqslant$ 65 km/h and occupant age $\leqslant$ 60 years, five cases of internal thoracic injury associated with rib fractures were observed, one of which was a flail chest.

In addition, it can be seen in this table that the probability of flail chest increases for V $>$ 65 km/h and for occupants over 60 years. These accident data are confirmed by other studies (12 to 16).

Since the predictive functions are used in the range of severity excluding extremely severe injuries (AIS = 4), the number of rib fractures will be a good indicator of thoracic injuries.

An advantage, with respect to AIS in general and thoracic AIS in particular, is that the number of rib fractures provides a more continuous scale. Consequently, we will attach greater attention to predictive functions using the number of rib fractures.

Remarks -

- A sternum was counted as one additional fracture. Clavicle fractures were not counted as supplementary fractures because the loading in this area results in further rib fractures.
- Generally, cadaver flail chest was considered to exist when at least 4 adjacent ribs had double fractures. This was defined by orthopaedic surgeons and pathologists.

DEFINITION OF A SINGLE PARAMETER REPRESENTATIVE OF THE ANTHROPOMETRY OF TEST CADAVERS Analysis is facilited by using a limited number of parameters. Accordingly, since we did not know, a priori, which anthropometrical data (or which combination thereof) would enable accurate characterization of test cadavers seated in car passenger compartments, we endeavoured, in an initial analysis, to determine a parameter that

Table 2 : Cadaver Thoracic Injuries - Cadaver Bone Characterization

	THORACIC INJURIES					BONE CHARACTERIZATION
Cadaver Ref.	TNF	STE	CLA	NRF	AIS	BCF
33	2	0	0	2	1	- 0.899
41	1	1	0	2	1	- 0.012
44	12	1	0	13	2	- 0.054
47	13	1	1	14	4	0.413
53	3	0	0	3	1	0.084
54	0	0	0	0	2	- 0.474
185	17	1	0	18	4	- 1.041
190	6	0	0	6	2	- 0.394
223	15	0	1	15	4	0.248
224	11	0	0	11	2	0.147
246	11	0	0	11	2	0.140
247	8	1	0	9	2	- 0.368
248	10	0	0	10	2	0.344
254	15	1	0	16	4	- 0.039
255	13	1	2	14	4	0.216
257	14	1	1	15	4	- 0.381
258	9	1	0	10	2	- 0.678
267	12	1	0	13	2	- 0.110
268	9	1	0	10	2	- 0.880
276	20	0	1	20	4	0.392
277	10	0	1	10	2	- 0.192
286	0	0	0	0	0	- 2.940
1152	5	1	0	6	2	0.172
1174	9	1	0	10	2	- 0.696
1242	17	1	0	18	4	0.421
1261	17	1	0	18	4	- 0.602
1352	9	0	0	9	2	- 0.206
1362	15	0	0	15	4	0.607
1482	8	0	0	8	2	0.406
1484	12	0	1	12	4	0.304
1841	20	1	1	21	4	0.843
3521	0	0	0	0	0	- 2.004
3532	12	1	0	13	2	- 0.559

Symbol meanings :

TNF : Total number of rib fractures
STE : Sternum fracture
CLA : Clavicle fracture
NRF : TNF + STE
AIS : Overall thoracic injury severity

Table 3 : Overview of Thoracic Injuries of 3-Point Belt-Wearers Involved in Real World Frontal Collisions, by Occupant Age and Impact Speed

Age Group (Number involved)	< 45 km/h			46 to 55 km/h			56 to 65 km/h			> 65 km/h		
	AIS 0/1 No RF	AIS 1/2 RF	AIS > 2	AIS 0/1 No RF	AIS 1/2 RF	AIS > 2	AIS 0/1 No RF	AIS 1/2 RF	AIS > 2	AIS 0/1 No RF	AIS 1/2 RF	AIS > 2
< 41 years (738)	608	10	1 (H+RF)	65	5	-	36	-	-	9	3	1 (LC)
41 to 50 years (147)	115	12	-	11	4	1 (FL)	2	1	1 (FL+H)	2	-	2 (FL)
51 to 60 years (119)	82	19	1 (H+RF)	7	4	-	2	-	1 (H+RF)	-	-	3
61 to 70 years (31)	21	6	1	1	-	-	1	-	-	-	-	1 (FL)
> 70 years (22)	15	5	-	-	-	2	-	-	-	-	-	-

Age Group (Number involved)	All speed			Percentage of injuries		
	AIS 0/1	AIS 1/2	AIS > 2	With RF	AIS > 2	F.L.
< 41 years (738)	718	18	2 (0 FL)	2,7	0,3	0
41 to 50 years (147)	126	17	4 (3 FL)	14,3	2,7	2
51 to 60 years (119)	91	23	5 (3 FL)	23,5	4,2	2,5
61 to 70 years (31)	23	6	2 (2 FL)	25,8	6,5	6,5
> 70 years (22)	15	5	2 (2 FL)	31,8	9,1	9,1

Symbols meanings :

H : Haemothorax
RF : Rib fracture
No RF : Number of occupants without rib fractures
FL : Flail chest
LC : Lung contusion

would summarize the six dimensional data on the human cadavers (set forth in Table 1), by means of factor analysis (17) of the correspondences (this is known as analysis by normed principal components).

The analysis finally showed that the four parameters used in the following function were sufficient to represent the anthropometry of the cadavers seated in the vehicle ; this function is termed A.P.B.O. (Anthropometrical Parameter for Belted Occupant).

The APBO values for cadavers used for the determination of the predictive functions of thoracic injuries are shown in Table 1.

Remark :

- For the 5th percentile female test dummy APBO = 1.7
- For the 50th percentile male test dummy APBO = 0.37
- For the 95th percentile male test dummy APBO = 1.89.

$$APBO = \frac{-0.75}{\sqrt{2.417}} \frac{HEP-66.83}{4.88\sqrt{4}} + \frac{-0.856}{\sqrt{2.417}} \frac{CTH-89.46}{8.08\sqrt{4}} + \frac{-0.732}{\sqrt{2.417}} \frac{ETH-20.58}{2.37\sqrt{4}} + \frac{0.766}{\sqrt{2.417}} \frac{TAI-1.67}{0.07\sqrt{4}}$$

PARAMETER REPRESENTATIVE OF THORAX IMPACT VIOLENCE - The diversity of the test conditions (vehicle velocity and deceleration, belt types, geometry of belt anchorage points, seat position occupied, etc...) made it necessary to define, as indicators of impact violence, physical parameters linked to the subject itself and independant of the vehicle environment.

For that purpose, from the available measurements (on T1, T4, T7 and T12) the following ten parameters were calculated.

- V : Integral of acceleration over time (m/s)
- M : Maximum acceleration (g)
- M3 : Level of acceleration exceeded during 3 ms (g)
- S : Severity index : integral of power 2.5 acceleration over time
- P1090 : Average power for overall impact $P1090 = 0.5x(80\%\ V)^2/T1090$ (watts/kg)
- P1050 : Average power during the severe period of the restraint $P1050 = 0.5x(40\%\ V)^2/T1050$ (watts/kg)
- PM : Peak power : maximum of the function (acceleration x instantaneous velocity) (watts/kg)
- B10 : Napierian log of V over 10 ms during which acceleration is greatest
- B20 : Napierian log of V over 20 ms during which acceleration is greatest
- J : Average jerk = $\frac{M\,3}{T1050}$ (g/s)

The calculated parameters are brought together in Table 4.

Nota bene

- B10, B20 and powers have sometimes been reported in the literature (18, 19) to describe thoracic injury tolerance, mainly in side impacts.

It was thought interesting to verify the relevance of these parameters in frontal collisions with belted occupants.

- Powers were expressed by unit of mass since this effective mass of the thorax, during impact, is unknown.
- PM is different of the product : (maximum acceleration x maximum velocity).
- P1090 is calculated from 10 % of V to 90 % of V (that is to say 80 % of V).
- T1090 is the duration between 10 % of V and 90 % of V (in seconds).
- P1050 is calculated from 10 % of V to 50 % of V.
- T1050 is the duration between 10 % of V and 50 % of V (in seconds).

RESULTS OF STATISTICAL ANALYSIS

CORRELATION BETWEEN THE SEVERITY OF THORACIC INJURIES AND CALCULATED PHYSICAL PARAMETERS - Tables 5 and 6 indicate the coefficient of correlation which, in our sample, links thoracic injury (AIS and number of rib fractures) to the 10 parameters used to represent impact violence to the thorax of test cadavers.

The results are classified according to the vertebrae on which acceleration was measured. The number of cadavers investigated, for each of the vertebrae on which acceleration was measured, is also indicated in Tables 5 and 6. Figures 1a and 3a illustrate the best results obtained with the AIS (with P1050 on T4). The results obtained at T4 with the acceleration peak exceeding 3 ms (M3) are shown in Figures 2a and 4a. This commonly used physical parameter is employed in the following chapters.

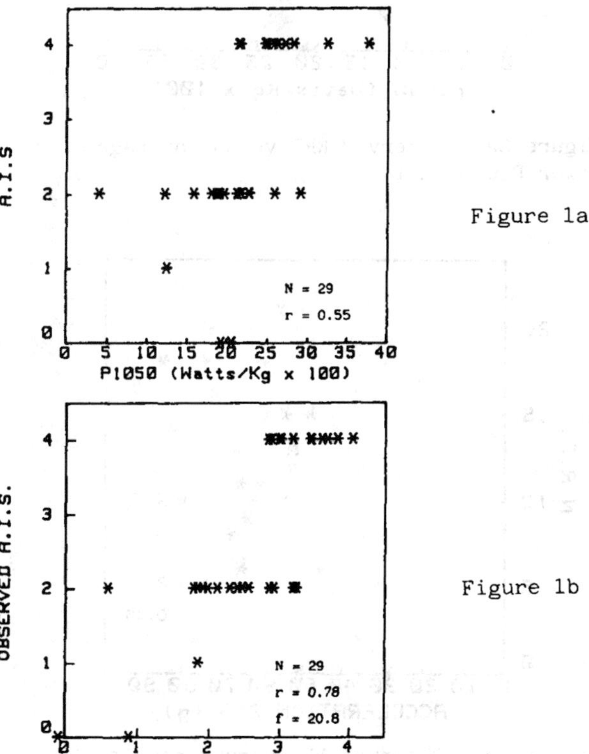

Figure 1a

Figure 1b

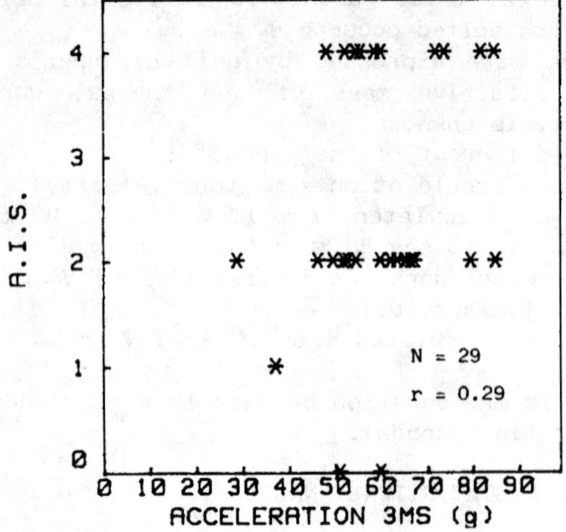

Figure 2a - Observed AIS versus resultant acceleration M3 on T4

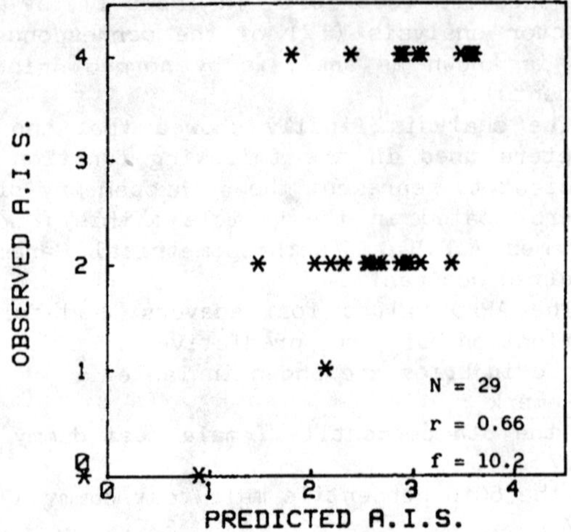

Figure 2b - Observed AIS versus predicted AIS using equation (6) on T4

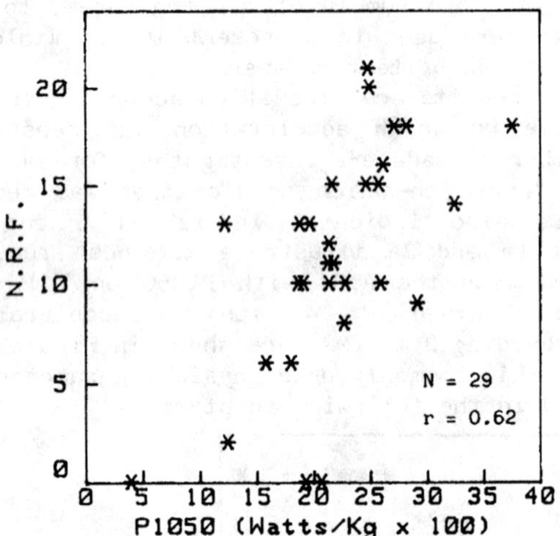

Figure 3a - Observed NRF versus average power P1050 on T4

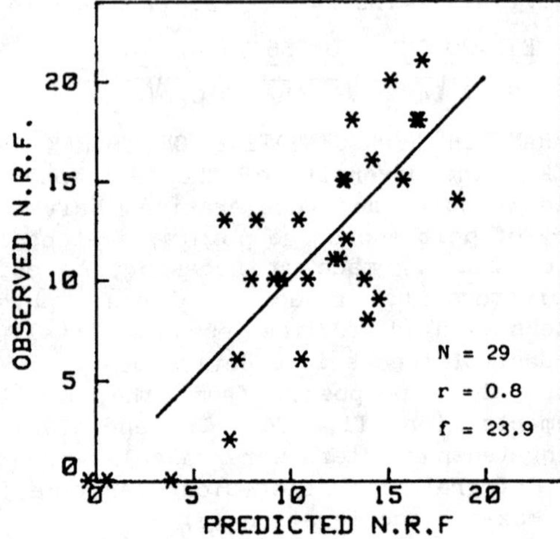

Figure 3b - Observed NRF versus predicted NRF using equation (1) on T4

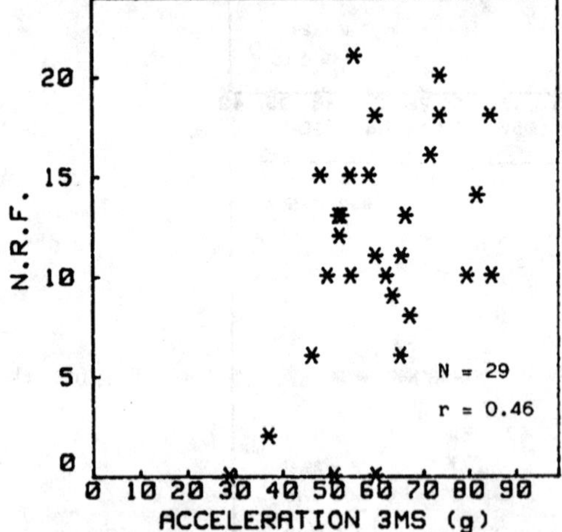

Figure 4a - Observed NRF versus resultant acceleration M3 on T4

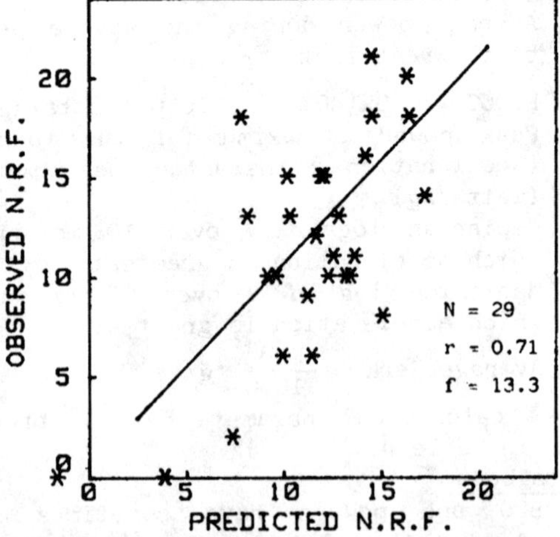

Figure 4b - Observed NRF versus predicted NRF using equation (5) on T4

Table 4 : Measurements Results - Calculated Physical Parameters

Location of the Measurement Points

Cadav. Ref.	T1				T4				T7				T12			
	P 1050	B 20	M 3	S	P 1050	B 20	M 3	S	P 1050	B 20	M 3	S	P 1050	B 20	M 3	S
33	*	*	*	*	*	*	*	*	*	*	*	*	*	*	*	*
41	*	*	*	*	1253.000	1.848	37.270	312.770	*	*	*	*	*	*	*	*
44	*	*	*	*	1228.000	2.153	52.900	494.520	*	*	*	*	*	*	*	*
47	*	*	*	*	*	*	*	*	*	*	*	*	*	*	*	*
53	*	*	*	*	*	*	*	*	*	*	*	*	*	*	*	*
54	*	*	*	*	404.000	1.584	28.940	89.010	*	*	*	*	*	*	*	*
185	2717.000	2.422	62.340	834.690	3763.000	2.456	60.100	993.750	3331.000	2.405	71.300	853.430	2754.000	2.395	72.610	830.150
190	1810.000	2.170	55.350	444.500	1586.000	2.264	65.340	574.330	1913.000	2.172	50.450	499.480	1105.000	1.829	38.450	214.150
223	2050.000	2.206	50.500	524.630	2165.000	2.259	54.690	571.080	*	*	*	*	1745.000	2.219	52.830	466.730
224	1382.000	2.028	46.880	331.920	2184.000	2.193	60.010	468.530	2267.000	2.299	64.170	517.590	2696.000	2.435	80.090	725.630
246	1941.000	2.209	51.360	491.700	2132.000	2.342	65.600	640.740	2478.000	2.457	67.900	755.530	2110.000	2.395	66.860	647.450
247	2507.000	2.250	59.020	610.410	2915.000	2.369	63.690	679.470	2639.000	2.418	70.770	707.400	2067.000	2.259	54.290	476.870
248	1700.000	2.001	45.900	327.120	2284.000	2.231	54.770	493.450	2102.000	2.243	52.360	503.520	1763.000	2.201	48.520	444.450
254	2135.000	2.298	62.100	535.320	2617.000	2.449	71.890	760.610	2822.000	2.577	82.120	957.040	3748.000	2.676	83.070	1140.130
255	2412.000	2.385	63.400	653.140	3254.000	2.628	82.000	1072.480	3325.000	2.646	93.730	1169.490	3745.000	2.597	82.410	1117.040
257	818.000	1.921	47.600	219.550	2576.000	2.358	58.740	759.930	3025.000	2.416	71.170	807.640	2786.000	2.456	86.030	821.790
258	1293.000	2.331	93.300	922.840	1918.000	2.362	79.750	735.250	2164.000	2.378	62.740	674.010	2760.000	2.355	60.620	701.210
267	1312.000	2.336	89.700	856.490	1974.000	2.297	66.530	623.440	2381.000	2.336	54.620	628.560	2196.000	2.365	63.240	626.160
268	1931.000	2.650	131.730	1649.880	2596.000	2.535	85.080	1169.440	2848.000	2.398	64.490	831.380	2760.000	2.349	63.210	720.920
276	2720.000	2.350	73.400	779.000	2496.000	2.274	73.890	618.860	3223.000	2.432	80.050	752.770	2664.000	2.354	62.060	600.320
277	1704.000	2.276	69.910	640.140	1871.000	2.179	49.770	557.470	2297.000	2.252	62.620	598.700	2759.000	2.343	62.910	658.980
286	888.000	1.854	39.700	208.060	1936.000	2.120	51.170	465.910	2867.000	2.360	61.290	778.790	2553.000	2.295	57.200	638.240
1152	*	*	*	*	1807.000	2.090	46.410	396.390	*	*	*	*	*	*	*	*
1174	*	*	*	*	2143.000	2.277	62.290	598.160	*	*	*	*	*	*	*	*
1242	*	*	*	*	2714.000	2.506	73.870	1012.680	*	*	*	*	*	*	*	*
1261	*	*	*	*	2820.000	2.613	84.850	1231.920	*	*	*	*	*	*	*	*
1352	*	*	*	*	*	*	*	*	2042.000	2.381	65.170	682.540	*	*	*	*
1362	*	*	*	*	2465.000	2.234	48.350	707.310	*	*	*	*	*	*	*	*
1482	*	*	*	*	2275.000	2.341	67.400	650.310	*	*	*	*	*	*	*	*
1484	*	*	*	*	2142.000	2.191	52.380	552.690	*	*	*	*	*	*	*	*
1841	*	*	*	*	2483.000	2.256	55.670	525.780	3361.000	2.424	63.890	793.060	*	*	*	*
3521	*	*	*	*	2073.000	2.272	60.020	597.220	3651.000	2.498	79.200	983.840	3005.000	2.410	71.080	759.040
3532	*	*	*	*	1881.000	2.207	52.060	523.870	1632.000	2.124	49.200	401.790	2218.000	2.298	62.470	588.550

Symbol meanings : See chapter concerning the parameters representative of thorax impact violence

Table 5 : Coefficients of Correlation Between Each Physical Parameter and Thoracic AIS

PHYSICAL PARAMETERS	T1 (16 cadavers)	T4 (29 cadavers)	T7 (19 cadavers)	T12 (18 cadavers)
	Location of the measurement points on spinal column			
V	0.065	0.305	- 0.022	0.108
M 3	- 0.028	0.297	0.422	0.366
M	0.084	0.296	0.467	0.305
S	0.064	0.473	0.301	0.357
P 1050	0.524	**0.550**	0.303	0.247
P 1090	0.189	0.480	- 0.062	0.157
P M	- 0.091	0.232	0.125	0.400
B 10	0.205	0.395	0.367	0.302
B 20	0.315	0.440	0.362	0.316
J	0.485	0.373	0.430	0.244

Table 6 : Coefficients of Correlation Between Each Physical Parameter and the Number of Rib Fractures (NRF)

PHYSICAL PARAMETERS	T1 (16 cadavers)	T4 (29 cadavers)	T7 (19 cadavers)	T12 (18 cadavers)
	Location of the measurement points on spinal column			
V	0.240	0.495	- 0.159	0.030
M 3	0.128	0.461	0.228	0.344
M	0.273	0.424	0.260	0.284
S	0.213	0.545	0.098	0.295
P 1050	0.542	**0.628**	0.200	0.198
P 1090	0.338	0.548	- 0.185	0.110
P M	- 0.019	0.312	- 0.056	0.298
B 10	0.332	0.579	0.177	0.310
B 20	0.423	0.600	0.338	0.346
J	0.492	0.490	0.338	0.274

Because of the low coefficients of correlation obtained, we decided to seek a new parameter, by means of combinations of the preceding ones, so as to obtain a better explanation of the injuries observed. To this end, we did factor analyses of the correspondences expressed in normed components for the four measurement points with the parameters defined above. These analyses showed that whatever the measurement point considered, the following parameters were the most relevant : P1050, M3, S and B20, and that a linear combination of these parameters did not make it possible to define a new parameter better correlated with the injuries, than each of these parameters taken separately.

DETERMINATION OF THORACIC INJURY PREDICTIVE FUNCTIONS - The general form of the predictive functions will be stated as follows :

AIS or NRF = K1 APBO + K2 BCF + K3 PP + C

in which :

NRF = number of rib fractures

APBO = anthropometrical parameter previously calculated for each cadaver

BCF = index for bone characterization of test cadavers

PP = physical parameter considered

and C, K1, K2 and K3 are constants calculated automatically by the statistical analysis program.

<u>Analysis of the Respective Influence of the Three Groups of Parameters used in the Predictive Functions for Thoracic Injury</u> - The purpose of this analysis was to appraise the respective effects of the three groups of parameters taken into account (i.e. APBO, BCF and PP shown above) in order to obtain the prediction of thoracic injuries, so as to reduce if possible the number of these parameters.

To this end, Student's test, which expresses the level of statistical significance between each parameter and the predicted value, was used. It then appeared that, whatever the function considered above, the APBO parameter (anthropometrical parameter) had a negligible influence on the prediction of thoracic injuries. It was therefore deleted from the preceding equations, which could then be stated as follows :

NRF = K1 BCF + K2 P1050 + K3 (1)
AIS = K'1 BCF + K'2 P1050 + K'3 (2)
NRF = K4 BCF + K5 S + K6 (3)
AIS = K'4 BCF + K'5 S + K'6 (4)
NRF = K7 BCF + K8 M3 + K9 (5)
AIS = K'7 BCF + K'8 M3 + K'9 (6)
NRF = K10 BCF + K11 B20 + K12 (7)
AIS = K'10 BCF + K'11 B20 + K'12 (8).

Tables 7 and 9 list the multiple correlation coefficients, <u>r</u>, and the Fischer coefficients, <u>f</u>, linking chest injuries (AIS and NRF) to bone condition (BCF) and to the various parameters selected to represent impact violence to the thorax.

In these tables :

P is the number of cadavers on which the relevant data were available,

r is the multiple correlation coefficient,

f is the Fischer coefficient.

The predictive functions are significant to 1/1000 :

when the Fischer coefficient, f, is $\geq$ 8.93 for 28 degrees of freedom (29 cadavers),

when the Fischer coefficient, f, is $\geq$ 10.39 for 18 degrees of freedom (19 cadavers),

when the Fischer coefficient, f, is $\geq$ 10.66 for 17 degrees of freedom (18 cadavers).

We also calculated, for each of the functions established, the confidence interval at 95 % :

$I = m \pm 1.96\,\sigma$; m is the mean value for the predicted number of rib fractures and σ is the residual variance. The I values are listed on Tables 7 and 9.

These data are classified according to the various measuring points.

<u>Analysis of Results Obtained with the AIS</u>

The predictive functions are expressed by the following equations :

$$AIS = 99.859 \times 10^{-5} \times P1050 + 89.128 \times 10^{-2} \times BCF + 60.476 \times 10^{-2} \quad (2)$$

$$AIS = 246.22 \times 10^{-5} \times S + 97.141 \times 10^{-2} \times BCF + 122.41 \times 10^{-2} \quad (4)$$

$$AIS = 2765.7 \times 10^{-5} \times M3 + 94.352 \times 10^{-2} \times BCF + 113.63 \times 10^{-2} \quad (6)$$

$$AIS = 249.27 \times 10^{-3} \times B20 + 91.771 \times 10^{-2} \times BCF - 284.98 \times 10^{-2} \quad (8).$$

The coefficients of correlation and Fischer coefficient f values are statistically significant, especially at T4 and T7.

Moreover, the use of Student's test (see Table 8) in each of the predictive functions established confirms that the influence of the physical parameters obtained at T4, particularly P1050, is the greatest.

Figure 1b illustrates the relation between the AIS observed on the test cadavers and the AIS predicted by function (2) using the BCF and P1050 which gives the best statistical results. The confidence interval at 95 % is + 1.5.

In spite of this, it can be seen in Figures 1a and 2a that, for average power (P1050), in the range of 2100 and 2900 watts/kg and for acceleration between 45 g and 85 g, injuries may be moderate (AIS = 2) or very severe (AIS = 4). It follows that the use of AIS raises problems due to its discontinuity.

Figure 2b illustrates the relation between the NRF observed on the test cadavers and the NRF predicted by function (6) using the BCF and M3.

<u>Analysis of Results Obtained with the Number of Rib Fractures</u> - The predictive functions are expressed by the following equations :

$$NRF = (53.901 \times 10^{-3})\ P1050 + (3.7347)\ BCF + 21.053 \times 10^{-2} \quad (1)$$

$$NRF = (13.088 \times 10^{-3})\ S + (4.1648)\ BCF + 3.6844 \quad (3)$$
$$NRF = (199.48 \times 10^{-3})\ M3 + (4.0195)\ BCF + 24.622 \times 10^{-3} \quad (5)$$
$$NRF = (16.077)\ B20 + (3.851)\ BCF - 24.399 \quad (7).$$

Although velocities obtained from spinal accelerometers are subject to directionnal as well as integration errors, the predictive functions have high statistical significance particularly at T4, T7 and T12 (see Table 9).

Whatever the measuring point considered, P1050 is the physical parameter which yields the best results.

The highest statistical significance values were obtained at T4, particularly with parameter P1050 ($r = 0.8$ and $f = 23.9$).

Moreover, the use of Student's test (Table 10) in each of the predictive functions established (Student's t allows appraisal of the influence of each parameter on the prediction) confirms the highest influence of the physical parameters obtained at T4, particularly P1050 which is the only physical parameter with a higher influence on the prediction than the BCF.

Figures 3a and 3b illustrate the correlation obtained with P1050.

COMMENTS - Analysis of the results in Table 9, also showed that as the measuring points become more distant, in a vertical plane, from the center of gravity of the thorax (as T1 and T12 in comparison with T7 and T4), the physical parameters have a lower influence on the prediction than BCF. The effects of the neck on T1 and of the pelvis on T12 may be observed.

Moreover, it may be noted that the confidence intervals obtained at 95 % are rather high, even at T4 (equation (1), $I = \pm 6.9$). This may be attributed to several factors :

- the relative position of the seat-belt strap with respect to the thorax of the cadavers ; this position varies depending on the position of seat-belt anchorage points and the height of the cadaver's bust,
- the movement of the strap on the thorax during impact,
- the non-linearity of rib fractures beyond a certain loading on the chest, with the appearance of posterior arch fractures and/or local thorax "collapse" (these injuries appear only at AIS = 4, i.e. flail chests).

The less good results using the acceleration peak exceeded during 3 ms (M3) can be partly explained by the fact that this parameter did not take into account either the energy absorbed by the thorax or the impact duration. Inversely, P1050 does take into account these two variables and does so during the most violent impact phase. Other parameters such as the speed of application of the impact forces to the thorax and the surface area receiving these forces represent other important variables needed to define thoracic tolerance. (These data have already been discussed in (32).) For belt wearers, these two last variables are approximately constant.

This is an additionnal reason explaining why P1050 is the best descriptor of overall sollicitation of the thorax when restrained by 3-point belt.

The set of results obtained in this study obviously calls for confirmation through a larger number of tests carried out with human cadavers of which the bone characterization is known ; however, on the basis of the tests carried out, it may be stated that :

- the predictive functions established have a very high statistical significance,
- among the physical parameters used, P1050 provides the best description of impact violence to the thorax, particularly at T4,
- characterization of cadaver bone condition (BCF) is necessary in interpreting thoracic injuries.

EXTRAPOLATION OF PREDICTIVE FUNCTIONS TO HIGHWAY ACCIDENT VICTIMS

The extrapolation of these predictive functions to living persons requires knowledge of their bone condition (BCF) and a prior check to ensure that the variables used in calculating test cadaver BCF are equally pertinent to bone condition among the living population exposed to the risk of traffic accidents.

To check this point, ribs were removed from 44 individuals (20) following their sudden death (victims of traffic accidents, suicides, etc...), prior to any degradation of their bone condition ; the ribs were tested as described in the reference publication (1) (mineralization and mechanical tests).

The variables used to calculate test cadaver BCF were obtained for the ribs of these 44 human subjects.

A factorial analysis of the correspondences of principal normed components was performed on the variables and showed that, on the factorial plane representative of overall rib resistance (see figure 9 in (1)), these variables had almost the same projections as those representing the overall rib resistance of the test cadavers. The BCF of each of the 44 human subjects representative of living road users was thus computed using the equation defined in (1).

The average age of these 44 subjects was 42 years (standard deviation = 21 years), and their average BCF was -1.2 (standard deviation : BCF = 1) :

minimum BCF = -3.45 for a subject 31 years old,
maximum BCF = 1.16 for a subject 77 years old.

Remark - It may be observed in the Table 2 that certain cadavers had a BCF located in the range of the 44 subjects representative of living road users.

PREDICTIVE FUNCTIONS OF THORACIC INJURY TO LIVING PERSONS - We have previously seen that several functions could be used to predict thoracic injuries. In what follows, we shall use only the function which provided the statistically most significant results, i.e. equation (1) with the average power during the severe period of the restraint (P1050). Since this

Table 7 : Relation Between the Observed Cadaver AIS and the Predicted Cadaver AIS

PREDICTIVE FUNCTIONS USED	Location of measuring point on dorsal vertebrae															
	T1				T4				T7				T12			
	P	r	f	I 95 %	P	r	f	I 95 %	P	r	f	I 95 %	P	r	f	I 95 %
(2)	16	0.64	4.6	± 1.9	29	0.78	20.8	± 1.5	19	0.78	12.8	± 1.6	18	0.72	8.3	± 1.8
(4)	16	0.55	2.9	± 2.1	29	0.77	19.2	± 1.6	19	0.77	12.4	± 1.6	18	0.75	9.6	± 1.7
(6)	16	0.55	2.8	± 2.1	29	0.66	10.2	± 1.8	19	0.76	11.4	± 1.6	18	0.72	8.4	± 1.9
(8)	16	0.58	3.4	± 2	29	0.72	14.4	± 1.7	19	0.74	9.8	± 1.7	18	0.70	7.4	± 1.9

Table 8 : Student t Relating Each Parameter to the Predicted AIS

Predictive Functions Used	Parameters	Location of the measurement point on the vertebrae			
		T1	T4	T7	T12
(2)	Γ 1050	1.5993	4.2352	2.7350	1.8164
	BCF	1.7969	4.6061	4.6734	3.8378
(4)	S	0.28126	3.9861	2.6445	2.2080
	BCF	2.3977	4.9078	4.5926	3.8713
(6)	M 3	- 0.1008	2.0393	2.4303	1.8673
	BCF	2.3901	4.0476	3.9926	3.5629
(8)	B 20	0.88374	3.1108	2.0378	1.5035
	BCF	2.2060	4.2791	3.8831	3.4460

Table 9 : Relation Between the Observed Cadaver Number of Rib Fractures and the Predicted Number of Rib Fractures

PREDICTIVE FUNCTIONS USED	Location of measuring point on dorsal vertebrae															
	T1				T4				T7				T12			
	P	r	f	I 95 %	P	r	f	I 95 %	P	r	f	I 95 %	P	r	f	I 95 %
(1)	16	0.70	6.3	± 7.1	29	**0.80**	**23.9**	**± 6.9**	19	0.79	14	± 7	18	0.76	10.1	± 7.2
(3)	16	0.65	4.9	± 7.5	29	0.78	20.6	± 7.3	19	0.75	10.8	± 7.5	18	0.77	11.1	± 7
(5)	16	0.63	4.4	± 7.3	29	0.71	13.3	± 8.2	19	0.75	10.3	± 7.6	18	0.76	10.8	± 7
(7)	16	0.68	5.8	± 7.3	29	0.79	22.1	± 7.1	19	0.74	10.1	± 7.7	18	0.76	10.7	± 7.1

Table 10 : Student t Relating Each Parameter to the Predicted Number of Rib Fractures

Predictive Functions Used	Parameters	Location of measuring point on spinal column			
		T1	T4	T7	T12
(1)	P 1050	1.661	**5.127**	2.199	1.665
	BCF	2.282	4.329	5.131	4.356
(3)	S	1.026	4.640	1.343	1.921
	BCF	2.991	4.608	4.617	4.355
(5)	M 3	0.630	3.363	1.164	1.844
	BCF	2.926	3.943	4.338	4.157
(7)	B 20	1.457	4.869	1.050	1.815
	BCF	2.705	4.358	4.333	4.130

parameter is largely unfamiliar and hardly ever used, we shall also present the results obtained with the maximum level of acceleration exceeded during 3 ms (M3) which is the most often used with test dummies to appraise thoracic protection, through equation (5).

Figures 5 and 6 illustrate the relation between the predicted number of rib fractures and the parameters P1050 and M3, depending on the BCF.

These figures provide representative results from cadaver experiments. Average BCF, characterizing the resistance of the thoracic cage of the population exposed to risk of accident, is BCF = -1.5.

This value is the average BCF of subsample of 44 subjects described above.

This choice was imposed by the low coefficient of correlation between age and BCF in the sample of 44 subjects (r = 0.52) which was inadequate to allow determination of BCF by age group for the population exposed to accident risk.

This does not mean that there is no relationship between BCF and the age of living occupants but that the available sample (44 subjects) is not large enough to establish such a relationship.

The sub-sample consists of 34 subjects of age ≤ 60 years :

average age = 33 years (standard deviation = 13),
average BCF = -1.5 (standard deviation = 0.85),
the minimum age of this population was 5 years,
the maximum age of this population was 60 years,
the minimum BCF for this population was BCF = -3.45 (age 31 years),
the maximum BCF for this population was BCF = -0.2 (age 43 years).

These values should be confirmed when a large enough number of BCF values characterizing the population exposed to the risk of traffic accidents is available.

Figures 5 and 6 made it possible to associate a value for the physical parameter under consideration to a given number of rib fractures. For instance, if it is considered that 8 rib fractures (without flail chest) is a reference injury level, it follows that the values for P1050 and M3 must not exceed 2500 watts/kg and 70 g, respectively. These values represent the most probable human chest tolerance of living belted persons if the maximum admissible injury severity is set at 8 rib fractures without flail chest (AIS = 2), from the analysis of present data.

COMMENTS

Concerning the choice of 8 rib fractures as an acceptable thoracic injury level - This value was determined by a team of orthopaedic surgeons and clinical pathologists, on the basis of their knowledge both of thoracic injuries observed on belt wearers and of potential associated functional problems. It should be pointed out here that even in the absence of flail chest, there may be more than 8 rib fractures, and that their location is of great importance.

Concerning the characterization of actual traffic accident victims - The establishment of an average BCF representative of the real world population exposed to risk must be based upon statistics. This has not been possible in the foregoing text, due to the inadequatly small number of available BCF data for the population involved in frontal collisions.

In addition, the following data relating to the age distribution of this population must also be borne in mind (21) :

90 % of those involved were between 10 and 60 years old,
80 % of those involved were between 10 and 50 years old,
63 % of those involved were between 10 and 40 years old.

Since average thoracic resistance increases when the age bracket is narrowed (its lower limit however being fixed at ten years), we may reasonnably assume that the average BCF = -1.5 used to established the tolerance threshold of the 10 to 60 age group, represents a very conservative value enabling an adequat level of thoracic protection for 90 % of living 3-point seat belt wearers in frontal collisions.

Furthermore, one is led to believe that, for the same BCF value, the thoracic resistance of the living population is greater than that of cadavers (especially because of their lack of muscular tonicity). One may therefore look upon the -1.5 BCF value as being particularly conservative.

DISCUSSION AND COMPARISON WITH DATA FROM THE LITERATURE

THORACIC INJURIES - If one wishes to establish human tolerances from experiments involving surrogates, especially human cadavers, it is first of all necessary to ensure that the injuries observed on the cadavers are of the same kind as those seen on real world accident victims who have been subjected to the same level of violence and who were rising the same type of protective devices.

For 3-point belt wearers, without a direct violent impact against the thoracic cage (for example, due to major intrusion of the passenger compartment), accident data show that the probability of injury of the intra-thoracic viscera is very low (11 to 16).

Table 3, from reference (11), shows this result : out of 984 belted occupants ≤ 60 years old (approximately 90 % of those involved), impact violence being ≤ 65 km/h, 60 occupants sustained thoracic injuries, 2 had a flail chest (AIS = 4) and 3 a haemothorax (AIS = 3) with rib fractures, without a flail chest.

We can conclude that injuries to the intra thoracic viscera consecutive to direct impact described by certain authors (22, 23, 24) are not representative of injuries sustained by 3-point belt wearers.

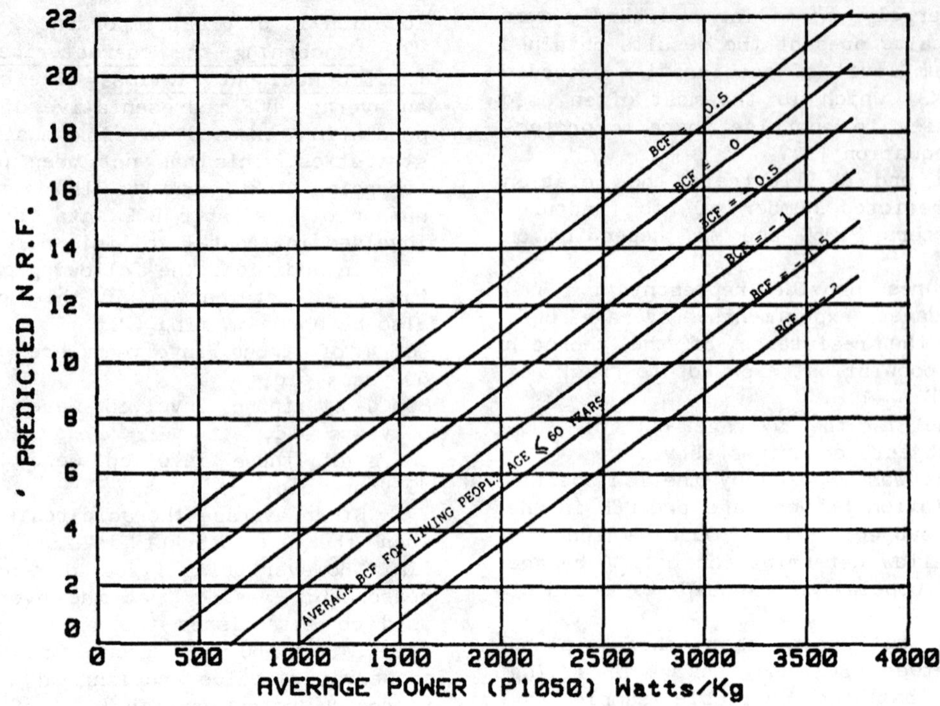

Figure 5 - Predicted number of rib fractures as a function of average power (P1050) measured on T4 and bone characterization (BCF)

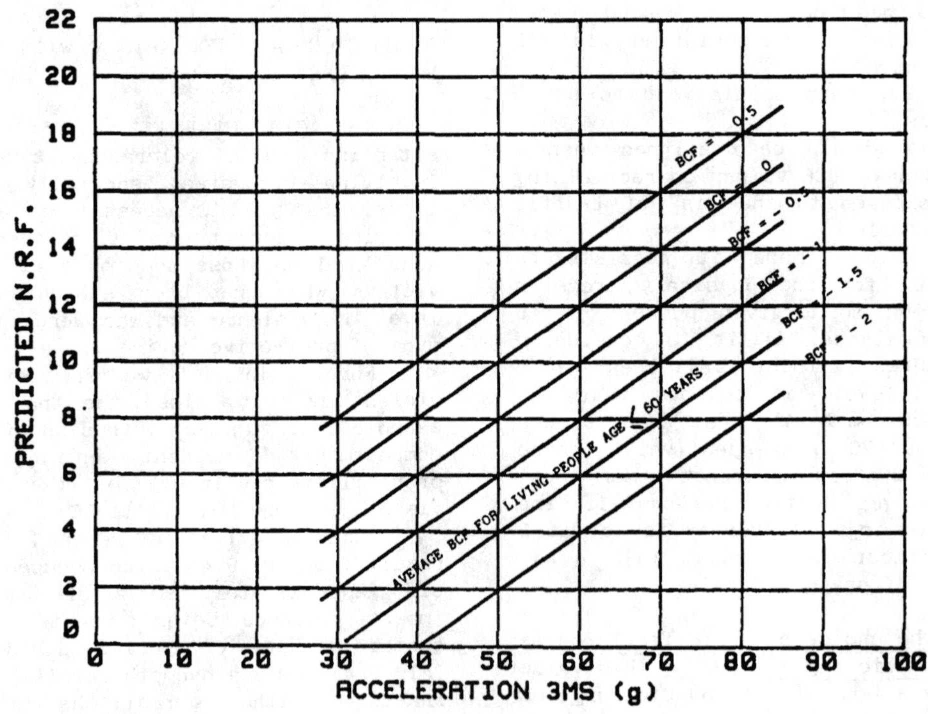

Figure 6 - Predicted number of rib fractures as a function of resultant acceleration 3 ms (M3) measured on T4 and bone characterization (BCF)

In consequence the number of rib and sternum fractures is a good indicator of thoracic injury severity for 3-point belt wearers.

This is especially true if injuries AIS ≤ 2 only are taken into account.

In addition, the similarity of injuries to accident victims and cadavers wearing 3-point belt being established, we can justifiably use cadaver experiment results obtained to determine thoracic tolerance for the living population, with the BCF method described above.

HUMAN TOLERANCE - The literature is full of data making it possible to evaluate the tolerance of the human thorax. Several physical parameters having this aim are used : deflexion measurements, forces and accelerations.

Deflexion Data (Antero-posterior) - Most data used relative deflexion (or compression) of the thorax from tests performed with impactors or pendulums striking the thorax at the sternum level (25, 26, 27).

Several overviews have been published (22, 23, 25). The most recent shows that, at high pendulum impact speed, the relative compression of the thorax is not sufficient to enable thoracic tolerance to be established (22).

This is so because of the visco-elastic characteristics of the intra-thoracic viscera. As we have however shown above, the type of impact is not representative of the way in which belt wearers' thorax is sollicited. The contact surface, the rate of speed increase of the sternum, impact duration, the energy dissipated in the thorax, etc... are amongst the parameters differentiating the consequences for belt wearers and those receiving a direct impact to the sternum.

Note - Abdominal injuries (liver, spleen, mesentery, etc...) are sometimes seen both on cadavers and accident victims wearing 3-point belts. These injuries are often the results of submarining and of forces applied by the lap belt and not by the torso webbing.

The data derived from this latter type of experiments can be used for unrestrained occupants but may on no account be transposed to 3-point belt wearers.

Nevertheless, it might have been thought that antero-posterior deflexion was a good indicator of thoracic injury severity amongst belt wearers especially if our interest was focused upon injuries below the threshold of tolerance (AIS < 4).

Few such data, unfortunately, are available because of the difficulties encountered in measuring deflexion.

In reference (2), based on 4 tests, it seems that a relative deflexion of about 30 % is near the level of human tolerance.

These results have shown the dominant effect of the bone condition of cadavers used. As an example, cadaver n° 225, tested at 50 km/h had ribs in poor condition and sustained a flail chest ; on the other hand, cadaver n° 258, tested at 65 km/h had rib condition average for the living population and received thoracic injuries at only AIS = 2.

Restraint Forces Data - Accident data have shown the influence of age and therefore of the characteristic of thoracic cage (28, 29) upon thoracic injury occurrence by using the breaking force of shock absorbing webbing located at the upper anchorage point as an indicator of forces exerted.

This force has often been used in cadaver experiments (2 to 10 and 30).

A synthesis of this was presented at the sixth E.S.V. conference (31) in which the force had been normalized, as if all the subjects had weighed 75 kg. A relationship was found connecting the severity of the thoracic injuries and the normalized force measured at the shoulder. Unfortunately the bone condition of the cadavers was unknown which made it impossible to transpose the results to the living population.

Moreover, for a given force applied perpendicularly to the thorax, the force measured in the strap can vary to an extremely wide extent depending on the positions of the thoracic strap anchorage points. One of the reasons accounting for the satisfactory correlation between injuries and normalized force measured above the shoulder was the large proportion of tests performed in the same test configuration (seat and position of belt anchorage points) (5, 6).

A more recent publication (2) used the resulting normalized force (as defined in (31)) applied perpendicularly to the thorax, obtained through dimensional film analysis of the positions of the thorax in relation to the belt. It was then possible to associate the value of 8.50 daN for 8 rib fractures of a human cadaver weighing 75 kg and having a "good" skeleton.

Although relevant to the establishment of thoracic tolerance, resultant force is difficult to evaluate because the relative position of strap and thorax is often very imprecise when the occupant is in a vehicle.

Another disadvantage is that this parameter cannot easily be transposed into a protection criterion measured on a dummy.

Acceleration Data on the Thorax Periphery - It had often been shown that the resultant acceleration peak (maximum or 3 ms) measured on a vertebra could not correctly describe the severity of thoracic injuries.

More recently, the so-called 12 accelerometers method developped for side impacts (18) has been used with restrained cadavers involved in frontal collision experiments (5).

We show below, in Table 11, the main results of this study.

9 tests were performed with cadavers within the age range between 19 and 39 years of the population at risk.

If we consider the resultant acceleration at T12 (which, it will be recalled, enables better prediction of thoracic injury than at T1 (see Table 9)) and the number of fractures observed on cadavers, we see that these results are distributed along the curve representing the

Table 11 : Kallieris data (5)

Run N°	Sex	Age	Weight (kg)	Number of Rib Fractures	Thoracic AIS	Resultant Acceleration (3 ms)(g) T1	T12
H 7915	M	22	58	2	2	38.2	-
H 7916	F	39	53	5	2	41.8	48.6
H 7917	M	18	77	0	0	44.6	41.2
H 7918	M	23	56	0	0	62.4*	41.6
H 8001	M	38	70	5	2	40.4	50.8
H 8002	F	32	61	3	2	42.6	54
H 8005	M	25	80	5	2	27.8	52.8
H 8006	F	34	48	4	2	36	38.2
H 8008	M	19	66	2	2	30.4	38.6
Average Value		28	63	3		38	45.7

* This value is considered doubtful by the author

population at risk (Fig. 7 coming from Fig. 6, with BCF = -1.5).

This result is very interesting although derived from T12. It confirms the accuracy of the predictive functions which were formerly based upon experiments performed with fairly old cadavers but whose bone condition compared with that of living people was known.

PROTECTION CRITERIA - The predictive functions established above have shown that, for a given value of the physical parameter, the number of rib fractures depends practically solely upon the cadaver bone condition (BCF). This means that, on the basis of the level reached by the physical parameter under consideration, we can predict the number of rib fractures which a living occupant of average bone resistance (BCF = -1.5) would have sustained in the same impact conditions.

In order to obtain the protection criterion corresponding to this injury prediction, we need to perform a dummy test which involves identical violence.

A sufficient number of dummy tests in cadaver test conditions will make it possible to establish the relationship between the prediction of injuries which living occupant would have sustained and the corresponding protection criteria observed on the dummy.

These data are not currently available. Nevertheless, for the time being, it is possible to use certain data already published.

Kallieris - He shows (5) of 5 tests performed with the Part 572 dummy in the same impact conditions as those of involving the 9 cadaver tests presented in Table 11.

The average acceleration peak value exceeded 3 ms is 37 g on the Part 572 dummy. We can see that the average values of the relationship

$$\frac{\text{cadaver peak acceleration (3 ms)}}{\text{dummy peak acceleration (3 ms)}}$$

are, respectively, 1 and 1.24 when acceleration at T1 and T12 are used.

If we hypothesise that the relationships are relatively constant whatever the impact conditions (which need to be demonstrated), we can transpose the level of physical parameters obtained with cadavers into data calculated on the Part 572 dummy.

Moreover, if we accept that the T4 measurements lie within those obtained between T1 and T12, we can predict the number of fractures which a living occupant with average bone condition (BCF = -1.5) would have sustained as a function of the value of the physical parameter calculated for the dummy on the basis of the relationship established above.

Figure 8 illustrates the results thus obtained.

If we accept that 8 fractures on a living occupant of average bone resistance represents the acceptable limit of thoracic injury severity, the resultant acceleration peak (3 ms) on a Part 572 dummy should be found to be within the range 56 g and 70 g.

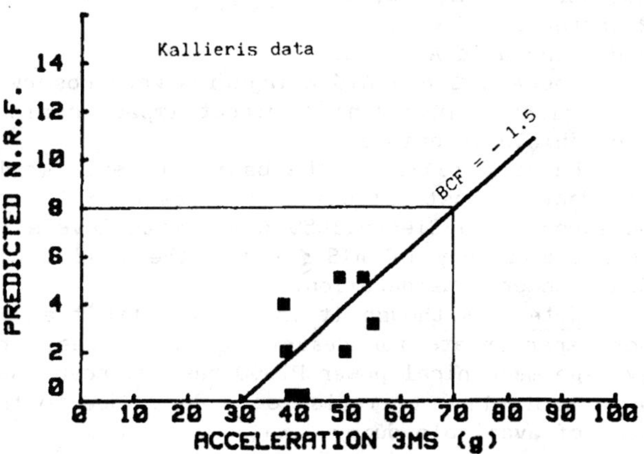

Figure 7 - Comparison between Kallieris data and predicted number of rib fractures for population at risk (BCF = - 1.5)

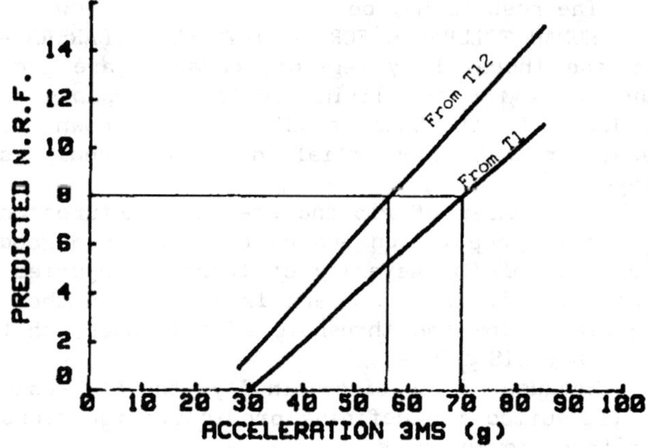

Figure 8 - Corridor of predicted number of rib fractures for population at risk (BCF = - 1.5) as a function of "Part 572" dummy acceleration

Other Bibliographical Data - Real-world accidents simulated by Saul (33) confirm the results obtained. The author even associates with an acceleration value (3 ms) of 38 g on the Part 572 dummy :
26 thoracic AIS = 0,
2 thoracic AIS = 1,
and 2 thoracic AIS = 2.

The AIS 1 and AIS 2 injuries were observed in accidents involving a direct impact against the vehicle interiors.

Patrick (34), on the basis of real- world accident reconstructions associates 60 g (3 ms) measured on a Sierra 1050 dummy with levels of thoracic injury of AIS $\leq$ 2 for the road accidents under consideration.

Note - Although it is a parameter that is more appropriate for describing tolerances, the average mechanical power P1050 has not been used as a protection criterion on a dummy due to the lack of available dummy data.

SUMMARY AND CONCLUSIONS

The set of results obtained requires confirmation through a larger number of experiments on dummies and on cadavers in the same test conditions and a better knowledge of the bone condition of the population exposed to the risk of traffic accidents. The results have their greatest validity for conditions approaching the experimental test conditions, i.e. impact velocities from 44 to 67 km/h and vehicle deceleration distances from 500 to 900 mm.

The results may be summarized as follows.

HUMAN TOLERANCE FOR 3-POINT BELT WEARERS - For the thorax body segment, cadavers are good surrogates for the living vehicle occupants on condition that bone condition is known (a parameter that is essential in the data analysis stage).

The number of rib and sternum fractures on the living population and on cadavers is a good indicator of the severity of thoracic injuries, especially if one's concern is limited to those injuries below the threshold of tolerance, that is to say AIS $\leq$ 2.

Because of its discontinuity, the AIS scale is ill-suited for defining predictive functions for thoracic injuries.

Anthropometrical variables have very little influence on injury.

Several physical parameters calculated from the deceleration pulse measured on these vertebrae may be used.

The most statistically significant predictive functions were obtained with T4 and T7 (mainly on T4).

The value for average power during the severe period of the restraint (P1050) yields the best results. If the value of 8 rib fractures without flail chest or internal lesions (very rare for this injury level) is considered to be the reference limit, the thoracic tolerance of living belted persons of age $<$ 60 years corresponds to the following values :
average power during the severe period of the restraint (P1050) = 2500 watts/kg (from T4),
maximum level of acceleration exceeded during 3 ms (M3) = 70 g (from T4).

PROTECTION CRITERIA ON THE PART 572 DUMMY - A value in the range 56 g and 70 g measured on a Part 572 dummy seems to correspond to an acceptable level of injury for the restrained population exposed to risk of accident.

These data confirm the relevance of the "60 g (3 ms)" criterion currently used for certain rule-making.

REFERENCES

(1) J. Sacreste, F. Brun-Cassan, A. Fayon, C. Tarriere, C. Got, A. Patel "Proposal for a thorax tolerance level in side impacts based on tests performed with cadavers having known bone condition" in Proceedings of 26th Stapp Conference, Ann Arbor - Michigan, October 1982.

(2) G. Walfisch, F. Chamouard, D. Lestrelin, C. Tarriere, C. Got, F. Guillon, A. Patel "Tolerance limits and mechanical characteristics of the human thorax in frontal and side impact and transposition of these characteristics into protection criteria" in Proceedings of IRCOBI Conference, Cologne - Germany, 1982.

(3) "The abbreviated injury scale", 1980 Revision, American Association for Automotive Medicine, Morton Grove - Illinois 60053 USA.

(4) G. Schmidt, D. Kallieris, J. Barz, R. Mattern "Results of 49 tests simulating frontal collisions of front seat passengers" in Proceedings of 18th Stapp Car Crash Conference, Ann Arbor - USA, 1974.

(5) D. Kallieris, H. Mellander, G. Schmidt, J. Barz, R. Mattern "Comparison between frontal impact tests with cadavers and dummies in a simulated true car restraint environment" in Proceedings of 26th Stapp Conference, Ann Arbor - Michigan - USA.

(6) G. Schmidt, Kallieris and Al. "Neck and thorax tolerance level of belt protected occupants" in Proceedings of 19th Stapp Car Crash Conference, San Diego - USA, 1975.

(7) L.M. PATRICK, R.S. LEVINE "Injury to unembalmed belted cadavers in simulated collisions" in Proceedings of 19th Stapp Car Crash Conference, San Diego - USA, 1975.

(8) J.R. Cromack, H.M. Ziperman "Three point belt induced injuries : a comparison between laboratory surrogates and real world accident victims" in Proceedings of 19th Stapp Conference, San Diego - USA, 1975.

(9) "Quantification of thoracic response and injury", Monthly Process Reports of Contract, N° DOT HS 4-00921, National Highway Traffic Safety Administration (N.H.T.S.A.), Department of Transportation, Washington DC - USA.

(10) "Sled tests of 3-point systems including air belt restraints, final report" Contract DOT-HS-5-01017, National Highway Traffic Accident Safety Administration, Washington DC - USA, 1976.

(11) J.Y. Foret-Bruno, G. Faverjon, F. Hartemann, C. Tarriere, C. Got, A. Patel "Description des blessures thoraciques observées sur un échantillon de 1057 porteurs de ceintures 3 points impliqués dans des collisions frontales", Unpublished data of PEUGEOT SA/RENAULT accidentological investigations, Laboratory of Biomechanics, April 1985.

(12) Nils Bohlin "A statistical analysis of 28000 accident cases with emphasis on occupant restraint value", Paper 670925, 11th Stapp Car Crash Conference, 1967.

(13) N. Bohlin, H. Norin, A. Anderson "A statistical traffic accident analysis" AB Volvo Car Division, IV E.S.V. Conference, Kyoto, March 1973.

(14) W.D. Nelson "Lap-shoulder restraint effectiveness in the United States", Paper 71-0077, SAE Automotive Engineering Congress, Detroit - USA, January 1971.

(15) W.D. Nelson "Restraint system effectiveness" 15th American Association of Automotive Conference, October 1971.

(16) G. Grime "Accidents and injuries to car occupants wearing safety belts" in Automotive Engineer, July 1968.

(17) J.P. Benzecri et Coll. "L'analyse des données, Tome 1 : la taxinomie ; Tome 2 : l'analyse factorielle des correspondances", Dunod Editeur.

(18) D.M. Robbins, R.J. Lehman, HSRI University of Michigan "Prediction of thoracic injuries as a function of occupant kinematics" in Proceedings of 7th E.S.V. Conference, Paris - France, 1979.

(19) R. Morgan, Hal P. Waters, NHTSA U.S. Department of Transportation "Comparison of two promising side impact dummies", 8th International Technical Conference on Experimental Safety Vehicles, Wolfsburg, October 1980.

(20) F. Chamouard, G. Walfisch, C. Tarriere, C. Got, A. Patel "Bone characterization of living persons exposed to actual traffic accidents", Unpublished data of PEUGEOT SA/RENAULT Laboratory of Biomechanics, April 1985.

(21) "Statistical analysis of french actual accidents", SETRA - FRANCE, 1984.

(22) David. C. Viano and Ian V. Lau, Biomedical Science Department, General Motors Research Laboratories, Warren - Mi 48090-9058 "Thoracic impact : A viscous tolerance criterion", 10th E.S.V. Conference, Oxford - England, June 1985.

(23) David. C. Viano, Ph. D., Biomedical Science Department, General Motors Research Laboratories, Warren - Michigan 48090 USA, "Thoracic injury potential" in Proceedings of IIIrd IRCOBI Conference, Lyon - France, Sept. 1978.

(24) J.P. Verriest and A. Chapon, ONSER "Validity of thoracic injury criteria based on the number of rib fractures", 10th ESV Conference, Oxford - England, June 1985.

(25) R.F. Neathery, C.K. Kroell, H.J. Mertz "Prediction of thoracic injury from dummy responses" in Proceedings of 19th Stapp Car Crash Conference SAE, Warrendale PA, 1975.

(26) C.K. Kroell, D.C. Schneider, A.M. Nahum "Impact tolerance and response of the human thorax II" in Proceedings of 18th Stapp Car Crash Conference 74 1187, SAE, Warrendale PA, 1974.

(27) A.M. Nahum, D.C. Schneider, C.K. Kroell "Cadaver skeletal response to blunt trauma" in Proceedings of 19th Stapp Car Crash Conference 75 1150, SAE, Warrendale PA, 1975.

(28) J.Y. Foret-Bruno, F. Hartemann, Ch. Thomas, A. Fayon, C. Tarriere, C. Got, A. Patel "Correlation between thoracic lesions and force values measured at the shoulder of 92 belted occupants involved in real accidents" in Proceedings of 22nd Stapp Car Crash Conference, Ann Arbor - Michigan - USA, 1978.

(29) J. Sacreste, G. Walfisch, A. Fayon, Ch. Thomas, C. Tarriere "Tolérance thoracique des porteurs de ceintures de sécurité" in Journal de Traumatologie, T1, n° 2, Masson, Paris, 1980.

(30) A. Fayon, C. Tarriere, G. Walfisch, C. Got, A. Patel "Thorax of 3-point belt wearers during a crash - Experiments with cadavers" in Proceedings of 19th Stapp Conference, San Diego - California, November 1975.

(31) R.M. Eppinger "Prediction of thoracic injuries using measurable experimental cadavers" in Proceedings of 6th E.S.V. Conference, Washington - USA, October 1976.

(32) Rolf. H. Eppinger, Jeffrey H. Marcus, US Department of Transportation, National Highway Traffic Safety Administration, Washington DC 20590 "Production of injury in blunt frontal impact" in Proceedings of 10th E.S.V. Conference, Oxford - England, July 1-5 1985.

(33) Roger A. Saul, Lisa K. Sullivan, Vehicle Research and Test Center, Jeffrey H. Marcus, Richard M. Morgan "Comparison of current anthropomorphic test devices in a three-point belt restraint system" in Proceedings of 27th Stapp Car Crash Conference, San Diego - California - USA, October 1983.

(34) L.M. Patrick, Wayne State University, A. Anderson, AB Volvo "Three-point harness accident and laboratory data comparison" in Proceedings of 18th Stapp Car Crash Conference, Ann Arbor - Michigan - USA, December 1974.

791005

Response of Belt Restrained Subjects in Simulated Lateral Impact

John D. Horsch,
Dennis C. Schneider, and
Charles K. Kroell
Biomedical Science Department
GM Research Laboratories
Warren, MI

Frank D. Raasch
Department of Pathology
University of California
San Diego, CA

Abstract

Far-side lateral impacts were simulated using a Part 572 dummy and human cadavers to compare responses for several belt restraint configurations. Sled tests were conducted having a velocity change of 35 km/hr at a 10 g deceleration level. It was estimated from field data that a 35 km/hr velocity change of the laterally struck vehicle represents about an 80th percentile level for injury-producing lateral collisions.

Subjects restrained by a three-point belt system with an outboard anchored diagonal shoulder belt (i.e., positioned over the shoulder opposite the side of impact) rotated out of the shoulder belt and onto the seat. The subject received some lateral restraint due to interaction with the shoulder belt and seatback. The subjects restrained by a three-point belt system with an inboard anchored diagonal shoulder belt (i.e., positioned over the shoulder on the side of impact) remained essentially upright due to shoulder belt interaction with the neck and/or head. Kinematic responses of the Part 572 dummy were generally similar to those of the cadaver subjects.

Injuries were found in cadavers restrained by both shoulder belt configurations, but were more extensive to the cervical region for those subjects receiving direct neck and/or head loading from the belt. However, limitations in the cadaver model and test environment, as well as the preliminary nature of the experiments do not permit definitive conclusions on the significance or applicability of the injury data obtained to real accident situations at this time.

OVERVIEW

Field studies have found that belt restraint systems are effective in reducing injury in lateral collisions [1-5], in part due to reduction of occupant ejection. For an occupant on the impacted side (near side) of the vehicle, there is little space for lateral displacement before contact with the side (interior) of the vehicle. However, for occupants more removed from the impact side of the vehicle, such as a far side or center occupant, there is in general appreciable lateral space available for displacement of the body. In this case, a belt restraint system might be effective in resisting lateral displacement.

This study is a preliminary investigation of belt restraint performance in lateral impacts where significant space for lateral displacement is available. Either an anthropomorphic test dummy or a human cadaver* (both embalmed and unembalmed were used) was positioned as a front seat occupant and was subjected to simulated lateral far side impact on a decelerator type crash simulator at the General Motors Research Laboratories (dummy only) or the University of California at San Diego (dummy and cadaver). These subjects were restrained by a three-point belt configuration, the diagonal upper strap having either an outboard or inboard anchor, or by a lap belt only (see Figure 1).

The purpose of these experiments was to investigate the mechanics of lateral impact restraint by these different belt configurations, (i.e., magnitude of

*The experimental use of these subjects complied with guidelines established by the U.S. Public Health Service and recommended by the National Academy of Sciences/ National Research Council.

restraining forces, body regions loaded, and occupant kinematics). In the laboratory experiments, a simple open test fixture was mounted to the sled structure. It consisted of an automotive bench seat and a flat footrest with a simulated tunnel so that only a limited number of structures other than the belt restraint were available to interact with a single occupant surrogate during the deceleration exposure.

According to reported field studies [1-9] lateral collisions represent a wide range of impact conditions. However, this preliminary investigation was restricted to a single impact angle (90° from frontal) and impact velocity (35 km/hr). The lateral impact simulated in this study is representative of a severe impact where a restraint system might provide significant occupant protection. Investigation of other impact conditions is necessary to apply the results of this study to the general spectrum of side impact field accidents.

Unembalmed and embalmed human cadavers were tested to provide a range of pretest musculoskeletal condition and impact response. The embalmed subjects were not "limbered" prior to testing except for that necessary to allow positioning in the fixture. Necropsies were preformed on the cadaver subjects.

Lateral impact experiments, conducted at lower velocity changes, have been reported with human volunteer subjects restrained by four-point parallel torso and lap straps by Zaborowski [18] and Ewing et al [19 and 20]. The well controlled and instrumented exposures reported by Ewing et al provided additional lateral restraint to the torso by a lightly padded board. Although these studies involve substantially different experimental methodologies than the present study, precluding a direct comparison of the two, they do provide voluntary response data on the tonic human neck subjected to sub-injury lateral inertial loading by the head.

METHODOLOGY

TEST ENVIRONMENT - The test fixture (Figure 1) consisted of an automotive front bench seat, fixed anchor supports for the belts, a flat floor pan with a rectangular-sectioned simulated tunnel, and a lateral support on the side of the seat opposite the impact to maintain the surrogate's initial position prior to sled deceleration. These components were mounted on the sled such that the acceleration vector was lateral with respect to the seat, fixture and occupant. The "open" fixture provided good visibility for high speed photography permitting three dimensional analysis of the belt interaction with the occupant. A bench seat for a full sized (four door) 1978 passenger car was used to minimize potential lateral restraint by the seat (as compared to a contoured bucket seat) and to provide a somewhat continuous lateral surface.

The subject was initially positioned on the far side of the seat (opposite the impacted side of the sled), so that the impact tended to move the subject towards the impact side of the seat (see Figure 1). The subjects were positioned in a conventional seated attitude, and were placed against the lateral acceleration support which prevented the head, thorax and thigh from moving prior to sled deceleration.

A three-point belt configuration having a diagonal shoulder strap was used for all tests except experiments 1137 and 1151 for which only a lap belt was used. The three point configurations were formed by a separate belt segment from each of the two floor anchors and a shoulder belt anchor, the three segments being attached together at a point laterally in line with the subject's hip joint on the side opposite from the shoulder belt anchor. All tests used the same floor anchor locations, which were chosen to be compatible with the seat. However, various shoulder belt anchor locations were used. These locations consisted of (see Figure 1):

Outboard (Opposite impact side)--
- Simulated typical in-vehicle location for the seat.

Inboard (Impact side)--
- Seat top height ("low") location (not attached to the seat);
- Roof height ("high") location.

Polyester automotive belt webbing was used, being attached to the anchorage hardware such that slip or "payout" of webbing did not occur. The belts were subjectively adjusted "snug" such that they were not under significant tension nor did they have free slack. Belt tension tranducers were located at or near the belt anchors for all tests.

IMPACT CONDITIONS - Tests were conducted at nominal sled conditions of: a velocity change of 35 km/hr; a constant 10 g deceleration level and a deceleration pulse duration of 110 ms. Sled deceleration response is shown in Figures 3 to 12 for typical tests.

A velocity change of 35 km/hr was chosen to simulate a severe lateral impact and is compared to field data reported by Hartemann et al. [1] in Figure 2. Their reported estimate of the velocity change of the laterally

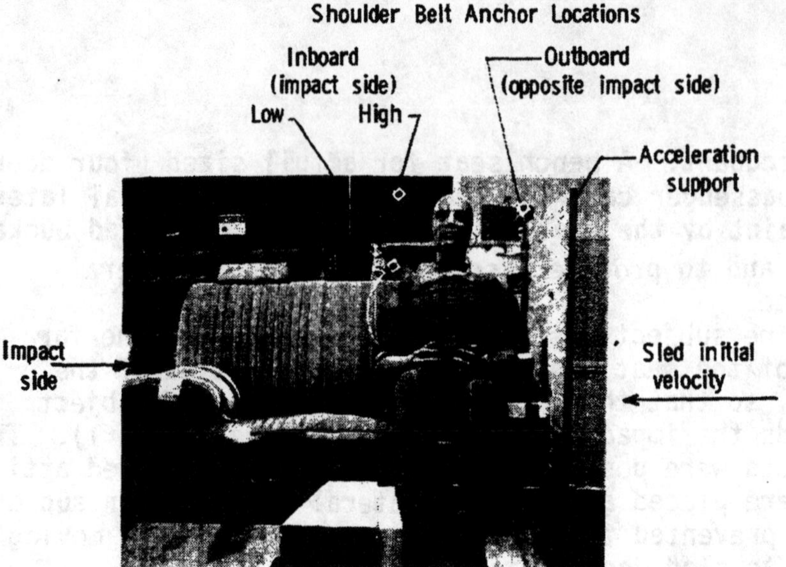

Fig. 1 - Test fixture mounted on the GMR crash simulator showing the bench seat, flat floor with tunnel and lateral acceleration support. The Part 572 dummy subject is restrained by a lap belt. Approximate shoulder belt anchor locations are indicated

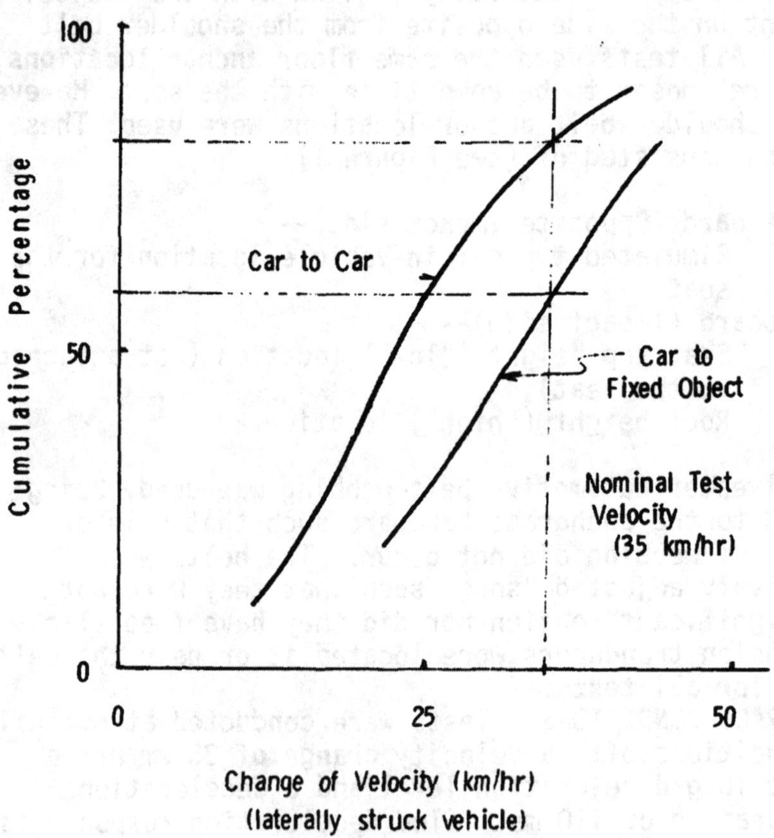

Fig. 2 - Comparison of the 35 Km/h test velocity to data reported by Hartemann et al. [1]

struck vehicle for 296 lateral impacts involving an injured occupant is plotted in Figure 2 as a cumulative percentage. A velocity change of 35 km/hr (for the laterally struck vehicle) represents about an 83 percentile severity level compared to their car to car data and about a 59 percentile severity level compared to their car to fixed object data (about 20% of the sample).

SUBJECT PREPARATION - The Part 572 dummy was adjusted with a nominal "1 g" adjustment of the knee, hip, shoulder, and elbow joints. The arm and leg twist joints were tight. Typical head and thorax triaxial accelerometers were installed to monitor response dynamics.

All tests using cadaver subjects were conducted at the University of California at San Diego Medical School. These cadaver subjects were stored at 3°C until approximately six hours prior to the test. Unembalmed subjects were allowed to pass through the characteristic stages of rigor mortis, and therefore an essentially flaccid state was realized for all tests. The pre-test preparation for the two embalmed subjects required manipulation of the extremity joints to permit maintenance of the desired seating attitude. However, the head/neck region was not flexed, and the tissue stiffness induced by the preservative chemicals was sufficient to maintain an upright head position. Care was exercised in pre-test handling of the subjects to prevent abnormal loads or displacements on individual body segments. Table 1 contains descriptive information for the cadaver subjects.

POST TEST NECROPSY - The post test necropsy procedure for the cadaver subjects included dissection and examination of the anterior thoracic wall, contents of the thorax and abdomen, cervical spine, thoracic and lumbar spine, and pelvic structure, though all areas were not examined for every test specimen (see Table 1). Traditional postmortem examination techniques were employed in the thoracico-abdominal and pelvic dissections. Examination of the cervical spine followed portions of the procedure recommended by Noguchi et al [21]. A superior-inferior incision was made along the mid-line of the neck and the overlying tissues reflected to expose the anterior neck organs. After inspection, the structures were reflected and the anterior portion of the vertebral column was examined. The subject was then placed in a prone position and a similar approach was utilized to expose the posterior aspect of the cervical spine. The neural arches were removed and the underlying dura examined. Opening of the dura permitted access to the spinal cord. For some subjects (see Table 1), an additional procedure was used. The cervical spine was removed *en bloc* from C-1 to C-7 and sawn in the midsagittal plane to permit viewing of the cross sections of the intervertebral discs. The dissection procedure employed in inspecting the thoracic and lumbar vertebral column was similar to that described above.

Control cadaver subjects were necropsied as part of the procedure. These control subjects received the same pre-test protocol as exposed specimens except that they were not placed on the sled fixture. The investigating pathologist performed the necropsy without being aware that these specimens had not received an acceleration exposure.

RESULTS/DISCUSSION

Test results are listed in Tables 1 and 2. Tests are grouped into those having similar belt configurations. Belt configurations differed only in the shoulder belt anchor location (or lack of shoulder belt) and are specified according to this location. The approximate shoulder belt locations are indicated in Figure 1 and can be seen in sequence photos for selected tests in Figures 3 to 12. Tests with the high inboard shoulder belt anchor were not entirely similar due to variations of subject size, seated posture and the fore-aft location of the anchor. In addition to the sequence photos which show surrogate kinematics, the corresponding transducer responses are also shown in Figures 3 to 12.

Peak belt forces are listed in Table 2 for each of the belt segments. A peak resultant belt force acting on the upper body was estimated by film analysis of occupant-belt kinematics and measured shoulder belt tension. This estimate was made by measuring the directions of the shoulder belt from body contact (i.e., the two tangent points on the upper body) and computing a resultant force vector. However, assigning the measured upper shoulder belt tension to both belt segments introduces some uncertainty in this calculation, since the actual tension in the lower end of the shoulder belt was not determined.

Responses of head HIC index and 3 ms g level and thorax GSI and 3 ms g level, all based upon triaxial linear acceleration, are given in Table 2 for the Part 572 dummy when measured.

COMPARISON OF KINEMATICS FOR VARIOUS RESTRAINT CONFIGURATIONS - Subjects having an outboard anchored shoulder belt (the belted shoulder opposite the impact side), Figures 3 & 4, exhibited significantly different kinematics than did subjects having an inboard anchored shoulder belt (the belted shoulder on the impact side), Figures 6-12. Dummy and cadaver subjects restrained by

Table 1 DESCRIPTION OF CADAVER SUBJECTS AND SUMMARY OF NECROPSY FINDINGS

Shoulder Belt Anchor Location[1]		Test No. --- Subj.[2]	Age (Year) --- Weight (kg) --- Height (cm)	Cause of Death --- Comments	SUMMARY OF NECROPSY FINDINGS						
					Pelvis	Abdomen	Thorax	Thoracic, Lumbar Spines	Neck		
									Skeletal	Ligamentous	Cord/Other
INBOARD (IMPACT SIDE)	LOW	048 FM	65/76.8/183	Cardio-respiratory arrest	-[3]	-[3]	-[3]	-[3]	Separation disc C5-6 complete--- Separation disc C6-7 partial--- Fx. trans. process C5 rt.--- Fx. trans. process C6 bilateral [4]	Torn post. long. lig. 75% of width at C5-6	None
INBOARD (IMPACT SIDE)	LOW	051 FM	56/91.8/180	Septicemia	-[3]	None	None	-[3]	Complete separation disc C6-7 with dislocation [4]	Torn ant. long. lig. at C6--- Torn interspinal lig. at C6-7	6 cm tear dura--- Transection cord at C6---Severed rt. vertebral artery at C6
INBOARD (IMPACT SIDE)	LOW	052 EM	59/83.2/173	Cardiac arrythmia Embalmed	-[3]	-[3]	-[3]	-[3]	Separation disc C6-7 complete--- Separation disc C5-6 partial [4]	None	None
INBOARD (IMPACT SIDE)	LOW	056 FM	23/47.7/170	Gunshot --- Weight loss due to cystic fibrosis	None	None	None	-[3]	None	None	None
INBOARD (IMPACT SIDE)	HIGH	049 FM	67/64.1/183	Hepatic failure	Linear fracture left acetabulum	-[3]	Fractured sternum	-[3]	Separation of lower C6 at disc margin complete--- Fx. of lat. art. facets & arch C6--- Fx. arch & spinous process C5 [4]	Torn ant. long. lig. at C6-7--- Torn dorsal lig. at C6	Two tears dura at C6--- Marked local softening cord C6
INBOARD (IMPACT SIDE)	HIGH	053 EM	59/75.0/189	Adenocarcinoma of Lungs --- Embalmed	-[3]	-[3]	-[3]	-[3]	Linear fracture C2--- Separation disc. C6-7--- Fracture C6	Loosening interspinous lig. C1-C2---Stretched ant. long. lig. C6--- Torn interspinal lig C6--- Torn post. long. lig. C6	1.3 cm tear in dura
OUTBOARD (OPPOSITE IMPACT SIDE)		054 FM	74/81.8/185	Cardio-respiratory Arrest--- Ribs easily broken Sternocleidomastoid muscle absent, rt.	None	-[3]	Fractured ribs* 1,3,4 left	-[3]	Partial separation disc C6-7--- Disc cleft[5] C5-6	None	None
OUTBOARD (OPPOSITE IMPACT SIDE)		057 FM	67/64.6/173	Cerebral Anoxia Ilium bone thin CPR prior to death	Disp. Fx.[6] rt. ilium*	None	Fx. sternum* Fx. ribs: 4,5 r & l; 6r	-[3]	Disc clefts[5]: C5-6, C6-7	None	None
OUTBOARD (OPPOSITE IMPACT SIDE)		058 FM	59/68.2/183	Metastatic melanoma	Disp. Fx. rt. ilium	None	None	None	Disc cleft[5]: C6-7	None	None
OUTBOARD (OPPOSITE IMPACT SIDE)		059 FM	65/65.9/173	Myocardial infarct CPR prior to death	None	None	Fractured sternum*	Traverse Fx. T1	Disc clefts[5]: C3-4, C4-5, C5-6, C6-7	None	None
NOT EXPOSED		Control FF	65/53.6/157	Congestive heart failure	-[3]	None	None	-[3]	Disc cleft[5]: C5-6	None	None
NOT EXPOSED		Control FM	69/46.8/168	Carcinoma of Stomach	-[3]	None	None	-[3]	Disc clefts[5]: C3-4, C4-5, C5-6, C6-7	None	None

[1]See text, Figure 1
[2]FM - Unembalmed male
EM - Embalmed male
FF - Unembalmed female

[3]Region not examined for that subject.
[4]Midsagittal section of spine not done.

[5]Observed only after midsagittal sectioning of spine.
*See comment for that subject.

[6]Fx. - Fracture.

Table 2
SUMMARY OF TEST CONDITIONS AND RESPONSES

Shoulder Belt Anchor Location[1]		Subject[3]		Test No.[4]	Sled velocity change (km/hr)	Peak Belt Forces (kN)				Triax. Result. Accel. (g)			
						Inboard Lap	Outboard Lap	Shoulder	Est. Result.[6]	Head[7] HIC	Head[7] 3 ms	Thorax[9] GSI	Thorax[9] 3 ms
Inboard (on impact side)	Low[2]	572 dummy		1065[5]	37	2.75	8.65	4.25	5.6	172	31	33	38
				047	36	2.10	5.90	4.90	6.7	-	-	-	-
		Cad.	FM	048[5]	37	2.70	5.70	3.95	5.9	-	-	-	-
			FM	051	34	2.00	7.45	4.10	6.5	-	-	-	-
			FM	056[5]	34	1.30	3.55	2.70	3.8	-	-	-	-
			EM	052	33	2.80	6.45	5.25	7.8	-	-	-	-
	High[2]	572 dummy		1138[5]	35	3.35	11.10	5.00	5.7	137	27	122	27
				1150	35	5.20	10.20	5.00	5.7	171	31	164	30
	High[2]	572 dummy		1063[5]	37	3.90	11.00	4.05	3.9	195	36	397	47
				046	38	2.25	8.60	3.90	3.8	-	-	-	-
		Cad.	EM	053[5]	35	2.40	6.30	3.90	3.2	-	-	-	-
	High[2]	Cad.	FM	049[5]	35	1.85	6.40	4.15	4.5	-	-	-	-
Outboard (opposite impact side)		572 dummy		1135[5]	36	3.45	4.25	2.60	2.3	455[8]	42[8]	66	23
				1152	35	4.05	4.30	3.60	3.0	254	33	59	19
		Cad.	FM	054[5]	37	3.25	2.55	4.00	-	-	-	-	-
			FM	057	35	3.65	2.25	4.10	2.5	-	-	-	-
			FM	058	35	3.50	2.50	3.35	-	-	-	-	-
			FM	059	35	1.90	2.85	0.90	-	-	-	-	-
None (Lap belt only)		572 dummy		1137[5]	35	3.40	6.55	None	0	1419[8]	130[8]	112	28
				1151	35	3.30	7.45		0	1266	93	120	30

[1]See Text, Figure 1.
[2]See Text, Figure 13.
[3]EM - embalmed male.
FM - unembalmed male.

[4]Test No.<100 conducted on USCD sled.
No head or thorax accelerometers.
Test No>1000 conducted on GMR sled.
[5]Indicates sequence photos and transducer responses given in Figures 3 to 12.

[6]Estimated resultant force of the shoulder belt on the upper body.
[7]SAE J211a class 1000 frequency response
[8]Head/arm impact significant
[9]SAE J211a class 180 frequency response.

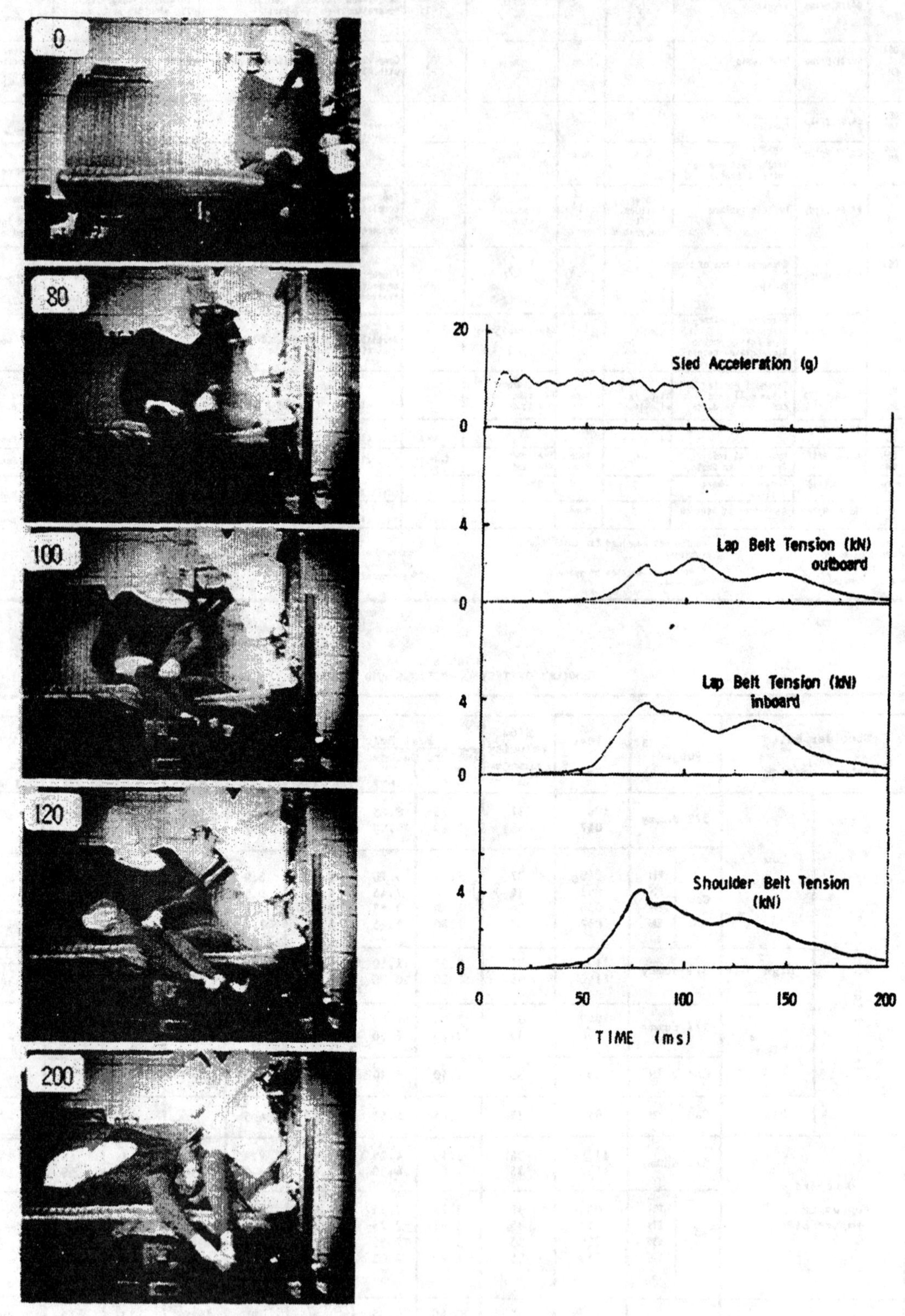

Fig. 3 - Sequence photographs and responses for test 057 -- outboard shoulder belt -- unembalmed cadaver subject

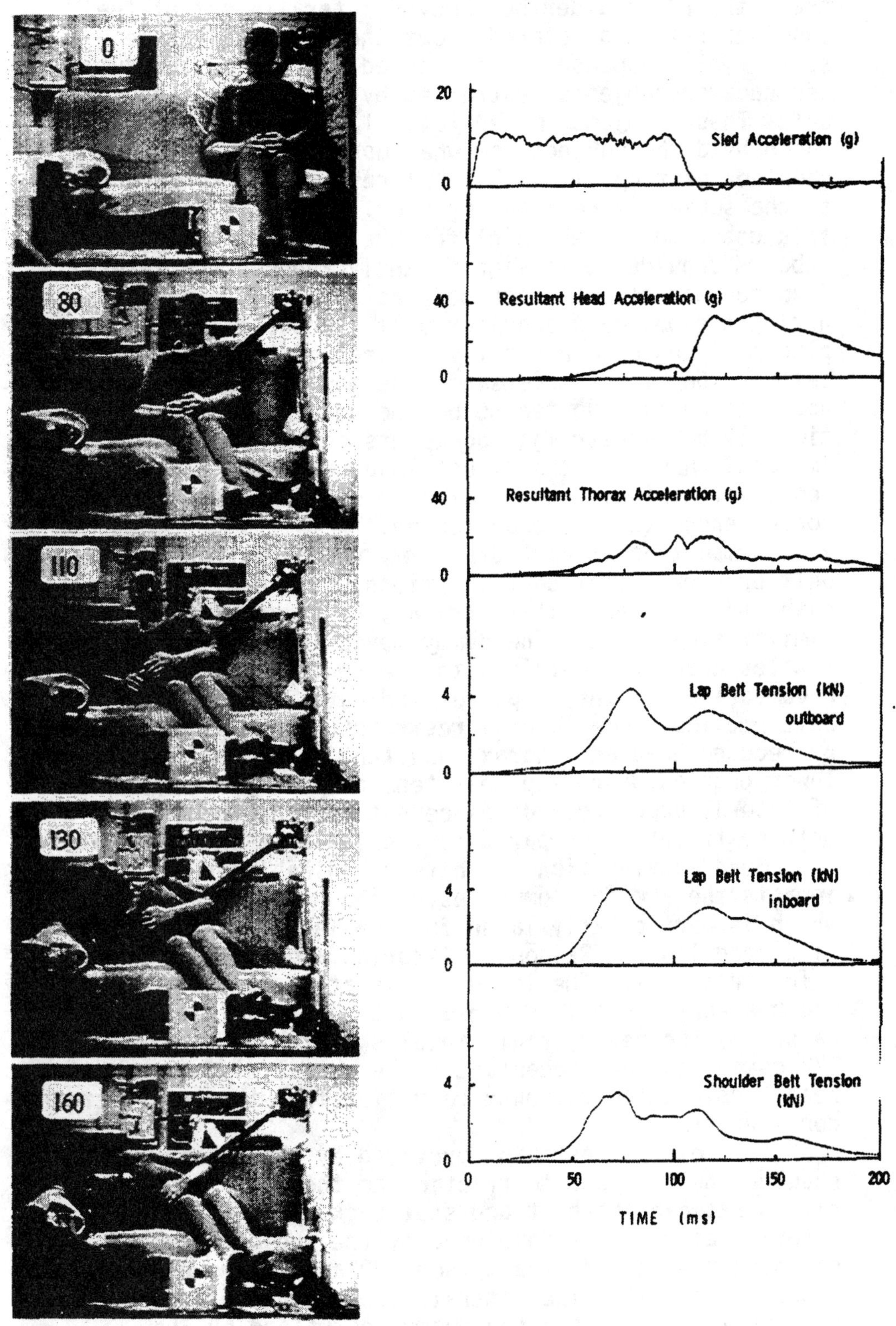

Fig. 4 - Sequence photographs and responses for test 1152 -- outboard shoulder belt -- Part 572 dummy

the outboard shoulder belt moved laterally out of the shoulder belt and rotated about the pelvis onto the seat. This response is contrasted to that of the dummy and cadaver subjects restrained by the inboard shoulder belt. These subjects moved laterally into the belt, which held the subject somewhat upright because the shoulder belt applied a lateral restraining force primarily to the subject's neck and/or head. As might be expected, this upper torso restraint for the subjects having the inboard shoulder belt significantly reduced the lateral displacement of the upper body as compared to subjects having the outboard shoulder belt. Reduced lateral displacement could be a positive aspect of the shoulder belt on the impact side since the most frequently injured body region is reported to be the head [3,6] for conventionally belt-restrained occupants in far side lateral impacts. However, the direct loading of the neck may represent a significant challenge with undetermined consequences to this body region.

Comparison of Part 572 dummy responses for the lap belt only restraint configurations, Figure 5, with the outboard shoulder belt restraint, Figure 4, indicates similar kinematics. The dummy moves laterally and rotates about the pelvis onto the seat. Interestingly, a comparison of the responses indicates that the shoulder belt provides some lateral restraint. This is demonstrated by reduced head and thorax resultant accelerations, by a lower peak outboard lap belt tension and by a reduction of lateral upper body displacement for the three point belt restraint as compared to the lap belt only restraint.

Head acceleration exhibits a "spike" when the head impacts the arm for dummy tests 1135, 1137 and 1151, which is particularly large for the lap belt only tests (1137 and 1151). Response histories are shown in Figure 5 for test 1151. The large acceleration "spike", due to the head-arm impact, is presumed to be an artifact caused by the heavy, rigid metal skeleton of the Part 572 dummy and thus accentuates the actual differences of head acceleration response resulting from the two restraint configurations.

Some of the restraint provided by the outboard shoulder belt is due to friction as the upper body slides between the belt and seat back. Additional lateral restraint is developed by the subject's elbow catching the shoulder belt (see 100 and 130 ms photos, Figures 3 and 4). The lateral displacement of the upper body is 11 cm less for the dummy restrained by the outboard shoulder belt than for the lap belt only. This can be seen by comparing tests 1152 and 1151, Figures 4 and 5, respectively. The restraint provided by the outboard shoulder belt is probably sensitive to the

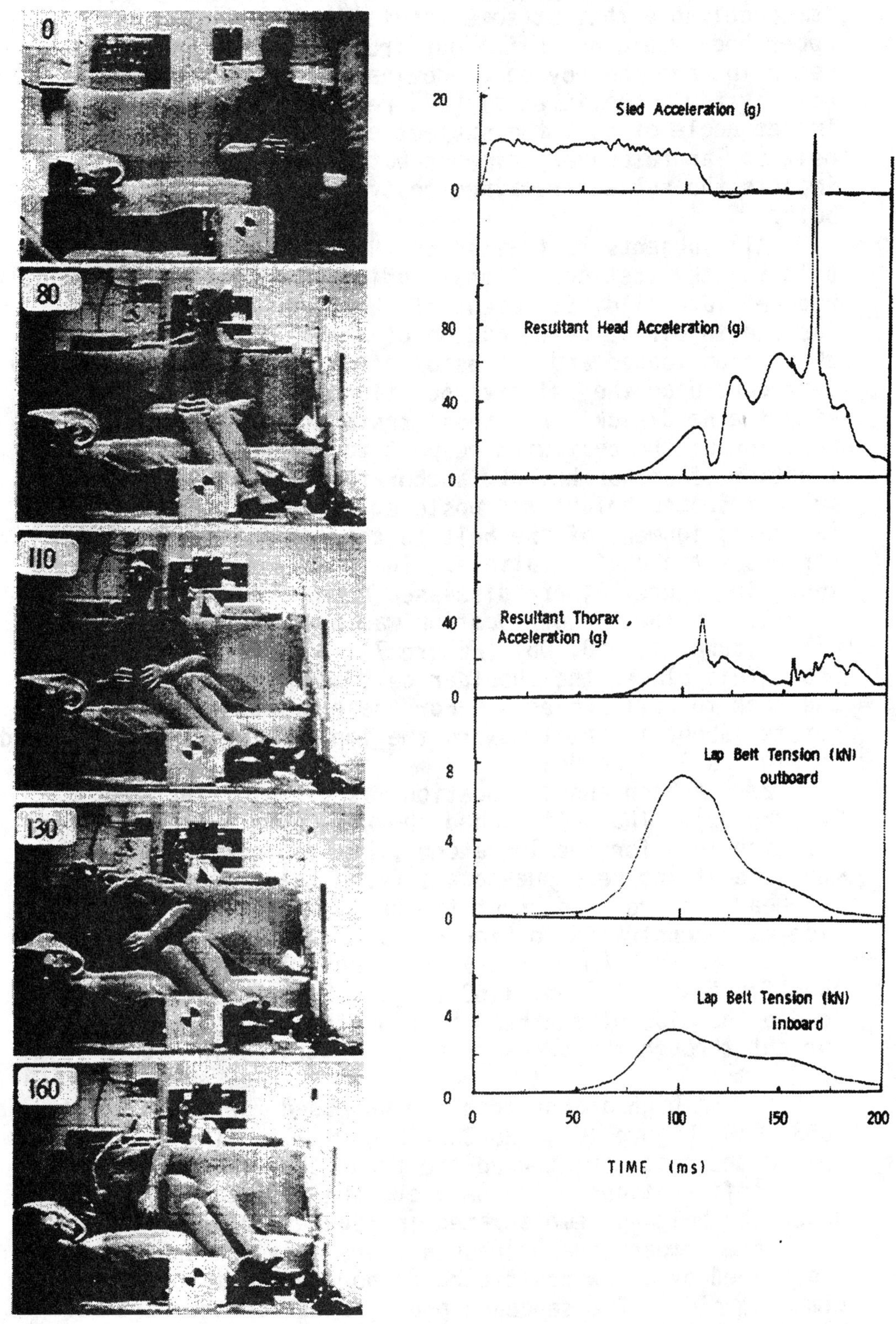

Fig. 5 - Sequence photographs and responses for test 1151 -- lap belt -- Part 572 dummy

level of belt slack/tension and to the impact angle. It is conceivable that at some frontal impact angle the upper body would not slide out from under the belt restraint and thereby be predominantly restrained by the belt system. Adomit et al [22] reported that at an impact angle of 58°, the subject escaped the shoulder belt in laboratory experiments but at more frontal impacts the subject remained restrained by the shoulder belt.

All subjects restrained by the inboard shoulder belt for the test conditions studied, exhibited significantly reduced lateral displacement of the upper body. This was due to direct belt loading of the neck/head. However, the region loaded and the head/neck kinematics were strongly dependent upon the relative position of the belt along side the head/neck. For these tests, the relative position of the belt with respect to the head/neck was a function of anchor height, anchor fore-aft position, and subject seated height and posture. Figure 13 shows the lateral alignment of the belt to the subject's head for various setup configurations. These alignments, as shown in Figure 13, are discussed below:

1. A low anchor location was used for tests 1065, 047 (Figure 6), 048, 051 (Figure 7), 052 (Figure 8), and 056. This placed the shoulder belt along the base of the neck for all subjects. For these tests the head rotated about the belt toward the impact side as demonstrated in Figures 6, 7, and 8.

2. A high anchor location was used for test 049 (Figure 12). The belt loaded the neck at a higher location than for the low anchor, but did not contact the side of the head due to a forward head position. The head also rotated about the belt, toward the impact side as demonstrated in Figure 12.

3. A high (more rearward) anchor location was used for tests 1138 and 1150 (Figure 9). The belt loaded the side of the head holding the head somewhat upright through the exposure as demonstrated in Figure 9.

4. A high anchor location was used for tests 1063, 046 (Figure 10), and 053 (Figure 11). The belt loaded the head more toward the front than the other test configurations. The head eventually rotated from under the belt as demonstrated in Figures 10 and 11.

Some comparisons of response are made for subjects restrained by a low positioned anchor on the inboard (impact) side. The sequence photographs for the Part 572 dummy (test 047, Figure 6), an unembalmed cadaver (test 051, Figure 7), and an embalmed cadaver (test 052, Figure 8) indicate a significant difference in head/neck rebound motion. The unembalmed cadaver's head (Figure 7) rotated over the belt and remained near the belted shoulder. The embalmed cadaver's head (Figure 8) rotated over the belt in a fashion similar to the unembalmed cadaver, but the head rebounded to an upright posture, remaining in this upright posture. This indicates significant restoring forces, presumably due to the tissue stiffness induced by applied chemical preservation. The Part 572 dummy's head (Figure 6) rotated over the belt similar to that of the cadaver subjects, then rebounded back beyond the original upright position.

The subject for test 056 was in some respects different from the other exposed cadavers -- age, weight, height, injury, and peak belt loads. A plot of estimated peak neck loading by the shoulder belt vs. subject weight is shown in Figure 14 for subjects restrained by a low inboard anchored shoulder belt. This plot suggests that the lower estimated peak neck loading for the 23 year old cadaver subject of test 056 was primarily due to a significantly lower subject mass than for the other subjects tested with the low inboard shoulder belt anchor.

NECROPSY FINDINGS - Although the human cadaver model has been reported to be more vulnerable to impact injury than living humans [10-15], a detailed examination for injuries was conducted to identify acute trauma, which might assist in identifying potential mechanisms of injury associated with lateral deceleration forces. It should be noted that not all specimens were examined to the same extent (see Table 1). The neck region, however, was thoroughly examined in all cases due to characteristics of the deceleration forces.

Fracture of the pelvis was found in three of the six subjects examined. These fractures are thought to be due to loading by the lap belt. No abdominal injuries were found for the five subjects examined.

Fractures of the ribs and/or sternum were found for four of the seven subjects examined. However, two of these subjects (057 and 059) had received CPR prior to death and this was quite possibly the cause of some or all of the musculoskeletal injury in these two cases. For another subject (054), the examiner noted that the bones were very brittle and easily broken.

Observed neck injuries were confined to the cervical spine and related ligamentous structures and not to the overlying soft tissues of the cervical region. Greater cervical damage was found in subjects receiving a direct neck/head loading by the inboard (impact side) upper torso belt than in subjects restrained by the outboard (opposite impact side) shoulder belt. Interestingly,

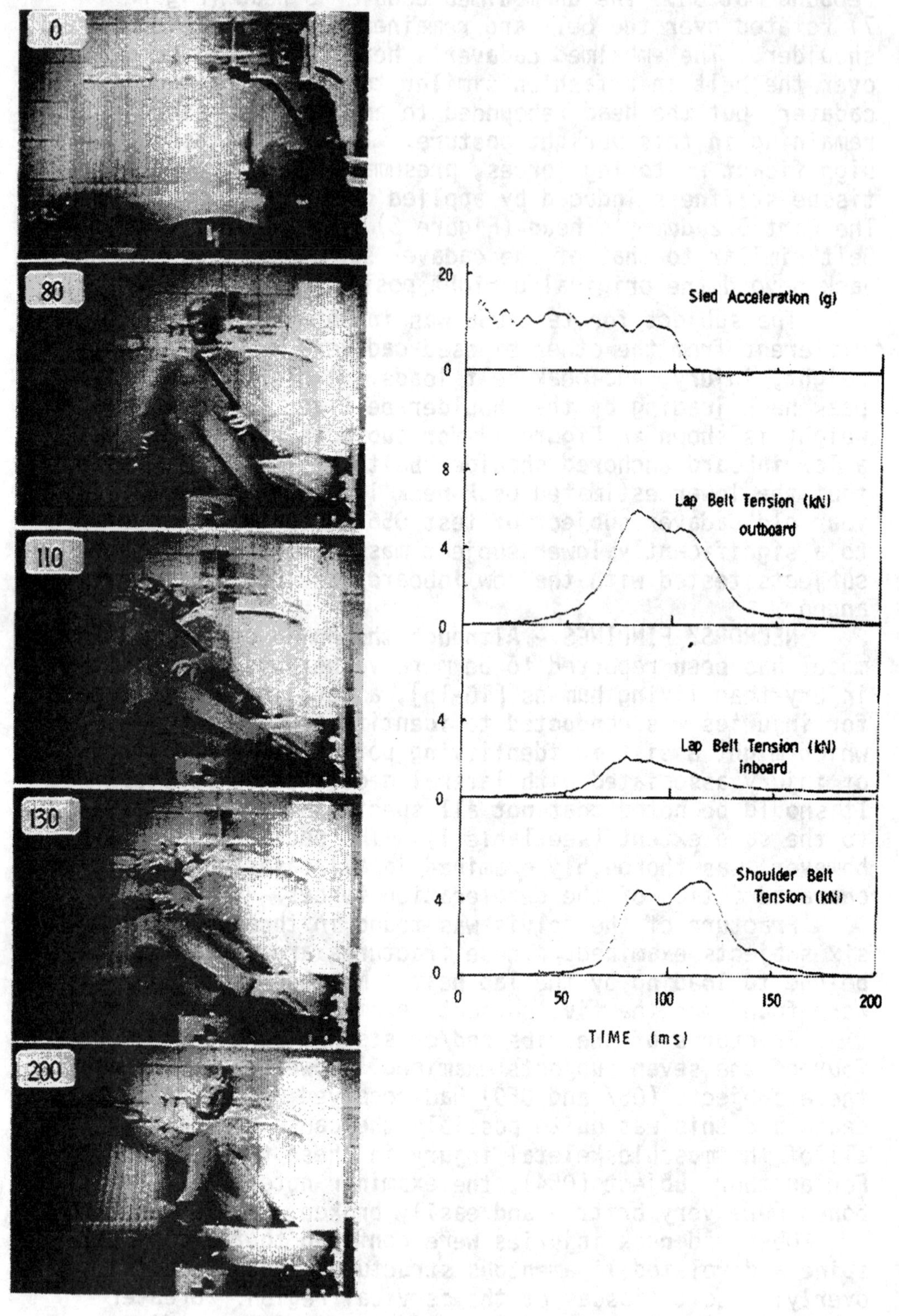

Fig. 6 - Sequence photographs and responses for test 047 -- low inboard shoulder belt -- Part 572 dummy

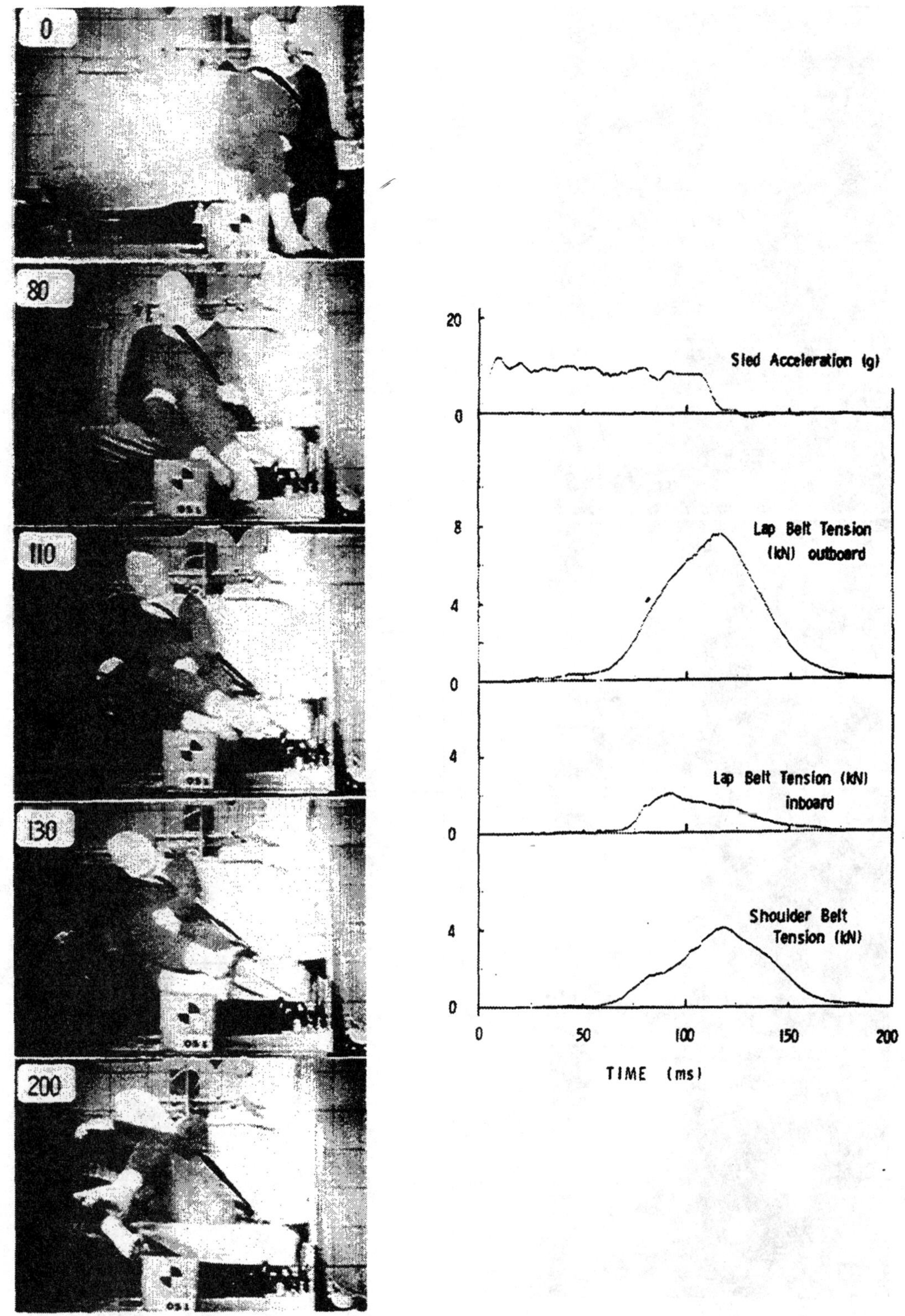

Fig. 7 - Sequence photographs and responses for test 051 -- low inboard shoulder belt -- unembalmed human cadaver

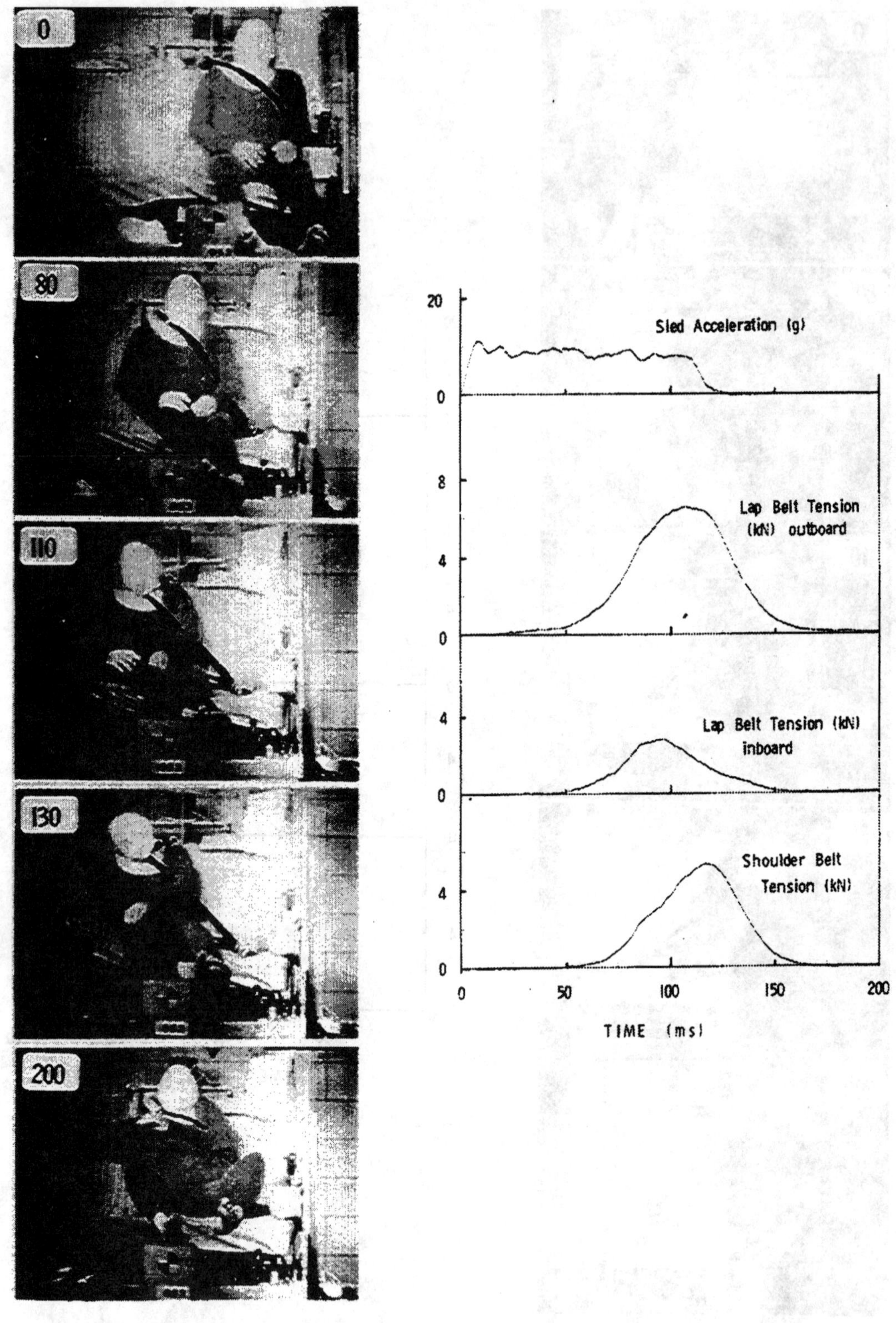

Fig. 8 - Sequence photographs and responses for test 052 -- low inboard shoulder belt -- embalmed human cadaver

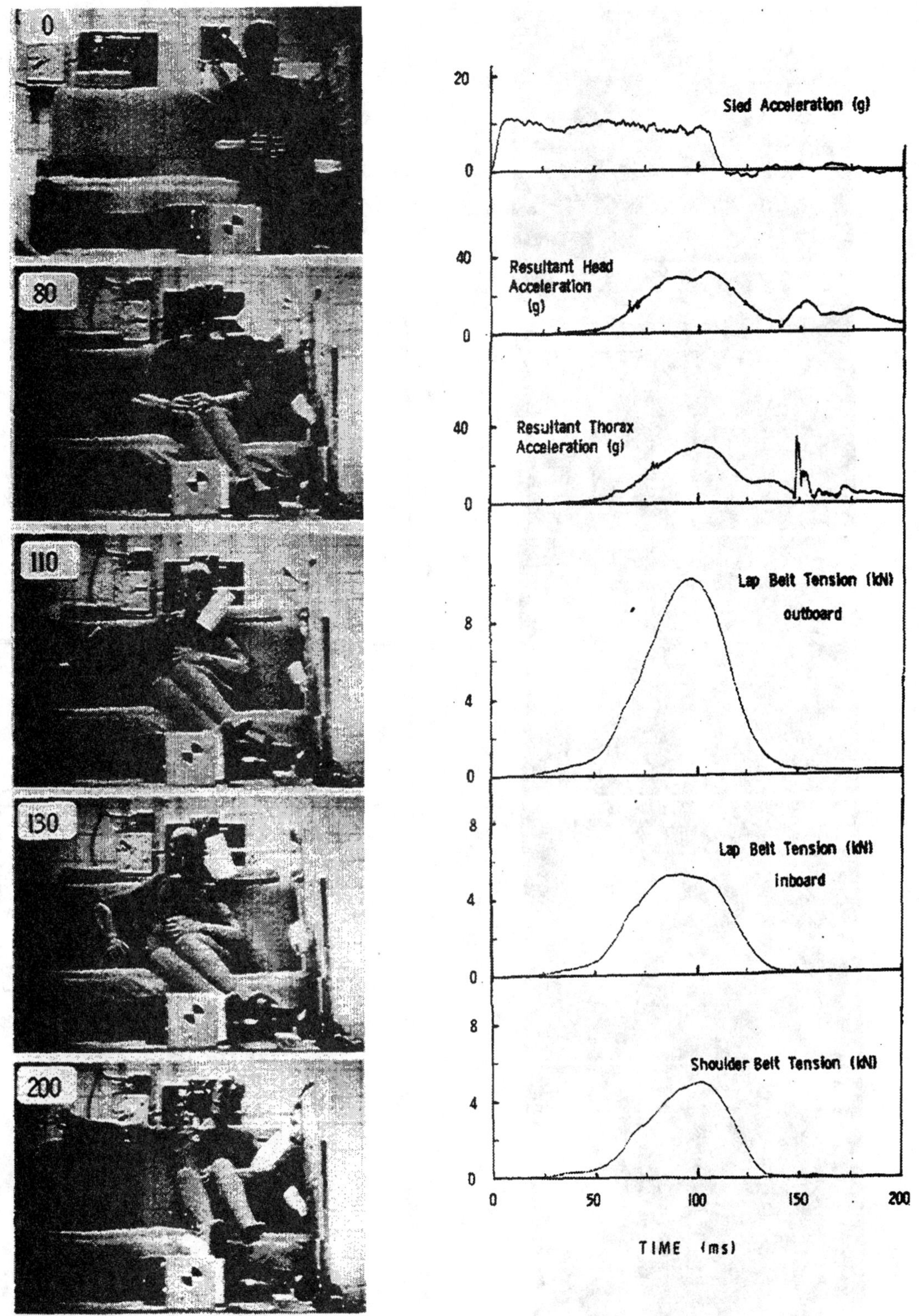

Fig. 9 - Sequence photographs and responses for test 1150 -- high inboard shoulder belt -- Part 572 dummy

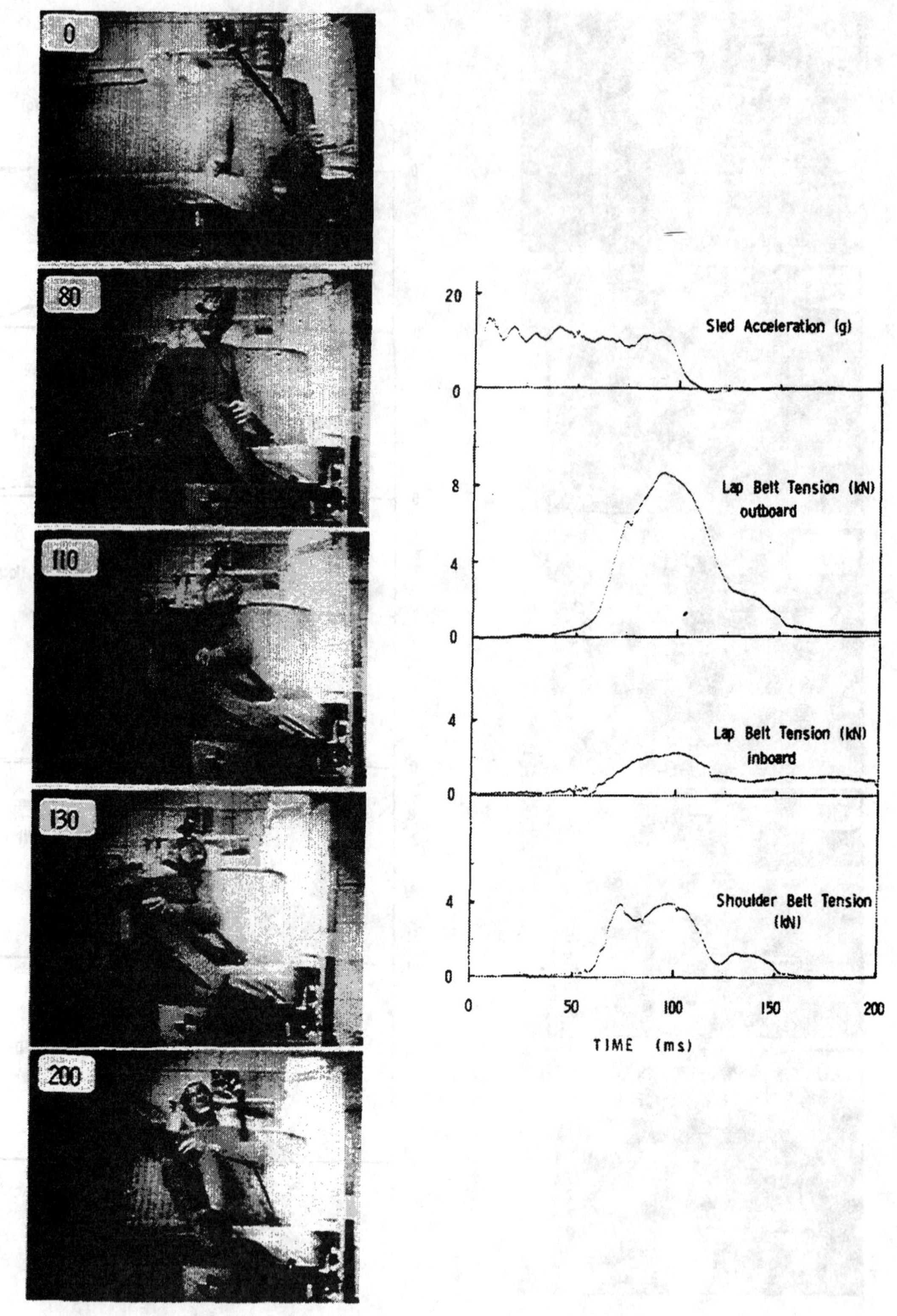

Fig. 10 - Sequence photographs and responses for test 046 -- high inboard shoulder belt -- Part 572 dummy

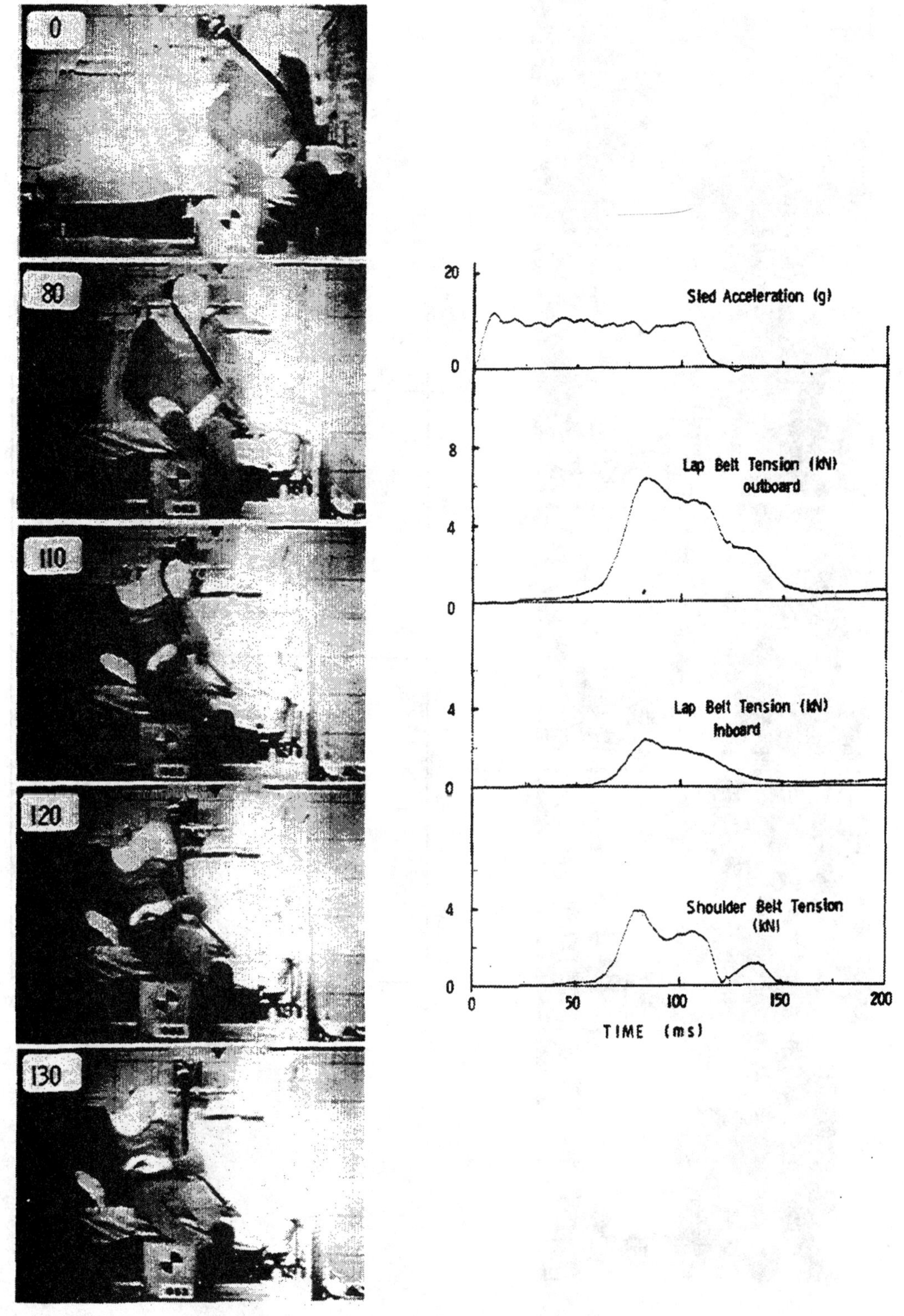

Fig. 11 - Sequence photographs and responses for test 053 -- high inboard shoulder belt -- embalmed human cadaver

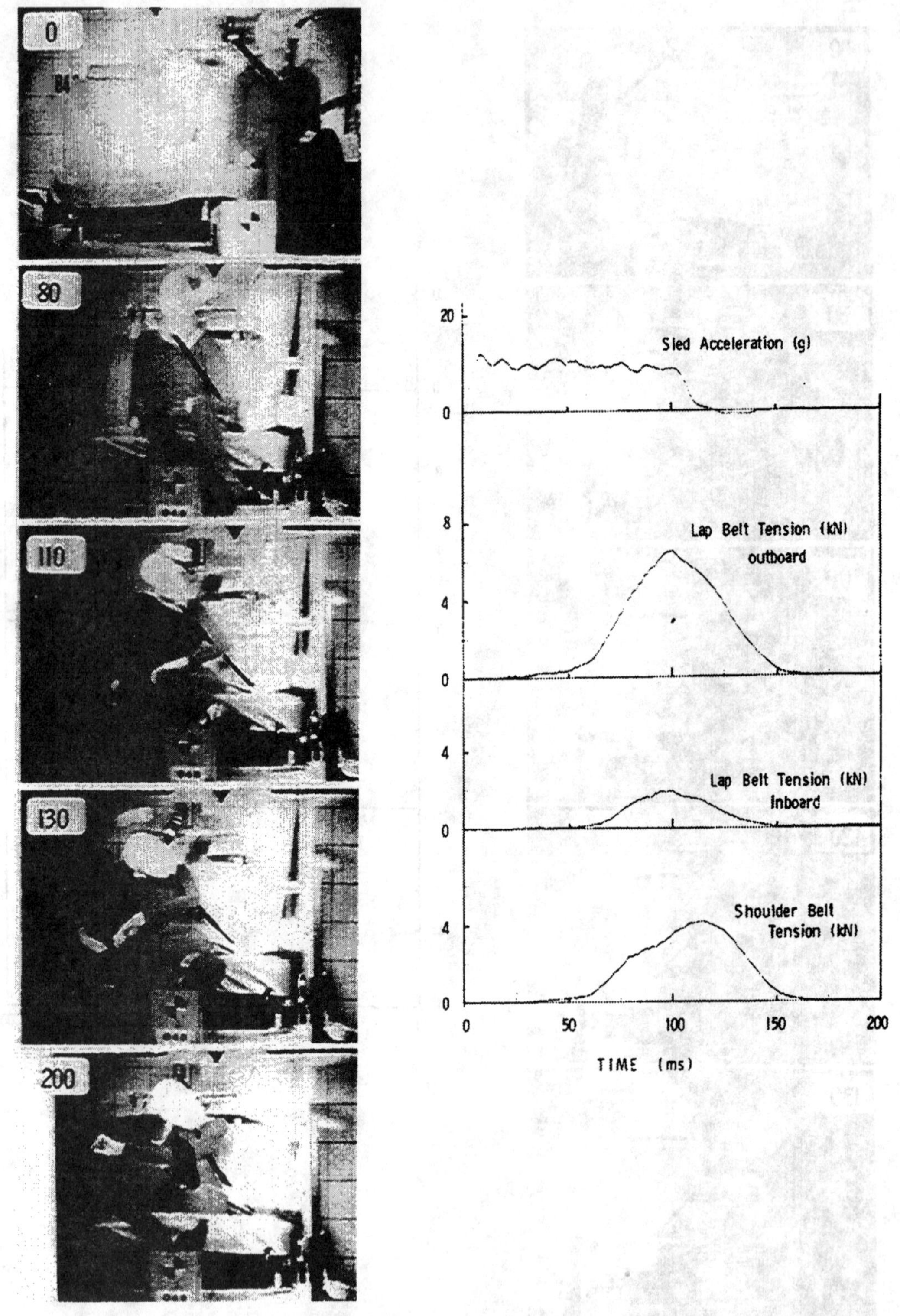

Fig. 12 - Sequence photographs and responses for test 049 -- high inboard shoulder belt -- subject's head initially forward -- unembalmed human cadaver

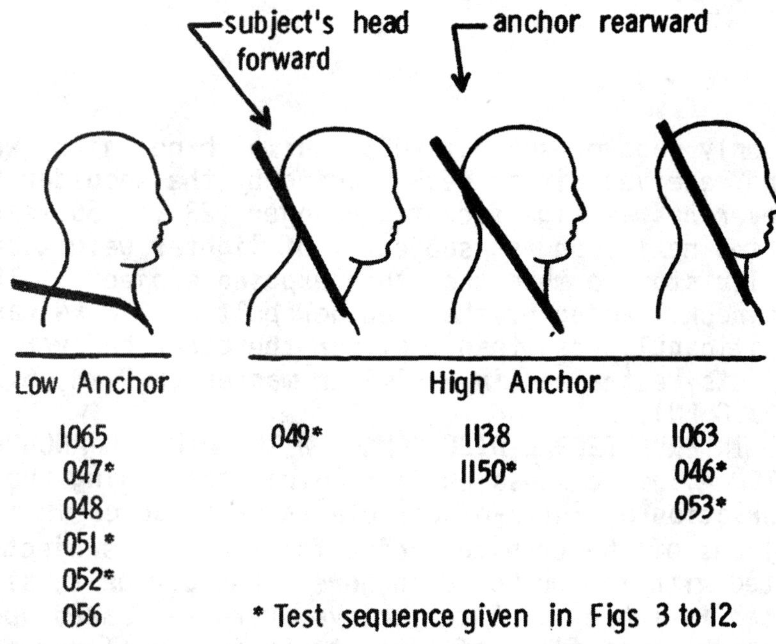

Fig. 13 - Schematic of the shoulder belt lateral alignment to the head/neck for the various inboard shoulder belt anchor locations

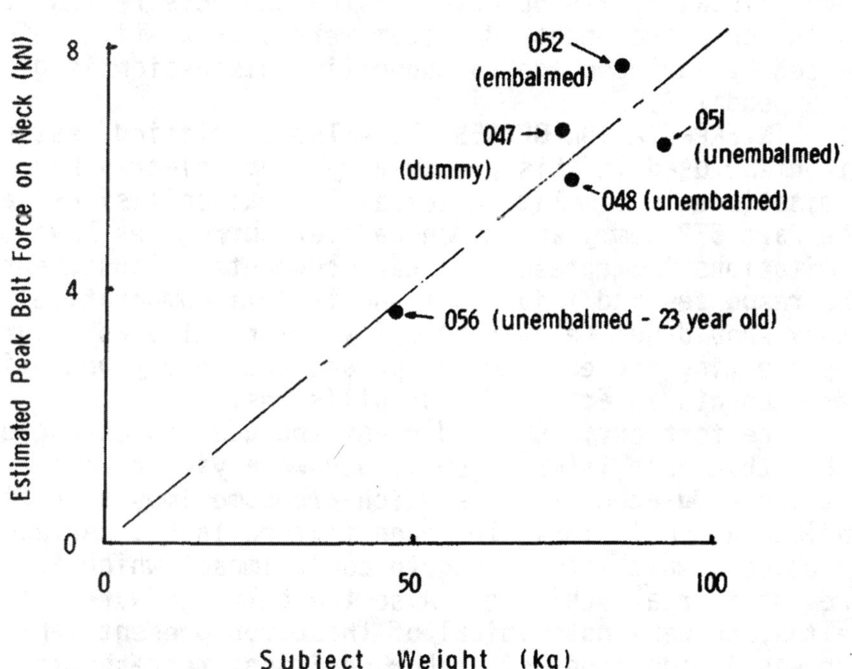

Fig. 14 - Estimated peak resultant shoulder belt load acting on the neck as a function of subject weight for the inboard shoulder belt with low anchor position

the only cadaver subject (056) in which no injury was found received direct neck loading by the shoulder belt. However he was significantly younger (23 vs. 56 years for the next youngest subject), of lighter weight, and smaller stature than the other exposed subjects. The peak neck loading by the shoulder belt of 3.8 kN was significantly less than that for the other cadaver subjects restrained in a similar manner (5.9 kN, 6.5 kN, and 7.8 kN).

INTERVERTEBRAL DISC DAMAGE WITH OUTBOARD-ANCHORED SYSTEM - Some discussion is required concerning the observation of intra-discal clefts or fissures in midsagittal sections of the cervical spine for the four subjects tested with the outboard anchored shoulder belt, as listed in Table 1. A subjective appraisal based upon the high speed films of these tests is one of rather innocuous kinematics -- a smooth roll-out from under the shoulder belt, with little relative angular displacement between head and torso. Each of the specimens so tested, however, revealed the intra-discal clefts at necropsy (Figure 15A). Subsequently, two untested control specimens were dissected, and very similar findings were detected in both cases (Figure 15B). It is thus possible that these discal clefts observed in the subjects restrained by the outboard-anchored system were pre-existing defects caused by advanced age (a supporting discussion is given in Appendix I).

INTERPRETATION OF RESULTS - The simplified test environment used in this preliminary study clearly has limitations in simulating actual lateral collisions, and the Part 572 dummy and human cadaver surrogates have limitations in representing car occupants. Consequently, the responses and injuries found in this comparative study should not be automatically interpreted as accurately representing the expected responses and injury potential of occupants in actual field collisions.

The test environment did not include impact angles other than pure lateral (90°); nor were yaw or roll motions simulated, factors which are sometimes significant in lateral collisions. The open fixture lacked surfaces or objects which the surrogate could impact which are present in real vehicles. Also the belt hardware and belt slack were not typical of those for present vehicles; nor was a wide range of anchor positions representing actual vehicles investigated.

Previous investigators who compared injuries of belt-restrained human cadavers in laboratory tests with field accident data [10-15], concluded that injuries to cadaver subjects were excessive. The cervical spine was one of the body regions injured. It is likewise believed

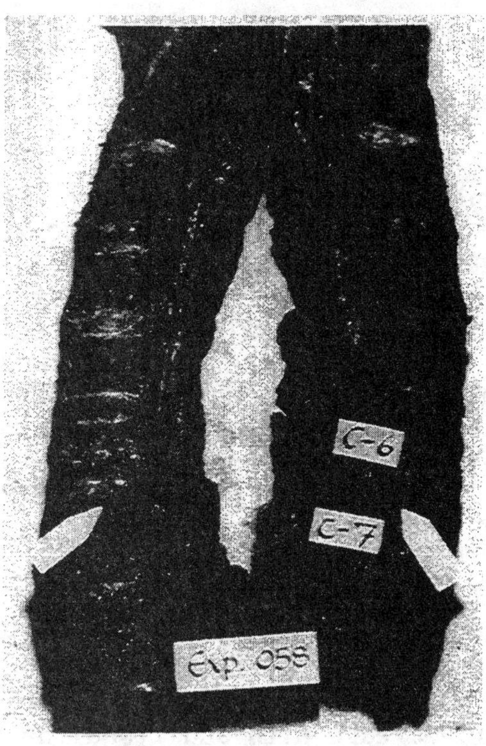

Fig. 15a - Midsagittal section of cervical spine for exposure 058 showing intra-discal cleft C6-7. Unembalmed human cadaver restrained by an outboard shoulder belt

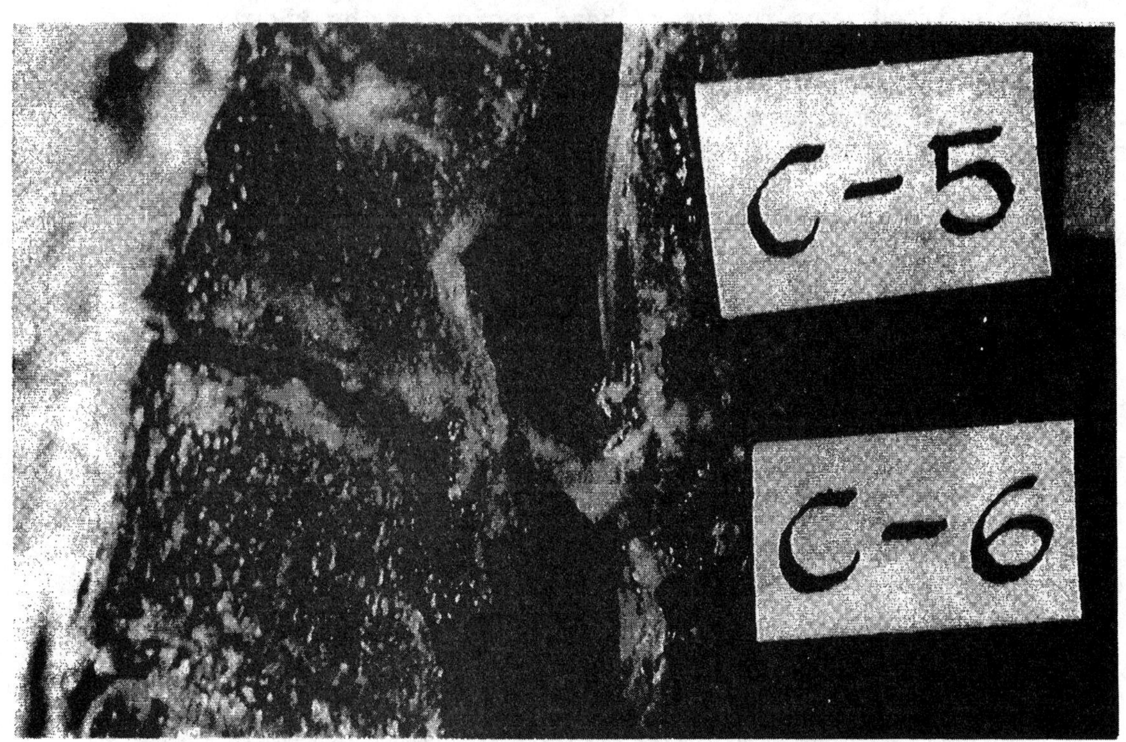

Fig. 15b - Midsagittal section of cervical spine for unexposed control subject (FF) showing intra-discal cleft C5-6. Unembalmed human cadaver

that the injuries observed in the cadaver tests in this study are also excessive when compared to the injury risk for similarly exposed car occupants and should not be automatically applied to represent that of car occupants.

In spite of these acknowledged limitations, the comparative interaction of the surrogate and restraint is believed to have been adequately simulated in terms of basic kinematics and load distribution to the body for the specific test conditions investigated. The generally similar kinematics of the Part 572 dummy, embalmed cadaver, and unembalmed cadaver surrogates strengthens this conclusion.

The results have shown that an outboard shoulder belt provides measurable restraint to the upper body in a far-side lateral impact. It is speculated that this restraint will increase as the impact becomes more frontal [22] or decrease with additional belt slack. The results have also demonstrated that an inboard shoulder belt significantly reduces lateral displacements of the upper body in far side lateral impact, which could be a positive aspect; however, the restraining load is applied primarily to the neck and/or head region. It may be speculated that if the restraining force could be shifted from the neck/ head region to the thorax, this would reduce the challenge to the cervical region and, at the same time, provide protection against head injury from impact with the vehicle far side interior, the adjacently seated occupants, or other objects.

SUMMARY/CONCLUSIONS

1. All belt systems investigated in this study provided lateral restraining forces in the experimental far side lateral impact simulation. However, the magnitude of the forces, the body regions exposed to belt loading, and the surrogate kinematics were significantly dependent upon the specific restraint configuration.

2. The lap portion of the belt systems provided lateral restraining forces on the pelvis for all belt configurations tested. However, the restraint of the upper torso was highly dependent upon the location of the shoulder belt relative to the side of impact.

   a. For the belted shoulder on the side of impact (inboard anchored), the shoulder belt substantially reduced lateral motion of the upper torso by directly loading the neck and/or head. The height of the shoulder belt anchor significantly affected the specific area of the neck/head loading and therefore, the head/neck kinematics.

   b. For the belted shoulder opposite the side of

impact (outboard anchored), the upper torso moved laterally between the shoulder belt and seat and rotated onto the adjacent seat structure. Some lateral upper torso restraint occurred by interaction with the shoulder belt and seat back as compared to no upper torso restraint for a lap belt only configuration.

3. The general surrogate kinematics were similar for the Part 572 dummy and both human cadaver preparations (unembalmed and embalmed) for similar test conditions.

4. Injuries were found in the cadaver subjects restrained by all shoulder belt configurations. The cervical region was more extensively damaged for those cadavers restrained by the shoulder belt anchored on the impact (inboard anchored) side.

5. It must be emphasized that the experimental test environment and surrogate models used in this study have significant inherent limitations in terms of simulating field accident conditions. However, it is believed that the basic restraint kinematic principles involving the interaction of occupant and belt restraint system provide basic response information for preliminary assessments of lateral collisions under field situations if the underlying qualifications are considered. The real life significance of the reported injury findings cannot be directly evaluated at this time. It is believed that the observed injury to the cadaver test subjects is excessive compared to an injury potential for similarly exposed car occupants.

REFERENCES

(1) F. Hartemann, et al, "Occupant Protection in Lateral Impacts." 20th Stapp Car Crash Conference, October 1976, Paper No. 760806.

(2) H. Appel et al, "Accident Analysis of Vehicle Side Collisions." 1977 International Conference on Impact Trauma, Berlin, Sept. 1977.

(3) H. Nielsen et al, "Lesions in Belted Car Riders from Oblique and Lateral Impacts." 1977 International Conference on Impact Trauma, Berlin, Sept. 1977.

(4) K. Langwieder, "Load of Car Structure and Injuries of Occupants in Side Impact Car Collisions with Trucks and Objects." 1977 International Conference on Impact Trauma, Berlin, Sept. 1977.

(5) D. Huelke et al, "The Effectiveness of Belt Systems in Frontal and Rollover Crashes." SAE March 1977, Paper No. 770148.

(6) F. Walz et al, "Belted Occupants in Oblique and Side Impacts." 1977 International Conference on

Impact Trauma, Berlin, Sept. 1977.

(7) P. Lachmann et al, "Car to Car Side Collision--A Comparison of Accident Analysis and Research Work." 1977 International Conference on Impact Trauma, Berlin, Sept. 1977.

(8) M. Danner et al, "Car/Vehicle Side Impacts--A Study of Accident Characteristics and Occupant Injuries." 6th International ESV Conference, October 1976.

(9) D. Cesari et al, "Biomechanical Study of Side Impact Accidents." 5th International ESV Conference, June 1974.

(10) J. Cromack et al, "Three-Point Belt Induced Injuries: A Comparison Between Laboratory Surrogates and Real World Accident Victims." 19th Stapp Car Crash Conference, November 1975. Paper No. 751141.

(11) L. M. Patrick et al, "Injury to Unembalmed Belted Cadavers in Simulated Collisions." 19th Stapp Car Crash Conference, November 1975, Paper No. 751144.

(12) R. Levine et al, "Effect of Quadricaps Function on Submarining." 22nd AAAM, 1978.

(13) J. Y. Foret-Bruno et al, "Correlation Between Thoracic Lesions and Force Values Measured at the Shoulder of 92 Belted Occupants Involved in Real Accidents." 22nd Stapp Car Crash Conference, October 1978, Paper No. 780892.

(14) A. Fayon et al, "Thorax of 3-Point Belt Wearers During a Crash (Experiments with Cadavers)." 19th Stapp Car Crash Conference, November 1975, Paper No. 751148.

(15) G. Schmidt et al, "Neck and Thorax Tolerance Levels of Belt-Protected Occupants in Head-On Collisions." 19th Stapp Car Crash Conference, November 1975, Paper No. 751149.

(16) R. Eppinger, "Prediction of Thoracic Injury Using Measurable Experimental Parameters." 6th Conference On Experimental Safety Vehicles, 1978.

(17) N. Alem et al, "Whole-Body Human Surrogate Response to Three-Point Harness Restraint." 22nd Stapp Car Crash Conference, October 1978, Paper No. 780895.

(18) A. Zabarowski, "Lateral Impact Studies Lap Belt Shoulder Harness Investigations." 9th Stapp Car Crash Conference, October 1965.

(19) C. Ewing et al, "Dynamic Response of the Human Head and Neck to +Gy Impact Acceleration." 21st Stapp Car Crash Conference, Octover 1977, Paper No. 770928.

(20) C. Ewing et al, "Effect of Initial Position on the Human Head and Neck Response to +y Impact Acceleration." 22nd Stapp Car Crash Conference, October 1978, Paper No. 780888.

(21) T. Noguchi et al, "Neck Injury Assessment

Protocol." Final Report DOT-HS-803 287, March 1978.

(22) D. Adomeit et al, "Expected Belt-Specific Injury Patterns Depend on the Angle of Impact." 1977 International Conference on Impact Trauma, Berlin, Sept. 1977.

(23) R. Walmsley, "The Development and Growth of the Intervertebral Disc." Edinburgh Medical Journal, August 1953, pp. 341-364.

(24) A. Peacock, "Observations on the Postnatal Structure of the Intervertebral Disc in Man." Journal of Anatomy, Vol. 86, Part 2, 1952, pp. 162-179.

(25) C. Hirsch and F. Schajowicz, Studies on Structural Changes in the Lumbar Annulus Fibrosus." Acta Orthopaed., Scand. 22:184, 1953.

APPENDIX I

DISCUSSION OF THE EFFECT OF AGE ON THE STRUCTURE OF THE INTERVERTEBRAL DISC - The intervertebral disc is a unique, highly complex anatomical structure which changes considerably with age. According to Walmsley [23], "...it undergoes more obvious structural change during pre- and post-natal life than any other joint in the body. At birth the outer portion, or annulus fibrosis, and inner portion, or nucleus pulposus, are structurally very distinct, the former being a highly intricate, although organized network of distinct collagen fibers and fibrocartilage and the latter a mucoid, or gelatinous, mass containing scattered cartilage cells.

As described by Peacock [24], structural changes occur with age in both regions, but particularly in the nucleus, where, by the second decade of life, a fibrocartilage network has begun to develop within the gelatinous substance, leading to a firmer consistency. As aging continues, dessication occurs, and the fibrocartilage mesh increases in density. The compositional and structural differences between nucleus and annulus become less distinct, and macroscopic discrimination becomes difficult by the seventh decade. Fiber degeneration eventually occurs, leading to the formation of fissures, or clefts, within the body of the disc. This is apparently characteristic of advancing age, but cases of gross cleft formation have been reported in cervical specimens representing the third decade of life (20-24 years). One theory for explaining early cleft formation in the cervical discs is that this is a response to functional mobility demands upon the cervical column. In any event, the foregoing structural changes, extensive as they may be, have been reported as characteristic of otherwise normal but aging tissue. Pathological changes, such as hemorrhage and pigmentation are also reported to occur with increasing frequency in aging discs.

Hirsch and Schajowicz [25] have likewise reported structural anomalies of intervertebral discs. In a study of 120 lumbar spine autopsy specimens representing nine decades of life, they found "concentric cracks or cavities" between the lamellae of the annulus fibrosis. These were observed in specimens as young as fifteen years and with increasing frequency in older subjects. Such defects were regarded as "physiological phenomena", generally without clinical significance. On the other hand, particularly in the 40-50 year age group, numerous instances of "radiating ruptures" in the annulus were observed; and the authors offer support for a relationship between such defects and the clinical picture of lumbago, or low back pain.

A relatively high incidence of intra-discal structural defects, especially with increasing age, is thus well documented in the literature. This, together with the finding of similar defects in our untested controls and the absence of other overt cervical damage in 3 of the 4 subjects restrained by the outboard anchored shoulder belt provides the basis for assuming that the lesions found by sectioning of the spine in these subjects may have been pre-existing and not, in fact, produced by the test exposure.

801310

Occupant Dynamics as a Function of Impact Angle and Belt Restraint

John D. Horsch
Biomedical Science Department
General Motors Research Lab.

Abstract

Sled tests were conducted to investigate the dynamics of a Part 572 dummy as a function of the belt restraint configuration and impact direction. The tests involved a 35 km/h velocity change and 10 g deceleration. An "opened" fixture, free of intervening surfaces, was oriented from frontal (0°), through oblique (±30°, ±45°, ±60°), to full lateral (±90°).

Restraint by only a lap belt resulted in the dummy's upper body rotating about the lap belt and continuing in the direction of sled deceleration. Restraint by a lap-shoulder belt greatly reduced upper-body displacement. However, the displacement and body loading were strongly dependent on the direction of deceleration, i.e., the orientation of the belt relative to the impact direction.

When the belted shoulder was opposite the impact (0° to +90°), the belt retained the upper body for impact angles of 0° to 45°. Although the upper body escaped from the shoulder belt from 60° to 90°, significant kinetic energy was removed from the upper body before escape, even for full lateral deceleration.

When the belted shoulder was on the impact side (0° to -90°), the upper body was restrained for all impact angles. However, the shoulder belt acted directly on the neck with increasing load as the impact became more lateral. Addition of lateral torso restraint, such as with a winged seat greatly reduced the loading of the neck by the shoulder belt for all impact angles.

OVERVIEW

Field accident studies [1-13] indicate that belt restraint effectively reduces injury over a wide range of crash configurations including frontal, oblique, lateral, and rollover accidents. One important function of belt restraint is to restrict the displacement of the occupant relative to the vehicle. Generally, this reduces the severity of impact with the vehicle interior, or even exterior to the vehicle, in the case of potential ejection.

The present study continues our previous work on belt-restrained subjects in lateral impact [14] and examines basic principles of occupant interaction with various belt-restraint configurations as a function of impact angle. The basic test environment was free of intervening surfaces other than the belt restraint, seat, and floor. Thus, the test conditions would more nearly simulate an occupant seated on the side of the vehicle opposite the impact than an occupant seated on the impacted side in the proximity of the door (except for the case of an opened door in which the belt restraint might act to restrict ejection). However, interpretation of the test results are not limited to the scope of present belt restraint systems which provide an outboard anchored shoulder belt for outboard located seating positions. The test conditions can also be viewed as simulating such restraint configurations as the shoulder-belt anchor on the inboard side of the occupant or a center-seated occupant restrained by a lap-shoulder belt.

METHODOLOGY

TEST ENVIRONMENT - The majority of tests were conducted with a test fixture (Figure 1) that has been previously described [14]. It consisted of an automotive front bench seat, fixed anchor supports for the belts, and a flat floor pan with a simulated rectangular tunnel. Lateral support on the side of the seat opposite the impact was used to maintain he dummy's position prior to sled deceleration. The fixture could be oriented at various angles relative to the sled acceleration vector. The "open" fixture provided good visibility for high-speed photography permitting analysis of the belt interaction with the occupant. A bench seat was used to minimize potential lateral restraint by the seat (as compared to a contoured bucket seat) and to provide a continuous lateral surface.

A lap belt or lap-shoulder belt was used in each test. For tests which used the bench-seat fixture, the lap belt was formed by a single segment attached to two floor anchors. The lap-shoulder system was formed by a separate belt segment from each of the two floor anchors and a shoulder-belt anchor. The three segments were attached together at a point laterally in line with the subject's hip joint on the side opposite the shoulder-belt anchor. The anchor locations were compatible with the seat. Polyester automotive belt webbing was attached to the anchorage hardware such that no slip or "payout" of webbing occurred. The belts were adjusted "snug," i.e., without significant tension or free slack. Belt tension transucers were located near the belt anchors.

In addition to the lap-shoulder belt restraint, lateral restraint for the thorax or pelvis was used in some of the tests. A simulated console was used for inboard lateral restraint of the pelvis and legs. The top of the "console" was in line with the seated height of the pelvis. A post, rigidly mounted to the sled frame, and covered by 50 mm of crushable foam (Fig. 2), provided inboard lateral restraint of the thorax.

Six additional tests explored the concept of a winged seat with a different fixture (Figure 3). Matched test pairs were conducted at -45° and -90° impact orientations. For each pair, one test was conducted with an automotive bucket seat, the other with an inboard wing attached to the seat. This wing, of a tubular construction, was welded to the seat-back frame and covered with foam. This was the only seat modification. The fixture had a simulated console and tunnel as well as a compatible belt system. An inertial retractor was used at the outboard floor anchor. A continuous length of webbing passed from a rigid attachment at the inboard floor anchor, across the lap to a webbing transfer ring fastened to the webbing from the retractor, and diagonally across the thorax to a rigid shoulder-belt anchorage. The shoulder-belt anchor location approximated a roof height and centerline installation for a two-passenger vehicle. The shoulder-belt anchor was 160 mm more lateral of the occupant than in the bench-seat tests.

IMPACT CONDITIONS - Tests were conducted on a decelerator sled at a velocity change of 35 km/h and 10 g deceleration level. The deceleration pulse was 110 ms in duration.

A 180° range of impact angles was used: 0° (frontal); ±30°; ±60°; and ± 90° (full lateral). The impact angle is defined as the difference between the sled deceleration vector and the angle of the fixture on the sled (as referenced to the front-facing direction of the seat). Instead of rotating the fixture on the sled 180°, the shoulder-belt anchor was moved from the outboard location to the inboard location, thus requiring a 90° range of fixture orientation. An impact on the opposite side of the belted shoulder (outboard) is referred to as a positive impact angle. An impact on the same side as the belted shoulder (inboard) is referred to as a negative angle. A free object would move

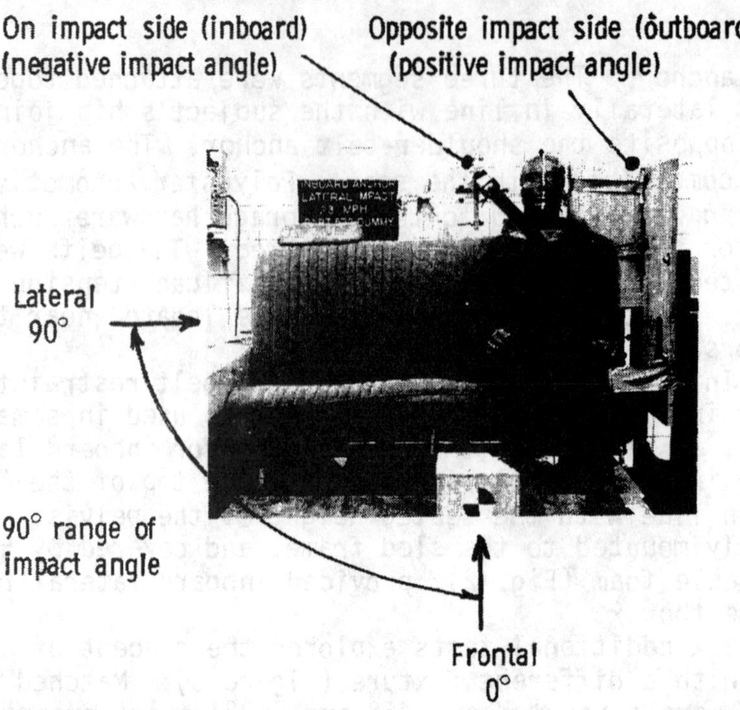

Fig. 1 - Bench-seat test fixture mounted on the GMR crash simulator. Shoulder-belt anchor locations are indicated. The Part 572 dummy is shown restrained by an inboard-anchored shoulder belt

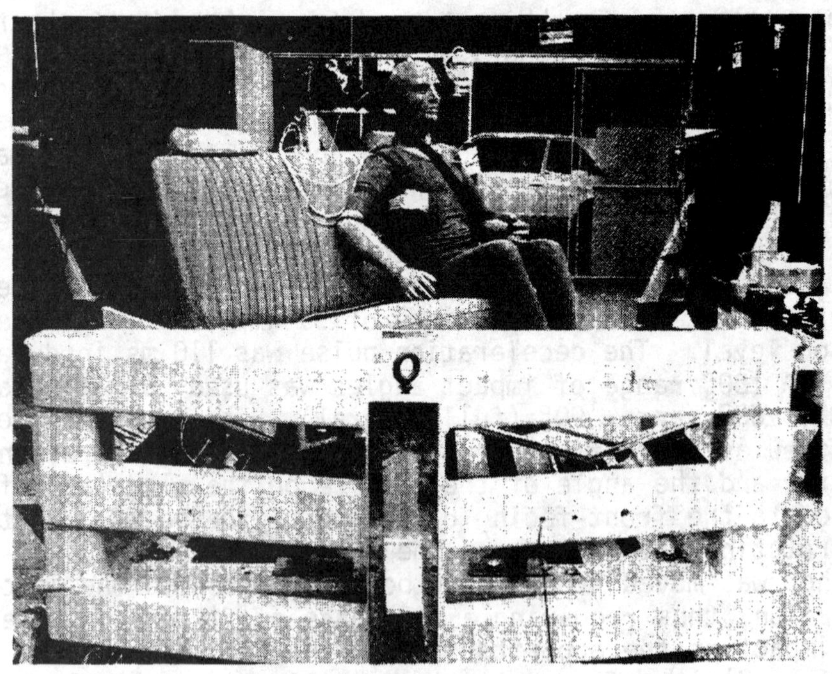

Fig. 2 - Inboard-anchored shoulder belt with inboard thoracic restraint (rigid post attached to fixture with crushable 50 mm foam pad)

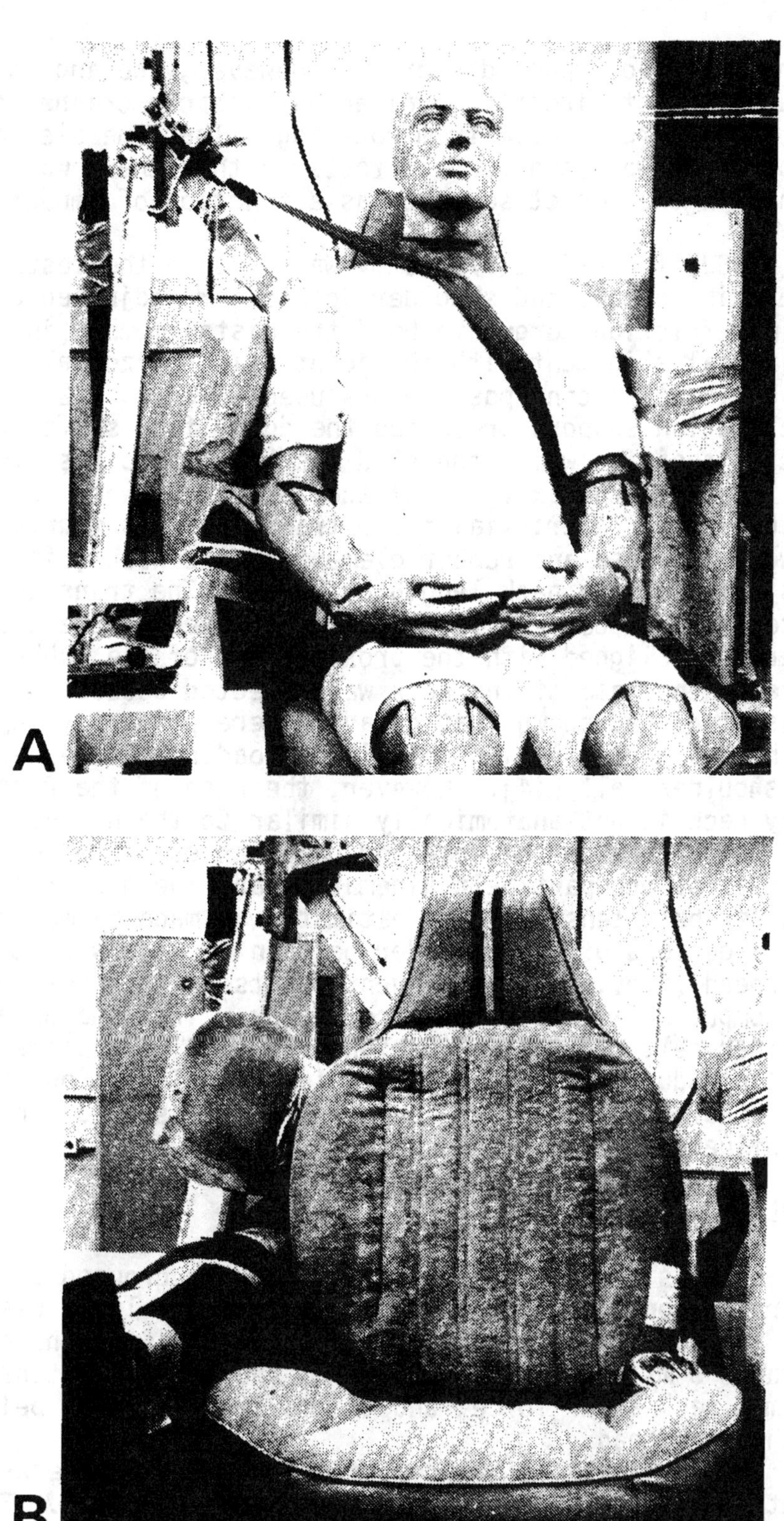

Fig. 3A&B - Seat wing shown attached to seat back

along the sled impact direction. However, the angular sled impacts do not directly model actual vehicle crashes because the sled is restricted from rotating. The same sled deceleration pulse was used for all tests so that there was no change in the impact severity as a function of impact angle.

SUBJECT - A Part 572 dummy was used in the tests. The knee, hip, elbow, and shoulder joints were adjusted at the maximum friction torque to hold the distal member in a horizontal alignment with the joint axis horizontal. A conventional seating posture was used (Figures 1,2,3). Acceleration support prevented the dummy from shifting the initial position while the sled was brought to test velocity.

Head and thorax triaxial accelerometers were installed in the dummy. A triaxial force transducer was mounted between the base and rubber element of the neck (Figure 4). This extended the neck length by 25 mm. The transducer was protected by a cover. The measurement axes of the transducer were aligned with the proximal end of the rubber neck. Force at the base of the neck was selected because a previous study had shown that cadavers were primarily injured in the C-5 to C-7 region for direct loading of the neck by the shoulder belt [14]. However, the base of the Part 572 dummy neck is not anatomically similar to the human.

Belt force on the neck and head was estimated by analysis of occupant/belt kinematics and the measured shoulder-belt tension. The estimate was made by measuring the directions of the shoulder belt in contact with the neck/head (i.e., the two tangent points on the neck/head) and computing a resultant force vector. The same upper shoulder-belt tension was assigned to both belt segments. This introduces some uncertainty in the calculation, since the actual tension in the lower end of the shoulder belt may be influenced by frictional effects.

RESULTS

Results from the bench-seat tests are grouped by similar restraint configuration in Table 1. Table 2 lists results from the bucket-seat tests. Responses consist of head HIC, thoracic 3-ms acceleration, maximum shoulder belt tension, maximum neck shear force, and the maximum external belt load on the neck/head.

The horizontal displacement of the neck was determined relative to the fixture from the high-speed movies. This displacement is shown in Figure 5 for selected bench-seat tests at impact angles of 30°, 45°, 60°, and 90°, and in Figure 6 for the bucket-seat tests at -45° and -90°. The maximum displacement is listed in Tables 1 and 2. The base of the neck was chosen as a representative point to compare

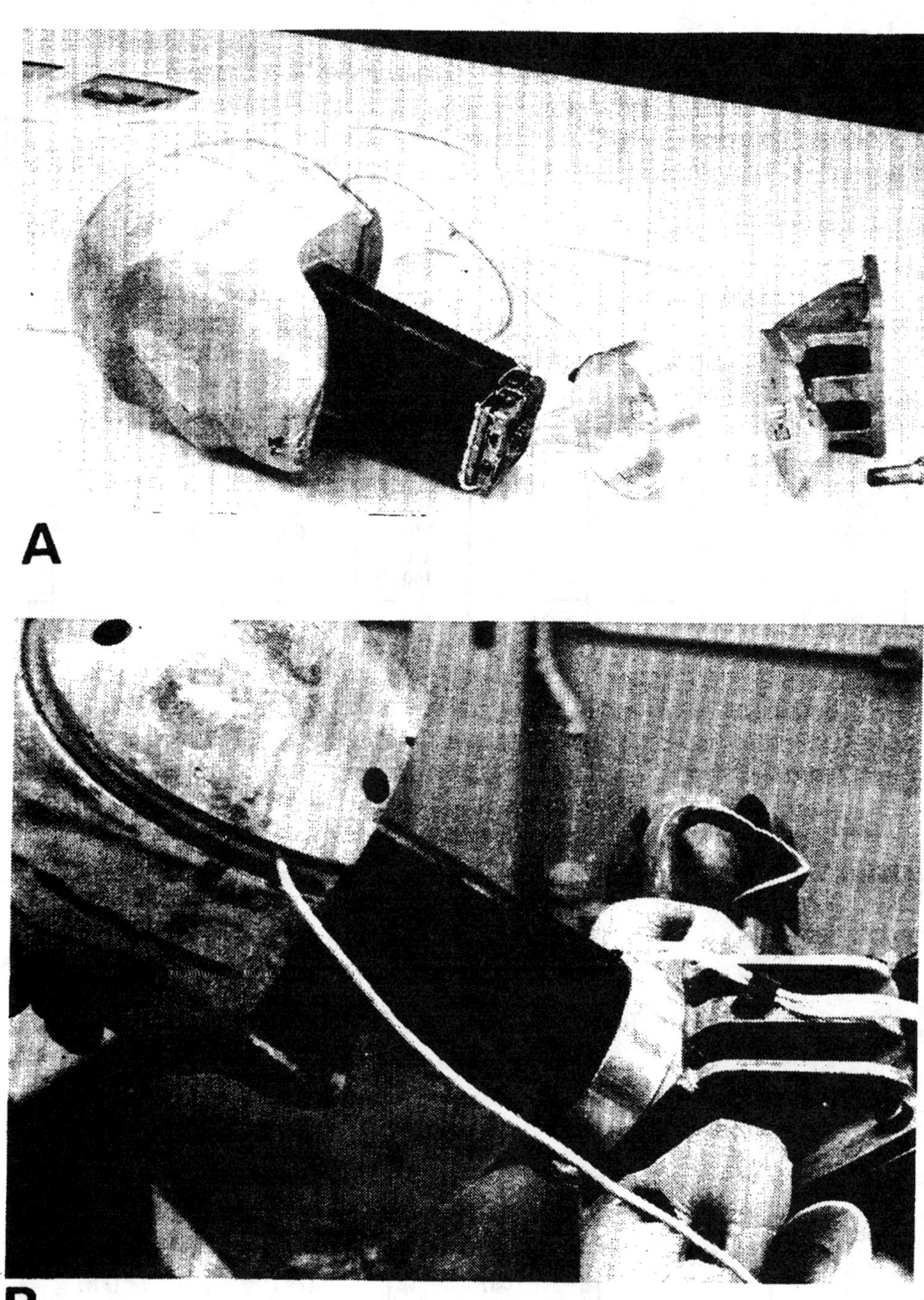

Fig. 4A&B - Installation of triaxial force transducer in the Part 572 dummy neck base

TABLE 1

Summary of Tests & Results Using Bench-Seat Fixture

Part 572 Dummy - 35 km/h

Restraint	Impact Angle	Maximum Neck-Base Horizontal Displ.(cm)**	Thorax 3-ms accel. (g)	Head HIC Index	Maximum Shoulder-Belt Tension (kN)	Maximum Neck-Base Shear (kN)	Maximum Shoulder-Belt Force on Neck/Head (kN)
Lap Belt	90°*	71	30	1270***	-	2.6	-
	90°*	N.D.	28	1420***	-	1.4	-
	60°	72	27	580	-	1.3	-
	45°	69	27	620	-	2.0	-
Outboard Shoulder Belt	+ 90°*	61	19	250	3.6	N.D.	0
	+ 90°*	N.D.	23	460***	2.6	0.8	0
	+ 60°	50	16	N.D.	5.4	0.8	0
	+ 45°	35	21	100	6.1	1.3	0
	+ 30°	22	22	N.D.	5.6	1.8	0
	0°	14	26	150	6.4	1.2	0.6
Inboard Shoulder Belt	- 90°*	28	30	170	5.0	2.5	5.7
	- 90°*	N.D.	27	140	5.0	2.4	5.7
	- 60°	22	24	120	5.9	2.0	5.0
	- 45°	16	23	110	6.0	1.4	3.3
	- 30°	13	21	N.D.	5.6	1.5	3.3
	0°	15	27	140	6.1	0.9	1.2
With Thoracic Support	- 90°	12	21	120	2.0	0.7	0.9
	- 60°	8	25	110	3.5	0.7	1.0
	- 45°	5	23	110	4.4	1.0	1.1
	- 30°	6	26	N.D.	6.0	0.6	1.1
With Lateral Lower Body Support	- 90°	26	32	128	3.33	2.4	3.5

* Previously reported in Reference [14]
** Does not include "rebound"
***Head-to-arm impact
N.D. - No Data

TABLE 2

Summary of Tests & Results Using Bucket Seat Fixture

Part 572 Dummy - 35 km/h

Impact Angle	Bucket Seat	Maximum Neck-Base Horizontal Displacement (cm)	Maximum Shoulder-Belt Tension (kN)	Maximum Shoulder-Belt Load on Neck/Head (kN)
-90°	Std.	47	3.1	4.5
	Wing	28	1.3	0.3
-45°	Std.	31	4.4	4.4
	Wing	23	4.0	1.4
-45°*	Std.	40	5.3	4.1
	Wing	31	4.2	1.1

*Shoulder-belt anchor 250 mm forward

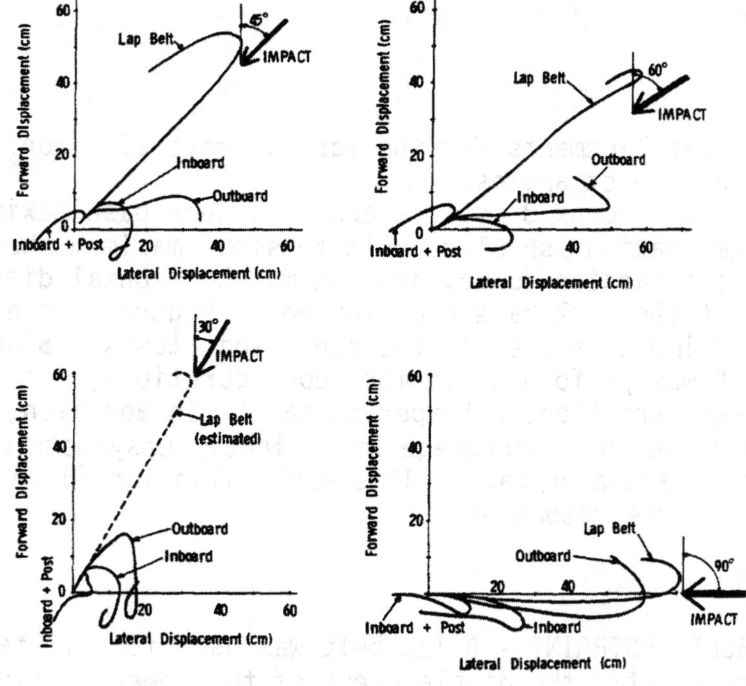

Fig. 5 - Overhead view of the neck-base displacement. Bench-seat tests. The horizontal coordinate system is relative to the seat, the origin being the initial location of the neck base. Inboard and outboard refer to shoulder belt anchor location, on the impact side and opposite the impact side, respectively

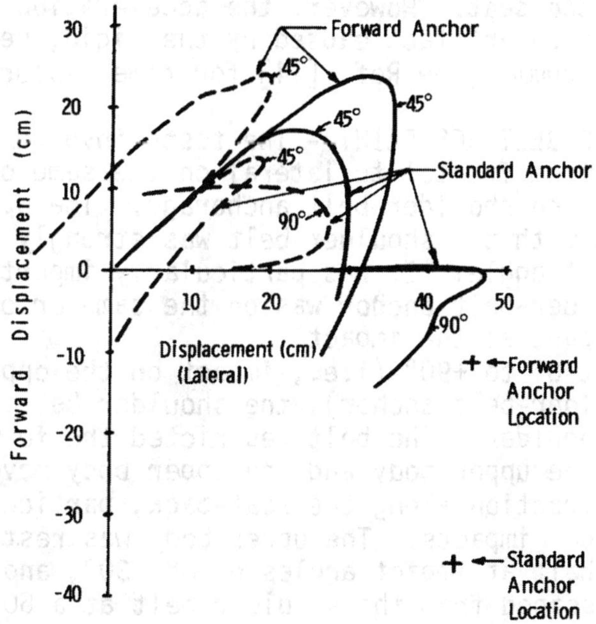

Fig. 6 - Overhead view of the neck base displacement. Bucket-seat tests. The horizontal coordinate system is relative to the seat, the origin being the initial location of the neck base. Trajectories for tests having a standard seat are shown solid -- trajectories for tests having a winged seat are shown dashed

upper body displacements for the various restraint configurations and impact angles.

Head HIC, thorax 3 ms acceleration, neck-base maximum shear force, maximum shoulder-belt tension, maximum shoulder-belt/neck interaction force, and maximum horizontal displacement of the neck base are plotted in Figure 7 as a function of impact angle for the bench-seat tests. Since a single test was performed for most configurations, some of the observed variation is "experimental." In addition, the inboard shoulder-belt anchorage was slightly unsymmetric with the outboard anchorage. This would also introduce a difference in the responses.

DISCUSSION

LAP-BELT RESTRAINT - A lap belt was used for all tests. The belt restricted the displacement of the lower portion of the dummy in the direction of deceleration. In the absence of a shoulder belt, the upper body pivoted about the lap belt in the direction of sled motion. The displacement of the head, neck, and thorax was largest with only lap-belt restraint of the restraint configurations tested (Figures 5 & 7). The head HIC and thorax 3-ms acceleration were also greatest for lap-belt restraint (Figure 7), especially HIC at ±90° impact because the head struck the arm during upper body loading of the seat. However, the acceleration "spike" is believed to be an artifact caused by the rigid, heavy metal arm of the dummy (see Ref. [14] for time history of test).

LAP-SHOULDER BELT RESTRAINT - The tests involve a range in impact angle from frontal to lateral on the same or opposite side of the shoulder-belt anchorage. The upper body interaction with the shoulder belt was strongly dependent on the impact angle. It was particularly important whether the shoulder-belt anchor was on the same or opposite side of the occupant as the impact.

For tests at 0° to +90° (i.e., impact on the opposite side as the shoulder-belt anchor), the shoulder belt loaded the thorax and shoulder. The belt restricted the forward displacement of the upper body and the upper body moved more in the lateral direction along the seat-back, particularly for the 45° and 60° impacts. The upper body was restrained by the shoulder belt at impact angles of 0°, 30°, and 45°. The upper body escaped from the shoulder belt at a 60° impact. However, most of the upper body kinetic energy was dissipated, resulting in little motion of the upper body after escape. A 56° angle was found by Adomeit [15] at which the upper body escaped from the shoulder belt. The upper body escaped from the shoulder belt with some reduction of kinetic energy for a 90° impact. In comparison with

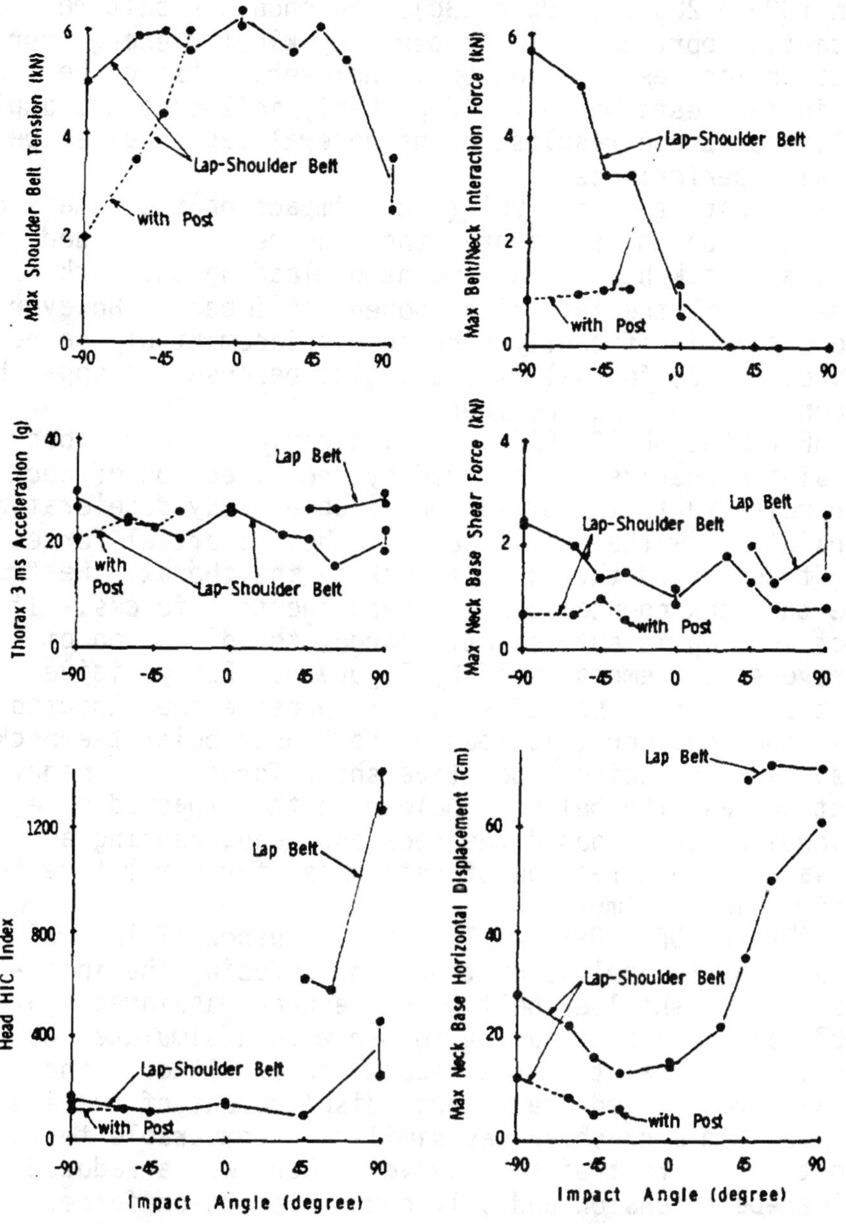

Fig. 7 - Various responses as a function of impact angle and restraint configuration

lap-belt only restraint, (Table 1) there is less upper body horizontal displacement (0.7 m vs. 0.6 m), a lower HIC (1266 & 1419 vs. 254 & 455), and a lower thoracic 3-ms acceleration (30g & 28g vs. 19g & 23g). The shoulder belt removes a substantial portion of the upper body kinetic energy for all impact angles tested, even 90°. However, a "snug" belt was used in the tests which would probably influence the applicability of these results to the general lap-shoulder belt restraint performance.

For test at 0° to -90° (i.e., impact on the same side as the shoulder-belt anchor), the shoulder belt loaded the thorax and neck/head. The amount of load on the neck increased with the lateral component of impact. However, there was a significantly reduced displacement of the neck (Figures 5 & 7) for all impact angles because the upper body was contained by the shoulder belt.

NECK-BASE SHEAR FORCE - An indication of upper body restraint mechanics is provided by the direction of neck shear relative to the direction of upper-body deceleration. Restraint below the neck results in head inertial forces being transmitted through the neck to the thorax. Restraint above the neck counteracts the head inertial forces. If the direct loading is sufficiently large, the direction of shear can reverse as demonstrated by Figure 8. For positive impact angles (the belted shoulder opposite the impacted side), the shoulder belt loaded the thorax below the neck transducer, producing a positive shear force. For negative impact angles (the belted shoulder on the impacted side), the shoulder belt loaded the neck and head, causing a reversal of the direction of neck shear force relative to the direction of impact.

LATERAL TORSO RESTRAINT - The influence of lateral restraint of the pelvis or thorax on reducing the interaction of the shoulder belt with the neck was investigated. A -90° lateral impact was performed with a simulated console for pelvic restraint on the impact side. Although the restraint restricted the lateral displacement of the lower body, maximum neck shear was similar to comparable tests without pelvic restraint. However, there was a reduced shoulder-belt tension and belt/neck interaction force.

Lateral support of the thorax on the impact side (see Fig. 2) was placed under the arm in contact with the dummy. This rigid support (an experimental device not intended as a vehicle restraint component) strongly influenced the upper body interaction with the shoulder belt for all impact angles tested (Figures 5 & 7) from -30° to -90°. The trajectory of the neck was altered by the thoracic support for the lap-shoulder belt restraint, primarily the lateral component of upper body displacement was restricted. The shoulder belt prevented the upper body from moving around the thoracic support. Obviously, "stiffness" or "give" in the belt system would influence the minimum size of the "wing" to prevent motion of the occupant around the support. When the rigid lateral thoracic support was used with the lap-shoulder-belt system, direct loading of the neck and neck shear were greatly reduced for impact angles of -30° to -90°. Neck shear force changed direction by introduction of the lateral thoracic support for these impact angles of -30° to -90°, Figure 8. However, head HIC and thorax 3 ms acceleration were similar for tests with and without thoracic restraint. Thus, head and chest acceleration are not sensitive indicators of the large changes in dummy kinematics and restraint load distribution in these tests.

WINGED-SEAT TESTS - The results from our preliminary tests with thoracic support suggested that a seat wing might substantially influence the performance of a lap-shoulder belt system in oblique and lateral impacts. However, the lateral support tested earlier was quite rigid and idealized. Additional tests were conducted with a more realistic thoracic restraint, a wing attached to the seat back. Oblique (-45°) and lateral (-90°) impact tests were conducted with a lap-shoulder-belted dummy on an automotive bucket seat (Figure 3). A belt system with webbing retractor was used so that elongation of the belt system could influence the performance of the wing. Comparison of paired tests with and without the seat wing demonstrated that the wing significantly influenced the response of the dummy. Displacement of the neck and interaction between belt and neck were substantially modified by the presence of the seat wing (see Figure 6 and Table 2). Neck shear loads are not reported for the winged-seat tests because the shoulder belt contacted the neck below the sensitive element in the neck transducer, and would not indicate the actual neck interaction with the shoulder belt. Comparable frames from high-speed movies (Figure 9) show kinematic differences between a -90° impact with and without the seat wing. Comfort, convenience and occupant size evaluations were not considered in this study.

SUMMARY

Although our laboratory tests involve a simple impact environment, an idealized belt system, a mechanical surrogate and no vehicle interior components, the results provide insight into occupant dynamics as a function of belt restraint and direction of deceleration. The lap belt restricts displacement of the lower body. However, in the absence of a shoulder belt the upper body pivots about the belt in the direction of sled deceleration. A "snug" shoulder belt effectively restricts upper body displacement

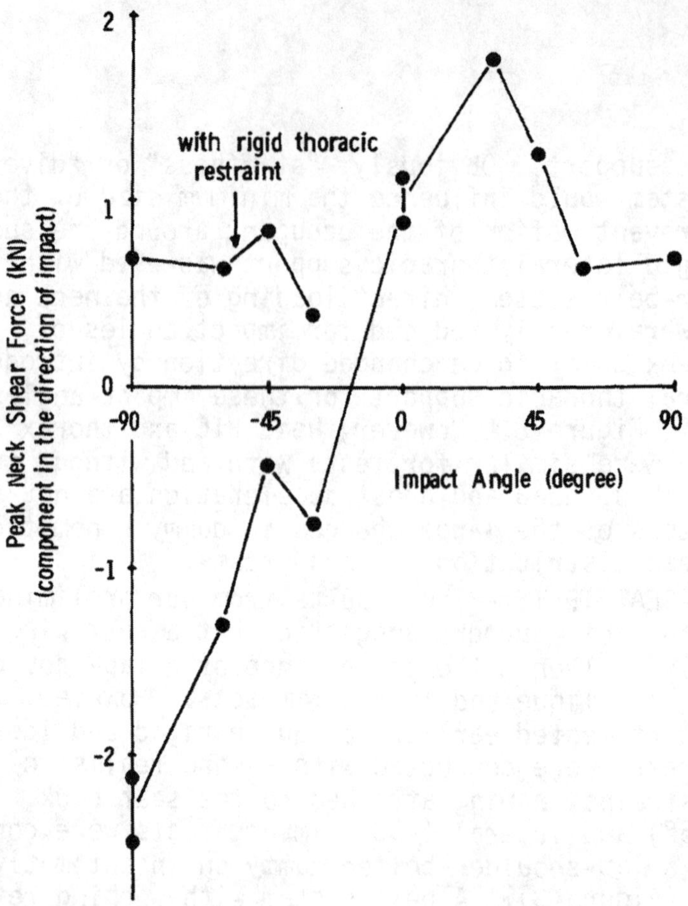

Fig. 8 - Estimated maximum neck-base shear-force component in the direction of impact (anatomical coordinates related to the initial seated position) as a function of impact angle for a lap-shoulder-belt restrained Part 572 dummy

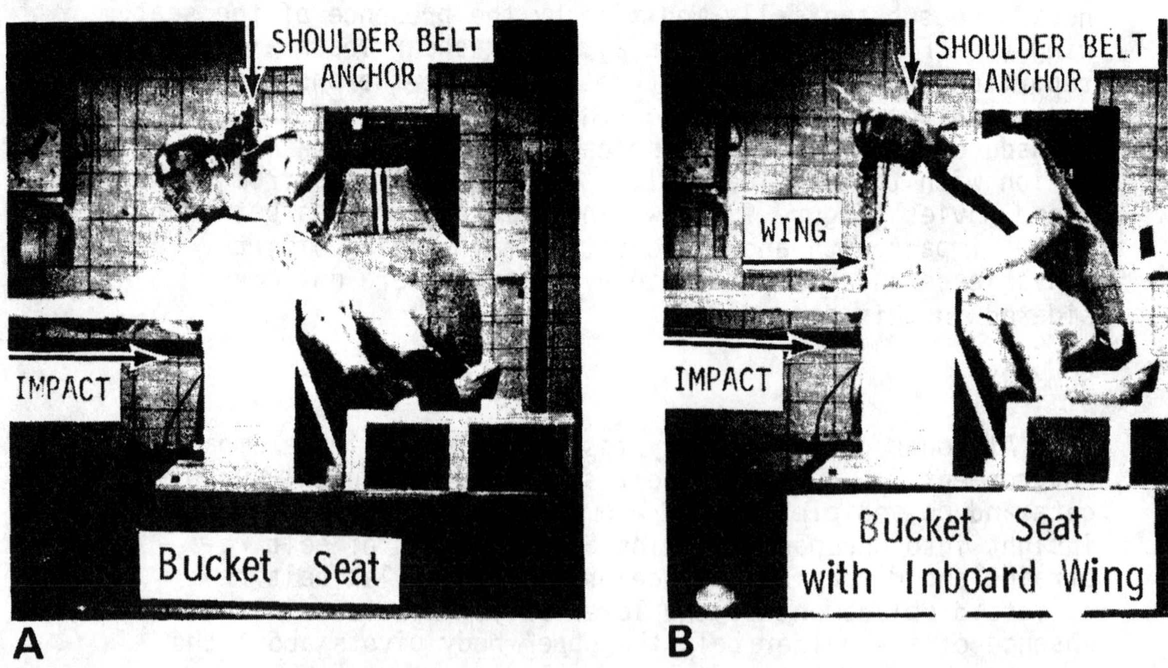

Fig. 9A&B - Comparison of Part 572 dummy response at time of maximum belt/neck interaction force for lateral (-90°) impact

for all impact directions from frontal to lateral (±90°). The greatest excursion of occupant's upper body occurred in far-side lateral impact (90°) as the upper body moved out of the shoulder belt. However, the restraint system reduced the kinetic energy and displacement of the upper body. Lateral impacts which displaced the upper body toward the shoulder-belt anchor (-30° to -90°) resulted in direct loading of the head and neck. The neck load increased in more lateral impacts. Lateral support for the thorax, e.g., padded post or a winged seat, restricted upper body displacements and significantly reduced the shoulder belt loading of the neck.

Interpretation of the results is not necessarily limited to the conventional restraint configuration of an outboard occupant with an outboard-anchored shoulder belt. The results are applicable to an outboard occupant having an inboard-anchored shoulder belt or a center occupant having a lap-shoulder belt restraint.

REFERENCES

1. D. Huelke et al, "The Effectiveness of Belt Systems in Frontal and Rollover Crashes." SAE March 1977, Paper No. 770148.
2. D. Huelke et al, "Effectiveness of Current and Future Restraint Systems in Fatal and Serious Injury Automobile Crashes." SAE February 1979, Paper No. 790323.
3. F. Hartemann et al, "Belted or Not Belted: The Only Difference Between Two Matched Samples of 200 Car Occupants." 21st Stapp Car Crash Conference, October 1977, Paper No. 770917.
4. C. Tarrierre, "The Efficiency of Three Point Belts in Real Accidents." 4th International Technical Conference on Experimental Safety Vehicles, March 1973.
5. F. Hartemann et al, "Occupant Protection in Lateral Impacts." 20th Stapp Car Crash Conference, October 1976, Paper No. 760806.
6. A. Rininger et al, "Lap-Shoulder Belt Effectiveness." 20th Conference American Association for Automotive Medicine, November 1976.
7. H. Appel et al, "Accident Analysis of Vehicle Side Collisions." 1977 International Conference on Impact Trauma, Berlin, September 1977.
8. J. Marsh et al, "Injury Patterns by Restraint Usage in 1973 and 1974 Passenger Cars." 19th Stapp Car Crash Conference, November 1975.
9. J. Danner, "Accident and Injury Characteristics in Side-Collisions and Protection Criteria in Respect of Belted Occupants." 21st Stapp Car Crash Conference, October 1977, Paper No. 770917.

10. H. Nielsen et al, "Lesions in Belted Car Riders from Oblique and Lateral Impacts." 1977 International Conference on Impact Trauma, Berlin, September 1977.

11. J. Shanks et al, "Injury Mechanisms to Fully Restrained Occupants." 23rd Stapp Car Crash Conference, October 1979, Paper No. 791003.

12. P. Niederer et al, "Adverse Effects of Seat Belts and Causes of Belt Failures in Severe Car Accidents in Switzerland During 1976." 21st Stapp Car Crash Conference, October 1977, Paper No. 770916.

13. F. Walz et al, "Belted Occupants in Oblique and Side Impacts." 1977 International Conference on Impact Trauma, Berlin, September 1977.

14. J. Horsch et al, "Response of Belt Restrained Subjects in Simulated Lateral Impact." 23rd Stapp Car Crash Conference, October 1979, Paper No. 791005.

15. D. Adomeit et al, "Expected Belt-Specific Injury Patterns Depend on the Angle of Impact." 1977 International Conference on Impact Trauma, September 1977.

872215

Chest Compression Response of a Modified Hybrid III with Different Restraint Systems

Egon Katz, Lothar Grosch, and Lothar Kassing
Daimler-Benz AG

ABSTRACT

Distribution of load has a major influence on type and severity of chest injuries. The introduction of the Hybrid III dummy into crash testing along with the requirement to measure sternum deflection for injury assessment has brought about the need to evaluate how well its thorax senses various loading conditions. Tests revealed that different load distributions, i.e. due to a diagonal shoulder belt or an airbag, did not produce the expected chest deflection patterns. This appears to be the result of both a relatively stiff sternum assembly and a nonsuitable design of the clavicle. Chest compression responses became more realistically when both, sternum and clavicle of the Hybrid II were mounted into the Hybrid III's thorax. As a consequence, this study suggests that the thorax of the Hybrid III dummy must be improved.

TESTS WITH BOTH CADAVERS AND ANIMALS confirmed that the chest acceleration alone is a poor indicator of chest injury potential and that the chest compression and other parameters derived from compression are superior predictors (1-7)*. Furthermore, several studies with Hybrid II dummies have demonstrated that a realistic evaluation of the effectiveness of restraint systems such as airbags, safety belts or combined systems should be based on individual measurements of rib deflection as accomplished by a system of strain gauges (8, 9). Results achieved with this technique (Fig. 1) support our experience from real world accidents of improved protection potential through the use of airbag systems (Fig. 2).

* Numbers in parantheses designate references at end of paper

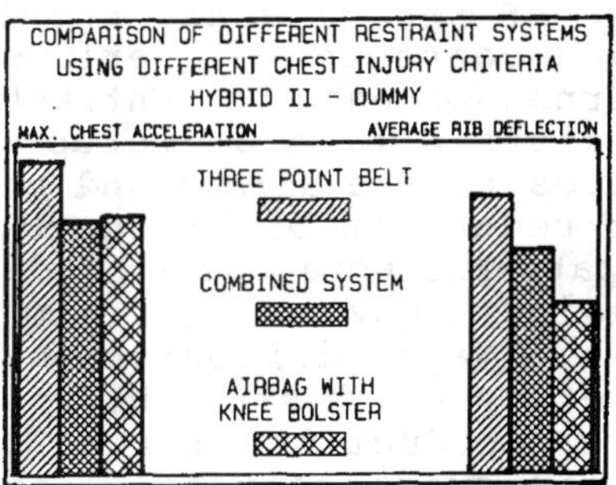

Figure 1 - Evaluation of chest injury risk using Hybrid II

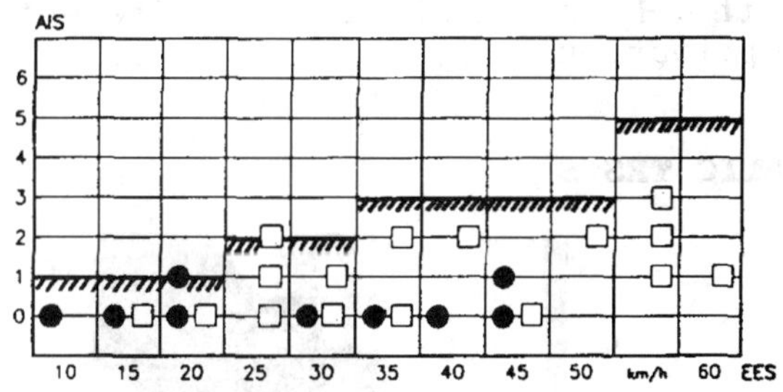

three point safety belt only
● airbag only
□ combined system

Figure 2 - Thoracic injuries in real world accidents with different restraint systems

Furthermore, it is interesting to see that the combined system does reduce chest injury severity, however, it does not prevent typical shoulder belt induced chest injuries.

As is evident from sled tests with cadavers conducted by the Institute of Forensic Medicine in Heidelberg (Fig. 3), high local forces due to shoulder belt loading frequently do cause multiple rib

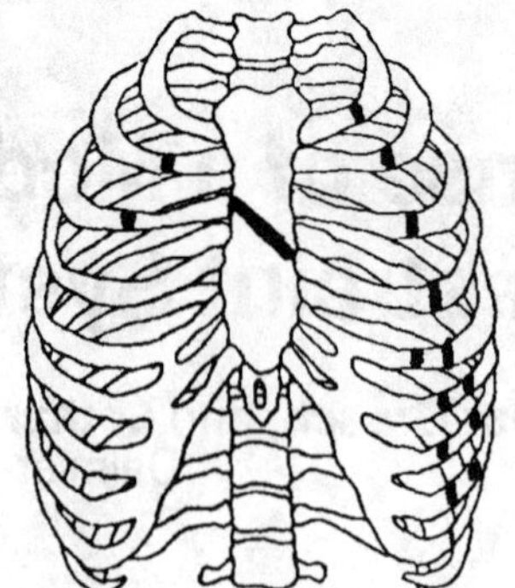

local pressure measured with Fuji-foil:
1-10 N/mm2
(peak values on the sternum and on the lowermost ribs)
pressure with airbag less than 1 N/mm2)

Figure 3 - multiple rib fractures in a cadaver test due to shoulder belt loading

fractures or even sternum fractures along the path of the diagonal shoulder belt and, as a consequence, serious injuries to internal organs. In contrast to this, by the distribution of forces over the wider area of the airbag and the corresponding reduction of pressure, no such damage patterns were observed due to airbag loadings. Obviously, distribution of load has a major influence on both type and severity of chest injuries. The projected introduction of the Hybrid III dummy into crash testing along with the requirement to measure sternum deflection for injury assessment has brought about the need to determine how well the thorax of the Hybrid III senses different load distributions.

STATIC TESTS

Figure 4 - Ribcage of the Hybrid III loaded with airbag

First of all, the ribcage was loaded in static tests (Fig. 4) with either a shoulder belt or a driver airbag (filled with foam). The belt was located similar to its position on a passenger car occupant. The compression was produced by pressing the ribcage into the restraint system until a force level of 2 kN was reached at the dummy's spine. Deflection measurements (Fig. 5) were taken between the spine and each single rib (70 mm to the left (d_1) and right (d_2) side of the center line). In addition to the 12 direct deflection measurements, each rib deflection (center of sternum relative to spine) was calculated from the strains (S_1, S_2) measured on the ribs on both sides close to the spine (9).

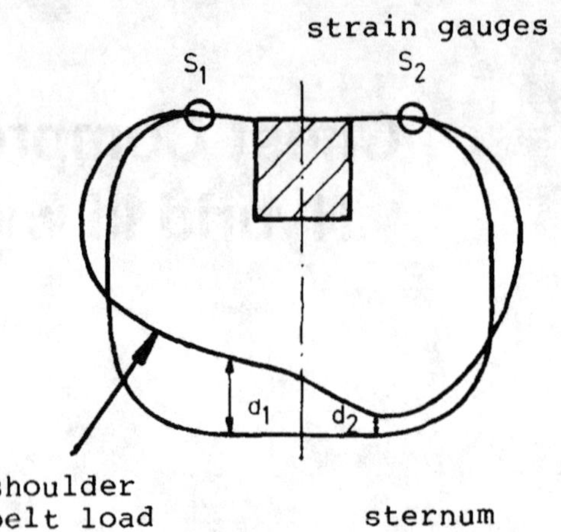

Figure 5 - Deflection and strain measurement - location of strain gauges

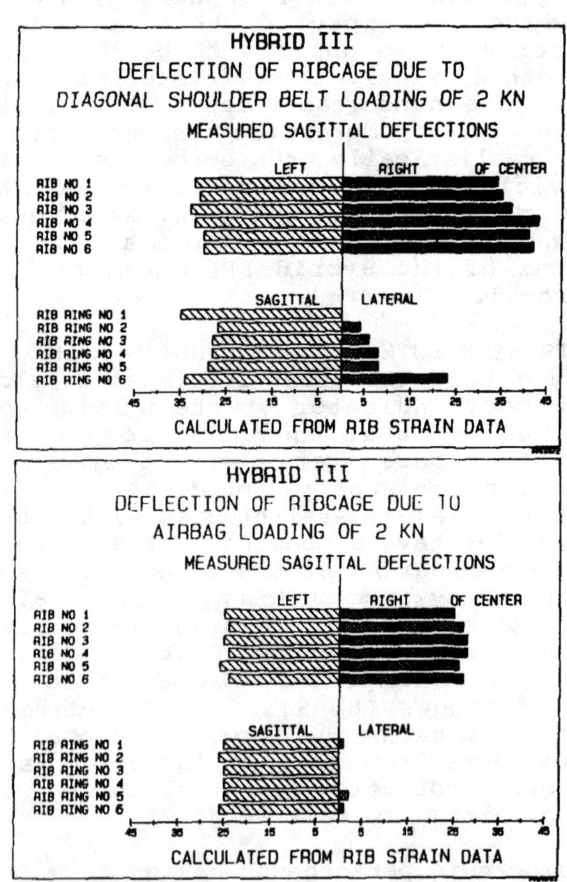

Figure 6 - Deflection of ribcage due to different restraint systems

Of course, static tests do not allow any conclusions about the stiffness of the restraint system. The same accounts for the dynamic damping effects built into the Hybrid III dummy ribs. However, the results summarized in Fig. 6 indicated that different load distributions did not produce the expected different chest deflection patterns. In comparison to the Hybrid II (10), where the path of the shoulder belt is distinctively reflected in the deflection pattern of the ribs, the deflections of the Hybrid III were nearly as uniform as with the airbag. Obviously, the sternum is so stiff in comparison to the ribs that the ribcage is only sensitive to the total force applied and individual rib deflection due to local forces cannot sufficiently occur. Furthermore, in comparison to the dominating sagittal deflections, lateral loadings may be underestimated with the Hybrid III. As a consequence, the sternum of the Hybrid III was exchanged by that of the Hybrid II, which is considered to be softer.

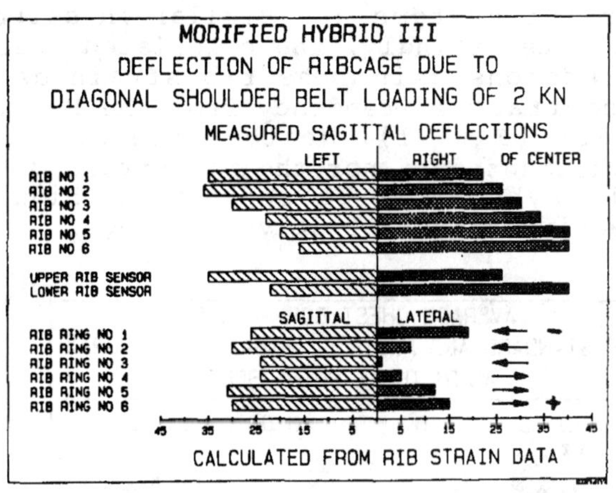

Figure 7 - Deflection of the Hybrid III's ribcage with the soft sternum assembly of the Hybrid II

The results (Fig. 7) reveal, that the thorax with the soft sternum assembly experienced deflection patterns similar to those of the Hybrid II, i.e. other than with the stiff sternum, the path of the diagonal shoulder belt was distinctively reflected in the rib deflections. Furthermore, lateral deflections were larger, which is particularly true for the lowermost ribs. It appears, because both local deflection and lateral loadings are sensed more distinctively, that the soft sternum is more suitable for monitoring different load distributions. It is interesting to note that the airbag produced the same uniform deflections with each of the sternum assemblies, that is the soft sternum does not change the stiffness of the thorax.

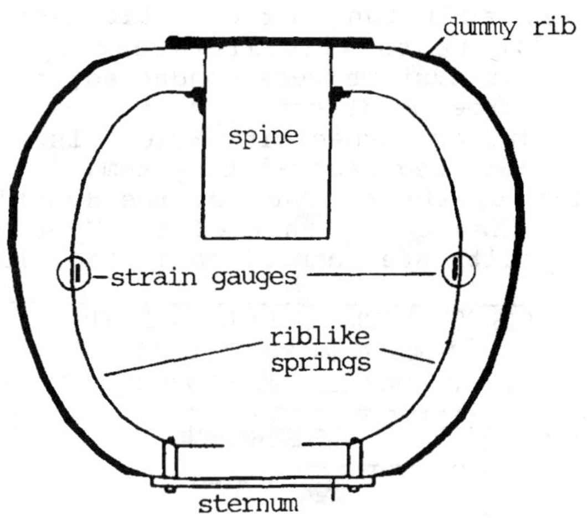

Figure 8 - Measuring technique with rib deflection sensors

Since the ribs of the Hybrid III, although the tolerance level of 2 inch for sternum displacement was not exceeded, took plastic deformations during the test program and had to be exchanged, a new technique to measure rib deflections was tested, where the strain gauges were not directly attached on the dummy's ribs. As shown in Fig. 8, extremely thin and soft springs similar in shape as a rib were attached between the sternum and the spine symmetrically on both sides of the center line - an upper pair on the second rib and a lower one on the fifth, each of them provided with strain gauges. From the measured strains, individual rib dedeflection can be calculated (9).

Particularly interesting is the distinctive belt load concentration on the clavicle. In contrast to the Hybrid II (10), without the clavicle mounted, the rib deflections are significantly larger due to shoulder belt loadings (Fig. 9).

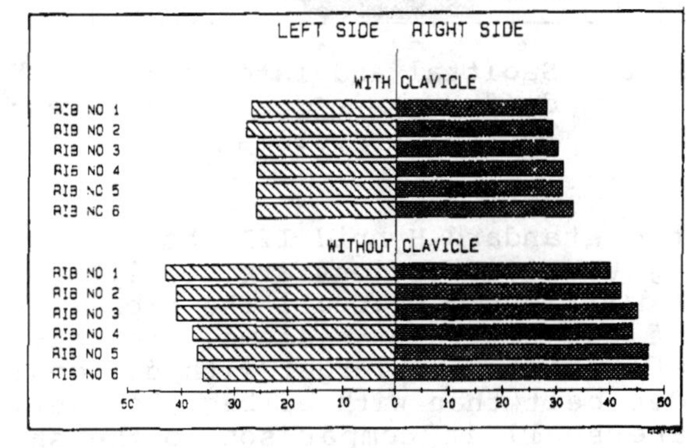

Figure 9 - Deflection of the Hybrid III's ribcage - with clavicle - without clavicle

SLED TESTS

In addition to the static tests, 30 mph sled tests with different types of restraint systems were conducted, namely:

- three point safety belt,
- driver airbag with knee bolster,
- combined restraint system.

At first, the sternum of the Hybrid III was exchanged by that of the Hybrid II; the results are summarized in Fig. 10:

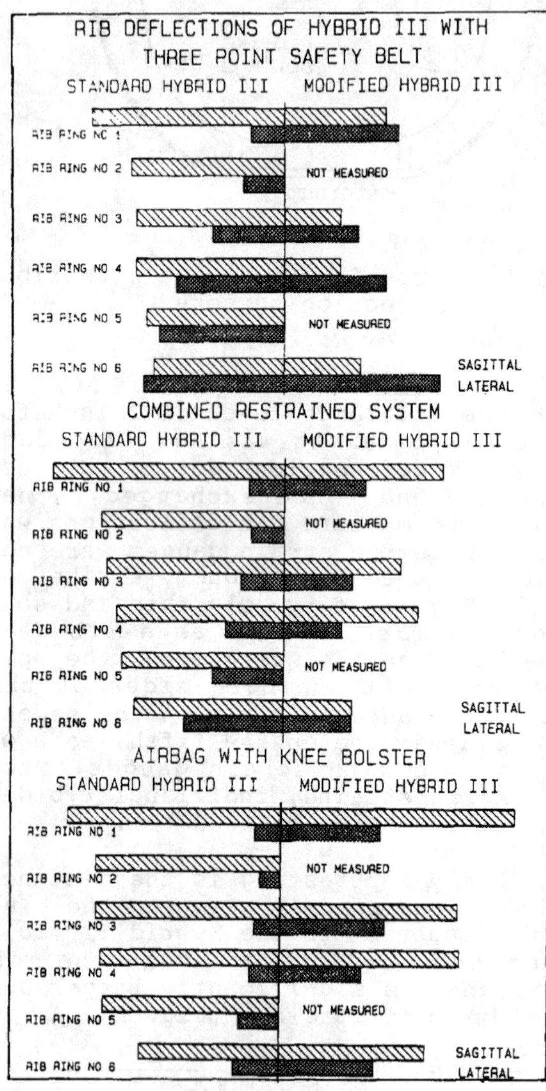

Figure 10 - Sagittal and lateral rib deflections due to different restraint systems

With the standard Hybrid III thorax, the sagittal deflections are nearly uniformly distributed over all the ribs with each of the restraint systems. The lateral deflections are larger with diagonal shoulder belt than with airbag, however, they are small in comparison to the sagittal components. Using the soft sternum assembly, both sagittal and lateral components measured with the three point belt are similar to those of the Hybrid II (10). However, it appears that the lateral deflections measured with the airbag, are unrealistically large. Further studies are necessary.

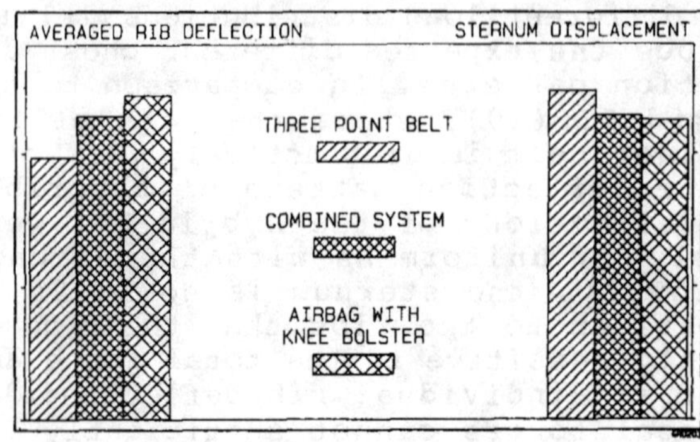

Figure 11 - Chest deflection of standard Hybrid III with different restraint systems

The sternum displacement values, measured with the centrally mounted internal gauge (Fig. 11), indicate that the evaluated restraint systems are similar in stiffness. Interestingly, the calculated chest compressions using the rib strain data are contrary in tendency from the Hybrid II data (Fig. 1). Neither the sternum displacement data nor the rib strain data of the Hybrid III do reflect the highly protective effect of driver airbags.

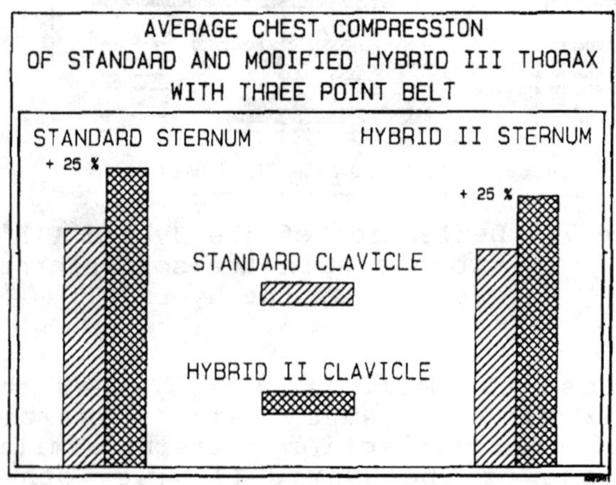

Figure 12 - Average chest deflections with different clavicles, measured with the riblike springs (Fig. 8)

With the Hybrid II's clavicle mounted on the Hybrid III's thorax (Fig. 12), the chest compressions due to shoulder belt loadings are significantly larger, with both, the standard and the soft sternum assembly. It appears that the belt load concentration on the Hybrid III's clavicle is too high and that the clavicle of the Hybrid II is more suitable.

SUMMARY AND CONCLUSIONS

Both the total stiffness of a restraint system relatively to that of the thorax and the distribution of load are important factors in preventing chest injuries on a passenger car occupant. However, in our experience of real world accidents, high local forces are dominant in producing injuries such as multiple rib fractures and serious injuries to internal organs. Therefore, the compression of a dummy's ribcage as a tool for evaluating restraint systems should reflect both, the stiffness of the restraint system and the distribution of load.

However, individual measurement of each rib deflection during static and dynamic tests with different restraint systems revealed that the ribcage of the Hybrid III, although superior in biofidelity, did not produce the expected deflection patterns when loaded with a diagonal shoulder belt. It appears, because of the stiff design of the sternum, that injury producing local forces are not sufficiently monitored. In contrast to this, the thorax of the Hybrid II, although it is considered to be too stiff, is obviously more suitable for monitoring different load distributions (10). As was shown, a softer sternum assembly, similar to that of the Hybrid II, appears to be more suitable. The same accounts for the Hybrid III's clavicle.

As illustrated in Fig. 13, different tolerance levels for chest deflection for seat belts on the one hand and airbags on the other hand, as required in FMVSS 208, do reflect the highly protective effect of the airbag. However, the question arises which tolerance level is valid for a combined system - generally 3 inch or

- 3 inch for a stiff airbag combined to a soft belt,
- 2 inch for a soft airbag in combination with a stiff belt?

Therefore, we would prefer realistic chest compressions with different restraint systems instead of different tolerance levels. As a consequence, this study suggests that the thorax of the Hybrid III dummy must be improved.

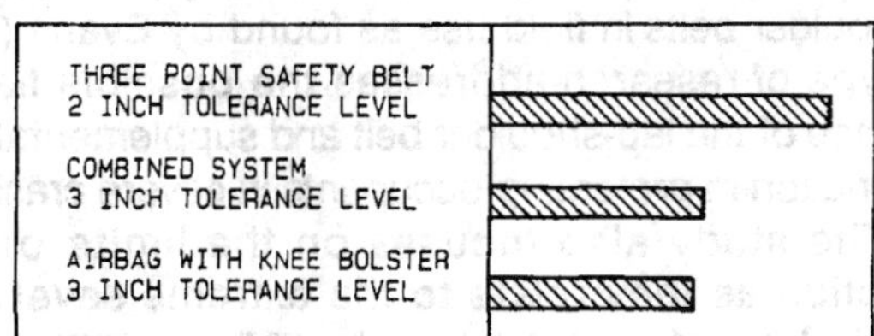

Figure 13 - Sternum deflections (Fig. 11) related to the corresponding tolerance levels in FMVSS 208

REFERENCES

(1) Viano, D., Lau, I.: Thoracic Impact: A Viscous Tolerance Criterion. 10th ITCESV, Oxford (1985).

(2) Eppinger, R., Marcus, J.H.: Production of Blunt Frontal Impact. 10th ITCESV, Oxford (1985).

(3) Kroell, C.K.: Thoracic Responses to Blunt Frontal Loading. SAE Special Publication P-67, Warrendale (1976).

(4) Lobdell, T.E., et al.: Human Torso Response to Blunt Trauma, Human Impact Response, Measurement and Simulation. Plenum Press, New York - London (1973).

(5) Fayon, A., Tarriere, C., et al.: Thorax of 3-Point-Belt Wearers During a Crash. SAE 751148. Proc. 19th Stapp Car Crash Conf., San Diego (1975).

(6) Walfisch, G., et al.: Tolerance Limits and Mechanical Characteristics of Human Thorax in Front and Side Impact and Transportation of these Characteristics into Protection Criteria. Proc. 7th IRCOBI Conf., Cologne (1982).

(7) Viano, D.: Chest Anatomy, Types and Mechanics of Injury, Tolerance Criteria and Units, and Injury Factors. Symp. on Biomechanics of Impact Trauma, San Diego (1983).

(8) Grösch, L.: New Injury Criteria for Combined Restraint Systems. 10th ITCESV, Oxford (1985).

(9) Grösch, L., Katz, E., Kassing, L.: New Measurement Methods to Assess the Improved Injury Protection of Airbag Systems. 30th Ann. Proc. of AAAM, Montreal (1986).

(10) Grösch, L., Katz, E., Kassing, L.: Chest Compression Response of Hybrid III with Combined Restraint Systems. 11th ITCESV, Washington (1987).

922523

Crash Injury Prevention: A Case Study of Fatal Crashes of Lap-Shoulder Belted Occupants

David C. Viano
General Motors Research and Environmental Staff

ABSTRACT

A case study was conducted of 123 crashes involving 144 fatally injured lap-shoulder belted front-seat occupants. The crashes occurred throughout the United States in 1985-86 and involved 97 driver and 47 right-front passenger deaths in new vehicles. A judgment was made by consensus of a safety panel on the potential for saving the victim's life by the addition of safety technology. Supplemental airbags provided the greatest potential for improving the life-saving effectiveness of current lap-shoulder belts. Overall, airbags may have prevented 12% of the belted occupant fatalities and 27% of the deaths in frontal crashes. The benefit of supplemental airbags was greater for the right-front passenger, in part, because of more females and occupants over 60 years of age in that seating position. A majority (68%) of the belted fatalities were judged unpreventable by reasonable restraint or vehicle modifications. This level is indicative of the extreme severity of many of the fatal crashes involving extensive vehicle damage and forces on the occupant compartment, unusual crash configurations and causes of death, and unique situations related to seating position and crash dynamics. Using published levels of belt effectiveness, 50% of all fatalities may not be preventable by the use of lap-shoulder belts, supplemental airbags, and practicable changes in crashworthiness systems. The study addressed only the potential benefit of supplemental-restraint components and airbags and did not consider possible adverse effects which might occur in real-world crashes.

INTRODUCTION

The choice of occupant restraint systems for passenger vehicles has been changing rapidly with the phase-in of passive restraint requirements. However, there are many factors involved in the ultimate choice including safety, comfort, convenience, cost and acceptability. A key factor seems to be the fatality reduction potential of a system and its components. Current highway crashes provide excellent data to analyze the potential of safety systems to prevent fatal injuries. For this reason, a panel reviewed fatal accidents involving lap-shoulder belted front-seat occupants of current model vehicles.

A panel of experienced safety researchers conducted a case-by-case evaluation of the field accident data. The data reflect the fatal crash experience of new-car owners who are restrained at the time of the impact. This is a specific cohort of drivers but a group that is older and selectively safer because of the voluntary use of safety belts. It is, however, an important subset of road users, since analysis of this group's crash experience provides insights on additional protection of this upper socioeconomic segment of society.

Specific circumstances and details of each belted fatality were evaluated and a consensus judgment reached on the possibility of preventing the fatality by the addition of a supplemental airbag, restraint feature such as belt pretensioner or grabber, interior padding, and seat or vehicle structural modification. These judgments were used to estimate the effectiveness of a combined lap-shoulder belt and supplemental airbag for the driver and right-front passenger.

This case study used an approach that is similar to that used in previous investigations of the fatality prevention effectiveness of lap-shoulder belts for unrestrained occupants (Huelke et al. 1979, Wilson et al. 1973, Pursel et al. 1978, Hartemann et al. 1977) and the potential benefits of supplemental restraint components for occupants wearing safety belts (Danner et al. 1984, Mulligan et al. 1987, Gloyns et al. 1989). Judgments by safety panels have been helpful in projecting the potential of restraint systems in preventing injuries. This is seen by comparing the projections of Wilson et al. (1973) and Huelke et al. (1979) with the actual safety performance of lap-shoulder belts in field use as found by Evans (1986a). This type of research addresses the possible field performance of the lap-shoulder belt and supplemental airbag combination in protecting occupants in severe crashes.

The study also focuses on the limits of crash protection as they relate to the extreme severity and unsurvivability of many fatal crashes (Viano, 1988a, 1988b, Thomas et al. 1990). This perspective is important as emphasis is increasing on crash avoidance. Driving aids, driver monitoring, vehicle controls, and collision warning systems are crash avoidance approaches that represent a next generation of major highway safety improvements (Viano 1988a).

METHODOLOGY

FATAL ACCIDENT DATA - One hundred and twenty three accidents were selected on the basis of a fatality of a front-seated occupant restrained by a lap-shoulder safety belt. The cases were part of the Motors Insurance Corporation (MIC) files involving a $10,000 insurance policy payment from an incentive program in 1985-86 for safety belt wearing. The policy was provided with all vehicles sold over an eighteen month period. These cases represent all of the fatal crashes of lap-shoulder belted occupants, and they were particularly well documented by insurance people and safety engineers.

A Collision Performance and Injury Report (CPIR) was filled out for each accident by a trained insurance adjuster investigating the claim. The cases were limited to fatal injury accidents involving a 1985-86 model vehicle. Included in the CPIR is information on environmental conditions, locality, collision details, vehicle damage, occupant statistics, occupant injuries and other pertinent crash data. This information is compiled from police reports, medical reports, witness accounts and measurements taken by trained personnel of the vehicle and the accident scene. These files usually include 35 mm photographic slides of the case vehicle and accident scene.

DETERMINATION OF PRE-CRASH FACTORS - The driving circumstances prior to the accident were determined by review of each file. In some cases, only a single condition could be identified as contributing to the accident; however, in others a total of two or three conditions were involved in the sequence of events preceding the crash. The following is a list of the conditions identified as factors in the fatal crashes involving safety belted occupants: 1) Traffic Violation, 2) Excessive Speed, 3) Intoxication/Drug Use, 4) Physical Condition, 5) Loss of Control, 6) Driver Error, 7) Inattention, 8) Weather/Road Condition, 9) Nothing Driver Could Do, 10) Unknown.

OCCUPANT INFORMATION AND CRASH TYPES - Data from each crash was summarized by the principal location of impact on the case vehicle and object struck. There were 144 fatally injured occupants who were lap-shoulder belted at the time of the accident. Occupant data was summarized by sex, age, seating position and time-of-death. The category "dead at scene" included all victims identified in the police report or accident investigation as being killed by the crash. In the other cases, victims survived the crash, were transported to hospital, and died within 45 days. Mechanism of death was determined for each victim.

METHOD OF EVALUATING FATALITY PREVENTION - A panel of experienced safety researchers performed the analysis of each crash and fatality. The panel included individuals with experience in design, development and testing of the vehicle crashworthiness, occupant restraints and crash protection systems. These individuals worked together and analyzed each fatality.

Relationship to Other Studies - The research approach in this study is similar to that used in previous case studies of seriously injured crash victims (Huelke et al. 1979, Wilson et al. 1973, Pursel et al. 1978). Although the previous studies are similar, this investigation deals only with belt restrained occupant fatalities so comparisons with earlier results from unbelted fatalities are not possible.

The Panel Evaluation - The panel met monthly during 1988 for about three hours to review about 25 cases. For each case, the following information was presented by an engineer involved in collecting the case information. Frequently, the engineer had personally investigated the crashed vehicle. Each case included: 1) A sketch of the accident site and vehicle trajectories leading up to and following the crash, 2) Slides of the accident scene and crashed vehicle(s), 3) A description of the accident circumstances, 4) A discussion of occupant injuries, at scene emergency medical services, and subsequent medical treatment.

For each case the significant details were discussed about the accident site, vehicle trajectories and causes of the crash. The 35 mm color slides of the vehicle exterior and interior were discussed as they pertained to an understanding of occupant injury mechanisms and possibilities to prevent the fatality.

A group discussion followed the review of the crash information and focused on the pros and cons of various approaches to potentially saving the victim's life. The diverse experience and knowledge of the panel provided unique insights into the causes of the fatality and potential for injury prevention. The group reached consensus on potential interventions although there were differing views voiced sometimes during the discussion. The agreed upon recommendations and conclusions were recorded as the consensus judgment of the panel.

Approaches Considered - As in any case investigation of this type, there are many accident variables. A systematic approach was taken for possible vehicle improvements. A questionnaire was developed and tested on several cases not included in the actual study. The system was set up to produce statistical information.

We used a two-level system of reporting judgments. The first level involved the type of intervention and the second the level of significance of the intervention in potentially preventing the fatality. We categorized four different types of intervention using the following scheme:

Accident Avoidance	Vehicle Changes
Improved Handling	Body Structure
Improved Braking	Interior Padding
Warning System	"Wing" Seat
Other	Stronger Seat
	Other

Restraint System Changes	Probably No Help
Airbag or Inflatable Restraint	
Better Restraint System	
Belt Webbing Grabber	
Belt Pretensioner	
Belt Load Limiter	
Other	

Levels of Significance - The classifications are fairly broad and incorporate a relatively wide range of injury prevention conditions. For each variable considered possibly helpful in preventing the fatality if available and used in the crash, a level of significance of injury prevention was assigned. The approach related to the proportion of cases in which the intervention was judged to be effective: 1) probably no help (0%), 2) maybe preventable (25%), 3) probably preventable (50%), and, 4) definitely preventable (100%). If an intervention was considered helpful in preventing a fatality, the panel reached consensus on the significance of the intervention to influence injury outcome.

DETERMINING RESTRAINT EFFECTIVENESS - Since the cases involved only fatalities of lap-shoulder belted occupants, we do not know the number of accidents in which fatal injury was prevented by safety belt use. It is possible, however, to estimate the total population of crashes by using published levels of safety belt effectiveness for the driver and right-front passenger seating positions.

Let N represent the total number of belted occupants involved in the population of serious crashes in which we have collected all of the fatal injuries (Figure 1). If D is the number of belted occupants that died in these crashes, the number N of involved occupants can be calculated using the published effectiveness of lap-shoulder belt use (E_B) in preventing fatal injuries. However this calculation assumes the published belt effectiveness level is representative of our sample of crashes:

$$N = \frac{D}{1-E_B}. \qquad (1)$$

The number of occupants saved S by safety belt use can be calculated from:

$$S = N-D = \frac{E_B}{1-E_B} D = E_B N. \qquad (2)$$

Now, let S_+ represent the number of occupants saved by the addition of a supplemental airbag, restraint technology, or vehicle modification. The percent of occupants saved $\%_S$ is:

$$\%_S = \frac{S_+}{D}. \qquad (3)$$

The level of effectiveness E of the combination safety belt and additional technology is given by:

$$E = E_B + \frac{S_+}{N} = E_B + \%_S (1 - E_B). \qquad (4)$$

The percent improvement $\%_+$ in fatality prevention or crash safety can be calculated using:

$$\%_+ = \frac{S_+}{S} = \%_S \frac{1-E_B}{E_B}. \qquad (5)$$

RE-EVALUATION OF RANDOMLY SELECTED CASES - The conclusions of this study are, in general, unbiased since all of the accidents in the MIC files for the 1985-86 time period were considered sequentially. However, the vehicles were General Motors' vehicles of the 1985-86 model years. The distribution of vehicles included: 30% small-compact cars, 33% mid-sized cars, 15% sports cars, 10% large-luxury cars, and 12% light trucks and vans.

An evaluation of the repeatability of the panel's consensus on injury prevention was performed by the group at the completion of the cases. It involved a series of 15 cases selected at random. In this re-analysis, 10 of 15 cases were identical in judgment and the other cases were within one level of significance and all involved judgments of either probably no help or maybe, the lowest levels of possible intervention. This indicated acceptable repeatability.

SPECIAL CASES - Although each crash was unique, six were considered to have special circumstances that made them particularly not amenable to fatality intervention. In one case, the victim drowned in a pond as the vehicle was submerged. In another case, the impacting car virtually landed on top of the victim's vehicle resulting in immediate decapitation. In two cases there was probable heart failure prior to the accident. There were possibly two cases of suicide.

RESULTS

Table 1 provides the distribution of fatal crashes by location of principal impact on the case vehicle. About half of the crashes involve deformation of the front of the case vehicle while about 35% involve side impact loading. This distribution of crashes is typical of national statistics from the Fatal Accident Reporting System (Figure 2) except for a lower incidence of rollover crashes and impacts involving damage to the top or roof of the vehicle (NHTSA 1988).

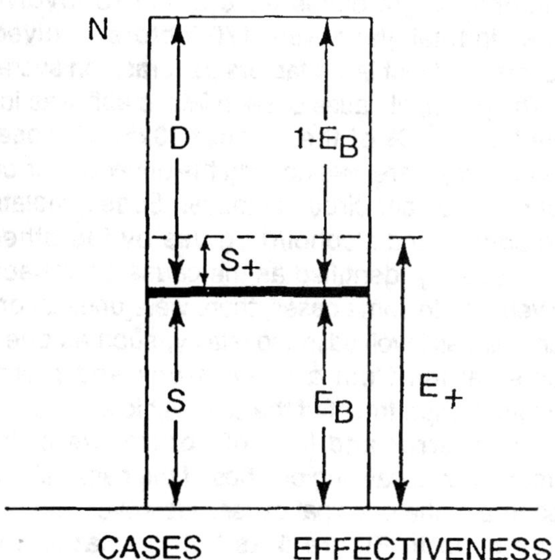

Figure 1: Number of occupants involved and effectiveness of occupant restraints for a population of serious crashes involving lap-shoulder belted occupants.

Table 2

Distribution of Fatal Crashes by the Type of Object Involved with the Case Vehicle

Object Struck	Cases (Frequency)
Car	55 (44.7%)
Pickup	16 (13.0%)
Van	3 (2.4%)
Tractor-Trailer	15 (12.2%)
Heavy Truck	5 (4.1%)
Train	3 (2.4%)
Bus	1 (0.8%)
Fixed Object	7 (5.7%)
Pole/Tree	9 (7.3%)
Guardrail/Wall	4 (3.3%)
Embankment/Ditch	4 (3.3%)
Water	1 (0.8%)
Total	123 (100%)

Table 1

Distribution of Fatal Crashes by Location of Principal Impact on the Case Vehicle

Point of Principal Impact	Cases (Frequency)
Front	64 (52.0%)
Side	
Driver	18 (14.6%)
Passenger	25 (20.3%)
Rear	6 (4.9%)
Rollover	4 (3.3%)
Top	2 (1.6%)
Noncrash	1 (0.8%)
Unknown	1 (0.8%)
TOTAL	123 (100%)

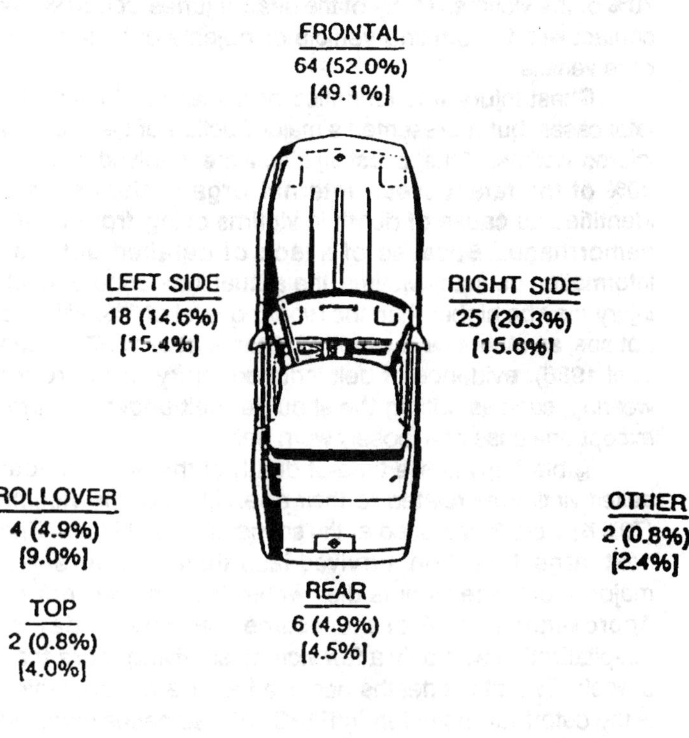

Figure 2: Distribution of 123 fatal crashes involving lap-shoulder belted, front-seated occupants by location of principal impact on the case vehicle (one unknown and one noncrash not shown). Frequencies in brackets are from the 1987 FARS (NHTSA 1988).

Table 2 describes the distribution of fatal crashes by the type of object struck. For this data set, a majority of crashes were multi-vehicle impacts. Approximately 80% of the crashes involved another moving vehicle, 60% where passenger vehicles, pick-up trucks or vans. About 20% of the multi-vehicle crashes involved a heavy truck or vehicle. This level of multi-vehicle involvement is higher than the national average, which was 54% in FARS for first harmful event in passenger car fatalities. The MIC fatal crashes involve fewer single-vehicle accidents.

Table 3 shows the distribution of fatally injured occupants by sex and seating position. Two-thirds of the victims were drivers. This is slightly lower than the 75% involvement in FARS. There was an approximately equal number of belted women and men killed in the crashes, and this is significantly different from the over-involvement of male victims in FARS (75% v 25%). The MIC crashes involve more female and right-front passenger victims than the average national fatal crash.

Injuries to the head and brain represented the most frequent cause of death to the accident victims (Table 4). Since the majority of cases classified as dying from multiple injuries involved head injury and major chest or abdominal trauma, serious head injuries occurred in about 70% of the victims. Many of the head injuries occurred by contact with the striking vehicle or objects outside of the case vehicle.

Chest injury was identified principally in 11% of the fatal cases, but represented a major fraction of the multiply injured victims. Thus, chest injuries were involved in up to 30% of the fatal cases. Internal organ injuries were identified as cause of death in victims dying from major hemorrhage. Because of a lack of detailed autopsy information on many victims, the actual incidence of neck injury may be higher than the reported level of 7%. We did not see, as others have reported (States et al 1987, Green et al 1986), evidence of belt induced injury or improper wearing, such as putting the shoulder belt under the arm, except one case of a loosely worn belt.

Table 5 gives the time-of-death of the lap-shoulder belted victims as related to their age. Most of the victims (70.8%) were found dead at the scene of the accident. The next largest fraction survived less than 24 hours -- a majority of these victims died within the first two hours. Approximately 12% of the victims died after a day of hospitalization with only a few victims surviving more than a week. Two of the deaths occurred after a month, which is the cutoff for inclusion in FARS. These cases involved significant medical complications over the treatment period.

Table 6 summarizes the factors in the accident report related to the cause of the fatal crash. In a majority of cases (65.5%), only one factor was identified as being involved in crash causation. In 33 cases (27.7%) two factors were identified as precipitating the crash. Most often, loss of control was related to evidence of excessive speed, driver inattention or poor weather and road conditions. Eight cases were found to involve three factors. In total, there were 170 factors involved in the accidents, and that is 1.4 factors per crash on average.

The principal cause of each fatal crash was identified (Table 6). In 31% of the crashes (32% of those with a known cause) there was nothing the driver did or could do about the accident circumstances. Speed, violation of a traffic control and alcohol/drug use by the other driver were frequently identified as the cause of impact of the case vehicle. In some cases, there were unusual or special circumstances involved in the fatality, such as one case in which steel fell from a heavy truck and pierced the occupant compartment of the case vehicle.

Driver error and loss of control were the most common factors causing crashes of the case vehicle. This is also true of the principal cause. However, many factors can be seen in Figure 3 as being related to crash causation. In over two-thirds of the crashes, a driver-related cause was identified. Some of the crashes involved evidence of aggressive driving behavior such as excessive speed, a traffic violation, or intoxication; and the remainder involved an apparent, inadvertent driving mistake.

Table 7 summarizes the judgments of the panel on the more obvious cases for crash avoidance or additional crash protection to prevent the lap-shoulder belted fatalities. Three levels of potential intervention were used: definite help, probable help, and maybe help in preventing the fatality. The judgments were based on fatally injured occupants so the sum for all of the cells is 139. In a majority of the crashes (90/139 = 64.7%) neither improvements in avoidance or protection were judged to be helpful in saving the victims' life. There were seven cases (5.0%) where the fatality was judged to be definitely preventable, 22 cases (15.8%) where the death was probably preventable, and 20 cases (14.4%) where help maybe was possible by crash avoidance approaches. Forty-five of the 49 lives potentially savable were related to additional crash protection technologies.

For the cases of crash avoidance, consideration was limited to the more clear-cut approaches that may have been helpful once the string of events leading to the crash, per se, had initiated. There were 14 cases in which accident avoidance technologies were judged to have potentially helped prevent the fatalities of the lap-shoulder belted occupants. The one case of definite intervention involved improved braking such as anti-lock brakes. A 21 year old female driver was speeding on a wet road surface and lost control of the vehicle while applying the brakes. The vehicle went off the road and struck a pole head on.

Three of the five cases of probable accident avoidance involved warning systems such as radar alert of an obstacle in the path of the vehicle. In one case, an elderly man and wife were driving late at night on a two-lane rural road in foggy conditions. They drove into the side of a train passing in front of their vehicle. This is also a

Table 3

Sex and Seating Position of the Fatally Injured Front-Seat Occupants

	Male	Female	Total
Driver	58	39	97 (67.4%)
Passenger	16	31	47 (32.6%)
Total	74 (51.4%)	70 (48.6%)	144 (100%)

Table 4

Principal Cause-of-Death for the Belted Victims

Cause of Death	Incidence (Frequency)
Head/Brain	57 (39.6%)
Multiple Injuries	42 (29.2%)
Chest	16 (11.1%)
Internal	10 (6.9%)
Neck	10 (6.9%)
Burn	4 (2.8%)
Drown	1 (0.7%)
Unknown	4 (2.8%)
Total	144 (100%)

Table 5

Time-of-Death for the Belted Victims by Their Age

Age (yr)	Dead at Scene	Time of Death 1-24 hr	1-7 days	1-4 weeks	Month +	Total
0-9	1	1	2	0	0	4 (2.8%)
10-19	6	1	0	0	0	7 (4.9%)
20-29	25	9	2	0	0	36 (25.0%)
30-39	16	5	2	1	1	25 (17.4%)
40-49	9	1	1	1	0	12 (8.3%)
50-59	6	3	0	0	0	9 (6.3%)
60-69	20	3	2	3	1	29 (20.1%)
70-79	17	1	0	1	0	19 (13.2%)
80-89	2	0	1	0	0	3 (2.1%)
	102 (70.8%)	24 (16.7%)	10 (6.9%)	6 (4.2%)	2 (1.4%)	144 (100%)

Table 6

Factors Causing Fatal Accidents
Based on Evaluation of Crash Reports
(Reprinted from Viano (1990) with Permission)

Crash Causation Factors	Factors Involved One	Two	Three	Total	Principal Cause
Traffic Violation	6	8	3	17(14.3%)	8(6.7%)
Excessive Speed	-	11	5	16(13.4%)	9(7.6%)
Intoxication	1	-	1	2(1.7%)	1(0.8%)
Physical Condition	2	-	-	2(1.7%)	2(1.7%)
Loss of Control	9	22	6	37(31.1%)	21(17.6%)
Driver Error	13	6	4	23(19.3%)	17(14.3%)
Inattention	4	10	3	17(14.3%)	13(10.9%)
Weather/Road Condition	2	9	2	13(10.9%)	7(5.9%)
Nothing Driver Could Do	37	-	-	37(31.1%)	37(31.1%)
Unknown	4	-	-	4(3.4%)	4(3.4%)
	78	66	24	170(143%)	119(100%)

Four cases were excluded from consideration because of insufficient information to render a judgment about the crash.

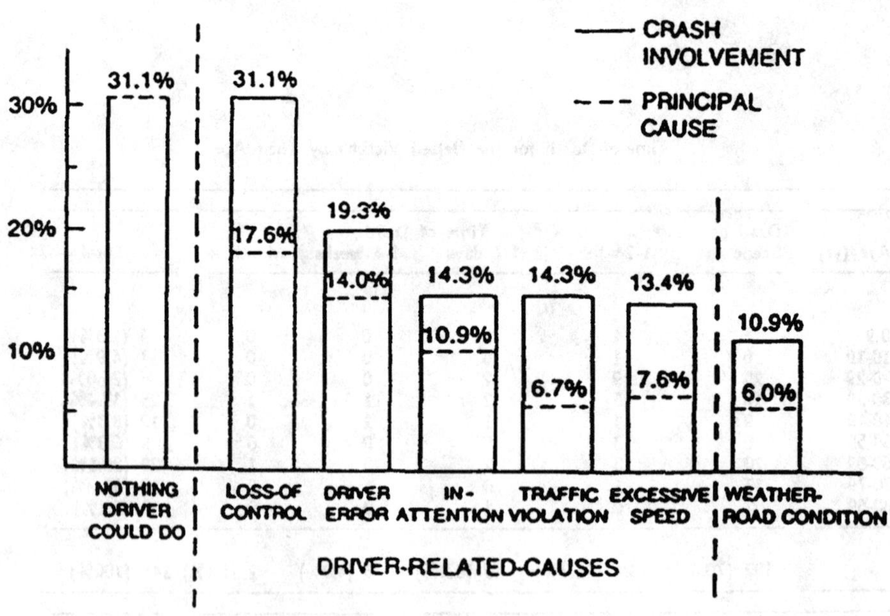

Figure 3: Frequency of involvement of factors in causing the fatal crash and principal cause of the accident.

Table 7

Judgment of Technologies to have Potentially Saved the Life of the Belted Victim Either by Additional Crash Protection or Crash Avoidance

Crash Avoidance	Crash Protection Definite	Probable	Maybe	No
Definite	-	-	-	1
Probable	-	2	2	1
Maybe	2	3	1	2
No	4	14	17	90

Table 8

Fatal Cases Judged to Have Been Potentially Preventable by Additional Crash Protection

	Definite	Probable	Maybe
All Cases	6	19	20
Airbag Cases	6 (2/4)[1]	17 (19)[2]	10
Airbag Only	2	6	6
Seat Wing/Structure	-	- (1)	1
Belt Pretensioner	- (1/3)	1 (11)	3
Belt Grabber	-	- (1)	-
Load Limiter	-	- (2)	-
Interior EA	- (-/1)	-	-
Vehicle Structure	- (1/-)	1 (2)	1
Non-Airbag Cases[4]	-	2	10
Seat Structure/Wing	-	1 (2)[3]	7
Interior EA	-	1	3

[1] These cases of definite airbag help include the judgment in parenthesis of probable/maybe help with belt system or vehicle improvements.

[2] These cases of probable airbag help include the judgment in parenthesis of maybe help with belt system or vehicle improvements.

[3] Includes judgment of additional life-saving potential maybe by a seat wing and improved interior energy absorption.

case in which a driver and passenger airbag were judged to have probably prevented the fatalities. In another case of probable intervention, a 15 year old female passenger was killed in a crash on a foggy road. The driver stopped and proceeded to make a U-turn in front of oncoming traffic. A warning system may have alerted the driver and averted the crash. In the remaining cases, improved handling and braking were judged as maybe helping to prevent the fatal crash. Most of these cases were also judged to have been helped by an airbag, or an improved restraint or seat structure.

Table 8 lists the 45 fatalities judged potentially preventable by additional crash protection. The principal approach to saving the belted victim's life was by the use of a supplemental airbag. The six cases judged to be definitely preventable were related to the use of a supplemental airbag. Five of the six victims were women and three were over 60 years old. Five of the victims survived 4-45 days.

The 19 cases of probable injury prevention involved 17 (89.5%) situations in which an airbag was judged to have been helpful. In these cases 9 (52.9%) involved women, 10 (58.8%) the right-front seating position and 9 (52.9%) occupants over 60 years old. The two cases of probable prevention without an airbag involved side impacts in which improved energy absorption of the interior probably would have helped prevent the fatality of a near-side seated occupant. In the other case, a winged seat would have helped minimize lateral displacement of a far-side seated occupant. The maybe help cases had an equal number of accidents in which an airbag, seat restraint or interior technology might have helped prevent the fatality. In general, all cases in which the airbag was considered a potential intervention were frontal impacts, whereas the cases of seat, restraint or interior-related interventions were essentially side-impact crashes.

There were 17 accidents in which the case vehicle was struck by a tractor-trailer. In these cases, 14 (82.4%) of the fatalities were judged to be unpreventable. In the 9 cases of heavy truck, bus or train involvement, 7 of the 12 occupant fatalities (58.3%) were judged to be unpreventable. We did not see a recurring incidence of under-run in the heavy vehicle crashes, although extensive vehicle damage was frequent. In the few frontal impacts where intervention was judged possible, the airbag was the most frequently sighted approach to preventing the fatality.

Table 9 summarizes the improvements in fatality prevention effectiveness that are possible by enhancing or augmenting the safety belts. Lap-shoulder belts of the type in 1985-86 products have a 42% effectiveness in preventing fatal injury for drivers and 39% effectiveness for right-front passenger. The calculation of effectiveness is based on analysis of FARS data using the method of double comparison developed by Evans (1986a, 1986b). The effectiveness level implies that out of 100 fatally injured but unrestrained drivers and passengers, lap-shoulder belt use would have prevented 42 driver fatalities and 39 right-front passenger deaths.

If all six of the fatalities in the definitely saveable category had been protected by an airbag, the effectiveness of the lap-shoulder belts and supplemental airbag would be 44% for the driver and 42% for the passenger. This calculation uses equations 1-5 and is based on 4 driver and 2 passenger fatalities definitely preventable by airbag use. If all of the definite and probable cases had been saved by a supplemental airbag, the combined effectiveness would be 49% and 55% respectively.

The projected effectiveness of the combined lap-shoulder belt and supplemental airbag system is 47% for the driver and 50% for the right-front passenger. This represents a potential saving of all of the lives judged definitely preventable, 50% of the cases judged probable and 25% of the cases identified as maybe preventable. The greatest effectiveness of the supplemental airbag was found for the right-front passenger seating position and involved preventing fatalities of female passengers -- female drivers were also the principal benefactor of the supplemental airbag. Thus, supplemental airbags would help prevent 10% of belted driver fatalities and 17% of belted passenger fatalities. This would potentially increase the safety effectiveness of lap-shoulder belts by 12% and 28% respectively. Seats and interior modifications were judged to have only a minor benefit in improving restraint effectiveness.

The effectiveness of lap-shoulder belts combined with airbags and seat, restraint and interior improvements are also given in Table 9. The projected maximum restraint effectiveness for all of the possible improvements is 47% (with a range in projection of 4% - 27%) for the driver and 51% (with a range in projection of 4% - 43%) for the right-front passenger.

Figure 4 summarizes the projected estimates of restraint effectiveness with supplemental airbags. It is based on the number of occupants fatally injured plus the estimated number of occupants who were saved by belt use. The 67 drivers saved fatality by use of the available lap-shoulder belts is based on 42% effectiveness (Evans 1986a, 1986b). Similar calculations were used for the passenger seating position. Based on the projected effectiveness with the combination lap-shoulder belt and supplemental airbag, 76 out of the 160 drivers would have been saved by the use of combination safety belts and airbag. This represents a potential safety effectiveness level of 47% in preventing fatal injury.

Figure 4 also indicates that 84 fatalities with safety belt and airbag use are unpreventable. This leads to 53% of the fatalities being unpreventable by the use of lap-shoulder belts and airbags. For the right-front seating position, the projected level of lives saved is 38 out of 76 and the projected effectiveness is 50%. The combined effectiveness for both seating positions is 48%, indicating that 52% of occupant fatalities are unpreventable by the combination of belts and airbags. When all potential changes in occupant protection are considered, 20 of the 139 fatalities (14%) are projected to have been pre-

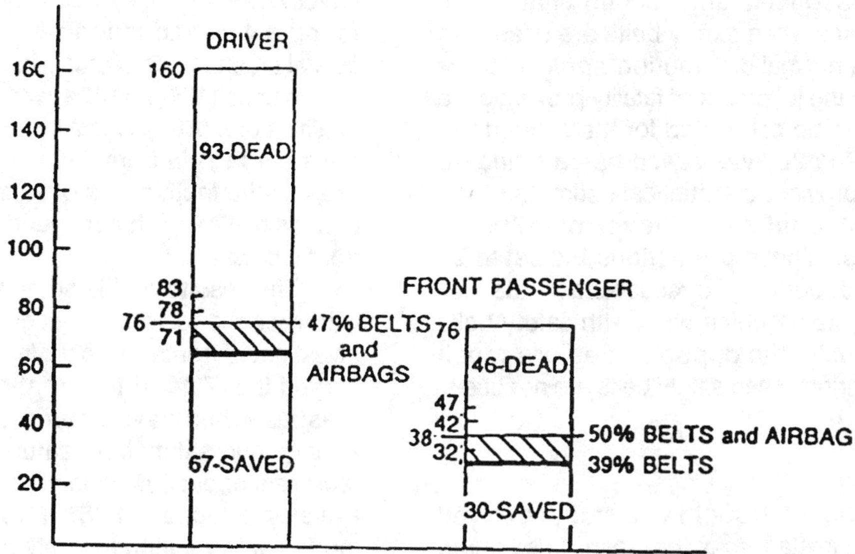

Figure 4: Number of lap-shoulder belted drivers and passengers involved in severe crashes in which 93 driver and 46 passengers were fatally injured. Shown are the potential lives saved and restraint effectiveness levels for safety belts and supplemental airbags in both seating positions based on the case analysis in this study.

Table 9

Projection of Effectiveness of Safety Belts Combined with Supplemental Airbags and Vehicle Interior Improvements Based on the Judgment of Preventable Fatalities

	Occupants Helped Number	Percent	Percent Effectiveness	Percent Improvement
Belts Only (Evans 1986)			42/39	
Supplemental Airbag				
Definite	4/2	4/4	44/42	5/8
Def + Probable	11/12	12/26	49/55	17/41
Def + Prob + Maybe	16/17	17/37	52/62	24/59
** Projection **	9/8	10/17	47/50	12/28
Seats + Interior				
Prob	1/1	1/2	43/41	2/5
Prob + Maybe	9/3	10/7	48/43	14/10
** Projection **	3/1	3/2	44/41	5/5
Airbag + Seats + Interior				
Definite	4/2	4/4	44/42	5/8
Def + Probable	12/13	13/28	49/57	17/46
Def + Prob + Maybe	25/20	27/43	58/66	38/69
** Projection **	11/9	12/20	47/51	12/31

Calculations are based on 93 drivers and 46 front-right passengers fatally injured in the crashes. The projection is based on preventing the fatality of all those judged in the definite category, half of those in the probable category and a quarter of those in the maybe category.

ventable. Based on this analysis, 50% of all fatalities are unpreventable by reasonable engineering changes in crashworthiness systems when safety belts are used with airbags. Assuming a normal distribution applies to the cases analyzed and in the judgment of fatality prevention, a standard deviation can be calculated for the estimate of fatality prevention. The 20 lives saved has a standard deviation of 4.5. This provides a statistical estimate of (50 ± 1.5)% and a 95% confidence level of 47% for unpreventable deaths. These projections indicated as much as a 6% further opportunity to reduce crash fatalities by crashworthiness systems which work with safety belts and airbags. Obviously, the opportunities are much greater for friendly interiors when safety belts are not used.

DISCUSSION

Each passenger-car fatality in this study involved unique events and circumstances of the crash. In addition, there were many contributing factors to occupant fatality such as human tolerance to the various loadings, velocities of body impacts in the vehicle and with exterior objects, and deformations of the occupant compartment. Each accident is further complicated by a variety in size, seating position, sex and belt wearing habit of a particular occupant. Thus, there are many factors that ultimately influence the overall effectiveness of a restraint system in actual crashes. It is possible, however, to list the significant factors that influence fatalities:

- Interior compartment intrusion, particularly in relation to the seating position of the restrained occupant.
- Crash severity, often time involving very high-speed crashes and impacts with large, heavy vehicles.
- Age of the restrained occupant, because older occupants have a lower tolerance to impact forces and greater vulnerability to injury.
- Health status of the occupant, including pre-morbid conditions, medications, and drug or alcohol use.
- Occupant height, weight, and seating position, which influence the potential for impacts on interior components and structures.
- Safety belt wearing that doesn't include positioning the lap belt low on the hips and shoulder belt tightly across the chest.
- Other occupants in the vehicle, effecting the restrained passenger, or loose objects in the vehicle, striking the restrained occupant.
- Non-collision factors including pre-crash cardiac arrest, drowning, fire and suicide.

This study was completed before the publication of an analysis of field accident data by IIHS. Zador and Ciccone (1991) evaluated driver airbag effectiveness and found a 9% reduction in overall fatality risk with supplemental driver airbags. This analysis projected a 10% reduction, but both studies result in an overall 47% effectiveness of safety belts and driver airbags. IIHS also found a 15% reduction with a supplemental airbag in frontal crashes. This study identified a 25% reduction in fatal injuries (17/64 = 27% with a range of 6-33 prevented fatalities or a 9%-52% fatality reduction) with supplemental driver airbags in frontal crashes. This would result in a projected effectiveness of safety belts and supplemental airbags of 57% (with a range in projection of 47%-71%) in frontal crashes.

The results of this study are in general agreement with recent accident investigations. Gloyns et al. (1989) investigated frontal impacts in the United Kingdom. They judged that 70% of the front-seated fatalities in frontal crashes could have been prevented by supplemental airbags and safety belt features. Since frontal crashes represent about 50% of fatal accidents, their study implies a fatality reduction of 35% (70% times 50%), for all crash modes, if the enhancements have minimal effect in non-frontal crashes. However, the 35% level is at the high end of our range of possible safety improvements with supplemental airbags and other technologies.

Side impacts and rollovers reduce the overall safety benefit of airbags because airbags are not effective in those crash types, whereas, safety belts are most effective in rollover crashes (Huelke et al. 1977, Evans 1988). Clearly, the most effective strategy for occupant protection is a system of technologies, and an approach in which a number of technologies work together enhancing protection in the diversity of real-world crashes.

The higher frequency of multi-vehicle crashes in this study than in national statistics may be explained, in part, by the higher age of the victims. In a recent study by Viano et al. (1990), older occupants were overrepresented in multi-vehicle side impacts at intersections. This crash type is a major cause of serious and fatal injury. The study also showed a lower involvement of older occupants in single vehicle impacts of fixed objects. The higher victim age may also influence the calculated effectiveness of restraints since human tolerance plays a key role in the potential for fatality prevention and the opportunities for new life-saving technologies.

The greater involvement of older victims and the consequences of more frequent multi-vehicle crashes would also influence any comparative study of vehicle crashworthiness. A population of crashes with differing vulnerabilities for occupant injury in a class of vehicles would complicate the analysis of vehicle factors in contributing to overall safety. There may be a need for further development of exposure methods to provide meaningful evaluations of the underlying crash safety of a vehicle. This could be particularly true even if the overall involvement rate in, for example, side impacts were similar since the decreasing rate of involvement of older occupants in single vehicle crashes may be offset by an increasing incidence in multi-vehicle side-impact accidents.

This study provides evidence for continued emphasis on improving the head protection since brain injury was the principal cause of death in 40% of the fatalities and a major factor in up to 30% of the fatalities due to multiple injuries. Thus, as many as 70% of the belted occupant deaths may be attributed to head impact and brain injury. Interestingly, many of the head injuries were due to impact with objects outside of the vehicle, either fixed objects or the impacting vehicle. This is consistent with the similar observations in other case studies (Gloyns et al. 1989, Mulligan et al. 1989, Rutherford et al. 1985).

Many of the fatal head injury cases involved partial loading of the front structure of the impacted vehicle. This concentrates forces on a portion of the front end and caused extensive deformation of the occupant compartment at the point of loading (Evans, Frick 1988). This type of injury mechanism has led some to consider an off-set frontal crash test to evaluate vehicle crashworthiness and injury risks for belt restrained occupants (Thomas et al. 1989, Grosch 1989).

A majority (70%) of the victims were killed in the crash or died prior to emergency assistance at the accident scene. In many of these fatalities there were multiple causes of death in very severe accident situations. This, in part, explains the frequent judgment from the expert panel that there was nothing that could be reasonably done to save the life of the victim. Nearly two-thirds of the fatalities (94/139 = 68%) were judged to be unpreventable by changes in crash protection systems. This attests to the extreme violence of most of the fatal crashes which often times involved extensive forces on the vehicle and passenger compartment, very high-speed impact with heavy vehicles or fixed objects, and unusual circumstances associated with the fatal injury.

The MIC accident investigation did not focus on alcohol or drug involvement as a cause of fatal crashes. This must be the case since the rate of reported involvement in these accidents is unrealistically low. Therefore, this study does not adequately address the role of alcohol and drug impairment as a critical factor in the causation of fatal crashes. However, we would expect a lower involvement than the national average due to the older population of belted occupants involved in these particular accidents. A lower alcohol involvement has been observed in recent case studies of multi-vehicle side-impact fatalities (Viano et al. 1990 and Mulligan et al. 1989).

In this study, over two-thirds of the fatal crashes were attributed to a driver-related cause. Some involved "aggressive" driving of the case vehicle, including excessive speed and traffic violations. These cases are related to the driver "pushing" the vehicle to and beyond their ability to recover, sometimes under adverse weather and road conditions. The remainder of the cases were related to an inadvertent mistake during reasonably good driving. The precipitating factor was usually inattentiveness or a driving error leading to loss of vehicle control. In 31% of the crashes, there was nothing the driver did or could do about the fatal crash. In many of these cases, the driver of the other vehicle was responsible for the accident by "aggressive" driving.

The study has only partly addressed accident avoidance approaches aiding drivers to prevent a crash. The most frequently identified technologies that may aid the driver were an obstacle detection and warning system to alert the driver of a fixed object in the path and anti-lock brakes to aid handling and braking during emergency maneuvers and stopping. However, other technologies, such as lane detection to warn the driver of wandering from the road surface, were not considered by the panel and may have potential advantages.

One important area of future advances in safety will involve crash avoidance and warning. This is underscored by the recurring finding that the combination lap-shoulder belt and supplemental airbag effectively approach a limit of protection during a crash, per se (Viano, 1988a, Thomas et al. 1990). This study has shown that belt use, airbags, and practical engineering changes in the restraint, interior, and vehicle structures can prevent only 50% of crash fatalities. This again is due to the overwhelming forces on the occupant compartment and frequent high-speed impacts with heavy vehicles or fixed objects. Obviously, crash avoidance is a principal approach to preventing these fatalities.

This study identified a 10%-17% (with a range in projection of 4%-37%) potential benefit in fatality reduction by the use of airbags with lap-shoulder belts for occupant protection. However, the estimate should be considered an upper bound for the actual performance of the combination because:

- Our sample involved only new car owners who were older and more safety conscious, as evidenced by voluntary belt use,
- The higher involvement in multi-vehicle crashes includes crashes where the airbag is reported to have a higher safety effectiveness than in single vehicle crashes (Evans, 1989),
- We looked only at the life-saving effectiveness of supplemental systems and did not consider that an additional component may alter occupant trajectories and crash events in some ways that will inadvertently result in fatal injuries where otherwise the occupant would survive,
- We did not consider the possibility that the additional safety components may contribute directly to occupant injury and death (Horsch et al. 1990), and,
- As technologies are introduced into new cars they are first owned by more safety conscious drivers and then pass to younger drivers as second hand vehicles. The riskier single vehicle exposures of younger drivers may off-set the additional safety features (Evans, 1987, 1989).

This study has identified a small potential for restraint system enhancements and other interior energy-absorbing improvements. Even though the supplemental restraint features may provide up to a 2-3% reduction in fatalities, they still represent a worthwhile area of investigation since several hundred fatalities may be prevented annually in United States traffic accidents.

Advances in crash protection may be realized by a systematic approach to safety design which includes an understanding of the distribution in crash types, severities, occupant tolerances and injury characteristics (Horsch 1987, Viano 1987, Langwieder et al. 1979, Searle et al. 1978, Lau and Viano, 1986, Lau et al. 1987, Viano and Lau, 1988). However, the potential for crashworthiness gains, once safety belt use and supplemental airbags are prevalent, is limited. The vehicle-related opportunities for major reductions in highway accident injury probably involve crash avoidance and warning which reduce the frequency or minimize the severity of high-speed and property damage crashes.

REFERENCES

1. Danner, M.; Langwieder, K.; and Hummel, T. The effect of restraint systems and possibilities of future improvements derived from real-life accidents. SAE Technical Paper #840394, Society of Automotive Engineers, Warrendale, PA; 1984.
2. Evans, L. The effectiveness of safety belts in preventing fatalities. *Acci Anal Prev* 18:229-241; 1986a.
3. Evans, L. Double pair comparison -- a new method to determine how occupant characteristics affect fatality risk in traffic crashes. *Acci Anal Prev* 18:217-227; 1986b.
4. Evans, L. Young driver involvement in severe car crashes. *Alcohol, Drugs and Driving* 3(3-4):63-78, 1987.
5. Evans, L. Restraint effectiveness occupant ejection from cars, and fatality reductions. General Motors Research Report GMR-6398, General Motors Research Laboratories, Warren, MI; 1988.
6. Evans, L.; and Frick, M. Seating position in cars and fatality risk. *Am J Pub Health* 78:1456-1458; 1988.
7. Evans, L. Airbag effectiveness in preventing fatalities predicted according to type of crash, driver age, and blood alcohol concentration. In the 33rd Annual Proceedings of the Association for the Advancement of Automotive Medicine, pp. 307-322, AAAM, Des Plaines, IL, 1989.
8. Gloyns, P.F.; and Jones, I.S. Characteristics of fatal frontal impacts and future countermeasures in Great Britain. Insurance Institute for Highway Safety Report; 1989.
9. Green, R.N.; German, A.; Gorski, Z.M. et al. Improper use of occupant restraints: case studies from real-world collisions. In Proceedings of the 30th Conference of the American Association for Automotive Medicine, pp. 423-438. American Association for Automotive Medicine, Des Plains, IL; 1986.
10. Grosch, L.; Baumann, K.H.; Holtze, H.; and Schwede, W. Safety performance of passenger cars designed to accommodate frontal impacts with partial barrier overlap. In Automotive Frontal Impacts SP-782, pp. 29-36, SAE Technical Paper #890748, Society of Automotive Engineers, Warrendale, PA; 1989.
11. Hartemann, F.; Thomas, C.; Henry, C.; et al. Belted or not belted: the only difference between two matched samples of 200 car occupants. In Proceedings of the 21st Stapp Car Crash Conference, pp. 95-150. Society of Automotive Engineers, Warrendale, PA; 1977.
12. Horsch, J.D. Evaluation of occupant protection from responses measured in laboratory tests. SAE Paper Technical #870222, Society of Automotive Engineers, Warrendale, PA; 1987.
13. Horsch, J.; Lau, I.V.; Andrzejak, D.; Viano, D.; Melvin, J.; Pearson, J.; Cok, D.; Miller, G.; Assessment of Airbag Deployment Loads. In Proceedings of the 34th Stapp Car Crash Conference, pp. 267-288, SAE Technical Paper #902324, Society of Automotive Engineers, Warrendale, PA, November, 1990.
14. Huelke, D.F.; Sherman, H.W.; Murphy, M.J.; Kaplan, R.J.; and Flora, J.D. Effectiveness of current and future restraint systems in fatal and serious injury automobile crashes. SAE Technical Paper #790323, Society of Automotive Engineers, Warrendale, PA; 1979.
15. Huelke, D.F.; Lawson, T.E.; Scott, R.; and Marsh, J.C. The effectiveness of belt systems in frontal and rollover crashes. SAE Technical Paper #770148, Society of Automotive Engineers, Warrendale, PA; 1977.
16. Huelke, D.F.; Sherman, H.W.; and Elliott, A.F. Clinical case studies of lap-shoulder belt effectiveness in passenger car crashes. SAE Technical Paper #840197, Society of Automotive Engineers, Warrendale, PA; 1984.
17. Huelke, D.F.; Compton, C.; Studer, R. Injury severity, ejection, and occupant contacts in passenger car rollover crashes. SAE Technical Paper #850336, Society of Automotive Engineers, Warrendale, PA; 1985.
18. Korner, J. A method for evaluating occupant protection by correlating accident data with laboratory test data. In Automotive Frontal Impacts SP-782, pp. 13-28, SAE Technical Paper #890747, Society of Automotive Engineers, Warrendale, PA; 1989.
19. Langwieder, K.; Danner, M.; Schmelzing, W.; Appel, H.; Kramer, F.; and Hofmann, J. Comparison of passenger injuries in frontal car collisions with dummy loadings in equivalent simulations. In Proceedings of the 23rd Stapp Car Crash Conference, pp. 199-232. SAE Technical Paper #791009, Society of Automotive Engineers, Warrendale, PA; 1979.
20. Lau, I.V.; and Viano, D.C. The viscous criterion - bases and applications of an injury severity index for soft tissues. In Proceedings of the 30th Stapp Car Crash Conference (P-189), pp. 123-142. SAE Technical Paper #861882, Society of Automotive Engineers, Warrendale, PA; 1986.

21. Lau, I.V.; and Horsch, J.D.; Andrzejak, D.V.; and Viano, D.C. Biomechanics of liver injury by steering wheel loading. *J Trauma* 27(3):225-235; 1987.

22. NHTSA, FMVSS 208 Regulatory Impact Analysis. National Highway Traffic Safety Administration, Department of Transportation, Washington, DC; 1984.

23. NHTSA, Fatal Accident Reporting System 1987. National Highway Traffic Safety Administration, Department of Transportation Report #807 360, Washington, DC; 1988.

24. Mulligan, G.; Dalkie, H.; and Redwood, A. An in-depth study of 121 fatal motor vehicle collisions involving 154 vehicle occupant fatalities; collision characteristics, causation and seat belt effectiveness. In Proceedings of the Canadian Multidisciplinary Road Safety Conference V; Calgary, Alberta; 1987.

25. Pursel, H.D.; Bryant, R.W.; Scheel, J.W.; and Yanik, A.J. Matching case methodology for measuring restraint effectiveness. SAE Technical Paper #780415, Society of Automotive Engineers, Warrendale, PA; 1978.

26. Rutherford, W.H.; Greenfield, T.; Hayes, H.R.M.; and Nelson, J.K. The medical effects of seat belt legislation in the United Kingdom. Department of Health and Social Security. Research Report No. 13. Her Majesty's Stationary Office, London.

27. Searle, J.A.; Bethell, J.; Baggaley, G. The variation of human tolerance to impact and its effect on the design and testing of automotive impact performance. In Proceedings of the 22nd Stapp Car Crash Conference (P-77), pp. 1-32. SAE Paper #780885, Society of Automotive Engineers, Warrendale, PA; 1978.

28. States, J.D.; Huelke, D.F.; Dance, M.; and Green, R.N. Fatal injuries caused by underarm use of shoulder belts. *J Trauma* 27(7):740-744; 1987.

29. Thomas, C.; Koltchakian, S.; Tarriere, C.; Got, C.; and Patel, A. Inadequacy of 0 degree barrier with frontal real-world accidents. In Proceedings of the 12th E.S.V. Conference, pp. xx-yy. NHTSA Paper #89-2a-0-002, National Highway Traffic Safety Administration, Washington, DC; 1989.

30. Thomas, C.; Koltchakian, S.; Tarriere, C.; et al. Primary safety priorities in view to technical feasibility limited to secondary automotive safety. In Proceedings of the 1990 FISITA Conference, Milan, Italy, 1990.

31. Viano, D.C. Evaluation of the benefit of energy-absorbing material for side impact protection: part ii. In Proceedings of the 31st Stapp Car Crash Conference, pp. 205-224. SAE Technical Paper #872213, SAE Transactions, vol. 96, Society of Automotive Engineers, Warrendale, PA; 1987.

32. Viano, D.C. Cause and control of automotive trauma. *Bull NY Acad Med*, Second Series, 64(5):376-421; 1988b.

33. Viano, D.C. Limits and challenges of crash protection. *Acci Anal Prev* 20(6):421-429; 1988a.

34. Viano, D.C.; and Lau, I.V. A viscous tolerance criterion for soft tissue injury assessment. *J Biomech* 21(5):387-399; 1988.

35. Viano, D.C.; Culver, C.C.; Evans, L.; Frick, M.; and Scott, R. Involvement of older drivers in multi-vehicle side impact crashes. *Acci Anal Prev* 22(2):177-188, 1990.

36. Viano, D.C. Effectiveness of safety belts and airbags in preventing fatal injury. In Frontal Crash Safety Technologies for the 90's (SP-852), pp. 159-171. SAE Technical Paper #910901, Society of Automotive Engineers, Warrendale, PA, 1991.

37. Wilson, R.A.; and Savage, C.M. Restraint system effectiveness - a study of fatal accidents. In Proceedings of Automotive Safety Engineering Seminar, pp. 27-39. General Motors Corporation, Automotive Safety Engineering, Environmental Activities Staff; Warren, MI; 1973.

38. Zador, P.L.; and Ciccone, M.A. Driver fatalities in frontal impacts: comparisons between cars with air bags and manual belts. Insurance Institute for Highway Safety, Arlington, VA, 1991.

RESTRAINED HYBRID III DUMMY-BASED CRITERIA FOR THORACIC HARD-TISSUE INJURY PREDICTION

Richard Kent, James Bolton, Jeff Crandall
University of Virginia

Priya Prasad, Guy Nusholtz, Harold Mertz
Alliance of Automobile Manufacturers

Dimitrios Kallieris
University of Heidelberg

ABSTRACT

Ninety-three sled tests (60 cadaver, 33 dummy) is used to assess the Hybrid III dummy and associated thoracic injury criteria to determine their ability to predict cadaver injury in matched impacts over a range of impact and restraint conditions. A statistical analysis is performed to evaluate the injury-predictive efficacy of the dummy-based maximum chest deflection, maximum chest acceleration, and CTI. A comparison of the selectivity and specificity of injury prediction models with each injury criterion reveals that the maximum dummy chest deflection is the best predictor of injury (C statistic = 0.8514) when used in an injury risk model including cadaver age, gender, and mass. Consideration of the maximum chest acceleration, either as an independent covariate or as part of CTI, is found to weaken the injury model. The restraint condition (belt-dominated loading, airbag-dominated loading, or combined loading), however, is found to be a more important injury predictor than any injury criterion.

Key Words: Injury criteria, cadavers, thorax, frontal impacts, restraint systems

THE DESIGN OF OCCUPANT RESTRAINT SYSTEMS is dictated to a large extent by the requirement to reduce thoracic injury risk. Restraint systems are assessed using a variety of tools. Human volunteer, cadaver, or animal tests may be performed to evaluate safety systems. These types of tests are expensive, laborious, and have limited repeatability, however; so anthropomorphic test devices (ATDs), or dummies, are often used as surrogates for humans. ATDs are instrumented to measure various mechanical parameters, including sternal displacement relative to the spine (chest deflection), acceleration measured at the chest center of gravity (cg), and others. Mechanical parameters, or combinations thereof, are correlated with the presence of injury and used as predictors of injury risk. Used in this manner, these mechanical parameters are referred to as injury criteria. Injury criteria are then often used in injury risk functions, which can incorporate other parameters affecting injury risk, such as occupant age (e.g., Funk et al. 2001). In order to be useful for injury prediction, an injury criterion or risk function must satisfy at least two requirements: it must differentiate injurious loading conditions from non-injurious loading conditions, and it must do this for the set of loading conditions that span the range of interest (including future advances in restraint system technology).

BACKGROUND – FACTORS AFFECTING THORACIC INJURY PREDICTION

The likelihood that a restrained (belt, airbag, or both) occupant will sustain a thoracic injury in a frontal impact depends on the characteristics of the collision (e.g., speed, occupant seating position), the characteristics of the restraint system (e.g., presence of a belt, presence of an airbag), and the characteristics of the occupant (e.g., size, age, physical condition). Characterization of injury risk in a

research or development test generally relies upon dummy-based mechanical parameters to represent the effect of the collision and restraint system characteristics. The purpose of this paper is to determine how well the Hybrid III 50th percentile male with a standard instrumentation package predicts cadaver hard tissue injury in a matched collision when the collision, restraint, and occupant characteristics are varied. A background discussion of these characteristics is therefore necessary.

EFFECT OF RESTRAINT CONDITION AND INJURY CRITERION: Thoracic injury criteria have been evaluated using a variety of loading conditions. Due to the low rates of seatbelt use in the 1960s and 1970s, early studies of thoracic injury and impact response focused on loading experienced by an unbelted subject (see Kroell 1994 for a review of thoracic impact studies performed prior to 1980). Characteristics of the thorax were developed using blunt objects representing a steering wheel hub or instrument panel, and loading rates were representative of those experienced by unrestrained occupants in severe collisions (e.g., Nahum et al. 1970, Kroell et al. 1971, Mertz and Gadd 1971, Kroell et al. 1974). As seatbelt use rates increased, it became necessary to evaluate thoracic response and injury criteria under localized belt impingement and at lower loading rates (e.g., Cromack and Ziperman 1975, Fayon et al. 1975, Mertz et al. 1991, Horsch et al. 1991, Crandall et al. 1994).

As airbag restraints became more common in the vehicle fleet, researchers recognized the need for a thoracic injury criterion appropriate for airbag loading and for combined belt-and-airbag loading. Recently, the National Highway Traffic Safety Administration (NHTSA) proposed the combined thoracic criterion CTI for the evaluation of diverse restraint conditions (Eppinger et al. 1999). CTI results from the finding that a two-parameter logistic regression model using the cadaver-based T1 peak acceleration and the maximum chest deflection as covariates correlated well with the probability of an AIS 3 or higher injury (defined by the number and distribution of rib fractures sustained by the cadaver). CTI is a sum of normalized maximum chest cg acceleration and normalized maximum chest deflection:

$$CTI = A_{max}/A_{int} + C_{max}/C_{int} \qquad [1]$$

where A_{int} and C_{int} are normalization constants defined for various dummy sizes.

EFFECT OF OCCUPANT CHARACTERISTICS (SIZE, GENDER, AGE): In addition to restraint condition, several characteristics of the occupant can affect the specific injury tolerance and the efficacy of an injury criterion for that occupant. Hard tissue injury tolerance is generally found to decrease with age and (for some types of loading) to increase with occupant size. Independently of size, females are often assumed to have lower hard tissue injury tolerance.

Effect of Occupant Age: Stein and Granik (1976) performed bending tests on three ribs from each of 79 human donors having an age range from 27 years to 83 years. They found a strong inverse correlation between breaking force and donor age at death ($p<0.001$). Those authors concluded that, like long bones, ribs apparently undergo progressive circumendosteal resorption with advancing age; but, unlike long bones, ribs show no evidence of continued subperiosteal apposition. This results in a general decrease in the percent of the rib cross-section that is cortical bone.

From a search of the available literature, it appears that the most comprehensive study of the effect of occupant age on thoracic hard tissue impact injury tolerance was presented by Zhou et al. (1996). They found that the reduction in thoracic injury tolerance for a senescent group (age 66-85) compared to a baseline younger group (age 16-35) was strongly dependent on the restraint condition: for belt loading, tolerance decreased by 72% while blunt load tolerance decreased by only 21% (frontal) or 27% (lateral).

Effect of Occupant Gender and Occupant Size: Studies often find females to have intrinsically weaker bones than males (e.g., Wertheim 1847, Rauber 1876, Burghele and Schuller 1970), but this finding is not universal. For example, Lindahl and Lindgren (1967) found no significant gender difference in ultimate tensile strength, mean deformation at failure, limit of proportionality, or the modulus of elasticity of human cortical bone. Yamada (1970) also found no significant gender differences in tensile strength or percent elongation in cortical bone. The data for compressive, bending, and torsional loading of cortical bone have been similarly mixed with respect to the effect of gender, as have the data for human cancellous bone (e.g., Messerer 1880, Evans and King 1961, Sonoda 1962, Chalmers and Weaver 1966).

Studies of gender influences on bone mineral density (BMD) for males and pre-menopausal females have likewise been mixed (e.g., Parsons et al. 1996, Slosman et al. 1994, Kelly et al. 1990). Henry and Eastell (2000) compared peak bone mineral density in male and female adults ages 20-37. Uncorrected dual-energy X-ray absorptiometry (DEXA) measures of femoral neck bone indicated higher BMD values for male subjects. After correcting for differences in skeletal size, however, these authors found no significant difference in femoral neck bone mineral apparent density (BMAD, an estimated volumetric bone density which attempts to normalize bone mineral density measurements for bone size) between male and female subjects. Lumbar spinal vertebra BMAD was actually significantly ($p < 0.0001$) greater for the female subjects in this pre-menopausal population. Studies have also evaluated the effect of bone apposition on the observed gender differences in fracture rates. Through the use of biochemical markers of bone formation, researchers have found greater bone formation in men (Resch et al. 1994, Vanderschueren et al. 1990, Duda et al. 1988) and in women (Epstein et al. 1994).

Based on these inconsistent findings, hard tissue injury tolerance differences observed between male and female specimens may be due to factors other than intrinsic differences in material strength; non-senescent males and females may not have material differences in hard tissue injury tolerance. As females age, however, the release of menopausal hormones does result in decreased bone mineral density compared to males, so age-dependent decreases in hard tissue injury tolerance are more pronounced for women than for men.

The effect of occupant size on thoracic injury tolerance to frontal impact restraint loading is not clear. Larger occupants have larger bony structures, and thus are capable of bearing greater forces than smaller occupants. In the case of non-inertial loading, the increased structural strength of larger occupants is manifest as an occupant size-dependent injury tolerance. In the case of non-force-limiting restraint loading in a frontal impact, however, the magnitude of force applied to the occupant is a monotonically increasing function of the occupant's thoracic mass. The compelling question is whether bony structural strength increases with occupant size at a greater or lesser rate than the inertial forces the structures must bear. At present, the authors know of no study that evaluates this relationship for the case of concentrated and distributed loading of the thorax during a frontal impact.

The estimation of injury risk requires an understanding of how all of the factors discussed above affect injury potential and also how they affect the predictive ability of an injury criterion. A statistical analysis of cadaver test data is a potential way to reach this understanding since regression techniques allow for the analysis of factors individually as well as factor interactions. Given the large number of factors that must be considered, however, the dataset must have certain characteristics. First, it must include cadavers that span the age and mass ranges of interest and both genders must be represented. Second, a range of restraint conditions must be considered so that the effect of load distribution can be evaluated. In addition, because thoracic injury data are censored, each age subset, mass subset, gender subset, and restraint subset must contain both injury and non-injury tests.

EXISTING CADAVER DATA

The existing data on cadaver frontal impact loading can be divided into three main groups: blunt impactor tests (e.g., Kroell 1994), seatbelt table tests (e.g., Cesari and Bouquet 1994), and sled tests (see Kuppa and Eppinger 1998 for a summary and for additional references). The blunt impactor tests were not designed to evaluate the effect of belt, airbag, and combined belt-and-airbag loading. In addition, the blunt impactor test setup appears to differ from a vehicle environment sufficiently to preclude the evaluation of combined thoracic injury criteria like CTI. The seatbelt table tests, while useful for evaluating the case of a concentrated load, does not contain any distributed-load tests for comparison. Further, the seatbelt table setup, which includes a reaction surface at the spine, precludes evaluation of any chest acceleration-dependent criterion.

The existing database of sled tests, therefore, has the greatest potential to elucidate the effects discussed in the Background section. The database contains numerous subjects over wide age and mass ranges. In addition, belt-dominated loading, airbag-dominated loading, and combined restraint loading test conditions are included. The sled-test database is, however, also subject to limitations. As discussed by Prasad (1999), Hassan and Nusholtz (2000), and Kent et al. (2000), these limitations include artifactual kinematic measurements that occurred when the cadaver's head struck the windshield or windshield header, resulting in large acceleration peaks transmitted through the spine to the cadaver's thorax. This phenomenon was observed for tests performed with the airbag-only restraint conditions, but was not observed for belted occupants. As a result, for the purpose of a statistical analysis of restraint condition effects, these tests introduce a systematic error into the database.

In addition, the instrumentation used in the cadaver sled tests introduced systematic errors into the database. Accelerometers were used to measured acceleration at several locations on the cadavers, including the first thoracic vertebra (T1). In most of the cadaver tests, the T1 acceleration was considered to be analogous to, or at least representative of, the chest cg acceleration. Subsequent cadaver tests, which included accelerometers at both T1 and at the eighth thoracic vertebra (T8), revealed that, for some conditions, the T1 acceleration does not necessarily represent the chest cg acceleration (Shaw et al. 2001). Because T1 acceleration magnitude is a reasonable representation of chest cg acceleration magnitude for some conditions but not for others, the use of T1 acceleration is a systematic error in the development of dummy-based injury criteria.

The use of chestbands is another source of uncertainty. As discussed by Bass et al. (2000), Shaw et al. (1999) and Hagedorn and Burton (1993), chestbands are subject to errors under the localized loading caused by seatbelts. The chestbands are, however, considered to be accurate indicators of thoracic deformations under the distributed loading caused by an airbag. This restraint-dependent measurement accuracy is another potential source of systematic error in the database.

REFINEMENT AND EXPANSION OF EXISTING SLED TEST DATABASE

Despite these limitations, the existing cadaver sled test database does contain the most diverse and detailed experiments available for the evaluation of injury criteria efficacy, so means should be sought that will allow the use of these data. The largest source of systematic error, the use of cadaver-mounted instrumentation, can be reduced through the use of matched Hybrid III dummy tests (performed under identical sets of circumstances – sled buck, restraint system, sled acceleration pulse). The use of a Hybrid III also contains inherent limitations. For example, while the dummy has biofidelity for mid-sternal blunt impacts, it has been shown to be too stiff statically (Lau and Viano 1988, Horsch et al. 1990) and at loading rates observed in belt tests (Backaitis and St-Laurent 1986, Cesari and Bouquest 1990). From a pragmatic standpoint, however, regardless of dummy biofidelity, a correlation between dummy measures and injury outcome is desired since the dummy is used to evaluate vehicle restraint design in research tests as well as in compliance tests and new car assessment tests. If a criterion can be identified that allows the existing dummy to predict injury risk for restraint and impact conditions that span the range of conditions in which the dummy is to be used, then the dummy's lack of biofidelity decreases in importance.

In addition to the improved repeatability and reliability of dummy measures compared to cadaver-based measures, this dummy-based approach has an additional advantage: several of the cadaver tests that had been identified by Prasad (1999) as having erroneous cadaver-based chest deflection measurements can be reincorporated into the dataset, since the only cadaver-based data of interest are the anthropometry, gender, age, and injury outcome.

EXISTING CADAVER SLED TEST DATABASE AND NEW CORRESPONDING DUMMY TESTS: The database of cadaver sled tests presented by Kuppa and Eppinger (1998), was used as a starting point. For each set of conditions tested using a cadaver at the University of Virginia (UVA), one or more corresponding 50th percentile male Hybrid III tests were performed with a nominally identical sled buck, sled acceleration pulse, and restraint condition. Standard Hybrid III instrumentation was used, including chest cg acceleration and mid-sternal chest deflection.

Analysis of high-speed film and instrumentation signals was used to identify several tests that, for a variety of reasons, were unacceptable for this study. The tests UVA103, UVA250, and UVA259, recommended for removal by Prasad, were included back into the dataset since the allegedly erroneous cadaver-based measurements were not used. Two other tests (UVA96 and UVA97), also recommended for removal by Prasad, were not reincorporated into the dataset because both the dummy and cadaver exhibited significant head strikes on the windshield header, which generated artifactual acceleration spikes and also resulted in a significant portion of the subject's momentum being arrested through the head/neck complex rather than through the thorax and restraint system.

Unacceptable tests also included those that had a loss of dummy instrumentation or those that exhibited significant differences between dummy kinematics and cadaver kinematics. In some tests, differences in spinal and thoracic compliance resulted in the cadaver loading the steering wheel heavily through the airbag while the dummy did not. In these cases, the tests were excluded from analysis because the kinematics were sufficiently different to cast doubt on the dummy's ability to represent cadaver injury potential. In three cadaver tests (UVA356-UVA358, airbag, no belt), the unbelted cadaver sustained rib fractures from steering wheel contact through the airbag. These tests were included in the dataset, however, since the dummy exhibited a similar interaction with the steering wheel. Except for the cases identified by Prasad and the cases of obviously artifactual chest acceleration peaks, film analyses of both the existing sled tests and the new sled tests were blinded. In other words, the restraint system and measured dummy responses were not used to target tests for film analysis. The comparison of dummy kinematics and cadaver kinematics was the only factor used to determine inclusion.

After the performance of dummy tests to match all UVA cadaver tests and the removal of several sets of tests for the reasons discussed above, the dataset of UVA tests was pared to 33 cadaver tests, each having a corresponding dummy test. The literature was searched to find additional cadaver tests having 50th percentile male Hybrid III dummies tested in identical conditions. Five such cadaver tests, presented by Yoganandan et al. (1991) of the Medical College of Wisconsin (MCW), were added to the UVA database. The resulting dataset of existing cadaver tests contains 38 cadaver tests, all of which have corresponding Hybrid III tests, and all of which exhibit similar kinematics between cadaver and dummy. In many cases, more than one cadaver was tested at a given set of conditions. In these cases, often only a single dummy was tested at that condition. Details of the existing cadaver tests and the corresponding dummy tests are presented in Table 1.

NEW CADAVER TESTS AND CORRESPONDING DUMMY TESTS: To expand the scope of the existing database, an additional 18 cadaver tests and 11 corresponding Hybrid III tests were performed at UVA and 4 cadaver tests with 3 corresponding Hybrid III tests were performed at the University of Heidelberg (UH) (Table 2). Seven of the dummy tests and eleven of the cadaver tests performed at UVA involved subjects seated in the RFP (designated by a "P" in the test number) (Kent et al. 2001). All tests were performed in vehicle bucks representing contemporary mid-sized sedans. Sled deceleration pulses were based on full-vehicle barrier tests.

FINAL DATABASE: The database of new cadaver tests and dummy tests was combined with the database of existing tests. This final database contains 60 cadaver tests, all of which have corresponding Hybrid III dummy tests, for a total of 93 sled tests. Restraint conditions include two-point shoulder belt systems, three-point lap-and-shoulder (L/S) belt systems, airbag-only systems, standard L/S-and-airbag systems, and force-limiting (F/L) belt-and-airbag systems. All airbags are of the head-and-torso type as opposed to the smaller bags common in Europe. For the purpose of evaluation, these restraints were divided into three groups, based on the thoracic loading (Table 3):

1. **Restraint A** – belt-dominated loading (all cases with a torso belt and without an airbag – this includes both two-point shoulder belt systems and three-point systems),
2. **Restraint B** – airbag-dominated loading (this includes cases with an airbag and without a torso belt), and
3. **Restraint C** – combined loading (this includes all cases with an airbag and a shoulder belt, regardless of whether the belt was force-limiting or standard).

Both male (n = 43) and female (n = 17) cadavers are included in the final database, the age-at-death range is 24 years to 72 years (mean 55, standard deviation 12.1), and the mass range is 47 kg to 117 kg (mean 71, standard deviation 15.8).

The test conditions for the new tests were chosen in an attempt to fill voids in the existing database (Table 3 – bold boxes indicate deficient areas in the dataset). Specifically, RFP tests, combined loading tests, and non-injurious tests were the focus. There are, however, still deficiencies. First, more non-injurious tests are needed, especially with female subjects and the belt-only restraint condition. Even with the new tests aimed at non-injurious loading, there are only 16 non-injury tests in the database. Only two of these 16 tests involved female subjects, and only two involved a belt-dominated restraint condition.

Table 1 –Database of Acceptable Existing Cadaver Tests with Corresponding Dummy Tests

Test No.	ID	Age	Gender	Weight (kg)	Height (cm)	Sled ΔV (km/h)	Restraint	No. of Rib Fx	Thoracic MAIS	Dummy Peak Chest Accel. (g)	Dummy Peak Slider Defl. (mm)	Dummy CTI
Tests Performed at UVA												
52	**Hybrid III**					34.8	2-pt			33	50	0.85
53	91-W-03	61	F	61	153	34.9	2-pt	19	4			
54	**Hybrid III**					38.1	2-pt			35	74	1.11
55	91-EF-10	62	F	91	176	36.9	2-pt	11	3			
101	**Hybrid III**					33.8	2-pt			30	57	0.88
102	92-UM-23	60	M	95	176	33.1	2-pt	19	4			
103	92-EM-21	57	M	103	179	32.5	2-pt	13	4			
104	92-EF-25	66	F	105	179	32.3	2-pt	11	4			
112	**Hybrid III**					46.8	2-pt			36	75	1.13
113	92-EF-22	24	F	57	159	47.3	2-pt	12	4			
114	92-FF-24	60	F	65	164	47.0	2-pt	27	4			
172	**Hybrid III**					24.6	L/S			25	56	0.82
173	92-EM-33	61	M	72	167	24.6	L/S	9	3			
174	93-EF-35	57	F	62	168	25.9	L/S	12	3			
175	93-EM-27	58	M	117	185	25.7	L/S	3	2			
222	**Hybrid III**					53.4	2-pt			45	64	1.12
223	93-EM-32	51	M	61	169	55.0	2-pt	13	4			
224	93-EM-26	58	M	65	175	54.4	2-pt	16	4			
225	93-EM-37	36	M	68	177	53.9	2-pt	16	4			
226	**Hybrid III**					53.3	2-pt			41	64	1.08
227	93-FM-29	53	M	70	165	53.5	2-pt	12	3			
228	92-FM-18	47	M	85	177	54.7	2-pt	16	4			
229	93-EM-36	37	M	60	183	54.0	2-pt	17	4			
249	**Hybrid III**					54.1	2-pt			45	57	1.05
250	94-EM-39	39	M	50	177	54.9	2-pt	12	4			
251	**Hybrid III**					58.9	2-pt			52	59	1.15
252	94-EM-38	37	M	67	177	58.2	2-pt	17	4			
256	**Hybrid III**					56.5	2-pt			46	52	1.02
257	94-EM-40	33	M	112	179	56.7	2-pt	10	4			
258	94-EM-43	69	M	64	178	55.3	2-pt	14	4			
259	94-EF-41	64	F	77	163	56.4	2-pt	15	4			
293	**Hybrid III**					56.8	L/S			47	37	0.88
294	94-EF-44	68	F	55	148	56.8	L/S	10	4			
295	94-EM-42	57	M	104	187	56.8	L/S	15	4			
296	94-EM-45	59	M	74	181	59.8	L/S	26	4			
302	**Hybrid III**					57.5	L/S and AB			51	31	0.86
303	93-EM-34	64	M	50	154	57.5	L/S and AB	4	2			
304	94-EM-47	65	M	57	168	59.4	L/S and AB	15	4			
305	94-EF-48	66	F	58	161	59.4	L/S and AB	12	4			
332	**Hybrid III**					58.2	F/L and AB			37	30	0.70
333	95-EM-51	51	M			58.2	F/L and AB	6	3			
334	94-EM-49	49	M			58.2	F/L and AB	5	3			
335	95-EM-50	50	M			58.6	F/L and AB	2	2			
355	**Hybrid III**					55.7	AB only			46	58	1.08
356	95-EM-52	64	M	73.9	176	57.2	AB only	30	4			
357	96-EM-55	48	M			57.2	AB only	19	4			
358	96-EM-56	40	M			59.0	AB only	17	4			
Tests Published by Yoganandan et al. (1991)												
1	**Hybrid III**					50.8	L/S			51	34	0.90
M1	Cadaver	58	M	82	180	49.0	L/S	10	4			
2	**Hybrid III**					49.7	L/S			44	39	0.87
D2	Cadaver	57	M	73	178	49.0	L/S	10	4			
3	**Hybrid III**					50.0	L/S			43	40	0.87
V3	Cadaver	66	M	77	178	50.0	L/S	7	3			
4	**Hybrid III**					51.8	L/S			40	29	0.73
F4	Cadaver	58	M	70	178	50.0	L/S	14	4			
5	**Hybrid III**					49.0	L/S			39	27	0.70
A5	Cadaver	67	M	73	175	50.0	L/S	12	3			

Restraint Codes: 2-pt – shoulder belt only, L/S – lap and shoulder belt, F/L – force-limiting L/S, AB – airbag.

Table 2 – Database of New Cadaver Sled Tests with Corresponding Dummy Tests

Tests Performed at UVA

Test No.	ID	Age	Gender	Weight (kg)	Height (cm)	Sled ΔV (km/h)	Restraint	No. of Rib Fx	Thoracic MAIS	Dummy Peak Chest Accel. (g)	Dummy Peak Slider Defl. (mm)	Dummy CTI
410	**Hybrid III**					57.0	F/L and AB			32	25	0.60
411	96-EM-57	60	M	Unk.	Unk.	57.5	F/L and AB	19	4			
412	96-EM-60	70	M	Unk.	Unk.	56.8	F/L and AB	14	4			
532	**Hybrid III**					48.6	F/L and AB			27	27	0.56
537	**Hybrid III**					48.9	F/L and AB			27	29	0.58
538	**Hybrid III**					48.1	F/L and AB			25	26	0.53
533	99-EF-104	67	F	64	163	48.6	F/L and AB	1	1			
534	97-EM-76	47	M	51	170	48.4	F/L and AB	4	3			
535	98-EF-95	57	F	53	163	48.1	F/L and AB	16	4			
544	97-EF-83	59	F	56	168	49.2	F/L and AB	9	4			
545	99-EM-103	67	M	74	176	48.1	F/L and AB	3	2			
571P	**Hybrid III**					47.6	F/L and AB			26	29	0.57
572P	**Hybrid III**					48.1	F/L and AB			28	28	0.58
576P	**Hybrid III**					48.1	F/L and AB			28	30	0.60
577P	99-FM-111	57	M	70	174	47.4	F/L and AB	0	0			
578P	99-FF-107	69	F	52.5	155	47.6	F/L and AB	4	3			
579P	99-FF-106	72	F	59	156	47.6	F/L and AB	11	4			
580P	99-FF-105	57	F	57	177	47.6	F/L and AB	0	0			
648P	**Hybrid III**					48.6	AB only			44	19	0.67
649P	**Hybrid III**					47.6	AB only			44	20	0.68
650P	00-FRM-124	40	M	47	150	48.9	AB only	4	2			
651P	00-FRM-121	70	M	57	176	48.6	AB only	0	0			
652P	00-FRM-118	46	M	74	175	49.7	AB only	0	0			
663P	**Hybrid III**					48.0	L/S and AB			45	40	0.89
664P	**Hybrid III**					48.0	L/S and AB			43	40	0.87
665P	99-FRM-112	55	M	85.3	176	48.0	L/S and AB	3	2			
666P	99-FRM-115	69	M	83.9	176	48.0	L/S and AB	3	2			
667P	00-FRF-120	59	F	79.4	161	48.0	L/S and AB	13	4			
668P	00-FRF-127	54	F	55.3	162	48.0	L/S and AB	23	4			
Tests Performed at the University of Heidelberg												
9001D	**Hybrid III**					47.0	L/S			43	22	0.69
9002D	**Hybrid III**					48.0	L/S			45	28	0.77
9013C	Cadaver	34	M	71	180	48.0	L/S	0	1			
9003D	**Hybrid III**					48.0	AB only			57	13	0.76
9014C	Cadaver	31	M	70	170	47.0	AB only	0	0			
9207C	Cadaver	25	M	74	184	49.0	AB only	0	0			
9212C	Cadaver	38	M	79	174	47.0	AB only	0	0			

Note: "P" in the test number represents tests performed with subjects in the right-front passenger position.

Another area for additional testing is the airbag-only restraint condition. Most of the original Kuppa and Eppinger (1998) tests that had to be removed from the database were performed with this restraint condition. This is a result of the relatively uncontrolled occupant motions that occur when a belt restraint is not used. Based on film analysis of the rejected tests, the dummy often does not represent the cadaver kinematics well in the absence of a belt and, as a result, many of these tests could not be used for the present study. The final database contains only nine airbag-only tests.

The database is also inadequate in the number of young cadavers. While there are obvious difficulties with obtaining young cadavers, age is known to be a significant predictor of injury risk. Additionally, the population of cadavers that have been tested under combined loading has a younger mean age and a lesser mean mass than the populations tested under the other two restraint conditions. These differences are not significant (≤ approximately one standard deviation), but may introduce bias into the restraint-dependency of the dummy measures.

Finally, the ratios of male-to-female cadavers are not equal among the three restraint conditions. In fact, there are no females tested in the airbag-only configuration. While the gender ratio of occupants receiving AIS 3+ chest injuries in the field is on the order of 1/1, the ratio in the sled test database is 43 males to 17 females.

Table 3 – Summary of Cadaver Database

	Existing database	New tests	Final Database
Total cadavers with a matching Hybrid III test	38	22	60
Number of these cadavers with AIS 3+ injury	35	9	44
Number of these cadavers w/o AIS 3+ injury	3	13	16
Tests with belt-dominated thoracic loading	29	1	30
Belt-dominated with AIS 3+ injury	28	0	28
Belt-dominated w/o AIS 3+ injury	1	1	2
Tests with airbag-dominated thoracic loading	3	6	9
Airbag-dominated with AIS 3+ injury	3	0	3
Airbag-dominated w/o AIS 3+ injury	0	6	6
Tests with combined thoracic loading	6	15	21
Combined with AIS 3+ injury	4	9	13
Combined w/o AIS 3+ injury	2	6	8
Females	9	8	17
Females with AIS 3+ injury	9	6	15
Females w/o AIS 3+ injury	0	2	2
Males	29	14	43
Males with AIS 3+ injury	26	3	29
Males w/o AIS 3+ injury	3	11	14
Under 56	15	9	24
Under 56 with AIS 3+ injury	14	2	16
Under 56 w/o AIS 3+ injury	1	7	8
Over 55	23	13	36
Over 55 with AIS 3+ injury	21	7	28
Over 55 w/o AIS 3+ injury	2	6	8
Driver-side tests	33	11	44
Driver-side tests with AIS 3+ injury	30	5	35
Driver-side tests w/o AIS 3+ injury	3	6	9
Passenger-side tests	5	11	16
Passenger-side tests with AIS 3+ injury	5	4	9
Passenger-side tests w/o AIS 3+ injury	0	7	7
Average age - belt-dominated tests	54.5	34.0	53.8
Standard deviation (years)	11.7	0.0	12.1
Average age - airbag-dominated tests	50.7	41.7	44.7
Standard deviation (years)	12.2	15.7	14.5
Average age - combined-loading tests	57.5	61.3	60.2
Standard deviation (years)	8.3	7.3	7.6
Average mass (kg) - belt-dominated tests	76.4	71.0	76.2
Standard deviation (kg)	18.0	0.0	17.7
Average mass (kg) - airbag-dominated tests	78.3	66.8	70.7
Standard deviation (kg)	3.8	12.3	11.4
Average mass (kg) - combined-loading tests	62.3	65.0	64.0
Standard deviation (kg)	9.9	12.3	11.4
Male/female ratio in belt-dominated tests	21/8	1/0	22/8
Male/female ratio in airbag-dominated tests	3/0	6/0	9/0
Male/female ratio in combined-loading tests	5/1	7/8	12/9

STATISTICAL ANALYSIS OF SLED TESTS (FINAL DATABASE)

METHODS: The data were analyzed by the generalized estimating equation method (GEE) (Liang and Zeger 1986). The outcome was a binary response, with cadaver thoracic AIS scores of 3 or greater considered as injury (Y=1) and scores <3 considered as non-injury (Y=0). Dummy-based injury criteria (A_{max}, C_{max}, and CTI), as well as characteristics of the cadaver [age (years), gender, and body mass (kg)] were treated as the set of potential predictors for thoracic injury. The data were deemed insufficient for an analysis of any interaction terms. The binary outcome was modeled by a generalized linear model with a logit link function. Model coefficients, β_i, were estimated by maximum likelihood and the variance-covariance parameters were estimated by the Huber-White estimator (Huber 1967, White 1982). The dummy test identification number was treated as a clustering factor, since in several instances multiple cadaver tests were performed under the same test configuration that was utilized during a single dummy test.

As a modeling strategy we fit 6 GEE models. Three of the models were specified so that only one of the dummy-based injury criteria was included along with the predictors associated with the cadaver (age, gender, and body mass). Model 1 used C_{max}, model 2 used A_{max}, and model 3 used CTI. Model 4 used both C_{max} and A_{max}. A baseline model (model 5) was developed, which included only the predictors associated with the cadaver (i.e., no dummy measures were included). Model 6 included C_{max} as well as consideration of the restraint condition (A,B,C).

The relative ranking of the predictors was based on a standardized version of the Wald chi-square statistic. By subtracting from the observed Wald chi-square statistic the total number of degrees of freedom associated with the statistic, the standardized value represents the difference between the observed chi-square statistic and the expected chi-square statistic under the null hypothesis of no association.

The comparison of the overall performance of the models was based on an internal bootstrap model validation (Efron and Tibshirani 1993). The bootstrap validation is a re-sampling procedure that corrects for the optimism (bias) in the observed value of model performance indices. Optimism is induced into the estimate of the model's overall performance as a consequence of utilizing the data at hand to estimate the model parameters. The bootstrap-adjusted index is bias-corrected and provides a better estimate of the model's performance when applied to a new sample of data.

The performance measure of interest in this analysis is the C statistic (Harrell 2001). The C statistic is an overall measure of the model's sensitivity and specificity with respect to discriminating between the subjects who fall into the category Y=1 and those who fall into the category Y=0. The C statistic can be interpreted in the following manner. For any random pair of subjects, one of which has realized outcome Y=1 and the other which has realized outcome Y=0, we would expect with probability C that the model-based prediction for the subject with realized outcome Y=1 to be greater in magnitude than the model-based prediction for the subject with realized outcome Y=0. For a set of predictors that produce a model C statistic of 0.50, the utility of the model is no better than the flip of a fair coin, while perfect discrimination produces a C statistic of 1.0. A set of predictors that produces a C statistic of 0.70 or greater is considered to have utility as a predictive tool (Harrell 2001).

RESULTS: Model 1, which includes C_{max} as a predictor, is a significant model of injury probability ($p < 0.0001$) (Table 4) and C_{max} is the most important covariate in that model (Figure 1). In contrast, model two, which includes A_{max} as the dummy measure, is not a significant model ($p = 0.0978$) and A_{max} is not a significant predictor ($p = 0.68$) in the model, ranking least important of the four covariates. Model 3, which includes CTI, is a significant predictor of injury probability ($p = 0.0004$) and CTI is the most important covariate in the model. To understand the importance of each component of CTI, consider model 4, which includes C_{max} and A_{max} as separate predictors. This model is a significant predictor of injury ($p < 0.0001$), but A_{max} is not a significant covariate ($p = 0.22$) and, in terms of relative ranking of predictors, ranks fourth out of the five predictors used in the model.

Table 4 – Comparison of Sled Test Injury Prediction Models

a) Model 1 ($p < 0.0001$)			b) Model 2 ($p = 0.0978$)			c) Model 3 ($p = 0.0004$)		
	β_i	p		β_i	p		β_i	p
Intercept	-0.6683	0.86	Intercept	-0.1733	0.95	Intercept	-6.5546	0.10
Gender	-0.8265	0.25	Gender	-1.4781	0.02	Gender	-2.2333	0.01
Mass (kg)	-0.05114	0.19	Mass (kg)	0.01088	0.64	Mass (kg)	-0.01867	0.52
Age (years)	-0.0009334	0.98	Age (years)	0.01540	0.60	Age (years)	0.03691	0.33
C_{max}	0.1652	0.00	A_{max}	0.01687	0.68	CTI	10.7951	0.00
Model C statistic = 0.8514			Model C statistic = 0.5484			Model C statistic = 0.8292		

d) Model 4 ($p < 0.0001$)			e) Model 5 ($p = 0.2516$)		
	β_i	p		β_i	p
Intercept	-2.6097	0.55	Intercept	0.6891	0.78
Gender	-1.3215	0.07	Gender	-1.3554	0.05
Mass (kg)	-0.05321	0.19	Mass (kg)	0.00976	0.67
Age (years)	0.004544	0.91	Age (years)	0.01173	0.71
C_{max}	0.1713	0.00			
A_{max}	0.05367	0.22			
Model C statistic = 0.8479			Model C statistic = 0.5776		

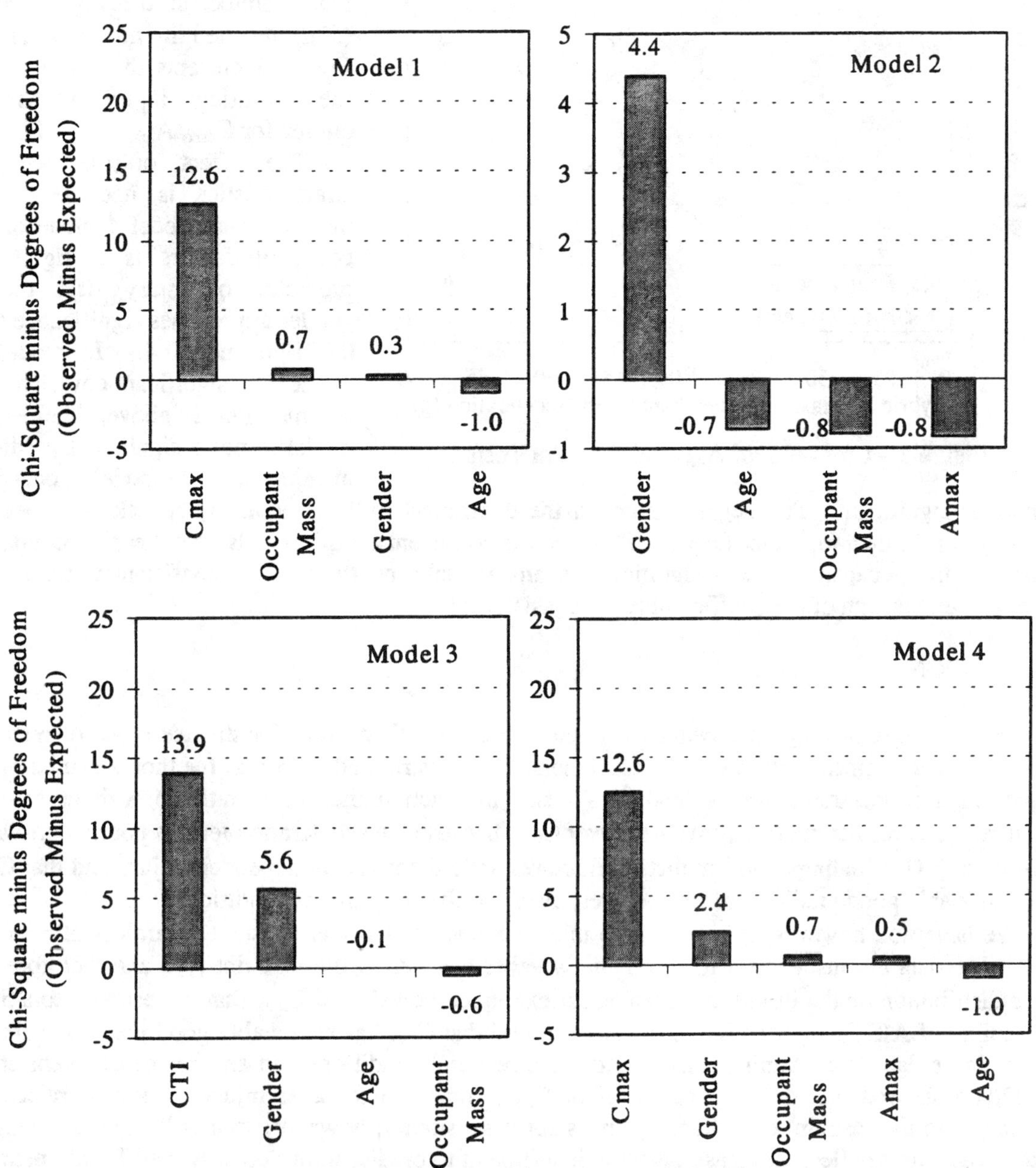

Figure 1 – Relative Ranking of Predictors Using Four Models of Sled Test Data

Since all models were identical except for the dummy criteria used as predictors, a comparison of models provides a means of comparing the efficacy of the various criteria. The C statistic is highest for model 1, indicating that, of the three criteria analyzed, C_{max} has the best injury sensitivity and specificity. The C statistic is lowest for model 2 (A_{max}) (Table 4). In fact, the addition of A_{max} in the model actually decreased the C statistic compared to model 5, which included no consideration of the dummy response. As expected based on this finding, the addition of A_{max}, either through the use of CTI (model 3) or the use of A_{max} as an additional covariate (model 4), decreased the C statistic relative to model 1. This lack of correlation between injury and A_{max} is illustrated in Figure 2. In this cross-plot, there is a clear relationship between C_{max} and the presence of 7 or more rib fractures. All tests with C_{max} below 25 mm resulted in fewer than 7 fractures. There is a transition zone between 25 mm and 40 mm in which some cadavers sustained greater than 7 fractures and some did not. Above 40 mm, all cadavers but one sustained greater than 7 fractures. A_{max} does not exhibit any

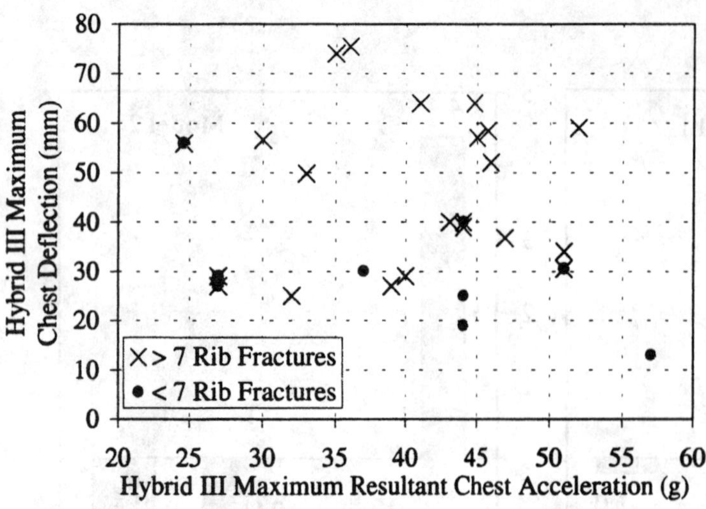

Figure 2 – Cross-plot of A_{max} and C_{max}, All Tests

relationship with the presence of 7 or more fractures. In fact, the test that resulted in the highest A_{max} (57g) resulted in no rib fractures. Figure 3 presents the test data and the resulting injury probability curves for C_{max}, A_{max}, and CTI.

The effect of the occupant characteristics is not clear. In model 1 and model 4, none of the occupant factors is a significant predictor of injury risk, though gender approaches significance ($p = 0.07$) in model 4. In model 2, gender is a significant covariate but, as mentioned above, the entire model is not a significant predictor of injury risk. In model 3, occupant gender is significant. The negative sign on the coefficient indicates that injury risk is higher for females ($x = 0$) than for males ($x = 1$). This trend is consistent for all models. Neither the occupant's mass nor the occupant's age is significant in any model, and there is no consistent trend among models (i.e., the sign of the coefficient is not consistent).

DISCUSSION

Our finding that C_{max} is a better rib fracture predictor than A_{max} for diverse types of restraint loading is in agreement with previous researchers. Well-distributed forces on the thorax can be much higher than concentrated forces (and thus result in much higher accelerations) without causing fractures. CTI results in no improvement over C_{max} because the acceleration term is poorly correlated with injury. Our findings confirm that rib fractures are a deformation-dependent injury and that C_{max} is a reasonably good predictor of rib fractures, even for diverse restraint conditions.

We have not, however, shown that the injurious level of C_{max} is constant regardless of restraint type. The data are insufficient to determine whether changes in the restraint type and therefore the force distribution on the thorax can change, for example, the value of C_{max} that corresponds to a 50% probability of AIS 3+ injury. We have demonstrated that C_{max} has reasonably good predictive ability for an entire dataset containing tests with diverse restraint conditions. In another paper (Kent et al. 2001), we showed that the injurious level of C_{max}, compared to other injury criteria, is relatively insensitive to the restraint condition. There is some sensitivity, however, so it is likely that a single, mid-sternal chest deflection measurement is insufficient to predict injury equally well for all restraint conditions. Advanced dummies, such as the Test device for Human Occupant Restraint (THOR), capable of measuring chest deformation at numerous locations, may allow for the identification of generally-applicable injury criteria that are not dependent on the restraint condition. The current measurement capabilities of the Hybrid III with its standard instrumentation package may not be sufficient for interpreting injury risk without consideration of the restraint. In support of this statement, consider another GEE model (model 6). In this model, the restraint condition (A,B,C) and the occupant's seating position (driver = 1, RFP = 0) are included as covariates in addition to C_{max} and the occupant characteristics age, gender, and mass. Despite the fact that C_{max} was the best of the three dummy criteria evaluated, the restraint condition is the dominant covariate in terms of injury prediction (Figure 4) and the C statistic (0.8789) is greater than the C statistic for model 1, indicating that consideration of the restraint improves model sensitivity and specificity. In this dataset, regardless of the dummy measurements, the belt-only restraint condition generated the greatest risk of injury, followed by the combined restraint condition. Due to the lack of concentrated belt loading, the airbag-only restraint condition was the least injurious.

The fact that no occupant characteristic was found to be a clear predictor of injury risk is an indication of the database's limitations. Of the three characteristics analyzed (gender, age, mass), age should be the strongest predictor of injury risk. Injury tolerance reductions with age are well documented. In this dataset, however, gender emerged as the most important occupant characteristic; it was the only characteristic to exhibit a consistent trend for all models and was the only characteristic to be a significant injury predictor in any model. Once gender-specific differences in age and size are accounted for, however, gender *per se* does not have a strong biomechanical basis as an indicator of hard tissue injury tolerance for a non-senescent population. The mean age of the subjects used in this study is over 50 years, however, so the observed gender effect is a reflection of the age-dependent differences in bone mineral density reduction.

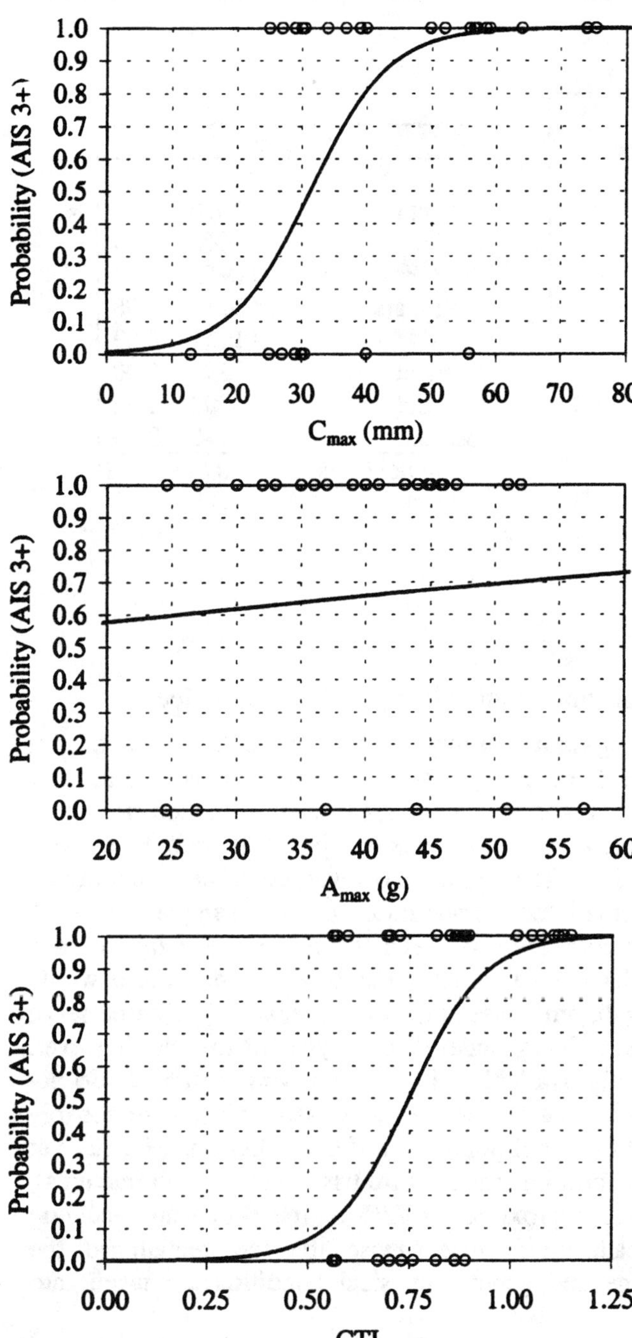

Figure 3 – Sled Test Model Results Showing Cadaver Injury Probability as a Function of Dummy-Measured C_{max}, CTI, and A_{max} for a 55 year-old, 71 kg, male (dataset mean age and mass)

The injury probability functions developed here are not applicable to a population of living humans because cadavers are generally more frail than the living, even for subjects of the same age. At present, we are aware of no studies that present a methodology for adjusting cadaver-based injury risk for application to the general population when diverse restraint conditions are considered. As a result, rather than predictors of living human injury risk, the statistical models presented here are useful primarily for evaluating the efficacy of the various dummy criteria and, to a lesser extent, for evaluating how critical values of the criteria change as a function of the occupant's age, gender, and mass. As a follow-up to this study, we will attempt to correlate dummy measures with injury risk to a population of living humans and in that way quantify the difference between living-human and cadaver injury risk.

In the tests presented here, the seatbelt was often the cause of rib fractures. This restraint dependence does not mimic the field experience of living humans. In the field, the addition of a belt restraint, rather than increasing thoracic injury risk, often increases occupant containment and precludes injurious strikes against hard vehicle components. Occupants in the field are also generally younger and less frail than cadavers used in sled tests and are thus less susceptible to rib fractures from belt loading (e.g., Martinez et al. 1994). The cadavers, then, may be representative of a population of elderly living humans, but exhibit different restraint-dependent injury trends than the general population of occupants. While it is important to consider the increased rib fracture risk for the elderly and the increased morbidity and mortality associated with rib fractures in the elderly, the use of cadavers to generate thoracic injury criteria may result in over-prediction of the injury-causing potential of belt loading and thereby result in sub-optimal apportioning of belt and airbag loading for the whole population.

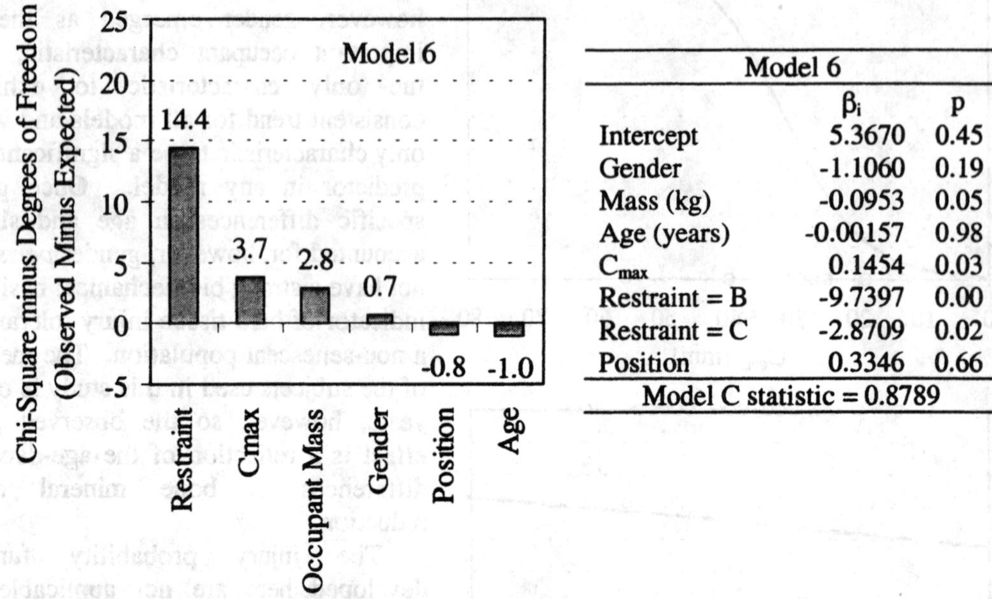

Model 6		
	β_i	p
Intercept	5.3670	0.45
Gender	-1.1060	0.19
Mass (kg)	-0.0953	0.05
Age (years)	-0.00157	0.98
C_{max}	0.1454	0.03
Restraint = B	-9.7397	0.00
Restraint = C	-2.8709	0.02
Position	0.3346	0.66
Model C statistic = 0.8789		

Figure 4 – Inclusion of Restraint and Seating Position in Injury Prediction Model

The use of cadaver injury to represent living-human injury is therefore a limitation of this analysis. In this, as in any cadaver-based dataset, the dominant thoracic injury type is rib fractures. The presence of multiple rib fractures is often associated with significant intrathoracic, intra-abdominal, and extra-cavity trauma (e.g., Pattimore et al. 1992). These injuries, therefore, are good indicators of the magnitude of overall thoracic trauma. They do not, however, correlate with all types of thoracic injuries sustained by the living, particularly soft-tissue injuries. For example, Lee et al. (1997) found that there is no clinically relevant correlation between thoracic skeletal injuries and acute traumatic aortic tear. It may be, therefore, that even a complete database of cadaver tests would not indicate the same injury risk factors, including dummy-based injury criteria, as a field-based set of data. For the purpose of evaluating injury criteria efficacy under diverse types of loading, however, cadaver testing is the most viable method currently available. In addition, while cadavers do not sustain soft tissue injuries representative of *in vivo* subjects, cadaver hard tissue injury is reasonably representative of the hard tissue injuries sustained by a frail population of living humans (Walfisch et al. 1985). Since approximately 61% of all AIS 2+ chest injuries are hard tissue injuries (rib fractures), and the maximum AIS is defined by fractures for approximately 72% of the occupants receiving MAIS 2+ chest injuries (Crandall et al. 2000), cadavers may be a reasonable representation of the living human population once differences in age and general physical condition are taken into account.

Injury criteria developed to predict rib fracture risk are an essential part of any vehicle design, assessment, or compliance testing effort. As optimized restraints and dynamically adaptable restraints are developed, however, it will become necessary to quantify the differences between cadaver-based injury and living-human injury since the use of a cadaver, as a conservative limit for the living human, is insufficient for robust optimization.

The finding that chest acceleration does not provide hard tissue injury-predictive value should not be interpreted to mean that the occupant's acceleration should not be limited in a crash. The human body does have an acceleration tolerance limit, but the types of injuries sustained at accelerations above this limit are not necessarily hard tissue injuries. Loss of consciousness, hemorrhaging, and other *in vivo* injuries, which have been associated with acceleration, currently cannot be evaluated using a cadaver model.

CONCLUSIONS

This paper presents analyses of 93 human cadaver and 50th percentile male Hybrid III dummy sled tests. Additional analysis, including more cadaver tests, is needed before a complete understanding of thoracic injury criteria efficacy and restraint dependence can be achieved. Even with the inclusion of 18 new cadaver tests and corresponding dummy tests, the sled test database is missing critical data points. The most important areas for additional testing are non-injury tests, tests with female subjects, tests with occupants in the RFP position, and tests with younger subjects.

The methodology outlined here eliminates some of the problems associated with thoracic injury criteria development from cadaver sled tests. The use of dummy measures collected using identical test conditions is of particular utility since it removes the largest source of systematic error (cadaver-mounted instrumentation) associated with cadaver sled tests. With this approach, it is critical to evaluate, using film and instrumentation, the dummy and cadaver kinematics to ensure that, first, the thoracic measures are due to thoracic loading rather than an unrelated artifact (e.g., a head strike) and, second, that the dummy's kinematics are similar to the cadaver's.

Relative ranking of injury prediction model covariates indicates that the injury-predictive effect of the restraint condition dominates that of the maximum dummy chest deflection or of any other dummy measure. Regardless of the dummy measures, the cadavers in this sample were more likely to be injured with a belt-dominated restraint condition than with a combined-loading restraint condition or an airbag-dominated restraint condition.

When the restraint condition was not considered, C_{max} was the best injury predictor of the three criteria considered. Consideration of the maximum chest acceleration, either as an independent covariate or as part of CTI, was found to weaken the injury model. The model that included A_{max} and not C_{max} was slightly worse than a model that included consideration of no dummy measure. This study is insufficient to exclude A_{max} as an injury criterion, however, since soft tissue injuries, particularly aortic injuries, which may be related to chest acceleration, are not considered.

Important caveats for cadaver test interpretation are given. In the laboratory, rib fracture risk increases when a belt is added to the restraint system because the belt results in concentrated force on the chest. In the field, the use of a belt controls occupant kinematics and often prevents injury-causing contacts against interior components. The validity of using cadaver-based injury criteria and risk functions to evaluate injury risk for a population of living, younger, and more robust occupants is currently not known and quantification of the differences between the living and the dead is an important area for research.

ACKNOWLEDGEMENTS

This research was sponsored in part by the Alliance of Automobile Manufacturers. This study also used tests performed under contract to the U.S. Department of Transportation, National Highway Traffic Safety Administration. The views expressed herein do not necessarily represent the views of the funding organizations. We gratefully acknowledge Jim Patrie of the University of Virginia Department of Health Evaluation Sciences, whose assistance with the statistical modeling was invaluable. Ricky Bryant of the UVA Automobile Safety Laboratory put in numerous hours preparing the cadavers and assisting in the cadaver and the dummy sled tests.

REFERENCES

The Abbreviated Injury Scale, 1990 Revision, Update 98, Association for the Advancement of Automotive Medicine, Des Plaines, Illinois, 1998.

Backaitis, S. and St-Laurent, A., Chest deflection characteristics of volunteers and Hybrid III dummies. Proc. 30th Stapp Car Crash Conference, pp. 157-166, 1986.

Bass, C., Wang, C., Crandall, J., Error analysis of curvature-based contour measurement devices. Paper number 2000-01-0054, Society of Automotive Engineers, Warrendale, Pennsylvania, 2000.

Burghele, N., Schuller, K., Die festigkeit der knochen kalkaneus und astragallus. *Zeitschrift für Orthopädie und ihre Grenzgebiete*, 107(3):447-461, 1970.

Butcher, J., Shaw, G., Bass, C., Kent, R., Crandall, J., Displacement measurements in the Hybrid III chest. Paper No. 2001-01-249, Society of Automotive Engineers, Warrendale, PA, 2001.

Calabrisi, P. and Smith, F., The dffects of embalming on the compressive strength of a few specimens of compact human bone. Naval Medical Research Institute, Project NM 001 056.02, 1951.

Cesari, D. and Bouquet, R., Behavior of human surrogates under belt loading. Proc. 34th Stapp Car Crash Conference, pp. 73-82, 1990.

Cesari, D. and Bouquet, R., Comparison of Hybrid III and human cadaver thoracic deformations. Paper 942209 Proc. 38th Stapp Car Crash Conference, 1994.

Chalmers, J. and Weaver, J., Cancellous bone: Its strength and changes with aging and an evaluation of some methods for measuring its mineral content. I. Age changes in cancellous bone. J. Bone Joint Surg 48-A(2):289-298, 1966.

Crandall, J., et al., A comparison of two and three point restraint systems. Joint Session of AAAM and International Research Council on the Biomechanics of Impact Conference, Lyon, France, September, 1994.

Crandall, J., Kent, R., Fertile, J., Martin, P., Mastrangelo, C., Rib fracture patterns and radiologic detection – a restraint-based comparison. *Proc. 44th Annual Scientific Conference of the Association for the Advancement of Automotive Medicine*, Chicago, Illinois, October, 2000.

Cromack, J and Ziperman, H., Three-point belt induced injuries: a comparison between laboratory surrogates and real world accident victims. Paper number 751141, Society of Automotive Engineers, Warrendale, Pennsylvania, 1975.

Duda, J., O'Brien, J., Katzmann, J., Peterson, M., Mann, K., Riggs, B., Concurrent assays of circulating bone gla-protein and bone alkaline phosphatase: effects of sex, age, and metabolic bone disease. J Clin Endcrinol Metab, 66:951-957, 1988.

Efron, B. Tibshirani, R.J., An Introduction to the Bootstrap. Chapman & Hall. NY, 1993.

Eppinger, R., Prediction of thoracic injury using measurable experimental parameters. Proc. 6th International Conference on Experimental Safety Vehicles, SAE-766073, pp. 770-780, 1976.

Eppinger, R., Morgan, R., Marcus, J., Side impact data analysis. Proc. 9th International Conference on Experimental Safety Vehicles, pp. 244-250, Kyoto, Japan, 1982.

Eppinger, R., On the development of a deformation measurement system and its application toward developing mechanically based injury indices. Paper 892426 Proc. 33rd Stapp Car Crash Conference, 1989.

Eppinger, R., et al., Development of improved injury criteria for the assessment of advanced automotive restraint systems – II. National Highway Traffic Safety Administration, U.S. Department of Transportation, November, 1999.

Epstein, S., McClintock, R., Bryce, G., Poser, J., Johnston, C., Hui, S., Differences in serum bone gla protein with age and sex. Lancet I:307-310, 1984.

Evans, F. and King, A., Regional differences in some physical properties of human spongy bone. in Evans, F. (ed.): Biomechanical Studies of the Musculoskeletal System. Charles C. Thomas pub., Springfield, IL, 1961.

Evans, F. Mechanical Properties of Bone. Charles C. Thomas publisher, Springfield, IL, 1975.

Fayon, A., Tarriere, C., Walfisch, G., Got, C., Patel, A., Thorax of 3-point belt wearers during a crash (experiments with cadavers). Paper number 751148, Society of Automotive Engineers, Warrendale, Pennsylvania, 1975.

Funk, J., Crandall, J., Tourret, L, MacMahon, C., Bass, C., Khaewpong, N., Eppinger, R., The effect of active muscle tension on the axial injury tolerance of the human foot/ankle complex. Paper number 01-S1-O-237, Proc. 17th International Technical Conference on the Enhanced Safety of Vehicles (ESV), Amsterdam, The Netherlands, 2001.

Hagedorn, A. and Burton, R., Evaluation of chestband responses in a dynamic test environment. Proc. 22nd Annual Workshop on Human Subjects for Biomechanical Research, pp. 83-101, 1993.

Harrell, F., Regression Modeling Strategies with Applications to Linear Models, Logistic Regression, and Survival Analysis. Springer, NY, 2001.

Hassan, J. and Nuscholtz, G., Development of a combined thoracic injury criterion – a revisit. Paper number 2000-01-0158, Society of Automotive Engineers, Warrendale, Pennsylvania, 2000.

Henry, Y. and Eastell, R., Ethnic and gender differences in bone mineral density and bone turnover in young adults: effect of bone size. Osteoporos Int, 11:512-517, 2000.

Horsch, J., Lau, I., Andrzejak, D., Viano, D., Melvin, J., Pearson, J., Cok, D., Miller, G., Assessment of air bag deployment loads. Proc. 34th Stapp Car Crash Conference, pp. 267-288, 1990.

Horsch, J. D. et al., Thoracic injury assessment of belt restraint systems based on Hybrid III chest compression. Paper 912895, *Proc. 35th Stapp Car Crash Conference*, 1991.

Huber, P., The behavior of maximum likelihood estimators under nonstandard conditions. Proceedings of the Fifth Symposium on Mathematical Statistics and Probability, Vol. 1. University of California Press, Berkeley, 221-223, 1967.

Kelly, P., Twomey, L, Sambrook, P., Eisman, J., Sex differences in peak adult bone mineral density. J. Bone Miner Res, 5:1169-1175, 1990.

Kent, R., Crandall, J., Bolton, J., Duma, S., Driver and right-front passenger restraint system interaction, injury potential, and thoracic injury prediction. *Proc. 44th Annual Scientific Conference of the Association for the Advancement of Automotive Medicine*, Chicago, Illinois, October, 2000.

Kent, R., Crandall, J., Bolton, J., Prasad, P., Nusholtz, G., Mertz, H., Influence of restraint condition on the efficacy of mechanical thoracic injury predictors. Stapp car crash journal, vol. 45 (abstract accepted), 2001.

Kroell, C., Schneider, D., Nahum, A., Impact tolerance and response of the human thorax. Paper number 710851, Society of Automotive Engineers, Warrendale, Pennsylvania, 1971.

Kroell, C., Schneider, D., Nahum, A., Impact tolerance and response of the human thorax II. Paper number 741187, Society of Automotive Engineers, Warrendale, Pennsylvania, 1974.

Kroell, C., Thoracic response to blunt frontal loading. in Biomechanics of Impact Injury and Injury Tolerances of the Thorax-Shoulder Complex, Backaitis, S. ed., Society of Automotive Engineers publication PT-45, 1994.

Kuppa, S. and Eppinger, R., Development of an improved thoracic injury criterion. Paper 983153, *Proc. 42nd Stapp Car Crash Conference*, 1998.

Lau, I. and Viano, D., How and when blunt injury occurs: implications to frontal and side impact protection. Paper number 881714, Proc. 32nd Stapp Car Crash Conference, pp. 81-100, 1988.

Lee, J., Harris, J., Duke, J., Williams, S., Noncorrelation between thoracic skeletal injuries and acute traumatic aortic tear. J. Trauma 43(3):400-404, 1997.

Liang, K-Y and Zeger, S.L., Longitudinal data analysis using the generalized linear model. *Biometrika* 73:13-21, 1986.

Lindahl, O. and Lindgren, A., Cortical bone in man. II. variation in tensile strength with age and sex. Acta Orthop Scand, 38:141-147, 1967.

Lindahl, O. and Lindgren, A., Cortical bone in man. III. variation of compressive strength with age and sex. Acta Orthop Scand, 39:129-135, 1968.

Marcus, J., Morgan, R., Eppinger, R., Human response to and injury from lateral impact. Paper Number 831634, Proc. 27th Stapp Car Crash Conference, pp. 419-432, 1983.

Martinez, R., Sharieff, G., Hooper, J., Three-point restraints as a risk factor for chest injury in the elderly. J. Trauma 37(6):980-984, 1994.

Mertz, H., Gadd, C., Thoracic tolerance to whole-body deceleration. Paper number 710852, Society of Automotive Engineers, Warrendale, Pennsylvania, 1971.

Mertz, H., et al., Hybrid III sternal deflection associated with thoracic injury severities of occupants restrained with force-limiting shoulder belts. Paper 910812, Society of Automotive Engineers International Congress and Exposition, February, 1991.

Messerer, O., Uber elasticitat und gestigkeit der menschlichen knochen. Verlag der J. G. Cotta'schen Buchhandlung, Stuttgart, Germany, 1880.

Morgan, R., Eppinger, R., Haffner, M., Yoganandan, N., Pintar, F., Sances, A., Crandall, J., Pilkey, W., Klopp, G., Dallieris, D., Miltner, E., Mattern, R., Kuppa, S., Sharpless, C., Thoracic trauma assessment formulations for restrained drivers in simulated frontal impacts. Paper number 942206, Society of Automotive Engineers, Warrendale, Pennsylvania, 1994.

Nahum, A, Gadd, C., Schneider, D., Kroell, D., Deflections of the human thorax under sternal impact. Paper number 700400, Society of Automotive Engineers, Warrendale, Pennsylvania, 1970.

Parsons, T., Prentice, A., Smith, E., Cole, T., Compston, J., Bone mineral mass consolidation in young British adults. J. Bone Miner Res, 11:264-274, 1996.

Pattimore, D., Thomas, P., Dave, S., Torso injury patterns and mechanisms in car crashes: an additional diagnostic tool. Injury: the British Journal of Accident Surgery 23(2): 123-126, 1992.

Prasad, P., Biomechanical basis for injury criteria used in crashworthiness regulations. Proc. International Research Conference on the Biomechanics of Impact (IRCOBI), Sitges, Spain, 1999.

Rauber, A., Elasticitat und Festigkeit der Knochen. Leipzig, Wilhelm Engelmann, 1876.

Resch, H., Pietschmann, P., Kudlacak, S., et al., Influence of sex and age on biochemical bone metabolism parameters. Miner Electrolyte Metab 20:117-121, 1994.

Shaw, C., Wang, C., Bolton, J., Bass, C., Crandall, J., Butcher, J., Khaewpong, N., Sun, E., Nguyen, T., Chestband performance assessment using quasistatic tests. Society of Automotive Engineers, Warrendale, PA, 2000.

Shaw, G., Crandall, J., Butcher, J., Biofidelity evaluation of the THOR advanced frontal crash test dummy. Proc. International Research Conference on the Biomechanics of Impact (IRCOBI), Montpellier, France, 2000.

Shaw, C., Kent, R., Sieveka, E., Crandall, J. Spinal kinematics of restrained occupants in frontal impacts. accepted for publication at the 2001 Conference of the International Research Council on the Biomechanics of Impact (IRCOBI), Isle of Man, 2001.

Slosman, D., Rizzoli, R., Pichard, C., Donath, A, Bonjour, J-P., Longitudinal measurement of regional and whole body bone mass in young healthy adults. Osteoporos Int, 4:185-190, 1994.

Sonoda, T., Studies on the strength for compression, tension, and torsion of the human vertebral column. J. Kyoto Pref. Med. Univ., 71(9):659-702, 1962.

Stein, I. and Granik, G., Rib structure and bending strength: an autopsy study. Calcif. Tissue Research, 20:61-73, 1976.

Vanderschueren, D., Gevers, G., Raymaekers, G., Devos, P., Dequeker, J., Sex- and age-related changes in bone and serum osteocalcin. Calcif Tissue Int, 46:179-182, 1990.

Walfisch, G., Chamouard, F., Lestrelin, D., et al. Predictive functions for thoracic injuries to belt wearers in frontal collisions and their conversion into protection criteria. Paper 851722, Proc. 29th Stapp Car Crash Conference, 1985.

Wertheim, M., Memoire sur l'elasticite et la cohesion des principaux tissus du corps humain. Ann. Chim Phys, 21:385-414, 1847.

White, H., Maximum likelihood estimation of misspecified models. *Econometrica.* 50:1-25, 1982.

Yamada, H., Strength of Biological Materials, ed. Evans, F., The Williams and Wilkins Company, Baltimore, MD, 1970.

Yoganandan, N., Skrade, D., Pintar, F., Reinartz, J., Sances, A., Thoracic deformation contours in a frontal impact. Paper number 912891, Society of Automotive Engineers, Warrendale, PA, 1991.

Yoganandan, N., Morgan, R., Eppinger, R. Pintar, F., Sances, A., and Williams, A., Mechanisms of thoracic injury in frontal impact. *J. of Biomech. Engineering.* **118**(4):595-597, 1996.

Zhou, Q., Rouhana, S., Melvin, J., Age effects on thoracic injury tolerance. Paper number 962421, Stapp Car Crash Conference, 1996.

902329

Comparison and Realism of Crash Simulation Tests and Real Accident Situations for the Biomechanical Movements in Car Collisions

Dietmar Otte
Accident Research Unit
Medical University Hannover

Abstract

In this study, the accident simulation tests, the so-called crash tests, enforced by legislation and put into praxis are evaluated in regard of their conformity with reality. They are based on accident analyses from investigations at the place of accident, which are carried out by a scientifically trained team which documentates details of the accident event. 826 cars involved in traffic accidents with cars, trucks and other objects, in the greater vicinity of Hannover (FRG) were at our disposal for evaluation purposes. The study clearly reveals that impact simulations like those carried out at present, cover only approximately 34% of all situations of road traffic. This conclusion is derived from a comparison of accident framework conditions like overlapping degree, impact impulse direction and impact situation. For the frontal impact the offset impact, with two-third degree of overlapping, without rail-bound lead should be favoured. The lateral collision should be simulated with oblique and eccentric impacting barriers from the front.

INTRODUCTION AND OBJECTIVE

With increasing traffic and rising accident figures, the question about safety in cars becomes more and more justified. Minimum measures for vehicle safety are enforced by law. These safety checks include the so-called crash tests. These provide the same safety standards for different vehicles under defined test conditions. However, as no generally applicable test procedure exists for every collision type, only a selection of tests are carried out. At present, this collection includes test procedures for occupant protection in the case of:

frontal impact
lateral impact
special accident situations

At present there are crash tests specified by the authorities which vary from country to country, but still show some similarities. In this study, mainly the tests which are enforced by law in the European Community (EG), the Economic Comission for Europe (ECE), and in the United States of America are discussed.

The objective of this survey is on the one hand to find out how far the events of the real traffic accident are demonstrated by these crash tests, and on the other hand, in which points the crash tests could contribute to still more safety, beyond the requirements of the real traffic accident situation.

The following special questions, therefore, emerge from the task:

- To which percentage are the real accident situations covered by crash tests, and how many are not?
- Are the speeds in the ordered tests comparable with those in the real accident events, under similar collision conditions?
- Do the actual impact impulse angles which occur conform with the respective angles in the test?
- Do the collision opponents conform with reality, are they to be seen as rigid or movable, and which geometrical shapes do the accident opponents have ?
- To which amount does the vehicle surface cover that of the opposing vehicle ?
- Which collision situations are the most dangerous for passengers ?

OVERVIEW OF LEGAL RESTRICTIONS FOR THE ACCIDENT BEHAVIOUR OF CARS

For the European Community the vehicle safety regulations are stipulated in

74/60/EWG interior
74/297/EWG steering wheel
74/408/EWG seat + adjustment

74/433/EWG	outer car shape/edges
76/115/EWG	seat belt anchorage
77/541/EWG	seat belts
78/937/EWG	head rests

Crash tests are demanded in the ECE reglements Nos. 32, 33 and 34/1/. These regulations stipulate test rulings for frontal and rear impacts (fig. 1). Tests concerning the crash behaviour of the lateral vehicle side are not yet enforced by law in the ECE reglements. They are, however intended to be included in the rules in future. Up to now, no details have been defined regarding these tests.

A fundamental difference between the US Regulations (FMVSS 208, 213 and 301-75) and the EC regulations lies in the fact that in the US frontal regulations dummies are included and that a lateral crash test without dummies are already included.

legal regulation	frontal collision	lateral impact	rear collision
ECE	33 without 34 dummies	-	32 34
FMVSS	208 with 301-75 dummies	208 301-75	301-75

fig. 1: Valid ECE and FMVSS regulations for different collision constellations

A side impact test procedure with dummies is in discussion in Europe (ECE: TRANS/SC1/WP23/GRSP/R48) and in the US (NRP).
Both tests differ in dummy, barrier face and impact angle [2]. The majority of vehicle manufacturers carry out further crash tests, beyond those enforced by law. Some of the European manufacturers have carried out tests in concerted actions, beside the legally enforced ones (3).

An overview of the various crash-test regulations is given by Brandsch (4) and shown in fig. 2.

<u>0 degree barrier impact</u> - A standard test for simulation of relatively serious frontal collisions is the frontal crash at a speed of approximately 50 km/h against a rigid barrier. As far as the stress on vehicles and passengers is concerned, this test is similar to a frontal collision of two equally heavy vehicles with an identical frontal structure and each with a speed of 50 km/h. This means that the relative impact speed at the time of collision was 100 km/h.
Contrary to other impact tests, with an only partly overlapping of the impact area, the whole of the vehicle front is involved symetrically.

<u>Frontal post impact</u> - A much more unfavourable constellation for the vehicle front is caused by a frontal post impact. With a similar amount of kinetic energy as in a 0-degree-barrier impact, the deformation forces are transmitted to a smaller area, depending on the post diameter.

<u>30 degree barrier impact</u> - In an impact to a 30 degree barrier, the vehicle front is only involved with one side. A part of the kinetic energy is used up by friction, rotation and translation by gliding off the vehicle. As a result, the energy transfered to the front structure is lower than in the 0-degree barrier impact.

<u>Offset barrier impact</u> - Contrary to the standard test with complete overlapping, only one half of the frontal structure is involved. The test vehicle is marched on a rail system, by which no rotation is possible. Whereas a deformation of the passenger compartment must only be expected in serious frontal collisions, in lateral collisions with an impact point in the door region, the door is moved toward the passengers, even at lower impact speeds. Due to this fact, the survival space in the passenger compartment is reduced.

<u>Lateral impact with the deformable barrier, 90 degrees</u> - For the simulation of lateral vehicle collisions, an expedient is used. The rocker-panel is hit only slightly. With approximately 300 mm distance of the lower edge of the deformation element measured from the ground. The barrier is set to 90 degrees to the vehicle longitudinal axis.
The stiffness of the deformation element is orientated on the mean European car.

<u>Lateral impact with a deformable barrier in a crabbed mode</u> - In this test configuration, a stationary vehicle is also hit at an angle of 90 degrees to the vehicle longitudinal direction, but the impact direction is arranged at an angle of 27 degrees. By the oblique directed wheels at this angle, the approach of the barrier takes place in backward movement. For this test arrangement consideration is given to the fact that in the majority of real accidents not only the impacting but also the impacted vehicle is moving. When the impact speed is simulated vertically to the longitudinal vehicle axis with 30 mph and the speed of the impacted vehicle with 15 mph, the vector sum shows the collision speed v=34 mph and the oblique impact direction at an angle of 27 degrees. This is the US proposal for side impact test procedures.

<u>Lateral impact against a post</u> - A vehicle is moved laterally against a post with a diameter of 300 mm which is supported by a block. The vehicle is moved on wheels at the side. By dividing the post into three areas, it is possible to regard the roof, door and rocker panel/floor structures seperately.

<u>Lateral impact against a rigid vehicle contour</u> - In this case, a rigid body mounted in front of the block has the simplified form of a vehicle front. The test vehicle, which is mounted on side wheels, is moved towards the vehicle contour at an angle of 90 degrees. The impact point is chosen so that the centre of gravity of the vehicles longitudinal axis meets with the perpendicular centre line of the contour.
Some vehicle manufacturers make further tests,

such as:

Roll-over of the vehicle - The vehicle is placed on a carriage in a slanted position. The carriage is set in motion and then braked abruptly. Because of the mass inertia, the vehicle performs a sideward tilt movement.

DESCRIPTION OF THE ACCIDENT SITUATION

CHOICE OF COMPARISON VARIABLES - Not so much the accidents, but far more the single car collisions and other conditions are of interest when comparing accident situations under crash-test conditions. A basis can be found in accidents from investigations at the site of the accident, which are documented by a research team and scientifically analysed.

METHOD OF ACCIDENT INVESTIGATION - In the greater vicinity of Hannover (West Germany) an interdisciplinary team of medical doctors and engineers drives to the site of the accident immediately after it has occured. They are informed by the headquarters of the police and fire-department. The team drives to the site of the accident, which is chosen from a statistical spot-check plan, in vehicles equipped with siren and blue-light. The operational area has a radius of about 100 km. Among others, the team investigates accident traces, vehicle deformations and contact areas with persons involved. True-to-scale pictures are made by use of a stereoscopic measuring camera, which enables a reconstruction and analysis of the accident and collision phases. A detailed description can be found in Otte /5/ and Dilling (6). The injuries are described in detail according to type, localisation and severity AIS (American Association for Automotive Medicine (7)). The accident data collection is made on the basis of a statistical spot-check plan to achieve representative accident evaluation results. For this, among other methods, the cases documented by the team are compared with the official police investigation results. The problem in answering these questions, is that it is difficult to achieve a scaled determination from the relatively exactly known, usually metric sizes and quantities of the framework conditions of the accident analysis. The detailed documentation of the investigation, which contains data of the accident reconstruction and true-to-scale drawings of the accident traces and vehicle deformations as well as photos of the vehicle deformations, is of advantage in this case.

DATA - For this study, accidents for the time from January 1985 to May 1987 were investigated. During that time, 1192 cases were covered. In 512 cases a total of 826 cars collided with another car, utility vehicle or an object respectively. Collisions with pedestrians and two-wheelers, or rollover accidents were not included. During an accident, a vehicle can be involved in more than one collision. In the investigated collective, up to seven collisions were established for one vehicle. From the distribution of collisions shown in fig. 3 an almost identical situation can be recognized for car against car and car against utility vehicle collisions. Approximately 84% of these are single collisions. The distribution of car against object collisions reveals with 48.1% distinctly fewer single collisions, but quite frequently secondary, third and fourth collisions. The frequency of multiple collisions in an impact with an object is understandable from the subsequent run-out movements during which additional collisions occur, with other traffic participants or other objects

IMPACT REGIONS - When looking at the accident situations in frontal, rear or lateral impacts (fig. 4) it becomes evident that for all collision types, i.e. car against car, car against truck, car against object, the frontal impact is the most frequent. In car with car collisions 52.2% occur frontally, 32% laterally, and 15.6% to the rear region. For collisions of cars with utility vehicles, the difference in frequency between the impact regions is not so distinct. In collisions of utility vehicles the relatively high proportion of laterally hit vehicles (43.2%) is striking. It was established that in the accident situation 50% of all vehicles are hit frontally in a collision (fig. 5). Lateral collisions are represemted with 36.1% of all cases, while rear collisions are established in 13.5%. It also became apparent that the left-side impact is more frequent than the right-sided.

IMPULSE ANGLE - As far as the load for passengers is concerned, physical rules of moving and pushed objects are the decisive factors. Passengers are subjected to relative kinematics which are effected by the impulse angle of the impact transmittance to the vehicle. In this accident analysis, the impulse angles for frontal collisions reveal an accumulation of approximately 180 degrees which corresponds with an impact direction directly from the front, parallel to the longitudinal axis of the vehicle. Almost two-thirds (63.8%) of all frontal collisions occur under an impulse angle in the region of 166 to 195 degrees, i.e. 180 +/- 15 degrees. Impacts oblique from the left dominate. However, 44.2% occured at an angle of 166 to 180 degrees. The exact 180 degrees collision can be defined with 23.7% of all frontal collisions. For rear collisions, an accumulation of 0 +- 15 degrees impulse angle can be observed. 78% of all rear collisions occur at this angle. In lateral collisions, such a small-angled region is not so apparent. Here the frontal to right-angled impacts from the side are very frequent (left 31.9%, for 106 to 135 degrees, 26.3% for 135 to 165 degrees, right 35% for 226 to 255 degrees, 19.2% for 195 to 225 degrees).

DELTA-V DISTRIBUTION - A measure for the force on passengers during the accident phase is the vehicle deceleration. As this is difficult to establish in accident analyses, for this study we used the speed change Delta-v (6) and for each respective collision type marked in figure 7.

80% of all collisions are established with Delta-v levels of maximum 25 to 33 km/h, depending on the impact situation.

A comparison of the impact regions rarely reveals Delta-v values above 50 km/h for rear collisions. In lateral and frontal collisions, however, this level is frequently exceeded. (Delta-v > 50: front 4.6%, side 2.3% and rear 0.1%).

OVERLAPPING GRADE - In figure 8, the percentual distribution of overlapping of the vehicle exteriors in collisions of car against car is shown. An overlapping of 100% could only be observed in 21.15% of all cases. Half of all frontal collisions show an overlappig of less than 50%. The diagram shows two maximums which look like so called bath-tub curve, with an often smaller and often higher overlapping. The first maximum shows an overlapping of 15% (15.4%), the second and bigger one a 100% overlapping (21.1%).

MULTIPLE COLLISIONS - Approximately 15% of all accidents consist of multiple collisions. This multiple burden for vehicles and passengers does not only result from the variously isolated single forces in the different impact impulses, but also from the different impulse directions and impact regions with resulting relative movements of the occupants.

Within the framework of this study, it is interesting to know whether and in how many cases the subsequent collisions reach a higher Delta-v level than each of the previous ones. Fig. 9 shows a comparison between the Delta-v values of the first and the second collisions.

Of the n=285 collisions n=123 (43.2%) have a Delta-v value which is just as high or even higher than that of the previous collision. With increasing Delta-v, the number of accidents in which the secondary collision has an at least equal Delta-v value as the first one, decreases with rising Delta-v value of the first collision. This fact proves that it is not always sufficient to look at just one collision situation only.

Side impact situations often occur after frontal collisions, and after a primary lateral impact, lateral collisions often occur. After a rear impact, frontal collisions are often subsequent impacts.

ACCIDENT SEQUENCES

In order to ascertain the injury risk for passengers in different collision situations, the maximum injury severity degree MAIS for the most frequent impulse angles, in relation to the Delta-v levels, is used. This evaluation was carried out for cars which were involved in one collision only. We used this method because an association to the injury cause for each collision was not always possible beyond all doubt.

TOTAL INJURY SEVERITY - The total injury severity grade MAIS (maximum AIS) for one accident victim is established from the maximum single injury severity grade of the various body regions.

In frontal collisions, below Delta-v level to 30 km/h, more than half the proportion of passengers remain uninjured (62.4%), and only a negligible proportion of them (1.8%) are seriously injured, i.e. they suffer MAIS above 2. This picture emerges for frontal collisions with an impact impulse angle of 135 to 225 degrees (fig. 10).

With increasing Delta-v level, the rate of slightly and not injured decreases. Above Delta-v level 50 km/h, the injury severity increases rapidly. At a speed level of 31 to 50 km/h only 4.8% seriously and fatally injured (MAIS > 2) are established, but with Delta-v level 51 to 70 km/h, there were already 18.8% injuries of MAIS > 2, which amounts to 53.7% at Delta-v above 70 km/h. Oblique impact-direction effects are partly of a more serious and partly of a lighter type, as far as injury sequences are concerned. In angles of 135 to 165 degrees and also of 195 to 225 degrees, there are clearly fewer, and with higher Delta-v levels sometimes fewer seriously injured (no diagram), depending on the relative movement of the passengers.

In lateral collisions, passengers sitting on the impact side are especially exposed to injuries, at an angle of 75 to 165 degrees. Up to Delta-v level of 30 km/h, 50.6% of the passengers on the impact side, but 71.2% of the passengers sitting impact-averted remained uninjured. In speeds of 31 to 50 km/h, only 21.9% of those sitting on the impact side (fig.12) and 28.6% of the impact-averted seated passengers were uninjured. Critically or fatally injured (MAIS 5/6) are found exclusively in impact-nearside positions. An injury-reducing function must be attributed to the seatbelt in connection with oblique frontal impact impulse directions.

HEAD INJURIES - With higher Delta-v levels, the injury severity for the head (fig. 11) also increases, but serious injuries occur in frontal collisions significantly only above Delta-v level of 50 km/h. With Delta-v levels up to 30 km/h, no head injuries occured in 84.1% of the cases. In a frontal collision with an impact angle of 135 to 225 degree, 50.9% of the head injuries occur by contact to the steering-wheel.

Literally no difference was observed for the frequency of head injuries in a lateral impact, between impact-nearside and opposite side seated passengers, not even with higher Delta-v levels. Up to 30 km/h, however, and in impact-averted positions, 83% of the passengers had no head injuries, in comparison to 70.9% on the impact side. The injury causes in lateral impacts are mainly the side windows (20.8%) and the side structures (30.8%), especially the *a*-pillars and door frames.

THORAX INJURIES - Serious thorax injuries only occur in frontal collisions at speed absorbing levels above Delta-v 50 km/h. With Delta-v level up to 70 km/h (fig 12) for instance, 47,7% of all front-seat passengers remained uninjured. 41.9% incure injuries of AIS 1/2, 10.4% suffered serious injuries of AIS 3/4.

Very serious or fatal injuries respectively are only registered at Delta-v levels above 70 km/h. Three quarters of all thorax injuries are caused by the seatbelt (70.5%). But 17.3% of these injuries were caused by the steering-wheel too, but mainly only in connection with high impact forces.

The single-case analysis established, that in lateral collisions, due to the increasingly frequent oblique impact transmittance and usually direct intrusion of the passenger compartment, the upper part of the body of passengers sitting on the impact-opposite side is moved toward the impact side. There they impact the steering-wheel (15.3%), the seats (15.5%) or are obstructed by the seatbelt (40.7%). On the other hand, the thorax, although under pressure on the impact side, is apparently well protected by the seat-belt strap and contacts the relatively large deformation region.

INJURIES TO THE PELVIS AND ABDOMEN - Serious pelvis and abdominal injuries only occur above Delta-v levels of 50 km/h (fig. 13). The predominant proportion of pelvis and abdominal injuries is caused by the seatbelt (63.6% for frontal collisions, 37% by the nearside impacted and 42.1% impact-opposite side seated passengers). As a rule, the injuries conform with AIS 1.

INJURIES TO THE LOWER EXTREMETIES - In frontal collisions and Delta-v level of 31 to 50 km/h, serious leg injuries occur to only 2.9% (fig. 14). Injuries of AIS >2 are mostly third degree open fractures with substantial soft-part lesions. With Delta-v values above 70 km/h, the proportion already concerns almost 50% of all belt-wearing front-seat passengers. Such injuries are observed in lateral collisions, exclusively with passengers sitting in nearside impact position. The leg injuries in the foot space of belt-wearing persons are mostly caused by shifting of front structures in frontal collisions, or side deformations in lateral impacts respectively.

SUMMARY OF COGNITIONS AND COMPARISON OF REAL ACCIDENT SITUATIONS UNDER CRASH CONDITIONS

Crash tests are carried out at certain speeds and a specially defined constellation, in order to simulate the forces to vehicle and passengers. The passengers are replaced by so-called dummies. These are equipped with measurement readers which as a rule record accelerations and decelerations. There are already first deviations visible of the simulations tests from the accident reality, which must be seen in the light of comparison between persons and dummies (7). Comparative tests can be carried out in laboratories (8).

The vehicle should be primarely constructed in such way that the forces, which occur during accidents are passed in reduced form to the passengers. The correct interpretation must be verified in the accident account. For this purpose, accidents must be evaluated and the applied simulation procedures evaluated in comparison with the accident situation.

This study clearly justifies the comprehensive evaluation that 34% is the smallest part of real road-traffic accident situation covered by a test. Test procedures can only cover a selection of possible collision configurations. They should however include those which occur most frequently and those with the most serious injuries.

The evaluation also reveals that 44.7% of all frontal and even 82.9% of all lateral car collisions discussed in the crash simulations do not conform with reality.

In lateral collisions, these are mainly collisions outside the compartment, which were not taken into consideration. 31% of the not considered collisions occur to the lateral front, 15% lateral to the rear.

In frontal collisions, among others, eccentric post collisions (approx. 26%) are not considered. Beyond that, it is shown that the impulse angles, the overlapping grades and the Delta-v values established in the study do not conform with reality. It is further a fact that all test procedures do not permit real reproductions of interactions by the collision partners. Test procedures are isolated procedures for the test vehicles. 68% of frontal collisions occur, however, to another car. The deformation conduct of the tested car surely depends on the deformation characteristics of the collision partner and thus does the injury possibility. Vehicle versus vehicle tests should be given preference.

Individually, the following important essential framework conditions of the real accident event appear to be important factors for simmulations:

FRONTAL COLLISIONS - Zero-degree barriers and offset barriers are at present included in the standard tests of frontal collisions. It can be stated that with this method approximately 21.4% of all real accident situations are taken into consideration. 41.6% of all frontal car collisions are covered (fig. 15). Up to now, tests were carried out at a speed level of approximately 50 km/h. The evaluations established that Delta-v values above 50 km/h occur in 4.6% of all collisions. As far as this fact is concerned, the tests include a substantial proportion of the real traffic accidents. As a rule, serious injuries have only been observed above these values. The crash tests and the real accidents differ in two vital points:

First of all, only 25% of the occuring impulse angles of the accidents are 180 degrees, the 90% region is orientated more to the left, at an angle of 180 +/- 35 degrees. Apart from the 0-degree impact it also seems sensible to carry out tests in which an impact impulse effects the vehicle at an angle of approximately 15 degrees from the front left. This appears to be efficiently simulated by the 30 degree barrier and the offset impact.

Secondly, the overlapping grade is not

basically always 100%. An overlapping of 100% is established in only 21% of all cases. It seems to be sensible to carry out further tests with an overlapping of 20 to 30%, beside those carried out up to now. This way it becomes understandable, why in an the offset impact, 15.8% of all frontal collisions are already covered (fig. 15).

Thirdly, in addition to an 0-grade impact, in which a central thrust impulse is simulated, a simulation with eccentric impact conditions must be tried. In real accidents, an overlapping of up to two-thirds of the vehicle front often occurs.

	impulse angle	
	180 +/- 15 degree	oblique
overlapping		
full face	56.8%	34.6%
two thirds	26.3%	26.9%
one third	16.9%	38.5%

This means that in connection with frequently occuring oblique impulse effect directions the thrusting force of a lever arm toward the center of gravity results in a rotation of the vehicle. As a rule injuries to passengers are caused by vehicle parts in the path of movement of their body regions (see tables 10 to 14). A rail-bound leading in an offset impact does therefore not conform with reality.

Approximately 50% of all frontal collisions have an overlapping grade between 66 and 100%, and 27% a grade of 33 to 66%. For this reason, the offset impact simulation should find preference, without a forced overlapping impact of the vehicle of less than two-third.

REAR COLLISIONS - This collision type is the smallest of the discussed groups. The tests enforced up to now are not exclusively carried out for the safety of passengers, but also for secondary interests, like for instance the fire protection.

The rear collision tests are carried out at a speed of 38 km/h. This Delta-v value is exceeded in the selected collective in 3.8% of all traffic accidents. This should be sufficient for all crash tests as the highest occuring injury severity grade is MAIS 4 for rear collisions. In rear collisions, 56% of the impulse angles are 0 degrees, so that an additional test seems sensible, and sufficient without an oblique position.

LATERAL COLLISIONS - In view of the evaluations made, and the fact that lateral collisions represent one third of all collisions, the author recommends crash tests, which establish the lateral deformation behaviour or the deceleration values respectively. These tests are sensible in view of the fact that the vehicle side is the region of the vehicle, which due to the limited deformation path, offers the lowest possible protection to passengers. It is noticeable for lateral collisions, that the proportion of uninjured is clearly smaller in comparison with frontal collisions (uninjured: frontal 31.6%, lateral 19.7%). Object accidents are dangerous, especially for the lateral region, in view of the intensive penetration. In lateral accidents, the passenger compartment is damaged to an amount of approximately 60%.

In the tests carried out up to now, a stationary vehicle is hit laterally and decreases its initial velocity which was built up by the impact, in the run-out phase. In reality, the vehicle has an initial velocity, it is hit laterally and moves after the accident with a speed changed in amount and direction. The Delta-v value together with the impulse angle demonstrates the impact impulse to the vehicle. The evaluation of the impulse revealed that, on average, the vehicles are impacted at an angle of 115 degrees. This means that often the real impact occures from the front, at an angle of 65 degrees. This angle is valid for the right-sided as well as the left-sided collisions. The US-test construction with an oblique impact barrier can realise this better. This way, approximately 10% of all lateral collisions are taken into consideration (fig. 15). A reality-conform test should be simulated in such way that the vehicle is not hit directly at its center of gravity in the region of the passenger compartment, but rather further front. The respective impulse angle should be 115 degrees, which means coming obliquely from the front at an angle of 65 degrees. The intensive impact results in a rotation of the vehicle around its center of gravity. In order to include 90% of all Delta-v values of this collision type, the impact should correspond to a Delta-v value of 40 km/h. This is significantly higher than the tests discussed presently with impact speeds of 50 km/h and defined barrier (v ≈ 20 km/h).

REFERENCES

(1) EWG/ECE
ECE-Richtlinien der Europäischen Wirtschaftsgem. für Strassenfahrzeuge Kirchbaum-Verlag, Bad Godesberg/FRG

(2) Friedel, B.
Seitenschutz von Pkw - Stand der Überlegungen zum Aufprall-Prüfverfahren in Europa und USA
Verlag TÜV Rheinland 'Kolloquium Unfallschutz der Fahrzeuginsassen in Europa', 113 - 133, 1988

(3) Kallina, I.; Zeidler,F.
Strategien zur Erhöhung der Sicherheit von Pkw-Insassen
Verlag TÜV Rheinland 'Kolloquium Unfallschutz der Fahrzeuginsassen in Europa', 13 - 31, 1988

(4) Brandsch, H.; Hoefs, R.
Über das Deformationsverhalten von Fahrzeugstrukturen
Zeitschrift Verkehrsunfall und Fahrzeugtechnik 2, 51-56, FRG (1990)

(5) Otte, D. et al.
Erhebungen am Unfallort
Unfall- und Sicherheitsforschung, Heft 37, Bundesanstalt für Strassenwesen Bergisch-Gladbach/FRG (1982)

6) Dilling, J.; Otte, D.
Die Bedeutung örtlicher Unfallerhebungen im Rahmen der Unfallforschung
Unfall- und Sicherheitsforschung Strassenverkehr, Heft 56, Bundesanstalt für Strassenwesen, 59-68, Bergisch-Gladbach/FRG (1986)

(7) American Association for Automotive Medicine - The Abbreviated Injury Scale - Revision 85
Americ. Ass. f. Autom. Med., Morton Grove, Illinois/USA (1985)

(8) Svens, M.J.; Wiedel, J.F.; Cheng, P.H.; Marandi, M.
The Vocabulary of Accident Reconstruction
SAE paper 89 06 34 in SP 777, 9-28, Warrendale/USA, (1989)

(9) Kallieris, D.; Mattern, R.
Belastbarkeitsgrenze und Verletzungsmechanik des angegurteten Fahrzeuginsassen beim Seitanprall
FAT-Schriftenreihe 36, Forsch.vereinigung Frankfurt/FRG (1984)

(10) Kallieris, D.; Schmidt, G.; Mattern, R.
Vertebral Column Injuries in 90-degrees Collisions, a study with post-mortem human subjects Proc. Ircobi, 189-292, Birmingham/UK (1987)

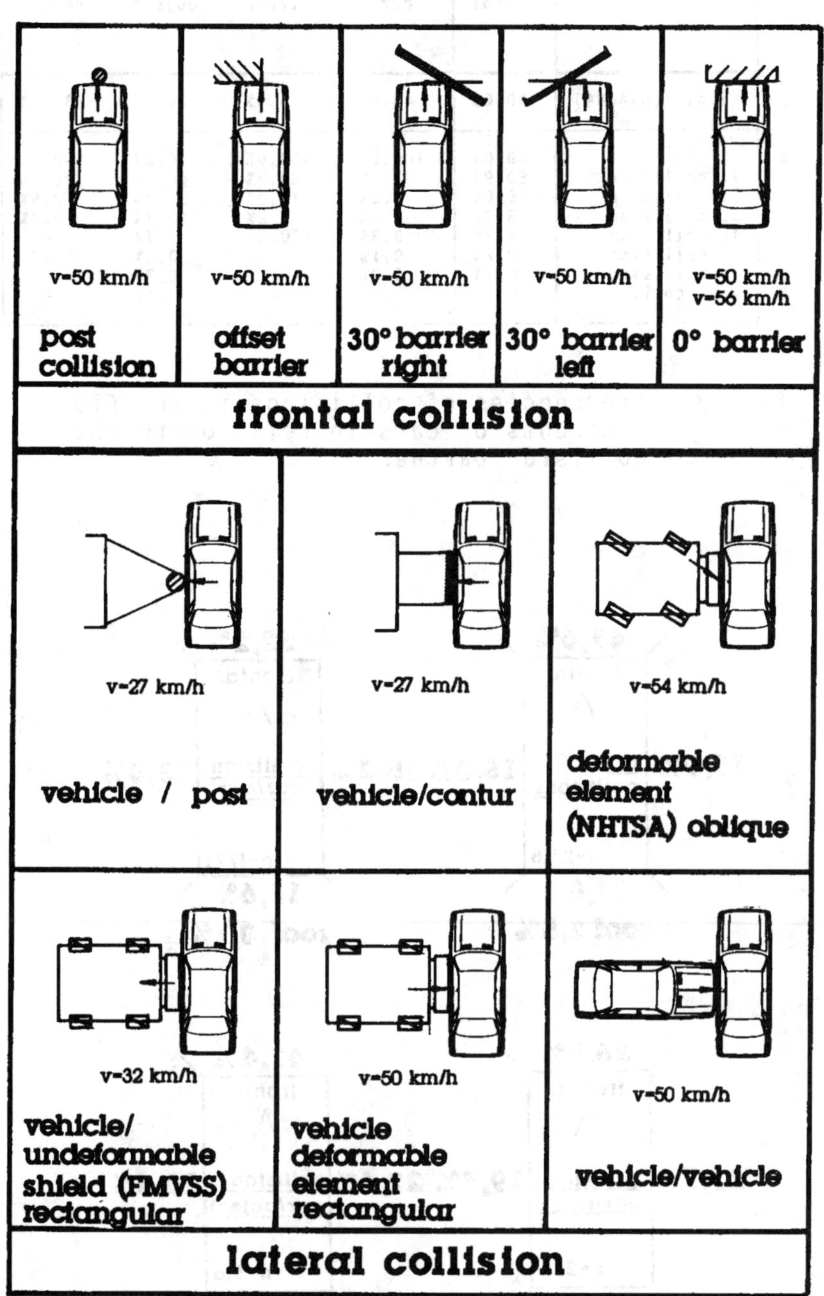

fig. 2: Impact situations in simulation tests for frontal and lateral collisions (Brandsch 1)

	total	collision partner of car			
		car	truck	object	pedes-trian/ cyclist
total (n=4084)	100,0%	43,4%	5,4%	20,0%	31,1%
	100,0%	100,0%	100,0%	100,0%	100,0%
1. collision	80,9%	83,9%	84,3%	48,1%	96,7%
2. collision	15,0%	14,0%	11,8%	37,6%	2,5%
3. collision	3,2%	1,6%	3,2%	10,4%	0,7%
4. collision	0,7%	0,3%	0,3%	2,7%	-
5. collision	0,2%	0,1%	-	0,8%	-
6. collision	0,1%	0,0%	-	0,3%	-
> 6. coll.	-	-	-	0,1%	-

fig. 3: frequencies of collisions in traffic accidents of cars in relation to the collision partner

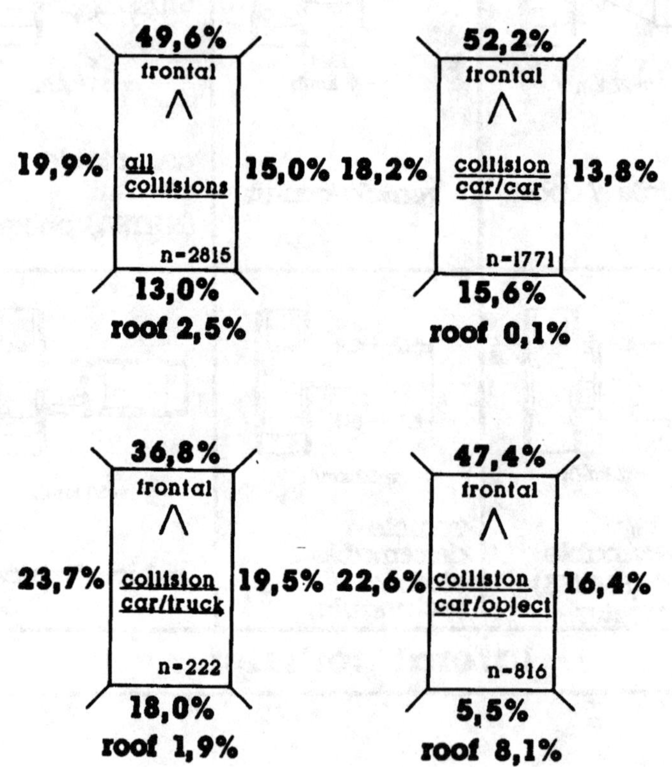

fig. 4: frequencies of impact regions for the different kinds of collision constellations in car accidents

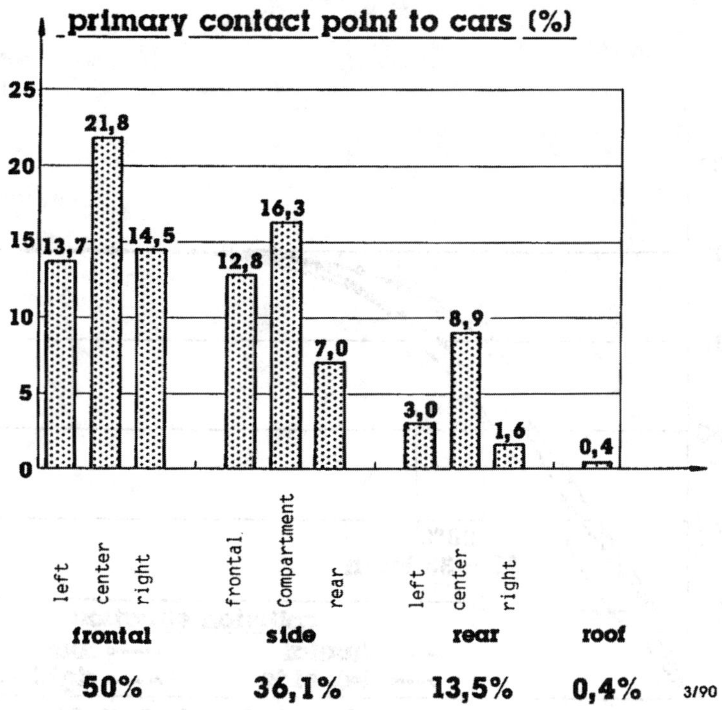

fig. 5: Primary contact points to cars

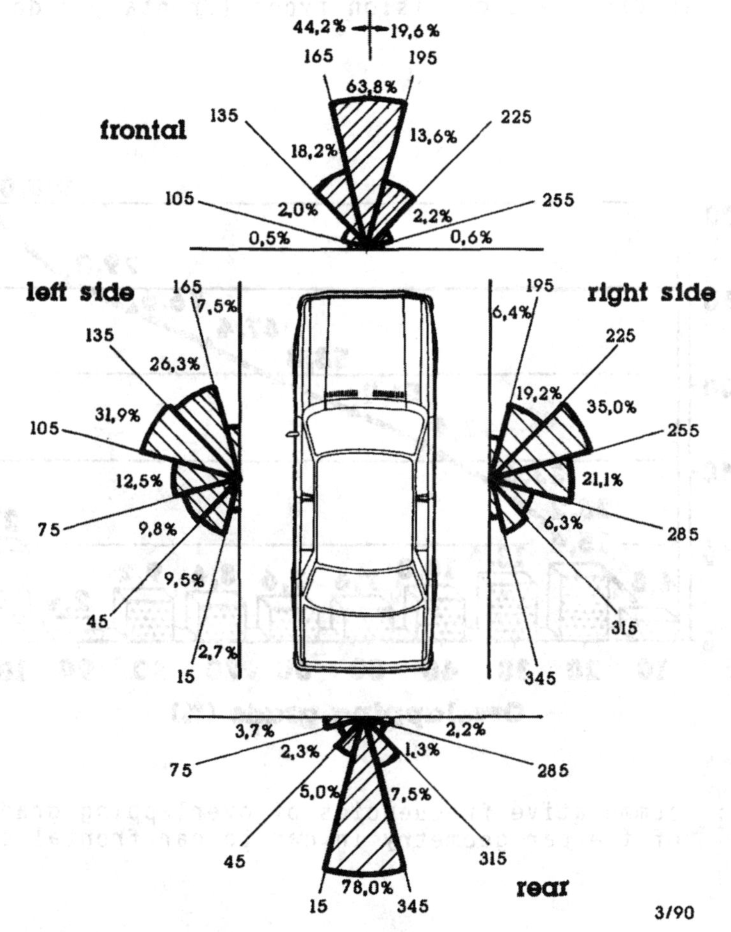

fig. 6: distribution of impulse angles and frequencies for car accidents, pointed out are the most common angle areas

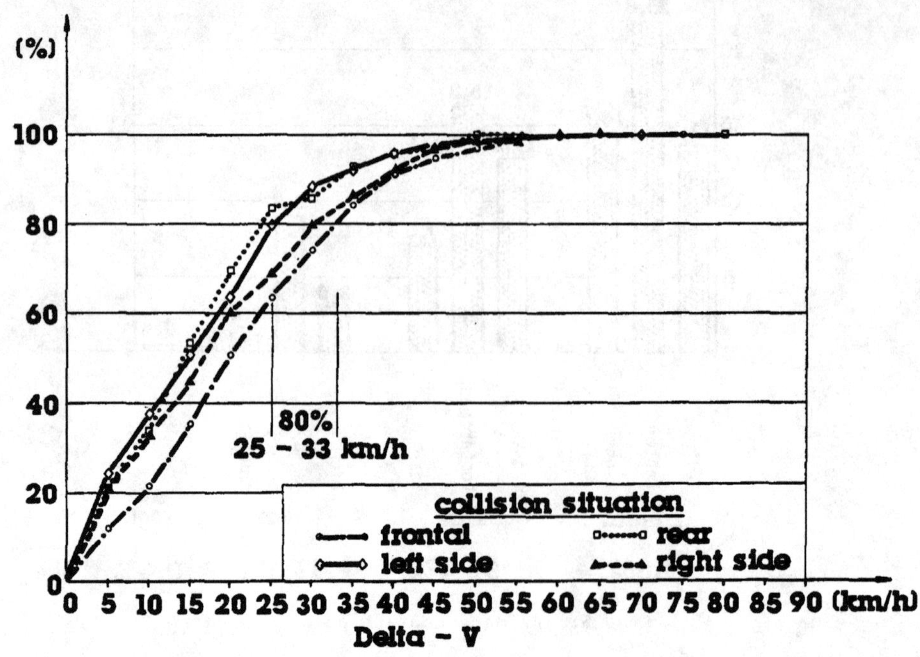

fig. 7: cummulative frequencies of delta-v-distribution in different collision types (frontal, side, rear)

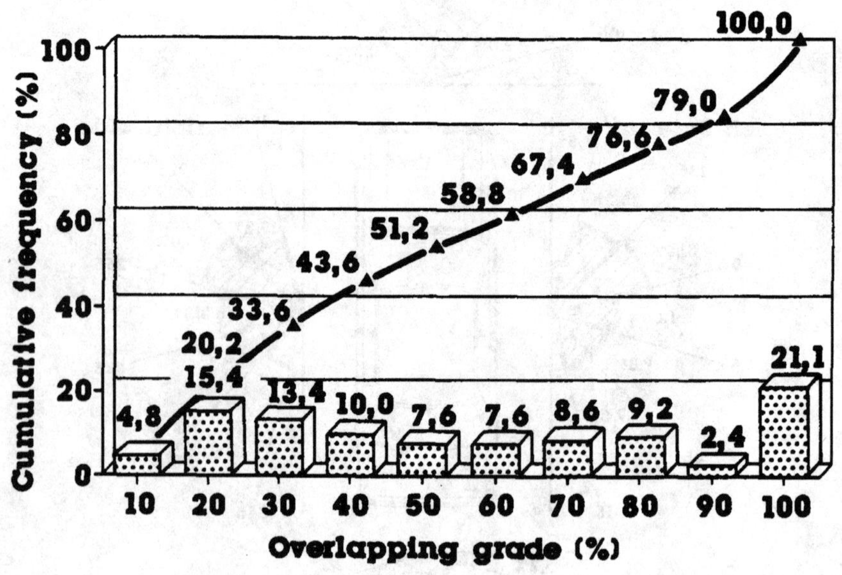

fig. 8: cummulative frequencies of overlapping grades of the car geometry in car to car frontal collisions

	total	delta-v1						
		1 -10	11-20	21-30	31-40	41-50	51-60	> 60
total (n=123)	100,0%	48,9%	27,0%	11,6%	0,6%	1,0%	-	-
delta-v2								
1 - 10	23,8%	17,7%	-	-	-	-	-	-
11 - 20	31,5%	13,3%	15,9%	-	-	-	-	-
21 - 30	24,6%	9,1%	4,9%	8,5%	-	-	-	-
31 - 40	10,3%	4,4%	3,2%	2,3%	0,4%	-	-	-
41 - 50	4,8%	2,5%	0,6%	0,8%	-	1,0%	-	-
51 - 60	3,3%	1,2%	1,6%	-	-	-	-	-
> 60	1,7%	0,7%	0,8%	-	0,2%	-	-	-

	total	primary collision				
		frontal	right side	rear	left side	roof
total (n=158)	100,0%	37,3%	18,4%	25,3%	18,4%	-
secondary coll.						
frontal	42,4%	7,6%	5,7%	22,2%	6,3%	-
right side	22,2%	10,8%	7,0%	1,9%	1,9%	-
read	11,4%	6,3%	1,9%	0,0%	3,2%	-
left side	22,8%	12,0%	3,2%	0,6%	7,0%	-
roof	1,2%	1,2%	0,6%	-	-	-

fig. 9: comparision of delta-v-distribution and impact areas in the primary and secondary collisions of cars (not included predestrins, cyclysts, rollovers)

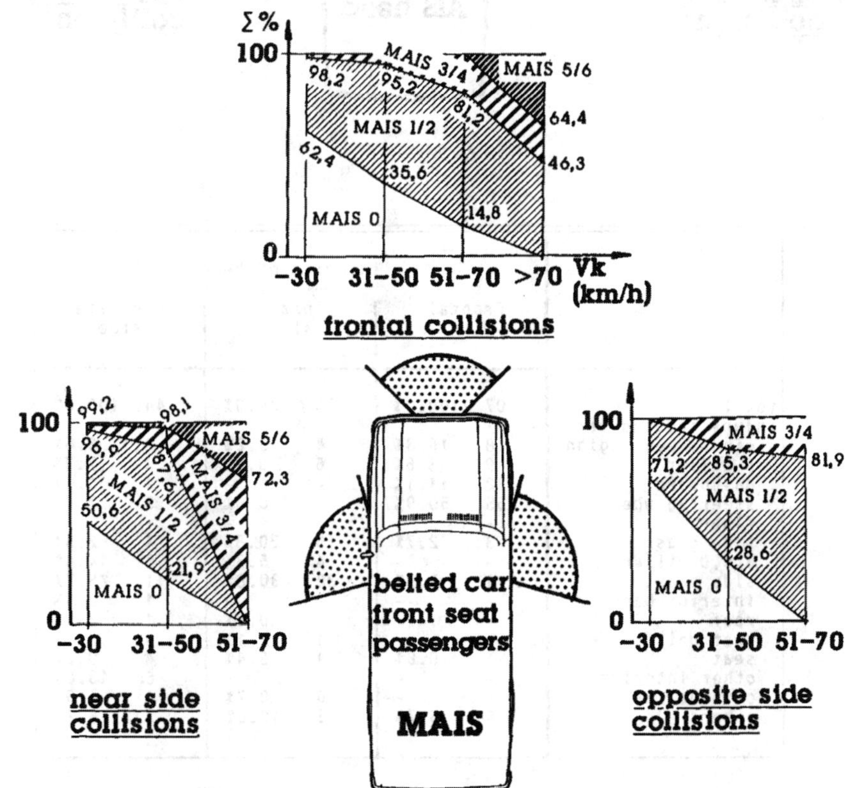

fig. 10: Injury severity grades MAIS for selected collision situations and injury causing parts for this observed injured collective

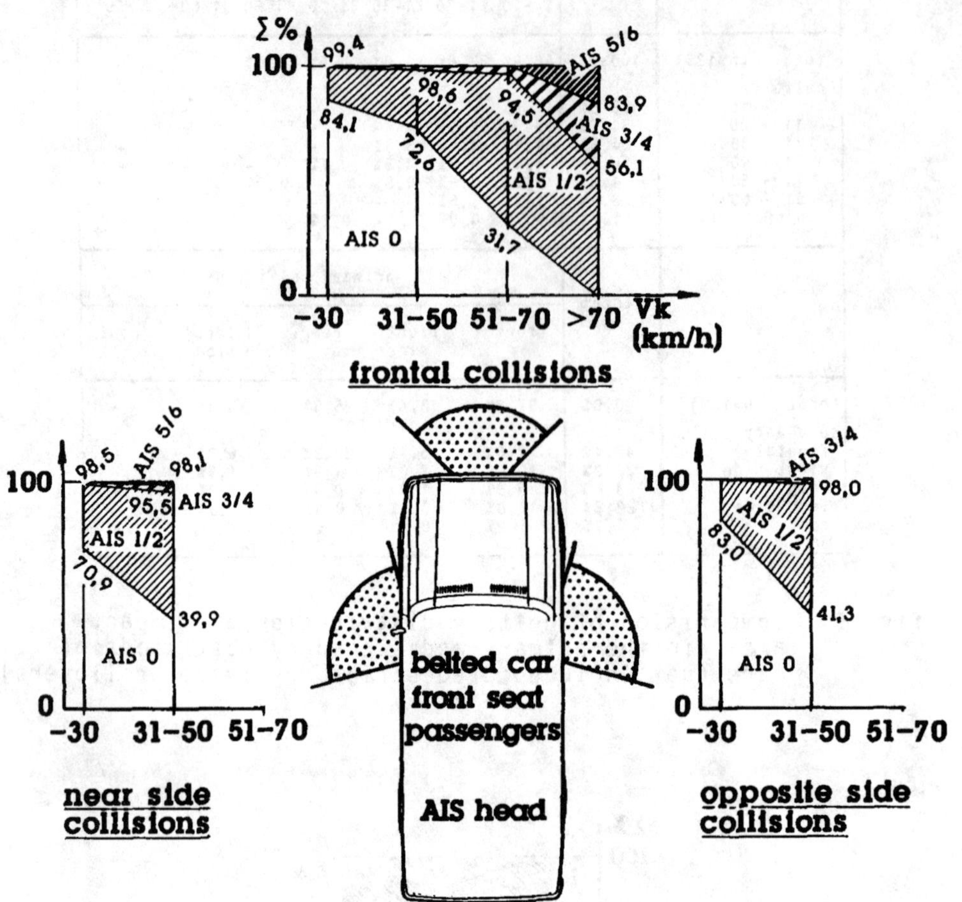

	injuries of head					
	frontal		near side		opposite side	
total	107	100.0%	70	100.0%	44	100.0%
injury causing part						
windscreen region	18	16.3%	4	5.2%	2	5.6%
A-pillar	9	8.6%	6	9.1%	3	7.3%
dashboard	12	11.1%	-	-	1	1.6%
steering wheel	55	50.9%	1	0.8%	3	7.4%
floor	-	-	-	-	-	-
side glass	3	2.7%	15	20.8%	1	2.5%
B,C,D-pillar	-	-	4	5.3%	6	14.1%
side panel	4	3.9%	22	30.8%	11	25.7%
interior back	-	-	-	-	1	2.5%
roof	-	-	1	0.8%	1	2.6%
seat belt	-	-	1	2.1%	-	-
seat	1	0.8%	4	5.4%	4	9.2%
other interior	-	-	-	-	6	13.6%
outside car	-	-	0	0.7%	1	1.6%
others	6	5.7%	13	19.0%	3	6.2%

fig. 11: Injury severity grades AIS-head for selected collision situations and injury causing parts for this observed injured collective

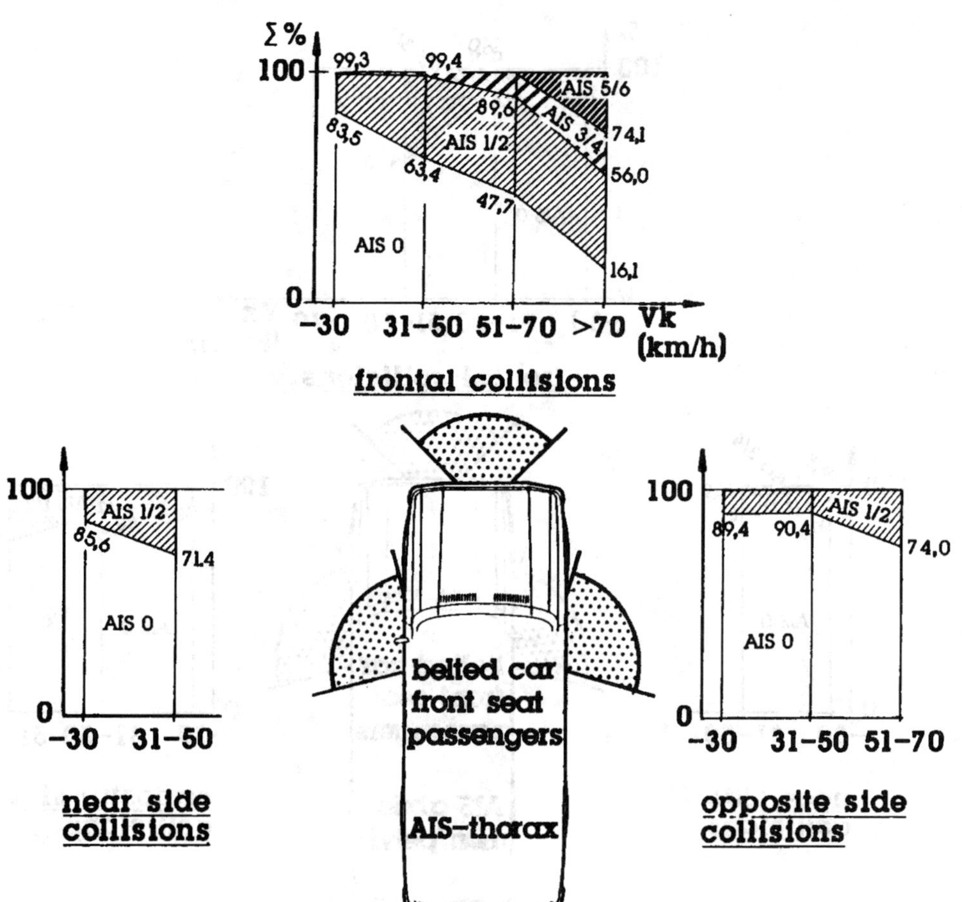

	injuries of thorax					
	frontal		near side		opposite side	
total	84	100.0%	29	100.0%	14	100.0%
injury causing part						
windscreen region	-	-	-	-	-	-
A-pillar	-	-	-	-	-	-
dashboard	3	4.0%	1	1.8%	1	4.3%
steering wheel	15	17.3%	1	3.2%	2	15.3%
floor	-	-	-	-	-	-
side glass	-	-	0	0.8%	-	-
B,C,D-pillar	-	-	-	-	-	-
side panel	3	4.1%	15	49.3%	2	11.6%
interior back	-	-	-	-	-	-
roof	-	-	-	-	-	-
seat belt	59	70.5%	12	42.1%	6	40.7%
seat	1	1.8%	1	1.9%	2	15.5%
other interior	1	1.0%	-	-	-	-
outside car	-	-	0	0.8%	1	8.3%
others	1	1.4%	-	-	1	4.3%

fig. 12: Injury severity grades AIS-thorax for selected collision situations and injury causing parts for this observed injured collective

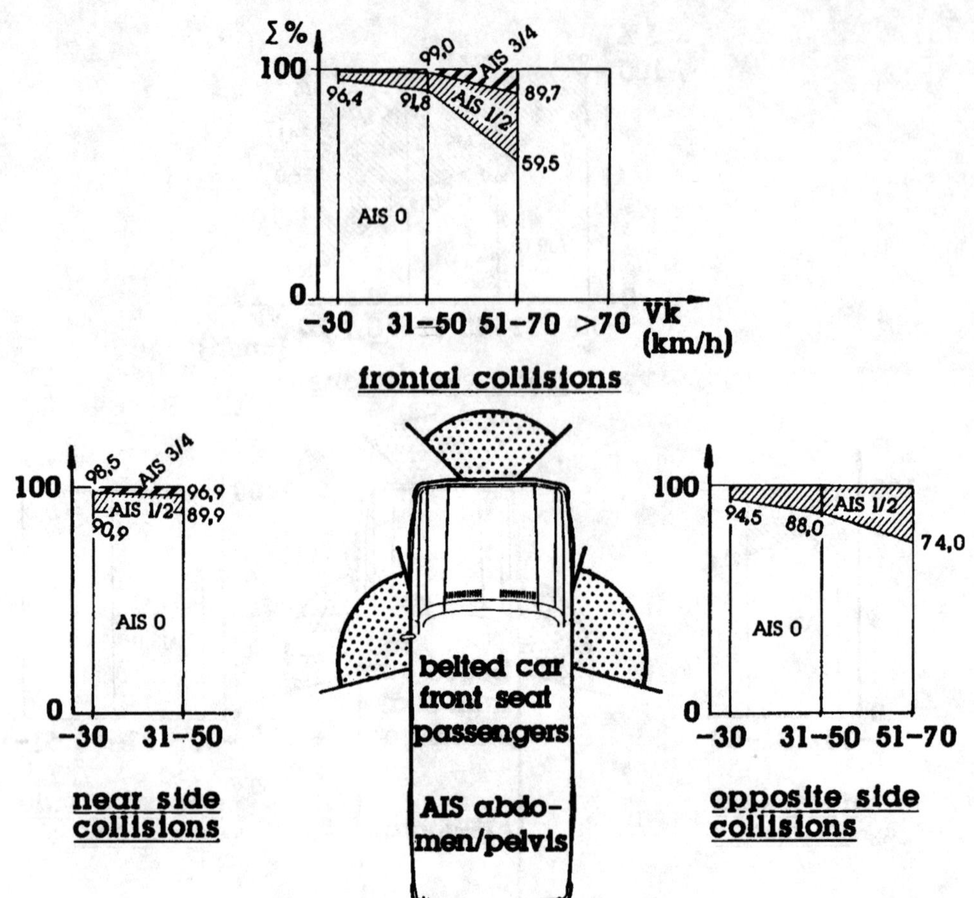

	injuries of abdomen and pelvis					
	frontal		near side		opposite side	
total	28	100.0%	15	100.0%	6	100.0%
injury causing part						
windscreen region	-	-	-	-	-	-
A-pillar	-	-	-	-	-	-
dashboard	2	5.5%	-	-	1	20.7%
steering wheel	2	5.7%	0	1.6%	-	-
floor	0	1.8%	-	-	-	-
side glass	-	-	-	-	-	-
B,C,D-pillar	-	-	-	-	0	8.3%
side panel	4	15.6%	7	45.2%	-	-
interior back	-	-	1	4.4%	1	11.1%
roof	-	-	-	-	-	-
seat belt	18	63.6%	6	37.0%	2	42.1%
seat	0	1.1%	1	5.5%	0	7.5%
other interior	0	0.9%	0	3.2%	-	-
outside car	-	-	0	3.2%	-	-
others	2	5.9%	-	-	1	10.4%

fig. 13: Injury severity grades AIS-pelvis/abdomen for selected collision situations and Injury causing parts for this observed injured collective

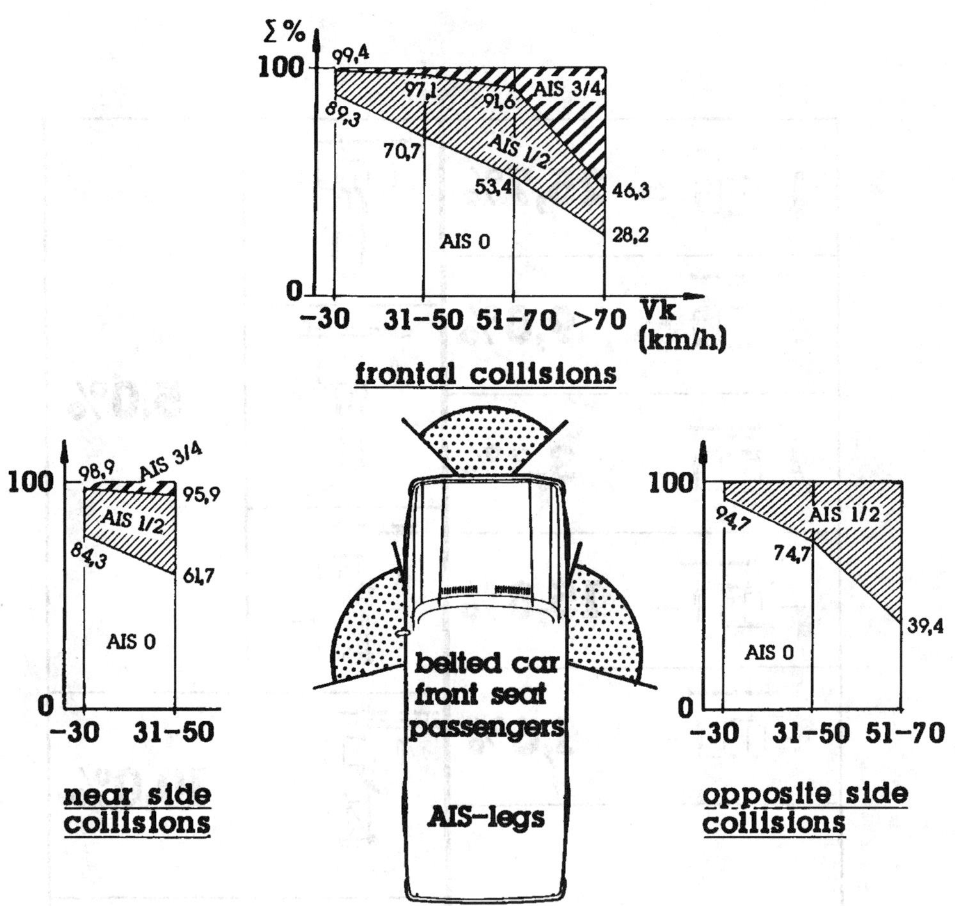

	injuries of lower extremities					
	frontal		near side		opposite side	
total	105	100.0%	41	100.0%	22	100.0%
injury causing part						
windscreen region	-	-	-	-	-	-
A-pillar	-	-	-	-	1	2.6%
dashboard	71	67.7%	6	14.8%	7	32.5%
steering wheel	-	-	0	1.2%	-	-
floor	29	27.5%	15	36.0%	6	24.8%
side glass	-	-	-	-	-	-
B,C,D-pillar	-	-	-	-	-	-
side panel	3	2.6%	11	26.5%	2	11.0%
interior back	-	-	-	-	1	6.6%
roof	-	-	-	-	-	-
seat belt	2	2.1%	-	-	0	2.1%
seat	-	-	5	13.1%	3	14.8%
other interior	-	-	2	4.3%	-	-
outside car	-	-	2	4.1%	1	3.2%
others	-	-	-	-	1	2.4%

fig. 14: Injury severity grades AIS-legs for selected collision situations and Injury causing parts for this observed injured collective

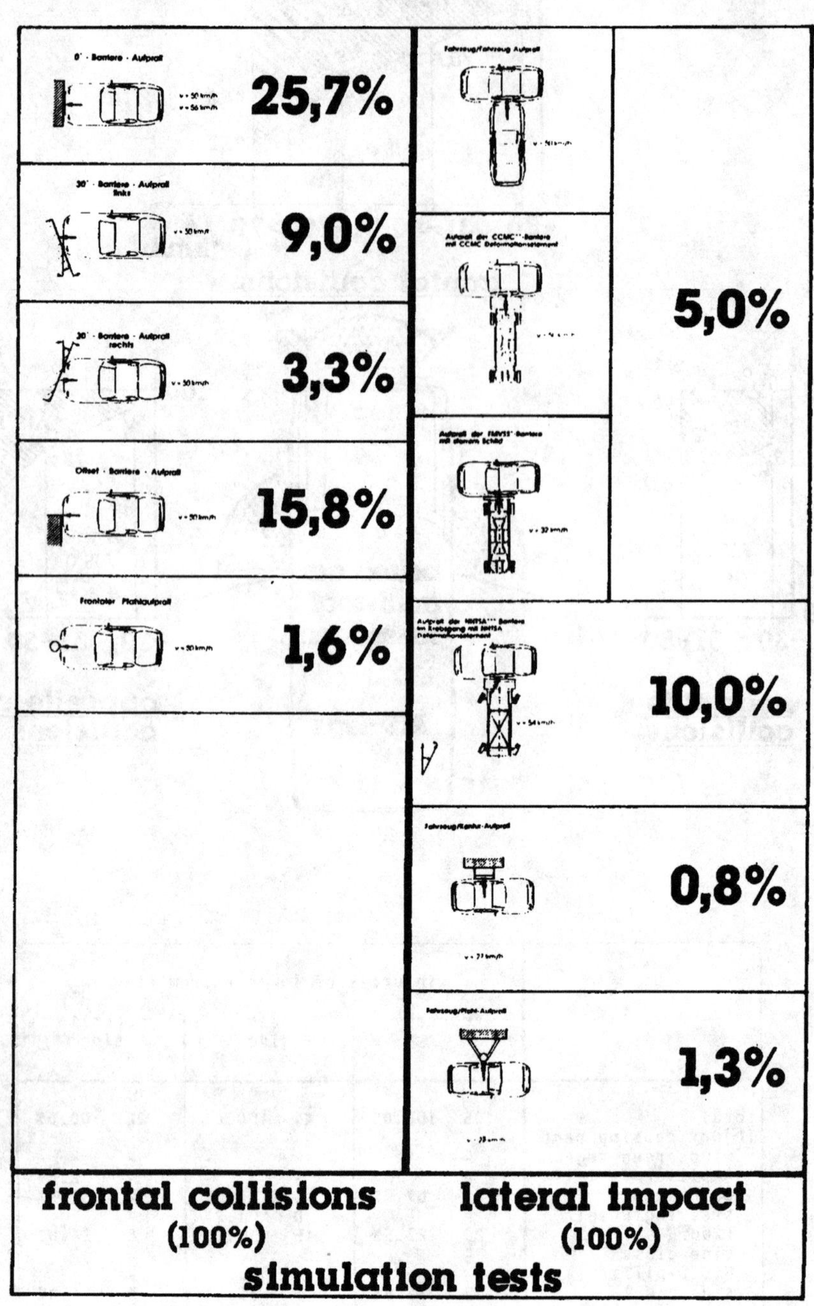

fig. 15: comparision of simulation test procedures and the confirmation of real accident situations

902328

Assessing the Safety Performance of Occupant Restraint Systems

David C. Viano and Sudhakar Arepally
General Motors Research Laboratories
Biomedical Science Dept.
Warren, MI

ABSTRACT

The purpose of this study was to investigate approaches evaluating the performance of safety systems in crash tests and by analytical simulations. The study was motivated by the need to consider the adequacy of injury criteria and tolerance levels in FMVSS 208 measuring safety performance of restraint systems and supplements. The study also focused on additional biomechanical criteria and performance measures which may augment FMVSS 208 criteria and alternative ways to evaluate dummy responses rather than by comparison to a tolerance level.

Additional analysis was conducted of dummy responses from barrier crash and sled tests to gain further information on the performance of restraint systems. The analysis resulted in a new computer program which determined several motion and velocity criteria from measurements made in crash tests. These data provide new insights on restraint performance including torso angle change, the effectiveness of occupant restraint from velocity buildup -- called the Restraint Quotient -- and forward displacement and rebound of the chest and pelvis. The various approaches are used to discuss Hybrid III responses from 24 barrier crash and 23 Hyge sled tests.

Four additional experiments were critically evaluated to validate the new analysis. They involved a lap-shoulder belted Hybrid III female dummy with the lap-belt positioned low on the pelvis for restraint or pre-positioned on the abdomen for submarining. By comparing calculated and film analyzed data, the Restraint Quotient program was found to predict accurately those test results.

Numbers relating to FMVSS 208 criteria values were reduced 15%-25% by lap-belt submarining over lap-belt restraint. The addition of an abdominal injury criterion and use of a frangible abdomen helped predict injury risk to the abdomen and an overall higher risk of injury with lap-belt submarining. The use of motion and velocity criteria also identified poorer performance with lap-belt submarining.

Several approaches were reviewed to interpret the overall performance and injury risks of safety systems in a crash. The preferred method was found to be injury risk assessment which represents dummy responses as injury risks using Logist probability functions. This enables injury in each body region to be assessed by the maximum risk from applicable biomechanical criteria. Whole body injury is determined by summing the individual risks from each body region.

The motion and velocity criteria determined from the Restraint Quotient program were useful in complementing the biomechanical criteria to assess injury. They may be helpful in an overall determination of the crash performance of restraint systems, supplements and enhancements.

THERE HAVE BEEN RAPID CHANGES in the type of restraint systems in passenger vehicles with the phase-in of passive safety requirements. While each system meets the response limits of FMVSS 208 in a barrier crash test, there is an emerging understanding that this represents only a basic requirement for safety assessment in a crash (1). Based on a review of the methods to assess the effectiveness of safety belt restraint, the following requirements are considered important to the overall quality of occupant restraint:

- Maintain FMVSS 208 responses below required limits.

- Provide primary restraining forces on the boney structures of the pelvis, upper thorax and shoulder of the belted occupant.

- Minimize loads on the compliant regions of the abdomen and lower thorax, thus limiting abdominal compression.

- Control forward excursion of the lower extremity and minimize rearward rotation of the pelvis.

- Ensure a slightly forward rotation of the torso to load the shoulder and upper thorax.

- Minimize contact HIC and facial loads.

- Minimize neck shear, bending and axial responses.

- Reduce chest compression, Viscous response and lateral shear deformation, and

- Minimize femoral bending, knee shear and tibial-ankle loads.

The above requirements represent a complicated and inter-related set of responses that need to be balanced systematically to ensure "overall" occupant restraint in a crash and minimize injury risks in critical body regions. This is particularly important when the distribution in age, tolerance and seating position for the whole family of occupants is considered in the development of new restraint systems.

For example, the current Part 572 dummy does not assess abdominal injury with lap-belt loading so that a system may adequately satisfy federal requirements while involving submarining and abdominal injury risk. In addition, occupant kinematics can include rearward rotation of the upper torso which may lower HIC and thoracic spinal acceleration but increase lower thoracic loading by the shoulder belt. This may inadvertently result in rib cage fractures, particularly in older occupants. Thus, as further refinements and advances in restraint systems are considered for use in passenger vehicles, a sufficient set of performance requirements should be utilized to help strive for reasonable safety with different systems. This will involve considering individual responses as part of the overall safety performance.

This paper is part of a three-part program to improve the assessment of occupant protection systems and the evaluation of restraint enhancements which may provide complementary protection. The research includes: 1) improvements in dummy injury criteria and assessment procedures to adequately address injury risk to various body regions (2-4), 2) attention to the biomechanical quality of restraints including motion sequence criteria (5-7) and injury biomechanics parameters (8-10), and 3) additional analysis of current test responses to better interpret restraint performance, including the influence of the distribution in crash types and severities, occupant age and tolerances, and injury mechanisms (11,12).

The study provides the rationale and procedures to determine motion sequence properties which are an important aspect in the evaluation of restraint performance. These data can be obtained from current dummy responses in barrier crash or sled tests of safety systems. They provide complementary information on the amount of restraint provided. The relative velocity of the occupant with respect to the interior, rebound, kinetic energy developed, and the forward displacement and posture of the occupant during restraint are evaluated and discussed in terms of potential injuries in a crash. In this study, five parameters of belt restraint quality are analyzed from current safety test data:

- Velocity of the occupant with respect to the interior.

- Rebound velocity as a fraction of the maximum forward velocity.

- Kinetic energy of the occupant with respect to the interior.

- Angle change of the upper torso during restraint.

- Displacement of the pelvis.

It is the intention of this study to show how these values relate to the nine restraint requirements listed above.

The lower the velocity of the occupant with respect to the interior, the greater the restraint and the lower the energy that must be managed. If the relative velocity is zero, the occupant would, in effect, be glued to the vehicle structure and experience the same deceleration as the passenger compartment. In contrast, an unbelted occupant builds up speed with respect to the interior as the vehicle stops by front-end crush. This can result in occupant impact with the interior at speeds approaching the vehicle impact velocity or possibly higher if interior impact occurs during rebound.

In a frontal barrier crash, an unbelted occupant can develop speeds that approach the initial impact speed of 30 mph (13.4 m/s) particularly in the passenger seating position since the steering system provides ridedown for the driver. The potential range in velocity build-up is zero to approximately 13.4 m/s. This produces a greater range in occupant kinetic energy because energy is related to the square of velocity. Thus, a 75 kg occupant may develop 0 to 6,733J of kinetic energy with respect to the passenger compartment.

If the occupant's instantaneous velocity in the vehicle is determined as a fraction of the overall change in velocity of the passenger compartment, the ratio is called the Restraint

Quotient (RQ). It normally varies between 0 and 1. The lower the RQ, the greater the restraint in a crash.

Another factor in occupant restraint is the angle change of the upper body during belt loading. For greater protection, the torso angle (θ) should change so that the spine is upright or slightly forward to ensure that belt load is on the shoulder. This applies force to the strong skeletal regions of the shoulder and upper chest.

At this time, the torso angle is not routinely reported in crash tests; yet, it provides information on the overall quality of restraint performance in loading the shoulder and upper chest. Poor quality restraint can involve a rearward orientation of the upper body with belt loads directed on the lower rib cage and abdomen. This can result in rib fractures and internal injury, and increases the risk of submarining. The torso angle is particularly important in properly restraining older occupants whose rib cage tolerance is lower than younger passengers. However, the need to manage the torso angle must be considered in context with the potential for head impact with the steering assembly.

This study provides an approach to obtaining new information from currently measured responses in dummy tests. It involves additional analysis of test results which may be helpful to making decisions on alternative safety systems and supplemental components. It provides data on the degree of restraint in terms of occupant velocity in the interior. This involves kinetic energy to be managed by occupant loadings. In addition, the angle change of the upper body is an indirect measure of the concentration of belt loading on the shoulder and away from the more compliant regions of the lower chest and abdomen.

METHODOLOGY

The following accelerations are used to determine the Restraint Quotient (RQ) and torso angle change (θ):

1) x- and z-axis of thoracic spinal acceleration;

2) x- and z-axis of pelvic spinal acceleration; and,

3) x-axis sled or passenger compartment acceleration.

The accelerations are routinely measured in barrier and sled tests. Accelerometers in the dummy are fixed to the AP and SI axis and move with the dummy during a crash. A computer analysis of the accelerations is used to generate properties of safety belt restraint.

The following assumptions are made and represent a first-order approximation based on dummy and vehicle responses through maximum occupant restraint in a crash test:

1) the resultant biaxial acceleration of the chest and pelvis is co-linear with the vehicle deceleration axis;

2) the lateral components of chest and pelvic acceleration can be neglected for this analysis;

3) the accelerometers are accurate enough and the responses do not involve sharp impacts so they may be double integrated to approximate occupant displacements; and,

4) the connection between the spinal and pelvic accelerometers is a rigid link.

Obviously, the lumbar spine is flexible and bends during restraint as much as 25°-40° in the Hybrid III dummy and the lap belt causes downward motion of the body but the effects are considered second order factors to the calculation of torso angle change. Using these approximations, the following calculations are made with respect to an inertial reference frame and a moving coordinate frame fixed to the passenger compartment. The calculations determine occupant dynamics and kinematics with respect to the vehicle interior. The resultant accelerations of the chest x_c and pelvis x_p are computed from the x-axis (AP direction of the dummy) and z-axis (SI direction) responses as a function of time:

$$\ddot{x}_c = (\ddot{x}^2_{cx} + \ddot{x}^2_{cz})^{1/2}, \quad \ddot{x}_p = (\ddot{x}^2_{px} + \ddot{x}^2_{pz})^{1/2}. \qquad (1)$$

The deceleration of the occupant compartment or sled (x_v) represents the average deceleration of the passenger compartment. For this analysis, data are filtered using SAE 180 channel class. The velocity of the dummy and vehicle are determined by standard numerical integration:

$$\dot{x}_c = \int \ddot{x}_c dt, \quad \dot{x}_p = \int \ddot{x}_p dt, \quad \dot{x}_v = \int \ddot{x}_v dt. \qquad (2)$$

These responses are related to the inertial reference frame. The velocity of the occupant with respect to the moving vehicle reference frame are:

$$v_c = \dot{x}_v - \dot{x}_c, \quad v_p = \dot{x}_v - \dot{x}_p. \qquad (3)$$

Displacements of the dummy chest and pelvis are shown in Figure 1. There are inherent problems in reliably determining velocity and displacement from integration of accelerometers. However, the placement in these tests on the dummy spine and pelvis reduces the potential for sharp spikes in acceleration and rapid changes in position, which are primary factors in increasing errors. Displacements in the vehicle are determined by integration of Equation 3:

$$d_c = \int v_c dt,$$
$$d_p = \int v_p dt. \qquad (4)$$

Occupant restraint during vehicle deceleration is measured as the relative velocity of the occupant in the passenger compartment divided by the maximum velocity change of the vehicle or sled. This parameter is a function of time and is called the Restraint Quotient (RQ). It varies between 0 -- assuming the occupant is glued to the passenger compartment and does not develop velocity -- and 1 -- assuming the occupant attains the total velocity change in the vehicle before impacting the interior. The Restraint Quotient is defined as:

$$RQ_c = \frac{v_c}{(\dot{x}_v)_{max}},$$
$$RQ_p = \frac{v_p}{(\dot{x}_v)_{max}}. \qquad (5)$$

The angular velocity of the upper body of the dummy is calculated using the relative velocity between the chest and pelvis:

$$\omega = \frac{v_c - v_p}{D}, \qquad (6)$$

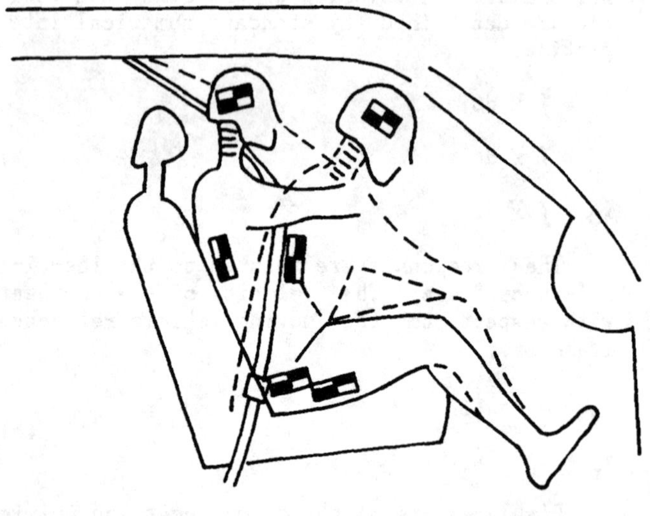

Figure 1: Motion of the occupant during a crash demonstrating the displacement of the chest and pelvis with belt restraint.

where D is the distance between the chest and pelvic accelerometer packages. For the 50th percentile male Hybrid III dummy the distance is D = 35 cm and for the 5th percentile female Hybrid III dummy it is D = 31 cm.

A positive angular velocity implies forward rotation of the upper torso, whereas a negative angular velocity represents rearward rotation of the upper body. The change in angle of the torso (θ) is found by integration of Equation 6:

$$\theta = \int \omega \, dt. \qquad (7)$$

Other parameters can be calculated from these responses. The term ridedown implies that the occupant benefits from restraining forces that act during front-end crush of the vehicle. This relates to an increased stopping distance for the occupant in the inertial reference frame and is associated with lower deceleration.

For the purposes of quantifying ridedown in barrier and sled crash tests, it is defined as starting when the occupant experiences restraining forces sufficient to reduce chest velocity in the interior more than 10% of the maximum velocity change of the vehicle. Ridedown continues until the end of vehicle deceleration. This is the point in which vehicle deceleration passes through zero, which separates front-end crush from rebound.

The duration of vehicle deceleration is defined as the time of positive deceleration of the passenger compartment in a barrier crash. Rebound is experienced by the vehicle and occupant. For the purpose of quantifying rebound for the occupant, it is measured as a ratio of the maximum rearward velocity normalized by the maximum forward velocity of the occupant in the interior:

$$RB_c = \frac{\max(-v_c)}{\max(+v_c)},$$
$$RB_p = \frac{\max(-v_p)}{\max(+v_p)}. \qquad (8)$$

Certain other parameters are reported from barrier crash or sled tests because they relate to occupant protection. For example, a relative kinetic energy is calculated using the maximum occupant velocity normalized by a velocity of 5 m/s. This kinetic energy factor (E) is:

$$E_c = \frac{\max(+v_c)^2}{25},$$
$$E_p = \frac{\max(+v_p)^2}{25}. \qquad (9)$$

The results of the Restraint Quotient computer analysis are presented in a one-page format with five separate plots as shown in Figure 2. This is an example of restraint performance in a 30 mph barrier crash with a lap-shoulder belted Hybrid III dummy in the passenger seating position. The responses include from top to bottom: 1) occupant compartment deceleration ($\ddot{x}_v$) and chest and pelvic acceleration ($\ddot{x}_c$ and $\ddot{x}_p$), 2) vehicle velocity change ($\dot{x}_v$) and occupant velocities in the vehicle (v_c and v_p), 3) displacement of the occupant in the vehicle (d_c and d_p), 4) the Restraint Quotient of the chest and pelvis (RQ_c and RQ_p) and 5) the torso angle change (θ).

Currently, the recline angle of automotive seatbacks is 26°, but the dummy doesn't typically have that great of a recline angle in the seat. Dummies in this study were usually seated with an initial reclining angle of 15°, so a positive torso angle change of 15° brings the torso into an upright (vertical) posture. Angle changes greater than 15° represent a forward rotation of the upper body beyond vertical and more belt loading into the shoulder region.

Validation Tests: The validity of the assumptions in the computer analysis was checked by analysis of a series of sled tests that had been conducted for another study (13). The experiments involved a lap-shoulder belted 5th percentile Hybrid III female dummy on a conventional bucket seat. Using a standard lap-shoulder belt system, two test conditions were simulated: one involved positioning the lap-belt low on the pelvis to ensure proper belt restraint and another pre-positioning the belt on the abdomen resulting in submarining. Each test condition was repeated twice to compare the calculations of occupant dynamics and kinematics against data obtained from high-speed movies of the sled test.

Barrier Crash Tests: The Restraint Quotient computer analysis was further evaluated on a series of 30 mph barrier crash tests conducted at the General Motors Proving Grounds during routine product development of several model passenger cars. The Hybrid III 50th percentile male dummy was used in the driver and right-passenger seating position. The following test conditions were evaluated:

- Unbelted occupants
- Manual lap-shoulder belted occupants
- Automatic lap-shoulder belted occupants
- Manual lap-shoulder belted driver with supplemental airbag

These tests routinely involve a full complement of dummy instrumentation so that the chest and pelvic accelerations were available for processing through the Restraint Quotient program. There was a similar mix of vehicle types in each restraint category so the responses are comparable between restraint type.

Sled Tests: Hyge sled tests were conducted in the Biomedical Science Department using an automotive body buck with standard interior including bucket seat, steering system and instrument panel components. The Hybrid III 50th percentile male dummy was used in various restraint configurations at increasing exposure severities. The combination of test conditions and velocities evaluated in this series of experiments included:

Restraint System

- Unbelted Driver
- Lap-Shoulder Belted Driver
- Lap-Shoulder Belted Driver with Supplemental Airbag
- Unbelted Driver with Supplemental Airbag
- Lap-Shoulder Belted Passenger

Test Velocity

- 14 mph (6.3 m/s)
- 20 mph (8.9 m/s)
- 33 mph (14.7 m/s)
- 38 mph (17.3 m/s)

The dummy was instrumented with sufficient channels to process the data through the Restraint Quotient computer program. In addition, the dummy was instrumented with advanced injury assessment transducers to compare other biomechanical measures of safety belt and restraint component performance. High-speed photographs of the experiments were also taken.

Statistical Analysis: The response data with different types of occupant restraints were compared using standard statistical tests in SAS. The analysis included a t-test to determine the significance of response differences between restraint types.

RESULTS

Validation of Restraint Quotient Calculations with Film Responses: Table 1 provides response data from two tests with lap-belt restraint and submarining using the Hybrid III 5th-percentile female dummy. Three out of four comparisons of chest and pelvic displacement by the Restraint Quotient program and direct film analysis were within 2%, at 100 ms. This degree of accuracy is not expected because of variability in response measurements, high-speed photography and film analysis. Accuracy within 10% would be acceptable for this type of analysis. The chest displacement with lap-belt submarining was 18% lower by film analysis than by calculation. The principal direction of rotation of the upper torso was accurately indicated by the RQ analysis and within 12% for the cases of lap-belt loading. For lap-belt submarining, the rearward angle calculated by RQ was significantly smaller than identified by film analysis.

Figure 2 shows examples of the one-page output from the Restraint Quotient program for lap-belt restraint and submarining. With lap-belt restraint, the torso angle increases as chest displacement is greater than pelvic dis-

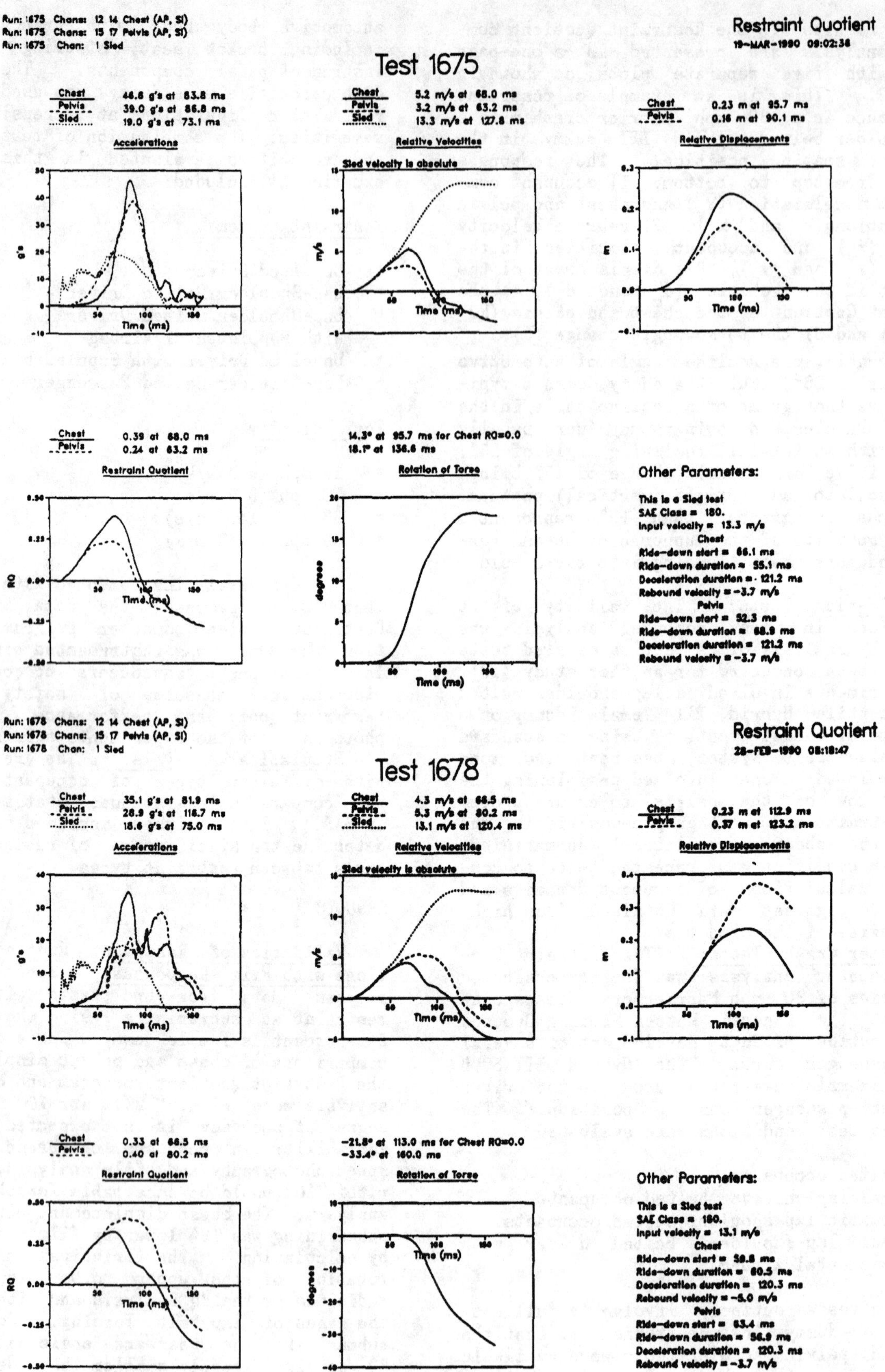

Figure 2: Examples of calculations in the Restraint Quotient program for a Hyge sled test with a 5th percentile female Hybrid III dummy experiencing lap-belt restraint (#1675) or lap-belt submarining (#1678).

placement. In contrast, lap-belt submarining involves a rearward rotation of the upper torso as the pelvis leads the chest in forward displacement. For the case of lap-belt restraint, there is tighter coupling and greater restraint of the pelvis than chest as indicated by a larger RQ_c than RQ_p. In contrast, lap-belt submarining involves a lower chest RQ_c than pelvis RQ_p.

Figure 3 compares the calculated and film analyzed chest and pelvic displacement and torso angle for lap-belt restraint and submarining. The data show acceptable similarity in the predicted and observed responses except for the lag in computed torso angle from the film response with lap-belt submarining.

Figure 4 shows the superimposed plot of Restraint Quotient with the corresponding shoulder or lap-belt load. The responses describe some of the dynamics of safety belt restraint and the Restraint Quotient. For example, with lap-belt restraint, RQ_c increases in magnitude until slightly after the sharp rise in shoulder belt load. This shows that shoulder-belt loading produces upper torso restraint and limits velocity build-up of the chest with respect to the interior.

The peak in RQ_c occurs between the onset and peak in shoulder belt loading and returns to zero slightly after the peak shoulder belt load. The effect of the shoulder belt on chest kinematics are illustrated by a reduction in RQ_c (indicating greater restraint) which coincides with an increasing shoulder belt load. This indicates that the peak shoulder belt load occurs at about the same time as the maximum forward excursion (and velocity) of the chest. Similar dynamics can be observed in the lap-belt submarining responses.

In summary, there is similarity between the computed and film analyzed displacements, in spite of known difficulties in using acceleration to determine velocity and displacement. The greatest differences were observed with the displacement response of the chest and torso angle for lap-belt submarining. For lap-belt restraint, there was excellent agreement between all calculated and film analyzed responses.

<u>Comparing Lap-Belt Restraint and Submarining Responses:</u> Lap-belt submarining reduced head HIC, chest compression, and chest and pelvic acceleration; but, there was an increase in peak head acceleration. This implies that lap-belt submarining may improve FMVSS 208 performance in some situations.

RQ_c was higher than the pelvic response with lap-belt restraint (Table 1: 0.38 v 0.24 or 58% higher). Lap-belt submarining involves a much higher RQ_p (62% increase) indicating a lower resistance to pelvic displacement. Lap-belt submarining also involved lower lap-belt loads (20% and 38% reduction respectively) without significantly affecting the peak shoulder belt tension.

The upper torso angle is positive with lap-belt restraint indicating forward rotation of the upper body. This resulted in a 15°-17° forward rotation which slightly exceeds an upright angle and involves belt loading on the shoulder and upper chest structure of the dummy at the peak in chest RQ_c. In contrast, lap-belt submarining involves a rearward rotation of the upper torso by 15°-20°. This increases the rearward torso angle to greater than -30° with respect to vertical.

Pelvic displacement is significantly larger with lap-belt submarining than restraint. It involves an increase in forward displacement of 14 cm (92% increase) because of belt compression of the abdomen. Chest displacement is slightly reduced as rearward torso rotation counteracts the forward displacement due to shoulder belt restraint. As indicated previously, the FMVSS 208 criteria of head HIC and chest acceleration are reduced by lap-belt submarining. There is also lower chest compression and pelvic acceleration. Lap-belt restraint results in higher chest velocities in the crash (5.1 m/s v 4.5 m/s, or 13% increase) than with lap-belt submarining. This results in slightly lower values of the RQ_c and higher levels of RQ_p with submarining.

With lap-belt submarining the film value of torso rotation is larger than computed with the RQ program. One reason is that the pelvis has a larger forward displacement with lap-belt submarining, whereas the chest is caused to rotate rearward and downward as the upper body pivots about the pelvis. Another assumption in the RQ program is that chest and pelvic accelerometers are attached by a rigid link. Since the RQ calculation assumes a horizontal displacement of the chest and pelvis, and there is a downward motion of the chest, a smaller negative angle change will be calculated by the RQ program due to the assumption of horizontal displacement of both components of displacement.

The peak resultant head acceleration increased by 22% with lap-belt submarining. This increase occurs in contrast to significant reductions in other head and chest responses. Figure 5 compares the resultant, longitudinal (AP) direction, and vertical (SI) direction accelerations for lap-belt restraint and submarining. The higher accelerations with lap-belt submarining are due to a larger longitudinal (AP) component of head acceleration near the end of shoulder belt loading. This is as the head is being arrested from its sweeping motion during shoulder belt restraint and as the head is facing downward. This can be seen by the large AP component of acceleration after the peak in vertical response which is typical of the phase of maximum torso restraint where the head is horizontal, experiencing maximum deceleration through the restraint system, and undergoing a large SI component of acceleration.

<u>Comparing Unbelted Versus Lap-Shoulder Belted Responses in Barrier Crashes:</u> Table 2 provides data from 30 mph barrier crash tests using Hybrid III dummies in the driver and

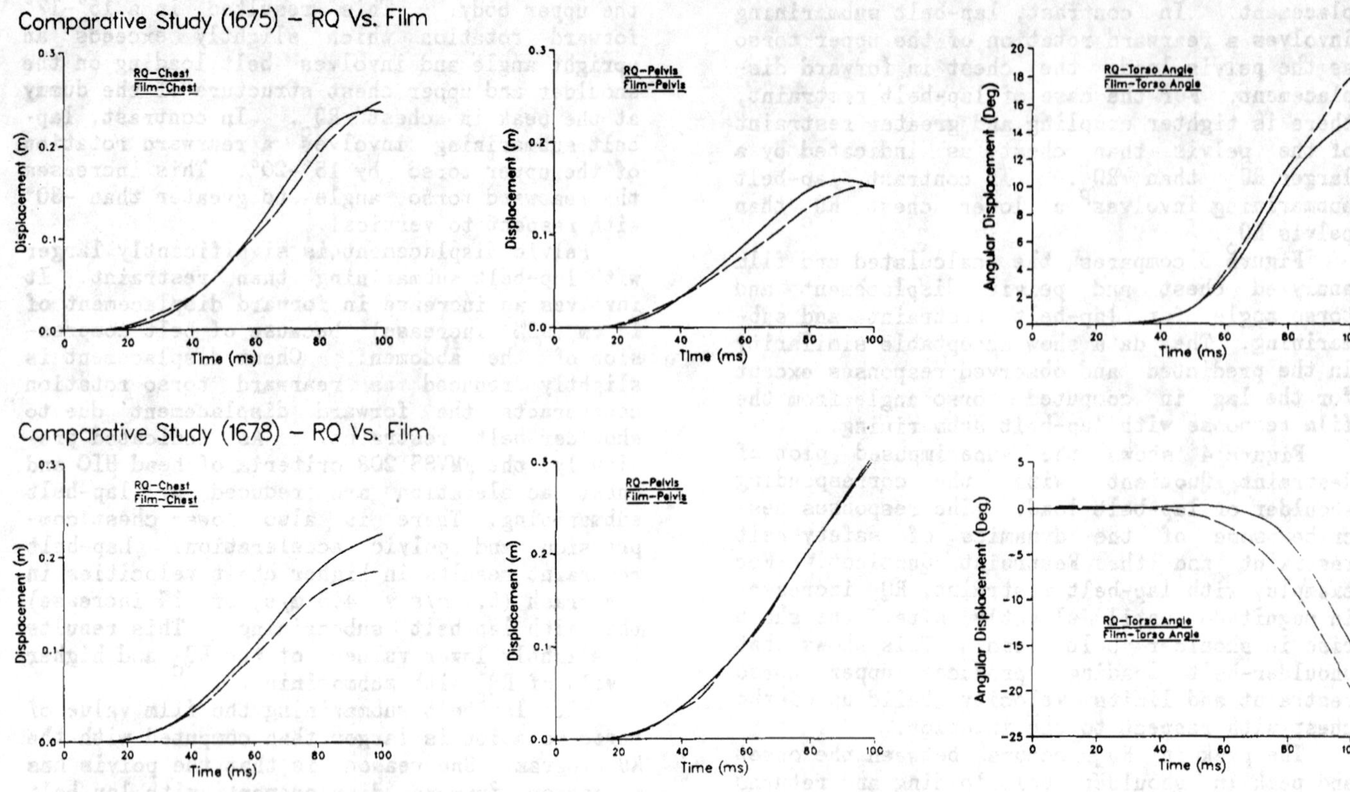

Figure 3: Comparison of calculated and film analyzed displacements of the chest and pelvis and torso angle change for tests #1675 and #1678.

Run No 1675 HEAD RESL. ACC. HEAD LONG ACCELERATION HEAD VERT ACCELERATION

Run No 1678 HEAD RESL. ACC. HEAD LONG ACCELERATION HEAD VERT ACCELERATION

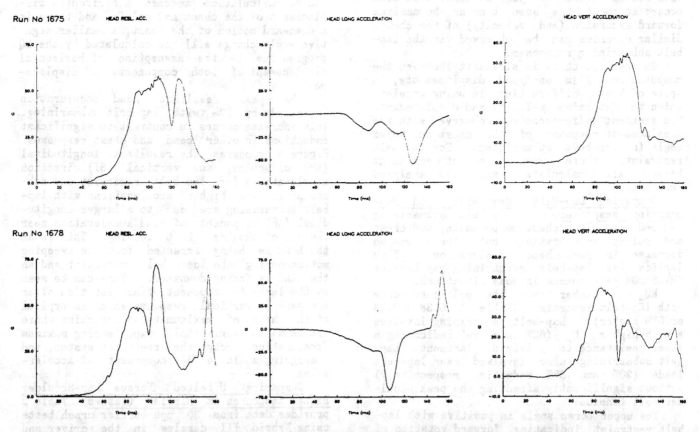

Figure 5: Resultant and components of head acceleration for tests #1675 and #1678.

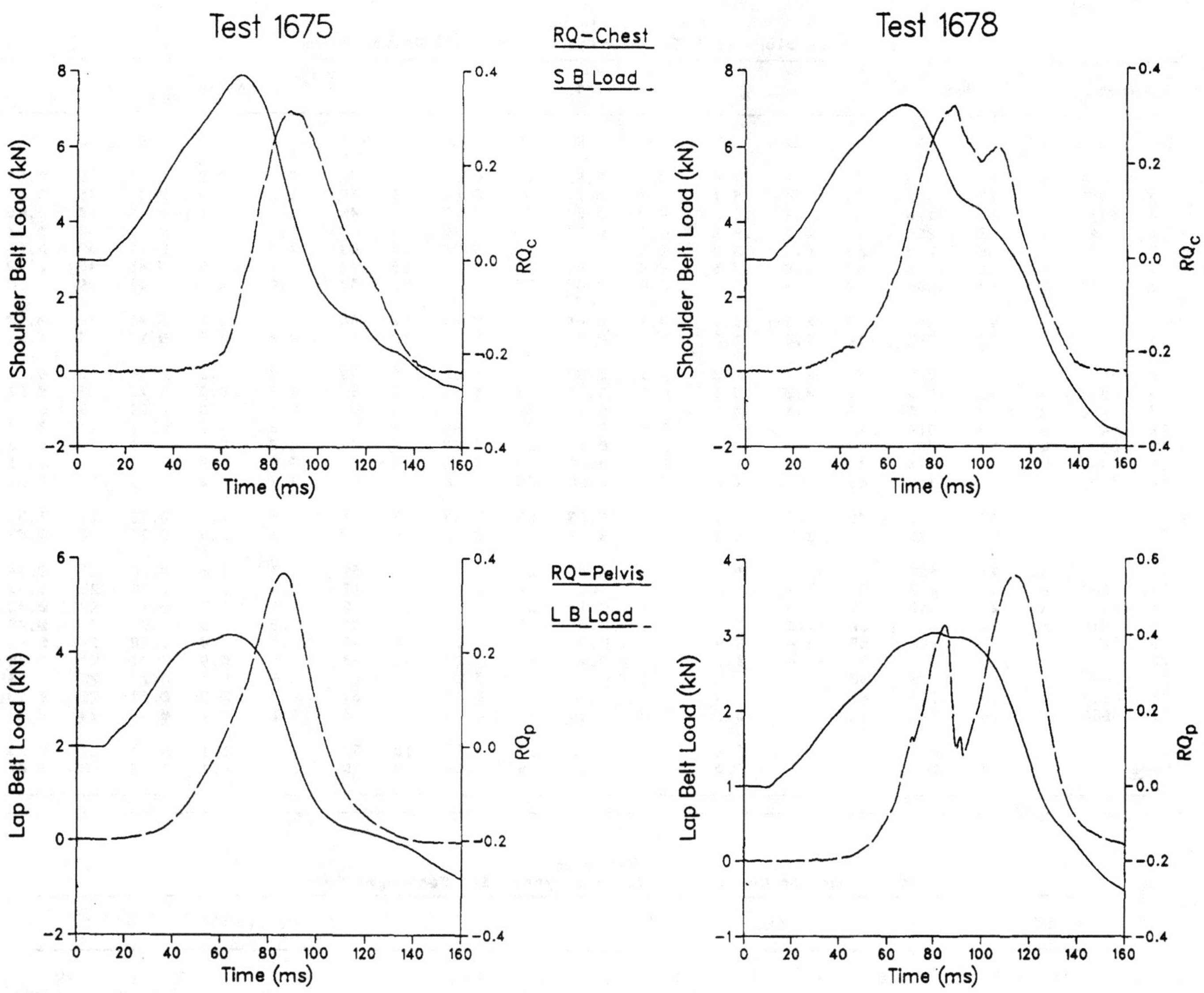

Figure 4: Shoulder and lap-belt loads with corresponding Restraint Quotient response of the chest and pelvis, respectively, for test #1675 and #1678.

Table 1

Hyge Sled Test Results Using a Hybrid III 5th-Percentile Female Dummy with Frangible Abdomen Experiencing Lap-Belt Restraint or Submarining in a Lap-Shoulder Belt System

Test Number	Head HIC	Head Acc (g)	Chest Compression (mm)	Acceleration (g) Chest	Acceleration (g) Pelvis	Displacement (cm) Chest RQ	Displacement (cm) Chest Film	Displacement (cm) Pelvis RQ	Displacement (cm) Pelvis Film	Restraint Quotient Chest	Restraint Quotient Pelvis	Torso Angle (°) RQ	Torso Angle (°) Film	Belt Loads (kN) Shoulder	Belt Loads (kN) Inboard Lap	Belt Loads (kN) Outboard Lap
1675	863	58.1	18.9	46.8	39.0	23.0	22.5	16.0	15.4	0.39	0.24	14.6	17.8	6.90	8.52	5.64
1676	889	61.0	23.8	48.4	39.2	22.0	21.8	15.0	15.0	0.37	0.24	15.6	16.0	7.20	8.88	5.74
Avg.	876	59.6	21.4	47.6	39.1	22.5	22.2	15.5	15.2	0.38	0.24	15.1	16.9	7.05	8.70	5.69
SD	18	2.0	3.5	1.1	0.1	0.7	0.5	0.7	0.3	0.01	0	0.7	1.3	.21	.25	.07
1678	706	72.0	17.4	35.1	28.9	23.0	18.5	30.0	30.0	0.33	0.40	-15.0	-19.6	7.00	6.84	3.76
1680	771	73.0	20.4	38.5	28.7	23.0	19.0	27.5	28.5	0.35	0.38	-8.2	-16.2	7.43	7.00	3.32
Avg.	739	72.5	18.8	36.8	28.6	23.0	18.8	28.8	29.3	0.34	0.39	-11.6	-17.9	7.22	6.92	3.54
SD	46	0.7	2.0	2.4	0.5	0.	.4	1.8	1.0	.01	0.01	(4.8)	(2.4)	0.30	.11	.31
% Diff.	-15.6%	+21.6%	-12.1%	-22.7%	-26.9%	+2.2%	-15.3%	+85.8%	+92.8%	-10.5%	+62.5%	-	-	+2.4%	-20.4%	-37.8%

Displacement and torso angle values compared at 100 ms.

Table 2a
30 mph Barrier Crash Tests Using a Hybrid III Driver Dummy

Test			Head		Chest								Pelvis					
OBS	RUN	RES	G	HIC	G	C	V	E	R	D	RQ	TH	G	V	E	R	D	RQ
1	6386	U	64	610	76	-	7.6	2.3	0.91	29	0.51	-8	73	8.0	2.6	0.24	35	0.54
2	6387	U	67	490	88	-	9.1	3.3	0.74	60	0.83	25	55	9.1	3.3	0.31	46	0.61
3	9411	U	89	1110	108	-	10.3	4.2	0.64	48	0.77	21	46	8.7	3.0	0.46	36	0.65
4	6723	U	82	820	79	4.3	10.5	4.4	0.37	50	0.70	26	46	7.1	2.0	0.49	34	0.47
5	6736	U	96	910	107	5.3	10.7	4.6	0.34	53	0.69	25	53	7.7	2.4	0.35	38	0.50
6	9755	U	125	1870	98	7.8	10.5	4.4	0.65	53	0.70	15	43	9.0	3.2	0.41	45	0.60
7	10023	U	93	1050	109	-	11.8	5.6	0.82	56	0.79	27	68	9.0	3.2	0.51	39	0.60
Mean			88	980	95	5.8	10.1	4.1	0.64	50	0.71	19	55	8.4	2.8	0.40	39	0.57
SD			20	451	14	1.8	1.3	1.0	0.22	10	0.10	12	12	0.8	0.5	0.10	5	0.07
8	6242	M	113	780	40	3.7	5.3	1.1	0.83	24	0.35	12	59	4.6	0.9	0.61	17	0.30
9	6244	M	92	630	37	4.0	4.6	0.9	1.13	21	0.31	10	50	4.0	0.6	0.77	15	0.27
10	6656	M	90	670	46	6.1	5.9	1.4	0.91	30	0.39	3	50	6.4	1.6	0.68	28	0.42
11	6409	M	92	810	40	4.3	5.1	1.0	1.17	26	0.34	-4	51	6.2	1.5	0.71	28	0.42
12	6694	M	54	630	47	-	5.3	1.1	0.51	34	0.33	-9	74	6.9	1.9	1.3	40	0.43
13	6695	M	70	670	42	-	5.3	1.1	0.69	32	0.33	15	55	5.0	1.0	0.34	24	0.31
14	6696	M	78	710	42	-	4.6	0.9	0.98	30	0.28	13	43	4.7	0.9	0.53	23	0.29
Mean			84	700	42	4.5	5.2	1.1	0.89	28	0.33	6	54	5.4	1.2	0.71	25	0.35
SD			19	71	4	1.1	0.5	0.2	0.24	5	0.03	9	10	1.1	0.5	0.30	8	0.07
15	6639	A	71	620	34	-	5.4	1.2	0.65	26	0.35	14	50	5.1	1.0	0.35	20	0.34
16	6640	A	62	470	37	-	5.3	1.1	0.9	27	0.36	8	47	5.5	1.2	0.71	23	0.37
17	6693	A	48	410	52	-	7.2	2.1	0.37	31	0.47	19	72	6.1	1.5	0.47	20	0.40
18	6816	A	72	770	50	4.0	7.6	2.3	0.68	35	0.51	12	61	6.9	1.9	0.87	29	0.46
19	6472	A	85	880	47	-	5.6	1.3	0.69	34	0.35	12	48	5.0	1.0	0.28	27	0.31
20	6470	A	70	690	35	-	6.2	1.6	0.47	41	0.39	33	46	4.8	0.9	0.41	24	0.30
21	6536	A	68	580	40	-	4.5	0.9	0.97	29	0.30	16	43	3.8	0.6	0.74	20	0.25
22	6544	A	69	730	45	-	4.9	1.0	0.94	32	0.31	16	51	4.7	0.9	0.42	23	0.29
23	6555	A	75	800	44	-	5.1	1.0	0.84	31	0.32	12	45	4.1	0.7	0.54	24	0.27
Mean			69	661	43	-	5.8	1.4	0.72	32	0.37	16	51	5.1	1.1	0.53	23	0.33
SD			10	155	7	-	1.0	0.5	0.21	5	0.07	9	9	1.0	0.4	0.20	3	0.07

Table 2b
30 mph Barrier Crash Tests Using a Hybrid III Passenger Dummy

	Head		Chest								Pelvis					
OBS	G	HIC	G	C	V	E	R	D	RQ	TH	G	V	E	R	D	RQ
1	65	440	80	1.4	12.2	6.0	0.57	63	0.82	32	34	7.6	2.3	0.53	44	0.51
2	94	1280	97	3.6	12.9	6.7	0.46	67	0.86	25	32	9.2	3.4	0.36	52	0.61
3	93	1050	104		10.6	4.5	0.81	58	0.83	22	36	8.8	3.1	0.38	45	0.69
4	83	660	79	1.5	11.8	5.6	0.57	62	0.78	20	29	7.9	2.5	0.16	51	0.52
5	76	720	100	2.3	12.0	5.8	0.55	62	0.78	21	42	8.3	2.8	0.24	50	0.54
6	96	920	59	2.9	11.4	5.2	0.57	62	0.82	20	32	8.4	2.8	0.41	50	0.61
7	83	610	87		12.2	6.0	0.71	67	0.89	38	79	9.0	3.2	0.54	44	0.65
Mean	84	811	87	2.3	11.9	5.7	0.61	63	0.83	25	41	8.5	2.9	0.37	48	0.59
SD	11	288	16	0.9	0.7	0.7	0.12	3	0.04	7	17	0.6	0.4	0.14	4	0.07
8	66	970	40	3.8	5.5	1.2	0.96	24	0.36	8	49	5.1	1.0	0.70	20	0.33
9	60	850	40	3.9	5.1	1.0	1.21	23	0.35	12	60	4.9	1.0	1.06	17	0.33
10	52	760	42	5.1	6.0	1.4	1.11	31	0.39	-2	47	6.9	1.9	0.27	33	0.45
11	51	490	37	4.4	4.7	0.9	1.27	25	0.32	-5	49	6.0	1.4	0.76	28	0.40
12	39	470	31		5.2	1.1	0.50	34	0.32	6	40	5.4	1.2	0.31	30	0.34
13	47	600	33		4.8	0.9	0.83	31	0.3	-1	40	5.7	1.3	0.29	32	0.35
14	45	560	35		4.7	0.9	1.0	32	0.29	1	43	6.1	1.5	0.49	31	0.37
Mean	51	671	37	4.3	5.1	1.0	0.98	29	0.33	3	47	5.7	1.3	0.55	27	0.37
SD	9	191	4	0.6	0.5	0.2	0.26	4	0.04	6	7	0.7	0.3	0.30	7	0.04
15	51	520	33		5.3	1.1	0.66	27	0.35	15	56	4.7	0.9	0.59	19	0.31
16	42	450	32		5.6	1.3	0.69	31	0.38	22	59	5.2	1.1	0.92	21	0.35
17	61	520	44		6.3	1.6	0.66	27	0.41	-3	54	7.4	2.2	0.28	29	0.48
18	62	830	44	3.1	5.7	1.3	0.53	25	0.38	1	63	6.2	1.5	0.37	24	0.41
19	53	660	36		5.0	1.0	1.06	31	0.31	2	42	5.1	1.0	0.41	30	0.32
20	50	600	32		5.4	1.2	0.68	36	0.33	3	36	6.2	1.6	0.21	34	0.38
21	43	490	31		4.7	0.9	0.70	31	0.31	14	43	4.1	0.7	0.78	23	0.27
22	48	560	35		5.2	1.1	0.88	35	0.33	8	40	5.7	1.3	0.49	30	0.36
23	46	520	45		4.8	0.9	2.25	23	0.3	1	45	4.9	1.0	0.28	27	0.32
Mean	51	572	38	-	5.3	1.1	0.09	30	0.34	7	49	5.5	1.1	0.48	26	0.36
SD	7	114	6	-	0.5	0.2	0.53	4	0.04	8	10	1.0	0.3	0.24	5	0.06

passenger seating positions. The test data were processed using the Restraint Quotient program and the results are summarized for three conditions for the two seating positions: 1) unbelted (U), 2) manual lap-shoulder belt restrained (M), and 3) automatic (passive) lap-shoulder belt restrained (A). While the Restraint Quotient program was developed to evaluate belt and airbag systems, the analysis can be applied to unbelted occupants for comparative purposes.

The head HIC and chest acceleration were significantly reduced by safety belt use in the driver and passenger seating positions. There was a 31% reduction in HIC (980 v 678, $p < .002$) in the driver position and a 24% reduction (811 v 616, $p < .05$) for the passenger. There was a larger reduction in chest acceleration averaging 56% for the driver (95 v 42, $p < .001$) and 57% for the passenger (86 v 37, $p < .001$).

The use of safety belts significantly reduced chest velocity with respect to the interior from an average of 10.1 m/s for the unbelted driver to 5.5 m/s for the belted driver (a 45.5% reduction, $p < .001$). There was a 11.9 m/s velocity for the unbelted passenger, which was reduced to 5.3 m/s with belt use (a 55.5% reduction, $p < .0001$). These changes resulted in a statistically significant reduction in the Restraint Quotient with safety belt use. It reduced RQ_c from 0.71 for the unbelted driver to 0.36 (a 49.3% reduction, $p < .001$) and from 0.83 for the passenger to 0.34 with belt use (a 59.0% reduction, $p < .0001$).

Because of greater pelvic than chest restraint for the unbelted dummies in both seating positions by early knee contact (RQ_p = .56 v .71 for the driver and RQ_p = .59 v .83 for the passenger), the upper torso rotated forward in the crash more in the unbelted test condition than with belt use. There was no statistical difference in torso angle for the unbelted and belted driver (θ = 18.6° v 11°, N.S.). However, the change in torso angle was statistically different in the passenger seating position (θ = 25.2° v 5°, $p < .0001$).

Manual vs Automatic Lap-Shoulder Belt Responses: Although there were slight differences in average responses between the manual and automatic lap-shoulder belt tests, there was no statistical significance in the differences in 31 out of 32 comparisons. In the one case where there was a statistical difference, the driver's torso angle was 5.7° with manual safety belts and 15.6° with the automatic belts ($p < .05$). A similar effect was not seen in torso angle in the passenger seating position. The data indicate comparable performance between the two restraint systems.

All Belted Driver vs Belted Passenger Responses: In general, the biomechanical and kinematic responses were similar in the two seating positions with safety belt use. There were three response comparisons out of 32 that were statistically different. The peak head acceleration was 76 g for the driver and 51 g for the passenger ($p < .0001$), the peak chest acceleration was 42 g for the driver and 37 g for the passenger ($p < .0005$) and the torso angle was greater for the driver than passenger (11° v 5°, $p < .05$).

Figure 6a-d shows examples of the Restraint Quotient analysis for unbelted and belted occupants in 30 mph barrier crash tests. In Figure 6a, chest velocity increased to 9.7 m/s resulting in an RQ_c = 0.61. Because of greater pelvic restraint by knee contact (RQ_p = 0.37), the peak velocity of the pelvis was 40% lower than the chest. This resulted in a 22° forward rotation of the torso. In contrast, Figure 6b shows a manual lap-shoulder belted driver. The chest velocity increased to 4.6 m/s and resulted in an RQ_c = 0.28 as well as a significantly lower forward displacement of the chest than occurred with the unbelted occupant (30 cm v 64 cm, $p < 0.0001$).

Figure 6c shows barrier test results for an automatic lap-shoulder belted driver. The chest developed slightly higher velocity than with manual belts and an RQ_c = 0.32 with a similar forward displacement and torso rotation. Figure 6d shows a lap-shoulder belted driver with supplemental air bag. The results of the Restraint Quotient analysis yielded an RQ_c = 0.28, forward displacement of 27 cm, and a lower rebound velocity than observed in other barrier tests analyzed.

Hyge Sled Tests: Table 3 provides Hyge sled test data for various exposure severities and restraint conditions. This matrix of tests provides a rich set of biomechanical responses to evaluate and compare trends in the 17 parameters determined for each experiment. Figures 7a-d shows examples of the Restraint Quotient analysis for a selection of test conditions showing changes in responses with crash severity. At the highest crash severity, there is an increase in head responses for the lap-shoulder belted occupant, but a dissimilar effect with the use of a supplemental airbag. In contrast, chest deflection and Viscous response remain low for belt restrained occupants irrespective of impact severity.

DISCUSSION

One of the key points addressed in this analysis is whether the motion sequence criteria espoused by Adomeit (5-7), velocity criteria presented here and other biomechanical measures (2,8) provide necessary and sufficient conditions to ensure a quality restraint system. It is not possible to answer the question fully in this paper but the criteria are probably not necessary to ensure the adequacy of an occupant safety system. Consider a radically different concept of occupant protection which focuses on, for example, rotation of the occupant's seat to place the individual in a spine horizontal position during a crash. This concept would necessarily violate many of the motion sequence criteria yet may be a viable -- although hypothetical -- system.

Based on the current concepts of occupant restraint, motion and velocity criteria seem to

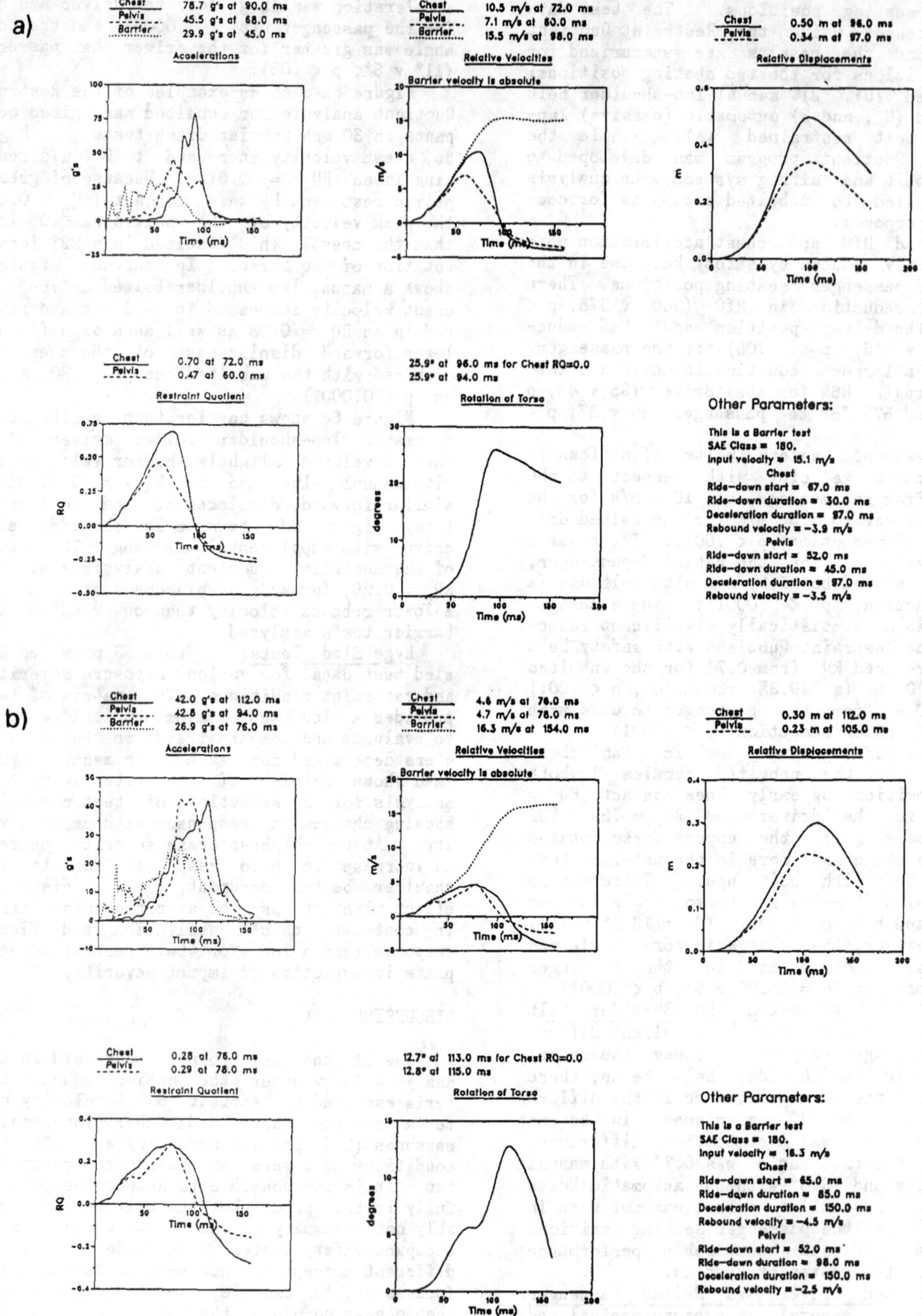

Figure 6: Examples of Restraint Quotient analysis of 30 mph barrier crash data for: a) an unbelted driver, b) a manual lap-shoulder belted driver, c) an automatic lap-shoulder belted driver, and d) a lap-shoulder belted driver with supplemental airbag.

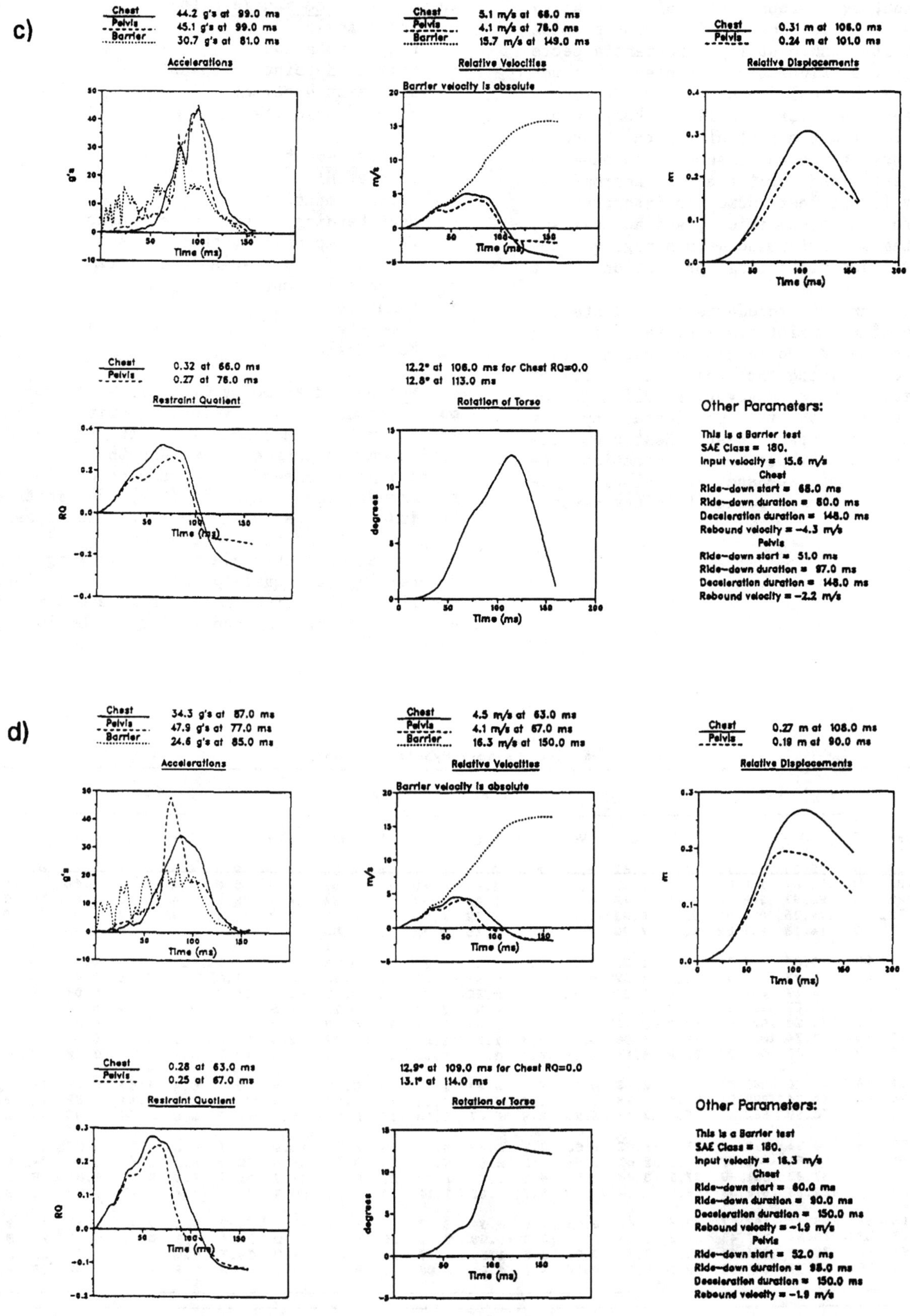
c)
Chest 44.2 g's at 99.0 ms
Pelvis 45.1 g's at 99.0 ms
Barrier 30.7 g's at 81.0 ms
Accelerations
g's
Time (ms)
Chest 5.1 m/s at 68.0 ms
Pelvis 4.1 m/s at 78.0 ms
Barrier 15.7 m/s at 149.0 ms
Relative Velocities
Barrier velocity is absolute
m/s
Time (ms)
Chest 0.31 m at 106.0 ms
Pelvis 0.24 m at 101.0 ms
Relative Displacements
m
Time (ms)
Chest 0.32 at 66.0 ms
Pelvis 0.27 at 76.0 ms
Restraint Quotient
RQ
Time (ms)
12.2° at 106.0 ms for Chest RQ=0.0
12.8° at 113.0 ms
Rotation of Torso
degrees
Time (ms)
Other Parameters:
This is a Barrier test
SAE Class = 180.
Input velocity = 15.6 m/s
Chest
Ride-down start = 68.0 ms
Ride-down duration = 80.0 ms
Deceleration duration = 148.0 ms
Rebound velocity = -4.3 m/s
Pelvis
Ride-down start = 51.0 ms
Ride-down duration = 97.0 ms
Deceleration duration = 148.0 ms
Rebound velocity = -2.2 m/s
d)
Chest 34.3 g's at 87.0 ms
Pelvis 47.9 g's at 77.0 ms
Barrier 24.6 g's at 85.0 ms
Accelerations
g's
Time (ms)
Chest 4.5 m/s at 63.0 ms
Pelvis 4.1 m/s at 67.0 ms
Barrier 16.3 m/s at 150.0 ms
Relative Velocities
Barrier velocity is absolute
m/s
Time (ms)
Chest 0.27 m at 108.0 ms
Pelvis 0.19 m at 90.0 ms
Relative Displacements
m
Time (ms)
Chest 0.28 at 63.0 ms
Pelvis 0.25 at 67.0 ms
Restraint Quotient
RQ
Time (ms)
12.9° at 109.0 ms for Chest RQ=0.0
13.1° at 114.0 ms
Rotation of Torso
degrees
Time (ms)
Other Parameters:
This is a Barrier test
SAE Class = 180.
Input velocity = 16.3 m/s
Chest
Ride-down start = 60.0 ms
Ride-down duration = 90.0 ms
Deceleration duration = 150.0 ms
Rebound velocity = -1.9 m/s
Pelvis
Ride-down start = 52.0 ms
Ride-down duration = 98.0 ms
Deceleration duration = 150.0 ms
Rebound velocity = -1.9 m/s

be complementary properties of a quality restraint system in comparative tests because they represent fundamentally important aspects of an occupant's kinematic response. The use of other biomechanical responses of the occupant to assess the risk of critical body injury is another important aspect of restraint evaluation. Injury assessment should be based on the best supported criteria and appropriate biofidelity in the test dummy to interpret the potential for occupant injury, and an analysis of the system's performance in managing occupant energy and enhancing protection in a crash.

Based on current procedures to evaluate the performance of restraint systems, the following criteria are considered a comprehensive set of properties determining the relative quality of restraint function. However, specific guidelines may depend on the particular type of safety system and crash test. Restraint criteria involve two types of information, one dealing with motion sequence and velocity criteria, and the other, occupant injury assessment:

<u>Motion Sequence and Velocity Criteria</u>

- Torso angle
- Pelvic rotation
- Pelvic (H-point) displacement
- Restraint Quotient
- Rebound velocity

<u>Injury Assessment</u>

- Contact HIC
- Facial impact force
- Neck bending, shear and extension
- Chest compression, shear, Viscous response, and spinal acceleration
- Abdominal compression
- Femur load
- Knee shear
- Foot-ankle load

<u>Motion and Velocity Criteria</u>: In terms of motion sequence and velocity criteria, there isn't much biomechanical data to validate limits on the parameters. For the purposes of comparing responses to a reference value, the recommendations of Adomeit (5-7) and limits set by judgment are used to normalize observed responses from individual tests. If each motion sequence criteria is equally weighted in determining the quality of belt restraint, a sum of the six criteria can be made determining a single performance index for the Motion Criteria (MC).

Table 3
Hyge Sled Tests Using a Hybrid III Dummy

Test				Head		Chest									Pelvis					
OBS	RUN	Speed	RES	G	HIC	G	C	VC	V	E	R	D	RQ	TH	G	V	E	R	D	RQ
1	1511	6.26	U	20.98	22.15	13.7	1.62	0.05	3.5	0.49	0.34	26	0.58	-12.9	9.3	4.3	0.74	0.00	41	0.71
2	1518	8.94	U	102.07	195.69	27.9	3.37	0.15	5.2	1.08	0.48	36	0.57	17.0	16.0	6.1	1.49	0.00	67	0.67
3	1345	14.22	U	96.97	348.04	41.0	5.70	0.78	7.2	2.07	0.64	47	0.54	44.5	58.2	5.4	1.16	1.04	26	0.41
4	1320	14.22	U	214.35	889.88	83.0	5.93	1.43	6.7	1.79	0.73	42	0.47	41.3	71.4	5.3	1.12	0.83	21	0.37
5	1336	17.17	U	114.13	897.52	96.7	7.20	3.48	9.2	3.38	0.72	46	0.56	39.5	106.4	6.9	1.90	0.96	27	0.42
6	1510	6.26	BD	16.30	37.62	14.9	1.05	0.02	2.2	0.19	1.09	16	0.36	-18.4	5.9	4.3	0.74	0.00	49	0.71
7	1512	8.94	BD	28.33	137.95	27.4	1.52	0.04	4.5	0.81	0.35	29	0.43	-33.1	11.8	7.1	2.01	0.00	90	0.67
8	1310	14.75	BD	38.17	354.59	32.0	2.24	0.05	3.6	0.52	1.03	14	0.25	-1.5	36.2	3.3	0.43	0.64	15	0.24
9	1318	14.75	BD	66.57	398.81	34.7	2.52	0.08	3.8	0.57	0.84	18	0.27	0.6	39.0	3.4	0.46	0.68	17	0.24
10	1319	17.32	BD	148.26	1028.95	41.6	2.76	0.13	4.6	0.84	1.24	17	0.28	-0.3	40.4	4.1	0.67	0.88	17	0.25
11	1340	17.32	BD	131.74	1021.19	49.0	4.06	0.20	5.4	1.16	1.04	23	0.32	2.9	65.1	5.1	1.04	0.69	22	0.31
12	1311	17.32	BD	132.03	646.07	38.0	3.17	0.28	4.2	0.70	0.95	16	0.25	-1.0	42.7	4.0	0.64	0.62	17	0.24
13	1516	8.94	BA	27.25	101.03	25.9	2.43	0.08	2.9	0.34	1.24	16	0.32	-21.8	9.7	6.4	1.64	0.00	75	0.71
14	1316	14.94	BA	44.64	329.96	32.6	3.24	0.15	3.2	0.41	0.93	12	0.22	0.4	33.9	3.2	0.41	0.62	12	0.22
15	1317	17.14	BA	56.89	552.89	40.8	3.16	0.23	3.8	0.57	1.13	15	0.23	2.8	40.9	3.8	0.58	0.89	13	0.23
16	1511	6.26	A	20.98	22.15	13.7	1.62	0.05	3.5	0.49	0.34	26	0.58	-12.9	9.3	4.3	0.74	0.00	41	0.71
17	1517	,8.94	A	21.51	62.89	25.1	2.61	0.03	3.9	0.61	0.69	23	0.43	-24.6	10.2	6.0	1.44	0.00	74	0.66
18	1314	14.81	A	47.06	351.99	37.5	3.40	0.55	4.1	0.67	1.02	21	0.29	11.7	32.3	3.9	0.61	0.64	15	0.27
19	1315	17.03	A	132.36	578.99	58.5	-	-	5.7	1.29	1.12	30	0.35	20.1	47.3	4.4	0.77	0.89	19	0.27
20	1530	8.94	BP	26.74	163.67	32.5	2.40	0.08	2.9	0.34	2.17	15	0.31	-2.6	23.6	3.2	0.41	1.47	19	0.34
21	1309	14.97	BP	46.04	347.43	31.2	2.54	0.10	3.8	0.58	0.79	19	0.26	0	32.4	3.5	0.49	0.57	19	0.24
22	1342	17.51	BP	73.36	1254.64	59.7	3.71	0.25	6.5	1.69	0.87	29	0.39	2.9	59.7	6.4	1.64	0.78	28	0.39
23	1308	17.51	BP	55.77	651.93	39.7	3.12	0.10	4.5	0.81	0.60	21	0.26	2.1	47.0	4.3	0.74	0.37	20	0.25

U Unbelted Driver
BD Lap-Shoulder Belted Driver
BA Lap-Shoulder Belted and Supplemental Driver Airbag
A Driver Airbag Only
BP Lap-Shoulder Belted Passenger

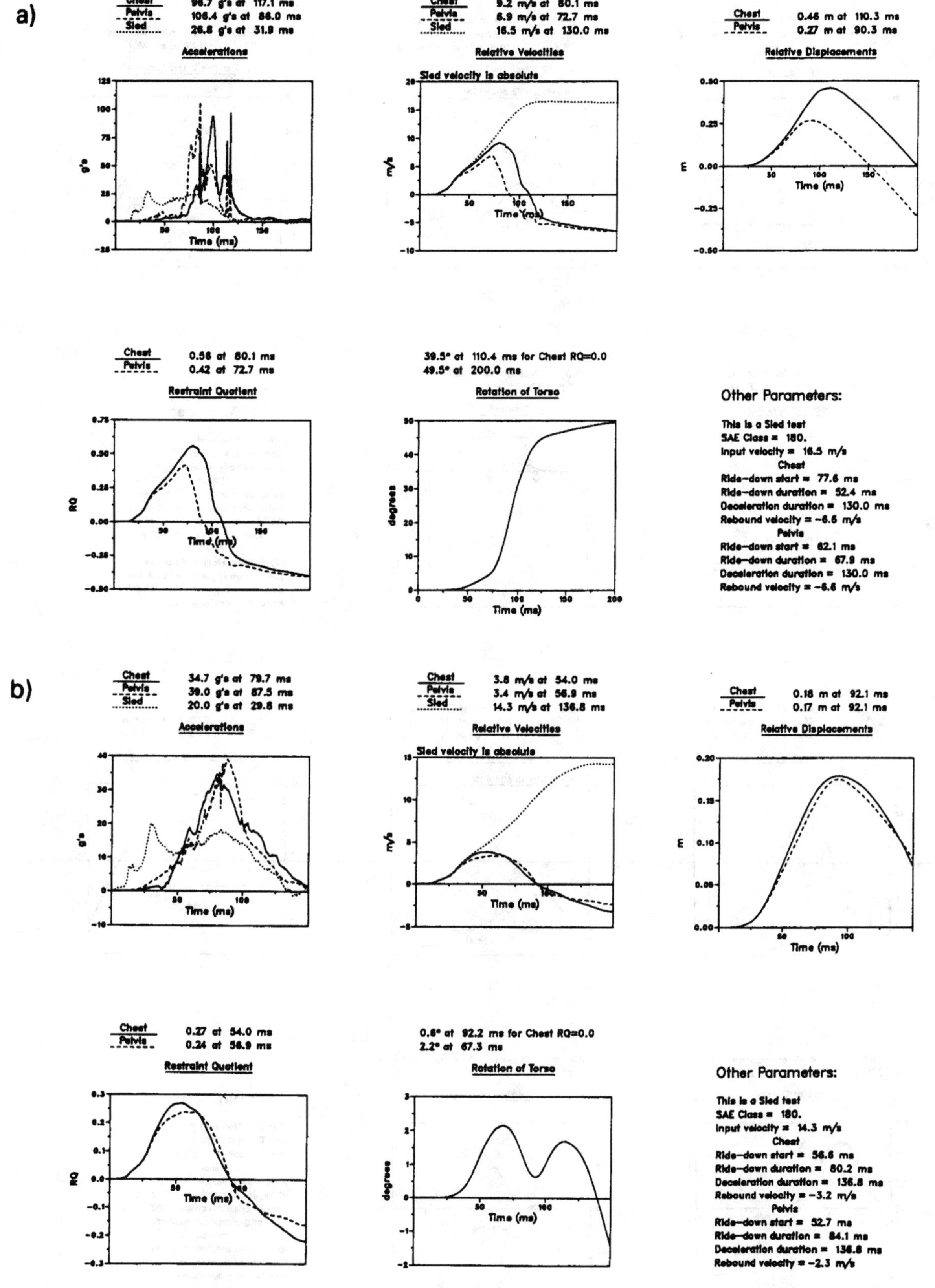

Figure 7: Examples of Restraint Quotient analysis of Hyge sled tests in a body buck using: a) an unbelted driver in a 17.2 m/s test, b) a lap-shoulder belted driver in a 14.8 m/s test, c) a lap-shoulder belted driver in a 17.3 m/s test, and d) a lap-shoulder belted driver with supplemental airbag in a 17.1 m/s test.

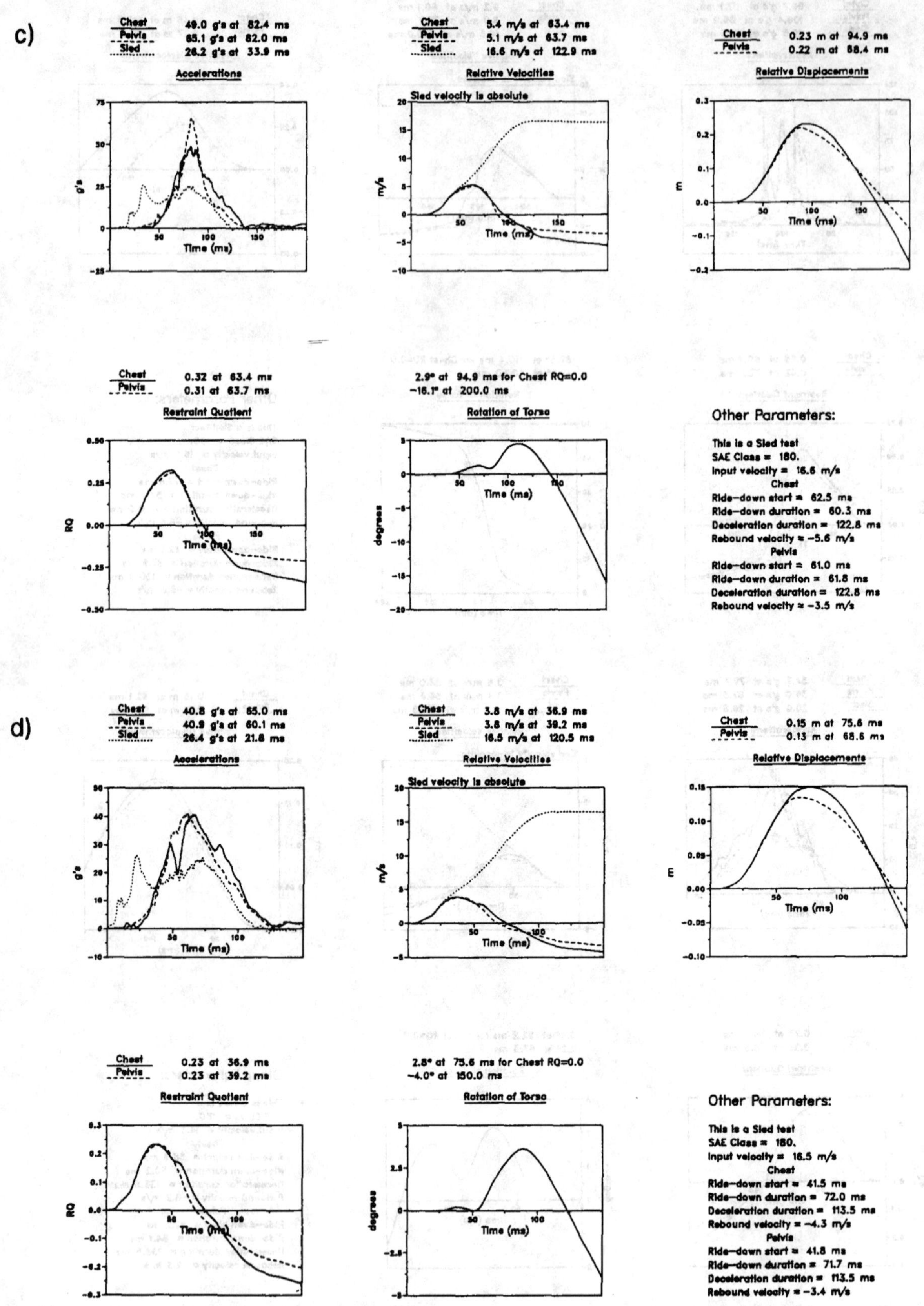

c)
Chest 49.0 g's at 82.4 ms
Pelvis 65.1 g's at 82.0 ms
Sled 26.2 g's at 33.9 ms
Accelerations
g's
Time (ms)
Chest 5.4 m/s at 63.4 ms
Pelvis 5.1 m/s at 63.7 ms
Sled 16.6 m/s at 122.9 ms
Relative Velocities
Sled velocity is absolute
m/s
Time (ms)
Chest 0.23 m at 94.9 ms
Pelvis 0.22 m at 88.4 ms
Relative Displacements
m
Time (ms)
Chest 0.32 at 63.4 ms
Pelvis 0.31 at 63.7 ms
Restraint Quotient
RQ
Time (ms)
2.9° at 94.9 ms for Chest RQ=0.0
-16.1° at 200.0 ms
Rotation of Torso
degrees
Time (ms)
Other Parameters:
This is a Sled test
SAE Class = 180.
Input velocity = 16.6 m/s
Chest
Ride-down start = 62.5 ms
Ride-down duration = 60.3 ms
Deceleration duration = 122.8 ms
Rebound velocity = -5.6 m/s
Pelvis
Ride-down start = 61.0 ms
Ride-down duration = 61.8 ms
Deceleration duration = 122.8 ms
Rebound velocity = -3.5 m/s
d)
Chest 40.8 g's at 65.0 ms
Pelvis 40.9 g's at 60.1 ms
Sled 26.4 g's at 21.8 ms
Accelerations
g's
Time (ms)
Chest 3.8 m/s at 36.9 ms
Pelvis 3.8 m/s at 39.2 ms
Sled 16.5 m/s at 120.5 ms
Relative Velocities
Sled velocity is absolute
m/s
Time (ms)
Chest 0.15 m at 75.6 ms
Pelvis 0.13 m at 68.6 ms
Relative Displacements
m
Time (ms)
Chest 0.23 at 36.9 ms
Pelvis 0.23 at 39.2 ms
Restraint Quotient
RQ
Time (ms)
2.8° at 75.6 ms for Chest RQ=0.0
-4.0° at 150.0 ms
Rotation of Torso
degrees
Time (ms)
Other Parameters:
This is a Sled test
SAE Class = 180.
Input velocity = 16.5 m/s
Chest
Ride-down start = 41.5 ms
Ride-down duration = 72.0 ms
Deceleration duration = 113.5 ms
Rebound velocity = -4.3 m/s
Pelvis
Ride-down start = 41.8 ms
Ride-down duration = 71.7 ms
Deceleration duration = 113.5 ms
Rebound velocity = -3.4 m/s

Since four of the six criteria are specified by the Restraint Quotient program, the following subset of responses is used for comparative purposes in this study:

$$MC_1 = \left[\frac{30° - \theta}{30°}\right] \text{torso angle} + \left[\frac{RQ_c}{.75}\right] \text{chest restraint quotient} + \left[\frac{V_R}{5\ m/s}\right] \text{rebound velocity} + \left[\frac{d_{ph}}{25\ cm}\right] \text{H-pt hor. disp.} \quad (10)$$

The two other criteria are:

$$MC_2 = \left[\frac{d_{pv}}{5\ cm}\right] \text{H-pt ver. disp.} + \left[\frac{30° - \phi}{30°}\right] \text{pelvic rotation} \quad (11)$$

so that the combined criteria becomes $MC = MC_1 + MC_2$.

Comparing the results of MC_1, for experiments involving lap-belt restraint and submarining, yield the following calculation for lap-belt restraint (Test 1675, Table 1):

$$MC_1 = \left[\frac{30° - 14.8°}{30°}\right] + \left[\frac{.39}{.75}\right] + \left[\frac{3.7}{5.0}\right] + \left[\frac{16}{25}\right]$$

$$MC_1 = 2.41$$

and for lap-belt submarining (Test 1678):

$$MC_1 = \left[\frac{30° + 22.5°}{30°}\right] + \left[\frac{.33}{.75}\right] + \left[\frac{5.0}{5.0}\right] + \left[\frac{37}{25}\right]$$

$$MC_1 = 4.67$$

Lap-belt submarining increases the Motion Sequence Criteria 94% above the level with lap-belt restraint. This increase is in spite of reduced FMVSS 208 numbers with submarining. The occupant kinematics shown in Figure 8 involve significant lap-belt loading on the abdomen with submarining. The confirmation of potential injury can now be made in such tests by the use of a frangible abdomen in the dummy (2). The average abdominal crush in comparable tests was 8.5 cm with lap-belt submarining and 0.0 cm with lap-belt restraint. The crush can be interpreted as a probability of occupant injury. These submarining tests resulted in a high probability of severe abdominal injury ($p > 48\%$), whereas lap-belt restraint involved no serious injury risk to the abdomen.

Weighted Tolerance Criteria: In terms of injury assessment, each biomechanical criterion has a tolerance level which is based on an analysis of injury from a series of impact exposures. A common approach to setting tolerance has been to select the response at which about 25% of the tested population experience serious injury. This approach is consistent with current tolerance levels in Federal Motor Vehicle Safety Standards and those set for chest compression and Viscous response (9). However, the various criteria do not have the same significance in terms of the actual occurrence of occupant injury in real-world crashes.

It is possible to use the distribution in human injury to weight the relative significance of individual responses from dummy tests. In this way, the importance of head injury is emphasized. This is important because of the risks of permanent disability and death associated with central nervous system trauma. One approach normalizes each response by its tolerance value and then multiplies it by a weighting factor which, when summed over all injury measures, provides a single value for overall injury risk. The approach helps balance the interpretation of various responses based on current occupant injury patterns. The approach can also be extended to injury disability. This helps avoid emphasis on a single response or parameter at the expense of other injury problems.

Table 4 gives the distribution in occupant injury by body region from two sources for unbelted and lap-shoulder belted occupants (14,15). Much of the field injury data isn't collected in a manner that is identical to the available measurements made in test dummies, and there are differences in injury severity and weighting that complicate direct comparison. However, it is possible to provide a reasonable weighting based on the injury data. For simplicity, the weighting is assumed equal for belted and unbelted occupants, and for occupants in the driver and passenger seating position. Obviously, as more refinements are made in the approach, separate procedures can be considered.

Table 5 provides a possible approach to determining occupant protection from dummy responses in crash tests. Since several responses can be measured in a single body region, the approach equally weights each response within a category so that the sum is representative of the overall significance of that body region to the current distribution of human injury in automotive crashes. Recognizing the wide distribution of crash and injury types involving occupants and injury, this is a simple approach; but, it does provide a way to balance interpretations of test data.

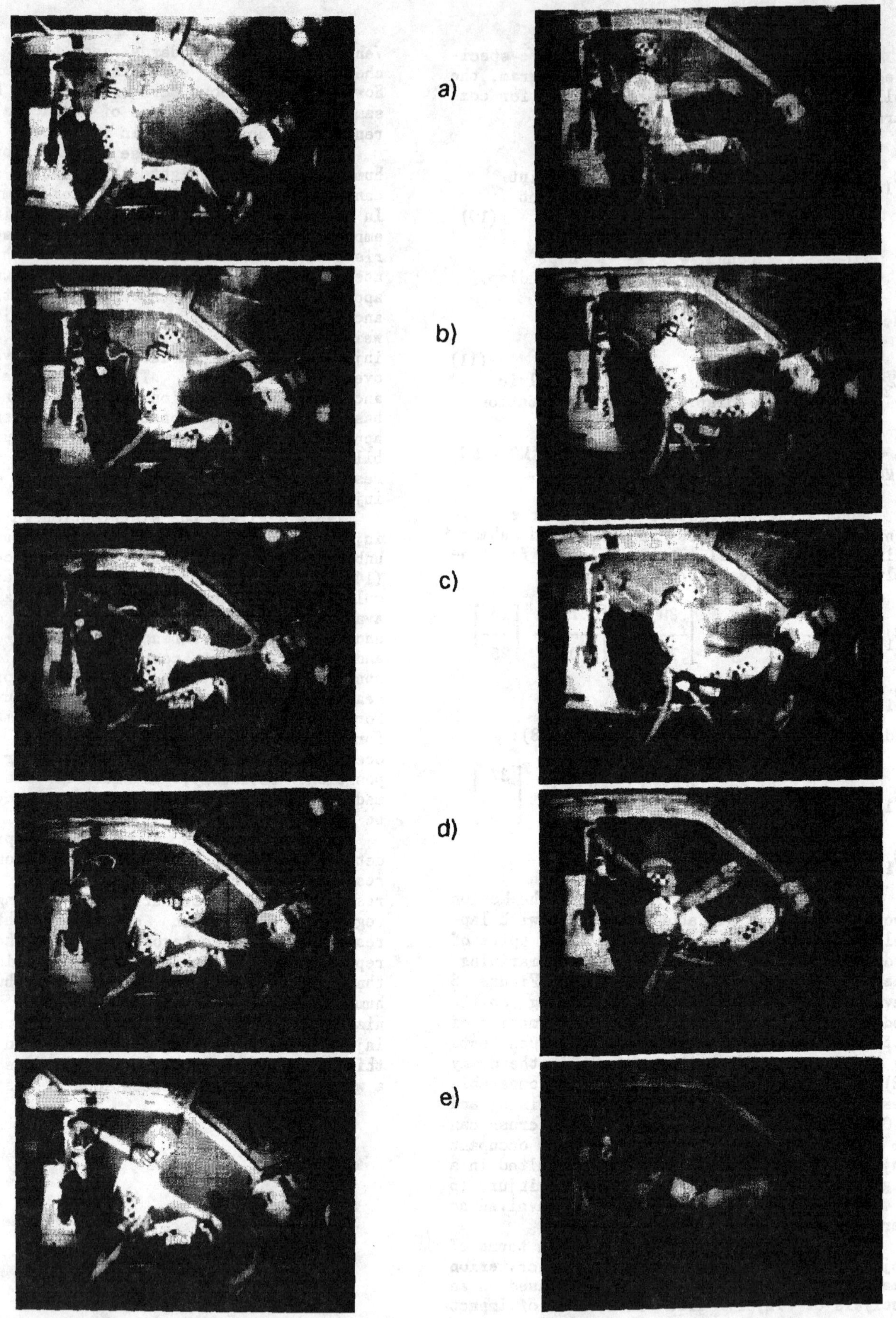

Figure 8: Sequence photographs from the high-speed movies of sled tests with lap-belt restraint (test #1675 - left sequence) and lap-belt submarining (test #1678 - right sequence).

Table 4

Distribution in Occupant Injury for Unrestrained and Safety Belted Front Seat Occupants

Body Region	Unrestrained		Lap-Shoulder Belted	
	USA[1]	Sweden[2]	USA	Sweden
Head	43.2	31.7/20.7	73.1	22.3/19.7
Face	8.5	9.2/10.3	11.9	15.2/7.5
Neck	9.3	-	1.9	-
Chest	22.1	15.5/21.0	4.1	23.3/27.9
Abdomen	6.6	18.2/27.4	1.7	18.4/26.1
Extremities	6.2	21.2/19.3	7.0	20.2/18.3
Other	4.3	4.2/1.3	0.5	0.7/0.4
	5,553		258	

1 Fraction of occupant injury harm.

2 Fraction of injury for occupants with ISS 11+. Driver/right-front passenger injuries.

Table 5

Injury Criteria, Tolerances and Logist Parameters for Injury Risk Assessment

Body Region	Injury Significance	FMVSS 208	Criterion	Tolerance	Risk Function α	β
Head	40%	60%	HIC	1000	5.02	0.00351
			A_{3ms}	>100	-	-
Face	10%	-	F	-	-	-
Neck	10%	-	Faxial	1100 N, 45 ms +	-	-
			Fshear	1100 N, 45 ms +	-	-
			Mextension	57 Nm	-	-
			Mflexion	190 Nm	-	-
Chest	20%	35%	A_{3ms}	60 g	5.55	0.0693
			C	34%	10.49	0.277
			VC	1.0 m/s	11.42	11.56
Abdomen	15%	-	C	48%	5.15	0.108
			VC	-	-	-
Extremity	5%	5%	F	10 kN	7.59	0.660
			A_{3ms}	-	-	-

Injury Criteria (IC) can be combined into a single parameter to assess overall safety performance. By using the data in Table 5:

$$
\begin{aligned}
IC = {} & \frac{0.40}{2}\left[\frac{HIC}{1000} + \frac{A_{3ms}}{(A_{3ms})_T}\right]_{\text{Head}} \\
& + 0.10\left[\frac{F}{(F)_T}\right]_{\text{Face}} \\
& + \frac{0.10}{4}\left[\frac{F_{ax}}{(F_{ax})_T} + \frac{F_{sh}}{(F_{sh})_T} + \frac{M_f}{(M_f)_T} + \frac{M_e}{(M_e)_T}\right]_{\text{Neck}} \\
& + \frac{0.20}{3}\left[\frac{C}{(C)_T} + \frac{VC}{(VC)_T} + \frac{A_{3ms}}{60g}\right]_{\text{Chest}} \\
& + 0.15\left[\frac{C}{(C)_T}\right]_{\text{Abdomen}} \\
& + \frac{0.05}{3}\left[\frac{F_l}{10\text{ kN}} + \frac{F_r}{10\text{ kN}} + \frac{(A_{3ms})}{(A_{3ms})_T}\right]_{\text{Extremity}}
\end{aligned}
\qquad (12)
$$

In terms of FMVSS 208 criteria, Equation (12) reduces to:

$$
\begin{aligned}
IC_{208} = {} & 0.60\left[\frac{HIC}{1000}\right]_{\text{Head}} + 0.35\left[\frac{A_{3ms}}{60\text{ g}}\right]_{\text{Chest}} \\
& + \frac{0.05}{2}\left[\frac{F_l}{10\text{ kN}} + \frac{F_r}{10\text{ kN}}\right]_{\text{Extremity}}
\end{aligned}
\qquad (13)
$$

It is possible to use the terms in Equation (13) for tests 1675 and 1678 involving lap-belt restraint and submarining. This yields a value of $IC_{208} = 0.793$ for Test 1675 and $IC_{208} = 0.629$ for Test 1678 (Table 6). On this basis lap-belt submarining could be interpreted as lowering injury risk.

If the results of frangible abdomen crush are used to reanalyze occupant injury, another term can be added to Equation (13) for abdominal injury assessment. This yields a different interpretation of injury risk, since:

$$
\begin{aligned}
IC_{208*} = {} & 0.60\left[\frac{HIC}{1000}\right]_{\text{Head}} \\
& + 0.20\left[\frac{A_{3ms}}{60\text{ g}}\right]_{\text{Chest}} + 0.15\left[\frac{C}{48\%}\right]_{\text{Abdomen}} \\
& + \frac{0.05}{2}\left[\frac{F_l}{10\text{ kN}} + \frac{F_r}{10\text{ kN}}\right]_{\text{Extremity}}
\end{aligned}
\qquad (14)
$$

and $IC_{208*} = 0.676$ for Test 1675 and $IC_{208*} = 0.688$ for 1678 (Table 7). Emphasis on abdominal loading now indicates greater injury risk with submarining. An increased injury risk is in agreement with the results of the motion sequence criteria, but the differences are small and don't seem to adequately quantify comparative risks of serious injury.

Injury Risk Assessment: A more informative and the preferred approach for assessing occupant protection involves interpreting dummy responses using an injury risk function which is based on Logist probability analysis of biomechanical data. This enables injury risk to be assessed in each body region.

Table 6 shows the situation using FMVSS 208 responses. Since only one response is measured for the head and chest, injury risk for these body regions can be directly calculated from the probability function in Logist using $p(x) = [1 + \exp(\alpha - \beta x)]^{-1}$. This function relates the probability of injury to a dummy response x using two parameters α and β, which are determined from the best fit using a sigmoidal relationship to approximate biomechanical data on impact injury.

In terms of interpreting risks for the extremities, where two independent leg responses are measured for each test, the overall risk is the sum of the two risks. Using this approach and the FMVSS 208 responses, the lap-belt restraint test results in a risk of p = 21.2% probability of serious injury and lap-belt submarining a lower risk of p = 11.6%.

If abdominal injury assessment is included in the interpretation of performance, the risk of serious injury with lap-belt restraint is identical, at p = 21.2%, whereas there is a 5-fold increase in perceived risk of injury to p = 59.6% with lap-belt submarining (see Table 7). This is due to significant crush of the frangible abdomen.

Augmenting FMVSS 208 responses seems to provide a more relevant interpretation of restraint performance since risk is reported as a probability for critical body regions. The overall risk of injury is determined by the sum of individual risk. Since individual risk varies from zero to one, the overall risk of injury can exceed p = 100%, but that would be consistent with some crash victims experiencing multiple injuries or causes of death. In addition, any discussion of trade-offs between responses is aided by this approach because of a more specific relation to potential injury in specific body regions.

A strength of injury risk assessment is the form of the sigmoidal function that relates a dummy response to injury risk and setting human tolerance at or below a 25% probability of injury. Tolerance levels are typically set in the low-risk region of the sigmoidal response (Figure 9). Thus, maximum dummy responses below tolerance involve a gradual change in risk with responses in the low-risk region of the sigmoidal function.

Table 6
FMVSS 208 Criteria to Assess Lap-Belt Restraint and Submarining

Body Region	Injury Significance	Criterion	Tolerance	Risk Function α	β
Head	60%	HIC	1000	5.02	0.00351
Chest	35%	A_{3ms}	60 g	5.55	0.0693
Femurs	5%	F	10 kN	7.59	0.660

Evaluation of Lap-Belt Restraint Responses and Injury Risk

Body Region	Criterion	Test Value 1675	Weighted Criterion	Injury Probability
Head	HIC	863	0.520	0.120
Chest	A_{3ms}	46.8	0.273	0.091
Femurs	F	0.65/0.82	0.000	0.001
		Total	0.793	0.212

Evaluation of Lap-Belt Submarining Response and Injury Risk

Body Region	Criterion	Test Value 1678	Weighted Criterion	Injury Probability
Head	HIC	706	0.424	0.073
Chest	A_{3ms}	35.1	0.205	0.042
Femurs	F	0.71/0.63	0.0	0.001
		Total	0.629	0.116

Table 7
FMVSS 208 Criteria Augmented by Abdominal Injury Risk to Assess Lap-Belt Restraint and Submarining

Body Region	Injury Significance	Criterion	Tolerance	Risk Function α	β
Head	60%	HIC	1000	5.02	0.00351
Chest	20%	A_{3ms}	60 g	5.55	0.0693
Abdomen	15%	C	48%	5.15	0.108
Femurs	5%	F	10 kN	7.59	0.660

Evaluation of Lap-Belt Restraint Responses and Injury Risk

Body Region	Criterion	Test Value 1675	Weighted Criterion	Injury Probability
Head	HIC	863	0.520	0.120
Chest	A_{3ms}	46.8	0.156	0.091
Abdomen	C	0	0.000	0.000
Femurs	F	0.65/0.82	0.000	0.001
		Total	0.676	0.212

Evaluation of Lap-Belt Submarining Responses and Injury Risk

Body Region	Criterion	Test Value 1678	Weighted Criterion	Injury Probability
Head	HIC	706	0.424	0.073
Chest	A_{3ms}	35.1	0.117	0.042
Abdomen	C	47%	0.147	0.480
Femurs	F	0.71/0.63	0.0	0.001
		Total	0.688	0.596

However, if a response approaches or is above tolerance, slight differences in response involve major changes in perceived risk because they occur in the transition region from low to high risk. This increasingly penalizes responses near or above tolerance, until responses are so large that further increases do not significantly modify the nearly absolute probability of injury occurrence. For cases where several responses are made in the same body region, such as in the chest where acceleration, compression and the Viscous response are evaluated, injury risk can be interpreted as the greatest risk from any of the applicable criteria.

Injury risk assessment was used on the average values from the unbelted and manual belted dummies in barrier crash tests, and an additional test involving belt and airbag restraint. Figure 10 shows the FMVSS 208 criteria. Figure 11 shows the overall risk of serious injury based on the analysis in Table 8. There is a 116.8% risk (range 64.6%-209.8% based on one standard deviation in test responses) of serious injury to the head, chest and legs. This is based on a 90.9% (55.7%-138.2%) probability of AIS 4+ injury to the head and chest, and a 28.9% (8.9%-71.6%) probability of AIS 3+ injury to the legs. Safety belts reduce the risk of injury to 14.7% (range 11.1%-19.6%), an 87% reduction from levels with an unbelted driver. The risk is further reduced to 8.0% with the combination of belts and airbag restraint.

Although the above is an informative example of the risk assessment approach, it also shows the uncertainty due to variability in dummy responses. If the confidence interval in the Logist function for injury probability is also considered, the risk range would be larger. However, this seems to be the appropriate way to consider the significance of differences in restraint performance. This approach takes advantage of the sigmoidal form of the injury probability function since dummy responses in the transition region are penalized in comparing occupant protection. Furthermore, the analysis shows that there is injury risk even with belt and airbag restraint.

While injury risk assessment seems to be the most meaningful approach to evaluate occupant restraints, it needs further evaluation, review, and discussion particularly in terms of the Logist parameters in determining the actual distribution of human injury, the adequacy of biomechanical data in quantifying occupant injury, the biofidelity of dummy responses in mimicking human response and validity of measurement approaches. The eventual approach should probably involve a sufficiently instrumented dummy to assess responses in the key body regions and use of appropriate parameters defining the injury probability functions from consensus review of biomechanical data.

Generalizing the Approaches: The Restraint Quotient is different for the chest in barrier and Hyge sled tests with similar restraint conditions (Figure 12). Higher values of RQ_c imply a greater chest velocity with respect to the interior in barrier tests than in comparable Hyge sled experiments. However, both crash types show a 48%-54% reduction in RQ_c with belted versus unbelted occupants. The combination of belts and driver airbag further reduced RQ_c by 15% over belt only restraint. While higher values of RQ_c were observed with airbag only restraint in Hyge sled tests, this type of evaluation involves only frontal crash analysis of restraint performance.

In terms of injury and motion sequence criteria, it is certainly advisable to extend the restraint criteria to consider non-frontal crash conditions and a range in crash severities and circumstances. This would probably involve an envelope of motion limits in the forward and lateral directions of occupant motion. Some work has already been completed on the oblique response of test dummies with safety belt restraints which have demonstrated the value of limiting lateral displacements as an approach to improved occupant protection.

In addition to more global motion sequence criteria, there would be a need to introduce lateral components of injury assessment. This would eventually lead to setting omni-directional restraint criteria which would be compatible with developing procedures and criteria in side impact crashes. This would build upon the work of Horsch (16, 17) and Culver et al (18) for oblique responses of belted dummies and the more recent work on pure lateral impact injury. Obviously, this is a future goal that would require an omni-directional dummy with the interpretation of injury risks and motion sequences based on a range of impact directions and severities occurring in motor vehicle crashes.

The Restraint Quotient concepts lose relevance when the occupant compartment undergoes deformations such that the mean acceleration of the passenger compartment is not a sufficient measure of the impact exposure. This is the case in severe side impacts because the door is rapidly accelerated in the crash and impacts the occupant at high velocity while the passenger compartment experiences a more gradual velocity change. In this situation, the injury risk associated with high-speed impact with the door is based on the Viscous response of the chest and the abdomen. This crash situation also points out the significance of the point of contact and passenger compartment deformation on potential occupant injury (19-21).

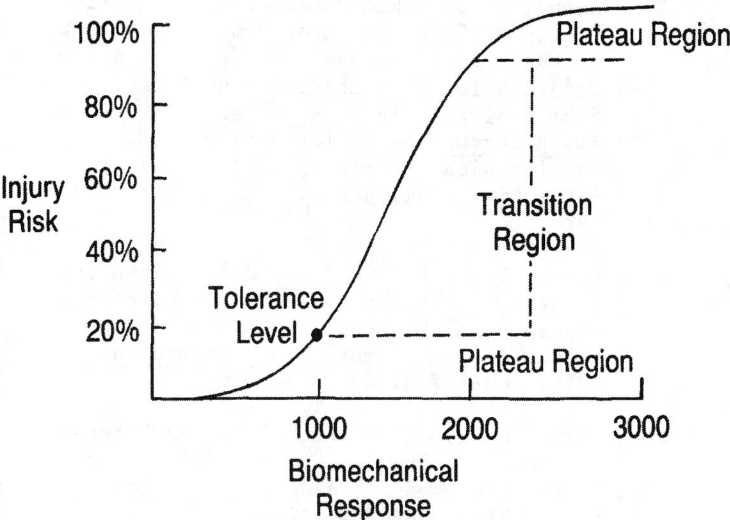

Figure 9: Example of an injury risk function based on HIC and serious head injury. The sigmoidal function is from Logist analysis of biomechanical data.

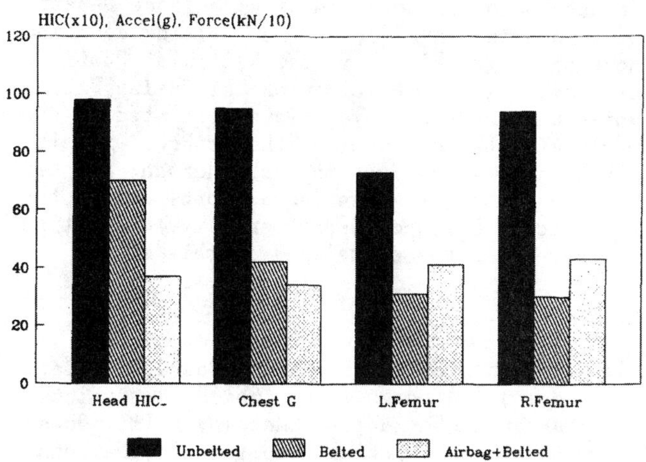

Figure 10: Comparison of FMVSS 208 responses for barrier crash tests with the Hybrid III dummy.

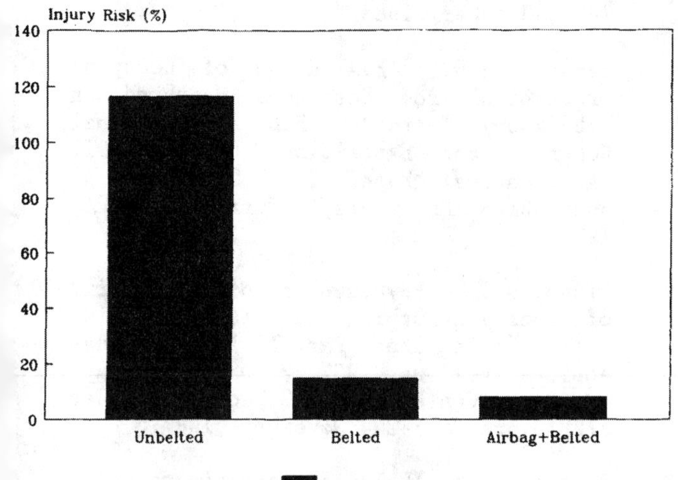

Figure 11: Overall risk of injury based on barrier crash test results.

Table 8

Injury Risk Assessment of Barrier Crash Data

	HIC		Injury Risk	
	Mean	Range	Mean	Range
Unbelted	980	529-1431	17.1	4.1-50.1
Belted	700	629-771	7.2	5.7-9.0
Airbag & Belts	370	-	2.4	-

	Chest G		Injury Risk	
	Mean	Range	Mean	Range
Unbelted	95	81-109	73.8	51.6-88.1
Belted	42	42-46	6.7	5.1-8.6
Airbag & Belts	34	-	3.9	-

	Femur kN		Injury Risk	
	Mean	Range	Mean	Range
Unbelted L	7.3	4.4-10.2	5.9	0.9-29.8
R	9.4	7.8-11.0	20.0	8.0-41.8
Belted L	3.1	1.8-4.4	0.4	0.2-0.9
R	3.0	1.3-4.7	0.4	0.1-1.1
Airbag & Belts L	4.1	-	0.8	-
R	4.3	-	0.9	-

ACKNOWLEDGEMENTS

We would like to recognize the assistance of Kenneth Baron of the Biomedical Science Department for his involvement in writing code and developing graphics for the Restraint Quotient program. The participation of Milan Patel, while a summer employe, is appreciated in the early efforts to develop the method. Special thanks are also due to John Horsch, Stephen Rouhana and the engineering support team in the Biomedical Science Department for conducting the Hyge sled tests analyzed in this study.

REFERENCES

1. Transportation Research Board, "Safety Belts, Airbags and Child Restraints: Research to Address Emerging Policy Questions." Special Report 224, National Research Council, Washington, DC, 1989.

2. Rouhana, S.W., Viano, D.C., Jedrzejczak, E.A., and McCleary, J.D., "Assessing Submarining and Abdominal Injury Risk in the Hybrid III Family of Dummies." In Proceedings of the 33rd Stapp Car Crash Conference, pp. 257-280, SAE Technical Paper #892440, Society of Automotive Engineers, Warrendale, PA, 1989.

3. Rouhana, S.W. and Horsch, J.D., "Assessment of Lap-Shoulder Belt Restraint Performance in Laboratory Testing." In Proceedings of the 33rd Stapp Car Crash Conference, pp. 243-256, SAE Technical Paper #892439, Society of Automotive Engineers, Warrendale, PA, 1989.

4. Horsch, J.D. and Hering, W.E., "A Kinematic Analysis of Lap-Belt Submarining for Test Dummies." In Proceedings of the 33rd Stapp Car Crash Conference, pp. 281-288, SAE Technical Paper #892441, Society of Automotive Engineers, Warrendale, PA, 1989.

5. Adomeit, D., "Seat Design - A Significant Factor for Safety Belt Effectiveness." In Proceedings of the 23rd Stapp Car Crash Conference, pp. 39-68, SAE Technical Paper #791004, Society of Automotive Engineers, Warrendale, PA 1979.

6. Adomeit, D., "Evaluation Methods for the Biomechanical Quality of Restraint Systems During Frontal Impact." In Proceedings of the 21st Stapp Car Crash Conference, pp. 913-932, SAE Technical Paper #770936, Society of Automotive Engineers, Warrendale, PA 1977.

7. Adomeit, D. and Heger, A., "Motion Sequence Criteria and Design Proposals for Restraint Devices in Order to Avoid Unfavorable Biomechanic Conditions and Submarining." In Proceedings of the 19th Stapp Car Crash Conference, pp. 139-165, SAE Technical Paper #751146, Society of Automotive Engineers, Warrendale, PA, 1975.

8. Viano, D.C., A.I. King, J.W. Melvin and K. Weber, "Injury Biomechanics Research: An Essential Element in the Prevention of Trauma." Journal of Biomechanics, 22(5):403-417, 1989.

9. Viano, D.C., and Lau, I.V., "A Viscous Tolerance Criterion for Soft Tissue Injury Assessment." Journal of Biomechanics, 21(5):387-399, 1988.

10. Viano, D.C., "Biomechanics of Head Injury: Toward a Theory Linking Head Dynamics Motion, Brain Tissue Deformation and Neural Trauma." SAE Transactions, vol. 97, 1988, 32nd Stapp Car Crash Conference, SAE Technical Paper #881708, pp. 1-20, October, 1988.

11. Horsch, J.D., "Evaluation of Occupant Protection from Responses Measured in Laboratory Tests." SAE International Congress and Exposition, Detroit, MI. SAE Technical Paper #870222, Society of Automotive Engineers, Warrendale, PA, 1987.

12. Viano, D.C., "Evaluation of the Benefit of Energy-Absorbing Material for Side Impact Protection: Part II." SAE Transactions, vol. 96, 1987, P-202, 31st Stapp Car Crash Conference, SAE Technical paper 872213, pp. 205-224 November, 1987.

13. Rouhana, S.W. Personal communication.

14. Carsten, O., "Relationship of Accident Type to Occupant Injuries." UMTRI-86-15, Final Report, The University of Michigan Transportation Research Institute, Ann Arbor, MI, 1986.

15. Nygren, A., "Injuries to Car Occupants - Some Aspects of the Interior Safety of Cars." Acta Oto-Laryngologica, Supplement 395, ISSN 0365-5237, The Almqvist & Wiksell Periodical Company, Stockholm, Sweden.

16. Horsch, J.D., "Occupant Dynamics as a Function of Impact Angle and Belt Restraint." In Proceedings of the 24th Stapp Car Crash Conference, pp. 417-438, SAE Technical Paper #801310, Society of Automotive Engineers, Warrendale, PA, 1980.

17. Horsch, J.D., Schneider, D.C., Kroell, C.K., and Raasch, F.D., "Response of Belt-Restrained Subjects in Simulated Lateral Impacts." In Proceedings of the 23rd Stapp Car Crash Conference, pp. 71-103, SAE Technical Paper #791005, Society of Automotive Engineers, Warrendale, PA, 1979.

18. Culver, C.C. and Viano, D.C., "Influence of Lateral Restraint on Occupant Interaction with a Shoulder Belt or Preinflated Air Bag in Oblique Impacts." In SAE Transactions, vol. 90, SAE Technical Paper #810370, Society of Automotive Engineers, Warrendale, PA, 1981.

19. Evans, L. and Frick, M., "Seating Position in Cars and Fatality Risk." Am J Pub Health 78:1456-1458, 1988.

20. Thomas, C., Koltchakian, S., Tarriere, C., et al, "Primary Safety Priorities in View to Technical Feasibility Limited to Secondary Automotive Safety." In Proceedings of the 1990 FISITA Conference, Milan, Italy, 1990.

21. Grosch, L., Baumann, K.H., Holtze, H. and Schwede, W., "Safety Performance of Passenger Cars Designed to Accommodate Frontal Impacts with Partial Barrier Overlap." In Automotive Frontal Impacts SP-782, pp. 29-36, SAE Technical Paper #890748, Society of Automotive Engineers, Warrendale, PA, 1989.

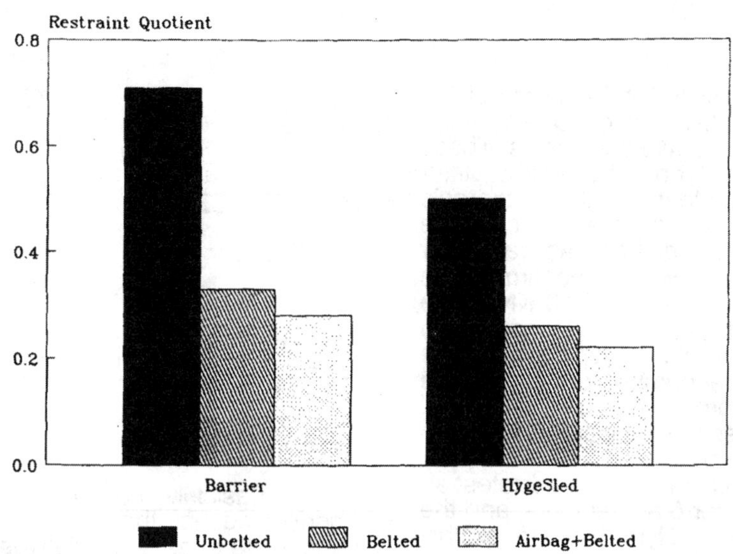

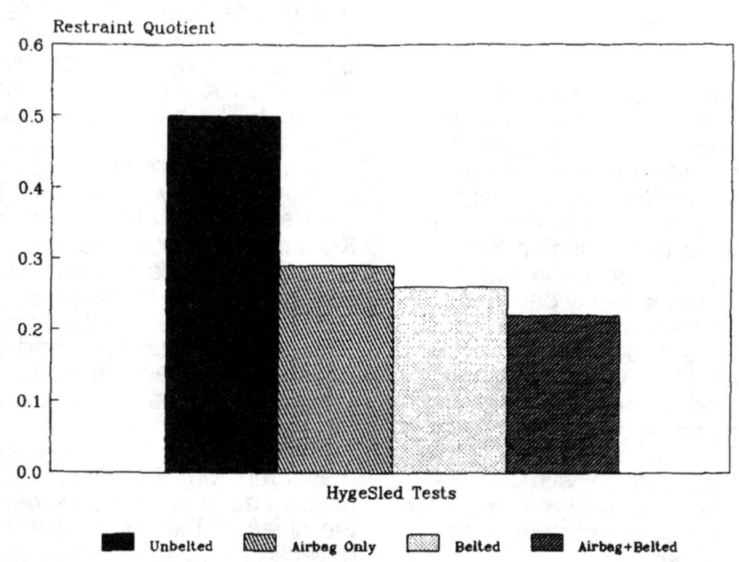

Figure 12: RQ_c for barrier and Hyge sled tests with various restraint conditions (data from Tables 2 and 3).

902310

Behaviour of Human Surrogates Thorax under Belt Loading

Dominique Cesari and Robert Bouquet
INRETS Laboratoire des Chocs et de Biomécanique

Abstract

To estimate the behaviour of the thorax of the human cadaver and Hybrid III a total of 33 belt impact tests were performed with the two surrogates. These tests have shown that the Hybrid III thorax is stiffer than that of the cadaver and that the internal thoracic deflection transducer may not necessarily record the maximum thoracic deflection. The belt load was lower value with the cadavers, which confirms the differences in stiffness. A belt force of 10 KN in the cadaver tests was associated with an average of 6 rib fractures. If we consider the relationship between the thoracic deflection and the number of rib fracture cadavers showing 5 or more rib fractures sustained an external thoracic deflection at least of 7.5 cm measured at the mid sternum.

The analysis of V*C parameter indicates an average V*C value of 0.77 for 6 rib fractures, and the values of V*C measured on Hybrid III are sligthly lower than those of cadaver tests.

THE response of the human thorax to impact is a biomechanical research problem which has been approached several years ago before the common use of belt restraint systems. Most of these studies (1,2.) used a disc impactor loading dynamically a human thorax. These tests increased our understanding of the behaviour of the human thorax to impact, but they do not represent the loading process involved with the use of the safety belt.

More recently a study comparing the thoracic deflection of human volunteers and Hybrid III under belt loading was performed (3,4). This study was limited to non injury levels, and to complete its results a test programme including cadaver and Hybrid III tests performed in the same conditions was conducted by INRETS in cooperation with NHTSA and Transport Canada. For that programme 20 Hybrid III and 13 cadaver tests were performed using the test appartus indicated in Figure 1.

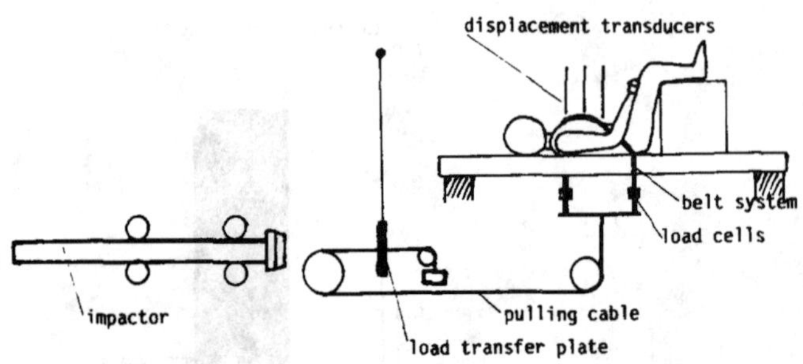

FIGURE 1 : TEST SET UP

TEST METHOD

Basically cadaver and dummy tests were performed in the same conditions: the human surrogate to be tested is was laying supine on a rigid flat table. A seat belt strap is placed across the torso in a geometrical layout similar to that of a car driver wearing a shoulder belt. The two extremities of the belt pass through the table over low friction supports, and are attached to an horizontal rod, as indicated in Figure 1. This bar is pulled down by a cable activated by a dynamic impactor. The impactor consists of a rigid tube propelled by rubber springs impacting at a constant velocity a plate attached to the cable pulling on the belt. The mass of the impactor tube was 22.4 Kg for some tests, and 76.1 Kg for the others : this corresponds to an impact speed almost divided by two for the same impact energy.

During the test the impactor speed prior to the impact was measured, and the impactor speed is equal to the speed at which the belt is pulled. The tensile belt forces are measured on each side of the torso.

Besides the internal thoracic deflection measured on the Hybrid III thorax, 10 external displacement transducers were placed above and on the sides of the torso ; for the last 6 cadaver tests deflection measurement at point #9 was replaced by a measurement at point #11, as previous tests had shown that there was almost no horizontal deformation on the right side of the thorax. Three transducer rods were attached to the belt, and the 7 others to the rib cage through the flesh. Fig 2 gives the location and the direction of the thoracic displacement transducers.

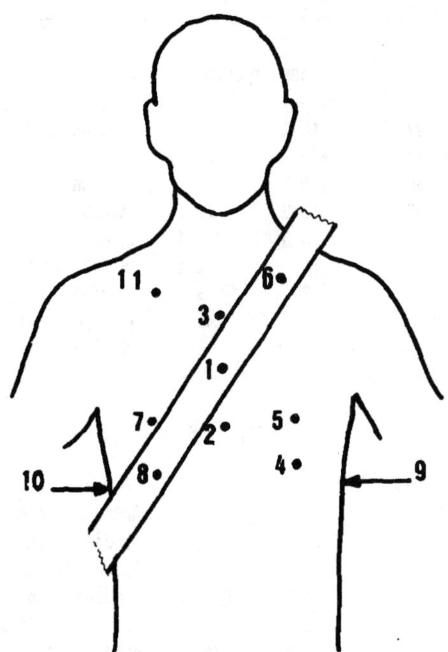

FIG. 2 : Location of thoracic deflection transducers

HYBRID III TEST RESULTS.

Twenty dynamic tests were performed on a Hybrid III dummy : 14 using a low mass (22.4 Kg) impactor and 6 with a heavy (76.1 Kg) impactor.

ANALYSIS OF THORACIC DEFLECTION - Points #1 (mid sternum) and 8 are the most deformed parts of the thorax, and deflection at point 1 is more interesting as it is located on the belt where it crosses the sternum. The deflection at this point is always higher than the internal deflection recorded through the original Hybrid III thoracic potentiometer, as shown in Figure 3. The difference between internal and external deflection is in the range of 1 to 2 cm in almost all the tests.

This difference is mainly due to the crush of the thorax flesh above the sternum which is about 2 cm thick. Low speed (high mass impactor) tests correspond to a lower difference between external and internal deflections.

There are three points on the sternum (upper, middle and lower) at which the deflection is measured (respectively points #2,1, and 3). In most cases the lower sternum deflection is slightly higher than that of the upper sternum. The internal deflection is of the same order of magnitude as deflections #2 and #3.

Deflections #9 and 10 are measured horizontally at the level of the 8th ribs : the deformation measured at these two points is an extension, which is a consequence of the antero-posterior compression. However this extension is very small especially, on the left side.

ANALYSIS OF BELT FORCE VALUES - The tension forces in the belt were measured at the two extremities above the pulling bar. The values were almost the same at each extremity and so it was possible to add the two values to determine the total belt force.

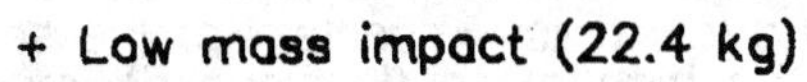

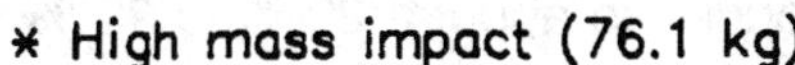

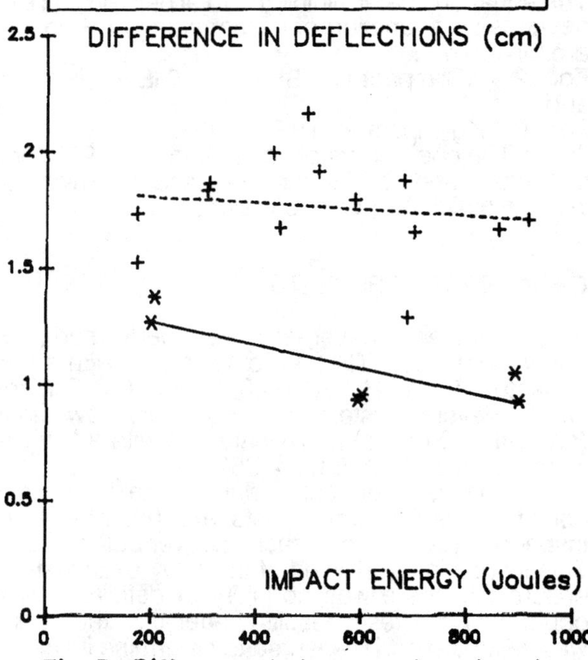

Fig 3: Difference between external and internal Hybrid III thoracic deflection.

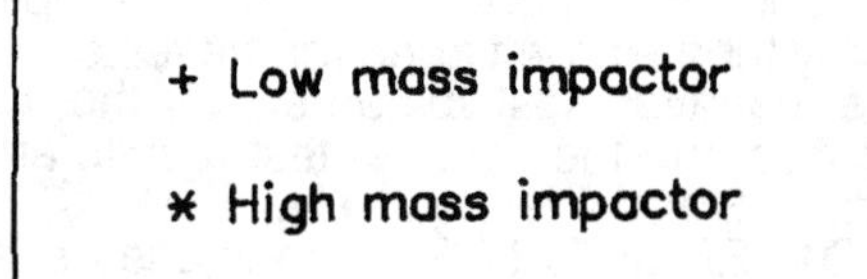

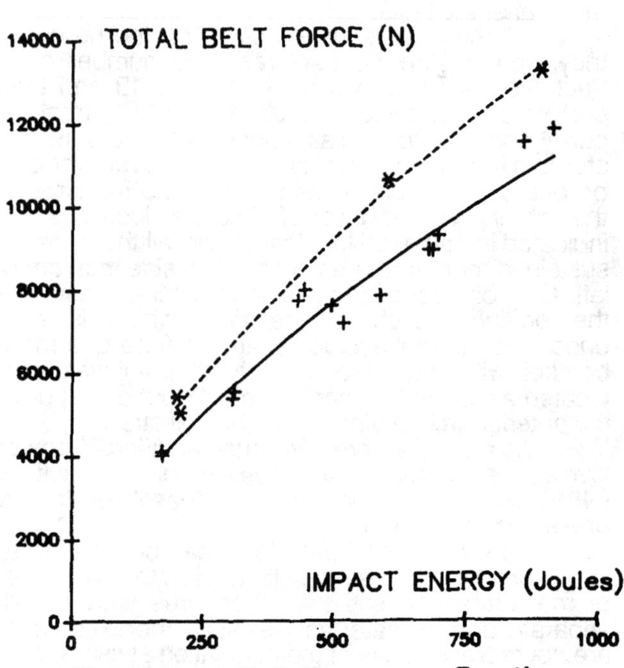

Fig 4: Total Belt Force as a Function of Impact Energy (H III)

Figure 4 shows the total belt force for high mass and low mass impacts. For the same impact energy the high mass (low speed) impactor tests corresponded to about a 20 % to 25 % higher total belt force.

The equations determining the total belt force (TBF) in newtons as a function of impact energy (IE) in joules are the following.

For 22.4 KG impactor : $TBF = 165.6\ IE^{0.618}$

and

For 76.1 Kg impactor : $TBF = 183\ IE^{0.632}$

The coefficients of correlation are 0.995 for the first case and 0.903 for the second which both indicate a high degree of correlation.

CADAVER TEST RESULTS

Thirteen cadaver tests were performed (7 with a low mass -tests THC 11 to 17- and 6 with a heavy impactor - tests THC 18 to 20, 62, 65, 69). The three last cadavers sustained a no injury low energy (around 100 joules) preliminary test with a low mass impactor (tests THC 61, 64, 68).

The cadaver tests were made in the same conditions as the dummy tests, except that a detailed inspection was made of each cadaver before the test. This inspection consisted of an X ray examination to ensure that there was no injury or deficiency which could alter the tests results ; after that the cadaver was weighted and measured to determine its detailed anthropometry. Before the cadaver was placed on the table a tracheotomy tube was installed. When the cadaver was in the test position the lungswere inflated to restore the abdominal organs to their correct positions and the normal thoracic shape. The tracheotomy tube was left open for the tests.

The cadaver test parameters, the cadaver characteristics and the cadaver test results are listed in appendices 3 and 4 respectively.

CADAVER INJURIES - The injuries sustained by the cadaver were recorded during the autopsy made after the tests.

All the injuries were located on the rib cage ; they were mainly rib fractures. The number of rib fractures varied from 0 (one case) to 19 and three cadavers showed multiple rib fractures corresponding to a flail chest (#14, 20, 65). A sternum fracture was found on two cadavers, isolated on one and associated with multiple rib fractures on the other. The details of fracture locations are indicated in Table 1. Note that almost all the cadavers sustained more fractures on the right side than on the left. Only one cadaver sustained fractures located on the posterior arch of the ribs, which is quite uncommon in real accidents, and may be due to the contact with the table. All the other injuries were located either on the median arch (7 out of 10) or on the anterior arch, mainly under the belt strap.

As the tests were performed at different energy levels, we compared the number of rib fractures (NRF) with the impact energy. These results are presented In Figure 5.

The results of the 13 tests can be considered in two groups : 10 tests results were in the same area of the figure whereas the other three were clearly separate and indicate less resistant thoraxes which are not representative of the population at risk. Taking into account the 10 other tests it is possible to draw a linear relationship between NRF and the impact energy. That relationship allows us to determine the

TABLE 1 - Cadaver Injuries

TEST N°	AGE	FRACTURES Right rib	Left rib	POSITION Arch/Sternum	NRF
		Low mass impactor			
THC 11	47	3 to 10	-	Median arch	8
THC 12	17	-	-		0
THC 13	86	5 & 7	-	Median arch	2
THC 14	69	2 to 8	2 to 5	Median arch	17
		1 to 3	2 & 4	Posterior arch	
THC 15	60	8	3 & 4	Median arch	3
THC 16	59	7	3	Median arch	4
		-	2 & 3	Anterior arch	
THC 17	71	4 to 10	-	Median arch	7
		High mass impactor			
THC 18	67	6 to 10	-	Median arch	6
		-	2	Anterior arch	
THC 19	83	5 to 8	-	Median arch	4
THC 20	70	2 to 7	2	Anterior arch	18
		2 to 9	3,8 &10	Median arch	
THC 62	72	6 to 8	1	Anterior arch	4
THC 65	71	5 to 9		Anterior arch	10
		6 to 9		Median arch	
		1		Sternum	
THC 69	40	1		Sternum	1

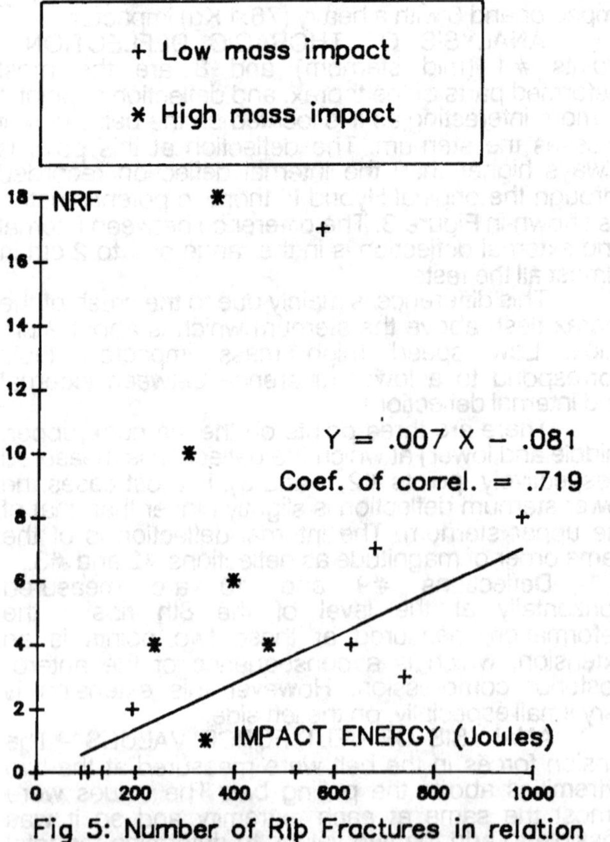

Fig 5: Number of Rib Fractures in relation to Impact Energy

value of kinetic energy needed to produce a specific number of rib fractures : an impact energy of 830 joules correspond to a value of 6 rib fractures which is the proposed acceptable limit for thoracic injuries (6).

THORACIC DEFLECTION - If we consider the thorax deformation, transducer #8 recorded the most important deflection. This point is located on the belt where it crosses the floating rib on the right side which is a relatively soft part of the thorax.

Point #1, located on the belt at the intersection with the sternum is also greatly deflected, more than the two other points located on the sternum, mainly because the crushed flesh under the belt increases the deflection.

BELT FORCES AND RIB FRACTURES - During the tests, the belt force was measured at the two extremities of the thoracic belt. Once again, the test results indicated similar values at both extremities for the same test and that enable us to use the total belt force value in our analysis of thorax behaviour. Tthe belt force has been computed for all the 13 cadaver tests.

Figure 6 shows the relationship between the number of rib fractures (NRF) and the total belt force.

There are clearly two tests results out of the "cloud" constituted by the other tests. In these two tests, the cadavers sustained a large number of rib fractures, whereas the belt force is not higher than for other tests with a small number of rib fractures. In the average the limit of NRF = 6 is about 10 KN total belt force. However there is a variation of the number of rib fractures of more than 1 to 2 for the same value of belt force.

One cadaver was much younger than all the others, and for that subject a total belt force of 7.5 KN did not induce any rib fractures.

VISCOUS CRITERION AND RIB FRACTURES - V*C is an injury predicting parameter based on a biomechanical analysis of thoracic impacts (5) ; it takes into account the non elastic response of human tissues to impact and is a viscous parameter.

It is defined as $V^*C = \frac{1}{MTT} \cdot \max [V(t) \cdot C(t)]$

In which MTT is the thoracic thickness
V(t) is the time/velocity history
C (t) is the time/deflection history

Figure 7 shows the relationship between viscous criterion values and the number of rib fractures. The viscous criterion values are determined at # point 1, which is the mid-sternum. This location is under the belt and corresponds to the internal measurement point on the Hybrid III.

This figure indicates a linear relationship between the number of rib fractures (nrf) and the V*C values. If we consider the limit of 6 rib fractures the corresponding value of V*C would be 0.77. Three tests with abnormal results (2 with a large number of rib fractures and 1 without rib fracture corresponding to a young subject out of the sample characteristics) were excluded.

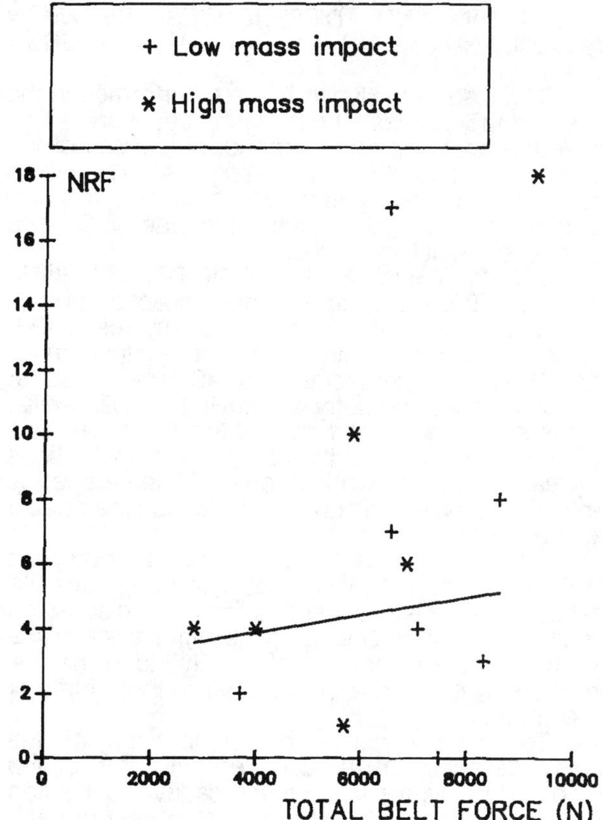

Fig 6: Number of Rib Fractures in relation to the Total Belt Force

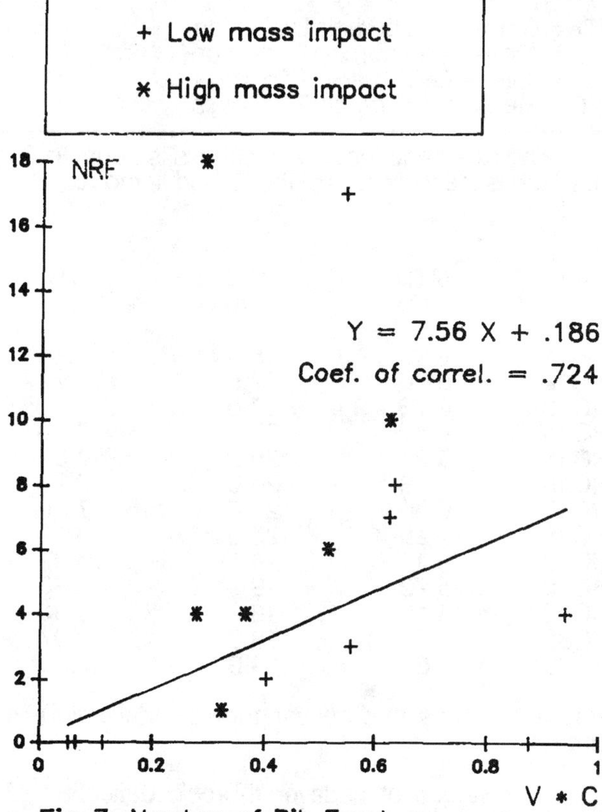

Fig 7: Number of Rib Fractures

SIMILARITIES AND DIFFERENCES BETWEEN HYBRID III AND CADAVER THORAX BEHAVIOURS.

The tests with Hybrid III were performed in the same conditions as those with cadavers. This enabled us to compare the impact responses of the thoraxes of two human surrogates. The main parameters to be compared are the thoracic deflection, the speed of deflection, the V*C, and differences in belt force values.

COMPARISON OF THORACIC DEFLECTIONS - On Figure 8 are plotted the time/deflection curves for a cadaver test (#17) and a dummy test (#22) performed in the same conditions : low mass impactor, impact speed equal to 7.78 m/s. Moreover the cadaver anthropometry is close to 50th percentile, and the injuries sustained are not too extensive. This figure shows clearly that thoracic cadaver deflections are greater than the Hybrid III ones. At some specific points, the cadaver deflection is more than the double that of Hybrid III.

On both models points #1 (mid sternum/belt strap) and point #8 (twelfth rib/belt strap) are the most deflected. It is proposed to focus the discussion on point #1 which is located on the symmetrical plane and above the attachement of the Hybrid III internal deflection transducer, approximately in the middle of the sternum.

To make possible the comparison of the thoracic deflection of the cadavers with that of the Hybrid III dummy, the value of the cadaver deflection is corrected according to the anthropometry of each cadaver as follows :

$$CTD = MTD \cdot \frac{RTT}{MTT}$$

In which :
CTD = Corrected Thoracic Deflection
MTD = Measured Thoracic Deflection
RTT = Reference Thoracic Thickness
MTT = Measured Thoracic Thickness

The reference thoracic thickness is equal to 26 cm, which is the thickness of the Hybrid III thorax.

Test#	MTD	MTT (cm)	CTD
THC11	6.13	18.0	8.85
THC12	6.20	17.5	9.21
THC13	4.26	17.0	6.52
THC14	6.55	22.0	7.74
THC15	5.82	20.0	7.57
THC16	7.08	20.0	9.20
THC17	6.30	21.0	7.80
THC18	7.25	20.0	9.42
THC19	6.11	21.5	7.39
THC20	5.72	19.0	7.83
THC62	4.10	18.0	5.92
THC65	6.49	18.0	9,37
THC69	5.81	19.0	7.95

Table 2 : Values of Cadaver thoracic deflection at point #1

The values of cadaver thoracic deflection of point #1 normalized as indicated above are listed in Table 2.

Comparison of cadaver and Hybrid III thoracic deflections shows that the dummy thorax is stiffer than that of the cadaver : for the same impact energy the cadaver corrected thoracic deflection is increased by 50 % to 100 %, i.e. 3 to 4 cm for impact energy between 200 and 1000 Joules, as shown by Figure 9.

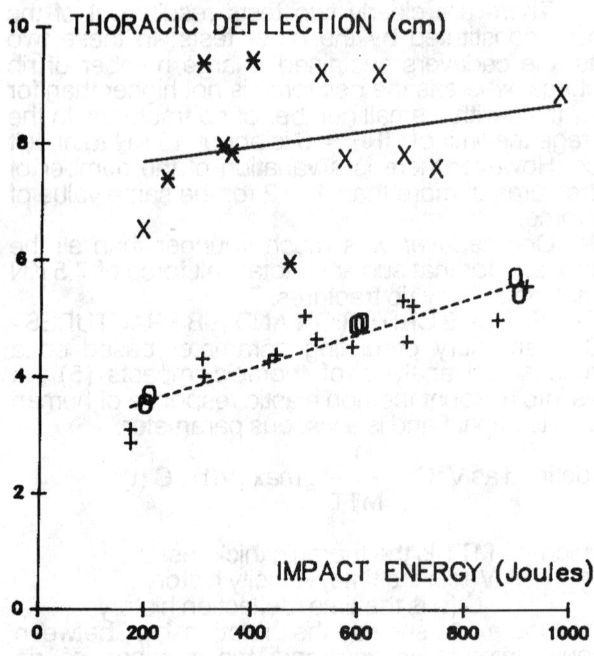

Fig 9: Comparison of thoracic deflection

The sample of cadaver tests can be devided in two groups according to the number of rib fractures : those with 4 or less rib fractures and those with 5 or more rib fractures.

For the first group (NRF ≤ 4) the maximum external corrected thoracic deflection measured at point 1 is below 8 cm whereas it is higher than 7.5. cm for the sub sample of 5 or more rib fractures. This limit of 7.5. to 8 cm on human subjects seems to be the tolerance limit, and the corresponding Hybrid III deflections would be 4 to 4.5 cm (external) and 3 to 3.5 cm (internal).

COMPARISON OF VISCOUS CRITERION VALUES - On all Hybrid III and cadaver tests a V*C value was computed taking into account deflection of point #1, which is derivated to determine the speed of deflection.

The values of V*C versus impact energy are presented in Figure 10. This figure shows that for the same human surrogate, V*C values of high mass impactor tests are below those of low mass. This could be expected as the V*C is directly influenced by the impact speed, and a low mass impactor test corresponds to a higher impact speed for a specific impact kinetic energy.

X Cadaver low mass impact
* Cadaver high mass impact
+ Hybrid III low mass impact
0 Hybrid III high mass impact

Cad : Y = .739 X + 139.3
Coef. of correl. = .707
H III : Y = .572 X + 42.7
Coef. of correl. = .803

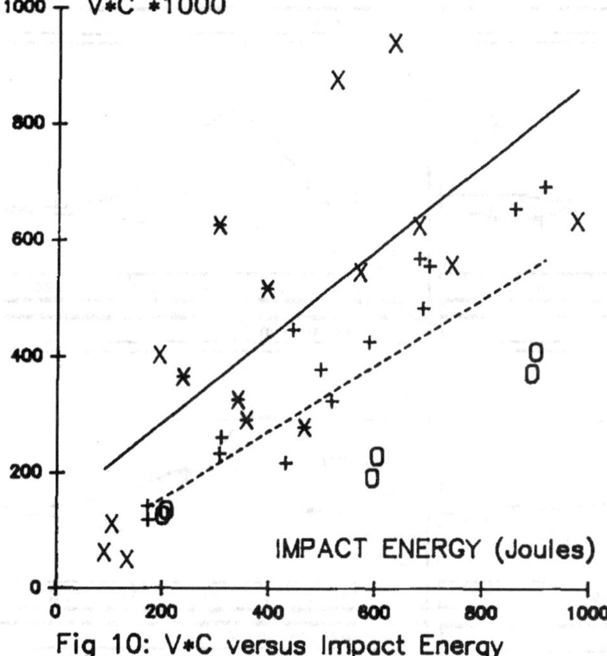

Fig 10: V*C versus Impact Energy

Figure 10 shows a clear relationship between V*C and impact speed for Hybrid III, almost independently from the impactor mass, whereas the results for the cadavers are more widely distributed especially for high values of impact speed. It is also noted that the Hybrid III thorax is more sensitive to V*C variation than was the case with the cadavers.

Comparison of results shown on figures 10 and 11 indicates a good correlation between V*C and impact energy, and between V*C and impact speed for both cadaver and Hybrid III. Always Hybrid III correspond to a lower V*C value than that of cadaver tests. The averaged difference is in the range of 0.15 to 0.3 in the range of the tests performed for this study.

If we consider the limit of 6 rib fractures, the cadaver V*C value is 0.77 and the corresponding Hybrid III values would be 0.62.

X Cadaver low mass impact
* Cadaver high mass impact
+ Hybrid III low mass impact
0 Hybrid III high mass impact

Cad : Y = 75.7 X + 80.2
Coef. of correl. = .687
H III : Y = 80.4 X − 114.5
Coef. of correl. = .900

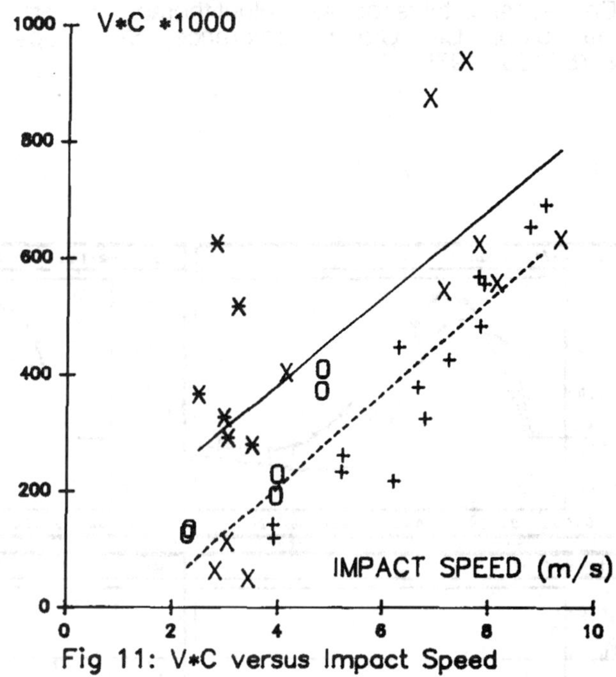

Fig 11: V*C versus Impact Speed

CONCLUSION

This research programme was facilitated a better understanding of the behaviour of the human thorax under belt loading. The test results provide data concerning human tolerance to belt loading : 6 rib fractures correspond to a belt load of about 10 KN or an impact energy of 830 Joules.

Comparisons between the responses of the cadavers and the Hybrid III dummy allow us to establish the biofidelity of the dummy thorax in terms of impact response and help us to determine the transfer functions from the human to the Hybrid III dummy. This analysis is the first step of a more complete programme in which low energy impact cadaver tests and static cadaver and Hybrid III belt tests will be performed in order to compare the responses of volunteers, cadavers, and the Hybrid III in different test conditions.

The V*C parameter value for the mid-sternum is related to the input data as well for the cadaver as for the Hybrid III. However the Hybrid III V*C values are lower than those of cadavers.

ACKNOWLEDGEMENT

The results presented in this paper are obtained from a joint programme between NHTSA, Transport Canada and INRETS. The authors would like to thank MM. Stan BACKAITIS and Dainius DALMOTAS who organized the technical supports for this research programme. A special thanks is given to Michelle RAMET, research director in our laboratory, and to the Lyons Medical University and INRETS persons who provided essential assistance during the preparation, the progress and the analysis of the tests.

REFERENCES

1. NAHUM A.M., SCHNEIDER D.C., KROELL C.K. "Cadaver skeletal response to blunt thoracic impact" 19th Stapp Car Crash Conference, SAE paper N° 751150 - 1975

2. KROELL C.K. "Thoracic responses to blunt frontal loading" SAE Publication P 67, 1976

3. L'ABBE R.J., DAINTY D.A., NEWMAN J.A. "An experimental analysis of thoracic deflection response to belt loading" 7th IRCOBI Conference, 1982

4. BACKAITIS S.H., ST LAURENT A "Chest deflection characteristics of volunteers and Hybrid III" 30th Stapp Car Crash Conference, SAE paper n° 861884 - 1986

5. VIANO D.C., LAU I.V. "Thoracic impact - A viscous tolerance criteria" GM Report N° GMR 5086, June 1985

6. FAYON A., TARRIERE C., WALFISCH G. "Thorax of 3 point belt wearers during a crash" 19th Stapp Car Crash Conference, SAE paper n° 751148 - 1975

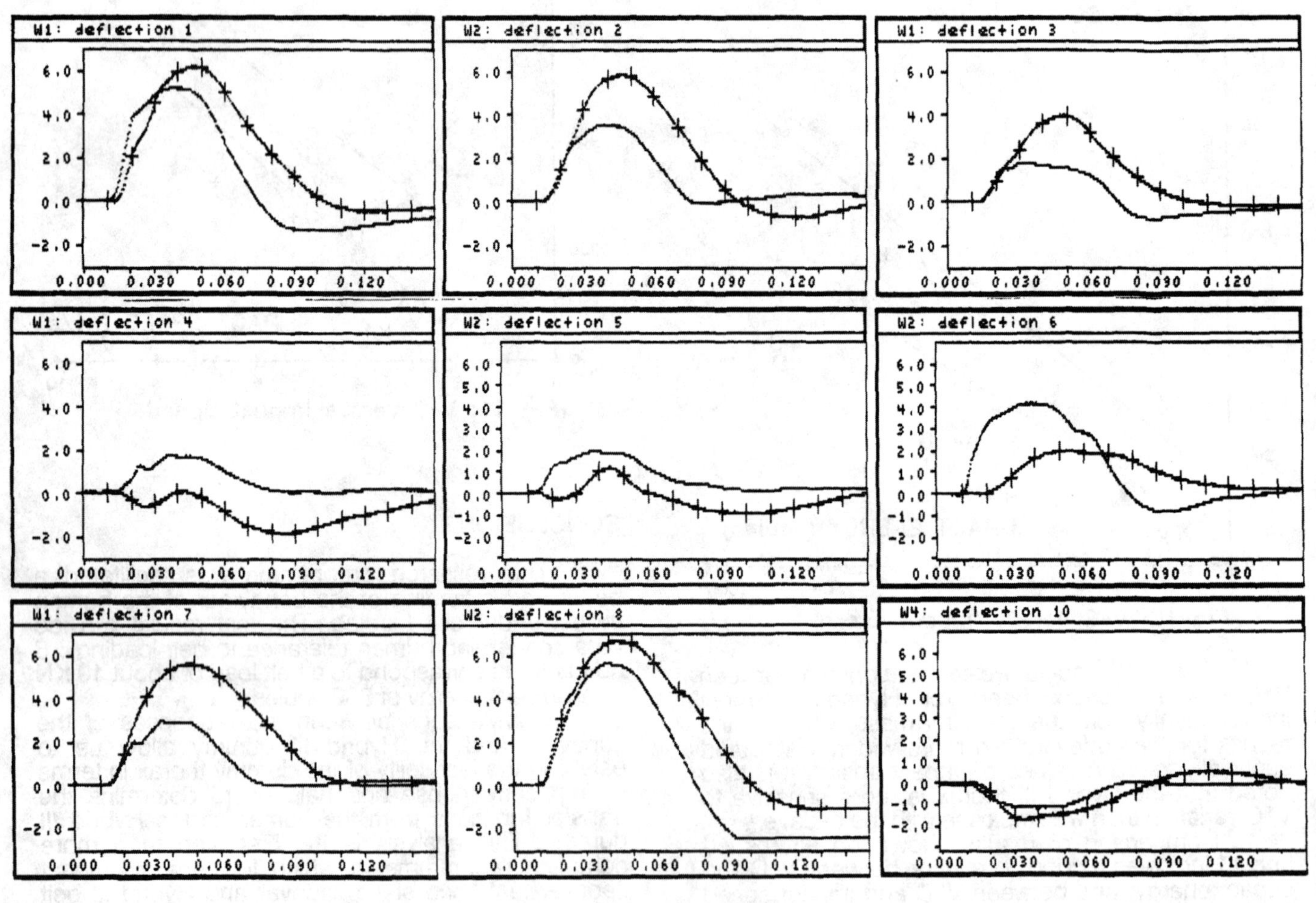

FIGURE 7 : DEFLECTION/TIME HISTORIES OF TESTS 17 AND 22

CHAPTER 3:

FIELD EFFECTIVENESS

670925

A Statistical Analysis of 28,000 Accident Cases with Emphasis on Occupant Restraint Value

N. I. Bohlin
Passenger Car Engineering Dept., AB Volvo

Abstract

The value of the three-point safety belt has been evaluated by a statistical analysis of more than 28,000 accident cases, which concerned mainly two cars only and in which 37,511 unbelted and belted front-seat occupants were involved. The safety harness concerned is the Volvo three-point combined lap and upper torso harness with a so called slip-joint. The average injury-reducing effect of the harness proved to vary between 40 and 90%, depending on the speed at which the accident occurred or the type of injury. Unbelted occupants sustained fatal injuries throughout the whole speed scale, whereas none of the belted occupants was fatally injured at accident speeds below 60 mph. Slight injuries only, mostly single rib cracks, bruises, etc., caused by the safety belt were reported in some cases. The three-point belt proved to be fully effective against ejection out of the car. Almost all cars involved were equipped with safety belts, of which, however, only 26% on an average were used. The frequency of use increased with the age of the occupants.

THE VALUE AND EFFECTIVENESS of occupant restraints, especially the upper torso restraint, in car accidents have been discussed and investigated in different ways. The investigations carried out have mainly concerned experimental studies in dynamic sled runs or full-scale car barrier crashes where anthropomorphic dummies have been used. In very few cases a human cadaver has been substituted for the dummy. Further mathematical models and computer programs have been the basis for the evaluation of the restraint value. Most of the investigations, which are based on data from real accidents, are considered to have one or more of the following deficiencies of greatest importance for the conclusions derived:

1. Insufficient and nonrepresentative number of cases.
2. The "car" involved has meant the smallest European vehicle, as well as the big American one — an uncontrolled parameter.
3. Insufficient consideration to the influence of the various systems, makes, and installations of belts or harnesses involved.

The primary purpose of this statistical analysis is to evaluate the effectiveness of the three-point upper torso shoulder restraint without the shortcomings mentioned above. It is considered to be a matter of course that findings and experiences derived from a representative and extensive number of real accident cases are superior to and more reliable than those of any experimental or theoretical study. To the best of the author's knowledge, the comprehensive investigation made is unique in regard to focusing attention on the combination of a specific car and a particular type of belt (Fig. 1). The cars involved throughout the collected material are of the same make and consist in practice of just two models (the Volvo P-12 and P-11) with about equal size and weight. The safety belt is the Volvo three-point safety harness. This harness is

Fig. 1 — The two car-models concerned in the analysis

classified as type 2 in SAE J4c and Federal Register 11528 and has a combined lap and upper torso strap with a so-called slip-joint in conjunction with the buckle, which is attachable to the transmission tunnel without an intermediate short strap (Fig. 1). The fact that Volvo cars, which are not subjected to model or body changes too often, have had the type of belt mentioned as standard on the front seats since 1959, created an essential ground of the investigation.

Method of Collecting Data

The Volvo "five-year guarantee" in Scandinavia made the collecting of data possible. This guarantee means briefly that every Volvo car is guaranteed by the company against damage in case of an accident during the first 5 years after delivery. The repair cost exceeding $80[1] is free for the car owner and when the car is written off — normally when the repair cost is calculated to exceed $800-1000 — it is replaced by the company with a new vehicle. Consequently, all accidents in which Volvo cars are involved and the repair cost is higher than $80 have to be reported to the company. The investigation was limited to Sweden and during the data collecting period of 1 year from March 28, 1965 to March 28, 1966 the average number of Volvo cars under obligation to report was 297,000 in Sweden. The average total number of Volvo cars on Swedish roads during this period was about 430,000.

A questionnaire or report form suitable for computer programming was to be filled in by the car owner or the driver after an accident. A translated sample of the form is shown in Fig. 2. The form comprises a total of 80 questions dealing with road and weather conditions, type of accident, speed, the general and also detailed behavior of the car, the use and type of safety belt, injuries sustained, and interior objects producing injury. Most of the questions provided data which were essential for this analysis. The return of the answers requested was not compulsory — but emphasized by the department involved — and had nothing to do with the ordinary regulations of the insurance or the financial side of the case. Selected people at the Volvo subdealers, who collected the forms within their dealer districts, were trained to give advice and instructions about the filling in of the form. The forms were then collected by the company, coded, checked, and finally transferred into punch cards. Severe accidents or other special cases were subjected to detailed investigation, contacts with the police, hospitals, etc.

Basic Collected Data

As a concentrated survey of more than 1 million facts obtained, some basic figures are given:

37,761 accidents (equal to the guarantee cases) were reported. In 28,780 of the cases (76%) the requested forms were satisfactorily filled in and recorded. These cases involved a total of 42,813 car occupants, consisting of 28,780 drivers, 8731 front-seat passengers, and 5302 rear-seat occupants. In 1803 cases (6.3%) 2445 occupants (5.7%) sustained injuries, from slight to fatal.

Models of cars involved were substantially just two: Volvo Model P-12 in 16,076 cases and Model P-11 in 12,610 cases: only 94 cases concerned the Volvo sports car P-1800.

[1] All monetary figures are in United States dollars.

AB VOLVO

Box 382, Göteborg 1

INFORMATION
ABOUT TRAFFIC ACCIDENT
Date

Record (1-5) 9 1845

To be filled in by Volvo 51 06 11

Note
The purpose of the information is to create a basis for the continuous development work of increasing interior safety in passenger cars to prevent injuries.
The information does not concern the insurance company or the investigation made by the police.
The information is completely confidential.
Information from a minor accident is as valuable as that from a severe one.

Owner or information source
Name N.N — Address Ljungby — Phone no 12490

The vehicle concerned
Type PV 544 Sport — Year model 66 — Chassis no 437026 — Colour Red — Code 02 6-7

1 [X] L.H. drive — 2 [ ] R.H. drive 8

Information about the accident
Where you in the car when the accident occurred? 1 [X] Yes 2 [ ] No — If yes 1 [ ] Driver 2 [ ] Passenger 9-10

Date of accident 9.12.1965 — Place of accident JÖNKÖPING

The accident occurred: 1 [ ] Straight road speed limit — 2 [ ] Straight road no speed limit — 3 [ ] Winding road speed limit — 4 [X] Winding road no speed limit 11

1 [X] In daylight — 2 [ ] In darkness — 3 [ ] At dusk or dawn 12

Weather conditions. Mark x in only one square: 1 [ ] Rain — 2 [X] Snow — 3 [ ] Fog — 4 [ ] Fine weather 13

Road surface: 1 [ ] Gravel — 2 [ ] Oil gravel — 3 [X] Asphalt — 4 [ ] Concrete — 5 [ ] Pavé — 6 [ ] Wood - for example a wood - bridge 14

Road conditions: 1 [ ] Not slippery — 2 [X] Slippery because of ice or snow — 3 [ ] Slippery because of rain — 4 [ ] Slippery because of clay 15

The accident was preceded by: 1 [ ] Braking with skidding — 2 [ ] Braking without skidding — 3 [X] Skidding without braking — 4 [ ] Neither skidding nor braking 16

Main type of accident: 1 [X] Head-on collision — 1 [ ] Side impact — 1 [ ] Rear impact 17-19
1 [ ] Roll over — 2 [ ] Spin — 1 [ ] Off the road — 2 [ ] Other type — Specify

20-21

What object did you run into? Mark x in only one square: 1 [ ] Passenger car colour — 2 [X] Truck Bus — 3 [ ] Off-road obstacle — 4 [ ] Tree, railing etc.
5 [ ] Rock face large stone — 6 [ ] Tractor — 7 [ ] Motorcycle scooter — 8 [ ] Other object. Specify 22

Your estimated speed at the time of the accident? Mark x in only one square: 1 [ ] Stationary — 2 [ ] Under 10 m.p.h. — 3 [ ] 10-25 m.p.h. — 4 [X] About 30 m.p.h.
5 [ ] About 40 m.p.h. — 6 [ ] 45-55 m.p.h. — 7 [ ] About 60 m.p.h. — 8 [ ] Over 60 m.p.h. 23

Position of car after the accident: 1 [ ] Lying on its roof — 2 [ ] Lying on its side — 3 [X] Standing on its wheels 24

Did your car catch fire? 1 [ ] Yes — 2 [X] No 25

Did any seat loosen? 1 [ ] Both front seats — 2 [ ] Left front seat — 3 [ ] Right front seat — 4 [X] No 26

Did either front door open? 1 [ ] Both doors — 2 [ ] Left door — 3 [ ] Right door — 4 [X] No 27

Did either rear door open? 1 [ ] Both doors — 2 [ ] Left door — 3 [ ] Right door — 4 [ ] No 28

Tyres: 1 [ ] Standard, fitted at factory — 2 [ ] Winter tyres on front and/or rear wheels — 3 [ ] Studded tyres on front and/or rear wheels — 4 [X] Other spec. tyres — Type of spec. tyres Pirelli 29

Deformed body parts
Mark on two figures at the right any essentially deformed parts. As an example see markings on figure at left.

General about safety belts (Mark x in only one square)
If safety belts were used at the time of the accident, did they: 1 [ ] Save life — 2 [X] Prevent or minimize injuries — 3 [ ] Prove useless — 4 [ ] Cause injuries 30

If safety belts were not used at the time of the accident, could they, if used: 1 [ ] Have saved life — 2 [ ] Have prevented or minimized injuries — 3 [ ] Have been useless — 4 [ ] Have caused injury 31

If you drive cars fitted with safety belts, do you use them: 1 [ ] Always — 2 [X] Only on long trips — 3 [ ] Rarely or never 32

If "rarely or never" please state under heading "other valuable information" overleaf

Fig. 2 — Sample of questionnaire - Part I

P.T.O

List of occupants	If you cannot or prefer not to provide this information mark with a cross instead of a name but mention the approximate age. The letters refer to position in car. See sketch below.					
	Name	Age	Phone no	Address	Code	
A	N.N	32	384...	xx	7	33
B	G.K	28	289...	xx	7	34
C						35
D						36
E	K.P	38		xx	7	37

Type of safety belts, if fitted, use by passengers

	A	B	C	D	E
Volvo belt - 3-point (type 2) (lap/diagonal) (Mark with x here) →	38 1 X	44 1 X	50 1	56 1	
Safety belt (diagonal type)	2	2	2	2	
Other type? Specify	3	3	3	3	
Safety belt not fitted	4	4	4	4	
Was the belt worn at time of accident? (x = yes)	39 1 X	45 1	51 1	57 1	
Was the belt difficult to release after the accident? (x = yes)	40 1	46 1	52 1	58 1	
Did the belt cause any injuries? (x = yes)	41 1	47 1	53 1	59 1	

If injuries occurred due to impact against the interior of the car, mark which of the objects below caused the injuries

64 [X] Windscreen	73 [ ] Window regulator or door handle
65 [ ] Steering wheel or column	74 [ ] Rear view mirror
66 [ ] Dashboard upper part	75 [ ] Front seat rear upper part
67 [ ] Dashboard lower part	76 [ ] Front seat rear lower part
68 [ ] Roof over windscreen	77 [ ] Windscreen side pillars
69 [ ] Ignition key	78 [ ] Door pillars
70 [ ] Roof	79 [ ] Glass splinters
71 [ ] Armrest	
72 [X] Other objects. Specify **Camera**	

If the belt caused injuries, specify

Injuries (Mark the degree of injury with x in only one square and group per person)

	A	B	C	D	E
Fatal injuries	42 1	48 1 X	54 1	60 1	62 1
Severe injuries (requiring hospital care)	2	2	2	2	2
Minor injuries	3 X	3	3	3	3
No injuries	4	4	4	4	4 X
Mark with x an occupant thrown out of the car or occupant sliding out of the belt completely or almost completely	43 1	49 1	55 1	61 1	63 1

Type of injuries (reference to hospital, doctor etc.)

Dr. Nilsson, The hospital of Jönköping

A: Cut on head

B: Damaged kidney, fract. pelvis

B died in hospital

Other valuable information — Information, viewpoints, suggestions you consider to be of value — Code 9 80

Volvo notes — $ 1950

Fig. 2 — Sample of questionnaire - Part II

The type of safety belt concerned in the belt cases was the three-point safety harness mentioned to an extent of 8992 in the front seats. In 574 cases only the single diagonal two-point upper torso belt was used. In three cases other types of belt were involved.

The main types of accident found are listed in Table 1. The main accident types only are given. It must, however, be understood that in most cases there was a complex combination of two or more of the cited types.

The statistical analysis, carried out in cooperation with experts in the field outside the company, was primarily concentrated to the injuries sustained by the front-seat occupants and to the injury-reducing effect of the safety belt. Therefore, all the statistical figures — unless specified otherwise — are based on the numbers of front-seat occupant (37,511) and not on the numbers of recorded accident cases (28,780). One accident case often means two occupant cases.

Table 1 – Types of Accidents

	Number of cases	%
Frontal impact	10,284	35.7
Side impact	9,641	33.5
Off-road	4,958	17.2
Roll-over	1,406	4.9
Rear-end impact and other types	2,491	8.7

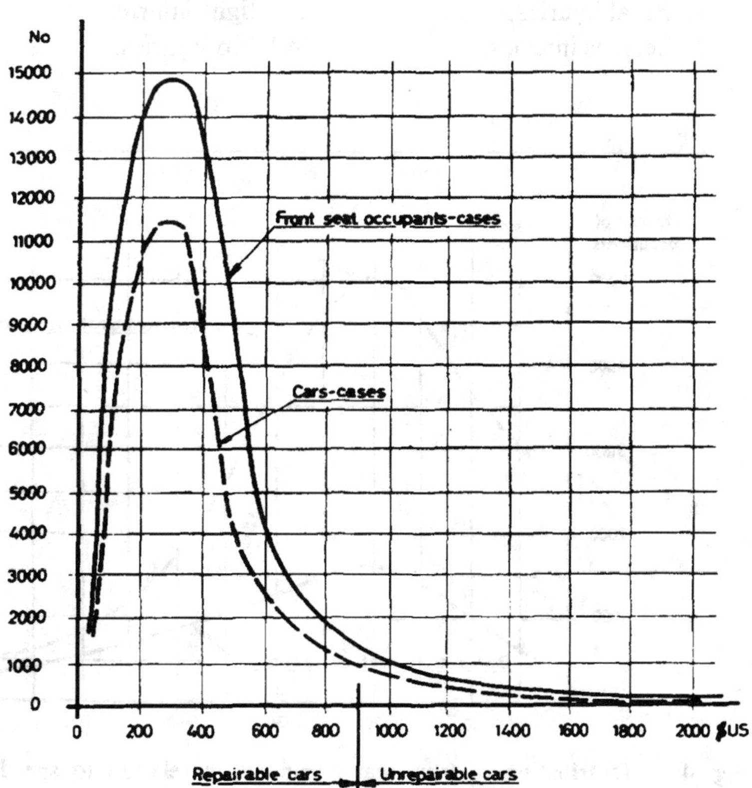

Fig. 3 — Number of cases in relation to repair cost (actual numbers)

Degree of Severity of Accident — This is considered to be a factor of importance for an adequate evaluation. Adopting the rule of experience that the repair cost of the damaged car as well as the speed at the time of the accident in general have certain correlations with the severity of the accident; these two factors are used as a kind of unit of measure for the accident severity.

Accidents with repair cost lower than $80 are not included. That means that all minor accidents are deleted. The distribution of the number of (occupant) cases along the repair cost scale is given in Fig. 3. The quantity representing a repair cost of $200-400 is most frequent. The recorded costs are mainly true costs, but calculated for values above $800-1000 because the damaged vehicle is then considered unrepairable and substituted by a new car.

The repair cost is not recorded in a number of cases for different reasons, delayed information, etc. Some investigations concerning the unrecorded repair cost cases have clearly indicated, however, that those cases usually concern badly damaged cars with high repair costs. Therefore, it is considered that the curve in Fig. 3 should be somewhat higher in the higher cost region to be representative.

Fig. 4 shows the distribution of front-seat occupants (belted and unbelted) in relation to speed at the moment of the accident. Cases within the speed range of 15-30 mph are significantly dominant in both groups. With due respect for the fact that people, for understandable reasons, are normally disposed to report a lower speed than the true one in an accident, it can be assumed that the curve in question should be transferred to rather higher speeds in general.

Injuries Sustained — The occupants' injuries were marked on the form by the car owner (driver). The injuries are divided into three (four) groups:

1. Fatal injuries.
2. Serious injuries.
3. Slight injuries.
(4.) No injuries.

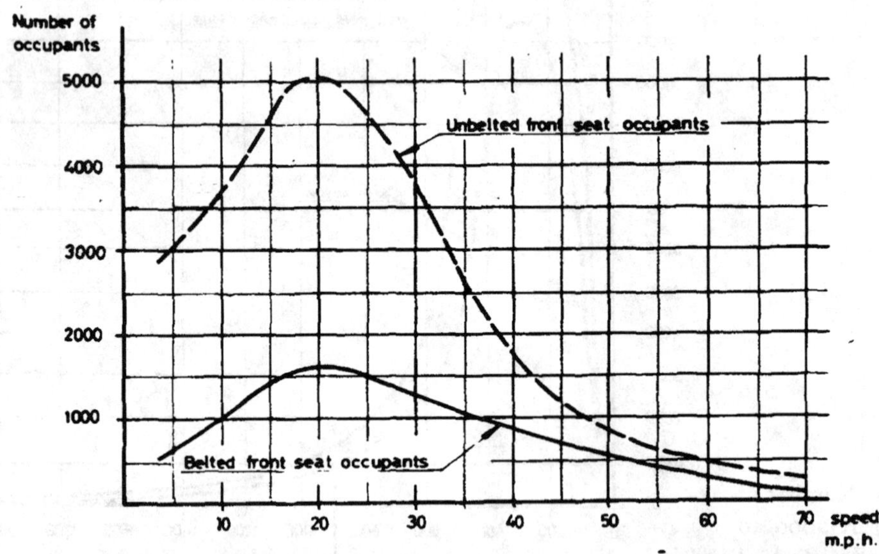

Fig. 4 — Distribution of front-seat occupants in relation to speed (actual numbers in the investigation)

The accuracy and reliability of this manner of data collecting depend on the reporter's judgment and understanding. In the fatal and no injuries cases, the criterium is undoubtedly clear. The "serious injuries" are defined as needing treatment in hospital and it was found by checking details of injuries reported, especially serious ones, that they were mostly obtained from doctors.

To avoid the influence of possible errors in the two middle groups of the injury rating, all main conclusions pertaining to the protective effect of the safety belt are referred to two (three) degrees only:

1. Fatal injuries.
2. Slight to serious injuries (nonfatal).
(3.) No injuries.

Some figures and tables, however, in which data are more extensively given for the convenience of the reader, comprise the four degrees of injury collected.

In order to provide the possibility of counteracting the error risk mentioned above and to obtain a more reliable and uniform graduation of the resulting injuries, the following graduation system was adopted. In nearly all cases the specific injuries were reported in great detail and described in addition to the injury rating on the form.

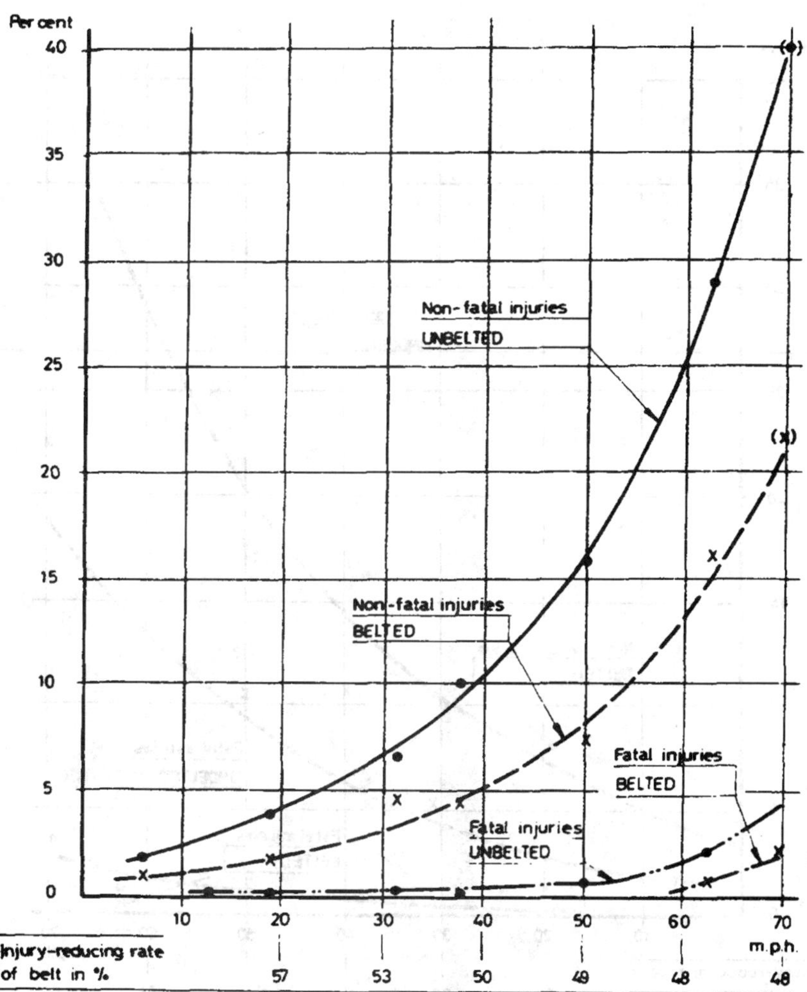

Fig. 5 — Frequency of injuries sustained by drivers in relation to accident speed

Each form covering an injured occupant was carefully studied and the cited injuries were then manually recorded, listed, and referred to as serious or slight in cooperation with a physician. Thus, graduation of injury is used in some parts of the analysis, for example, frequency of main injuries to the occupants and distribution of injuries on the body.

All statistical figures are given in numbers per 10,000 unless otherwise specified. (Note: The cars involved have left-hand steering and the accidents occurred in left-hand traffic.)

Table 2 shows the number of injuries sustained by the drivers and front-seat passengers. It is both misleading and inadequate to derive more than preliminary conclusions about the value of the safety belt from these average survey figures. The frequency of injury, as well as the degree of its severity, must be related to the speed at which the accident occurred, which could be considered as a relevant measure of the severity of the accident.

The frequency of injuries related to accident speed is graphically shown in Figs. 5 and 6. For reasons previously mentioned, the injuries here are divided into two classes only: fatal and nonfatal equaling all injuries from slight to very serious ones. The

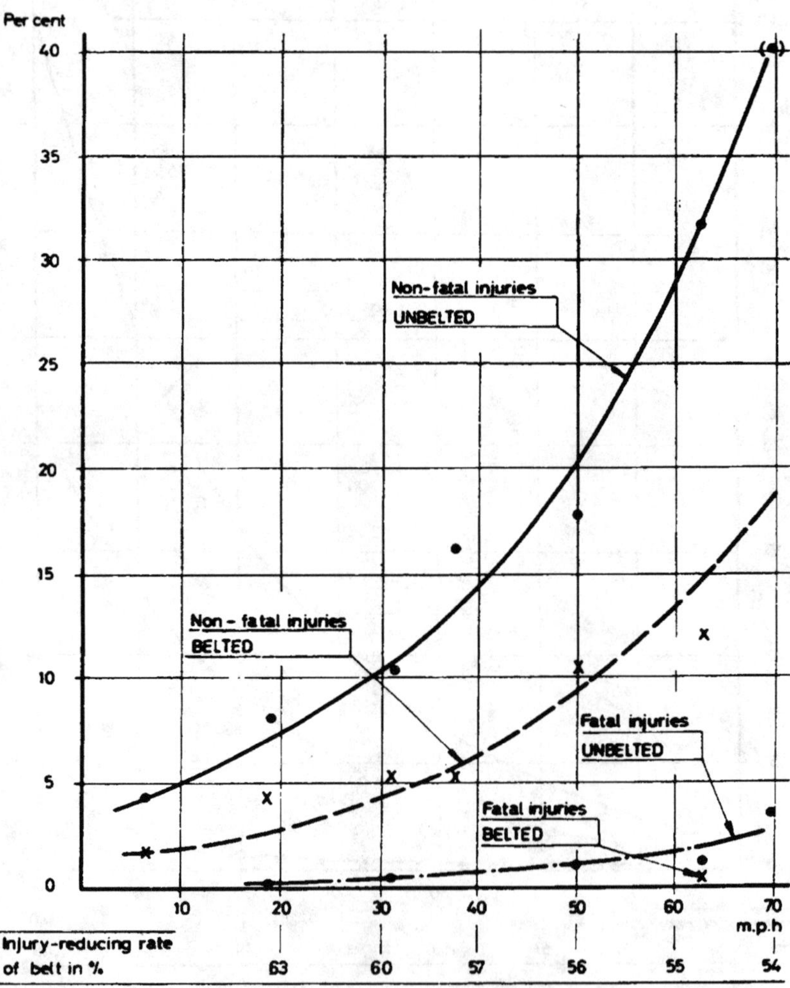

Fig. 6 — Frequency of injuries sustained by front-seat passengers in relation to accident speed

number of injuries increases overall rapidly at increasing speed. The frequency at 50 mph is about 2.5 times the frequency at 30 mph for the driver (about twice as high for the passenger). The unbelted front-seat passenger sustained injuries more often than the unbelted driver, as much as 50% more at 30 mph, about 22% more at 60 mph.

The fatal injuries occurred in the belted cases at accident speeds exceeding 60 mph. The fatalities in the unbelted cases were spread over the whole speed scale, starting as low as 12 mph.

The injury reducing rate of the belt is calculated at various speeds (Figs. 5 and 6). With regard to nonfatal injuries, the reducing rate in per cent of the driver's belt varies from about 57% at lower speeds to 48% at higher speeds. The front-seat passenger's belt shows somewhat higher rates (63 to 55%, respectively). Considering the fatal injuries, the bold conclusion can be derived that the safety belt offers "full" protection

Table 2 — Number of Injuries Sustained, per 10,000
(Numbers in parentheses show actual number of cases)

	Fatal	Serious	Slight
Drivers			
Unbelted	17.2 (37)	123 (263)	388 (835)
Belted	2.9 (2)	74 (51)	255 (175)
Front-Seat Passengers			
Unbelted	18.6 (12)	249 (160)	682 (439)
Belted	3.7 (1)	82 (22)	404 (109)

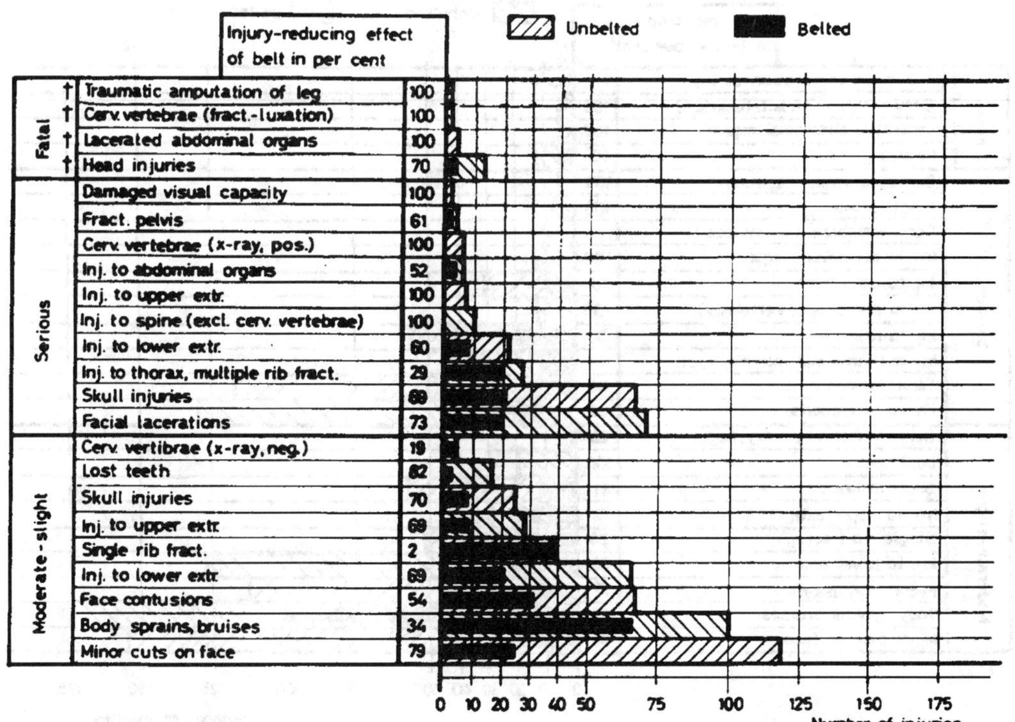

Fig. 7 — Frequency of injuries to driver (cases per 10,000)

against fatal injury up to accident speeds of about 60 mph.

The types of frequent injuries sustained by the occupants were derived and analyzed from each report form, as previously mentioned. A graphical survey, which tells the frequency of specific injuries sustained by the unbelted driver as well as his belted colleague, is presented in Fig. 7. Corresponding figures for the front-seat passenger are in Fig. 8. Again, the unbelted front-seat passenger is substantially more subject to injuries in relation to the driver. The high frequency — up to twice as high — for skull, facial, and thorax injuries should be particularly noted. The increased frequency of leg injuries is proved too. No remarkable corresponding difference between the belted driver and the belted passenger can be pointed out. The unbelted driver's lower injury rate related to the passenger's might be referred to the steering wheel, which offers some "support" at lower speeds. Again, the value of the safety belt in decreasing the injury rate is high. The reduction per cent with respect to the specified injuries is included in the survey as another adequate value rating of the safety belt. It is to be especially noted that the reduction percentage is particularly high in the case of fatal and serious injuries. Further, the number of head and thorax injuries are reduced by 60-85% when the belt is used. "Single rib fracture" injury has, however, a comparatively low reduction per cent (28% for front-seat passenger). This will be discussed further in the following paragraphs. The injury reduction percentage of the passenger's belt is generally much higher than that of the driver's belt. This advantage for the passenger can certainly be referred to the greater clearance between him and the parts of the car in front of him.

In quite a number of the belt cases, it was recorded that the belt was worn too

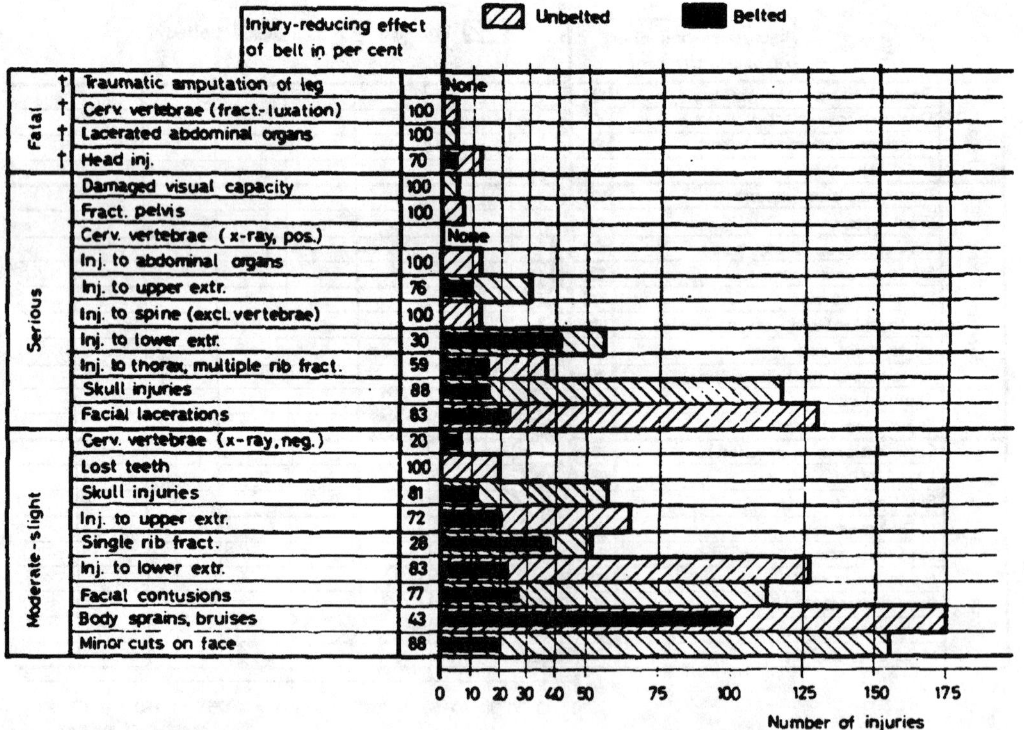

Fig. 8 — Frequency of injuries to front-seat passenger (cases per 10,000)

Table 3 — Injuries Caused by Seat Belts
(Actual number of cases)

	Driver	Front Seat Passenger
Slight single rib fracture (no hospital treatment)	10[a]	10
Multiple rib fracture	5[a]	1
Pain in chest and bruises	6	5
Slight to moderate fracture to chest bone	3[a]	1
Bruises and soreness in lap area	3	2
Bruises and soreness in shoulder area	1	2
Fracture of collar bone	3	2
Slight superficial injury to neck due to webbing	1	1
Soreness in back, not defined	2	–
Slight injury, not defined	–	1
Total	34	25

[a] It was reported in two cases that the belt was worn too loosely.

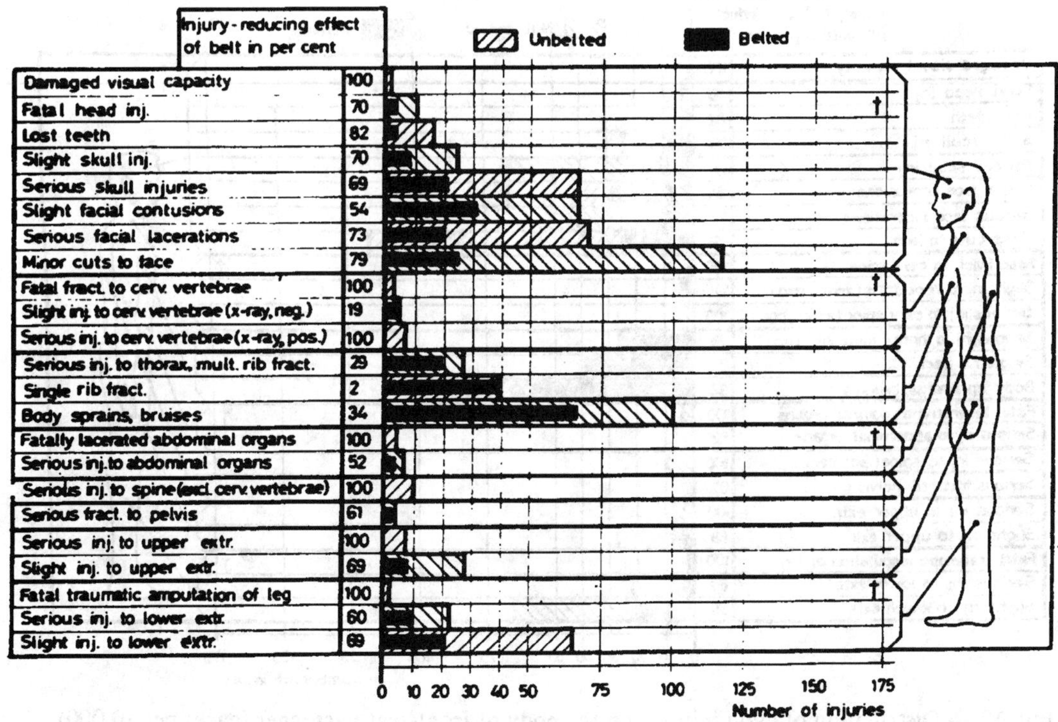

Fig. 9 — Distribution of frequent main injuries on the driver's body (cases per 10,000)

loosely, which has contributed to decreased effectiveness of the belt. Injuries sustained when belts were used, therefore, could be referred partly to such factors. There are good grounds for assuming that the injury reduction value of the belt would have become still higher if the belts had been worn snugly or better adjusted throughout. As a matter of fact, in some belt cases the belt was apparently so poorly adjusted that it would be more accurate to consider them as unbelted cases.

Injuries caused by the safety belt were reported in 34 cases for drivers and in 25 cases for the front-seat passenger. The types and frequencies of injuries caused by belts are listed in Table 3.

"Multiple rib fracture" and "fracture to chest bone" injuries occurred significantly more often to the belted driver than to the belted passenger. In view of the additional information in some of these cases that the belt was not properly tightened, it could be assumed that some of those injuries were caused not by the pressure of the belt, but by impact against the steering wheel due to decreased effective restraint of the belt. Slight single rib fracture was found to be the most frequent injury to both the driver and his passenger caused by the belts. These injuries did not, however, need hospitalization. Collar bone fracture occurred on the shoulder strap side. No cervical vertebrae injuries caused by the belt were reported. Most injuries reported to be caused by the belt occurred in conjunction with seriously injured unbelted occupants and in accidents at high speeds.

The distribution of injuries on the occupant's body is illustrated in Fig. 9 (driver) and Fig. 10 (front-seat passenger). Skull and head injuries dominate, but were effectively counteracted by the safety belt.

Ejection of front-seat occupants out of the car happened in 159 unbelted cases, and

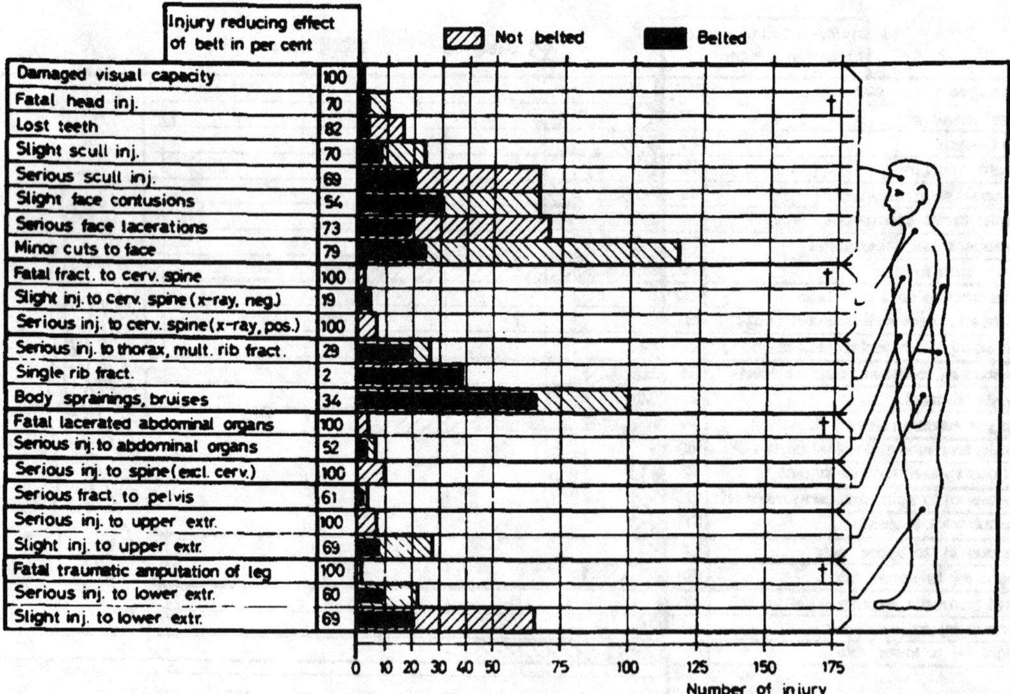

Fig. 10 — Distribution of main injuries on the body of front-seat passenger (cases per 10,000)

possibly in one single belt case. Ejection of unbelted occupants was reported in conjunction with fatal injuries in 10 cases out of a total of 49 fatal cases and serious injuries in 59 cases out of a total of 423 cases. In most cases, it was not possible to obtain positive confirmation that the ejection had caused the fatal and serious injuries. In 3 of the 10 fatal cases, however, it could be determined for good reasons that ejection was the primary cause. Even with this cautious derivation of cause, *the risk of fatal injury increases more than 10 times in the case of ejection* compared to nonejection.

In one single belt case, partial ejection might have occurred. Unfortunately, the investigation in question did not definitely clarify what really happened. The driver, with his small children 3 and 7 years old in the rear seat, could not positively recognize how he was positioned or when and how he released his belt at the moment after the accident, which was a severe roll-over at 50 mph. He thought he had been partly outside the car and reported accordingly on the form. The man sustained moderate to serious injuries to pelvis and leg and also fracture of the arm.

Analysis of Cases with Two Front-Seat Occupants

To analyze the injury-reducing effect in more detail and with other aspects the investigation material was separated into the following four groups (passenger means front-seat passenger and the belt is the three-point shoulder harness):

1. Unbelted driver in relation to unbelted passenger.
2. Belted driver in relation to belted passenger.
3. Unbelted driver in relation to belted passenger.
4. Belted driver in relation to unbelted passenger.

The analysis of the combinations cited concerns all cases with occupants in both the front seats at least.

The distribution of the actual number of cases among the groups mentioned was:

Group 1 — 6106 cases (67.6%). Group 3 — 275 cases (3.1%).
Group 2 — 2413 cases (26.6%). Group 4 — 245 cases (2.7%).
Total 9039 cases

It is to be noted that the first two groups (both driver and front-seat passenger are equally "equipped" — either belted or unbelted) represent the most substantial part of the material. The explanation might be that the driver's (or the passenger's?) "opinion" or habit to be belted or unbelted strongly influences his front-seat passenger.

The distribution of injuries in Groups 1 and 2 is shown in Table 4. It is concluded that the passenger's seat, in general, is more exposed to injury-producing conditions than the driver's seat. In adding the numbers of all injuries from fatal to slight, one gets an average relation factor for Group 1 of 1.4 (555/395) and for Group 2 of 1.44.

From this relation factor one might be inclined to conclude that the belt in that respect does not have any favorable effect. The influence of speed, however, must be considered. Limited studies pertaining to the influence of speed indicate that the "belted passenger's seat" from an injury-producing point of view is significantly more exposed than the "belted driver's seat" at low and moderate speeds, but less exposed at higher speeds (Fig. 14). One explanation for this could be that the driver has some advantage from the steering wheel in moderate-speed accidents, but in high-speed impacts, when the (front of the) car generally is badly damaged and the steering wheel is often displaced rearwards, the clearance to vehicle structure is insufficient to give the driver's belt full effectiveness.

The 15 drivers killed in relation to (only) 10 passengers killed in Group 1 might seem to be somewhat controversial in view of the "relation factor" mentioned above. However, these 15 fatal cases included 4 cases of door opening with subsequent ejection (1 case only for passenger's seat) and 1 case where the driver was just moderately injured in the accident (one of his legs was broken) but due to an uncommon sequel he died a few days later in hospital.

To find the average injury-reducing effect of the shoulder belt (in addition to the basic figures in the main report) Groups 1 and 2 were compared, and the effect mentioned was found to be for

Driver's seat - 54%

Passenger's seat - 52%

Table 5 shows the distribution of injuries in Groups 3 and 4. In these groups the total number of cases as well as the injury cases is considered too small to permit more than general conclusions. The figures on an average clearly indicate, however, the positive effect of the belt.

To provide further information about the accident history, each of the 62 injury cases in Group 3 and all of the 43 cases in Group 4 were studied in detail from the

Table 4—Distribution of Injuries of Front-Seat Occupants, Groups 1 and 2 (Actual numbers)

Group	Classification of Injury	No. Injured	Total Injured
1	Unbelted Driver		
	Fatally injured	15	395
	Seriously injured	89	
	Slightly injured	291	
	Uninjured	5711	
	Unbelted Passenger		
	Fatally injured	10	555
	Seriously injured	143	
	Slightly injured	402	
	Uninjured	5551	
2	Belted Driver		
	Fatally injured	–	72
	Seriously injured	16	
	Slightly injured	56	
	Uninjured	2341	
	Belted Passenger		
	Fatally injured	1	104
	Seriously injured	17	
	Slightly injured	86	
	Uninjured	2309	

accident forms. The following was revealed.

Group 3 — In the only fatal cases recorded, the unbelted driver sustained fatal injuries to the chest and head when thrown against the steering wheel and the header above the windshield in a head-on collision at a speed of about 50 mph. The belted passenger sustained no injuries. One unbelted passenger in the rear seat was seriously injured.

The Group 3 material comprises 14 further cases of frontal impact in which the unbelted drivers were seriously injured in 6 cases, whereas the belted passengers were seriously injured in only 2 cases.

The driver's door opened in 7 cases in which 2 unbelted drivers were ejected out of the vehicle and thereby seriously injured. The passenger's door opened in five cases, but there was no ejection of the belted passenger.

Side impacts and other types of untypical frontal accidents happened in a total of 16 cases. Rear-end impact occurred in two cases.

Group 4 — Details from the two fatal accidents recorded in Group 4 can be reported in one case. There was a frontal impact at about 30 mph where the belted driver sustained a slight cut (broken glass) on his head, whereas the unbelted front-seat passenger was fatally injured (damaged kidney, broken pelvis). One occupant centrally located in the rear seat was uninjured.

The accident types included 15 frontal impacts, 4 side impacts (of which the driver's side was hit in 3 cases), and 3 rear-end impacts.

In 11 out of the 15 frontal impacts the belted driver was less injured than the unbelted passenger. In three cases the reported degree was equal (slight for both the occupants) and in one case the driver sustained slight injury, whereas the passenger was uninjured. That single case concerned an accident where the vehicle was hit in the front end, about 30 deg to the left, at a speed of about 25 mph. The driver sustained slight bruises on his nose and his knee.

Door opening occurred in four cases (in side impacts and roll-overs) on the passenger's side causing the ejection of one unbelted passenger, who was seriously injured on his head and his back. Driver's door opened in two cases, but there was no ejection of the belted driver.

Unbelted Rear-Seat Passenger in Relation to Shoulder-Belted Front-Seat Occupant in Frontal Impacts

This part of the analysis concerns the possible chances that an unbelted rear-seat passenger in a frontal impact is thrown forward against the shoulder-belted occupant in front of him, thereby causing injuries (head against head) to either of the "colliding" occupants.

In the 28,780 accident cases there were 5039 rear-seat passengers involved, of whom 2178 were involved in frontal impact accidents. Only three of those rear-seat passengers were belted and none were injured. The type of belt was the single diagonal shoulder belt. The unbelted 2175 rear-seat occupants were positioned in the car as follows:

Rear right (behind the front-seat passenger)	1043
Rear left (behind the driver)	747
Rear middle	385

Injuries sustained by the rear-seat occupants (2175) in frontal impacts are shown in Table 6.

The number of cases to be further studied to answer the question mentioned above is thus: 52 + 35 + 23 = 110, where injuries of all degrees were sustained, of which 43 were fatal or serious injuries.

The criterium of the cases to be finally concerned was settled to be the combination of the following four conditions:

1. Frontal or mainly frontal impact.
2. Either driver or front-seat passenger, or both, belted.

Table 5 – Distribution of Injuries of Front-Seat Occupants, Groups 3 and 4 (Note: Numbers of injuries are too small to justify a significant classification)

Group	Classification of Injury	No. Injured	Group	Classification of Injury	No. Injured
3	**Unbelted Driver**			**Belted Driver**	
	Injured, total	34	4	Note: See Group 2, Table 4	
	Uninjured	241			
	Belted Passenger			Injured, total	17
				Uninjured	228
	Injured, total	28		**Unbelted Passenger**	
	Uninjured	247		Injured, total	26
				Uninjured	219

Table 6 – Distribution of Injuries of Rear-Seat Occupants (Actual numbers)

	No. of Injuries	Total Injured
Rear Right Occupants		
Fatal injury	1	
Serious injury	22	52
Slight injury	29	
No injury	991	
Rear Left Occupants		
Fatal injury	1	
Serious injury	10	35
Slight injury	24	
No injury	712	
Rear Middle Occupants		
Fatal injury	3	
Serious injury	6	23
Slight injury	14	
No injury	362	

3. At least one rear-seat occupant belted or unbelted.

4. Either the belted front-seat occupant or the rear-seat occupant, or both, more than slightly injured.

It is considered to be reasonable to exclude the "slight injury" cases (that is, no hospital treatment) because a possible head-to-head collision would certainly have more serious consequences.

Eleven cases complied with the earlier conditions mentioned. Each of these is reported in more detail below:

1. Speed about 60 mph, impact against another vehicle. The belted front-seat occupant sustained some forehead injuries (side door structure) and fracture of one of his feet. The unbelted right rear-seat passenger sustained fracture to his jawbone when thrown against the upper part of the front seat.

2. Speed about 50 mph, impact against a truck. Belted front-seat occupant sustained no injuries, whereas the unbelted rear-seat passenger sustained a skull fracture when striking the roof structure.

3. (Frontal) impact followed by rear-end impact. Speed not reported. The belted driver was not injured, the unrestrained rear-seat occupant sustained slight-to-moderate injuries to face when hitting front-seat structure.

4. Impact against another vehicle at a speed of 37 mph. The belted driver sustained rib fracture. One of the two unbelted occupants in rear seat was seriously injured on his right leg, and the other slightly injured from upper part of front seat.

5. Mainly frontal impact (slightly from the right) against a bulldozer at a speed of about 25 mph. The two belted occupants of the front seats sustained slight undefined injuries, the single rear-seat occupant sustained a broken arm when thrown against the seat and door structure.

6. (Frontal) impact followed by roll-over at a speed of more than 60 mph. The vehicle was completely wrecked. The belted driver as well as the belted passenger in the front seat sustained fracture of collar bone and rib fracture. The single rear-seat occupant behind the right front seat was seriously injured in the back and sustained fractures of vertebrae. He was thrown out of the vehicle through the rear window at the third roll-over, which followed the impact.

7. Impact against another vehicle, speed about 25 mph. Belted driver sustained single rib fracture, belted passenger was uninjured. The unbelted and left positioned rear-seat occupant smashed his lips and teeth against front-seat structure.

8. Impact against an elk at a speed of 50 mph. The belted driver sustained rib fracture by striking the steering wheel in spite of the belt, cuts from glass on his right hand. The belted passenger next to him sustained slight cuts on his hands from glass. One of the two unbelted rear-seat occupants sustained a broken nose and cut on lower part of jaw from seat structure. The other had slight, but undefined injuries.

9. Impact against a car, speed 25 mph. Belted driver sustained slight injuries, and his belted passenger sustained cuts on face from glass and fracture of his right lower arm. The occupants in the rear seats were both slightly to moderately injured (cuts on the legs and soreness in abdominal area).

10. In this accident the driver fell asleep and the vehicle impacted a tree at a speed of more than 60 mph. The car was wrecked. The belted driver was killed due to smashed thorax and skull fracture. The driver's seat was torn loose and the belt webbing burst, probably caused by the impact created by the unbelted rear-seat

occupant, when he was thrown forward against the driver's seat. At the moment of the accident he was lying asleep in the rear seat. The passenger sustained mainly just a leg fracture.

11. Impact against a passenger car, speed above 60 mph, followed by roll-over and side impact. Both the driver and the front-seat passenger were belted. Driver seriously injured: rib fracture, moderate concussion, and contusion to one of his thighs. Front passenger sustained fatal skull fracture when he was thrown against the door structure. The single rear-seat passenger positioned to the right sustained no head injuries, but serious injuries to thorax and fracture of thigh bone.

With respect to the comparatively high number of cases of the total material analyzed, it might be concluded with good reason that: The risk of a "collision" between the head of the shoulder belted front-seat occupant and the head of the unbelted rear-seat passenger is minor. Anyhow, the occurrence of that risk is not confirmed by the analysis.

Use of Safety Belt

More than 98% of the cars involved in the reported accidents were equipped with safety belts on the front seats. That means that almost every driver and front-seat passenger had the chance to be restrained.

The analysis revealed, however, that the safety belts were used on an average only by about 25% (6870) drivers and by about 30% (2699) of the front-seat passengers. As shown in Fig. 11, the "use per cent" is found to increase, however, at increasing accident speed. Is the driver who fastens his belt a high-speed traveler, or do people prefer to be belted during (longer) trips which result in higher speeds?

The use per cent of the front-seat passenger increases gradually with the speed, whereas that of the driver rises significantly steeply at 30 mph. The explanation for this could be that in Sweden the speed limit in city areas is generally 30 mph. Do longer trips in unrestricted areas mean more belt users? This theory is partly

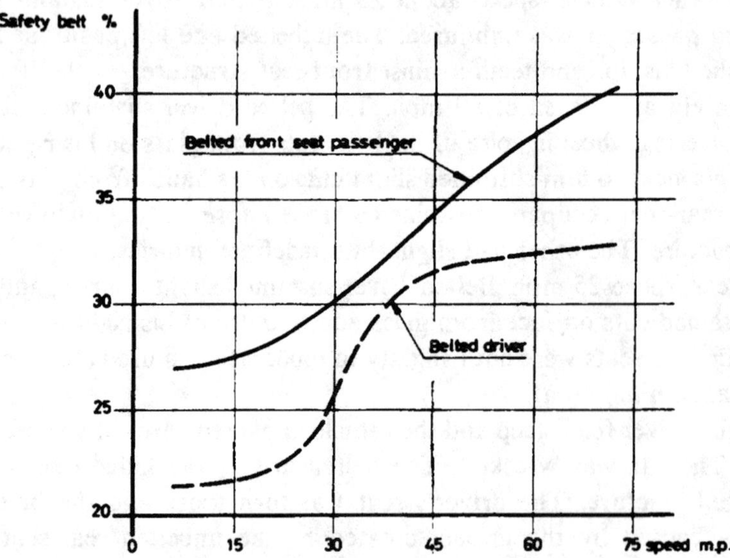

Fig. 11 — Use of seat belt in relation to accident speed

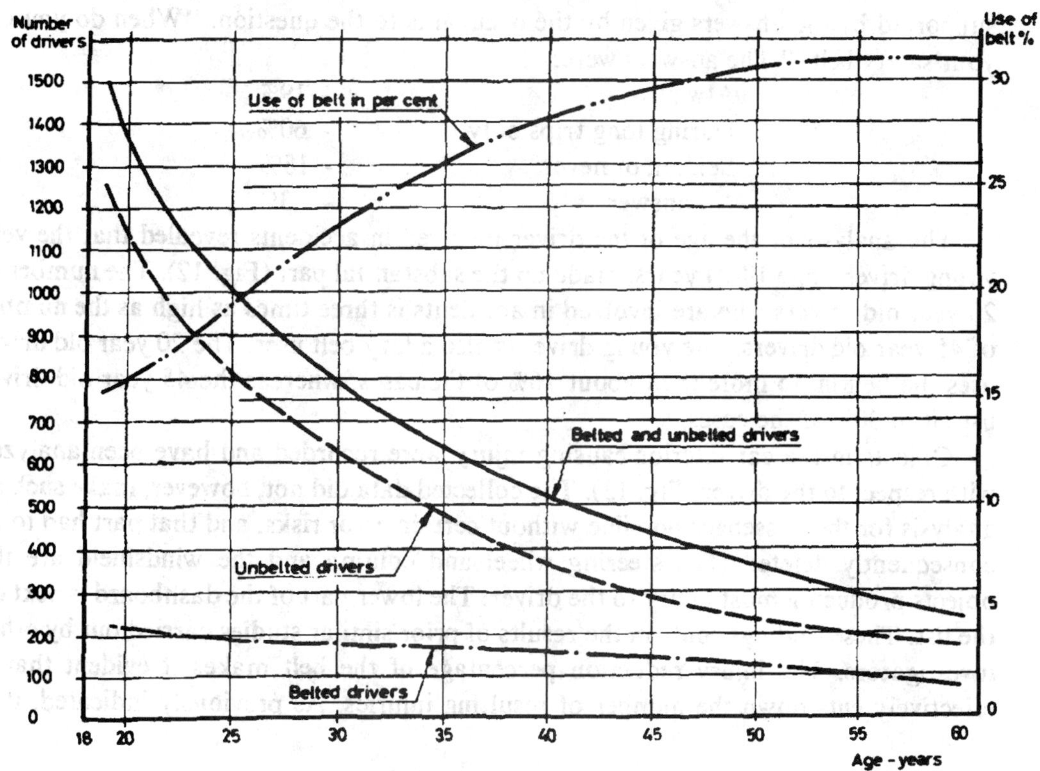

Fig. 12 — Frequency of drivers related to age

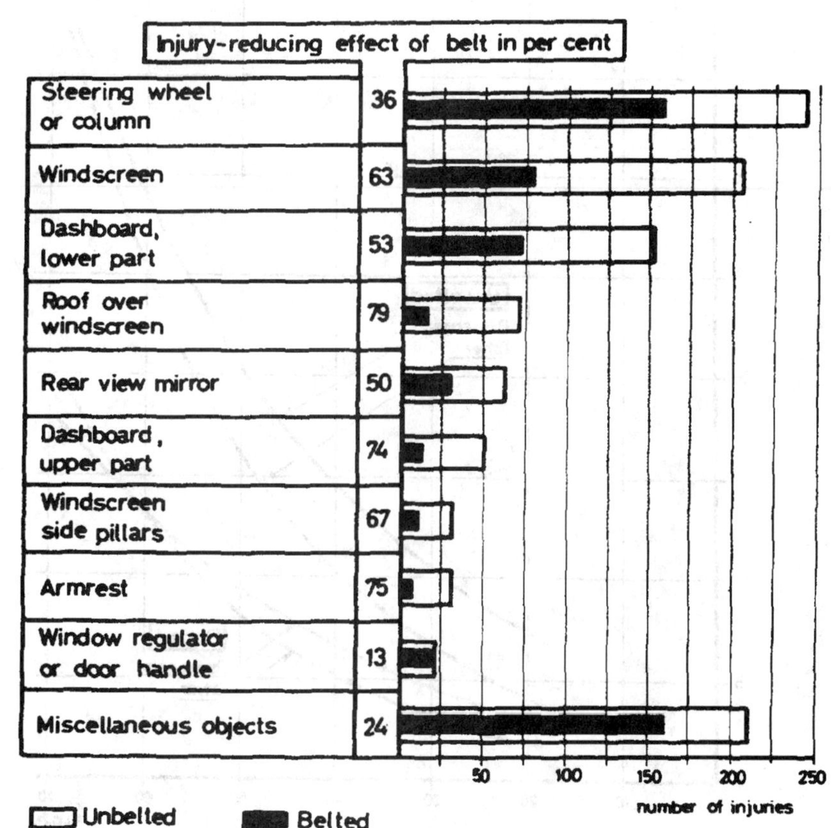

Fig. 13 — Injury - producing objects in car interior (cases per 10,000 drivers)

supported by the answers given by the occupants to the question: "When do you use your safety belt?" The answers were:

Always	- 19%
During long trips only	- 60%
Seldom or never	- 18%
No answer	- 3%

The analysis of the age of the driver involved in accidents revealed that the very young drivers, age 18-24 years, made up the substantial part (Fig. 12). The number of 20 year old drivers who are involved in accidents is three times as high as the number of 45 year old drivers. The young driver is also a lazy belt user. The 20 year old driver uses the belt in no more than about 16% of the cases, whereas the 45 year old driver uses it in 30% of the cases.

Objects in the car interior causing injury were recorded and have been analyzed with respect to the driver (Fig. 13). The collected data did not, however, make such an analysis for the passenger possible without certain error risks, and that part had to be consequently deleted. The steering wheel and column and the windshield are the objects producing most injury to the driver. The lower part of the dashboard is next on the list. These findings confirm the results of prior similar studies carried out by other investigators. The injury reduction percentage of the belt makes it evident that it effectively cuts down the number of resulting injuries. As previously indicated, it is

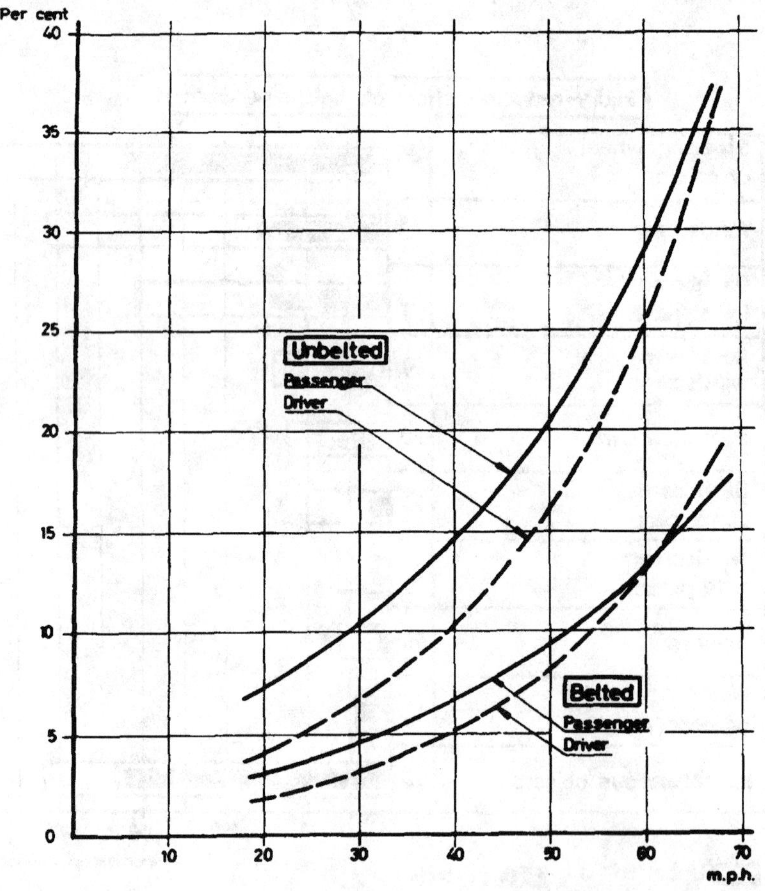

Fig. 14 — Frequency of injuries, unbelted and belted front-seat occupants

considered that the poorly adjusted belt offers reduced protection, especially with respect to contact with closely located objects, for example, the steering wheel. The snugly worn belt should certainly have given even more favorable reduction figures.

Summary and Conclusions

From the analysis carried out, it is concluded that the three-point combined lap and diagonal shoulder harness with slip-joint:

1. Reduces substantially the frequency of injury of all kinds and the frequency of certain injuries (particularly concerning the head and upper torso) very substantially.
2. Offers effective protection against ejection.
3. Does not cause any serious injury to the wearer, but in some severe accidents slight injuries such as cracks in single rib, bruises, etc.

To improve the effect of the safety belt even more, it is strongly considered that particular emphasis should be given to factors which will contribute to more frequent use of the belt and to easier adjustment of the belt.

Acknowledgment

The author wants to thank all those who have contributed to the collection of the data or have otherwise assisted. He and his company owe all the hundreds of Volvo dealers in Sweden particular thanks for their positive cooperation.

790323

Effectiveness of Current and Future Restraint Systems in Fatal and Serious Injury Automobile Crashes. Data from On-Scene Field Accident Investigations

Donald F. Huelke
Univ. of Michigan Medical School
Harold W. Sherman, Michael J. Murphy,
Richard J. Kaplan, and Jerry D. Flora
Highway Safety Research Inst., Univ. of Michigan

Estimates of the effectiveness of lap belts, lap-shoulder belts, air bag restraints, and passive belt systems has generally been made from mass data statistics as examplified by NHTSA activities (1).* In 1972, Wilson and Savage reviewed 706 fatalities evaluating various occupant restraint systems as to their fatality reducing potential (2). They reviewed field accident data of fatal automobile crashes, including photographs, medical and/or autopsy reports and other necessary information. In this study we reviewed the data on fatal cases that were investigated at the scene of the crash, in Washtenaw County, Michigan. Most of the fatal crashes were at high speed in the rural areas. Additionally the more serious injury crash cases that we investigated were also reviewed in order to make estimates of effectiveness of the various kinds of occupant restraint systems.

*Numbers in parentheses designate References at the end of paper.

DATA BASE

There were 101 front seat occupant fatalities in 80 crashes retrieved from our files, crashes that occurred between January 1, 1973 and December 31, 1977. All fatal passenger car cases were reviewed; fatalities in vans, pickup trucks, larger trucks, and rear-seat automobile fatalities, were excluded. Of the 101 fatalities, four were wearing belts. One lap belted passenger died of thoracic injuries from impact to the right door. A lap belted driver had fatal head and thoracic injuries from contact with the left door and the striking car. A lap-shoulder belted passenger died from head injuries sustained by impacting the left door; another passenger had fatal head injuries from roof impact. In these cases all estimators agreed that none of the restraint systems would have prevented these fatilities.

Make, model and year of car, crash configuration and major damage areas of the cars,

ABSTRACT

Data from 101 front seat automobile occupant fatality crashes that the authors had investigated were reviewed along with 70 front seat automobile occupants who had the more severe (AIS 3, 4, or 5) level injuries who did not die. The effectiveness of the lap belt alone, lap-shoulder belt, air bag alone, air bag with lap belt, and the passive shoulder belt were made. The estimates reveal that none of the restraints would have prevented 42 to 51 of the fatalities. The air bag with lap belt, and the lap-shoulder belt system, have the highest effectiveness for reducing fatalities (AB+LB, 34%; LB+SH, 32%). The air bag with lap belt has an effectiveness of 68% in reducing the more serious injuries with the lap-shoulder belt nearly as equal (64%). NHTSA's fatality reduction estimates are excessively high and overly optomistic compared to ours, but theirs are noticeably lower for serious injury reduction than are ours. Comparisons with other restraint effectiveness studies are also made.

determined from a review of the police reports, photographs, investigative reports, and field notes were tabulated (Appendix A). Using a variety of medical sources for the injury descriptions, including hospital records, autopsy reports, body examinations (where no autopsy was performed but the cause of death was obvious), and death certificates, the cause of death was listed along with the fatal producing impact areas.

Independently three of the authors (DFH, HWS, MJM)-"estimators"-made an assessment of the effectiveness of each of five types of restraint systems. The systems evaluated are:

1. lap belt alone
2. lap-shoulder belt
3. air bag alone
4. air bag with lap belt
5. passive belt (shoulder portion only)

All estimators are investigators experienced in injury causation mechanisms and have investigated crashes involving unrestrained, lap, and lap-shoulder belted occupants.

Where the investigators felt that there was insufficient data for restraint system effectiveness evaluation (mainly incomplete medical data), the cases are included, but effectiveness is listed as "unestimatable". In the few cases of pre-crash "possible heart attack", it was assumed that the heart condition would not have been a complication for survival and thus are included. In some cases the restraint system was recorded as "not applicable (NA). These involved front center passengers, or an occupant of a convertible (no shoulder belt available).

THE MORE SERIOUS INJURIES

During the same time period (1973-1977), the authors investigated car crashes where front seat occupants sustained injuries of a more severe nature according to the Abbreviated Injury Scale (AIS). The AIS groups injuries into categories of minor-AIS 1, moderate-AIS 2, serious-AIS 3, severe-AIS 4, and critical-AIS 5 (3). Determination of the restraint effectiveness to reduce the overall injury to, at most, an AIS 2 (moderate injury) level was made independently by each investigator. In this serious injury group there were 70 front seat occupants with AIS 3, 4, or 5 level injuries including 52 drivers, 4 front center and 14 front right occupants (Appendix B). None of these occupants died.

In these crashes with the 70 seriously injured there were 14 other front seat occupants most (11) having minor injuries. All were unrestrained except one lap-shoulder belted driver. Of the three with more than minor injuries two unbelted occupants had moderate level injuries (AIS-2) and one other unbelted passenger was killed. This fatality is included in the 101 fatal series.

ASSUMPTIONS

The lap belt and lap-shoulder belt types assumed are those that have been standard equipment in current or past model cars. The air bag restraint systems were envisioned to be of two types, i.e., (1) mounted in the steering wheel hub, as in the General Motors ACRS, 1973 Chevrolets, and in those GM cars that were of the optionally equipped Volkswagon Rabbit and the Chevrolet Chevette. It was assumed that the air bag restraint system would deploy in each crash. Further, it was assumed that the various restraint systems themselves would not have caused any significant injuries.
The passive belt system considered was like that of the optionally equipped Volkswagon Rabbit. It was assumed that the air bag restraint system would deploy in each crash. Further, it was assumed that the various restraint systems themselves would not have caused any significant injuries.

RATING METHOD

The probability of fatality or injury reduction for the various restraint systems for each front seat fatality is as follows:

0% -essentially no chance for survival
10%,20%,30%-survival possible, but unlikely
40%,50%,60%-survival definitely possible
70%,80%,90%-survival very likely
100% -survival essentially certain

Each estimator reviewed the cases independently of the other two, and assigned a fatality or injury reduction percentage to each individual for each type of restraint system. All three estimators then met collectively to review those cases where there were noticeable rating differences in the potential restraint system effectiveness. These were reviewed with respect to the crash type, severity, injuries, and injury causation. Each estimator stated his reasons for his opinion, and following group discussion of the crash and of injury causation, some reevaluations were made, based on new or previously unnoted data. Oftentimes, following these reviews, opinions were unchanged. There were 303 estimates made on each restraint system for fatalities (101 fatalities x 3 estimators) and 210 estimates on each system for the more serious injury group (70 x 3 estimators).

By summing each estimators effectiveness percentages on each restraint system and dividing by the total number of occupants (omitting those of "unestimatable or not applicable") the range of potential effectiveness of each system was determined. A total sum of effectiveness percentages of the three estimators on each system divided by the total number of occupants (omitting the unestimatable and not applicable") gave the "average of potential effectiveness".

FATALITIES

UNSURVIVABLE ACCIDENTS-Of the 101 front seat automobile fatalities all 3 estimators were in unanimous agreement that 42 fatalities were non-survivable, i.e., none of the restraint systems would have prevented the occupant from being killed.

Additionally, there were 5 fatalities where 2 of the 3 estimators concluded that each of the restraint systems would not have prevented the fatality with the third estimator opining that either the data were "unestimatable" or gave a very low percentage survivability for some of the systems.

Also there were 4 additional fatalities where one of the three estimators concluded that there was no possibility of survival with the others giving a minimal chance of survival.

Thus, no survival, or very limited chances of survival, was found in 42 to 51 of the 101 fatalities.

RESTRAINT EFFECTIVENESS

Lap Belts - The potential effectiveness of the lap belt restraint system is rated the lowest of all the systems. With the severe nature of most of the crashes reviewed and the relative infrequency of ejection, 67% of the 303 estimates indicated no chance of survival had the lap belt alone been worn (Table 1). Furthermore the estimators determined that 15% of the fatalities may have survived (10-30% category) with possible to probable survival (40-90%) at the 12% level. Only 3% of the estimates indicated certain survival. The range of probable effectiveness for lap belts is 9.2 - 15.9% (Ave. 12.8%).

Lap-Shoulder Belts - In the series of fatalities, 48% would not have survived by the use of lap-shoulder belts (Table 1). It was estimated that 8% would definately have survived and 19% most probably would have survived had lap-shoulder belts been worn. The range of potential effectiveness is 30.6 - 32.4% (Ave. 32.1%).

TABLE 1

COMPARATIVE EFFECTIVENESS OF VARIOUS RESTRAINT SYSTEMS FOR FATALITY REDUCTION

	PROBABILITY OF FATILITY REDUCTION						
Restraint Types	0% (%)	10-30% (%)	40-60% (%)	70-90% (%)	100% (%)	Unest.* (%)	NA* (%)
Lap Belt	67	15	9	3	3	2	1
Lap-Shoulder Belt	48	8	11	19	8	2	4
Air Bag alone	59	8	10	14	7	2	--
Air Bag + Lap Belt	50	9	7	15	16	2	1
Passive Belt	49	12	12	14	7	2	4

*Unest. - unestimatable; NA - not applicable

Air Bag Alone - Estimator certainty of air bag restraint effectiveness (100%) was 7% with probable survival at 14% (Table 1). Non-survivability averaged 59%. The range of potential effectiveness is 23.2 - 27.1% (Ave. 25.1%).

Air Bag plus Lap Belt - Of those killed, it was estimated that 16% would definitely have survived had the air bag deployed and the occupant worn a lap belt (Table 1). Additionally, 15% would probably have survived, yet half (50%) of the fatal front seat occupants would not have survived with the air bag and a lap belt system. The range of potential effectiveness for this system is 32.6 - 35.3% (Ave. 34.2%).

Passive Belt - Of those killed it was estimated that 7% definitely would have survived, and that 14% probably would have survived with a passive shoulder belt (Table 1). Half would not have survived. The range of potential effectiveness is 26.8 - 29.6% (Ave. 28.0%).

RESTRAINT EFFECTIVENESS COMPARISON

Table 1 shows the effectiveness of the five restraint systems by category of survival probability. Note that for the lap-shoulder belts, passive belt, and air bag with lap belt, half of the occupants would not have survived (lap belt non-survivability was 67%). More occupants would not have survived by the use of the air bag alone than by the lap-shoulder belt, passive belt or air bag with lap belt.

At the other end of the scale, certain survival (100%) is essentially the same for lap-shoulder belts, air bag alone, and passive belts. However, combining the air bag with the lap belt doubles the frequency of assured survival to 16%. "Survival essentially certain" frequencies show that the air bag alone is no more effective than the full usage of the lap-shoulder belt or the passive belt.

Survival very likely (70-90%) is the same for the air bag, with or without belts, or the passive belt systems (Table 1). In this category the full use of lap-shoulder belts was most often indicated for its effectiveness above all others.

When the "survival very likely" and "survival essentially certain" categories (70-100%) are combined, the air bag with lap belt has the highest probable effectiveness level (31%) followed closely by the lap-shoulder belt system (27%). Air bag alone or passive belts are at the 21% effectiveness level.

All fatal crashes in our series occurred in a rural environment with many of the crashes at high speed, involving marked intrusion and compromise of the occupant space. Our results are similar to the findings of Wilson and Savage in their study of 706 fatalities (Table 2). Our estimates of lap-shoulder belt effectiveness is basically identical to their results.

Our lap belt potential effectiveness is less (13 vs. 17%), with our estimates of air bag alone, and air bag plus lap belt, being higher (25 vs. 18%; 34 vs. 29%). However these differences are not statistically significant. (X^2(df=3)=2.98, p=.38).

Our review of fatal crashes indicates that the air bag with lap belt has the highest potential effectiveness averaging 34%, closely followed by the lap-shoulder belt system (32%). Average effectiveness for the air bag alone is 29%.

Even with our familiarity with the cases reviewed we could not estimate potential restraint system effectiveness in 2% of the cases. Additionally, we found that 1-4% of the systems were "not applicable" to specific cases.

POTENTIAL SURVIVAL BY SEATING POSITION

There is very little difference in the average potential restraint system effectiveness between drivers and passengers among all systems except the lap belt (Table 3). Drivers are more often protected by the lap belt than are passengers by a factor of 50%; however, lap belt protection in this study gave the lowest potential effectiveness of all restraint systems.

POTENTIAL SURVIVAL BY IMPACT LOCATION

Because of the high percentage of drivers, restraint systems are more effective in far (right) side impacts than in "near side" impacts (Table 4). In right side impacts the air bag plus lap belt is potentially the most effective, followed by the air bag alone and lap-shoulder belts mainly affecting drivers in these far side impacts (Table 4).

In near (left) side impacts the potential effectiveness for most all systems is lower than that for far (right) side impacts. The vulnerability of the driver in these "near side" impacts is reflected in the comparatively low effectiveness of the air bag alone or with a lap belt. In these near (left) side crashes the lap-shoulder belt system is twice as effective as the air bag alone, and 20% more effective than the air bag with lap belt.

TABLE 2

COMPARISON OF POTENTIAL EFFECTIVENESS OF VARIOUS RESTRAINT SYSTEMS FOR FATALITY REDUCTION

Restraint Type	This Study		Wilson & Savage
	Ave.	Range	
Lap Belt	12.8%	(9.2-15.9%)	17%
Lap-Shoulder Belt	32.1%	(30.6-32.4%)	31%
Air Bag alone	25.1%	(23.2-27.1%)	18%
Air Bag + Lap Belt	34.2%	(32.6-35.3%)	29%
Passive Belt	28.0%	(26.8-29.6%)	Not Estimated

In frontal crashes, the air bag with lap belts is more effective than the air bag alone, the passive belt, or the lap-shoulder belt.

It is possible that our estimates on the air bag alone and air bag plus lap belt are high. Our conclusions on these systems were based on the assumptions that the air bag system would always deploy and offer the desired protection, especially in cases where the right side of the instrument panel was significantly deformed, twisted, and in general, out of its design position. Bending of the instrument panel due to right side occupant compartment compromise may prevent deployment, or tear an inflating or inflated bag, thus decreasing occupant protective levels. Only laboratory tests and/or real world crash data of air bag equipped cars will prove the effectiveness of the system in severe right side impacts.

In the "other impact cases", mainly rollovers, the effectiveness of the belt system is obvious, for the air bag alone has but a 4% average potential effectiveness level, whereas all belt systems are near the 30% effectiveness level.

PREVIOUS FINDINGS FROM SAME GEOGRAPHICAL AREA

A study of fatal automobile accidents was conducted in 1961-65 in the same geographical area where the crashes in this study occurred (4). In that study there were 137 crashes in which 177 occupants were killed. Of the 177 individuals, 154 were front seat occupants. It was estimated that of the front seat occupants, belt systems would potentially have saved 56% (40% lap belts, 56% with lap-shoulder belts) with 37% not having survived even if belts had been worn.

The apparent discrepancy between the potential effectiveness of the lap belt in the current study (13%) and the relatively high percentage of estimated lap belt effectiveness in the older study (40%) is probably due to the different frequencies of injury mechanisms. In the 1961-65 study, there were 43 front seat occupants who were ejected; it was estimated that 81% of them would not have been killed had they been using the lap belt, and an additional

TABLE 3

AVERAGE POTENTIAL RESTRAINT SYSTEM EFFECTIVENESS FOR FATALITY REDUCTION

Restraint Types	All Front Seat Occupants	Drivers	Passengers
	(%)		
Lap Belt	13	14	9
Lap-Shoulder Belt	32	31	31
Air Bag alone	25	24	27
Air Bag + Lap Belt	34	34	32
Passive Belt	28	28	26

7% had the shoulder belt been available and worn. Thus, the high frequency of potential survivability with the lap belt in the 1961-65 study was due to the potential to reduce ejection frequency. In the current study, about 11% of the front seat occupants were ejected; thus there are more people remaining within the confines of the occupant compartment which probably had an affect on the relative effectiveness of the lap belts. Improved latch strikers, hinges and door reinforcement may be factors for increased occupant containment.

However, in those types of crashes where ejection would be most likely to occur, i.e., rollovers, we find in the present study that belt systems are relatively high in average potential effectiveness (Table 4).

ESTIMATES OF LIVES SAVED

If our estimates of potential restraint system effectiveness are representative, they then can be used to determine the number of front seat lives expected to be saved in various crash modes (Table 4). There are 27,200 front seat automobile occupant fatalities for the year 1975 (1).

In the federal report on fatal crashes "Fatal Accident Reporting Systems - 1976 Annual Report" (5) are illustrated percentages of impact locations for single and multiple car crashes (Fig. 3.6, p. 67 and Fig. 4.6, p. 85). Averaging these impact locations and combining them for left side (8, 9, 10 o'clock impacts, 14.2%), right side (2, 3, 4 o'clock impacts, 16.8%), and frontal crashes (11,12, 1 o'clock impacts, 60.0%) restraint systems effectiveness in terms of total number of lives saved nationally can be calculated.

When the number of front seat occupant fatalities, multiplied by impact location effectiveness estimates, and by the frequency of impacts at that location (FARS Report), the total number of occupants potentially saved in the United States can be estimated.

Number of Lives Saved Nationally	=	Number of Drivers (or Passengers)	X	Impact Location Effectiveness	X	Impact Location Frequency

How many lives then could be saved? All those who were not listed as "essentially no chance for survival" would be the maximum number possibly saved. Only those in the "survival very likely" or "essentially certain"

TABLE 4

ESTIMATED NUMBER OF LIVES SAVED ANNUALLY IN USA BY VARIOUS RESTRAINT SYSTEMS AND IMPACT LOCATIONS*+

Restraint Types	Lt. Side** Impacts		Rt. Side Impacts		Frontal Impacts		Other Impacts		Total
	%++	No.	%	No.	%	No.	%	No.	No.
Lap Belt	10	386	14	640	10	1,632	27	661	3,319
Lap-Shoulder Belt	20	772	22	1,005	37	6,038	36	881	8,696
Air Bag alone	9	348	22	1,005	37	6,038	4	98	7,372
Air Bag + Lap Belt	16	618	28	1,280	43	7,018	29	710	9,626
Passive Belt	15	579	18	823	35	5,712	32	783	7,897

*Based on 27,200 front seat automobile fatalities (see reference No. 1)

+27,200 fatalities x frequency of impacts x potential restraint effectiveness

**Frequency of impacts: left side (8, 9, 10 o'clock) = 14.2%; right side (2, 3, 4 o'clock) = 16.8%; frontal (11, 12, 1 o'clock) = 60.0%; others (rollovers, rear impacts) = 9.0%.

++Average potential restraint system effectiveness.

categories would be the minimum number. Realistically, however, we believe that the actual number of lives saved would be between the maximum and minimum figures, including the "survival definitely possible, very likely, and essentially certain". The estimated number of lives saved for the various restraint systems is shown in Table 5.

EQUIVALENT EFFECTIVENESS OF RESTRAINT SYSTEMS

Using the average potential effectiveness for each system, our estimates of average restraint effectiveness (Table 5) by comparative combinations of restraint systems indicates:

1) Air bag alone = 78% lap-shoulder belt usage
2) Air bag alone = 89% passive belt usage
3) Air bag plus 25% lap belt usage = 85% lap-shoulder belt usage
4) Air bag plus 20% lap belt usage = 84% lap-shoulder belt usage
5) 100% passive belt usage = 87% lap-shoulder belt usage

TABLE 5

ESTIMATED RANGE OF THE ANNUAL NUMBER OF LIVES SAVED NATIONALLY BY RESTRAINT SYSTEMS

Restraint Types*	Minimum+		Maximum**		Average++	
	%	No.	%	No.	%	No.
Lap Belt	6	1,632	32	8,704	13	3,536
Lap-Shoulder Belt	27	7,344	51	13,872	32	8,704
Air Bag alone	21	5,712	41	11,152	25	6,800
Air Bag + Lap Belt	31	8,432	50	13,600	34	9,248
Passive Belt	21	5,712	50	13,600	28	7,616

*Assumes 100% belt usage or 100% air cushion deployment.

**Maximum Number = (100 - % Non-Survival) x 27,200.

+Minimum Number = Combined percentages of "Survival very likely and survival essentially certain" (Table 1) x 27,200.

++Average potential effectiveness - Table 2.

COMPARISON WITH FEDERAL DATA

NATIONAL POTENTIAL RESTRAINT EFFECTIVENESS- NHTSA has indicated 27,200 as the total front seat car occupant fatalities (1). Using this number, the minimal estimate of lives saved, if only the "survival very certain or essentially certain" percentages are used, shows the air bag plus lap belt as most effective followed by the lap-shoulder belt (Table 5). The maximum estimate of lives saved nationally would be 13,872 by full usage of the lap-shoulder belts. These include all fatalities not classified as "no chance for survival". The air bag plus 100% lap belt usage, or 100% passive belt usage, would maximally save 13,600.

Realistically, however, the average potential effectiveness values would be more accurate saving 9,248 lives by the air bag plus 100% lap belt usage, and 8,704 by 100% lap-shoulder belt usage (Table 5).

Table 6 compares our potential restraint system effectiveness benefits with those of NHTSA.

Assuming 100% usage, our lap belt effectiveness is only 1/3 that of NHTSA's.

Assuming 100% usage, our lap-shoulder belt effectiveness is 53% of the NHTSA estimate.

Our potential effectiveness for the air bag alone is 37% less than that of NHTSA.

Assuming 100% lap belt usage with the air bag, our potential effectiveness is about 48% less than NHTSA's.

Our best estimate for the most effective restraint system for all types of fatal crashes is the air bag with 100% belt usage. Our estimate is that the air bag alone would save 6,800 lives, significantly less than the calculated NHTSA estimate of 10,700. Lap-shoulder belts (100% usage) could save 8,704 lives, 47% less than NHTSA's estimate. It is fair to assume that some of the 27,200 fatalities were belted (lap or lap-shoulder) and therefore our estimates of restraint effectiveness nationally would be somewhat high.

Thus, there is a noticeable difference between the estimates of NHTSA and ours. NHTSA indicates that lap-shoulder belt usage would save 16,300 lives, i.e., 60% of all front-seat fatalities. With the air bag plus 40% lap belt usage their estimate is 50% survival. This seems to us to be an excessively high estimate and overly optimistic for we have found that in the fatal crashes investigated at least 50% of those who died could not have survived the crash by the use of any of the restraint systems reviewed.

An alternative method of computing the potential fatality reduction is to use the probabilities derived from the combined three-judge estimates. Using each interval (10%, 20%, etc.) as the probability, and multiplying by the number of cases in each interval and by the 27,200 front seat fatalities, yields the effectiveness of each restraint system. These figures appear in the third column of Table 6. Note that these figures are even more divergent from NHTSA's numbers than are our average estimates.

RESTRAINT EFFECTIVENESS - OTHER STUDIES

There have been numerous recent studies on the effectiveness of lap and lap-shoulder belts using mass accident statistics (6-10). These studies have compared injury severity and fatalities of individuals who were unrestrained, lap belted, or lap-shoulder belted. Rininger and Boak (6) studied front-seat occupants injured and killed in full size 1974-75 General Motors cars involved in towaway accidents. They found that 77% of the fatalities in their study would have been prevented by the use of lap-shoulder belts. Similarily, a study by Scott, et al (8), of the Highway Safety Research Institute (RSEP Study) (1973-75 American made passenger cars) showed that the lap and shoulder belts are 62% effective in reducing the incidence of fatalities. Renfurt, et al (9) from the Highway Safety Research Center, indicated that lap-shoulder belts were 55% effective in reducing fatal injuries. Backstrom, et al, Bohlin, et al, MacKay, and Tarriere (10, 11, 12, 13) have indicated that conventional three point belts prevent serious and fatal injuries to front seat occupants at about the 55% level. Additionally, a paper by Huelke, et al (14) showed that there is a decrease in fatalities by 57-91% depending on the type of crash and data base used. All of these studies are in relative agreement; they generally indicate an effectiveness estimate of the lap-shoulder belt system twice that which we have in our current study.

Overall reduction in front seat fatalities estimated from the present study is lower than that suggested in NHTSA air bag effectiveness presentations, but agrees well with the study by Wilson & Savage (2). One possible reason is a difference in the population studied - our data representing a relatively rural high speed environment, i.e., rural accidents may be a lot more unsurvivable than, say, half urban - half rural. Most of the other studies mentioned compared mass data of occupants who had been unrestrained, lap belted or lap-shoulder belted at the time of the crash. In the other studies, there were specific criteria for data inclusion with some of the data being weighed and analysis techniques used which were designed to avoid the biasing effect of several study variables. Our study is a self-selecting sample tending to be biased toward non-survival -a special class of crashes. However it may better represent the "real world" of fatal crashes and be most appropriate for comparison with NHTSA's numbers.

TABLE 6

ESTIMATED LIVES SAVED BY RESTRAINT SYSTEM TYPE

Restraint Types	This Study			NHTSA*	
	Number**			No.	%+++
	Range	Ave.	Prob. Est.***		
Lap Belt (100% usage)	1,632 - 3,704	3,536	3436	10,900	40
Lap-Shoulder Belt (100% usage)	7,344 - 13,872	8,704	8278	16,300	60
Air Bag alone (100% deployment)	5,712 - 11,152	6,800	6909	10,700+	39
Air Bag + Lap Belt (100% lap belt usage)	8,432 - 13,600	9,248	9112	17,700+	65
Passive Belt (100% usage)	5,712 - 18,600	7,616	7371	13,400++	50

*Table A6, p.A-8: The Secretary's Decision Concerning Motor Vehicle Occupant Crash Protection. DOT, Washington, D.C., December 6, 1976.

**Minimum to Optimum number of lives saved of 27,200 front seat car occupant fatalities. See Table 5.

***Probability Estimates. See discussion in text.

+Our estimate from NHTSA data, where air bag with 20% lap belt usage = 12,100; with 40% usage = 13,500; thus, 10,500 without belts and 17,700 with 100% lap belt usage.

++Our estimate from NHTSA data, where passive belt usage of 60% = 9,800; with 70% usage = 10,700; thus 100% = 13,400.

+++Calculated percentages from NHTSA data - Table A6, p.A-8. See * above.

REASONS FOR DIFFERENCES IN RESULTS BETWEEN STUDIES

The actual reasons for the large difference in the restraint effectiveness (of preventing fatalities) estimated in this study and those found in other studies cannot be specifically identified. The differences are likely due at least in part to the differences in study methods. The present study investigated crashe in which at least one person was killed. These crashes tended to be quite severe and were found to be mostly high-speed, rural crashes. Other studies have estimated restraint effectiveness based on fatality (or injury) rates among occupants with different restraints in a large set of crashes. As a result, the crashes investigated in other studies tended to be of lesser severity on the average. In addition, some studies involved primarily crashes in urban or suburban areas, which are typically at lower speeds than those in rural areas. Thus the self selecting nature of a set of fatal accidents is likely to have played a role in the difference in the estimated restraint effectiveness.

A second possible reason for the difference in effectiveness estimates is that restraint effectiveness may be a function of crash severity (as measured by Δ V, speed at impact, vehicle deformation index, etc.). If this is the case, then restraint effectiveness is likely to be a decreasing function of crash severity at least after some threshold. The selection of a set of fatal crashes, with its assumed selection of high severity crashes would then have a much higher proportion of crashes which are so severe that survival is unlikely with any restraint system so that the effectiveness of restraints is likely to be quite small.

Finally, this study has found that about half of the fatalities were in crashes so severe that none of the restraint systems studied would have increased the probability of survival. This may imply that some fraction--about 40-50% of fatalities would not have been eliminated by any of the restraint systems. If this speculation is the case, then the previous restraint effectiveness estimates in other studies only apply to the "survivable" crashes, not to all fatal crashes. If the previous effectiveness estimates were applied to the, say 55%, of fatalities which occurred in "survivable" crashes, then the estimated savings would be similar. Thus much of the difference in estimated effectiveness rates can be explained if the effectiveness is assumed to depend on the severity of the crash and if it is assumed that a large percentage (40-50%) of the fatalities occur in crashes so severe as to be "unsurvivable."

THE MORE SERIOUS INJURIES

NON-REDUCTION OF INJURY LEVELS BY RESTRAINT SYSTEMS--Of the 70 occupants in this group all three estimators were in unanimous agreement that 11 individuals would not have benefited from any of the restraint systems to reduce their overall level of injury severity. In addition, two of the three estimators opined that none of the restraints would have been of benefit, with the third estimator giving a low rating for one or two of the restraint systems on 8 other injured occupants. Thus 16%-27% of these front seat occupants would not have benefited, or have had little benefit, from the various restraint systems.

RESTRAINT EFFECTIVENESS

Lap Belts - As in the fatal series, the lap belt has the lowest potential effectiveness for injury reduction of all systems rated (Tables 7 & 8). However, the lap belt is significantly more effective in reducing the more severe injuries than it is in reducing fatalities. The estimate potential effectiveness at the 70-100% level is 34%. The average potential effectiveness is 39% (Range: 33-46%).

Lap-Shoulder Belts-In this series of front seat occupants with the more serious injuries, one of four would not have had their injury severity reduced by the use of the lap-shoulder belts. Yet the estimators opined that lap-shoulder belts have an effectiveness level of 64% (Range: 64-65%).

Air Bag Alone-Estimator certainty (the 100% level) of air bag effectiveness is 42%. However it was estimated that one-third of the front seat occupants would not have benefited from the air bag alone. The range of potential effectiveness is 57-62% (average 58%).

Air Bag plus Lap Belt-This type of restraint system, as in the fatal series, was indicated to have the greatest potential effectiveness. More than half (55%) of the more seriously injured definitely would have benefited from the air bag and lap belt combination with an additonal 10% most probably having received injury reduction via this system. On the other hand, fully one in four would not have received any benefit. The range of potential effectiveness is 68-69% (average 68%).

Passive Belt- Estimators indicate that one in four front seat occupants would not have benefited from passive belt usage. However, 30%

TABLE 7

POTENTIAL RESTRAINT EFFECTIVENESS FOR SERIOUS INJURY REDUCTION

Restraint Type	All Front Seat Occupants*		Drivers	Passengers
	Ave.(%)	(range)	%	%
Lap Belt	39	(33-46)	43	25
Lap Shoulder Belt	64	(64-65)	70	44
Air Bag alone	58	(57-62)	58	59
Air Bag & Lap Belt	68	(68-69)	70	62
Passive Belt	58	(53-61)	61	45

*Front seat occupants = 70 (Drivers - 52, Passengers - 18

definitely would have benefited with an additional 20% "probably have benefited" from these belts. The range of potential effectiveness is 53-61% (average 58%).

DRIVER AND PASSENGER DIFFERENCES

Data in Table 7 indicate that drivers receive markedly greater benefits from all restraint systems, except the air bag alone, than do passengers.

RESTRAINT EFFECTIVENESS COMPARISON

Table 8 indicates the potential levels of effectiveness of the various restraint systems. Even when upper torso protection is provided (shoulder belts, air bags) approximately one of four front seat occupants would not have had a reduction of their overall injury severity. The air bag alone was estimated to be non-effective in 34% of cases, greater than all other systems except the lap belt which is 41%.

The injury reduction effectiveness "essentially certain" is at the 42-43% level for lap-shoulder belts and the air bag alone, and the 55% level for the air bag and lap belt system. Passive belts, or the lap belt alone is lower (30%; 16%).

When the categories of injury reduction "essentially certain and very likely" are combined, then all systems with upper torso restraint are at or above the 50% effectiveness level with the lap belt alone being at the 34% level.

RESTRAINT SYSTEM EFFECTIVENESS BY IMPACT TYPES

All restraints are most effective in the frontal crash type and are less effective in near or far side impacts (Table 9). In the "other impacts"-primarily rollovers-the lap belt is extremely effective (61%) as is the lap-shoulder belt system. However, the passive systems (bags or belts) are less effective than on the average. There are no national data available on serious injuries by impact location with which we can compare our data.

INJURY REDUCTION BY BODY AREAS

A review of the body areas injured at the AIS 3-5 level indicates that the head and face, or the chest regions, were most often injured (Table 10) agreeing with other studies (15, 16, 17). Single body areas with the more severe injuries predominate, although 19 of the 70 occupants had injuries at the minor or moderate (AIS 1 or 2) level of severity in the same or in some other body areas.

ESTIMATES OF INJURY REDUCTION

The maximum, and minimum numbers of serious injuries reduced and the average potential restraint system effectiveness are shown in Table 11. The technique for estimating maximum and minimum numbers was discussed previously under

TABLE 8

COMPARATIVE RESTRAINT SYSTEM EFFECTIVENESS FOR SERIOUS INJURY REDUCTION

PROBABILITY OF INJURY REDUCTION

Restraint Type	0 %	10-30 %	40-60 %	70-90 %	100 %	Unest* %	N.A.* %
Lap Belt	41	10	10	18	16	3	1
Lap-Shoulder Belt	24	4	4	16	43	3	6
Air Bag alone	34	1	5	15	42	3	-
Air Bag & Lap Belt	26	2	2	10	55	3	1
Passive Belt	26	5	10	20	30	3	6

*Unest. - unestimatable; N.A. - not applicable

TABLE 9

POTENTIAL RESTRAINT SYSTEM EFFECTIVENESS
FOR
SERIOUS INJURY REDUCTION BY IMPACT DIRECTION

Restraint Types	Frontal Impacts	Lt. Side Impacts*	Rt. Side Impacts	Other Impacts
	% (Range)	% (Range)	% (Range)	% (Range)
Lap Belt	39 (30-48)	9 (3-12)	36 (33-41)	61 (50-69)
Lap-Shoulder Belt	68 (68-69)	51 (43-57)	52 (44-60)	67 (64-70)
Air Bag	71 (69-74)	28 (25-30)	56 (53-59)	30 (24-37)
Air Bag & Lap Belt	77 (75-80)	32 (25-40)	63 (62-63)	64 (56-69)
Passive	64 (64-65)	41 (25-52	49 (40-59)	48 (30-67)

*Frequency of impacts: left side (8, 9, 10 o'clock) = 9%; right side (2, 3, 4 o'clock) = 19%; frontal (11, 12, 1 o'clock) = 57%; others (primarily rollovers) = 16%.

the 'Estimates of Lives Saved' section.

COMPARISON WITH NHTSA DATA AND OTHER STUDIES

Our occupant numbers (Table 11) represents the estimates of the overall injury reduction from the more severe levels. NHTSA personnel have indicated that "prevented", as used in the Secretary's decision report, actually means "reduced in severity" (personal communication to DFH-6/78). In general our injury average reduction numbers agree with NHTSA's for 100% usage of the lap, and the lap-shoulder belt systems. Our estimates for the air bag alone (100% deployment) is 80% greater than NHTSA's estimates; our air bag plus lap belt estimate is 36% greater than their's. Our passive belt estimate is 26% greater.

Using the alternative method described in the discussion of Table 6, the estimates for this study are not significantly changed, except for the air bag plus lap belt system estimates.

In their study Rininger & Boak (6) and Reinfurt, et al (9) concluded that lap-shoulder belts are 57% effective in reducing injuries of AIS 2 or greater. Scott, et al (8) indicated an effectiveness of 42% for these restraint systems at the AIS 2 or greater level, and Huelke, et al (14) noted a 48-51% reduction in AIS 3 or greater injuries with lap-shoulder belts depending on the type of crash. Other authors (9, 10, 11, 12) indicate at least a 55% effectiveness level of lap-shoulder belts for the prevention of serious and fatal injuries. MacKay (15) questioned the effectiveness of air bags because of intrusion into the passenger compartment. His data indicate a decreasing effectiveness for both the air bag and the lap-shoulder belt restraint systems with increasing injury severity.

TABLE 10

BODY AREAS WITH THE MORE SEVERE INJURIES

Body Area	No.	Body Area	No.
Head & Face	19	Head, Chest &:	
Neck	2	Neck	1
		Abdomen	1
Chest & Thoracic Spine	16	Upper Ext.	2
		Lower Ext.	1
Abdomen & Lumbar Spine	3	Head, Abdomen, Pelvis &:	
Upper Ext.	5	Chest	1
		Lower Ext.	1
Lower Ext.	6		
		Chest &:	
Head &:		Neck	1
Chest	2	Upper Ext.	2
Abdomen	1	Abdomen	1
Lower Ext.	1	Abd. & Upper Ext.	2
Upper Ext.	2		

TABLE 11

BENEFITS OF RESTRAINT SYSTEMS FOR SERIOUS INJURY REDUCTION

Estimated Number of Injuries Reduced*

Restraint Types**	Minimum***	Maximum+	Average++	Prob. Est.+++	NHTSA****
Lap Belt	24,038	41,713	27,573	27,119	22,840
Lap-Shoulder Belt	41,713	53,732	45,248	43,943	41,876
Air Bag alone	40,299	46,662	41,006	42,524	22,840
Air Bag + Lap Belt	45,955	52,318	48,076	44,425	35,238
Passive Belt	35,350	52,318	41,006	40,432	32,630

*The Secretary's decision concerning motor vehicle occupant crash protection. D.O.T., Washington, D.C., December 6, 1976. (Table A1, p. A-2) This table indicates that 70,700 front seat occupants are injured at the AIS 3-5 level (AIS-3 = 54,400; AIS-4 = 13,600; AIS-5 = 2,700).

**Assumes 100% belt usage or 100% air bag deployment.

***Minimum Number - Combined percentages of "Survival very likely and survival essentially certain" (Table 8) x 70,700.

+Maximum Number = (100 - % Non-Survival)x 70,700.

++ Average potential effectiveness - Table 7.

+++Probability estimates - derived by the techniques discussed in the text for Table 6.

****The Secretary's decision cited above. Effectiveness levels for the various restraints systems at different AIS levels are presented. (Table A4, p.A-6) Using these percentages and the number of injured from TA1, p.A-2 gives the benefits in numbers of injuries reduced.

REFERENCES

1. The Secretary's Decision Concerning Motor Vehicle Occupant Crash Protection. December 6, 1976.

2. R.A. Wilson, and C.M. Savage: Restraint System Effectiveness - A Study of Fatal Accidents. Proceedings, Automotive Safety Engineering Seminar, p. 27-39, June 20-21, 1973.

3. The abbreviated Injury Scale, 1976 Revision. Available from: AAAM, P.O. Box 222, Morton Grove, Illinois, 60053.

4. D.F. Huelke and P.W. Gikas: Causes of Deaths in Automobile Accidents. Final Report, University of Michigan, April, 1966.

5. Fatal Accident Reporting System - 1976 Annual Report. November, 1977.

6. A.R. Rininger and R.W. Boak: Lap-Shoulder Belt Effectiveness. Proceedings of the 20th Conference, American Association for Automotive Medicine, Atlanta, Georgia p. 262-279, November 1-3, 1976.

7. B.J. Campbell: Oral Comments. Presented at the SAE Automotive Engineering and Manufacturing Meeting, Detroit, Michigan, October 13-17, 1975.

8. R.E. Scott, J.D. Flora, and J.C. Marsh: An Evaluation of the 1974 and 1975 Restraint Systems. Highway Safety Research Institute, University of Michigan, Ann Arbor, Michigan, May, 1976.

9. D.W. Reinfurt, C.Z. Silva, and A.F. Seila: A Statistical Analysis of Seat Belt Effectiveness in 1973-75 Model Cars Involved in Towaway Crashes. Highway Safety Research Center, University of North Carolina, May, 1976.

10. C.G. Backstrom, C.E. Anderson, E. Forsman, and L.E. Nillson: Accidents in Sweden With Saab #99. Saab-Scania, 1973.

11. N. Bohlin, H. Norin, and A. Andersson: A Statistical Traffic Accident Analysis Referring to Occupant Restraint Value and Crash Safety Requirements for the Experimental Safety Car. Proc. 4th International Technical Conference on Experimental Safety Vehicles. Kyoto, Japan p. 359-392, 1973.

12. C. Tarrierre: The Efficiency of Three Point Belts in Real Accidents. Committee of Common Market Constructors. Proc. 4th International Technical Conference on Experimental Safety Vehicles. Kyoto, Japan p. 607-620, 1973.

13. G.M. MacKay: Some Cost Benefit Considerations of Car Occupant Restraint Systems. Technical Aspects of Road Safety #59, p. 4.1-4.11, September, 1974.

14. D.F. Huelke, T.E. Lawson, R. Scott, and J.C. Marsh: The Effectiveness of Belt Systems in Frontal and Rollover Crashes. SAE International Automotive Engineering Congress and Exposition, Cobo Hall, Detroit, February 28 - March 4, 1977.

15. G.M.MacKay: Airbag Effectiveness-A Case for the Compulsory Use of Seat Belts. FICITA, X1VTH Congress p. 3/66-3/70,London 1972.

COMPARATIVE STUDY OF 1624 BELTED AND 3242 NON-BELTED OCCUPANTS: RESULTS ON THE EFFECTIVENESS OF SEAT BELTS

C. Thomas, G. Faverjon, C. Henry, C. Tarrière
Laboratoire de Physiologie et de Biomécanique de l'Association
Peugeot S.A.-Renault
18, rue des Fauvelles,
92250 La Garenne-Colombes
France

C. Got, A. Patel,
Institut de Recherches Orthopédiques
Hôpital Raymond Poincaré
92380 Garches

ABSTRACT

Since 1970, a team of doctors and technicians, in collaboration with the police, has studied 3138 vehicles involved in accidents. The medical and technical data gathered has enabled exact determination of the circumstances and the configurations of collisions and identification of the causes of minor, serious and fatal injuries suffered by each of the occupants transported to the hospital. The material indexes dependent on the belt itself allowed pinpointing the cases in which the belt was worn effectively, thereby eliminating all risk of error in this fundamental point.

If the effectiveness of the 3-point belt is easy to measure in the experimental conditions to which the dummies are subjected, the prediction of the number of lives that the belt can spare on the road requires delicate handling. Real accidents are extremely varied and not all of them can be reproduced by the principal tests. The accident scenes must be observed not only to verify the results of these tests, but to prescribe new ones as well. The protection afforded by the safety belt is compared for drivers and front seat passengers.

In frontal impact, the effectiveness of the belt is evaluated:

- in terms of the force of the impacts characterized by the speed variation of the occupant (ΔV) on the one hand, and by the average deceleration of the vehicle (γaverage), on the other hand;
- by body regions.

Finally, the role of the belt is analyzed in lateral impact and in rollover.

THE METHOD OF ANALYSIS of real accidents used since 1970 in more than 2300 accidents is described as follows: all automobile accidents occurring in a zone situated to the west of Paris which involved bodily injuries are the object of a preliminary description by qualified police officers recorded on a special form and accompanied by photographs taken at the scene.

The information concerning the occupants involved in an

accident but not injured, or those whose condition does not warrant hospital examination or treatment, is collected at this stage of the investigation. This information is essential in judging the effectiveness of the belt. The doctors record their observations in a formula developed for this study. The severity of the injuries is recorded on a severity index for each body region according to the injury severity scale, Abbreviated Injury Scale (AIS)(1). Finally, each vehicle involved in the accident is analyzed and photographed at the garage. Next, there is a technical analysis of the reaction of the structure and of the bodywork, and, consulting the medical form, an examination of the impact outlines left by the occupants, conducted to determine the cause of their injuries. This examination results in reconstructing the trajectory of each of the occupants during impact. The participation of specialized engineers and the opinion of the doctors is sometimes necessary to eliminate certain doubts as to cause of a particular injury. It is during the course of this technical assessment that the indexes for wearing belts are investigated.

The violence of the frontal impact is evaluated with the aid of a method established in 1973 by Renault's Research and Development Department, and subsequently adopted by a growing number of foreign teams. The method allows calculation of:

- the speed variation suffered by the occupant during the critical phase of impact, that is during the vehicle's contact with the obstacle,
- the average deceleration of the vehicle during this contact phase (Fig. 1).

Both measurements are necessary: it is understood that of two occupants having been subjected to a speed variation of 50 km/hr, the one most severely exposed is the one for which this speed variation occurred in the shortest amount of time. Each vehicle involved in an accident is compared to a series of vehicles of the same make and type which were tested in various impact configurations at different known speeds. The comparison of the deformations is the connecting criteria with one of the reference vehicles. Such is the source of the basis of calculation of the violence of real impacts. It is to be noted that two circumstances often reduce this violence: the braking that precedes the accident in approximately 60% of the cases, and the breaking free of the vehicle which, after being impacted, sometimes travels considerable distance before becoming immobile. In the latter case, the initial speed of the vehicles is not completely halted by the impact, but only reduced; the reduction to zero follows during a fairly long time period during which the occupant runs almost no risk. Braking and the breaking free of the vehicle explain why the speed variations measured in accidents (and for which values will be provided later on) are considerably removed from speeds driven on the road, which are much higher.

The belts worn are 3-point anchorage systems with or without retractors. The proof of wearing the belt is furnished by observation of the deterioration of the belt's webbing and of its returns to the anchorages due to efforts to restrain the occupant (Fig. 2). These impressions are even more subtle when the force of the impact is higher. The types of belts encountered in the investigation are distributed in the following manner:

- 3 anchorage points: . without retractor: 1461, with retractor: 203
- harness: 1
- undetermined: 3

In this study we will disregard the several 2 anchorage point belts.

The review of possible impact points of the occupant against interior elements of the passenger compartment enables reconstructing the trajectory during impact. The analyst determines from examining the belt, in addition, the indexes that serve to indicate the adjustment adopted by the wearer in the case of manually adjustable belts.

The size and weight of the individual involved in the accident are known, as well as the initial position of the seat as it may be readjusted. All the indexes taken together are very informative as to the more or less correct manner in which the belt was worn at the moment of impact.

In approximately 65 % of the cases, the manually adjustable belts were adjusted correctly; i.e., the displacement of the head and shoulders of the occupants does not surpass 10 cm before the tensing of the webbing. To measure the effectiveness of the belt from a sampling of accidents implies linking of certain conditions without which the proceeding can result in incorrect conclusions. It is not sufficient to observe that belted car occupants are less severely injured than others to legitimately attribute this difference to the belt. The difference can have other causes which explain completely or in part:

- the impacts in which those wearing the belt are involved can be, on the average, less violent than the others,
- the occupants wearing belts have possibly suffered certain types of dangerous collisions - lateral collisions in particular - in a smaller proportion than those not wearing belts; but, since the severity of this category of accidents is particularly high, the disproportion would suffice to justify the fact that the bodily injuries caused by the impacts they were subjected to are less serious,
- the age distribution of the two groups compared is perhaps disadvantageous to those people without a belt. In the hypothesis where they would be older they would suffer a handicap - verified experimentally - of less resistance to impacts which would take into account only this factor in examining the greater severity of their condition.

The above influences, among others, must be controlled. If it appears that the two groups have not been subjected equally to the influences, corrections are imperative before undertaking quantification of the effects of the factor studied.

Undoubtedly the size of the sample available here presents a significant bias, but we would not know how to eliminate it in order to verify it.

The percentage distribution of impacts occurring to front seat occupants, with and without belts, by types of impacts and obstacles, is provided in Table 1.

Table 1 - Percentage Distribution of Impacts Occurring to Front Seat Occupants (1624 Belted (B) and 3242 Non-Belted (N)) by Types of Impacts and Obstacles

Types of Impacts		Fixed Objects	Cars	Heavy Trucks	Miscellaneous	Total
		Types of Obstacles				
Frontal	B	8,6	39,0	6,5	-	54,1
	N	14,2	39,5	5,2	-	58,9
Lateral	B	2,8	9,1	1,9	-	13,8
	N	5,0	10,8	2,1	-	17,9
Rear	B	0,6	15,2	2,5	-	18,3
	N	0,6	9,7	1,1	-	11,4
Rollover	B	-	-	-	12,9	12,9
	N	-	-	-	10,8	10,8
Unclassified	B	-	-	-	0,9	0,9
	N	-	-	-	1,0	1,0
Total	B	12,0	63,3	10,9	13,8	100
	N	19,8	60,0	8,4	11,8	100

One can note an under-representation of the belted people in frontal and lateral collisions against fixed obstacles and an over-representation in frontal and rear crashes against (or by) heavy trucks and also, in cases of rear collisions by other cars.

The distributions of age of belted and non-belted people are given in percentage in Table 2.

Table 2 - Distribution in Percentage of Age Classes of Belted and Non-Belted

Age (years)	Belted	Non-Belted
<25	15,7	23,2
25 - 49	41,6	43,5
$\geqslant 50$	14,6	12,0
Unknown	28,1	21,3

In general, the age of the uninjured people is difficult to determine, at least to conduct particular measurements, which explains the fairly large proportion of unknown ages. In any case, it is verified that those with a belt are not systematically given the advantage of a younger average age than the others. Even if all the occupants whose age is unknown were less than 25 years old, this would not unbalance the distributions.

Definitely, this heterogeneity, all relative, in the frequency of involvement by types of impact and obstacles, as well as by class of age, justifies specific analyses bearing on the two matched samples.

They must be composed in such a way so that for each occupant wearing a belt, there is a corresponding occupant not wearing a belt, being equivalent to each other in terms of the following factors:

- type of impact
- seating position of the occupant
- age
- direction of impact
- violence of impact (ΔV, γ average)
- intrusion
- overload

This method of comparison enables an evaluation of the performance of the 3-point belt with maximum precision. It has already been applied in frontal impact in 100 pairs of occupants wearing and not wearing belts (4) taken from a sample of the more than 3000 vehicles that are the subject of this analysis.

RESULTS

EFFECTIVENESS OF THE BELT IN ALL TYPES OF ACCIDENTS - The severity of injuries to drivers and front seat passengers, with and without a belt is given in Table 3:

Table 3 - Severity of Injuries to Drivers and Front Seat Passengers - All Types of Impacts

	Drivers		Front Seat Passengers	
O.AIS	Belted	Non-Belted	Belted	Non-Belted
0	384	424	172	155
1-2	536	1237	364	795
3-4-5	77	267	40	123
6	35	149	13	80
	1032	2077	589	1153

In the 3138 vehicles involved in an accident:

- the proportion killed among the 1032 belted drivers was divided by 2.11
- the proportion killed for the 589 passengers wearing belts was divided by 3.10

The chances of the driver wearing a belt being uninjured are multiplied by 1.8, the chances of passengers wearing belts being uninjured are multiplied by 2.3

EFFECTIVENESS OF THE BELT IN FRONTAL IMPACT - The risks against which the belt protects are of two types: ejection, and projection against the windshield. In frontal impact, of 1000 non-belted front seat occupants, 6.2 are killed as a result of ejection. Projection against the windshield, its pillars, etc. is the most serious risk to which those involved in frontal impact are subjected. Out of 1810 frontal impacts, the following mortality rates (no. killed x 100/no. involved) are registered according to seat occupied and the wearing/not wearing of safety belts:

	Non-Belted	Belted
Drivers	6.00	2.83
Front Passengers	6.39	2.54

The risk of death by projection against the pillars of the passenger compartment is divided by more than 2 for drivers and the front seat passengers.

Nonetheless, the protection of the passenger is slightly better than that of the driver. This stems from the fact that the driver sometimes impacts against the entire steering column, the aggressivity of which is even greater on a certain number of models of older design, and is also due equally to the fact that the frontal impacts against other vehicles are more often originated on the left side of the vehicles; when they are very violent they can cause collapse of the passenger compartment: the driver does not benefit from, in this case, sufficient absorption distance provided by the belt.

These results take into account all frontal impacts of the sample, including those for which the violence of the impact could not be evaluated.

But, just as measurement of the effectiveness in all accidents has no value unless the various categories of accidents are represented in proportions that do not differ significantly for those wearing belts and those not wearing belts, likewise, an analysis by type of impact merits being conducted after verification of the fact that the belted occupants suffered impacts comparable in violence to unbelted occupants.

Table 4 demonstrates that the statistical distribution of impacts by degree of violence is similar for belted and nonbelted occupants. We must specify that occupants involved in frontal impact, but for which the violence of impact could not be estimated (such as impacts against heavy duty trucks) are no longer taken into account and are consequently eliminated from the rest of the analysis.

Examination of body regions reveals that the HEAD segment (including the face) is the best protected body area thanks to the belt.

The degree of severity of injuries to the head is given in Table 5 for drivers and passengers with or without a belt.

The reduced risk of severe or fatal injuries to the head is very important for front seat passengers wearing belts. Only one head AIS $\geqslant$ 3 out of 146 involvements is recorded, compared to 16 out of 281 front passengers not wearing belts.

In spite of the 15.5 % head-against-steering-wheel contacts of drivers wearing belts (fig. 3), the reduction of risk of an AIS $\geqslant$ 3 injury to the head is 3.7. The NECK incurs serious injuries in 2 front occupants wearing belts out of 407[x]

---

[x] The differences observed in comparison with Table 4 stem from the fact that the injuries are unknown for a small number of victims.

Table 4

Percentage Distribution of Belted and Non-Belted Passengers by Classification of the Violence of Frontal Impact.

VIOLENCE OF IMPACT			
$\Delta V < 25$ km/hr or γ average < 5 g	25 km/hr $\leqslant \Delta V \leqslant 55$ km/hr and 5 g $\leqslant \gamma$ average $\leqslant$ 11 g	$\Delta V > 55$ km/hr and γ average $>$ 11 g	
49.0 %	47.9 %	3.1 %	100 %
54.0 %	42.3 %	3.7 %	100 %

Top row of percentages shows the Belted (N = 413)
Bottom row shows the Non-Belted (N = 871)

Table 5 - Degree of Severity of Injuries to the Head of Drivers and Passengers with or Without Belts

A.I.S.

Drivers	0	1	2	3	4	5	TOTAL
Belted	167	59	27	5	0	5	263
%	63.5	22.4	10.3	1.9	0	1.9	100 %
Non-Belted	180	225	109	16	4	17	551
%	32.7	40.8	19.8	2.9	0.7	3.1	100 %
Front Seat Passengers							
Belted	111	30	4	1	0	0	146
%	76	20.6	2.7	0.7	0	0	100 %
Non-Belted	56	147	62	8	1	7	281
%	19.9	52.3	22	2.9	0.4	2.5	100 %

and in 3 front occupants not wearing belts out of 813. Therefore, the belt does not cause cervical (neck-related) injuries. These injuries only occur through the effect of head impact during the course of forward movement of the upper torso in very violent impacts.

The THORAX is, with the pelvis, a susceptible body region with the belt since it is through these body parts that the main restraint force passes. Therefore, it can happen that an occupant suffers several fractured ribs when the speed variation is considerable and the deceleration level high. This problem presents itself even more sharply when the belted victim is older.

When an irreversible injury to the thorax occurs, it is most often a result of the crushing of the occupant against the front part of a completely collapsed passenger compartment.

Table 6 shows the degree of severity of thoracic injuries among drivers and front seat passengers.

Table 6 - Degrees of Thoracic Injuries

	AIS					
	0	1	2	3	4	5
Drivers						
Belted	176	74	8	3	0	2
Non-Belted	390	101	19	15	4	8
Front Seat Passengers						
Belted	82	47	10	5	0	2
Non-Belted	228	33	8	0	5	2

The PELVIC fractures that a considerable number of occupants not wearing safety belts suffer are reduced in large proportions by this.

Table 7 - Degrees of Pelvic Injuries

	AIS					
	0	1	2	3	4	5
Drivers						
Belted	239	17	3	3	0	0
Non-Belted	512	4	1	9	11	0
Passengers						
Belted	135	9	1	1	0	0
Non-Belted	253	4	4	11	4	0

The fractures with displacement (AIS 3 and more) that occur among 1 % of those wearing belts and 4.3 % of those not wearing belts occur following an excessive force transmitted by the knee-femur segment after an impact below the dashboard. The reduction of risk with the belt is appreciable: it is divided by more than 4.

The ABDOMEN is exposed to the risk - eliminated in vehicles of recent design - called "submarining". For models where the installation of the belts still leaves something to be desired, the sub-abdominal webbing which should normally catch hold and sustain its support on the pelvic bones, rises again during the movement the occupant makes toward the front and compresses the abdominal wall; severe injuries can result. This mechanism, well-researched today, is avoided by a much more advanced positioning of the anchorage between the seats. It is to be noted that, in spite of these submarining injuries, the risk of abdominal injuries among those wearing belts is not higher that for those not wearing belts (see Table 8).

Table 8 - Degree of abdominal injuries

	AIS					
	0	1	2	3	4	5
Drivers						
Belted	244	12	0	0	5	1
Non-Belted	491	25	1	1	12	7
Passengers						
Belted	135	7	0	2	2	0
Non-Belted	268	4	0	1	3	0

The effectiveness of the belt in frontal impacts has LIMITS. The limits are a function of two factors: those due to the designs of the vehicles and those due to the impact resistance of the human organism.

It happens that these limits are attained almost simultaneously, and it is not by chance. Since the time that vehicles have been equipped with belts, it has been necessary to produce a deformable front structure in order to maximize the distance by which the occupant reduces his speed, as was noted previously. However, there are several ways to adapt the structure to this requirement. It can be calculated in such a way that the maximum deformation is attained, without intrusion in the passenger compartment, at a speed of impact, against a stiff wall, of 80 km/hr, for example. The threshold can also be set slightly lower, e.g. 55 km/hr. The solutions and their consequences for the occupants will be very different. If the speed is fixed at 80 km/hr, the structure will have to be of considerable stiffness given the considerable quantity of energy to dissipate in the short useable distance of the vehicle; in the same way the stiffness of the belt will have to be sufficient so that its elongation does not allow impact of the occupant against the passenger compartment. If these conditions are fulfilled, it will be demonstrated that not only is the impact intolerable for the human organism, but also that in 55 km/hr tests, the rigidity of the structure and that of the belt is too high. The impact will then provoke a slight deformation and slight elongation of the belt; the deceleration will be much more acute and will expose the most fragile occupants to the risk of fatal injuries. Therefore a reference speed as high as possible, but satisfying the following requirements, must be chosen:

- assure the survival of a belted occupant at this speed,
- not disadvantage occupants subjected to impacts occurring at lower speeds,
- not render the vehicle more aggressive with regard to the others, particularly in frontal-lateral collisions; in fact, the severity of lateral impact depends for the most part on the architecture and the rigidity of the front of the impacting vehicle.

These requirements are incorporated in present production vehicles.

96 to 98 % of frontal bodily impacts, i.e., causing, at least, injury to an unbelted passenger, occur at a violence that does not surpass 55 km/hr speed variation and 11 g of average deceleration; this is just about the reference speed chosen for impact tests that verify the chances of survival are then maximum and which structure in the conditions most often encountered in real accidents. It is understood then that wearing the safety belt is not a guarantee of protection against any violence of impact. Knowing equally that resistance to impact varies greatly according to individuals - age in particular greatly increases the fragility of the skeleton - it is not surprising that the effectiveness of the belt decreases as violence of impact increases. Table 9 makes this evident.

Table 9 - Comparative Effectiveness of the Belt in Reduction of Fatalities According to Degree of Violence

	Degree of Violence		
	$\Delta V < 25$km/hr or γ average < 5g	25km/hr $\leqslant \Delta V \leqslant$ 55km/hr and 5g $< \gamma$average $<$ 11g	$\Delta V >$ 55km/hr and γaverage $>$ 11g
Belted			
Killed	0	4	4
Involved	202	198	13
Killed/100 Involved	0%	2%	31%
Non-Belted			
Killed	2	44	15
Involved	471	368	32
Killed/100 Involved	0.5%	12%	47%
Effectiveness of the Belt	absolute	risk of death divided by 6	risk of death divided by 1.5

EFFECTIVENESS OF THE BELT IN LATERAL IMPACT - 531 vehicles analyzed in the investigation suffered a lateral impact. 27 % of the front seat occupants were wearing safety belts. The statistics are given in Table 10 as a function of the incidence of ejection and wearing or not wearing of safety belts.

Table 10 - Effectiveness of the Belt in Lateral Impact

		Belted	Non-Belted
Ejected:	- killed	0	24
	- involved	0	83
	- killed/100 involved	0%	29%
Not Ejected:	-killed	13	35
	-involved	224	499
	-killed/100 involved	5.8%	7%
Total:	- killed	13	59
	- involved	224	582
	- killed/100 involved	5.8%	10.1%

Wearing the belt therefore divides the risk of being killed by 1.7, essentially because the belt prevents ejection.

If the total of risks of being killed or severely injured is considered, the belt divides the risk by 2. The belt prevents numerous severe injuries following the projection of occupants situated on the side opposite the impacted side.

EFFECTIVENESS OF THE BELT IN ROLLOVER - Ejection is the primary risk against which the safety belt protects. Risk diminishes, as in other types of impact, not only due to safety belts, but also is attributable to a new system of door locking which has been installed in vehicles in France since 1972, and which considerably reduces the risk of opening. As shown in Table 11, the mortality rate for those not wearing belts is five times higher than for those wearing belts.

If the death rate for those wearing seat belts is slightly higher than that for the non-belted remained aboard the vehicle, it is obviously due to the fact that rollovers involving belted occupants cover the entire range of violence. However, non-belted occupants who were not ejected no doubt underwent less serious rollovers.

CONCLUSIONS

The two-disciplinary survey constitutes an excellent tool in evaluating the effectiveness of a protection system provided however, that the sample is representative of all the serious and fatal accidents occurring in the country. As concerns evaluation of the effectiveness of the seat belt, the following are necessary:

- whether or not the seat belt was worn must be determined by a meticulous analysis of the state of the seat belt

Table 11 - Mortality Rate Observed in Rollover Among Those Wearing/Not Wearing Belts According to Occurance of Ejection

		Belted	Non-Belted
Ejected:	- killed	-	31
	- involved	-	87
	- killed/100 involved	-	35.6 %
Not Ejected:	- killed	4	2
	- involved	209	262
	- killed/100 involved	1.9%	0.7%
Total:	- killed	4	33
	- involved	209	349
	- killed/100 involved	1.9%	9.5%

and the vehicle, and not on declarations or reports alone,

- all seriously damaged vehicles must be considered including those which did not involve body injuries.

The analysis of 3138 damaged vehicles examined since 1970 comprising 1624 belted and 3242 non-belted front-seat occupants gave the following results clearly establishing the effectiveness of wearing seat belts.

1 - Wearing a seat belt did not penalize any of the 1624 belted occupants studied in our survey. This fact results from a case by case examination of each of the belted occupants.

2 - For the 3138 damaged vehicles, the proportion of fatalities among the 1032 belted drivers is divided by 2.1; the proportion of fatalities among the 589 belted passengers is divided by 3.1. The chances of being uninjured for belted drivers are multiplied by 1.8; the chances of being uninjured for belted passengers are multiplied by 2.3.

3 - The effectiveness of wearing a seat belt is shown clearly when the different types of collisions are distinguished. In all cases, it remains high, as can be seen in the following table:

Type of Collision	% of Cases	Division of Risk of Being Killed
Frontal	58	2.3
Side-Impact	17	1.7
Rear	13	2.1
Rollover	11	5.0
Others	1	1.3

Because it prevents ejection, wearing seat belts divides the risk of being killed in rollovers by five.

For frontal collisions, which are the most frequent, the protection afforded to the occupants is high, since:

- the risk of serious or fatal head injuries (AIS $\geqslant$3) for belted front-seat passengers is considerably reduced in comparison with non-belted occupants (respectively 1 case/146 and 16 cases/281).

- for belted occupants, protection afforded by the seat

belt decreases as the violence of the collision increases. However, wearing the belt still divided the risk of fatality by 1.5 in the most violent collisions ($\Delta V > 55$ km/hr and average $\gamma > 11$ g).

REFERENCES

1. J.D. States, D.F. Huelke and L.N. Hames, "1974 AMA-AAAM Revision of the Abbreviated Injury Scale (AIS)", Proceedings of 18th A.A.A.M. Conference, 1974.

2. P. Ventre, J. Provensal, "Proposition d'une méthode d'analyse et de classification des sévérités de collision en accidents réels", Proceedings of 1st IRCOBI Conference, Amsterdam, June 26-27, 1973.

3. C. Tarrière, A. Fayon, F. Hartemann, P. Ventre, "The Contribution of Physical Analysis of Accidents Towards Interpretation of Severe Traffic Trauma", Proceedings of 19th Stapp Car Crash Conference, San Diego, California, Nov. 1975, Edited S.A.E., New-York.

4. F. Hartemann, C. Thomas, C. Henry, J.Y. Foret-Bruno, G. Faverjon, C. Tarrière, C. Got, A. Patel, "Belted or Not Belted: the Only Difference Between Two Matched Samples of 200 Car Occupants", Proceedings of the 21st Stapp Car Crash Conference, SAE paper 770 917, S.A.E. Inc., New-York, 1977.

5. J.Y. Foret-Bruno, F. Hartemann, C. Thomas, A. Fayon, C. Tarrière, C. Got, A. Patel, "Correlation Between Thoracic Lesions and Force Values Measured at the Shoulder of 92 Belted Occupants Involved in Real Accidents", Proceedings of 22nd Stapp Car Crash Conference, S.A.E. Paper 780 892, S.A.E. Inc., New-York, 1978.

6. C. Thomas, J.Y. Foret-Bruno, F. Hartemann, C. Tarrière, C. Got, A. Patel, "Influence of Age and Restraint Force Value on the Seriousness of Thoracic Injuries Sustained by Belted Occupants in Real Accidents", Proceedings of 4th IRCOBI Conference, Gõteborg, Sweden, 5/7 Sept. 1979.

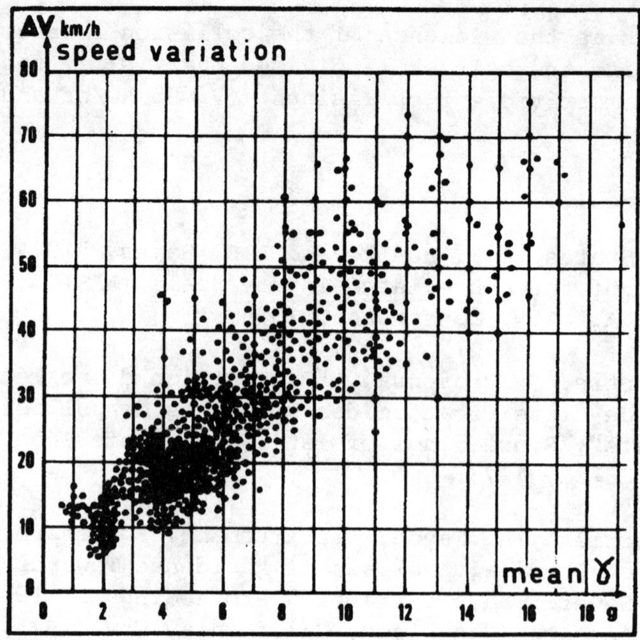

Figure 1 - Diagram of frontal Car Crashes Violence

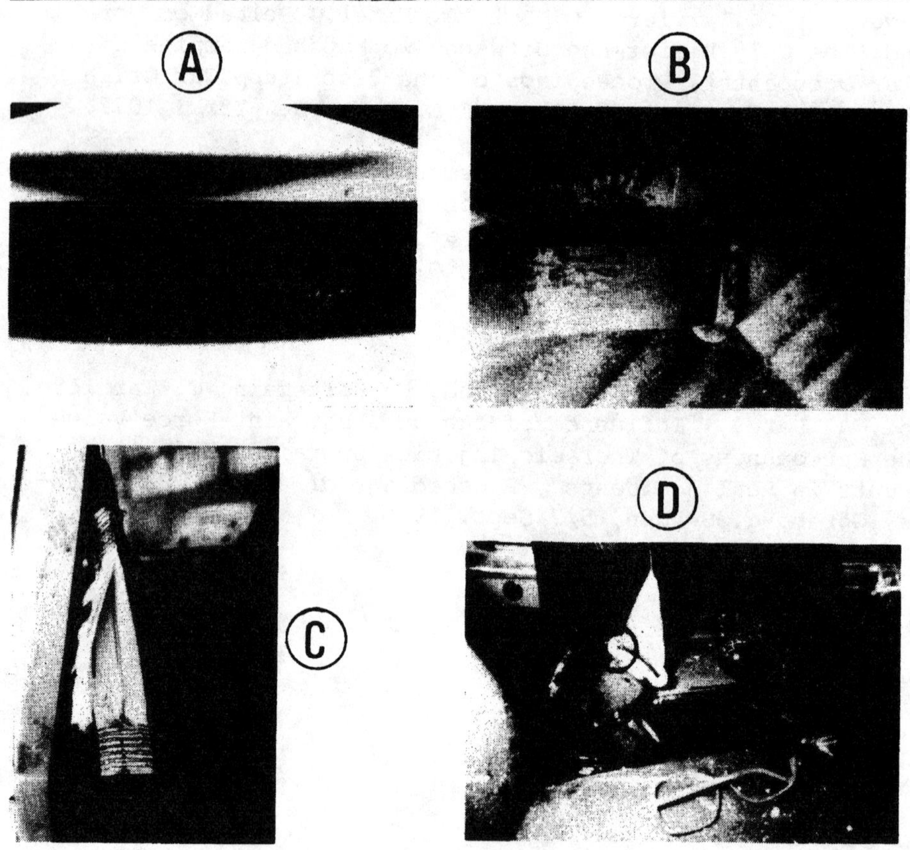

Figure 2 - Examples of Marks Proving that the Occupant Was Belted
(a) Abrasions on Running Loop Coating
(b) Webbing Abrasions
(c) Tear Webbing After Impact
(d) Damages to Running Loop Protection

DRIVERS

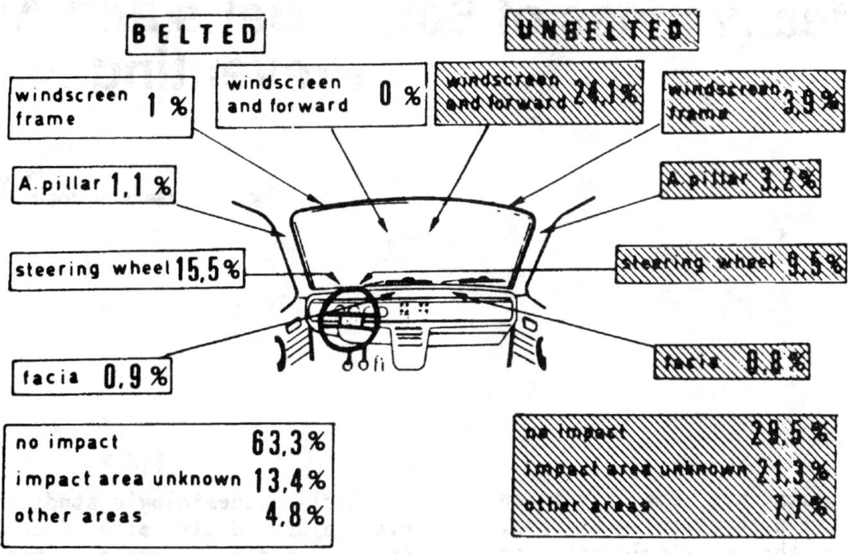

FRONT PASSENGERS

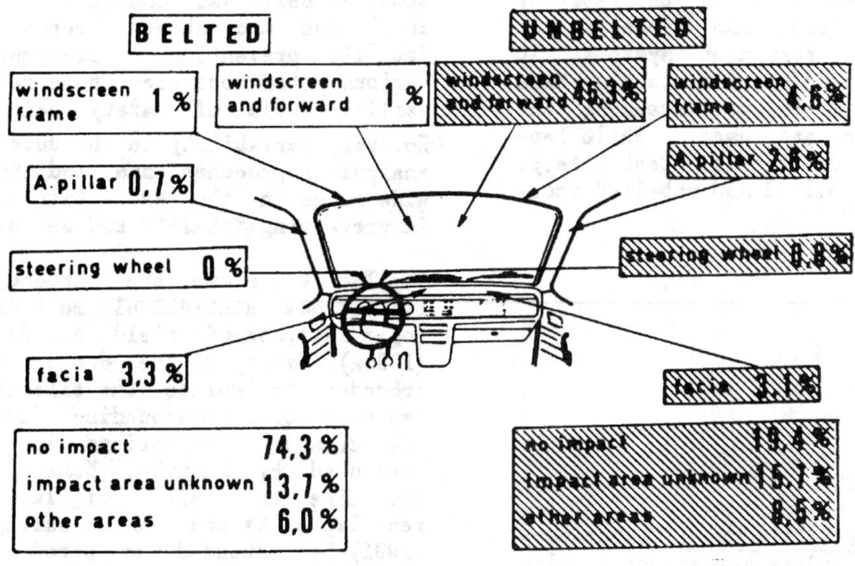

Figure 3 - Frequency of Drivers and Front Passengers Head Contacts With the Most Important Parts of the Passenger Compartment in Frontal Collisions. Comparison Between Belted and non-Belted

910901

Effectiveness of Safety Belts and Airbags in Preventing Fatal Injury

David C. Viano
Biomedical Science Dept.
General Motors Research Laboratories
Warren, MI

ABSTRACT

Airbags and safety belts are now viewed as complements for occupant protection in a crash. There is also a view that no single solution exists to ensure safety and that a system of protective technologies is needed to maximize safety in the wide variety of real automotive crashes. This paper compares the fatality prevention effectiveness. and biomechanical principles of occupant restraint systems. It focuses on the effectiveness of various systems in preventing fatal injury assuming the restraint is available and used. While lap-shoulder belts provide the greatest safety, airbags protect both belted and unbelted occupants.

Early epidemiologic studies of motor vehicle injury dealt with fleet evaluations of interior safety features, including the energy absorbing steering system, high penetration resistant windshields, and side-guard door beams. The first comprehensive study of lap-shoulder belts was conducted by Bohlin (1967) in Sweden and showed impressive injury and fatality prevention. Subsequent studies in various other countries have substantiated the earlier levels of safety belt effectiveness. However, variability in the underlying data and analysis approaches has led to a relatively wide range in the estimation of effectiveness in preventing fatality and serious injury.

The 1980's saw the introduction of sophisticated new statistical methods for epidemiologic analyses of field accident data. Evans (1986a) developed the double pair comparison procedure to isolate the effectiveness of belt use from other confounding factors in automotive crashes. Variations in the method have been used by Partyka, Kahane and others to determine the safety of lap-shoulder belts, rear lap belts and child safety seats. Evans (1991) has extended the approach to investigate the effects of alcohol use, occupant age, seating position, and direction of crash on fatality risks. The methodology has enabled accurate quantification of factors influencing crash injuries.

RESTRAINT EFFECTIVENESS

Estimates of Effectiveness: Laboratory tests involving dummy injury assessment provide an objective evaluation of safety systems, but not an accurate estimate of restraint effectiveness in saving lives and preventing injury. In part, this is due to the limited type of crash testing conducted in relation to the wide range of real world crashes and the evolution of test dummies and injury criteria in simulating the responses of real occupants. While significant improvements have been made in the biofidelity of dummies, understanding of biomechanics, and validity of injury criteria that enable laboratory tests to better predict restraint effectiveness and better related to real-world safety performance, the most objective assessment of the effectiveness of occupant restraints is by analysis of real-world crashes.

Statistical Analysis of Restraints in Fatal Crashes: Evans (1986a) developed the double paired comparison method to determine the effectiveness of occupant restraint as a function of seating position and crash direction. The method compares the number of fatalities to either of two occupants under two conditions, such as restrained or unrestrained, driver or passenger, and ejected or non-ejected. One of the occupants serves a normalizing or exposure estimating role for the frequency of fatality of the other under a particular crash situation. This forms the basis for the estimation of restraint effectiveness.

For example, the effectiveness of safety belt use by the right-front passenger (RFP) is determined by the double pair comparison using two sets of fatal crashes. The first set consists of a restrained RFP and an unrestrained driver, at least one of whom is killed. From the numbers of RFP and driver fatalities, a restrained RFP to unrestrained driver fatality ratio is calculated. From the second set of data on unrestrained RFPs and unrestrained drivers, an unrestrained RFP to unrestrained driver fatality ratio is similarly estimated. When the first fatality ratio is divided by the second, it gives the probability that a restrained RFP is killed compared to that of an unrestrained RFP in actual traffic accidents. This is the estimate of restraint system effectiveness.

Restraint Effectiveness: Determining the fatality prevention effectiveness of lap-shoulder belts was one of the first applications of the double pair comparison (Evans 1986b). Lap-shoulder belts were shown to be (41 ± 4)% effective in preventing fatality for front-seated occupants. They are (42 ± 4)% effective for drivers and (39 ± 4)% for right-front passengers (Table 1). The overall effectiveness was estimated at 43% by combining the average from Evans with those of NHTSA and others estimating 40-50% effectiveness.

Subsequently, the effectiveness of lap-shoulder belts was determined as a function of the direction of impact (Evans 1988c), including contributions from reducing ejection (Evans and Frick 1989). This type of analysis (Figure 1) helps differentiate two essential components of occupant protection with safety belt use. One is protection against ejection and is primarily due to the lap portion of the belt system. The other is mitigation of interior impact and is largely contributed by upper body restraint from the shoulder harness.

The relative safety contribution by reducing ejection with belt use significantly depends on the type of crash and is highest in primarily rollover accidents. Safety belt use is most effective in preventing driver fatality in crashes where rollover is the first harmful event and where occupant containment is a key feature of safety performance. The lap-shoulder belt system is least effective in preventing driver fatality in left-side impacts as the principal point of vehicle impact and deformation of the occupant compartment are critical factors in increasing fatality risk for an occupant.

The difference between overall lap-shoulder belt effectiveness and ejection reduction was used by Evans (1988c) to determine the safety benefit of a driver and passenger airbag system. The analysis inferred the fatality reduction effectiveness of the airbag only system as a component of impact mitigation and estimated an (18 ± 4)% effectiveness for the unbelted driver and (13 ± 4)% for the unbelted right-front passenger. This level compares favorably with the results of an expert judgment of fatality prevention potential of airbag restraints (Wilson and Savage 1973).

In a further study, Evans (1988a) determined an overall (18 ± 9)% effectiveness of lap-belt use by rear-seated occupants. Most of the safety benefit is from anti-ejection since this level compares favorably with the 17-19% effectiveness of lap-shoulder belts in preventing ejection. A major component of lap-belt effectiveness is occupant containment in the vehicle. For primarily frontal crashes, the effectiveness of lap-belt use by rear seated occupants has a larger variability and is relatively low for impact mitigation.

Lap-shoulder belts for rear outboard occupants are available in most new passenger vehicles and will increase in the vehicle fleet. However, there isn't sufficient crash injury data to conduct a statistical analysis of effectiveness. It is possible to use the understandings of restraint effectiveness in other seating positions to make a judgment about the level of impact mitigation and ejection prevention. Lap-shoulder belts are estimated to be 27% effective in preventing fatal injury in rear seats. This is essentially due to impact mitigation improvements.

Table 2 summarizes the available estimates of belt restraint and airbag effectiveness in preventing occupant fatalities. The results show the substantial effectiveness of lap-shoulder belts. They are clearly the principal safety feature in protecting occupants in severe crashes. Since airbags and other interior components do not hold the occupant in a seating position, they have lower overall effectiveness because of a much lower effectiveness in reducing ejection. Their contribution to overall crash protection is essentially limited to impact mitigation which is about half the overall benefit of lap-shoulder belt use.

The overall safety benefit of the combination of lap-shoulder belt use and airbag has not been determined from field accident data. However, it is possible to estimate the effectiveness by considering the current safety effectiveness studies and the frequency of unsurvivable crashes. Based on analysis of fatal crashes to unbelted occupants, Huelke, Sherman and Murphy (1979) estimated that approximately 50% of the crash fatalities are unpreventable by currently available occupant restraints. The limit, in part, reflects the

Table 1

Effectiveness of Occupant Protection Systems
(Adapted from Evans with additional estimates added)

Safety System	Impact Mitigation	Ejection Prevention	Overall Effectiveness
Driver			
Lap-Shoulder Belt	(23 ± 4)%	(19 ± 1)%	(42 ± 4)%
Airbag	(18 ± 4)%	-	(18 ± 4)%
EA Steering System	6%	-	(6 ± 3)%
Lap-Shoulder Belt (Plus Air Bag)	[27%]	[19%]	[46%]*
Right Front Passenger			
Lap-Shoulder Belt	(22 ± 4)%	(17 ± 1)%	(39 ± 4)%
Airbag	(13 ± 4)%	-	(13 ± 4)%
Friendly Interior	[<6%]	-	[<6%]
Lap-Shoulder Belt (Plus Airbag)	[26%]	[17%]	[43%]
Rear-Seat Passenger			
Unbelted Rear Versus Front-Seat Position	-	-	(26 ± 2)%
Lap-Belt	(1 ± 9)%	(17 ± 1)%	(18 ± 9)%
Lap-Shoulder Belt	[10%]	[17%]	[27%]

*[] New estimates of restraint effectiveness based on judgement.

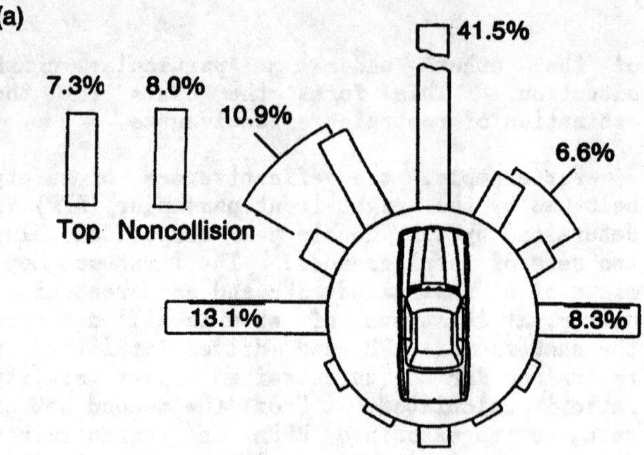

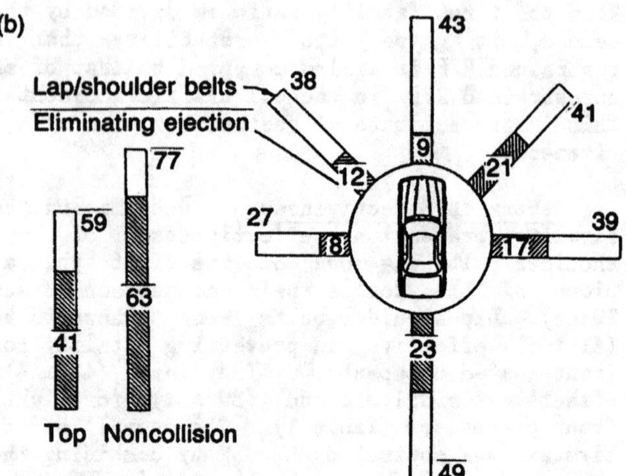

Figure 1: (a) Distribution of driver deaths by principal impact point and (b) effectivenes of lap/shoulder belts in preventing driver fatalities. The fraction of fatali ties prevented by eliminating ejection is shown in the hashed portion of the ba according to impact direction. For example, in frontal (12 o'clock) crashes lap/shoulder belts prevent 43% of driver fatalities; 9% of this is due to elimin ating ejection, so that 34% is due to interior impact reduction (from Evans 1988 with permission).

severity of many fatal crashes which may involve extreme vehicle damage and forces on the passenger compartment, unusual crash configurations and causes of death, and unique situations associated with particular seating positions and crash dynamics.

By comparing the level of unpreventable fatalities with the overall protective effect of lap-shoulder belts, Viano (1988a) found a maximum of 7% additional fatality prevention may occur with a combination of restraints or additional supplemental features. Based on a similar review of fatalities to lap-shoulder belted front-seated occupants, a potential 3-5% additional fatality prevention is estimated with a combination of lap-shoulder belt and airbag system.

Using a 4% additional benefit to safety belt wearers from airbags, there is an overall effectiveness of 46% for the belted driver with supplemental airbag (Figure 2) and 43% for the right-front passenger. However, this estimate is based on expert judgment and not statistical methods applied to crash injury data.

The 4% increment in effectiveness with the airbag supplementing lap-shoulder belt use is consistent with a 5% additional benefit determined by NHTSA (1984). The fact that a lap-shoulder belt and airbag are only 43% to 46% effective in preventing fatality underscores that absolute protection is not achievable by occupant restraints and that injury and fatality will continue to occur to belt wearers despite a significant overall net safety gain by restraint usage, even if augmented by airbags. This is because of accident severity, configuration, and limits of human tolerance.

Fatality risk depends on the particular occupant seating position, the principal impact point and the proximity of passenger compartment crush. Evans and Frick (1988) have shown (Figure 3) more than a 7 to 1 increase in fatality risk for a right-front passenger in right-side (nearside) versus a left-side (far-side) impact.

A better appreciation of restraint effectiveness estimates may be possible by a fuller understanding of the biomechanics of restraint systems in reducing impact forces and controlling occupant kinematics over the wide variety of real-world crash. The following section expands on a recent review by Viano (1988b) and addresses key features of occupant restraint performance in frontal and rollover accidents.

RESTRAINT BIOMECHANICS AND PERFORMANCE

Driver Airbag: During the development of crash protection systems for automobile occupants in the early 1960's, many concepts for energy absorption and load distribution were conceived and evaluated. A driver airbag took advantage of rapid filling of a concealed bag to provide a cushion in front of an occupant in a crash (see review of SAE publications in Viano 1988c). This provided a large area to gradually decelerate the driver. However, it was necessary to provide lower torso restraint through either a lap-belt or knee bolster to prevent the driver from submarining the airbag (Figure 4a and b) and experiencing higher forces on interior impact in a frontal crash.

Although compressed gas in a cylinder was one of the early concepts, the eventual production system took advantage of a relatively small mass (70-100 g) of sodium azide. The chemical can rapidly inflate a 80-100 l driver airbag by ignition and conversion to harmless nitrogen gas when sensors detected a severe frontal crash. The airbag is stored inconspicuously in the interior and deployed only during a severe crash, sensors are located in the engine compartment to detect rapid decelerations of the front-end during a crash and an electronic system is used to monitor and initiate deployment of the airbag.

Production airbag systems were available from General Motors in 1974. However, the customer demand never developed for the safety option so the program was cancelled in 1976 after selling only 10,321 vehicles. Mertz (1988) recently published an analysis of the field performance of the airbag based on the work of Pursel et al (1978) using comparable non-airbag crashes. The driver airbag system was found to be 21% effective in preventing AIS 3+ injuries (16% for the passenger system), but was -34% ineffective for AIS 2+ passenger injury. However, the matched crash analysis only allows the determination of effectiveness in deployment accidents. The overall effectiveness of a driver airbag in typical crashes would be approximately 9% in preventing AIS 3+ driver injuries, assuming deployment accidents represent 45% of all severe crashes.

Many of the airbag inflation injuries occurred in moderate severity crashes without distortion of the occupant compartment. They were considered "unexpected" and caused by the occupant being near the cushion at the time of deployment. Horsch and Culver (1979) and more recently Horsch et al (1990) have observed significant forces adjacent to an airbag if the normal path of inflation is blocked by an occupant. Thus, the deployment, per se, of an airbag has the potential to seriously injury or kill, independent of crash severity.

The use of lap-shoulder belts helps minimize the risk of airbag injury, but there is a potential consequence of public misperception of driver airbag protection. Some car occupants may not recognize that the steering wheel airbag is only a supplement, and not an alternative, to the primary occupant protection system in the vehicle, the lap-shoulder belts.

Table 2

Effectiveness of Occupant Restraints

	Driver	Right Front Passenger	Rear Passenger
Lap-Shoulder Belts	(42 ± 4)%	(39 ± 4)%	[27%]
Airbag Only	(18 ± 4)%	(13 ± 4)%	--
Belts and Airbag	[46%]	[43%]	--
Lap Belt	--	--	(18 ± 9)%

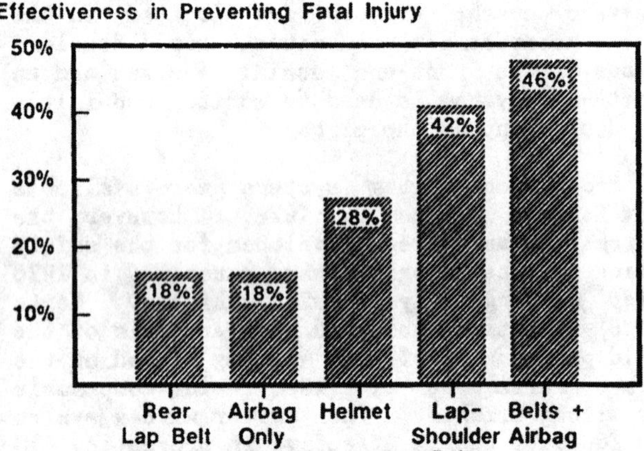

Figure 2: Effectiveness of safety devices in preventing fatal driver injury in severe motor vehicle crashes and motorcyclist injury with helmet use (from Viano 1990 with permission).

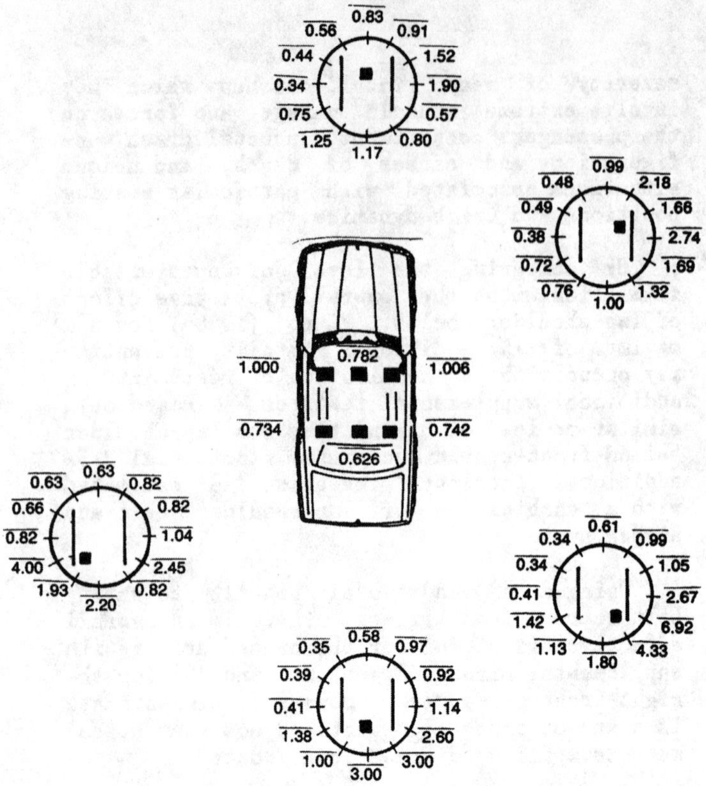

Figure 3: Relative fatality risk to passengers in different seating positions in relation to that of the driver set at 1.000 and fatality risk for passengers in various car seating positions relative to the principal direction of impact (based on Evans and Frick 1988 with permission).

If drivers fail to buckle up, they significantly reduce their driving safety, particularly in side impact and rollover crashes where the lack of belt restraint in the seat may lead to interior impact or ejection. Evans (1989) calculated a 41% increase in fatality risk by drivers ceasing to wear the lap-shoulder belt because they have a supplemental bag system. Fortunately, current belt use surveys are not finding a lower rate of belt use in airbag equipped vehicles.

Lap-belts: An early use of the lap-belt was to complement crash protection of the chest by a driver airbag or the energy absorbing (EA) steering system, another 1960's interior safety concept. The belt controlled forward excursion of the lower torso (Figure 4c) by restraining the pelvis and maintained an upright posture of the occupant (Horsch, Peterson and Viano, 1982).

In the absence of pelvic restraint, loads would be applied through the knees and seat pan to restrain the occupant and control the upright posture of the driver. These loads are necessary to take advantage of the cushioning affect of the airbag or EA steering column on the upper body. The lap-belt also complemented early designs for passenger protection by a padded dash board and found their way into rear seating positions to supplement padded seat backs (Figure 4d).

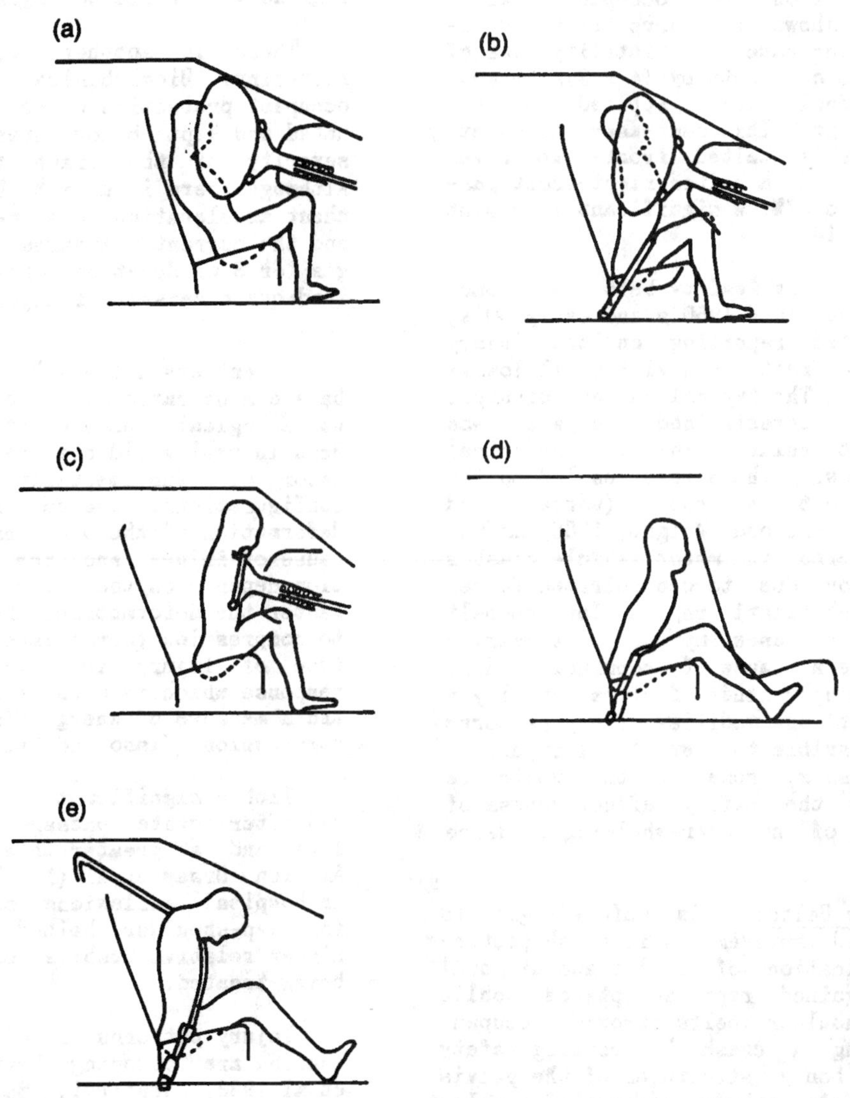

Figure 4: Kinematics of a driver restrained by (a) a steering wheel airbag where loads act on the upper body but not on the lower extremity, (b) the combination of lap belt and airbag giving pelvic and upper body restraint, (c) the energy absorbing steering system and knee bolster, and passenger restrained by (d) a lap belt in the rear seat and (e) a lap-shoulder belt in the right-front seating position.

Although much of the development work on restraints has involved frontal barrier and sled testing, safety engineers recognised the importance of the lap-belt in preventing ejection, which had been identified as the leading cause of fatality in the mid-1950's by crash investigations. Thus, occupant containment within the vehicle during a crash was a principal benefit of lap-belt usage. Since lap-belt restraint was only one part of the systems engineering for occupant protection by friendly interior components, development work focused on assuring that the various safety components worked together as a system in a crash.

During testing, it also became apparent that lap-belt use also reduced the potential for rear-seat occupant loading on the front-seat back. This is particularly important for belt restrained front-seat occupants, since Park (1987) has shown that unrestrained rear-seat occupants increase the fatality risk of belted front-seat occupants by (4 ± 2)% because of the additional load applied on the restrained occupant. This decreases the safety effectiveness for a belted front-seat driver from 42% to 40% and a belted right-front passenger from 39% to 37%, a significant increment in crash injury risk.

With the advent of safety belt use in passenger cars in the late 1950's and early 60's, physicians started reporting on the injury patterns of belt restrained victims (Kulowski and Rost, 1956). The typical pattern of upper body injury to unrestrained occupants was replaced by belt related injury to abdominal organs and tissues. These injuries led to the phrase "seat belt syndrome" (Garrett and Braunstein 1962, Fish and Wright, 1965) as the new injury patterns in motor vehicle crashes received attention due to concentrated forces on the lower abdominal region for lap-belt wearers. In many cases improper belt wearing was identified as a cause of abdominal loading and injury, although other factors may play a role. Since belt use modifies injury patterns and it is not possible to prevent all injury to occupants in crashes, some in the public are suspicious about the safety effectiveness of belts, in spite of now over-whelming evidence of benefit.

Lap-Shoulder Belts: As safety engineers pursued continued improvements in crash protection, the combination of a lap and diagonal shoulder belt gained rapid acceptance (Bohlin 1967). Lap-shoulder belts provide occupant restraint during a crash by routing safety belts over the boney structures of the pelvis and shoulder. This takes advantage of a relatively high tolerance to impact forces for these regions of the skeleton and avoids concentrating load on the more compliant abdominal and thoracic regions. Control of occupant kinematics helps insure maximum protection by belt restraints.

The fundamentals of a high quality belt restraint system involve kinematic controls (Adomeit and Heger 1975, Adomeit 1977, 1979, Viano and Arepally, 1990) which maintain the lap portion of the belt low on the pelvis through adequate seat cushion support. This minimises pelvic rotation and reduces the tendency for the lap-belt to slide off the illium and directly load the abdomen. Forward rotation of the upper torso of slightly greater than 90° upright posture (Figure 4e) directs a major portion of the upper torso restraint into the shoulder and upper chest region. This reduces loads on the more compliant areas of the lower rib cage. Each of the kinematic controls helps direct forces onto the skeletal structures above and below the center of gravity of the torso thus balancing the restraint and keeping loads away from more compliant body regions vulnerable to injury.

There is another component of quality restraint. Biomechanical responses related to occupant protection assessment need to be evaluated and kept below human tolerance for the severity of the crash test (Horsch 1987). Although there is a rich history in the use of chest acceleration as a measure of injury risk and the current standards require less than 60 g's for 3 ms duration, chest acceleration is an inadequate measure of injury risk.

Recent evaluations by Groesch et al (1986) have demonstrated that chest acceleration is not a logical measure of restraint effectiveness in real-world crashes or an adequate indicator of fundamentally different restraint configurations. The current evidence points to deformation of the body and its organs as the cause of injury, and the tolerance to deformation depends on the velocity of loading. The faster the deformation, the lower the tolerance to compression (normalized deformation). This type of injury is related to the Viscous response which is a cause of soft tissue injury and a measure of energy dissipated during rapid compression (Viano and Lau 1988).

With a significant increase in safety belt use after state passage of mandatory wearing laws and a greater health consciousness in America, Orsay et al (1988) has seen reductions in hospital admissions and injury severities for lap-shoulder belted occupants, even as higher relative numbers of belted victims are being treated.

Injury patterns associated with safety belt wearing are becoming better understood (Denis et al 1983, Arajarvi, Santavirta and Tolonen, 1987, Banerjee 1989). The potential for improper use has also continued with the advent of lap-shoulder belt systems, particularly placement of the shoulder harness under the arm (States et al 1987) and wearing of the lap-belt high on the abdomen with poor seating posture.

An improvement in passenger safety has been made by adding a shoulder belt to the out-board rear-seat lap-belt. This further improves the safety of rear-seat occupants, which is otherwise similar to belted front-seat occupants because of the inherently greater safety of rear seating positions (Evans 1988a). That is, the restraint effectiveness of lap-shoulder belt use by front-seat occupants is similar to lap-belt use by rear-seat occupants (39% effectiveness for lap-belted rear-seat occupants versus 43% for a lap-shoulder belted driver and 39% for the right-front passenger). Thus, combining the shoulder harness to the rear-seat lap-belt should increase effectiveness by mitigating interior contacts of the upper body. This should increase the safety of belted rear-seat occupants to an estimated level of 27% with an overall gain in safety over those belted in the front seat.

An analysis of rollover crashes by Huelke, Compton and Studer (1985) determined that safety belt use virtually eliminated the risk of paralysing cervical injury and ejection. This observation is consistent with Evans' (1986b) finding that belt restraints are most effective in rollover crashes, particularly where rollover is the first harm event. In such crashes, lap-shoulder belts are (82 ± 5)% effective in preventing driver fatalities. Sixty-four percent (64%) of the overall effectiveness is by preventing ejection with the remaining 18% effectiveness by reducing forces due to interior impacts.

In contrast, unrestrained occupants in rollovers are subjected to a series of impacts and potential ejection during a rollover (Viano 1990). In a simulation of experimental rollovers by Robbins and Viano (1984), the complex kinematics of an unbelted occupant were studied using several crash scenarios, including complete occupant containment, and subsequent driver or passenger door opening (Figure 5). The potential for serious injury by interior impacts or ejection were apparent in the rollover sequences for the unrestrained driver.

PERCEPTIONS OF OCCUPANT RESTRAINTS

Interestingly, the overall fatality prevention effectiveness of a driver airbag without safety belt use and lap-belt use in the rear is similar. Both systems provide an 18% reduction in fatal crash injury risk. This level of effectiveness is far greater than built-in safety devices, such as the EA steering column at 6% effectiveness, and high penetration resistant windshields, side door-beams, and interior padding which have lower effectiveness levels. Therefore, airbags and rear-seat lap-belts are effective automotive safety features and are second only to front-seat lap-shoulder belts which have a 41% safety effectiveness. However, this is the effectiveness when the airbag is available and the lap belt is used. It also does not consider the effects of occupancy rates on overall injury prevention.

Driver airbags and rear-seat lap-belts provide protection in different crash types. Airbag effectiveness is essentially due to interior impact mitigation in primarily frontal crashes. It has minimal effectiveness in lateral or rollover crashes. In contrast, rear-seat lap-belt effectiveness is essentially due to ejection prevention in primarily non-frontal crashes such as rollovers, since it is less effective in reducing impact forces in frontal crashes. Rear-seat lap-belts are currently available in virtually all passenger cars and only require buckling by an occupant to be effective. On the other hand, driver airbags are available in much fewer vehicles but provide crash protection independent of any action by the occupant.

Both systems have the possibility of modifying injury in particular situations. The high energy release of an airbag may injure an occupant against the system at the instant of deployment. Blocking the path of deployment increases pressures in the cushion during gas generation and develops high forces on the occupant. Since the force occurs with high velocity, there is a risk of injury by a Viscous mechanism.

Placing the lap-belt on the abdomen as compared to correctly on the pelvis does not direct restraining loads through the skeleton in severe frontal crashes. Abdominal loading and deformation occurs as forces in the lap-belt restrain forward excursion of the lower body. This situation, and another in which the lap-belt slips above the pelvis during a crash, may result in abdominal organ injury and hemorrhage by submarining pelvic restraint.

Since there is a general lack of information on the technical aspects, performance, and efficacy of safety systems, the public has developed an exaggerated perception of the safety effectiveness of a driver airbag and insecurities about rear-seat lap-belt use. Many falsely believe that airbags are safer than safety belts. The facts about safety systems need to be accurately covered in the news and conveyed to the public to gain understanding.

In addition, there should be a better awareness that safety devices cannot provide full or complete protection, function in a specific range of crash types, and may be associated with injury in particular crash situations. Misleading information or rumor about either safety benefits of crash protection components or injury risks, which may be infrequent in comparison to overall safety benefit, may cause occupants to reduce their overall

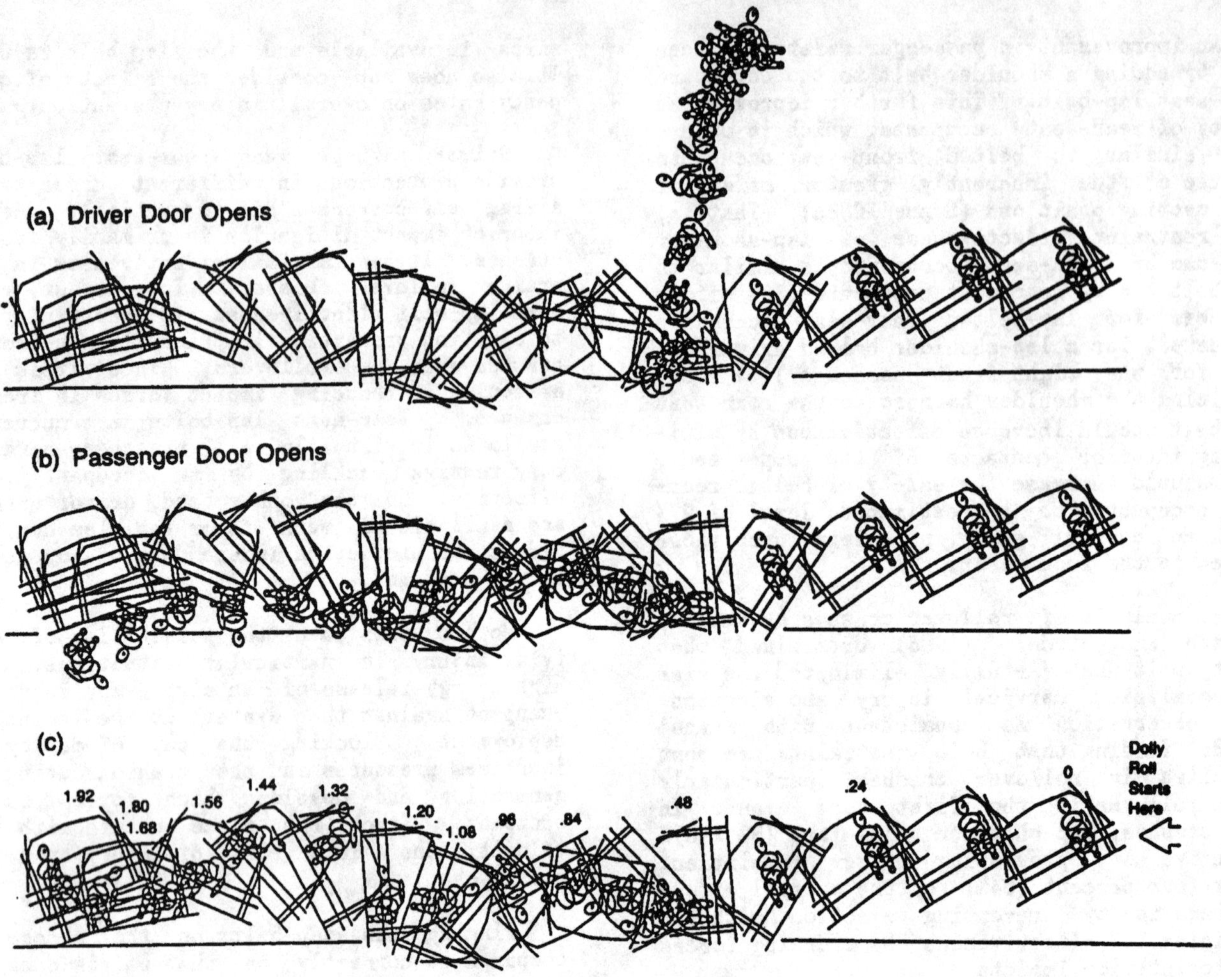

Figure 5: Kinematics of an unrestrained driver in a rollover crash in which (a) the driver door opens and the occupant ejects and is thrown upward, (b) the passenger door opens, the driver ejects and is crushed between the rolling car and the road, and (c) the occupant is contained in the vehicle (from Robbins and Viano 1984 with permission).

driving safety by failing to properly use inherently safe technologies to protect themselves and their families.

REFERENCES

1. Adomeit, D. and Heger, A., "Motion Sequence Criteria and Design Proposals for Restraint Devices in Order to Avoid Unfavorable Biomechanic Conditions and Submarining." In Proceedings of the 19th Stapp Car Crash Conference, 139-166, SAE Technical Paper #751146, Society of Automotive Engineers, Warrendale, PA, 1975.
2. Adomeit, D., "Evaluation Methods for the Biomechanical Quality of Restraint Systems During Frontal Impact." In Proceedings of the 21th Stapp Car Crash Conference, 911-932, SAE Technical Paper #770936, Society of Automotive Engineers, Warrendale, PA, 1977.
3. Adomeit, D., "Seat Design -- A Significant Factor for Safety Belt Effectiveness." In Proceedings of the 23rd Stapp Car Crash Conference, 39-68, SAE Technical Paper #791004, Society of Automotive Engineers, Warrendale, PA, 1979.
4. Arajarvi, E., Santavirta, S. and Tolonen, J., "Abdominal Injuries Sustained in Severe Traffic Accidents by Seat Belt Wearers." Journal of Trauma 27(4):393-397, 1987.
5. Banerjee, A., "Seat Belts and Injury Patterns: Evolution and Present Perspectives." Postgraduate Medical Journal, 65:199-204, 1989.
6. Bohlin, N.I., "A Statistical Analysis of 28,000 Accident Cases with Emphasis on Occupant Restraint Value." In Proceedings of the 11th Annual Stapp Car Crash Confer-

ence, University of California, Los Angeles, CA, October 10-11, 1967. SAE Technical Paper #670925, Warrendale, PA, Society of Automotive Engineers, 455-478, 1967.

7. Denis, R., Allard, M., Atlas, H. and Farkouh, E., "Changing Trends with Abdominal Injury in Seat Belt Wearers." Journal of Trauma 23(11):1007-1008, 1983.
8. Evans, L., "The Effectiveness of Safety Belts in Preventing Fatalities." Accident Analysis and Prevention 18:229:241, 1986b.
9. Evans, L., "Double Pair Comparison -- A New Method to Determine How Occupant Characteristics Affect Fatality in Risk in Traffic Crashes." Accident Analysis and Prevention 18:217-227, 1986a.
10. Evans, L., "Passive Compared to Active Approaches to Reducing Occupant Fatalities." GM Research Publication GMR-6596, Experimental Safety Vehicles paper #ESV 89-5B-0-005, Goteborg, Sweden, May, 1989.
11. Evans, L. and Frick, M.C., "Potential Fatality Reductions Through Eliminating Occupant Ejection from Cars." Accident Analysis and Prevention, 22(2):169-182, 1989.
12. Evans, L., "Rear Seat Restraint System Effectiveness in Preventing Fatalities." Accident Analysis and Prevention 20:129-136, 1988a.
13. Evans, L., "Occupant Protection Device Effectiveness in Preventing Fatalities." In Proceedings of the 11th International Technical Conference on Experimental Safety Vehicles, Washington, DC, May 12-15, 1987, U.S. Department of Transportation, National Highway Traffic Safety Administration, DOT HS 807 233, pp. 220-227, 1988b.
14. Evans, L., "Restraint Effectiveness Occupant Ejection from Cars, and Fatality Reductions." General Motors Research Report #GMR-6398, General Motors Research Laboratories, Warren, MI, 1988c.
15. Evans, L. and Frick, M., "Seating Position in Cars and Fatality Risk." American Journal of Public Health, 78:1456-1458, 1988.
16. Evans, L., Traffic Safety and the Driver. Van Nostrad Reinhold, January, 1991.
17. Fish, J. and Wright, R.H., "The Seat Belt Syndrome -- Does It Exist?" Journal of Trauma, 5:746-750, 1965.
18. Garrett, J.W. and Braunstein, P.W., "The Seat Belt Syndrome." Journal of Trauma 2:220-238, 1962.
19. Groesch, L., Katz, E., Marwitz, H., Kassing, L., "New Measurement Methods to Assess the Improved Injury Protection of Airbag Systems." In Proceedings of the 30th Annual Conference of the American Association for Automotive Medicine, Association for the Advancement of Automotive Medicine; Des Plains, IL, 235-246, 1986.
20. Horsch, J.D. and Culver, C.C., "A Study of Driver Interactions with an Inflating Air Cushion." In Proceedings of the 23rd Stapp Car Crash Conference, SAE Technical Paper #791029, Society of Automotive Engineers, Warrendale, PA, 1979.
21. Horsch, J.D., Petersen, K.R. and Viano, D.C., "Laboratory Study of Factors Influencing the Performance of Energy Absorbing Steering Systems." SAE Transactions, Vol. 91, 1982, SAE Technical Paper #820475, in SAE Special Publication 507, Occupant Interaction with the Energy Absorbing Steering System, pp. 51-63, 1982.
22. Horsch, J.D., "Evaluation of Occupant Protection From Responses Measured in Laboratory Tests." SAE International Congress and Exposition, Detroit, MI, February 23-27, 1987. SAE Paper #870222. Society of Automotive Engineers, Warrendale, PA, 1987.
23. Horsch, J., Lau, I., Andrzejak, D., Viano, D., Melvin, J., Pearson, J., Cok, D., Miller, G., "Assessment of Airbag Deployment Loads." In the Proceedings of the 34th Stapp Car Crash Conference, pp. 276-288, SAE Technical Paper #902324, November 1990.
24. Huelke, D.F., Sherman, H.W., Murphy, M.J., "Effectiveness of Current and Future Restraint Systems in Fatal and Serious Injury Automobile Crashes." SAE Technical Paper #790323. Warrendale, PA, Society of Automotive Engineers, 1979.
25. Huelke, D.F., Compton, C. and Studer, R., "Injury Severity, Ejection, and Occupant Contacts in Passenger Car Rollover Crashes." SAE Paper #850336, Society of Automotive Engineers, Warrendale, PA, 1985.
26. Kulowski, J. and Rost, W.B., "Intra-Abdominal Injuries from Safety Belt in Auto Accident." Archives of Surgery, 73:970-971, 1956.
27. Mertz, H.J., "Restraint Performance of the 1973-76 GM Air Cushion Restraint System." SAE Technical Paper #880400, 1988:61-72, Society of Automotive Engineers, Warrendale, PA.
28. National Highway Traffic Safety Administration, "FMVSS 208 Regulatory Impact Analysis." Department of Transportation, 1984.
29. Orsay, E.M., Turnbull, T.L., Dunne, M., Barrett, J., Langenberg, P., and Orsay, C.P., "Prospective Study of the Effect of Safety Belts on Morbidity and Health Care Costs in Motor-Vehicle Accidents." Journal of American Medical Association 260(24):3598-3603, 1988.
30. Park, S., "The Influence of Rear-Seat Occupants on Front-Seat Occupant Fatalities: The Unbelted Case." General Motors Research Laboratories Research Publication GMR-5664, January 8, 1987.

31. Pursel, H.D., Bryant, R.W., Scheel, J.W., and Yanik, A.J., "Matching Case Methodology for Measuring Restraint Effectiveness." SAE Technical Paper #780415, Society of Automotive Engineers, Warrendale, PA, February, 1978.
32. Robbins, D.H., and Viano, D.C., "MVMA-2D Modeling of Occupant Kinematics in Rollovers." SAE Transactions, Vol. 93, 1984. In Mathematical Simulation of Occupant and Vehicle Kinematics, (P-146), 65-77, SAE Technical Paper #840860, Society of Automotive Engineers, Warrendale, PA, 1984.
33. States, J.D., Huelke, D.F., Dance, M. and Green, R.N., "Fatal Injuries Caused by Underarm Use of Shoulder Belts." Journal of Trauma 27(7):740-, 1987.
34. Viano, D.C., "Limits and Challenges of Crash Protection." Accident Analysis and Prevention, 20(6):421-429, 1988a.
35. Viano, D.C., "Cause and Control of Automotive Trauma." Bulletin of the New York Academy of Medicine, Second Series, 64(5):376-421, 1988b.
36. Viano, D.C., (editor), SAE Passenger Car Inflatable Restraint Systems: A Compendium of Published Safety Research, SAE Progress in Technology PT-31, Society of Automotive Engineers, Warrendale, PA, 1988c.
37. Viano, D.C. and Lau, I.V., "A Viscous Tolerance Criterion for Soft Tissue Injury Assessment." Journal of Biomechanics, 21(5):387-399, 1988.
38. Viano, D.C., Arepally, S., "Assessing the Safety Performance of Occupant Restraint Systems." In the Proceedings of the 34th Stapp Car Crash Conference, pp. 301-328, SAE Technical Paper #902328, November, 1990.
39. Viano, D.C., "Cause and Control of Spinal Cord Injury in Automotive Crashes." G. Heiner Sell Lecture. Submitted to Paraplegia, 1990 and available until publication as GMR-6885, 12/7/89.
40. Viano, D.C., "Testimony Before the United States Senate Committee on Environment and Public Works Subcommittee on Water Resources, Transportation, and Infrastructure Concerning Senate Bill S.1007 The National Highway Fatality and Injury Reduction Act of 1989." October 17, 1989.
41. Wilson, R. A. and Savage, C.M., "Restraint System Effectiveness - A Study of Fatal Accidents." In the Proceedings of Automotive Safety Engineering Seminar, Warren, MI, General Motors Corporation, Automotive Safety Engineering, Environmental Activities Staff; 27-39. 1973.

REDUCTION IN SEAT BELT EFFECTIVENESS DUE TO MISUSE

E.G. Janssen, C.G. Huijskens, M.C. Beusenberg
TNO Road-Vehicles Research Institute
Crash Safety Research Centre
Delft, The Netherlands

Abstract

The wearing rate of standard 3-point seat belts by front seat car occupants in the Netherlands is about 70%. Field studies have shown that about 1/3 are used incorrectly. So approximately 50% of the front seat occupants are not or insufficiently protected by their seat belts.
The influence of incorrect use or misuse on the seat belt effectiveness has been studied by the TNO Crash-Safety Research Centre in an extensive research programme. Eight full-scale tests have been performed with a car body mounted on an impact sled. The influence of variations in seat position and belt routing on adult and child dummy responses are studied. The effect of an additional driver airbag system on the dummy loadings is also analyzed in this programme.
The results of this test programme are presented in terms of Ride-Down-Effect (RDE), dummy velocity-displacement curves, dummy contact with the car interior and injury criteria. The injury criteria of the individual body parts are combined into one overall value. Injury probability analysis has been used to evaluate the effect of misuse configurations.
It is concluded that certain misuse configurations extremely reduce seat belt effectiveness. A clear understanding of the type of misuse and the effect on occupant injuries have important implications for the design of protection systems, community actions (e.g. governmental campaigns) and legislation.

INTRODUCTION

Seat belts were introduced some 30 years ago to protect car occupants in a frontal crash. The belt should prevent contact of the occupant with the car interior. Furthermore, the belt should decelerate the occupant optimally, by using the crash deformation of the car and by using the elongation of the belt.
Since June 1975 the wearing of seat belts by front seat occupants in passenger cars is mandatory in the Netherlands. National belt use rates increased from around a 25% level in 1974 to around 50% in 1975. Since then, no steep increase has taken place, despite several public campaigns. Seat belt use has stabilized at around 60% inside built-up areas and around 80% outside built-up areas for the past few years [1]. This is relatively low compared with other European countries like Germany and the United Kingdom.
Besides this low wearing rate, it is know from literature that seat belts are not always worn correctly, which will probably influence their effectiveness in crashes. Often mentioned as incorrect use or misuse of the seat belt system are [2, 3]: excessive slack, a wrong position or routing of the belt, a wrong position of the seat or a wrong body attitude. Appel et al. [2]

found that misuse contributed for 35% to the 'failures' resulting in injuries to occupants involved in a frontal crash. Excessive slack in the shoulderpart and/or lappart of the belt system was frequently observed, however serious injuries were mainly caused by an incorrectly positioned lappart.
Green et al. [3, 4] described several cases of misuse and the resulting injuries. As examples of misuse two cases are mentioned here, one case in which the shoulderpart was positioned under the arm-pit and the other one where the seat back was placed in a sleeping position. Niederer et al. [5] found that misuse strongly influences the effectiveness of seat belts; a large amount of slack considerably increases the overall injury severity.

How many car occupants in the Netherlands are sufficiently protected by seat belts taking into account that the wearing rate is only 70% and among these there could be a considerable number of incorrectly used or even misused belt systems? In order to obtain an insight in this problem, two reseach programmes have been carried out recently in the Netherlands. A field study has been carried out by the SWOV Institute for Road Safety Research, aimed at investigating the frequency of misuse of adult seat belts and child restraint systems (see also [6]). An experimental test programme has been performed by the TNO Crash-Safety Research Centre to study the effect of misuse on the occupant kinematics and injury severity. The effect of an additional driver airbag system was also analyzed in this programma. A second purpose of the study was to obtain high speed films for public campaigns with respect to misuse.

This paper describes the experimental research programme, after a short presentation of the field study.

FIELD STUDY

The methodology of the Dutch field study aimed at investigating the frequency of certain types of misuse of restraint systems and results of this study specifically concerning child restraint systems are described by Schoon et al. [6].
Table 1 shows the results for the front seat car occupants with respect to correct use, incorrect use and misuse of the standard 3-point seat belt. In 35% of all cases the seat belt was worn such that the researchers judged this as misuse. Specific aspects of interest to be judged in this field study were; routing of the belt system (e.g. lappart on abdomen, twisted belt, belt under arm-pit, belt close to neck) and seating position of the occupant (e.g. forward vs. rearward position of the seat, seat back angle).

Table 1 *Correct use, incorrect use and misuse of seat belts by front seat occupants [7].*

	Correct		Incorrect		Misuse		Total	
	n	%	n	%	n	%	n	%
Drivers	122	55	37	17	63	28	222	100
Passengers	64	40	26	16	69	43	159	100
Total	186	49	63	17	132	35	381	100

Every aspect was judged in terms of correct use or not (0 or 1 point) and then weighted by a predefined factor, based on engineering judgement with respect to the probability of serious injuries. The scores were then added up and a final assessment was obtained in terms of correct use, incorrect use or misuse [7]. One aspect or a combination of aspects can therefore result in a total score of 'misuse'.

The frequency of specific aspects found in this field study are:

- shoulderpart under arm-pit (2% of all drivers and 2% of all passengers);
- shoulderpart behind back of occupant (0.5% of all drivers; 1 case);
- shoulderpart on 'arm' (9% of all drivers and 21% of all passengers);
- shoulderpart close to neck (8% of all drivers and 8% of all passengers;
- twisted belt (20% of all drivers and 23% of all passengers);
- excessive slack in belt system (8% of all drivers and 9% of all passengers);
- hippart positioned on abdomen (4% of all drivers and 9% of all passengers);
- large seat back angle (6% of all drivers and 9% of all passengers).
- seat positioned far rearward (9% of all drivers and 17% of all passengers);
- not optimal position of adjustable anchorage point B-pillar (19% of all drivers and 38% of all passengers).

It seems obvious from Table 1 that more drivers than passengers are using the seat belt correctly.

Since more than 1/3 of the front seat occupants in the Netherlands using a seat belt are not using the belt system correctly (in combination with the car seat) and since the wearing rate in the Netherlands for these car occupants is approximately 70%, it can be concluded that more then 50% is insufficiently protected.

The misuse of seat belts if even more stimulated by public advertisements in which a comfort-clip is promoted, which avoids the "tightning" effect of the belt during long trips (see Figure 1). The slack in the shoulderpart is increased by the user and then the clip is clamped on the belt, blocking the function of the pillar-loop and the retractor.

Until 1992, another 'misuse' problem in the Netherlands was the fact that children under the age of 12 years, sitting on the front passenger seat, were allowed to use a standard 3-point adult belt as lapbelt by wearing the shoulderpart behind the back of the child. This was advised by the Dutch Government in favour of a situation where the shoulderpart would penetrate the neck. This and other misuse aspects have been evaluated by TNO in an experimental test programme.

Autogordelklemmen
Hiermee regelt u zelf de strakheid van uw autogordel.
Vooral een uitkomst bij langere autoritten; knellen van
de gordel behoort zo tot het verleden.
8.95

Figure 1 *Example of advertisement for 'comfort-clip'.*

TEST PROGRAMME

Introduction

Three series of sled tests have been performed; one series with correctly used seat belts (called standard tests or C-tests), one series with a misuse configuration (M-tests) and one series with an airbag system installed (A-tests). The test set-up is described below in terms of used vehicle and dummies, as well as seat and restraint position. Table 2 gives a summary of the complete test programme.

Table 2 *Variations in the test programme.*

Testno.	Driver		
	Dummy	Belt routing	Seat position
C-1	Hybrid II	Correct	Standard
C-2	Hybrid III	Correct	Standard
C-3	TNO-10	Correct	Standard
M-1	Hybrid III	Correct	Rearward
M-2	Hybrid III	Under arm-pit	Standard
M-3	Hybrid III	Behind back dummy	Standard
A-1	Hybrid II	Correct + airbag	Standard
A-2	Hybrid II	No belt + airbag	Standard
	Passenger		
C-1	Hybrid III	Correct	Standard
C-2	Hybrid II	Correct	Standard
C-3	TNO P3/4[2]	Correct + CRS[1]	Rearward
M-1	Hybrid II	Correct	Back angle 40^0
M-2	Hybrid II	Behind back dummy	Standard
M-3	TNO P6[2]	Behind back dummy	Standard

1 Rear ward facing Child Restraint System
2 Child dummy

Vehicle

A reinforced body of a European passenger car was rigidly mounted on a moving barrier (i.e. sled). Figure 2 shows the test set-up. The sled was decelerated using crumple tubes, where the deceleration characteristics of the passenger car in a 56 km/h NCAP test were used as test condition in this programme. Standard interior components, such as seat belts, front seats and steering wheel of the passenger car were used. They were replaced by new ones after each test. In testno. A-1 and A-2 a steering wheel, including an airbag system, and a kneebar from another European passenger car were used. Sled, as well as vehicle body accelerations were measured. In the airbag tests also the steering column accelerations were recorded.
The tests were filmed using 8 high speed film camera's.

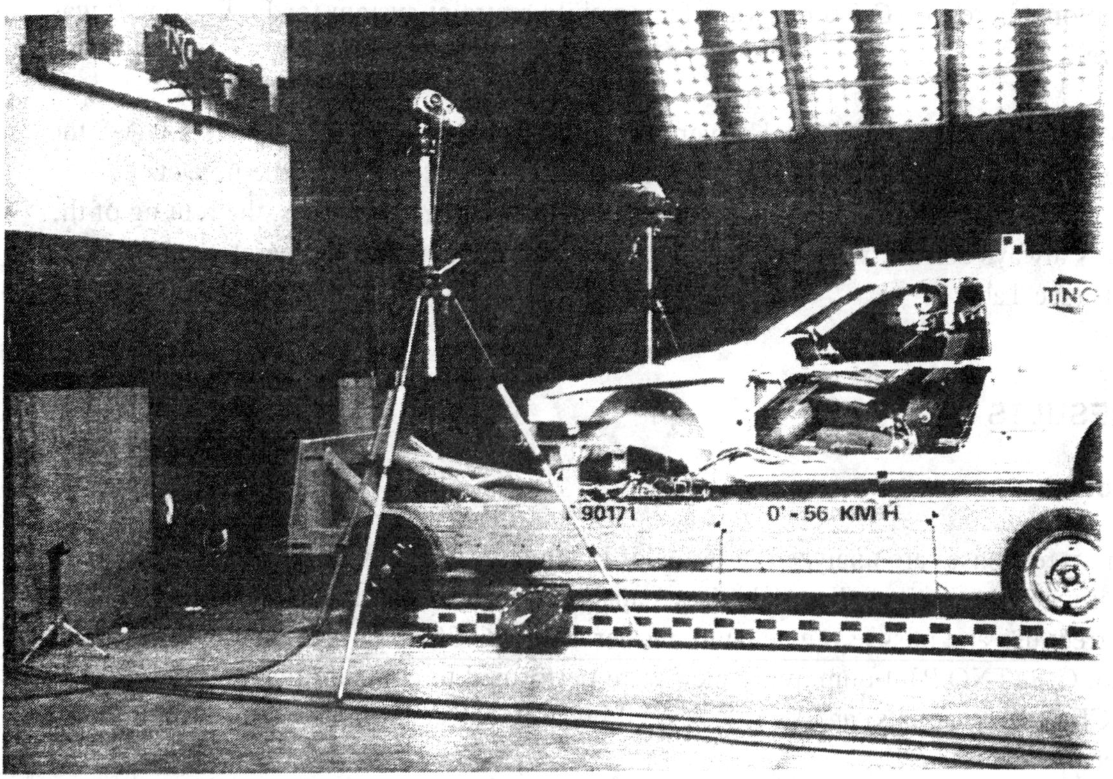

Figure 2 *Test set-up showing car body mounted on impact sled.*

Dummies

Hybrid II and Hybrid III dummies were used as driver and/or passenger in this programme. Head, chest and pelvis accelerations, femur forces, neck forces and moments (only Hybrid III) were recorded.
For comparison reasons the TNO-10 dummy, described in ECE Regulation 16 (seat belts), was used as driver in one test. The TNO P6 and P3/4 child dummies, described in ECE Regulation 44 (child restraint systems) were used as front seat passengers in two tests. Head and chest accelerations were recorded with these three TNO dummies.

Seat position and restraint system

Two types of misuse have been studied:

- a non-optimal seat position creating 'space' between the shoulder and the belt;
- a non-correct routing of the belt by positioning the belt under the arm-pit or behind the back of the dummy.

The driver seat was adjusted in the middle (fore-after) and lowest position (top-down), while the passenger seat was placed in the middle/middle position. In testno. M-1 and C-3 the driver seat, respectively the passenger seat was placed in the most rearward position (see Table 2). The standard seat back angle was 25° for both driver and passenger seat, except in testno. M-1 were the passenger seat back angle was 40°.

In all tests the standard 3-point belt with retractor was used. Belt forces were measured at three locations. In testno. C-3 a rearward-facing child restraint system for ECE group 0 was used as well.

In testno. A-1 and A-2 an additional driver airbag and kneebar were used. In testno. A-2 the belt system was not used. The airbag was externally triggered 10 ms after the impact.

In the A- and C-tests the position of the seat belt was correct. In the M-tests, the routing of the lappart was always correct, while the routing of the shoulderpart was deliberately changed in some tests (see Table 2).

TEST RESULTS

Introduction

The dummy responses are analyzed in terms of kinematics (i.e. motion and velocity) and in terms of injury criteria (including risk). The capability of these parameters to distinguish between correct use and misuse is evaluated. Driver testno. C-3 (TNO-10 dummy) and passenger testno. C-3 (TNO P3/4 dummy), which were performed to compare the results also with that of ECE-R16 (seat belts) and ECE-R44 (child restraint systems) respectively, are not presented here.

The results of the M-series and A-series are always compared with that test of the C-series in which the same dummy type (i.e. Hybrid II or Hybrid III) was used.

Kinematics

Ride-Down-Effect

The belt system should decelerate the occupant smoothly by using the crash deformation path of the car. The so-called Ride-Down-Effect (RDE) has been calculated to assess the amount of the car deceleration shared by the occupant (see also ref.[8]):

$$RDE = (S_V - S_r) \times 100\% / S_V$$

S_V = maximum outer deformation path of car

S_r = deformation path of car up to the time t_r at which the restraint system comes into effect

To determine the Ride-Down-Effect, the resulting thorax deceleration is required, as well as the time function of the car deformation path which can be determined by double integration of the car deceleration (see Figure 3).
To determine the time t_r, at which the retractor has blocked and the slack in the belt has been taken up, a straight line is placed on the rising curve of the resulting thorax acceleration. The line connects the points on the curve representing 25% and 75% of the (first) peak acceleration (see Figure 3). The intersection of the line with the time axis marks the point in time t_r from which the restraint system is assumed to take effect. At this time, the car has already passed through a deformation path of S_r. This path is substracted from the maximum dynamic deformation path and thus describes the percentage of the car deceleration shared by the occupant. According to this calculation RDE=100% means that the occupant is decelerated immediately, without belt slack etcetera, while RDE=0% indicates that the occupant is not decelerated until the maximum dynamic car deformation path has been reached.

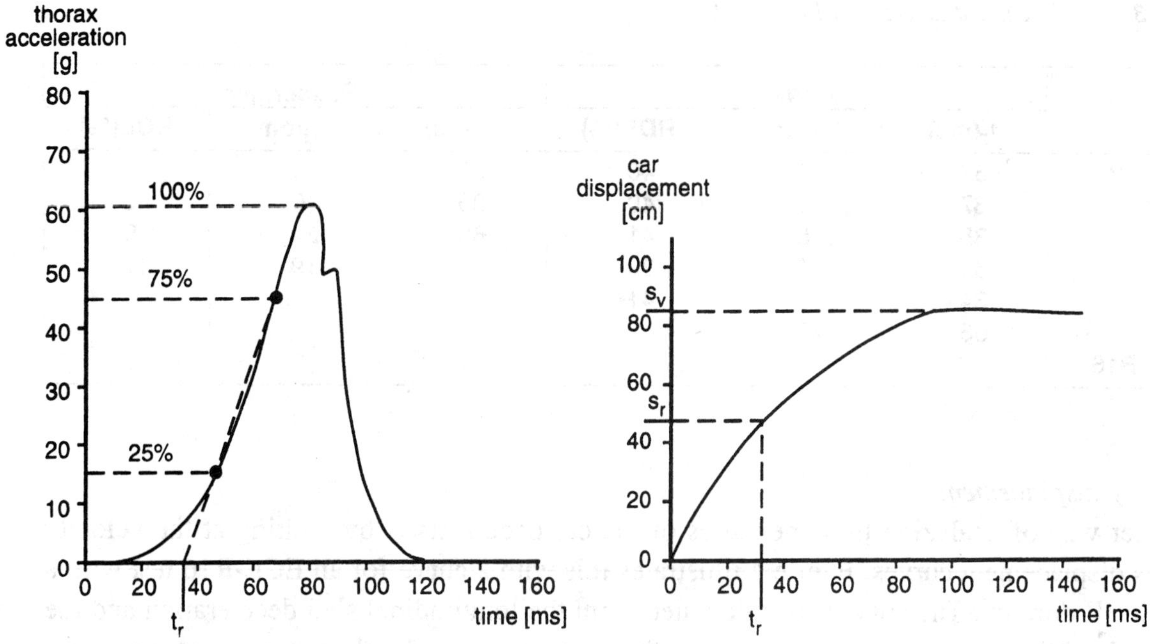

Figure 3 *Resultant thorax acceleration versus time showing the definition of t_r and the car displacement versus time showing the definition of S_r and S_v.*

Table 3 summarizes the calculated RDE values for the tests evaluated here. S_d is the dummy thorax displacement relative to the car at time t_r, calculated from the double integration of the longitudinal thorax acceleration minus S_r. For comparison reasons the RDE calculation of this seat belt system in a standard ECE Regulation 16 test using a Hybrid III dummy is also presented in Table 3. Here the value is somewhat higher than that in the car sled tests with the correctly used belt system (C-series). So an RDE value of 55% obviously indicates a good effectiveness of this belt system.
From Table 3 it seems that the position of the seat and the seat back do not influence RDE, the results of testno. M-1 are similar to that of C-1 and C-2. Positioning of the shoulderpart of the adult belt under the arm-pit (testno. M-2 driver) also appears to have no influence on RDE. However, positioning of the shoulderpart of the belt behind the back of the driver (testno. M-3) or the passenger (testno. M-2) considerably reduces the calculated RDE; from

40% to 20% and from 40% to 8% respectively. The RDE value of the driver is somewhat higher than that of the passenger in this misuse configuration, because the thorax of the driver is partly decelerated by the steering wheel. The 6-year child passenger wearing the belt behind the back shows a relatively low RDE-value.
Introducing an airbag besides wearing the seat belt (testno. A-1) seems not to influence the RDE. However, when the airbag 'replaces' the safety belt (testno. A-2), the RDE drops to 7%. This means that the car is almost stopped when the occupant impacts the airbag, with a relative high impact velocity.

If the time t_r from which the restraint system takes effect is very late, so the RDE-value is low, it can be expected that the forward motion of the thorax relative to the car is large. This is illustrated in Table 3 by the distance S_d. If this displacement is too large, the thorax and/or the head will contact the car-interior, which could lead to serious injuries.

Table 3 *Calculated Ride-Down-Effect RDE.*

Testno.	Driver			Passenger		
	t_r(ms)	S_d(cm)	RDE(%)	t_r(ms)	S_d(cm)	RDE(%)
C-1/C-2	37	5	40	36	5	40
M-1	37	6	40	36	5	40
M-2	35	5	41	69	24	8
M-3	54	13	20	59	19	17
A-1	34	5	41			
A-2	66	27	7			
ECE-R16				14	1	55

Velocity-displacement

Another way of analyzing the kinematics of the car occupants is by looking at the velocity versus displacement curves. Figure 4 illustrates this with a curve for an ECE-R16 test with a Hybrid III dummy. The curves are calculated from the longitudinal sled deceleration and the longitudinal thorax acceleration. The earlier the thorax starts to decelerate the more the thorax uses the car deformation path. The time-points t_r on both curves are connected by a line. The horizontal component of this line represents the relative thorax-to-car displacement S_d and the vertical component represents the relative thorax-to-car velocity. It can be seen from Figure 4 that the relative thorax displacement at time t_r is approximately 1 cm (see also Table 3). The maximum thorax displacement is approximately 21 cm more than the maximum sled displacement. The RDE-value is also given in the sled velocity-displacement curve. The lower the RDE-value, the larger the thorax displacement before the thorax velocity starts to drop. As an example, also a situation without a seat belt is shown in Figure 4.

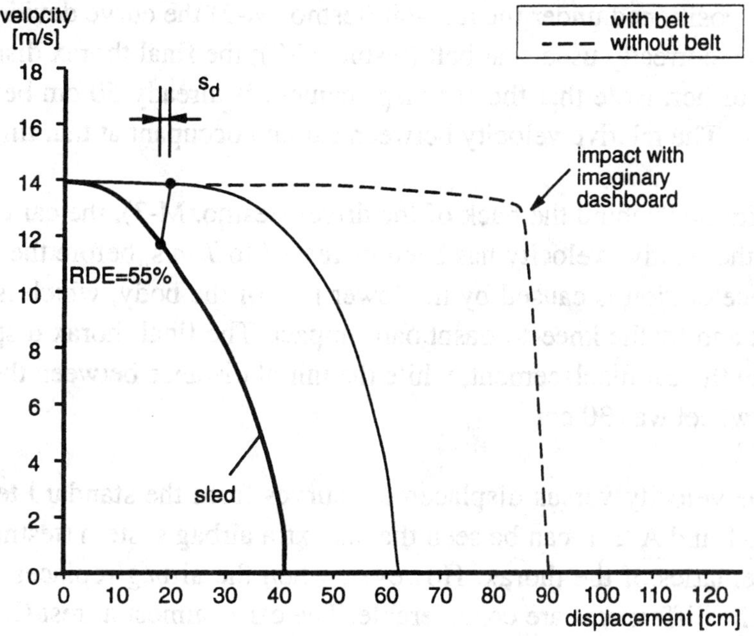

Figure 4 *Velocity versus displacement of sled and Hybrid III thorax in ECE-R16 sled test. An imaginary curve for a dummy without a seat belt impacting a dashboard is also shown.*

Figure 5 shows the velocity-displacement curves for the car-body mounted on the sled and the driver in testnumber C-2, M-1, M-2 and M-3. The t_r-lines and RDE-values are also presented in this figure. From this figure it appears that the seat position (testno. M-1) has practically no influence on the kinematic results.

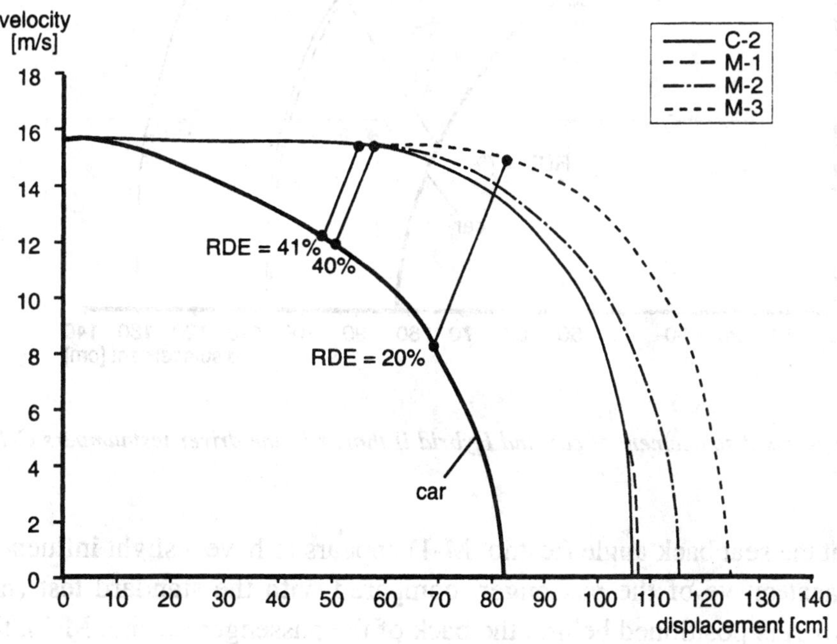

Figure 5 *Velocity versus displacement of car and Hybrid III thorax in the driver testnumbers C-2, M-1, M-2 and M-3.*

With the shoulderpart positioned under the arm-pit (testno. M-2) the curve devides somewhat from the curve with the correctly used seat belt (testno. C-2); the final thorax displacement is approximately 9 cm further. Note that the car displacement is already 50 cm before the belt system becomes active. The relative velocity between car and occupant at that time is approximately 3.5 m/s.
With the seatbelt positioned behind the back of the driver (testno. M-3), the car displacement is already 70 cm and the relative velocity has been increased to 7 m/s, before the thorax starts to decelerate. This deceleration is caused by the lower part of the body, which is stopped by the lappart of the belt and by the knee-to-dashboard impact. The final thorax displacement is some 42 cm more than the car displacement, while the initial distance between thorax and the centre of the steering wheel was 30 cm.

Figure 6 compares the velocity versus displacement curves from the standard test (no. C-1) and the airbag tests A-1 and A-2. It can be seen that an extra airbag system (testno. A-1) does not influence the kinematics of the thorax. However, when the airbag replaces the seat belt system (testno. A-2) the differences are considerable. The car is almost at rest (RDE-value of 7%) before the thorax impacts the airbag. The relative velocity between car and occupant is then very high (i.e. 8.5 m/s).

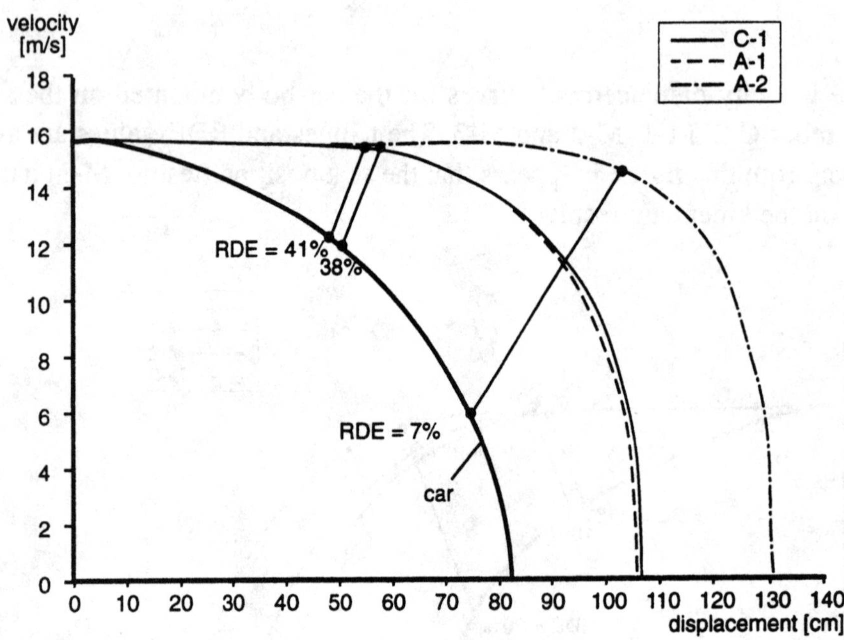

Figure 6 *Velocity versus displacement of car and Hybrid II thorax in the driver testnumbers C-1, A-1 and A-2.*

Figure 7 shows that the seat back angle (testno. M-1) appears to have a slight influence on the velocity-displacement curve of the passenger, compared with the standard test (no. C-2). When the shoulderpart is positioned behind the back of the passenger (testno. M-2), the car is almost at rest (80 cm displacement) before the thorax is decelerated by the lower body. The RDE-value is extremely low (i.e. 7%). The maximum thorax displacement is almost 70 cm more than that of the car, while the initial distance between thorax and dashboard was approximately 60 cm. However, the car displacement is practically linear, while the dummy thorax follows an arc.

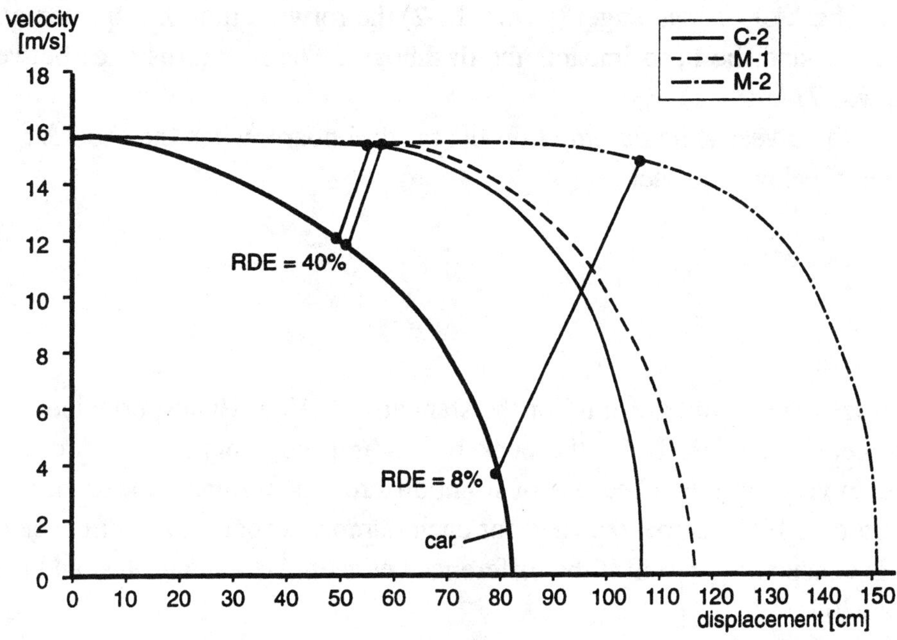

Figure 7 *Velocity versus displacement of car and Hybrid II thorax in the passenger testnumbers C-2, M-1 and M-2.*

Contacts with interior

The relatively large forward displacement of the thorax in some tests, has resulted in contact of the head and/or thorax with the car interior.

With a correctly used seat belt the face of the driver impacts the relatively hard centre of the steering wheel in these 56 km/h sled tests (note: no intrusion of steering column or dashboard occurs). The forehead of the driver impacts the same point on the steering wheel if the seat is positioned rearwards (testno. M-1). With the seat belt under the arm-pit (testno. M-2) the forward displacement of the thorax is larger than that in the standard test (see Figure 5); the thorax impacts the lower ring of the steering wheel. The forward head excursion is also larger; the face impacts the upper ring of the steering wheel slightly before impacting the dashboard behind the steering wheel. With the seat belt positioned behind the back of the driver (testno. M-3), the kinematics even look worse. The lower ring of the steering wheel penetrates the soft abdomen, the thorax impacts the whole steering wheel and the head of the driver is lashed over the steering wheel onto the dashboard. A considerable dynamic deformation of the dashboard can be seen on the high speed films. The initial distance between the head and the upper ring of the steering wheel was 44 cm.

With an additional driver airbag system (testno. A-1), the head contacts the airbag rather than the steering wheel. When the seat belt is not used, but only an airbag system (testno. A-2), the thorax is stopped by the airbag instead of by the belt. The head slides over the airbag and impacts the windscreen.

With a correctly used seat belt, the head of the passenger will not contact the car interior in these 56 km/h sled tests (note: no intrusion of the dashboard occurs). However, the chin im-

pacts the thorax with considerable violence. The forward thorax displacement is slightly larger than that of the standard test, when the passenger seat back is inclined backwards (see Figure 7). However, the head still does not contact the car interior. With the seat belt positioned behind the back of the passenger (testno. M-2) the forward thorax displacement is very large (see Figure 7) and the head impacts the dashboard. The initial distance between head and dashboard was 71 cm.
Even the head of the 6-year child dummy impacts the dashboard, when the shoulderpart of the belt system is worn behind the back.

Injury criteria

Maximum values
Table 4 summarizes the dummy results for the standard FMVSS 208 criteria and some additional measurements. The results for the neck forces and moments are not presented here, since they seem to vary too much because of slight differences in impact location.
Since a variation of ± 10% can be expected for each dummy response, an influence of a misuse parameter is considered present if the difference between the standard test and the misuse test is more than 20%.

Table 4 *Maximum dummy responses.*

	Testnumber						
Dummy responses	**C-1**	**C-2**	**M-1**	**M-2**	**M-3**	**A-1**	**A-2**
Driver head							
HIC	991	1595	1024	724	2003	193	673
3 ms max accel [g]	93	114	84	71	120	42	59
Driver thorax							
3 ms max accel [g]	50	52	49	42	61	51	71
max deflection [mm]	--	42	35	54	28	--	--
Driver pelvis							
3 ms max accel [g]	60	56	53	57	70	60	62
Driver femur							
max force left [kN]	1.5	1.9	4.4	1.3	2.5	2.9	7.6
max force right[kN]	1.2	0.7	1.7	1.9	2.9	3.8	5.6
Passenger head							
HIC	1246	887	1328	6120	NA[1)]		
3 ms max accel [g]	87	72	85	213	166		
Passenger thorax							
3 ms max accel [g]	48	45	38	92	49		
max deflection [mm]	43	--	--	--	--		
Passenger pelvis							
3 ms max accel [g]	64	58	60	65	--		
Passenger femur							
max force left [kN]	1.9	2.0	2.8	2.0	--		
max force right[kN]	1.0	1.7	2.2	2.1	--		

1) Not available

Figure 8 shows the 3 ms maximum resultant head acceleration for the driver in testnumbers C-2 (correct use), M-1 (seat rearward), M-2 (belt under arm-pit) and M-3 (belt behind back). The results of the M-testseries are presented as percentage of the C-2 result, which is defined as 100%. It can be seen that the results of testno. M-1 and M-2 are considerable lower than 100%, while the M-3 result is similar to that of testno. C-2. The HIC-values presented in Table 4 show a similar trend; for testno. M-1 and M-2 lower values are obtained and the HIC-value in testno. M-3 is significantly higher.
Figure 9 shows the results for the passenger head in testnumbers M-1 (angled seat back) and M-2 (belt behind back) as percentage of the result in testno. C-2 (correct use). The 3 ms maximum acceleration in testno. M-1 is within the ± 20% corridor, while the M-2 result is extremely higher than that of the standard test. The HIC-values presented in Table 4 indicate a significantly higher result in testno. M-1 and M-2 compared with the standard test.
The head injury criteria seem to indicate extreme misuse of the belt system, but not all misuse configurations can be assessed by looking at the head injury criteria. For some driver misuse configurations even a lower value is shown.

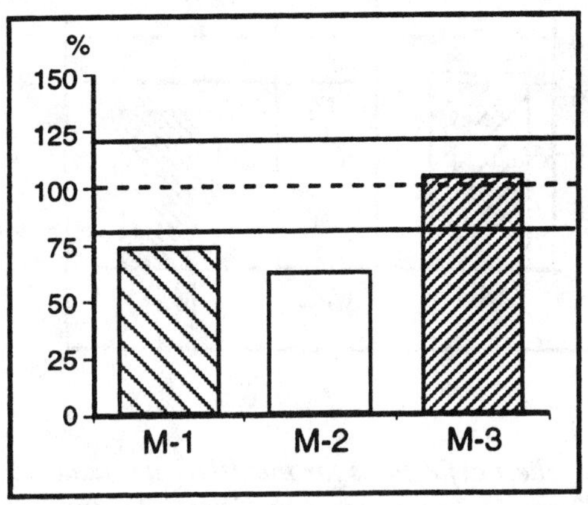

Figure 8
3ms maximum resultant head acceleration for the driver testnumbers M-1, M-2 and M-3 as percentage of the C-2 result

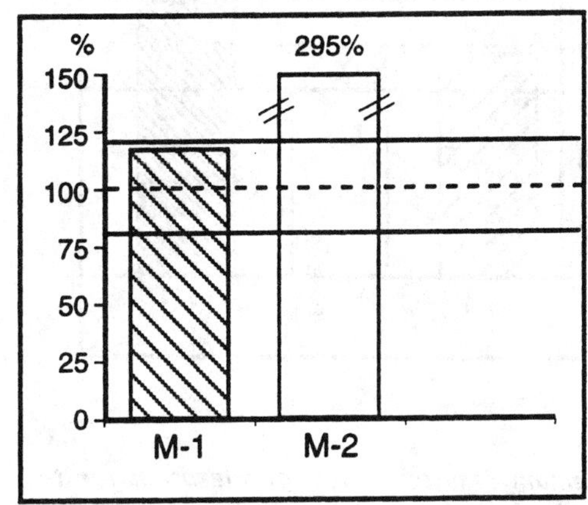

Figure 9
3ms maximum resultant head acceleration for the passenger testnumbers M-1 and M-2 as percentage of the C-2 result.

Figure 10 shows the 3 ms maximum resultant chest acceleration of the driver in the standard test and the 3 misuse tests. All results are within the ± 20% corridor. The two misuse configurations "belt under arm-pit" (testno. M-2) and "belt behind back" (testno. M-3) indicate a different trend; a lower respectively a higher 3ms maximum chest acceleration compared with the standard test. The maximum chest deflections, shown in Figure 11, indicate an opposite trend; a much higher deflection in testno. M-2 and a much lower in testno. M-3.
Figure 12 shows the 3ms maximum resultant chest acceleration for the passenger in the standard test and 2 misuse tests. The result of testno. M-1 is within the ± 20% corridor, while the result for testno. M-2 (belt behind back) is considerably higher.
Extreme misuse configurations are seen by the chest injury criteria, however with different outputs. Sometimes a lower maximum is found for one criterion, while another shows a higher value for that specific test.

The 3 ms maximum pelvis accelerations are not strongly influenced by the misuse configurations tested here (see Table 4). Table 4 shows that the femur forces are to some extend influenced by the seat position, however they are far below the FMVSS 208 tolerance criterion of 10 kN.

Figure 13 shows that the 3 ms maximum resultant head acceleration in the two airbag tests is much lower than that of the standard test (C-1). The HIC-values presented in Table 4 indicate the same trend. Figure 14 shows that the 3 ms maximum resultant chest acceleration is not influenced by the additional airbag (testno. A-1), while this parameter is much higher for the test with an airbag and without seat belts (testno. A-2). The reinforced dashboard with knee-bar in testno. A-1 shows higher femur forces than in the standard test, and much higher forces in the test without seat belt system (testno. A-2).

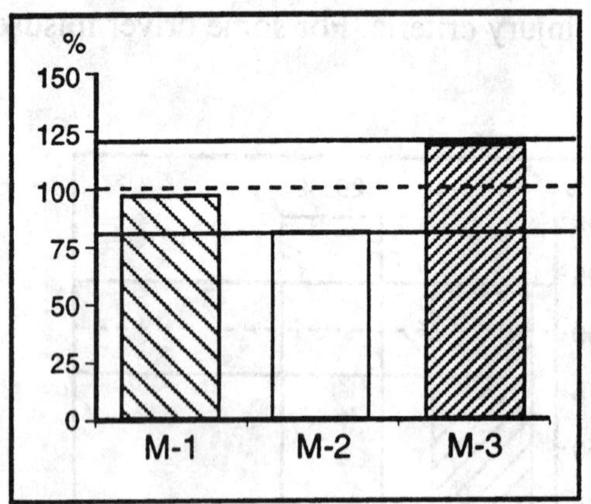

Figure 10
3ms maximum resultant chest acceleration for the driver testnumbers M-1, M-2 and M-3 as percentage of the C-2 result.

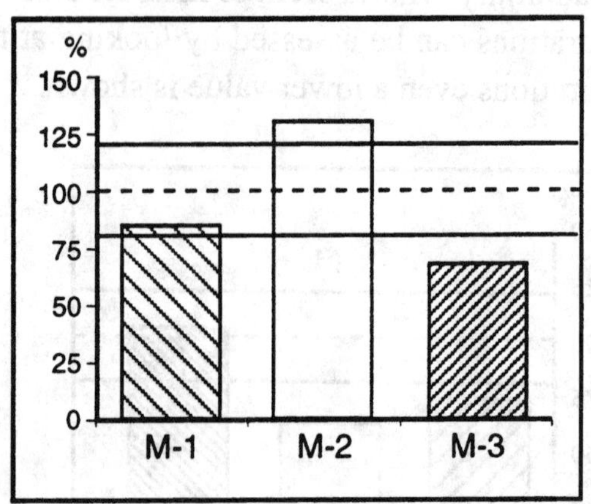

Figure 11
Maximum chest deflections for the driver testnumbers M-1, M-2 and M-3 as percentage of the C-2 result.

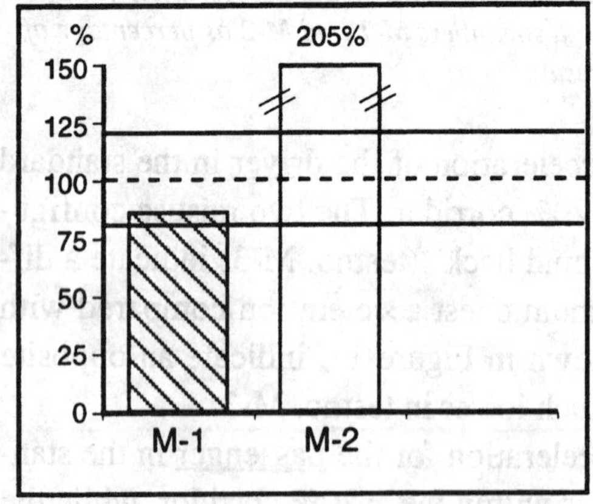

Figure 12
3ms maximum resultant chest acceleration for the passenger testnumbers M-1 and M-2 as percentage of the C-2 result.

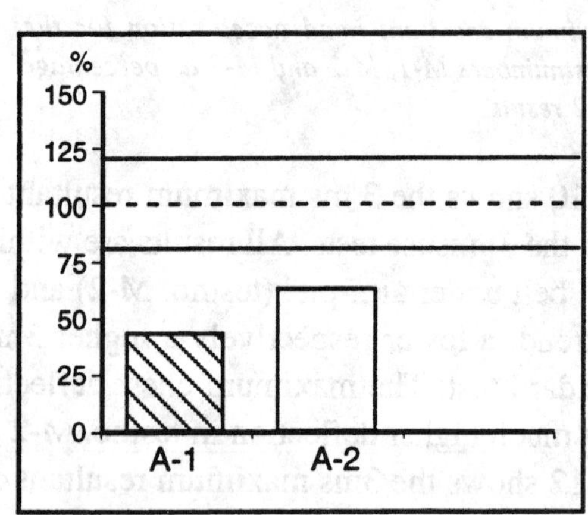

Figure 13
3ms maximum resultant head acceleration for the driver testnumbers A-1 and A-2 as percentage of the C-1 result.

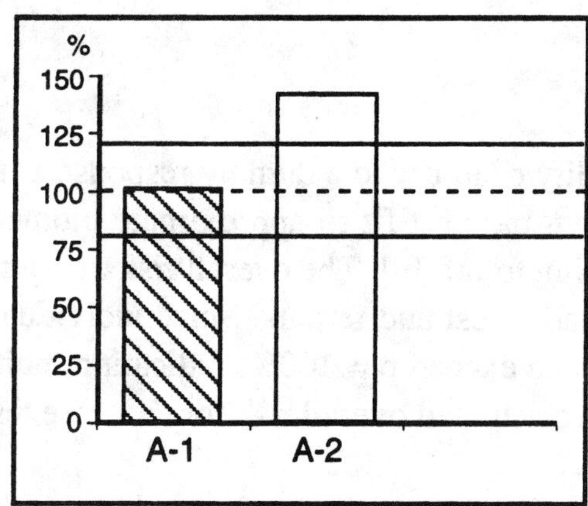

Figure 14 *3ms maximum resultant chest acceleration for the driver testnumbers A-1 and A-2 as percentage of the C-1 result.*

Single parameter

In [9] a method is described to calculate a single value for overall injury risk by normalizing each response by its tolerance value and then multiply it by a weighting factor. This factor is 60% for the head response (HIC= 1000), 35% for the chest (A_{3ms}= 60 g) and 5% for the lower extremities (F= 10 kN) in terms of FMVSS 208 citeria. So the injury criteria can be combined in a single parameter to assess overall safety performance called IC_{208}:

$$IC_{208} = 0.60*[HIC/1000] + 0.35*[A_{3ms}/60g] + (0.05/2)*[(F_l/10kN) + (F_r/10kN)]$$

Table 5 shows the results of this calculation for the three test series. It appears that the driver misuse configurations M-1 and M-2 show a decreasing IC_{208}-value compared to the standard test. This is caused by the lower HIC-values in these tests. Only testno. M-3 (belt behind back) indicates a higher injury risk. For the passenger the misuse configurations show an increasing IC_{208}-value, especially when the belt is positioned behind the back (testno. M-2). With an additional driver airbag (testno. A-1) the HIC-value is decreased considerably, resulting in a 50% lower IC_{208}-value. However, when the belt system is not used (testno. A-2), the airbag alone indicates the same risk as in the standard test.

Table 5 *Combined injury criteria IC_{208}.*

	Testnumber						
Occupant	**C-1**	**C-2**	**M-1**	**M-2**	**M-3**	**A-1**	**A-2**
Driver	0.89	1.27	0.92	0.69	1.57	0.43	0.85
Passenger	1.04	0.80	1.03	4.22			

Probability function

A more preferred approach for assessing occupant protection involves interpreting dummy responses using an injury risk function based on probability analysis of biomechanical data [9].

The probability function is non-linear, with a strongly increasing injury risk above the tolerance level:

$$p(x) = [1 + \exp(\alpha - \beta x)]^{-1}$$

This function relates the probability of injury to a dummy response x using two parameters α and ß, which are determined from the best fit to approximate biomechanical data. Table 6 presents these parameters according to ref. [9]. The overall risk of injury is determined by the sum of the individual risks of head, chest and femurs. Since individual risk varies from zero to one, the overall risk of injury can exceed p = 100%, indicating multiple injuries or causes of death. Table 7 summarizes the calculated overall risk for the three test series.

The probability of serious injuries (i.e. AIS 4+ for head and chest, AIS 3+ for femurs) for the driver in the standard test (C-2) is large (77%) due to the severe head to mid-steering wheel impact. The M-series shows the same trend as presented above for the IC-value; decreasing for testno. M-1 and M-2, increasing (above 100%) for testno. M-3.
The standard test for the passenger (testno. C-2) appears to have a much lower probability of serious injuries (21%) compared with that of the standard driver test. The passenger misuse configurations show a higher probability of serious injuries; 47% and 170% for testno. M-1 and M-2 respectively.

The test with the additional airbag system (testno. A-1) indicates a 50% lower probability of serious injuries compared with the test using seat belts only (testno. C-1). When the seat belts are not used, the airbag alone is not able to keep p(x) at the same level as in the standard test; the injury risk is much higher (see Table 7).

Table 6 ***Parameters for injury risk function [9].***

Body region	Parameters	
	α	β
Head	5.02	0.00351
Chest	5.55	0.0693
Femurs	7.59	0.660

Table 7 ***Injury risk using probability function.***

Occupant	Testnumber						
	C-1	C-2	M-1	M-2	M-3	A-1	A-2
Driver	0.29	0.77	0.31	0.15	1.10	0.14	0.51
Passenger	0.44	0.21	0.47	1.70			

DISCUSSION AND CONCLUSIONS

Car seat belt use in the Netherlands has stabilized at around 60% inside built-up areas and around 80% outside built-up areas, despite several mass public campaigns. A recent field study has shown that 35% of the car drivers and front seat passengers wearing a seat belt appear to 'misuse' the belt system. This means that more than 50% of the front seat occupants in the Netherlands are not or insufficiently protected by the seat belt. Since the misuse frequency of child restraint systems is even 70% [6], public campaigns should focus more on the correct use of restraint systems. Furthermore current laws should be extended by an improved description of 'wearing the seat belt' and by forbidding the use of 'comfort-clips'.

The TNO Crash-Safety Research Centre has conducted a series of sled tests to evaluate the influence of certain type of misuse configurations on the dummy kinematics and injury criteria. Two types of misuse have been studied:

- a non-optimal seat position creating 'space' between the shoulder and the belt;
- a non-correct routing of the belt by positioning it under the arm-pit or behind the back of the dummy.

In the first type, seen in 15% of the cases for drivers and in 26% of the cases for passengers [7], the upper torso is not immediately restrained by the belt. The amount of the car deceleration shared by the occupant is called the Ride-Down-Effect (RDE). It appears that the RDE in these 56 km/h sled tests is similar for correctly used seat belts and misuse of the first type (i.e. RDE=40%). So RDE can not differentiate between correctly used belt systems and incorrectly used belt systems.
The second type of misuse is seen in approximately 2% of the cases for drivers as well as passengers [7]. The RDE seems to decrease considerably (i.e. from 40% to 8-20%) when the shoulderpart of the belt is positioned behind the back of the dummy. RDE appears not to be influenced when the shoulderpart is worn under the arm-pit, obviously a 'quick' deceleration of the upper torso is still present. However, analysis of the high-speed films have indicated that this misuse configuration can result in severe head-to-dashboard impacts. It seems that RDE can only indicate very severe misuse configurations, like wearing the shoulderpart behind the back.

In [9] a so-called Restraint Quotient is presented, not based on relative displacement like RDE, but on the ratio between relative velocity of the occupant and maximum velocity change of the car. This ratio has not been calculated here, but the kinematics of the occupants were studied by analysis of velocity-displacement curves. It appears that large differences between car and occupant can be found with respect to displacement and velocity, especially for the misuse configurations of the second type. This means that the effectiveness of the belt system, smoothly decelerating the occupant by effectively using the car's crash zone, is reduced. The second function of the belt system, avoiding occupant contact with the car interior, was also analyzed. Extreme head impact locations can be seen in the tests with the shoulderpart of the belt under the arm-pit or behind the back of the driver. In these tests also chest-to-steering wheel impacts occured. In the passenger tests head-to-dashboard impacts occured in the tests where the belt was positioned behind the back of the adult or 6-year child dummy. So analysis of the motion of occupants and the (possible) impact locations with the car interior can indicate gross misuse configurations. It should be noted here that since April

1992 in the Netherlands children under the age of 12 years are not allowed to sit on the front seat without a child restraint system. Children between 3 and 12 years are still allowed to use the 3-point adult belt as lapbelt, but only on the rear seats and only if no child restraint system is available.
Analysis of the injury severity in this test programme is to some extend influenced by the limited number of tests, and especially of repeated tests. However, some general conclusions are given here. The HIC is strongly influenced by the stiffness of the impact location and could show a low value in a misuse configuration. When no head impacts occurs, for instance in 'slight' misuse configurations, HIC can not be used. The chest injury criteria (i.e. 3ms maximum acceleration and maximum deflection) sometimes show an opposite trend; a low value for one parameter and a high for the other. Furthermore, sometimes lower values are found in a misuse configuration compared with the correct use test.
Combination of the injury criteria in a single parameter shows good results for the passenger tests; higher values in misuse configurations. The results of this calculation for the driver tests appear to be too much influenced by the HIC-value, which depends too much on 'coincidental' head impacts. Similar results are found when the injury probability is calculated. However, a considerable difference was found between the combined criteria calculation and the injury probability calculation in the airbag only test; a similar value respectively a much higher risk compared with the standard belt test.
An important disadvantage in the use of injury criteria and injury risk functions is the limited instrumentation available in the current frontal dummies. The Hybrid III dummy offers more possibilities in this respect than the Hybrid II dummy. Comparison of the 3 ms maximum accelerations of both dummies as driver and passenger in the standard tests C-1 and C-2, shows that the thorax and pelvis responses are more or less identical (i.e. 4-11% differences). The 3 ms maximum head accelerations of the Hybrid III are 20-23% higher and the HIC is 40-61% higher than that of the Hybrid II dummy in similar tests.

Analysis of the kinematics and injury criteria of the driver and passenger dummies in these 56 km/h sled tests, showed that extreme misuse of the 3-point seat belt can significantly reduce the effectiveness of the belt system. Not all misuse configurations could be well identified. Therefore more sophisticated dummies and analysis methods seem necessary.
Several technical solutions can avoid or decrease (the effect of) misuse of seat belts, for instance adjustable anchorage points, integrated belt systems, pretensioners and automatic restraint systems. Airbags can reduce the severity of the head-to-steering wheel impact in a severe crash with a correctly used belt system, but also in a moderate crash with an incorrectly used belt system. The current research programme has also shown that an airbag should not be used instead of the 3-point seat belt, but in combination with this system.

ACKNOWLEDGEMENTS

The research programme was sponsored by the Dutch Ministry for Transport and Public Works.

RERERENCES

1. Hagenzieker, M.P.:
'Strategies to increase the use of restraint systems.' Proceedings of a Workshop at the VTI-TRB International Conference Traffic Safety on Two Continents, Gothenburg, September 1991. SWOV reportno. R-91-60.

2. Appel, H. et al.:
'Misfunctions of Safety Belts Unavoidable and Avoidable Injuries'. Proceedings IIIrd International Meeting on the Simulation and Reconstruction of Impacts in Collisions, Lyon, September 1978.

3. Green, R.N. et al.:
'Misuse of Three-point Occupant Restraints in Real-World Collisions'. Proceedings 1987 International IRCOBI Conference on the Biomechanics of Impacts, Birmingham, September 1987.

4. Green, R.N. et al.:
'Abdominal Injuries Associated with the Use of Rear-Seat Lap Belts in Real-World Collisisons'. Proceedings 1987 International IRCOBI Conference on the Biomechanics of Impacts, Birmingham, September 1987.

5. Niederer, P. et al.:
'Adverse Effects of Seat Belts and Causes of Belt Failures in Severe Car Accidents in Switzerland During 1976'. Proceedings 21st Stapp Car Crash Conference, New Orleans, October 1977.

6. Schoon, C.C., Huijskens, C.G. and A.H. Heijkamp:
'Misuse of Restraint Systems for Childs in the Netherlands'. To be presented at 1992 International IRCOBI Conference on the Biomechanics of Impacts, Verona, September 1992.

7. Schoon, C.C. et al.:
'Onderzoek naar verkeerd gebruik van autogordels en kinderzitjes'. SWOV reportno. R-91-88, Leidschendam, 1991.

8. Lutter, G., Kramer, F. and H.Appel:
'Evaluation of Child Safety Systems on the basis of Suitable Assessment Criteria'. Proceedings 1991 International IRCOBI Conference on the Biomechanics of Impacts, Berlin, September 1991.

9. Viano, D.C. and S. Arepally:
'Assessing the Safety Performance of Occupant Restraint Systems'. Proceedings 34th Stapp Car Crash Conference, Orlando, November 1990.

Correlation of Vehicle Performance in the New Car Assessment Program with Fatality Risk in Actual Head-On Collisions

Charles J. Kahane
James R. Hackney
Alan M. Berkowitz
National Highway Traffic Safety Administration
United States

Paper No. 94-S8-O-11

ABSTRACT

The New Car Assessment Program (NCAP) has gauged the performance of vehicles in frontal impact tests since model year 1979. NCAP test speeds and impact locations closely resemble the conditions in a large proportion of actual frontal crashes that result in fatalities or serious injuries. The relationship between NCAP test scores and actual fatality risk on the road was studied. Head-on collisions between two 1979-91 passenger cars in which both drivers wore safety belts were selected from the 1978-92 Fatal Accident Reporting System. There were 396 collisions (792 cars) in which both cars were identical with or very similar to vehicles which had been tested in NCAP. In the analyses, adjustments were made for the relative weights of the cars, and for the age and sex of the drivers.

There are statistically significant correlations between NCAP scores for head injury, chest acceleration and femur loading and the actual fatality risk of belted drivers. In a head-on collision between a car with good NCAP score and a car of equal weight with a poor score, the driver of the car with the better NCAP score has, on average, a 15 to 25 percent lower risk of fatal injury. Cars built from 1979 through 1982 had, on the average, the poorest NCAP scores. Test performance improved substantially from 1983 onwards. In parallel, fatality risk for belted drivers in actual head-on collisions decreased by 20 to 25 percent in model years 1979-91, with the largest decreases just after 1982. The paper concludes with a survey of possible future goals for NCAP.

INTRODUCTION

The Fiscal Year 1992 Report of the Appropriations Committee of the United States Senate and the Senate-House Conference Committee Report required the National Highway Traffic Safety Administration (NHTSA) to implement improved methods to inform consumers of the comparative levels of safety of passenger vehicles as measured in the New Car Assessment Program (NCAP) and to examine and study the results of previous model year NCAP results to determine the validity of these test data in predicting actual on-the-road injuries and fatalities.[1] In December 1993, NHTSA presented a report to Congress that responded to these requirements.[2] NHTSA issued a technical report in January 1994, which analyzed the correlation of NCAP performance with fatality risk in actual head-on collisions.[3] Analyses, findings and conclusions of the Congressional and technical reports are summarized in this paper.

Brief History of the New Car Assessment Program

In 1978, NCAP was initiated with the primary purpose of partially fulfilling one of the requirements of Title II of the Motor Vehicle Information and Cost Savings Act of 1973.[4] The purpose of this requirement was to provide consumers with a measure of relative crashworthiness of passenger motor vehicles. NHTSA concluded that by using existing technical approaches, safety information on the relative crashworthiness that vehicles provide in frontal crashes could be developed. This provided consumers with important information to aid them in their vehicle purchase decisions. The ultimate goal of NCAP was to improve occupant safety by providing market incentives for vehicle manufacturers to voluntarily design better crashworthiness into their vehicles, rather than by regulatory directives.

In this program, vehicles are subjected to a frontal crash test. The vehicles are towed head-on into a fixed, rigid barrier at 35 mph. Each vehicle carries two instrumented anthropomorphic test devices (dummies) that simulate 50th percentile adult males. These dummies are located in the front driver and front-right passenger seats and are restrained by the vehicle's safety belts and air bags, if available. During the crash, measurements are taken from each dummy's head, chest, and upper legs. These measurements are used as surrogates for the likelihood of serious injury and, thereby, the relative crashworthiness of the vehicle in a severe frontal impact.

The testing protocol used by NCAP is based on years of development work conducted by NHTSA, the automobile industry, and others to create the test devices and test procedures used in determining compliance with Federal Motor Vehicle Safety Standard (FMVSS) No. 208, "Occupant

Crash Protection."[5] This standard requires that certain injury criteria, as measured by the dummies, not be exceeded in a 30-mph frontal crash test. The injury criteria apply to the head (as measured by a composite of acceleration values known as the Head Injury Criterion or HIC), chest (as measured by a chest deceleration value known as chest G), and upper legs (as measured by compressive forces on each of the femur bones). These criteria are used to assess the performance of the vehicles tested in the NCAP.

The NCAP crash tests are conducted at 35 mph in order to provide a level of impact severity sufficiently higher than the FMVSS 208 requirement at 30 mph so that differences in frontal crashworthiness performance among vehicles can be more readily observed. Since kinetic energy is proportional to the square of the velocity, there is 36 percent more kinetic energy in a 35-mph crash than one at 30 mph. Another measure of severity in a frontal, fixed barrier test is the total instantaneous change in velocity of the vehicle (known as delta V), including the rebound from the barrier. In the 35-mph NCAP test, the average delta V is 40 mph, including the rebound velocity from the barrier. In a 30-mph test, the average delta V is 33 mph.

From an analysis of the National Accident Sampling System's (NASS) files, the relationships of delta V to injury and fatalities have been developed for passenger car drivers restrained by available belt systems (no air bag equipped vehicles are included).[6] These data are shown in Figures 1 and 2. Curves are given for Abbreviated Injury Scale (AIS) 3 and greater injuries, AIS $\geq$ 4 injuries, and fatalities.[7] AIS 3 injuries are serious but often not life-threatening with emergency care. AIS $\geq$ 4 injuries are severe and life-threatening. AIS $\geq$ 4 injuries to the head may include severe skull fractures and/or brain injury. AIS $\geq$ 4 injuries to the thorax may include severe damage to the lungs, torn aortas, or massive collapse of the rib structure.

The NASS data indicate that the fatality and injury rates for restrained, front-seat drivers are several times greater in a crash with a 40-mph delta V than in a crash with only a 33-mph delta V (See Figure 1). The NASS files also show that approximately 50 percent of the life-threatening injuries and nearly 80 percent of the fatalities of restrained drivers in frontal collisions occur in crashes with a delta V greater than 33 mph (See Figure 2). As in the real-world crashes, the injury data obtained in the 35-mph crash tests show a much greater injury potential and a much greater spread among the safety performance measures of various vehicles than observed in the 30-mph crash tests.

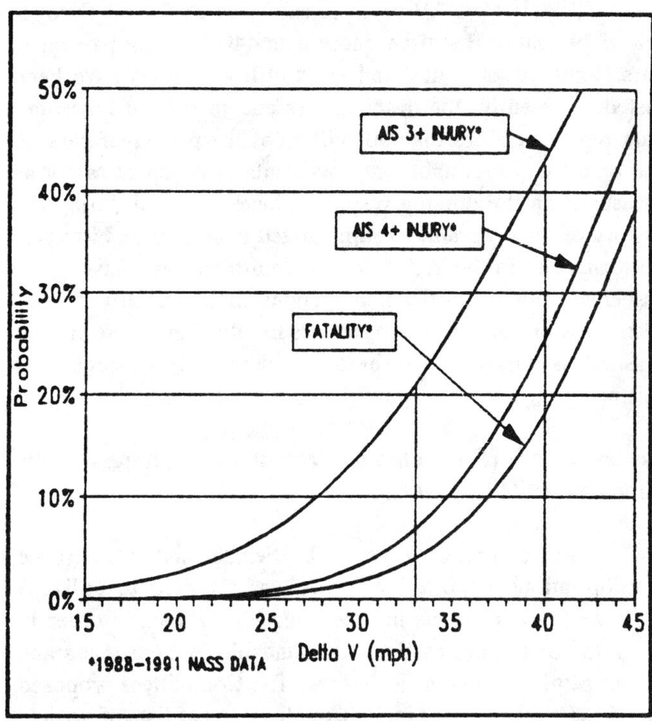

Figure 1. Estimated Probabilities of Injury and Fatality for Restrained Drivers in Frontal Collisions.

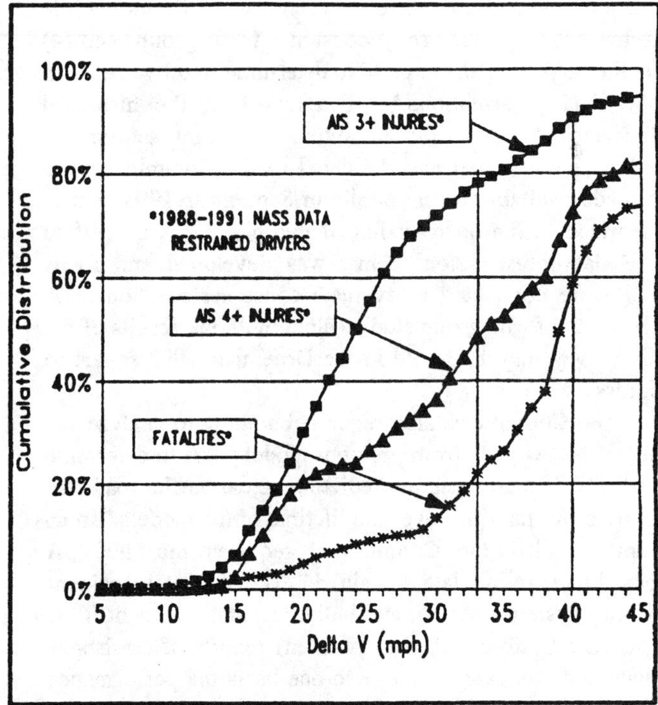

Figure 2. Cumulative Distribution of Injuries and Fatalities for Restrained Drivers in Frontal Crashes.

The first NCAP press release was issued on October 16, 1979. Since that time, more than 440 different passenger cars, light trucks, vans, and sport utility vehicles have been tested. Presently, the tested makes and models of passenger cars represent more than 50 million of the passenger cars on the road today. Notable improvements in occupant safety as measured by the dummy responses have occurred during the history of the program, as summarized in section on historical performance in NCAP. These improvements have been associated with significant reductions in the fatality risk of restrained drivers of passenger cars involved in severe frontal crashes, as will be shown in the accident analysis sections of this paper.

Review of NHTSA's Plan as Proposed to Congress in the February 1992 Report

In the Fiscal Year 1992 Senate and Conference Appropriations Reports,[1] NHTSA was required to utilize a variety of new methods in presenting NCAP data in order to make the data more easily understandable by consumers and more useful as a market incentive. The Committees proposed that these methods may include publications of lists of vehicle models performing best and worst on different injury criteria, lists of vehicle models with the highest and the lowest HIC, lists of vehicle models in rank order of their performance on NCAP tests, and the historical performance of different automobile manufacturers on NCAP tests.

In response, NHTSA proposed to: (1) develop a report of the historical performance of the different automobile manufacturers in NCAP, (2) analyze the NCAP data base and determine an appropriate format for presenting the various suggestions for new lists, (3) evaluate the potential impact of these presentation methods on the car-buying public and evaluate the vehicle safety needs and choices of the automobile consumers through the use of consumer focus groups and (4) enlist the help of media experts to determine improvements in NCAP data presentations.[8] The report of the historical performance of the different automobile manufacturers in NCAP was completed and delivered to the Committees and then made available to the public in September 1993.[9] It is summarized in the next section of this paper. A simplified NCAP data presentation format was developed and focus groups were conducted to evaluate consumer reactions. A review of the focus group studies along with the results of the media survey may be found in the December 1993 report to Congress.[2]

The Committees also requested a study to analyze the results of NCAP data from previous model years to determine the validity of these tests in predicting actual on-the-road risk of injuries and fatalities over the lifetime of the models. In an attempt to fulfill the Committees' requirements, NHTSA proposed to examine data contained in NASS, Fatal Accident Reporting System (FARS), and individual state accident files, and analyze "hard-copy" (i.e., written) reports of crashes to evaluate and compare on a one-to-one basis the performance of specific models which have been tested in NCAP and also have been involved in high-severity frontal impacts on the highway. Those studies are summarized in this paper.

HISTORICAL PERFORMANCE OF DIFFERENT AUTO MANUFACTURERS IN NCAP

In the September 1993 historical NCAP report, trends of improved vehicle safety performance as measured by NCAP were provided.[9] These trends, based on the dummy HIC and chest G responses are shown in Figure 3 for all tests of passenger cars that have been conducted through MY 1993. The average values for the dummy response parameters are given for each model year. Also, the averages for the fleet of NCAP-tested passenger cars, as determined from vehicle registrations, are shown for each year. (Note: After the first year of NCAP testing, this fleet included approximately two million of the passenger cars on the road. At the conclusion of the MY 1992 NCAP testing, this fleet included over 52 million of the registered passenger cars. The file has not yet been updated with vehicle registrations for MY 1993. Therefore, weighted values are only available through MY 1992.) Conspicuous downward trends are shown for each of the injury parameters.

In Tables 1 and 2, summary information from the September historical report on the different motor vehicle manufacturers is given. These data include: the number of vehicles which have been tested, the percentage of vehicles which have met FMVSS 208 requirements (HIC's not exceeding 1,000, chest G's not exceeding 60, and femur loads not exceeding 2,250) in the higher-speed NCAP tests, and overall average values for the driver HIC, passenger HIC, driver chest G, and passenger chest G. For passenger cars, where adequate data exist, this information also is given for two time periods, MY 1979 through MY 1986 and MY 1987 through MY 1993. The phase-in of the automatic occupant protection requirements of FMVSS 208, beginning in MY 1987 led to extensive use of air bags as supplemental restraints, which further improved the safety performance of passenger cars in NCAP.

Substantial reductions in average driver HIC and passenger HIC values have occurred in MY 1987 through 1993 passenger cars when compared to MY 1979 through 1986 passenger cars. The average driver HIC values along with these reductions for the 6 major manufacturers are graphically shown in Figure 4.

A much higher percentage of passenger cars are now meeting the requirements of FMVSS 208 at the higher NCAP crash speed. Almost 80 percent of the passenger cars tested in NCAP during 1993 met the FMVSS 208 requirements. These historical records and the trends shown in Figure 3, indicate that:

- The vehicle manufacturers have the knowledge and capability to design passenger cars that achieve excellent scores in the severe 35-mph crash test if all occupant protection systems are used
- With the phase-in of automatic occupant protection beginning with MY 1987, the vehicle manufacturers notably improved occupant protection in 35 mph crashes as measured by the dummy responses.

TABLE 1. NCAP - SUMMARY DATA ON PASSENGER CARS

MANUFACTURER	NO. OF CARS TESTED		% MEETING FMVSS NO. 208 CRITERIA			DRIVER HIC AVERAGE			PASSENGER HIC AVERAGE			DRIVER CHEST G AVERAGE			PASSENGER CHEST G AVERAGE		
	MODEL YEARS		MODEL YEARS			MODEL YEARS			MODEL YEARS			MODEL YEARS			MODEL YEARS		
	ALL	87-93	ALL	79-86	87-93	ALL	79-86	87-93	ALL	79-86	87-93	ALL	79-86	87-93	ALL	79-86	87-93
GM	71	33	59	61	58	858	897	812	806	802	811	46	44	48	40	39	42
FORD	51	22	48	19	89	920	1090	693	796	1018	500	52	55	47	44	47	41
CHRYSLER	44	20	48	38	61	969	1111	799	974	1069	853	50	51	48	44	43	45
TOYOTA	29	13	62	62	62	883	910	849	753	853	631	50	50	51	47	48	44
NISSAN	25	15	40	20	53	982	1142	874	939	1301	697	53	56	51	46	50	43
HONDA	28	17	69	50	81	909	1176	736	795	1016	652	49	49	49	41	38	43
VOLKSWAGEN	17	6	19	10	33	1136	1250	945	958	911	1035	53	54	52	45	44	45
MAZDA	12	7	58	0	100	881	1065	750	1012	1445	703	55	60	51	48	49	48
MITSUBISHI	10	7	78	67	83	891	879	897	830	1168	685	54	62	50	44	45	44
PEUGEOT/RENAU	13	4	0	0	0	1906	1957	1793	1866	2011	1577	59	58	60	49	47	52
VOLVO	7	2	86	80	100	742	879	400	700	724	640	41	42	40	39	39	40
HYUNDAI	8	7	25	0	29	888	1000	871	971	2662	729	56	73	53	45	55	44
ISUZU	5	2	0	0	0	1570	1821	1194	1523	1711	1240	47	42	54	48	47	48
SUBARU	8	4	38	25	50	1055	1230	880	988	1293	682	53	54	51	46	49	43
MERCEDES	3	1	33	0	100	984	1076	800	979	1052	833	59	58	60	49	44	58
SAAB	5	3	40	0	67	658	754	594	1029	1304	846	48	55	43	38	40	37
BMW	3	2	33	0	50	1093	1539	870	622	547	698	49	42	52	40	39	40
TOTAL	339	165	50	37	63	967	1101	826	905	1055	746	50	51	49	44	44	44

TABLE 2. NCAP - SUMMARY DATA ON LIGHT TRUCKS, VANS & SPORT UTILITY VEHICLES (LTVS)

MANUFACTURER	NO. OF LTVS TESTED	% MEETING FMVSS NO. 208	DRIVER HIC AVERAGE	PASSENGER HIC AVERAGE	DRIVER CHEST G AVERAGE	PASSENGER CHEST G AVERAGE
		MODEL YEARS	MODEL YEARS	MODEL YEARS	MODEL YEARS	MODEL YEARS
		ALL	ALL	ALL	ALL	ALL
GM	21	29	1274	1215	60	49
FORD	17	44	1124	901	52	47
CHRYSLER	18	44	857	1005	51	45
TOYOTA	12	8	1250	828	55	50
NISSAN	8	38	1080	810	54	46
VOLKSWAGEN	3	0	1507	874	56	49
MAZDA	3	33	1002	857	55	48
MITSUBISHI	6	50	1203	976	52	54
ISUZU	10	10	1282	1207	61	59
SUZUKI	3	33	1214	1548	62	53
TOTAL	101	31	1150	1020	55	49

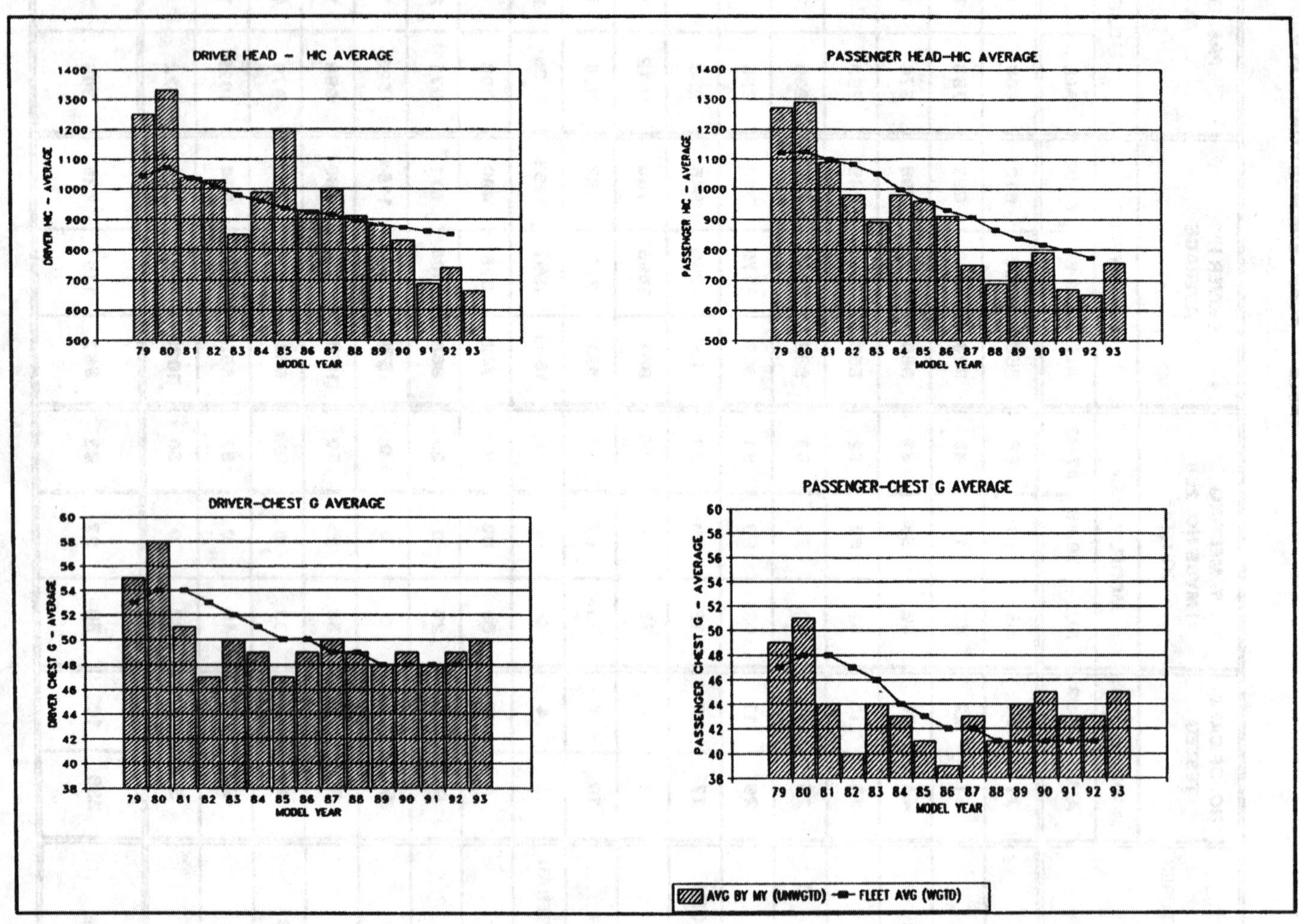

Figure 3. NCAP Dummy Response Trends for Passenger Cars

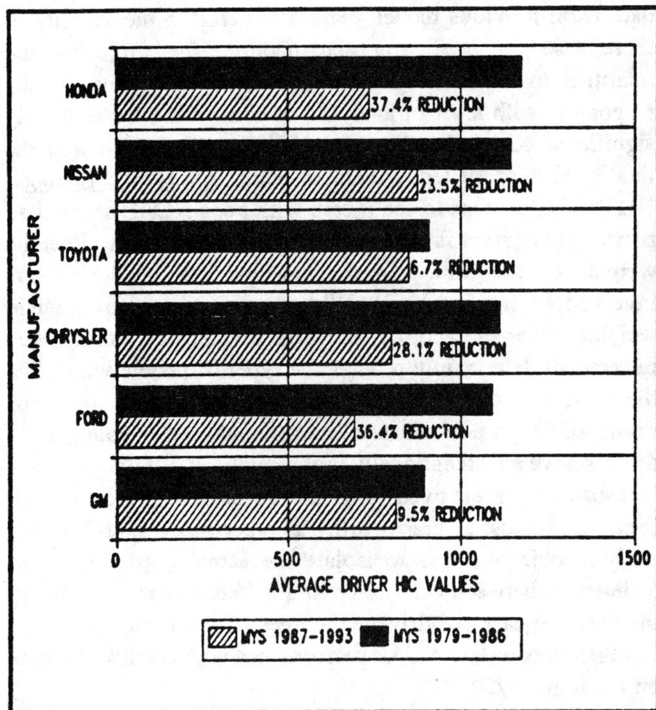

Figure 4. Average NCAP Driver HIC Values with the Percentage Reduction when Comparing MY 1987-1993 Passenger Cars to MY 1979-1986 Passenger Cars.

CHARACTERISTICS OF FRONTAL CRASHES IN THE NATIONAL ACCIDENT SAMPLING SYSTEM

In response to the Congressional Committees' request to correlate NCAP test results with fatality and injury risk in actual crashes, NHTSA examined data contained in State accident files, the National Accident Sampling System (NASS),[6] and the Fatal Accident Reporting System (FARS).[10] State files have large samples, but are of limited utility because they do not identify which injuries are serious or life-threatening. Also, the accuracy of belt-use reporting is questionable in some cases. NASS contains detailed information on occupant injuries and belt use, but the number of serious-injury crashes on the file is insufficient to analyze the correlation of injury risk with NCAP test results. On the other hand, the NASS data are quite useful for tabulating frequency distributions of severe frontal crashes and comparing the characteristics of actual crashes to NCAP test conditions. Two of the more important crash parameters for frontal crashes are the change in velocity (delta V) which occurs during the impact and the impact configuration. As previously noted, the NCAP tests result in delta Vs of approximately 40 mph and the NCAP crash configuration is a full-frontal barrier impact.

Crash Severity

Figure 2 graphs the cumulative distributions of injuries and fatalities by delta V, as found in the NASS file for restrained drivers in frontal towaway crashes. These data indicate that almost 60 percent of the fatalities and approximately 90 percent of the serious injuries for restrained drivers occur below the NCAP delta V of 40 mph. Assuming that NCAP results reflect the relative potential safety that a vehicle provides for belted occupants within 5 mph of the NCAP delta V (i.e., the NCAP data are applicable from 35- to 45-mph delta V), nearly 50 percent of the fatalities occur within this range.

Crash Type

The NCAP test configuration is based on FMVSS 208. This configuration is a full-frontal crash into a fixed-rigid barrier. This is approximately the same as two similar vehicles colliding head-on. Such collisions result in extensive damage across the full front of the vehicle and expose the occupants to high forces which must be effectively controlled by the restraint systems and the gradual deformation of the vehicle structure in order to prevent serious or fatal injury.

In Figures 5 and 6, NASS data provide insight into the relationship of real-world crash configurations to this laboratory test condition. In Figure 5, it is seen that more than 70 percent of the real-world frontal crashes which result in AIS 3 or greater injuries have a direction of force of 12 o'clock or head-on. In Figure 6, it is shown that 54 percent of the frontal crashes have induced or direct damage across the full front of the vehicle and another 27 percent have induced or direct damage which extends two-thirds of the way across the front of the vehicle.

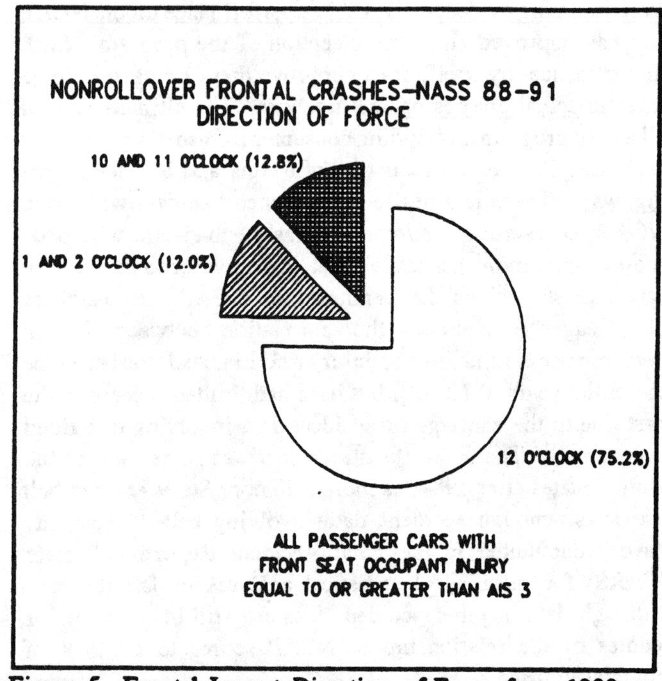

Figure 5. Frontal Impact Direction of Force from 1988-1991 NASS - Retrained and Unrestrained Front Seat Occupants

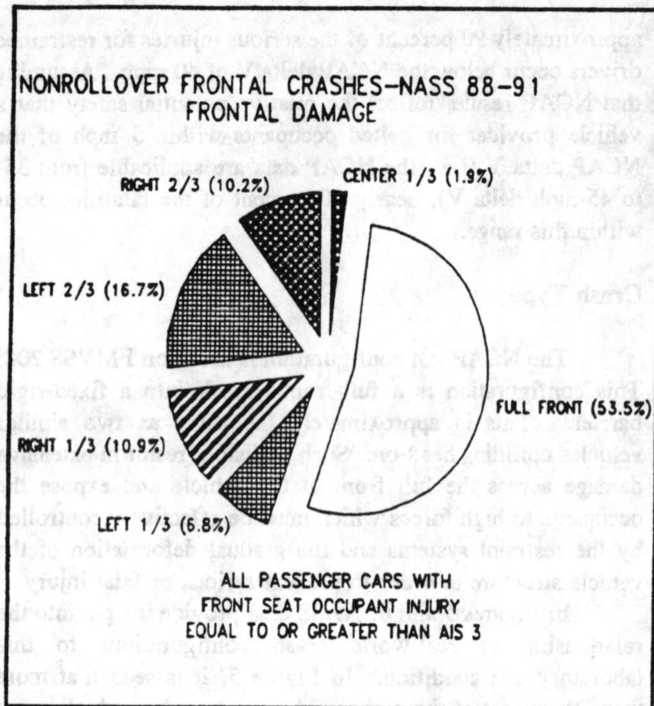

Figure 6. Frontal Impact Damage Pattern from 1988-1991 NASS - Restrained and Unrestrained Front Seat Occupants

These NASS data indicate that the FMVSS 208 and NCAP test configurations reflect closely the real-world frontal crash configurations which result in the largest number of serious injuries and fatalities.

CORRELATION OF NCAP SCORES WITH FATALITY RISK IN ACTUAL HEAD-ON COLLISIONS

The historical trends presented earlier in this paper are overwhelming evidence that vehicle performance on the NCAP test has improved since the inception of the program. Still, that evidence, by itself, does not prove that cars have become safer in actual crashes on the highway. The ultimate goal of all safety programs, including consumer information programs such as NCAP is the reduction of deaths and injuries on the highway. There is a desire for evidence that cars with good NCAP scores are safer in actual crashes than cars with poor scores, and, more generally, that cars have become safer in actual crashes since the beginning of NCAP. Researchers have eagerly explored the correlation between NCAP performance and fatality or injury risk in actual crashes since the initial years of NCAP, but have had limited success in the past due to the shortage of accident data involving restrained occupants.[11] Thanks to the steady increase in belt use in the United States after 1984, as more and more States enacted belt use laws, enough accident data involving belted occupants have accumulated in the Fatal Accident Reporting System (FARS) for meaningful statistical analyses of fatality risk, although data in other accident files are still insufficient for studies of the relationship of NCAP scores to the risk of nonfatal injuries.

The Appropriations Act for Fiscal Year 1992 directed "NHTSA to provide a study to the House and Senate Committees on Appropriations comparing the results of NCAP data from previous model years to determine the validity of these tests in predicting actual on-the-road injuries and fatalities over the lifetime of the models."[1] The agency responded with a set of analyses demonstrating a statistically significant correlation between NCAP performance and the fatality risk of belted drivers in actual head-on collisions.[3] NHTSA's goal was to see if cars with poor NCAP scores had more belted-driver fatalities than would be expected, given the weights of the cars, and the age and sex of the drivers involved in the crashes. Without adjustment for vehicle weight, driver age and sex, the large diversity of fatality rates in accident data mainly reflects the types of people who drive the cars, not the actual crashworthiness of the cars. For example, "high-performance" cars popular with young male drivers have an exceptionally high frequency of fatal crashes - because they are driven in an unsafe manner - even though they may be just as crashworthy as other models. NHTSA's analysis objective was to isolate the actual crashworthiness differences between cars, removing differences attributable to the way the cars are driven, the ages of the occupants, etc., and then to correlate NCAP performance with crashworthiness on the highway.

Analysis Overview

Since NCAP is a frontal impact test involving dummies protected by safety belts, the agency limited the accident data to frontal crashes involving belted occupants. However, NHTSA did not consider all types of frontal crashes, but further limited the data to head-on collisions between two passenger cars, each with a belted driver, which resulted in a fatality to one or to both of the drivers. A head-on collision is a special type of highway crash ideally suited for studying frontal crashworthiness differences between two cars. Both cars are in essentially the same frontal collision. It doesn't matter if one of them had a "safe" driver and the other, an "unsafe" driver; at the moment they collide head-on, how safely they were driving before the crash is nearly irrelevant to what happens in the crash. Which driver dies and which survives depends primarily on the intrinsic relative crashworthiness of the two cars, their relative weights, and the age and sex (vulnerability to injury) of the two drivers.[12]

If car 1 and car 2 weigh exactly the same, and both drivers are the same age and sex, the likelihood of a driver fatality in a head-on collision would be expected to be equal in car 1 and car 2. If car 1 and car 2 have different weights, etc., it is still possible to predict the expected fatality risk for each driver in a head-on collision between these two cars. The relationship between fatality risk and vehicle weight, driver age and sex is calibrated from the accident data by a logistic regression.[13] The expected fatality risk for driver 1 is

$$\frac{\exp[.616 - 5.427(\log W_1 - \log W_2) + .0531(A_1 - A_2) + .34(F_1 - F_2)]}{1 + \exp[.616 - 5.427(\log W_1 - \log W_2) + .0531(A_1 - A_2) + .34(F_1 - F_2)]}$$

where W_1 is the curb weight of car 1, A_1 is the age of driver 1 and F_1 is 1 if driver 1 is female, 0 if male. The expected fatality risk for driver 2 is

$$\frac{\exp[.616 + 5.427(\log W_1 - \log W_2) - .0531(A_1 - A_2) - .34(F_1 - F_2)]}{1 + \exp[.616 + 5.427(\log W_1 - \log W_2) - .0531(A_1 - A_2) - .34(F_1 - F_2)]}$$

These formulas measure the relative vulnerability to fatal injury of the two drivers, given that their cars had a head-on collision. The risk is greater in the lighter car than the heavier car, and the older or female driver is more vulnerable to injury than the younger or male driver. The formulas do not address the propensity of cars to get involved in head-on collisions as a function of driver age, sex, etc. For example, given 100 fatal head-on collisions between 3000 pound cars driven by belted, 20-year-old males and 2500 pound cars driven by belted, 50-year-old females, the formulas predict 9 deaths among the young males in the heavier cars and 97 deaths among the older females in the lighter cars (for a total of 106 fatalities in the 100 collisions, since some of them resulted in fatalities to both drivers).[14]

Cars with average crashworthiness capabilities will experience an actual number of fatalities very close to what is predicted by these formulas, which are calibrated from the collision experience of production vehicles. If a group of cars, however, consistently experiences more fatalities than expected in their head-on collisions, then the empirical evidence suggests that this group of cars is less crashworthy than the average car of similar mass. The gist of the analyses is to see if groups of cars with poor NCAP scores have significantly more belted-driver fatalities per 100 actual head-on collisions than expected (and there are several ways to define a "poor" score). The analyses measure the reduction in fatality risk, in actual head-on collisions, for a car with good NCAP scores relative to a car with poor NCAP scores. They measure the overall reduction in fatality risk, for belted drivers in head-on collisions, since model year 1979, when NCAP testing began, until 1991, the latest model year for which substantial accident data were available as of mid-1993.

The analyses require a data file of actual head-on collisions, with both drivers belted, resulting in a fatality to at least one of the drivers, indicating, for both cars, the curb weight, the driver's age and sex, and the HIC, chest G's and femur loads that were recorded for the driver dummy when that car was tested in NCAP. FARS, a census of fatal crashes in the United States, from 1978 through mid-1992, provided the basic accident data for the study.[10] Accurate curb weights are indispensable, because the relative fatality risk for two vehicles in a head-on collision is highly sensitive to the relative weight. The FARS data were supplemented with accurate curb weights, derived from weights specified in R. L. Polk's National Vehicle Population Profile[15] as well as actual weights measured in NHTSA FMVSS compliance tests on production vehicles.[16] Insufficient NCAP and FARS data were available to include light trucks, vans or sport utility vehicles in the analyses. Thus, the study is limited to collisions between two 1979-91 passenger cars.

NHTSA staff reviewed the cars involved in head-on collisions on FARS and identified, where possible, the NCAP test car that came closest to matching the FARS case. They found 396 head-on collisions, involving 792 cars, in which both drivers were belted and both cars match up acceptably with an NCAP case: (1) The make-models on FARS and NCAP are identical or true "corporate cousins" (e.g., Dodge Omni and Plymouth Horizon). (2) The model years on FARS and NCAP are identical, or the FARS model year is later than the NCAP model year, but that model was basically unchanged during the intervening years. The FARS cases were supplemented with the matching NCAP test results for each car. The sample is large enough for a statistical analysis of NCAP scores and fatality risk.[17]

Whereas FARS data can be used to distinguish head-on collisions from other crashes, they currently do not identify the impact speeds in the collisions or the exact alignment of the vehicles. FARS data do not single out those head-on collisions that would essentially duplicate an NCAP test: perfectly aligned collisions of two nearly identical cars, with minimal offset, a closing speed close to 70 mph. (Nevertheless, the NASS data showed that many fatal frontal crashes extensively resemble an NCAP test: approximately 50 percent of the fatal frontal crashes for restrained drivers occur within 5 mph of the NCAP delta V, and most severe frontal crashes involve damage across a large portion of the front of the vehicle.) Other major differences between NCAP tests and actual crashes include:

- Differences between the physical characteristics of the human driver population and the anthropomorphic dummy (the dummy represents a 50th percentile male)
- Variations in the vulnerability to fatal injury due to age and sex[18]
- Location of the fatal lesions (injury parameters are measured only in the head, chest, and femurs of the dummies in NCAP, and not directly on the neck or abdomen)

It is inappropriate to expect perfect correlation between NCAP test results and actual fatality risk in the full range of head-on collisions represented in the FARS sample. Moreover, if there is any significant correlation between the two, it would suggest that the NCAP scores say something about actual crashworthiness in a range of crashes that goes far beyond the specific type tested in NCAP. It would also uphold the premise that vehicles which meet the FMVSS 208 criteria in crash tests are safer in actual crash tests than vehicles that do not meet these criteria.

Correlation of NCAP Scores and Fatality Risk

The goal of the analysis is to test if cars with poor scores on the NCAP test have higher fatality risk for belted drivers, in actual head-on collisions, than cars with good or acceptable scores. There are many ways to define "poor" and "good" scores and measure the difference in fatality risk. All of the methods tried out by NHTSA staff demonstrate a statistically significant relationship between NCAP scores and actual fatality risk, as shown in Table 3.

Table 3

Collisions of Cars with "Good" NCAP Scores into Cars with "Poor" NCAP Scores
(N of crashes approximately 120 in each analysis)

		Performance in Actual Crashes	
"Good" NCAP Performance	"Poor" NCAP Performance	N of Crashes	Fatality Reduction for Good Car (%)
Chest g's $\leq$ 56	Chest g's > 56	125	19*
HIC $\leq$ 1000	HIC > 1200	113	14*
L Femur $\leq$ 1600 AND R Femur $\leq$ 1600 AND L+R Femur $\leq$ 2600	L Femur > 1600 OR R Femur > 1600 OR L+R Femur > 2600	132	20**
HIC $\leq$ 1100 AND Chest g's $\leq$ 60	HIC > 1300 OR Chest g's > 60	125	19*
Chest g's $\leq$ 56 AND L Femur $\leq$ 1400 AND R Femur $\leq$ 1400 AND L+R Femur $\leq$ 2400	Chest g's > 60 OR L Femur > 1700 OR R Femur > 1700 OR L+R Femur > 2700	134	22**
HIC $\leq$ 900 AND L Femur $\leq$ 1400 AND R Femur $\leq$ 1400 AND L+R Femur $\leq$ 2400	HIC > 1300 OR L Femur > 1700 OR R Femur > 1700 OR L+R Femur > 2700	121	19*
HIC $\leq$ 900 AND Chest g's $\leq$ 56 AND L Femur $\leq$ 1400 AND R Femur $\leq$ 1400 AND L+R Femur $\leq$ 2400	HIC > 1300 OR Chest g's > 60 OR L Femur > 1700 OR R Femur > 1700 OR L+R Femur > 2700	118	21**
NCAPINJ $\leq$ .6	NCAPINJ > .6	117	26**

*Statistically significant at the .05 level
**Statistically significant at the .01 level

A straightforward way to delineate "poor" and "good" scores is to partition the cars based on their NCAP score for a single body region - chest G's, HIC or femur load - and to consider only a subset of the 392 head-on crashes where one car has a score in the "poor" range and the other car has a score in a good or acceptable range. This subset should contain approximately 120 crashes, which is equivalent to defining the worst 20 percent of cars as "poor" performers and the remaining 80 percent as good or acceptable. Do the cars with the poor NCAP scores have significantly more driver fatalities than expected?

When chest G's are used to partition the cars into acceptable and poor performance groups, the cars with high chest G's almost always have significantly more fatalities than the cars with acceptable chest G's.[19] For example, there are 125 actual head-on collisions (both drivers belted) in which one of models had more than 56 chest G's for the driver when it was tested in NCAP, and the other had 56 g's or less. In the 125 cars with chest G's > 56, 80 drivers died, whereas only 68.2 fatalities were expected, based on car weight, driver age and sex. In the 125 cars with chest G's $\leq$ 56, there were 74 actual and 77.6 expected driver fatalities. That is a statistically significant fatality reduction of

$$1 - [(74/80) / (77.6/68.2)] = 19 \text{ percent}$$

for the cars with the lower chest G's.

The statistical significance of this difference in fatality risk is tested by examining a variable, RELEXP, which is computed for each collision. If E_1 is the expected probability of a fatality in the low-chest G car, based on the formula using car weight, driver age and sex, and E_2 is the expected probability in the high-chest G car, while A_1 and A_2 are the actual outcome in each car (1 if the driver died, 0 if the driver survived),

$$\text{RELEXP} = (A_1 - E_1) - (A_2 - E_2)$$

measures actual performance "relative to expectations." It can range from -2 to +2. The more negative it is, the better the low-chest G car did, relative to expectations. If both groups of cars were equally safe, the average value of RELEXP should be close to zero. But in these 125 crashes, the average value of RELEXP is significantly less than zero ($t = 2.32$, $p < .05$), which means that the fatality reduction for the low-chest G cars is statistically significant.[20]

The relationship between chest G's on the NCAP test and fatality risk over the range of head-on collisions experienced on the highway, although statistically significant, is not perfect. Merely having the lower NCAP score of the two cars in the collision does not guarantee survival, even if the two cars are of the same weight and the drivers of the same age and sex. Yet, on the average, in collisions between cars with $\leq$ 56 chest G's on NCAP and cars with > 56 chest G's, the driver of the car with the better NCAP score had 19 percent less fatality risk than the driver of the car with the poorer NCAP score, after controlling for weight, age and sex.

Fifty-six chest G's are just one possible boundary value between "good" and "poor" performance. The fatality reduction for "good" performers can be magnified by using a higher boundary value or by replacing a single boundary value with a gap, putting some distance between the "good" and the "poor" groups. For example, in collisions of cars with chest G's $\leq$ 60 into cars with chest G's > 60 (the pass-fail criterion in FMVSS 208), the fatality reduction in the "good" performers is 24 percent. However, there are only 92 crashes meeting those criteria. Many other boundary values between low and high chest G's will also produce statistically significant fatality reductions for the group with low chest G's, but the boundary value of 56 maximizes the fatality reduction for an accident sample close to 120 crashes.

The Head Injury Criterion (HIC) can be used to partition the cars into two performance groups. In 113 head-on collisions between a car with HIC $\leq$ 1000 on the NCAP test and a car with HIC > 1200, the fatality risk was a statistically significant 14 percent lower in the cars with HIC $\leq$ 1000. The femur loads measured on the NCAP tests can also, by themselves, differentiate safer from less safe cars. The "good" performers are defined to be the cars with $\leq$ 1600 pounds on each leg, and the sum of the two loads $\leq$

2600 pounds. The "poor" performers are those with > 1600 pounds on either leg, or a sum > 2600 pounds. In 132 head-on collisions, the fatality reduction for the "good" NCAP femur load performers was a statistically significant 20 percent.[21]

One reason that chest G's, HIC and femur load all "work" by themselves is that the three NCAP test measurements are not independent observations on isolated body regions. Cars with intuitively excellent safety design tend to have low scores on all parameters, while cars with crashworthiness problems tend to have high scores on one or more parameters, but it is not always predictable which one. Still, the reasons for the significant correlation between NCAP femur load and actual fatality risk are not completely understood at this time, since injuries to the lower extremities, by themselves, are generally not fatal.

Any two NCAP parameters, working together, can do an even more reliable job than any single parameter. In 125 actual head-on collisions between cars with driver HIC ≤ 1100 and chest G's ≤ 60 on the NCAP test and cars with either HIC > 1300 or chest G's > 60, the fatality risk was a statistically significant 19 percent lower in the cars with low HIC and chest G's. Table 3 shows how chest G's and femur load, or HIC and femur load can be used to partition the cars, with statistically significant 19-22 percent fatality reductions for the "good" performers, in samples of 121-134 crashes.[22]

NHTSA's Report to Congress highlighted three analyses in which cars were partitioned into good and poor performance groups according to the HIC and chest G's.[2] In Case I, the "good" car has to meet the FMVSS 208 criteria (HIC ≤ 1000 and chest G's ≤ 60) while the "poor" car has to fail at least one criterion (HIC > 1000 or chest G's > 60). Cases II and III place a gap between "good" and "poor" performance. In Case II, the "good" car has to meet the FMVSS 208 criteria for HIC and chest G's while the "poor" car has to have HIC > 1200 or chest G's > 70. In Case III, the "good" car must have HIC ≤ 900 and chest G's ≤ 56, while the poor car has HIC > 1250 or chest G's > 65. (These three cases are of particular interest because they were used in a parallel analysis of fixed-object impacts, that is summarized in the next section of this paper.) Table 4 shows the actual and expected fatalities in the three cases, as well as the average vehicle weight and driver age. The "unadjusted fatality reduction" (simple difference of actual fatalities in the "good" vs. the "poor" cars) is 20 percent in Case I, 30 percent in Case II and 33 percent in Case III. After adjusting for car weight, driver age and sex, the fatality reductions are 14, 19 and 27 percent, respectively. The adjusted fatality reductions are not too much smaller than the unadjusted reductions, because in these particular analyses, the "good" cars have a modest weight advantage over the "poor" cars, as shown in Table 4, but the advantage is partly offset because the drivers are slightly younger in the "poor" cars.

Table 4

Collisions of Cars with "Good" NCAP HIC and Chest G's into Cars with "Poor" HIC or Chest G's

	CASE I	CASE II	CASE III
Definition of "good" car: HIC ≤	1000	1000	900
AND Chest G's ≤	60	60	56
Definition of "poor" car: HIC >	1000	1200	1250
OR Chest G's >	60	70	65
Average car weight:			
"Good" cars	2920	2941	2944
"Poor" cars	2769	2769	2761
Average driver age:			
"Good" cars	43.7	42.2	46.4
"Poor" cars	41.1	41.0	43.5
N of head-on collisions	170	104	81
Actual driver fatalities			
In the "good" cars	89	50	39
In the "poor" cars	111	71	58
Expected driver fatalities			
In the "good" cars	96.2	56.8	45.8
In the "poor" cars	103.8	65.2	49.9
"Unadjusted" fatality reduction for the good cars (%)*	20	30	33
"Adjusted" fatality reduction for the good cars (%)**	14	19	27

*e.g., in Case I, the unadjusted reduction is 1 - (89/111).

**e.g., in Case I, the reduction is 1 - [(89/111) / (96.2/103.8)] after adjusting for car weight, driver age and sex.

NCAP scores for all three body regions, with an independent "pass-fail" criterion on each score, work about as well as scores for any two body regions. "Good" performance could be defined as HIC ≤ 900 and chest G's ≤ 56 and femur load ≤ 1400 on each leg and ≤ 2400, total, while HIC > 1300 or chest G's > 60 or femur load > 1700 on either leg or > 2700, total defines "poor" performance. The fatality risk in 118 actual head-on collisions between a good and a poor NCAP performer is a statistically significant 21 percent lower for the drivers of the cars with good NCAP scores, after controlling for vehicle weight, driver age and sex. These criteria can be varied by a moderate amount and the fatality reduction for the "good" performers will still be statistically significant, as long as the HIC cutoff is reasonably close to or slightly above the FMVSS 208 value of 1000, the chest G cutoff is not far from the FMVSS 208 value of 60 g's, and the femur load cutoff ranges from about 1400 pounds up to the FMVSS 208 value of 2250 pounds.[23]

A highly efficient way to use the NCAP scores for the three body regions, however, is to combine them into a single composite score, wherein excellent performance on two body regions might compensate for moderately poor performance on the third. The composite score could be some type of weighted or unweighted average of the scores for the various body regions. For example, a weighted average measure of NCAP performance, NCAPINJ, was derived by a two-step process. First, the actual NCAP results for the driver dummy were transformed to logistic injury probabilities, HEADINJ, CHESTINJ, LFEMURINJ and RFEMURINJ, each ranging from 0 to 1.[24] The weighted average

$$\text{NCAPINJ} = .21\ \text{HEADINJ} + 2.7\ \text{CHESTINJ} + 1.5\ (\text{LFEMURINJ} + \text{RFEMURINJ})$$

has the empirically strongest relationship with fatality risk for belted drivers in the specific data set of actual head-on collisions described above (396 collisions, 792 cars). The accident data include 117 head-on collisions of a car with NCAPINJ ≤ 0.6 into a car with NCAPINJ > 0.6. Fatality risk is a statistically significant 26 percent lower in the cars with NCAPINJ ≤ 0.6 (t for RELEXP is 3.22, $p < .01$). Since NCAPINJ is a weighted sum of NCAP scores for all of the body regions, the cars with NCAPINJ ≤ 0.6 have, on the average, substantially lower HIC, chest G's and femur loads than cars with NCAPINJ > 0.6.[25]

The statistically significant relationship between NCAP performance and actual fatality risk is not limited to head-on collisions in which one vehicle is specifically a "good" performer and the other has "poor" NCAP scores. There is also a strong relationship in the full data set of 392 head-on collisions, which includes crashes between two "good" performers or two "poor" performers. NCAPINJ was defined in a way that makes it easy to test if it is correlated with fatality risk. DELNCAP = $\text{NCAPINJ}_1 - \text{NCAPINJ}_2$, the relative NCAP score for the two vehicles in the crash, and RELEXP, the measure of actual fatality risk defined above, have a correlation coefficient of .166, which is significant at the .001 level in a sample of 392 crashes. In other words, the higher the NCAP score for car 1 relative to car 2, the higher the fatality risk for driver 1 relative to driver 2, after adjusting for car weight, driver age and sex.[26]

The purpose of defining NCAPINJ was to illustrate the strength of the overall relationship between NCAP performance and fatality risk. However, NCAPINJ is not a "magic bullet" or "ideal" way to combine the NCAP scores, resulting in far higher correlations than other methods. Many other weighted averages, or even an unweighted sum of the logistic injury probabilities, work almost as well for differentiating the safer from the less safe cars on the principal accident data set. On a more restricted alternative accident data set of 310 collisions and 620 cars, where the FARS vehicles are also required to have the same number of doors as their matching NCAP test vehicles, NCAPINJ is not the optimum weighted average (although it comes close to the optimum), and it is only slightly more correlated with fatality risk than an unweighted sum of the logistic injury probabilities. Moreover, on this alternative data set, HIC and femur load have about equally strong correlation with fatality risk.[27]

Improvements in Actual Crashworthiness and NCAP Performance during 1979-91

It was shown above that the performance of passenger cars on the NCAP test has greatly improved since the program was initiated in 1979. Has the historical trend of better performance on the NCAP test been matched by a reduction in the actual fatality risk of belted drivers in head-on collisions?

In general, it is not easy to compare the crashworthiness of cars of different model years. Fatality rates per 100 million vehicle miles have been declining for a long time. In any given year, the fatality rate per 100 million miles or per 100 crashes is lower for new cars than for old cars. Both trends create the impression that "cars are getting safer all the time," but, in fact, the declines in fatality rates to a large extent reflect changes in driving behavior, roadway environments, demographics or accident-reporting practices. A head-on collision between cars of two different model years, however, reveals their relative crashworthiness. Both cars are in essentially the same frontal collision, on the same road, in the same year, on the same accident report. The behavior of each driver, prior to the impact, has little effect on who dies during the impact. After adjustment for differences in car weight, driver age and sex, the model year with more survivors is more crashworthy. For example, a similar analysis of head-on collisions involving two unrestrained drivers found little change in fatality risk between model years 1970 and 1984.[28]

There have been 241 actual head-on collisions between a model year 1979-82 car and a model year 1983-91 car, in which both drivers were belted. These collisions allow a comparison of cars built during the first four years of NCAP to subsequent cars, where manufacturers have had time to build in safety improvements. In the 241 older cars, 146 drivers died, whereas only 126.6 fatalities were expected, based on car weight, driver age and sex. In the newer cars, there were 132 actual and 147.1 expected driver fatalities. For the 1983-91 cars, that is a fatality reduction of

$$1 - [(132/146) / (147.1/126.6)] = 22 \text{ percent}$$

and it is statistically significant (t for RELEXP is 3.43, $p < .01$).[29]

A more generalized analysis, which allows a larger sample size of 1189 crashes, applies to head-on collisions in which the "case" vehicle of interest is a 1979-91 car that matches up with an NCAP test, whose driver wore belts, but the "other" vehicle in the crash can be any 1976-91 passenger car with a belted driver. For any subset of crashes, the actual fatalities are tallied in the "case" and "other" vehicles, and so are the expected fatalities (based on the relative weights of the two cars, and the age and sex of the two drivers). The fatality risk index for the "case" vehicles is

$$100\ [(\text{actual}_{\text{case}} / \text{actual}_{\text{other}}) / (\text{expected}_{\text{case}} / \text{expected}_{\text{other}})]$$

The lower the risk index, the more crashworthy the car (100 = average). The actual fatality risk indices can be compared in three model-year groups, 1979-82, 1983-86 and 1987-91. So can the NCAP test performance, as measured by a composite score such as NCAPINJ, by the average values of the actual NCAP parameters for the three body regions, or by the average value of the joint probability of AIS $\geq$ 4 injury to the head or chest:

	Model Years		
	1979-82	1983-86	1987-91
Fatality risk index in actual head-on collisions	119	95	91
Average value of NCAPINJ	.59	.40	.37
Percent of cars with NCAPINJ > 0.6	49	14	9
Average HIC	1052	915	827
Average chest G's	54.9	46.8	46.5
Joint head-chest injury probability	.32	.25	.22
Average left femur load	928	883	1002
Average right femur load	1079	784	1018

The trends in the actual fatality risk and the average value of NCAPINJ are almost identical. The risk index decreased by a statistically significant 20 percent from 1979-82 to 1983-86, and by another 4 percent from then until 1987-91 (nonsignificant). In all, the actual fatality risk for belted drivers in head-on collisions decreased by a statistically significant 24 percent from model years 1979-82 to 1987-91. A composite NCAP score, such as NCAPINJ, nicely portrays the improvement in NCAP performance over time. Parallel to the reduction in the fatality risk index, NCAPINJ greatly improved from an average of 0.59 in model years 1979-82 to 0.40 in 1983-86, with an additional, modest improvement to 0.37 in 1987-91. If NCAPINJ = 0.6 is defined as the limit of "acceptable" NCAP performance, the passenger car fleet has truly progressed since the inception of NCAP: initially, 49 percent of the cars had NCAPINJ > 0.6, but that decreased to 14 percent in 1983-86 and 9 percent in 1987-91. Average HIC and chest G's declined substantially during the NCAP era; average femur loads stayed about the same, but well below the 2250 pounds permitted in FMVSS 208.[30]

A final question is whether the correlation of actual fatality risk with NCAP performance is merely a coincidence, in the sense that fatality risk and NCAP scores both decreased during model years 1979-91, and thus became correlated with one another just because both are correlated with model year. The earlier analysis of collisions of cars with NCAPINJ $\leq$ .6 into cars with NCAPINJ > .6 was rerun, limited to crashes in which the "good" and the "poor" NCAP performers had similar model years ($-5 \leq MY_{GOOD} - MY_{POOR} \leq 3$). In these 61 crashes, where the average model year for the "good" and "poor" NCAP performers was equal, and the average car weight, driver age and sex nearly equal, the fatality risk was a statistically significant 32 percent lower in the "good" NCAP performers than in the poor performers (t for RELEXP is 3.03, $p < .01$). This result and other, similar analyses confirm that the strong association between NCAP performance and actual fatality risk in head-on collisions is quite independent of model year.[31]

Principal Findings, Conclusions and Caveats for the Analysis of Fatality Risk in Head-On Collisions

- There is a statistically significant correlation between the performance of passenger cars on the NCAP test and the fatality risk of belted drivers in actual head-on collisions. Since many head-on collisions differ substantially from NCAP test conditions, this suggests NCAP scores are correlated with actual crashworthiness in a wide range of crashes.
- In a head-on collision between a car with "acceptable" NCAP performance and a car of equal mass with "poor" performance, the driver of the "good" car has, on the average, about 15-25 percent lower fatality risk.
- A highly effective way to differentiate "good" from "poor" NCAP performance is by a single, composite NCAP score, such as a weighted combination of the scores for the three body regions. However, even the NCAP score for any single body region can be used to partition the fleet so that the cars with "good" scores have significantly lower fatality risk than the cars with "poor" scores. The borderline between "good" and "poor" NCAP scores that optimizes the differences in actual fatality risk is close to the FMVSS 208 criteria for each of the three body regions.
- NCAP scores have improved steadily since the inception of the program in 1979, with the greatest improvement in the early years. By now, most passenger cars meet the FMVSS 208 criteria in the 35 mph NCAP test. This achievement has been paralleled by a 20-25 percent reduction of fatality risk for belted drivers in actual head-on collisions in model years 1979-91, with the largest decreases during the early 1980's.
- This is a statistical study and it is not appropriate for conclusions about cause and effect. It shows that passenger cars became significantly safer in head-on collisions during 1979-91, as NCAP scores improved. It does not prove that the NCAP program was the stimulus for each of the vehicle modifications that saved lives during 1979-91. (For example, the automatic protection requirement of FMVSS 208 was another important stimulus.)
- The correlation between NCAP scores and actual fatality risk is statistically significant, but it is far from perfect. On the whole, cars with poor NCAP scores have higher-than-average fatality risk in head-on collisions, but there is no guarantee that every specific make-model with poor NCAP scores necessarily has higher fatality risk than the average car. Conversely, there is no guarantee that a specific model with average or even excellent scores necessarily has average or lower-than-average fatality risk in head-on collisions.
- The data show that cars with poor NCAP scores (e.g., above the FMVSS 208 criteria) have significantly elevated fatality risk in head-on collisions, but they do not show a significant difference between the fatality risk of cars with exceptionally good NCAP performance and those with merely average performance.

ANALYSIS OF FATAL CAR-TO-FIXED OBJECT FRONTAL COLLISIONS

Concurrent with the analysis of head-on collisions, a more generalized study of FARS was conducted to determine if cars with "good" NCAP scores also had lower-than-average fatality risk in frontal crashes other than the car-to-car head-on collisions. In this analysis, the fatality rate for restrained drivers in frontal fixed-object collisions, per million vehicle years, was compared for cars with "good" NCAP scores and cars with poor scores.

The study is based on 1979-90 FARS data on MY 1979-90 passenger cars.[10] The first step in the analysis was to identify the cars in FARS for which NCAP test results were available. The make-model and number of doors had to match exactly in FARS and NCAP. The model years of the FARS and NCAP could be identical, or the FARS model year could be later than the NCAP model year, if that make-model was basically unchanged during the intervening years.[32] For these applicable vehicles, records of restrained drivers killed in single-vehicle frontal, fixed-object collisions were extracted from FARS. Exposure data (vehicle registration years) were obtained from R. L. Polk's National Vehicle Population Profile.[15] The number of exposure years from the Polk file was multiplied by the on-the-road belt use rate for drivers,[33] to obtain the "restrained exposure years." The frontal, fixed-object fatality rates per million exposure years (for restrained drivers) was computed and compared for cars with good and poor NCAP performance, where the definitions of "good" and "poor" are based on HIC and chest G's, as in Case I, Case II and Case III of Table 4.

In Case I, the "good" car has to meet the FMVSS 208 criteria (HIC $\leq$ 1000 and chest G's $\leq$ 60) while the "poor" car has to fail at least one criterion (HIC > 1000 or chest G's > 60). Cases II and III place a gap between "good" and "poor" performance. In Case II, the "good" car has to meet the FMVSS 208 criteria for HIC and chest G's while the "poor" car has to have HIC > 1200 or chest G's > 70. In Case III, the "good" car must have HIC $\leq$ 900 and chest G's $\leq$ 56, while the poor car has HIC > 1250 or chest G's > 65.

The results for frontal, fixed-object crashes are shown in Table 5 along with the average vehicle test weight, drivers' HICs, and drivers' chest G's from NCAP. The last row of Table 5 shows the reduction in the fatality rate, for "good" relative to "poor" NCAP performers. The fatality reduction for good NCAP performance is 19 percent in Case I, 22 percent in Case II and 36 percent in Case III. Unlike the analysis of head-on collisions, the fatality rates here have not been adjusted for differences in car weight, driver age or sex. Also unlike the head-on collisions, these data do not "self-adjust" for differences in crash-involvement propensities. Nevertheless, as was noted in Table 4, there is, on the average, little difference in the vehicle weights and driver ages of "good" and "poor" NCAP performers. In the analyses of head-on collisions in Table 4, both "adjusted" and "unadjusted" fatality reductions were computed for the good NCAP performers, and they were not too far apart. The fatality reductions in Table 5 must be compared to the "unadjusted" reductions in Table 4.

Table 5

Summary of Real-World NCAP Effects Based on FARS Analysis of Car-to-Fixed Object Frontal Collisions

Parameter	Group No.	Case I*	Case II*	Case III*
Average vehicle NCAP test weight	1	3183	3183	3150
	2	3197	3180	3202
Average drivers' HIC from NCAP	1	722	722	676
	2	1315	1614	1435
Average drivers' chest G's from NCAP	1	45	45	44
	2	58	58	62
Reduction in fatality rate - cars in group 1 vs. cars in group 2 - actual FARS data		19.2 %	21.8 %	35.7 %

*Case I - Cars in Group 1 have HIC values $\leq$ 1000 and chest G's $\leq$ 60 in the NCAP tests. Cars in Group 2 have HIC > 1000 and/or chest G's > 60 in the NCAP tests.

*Case II - Cars in Group 1 have HIC values $\leq$ 1000 and chest G's $\leq$ 60 in the NCAP tests. Cars in Group 2 have HIC > 1200 and/or chest G's > 70 in the NCAP tests.

*Case III - Cars in Group 1 have HIC values $\leq$ 900 and chest G's $\leq$ 56 in the NCAP tests. Cars in Group 2 have HIC > 1250 and/or chest G's > 62 in the NCAP tests.

Figure 7 reveals remarkable similarity in the results for head-on and fixed-object collisions. In Case I, the unadjusted fatality reduction for the good NCAP performers was 20 percent in the head-on collisions and 19 percent in the fixed-object impacts. In Case II, the reductions were 30 and 22 percent. In Case III, where the definitions of a "good" and a "poor" car straddled the FMVSS 208 criteria and produce the largest relative fatality reduction, that reduction is 33 percent in the head-on collisions and 36 percent in the fixed-object crashes. NHTSA is contemplating more detailed analyses of single vehicle crashes, including methods to control for driver age, etc., and to test the statistical significance of differences in fatality rates.[34] For now, these initial analyses show consistency with the results in head-on collisions and suggest that the findings in car-to-car head-on crashes may also be applicable to these other frontal crashes.

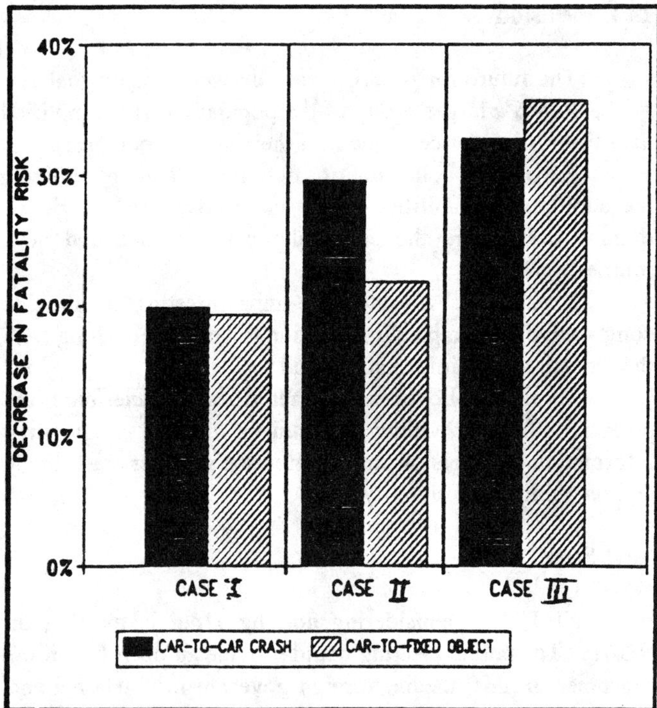

Figure 7. Comparison of the Decrease in Fatality Risks for "Good" Performing Cars in NCAP in Car-to-Car and Car-to-Fixed Object Collisions

THE FUTURE FOR NCAP

Make NCAP Easy to Understand

NCAP has produced extensive frontal crash test information for use by consumers and the media. However, as noted in NHTSA's Focus Group Study and Media Survey, this information has been difficult for some consumers to understand and the media to use.[35] NHTSA's first step toward the goal of reaching a larger group of the population has been to simplify the data in order to assist consumers in making their vehicle purchase decision.[36]

The primary element for Fiscal Year 1994 will be a consumer brochure in a computerized format. This will permit easy updating. The format will also be adaptable to print media requirements. The brochure will utilize an easy to read and simple presentation technique. It will contain a description of NCAP and the comparative results from the vehicle tests.

Expand the Usefulness of NCAP

NCAP has evolved into a real catalyst in the automobile market place. Consumer enlightening publications highlight crash test results as an important ingredient to consider in the vehicle selection process. As explained at the beginning of this paper, the overall trend of the NCAP test results indicate the favorable influence the program has had on motivating the manufacturers to improve restraint systems, steering assemblies, and structural crash characteristics of many of their products. The accident analyses highlighted the significance of these improvements as shown, statistically, in the reduction of fatality risks for restrained occupants in the "good" performing passenger cars. In addition, NCAP continues to be a main source of research and engineering data for use by NHTSA and others in directing research programs and analyzing safety problems. With the exclusive use of the Hybrid III dummy in the NCAP frontal tests, NHTSA will expand the collection of safety information by utilizing the additional capabilities of the more advanced dummy to measure the potential for lower limb and neck injuries. From these perspectives, the frontal crash testing of NCAP has been and continues to be successful.

The focus group recommendations critically pointed out that NCAP provides information for frontal crashes only. Although the frontal crashes account for the highest percentage of fatalities, as shown in Figure 8, side crashes and rollovers are also very prominent crash modes. Almost 8,000 fatalities occurred in side crashes in 1991 and more than 9,000 fatalities occurred in rollover crashes. The focus group study indicates that consumers desire overall safety information on vehicles. In essence, NHTSA needs to expand the crash modes covered by NCAP.

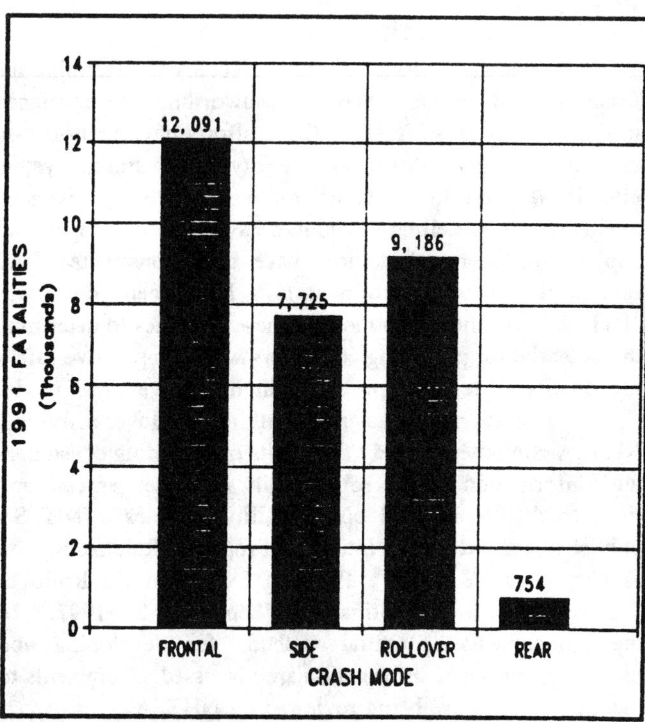

Figure 8. 1991 Fatalities occurring in Frontal, Side, Rollover, and Rear Crash Modes - Passenger Cars and Light Trucks.

The enactment of the upgraded side-impact protection standard, beginning with MY 1994 passenger cars, has provided the opportunity to expand NCAP into side-impact protection.[37] The expansion of NCAP into side-impact protection has the potential for improving occupant protection well above that required in the applicable standard if the vehicle manufacturers, which have been responsive to the frontal NCAP test results, are equally responsive to such a program in side-impact testing. As in the frontal NCAP, a side-impact NCAP would provide an engineering data base which can be used to inform consumers of relative vehicle crashworthiness performance. That data base can also serve as a basis for further research and additional studies in side-impact.

Side Impact NCAP

In Fiscal Years 1992 and 1993, Congress provided funds as requested by NHTSA to conduct a study to develop the requirements and procedures for the possible expansion of NCAP into side-impact protection. This two-year study included a pilot crash testing program to determine an NCAP crash severity level, to assure that testing, instrumentation, and test device performance are consistent. The results from this program support the feasibility of a side-impact NCAP which could provide comparative results to consumers. If Congressional funding is provided, side-impact NCAP tests would be conducted on passenger cars and the information would be provided to consumers along with the frontal NCAP information. Initiation of this side-impact NCAP would provide consumers with comparative safety data on two of the most important crash modes.

Rollover Testing

Research efforts continue in NHTSA to determine the feasibility of determining vehicle crashworthiness performance in the rollover crash mode. These efforts have focussed on evaluating vehicle structural integrity and restraint system effectiveness during dynamic rollover events. Advanced mathematical modelling techniques have been developed and applied, rollover test devices have been constructed, and several demonstration rollover tests have been conducted. NHTSA will continue to monitor these activities to determine the potential for providing consumers with comparative safety information on levels of protection in the rollover crash mode.

In addition to these crashworthiness rollover activities, NHTSA continues to study the merits of providing consumers with information on the roll stability of passenger cars and light trucks, vans, and sports utility vehicles. NHTSA published an Advance Notice of Proposed Rulemaking on January 3, 1992[38] and a Planning Document for Rollover Prevention and Injury Mitigation on September 23, 1992.[39] In these documents, potential methods for developing and providing consumer information are discussed. Comments to these documents are being reviewed by NHTSA.

In Conclusion

The future for NCAP includes several major goals:

- Reach a larger group of the population with simplified data that will assist consumers in their vehicle purchases,
- Expand the collection of safety information by utilizing the additional capabilities of the more advanced Hybrid III dummy to measure the potential for lower limb and neck injuries,
- Expand NCAP into side-impact testing to provide comparative side impact information to consumers along with the frontal NCAP information, and
- Monitor rollover safety activities to determine the potential for providing consumers with comparative information on levels of protection in the rollover crash mode and on vehicle roll stability.

Next Steps

NHTSA is considering holding a public meeting on NCAP. The public meeting could provide an open forum for consumer groups, media, foreign governments, national and international safety organizations, and motor vehicle manufacturers to discuss the above NCAP goals. A *Federal Register* notice was issued on January 4, 1994, requesting comments on the possibility of convening such a meeting and on the scope of materials which might be discussed; NHTSA has received the comments and is reviewing them.[40]

REFERENCES

1) *Department of Transportation and Related Agencies Appropriations Act for 1992*, Conference Report No. 102-243, United States House of Representatives, Washington, October 7, 1991.

2) *New Car Assessment Program: Response to the NCAP FY 1992 Congressional Requirements*, Report to the Congress, National Highway Traffic Safety Administration, Washington, 1993.

3) Kahane, Charles J., *Correlation of NCAP Performance with Fatality Risk in Actual Head-On Collisions*, Report No. DOT HS 808 061, National Highway Traffic Safety Administration, Washington, 1994.

4) *The Motor Vehicle Information and Cost Savings Act of 1973*, Public Law 92-513, as amended, 15 *United States Code* 1901-2012.

5) *Federal Register* 49 (17 July 1984): 28962. Also, *Code of Federal Regulations*, Title 49, Part 571.208, General Printing Office, Washington, 1992.

6) *National Accident Sampling System 1986*, Report No. DOT HS 807 296, National Highway Traffic Safety Administration, Washington, 1988.

7) *The Abbreviated Injury Scale (AIS) - 1990 Revision*, American Association for Automotive Medicine, Des Plaines, IL, 1990.

8) *New Car Assessment Program: Plan for Responding to the NCAP FY 1992 Congressional Directives*, Report to the Congress, National Highway Traffic Safety Administration, Washington, 1992.

9) *Historical Performance of Different Auto Manufacturers in the New Car Assessment Program Tests*, Report to the Congress, National Highway Traffic Safety Administration, Washington, 1993.

10) *Fatal Accident Reporting System, 1981 Coding and Validation Manual*, National Highway Traffic Safety Administration, Washington, 1980. *Fatal Accident Reporting System, 1987 Coding and Validation Manual*, National Highway Traffic Safety Administration, Washington, 1986. *Fatal Accident Reporting System, 1991 Coding and Validation Manual*, National Highway Traffic Safety Administration, Washington, 1990.

11) *Comparison of New Car Assessment Program Crash Test Results with Real-World Crash Data: Summary and Analysis of Existing Studies*, Submission to Docket No. 79-17-GR-060, National Highway Traffic Safety Administration, Washington, 1988.

12) Previous analyses of relative crashworthiness based on head-on collisions include Zador, Paul L., Jones, Ian S., and Ginsburg, Marvin, "Fatal Front-to-Front Car Collisions and the Results of 35 mph Frontal Barrier Impacts," in *Proceedings of the Twenty-Eighth Conference of the American Association for Automotive Medicine*, American Association for Automotive Medicine, Morton Grove, IL, 1984; and Kahane, Charles J, *An Evaluation of Occupant Protection in Frontal Interior Impact for Unrestrained Front Seat Occupants of Cars and Light Trucks*, Report No. DOT HS 807 203, National Highway Traffic Safety Administration, Washington, 1988, pp. 111-140.

13) Joyner, Stephenie P., ed, *SUGI Supplemental Library User's Guide, 1983 Edition*, SAS Institute Inc., Cary, NC, 1983.

14) Kahane (1994), op. cit., pp. 7-12 and 60-61.

15) *National Vehicle Population Profile*, R. L. Polk, Detroit, Annual Publication.

16) Kahane (1994), op. cit., pp. 19-27.

17) Ibid., pp. 27-51.

18) Evans, Leonard, *Traffic Safety and the Driver*, Van Nostrand Reinhold, New York, 1991, pp. 19-28.

19) Kahane (1994), op. cit., pp. 86-88.

20) Ibid., pp. 60-63.

21) Ibid., pp. 89-98.

22) Ibid., pp. 98-103.

23) Ibid., pp. 103-106.

24) Hackney, James, "The Effects of FMVSS No. 208 and NCAP on Safety, as Determined from Crash Test Results," in *Thirteenth International Technical Conference on Experimental Safety Vehicles*, National Highway Traffic Safety Administration, Washington, 1992. These functions are

HEADINJ = 1 / [1 + exp(5.02 - .00351 HIC)]
CHESTINJ = 1 / [1 + exp(5.55 - .06930 chest g's)]
LFEMURINJ = 1 / [1 + exp(7.59 - .00294 left femur load)]
RFEMURINJ = 1 / [1 + exp(7.59 - .00294 right femur load)]

They measure the probability of AIS $\geq$ 4 head and chest injury and AIS $\geq$ 3 leg injury, as a function of HIC, chest g's and femur load. They are based on research in Prasad, Priyaranjan, and Mertz, Harold J., *The Position of the United States Delegation to the ISO Working Group 6 on the Use of HIC in the Automotive Environment*, SAE Paper No. 851246, Society of Automotive Engineers, Warrendale, PA, 1985; and Viano, David C., and Arepally, Sudhakar, *Assessing the Safety Performance of Occupant Restraint Systems*, Report No. GMR-7093, General Motors Research Laboratories, Warren, MI, 1990.

25) Kahane (1994), op. cit., pp. 53-60 and 79-86.

26) Ibid., pp. 63-64.

27) Ibid., pp. 49-53, 76-77 and 110-115.

28) Kahane, Charles J, *An Evaluation of Occupant Protection in Frontal Interior Impact for Unrestrained Front Seat Occupants of Cars and Light Trucks*, Report No. DOT HS 807 203, National Highway Traffic Safety Administration, Washington, 1988, pp. 111-140.

29) Kahane (1994), op. cit., pp. 129-135.

30) Ibid., pp. 135-142. The joint probability of AIS $\geq$ 4 head or chest injury is 1 - [(1 - HEADINJ)(1 - CHESTINJ)], where HEADINJ and CHESTINJ are the logistic injury probability functions defined from HIC and chest G's in footnote 24.

31) Ibid., pp. 142-147.

32) *NCAP Test Results for Drivers of Passenger Cars, 1979-91*, Submission to NHTSA Docket No. 79-17-N40-002, National Highway Traffic Safety Administration, Washington, 1994.

33) Safety belt use as a function of vehicle make, vehicle age and calendar year is reported in *Restraint System Usage in the Traffic Population*, National Highway Traffic Safety Administration, Washington, Annual Publication (1982-90); and Phillips, Benjamin M., *Safety Belt Usage among Drivers*, Report No. DOT HS 805 398, National Highway Traffic Safety Administration, Washington, 1980.

34) A regression analysis on fatality rates per million vehicle years, controlling for driver age and other factors, is presented in *Evaluation of the Effectiveness of Occupant Protection: Federal Motor Vehicle Safety Standard 208*, Report No. DOT HS 807 843, National Highway Traffic Safety Administration, Washington, 1992, pp. 25-34 and 39-41.

35) *New Car Assessment Program: Response to the NCAP FY 1992 Congressional Requirements*, Report to the Congress, National Highway Traffic Safety Administration, Washington, 1993, Sections 2A, 2B and 2C.

36) *NHTSA Releases First 1994 Crash Test Results in a New Format*, Press release 59-93, National Highway Traffic Safety Administration, Washington, December 14, 1993.

37) *Code of Federal Regulations*, Title 49, Part 571.214, General Printing Office, Washington, 1992.

38) *Federal Register* 57 (3 January 1992): 242.

39) *Planning Document for Rollover Prevention and Injury Mitigation*, Submission to NHTSA Docket No. 91-68-N02-001, National Highway Traffic Safety Administration, Washington, 1992.

40) *Federal Register* 59 (3 January 1994): 104.

CHAPTER 4:

INJURIES OF BELTED OCCUPANTS

780892

Correlation Between Thoracic Lesions and Force Values Measured at the Shoulder of 92 Belted Occupants Involved in Real Accidents

J. Y. Foret-Bruno,
F. Hartemann,
C. Thomas,
A. Fayon, and
C. Tarrière
Laboratoire de Physiologie et de Biomécanique PSA/RNUR,
La Garenne–Colombes (France)
C. Got and
A. Patel
IRO/IRBA, Institut de Recherches Orthopédiques de l'Hôpital Raymond Poincaré,
Garches (France)

Abstract

The 3-point static belts that are installed in Renault and Peugeot vehicles are equipped with a force limiter near the upper anchorage. This system is made up of several bands of textiles that tear successively for the increasing levels of force exerted by the occupant. One can thus associate, for each person in the accident, the degree of the thoracic AIS and the value of the support force, expressed in daN.

This relationship is established for 92 belted occupants who were involved in frontal impacts. In addition it is indicated which are the distributions of impact violence parameters incurred and which are the distributions of ages in order to determine the statistical meaning of the required results.

The levels of tolerance observed in this sample are compared to thoracic injuries observed on belted cadavers exposed to equivalent violent impacts.

LIMITATION OF THE LOAD ON THE SHOULDER is one of the "Injury Criteria" that it is planned to apply to dummies restrained by a 3-points belt.

While this criterion offers the advantage of simplicity, it nonetheless raises two crucial questions, namely:

- is it a good indicator of the risk of injuries to the thorax?

- what limit should be ascribed to it to ensure satisfactory thorax protection for occupants exposed to real-life accidents?

The occurence mechanism of the injuries caused by retention by the chest webbing of a 3-points seat-belt during very violent impacts is still only poorly understood. As regards rib fractures, it is observed that certain take place on the path covered by the belt, which suggests a shear phenomenon, while others occur away from the chest webbing, which suggests an explanation by flexion. If the ribs cede to flexion, a criterion based on the load applied during the time would at first sight seem more pertinent than a mere maximum load.

In addition, the angles inscribed by the part of the webbing above the shoulder with the thorax and the longitudinal axis of the vehicle cause equal tensions measured in the belts to have highly different effects on the thorax. This is why the only criterion for load at the shoulder is a priori a very coarse one in theory. Nonetheless, in practice, injuries appear in a fairly confined modular variation of the speed (ΔV) and mean deceleration (mean γ), during the violent phase of the accident; the varying degree of advancement of the position of the occupant in the car reduces the differences between the various webbing/vehicle angles depending on the vehicle considered. The conditions obtained with the appearance of injuries are similar enough to justify a study based on the maximum load on the shoulder, even if this criterion remains too approximate from the standpoint of homologation.

As to definition of the acceptable load limit, and without prejudgment of the measuring conditions, it must refer to tests performed on a population of subjects representative of the occupants of the vehicles which undergo the accidents.

Reconstitutions of real-life accidents carried out using dummies by L.M. Patrick (2)[x] enabled him to set the tolerance threshold (50 % of passengers with an AIS $\geqslant 3$) at a load of 1930 lbs measured above the shoulder.

Tests using fresh cadavers are another way of acquiring data. They offer the advantage of being capable of exploring a very wide range of impact violence for well-controlled parameters, though the results they yield must be corrected to allow for the brittleness of the cadaver compared to that of a live subject. The reconstitution of real-life accidents using cadavers is one method of acquiring correction bases. For instance, A. Fayon was able to compare for the same retention forces on the shoulder the number of fractures observed on fresh cadavers and on the occupants of cars involved in accidents (3).

Three frontal impact reconstitutions involving 4 belted occupants gave the results indicated below (Table 1). These verify the assumption that the fresh cadaver has less tolerance than a live subject of the same age and situate at a much higher level than 1300 lbs the shoulder retention force at which the initial rib fracture will occur.

There also exists another source of information concerning the tolerances of live subjects; this stems from accidents which occurred to persons wearing 3 static point belts of the type equipping Peugeot and Renault cars which came off the production line between 1970 and 1977 and were sold in France. These belts comprise a load limiter located between the shoulder and the top anchorage point. Examination of this system on a crashed vehicle provided knowledge of the value of the load supported by the occupant and enabled it to be considered in relation to the rib injuries, if any, noted in the hospital. This analysis was carried out in the framework of the multi-disciplinary survey on passenger car accidents conducted by the Association Peugeot-Renault since 1970.

The impact violence parameters (ΔV and mean γ) have also been evaluated in accordance with the method already described (C. Tarrière et al., 6). It is therefore possible to estimate the probabilities of attaining or exceeding any given level of force in real-life head-on impacts and ascribe to these probabilities a risk of thoracic injuries.

METHODOLOGY

THREE DIFFERENT LOAD LIMITER SYSTEMS were fitted in turn to 3 points static belts on Peugeot and Renault cars on sale in France since 1970.

1) Load limiter with 5 unequal length ribbons (Type A, Figures 1 and 2).

The mean values of the rupturing load from the 1st to the 5th ribbon are as follows:

---

x Numbers in parentheses designate References at the end of paper.

	1st ribbon	2nd ribbon	3rd ribbon	4th ribbon	5th ribbon
Rupturing load (daN)	210	385	325	400	440

These values were established on the basis of 10 dynamic tests.

2) Load limiter with 5 equal lengths (Type B, Figures 1 and 2).

Here, the mean values are distinctly higher:

	1st ribbon	2nd ribbon	3rd ribbon	4th ribbon	5th ribbon
Rupturing load (daN)	740	720	790	800	800

3) Tear-webbing (Type C, Figures 1 and 2).
Triggering peak: 550 daN
Mean steady-state value: 410 daN

MEASUREMENT OF THE LOAD for a belted person involved in an accident is established in the light of the above values. However, in cases where a Type A load limiter has 3 broken ribbons, the fracture breaking effort of the 2nd strand, which is higher (385 daN), is retained. In the case of a type C limiter, the starting peak value is adopted. The values indicated in the remainder of this study are corrected for the weight of the occupant, in accordance with a formula proposed by Eppinger (4) following dimensional analysis considerations, namely:

$$F_{\text{corrected shoulder}} = F_{\text{measured shoulder}} \left[\frac{75\text{kg}}{\text{weight of occupant}} \right]^{\frac{2}{3}}$$

92 BELTED OCCUPANTS, involved in head-on collisions, were subjected to determination of the load on the shoulder. They satisfy the following conditions:

- at least one ribbon broken, whether for a type A or type B limiter, or starting of tears in the type C limiter.
- Age and weight are known.
- No contact of the thorax against an element other than the seat webbing itself in occupants suffering from rib cage injuries.

RESULTS

CHARACTERISTICS OF SAMPLE STUDIED - The 92 occupants break down in accordance with 3 load limiter types, namely:

Table 1 - Forecasted and Observed Rib Fractures in Accident Reconstruction.

Occupant No	Shoulder Belt Load (daN)	Weight (kg)	Age	Corrected shoulder belt load (1)	Fractures Observed on cadavers	Fractures Forecasted by Eppinger (2)	Fractures Observed on the occupants
1	460	48	54	620 daN (1370 lbs)	10	4-5	0
2	520	68	55	550 daN (1200 lbs)	7	3	0
3	750	68	53	800 daN (1770 lbs)	17	6	5
4	860	66	29	935 daN (2050 lbs)	9	6	1

(1) Corrected shoulder belt load: measured shoulder belt load x $\left(\frac{\text{Dummy 50 \% weight}}{\text{actual weight}}\right)^{2/3}$

(2) Forecasted for a cadaver of same age than the occupant.
Forecasted rib fractures = -3 (2.10^{-4} x corrected shoulder load x age)

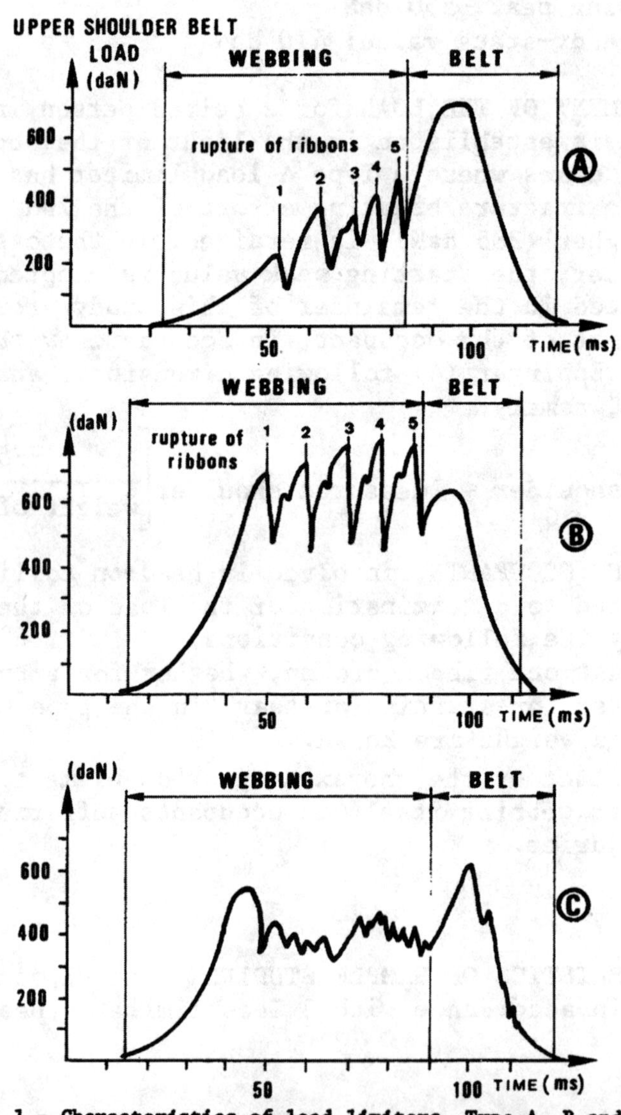

Fig. 1 - Characteristics of load limiters, Type A, B and C

- Type A: 82 cases, of which 23 caused 5 ribbons to break. These 23 occupants may therefore have withstood a load of over 440 daN, corresponding to fracture of the 5th ribbon.

- Type B: 4 cases, of which 1 broke the 5 ribbons of the device.

- Type C: 6 cases. None tore the tear webbing throughout its length.

The list of cases is given in Annex 1. The values of the load which feature in it have not been corrected.

Age breakdown

	<35	35-45	45-55	>55 years	
Number of cases	56	15	17	5	92
%	60	16	18	6	100

In comparison to the age distribution of the 394,792 car occupants involved in accidents in France in 1975 (last year for which national statistics are available), one can merely remark a slight under-representation of people over 55 years old in our sample (see Table 2).

The violence of the impacts in which the 92 occupants were involved differs considerably with the age group: those less than 35 years old are all the more over-represented, the more violent the impact, whereas the proportion of occupants from 35 to 55 years old falls off (Table 3). Persons over 55 years old are represented in the three classes of violence in proportions not significantly different from their total representation. This distinct tendency of young drivers or front passengers to be involved in impacts which are more violent than those of their elders is again to be found in the overall population involved in head-on impacts in the Peugeot-Renault survey. This results from the combined effects of faster driving and the possession of lighter vehicles, at a disadvantage in impacts between vehicles. The means of the weights of vehicles occupied by the three age groups in the sample work out as follows:

35 years : mean mass of vehicle : 867 kg
35 to 55 years: mean mass of vehicle : 944 kg
55 years : mean mass of vehicle : 908 kg

(2% of the vehicles in the total number of cars involved in accidents weight over 1350 kg and 2% less than 550 kg).

In order to decide on a maximum value for the shoulder load, assuming the existence of regulations on this criterion, one will therefore have to allow for the fact that too low a threshold, which would be advantageous to the more

Before Impact After Impact

type A

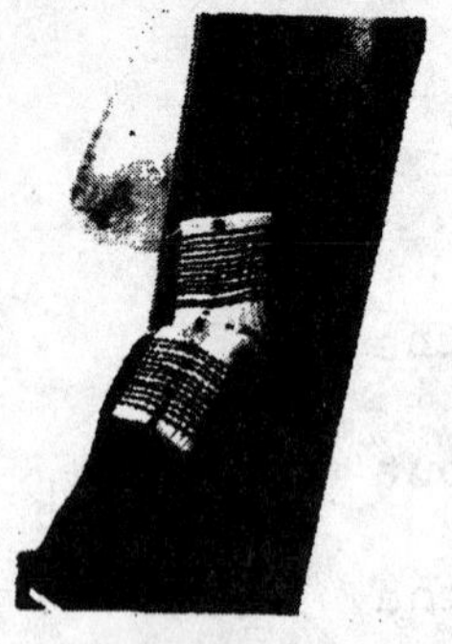

type B

type C

Fig. 2 - Load limiters Type A, B and C before and after being used in actual accidents

Table 2 - Comparative Age Distributions in Present Sample and in the Whole Car Occupants Population at Risk, France, 1975.

	Age				
	<35	35 - 45	45 - 55	>55	All
Sample (N = 92)	60%	16%	18%	6%	100%
Car occupants involved in accidents - France (N = 394.792)	59%	17%	13%	11%	100%

Table 3 - Age Versus Impact Violence.

ΔV		Age <35	36 - 55	>55	Total	N
>55 km/h	(1)	75%	12.5%	12.5%	100%	8
	(2)	69%	21%	10%	100%	52
35 - 55 km/h	(1)	67%	33%	---	100%	27
	(2)	60%	30%	10%	100%	207
<35 km/h	(1)	40%	50%	10%	100%	40
	(2)	56%	36%	8%	100%	469
Total	(1)	53%	40%	7%	100%	75
	(2)	58%	33%	9%	100%	728

(1) Sample of 75 belted front seat occupants

(2) Whole population of front seat occupants involved in frontal collisions (N = 728).

elderly persons involved in impact of moderate violence, would unduly penalize the younger population. This high tolerance level of the latter as qualified by the observations which are to follow, should be turned to the best possible advantage in order to reduce their risks of injury in the violent impacts where they are widely represented.

VALUES OF THE LOAD ON THE SHOULDER AND RESPONSES AT THE THORAX - Occupants less than 30 years old can withstand up to 740 daN without any injuries to the thorax. Between 740 and 900 daN, 3 cases of an AIS at the thorax of less than 3 were recorded, and one involving 5 rib fractures (Figure 3).

For occupants over 50 years old, the risk of rib or sternum fracture rises at a much lower level.

Too few people from 30 to 50 years old subjected to a shoulder load of over 500 daN were available for a threshold value to be indicated.

It will be noted that the frequency of injuries to the thorax is relatively low, although the loads are considerable in a great many cases.

The frequency is in the same order of magnitude as that observed for all occupants involved in car accidents wearing seat belts with a load limiter (Table 4).

COMPARISON AGAINST THE RESPONSES OF THE THORAX IN CADAVERS - For values of the load on the shoulder identical to those measured on the persons of this sample (corrected for their weight), the thorax responses expected on cadavers of the same age as the live occupants are shown on Figure 4. The tolerances of the live occupants are much higher. The difference is of 3 to 5 rib fractures in the majority of cases.

RISK OF EXCEEDING THE TOLERANCES IN REAL LIFE-ACCIDENTS- On the ΔV, γ diagram presented below (Figure 5), one can see for a given impact violence the corresponding values of the load observed (corrected). Their scatter (as a result of various factors: overloading by an occupant being projected toward the rear, the degree of slackness with which the belt is worn, the deceleration law of the vehicle, the geometry of the belt) means that the values expected can be situated only approximately with respect to the various impact violence levels. It can be considered that for wearers of 3 point seat belts as a whole, these belts being equipped with a load limiter, the value of 700 daN is reached when the parameters ΔV and mean γ exceed 55 km/hr and/or 13 g, in the absence of any overload by the rear.

The diagram shown on Figure 6 makes it possible to situate the present sample with relation to a population representative of head-on impacts which occured on the French national highway system.

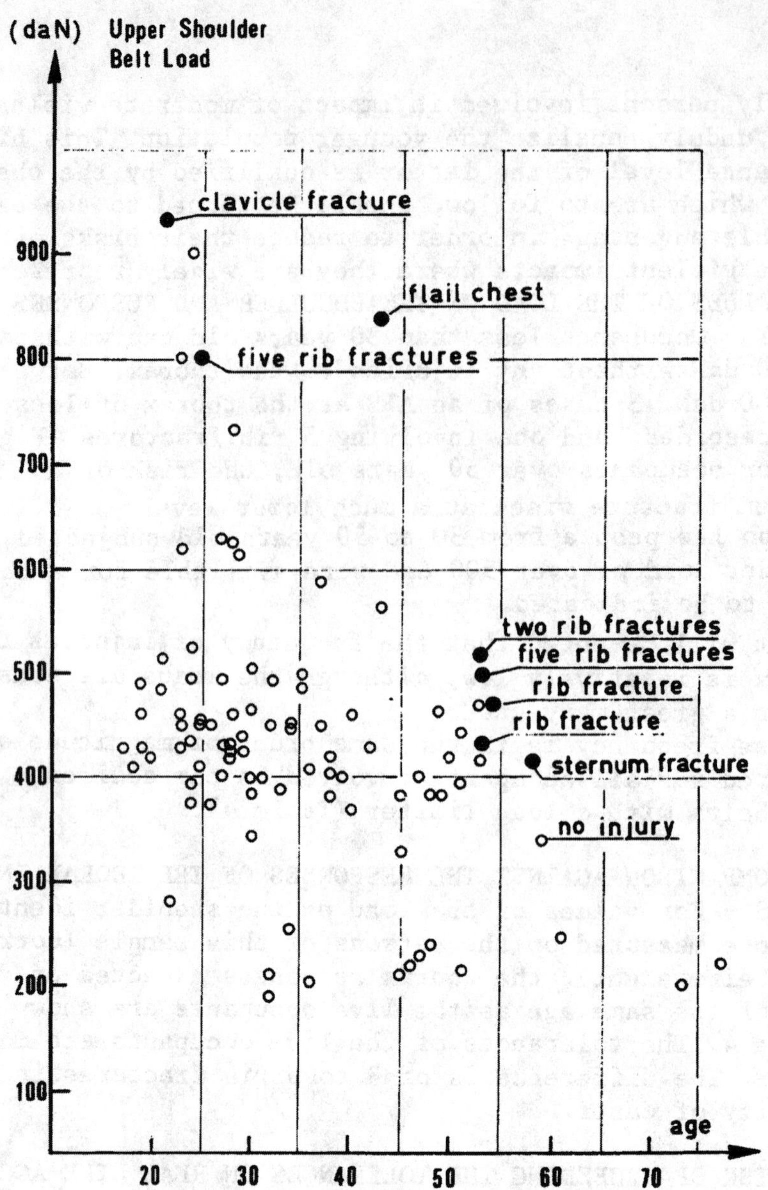

Fig. 3 - Shoulder belt load versus age for 92 occupants involved in frontal collisions

Table 4 - AIS Thorax Distribution in Present Sample and in the Whole Sample of Occupants Wearing Belt Equipped with Force Limiter. Peugeot-Renault Accident Investigation.

AIS Thorax	Present Sample	Whole Sample
0 - 1	91.3 %	90.5 %
2	4.3 %	4.0 %
3	3.3 %	3.6 %
4	1.1 %	1.9 %
Total:	100 %	100 %
	(N = 92)	(N = 251)

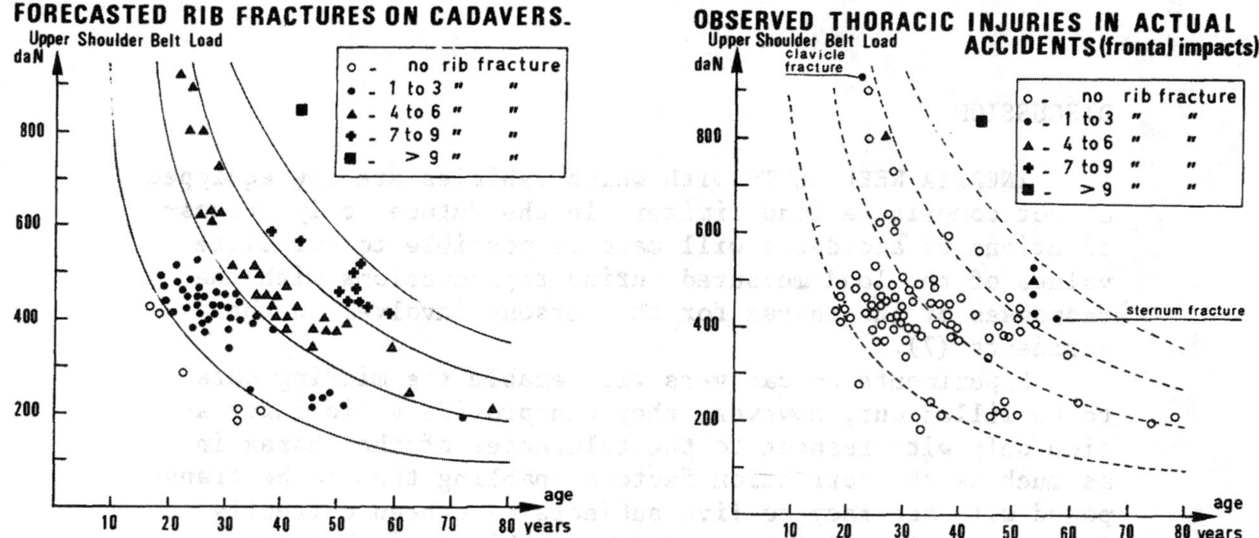

Fig. 4 - Above: Forecasted rib fractures for cadavers,
Below: Thoracic injuries in actual accidents

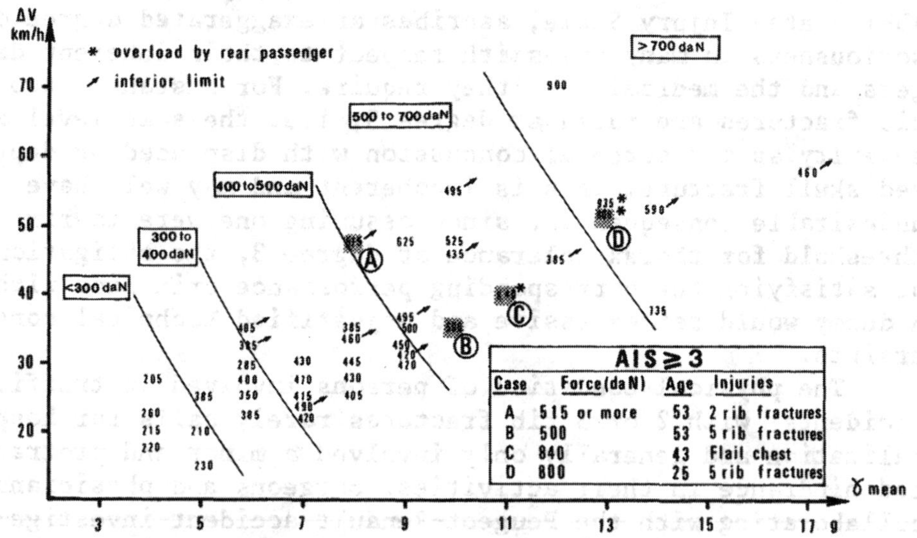

Case	Force(daN)	Age	Injuries
A	515 or more	53	2 rib fractures
B	500	53	5 rib fractures
C	840	43	Flail chest
D	800	25	5 rib fractures

Fig. 5 - Velocity change versus mean deceleration. Shoulder belt load values for 42 occupants involved in actual accidents

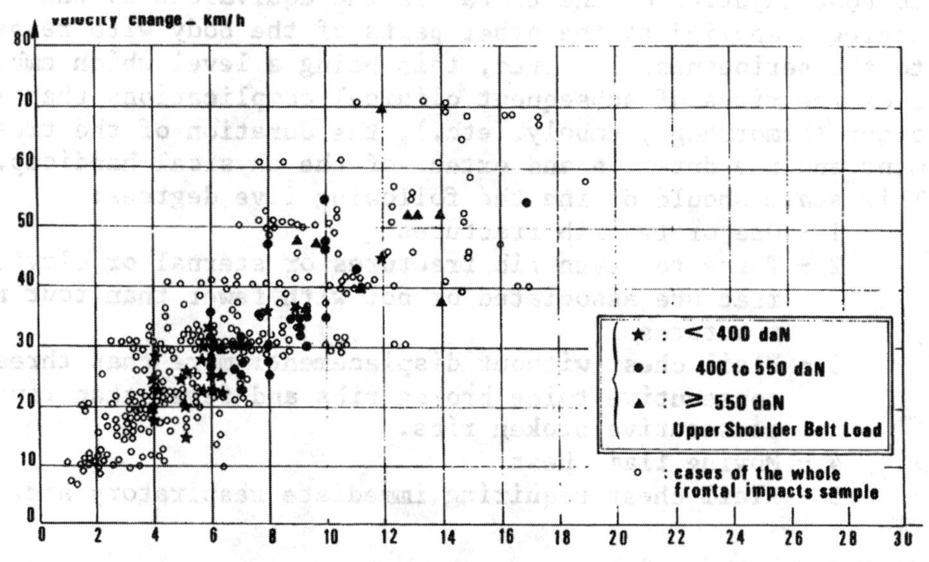

Fig. 6 - Velocity change versus mean deceleration for 60 frontal impacts of the present sample compared with the whole frontal impacts sample. Peugeot-Renault Accident Investigation

DISCUSSION

INERTIA REEL BELTS with which vehicles are now equipped do not comprise a load limiter. In the future, only reconstitutions of accidents will make it possible to correlate values of the load measured during reproductions with the responses of the thorax for the persons involved in the accidents (7).

Experiments on cadavers will enable the missing data to be filled out; however, they can provide valid information only with respect to the tolerances of the thorax in as much as the correction factors enabling them to be transposed with accuracy to live subjects have been correctly determined. An original research on this topic has been performed by A. Fayon and J. Sacreste (8).

THE SCALE FOR RATING THORAX INJURIES, proposed in the Abbreviated Injury Scale, ascribes an exaggerated degree of seriousness in many cases with respect to their inherent dangers and the medical care they require. For instance, two rib fractures are rated at degree 3, i.e. the same level as severity as for cerebral concussion with displaced or depressed skull fracture. This is incoherent and may well have undesirable consequences, since assuming one were to fix the threshold for thorax tolerance at degree 3, the obligation of satisfying the corresponding performance criterion with a dummy would set excessive and unjustified technical constraints.

The physical condition of persons involved in traffic accidents with 2 or 3 rib fractures rarely calls for hospitalization and generally only involved a minor and protracted hindrance in their activities. Surgeons and physicians collaborating with the Peugeot-Renault accident investigation teams suggest that a thoracic injury scale has to be established in such a way that the degree 3 that it ascribes to bone injuries of the thorax is the equivalent to the degree 3 applied to the other parts of the body with respect to the seriousness involved, this being a level which embodies the risks of subsequent clinical complications that may occur (hemorrhage, emboly, etc.), the duration of the treatment and the duration and extent of the physical handicap. This scale should define the following five degrees:

1 - One or two rib fractures
2 - Three to seven rib fractures or sternal or clavicle fracture associated or not with fewer than four rib fractures
3 - Flail chest without displacement: more than three consecutive twice broken ribs and fewer than five consecutive broken ribs.
4 - Moving flail chest
5 - Flail chest requiring immediate respiratory aid.

The use of this scale would reduce the thorax injuries of the 92 occupants presented in Figure 3 to a single case of AIS 3 and would lead to situating the load threshold values in the following brackets:

<31 years: 850 to 900 daN
35 to 50 years: 600 to 800 daN
>50 years: 500 to 600 daN

These loads being proposed as the limiting values above which one person out of two may suffer injuries of AIS 3 or more in the scale proposed.

THERE IS THE ADVANTAGE TO BE GAINED FROM CHOOSING A RESTRAINT LOAD as high as possible with a view to regulation of the shoulder load. Should the level be too low, one may apprehend that for an impact a little more severe than that for which this level has been respected, the retention would become unduly sudden and be fatal to the occupants who might have been saved by a higher maximum load.

Indeed, a higher retention load enables more kinetic energy of the occupant to be dissipated into the retention system. As a result of the phenomenon of relative coupling between this occupant and the structure of the vehicle more energy is also dissipated into the structure of this vehicle.

Let us assume that for every extra Joule dissipated into the belt, there corresponds 1 additional Joule into the structure with an adequate restraint system.

Let us likewise assume that one can increase by 325 daN the retention load at the shoulder over a length of 0.2 m in the vehicle (325 daN correspond to 1930 lbs less 1200 lbs).

Lastly, let us assume that shoulder retention involves 40 % of the total weight of the occupant retained and that increases in the other retention loads (for instance pelvis) remains possible from the tolerance standpoint.

A simple calculation based on a 50 km/hr reference speed shows that one can dissipate the additional energy of an occupant at up to 10 km/hr more speed.

The interest of these 10 km/hr is revealed when one situates them on a curve of cumulative mortality occurences: in the neighborhood of a ΔV of 50 km/hr, 10 km/hr correspond to about 20 % of those killed.

CONCLUSIONS

1 - An examination of the load limiters fitted to three point static belts of crashed vehicles has made it possible to note the following on a sample of 92 persons:

- that occupants less than 30 years old can withstand up to 740 daN of shoulder load without any chest injuries (load corrected in terms of their weight);

- that over 50 years old, rib fractures are observed at a much lower level of load, but this category of front seat car occupant is rarely involved in violent accidents.

It did not prove possible to establish the curve of deterioration of tolerances between the ages of 30 and 50 through lack of any implication of persons in this age group in fairly violent impacts.

2 - The rating of the injuries by the Abbreviated Injury Scale exaggerates the seriousness of a considerable number of cases by ascribing a rating of 3 after two rib fractures. The use of a scale coherent with that of the other regions of the body would, were an AIS of 3 to continue to be taken as reference for the definition of injury criteria, make it possible to adjust better the level of performance to be reached during regulatory tests with respect to the real risks run by the occupants of the accident vehicles.

3 - The underestimation of thorax tolerances of the live subjects by tests made on fresh cadavers is verified. The present study suggests that this underestimation is of about 3 to 5 rib fractures.

4 - Occupants of vehicles less than 35 years old are by far the most widely represented in violent accidents. This requires confirmation for high traffic density countries as a whole.

There stems from this the need to set the tolerance threshold fairly high in order to save the greatest possible number of persons exposed to risk.

5 - The development of reconstitutions of real-life accidents should make it possible to gain data which is at present still lacking in order to establish the relevance between the thorax responses of live patients and those of cadavers or dummies. These reconstitutions would also have the additional advantage of subjecting subjects to realistic impact violences, which is not always the case in tests with cadavers, certain of which far exceed the deceleration levels attained in even the most violent real-life accidents.

REFERENCES

1. A. Fayon et Al.: "Thorax of 3-pt Belt Wearers During a Crash", 19th Stapp Car Crash Conference.

2. L.M. Patrick et Al.:"Three-Point Harness Accident and Laboratory Data Comparison", 18th Stapp Car Crash Conference.

3. "Supplemental Comments from Peugeot-Renault on the Draft Injury Criterion: 1200 lbs for Torso Belt Load" (Docket 74-14; Notice 7).

4. R.H. Eppinger: "Prediction of Thoracic Injury Using Measurable Experimental Parameters", 6th Conference on Experimental Safety Vehicles.

5. G. Schmidt et Al.: "Neck and Thorax Tolerance Levels of Belt Protected Occupants in Head-on Collisions", 19th Stapp Car Crash Conference.

6. C. Tarrière et Al.: "The Contribution of Physical Analysis of Accidents Towards Interpretation of Severe Traffic Trauma", 19th Stapp Car Crash Conference, SAE paper n° 761 176.

7. M. Balthazard: "Complete Analysis of a Head-on Collision Reconstructed with Dummies and Fresh Cadavers", IRCOBI Conference, Lyon, France, Sept. 1978.

8. A. Fayon et Al.: "Methods for Improving the Backing-up of Conclusions Reached on the Basis of Accident Reconstructions with a Limited Number of Instrumented Cadavers". IRCOBI Conference, Lyon, France, Sept. 1978.

INFLUENCE OF AGE AND RESTRAINT FORCE VALUE ON THE SERIOUSNESS OF THORACIC INJURIES SUSTAINED BY BELTED OCCUPANTS IN REAL ACCIDENTS.

by

C. Thomas, J.Y. Foret-Bruno, F. Hartemann, C. Tarrière

Biomechanical and Physiological Laboratory of the Peugeot-Renault Association
92250 - La Garenne-Colombes, France

C. Got, A. Patel

IRO/IRBA - Institute for Orthopaedic Research, R. Poincaré Hospital
92380 - Garches, France

Three studies published over recent years deal with the force at shoulder level of belted occupants in frontal collisions:

- L.M. Patrick (1)[x], following reconstitutions of real accidents using dummies, determines the tolerance threshold (50 % of occupants having an AIS 3) at 858 daN (1930 lbs) measured above the shoulder.

- N.H.T.S.A. (2), based on research investigations conducted with cadavers, has suggested a shoulder force limit of 534 daN (1200 lbs).

- J.Y. Foret-Bruno (3) has analysed 92 belted occupants involved in frontal collision, having a force limiting device between the shoulder and the upper anchorage point of the shoulder belt. He observes that the less than 30 year olds withstand a shoulder force of 740 daN with no thoracic injury, whereas beyond 50 years old, one belted occupant out of two risks one or more thoracic fractures at the 450 daN threshold. Finally, he verifies the under-estimation for living people by tests conducted with cadavers.

The present research investigation is also based on data acquired from observing force limiting devices fitted in Peugeot-Renault 3-point static belts. It incorporates 11 new cases. Stress is laid on those shoulder force thresholds where the first thoracic fracture appears, according to age.

Above all, the study attempts to quantify per age group the risk of thoracic fracture due to the belt for the front occupants involved in real frontal collisions, the violence of which is evaluated by the variation in velocity of the front occupant, and the mean deceleration of car mean γ.

The plan is as follows:

1 - Shoulder force values and assessment of thorax according to age,

2 - Shoulder force values according to violence of frontal collision (diagram ΔV, mean γ),

---

(x) - Parentheses refer to the bibliography at end of paper.

3 - Risks, per age group, of going beyond the force threshold at which belt-induced rib fracture appears in real frontal collisions.

Verification on a sample of 227 belted occupants.

DEVELOPMENTS AND RESULTS

1. SHOULDER LOAD VALUES AND ASSESSMENT FOR THORAX ACCORDING TO AGE

Method: The shoulder load level for living subjects is supplied us from accidents involving wearers of static, 3-point belts as fitted to Peugeot and Renault standard production cars sold in France, from 1970 to 1977.

These belts have a load-limiting device located between the shoulder and the upper anchorage point.

Three different load-limiting systems were fitted successively to the cars as shown in the photos (figure 1), which exhibit them before and after the impact.

Type A - Load limiter with 5 different lengths of band: the mean load values from 1st. to 5th. bands are as follows:

Band No	1st	2nd	3rd	4th	5th
Breaking load (in daN)	210	385	325	400	440

Type B - Load limiter with 5 bands of equal length: the mean values are considerably higher:

Band No	1st	2nd	3rd	4th	5th
Breaking load (in daN)	740	720	750	800	800

Type C - Tear webbing: triggering peak: 550 daN.

Important remark: when all the shock-absorbing material has been used (5 broken bands for types A and B, completely ripped tear webbing for type C), the load undergone by the shoulder was at least equal to the maximum value stated for the type of load-limiting device in question. The load value taken into account is not the exact value but a "threshold" value.

LOAD MEASUREMENT for a belted occupant in an accident is conducted taking account of the above-mentioned values. The values shown henceforth in this study are corrected in accordance with occupant mass, in compliance with the equation suggested by Eppinger (4) and which we would state once again:

$$F_{\text{corrected shoulder}} = F_{\text{measured shoulder}} \left(\frac{75 \text{ kg}}{\text{occupant mass}}\right)^{2/3}$$

103 belted occupants involved in frontal collisions had their shoulder forces determined. The following conditions were satisfied:

- at least one band broken, whether it was a type A or B load limiter, or beginning of tearing of type C limiter,
- age and weight known,

- no thorax contact with an element other than the belt itself for occupants suffering thorax injuries.

Collating the force value sustained by the occupant and the assessment of thorax injuries if there are any according to the age enables the rib fracture risk threshold to be defined per age group.

RESULTS - Figure 2 shows the corrected shoulder force according to age for the 103 belted occupants having a material shock-absorber, involved in a frontal collision. Possible cases of rib fractures are also shown.

The 24 triangles indicate that the initial measured force value is caused by the material shock-absorber the 5 bands of which have broken. These belts have therefore withstood a greater force than the breaking load of the 5th. band.

The dash line connects those belted occupants who, in each age group, have the highest shoulder force with no fracture of the thoracic segment. This line separates, as best it can, the 11 cases with rib fracture from the 92 cases with no fracture. However, 2 of the 4 cases with fracture are situated below this line.

The first case (25 years old, 840 daN) sustained 5 rib fractures. In the real accident, he broke the 5 material shock-absorber bands (type A). This collision was accurately reconstituted using cadavers. The corrected shoulder load was 840 daN. However, the cadaver seated behind the subject was not so heavy as in the real accident. Deformation of the upper anchorage point was less considerable in the reconstitution than in the real accident. These indications make us think that the load of 840 daN is the lower shoulder load limit effectively withstood by this young belted occupant.

In the second case (54 years old - 440 daN), the corrected load is exact. The case of the belted occupant of same age in the neighbourhood of this observation (54 years old, 470 daN) shows that we are in an area where, for similar age and loads, the thoracic response is the result, among others, of the difference in tolerances between individuals.

Among the 11 fracture cases, 5 underwent the harmful overload of the rear adult occupant. These were:

- the 3 belted occupants under 45 years of age,
- 2 cases of flail chest, the victims being 75 and 42 years old.

It will be seen in what follows that it is probable that with no overloading due to the rear occupant, the 3 victims less than 45 years old would not have run the risk of rib fracture, seeing the violence of the collisions in which they were involved.

Age	Corrected load (daN)	ΔV (km/hr)	γ Mean (g)	Probable load area (x) (daN)
21	930) lower	53	13	650 - 900
24	800) limit	53	13	650 - 900
42	840	40	11	500 - 650

FRACTURE THRESHOLDS FOR LIVE, BELTED OCCUPANTS ACCORDING TO AGE - They can be situated as follows:

< 35 years old:	900 daN	55 to 64 years old:	350 daN
35 to 44 years old:	650 daN	≥ 65 years old:	250 daN
45 to 54 years old:	500 daN		

(x) see figure 3

The reliability of such an estimate resides in the number of observations in the neighbourhood of the load threshold suggested for a given age group. The highest uncertainty is for the 35-44 years old age group.

One must be careful of the fact that they are thresholds for the appearance of just one rib fracture and not flail chest (at least 3 consecutive double rib fractures), which situates the real limit of thoracic tolerance. For information, we would mention that Foret-Bruno (3), based on Eppinger's (4) results obtained with cadavers, shows how wide the difference in load is between the moment 1st. and 7th. rib fractures appear depending on age. The difference is 370 daN for 35 year-old, 290 daN for 45 year-old, 240 daN for 55 year-old and 180 daN dor 65 year-old.

Let us now see to what collision violences correspond the load thresholds per age group.

2. SHOULDER LOAD VALUES VERSUS FRONTAL COLLISION VIOLENCE (DIAGRAM ΔV, MEAN γ)

Method: The frontal collision violence parameters (ΔV and mean γ) are estimated in accordance with a method already stated (TARRIERE et al. (5)). We now want to find at what collision violences shown in the ΔV, mean γ diagram correspond the different shoulder load values related to the appearance of rib fracture for each of the age groups under consideration.

To draw up such a graph requires each observation used to satisfy further conditions.

- The measured value must be exact. We must therefore reject those belted occupants for which all the breaking capacity of the material shock absorber has been used, and who may have sustained a load greater than the maximum breaking threshold.
- ΔV and mean γ must be known.
- The belted occupants under consideration should not have suffered overloading due to rear occupant, sub-marining, wearing a manifestly too-loose static belt.

These conditions reduce from 103 to 29 the number of belted occupants that can be used for the graph.

Taking account of the end we are working towards, it has become necessary to resort to similar results acquired in experiments provided that the violence of the collisions is comparable to that observed in real accidents. The following were used:

- 12 cadavers stemming from collisions conducted by the Peugeot-Renault (6) Laboratory,
- 62 dummies stemming from collisions conducted by the Renault and Peugeot test departments (not published),
- 54 dummies stemming from tests conducted by Patrick (1),
- 4 results acquired from an American volunteer with no belt pre-tensioning (7).

The estimation of the collision violence shown in the ΔV, mean γ diagram, and corresponding to the different shoulder load limits per age group, is in short effected on a sample of 161 belted occupants.

RESULTS - Figure 3, "ΔV, mean γ" diagram shows the shoulder load (corrected values) for the 161 belted occupants. The violence of the previously stated load thresholds is as follows:

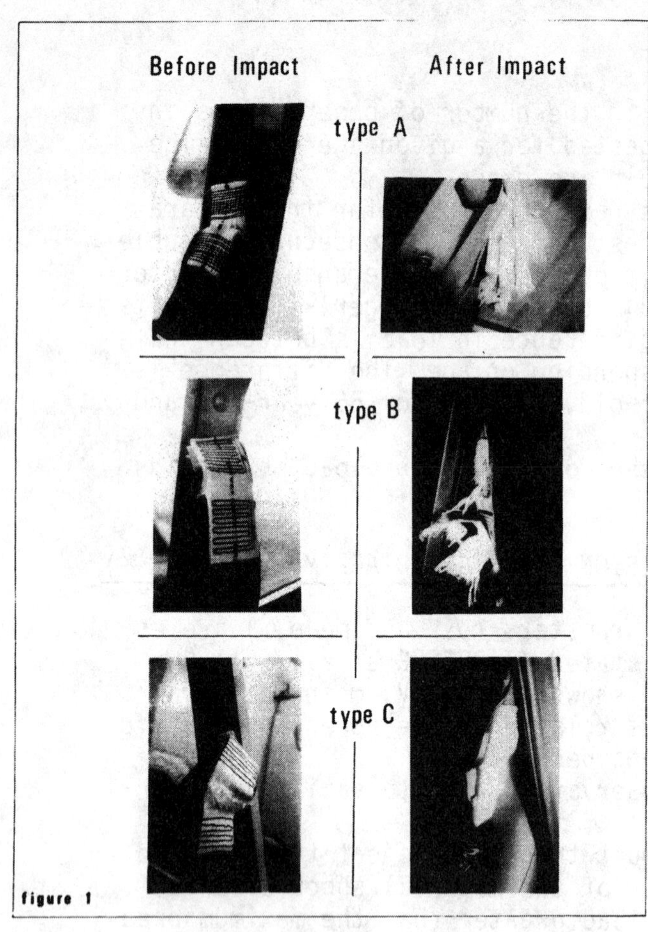

figure 1

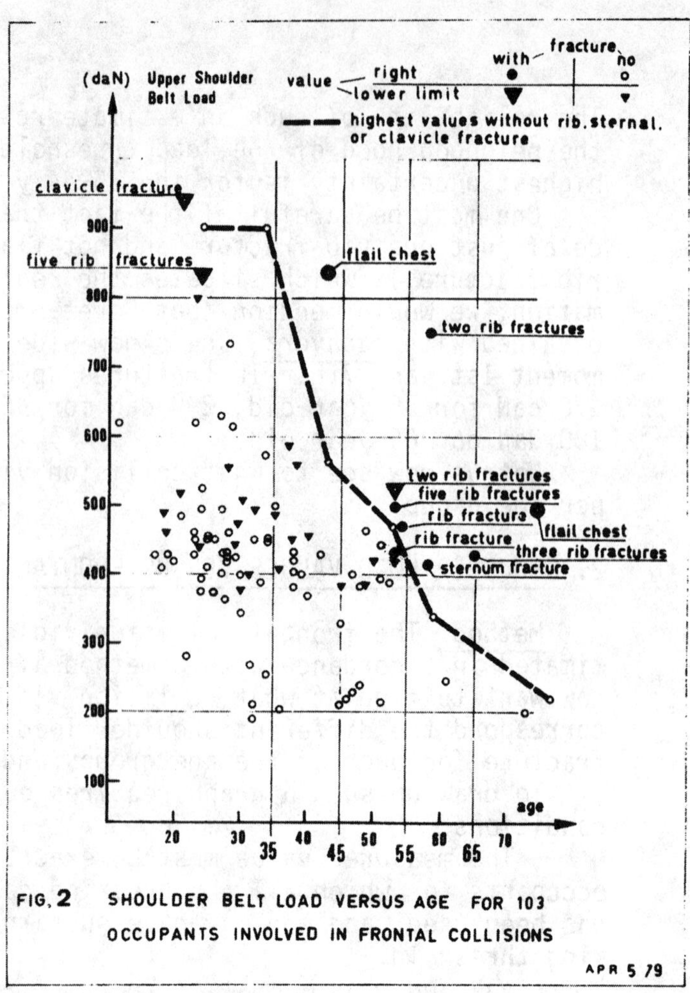

FIG. 2 SHOULDER BELT LOAD VERSUS AGE FOR 103 OCCUPANTS INVOLVED IN FRONTAL COLLISIONS

APR 5 79

fig.3: CORRECTED* SHOULDER BELT LOAD VERSUS "DELTA V" & MEAN GAMMA

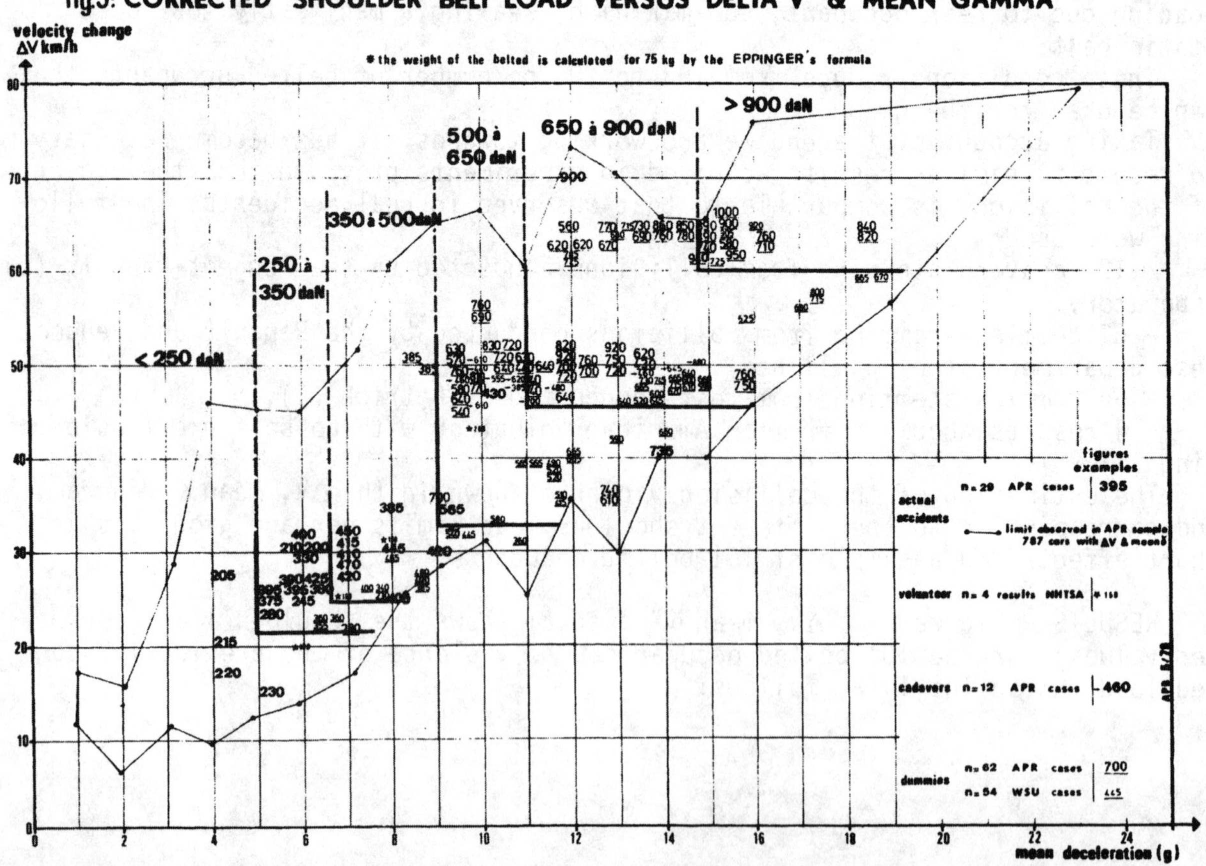

Load threshold (daN)	Mean collision violence ΔV(km/hr)	γ (g)
900	$\geqslant$60	$\geqslant$ 14.7
650	$\geqslant$45	$\geqslant$ 11
500	$\geqslant$34	$\geqslant$ 9
350	$\geqslant$25	$\geqslant$ 6.5
250	$\geqslant$22	$\geqslant$ 5

It is obvious that a greater number of observations would have been preferable and that, consequently, these results should be regarded cautiously.

In spite of considerable scattering about the class, each load class counts practically one case out of two among the belted occupants, the shoulder load of which corresponds to this class as shown in the following table:

Load class (daN)	Number satisfying load class (N=161)	Number per load class (daN)					
		250	250 to 349	350 to 499	500 to 649	650 to 899	900
250	11	5	3	3	0	0	0
250 to 349	5	0	3	1	1	0	0
350 to 499	38	0	10	18	7	3	0
500 to 649	40	0	0	2	21	16	1
650 to 899	58	0	0	0	13	37	8
900	9	0	0	0	0	3	6

Scattering observed on the corrected shoulder load value can be explained here by four series of factors that it has not been possible to take into account up till now in our accident investigation. They concern:

- the car: deceleration profile of car (which varies furthermore depending on the type of obstacle and the velocity),
  the geometry of belt anchorage points,
- the classic "3-point" belt (static/reel):
  . how well belt is worn (even if considered correct) (importance of "belted occupant/car" loop);
  . technical characteristics of belt (stretch, dynamic response curves).
- the belted occupant:
  . morphology of belted occupant (weight being equal),
  . thorax strength.
- the obstacle:
  . strength of obstacle (already stated since it is taken into account in the car deceleration profile),
  . collision angle (and angle of belted occupant displacement-thorax sliding under shoulder belt).

The overload due to the rear occupants, "submarining" and intrusion phenomena also affect shoulder load, but these cases have been excluded from the present analysis.

Let us now examine the risks of going beyond the load thresholds (for rib fracture) in real accidents per age group.

3. RISKS PER AGE GROUP OF EXCEEDING THE LOAD THRESHOLD AT WHICH BELT-INDUCED RIB FRACTURE APPEARS IN REAL FRONTAL COLLISION.

Method: for each age group, the probability is calculated of exceeding the risk of a simple rib fracture by referring to the occupant accident frequency (in each age group) having a greater or lower violence than that which causes the critical restraint load for the age group they belong to.

For this evaluation, we have 1134 front occupants at our disposal (whose ages are known) involved in frontal collision and classified in ΔV, mean γ. The cases issue from the multipurpose investigation conducted since 1970 by the Peugeot-Renault Association (8).

This risk estimation is verified on the sub-sample of 227 belted occupants who have suffered neither notable intrusion nor rear occupant overload.

The thresholds of initial rib fracture per age group in the "ΔV, mean γ" diagram were defined in the previous chapter for belted occupants weighing 75kg. These thresholds must be corrected to take account of the real average weight of the front occupants per age group (correction made using the 2402 front occupants in our investigation, of whom we know the age and weight).

The shoulder load correction is made by applying the Eppinger equation.

Finally, the new thresholds are defined in the " ΔV, mean γ " diagram by interpolating with relation to the results obtained for 75 kg.

The results are shown in the following table:

Age group (years)	Average weight (kg)	Shoulder load (daN) 75 kg	Shoulder load (daN) for average weight	Threshold in " ΔV, mean γ "	
< 35	62,8	900	800	55	13,5
35 - 44	66,1	650	600	42	10,5
45 - 54	67,3	500	465	32	8,5
55 - 64	68,3	350	330	24	6,2
≥ 65	67,5	250	235	21	4,9

It can be seen, for example, that in a real frontal collision, the ΔV limit 55 km/hr, and mean deceleration 13.5 g is a limit not to be exceeded for young people less than 35 without running the risk of a belt-induced rib fracture.

We are now trying to find the theoretical frequency of the belt-induced rib fracture risk per age group for those involved in real frontal collision.

RESULTS - The Peugeot-Renault "accident" investigation includes 1134 front seat occupants, whether belted or not, whose age and frontal collision violence (ΔV, mean γ) are known.

The situation of each of the 1134 occupants is compared with regard to age and frontal collision violence, with the limit of violence (related to age) corresponding to initial, belt induced rib fracture.

The results are shown in the following table:

Age Group (years)	Percentage involved	Relationship between those exceeding the threshold (of initial belt-induced rib fracture) with all those involved. - Involved -	- Percentage -
< 35	59 %	24/666	4 %
35 - 44	18 %	20/198	10 %
45 - 54	13 %	38/151	25 %
55 - 64	6 %	22/70	31 %
≥ 65	4 %	28/49	57 %
Total:	100 %	132/1134	12 %

It appears that 88 % of front occupants involved in a frontal collision do not run the risk of belt-induced thorax fracture. This risk varies considerably depending on age.

The risk is only 4 % for front occupants less than 35, who represent 60 % of front seat occupants. Over 65, one occupant out of two risks one or more thoracic fractures. Luckily, they represent only 4 % of front occupants.

One must be careful of the fact that too-low a shoulder load limitation obtained by increasing the stopping distance would penalize the more tolerant belted occupants. The latter, who are by far the greatest number, would then be exposed to more frequent impacts with the passenger compartment units.

Sub-sample of belted occupants and their own - The risk of rib fracture is verified versus age group for the 227 belted front occupants having sustained no considerable intrusion, no rear occupant overload, and involved in a frontal collision classified as "ΔV, mean γ ".

The following table gives the results of this analysis.

Belted occupants involved in a frontal collision less violent than threshold: No of cases with fracture	No of involved	(%)	. Age group . Corrected load threshold for average mass of age group . Limit in diagram ΔV, mean γ .	Belted occupants involved in frontal collision above threshold: No of cases with fracture	No of involved	(%)
3	124	2 %	< 35, 800 daN, 55 km/hr 13,5 g	1	5	20 %
2	40	5 %	35-44, 600 daN, 42 km/hr, 10,5 g	0	4	0 %
4	31	13 %	45-54, 465 daN, 32 km/hr, 8,5 g	5	8	63%
1	9	11 %	55-64, 330 daN, 24 km/hr, 6,2 g	7	8	88%
0	7	0 %	≥ 65, 235 daN, 21 km/hr 4,9 g	3	7	43%
10	211	5 %	TOTAL	16	32	50%

Out of the 227 belted occupants, only 26 (i.e. 11 %) sustained one or more thoracic fractures. Although women represent only 35 % of those involved in front seats, and although, on an average, they weigh less than men, we observe that they are over-represented among the 26 belted occupants sustaining rib fracture (16/26, i.e. 62 %). Belted occupants involved in collisions less violent than the thresholds corresponding to their age, have one or more rib fractures in the proportion 10/211 (i.e. 5 %).

On the other hand, when the threshold is exceeded, the proportion rises to 16/32 (i.e. 50 %).

The ideal theoretical proportions from 0 to 100 %, depending on whether the defined threshold is exceeded or not, are not obviously verified.

There is no need for concern in as much as scattering is caused by:

- small sample size,
- differences in weight with relation to the average weight of each age group,
- a probable smaller tolerance for women,
- restraint because of knee contact.

This partly explains why 8 belted occupants out of 9 aged less than 45 suffered no rib fracture although the corresponding threshold for their age was exceeded.

Only two cases of flail chest were recorded out of the 227 belted occupants studied (i.e. 1 %). These were two belted occupants over 55 involved in serious collisions (40 km/hr - 11 g for one, 60 km/hr, 15 g for the other).

DISCUSSION

1 - Specifying a correct parameter for the thorax is an important problem for biomechanics. All occupant protection systems in frontal collisions, whether they be safety belts, air-bags, steering wheel or dash-board, apply forces directly to the thorax. If the head doesn't hit anything, the critical part is the thorax.

Specialists agree on one point. It is desirable for the thorax protection criteria to be measured on the dummy itself and to be independant of the restraint system used.

This is not the case for shoulder load (or load at upper belt anchorage point) which is closely related to 3-point belt technology (static or reel), having a 5 to 6 cm wide shoulder belt.

The results related to the shoulder force are not applicable to wide or inflatable shoulder belts, and even less to air-bags. As they are "pretension" belts, only those data relating to the shoulder loads corresponding to the initial thoracic fracture thresholds according to age remain valid (figure 1). Pretension belts lower the level of shoulder load for a given set of violence and weight conditions.

Shoulder load, or upper belt anchorage point load, is related to the 3-point belt technology. Not being measured on the dummy, this parameter cannot be retained as a thorax protection criterion.

It remains with biomechanics to define the parameter which most accurately reflects human response, taking into account the observations for live belted occupants. The present proposals concern the total load, the load per unit area, the deflection or any index derived from the deceleration/time curve recorded with an accelerometer located in dummy thorax.

Reconstituting real accidents using dummies and cadavers, plus the work conducted on characterizing the cadaver bone condition, should enable the required correspondences between the thoracic responses of live occupants, cadavers and dummies to be established.

2 - Stating injuries using the Abbreviated Injury Scale (9) exagerates the seriousness of a large number of cases by assigning a degree 3 to 3 rib fractures (or sternum or clavicle fractures). Out of 22 belted occupants showing only one or two thoracic fractures and coded AIS 3, 9 were not admitted to hospital, 12 were admitted to hospital for less than 9 days and 1 stayed in hospital for 18 days. The arithmetical mean is 3 days in hospital. The injuries, from the standpoint of seriousness expressed by AIS, showed the same severeness of injury as those who suffered head injury with displaced fracture of the skull, the therapeutic difficulties of which we know of, as well as the risk of after-effects. This lack of consistency means general overestimation of the seriousness of one or two rib fractures. It is proposed to review this point and to replace it with the following scale:

Allowable levels:	AIS 1 - One or two rib fractures
	AIS 2 - From 3 to 7 rib fractures, or fracture of the sternum or the clavicle, associated or not with a number of ribs lower than 4.
Levels at which protection is sought:	AIS 3 - Flail chest with no displacement. At least 3 consecutive double rib fractures or more than 7 rib fractures.
	AIS 4 - Flail chest with displacement.
	AIS 5 - Flail chest requiring respiratory aid.

CONCLUSIONS

1 - The limit shoulder load withstood by the belts corresponding to the <u>absence of thoracic fracture</u> (fitted with classic "3-point" belts) varies depending on age. The estimate of the limit load (corrected for a weight of 75kg) per age group is as follows: $<$35 years old: 900 daN, 35 to 44: 650 daN, 45 to 54: 500 daN, 55 to 64: 350 daN, $\geqslant$ 65: 250 daN.

2 - Seeing the average weight per age group, we can estimate that the risk of initial, belt-induced thoracic fractureoccurs, at the present time, at the following (ΔV, mean γ) violence thresholds: $<$ 35 years old: 55 km/hr and 13,5 g, 35 to 44: 42 km/hr and 10,5 g, 45 to 54: 32 km/hr and 8,5 g, 55 to 64: 24 km/hr and 6,2 g, $\geqslant$ 65: 21 km/hr and 4,9 g.

3 - The percentage of front occupants involved in frontal collision per age group is as follows: $<$35 years old: 59 %, 35 to 44: 18 %, 45 to 54: 13 %, 55 to 64: 6 %, $\geqslant$65: 4 %.

4 - 88 % of front occupants, whether belted or not, are involved in frontal collisions the violence of which is lower than the different thresholds (related to age) of initial, belt induced thoracic fracture (in the absence of overloading by rear occupant).

It should be underlined that front occupants less than 35 years old not attaining the threshold of initial rib fracture (800 daN for 62,8 kg, or again 900 daN for 75 kg, i.e. 55 km/hr, 13.5 g) represent on their own 55 % of all front occupants involved in frontal collisions.

5 - The risk of one or more belt-induced thoracic fractures varies considerably depending on age. The probability is 4 % for the less than 35 year olds, 10 % from 33 to 44, 25 % from 45 to 54, 31 % from 55 to 64, 57 % for the 65 year olds and more.

6 - Belt-induced flail chest is rare (1%). Out of 227 belted front seat occupants involved in frontal collisions rated as " ΔV, mean γ " and not ha-

ving suffered any notable intrusion or overloading by rear passenger, only two cases of flail chest were observed for persons age more than 55.

7 - An injury criterion based on "shoulder force or upper belt anchorage point force" is closely related to the technology of 3-point belts having a 5 to 6 cm wide shoulder belt. Such a criterion is not applicable to the other restraint systems. Direct measurement on dummy should be preferred as a criterion.

8 - Reconstituting real accidents using dummies and cadavers, plus the work conducted on characterizing the cadaver bone condition, should enable the required correspondences between the thoracic responses of live occupants, cadavers and dummies to be established.

9 - Stating thoracic bone injury with the Abbreviated Injury Scale exagerates the seriousness of a large number of cases by assigning a degree 3 to 3 rib fractures. A scale is proposed whereby degree 3 would be attained for flail chest without displacement or more than 7 rib fractures.

ALL IN ALL, the risk of rib fracture by the shoulder belt is low, in the region of 12 %. The risk is only 1 % for these injuries to be serious enough to achieve flail chest which alone shows any real character of seriousness.

Flail chest is sustained by old people whose tolerance is three times less than for young people (measurement of shoulder load). To protect would require very low shoulder load levels, which would benefit in no way the large majority of belts as young people withstand the impact much better.

BIBLIOGRAPHY

1. L.M. Patrick, N. Bohlin and A. Andersson: "Three Point Harness Accident and Laboratory Data Comparison", 18th Stapp Car Crash Conference.

2. N.H.T.S.A.: "Draft Injury Criterion: 1200 lbs for Torso Belt Load", docket 74-14, Notice 7.

3. J.Y. Foret-Bruno, F. Hartemann, C. Thomas, A. Fayon, C. Tarrière, C. Got and A. Patel: "Correlation between Thoracic Lesions and Force Values Measured at the Shoulder of 92 Belted Occupants Involved in Real Accidents", 22nd Stapp Car Crash Conference.

4. R.H. Eppinger: "Prediction of Thoracic Injury Using Measurable Experimental Parameters", 6th Conference on Experimental Safety Vehicles.

5. C. Tarrière, A. Fayon, F. Hartemann and P. Ventre: "The Contribution of Physical Analysis of Accidents Towards Interpretation of Severe Traffic Trauma", 19th Stapp Car Crash Conference.

6. A. Fayon, C. Tarrière, G. Walfisch, C. Got and A. Patel: "Synthèse des résultats et conclusions d'une série d'essais de ceintures de sécurité retenant des cadavres", 2nd IRCOBI.

7. T.H. Glenn: "Anthropomorphic Dummy and Human Volunteer Tests of Advanced and/or Passive Belt Restraint Systems", 3rd International Conference on Occupant Protection.

8. F. Hartemann, C. Thomas, C. Henry, J.Y. Foret-Bruno, G. Faverjon, C. Tarrière, C. Got and A. Patel: "Belted or not Belted: the Only Difference Between Two Matched Samples of 200 Car Occupants", 21st Stapp Car Crash Conference.

9. JOINT COMMITTEE ON INJURY SCALING OF SAE, AMA AND AAAM: "The Abbreviated Injury Scale (1975 revision)., 19th A.A.A.M.

Injuries to Restrained Car Occupants; What are the Outstanding Problems?

P.L. Harms, M. Renouf,
Department of Transport,
P.D. Thomas, M. Bradford,
Institute for Consumer Ergonomics,
United Kingdom

Abstract

This Paper reports on the results of a comprehensive crash-injury investigation currently underway in England and has principally addressed the remaining injury problems associated with restrained occupants.

As a result of legislation introduced in 1983, seat belt wearing rates are of the order of 95 per cent for front seat occupants. The fitment and therefore the wearing rate for rear seat occupants is low but it is hoped that compulsory fitting of rear belts in new cars sold after April 1987 will lead to a corresponding reduction in rear seat occupant casualties.

The head and chest are seen as vulnerable areas requiring added protection, principally due to steering wheel contact (drivers) in frontal impacts and intruding objects with side impacts. The majority of struck-side occupant injuries for both chest and abdomen are due to door contact usually supported by an external object. Footwell intrusion is seen as a major source of lower limb injuries.

Impact zones and collision speeds have shown the contact areas to be considered for type approval testing. The majority of accident-involved cars are impacted by other vehicles which confirms that the vehicle structure and any modification to it still has a major part to play in occupant protection. A feature of the study is the estimation of velocity change in most of the impacts, the resulting indications being valuable for the selection of test conditions for regulatory tests.

The correct use of padded structures, particularly to the steering wheel, together with seat belt pretensioners could further assist in the reduction of casualties amongst the restrained car occupant population.

Introduction

After considerable deliberations by experts followed by much debate in Parliament, the seat belt law came into force in the United Kingdom on 31 January 1983. It is the personal responsibility of all *front* seat occupants over 14 years of age (with very few exceptions) to wear an approved restraint whilst travelling in cars and light vans. It is also the legal responsibility of the driver to ensure that children under 14 years old wear an approved front seat child restraint or an adult belt when travelling in the front passenger seat. Children under one year old must be in an approved child restraint designed for the child's age and weight. This law resulted in an initial saving of more than 200 deaths and 7,000 serious injuries per year compared to the pre-legislation period (1). Since the introduction of the law, compliance has remained at about 95 per cent. All cars sold from 1 April 1987 must have adult belts fitted in the rear but there is no legislation at the moment to cover their compulsory wearing. Approved child restraints are available for use in the front or rear and their fitting and use is encouraged in various ways.

A detailed clinical study was carried out by Rutherford et al (2) to assess the effectiveness of the seat belt legislation in injury terms. Casualty information was obtained from 14 participating hospitals one year before and one year after the legislation date of 31 January 1983. The study was organized as a series of hypotheses predicting changes in injury patterns and other key factors. The most important of these are presented in Table 1 and show that the hypotheses were confirmed in most cases.

Skull and facial fractures did increase slightly for drivers and the present study explores this further.

National Data

The police are usually called to the scene of a road traffic accident where there is injury, allegations of traffic offences and/or in cases of a potential traffic hazard. Frequently all three factors are present. These are not hard and fast criteria and sometimes depend upon police resources available at the time and other factors. When a police accident report is completed it is used for (a), their own purposes and (b), the compilation of national accident statistics (3). Only injury accidents are used in the latter and information on 'damage only' impacts is not available nationally. However information on such accidents reported to the police is often available at local level.

The injury severity of any casualty and therefore the accident severity is assessed by the police officer based on medical evidence available to him/her at the time. The categories of 'Fatal', 'Serious' and 'Slight' injury are used, working to agreed criteria. These assessments tend to vary slightly between officers and the descriptors cannot always relate to subsequent clinical findings (with the possible exception of fatalities). However, they are accepted as being adequate for national accident data which forms a large databank whose output is particularly useful for monitoring accident trends. Due to the relatively large number of cars in use, it is not surprising that these are involved in more road traffic accidents than any other class of vehicle. This results in a correspondingly higher proportion of car occupant casualties than any other road user group (Table 2).

National figures also show that the majority of pedestrian and motorcyclist accidents involve a car. Potential engineering solutions aimed at reducing casualties amongst these two road user groups will be presented by TRRL at this Conference.

It is interesting to note that, despite an increase in traffic over recent years, there has been no corresponding increase in casualty rates when related to cars and drivers (Figure 1).

However in spite of the above trends and the success of front seat restraint legislation, there is still cause for concern that over 2,000 car occupants are killed and over 27,000 are seriously injured each year. This Paper considers the detailed injury mechanisms and related factors from a sample of car accidents investigated as part of a comprehensive crash-injury study.

Study Background

As an essential part of an initiative on vehicle safety, the Transport and Road Research Laboratory

Table 1. Medical effects of seat belt legislation (Rutherford et al).

Factor	Prediction (↘ Decrease ↗ Increase)	Outcome		
		Drivers (%)	F.S.P. (%)	Confirmed
Out-Patients	↘	-10	-22	Yes
In-Patients	↘	-23	-43	Yes
Bed Days	↘	-27	-35	Yes
Severe Injuries	↘	-20	-24	Yes
Lung Injuries	↘	-33	-58	Yes
Sprained Necks	↗	+22	+ 8	Yes
# Sternum	↗	+24	+14	Yes
Brain Injuries	↘	-30	-57	Yes
Skull Fractures	↘	+ 8	-71	No
Minor Facial	↘	-44	-63	Yes
Eye Injuries	↘	-38	-40	Yes
Facial #	↘	+10	-46	Not Sig. Overall

Table 2. Casualties in Great Britain 1985.

Severity \ Road User Group	Car Occupants N (%)	Pedestrians N (%)	TWMV* N (%)	All Others N (%)	All Casualties N (%)
Killed	2061 (40)	1789 (35)	796 (15)	519 (10)	5165 (100)
All Severities	149452 (47)	61390 (19)	56591 (18)	50091 (16)	317524 (100)

*Two Wheel Motor Vehicles (Riders and Passengers)

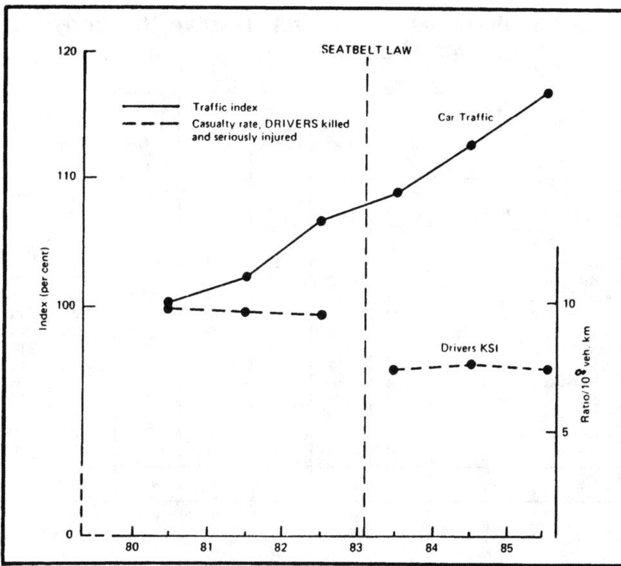

Figure 1. Traffic index and car driver casualty rate. Rel. to 1980 (Great Britain)

(TRRL), acting on behalf of the Department of Transport, are principal sponsors of a car occupant crash-injury research programme. The work is actively supported by the Rover Group, Jaguar Cars and also the Ford Motor Company acting as co-sponsors. The majority of investigations are carried out under contract by the Accident Research Units (ARU) of the University of Birmingham and the Institute for Consumer Ergonomics at Loughborough, with both Units analysing the data. In addition, the Department of Transport's own Vehicle Examiners play an important role in data collection and assessment. Altogether, information on approximately 650 accidents/year is sampled from various locations throughout England. Scientific staff at TRRL are involved in project management, data verification and analyses. All investigators work to standard data forms and collection procedures which have been fully described elsewhere (4,5). All data elements are computer codeable and stored electronically at the ARU's and TRRL. Data consistency checks are carried out by specialist TRRL staff in order to enhance data quality and check for possible errors, the latter being corrected in co-operation with ARU staff. After removing identifying features, all principal users of the data receive complete copies of the appropriate data sheets together with a set of photographic slides. Part analyses from earlier periods of this study have already been reported (6,7). However, the results presented in this Paper will be the first time that a more comprehensive database has been analysed. The study is on-going and subsequent papers and presentations will be able to call on a growing and enhanced database.

The Basic Data

Detailed information on 1,618 vehicles and 2,720 occupants collected between January 1984 and June 1986 is currently available for computerized analysis. To be considered for investigation, a 'case' vehicle must be less than 6 years old at the time of the accident and sufficiently disabled to be towed from the scene. The accident would also have been reported to the local police. Resources do not permit the investigation of every car accident within a given catchment area so the accidents are further stratified according to the police assessment of severity previously described. Using these categories, investigating teams endeavour to cover all fatalities, a high proportion of 'police serious' and lesser proportions of 'police slight' and 'police non-injury'. By their nature the latter category are bound to be underrepresented when compared to the possible numbers involved—particularly as only a proportion are reported to the police. For this reason damage-only accidents have not been used in this particular paper. The non-fatal accidents are chosen at random from police information. Weighting factors are computed to relate numbers of accidents investigated in a given geographical area to those reported to the police. Data 'weighting' is an accepted statistical technique for estimating the total population who may be involved in any particular study. A major advantage is that weighted numbers enable a true balance to be made between frequencies of accidents of different severities. A minor disadvantage of weighting is the possibility of spurious affects when applied to small numbers and consequently firm conclusions cannot be drawn when this occurs. Weighted and unweighted data are also compared for similarity of trends. Rounding off numbers and percentages will lead to minor differences in totals in several of the analyses. It is not possible to compare all parameters in the weighted sample with those in national accident statistics due to restrictions on the latter. However comparisons have been made of such factors as accident severity, casualty severity and seating position (Tables 3, 4 and 5).

The ratio between national data and the weighted sample shows close agreement and similar trends. Any minor differences could be due to the fact that a 'case' vehicle is less than 6 years old and has been

Table 3. Accident sample distribution (cars less than 6 years old).

Maximum Injury Severity/Vehicle	Unweighted Sample N	(%)	Weighted Sample N	(%)	National Data N	(%)
Fatal	97	(10)	125	(2)	1048	(2)
Serious	508	(50)	1392	(27)	11771	(25)
Slight	401	(40)	3607	(71)	33777	(72)
All Severities	1006	(100)	5124	(100)	46596	(100)

Table 4. Percentage age/seating position distribution—casualties in cars (Known values—all severities).

Position \ Age	0–4	5–9	10–16	17–24	25–34	35–44	45–54	55-59	60–64	65+	N.K.
Drivers											
National	–	–	<1	33	24	18	11	4	3	6	1
Weighted Sample	–	–	<1	23	24	18	10	6	5	4	9
Front Pass.											
National	<1	<1	6	35	17	11	9	4	4	9	4
Weighted Sample	<1	-	5	23	16	14	10	4	7	7	14
Rear Pass.											
Nationl	10	12	17	31	9	5	5	2	2	6	1
Weighted Sample	10	12	10	24	9	4	3	3	3	4	18

towed away whereas national figures include vehicles of all ages, not necessarily towed away. There may be other minor differences, missing, incorrect data but overall it is considered that the weighted sample is fairly representative of national data. Thus the weighted number of cars is 5,131 and this is used for analysis.

Impact Configurations

The majority of the vehicles in the sample were saloon or hatchback cars reflecting the popularity of this body style on British roads. The majority were struck by other vehicles—principally cars. Roadside furniture of various types was also struck (Table 6).

In many of these impacts, the car structures together with in-built safety features have a major part to play in occupant protection and may possibly be further improved.

Principal direction of force (PDOF) was recorded using the Collision Deformation Classification system (CDC) (8) and shows that the majority (67 per cent) of impacts were to the front of the vehicle and 23 per cent to the side—more to the driver's off-side (Table 7). These proportions agree with national data. A

Table 5. Percentage comparisons—casualties in car accidents.

Severity/Position	National Data (%)	Weighted Sample (%)
Slight Injury*		
Drivers	57	54
Front Passenger	25	25
Rear Passenger	18	21
Serious Injury*		
Drivers	58	55
Front Passenger	24	22
Rear Passenger	18	19
Fatal Injury*		
Drivers	62	64
Front Passenger	22	22
Rear Passenger	16	14
All Severities*		
Drivers	57	56
Front Passenger	25	24
Rear Passenger	18	20

*Using Police Severity Criteria - see text

Table 6. Principal object hit relative to bodystyle (weighted data).

Object hit \ Body style	Saloon	Hatch-back	Estate	Convert-ible, top on	Light Goods Multi-purpose	3 Wheeler	Total
Car	1145	1720	192	16	20	0	3093
Light Goods Vehicle	65	121	2	0	0	10	198
Heavy Goods Vehicle	202	217	40	0	1	0	460
Public Service Vehicle	16	70	0	3	0	0	89
Other Vehicle	7	55	15	0	0	0	77
Pedal - Motorcycle	6	22	0	0	0	0	28
Sign, Post or Pole	123	171	27	5	17	0	343
Crash Barrier	3	50	0	0	0	0	52
Wall	41	108	5	0	0	0	154
Road Furniture	72	32	17	0	0	0	122
Tree	56	52	3	1	0	0	112
Ditch/Natural Object	18	58	5	0	8	0	89
Roll over - no impact	35	75	3	0	2	0	146
Roll over + Sig. impact	54	52	5	0	0	0	115
Other	7	37	0	0	0	0	44
Not known	12	6	0	0	0	0	18
Total	1863	2860	314	36	49	10	5131

more detailed analysis related to impact direction and location will be presented in later Sections.

Where possible, change of speed on impact (△V) estimates were computed using the CRASH 3 routine (9). Values were available for 49 per cent of the weighted sample (Figure 2), the majority being in the 11-40 km/h band with a peak between 20 and 30 km/h. It is presumed that low speed impacts would either not have been reported to the police or not sampled.

Restraint Use Related to Seating Position and Age

The majority of vehicles were fitted with three point inertia reel belts in the front seating positions. Less than one per cent of adult restraints were fitted in the rear. The distribution of restraint use related to seating position is give in Table 8. The 'Used' category indicates that there are witness marks on the belt or webbing and/or appropriate injuries to the wearer. 'Used-Unproven' is where an occupant states that he or she wore the belt at the time of the accident (in response to a questionnaire) but there is no visual or medical evidence to support this. The latter two

Table 7. Principal direction of force (PDOF) related to struck side (vehicles—weighted data).

PDOF \ Struck side	Front	Right Side	Left Side	Rear	Top	Under	Unclass.	Total
Rollover	7	23	22	15	123	4	65	259
1 o'clock	615	71	0	0	0	0	0	686
2 o'clock	115	112	0	0	0	5	0	233
3 o'clock	12	315	0	0	0	0	0	327
4 o'clock	0	66	0	5	0	0	0	71
5 o'clock	0	75	0	71	0	0	0	147
6 o'clock	0	2	0	225	0	0	0	226
7 o'clock	0	0	21	5	0	0	0	26
8 o'clock	0	0	42	0	0	0	0	42
9 o'clock	0	0	182	0	0	0	0	182
10 o'clock	107	0	157	0	0	0	0	264
11 o'clock	460	0	39	0	3	0	0	502
12 o'clock	2104	52	7	0	3	0	0	2168
Total	3420	716	470	320	128	9	65	5131

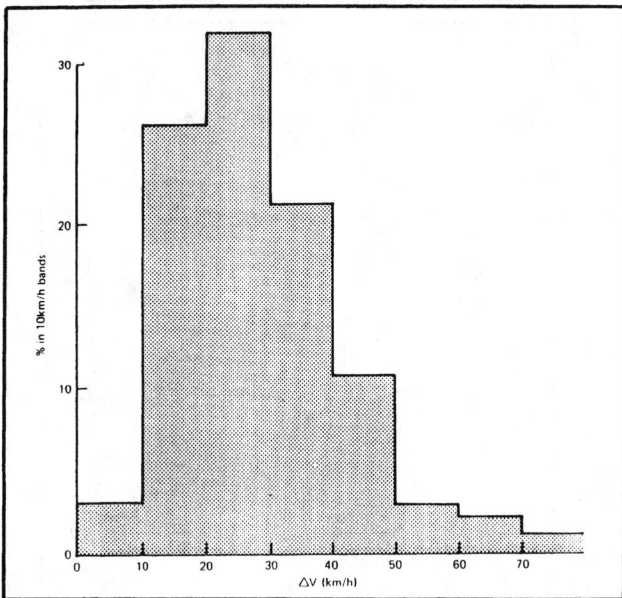

Figure 2. Distribution of velocity change on impact (ΔV). Cars in injury accidents (weighted data)

categories will be combined for some later analyses. Some of the vehicles involved were stationary with the drivers seating position unoccupied. This would account for a slight difference between the number of drivers and number of vehicles, the difference being magnified when weighted.

In addition to the percentage distribution given in Table 8, the data also shows an estimated 93 per cent of drivers and 91 per cent of front seat passengers wore restraints, ignoring 'Not Known' values. These figures indicate a slightly lower use rate than the national average of 95 per cent. The data also shows that only 4 per cent of the rear seat occupants wore restraints. Age related to restraint use shows that the young are at risk (Table 9).

Analysis of restraint use by age shows that most of the unrestrained were in the rear seats although a significant proportion were front seat occupants, particularly those in the middle years age group (Table 10).

The data shows the increase in casualty involvement in the 17-24 year age group. This particular age group ranks high on the national road traffic casualty scale and almost 80 per cent of their accidental deaths are due to road traffic accidents.

Table 8. Restraint use and seating position in injury accidents (weighted data).

Belt Use \ Position	Driver N	Driver (%)	Front Pass N	Front Pass (%)	Rear Pass N	Rear Pass (%)	N.K. N	Total N	Total (%)
Used	3412)	(82)	1277)	(80)	55)	(4)	2	4746)	(66)
Used-Unproven	750)		434)		17)		0	1202)	
Not used	306	(6)	164	(8)	1692)	(94)	10	2172	(24)
Not known	620	(12)	258	(12)	36	(2)	4	918	(10)
Total	5089	(100)	2133	(100)	1801	(100)	16	9038	(100)

Table 9. Age related to restraint use—all seating positions (weighted data).

Belt Use \ Age	0–4 %	5–9 %	10–16 %	17–24 %	25–34 %	35–44 %	45–54 %	55–59 %	60–64 %	65+ %	Age N.K. %	Total [N]
Used	11	<1	28	56	58	60	57	54	53	45	51	4746
Used-Unproven	4	–	4	8	13	23	21	12	23	11	9	1202
Not used	85	96	67	26	15	8	12	24	20	28	28	2172
Not known	–	3	2	10	14	9	10	10	4	16	11	918
Total [N]	186	208	291	2105	1721	1352	782	440	436	484	1033	9038

Table 10. Unrestrained car occupants in injury accidents percentage per age group (weighted data).

Position \ Age	0–4 %	5–9 %	10–16 %	17–24 %	25–34 %	35–44 %	45–54 %	55–50 %	60–64 %	65+ %	N.K. %	Total N
Driver	–	–	1	13	35	39	26	43	5	12	3	306
Front Pass	–	–	2	7	1	5	25	2	38	26	8	164
Rear Pass	100	100	97	79	64	57	49	56	58	63	87	1692
Other /NK	–	–	–	<1	–	–	–	–	–	–	2	10
Total [N]	158	200	195	545	258	102	95	105	88	137	287	2172

Frontal Impacts

Table 7 showed that the majority of impacts were to the front. In order to eliminate end swipe accidents for this part of the analysis, only impacts of between 11 and 01 clock direction will be considered. This represents 62 per cent of the weighted sample.

Table 11 shows the CDC vehicle front body region impacted in relation to object hit. Collisions with other vehicles, particularly cars, still predominate. Two-thirds of the frontal impacts involved the car centre and the offside front was struck more frequently than the nearside. Table 8 showed that only 4 per cent of rear seat occupants were restrained in the accidents. Therefore, in order to give more precise definitions of injuries and mechanisms, only restrained front seat occupants will be considered in this part of the analysis. Two accidents involving total fire

Table 11. Principal object hit relative to vehicle body zone cars in injury accidents—frontal impacts (weighted data).

Object hit \ Part Impacted	Left	Centre	Right	Left + Centre	Right + Centre	Full Width	Total
Car	288	14	346	234	545	585	2013
Light Goods Vehicle	22	0	33	13	18	36	121
Heavy Goods Vehicle	16	9	76	34	74	71	280
Public Service Vehicle	10	0	5	2	4	17	38
Other Vehicle	48	0	1	1	0	20	70
Sign, Post or Pole	59	55	59	34	24	0	232
Crash Barrier	6	0	6	0	0	22	34
Wall	27	5	30	8	14	44	128
Road Furniture	6	19	53	1	0	3	82
Tree	17	12	13	2	14	10	67
Ditch/Natural Object	8	0	8	2	0	47	83
Other	2	2	5	6	12	16	42
Total	507	116	636	338	704	870	3171

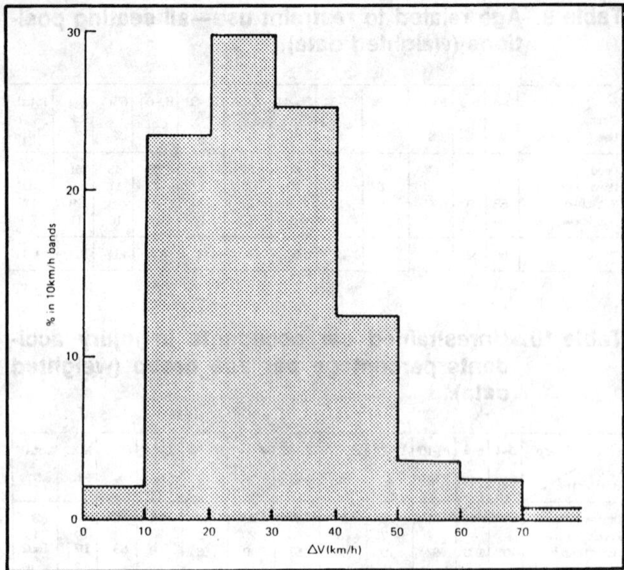

Figure 3. Distribution of velocity change on impact (△V). Cars in injury accidents—frontal impacts (weighted data)

loss with corresponding fatal injuries, 5 light goods and 3 fibre glass body vehicles have also been eliminated. With this slightly reduced sample, △V values are shown in Figure 3 and again indicate predominant values in the 11-40 km/h bands.

Injury Severity

One-third of the front seat occupants were uninjured and approximately half only had injuries of MAIS = 1 which strongly suggests that seatbelts play a major role in injury reduction (Table 12).

The work by Hobbs and Mills (10) related probability of injury to changes in △V and also indicated the injury shift between the restrained and unrestrained front seat occupant. Relating injury severity to impact severity for restrained Drivers and Front Seat Passengers in frontal collisions shows that MAIS 3 injuries occur from approximately 45 km/h and also shows little difference exists between these two groups of front seat occupants until higher values of △V are reached (Figure 4).

The importance of injuries to each body region can be assessed by several methods. The proportion of occupants with an injury to a particular region is a measure of injury incidence. Using this technique the lower limbs are the most frequently injured (81 per cent) followed by the head (52 per cent), as shown in Table 13.

A further measure is to only consider severe injuries ie, AIS 3+. (195 front occupants). A third method suggested by Malliaris (11) is to further weight the AIS values by a 'harm' factor. This is based on injury cost which would vary between countries for a variety of economic reasons. However, proportions are likely to be similar and will serve in this instance to illustrate the point. Using the HARM scale an AIS 6 injury has a weight 378 times greater than an AIS 1 injury (Table 14).

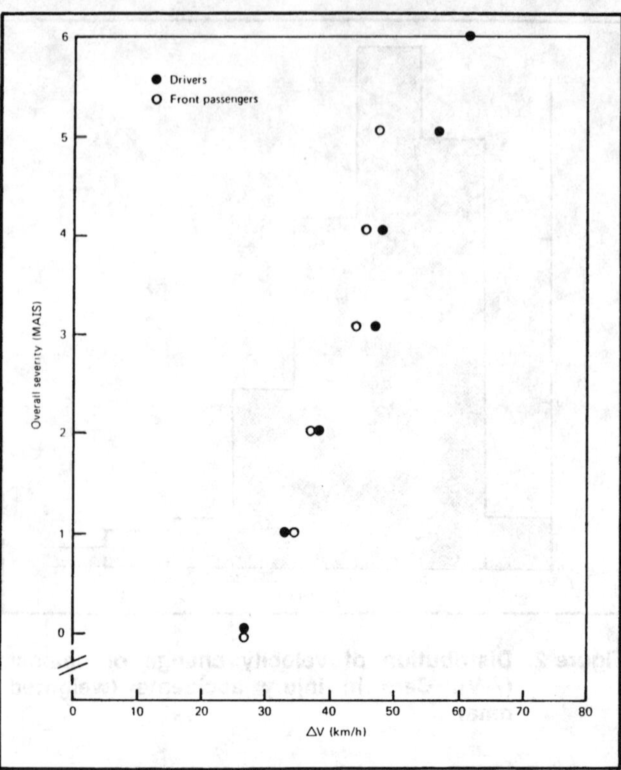

Figure 4. Max. injury severity vs. mean impact severity. Restrained front occs. of cars in frontal impacts

Table 12. Overall injury severity distribution (%)—restrained front occupants in frontal collisions (weighted data—injury accidents).

Position \ MAIS	Uninjured	1	2	3	4	5	6	Total N (%)
Driver	31	45	16	4	<1	<1	<1	2602(100)
Front Pass	35	51	11	2	<1	<1	-	1092(100)
Total N (%)	1188(32)	1731(47)	545(15)	133(4)	30(<1)	23(<1)	9(<1)	3694 (100)

Table 13. Maximum AIS for body regions of restrained front occupants frontal impacts (weighted data).

Body region \ Max AIS	1	2	3	4	5	6	All injuries N	(%)*
Head/face	990	252	24	23	11	4	1304	(52)
Neck	156	11	16	-	3	3	189	(8)
Upper limbs	801	138	47	-	-	-	986	(39)
Chest-Upper back	978	180	43	8	6	4	121	(49)
Abdomen-Lower back	245	4	17	12	10	-	288	(11)
Pelvis and Hips	410	22	13	-	-	-	445	(18)
Lower limbs	1345	178	52	-	-	-	2020	(81)

* Expressed as a percentage of 2506 injured, front occupants (weighted)
NB: Values are not mutually exclusive.

Table 14. Harm weighting factors based on injury costs (Malliaris).

AIS value	1	2	3	4	5	6
HARM weighting factor	0.7	3.0	9.2	56.7	232.5	264.9

Table 15 summarises the ranking of body areas using the three methods. It is considered that the HARM system is the most useful rank order as it is the only scale accommodating both frequency and severity of injury.

Table 15. Rank order of body area importance by three methods restrained front occupants in frontal impacts (weighted data).

Rank Order	All Injuries Body Region	(%)	Injuries AIS 3+ Body Region	(%)	Total harm Body Region	(%)
1	Lower limbs	(81)	Head/Face	(32)	Head/Face	(33)
2	Head/Face	(52)	Chest-Upper back	(31)	Chest-Upper back	(23)
3	Chest-Upper back	(49)	Lower limbs	(27)	Abdomen-Lower back	(17)
4	Upper limbs	(39)	Upper limbs	(24)	Lower limbs	(10)
5	Pelvis and Hips	(18)	Abdomen-Lower back	(20)	Neck	(9)
6	Abdomen-Lower back	(11)	Neck	(11)	Upper limbs	(7)
7	Neck	(8)	Pelvis and Hips	(7)	Pelvis and Hips	(2)

Based upon weighted data both the serious injury and the HARM scale suggest that the priorities are to reduce head and chest injuries. It should be noted that percentage totals exceed 100 as injuries are not mutually exclusive.

Injury Mechanisms

Contact locations giving rise to AIS 2+ injuries to a vulnerable body regions show that the steering wheel has a major involvement (Table 16).

Table 16. Restrained front occupants in frontal collisions contact locations for vulnerable body regions (AIS≥2) (weighted data).

Contact locations \ Body Region AIS	Head/Face		Neck		Chest-Upper back		Abdomen-Lower back		Total	
	2-3	4-6	2-3	4-6	2-3	4-6	2-3	4-6	N	(%)
Steering wheel	144	15	–	3	22	8	4	6	202	(34)
Belt Webbing	–	–	–	–	178	2	5	4	189	(32)
Other vehicle	14	9	–	1	2	2	–	2	30	(5)
'A' Pillar	19	1	1	–	4	3	–	–	28	(5)
Windscreen, surround	32	–	–	–	–	–	–	–	32	(5)
Fascia	13	–	–	–	7	–	–	2	22	(4)
Other Occupant	6	1	1	–	1	–	6	1	16	(3)
Bonnet	18	2	–	–	–	–	–	–	20	(3)
Roof	6	3	4	–	–	–	–	–	13	(2)
Glass	10	–	–	–	–	–	–	–	10	(2)
Front Header Zone	10	–	1	–	–	–	–	–	11	(2)
Own Seat	1	–	–	–	6	–	–	–	7	(1)
Other	6	–	–	–	3	–	1	4	14	(2)
								Total	594	(100)

(NB. Values are not mutually exclusive)

The above values are not mutually exclusive in that it is possible to obtain a chest *and* head injury from say, a steering wheel contact in higher energy impacts. The other vehicle intruding into the occupant space as well as the 'A' pillar and windscreen/surround were also responsible for a significant number of head injuries.

Seat belt webbing was responsible for a high number of AIS 2-3 chest injuries. These were frequently minor rib/sternum fractures and anticipated (Table 1). They could almost be considered a 'trade-off' in relation to serious and fatal injuries occurring to the unrestrained in similar accidents. There is a possibility that seat belt pre-tensioners could reduce a number of the head/chest steering wheel related and also seat belt induced injuries.

Referring to non-vulnerable body regions, foot-well intrusion was responsible for the majority of injuries to the lower limbs (Table 17). Intrusion related injuries will be dealt with in a later Section. The fascia area also featured in lower limb injury causation.

A further area of concern is neck sprains. These are not life-threatening and often present sometime after the accident. They sometimes cause severe discomfort and have been known to persist for considerable periods of time. They have been coded as 'Other' injuries in this particular study and an estimated 653 such lesions (11 per cent) occurred to restrained front seat occupants in frontal impacts. At the moment there is insufficient data to assess the effectiveness of head restraints in reducing this type of injury. However, it is hoped to carry out such an analysis at a later date.

Intrusion Related Injuries

The study notes whether passenger compartment intrusion was a factor causing injury. This is a difficult area as it could be argued that a particular body region may have struck the vehicle interior anyway without intrusion taking place. A further factor is that the body region concerned did not have to travel so far before impact which may have lessened the injury severity in some cases. However previous information (Table 17) indicated that foot-well certainly caused a high percentage of lower limb

Table 17. Restrained front occupants in frontal collisions contact locations for non-vulnerable body regions (AIS 2-3) (weighted data).

Vehicle Location \ Body Region	Upper limbs	Lower limbs	Hips/ Pelvis	Total N	(%)
Pedals/Intrusion	–	206	3	209	(44)
Fascia	18	61	1	80	(17)
Belt Webbing	41	–	–	41	(9)
'A' Pillar	18	8	2	28	(6)
Glass	18	–	–	18	(4)
Steering wheel	16	–	–	16	(3)
Column area	–	14	1	15	(3)
Door	11	–	2	13	(3)
Front firewall	–	6	3	9	(2)
Loose object	8	–	–	8	(2)
Car centre	–	5	3	8	(2)
Bracing	3	–	3	6	(1)
Other	16	6	1	23	(5)
			Total	474	(100)

(NB. Values are not mutually exclusive)

injuries. Another side of the argument is that any decrease in occupant space will tend to negate any safety features which may have been designed to operate with an optimum size occupant compartment and the structures/components associated with it. Also the effects of body shell distortion arising out of intrusion cannot be ignored.

Although some form of energy absorption exists on the steering columns, the intruding steering wheel played a major part in injury causation, particularly to the head/face and chest (Table 18). Impact distortion of the vehicle front and firewall area may have contributed to steering assembly movement.

The intrusion of the other vehicle involved in the accident was responsible for the next highest group of injuries, principally to the head area and usually via the glazed sections of the vehicle.

Footwell intrusion and pedal involvement featured highly in relation to the lower limb injuries (Table 19).

Intrusion in this area is likely to be an indirect result of engine compartment distortion, sometimes involving the front road wheel assembly.

General Discussion—Restrained Front Occupants In Frontal Impacts

The current study shows that approximately 67 per cent of car accidents are frontal, the striking object principally being another vehicle and to a less extent roadside furniture. The impact zone involves either the centre or the full width of the car front. The majority of the impacts in this study occur in the 11-40 km/h$\triangle$V range, peaking in between 20-30 km/h.

Using either 'harm' or severe injury ranking order, the head and chest are body regions where further protection is needed. The steering wheel ranks highly as causing injury to these two body regions, either as a direct contact with the trunk and head going forward following impact or as the result of steering mechanism intruding into the passenger area. Forward body movement is also likely to be a related feature and could possibly be countered by seat belt pretensioners. Seat belt induced injuries occur to the chest, principally bruising and simple fractures. Although these are still undesirable injuries, they could be considered as a 'trade-off' relative to the more serious or fatal injuries that could occur to unrestrained occupants in similar collisions.

The area of concern for the non-vulnerable body regions is the high proportion of lower limb injuries resulting from footwell intrusion. However the introduction of anti-intrusion measures into car occupant compartments must also take account of high peak decelerations that may occur to the occupants unless energy absorbing countermeasures are introduced at the same time. This point has been well recognized in the work by Hobbs et al and reported at this Conference (12). Sufficient energy absorption must be provided ahead of the passenger compartment. However, with a high percentage of smaller cars, the depth available for such absorption is limited by physical constraints within the engine compartment and surrounding areas. In this case the dynamic stiffness of the absorbing structure must be high leading to an even stiffer passenger compartment to ensure that the latter does not collapse. In an experimental design, an Austin Metro has been modified to take account of these guidelines and will be fully reported at this Conference (12).

There is concern that only an estimated 4 per cent of the rear seat occupants were restrained in the accident sample. This situation should improve with the compulsory fitting of adult rear belts in cars. However it is too early to judge the likely extent of their wearing rate and child restraint use must be seen as a feature of this. It has been suggested that bench-type rear seats may present problems in relation to belt mounting points and the lie of the webbing across the body. Collapsible seat backs, either full width or split, as in the hatchback/estate type of car,

Table 18. Restrained front occupants in frontal collisions intrusion contacts for vulnerable body regions (AIS ≥2) (weighted data).

Location \ Body Region	Head/face		Neck		Chest-Upper back		Abdomen-Lower back		Total	
AIS	2-3	4-6	2-3	4-6	2-3	4-6	2-3	4-6	N	(%)
Steering wheel	82	10	–	2	17	7	4	6	128	(51)
Other vehicle	14	9	–	1	2	2	–	2	30	(12)
'A' Pillar	17	1	–	–	3	3	–	–	24	(10)
Windscreen, surround	19	–	–	–	–	–	–	–	19	(8)
Roof	6	3	4	–	–	–	–	–	13	(5)
Bonnet	10	2	–	–	–	–	–	–	12	(5)
Fascia	6	–	–	–	6	–	–	–	12	(5)
Front header area	4	–	1	–	–	–	–	–	5	(2)
Door	–	–	–	–	2	–	–	2	4	(2)
Other	1	–	–	–	2	–	–	–	3	(1)
								Total	250	(100)

(NB. Values are not mutually exclusive)

Table 19. Restrained front occupants in frontal collisions intrusion contacts for non-vulnerable body regions (AIS 2-3) (weighted data).

Location \ Body Region	Upper Limbs	Pelvis/Hips	Lower Limbs	Total N	(%)
Footwell intrusion	–	3	128	131	(51)
Pedals	–	–	27	27	(11)
Fascia/Parcel shelf	12	–	14	26	(10)
'A' Pillar	6	2	8	16	(6)
Steering column area	–	1	13	14	(5)
Door	9	2	–	11	(4)
Steering wheel	9	–	–	9	(3)
Front Firewall	–	3	3	6	(2)
Other Vehicles	3	1	–	4	(1)
Other	6	–	5	11	(4)
			Total	255	(100)

(NB. Values are not mutually exclusive)

are additional difficulties. These points will be monitored as rear belt usage increases.

Side Impact Characteristics

Improvements in side impact protection have been suggested by Hobbs et al and reported at this Conference (13). A standard hatchback car has been modified and is currently on display.

Vehicles that sustained their most severe impact on the right side with a direction force of 2, 3 or 4 o'clock or to the left side with a direction force of 8, 9 or 10 o'clock were defined as side impacts. Within the population of 51 31 vehicles there were 831 (17%) that sustained a side collision as the impact that dissipated the most energy within the vehicle structure. These numbers were based on an original unweighted sample of 193 vehicles. Of these 831 cars, 29 (4%) were involved in fatal accidents, 277 (33%) were in serious injury accidents and 525 (63%) in slight injury accidents. These severities were determined using the UK Police classification system.

Directions of impact

The most common directions of force were the perpendicular ie, 3 o'clock and 9 o'clock, 482 (58%) of the 831 vehicles receiving a side-impact. There were 501 vehicles (60%) struck on the right (driver's) side at ± one clock point either side of perpendicular.

The direction of force of side impacts is described for each injury severity in Table 20 but expressed in terms of the angle from straight ahead. Amongst those vehicles in which the maximum injury severity was AIS 1 or AIS 2/3 the most common impact direction was perpendicular to the vehicle in 368 (60%) and 105 (53%) cases respectively. To simulate accidents of these severities an impact would correspond well to reality if the direction of force were perpendicular. If an impact were to attempt to reproduce the most severe type of side impact in which AIS 4+ injuries are found only 10 (33%) would be simulated by a perpendicular impact. A larger proportion (66%) of fatal impacts would be simulated in an impact with a direction of force of ± 60° from straight ahead.

Table 20. Direction of force for each injury severity (°from straight ahead): all side impact vehicles.

Directions from Straight Ahead(°)	Maximum AIS in Vehicle		
	0/1	2/3	4-6
± 60°	172 (28%)	74 (37%)	20 (67%)
± 90°	368 (60%)	105 (53%)	10 (33%)
± 120°	77 (12%)	19 (10%)	0
Total	617(100%)	198(100%)	30(100%)

Table 21. Location of damage along car side, all vehicles with restrained occupant.

Position	AIS of most severe injuries			
	1	2-3	4-6	All known severities
Front alone	143 (34%)	62 (26%)	0	205 (30%)
Front and passenger compartment	110 (26%)	54 (23%)	15 (45%)	179 (26%)
Passenger compartment alone	91 (22%)	74 (31%)	15 (45%)	180 (26%)
Passenger compartment and rear	48 (11%)	44 (19%)	1 (3%)	92 (13%)
Rear alone	19 (4%)	0	0	19 (3%)
Whole side	12 (3%)	2 (1%)	2 (6%)	16 (2%)
Total population estimate	423(100%)	236(100%)	33(100%)	692(100%)

Damage location

The collision deformation classification (CDC) was used to describe the nature of the direct contact on the vehicle. The location of the damage along the side of the car is shown in Table 2 for all impacts, and for each severity, classified using the fourth character of CDC.

Amongst all severities of accident there were 451 (65%) cars with the direct crush overlapping the passenger compartment. There were therefore 241 (35%) cars in side impacts where there was no overlap of the passenger compartment by the striking object. However, these vehicles tended to be involved in the less severe accidents. Amongst those 423 vehicles in which only AIS 1 injuries occurred, 162 (38%) did not have direct contact over the passenger compartment while of the 236 involving AIS 2 or 3 injuries, 62 (26%) did not. Of the 33 cars where the maximum severity was AIS 4+ all had direct contact over the passenger compartment. When the passenger compart-

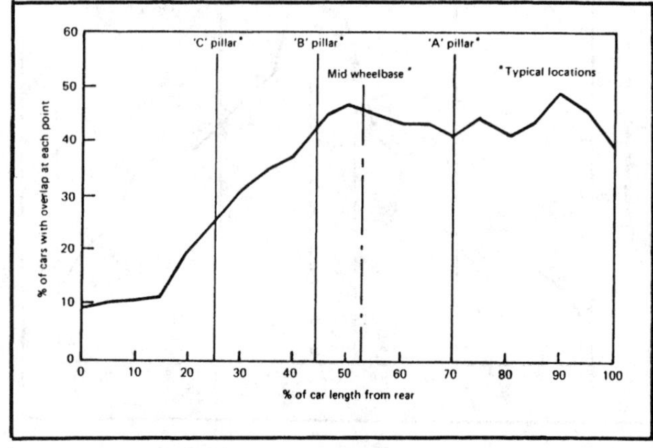

Figure 5. Probability of direct contact, all side impacts (all injury severities—weighted data)

ment is not directly struck the resulting injuries only rarely become serious or fatal. It can also be concluded that measures that strengthen the passenger compartment to reduce the effects of intrusion are unlikely to benefit the occupants of 26% of the cars involving AIS 2-3 injuries and the 38% involving AIS 1 injuries. Such measures might, however, be expected to benefit the remainder of the occupants of cars in each category.

The location of the direct contact along the side of all cars is shown in Figure 5. It is rare that cars in which an occupant is injured is struck towards the rear. Less than 10% of these cars have direct contact within the rear 15% of the car side. The front 60% of the car side is most likely to be involved with typically 40% of cars being struck in this area. The positions of the A and B pillars, the mid point of the wheelbase and the rearmost part of the passenger compartment were measured in a small group of cars—large and small hatchbacks and saloons. The location of these points was found to vary only a little and their average positions have therefore been superimposed upon the graphs as a guide.

The probability of direct contact along the car side for fatal side impacts is shown in Figure 6 with typical positions of car structures indicated. As shown, the zone most commonly involved in direct contact is narrowly concentrated between the 'B' pillar and slightly forward of the wheelbase centre. This corresponds to 20 per cent of the car length. The part most frequently directly contacted is the area between the 'B' pillar and wheelbase centre where 70 per cent of side impacted vehicles in which someone is killed is struck.

Proposals have been made concerning a legislative side impact performance test, these frequently involve

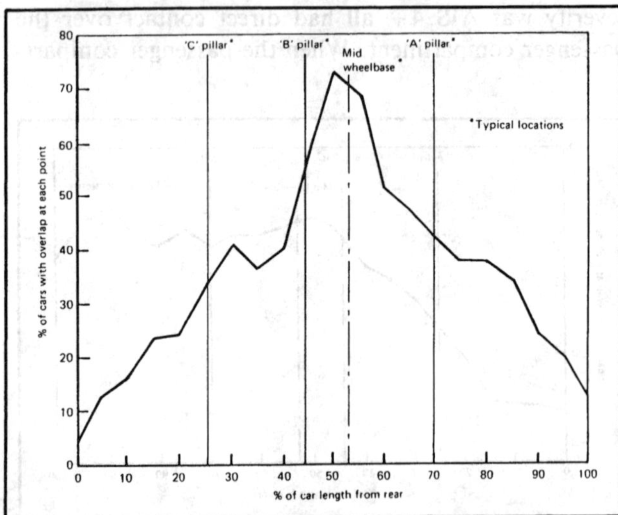

Figure 6. Probability of direct contact, all side impacts (fatalities—weighted data)

a mobile barrier. The proposed width of the EEVC barrier is 150cm, and the NHTSA barrier is 168cm. These correspond to 41% and 36% of a typical car length. Both barriers fit the direct contact distributions of side impacts of all severities fairly closely. However, both barriers appear to be double the length necessary for a close fit to the fatal side impact damage distribution.

Impact Severity

The CRASH 3 computer program was used to estimate the speed change within the impact. The values for 383 were calculated, the remainder violated at least one of the assumptions of CRASH 3. The known delta-v of the cars sustaining a side impact is shown in Figure 7. The mean value of delta-v for the whole population is 25 km/h, for AIS 4+ impacts it is 50 km/h, 31 km/h for AIS 2-3 and 22 km/h for AIS 1 impacts.

If a test impact were to attempt to reproduce the most common side impacts, it is recommended that the delta-v for the side impact vehicle should be 25 km/h. If the test impact were intended to reduce the

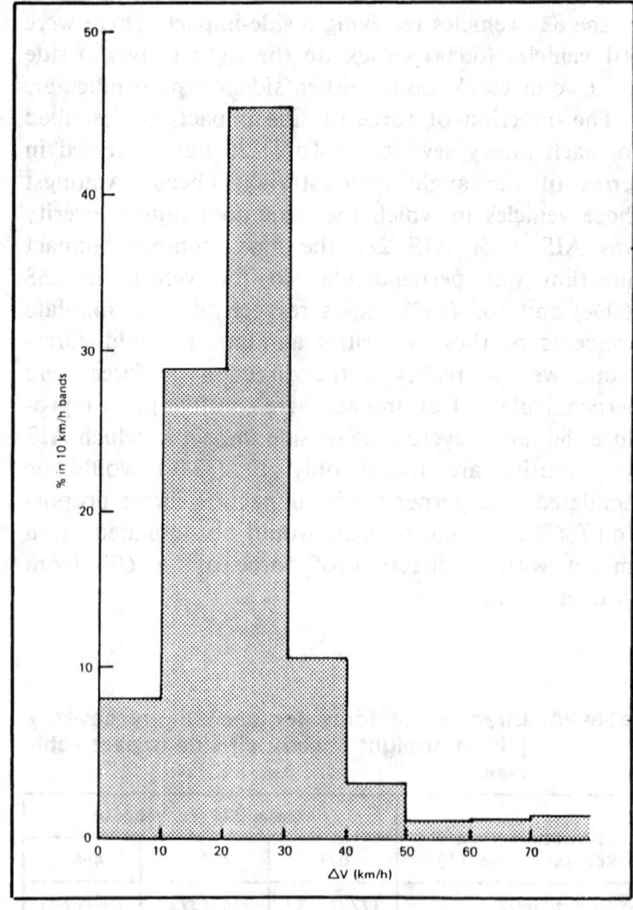

Figure 7. Distribution of velocity change on impact (ΔV). Cars in injury accidents—side impacts (weighted data)

Table 22. Seating positions of restrained occupants maximum AIS = 1+.

	Struck Side	Non-struck Side
Front side	304 (74%)	176 (61%)
Front left	104 (25%)	114 (39%)
Rear right	2 (0.5%)	0
Rear left	0	1 (0.5%)
Total	410 (100%)	292 (100%)

incidence of the most severe injuries it should be performed at a delta-v of at least 50 km/h although some reduction in fatalities might be expected with a test involving a lower delta-v.

Occupant Characteristics

There were 701 injured occupants in the 692 cars involved in a side impact who were also restrained. Their seating positions are shown in Table 22.

Since there is no legal requirement in the UK that rear seat occupants be restrained, 99% of those examined in this analysis are front seat occupants. The severity of the injuries all 701 sustained is shown in Table 23 for the Police severity and Table 24 for the maximum AIS for struck side and non-struck side occupants separately.

Table 23 shows that 42% of struck side occupants were in the serious or fatal categories compared with 25% of the non-struck side occupants. This pattern is repeated in Table 24, 15% of struck side occupants sustained injuries of AIS 3 or above compared with 7% of non-struck side occupants. These differences cannot automatically be attributed to any additional protection non-struck side occupants may receive as these occupants happened to be passengers in vehicles with lower impact severities than struck side occupants as Table 25 shows.

There were 171 struck side occupants with a known delta-v and 167 non-struck side occupants. Above 30 km/h the two distributions are closely similar but significant differences lie in the lower bands. 85 (50%) of the struck side occupants were in vehicles with a delta-v of 20-29 km/h compared with only 63 (38%) of the non-struck side occupants. Also 66 (39%) of non-struck side occupants had a vehicle delta-v of 10-19 km/h compared with 42 (24%) of struck side occupants.

Table 23. Police severity of restrained occupants with maximum AIS = 1+.

Severity	Struck side	Non-struck side
Uninjured	11 (3%)	7 (3%)
Slight	229 (56%)	214 (73%)
Serious	158 (39%)	66 (23%)
Fatal	12 (3%)	4 (2%)
Total	410(100%)	292(100%)

Table 24. Maximum AIS of restrained occupants with injuries AIS = 1+.

Maximum AIS	Struck side	Non-struck side
1	215 (53%)	218 (75%)
2	132 (32%)	50 (17%)
3	42 (10%)	10 (4%)
4	7 (2%)	4 (1%)
5	8 (2%)	8 (3%)
6	6 (2%)	1(0.4%)
Total	410(100%)	292(100%)

Struck side occupant injuries

Table 26 shows the distribution of the severities of injury for each body region of the 294 struck-side occupants. Also shown in Table 26 is the contribution towards the total HARM for each body region which was calculated using the weightings derived by Malliaris shown in Table 14.

The importance of injuries to each body area can be assessed using several methods. The proportion of occupants with an injury in a particular body region is a measure of incidence of injury. Using this technique the head is the most frequently injured, 283 (69%) of the 410 of struck-side occupants sustained a head injury.

When the number of occupants with an injury of AIS 3 or above is expressed as a proportion of all injuries, this produced a measure of the severity of those injuries that do occur. Under this measure the abdomen is most frequently the body area sustaining severe injuries, 52% of the injuries found in the

Table 25. Delta-v of struck side and non-struck side occupants vehicles where known (km/h).

Delta-v	Struck side	Non-struck side
1 - 9	8 (5%)	2 (1%)
10 - 19	42 (24%)	66 (39%)
20 - 29	85 (50%)	63 (38%)
30 - 39	21 (12%)	24 (14%)
40 - 49	9 (5%)	6 (4%)
50 - 59	1(0.6%)	2 (1%)
60 - 69	3 (2%)	1(0.8%)
70+	3 (2%)	4 (2%)
Total	171(100%)	167(100%)

Table 26. Distribution of maximum AIS within each body area related to harm for struck side occupants. Numbers of injuries and proportions of known severities.

Body Area		Struck Side of Occupants								
		Maximum AIS							Amount of Harm	
		0	1	2	3	4	5	6	No's	%
Head/face	No.	127	143	116	12	6	3	3	2391	33
	%	31	35	28	3	1	0.8	0.8		
Neck	No.	332	76	0	1	0	0	0	50	0.7
	%	81	19		0.3					
Chest	No.	245	123	12	19	1	5	5	2840	39
	%	60	30	3	5	0.2	1	1		
Abdomen	No.	364	22	0	14	5	5	NA	1590	22
	%	89	5		4	1	1			
Arms	No.	255	102	51	1	NA	NA	NA	231	3
	%	62	25	12	0.3					
Legs	No.	276	115	9	10	0	NA	NA	200	3
	%	67	28	2	3					
Total Harm									7302	100%

abdomen were AIS 3 or above, a substantially higher proportion than any other body region.

A third measure is to calculate the contribution to the total value of 'harm' found within the side impact population of injuries to each body area. Using these factors chest injuries are the most expensive and contribute 39% of the total harm.

Table 27 summarises the rankings of body areas using each of these three methods. It is considered that the most useful rank order is produced using 'harm' as it is the only scale that reflects both the frequency and severity of injury.

Table 27 shows that chest injuries are frequent, being sustained by 40% of struck side occupants, they are also severe with 18% of injuries being of AIS 3+. Additionally Table 26 shows that 10 out of the 30 AIS 3+ chest injuries have a very heavy weighting of harm. These 3 factors combine to make the chest the most significant in terms of harm.

Table 27 also shows that head/face injuries are very common but seldom severe whilst abdominal injuries are rare but frequently severe. However, it is considered likely that protection against chest injuries is likely to reduce abdominal injuries due to the coupled mechanisms of injury. It should be noted that for the purposes of this analysis the abdomen includes the pelvis and hip joints.

The maximum AIS possible for arm and leg injuries is 4 although such scores are rare. The contribution towards the total value of harm of each body area is therefore low at 3% each.

Finally Table 27 shows that neck injuries are the least commonly injured in struck side impacts and are rarely severe. Neck injuries only contribute 0.7% to the total harm. The harm scale suggests that the reduction of the chest injuries of struck side occupants should be the principle objective of side impact protection. Further objectives should be the reduction of head/face and abdominal injuries in that order.

Table 27. Rank order of body area importance using frequency of injury, frequency of AIS 3+ injury and harm—struck side occupants.

Rank	Body Region and % Injured		Body Region and % of Injuries AIS 3+		Body Region and % of Total Harm	
1	Head/face	69%	Abdomen	52%	Chest	39%
2	Chest	40%	Chest	18%	Head/face	33%
3	Arms	38%	Head/face	8%	Abdomen	22%
4	Legs	33%	Legs	7%	Legs	3%
5	Neck	19%	Neck	1%	Arms	3%
6	Abdomen	11%	Arms	0.6%	Neck	0.7%

Table 28. Distribution of maximum AIS within each body area related to harm for non struck side occupants. Numbers of injuries and proportions of known severities.

Body Area		Non-Struck Side of Occupants								
		Maximum AIS							Amount of Harm	
		0	1	2	3	4	5	6	No's	%
Head/face	No.	210	52	14	6	1	8	1	2315	76
	%	72	18	5	2	0.3	3	0.4		
Neck	No.	275	14	0	3	0	0	0	37	1
	%	94	5		1					
Chest	No.	185	96	10	0	0	0	0	97	3
	%	63	33	4						
Abdomen	No.	271	9	6	3	4	0	NA	279	9
	%	93	3	2	0.9	1				
Arms	No.	188	77	26	1	NA	NA	NA	141	5
	%	65	26	9	0.3					
Legs	No.	77	211	1	2	0	NA	NA	169	6
	%	27	72	0.4	0.7					
Total Harm									3038	100%

Non-struck Side Occupant Injuries

The pattern of injuries of the 292 non-struck side occupants in the population is shown in Table 28. It is rare for an injury above AIS 2 to occur in any body region. The rank order of the body areas was calculated using the methods previously applied, and the results are shown in Table 29. This table shows that the greatest contribution to the total harm by far is from head/face injuries. The contribution is 76% and is a result of a 27% incidence rate within the

Table 29. Rank order of body area importance using frequency of injury, frequency of AIS 3+ injury and harm: non-struck side.

Rank	Body Region and % Injured		Body Region and % of Injuries AIS 3+		Body Region and % of Total Harm	
1	Legs	73%	Abdomen	33%	Head/face	76%
2	Chest	37%	Head/face	20%	Abdomen	9%
3	Arms	35%	Neck	6%	Legs	6%
4	Head/face	28%	Arms	1%	Arms	5%
5	Abdomen	7%	Legs	1%	Chest	3%
6	Neck	6%	Chest	0%	Neck	1%

population and a high proportion of AIS 3+ injuries. Below head/face the abdomen accounted for 9% of the total harm, these injuries being extremely severe amongst non-struck side occupants but only rarely occurring.

The remaining 4 body areas each contributed only a little to the total harm. AIS 3+ injuries were rarely found to occur to the legs, chest or arms. They occurred in 6% of neck injuries but only 6% of non-struck side occupants sustained a neck injury. These 4 body regions accounted for only 15% of the total harm.

Struck and Non-struck Side Occupants Compared

Tables 26 and 28 showed that the total harm to struck and non-struck side occupants was 7302 and 3038 'harm' units respectively. When all body areas of *all* occupants are considered together, the greatest contribution to the total value of 10340 harm units was 2840 harm units from struck side occupants chest injuries. The head/face injuries of struck side occupants followed, this accounting for 2391 harm units. The third highest contribution however was 2315 from the head injuries of non-struck side occupants. Inspection of Tables 26 and 28 shows that 9 non-struck side occupants sustained AIS 5 or 6 head injuries compared with only 6 struck side occupants. The weighting factors of harm give this difference of 3 injuries a value of 633 harm units.

Contact Analysis

As part of the procedure of vehicle examination in the study, the interior was inspected for evidence of occupant contact. This information was later combined with a detailed description of the injuries sustained, coded using AIS. Modern cars tend to be built using materials that do not always show contact evidence particularly if the contact is light. It is therefore not always possible to determine the part of the car that caused the injury and a "not known" code is used in these circumstances.

The coding procedure used allows up to 2 contacts to be associated with each body injury. Also the gross body areas used for this analysis may be constructed of several smaller units. For example, when injuries to the body area "chest" are initially coded they are categorized as being sustained either to the upper back and spine or to the remaining part of the chest. There are therefore potentially 4 contact codes that may be associated with injuries to the body area "chest". For these reasons the numbers of contacts in the tables presented are likely to be different from the numbers of injuries in each body area.

Injuries are only seldom associated with 2 contacts but when this occurs there is no discrimination as to which is the more severe. For example, a common contact in a side impact is with an intruding door supported by a striking vehicle. Both door and vehicle would be given a contact code.

Similarly an occupant's head may break the side window and strike a tree, both window and tree would be coded and would feature in the tables of contacts presented in this analysis. It would not be true to assume that both contacts played an equal role in the causation of the injury.

When the contacts that cause injuries of any severity are examined, this effect is not relevant as every contact caused an injury of some severity. However, amongst the contacts that caused the more severe injuries, the presence of less severe contracts may cloud any conclusions drawn. For example, Table 30 shows both side glass and intruding external objects to be associated with AIS 3+ head injuries. These areas of possible misinterpretation are highlighted in the text.

For each body area the tables presented show the most common contacts that together account for at least 67% of all identified contacts.

Head and Face Injuries

Of the 410 struck side occupants, 283 sustained a classifiable injury to the head or face and when measured using harm this body area ranked second. Of these occupants, the vast majority, 259 (92%), received minor to moderate (AIS 1 and AIS 2) injuries, whilst the remaining 24 (8%) sustained more severe AIS 3+ injuries (see Table 26). Table 30 shows the frequency of contacts causing injuries of any severity and AIS 3+ injuries respectively.

When considering the contacts causing injuries of any severity, the most commonly occurring were with glazing materials; there were 146 (32%) contacts from this source. The steering wheel was contacted 47 times. All of these occupants involved except one were involved in multiple impacts, one of which was a

Table 30. Contacts causing the head and face injuries of restrained struck side occupants.

Struck Side Occupants					
All Head/Face Injuries			AIS 3+ Head/Face Injuries		
Contact Source	No. of Contacts	Relative Frequency %	Contact Source	No. of Contacts	Relative Frequency %
Glazing materials	146	32	Intruding external object	28	52
Steering wheel	47	15	Side window glass	17	31
Intruding external object	40	12	Other known contacts	19	17
Side header rail	32	10			
Other known contacts	57	31			
Total	322	100%	Total	54	100%

frontal. The next most frequent was from intruding external objects accounting for 40 (12%) of the contacts. (This includes such objects as other vehicles, roadside furniture etc, without the occupant being ejected. There were 32 (10%) contacts with the side header rail. These contact sources accounted for 79% of the known injury causes.

If the contacts causing the AIS 1 and AIS 2 injuries within this group are excluded then there are 54 remaining contacts causing AIS 3+ injuries; the majority, 28 (52%) were from intruding external objects and 17 (31%) were from side window glass. These 2 contacts accounted for a total of 73% of the known injury causes. Because more than one object can be contacted by the same body area, the raw data was examined revealing that those occupants sustaining AIS 3+ injuries from side window glass also made contact with an intruding external object. Therefore, the most significant cause of serious to fatal head injuries for struck side occupants is from intruding external objects.

Table 28 showed that only 82 non-struck side occupants sustained head or face injuries and of these 66 (80%) had AIS 1 or AIS 2 injuries whilst 16 (20%) sustained AIS 3+ injuries. The analysis also shows that there were slightly less than one third the number of non-struck side occupants receiving a head or face injury when compared with those on the struck side. However, in terms of harm, they were the most important body area to be injured by non-struck side occupants.

The contacts causing these injuries at any severity level is shown in Table 31 where it can be seen that, as for the struck side occupants, glazing materials are the most common source of injury (27%). However the second most common cause is quite different, coming from contact with another occupant (20%) rather than from an intruding external object. There were 10 (16%) contacts with the facia and 7 (11%) cases of injury without contact. These contacts account for 66% of all known sources.

Table 31 also shows that AIS 3+ injuries occur most commonly from contact with a door (24%), other occupants (20%) and then equally frequent from A-pillar (17%) and rear windows. It should be noted however, that the total number of non-struck side occupants sustaining AIS 3+ injuries was small, there were only 13 contacts.

Table 31. Contacts causing the head and face injuries of restrained non-struck side occupants.

Non-Struck Side Occupants					
All Head/Face Injuries			AIS 3+ Head/Face Injuries		
Contact Source	No. of Contacts	Relative Frequency %	Contact Source	No. of Contacts	Relative Frequency %
Glazing materials	17	27	Door	3	24
Other occupant	12	20	Other occupant	3	20
Facia	10	16	A-pillar	2	17
Injury without contact	7	11	Rear window	2	17
Other known contacts	15	26	Other known contacts	3	20
Total	61	100%	Total	13	100%

Neck Injuries

Table 26 showed that only 77 (19%) of the 410 struck side occupants received a neck injury. Of these, 76 (97%) received AIS 1 injuries and only 1 sustained an AIS 3 injury. The low numbers of neck injuries is also reflected in Table 27 where they are shown as the least important in terms of harm, accounting for only 0.7 of the total harm distribution among all body areas.

Table 32 shows that by far the most common cause was non-contact injuries (ie deceleration without contact); a total of 55 contacts 36 (65%), were of this type. The single AIS 3 injury was caused by an unidentified contact.

Table 28 showed that a lesser proportion of non-struck side occupants sustained a neck injury. There were 17 such casualties representing only 6% of non-struck side occupants. Neck injuries were also the least important in terms of harm. Of the 17 neck injuries, 14 (82%) were AIS 1 and 3 were AIS 3. The contacts causing these injuries are shown in Table 33 and the findings are similar to those for struck side occupants with the vast majority sustaining their injuries as a result of deceleration without contact. The contacts causing the AIS 3 neck injuries were from an unknown source.

Chest Injuries

It can be seen from Table 27 that chest injuries for struck side occupants are the most important in terms of harm, accounting for 39% of the total harm distributed amongst all the body areas. 165 struck side occupants received classifiable injuries and of those

Table 32. Contacts causing the neck injuries of restrained struck side occupants.

Struck Side Occupants					
All Neck Injuries			AIS 3+ Neck Injuries		
Contact Source	No. of Contacts	Relative Frequency %	Contact Source	No. of Contacts	Relative Frequency %
Non-contact injury	36	65	Non-identified	0	
Other occupant	8	15			
Other known contacts	11	20			
Total	55	100%	Total	0	

Table 33. Contacts causing the neck injuries of restrained non-struck side occupants.

Non-Struck Side Occupants					
All Neck Injuries			AIS 3+ Neck Injuries		
Contact Source	No. of Contacts	Relative Frequency %	Contact Source	No. of Contacts	Relative Frequency %
Non contact injury	5	83	None identified		
Other known contacts	1	17			
Total	6	100%	Total	0	

Table 35. Contacts causing chest injuries of restrained non-struck side occupants.

Non-Struck Side Occupants					
All Chest Injuries			AIS 3+ Chest Injuries		
Contact Source	No. of Contacts	Relative Frequency %	Contact Source	No. of Contacts	Relative Frequency %
Seat belt webbing	71	78	No injuries	0	0
Other known contacts	19	22			
Total	90	100%	Total	0	

125 (76%) were AIS 1 or AIS 2. The remaining 40 (14%) were AIS 3+ injuries (see Table 26).

Table 34 shows the frequency of contacts occurring for any severity of chest injury to struck side occupants. From a total of 172 contacts, 84 (48%) were from seat belt webbing closely followed by the door 75 (43%) as the next most common. These 2 contact sources accounted for 91 of all the known injuries. If all the AIS 1 and AIS 2 contacts are excluded then the most common frequent contact causing an AIS 3+ injury is the door (66%).

A total of 107 non-struck side occupants received chest injuries and all were either AIS 1 or AIS 2. Harm ranks these injuries fifth, only 2% lower than arm injuries but significantly lower than the head injuries for non-struck side occupants.

Table 35 shows that the most common cause of chest injuries for non-struck side occupants is seat belt webbing. Of the 90 contacts the vast majority 71 (78%) were from this source.

Abdomen Injuries

Table 26 showed that 46 struck side occupants received injuries to the abdomen representing 11% of the total population of struck side occupants. These injuries ranked third measured using harm (22%). Twenty-two (48%) received AIS 1 or AIS 2 injuries and the remaining 24 (52%) received more serious AIS 3+ injuries. The contacts causing these injuries are shown in Table 36.

When considering the causes of abdomen injuries for any severity the door and door furniture are by far the most common (70%). Of these 101 contacts 82 (59%) were caused by the door, 15 (11%) by the door handle and 4 (3%) by the door bin. Amongst those with AIS 3+ injuries there were 33 (57%) door contacts and 17 (30%) contacts with intruding external objects.

Although the door is shown to be the most important cause of both slight and severe injuries to the abdomen, it is quite likely that any one occupant could have received the same injury by contacting both the door and the intruding external object, if that object was supporting the door at the time of contact. Further investigation of the raw data revealed that this was the case for the 17 AIS 3+ contacts, thus; intruding external objects such as other vehicles and roadside furniture, were by far the most injurious abdomen contact for struck side occupants.

Table 28 showed that 22 (7%) of the non-struck side occupants received an injury to the abdomen. However, the majority 15 (68%) were slight to moderate AIS 1 and AIS 2 injuries, with 7 (32%) AIS 3+ injury. Measured using harm these injuries ranked second (9%), only 3 more than leg injuries, but they were substantially less important than head injuries for non-struck side occupants. The analysis shows, therefore, that the population by far the most likely to receive a serious abdominal injury are struck side occupants.

Table 34. Contacts causing the chest injuries of restrained struck side occupants.

Struck Side Occupants					
All Chest Injuries			AIS 3+ Chest Injuries		
Contact Source	No. of Contacts	Relative Frequency %	Contact Source	No. of Contacts	Relative Frequency %
Seat belt webbing	84	48	Door	25	66
Door	75	43	Other known contacts	11	34
Other known contacts	13	9			
Total	172	100%	Total	36	100%

Table 36. Contacts causing the abdomen injuries of restrained struck side occupants.

Struck Side Occupants					
All Abdomen Injuries			AIS 3+ Abdomen Injuries		
Contact Source	No. of Contacts	Relative Frequency %	Contact Source	No. of Contacts	Relative Frequency %
Door and door furniture	101	73	Door and door furniture	33	57
Other known contacts	36	27	Intruding external object	17	30
			Other known contacts	7	13
Total	137	100%	Total	57	100%

Table 37. Contact causing the abdomen injuries of restrained non-struck side occupants.

Non-Struck Side Occupants					
All Abdomen Injuries			AIS 3+ Abdomen Injuries		
Contact Source	No. of Contacts	Relative Frequency %	Contact Source	No. of Contacts	Relative Frequency %
Seat belt webbing	43	70	Seat belt webbing	4	42
Other known contacts	18	30	Other occupant	3	34
			Other known contacts	2	24
Total	61	100%	Total	9	100%

The contacts causing the abdomen injuries to non-struck side occupants are shown in Table 37. Only one contact type, seat belt webbing accounted for 70% of all the non-struck side occupants' abdomen injuries. The belt (42%) and other occupant contacts (34%) were the most frequent source of AIS 3+ injuries.

Arm Injuries

Table 26 showed that 155 of the 410 struck side occupants had an injury to the arm with a classifiable AIS. Arm injuries, however, ranked fifth when measured using harm. The contacts by the arms of these 155 occupants are shown in Table 38. The most common contact was with the door, occurring 71 (49%) times.

Table 26 showed only one occupant sustaining AIS 3 arm injuries so for this region the contacts causing AIS 2 and 3 injuries are shown separately also in Table 38. This table shows that of the more serious injuries to arms 48 (71%) also arise from door contact.

Table 39 shows the sources of arm injuries amongst non-struck side occupants. There were 104 such occupants with an arm injury but only 50 contacts were identified. The most common contact to cause an injury of any severity was with another occupant accounting for 22 (43%) of all contacts. Other relatively frequent contacts were flying glass and clothing. The low total number of contacts means it is difficult to place a reliable rank order on these contact regions. There were 18 identified arm contacts causing an AIS 2+ injury, all were with another occupant.

Table 38. Contacts causing the arm injuries of restrained struck side occupants.

Struck Side Occupants					
All Arm Injuries			AIS 3+ Arm Injuries		
Contact Source	No. of Contacts	Relative Frequency %	Contact Source	No. of Contacts	Relative Frequency %
Door	71	49	Door	48	71
Glazing materials	38	26	Other known contacts	20	29
Other known contacts	36	25			
Total	145	100%	Total	68	100%

Table 39. Contacts causing the arm injuries of restrained non-struck occupants.

Non-Struck Side Occupants					
All Arm Injuries			AIS 2+ Arm Injuries		
Contact Source	No. of Contacts	Relative Frequency %	Contact Source	No. of Contacts	Relative Frequency %
Other occupant materials	22	43	Other occupant	18	100
Flying glass	7	13			
Clothing	6	12			
Other known contacts	15	33			
Total	50	100%	Total	18	100%

Leg Injuries

Leg injuries ranked fourth in Table 27 when harm was considered although they are relatively frequent, 33% of struck side occupants sustaining these injuries. Table 40 shows the contacts causing the leg injuries of struck side occupants for injuries of all severities. The contacts for AIS 2+ leg injuries are also shown (AIS 3 leg injuries being rare).

The contacts causing leg injuries were the same for both groups of severities. The most common contact was the door accounting for 155 (72%) of injuries of any severity and 20 (49%) of AIS 2+ injuries.

Amongst AIS 2+ injuries the second most common was the intruding footwell and pedals, accounting for 15 (36%) of contacts.

Table 28 showed that leg injuries were sustained by 215 (73%) of all non-struck side occupants. Of these 211 (98%) were AIS 1 but leg injuries, while contributing only 6% of the total harm, ranked second. Table 41 shows the contacts that were associated with these injuries. There were 227 identified contacts associated with injuries of any severity. The most common two were with the centre console and the lower facia. These two contacts occurred 68 (30%)

Table 40. Contacts causing the leg injuries of restrained struck side occupants.

Struck Side Occupants					
All Leg Injuries			AIS 2+ Leg Injuries		
Contact Source	No. of Contacts	Relative Frequency %	Contact Source	No. of Contacts	Relative Frequency %
Door	155	72	Door	20	49
Other known contacts	60	28	Intruding footwell	15	36
			Other known contacts	6	15
Total	215	100%	Total	41	100%

Table 41. Contacts causing the leg injuries of restrained non-struck side occupants.

Non-Struck Side Occupants					
All Leg Injuries			AIS 2+ Leg Injuries		
Contact Source	No. of Contacts	Relative Frequency %	Contact Source	No. of Contacts	Relative Frequency %
Centre console	68	30	Gear lever	1	50
Lower facia	64	28	Self	1	50
Pedals and bracketry	23	10			
Other known contacts	72	32			
Total	227	100%	Total	2	100%

and 64 (28%) times respectively. There were also 23 (10%) contacts with the pedals and bracketry.

There were only 2 identified contacts associated with the more severe injuries, these were with the gear lever and another part of the occupant.

Other Impact Configurations

Table 5 showed that 6 per cent of the impacts were to the rear and 4 per cent involved a rollover. Tables 42 and 43 show the injury severity distribution for restrained front seat occupants where the vehicles have been involved in these two impact configurations.

This shows the outcome in rear impacts is less severe than for frontal and side accidents ie, two-thirds were uninjured and a third only sustained an injury of MAIS 1. The estimated mean $\triangle V$ was 26 km/h.

Not surprisingly, only a third of the occupants were uninjured in rollovers and half had a maximum severity level of MAIS 1. Multiple contact must also be a feature of this type of impact as the restraint may not be performing in its normal mode of operation and webbing could be slack. It is often difficult to ascribe contact points for the injuries sustained to occupants in this type of accident. It is conceivable that belt pre-tensioners may assist in reducing injuries provided they can be initiated in this type of impact. Selective padding will also assist when further information is available on the structure contacted.

Table 42. Maximum injury severity distribution—restrained front occupants in rear impacts (weighted data)..

Severity / Position	Uninjured (%)	MAIS 1 (%)	MAIS 2-3 (%)	MAIS 4-6 (%)	Total N (%)
Driver	(64)	(33)	(2)	<1	481 (100)
Front Passenger	(64)	(31)	(5)	-	104 (100)
Total N (%)	378 (65)	190 (32)	15 (3)	2 (<1)	585 (100)

Table 43. Maximum injury severity distribution.

Severity / Position	Uninjured (%)	MAIS 1 (%)	MAIS 2-3 (%)	MAIS 4-6 (%)	Total N (%)
Driver	(29)	(52)	(16)	<1	256 (100)
Front Passenger	(32)	(52)	(13)	4	104 (100)
Total N (%)	108 (30)	188 (52)	54 (15)	6 (2)	359 (100)

Conclusions

General

This Paper has reported on the results of a comprehensive crash-injury study currently underway in England. Weighted data has been used throughout and compared with national accident figures where appropriate. The following is concluded:

The study sample is reasonably representative of national accident data.

Although compulsory seat belt legislation has brought about a wearing rate of approximately 95 per cent amongst front seat car/van occupants, there is still cause for concern that over 2,000 occupants are still being killed and 27,000 seriously injured per year. However the driver casualty rate is not increasing at the same rate as traffic volume. An estimated 94 per cent of the rear seat occupants in the sample were unrestrained. It is hoped that the introduction of compulsory fitting of rear belts in new cars from April 1987 will encourage their use and reduce mortality and injury amongst rear seat occupants. Problems that may arise in respect of belt mounting and body fit will have to be monitored as usage increases.

An analysis of the objects impacted shows that the majority are struck by other cars. This fact, and also the other structures impacted indicates that the car structure and any improvements to it, still have a major part to play in occupant protection.

The majority (67 per cent) of impacts were found to be to the front of the vehicle with a significant number (23 per cent) to the side—with rather more to the driver's side than to the nearside.

Young children featured highly amongst the unrestrained occupants.

Frontal Impacts

Two thirds of the impacts in this direction involved the front centre with a high proportion extending across the whole front. Corner impacts also occurred as an important sub-group.

Changes of speed on impact ($\triangle V$) indicates the majority were in the 11-40 km/h band.

Injury severities of MAIS 3+ to restrained occupants are likely to result from impacts in excess of 45 km/h. (△V).

Assessment of injury severity shows the head and chest to be the most vulnerable areas with contact to the steering wheel featuring highly as an injury mechanism - either as direct contact or when the wheel itself has intruded into the passenger compartment. Contact with the 'A' pillar and by the other vehicle in the accident also produced a number of injuries. It is reasoned that seat belt pre-tensioners could greatly assist in preventing contact or intrusion-related injuries as described.

Footwell intrusion was responsible for 51 per cent of lower limb injuries. Although not life threatening, this type of injury could lead to prolonged recovery times and permanent disability in later life.

The effect of head restraints in reducing neck sprains amongst restrained occupants in frontal collisions requires monitoring as the introduction of such restraints increases.

Side Impacts

To reproduce the characteristics of side impacts of all severities of injury, a test would be perpendicular to the side of the car involving A and B pillars at a △V of 21-28 km/h.

To reproduce fatal side impacts a test could be angled 30° forward of perpendicular, with an overlap of 20 per cent of the car length between B pillar and wheelbase center at an average △V of 51 km/h. This represents a balance between fatal impacts of much lower speeds and very severe impacts in which the car was completely destroyed. It is not feasible to build in protection for the latter but might be realistic to attempt to prevent fatalities that occur at the lower edge of the speed range.

Within the population studied, restrained non-struck side occupants were less severely injured than restrained struck side occupants. They did, however, tend to be involved in impacts with a lower △V.

The use of 'harm' measures suggests that priority should be given to the reduction of struck side occupants chest and head injuries, non-struck side occupants head injuries and struck side occupants abdomen injuries, in that order.

The most common contacts associated with struck side occupants AIS 3+ chest injuries are: side door (66 per cent), often supported by an intruding external object.

The most common contact associated with struck side occupants AIS 3+ head injuries is an intruding external object (51 per cent). The most common contacts associated with non-struck side occupants AIS 3+ head injuries are: door (24 per cent), other occupant (20 per cent).

The most common sources of struck side occupants AIS 3+ abdominal injuries are the side door (40 per cent), including those supported by an intruding external object (28 per cent).

Concluding remarks

Further analyses of the growing database derived from real-world accidents will enable detailed information to be presented to design engineers. This will take many forms and greatly assist legislators and others responsible for proposed type-approval test procedures.

This Paper represents a significant analysis of a comprehensive crash injury database which in turn represents the efforts of many people. Other analyses will doubtless be refined in the future with many detailed studies of particular aspects of the total problem.

Acknowledgements

A study as complex as injury investigation could not take place without the close co-operation of many people. The authors are indebted to the following:

Staff of the Accident Research Units at Birmingham and Loughborough.

Department of Transport Vehicle Examiners.

Co-sponsors Gaydon Technology and Associated Companies and also the Ford Motor Co.

The Chief Constables and staff of the Police Forces in whose areas the study has taken place.

The consultants and staff of participating hospitals.

H.M. Coroners and their officers.

Support staff associated with the ARU's and TRRL.

In particular the efforts of Jeff Meades and Yomi Otubushin. The former for his involvement with the project as a whole and particularly the computerized database. The latter for his assistance for preparing the side impact section of this paper.

References

1. Compulsory Seat Belt Wearing. Report by the Department of Transport. HMSO, London. October 1985.
2. Rutherford W. H. et al. The Medical Effects of Seat Belt Legislation in the United Kingdom. DHSS Research Report No. 13. HMSO, London, 1985.
3. Road Accidents in Great Britain. Published annually by HMSO, London.
4. MacKay G.M. et al. The Methodology of In-Depth Studies of Car Crashes in Great Britain. SAE Paper 850556. Detroit, 1985.
5. Otubushin Y. and M. Galer. Crashed Vehicle Examination Techniques. SAE Paper 860372. Detroit, 1986.
6. Galer M. et al. The Causes of Injury in Car Accidents—and Overview of a Major Study currently underway in Britain. Proceedings of the 10th Internation Conference on Experimental Safety Vehicles. Oxford, July 1985.
7. Ashton S.J. et al. The Effects of Mandatory Seat Belt Use in Great Britain. Proceedings of 10th International Conference on Experimental Safety Vehicles, Oxford, July 1985.
8. Collision Deformation Classification. Report of the Automotive Safety Committee. SAE Report J 224. March 1980.
9. Crash 3 User Guide and Technical Manual. US Department of Transportation, National Highway Traffic and Safety Administration. Washington, U.S.A.
10. Hobbs C.A. and P.J. Mills. Injury Probability for Car Occupants in Frontal and Side Impacts. Transport and Road Research Laboratory. Report LR 1124. Crowthorne, England. 1984.
11. Malliaris A.C., R. Hitchcock and M. Hansen (NHTSA). Harm Causation and Ranking in Car Crashes. SAE Paper 850090. February 1985.
12. Hobbs C.A., S. Penoyre and S.P.F. Petty. Progress towards to improving Car Occupant Protection in Frontal Impacts. Proceedings of the 11th International Conference on Experimental Safety Vehicles. Washington, U.S.A. May 1987.
13. HOBBS C.A. et al. Development of the European Side Impact Test Procedure and Related Improvements. Proceedings of the 11th International Conference on Experimental Safety Vehicles. Washington, U.S.A. May 1987.

 The work described in this Paper forms part of the programme of the Transport and Road Research Laboratory and the Paper is published by permission of Director.

INJURIES CAUSED BY SEAT BELT LOADS TO DRIVERS AND FRONT SEAT PASSENGERS

Julian Hill, Steven Parkin and Murray Mackay
Birmingham Accident Research Centre
The University of Birmingham
Birmingham, UK

ABSTRACT

This paper examines how the characteristics of front seat car occupants affected their vulnerability to non-minor injury (Abbreviated Injury Scale ≥2) from seat belt loads, and how the very different environments for drivers and front seat passengers (FSPs) affected injury rates. FSPs suffered belt injuries considerably more often than drivers. Occupants who suffered non-minor injuries from the belt tended to be older in both low and high speed crashes. Weight was seen to be a factor in belt injuries to drivers. Equivalent Test Speeds were calculated showing that torso injuries from belts rose in frequency with crash severity up to 55km/h. Above that speed belts were seen to be significantly less important causes of injuries to FSPs, while other causes were the more significant problems for drivers at 35 km/h and above.

The effectiveness of seat belts has long been established using various methods. These have focused on two areas. First, injury rates and injury severity rates have been compared before and after a legislative intervention. For example, Rutherford et al (1985) following the 1983 introduction of the front seat belt law in the United Kingdom, found in this before and after study that the number of patients taken to hospital was reduced by 15%, and the number of patients requiring admission to hospital wards was reduced by 25%. Within that study was a subsection dealing with fatalities seen at eight coroners' districts, in which it was found that the number of deaths fell by 25.7% when seat belt use in front seats rose from around 38% to some 90%. The second type of study, comparing 0% seat belt use to 100% seat belt use has recently been performed by Evans (1994). He suggested that many studies are fundamentally flawed due to selective recruitment, in that unbelted drivers tend to be riskier drivers, thereby introducing large biases. Evans removed such biases by using a matched pairs technique and found that the seat belt reduced driver fatality risk by 42 ± 4%. His paper also quantifies how seat belt effectiveness for fatality and general injury risk varies with crash severity, in that it approaches 100% effectiveness at low severity crashes, and offers reduced protection as crash severity increases. This point is illustrated in a study by Viano (1992) who considered how effective supplemental restraint devices (such as airbags) would have been in fatal crashes involving lap-shoulder belted occupants. Viano concluded that 68% of the belted occupants could not have been saved by additional technology due to the extreme severity of many of the crashes involving extensive vehicle damage and crush of the occupant compartment.

It is therefore clear that seat belts do not offer complete protection. Various papers also report on injuries produced by seat belt loading directly, although when considering the injuries produced by seat belts, it should be borne in mind that for the general population of occupants and crashes, lack of seat belt use results in higher levels of injury. In the great majority of cases, injuries from seat belts themselves are minor. That was the situation reported by Hill et al (1994) who found that 29.6% of belted front seat occupants on the United Kingdom Co-operative Crash Injury Study (CCIS) database sustained injuries from belt loads rated one on the Abbreviated Injury Scale (AIS). In contrast 7.1% of occupants sustained AIS ≥ 2 injuries from belt loads. Otte et al (1987) in a German study, found that in frontal impacts the seat belt was the contact source for 14.6% of the drivers' injuries, and 20.5% of the front seat passengers' injuries. They also found that the seat belt was the contact for 6% of struck side occupants' injuries, and 15.4% of the non-struck side occupants in lateral crashes. Hill et al (1994) reported that seat belts caused AIS ≥ 2 injuries in 14.6 % of fatally injured occupants, 9.9% in seriously injured occupants, and 2.4% in slightly injured occupants (severity determined according to the British Government Scheme described by H.M.S.O., 1991). Huelke et al (1993) reported on several case studies of intraabdominal injuries suffered by belted front seat occupants involved in frontal impacts with little intrusion and presented an extensive bibliography of clinical cases where seat belts have caused injury. The causes of the injuries were thought to be due to positioning of the belt on the belly wall, submarining, or inertial loading of the organs. Rib fractures from the seat belt were also cited with associated trauma to the liver or other upper abdominal organs. Huelke et al (1992) also reported on cervical fractures and fracture dislocations without head impact sustained by belted occupants, and thoracic and lumbar spine injuries associated with the lap-shoulder belt. These types of injuries are rare, as illustrated by the work of Augenstein et al (1995) who examined the frequency of different types of chest and abdominal injuries suffered by belted occupants contained on the National Highway Traffic Safety Administration's (NHTSA) National Accident Sampling System, Crashworthiness Data (NASS/CDS 1988-1992). The analysis showed that over three quarters of the injuries were to the skin or muscle and that over half of the remaining injuries were to the skeletal system (excluding the vertebrae). Pulmonary injuries occurred at a rate of 1.56 in every 100 injuries. Similarly the figures given for some other regions were: liver (1.05/100), kidneys (1.05/100), spleen (0.67/100), heart (0.89/100), spine (0.03/100), vertebrae (0.10/100).

This paper sets out to investigate injuries directly caused by seat belt loads to car front seat occupants. The paper addresses specific injury details and how they were influenced by occupant characteristics. Of particular interest were the two very different crash environments occupied by drivers and front seat passengers (FSPs).

MATERIALS AND METHODS

CRASH DATA - The Co-operative Crash Injury Study (CCIS) database was used for this analysis (Mackay, 1985; Hassan et al., 1995). This national study has been underway in the UK since December 1983. In-depth vehicle examinations are made at garages or scrap yards several days after the crash occurs. Medical details of car occupants are obtained from hospital records, coroners reports and questionnaires sent to survivors. Crashes are sampled according to a plan that ensures that a full range of accident injury severities are obtained equally from crashes occurring in rural and urban areas. The sample plan is weighted towards fatal and serious injury cases but also includes slight and damage only cases as defined by H.M.S.O. (1991).

The current analysis uses data for accidents that occurred from April 1992 to December 1995. Cases from before 1992 are not presently compatible with these later cases. From the entire data set (as defined above) a sub-sample of seat-belted front occupants was selected for this study as defined below.

FRONT SEAT OCCUPANTS - This study involves occupants selected according to the following criteria:-

1. Drivers and outboard-seated FSPs,
2. Seat-belted using a three-point (lap & diagonal) inertia-reel system,
3. Appropriately belted: occupants were excluded if known to be using belts in an inappropriate manner, if they were children, or if a child restraint system was used. After consideration of work by Stalnaker (1993), children were defined as occupants whose mass was < 35 kg, or if mass unknown < 13 years.

BELT INJURIES AND BODY REGIONS- Belt injuries were defined as injuries occurring under the lie of the seat belt webbing or buckle. Thus induced injuries from inertial effects, such as soft tissue neck sprains occurring because of non-contact neck loads are specifically excluded from this analysis. The torso was defined as the region at risk from belt injuries. It included neck, shoulders, thorax, thoracic spine, lumbar spine, abdomen and the lap region. Many injuries were AIS 1 and reported via questionnaires sent back from occupants. It was often impossible to determine if these injuries occurred at abdomen, pelvis or thigh level. Consequently, all such injuries were classified into the "lap region".

This paper concentrated on belt injuries only. While vehicles are increasingly fitted with airbags, belt pretensioners and grabbers they only constitute a small proportion of the current database. Such cases were not excluded from this study, but an investigation into effects due to supplementary technologies was beyond the scope of the present database.

OTHER METHODOLOGY - Frontal impact severities were rated via CRASH3 in terms of Equivalent Test Speeds (ETS; Mackay et al., 1985). Injuries were rated according to the Abbreviated Injury Scale (AIS; AAAM, 1990), with each occupant's overall severity represented by the Maximum AIS (MAIS).

Non-parametric statistical tests (chi-squared and median tests) were performed where appropriate. Results were said to be significant when tests gave probabilities (p) < 0.07. Significance levels and degrees of freedom (df) are given. Comparisons are made between occupants with MAIS 0-1 injuries and MAIS ≥2 injuries for various characteristics such as sitting position, age and gender.

RESULTS AND DISCUSSION

CCIS SUB-SAMPLE - Filtering occupants as described above produced a sub-sample of 2557 drivers and 759 FSPs. Table 1 shows a large age range at all seating positions including several under-age drivers.

BELT INJURIES - 721 drivers (28% of drivers in the sub-sample) and 294 FSPs (39%) sustained belt injuries rated AIS ≥1. That was a significant difference (df = 1, $p<0.07$). Body regions injured were largely those "intended" to bear

Table 1 - Characteristics and frequencies of Belted, Front Seat Occupants

Type of Occupant	Gender	n	%	Age (years) min.	median	max.	Age not known (n)
Driver	Male	1696	68%	14	33	89	188
	Female	788	32%	16	34	87	48
	Not known	73	-				68
	Overall	2557	100%	14	33	89	304
FSP	Male	309	41%	11	24	88	0
	Female	449	59%	9	35	92	2
	Not known	1	-				0
	Overall	759	100%	9	29	92	2

Table 2 - AIS ≥1 Belt Injury Rates by Body Region

Body Region	Drivers	FSPs	df & p for significant differences between occupant types
	(n=721)	(n=294)	
Neck	2.2%	3.1%	-
Shoulders	23.8%	20.8%	-
Thorax	77.2%	81.6%	-
Lap Region	29.5%	42.6%	df = 1, $p < 0.07$
Thoracic and Lumbar Spine	0.4%	1.4%	-

Table 3 - AIS ≥2 Belt Injuries

Injury	Drivers (n)			FSPs (n)		
	AIS			AIS		
	≥2	≥3	≥4	≥2	≥3	≥4
Neck						
nerves	1			-		
fracture or dislocation	1			3		
other c. spine injury	1			-		
Shoulders						
surface	-			1		
fracture or dislocation	21			14		
other bony or joint injuries	2			-		
Thorax						
aorta	1	1	1	-		
heart	1	1		1	1	1
haemo/pneumothorax	5	5	1	4	4	1
other internal organ injuries	8	7	3	4	4	1
sternum fractures	67			39		
2 to 3 rib fractures	9	1		9		
4 or more rib fractures	2	2	1	7	7	3
other skeletal injuries	2			1	1	
Abdomen						
bladder	2			-		
liver	3		2	2		1
spleen	5		1	1		
other internal organs	7		1	5		2
Thoracic spine fracture or dislocation	2		1	2		
Lumbar spine fracture or dislocation	1			1		

seat belt loads: shoulders, thorax and lap (Table 2). FSPs had many more injuries in the lap region (df = 1, $p<0.07$).

Injuries rated AIS ≥2 (non-minor injuries) were of greater concern (Table 3). The sternum was the most frequent fracture site and these injuries make up the greater part of all non-minor belt injuries. Fractured and dislocated shoulders were also found frequently.

IMPACT TYPES, STEERING WHEELS AND INTRUSION- Drivers and FSPs experienced, as expected, very different crash circumstances. A description of how two important factors, intrusion and steering wheel contacts, affected drivers and FSPs differently was desirable before considering specific belt injuries. Intruding structures such as roof, pillars or steering wheels were commonly seen to compromise the space occupied by drivers and passengers. Intrusion was defined as any deformation of passenger compartment or steering system ≥ 1cm toward an occupant. Occupants were more often than not involved in frontal impacts (58% overall, Figure 1). It can be seen that intrusion could not be overlooked in front impacts (particularly regarding drivers,) struck side and roll over impacts.

Frontal Impacts - Considering frontal impacts, it is well recognised that drivers often contact steering wheels whether

Figure 1 - Impact-type Distributions also showing the Occurrence of Head/Torso Injuries on Intruding Structures

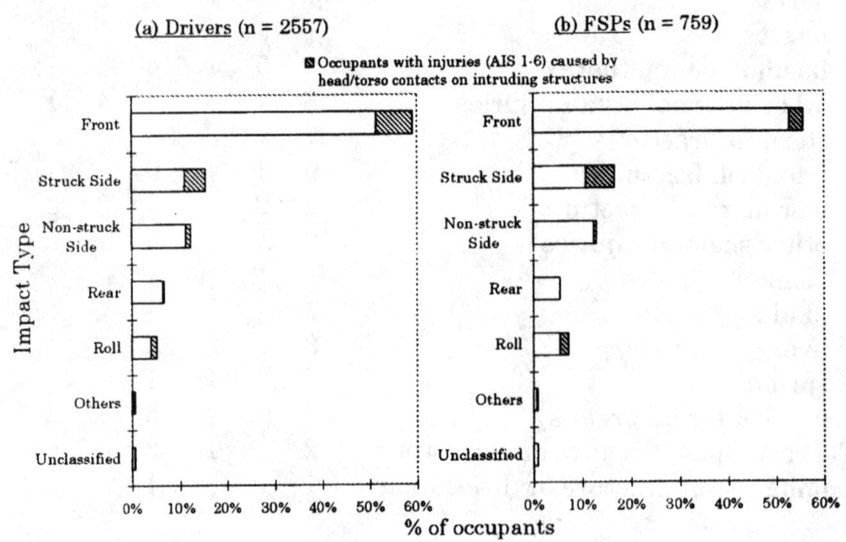

Figure 2 - Driver and FSP Head/Torso Injuries in Frontal Impacts versus ETS.

Rates are shown for:

i) all steering wheel contacts,

ii) all contacts on intruding structures (including intruding steering wheels).

(a) Drivers

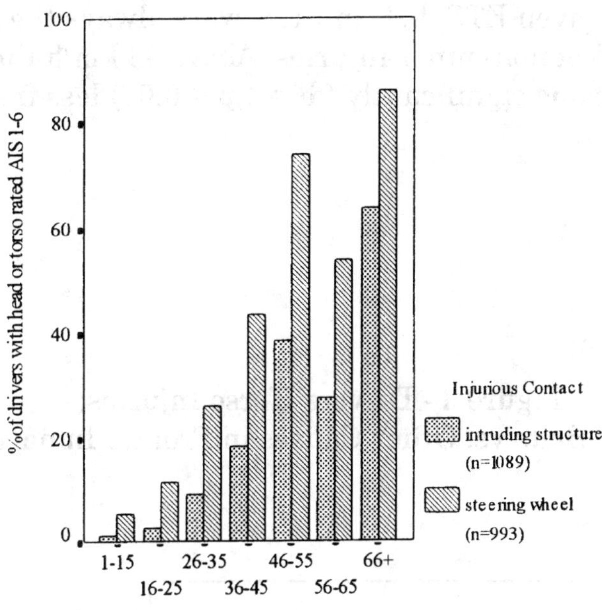

(b) FSPs

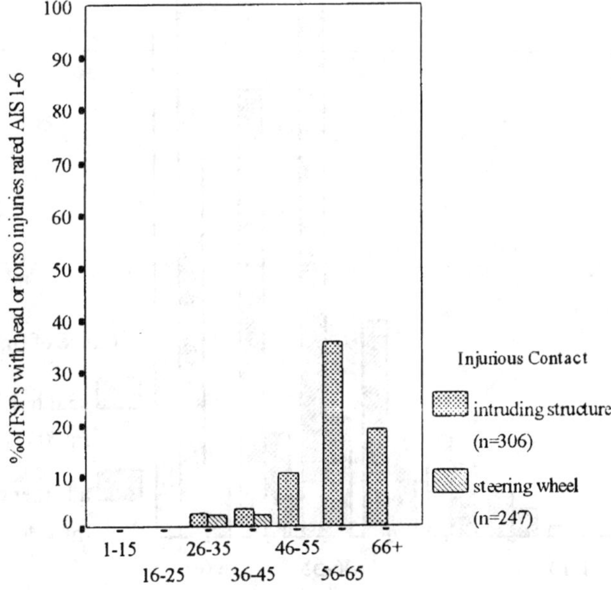

intruding or not. Figure 2a shows how steering wheels and intrusion (of various structures including steering wheels) came into contact with drivers with ever increasing frequency for greater values of ETS. In contrast, FSPs (Figure 2b) made far fewer such contacts with intrusion and, of course, hardly ever with steering wheels.

Belt injuries to drivers rated AIS ≥2 were compared with torso injuries from all other causes in frontal impacts (Figure 3.) For any given ETS, belt injuries were always the least likely cause of non-minor injuries. Above 35 km/h these injuries became significantly (df = 1,p < 0.07) less frequent

Figure 3 - Drivers' Torso Injuries:
Seat Belts vs. Other Causes in Frontal Impacts

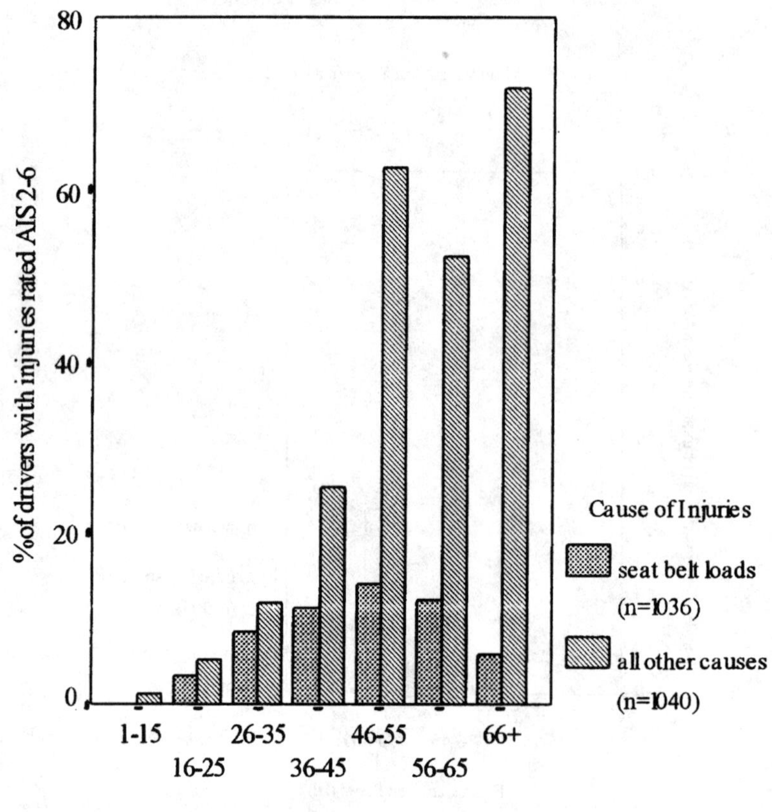

than other causes of injuries. Steering wheels were most likely to be involved in the cause of serious torso injuries with increasing ETS (Figure 4.)

For FSPs, the situation was seen to be very different. With less intrusion and virtually no steering wheel interactions, belt injuries appeared to dominate other injuries up to approximately 55 km/h (Figure 5). Other causes of injuries were significantly (df = 1, $p < 0.07$) more important than seat belt injuries above 55 km/h. It has already been noted that belt injuries were less important for drivers than steering wheel contacts above 35 km/h. These results show that seat belt loads are the greater problem for FSPs in the 35 to 55 km/h range of ETS while their drivers are more likely to be injured by steering wheels. This explains why the AIS ≥2 belt injury rates were 7.3% for drivers and 19.3% for FSPs over all front impact severities. The considerable difference in belt injury rates for drivers and FSPs was also due to their different characteristics.

Figure 4 - Drivers with Torso Injuries Rated AIS ≥2 Showing Steering Wheel Involvement in Frontal Impacts (n=173)

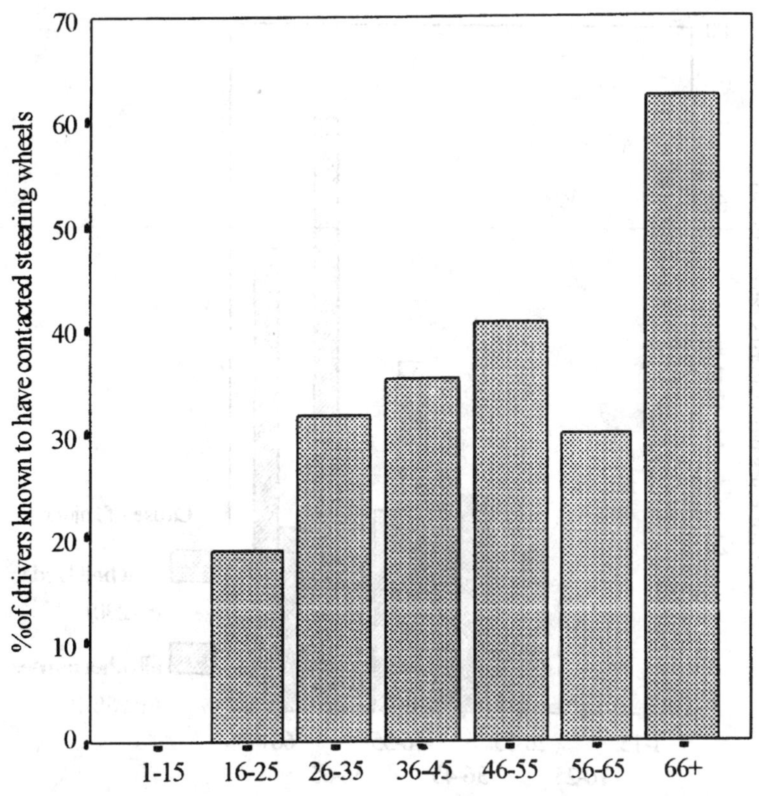

OCCUPANT CHARACTERISTICS IN FRONTAL IMPACTS - To address how occupant characteristics affect belt injury outcome, occupants known to have made head or torso contacts with intruding structures were excluded from the analysis. Seat belt performance will be least affected if only cases with no intrusion are considered. A small number of cases were noted with potentially confusing circumstances such as when slack had been introduced into the seat belt webbing, or when interaction occurred between occupants or luggage. While such cases would form an important part of a further detailed study, it was decided to exclude them before making further analyses in this study. Results were given for frontal impacts, the largest and most important group.

Figure 5 - FSPs' Torso Injuries:
Seat Belts vs. Other Causes in Frontal Impacts

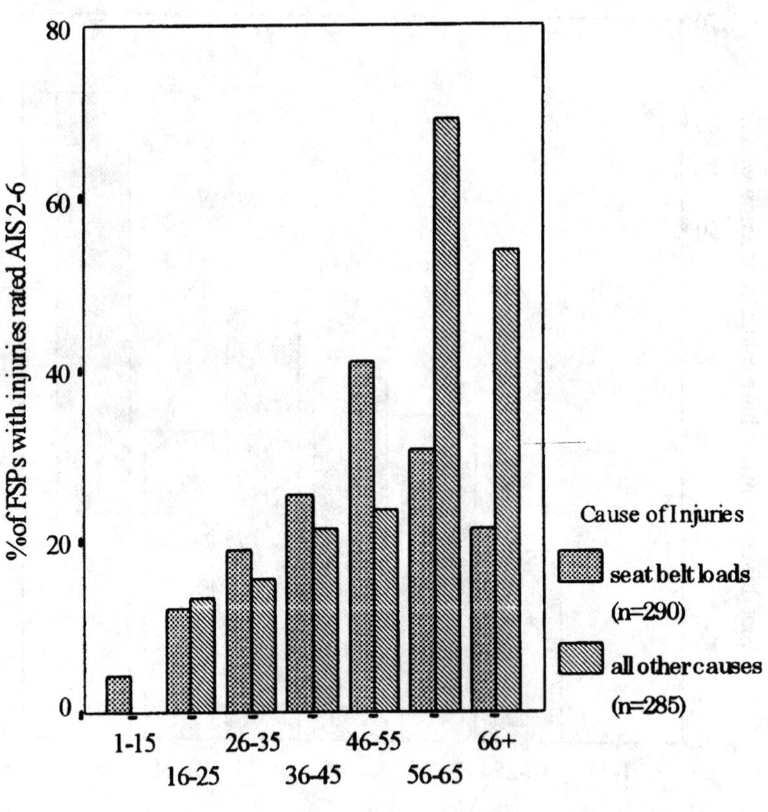

Table 4 - Median Age for Occupants in Frontal Crashes and ETS < 35 km/h

		Median Age (years)		
		Whole body, any cause MAIS 0-1	Belt Injuries AIS ≥2	df & p for significant difference between medians
Male	Driver	34	58	df = 1, p < 0.07
	FSP	28	55	-
Female	Driver	35	39	-
	FSP	32	68	df = 1, p < 0.07

Occupant Age - Details were available for age, for occupants with belt injuries, by gender and seating position. Frontal impacts were banded into ETS < 35 km/h (Table 4) and ETS ≥ 35 km/h (Table 5).

Injury severities were classified between AIS 0-1, and AIS ≥2. Occupants sustaining only MAIS 0-1 were included with injuries due to any cause and at any body region. Those occupants were compared with occupants sustaining AIS ≥2 belt injuries. Two initial observations were made concerning the median ages of occupants in the AIS 0-1 groups. Firstly, male and female FSPs were younger than drivers. Secondly, higher speed impacts involved younger occupants with the exception that female drivers had a median age of 35 years in both low and high speed crashes.

Considering belt injuries rated AIS ≥2, age clearly had a strong positive relationship (Table 4 and Table 5). Median ages for belt injured people were significantly (df = 1, p < 0.07) older for the following:

- all male drivers,
- male FSPs in crashes at ≥ 35km/h,
- all female FSPs.

Table 5 - Median Age for Occupants in Frontal Crashes and ETS ≥ 35km/h

		Median Age (years)		
		Whole body, any cause MAIS 0-1	Belt Injuries AIS ≥2	df & p for significant difference between medians
Male	Driver	29	36	df = 1, p < 0.07
	FSP	22	39	df = 1, p < 0.07
Female	Driver	35	46	-
	FSP	23	38	df = 1, p < 0.07

The median for female FSPs sustaining non-minor belt injuries was the greatest at 68 years in low speed crashes, and that was considerably older (36 years) than median for the minor injured group, 32 years. The higher ETS band allowed a greater range of impact severity, and it was most likely that increased crash severity would alone have been the dominant cause of injuries in a number of impacts.

Occupant Weight - Drivers with belt injuries (AIS ≥2) were lighter ($df = 1$, $p < 0.07$, Figure 6a). That was not so for FSPs (Figure 6b). The result for drivers was unexpected because light occupants will experience lower belt loads than

Figure 6 - Occupant Weight Distributions In All Types of Impact

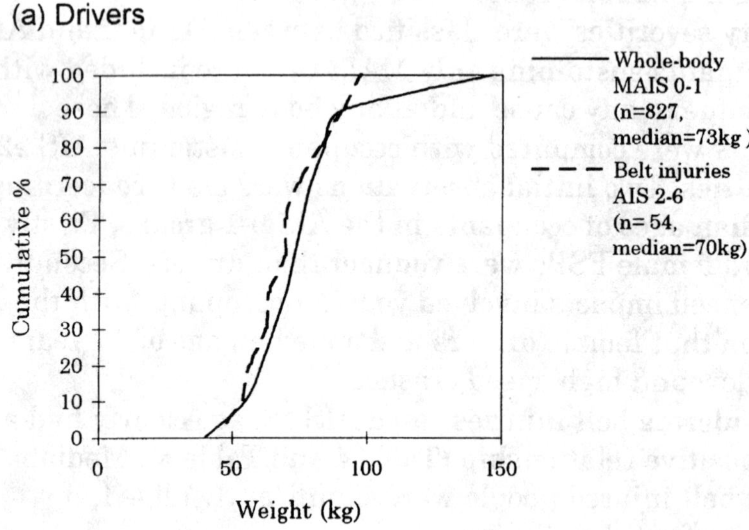

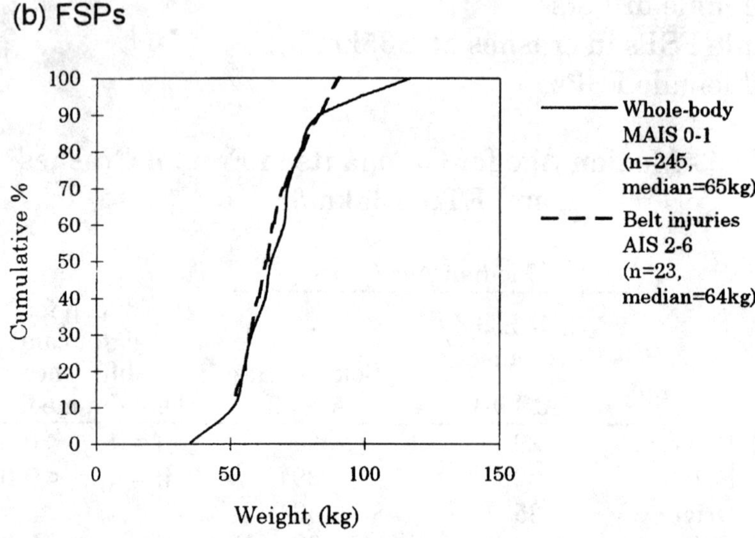

heavy occupants all other factors being equal. A detailed examination revealed these drivers to be mostly females and some ten years older on average. Of course females will typically be lighter in weight, and age was clearly related to risk of injury. Clearly a simple univariate observation, such as this based on weight alone, might be open to misinterpretation. However weight alone is the fundamental occupant characteristic influencing belt webbing loads. These distributions for belt-injured drivers and passengers centred at 70 and 64kg respectively, may describe optimum design parameters for occupants represented by this study to be at risk, and it is interesting to note how, for drivers but not passengers, that requirement may differ from the minor injury population.

CONCLUSIONS

Numerous prior studies have looked at the overall effectiveness of current seat belts and many have addressed the injuries which occur at the limiting condition, i.e. when a belted occupant is injured, what are those injuries? This study examines some of the characteristics of the occupants themselves which have a bearing on injury outcome. Thus age, gender and body weight are variables included in this analysis.

To get such data requires an in-depth database such as is provided by the CCIS procedures. This particular study uses age and gender as a surrogate for biomechanical tolerance variation and considers how these background factors affect outcome.

We show that gender differences between drivers and FSPs are important and that age particularly has a significant influence on injury outcome. This is especially true in moderate speed frontal crashes.

By limiting our analysis to injuries specific to the seat belt itself we illustrate how FSPs are injured from the belt more frequently in comparison to drivers because of steering wheel interaction.

Overall, FSPs suffered belt injuries at over twice the rate seen in drivers. Occupants who suffered serious injuries from the belt tended to be older than those who didn't, in both low and high speed crashes. Occupant weight was seen to be a factor relevant to belt injuries in drivers. Occupant weight and age distributions do not directly explain variations in tolerance to injuries from belt loads, but median values as given here currently offer the best surrogate parameters for biomechanical tolerance.

Torso injuries from belts were seen to rise in frequency with crash severity up to 55km/h. Above that speed belts were seen to be relatively less important causes of injuries to FSPs. The steering wheel was another important source of injuries for drivers. That and the greater rates of injuries associated with compartment intrusion seen at the drivers' seating positions were considered to be the main reasons why belt injuries were a significantly less frequent problem for drivers at speeds above 35 km/h.

A consequence for the design of restraint systems is to suggest that there is merit in having different characteristics for the restraints fitted for drivers in comparison to passengers. Passengers are more frequently female. Passengers who are seriously injured are female more frequently and older in comparison to drivers. The absence of the steering wheel allows larger ride-down distances and intruding structures for FSPs do not feature as a frequent injury source. Thus having "softer" belts for passengers in comparison to drivers would be beneficial.

Such specific tuning of restraint systems for individual sitting positions in the car is however just a precursor for intelligent restraints which conceptually could adjust to more specific disaggregation of occupant variability such as age, gender, weight and head position. The technology to achieve that is some few years away and as an interim measure, this study suggests that at least designing for the specifics of the two front seat positions and their occupants could have advantages.

ACKNOWLEDGEMENTS

The Co-operative Crash Injury Study is funded by the Department of Transport, Ford Motor Company Limited, Nissan Motor Company, Rover Group, Toyota Motor Company and Honda Motor Company. The project is managed by the Transport Research Laboratory.

REFERENCES

Association for the Advancement of Automotive Medicine. The Abbreviated Injury Scale (AIS) 1990 Revision. Des Plaines, IL, 1990.

Augenstein, J.S., Digges,, K.H., Lombardo, L.V., Perdeck, E.B., Stratton, J.E., Quigley, C.A., Malliaris, A.C., Byers, P.M., Nunez Jr, D.B., Zych, G.A., Andron, J.L., Craythorne, A.K and Young, P.E.; Chest and abdominal injuries suffered by restraint occupants; SAE International Congress and Exposition: Advances in occupant protection technologies for the mid-nineties: 37-44, 1995.

Evans, L.; Safety-belt effectiveness: The influence of crash severity and selective recruitment; Porch. 38th AAAM: 25-42; 1994

Hassan, A. M.; Hill, J. R.; Parkin, S.; Mackay, G. M. Secondary Safety Developments : Some Applications of Field Data. Autotech 1995; I Mech E; 1995.

Hill, J.R., Mackay, G.M and Morris A.P.; Chest and abdominal injuries caused by seat belt loading; Accident Analysis and Prevention, Vol. 26 (1): 11-26, 1994.

H.M.S.O. Road Accidents Great Britain 1990, The Casualty Report. London, U.K.; 1991.

Huelke, D.F., Mackay, G.M., Morris A.P. and Bradford, M.; Cervical fractures and fracture-dislocations without head impacts sustained by restrained occupants; Proc. 36th AAAM: 1-23; 1992.

Huelke, D.F., Mackay, G.M. and Morris A.P.; Intraabdominal injuries associated with lap-shoulder belt usage; SAE International Congress and Exposition: Frontal impact protection: Seat belts and air bags: 39-47, 1993.

Mackay, G. M.; Galer, M. D.; Ashton, S. J.; Thomas, P. D. The Methodology of In-depth Studies of Car Crashes in Britain. International Congress and Exposition; 1985; Detroit, Michigan, U.S.A. : S.A.E. 365-390; 1985.

Otte, D., Sudkamp, N and Appel, H.; Variations of injury patterns of seat-belt users; SAE International Congress and Exposition: Restraint: Technologies: Front seat occupant protection: 61-71; 1987.

Rutherford, W.M., Greenfield, T., Hayes, H.R.M. and Nelson, J.K.; The medical effects of seat belt legislation in the

United Kingdom; HMSO Research Report No. 13:ISBN 0113210396. 1985.

Stalnaker, R. L. Inconsistencies in State Laws and Federal Regulations Regarding Child Restraint Use in Automobiles. Child Occupant Protection. S.A.E.; SP-986 933087: 51-69; 1993.

Viano, D.C.; Crash injury prevention: A case study of fatal crashes of lap-shoulder belted occupants; Proc. 36th Stapp Car Crash Conf: 179-192; 1992.

ASSESSMENT OF MEASURES REDUCING RESIDUAL SEVERE AND FATAL INJURIES MAIS 3+ OF CAR OCCUPANTS

Dietmar Otte
Accident Research Unit, Medical University Hannover
Germany
Paper Number 96-S4-W-18

ABSTRACT

The Accident Research Unit at the Medical University Hannover carried out an in-depth investigation of road accidents in a statistical representative manner. Each year approx. 1000 accidents are collected on order of BASt. These data are a good tool for analysing the accident scene of severe car crashes, answering the question, how car crashes of severe and fatal injury occurence look like. While many safety measures for car occupants and test procedures for car developments exist, 1/3 of all severe and fatal injured persons in road accidents are car occupants. 5.5% of all car occupants involved in road accidents suffered injury severities of more or equal than MAIS 3, but it must be the aim to avoid these victims. For this purpose the accident situations were analysed in detail.

In the study the injury mechanisms are described, measures reducing these residual severe and fatal injuries assessed and demands for test procedures and car design formulated with regard to an optimized occupant protection.

INTRODUCTION

The amount of fatally injured car passengers has steadily reduced during the past years (figure 1). 8,989 killed car passengers were registered in Germany in 1970, compared to only 3,974 in 1994 (StBA-1). This means a reduction of 56%. The amount of severely injured has also reduced by 49% during the same period of time, but the amount of slightly injured has only reduced by 14%.

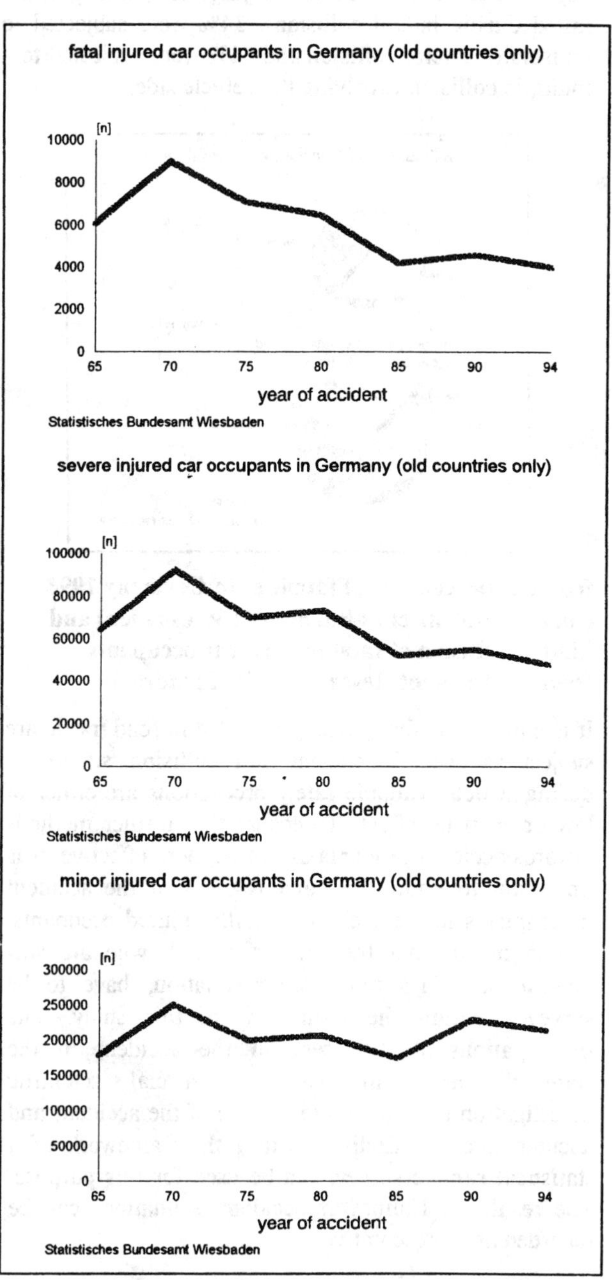

figure 1 injury situation of car occupants in Germany (old countries only) from 1965 to 1994

This has mainly been achieved by the development of vehicle safety measures such as the safety belt, the airbag and the crumple zone, as 50% of all collisions concerning injured passengers are frontal collisions, 32% are lateral collisions and 13% are rear collisions and most of the safety elements come into effect during frontal collisions. But figure 2 shows the collision situations of car drivers killed in accidents and it appears that 65% of all fatally injured car passengers are killed during lateral collisions, 34% were subjected to an isolated lateral collision and 31% were subjected to a multiple collision involving the vehicle side.

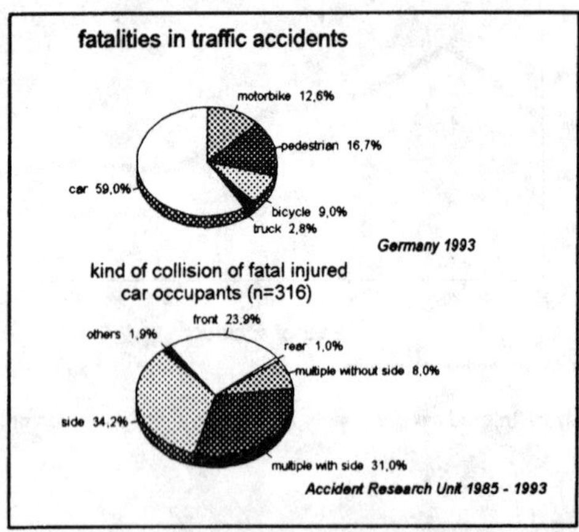

figure 2 percentage of fatalities in Germany 1993 (source: Statistisches Bundesamt Wiesbaden) and kind of collision of fatal injured car occupants (source: Accident Research Unit Hannover)

It is quite likely that passengers killed in road traffic are subject to special accident and collision situations, during which available safety precautions are either of low or even no effect. To ensure that further methods towards accident prophylaxis can be more effective, it is important to know the circumstances of the accident mechanisms for severely and fatally injured occupants. Which means that the severly injured, who are still present in todays road traffic situation, have to be surveyed within the framework of this study. The investigations at the scene of the accident in the Hannover area, in which a special scientific investigation team drives to the site of the accident and documentates its findings within the framework of a statistical sample survey, can be used for this purpose. The results of Hannovers accident evaluations can be regarded as representative.

THE INVESTIGATION METHOD

Since 1973, traffic accidents in the greater district of Hannover have been registered on the site and vehicle damages as well as injuries have been stored in a data bank, by order of the German Federal Highway Research Institute BAST. Since 1985 the accident documentation is based on a statistical random sample and the evaluation is carried out by means of a weighting method (Otte-2). The accidents evaluated in the course of the study were recorded between 1985 and 1994. The Abbreviated Injury Scale (AIS) is used to describe the accident severity (American Association of Automotive Medicine-3). Vehicle deformations are recorded and measured by photographic methods. The vehicle interior is searched for contact and collision points and deformations are represented by means of a computerized matrix system (Otte-4). An extensive reconstruction supplies details of the collision speed as well as the movement behaviour of the vehicles and the occupants. The speed variations caused by the collision delta-v calculated by a mathematical-physical collision analysis is used to assess the forces applied to the passengers. This enables a description of the injury mechanisms as well as the occured load conditions of the documentated cases within the framework of this study.

RESULTS OF THE STUDY

Assessment and Defintion of the Severely and Fatally Injured of MAIS 3+

The killed and severely injured persons found in official statistics are classified according to hospital treatment. According to the statistics, out-patients are regarded as „minor injured", in-patients are „seriously injured" and those who die within 30 days are regarded as „killed". The scientific approach, on the other hand, uses the AIS scaling method. This incorporates a numeric scale of the degrees 1 (minor injured) to 6 (fatally injured) using the graduations

AIS 1	minor
AIS 2	moderate
AIS 3	serious
AIS 4	severe
AIS 5	critical
AIS 6	maximum

This scale can be used for individual injuries as well as for total injuries of the body region and for the total injuries of the person as MAIS. But MAIS does not contain any information concerning the relevant status of the event of death. It is evident that all persons of MAIS 6 injured persons died, but also 41% of MAIS 5, 10% of MAIS 4 and even 2% of MAIS 3 injured persons die as well. A comparison of the official injury severity grades minor, serious and fatal with the AIS scale is possible (Otte-5). Fig. 3 shows a good correlation of about 80%, when

MAIS 1 = minor injured
MAIS 2-4 = seriously injured
MAIS 5/6 = most seriously/fatally injured

are defined. A 95% probability for exclusively serious and most serious injuries is possible by use of the injury severity degree MAIS 3 and above, described as **MAIS 3+** during the following.

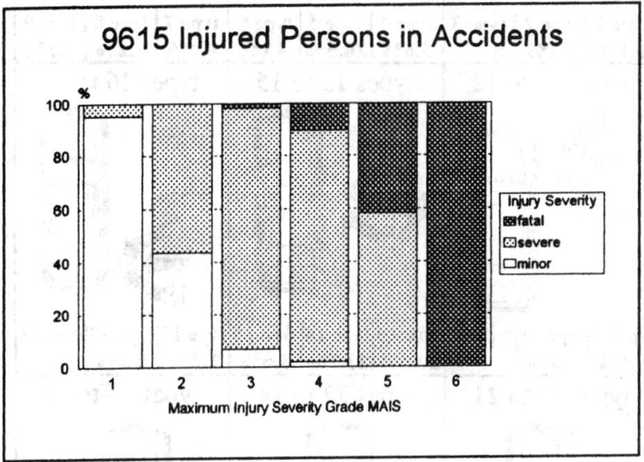

figure 3 correlation of MAIS and official injury severity grades

8,500 accidents were documentated within the framework of the Accident Investigations of Hannover during the 10 years between 1985 and 1994, these included 6,908 with car participation. 563 car passenger accidents of MAIS 3+ were registered. A total of 691 car passengers with MAIS 3+ injuries were recorded.

Fig. 4 shows the MAIS distribution of all injured car passengers, 80% suffered MAIS 1 (minor injured), only 5.5% of the involved suffered MAIS 3+ injuries, 25% of the MAIS 3+ patients died..

	total	fatalities
total	5165	70
maximum AIS		
MAIS 1	80.0%	0.0%
MAIS 2	14.5%	0.0%
MAIS 3	3.2%	3.4%
MAIS 4	0.9%	15.6%
MAIS 5	0.7%	52.5%
MAIS 6	0.7%	100.0%
MAIS 3+	5.5%	25.0%

figure 4 frequency of injury severity MAIS and percentage of fatalities for different MAIS grades

Population of the MAIS 3+ occupants

2/3 of the MAIS 3+ patients were drivers (66%), 21% were front-seat passengers and 10% were rear-seat passengers. 69% used a seat belt. 37% were up to 25 years of age, only 14% were older than 55. Which shows that high age, that leads to a reduced endurance limit, is not responsible for the severe injuries of car passengers, but rather that the corresponding accident situation causes the resulting injury. A comparison of the ages of all car passengers, including the non-injured, shows that 32% are up to 25 years of age and 12% are older than 55. But 43% of the MAIS 3+ patients had a height of more than 175 cm in relation to 30% of the MAIS 0 to 2 persons.

The accident and collision situation of the injured occupants MAIS 3+

25.5% of the cars with MAIS 3+ injured persons collided with another car, 22.9% collided with a pole and 36.8% suffered multiple collisions. When a closer look is taken at the multiple collisions in regard to the collision opponent (figure 5), car and pole collisions make up 2/3 of all collision situations of the severely injured. 56 % of the cars collided frontally and 36% laterally. This is a distribution, which is not differ from the situation of all accidents with injured people.

	n	%
total	252	100.0%
collision partner		
car	86	34.2%
truck to 7,5t	18	7.0%
truck > 7,5t	15	6.0%
pole	92	36.6%
other object	33	13.0%
two-wheeler	-	-
pedestrian	-	-
others, unknown	7	2.7%
impact area		
front	143	56.7%
side	90	35.6%
rear	5	1.8%
others	15	5.9%

figure 5 collision partners and impact areas of cars with MAIS 3+ occupants

The following collision types were established (figure 6 and figure 7). In the case of car to car collisions, the angle between the longitudinal axis of both vehicles was established and assigned to the collision points on the cars. For this reason every side of the vehicle is divided into 3 zones according to VDI (Vehicle Deformation Index - 6). The angle of the transmitted impulse was established and determined to the according impact points on the car for the representation of the collision types with poles.

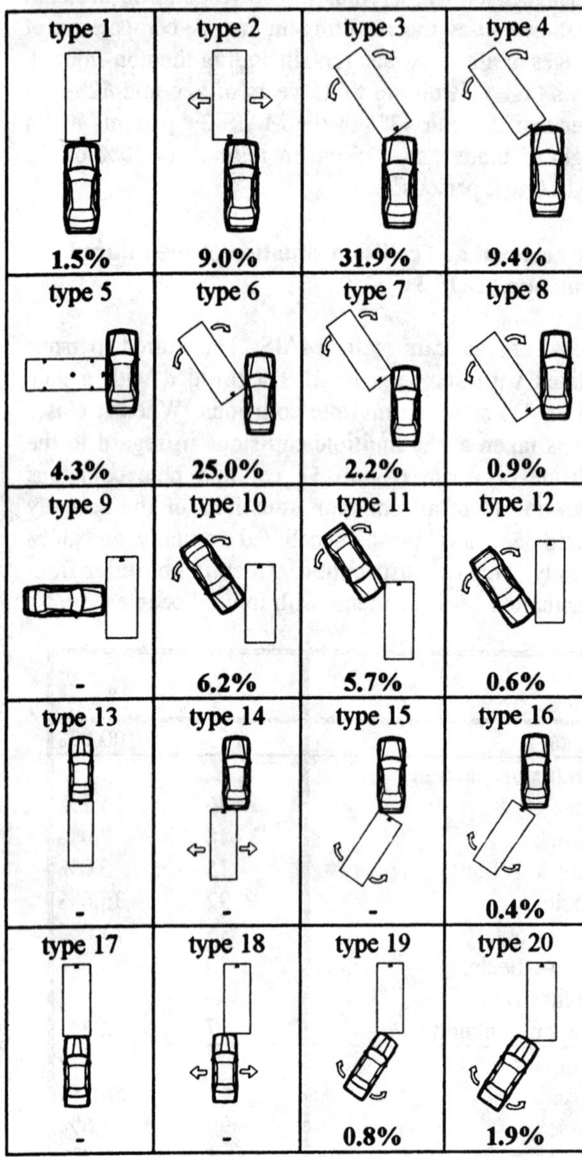

figure 6 definition of collision types in car to car accidents, the blacked car is the MAIS 3+ vehicle

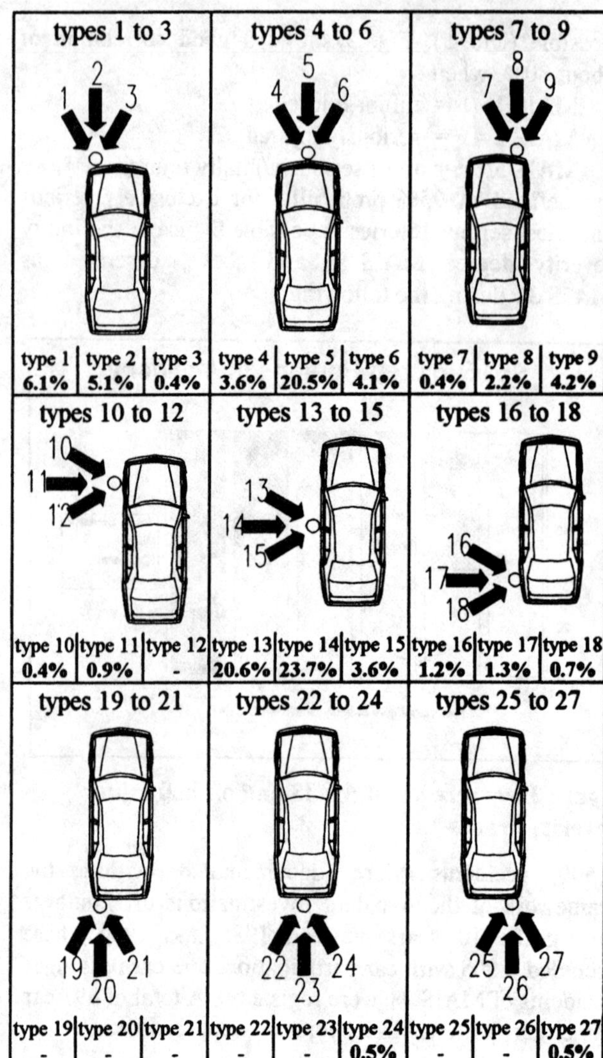

figure 7 definition of collision types in car to pole accidents

3/4 of the most seriously injured MAIS 3+ cases were assigned to frontal collisions at an oblique impact angle of the opposing car, where as in most cases the vehicle fronts partly covered each other, as well as oblique lateral impacts against the compartment area (see the following sketches)

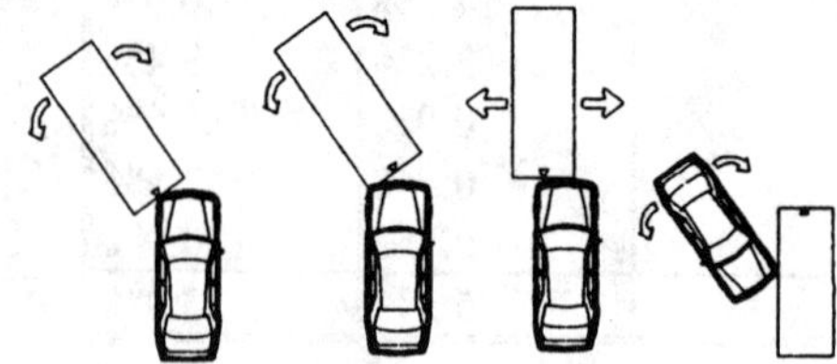

As the main impact constellation with poles, 3 different collision types were established,

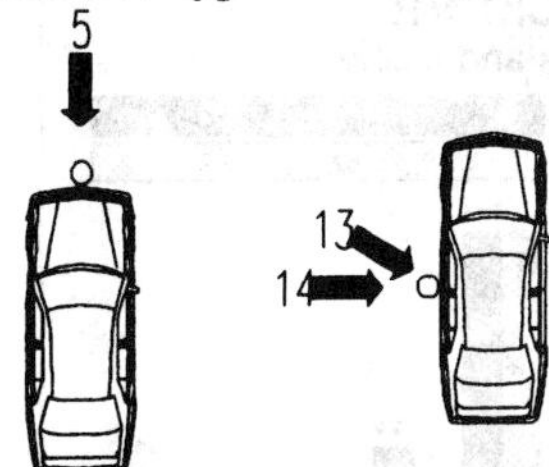

2/3 of the most seriously injured occupants were registered in these 3 main collision types. These are a pole impact against the mid of the car front as well as the rectangular and oblique impact against the lateral part of the compartment.

INJURY PATTERNS OF THE MAIS 3+ CASES

The injury cases with an injury severity grade of MAIS 3+ or higher usually suffered injuries of various body regions. 37% are regarded as so-called polytraumatised, which according to the definition, influences at least 3 different body regions, each injury being so severe that it can influence the occurance of dying (Heberer - 7). The AIS 3+ injuries were located on the head (35.1%), thorax (41.5%), abdomen (20.5%) and legs (38.9%). Figure 8 shows that patients who suffered head injuries of the severity of AIS 3+ also suffered AIS 2 and higher injuries to the thorax (55.2%), abdomen (30.8%) and legs (45.6%). 62.2% of those with serious thorax injuries of AIS 3+ also suffered head injuries of AIS 2 and above. 61.5% of those with leg injuries of AIS 3+ suffered head injuries of AIS 2+ and above. Which shows that head injuries were especially frequent among the MAIS 3+ injured. Only 20% of the passengers suffered no head injuries.

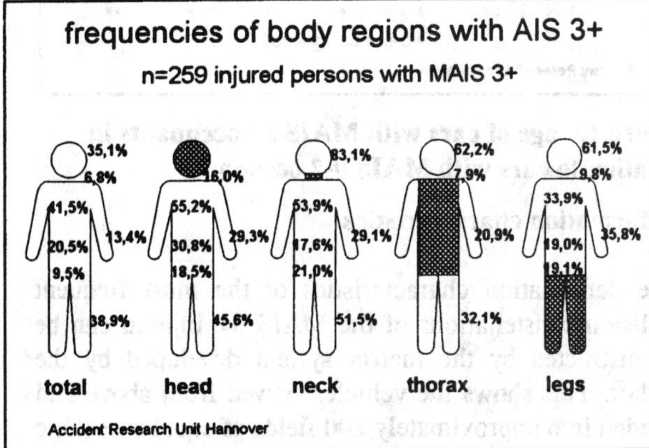

figure 8 frequency of body regions with MAIS 3+ in total (100% all AIS 3+ injuries) and for different body regions with MAIS 3+ (100% each blacked body region)

The most relevant injuries are brain traumata, which suffered 48.9% of the MAIS 3+ patients (figure 9). 21.5% occured fractures of the mid facial bones. It is generally noticed, that 1/5 of the MAIS 3+ patients suffered fractures to the spine. This pattern can be explained as bending and sharing load to the whole body and identify the body movement of the belted occupants, 21.2% of the belted MAIS 3+ patients suffered spinal injuries in frontal collisions.

		impact area	
	total	front	side
total (n)	259	149	89
skull fracture	9.6%	8.6%	10.1%
facial fracture	21.5%	30.1%	7.7%
fracture of base of skull	6.3%	6.7%	4.2%
brain injury	48.9%	44.1%	53.7%
spine fracture	19.9%	21.2%	18.1%
rib fracture (> 3 ribs)	19.0%	16.3%	21.5%
organ injury thorax	23.5%	19.4%	24.3%
intra-abdominal injury	16.7%	14.2%	17.5%
pelvic fracture	16.0%	12.0%	24.2%
organ injury pelvis	0.5%	0.3%	1.0%
closed fracture of upper leg	25.5%	32.1%	19.3%
open fracture of upper leg	4.5%	7.0%	1.1%
fracture of knee	7.1%	10.0%	1.0%
closed fracture of lower leg	9.4%	11.4%	7.2%
open fracture of lower leg	6.6%	8.1%	4.7%
fracture of foot or ankle joint	13.4%	15.6%	8.0%
fracture of upper extremities	25.5%	29.8%	15.8%
fracture of shoulder	9.0%	7.4%	11.9%

figure 9 frequencies of common injuries of MAIS 3+ occupants

In lateral collisions severe brain injuries as well as serial fractures of the ribs, multiple thoracic lesions and pelvis fractures occur very often. In frontal collisions thigh fractures could be established in 32% of the MAIS 3+ patients. This explains itself by the high deformation and load transmission about the dashboard.

BASIC CIRCUMSTANCES OF THE ACCIDENT

Level of accident severity

The data reveals that the passengers of cars with MAIS 3+ injuries are usually prone to extence accident severity (figure 10). The change of speed due to the collision „delta-v“ is an dominant accident severity parameter.

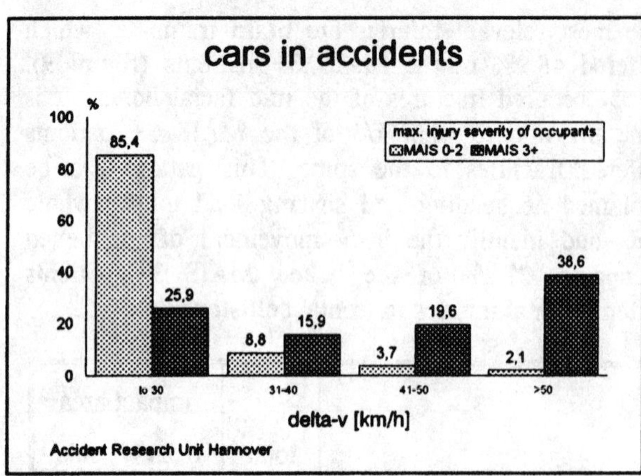

figure 10 distribution of delta-v for cars MAIS 3+ occupants in relation with MAIS 1/2 occupants

Where as accidents which lead to minor injuries have delta-v values of 85.4% up to 30 km/h, raising only by 2.1% at speeds above 50 km/h, only 25.9% of the car passengers with MAIS 3+ injuries had delta-v values of up to 30 km/h, but 38.6% even recorded values above 50 km/h.

An influence of the vehicle weight on the resulting injury severity could not be established. Roughly the same distribution of vehicle crashweight was established for cars with MAIS 3+ injured as for cars with minor injured passengers (figure 11). But as the impulse examination interrelates the mass and delta-v, it is significantly proved that with higher delta-v values the occurrence for resulting injury severity grade is higher too.

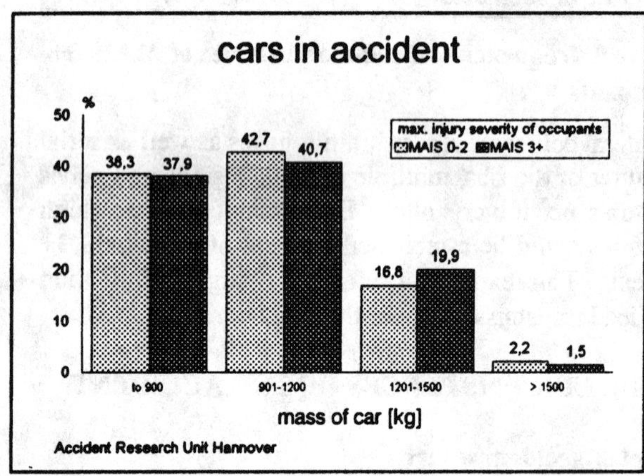

figure 11 mass classes of cars with MAIS 3+ occupants in relation to those with MAIS 0-2 occupants

The mass proportions of the vehicles colliding together were also analysed. It could be established, that for 77.2% of the MAIS 3+ vehicles a mass relation factor above 1 was often calculated (figure 12)

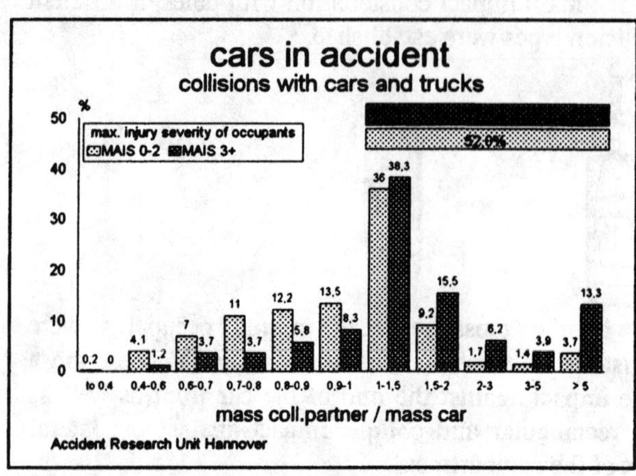

figure 12 mass ration (mass collision partners / mass of car with MAIS) for occupants with MAIS 3+ in relation to MAIS 0-2

An analysis of the age of the vehicle as a possible influential parameter only lead to a slight difference in the comparison of severe and minor injuries. Where as 59.5% of the cars with MAIS 3+ injured were more than 6 years old, 53.4% of the cars with minor injured were of the same age (figure 13).

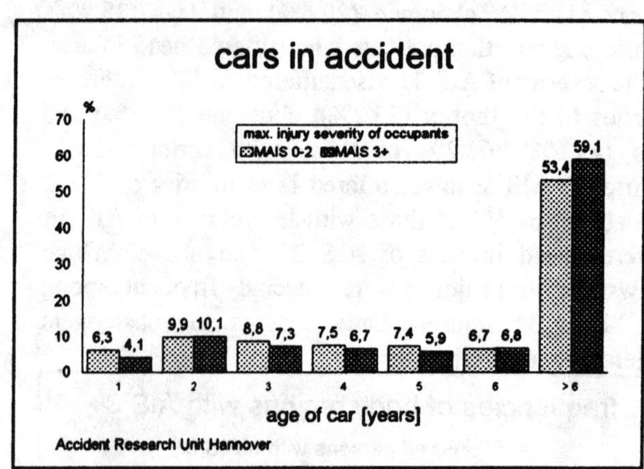

figure 13 age of cars with MAIS 3+ occupants in relation to cars with MAIS 0-2 occupants

Deformation characteristics

The deformation characteristics of the most frequent collision constellations of the MAIS 3+ injured can be reconstructed by the matrix system developed by the author. This shows the vehicle, viewed from above and divided into approximately 200 fields of equal size (Otte - 4). The accident documentation records the deformated zones by true-to-scale marking of the corresponding fields and this reproduction of the deformation pattern of every accident is stored in a computer. This means that it is possible to reproduce the deformation pattern of every case and at free choice by means of the accumulated addition of all the damaged zones as procentual frequency of the damaged zones.

The figures 14 and 15 reproduce the deformation patterns of car to car collisions involving an oblique frontal impact as well as car to pole collisions to the mid of the front as well as involving the passenger compartment in lateral collisions. It can be seen that in frontal collisions the deformation often reaches as far as the passenger compartment. 69% of the cars with MAIS 3+ injured were damaged by an intrusion of the compartment in frontal collisions, 86% in lateral collisions. 1/3 of the lateral impacts lead to deformations of more than 40% of the vehicle breath.

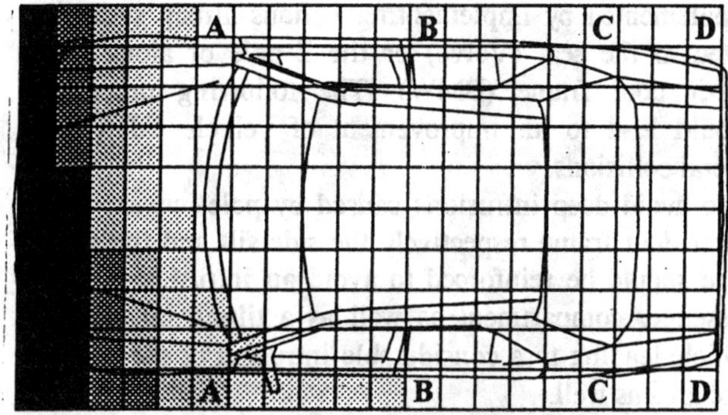

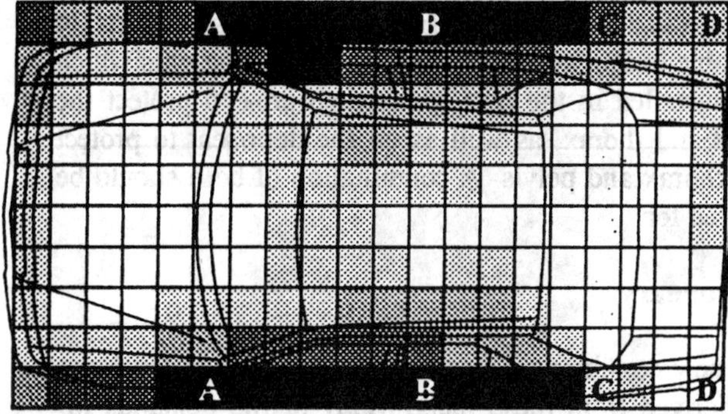

figure 14 <u>car to car accidents</u>, upper matrix collision type 3 and lower matrix collision type 6 - reference figure 6

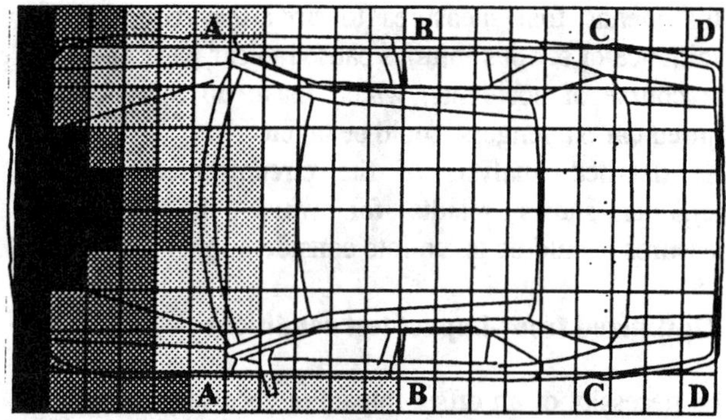

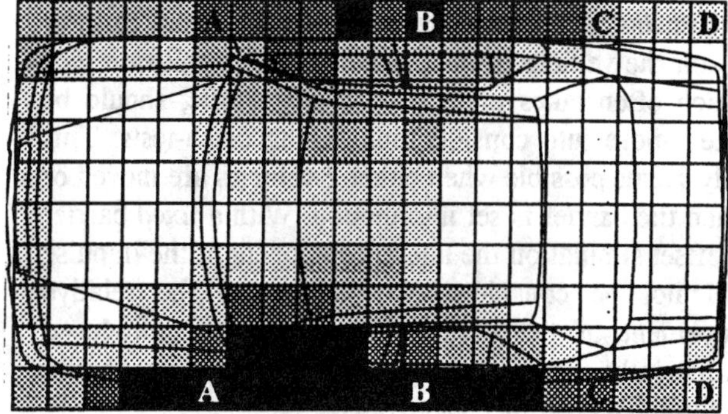

figure 15 <u>car to pole accidents</u>, upper matrix collision type 5 and lower matrix collision types 13 and 14 - reference figure 7

CONCLUSIONS

The detailed analysis of the car accidents, by which a passenger was severely injured with severity grades MAIS 3 and higher (MAIS 3+) shows that special collision situations were always in effect.

These are for instance

- Frontal collisions involving an oblique impact of the opposing car as well as a mid frontal impact with a pole

as well as

- the oblique lateral impact of a car as well as the rectangular and oblique impact of a pole against the compartment side

Deep intrusions as well as high delta-v values are dominant for the accident conditions. MAIS 3+ passengers are often polytraumatised and suffer serious injuries of the head, thorax and legs. The head is almost always injured and this is usually responsible for the resulting trauma consequences. The half of MAIS 3+ patients suffered brain traumata, a quarter died. There is no significant influence concerning the age of persons or vehicles, but 43% of the MAIS 3+ occupants had a body height of more than 175 cm, compared to 30% only in the group of persons with MAIS 0 to 2.

This means that measures towards the reduction or avoidance of serious injuries caused to car passengers in the course of accidents, which make up 5.5% of all injured car passengers, could be taken.
The detailed analysis of the circumstances of the accidents shows results for which the following measures should be taken into consideration:

1 Optimized front-impact test conditions

The necessity of an offset-impact is necessary to reduce the amount of severely injured car passengers. The circumstances of oblique collisions between cars by which the vehicle fronts partly overlap, a collision type which often takes place in everyday traffic, should be taken more into consideration during crash-tests. This only seems possible when either 2 vehicles are moved or when the barrier is set into motion. With a fixed barrier of offset simulation the lateral component of the impulse will not be considered in reality and the relative movement of the occupants will not be reproduced realistically. The seating position of the dummy should be foremost position.

- A crash test car against a pole, which collides mid-frontally, is also necessary. Oblique impulse angles are not necessary in this case.

2 Optimized lateral impact test conditions

- A barrier, moved under an oblique direction against the lateral compartment, should be used for crash-test conditions of a simulated collision of a car against the side of another car, where as the front corner of one car impacts the side of the other car first, which is not a crabbed impact of the whole front against the side as described in the US-test.
- The lower edge of the barrier should be positioned as high as possible in both the EC as well as US test, as the real accident situation shows that the impact height always exceeds the side sill, thus causing a deep intrusion of the passenger door.
- A single test condition for the simulation of lateral car collisions is not sufficient. More attention should be paid to the frequency and severity of object impacts, especially those involving poles.
- To enable accident conform test conditions of an impact with an object, the pole should impact the car between the A and B-pillar of the vehicle directly in front of the pelvis of the passenger. The impact direction should incorporate an oblique component from the front.
- The height of a pole-barrier can be as high as the vehicle from the door-base to the roof.

3 Optimized vehicle safety for frontal collisions

The present safety fittings of the vehicle such as the seat belt, airbag and construction measures of the car seem to be sufficient and are demonstrated in the detailed study. A few optimizations seem sensible
- in the foot region to avoid foot fractures
- in the compartment structure to avoid intrusions of the passenger compartment

4 Optimized vehicle safety for lateral collisions

- Safety precautions for lateral collisions can be supplemented by implementing various airbag systems fitted to the seat (Volvo) or the B-pole or along the upper door frame (BMW). The following proposal should lead to an improvement of vehicle safety in lateral collisions.
- To avoid deep intrusions caused by pole-impacts the lower door frame respectively the side sill and the roof edge should be reinforced to avoid an intrusion of the passenger compartment as well as a tilting up of the vehicle leading to a considerable intrusion of the upper roof area as well.
- Passenger impacts in the region of the compartment sides should be made less severe by installations of suitable airbag systems. These should be systems which are installed in the post/windscreen area to protect the head and thorax, as well as on the door/seat to protect the thorax and pelvis. A combination of both should be strived for.

5 Summary

No new cognitions beyond todays knowledge level can be derived from these facts. Many of the demands for measures of injury reduction mentioned have already been proposed in earlier publications (e.g. Otte-8,9), but they gain more importance by the extensive detailed representation and analysis of the injury and collision situation of the severly injured MAIS 3+ found in todays traffic scene. It can be suggested once again that the two aims

stable passenger compartment
energy absorption

although their safety strategy aspect may differ, should be used mutually in the concept.

The aspects of compatibility should be further pursued, as the study revealed that many of the MAIS 3+ injured were hit by vehicles of larger mass.
Softer front structures of the heavier impacting vehicle could lead to better EES relation.

On the whole, an intrusion must not be regarded as negative, as a reduction of kinetic energy by means of deformation crumpling is needed to reduce impact. But lateral impacts do not offer this facility, which means that injury sources can be found in todays vehicles, caused by relatively uncontrolled characteristics of the deformation leading to sharp-edged parts, as can be derived from the documented accidents. A reduction of injuries could be achieved by a behaviour leading to a defined, wide area surface pressure. The offset and lateral impact test conditions (ETSC - 10) which are to become obligation Europe-wide from 1998 onwards fulfil a few of these demands.

REFERENCES:

(1) StBA
Verkehrsunfälle in der Bundesrepublik Deutschland
Statistisches Bundesamt Wiesbaden, Fachserie 8, Reihe 7, Verlag Metzler-Poeschel, 1994

(2) Otte, D.:
The Accident Research Unit Hannover as Example for Importance and Benefit of Existing In depth Investigations
SAE-Paper 940712, SP 1042, Detroit/USA, 1994

(3) AIS - The Abbreviated Injury Scale - Revision 90
American Association for Automotive Medicine, Morton Grove, Illinois/USA, 1990

(4) Otte, D.:
Deformation Characteristics and Occupant Load Capacity in Lateral Impacts for Nearside Belted Front Car Passengers
SAE-Paper 933126, 37th STAPP Conference, P 269, 185-197, San Antonio/USA, 1993

(5) Otte, D.:
Injury Scaling: from lesion assessment to passive safety improvement
Round Table, Inst. Legal Medicine, University Verona/Italy, 1995

(6) Vehicle Deformation Index
Collision Deformation Classification - CDC
SAE Recommended Practice, SAE Handbook SAE J 224 a, Society of Automotice Engineers, Inc., Warrendale/USA, p. 34-109, 1979

(7) Heberer, G.; Köle, W.; Tscherne, H.:
Lehrbuch der Chirurgie
Springer-Verlag, Berlin/Germany, 786, 1977

(8) Otte, D.:
Importance of existing car safety concepts for the accident behaviour and demands to future developments
Vortrag 26. ISATA-Konferenz, Aachen, 13-17.09.93, Proc. Road and Vehicle Safety, Paper No. 93FO26, 49-62, 1993

(9) Otte, D.; Pohlemann, T.:
Injury mechanisms for pelvis fractures of nearside occupants in lateral car impacts and influences of deformation characteristics
Vortrag und Proc. 1995 International IRCOBI Conference, Brunnen/Schweiz, 1995

(10) ETSC
Front and side impacts proposals „Fail to optimize safety"
Crash ETSC Newsletter on European Vehicle Crash Protection, 1. Edition, European Transport Safety Council, March 1995

INJURIES TO RESTRAINED OCCUPANTS IN FAR-SIDE CRASHES

Kennerly Digges
The Automotive Safety Research Institute
Charlottesville, Virginia, USA

Dainius Dalmotas
Transport Canada
Ottawa, Canada
Paper Number 351

ABSTRACT

Occupants exposed to far-side crashes are those seated on the side of the vehicle opposite the struck side. This study uses the NASS/CDS 1988-98 to determine distributions of serious injuries among restrained occupants exposed to far-side crashes and the sources of the injuries. Vehicle-to-vehicle crash tests were conducted to study dummy kinematics.

The NASS/CDS indicated that the head accounted for 45% of the MAIS 4+ injuries in far-side collisions and the chest/abdomen accounted for 39%. The opposite-side interior was the most frequent contact associated with driver AIS 3+ injuries (26.9%). The safety belt was second, accounting for 20.8%.

Vehicle-to-vehicle side impact tests with a 60 degree crash vector indicated that different safety belt designs resulted in different amounts of head excursion for the far side Hybrid III dummy. For all three point belt systems tested, the shoulder belt was ineffective in preventing large amounts of head excursion. Restraint was achieved by the lap belt loading the abdomen. A single retractor design with low friction sliding latch plate permitted the greatest head excursion in the far side crash tests. A dual retractor system with a fixed latch plate permitted the least.

INTRODUCTION

Test procedures required by present safety standards for side crashes require the crash dummies to be located on the side of the vehicle closest to the impact. Far-side occupants, those located on the side opposite the impact, are not included. Studies of injuries in far-side crashes may assist in identifying safety systems and test procedures to further improve occupant safety.

An objective of this study was to examine injury patterns for far-side front seat occupants in side collisions and to conduct crash testing to better understand the occupant kinematics that cause the most frequent injuries. The study is a continuation of the analysis that incorporated data from the Miami School of Medicine's William Lehman Injury Research Center (Augenstein, 2000). Previous authors have also investigated the far side injury problem.

Mackay (1991) examined 193 crashes with belt restrained far-side occupants during the period 1983-1989. The 193 cases contained 150 AIS 2 injuries and 15 AIS 3+ injuries. Among those with AIS 2+ head injuries, 35% came out of the shoulder belt. For those with AIS 2+ abdominal injuries, 72% were from contact with the safety belt. Contact with the belt system was the most frequent source of chest injuries (59%).

Frampton (1998) studied 295 crashes with belt restrained far-side occupants between June 1992 and April 1996. These cases included 46 MAIS 2 and 33 MAIS 3+ injuries. The MAIS 2 median deltaV was 25 km/hr. The median MAIS 3+ deltaV was 35 km/hr. Frampton found that MAIS 2+ injury rates were higher in perpendicular crashes than in oblique crashes.

Thomas (1999) investigated a sample of 474 side crashes between 1992 and 1998. The cases contained 226 MAIS 3+ survivors, and 188 fatalities. Thirty-four percent of the MAIS 3+ survivors and 32% of the fatalities were seated on the non-struck side. The distribution of injuries by body region for the 21 MAIS 3+ survivors on the non-struck side and without interaction with other occupants were: Head – 52%; Neck – 14%; Thorax – 19%; Upper Extremity – 38%; Pelvis – 14%, and Lower Extremity 14%. For the fatalities the equivalent : Head – 68%; Neck – 18%; Thorax – 86%; Abdomen – 41%; Upper Extremity – 9%; Pelvis – 9%, and Lower Extremity 18%.

Fildes (1991) examined injuries sustained in side collisions by drivers in Australia. The study was based on the Monash University crashed vehicle file consisting of 227 vehicles and 267 patients from crashes that occurred in Victoria during 1989 and 1990. The file contained 572 variables to describe the crash and the occupant. Fildes found that the injury rate of AIS 2+ head injuries was twice as high in far-side impacts as in near-side impacts. In far-side impacts, head and chest injury rates were about equal. The four most frequent sources of injuries were the instrument panel, the roof, the door panel, and the other occupant. The injury rate due to seat belts was about half that of the four most frequent sources.

METHODOLOGY AND DATABASES

In this study, National Automotive Sampling System/ Crashworthiness Data System (NASS/CDS) for the years 1988 to 1998 was used to examine the distribution of injuries and injuring contacts for belted occupants in far-side impacts.

The National Highway Traffic Safety Administration (NHTSA) maintains the NASS/CDS database of vehicle crashes in the United States. The NASS/CDS is a stratified sample of light vehicles involved in highway crashes that were reported by the police and involved sufficient damage that one vehicle was towed from the crash scene. The database was compiled between 1988 and 1998 and has been used extensively by NHTSA and others to assess the effectiveness of safety systems in reducing casualties in the crashes that occur on US highways.

INJURIES IN FAR-SIDE CRASHES BASED ON NASS/CDS

In the NASS/CDS data query, far-side occupants were defined as drivers in vehicles with right side damage and principal direction of force in the 1 to 5 o'clock direction or right front passengers in vehicles with left side damage and principal direction of force in the 7 to 11 o'clock direction.

In addition, the following restrictions were imposed: belted occupants only, age 16 or older, and no subsequent rollover of the struck vehicle. The data set contained 4696 cases – 3576 drivers and 1120 right front passengers. Of this driver population, 286 occupants had serious (MAIS 3 or greater) injuries with an aggregate of 776 AIS 3+ injuries.

Table 1 shows the MAIS 2, 3 and 4+ distribution by body region for belted occupants in far-side crashes. Each NASS/CDS case contains a weighting factor that is used by the NHTSA to extrapolate the individual cases to the national numbers. The distributions in Table 1 show both weighted and unweighed populations. The average weighting factor is also shown. In NASS/CDS, crashes with higher weighting factors have severities that occur more frequently in the population. The lower weighting factor cases tend to be in the more severe crashes.

All skull, brain and facial injuries were classified as head injuries. Injuries to the chest and abdomen were classified as trunk injuries. Injuries to the pelvis were classified as lower extremity injuries and shoulder injuries are included in the upper extremity category. Based on weighted data, head injuries accounted for 39% of the AIS 2 injuries, 24% of the AIS 3 injuries, and 45% of the AIS 4+ injuries. Spinal injuries contributed another 16% of the weighted AIS 4+ injuries. Chest/abdominal injuries dominated the AIS 3 injuries with 58%.

Table 1.
MAIS 2+ Injury Distribution for Belted Front Seat Occupants in Far-Side Crashes by Body Region, NASS/CDS 1988-1998

MAIS 2

Body	No.	Unwgt	Wgt	Ave Wgt
Head	205	44%	39%	183
Trunk	89	19%	24%	256
Lower X	88	19%	18%	199
Upper X	79	17%	19%	226
Total	461	100%	100%	208

MAIS 3

Body	No.	Unwgt	Wgt	Ave Wgt
Head	34	23%	24%	168
Trunk	62	41%	58%	223
Lower X	33	22%	7%	54
Upper X	21	14%	11%	126
Total	150	100%	100%	160

MAIS 4+

Body	No.	Unwgt	Wgt	Ave Wgt
Head	77	57%	45%	52
Spine	7	5%	16%	204
Trunk	52	38%	39%	66
Total	136	100%	100%	65

Table 2 shows the distribution of AIS 3+ injuries by injuring contact. The columns in Table 2 are similar to Table 1. The AIS 3+ injury distribution by body region and injuring contact is displayed in Table 3. Body regions and contacts that constituted less than 2% of the AIS 3+ weighted injuries were not included in Table 3. In Tables 2 and 3, the Side Interior category includes all interior side surfaces of the vehicle above the floor and below the roof.

Table 2.
AIS 3+ Injury Distribution for Belted Front Seat Occupants in Far-Side Crashes by Injuring Contact, NASS/CDS 1988-1998

Contact	No.	Unwgt	Wgt	Ave Wgt
Far Side Interior	245	31.5%	26.9%	70
Safety Belt	75	9.7%	20.8%	178
Roof	57	7.3%	12.2%	137
All Other	90	11.6%	8.6%	61
Seat	43	5.5%	7.5%	111
Near Side Interior	39	5.0%	7.0%	116
Non Contact	88	11.3%	6.5%	48
Dashboard	55	7.1%	5.2%	60
Other Occupant	58	7.5%	2.9%	178
Steering System	27	3.5%	2.4%	58
Raw No.	777	100.0%	100.0%	83

Table 3.
AIS 3+ Injury Distribution for Belted Front Seat Occupants in Far-Side Crashes by Body Region and Injuring Contact, Weighted NASS/CDS 1988-1998

Body Region	Injuring Contact	Weighted
Trunk	Belt	20.6%
Trunk	Right Side Interior	11.8%
Head	Right Side Interior	11.4%
Head	Roof	10.2%
Head	Left Side Interior	6.1%
Trunk	Seat	5.4%
Trunk	Other Occupant	2.3%
Trunk	Non Contact	2.2%
Head	Non Contact	2.1%
Head	Seat	2.1%
Spine	Roof	2.0%
Head	Dash	1.9%
Trunk	Dash	1.6%

Table 3 provides additional insights into injury mechanisms. The seat belt to trunk contact accounts for virtually all AIS 3+ Seat Belt contact injuries. The Side Interior (generally the opposite side) contacts account for a large fraction of the head injuries. The Roof contact is another significant source of head injuries.

CRASH SEVERITY

NASS/CDS 1988-1998 contains 150 cases of MAIS 3+ injured far side occupants with known delta-V. The distributions of lateral and total delta-V for these cases are shown in Figure 1. Approximately 50% of the MAIS 3+ crashes occur at lateral delta-V less than 30 kph. and 85% occur at lateral delta-V less than 50 kph.

A search of the NHTSA/FHWA crash test database maintained by the National Crash Analysis Center, at George Washington University disclosed one available crash test involving a far side dummy. The test was of a full size 1988 Chevrolet pickup impacted at 31 kph by a movable barrier with a rigid face. The direction of impact was 9 o'clock and the test dummy was a 50% male Hybrid II. The restraint system had a fixed latch plate, and both the shoulder and lap belt were attached to a retractor. In the test, the shoulder belt provided virtually no restraint. Upon the onset of side acceleration, the dummy immediately slid out of the shoulder belt, and was restrained only by the lap belt.

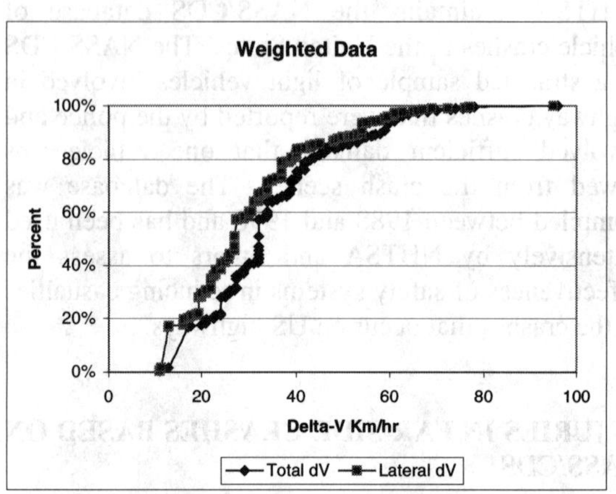

Figure 1. Crash Severity for Belted Far Side Front Seat Occupants with MAIS 3+ Injuries, NASS/CDS 1988-98.

As indicated in Figure 1, far side crashes with MAIS 3+ injuries frequently involve both lateral and longitudinal components of acceleration. It was postulated that a frontal acceleration component would improve the performance of the shoulder belt system. This possibility was explored in crash tests conducted for this study.

CRASH TEST RESULTS

Crash tests were conducted to study the dummy kinematics in far side impacts. The baseline test was a vehicle-to-vehicle side crash with a lateral delta-V of approximately 50 kph. The crash direction was 60 degrees relative to the centerline of the struck vehicle. A 1988 Chevrolet pickup, similar to the one in the NHTSA test was used as the test vehicle. It was impacted at the occupant compartment by a full size passenger car. A belted Hybrid III 50% male dummy was seated on the far-side of the impact. The restraint system was the same as in the 90 degree NHTSA test described earlier.

In the crash, both the lateral and longitudinal components of acceleration reached a peak of around 30g during the initial 12 ms. After 12 ms., the longitudinal acceleration , rapidly decreased, while the lateral acceleration continued for about 80 ms. Maximum lateral acceleration during the 12 to 80 ms. period was 27 g..

The load vs. time for the lap and shoulder belt are shown in Figure 2. The shoulder belt was loaded during the initial 90 ms. of the crash. After 90 ms. the load decreased and the lap belt load continued to increase. The maximum shoulder belt load was 2256N at 90.4ms. The maximum lap belt load was 2882N at 143 ms.

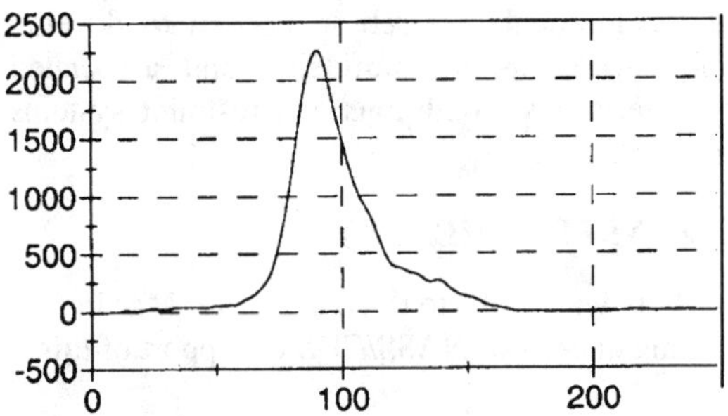

Shoulder Belt Force (Newtons) vs. time, ms.

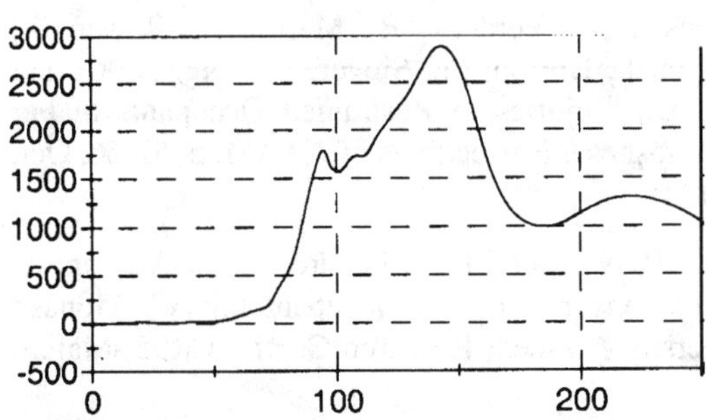

Lap Belt Force (Newtons) vs. time, ms.

Figure 2. Belt Forces in Vehicle-to-Vehicle Far Side Crash Test

The dummy injury readings were all low. The HIC was 194; the max chest g was 16.7; the max pelvic g's were 21, and the neck Nij was 0.3. None of these readings indicated injuries due to belt loading or head contact as expected based on the injury frequencies listed in Tables 1, 2 and 3. However, the injuries from opposite side head impact and trunk belt loading may not be adequately reflected by the conventional dummy readings. The maximum lap belt load and the extent of head excursion are more likely criteria for these types of injuries.

The position of the dummy at 130 ms after initial impact is shown in Figure 3. This position corresponds to the time of high lap belt loading.

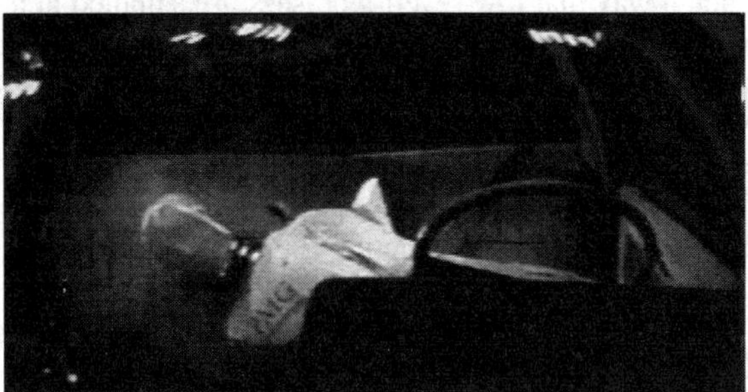

Figure 3. Dummy Position During Highest Belt Loading

A view from a second camera is shown in Figure 4. This view shows the configuration of the lap belt late in the crash at maximum head excursion. The lap belt, limits the head excursion by loading the abdomen. The shoulder belt loading is also across the abdomen.

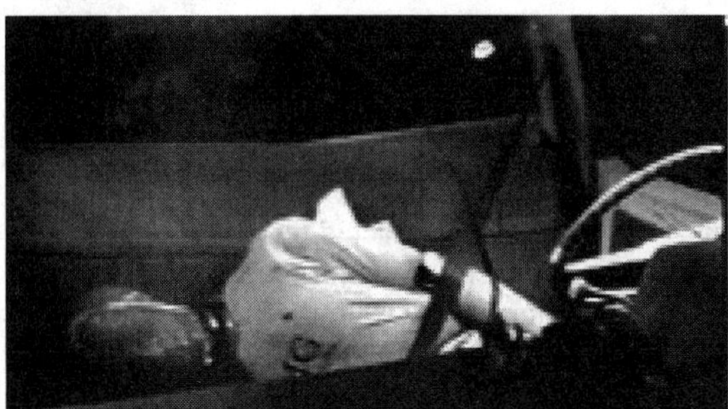

Figure 4. Dummy Position Showing Head Excursion

Identical crash tests were conducted for two additional OEM safety belt systems. The first was a belt system with a single retractor and latch ring plate that moved with low friction. This system permitted easy movement of the latch plate along the belt, even when the belt was buckled. A second system was a single retractor belt with a high friction latch ring. The crash test results showed that the low friction latch plate permitted the most head excursion, and the two retractor system discussed earlier permitted the least. All of the systems prevented the dummies from impacting the far side interior of the vehicle.

DISCUSSION

The NASS/CDS data for belted front seat occupants indicates that the safety belt accounts for 20.8% of AIS 3+ injuries in far-side crashes. Earlier studies by Mackay found that 59% of AIS 2+ chest injuries and 72% of abdominal injuries among belt restrained far-side occupants were from seat belt contacts (Mackay, 1991). The NASS/CDS data sample is stratified in such a way that low severity crashes are sampled at a much lower rate than the higher severity crashes. Table 2 shows that belt contacts carry a large average weighting factor, suggesting lower severity crashes, and therefore a smaller than average sample of cases. The NASS/CDS sample may not be sufficiently robust to adequately capture injury modes in the lower crash severity ranges.

The NASS/CDS indicated that the largest source of injury to belted occupants in far-side impacts was the Far Side Interior. For the weighted data the Far Side Interior percentage was 26.9% and for the unweighted data it was 31.5%. Fildes reported the door panel as one of the most frequent injury contacts for Australian far-side AIS 2+ injuries (Fildes, 1991). Mackay reported that the far-side occupant came out of the belt in 35% of the cases with AIS 2+ head injuries (Mackay, 1991). These studies suggest that opportunities exist for improving the occupant retention and chest/abdominal loading of safety belts in far-side crashes.

The crash testing of three different belt system designs showed differences in head excursion, but none of the dummies contacted the opposite side interior. The conventional injury measurements of a Hybrid III dummy were low in the 50 kph lateral delta-V crash tests conducted in this study. The failure to measure any injuries in the crash tests may indicate that the test configuration is not representative of crashes that produce severe injuries in the real world. However, the tests did indicate that the shoulder belt was largely ineffective in preventing large amounts of head excursion, even in far side crashes with longitudinal acceleration. Head contact with the opposite side interior would more likely occur if the vehicle interior were smaller (small car) and/or if significant occupant compartment intrusion had occurred.

Further, the tests showed lap belt loading of the abdominal region. The maximum belt force developed was lower than that permitted in a frontal crash test where the lap belt transmits the loading through the pelvic structure. Different injury criteria would be applicable to belt loading of the abdomen.

Finally, the tests may indicate the need for better fidelity of the dummy in far side impacts. The Hybrid III has biofidelity in a frontal crash, but no attempt has been made to evaluate its performance in a far side crash.

CONCLUSIONS

For belted occupants in far-side crashes, the most harmful injury source is the opposite side of the car (26.9%). The second most harmful injury source is the safety belt (20.8%). Impacts with the roof account for 12.2%. The contacts with the opposite side, roof and belt may be influenced by safety belt design. Crash tests indicate differences in the extent of restraint offered by different belt systems in far side crashes. In all three point belt systems tested, the shoulder belt was ineffective in preventing large amounts of head excursion. Restraint was achieved by the lap belt loading the abdomen. The baseline, belt system with a fixed latch plate and two retractors permitted the lowest head excursion of the systems tested. Additional research in needed to develop dummies with far-side biofidelity and associated injury criteria, test conditions and restraint systems for far side crashes.

ACKNOWLEDGEMENTS

The authors would like to thank Dr. Peter Martin for conducting queries of NASS/CDS in support of this study.

REFERENCES

Augenstein, J, Perdeck, E., Martin, P., Bowen, J., Stratton, J., Horton, M., Singer, M., Digges, K., and Steps, J., "Injuries to Restrained Occupants in Far Side Crashes", Proceedings of AAAM, p. 57-66, Oct, 2000.

Fildes, B. N., Lane, J. C., Lenard, J., and Vulcan, A. P., "Passenger Cars and Occupant Injury", Monash University Accident Research Centre, CR95, March, 1991.

Frampton, R., Brown, R., Thomas, P., and Fay, P., "The Importance of Non-struck-side Occupants in Collisions", Proceedings of the 42nd AAAM, p. 303-320, 1998.

Mackay, M., Parkin, S., Hill, J., and Munns, J., "Restrained Occupants on the Non-struck Side in Lateral Collisions", ", Proceedings of the 35th AAAM, p.119-131, 1991.

Thomas, P., and Frampton, R., "Injury Patterns In Side Collisions, A New Look with reference to current Test methods and Injury Criteria", 43rd Stapp Car Crash Conference, P-350, 99SC02, p. 13-24, October 1999.

905122

Females More Vulnerable than Males in Road Accidents

J. Y. Foret-Bruno, G. Faverjon, F. Brun-Cassan, C. Tarriere
Laboratoire de Physiologie et de Biomécanique Peugeot-Renault
C. Got, Institut de Recherches en Biomécanique et Accidentologie, Boulogne
A. Patel, F. Guillon, Institut de Recherches Anatomiques Chirurgicales, Garches

ABSTRACT

It is well known that the risks taken by drivers are different according to their sex. This clearly explains why, according to the official statistics, the fatality rate for male accident victims is always higher than that for female victims (due to overspeeding, drugs, alcohol, high annual mileage, etc.).
However, this situation may be changing, because J.C. Fell (1) and C.L. Popkin (2) observe in the USA an increasing number of young women under the influence of alcohol involved in fatal accidents.
If the risks taken by both sexes were to become identical, that is, if the drivers of both sexes were to become involved in accidents of identical violence and configuration, the severity for females would be far greater than for males of identical age, and the difference in severity would increase with age.
These conclusions are based on a comparative study of the risks of thorax injuries to restrained occupants of both sexes involved in frontal impacts and on bone resistance tests on cadavers. For example, it is observed in real accidents that, for an identical impact violence, restrained females reach their maximum thorax resistance sooner than males of the same age, even though they are lighter and the load applied to their thorax is smaller . Moreover, the seat position occupied (driver or passenger) has a major influence on the risk. It is also demonstrated that the curved showing the maximum forces (measured on the seat belt at shoulder level) sustainable according to age are similar to the curves showing the mean values of rib bending fracture stress for a sample of cadavers.
Extrapolation of an extra injury risk for the female thorax, assessed at 20% on average, to a global additional risk, does not seem as unrealistic as one might think, since Evans (3), using a statistical method on a large sample of accident victims, achieves a similar result for the fatality rates of accident victims of both sexes.

INTRODUCTION

Men sometimes speak of "wome weakness" in mocking or male chauvir terms. Are females really weaker than ma in biomechanical terms?

It is true that the female populatio does not (on average) achieve the sam performance levels as the male population i: athletic events where muscles play a role.

It is true also that the finer bones of the female skeleton imply lower bone resistance for an identical applied stress.

It may therefore be felt that such performances, which can be explained in mechanical terms, are quite simply proportional to the morphological features of the two populations, e.g., weight being influenced by muscle development and the skeleton. In what follows, of course, we are talking about the mean of each population, since there are great disparities from person to person.

However, if body mechanics are proportional to body morphology, it would then be natural, for women subjected to the same impact conditions as men, i.e., an impact in which the weight difference would be the only variable, to observe an equivalent level of risk.

Now in what follows, we demonstrate that this is not the case and that women do not have the same capacity as men to support their own weight without injury. The female population therefore seems more vulnerable and the term "women's weakness" seems to us, without any chauvinism, quite realistic.

ANALYSIS METHODS AND STUDY SAMPLES

To verify the above assertion, two methods can be used:

- A sophisticated statistical comparison on the basis of a huge sample to try and eliminate the various possible sources of bias (age, delta-V, seat position, seat belt usage). This is the method used by Evans (3), which gives interesting results rather similar to our conclusions. We shall mention this again later.

- A comparison of the risks for a "sensitive" body area in which the only factor involved is the weight of the occupant without consideration of occupant size or the part of the car impacted.

To do this, we examined the risks of thoracic injuries attributable to the safety belt for all front-seat occupants involved in frontal impacts. To avoid as many sources of bias as possible, we eliminated cases in which the occupants:
- either sustained an overload due to a back-seat occupant;
- or, sustained major intrusion preventing use of the restraining device without direct impact against the steering wheel or front panel.

The Accident Research Laboratory survey enabled us to study a sample of 1230 belted casualties in the circumstances mentioned above, of whom 109 suffered thoracic injuries attributable to the seat belt.

These observed injuries are mainly rib fractures, sometimes associated with internal injuries (hemothorax, pneumothorax for the most part).

To try and assess the difference in thorax tolerance between males and females, we then analyzed:
- a sub-sample of 295 occupants wearing a seat belt equipped with a textile shock absorber;
- a series of thorax bone resistance tests covering 245 cadavers and occupants who deceased suddenly in road accidents.

THORACIC INJURIES ATTRIBUTABLE TO THE SEAT BELT IN FRONTAL IMPACTS

Before studying the influence of the various parameters (age, delta-V, seat position, sex) it should be specified that internal thoracic injuries are generally associated with multiple rib fractures (Table 1). The rare cases of internal injuries without rib fractures are observed in young belted occupants involved in very severe impacts (delta-V > 65 km/h).

Table 1

	Number of rib fractures				TOTAL
	0	1 to 3	4 to 8	> 8	
Number of internal thoracic injuries/ number of accident victims	6/2038	3/122	8/32	9/16	170/2208
Percentage	0.3	2.5	25	56.2	7.7

INFLUENCE OF SPEED. The forces involved increase with the impact velocity, and it is normal to observe for a delta-V of between 25 and 65 km/h an increase in risk from 4.5 to 43 (Figure 1).

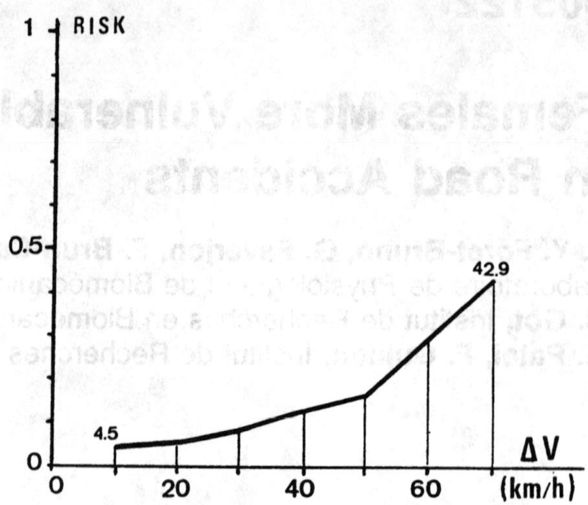

Figure 1. Risk of Rib fractures in Terms of delta-V

INFLUENCE OF AGE. Age, as we know, has a major influence on injury severity. Figure 2 shows that the thorax risk increases from 2.8 to 33 as occupant age increases from 30 to 65 or over.

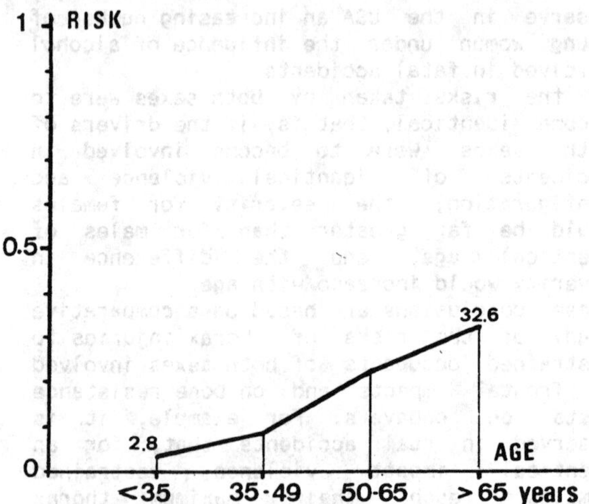

Figure 2. Risk of rib fractures according to age

INFLUENCE OF SEAT POSITION AND SEX. As we can observe, a belted driver and passenger in the same vehicle are not subjected to the same thorax risks. Due to the presence of the steering wheel and steering column, the driver is subjected to:
- some staying of the arms which may, at low and medium velocities, reduce belt stress;
- knee buttressing against the steering column, which reduces pelvis movement and accordingly reduces stress at the thorax level.

Belted Couples. We first compared belted couples in the same car in the impact conditions defined in paragraph 2. Other requirements were that:

- The occupants be of different sexes;
- One of the two occupants have a thoracic injury.

48 couples met these criteria, although in 90% of cases the women occupied the front passenger position. Nevertheless, the results are revealing (Figure 3):

- For identical age (± 1 year), a risk of thoracic injury (rib fracture or internal injury) to the female is observed in 82% of cases;
- An equivalent risk for male and female is reached when the man is 9 years older than the woman.

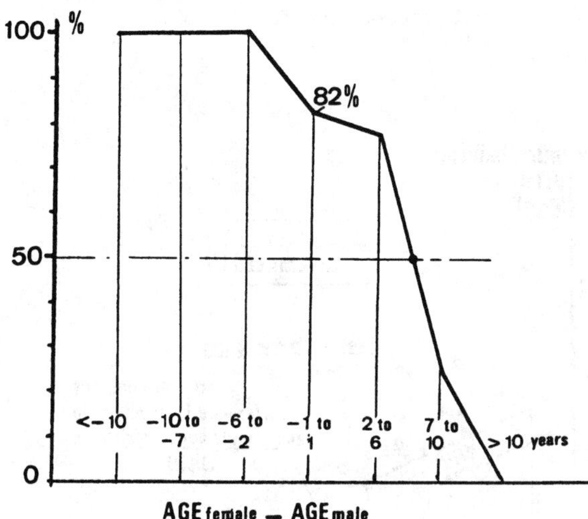

Figure 3. Percentage of females who sustained rib fractures for couples (belted males/females) in the same vehicle, in terms of age range observed for each couple

All Belted Occupants. To obtain a sufficiently large sample, we studied the risks for drivers and passengers according to sex for 3 speed classes (delta-V less than 25 km/h, from 26 to 45 km/h, and greater than 45 km/h).

Figure 4 shows that the risk of thoracic injuries for identical age are:

- always greater for females, whether they be the driver or passenger;

- always greater for the passenger than for the driver of identical sex;
- similar for a female driver and a male passenger.

Figure 5 shows, for the two extreme age classes, the greater risk for the elderly.

For males and females of each age class involved in impacts of identical violence, the calculated mean risks would be as follows:

- Male drivers 0.06
- Male passengers 0.11
- Female drivers 0.12
- Female passengers 0.17

It is immediately clear that a well designed kneestop for the passenger would greatly reduce the risk of rib fractures, not to mention reduced risk of abdominal injuries, which are also more frequent for passengers.

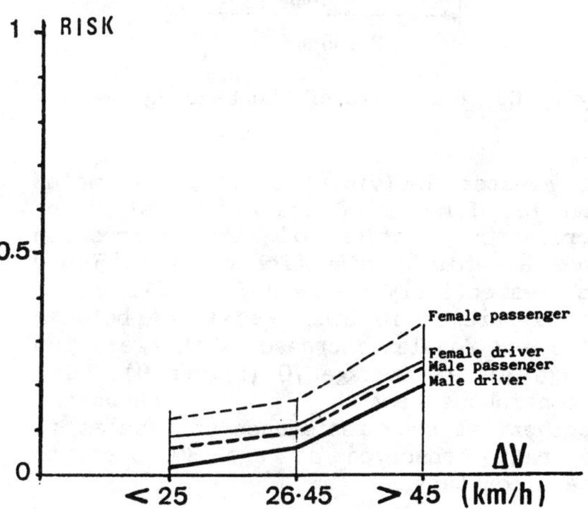

Figure 4. Risk of rib fractures versus delta-V according to sex and seat location

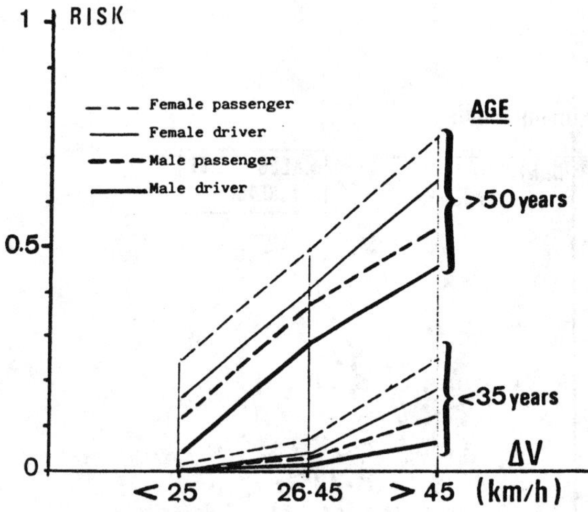

Figure 5. Risk of rib fractures versus delta-V according to sex, seat location and age

BONE RESISTANCE OF THE THORAX

RIB BENDING STRENGTH. For 245 cadavers, including 60 road fatalities, the bone resistance was tested by bending tests on the 6th right and left ribs by the procedure schematically represented in Figure 6. The bending strengths of these ribs show:

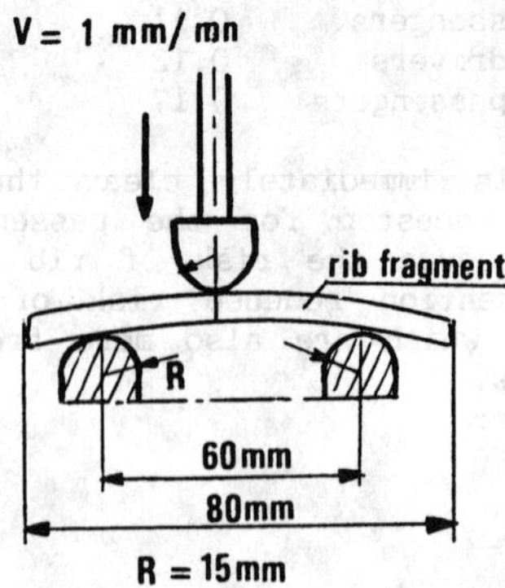

Figure 6. Principle of the bending tests

- The greater individual scatter for males than for females of identical age. The correlation factors of the regression lines are highly significant, at 0.45 and 0.67 respectively (Figures 7 and 8).
- The difference in bone resistance between males and females increases with age: 16% at age 20, 44% at age 70 (Figure 9). The discontinuous line (_ _ _ _) shows the hypothetical bone resistance of females if the ratio observed at age 30 were to remain constant.

These results, obtained by simple rib bending tests, can be explained by problems of osteoporosis observed in postmenopausal women (reduced bone resistance due to decalcification of the skeleton).

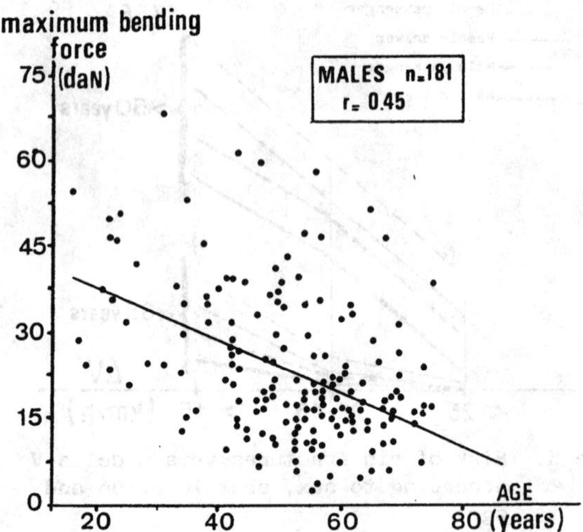

Figure 7. Maximum bending force of the sixth rib versus age for males

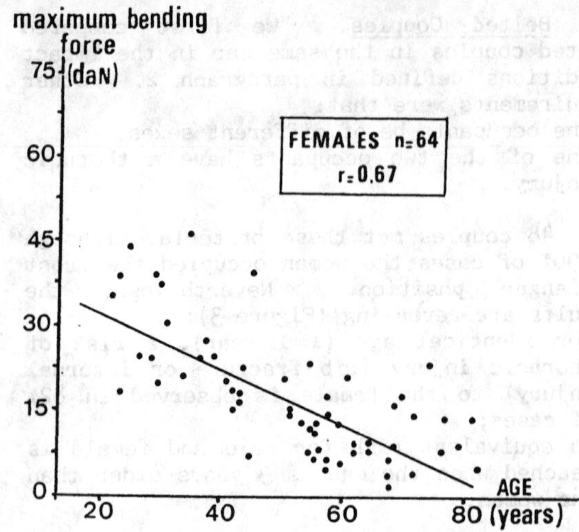

Figure 8. Maximum bending force of the sixth rib versus age for females

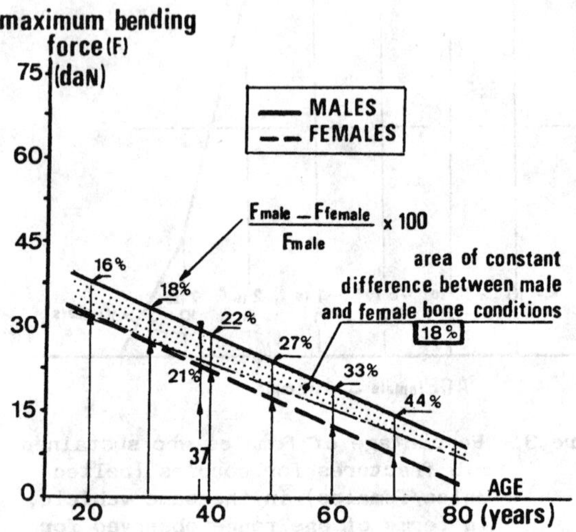

Figure 9. Bone condition according to sex and age

MAXIMUM FORCES SUSTAINABLE BY THE THORAX IN REAL ACCIDENTS. A publication by the Laboratory (5) gave, for 295 occupants belted by a 3-point seat belt equipped with a load limiter, the force levels sustained in frontal impacts versus age (Figure 10).

One immediately observes that the maximum forces sustainable without rib fracture are twice as high for those aged under 30 (800 to 900 daN) as for those aged over 50 (400 to 500 daN).

Hence, the observed ratio between these two age classes is similar to that defined for rib bending tests only (33 daN at age 25, 16 daN at age 60 on average).

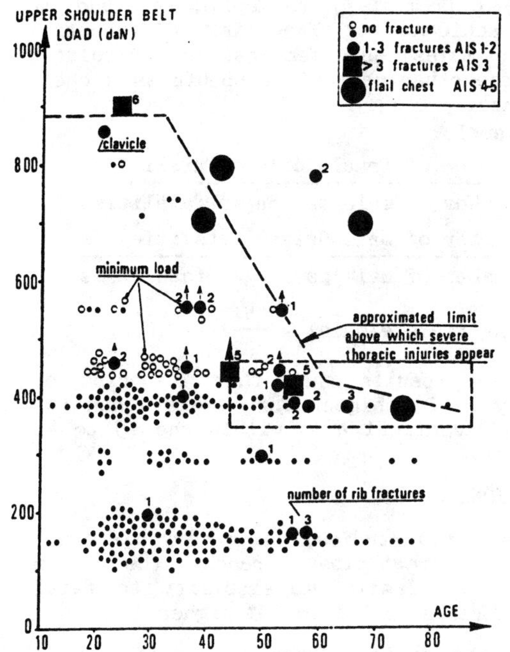

Figure 10. Shoulder belt load versus age for 295 occupants involved in frontal collisions and approximate limit above which severe thoracic injuries appear

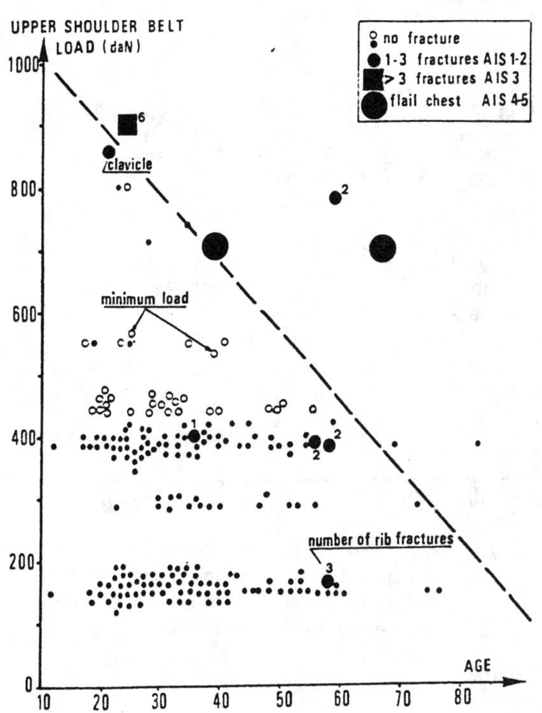

Figure 11. Shoulder belt load versus age for male occupants involved in frontal impacts and approximate limit above which severe thoracic injuries appear

The curves showing the maximum forces without rib fracture versus age for each of the sexes (for these 295 cases) are hard to calculate given the small numbers involved for the highest impact forces. However, one may note that in the range of forces between 380 and 440 daN, the risk of rib fractures is significantly higher for females in the over-45 age group.

- Males 2/19 11%
- Females 4/7 57%

In view of these observations, the lines obtained for static rib breaking forces can be extrapolated to lines indicating the maximum dynamic forces sustainable at the shoulder level prior to rib fracture occurrence, for both sexes. This extrapolation for the whole thorax does not, according to Figures 11 and 12, seem as unrealistic as might have been thought.

RIB FRACTURES AS A FUNCTION OF SHOULDER FORCE AND BONE RESISTANCE. For an identical dynamic force measured at the shoulder level in 45 frontal impact tests with belted cadavers, one observes that the greater the bone strength of a cadaver (static rib breaking force shown on the abscissa in Figure 13), the lower the risk of rib fractures.

These tests, which were performed in very different conditions (multiple anchorage locations) show that while the age factor is very important and in relative correlation to the maximum acceptable force, there is no doubt that a better correlation can be obtained for bone strength which, especially in the case of males, can differ greatly at an identical age.

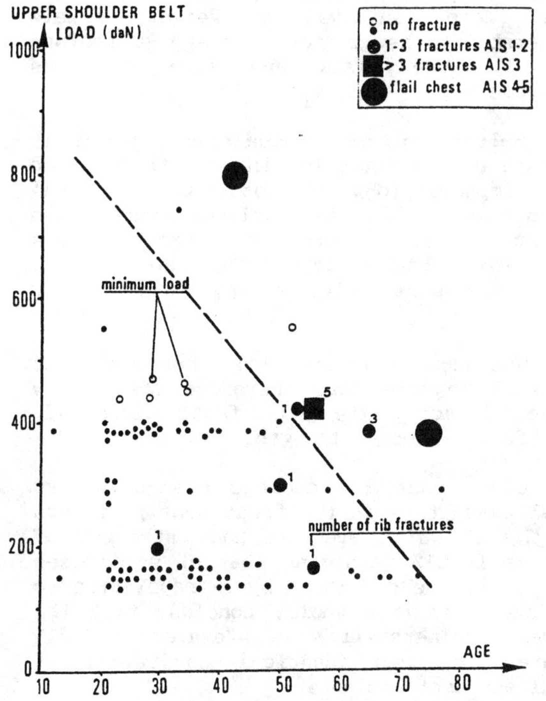

Figure 12. Shoulder belt load versus age for female occupants involved in frontal impacts and approximate limit above which severe thoracic injuries appear

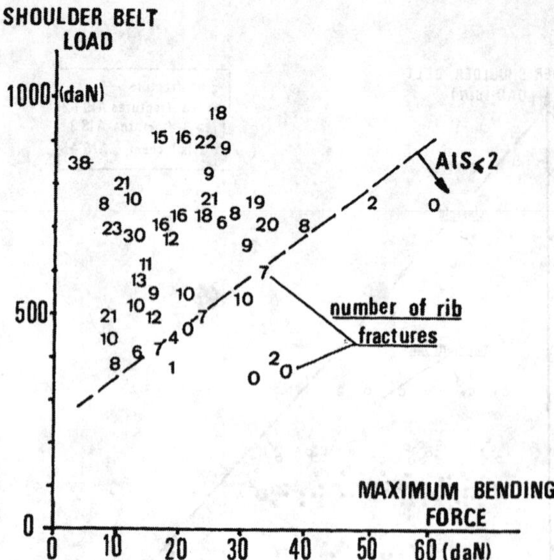

Figure 13. Number of rib fractures according to shoulder belt load and maximum bending force of the sixth rib for 45 tests with cadavers

DISCUSSION

The average weight of females is 18% to 20% less than that of males. This implies that in frontal impact the forces sustained at the thorax level by belted females are in the same ratio. Now, over age 30, it has been observed, as shown in Figure 9, that the bone resistance of females is over 18% less than that of males of identical age, and this percentage reaches 44% at age 70. For these precise frontal impact conditions, then, where the only factor involved is weight, female vulnerability is greater over age 30 (before age 30, the risks are more or less equivalent).

Belted fatal casualties generally sustain direct intrusion in both frontal and side impacts (85% of total belted fatal casualties). In these circumstances, the effect of the occupant's weight in his restraining device plays a far smaller role, since the seat belt is subjected to low stress.

Skeleton injuries and the resultant internal injuries are therefore created by direct impacts; the only factor involved, therefore, is bone strength.

Given that the average age of belted fatal casualties in the front seat positions is 37, at which age the bone strength of females is 21% less than that of males (see Figure 9), one can then, by extrapolation to the skeleton as a whole, conclude that the average vulnerability of females is 21% greater in such identical conditions of fatal accident violence.

By a statistical method, Evans has endeavoured to eliminate the sources of bias described at the start of this paper (violence, occupant's seat position, age) so as to compare the risk for both sexes.

This method is based on comparisons of double pairs of occupants (81,994 taken from the FARS from 1975 to 1983) involved in fatal accidents. The risk differential between males and females is calculated using the other pair of occupants as a check reference.

Example:

$$R1 = \frac{\text{Number of female driver fatalities}}{\text{Number of male passenger fatalities}}$$

$$R2 = \frac{\text{Number of male driver fatalities}}{\text{Number of male passenger fatalities}}$$

$$\text{Risk (female/male)} = \frac{R1}{R2}$$

These results show that the risk of fatality is 25% higher for females than for males of identical age, within the 15 to 45 age range.

CONCLUSIONS

These two studies, using very different methods, nevertheless reach the same conclusion: female vulnerability to fatal accident risks is 20% to 25% higher.

On the basis of our study, it can be concluded that this higher risk is attributable to the female skeleton's being less able to sustain the same stresses as males. Above all, though, one observes that female impact tolerance is less than one would expect on the basis of the proportional difference in mean weights between the two sexes.

If, as indicated in certain studies, the risks taken by females are gradually increasing and may well reach the risk level of males within a few years, then it will be seen that the female fatality rate will approach and subsequently exceed the male fatality rate.

REFERENCES

(1) "Alcohol Involvement Rates in Fatal Crashes, A Focus on Young Drivers and Female Drivers", James C. Fell, NHTSA Washington DC20950. Proceedings of the 31st American Association for Automotive Medicine, New Orleans, 1987.

(2) "Drinking and Driving by Young Females", Carol Lederhaus Popkin, HSRC University of North Carolina. Proceedings of the 33rd American Association for Automotive Medicine, Baltimore, 1989.

(3) "Fatal and Severe Crash Involvement Versus Driver Age and Sex", Leonard Evans, General Motors Research Laboratories. Proceedings of the 31st American Association for Automotive Medicine, New Orleans, 1987

(4) "Age and Sex Effects on Severe and Fatal Injury Rates in Traffic Crashes", Leonard Evans, General Motors Research Laboratory, SAE 885053.

(5) "Thoracic Deflections of Hybrid III. Dummy Response for Simulations of Real Accidents" J.Y. Foret-Bruno et al., Laboratory of Physiology and Biomechanics associated with Peugeot S.A./Renault, 12th ESV Conference, 1989.

CHAPTER 5:

SUBMARINING

751146

Motion Sequence Criteria and Design Proposals for Restraint Devices in Order to Avoid Unfavorable Biomechanic Conditions and Submarining

Dieter Adomeit and Alfred Heger
Institute of Automotive Engineering
Technische Universität Berlin

Abstract

Our aim is to show that today's safety standards (FMVSS 208, EC-Proposals) are inadequate in the present state to ensure optimum protection for belted passengers. These standards do not take into account motion sequence during impact. The postulated tolerance limits - HIC, SI, forward displacements etc. - cannot describe the dynamic behaviour adequately.

We emphasize the importance of motion sequence to ensure optimum biomechanic consitions, because motion sequence is the necessary prerequisite for any discussion about biomechanical tolerance limits.

First results of our current belt-accident-investigations indicate that this is an important problem. By applying experience from crash tests and accident investigations, we try to define simple and well controllable criteria for motion sequence. These criteria should not only guarantee favourable biomechanical loads for the belt-restrained passenger, but also give the possibility of simplifying data registration and calculation of results.

Our final aim is to reduce belt-specific injuries, since they are avoidable by using current technical capability.

The solution of these problems can be achieved by modifications of safety standards using defined motion sequence criteria. This would lead to further developments of the system comprising a 3-point-belt and the seat.

In this paper we introduce one possible development. Results of a number of crash tests show its efficiency.

BACKGROUND

BELT ACCIDENT STATISTICS - Accident investigations of 3-point belted occupants evaluate several belt-specific injury patterns (1 - 6)*. First of all it should be stated that these injuries are of essentially lower AIS than those of unbelted passengers in similar accidents. The injuries of the belted occupants should be differentiated as: 1. injuries directly caused by the belt, and 2. injuries occuring by contact to interior parts of the compartment in spite of belt usage.

Until now the results of accident investigations - in attempt to typify injury patterns and their causes - are rather rare and not uniform. None of these present investigations deals thoroughly with the often discussed problem of "submarining" and its typical injury patterns. This gives the impression that submarining is no problem today - submarining in its classical definition: the lap belt slides over iliac crest with lap belt forces effecting the internal abdominal organs during forward displacement of the lower torso. Patrick (5) concluded from his case studies that classical submarining has no statistical relevance. Contrary to that Bäckström (2) demonstrates a relative increase of abdominal injuries for belted passengers without further explanation. A review of the severe accidents investigated by Henderson (4) shows head injuries in addition to very frequent and severe abdominal injuries caused directly by the belt.

<u>Submarining Injuries: Not Only Abdominal Injuries</u> - The first evaluation of our own accident investigations on scene**confirm the

---

* Numbers in parentheses designate References at end of paper.

**Accident Investigation Program of the Institute of Automotive Engineering at the Technical University, Berlin, on behalf of BASt (federal road traffic institute)

tendency for belt-specific injury patterns, which can also be recognized in the above mentioned papers.

This injury pattern demonstrates that submarining or submarining-similar motion sequences are an important problem now. This type of motion produces injuries which are undoubtedly avoidable by using current technical resources. Failure to find abdominal injuries resulting from submarining in the usual sense has led to an underestimation of this problem.

Truly abdominal injuries are not frequently observed, but in case of occurance they are severe injuries (AIS 4-5) (4,1). Instead of abdominal injuries alone, we have observed other types of injuries giving unquestionable signs of submarining or submarining-similar motion sequences. Primarily injuries of the knees should be considered, being of special importance:

First, the injuries of the knees indicate - depending on the size of the passenger compartment - an unfavourable forward displacement of the lower torso;

second, the impact of the knees lowers the lap belt load and so prevents abdominal loading. Although submarining or a submarining-similar motion sequence occured, it is usually not discovered as the reason for the injury pattern. Further injuries following from this submarining-similar behaviour will be evaluated under "Injury Pattern of Submarining-Similar Motion Sequences".

<u>Motion Sequences</u> - It is well known that belt forces and dummy decelerations (HIC and SI) are influenced by the type of the motion sequence of the dummy in sled tests. In addition, some facts are known about the correlation between 3-point belt geometry, webbing stiffness, seat design and motion sequences.

In the following paragraphs we shall discuss three types of motion sequence with their special aspects.

Figures 1, 2 and 3 demonstrate the final positions of the dummy and the trajectories of the hip (H-point) and the shoulder points (S-point).

The torso angle α, defined as the angle between the H-S-plane and the horizontal plane, will prove later on to be an important parameter.

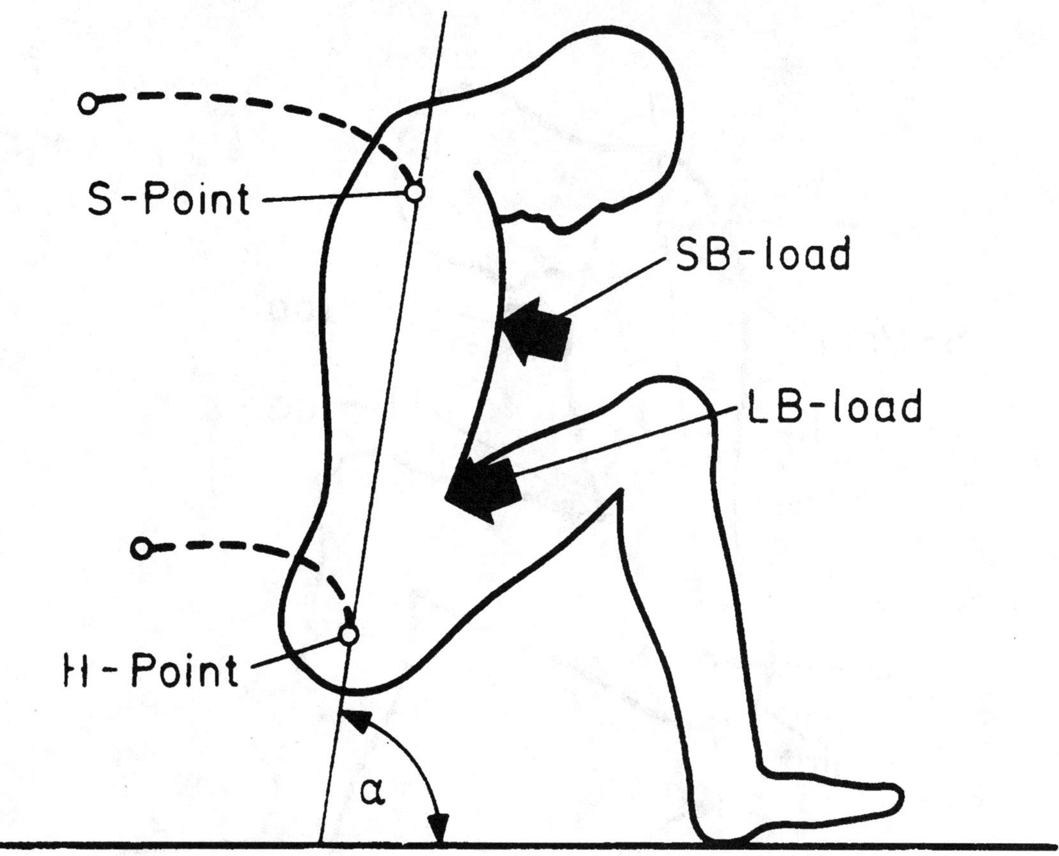

Fig. 1 - Trajectories and final position of 3-point-belted passenger without classical "submarining" seated on a soft spring-cushion-seat

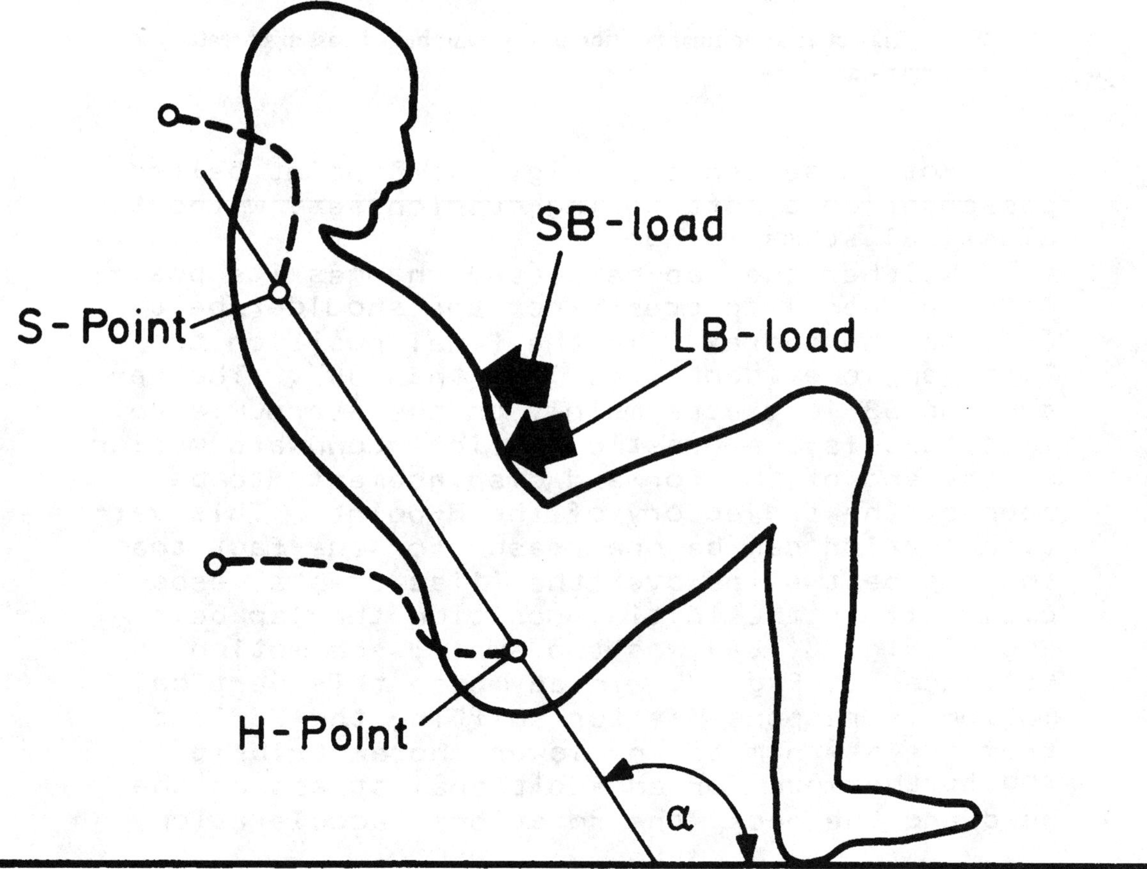

Fig. 2 - Trajectories and final position of a "submarining" 3-point-belted passenger seated on a soft spring-cushion-seat

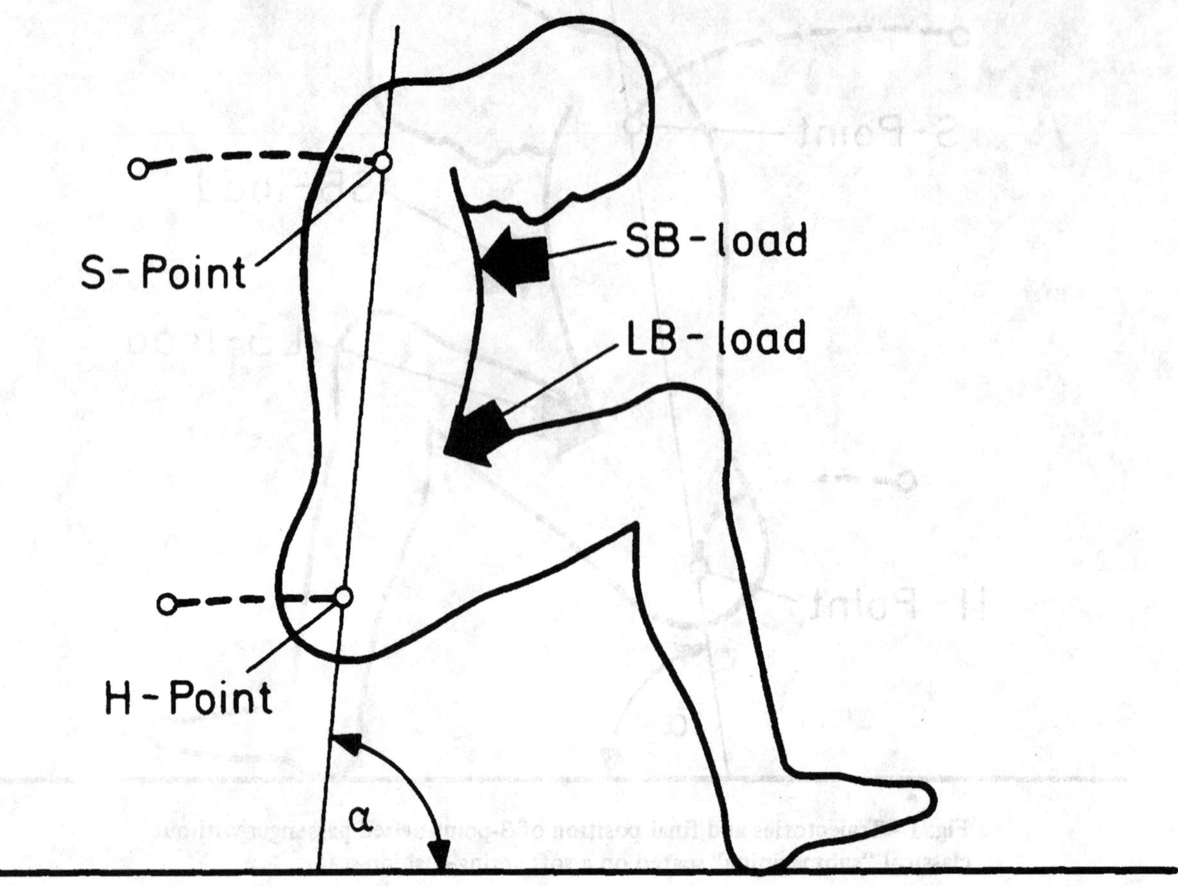

Fig. 3 - Trajectories and final position of a 3-point belted passenger seated on a rigid seat

Motion sequence I (Fig. 1): 3-point belted passenger on a soft spring-cushion seat without classical submarining:

Neither the lap-belt (LB) changes its position on the iliac crests nor the shoulder-belt (SB) on the thorax. In the final position the torso angle evidently is less than 90°. The resulting SB-load acts mainly on the sternum area. Important is the drastic vertical downward motion at the end of the forward displacement, to be seen by the trajectory of the H-point. This vertical motion can be one reason for the fact that the lap belt slips over the iliac crests, especially if in the initial position the lap belt angle β (Fig. 8, 8a) was too small; see motion sequence II, Fig. 2. In any case this vertical motion is responsible for shifting the SB-load from the sternum to the lower thorax (ribs 6-12) and furthermore for an additional stress on the head and the neck: the rotational acceleration of the head increases when the vertical motion of the H-point is abruptly stopped.

Data characteristics:

SI (chest): favourable
(depending on forward displacement of chest (t,s))

HIC (head) : problematic.

Motion sequence II (Fig. 2): 3-point belted passenger on a soft spring-cushion seat.

Fig. 2 shows the known motion sequence with submarining on a spring-cushion seat. Noteworthy are the trajectories of the H-point and, resulting from this, the movement of the S-point. The forward displacement of the hip is larger than the forward displacement of the chest, torso angle $\alpha > 90^{\circ}$. There are low bending moments affecting the neck and low rotational accelerations of the head. On the other hand, the known overload of the abdomen and the lower thoracic area caused by the lap belt sliding upward, has to be mentioned. Very important is strain of the thorax below the sternum (ribs 6-12) caused by the shoulder belt.

Data characteristics:

SI: favourable (!)
(depending on total displacement (t,s))

HIC: favourable (!) .

Motion sequence III (Fig. 3): 3-point belted passenger on a rigid seat.

Fig. 3 shows the motion sequence on a rigid test seat according to EC-proposals with 3-point belt.

There is no vertical movement of the H-point and there is no relative slip between the belt loops and the torso. Appropriate webbing being provided, forward displacement of the H-point, which relieves head and neck: no whiplash effect.

This sequence is characterized by favourable biomechanical conditions:

- advantageous loading of the thorax on the sternum area by the shoulder belt and of the lower torso on the iliac crests by the lap belt
- low forces affecting head and neck, upper and lower torso, provided there is an optimized forward displacement versus time.

Data characteristics:

SI: favourable
(depending on forward displacement (t,s))

HIC: favourable .

This description explains principally the importance of the motion sequence for advantageous biomechanical conditions.

In belt accident investigations the outer crash conditions (ETS, deceleration-level, -time, -direction, deformation of the compartment), the marks inside the compartment (marks on the webbing, deformation of the seat and the dashboard, etc.), and the injuries of the passenger are available for reconstruction of the passenger's motion sequence.

Two difficulties appear concerning today's belt accident investigations:

First, for demonstrating the severity of accidents usually only the ETS is applied, an inadequate criterion,

second, usually the marks of the interior compartment, describing the initial conditions relating to the passenger and to the geometry of the construction, were neglected. Nevertheless, it is wellknown that these conditions are of high influence on the passenger's dynamic behaviour.

Therefore, the conclusion concerning the correlation between type of loading and injuries is often questionable.

Injury Pattern of Submarining-Similar Motion Sequences - In our current accident analysis program we try to evaluate additional data in order to get more precise information on the severity of accidents. Further, we gather all nenessary data in order to reproduce the passenger compartment conditions, which are namely the initial position of the passenger, belt slack, belt geometry, and the construction of the seat. Analysis of our cases as well as reviews of the mentioned investigations (2,3,4,5) repeatedly show types of injuries, which can only be caused by submarining-similar motion sequences. They are induced by incorrectly adjusted restraint systems, often augmented by unfavourable geometric belt-seat conditions.

This "Submarining-Seat-Belt-Syndrome" exhibits combined or single injuries, which essentially are: lacerations and abrasions in the neck area, thoracic injuries mainly to the lower thorax (ribs 6-12), injuries to the abdomen and injuries to the knees.

- Neck injuries are caused by sliding of the neck along the stretched webbing,
- Thoracic injuries: a torso angle of more than 90° ($\alpha > 90^{o}$) leads to a resulting shoulder-belt load in the area of the ribs 6-12 (5), (Fig. 2,5),
- Abdominal injuries are caused by compression of internal organs after the lap belt slipped over the iliac crests,
- Knee injuries are caused by the impact of the knees against the lower panel after an excessive forward displacement of the H-point.

CRITERIA OF MOTION SEQUENCE CONCERNING BIOMECHANICS - The above description is confirmed by crash test experience, so that the standard criteria for the head, HIC, and the chest, SI (according to test conditions of FMVSS 208), do not permit any qualification to the type of passenger's strain and the "biomechanical quality" of the process. We even found favourable values of HIC and SI in submarining-similar motion sequences which means bad "biomechanical quality".

Further belt accident analyses show clearly the necessety of additional criteria describing limits for the motion sequence in addition to or even replacing some criteria used up to now.

Appropriate additional criteria seem to be the above mentioned trajectories of H-point and S-point as well as the deduced torso angle α during forward displacement (Fig. 1,2,3). The registration of these data can be achieved without any problem.

<u>Required Ranges for Criteria of Motion Sequence to Avoid Submarining</u> - Interpretation of Figures 1, 2 and 3 lead to certain ranges of the motion sequence criteria.

- Trajectory of H-point:

The main requirement is to avoid a vertical downward motion. The H-point has to remain on its initial level of h=0 during displacement, a gentle upward motion up to h=40 mm can be permitted (Fig. 4)

$$0 < h < 40 \text{ mm}.$$

Thereby, additional loads on the head-neck-region (rotational acceleration according to motion sequence I, Fig. 1) and in longitudinal di-

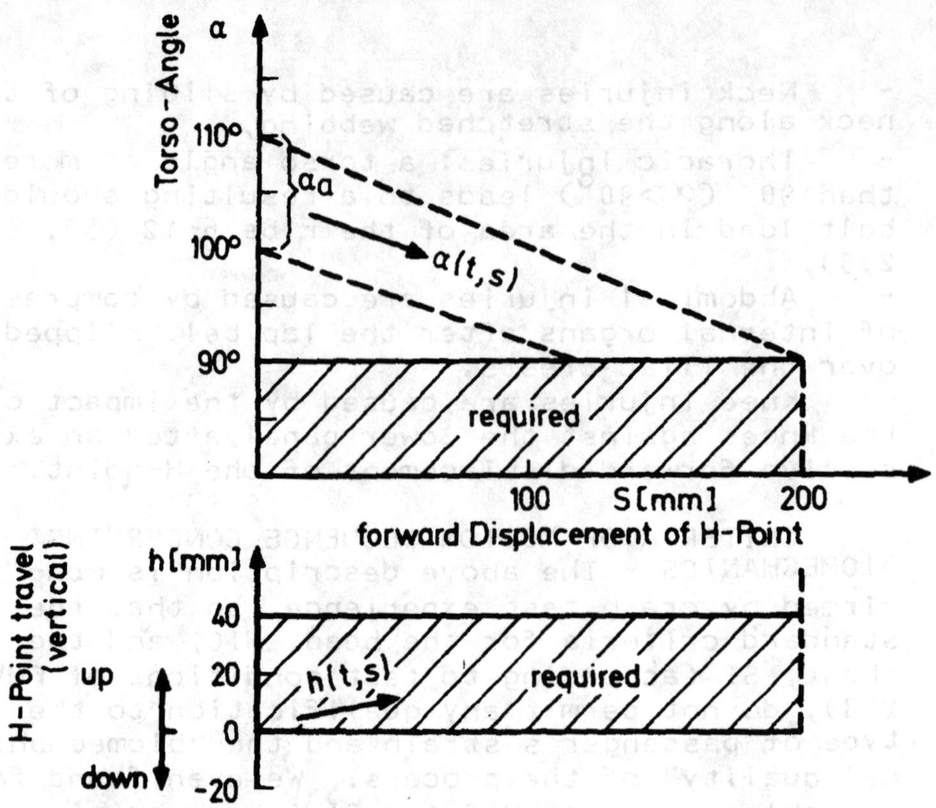

Fig. 4 - Required ranges of h- and α-values for optimum positioning of resultant belt load vectors

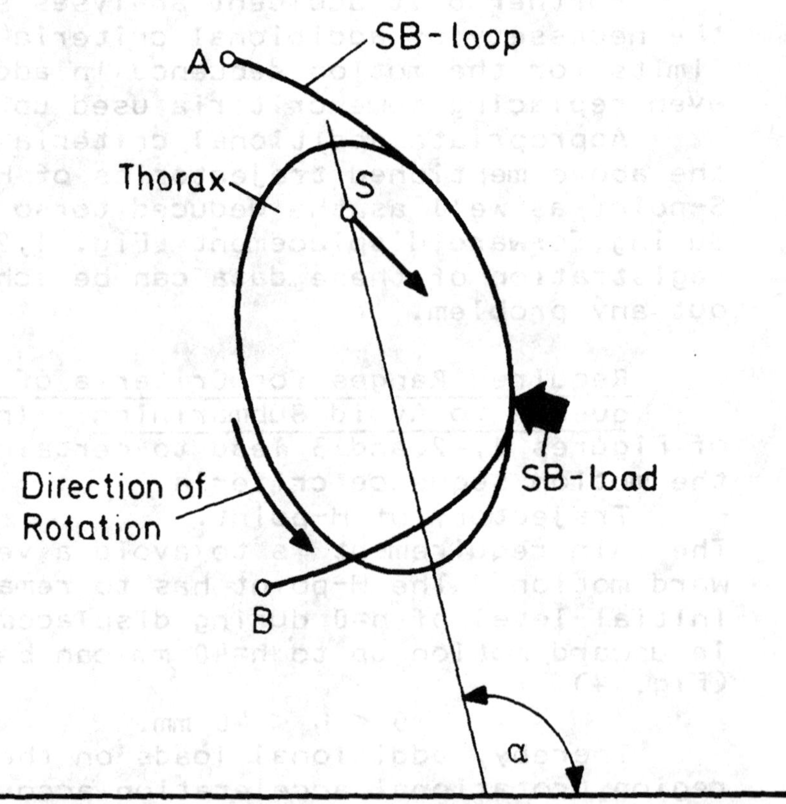

Fig. 5 - Sketch of thorax with shoulder belt (SB) for submarining-similar motion-sequence

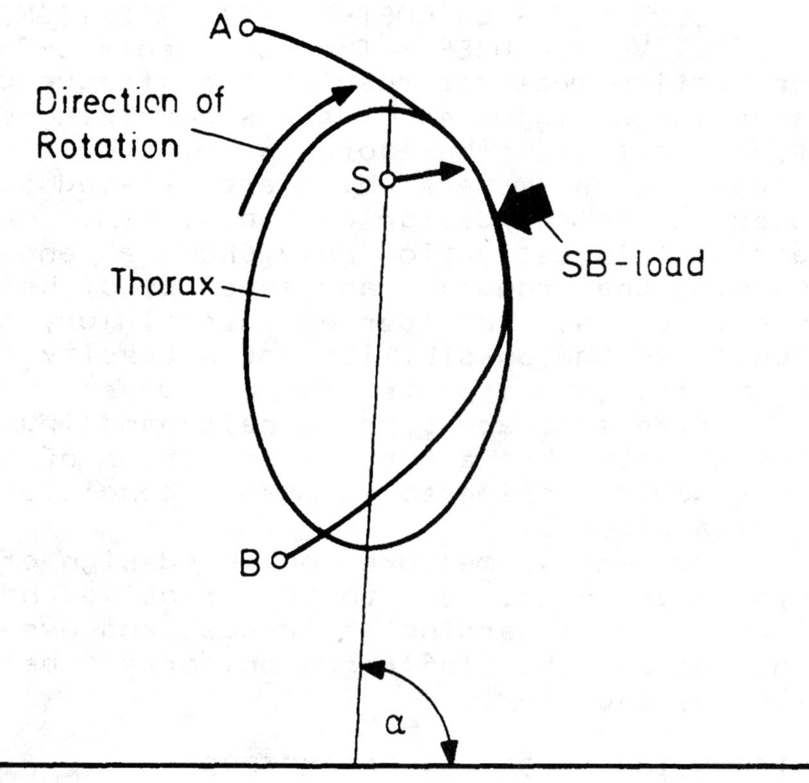

Fig. 6 - Sketch of thorax with shoulder belt (SB) for motion-sequence without any submarining

rection on the spine can be prevented. In the region of the ribs 6-12 the thorax is relieved of the vertical component of shoulder belt loads. A gentle upward motion would even shift the resulting shoulder belt force up to the sternum (Fig. 6 compared to Fig. 5). Thus at least one cause for submarining is eliminated.

- Torso angle α:

The main requirement is a decrease of the initial torso angle (α_a=100-115°) to a value of 90° or less at the moment of the maximum upper shoulder-belt force F_A (Fig. 1,3,6) and of the maximum forward displacement of the chest (F_A defined according to Fig. 7 and 12)

$$\alpha\,(F_{Amax}) \leq 90^\circ$$

This limit criterion guarantees that the resulting shoulder-belt loads are transferred to the sternum. Further, in connexion with an optimized chest deceleration versus time and displacement characteristic, a favourable strain of head and neck thus is achieved.

SOME NOTES ON COST-BENEFIT-RELATIONSHIP OF PROTECTIVE MEASURES - The cost-benefit-factor of protective measures results from frequency and severity of injuries of the submarining-seat-belt syndrome-type. Therefore, we must consider the causative parameters which are related to the user and to the design of the system. At present, accident investigation only shows a tendency concerning the frequency and severity of beltspecific injuries. Considering these injury causes we recognize the possibility and necessity for immediate additional protective measures:

Frequency and type of belt-handling mistakes demonstrate that a further education of the belt user would considerably improve conditions in real accidents.

Preventive measures on the design of the system belt-seat seem to be effective and very interesting regarding their cost and even regarding the possible influence on correct belt handling by the user.

DISCUSSION OF DESIGN PARAMETERS

Crash tests often assume ideal conditions concerning belt geometry and seat design (EC-standard) with its consequences on the motion sequence. Analysis of real accidents show important differences to these ideal test conditions and illustrate influences of the respective design parameters.

BELT GEOMETRY -

Positioning of A-Point - The positioning of the A-point is responsible for the location of the shoulder belt (SB) on the upper thorax. This position influences:

- the rotation of the torso around its vertical axis
- the endangerment of the neck by the stretched webbing (A" in Fig. 7)
- the possibility that the thorax will slip out of the SB-loop sideways (A' in Fig. 7)
- the application region of the resulting SB-load on the thorax.

Positioning of B-Point and the Length $\overline{VB}$ of the Buckle Part - The positioning of the connection of SB and LB, the V-point (Fig. 7,8), is determined by the location of the B-point and

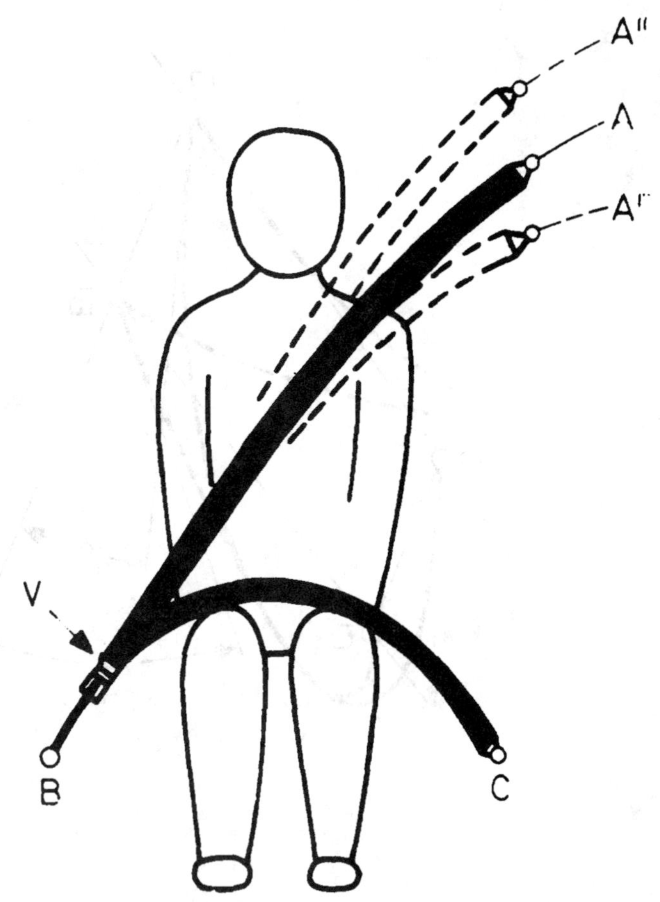

Fig. 7 - Definition of A-, B-, C- and V-Point and A-Point-Positioning

the length $\overline{VB}$ of the buckle part. This is responsible for the location of the SB on the lower thorax. A steep LB-angle β (see B' in Fig. 8a), principally desirable in order to avoid submarining, as well as a long buckle part $\overline{VB}$ (see V_{III} in Fig. 8) promote torso rotation around its vertical axis, since with this arrangement the SB-force F_A increases earlier than LB-loads. That means an excentric force is initially affecting the torso. Following Fig. 8 we can recognize a further influence of a over-long buckle part $\overline{VB}$: with an increase of the SB-force F_S the LB will be pulled up one-sided by the resulting vertical component F_{VIII}, often an initial reason for submarining-similar behaviour.

<u>Positioning of C-Point and LB-Angle β</u> - The location of the C-point determines the LB-angle β in connection with the position of B-point under consideration of positioning of B-point and the

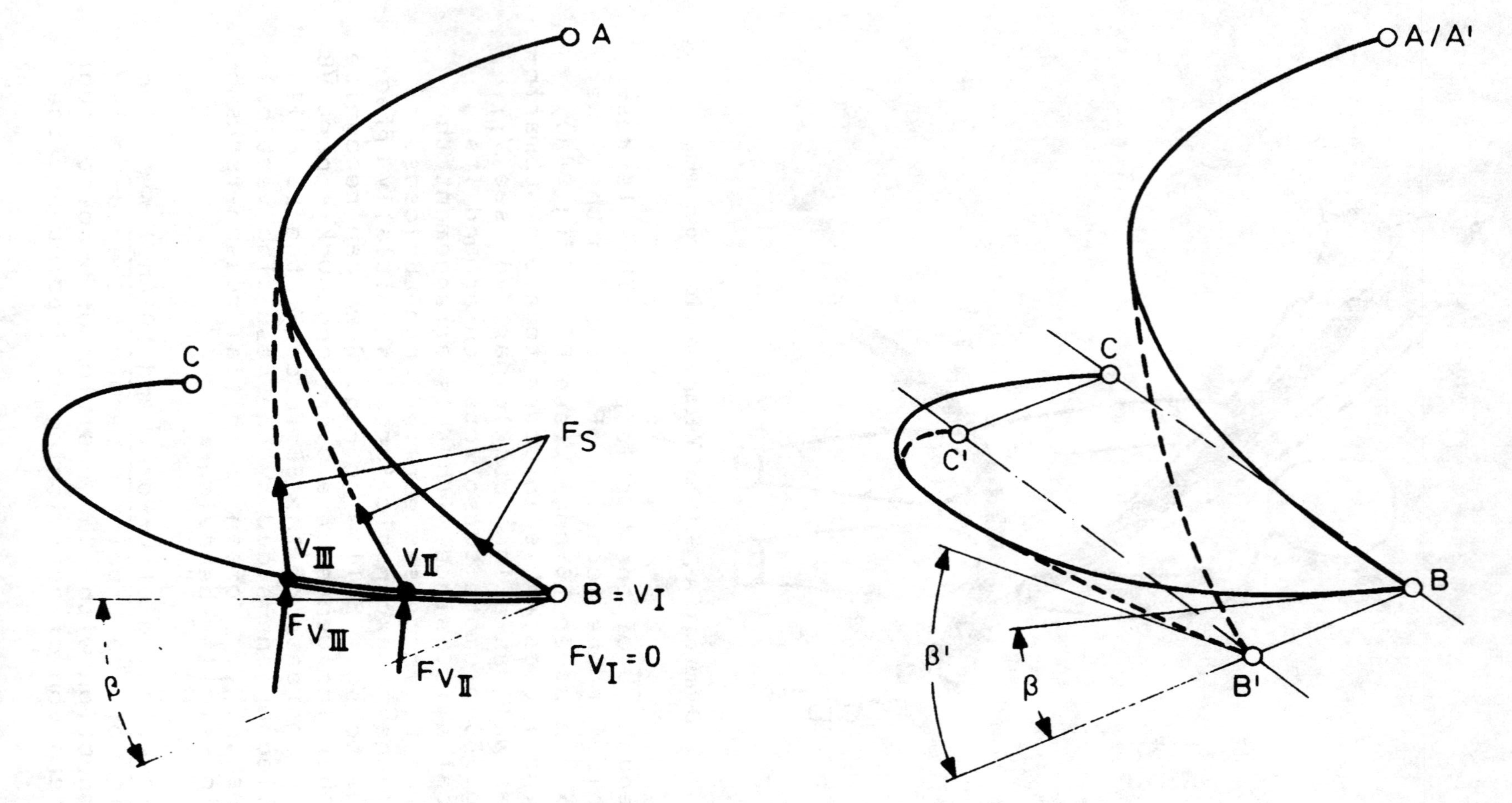

Fig. 8 - Influence of length VB on component F_v (F_v causes vertical displacement of lap belt)

Fig. 8A - Influence of B/C-Point-Positioning on angle β and shoulder belt geometry

length $\overline{VB}$ of the buckle part. The importance of β was often stated (6,7).

It is not possible to discuss peculiarities of β without regard to properties of the used seat (see "Influence of the Seat Cushion on the Downward Movement of the H-Point"). Principally there is no risk of LB-slipping over the iliac crests with steep LB-angles β in ranges about $\beta = 50^\circ - 70^\circ$. However, extremely large angles β produce disadvantages according to positioning of B-Point and the length $\overline{VB}$ of the buckle part (rotation of torso and delayed increase of LB-loads)

<u>Resulting Design Proposals for Belt Geometry</u>

The positioning of the anchorpoints A, B, and C should guarantee:

- a central location of the SB on clavicle and sternum
- a LB-angle β in a range of $\beta = 45^\circ - 50^\circ$ - limit for negligable torso rotation
- that the V-point is located close to the B-point - thereby less influence of SB on LB.

In order to achieve independance of the LB-angle β from passenger size and seat-adjustment we propose that the standards require the anchorpoints B and C be connected with the frame of the seat.

SEAT DESIGN -

<u>Influence of the Seat Cushion on the Downward Movement of the H-Point</u> - A vertical component F_{LBvert} of the LB-load F_{LB} results from the LB-angle β, the tangential iliac crest angle γ, a function of torso angle α ($\gamma(t) = f(\alpha)$, Fig. 9, 9a) acting on the iliac crest during forward displacement. F_{LBvert} presses the hip downward into the seat cushion. The absolute value of the downward motion of the H-point depends on the type of seat cushion, on the mentioned angles and on the level of F_{LB}.

The LB-angle β decreases coincident with the downward movement of the H-point. As a result the risk of LB-slip on the iliac crests increases - inducing submarining.

<u>Influence of Seat Cushion on Forward Displacement of H-point and Torso Angle α</u> - Elongation of LB and decrease of LB-angle β lead to H-point forward displacement. From the above

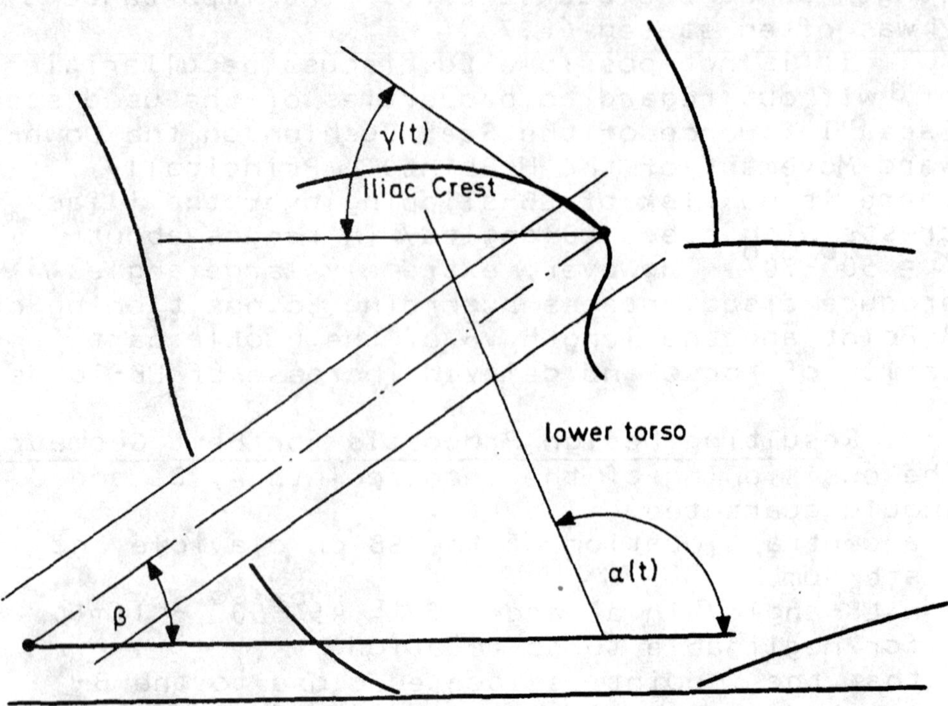

Fig. 9 - Relation between belt configuration, anatomical conditions and seating position

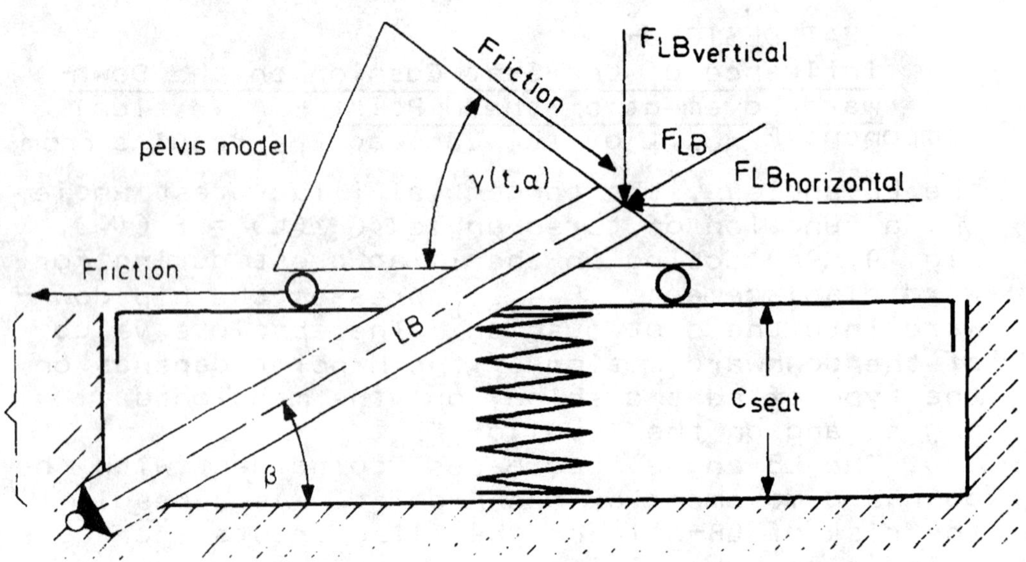

Fig. 9A - Lap belt forces resulting from pelvis and seat configuration, as transferred into a mechanical model

explained vertical H-point movement follows an additional decrease of LB-angle β. This is the reason why an additional forward displacement of the H-point occurs, but without remarkable energy absorption. The influence of H-point movement on torso angle α is obvious: if this forward displacement of the H-point is too great and too early compared to that of the S-point, then the required torso angle

$$\alpha(F_{Amax}) \leq 90^\circ$$

isn't attainable.

Consequences for Seat Designing - Our aim of limiting the vertical motion of H-point with the required range of the criterion $0<h<40$ mm has already been explained. LB-angle $\beta=45^\circ-50^\circ$ has been chosen according to "Resulting Design Proposals for Belt Geometry", which is a compromise in order to avoid submarining and torso rotation around the vertical axis. With this LB-angle the LB-loads will increase later than the SB-force F_A. We can compensate this annoying fact and reduce the restraining LB-loads on the iliac crests by inducing additional restraint forces into the ischia produced by the seat cushion.

The design problem is to ensure energy absorption during the compression of the seat cushion. The design of a seat with these properties is shown in Fig. 10: a sheet metal pan with a reinforced front wall serves as the frame for this seat. In order to fulfil special requirements the cushioning consists of different types of soft foam.

For the average requirements of seat comfort we chose a soft foam of a volumetric weight of 30 kg/m^3 in the area of vertical static and dynamic compression by the ischia (C_1 in Fig. 10). In the forward area of the seat pan we used a soft foam of 53 kg/m^3 (EA in Fig. 10). A soft foam of 28 kg/m^3 with a thickness of 30 mm covered the entire seating. This seating design does not negatively influence the seat comfort.

During crash conditions the lower pelvis and ischia compresses the EA-foam against the front wall of the seat pan on a large area and with respective relative velocity. The vertical component F_{LBvert} of the LB-loads (Fig. 9a) prevents the pelvis from sliding along the seat cushion

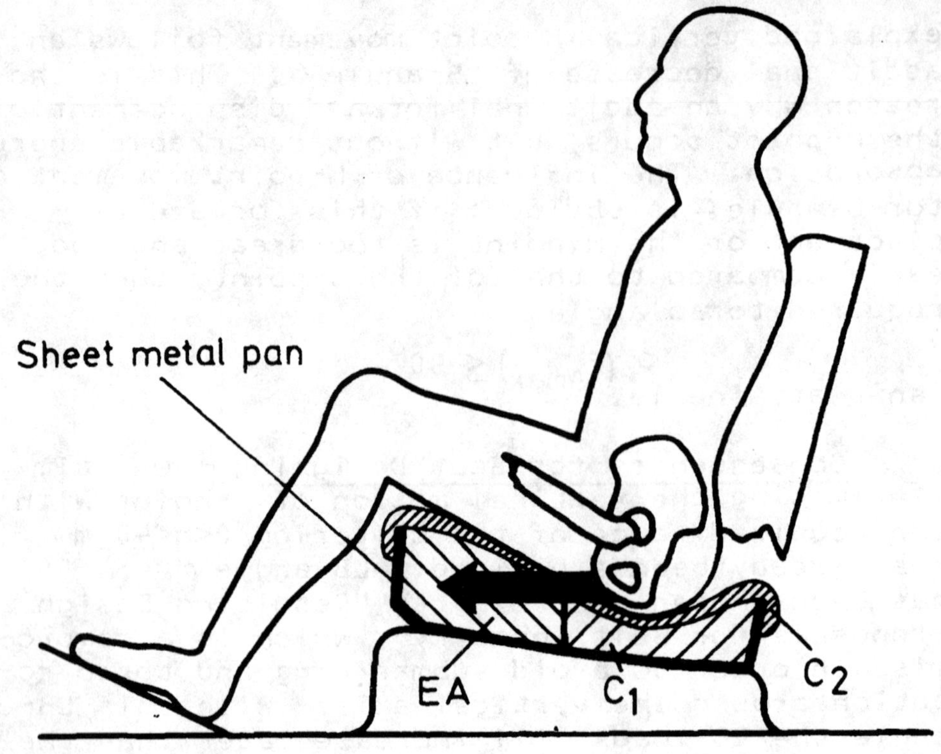

EA = energy absorbing foam ($53\ kgm^{-3}$)
C_1 = comfort foam ($30\ kgm^{-3}$)
C_2 = all over comfort foam ($28\ kgm^{-3}$)

Fig. 10 - Sketch of an energy absorbing seat

without compressing the EA-foam. In this case the LB functions chiefly in keeping the pelvis down.

CRASH TESTS AND TEST RESULTS

The influence of the discussed design parameters were examined in a crash-test series.

TEST SET-UP - Fig. 11 shows the simulated 50 km/h-wall impact deceleration, Fig. 12 is a sketch of the test sled. We used the Alderson dummy VIP 1030. The test sled was equipped with

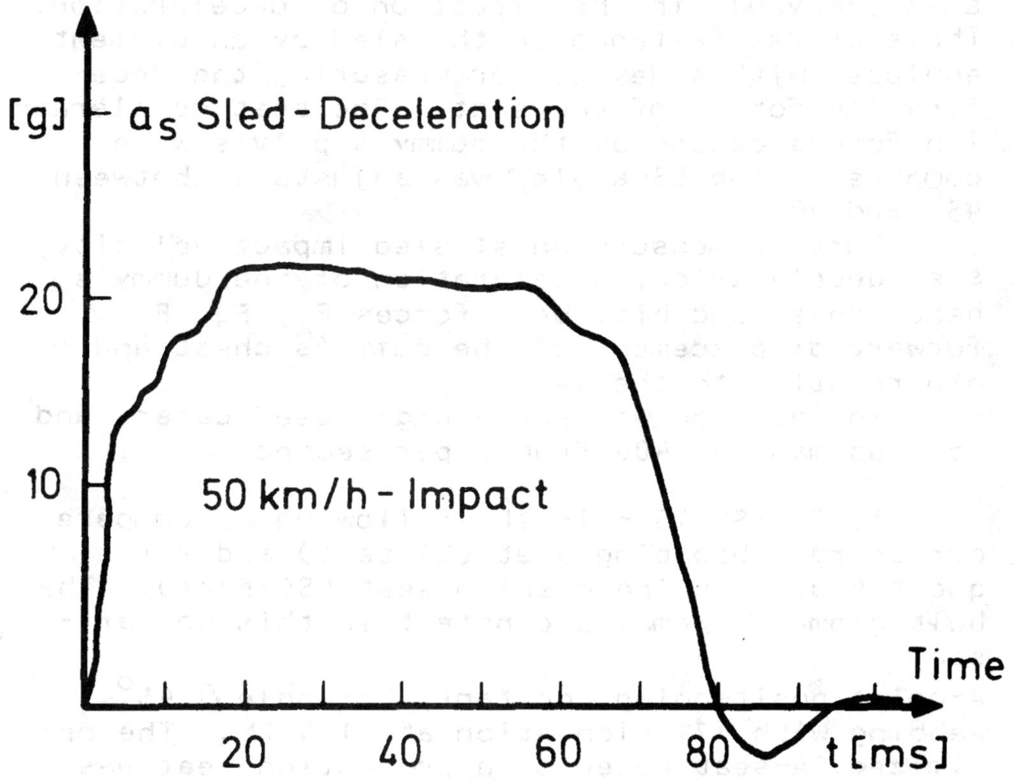

Fig. 11 - Deceleration-Time-History of Sled-Test

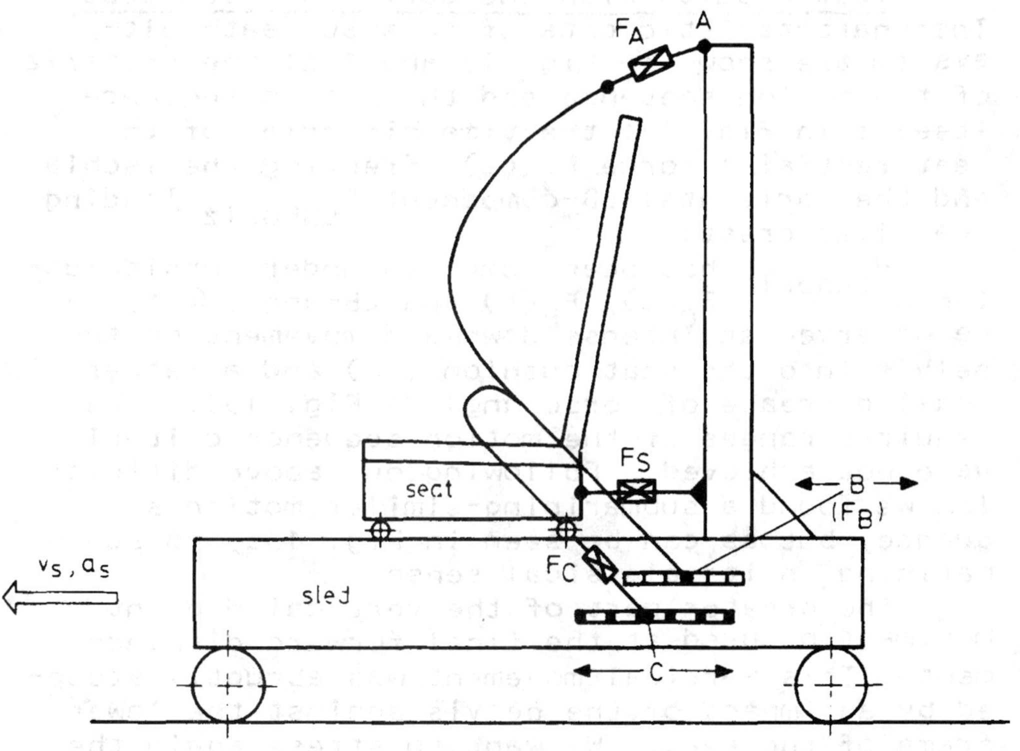

Fig. 12 - Sketch of test set-up

a seat movable in the direction of deceleration. The seat was fastened to the sled by an element equipped with a device for measuring the deceleration forces of the seat. The seat deceleration forces caused by the dummy's pelvis were computed. The LB-angle β was adjustable between 45° and 90°.

Further measurements: sled impact velocity, sled deceleration, deceleration of the dummy's head, chest and hip, belt forces F_A, F_B, F_C, forward displacement of the dummy's chest and hip relative to the seat.

In addition we used a high-speed camera and took as many as 400 frames per second.

TEST RESULTS - In the following we compare our energy absorbing seat (EA-seat) and a frequently used spring cushion seat (SC-seat). The belt geometry remains constant in this comparison:
A-point positioning constant, LB-angle $\beta=45^{\circ}$, webbing with 17% elongation at 11.5 kN. The optimized EA-seat based on a production seat was considerably modified. These modifications resulted from the current crash tests.

Test Results with the Spring Cushion Seat - The characteristic data of this SC-seat-belt-system are shown in Fig. 13 and 13a: the criteria of the motion sequence and the motion sequence itself; in Fig. 14: the time histories of the seat restraint force $F_{SD}(t)$ affecting the ischia and the horizontal LB-component $F_{LBhoriz}$ loading the iliac crests.

$F_{LBhoriz}$ has been computed under consideration of $F_C(t)$, $F_B(t)$ and LB-angle $\beta(t)$. We observed an intense downward movement on the pelvis into the seat cushion (-h) and a rather small decrease of torso angle α (Fig. 13). The required ranges of the motion sequence criteria were not achieved. Following our above difinition we found a submarining-similar motion sequence, but as can be seen in Fig. 13a, no submarining in its classical sense.

The greater part of the vertical H-point movement occured at the final forward displacement. This vertical movement was abruptly stopped by an impact of the pelvis against the lower frame of the seat. We want to stress again the

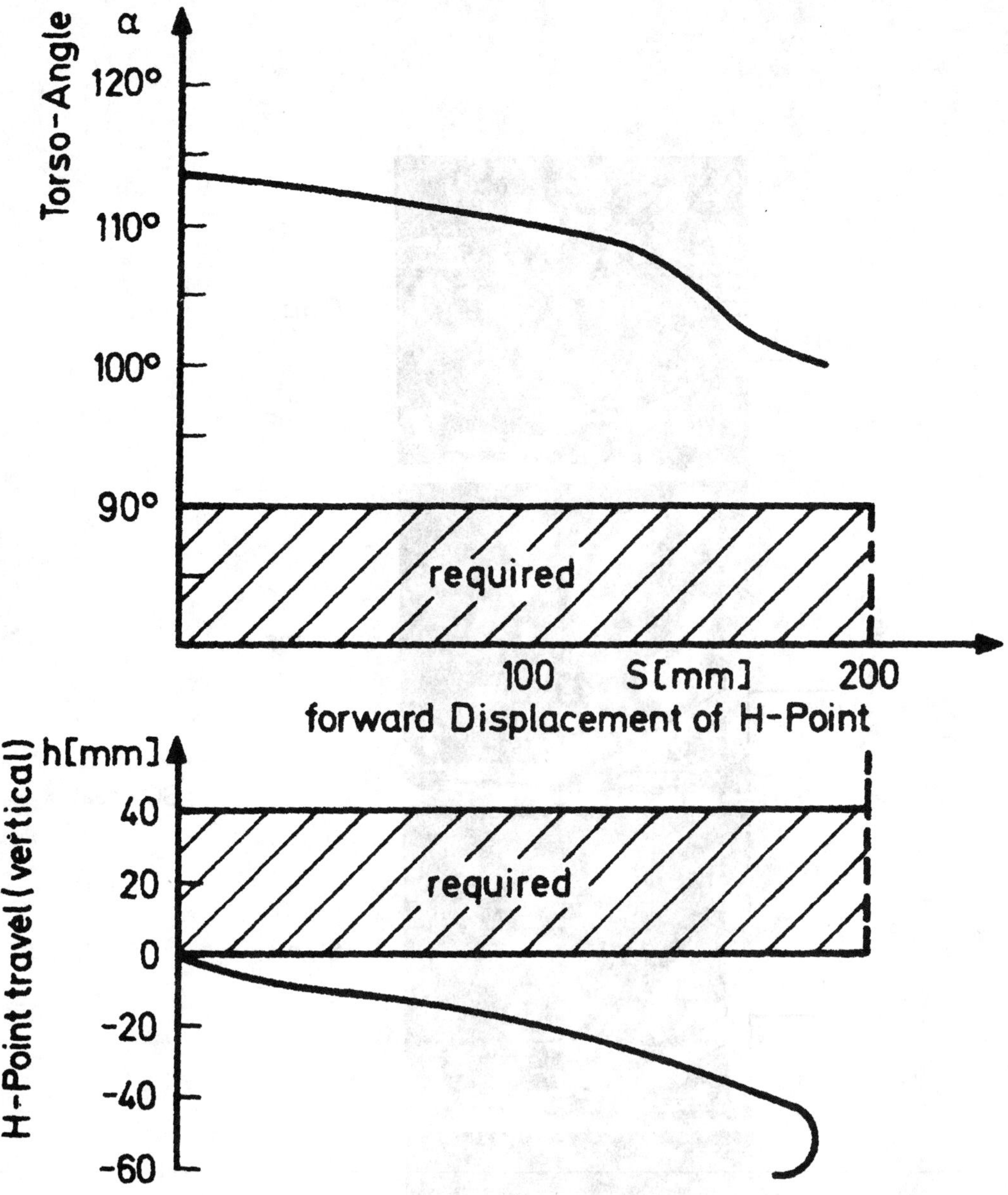

Fig. 13 - Motion-sequence-criteria h and α of tests with spring-cushion-seat (SC-seat)

fact of the resultant additional loads, explained by "Motion Sequences" and "Criteria of Motion Sequence Concerning Biomechanics". Torso angle α is a typical result of the forward displacement of the H-point versus time.

The main part of the lower torso restraint had to be brought about by the LB with its horizontal component $F_{LBhoriz}$ (Fig. 14, 9a).

The LB-load level was acceptable, however, the excessive forward displacement of the H-point

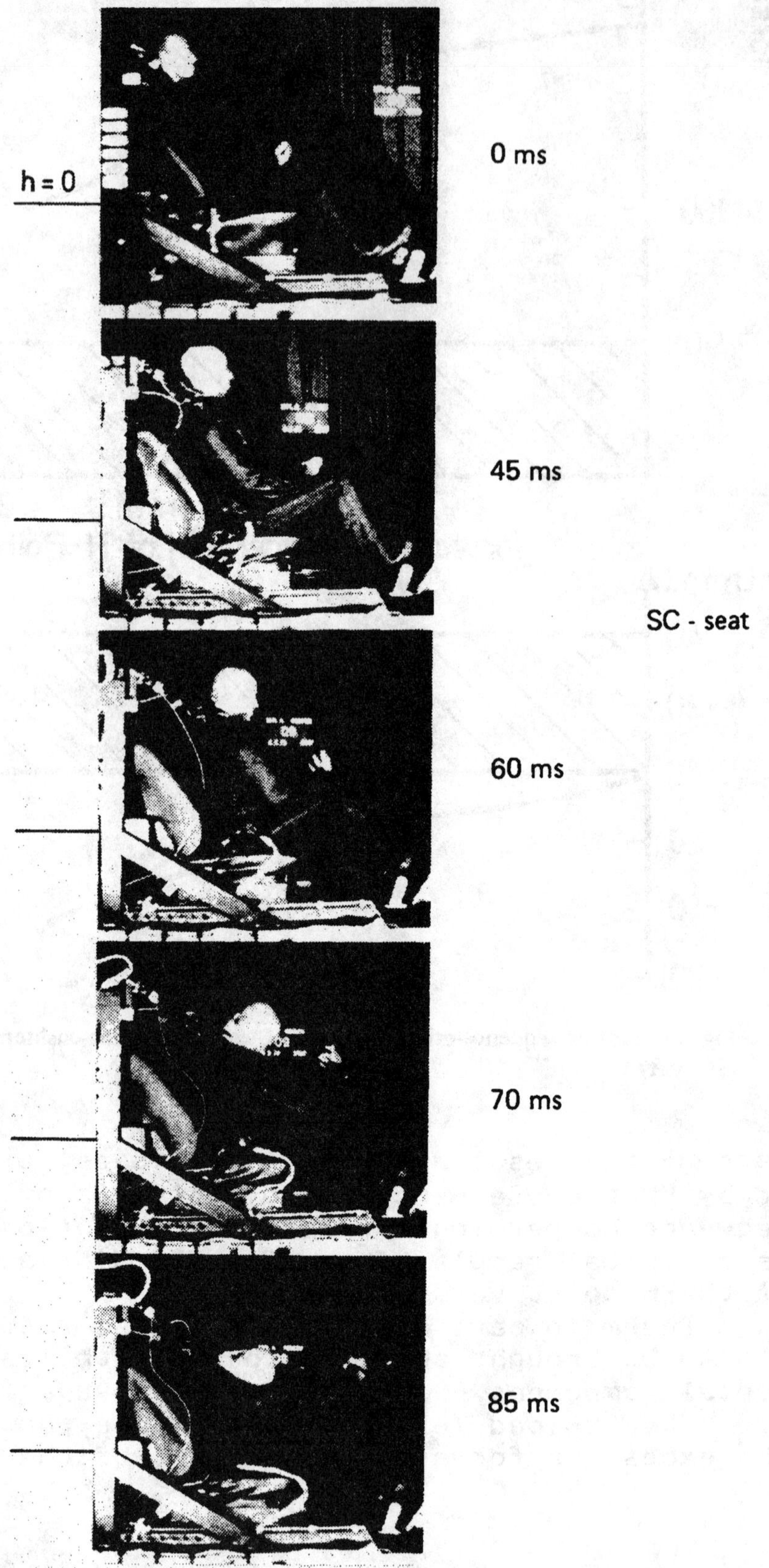

Fig. 13A - Motion-sequence of dummy seated on SC-seat

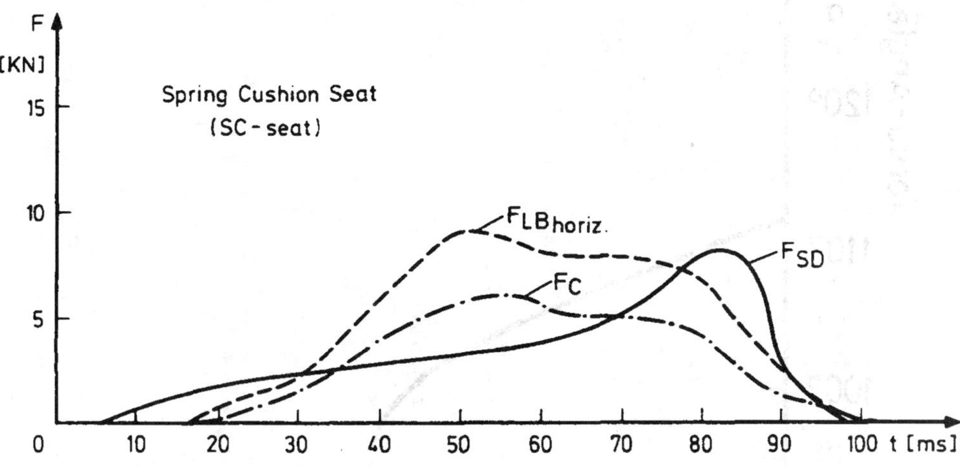

Fig. 14 - Time histories of seat-restraint-force F_{SD}, horizontal LB-restraint-$F_{LBhoriz}$ and force of anchorpoint C/F_c

shows that it is not possible to further reduce the LB-loads.

The torso angle α, in this case with a final value of about 100°, should have been brought down to about 90° by use of appropriate SB-webbing, but with the coincident risk of head- and chest-contact with elements of the inner compartment.

The unfavourable influences of the downward H-point movement (following "Motion Sequences" and "Required Ranges for Criteria of Motion Sequences to Avoid Submarining") persist without any restrictions.

Test Results with the Energy-Absorbing Seat

For our design the required ranges of the motion sequence criteria were clearly sustained and our EA-seat-belt-system does not show any type of submarining (Fig.15 and 15a).

We observed the expected slight forward displacement of the H-point accompanied by a slight upward motion. In spite of this LB-load did not significantly exceed the values of the SC-seat-belt-system. Therefore, further reduction of LB-load seems to be possible.

The main part of the restraining forces on the lower torso was produced by the EA-seat (see $F_{SD}(t)$ in Fig. 16). The line of the seat restraint force $F_{SD}(t)$ shows the efficiency of the EA-foam. The F_{SD} peak value of about 18 kN seems to be acceptable, because the seat restraint

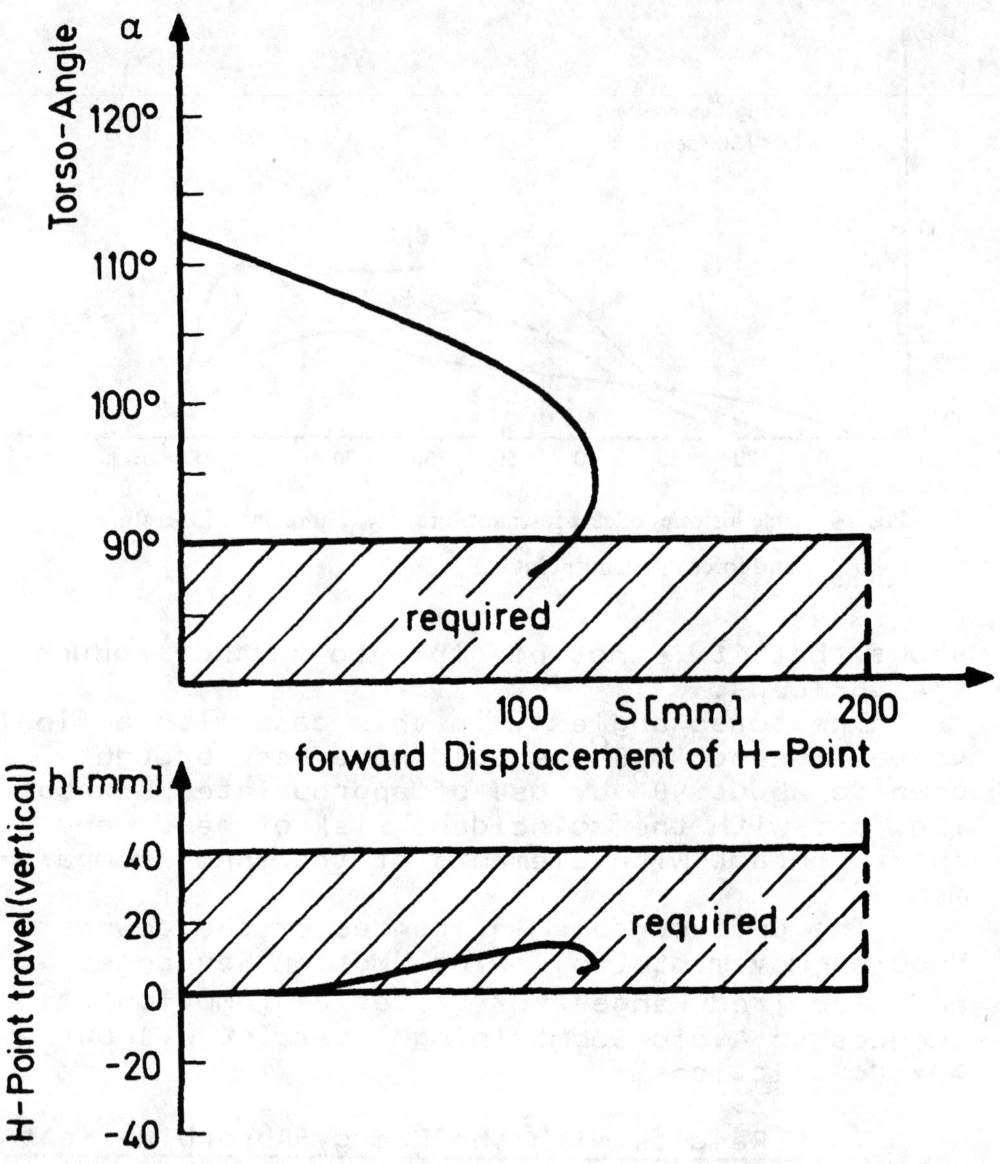

Fig. 15 - Motion-sequence-criteria h and α of tests with energy-absorbing-seat (EA-seat)

force was applied on a large area into the ischia region without local peak values.

The abrupt descent of the horizontal component of the LB-load $F_{LBhoriz}$ at the maximum seat restraint force F_{SD} is remarkable.

<u>Summary of the Sled-Test-Results</u> - The fact that we observed no classical submarining in the SC-seat-test-series results from ideal test conditions concerning belt geometry and belt slack. With the EA-seat it was possible to avoid a submarining-similar motion sequence according to

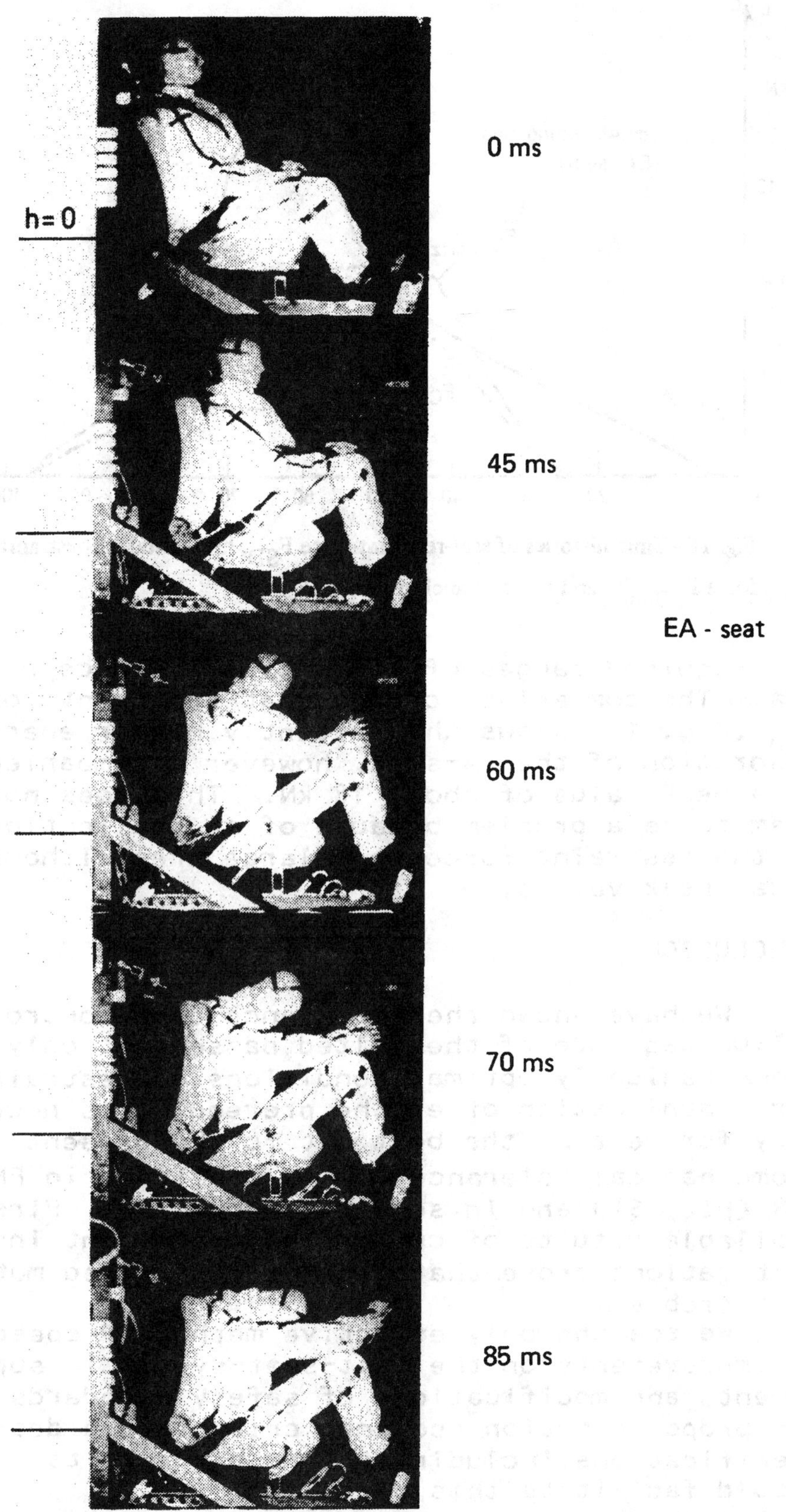

Fig. 15A - Motion-sequence of the dummy seated on the EA-seat

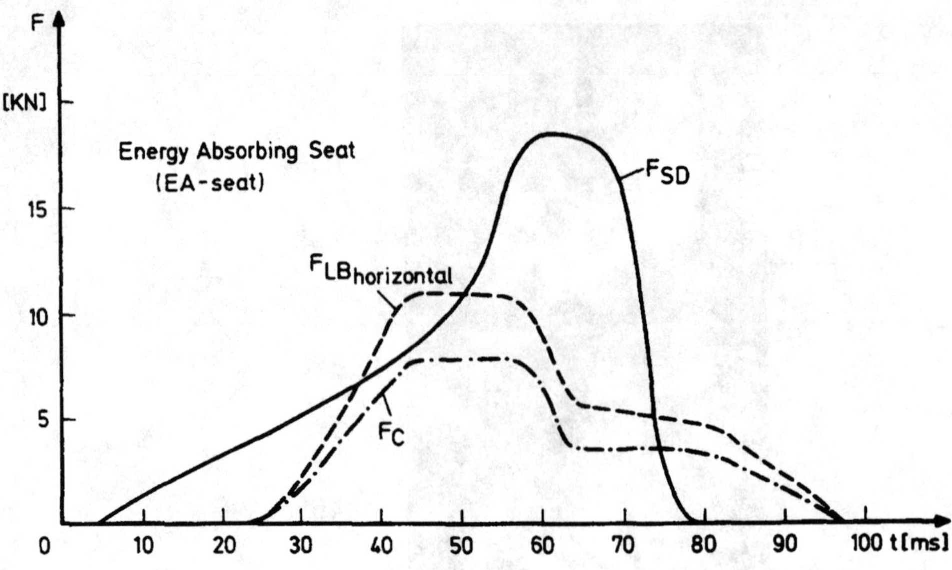

Fig. 16 - Time histories of seat-restraint-force F_{SD}, horizontal LB-restraint-force $F_{LBhoriz}$ and force of anchorpoint C/F_c

our required ranges of the motion sequence criteria. The comparison of the seat restraint forces F_{SD} (Fig. 17) shows the evidently higher energy absorption of the EA-seat, however accompanied by a peak value of about 18 kN. This does not seem to be a problem because of an application of the restraint force on a large area without local peak values.

CONCLUSION

We have shown the importance of a controlled motion sequence of the belted passenger: only biomechanically optimal conditions of restraint force application offer the prerequisites necessary for getting the best out fo the present biomechanical tolerance limits (included in FMVSS 208 (HIC, SI) and in some EC-proposals). First available results of current belt accident investigations prove that lack of controlled motion is a problem.

We see the only effective method of speeding up improvements on the belt-seat-system in supplements and modifications of safety standards. Our proposed motion sequence criteria and design specifications including their test results should facilitate this direction.

We believe that these criteria could replace without disadvantages the problematical Head-

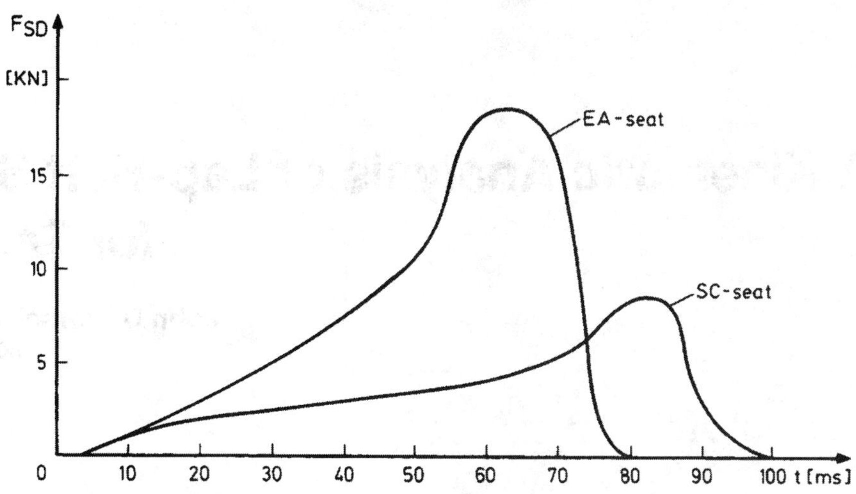

Fig. 17 - Seat-restraint-forces F_{SD} of a spring-cushion-seat and an energy-absorbing seat under similar crash-conditions

Injury-Criterion and Chest-Severity-Index, if accompanied by a limitation of the SB-force $F_A(t)$ and the defined displacement of the chest and hip.

We introduced one possible belt-restraint-seat-design with its favourable biomechanical properties. The sled test results explain the use and the effectiveness of the required ranges of the criteria of the motion sequence.

Our proposed design solution demonstrates the direction of necessary and possible improvements.

REFERENCES

1. W. Appel, D. Adomeit et al, "Verletzungen durch einen 3-Punkt-Automatik-Gurt." Monatszeitschrift für Unfallheilkunde 78, 1975.

2. C. G. Bäckström et al, "Untersuchung von Verkehrsunfällen." First part of a Saab Study.

3. N. I. Bohlin, "A Statistical Analysis of 28,000 Accident Cases with Emphasis on Occupant Restraint Value." Paper 670925, 11th Stapp Car Crash Conference.

4. J. M. Henderson et al, "Seat Belts - Limits of Protection: A Study of Fatal Injuries Among Belt Wearers." Paper 730964, 17th Stapp Car Crash Conference.

5. L. M. Patrick et al, "Three-Point Harness Accident and Laboratory Data Comparison." Paper 741181, 18th Stapp Car Crash Conference.

6. J. D. States, "Trauma Evaluation Needs." G.M. Symposium, Edit. Plenum Press, N.Y., 1973.

7. C. H. Tarriere, "Proposal for a Protection Criterion as Regards Abdominal Internal Organs." Proceedings of 17th AAAM

892441

A Kinematic Analysis of Lap-Belt Submarining for Test Dummies

John D. Horsch and William E. Hering
Biomedical Science Dept.
GM Research Labs.
Warren, MI

ABSTRACT

A kinematic view of the test dummy pelvis "unhooking" from the lap-belt was developed from a series of sled tests. The dynamics of the test resulted in reducing the vertical angle of the lap-belt and rearward rotation of the top of the pelvis. Both of these motions acted to "unhook" the belt from the pelvis. When a "critical" angle between the belt and pelvis was reached, the belt "slipped" from the pelvic spines and directly loaded the abdomen. In these tests, rearward rotation of the pelvis was a predominant mechanism.

The study also identified a threshold test severity. At test severities less than the threshold, the dummy did not submarine and at severities greater than the threshold the dummy submarined. The critical pelvis-to-belt slip angle and threshold test severity associated with the pelvis unhooking from the belt are parameters that can enhance assessment of submarining performance beyond a yes/no evaluation.

LABORATORY TESTING OF belt-restraint systems can, in some situations, result in the dummy pelvis "unhooking" from the lap-belt, with the belt directly loading the abdominal region. There are many issues related to assessment of belt restraint performance associated with this observation of direct abdominal loading. One group of issues is directed toward field relevance, another directed toward laboratory assessment of submarining, and still another is directed toward restraint performance for a range of dummy sizes and seating positions in laboratory tests.

Field relevance includes such considerations as: the frequency and severity of abdominal injuries due to belt loading observed in the field; and what factors are associated with abdominal injuries from belt loading. Laboratory assessment includes: detection of abdominal loading by the belt; the perceived risk of injury if direct abdominal loading by the belt restraint does occur; the biofidelity of the dummy response; and the relevance of the test conditions to field conditions [1-9].

The development of restraint systems which prevent the dummy pelvis unhooking from the lap-belt in laboratory tests presents different types of issues: the influence of restraint and test exposure parameters; the submarining tendency of various dummies; or test adjustments such as dummy or belt positioning. There have been a variety of recommendations in the literature for system configurations to reduce the tendency of belt slippage from the pelvis [10-15].

This paper summarizes test data that provides a basis for a kinematic analysis of laboratory data related to the dummy's pelvis unhooking from the lap-belt and provides suggestions for methods of performance analysis in addition to whether or not the dummy submarined.

METHOD

TEST ENVIRONMENT - The sled test fixture was developed to provide a simple, repeatable and reproducible test environment in which the submarining responses of various surrogates could be evaluated and compared. The fixture, Figure 1, consisted of a horizontal seat pan and a flat seat back having a rearward angle from vertical of 20°. The seat pan was 355 mm wide, sufficient to completely support the subject but narrow enough to prevent contact with the lap-belt. The seat pan was constructed from a flat, metal plate covered by a ribbed vibration isolation padding about 12 mm thick which was covered by vinyl seating material. The seat back was plywood covered by a 25 mm thick padding to reduce rebound forces. The sled fixture was based on the fixture used for belt restraint studies with human cadaver subjects at the University of Michigan (HSRI) [16].

The lap-belt was separate from the shoulder belt to eliminate effects of the shoulder-belt lifting the lap-belt. Lap-belt anchors were 381 mm laterally apart, allowing adjustment in the vertical location. Belt webbing was the same as used in the HSRI belt restraint tests [16]. Upper-body restraint was symetrical about the midsaggital plane to give a symetrical response and consisted of dual diagonal shoulder-belts crossing about mid- sternum. Belt upper anchor locations were typical of those used in the HSRI studies [16]. Belts were adjusted "snug" (approximately 45 N). The upper torso belt configuration was intended to minimize variations in shoulder belt/torso interactions which could influence lap belt interactions with the pelvis.

Tests with the Part 572 dummy were conducted using a square-wave (constant level) deceleration pulse with a 3 to 5 ms rise time. The stopping distance (dynamic crush) was proportional to sled velocity (10.4 mm /kmph).

Quasi-static tests to determine belt-to-pelvis slip angle were conducted with the fixture shown in Figure 2. The horizontal seat pan was covered by a flat plate supported on rollers to provide low seat-friction. The lap-belt configuration was the same as in the dynamic tests. Lap-belt forces were generated by pulling on the subject's legs. Testing was accomplished by applying the force to the legs and then rotating the pelvis slowly rearward while maintaining this force until the belt unhooked from the pelvis, directly loading the abdominal region.

The Part 572 dummy was dressed in cotton "thermal" underware, with the shirt tucked into the pants to provide a double layer of clothing under the lap-belt. The dummy was modified for a front mounted photographic target for setting the initial pelvis angular position and measuring the dynamic angular position from high-speed movies. The lap-belt was placed as low on the pelvis as practical and adjusted snug with an approximate 45 N preload.

RESULTS

SLED TEST RESULTS - A series of sled tests were conducted for each of three restraint configurations. The test configurations differed by the lap-belt angle ("steep" or "shallow") or the clothing on the dummy (cotton clothes or none). The purpose of these tests was to evaluate two hypotheses: 1) a "submarining" threshold severity concept, and 2) a threshold or critical belt-to-plevis slip angle.

Sled test results are provided in Table 1. The belt-to-pelvis angle, see Figure 3, is defined as the angle between the side view projection of the lap-belt and the pelvic target, directly determined from high-speed movies. The lap-belt had little spread laterally between the pelvis and anchors and thus the belt was nearly in the plane of the side projection. This differs greatly from some current belt configurations. The pelvic target was parallel to the lumbar spine mounting surface and located in the midsagittal plane.

Unhooking of the pelvis from the lap-belt results in a relative slip motion between the lap-belt and pelvis, referred to in this paper as "belt slip". Belt slip from the pelvis was strongly associated with the rearward rotation of the pelvis. The amount of rearward rotation was directly influenced by test severity.

QUASI-STATIC TESTS - Quasi-static tests were performed using the "shallow" lap-belt anchor location and clothing/no clothing combinations as in the sled tests. Table 2 provides a summary of test data in terms of the belt force (the sum of the right and left belt tensions) and the angle between the lap-belt and the pelvic target at the time of the belt slipping from the pelvis. The belt-to-pelvis angle at which the belt slips from the pelvis depends on belt force and dummy clothing.

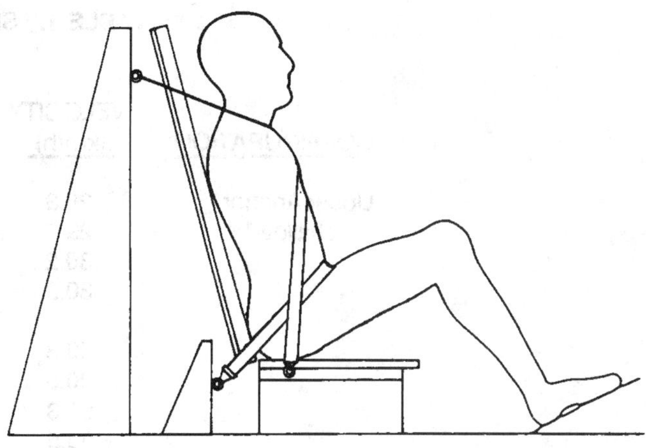

Figure 1. Schematic of the sled fixture.

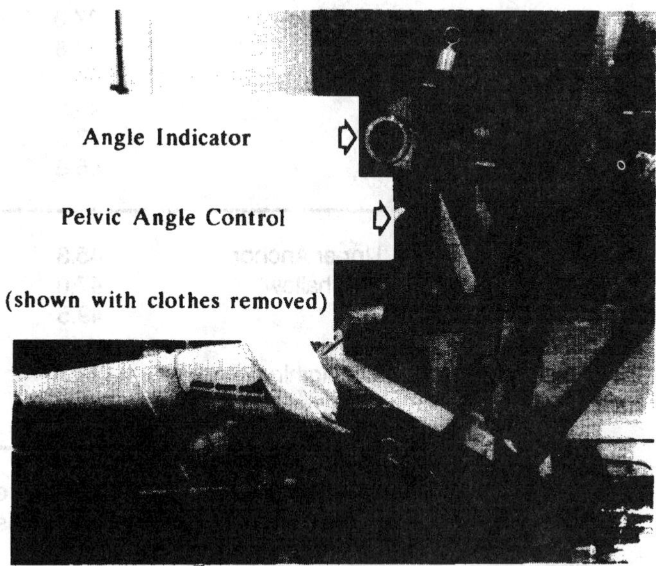

Figure 2. Quasi-static test method using the modified sled fixture. Load was developed by pulling on the legs.

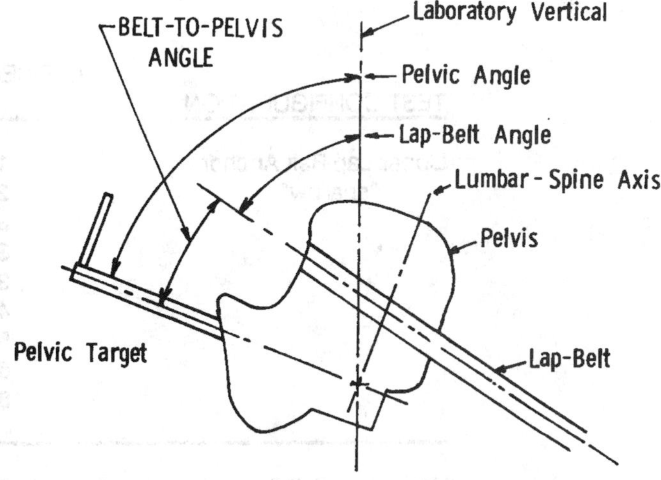

Figure 3. Definition of lab-belt and pelvic angles. Belt-to-pelvis angle is the lap-belt angle minus the pelvic angle (slightly negative as shown).

TABLE 1. SLED TEST RESULTS

CONFIGURATION	VELOCITY (km/h)	BELT-TO-PELVIS ANGLE* (degree)	SUBMARINING?
Upper Anchor "shallow"	26.3	21	NO
	29.7	28	NO
	30.2	29	NO
	30.2	31	NO
	30.3	29	YES
	30.5	30	YES
	33.3	30	YES
	33.5	31	YES
Lower Anchor "steep"	33.3	24	NO
	37.0	29	NO
	39.8	32	NO
	43.2	33	YES
	43.3	33	YES
	46.8	34	YES
Upper Anchor "shallow"	45.3	46	NO
	47.0	46	NO
	49.5	47	NO
No Clothing	53.7	47	YES
	53.8	51	YES

* At the time of belt slip or if the belt did not slip, the maximum observed in the test.

TABLE 2. QUASI-STATIC BELT-SLIP TESTS

TEST CONFIGURATION	LAP-BELT FORCE (kN)	BELT-TO-PELVIS SLIP ANGLE (degree)
Upper Lap-Belt Anchor "shallow"	1.31	23
	2.57	28
	3.10	30
	3.59	30
	3.86	32
	4.99	34
	5.16	35
	6.00	35
	8.01	38
Upper Lap-Belt Anchor "shallow" No Clothes	2.52	51

DISCUSSION

The belt-to-pelvis angle at which the belt slips off the pelvis and the threshold test severity provide measures of the tendency to submarine for a given situation. The sled test data from Table 1 is plotted in Figure 4. The belt-to-pelvis angle tabulated and plotted is at the time of slippage of the belt from the pelvis or the maximun observed, if the belt did not slip off the pelvis.

SEVERITY - One trend indicated in Figure 4 is the existence of a submarining threshold severity for each test configuration --- below this test severity (plotted as test velocity) the dummy did not submarine and above this threshold the dummy submarined. For example, making the lap-belt angle steeper increased the threshold submarining severity from 30 km/h to 42 km/h. Increasing the coefficient of friction between the belt and dummy (removing the cloths from the dummy) increased the threshold submarining severity from 30 km/h to 50 km/h at the shallower belt angle.

The submarining threshold concept suggests that for each test environment, subject, and restraint adjustment, a submarining tendency could be established and used for performance analysis based on test severity. However, this technique requires many tests to be used and limits its application as an analysis technique. It might be useful for comparing the submarining tendency among types of dummies (i.e., a 5th vs 50th dummy, a Hybrid III vs a Part 572 dummy).

BELT-TO-PELVIS "CRITICAL" SLIP ANGLE - A second trend indicated in Figure 4 is the correlation of the belt-to-pelvis angle with the belt slipping off the pelvis. The data suggests a "critical" angle at which submarining occurs -- the dummy will not submarine if this angle is not achieved and will submarine when it is. The higher coefficient of friction associated without clothing on the dummy indicates a larger "critical" belt-to-pelvis angle before the belt can slip off the pelvis. A theoretical model is presented in Figure 5 to help explain the tendency for a "critical" belt-to-pelvis angle for submarining. The belt will slip when the slip force is greater than the friction force: or $T \sin \theta > \mu T \cos \theta$; thus when $\tan \theta > \mu$. The "critical" belt-to-pelvis slip angle (θ) depends on the coefficient of friction (μ). Lifting of the lap-belt by the shoulder-belt reduces the "critical" slip angle (θ).

The simple belt slip model shown in Figure 5 assumes that the iliac spines are straight and that the force resisting slippage of the belt is due to friction (the normal force times the coefficient of friction). The belt will slip when the slip force exceeds the resisting force. The model indicates that the two key parameters are the belt-to-pelvis angle and the coefficient of friction. Of course shoulder belt lifting of the lap-belt would add to the slip force and the belt could slip off the pelvis at a smaller belt-to-pelvis angle. The model does not consider the "flesh" over the pelvic bones, the complex shape of the pelvis, or the three dimensional nature of the belt and pelvic geometry, some of these factors discussed by Leung, et al [3].

The sled tests listed in Table 1 were conducted, in part, to explore the concept of a "critical" belt-to-pelvis angle and to test the importance of the coefficient of friction suggested by the theoretical model. The test data does indicate a tendency for the dummy to submarine at a given belt-to-pelvis angle for each test configuration and demonstrates that increasing the coefficient of friction (by removing the clothes in these tests) does significantly increase the "critical" belt-to-pelvis slip angle (and thus decrease the tendency of the dummy to submarine).

Quasi-static tests were conducted to further explore if a well defined "critical" belt-to-pelvis slip angle exists as suggested by the sled test data and by the belt slip model. The "critical" belt-to-pelvis angle is plotted as a function of lap-belt force in Figure 6. The trends indicate that the belt is better hooked at higher loads (requires a greater belt-to-pelvis angle for the belt to slip from the pelvis). This is likely due to the simulated flesh having greater

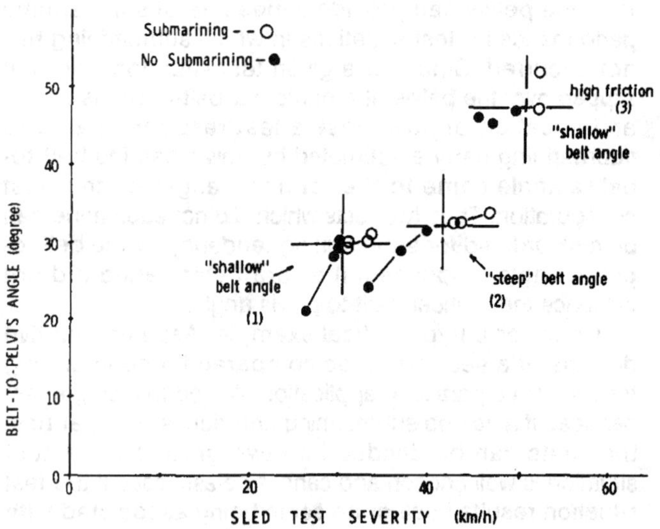

Figure 4. Sled test data for three test configurations. For each configuration there is a threshold test severity and a "critical" belt-to-pelvis angle at which the belt slips over the pelvis. Plotted belt-to-pelvis angle is the angle at the time of slip or maximum for no slip.

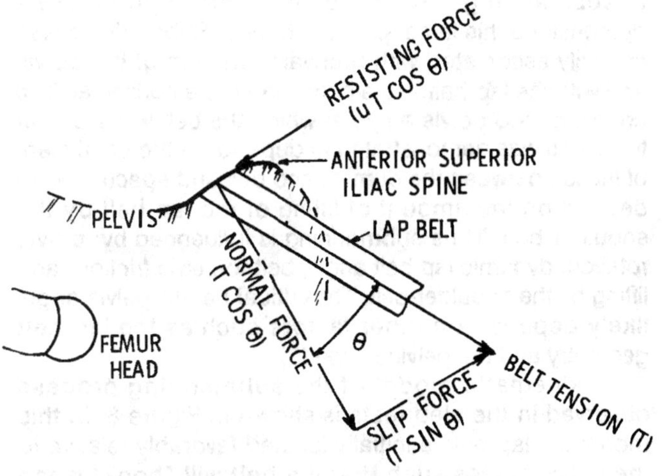

Figure 5. Theortical belt slip model.

compression at higher loads. The belt-to-pelvis angle at the time of submarining from the sled tests is also plotted in Figure 6 as a function of peak belt force (the belt force at the time of submarining is not available; importantly the belt tension at the time of submarining is not necessarily the maximum, see Figure 7). The data indicate that the belt will slip off the pelvis at a smaller angle dynamically than statically for the same belt load. This might be due to rate sensitivity of the dummy flesh not allowing deep deflection or a dynamic vs static friction difference. The small difference in the "critical" belt-to-pelvis angle between the "shallow" and "steep" lap-belt angles (Figure 4) seems to be explained by the greater belt tensions in the higher severity tests that are associated with submarining at the steeper belt angle.

The "critical" belt-to-pelvis angle at which the belt slips from the pelvis can provide a measure of submarining performance for test situations in which submarining has not occurred. Since for a given test situation, the belt slipped over the pelvis at a particular belt-to-pelvis angle, an indication of how close a test response came to submarining can be estimated by how close the belt-to-pelvis angle came to the "critical" angle for that test configuration. Thus two tests which did not submarine can be rank ordered for submarining tendency by the belt-to-pelvis angle response if the parameter varied did not influence the "critical" belt-to-pelvis angle.

Consider a hypothetical example. Assume that five designs of a seat are to be compared for submarining tendency in a particular application, limited to a single test per seat. If a yes/no submarining criterion is used, at best the seats can be divided into two groups if the test situation is well chosen and cannot be assessed if the test situation results in all or no seats being associated with submarining. If the test conditions are chosen such that none of the seats are associated with submarining, the amount of pelvic rotation or the belt-to-pelvis angle can be used to rank order the seats for submarining tendency in the test situation.

A KINEMATIC VIEW-POINT OF SUBMARINING - The test results and analysis show that lap-belt slippage off the pelvis was associated with a large change in the lap-belt-to-pelvis angle between the initial position and that at submarining. Test responses shown in Figure 7 demonstrate this large angular change. Submarining was strongly associated with rearward rotation of the pelvis and with the lap-belt angle becoming more horizontal. The critical belt-to-pelvis angle at which the belt will slip over the pelvis was demonstrated to depend on the coefficient of friction between the dummy and belt and speculated to depend on the amount of lifting of the lap-belt by the shoulder-belt. Thus submarining is influenced by: pelvic rotation, dynamic lap-belt angle, belt-to-pelvis friction, and lifting by the shoulder-belt. The critical belt-to-pelvis angle likely depends on other factors such as the lap-belt geometry and the pelvic shape [3].

A kinematic model of the submarining process observed in the sled tests is shown in Figure 8. In this model the lap-belt is initially located favorably relative to the pelvic spines such that the belt will "hook" when loaded. However the dynamics of the frontal crash test results in changes of the belt and pelvic angular positions

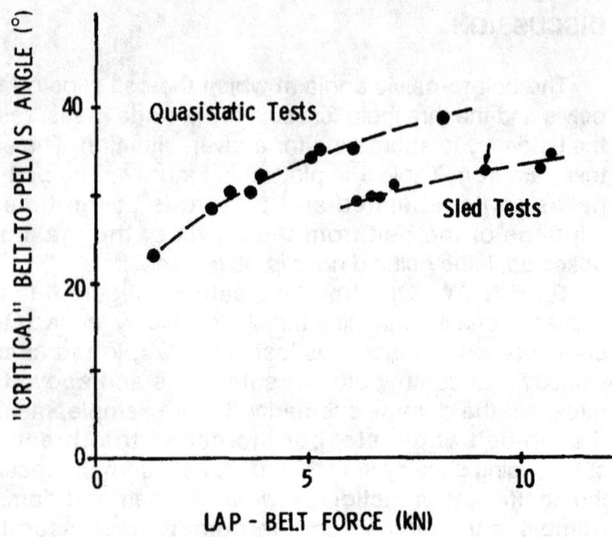

Figure 6. Relationship of the belt force acting on the pelvis with the "critical" belt-to pelvis angle at which the belt slips over the pelvis. Importantly, the sled test data is for maximum belt force, not necessarly the force at slip. Responses from test indicated by arrow are shown in Figure 7.

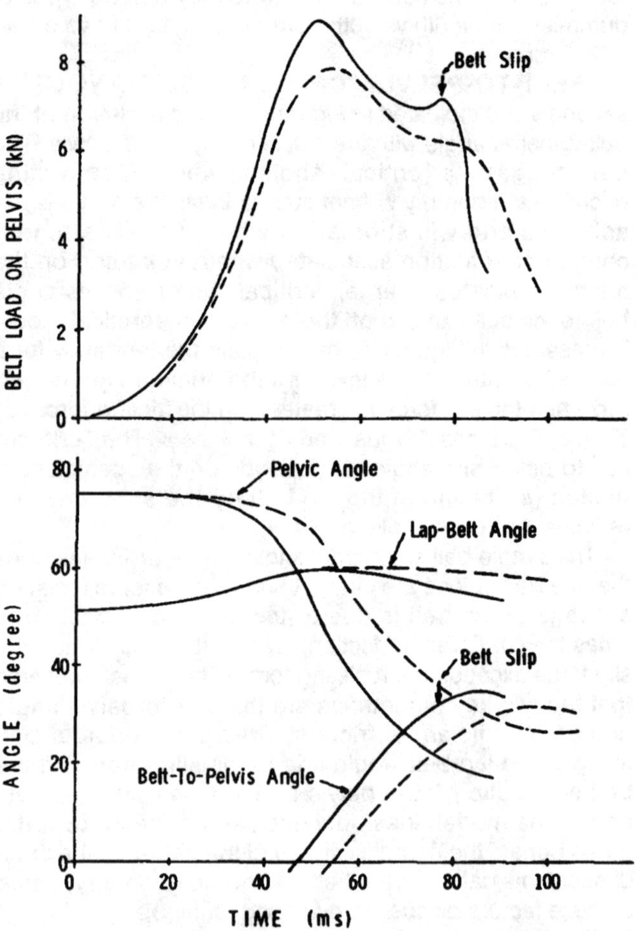

Figure 7. Test responses for the "steeper" lap-belt angle.
Solid line - belt slip occurred (43.2 km/h test)
Dashed line - belt did not slip (39.8 km/h test)

and tends toward a less favorable alignment of the belt and pelvis.

The dummy moves forward relative to the seat and inertially loads the lap and shoulder belts. The vertical component of the belt force acts to displace the dummy downward into the seat. Both the forward and downward motions of the dummy act to make the lap-belt angle more horizontal and thus in a direction closer to the "critical" belt-to-pelvis angle at which the belt will slip off the pelvis. The seat vertical stiffness, the lap-belt geometry, the lap-belt elongation, the seat resistance to forward motion of the pelvis, and lifting of the lap-belt by the shoulder-belt are factors which influence the dynamic change of the lap-belt angle. The foot or knee restraint and leg position will influence the force acting at the hip-joint.

The pelvis of the dummy rotates in response to a variety of forces and moments acting on the pelvis. The experimental study did not address these forces and moments. A list of some potential forces and moments acting on the dummy's pelvis is provided in Figure 8, which include the lap-belt force restraining the pelvis; the pelvis inertia; the leg forces and the moment transmitted through the hip-joint; the seat force acting on the pelvis; and the lumbar spine forces and moments acting on the pelvis. For example, moving the shoulder-belt junction with the lap-belt from "typical" to the lap-belt anchor strongly influenced the upper-torso kinematics and thus lumbar spine forces and moments on the pelvis which resulted in greatly different pelvic angles, Figure 9. Force components which act to rotate the pelvis forward improve belt retention on the dummy's pelvis.

The belt slip model suggested that the coefficient of friction between the dummy's pelvis and lap-belt would have a major influence on the belt slip from the pelvis -- confirmed by the experimental study. "Slippery" clothing on the dummy might increase the tendency of the dummy to submarine. "Holes" in the clothing such that the lap-belt could contact the skin might reduce the tendency of the dummy to submarine. The condition of the "flesh" covering the pelvic spines in terms of tears, splits, and repairs might also influence the tendency of the belt to slip off the dummy's pelvis.

Our kinematic analysis of submarining suggests evaluation methods for test responses relating to submarining in laboratory tests. Additionally, evaluating test responses in terms of the model might be helpful in determining effective design directions to influence submarining performance in laboratory tests. There are important differences in lap-belt geometry in some current restraint systems and that used in our study. The extrapolation of our results to vehicles or car occupants should consider differences between our test environment and vehicles in real crashes, and between the Part 572 dummy and car occupants. Methods to assess the potential injury risk associated with abdominal loading in a laboratory exposure would increase the objectivity of an extrapolation of laboratory results to real world issues.

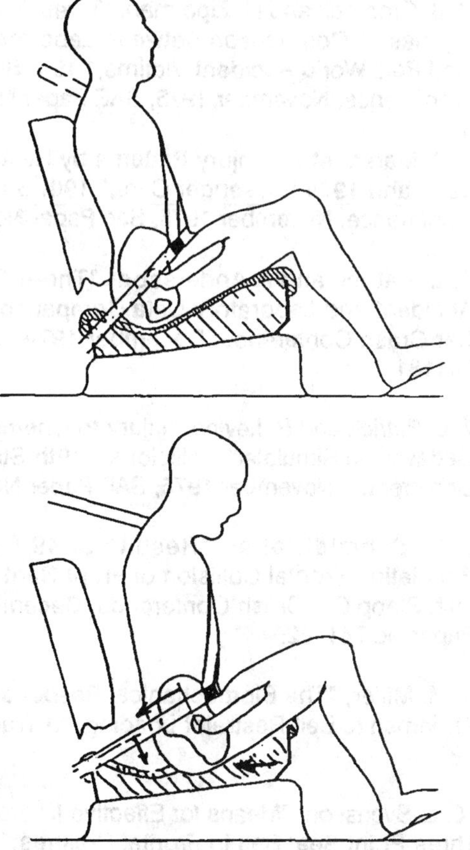

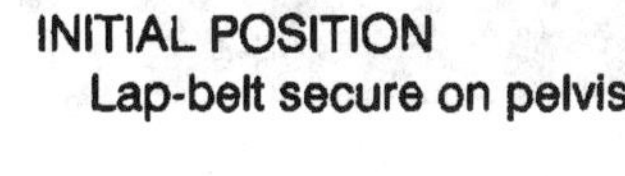

Figure 8. Schematic of lap-belt slip kinematics.

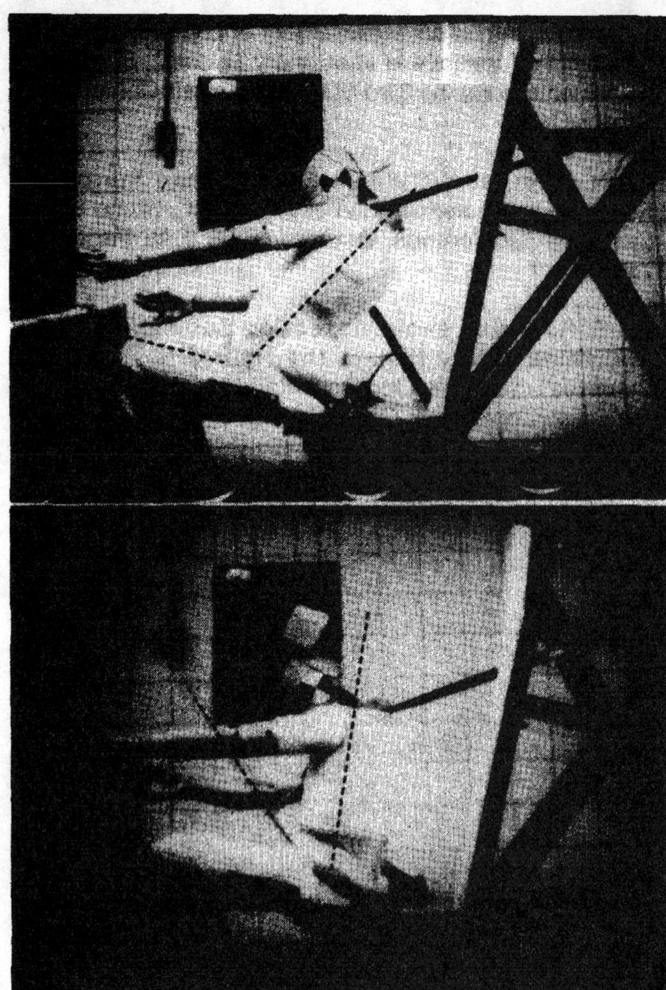

Figure 9. Demonstration of the strong influence of upper-torso restraint on pelvic rotation in 55 km/h sled test. Single upper-torso belt with the lower anchor at: (a) the lap-belt anchor or: (b) attached to the lap-belt near the dummy H-point. Pelvic target axis and thoracic alignment shown by dashed lines.

SUMMARY

A kinematic analysis was made of the dummy's pelvis "unhooking" from the lap belt, with the lab-belt slipping from the dummy's pelvic spines to directly load the abdominal region. In our laboratory tests, belt slippage was associated with large rearward rotation of the Part 572 pelvis, as much as 50° being observed. To a lesser extent belt slippage from the pelvis was associated with the lap-belt becoming more horizontal due to forward and downward motions of the pelvic spines.

There appeared to be a "critical" belt-to-pelvis angle for belt slip. At a lesser angle the belt did not slip, but did slip from the pelvis when the critical angle was reached. The coefficient of friction between the dummy and lap-belt influenced this critical belt slip angle. We speculated that lifting by the shoulder-belt could reduce the critical angle for belt slip from the pelvis.

A threshold test severity was observed for belt slip from the pelvis. Below this test severity, the belt did not slip and above this test severity the belt would slip from the dummy pelvis. Test parameters such as lap-belt angle and the friction between the lap-belt and the dummy strongly influenced the threshold test severity associated with submarining.

ACKNOWLEDGMENT - The authors thank Dave Viano, Steve Rouhana, and John Melvin for their technical review and comments, and Joseph McCleary and William Little for technical assistance in conducting the experiments.

REFERENCES

1. D. Dalmotas, "Mechanisms of Injury to Vehicle Occupants Restrained by Three-Point Seat Belts," 24th Stapp Car Crash Conference, October 1980, SAE Paper No 801311.

2. Y. Leung, et al, "A Comparison Between Part 572 Dummy and Human Subject in the Problem of Submarining," 23rd Stapp Car Crash Conference, October 1979, SAE Paper No 791026.

3. Y. Leung, et al, "Submarining Injuries of 3 Pt. Belted Occupants in Frontal Collisions -- Description, Mechanisms and Protection", 26th Stapp Car Crash Conference, October 1982, SAE Paper No 821158

4. J. Cromack and H. Ziperman, "Three-Point Belt Induced Injuries: A Comparison Between Laboratory Surrogates and Real World Accident Victims," 19th Stapp Car Crash Conference, November 1975, SAE Paper No 751141.

5. J. Marsh, et al, "Injury Patterns by Restraint Usage in 1973 and 1974 Passenger Cars," 19th Stapp Car Crash Conference, November 1975, Sae Paper No 751143.

6. L. Patrick and A Andersson, "Three-Point Harness Accident and Laboratory Data Comparison," 18th Stapp Car Crash Conference, December 1974, SAE Paper No 741181.

7. L. Patrick and R. Levine, "Injury to Unembalmed Belted Cadavers in Simulated Collisions," 19th Stapp Car Crash Conference, November 1975, SAE Paper No 751144.

8. G. Schmidt, et al, "Results of 49 Cadaver Tests Simulating Frontal Collision of Front Seat Passengers," 18th Stapp Car Crash Conference, December 1974, SAE Paper No 741182.

9. M. Miller, "The Biomechanical Response of the Lower Abdomen to Belt-Restraint Loading," J Trauma, in press, 1989

10. L. Svensson, "Means for Effective Improvement of the Three-Point Seat Belt in Frontal Crashes," 22ond Stapp Car Crash Conference, October 1978, SAE Paper No 780898.

11. D. Adomeit, "Seat Design - A Significant Factor for Safety Belt Effectiveness," 23rd Stapp Car Crash Conference, October 1979, SAE Paper No 791004.

12. D. Adomeit and A. Heger, "Motion Sequence Criteria and Design Proposals for Restraint Devices to Avoid Unfavorable Biomechanical Conditions and Submarining," 19th Stapp Car Crash Conference, November 1975, SAE Paper No 751146.

13. D. Adomeit, "Evaluation Methods for the Biomechanical Quality of Restraint Systems During Frontal Impact," 21st Stapp Car Crash Conference, October 1979, SAE Paper No 770936.

14. C. Freeman and D. Bacon, "The 3-Dimensional Trajectories of Dummy Car Occupants Restrained by Seat Belts in Crash Simulations," 32ond Stapp Car Crash Conference, October 1988, SAE Paper No 881727

15. S. Rouhana, et al, "Assessment of Lap-Shoulder Belt Restraint Performance in Laboratory Testing", Submitted for presentation in the 33rd Stapp Car Crash Conference, October 1989.

16. N. Alem, et al, "Whole-Body Human Surrogate Response to Three-Point Harness Restraint", 22ond Stapp Car Crash Conference, October, 1978, SAE Paper No 780895

17. S. Rouhana, et al, "Assessing Submarining and Abdominal Injury Risk in the Hybrid III Family of Dummies", Submitted for presentation in the 33rd Stapp Car Crash Conference, October 1989

821158

Submarining Injuries of 3 Pt. Belted Occupants in Frontal Collisions — Description, Mechanisms and Protection

Y. C. Leung, C. Tarrière, and D. Lestrelin
Laboratory of Physiology and Biomechanics
Peugeot-Renault Association (France)

J. Hureau
Faculty of Medicine, University of Paris
France

C. Got, F. Guillon, and A. Patel
I.R.B.A. Raymond Poincaré Hospital
Garches, France

ABSTRACT

Accidentological studies show, firstly, what kind of injuries are sustained by seat-belt wearers in frontal collisions, to abdomen, lumbar spine and lower members and, secondly, how to determine their frequencies and severities. Corresponding data are presented.

Then, a synthesis is made, in which the results of extensive cadaver testing - more than 300 human subjects - are examined with particular emphasis on the abdominal injuries, and on the association of injuries, such as lumbar spine injuries. Causation is particularly looked at. This experimental survey is completed by the results of specialized testing in abdominal tolerance when submarining occurs.

These two surveys enable the development of protection.

Finally, former attempts for defining an abdominal protection criterion are reviewed and a final definition for such a criterion is presented and justified.

ABDOMINAL PROTECTION IN AUTOMOBILE ACCIDENTS IS A SUBJECT WHICH HAS NOT BEEN FULLY EXPANDED UPON IN LITTERATURE ON ACCIDENTOLOGY AND BIOMECHANICS.

It is true that the head, thorax and lower members were the areas where those sustaining multiple injuries were the most frequently and most severely injured. Injuries in these areas were often evident (coma, breathing problems and fractures of long bones) whilst the possible share of abdominal injuries is more difficult to ascertain. In fatal accidents in particular, the autopsy is too often incomplete and only indicates the main injuries which are enough to account for the cause of death. Here again, abdominal injuries are often ignored in the same way as spinal injuries which require long and tedious research work before they can be detected.

In the 1970's, the wearing of seatbelts became general practice, especially in Europe. A new approach to regulation from the USA is gradually making headway. It is known as the global approach and sets out to appraise the secondary safety provided by a car, by simulating the most frequent and most deadly accidents and by measuring performance on one or more anthropomorphical dummies.

The first proposal for this type of regulation dates back to 1971. The American Administration wishes to appraise the protection provided by the vehicle in a frontal collision. The performance to be measured on the dummy concerns acceleration of the head and thorax and the load through the main axis of the thigh bone.

It is not surprising that the head, thorax and lower members are privileged areas we wish to protect. Indeed, methodological tools exist which we are not challenging. It is easy to measure the resultant acceleration of almost rigid models (head and thorax) by installing a three directional accelerometer at their centre of gravity.

Difficult research work is being conducted to determine the thresholds beyond which occupant protection cannot be ensured (to eliminate, for example, the risk of irreversible injuries). These thresholds must be based on human impact tolerance.

For 30 years, data have been progressively gathered and refined,
Curiously, the abdomen was overlooked. Until 1979, the first tolerance values (load-deflection) of a human abdomen loaded by a seatbelt in static test conditions was presented(1). Because the tolerance could not be measured by accelerometer techniques, a specific methodology was used to define abdominal tolerance and translated into a measurable performance on a dummy. To make the distinction between values which are specific to the human being and their applications to a dummy, we employ the term "tolerance" for the former and "protection criterion" for the latter. The protection criterion may be different if the dummy itself changes definition.

The purpose of this publication is to establish a consolidated version of the most recent work devoted to abdominal protection <u>in frontal</u>

collisions(xx) for belted occupants and to attempt to ascertain whether the protection criterion currently available would be sufficient to avert the great majority of abdominal injuries. Because submarining is the most involved mechanism, the analysis is enlarged to all injuries induced, by submarining that is to say lumbar, lower members, pelvic as well as abdominal injuries,

An analysis is presented for the abdominal injuries observed in a sample of 1542 subjects wearing 3-point seatbelts and involved in frontal impacts. This is followed by a corresponding analysis of abdominal injuries observed in the tested cadavers which were divided-up into two independent samples groups, one composed of 70 subjects, and the other 211 subjects, corresponding to the observations supplied by different research groups.

After having specified the relative importance of the risk of lumbar, lower members and abdominal injuries, the study goes on to analyze and consider the prevention of abdominal injuries caused by the mechanism the most often involved, well-known under the name of "submarining" (factors which influence the occurrence of submarining and in particular the definition of a protection criterion to eliminate the risk of submarining by means of principles of a well-designed vehicle).

By discussing all available information, it is possible to estimate the extent of protection provided by currently available criterion.

(x) Numbers between parentheses are references at the end of paper

(xx) Other publications cover a similar study on abdominal protection in side impacts,

SUBMARINING INJURIES IN REAL-LIFE ACCIDENTS

The accident survey file contains 1017 frontal impacts in which drivers wearing three-point seatbelts were involved.

The impacts are selected based on the following body injury criterion ; at least one of the occupants involved in the accident was taken to the hospital.

SEVERITY OF SUBMARINING INJURIES IN RELATION TO OVERALL RISK - The population of seriously injured among the 1017 drivers (M.AIS 3, 4 or 5) is slightly less than that observed among the 525 passengers. Respective proportions of those killed are not significantly different, (Table 1). (See also Appendix I)

From among the severely injured front occupants, 1 driver out of 5 and 1 passenger out of 3 sustained severe injuries to abdomen and dorso-lumbar spine fractures (Table2).

Table 2 - Proportion of AIS ≥ 3 to Abdomen and Dorso-Lumbar Column Among Severely Injured Drivers and Passengers (M.AIS ≥ 3) wearing 3-pt belt.

	Thoracic-lumbar spine and Abdomen AIS		
	<3	≥3	Total
Drivers	69 (79%)	18(21%)	87(100%)
Passengers	37 (64%)	21(36%)	58(100%)
Total	106 (73%)	39(27%)	145(100%)

SEVERITY OF ABDOMINAL INJURIES IN RELATION TO OTHER BODY AREAS - The abdomen, with the head and thorax, is, for belted drivers, the body area most exposed to critical injury (AIS ≥ 4). Injuries to lower members very largely dominate at level 3, (Table 3).

One of the advantages of the adoption of retractor belt systems which are fitted to cars produced since 1978 has been to reduce the slack with which static belts were habitually worn. Further, their geometrical lay-out in the car is a great improvement; the central mounting systems do, in particular, eliminate any further risk of initial contact of the adjusting buckle against the abdomen.

This progress accounts for the very great difference in the risk of abdominal injury between those wearing static and those wearing retractor belts (Table 4).

The improvement to seat belt has, therefore, approximately halved the share of abdominal injuries

However, they are still dangerous and some progress should be made. A precise description of the injuries and their most probable mechanisms may help to throw some light on the subject.

Table 1 - Severity of Frontal Impacts for Front, Belted (3-point) Drivers and Passengers.

	M.AIS			
	0-1-2	3-4-5	≥6	Total
Drivers	892 (87.8%)	87 (8.5%)	38 (3.7%)	1017 100%
Passengers	451 (86%)	58 (11%)	16 (3%)	525 100%

Table 3 - Severity of Injuries by Body Area Sustained by Belted Drivers and Passengers During Frontal Impact (Excluding Those Killed and not Subjected to Autopsy).

DRIVERS — Frontal Impacts (11.12.01 o'clock)

	Head/Face	Neck	Thorax	DL. Spine	Abdomen	Pelvis	L. Members
AIS - 0	651	917	762	968	948	938	670
1	225	75	207	24	31	40	229
2	96	1	16	6	3	6	50
3	16	4	2	1	4	15	50
4	2	1	11	-	11	-	-
5	9	1	1	-	2	-	-
	999	999	999	999	999	999	999

RIGHT FRONT PASSENGERS

	Head/Face	Neck	Thorax	DL. Spine	Abdomen	Pelvis	L. Members
AIS - 0	395	456	326	492	477	498	366
1	90	61	159	27	30	23	123
2	33	4	28	4	1	1	19
3	4	3	7	3	5	4	18
4	3	1	6	-	13	-	-
5	1	1	-	-	-	-	-
	526	526	526	526	526	526	526

Table 4 - Proportions (%) of Occupants Sustaining Severe Abdominal Injuries (AIS ≥ 3) and/or Dorso-Lumbar Fractures (AIS ≥ 3) Among the Seriously Injured Drivers and Passengers (M.AIS 3, 4 or 5).

	Static Belt	Retractor Belt
Drivers	22.6%	12 %
Passengers	37.5 %	16.7 %

THE RISK OF SUB-MARINING INJURIES INCREASES WITH THE VIOLENCE OF IMPACT - This relationship, which can be seen in Table 5 allows us to rule out the assumption, occasionally put forward, wheby the risk of injury to the abdomen is primarily governed by malfunctions or by the presence, in the population studied, of dangerous and badly designed systems, irrespective of the violence. The mechanisms which sometimes cause the abdomen to be subjected to excessive loading, resulting in injury, are not easy to describe for the accidentologist. A number of factors do, however, allow us to state, with a good degree of certainty, that each case matches up with one of the following configurations :

1- Restraint is achieved by the most resistant bone segments, including-and more especially- the pelvic region, but the abdominal mass undergoes a deceleration beyond the level of tolerance. It results in tears, wounds or rupture of viscera which certain authors thought could be described by specific features (2).

2 - A variant of these cases on which the seatbelt firmly fastens the pelvis and where we could still observe abdominal injuries, would correspond to another mechanism, that is a very important flexion of the trunk around the pelvis. This flexion can be clearly observed in experimental impacts (Figure 1), and the head can come into contact with the knees. A great kinematics was produced possibly within the certain car models, in which the superior anchorage was very behind that favored the sliding of the shoulder under the shoulder belt. Abdominal injuries would not be the result of the deceleration of the viscera in relation to the belt restrained skeletal frame, but would be induced by very high pressure in the viscera, and even compression between the trunk and thighs. The stout persons are more exposed than others to this type of compression. If here is a submarining process, the lap-belt still increases the pressure in viscera.

3 - The belt is worn in such a way that, when the slack is taken up, the lap-belt presses against the abdomen, penetrating through even as far as the spinal column when deceleration is high (Table 6). This second configuration, where play is a major factor, is frequent when the seat belt anchorage installation is faulty. Its consequences are aggravated when the restraint system on the oldest static belt models includes a buckle for adjusting belt tension, located on the end of a flexible stalk anchored to the centre of the car. This system has significantly deteriorated seat belt performance in numerous accidents which have occurred over the past ten years.

4 - The lap-belt is correctly positioned against the pelvic bone but, when the occupant is displaced forward, loading conditions of the restraint system are such that the lap-belt, under tension, runs up over the iliac crests and compresses the abdomen. This is submarining in the strict sense of the term.

The last two configurations have the most adverse effects on the abdomen. The injuries they cause have no specific typology whereby it would be possible, from a review of the medical file, to determine whether the case in point falls into configuration 3 or 4. Amongst the relevant indications capable of distinguishing between them, we can name the presence of cutaneous erosion in the pelvic region, particularly below or above the anterior superior iliac spines: the distance over which the belt shows signs of friction compared to occupant stopping distance.

In the last two configurations, the pressure exerted on the abdomen is the risk to be avoided. Whether it be at the just beginning of occupant displacement or after slipping from the pelvis, that in no way obviates the need to design restraint systems and their installation in such a way that restraint acts only against the pelvis throughout the whole of the occupant's deceleration phase. The share of the shoulder belt as a direct cause or aggravating factor in injuries to tissues or organs in the upper abdomen region (liver, spleen, in particular) is not negligible but the results of an analysis indicated in Table 7 show that, when it exists, it is most often accompaning the preponderant action from the lap-belt.

Table 5 - Proportion of Occupants With Severe Abdomen Injuries and Dorso-Lumbar Fractures and Proportion of Severely Injured (M. AIS ≥3) in Each ΔV Category.

ΔV (kph)	≤35	36-45	46-55	56-65	≥65
AIS abdomen ≥ 3	0.3%	2.4%	13.3%	13.5%	46.2%
M.AIS ≥ 3	2%	16.0%	48.0%	59.5%	92%

Likewise, drivers sometimes strike against the steering unit which contributes to the injuries or is the sole cause of them, but this is a rare occurrence, since we have seen (above) that the abdominal risk is lesser for driver than for passenger.

The 26 drivers and 21 passengers who sustained severe submarining injuries (AIS≥3) were equipped with seatbelts almost half of which had a geometrical defect and were worn too loose. The presence of one or other of these two factors or both combined, was noted in more than two out of three cases (Table 8).

CONCLUSIONS

1. The process described by the term of submarining of the lower part of the trunk (pelvis-abdomen-lumbar- spine and the lower thorax) under the lap-belt is a mechanism fairly frequently encountered among belted occupants in frontal impact.

Submarining is considered to be the mechanism most probably at cause in the vast majority of cases involving the seriously injured (AIS≥3) sustaining injuries to the abdomen, the lower members, the pelvis or the dorso-lumbar column.

One of the original aspects of this in-depth cases analysis is that it shows how submarining can cause severe injuries to the lower members (legs-knees-femurs), the pelvis and the dorso-lumbar column (fracture-compression of the first lumbar vertebraes or fracture of the transverse processes) as well as to the abdomen. There are besides often combinations of these different injuries.

Other injuries may more exceptionally be combined, such as wounds or injuries to the neck and occasionally rib fractures.

2. The lap-belt is certainly the cause of 68% (32/47) of cases of victims of abdominal (AIS ≥ 3) and/or dorso-lumbar column (AIS ≥ 2) injuries.

Table 6 - Dorso-Lumbar Spine Injuries (AIS $\geqslant$ 2) - N = 14 Cases

LEVEL OF FRACTURES	INJURIES' MECHANISMS: SUBMARINING			DIRECT REAR IMPACT (rear passenger or object)		
	Sure	Probable	Uncertain	Sure	Probable	Uncertain
D9 - D11			1*			
D 12	1	1	1			
L 1	4					
L 2	2					
Lumbar vertebrae process-transverse:	1	2*	1			
L5 process transverse:				1		

(*) aggravating influence of the rear passenger

Table 7

SYNTHESIS OF SUBMARINING - ANALYSIS OF INJURIES' MECHANISMS

PARAMETERS	LAP-BELT			SHOULDER BELT			OTHER DIRECT IMPACTS (steering-system for example)		
	Sure	Probable	Uncertain	Sure	Probable	Uncertain	Sure	Probable	Uncertain
POOR GEOMETRY OF THE RESTRAINT	14	6 43%(A)	1	1	0	2	0	0	0
BELT WORN WITH SLACK	13	4 36%	2	0	0	1	0	0	0
FRONT SEAT TRACK DAMAGE	13	5 38%	1	0	1	2	0	0	0
ADDITIONAL LOADING	7	3 21%	0	0	0	1	0	0	0
NECK CONTUSIONS OR ABRASIONS BY SHOULDER-BELT	6	3 19%	0	0	1	1	0	0	0
LOWER MEMBERS IMPACT AGAINST LOWER PANEL	30	9 83%	5	1	1	4	0	2	1
CRASH VIOLENCE $\Delta V \geqslant 50$ km/h mean $\gamma \geqslant 10$ g	19	6 58%	2	0	1	3	0	1	1
ABDOMINAL INJURIES SUB-MESOCOLIC	25	5 ·64%	2	0	1	3	0	1	0
ABDOMINAL INJURIES ABOVE MESOCOLIC	8	2 21%	1	1	1	3	0	1	0
DORSO-LUMBAR FRACTURES	8	3 23%	1	0	0	0	1	0	0
LOWER MEMBERS FRACTURES	17	4 45%	2	1	1	2	0	0	0
PELVIC INJURIES	3	1 8%	2			1	0	0	0
RIB FRACTURES	14	1 32%	1	1	0	1	0	0	0
AGGRAVATING INFLUENCE OF RIB FRACTURES	3	1 8%	0	1	0	0	0	0	0

IMPORTANT REMARKS:

(A) This percentage corresponding to the two first columns means that the parameter involved plays for x% of all cases where lap-belt submarining is sure or the most probable. For example, "Poor geometry..." plays in 20 cases out of 47 where submarining is sure or the most probable.

(B) The total in each ROW could overpass the number of involved people because, for some cases, two mechanisms could play simultaneously (ex.: lap-belt submarining and abdominal injury induced by shoulder-belt (rib fracture).

Its rôle is very probable in 19% of the other cases;only in 13% of cases it is absent or doubtful.

The rôle of the shoulder belt is unquestionable in 1 out of 47 cases as complementing the action of the lap-belt and is possible in one other case. In 4 cases there is some doubt.

3. It is difficult to strictly specify the mechanism causing abdominal injuries as it means taking into consideration all the medical, anthropometric and technical data in the accidentology files. Two analysis grids have been established and they are given in Appendices I and II.

Submarining is evident when contributory factors such as poor geometry of the restraint system and (or) slack in the belts (one or other or both these factors combined are present in 74% of cases) are observed with the following:

a) Seat-tracks damages in 38% of cases

b) Knee or leg impact under the instrument panel with very obvious deformation of the latter in 83% of cases.

c) Violent impact ($\Delta V \gg$ 50kph and $\gamma \gg$ 10g in 58% of cases where submarining is certain or very probable).

4. It should be noted that apart from submarining and the auxiliary rôle played by the shoulder belt, the responsability of the instrument panel and/or the steering unit is exceptional in the occurrence of abdominal or lumbar injuries (one probable case and one doubtful).

5. The additional load from the rear passengers is an aggravating factor (in 21% of cases where submarining is certain or probable).

6. Fractures of the pelvis are seldom but could be the result of submarining (8% of cases where submarining is certain or probable)whether by knee impact or by direct action of the lap-belt.

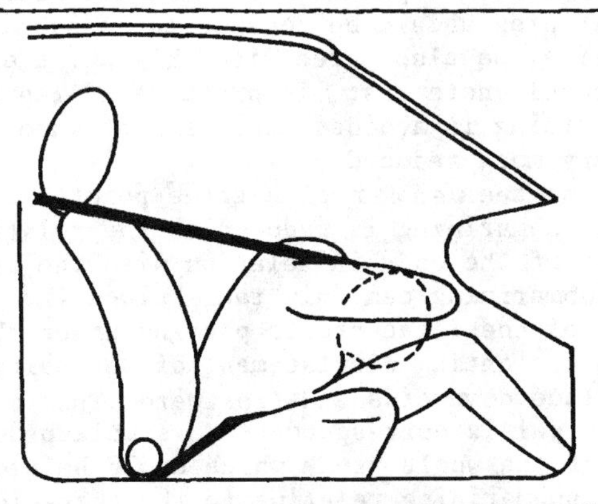

FIG. 1 . DIAGRAM OF A CADAVER TEST.

7. Another Sub-marining consequence already noted in a previous publication (2) is the great frequency (23%) of dorso-lumbar vertebraes fractures (D12-L1-L2) by direct impact of the lap belt or (and) hyperflexion of the trunk around the lap-belt penetrated in the abdomen.

8. The lower members injuries (legs,knees and femurs fractures) are not specific ; however they are observed in 45% of cases of sub-marining sure or most probable and could be considered as another consequence of this process.

ANALYSIS OF ABDOMINAL INJURIES OBSERVED IN EXPERIMENTS CONDUCTED ON HUMAN CADAVERS FOR 281 FRONTAL IMPACT TESTS.

Two test samples with cadavers simulating frontal collisions with three-point seatbelts provide a total of 281 observations. All the subjects were fresh cadavers conserved at a temperature of around 0° and experiments were conducted a few hours or days following their death.

An initial sample known as "APR" groups 70 subjects. A second one, the "Heidelberg" sample groups 211 subjects.

ABDOMINAL INJURIES OBSERVED IN THE "APR" SAMPLE - The most frequent injuries are fractures of the rib cage and litterature already exists on this subject.

In this paper, the injuries indicated are only the abdominal injuries or the lumbar vertebra injuries which can be produced by the same mechanism, in particular submarining.

Out of the 70 frontal impact tests using cadavers, there were 47 cases with injuries, 23 with lap-belt submarining and 24 without lap-belt submarining. The rest were cases where no injury was sustained and there was no evidence of submarining. The appended Table 9 illustrates which of the tested cadavers sustained only the injuries observed in the liver, the spleen, the lumbar vertebra and other abdominal injuries,including the intestines, mesentery, colon, iliac crest and abdominal muscles.Cases with only iliac crest fracture or abdominal muscle injury are alsopresented in this Table which contains 24 cases, 6 of which are non-submarining and 18 submarining.

Table 8 - Quality of Restraint System Geometry and Belt Wearing Among Victims of Abdominal injuries and Dorso-Lumbar Fractures

		POOR GEOMETRY		
		YES	NO	TOTAL
EXCESS SLACK	YES	8	8	16
	NO	12	13	25
	UNCERTAIN	3	3	6
	TOTAL	23	24	47

The following comments can be made :

1. No lumbar vertebra fracture was found in non-submarining cases but lumbar vertebra fractures occurred frequently in submarining cases (5 cases out of 14).

2. No injuries observed in intestine,colon, and mesentery were found in non-submarining cases (except for N° 25 who sustained a light mesentery injury). These injuries were frequently observed in the submarining cases (6 cases out of 17).

3. 5 drivers sustained liver injuries, 2 of which were non-submarining cases.

4. 2 passengers sustained liver injuries with submarining.

5. All four "spleen injury" cases were observed only on front passengers, 2 of which were non-submarining cases, three were submarining cases

ABDOMINAL INJURIES OBSERVED IN THE "HEIDELBERG" SAMPLE - An ISO document based on experimental data from the University of Heidelberg which have appeared in numerous past publications presents the following summary as shown in Table 10.(3)

Except for spleen injuries, there is no preferential distribution of injuries between drivers and passengers. Liver is injured in 16% of cases for the passenger and 3.5% for the driver. The mean sled deceleration was exceeding 16g.

These results confirm the absence of specificity regarding liver injuries which would, as for spleen injuries, seem to be sustained more often by the passenger. This does, therefore, consolidate the findings of the APR sample analysis.

DISCUSSION

1. Out of 23 cases of liver injuries,7 were drivers. 2 of them were non-submarining cases, 5 were submarining cases. This is not sufficient to ascertain that the shoulder belt is responsible for liver injuries in non-submarining cases, as we could see similar results for the front passenger. 16 cases of liver injuries were observed among front passengers (or right rear passengers),and they were submarining cases. The possible effect of the shoulder belt on liver injuries was relatively small in this group of the sample.

2. All of 5 spleen injury cases were observed in the front passengers. Two of them occurred in submarining cases.

3. As observed for the spleen, or for the liver, any sustained injury cannot be automatically due to the shoulder belt. It would be more suitable to say that the shoulder belt favours submarining on the buckle (interior)side. In certain cases, the shoulder belt could play a role in provoking such injuries without the occurrence of submarining.

4.No intestine,colon,mesentery injuries were found in non sub-marining cases (except one case where a light mesentery injury occured).

5.In our previous submarining studies (4), it was found that submarining began,in most case: on the interior side,for both cadaver and dummy tests.Two cases were found where submarining was limited to the interior side only, This illustrates the statement made in 3. to the effect that "the shoulder belt favours submarining on the buckle (interior) side".

CONCLUSIONS

1, Abdominal injuries occurred rarely in non submarining cases,

2, Dorso-lumbar vertebra fracture was never found in non-submarining cases, But lumbar vertebra fractures are often associated with submarining and, sometimes, without abdominal injury.

3, Spleen and lever injuries are possible consequences of lap-belt sub-marining, Liver injuries are not specific to anyone occupant seat. Spleen injury is found particularly to the right passenger (front or rear),

4, The possible effect of shoulder belt is rather low. The shoulder belt could favour the submarining process on the buckle interior side.

5, All sub-mesocolic injuries (except one) are consequences of lap-belt sub-marining.

DEFINITION OF SUBMARINING (MECHANISM)

A definition problem regarding the word"submarining" was experienced on several occasions these last years, in particular with I.S.O. (ISO TC22/SC12/GT 6),

Corresponding to a personal communication of G.M. NYQUIST, several types of submarining referred to by various authors in the past are :

- Lap-belt submarining
- Shoulder belt submarining
- Air bag submarining
- Instrument panel submarining.

He stressed the fact that abdominal injuries are not the only consequences of a submarining and that knee injuries due to impact following submarining should be considered as submarining injuries. We also agree with this point of view. One should note that, in practice if lap belt submarining is avoided, the risk of knee impact is very much reduced.

For the wearer of a three point safety belt, submarining is reduced to the relative movement of the body in relation with the belts. The submarining can only take place through release of the iliac crests passing under the lap belt, A slanting displacement of the pelvis takes place downwards and frontwards.This movement of the pelvis corresponds to a simultaneous movement of the whole trunk which could be named thoracic submarining relating to the thoracic belt. However, no thoracic syndrome specific to submarining has yet been described. Some cervical injuries could be due to this whole trunk submarining as we have seen in our accidentological sample.

So as not to complicate things, at least as regards the wearer of a three point safety belt, one should avoid talking of submarining regarding the body as a whole. To make things clear one should talk about pelvis submarining relating to the pelvis belt. This usually causes abdominal injuries, injuries of the dorso-lumbar vertebra, of the knee-femur-pelvis axis, all these injuries which can either be isolated or associated.

It is difficult to distinguish, within this category, the difference of the submarining occurred with the initial positioning of the lap-belt on abdomen and with the initial positioning on the pelvis. In real accidents, in most cases, it is impossible to know exactly the initial positioning. This depends on the care taken by the wearer in placing correctly the safety-belt (however does the wearer know that the lap-belt, sometimes called "abdominal", should really restrain the pelvis ?) The placing of the belt depends firstly on the restraint geometry but also of the posture and slack in the belt.

FACTORS INFLUENCING THE OCCURRENCE OF LAP-BELT SUBMARINING

"Submarining" is a complicated problem because its occurrence is associated with many parameters, for example the geometry of the seat-belt system, orientation of the pelvis, dynamic characteristics of the seat and vehicle, impact velocity, etc... These are the external parameters determined by test conditions. An internal parameter is the pelvis shape; the area in question is the upper half of the notch below the Anterior Superior Iliac Spine (A.S.I.S.). Here as in previous publications (4)(5), this pelvic area situated just below the ASIS is given the name "Sartorius" and has a length of A1A2 (see figure 2).

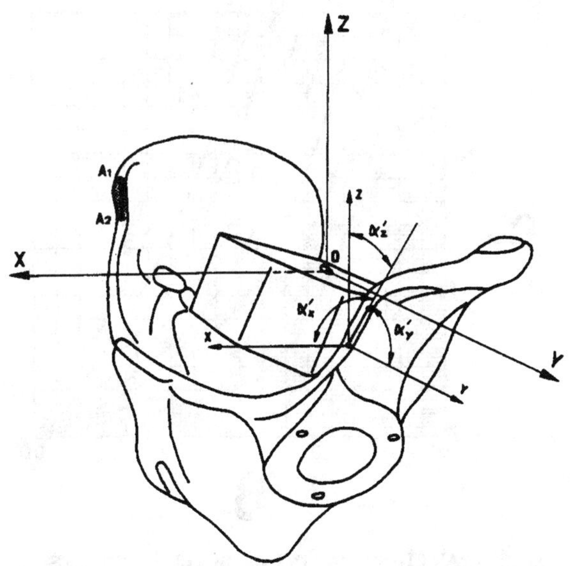

FIG. 2 . DIRECTION OF THE UPPER HALF NOTCH (A1A2)"SARTORIUS" DEFINED IN THREE DIMENSIONS.

In fact, to avoid any confusion, it is necessary to state that Sartorius is the original name given to the muscle attached to this referred bone area in Gray's Anatomy (pp. 228, 492).

INFLUENCE OF LAP-BELT ANGLES - During frontal impact, the load applied to the "Sartorius" on one side of the pelvis is the resultant of lap-belt tensions on the corresponding side. This load can be resolved into two components: one is parallel to the "Sartorius", the other is perpendicular. The parallel one orients downward and backward along the "Sartorius"; it can be defined as a load to keep the pelvis in a favorable position necessary to prevent the process of submarining. This load is related to the orientation of the "Sartorius" and the geometry of the lap-belt as well as to lap-belt tension. The coefficient of the load can be used as an indicator of the efficiency in reducing the submarining tendency. A brief technical term referring to "anti"-submarining scale was used in a previous paper (4).

If the orientation of the Sartorius is considered as a constant, the anti-submarining scale is only a function of the geometry of the lap-belt defined in three dimensions. Since some experimental results cannot be explained by the traditional geometry definition of the lap-belt given in two dimensions, a complete geometrical definition of the lap-belt determined in three dimensions was developed.

Thanks to the anti-submarining scale curves established on the basis of the present theory and the submarining tendency as a function of lap-belt angles $\beta 1$, $\beta 2$, determined in experimental results, an anti-submarining scale $\xi=0.63$ is proposed for a limit of submarining risk on lap-belt geometry. Hence, this paper shows that the graphs of the anti-submarining scales can be used for checking the submarining risk in the seat-belt system.

Theoretical study: using a geometrical model of the lap-belt during impact (Fig. 3), the following equations can be found:

$$F3 = F \cos\theta$$
$$F4 = F \sin\theta$$

where load F is the resultant of lap-belt tension applying to the "Sartorius" (A1A2) on one side.

When the magnitude of F is known, F3 and F4 depend only on the angle θ which is formed between the direction of Sartorius and the resultant load F. F3 is a load parallel to the Sartorius, preventing the lap-belt from riding-up over the A.S.I.S. and keeping the pelvis in a correct position. Therefore, it is a load that plays an important role in reducing the submarining tendency. In contrast, F4 is a load which rotates the pelvis backward and downward, i.e. it increases the risk of submarining. F3 could be used as an indicator of the efficiency in reducing the submarining tendency.

By a mathematical study (4), load F3 is given by another expression,

$$F3 = \xi F1$$

Lap-belt tension in this equation depends on collision velocity, seat and vehicle dynamic characteristics, mass of the belted occupants, etc... If F1 is determined, F3 is a function of the coefficient ξ. Since F3 is defined as an indicator of the efficiency in reducing the submarining tendency, the coefficient ξ can be designated as an anti-submarining scale indicating the efficiency to prevent the process of submarining for a given configuration.

The coefficient ξ depends, on one part, on the direction cosines of the "Sartorius" and, on the other part, on the lap-belt geometry designated by angles $\beta 1$, $\beta 2$. It is presented precisely by the following equation:

$$\xi = \cos\beta 1 . \cos\left[\tan^{-1}(\cos\beta 1 . \tan\beta 2)\right] . \cos\alpha'_x + \left[\cos 1/2\tan^{-1}(\cos\beta 1 . \tan\beta 2) - \sin 1/2\tan^{-1}(\cos\beta 1 . \tan\beta 2)\right]^2 . \cos\alpha'_y + \sin\beta 1 . \cos\left[\tan^{-1}(\cos\beta 1 . \tan\beta 2)\right] . \cos\alpha'_z$$

Since $\alpha'x$, $\alpha'y$ and $\alpha'z$ were three-dimensions defined previously for the "Sartorius" of an occupant within a car (5), anti-submarining scale could be determined as a function of lap-belt angles $\beta 1$, $\beta 2$.

By using above equation, the relationship between the coefficient ξ and the lap-belt angles ($\beta 1$, $\beta 2$) can be obtained, as illustrated in Figure 4 for human subject (male) and for Part 572 dummy (Figure 5).

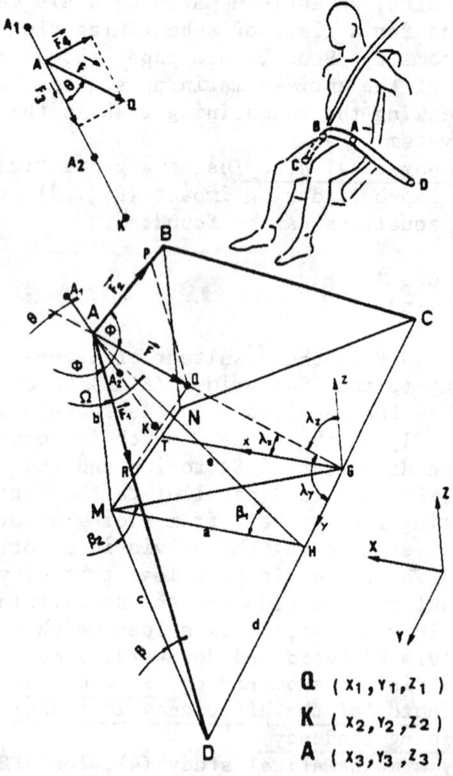

FIG. 3 . GEOMETRICAL MODEL OF THE LAP-BELT DURING IMPACT.

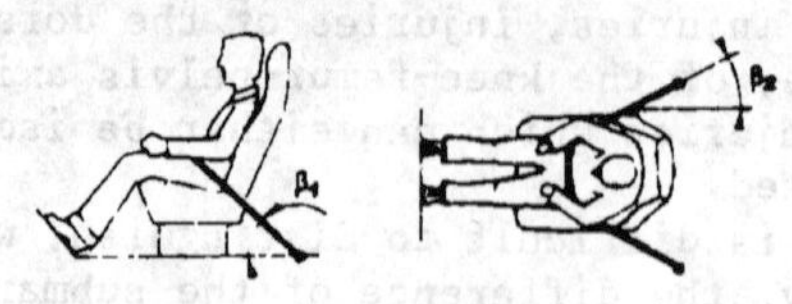

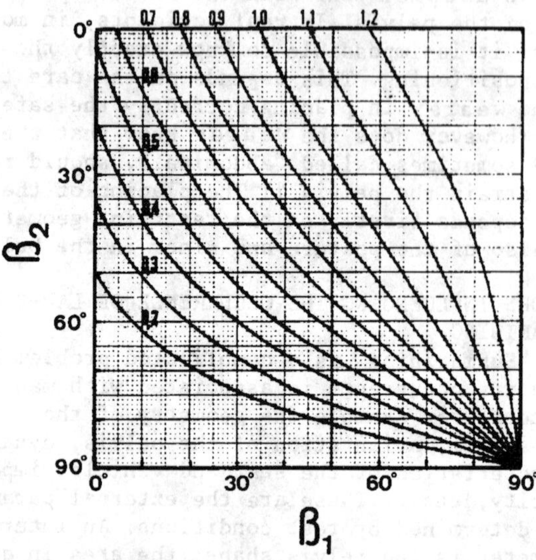

FIG. 4 . ANTI-SUBMARINING SCALE ξ CURVES AS A GUNCTION OF LAP-BELT ANGLES β1, β2 DERIVED FOR HUMAN SUBJECTS (MALE).

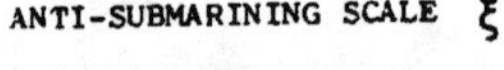

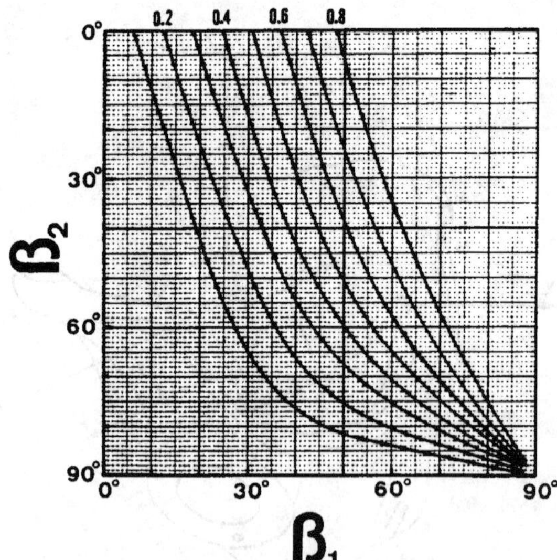

FIG. 5 . ANTI-SUBMARINING SCALE ξ CURVES AS A FUNCTION OF THE LAP-BELT ANGLES β1, β2 DERIVED FOR PART 572 DUMMY.

Experimental study - In order to put into practice and examine this theoretical study and the results, a series of sled tests was performed with human cadavers, the Part 572 dummy and the modified dummy. Three different bodies of European car models were used and mounted on the sled for the tests.

The lap-belt angles $\beta 1$, $\beta 2$ were measured with the tested subjects in the initial position, the test conditions and the complete experimental results were given in reference (4).

The details of the conservation of human cadavers can be consulted in reference (6). Briefly, the cadavers were fresh and non-embalmed. Death occurred less than 4 days before the test. They were put in a cold box and then taken out a few hours before the test.

Following the experimental results, submarining consequences in relation to the lap-belt angles $\beta 1$, $\beta 2$ are illustrated generally in Fig. 6 for the human subjects, Part 572 dummy and modified dummy.

Based on both the theoretical study and experimental results, the anti-submarining scale ξ can be traced between the groups of submarining data points corresponding to the 0.66 curve on Figure 4. It can be found in this figure: firstly that the submarining data points locate at the area which is formed by the smaller angles $\beta 1$ and the greater angles $\beta 2$ corresponding to the lower anti-submarining scale zone. Secondly, non-submarining data points locate inversely at the other area where $\beta 1$, $\beta 2$ are respectively greater and smaller, corresponding to the higher anti-submarining scale zone. Thirdly, two non-submarining data points for a cadaver test and a modified dummy test are situated in the submarining zone. This may be explained once more by the fact that the cadaver and the modified dummy are less inclined to submarine than the Part 572 dummy.

The figuration and location of the curve indicate that for the same anti-submarining scale, the decrease of $\beta 1$ should be associated with the decrease of $\beta 2$, i.e. if angle $\beta 1$ is favorable to submarining (smaller), it is necessary to associate a good angle $\beta 2$ (smaller) to reduce the submarining tendency if other conditions remain unchanged. This case corresponds to the highest data points (circle). On the other hand, while angle $\beta 1$ is great enough, according to previous beliefs, it seems to be unfavorable to submarining. Unfortunately, the submarining process occurred. In Figure 6, the three or four lowest black data points correspond to this case and it can be explained by the fact that these data points have greater angles $\beta 2$. Experimental results reveal that the geometrical definition of the lap-belt designated only by angle $\beta 1$ is insufficient to define its submarining tendency.

The traditional geometrical definition is used in two dimensions given as the angle formed by the lap-belt and the X-axis from side view, specified as angle $\beta 1$. A wide range of this angle had been proposed as an unfavorable angle to submarining by the previous researchers. There were, for example, 45-50° proposed by Haley (7), 45° by Patrick (8), 50-70° by Adomeit (9), Hontschik (10), and 69° by Billault (11). These reference angles are represented in Figure 6 for a comparison.

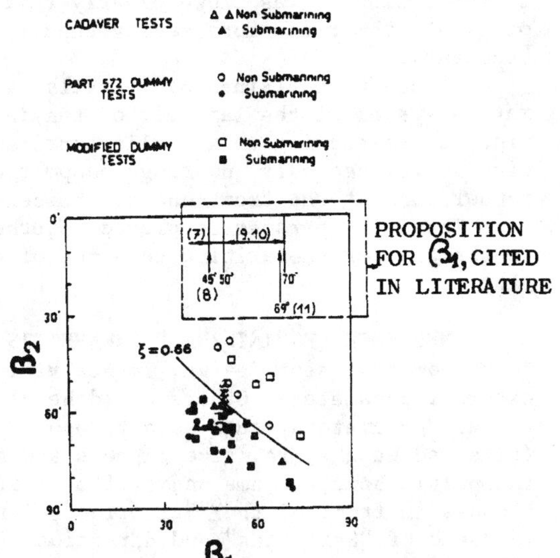

FIG. 6 . ANTI-SUBMARINING SCALE DETERMINED IN CADAVER, PART 572 AND MODIFIED DUMMY TESTS. THE RESULTS GIVEN IN THREE DIMENSIONS (β_1, β_2) ARE COMPARED WITH THOSE GIVEN IN TWO DIMENSIONS (β_1) PROPOSED IN LITERATURE.

Angles $\beta 1$ proposed in the above references cover almost all the submarining and non-submarining data points in the Figure 6. This variation was caused certainly by the variable angles $\beta 2$ which had been used but not observed by the authors of the references. Indeed, it is difficult to state what is the unfavorable angle $\beta 1$ to submarine if angle $\beta 2$ is not given.

Actually, the submarining tendency cannot be indicated when using the traditional definition of the lap-belt geometry as determined by angle $\beta 1$ alone. It is according to our theoretical and experimental study.

INFLUENCE OF SEAT DESIGN - Seat design is another parameter influencing the submarining consequence. Necessary motions of the pelvis are the forward and downward displacements which provoke probably a lap-belt submarining. To limit these two displacements, there is the conception for the creation of an "anti-submarining" seat.

Adomeit and Heger (9) proposed an energy-absorbing seat constructed with the front wall of the seat metal pan. On the forward area of the seat pan, a density of foam of 53 kg/m^3 was used (Figure 7).

Lundell et al. (12) presented a seat which had a contourned floor pan with a pronounced ridge at the front end. The seat cushion had a greater thickness which decreased gradually to the front edge of the cushion (Figure 8). In comparison with their flat seat tests, it showed that the new seat design gave both reduced injury criteria and low risk of submarining.

On the previous study (4), a comparison of submarining tendency was taken for the front occupant in the passenger seat and rear occupant in the bench. It was shown clearly that the rear occupant submarined more easily than the front occupants.

It can be explained by, firstly, a bad fixation system of the lap-belt on the interior side, corresponding to a smaller anti-submarining scale; secondly, no ridge support under the rubber foam at the front end of the bench.

These two parameters played together a role in the greater submarining tendency of rear occupant.

INFLUENCE OF "INTERNAL" PARAMETERS - Seat-Belt geometry, seat design, impact violence are external parameters. Others could be called "internal" parameters, such as anthropomorphic data (size and weight, relative bigness and shape of the pelvic bone, volume and stiffness of soft tissues in front of the "Sartorius", direction in space of "Sartorius" and direction of the pelvis in sitting position.

As indicated in the anti-submarining scale equation, if the lap-belt angles ($\beta 1$, $\beta 2$) are determined, we can see how the direction of "Sartorius" affects the submarining tendency. A comparison of the anti-submarining scales can be shown in Figure 9 and Figure 10 for human sub-

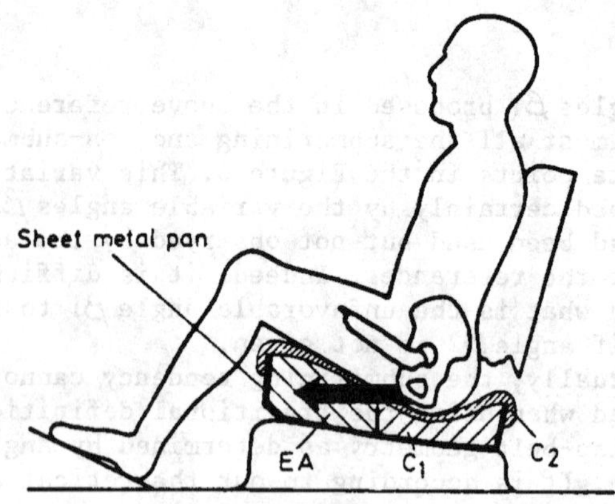

FIG. 7. DIAGRAM OF A SEAT DESIGN GIVEN BY ADOMEIT ET AL.(9).

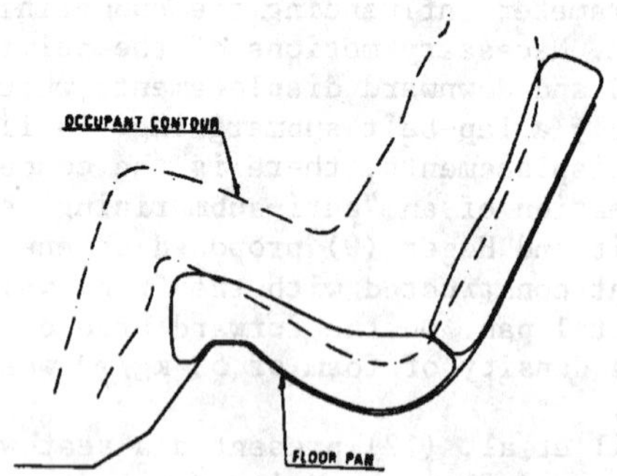

FIG. 8. DIAGRAM OF A SEAT DESIGN GIVEN BY LUNDELL ET AL.(12).

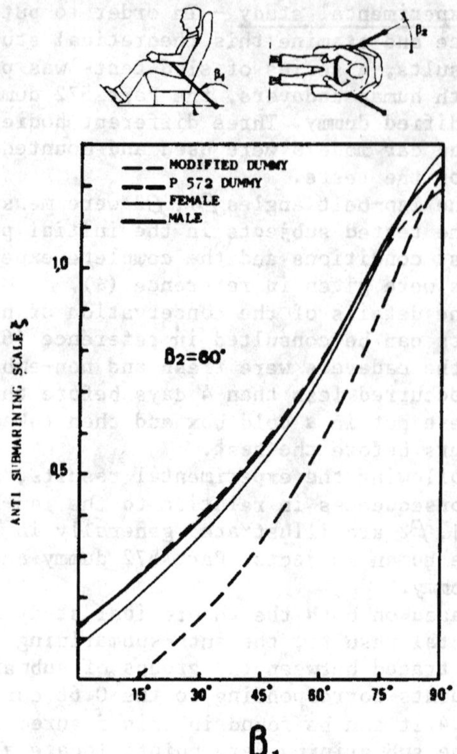

FIG. 9 . COMPARISON OF THE ANTI-SUBMARINING SCALE CURVES BETWEEN THE DIFFERENT SUBJECTS STUDIED AS A FUNCTION OF $\beta 1$ WHEN $\beta 2$ =60°.

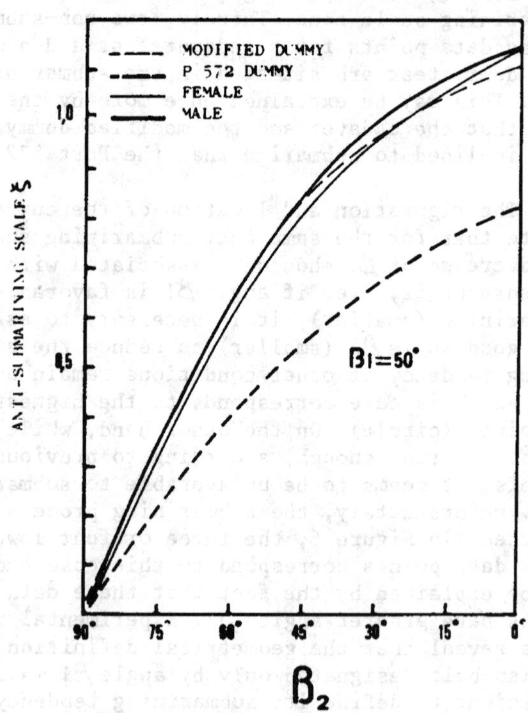

FIG. 10 . COMPARISON OF THE ANTI-SUBMARINING SCALE CURVES BETWEEN THE DIFFERENT SUBJECTS STUDIED AS A FUNCTION OF $\beta 2$ WHEN $\beta 1$ =50°.

jects (male and female), Part 572 dummy and modified dummy.

In (5), it was shown that the direction of the Sartorius varies with the sex of the individual and with the different pelvic models. For example, the angle $\alpha'x$ (Fig. 2) is the most critical angle for indicating the tendency to submarine. This angle varies from 85° for males to 79° for females and to 109° for Part 572 and 83° for the FAA-USAF-NHTSA pelvic model. The submarining tendency is stronger when this $\alpha'x$ is higher (4)(5).

Flexibility of the soft abdominal tissue is greater for human subjects than for Part 572 dummy (5). An attempt was made in order to have a more human-like dummy. A Part 572 dummy modified at the level of "Sartorius" and abdominal tissue was defined (Fig. 11) and evaluated (4).

The theoretical approach still applies to these evaluations and experiments already widely described (4) have shown that the frequency of submarining is very similar for the modified dummy and the cadavers (65 % and 62.5 % respectively), smaller than for Part 572 (75 %). It was also observed that submarining occurs in a shorter time for the Part 572 (67.8 ms against 74.5 and 73.5 ms for the human subject and the modified dummy). This constitutes another reference for the justification of submarining tendency.

The direction of the "Sartorius" depends not only on the pelvic shape of the individual but also on the orientation of the pelvis in sitting position. It is known that the lap-belt submarining performance depends significantly on pelvic orientation. The submarining tendency increases with the rearward rotation of the pelvis. This rotation was defined for a sitting position within a car in relation to a standing position. An angle of 36.7° was recommended by Nyquist (13); it was determined an X-ray-radiograms-study on the basis of two volunteers from his study (14) and 5 volunteers from reference (5).

STUDIES BY MATHEMATICAL MODEL OF THE INFLUENCE OF CERTAIN OCCUPANT PARAMETERS ON THE RISK OF SUBMARINING OCCURRING.

A study by mathematical model of occupant related factors was conducted in 1979 with the financial backing of the French Government, as part of the Programmed Thematic Action "Vehicle Safety - Submarining Criterion". The purpose of this study was to evaluate the influence of parameters such as stiffness of the lumbar column, hips and knees, or mass distribution adjacent to the pelvis, on the tendency of a dummy to submarine (15).

MODEL - The model used, "Prakimod", is a two-dimensional one. It has ten degrees of freedom and offers the advantage of being a sophisticated and tried method of simulating a belt restraint system (16).

REFERENCE TEST - An experimental test - 50th Percentile Part 572 dummy in front passenger position, restrained by a 3-point-inertia-reel-belt type on a sled (Renault 5 configuration without instrument panel), catapulted at 50 kph against a 30° inclined rigid wall - was reproduced by mathematical modelling, to act as a reference.

OUTPUT PARAMETERS - Actual submarining is not simulated by the model, but it does provide, at each moment, an output of the angle formed by the inboard and outboard webbings of the lap-belt with the pelvis projected along the sagittal plane (Figure 12). The maximum value of this angle during impact, together with its value when maximum forces in the lap-belt occur, make it possible to identify the risk of submarining. Maximum head, thorax and pelvis accelerations are also recorded as well as the maximum tensions of the different webbings and resultant forces applied to the dummy.

SIMULATIONS FOR COMPARISON PURPOSES - After the reference simulation, eleven other simulations were carried out using the same set of data but modidying each time the value of one of the input parameters to be tested.

RESULTS (Table 11) - Stiffening of the lumbar-column would greatly reduce the risk of submarining which is natural. A major increase in head accelerations would be observed simultaneously, along with a significant decrease in the longitudinal displacement of the centre of the head.

Stiffening of the hips generally heightens the risk of submarining, reduces head and pelvis accelerations and increases thorax accelerations. Moderate stiffness of the knees is of very little consequence.

Significant stiffening tends to slightly reduce the risk of submarining, as well as diminishing all maximum accelerations and belt forces. Transferring load from the thigh to the trunk tends to reduce the risk of submarining, especially if the load is transferred to the abdomen and not to the pelvis. Furthermore, a reduction in forward movement of the head and maximum acceleration of the pelvis and an increase in shoulder belt loading were recorded.

FINDINGS OF STUDY - Two possible orientations emerged from this study to reduce the tendency of the Part 572 dummy to submarine, independently of any change to the shape of the pelvis:

-Pending accurate accurate and reliable anthropometric data, displace load from the thighs to the pelvis, or better still, to the abdomen. This would entail modidying the dummy build.

-Or simply modify the adjustment of hip and knee friction torques, by requiring that these torques balance not the weight of the members as specified in Standard 208 (Figure 13A) but the weight of the whole body, realistically positioned in line with the working of the muscles of these members (Figure 13 B). This gives a smaller hip torque (61Nm against 100Nm for both hips), but a knee torque more than seven times greater (146Nm against 20Nm).

Two further simulations confirmed the accuracy of this solution regarding the change in output parameters (risk of submarining, accelerations..) but also showed that excessive blocking of the

Table 9 - Abdominal and spine injuries observed in frontal impact tests with cadavers (APR sample).

Test N°	Seat N°	Submarining	Liver	Spleen	Lumbar vertebra	Other abdominal injuries and pelvic bones fractures
2	1	yes	yes		L4 sacrum	colon, mesentery
4	1	no	yes			
6	2	yes				
8	2	no		yes		ilium, diaphragm
10	1	no	yes			ilium
12	2	yes				mesentery
16	2	no		yes		diaphragm
25	2	no				mesentery, ilium*
44	2	no				
127	4	yes			L1	
154	4	yes	yes		L2	iliac crest
170	1	yes			L4,L5	
182	2	yes			L1	
183	1	yes	yes			
184	1	yes	yes			
185	2	yes				mesentery
189	4	yes			L5,S1	abdominal muscle, mesentery
231	2	yes	yes	yes		mesentery
243	4	yes				colon, abdominal muscle
244	4	yes				left ilium crest
245	4	yes				abdominal muscle
246	4	yes			T6,T7	abdominal muscle
247	4	yes				abdominal muscle
255	2	yes		yes		

Seat N°. 1 : driver
Seat N°. 2 : front passenger
Seat N°. 4 : right rear passenger

(*) very light injury

Table 10 - Abdominal injuries observed in Frontal Impact tests with cadavers ("Heidelberg" sample).

	Mean $\gamma \leq 16$ g		Mean $\gamma > 16$ g	
	Drivers	**Passengers**	**Drivers**	**Passengers**
Size of sample	81	0	46	84
Liver	0	0	2	14
Spleen	0	0	0	5
Kidneys	1	0	1	3
Mesenterium	2	0	4	11
Intestines	1	0	3	4
Vessels	0	0	2	4

Table 11 - Part of Mathematical Simulation Results

	REFERENCE	HIP STIFFNESS DIVIDED BY 3	MULTIPLIED BY 3	MULTIPLIED BY 9	KNEE STIFFNESS DIVIDED BY 3	MULTIPLIED BY 3	MULTIPLIED BY 9	LUMBAR STIFFNESS DIVIDED BY 3	MULTIPLIED BY 3	MASSES DISPLACEMENT -4kg from upper legs +4kg onto pelvis	-4kg from upper legs +2kg onto pelvis +2kg onto lumbar segment	-4kg from upper legs +4kg onto lumbar segment
TEST No.	1	2	3	4	5	6	7	8	9	10	11	12
MAX. HEAD ACCELERATION (G)	71.00	70.38	57.51	53.59	71.12	70.64	67.85	42.16	75.50	72.04	64.18	50.81
HEAD G.S.I.	511.75	521.09	397.71	371.35	514.13	504.84	473.78	281.06	568.40	535.11	417.17	371.35
H.I.C.(*1)	408.17	417.15	327.41	298.28	410.38	401.64	373.87	242.50	465.93	416.61	380.61	315.55
MAX THORAX ACCELERATION (G)	46.32	47.75	47.14	50.02	46.64	45.37	42.37	49.01	53.10	46.55	49.27	51.35
THORAX G.S.I.	262.28	260.97	269.15	325.80	265.11	253.70	228.97	297.28	354.94	249.76	258.16	276.90
MAX PELVIS ACCELERATION (G)	71.30	74.27	70.36	63.21	71.42	70.88	69.53	69.83	69.27	68.90	67.40	65.97
MAX FORCE 1 (*2) (N)	12813	13266	13117	13509	12833	12750	12523	13082	13529	13251	13669	14025
MAX FORCE 2 (*2) (N)	9589	10163	9532	9937	9624	9482	9133	10974	10550	9961	10522	11148
MAX FORCE 3 (*2) (N)	15713	16177	15753	16329	15809	15423	14484	14569	17848	16089	15876	15590
AT TIME (ms)	50	50	52	63	50	51	51	66	51	50	50	50
θ ANGLE WHEN FORCE 3 IS MAX. (°)(*2)	127.97	124.73	132.67	138.03	128.13	128.93	127.14	158.15	123.30	126.42	125.64	124.88
MAX. θ ANGLE (°)	152.14	149.92	147.44	138.04	152.35	151.47	148.91	165.02	134.59	150.54	149.39	148.16
HEAD x-DISPLACEMENT (cm)	51.78	50.33	51.22	50.18	51.76	51.86	51.78	51.24	43.72	51.41	49.92	48.07

(*1) Not significant because of absence of head impact

(*2) see Figure 12

knees could significantly increase the horizontal displacement of the head.

-It can be clearly seen that the study to reduce the severity of head impacts cannot be carried out separately from the controlling of submarining.

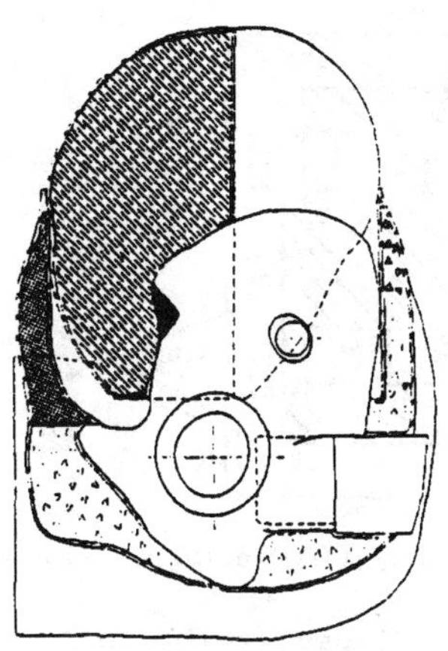

Silicone foam RTV 5370
Density 0.20 instead of 0.26

Silicone foam RTV 5370
Density 0.16 instead of 0.20

Removed portion of the pelvis

To provide a human like three dimentional direction of "Sartorius" in sitting posture

FIG. 11 . PELVIC MODIFICATIONS.

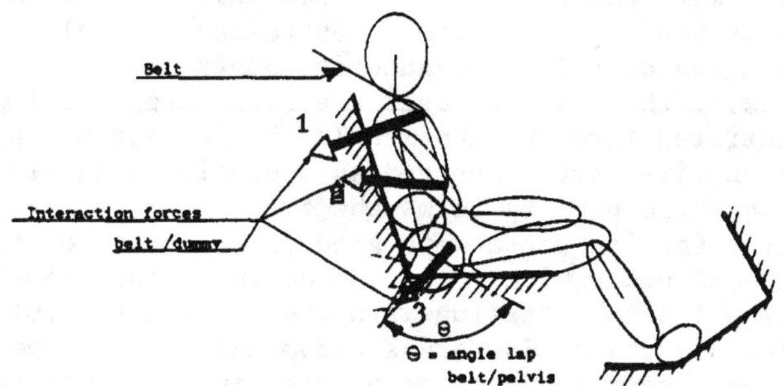

FIG. 12 . INITIAL POSITION OF THE DUMMY.

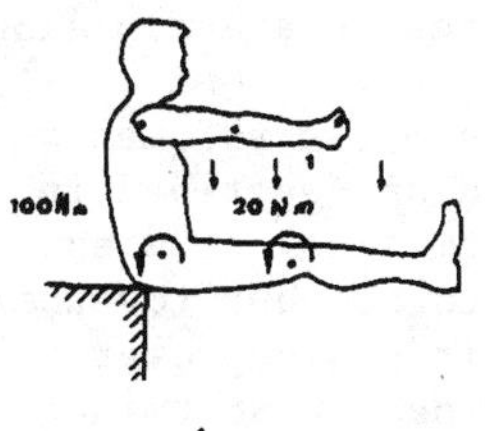

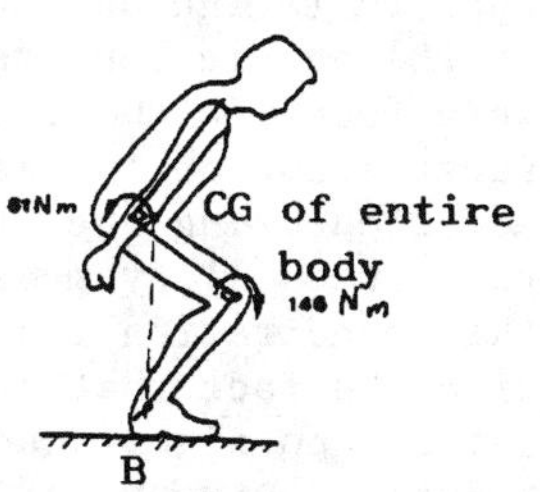

A
According to standard 208.

B
According to a more natural position.

FIG. 13 . TORQUES APPLYING TO DUMMY'S LOWER LIMB JOINTS.

DEFINITION OF AN ABDOMINAL PROTECTION CRITERIA

Past publications have described the complexity of the factors influencing the occurence of submarining and concluded that it was impossible to avert it by a single geometrical criterion as exists in Regulation 14,for example.That is why a great deal of work had been carried out with the object of seeking an abdominal protection criterion.

These different approaches have in common the determination to eliminate the risk of abdominal injuries induced by submarining under the lap-belt.

AN ACCOUNT OF THE DIFFERENT ATTEMPTS MADE TO DEFINE AN ABDOMINAL PROTECTION CRITERION

To our knowledge,the first attempt at defining a protection criterion for abdominal organs was published in 1973 (17).It stated that the "protection of abdominal will be satisfactorily guaranteed if any rising of strap -in dynamics-(during test) above anatomical reference marks materializing the limit compatible with a satisfactory rest on pelvis is forbidden"(Figure 14) "Compliance with the criterion is checked by cinematographic observation".

This first method successfully used in research could not be retained for the regulation test as the presence of doors -and unmodified doors-precludes filmed observation of the pelvis.

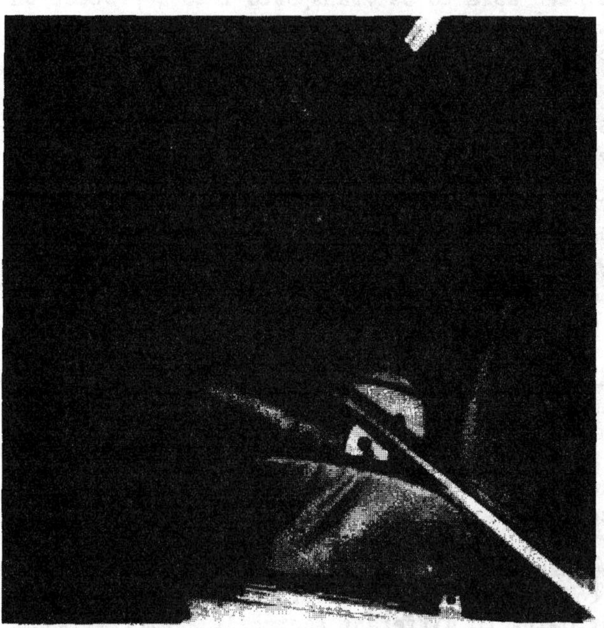

FIGURE 14.

The following year,in June 1974,at the ESV Conference in London,Citroen proposed fitting the dummy with optical transducerslocated on the pelvis beneath the Anterior Superior Iliac Spine (A.S.I.S.).These transducers would allow movement of the lap belt in relation to the A.S.I.S. to be detected. (18)

In the same year,a Ford patent dated 15th October 1974,proposed an almost identical system of detec

tion in which the optical transducer was replaced by force transducers. (19) .This solution was also to be retained by Citroen in 1977 (20).
A criticism which can be levelled against two above solutions is that abdominal danger cannot be stated to exist by the mere fact of the belt running up over the iliac crests.Indeed,the belt may be positioned against the abdomen at the end of impact without there being the risk of the occupant sustaining abdominal injury,if belt load is zero or low.Experimental work on human subject showed that certain cases of submarining proved to be of little danger provided that belt penetration into the abdomen as well as belt load -these two variables are obviously interdependent- were lower than a given limit.Moreover,the abdominal tolerance which justifies the abdominal protection criterion presented in following chapter is based on such values.
Three other ideas were formulated which would enable the detection of submarining:

-measure the angle of rotation of the pelvis (21)

-measure the pressure in the abdominal air bag with which dummy Part 572 is equipped (22)

-measure the pressure exerted on a set of pressure transducers arranged over the anterior face of the dummy (abdomen and thorax) (23).

These ideas are still in the early proposal stage. With the lack of experiments on dummies and references to the tolerance of human beings,they have not been able to be transposed into a protection criterion determining a limit which must not be exceeded if the non-occurence of abdominal injuries is to be guaranteed.

EXPERIMENTAL WORK CONDUCTED ON CADAVERS TO DEFINE THE ABDOMINAL TOLERANCE IN THE SUBMARINING CASE.

To define this abdominal tolerance in the event of submarining under the lap-portion of a three-point belt,10 cadaver tests were performed. Results have already been published as an ISO document(24).

METHODOLOGY USED IN CADAVER TESTS

Ten human cadaver tests were first and last performed with a current Renault 18 and Renault 20 car body mounted on a sled.The maximum sled deceleration and impact velocity were 21-30g and 50 kph.The fresh cadavers were placed on the front passenger seat (Nb 2) or on the rear bench seat at the right side (Nb 4).Three-point retractor belts were used.Because the car occupant is more likely to submarine in the rear seat than in the front seat,owing to different seat-belt geometry (4), most of the fresh human cadavers were placed on the rear bench seat at the right side (seat Nb 4). Eight cameras (velocities 500 or 1000 pictures/second) were used for this study.The positions of these cameras are shown in figure 15.
For two subjects (Nb 246 and 247),accelerations were recorded for the thorax at D1,D4,D7,D12,the ribs and the pelvis (on sacrum).No head deceleration recordings were taken because of recording capacity.Two shoulder belt loads (upper and lower) two lap-belt loads (inboard and outboard) were also recorded.
Autopsies on the human cadavers were carried out after tests by the specialist of IRBA under the direction of Pr. C.GOT.

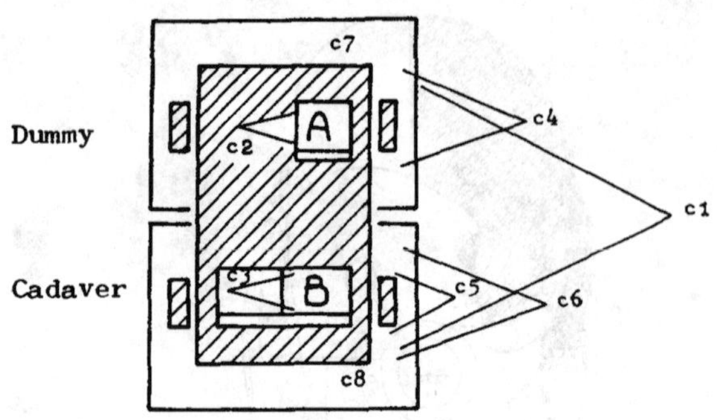

Cameras C1,C4,C5,C6:Placed at the right side of the sled.
Cameras C2,C3:Placed at the left side of the sled.
Cameras C7,C8:Placed at the top of the sled.

FIG. 15 . POSITION OF THE CAMERAS USED IN THE TESTS.

RESULTS AND DISCUSSION

The acceleration recordings are presented in Table 12.The anthropometrical data of the human cadavers and the results of the dissections are shown in Tables 13 and 14.
In five recently conducted tests ,Nb 244 and Nb 245 were normal cases of submarining.This means that the lap-belt firmly restrained the pelvic bone region below the Anterior Superior Iliac Spines.It then rode-up over the iliac crests and penetrated into the abdomen.In this case,the tension time curve appeared as a distinct "saddle shape".In precise terms,the first peak of lap-belt tension was always greater than that of the second peak which corresponded to a lower compression load penetrating into the abdomen(Fig.16). That is why no dangerous abdominal injuries were found.Heavy submarining occured in the remaining tests.The lap-belt rode-up rapidly over the ASIS and tension increased continuously(Fig.17).The tension was usually greater at the second peak applied to the abdomen than at the first peak, to the pelvic bone region.Therefore,abdomen AIS$\geqslant$3 were found in the cadavers.
Penetration of the lap-belt into the abdomen was measured using the same method as described in reference (1).By means of kinematical studies, this penetration was taken into account together with the fact that the lap-belt rode-up over ASIS. Because no significant difference in abdominal injuries between left and right hand sides was observed,either in the autopsies or in the real-life traffic accident,the severity of the abdominal injuries is specified for the whole abdomen as given by the anatomo-pathologists.The average lap-belt tension was taken from both sides.
In order to compare the dummy tests,lap-belt ten-

TABLE 12- DECELERATIONS RECORDED - TESTS WITH HUMAN CADAVERS

Test N°	Seat N°	Sled decel. γ (g)	Sled speed (km/h)	Thorax [γ (g)/SI] D1	D4	D7	D12	Sacrum γ(g)/SI	Shoulder belt (N) upper	lower	1st peak lap-belt (N) int.	ext.	2nd peak lap-belt (N) int.	ext.
127	4	23	50.9	-	-	-	-	86/1501	6800	8500	4000	9100	3900	9000
148	2	21	50.1	-	-	-	-	50/563	7700	2250	7600	4100	4200	2000
148	4	21	50	-	-	-	-	53/340	6000	5000	8600	4500	6500	2800
154	4	37	50.7		82/620	147/8064		71/779	6400	10050	5400	7400	5400	6700
182	2	24	47.5	39/171	36/76	81/555	107/437	45/272	5200	5500	6450	3900	7750	4300
243	4	27	47.4	-	-	-	-	93/471	4200	4600	600	700	4900	5100
244	4	34	50.1	-	-	-	-	63/334	5000	2800	3400	3900	1800	2200
245	4	22	49.8	-	-	-	-	52/859	5000	1800	3000	3700	2800	3700
246	4	30	50.1	56/521	82/678	81/774	79/637	153/1070	6600	4500	1500	4800	6000	7200
247	4	30	50.5	67/622	76/683	79/723	65/494	133/1067	8900	5800	5000	5200	6800	8600

TABLE 13 - NORMALIZED LAP-BELT TENSIONS FOR THE SECOND PEAK (AFTER SUBMARINING)

Test N°	Mass Kg	Coefficient for normalized tension	Average 2nd peak tension of 2 sides (N)	Normalized lap-belt tension (N)	Abdomen AIS	Relative bigness x 10^{-6}
127.4	41	1.49	6450	9610,5	3	10,2
148.2	59	1.17	3100	3627	0	13,9
148.4	67	1.08	4650	5022	0	13,2
154.4	42,5	1.46	6050	8833	5	8.5
182.2	62	1.14	6025	6868,5	3	11,4
243.4	74	1.01	5000	5050	4	14,5
244.4	54	1.24	2000	2480	0	12.0
245.4	62	1.14	3250	3705	1	16.0
246.4	52	1.28	6600	8448	3	11.5
247.4	58	1.19	7650	9104	4	13.3

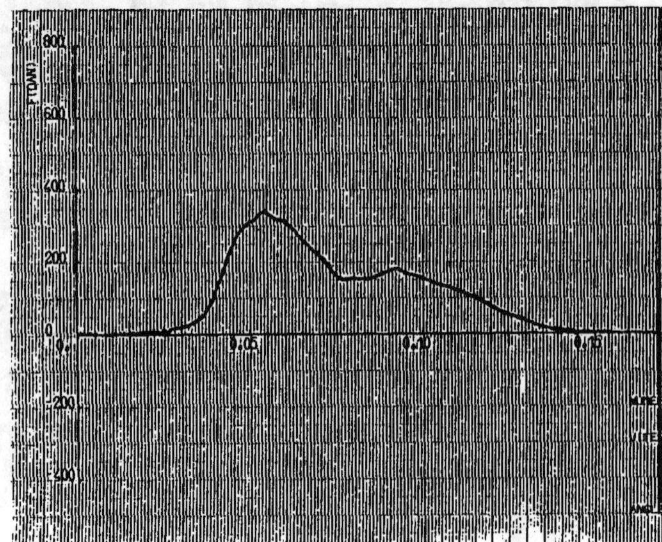

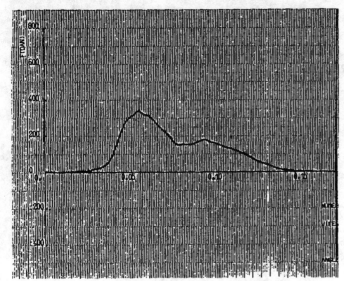

FIG. 16 . LAP-BELT TENSION/TIME HISTORY RECOR- -DED AT THE INTERIOR SIDE OF CADAVER N°244

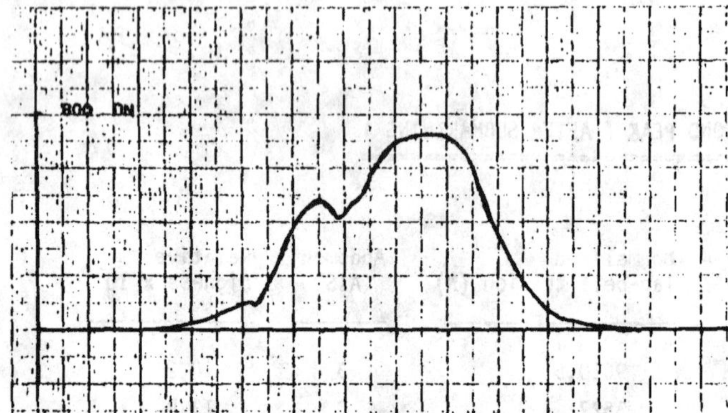

FIG. 17 . LAP-BELT TENSION/TIME HISTORY RECOR- -DED AT THE EXTERIOR SIDE OF CADAVER N°246.

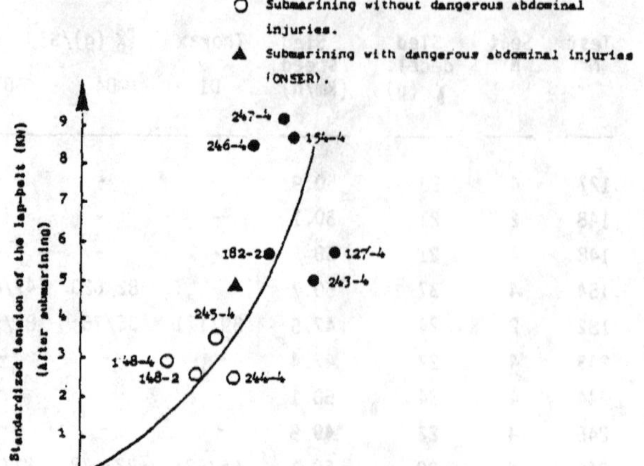

Test N°	Mass	Size	AIS
127-4	41kg	159cm	3
148-2	59kg	161cm	0
148-4	67kg	172cm	0
154-4	42,5kg	171cm	5
182-2	62kg	176cm	3
243-4	74kg	172cm	4
244-4	54kg	165cm	0
245-4	62kg	157cm	1
246-4	52kg	165cm	3
247-4	58kg	163cm	4

Previous results: 127-4 to 182-2. Recent results: 243-4 to 247-4.

FIG. 18 . ABDOMINAL TOLERANCE TO THE SUBMARINING.

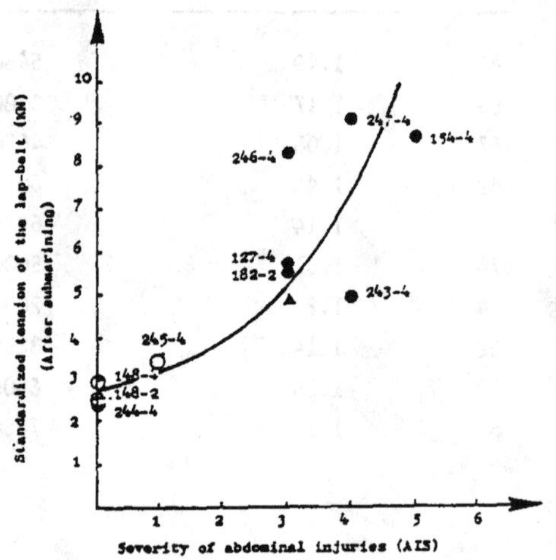

FIG. 19 . RELATIONSHIP BETWEEN THE STANDARDIZED TENSION OF THE LAP-BELT (AFTER SUBMARINING) AND THE ABDOMEN AIS.

sion recorded in the cadaver tests should be standardized with respect to the mass of the Part 572 dummy (75kg).A formula given by Eppinger is used in this study,which has been described in references (24)(25).

In figure 18,a parabolic curve can be drawn as a relationship between the standardized tension of the lap-belt and abdominal penetration.The scatter of the data points is due to the differences in the anthropometric data of the cadavers.A similar curve can be found also in figure 19 for the standardized lap-belt tension (after submarining) and the abdomen AIS.Previous results are also presented in these figures.It can be seen that Nb 243-4 is the lowest point (5KN) in the new group with dangerous injuries.This value is smaller than that (5.7KN) of Nb 182-2,the lowest one in the old group of previous results.One datum point (without test number) given by ONSER is also presented in these two figures.

Specific comments on the severity of abdominal injuries (AIS) concerning cadavers Nb244 and 245 are given below:

Nb 244 sustained only pelvic fracture with an AIS=3 but no abdominal injuries.An abdominal AIS=0 was given to this case.It implied that if no pelvic fracture has occured,the lap-belt would probably have penetrated into the abdomen,with greater force and caused abdominal injuries,which in real-life could be a dangerous situation.It is an ambigous case difficult to interpret and we prefer to avoid underestimating the abdominal risk.

-Nb 245 sustained abdominal AIS=1 and was the only subject situated between the group with dangerous injuries (AIS ⩾3) and the group with injuries (AIS=0).Since the severity of the abdominal injuries due to submarining is probably influenced by the "relative bigness" defined as "weight/height^3" (kg/cm^3) in reference (24),a further study of this factor is described here.The "relative bigness" of the human cadavers versus their abdomen AIS is illustrated in figure 20 where the previous results are also presented.Cadaver Nb 245 had the greatest "relative bigness" in all results.We made the assumption that this subject would sustain a greater AIS if his bigness had been so to speak,smaller.It may be for this reason that the only fissure of the abdominal adipose tissue was found in this subject,

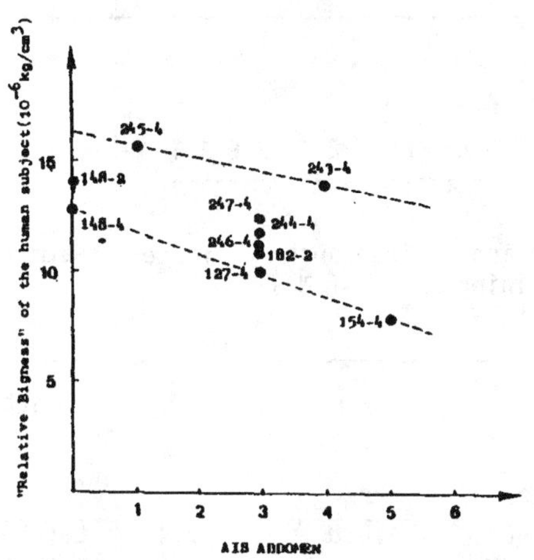

FIG. 20 . "RELATIVE BIGNESS" VS SEVERITY OF THE ABDOMINAL INJURIES (AIS).

The abdomen AIS is certainly influenced by other parameters,for example rib cage strength,which are not discussed in this study.In general,no big difference can be found between previous and recent results.But the limits between the groups with and without injuries are closer due to the recent results;therefore the latest determination of the abdominal tolerance will be more accurate. The determination of critical lap-belt tension should be considered more carefully.As just mentionned above,the case with Nb 244 cadaver produced no injuries,yet it is nevertheless a dangerous situation.On the other hand,Nb 243 (5KN) produced the lowest datum point in the group with injuries in recent results;Nb 182-2 (5.7KN) was the lowest one in previous results.Based on previous and recent available data points presented in figure 18 and figure 19,the critical standardized lap-belt tension was reconsidered.A value of 3 KN can be contemplated,which is smaller than the previous value of 4.4 KN which had been proposed(25,26,27).

EXPERIMENTAL WORK WITH DUMMIES INSTRUMENTED WITH "ILIAC CREST TRANSDUCERS"

SCHEDULE OF RECENT RESEARCH PROGRAMS FOR DUMMY TESTS

-First series - Dummy tests with a Renault 20 car body and seat -

The same car body and sled used in cadaver tests were employed for the dummy tests.The sled deceleration and impact velocity were similar to those registered in cadaver tests.The Part 572 dummy was equipped with short (40mm) or long (55mm) APR submarining transducers and placed in the front passenger seat (Nb 2) or in the rear bench seat on the right side (Nb 4).Three-point retractor seat belts (60mm width) and different positions of lap-belt were used in these tests.Standard recordings were made.Test conditions were similar or more severe with respect to test Nb 5 given in reference (28).

-Second series - Dummy sled tests with Renault 20 seat directly on the sled -

This series of tests was performed by T.N.O.The test conditions were similar to those used in reference (28).Some tests were performed under excessive test conditions (with the seat back set at an angle of about 30° from the vertical for example.Classical recordings were taken.

-Third series - Dummy sled tests with Citroen Visa seat directly on the sled -

This series of tests was the reconstruction of test NB 10 in reference (29).Briefly,the test conditions were:

-front passenger Citroën Visa seat
-Part 572 dummy without fore-arms
-3-pt static seat belts (50mm width)
-sled deceleration and impact velocity about 20g and 50kph respectively
-inclination of seat back 28°
-weakened seat frame
-slack of the lap-belt and shoulder belt about 25 mm

Recordings were made for pelvic deceleration, shoulder belt tensions (upper and lower) and lap-belt tensions (inboard and outboard).Citroën transducer and APR short or long transducers were used,

RESULTS

-First series - Dummy tests with Renault 20 seat,

The results of this series of tests are presented in Table 15.The first three tests (Nb 1013,1014, 1015) are cases of non-submarining with the lap-belt positioned correctly.Tests Nb 1018,1019 and 1021 are submarining cases with the lap-belt positioned 1 cm higher than in the first three tests. The compression loads recorded by the submarining transducer are in the range between 463N and 798N. Tests Nb 1016,1017 and 1020 are cases of excessive submarining in which the lap-belt rode up over the submarining transducers.Nb 1016 and 1017 are tests in seat Nb 2 with the worst position of the lap-belt,Nb 1020 is a test performed in seat Nb 4 with a correct position of the lap-belt and using short transducers (Figure 21).The recorded loads range from 805N to 1116N and it serves as an exam-

Table 14 - Injuries of the Tested Human Cadavers

Cadaver No.	Sex	Age	Size (m)	Weight (kg)	Injuries and Corresponding AIS: Thorax (Rib fractures)	Abdomen	
127-4	M	57	1.59	41	28 ribs, AIS 4	L 1 fracture	AIS 3
148-2	F	65	1.62	59	8 ribs, AIS 3		AIS 0
148-4	M	62	1.72	67	12 ribs AIS 3		AIS 0
154-4	M	63	1.71	42.5	40 ribs+sternum AIS 4	liver + L2 fracture	AIS 5
182-2	M	57	1.76	62	8 ribs AIS 3	L 1 fracture	AIS 3
243-4	M	61	1.72	74	7 ribs AIS 3	colon + break of rectus	AIS 4
244-4	F	57	1.65	54	24 ribs AIS 4	fractures of left ilium crest	AIS 3
245-5	M	56	1.57	62	7 ribs AIS 3	fissure of adipose tissue	AIS 1
246-4	M	62	1.65	52	12 ribs AIS 3/T6-T7 AIS 3	break of rectus	AIS 3
247-4	M	42	1.63	58	8 ribs+sternum AIS 3	break of rectus	AIS 4

TABLE 15 - DUMMY TESTS WITH RENAULT 20 SEAT

Test N°	Seat N°	Submarining transducer	Position of lap-belt with relation to ASIS from side view (cm)	Lap-belt tension after submarining Int (N)	Ext (N)	Transducer force (N) Int.	Ext.	Remarks
1013	2	short	+ 4	-	-	-	-	no submarining
1014	2	"	+ 3	-	-	-	-	" "
1015	2	long	+ 2	-	-	-	-	" "
1016	2	"	- 1	2200	2240	1030	994	Lap-belt rode-up over transducers
1017	2	"	+ 1	1940	1550	869	805	"
1018	2	"	0	1667	4520	798	614	Transducers worked correctly
1019	2	"	0	1117	1620	638	521	" "
1020	4	short	+ 4	5237		1116	-	lap-belt rode-up and touched direct submarining transducers
1021	2	"	+ 1	1122	-	463	-	transducers worked correctly

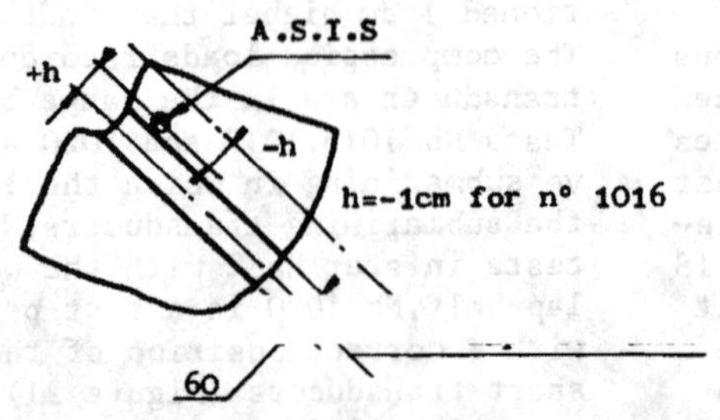

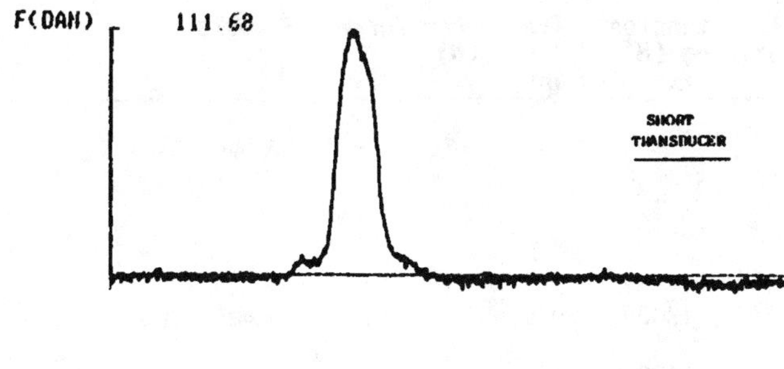

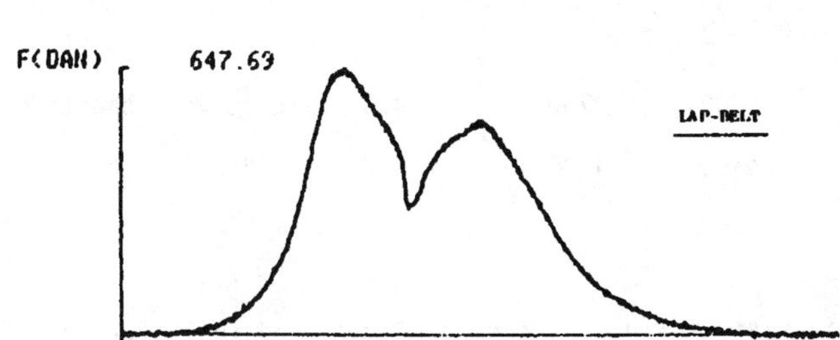

FIG. 21 . TENSION (T) AND FORCE (F) CURVES OF TEST N° 1020 (INTERIOR SIDE).

ple where the short transducer works correctly and records a greater load under test conditions more favorable to submarining in seat Nb 4 (rear bench) than in seat Nb 2 (front individual seat) (4).
In order to make a comparison with cadaver tests, a new research program is being prepared to perform the dummy tests under these same conditions. It can be recognized from the above analyses that the responses of the transducer to the submarining process were in accordance with the different test conditions.

-Second series - TNO sled dummy tests -

The results of this series of tests are presented in Table 16.No submarining occurred in the first three out of a total of nine tests,even with the lap-belt located 15 mm higher than the correct position.In the rest of the tests,a submarining phenomenon could only be obtained by placing sheets of double plastic foil between the dummy and the car seat in combination with a 25 mm "higher than correct" position of the belt on the pelvis.Excessive submarining in tests Nb 454,456 and 457 could only be obtained by additionally inclining the seat back to about 30° from the vertical.The excessive submarining cases all resulted in a short peak loading of the transducers (duration 6 ms),followed by an over-riding of the transducers.Consequently,the lap-belt was loaded again by the lumbar spine.In all tests,a reasonable symmetrical loading of left and right lap-belt and transducers was found.

-Third series - Dummy tests with Citroën Visa seat

In table 17 tests Nb 5363 and 5364 correspond to non-submarining cases.The lap-belts were positioned correctly.Tests Nb 5360,5361 and 5365 are normal submarining cases.Their descriptions are similar to the ones given for the first series of dummy tests.The positions of the lap-belts are relatively worse than in the non-submarining group.Both short and long transducers work correctly in these three tests.The long transducer is used in test Nb 5361;the lap-belt tension and transducer load as function of time for the outboard side are illustrated in Figure 22.

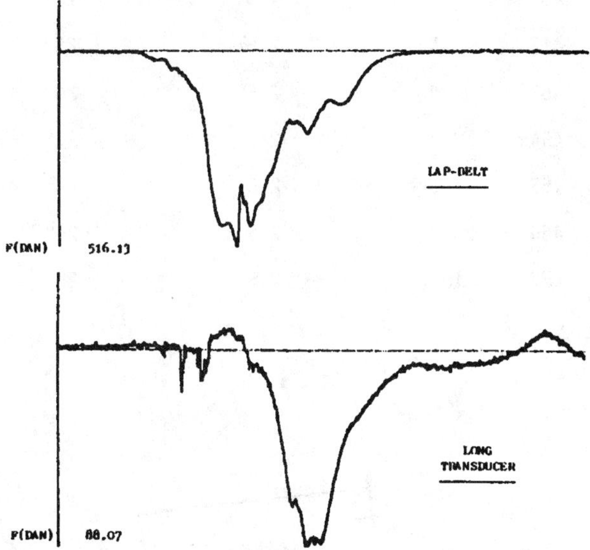

FIG. 22 . TENSION (T) AND FORCE (F) CURVES OF TEST N° 5361 (EXTERIOR SIDE).

Test Nb 5362 corresponds to a submarining case in which the lap-belt rides-up over the transducer. The position of the lap-belt was the least desirable one observed in this series of tests.However, a significant load of 2531N was recorded by the short transducer.This sample again illustrates how the submarining transducer can provide a correct response:when the test is performed under more severe submarining conditions,a greater compression load can be recorded by the transducer, so indicating that it would be a more dangerous submarining case.The lap-belt tension (t) and transducer load (f) for outboard side in the test Nb 5362 are illustrated in Figure 23.

RELATIONSHIP BETWEEN LAP-BELT TENSION AND THE SUBMARINING TRANSDUCER LOAD

All data points obtained from the recent results of dummy tests are illustrated in Figure 24 The two parameters are the maximum transducer load and the corresponding lap-belt tension at the same points in time.The previous results are also presented.A linear correlation between lap-belt tension and transducer load can be proposed.
Since the standardized critical lap-belt tension (3000N) is already determined in Figures 18 and 19 a corresponding transducer load of 800N can be found in Figure 24.
Based on previous and recent results obtained from cadaver and dummy tests,a submarining transducer load of 800N is proposed as an abdominal protection criterion in the dummy tests.

CONCLUSIONS

TABLE 16 - T.N.O. SLED TEST EVALUATION OF APR SUBMARINING TRANSDUCERS ON PART 572 DUMMY WITH RENAULT 20 SEAT

Test No.	Submarining transducer	lap-belt position upper ASIS (cm)	Seatback	Max. seat belt tension after submarining (N) Int.	Ext.	Transducer force (N) Int.	Ext.	Remarks
449	long	correct	25°	-	-	-	-	no submarining
450	"	1	25°		-	-	-	" "
451	"	1.5	25°	-	-	-	-	" "
452	"	2.5	25°	1500	1300	900	1000	submarining
453	"	2.5	25°	2000	1100	1130	1100	"
454	"	2.5	30°	2200	1000	1100	1000	excessive submarining
455	short	2.5	25°	-	-	-	-	no submarining
456 *	"	2.5	>25°	(?)	3200 (?)	750	450	excessive submarining
457	long	2.5	>25°	3600	2100	1100	820	" "

* In this test, there were the problems found in the recordings.

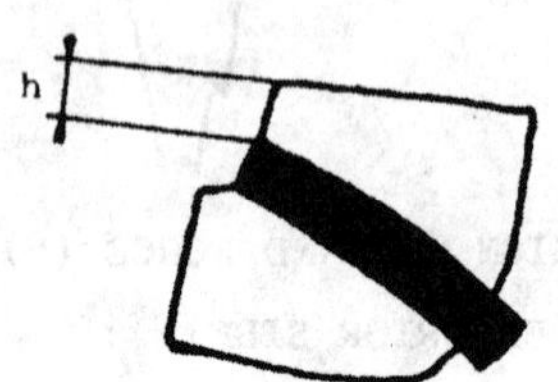

TABLE 17 - DUMMY TESTS WITH CITROEN VISA SEAT

Test No.	Seat No.	Submarining transducer	Position of lap-belt (cm) *	Seat	Lap-belt tension after submarining Int (N)	Ext (N)	Transducer force (N) Int.	Ext.	Remarks
5360	2	short	2	weaken	2627	1564	-	607	transducer worked
5361	2	"	1	"	-	2198	-	880	" "
5362	2	"	0	"	-	4050	-	2531	lap-belt rode-up over transducer
5363	2	"	3	weaken+flexible	-	-	-	-	no submarining
5364	2	long	3	" "	-	-	-	-	" "
5365	2	"	3	" "	4506	-	1239	-	transducer worked

* Definition is same as given in the diagram presented in Table 2.

1.A standardized critical lap-belt tension of 3000N is determined,based on available values from previous and recent submarining cadaver tests.
2.In human cadaver submarining tests,a relationship between abdominal injury (AIS) and "relative bigness" of the subject can be found:abdomen AIS increases as "relative bigness" decreases.

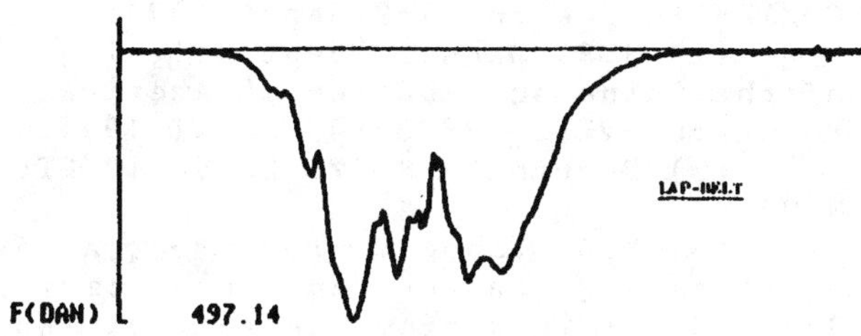

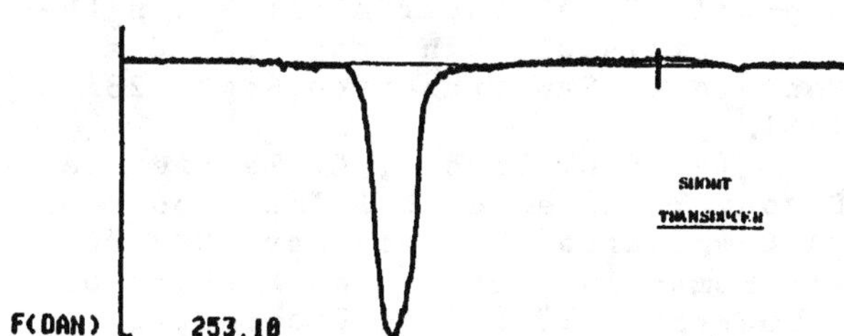

FIG. 23 . TENSION (T) AND FORCE (F) OF TEST N°5362 (EXTERIOR SIDE).

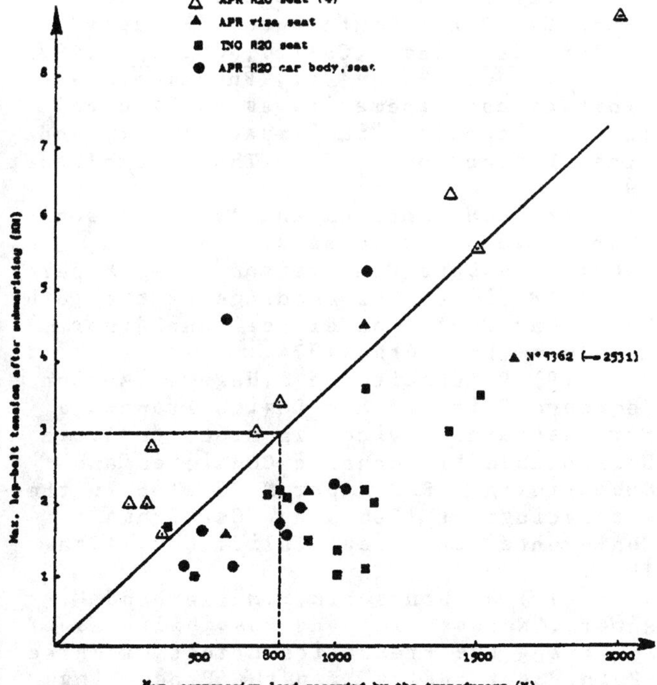

FIG. 24 . RELATIONSHIP BETWEEN THE LAP-BELT TENSION AND THE COMPRESSION LOAD RECORDED BY THE TRANSDUCERS.

3.With a dummy in the front passenger seat (Nb 2) and a correctly positioned lap-belt,non-submarining cases were usually obtained.
4.Submarining cases were obtained at the front passenger seat (Nb 2) by deliberately positioning the lap-belt at an excessive level on the abdomen or by increasing backrest rake,only weakening the seat frame.
5.In all submarining cases,the short (40 mm) and long (55 mm) transducers were able to detect the submarining phenomenon.
6.The values recorded by the short or long submarining transducers corresponded to the different conditions used in the dummy tests.
7.Based on recent critical lap-belt tension determined from the cadaver tests and,also the data points concerning a relationship between the lap-belt tension and the transducer load found in dummy tests,the abdominal protection criterion of 800N is proposed for dummy tests.
8.The short transducer can usually be used in submarining tests for the abdominal protection criterion;the long transducer can be used successfully in tests performed particularly under excessive submarining conditions.

GENERAL CONCLUSIONS

1.LAP-BELT SUBMARINING in real-life accidents is a process inducing severe injuries in a relatively high proportion of severe frontal crashes (27%) - but it is too often underestimated in previous studies-.Reasons of underestimation were for example that dorso-lumbar spine fractures,or lower members fractures,or above mesocolic injuries were not often considered as submarining consequences.Present findings constitute an attempt to clarify the whole pattern of such lap-belt submarining.
From a whole sample of 1423 three-point belted front occupants involved in 958 frontal car crashes,a sub-sample of 45 cars have been selected in which,at least one of the 77 front occupants sustained either a severe abdominal injury (AIS $\geqslant$3) or a dorso-lumbar column fracture (AIS$\geqslant$2). Among this survey sample,47 injured people sustained a lap-belt submarining which was sure or most probable (61%).These submarinings induced three main types of injuries:
-abdominal injuries(sub-mesocolic but also above mesocolic) (AIS$\geqslant$3)
-dorso-lumbar spine fractures (mainly T12, L1,L2) (AIS$\geqslant$2)
-lower members fractures (mainly legs,knees and femurs fractures) (AIS$\geqslant$2)

2.Among submarining cases,68% of cases of abdominal injuries (AIS$\geqslant$3) and (or) dorso lumbar spine fractures (AIS$\geqslant$2) were SURELY induced by the lap-belt section.The percentage reached 94% if we considered the cases where the lap-belt influence was sure or only MOST PROBABLE.

3.The influence of shoulder-belt plays only as an

aggravating factor - complementary to the lap-belt- for less than 10% of cases.

4. Poor geometry and (or) slackly worned belts were present in 74% of submarining cases.

5. Most of the submarining cases were observed in high violence crashes ($\Delta V \geqslant 50$ kph and mean $\gamma \geqslant$ 10g occured in 58% of cases).

6. In real-life accidents, the most frequent severe consequences of lap-belt submarining are according to a decreasing order:
   - sub-mesocolic injuries
   - lower members fractures
   - dorso-lumbar spine fractures
   - above-mesocolic injuries
   - pelvic fractures

7. Taking into account the severity of above-mesocolic injuries (liver and (or) spleen), it is noticeable that 3/5 of such victims sustained a lap-belt submarining which is sure or most probable (of course these lesions are often associated to sub-mesocolic injuries).

8. In cadaver tests (with blood pressure restored at normal level), submarining process is checked by special in-board camera which shows then all these above described injuries which could be induced by lap-belt submarining, even for above-mesocolic injuries (liver or spleen injuries) or dorso-lumbar spine fractures. These experiments also confirm the possible aggravating influence of shoulder-belt section.

9. Abdominal or dorso-lumbar spine TOLERANCE to lap-belt submarining is low. Injuries are observed for lap-belt tension higher than 3000N.

10. A PROTECTION CRITERION - ANTI-LAP-BELT SUBMARINING CRITERION - has been proposed. It consists in the record of lap-belt loading against specific ILIAC-CREST TRANSDUCERS symetrically installed on the pelvis of the dummy. Based on recent critical lap-belt tensions determined from specific cadaver tests and - also - the data points concerning a relationship between the lap-belt tensions and the Iliac-Crest-Transducer loads found in dummy tests, the LAP-BELT-SUBMARINING PROTECTION CRITERION of 800N is proposed for dummy test.

REFERENCES

(1) G. Walfisch, A. Fayon, Y.C. Leung, C. Tarriere; C. Got, A. Patel "Synthesis of Abdominal Injuries in Frontal Collisions with Belt-Wearing cadavers Compared with Injuries Sustained by Real-Life Accident Victims. Problems of Simulation with Dummies and Protection Criteria." in Proceedings of IRCOBI GOETEBORG, Sweden, 7-9 Sept. 1979.

(2) J.S. Dehner, "Seat Belt Injuries of the Spine and Abdomen ." American J. Roentgen , VIII, PP833-843, April 1971.

(3) Document ISO/TC 22/SC 12/GT6 N107.

(4) Y.C. Leung, C. Tarrière, A. Fayon P.Mairesse, P. Banzet. "An Anti-Submarining Scale Determined from Theoretical and Experimental Studies Using Three-Dimensional Geometrical Definition of the Lap-Belt " SAE Paper n°811020, in the Proceedings of 25th Stapp Car Crash Conference, San Francisco, Sept. 28-30, 1981.

(5) Y.C. Leung , C. Tarrière, A. Fayon, P. Mairesse, A. Delmas and P. Banzet, "A Comparaison Between Part 572 Dummy and Human Subject in the Problem of Submarining ." in the Proceedings of 23rd Stapp Car Crash Conference, San Diego Calif., Oct. 17/19, 1979, SAE Transaction Paper N°791.026.

(6) A. Fayon , C. Tarrière, G. Walfisch C. Got, A. Patel "Thorax of 3-Point Belt Wearers During a Crash (Experiments with Cadavers)" in the Proceedings of 19th Stapp Car Crash Conference, SAE paper 751148, San Diego , Calif., Nov. 17/19, 1975.

(7) J.L. Haley Jr., "Fundamentals of Kinetics and Kinematics as Applied to Injury Reduction "in "Impact Injury and Crash Protection ", C.C. Thomas Publisher 1970.

(8) L.M. Patrick and A. Andersson "Three -Point Harness Accident and Laboratory Data Comparaison". SAE Paper N°741.181, in the Proceedings of the 18th Stapp Car Crash Conference , Ann Arbor Michigan , Dec. 4th , 1974.

(9) D. Adomeit and A. Heger , "Motion Sequence Criteria and Design Proposals for Restraint Devices in order to Avoid Unfavorable Biomechanic Condition and Submarining". SAE Paper N°751.146 In the Proceedings of 19th Stapp Car Crash Conference , San Diego, Calif. Nov. 17th 1975.

(10) H. Hontschik, E. Müller and G. Rüter , "Necessities and Possibilities of Improving the Protectice Effect of Three -Point Seat -Belts ", in the Proceedings of the 21st Stapp Car Crash Conference , New Orleans , Louisiana , Oct. 19/21st 1977.

(11) P. Billaut , C. Tisseron , M. Dejeammes, R. Biard, P. Cord, P. Jenoc, "The Inflatable Diagonal Belt "7th

International Technical Conference on the Experimental Safety Vehicles,Paris June 5/9 ,1979.

(12) B. Lundell,H.Mellander,I.Carlson ,"Safety Performance of a rear Seat Belt System with Optimized Seat Cushion Design",SAE Paper N°810.796,Passenger Car Meeting Dearborn ,Michigan ,June 8 12,1981

(13) G.W. Nyquist "Comparaison of Vehicule-Seated Volunteer Pelvic Orientations Determined by Leung et al, and by Nyquist et al .Document ISO /TC22/SC12/ WG5?April 10,1980.

(14) G.W. Nyquist et al . ,"Lumbar and Pelvic Orientations of the Vehicule Seated Volunteer" SAE 760821 ,20th Stapp Car Crash Conference 1976.

(15) D. Lestrelin "Etude par modele Mathematique de l'Influence de quelques Paramètres du Mannequin Part 572 sur sa Propension au sous-Marinage ".Rapport Interimaire N°2 Contrat N°78043 "Critères de Sous-Marinage" dans le Cadre des Actions Thematiques Programmees Françaises, 1979.

(16) D. Lestrelin ,A. Fayon , C. Tarrière "Development and use of a Mathematical Model Simulating a Traffic Accident Victim" Proceedings of 5th International IRCOBI Conference , Birmingham,Sept. 1980.

(17) C. Tarrière "Proposal for a Protection criterion as Regards Abdominal Internal Organs" P371 Proceedings of Conference of A.A.A.M., Oklahoma City ,Oklahoma,Nov.14/17,1973.

(18) M. Clavel "Restraint Systems Improvement " Proceedings of 5th Interna -tional Technical Conference on Experimental Safety Vehicles,London ,June 4-7 1974.

(19) R.P. Daniel "Test Dummy Submarining Indicate in United States Patent ,3.841.163 Oct. 15, 1974.

(20) Citroën,"Methode de Detection de Depassement des crêtes Iliaques" Document ISO/TC 22/SC 12/GT6(F3)21F,1977.

(21) D. Adomeit "Seat Design -A Significant Factor for Safety Belt Effec -tiveness " SAE Paper N°791004, in Proceedings of 23rd Stapp Car Crash Conference ,San Diego ,Calif. ,Oct.17-19 1979.

(22) Bröde,Personal Communication Nov.24,1980.

(23) Fiat "Development of a Device to Evaluate the Abdominal Injuries in Submarining",Document ISO /TC 22/SC 12/ GT 6(Italie 1)N 71,Sept.1980.

(24) Y.C. Leung ,C. Tarrière,J. Maltha "A Review for the Abdominal Protection Criterion" Document ISO/TC 22/ SC 12/WG6 N97 October 1981.

(25) "Experimental Elements for the Definition of Abdominal Protection Criterion in a Submarining Possibility" July 27th,1980. ISO/TC 22/SC 12/WG. 6 N°72

(26) "Proposal for an abdominal Protection Criterion ",March 1980. ISO/TC 22/SC 12/WG. 6 N°58

(27) Y.C. Leung ,P. Mairesse,P.Banzet "Submarining Criterion "Sept. 1980. ISO/TC 22/SC 12/WG N°77.

(28) R.L. Stalnaker "Submarining sled Testd Part 572: Pelvis with and without PSA/Renault Submarining Transducers and PSA/Renault modified Part 572 Pelvis with Submarining Transducers" TNO, October 8th,1980.

(29) M.Dejeammes,R.Biard,Y.Derrien "Factors influencing the estimation of Submarining on the Dummy". ISO/TC 22/SC 12/WG.6-96, August 1981.

892440

Assessing Submarining and Abdominal Injury Risk in the Hybrid III Family of Dummies

Stephen W. Rouhana, David C. Viano, Edward A. Jedrzejczak, and Joseph D. McCleary
Biomedical Science Dept.
General Motors Research Labs.

ABSTRACT

This paper details the development of an abdominal injury assessment device for loading due to belt restraint submarining in the Hybrid III family of dummies. The design concept and criteria, response criteria, choice of injury criterion, and validation are explained. Conclusions of this work are:

1) Abdominal injury assessment for belt loading due to submarining is now possible in the Hybrid III family of dummies.

2) The abdomen developed has biofidelity in its force-deflection characteristics for belt loading, is capable of detecting the occurrence of submarining, and can be used to determine the probability of abdominal injury when submarining occurs.

3) Installation of the abdomen in the Hybrid III dummy does not change the dummy kinematics when submarining does not occur.

4) When submarining does occur, the dummy kinematics are very similar to baseline Hybrid III kinematics, except for torso angle. The change in torso angle is or may be a result of a more human-like compliance in the abdomen.

A SIGNIFICANT AMOUNT OF RESEARCH HAS BEEN DONE to assess the effectiveness of safety belts. Evans [1] applied a double pair comparison method to the 1975-86 FARS data and determined an overall effectiveness of (42 ± 3)% in fatality prevention with lap-shoulder belt use by front seated occupants. His more recent analysis [2] indicates that the highest level of driver fatality prevention by belt wearing is (82 ± 5)% in crashes where rollover is the first harmful event. Protection of the driver by lap-shoulder belt use is lowest in left-side impacts (27 ± 17)%.

There are essentially two components to occupant protection by safety belt use. One is anti-ejection protection and is primarily due to the lap portion of the belt system [3]. The other is mitigation of interior impact and is largely contributed by restraint of the upper body from the shoulder harness.

CONCEPT OF EFFECTIVE SAFETY BELT RESTRAINT - A principal design feature of lap-shoulder belts is to provide controlled loading and occupant restraint during a crash by routing safety belts over the bony structures of the pelvis, upper thorax, and shoulder. This takes advantage of a relatively high tolerance to impact forces for these regions of the skeleton and avoids concentrating load on the more compliant abdominal and lower thoracic regions. Control of occupant kinematics helps ensure maximum protection by belt restraints.

Adomeit [4-6] first espoused the fundamentals of effective belt restraint. The kinematic criteria, shown in Figure 1, help maintain the lap-portion of the belt low on the pelvis. This minimizes pelvic rotation and reduces the tendency for the lap belt to slide off the ilium and directly load the lower abdomen. Kinematic controls contribute to directing restraining forces onto the skeletal structures and away from more compliant body regions which are more vulnerable to critical injury.

Another component of effective restraint is to ensure that the biomechanical responses measured in crash testing, and related to assessment of occupant protection, consider human tolerance. While the Hybrid III dummy is considered to have biofidelity in its chest deflection response, and is considered to be a suitable tool for assessing injury risk by the Viscous and compression criteria [7-10], there is no injury assessment capability currently available for its abdominal region. In addition, the dynamic location of the belts is frequently obscured from direct photographic observation, especially in full-vehicle barrier crash testing. These situations complicate abdominal injury assessment.

HUMAN TOLERANCE - The human body has significant tolerance to impact loading as evidenced by survival in extremely severe motor vehicle crashes. In spite of the compliance of the chest and abdomen, these body regions are capable of tolerating impact force because of elastic and viscous resistance to body deformation. The pioneering work of Gadd and Patrick [11] demonstrated a tolerance to concentrated loading of the sternum of 3.3 kN (740 lb) and a tolerance of 8.0 kN (1800 lb) for distributed loading on the chest and shoulder. This research underscored the importance of the shoulder region as a load path for occupant restraint.

More recent studies have shown the importance of the Viscous response as a mechanism of impact injury of the abdomen and chest [9,10,12,13,14], and of compression [7,8] as a mechanism of crushing injury of the chest. Impact force is generally an inadequate predictor of abdominal injury [15]. However, Rouhana [13] demonstrated that the product of force and compression is a good predictor of abdominal injury. Following a limited series of experiments [16] on upper abdominal injury by a shoulder belt, Miller [17] recently determined injury tolerances of lower abdominal organs for direct lap belt loading. A Viscous tolerance of VC = 1.4 m/s and compression tolerance of C = 48% was determined for a threshold risk of 25% probability of critical abdominal injury (AIS 4+). Miller also confirmed the adequacy of the force and compression product in prediction of abdominal injury from direct belt loading. Sacco, et al [18] recently showed a 10.5% probability of death given an AIS 4+ abdominal injury by analysis of trauma patients in the Multiple Trauma Outcome Study (MTOS). This establishes the Viscous and compression tolerance for lap belt loading of the lower abdomen at approximately 2.6% probability of death (25% risk of 10.5% fatality outcome).

LIMITS OF CRASH PROTECTION - A recent analysis by Viano [19] brought together the results of several studies on crash protection. One was a crash investigation study and analysis which determined that approximately 50% of fatalities to unrestrained occupants are not preventable by the use of lap-shoulder belts or lap belt and air bag combination. This limit, in part, reflects the severity of many fatal crashes which involve extreme vehicle damage and forces on the passenger compartment, unusual crash configurations and causes of death, and unique situations associated with particular seating positions and crash dynamics. This implies that

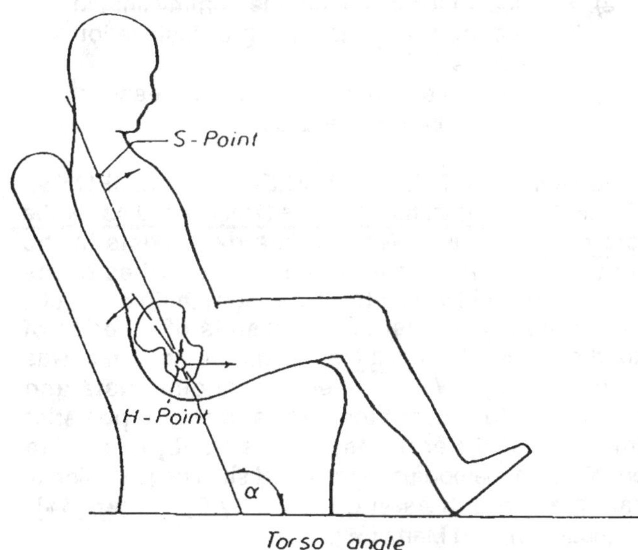

Figure 1. Adomeit's Motion Sequence Criteria
- H-point vertical disp. ≲ ±50 mm
- H-point horizontal disp. ≲ 250 mm
- Pelvic rotation < 30° CCW
- Torso axis ~90° at max. shoulder belt load

fatality risk depends on the particular occupant seating position with respect to the principal impact point. Rouhana and Foster [20] found a more than 19 to 1 increase, and Evans [21] found more than a 7 to 1 increase in fatality risk for a right-front passenger exposed to a right side (nearside) versus a left side (farside) impact in the NCSS and FARS data bases, respectively.

If the results of belt effectiveness and unpreventable deaths are compared, currently available lap-shoulder belt restraints provide a very high degree of effectiveness. There is approximately 7% additional safety potential achievable by the addition of interior safety features which supplement belt usage, such as air bags, restraint enhancements, and friendly interiors. However, the fact that lap-shoulder belts are only 42% effective in preventing fatalities underscores that absolute protection is not achievable by occupant restraints. Furthermore, it suggests that despite a significant overall net safety gain by restraint usage, injury and fatality will continue to occur to belt wearers even if belt restraints are supplemented by airbags.

BELT RELATED INJURIES - With the advent of lap belts in passenger cars in the late 1950's and early 60's, physicians started reporting new injury patterns for belt restrained victims. The typical pattern of facial and upper body injury to unrestrained occupants was replaced by belt related injury to abdominal organs and tissues [22-28]. These injuries led to the phrase "seat belt syndrome" [29-30] as the new injury patterns in motor vehicle crashes received attention due to concentrated forces on the lower abdominal region for lap belt wearers. In many cases improper belt wearing was identified as a cause of abdominal loading and injury [31]. The potential for improper use has continued with the advent of lap-shoulder belt systems, particularly placement of the shoulder harness under the arm [32] and wearing of the lap belt high on the abdomen with poor seating posture.

The most frequent belt related injuries are to the liver and spleen which are organs in the upper quadrant of the abdomen [33-35]. Injuries to lower abdominal tissues do occur [36,37] but at lower incidence suggesting a greater tolerance to abdominal compression. In some situations, upper abdominal injuries may be incorrectly assigned to the lap belt of the 3-point restraint system, when in fact the shoulder harness or steering wheel may have caused the injury. In very severe crashes, injuries may extend to the lumbar spine [38-39].

ABDOMINAL INJURY ASSESSMENT - With the introduction of lap-shoulder belts as standard equipment for front-seat occupants in the late 1960's, safety and design engineers saw the need to modify the test dummies and add new components to evaluate belt loading during crash testing. This work has continued as efforts have intensified to reduce the potential for belt-related injuries. There has also been continued progress in belt system design as injury mechanisms become better understood [40-45].

An indirect approach to abdominal injury assessment involved the insertion of load-measuring bolts at three levels in the ilium of the dummy pelvis to study the dynamics of lap belt loading and load transfer on the hip [46]. Later, the research group at Association Peugeot

Renault [47,48] added a load cell above the ilium which could react to belt loading if the restraint slipped off (above) the pelvis. Research has also focused on improving the anthropometry of the dummy pelvic structure as more detailed information on the human seating posture became available. Melvin and Weber [49] developed a fluid filled, abdominal "intrusion sensor" for the Part 572 3 year old child dummy. While the abdomen could discriminate among various restraint systems, they noted that they did not attempt to make the response of abdomen biomechanically realistic. In fact, none of the early efforts included biofidelity in the dummy abdomen or injury assessment using a valid biomechanical response and injury criterion [50].

The importance of abdominal biofidelity and injury assessment has been discussed recently [51-53] as greater belt wearing has been achieved in the United States and passive belt systems are being introduced. Our development of an injury assessing abdomen for belt restraint loading involved matching the force-deflection properties of the human abdomen, and knowledge of the criteria upon which to estimate the risk of injury. Once a design was conceived and built, we conducted tests to ensure that the modified dummy and abdominal components functioned according to the design and development criteria.

The same basic design will be used in each member of the Hybrid III family of dummies. Values for the 50th percentile male used in the development phase have been scaled to the 5th percentile female dummy and other dummy sizes.

Early in the project we decided to develop a frangible abdomen (literally, "one that breaks readily or easily"; Webster) using plastic or foam, an approach which proved to be successful. The remainder of this paper details our design, development, and validation of an abdominal injury assessment insert for the Hybrid III family of dummies.

METHODS

DESIGN CONCEPT AND CRITERIA FOR THE ABDOMEN - The conceptual design for the injury assessing abdomen is a frangible insert that possesses biofidelity in belt restraint loading and minimizes changes to the Hybrid III dummy. The design concept involves the removal of the current "gut sack" and chest deflection potentiometer. Both of these components are situated in the abdominal region of the dummy in the location chosen for this abdominal injury transducer. The design includes a reaction surface for the frangible insert which is attached to the pelvis, rather than the rib cage, so that the orientation of the abdomen during belt loading remains fixed relative to the pelvis. The frangible insert design has biofidelity and the capability for measuring an abdominal injury criterion. The final design ensures that the amount and the characteristics of the rotation of the lumbar spine in the Hybrid III are preserved.

The design criteria for the frangible abdomen were grouped into two categories, viz., those that apply to the material chosen for the abdomen, and those that apply to the interface between the abdomen and the dummy.

Material Criteria for the Abdominal Insert - The following criteria were developed and applied to the selection of the material used for injury assessment in the abdomen:

1) The force-deflection characteristics of the material must fall into established corridors.
2) When loading typical of submarining is encountered, the integrity of the material must not be compromised.
3) Restitution (or elastic spring-back) of the material does not occur, or if it does the amount of restitution is well characterized.
4) The material is easily machineable or moldable.
5) There is minimal variability between batches in production of this material.

Abdomen/Dummy Interface Criteria - The following criteria were developed for the attachment interface between the dummy and the abdomen:

1) There should be minimal modifications to Hybrid III family of dummies.
   a) Any mass and inertial changes should be well known so that reballasting of the dummies can be performed.
   b) There should be no effect on dummy posture.
   c) Performance of the dummy, as measured by dummy kinematics, should be unaffected when there is no submarining. Dummy kinematics may change when submarining occurs because the belt/dummy interaction will take place through the frangible abdomen which will be human-like and will differ in stiffness and crush performance from the existing Hybrid III abdomen.
2) Physical modifications to the dummy should not require major machining or installation procedures.
3) The frangible abdomen should be easy to install and remove after a test.

ABDOMINAL FORCE-DEFLECTION RESPONSE CRITERIA - Normalization of Force-Deflection Data for the Porcine Abdomen - While some data exists in the literature for the force-deflection properties of the abdomen, most of it is for blunt loading with rigid objects. Miller recently published the results of a series of experiments in which belt loading on the abdomen was examined [17]. We obtained the original data and normalized it to account for mass and antero-posterior dimensional differences between subjects. The normalization method used the equal stress/equal velocity scaling approach as suggested by Eppinger [54], Langhaar [55], and Mertz [56].

We chose to scale the force-deflection data, because it was necessary to preserve this characteristic in the design of the abdomen. Scaling was performed in a manner analogous to Mertz's scaling of force-time data, using a force scale factor, and a displacement scale factor. Because both time scales are identical, the time data did not have to be scaled. Scaling the force-deflection data in this manner is equivalent to scaling the force-time and displacement-time data separately, and then cross-plotting them.

The force and displacement are scaled according to equations (1) and (2).

$$F_s = \lambda_f * F_i \tag{1}$$

$$D_s = \lambda_d * D_i \tag{2}$$

where F_s = the normalized force (force on a standard subject),
F_i = the unnormalized force (force on the "ith" subject),
D_s = the normalized displacement,
D_i = the unnormalized displacement,
and λ_f, λ_d are determined by equations (3) and (4).

$$\lambda_f = \sqrt{\frac{M_s}{M_i}\frac{L_s}{L_i}} \tag{3}$$

$$\lambda_d = \sqrt{\frac{M_s}{M_i}\frac{L_i}{L_s}} \tag{4}$$

where M_s = the average mass of the the porcine subjects tested
M_i = the mass of the ith subject
L_s = the average antero-posterior depth at the 4th Lumbar vertebrae
L_i = the A-P depth at L4 of the ith subject

Determination of Abdominal Force-Deflection Properties of Porcine Cadavers - The only force-deflection data available was from live anesthetized porcine and human cadaver subjects. Therefore, a small number of experiments was carried out to determine the force-deflection characteristics of the porcine cadaver for comparison with the data from the live anesthetized porcine subjects.

Fifteen experiments were performed on 7 Landrace-Yorkshire crossbred males* with an average mass of 46.1 ± 1.3 kg (101.4 ± 2.9 lbs). The subjects were preanesthetized by intramuscular injection of Acetylpromazine Maleate (0.37 mg/kg) and Ketamine HCl (33 mg/kg). They were then induced into deep surgical anesthesia by inhalation of Nitrous Oxide (50% for induction) and Methoxyflurane (3% for induction). Euthanasia was administered by overdose of anesthesia which was increased to lethal levels. Death was assured by examination of heart rate, respiration rate and volume, and neurological signs.

The protocol followed in these experiments simulated that used in human cadaver experiments such as those done by Kroell [7]. Within 6 hours after euthanasia, each subject was placed in a refrigerator where it remained at 2 °C until tested. Refrigeration times ranged from 2 to 8 days. The subjects were removed from the refrigerator at 6:00 AM on the day of the test, and allowed to warm at room temperature for approximately 8 hours. Rectal temperatures at the time of test ranged from 8 to 12 °C (47 to 54 °F).

Most subjects were utilized in more than one test in an attempt to obtain as much biomechanical information as possible. As seen in the matrix of experiments (Table 1), multiple experiments on a single subject were performed in order of increasing velocity. The primary focus of these experiments was a determination of force-deflection curves at velocities similar to those used by Miller. A determination of abdominal injury in the porcine cadavers was also made for comparison with the data from live anesthetized porcine subjects.

Each subject was placed in dorsal recumbency on a plexiglas V-block support which was bolted to the platen of a hydraulic, materials testing machine (MTS). The V-block was positioned such that impact would occur approximately at the level of the fourth lumbar vertebrae. The impacting object was the same 381 mm (15") length of 5.0 cm wide polyester safety belt used by Miller. The belt was attached to a yoke through two pivots (Figure 2), and the yoke was attached to the actuator piston of the MTS through an inertially compensated biaxial load cell (GSE 3182).

Prior to impact, the piston was lowered to the point where all belt slack was removed, but negligible preload was developed. The MTS piston was then cycled through a programmed displacement stroke and velocity. Total axial force developed, belt force at the pivot points (measured using strain gauges), and piston stroke were recorded on FM tape. Velocities ranged from 0.2 to 5.3 m/s, and piston stroke ranged from 118 to 126 mm (45 to 69 % of the antero-posterior dimension of the subject).

An abdominal necropsy was performed after last test on each subject, with abdominal injuries recorded and assessed using the Abbreviated Injury Scale (AIS-85) [57].

Extrapolation of Porcine Data and Human Cadaver Data to Corridors for Living Humans - Our desire was to determine the force-deflection characteristics of living humans. It was unknown whether force-deflection properties are independent of species (Porcine versus

---

* The rationale and experimental protocol for the use of an animal model in this program have been reviewed by the Research Laboratories' Animal Research Committee. The research follows procedures outlined in the Guide for the Care and Use of Laboratory Animals, U. S. Department of Health and Human Services , Public Health Service, National Institutes of Health NIH Publication 85-23, Revised 1985, and Public Health Service Policy on Humane Care and Use of Laboratory Animals by Awardee Institutions, NIH Guide for Grants and Contracts, Vol. 14, No. 8, June 25, 1985, and complies with the provisions of the Animal Welfare Act of 1966 (P.L. 89-544), as amended in 1970 (P.L. 91-579)and 1976 (P.L. 94-270), and the Food Security Act of 1985 (P.L. 99-158).

TABLE 1. Matrix of Porcine Cadaver Experiments

Expt #	Subject #	AP Dimen (mm)	Velocity (m/s)	Compression (mm)	(%)	AIS
1	1	207	.23	126		
2	1	"	.84	123		
3	1	"	5.3	118	57.1	5
4	2	242	NA	NA		
5	2	"	NA	NA		
6	3	268	.22	122		
7	3	"	5.2	120	44.6	4
8	3	"	5.0	119		
9	3	"	4.6	120		
10	4	181	.59	124		
11	4	"	.85	125	68.8	0
12	5	211	.86	125		
13	5	"	.88	125	59.2	3
14	6	199	.86	124	62.3	2
15	7	183	5.2	118	64.3	4

Human), or if the force-deflection properties of living subjects should be equivalent to those of cadaver subjects (assuming that tissue properties may change after death). Therefore, the force-deflection curves from porcine cadavers were compared to the curves from human cadavers, and to the curves obtained for living anesthetized porcine subjects. Any observed differences could then be used to extrapolate human cadaver force-deflection data to living human force-deflection data.

CHOICE OF ABDOMINAL INJURY CRITERION FOR BELT LOADING - Previous work has shown that the risk of abdominal injury can be predicted using a number of different criteria. The viscous criterion has been shown to apply to the abdomen by Rouhana [12], Viano and Lau [9,10], and Stalnaker [58] for rigid loading, and recently by Miller [17] for belt loading. In addition, Rouhana showed that the product of maximum force and maximum compression was a good indicator of the probability of abdominal injury for human cadavers. Miller [17] has shown that this criterion also worked well in live anesthetized porcine subjects. Miller also proposed that compression alone is a good indicator of abdominal injury in belt loading to the abdomen. Tolerance values for compression are presented in Figure 3 [17].

Verriest, et al. [59] have shown that belt loading is a low velocity phenomenon (Figure 3) which occurs at around 3 m/s. Given this low velocity loading, maximum compression is the injury criterion of choice for assessment of abdominal injury from belt restraint submarining.

DESIGN VALIDATION - The design of the frangible abdomen was validated in material tests and sled tests.

Material Tests - All material tests were performed on the hydraulic testing system (MTS) described above. The material tests were performed to assess:

a) Force-deflection of the design - Samples of the frangible abdomen were placed in a Hybrid III dummy pelvis which had been modified to accommodate a support structure for the abdomen. The same belt yoke and protocol that was used to determine the porcine force-deflection data was used to develop and assess the force-deflection data for the frangible abdomen samples. The abdomen was loaded by a standard belt in what is an anterior to posterior direction, *in situ*, in the dummy pelvis. The force-deflection of the final design could then be compared to the established corridors.

b) Repeatability of the force-deflection curves - Five samples were tested at 1 m/s and five at 6 m/s using the protocol described in a). Each force-deflection curve was compared to others generated at the same velocity.

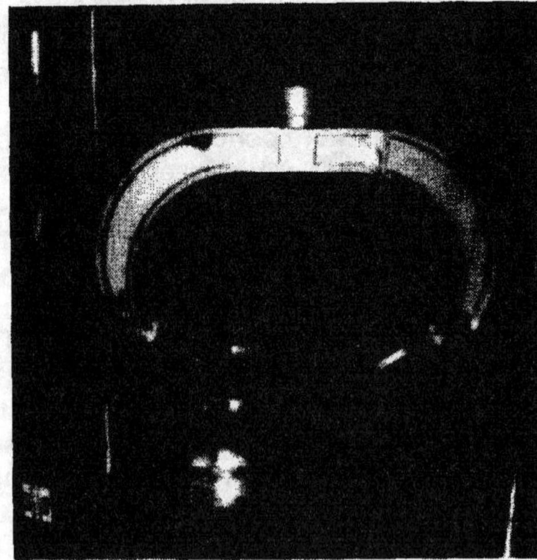

Figure 2. MTS impactor with yoke and belt used in live anesthetized and porcine cadaver experiments.

c) Rate dependence of the material chosen - Five samples were tested at 1 m/s and five at 6 m/s using the protocol described in a). The force-deflection curves generated at 1 m/s were then compared to those generated at 6 m/s.

d) Restitution of the material chosen - A small number of experiments were conducted during the development phase using a 51 x 127 mm (2 x 5 inch) rectangular aluminum impactor. Elliptical samples of the proposed material with a major axis of 279 mm (11.0") and a minor axis of 203 mm (8.0") were placed in a test fixture on the MTS platen. A programmed deformation-time history was then imposed on it. The restitution of the material chosen for the frangible abdomen could be determined by subtracting the amount of residual deformation of the material (measured after the test) from the amount of piston travel programmed. For example, if the programmed piston travel is 76 mm (3.0"), and the residual deformation is 70 mm (2.75"), then the restitution would be 6 mm (0.25").

e) Function and integrity of the material - The function of the material was assessed during the development phase by performing experiments as described in a), but varying the belt loading direction and amount of slack. The loading for all tests in this series was in an antero-posterior direction at a 30° angle to a transverse plane through the material, coupled with one of the following: no slack in the belt; 25 mm (1.0") of slack in the belt; or the initial position of the belt contact point either on an edge or in the center of the material. The structural integrity of the material was assessed post-loading. In addition, the "behavior" of the material (symmetric crush for symmetric load, etc.) was assessed.

f) Interaction between the rib cage and abdominal element - In sled or barrier tests in which submarining does not occur, a dummy can experience torso flexion (bending forward about the hip in which the front of the thorax (chest) moves toward the top of the knees). During torso flexion, the bottom of the rib cage interacts with any material in the area between the rib cage and the pelvis. Thus, if the kinematics of the dummy are to be preserved, the interaction of the bottom rib with the abdomen should be similar to current dummies.

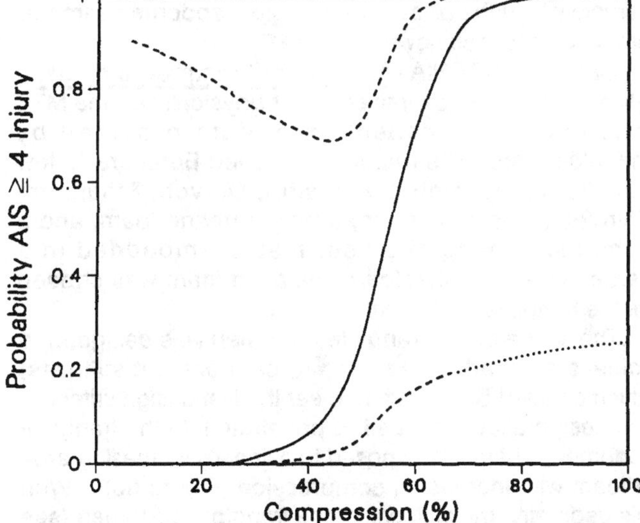

Figure 3. Probability of AIS ≥ 4 abdominal injury as a function of compression. (From logistic regression analysis of Miller data.)

To assess this interaction, MTS tests were performed using a simulated lower rib. This rib was made of a half steering wheel segment and was mounted to the hydraulic piston through the biaxial load cell described above. The force-deflection properties were recorded for the frangible abdomen, the abdomen from a Hybrid-III 5th percentile female, and both abdomens from a Hybrid III 50th percentile male (the abdomen with a clearance notch for the chest potentiometer arm, and the abdomen without the notch). Each abdomen was tested four times with vertical loading, and four times with the direction of loading at a 30° angle to the vertical (Figure 4). For each loading direction, 2 of the four tests were performed quasi-statically at .36 m/s and the other 2 were performed at 89 m/s.

Sled Tests - Sled tests were performed to assess the kinematics of the dummy and the function of the foam in actual use (indication of submarining, integrity after loading in a dummy, and ease of installation and handling.)

The sled tests were performed with a generic restraint system in a body buck on a Hyge Sled. The sled buck had a single bucket seat in the driver's position, and there was no instrument panel or steering system. The seat used in these tests was constructed of foam on a zig-zag spring suspension and sheet metal frame. The foam Indentation Load-Deflection characteristic was 240 N. The seat pan angle was 8° and the seat back angle was approximately 26° (not a reclining seat).

The dummy was in the driver's seat wearing a single retractor continuous loop lap/shoulder belt system with a cinchable latch plate. The retractor was on the shoulder portion of the belt system. Three types of tests were run which were classified by the intended degree of submarining:

a) no submarining - The lap portion of the belt system was positioned below the anterior superior iliac spines of the dummy pelvis and was cinched snug. The shoulder belt was pulled out of the retractor and allowed to retract until there was a "fistful" of slack in the belt (25-38 mm (1-1.5") of webbing). The retractor was locked before the test to avoid any changes in configuration.

b) incipient submarining - The lap portion of the belt system was positioned below the anterior superior iliac spines of the dummy pelvis and but was not cinched. Instead, the lap belt was left loose with 76 mm (3.0") of slack. The shoulder belt was "cinched" tight by pushing the back of the dummy into the seat back, retracting all shoulder belt slack, and locking the retractor

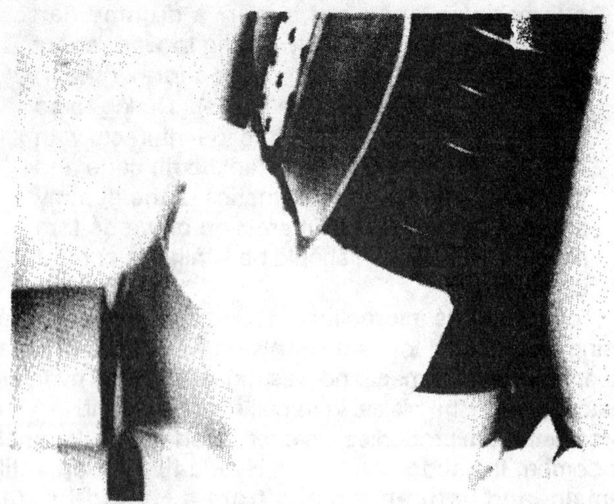

Figure 4a. Side view of Hybrid III showing position of rib cage relative to foam insert.

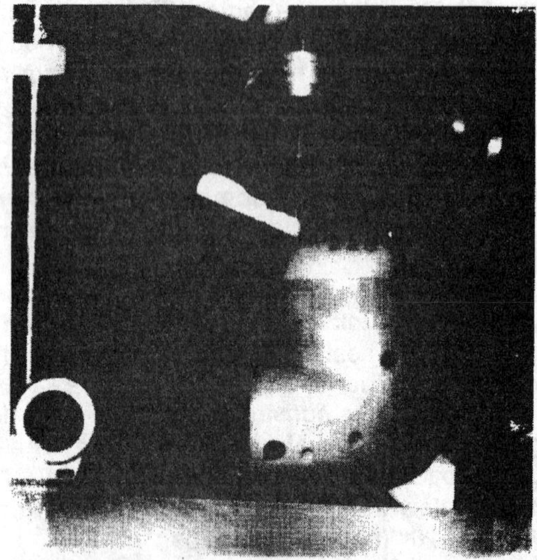

Figure 4b. Superior/inferior loading simulation.

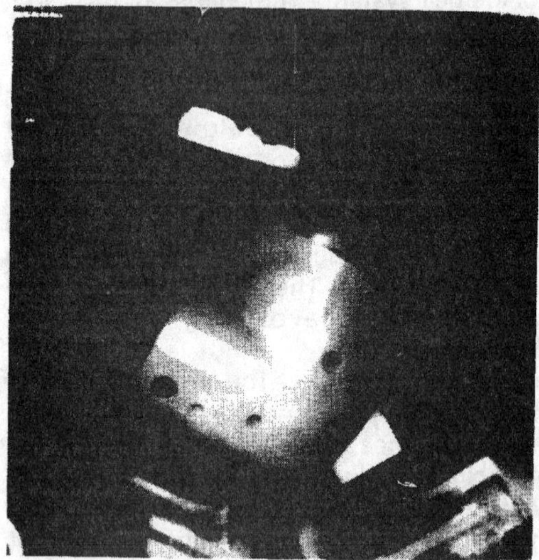

Figure 4c. Torso flexion loading simulation.

before releasing the dummy.

c) presubmarined - The lap portion of the belt system was positioned above the anterior superior iliac spines of the pelvis and lightly cinched without preloading the abdomen. The shoulder belt was also "cinched" into position by pushing the back of the dummy into the seat, retracting all slack, and locking the retractor before releasing the dummy.

All tests used a single 50th percentile male Hybrid III dummy. Shoulder belt loads were monitored using a GSE Belt Load Cell in a position midway between the guide loop and dummy shoulder. Inboard belt loads were also measured between the inboard anchor and the buckle. Sled parameters were set to provide a generic crash pulse with a 30 mph delta-v.

In all tests, high speed movies were taken with an onboard camera view from the driver's right side. The movies were used to compare baseline Hybrid III dummy kinematics to kinematics of the Hybrid III with the frangible abdomen. Dummy motion was assessed using Adomeit's kinematic criteria [4-6] and other criteria. Specifically, we monitored: the maximum excursion of the head in the x and y direction (by tracking the temporo-mandibular "joint" on the dummy with x forward/backward and y up/down); the angle of the neck at maximum head excursion (vertical was zero degrees, forward was increasing angle); the torso angle at maximum belt load (vertical was 90°, forward was decreasing angle); and the maximum excursion of the H-point in the x and y direction.

RESULTS

CONSTRUCTION OF DESIGN - The frangible abdomen is constructed of styrofoam floatation foam which is cut on a standard band saw. The shape of the foam after cutting is shown in Figures 5a-c. Since styrofoam floatation foam is extruded, the force-deflection properties may vary depending on the orientation of the sample within the billet. The frangible abdomen samples are machined as shown in Figure 6.

DESIGN CRITERIA - Material for Frangible Abdomen - Six materials were examined either physically on the MTS machine or by inspection of data provided by manufacturers. The materials included Betacore (a low density honeycomb composite), Dorvon, Ethafoam, Styrofoam floatation foam, a polyurethane foam, and a composite made of a fiber matrix embedded in a hardened resin. Styrofoam floatation foam was chosen for the frangible abdomen.

The shape of the frangible abdomen was designed to achieve a linearly increasing force (constant stiffness) which enabled Styrofoam to meet the first design criterion. The design allows the belt to penetrate into the frangible abdomen, where it is exposed to a linearly increasing area of foam with increasing compression (Figure 5c). With this geometry, the stiffness of the frangible abdomen (see Figures 16-18) is not significantly different from the porcine abdominal stiffness, at the .05 level of significance. The material and sled tests described below demonstrated that styrofoam also met design criteria 2-5.

Abdomen/Dummy Interface Criteria - The chest

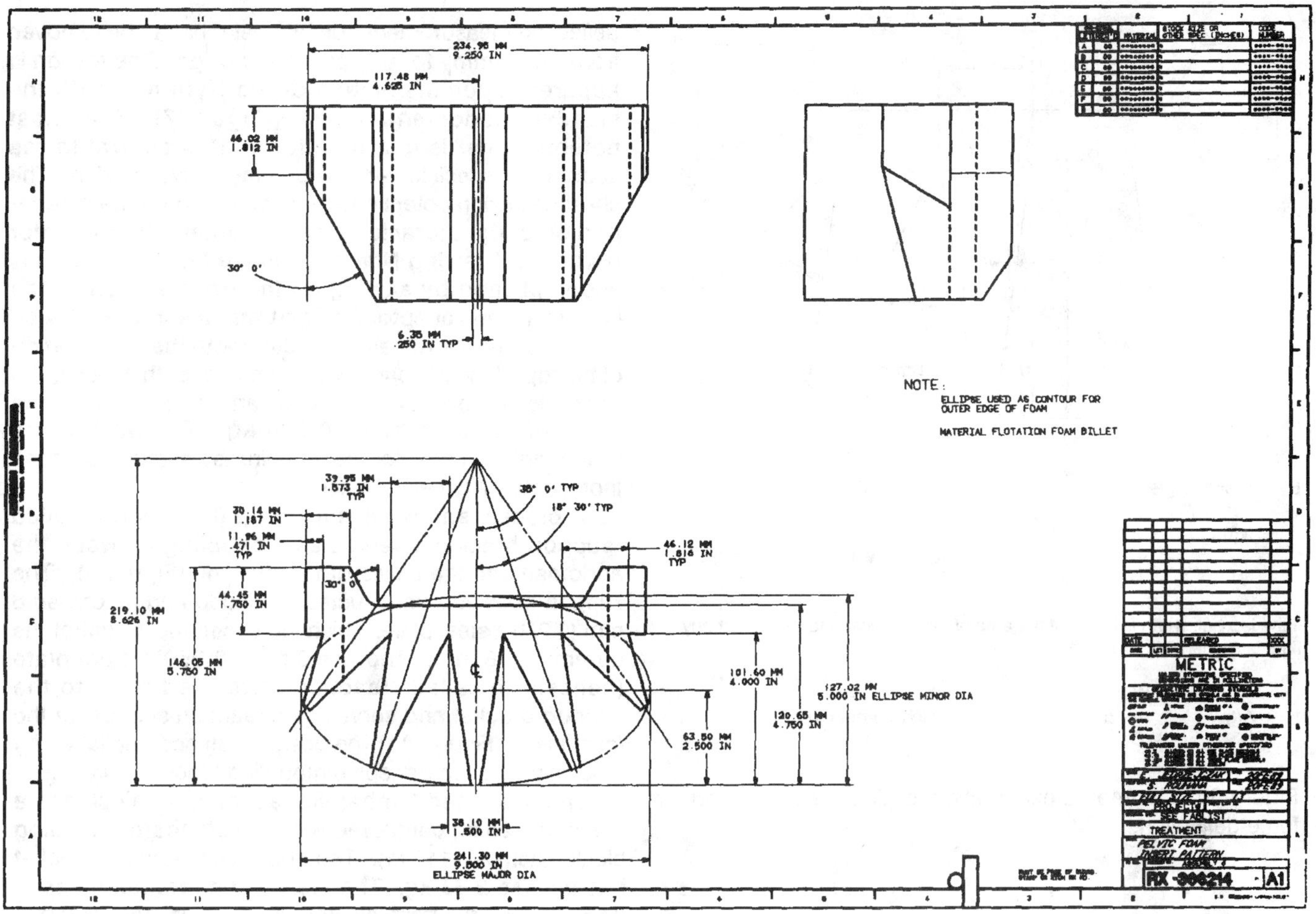

Figure 5a.

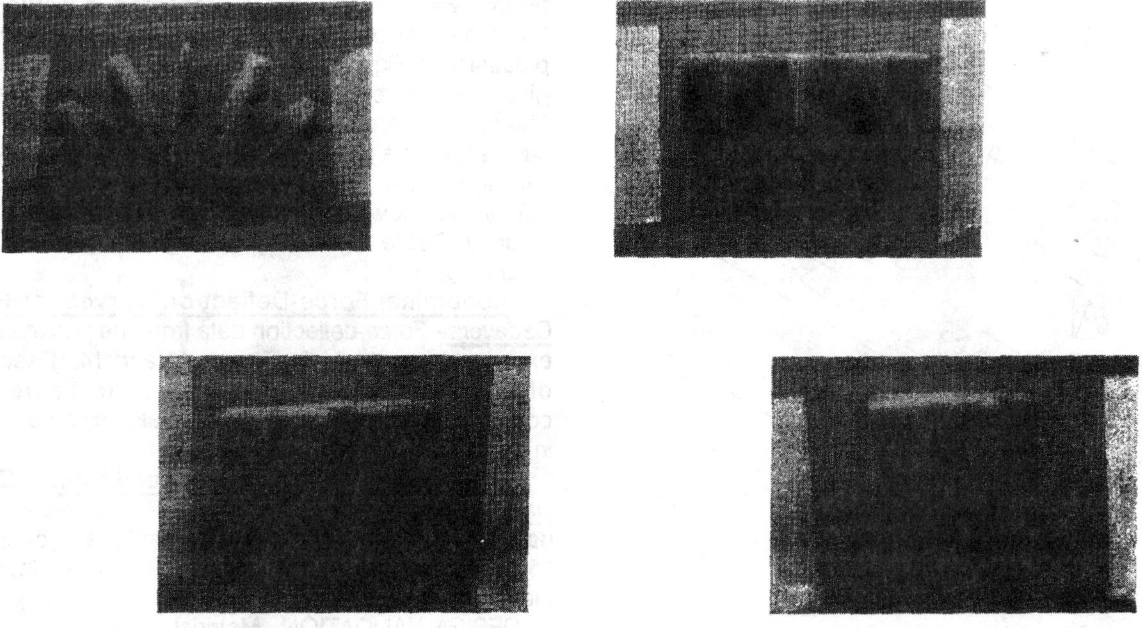

Figure 5b. Frangible abdomen foam insert.

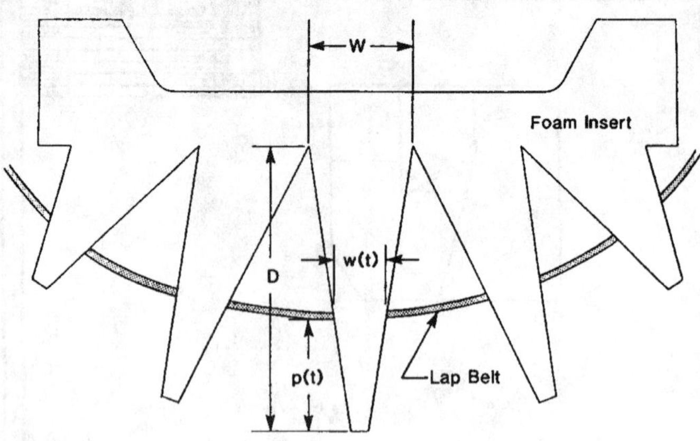

By similar triangles:

$$\frac{w(t)}{W} = \frac{p(t)}{D} \qquad \text{so,} \qquad w(t) = \frac{W}{D}\, p(t)$$

If the lap belt width is h, and if the area of foam exposed to the belt at any time is A, then $A(t) = h\,w(t)$, or

$$A(t) = \frac{Wh}{D}\, p(t) \qquad \text{which is } \textbf{linear} \text{ in } p(t).$$

Figure 5c. Geometric explanation for linearly increasing force-deflection.

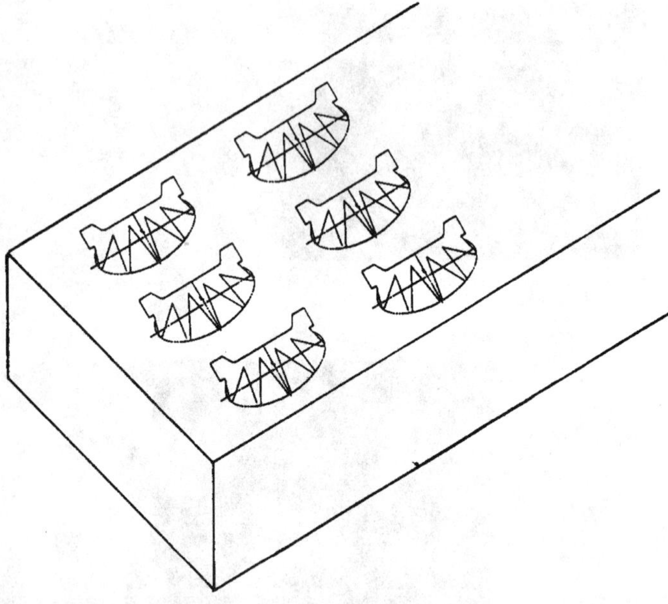

Figure 6. Location of foam sample within billet before cutting.

deflection measurement potentiometer had to be removed from the dummy for any practical design. The reason is apparent upon inspection of the Hybrid III with the standard abdomen removed (Figure 7). The chest potentiometer is in a location that is central to the abdomen, especially with the dummy in torso flexion. The chest deflection potentiometer is mounted on the anterior portion of the thoracic spine instrumentation adaptor (Hybrid III Drawing Number 78051-88). Its removal is accomplished by sawing off the front portion of the instrumentation adaptor (the part with the molded flexion stops) so that the remainder is flush with the front portion of the top of the lumbar spine. Removal of this piece, the chest deflection potentiometer, and arm, reduces the mass of the dummy by 0.845 kg. The part of the instrumentation adaptor that remains is used to mount the thoracic spine accelerometers.

A bracket shaped like a stirrup (henceforth, called "support bracket") served as a coupling between the abdomen and the pelvis of the dummy (Figure 8). The support bracket is constructed of welded and machined 6 mm (0.250") steel plate, which increases the height of the dummy by 6 mm. Another 6 mm (0.250") steel plate (henceforth called "reaction plate") is bolted to the support bracket and serves as a reaction surface for the foam insert (Figure 9). The design was accomplished by machining the lumbar spine mounting block (Figure 10).

The machined lumbar spine mounting block has a mass of 5.26 kg compared with the standard mounting block mass of 7.145 kg. The abdominal support bracket has a mass 2.63 kg. The standard abdomen mass is 0.656 kg, and the frangible abdomen insert mass is 0.062 kg. Then the net mass change of the Hybrid III dummy is a decrease of 0.694 kg.

Normalized Abdominal Force-Deflection Curves for Living Porcine Subjects - The zoometric data and test conditions from Miller's experiments, and the normalizing factors we determined are presented in Table 2. The force-deflection curves for the remaining tests are presented in Figures 11 and 12. Two separate graphs are given designated "higher velocity" and "lower velocity". The curves were separated because the higher velocity curves have a significantly greater stiffness than the lower velocity curves. Higher velocity curves were those above 5.8 m/s, and lower velocity curves were those below that velocity. Table 3 gives the average stiffness for the two different groups.

Abdominal Force-Deflection Curves for Porcine Cadavers - Force-deflection data from the porcine cadaver experiments is shown in Figures 13 and 14. Descriptions of the resulting injuries are given in Table 4, and comparison of stiffness with live anesthetized porcine data in Table 5.

Extrapolation of Porcine Data and Human Cadaver Data to Corridors for Living Humans - The force-deflection data from human cadaver experiments is shown Figure 15. Note that only low velocity data was used since this is the velocity range of belt loading [59].

DESIGN VALIDATION - Material Tests -

a) Force-deflection of final design - The force-deflection curves obtained from 5 tests of the frangible abdomen at 1 m/s are shown in

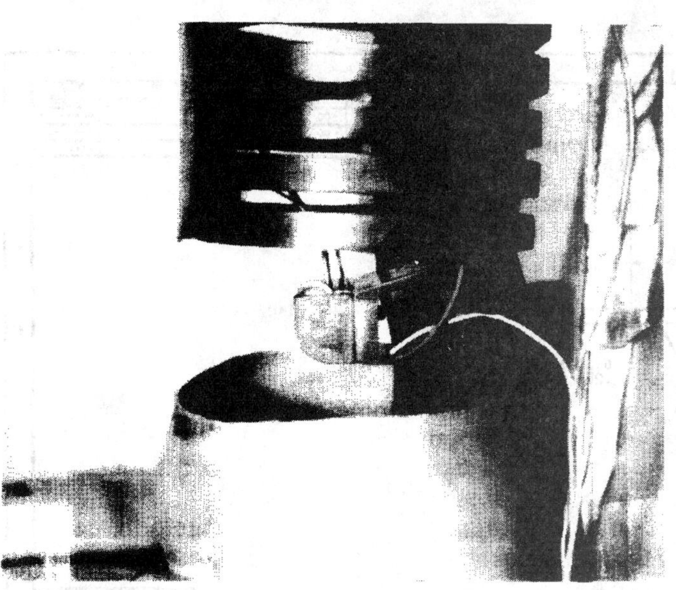

Figure 7a. Side view of Hybrid III dummy showing location of chest deflection potentiometer.

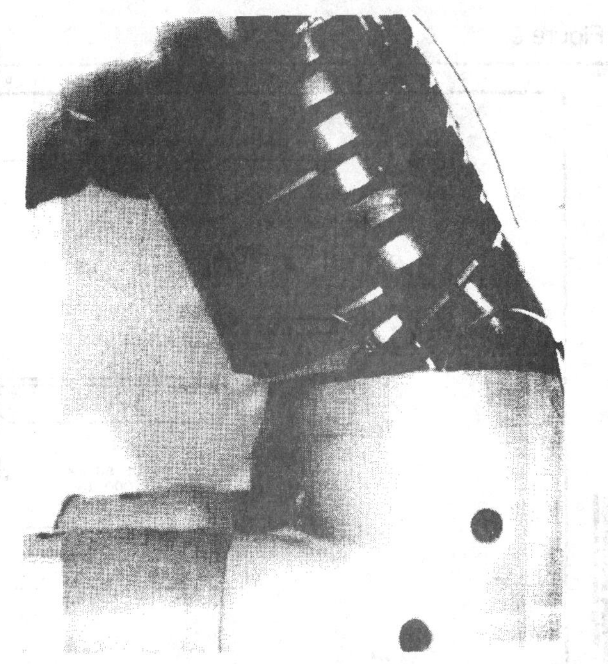

Figure 7b. Side view of Hybrid III dummy in torso flexion

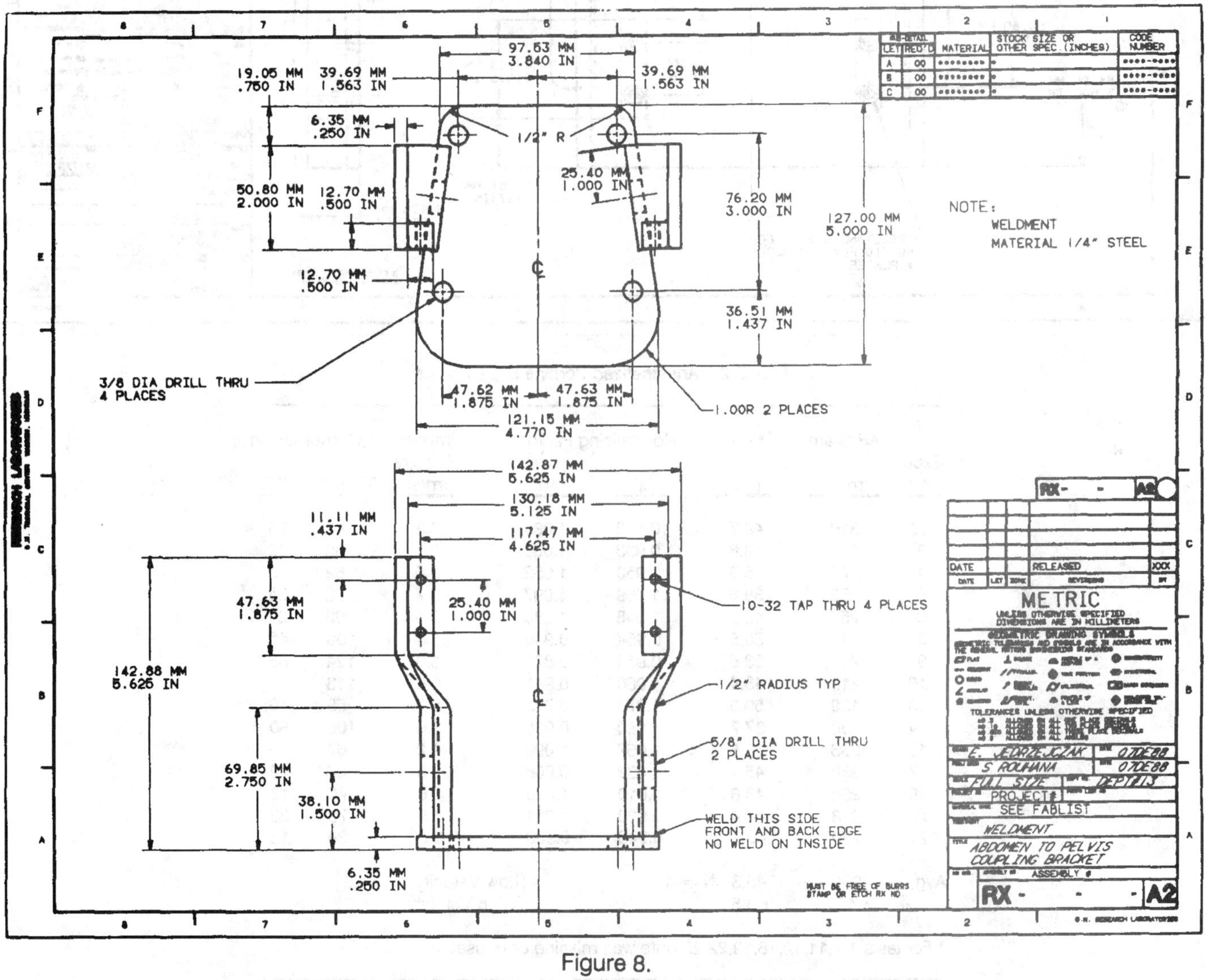

97.63 MM
3.840 IN
19.05 MM
.750 IN
39.69 MM
1.563 IN
39.69 MM
1.563 IN
6.35 MM
.250 IN
1/2" R
25.40 MM
1.000 IN
50.80 MM
2.000 IN
12.70 MM
.500 IN
76.20 MM
3.000 IN
127.00 MM
5.000 IN
12.70 MM
.500 IN
36.51 MM
1.437 IN
3/8 DIA DRILL THRU
4 PLACES
47.62 MM
1.875 IN
47.63 MM
1.875 IN
121.15 MM
4.770 IN
1.00R 2 PLACES

NOTE:
WELDMENT
MATERIAL 1/4" STEEL

142.87 MM
5.625 IN
130.18 MM
5.125 IN
117.47 MM
4.625 IN
11.11 MM
.437 IN
47.63 MM
1.875 IN
25.40 MM
1.000 IN
10-32 TAP THRU 4 PLACES
142.88 MM
5.625 IN
1/2" RADIUS TYP
5/8" DIA DRILL THRU
2 PLACES
69.85 MM
2.750 IN
38.10 MM
1.500 IN
WELD THIS SIDE
FRONT AND BACK EDGE
NO WELD ON INSIDE
6.35 MM
.250 IN
MUST BE FREE OF BURRS
STAMP OR ETCH RX NO

METRIC
UNLESS OTHERWISE SPECIFIED
DIMENSIONS ARE IN MILLIMETERS
PROJECT#
SEE FABLIST
WELDMENT
ABDOMEN TO PELVIS
COUPLING BRACKET
RX- - - A2

Figure 8.

Figure 9.

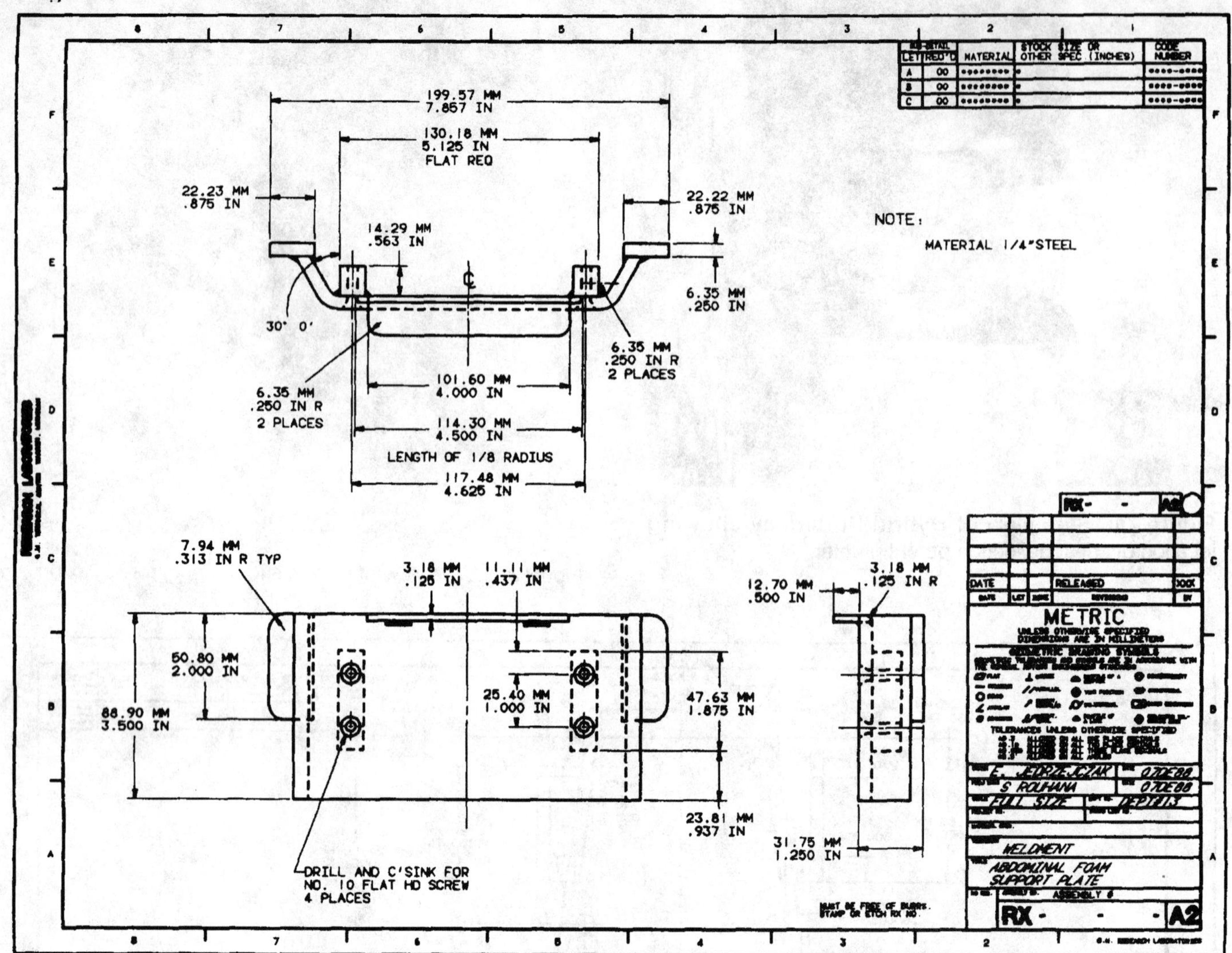

TABLE 2. Anesthetized Porcine Experiments

Expt. #	AP Diam (mm)	Mass (kg)	Normalizing Factors λf	Normalizing Factors λd	Impact Velocity (m/s)	Compression (mm)	Compression (%)
2*	310	48.7	0.858	1.083	3.3	62	20
3	285	44.8	0.933	1.083	3.5	62	22
4	272	45.3	0.950	1.052	3.5	62	23
5	252	38.6	1.069	1.097	3.7	62	25
6	253	40.0	1.048	1.080	3.5	62	24
8	242	50.5	0.954	0.940	5.0	109	45
9	221	53.6	0.969	0.872	6.0	124	56
10	218	45.4	1.060	0.941	6.6	113	52
13	138	50.0	1.270	0.713	2.7	55	40
14	180	37.7	1.280	0.938	3.0	108	60
15	253	38.6	1.067	1.099	4.2	67	26
17	230	45.4	1.032	0.966	2.5	27	12
18	250	43.6	1.010	1.028	3.8	45	18
20	278	46.4	0.929	1.051	5.5	73	26
21	250	49.7	0.946	0.963	3.6	29	12
Avg	238 ± 45.6	45.3 ± 4.5	N = 15		{Low Velocity: N = 13} {High Velocity: N = 2}		

* For tests 1,7,11,12,16,19,22,23 data was missing or unusable.

Figure 10.

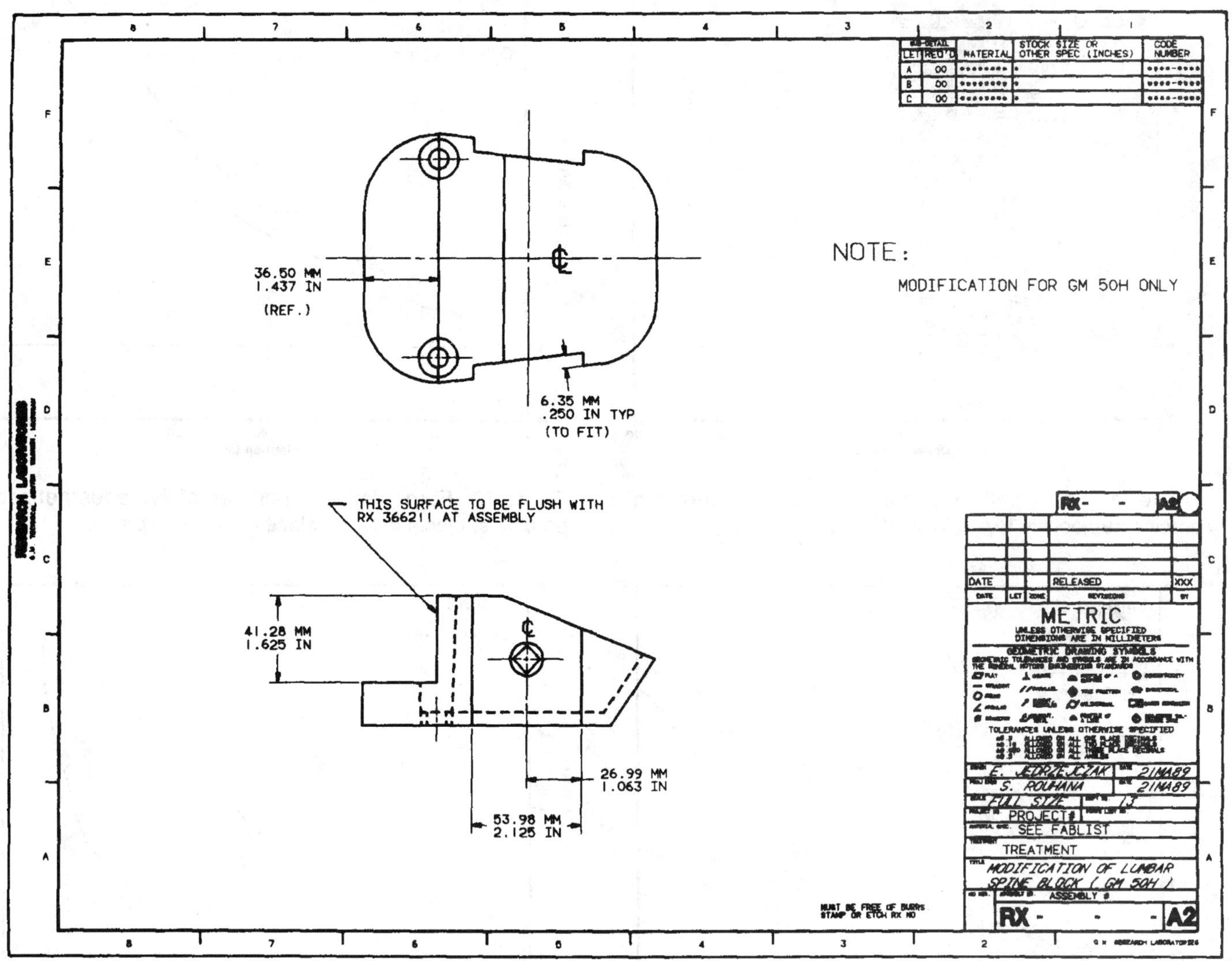

TABLE 3. Anesthetized Porcine Abdominal Stiffness

Experiment Series	N	Stiffness (N/mm)	Velocity (m/s)
Lower Velocity	13	23 ± 10	3.7 ± .84
Higher Velocity	2	63 ± 13	6.3 ± .42

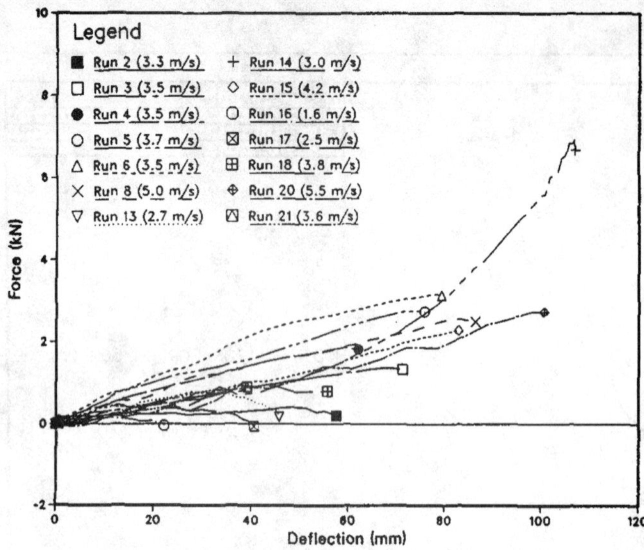

Figure 11. Force-deflection curves of live anesthetized porcine subjects in low velocity lap belt impacts.

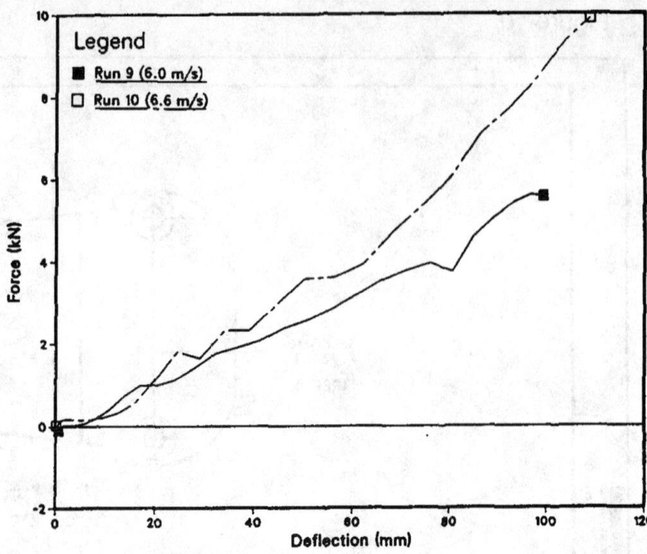

Figure 12. Force-deflection curves of live anesthetized porcine subjects in high velocity lap belt impacts.

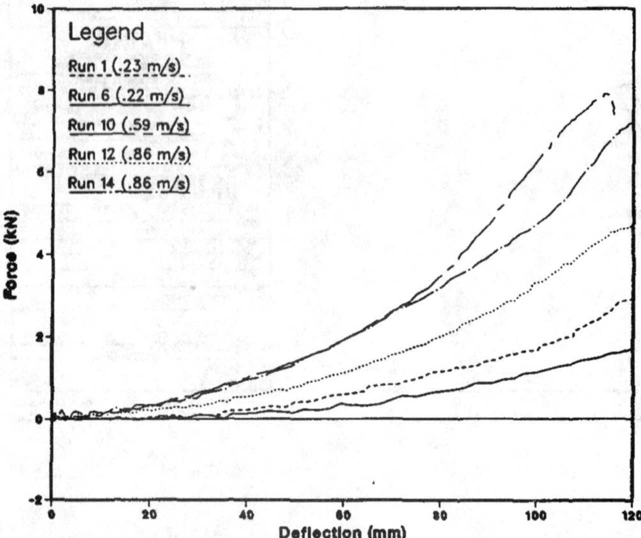

Figure 13. Normalized force-deflection curves for porcine cadaver subjects from low velocity lap belt impacts. (These are from the first impact to each subject.)

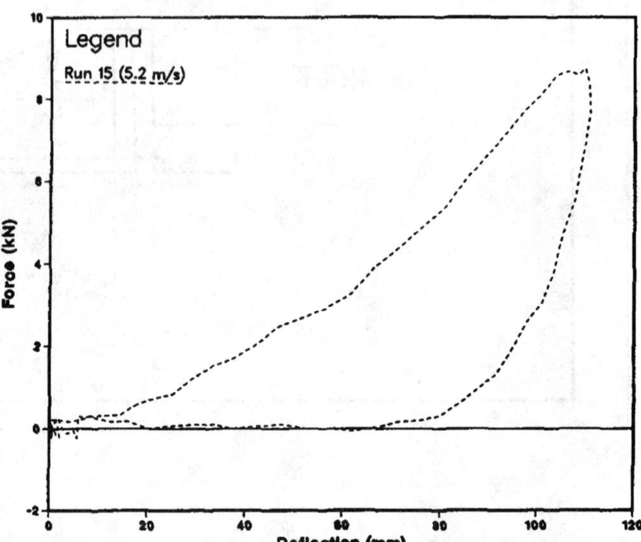

Figure 14. Normalized force-deflection curves for porcine cadaver subjects from high velocity lap belt impacts. (These are from the first impact to each subject.)

TABLE 4. Abdominal Injuries in Porcine Cadaver Experiments

Test	Injury Description	AIS
1	Large Intestine: Major Rupture	5
2	Large Intestine: Rupture with contam.	5
3	Large Intestine: Rupture	4
4	No Injury	0
5	Small Intestine: Rupture	3
6	Hemoperitoneum: Origin Unknown	2
7	Spiral Colon: Rupture	4

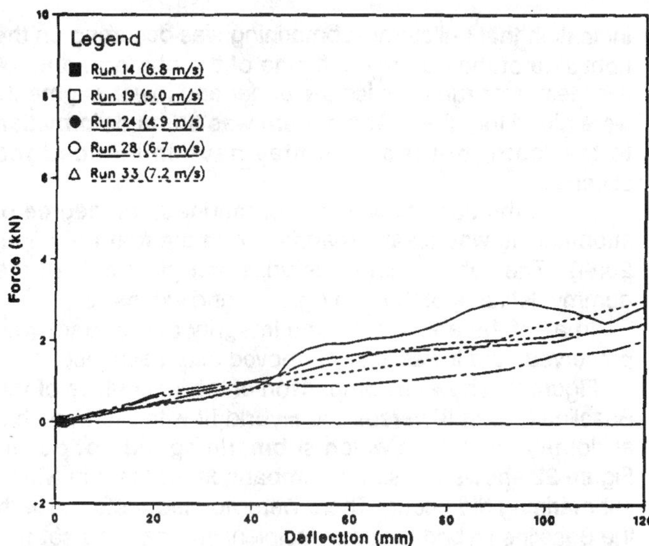

Figure 15. Force-deflection curves for human cadaver subjects, from Cavanaugh, et al.

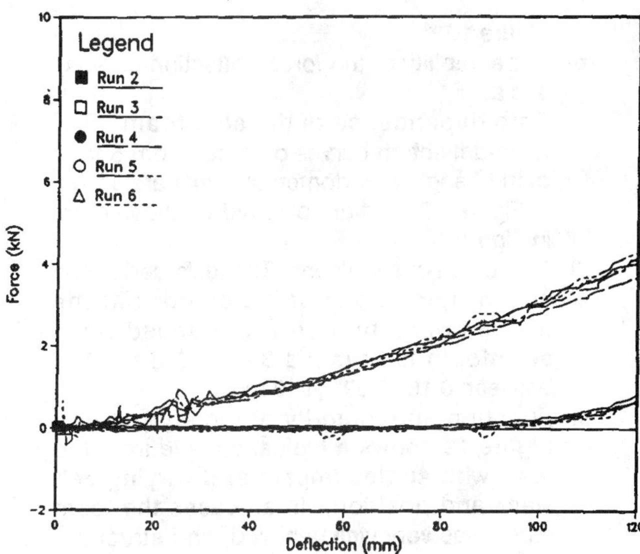

Figure 16. Force-deflection curves for frangible abdomen in 1 m/s lap belt impacts.

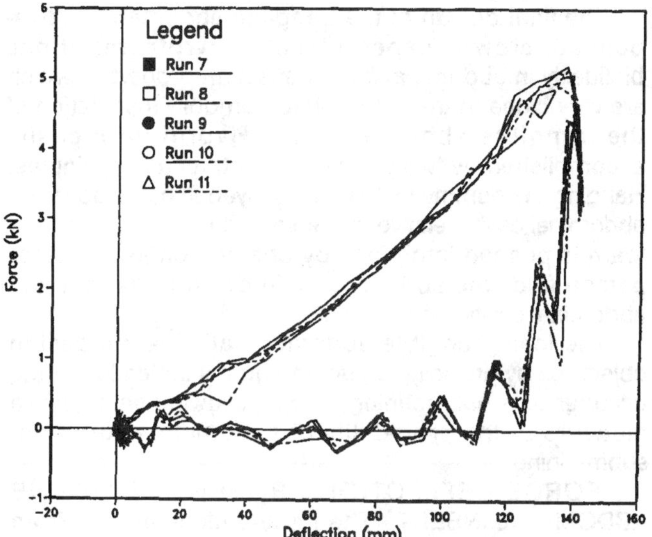

Figure 17. Force-deflection curves for frangible abdomen in 6 m/s lap belt impacts.

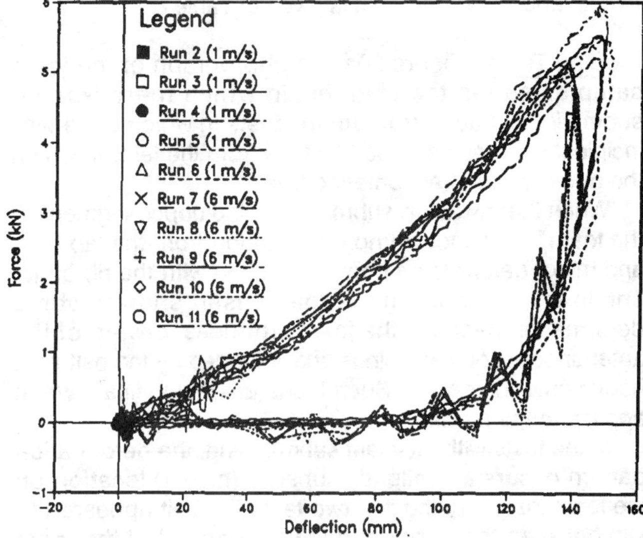

Figure 18. Overlay of 1 m/s and 6 m/s force-deflection curves for frangible abdomen.

TABLE 5. Stiffness of Porcine Cadavers
Low Velocity; First Test on Cadaver

Velocity (m/s)	Stiffness (N/mm)
0.23	12.8
0.22	27.8
0.59	28.4
0.86	22.0
0.86	59.2

Average Stiffness = 31 ± 9 N/mm (N=5)
Average Velocity = 0.55 ± .32 m/s

Figure 16.

b) Repeatability of the force-deflection curves - see a).
c) Rate dependence of the styrofoam - The force-deflection curves obtained from 5 tests of the frangible abdomen at 6 m/s are shown in Figure 17, and are overlaid on those of (a) in Figure 18.
d) Restitution of styrofoam - The tests performed with a rigid impactor indicated that the amount of restitution experienced when styrofoam is crushed 3.0" (76.2 mm) is between 0.16"-0.39" (4-10 mm).
e) Function and integrity of the styrofoam - Figure 19 shows a typical sample from the tests with angled impact and varying belt slack and position. In all cases the foam crush was very well behaved, and structural integrity of the undeformed foam was well maintained.
f) Interaction between the rib cage and the frangible abdomen - The stiffness of each abdominal element is given in Table 6.

Sled Tests - Figure 20a is a photograph of the foam sample used in the sled test in which there was no submarining, 20b-d are from the tests in which there was incipient submarining and 20e-f are from the tests in which the dummy was presubmarined.

When there was no submarining, the upper segment of the foam insert showed no deformation from the lap belt, and minor deformation from interaction with the rib cage. The lower segment of the foam insert showed some deformation because the foam protrudes forward of the anterior-superior iliac spines and is loaded by the belt as it hooks onto the pelvis. Such loads are insignificant when applied to the pelvis.

In the tests with incipient submarining, the deformation pattern occurs at a slightly superior (higher) location on the foam insert. In the first two tests the belt appeared to slip between the upper and lower segment of the foam insert. There is slight deformation to the insert in the first test, but, in the second test (Figure 20c) there is a clear indication that unilateral submarining was occurring on the right side of the dummy (left side of the photograph). A third test was run in which the upper and lower segments were glued together. Again there was some deformation to the foam, but the glue may have influenced the stiffness.

When the dummy was presubmarined, the degree of submarining was clearly manifested in the foam (Figures 20e-f). The submarining was unilateral (right side of the dummy, left side of the photograph) and extensive.

In all of the above tests the integrity of the foam was preserved, and it was easily removed after each test.

Figure 21 shows a comparison of the kinematics of the Baseline Hybrid III versus the Hybrid III with the frangible abdomen for tests in which submarining did not occur. Figure 22 shows the same comparison for tests in which submarining did occur. There were no sled tests run with the Baseline Hybrid III in the incipient submarining set up, so a comparison of those kinematics was not performed.

DISCUSSION

The final design of the frangible abdomen is a five-pointed, crown shaped piece of styrofoam. It has biofidelity in abdominal force-deflection properties which are very close to the established corridor. Installation of the frangible abdomen in the Hybrid III is easily accomplished by two people in less than a few minutes. Hanging the dummy by the cranial eyebolt helps open the abdominal cavity relative to the seated dummy. Then the foam is pushed into place by one person as the other person folds the abdominal skin down to expose the abdominal cavity.

The foam frangible abdomen satisfies the design objectives by providing an unambiguous indication of the occurrence of submarining, and by allowing an objective measure of the probability of abdominal injury from submarining.

FORCE-DEFLECTION PROPERTIES OF THE ABDOMINAL INSERT - The force-deflection data from each group of experiments was "self-normalized" using equal velocity/equal stress scaling. The purpose of the self-normalization was to account for mass and geometry

Figure 19. Crush of typical styrofoam sample in angled lap belt impact on MTS machine.

TABLE 6. Dummy Abdomen Stiffness Superior-Inferior Loading

Abdomen Type	Low Speed Stiffness (N/mm)	High Speed Stiffness (N/mm)
Hybrid III (old)	6.0	6.8
Hybrid III (new)	9.7	12.3
Frangible Abdomen	13.1	11.4
5th Female - HybIII	20.2	19.1

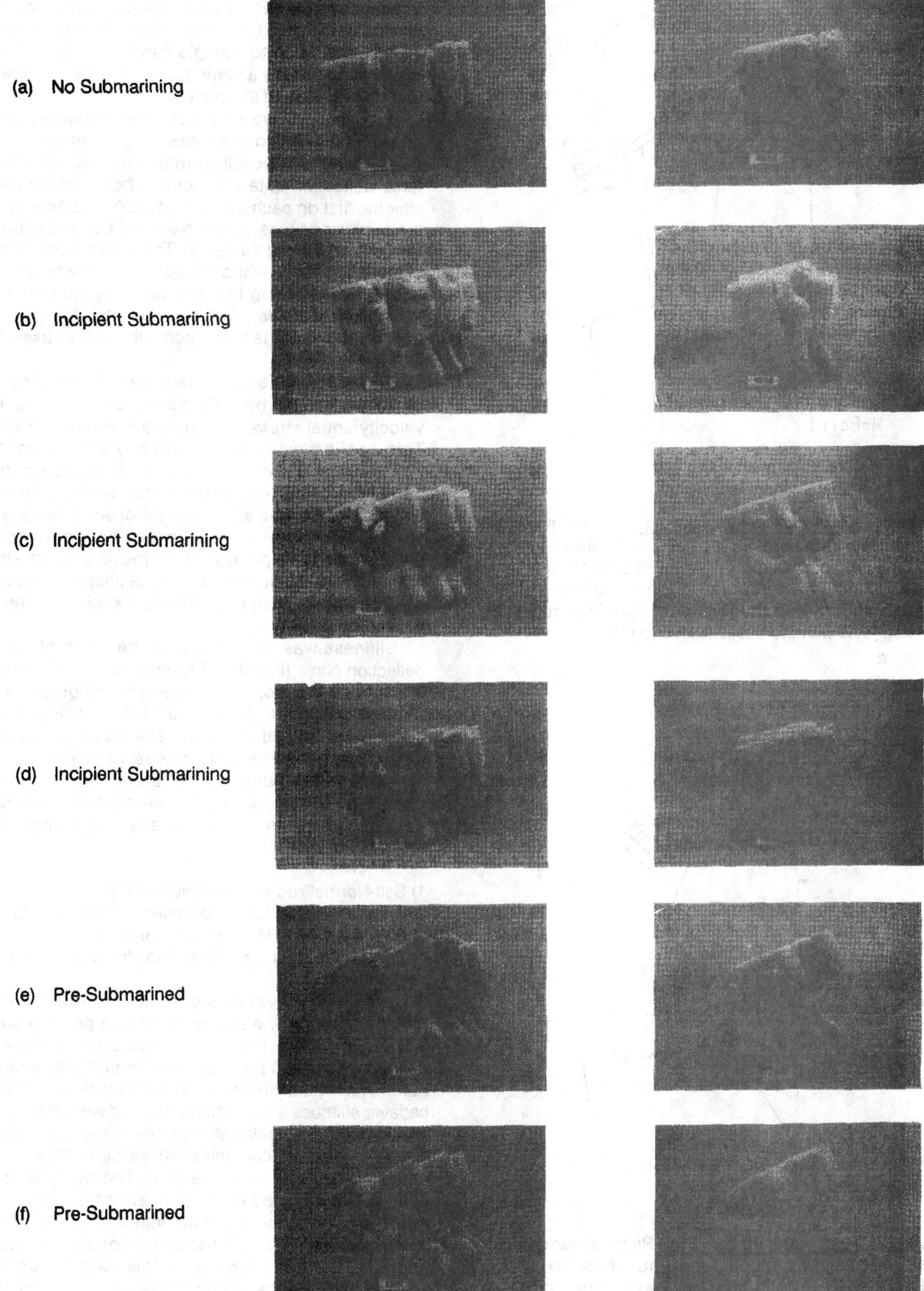

Figure 20. Crush pattern of frangible abdomen in various sled tests.

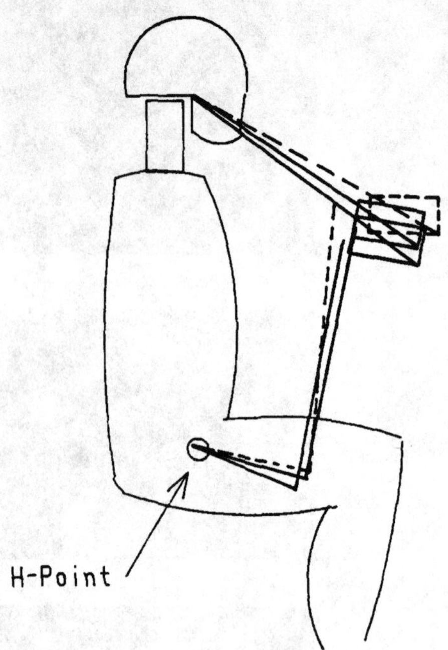

Figure 21. Tests with no submarining. Pictorial summary of kinematic analysis including maximum head excursion and neck angle, maximum H-point excursion and torso angle. (Dashed line represents one sled test with the frangible abdomen in the Hybrid III; solid lines represent two sled tests with the baseline Hybrid III.)

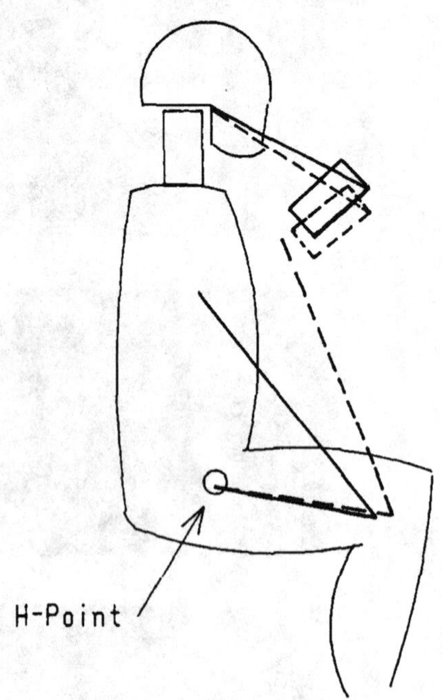

Figure 22. Tests with submarining. Pictorial summary of kinematic analysis including maximum head excursion and neck angle, maximum H-point excursion and torso angle. (Dashed line represents one sled test with the frangible abdomen in the Hybrid III; solid line represents one sled test with the baseline Hybrid III.)

differences between various subjects. While in principle normalization could make the data within each group of experiments fall onto a single curve, that rarely happens in practice. Instead a corridor is determined by the upper and lower bound of the curves.

As mentioned previously, several tests were performed on each porcine cadaver subject in an attempt to gain the most information possible from each test. However, the force-deflection data was not well behaved for the tests after the first on each subject. A likely explanation is that the cadaver tissues do not have the same restitution as the tissues of living subjects. These test were performed in rapid sequence without much time in between. It was noted during testing that the subjects did not return to their original shapes. Therefore, only the force-deflection data from the first test on each subject was used for this analysis.

The porcine cadaver data was also scaled to the Cavanaugh Human Cadaver data using equal velocity/equal stress scaling, and is shown in Figure 23. This scaling was mathematically equivalent to the normalization described in the previous paragraph, but allowed direct comparison between the human and porcine cadaver subjects. The use of equal velocity/equal stress scaling in this instance may not be appropriate, because the basic premise of geometric similitude may be violated. The amount by which the shape of the porcine subjects differs from that of human subjects determines the scaling error.

Stiffness was determined as the slope of the force-deflection curve (in units of N/mm), and these are shown in Table 7. The slopes were determined graphically, by visually fitting a straight line to the initial loading portion of each force-deflection curve. The mean and standard deviation of these stiffnesses for each group of experiments was determined, and the means between groups were compared using single factor Analysis of Variance (ANOVA). Specifically, we compared the stiffness between:

1) Self-Normalized Porcine Cadaver Data
2) Self-Normalized Live Anesthetized Cadaver Data
3) Self-Normalized Human Cadaver Data
4) Porcine Cadaver Data Scaled to Human Cadaver Data

At the .05 level of significance, there were no differences found: between the mean porcine cadaver stiffness and mean live anesthetized porcine stiffness; or between the mean porcine cadaver stiffness and mean human cadaver stiffness; or between mean scaled porcine cadaver stiffness and mean human cadaver stiffness.

Although the porcine stiffness data appears to be very similar the human cadaver stiffness data, they were not determined in the same manner. The porcine cadaver data was determined using standard safety belt material as the loading surface, but the human cadaver data was determined using a rigid bar as the loading surface. In addition, the human cadaver low velocity data was actually performed at a higher velocity range than these porcine experiments (Porcine low velocities were 1.6-5.5 m/s; Human low velocities were 4.9-7.2 m/s). While this may make the comparison less than ideal, there is no

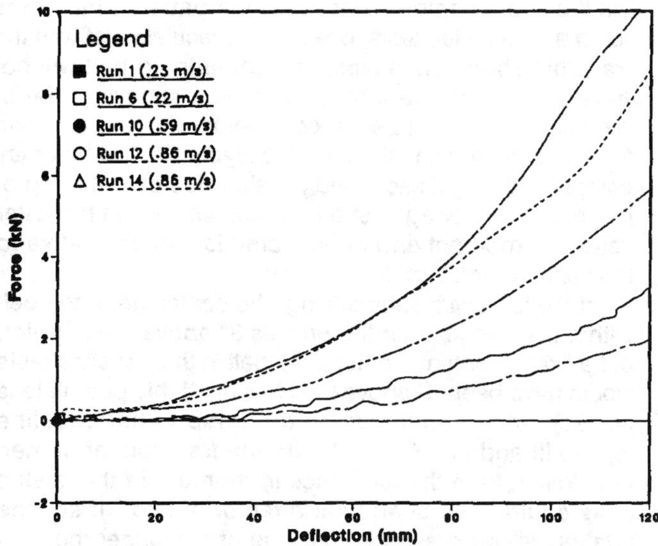

Figure 23. Force-deflection curves for porcine cadaver subjects normalized to human cadaver data.

other data available which gives the human cadaver stiffness with belt loading.

Given the above considerations, the target stiffness for the frangible abdomen was determined to be 23.0 N/mm.

The results from the MTS tests which were performed to assess interactions between the dummy ribs and abdomen are shown in Table 6. The frangible abdomen is stiffer in the superior/inferior direction than the old standard Hybrid III abdomen, about as stiff as the new Hybrid III abdomen with the notch for the chest deflection potentiometer arm, and less stiff than the Hybrid III 5th percentile female abdomen. With this relative stiffness, we do not expect unusual kinematics to result from interaction of the dummy ribs with the frangible abdomen.

REACTION SURFACE AND OTHER DUMMY MODIFICATIONS - Care was taken in the design of the support bracket and foam reaction plate to ensure that the ability of the dummy to flex forward, and sit "naturally" was maintained. A minimum clearance of 16 mm (0.625") between the flexible lumbar spine and the rigid reaction plate was designed after static dummy bending tests to ensure no contact during flexion. Sled tests with a prototype plate and bracket were performed which confirmed no interaction between the spine and plate.

TABLE 7. Stiffness Comparison

Anesthetized Porcine		**Porcine Cadavers**		**Porcine Scaled to Human**	
Test	Stiffness (N/mm)	Test	Stiffness (N/mm)	Test	Stiffness (N/mm)
2	8	1	25	1	26
3	19	6	21	6	19
4	30	10	41	10	52
5	37	12	26	12	28
6	40	14	41	14	52
8	29				
13	20				
14	21				
15	21				
17	8				
18	15				
20	25				
21	24				

Porcine Cadavers: Average (N = 5)
Stiffness: 31 ± 9 N/mm
Velocity: .55 ± .32 m/s

Porcine Scaled to Human: (N = 5)
35 ± 16 N/mm
.55 ± .32 m/s

Anesthetized Porcine: Average (N = 13)
Stiffness: 23 ± 10
Velocity: 3.7 ± .84 m/s

Human Cadaver Average Stiffness
From Cavanaugh (N = 5)

Stiffness = 23 N/mm
Velocity = 6.1 ± 1.1 m/s

As noted in the results, a height increase was introduced into the dummy by the support bracket. The support bracket raises the lumbar spine mounting block relative to the pelvis. This causes an offset between the H-point access hole and wrench socket which makes it difficult to set the H-point using standard tools. Since the lumbar spine mounting block must be machined for this design to work, a reasonable approach could be to incorporate the support bracket into the casting of the lumbar spine mounting block. This would eliminate both the 6 mm (0.250") height increase, and any difficulty setting the H-point.

With the addition of the support bracket, machining of the lumbar spine block, replacement of the standard abdomen, and removal of the chest deflection potentiometer, the net change in dummy mass is a decrease of .694 kg. This does not constitute a major modification to the Hybrid III since the dummy mass can be returned to its exact specified mass and inertia by addition of ballast material.

A important change with the frangible abdomen design is the removal of the chest deflection potentiometer. The ability to measure chest deflection and viscous response during impact at the same time as submarining injury potential is highly desirable. Use of high tension string potentiometers has recently been discussed as replacement technology for rotary chest potentiometer. The mounting locations under discussion are within the thoracic cavity, and as such would not interfere with the function of the frangible abdomen. If the rotary potentiometer is replaced with these advanced string potentiometers this concern would be alleviated.

Since these design changes have not been inspected by the NHTSA at this date, the Hybrid III dummy with the frangible abdomen cannot be used in certification tests. In addition, the force-deflection curves upon which the abdomen is based were generated using belt loading. They may also be applicable for impact with rigid objects, but further validation would be necessary. However, the reader is cautioned that impacts with rigid objects, such as those of unbelted occupants into steering wheels, may occur at higher velocities than belt loading. Then because of the rate sensitivity of the human abdomen, and the lack of rate sensitivity in the frangible abdomen, the stiffness of the frangible abdomen would have to be adjusted to a higher velocity stiffness to examine such phenomena.

DUMMY KINEMATICS WITH THE FRANGIBLE ABDOMEN - The kinematics of the dummy in tests without submarining appeared to be independent of whether or not the Frangible Abdomen was installed in the dummy. As can be seen in Figure 21, the head excursion, neck angle, torso angle, and H-point motion are all very similar in the baseline Hybrid III and the Hybrid III with the Frangible Abdomen installed.

When there was submarining, the head, neck, and H-point excursions were again independent of presence of the Frangible Abdomen (Figure 22). However, the torso angle was 16° greater in the Baseline Hybrid III, than in the Hybrid III with the frangible abdomen (129° vs 113°; where at 90° the dummy is sitting straight up, and at 180° it is lying flat on its back). A difference in torso angle during submarining might be expected when the frangible abdomen is in place because the reaction force of the lap belt in submarining acts directly upon the torso, probably via the lumbar spine. Based on the patterns left in the foam after the sled tests, one might speculate that with the frangible abdomen in place, a submarining belt will not have significant freedom to move in a superior/inferior direction because it becomes enveloped in the crushed foam. In contrast, in the baseline Hybrid III, the belt may compress the "gut sack" and possibly ride over the top of it to react directly against the lumbar spine. In the latter case, the moment arm on the torso is longer relative to the center of rotation (the H-point).

In the tests with submarining, the centerline of the belt within the frangible abdomen was 3" above the H-point, but a likely moment arm for a lap belt in the baseline tests would have been 6" above the H-point. If this postulate is correct, and if the forces on the torso in the baseline Hybrid III and the Hybrid III with the frangible abdomen are similar, then the difference in moment in these sled tests could have been near a factor of two. If so, the rotation will be greater in the case of the longer moment arm, viz. the Baseline Hybrid III.

POST-TEST CRUSH PATTERNS IN THE FRANGIBLE ABDOMEN - The frangible abdomen, as designed, functioned very well in bench test loading on the MTS and in sled tests. The foam crushed when loaded, but in a very well behaved manner. In particular, off-axis loading did not change the foam response, the crushed foam remained in one piece, it did not shatter or splinter, and was able to be removed without falling apart. The original design was limited to a two piece configuration as seen in Figures 20a-f because of the size of the billets available at that time. We now obtain billets twice as thick, which allow one piece construction (Figure 5a).

In the incipient submarining tests, the results were limited by the two piece design of the foam. While it was clear that submarining occurred, the exact amount of deformation of the foam was uncertain because part of the penetration of the belt into the abdomen occurred in between the two pieces of foam.

In the sled tests in which the dummy was presubmarined, and the tests in which no submarining took place, the ability of the foam to determine whether submarining had occurred was quite clear. Deformation to the foam above the location of the anterior-superior iliac spines indicated the occurrence of submarining.

INJURY ASSESSMENT WITH THE FRANGIBLE ABDOMEN - Because the force-deflection characteristics of the foam are known to match human force-deflection characteristics, the abdomen has biofidelity, and we can estimate the injury potential associated with the its deformation. For example, if the measured deformation of the abdomen after a test is 3.5" (89 mm), then when the elastic springback of the foam is accounted for, the maximum deformation during the test would be approximately 3.9" (99 mm) . A 50th percentile male has an abdominal depth of 8.1" (205 mm). The percent compression associated with 3.9" deformation is 48%. Using Figure 3 such belt loading would have a 25% probability of causing an AIS >= 4 abdominal injury. Thus, the frangible abdomen demonstrates the occurrence of submarining <u>and</u> the associated risk of injury.

The procedure for determining the probability of injury in a test with lap belt submarining is to remove the foam post-test and measure the maximum deformation of the middle three points. This deformation can be converted into a compression of the 50th percentile male abdomen by dividing by 206 mm (8.1"), and the probability of injury can be determined using Figure 3.

The frangible abdomen force-deflection characteristics were designed for low velocity loading (2-6 m/s). This is consistent with belt loading rates cited in the literature, but in higher velocity loading it could overestimate the probability of injury. Such an overestimate could occur because the stiffness of the human abdomen is rate sensitive. That is, at higher velocities, it takes more force to compress the human abdomen than at lower velocities. Therefore, if load from a higher velocity impact is applied to a dummy abdomen which is not rate sensitive and has been developed for a lower velocity impact, the loading will probably cause more deformation than it would have if the dummy abdomen was stiffer or rate sensitive. Thus, a dummy abdomen which is either stiffer, or one which is rate sensitive is desireable to study higher speed loading events that may take place (such as loading to the abdomen of an unbelted occupant).

Another consideration which indicates the need for a rate sensitive abdomen is that belt loading of an occupant in a vehicle has both loading and unloading phases. Near the end of the loading phase the rate of compression decreases, while the magnitude of the compression continues to increase, and the force may continue to increase. As the rate of compression decreases a rate sensitive abdomen would become less stiff, but a non rate sensitive abdomen would maintain a constant stiffness. Then the rate sensitive abdomen may compress more than the non rate sensitive abdomen. This indicates that the constant stiffness abdomen may underestimate the compression from actual belt loading.

HYBRID III FAMILY OF DUMMIES - Work continues in our laboratory to develop an abdomen that is rate sensitive, reusable, and gives a deformation time history. Concepts being explored include a fluid filled abdomen with advanced deflection sensors and a set of damped springs similar to Hybrid III ribs.

Figure 24. Support bracket and foam reaction plate for 5th percentile female Hybrid III dummy.

While this development was done with a 50th percentile male Hybrid III dummy, the same principles apply to other size dummies. At the time this project was started, the 5th percentile female dummy was in the process of being upgraded to Hybrid III status. Since no dummies were available we decided to work only on the 50th percentile dummy, and later scale the results down to the size of the 5th female. A foam reaction plate and support bracket have been designed for this dummy, and are shown in Figure 24. The lumbar spine block has been incorporated into the support bracket for simplicity. A need for submarining detection is not anticipated for the 95th percentile male dummy, but the principles and technology presented here apply as well.

The frangible abdomen has been transferred to the Safety Research and Development Laboratory at the Milford Proving Ground for use in sled and barrier testing. Biomedical Science will serve as the focal point for the results of these tests until a large enough data base has been developed to standardize the data analysis and procedures. Work is also proceeding on a version of frangible abdomen for the 5th percentile female Hybrid III dummy.

CONCLUSIONS

1) Abdominal injury assessment for belt loading due to submarining is now possible in the Hybrid III family of dummies.
2) The abdomen developed has biofidelity in its force-deflection characteristics for belt loading, is capable of detecting the occurrence of submarining, and can be used to determine the probability of abdominal injury when submarining occurs.
3) Installation of the abdomen in the Hybrid III dummy does not change the dummy kinematics when submarining does not occur.
4) When submarining does occur, the dummy kinematics are very similar to baseline Hybrid III kinematics, expect for torso angle. The change in torso angle is or may be a result of a more human-like compliance in the abdomen.

ACKNOWLEDGEMENTS

The authors acknowledge the contributions of many people that went into the successful execution of this project. In particular, we thank:
Dennis Andrzejak - for photographic work;
Joseph Balser - for technical discussions related to dummy development and use;
Howard Bender, Gerald Horn, Timothy Sorenson, Todd Townsend - for technical support;
Clyde Culver - for technical discussions;
Mary Foster - for computer support, assistance with figures, and paste-up of the final manuscript;
William Hering - for providing results of his own material tests on various foams;
John Horsh - for technical review;
Ahmed Kabir - for computer support;
John Melvin - for technical discussions, especially as

related to scaling and compression of a constant stiffness abdomen;
Mary Alice Miller - for providing raw force-deflection data from her initial porcine lap belt loading experiments;
Jeff Welch - for technical discussions related to foams.

REFERENCES

1. Evans, L., "Double Pair Comparison -- A New Method to Determine How Occupant Characteristics Affect Fatality Risk in Traffic Crashes." *Accid Anal & Prev* 18(3):217-227, 1986.

2. Evans, L., "Restraint Effectiveness, Occupant Ejection from Cars, and Fatality Reductions." General Motors Research Laboratories Report GMR-6398, September, 1988.

3. Evans, L., "Rear Seat Restraint System Effectiveness in Preventing Fatalities." *Accid Anal & Prev* 20(2):129-136, 1988.

4. Adomeit, D. and Heger, A., "Motion Sequence Criteria and Design Proposals for Restraint Devices in Order to Avoid Unfavorable Biomechanic Conditions and Submarining." In Proceedings of the 19th Stapp Car Crash Conference, SAE Technical Paper #751146, Warrendale, PA, 1975.

5. Adomeit, D., "Evaluation Methods for the Biomechanical Quality of Restraint Systems During Frontal Impact." In Proceedings of the 21st Stapp Car Crash Conference, SAE Technical Paper #770936, Warrendale, PA, 1977.

6 Adomeit, D., "Seat Design -- A Significant Factor for Safety Belt Effectiveness." In Proceedings of the 23rd Stapp Car Crash Conference, SAE Technical Paper #791004, Warrendale, PA, 1979.

7. Kroell, C.K., "Thoracic Response to Blunt Frontal Loading." In *The Human Thorax-Anatomy, Injury and Biomechanics,* SAE Publication P-67, pp.49-78, Warrendale, PA, 1976.

8. Neathery, R.F., Kroell, C.K., and Mertz, H.J., "Prediction of Thoracic Injury from Dummy Responses." In Proceedings of the 21st Stapp Car Crash Conference, SAE Technical Paper #751151, Warrendale, PA, 1975.

9. Viano, D.C. and Lau, I.V., "A Viscous Tolerance Criterion for Soft Tissue Injury Assessment." *J Biomech* 21(5):387-399, 1988.

10. Lau, I.V. and Viano, D.C., "The Viscous Criterion: Bases and Applications of an Injury Severity Index for Soft Tissue." SAE Transactions, vol. 95, 1986, P-189, In Proceedings of the 30th Stapp Car Crash Conference, pp. 123-142, SAE Technical Paper #861882, October, 1986.

11. Gadd, C.W. and Patrick, L.M., "Systems Versus Laboratory Impact Tests for Estimating Injury Hazard." SAE Technical Paper #680053, Society of Automotive Engineers, 1968.

12. Rouhana, S.W., Lau, I.V., and Ridella, S.A., "Influence of Velocity and Forced Compression on the Severity of Abdominal Injury in Blunt, Nonpenetrating Lateral Impact.", *J Trauma* 25(6):490-500, 1985.

13. Rouhana, S.W., "Abdominal Injury Prediction in Lateral Impact - An Analysis of the Biofidelity of the Euro-SID Abdomen.", In Proceedings of the 31st Stapp Car Crash Conference, pp.95-104, SAE Technical Paper #872203, Warrendale, PA, 1987.

14. Viano, D.C., "Cause and Control of Automotive Trauma." *Bulletin of the New York Academy of Medicine,* Second Series, 64(5):376-421, June, 1988.

15. Rouhana, S.W., Ridella, S.A., and Viano, D.C., "The Effect of Limiting Impact Force on Abdominal Injury: A Preliminary Study.", SAE Transactions, vol. 95, pp. 634-648, 1986, In Proceedings of the 30th Stapp Car Crash Conference, pp. 65-79, SAE Technical Paper #861879, Warrendale, PA, October, 1986.

16. Lau, V.K. and Viano, D.C., "An Experimental Study on Hepatic Injury from Belt Restraint Loading." *Avia Space Envir Med,* 52(10):611-617, October, 1981.

17. Miller, M.A., "The Biomechanical Response of the Lower Abdomen to Belt Restraint Loading." *J Trauma,* in press, 1989.

18. Sacco, W.J., Jameson, J.W., Copes, W.S. et al, "Progress Toward a New Injury Severity Characterization: Severity Profiles." In Computers in Biology and Medicine, December, 1988.

19. Viano, D.C., "Limits and Challenges of Crash Protection." *Accid Anal & Prev* 20(6):421-429, 1988.

20. Rouhana, S.W. and Foster, M.E., "Lateral Impact - An Analysis of the Statistics in the NCSS.", SAE Transactions, vol. 94, 1985, In Proceedings of the 29th Stapp Car Crash Conference, pp. 79-98, SAE Technical Paper #851727, Warrendale, PA, 1985.

21. Evans, L. and Frick, M., "Seating Position in Cars and Fatality Risk." General Motors Research Laboratories Report GMR-5911, July, 1987.

22. Kulowski, J. and Rost, W.B., "Intra-abdominal Injuries from Safety Belt in Auto Accident." *Arch Surg* 73:970-971, December, 1956.

23. Sube, J., Ziperman, H.H. and McIver, W.J., "Seat Belt Trauma to the Abdomen." *Amer J Surg* 3:346-350, March, 1967.

24. Porter, S.D. and Green, E.W., "Seat Belt Injuries." *Arch Surg* 96:242-246, February, 1968.

25. Mackay, G.M., "Abdominal Injuries to Restraint Front Seat Occupants in Frontal Collision." In Proceedings of the American Association for Automotive Medicine, pp. 146-148, 1982.

26. Ryan, P. and Ragazzon, R., "Abdominal Injuries in Survivors of Road Trauma Before and Since Seat-Belt Legislation in Victoria." *Aus N.Z. J Surg* 49(2):200-202, 1979.

27. Gallup, B.M., St-Laurent, A.M., Newman, J.A., "Abdominal Injuries to Restrained Front Seat Occupants in Frontal Collisions." In Proceedings of the 26th Annual American Association for Automotive Medicine Conference, Ottawa, Canada, October, 1982.

28. Dalmotas, D.J., "Mechanisms of Injury to Vehicle Occupants Restraints by Three Point Seat Belts." In Proceedings of the 24th Stapp Car Crash Conference, pp. 439-476, SAE Technical Paper #801311, Warrendale, PA, 1980.

29. Fish, J. and Wright, R.H., "The Seat Belt Syndrome -- Does it Exist?" *J Trauma,* 5:746-750, November, 1965.

30. Garrett, J.W. and Braunstein, P.W., "The Seat Belt Syndrome," *J Trauma,* 2:220-238, May, 1962.

31. Cocke, W.M. and Meyer, K.K., "Splenic Rupture Due to Improper Placement of Automobile Safety Belt," *JAMA,* 183:693, February, 1963.

32. States, J.D., Huelke, D.F., Dance, M., and Green, R.N., "Fatal Injuries Caused by Underarm Use of Shoulder Belts." *J Trauma,* 27(7):740-, 1987.

33. Trollope, M.L., Stalnaker, R.L., McElhaney, J.H., et al, "Mechanism of Injury in Blunt Abdominal Trauma." *J Trauma,* 13(11):962-970, 1973.

34. Huelke, D.F., Lawson, T.E., "Lower Torso Injuries and Automobile Seat Belts." Society of Automotive Engineers Technical Paper #760370, Warrendale, PA, 1978.

35. Dardik, H., Ibraham, M.I., "The Spectrum of Seat Belt Injuries." *Lawyer's Med J,* 6:50-75, 1977.

36. Williams, J.S., Lies, B.A., Hale, Jr., H.W., "The Automotive Safety Belt, in Saving a Life, May Produce Intra-Abdominal Injuries." *J Trauma,* 6:303, 1966.

37. Witte, C.L., "Mesentery and Bowel Injuries from Automotive Seat Belts." *Ann Surg,* 167:486, 1968.

38. Smith, W.S. and Kaufer, H., "Patterns and Mechanism of Lumbar Injuries Associated with Lap Safety Belt." *J Bone Joint Surg* 51-A, 239-254, 1969.

39. Dehner, J.S., "Seat Belt Injuries of the Spine and Abdomen," *Amer J Roentgen,* VIII:833-843, April, 1971.

40. Denis, R., Allard, M., Atlas, H. and Farkouh, E., "Changing Trends with Abdominal Injury in Seatbelt Wearers." *J Trauma,* 23(11):1007-1008, November, 1983.

41. Arajarvi, E., Santavirta, S. and Tolonen, J., "Abdominal Injuries Sustained in Severe Traffic Accidents by Seatbelt Wearers." *J Trauma,* 27(4):393-397, April, 1987.

42. Shanks, J.E. and Thompson, A.L., "Injury Mechanisms to Full Restrained Occupants." In Proceedings of the 23rd Stapp Car Crash Conference, SAE Technical Paper #791003, Warrendale, PA, 1979.

43. Dalmotas, D.J., "Mechanisms of Injury to Vehicle Occupants Restrained by Three-Point Seat Belts." In Proceedings of the 24th Stapp Car Crash Conference, SAE Technical Paper #801311, Warrendale, PA, 1980.

44. Society of Automotive Engineers, Advances in Belt Restraint Systems: Design, Performance and Usage, P-141, Society of Automotive Engineers, Warrendale, PA, 1984.

45. Society of Automotive Engineers, Passenger Comfort, Convenience and Safety: Test Tools and Procedures, P-174, Society of Automotive Engineers, Warrendale, PA, 1986.

46. Daniel, R.F., "Test Dummy Submarining Indicator System." United States Patent 3,841,163, October, 15, 1974.

47. Tarriere, C.H., "Proposal for a Protection Criterion as Regards Abdominal Internal Organs." In Proceedings of the 17th American Association for Automotive Medicine Conference, pp. 371-382, 1973.

48. Leung, Y.C., Tarriere, C., Fayon, A. et al, "A Comparison Between Part 572 Dummy and Human Subject in the Problem of Submarining." In Proceedings of the 23rd Stapp Car Crash Conference, pp. 677-720, SAE Technical Paper #791026, Warremdale, PA, 1979.

49 Melvin, J.W. and Weber, K., "Abdominal Intrusion Sensor for Evaluating Child Restraint Systems.", In Passenger Comfort, Convenience and Safety: Test Tools and Procedures, P-174, SAE Technical Paper #860370, Society of Automotive Engineers, Warrendale, PA, 1986.

50. DeJeammes, M., Biard, R., and Derrien, Y., "Factors Influencing the Estimation of Submarining on the Dummy." In Proceedings of the 25th Stapp Car Crash Conference, pp. 733-762, SAE Technical Paper #811021, Society of Automotive Engineers, Warrendale, PA, 1981.

51. Leung, Y.C., Tarriere, C., Lestrelin, D., et al. "Submarining Injuries of 3 Pt. Belted Occupants in Frontal Collisions - Description, Mechanisms and Protection." SAE Technical Paper #821158, pp. 173-205, Society of Automotive Engineers, Warrendale, PA, 1982.

52. Mooney, M.T. and Collins, J.A., "Abdominal Penetration Measurement Insert for the Hybrid III Dummy." SAE Technical Paper #860653, Society of Automotive Engineers, Warrendale, PA, 1986.

53. Biard, R., Cesari, D. and Derrien, Y., "Advisability and Reliability of Submarining Detection." In Restraint Technologies: Rear Seat Occupant Protection SP-691, pp. 27-38, SAE Technical Paper #870484, Society of Automotive Engineers, Warrendale, PA, 1987.

54. Eppinger, R.H., "Prediction of Thoracic Injury Using Measurable Experimental Parameters.", The 6th International Conference on Experimental Safety Vehicles, pp. 770-780, 1976.

55. Langhaar, H.L., Dimensional Analysis and Theory of Models, Wiley, New York, 1957.

56. Mertz, H.J., "A Procedure for Normalizing Impact Response Data.", SAE Technical Paper #840884, 1984.

57. American Association for Automotive Medicine, "The Abbreviated Injury Scale", 1985.

58. Stalnaker, R.L. and Ulman, M.S., "Abdominal Trauma - Review, Response, and Criteria.", In Proceedings of the 29th Stapp Car Crash Conference, pp. 1-16, SAE Technical Paper #851720, Warrendale, PA, 1985.

59. Verriest, J.P., Chapon, A., and Trauchessec, R., "Cinephotogrammetrical Study of Porcine Thoracic Response to Belt Applied Load in Frontal Impact - Comparison Between Living and Dead Subjects.", In Proceedings of the 25th Stapp Car Crash Conference, pp. 499-545, SAE Technical Paper #811015, Warrendale, PA, 1981.

902317

Assessing Submarining and Abdominal Injury Risk in the Hybrid III Family of Dummies: Part II — Development of the Small Female Frangible Abdomen

Stephen W. Rouhana, Edward A. Jedrzejczak, and Joseph D. McCleary
Biomedical Science Dept.
General Motors Research Laboratories

ABSTRACT

The Frangible Abdomen is a crushable Styrofoam insert for the abdominal region of the Hybrid III family of dummies, which has biofidelity, and assesses the occurrence of submarining and its risk of injury. It was first developed for the mid-sized male Hybrid III dummy. This paper describes the design of the Frangible Abdomen for the small female Hybrid III dummy, and how to use it to assess the occurrence and the risk of injury from submarining.

The force-deflection properties of the mid-sized male insert were scaled to the small female dimension using equal stress/equal velocity scaling. Sled tests were run to compare the kinematic and dynamic performance of the baseline small female Hybrid III dummy with the same dummy modified to incorporate the Frangible Abdomen. The kinematic and submarining performance of the small female Hybrid III dummy was unchanged by the addition of the Frangible Abdomen.

The Frangible Abdomen was easy to install and use, and had excellent repeatability. Injury assessment with the Frangible Abdomen is based on the depth of foam deformation. Tables and graphs are included which relate the risk of injury to the amount of crush. Issues regarding its use, handling, applicability, spinal injury, belt roping, and unilateral submarining are discussed. Appendices containing information on retrofitting existing dummies, calibration testing of Styrofoam, and with more detailed tabulations of sled test data are provided.

LAP BELT SUBMARINING can be defined as relative motion between the occupant and the lap belt such that the lap belt load shifts from the pelvic skeletal structure to the softer abdominal region. Submarining sometimes occurs during the development tests of a belt restraint system, but it can be difficult to objectively assess using a standard anthropomorphic test device (dummy) [1]*.

Previous authors have suggested a variety of dummy modifications to provide submarining assessment capability including: a pelvis, developed by Daniel, which is fitted with load bolts to measure belt force on the anterior iliac spines as a function of time [2], a strain gauged beam within the abdominal cavity, developed by APR [3,4], measurement of pelvic rotation using a five accelerometer array [1], and others [5-9]. While there has been some success with these methods, none of them provides the combination of biofidelity in the abdominal force-deflection response, objective indication of the occurrence of submarining, and more importantly, a quantitative estimate of the risk of injury to the abdominal organs.

Rouhana, et al. [10] recently proposed a method for submarining assessment using an abdominal insert which has biofidelity under belt loading conditions, and the capability to assess both the occurrence of submarining and the risk of injury associated with it. The insert, known as the Frangible Abdomen, was developed using the mid-sized male Hybrid III dummy, with the plan to implement a scaled version in the small female (5th percentile) Hybrid III dummy when it became available. The development of the Frangible Abdomen for the mid-sized (50th percentile) male dummy was described in [10].

The purpose of this paper is to document the extension of the Frangible Abdomen concept to the small female Hybrid III dummy, and to describe the method of assessing injury risk using the Frangible

---

*Numbers in square brackets indicate references which are listed at the end of the paper.

Abdomen. This report includes the scaling procedure used to obtain the stiffness corridor of the dummy abdomen, and details of the design, construction, and sled testing of the small female frangible abdomen. It also includes tables of data which relate the amount of deformation measured from the foam insert to the risk of abdominal injury associated with that deformation. The extension of this concept to the small female dummy is important in light of recent results which indicate that the old small female dummy (ARL VIP-5F) has a tendency to submarine more readily than the mid-sized male dummy [1].

DESCRIPTION OF THE FRANGIBLE ABDOMEN

The Frangible Abdomen is a crushable foam insert, which is shaped with 5 tapered points to achieve a force-deflection response similar to the human abdomen. When belt loading of the foam occurs, the foam crushes and remains crushed leaving a permanent record of the maximum penetration. Visual inspection of the foam determines whether submarining occurred, and if so, measurement of the depth of crushed foam allows one to make an objective estimate of the risk of occupant injury due to that loading.

The Frangible Abdomen consists of three main components: 1) a foam insert which serves as the deformation transducer, 2) a reaction plate for the foam insert, and 3) a support bracket which couples the reaction plate and insert to the pelvis.

1) The **foam insert** (Figure 1) is located in the abdominal cavity of the dummy (beneath the rib cage, in front of the lumbar spine). The geometry of the insert gives it a force-deflection response similar to a human abdomen in belt loading (biofidelity). On each side of the "central point" of foam (Figure 2, Top View) is a "major side point" (the larger point) and a minor side point (the smaller point).

2) The **reaction plate**, is located behind the foam insert in the dummy, and as the name implies, is the surface which supports the foam insert during belt loading (Figure 3).

3) The **support bracket** bolts to the pelvic structure to anchor the reaction plate (Figure 4), and serves as the attachment point for the lumbar spine in place of the standard lumbar spine attachment cylinder. The position of the reaction plate was chosen to reduce the possibility of interference with the lumbar spine when the dummy torso flexes forward, while at the same time allowing room for a 121mm (4.75") foam insert in front of the plate.

Prints of the various components of the small female Frangible Abdomen are given in Appendix I.

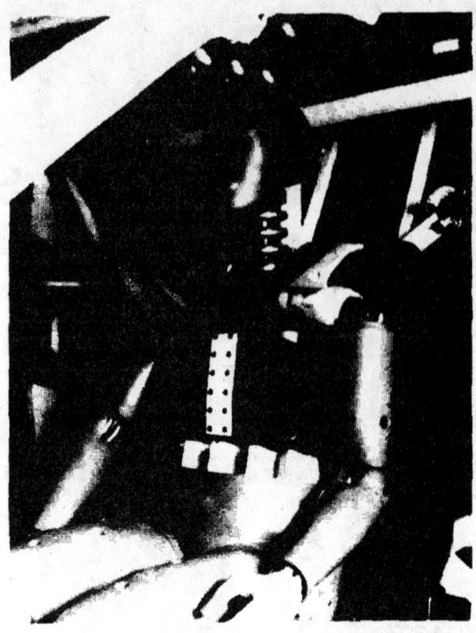

Figure 1. Frangible Abdomen in place in a Hybrid III mid-sized male dummy (beneath the rib cage, in front of the lumbar spine, behind the abdominal "skin", above the pelvis).

DUMMY MODIFICATIONS

As for the mid-sized male dummy [10], the rotary potentiometer for measuring chest deflection must be removed from the small female dummy for the installation of the Frangible Abdomen. Figure I.4, in Appendix I, shows the replacement for the bracket which holds the chest deflection potentiometer, and which joins the thoracic spine to the lumbar spine. The front edges of the upper lumbar spine attachment plate and the chest deflection potentiometer replacement plate are machined so that they are flush with one another, and are as far back in the dummy as possible (as close to the front surface of the lumbar spine as possible). The front two 1/4-20 "socket head cap" screws used to connect the lumbar and thoracic plates are replaced with 1/4-20 "button head socket cap" screws.

To measure chest deflection in these experiments we used a high tension string potentiometer (Space Age Controls Model 160-321H; 56oz. pull). The string pot body was mounted to the thoracic spine at a position just below the 3rd rib. The end of the string was connected to a point mount on the sternum approximately 25mm lateral of the mid-sagittal plane with the string perpendicular to the rib cage at the point of attachment.

The addition of the reaction plate and other attachment hardware to the small female dummy caused the weight to increase by 0.59kgf (1.3 lb.), from 4.81kgf (10.6 lb.) to 5.40kgf (11.9 lb.). The

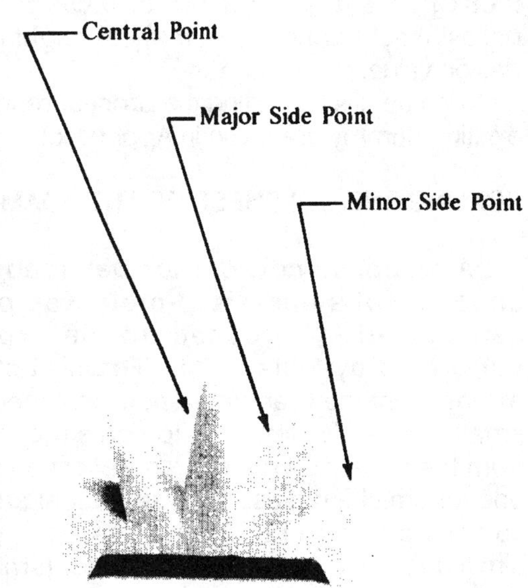

Figure 2. The foam insert in a) side view, b) front view, and c) top view. The tapered point geometry gives the insert a force-deflection response similar to a human abdomen in belt loading.

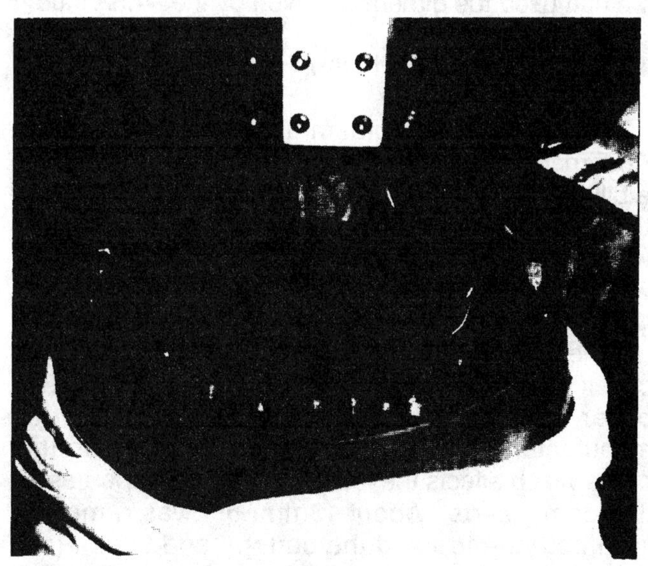

Figure 3. The reaction plate is located behind the foam insert in the dummy, and is the surface which supports the foam insert during belt loading.

Figure 4. The support bracket (right) bolts to the pelvic structure to anchor the reaction plate (left). It also serves as the attachment point for the lumbar spine in place of the standard lumbar spine attachment cylinder.

height of the dummy was unchanged by the modifications. Making the reaction plate out of aluminum would lower its weight by approximately 0.68kgf (1.5 lb.). Addition of 0.09kgf (0.2 lb.) of ballast would bring the dummy's weight back to the design value.

More details regarding the procedure to retrofit an existing dummy are given in Appendix I.

SCALING THE STIFFNESS OF THE FOAM INSERT

A response corridor for belt loading to the abdomen of a mid-sized male was previously established [10], based on the experiments performed by Miller [11]. The first step in the extension of the Frangible Abdomen concept to the small female dummy was to scale the mean value from the mid-sized male force-deflection corridor to one for small females. We used equal stress/equal velocity scaling [12,13] and assumed geometric similitude exists between the model (small female) and prototype (mid-sized male). Then the scale factor for length dimensions is the ratio of some characteristic length dimension in the model to the same length dimension in the prototype. That is,

$$\lambda_1 = \frac{l(\textit{small female})}{l(\textit{mid-sized male})}$$

where, λ_1 is the length scale factor, and l = the characteristic length in the model or the prototype.

With equal stress/equal velocity scaling, the force scales as λ_1^2 and the length scales as λ_1. Since the stiffness is the ratio of force over length, it also scales as λ_1. Then, the length scale factor can be applied to the mid-sized male abdominal stiffness to determine the abdominal stiffness of the small female.

The characteristic length chosen for this scaling was the distance between the 5th lumbar vertebra (L5) and the maximum abdominal protrusion. Two values were selected from the various anthropometric studies reported in the literature. A study done at UMTRI for the NHTSA sponsored Advanced Anthropomorphic Test Device project [14], provided a value of the characteristic length for subjects seated in automotive seats. A data base compiled by NASA on the anthropometry of military subjects provided a value of the characteristic length for standing subjects [15].

The distance between L5 and the maximum abdominal protrusion in the UMTRI study was 258mm for the mid-sized male, and 205mm for the small female vehicle occupants measured. Then, the UMTRI based scale factor for length and stiffness is equal to 0.795 (the small female abdominal stiffness should be 0.795 times the mid-sized male stiffness). Scaling the 23N/mm mid-sized male abdominal stiffness results in a value of 18N/mm for the small female abdominal stiffness.

The abdominal depth of standing subjects in the NASA data base was 208mm for the mid-sized male, and 180mm for the small female. These dimensions yield a NASA based scale factor of 0.866, and a scaled stiffness equal to 20N/mm.

While the UMTRI anthropometry is representative of seated automotive occupants, the NASA anthropometry more closely approximates the subject posture in the Miller experiments [11] which form the basis of the force-deflection corridor. Both NASA's and Miller's subjects had their torsos extended (in the anatomical sense), which shortens the abdominal depth by eliminating the paunch induced by the seated posture. But since NASA's subjects were standing and Miller's subjects were supine (on their backs), they may have different external abdominal dimensions because of movement of the highly mobile abdominal organs, toward the feet in NASA's subjects and towards the back in Miller's subjects.

The stiffness chosen for the small female Frangible Abdomen was the 18N/mm value based on the UMTRI anthropometric measurements. This value was used because it is representative of seated occupants and the UMTRI study served as the basis for the small female Hybrid III anthropometry. Since this value yields a lower stiffness after scaling, the abdomen developed will be more conservative than if we had used the dimension given by the NASA study.

MATERIAL TESTS ON FOAM FOR THE INSERTS

The flotation Styrofoam used for the Frangible Abdomen is obtained in a relatively large block called a billet (ours were 10"x20"x96"; Dow Chemical calls them Buoyancy Billets). The direction of crushing was found to affect the force-deflection properties as will be discussed later, so a billet coordinate system was set up as shown in Figure 5.

The outer surface of the billet of Styrofoam was cut off and the sides of the billet made parallel. The outer surface was removed, because in the manufacturing process a tough skin forms on the billets which affects the force-deflection properties of the foam inserts. About 13mm(.5") was removed from both the top and the bottom, and 25mm (1") from each side of the billet to remove the skin and square off the faces. Then, before any other cuts were made on the billet, a template of the top view of the Frangible Abdomen was used to draw the pattern on the top surface of the billet. The orientation of the inserts within the billet must be such that the back ends of the inserts are facing the edges of the billet (Figure 6). Drawing the insert pattern on the billet before cutting eliminates the possibility of making a mistake in orientation. Using this procedure more

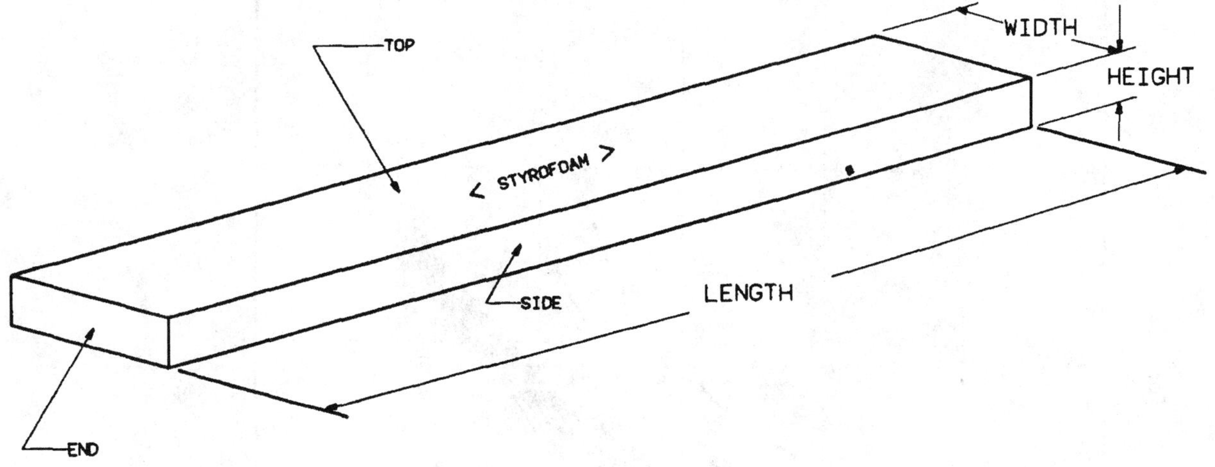

Figure 5. Styrofoam billet with coordinate system convention.

than 60 inserts were obtained from one 10"x20"x96" billet of Styrofoam.

Material tests on the foam inserts were performed using a Material Test System (MTS), which is a high speed hydraulic piston. One of two impact interfaces was attached to the piston. For Styrofoam calibration tests, a 25.4mm (1") diameter flat aluminum disk was used. For testing foam insert sample patterns, a belt-yoke fixture was used with standard automotive belt webbing as the impact interface. The belt-yoke setup was described in [10], and the reader is referred to that publication for further information on the test equipment and procedure.

The force-deflection curves of the samples were compared to the scaled force-deflection corridors. If the sample was too stiff or not stiff enough, the geometry of the sample was modified, and the tests were repeated. These iterations continued until the force-deflection corridor was matched by the Styrofoam, which established a shape for the small female Frangible Abdomen foam inserts.

The force-deflection curves for the small female Hybrid III dummy Frangible Abdomen are shown in Figure 7 with the scaled abdominal force-deflection corridor superimposed. The shape of the abdominal insert that gave this force-deflection response is shown in Appendix I, Figure I.1.

Like the insert for the mid-sized male dummy, the force-deflection of the small female insert is in good agreement with the established corridor and has excellent repeatability. The curves in Figure 7 represent 25 identically shaped inserts from different positions in 7 different billets of Styrofoam (4 different positions within each of 5 billets, 3 different positions in 1 billet, and 2 different positions in 1 billet). Thus, the Frangible Abdomen repeatability encompasses different samples with the same shape, different locations within a single billet, and different billets.

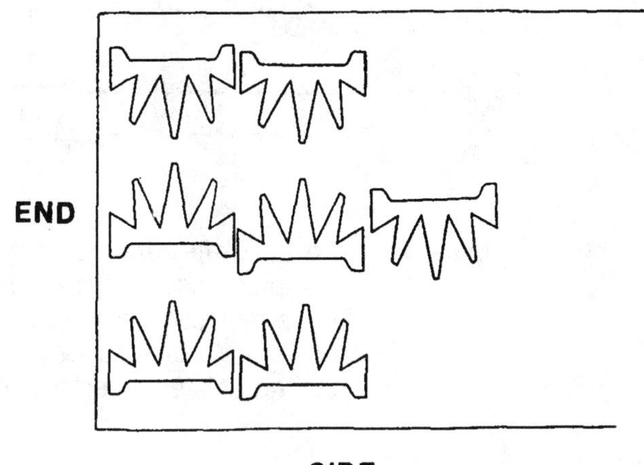

Figure 6. Top view of billet showing orientation of templates for cutting out foam inserts. Note that because of edge effects, inserts cut from the edges of a billet are cut with their back ends toward the sides of the billet (inserts cut from the middle can face either side). Also, the front to back axis of the insert is parallel to the side-to-side direction of the billet.

We developed a material test, to calibrate new billets of Styrofoam, which encompasses crushing a 76mm (3") cube of foam. The cubes for calibration testing were cut from the middle of a 76mm (3") wide slice off of either end of the billet (Figure 8). The aluminum disk compressed the cube in a side-to-side direction with respect to the billet coordinate system. The side-to-side direction was chosen for calibration because that is the direction of crush of the inserts machined from the billet. The typical crush distance

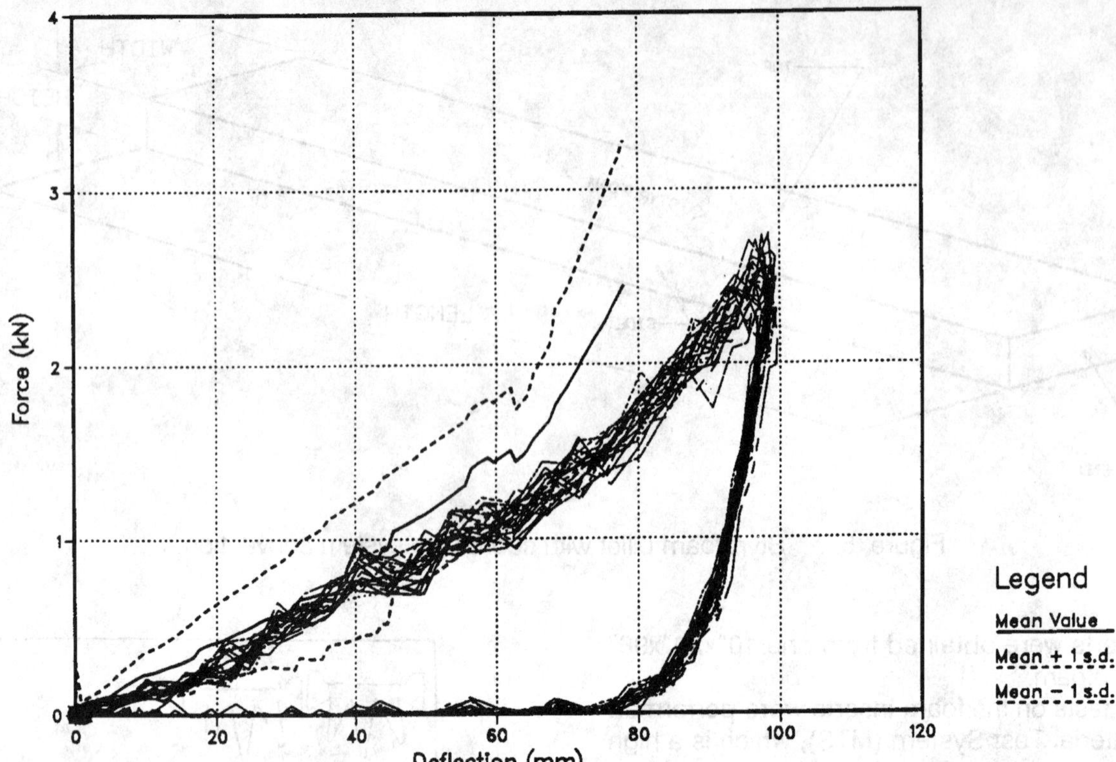

Figure 7. Force-deflection response of 25 small female Frangible Abdomen samples from 7 different billets of Styrofoam superimposed on the scaled force-deflection corridor. The corridor is the mean value, plus and minus one standard deviation, of the normalized force-deflection response from the lap belt tests performed by Miller [11].

we used was 51mm (2"), and tests were conducted both quasi-statically and dynamically. The Styrofoam samples we tested were not rate dependent. The force-deflection curves for 14 cubes of foam (one from each of 2 different positions in 7 different billets), tested in a side-to-side crush direction, are shown in Figure 9.

On the scale of the calibration cubes, the properties of Styrofoam flotation foam are not isotropic, nor homogeneous. Therefore, the force-deflection response in the cube calibration tests is a function of the position of the sample within the billet. The stiffness of the cube samples within 102mm (4") of the edge of the billet can be greater than those taken from the middle of the billet. Because of this anisotropy in the material response, it is recommended that the samples used for calibration check of the foam be taken from the middle region of the 76mm (3") slice of foam from the billet.

The foam properties appear to average out as the

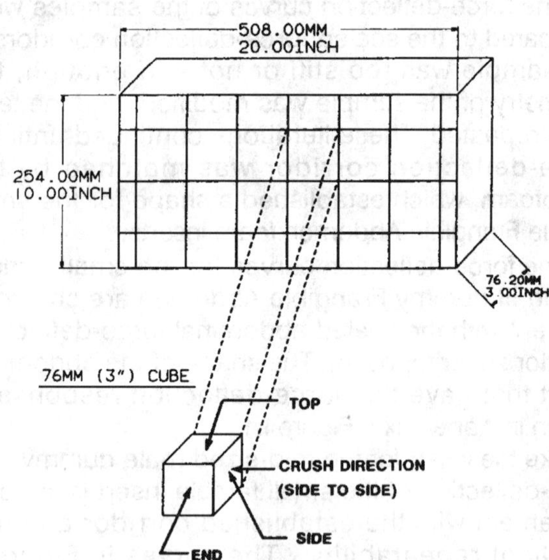

Figure 8. A 76mm thick slice of foam (shown in end view), from which cubes are cut for material testing of a billet. A 25.4mm diameter aluminum disk is used to crush the cubes to a depth of 51mm in a side-to-side direction.

sample size increases to the size of the Frangible Abdomen foam insert. However, the Frangible Abdomen foam inserts should be cut such that the back surfaces of the inserts face the edge of the billet rather than the middle of the billet (the tips of the points face the middle of the billet) (see Figure 6).

Shortly after we developed the prototype small female Frangible Abdomen insert, the maker of Styrofoam changed its formulation to replace the CFC blowing agent used in production. This change in formulation was reflected in different stiffness results on the level of the cubes and the small female Frangible Abdomen foam inserts, but did not affect the larger mid-sized male foam insert. When the cause for the difference in properties was determined, the small female insert pattern was modified to match the scaled force-deflection corridor again. This illustrates both a negative and a positive aspect of the Frangible Abdomen concept. On the negative side, there may be effects, on the force-deflection properties of different batches of Styrofoam, which necessitate the calibration tests. On the positive side, the design is robust enough such that the force-deflection can be adjusted easily by changing the pattern of the foam. Previous experience with different batches did not show different properties, so now that the Styrofoam formulation has stabilized again, there should be no need to alter the pattern of the foam insert, but calibration checks are of the utmost importance if biofidelity and accuracy in injury risk assessment are to be assured. Appendix II contains a discussion of considerations necessary for changing the shape of the foam insert.

SLED TESTS WITH THE FRANGIBLE ABDOMEN

A series of sled tests was performed using a HYGE sled with a body buck to compare the kinematics and dynamics of the baseline Hybrid III small female dummy with the kinematics and dynamics of the dummy with the Frangible Abdomen installed.

The sled tests reported here were conducted with the prototype small female Frangible Abdomen hardware. Analysis of the data showed that there was contact between the steel plate on top of the lumbar spine, and the Frangible Abdomen hardware which affected some of the dynamic measurements. The design was modified to reduce the possibility of contact. There was no evidence of contact in 6 out of 6 tests performed with the modified design. The more extensive tests of the prototype are reported here, because they clearly demonstrate that the Frangible Abdomen had no effect on kinematics or submarining performance of the dummy, and only limited effect on dynamics of the dummy. Our judgement is that the modifications made, which are

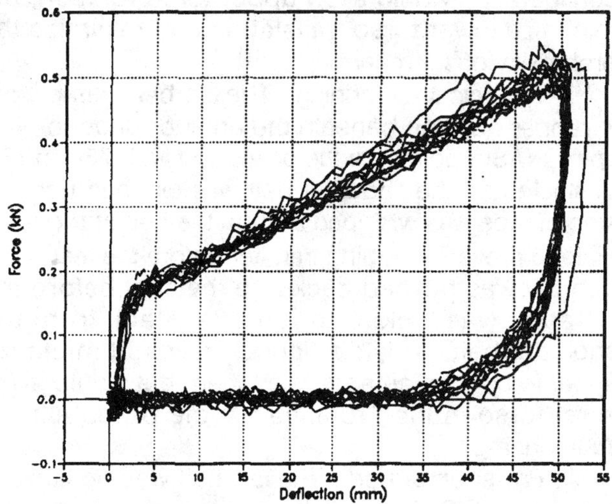

Figure 9. Force-deflection curves for 14 cubes of Styrofoam from 7 different billets, tested in a side-to-side crush direction.

included in the prints in Appendix I, should improve the agreement between the baseline and Frangible Abdomen dummy configurations.

The tests were run with the dummy seated on the driver's side of the buck in an automotive bucket seat. There was no dashboard or steering system in the sled buck. The structures to which the steering column and instrument panel normally bolt were present in the sled buck, but were padded with Rubatex R310V foam. The seat used in these tests was constructed of foam on a zig-zag spring suspension and sheet metal frame. The foam Indentation Load-Deflection characteristic was ≃240N. The seat pan angle was ≃8°, and the seat back angle was ≃26° (not a reclining seat). The seat was fixed 2 notches forward of the full rear position, and a block of Styrofoam was placed in front of each foot of the dummy to simulate the toe pan of the vehicle.

A standard automotive lap-shoulder belt was fastened around the dummy in one of three configurations (see below). The belt was a single retractor, continuous loop system with a locking latch plate. The retractor was mounted to the B-pillar of the sled buck, and was pre-locked. The dummy was pre-positioned by setting the head, H-point, and knee targets in locations marked by 3 laser beams fixed in the laboratory reference frame.

The belt system configurations were chosen to specifically vary the degree of submarining. All configurations used the same anchor locations. The configurations consisted of:

i) No submarining - The lap belt was cinched tight and 76mm (3") of slack was introduced into the shoulder belt before locking the retractor. This configuration would allow upper torso rotation, but minimize lower torso translation and minimize the probability of submarining.

ii) Dynamic submarining - The lap belt was placed in proper position beneath the anterior-superior iliac spines (ASIS points) of the pelvis, but with 76mm (3") of slack, and the shoulder belt was cinched tight. A strip of tape was wrapped around the belt at the latch plate to prevent webbing transfer before the test. The dummy was pushed back into the seat before the retractor was locked to remove slack from the shoulder belt. This configuration would minimize upper torso rotation and translation, but would allow lower torso motion to enhance the probability of submarining.

iii) Pre-submarined - The lap belt was positioned above the ASIS points of the pelvis and was cinched to just contact the abdomen. The shoulder belt was cinched tight before the retractor was locked. This configuration ensured that submarining would occur because the lap belt started out in contact with the abdomen.

Each test was run at 30mph using a sled pulse with a peak acceleration of ≈20g, as shown in Figure 10. A single small female Hybrid III dummy with or without the Frangible Abdomen was used in each test. The dummy instrumentation was used to make the following measurements:

- head acceleration (longitudinal, lateral, and vertical)
- neck force (longitudinal, lateral, and vertical)
- neck moment (about longitudinal, lateral, and vertical axes)
- chest acceleration (longitudinal, lateral, and vertical)
- chest deflection (using a string potentiometer)
- pelvis acceleration (longitudinal, lateral, and vertical)
- pelvis rotation (5 inline accelerometers)
- femur load (in right and left femur)

There were 3 experiments for each belt and dummy configuration. The dummy configurations included the baseline small female Hybrid III dummy and the small female Hybrid III dummy modified to include the Frangible Abdomen. Note that chest compression in the baseline dummy was measured using a string potentiometer as in the dummy with the Frangible Abdomen.

There were 18 sled tests in total (2 dummy configurations * 3 belt configurations * 3 tests each). The data from these tests was taken using a high speed digital data acquisition system which samples each channel at 10,000 samples per second. The data was stored on a VAX 6220 computer for subsequent analysis.

High speed movies were taken of each sled test for comparison of dummy kinematics. The motion of the head, H-point, and knee was tracked in the sagittal plane of the dummy for each test. In addition, the neck angle and torso angle were determined, for each test, at maximum head excursion and maximum torso excursion, respectively.

KINEMATICS

The kinematics of the dummy with and without the Frangible Abdomen were compared for the tests with each belt configuration. Installation of the Frangible Abdomen did not change the kinematics either qualitatively or quantitatively. A qualitative description of the kinematics follows. The quantitative analysis included comparison of maximum horizontal and vertical excursion of the H-point and head target, torso angle at maximum torso excursion, neck angle at maximum head excursion, time of submarining, time of rearward torso rotation, and time of knee contact with the front structure of the sled buck. Tables summarizing the quantitative data can be found in Appendix III.

NO SUBMARINING - Kinematics of the dummy with the Frangible Abdomen were the same as those of the baseline dummy. The dummy translated forward in an upright position, with a constant torso angle relative to the vertical, until the lap belt caught the pelvis and began to restrain it. As pelvic restraint began, the belt stayed low on the pelvis and the torso rotated forward until shoulder belt restraint began.

The feet of the dummy loaded the two Styrofoam blocks which comprised the sled fixture toe pan, but the knees of the dummy did not contact the front structures of the sled buck when there was no submarining. At maximum head excursion there was a slight twist of the head and torso in a counter

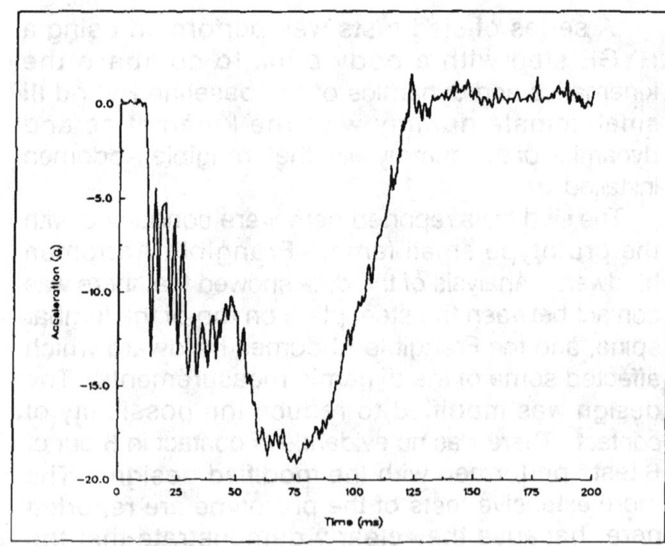

Figure 10. Sled pulse used in the sled testing to validate the small female Frangible Abdomen.

clockwise direction when viewed from above (the right shoulder moved forward relative to the left shoulder as if the dummy was going to rollout of the belt).

It is interesting to note that the side view close-up camera showed the lap belt roping while on the pelvis in one test. In belt roping, the belt curls up and presents a rope-like loading surface instead of the usual flat and wide loading surface.

DYNAMIC SUBMARINING - Kinematics of the dummy with the Frangible Abdomen were the same as those of the baseline dummy. The dummy translated forward in an upright position, with a constant torso angle relative to the vertical. The dummy initially moved forward with no torso rotation. The belt started out in contact with the pelvis beneath the ASIS points. As soon as the belt began to tighten, it started to migrate up the pelvic spine until it snapped over the ASIS points at 85-88ms, in the baseline tests and 84-96ms in the tests with the Frangible Abdomen. Submarining occurred just before knee contact with the front structures of the sled buck, while the dummy was in an upright position. The maximum abdominal penetration appeared to be at ≃112ms when rotation about the lap belt began as evidenced by rearward rotation of the torso and forward translation of the pelvis. There was never any forward rotation of the torso (flexion).

PRE-SUBMARINED - Kinematics of the dummy with the Frangible Abdomen were the same as those of the baseline dummy. The dummy translated forward in an upright position, with a constant torso angle relative to the vertical. At ≃62ms penetration of the belt into the abdomen began. At ≃102ms body began to rotate around the belt while the H-point continued forward (this may have been the time of maximum belt penetration and coincided with a second peak in the belt load which occurred after the initiation of submarining). Knee contact with the front structures of the sled buck occurred at ≃107ms, which also coincides with a maximum in head longitudinal acceleration. Head-to-seat back contact occurred at ≃154ms, and the dummy came to rest on its back looking up at the ceiling.

Belt penetration began in a direction perpendicular to the abdomen, but the dummy rotated around the lap belt so that the final belt angle was about 30° off the z-axis of the dummy's upper torso in an upward direction.

In summary, there were no significant differences between kinematics of the baseline dummy and the kinematics of the dummy with the Frangible Abdomen, for any configuration tested.

DYNAMICS

The dynamic data from the dummy with and without the Frangible Abdomen were compared for the tests with each belt configuration. Some of the responses were different in the dummy with the Frangible Abdomen. The analysis included comparison of head accelerations, neck forces and moments, chest accelerations and displacements, pelvis acceleration and rotation, femur loads, and crush of the Frangible Abdomen. Tables summarizing the dynamic data are presented in Appendix III.

NO SUBMARINING - There were differences between the vertical head acceleration, lateral neck force, and vertical neck force in the baseline dummy when compared with those values for the dummy with the Frangible Abdomen. The magnitudes of those 3 responses were greater in the dummy with the Frangible Abdomen than those in the baseline dummy (significant at the .05 level). There were no other significant differences.

DYNAMIC SUBMARINING - The chest displacement in the baseline dummy was different from that in the dummy with Frangible Abdomen at the .05 level of significance. There were no other significant differences.

PRE-SUBMARINED - There were differences between the longitudinal head acceleration, the lateral head acceleration, HIC, longitudinal neck force, vertical neck force, and vertical pelvis acceleration in the baseline dummy when compared to the dummy with Frangible Abdomen (significant at the .05 level). In all of these measurements except the lateral head acceleration and the vertical pelvis acceleration, the dummy with the Frangible Abdomen had lower values than the baseline dummy. There were no other significant differences.

ASSESSMENT OF THE OCCURRENCE OF SUBMARINING

Photographs of the Frangible Abdomen inserts after the sled experiments are shown in Figure 11. The inserts from the pre-submarined experiments show the most deformation, those from the dynamic submarining experiments show less deformation, and those from the experiments without submarining show the least deformation.

SUBMARINING - Lap belt submarining leaves a clear impression on the frangible abdomen foam insert (Figure 11a,b). The location, area, and depth of the deformation are indicators of the extent of the submarining. For example, one side of the foam may be crushed in case of unilateral (one-sided) submarining. In such a case, a front view of the foam will show a diagonal crush pattern, with the depth of penetration increasing from the non-submarined side of the pelvis to the submarined side. Alternatively, there can be bilateral submarining in which the crush pattern extends from one side of the foam to the other, and the crush is entirely above the anterior-

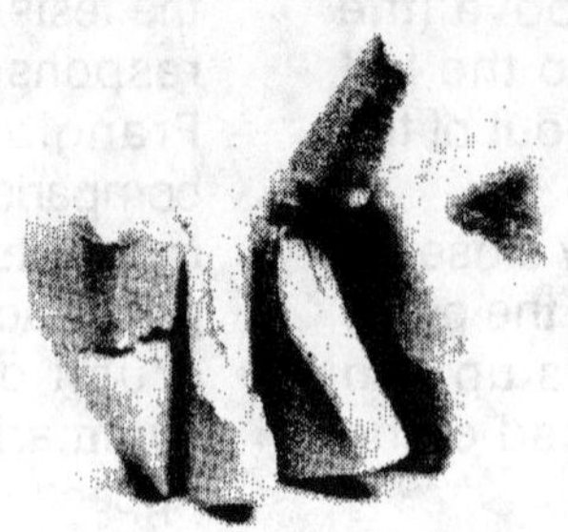

Figure 11a. Oblique and side views of Frangible Abdomen samples after sled testing in which the dummy was **pre-submarined**.

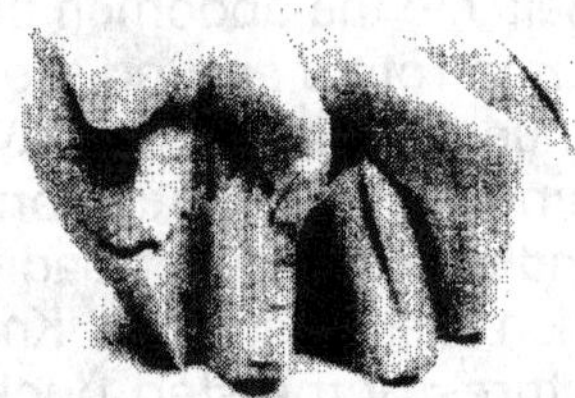

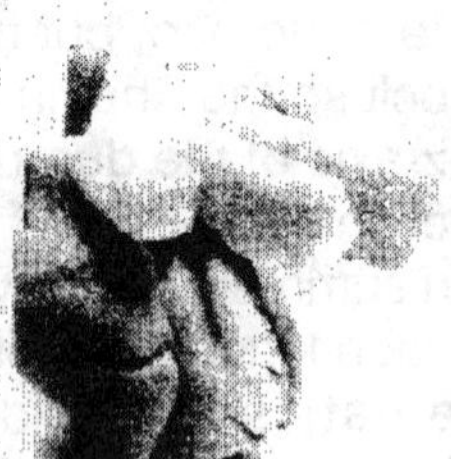

Figure 11b. Oblique and side views of Frangible Abdomen samples after sled testing in which there was **dynamic submarining**.

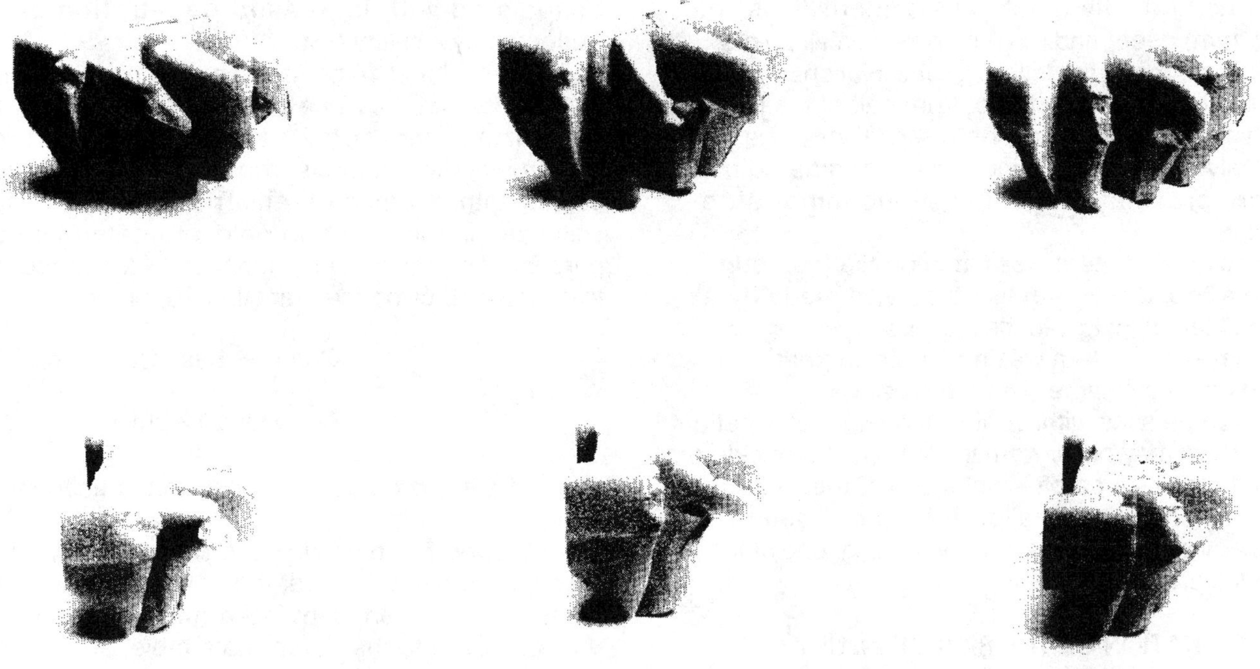

Figure 11c. Oblique and side views of Frangible Abdomen samples after sled testing in which there was **no submarining**.

superior iliac spines of the pelvis. The pattern of deformation can be as wide as the belt, or it can be narrow if the belt "ropes" (curls up) during the submarining event.

NO SUBMARINING - When there is no submarining, there can still be some crush of the foam insert (Figure 11c). The location and character of the crushed foam is the key to making a determination as to the occurrence of submarining. When submarining has not occurred, the crush appears as a 5 cm wide flattened zone along the bottom of the front portion of the foam insert mainly below the anterior-superior iliac spines of the pelvis. This occurs because, as in the human being, the abdomen of the dummy extends beyond the spines of the pelvis (see Figure 12). Then when the lap belt tightens up in a collision, even though it is located correctly on the pelvis, it does compress the abdomen. This compression however, is low on the pelvis, is limited by the pelvic spines, and is probably not a source of injury.

In the tests with no submarining, we observed deformation on the top surface of the foam insert. This deformation was caused by the lower rib of the dummy as the dummy torso flexed forward, and was typically 1/2"(12mm) deep at its maximum, with some deformation on each point. Top surface deformation

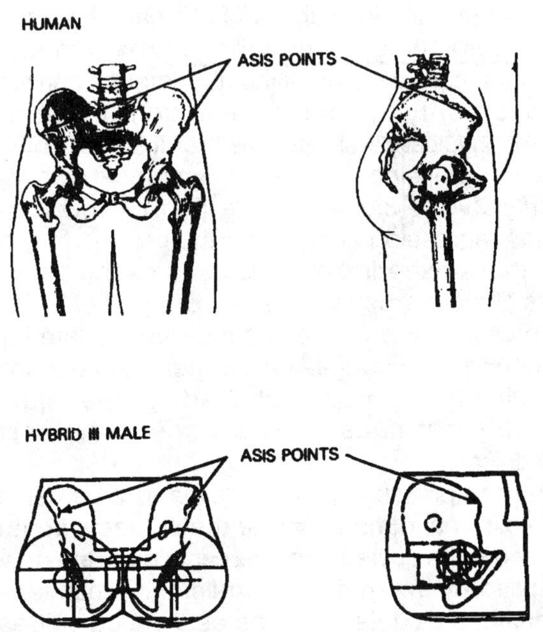

Figure 12. The Anterior Superior Iliac Spines (ASIS points) of the pelvis in relation the the overlying abdominal soft tissue.

was also observed in some submarining situations but was typically much less evident (less than 1/4"(6mm) deep) and did not appear on all points.

We also made the following observations:

- The crush was on the bottom half of the foam mainly below the anterior-superior iliac spines.
- Only the middle point and major side points were crushed (not either of the minor side points).
- The crush pattern was flat in profile (not round like a roped belt, not like a crevice made by a belt edge digging into the foam).
- The crush pattern was horizontal in front view, and about the same size as the belt width.

In some cases without submarining, deformation of the foam may occur from loading by the buckle or shoulder belt. A careful evaluation of the pattern of the crushed foam will allow the user to determine whether the crush is due to submarining, one of these other factors, or both.

DETERMINATION OF THE RISK OF INJURY

It is well known that the stiffness of the human abdomen is dependent on the velocity of loading. Experiments have shown that shoulder belt loading of the thorax is a low velocity event (about 3.3 m/s) [4]. The Frangible Abdomen has been designed to represent the stiffness of the human abdomen in this velocity range.

Experiments performed by Miller using live anesthetized porcine subjects [11], have shown that, in low velocity loading by belts ($v < 5$ m/s), the maximum compression of the abdomen is correlated to the severity of abdominal injury. Miller used several physical parameters as the dose variable in a logistic regression dose-response model with injury severity of AIS ≥ 3 or 4 as the response variable. The logistic regression analysis on the belt loading data from those experiments was used as the basis for tables of foam crush versus risk of injury. Maximum compression was chosen as the appropriate injury criterion for the Frangible Abdomen because in the low velocity regime of belt loading, the injury is probably not caused by a viscous or inertial mechanism.

The compression data was given as a percentage of the external dimensions of the subjects so it did not need to be scaled. The percent compression was multiplied by the mid-sized male and small female dimensions to determine the associated values of crush of the Frangible Abdomen for the central and major side points. These values were corrected for the effects of foam restitution (experiments have shown that the foam springs back slightly after being crushed [10] and the amount of restitution increases linearly as the crush depth increases). The results presented in Tables 1-4, for the mid-sized male and small female, demonstrate the increased risk of injury associated with increased penetration of the abdomen by a safety belt.

Figures 13 and 14 present the data on foam crush versus risk of injury in a different format for easier use. The anthropometric data for the external abdominal dimension is from the UMTRI study on seated anthropometry [14]. The abdominal depth was used in the equation below to determine the percent compression as a function of the amount of foam crushed using the Frangible Abdomen.:

$$\text{Compression} = \frac{\text{Crush} + \text{Rest. Corr. Fact.}}{\text{Abdominal Depth}}$$

where, Rest. Corr. Fact. = correction to account for restitution of the foam.

In theory, to determine the risk of injury when submarining has occurred, the depth of crush of the Frangible Abdomen foam insert must be measured and compared to the appropriate table. In practice, we used outside calipers to measure the depth of foam remaining (the depth of foam that has not been crushed), called the "remaining thickness", given in the first column of the tables. The remaining thickness is converted to the amount of foam crushed and the corresponding compression in the second and third columns, respectively.

The measurement of crush was made along the centerline of the point being measured (not perpendicular to the back of the foam insert) because it was assumed that the belt wraps around the circumference of the abdomen as submarining occurs. Then, the force from the belt would be directed radially in toward the spine.

The amount of crush was measured for the central point and both major side points, and the risk of injury was determined for each. The reason for the different tables is that although the risk of injury depends only on the amount of crush, the remaining thickness of foam measured is a function of the length of the points which is different for the central point compared to the side points.

The probability of AIS ≥3 or AIS ≥ 4 abdominal injury as determined from logistic regression of the Miller data is listed in the last two columns. For example, if the left major side point of a small female dummy has been crushed 3.0", the central point 3.5", and the right major side point 4.0", the probability of AIS ≥ 4 injury would be 18%, 49%, and 78%, respectively (from Tables 1&2). The probability of abdominal injury in the given sled test would then be given by the maximum of these three values (78%).

The risk of occupant injury was measured from each Frangible Abdomen used in these sled experiments and is given in Table 5. In this Table, only the maximum risk of injury is given. That is, the

TABLE 1
TABLE OF FRANGIBLE ABDOMEN CRUSH VERSUS PROBABILITY OF INJURY
Seated Anthropometry
Small Female Hybrid III Dummy
Central Point

REMAINING THICKNESS (") *	AMOUNT OF CRUSH (") **	CORRESPONDING COMPRESSION (%) ***	PROBABILITY OF SERIOUS ABDOMINAL INJURY (AIS ≥ 3) (%)	PROBABILITY OF SEVERE ABDOMINAL INJURY (AIS ≥ 4) (%)
4.75	0	0.0	0.6	0.00
4.50	0.25	5.1	1.0	0.01
4.25	0.50	8.2	1.4	0.02
4.00	0.75	11	1.9	0.03
3.75	1	14	2.6	0.06
3.50	1.25	17	3.5	0.10
3.25	1.50	20	4.8	0.18
3.00	1.75	24	7.2	0.40
2.75	2	27	9.6	0.64
2.50	2.25	30	13	1.1
2.25	2.50	33	17	1.9
2.00	2.75	36	22	3.3
1.75	3	40	30	6.7
1.50	3.25	43	37	11
1.25	3.50	47	48	21
1.00	3.75	50	56	31
0.75	4	54	66	49
0.25	4.50	60	79	74

TABLE 2
TABLE OF FRANGIBLE ABDOMEN CRUSH VERSUS PROBABILITY OF INJURY
Seated Anthropometry
Small Female Hybrid III Dummy
Major Side Points

REMAINING THICKNESS (") *	AMOUNT OF CRUSH (") **	CORRESPONDING COMPRESSION (%) ***	PROBABILITY OF SERIOUS ABDOMINAL INJURY (AIS ≥ 3) (%)	PROBABILITY OF SEVERE ABDOMINAL INJURY (AIS ≥ 4) (%)
4.00	0	0.0	0.6	0.00
3.75	0.25	5.1	1.0	0.01
3.50	0.50	8.2	1.4	0.02
3.25	0.75	11	1.9	0.03
3.00	1	14	2.6	0.06
2.75	1.25	17	3.5	0.10
2.50	1.50	20	4.8	0.18
2.25	1.75	24	7.2	0.40
2.00	2	27	9.6	0.64
1.75	2.25	30	13	1.1
1.50	2.50	33	17	1.9
1.25	2.75	36	22	3.3
1.00	3	40	30	6.7
0.75	3.25	43	37	11
0.50	3.50	47	48	21
0.25	3.75	50	56	31
0.00	4	54	66	49

* Measure with "outside calipers"

** CRUSH = Original Length of Point - Measured Thickness Remaining

$$\text{*** Corresponding Compression} = \frac{\text{Crush + Restitution Correction Factor}}{\text{Small Female Waist Depth****}}$$

**** Small Female Waist Depth = 8.1"(179 mm); Reference [9]

<u>Restitution Correction Factors</u>

Crush Depth (in)	0	1	2	2.5	3.0	3.5	4.0
Restitution (in)	0	.16	.16	.20	.25	.32	.35

TABLE 3
TABLE OF FRANGIBLE ABDOMEN CRUSH VERSUS PROBABILITY OF INJURY
Seated Anthropometry
Mid-sized Male Hybrid III Dummy
Central Point

REMAINING THICKNESS (") *	AMOUNT OF CRUSH (") **	CORRESPONDING COMPRESSION (%) ***	PROBABILITY OF SERIOUS ABDOMINAL INJURY (AIS ≥ 3) (%)	PROBABILITY OF SEVERE ABDOMINAL INJURY (AIS ≥ 4) (%)
4.75	0	0	0.58	0.00
4.50	0.25	4	0.89	0.01
4.25	0.50	7	1.2	0.02
4.00	0.75	9	1.5	0.02
3.75	1	11	2.0	0.04
3.50	1.25	14	2.5	0.06
3.25	1.50	16	3.3	0.09
3.00	1.75	19	4.3	0.14
2.75	2	21	5.5	0.23
2.50	2.25	24	7.1	0.36
2.25	2.50	27	9.4	0.61
2.00	2.75	29	12	0.96
1.75	3	32	16	1.7
1.50	3.25	35	20	2.6
1.25	3.50	38	25	4.6
1.00	3.75	40	31	7.1
0.75	4	43	38	11
0.25	4.50	48	51	24

TABLE 4
TABLE OF FRANGIBLE ABDOMEN CRUSH VERSUS PROBABILITY OF INJURY
Seated Anthropometry
Mid-sized Male Hybrid III Dummy
Major Side Points

REMAINING THICKNESS (") *	AMOUNT OF CRUSH (") **	CORRESPONDING COMPRESSION (%) ***	PROBABILITY OF SERIOUS ABDOMINAL INJURY (AIS ≥ 3) (%)	PROBABILITY OF SEVERE ABDOMINAL INJURY (AIS ≥ 4) (%)
4.5	0	0	0.58	0.00
4.25	0.25	4	0.89	0.01
4.0	0.50	7	1.2	0.02
3.75	0.75	9	1.5	0.02
3.50	1	11	2.0	0.04
3.25	1.25	14	2.5	0.06
3.0	1.50	16	3.3	0.09
2.75	1.75	19	4.3	0.14
2.50	2	21	5.5	0.23
2.25	2.25	24	7.1	0.36
2.00	2.50	27	9.4	0.61
1.75	2.75	29	12	0.96
1.5	3	32	16	1.7
1.25	3.25	35	20	2.6
1.00	3.50	38	25	4.6
0.75	3.75	40	31	7.1
0.50	4	43	38	11
0.00	4.50	48	51	24

* Measured Remaining Thickness is measured with "outside calipers"

** CRUSH = Length of Original Point - Measured Remaining Thickness

$$\text{*** Corresponding Compression} = \frac{\text{Crush + Restitution Correction Factor}}{\text{Mid-sized Male Waist Depth****}}$$

**** Mid-sized Male Waist Depth = 10.1"(203 mm); Reference [9]

<u>Restitution Correction Factors</u>

Crush Depth (in)	0	1	2	2.5	3.0	3.5	4.0
Restitution (in)	0	.16	.16	.20	.25	.32	.35

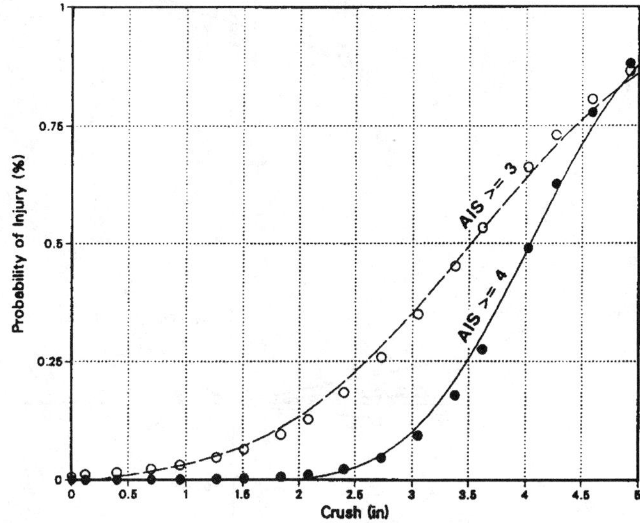

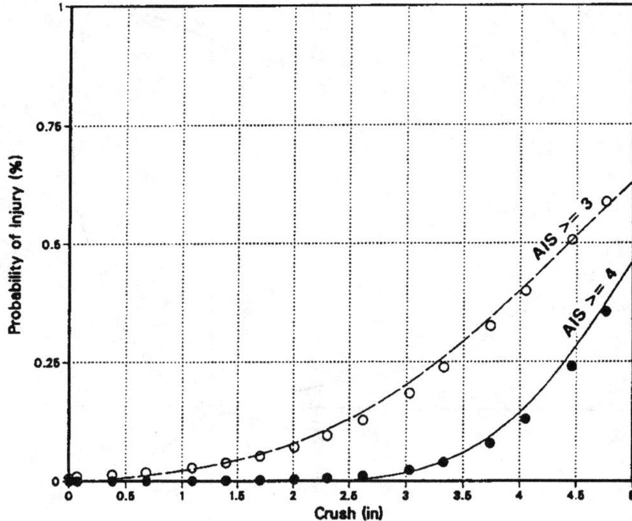

Figure 13. Probability of injury versus crush of the Frangible Abdomen for the central point of the small female Hybrid III dummy. Data based on UMTRI anthropometry.

Figure 14. Probability of injury versus crush of the Frangible Abdomen for the central point of the mid-sized male Hybrid III dummy. Data based on UMTRI anthropometry.

TABLE 5
RISK OF INJURY
AS DETERMINED BY THE FRANGIBLE ABDOMEN

TEST CONDITION	RISK OF INJURY
DYNAMIC SUBMARINING	63%
	56%
	63%
Mean Value ± s.d.	60.7 ± 4.0% (n=3)
PRE-SUBMARINED	59%
	82%
	78%
Mean Value ± s.d.	73.0 ± 12.3% (n=3)
Mean Value ± s.d. (without 1681)	80.0 ± 2.8% (n=2)

amount of crush of the 3 points on each insert was measured and converted to a risk of injury. The maximum risk was then listed.

DISCUSSION

KINEMATIC ANALYSIS - Figures 15a)-c) show graphically, to scale, the trend in H-point and head, maximum longitudinal and vertical excursion for the no submarining, dynamic submarining, and pre-submarined experiments. In these figures, the initial position of the head was placed at the head c.g., for simplicity. The straight lines which connect the initial and final positions of the head c.g. and H-point were drawn for ease of viewing and do not represent the trajectories of those targets.

As the degree of submarining increases, the forward and upward motion of the h-point increases, but the forward and downward motion of the head, which is typical of non-submarining kinematics, decreases (i.e., the upper body does not move as far forward). These results agree with intuition and experience, and both the baseline small female Hybrid III and the dummy with the Frangible Abdomen exhibit the same pattern.

In fact, there were no statistically significant differences in kinematic measures between the baseline dummy and the dummy with Frangible Abdomen in any of the experiments. Although not significant at the .05 level, in the pre-submarined experiments, there was near significance to the

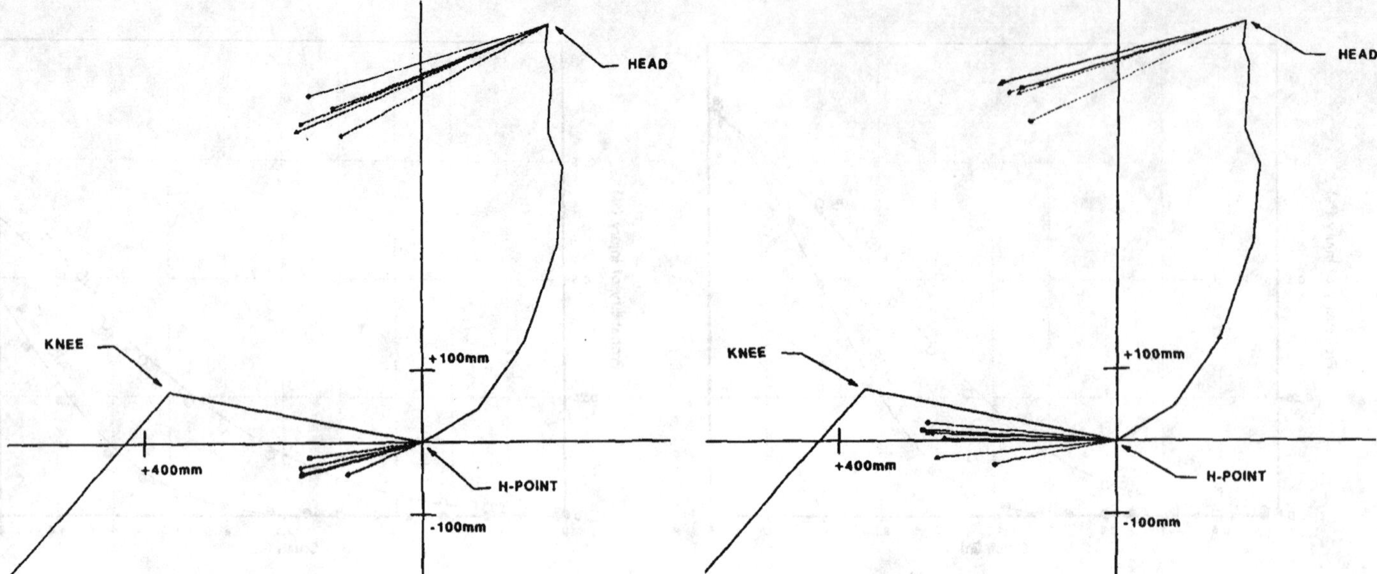

Figure 15a. Maximum head and H-point travel for the small female dummy with the Frangible Abdomen (solid lines) and without the Frangible Abdomen (dashed lines) when there was **no submarining**. (Note, the lines are not the actual trajectories, but were drawn to facilitate comparison.)

Figure 15b. Maximum head and H-point travel for the small female dummy with the Frangible Abdomen (solid lines) and without the Frangible Abdomen (dashed lines) when there was **dynamic submarining**. (Note, the lines are not the actual trajectories, but were drawn to facilitate comparison.)

difference in torso angles, with that for the dummy with the Frangible Abdomen being lower than that for the baseline dummy. A similar relationship was shown in the Frangible Abdomen tests for the 50th percentile Hybrid III dummy [10], which may reflect a difference in the interaction of the lap belt with the abdomen.

DYNAMIC ANALYSIS - There were some differences in the dynamic responses between the baseline dummy and the dummy with the Frangible Abdomen. The differences in the no submarining cases were most likely due to contact between the steel plate on top of the lumbar spine and foam reaction plate and support bracket. The results in this report are for the original reaction plate design. When the lumbar spine contact was discovered the reaction plate was modified as shown in Figure 16. Recent tests with the new reaction plate have not shown any evidence of contact which should minimize or eliminate the few differences in the dynamic data.

Most of the dynamic responses were unchanged in the other test configurations.
The differences that were observed were probably due to changes in the way that the belt interacts with the abdomen which should be more realistic with the Frangible Abdomen since it has biofidelity.

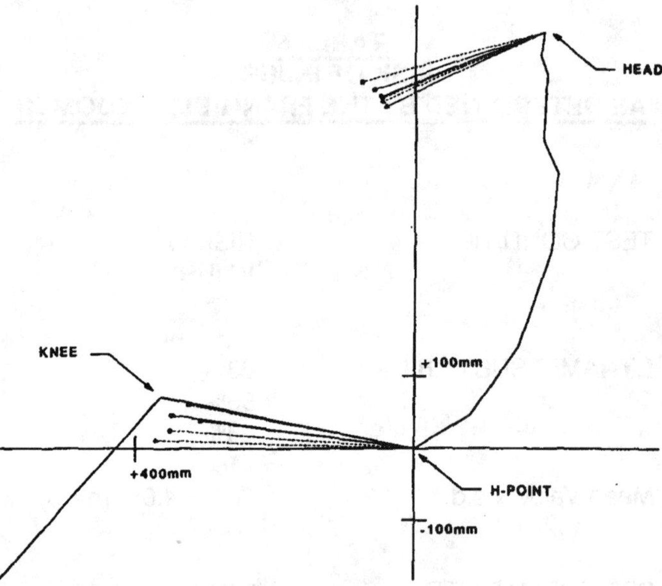

Figure 15c. Maximum head and H-point travel for the small female dummy with the Frangible Abdomen (solid lines) and without the Frangible Abdomen (dashed lines) when the dummy was **pre-submarined**. (Note, the lines are not the actual trajectories, but were drawn to facilitate comparison.)

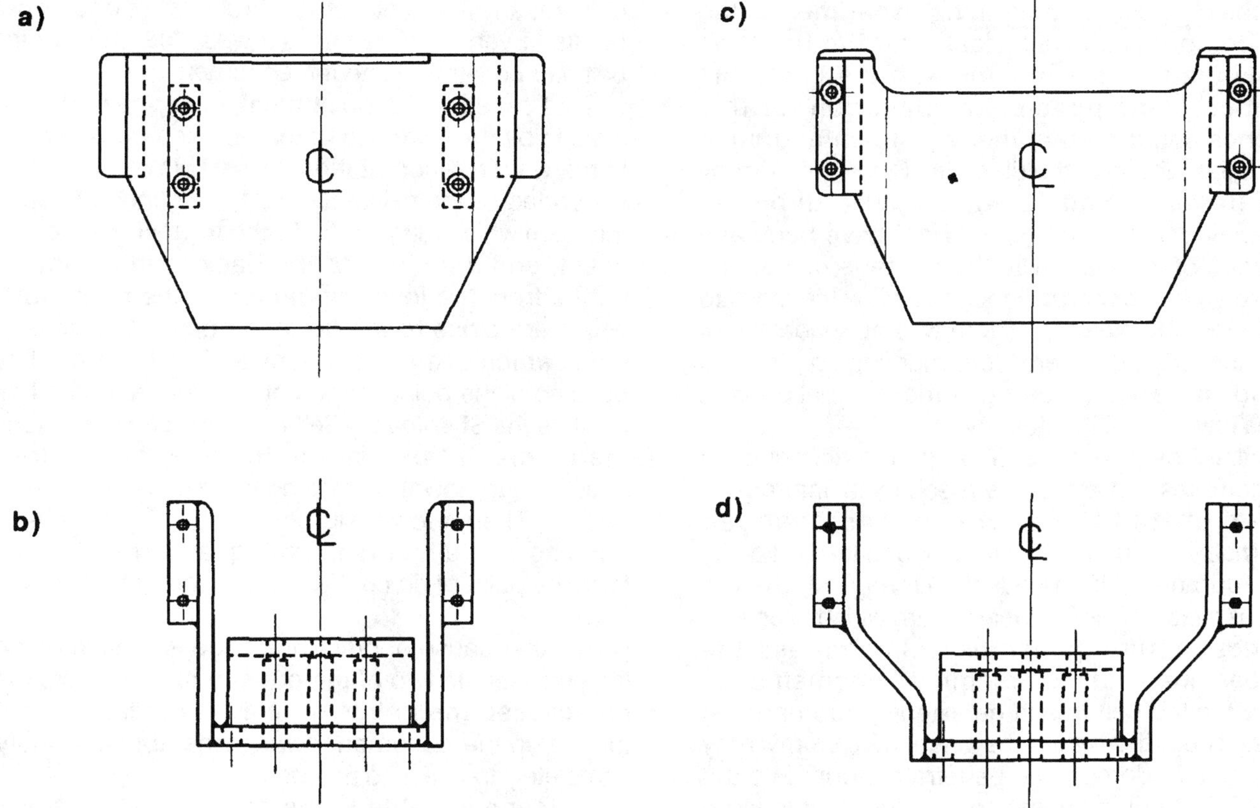

Figure 16. Reaction plate and support bracket used in the validation sled tests (a & b), compared with the reaction plate and support bracket after modifications to minimize contact with the lumbar spine (c & d).

For the no submarining case, there were differences in the vertical head accelerations, and in the lateral and vertical neck forces between the baseline dummy and the dummy with the Frangible Abdomen which were significant at the .05 level. In addition, while not significant at the .05 level, there were differences in HIC values between the baseline dummy and dummy with Frangible Abdomen and between tests with the same dummy (Appendix III). We believe the differences are the result of contact by the steel plate on top of the lumbar spine with the Frangible Abdomen reaction plate. The contact was evidenced after a number of tests were run as a dent in the ledge at the top of the reaction plate (Figure 16a) which keeps the foam from being pushed up into the dummy's chest. In addition to the physical marking, there were significant spikes in the vertical head acceleration, the vertical neck force, all three chest accelerations, and all three pelvis acceleration signals. The acceleration spikes were greatest in the vertical direction.

These observations are consistent with the theory that the lumbar spine was making contact with the top of the reaction plate. We did not expect this contact in the dynamic submarining or in the pre-submarined experiments because the torso does not flex (rotate forward) in those experiments. The pelvis may rotate backwards far enough to cause contact, but the forces involved in that interaction should be much lower than those involved when the upper torso of the dummy is abruptly stopped by the contact.

Modifications were made to the reaction plate and support bracket to eliminate or reduce this contact (Figure 16). In particular, the height of the reaction plate was reduced in the region of possible contact by 15.9mm (5/8"), and the ledge on the reaction plate was replaced by two tabs (Figure 16b), placed as far apart as possible to reduce the possibility of contact with the thoracic/lumbar junction. In addition, the wings of the support bracket were widened to clear the area between the lumbar spine and reaction plate of any possible obstructions (Figure 16c,d). The drawings in Appendix I (Figure I.2) are prints of the reaction plate and support bracket after the modifications.

Recent tests with the modified hardware has shown no further evidence of contact which should reduce or eliminate the differences in the dynamic data. As already discussed, the kinematics, and in particular, the submarining performance of the small female dummy were not affected by the addition of the Frangible Abdomen.

In the dynamic submarining experiments, the chest compression was greater for the baseline dummy than for the dummy with the Frangible Abdomen. One possible explanation for this difference could be that the Styrofoam abdomen restricts the motion of the rib cage. Since the dummy torso never flexed forward in the dynamic submarining tests, there should not have been any interaction of the foam with the rib cage unless the pelvis rotated backwards enough to allow the foam to get into the chest cavity. There was no evidence of that on the foam, but there were markings on the top of the foam inserts indicating some minimal contact with the lower rib of the dummy.

Another possible explanation for the difference in chest compression is that the mechanical interaction of the belt with the abdomen has changed (presumably it is more realistic owing to the biofidelity of the abdomen). In the baseline Hybrid III, the gut sack was made of vinyl covered foam which probably does not capture the lap belt. In contrast, the Styrofoam insert of the Frangible Abdomen does capture the lap belt to a large extent if submarining occurs. Then, the belt in the baseline dummy may slip over the abdomen and penetrate further into the dummy. That could cause the shoulder belt to wrap around the rib cage in a much different way than in the dummy with the Frangible Abdomen, and may be the cause of the increased compression. Chest compression was the only parameter, in the dynamic submarining experiments, which exhibited a significant difference between the baseline dummy and the dummy with a Frangible Abdomen.

In the pre-submarined configuration, the longitudinal head acceleration and neck force, vertical neck force, and HIC all decreased in the dummy with the Frangible Abdomen. One possible explanation for the decrease in these parameters is again related to the difference in the composition of the abdominal elements of the dummies. In the baseline dummy, if the lap belt were to slip over the top of the vinyl covered abdomen, it would move unimpeded until it struck the lumbar or thoracic spine, at which point it would abruptly apply restraining force to the dummy. In comparison, the dummy with the Frangible Abdomen tends to catch the lap belt as the foam insert crushes during submarining. Then, the restraining force in the dummy with the Frangible Abdomen, would be applied more gradually, possibly leading to lower accelerations in the longitudinal direction.

The reasons for the increased lateral head acceleration and vertical pelvis acceleration in the dummy with the Frangible Abdomen are not clear.

FRANGIBLE ABDOMEN REPEATABILITY AND USABILITY - The Frangible Abdomen was simple to install and remove in the dummy, and did not affect turn around time for testing. In some tests, a section of foam, above the belt crush, broke off of one or two points. Even in those cases it was clear how far the belt had penetrated into the abdomen.

One cause of the point breakage appeared to be loading by the lower rib of the rib cage possibly due to rearward rotation of the pelvis. This hypothesis is supported by the existence of: 1) tear lines on top of the foam which run parallel to the front surface of the chest, and are close to the back of the point; 2) marks from the lower rib on top of the foam point which are close to the tip of the point, 3) points of foam which are bent downward at the tip. The bending of the points apparently occurs after the belt crushes the Styrofoam. Belt crush probably weakens the points of the insert in the inferior/superior direction (up/down) by making them thin cantilevered beams. Then if lower rib contact occurs it causes bending with a moment arm equal to the distance from the point of rib contact to the point of maximum crush.

Another cause of point breakage is motion of the lap belt itself. In two of the pre-submarined tests the belt caused the points to tear off the insert as it moved up the abdomen towards the ribs, ultimately completely transecting the foam.

As seen in Table 5, the risk of injury in the 3 dynamic submarining experiments were very similar (± 7%). The risk of injury in the pre-submarined experiments had more variability (± 17%). The reason for the increased variability in the pre-submarined experiments unknown.

MEASUREMENT OF THE AMOUNT OF CRUSH - The amount of crush is measured along the centerline of the central or major side points. The percent compression has been calculated assuming that the distance from the tip of the point to the back of the dummy is a constant for a given dummy size and choice of data base. Because the points are of different lengths, the crush must be measured at each point and converted to probability of injury before making a judgement as to the maximum injury potential. It is incorrect to assume that the point which has crushed the closest to the back of the foam insert has the maximum risk of injury. As seen in Tables 1-4, the major side points have a lower risk of injury for the same amount of foam remaining uncrushed.

Miller's data was obtained with a relatively flat loading by the belt because of the short length of belt used and the yoke loading arrangement. During submarining in a vehicle, the belt will probably tend to wrap around the occupant resulting in loading from the sides as well as from the front. It is not clear what effect this might have on the resulting injury.

ANALYSIS METHOD - Using the injury risk curves in Figures 13 and 14, one can objectively compare different restraint systems using the Frangible Abdomen. If submarining occurs for a restraint system under development, the amount of crush recorded from the Frangible Abdomen could be plotted right on a copy of the appropriate injury risk curve (Figure 17). Then after a design change and sled test, if submarining has recurred, the amount of crush can be recorded on the same graph. In this way, the Frangible Abdomen can give design direction, by graphically demonstrating the improvement or degradation in performance attributable to a specific change, and by showing the magnitude of the difference. The user should consider the minimum error band associated with the injury risk to be the risk corresponding to the recorded crush value ± 0.25" (6.4mm), based on the range of values mentioned above.

INDETERMINATE RISK OF INJURY - We have observed one case, in the sled testing with the Frangible Abdomen, in which the submarining of the dummy was so severe that the belt ended up above the foam insert and hooked on the lower margin of the rib cage. As the dummy slipped under the belt there was significant deformation of the foam, but instead of a concentrated crush zone, a wavy fracture pattern appeared.

Cases like this in which the dummy slips underneath the belt with a relative motion that is along the abdomen instead of into the abdomen, may result in an indeterminate risk of injury, but will still provide an indication of belt/abdomen interaction. Film analysis of the dummy kinematics, where available, can be used to supplement the Frangible Abdomen results in this situation.

HANDLING - While not fragile, the Frangible Abdomen can be deformed during dummy placement in a sled fixture or barrier vehicle if precautions are not taken. Therefore, technical support personnel need to know that the purpose of the Frangible Abdomen is to measure crush of the foam. They should be cautioned to install the Frangible Abdomen in the dummy after it is placed in the test vehicle, and not to apply force to the abdominal area when handling the dummy to get the H-point in its proper location.

APPLICABILITY - The Frangible Abdomen is a frontal impact test device, whose side impact properties are uncalibrated. It was designed for frontal impact tests involving a 3-point belt.

Although it may also be applicable to lap belt-only systems, the Frangible Abdomen has not been tested with that belt configuration. The reaction plate contact in torso hyperflexion could be a major consideration in lap belt-only tests, because of the unrestricted motion of the upper torso. The reaction plate could possibly be modified to accommodate such testing, however, with a lap belt-only, the torso hyperflexion could result in shearing off of the front portion of the points. If submarining occurs it may or may not occur before interaction of the rib cage with the foam insert.

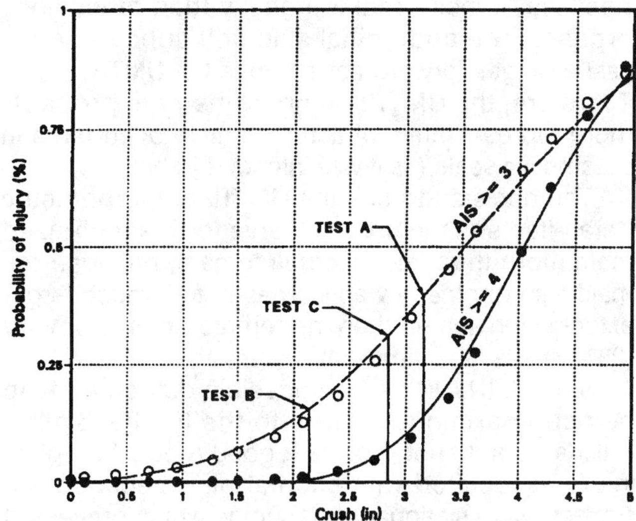

Figure 17. One method of comparing restraint system designs using the Frangible Abdomen.

The force-deflection corridor of the Frangible Abdomen represents the results of controlled low speed loading with a lap belt applied to anesthetized swine [11]. Since the human abdomen is rate sensitive, its stiffness increases in higher speed impacts such as those encountered by an unbelted occupant. However, since the Styrofoam Frangible Abdomen is not rate sensitive, its force-deflection response will not be stiff enough in unbelted dummy tests, and the resulting amount of crush will be an overestimate of what a human occupant would experience. Although the Frangible Abdomen would not be able to provide a precise risk of injury in those tests, it would still be useful as an indicator of contact by vehicle components and of the load distribution during that contact.

INJURY RISK CURVES & ANTHROPOMETRY - The choice of which anthropometric data to use can affect the injury risk curves because the injury risk may be a function of the posture of the occupant. The subjects in the experiments performed by Miller were tested in dorsal recumbency (on their backs) with the torso extended (in the anatomic sense). Then, the NASA data base discussed previously is a good representation of the external geometry of the subjects in these experiments, but possible differences in the position of the internal organs when comparing standing and supine subjects may reduce the quality of the representation.

Similarly, typical vehicle occupants are neither standing, nor supine, but are seated. A number of changes occur in abdominal anthropometry while seated as compared with standing. When seated there is a relative sagging of the abdominal organs because of their greater mobility than other body organs. The abdominal skin and subcutaneous tissue bulges forward as shown in the UMTRI study. Therefore, the UMTRI anthropometry is probably more representative of actual vehicle occupants in automotive seats (as it was intended to be).

The decision to use the UMTRI anthropometric data affects the injury risk curve for the mid-sized male more than for the small female, because the male anthropometry was increased to a much larger extent when seated than the female anthropometry (28% versus 14% increase).

SPINAL INJURY - One possible outcome of lap belt submarining is injury to the lumbar spine. Miller's tests do not provide a good model for spinal injury associated with submarining because the subjects were stationary and supine which prevented torso hyperflexion. Leung, et al. [7] have shown that spinal injury can occur with or without associated abdominal injury. Therefore, the risk determined by the Frangible Abdomen is for abdominal organs and does not account for injury to the spine.

BELT "ROPING" - Lap belt "roping" occurred in the sled tests, but probably not in Miller's tests in which the belt length was short in comparison to a standard belt, and the attachment to the yoke prevented the ends from roping. It is not clear what effect this might have on the test results. Since a roped belt presents a much smaller cross-sectional area to the abdomen, it may have penetrated more easily due to higher pressure (same force over smaller area). This is dependent upon when the belt actually does rope. Roping may occur as the belt penetrates or before penetration begins. In addition, a roped belt may result in injuries, tolerances, or both injuries and tolerances that are different from those due to submarining with a flat belt. In either case, the Frangible Abdomen should be more compliant if the belt ropes, and will therefore present a more conservative estimate of injury risk.

UNILATERAL SUBMARINING - When submarining occurs only on one side of the dummy, it is possible that the risk of injury is less than if the submarining was across the entire abdomen. This is because the mechanism of injury in submarining may involve pinching or crushing of the abdominal organs between the lap belt and lumbar spine with a resultant increase in intra-abdominal pressure. Since submarining is a low velocity event, it is unlikely that inertial or viscous mechanisms predominate. In unilateral submarining, the organs may be pushed backwards, but may not be pinched or crushed against the lumbar spine with the same amount of force as in bilateral submarining since the lumbar spine is along the midline of the body, and the load is being shared by one iliac spine (a pelvic "wing"). Alternatively, since the belt is in tension and it is around the circumference of the dummy abdomen, there will be components of force directed toward the lumbar spine at each point of belt contact on the submarined side.

Since we do not currently have any biomechanical data which examines this question, a conservative approach would be to assume that the risk of injury when submarining occurs on one side is the same as in the case of submarining across the entire abdomen, i.e., determined by the maximum crush of any point.

SUMMARY

The Frangible Abdomen is a Styrofoam insert for the abdominal region of the Hybrid III family of dummies which has biofidelity, and assesses the occurrence of submarining and its risk of injury. It is easy to install and use, and provides an objective assessment of belt restraint submarining performance. It was first developed for the mid-sized male Hybrid III dummy. This paper describes the design of the Frangible Abdomen for the small female Hybrid III dummy, and how to use it to assess the risk of injury from submarining.

The force-deflection properties of the mid-sized male insert were scaled to the small female dimension. Sled tests were run to compare the kinematic and dynamic performance of the baseline small female Hybrid III dummy with the same dummy modified to incorporate the Frangible Abdomen.

No significant kinematic differences exist, either qualitatively or quantitatively, between the baseline dummy and the dummy with the Frangible Abdomen.

The dynamic responses of the dummy with the prototype Frangible Abdomen differed in some cases from those of the baseline dummy. The differences were traced to contact between the steel plate on top of the lumbar spine and the steel reaction plate and support bracket for the styrofoam insert. The Frangible Abdomen hardware as specified in the mechanical drawings in Appendix I is a design which was modified to reduce possible interferences. The modifications made to the reaction plate and support bracket should make the dynamic performance of the dummy with the Frangible Abdomen much closer to the baseline dummy since further limited testing has shown the elimination of contact. More importantly, the kinematic and submarining performance of the small female Hybrid III dummy is unchanged by the addition of the Frangible Abdomen.

Injury assessment with the Frangible Abdomen is achieved by comparing the depth of foam deformation to tables provided which relate the risk of injury to the amount of crush.

ACKNOWLEDGEMENTS

The authors would like to thank Gerald Horn, Timothy Sorenson, Todd Townsend, and Frank Wood for technical assistance during these tests. In addition, we appreciate the many technical discussions with Clyde C. Culver (now deceased), whose advice and counsel are sorely missed, and we thank Joseph Balser, John D. Horsch, Michael Marshall, John W. Melvin, and David C. Viano for many fruitful technical discussions and for their editorial contributions.

REFERENCES

1. Rouhana, S.W., Horsch, J.D., and Kroell, C.K., "Assessment of Lap-Shoulder Belt Restraint Performance in Laboratory Testing", In Proceedings of the 33rd Stapp Car Crash Conference, pp. 243-256, SAE Technical Paper, Society of Automotive Engineers, #892439, Warrendale, PA, 1989.
2. Daniel, R.F., "Test Dummy Submarining Indicator System." United States Patent 3,841,163, October, 15, 1974.
3. Tarriere, C.H., "Proposal for a Protection Criterion as Regards Abdominal Internal Organs." In Proceedings of the 17th American Association for Automotive Medicine Conference, pp. 371-382, 1973.
4. Leung, Y.C., Tarriere, C., Fayon, A. et al, "A Comparison Between Part 572 Dummy and Human Subject in the Problem of Submarining." In Proceedings of the 23rd Stapp Car Crash Conference, pp. 677-720, SAE Technical Paper #791026, Society of Automotive Engineers, Warrendale, PA, 1979.
5. Melvin, J.W. and Weber, K., "Abdominal Intrusion Sensor for Evaluating Child Restraint Systems.", In Passenger Comfort, Convenience and Safety: Test Tools and Procedures, P-174, SAE Technical Paper #860370, Society of Automotive Engineers, Warrendale, PA, 1986.
6. DeJeammes, M., Biard, R., and Derrien, Y., "Factors Influencing the Estimation of Submarining on the Dummy." In Proceedings of the 25th Stapp Car Crash Conference, pp. 733-762, SAE Technical Paper #811021, Society of Automotive Engineers, Warrendale, PA, 1981.
7. Leung, Y.C., Tarriere, C., Lestrelin, D., et al. "Submarining Injuries of 3 Pt. Belted Occupants in Frontal Collisions - Description, Mechanisms and Protection." In Proceedings of the 26th Stapp Car Crash Conference, pp. 173-205, SAE Technical Paper #821158, Society of Automotive Engineers, Warrendale, PA, 1982.
8. Mooney, M.T. and Collins, J.A., "Abdominal Penetration Measurement Insert for the Hybrid III Dummy." SAE Technical Paper #860653, Society of Automotive Engineers, Warrendale, PA, 1986.
9. Biard, R., Cesari, D. and Derrien, Y., "Advisability and Reliability of Submarining Detection." In Restraint Technologies: Rear Seat Occupant Protection SP-691, pp. 27-38, SAE Technical Paper #870484, Society of Automotive Engineers, Warrendale, PA, 1987.
10. Rouhana, S.W., Viano, D.C., Jedrzejczak, E.A., and McCleary, J.D., "Assessing Submarining and Abdominal Injury Risk in the Hybrid III Family of Dummies", In Proceedings of the 33rd Stapp Car Crash Conference, pp. 257-280, SAE Technical Paper #811015, Society of Automotive Engineers, Warrendale, PA, 1989.
11. Miller, M.A., "The Biomechanical Response of the Lower Abdomen to Belt Restraint Loading", J. Trauma, Vol. 29(11), 1989.
12. Eppinger, R.H., "Prediction of Thoracic Injury Using Measurable Experimental Parameters.", In Proceedings of the 6th International Conference on Experimental Safety Vehicles, pp. 770-780, NHTSA, 1976.
13. Langhaar, H.L., Dimensional Analysis and Theory of Models, Wiley, New York, 1957.
14. Schneider, L.W., Robbins, D.H., Pflug, M.A., Snyder, R.G., "Development of Anthropometrically Based Design Specifications for an Advanced Adult Anthropomorphic Dummy Family", Final Report, DTNH22-80-C-07502, NHTSA, 1983.
15. "Anthropometric Source Book Volume II: A Handbook of Anthropometric Data", NASA Reference Publication 1024, pg. 94, 1978.

Measurement of submarining on Hybrid III 50° & 5° Percentile Dummies

Jérôme Uriot, Martin Page
Institut de Recherches Biomécaniques et Accidentologiques
Claude Tarrière
Département Biomédical de l'Automobile de RENAULT S.A.
Farid Bendjellal
Département d'étude et d'essais de sécurité de RENAULT S.A.
Frédéric Loiseleux
Institut de Recherches Anatomo-Chirurgicales et de Biomécanique Appliquée
Mick Saloum
First Technology Safety Systems
Philippe Fournier
P.S.A. Peugeot Citroën

Paper N° 94 S1 W 23

ABSTRACT

Instrumentation for the study of submarining on the Hybrid II test dummy has already been described in a previous paper. The instrumentation consisted basically of two 2D force transducers installed in the dummy's pelvis. This system has been evaluated, improved and installed on the 50th percentile Hybrid III dummy. The aim of this paper is to describe the general concept arrived at through this study, namely:

- Detection of movement of the lap-belt via a simple system installed in the iliac wing of the pelvis.
- Measurement of the amplitude of submarining via a 2D force transducer installed back from the iliac crests.

This work, carried out in conjunction with FTSS, required changes on the levels of the spinal/lumbar column connection and measurement of thoracic deflection. The assembly was tested in various impact configurations (subsystem and sled configuration tests). The corresponding results show that the system detects submarining well and does not affect the overall response of the instrumented dummy by comparison with a standard dummy.

The new system detecting seat-belt slippage was also installed on the 5th percentile Hybrid III dummy. Sled tests gave reliable recordings of the detection of submarining.

From the results available, it seems that the two versions presented (50th and 5th percentile) enable improved study of submarining and accordingly the optimization of occupant restraining systems.

INTRODUCTION

Instrumentation for the study of submarining has been developed at the Laboratory of Accidentology and Biomechanics P.S.A. Peugeot Citroën/RENAULT in conjunction with First Technology Safety Systems for the 50°P Hybrid II dummy. This instrumentation consists of two sensors measuring seat belt forces on the abdomen during submarining along two axes (anatomical X and Z). It is supplemented by gauges bonded onto the aluminium structure of the pelvis at the level of the iliac wings. These gauges detect the moment at which the seat belt passes from the iliac wings onto the abdomen. The benefit of equipping dummies with instrumentation of this type is to be able to study submarining precisely, in terms of duration extent and time of occurence, which is practically impossible with standard measurements or a film analysis.

The following study describes the adaptation and improvement of this instrumentation on the 50°P and 5°P Hybrid III dummies.

TECHNICAL DESCRIPTION OF THE SYSTEM

Sensor to detect seat belt sliding (SWING)

This sensor detects seat belt presence or not on the iliac wings. Installed in the left and right iliac wings, its role is to warn of any submarining and of when it occurs. This sensor is an improvement on the sensor developed for the 50°P Hybrid II by Bendjellal in 1989 (2).

Description - The SWING (Sensor Iliac Wing) can be likened to a flexible pin equipped with strain gauges (Figure 1).

Location - This sensor is mounted forcibly in a housing created in the iliac wing. This housing is extended by a groove to enable it to be deformed by seat belt forces, and this deformation has repercussions on the sensor (Figure 2). Two types of pelvis are equipped with this housing, the pelvis of the 50°P Hybrid III, and the new 5°P SAE version (Figure 3). The advantage of this type of location is that it does not alter the geometry of the iliac wing.

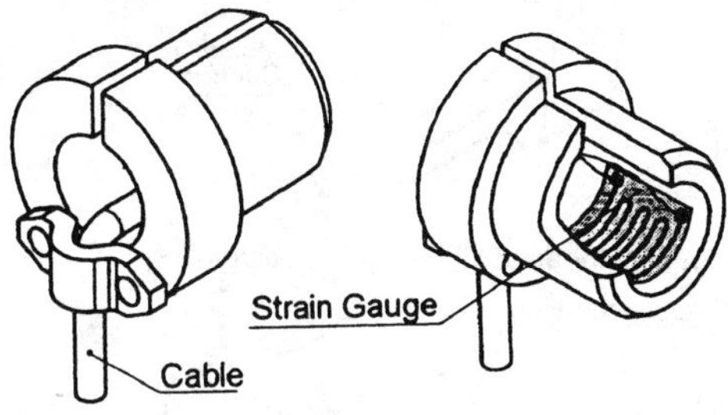

Figure 1 SWING sensor for detection of submarining

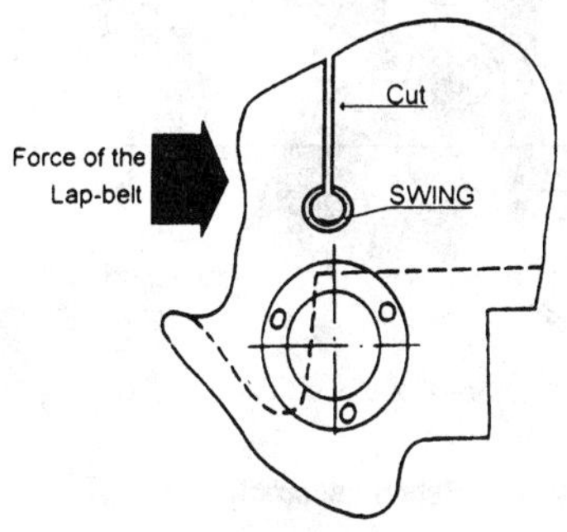

Figure 2 Operating principle of SWING sensor (Adaptation to the Hybrid III 50°P Pelvis)

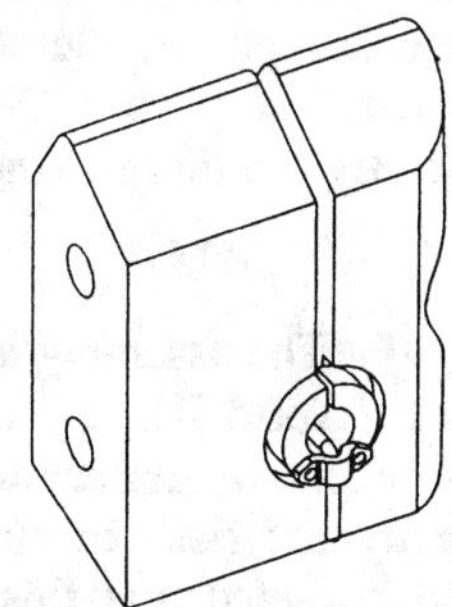

Figure 3 Location of SWING in the Hybrid IIII 5°P pelvis

CAP-SM2D, two-dimensional submarining sensor

This sensor was originally developed for 50°P H II, based on a sensor designed by Leung (1) in 1979 and improved by Bendjellal in 1989 (2), which measures the force exerted by the seat belt during the submarining phase. These sensors are used in pairs, and hence indicate the left and right submarining forces along anatomical X and Z.

Description - The submarining sensor consists of two elements:

- A sensor part formed of 2 beams equipped with strain gauges measuring the forces applied by the seat belt along the anatomical X and Z axes, and a protective aluminium casing (Figure 4).
- A support for sensor fastening in the dummy.

There are two types of support, one for the H II dummy and one for the 50° P H III dummy. The sensor part is identical in both dummies (Figure 5). The support/sensor assembly makes up the CAP-SM2D (Figure 6).

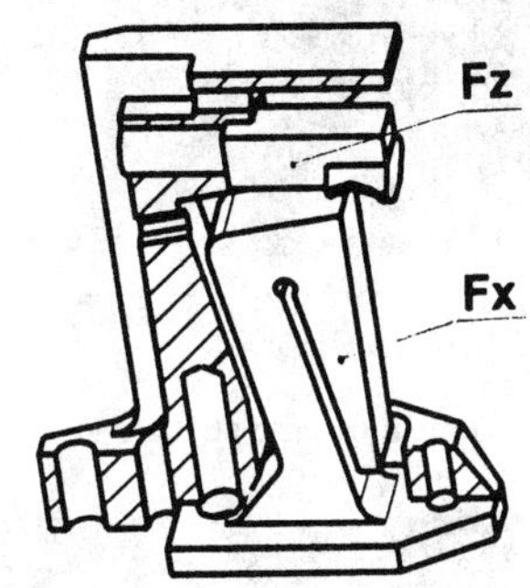

Figure 4 Submarining sensor

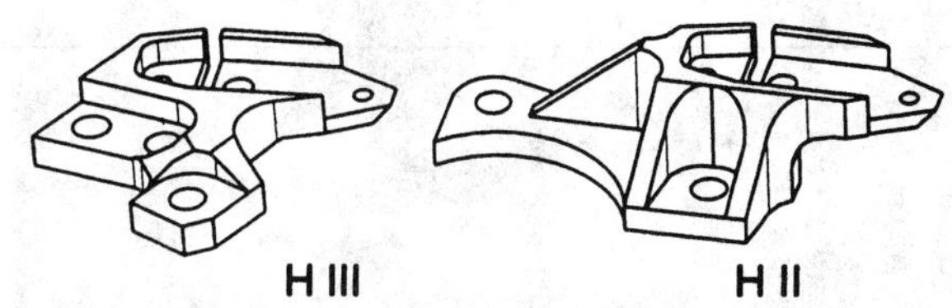

Figure 5 Support for the Hybrid III and 50°P Hybrid II

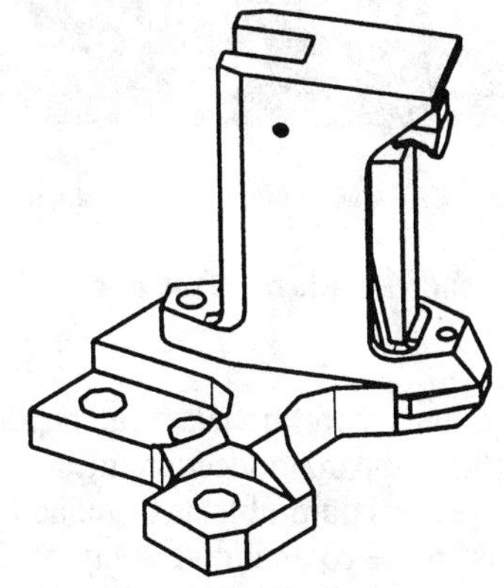

Figure 6 CAP-SM2D assembly

Location (Figure 7) - The CAP-SM2D can be installed at present on the H II and 50°P H III dummies, but not on the 5°P H III for reasons of size. It is installed on the lumbar column support, to which a slight modification is required. Moreover, the installation of the CAP-SM2D in the 50° P Hybrid III poses a problem of interaction between the protective casing of the standard deflection sensor and the casing of the CAP-SM2D, as shown in Figure 8. To solve this problem of size we have replaced the standard rotary deflection sensor (rod potentiometer) with a string potentiometer, which has enabled us to reduce the width of the spinal column/thoracic column interface.

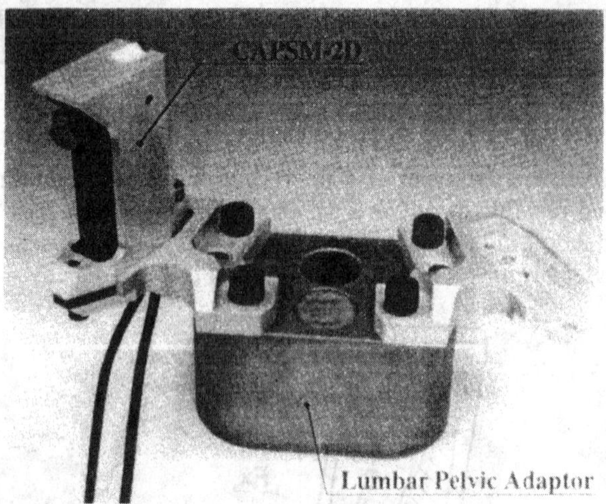

Figure 7 CAP-SM2D on the lumbar column support

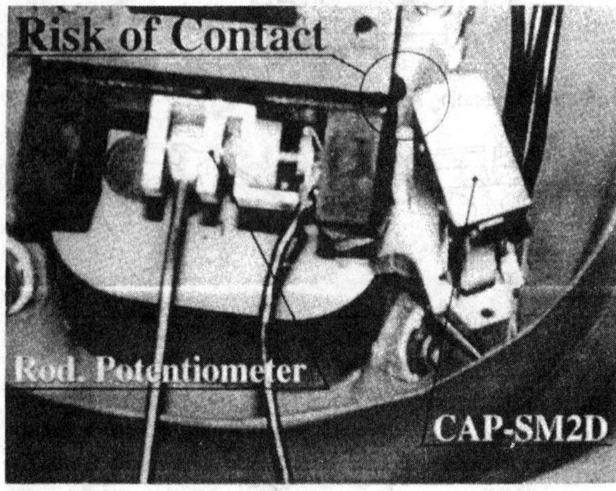

Figure 8 Contact between CAP-SM2D and rod potentiometer

Change made in the standard thoracic deflection measurement

Description - We have therefore taken the opportunity to try to improve the standard deflection measurement, especially since it poses certain problems of reliability. F. Bendjellal shows this in the comparative study performed for ACEA (3) in which it is apparent that in certain impact conditions the standard potentiometer does not always deliver a measurement, which is not the case for the string potentiometer.

A mounting bracket has therefore been developed to place this string measuring system inside the dummy, allowing the installation of 5 sensors (Figures 9 and 10):

- One string potentiometer (BIOSID type)
- Three single-axis acceleration sensors (ENDEVCO 7264 or ENTRAN)
- One angular velocity transducer MHD (ATA)

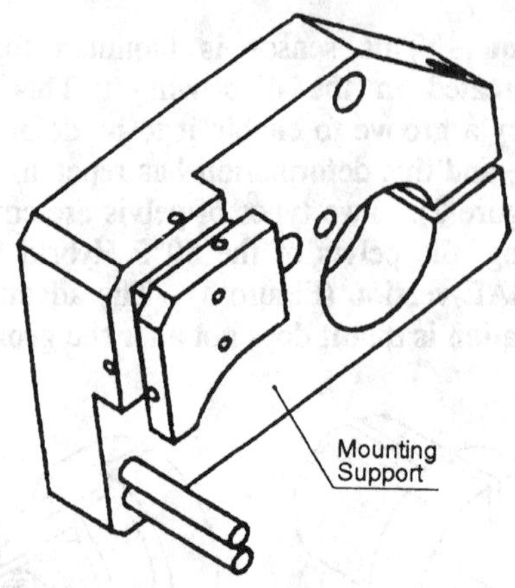

Figure 9 Bracket for the 50°P Hybrid III column

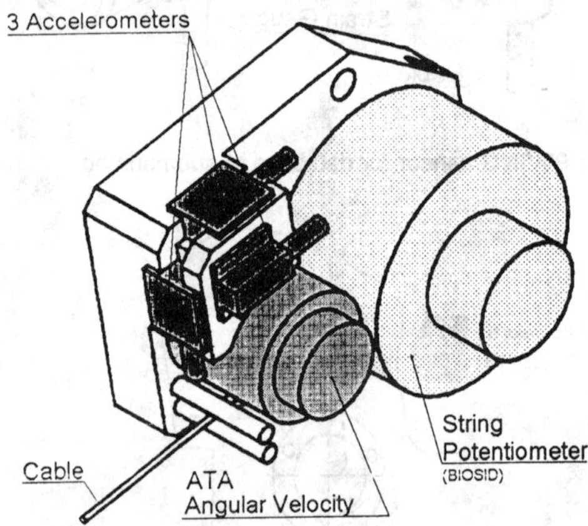

Figure 10 Thoracic instrumentation assembly

The string potentiometer replaces the standard rod potentiometer. The three acceleration sensors correspond to the measurements of thorax accelerations, and the angular speed transducer has been added to make it possible to measure directly the thorax kinematics during the impact.

Location (Figure 10) - The assembly is installed in a spinal column of a 50°P Hybrid III. By eliminating the standard rotary measurement of deflection, it has been possible to reduce greatly the risks of contact with the CAP-SM2D. It should be noted that this assembly can also be used in a conventional column.

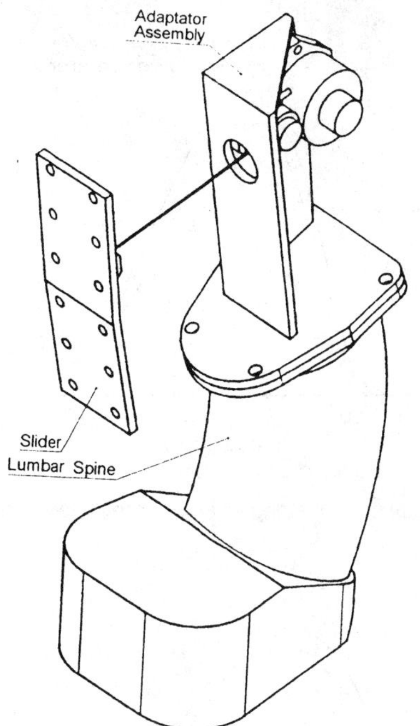

Figure 10 Modified assembly for the chest string potentiometer

Description of the overall chest-pelvis system

The overall system for the 50°P Hybrid III, shown in Figure 11, contains 7 extra measurement channels by comparison with conventional instrumentation.
Three instrumentation combinations are possible. These are:

- Prototype 50°P Hybrid III assembly equipped with two SWINGs, two CAP-SM2Ds and the string potentiometer.
- Prototype 50°P Hybrid III assembly equipped with two SWINGs
- Prototype 5°P Hybrid III assembly, new SAE version equipped with two SWINGs.

VALIDATION OF STRING POTENTIOMETER

These tests were performed at FTSS.

Description of tests (Figure 12)

A guided 5 kg mass of diameter 150 mm was dropped onto the thorax alone. Deflection was measured at the centre of impact. One thoraxe was used in these tests, equipped with the string potentiometer or a standard rod potentiometer. The measurements performed in these tests were:

- Impactor acceleration
- Thoracic deflection (string potentiometer or rod potentiometer)
- Two sternum accelerations
- Two sternum displacements (2 linear potentiometers)

8 tests were performed, at 2, 4, 6 and 7 m/s

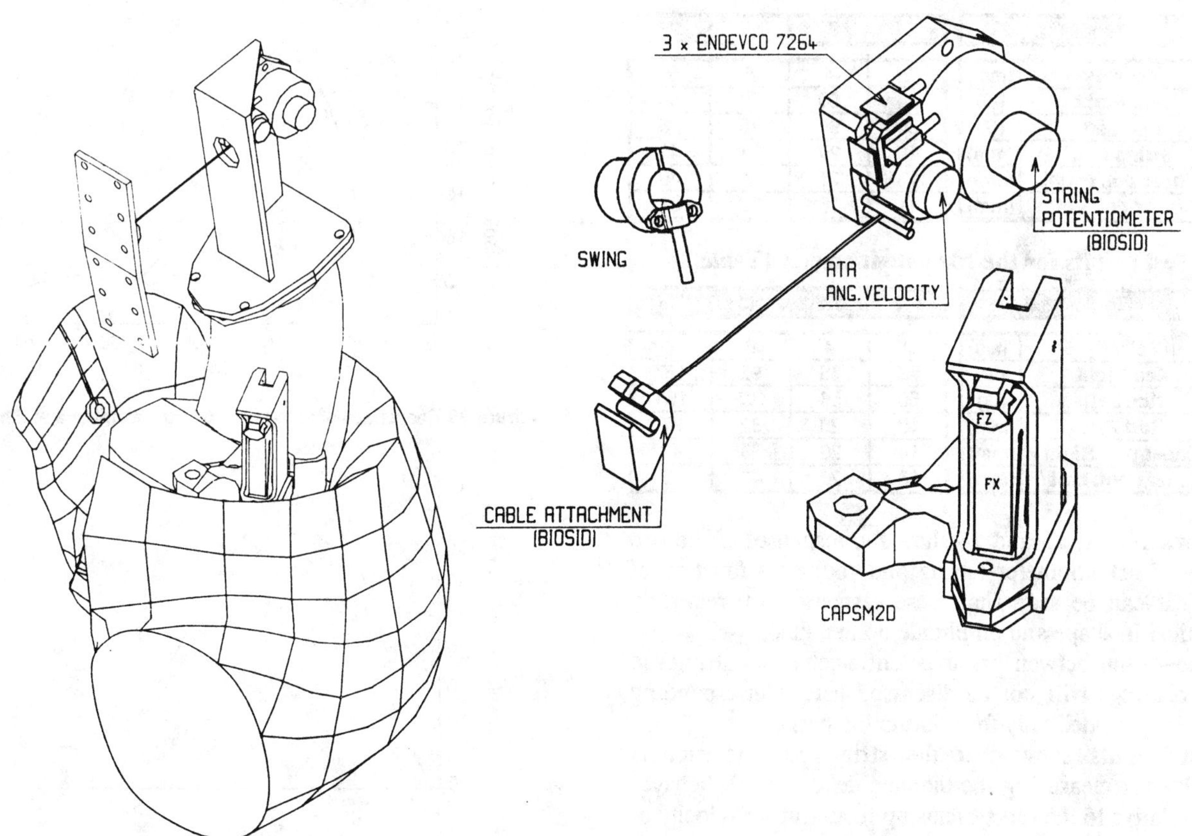

Figure 11 Overall chest-pelvis assembly for the 50°P Hybrid III dummy

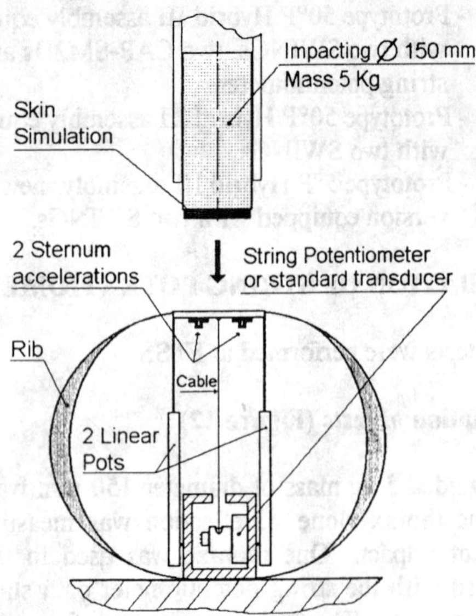

Figure 12 Drop test configuration

Analysis and discussion

8 tests were performed at 4 different speeds - 2, 4, 6 and 7 m/s - with the string potentiometer and the standard rod potentiometer. Tables 1 and 2 summarize the results obtained in these tests.

Test results for the string potentiometer (Table 1)

Test N°		S1	S2	S3	S4
Velocity	(m/s)	2	4	6	7
Acc Right	(G)	54	96	72	86
Acc Left	(G)	50	90	78	95
String Pot.	(mm)	10	20	31	42
Linear pot. Right	(mm)	13	21	31	49
Linear pot. Left	(mm)	11	18	32	51

Test results for the rod potentiometer (Table 2)

Test N°		R1	R2	R3	R4
Velocity	(m/s)	2	4	6	7
Acc Right	(G)	53	65	92	157
Acc Left	(G)	50	56	103	153
Rod Pot.	(mm)	10	20	32	37
Linear pot. Right	(mm)	14	20	33	40
Linear pot. Left	(mm)	13	20	43	35

Figures 13, 14, 15 and 16 show the responses of the two types of potentiometer (string and rod) as a fonction of time. It can be seen that these responses are relatively identical in shape and amplitude up to 6 m/s.

Comparaison betwen linear potentiometer and string/rod pot. readings will not be discussed here. Corresponding data are provided only for information purpose.

These results show that the string potentiometer is suitable for measuring the thoracic deflection. It behaves well relative to the rod systems up to an impact velocity of 6 m/s.

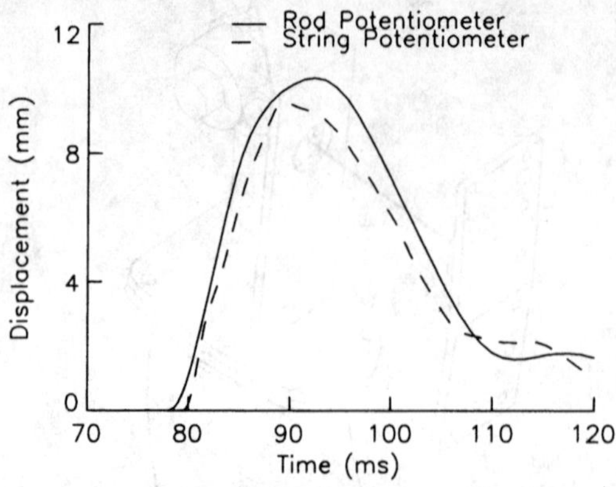

Figure 13 Responses of string and rod potentiometers at 2 m/s.

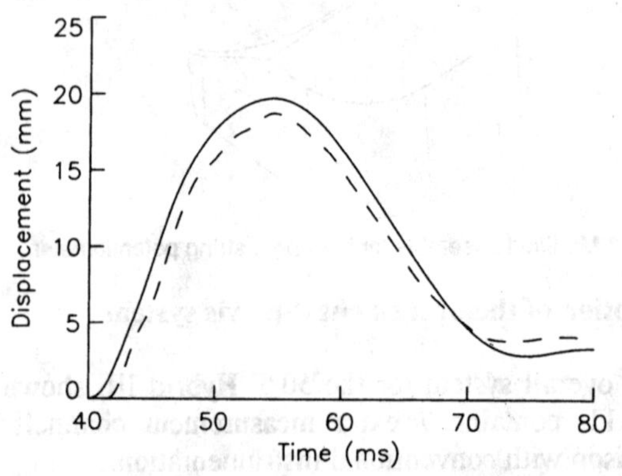

Figure 14 Responses of string and rod potentiometers at 4 m/s.

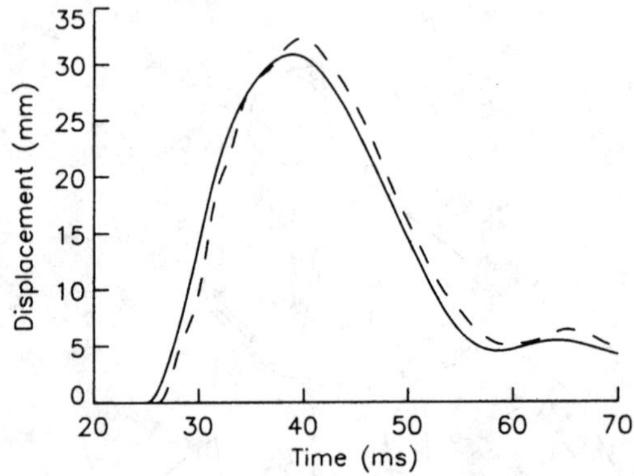

Figure 15 Responses of string and rod potentiometers at 6 m/s.

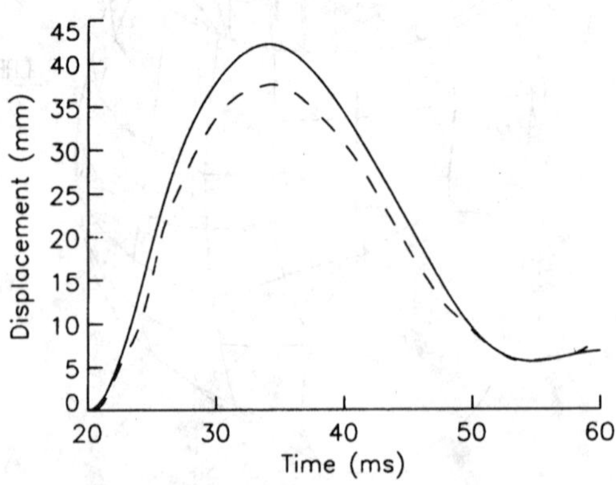

Figure 16 Responses of string and rod potentiometers at 7 m/s.

VALIDATION OF THE PROTOTYPE 50°P HYBRID III IN HYGE SLED TESTS.

To validate the prototype, we compared it with a standard dummy. To do this, tests were performed at P.S.A on an inverted catapult.

Description of the tests

The tests were performed on the seat cushion of a car body. The two dummies were positioned simultaneously as left and right back-seat passengers (Figures 17 and 18). For the tests with submarining, the restraint conditions were degraded by increasing the length of the belt buckle and placing it more horizontally (the same as for the lap belt) and by eliminating the anti-submarining boss. Two tests are described here, and the test conditions and the results obtained are summarized in Table 3.

Figure 17 Configuration of Hyge sled test, prior to impact

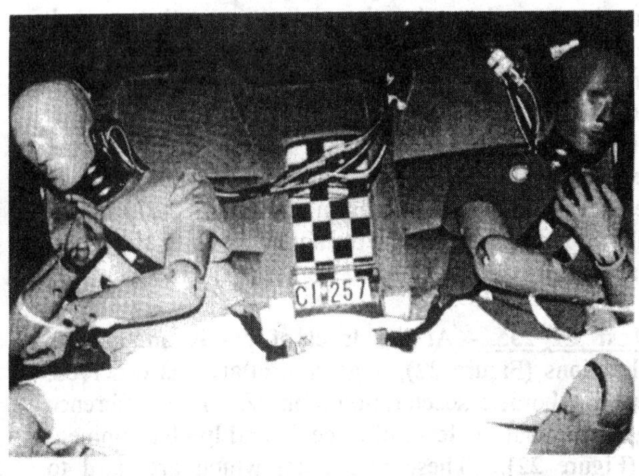

Figure 18 CI 257 configuration post impact

Analysis of prototype measurements

Summary table (Table 3)

Test No.		CI 257		CI 235	
Seat position		Left	Right	Left	Right
Dummy type		Stand.	Proto.	Stand.	Proto.
Thorax	X (G)	62	52	55.6	48.7
Acceleration	Z (G)	-20;18	-25;25	-27;22	-11;15
Pelvis	X (G)	35	28	41	/
Acceleration	Z (G)	30	27	23.7	26.1
Lumbar	Fx (daN)	170 -265	400 -310	246 -287	223 -211
column	Fz (daN)	50 -240	168 -170	-119 77.	-123 173
force	My (N.m)	57	66	38.2	24.6
Belt	Lap (daN)	900	850	766.3	720.1
Force	Shoulder (daN)	900	970	663.5	696.4
Submarining		Yes	Yes	No	No
Fx Right	(daN)	/	2	/	0
Fz Right	(daN)		1	/	0
Fx Left	(daN)	/	160	/	0
Fz Left	(daN)	/	160	/	0
Belt Buckle Angle	(Deg)	54	55	52	52
Belt Buckle length	(mm)	280	280	335	335
Lap belt Angle	(Deg)	60	57	70	70

<u>Response of the CAP-SM2D and SWING with submarining</u> - Using these sensors, submarining can be broken down into three phases in the CI 257 test (Figure 19).

- **Phase 1**: Pelvis movement forward over the seat belt, and loading of the iliac wings. A correlation is observed between the rise of the signals coming from the right and left SWING channels and the lap belt force.

- **Phase 2**: Slippage of the seat belt from the iliac wings onto the abdomen. This can be observed on the SWING channels by a drop in the signal, which takes place rapidly on the left side, and in two plateaux on the right side, before the drop in the seat belt force signal. In this specific case, the drop in the SWING signal was not very sudden, which indicates submarining of relatively low violence.

- **Phase 3**: Presence of the seat belt on the abdomen. The amplitude of this phenomenon is measured by the CAP-SM2D. Relatively slight submarining is measured on the left along anatomical X and Z, of approximately 160 daN. On the right side there is practically no submarining, which is consistent with the measurement on the right SWING channel.

These six measurement channels enabled us to describe accurately and in detail the submarining sustained by the dummy in test CI257. By cross-checking, which can be performed with the SWING channels, the CAP-SM2D and the lap belt force, it can be known with certainty whether submarining occurs and what is its amplitude.

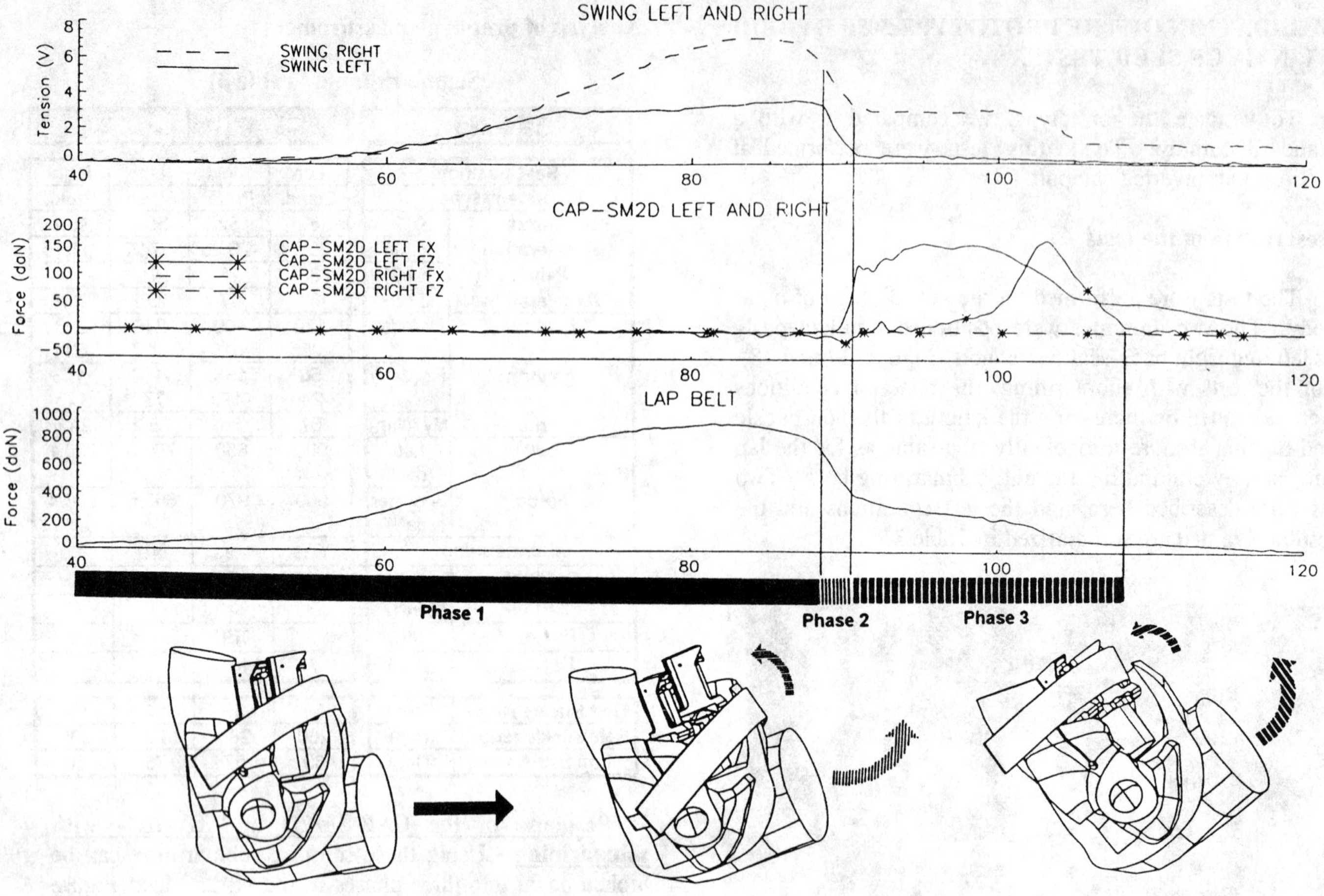

Figure 19 Submarining analysis based on the prototype channels in test CI257

Response of the CAP-SM2D and SWING without submarining - When no submarining occurs, as in test CI 235, the CAP-SM2D sensors detect nothing. The responses of the SWING sensors are correlated throughout the impact with the measurement of the lap belt force (Figure 20).

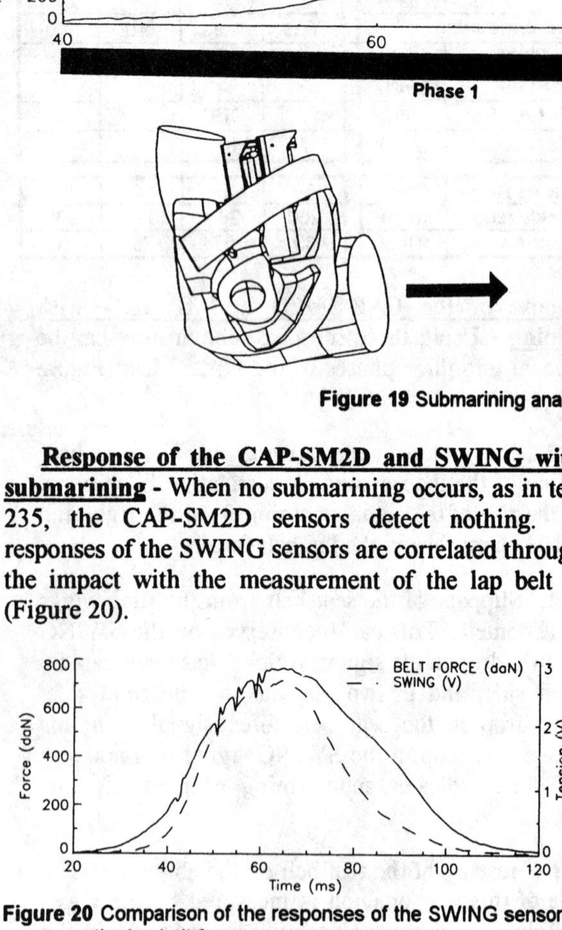

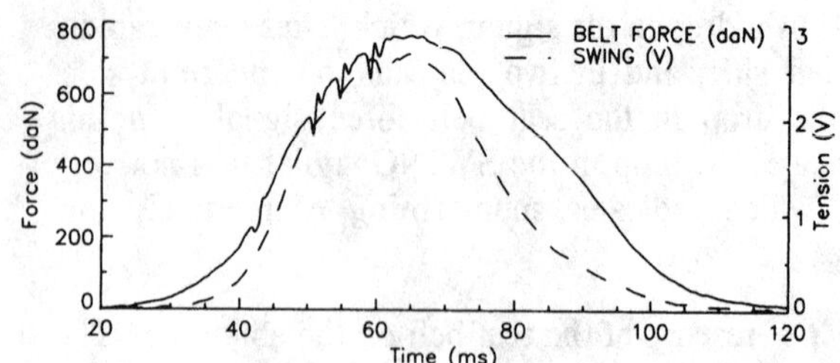

Figure 20 Comparison of the responses of the SWING sensors with the lap belt force.

Comparison of standard measurements

In these tests, we have compared the acceleration and kinematics of the thorax and pelvis, and lumbar forces, to evaluate the influence of the prototype instrumentation on the dummy responses.

Test CI 257 - At the level of pelvis and thorax accelerations (Figure 21) a slight difference is observed, which is accentuated around 60 ms, which corresponds to the start of submarining.

At the kinematic level (Figure 21), the comparison is good, even after 60 ms.

On the other hand, the lumbar measurements (Figure 21) diverge sharply above 70 ms. This is due to the design of the CAP-SM2D which prevents the seat belt from being supported by the ribs and lumbar column during submarining; the forces generated by the seat belt are therefore transmitted directly to the pelvis.

It may be pointed out that the movement of the seat belt onto the ribs and the column in a standard dummy interferes with the thoracic deflection measurement. The standard potentiometer, located at the base of the spinal column, sustains the seat belt force directly during submarining, and this can even damage the sensor, which gave unusual high deflections measured on the standard dummy. This phenomenon cannot occur with the string potentiometer, which is located higher in the thorax.

Test CI 235 - At the level of pelvis and thorax accelerations (Figure 22), a good similarity is observed, except for thoracic acceleration along Z. This difference is again found at the level of force Fz and lumbar moment My (Figure 22). These responses, which are hard to explain, do not seem to be due to the prototype instrumentation, because no physical contact was observed on the CAP-SM2D.

Moreover, no major differences are observed at the kinematic level (Figure 22).

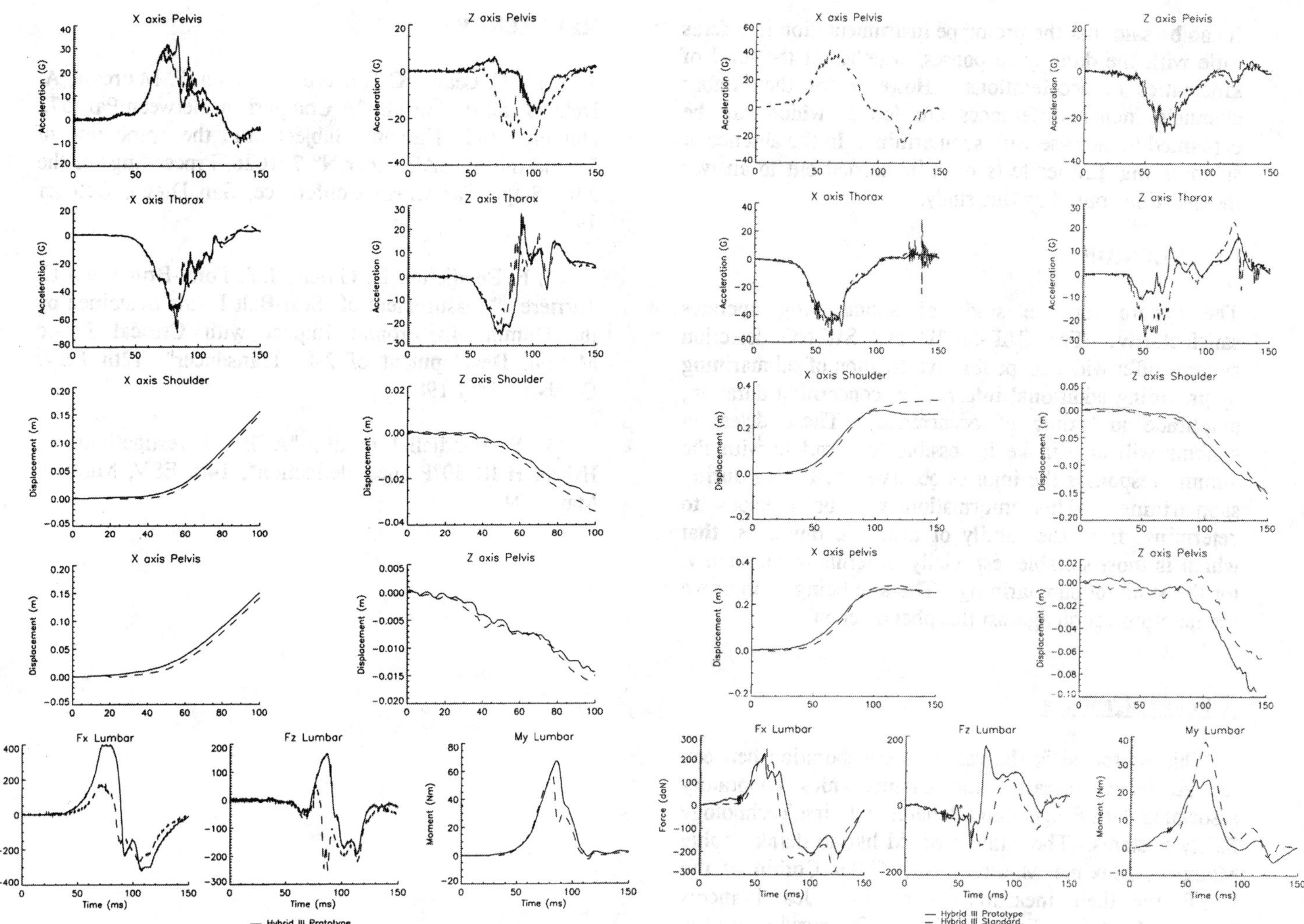

Figure 21 Comparison of thorax, pelvis and lumbar column responses for test CI 257 with submarining.

Figure 22 Comparison of thorax, pelvis and lumbar column responses for test CI 235 without submarining.

It can be said that the prototype instrumentation interferes little with the dummy responses, whether at the level of kinematics or accelerations. However for the lumbar channels, major differences are found, which can be explained in the case with submarining. In the absence of submarining, further tests must be carried out to answer the questions raised by this study.

CONCLUSION

The system for the study of submarining operates satisfactorily. The CAP-SM2D and SWING detection systems offer wider scope for investigation of submarining by providing additional information concerning duration, amplitude and time of occurrence. These detection systems will also make it possible to correlate with the dummy responses the injuries observed on PMHS during submarining. This information will be precious to determine, from the family of available dummies, that which is most suitable, especially in terms of biofidelity, for the study of submarining. The aim being to improve occupantprotection against this phenomenon.

Acknowledgements

This prototype is the result of collaboration between the Accident Research and Biomechanics Laboratory associated with Peugeot and Renault and First Technology Safety Systems. The authors would like to thank Sophie Mairesse, Maxime Moutreuil and Gilles Corbin of the L.A.B for their measurement services, Jean-François Huere and François Laurent of P.S.A for carrying out the catapult tests, Muir Parker, Gerald S. Lock and Joshua Y. Zhu of F.T.S.S. for carrying out the deflection tests and providing prototype parts, Craig Morgan of Denton Corp. for his technical assistance and collaboration in manufacturing of the SWING sensor, and all those who contributed to the completion of this project.

REFERENCES

(1) Y.C. Leung, C.Tarrière, A. Fayon, P. Mairesse, A. Delmas and P. Banzet, "A Comparison Between Part 572 Dummy and Human Subject in the problem of Submarining", SAE paper N° 791026, Proceedings of the 23rd Stapp Car Crash Conference, San Diego, October 1979

(2) F. Bendjellal, D. Gillet , J.Y. Foret-Bruno and C. Tarrière, "Measurement of Seat-Belt Loads Sustained by the Dummy in Frontal Impact with Critical Pelvic Motion: Development of 2-D Transducer", 12th ESV, Göteborg, May 1989.

(3) F. Bendjellal et al., "ACEA Investigations on Hybrid H III 50°P chest deflection", 14th ESV, Munich, May 1994.

SECTION 2:

SYSTEM ENHANCEMENTS/FEATURES

CHAPTER 6:

PRETENSIONERS

840395

Three-Point Belt Improvements for Increased Occupant Protection

Jüergen E. Mitzkus and Heinz Eyrainer
TRW Repa GmbH
Alfdorf/West Germany

ABSTRACT

Standard three-point belts offer a maximum of comfort and convenience. We have developed various systems designed to improve protection performance as well. This paper describes two of these systems, presents all necessary data on test equipment and procedure, and discusses test results. These two systems are, respectively:

The PYROTECHNIC PRETENSIONER, which retracts and preloads the belt by rotating the retractor spindle, thereby eliminating potential belt slack. The relations between actuation time, pretensioning, and improved occupant protection are shown.

The WEBBING CLAMP to eliminate the spool-out effect of the webbing stored on the retractor spindle, in two different versions:

- for clamping at the retractor, or
- for clamping at the D-ring.

The main advantage of these systems is the reduction in occupant forward displacement. The danger of impact is appreciably reduced.

DEMANDS REGARDING THE PASSENGER RESTRAINT SYSTEM "SEAT BELT" have been increased continuously. On the one hand, it is important that comfort and convenience are assured so that seat belts are more widely used. (1)* On the other hand, constant improvement of protection performance is required.

For reasons of competition U.S. automobile manufacturers endeavor to meet the recommended protection goals set forth by FMVSS 208 even in the 35 mph crash. In Europe, where a high belt usage rate is a direct consequence of traffic legislation (in most countries), analyses of traffic accidents have stimulated attempts to improve passenger protection. Investigations for instance by Appel/Vu-Han (2) and Walz (3) have shown that automobile occupants can suffer injuries in spite of having fastened their seat belt. Head, chest, and abdomen are the areas most likely to sustain serious injuries. Head and chest injuries, caused mainly by impact against the front structures of the car, are a function of <u>occupant forward displacement</u>. Abdominal Injuries are caused by submarining, i.e. by the belt slipping over the pelvic bone and applying load to the abdomen as the pelvis slides forward in a downward motion. This effect is caused not only by seats that are too soft and by unfavorable belt geometry but also by excessive <u>belt slack</u>.

We as a manufacturer of seat belts have developed auxiliary seat belt components for the elimination of belt slack and the reduction of passenger forward displacement. The effectiveness of pretensioner systems has been borne out in a number of investigations (4, to 9). This paper deals with the application of our seat belt pretensioning and webbing clamping devices with special reference to three-point belt systems and records their performance in 35 mph frontal barrier crash experiments.

DESCRIPTION OF BELT RESTRAINT SYSTEMS

SYSTEM A - STANDARD THREE-POINT BELT (Fig.1)

The Standard three-point belt with emergency locking retractor (3PB/ELR) serves as reference basis for all further systems. It is equipped with a standard polyester webbing with an elongation of 6 % and a total length of 3300 mm. 500 mm of webbing is stored on the spindle. These figures apply for all tests.

* Numbers in parentheses designate References at end of paper.

SYSTEM B1 - THREE-POINT BELT WITH RETRACTOR-MOUNTED WEBBING CLAMP (Fig. 2) - In order to eliminate the spool-out effect the webbing is clamped at the retractor. In other respects the construction of the belt corresponds to System A.

This clamping device is mounted on a standard emergency-locking retractor. At the present only prototype samples are available (Fig. 3). Since the forces acting on the retractor are greatly reduced by the clamping procedure, a lightweight retractor might be used in future applications.

The system of clamping the webbing at the retractor is presented in Fig. 3. The advantage of this version is the especially soft clamping action. In addition to the actual clamping between the eccentric clamping roller ④ and the clamping bar ① there is wrap-around friction between the webbing and the rubber-coated clamping roller sleeve ③. This clamping roller sleeve, mounted at each end in ball bearings, bends under belt load and is pressed against the rigid clamping roller ④.

SYSTEM B2 - THREE-POINT BELT WITH WEBBING CLAMP AT D-RING (Fig. 2) - A second version of a webbing clamping device is the combination with the D-ring. In other respects the construction of the belt corresponds to System A.

This webbing clamp at the D-ring is likewise available only as a prototype sample. The system is displayed in Fig. 4. With the D-ring ① under load, the sliding plate ② is pressed against the eccentric clamping roller ③. The rubber-coated clamping roller forces the webbing against the clamping bar ④ and secures it by friction and pressure.

This clamping device has no adverse effects on the wearing convenience of the belt since there are no additional friction points or deflections.

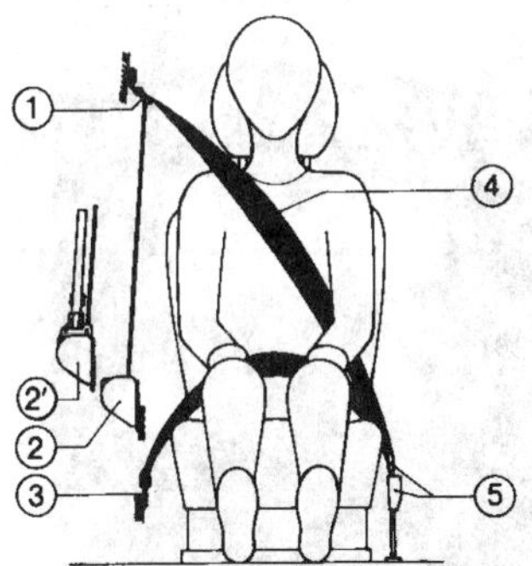

Fig. 1 - SYSTEM A: Standard three-point belt with ELR
SYSTEM C: Three-point belt with ELR and pretensioner

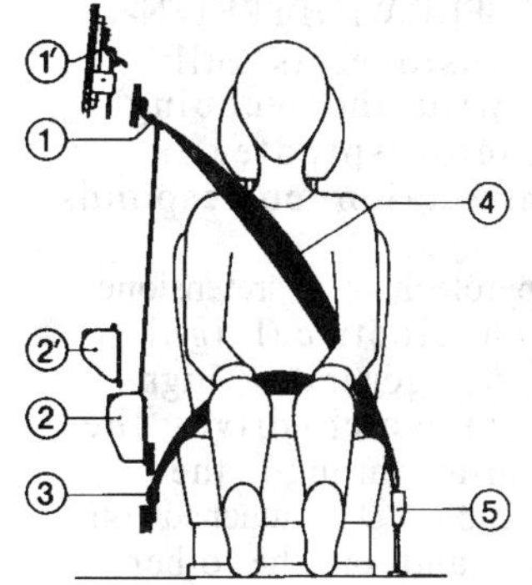

Fig. 2 - SYSTEM B1 and B2
Three-point belt with ELR and webbing clamp at the retractor or at the D-ring, respectively

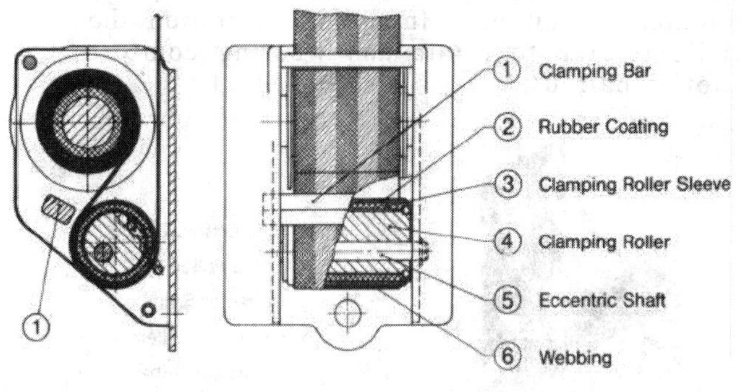

Fig. 3 - Retractor-mounted webbing clamp

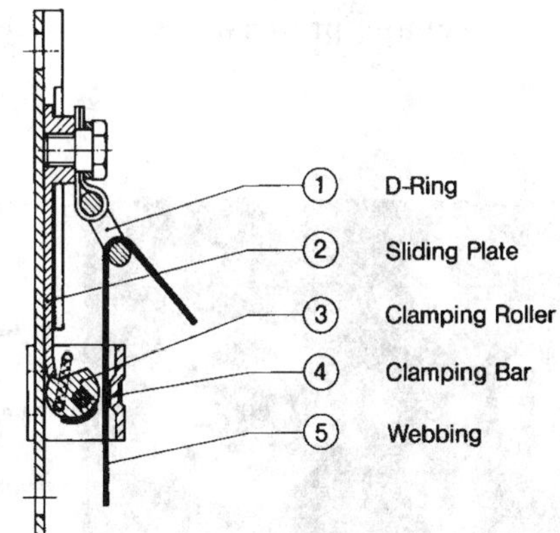

Fig. 4 - Webbing clamp integrated in D-ring

SYSTEM C - THREE-POINT BELT WITH PRETENSIONER (Fig. 1) - The pretensioner is fully integrated in the retractor, with the webbing being retracted by the retractor spindle. In other respects the belt construction corresponds in System A.

The layout of the pyrotechnic pretensioner is displayed in Fig. 5. Upon electrical ignition the pyrotechnic charge ① generates high-pressure gas (approx. 500 bar) which drives the pretensioner piston ⑦ upwards through the aluminum tube ⑤ . A steel cable ⑧ attached on the one end to the piston ⑦ and on the other end to the cable pulley ② , converts the linear motion into rotary motion. The three clamping rollers ④ are driven over sloping ramps against the retractor spindle ③ and are firmly pressed into the zinc material of the spindle (Fig. 6). This positive engagement provides a direct connection between the pretensioning element and the retractor which enables the retractor spindle to retract the webbing ⑥ . The system under test is designed for a maximum webbing retraction of 180 mm. In normal operation the spindle is free to rotate and wearing comfort is not impaired.

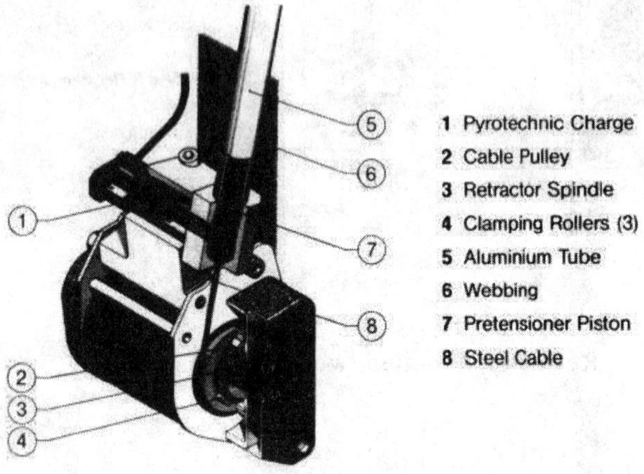

Fig. 5 - Pyrotechnic pretensioner

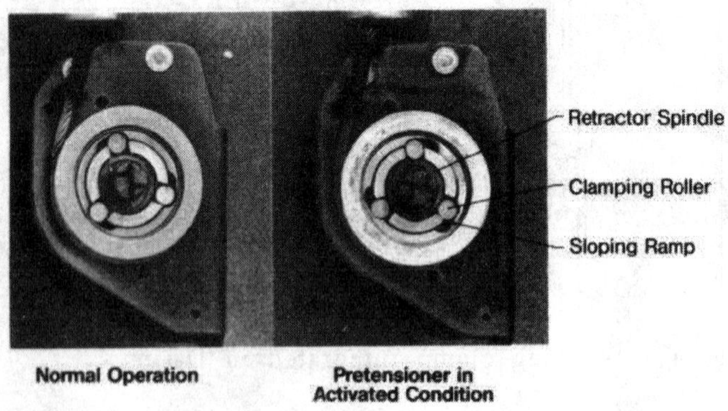

Fig. 6 - Pretensioner clutch system

This pyrotechnic pretensioner is commercially available and is being used in Daimler Benz automobiles.

The pretensioner is activated electronically by sensors registering acceleration as a function of time. These sensors have been developed by Bosch and by Radio-Becker.

SYSTEM D - IDEALIZED THREE-POINT BELT - The system idealized with respect to its protection performance is designed as a static three-point belt (Fig. 7) with a webbing of 3 % elongation (under 11,300 N) and pretensioned with a force of 300 N in the lap and shoulder area. The upper belt anchorage incorporates a deformation element (sheet metal strip) limiting shoulder belt loads to 6400 N. This arrangement allows the occupant to participate in vehicle deceleration at a very early stage, while the energy is transformed to an optimum degree, making use of an extensive forward displacement.

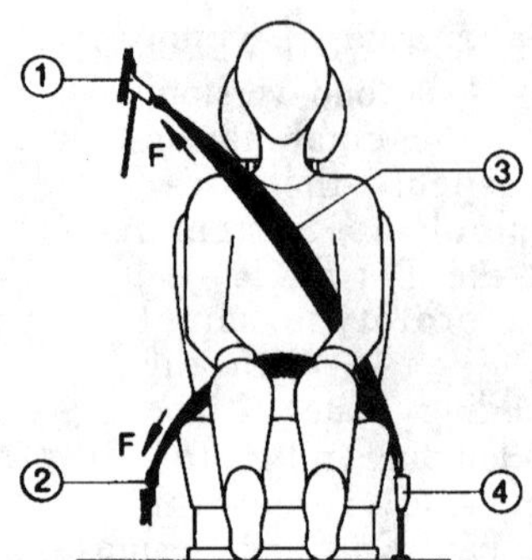

Fig. 7 - SYSTEM D - Idealized three-point belt

TEST IMPLEMENTATION

TESTING EQUIPMENT - The sled in our crash track IS accelerated over a distance of 40 m by means of rubber bungies. Deceleration is effected by a steel plate brake at the barrier (Fig. 8).

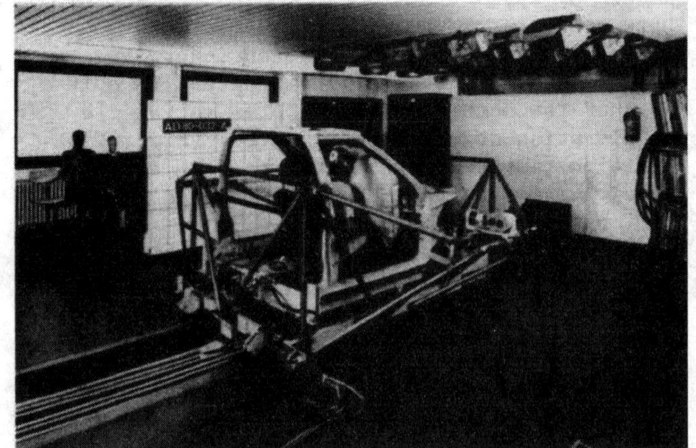

Fig. 8 - Crash track

Size and quantity of the steel plates are variable so that vehicle deceleration performance may be readily simulated.

The data obtained are transmitted by a modern PCM unit and stored and processed by computer in on-line operation (Fig. 9).

TEST CONFIGURATION - The sled (Fig. 10) was equipped with a standard vehicle seat (a new one for each test) and a rigid framework for the installation of the seat belt. The belt geometry was equivalent to that of a subcompact car. The tests were performed without dashboard on the passenger side. Thus there was no head impact.

TEST PARAMETERS AND EVALUATION

Impact speed:	35 ± 0.3 mph (56.3 km/h)
Deformation distance:	650 ± 25 mm
Deceleration pulse:	corresponding to a subcompact car as shown in Fig. 11
Anthropometric dummy:	Humanoid, 50th percentile male, as per Part 572

Dummy motions and forward displacements were recorded by high speed cameras. Evaluation and assessment of the injury values were made in accordance with FMVSS 208.

TEST RESULTS

Detailed results of all tests are contained in Table 1 (Annex 1).

COMPARISON OF WEBBING CLAMPING AND PRETENSIONING TO STANDARD THREE-POINT BELTS - The main advantage of a clamping device is elimination of the spool-out effect of the webbing stored on the retractor spindle (approx. 60 - 70 mm with 500 mm of webbing on the spindle). Both webbing clamps - the one at the retractor (Bl) as well as the one at the D-ring (B2) - clamp the webbing in a crash equally fast and permit a webbing extraction of approx. 15 to 30 mm. As there is only minor difference in results between Bl and B2 they may be considered as being equal for the purpose of this comparison, although the elongation-effective length of the webbing varies by approx.600 mm, i.e. the distance between retractor and D-ring.

The advantage of the pretensioner lies in the rapid pretensioning of the entire belt so that the occupant participates in vehicle deceleration at a very early stage (Fig. 12). Ignition time was established as t_1 = 10 ms, as explained in the section "Influence of Pretensioner Ignition Time".

Fig. 13 clearly illustrates the improved protection capability of the above systems as compared to the standard three-point belt, with the pretensioner showing the best results with regard to head and chest loads.

The clamping systems are slightly superior to the others with regard to head forward displacement (Fig. 14), though the difference here between clamping and pretensioning may be considered negligible.

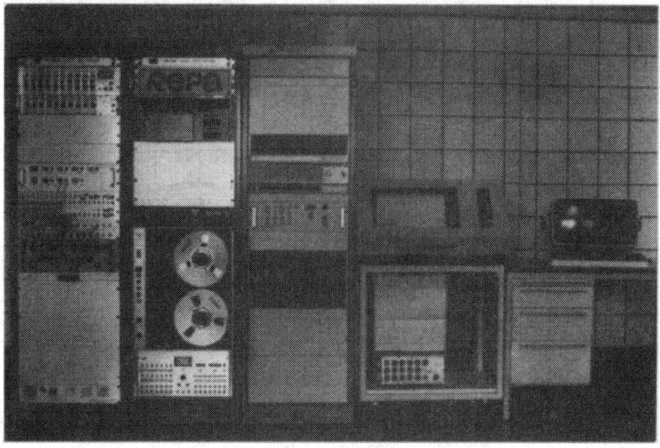

Fig. 9 - Data recording and processing unit

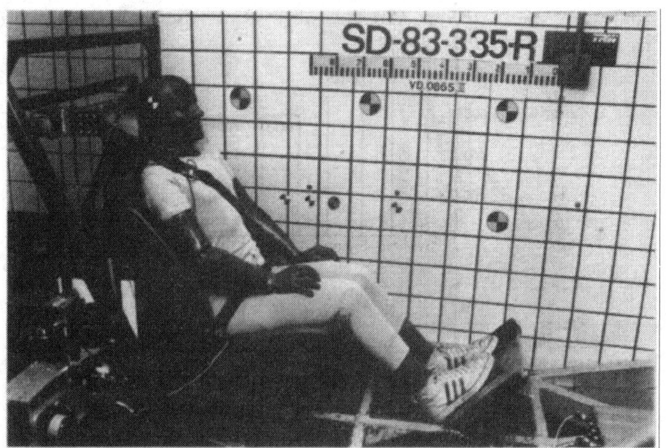

Fig. 10 - Test configuration

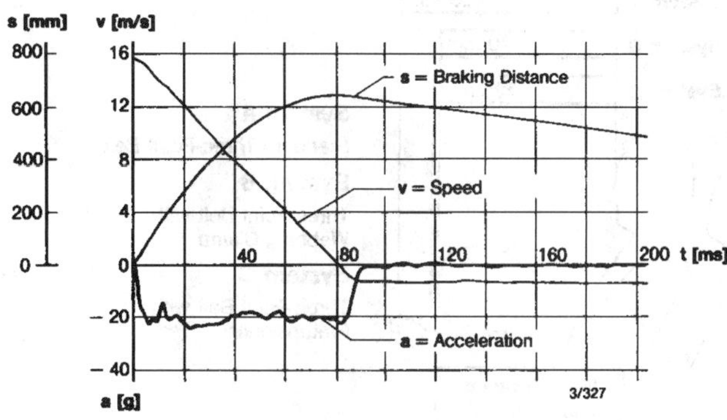

Fig. 11 - Sled deceleration pulse

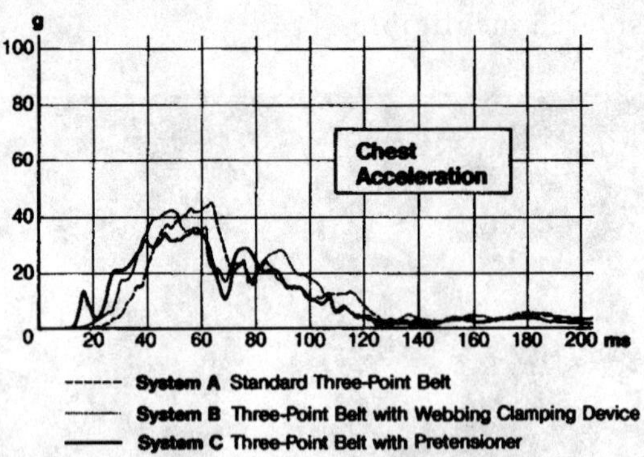

Fig. 12 - Typical chest acceleration curves

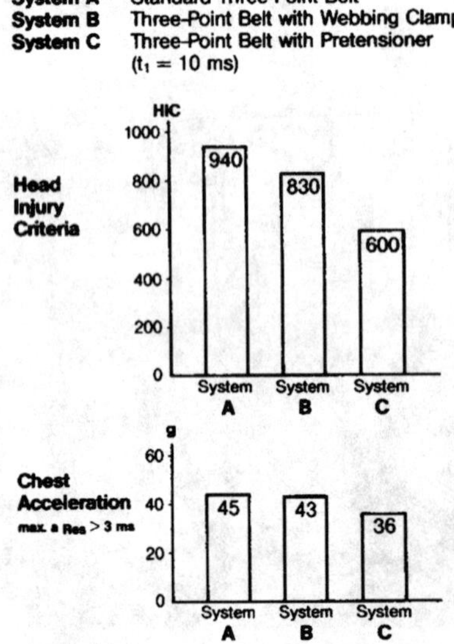

Fig. 13 - Comparison of injury values

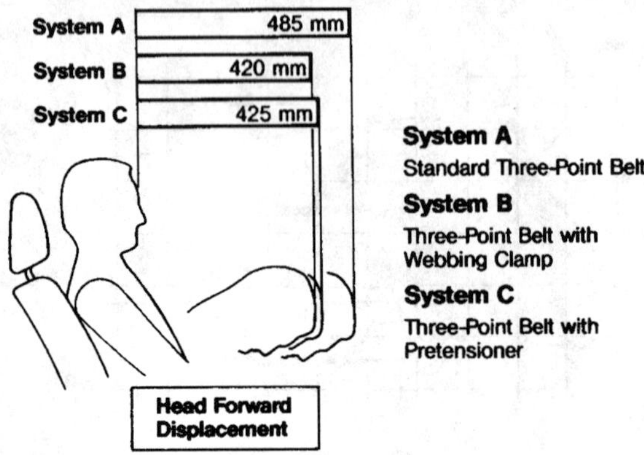

Fig. 14 - Comparison of head forward displacement values

WEBBING ELONGATION - In order to determine the influence of webbing elongation two types of polyester webbing with the extreme elongation limits of 6 % and 17 % under 1130 daN were used.

In all the systems tested, high webbing elongation resulted in reduced chest acceleration values while the head forward displacement values were increased (Fig. 15). This correlation is explained by the physical law that when energy absorption remains constant, the resultant forces, i.e. acceleration values, are reduced when distributed over a longer distance, though this law is influenced by the differences in load onset time and load onset rate.

No definite conclusions may be drawn with regard to HIC value. The variance is so great that an increased number of tests would have been required.

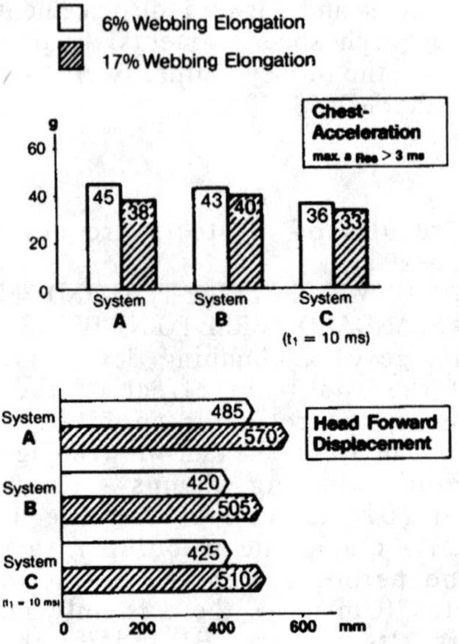

Fig. 15 - Chest acceleration and head forward displacement at different webbing elongation limits

BELT SLACK - The standard tests were performed with the seat belt properly fastened. In actual driving conditions, however, there is always a certain amount of belt slack - because of negligent usage of the belt, because the belt is being caught by a vehicle part, because of heavy winter clothes, or because of the tension eliminator at the retractor as used in the United States.

In these cases the pretensioner is especially advantageous. For the relevant tests a layer of foam material (40 mm) was inserted between the dummy and the belt in the chest area and in the pelvic area, simulating a belt slack of 50 mm and 70 mm respectively.

As compared to the standard three-point belt with equal belt slack the pretensioner reduces the HIC value by approx. 50 %, chest acceleration by approx. 20 %, and head forward displacement by approx. 15 % - as illustrated in Figures 16, 17, and 18.

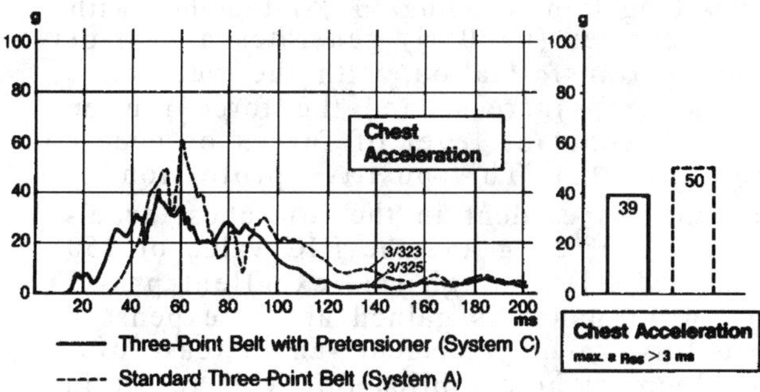

Fig. 16 - Chest acceleration values - 3PB/ELR with and without pretensioner, with a belt slack of 120 mm each

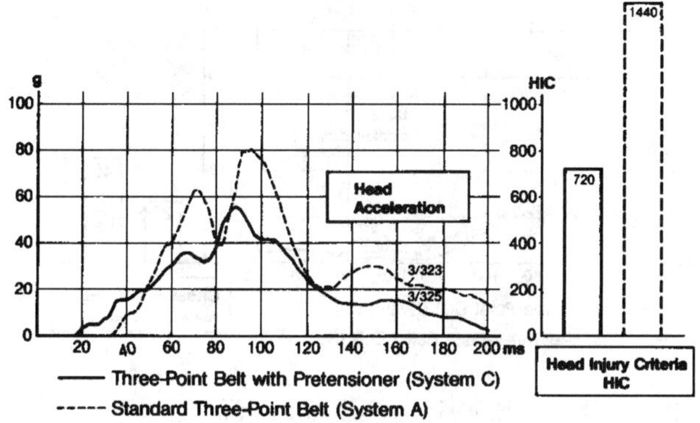

Fig.17- Head acceleration and HIC values - 3PB/ELR with and without pretensioner, with a belt slack of 120 mm each

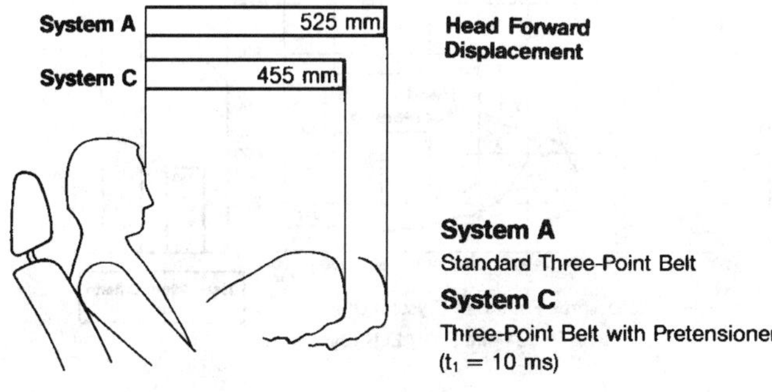

Fig. 18 - Head Forward displacement values - 3PB/ELR with and without pretensioner, with a belt slack of 120 mm each

INFLUENCE OF PRETENSIONER IGNITION TIME - The performance of the pretensioner is vitally influenced by the ignition time (t_1) after crash onset. The sooner the ignition the better the results; with a delayed ignition of t_1 = 30 ms the pretensioning effect is virtually nullified and the results are similar to those of a standard three-point belt - as shown in Fig. 19 and 20.

The lower limit of t_1 is determined by actual driving conditions (bumpy roads, curbstones, etc.) where the actuation of the pretensioner without a crash situation is to be avoided. By chosing t_1 to be 10 ms all tests to date have been based on a realistic value corresponding to the driving and crash performance of a subcompact car.

With an extended t_1 value the length of webbing retracted by the pretensioner is reduced (Fig. 21). This results from the fact that the webbing is retracted against the weight of the occupant moving forward only to the point where an equilibrium of forces in the belt is reached. The belt slack and part of the spool-out of effect are eliminated. This is followed by webbing extraction at the retractor. These motion sequences are demonstrated in Fig. 21. A comparison of webbing movement with and without pretensioner (t_1 = 10 ms) shows an effective different-e of 90 mm of extraction, which is appreciable. Among other things, this reduces forward movement of the occupant. For a brief moment belt forces generated by pretensioning amount to approx. 3500 N and then drop to a level of approx. 500 N (Fig. 22).

The pretensioner is incapable of pulling the occupant backwards. Its working energy (1 gram of pyrotechnic charge) is insufficient for more than a tensioning of the belt.

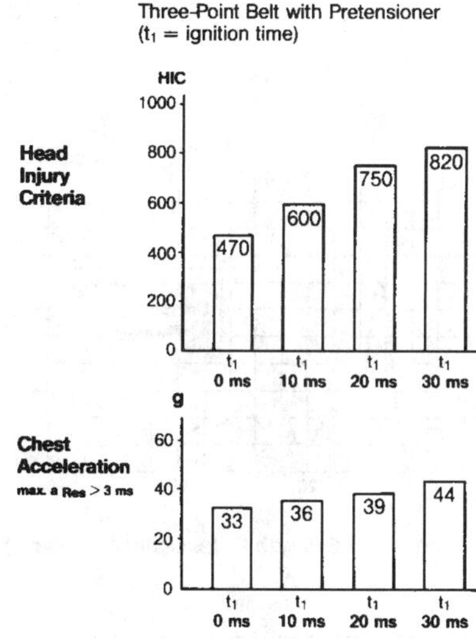

Fig. 19 - HIC and chest acceleration values as function of pretensioner ignition time

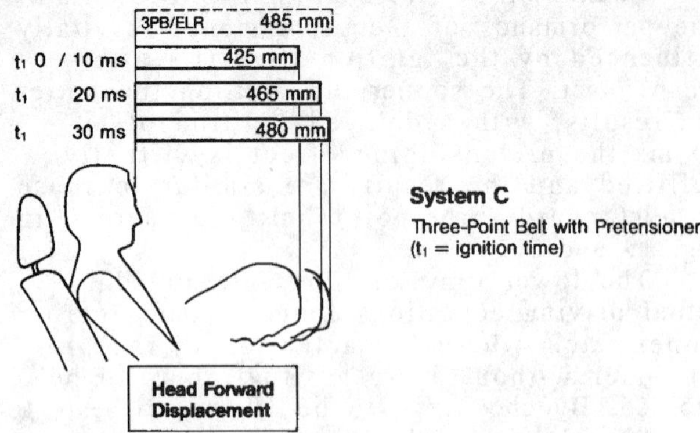

Fig. 20 - Head forward displacement values as function of pretensioner ignition time

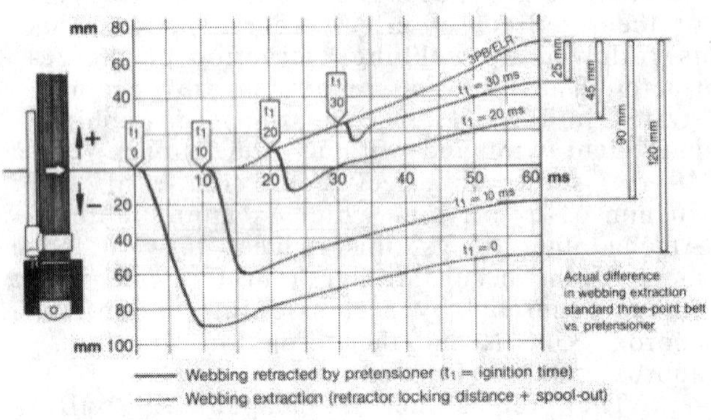

Fig. 21 - Webbing movement at the retract-or as function of pretensioner ignition time

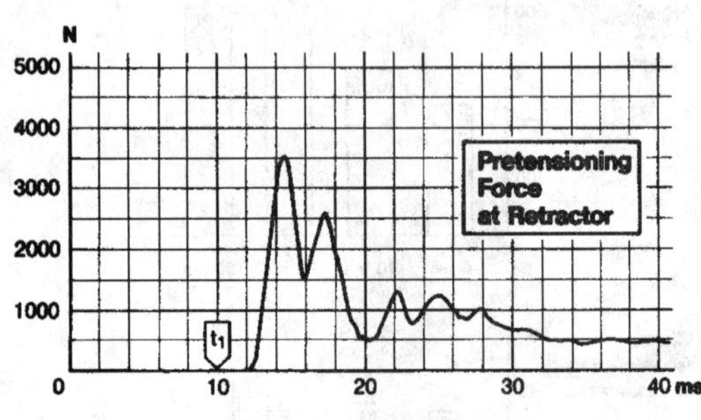

Fig. 22 - Pretensioner retraction forces, as measured in a static test

SYSTEM D - IDEALIZED BELT RESTRAINT SYSTEM- This system shows the maximum improvement potential of a three-point belt restraint system. By virtue of the pretensioning of the belt the occupant takes part in vehicle deceleration at a very early stage. Chest acceleration increase begins after a mere 10 ms as compared to 27 ms with the standard three-point belt (Fig. 23). The low-elongation webbing (3 %) together with the force limiter (6400 N) generates a near-perfect energy transformation, with the webbing causing a steep increase and the force limiter affording a constant level of forces or acceleration (Fig. 23). This positive protection performance is evident in the low chest acceleration value of 27 g and the HIC value of 350 (Fig. 23 and 24), though this excellent protection performance is gained at the expense of head forward displacement (an increase of 30 %). In most automobiles this will result in head impact.

The auxiliary seat belt devices pretensioner and webbing clamp offer the material for a compromise between idealized and practicable seat belt systems.

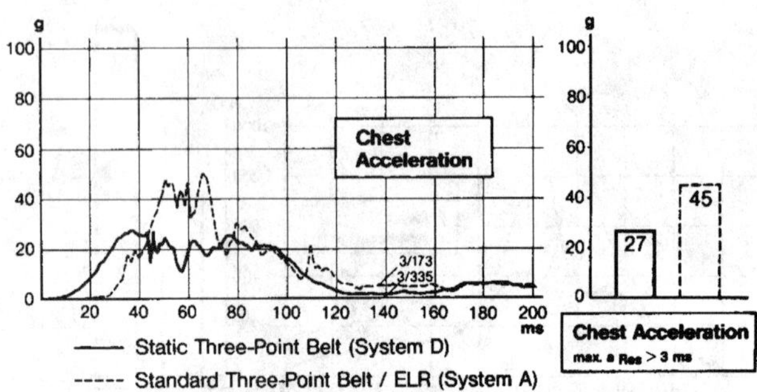

Fig. 23 - Chest acceleration curves - Standard three-point belt vs. idealized belt system

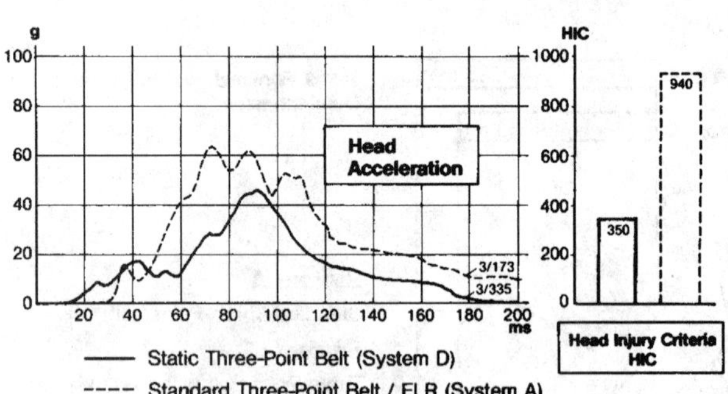

Fig. 24- Head acceleration curves & HIC data- Standard three-point belt vs. idealized belt system

CONCLUSIONS

- The main advantage of the webbing clamping device, regardless of whether it is mounted at the retractor or at the D-ring, lies in the reduction of occupant forward displacement. This can help to avoid head impact or chest contact with front structures of the car. The performance of the three-point belt with clamping system is similar to that of a well-fastened static three-point belt, though it offers the fastening convenience and wearing comfort desired in cars of today. It is also possible to integrate the webbing clamp in a height-adjustable D-ring. The shoulder belt may thus be adapted to the size of the occupant and the belt geometry improved.
- This investigation clearly shows the effectiveness of the pretensioner. In addition to reduced occupant forward displacement - similar to the webbing clamping system - head and chest injuries are considerably reduced. In the case of the preloaded belt (allowing for a realistic 120 mm of belt slack) the HIC value is reduced by 50 % as compared to standard three-point belts.
- Pretensioning performance is very critically influenced by the actuation time after crash onset. The sooner the ignition the greater the effect. An ignition time of 30 ms, slow with respect to the specified deceleration pulse, offers almost no improvement over standard three-point belts. Optimum sensor adjustment is thus of vital importance. In case of inadvertent ignition the belted occupant will not be hurt. The pretensioning action itself as well as the resultant belt loads (approx. 500 N) acting on the occupant are easily tolerable. This has been tested on living persons. Simulated faulty ignitions in road tests have not resulted in dangerous situations.
- In addition to the advantages quantifiable in crash tests the pretensioner shows very positive performance in real-life accidents: By eliminating belt slack in the lap portion of the seat belt, submarining and resultant abdominal injury may be prevented.
- A very interesting solution with regard to comfort as well as to safety would be a combination of the pretensioner described here and the tension eliminator as used in the United States. The belt slack inherent in the latter system could be eliminated completely, allowing the full protection potential of the belt system to take effect.
- An unrealistic seat belt construction,the idealized System D, was investigated for the purpose of determining its performance limits. We could imagine that a combination of pretensioner, webbing clamp, and force limiter, properly coordinated and adapted, would provide the requisite for an equally effective belt system. We might point out that chest injury values could further be improved by increased webbing elongation, provided there is sufficient space for unrestricted occupant forward displacement.
- Finally, the commercial aspect. The costs of an advanced seat belt system with pyrotechnic pretensioner are about twice those of a standard three-point belt, not including the ignition sensor.

REFERENCES

1. H. JOHANNESSEN, T. H. VOS
"The Changing Shape of Seat Belt Systems"
SAE Paper No. 820796, June 1982

2. H.APPEL, V. VU-HAN
"Schutzwirkung von Sicherheitsgurten - Band 3 - Auswertung von Gurtunfällen"
Bundesanstalt für Straßenwesen, Köln/W.Germany 1978

3. F. WALZ, P. NIEDERER and al.
"Analyse des accidents par rapport aux ceintures de sécurité"
Départment Fédérale de Justice et Police, Berne, 1977

4. R. WEISSNER
"A Comparison of Advanced Belt Systems Regarding Their Effectiveness"
SAE Paper No. 780414, March 1978

5. G. HOFFMANN, W. KÖPKE, DR. R. WEISSNER
"The Influence of Preloading on the Efficiency of Belt Systems"
ISATA, Wolfsburg/West Germany, Sept. 1978

6. G. HOFFMANN, R. WEISSNER
"The Influence of the Delay Time on the Efficiency of Preloaded Belt Systems"
ISATA, Graz/Austria, Sept. 1979

7. W. REIDELBACH, H. SCHOLZ
"Advanced Restraint System Concepts"
SAE Paper No. 790321, March 1979

8. M. DEJEAMMES, R.BIARD, Y. DERRIEN
"Improvement of the Automatic Belt: The Possible Efficiency of the Pyrotechnical Retractor"
Lab. des Chocs et de Biomécanique - ONSER

9. M. J. WALSH, B. J. KELLEHER
"Development of a Preloaded, Force-Limited Passive Belt System for Small Cars"
SAE Paper No. 800300, Feb. 1980

A Mechanical Buckle Pretensioner to Improve a Three Point Seat Belt

Yngve Håland and Torbjörn Skånberg,
Electrolux Autoliv AB,
Sweden

Abstract

The occupant protection in a car can be improved by supplementing the three point seat belt with an efficient pretensioner. This is of special importance if there is a lot of slack in the belt caused by e.g. thick clothes.

This paper describes a mechanical pretensioner which has been in production since January 1989. It acts on the seat belt buckle. At the very beginning of a crash the buckle is rapidly pulled down, thereby pretensioning both the lap- and chest belt parts.

The pretensioner comprises a preloaded torsional spring, which acts upon the buckle and a sensor mechanism for triggering. The design, performance and effectiveness of the buckle pretensioner in a three point belt system with and without slack is presented. The main advantages are reduction of occupant forward displacement, lower impact speed into the steering wheel, less head injury criteria (HIC) and less risk of submarining.

Introduction

The three point seat belt is an efficient restraint system in cars. There are however means to further improve it. One problem is that excessive belt slack can cause unnecessary injuries to the head, chest and abdomen. A pretensioning device which acts very early in a crash, before an unrestrained mass has moved more than 20 mm, and that is strong enough (1000 N), will reduce the risk of these injuries.

Our company, a manufacturer of seat belts, has developed a low cost mechanical pretensioner which works by rapidly pulling down the seat belt buckle. The pretensioner has been in production since January 1989.

This paper gives a description of the buckle pretensioner, how the triggering is set and results from 35 mph tests.

Background

During the initial phase of an impact, when the front zone of the car is being deformed, a person restrained by a standard belt system initially continues at an unchanged speed. The distance the person travels before the belt starts to absorb energy and decelerates him depends partly on the locking distance of the retractor and the so called filmspool effect, but is mainly dependent on the slack of the belt. The amount of slack depends on several factors, for example the type of clothing worn by the person, the beltgeometry and seat position. A way of eliminating slack is to equip the belt with a pretensioner. There are different ways of pretensioning. You can for example retract the belt webbing by rewinding the retractor. This is often done by pyrotechnical means. An other way is to pull the buckle downwards towards the buckle attachment point. The advantage of pretensioning at the buckle is that you simultaneously pull in the diagonal and the lap part of the belt.

Description of the Buckle Pretensioner

The buckle pretensioner is an all mechanical solution bolted into the seat at the standard buckle attachment point. The pretensioner assembly is shown in figure 1. It consists of three main parts, namely the sensor, the pulling and the locking parts. They are separately described below.

Figure 1. Mechanical buckle pretensioner.

Sensor part

The sensor is fully mechanical and is integrated with the mechanism that makes the pulling. It consists of five parts.

The different parts are configured in a way that the torque from the tensioned torsion bar spring creates the following arrangement of reaction forces acting upon the lever.

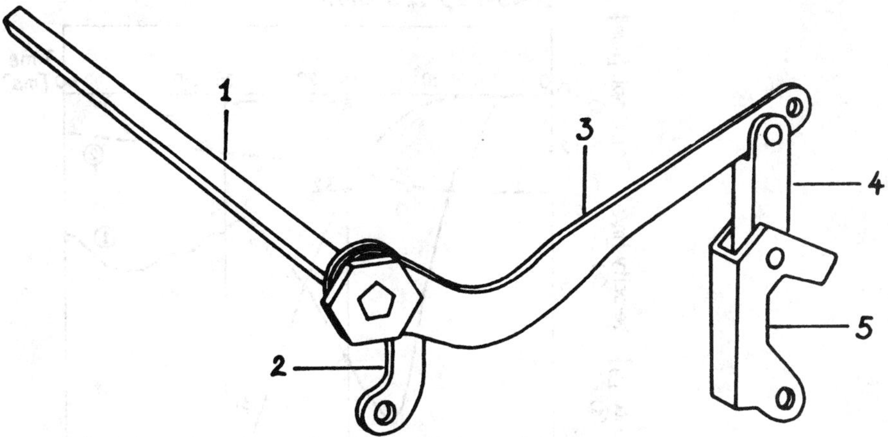

Figure 2. Sensor part.

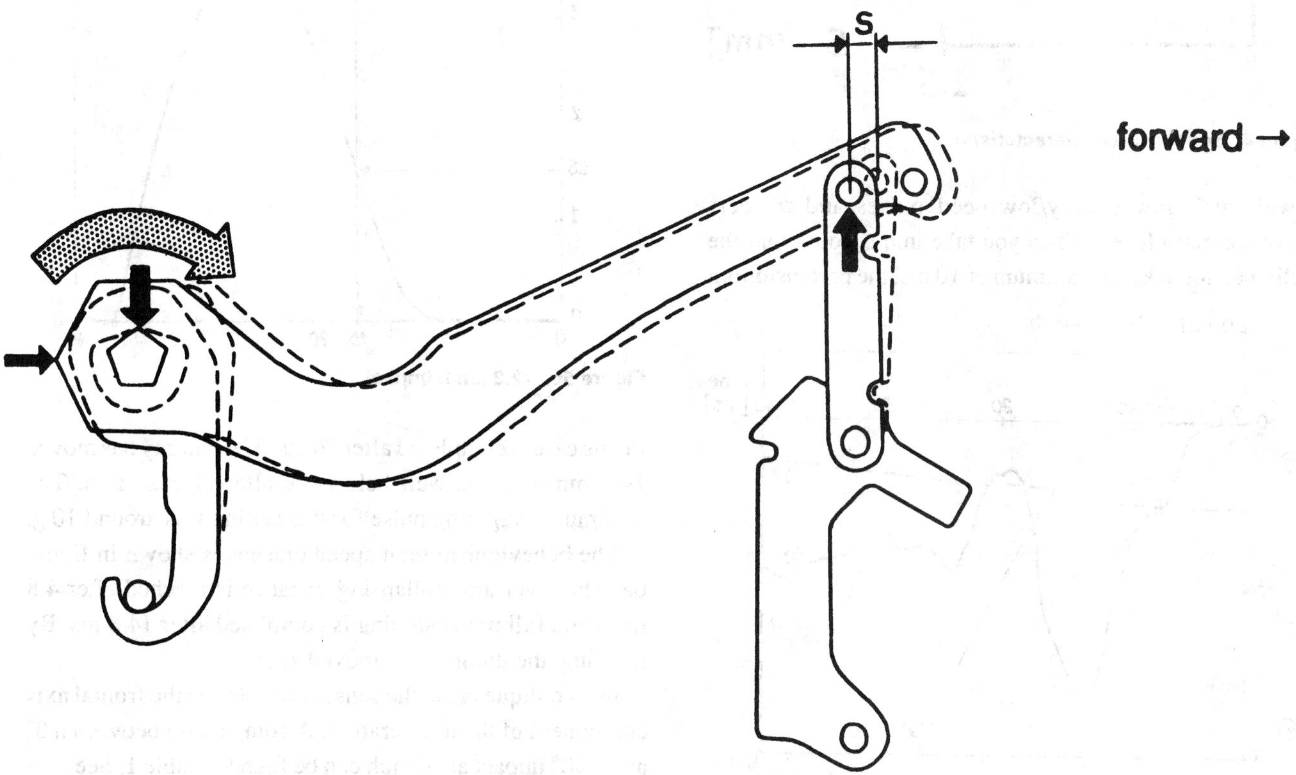

Figure 3. Reaction forces.

The lever arm is striving to move backwards but is prevented from doing so and is only allowed to move forward. The force needed to move the lever arm forward is in theory constant, but taking into account the bending of the torsion bar and some friction the force actually increases slightly. If the lever arm is moved forward a distance S mm the two legs, upper and lower, are positioned in a straight line and for an additional motion they will collapse. The pretensioner has then triggered and the pretensioning force is released to pull the buckle downwards. The sensor characteristics can easily be altered to suit various carpulses, only by changing the geometry of the legs.

A typical sensor force characteristic is shown in figure 4.

The characteristic is chosen so that pretensioning will take place only when needed avoiding inadvertant triggering. A complete pretensioning has to be completed before the person (dummy) has moved 20 mm forward relative to the car during the impact. The low speed impact behaviour is shown by the two figures 5a and 5b. By following the full drawn line it can be seen that for a 9.3 km/h impact speed the lever arm doesn't reach the collapsing distance. It only moves 0.9 mm whilst at 12.3 km/h it collapses after 16 ms. The dummy (an unrestrained mass) has by this time moved only 3.2 mm. Thus it is evident that the sensor characteristic

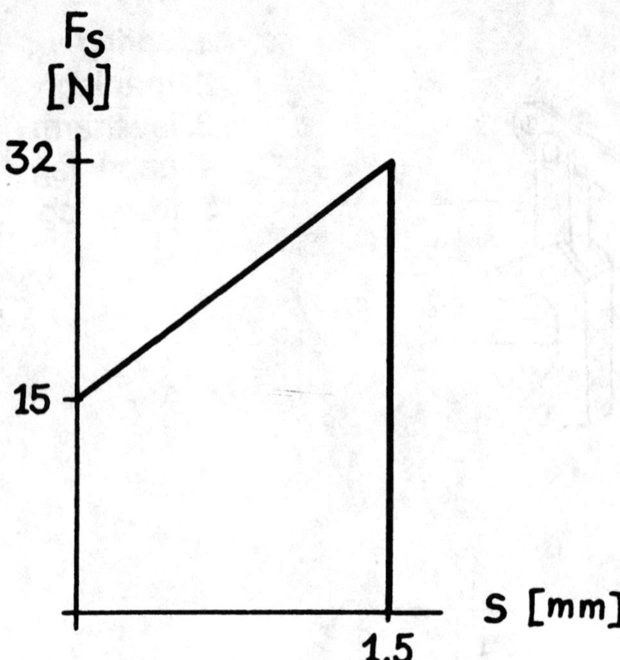

Figure 4. Sensor force characteristic.

"swallows" low energy/lowspeed pulses and triggers above a certain level. When you take into account that the pulling action takes a maximum of 10 ms, the pretensioning

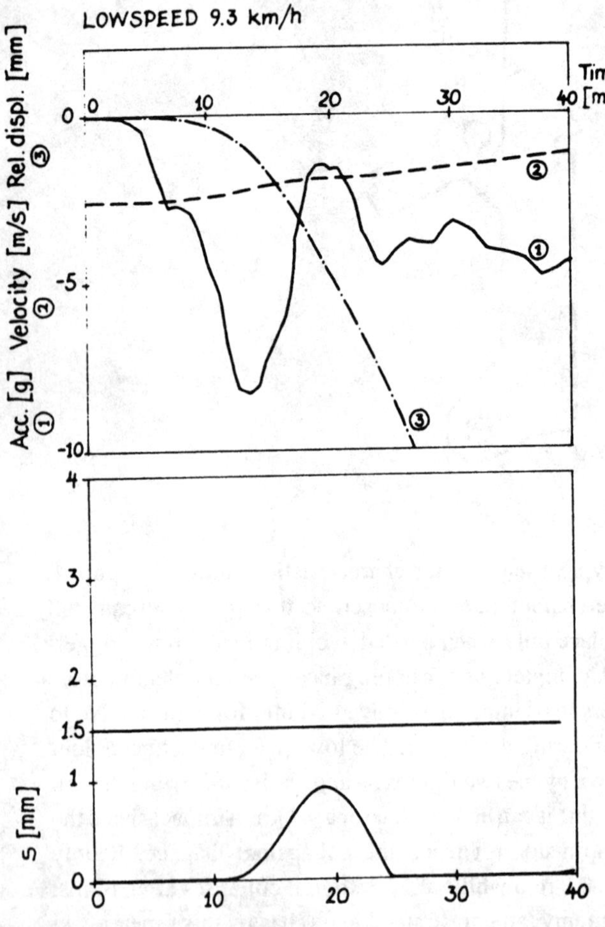

Figure 5a. 9.3 km/h impact.

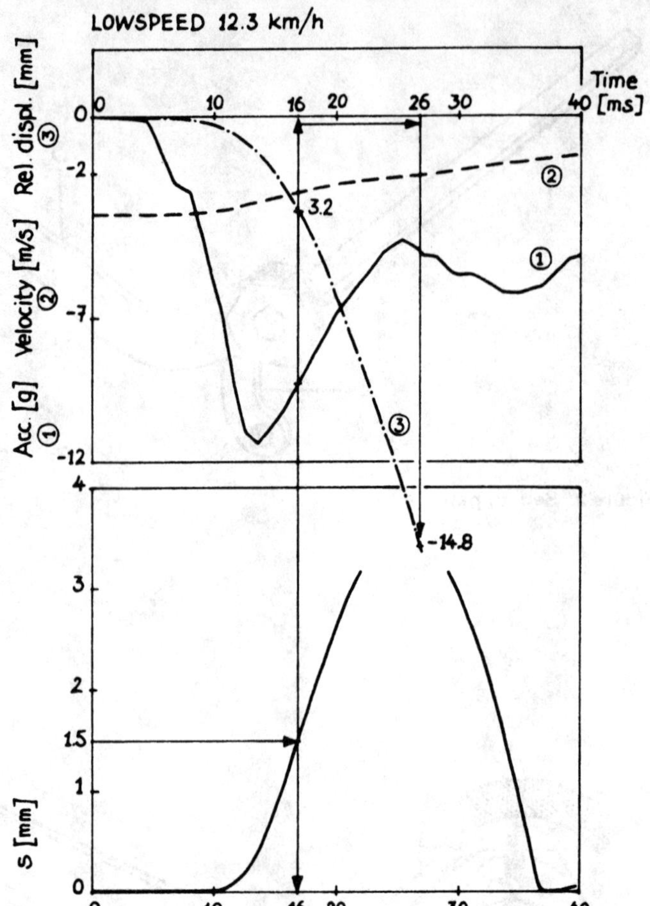

Figure 5b. 12.3 km/h impact.

in this case is completed after 26 ms. The dummy has moved 14.8 mm—this is well below the allowable 20 mm. The minimum triggering pulse for this setting was around 10 g.

The behaviour in high speed crashes is shown in figure 6a. The lever arm collapsing distance is reached after 4.8 ms—and full pretensioning is completed after 14.8 ms. By this time the dummy has moved 16.5 mm.

In an oblique crash the sensor only senses the frontal axis component of the deceleration. A comparison between a 0° and a 30° impact at 30 mph can be found in table 1. See also figures 6a and 6b respectively.

Table 1. Comparison of triggering in 0° and 30° frontal 30 mph impact.

	Triggering Time (ms)	Time to full Pretensioning (ms)	Dummy Motion (mm)
0°	5	15	17
30°	19	29	18

Pulling part

The pulling part consists of a spring and a lever arm.

When the sensor part has collapsed the spring is free to pull the buckle downwards. The nominal torque of the

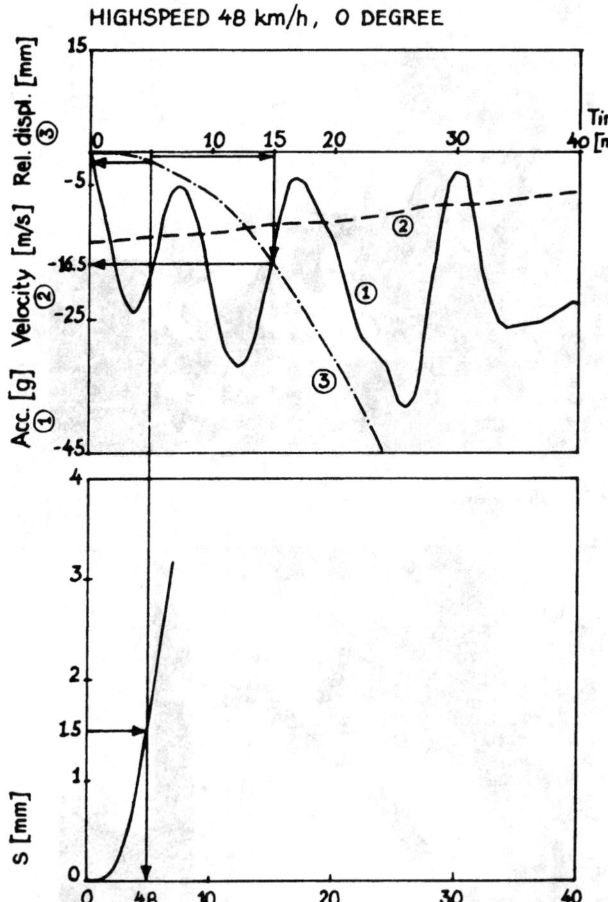

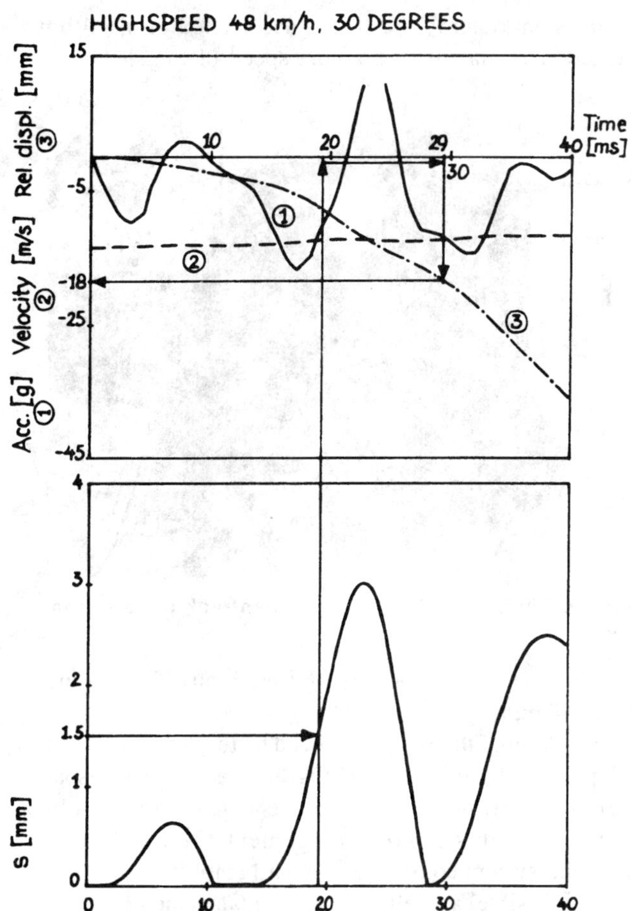

Figure 6a. 48.4 km/h 0° impact.

Figure 6b. 48.7 km/h 30° impact.

spring is 150Nm and the pulling force 1000 N. This force decreases to 40% when the pulling operation is complete. The force vs time characteristic is shown below.

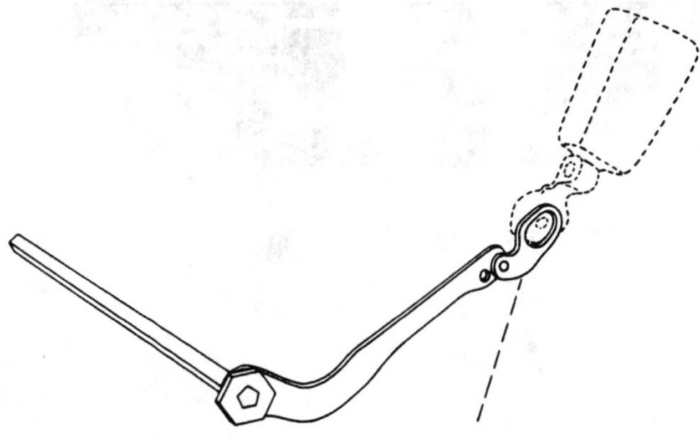

Figure 7. Pulling part.

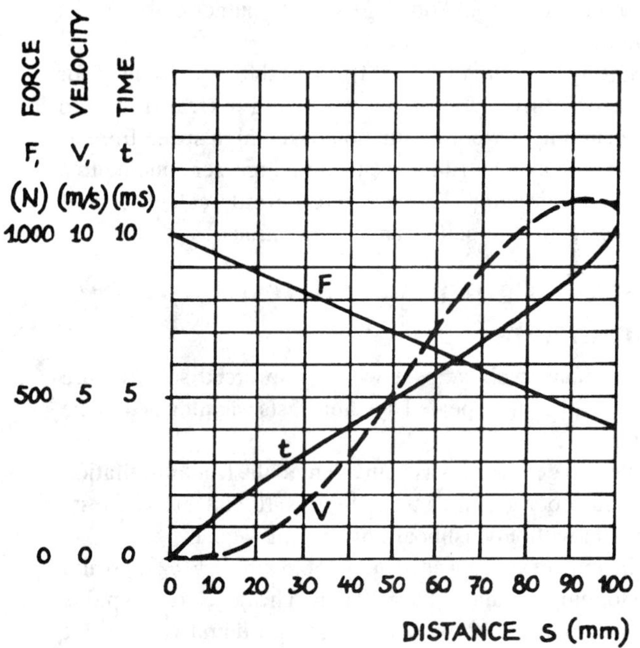

Figure 8. Force and torque characteristics.

Locking part

The occupant restraint load is taken by the locking plate, which has a toothed element. It must be able to take the load imparted by a Hybrid II 95th percentile male dummy in a 40 mph impact. The locking plate is directly bolted into the seat. The load is thereby transferred to the normal seat attachment point.

Test Equipment

The sled crash track facility at the Electrolux Autoliv R&D centre in Vågårda Sweden uses a hydraulic transmission system. The sled is accelerated over a 40 m distance. The maximum possible speed of the sled is 50

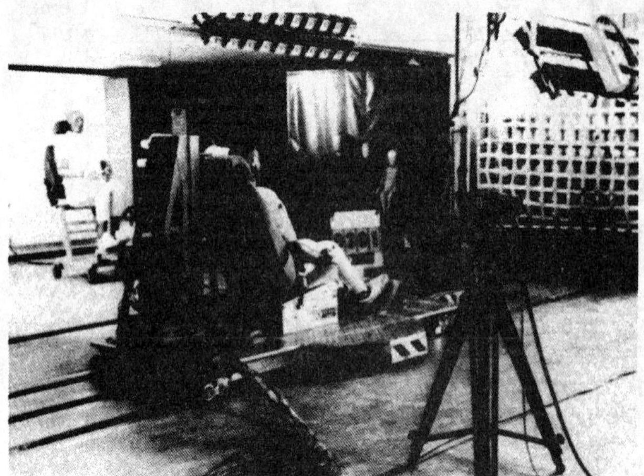

Figure 9. The Electrolux Autoliv crashtrack in Vårgårda, Sweden.

mph. Max weights that can be pulled are 1 ton at 50 mph and 2 ton at 40 mph.

The braking of the sled is achieved by the deformation of steel plates of different lengths and cross sections at different positions. Any carpulse can be accurately simulated by this brake arrangement. The sled data acquisition system consists of onboard amplifiers (max 54 channels possible) as well as an onboard data memory. The data are transferred after the test to a mini computer for processing. The amplifiers have a basic filtering of CFC 1000 and additional filtering is made numerically by the computer.

High speed cameras (max 3) are used for recording of for example dummy motion, webbing extraction and pretensioning action. Evaluation of the high speed films is undertaken using a filmscreen with a digitizer coupled to a personal computer. The software permits calculation of displacements, velocities and accelerations.

Test Configuration and Test Parameters

A standard vehicle seat with a reinforced substructure able to withstand repeated dynamic tests was attached to the sled.

The belt geometry was equivalent to the front installation of a four door sedan car. See figures 10 and 11. The tests were run without dashboard or steering wheel.

The impact speed was 35 mph (56,3 km/h) ± 0,2 mph and the stopping distance 585 ± 20 mm. The deceleration pulse was acc. to figure 12. An antropomorphic dummy, Hybrid II 50th percentile male, was used. The evaluation of test data was made in accordance with FMVSS 208.

To simulate belt slack a foam material of 30 mm thickness was inserted between the webbing and the dummy in the chest area and with two layers (60 mm) in the pelvis/abdomen area. The characteristic of the 30 mm foam was a

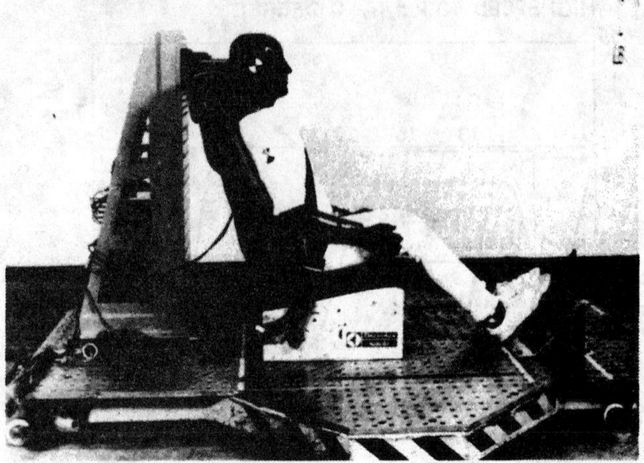

Figure 10. Side view.

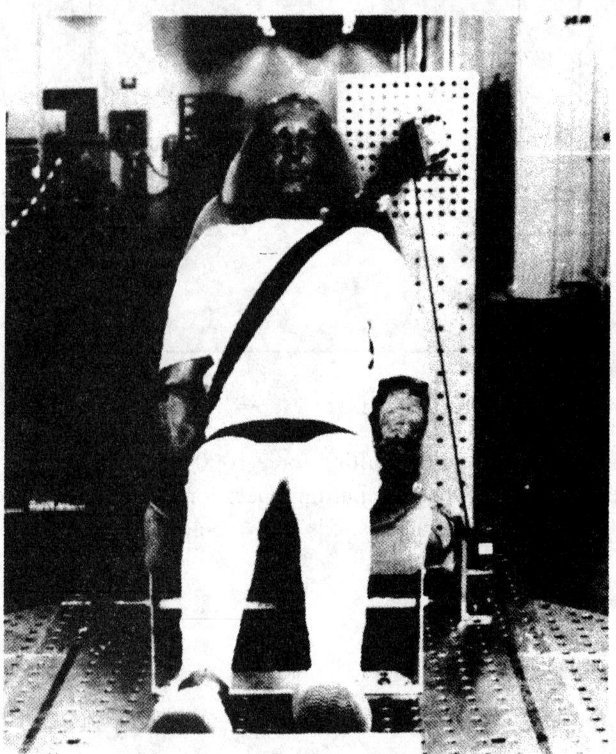

Figure 11. Front view.

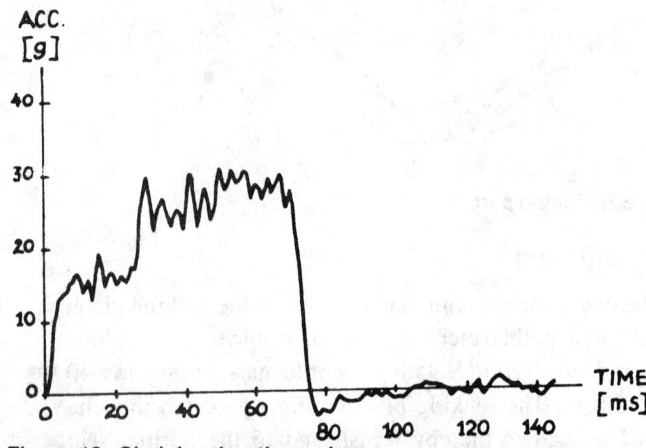

Figure 12. Sled deceleration pulse.

4.6 N/cm pressure resistance at a 40% compression. A total of about 120 mm of slack was created.

Test Results

The test results presented here are for 35 mph tests. The positive effect of the mechanical buckle pretensioner is increased at this impact speed compared to 30 mph. However some typical figures from 30 mph tests are mentioned. With the same seat configuration and belt geometry as in the 35 mph tests the use of a pretensioner resulted in a typical reduction of HIC of about 25%. This was applicable to belts both with (120 mm) slack and without. The reduction of maximum chest acceleration (>3 ms) was about 4 g in both cases.

Forward displacement

The displacements of head, chest and pelvis can be seen in the table below. The time to trigger the buckle pretensioner was 9–10 ms. The pulling down distance of the buckle was 35 mm for the belt with no slack and 65 mm for the belt with 120 mm slack.

Table 2. Head, chest and pelvis displacement.

	Head Forward Displacement (mm)	Chest Forward Displacement (mm)	Pelvis Forward Displacement (mm)
Std. three point belt. No slack.	600	350	265
Three point belt with buckle pret. No slack.	570	315	180
Std three point belt. 120 mm slack.	630	400	310
Three point belt with buckle pret. 120 mm slack.	590	325	220

The trajectory of the head for belts both without slack and with slack can be seen in figures 13 and 14.

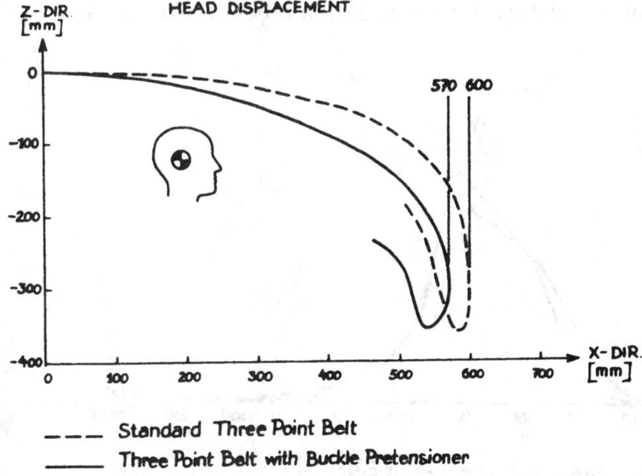

Figure 13. Head displacement. Three point belt without slack.

As expected the reduction in forward displacement is larger for a belt system with slack than for a corresponding one without slack. The reduction is also larger for the pelvis than for the head. For belts with 120mm slack the head forward displacement is reduced by 40mm from 630 to 590 mm and the pelvis forward displacement by 90 mm from 310 to 220 mm. This latter figure indicates how effective the buckle pretensioner is in preventing submarining.

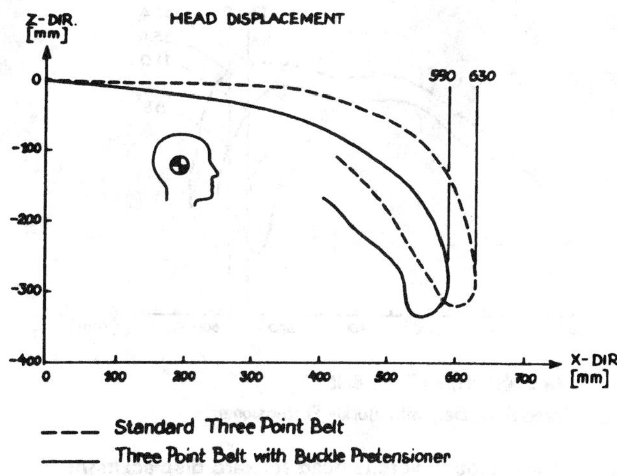

Figure 14. Head displacement. Three point belt with 120 mm slack.

As far as the head is concerned it is not only the forward displacement that is of interest but also the speed when for instance impacting the upper rim of the steering wheel. Figures 15 and 16 show the velocity of the head versus forward displacement.

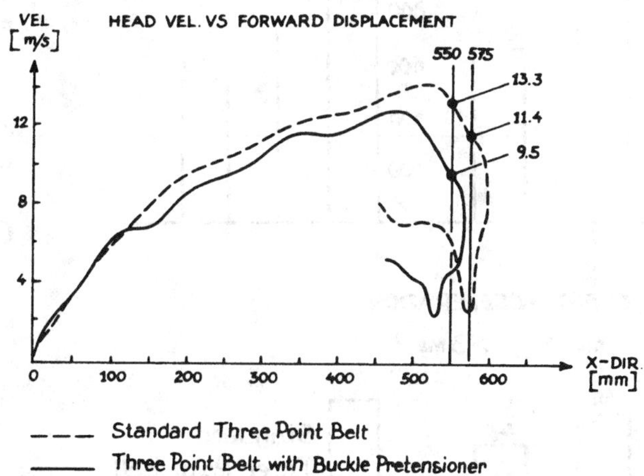

Figure 15. Head velocity versus head forward displacement. Belts without slack.

If the head impacts the steering wheel after a 550 mm forward movement the velocity is 13.3 m/s for the standard three point belt with no slack and only 9.5 m/s for the belt with a buckle pretensioner. If the steering wheel is another 25 mm forward (at 575 mm) the impacting velocity is 11.4 m/s for the standard belt whilst head contact will not occur in the case of a belt with a pretensioner. The test results for belts with slack show the same effect. The buckle pretensioner is therefore effective in reducing the impact of the head into the steering wheel.

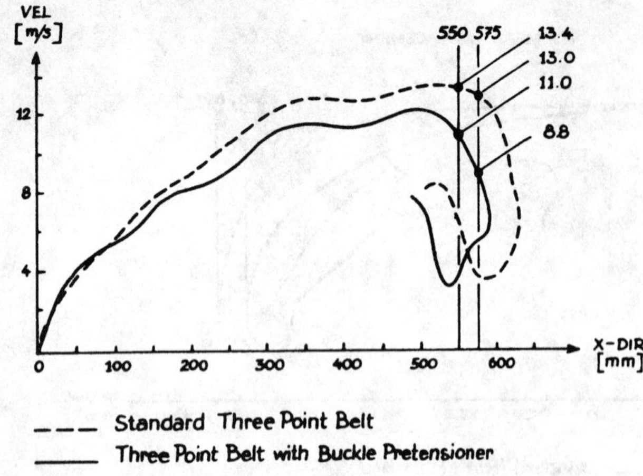

Figure 16. Head velocity versus head forward displacement. Belts with 120 mm slack.

Head injury criteria and chest acceleration

Figures 17 and 18 show values for head injury criteria (both HIC and HIC_{36}) and chest acceleration for the different cases.

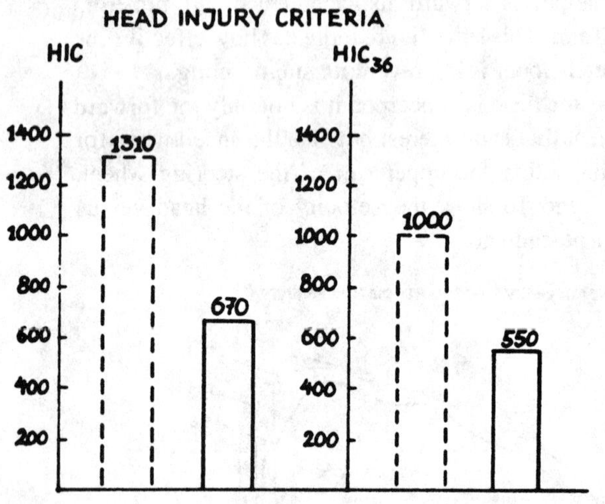

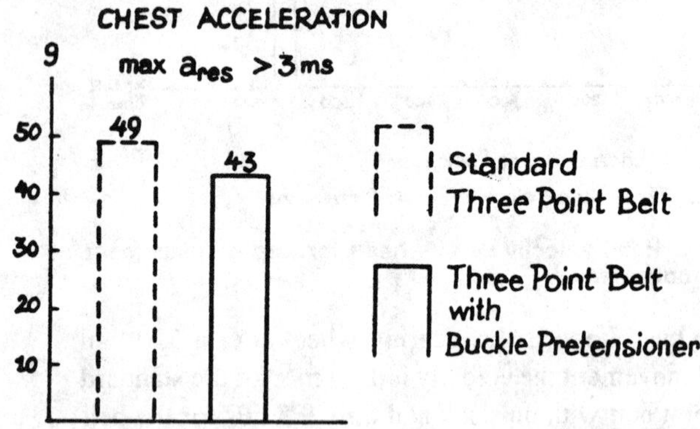

Figure 17. HIC and chest acceleration. Belts without slack.

The reduction of HIC is considerable—about 50% in all four cases. The largest reduction, from HIC 1880 to 870 (–54%), was achieved with the belt with 120 mm slack. The corresponding figures for HIC_{36} were 1220 and 670 respectively.

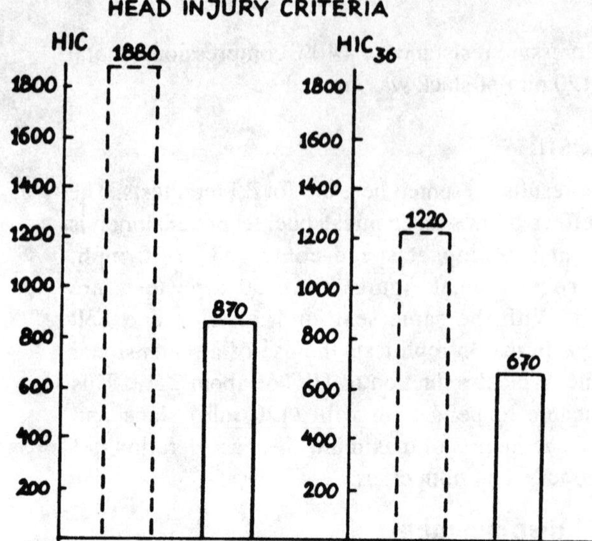

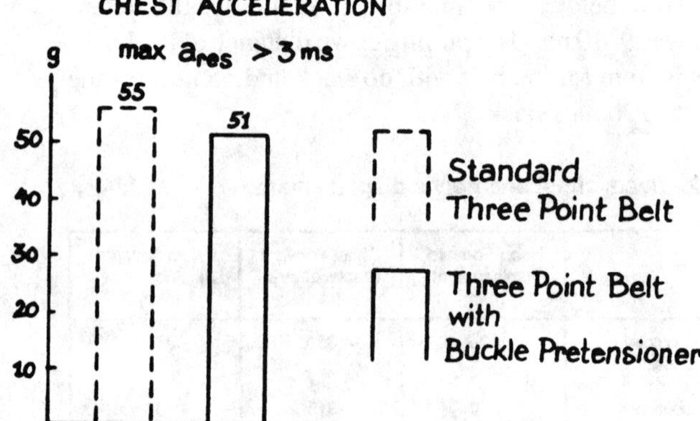

Figure 18. HIC and chest acceleration. Belts with 120 mm slack.

There is also a significant reduction of max chest acceleration (>3 ms). The reduction was from 49 g to 43 g for belts with no slack and from 55 g to 51 g for belts with slack.

Chest belt force

The curves for the chest belt forces in figures 19 and 20 show that the dummy is arrested earlier (about 8–10 ms) in

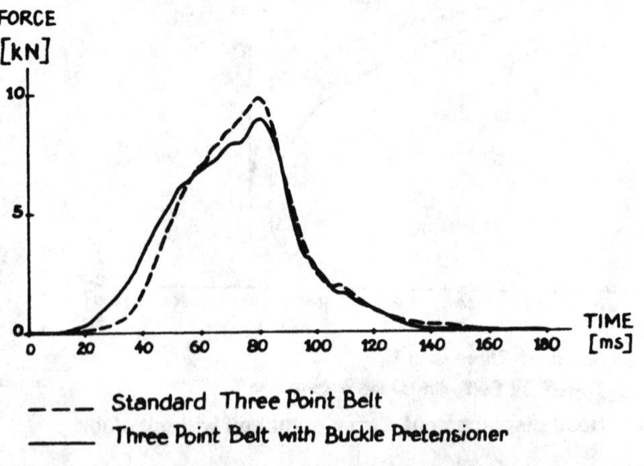

Figure 19. Chest belt force. Belt without slack.

belts where the buckle pretensioner has worked. The slope of the curves are lower and there is a reduction in peak values.

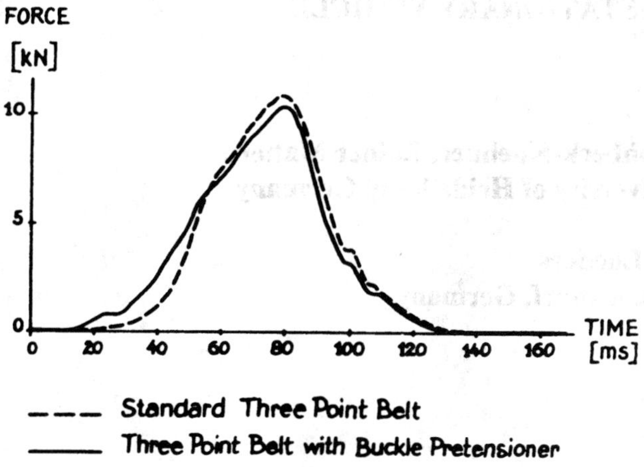

Figure 20. Chest belt force. Belts with 120 mm slack.

Summary and Conclusions

The mechanical buckle pretensioner acts with a high initial spring force of 1000 N. It can trigger and complete the pretensioning operation before an unrestrained mass has moved 20 mm forward. The sensor threshold level can be set so that inadvertant triggering can be avoided.

Test result show that the use of a buckle pretensioner will mean a less severe head impact with the steering wheel. The HIC value can be reduced significantly. In 35 mph barrier tests a reduction of HIC by 50% to less than 1000 was achieved with 120 mm of initial slack in the belt.

The pelvis displacement was decreased by up to 90 mm, thereby reducing the risk of submarining.

The belt with a pretensioner is loaded quicker (about 10 ms). This means lower peakloads and reduced maximum chest accelerations.

It can be concluded that the mechanical buckle pretensioner is a cost effective device to supplement the standard three point belt.

VOLUNTEER TESTS ON HUMAN TOLERANCE LEVELS OF PRETENSION FOR REVERSIBLE SEATBELT TENSIONERS IN THE PRE-CRASH-PHASE

PHASE I RESULTS: TESTS USING A STATIONARY VEHICLE

Bernd Lorenz, Dimitrios Kallieris, Peter Strohbeck-Kuehner, Rainer Mattern
Institute of Legal and Traffic Medicine, University of Heidelberg, Germany

Uwe Class, Michael Lueders
TRW Occupant Safety Systems, Alfdorf, Germany

ABSTRACT

It is the aim of this current study to define the maximum force of seatbelt pretension in the pre-crash-phase tolerable for a car passenger. This is attempted by volunteer tests using a car fitted with a prototype of a reversible system for belt pretension. The volunteers (14 f, 10 m, aged 16 to 73) represent a broad spectrum of car-users.

Up to now 64 tests were conducted in a stationary vehicle to determine the tolerable strain especially under Out of Position (OOP) conditions.

The head acceleration measured through accelerometers which were mounted on individually fitted dental adapters, was rather low in all tests with some increase for the OOP-experiments (a_{max}= 2,9 g).

Belt forces were 0,16 kN < F_{Bmax} < 0,29 kN.

Under the present test set up and conditions, the loadings were assessed by the test persons as tolerable and acceptable. In general, the belt forces measured with shorter and lightweight persons were higher than those measured with tall and heavier persons. This can be improved by a special algorithm e.g. in connection with a weight-sensor in the car seat so as to better adapt the system to the anthropometric parameters of the occupant.

Key words: reversible seatbelt tensioner, safety belts, volunteers, tolerances, soft tissue

ADVANCED RESTRAINT SYSTEMS are being intensively researched on, in order to improve the safety of car passengers. Reversible systems are the first step to integrate seatbelts into active safety of cars (pre-crash phase). The reversible seatbelt tensioner system is developed to reduce slack of the seatbelt system in critical situations of the vehicle. Further a better positioning of the occupant is achieved and Out of Position (OOP) situations can be avoided.

It is the aim of the current study to define the maximum force of belt pretension in the pre-crash-phase tolerable for a car passenger. This is attempted by volunteer tests using a car fitted with a prototype of a reversible system for belt pretension. The reversible belt tensioner is mounted at the B-pillar. Presently, the reversible tensioner is limited to the maximum belt forces measured by the performance of the motor available.

This study consists of two major parts. In phase 1, tests were conducted in a stationary vehicle to determine the tolerable strain especially under OOP conditions. Phase 2 consists of driving tests simulating emergency situations (in progress).

Several studies have been published on related subjects. There are some studies on effects and injury mechanisms of restraint systems, i.e. 3-point-belts and air bags in real collisions and under laboratory conditions (p. ex. Kallieris et al. [5]-[8]; see: References). However, no comparable studies were found on the effects of reversible seatbelt tensioners, that, on one hand, generate distinctly lower belt forces and belt velocities, but on the other hand, show performance characteristics over a much longer period of time in contrast to the pyrotechnic seatbelt tensioners, which ignite only on the occasion of a collision. In the meantime, some studies have been published on the problems of OOP during extreme driving manoeuvres, i.e. ABS braking and simultaneous evasive action. These studies emphasize the necessity for reducing the forward displacement of the passenger so as to take the passenger back to a defined better or ideal position (p. ex. Kümpfbeck et al. (1999, [9]) and Zuppichini et al. (1997, [24])).

METHOD

THE GROUP OF RESPONSIBLE TEST SUBJECTS: Fourteen female (aged 16-73 yrs) and ten male (aged 32-62 yrs.) persons, with two exceptions all members of the institute, volunteered for the experiments (cf. table in the appendix). Each participant received detailed instructions on the study and course of the experiment according to suggestions from the ethics commission. Participation was entirely voluntary and each test subject took responsibility for their own tests, continuing or discontinuing the experiment at any given moment without showing cause or incurring any disadvantage whatever.

KIND AND NUMBER OF TESTS: Each test subject would, generally, participate in at least two tests. Ten pre-tests were designed to give a rough idea of the involved forces and to test the measurement technology including its periphery.

A total of 64 tests (cf. tabular overview in the appendix) has so far been conducted. Wherever possible, each test subject would participate in one test for normal or ideal seating position („In Position" IP) and in one for an Out Of Position-situation (OOP).

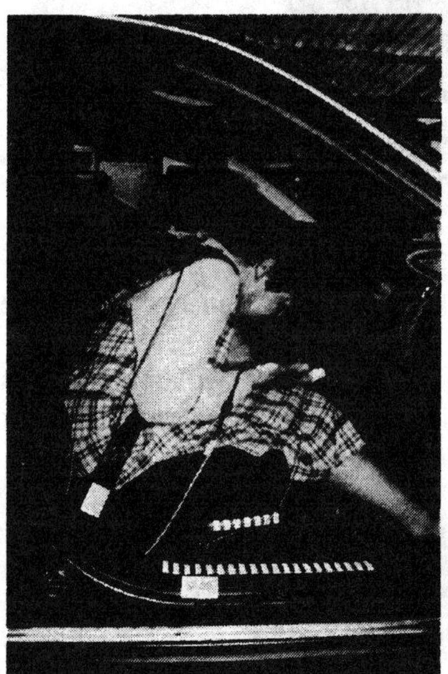

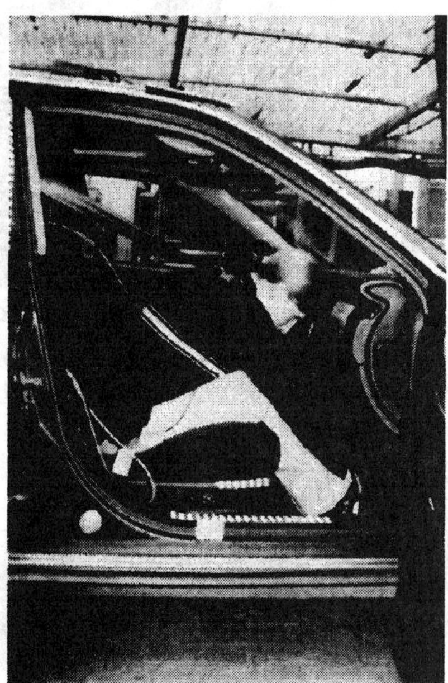

Figs. 1+2: Examples of OOP-situations (V 39, V41)

For the first IP-test, each test subject could adjust the seat to a subjectively comfortable and relaxed seating position. The OOP-situation was also simulated according to personal preferences, ideally leaving the person doubled over with the head close to the dashboard (resembling the position while tying shoe-laces or reaching for a handbag, cf. Figs. 1 and 2).

MEASUREMENT

Points of measurement on the volunteers

Head acceleration was taken using individually fitted (if accepted by the volunteers) dental adapters (cf. Figs. 3 and 4) on which three uniaxial accelerometers were mounted. The dental adapters are similar to those used by Lorenz et al. (1999).

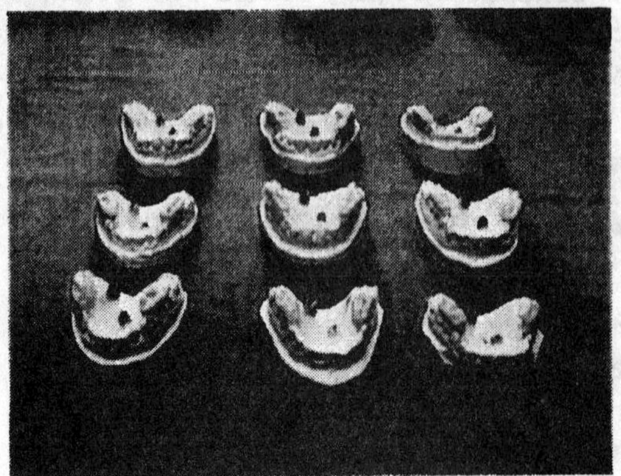

Fig. 3: impressions of the upper jaw with adapters

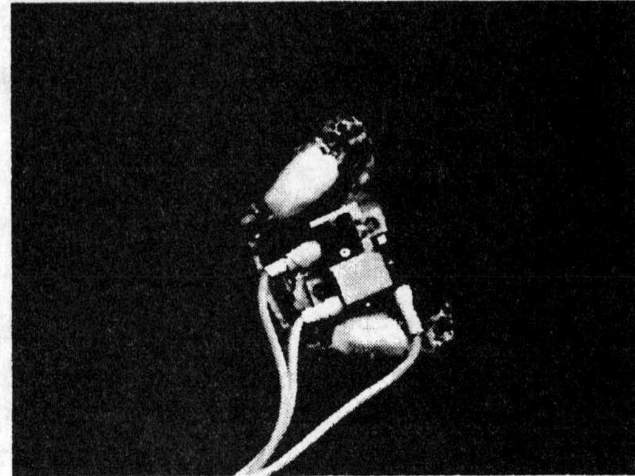

Fig. 4: Adapter with accelerometers

As parameters for physiological refernce, measurements of muscle tonus, skin surface resistance and pulse frequency were taken using a compact measurement equipment named BIOPAC developed by Dr. Maus Elektronik, Frankenthal (cf. Fig. 5)

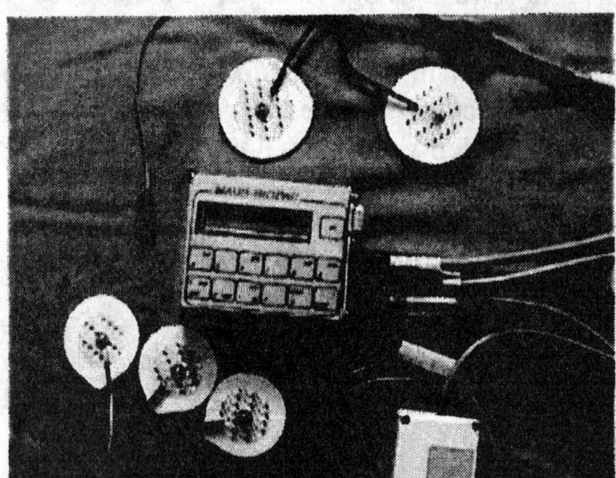

Fig. 5: BIOPAC with sensors for measuring muscle tonus, skin surface resistance, pulse frequency

Signals of muscle tonus were derived through three adhesive electrodes (two active, one passive, cf. Fig. 6) connected to the BIOPAC and attached to the skin of the test persons on the left shoulder next to the cervical vertebrae (trapezius muscle).

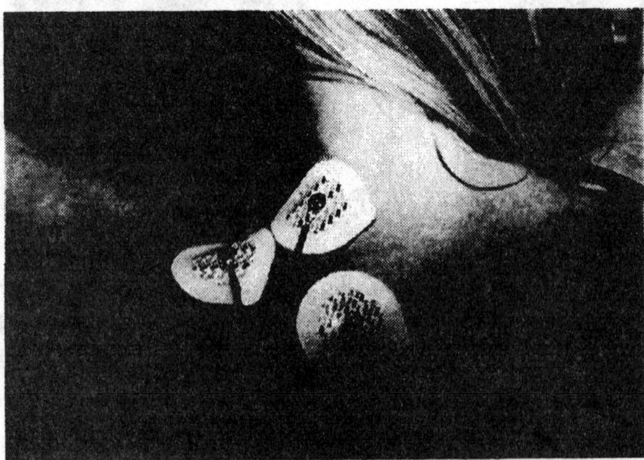

Fig. 6: Adhesive electrodes for measuring muscle tonus in position

Skin resistance was measured with two electrodes on the palms of the right hand.

Measuring pulse frequency on the right ear lobe turned out the be impracticable and unreliable, later infra-red measurements were carried out on the right arm.

The BIOPAC was modified several times throughout the current series of tests by Dr. Maus Elektronik (p. ex. Adjustment of sensitivity for muscle tonus measurements)

Measurement on the reversible belt tensioner system

Belt forces were measured through a belt force transducer attached to the seat belt at shoulder of the test person.

Belt movement was derived from the analysis of high speed video films.

For tests No. 59 to 64 an opto-electronic belt movement sensor was at our disposal (resolution 0,25 mm).

Belt velocity was calculated from the measured belt movement.

A measuring shunt for taking the motor current intensity was also available from test 59 onwards.

DOCUMENTATION: The tests were documented using a high-speed video camera (500 colour expositions per second) and standard video cameras.

Additional photographs were taken immediately before the tests to document the seating position.

PSYCHOLOGICAL ASSESSMENT AND PHYSIOLOGICAL PARAMETERS: Subjective statements whether a particular stimulus is rated as pleasant, neutral, or unpleasant not only depend on the stimulus itself but also on the person´s expectation, psychological well-being, and activation since these factors have an impact on the individual´s threshold of perception. Hence, in studies like this, it is necessary to consider that rather anxious persons have negative expectations that can lead to a lower threshold of perception and, consequently, to a more intense perception.

In order to control these effects in the present study, state of anxiety and subjective activation of the test subjects were assessed using the short form of the General Activation-High Anxiety-State-Scale (GA-HA-State-KF; Wieland-Eckelmann, 1992; a German adaption of the Activation-Deactivation-Check-List by Thayer, 1967). This self-report check-list consists of eight adjectives representing two sub-scales: „General Activation“ (GA) and „High Anxiety“ (HA). Immediately before the test, test persons were asked to fill in the check-list on how they actually feel.

In order to investigate potential effects of the test persons´ anxiety and activation on the rating of the present tests, the results obtained through the GA-HA-State-KF check-list of the present study were compared with the results of two other samples obtained in other settings using the same check-list. One sample to be compared were 180 persons undergoing a medical-psychological driver selection setting (sample I). The other sample consisted of 48 volunteers participating in an experiment on the impact of music on attention (sample II).

INTERVIEW AND PHYSICAL EXAMINATION: The test subjects were questioned concerning their personal impressions and examined physically by a medical doctor if necessary immediatly after each test. The interviews were documented on video and minuted.

The medical findings, if any, were documented.

RESULTS AND DISCUSSION

The results evaluated so far (belt forces, belt movement, belt velocity and head acceleration) are available in tabular form (maximum values) in the appendix. The table also includes the relevant anthropometric data for each test person.

The head acceleration was rather low in all tests with some increase for the OOP-experiments ($a_{cresmax}$= 2,9 g), which is why it was not recorded during later tests.

Belt forces were 0,16 kN < F_{Bmax} < 0,29 kN.

Fig. 7 shows a typical belt force time history for an OOP-test including the head acceleration time history of the test with the highest measured head accelerations. Fig. 8 shows the plots for a IP-test.

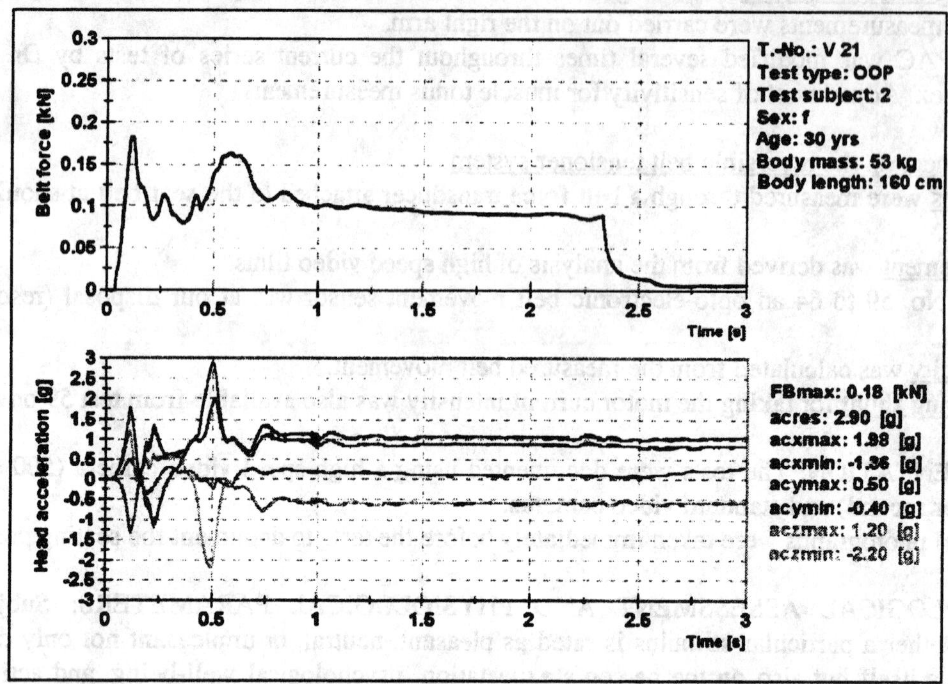

Fig. 7: Belt force and head acceleration of an OOP-test

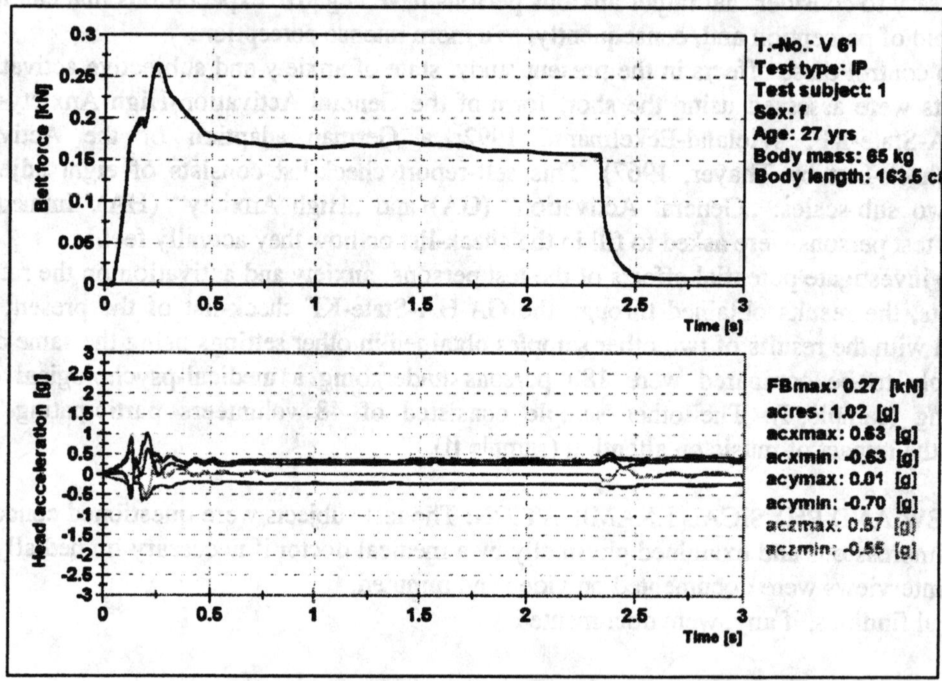

Fig. 8: Belt force and head acceleration of an IP-test

In general, it proved to be impossible to avoid a certain anticipation of test subjects, despite all efforts to the contrary in creating a relaxed and comfortable situation. This impression was also supported by physiological data (with reservation as to future data evaluation).

Only in one experiment (test subject 4), when, due to a defect in the switch, the tensioner was tightened inadvertently a second time was the test subject genuinely surprised while he was removing the dental adapter and electrodes. There were, however, no ill effects.

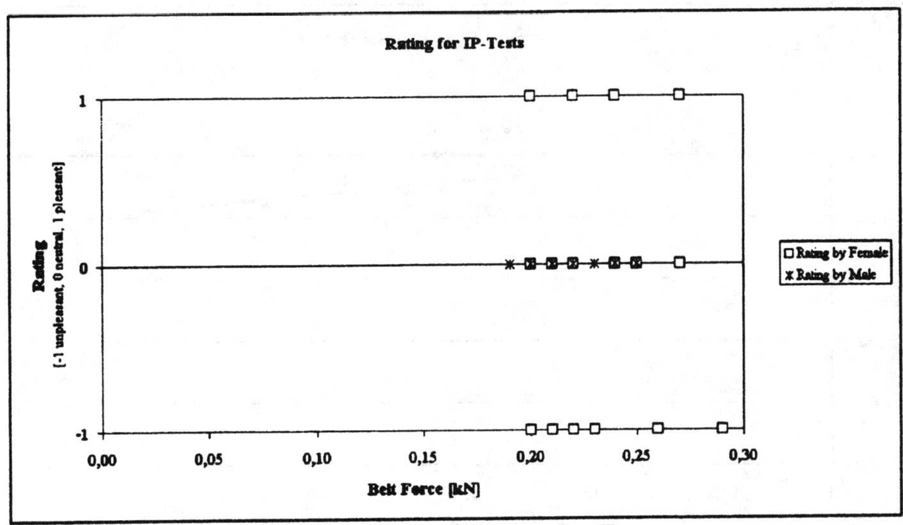

Fig. 9: Rating of volunteers for IP-tests

The male test persons suffered no discomfort during either the IP or with one exception the OOP tests (cf. Figs. 9 and 10). The system was felt to be neither particularly pleasant or unpleasant, at most as unproblematic and well tolerable. Two of the bigger and heavier male test subjects (numbers 6 and 20, 193 cm, 93 kg and 183 cm, 120 kg, respectively) doubted, however, that the belt force would be sufficient to pull them back from an OOP-position during braking.

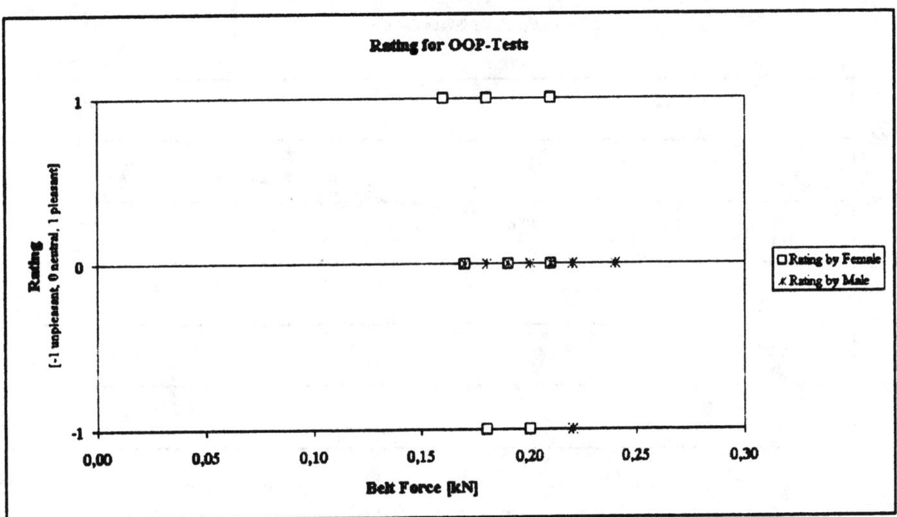

Fig. 10: Rating of volunteers for OOP-tests

Belt force tended to be higher in IP-situations (cf. Fig. 11) especially for lightweight and shorter persons. The highest measurement occurred in normal seating position during test 10, test subject 7, a female of 57 kg and 38 years old. The seat had been moved as far back as possible. The test subject criticized the system after the test as unpleasant and complained of the feeling of high pressure exerted on the thorax by the seat belt („thorax is compressed"). There were, however, neither lasting complaints nor medical findings on the skin.

After analysing high speed video films and comparing data from several tests, an assumption was made that the force of belt pre-tension was decicively influenced not only by seating position, body mass and height, but also by the clothes due to friction. Therefore, the test person 7 (f) was asked to repeat the test wearing different clothes (tests 45 and 46) and the same clothes as in the tests 10 and 15 (tests 47 and 48). The forces measured with clothes made out of synthetical fibres (95% polyamid and 5% elasthan) were higher than those measured with other fabrics.

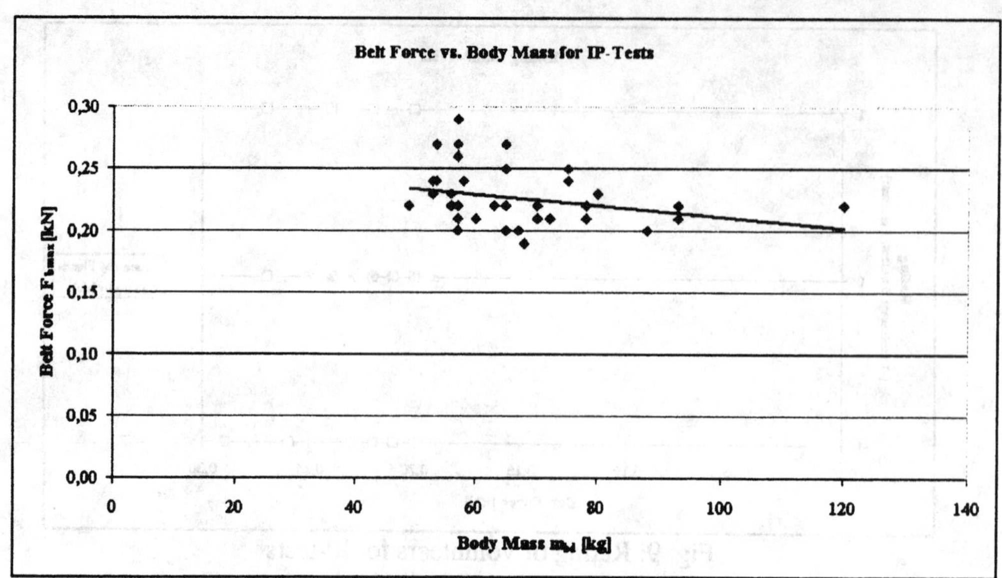

Fig. 11: Belt force vs. body mass for IP-tests

Belt force tended to be lower in OOP-situations (cf. Fig. 12), and subjective assessment of the system for OOP-situations also provided one unexpected result. Four female test subjects (7, 10, 14, 15) claimed the feeling of being pulled back from the OOP-position to be very pleasant.

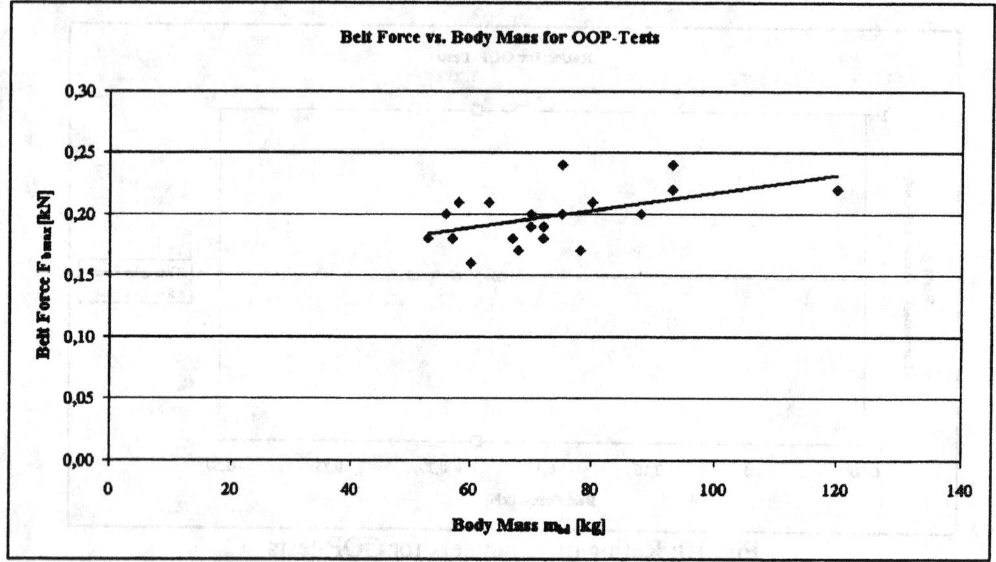

Fig. 12: Belt force vs. body mass for OOP-tests

One further unexpected result occurred for tests (40, 41) with a 16-year-old test subject (19), who already reported some slight pain on the clavicle even after the first test. She insisted on another (OOP) test, a medical examination immediately afterwards showed distinct redness of the skin. On the following day, both redness and pain had subsided entirely, but the impression of the belt on the skin was still clearly visible.

Although the test subject had earlier witnessed other tests and was in anticipation, she showed clear signs of being startled on the release of the system both times. It should be mentioned that her clavicles protrude slightly more than usual.

In order to verify this result, it was searched for more persons with similar protruding clavicles among the institute members. Two test subjects (23 and 24) were found and volunteered for the tests. They did not know what the aim of the tests was. Both test persons have not reported any problems in the area of clavicles; however, they rated the tests as rather unpleasant.

The oldest test subject (72 yrs, subject 16) claimed to have suffered no discomfort during either IP- or OOP-test (cf. Fig. 1). However, she, too, was clearly startled in both tests, notwithstanding her anticipation.

Fig. 13 shows the time histories of belt force, belt movement, belt velocity and motor current intensity for an IP-test with test subject 13.

The opto-electronic belt movement sensor was at our disposal only for the last IP-tests (tests No. V 59 to V 64). The measured belt movement is about 80 to 96 mm and similar to those values for IP-tests derived from high speed film analysis (80 to 105 mm).

The calculated belt velocity was about 1 m per sec.

The belt movement for the OOP-tests depends on the seating position respectively how far the test subjects were bent forward. The range was from 105 to 500 mm.

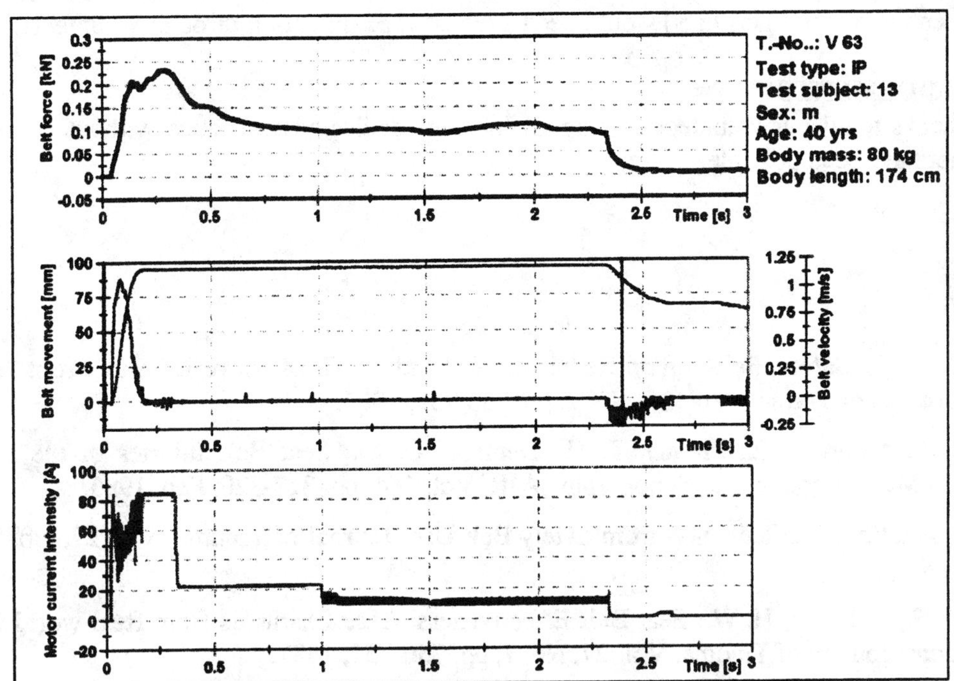

Fig. 13: Time histories of belt force, belt movement, belt velocity and motor current intensity

Out of the 64 tests, 11 tests were rated by test persons as pleasant, 40 as neutral (neither pleasant nor unpleasant), and 13 as unpleasant.

Regarding the effect that anxiety and activation might have had on the rating of the tests, the present study and samples I and II were compared. The mean of the sum-scores of the anxiety check-list was with 6.51 significantly lower ($p<0.01$) in the present study than the mean of sample I (8.87). In contrast, there was no difference to sample II (mean 7.08). As far as activation was concerned, the data of the persons in the present study were similar to those in sample I. In contrast, the test subjects of the present study reported a lower subjective activation than the volunteers of sample II.

These results show that the test subjects of the present study have not reacted to the test with any remarkable anxiety or increased activation. Consequently, the risk that the assessment of the stimulus was influenced by changes in the psychological well-being of the test subjects caused by a test situation can be rated as rather low.

Furthermore, it was analysed whether the test subjects who have rated the test as „unpleasant" have, at the same time, reported increased anxiety or activation. Those test persons who have described the test as „unpleasant" have shown a higher mean of state anxiety (7.37) than those who have assessed the test as „neutral" or „pleasant" (mean 6.32). In contrast, as far as the subjective activation was concerned, there have not been noticed any considerable differences in the means. Due to the small sample size of test subjects who have described the test as „unpleasant", no further statistical analysis was performed.

CONCLUSIONS

Under the present test set up and conditions, the loadings applied by the prototype of the reversible belt tensioner were assessed by the test persons as tolerable and acceptable. However, risk groups (e.g. pregnant women or persons with osteoporosis) would need a separate assessment of their potential injury risk.

In general, the belt forces measured with lightweight and shorter persons were higher than those measured with tall and heavier persons. This could be improved by a special algorithm e.g. in connection with a weight-sensor in the car seat so as to better adapt the system to the anthropometric parameters of the occupant.

The represented results of the first phase of the project still have to be validated through driving tests that are already in preparation. Furthermore, the final version of the reversible belt tensioner (changed trigger and motor algorithms) would need a repeated examination through volunteers.

ACKNOWLEDGEMENTS

Special thanks to all our volunteer colleagues for their dedicated assistance without which this study would not have been possible.

REFERENCES

[1] Campbell, B. J.: Safety Belt Injury Reduction Related to Crash Severity and Front Seated Position. Journal of Trauma, Vol. 27, No. 7, pp. 733-739, 1987.

[2] DiPiro, P. J.; Meyer, J. E.; Frenna, T. H.; Denison, C. M.: Seat Belt Injuries of the Breast: Findings on Mammography and Sonography. AJR, Vol. 164, pp. 317-320, Feb. 1995.

[3] Evans, L.: Fatality Risk Reduction from Safety Belt Use. Journal of Trauma, Vol. 27, No. 7, pp. 746-749, 1987.

[4] Huelke, D. F.; Sherman, H. W.: Seat Belt Effectiveness: Case Examples from Real-world Crash Investigations. Journal of Trauma, Vol. 27, No. 7, pp. 750-753, 1987.

[5] Kallieris, D.; Mellander, H.; Schmidt, G.; Barz, J.; Mattern, R. Comparison Between Frontal Impact Tests with Cadavers and Dummies in a Simulated True Car Restrained Environment. Paper-No. 821170, Proc. 26. STAPP Car Crash Conf., P-113, pp.. 353-367, SAE, Ann Arbor, Mich.; USA, 1982.

[6] Kallieris, D.; Rizzetti, A.; Mattern, R.: The Biofidelity of Hybrid III Dummies. Proc. IRCOBI Conf., pp. 135-154, Brunnen, Switzerland, 1995.

[7] Kallieris, D.; Rizzetti, A.; Mattern, R.; Morgan, R.; Eppinger, R.; Keenan, L.: On the Synergism of the Driver Air Bag and the 3-Point Belt in Frontal Collisions. Paper No. 952700, Proc. 39. STAPP Car Crash Conf., P-299, SAE, San Diego, USA, 1995.

[8] Kallieris, D.; Del Conte-Zerial, P.; Rizzetti, A.; Mattern, R.: Prediction of Thoracic Injuries in Frontal Collisions. 16. ESV-Conference, P- 98-S7-O-04, pp. 1-14, Windsor, Canada, June 1-4, 1998.

[9] Kümpfbeck, M.; Oertel, D.; Pilatus, K.: Occupant Kinematics During Emergency Breaking - An Investigation with Regard to the Out of Position issue for Passengers. Proc. IRCOBI Conf., pp. 443-454, Sitges, Spain, 1999.

[10] Leung, Y. C.; Tarrière, C.; Fayon, A.; mairesee, P.; Delmas, A.: A Comparison between Part 572 Dummy and Human Subject in the Problem of Submarining. SAE Paper-No. 791026. 23. Stapp Car Crash Conf., 677-719, 1979.

[11] Leung, Y. C.; Tarrière, C.; Lestrelin, D.; Got, C.; Guillon, C.; Patel, A.; Hureau, J.: Submarining Injuries of 3 Pt. Belted Occupants in Frontal Collisions - Description, Mechanism and Protection. Proc. 26. STAPP Car Crash Conf., P-113, pp. 173-205, SAE Paper No. 821158, Ann Arbor, Mi., USA, 1982.

[12] Lorenz, B.; Kallieris, D.; Mattern, R.; Wienecke, P.: Volunteer and Dummy Tests for the Validation of Biomechanical Assessment Criteria under the Euro-NCAP: Proc. IRCOBI Conf., pp. 385-404, Sitges, Spain, 1999.

[13] Martinez, R.; Sharieff, G.; Hooper, J.: Three-Point Restraints as a Risk Factor for Chest Injury in the Elderly. The Journal of Trauma, Vol. 37, No. 6, pp. 980-984, 1994.

[14] Meßner, G.; Hübner, W.: Unfallforschung: Konsequenzen für die Auslegung von Rückhaltesystemen. Verkehrsunfall und Fahrzeugtechnik, Heft 7/8, S. 186-190, 1997.

[15] Otte, D.; Richter, M.: Häufigkeit und Mechanismen von Armverletzungen bei gurtgeschützten Pkw-Fahrern. Verkehrsunfall und Fahrzeugtechnik, Heft 1, S. 2-9, 1999.

[16] Patrick, L. M.; Trosien, K. R.: Volunteer, Anthropometric Dummy, and Cadver responses with Three and Four Point Restraints. Automotive Engineering Congress, SAE Paper-No. 710079, 1971.

[17] Sato, T. B.: Effects of Seat Belts and Injuries Resulting from Improper Use. Journal of Trauma, Vol. 27, No. 7, pp. 754-7758, 1987.

[18] Schmidt, G. Injury Patterns of Fatally Impacted Car Front Passangers in Regard to the Three-Point Seat Belt and Related Occupant Injury Modes. Int. Congress & Exposition, Detroit, Mich., USA, Paper-No. 870225, SAE, Feb. 23-27, 1987

[19] Siegel, J. H.; Mason-Gonzalez, S.; Dischinger, P.; Cushing, B.; Read, K.; Robinson, R.; Smialek, J.; Heatfield, B.; Hill, W.; Bents, F.; Jackson, J.; Livingston, D.; Clark, C. C. Safety Belt Restraints and Compartment Intrusions in Frontal and Lateral Motor Vehicle Crashes: Mechanisms of Injuries, Complications, and Acute Care Costs. The Journal of Trauma, Vol. 5, No. 34, pp. 736-759, 1993.

[20] Sleet, D. A.: Motor Vehicle Trauma and Safety Belt Use in the Context of Public Health Priorities. Journal of Trauma, Vol. 27, No. 7, pp. 695-702, 1987.

[21] States, J. D.; Huelke, D. F.; Dance, M.; Green, R. N.: Fatal Injuries Caused by Underarm Use of Shoulder Belts. Journal of Trauma, Vol. 27, No. 7, pp. 740-745, 1987.

[22] Thayer, R.E.: Measurement of Activation Through Self-Report. Psychological Reports 20, pp. 663-678, 1967.

[23] Wieland-Eckelmann, R.: Kognition, Emotion und psychische Beanspruchung. Hogrefe, Göttingen, 1992.

[24] Zuppichini, F.; Bigi, D.; Bachi, E.; Adamich, A. de: How Drivers Behave in Pre-Impact Emergency Situations. Proc. IRCOBI Conf., pp. 433-441, Hannover, Germany, 1997.

CHAPTER 7:

LOAD LIMITERS, WEB CLAMPS AND LOCKS

870328

Items of an Engineering Program on an Advanced Web-Clamp Device

Dieter Adomeit
Adomeit Automotive Engineering
Berlin, West Germany
Werner Balser
Hamburg, West Germany

ABSTRACT

The objective of this paper is to introduce a web-clamp concept which can be adapted to emergency locking retractors (ELR's) or entirely replace these current seat belt locking devices in order to enhance the safety of car passengers in collision-accidents.

Different types of web-clamp systems which demonstrate the discussed clamping concept are demonstrated and evaluated in comparison with the current ELR, as well as the pyrotechnically pre-loaded ELR system.

The theory of the clamping concept as developed here, as well as different designs with correlating performance data (safety, comfort), will be discussed.

It will be shown that a well-adjusted restraint system based on this web-clamp concept can reach a safety and comfort level close to pyrotechnically pre-loaded ELR systems.

THE LATEST GENERATION of duo-sensitive emergency locking retractors (ELR's) reached a significant level of safety performance. Accuracy of vehicle and webbing sensitivity increased, and locking distance was reduced. Futhermore, impressive weight reduction and even certain production-cost reduction can be observed.

However, principle disadvantages due to the locking concept of these systems have not been eliminated. These are:

- delayed restraint force onset due to necessary web-extraction for activating the locking mechanism
- the film spool-effect, leading to further passenger displacement on an ineffectively low restraint-force level.

To overcome these disadvantages of standard ELR systems, low-elongation webbing is used to control and limit the passenger's forward displacement. This causes the following effects on the passenger:

- high restraint force level during the final third of displacement
- unnecessary high passenger loadings by the restraint system (HIC, decelerations, etc.)

In terms of crash-test data, disadvantages of the standard ELR can be characterised as:

- restraint activation delayed in comparison to beginning of passenger displacement after t_0
- low onset rate of restraint forces during the first two thirds of passenger displacement

Since further reduction of passenger loading during the crash phase cannot be realized by increasing displacement, due to space limits, there is obviously only one measure for more favourable restraint conditions:

To obtain forward displacement at reasonable restraint force levels immediately after beginning of deceleration of the passenger compartment (1).* This means earlier and faster locking and more rigid behaviour of the restraint system in the restraint force onset phase.

By this measure, energy-absorbing effectiveness of passenger forward displacement can be increased, resulting in turn in lower restraint forces for an equal amount of kinetic energy.

Pyrotechnically pre-loaded seatbelt systems represent the latest effort to approach this aim, even in serial production.

However, the need for electronic sensor techniques, wiring, and partial redundance of the system for quality and safety leads to comparably high costs.

---

* Numbers in parentheses designate References at end of paper.

Web-Clamp Systems, well-designed and effectively incorporated into the restraint system, may be an alternative, directed to the same aim, however, at comparably low additional costs and only minor increase of weight. It is well known, however, that excessive belt slack caused by passenger handling and usage failures cannot be eliminated by this device.

Therefore, a development program was started for studying design and production needs of a web-clamp system in relation to the possible increase of safety. Assumption has been made of correct belting conditions, following ECE/FMVSS 209 specifications (2).

Aims of this web-clamp development and design program (*) were, in detail, as follows:

- to minimize the delay until locking and, by this, reduce webbing extraction
- to eliminate the film-spool effect
- to enhance current handling and wearing comfort
- to keep production costs within a range of actual ELR systems.

CONCEPT OF WEB-CLAMP AS DEVELOPED HERE

Web-clamp patents and web-clamp systems handmade for testing more or less regularly use the wedge-concept for clamping and locking the webbing.

Without going into more details, the following performance limits and failure-modes were observed in this type of system observed: to date:

- release after accidental locking was critical in normal use
- webbing tear-off occured in an uncontrolled manner under crash-conditions, due to shear forces acting on the contact surface
- uncontrolled slip during webbing extraction could be measured under normal and shear force, until final locking.

These decisive concept-related disadvantages often led to the termination of web-clamp research.

The web-clamp system as developed here follows a completely different concept. It basically, a simple lever system. Geometry and kinematics of the lever-housing arrangement result in pure normal-force loading of the webbing in locking conditions.

THEORY OF THE WEB-CLAMP GEOMETRY - The principle of the web-clamp restraint system is that of a pure lever concept without clamping-force amplification during clamping action, as occurs with the employment of clamping wedges.

(*) Different projects in question were adviced, supported and/or sponsored by
- Allied Corporation / formerly the Klippan Group Europe
- AUDI, BMW, Daimler Benz, Opel, Porsche, Renault/Peugeot, Volkswagen, Volvo
- Technical University Berlin

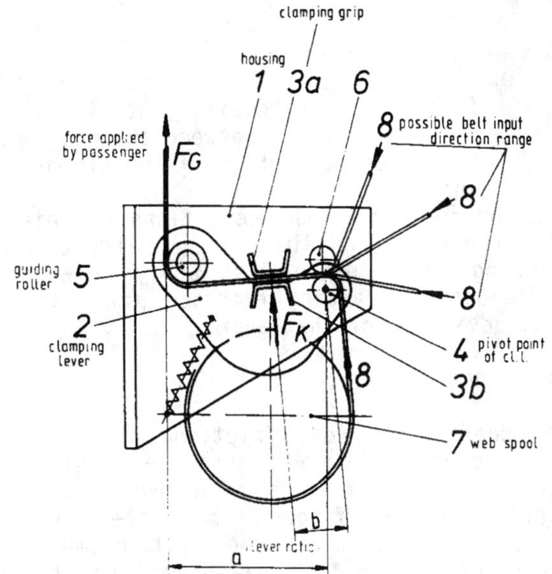

Fig. 1 - Basic concept of the lever web-clamp system

Fig. 1 elaborates the simple kinematic relations involved here, as described below:

One clamping grip 3a is immovably installed in the housing 1. At a pivot point 4, a clamping lever 2 is mounted in the housing 1 so as to rotate with the movable clamping grip 3b. The web belt runs into the housing either from a roller 7 through the guide roller 4, or from any other direktion 8, through the guide rollers 4 or 6, between the clamping grips 3a and 3b, and then through the housing to another guide roller 5. From roller 5, the belt runs at a return angle of approximately 60 ... 120° to the D-ring fitting on pillar B of the vehicle.

The following are represented for the arrangement in ist clamped state, in accordance with Fig. 1:

- the belt force F_G
- the operating lever a
- the clamping force F_K
- the clamping lever b.

The belt force F_G applied by the passenger creates a clamping force F_K in accordance with the following equation:

$$F_G \cdot a = F_K \cdot b$$

$$F_K = \frac{a}{b} \cdot F_G \qquad (1)$$

If assumption is made of an otherwise frictionless system at the points of belt reversal, then the following equation, furthermore, will apply for secure holding of the belt:

$$(2) \qquad F_K \cdot \mu - F_G$$

(where μ is the coefficient of sliding friction between the web belt and the surface of the clamping grip)

In order to fulfill the conditions of this inequality for secure holding of the belt, then the following relationship must apply between the coefficient of clamping friction μ and the mechanical advantage of the lever:

$$(3) \qquad \mu - \frac{b}{a}$$

The coefficient of friction, however, should be coordinated with the mechanical advantage of the lever in such a manner that unnecessarily great forces from the belt material do not result at the belt clamping point 3a/3b as a result of clamping force F_K.

For series employment in accordance with cycle testing stipulated by ECE or FMVSS, adjustment should be performed so as to come "as close as possible" to the following relationship:

$$\mu = \frac{b}{a}$$

The following key words can express the advantages of this web-clamp system:

- Clamping of the web belt takes place at the clamping point with practically purely <u>normal force</u> (compressive force), without the action of shearing forces on the belt material. As a result, tearing of the belt can be reliably prevented under all known conditions of employment.
- The web belt is less harshly stressed at the point of clamping, since at all times only that clamping force is applied to it which is required for withstanding the originating belt force F_G.

  Uncontrolled self-amplification effects to not arise; i.e., the following apply <u>with an always-constant geometrically defined mechanical advantage:</u>
  - -- small belt forces F_G result in small clamping forces, leading to a remarkable safety against damage of webbing under misuse-conditions
  - -- great belt forces F_G result in great clamping forces, producing a pure YES (lock) / NO (open) - operation under any level of originating belt force F_G
- When the clamping lever is open, the web belt passes practically without friction through the clamping point (a prerequisite for wearing comfort).
- Even after application of great loading, opening the belt after the clamping action can always be performed without difficulty by only the small spring forces involved.

DESIGN FEATURES OF WEB - CLAMP (BASIC SYSTEM) - A first design of this web-clamp system was carried out closely according to the above-discussed web-clamp concept. Fig. 2 demonstrates this. For easy correlation to Fig. 1, the same numbers were used to equivalent parts:

It shows clamping lever 2 pivoting at spool-pin 4. Lever 2 carries clamping grip 3b, which is able to grip webbing on the surface of clamping grip 3a, mounted in housing 1, after lever 2 has been lifted by spedified locking criteria.

Displacement (lift) of guiding roller 5 at least leads to the only webbing motion in the direction of the belted passenger, until an abrupt non slip-stop of the webbing occurs. This is due to the geometry of the system, under pure normal-force pressure on the webbing, without belt-wearing sheer-stresses, as mentioned above.

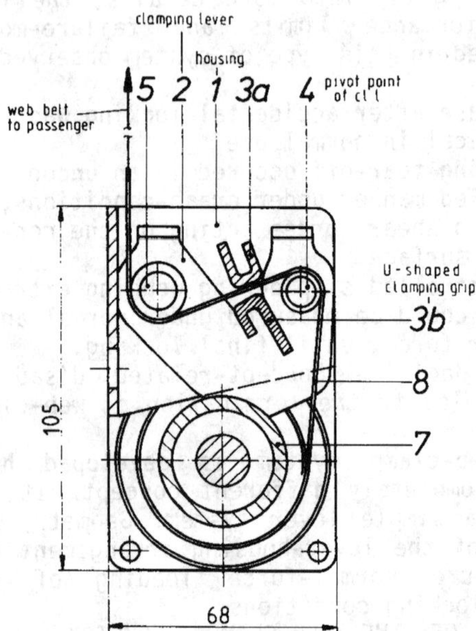

Fig. 2 - Standard-design arrangement of a web-clamp system, according to basic concept

To reach a favorable compression distribution on the webbing in the clamping area under high loads, the clamping grips are canted about 6 mm toward the centerline of the belt. In this it is guaranteed that no clamping pressure acts on the edges of the webbing under any condition.

Clamping pressure on the edges would cause initial tear-off of the belt and, as a result, total failure of the restraint system. With this web-clamp basic system, different types of test programs were carried out, serving as basis for the per-for-mance data, represented here.

PERFORMANCE DATA

Test models were developed into the initial automatic web-clamp system, as shown in section in Fig. 2, for work performed in different test series. The following tests were performed in this series:

- Cycle tests in accordance with ECE R 16, with subsequent R 16 crash tests with a 95 % dummy
- Crash tests oriented to ECE R 16, with an ECE R 16 seat and dummy
- Full seize casco crash test (fully equipped interior).

These test were designed with the objective of further elaborating the characteristic advantages of the web-clamp system in comparison with conventional automatic restraint devices (ELR's) or pyrotechnically pre-loaded restraint systems.

RESULTS OF CYCLE TESTING - ECE R 16 cycle tests (2) were successfully conducted. The belt webbing was subjected to loading by the clamping action involved, in regular, cyclical intervals within a narrowly resticted range. After being subjected to 70,000 test cycles (14,000 clampings), the same system successfully passed the ECE R 16 crash test.

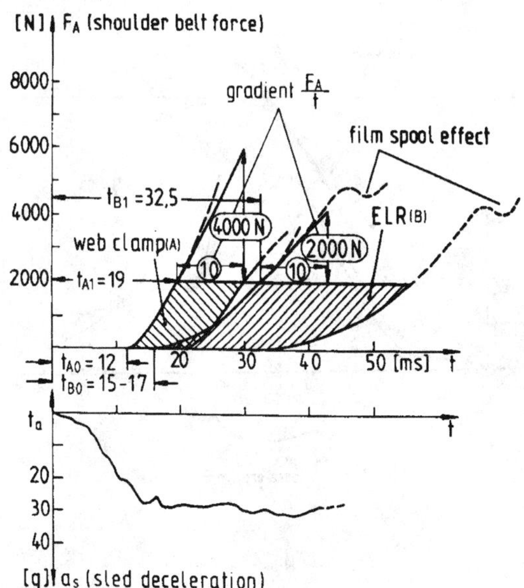

Fig. 3 - Range of delays and onset characteristics of shoulder belt forces comparing
- current ELR's
- web-clamp system

in a crash test series

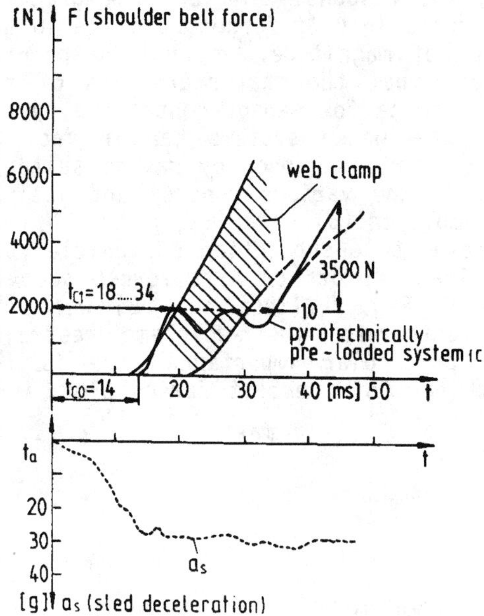

Fig. 4 - Delay and onset of shoulder belt force of a pyrotechnically pre-loaded ELR system in comparison to web-clamp response

SPECIAL ASPECTS OF CRASH TEST RESULTS (ECE R 16) - The crash tests consisted of a test series with conventional automatic belt restraint systems (ELR's), and of a series of tests with the web-clamp restraint system. The clamping system was tested with different versions of the clamping grips, as well as with various types of web belts.

Fig. 3 and 4 show the restraint-system reaction time and the characteristics of the increase in belt force, in comparative testing between the following:

- the web-clamp restraint system
- standard automatic belt restraint systems (ELR).
- pyrotechnically pre-loaded ELR-system (Fig. 4 only)

The superior features of the web-clamp restraint system can be seen in the following factors:

- a significant reduction in the response time t_0 from about 30 ms for standard automatic belt restraint systems to 15... ... 20 ms for the web-clamp.
- above all: the clearly greater slope, following response, of the belt force curve, with complete prevention of spool slippage in the belt. Such slippage, of course, is responsible for the introduction of additional slack into currently employed automatic restraint systems (see Fig. 3).

In contrast to sensor-controlled, pyrotechnically pre-loaded standard automatic-retraction restraint systems (shown in Fig. 4), the superior characteristics of the web-clamp restraint system are obvious: under laboratory conditions, the response time and the gradient for the increase in belt forces both lie in similar orders of magnitude. It must be pointed out, however, that the characteristics of the rise in belt force for sensor-controlled, pyrotechnically pre-loaded systems can in fact be modified over certain ranges by making suitable adjustments in the various control and activation units employed (3).

With respect to subjection of vehicle passengers to loads, plots of belt forces such as that shown in Fig. 5 (test no. 08) were observed in tests with the web-clamp restraint system. Of particular importance here is the "filled-out" form of the belt forces plot, with a modest peak value of $F_{A\,max}$ = 6,5 kN.

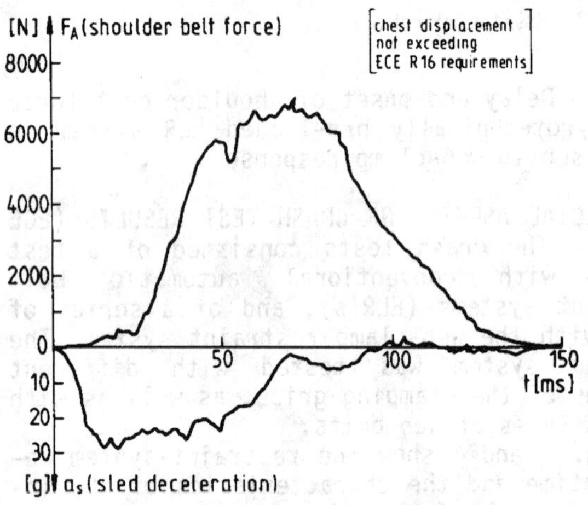

Fig. 5 - Typical plot of a shoulder belt force history of the web-clamp restraint in 30 mph-crash test (ECE R 16-Spec.)

This favorable plot was obtained by a combination of the automatic web-clamp restraint system with an especially adjusted belt material, with an elongation of ε = 25 ... 28 % for the range of 5,000 ... 10,000 N loading (4). Even with this belt material, the limits of forward passenger displacement as stipulated in ECE R 16 were clearly ovserved.

With the standard automatic-retraction restraint systems (ELRs) tested under identical crash conditions, results were observed of 9,500 ... 11,500 N, with slightly greater forward displacement of the dummy thorax. The belt material used here exhibited an elongation of 10 % for tensile force of 11,500 N.

COMPARISON OF WEB-CLAMP AND PYROTECHNICALLY PRE-LOADED ELR IN A CASCO CRASH TEST. - Further crash test series were carried out incorporating

- the casco of a medium seize vehicle with fully equipped interior (ex steering system)
  (Deceleration generated by a Bendix-sled-system)
- passenger and driver 50 Percentile Part 572 - Hybrid II - Dummies
  -- one restrained by web-clamp
  -- one restrained by pre-loaded ELR

All FMVSS 208-specified data (2) were computed. Specific means were provided for high-speed-film (1,000 frames/s) and trajectory analysis.

Again, and similar to the previous test series discussion, we only want to point out specific aspects of this comparison to characterise web-clamp performance.

Related to these test series we want to show and emphasize the motion sequences of the dummies (5). The plots of

- chest-deceleration history
- displacement trajectories of head, shoulder, hip and knee points

of one test run were worked out to show typical trends.

In order to compare both systems, we primarily examine at the data output of the chest. Corresponding values of displacements as well as accelerations of chest are shown in Fig. 6, resulting from the same crash test with both systems.

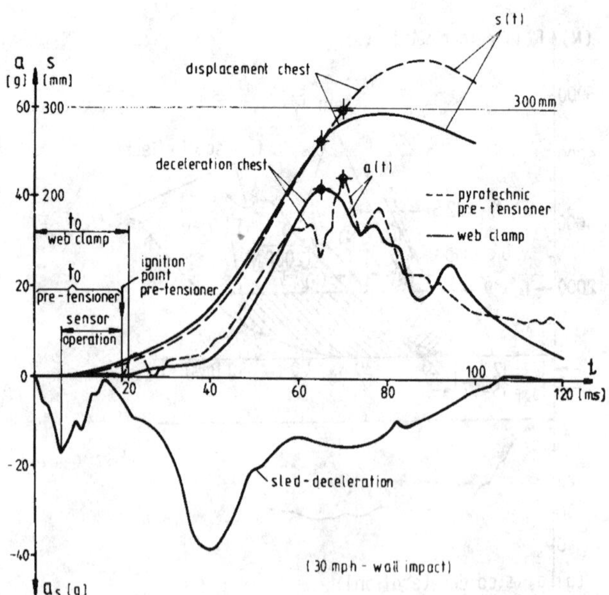

Fig. 6 - Comparison of chest displacement and chest loading of dummies restrained by pyrotechnic pre-tensioner or web-clamp resp.

The following can be seen from Fig. 6: within 20 ms the pyrotecnically pre-loaded restraint system is activated by the electrical signal of the sensor and starts the tension of the webbing, showing a maximum of acceleration sequence after an additional 3 ms (with loading of the restraint system after a total of 27 ms) At the same time - 20 ms after crash - the displacement of the dummy has activated the web-clamp system as discussed. The energy absorption of the pyrotechnically pre-loaded system is slightly more efficient in the beginning, caused by the fully eliminated slack of the system. After 60 ms the energy absorption of the pyrotechnic system will be hindered by the typical film spool effect of an ELR, which leads to a higher displacement compared to the web-clamp system.
in other words, due to the stiff structure of the web-clamp, the efficiency of this system is higher in terms of energy absorption. The maxima of accelerations are of about the same magnitude. Nevertheless, displacement of chest is 30 mm less with the web-clamp system.

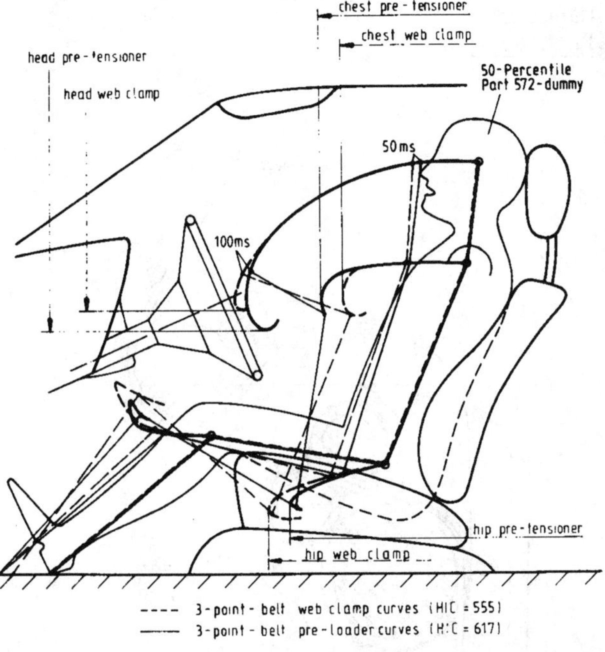

Fig. 7 - Comparison of trajectories of dummies restrained by web-clamp or pyrotechnic pre-tensioner system resp.in 30 mph-crash test

Fig. 7 gives informations about the related trajectories of the dummies. It is significant that the dummy restrained by the pyrotechnically pre-loaded system moves with a smaller resulting lumbar spine angle. Forward displacement of the hip. The reason seems to be that the tensioner device is able to eliminate even the slack in the lap belt portion.

The following table elaborates on the figures derived from the film analysis of motion sequences of the dummies:

Data of trajectories of dummies subjected to 30 mph - crash test:

	pyrotechnically pre-loaded ELR	web-clamp system
displacements (mm) at t = 50 milliseconds		
head	139	157
chest	135	142
hip	107	132
displacements (mm) at t = 100 millisec.		
head	535	550
chest	328	285
hip	214	264
displacements (mm) at a_{max} (t) of chest		
head	539	552
chest	296	268
hip	139	207
displacements (mm), Maxima		
head	546	564
chest	346	292
hip	239	268

The hip point of the web-clamp protected dummy travels farther forward. However - as exspected - the hip deceleration maximum is lower in this case. The deeper plunge of the web-clamp restrained hip into the seating structure and, consequently, the higher energy absorption rate of the seat in this case, reduced loading of the lap belt. As a result we found that the efficiency of the web-clamp system with respect to energy absorption is more favorable, taking the relation of accelerations and maximum travel of hip and chest into account.

Trajectories of the heads of both dummies were quite similar. HIC - values - 617 for the pyrotecnically pre-loaded restraint system and 555 for the web-clamp system - were not remarkably different. More rotational movement of the head of the pre-tensioner-restrained dummy can be marked, leading to unfavorable head contact to the rim of the steering wheel, and resulting in more neck bending.

Summerizing this specific test analysis, one can observe the difference in the balance of both systems, the interworking of correlating components of the complete restraint system:

- seat belt/seat belt geometry
- locking device/kind of integration into the restraint system
- seating structure
- steering system
- lower dashboard structure

Current knowledge and experiences allow a detailed adjustment of these components to the achieve optimum from the restraint system and the car compartment involved.

In this respect, we want to state here that a web-clamp equipped restraint system fitted to an optimally adjusted interior envirement may result in significant increase of passenger protection.

TYPES AND CURRENT DEVELOPMENT TRENDS FOR THE WEB-CLAMP RESTRAINT SYSTEM

Past as well as current project development trends as carried out for the web-clamp restraint system can be divided into two different directions:

Web-clamp devices

- to replace the loop-around belt-fitting and to lock at this point, close to the belted passenger,

  the D-ring type

- to replace, or to be adapted to, a standard ELR,

  the package type.

Web-clamp types of both directions show specific advantages and disadvantages on at least equal safety level with respect to the acceptance and adaptation to current car bodies.

TYPES OF WEB-CLAMP SYSTEMS

D-ring type

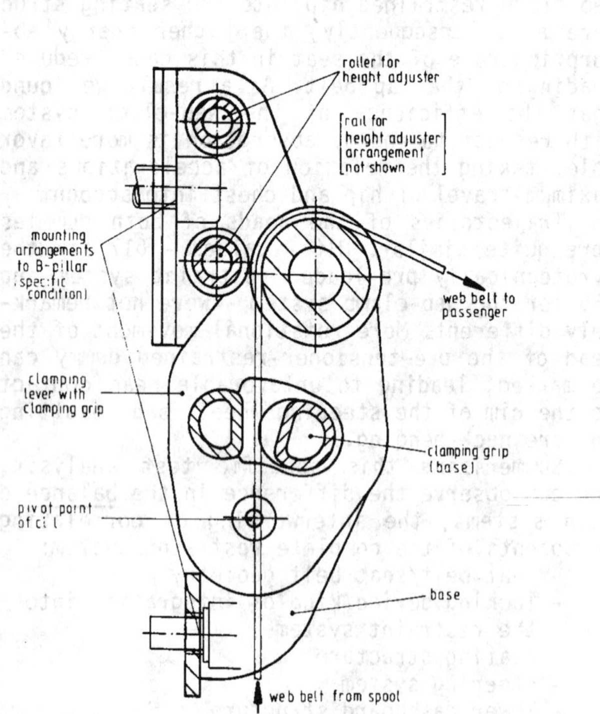

Fig. 8 - D-ring type web-clamp device

Fig. 8 shows a section view of the structural design, as well as the belt arrangement, for the D-ring type which is designed to replace a conventional loop-around belt fitting. This web-clamp restraint system can be provided in versions with or without a height adjustment feature. We consider this type to represent an independent direction of development with particular design problems. The following represents a tabular listing of advantages and disadvantages, with simple delineation of the types involved:

Advantages:
- great comfort
- small number of parts
- fast, exact locking of the system
- "hard" response of the restraint system

Disadvantages:
- greater space requirements in the upper area of the B-pillar/adaptation problems
- requirement of coordinating the design of the vehicle body system with the design of the automatic web-clamp restraint system

Adaptation to any current or new car structure (upper B-pillar) necessitates a close cooperation with the body-design department of the car manufacturer.

Package type

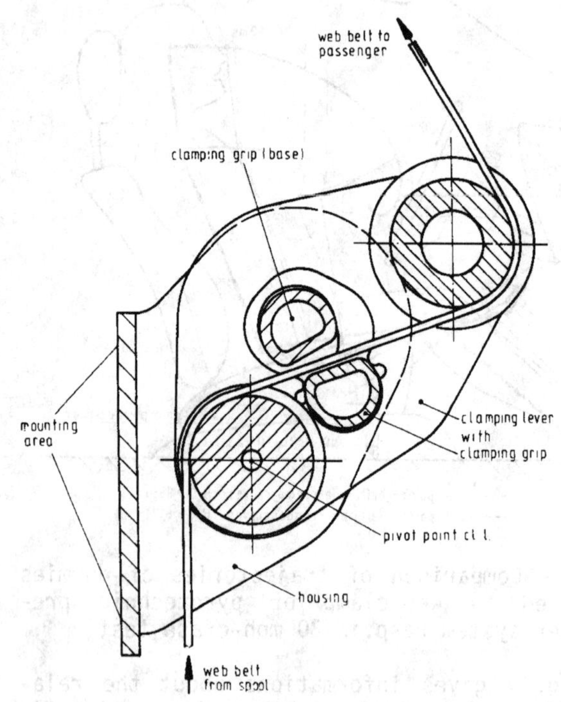

Fig. 9 - Package type web-clamp device (comfort and weight optimized)

Fig. 9 shows a section view of a standard-type automatic web-clamp restraint system. The belt-reversal arrangement necessary for activation of the locking system is not shown.

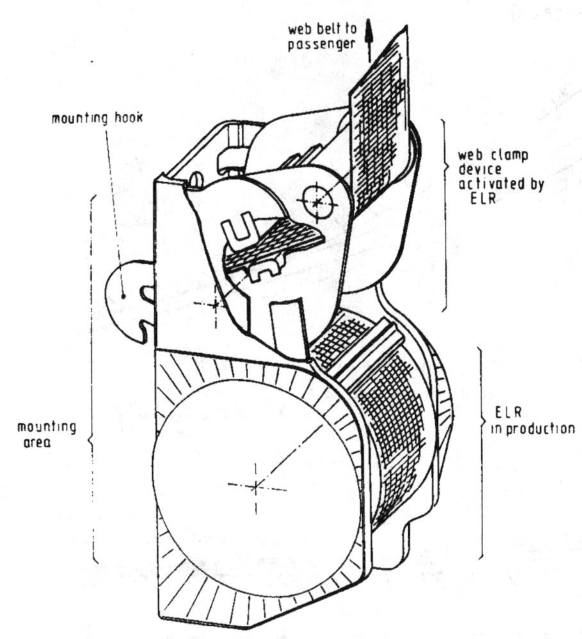

Fig. 10 - Package of web-clamp / ELR

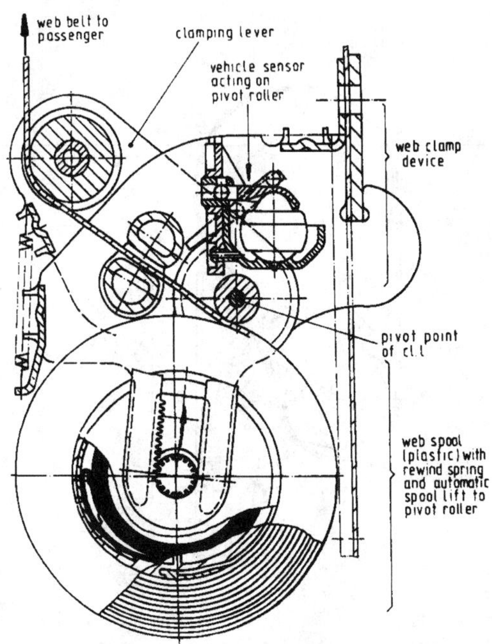

Fig. 11 - Integral automatic web-clamp system (weight and comfort optimized) (Version A)

Fig. 10 shows the assembly arrangement of the standard-type web-clamp system, with a currently conventional automatic retraction device (which in this case performs activation of the clamping system). This web-clamp ELR package restraint system is designed for conventional installation in the B-pillar foot area.

Advantages:
- safety performance
- easy adaptation within the space of current B-pillar

Disadvantages:
- reasonable comfort only
- doubling of load carrying parts (costs)
- size and weight

DIRECTIONS AND RESULTS OF ACTUAL DEVELOPMENTS - Effective designs have recently been developed for various applications. Both the crash-test results as well as comfort investigations have been implemented to achieve a high level of standards. The objective of work performed along these lines of development is to incorporate all results from these programs into final development of a so-called <u>integral automatic web-clamp restraint system</u>. A further step will include adaption of a pyrotechnic pre-loading device for belt tensioning.

Fig. 11 shows a concept for an integral automatic web-clamp restraint system in which a vehicle-sensitive activation system acts directly on the locking rocker of the clamping systems. The belt take-up reel as well as the associated mechanical parts are made of plastic in the integral system, since this reel device is not subjected to restraint forces.

Objectives and advantages of the integral automatic web-clamp restraint system are as follows:

- fast and simple belt locking system, with duo-sensitive activation
- comfort on the level of the present conventional restraint systems with automatic retraction
- belt take-up reel executed in plastic (for savings in weight and costs)
- the small number of simple parts with less demensional tolerance needs due to concept-related permanent synchronisation of locking procedure
- safety above the level of current legal stipulations and customer requirements
- easy adaptation to different mounting conditions.

The main advantage of the integral system over the D-ring type consists in the fact that it can be implemented in presently operating and future vehicles without basic modifications to the restraint system.

Fig. 12 gives the package dimensions which are presently possible in accordance with the present state of development of an integral version of the automatic web-clamp restraint system.

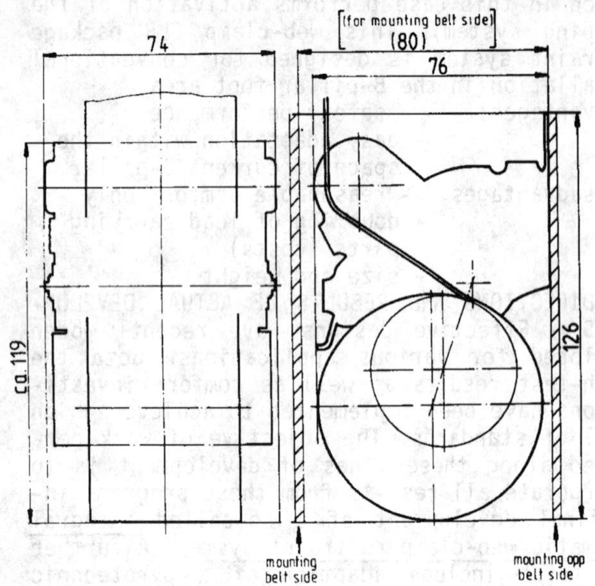

Fig. 12 - Mounting dimensions and conditions of integral automatic web-clamp system (Version A)

Wearing comfort of integral web-clamp

Owing to the additional belt-reversal points required by the design principle involved with the standard-type automatic web-clamp restraint system, the problem of wearing comfort arises. To be sure, extremely small bearing diameters on the take-up reel and on the first belt-reversal reel are possible owing to the fact that the mechanical reel mechanism is not subject to loads imposed by crashes. In the extraction/retraction hysteresis, however, the additional effort required for flexure of the belt is constantly noticeable. Investigations have been conducted toward minimizing the work losses associated with this phenomenon. The following influencing factors have been examined:

- the diameter of the belt-reversal reel
- the angle of belt reversal
- the overall geometry of the belt sequence, running from the reel to the D-ring.

We have since arrived at an arrangement, shown in Fig. 14, in which geometry and installed size have been optimized, resulting in extraction/retraction behavior of the integral web-clamp restraint system which is on the level of currently available standard restraint belts with automatic retraction.

Fig. 13 represents an extraction/retraction curve within representative requirement limits reached by this version as per Fig. 14.

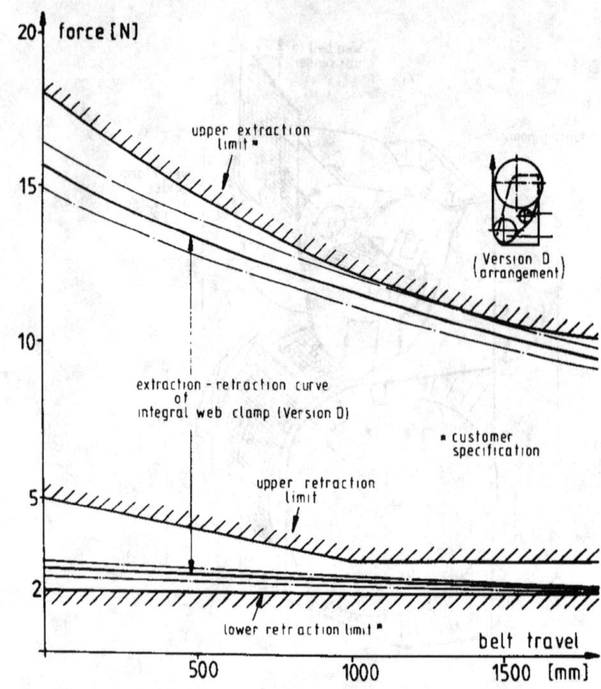

Fig. 13 - Comfort: Extraction-Retraction diagram of integral automatic web-clamp (Version D)

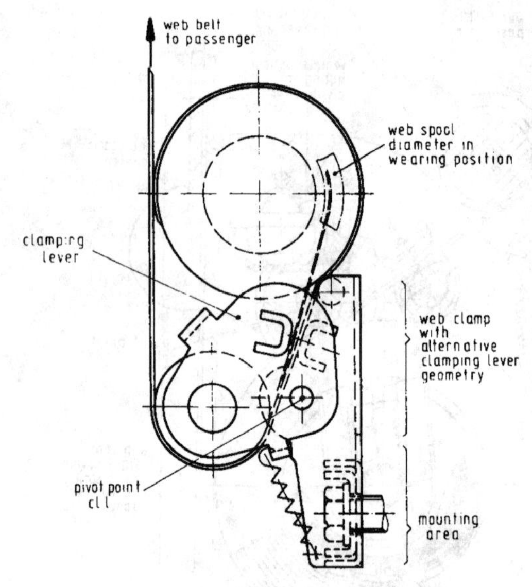

Fig. 14 - Integral automatic web-clamp system (Version D, for specific mounting conditions)

In addition to the geometry of the integral system and how it is determined by comfort requirements, the optimization of the installed size and weight characteristics is also influenced by the specific installation condition in the vehicle involved. The required belt reel diameter, as well as the enclosure size as determined by the mounting dimensions, both dictate the package dimensions involved in each case of installation. The design of the belt take-up reel and activation unit as a non-load-bearing system (made of plastic parts) enables savings in weight and expense which can be employed to compensate for the costs and the weight of the expensive load-bearing parts of the clamping system.

Offical approval - As far as legal currently valid regulations are concerned (ECE/FMVSS) there are no additional obstacles for official approval in accordance with presently valid legal stipulations, if the clamping system and the belt take-up reel unit form a package unit. In this case definitions and approval test procedures can obviously be applied to the web-clamp system.

CONCLUSIONS

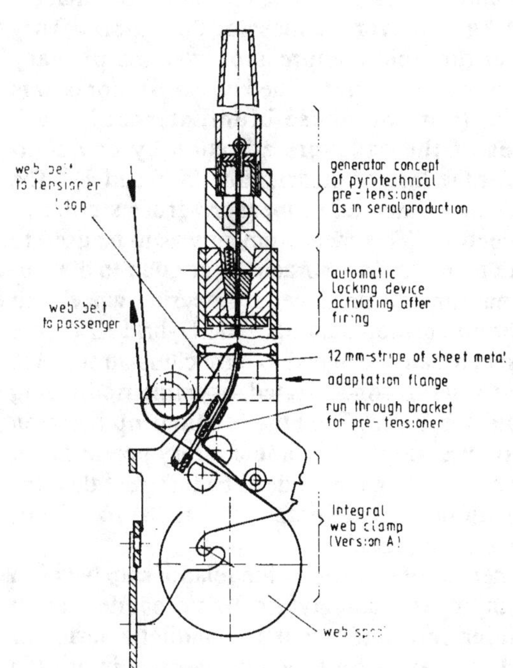

Fig. 15 - Integral web-clamp with pyrotechnical pre-tensioner device

The web-clamp concept as discussed here offers a significant increase of safety performance as summarized in the following:

- Restraint loading can be reduced within existing space limits
- Displacements of passenger can be reduced on reasonable restraint force levels, if needed.

Safety level is of a magnitude similar to that of pyrotechnically pre-loaded ELR systems, but at much less cost,

- if the web-clamp system is installed within a well balanced environment regarding seat, belt geometry and steering system
- if passenger misuse (excessive slack) is avoided.

To overcome even this final problem a combination of the web-clamp with a pyrotechnically pre-loaded system can be considered as a conclusive and logical step.

Fig. 15 may demonstrate a design of such a combination. One can easily observe a further gain that - adverse to the problems arising with the adaptation of pre-loaders to standard ELRs - this web-clamp concept does not need any structural reinforcements to take the additional stress of the pre-tensioning process.

REFERENCES

1. Bundesminister für Forschung und Technologie BRD, "Technologien für den Straßenverkehr", Bonn 1976.

2. - ECE Reglement No. 16, E/ECE/Trans/505
   - Federal Motor Vehicle Safety Standard No. 208, "Occupant Protection in Frontal Impact."

3. J.E. Mitzkus and H. Eyrainer, "Three-Point Belt Improvements for Increased Occupant Protection." SAE Paper No. 840395.

4. J. Takada, "Development of Energy-Absorbing Safety Belt Webbing." SAE Paper No. 740581.

5. D. Adomeit, "Evaluation Methods for the Biomechanical Quality of Restraint Systems During Frontal Impact." STAPP Paper No. 770936.

6. D. Adomeit, "Seat Design - A Significant Factor for Safety Belt Effectiveness." STAPP Paper No. 791004

910812

Hybrid III Sternal Deflection Associated with Thoracic Injury Severities of Occupants Restrained with Force-Limiting Shoulder Belts

Harold J. Mertz, John D. Horsch, and Gerald Horn
General Motors Corp.

Richard W. Lowne
Transport and Road Research Laboratory

ABSTRACT

A relationship between the risk of significant thoracic injury (AIS ≥ 3) and Hybrid III dummy sternal deflection for shoulder belt loading is developed. This relationship is based on an analysis of the Association Peugeot-Renault accident data of 386 occupants who were restrained by three-point belt systems that used a shoulder belt with a force-limiting element. For 342 of these occupants, the magnitude of the shoulder belt force could be estimated with various degrees of certainty from the amount of force-limiting band ripping. Hyge sled tests were conducted with a Hybrid III dummy to reproduce the various degrees of band tearing. The resulting Hybrid III sternal deflections were correlated to the frequencies of AIS ≥ 3 thoracic injury observed for similar band tearing in the field accident data. This analysis indicates that for shoulder belt loading a Hybrid III sternal deflection of 50 mm corresponds to a 40 to 50% risk of an AIS ≥ 3 thoracic injury.

THE PRIMARY DESIGN OBJECTIVE for the Hybrid III thorax was to develop a structure that could be used to assess the restraint efficacy of energy absorbing steering assemblies. Gadd and Patrick (1)* conducted tests of energy absorbing steering columns using cadavers. They noted that the total tolerable restraint load applied to the torso could be more than doubled if the wheel was designed so that part of the restraint load was applied to the shoulder structure by the steering wheel rim. This led to the understanding of how the total upper torso restraint load could be increased without increasing the risk of thoracic injury by increasing the load transmitted to the shoulder structure. Nahum, Kroell et al (2, 3, and 4) conducted cadaver tests to determine thoracic tolerance to sternal loading. They chose to impact the cadavers at mid sternum using a rigid face impactor. The impactor face was 152 mm in diameter which represented the area of a hub of a steering wheel used by Gadd and Patrick (1). The mass of the impactor was 23.4 kg, the average mass of the thorax. They found that thoracic compression was the primary thoracic injury parameter. The Hybrid III thorax was developed to mimic these blunt, sternal impact responses of the cadavers adjusted by 667 N to account for the lack of muscle tone (5, 6 and 7). The chest was instrumented to measure gross sternal to spine deflection. This measurement was to be used to assess the potential for thoracic injury due to distributed sternal impacts produced by steering wheel hub loading or air cushion loading. The Hybrid III thorax was not specifically designed to mimic human thoracic response for more concentrated, asymmetric loading of a shoulder belt. Thus the relationship between sternal compression and thoracic injury potential for shoulder belt loading may indeed be different than the relationship noted by Neathery et al (8) for blunt, frontal sternal impacts.

One method of estimating the relationship between thoracic injury risk and Hybrid III sternal deflection for shoulder belt loadings is to conduct a series of simulated car crashes for which the thoracic injuries to shoulder belted occupants are known. A major difficulty with this method has been the lack of exposure details for the car crashes. An important set of car crash data has been published by Foret-Bruno et al (9). They investigated accidents of Renault and Peugeot vehicles that were equipped with force limiters in the shoulder belt webbings. The force limiting elements consisted of a series of stitched webbing bands of different lengths. The bands were designed to tear at specified force levels. When the shortest band tore, the belt load was transmitted through the next shortest band. Published accident data are available on the level of band tearing and the corresponding thoracic

*Numbers in parentheses designate references listed at end of paper.

injury for 386 occupants (9, 10 and 11). This is the set of data that will be used to develop a relationship between thoracic injury due to shoulder belt loading and Hybrid III sternal deflection.

Association Peugeot-Renault (APR) provided force-limiting elements for simulating various belt loadings with a Hybrid III dummy. This report provides a summary of Hyge sled tests of these force-limiting elements conducted using a Hybrid III dummy, an analysis of the field accident data for belt systems using force limiters, and the development of a relationship between Hybrid III sternal deflection and the risk of thoracic injury for shoulder belt loading.

HYBRID III SLED TESTS

TEST CONDITIONS - The sled fixture consisted of a "hard" seat, toe pan, and belt anchor supports. The "hard" seat contained a rigid metal plate under the cushion, which provides a reusable and repeatable seat resulting in a nominal horizontal pelvic motion. The selection of anchor locations was based on a tabulation of anchor locations for selected crashes investigated by APR, Figure 1. Dimensions of our test configuration with the toe pan and "hard" seat location are also provided. The belt restraint used current automotive webbing, fixed at all three anchors (no

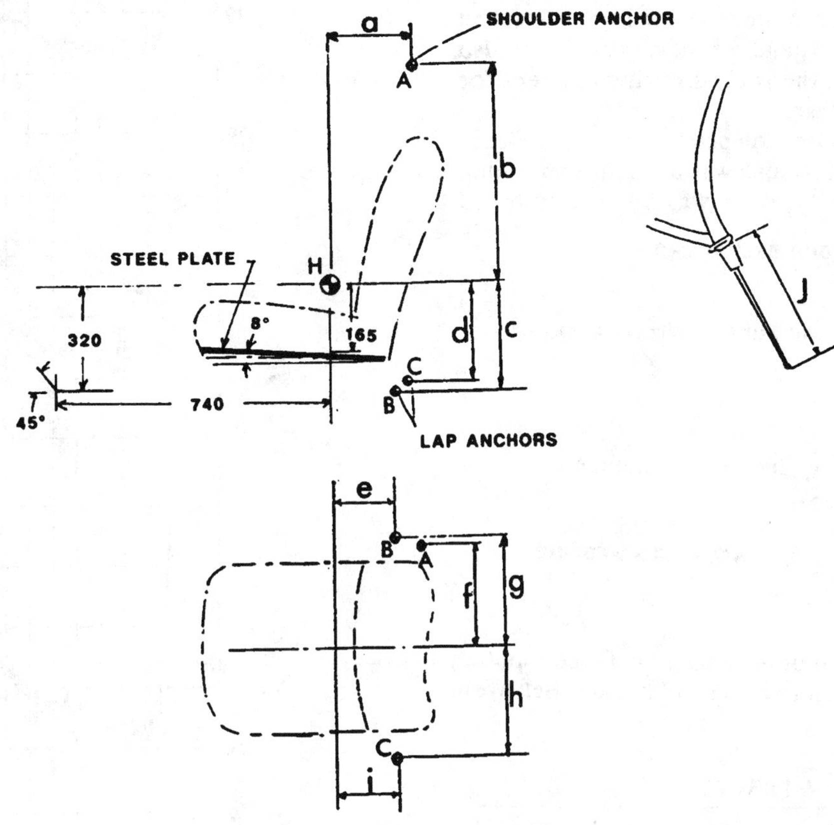

	CASES	APR DATA 2091	2165	3168	3805	4061	4549	SLED TESTS
D	a	220	430	220	125	500	205	270
I	b	600	600	600	600	600	600	600
M	c	220	220	220	270	220	270	230
E	d	270	300	270	310	300	310	280
N	e	170	150	170	100	220	180	170
S	f	270	250	270	270	250	270	260
I	g	350	250	350	280	250	280	280
O	h	300	300	300	300	300	300	300
N	i	170	150	170	100	220	180	170
S	j	310	360	310	340	360	340	380

Dimensions in mm; A, B, C are belt anchor locations.

Figure 1 - Dimensions Used in the Sled Tests Compared with Vehicle Dimensions Provided by APR.

retractors), and had a locking latch-plate. The force-limiting element was located at the shoulder belt anchor, Figure 2. Adjustment of the belt restraint was "snug" for the lap belt and 25 mm slack in the shoulder belt. Placement of the belt on the shoulder was distal to the neck belt-guard. Dummy positioning included a pelvic angle of 22 degrees and head X-axis horizontal.

The force-limiting elements (Type B) provided by APR have 5 individual bands which are loaded in sequence such that when the first loaded band is torn, the load shifts to the next band. Figure 3 provides published APR data for the Type B element tested, indicating the force limit associated with tearing of each band (9).

A range of test severities was used which resulted in the tearing of 2, 4, and 5 bands. Tests were also conducted without the force-limiting element for comparison of responses.

Generic sled acceleration pulses were used, Figure 4. Since shoulder belt load was to be the controlling parameter of severity, there was no need to try to

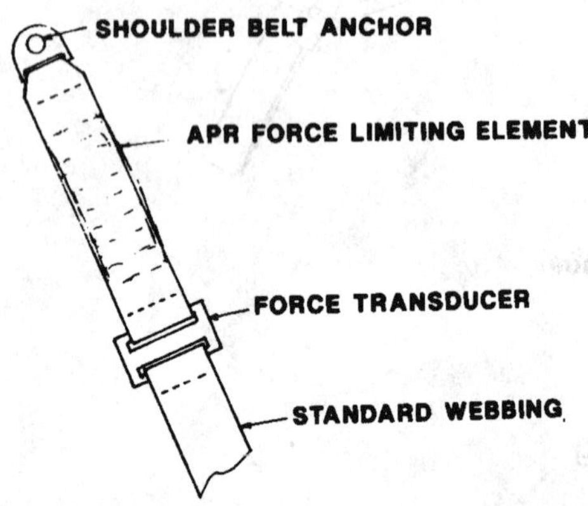

Figure 2 - Method of Attaching the Force Limiting Element to Standard Shoulder Belt Webbing.

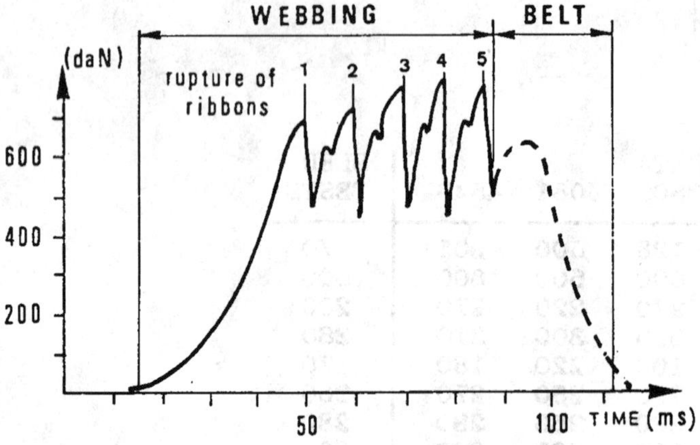

Figure 3 - APR Data Characterizing the Dynamic Load Response of a Type B Force Limiting Element (9).

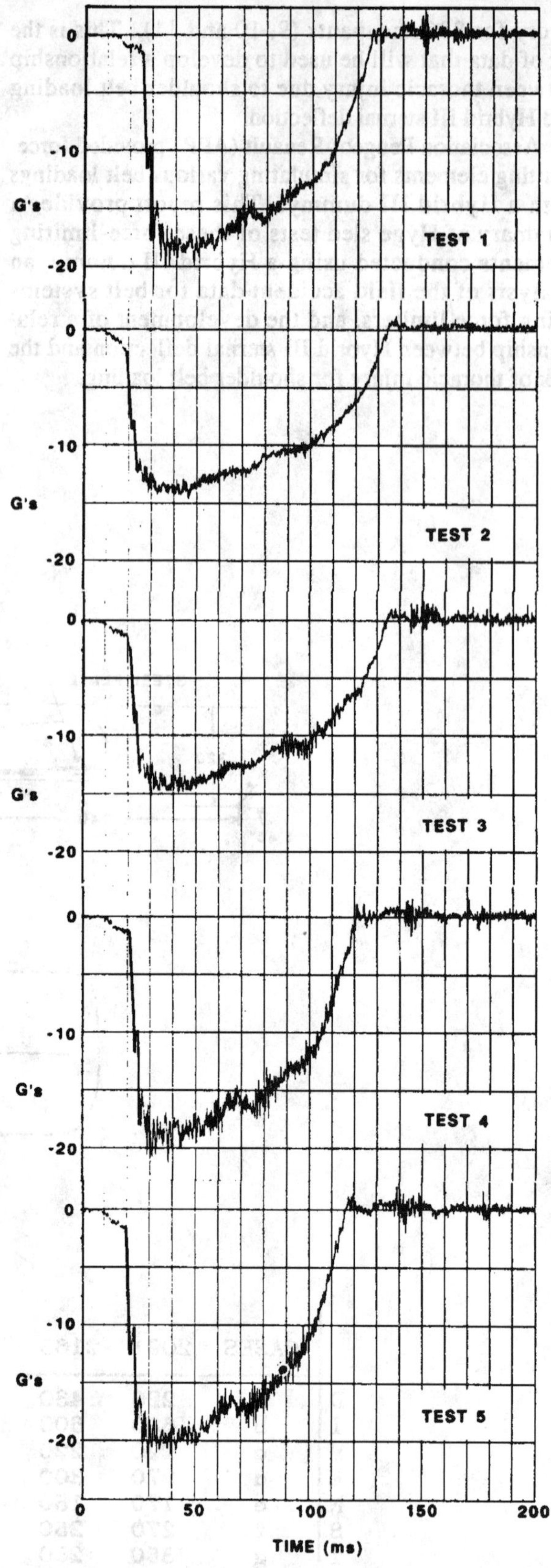

Figure 4 - Acceleration Profiles of Sled Tests. Profiles Chosen to Produce Prescribed Amount of Band Tearing. Filter Frequency of 2000 Hz.

duplicate any specific collision pulse. The pulse duration was constrained to 100-120 ms. Increase in test severity was obtained by increasing the level of sled acceleration. The test conditions are provided in Table 1.

Instrumentation included Hybrid III sternal deflection, thoracic acceleration, and shoulder belt tension. The shoulder belt tension transducer location is shown in Figure 2. Data plots are not filtered and have a nominal frequency response associated with the 2000 Hz anti- aliasing filter to better illustrate responses related to rapid release of load due to stitching tearing.

TEST RESULTS - Plots of shoulder belt force and sternal deflection as functions of time are provided in Figures 5 and 6, respectively. The tearing of each band is clearly evident by a sudden drop in shoulder belt load with a lesser influence on the sternal deflection.

Belt tensions for tearing of each band and the corresponding sternal deflections are provided in Table 1. Belt force required to tear the bands ranged from 5.5 to 6.9 kN with a trend that "early" tearings were at a lower force level. Sternal deflection associated with band tearing ranged from 26 mm to 35 mm. Most of the variation of sternal deflection is due to the variation in belt loads that caused band tearing.

Figure 7 contains cross-plots of belt force with sternal deflection. Note the linear relationship between shoulder belt loading force and Hybrid III sternal deflection. Figure 8 is a plot of Hybrid III sternal deflection corresponding to the belt force that produced band tearing; i.e., the data from Table 1. Forty millimeters of sternal deflection correspond to 8 kN of belt tension, a slope of 5 mm/kN. For this belt geometry, this slope is essentially constant over the range of sled test severities used and for shoulder belts with or without the load limiting element. This observation implies that if the shoulder belt load is known for a given accident then the Hybrid III sternal deflection can be estimated. This is the relationship that is needed in order to correlate Hybrid III sternal deflection with thoracic injuries experienced by car occupants.

In all five tests, the rate of sternal deflection was less than 3 m/s resulting in viscous criterion levels of less than 0.25. Thoracic injury due to rate of loading associated with the belt loading or band tearing is not a primary concern with this belt system.

FIELD ACCIDENT DATA

The field accident data used in the analysis were compiled from the studies of Foret-Bruno et al (9 and 10), case histories of seven accidents that were reviewed by Working Group 6 of ISO/TC22/SC12 (11) and personal communications with Mr. Foret-Bruno. The cars were 1970-1977 Peugeots or Renaults that were equipped with front seat three-point belts with a force-limiting shoulder belt. Three types of force limiters were used and are denoted as A, B and C. The multidisciplinary team from the Laboratory of Physiology and Biomechanics of the Association Peugeot/Renault has investigated accidents where 427 occupants were wearing a three-point belt with a force limiter in the shoulder belt. Foret-Bruno et al (10) have provided data for the accidents of 386 occupants for which the age of the occupant is known. Table 2 provides a summary of accident data for these 386 occupants. Note that 6 occupants (1.6%) suffered AIS = 3 thoracic injuries and 4 occupants (1%) suffered AIS ≥ 4 thoracic injuries. Table 3 is a summary of data for 15 occupants involved in accidents wearing a Type B load limiter. This Table contains the data for Type B accidents where part or

TABLE 1 - Summary of Sled Test Conditions and Results

Run No.	Belt Type	Sled Kinematics Acc. (G)	Vel. (km/h)	Stitch Bands Torn	Tear Force (kN) / Sternal Def. (mm)				
1	STD	18.0	51	-	-				
2	STD	14.0	44	-	-				
3	B	14.5	45	2	6.8 / 32	6.9 / 33			
4	B	18.5	52	4	5.6 / 26	5.9 / 27	7.2 / 35	6.9 / 34	
5	B	19.0	53	5	5.5 / 28	5.7 / 29	5.8 / 29	6.3 / 29	6.3 / 29

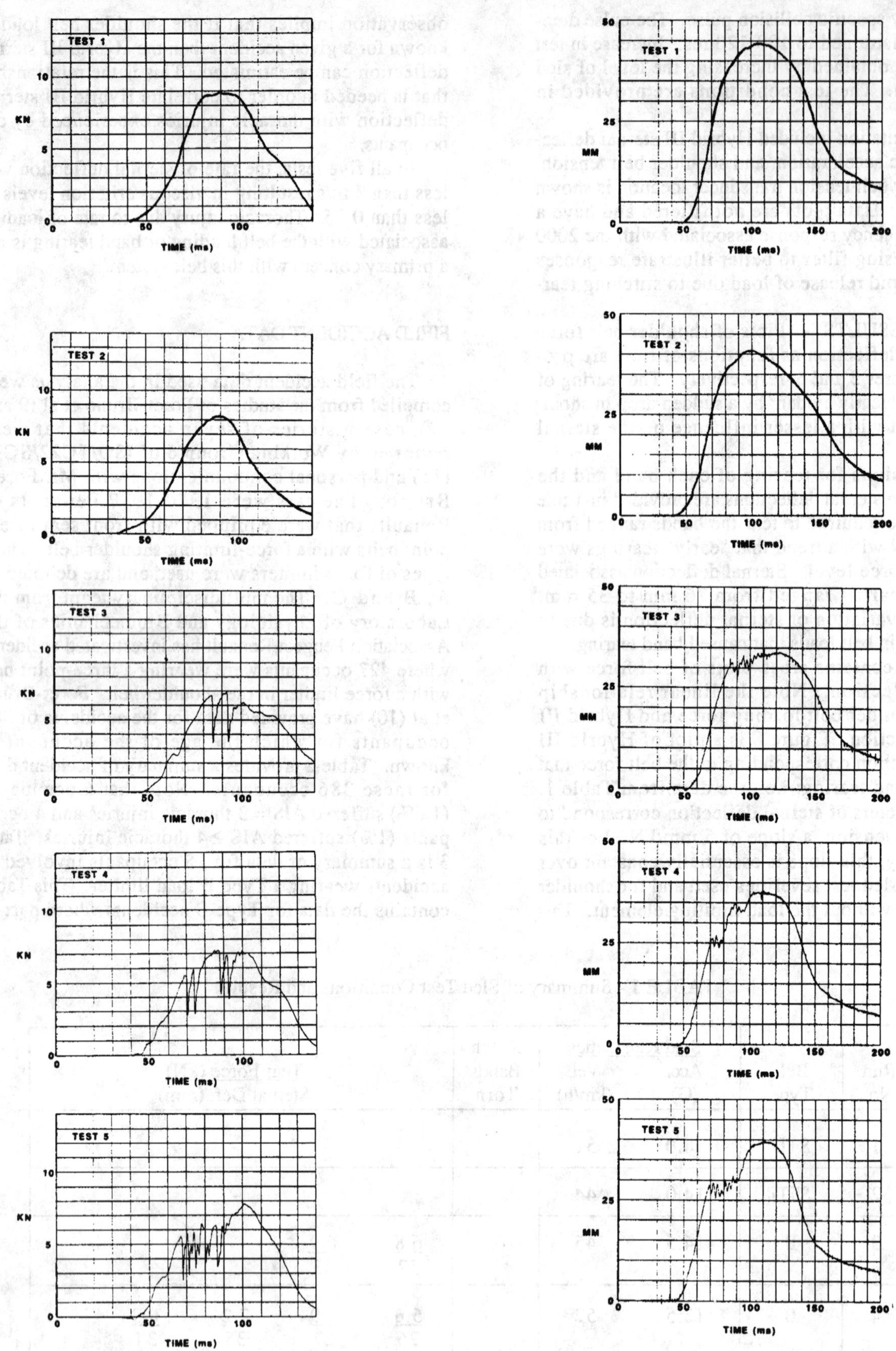

Figure 5 - Shoulder Belt Loadings from Sled Tests. Filter Frequency of 2000 Hz.

Figure 6 - Hybrid III Sternal Deflections from Sled Tests. Filter Frequency of 2000 Hz.

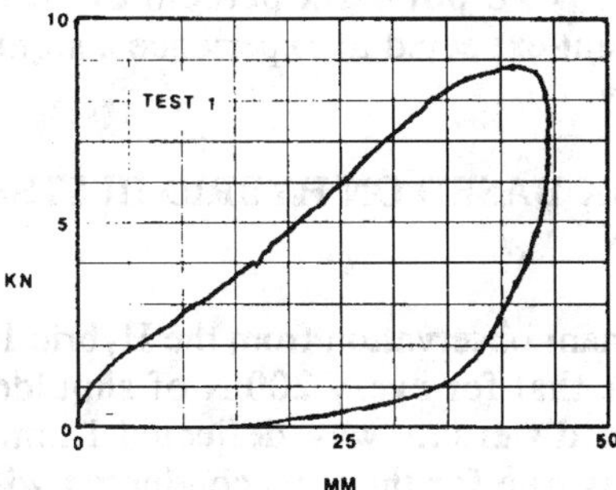

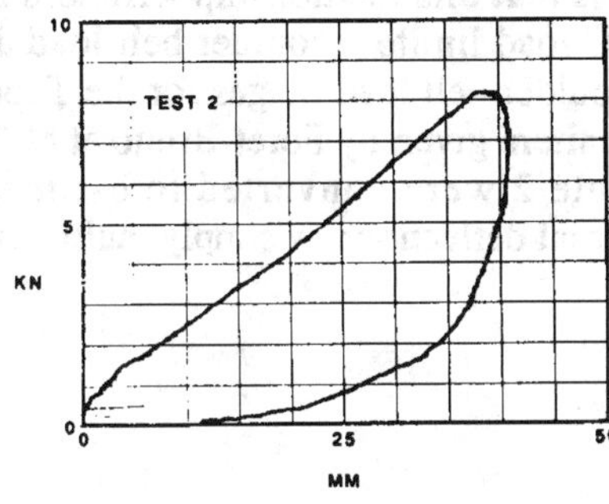

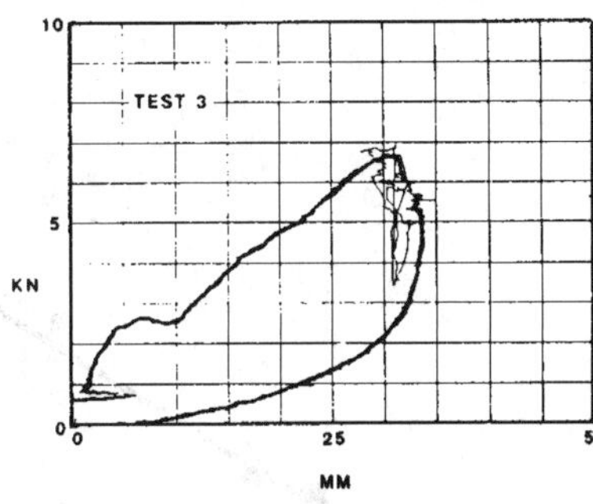

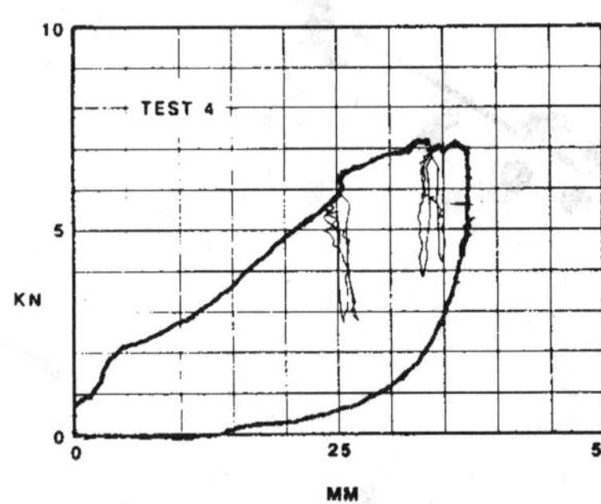

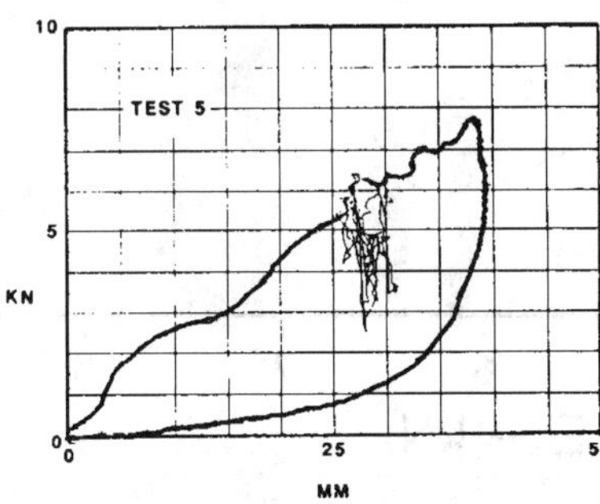

Figure 7 - Crossplots of Shoulder Belt Force vs. Hybrid III Sternal Deflection from Sled Tests.

all of the bands tore and accidents where the bands did not tear, but an occupant suffered rib fracture.

TYPE "A" LOAD LIMITER - The Type A load limiter had five discrete bands. Each band tore at a prescribed load level. The possible shoulder belt load range for tearing of each band is given in Table 2 and are based on the data given by Foret-Bruno (9). For example, if none of the bands tore, then the shoulder belt load could be any value from zero to 2.1 kN which is the average load required to tear the first band. If the first band tore, then the shoulder belt load could be any value between 2.1 kN which is the load required to tear the first band and 3.85 kN which is the load required to tear the second band. If the second band tore, then the shoulder belt load must have been 3.85 kN which is the load required to tear the second band. A higher load is not possible since the load required to tear the third band is only 3.25 kN. The shoulder belt load ranges for the tearing of bands 3 and 4 were deduced in a similar fashion. If all 5 bands tore, then the shoulder belt load can be any value greater than 4.4 kN.

There were 273 occupants involved in accidents with Type A force limiters. One hundred forty-nine occupants (54.6%) experienced shoulder belt loads between zero and 3.85 kN without any thoracic injury greater than AIS = 2. Ninety-one occupants (33.3%) experienced shoulder belt loads between 3.85 kN and 4.40 kN. Two of these occupants (2.2%) experienced thoracic injuries greater than AIS = 2. Thirty-three occupants (12.1%) tore all five bands. The shoulder belt loads for these occupants are not known; however, in all cases the load exceeded 4.4 kN. Of these occupants, two occupants (6.1%) experienced thoracic injuries greater than AIS = 2.

TYPE "B" LOAD LIMITER - The Type B load limiter also had five discrete bands of stitching, but the tearing loads for each band were approximately twice the tearing loads of the Type A. The load ranges for each band are based on the data given by Foret-Bruno et al (9) and were calculated in the same manner as was indicated for the Type A load limiter. Although not specified in References 9 or 10, Foret-Bruno has noted that there were 79 occupants involved in accidents with Type B load limiters. Seventy-two occupants (91.1%) experienced shoulder belt loads between zero and 7.4 kN. Three of these occupants (4.5%) suffered thoracic injuries which were classified as AIS greater than 2. Two of these occupants, a 39 year old male and a 67 year old male (Table 3), suffered flail chests (AIS = 5). Six occupants (7.6%) experienced shoulder belt loads ranging from 7.4 kN to 8.0 kN. One of these occupants (16.7%), a 43 year old female (Table 3), suffered a flail chest (AIS = 5). One occupant (1.3%), a 23 year old male (Table 3), experienced a shoulder belt load greater than 8.0 kN. He suffered a chest contusion classified as AIS = 1.

TYPE "C" LOAD LIMITER - The Type C load limiter had a single band of stitching that began to tear at a shoulder belt load of 5.5 kN (9). Although not noted in Reference 9 or 10, Foret-Bruno has indicated that thirty-four occupants were involved in accidents with this type of load limiter. Twenty-one occupants (61.8%) had no tearing of the stitching. Three occupants (8.8%) had partial tearing of the stitching so their shoulder belt load was limited to 5.5 kN. Ten occupants (29.4%) had complete tearing of the stitching. Their shoulder belt loads exceeded 5.5 kN, but the exact values are not known. None of the occupants involved in accidents with the Type C load limiter experienced thoracic injuries classified as AIS 3 or greater. This lack of an occupant with a significant thoracic injury is due to two facts. Most of the 386 accidents were not severe since only 10 occupants (2.6%) suffered AIS ≥ 3 thoracic injuries. There were only 34 occupants involved in accidents with Type C load limiters. Two point six percent of 34 is less than 1 occupant expected to experience a significant thoracic injury.

INJURY RISK BASED ON HYBRID III STERNAL DEFLECTION

An important observation from the Hybrid III sled test results is that for every 200 N of shoulder belt load the Hybrid sternum was deflected 1 mm. This relationship is true for the tests conducted with and without the Type B load limiter and for two levels of simulated collision environments. The implication of these results is that this relationship will hold for the Type A and C load limiter shoulder belt load data as well. The shoulder belt load ranges for the Type A, B and C load limiters given by Foret-Bruno et al (9) and listed in Table 2 were converted to estimates of Hybrid III sternal deflections by simply multiplying the

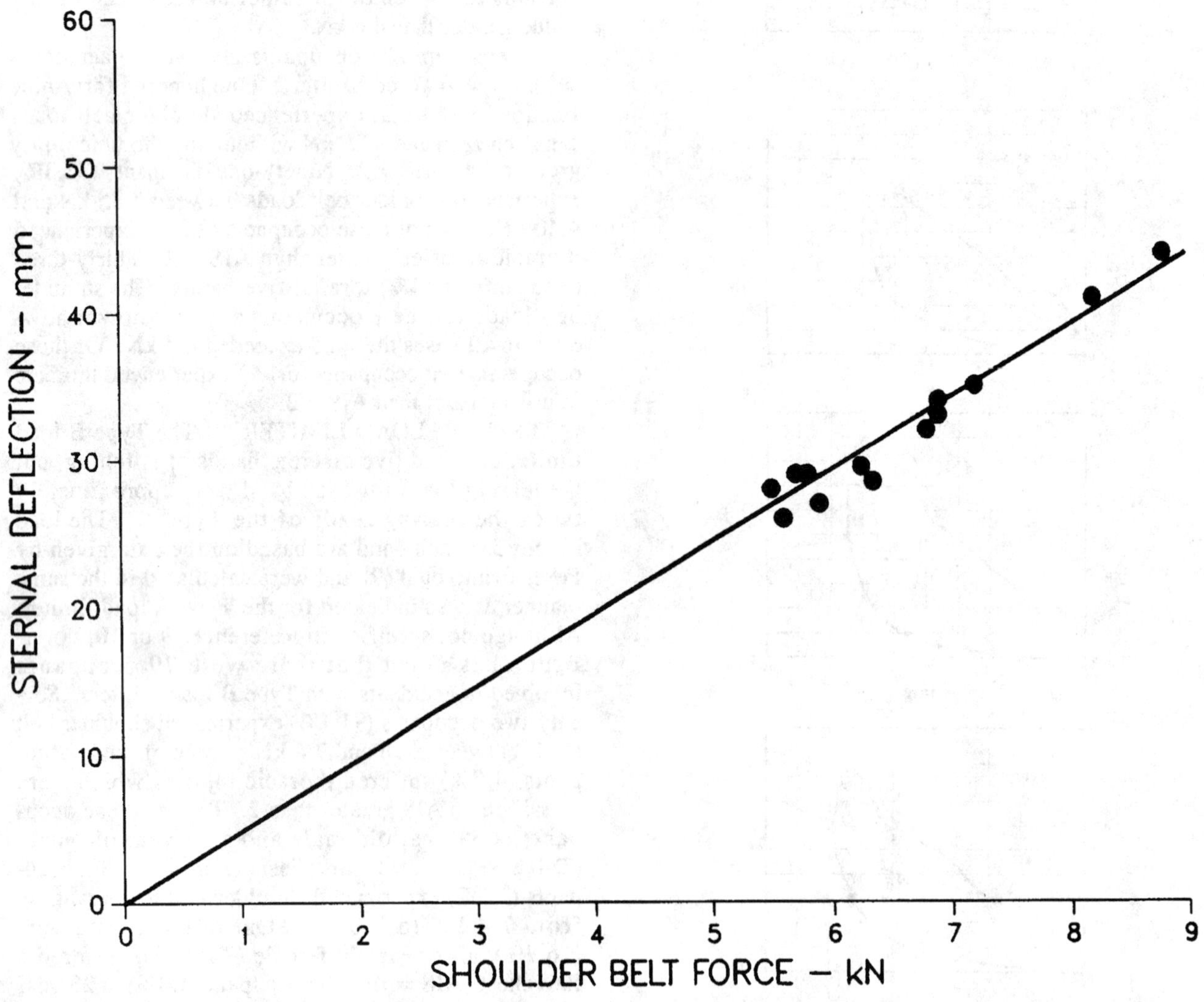

Figure 8 - Correlation Between Hybrid III Sternal Deflection and Shoulder Belt Force Corresponding to Band Tearing.

TABLE 2 - Data Summary of Accidents Involving Load Limiting Shoulder Belts. Shoulder Belt Load Ranges Taken from Foret-Bruno et al (9).

Type A Load Limiter

Stitch Bands Torn	Load Range (kN)	No. Occupants with Thoracic Injury			No. Cases
		AIS = 0 to 2	AIS = 3	AIS = 4 to 6	
0	0 - 2.10	123	0	0	123
1	2.10 - 3.85	26	0	0	26
2	3.85				
3	3.85 - 4.00	89	1	1	91
4	4.00 - 4.40				
5	> 4.40	31	2	0	33
				Total	273

Type B Load Limiter

Stitch Bands Torn	Load Range (kN)	No. Occupants with Thoracic Injury			No. Cases
		AIS = 0 to 2	AIS = 3	AIS = 4 to 6	
0	0 - 7.40	67	3	2	72
1	7.40	2	0	0	2
2	7.40 - 7.90	1	0	0	1
3	7.90 - 8.00	1	0	0	1
4	8.00	1	0	1	2
5	> 8.00	1	0	0	1
				Total	79

Type C Load Limiter

Torn Stitching	Load Range (kN)	No. Occupants with Thoracic Injury			No. Cases
		AIS = 0 to 2	AIS = 3	AIS = 4 to 6	
None	0 - 5.50	21	0	0	21
Partial	5.50	3	0	0	3
Complete	> 5.50	10	0	0	10
				Total	34

load values by 5 mm/kN. These estimated Hybrid III sternal deflections and the corresponding thoracic injury data are given in Table 4. The data for tearing 1 to 4 bands of the Type B load limiter have been grouped together for a Hybrid III sternal deflection range of 37 to 40 mm. The accident data observations for those occupants where all the bands or stitching tore are not given in Table 4. These data cannot be used in the injury risk analysis, since the shoulder load was not bounded and consequently the Hybrid III sternal deflection is not bounded.

SENSITIVITY ANALYSIS - The sensitivities of the sample sizes of the field accident data that were used to calculate the various injury risk values given in Table 4 can be determined as follows. For example, there were 123 occupants restrained by a Type A force limiter where no bands were torn. Since none of these occupants experienced an AIS ≥ 3 thoracic injury, a definitive injury risk value can not be calculated. However, an upper bound of the risk can be calculated by assuming that the next occupant involved in an accident where there is no tearing of a band of a Type A load limiter would experience an AIS ≥ 3 thoracic injury. If such were the case, then the risk of AIS ≥ 3 thoracic injury would be 1 in 124 occupants or 0.8% which is an upper bound. A lower bound of

TABLE 3 - Summary of Accident Data for Type B Load Limiter. All Cases Where Bands Broke. All Cases Where Bands Did Not Break, But There Was Rib Fracture.

Stitch Bands Torn	Occupant Information						Loading By Rear Occupant
	Age	Sex	Ht. (mm)	Mass (kg)	No. Rib Fracture	Thorax AIS	
0	31	F	-	-	1	1	Yes
0	68	M	-	-	1	1	No
0	72	F	-	-	2	1	No
0	52	F	-	-	4	3	No
0	33	F	-	-	4	3	Yes
0	56	F	-	-	4	3	Yes
0	67	M	-	-	14	5	Yes
0	39	M	-	-	15	5	Yes
1	28	M	180	76	0	1	No
1	34	F	157	55	0	1	Yes
2	33	M	185	92	0	1	No
3	59	M	175	80	2	1	No
4	24	M	176	63	0	0	No
4	43	F	160	70	14	5	-
5	23	M	188	75	0	1	Yes

risk can be calculated by assuming that this next occupant does not experience an AIS ≥ 3 thoracic injury. For this case, the injury risk would be zero in 124 occupants or 0.0% which is a lower bound. This procedure for determining upper and lower bounds was applied to all the accident data sets given in Table 4. The results of these calculations are summarized in Table 5.

INJURY RISK CURVES - Two methods were used to obtain curves that can be used to estimate the risk of significant thoracic injury due to shoulder belt loading as a function of Hybrid III sternal deflection. In the first method, the mean Hybrid III sternal deflections and associated risks for AIS ≥ 3 and AIS ≥ 4 thoracic injury given in Table 4 were subjected to Probit analyses (13, 14). The resulting risk curves for AIS ≥ 3 and AIS ≥ 4 thoracic injury are shown in Figure 9. Figure 10 shows the AIS ≥ 3 risk curve and its associated 95% confidence limits. These confidence limits reflect the spread in the upper and lower bounds of the risk values given in Table 5; that is, the curve is well determined for Hybrid III sternal deflections of 20 mm or less with greater uncertainty for larger chest deflections.

A second approach for obtaining an injury risk curve is to use only those data points from Table 4 for which the Hybrid III sternal deflection is closely bounded and the injury risk is also closely bounded. Only two Hybrid III sternal deflection ranges (19.3 to 22.0 mm and 37.0 to 40.0 mm) meet these requirements. The average sternal deflections for these ranges (20.7 mm and 38.5 mm) and their corresponding risks of AIS ≥ 3 thoracic injury (2.2% and 16.7%) were plotted on normal distribution graph paper, Figure 11. A straight line drawn through these two points provide a risk curve of AIS ≥ 3 thoracic injury as a function of Hybrid III sternal deflection for shoulder belt loading. Also shown on Figure 11 are dashed lines representing estimates of the upper and lower bounds of the injury risk curve. These boundaries were determined by plotting the upper and lower bounds of the injury risk values for mean Hybrid III sternal deflections of 20.7 mm and 38.5 mm given in Table 5. The dashed lines were drawn through the upper bounds and lower bounds of these mean sternal deflection levels. Note that the risk curve is well bounded from below and is quite well determined for sternal deflections of less than 20 mm.

TABLE 4 - Estimated Hybrid III Sternal Deflections Produced by Shoulder Belt Loading Associated with Risks of Various Thoracic Injury Severity Levels. Shoulder Belt Load Ranges and Occupant Injury Data Taken from Foret-Bruno et al (9).

Force Limiter Type	Stitch Bands Torn	Load Range (kN)	Hybrid III Sternal Defl Range (mm)	Hybrid III Sternal Defl Mean (mm)	Total Occupants	Thoracic Injury AIS ≥ 3 No.	AIS ≥ 3 %	AIS ≥ 4 No.	AIS ≥ 4 %
A	None	0 - 2.10	0 - 10.5	5.3	123	0	0.0	0	0.0
A	1	2.10 - 3.85	10.5 - 19.3	14.9	26	0	0.0	0	0.0
A	2 to 4	3.85 - 4.40	19.3 - 22.0	20.7	91	2	2.2	1	1.1
C	None	0 - 5.50	0 - 27.5	13.8	21	0	0.0	0	0.0
C	Partial	5.50	27.5	27.5	3	0	0.0	0	0.0
B	None	0 - 7.40	0 - 37.0	18.5	72	5	6.9	2	2.8
B	1 to 4	7.40 - 8.00	37.0 - 40.0	38.5	6	1	16.7	1	16.7

TABLE 5 - Bounds of Risk of Significant Thoracic Injury Due to Shoulder Belt Loading for Various Hybrid III Sternal Deflection Levels.

Hybrid III Sternal Defl.		Bounds of Risk of Thoracic Injury (%)			
Range (mm)	Mean (mm)	AIS ≥ 3		AIS ≥ 4	
		Lower	Upper	Lower	Upper
0 - 10.5	5.3	0.0	0.8	0.0	0.8
10.5 - 19.3	14.9	0.0	3.7	0.0	3.7
19.3 - 22.0	20.7	2.2	3.3	1.1	2.2
0 - 27.5	13.8	0.0	4.5	0.0	4.5
27.5	27.5	0.0	25.0	0.0	25.0
0 - 37.0	18.5	6.8	8.2	2.7	4.1
37.0 - 40.0	38.5	14.3	28.6	14.3	28.6

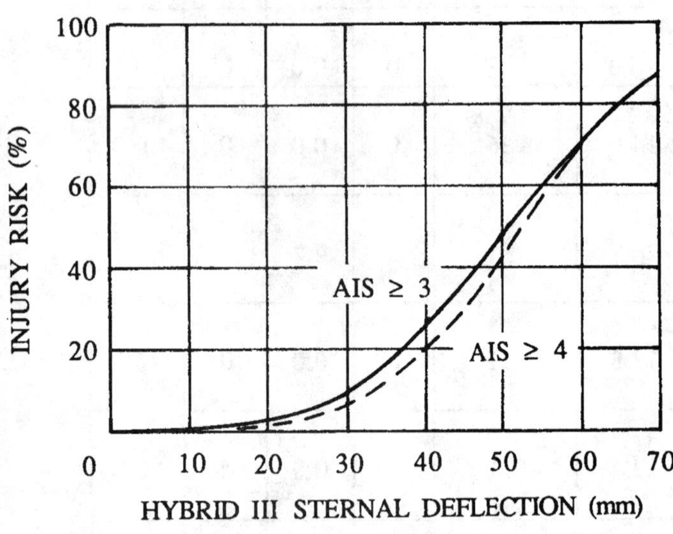

Figure 9 - Risks of AIS ≥ 3 and AIS ≥ 4 Thoracic Injury as a Function of Hybrid III Sternal Deflection for Shoulder Belt Loading. Risk Curves Determined by Probit Analysis.

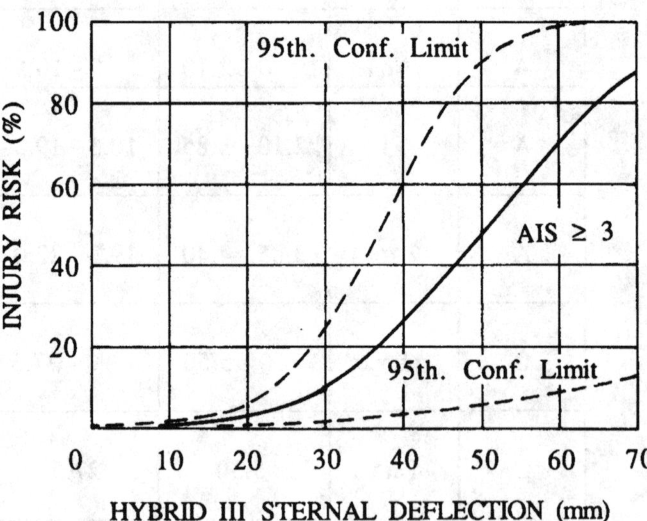

Figure 10 - Injury Risk Curve and Associated 95% Confidence Bands for AIS ≥ 3 Thoracic Injury Based on Probit Analysis.

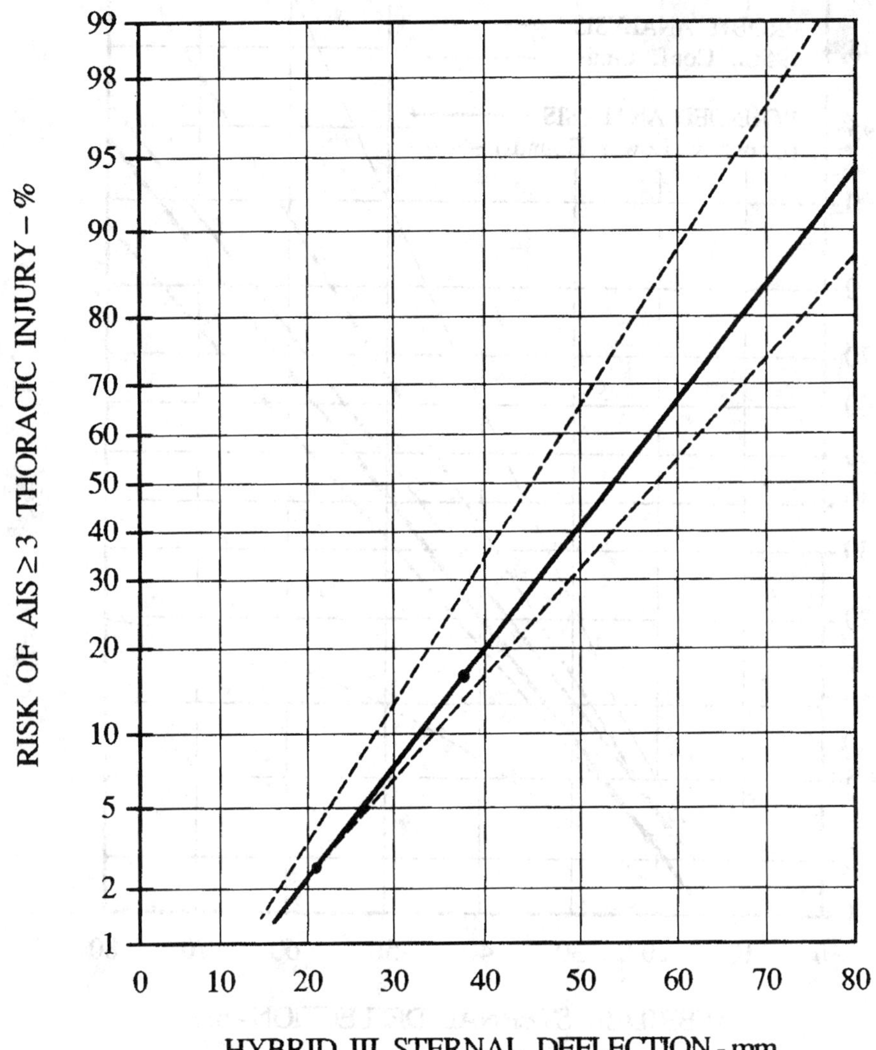

Figure 11 - Risk of AIS ≥ 3 Thoracic Injury Due to Shoulder Belt Loading as a Function of Hybrid III Sternal Deflection. Risk Curve Determined Using Closely Bounded Data. Dash Lines are Bounds Due to Sensitivity of Accident Sample Size.

Additional field data are needed for the 38.5 mm deflection level to decrease the sample size sensitivity of the upper bound. A risk curve for AIS ≥ 4 thoracic injury was not plotted using this method because there were only four such injury observations in the accident sample given in Table 4.

Figure 12 compares the injury risk curves for AIS ≥ 3 thoracic injury calculated by the two methods. Note the close agreement of the two curves, especially at low and high injury risk levels where the differences between the curves are less than one percent. Both curves lie within the 95% confidence bands determined by the Probit analysis. They also lie within the sample size sensitivity bounds for Hybrid III sternal deflections greater than 20 mm. For Hybrid III sternal deflections of less than 20 mm, the Probit curve lies slightly outside of sample size sensitivity bounds probably because of the greater uncertainty of the data used in the Probit analysis. Both curves are in excellent agreement and well determined for Hybrid III sternal deflections corresponding to low injury risks.

DISCUSSION

Foret-Bruno et al (10) conducted a series of Hybrid III sled tests at UTAC using the Type B load limiter. In the UTAC tests, the average Hybrid III sternal deflection during tearing of the bands was 60 mm (Test 1943) compared to 29 mm (Run 5) in our tests. Upon review of the UTAC data, an error was found in the sternal deflection measurements. To confirm this finding two additional sled tests were conducted. The corrected results along with the results of the two additional tests have been summa-

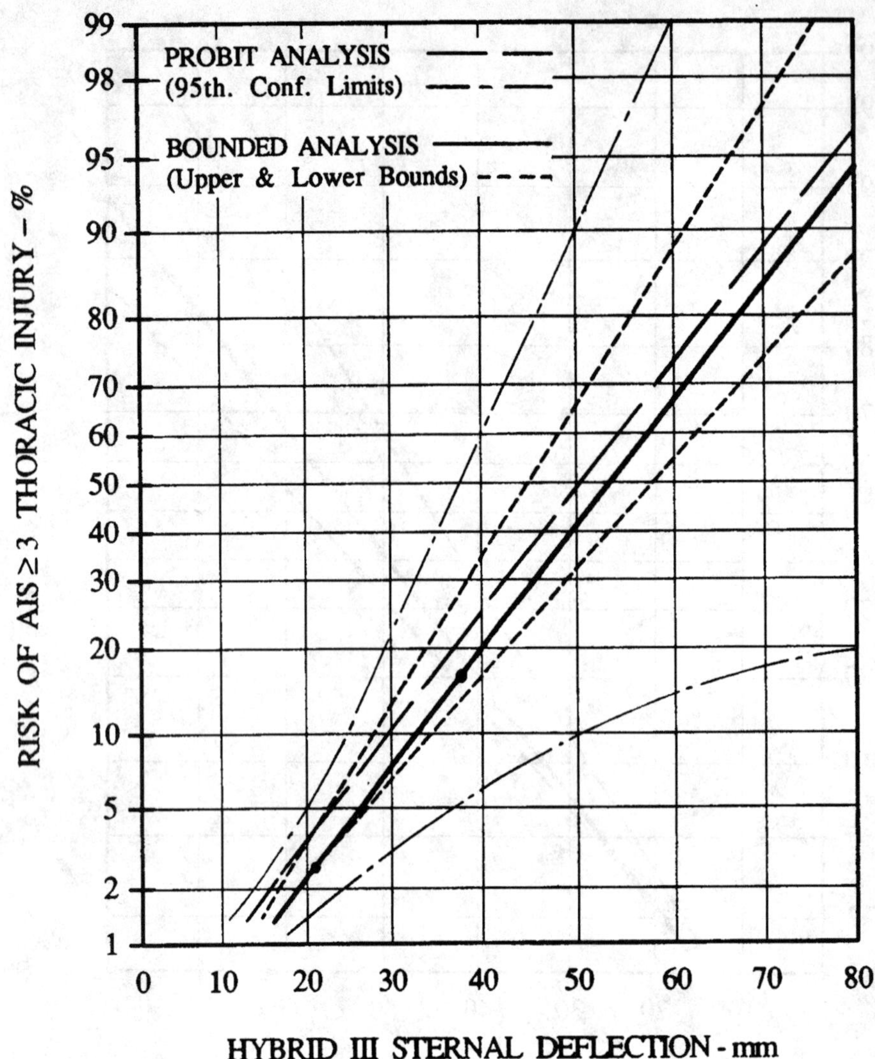

Figure 12 - Comparison of Injury Risk Curves Determined by Probit Analysis and Bounded Sample Analysis.

rized by Foret-Bruno and Bendjellal (12). Hybrid III sternal deflections for their corrected results ranged from 30 to 43 mm during band tearing which compare favorably with our sternal deflections which ranged from 26 to 35 mm. Based on these corrected results the Foret-Bruno et al (10) conclusion relative to tolerable sternal deflection for young people should be restated as, "a load of 8.0 kN corresponding to a sternal deflection of 40 to 50 mm for the Hybrid III can be withstood without thoracic injury by young adults in the population". This conclusion is now in agreement with the injury risk curves shown in Figure 12 which indicate that sternal deflections of 40 to 50 mm of the Hybrid III dummy produced by shoulder belt loading pose a risk of AIS ≥ 3 thoracic injury to 20 to 50 percent of the population.

Foret-Bruno et al (10) have eliminated from their analysis the accident cases where a rear seat occupant loaded the front seat occupant who was wearing the three-point belt equipped with the shoulder belt load limiter. We chose to include these data since the shoulder harness limits the loading. The loading of the rear passenger is simulated by simply increasing the level of the sled pulse to produce the observed level of band tearing. Foret-Bruno et al (10) also chose to segregate the field accident by age groups. We chose to do our risk analysis for the population as a whole. Obviously, people with lower bone strength, either due to the effects of aging, health or physical geometric size, will have a greater risk of experiencing a significant thoracic injury for a given shoulder belt load. It is because of these types of factors that we have the risk curves shown in Figure 12.

The field accident data for the Type A, B and C load limiters given in Table 2 can be used to estimate the collision severity distribution of real world accidents as measured by Hybrid II sternal deflection. Note that for the Type A load limiter, 123 of the 273 occupants (45.1%) had shoulder belt loadings that would produce Hybrid III sternal deflections of less than 10.5 mm. For Type B load limiter, 72 of 79 occupants (91.1%) experienced shoulder belt loadings that would have produced Hybrid III sternal deflections of less than 37 mm. For Type C limiter, 21 of 34 occupants (61.8%) had shoulder belt loadings corresponding to Hybrid III sternal deflections of less than 27.5 mm. Using these values, a cumulative frequency distribution curve of collision severity of the accidents involving occupants wearing three-point harnesses that were investigated by APR can be plotted, Figure 13. FMVSS 208 barrier tests of cars equipped with three-point belt systems can produce shoulder belt loads ranging from 9 to 10 kN and corresponding Hybrid III sternal deflections ranging from 45 to 50 mm. For example, Run 1 with standard webbing (Table 1) was slightly less severe than a 30 mph FMVSS 208 barrier test. It produced a peak belt load of 8.9 kN and Hybrid III sternal deflection of 44 mm. From Figure 12, Hybrid III sternal deflections of 45 to 50 mm correspond to a 30 to 50% risk of AIS ≥ 3 injury. Note from Figure 13 that less than 5% of the field accidents investigated by APR would produce a Hybrid III sternal deflection of 45 mm or greater. This implies that very few of the accidents investigated are as severe as a FMVSS 30 mph barrier test. Also note that the average risk of an AIS ≥ 3 thoracic injury for the APR accident sample is 10 in 386 or 2.6%. From Figure 12, this risk level corresponds to a Hybrid III sternal deflection of 17 to 21 mm which is much less than the deflection for a FMVSS 208 barrier test. Clearly, it would be inappropriate to associate a Hybrid III sternal deflection that is measured in a FMVSS 208 frontal barrier test with the average risk of AIS ≥ 3 thoracic injury for the APR accident sample.

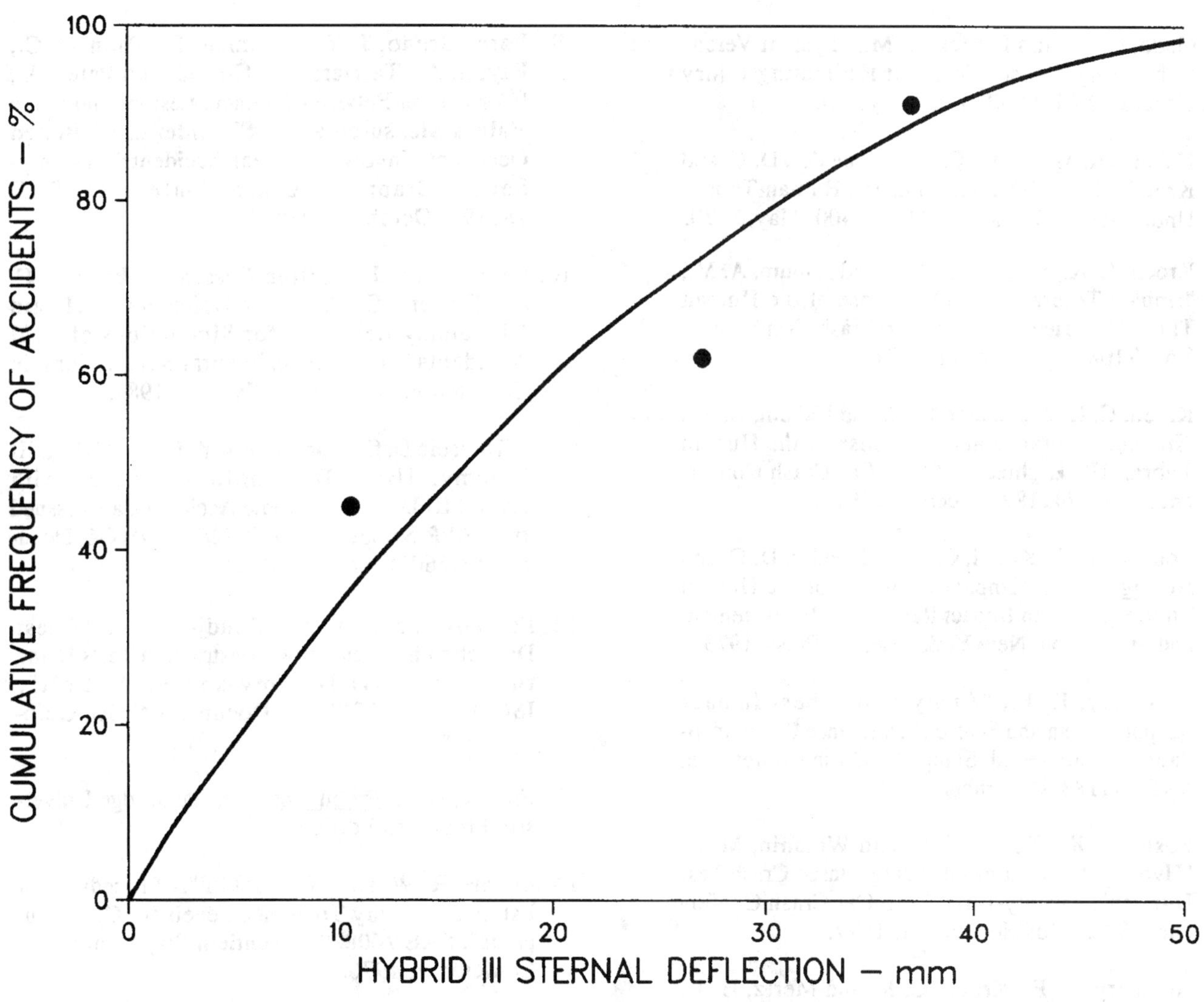

Figure 13 - Cumulative Frequency of Accidents Involving 3-Point Restrained Occupants Investigated by APR as a Function of Equivalent Hybrid III Sternal Deflection.

SUMMARY

Field accidents data obtained by the Association Peugeot-Renault of 386 occupants who were involved in frontal collisions while being restrained by three-point belt systems that used a shoulder belt with a force-limiting element were analyzed. For 342 of these occupants, the magnitude of the shoulder belt force could be estimated with various degrees of certainty from the amount of force-limiting band ripping. These data were used to calculate the risks of AIS ≥ 3 and AIS ≥ 4 thoracic injury for the various peak shoulder belt loads. Hyge sled tests were conducted with a Hybrid III dummy to reproduce the various degrees of band tearing. The results of these tests indicated that Hybrid III sternal deflection was well correlated with shoulder belt load. Every 5 mm of Hybrid III sternal deflection corresponded to 1 kN of shoulder belt load. This relationship was used to convert the estimated shoulder belt loads of the field accident data to equivalent Hybrid III sternal deflections. Statistical methods were used to develop relationships between Hybrid III sternal deflections and the risks of AIS ≥ 3 or AIS ≥ 4 thoracic injury produced by shoulder belt loading. This analysis indicates that for shoulder belt loading of the Hybrid III thorax a sternal deflection of 50 mm corresponds to a 40 to 50% risk of an AIS ≥ 3 thoracic injury.

ACKNOWLEDGMENTS

The authors wish to acknowledge the cooperation of the following personnel of the Association of Peugeot-Renault: Dr. Tarriere for providing the force limiters used in our test program, Mr. Bendjellal who participated in a joint review of the APR and our sled test data, and Mr. Foret-Bruno for providing additional details of the APR field accident data. The contribution from Mr. Lowne is published with permission of the Director of TRRL.

REFERENCES

1. Gadd, C. W. and Patrick, L. M., "System Versus Laboratory Impact Tests for Estimating Injury Hazard", SAE 680053, January, 1968.

2. Nahum, A. M., Gadd, C. W., Schneider, D. C. and Kroell, C. K., "Deflection of the Human Thorax Under Sternal Impact", SAE 700400, May, 1970.

3. Kroell, C. K., Schneider, D. C. and Nahum, A. M., "Impact Tolerance and Response of the Human Thorax", Fifteenth Stapp Car Crash Conference, SAE 710851, November, 1971.

4. Kroell, C. K., Schneider, D. C. and Nahum, A. M., "Impact Tolerance and Response of the Human Thorax-II", Eighteenth Stapp Car Crash Conference, SAE 741187, December, 1974.

5. Lobdell, T. E., Kroell, C. K., Schneider, D. C. and Hering, W. E., "Impact Response of the Human Thorax", Human Impact Response - Measurement and Simulation, New York: Plenum Press, 1973.

6. Neathery, R. F., "Analysis of Chest Impact Response Data and Scaled Performance Recommendations", Eighteenth Stapp Car Crash Conference, SAE 741188, December, 1974.

7. Foster, J. K., Kortge, J. O. and Wolanin, M. J., "Hybrid III - A Biomechanically-Based Crash Test Dummy", Twenty-First Stapp Car Crash Conference, SAE 770938, October, 1977.

8. Neathery, R. F., Kroell, C. K. and Mertz, H. J., "Prediction of Thoracic Injury from Dummy Responses", Nineteenth Stapp Car Crash Conference, SAE 751151, November, 1975.

9. Foret-Bruno, J. Y., Hartman, F., Thomas, C., Fayon, A., Tarriere, C., Got, C. and Patel, A., "Correlation Between Thoracic Lesions and Force Values Measured at the Shoulder of 92 Belted Occupants Involved in Real Accidents", Twenty-Second Stapp Car Crash Conference, SAE 780892, October, 1978.

10. Foret-Bruno, J. Y., Brun-Cassan, F., Brigout, C. and Tarriere, C., "Thoracic Deflection of Hybrid III Dummy Response for Simulations of Real Accidents", 12th Experimental Safety Vehicle Conference, Gothenburg, Sweden, 1989.

11. " Thoracic Deflection Limit with Belted Hybrid III Dummy: Useful Data for Reconstruction with Hybrid III Dummy of Some Accident Cases Issued from APR Sample", ISO/TC22/SC12/WG6, Document N260, October, 1987.

12. Foret-Bruno, J. Y. and Bendjellal, F., "Chest Deflection in Accident Reconstruction Tests Using the Hybrid III Dummy and E. A. Belt", ISO/TC22/SC12/WG6, Document N313, October, 1990.

13. Finney, D. J., Probit Analysis, Cambridge University Press, 3rd Edition, 1971.

14. Lowne, R. W. and Wall, J. G., "A Procedure for Estimating Injury Tolerance Levels for Car Occupants", SAE 760820, Twentieth Stapp Car Crash Conference, 1976.

973333

The Programmed Restraint System — A Lesson from Accidentology

Farid Bendjellal, Gilbert Walfisch, and Christian Steyer
Safety Engineering Department, Renault

Philippe Ventre
Engineering Division, Renault

Jean-Yves Forêt Bruno
PSA - Renault

Xavier Trosseille
Département Biomédicale de l'Automobile, Renault

Jean-Pierre Lassau
Institut d'Anatomie de l'UER Biomédicale des Saints Pères

ABSTRACT

Accident studies show that frontal collisions, both as regards the number of people killed and those seriously-injured, are by far the type of crash with the most serious consequences. In order to improve this situation, it is necessary to ensure that the means used to restrain occupants work as efficiently as possible, whilst preserving the occupant compartment and thus by eliminating intrusion on the occupant restrained by seat-belts and pretensioners.

In frontal collisions where vehicle intrusion is minor, the main lesions caused to occupantss are thoracic, mainly rib fractures resulting from the seat-belt. In collisions where intrusion is substantial, the lower members are particularly vulnerable. In the coming years, we will see developments which include more solidly-built cars, as offset crash test procedures are widely used to evaluate the passive safety of production vehicles. If nothing is done in order to limit the restraining forces, occupant lesions will be mainly thoracic, since airbags will have reduced the risk for the head, and lesions of the lower members, linked to intrusion, will be limited.

In order to reduce loads on occupants restrained by seat-belts, it has become necessary to work on an optimized limitation of the restraining forces, while taking account of the broadest possible population, especially elderly people. A first step in this reduction was taken in 1995 with the introduction of the first-generation Programmed Restraint System (PRS), with a seat-belt force threshold of 6 kN; thirty seven frontal accident cases involving this type of restraint were investigated. The corresponding data, crash severities and occupant injuries, are reported in this paper. Analysis of these data combined with findings from the University of Heidelberg / NHTSA study, shows that it is necessary to go a step further by reducing the shoulder belt force to 4 kN. As this objective cannot be achieved with a standard restraint system, it was necessary to redesign the airbag and its operating mode, that is, a new seat-belt + airbag combination called PRS II.

This paper summarizes the data obtained with the 6 kN load limiter restraint in real-world collisions. A description of the new system is given and its performance in offset crash situation with respect to a European standard belt + air bag system is discussed. The paper provides data on the validation of the PRS II in various frontal collisions and static out-of-position (OOP) tests.

INTRODUCTION

IMPORTANCE OF FRONTAL COLLISIONS. Detailed analyses of all fatal accident reports in France in 1990 and of the accidentology file of the PSA/Renault Laboratory enabled to determine the distribution of fatalities and seriously injured occupants with respect to collision configurations. The percentages related to frontal impact are respectively 50% and 70%, as shown in Figure 1; illustrating the predominant role of this crash configuration on occupant injuries. In order to assess the distribution of lesions in frontal collisions as regards the main body segments, an analysis was conducted on 100 belted front seat occupants taking into consideration serious injuries. Figure 2 presents the distribution of AIS ≥3 injuries for the head, the thorax, the abdomen and the lower members. It can be observed that the thoracic risk is highest for the passenger, and secondly, for the driver. For the latter, injuries to the lower members constitute the most frequent risk. Since 1992, improvements have been noted in Europe in cars as regards the resistance of the passenger compartment, especially the reduction in intrusion. In addition the majority of cars are today equipped with belt pretensioners. The combination of these improvements

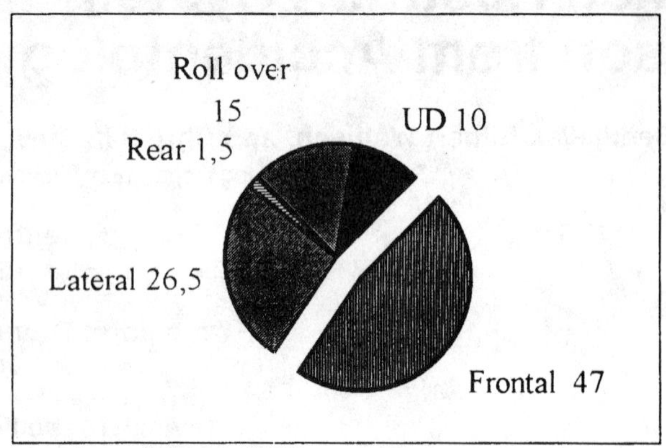

Number of killed: 6000
UD means undetermined type of collisions

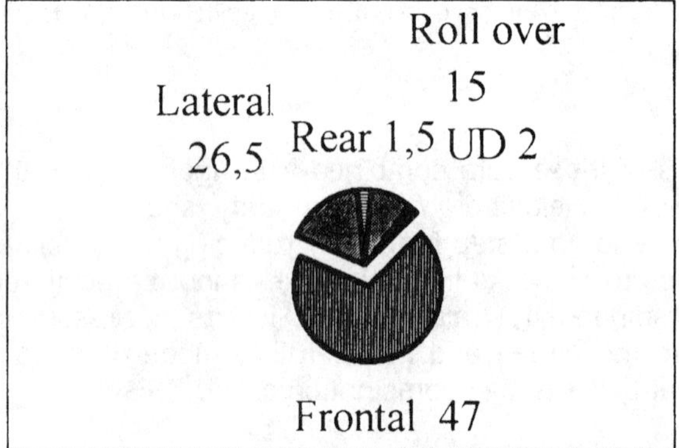

Number of severely injured: 18000

Figure 1 : Predominance of frontal collisions in terms of fatalities and severely injuried occupants from LAB PSA/Renault accident database. UD means undetermined cases.

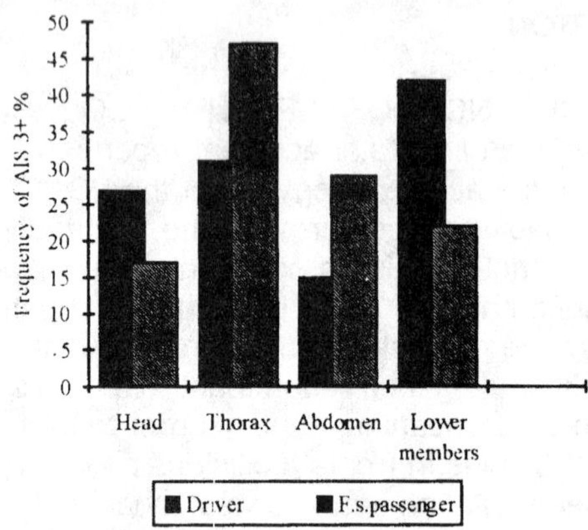

Figure 2 : Distribution of severe injuries per body regions for 100 injured occupant (AIS ≥ 3) in frontal collisions, occupant belted. Driver and front seat passengers.

would suggest a certain benefit in reducing the severity of injuries to the occupant. To assess this hypothesis two accident files, including belted drivers involved in frontal collisions, were selected. The first file (A) comprises 2000 vehicles manufactured before 1991 and with no belt pretensioners in the restraint system. The second file (B) includes 160 vehicles, manufactured since 1992, all equipped with belt pretensioners and structural reinforcements. The two files were compared taking into account the frequency of moderate to serious injuries, AIS ≥ 2 , corresponding to the main body segments, as shown in Table 1. When comparing files A and B, a tendency in the reduction of injury frequence is observed for the head, the abdomen, the lower limbs. For the thoracic segment an opposite trend appears with an increase of risk. As this tendency to reduce intrusion will continue and, as airbags will become more widespread in Europe, one may expect gains as regards the risk of injuries to the head and lower members, and abdominal risks will be maintained. For the thorax, there will be increased risk since rigidifying the structure will result in a direct increase in restraining forces on the occupant.

The study presented in this paper was initiated in order to address this rising risk of chest injuries.

LIMITATION OF THORACIC RISK LINKED TO SEAT-BELT

BELT INDUCED INJURIES AND OCCUPANT AGE - The 3-point seat-belt was designed to protect the occupant as regards contact with the passenger compartment and to avoid his being thrown out of the vehicle. In order to provide this protection, the seat-belt exerts substantial and localized forces on the thoracic cavity. These forces, which may reach 10 kN, generate broken ribs which may or may not be combined with internal lesions of the thorax. The first relationship between seat-belt tension and the associated thoracic risk level was established by J.Y. Forêt Bruno in 1978 (1) based on an analysis of 90 accident cases. The vehicles in question, sold in France in the 1970's, were equipped with 3-point static seat-belts in the front seats with a load limiter located in the belt webbing between the occupant's shoulder and the upper anchorage point. The load limitation was obtained by tearing of the stitching which was used to sew loops in the webbing. In case of impact the stitching tore, thus allowing more webbing from the loop; as a consequence the torso can move relative to the vehicle at a controlled load level. A view of such a load limiter before and after impact is shown in Figure 3a and its force-time response in dynamic test is illustrated in Figure 3b. In 1989, other cases were added to this investigation, bringing the total of this database up to 290 accidents (2). The key point of this unique database is the possibility of showing a relationship between the seat-belt tension exerted on the occupant, his age and the type of resulting lesions; this relationship, reproduced from Foret Bruno study (2), is given in Figure 4. This data clearly shows that thoracic

Table 1: Risk of AIS ≥ 2 injuries in frontal collisions involving belted drivers. Comparison of 2 accident samples with cars manufactured before 1991 (A) and with cars manufactured since 1992 (B)

Body segments	Accident Sample A (before 1991). Belt restraint without pretensioners. Drivers Frequency of AIS ≥ 2 injuries in %	Accident Sample B (After 1992). Belt restraint with pretensioners. Drivers Frequency of AIS ≥ 2 injuries in %
Head	23	13
Thorax	12	21
Abdomen	4	2
Lower Members Femur and knee	12	7
Lower Members Tibia and Foot	15	10

risk among occupants restrained by seat-belts increases with age and that a shoulder belt force of 8 to 9 kN may induce a high risk for the chest.

a) Load limiter before and after impact

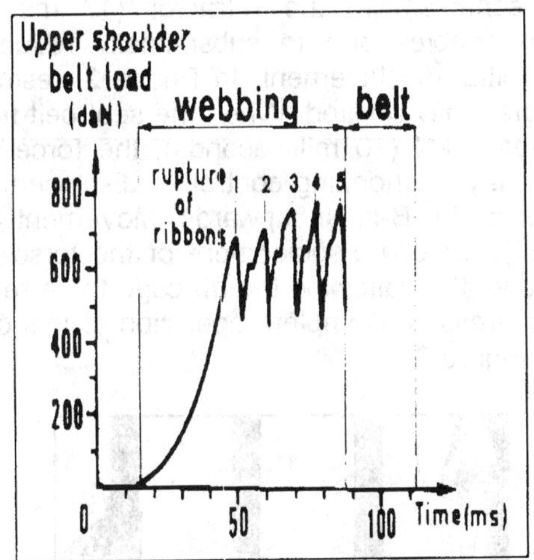

b) Response of the load limiter in dynamic test

Figure 3: A load limiter installed in cars sold in France between 1970 and 1977.

Real-world Accident with recent - Is this risk still valide with present cars ? To answer this question, at least from qualitative point of view, four examples of recent vehicles involved in frontal crashes have been analyzed. The particularity of these cases is the low intrusion - from 0 to 10 cm - noted opposite the occupant. Another common point: the restraint system comprised of a standard seat-belt with pretensioner. A photographic summary of each accident is presented in the following, including a description of occupant thoracic injuries.

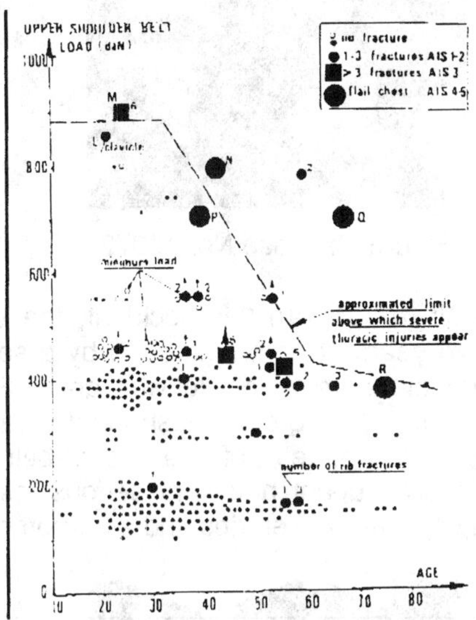

Figure 4: Relationship between shoulder belt tension, age of occupant and injury severity to the chest. Reproduced from (2).

Case N° 11975 - This involves a crash offset to the left, with a 2-cm dashboard backward displacement. The driver, 55 years of age, was restrained by a seat-belt. The pretensioner and the airbag functioned properly. The violence of the impact is estimated at 55 km/h in terms of Equivalent Energy Speed (EES); the occupant suffered broken ribs (AIS 3).

Figure 5a: Case No. 11975

Case N° 11972 -This involves a frontal crash without intrusion, with two occupantss restrained by seat-belts in the front seats in the vehicle under consideration. The EES is estimated at 50 km/h. The driver, whose protection system included an airbag, had serious lesions and 4 broken ribs. The passenger, 54 years of age, with no airbag, also had an AIS 3 at the level of the thorax.

Figure 5b: Case No. 11972

Case N° 10851 - In this accident, the vehicle, with a driver 48 years of age, restrained by a seat-belt, struck the side of another vehicle with an EES of 50 km/h. There is a 10-cm dashboard backward displacement. In spite of a seat-belt+airbag combination, the occupant had very serious lesions at the thorax (AIS4), with broken ribs and a hemothorax.

Figure 5c: Case No. 10851

Case No. 11975- The vehicle was involved in an offset crash, with an EES of 55 km/h without intrusion on the passenger side. The assessment of chest lesions for this passenger, 53 years of age, is severe as regards the thorax with an AIS level of 4.

Figure 5d: Case No. 11975

LIMITATION OF THORACIC RISK LINKED TO SEAT-BELT - The data discussed in the previous sections and these accident cases show the necessity of reducing seat-belt tension forces in frontal crashes. An initial stage, consisting of limiting this force to 6 kN, was carried out in 1995 on Renault vehicles with the introduction of the PRS system (Programmed Restraint System). This system is comprised of a pretensioner pyrotechnic buckle, a retractor webbing clamp and a steel part, fastened between the retractor and the seat-belt anchoring point as shown in Figure 6. This part, designed to deform at a given level of force, acts like a force limiter.

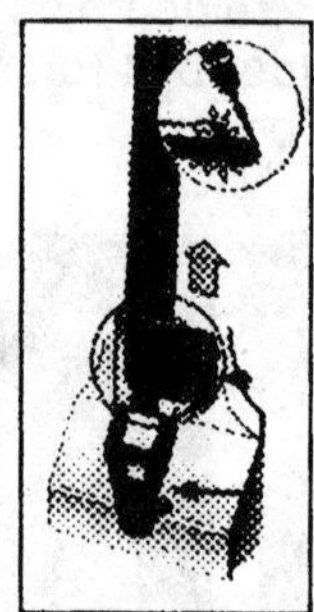

Figure 6: The Programmed Restraint System installed in Renault cars since 1995 (6 kN shoulder belt load limiter)

The system's operating method includes 3 phases: at the beginning of the impact (15 milliseconds), the buckle pretensioner triggers in order to take up the seat-belt/occupant slack. The occupant's coupling is increased in this phase with the action of the strap blocking mechanism in the retractor (17 ms). This combination enables one to substantially reduce the occupant's initial displacement. In Phase 2, restraining forces are gradually applied. When the seat-belt tension level reaches 6 kN (70 milliseconds), the force-limiter comes into play, authorizing controlled displacement of the retractor in the B-pillar, upwards. Movement of the retractor will enable a displacement of the torso under controlled load, thus allowing the rib cage to be relieved of seat-belt stresses. Complete operation of this device is shown in Figure 7.

Figure 7: The PRS operating mode - Phase 1 Initial part of the crash and belmt pretension activation, Phase 2 Action of the webbing clamp, Phase 3 Load limiter activation, Phase 4 End of impact

<u>Behaviour of the Programmed Restraint System in Real - World Accidents</u> - To date, 41 accident cases related to frontal collisions with cars equipped with this device have been investigated since 1995. Thirty seven cases were selected as the necessary data, i.e. collision information and a detailed medical report on occupant injury, were completed. The main parameters of this sample are summarized in Figures 8a through 8d and detailed findings are presented in Table A1 in the appendix. Age distribution of occupants ranges from 17 years to 72 years, with 11 cases (30%) with age < 25 years, 7 cases (19%) with age ranging from 26 to 35 years, 3 cases (8%) between 36 and 45 years, 8 cases (21.5%) between 46 and 55 years, and 8 cases (21.5%) with age > 56 years. The severity of the collisions, expressed in terms of EES, ranges from 35 km/h to 75 km/h. Nearly half of this sample (48.6%) corresponds to a severity which is superior to EES of 55 km/h.

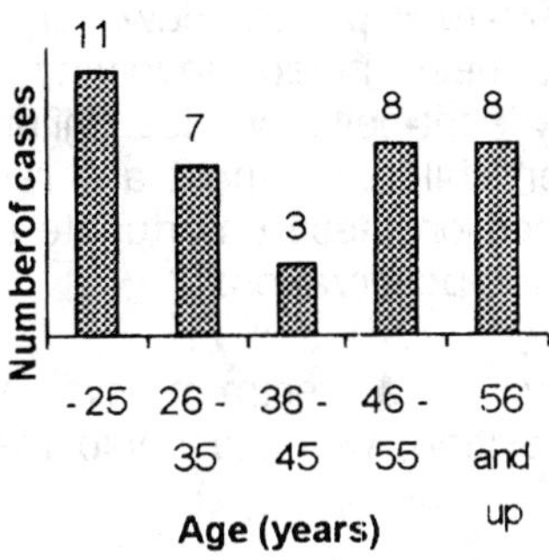

Figure 8a : Age distribution in the 37 accident cases with PRS

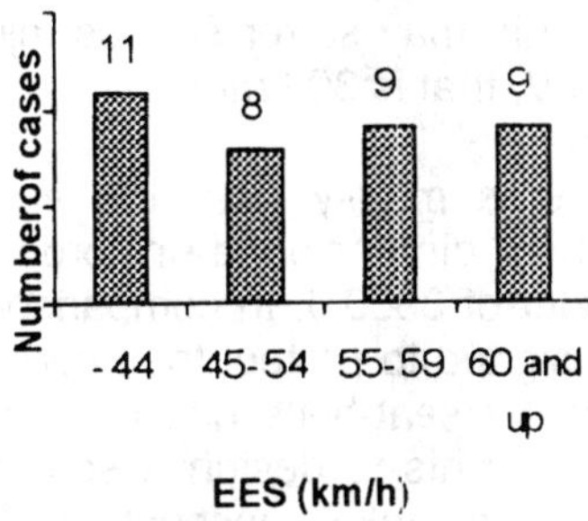

Figure 8b: Collisions' severity in the 37 accident cases with PRS

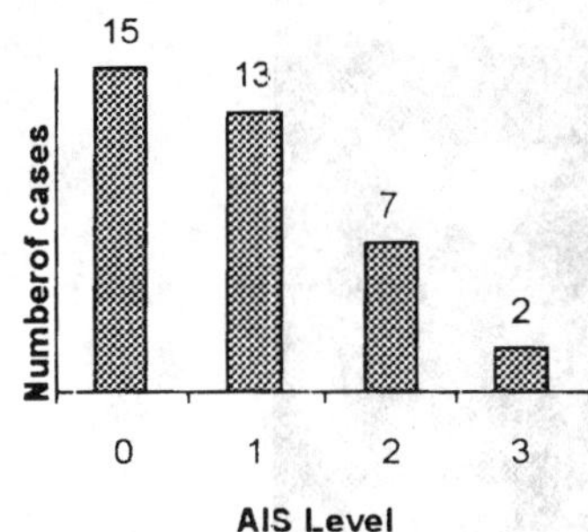

Figure 8c: Maximum thoracic AIS in the 37 accident cases with PRS

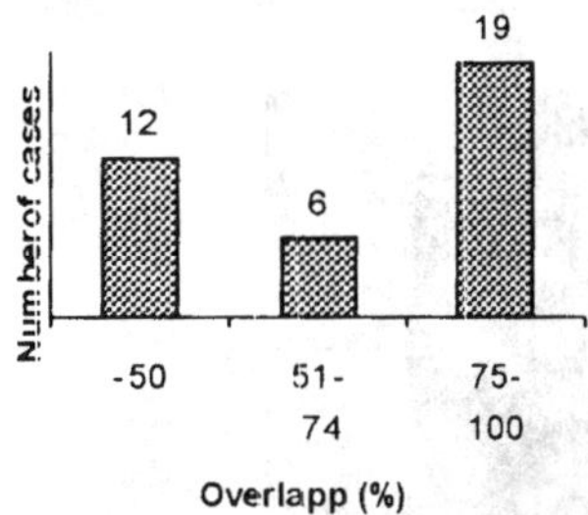

Figure 8d: Range of overlapp in the 37 accident cases with PRS

Regarding the injury severity for the thorax, only 2 cases (5.4%) are related to an AIS level of 3. These are cases No.11419 and No.12046; in the first case the driver, a 72 old male sustained 3 right rib fractures and lung contusion. The car was involved in an offset collision to the left, with an overlap of 85% and with an EES of 50-55 km/h. Except fractures to the left metatarse, no other injuries were found. In the second case two front seat occupant were involved; a 58 years old male in the driver position and a 60 years old female in the passenger position. The driver sustained a fracture to the sternum (AIS 2) and the passenger had 4 left rib fractures (AIS 3). In both accident the shoulder belt load, estimated from the PRS deployment, was 6kN. The other thoracic AIS levels observed for the rest of the sample are AIS 2 with 7 cases (19%), AIS 1 with 13 cases (35%) and AIS 0 with 15 cases (40%). Regarding the overlap distribution among these accident cases, half of the sample corresponds to offset configuration with an overlap below 74% and the other half is close to a full barrier test. An illustration of one accident, case No.1241 in Table A1, is given in Figure 9 with photographs of the car deformation and the PRS deployment.

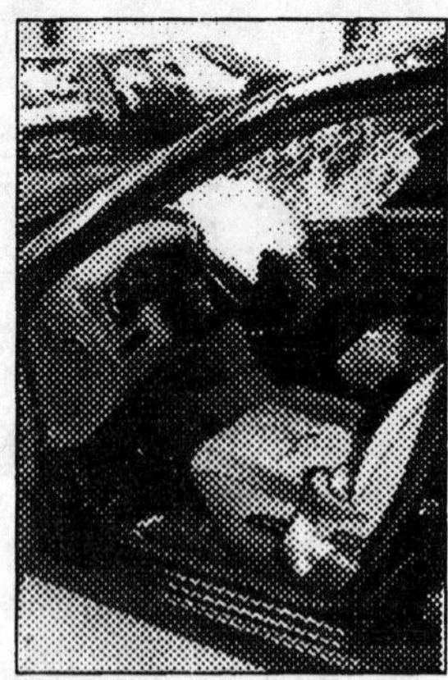

Figure 9a: Illustration of car deformation in a frontal collision. Case No.12041

Figure 9b: Illustration of PRS actual deployment in a frontal collision. Case No.12041

Belt limitation threshold - The accident cases presented in the previous section and the data contained in Table A1 show that a threshold of 6 kN for belt load limitation is not sufficient to prevent a risk of serious injury to the thorax as 2 cases with occupants having sustained an AIS 3 level were found. This observation is consistent with the data from Forêt Bruno (2) published in 1989 . It is therefore necessary to go a step further in the reduction of shoulder belt load. As this reduction will result in an increase in excursions of the head and thorax; it is therefore essential that with this kind of seat-belt, it is necessary to combine the pretensioner and, quite obviously, the airbag. The combination of an air bag and a 3-point belt restraint was discussed in various publications among them are the paper from Kompass in 1994 (3), the study of Kallieris et al. in 1995 (4) and Mertz et al. investigation in 1995 (5). According to the data discussed in Kallieris paper (4)and the mathematical simulation investigated by NHTSA (4) for a variety of crash conditions (frontal and rollover), crash severities and occupant sizes (5°, 50° a,d 95° percentiles) a threshold of 4 kN for the shoulder belt load limitation appears to be suitable for reducing the risk for thoracic injury, without negative consequences on other injury measurements. Therefore, a 4 kN load limitation threshold was chosen for the belt system.

Whilst working in the same thorax/steering wheel distance deceleration space for a 50° percentile of approximatly 300 to 350 mm - in this case, it is necessary for the airbag to play an important role by taking part of the thoracic restraint. Before pursuing this approach, it is useful to assess the role of the airbag in reducing injuries in the USA and in France.

Air bag accident data in the USA and in France - When the FMMVSS 208 was introduced in the USA in the beginning of the 1980's, according to investigations carried out by NHTSA, most people did not use seat-belts. The percentage of people wearing seat-belts at that time was on the order of 15%; this would suggest the necessity of including occupant protection by means of a restraint system independent of the seat-belt. The physics of vehicle impacts, 50 km/h against a rigid barrier, with 50° percentile dummies not restrained by a seat-belt, imposed de facto paddings or knee plates for the protection of the femurs and knees, and the airbag for protection of the upper part of the body. The assessment of such a restraint system combined with the seat-belt is quite positive, with more than 1500 lives saved (6) during the 1990-1996 period. However, cases of fatal accidents have been noted, involving either adults not restrained by seat-belts or else children in rearfacing seats, or even children without any restraint system whatsoever. Based on these unfortunate cases, one can make the following observations:

- Incompatibility of a restraint system for children in rearfacing position, with the deployment of the passenger airbag.

- The concern of US manufacturers in the 1970's with respect to risks linked to the aggressiveness of an FMVSS 208 airbag, for occupants near the steering wheel or the dashboard are confirmed.

- Smaller occupants may suffer serious injuries in crashes with low Delta V, that is 30 km/h.

This problem stems mainly from the energy parameters of the airbag dimensioned in order to absorb energy on the order of 3000 J. In comparison, a Eurobag or "facebag" designed to protect the head of a 50° percentile restrained by seat-belts has an energy potential of 200 J. If one wants to design a seat-belt-airbag restraint system which takes account of OOP situations, it is therefore necessary to explore other possibilities.

Current situation in France - Out of the total number of automobiles in France — some 25 million — only 2 to 3% of vehicles are equipped with driver airbags. We lack data on airbag efficiency in Europe, since the target survey files remain statistically low in comparison with the USA, only 100 cases have been studied in France by the Laboratoire d'Accidentologie et de Biomécanique Peugeot-Renault; 75 cases involved frontal collisions with belted drivers.

In Figure 10 a risk comparison for the head, with and without airbags, is given. For the 25 to 45 km/h speed range, one notes moderate lesions (11%) for cases with no air bag as opposed to 0% for cases with air bag. For the 46 to 65 km/h speed range, the frequency of AIS ≥ 2 is 40% without air bag and only 14% with air bag. No facial fractures were observed with air bag, whereas half of the sample without air bag represents facial fractures. The tendancy of air bag to improve head protection is confirmed.

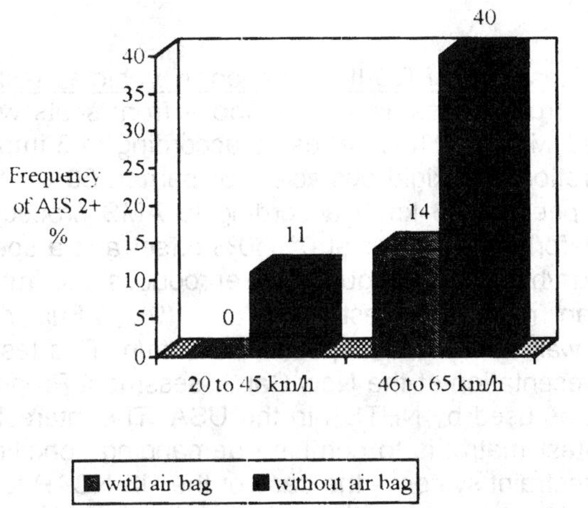

Figure 10: Accident survey with frontal collisions involving occupants with 3 points belt + Air bag restraint system. Frequency of AIS 2+ injuries to the head. All cases with Eurobag type of air bag.

SPECIFICATIONS FOR AN OPTIMIZED SEAT-BELT + AIR BAG RESTRAINT SYSTEM - The basic principle is that occupant restraint energy must be managed, whilst complying with human tolerance limits. In this context, the thoracic cavity is more tolerant to distributed pressure (air bag) than to a very localized pressure (belt). With the same stopping distance for the occupant in the vehicle, it is possible for the airbag to take a part of the seat-belt forces.

Once the basic elements of the seat-belt, that is, pyrotechnic pretensioner and force-limitation, have been determined, it is now necessary to define the airbag characteristics. The corresponding specification is based on 2 separate parts: to contribute actively to restraining the occupant, and to control the aggressiveness of the deployment of the airbag. This results in the 3 following main functions:

•The airbag must inflate very early on in the impact and "wait for" the occupant's contact; this is the anticipation function, analogous to the effect of a pretensioner on the PRS seat-belt.

•Having a law of force which is as constant as possible. This is equivalent to controlling the pressure in the airbag and the force exerted on the occupant. This is similar to the action of the force-limiter in the PRS seat-belt.

•These two functions result in an increase in the generator power in relation to the Eurobag. In order to control the bag aggressiveness in OOP situations, it is necessary to compensate this through a more elaborate airbag-folding strategy, in order to reduce the punch out transmitted to the occupant. This objective results in a deployment mode distributed in 3 directions: first downwards and sideways and then toward the occupant.

Based on these elements, a new airbag has been developed in the frame of the new system called the PRS II.

DEVELOPMENT OF THE PRS-II DESCRIPTION AND VALIDATION

The system comprises 3 main components. these are the pretensioner, the belt load limiter and the air bag.

The pretensioner - This is a device which enables the seat-belt strap to be drawn taut very quickly at the initial moment of impact. For the PRS-II system, and given the experience acquired on Renault vehicles since 1992, a pyrotechnic buckle pretensioner has again been selected, especially for its efficiency with respect to submarining. In 4 milliseconds, it enables to take up the seat-belt slack and secure the occupant to the seat.

The seat-belt force-limiter - The force limitation function is located at the core of the retractor with a torsion bar whose plastic deformation comes into play as soon as the seat-belt force at the shoulder reaches 4 kN. For this function an another option is to use the deformable steel plate, as in the PRS-I generation, providing a sufficient space in the B-pillar packaging.

The airbag - The airbag is a 60 liters bag with a pressure limitation function and a folding which allows a deployment from top to bottom and to the sides. As opposite to Eurobag, this bag is defined to protect the head and the thorax. There are different ways to control the pressure of the air bag; the system described here refers to a set of vents in a row, contained in a meltable seam. One the air bag is full and the vents closed At a given pressure, the seam tears and the vents open successively. The restraint force acting on the occupant from the air bag is thus controlled.

Development of the PRS II - After a computer simulation phase, the opening pressure of the airbag vents has been validated during tests using a free fall pendulum system. At the same time, the seat-belt force limiter was developed. Then, sled tests were conducted in order to fine-tune the system's characteristics. The validation program also included static tests in OOP, according to ISO recommendations (7), and crash tests with vehicles.

Figure 11 provides a description of PRS-II components. The operating phases of the system, as obtained in a 50% offset rigid barrier test, are illustrated in the same figure where the 4 upper sequences indicate the air bag work and the 4 lower sequences relate to the belt actions. Sequence 1 in Figure 11 represents the firing of the belt pretensioner at 12 ms followed by the start of air bag deployment at 15 ms. Note that once the air bag deployment is achieved (sequence 2) , the vents are still closed; the air bag is wainting for the occupant. When the thorax contacts the air bag, in sequence 3, the seam covering the vents starts to tear, thus liberating the first vent. The bag pressure is now under control; in sequence 4 the belt load limiter function starts to work in conjunction with the opening of the remaining vents in the air bag: with this last sequence the thoracic restraint loads are controlled thorough the impact duration.

Figure 11: PRS II- Principle of belt load limiter and air bag actions in a frontal test

Actual driver motion sequences as obtained in a frontal offset test are provided in Figure A1 in the Appendix. In this test a car of 1200 kg mass, with 2 front seat occupants (Hybrid III 50°percentile), was subjected to a 56 km/h rigid barrier test with 50% overlap. The sequences in Figure A1 refers to the mains events of the crash and in particular to the PRS II work.

COMPARAISON OF PRS II WITH A CONVENTIONAL RESTRAINT SYSTEM - Various mathematical simulations and sled tests were conducted in order to assess the PRS-II performances in frontal collisions. In addition two crash tests with the same vehicle model (mass of the vehicle 1200 kg) were performed; the test configuration corresponds to a 50% offset rigid barrier test at 56 km/h. One of the vehicle was equipped with a conventional belt + air bag system; the belt included a pyrotechnic buckle pretensioner and the air bag was of Eurobag type(volume of 45 liters). The other vehicle had the PRS-II system. In the front seats of both vehicles instrumented Hybrid III 50° dummies were installed. The results from both tests are illustrated in Table 2 and time-histories for the head acceleration, the chest acceleration and the shoulder belt load are provided in Figure 12. With the PRS-II the head HIC and 3ms acceleration are reduced, respectively 75% and 55%; the neck shearing force is also reduced respectively 60% for -Fx and 57% for +Fx. The neck extension moment is increased with the PRS II with a maximum of 35 Nm as opposed to 11 Nm with the conventional system. The shoulder belt load reduction with the PRS-II is significant -55%, as a direct result of the combined work of the belt and the airbag. The thoracic acceleration is also reduced but the amount of reduction (24%) is smaller than those observed with the other criteria. This last result shows that 1) the occupant stopping distance is the same for the 2 systems we are comparing and 2) the energy distribution on the thorax is spread differently with the PRS-II. Chest injury parameters, such as the chest deflection and the VC, cannot be compared as the data corresponding to the conventionnal system (with the same vehicle) are not available. The maximum chest deflection and VC with the PRS II are 25 mm and 0.09 m/s. Compared to the conventionel system the PRS II allowed an increased x-displacement of the chest (+60mm).

Results of PRS-II validation in vehicles tests - Vehicles from the same model whose front seats were equipped with PRS II were tested according to 3 impact configurations : **1**.Rigid obstacle, 15° barrier, 50% offset and a speed of 56 km/h according to AMS procedure (8), **2**.Deformable barrier at 0°, 40% offset and a speed of 56 km/h. This configuration reproduces the future European regulatory test, ECE 94 (9), **3**.Full rigid barrier, wall at 0°, and a speed of 56 km/h. This test is the representation of the New Car Assessment Program (NCAP) as used by NHTSA in the USA. The interest of such a test matrix is to combine demanding conditions for the restraint system - the case of the US NCAP test - and for the structure of the vehicle with the other two offset crashes. The first offset test condition allows to assess both the structure of the vehicle and the restraint system. The test according to the procedure defined by the EEVC (ECE 94) is a special case, since this configuration enables to simulate a car to car collision and also to judge the quality of the triggering system for the belt restraint and the airbag restraint, in particular as the first part of the crash is soft compared to the two other test configurations.

The results of these tests are documented in Table 3, which includes the resulting accelerations of the head and thorax, the Head Injury Criterion (HIC 36 ms), the upper neck shear force, the upper neck extension moment, and the shoulder belt tension, the chest acceleration, the chest deflection and VC. All the maximum values refer to measurements obtained from Hybrid III 50° percentile dummy, for both the driver and passenger.

The PRS-II system behaved well in all 3 configurations, both belt load limiter and air bag pressure limiter worked. The shoulder belt tension was between 3.9 kN and 4.8 kN. The lowest value was recorded for this parameter was obtained in the EEVC test (for he driver) and the highest value in the US NCAP test (for the passenger). This difference is due to the friction in the D-ring. As this friction is directly related to the dummy forward displacement, its effect on the shoulder peak load is more pronounced for the passenger. Chest accelerations were all below 46 G; this result indicates no chest to steering wheel contact. Neither head to steering wheel contact was observed as illustrated by the low values recorded with the HIC acceleration - between 24 G and 53 G. Chest deflections ranges from 15 mm to 40 mm; the lowest value was obtained in the EEVC test for the passenger and the highest in the NCAP test for both the driver and the passenger.

VC values were between 0.03 m/s and 0.64 m/s. Both head and chest accelerations and also chest deflections and VC's ensure that the use of belt load limitation, in the test conditions discribed here, combined with air bag pressure control has no negative effects on injury measures.

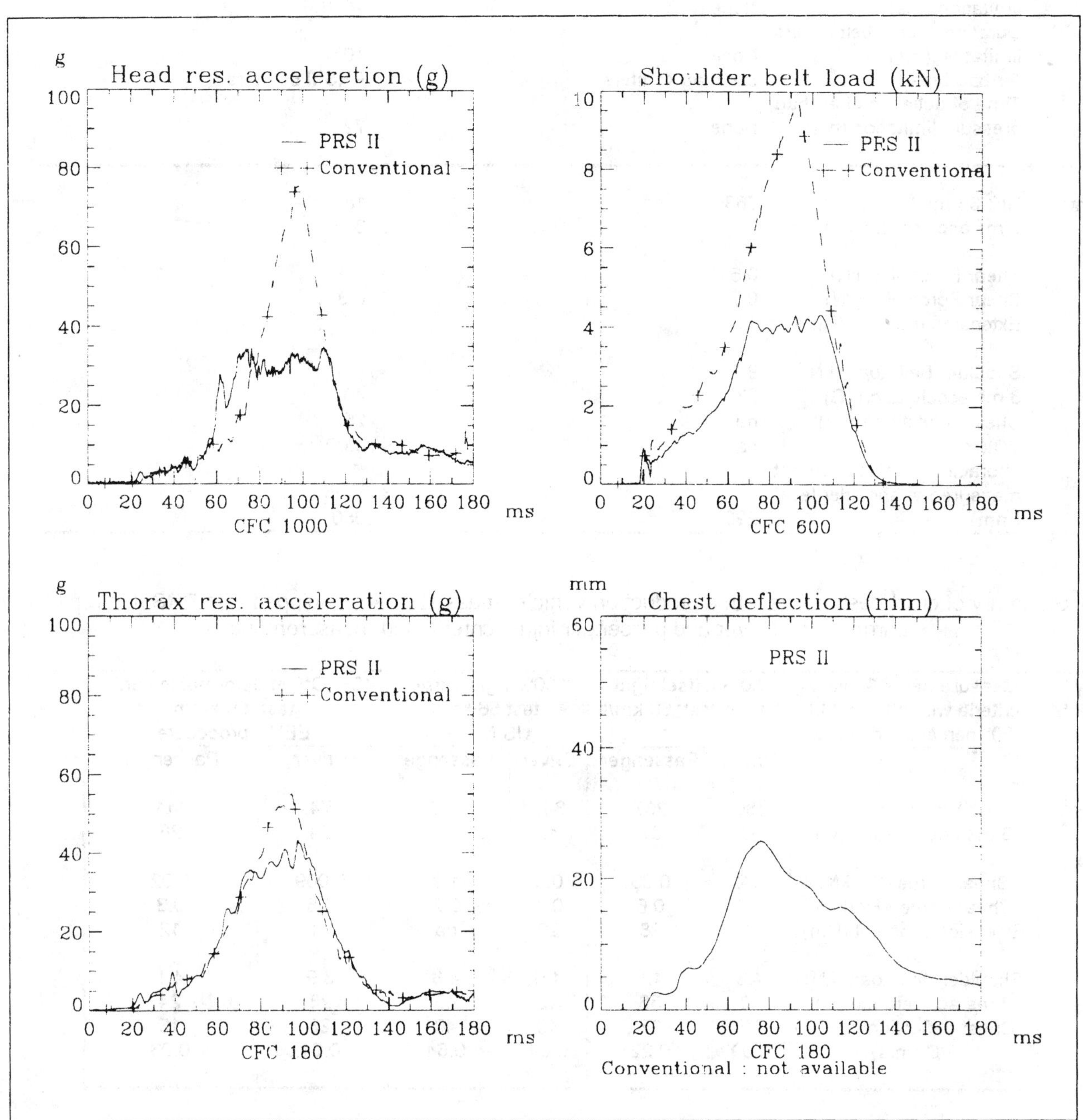

Figure 12: Comparison of PRS II responses with those of a conventional belt + air bag system. Driver data as obtained from a 56 km/h offset barrier test (same vehicle used in both tests).

Table 2: 50% offset rigid barrier test at 56 km/h. Comparison of PRS II responses with those of a conventional belt + air bag system - Same vehicle used in both tests, driver data.

	Measurements & injury criteria with a Hybrid III 50° percentile dummy	50% Offset rigid barrier test, 56 km/h with a conventional restraint system	50% Offset rigid barrier test, 56 km/h with the PRS II
Restraint system	Buckle pretensioner activation time (ms)	18	16
	Belt pretension (mm)	49	49
	Initiation of belt load limitation (ms)	None	70
	Duration of belt load limitation (ms)	None	40
	Air bag type	Eurobag 45 liters	PRS II 60 liters
	Time of actiation of air bag pressure limitation (ms)	None	72
Body segments Head	HIC 36 ms	763	186
	3 ms acceleration (G)	74	33
Neck	Shear Force -Fx (kN)	0.5	0.2
	Shear Force +Fx (kN)	0.7	0.3
	Extension moment (Nm)	11	35
Thorax	Shoulder Belt Load (kN)	9.7	4.3
	3 ms acceleration (G)	53	40
	Chest deflection (mm)	na	25
	VC (m/s)	na	0.09
	Thoracic X-displacement measured at shoulder level (mm)	290	350

Table 3: Summary of crash test results with a production vehicle (mass 1200 kg) equipped with PRS II in offset and full barrier tests. Driver and passenger injury criteria and measurements.

Body segments	Measurements & injury criteria with a Hybrid III 50° percentile dummy	50% Offset rigid barrier test, 56 km/h		100% rigid barrier test,56 kmh US NCAP		40% Offset deformable barrier test, 56 km/h EEVC procedure	
		Driver	Passenger	Driver	Passenger	Driver	Passenger
Head	HIC 36 ms	186	257	347	519	74	111
	3 ms acceleration (G)	33	37	45	53	24	28
Neck	Shear Force -Fx (kN)	0.2	0.03	0.5	1.2	0.009	0.02
	Shear Force +Fx (kN)	0.3	0.6	0.4	0.2	0.5	0.3
	Extension moment (Nm)	35	28	29	na	11	12
Thorax	Shoulder Belt Load (kN)	4.3	4.5	4.6	4.8	3.9	4.1
	3 ms acceleration (G)	40	36	42	45	23	24
	Chest deflection (mm)	25	27	40	40	23	15
	VC (m/s)	0.09	0.22	0.64	0.64	0.01	0.03

CONSIDERATION OF NECK SECONDARY RISK IN OOP - In the absence of reference data published for adults subjected to airbag-generated loads, a test protocol with Post Mortem Human Subject (PMHS) has been prepared. The objective is to verify the harmlessness of the new airbag as far as the neck region is concerned and to acquire biomechanical data in this field. Tests in the same conditions were performed with the Hybrid III 50° dummy. Only two tests have been conducted so fare with PMHS and the Hybrid III and their preliminary results are reported in this section.

Equipment and method - The subject is seated on a standard seat, leaning over the steering wheel. Two occupant positions relative to the steering wheel were used. In the first position (Figure 13a), the forehead rests on the steering wheel rim and the chin is on top of the airbag module. The horizontal distance from the chest to the rim was 35 mm. In the second position, the nose was located on the module and the chest to rim distance was 100 mm, as shown in Figure 13b. Once the subject is in place, the airbag is triggered. The airbag/subject contact is recorded by high speed cameras and a set of measurements.

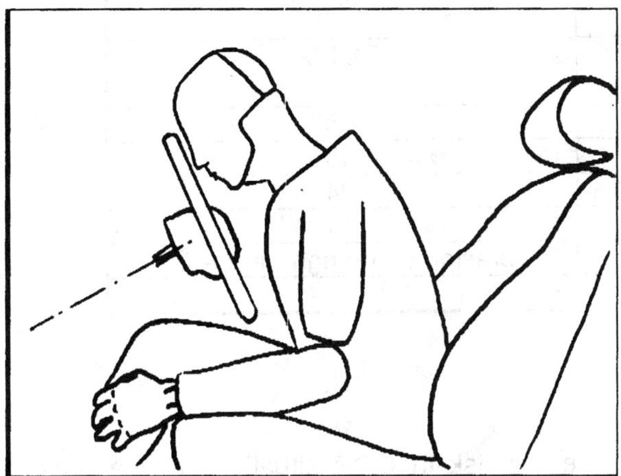

Figure 13a: Occupant relative position to steering whell in static air bag deployment OOP tests. Position forehead on rim.

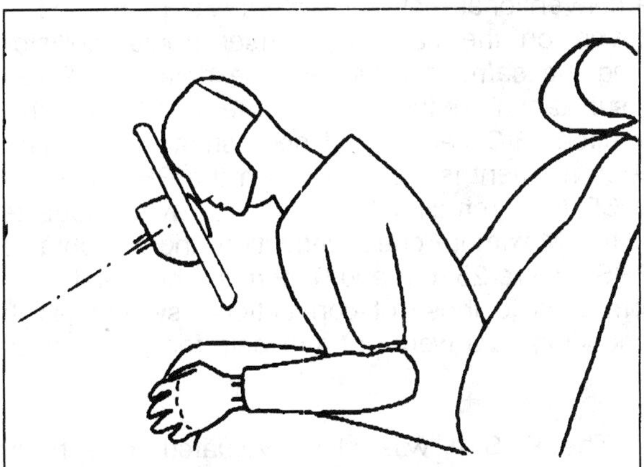

Figure 13b: Occupant relative position to steering whell in static air bag deployment OOP tests. Position nose on module.

Preparation of subjects - The subjects are fresh cadavers from the Anatomical Donors' service of the Anatomical Laboratory of the Saints-Pères Faculty, Paris V. The vascular system is re-pressurized (bi-carotid injection) and the pulmonary volume is re-established.

Instrumentation - The subjects are equipped with a device which measures the linear and angular accelerations of the head, fastened to the occiput. The T4, T12 vertebrae, as well as the sacrum, are connected to accelerometers. Thoracic deflection is measured with a chest band at level T5. Blood and vascular pressures are measured, as is the airbag pressure. For the Hybrid III measurements comprised neck loads, head and thorax acceleration, head and thorax angular velocity, chest deflection (potentiometer measurement). The Hybrid III was equipped with the neck shield.

Test results - Table 4 gives a summary of the Hybrid III tests and the autopsy findings from the PMHS tests. For the Hybrid III, neck positive Fx are 0.1 kN and 1.1 kN for the forehead/rim position and nose/module position. Maximum neck extension moment are respectively 40 Nm and 27 Nm and neck tension forces are 1.8 kN and 1.2 kN. All neck measurements are below Injury Assessment reference Values (12). Chest deflections reached values of 18.5 mm and 13 mm.

For the PMHS tests one subject was tested in each configuration. No macroscopic lesion was noted during the autopsy, except for a slight burn on the forearm of the first subject (AIS 1). The pressure limiting function in the air bag worked in both tests. Both dummy and PMHS tests are encouraging though not yet conclusive. Further work is needed, in particular with the Hybrid III 5° percentile dummy.

SUMMARY AND CONCLUSION

This study was initiated to address the rising risk of belt induced chest injuries in frontal impact. The starting point was the analysis of 290 frontal accident cases with vehicles that were equipped in France in the 1970's with a belt load limiter in front seats. The load limiter was based on a tear-webbing principle and was located near the upper belt anchorage point. This data base shows that older people (≥ 50 years) may sustain severe chest injuries. Based on this experience a program was initiated at Renault with a view to reduce the shoulder belt load. In 1995, a belt restraint system called PRS was introduced; its comprises a combination of a pyrotechnic pretensioner located at the buckle, a clamp retractor and a steel part attached to the retractor and to the belt anchorage point. This steel part, designed to deform at a given load, acts as a load limiter. This allowed to control the shoulder belt load at a 6 kN level. Accident cases involving this type of restraint were collected and analyzed; in particular the behavior of the belt load limiter was investigated in relation with occupant injuries. The data from 37 cases with belted front seat occupants, are reported in this paper. Crash severities ranged from 35 km/h to 75 km/h. A significant part of this sample, 27% of occupants with age ≥ 50 years, sustained minor to moderate chest injuries. The combination of belt pretension and a 6 kN belt load limitation appears to

Table 4: Static air bag deployment tests with the PRS-II air bag. Hybrid III 50° and PMHS test results.

Measurements	Units	HIII 50° responses-Position Nose on Module	HIII 50° responses-Position Forehead on rim
Chest to lower rim distance	mm	100	35
Neck Upper Fx/-Fx	kN	1.1/0.1	0.1/0.7
Neck Upper Fz Tension/Compression	kN	1.28/0.1	1.8/0.02
Neck Upper Moment Flex/extension	Nm	78/27	4/40
Head x-acceleration	G	64	39
Head y-Ang velocity	rad/s	28	16
Thorax x-acceleration	G	13	16
Thorax y-Ang velocity	rad/s	7	10
Chest deflection	mm	13	18
Corresponding PMHS Tests		MS 501	MS 500
Age and Height	years/cm	73/176	54/172
Mass	kg	96	68
Head 3ms acceleration	G	81	46
Thorax 3 ms acc. T4	G	24	24
Thorax 3 ms acc. T12	G	9	14
Injuries	-	none	Abrasion to the right arm AIS 1

have benefits in reducing thoracic loads from the belt for this population; the 6 kN level is however not sufficient to cover the whole population. Thus, a further step in reducing the shoulder belt load is necessary. As this reduction will involve increased excursions of the head and the thorax, the belt load load limitation has to be combined with an air bag.

The combination of an air bag and a 3-point belt restraint was discussed in various publications among them are the paper from Kompass in 1994 (3), the study of Kallieris et al. in 1995 (4) and Mertz et al. investigation in 1995 (5). According to the data discussed in Kallieris paper (4)and the mathematical simulation investigated by NHTSA (4) for a variety of crash conditions (frontal and rollover), crash severities and occupant sizes (5°, 50° a,d 95° percentiles) a threshold of 4 kN for the shoulder belt load limitation appears to be suitable for reducing the risk for thoracic injury, without negative consequences on other injury measurements.

From the experience acquired with the PRS a new approach in the occupant restraint system was developed. The PRS-II combines a pyrotechnic buckle pretensioner with a 4 kN belt load limiter and an air bag specially designed with respect to 2 key factors: a deployment to the sides and from top to bottom in order to reduce the risk in OOP situations and a pressure control which operates when a certain load is applied by the thorax. One the major concern with the belt load limitation was the possibility to increase the injury risk for the head and for the thorax. A comparison of PRS II with a conventional restraint system was performed. for the driver, on the basis of offset frontal collisions involving the same car model. The data with PRS II show substantial reductions for the head and chest acceleration, HIC values and neck shear forces. Neck extension moment is increased with the PRS II but the value, 35 Nm, remains below the 57 Nm suggested IARV (10). Maximum chest deflection and VC obtained with PRS II were 25 mm and 0.09 m/s. These data were not compared to those of conventional system. as the corresponding data were not available for the same car model.

The PRS II was also evaluated in 3 frontal collisions: offset rigid barrier test at 56 km/h, offset deformable barrier test at 56 km/h (EEVC frontal impact test procedure), and in full rigid barrier test at 56 km/h (NHTSA frontal NCAP test). In the test conditions described here, the combination of a 4 kN belt load limitation with the pretensioner and air bag pressure control has no negative effects on injury measures.

A preliminary investigation to evaluate the risk of neck injury with the PRS II air bag was performed, using the Hybrid III 50° percentile dummies and 2 subjects (PMHS). Two occupant/steering wheel positions, i.e. forehead on rim and nose on module, were used. The air bag pressure control worked in all tests. For the Hybrid III all neck loads were below the suggested IARV. For the PMHS, no lesions were found during the autopsy, except for a slight burn on the forearm for one subject (AIS 1). These data appears to be encouraging at this stage; further investigations with the Hybrid 5° percentile and the analysis of the PMHS test data remain to be performed. In addition further analysis of the present and future accident cases with the PRS 6 kN, will allow to establish an injury risk curve for the thorax.

ACKNOWLEDGMENTS

The PRS II principle was initiated by the Renault Safety Eng. Dept. Many people were involved in its development in particular vehicle plateforms teams and advanced research groups. The authors would to thank Michel Kozireff and his team from Autoliv France for their important contribution. Many thanks to the Faculté des Saints Pères and to the CEESAR for the biomechanical evaluation. This study was made possible thanks to the sled and crash tests performed at the Renault Lardy Test Center.

Views and opinions expressed here are those of the authors and not necessarily those of Renault S.A.

REFERENCES

1- J.Y.Forêt-Bruno et all.: »Correlation Between Thoracic Lesions and Force Values Measured at Shoulder of 92 Belted Occupants Involved in Real Accidents ». SAE Paper No.780892, Stapp Car Crash Conference, USA, 1978.

2- J.Y.Forêt-Bruno, F.Brun-Cassan, C.Brigout, C.Tarrière : Thoracic Deflection of Hybrid III: Dummy Responses for Simulation of Real Accidents ». In proceedings of the 12th International Technical Conference on Experimental Safety Vehicles. Goteborg, Sweden, May 1989.

3- K.Kompass: « Opportunities and Limits of an Air bag Optimisation Based on the Passie Requirements of Standard 208. Paper No. 94-S4-O-08, 14th ESV Conference. Munich, Germany.

4- D.Kallieris, A.Rizzeti, R.Mattern, R.Morgan, R.Eppinger and L.Keenan : »On the Synergism of the Driver Air bag and the 3-point Belt in Frontal Collisions ».SAE Paper No.952700, Proceedings of Stapp Car Crash Conference. USA, 1995.

5- H.J.Mertz, J.E.Williamson and D.A.Lugt « The Effect of Limiting Shoulder Belt Load with Air Bag Restraint ». SAE Paper No.950886, International Congress and Exposition. Detroit, USA, 1995.

6- ISO/TC22/SC10/WG3: « NHTSA Communication on Air Bag Related Accidents ». Meeting in Delft, The Nederlands, May, 1997.

7- ISO/DTR 10982 - Road Vehicles - Test Procedures for Evaluating Out of Position Vehicle Occupant Interactions with Deploying Air Bags. November, 1995.

8- Auto Motor und Sport - Mai 1997 Report - Germany

9- »Uniform Provisions Concerning the Approval of Vehicles with Regard to the Protection of the Occupants in the Event of Frontal Collision » Official Journal of European Communities, Brussels, July 1996.

10- H.J.Mertz : « Anthropomorphic Test Devices « .Accidental Injury - Biomechanics and Prevention, A.M.Nahum and J.M.Melvin, eds, Springer - Verlag , New York, 1993.

983166

Thoracic Injury Risk in Frontal Car Crashes with Occupant Restrained with Belt Load Limiter

J-Y. Foret-Bruno, X. Trosseille and J-Y. Le Coz
Lab Renault / PSA Peugeot-Citroën

F. Bendjellal and C. Steyer
Safety Engineering Department, Renault

T. Phalempin, D. Villeforceix, P. Dandres and C. Got
CEESAR

ABSTRACT

In France, as in other countries, accident research studies show that the greatest proportion of restrained occupants sustaining severe injuries and fatalities are involved in frontal impact (70% and 50% respectively). In severe frontal impacts with restraint occupants and where intrusion is not preponderant, the oldest occupants very often sustain severe thoracic injuries due to the seat belt. In the seventies, a few cars were equipped in France with load limiters and it was thereby possible to observe a relationship between the force applied and the occupant's age with regard to this thoracic risk.

The reduction of intrusion for the most violent frontal impacts, through optimization of car deformation, usually translates into an increase in restraint forces and hence thoracic risks with a conventional retractor seat belt for a given impact violence. It is therefore essential to limit restraint forces with a seat belt to reduce the number of road casualties, especially for the most elderly.

In order to address the thoracic risk in frontal impact, a restraint system combining belt load limitation and pyrotechnic belt pretension, known as the Programmed Restraint System (PRS), has been fitted to Renault cars since 1995. The belt load limiter, which is a steel part designed to deform at a given shoulder force threshold, i.e. 6 kN, is fastened between the retractor and the lower anchorage point of the belt.

From static and dynamic tests performed with the load limiter, it is possible to determine the shoulder belt force applied to the occupant from the amount of load limiter deformation. Eighty-nine accident cases, with EES's ranging (Equivalent Energy Speed) from 40 to 80 km/h and involving frontal collisions with cars equipped with the PRS, are reported in this paper.

The purpose of the present study, based on this accident file and also on another accident file with textile limiters, is to establish, for belted occupants, the thoracic injury risk as a function of occupant age and the load applied at shoulder level. One observes that for 50% of thorax risk of AIS3+, the force for all ages together is 6.9 kN. Occupant's age is a very important factor. These results are obtained for 256 occupants, where the age's distribution is similar to that of front seat occupants of the French accident file. Despite occupant's sizes and belt geometries in this accident sample, the shoulder belt load appears to be in accordance with the occurrence of chest injuries.

A comparison of results obtained in real-world accidents and the tests performed with 209 PMHS on Laboratory (Post Mortem Human Subject) by Heidelberg University is also described in this paper. Both databases show the same trend for the thoracic risk according to age, but the force for a given risk and a given age is 2 kN lower than that for occupants in real-world accidents.

A relationship between the Hybrid III thoracic injury measurements and the shoulder belt load is also investigated in this paper. For a given shoulder belt load in crash tests performed in cars identical to those in the accident analyzed with PRS, the sternal deflection scatter does not allow to satisfy evaluation of the thoracic risk. These results are discussed and compared with several papers. It is suggested to include the shoulder belt load measurement in homologation and rating tests for a better assessment of thorax protection.

This study confirms that a 6 kN force level is not sufficient to protect a larger proportion of the population. A belt load limitation of 4 kN, combined with a specifically design airbag, would make it possible to protect 95% of those involved in frontal impacts from thorax injuries of AIS3+.

LOAD LIMITERS IN REAL-WORLD ACCIDENTS

TEXTILE LIMITERS IN THE 1970'S – The first relationship between seat-belt tension and the associated thoracic risk level was established in 1978 (1) and was based on an analysis of 90 accident cases. The vehicles analyzed, sold in France in the 1970's, were equipped with 3-point static seat-belts in the front seats with a load limiter located in the belt webbing between the occupant's shoulder and the upper anchorage point. The load limitation was achieved by tearing of the stitching which was used to sew loops in the webbing. In case of impact the stitching tore, thus allowing more webbing from the loop; as a consequence the thorax was able to move relative to the vehicle at a controlled load level.

In 1989, other cases were added to this investigation, bringing the total of this database up to 290 accidents (2). The key feature of this unique database is the possibility of showing a relationship between the seat-belt tension exerted on the occupant, his (or her) age and the type of resulting lesions.

These data clearly shows that thoracic risk among occupants restrained by seat belts increases with age and that a shoulder belt load of 8 to 9 kN may induce a high risk for the chest, especially in elderly people.

This sample, which consisted of relatively few occupants for whom the forces applied were high, made it possible to obtain a trend but limited the potential for establishing reliable curves of thoracic risks according to age. The delta V's were 40 km/h on average, and there were only three occupants with AIS 3+ thorax injuries for which the force sustained was known precisely. The 3-point retractor seat belt replaced this static seat belt in 1976, and those textile load limiters were then abandoned.

THE NEW LOAD LIMITER : THE PRS – To reduce restraint forces and consequently the number of thoracic injuries in frontal impact, Renault in 1995 again equipped its cars with a load limiter based on a different concept, located at the retractor level, the PRS 1 (Programmed Restraint System, first generation).

A special accident investigation program was set up throughout France in order to examine the behavior of this restraint system in real-world accidents, in relation with occupant exposure. In this new sample, the frontal impacts analyzed are much more severe than in the previous sample with textile limiters, since the mean of the delta V values is approximately 55 km/h.

This new load limiter was initially designed to limit forces to 6 kN, and nearly all the cars studied in this paper were equipped with this PRS 1. In a second stage, the belt force is limited to 4 kN, but the occupants have an airbag developed specially to provide optimum force distribution over the thorax and protect the head in excursions which are obviously greater (PRS 2). This PRS 2 came on the market in 1998 and only two cases have been analyzed.

This system (PRS 1) consists of a pretensioner pyrotechnic buckle, a retractor webbing clamp and a steel part, fastened between the retractor and the seat-belt anchoring point as shown in Figure 1. This part, designed to deform at a given level of force, acts as a load limiter.

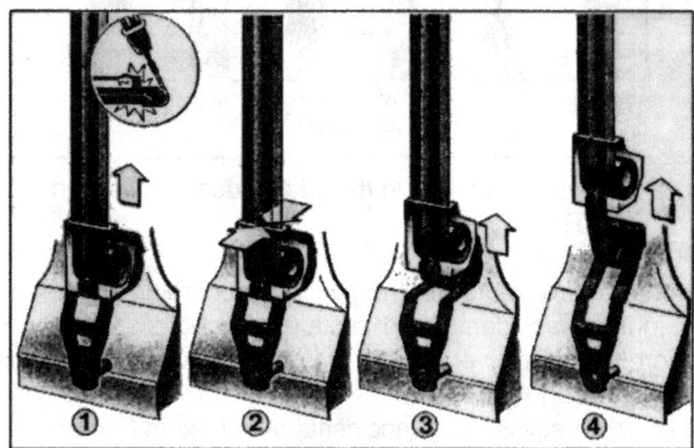

Figure 1. Principle of PRS 1 operation - Phase 1 : Initial part of the crash and belt pretension activation. Phase 2 : Action of the webbing clamp. Phase 3 : Load limiter activation. Phase 4 : End of impact.

A detailed description of the operation of these PRS's was given by F. Bendjellal in a paper presented in 1997 (3).

In this paper we will first describe the accident sample (89 cases) with PRS 1 and establish the thoracic injury risk due to the belt. Further analysis of this risk is proposed with enlarged database, where the accident sample with PRS 1 is combined to that with the textile load limiter , bringing the total number of cases to 256. A discussion of the findings in relation to those from Heidelberg (4) and Mertz (5) is provided.

DESCRIPTION OF ACCIDENT SAMPLE

To date, 89 accident cases relating to frontal collisions with cars equipped with this device have been investigated since 1995.

The main parameters of this sample are summarized in Figures 2a through 2c and detailed findings are presented in Table IA in the appendix for 51 drivers and 38 right front passengers. In this databank, only 15 cases are observed with airbag.

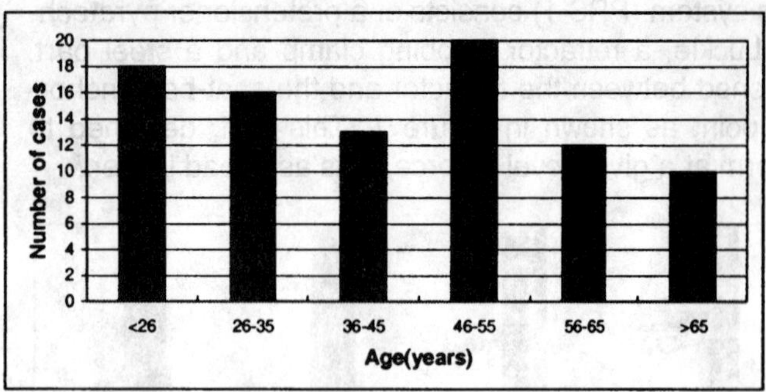

Figure 2a. Age distribution in the 89 accident cases with PRS

To better approximate the thoracic risk according to age, a majority of accidents with occupants aged more than 40 were selected everywhere in France, since the risks for the lower age groups are very slight. Here the median age is 42, versus 37 in all accidents observed in France.

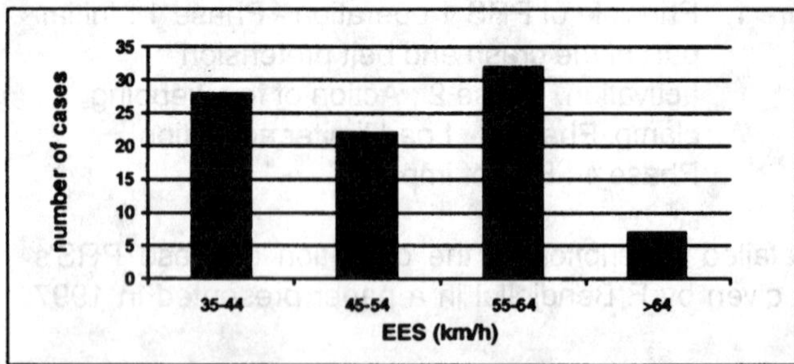

Figure 2b. Collision severity in the 89 accident cases with PRS

The accidents analyzed are also of a severity well above the national average, to be able to approximate the thoracic injury risk which is obviously higher at the highest speeds. In this sample, the median of the EES's (and the delta V's) is 53 km/h in impacts which are usually offset.

The AIS 2+ head injury risks in this sample with load limiter (PRS 1) are not different as regard to a sample without load limiter.

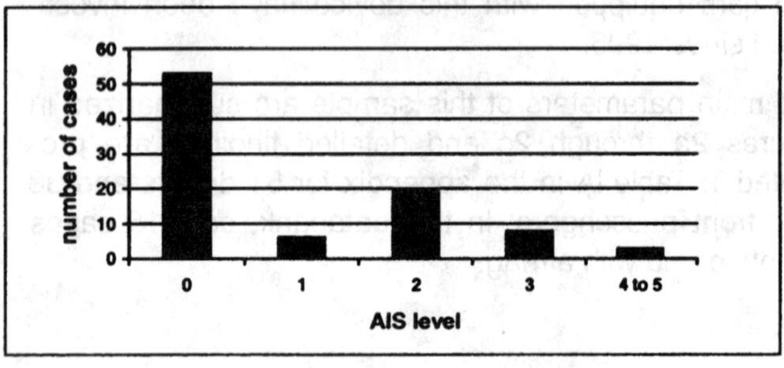

Figure 2c. Maximum thoracic AIS in the 89 accident cases with PRS

The thoracic injuries according to AIS taken into account hereafter are as follows :

- AIS 1 thoracic injury cases are solely occupants having a rib fracture detected by X-ray, and cases in which the AIS 1 is attributable to a mere pain in the thorax are therefore not counted.
- AIS 2 injury cases are fractures of two or three ribs (AIS code) and/or the sternum, plus fractures of the clavicle due to the seat belt.
- AIS 3+ injuries are as defined by the international AIS code (6).

The three thoracic injuries of AIS 4 and 5 are observed for maximum forces of 6 kN for front-seat passengers. These cases are as follows :

- two occupants aged 67 and 88 who died of major thoracic injuries (the restrained front-seat passenger aged 67 was hemiplegic, so was his tolerance reduced?)
- one occupant aged 58 suffering multiple rib fractures associated with a bilateral pneumothorax.

The severity of the accidents in which these occupants were involved can be observed in this graph 2c, since 30 of the 89 occupants in this sample show thorax injuries of level 2+ and 11 of level AIS 3+.

Nearly all of the forces measured at the shoulder do not exceed 6 kN with the PRS, and it can be seen immediately that this level of force is not yet optimized to ensure maximum protection for a large number of occupants.

CASE EXAMPLES

The following three accident cases show the benefits of this limiter but also its limitations for protection of the most elderly occupants.

In some cases, for forces close to 6 kN and for elderly occupants, one observes moderate thoracic injuries.

EXAMPLE 1 – Accident case 12092, the front female passenger aged 71 has a fracture of the right clavicle and a fracture of the left iliac wing for a plateau force of 5.5 kN with tearing of the PRS over 65 mm (photos 1 and 2).

EXAMPLE 2 – In this second accident (case 12041), the driver, aged 52 with airbag, is unharmed for a plateau force of 5.5 k N with tearing of the PRS over 200 mm (photos 3 and 4).

Without these limiters, it can be assumed here that the forces applied at the shoulder level would have been close to 9 to 10 kN, as observed in the first cars not equipped with PRS's tested with dummies. The few injuries observed here can be attributed to the beneficial presence of the PRS.

EXAMPLE 3 – It can also be seen, in case 12438, that the tolerable limits are exceeded for the two front occupants who sustained a force level of 6 kN (see photos 5 and 6) :

- the driver aged 67 suffers 6 rib fractures for tearing of the PRS over 140 mm;
- the front female passenger aged 88 dies of major thoracic injuries for tearing of the PRS over 160 mm.

photo 1. case number 12092

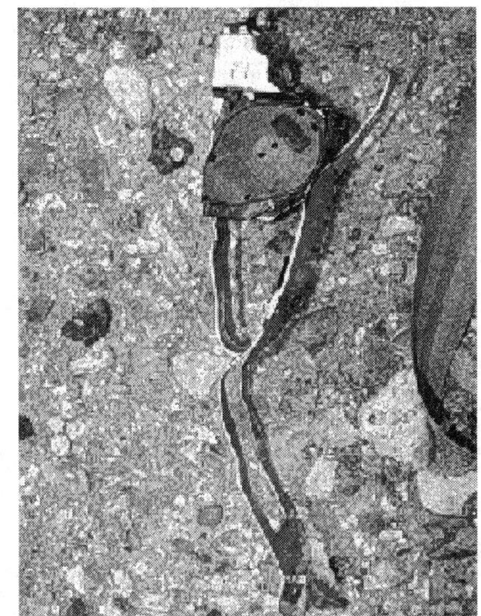
photo 2. case number 12092

photo 3. case number 12041

photo 4. case number 12041

photo 5. case number 12438

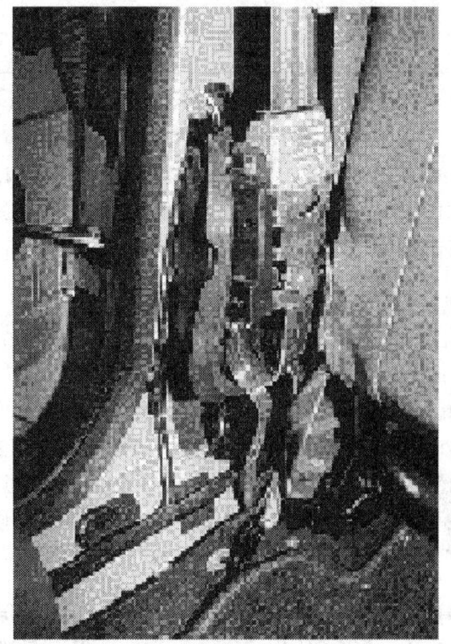
photo 6. case number 12438

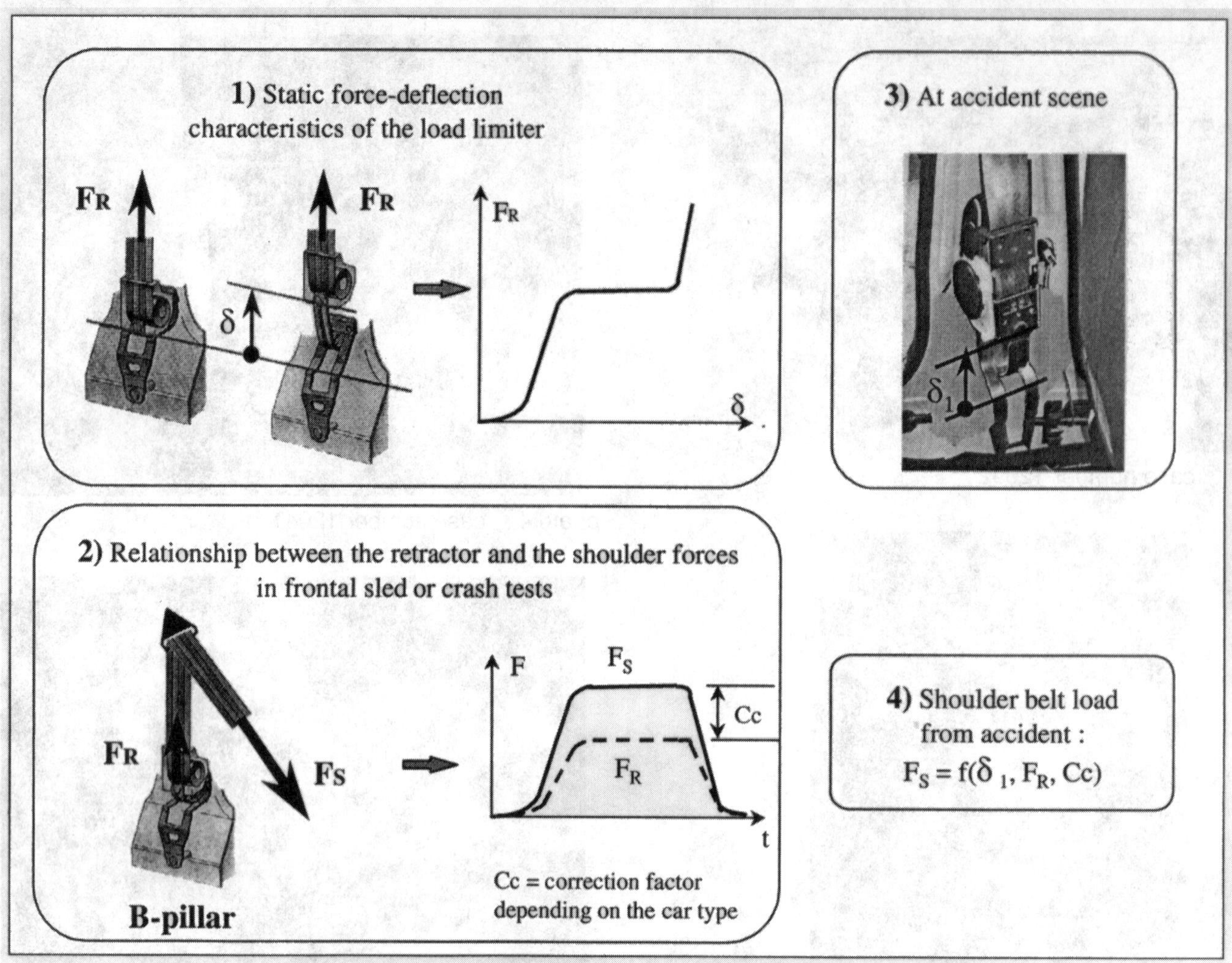

Figure 3A. How shoulder belt load was obtained from accident.

INFLUENCE OF LOAD ON THE OCCURRENCE OF RIB FRACTURES

FORCE MEASURED ACCORDING TO DEPLOYMENT OF THE PRS : – For nearly all the cars analyzed, the PRS plateau forces range between 5 and 6 kN, and for one specific car the plateau force is 7.5 kN (coupe and convertible), but this car represents only 5% of this sample.

Before reaching this plateau force, a force increase is measured which results in deployment of the PRS and the slope of which is variable for the various PRS's. At 30 mm of deployment, this force has then reached its plateau value and remains stable until it reaches the thrust stop, which is a very rare occurrence. This thrust stop acts only after 260 mm deployment of the PRS.

Numerous dynamic and static tests have enabled a relationship to be established between this deployment measured statically on an accidented car and the shoulder load sustained by the occupant. Forces were measured in crash tests forward and backward of the D-ring (figure 3A). This allows to establish between forces at shoulder level and forces at retractor level established from PRS deployments. The curves presented in figure 3B from PRS-1 have been corrected for the friction forces at the D-ring and therefore correspond to shoulder belt forces. One can see the various curves used to evaluate these loads for the first centimeters of initial deployment (before the plateau forces) of four different PRS's. According to various PRS's, the minimal force values just before the plateau forces, are function of the form of the steel part.

The minimum levels obtained from tests, to damage the steel part are 3.5 kN for PRS model A, 3.9 for PRS model B, 4.5 for PRS model C and 5 kN for PRS model D. It must be born in mind that there are cases without any PRS' steel part damage. Of course, the shoulder belt load value are below the previous mentioned levels, but is not know precisely. No rib fracture are recorded in those cases which are not included in the present study.

It can be observed from figure 3 that for 3 PRS' versions, the plateau forces is early reached between 5 and 10 mm, but much later (30 mm) for the fourth PRS' version.

80% of these limiters reached the plateau force and 13 show deployments of more than 100 mm.

It is certain that the greater the deployment of the limiter during application of the plateau force, the greater would have been the force which would have been sustained without this limiter.

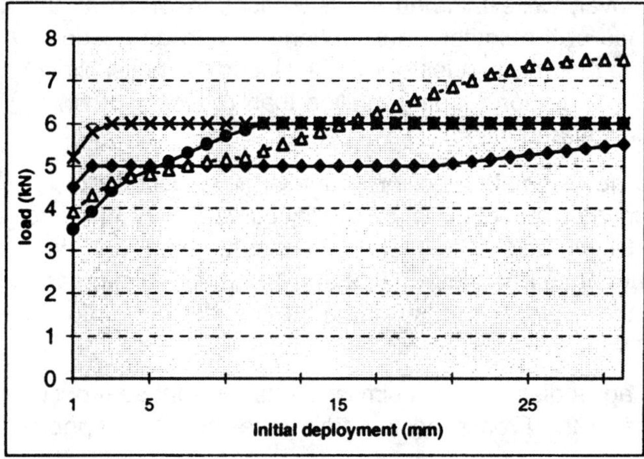

Figure 3B. Phase of initial deployment before plateau force for the 4 various PRS's

RISK ANALYSIS WITH PRS – The 89 cases of this sample are presented in Figure 4 according to age, force and thoracic AIS level. Severe thoracic injuries (AIS 3+) are observed in most cases for the most elderly, over 55 years old. But one also observes that thoracic injuries of AIS level 2 are not sustained solely by the most elderly, since sternum and clavicle fractures also occur for occupants aged less than 40.

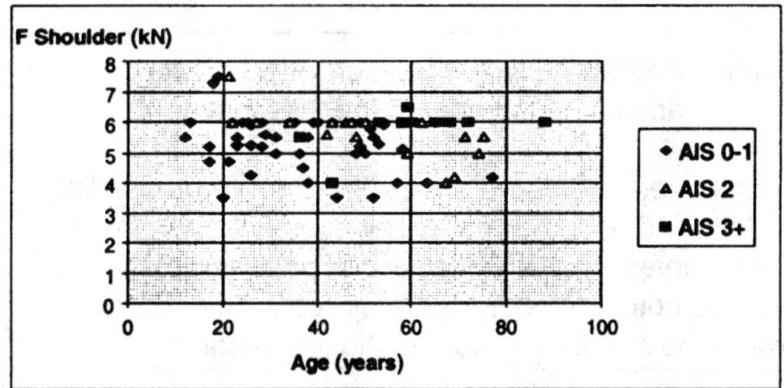

Figure 4. Age, force and thorax AIS for the 89 cases

Figure 5 gives, for 50% of AIS 2+ and AIS3+ risks respectively, shoulder load values of 6 and 6.3 kN. The method used here to establish this curve is the certainty method described by Mertz (7).

The difference observed between these two curves is very slight, but it can undoubtedly be explained by the fact that between the appearance of a thoracic injury of AIS level 2 (fracture of the sternum, the clavicle or two or three ribs) and a level 3 injury (more than three ribs), the additional force applied does not need to be very great. In this sample, the average age is 42 and the standard deviation is 17 years.

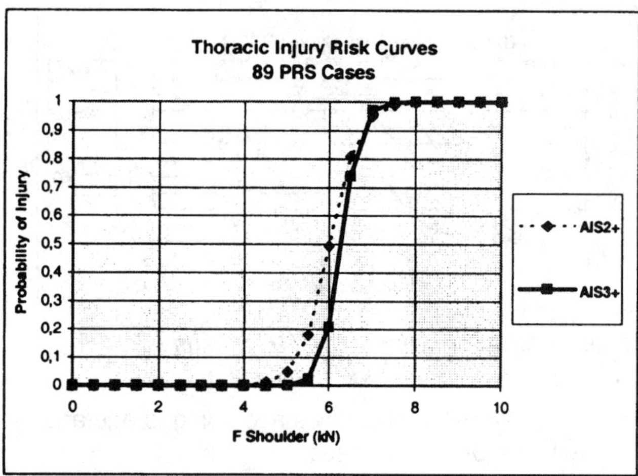

Figure 5. AIS 2+ and AIS 3+ thoracic risks according to shoulder load

But these risk probabilities obtained here for all the occupants do not show the preponderant influence of age on the occurrence of severe thorax injuries.

Now, to better assess the thoracic risk with these PRS's, we selected more especially cases with elderly occupants. It is therefore essential to give by age the thoracic risks with seat belt. We use the certainty method applied to age and load.

For a prescribed level of shoulder forces and age, we make a group (the certainty group) including only specimens where it is known for sure that specimens would have experience chest injuries or no chest injuries at this level of shoulder forces and age. The probability of chest injuries at the given level of shoulder forces and age is then estimated by calculating the ratio of the number of specimens that would have experience chest injuries to the total number of specimens in the certainty group.

To define if a specimen would have experienced or not experienced chest injuries at a given level of forces and age, we assume that if someone sustains chest injuries at a given level of force, he will also sustain at least the same chest injuries at a higher force level. Likewise, an older subject is supposed to sustain at least the same chest injuries. On the contrary, we assume that if someone of a given age doesn't sustain injury at a force level, he will not experience injuries at a lower level of forces, the same than for a younger subject at the same level of forces. The values obtained are then analyzed numerically using Logistic analysis.

Figures 6 and 7 illustrate very great differences of risks according to age. Hence, one observes deviations of 5 kN between occupants aged 80 and those aged 20, for a given risk equivalent to 50% of AIS 2+ cases.

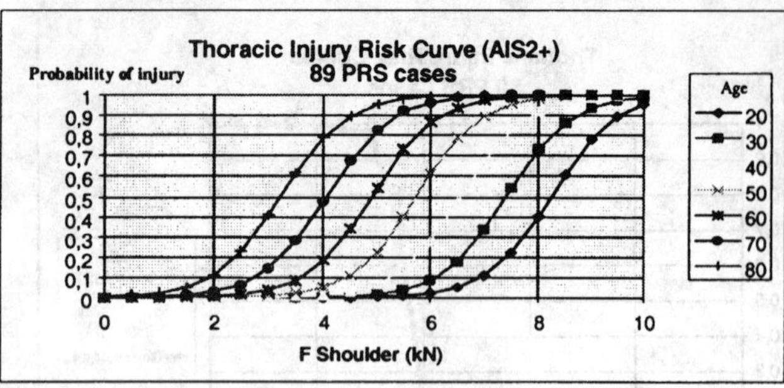

Proba (AIS2+)=1 / (1 + exp (16.6 - age / 7.1 - F shoulder / 59.7))

Figure 6. AIS2+ risk probabilities according to age and force for the 89 cases

Concerning AIS 3+ risks, one observes that the 50% deviations between the youngest and most elderly cases considered here are 7 kN. But it is true that for the youngest, the risk calculations are here extrapolated from the other age groups, due to the absence of AIS 3+ cases for those aged less than 30.

Evaluation of the thoracic risk for all ages together does not therefore provide a clear view of this tolerance scatter between extreme age groups.

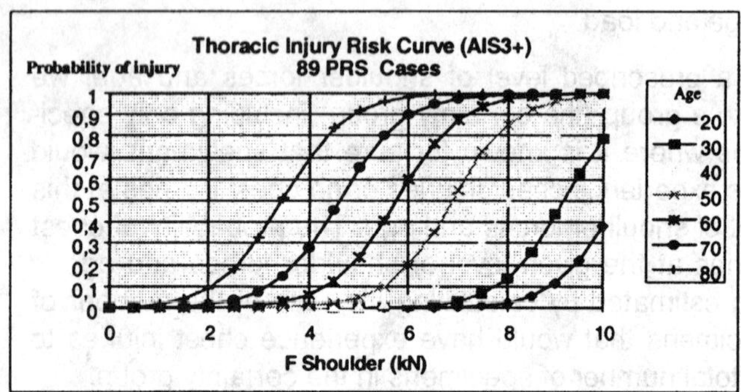

Proba (AIS3+)=1 / (1 + exp (19.8 - age / 5.6 - F shoulder / 63.4))

Figure 7. AIS3+ risk probabilities according to age and force for the 89 cases

RISK ANALYSIS WITH ALL LOAD LIMITERS – To obtain a more statistically significant sample, we now include in cases with PRS all the cases with textile load limiters already described previously, and for which we know the exact force at the shoulder level (see table 1B in the appendix : 167 textile limiters).

For this purpose, we eliminated from this textile limiters sample all the cases in which no strand of the limiter was torn because it is impossible to know exactly the shoulder belt load. For the textile load limiter, we didn't found any way to analyse the first real damage. It is very different with PRS's metal-deforming. The obtained information is only « it's all or nothing » for textile limiters.

Moreover, we eliminated the few cases in which all the strands of the limiter were broken but with an occupant having thoracic injuries of AIS 2+, due to the possibility of a far greater force being applied than the force at which the last strand failed.

This new sample therefore contains 256 occupants for which we plot the same probability curves for AIS 2+ and AIS 3+ thoracic risks, first for all ages together and then according to age (figures 8, 9 and 10). In this larger sample, the average age is 37 and the standard deviation is 15 years.

The age's distribution is similar to that of front seat occupants of the French accident file (see table 2 in appendix).

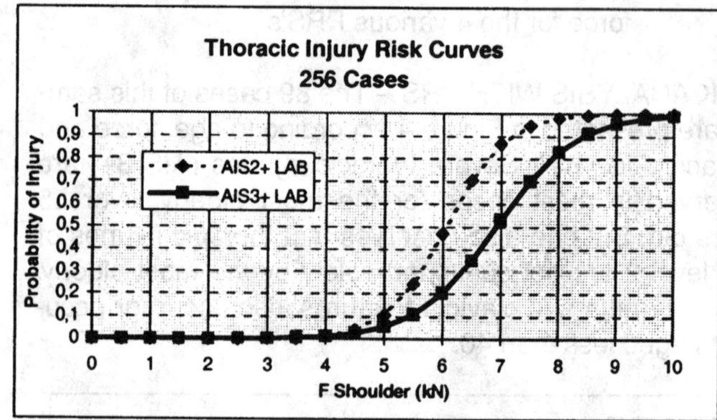

Figure 8. Probabilities of AIS2+ and AIS3+ risks according to force for the 256 cases

One observes, in figures 5 and 8 giving the probabilities of AIS 2+ and AIS 3+ risks, for all ages together, that the AIS 2+ curves are very similar for both samples. Likewise, according to age, there is very little difference between the AIS 2+ risk curves (figures 6 and 9).

For the AIS 3+ risk curves, one notes that the value corresponding to a 50% risk diverges, especially for the highest risk probabilities. At 50%, the risk is 6.9 kN instead of 6.3 kN in Figure 5. This slight difference as regards AIS 3+ can be partly explained by the younger sample age for the 256 cases (combined sample), with an average age of 37, compared with age 42 in the sample of 89 cases with PRS 1.

As regards the probabilities of AIS 3+ risks according to age, in figures 7 and 10, one notes a slight difference especially for risks to the youngest occupants. The deviation at 50% of risks for these youngest occupants does not, however, exceed 1 kN, i.e. 9.2 kN for the 256 cases and 10.2 kN for the 89 cases. For the most elderly, no difference is observed.

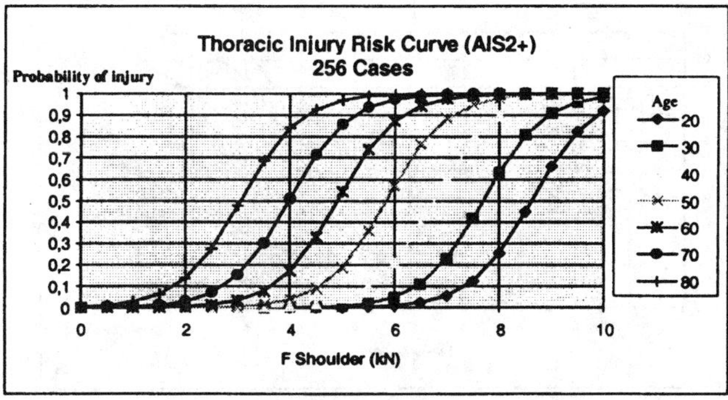

Proba (AIS2+)=1 / (1 + exp (18.3 - age / 6.2 - F shoulder / 57.3))

Figure 9. Probabilities of AIS2+ risk according to age and force for the 256 cases

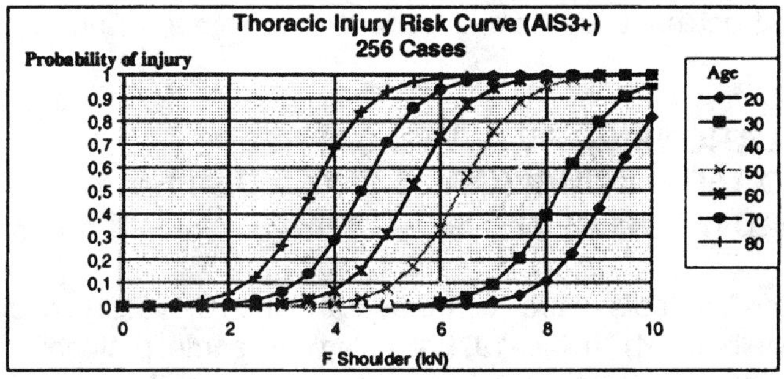

Proba (AIS3+)=1 / (1 + exp (19.9 - age / 5.9 - F shoulder / 55.7))

Figure 10. Probabilities of AIS3+ risk according to age and force for the 256 cases

From these values, calculated for 25%, 50% and 75% of AIS 2+ and AIS 3+ risks, it is then possible to determine the tolerance limits according to age and shoulder belt force on the same Figures 11A and 11B.

The force deviations for two identical risk probabilities (e.g. 25% of AIS 2+ and 25% of AIS 3+) and for a given age are 0.6 kN and are equivalent to a transition from 25% to 50% for a given risk.

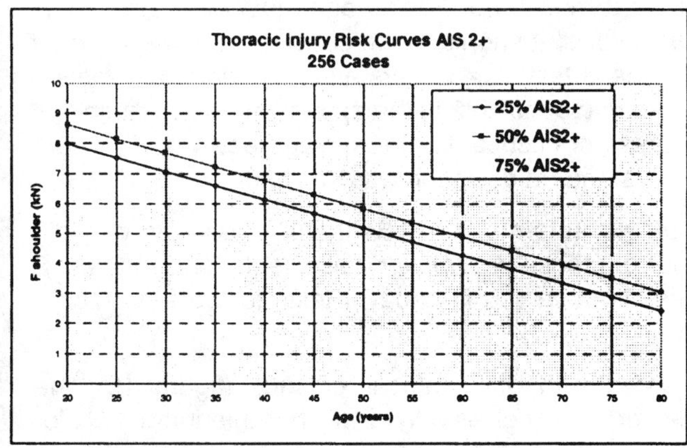

Figure 11A.Thoracic risks of AIS 2+ as a function of shoulder belt load and occupant's age

Occupants aged more than 80 with seat belt still show a significant risk for a force of 4 kN. Their probability of AIS 3+ thoracic risk is between 50% and 75%, but this population's involvement in accidents nevertheless remains very slight, as we shall see further on.

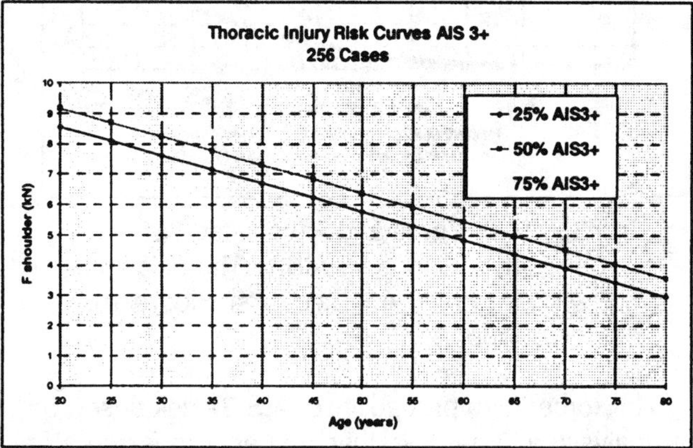

Figure 11B.Thoracic risks of AIS 3+ as a function of shoulder belt load and occupant's age

SHOULDER BELT LOAD AS THORACIC RISK INDICATOR

In 1991, Mertz (5) established a relation between the deflection values obtained on Hybrid III 50e percentile and the forces measured on the seat belt in five sled tests with the same limiters as in certain real-world accidents studied by the LAB and in the same anchorage system location and impact violence conditions.

These tests made it possible to draw a relationship between shoulder belt force and chest deflection (Figure 12). Hence, for a 10 mm increase in deflection, an additional shoulder belt force of 2 kN was measured.

Mertz then performed a statistical analysis on the probabilities of AIS 3+ risks, based on all the LAB's accident cases with textile load limiters. Figure 13 below, taken from his paper, gave the thorax AIS3+ risk according to deflection for this LAB sample. Knowing that in sled tests deflection is very directly related to force, one could therefore associate with a 50% risk of AIS 3+ a force of 10 kN, but also in this case a 50 mm deflection.

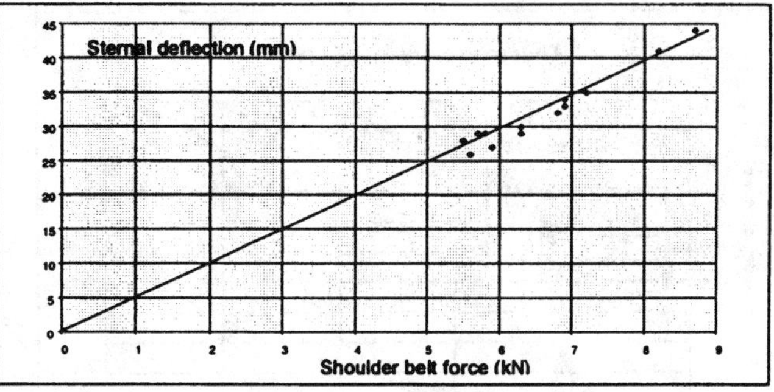

Figure 12. Correlation Between Hybrid III Sternal Deflection and Shoulder Belt ForceCorresponding to Band Tearing (from Figure 8, Mertz paper 5)

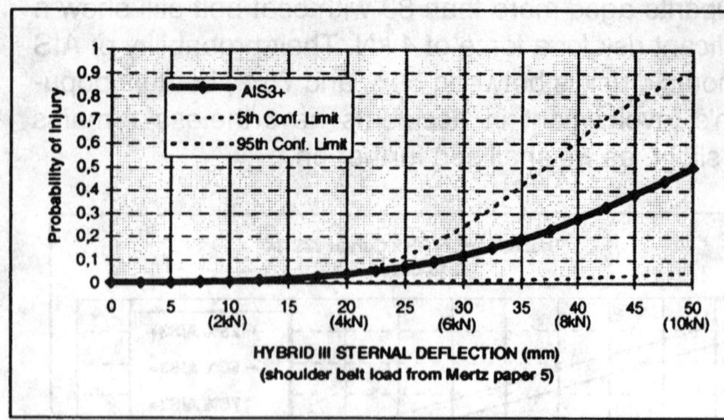

Figure 13. Injury Risk Curve and Associated 95% Confidence Bands for AIS3+ Thoracic Injury Based on Probit Analysis (from Figure 10, Mertz paper 5)

This 10 kN force, for a probability of AIS 3+ risk based on probit analysis, all ages together, is therefore far greater than that observed in our sample which is 6.9 kN and based on certainty method (deduced from Figure 8). It is true that the confidence limits given by Mertz for deflection values of 50 mm are very wide on this figure, due to the limited number of occupants on which the forces applied were high, and hence the number of severe thorax injuries.

The new database presented here, which is larger, but especially more severe in terms of impact violence and hence force sustained (shoulder loads between 5 and 6kN) and associated risk, allows curves to be produced as shown in Figure 14, which are very different from those given by Mertz for probabilities of AIS 3+ risks.

It can also be verified in Figure 15 that the confidence limits are far more reliable for our curve of AIS 3+ risks, compared with Mertz's curve. It is also true that the statistic methods used are different.

Both curves are similar up to 5 or 6 kN of force and then diverge very sharply, which is not surprising given the very small number of cases with forces exceeding 5 kN (6 cases) for Mertz to calculate reliable risk probabilities beyond these force levels. Our old sample, which was used as a basis for his evaluation, was obviously not sufficiently large for high shoulder belt loads.

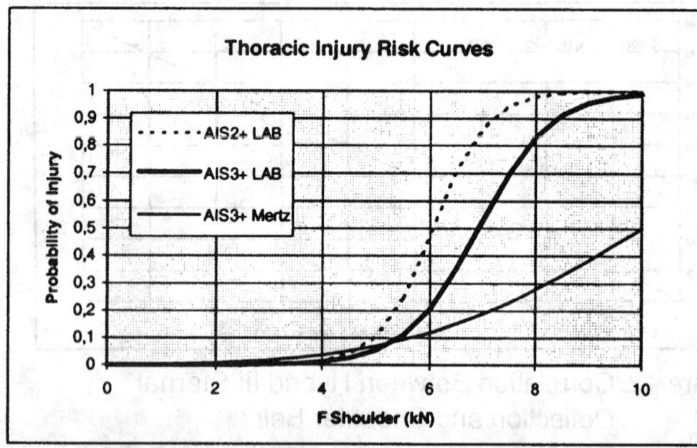

Figure 14. Injury Risk Curve for AIS 2+ and AIS3+ Thoracic Injury (from Mertz and LAB)

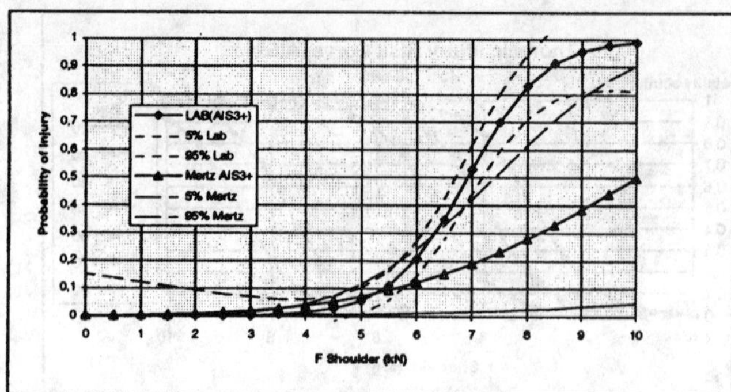

Figure 15. Injury Risk Curve and Associated 95% Confidence Bands for AIS3+ Thoracic Injury (from Mertz and LAB)

But first, it is interesting to compare an analysis of the same type on the Heidelberg cadaver tests with our real-world accident results with regard to risks according to age.

RELATIONSHIP BETWEEN CADAVER THORACIC INJURIES AND SHOULDER BELT LOAD (FROM HEIDELBERG TESTS)

209 sled tests were performed by the University of Heidelberg (4) in the 1970's, using the same protocol, that is to say that the same seat model type, the same anchorage point and the same seat belt model type were used for all of 209 tests. Only the speed (between 30 to 60 km/h) and the deceleration (average sled deceleration : between 9 to 25g with trapezoidal form) pattern were varied to obtain a wide range of impact and restraint force violence.

The object of the research project was to determine the loadability limits of the vehicle occupant belted in the seat and to analyse the injury mechanisms. The loadability limit can be defined by physical parameters and medical findings. The final report was published in 1978.

The tests enable the same risk curves to be defined as for real-world accidents, since the forces at the shoulder level were measured. For all ages together, the results are as follows. It should be emphasized that the average age in these tests is 41 years and the standard deviation is 17 years (see table 2 in the appendix). In our sample of 256 cases of occupants, the average age was 37 years and the standard deviation was 15 years.

The results in terms of AIS 3+ thoracic risks are very different from those of real-world accidents, since almost 2 kN less is observed in cadaver tests for 50% of AIS 3+ risks.

Compared with our real car crashes (figure 10), the Heidelberg study gives a load which is inferior of 4 kN for the same risk and for seventy year old occupants. For the youngest occupants, the difference in load (50% of AIS 3+) is only 1.5 kN.

This very large difference can be explained only by weaker bone resistance of the cadavers used for these tests, particularly for the oldest, or would an autopsy enable more rib fractures to be visualized than with mere X-rays, as written by Morgan (8) and Yoganandan (9) ? It may be felt that both hypotheses are possible.

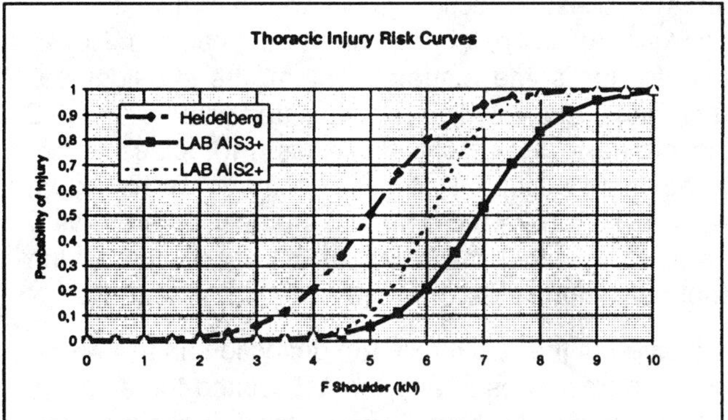

Figure 16. AIS 3+ risk according to shoulder belt force for the Heidelberg tests

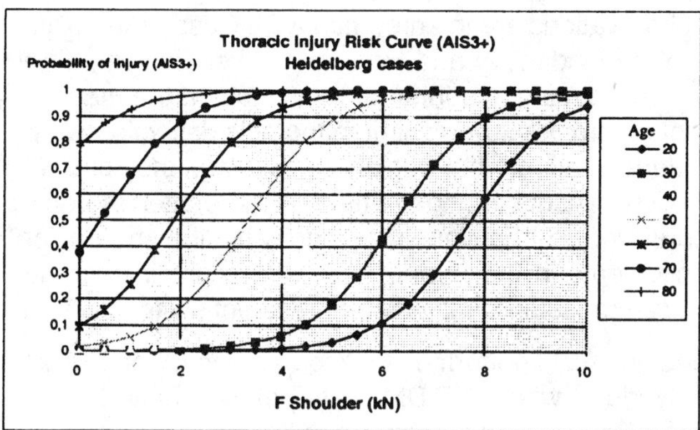

Proba (AIS3+)=1 / (1 + exp (13.1 - age / 5.6 - F shoulder / 81.1))

Figure 17. AIS 3+ risk according to shoulder belt force and age for the Heidelberg tests

We can now consider what are the deflection and force values obtained for various crash tests, offset or not, with cars of the same types as those studied in real-world accidents with PRS.

RELATIONSHIP BETWEEN HYBRID 3 THORACIC INJURY MEASUREMENTS AND SHOULDER BELT LOAD

In real-world accidents, these limiters are the only way of getting of the shoulder belt forces exerted on the belted occupants involved in a frontal impact. It was therefore interesting to compare these accidents with crash tests performed with similar cars to compare the thoracic criteria (deflection and 3 ms acceleration) with the shoulder belt loads measured on dummies.

These tests, performed during development of the PRS, therefore include certain cars not equipped with these PRS systems and, of course, a large very proportion of offset crashes (40% to 50% offset).

All the tests are performed with HIII dummies, 50e percentile in the driver and front-seat passenger positions, which are differentiated in figures 18 and 19.

These various tests show that, for a given seat-belt restraint force, there are major scatters for the two thoracic criteria. It seems, in light of these figures, that it is hard to predict a thoracic risk in a test based solely on deflection or acceleration, knowing that there is a good correlation between the force measured on the shoulder and the risk of rib fractures.

On Figure 18, moreover, one can see the regression line of Figure 10 deduced by Mertz from these sled tests. Why so much scatter for the two criteria for a given shoulder belt load in all the tests?

In all these tests, one usually observes that the maximum deflection values are obtained in full barrier tests with a 100% overlap, and that the minimum deflection values are more frequently observed in offset crashes, for a given shoulder belt load.

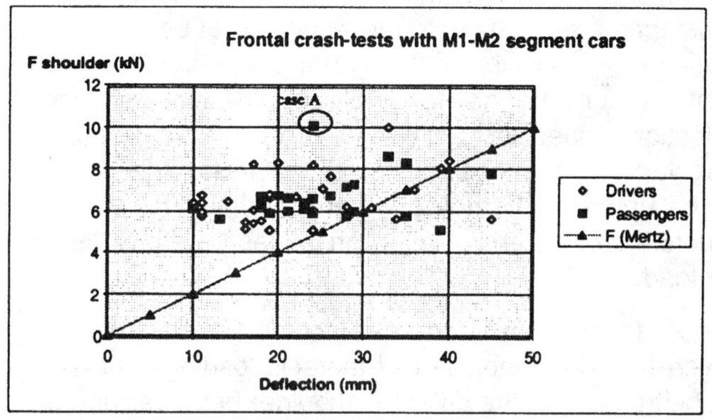

Figure 18. Shoulder belt load and chest deflection - Mertz relationship Renault test results with front occupant restrained by 3 point-belt with Eurobag. This type of airbag doesn't involve the chest

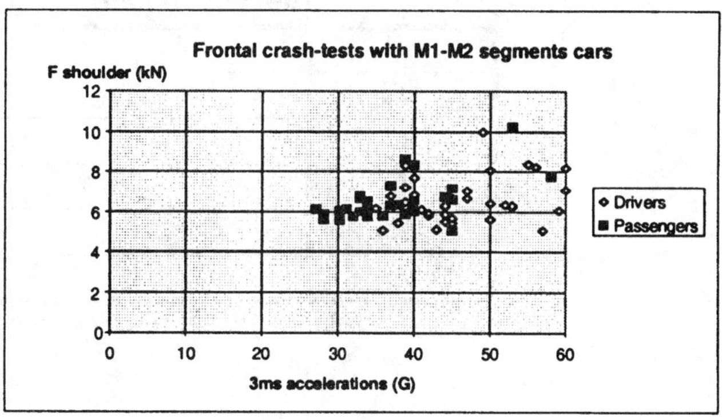

Figure 19. Shoulder belt load and 3 ms chest acceleration

DISCUSSION

The main purpose of this paper was to establish a relation between the shoulder belt load derived from real world accidents and the thoracic injury risk, as the accident data file includes cars equipped with a belt restraint allowing shoulder belt load to be determined. The results obtained allow to express the probability of thoracic AIS 3+ injuries as a function of maximum shoulder belt load. These results may not apply to all belt geometries and occupant sizes, but they can be considered as valuable basis to improve thoracic injury assessment . Chest deflection is certainly an appropriate injury criterion to assess the thoracic risk, but its present use with the Hybrid III dummy shows some limitations that are discussed in the following.

The five sled tests performed by Mertz (5) to determine the sternal deflection according to force are carried out with identical anchorage system positions, with the same speed and deceleration, on rigid seats and, of course, strictly identical dummy positions. These conditions limit to a minimum the scatter of seat belt positioning on the thorax. also due to very pure dummy trajectories.

The same is certainly not true with offset crashes which involve slightly oblique dummy trajectories. In such circumstances, these trajectories result in seat belt slippage relative to the position of the sternal deflection measurement sensor and consequently very different sternal deflection values for a given shoulder belt load. Moreover, very different positionings of the anchorage system for the upper anchorage point of the shoulder belt also influence these sternal deflection values for a given seat belt load.

Indeed, for a given force sustained at the clavicle level (which is a good indicator of thoracic loading, whatever the belt position), the different shoulder belt positions on the thorax influence the deflection measured at the sternum deflection sensor. So, the sternum deflection sensor alone is certainly not sufficient to establish, in all test configurations, the real thoracic risk for the whole thorax.

According to the data presented, in particular in figure 18, a restraint system with no belt load limitation may give an acceptable chest deflection value, i.e. 23 mm which indicates 5% of AIS 3+ risk based on Mertz (case A). Nevertheless for the same restraint system, the shoulder belt load reaches 10 kN, which is associated with 98% of AIS 3+ according to the real world cases, regardless of belt geometry and occupant size. We observe in this figure several cases where the potential thoracic risks are very different using either the shoulder belt load value or the sternal deflection value.

The location of the real maximum deflection (in real-world crashes or crashs tests) and the associated thoracic risk is not systematically at the sternum level. It explains certainly on crash tests, the different values of thoracic risks when using either the shoulder belt load value or the sternal deflection value.

Our observations are in agreement with Cesari (10), Katz (11), Yoganandan (9) and Horsch (12) who indicate that under belt loading, the internal transducer in the manikin did not record the true maximum thoracic deflections and that depending on the region of loading, the internal deflection transducer underestimated more often thoracic compressions. Moreover, they indicate the need for improvements in the Hybrid III thorax to obtain realistic chest compressions with different restraint systems.

Indeed, the tests performed by Yoganandan (9) on PMHS and Hybrid III with 2 EPIDM (External Peripheral Instrument for Deformation Measurement) show that maximum deflections generally do not occur at sternum level as indicated in the following tables 3a and 3b :

Table 3a. Summary of Chest Deflection for Specimen Tests (reproduced from Yoganandan paper 9, table 3a)

specimen	UPPER EPIDM		LOWER EPIDM	
	maximum déflection (cm)	location* (cm)	maximum déflection (cm)	location* (cm)
M1	2.7	0	12.3	114
D2	5.1	3.8	11.4	6.4
V3	-	-	6.2	0
F4	9.1	3.8	2.6	3.2
A5	8.1	5.1	7.1	10.2
J6	6.2	10.2	6.5	12.7

* distance measured clockwise from sternum as viewed from top

Table 3b. Summary of Chest Deflection for Manikin Tests (reproduced from Yoganandan paper 9, table 3b)

TEST N°	shoulder belt load (kN)	internal chest pot (cm)	UPPER EPIDM maximum déflection (cm)	UPPER EPIDM location* (cm)	LOWER EPIDM Maximum déflection (cm)	LOWER EPIDM location* (cm)
1	7.9	3.4	3.2	2.5	2.9	11.4
2	6.9	3.9	-	-	3.9	6.4
3	8.6	4.0	-	-	5.7	2.5
4	7.6	2.9	3.3	5.1	3.8	5.1
5	7.8	2.7	4.3	3.2	5.8	8.3
6	7.9	-	6.7	3.2	3.7	5.1

* distance measured clockwise from sternum as viewed from top

Moreover, if we compare the potential thoracic risks between tests with PMHS and identical tests with Hybrid III, we observe significant differences :

- For each PMHS, the severe thoracic injuries observed (table 3c) are in accordance with the probabilities (AIS 3+) of thoracic injury predicted according to shoulder belt load (here, between 6.1 and 8.8 kN) from figures 10 (according to age classes) and 8 (all ages).
- For Hybrid III, the deflections measured with the sternal sensor (shoulder belt loads between 6.9 and 8.6 kN) are contained between 2.7 and 4 centimeters and correspond to probabilities of thoracic injury contained only between 0.18 and 0.35 from figure 13.

Table 3c. Thoracic Risk Estimations using risk curves from figure 8 and figure 10 (Specimen tests from Yoganandan paper 9)

specimen and age	shoulder belt load (kN)	AIS thorax	probability AIS3+ according to age (from LAB figure 10)	probability AIS3+ for all ages (from LAB figure 8)
M1-58	7.3	4	0.95	0.6
D2-57	6.1	4	0.7	0.25
V3-66	8.8	3	1	0.95
F4-58	7.3	3	0.95	0.6
A5-67	6.6	3	0.95	0.4
J6-44	8,2	4	0.9	0.85

As a consequence, and according to Yogonandan data, it can be noted that the present thoracic injury criterion as used with the Hybrid III on the basis of sternal deflection is not sufficient. So, it is difficult to assert that this deflection value alone, measured at the sternum level, predicts the real thoracic risk for the whole chest, for all anchorage locations and occupant sizes.

One solution to address this issue would be to measure chest deflection in several locations. Another solution, which will not affect the present dummy design, is to use shoulder belt load in addition to sternal deflection. Injury risk curves provided here can be considered as they are based on real world conditions, i.e. with living people involved in real crashs. In addition, despite occupants' sizes and belt geometries in the accident sample discussed here, the shoulder belt load appears to be in accordance with the occurrence of chest injuries.

THE NEED FOR A 4 KN BELT LOAD LIMITATION

Accident data discussed here show that a threshold of 6 kN for belt load limitation is not sufficient to prevent a risk of serious injury to the thorax. It is therefore necessary to go a step further in reducing the shoulder belt load. As this reduction will result in an increase in excursions of the head and thorax, it is therefore essential to combined with this kind of seat belt a pretensioner and, of course, an airbag. The combination of an airbag and a 3-point belt restraint is discussed in various publications, including the paper by Kompass in 1994 (13), the study by Kallieris et al in 1995 (14), and the investigation by Mertz et al in 1995 (15). According to the data discussed in the paper by Kallieris (14) and the mathematical simulation investigated by NHTSA (16) for a variety of crash conditions (frontal and rollover), crash severities and occupant sizes (5, 50 and 95 percentiles), a threshold of 4 kN for the shoulder belt load limitation appears to be suitable to reduce the risk of thoracic injury without negative consequences on other injury measures.

Due to the effect of offset test configurations, the car structures are more and more stiffer especially in Europe and the car-to-car compatibility criteria is leading also to increase of the stiffness of small cars (17), this belt load limitation appears to be inevitable in order to balance occupant risks related to intrusion and deceleration.

Considering a belt load limitation of 4 kN, the population at risk involved in accidents is very broad, since the over-80's represent only 1.4% of those involved, 1.8% of severe injuries and 4.2% of fatalities (Table 2 in the appendix). One can also see all the benefits of reducing the force level from 6 kN to 4 kN, given the number of occupants involved aged over 50, who account for 21.8% and 30.8% of fatalities.

SUMMARY AND CONCLUSION

This study was initiated to address the rising risk of belt-induced chest injuries in frontal impact. The starting point was the analysis of 290 frontal accident cases with vehicles that were equipped in France in the 1970's with a belt load limiter in the front seat positions. The load limiter was based on a tear-webbing principle and was located near the upper belt anchorage point. This database shows that older people (age Š 50) may sustain severe chest injuries. Based on this experience a program was initiated at Renault with a view to reducing the shoulder belt load.

In 1995, a belt restraint system called PRS was introduced; comprising a combination of a pyrotechnic pretensioner located at the buckle, a clamp retractor and a steel part attached to the retractor and to the belt anchorage point. This steel part, designed to deform at a given load, acts as a load limiter. This allowed the shoulder belt load to be controlled at a level of 6 kN.

Accident cases involving this type of restraint were collected and analyzed, and in particular the behavior of the belt load limiter was investigated in relation to occupant injuries. The data from 89 cases with belted front seat occupants are reported in this paper. Crash velocities ranged from 40 km/h to 80 km/h.

The probabilities of AIS 2+ and AIS 3+ thorax risks according to age and force sustained at the shoulder level are calculated based on all these 89 cases. In a second stage, a larger accident sample (256 cases), including the textile load limiters and the PRS1 limiters, is considered.

One observes that for 50% of thorax risks of AIS 3+, the force for all ages together is 6.9 kN. Occupant's age is also a very important factor. Other factors such as belt attachment geometries and occupant size may have an influence on the injury risk curves presented in Figures 5 through 10. The authors have no evidence to establish this influence.

This 6.9 kN force threshold is smaller than that given by Mertz in 1991 (10 kN), who, at the time, used for his evaluation the LAB data with textile force limiters. The limited number of severe impacts and cases with severe thoracic injuries combined with known loads with precision (3 cases with AIS3+ among 386 total cases) at the time, easily explains this difference. The risks observed with this large number of additional severe cases are therefore now far more reliable.

Analysis of the same thoracic risks based on the tests performed on cadavers by the University of Heidelberg shows this same increase in risk according to age. Both databases show the same trend for the thoracic risk according to age, but the force for a given risk and a given age is here 2 kN lower than that for occupants with force limiters involved in real-world accidents.

The relation between shoulder belt load and sternal deflection is not good for all the crash tests performed in cars identical to those in the accidents analyzed with PRS, no doubt due to offset crashes which exert very different loads on the sensor for a given shoulder belt load. This deflection scatter measured only at the sternum level makes is difficult to satisfy evaluation of the thoracic risk in frontal crashes. So, it seems difficult to assert that this deflection value alone, measured at the sternum level, predicts the real thoracic risk for the whole chest, for all anchorage locations and occupant sizes.

This study verifies the fact that gains in terms of victims can be achieved with this 6 kN force level, but that this force level must be reduced further to protect a greater proportion of the population. It may be noted that a force limitation of 4 kN would make it possible to protect 95% of those involved in frontal impacts from thorax injuries of AIS3+. Such, a restraint system has to include a specifically design airbag which has to absorb a significant part of restraint energy. A detailed description of this system is provided in paper 3.

The study highlights :

- The reduction of chest loadings due to the belt appears to be essential for improving car occupant protection.
- As car structures are more and more stiffer especially in Europe (effect of the offset test configuration), the belt load limitation appears to be inevitable in order to balance occupant risks related to both intrusion and deceleration.
- Car-to-car compatibility criteria is leading to an increase of the stiffness of small cars. Controlling

occupant loads of the restraint system should be consequently necessary.

- For a better assessment of chest protection, the authors are looking for ways to implement the measurement of shoulder belt loads into European rating crash tests and European regulatory tests.

REFERENCES

1. J.Y Foret-Bruno et al. : « Correlation Between Thoracic Lesions and Force Values Measured at Shoulder of 92 Belted Occupants Involved in Real Accidents ». SAE Paper No.780892, Proceedings of Stapp Car Crash Conference, USA, 1978.
2. J.Y Foret-Bruno, F. Brun-Cassan, C. Brigout, C. Tarrière : « Thoracic Deflection of Hybrid III : Dummy Responses for Simulation of Real Accidents » . In proceedings of the 12th International Technical Conference on Experimental Safety Vehicles. Goteborg, Sweden, May 1989.
3. F. Bendjellal and all : « The Programmed Restraint System- A Lesson from Accidentology » SAE Paper No.973333, Proceedings of Stapp Car Crash Conference, 1997
4. Institute of Forensic Medicine, Heidelberg : « Biomechanics- Determination of the mechanical loadability limits of the occupants of a motor vehicle », final report at end of project N° 3906, 1978
5. H. Mertz, J. Horsch, G . Horn, R. Lowne : « Hybrid III Sternal Deflection Associated with Thoracic Injury Severities of Occupants Restrained with Force Limiting Shoulder Belts »; SAE Paper No.910812
6. The Abbreviated Injury Scale, 1990 Revision American Association for the Advancement of Automotive Medicine, IL, 1990.
7. H. Mertz, P. Prasad, G. Nusholtz : « Head Injury Risk Assesment for Forehead Impacts », International Congress and Exposition, Detroit, Michigan, SAE paper N° 960099, February 26- March 1, 1996
8. R. Morgan and all : « Thoracic Trauma Assessment Formulations for Restrained Drivers in Simulated Frontal Impacts » SAE Paper 942206, Proceedings of Stapp Car Crash Conference, 1994
9. N. Yoganandan, D. Skrade, F. Pintar, J.Reinartz, A. Sances : « Thoracic Deformation Contours in a Frontal Impact », SAE Paper No.912891, Proceedings of Stapp Car Crash Conference, 1991
10. D. Cesari, R. Bouquet : « Behaviour of Human Surrogates Thorax under Belt Loading », Proceedings of 34 th Stapp Car Crash Conference, 1990, pp 73-81
11. E. Katz, L.Grosch, L.Kassing : « Chest Compression Response of a Modified Hybrid III Dummy Rib Cage Features », Proceedings of 31 th Stapp Car Crash Conference, 1987, pp 245-249
12. J. Horsch, J. Melvin, D. Viano, H. Mertz : « Thoracic Injury Assessment of Belt Restraint Systems Based on Hybrid III Compression », Proceedings of 35 th Stapp Car Crash Conference, 1991, pp 85-108
13. K. Kompass : « Opportunities and Limits of an Air bag Optimization Based on the Passive Requirements of Standard 208 ». Paper No. 94-S4-O-08, 14th ESV Conference. Munich , Germany.
14. D. Kallieris, A. Rizzeti, R. Mattern, R. Morgan, R. Eppinger and L. Keenan : « On the Synergism of the Driver Air bag and the 3-point Belt in Frontal Collisions ».SAE Paper No.952700, Proceedings of Stapp Car Crash Conference, USA, 1995.
15. H.J. Mertz, J.E. Williamson and D.A. Lugt : « The Effect of Limiting Shoulder Belt Load with Air Bag Restraint ». SAE Paper No.9508861 International Congress and Exposition. Detroit, USA, 1995.
16. ISO/TC22/SC10/WG3 : « NHTSA Communication on Air Bag Related Accidents ». Meeting in Delft, The Nederlands, May, 1997.
17. C. Steyer, M.Delhommeau, P.Delannoy : « Proposal to Improve Compatibility in Head on Collisions », Paper No. 98-S3-O-05, 16th ESV Conference. Windsor , Canada.

Table 1A. CASES WITH PRS1

case number	place	Age/Sex	airbag	EES	chest injuries	MAIS	chest AIS*	other injuries AIS>2	shoulder load in kN
10295	D	48M		55		3	0	2 : head; 3 : radius cubitus, tibia	5,20
11173	D	69 F		40-45	sternum fracture	2	2		4,20
11173	RFP	77M		40-45	11 th left rib fracture	1	1		4,20
11349	D	50M		55-60	sternum fracture and 6th+7th rib fractures	2	2		6,00
11393	D	27M		55	pain chest	2	0	2 : foot	6,00
11393	RFP	25 F		55	pain chest	1	0		6,00
11419	D	72M		50-55	3 right rib fractures, lung contusion	3	3	2 : foot	6,00
11436	RFP	37 F		40-45	2th right rib fracture and 10th left rib fracture	1	1		4,50
11483	RFP	19 F	YES	75-80		3	0	3 : head	7,50
11528	D	25M		55	pain chest	1	0		6,00
11530	D	27M		45	clavicle fracture and pain chest	2	2		6,00
11530	RFP	28M		45	12th left rib fracture	1	1		6,00
11535	D	40M		60-65		2	0	2 : foot	6,00
11547	D	22 F		45	sternum fracture	2	2	2 : face	6,00
11547	RFP	29 F		45	10th left rib fracture	1	1		5,60
11549	D	23M		55-60		1	0		5,25
11594	D	50M		40	pain chest	1	0		5,00
11733	D	54M		35		1	0		5,90
11733	RFP	17M		35		1	0		5,20
11741	D	25M		55		0	0		6,00
11741	RFP	26 F		55		2	0	2 : ankle	4,25
11853	D	20 F		40-45	pain chest	1	0		3,50
11853	RFP	21 F		42		0	0		4,70
11910	D	62M		45-50	sternum fracture	2	2		6,00
11914	RFP	25 F		45-50	pain chest	1	0		6,00
11928	D	26M	YES	55-60		1	0		5,25
11977	RFP	58M	YES	35	pain chest	1	0		5,10
12041	D	52M	YES	60-65		0	0		5,50
12046	D	58M		60-65	sternum fracture	2	2		6,00
12046	RFP	60 F		60-65	4 left rib fractures	3	3		6,00
12051	D	43M		40-45	sternum fracture	2	2		6,00
12069	D	21M		60	clavicle fracture	3	2	2 : head,pelvis; 3 : abdomen	7,50
12069	RFP	28M		60		3	0	2 : head. 3 : femur+ radius	6,00
12076	D	44 F		40-45		1	0		3,50
12091	D	61 F		60-65	9th right rib fracture	2	1	2 : radius	6,00
12092	RFP	71 F		55-60	clavicle fracture	3	2	3 : pelvis, radius, cubitus	5,50
12093	D	54 F		35-40		2	0	2 : patella	6,00
12093	RFP	59 F		35-40	sternum fracture	2	2		5,00
12102	RFP	50 F		50	clavicle fracture+ one left rib fracture	2	2	2 : foot	6,00
12120	D	47M		60		3	0	2 : ankle; 3 : femur	6,00
12154	D	59M		50	sternum fracture+ lung contusion	3	3		6,50
12154	RFP	18 F		50		1	0		7,30
12166	D	48 F		55	clavicle fracture+ 1th left rib fracture	2	2	2 : lumbar spine and feet	5,50
12196	D	34 F	YES	45		1	0		6,00
12196	RFP	34 F		45	sternum farcture and clavicle fracture	2	2		6,00
12210	D	38 F	YES	50-55		1	0		4,00
12408	D	28M		40		1	0		5,20
12409	D	46M		40	sternum fracture	2	2		6,00

case number	place	Age/Sex	airbag	EES	chest injuries	MAIS	chest AIS*	other injuries AIS>2	shoulder load in kN
10295	D	48M		55		3	0	2 : head; 3 : radius cubitus, tibia	5,20
11173	D	69 F		40-45	sternum fracture	2	2		4,20
11173	RFP	77M		40-45	11 th left rib fracture	1	1		4,20
11349	D	50M		55-60	sternum fracture and 6th+7th rib fractures	2	2		6,00
11393	D	27M		55	pain chest	2	0	2 : foot	6,00
11393	RFP	25 F		55	pain chest	1	0		6,00
11419	D	72M		50-55	3 right rib fractures, lung contusion	3	3	2 : foot	6,00
11436	RFP	37 F		40-45	2th right rib fracture and 10th left rib fracture	1	1		4,50
11483	RFP	19 F	YES	75-80		3	0	3 : head	7,50
11528	D	25M		55	pain chest	1	0		6,00
11530	D	27M		45	clavicle fracture and pain chest	2	2		6,00
11530	RFP	28M		45	12th left rib fracture	1	1		6,00
11535	D	40M		60-65		2	0	2 : foot	6,00
11547	D	22 F		45	sternum fracture	2	2	2 : face	6,00
11547	RFP	29 F		45	10th left rib fracture	1	1		5,60
11549	D	23M		55-60		1	0		5,25
11594	D	50M		40	pain chest	1	0		5,00
11733	D	54M		35		1	0		5,90
11733	RFP	17M		35		1	0		5,20
11741	D	25M		55		0	0		6,00
11741	RFP	26 F		55		2	0	2 : ankle	4,25
11853	D	20 F		40-45	pain chest	1	0		3,50
11853	RFP	21 F		42		0	0		4,70
11910	D	62M		45-50	sternum fracture	2	2		6,00
11914	RFP	25 F		45-50	pain chest	1	0		6,00
11928	D	26M	YES	55-60		1	0		5,25
11977	RFP	58M	YES	35	pain chest	1	0		5,10
12041	D	52M	YES	60-65		0	0		5,50
12046	D	58M		60-65	sternum fracture	2	2		6,00
12046	RFP	60 F		60-65	4 left rib fractures	3	3		6,00
12051	D	43M		40-45	sternum fracture	2	2		6,00
12069	D	21M		60	clavicle fracture	3	2	2 : head,pelvis; 3 : abdomen	7,50
12069	RFP	28M		60		3	0	2 : head. 3 : femur+ radius	6,00
12076	D	44 F		40-45		1	0		3,50
12001	D	61 F		60-65	9th right rib fracture	2	1	2 : radius	6,00
12092	RFP	71 F		55-60	clavicle fracture	3	2	3 : pelvis, radius, cubitus	5,50
12093	D	54 F		35-40		2	0	2 : patella	6,00
12093	RFP	59 F		35-40	sternum fracture	2	2		5,00
12102	RFP	50 F		50	clavicle fracture+ one left rib fracture	2	2	2 : foot	6,00
12120	D	47M		60		3	0	2 : ankle; 3 : femur	6,00
12154	D	59M		50	sternum fracture+ lung contusion	3	3		6,50
12154	RFP	18 F		50		1	0		7,30
12166	D	48 F		55	clavicle fracture+ 1th left rib fracture	2	2	2 : lumbar spine and feet	5,50
12196	D	34 F	YES	45		1	0		6,00
12196	RFP	34 F		45	sternum farcture and clavicle fracture	2	2		6,00
12210	D	38 F	YES	50-55		1	0		4,00
12408	D	28M		40		1	0		5,20
12409	D	46M		40	sternum fracture	2	2		6,00

Table 1B. CASES WITH TEXTILE LIMITERS

case number	place	age	EES	type of limiter	number of broken band(s)	chest injuries	MAIS	chest AIS*	other injuries AIS >2	shoulder load in kN
510	D	33	60	A	5	pain chest	3	0	3 : pelvis and femur fracture	4,40
755	D	29	50	A	5	pain chest	1	0		4,60
755	RFP	24	50	A	5	pain chest	2	0	2 : head	5,20
763	D	42	50	A	4		2	0	2 : pelvis	4,20
779	D	59	50	A	4	pain chest	3	0	3 : lower member; 2 : head	4,20
870	RFP	35	30	A	3		1	0		4,00
877	D	51	30	A	2	pain chest	2	0	2 : lower member	3,80
962	D	56	20	A	1	pain chest	1	0		2,10
966	D	40	60	A	5	pain chest	2	0	2 : head	4,40
1000	D	36	40	A	5		1	0		4,40
1034	D	41	20	A	2		0	0		3,70
1034	RFP	42	20	A	2		1	0		3,70
1039	D	35	40	A	5		1	0		4,50
1187	RFP	23	30	A	2	pain chest	1	0		3,60
1210	D	35	30	A	3		0	0		3,90
1368	RFP	30	55	A	2		6	0	6 : head	3,70
1389	D	40	20	A	1		0	0		2,10
1411	D	24	20	A	2		0	0		3,80
1411	RFP	25	20	A	2	pain chest	1	0		3,80
1452	D	36	30	A	1	pain chest	1	0		2,10
1479	D	53	40	A	4	1 rib fracture	2	1	2 : head	4,30
1497	RFP	34	60	A	4		3	0	3 : lower member	4,20
1505	D	35	40	A	5	pain chest	2	0	2 : head	4,40
1514	D	49	25	A	2		0	0		3,70
1514	RFP	50	25	A	3	pain chest	1	0		3,90
1547	D	32	20	A	1		2	0	2 : neck	2,00
1628	D	51	20	A	1		0	0		2,10
1639	RFP	77	20	A	1	pain chest	1	0		2,10
1791	D	25	40	A	2		0	0		3,70
1791	RFP	28	40	A	2		0	0		3,60
1805	D	48	30	A	2		1	0		3,70
1806	D	38	25	A	2		1	0		3,70
1811	D	26	20	A	2		1	0		3,60
1837	RFP	58	20	A	2	sternum fracture	2	2		3,70
2056	D	32	30	A	3		0	0		4,00
2091	D	28	40	B	1	pain chest	1	0		7,20
2112	RFP	53	30	A	4	5 left rib fractures	3	3		4,20
2165	RFP	23		B	5	pain chest	2	0	2 : head	8,00
2168	D	22	20	A	1		1	0		2,10
2214	D	19	40	A	5		0	0		4,40
2214	RFP	21	40	A	5		0	0		4,50
2254	D	50	30	A	5		0	0		4,40
2266	D	32	30	A	1		1	0		2,10
2272	D	24	30	A	3		0	0		4,00
2277	D	34	30	A	3	pain chest	2	0	2 : head	3,90
2283	D	51	20	A	2		1	0		3,70

case number	place	age	EES	type of limiter	number of broken band(s)	chest injuries	MAIS	chest AIS*	other injuries AIS >2	shoulder load in kN
2308	D	30	50	A	5	pain chest	1	0		4,50
2342	D	45	30	A	4		0	0		4,20
2349	D	45	?	B	1		3	0	3 : upper mamber; 2 : head	7,30
2356	D	29	20	A	5		1	0		4,40
2460	D	52	40	A	1	pain chest	1	0		2,10
2500	D	21	55	B	5	clavicle fracture	3	2	3 : head; 2 : lower member	8,60
2500	RFP	25	55	B	5	6 left rib fractures	4	3	4 : abdomen	9,00
2504	D	55	40	A	5		3	0	2 : head, lower member; 3 : neck	4,40
2508	RFP	26	30	A	2	pain chest	1	0		3,70
2526	D	19	30	A	2		0	0		3,80
2526	RFP	20	30	A	5		0	0		4,40
2604	D	53	20	A	2	1 rib fracture	1	1		3,70
2617	D	47	15	A	1	pain chest	1	0		2,10
2666	D	32	20	A	1		1	0		2,20
2725	D	25	60	A	4	pain chest	2	0	2 : head	4,20
2725	RFP	32	60	A	5	pain chest	4	0	4 : abdomen; 2 : head	4,40
2733	D	24	20	A	2	pain chest	1	0		3,70
2737	D	21	60	A	2	pain chest	3	0	3 : head and lower member	3,60
2758	D	27	30	A	3		2	0	2 : upper member	3,90
2758	RFP	28	30	A	3		1	0		4,00
2883	D	28	40	A	3	pain chest	3	0	3 : lower member	3,90
2946	D	27	50	A	2		3	0	2 : head, lower member; 3 upper memb.	3,70
2963	RFP	43	40	B	4	flail chest	6	5		8,00
2969	D	28	30	A	3	pain chest	1	0		3,85
2969	RFP	30	30	A	5	pain chest	1	0		4,45
2977	RFP	17	50	A	2		2	0	2 : head	3,70
3020	D	35	50	C	all	pain chest	2	0	2 : head and upper member	5,50
3035	D	46	30	A	1		0	0		2,10
3076	D	38	30	A	4		1	0		4,20
3098	D	30	30	A	5		1	0		4,60
3168	D	24	70	B	4		3	0	3 : femur fracture; 2 : head	8,00
3205	RFP	39	20	A	3	pain chest	1	0		4,00
3205	D	40	20	A	3		0	0		3,90
3207	D	23	30	A	2		1	0		3,80
3253	D	29	30	A	2		0	0		3,80
3279	D	18	50	A	2		1	0		3,70
3279	RFP	19	50	A	2		3	0	3 : femur fracture; 2 : upper member	3,70
3337	D	46	50	A	2		3	0	3 : pelvis fracture	3,80
3385	D	22	20	A	1	pain chest	1	0		2,20
3407	D	38	30	A	2		0	0		3,70
3409	D	33	30	A	3	pain chest	1	0		3,90
3425	D	47	30	A	2		1	0		3,70
3493	D	73	30	A	1		0	0		2,10
3505	D	48	30	A	1	pain chest	1	0		2,10
3517	D	49	30	A	5	pain chest	1	0		4,40
3525	RFP	37	50	A	5		3	0	3 : upper member; 2 : lower member	4,40
3542	D	30	30	A	1	pain chest	1	0		2,10

Table 1B. CASES WITH TEXTILE LIMITERS

case number	place	age	EES	type of limiter	number of broken band(s)	chest injuries	MAIS	chest AIS*	other injuries AIS >2	shoulder load in kN
3656	D	39	60	C	100mm		4	0	4 : abdomen; 3 : upper member; 2 : head	5,50
3666	RFP	75	40	A	3	flail chest	6	5		3,90
3666	D	38	40	A	4	pain chest	2	0	2 : head	4,20
3671	RFP	58	15	A	1		0	0		2,10
3776	RFP	12	30	A	3		1	0		3,90
3792	D	28	40	A	2		1	0		3,70
3805	RFP	59	50	B	3	2 rib fractures	2	2		7,80
3853	D	21	50	A	5		2	0	2 : upper member	4,40
3889	D	31	50	A	3		3	0	3 : lower member; 2head	3,90
3889	RFP	51	50	A	3		4	1	4 : abdomen; 3 : upper member; 2 : head	3,90
3979	D	30	20	A	1		0	0		2,20
4026	D	28	40	A	5		1	0		4,40
4061	D	34	40	B	1	pain chest	2	0	2 : head	7,20
4084	RFP	65	20	A	3	3 rib fractures	2	2		3,90
4085	RFP	37	30	A	2		0	0		3,70
4193	D	22	30	A	3	pain chest	1	0		3,90
4197	D	39	40	A	3	pain chest	2	0	2 : head	3,90
4218	D	22	50	A	5		1	0		4,50
4225	D	37	60	A	1	pain chest	2	0	2 : head and abdomen	2,10
4284	D	31	40	A	2		1	0		3,80
4291	D	19	55	A	3		2	0	2 : head and lower member	3,90
4301	D	53	50	C	70mm	1 rib fracture	1	1		5,50
4301	RFP	52	50	C	80mm	pain chest	4	0	4 : abdomen; 2;upper member	5,50
4421	D	22	40	A	4		1	0		4,20
4452	RFP	26	20	A	2		0	0		3,80
4453	D	21	40	A	3		2	0	2 : head	3,90
4453	RFP	26	40	A	3		2	0	2 : lower member	3,90
4458	D	22	40	A	5		2	0	2 : head	4,40
4516	D	26	40	A	4	pain chest	1	0		4,20
4525	D	55	55	A	2		3	0	3 : upper and lower member	3,70
4527	D	19	20	C	30mm		0	0		5,50
4549	D	33	40	B	2	pain chest	1	0		7,40
4553	D	21	30	A	2		1	0		3,70
4692	D	35	30	A	1		0	0		2,10
4741	D	33	20	A	1		1	0		2,10
4793	D	31	55	A	5		1	0		4,40
4843	D	23	55	C	100mm		2	0	2 : lower member	5,50
4843	RFP	25	55	C	100mm		4	0	4 : abdomen	5,50
4866	D	23	30	A	2		0	0		3,70
5029	D	45	35	A	3		0	0		3,90
5037	D	38	25	A	1		1	0		2,10
5041	D	55	35	A	3		0	0		3,90
5074	RFP	46	30	A	2	pain chest	2	0	2 : upper member	3,70
5106	RFP	18	40	C	all		1	0		5,50
5224	D	20	45	C	12mm		3	0	3 : lower member	5,50
5244	D	33	25	A	2		1	0		3,80
5244	RFP	30	25	A	5		0	0		4,40
5272	D	24	45	A	3		2	0	2 : upper member	3,90

case number	place	age	EES	type of limiter	number of broken band(s)	chest injuries	MAIS	chest AIS*	other injuries AIS >2	shoulder load in kN
5296	D	83	35	A	2	pain chest	1	0		3,70
5368	D	15	25	A	2		1	0		3,70
5368	RFP	23	25	A	2	pain chest	1	0		3,75
5388	D	33	55	A	2		2	0	2 : lower member	3,70
5388	RFP	24	55	A	3		3	1	3 : lower member	3,85
5471	D	38	25	A	2		0	0		3,60
5521	D	36	35	C	80mm	sternum fract+1 rib fr	2	2		5,50
5597	RFP	43	40	A	3	pain chest	1	0		3,90
5602	D	50	25	A	1	1rib fracture	1	1		2,10
5670	D	69	45	A	3	pain chest	3	0	3 : head	3,90
5768	D	48	30	A	3	pain chest	1	0		3,90
5803	D	25	50	C	20mm		2	0	2 : head	5,50
5965	D	17	35	A	2		1	0		3,80
5965	RFP	19	35	A	3		0	0		3,90
6147	D	21	25	A	1	pain chest	1	0		2,10
6245	D	32	20	A	3	pain chest	1	0		3,90
6578	D	21	25	A	2		0	0		3,80
6628	D	31	30	A	2		1	0		3,70
6675	RFP	29	25	A	2	pain chest	1	0		3,70
6891	D	25	40	A	5	pain chest	2	0	2 : head	4,40
7004	D	39	45	C	all	pain chest	1	0		5,50
7492	D	21	35	A	1		0	0		2,20
7492	RFP	21	35	A	1		0	0		2,00
7734	RFP	35	35	A	3	pain chest	1	0		3,85
8615	D	18	75	A	5	pain chest	5	0	5 : abdomen; 3 : lower memb; 2 : head	4,40

Table 1. Age classes for the 3 samples and comparison with front-seat occupants involved in France according to injury severity (1996)

age classes	sample 89 cases PRS1	256 cases all force limiters	209 Heidelberg tests	involved in France	seriously injured in France	fatalities in France
< 30 ans	32	41	30	40	42	36
31-40	15	23	20	21	19	18
41-50	19	15	19	17	15	16
51-60	18	13	17	10	10	10
61-70	9	4	9	7	7	9
71-80	6	3	4	4	5	7
>80	1	1	1	1	2	4
total	100	100	100	100	100	100
mean age	42	37	41	38	40	42

CHAPTER 8:

SMART RESTRAINTS

S9-O-13

Frontal Impact Protection Requires a Whole Safety System Integration

C. Tarriere, C. Thomas, X. Trosseille
Renault

Abstract

Beyond the generalization of the belt wearing, the improvement of the frontal impact protection is one of the most efficient action to reduce the number of the severe road victims. However, the attempt to evaluate the potential gains shows some important limitations to this efficiency and indicates the necessity of complementary actions. Among them:

- the front-end of the trucks needs to be modified to avoid underide and too severe decelerations of car occupants,
- due to the interaction between the protection in frontal and in lateral impact, the gain in frontal could be lost by an increase of the aggressiveness of the impacting car in side collisions,
- in car-to-car head-on collisions, the gain would be reduced by the increasing aggressiveness of the heavier car.

The author presents the quantification of the expected gains for the most prioritary countermeasures, discusses the major interactions between them, and tries to define the required conditions to optimize the whole safety system.

Introduction

Achieving the most substantial benefits in secondary safety requires a two-pronged approach aimed both at improving the performance of the "structure/restraint systems" combination in asymmetric frontal impacts for a broad-enough range of velocities (delta-V from 55 to 60 km/h), and at reducing frontal aggressiveness against other vehicles, particularly in the case of side-impact crashes.

In addition, such an approach cannot ignore the inescapable fact that cars are getting lighter in order to cope with environmental demands (better fuel economy, less CO2 and pollution).

In the final analysis, the search for safety benefits will necessarily involve a broad, systemic approach based on the following indicators:

- protection criteria, measured on instrumented dummies,
- criteria for reducing aggressiveness, measured on a dynamometric barrier,
- and criteria for reducing fuel consumption and the production of C02, which will need to take into account a reduction in the power/weight ratio in order to control and then reduce the death rate in single-vehicle crashes.

Why Is Improving Protection Against Frontal Impact the Number 1 Priority?

Bureaucratic Truth vs. Scientific Proof

There are already so many regulations on frontal impact, some say, that the new priority should be introducing regulations aimed at improving side impact protection. This line of thinking is purely bureaucratic and simply ignores the scientific facts. According to our own assessment, the benefits to be expected from the side impact regulations that are being contemplated in Europe—and that were recently introduced in the U.S.—would be 6 times less effective than those expected from improvements in frontal impact protection.

Conditions Required for Achieving the Best Gains in Frontal Impact

Selecting a test that matches as closely as possible the deadliest front impact configurations. Such a test will be deemed representative of global frontal impacts if its configuration agrees with the characteristics of real-life accidents, therefore ensuring that test results will eventually translate into better highway safety.

Investigating actual collisions provides a basis for designing a suitable configuration with the most appropriate velocity.

The asymmetric configuration:

This configuration is the one used most widely in studies relating to overlap, deformation and resultant trajectories.

Based on 413 cars involved in front-end crashes (all severities of injuries combined) that were investigated in the region of Hanover (Germany), Otte (1) reports that 79% of them showed no evenly-distributed frontal deformation. The same author notes that in frontal impact the resultant trajectory of the forces is exactly parallel with the longitudinal axis of the car in only 23.7% of the cases.

Zeidler et al. (2) reports that 84% of the 822 cars they analyzed that were involved in injury-inducing frontal crashes were not symmetrically and evenly deformed along the front end.

Gloyns et al. (3) reports that 90% of the fatal frontal crashes they investigated could not be represented by the full frontal barrier test.

Lastly, the results from a survey conducted by the Peugeot-Renault Association involving 1831 belted front seat occupants confirm these findings. Figure 1 gives the state of these occupants according to the type of frontal overlap with the obstacle and the type of deformation for one sample, all brands and models combined. The number of occupants was reduced to 1000 for easier reading. Only 23 out of the 126 fatally and seriously injured (MAIS 3+)—i.e. 18%—had the front ends of their cars evenly and symmetrically deformed. Before making any comparisons with the full frontal barrier test, it may be useful consider the mean acceleration sustained by these occupants. For 7 out of the 23 seriously injured reported under these conditions, the obstacle crashed against most often was the door of another car. In these cases, the estimated mean acceleration is much below that recorded in tests against an inflexible barrier. These cases appear in the "Symmetric (low deceleration)" category in Figure 1.

		SEVERITY OF INJURIES FOR 1,831 BELTED FRONT OCCUPANTS (N = 1,831 occupants reduced to 1,000 for this table)				
		M.AIS 0	M.AIS 1-2	M.AIS 3-4-5	Killed	TOTAL
(1/4 track)	All cases	64	79	9	2	154
(1/3 track)	All cases	54	54	9	2	119
(1/2 track)	Rectangular deformation	6	7	1	-	14
	Oblique deformation	59	102	18	6	185
(2/3 track)	Rectangular deformation	9	10	2	1	22
	Oblique deformation	52	87	21	7	167
(distributed)	Rectangular deformation (low mean acceleration)	48	54	6	1	109
	Rectangular deformation (high mean acceleration -impacts related to 0° test)	9	33	13	3	58
	Oblique deformation (impacts related to 30°test)	13	27	10	4	54
(pinpoint)	All cases	10	10	2	-	22
	Overhanging obstacle with passenger compartment intrusion (all types of overlap)	5	9	2	4	20
	Above or below the frame No passenger compartment intrusion (all types of overlap)	38	35	3	-	76
	TOTAL	367	507	96	30	1000

Figure 1. Frontal Real-World Accidents—M.AIS of Belted Front Occupants According to Type of Overlap and Type of Deformation

All in all, the full frontal barrier test represents only 13% (16/126) of the belted front seat occupants who were seriously or fatally injured in frontal crashes.

Figure 1 also shows the types of overlap and deformation observed in the other cases. The data show that the deformation equivalent to the 30° angled barrier test resulting from frontal impact with a 50 to 100% overlap covers:

- 41% of the belted front seat occupants, all injuries combined,
- 52% (66/126) of the seriously and fatally injured,
- and 57% (17/30) of the fatalities alone.

Other tests such as the corner impact (1/2 overlap) can also cause asymmetric front-end deformations, but at a right angle. However, this type of deformation is rarely observed in actual crashes.

Selecting a priority-type asymmetric frontal test configuration—such as the 30° angled barrier on the driver's side—also requires estimating the force of the impact sustained by the belted occupants.

Force of the impact:

The force involved in actual frontal crashes is traditionally estimated in terms of the instantaneous velocity variation of the occupant (delta-V) and the mean acceleration of the vehicle (Am).

Figure 2 shows the cumulative delta-V percentages for belted front seat occupants that were not subjected to additional loading from rear seat passengers, according to various severities of injury. A delta-V value of 53 km/h covers half of the 169 severely and fatally injured for whom the force of the impact could be evaluated.

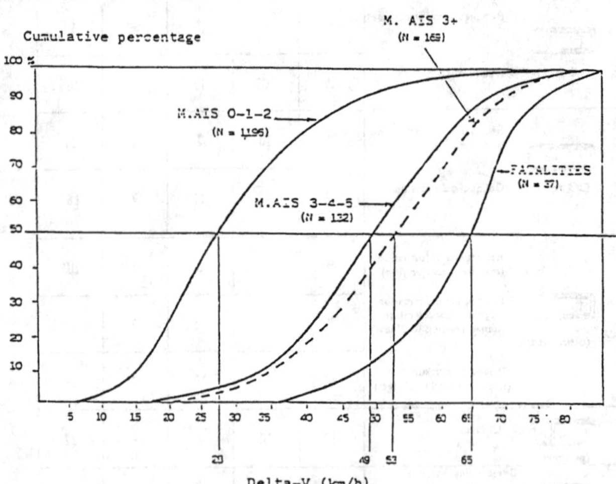

Figure 2. Frontal Impacts—Cumulative Percentages of Belted Front Occupants (Without Rear Occupant Overload) According to Delta-V and Severity of Injuries (N=1,365)

Another way to quantify the most representative velocity consists in examining the overall severity of the lesions in each delta-V category (see Figure 3). The data show that one out of two belted front seat occupants is

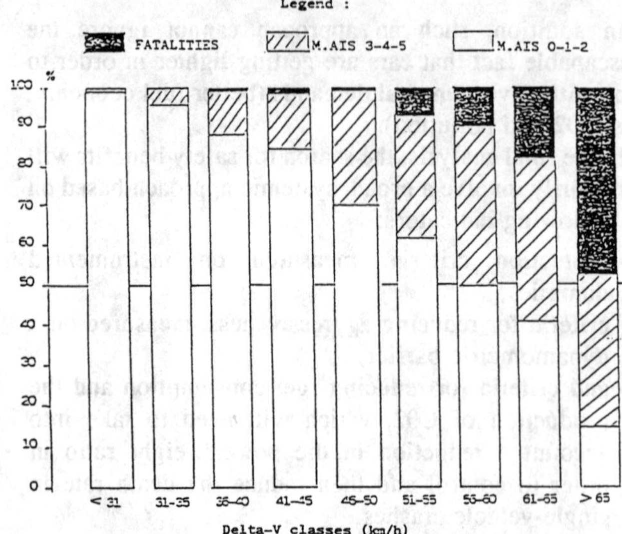

Figure 3. Frontal Impacts—Breakdown of Severity of Injuries for Belted Front Occupants (Without Rear Occupant Over Load) by Delta-V Classes

seriously or fatally injured in the delta-V category ranging between 56 and 60 km/h.

Thomas et al. (4) investigated the estimated mean acceleration in actual frontal crashes equivalent to the 30° angled barrier test with delta-V values on the order of 55 km/h. Based on 320 actual frontal crashes that could be assimilated to that test, the data show that the majority of the mean acceleration values for delta-V values between 50 and 60 km/h fall between 10 and 13 g (4). Most of the mean acceleration values measured in 30° barrier tests for this same range of velocities fall between 11 and 15 g (4). One may conclude that the 30° barrier test simulates rather faithfully the mean acceleration sustained by car occupants involved in crashes between 50 and 60 km/h.

In addition to delta-V and mean acceleration, intrusion is yet another factor that may bear on the risk incurred by belted front seat occupants. Table 1 shows the distribution of 403 such occupants broken down into delta-V categories according to the intrusion level inside the passenger compartment. The data show that for delta-V values ranging from 51 to 60 km/h, intrusion involves one fourth (25/101) of the belted front seat occupants, and most of all, 46% (17/37) of the seriously and fatally injured.

Table 1. Distribution of 403 Belted Occupants by Classes of Delta-V in Relation to the Intrusion

Level of Intrusion (inward displacement of the lower edge of the windshield)	Overall severity	Delta-V (in km/h)			
		40-50	51-60	61-70	TOTAL
Null or moderate (< 250 mm)	M.AIS 0-1-2	208	56	11	275
	M.AIS 3-4-5	26	19	19	64
	Killed	1	2	2	4
	Total	235	32	32	343
Critical (> 250 mm)	M.AIS 0-1-2	7	8	2	17
	M.AIS 3-4-5	7	13	7	27
	Killed	0	4	12	16
	Total	14	25	21	60

Figure 4 provides a distribution of the severities of injury according to the delta-V, mean acceleration and intrusion for actual crashes equivalent to the 30° angled barrier test, although no information is given on the resultant trajectory of the occupant.

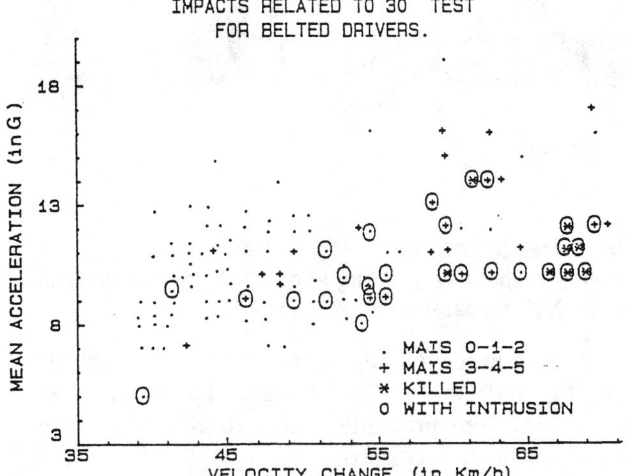

Figure 4. M.AIS of Belted Drivers According to Velocity Change, Mean Acceleration of Cars and Intrusion for Frontal Real-World Accidents Related to the 30° Angled Barrier Test

Otte (1) reports that the mean resultant of the forces is located to the left of the longitudinal axis, forming an angle estimated to be about 5°. The frequency of head impacts against the steering wheel also provides some indication as to the drivers' trajectories.

P. Thomas (5) observes that for values of delta-V between 50 and 59 km/h, 56% (56/99) of the belted drivers experience head/wheel impact. G. Walfisch et al. (6) report a similar frequency in the same delta-V category, based on the crash sample investigated by the PSA/Renault Association.

From an experimental standpoint, these authors note that in the delta-V category from 46 to 60 km/h, the impact of the head of the belted dummy driver is almost systematic in 0° or 50% offset tests, which is not the case in real life.

At any rate, the problem is elsewhere. In frontal crash tests—run as part of car design, regulation compliance or even car rating—the reference to a single head-wheel impact is unacceptable. In real life, there are not one but many conditions under which head-wheel impacts can occur, depending on the size of the occupant, the kinematics of the occupant, the characteristics of the frontal crash, the way the steering column moves, and the exact location of the impact against the steering wheel.

Therefore, a specific test needs to be designed that can measure the protection to the head (face, brain, skull, neck) provided by the steering wheel, and at various locations thereof. Such a test is the necessary complement to any overall frontal barrier test, whatever that test may be.

In the last analysis, any representative frontal test should restitute the parameters that govern the risks incurred by the maximum number of belted front seat occupants, i.e. it needs to provide for:

- a delta-V of at least 55 km/h,
- a mean acceleration on the order of 13 g,
- an asymmetric deformation of the front end of the car giving rise to a potential risk of intrusion on the driver's side,
- and a slightly oblique occupant trajectory that provides for the possibility that the head of the driver dummy will hit the left hand side of the steering wheel.

It should be specified however that such a test will not, of course, be sufficient to predict how safe a car is in all frontal crash situations. The test described above only seems the one most appropriate at present to cover close to half the seriously and fatally injured belted front seat occupants of cars involved in frontal crashes.

Selecting the tool for predicting the risk of lesions. This means designing a biofaithful dummy that can help scientists avoid making errors when they interpret test results. The base for such a dummy already exists: Hybrid III. It has a thorax, neck and head that are much improved compared with its predecessor's—Hybrid II—and that gives dynamic responses which are more closely related to those of human beings.

But although Hybrid III represents a significant step forward, further improvements are still needed in the following areas:

- collar bone stiffness,
- face biofidelity,
- pelvic behavior in submarining,
- chin/sternum contact,
- and adding asternal rib simulation.

In addition to progress in terms of biofidelity, Hybrid III makes it possible to measure several parameters that can be used to assess the risk of lesions in various body regions. A thorough analysis of all the parameters—both technical and medical—for a large number of actual crashes can help scientists clearly specify the risks facing car occupants: Accidentology can thus determine the types of impact encountered, their force, the body regions that are most at risk, and the types of lesions that are observed.

Based on the data, the role of biomechanics is to translate the physical parameters measured on dummies into protection criteria, and to set the various thresholds beyond which lesions may occur. That is the only way the risk of bodily injuries to the occupants involved in actual crashes can be predicted. Table 2 sums up the criteria proposed for frontal impact.

Table 2. Criteria Synthesis Suggested for the Frontal Test

	US CRITERIA	RENAULT CRITERIA : Proposal
SKULL	HIC < 1000	HIC<1000 If direct head impact. To be calculated during head contact
BRAIN		w<25000 rd/s2
FACE		F1<500 daN (forehead on rim) F2<250 daN (below forehead, with deformable face and on a large area)
NECK		My<250 Nm) Fx <250 daN) without head impact with impact : under determination
CHEST	a < 60 g	Deflection < 50 mm a < 60 g
ABDOMEN		F < 150 daN in case submarining is detected
FEMUR	F < 1000 daN	F < 1000 daN

Why Is It Important Both to Improve Protection Against Frontal Impact and to Reduce Frontal Aggressiveness Against Other Vehicles, Particularly in Side-Impact Crashes?

At present, car safety policies vary from country to country, but the one feature they share is that they only address the issue of protection inside the vehicle being certified. they ignore the risks that such a vehicle may pose to other users.

This void is that much more serious that there is no other truly effective means of protecting these users. Accordingly, a car's frontal aggressiveness cannot be effectively compensated by a regulation aimed at making an impact against for instance the doors of another vehicle tolerable by its occupants. Of course, the overall number of injuries can be lowered by making specific design changes to the side of cars, but the effectiveness of such changes is too limited to affect the number of serious injuries and fatalities (7, 8). Effective protective measures—measures that could cut down that number in half—are technically and economically unfeasible, as shown by the evaluation program for the Renault COVER (9 to 11). Naturally, such inability to compensate reaches 100% in the case of road users that are most vulnerable, i.e. pedestrians and two-wheelers.

Charges against the aggressiveness of certain cars were levelled over fifteen years ago (12 to 17). Its components were identified: weight, architecture and stiffness. Spectacular demonstrations were made, such as Philippe VENTRE' s to the 3rd International Conference on ESV (18) in a communication entitled: "A Homogeneous Safety in a Heterogeneous Fleet." The author clearly demonstrated that a small Renault 5 weighing 660 kg could offer the same protection as that of a standard-sized American car weighing twice as much, but whose aggressiveness would have been reduced. The case was made using a front-end collision between the two vehicles (Figure 5).

To date, such studies have not prompted any new regulatory developments. Could it be for lack of quantifying the influence on highway fatalities? If so, the void is now filled: we know for instance that the risk is 5 times higher in the most aggressive car category (over 1000 kg) than in the least aggressive one (under 800 kg) (Figure 6). And these are but averages by weight category. The risk runs probably from 1 to 20 between the least and the most aggressive car.

Figure 5. Head-On Crash Test Between Two Cars Weighing Respectively 660 kg and 1320 kg Running Each at 70 km/h (Source: from Ventre (18))

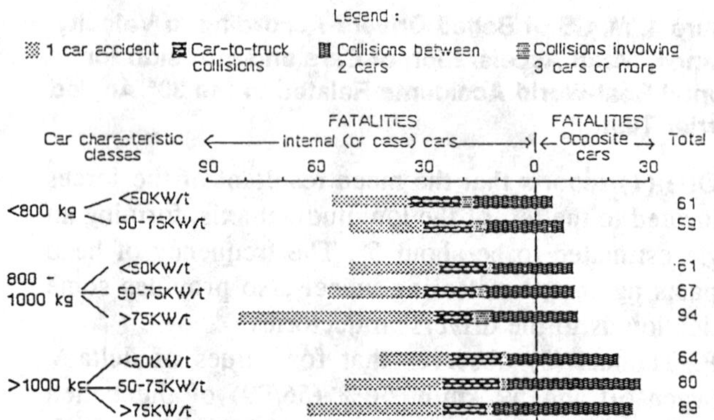

Figure 6. Distribution of Fatalities Among 1,000 Drivers in Each Car Class of Weight and "Power/Weight" Ratio According to Obstacle Type (Source: Sample of 41,944 Drivers, France 1989, Gendarmerie File)

The stakes are high, even higher when pedestrians and two-wheelers are taken into account—and justifiably so—since here again the risk varies in relation with the aggressiveness of the vehicle (Figure 7).

It is often said, particularly in the United States, that safety inside a car drops as the car gets lighter. This indeed complies with the laws of mechanics. The velocity variation of car occupants wearing seat belts anchored to the non deformed section of their cars is proportional to weights ratio. For example comparing two cars involved in a frontal collision, one weighing 700 kg and the other 1400 kg, with a closing speed of 100 km/h, delta-V will be 67 km/h for the lighter car and 33 km/h for the heavier one. Turning now to the statistics on actual crashes, the death rate is 25 per thousand drivers of cars under 800 kg, compared to 10 for cars over 1000 kg. Therefore, the death rate indeed increases as cars get lighter, but the ratio is still below 2.5. Accordingly, the

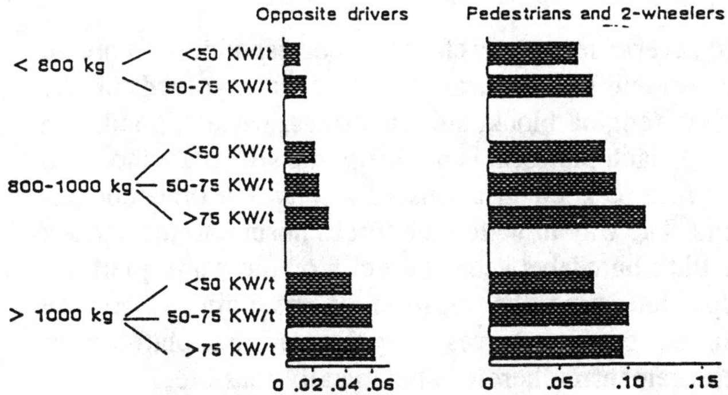

Figure 7. Fatality Rates of Opposite Car Drivers and Pedestrian or Two-Wheelers According to the Striking Car Weight Class and "Power/Weight" Ratio (Source: Sample of 50,710 Car Drivers, France 1989, Gendarmerie File)

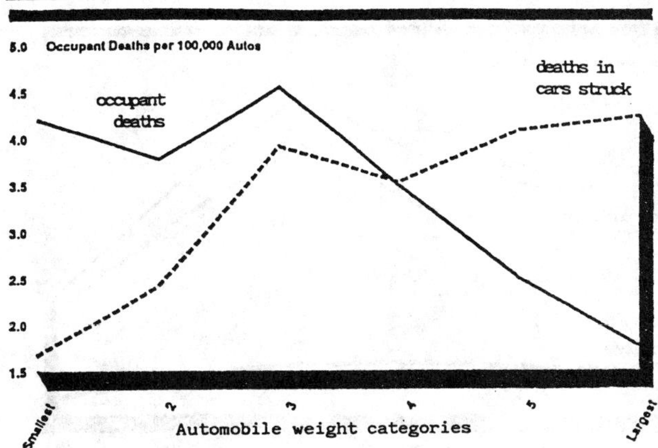

Figure 8. Deaths in Two Cars Accidents—USA, FARS 1975-1988 File (Source: From Chelimsky (19))

increased safety inside the heavier cars does not offset, by far, the risk they pose to the outside since, as indicated above, the corresponding ratio is 1 to 5 when comparing the average for cars under 800 kg and that for cars over 1000 kg.

Figure 6 clearly illustrates the results drawn from a data base on 40,000 drivers involved in injury-inducing traffic accidents which occurred in France in 1989. Our findings are in good agreement with those published recently in the United States (19). The author, E. Chelimsky, points out that " ... the safety provided to the occupants of a large car must be considered together with the risk posed by that same car to passengers in other automobiles.... (I)t is not true that cars become more dangerous simply by getting lighter."

Our findings are in full agreement with C. Thomas's paper (20) presented and discussed 6 months earlier at the AAAM Conference held in Scottsdale (USA) in October 1990. The author shows that although the death rate is higher inside lighter-than-average cars in the French automobile fleet, overall, the heavier cars are more dangerous. The number of victims in the struck cars (only 2-car collisions are studied) increases faster with weight than their number decreases inside these same vehicles.

The results obtained in the US regarding only 2-car accidents relate to crashes that occurred in the 1978-1988 period. These crashes are broken down into six equal-sized categories of car weights, from category 1 (smallest cars) to category 6 (largest cars).

It is noteworthy that the risk inside the cars (as opposed to the risk outside, in cars that are struck) is highest in the 3rd weight category, and not in the two lightest car categories (Figure 8).

The same figure also shows (dashed line) the effect of car size as an initiator of force. The risk of fatalities in the "other" car increases dramatically starting from the 3rd heaviest car category.

It is clear that when you add the two categories of fatalities (inside and outside) in each weight category, the lightest car category is the least dangerous overall. For the rest, the results differ slightly from those recorded in France, and such for various reasons including the lower seat belt use rate in the US.

Thus, on the basis of two very different samples, the French one representing 90% of all traffic injuries sustained in that country, one can better understand why there is an imperative need to tie in the search for better frontal protection with a limit on the aggressiveness of vehicles. Any improvement to frontal protection alone is likely to cause more heterogeneity in the fleet, particularly between the heaviest cars—that would be made even stiffer to meet the new requirements—and the existing fleet. Accordingly, there is an urgent need to introduce objective means of measuring aggressiveness, such as those proposed hereinbelow.

Analysis of the Influence of the Power/Weight Ratio

In this comparative study of risk in each weight category of cars in the fleet, there is yet another important parameter beside weight: the power/weight ratio.

The above-mentioned US study also investigates one-car accidents (Figure 9). For this type of crashes the sample only includes recent—one to two-year old—passenger cars involved in fatal accidents from 1986 to 1988 grouped in the same six weight categories as above.

Here, the most dangerous cars are in the fourth weight category, where the risk is much higher than that in the three lower weight categories. The risk is lowest in the sixth category, which is comprised of the heaviest cars. These cars also have the lowest risk of rollover.

The author offers no explanation for the fact that cars in the fourth category are more dangerous.

Referring to our own results, we submit that an additional parameter plays an important role in one-car accidents: the power/weight ratio.

The influence of the power/weight ratio is felt across all the weight categories (see Figure 6). For a same category of power to the ton, this influence tops out in the

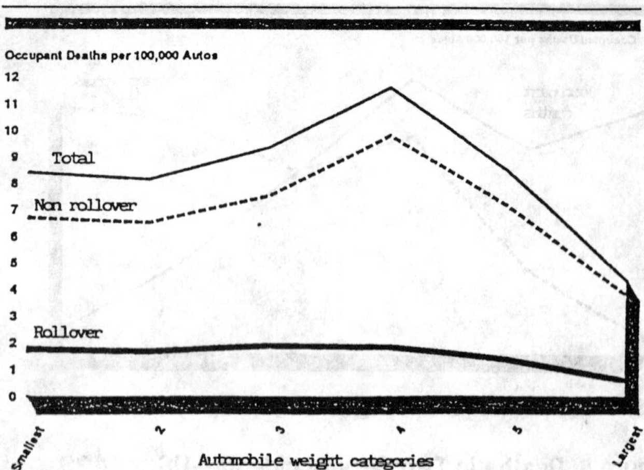

Figure 9. Deaths in One Car Accidents—USA, FARS 1975-1988 File (Source: From Chelimsky (19))

cars of average or below average weight. For example in cars weighing between 800 and 1000 kg with a power/weight ratio of 75 kW/ton, there is a sizable excess risk that represents 66% of the "inside" fatalities and 59% of all the fatalities ("inside" + "outside") associated with that category. Loss of control is the major cause, particularly going off the road round a curve and crashing against a fixed obstacle. But the power/weight ratio also translates into more aggressiveness toward the most vulnerable users (Figure 7).

One can see how important it will be to limit the power/weight ratio in the near future, when the average weight of cars tends to drop in order to meet environmental requirements.

Proposals on How to Estimate the Aggressiveness of Car Structures

Several approaches can be imagined to estimate the aggressiveness of cars. But rather than choose a specific test, we suggest that the most cost-effective solution consists in tying in this estimate with the future frontal impact test.

Indeed, would it not suffice to measure the forces applied to the barrier in dynamic tests to get a good description of their maximum values and distribution? The side members and the engine/gearbox unit transmit the brunt of the forces, and it needs to be accurately recorded by the selected frontal impact test configuration. Among the asymmetrical impacts, the so-called "overlap higher than 1/3" impacts do not adequately account for the engine/gearbox unit forces. However, the asymmetrical impact against a 30° angle barrier, shown above to be quite representative, would be acceptable to estimate the aggressiveness involving the quasi instantaneous impact of one side member and the engine/gearbox unit and, at high speed, of the whole front end of the car.

How feasible is estimating a car's aggressiveness using the 30° angle barrier test? Instrumenting the 30° angle barrier using force cells has already been done (21). Measurement platforms divide up the barrier area into several sectors, each sector corresponding to one of the vehicle's structural components involved in the impact (engine block, side members, wheels, hood, and so on); each platform is made up of a stiff plate mounted onto four piezoelectric sensors with two or three components. The way in which the forces normal to the surface are distributed between the cells of the same platform helps determine the location of the point where the resultant of these forces is applied on that platform. A measurement is therefore technically feasible.

What needs to be agreed on next is how many cells there should be and how they should be distributed. Indeed, the spatial measurement of the forces must be accurate enough so that the authorized limits—which are yet to be determined—depending on the location are compatible with an actual decrease in side impact aggressiveness.

Some may think that running additional tests is not cost effective so long as the full front impact test remains a mandatory requirement.

With respect to this type of impact, we have a picture of the French car fleet in the 80s as given by the moving deformable barrier test proposed by UTAC (22). For impacts at 35 km/h covering 20 representative vehicles, mean overall stiffness values ranging from 500 kN/m to 5000 kN/m (!) and peak forces from 179 kN to 510 kN were recorded. In a first step, it would be interesting to get a new picture of the fleet in 1991. This picture would help define specifications based on the least aggressive vehicle, making sure that the vehicle that is selected offers good protection against frontal impacts.

However, such initial approach to the 0° barrier test should not prevent us from considering the feasibility of estimating aggressiveness using the 30° angle barrier test, a potentially fruitful path for the future.

Conclusion

In the light of these findings, it should be obvious that the problems at hand are highly complex.

Gone are the times when simplistic proposals such as "Improving vehicle safety merely requires better side impact protection" or "Improving vehicle safety merely requires tougher regulations in terms of passive safety, and running tests at a higher speed" could be made.

Before making users pay for implementing such proposals, let's first ascertain what benefit they would derive in terms of a diminished risk on the road.

As concerns protection against side impacts for instance, our estimates show that the benefit would indeed be very small, due to the shape—thinness of the side wall—and to the highly unfavorable ratio between the stiffness of the front end of cars and that of the side wall.

Increasing protection against frontal impacts is a much higher priority since the expected benefits are five times higher than in side impacts. Nevertheless, such benefits should be attained using the "softest" methods, i.e. those that improve protection inside the vehicle without increasing its aggressiveness outside. This means putting a priority on optimizing the effectiveness of protection systems inside the passenger compartment.

Such optimization should be applied to each seat in the vehicle: indeed, the requirements are different for the driver, the front seat passenger and the rear seat passengers depending on whether they are seated next to a door or in the middle, and whether they are adults or children. And for the latter the demands are still different for each age group.

There is still a lot of work to be done in this area, and the "COVER" experimental vehicle presented by RENAULT provides many useful applications.

Progress in the area of frontal impacts can and must be sought by improving the behavior of car structure. This does not mean stiffening, as stiffening fails to deliver better overall protection. Recent examples show that cars that are too stiff transmit forces to the occupants through the restraint system, forces that exceed human tolerance particularly in the thoracic region.

To improve structural behavior means getting better control over this parameter with a view toward maximizing the dissipation of energy while at the same time avoiding unacceptable deformation of the passenger compartment. The criteria measured on instrumented dummies can be used to predict the risk of injury, and hence to assess any improvement to the "structure-restraint system" couple.

It should also be stressed that the damage to a car (its deformation) does not provide—in and by itself—any pertinent measurement of safety.

But there is yet another imperious reason not to increase structural stiffness. In addition to the risk stiffness poses to the occupants in crashes against fixed obstacles—i.e. 39% of all the frontal impacts involving fatalities in France—it also constitutes a considerable danger on the road by increasing the aggressiveness of striking vehicles. Let's not forget that any car on the road is as much a potentially struck vehicle than a potentially striking one.

Limiting front end aggressiveness is an imperious necessity. This objective can only be reached if one acts simultaneously to improve protection against frontal impacts—a desirable goal—and to limit aggressiveness.

The aggressiveness of vehicles is highly dependent on their weight. One can easily conceive how dangerous it would be to let the heterogeneity of cars on the road in terms of weight and stiffness further increase without controlling their aggressiveness.

Studies have also shown that the weight-safety relationship is not a linear one. There is an excess risk related to the power/weight ratio that comes into play independently and that primarily manifests itself in the form of loss of control of the car in a curve (one car accidents).

In the final analysis, it is becoming ever more obvious that safety can only be served by a systemic approach.

Frontal impact protection of course remains the number one priority in passive safety, but it needs to be considered by curbing the evolution of weight and by limiting the aggressiveness of cars against one another.

Overall highway safety will advance as the heterogeneity of the fleet decreases. This goal has to be reached in the context of an inescapable drop in weight in order to lower both fuel consumption and pollution.

The stakes thus appear in an overall perspective where safety imperatives are confronted with those of the environment.

Safety benefits can be expected in spite of—or thanks to—lighter cars, but two conditions need to be fulfilled:

- fleet heterogeneity must be fought against and reduced,
- and power must decrease as cars become lighter. In other words, the power/weight ratio must not increase, it must go down.

Fulfilling these conditions requires a clearly stated political will. Indeed, the natural trend in the past decades has been toward more power and heavier cars, and specifically, more weight in cars that are among the heaviest in the existing fleet.

Finally, it is on purpose that this report has only addressed those issues that concern secondary safety. In fact, safety as a whole is a much more complex matter. A coherent road safety policy cannot avoid comparing the cost effectiveness of the various proposals being made, according to whether they relate to injury prevention (so-called passive or secondary safety) or accident prevention (so-called active or primary safety). The fact that protection against side impacts is not very effective does not mean that nothing should be done. It only shows that lives should be saved by other, more cost effective means, i.e. crash avoidance. Measures of the highest priority need to be taken to improve the road infrastructure. For instance:

- a traffic circle in place of an intersection cuts down the risk of accidents involving serious injuries twenty-fold;
- and putting up guardrails along roads lined with trees would likely cut almost in half the number of killed against trees recorded in FRANCE in 1990 by the state police force.

Accident prevention will account for the great majority of the lives saved in the future (23, 24). Broadening the analysis to all the safety issues and integrating them into a comprehensive system should not be used as an alibi to do nothing. Measures designed to improve frontal impact protection constitute a priority that is that much less arguable that they form an overall system coupled with support measures that give it a clearly established coherence and effectiveness.

References

1. D. Otte, "Comparison and Realism of Crash Simulation Test and Real Accidents Situations for the Biomechanical Movements in Car Collisions" SAE paper 90 23 29.
2. F. Zeidler, H. H. Schreier and R. Stadelmann, "Accident Research and Accident Reconstruction by the EES-Accident Reconstruction Method" Proceedings of SAE Congress, Detroit, Michigan, February 25-March 1, 1985 P. 159. SAE Paper 850256.
3. P.F. Gloyns, S.J. Rattenbury and I.S. Jones, "Characteristics of Fatal Frontal Impacts and Future Countermeasures in Great Britain" Proceedings of 12th ESV Conference, Goteborg-Sweden May 29-June 1, 1989.
4. C. Thomas, S. Koltchakian, C. Tarriere, C. Got and A. Patel, "Inadequacy of A 0° Degree Barrier with Frontal Real-World Accidents" Proceedings of 12th ESV Conference, Goteborg-Sweden, May 29-June 1, 1989.
5. P. Thomas, "Head and Torso Injuries to Restrained Drivers From the Steering System" Proceedings of IRCOBI Conference, Birmingham-U.K., September 8, 9, 10, 1987.
6. G. Walfisch, D. Pouget, C. Thomas and C. Tarriere, "Head Risks in Frontal Impacts: Similarities and Differences Between Tests and Real-Life Situations" Proceedings of 12th ESV Conference, Goteborg-Sweden, May 29-June 1, 1989.
7. D. Viano, "Estimates of Fatal Chest and Abdominal Injury Prevention in Side-Impact Crashes" Journal of Safety Research, Vol. 20, pp 145-152, 1989.
8. C. Henry, C. Thomas and C. Tarriere, "Side Impacts: Expected Benefits of Planned Standards" Proceedings of 12th International ESV Conference, Goteborg, Sweden, May 29-June 1, 1989.
9. N. Casadei, "VSS—Safety Synthesis Vehicle" Proceedings of 13th International ESV Conference, Paris-France, November 4-7, 1991.
10. G. Walfisch, "Renault VSS Safety Vehicle: Occupant Safety in Frontal Impacts" Proceedings of 13th International ESV Conference, Paris-France, November 4-7, 1991.
11. J. Rio, "Renault VSS Safety Vehicle: Occupant Safety in Lateral Impacts" Proceedings of 13th International ESV Conference Paris-France, November 4-7, 1991.
12. G. Chillon, "The Importance of Vehicle Aggressiveness in the Case of a Transversal Impact" Proceedings of 1st International ESV Conference, January 25-27, 1971.
13. H. Appel, "Optimum Deformation Characteristics for Front, Rear and Side Structures of Motor Vehicles in Mixed Traffic."
14. E. Chandler, "Car-to-Car Compatibility" Proceedings of 4th International ESV Conference, Kyoto, Japan, March 13-16, 1973.
15. P. Ventre, "Proposal for Test Evaluation of Compatibility Between Very Different Passenger Cars" Proceedings of 4th International ESV Conference, Kyoto, Japan, March 13-16, 1973.
16. U. Seiffert, "Compatibility on the Road" Proceedings of 5th International ESV Conference, London, England, June 4-7, 1974.
17. P. Ventre, "Compatibility Between Vehicles in Frontal and Semi-Frontal Collisions" Proceedings of 5th International ESV Conference, London, England, June 4-7, 1974.
18. P. Ventre, "Homogeneous Safety Amid Heterogeneous Car Population?" Proceedings of 3rd International ESV Conference, Washington, D.C. May 30-June 2, 1972.
19. E. Chelimsky, "Automobile Weight and Safety" Statement to the Senate Committee on Commerce, Science and Transportation, Subcommittee on Consumer, Washington D.C., April 11, 1991. TECH-91-229.
20. C. Thomas, G. Faverjon, C. Henry, J.Y. Le Coz, C. Got and A. Patel, "The Problem of Compatibility in Car-to-Car" Proceedings of 34th AAAM Conference, Scottsdale, Arizona, October 1-3, 1990.
21. Kistler, Technical information: crash Dynamometer System.
22. E. Chapoux, J.C. Jolys, R. Dargaud, "An Approach for the Design of a Deformable Mobile Barrier to Evaluate the Protection Afforded to Occupants of a Passenger Car Involved in a Side Collision" Proceedings of IXth ESV, Kyoto, Japan, 1982.
23. D.C. Viano, "Limits and Challenges of Crash Protection" In Accident Analysis and Prevention, Vol. 20, No 6, pp 421-429, 1988.
24. C. Thomas, S. Koltchakian, C. Tarriere, B. Tarriere, C. Got and A. Patel "Primary Safety Priorities in View to Technical Feasibility Limits to Secondary Automotive Safety" Proceedings of 23rd FISITA Congress, Torino, Italy, 7-11 May 1990.

970776

Adaptive Airbag-Belt-Restraints - An Analysis of Biomechanical Benefits

H.-D Adomeit, O.Wils, and A. Heym
PETRI AG Engineering Center for Automotive Safety

ABSTRACT

Current restraint systems with pretensioner, belt force limiter and airbag are not only designed for a crash test such as FMVSS 208. Automobile manufacturers carry out test configurations over and above these parameters to check the effectiveness of restraint systems, for instance: NCAP, frontal crash with 40% offset against a deformable barrier, which is a provision of the ECE-R 94 and a binding prescription for all new European vehicles as of October 1998. However it is not possible to cover the entire spectrum of accident configurations found in real-world accident occurrences.

In the development of restraint systems with reference to aspects of occupant biomechanics, the possibilities on the testing side for reproducing reality are particularly limited. Aside from the 50%-HYBRID III dummy as occupant of average size and weight, only the two extremes of the occupant - 5% female dummy and 95% male dummy - are used.

This paper defines the concept of *Adaptability of the restraint system*, that is the adaptability of its protection performance to various different possible initial parameters such as type and severity of accident, and the occupancy/car interior/ constellation. The adaptation requirement for the restraint system is determined in an occupancy/car interior model with simulation for individual crash occupancy combinations in a frontal crash. With the help of selected crash configurations with high adaptation requirements on the restraint system (RS), the potential of such restraint systems with regard to the increase in occupant protection is demonstrated. The question of the value of adaptability should be discussed in comparison to a restraint system which has been alternatively further optimized by primary measures.

INTRODUCTION

The current discussion concerning the optimization of occupant protection systems in motor vehicles is governed by the concept of *adaptability*.

Adaptable should mean that the restraint system adjusts itself individually to the initial crash parameters such as type of crash and speed, which determine the severity of the accident, as well as to the biomechanical conditions regarding the occupant and his particular sitting position. This means that for all crash configurations in combination with all possible occupant positions and proportions, the restraint system must *always* guarantee an *early* and *appropriate* introduction of restraining forces on the occupants, so that the injury-causing force load inflicted on the occupants is minimized.

The qualitative evaluation of the effectiveness of restraining forces is expressed as *Biomechanical Quality* [ADOM 93]. The restraining forces should only be applied to those portions of the occupant skeletal frame capable of bearing loads. In doing so, relative movements between individual body parts should be avoided if at all possible, so that the resulting coupling forces are minimized.

An adaptive or intelligent restraint system (RS) must fulfill the following prerequisites:

1. Crash sizes and biomechanical sizes must able to be reliably measured by sensors and immune to interference.
2. The RS must be able to differ its performance output in the subsystems safety belt or airbag system.

The RS must react to the initial parameters of the accident with the manipulatable variables of its subsystems, usually before the interaction with occupants; for future systems it is theoretically conceivable that the RS would also operate adaptively during the interaction.

MAIN FUNCTION OF RESTRAINING SYSTEMS IN AUTOMOTIVE SAFETY

Occupant restraint systems are designed to couple the occupants with the vehicle by a defined force forward displacement characteristic during the crash. The occupant load on the individual bodily parts should be kept as low as possible. Prerequisite for minimizing the occupant loads is on the one hand the ideal direction of the initial stress inflicted on those parts of the human skeletal structure capable of load-bearing, and on the other hand the ideal and largest possible surface distribution of restraining forces on the body. The coupling loads between the pelvis and the thorax as well as between thorax and head must remain as small as possible. The directions of the restraining forces, especially that of the airbag, particularly determine the flexion and extension moments in the neck (see Figure 1), and also in the lumbar spine.

DEVELOPMENT OF RESTRAINING SYSTEMS

Nowadays in the development of restraint systems with occupant loads one is often satisfied simply if the legally binding protection criteria are fulfilled. The restraining force characteristic or the occupant kinematics resulting from the kind of loads inflicted are not evaluated!

DEVELOPMENTAL TENDENCY OF "SMART RESTRAINT" - For those occupant crash configurations which cannot be covered by a laboratory test (FMVSS 208 or NCAP) for example because they demonstrate extremely severe accidents or extreme occupant behavior of some sort, complex control techniques are developed to design the restraint system in such a way to encompass such situations in an adaptive manner. However due to the rarity of such severe accidents in real life, the enormous development costs for such complex control technology seems questionable compared to the biomechanical benefits (only 4% of all frontal collisions have a Δv > 56 kph [SEIF 93]; see Figure 2). An adjusted inflator performance of the airbag in cases of increased accident severity could contribute to an improvement in safety. The benefits arise from the reduction in thorax load, since in this case, in contrast to the currently prevalent inflator designs, a breakthrough of the airbag is prevented, hence hindering the crash of an occupant on the contact system of the vehicle. Compared with an airbag design with the current 100% inflator performance, a inflator stage adjusted for lesser accident severity will always mean an increase in occupant forward displacement [PETR 96].

The complex control techniques play a particularly large role in out-of-position (OOP) situations, however, or inflation-induced injuries. These are situations in which at the onset of the accident the occupant is in the immediate vicinity of the airbag module, due to faulty behavior or - without being secured by the belt - due to a preceding hard braking. In a reliably sensored OOP situation, the airbag could be switched off. The firing of a reduced inflator stage in an out-of-position situation also always means increased forward displacement, and depends on the occupant energy or accident severity, so that it is practically impossible to demonstrate any residual benefit for occupant safety.

DEVELOPMENTAL PHILOSOPHY OF "REAL SMART PROTECTION" - Another approach in the further development of restraint systems for reducing the occupant load and optimizing the biomechanically compatible restraining force introduction is the increase in effectiveness of the restraint system. Increasing the effectiveness pursues the main goal of significantly lessening the occupant load during forward displacement by an early and rapid rise in the restraining forces on the occupant. In physical terms, this involves a lesser relative speed between the occupant and the vehicle. The loads on the occupant during the ride-down phase are at a considerably lower level than with a conventional, less effective restraining characteristic.

Aside from a reduction of occupant load, the occupant kinematics must also be controlled. The suitable introduction of the restraining forces of the airbag on the body of the occupant can prevent injuries.

The reduction of the load peak values as well as the reduction of occupant forward displacement build a potential which makes more complex restraint systems with adaptability for statistically unimportant individual cases superfluous.

For the great majority of all occupant crash configurations in frontal collisions in real life accident occurrences [SEIF 92], taking advantage of the system-specific potentials of current restraint systems offers a significantly higher degree of occupant protection.

Despite the rise in effectiveness of restraint systems, provisions must nevertheless be made for OOP situations with an additional control technique, at least for the airbag.

METHODOLOGY OF ANALYSIS WITH OCCUPANT CRASH SIMULATION MODEL

Here the goal is to investigate the requirements for adaptability in the restraint system for different initial crash parameters (see Figure 3). We employ a validated occupant crash simulation model with the following configuration:

- occupant - 50% HYBRID III dummy.
- RS - 3-point belt with constant belt geometry;
  pretensioner;
  2 kN belt force limiter in shoulder belt;
  - driver airbag (67 Liter) with a venting outlet of 55 mm diameter;
  - rigid steering wheel.
- Passenger compartment - rigid seat shell;
  - windshield;
  - bottom plate;
  - leg support.
- Crash characteristics
  - medium-class vehicle at
  - FMVSS 208 crash.

In order to exclude pelvic load induced by the bolster, a knee bolster was not modeled. The airbag dominates in the restraint system here, due to the 67 liter airbag and the low belt load level, in order to take advantage of additional protective characteristics of the airbag [ADOM 94;SCHE 96]. The contours of the car interior and the belt geometry both correspond to the medium-class vehicle being tested. The seat is designed as rigid seat shell to control the pelvic movement. Pelvic kinematics, rotation and vertical travel, are hence defined. The steering system is rigid and cannot intrude into the passenger compartment. The influence of forward displacement or the addition of steering column and steering wheel deformations on the occupant load are thus eliminated. This all serves to foster the reproducibility of the test conditions.

The occupant crash configurations tested with this simulation model are depicted in Table 1. In the variation runs 1 to 11 one parameter was changed respective to the base case (FMVSS 208); in the mathematical model runs 12 to 17 (40%-offset, deformable barrier, 50 kph) and 19 to 23 (0°-rigid barrier, 100% overlap, 30 kph) the occupant parameters were modified.

In all parameter variations of the simulation model a 3-point belt with pretensioner and belt force limiter were employed in order to maintain reproducible results in the simulation.

The main parameter used here in the evaluation is the thorax restraining force in x-direction via the thorax forward displacement - also here only the x-component. The path of force characteristics of these parameter variations are depicted in Figures 4 to 9.

DEMAND FOR ADAPTATION OF RESTRAINT SYSTEM PERFORMANCE

DEFINITION - The need or demand of a restraint system to be adaptive has to be identified by the spread of the occupant's load, dependent on input parameters of accident severity or the occupant's characteristics. The more the loads differ within the range of injury criteria under different test conditions or under real world accident conditions - or even exceed injury criteria under certain circumstances - the more we need active restraint system adjustments related to input parameters: in other words, adaptation of restraint system. The aim of adaptation is to ensure for any accident condition sufficient reserves in restraint system performance - the loads to be transferred onto the occupant's body - with respect to defined injury criteria.

GRADE OF ADAPTATION

It is of highest interest to identify extreme input parameters resulting in excessive deviations of the occupant's restraint loads. In addition statistical analysis has to be applied in order to judge the relevance for cost-benefit considerations.

Our method here is to evaluate output parameters loads, travel, and kinetic energy of the occupant. We make use of single parameters or combined-weighted parameters to at least measure differences of occupant loads in the matrix of crash conditions.

The characteristic value, which should evaluate the adaptation requirement of the restraint system according to the different input parameters, is the quotient from the real and ideal a_{3ms}-values (when one ideally assumes a constant path of force characteristic of the thorax restraining forces; see Figure 11) as well as the relationship between the ideal and real load level (see Figure 10). The forward displacement is likewise taken into consideration. In this case the degree of risk for contact between the occupant and the steering wheel calls for the need for adaptation, increasing as the maximal forward displacement is further exploited.

At this point it is clear that forward displacement with respect to contact risks for not-in-position occupants must then be newly assessed when the accompanying occupant loads lie within the tolerance limits.

RESULTS - This evaluation model was used to develop the table of results depicted in Figure 10. The need for adaptability for the individual crash configurations can be summarized as follows:

- The adaptability requirement is very strongly dependent on the type of crash (see Figure 10, Variations B,1,2).
- Of equal importance as accident parameter is the collision speed (Variations 3-5).
- The greatest requirement for adaptability exists in very severe accidents, such as the 60 kph wall crash (Variation 5); especially when forward displacement is taken into account (see Figure 10).
- The adaptation requirement for occupant parameters is slight. For 0°-wall crash the adaptation requirement for occupant parameters (Variations 6-11) is higher than for 40% offset crashes (Variations 12-17) and 30°-crash (Variations 18-23).
- The parameter "occupant size" has no significant influence on the adaptation requirement.
- The adaptation requirement for the 40% offset crash is particularly slight, since in this case the thorax of the occupant is subjected to a slight, almost constant load by the restraint system, caused by an extremely soft vehicle impulse. The restraint system reacts here with a nearly ideal constant low load level.

ADAPTATION MEASURES IN SO-CALLED "SMART-RESTRAINTS"

An example of a technical development being currently discussed for an adaptive restraint system is the multiple stage inflator. In order to check the effectiveness of this proposal, those of our simulation runs were selected which demonstrate the largest adaptation requirement. The speed is examined for a 0°-wall crash with full coverage, at the extremes 30 kph and 60 kph. The type of crash will be examined, from 0°-wall crash and 40% offset crash against a deformable barrier.

The gas mass flow in a well-known inflator was scaled to produce a multiple stage inflator as model. A reliable and robust sensoring of the individual frontal crash types was assumed for the simulation runs considered here. For less severe accidents, a reliable sensoring is particularly difficult, since the mass and rigidity of an obstacle or another car can under some circumstances be recognized too late. The firing of a inflator stage with reduced mass flow has fatal consequences when an accident occurs against a more weightier counterpart, and due to the nature of its front end characteristic, the sensor system initially identifies an only slight accident.

In the simulation of the 30 kph crash, a mass flow of 100% and 75% was set, whereas in the 60 kph crash simulation the mass flow of the inflator was set at 100% and 110%. For the 40% offset crash against a deformable barrier, a mass flow of 100% and 117% was assumed (see Figure 15).

RESULTS - In the cases examined, the implementation of a multiple stage inflator as most important component of a so-called smart restraint system revealed that this design had only slight potential for reducing load. In a 30 kph crash the firing of a inflator for a reduced stage of mass flow does not result in any appreciable change in the occupant loads. However, an increase in forward displacement could be registered. Hence the risk increases, for the head in particular, of hitting the steering wheel or dashboard (see Figure 13). For the 60 kph crash the firing of two inflator stages - resulting together in a mass flow of 110% compared to the current 100% stage - there is no reduction of the thorax restraining forces compared to the current RS. The forward displacement, on the other hand, is slightly reduced (see Figure 14). For the ODB (offset deformable barrier) case with firing of two inflator stages delivering a total 117% mass flow, likewise no reduction of the thorax load ensues. Here also a reduction in the forward displacement can be observed (see Figure 15).

These results demonstrate that there is no recognizable advantage in this measure, especially since the cases treated here have statistically only a slight relevance in real life accident occurrences.

"REAL-SMART" APPROACH OF RESTRAINT SYSTEM OPTIMIZATION

The negligible or completely missing potential for reducing occupant loads by employing a multiple stage inflator can hardly justify the high costs of developing such a system. Hence the question: Do we need a restraint system with adaptive characteristics or is it a better solution to develop a restraint system with all told increased effectiveness - lower loads and even, in some circumstances, with reduced forward displacement?

Based on the purely physical properties which offer such untapped potential in current restraint systems, high effectiveness attained by earlier controlled restraint of the occupant (see Figure 12) emerges as an undeniably important core measure.

Compared to the current situation we need improved gas inflators (see Figure 16) for such an effective restraint of the occupant, as well as improved and adjusted techniques in belt force limiters. Furthermore a "Real Smart" venting concept for the airbag is called for, since the occupant loads are derived from the interaction of the occupant diving into the deployed airbag. In order to allow the airbag restraining forces to rapidly take effect on the occupant, a part of the gas mass in the airbag is released as of a certain internal pressure, by means of a pressure-dependent vent opening - below this pressure, only the fabric porosity is responsible for the loss of gas mass. The limiting pressure value should only be reached once the occupant has fallen into the airbag. The effect of these techniques in increasing airbag efficacy is described in the simulation model in Figure 18. Aside from a significant reduction in the maximal load, even the forward displacement is reduced.

The minimization of forward displacement, when it goes hand-in-hand with a decrease in load, *must be seen in a new light when considering OOP accidents.*

Up until now it was the declared goal of development engineers for restraint systems to make full use of the entire available forward displacement travel. With the previously only limited performance of the restraint systems, this usually was equivalent to keeping the load down to sufficiently low levels. With highly effective future systems the free travel path of the occupant will be transformed into a "valuable asset". This value namely includes the risk potential of the restraint system within the ranges of occupant parameter mass, sitting position, and sitting geometry up to the case of OOP conditions.

RESULT - The "Real Smart" measures to increase effectiveness of the RS demonstrate significant reductions in maximal loads and forward displacement in the observed crash constellations. A multiple stage inflator cannot equal this performance in an adaptive RS for physical reasons.

Equally low loads are attained with "Real Smart" for the 30 kph crash, although the forward displacement is reduced (see Figure 19). In the ODB crash the forward displacement is also reduced with the same maximal thorax restraining force (see Figure 21). In both cases the risk of contact of the occupant with the steering wheel or dashboard is hence reduced. For the 60 kph crash a drastic reduction of the maximal force is achieved, compared to the firing of two inflator stages with a total of 110% mass flow (see Figure 20).

The advantages in terms of occupant loads and biomechanical quality, as well as further potentials for increasing occupant protection, in employing such a "Real Smart" concept compared to the so-called "smart restraint" system with multiple stage inflator, are easily recognizable.

The "Real Smart" aspect of such a system is the highly developed sensor system. This has to better identify accident severity and accident type, in order to better adjust the initial occupant-RS interaction in the given system-specific window of time.

"Real Smart" must be an advanced inflator concept which offers a gas mass flow that supports the pressure-controlled high-performance airbag.
And finally, "Real Smart" must include the secondary car interior sensory system, which has to be in the position to switch into the restraint system in catastrophic constellations, namely in occupant out-of-position situations.

NEED FOR ADAPTATION OF RESTRAINT SYSTEM WITH OOP-OCCUPANT

For an out-of-position occupant, driver or passenger positioned close by or contacting the airbag module, adverse effects of the restraint system have to be faced. Sensing this dangerous occupant position before or early in crash seems very advisable for the near future in making decisions within the restraint system with regard to performance. This is valid for so called "smart restraints" as well as our "Real Smart" restraint approach. There are however quite a number of design features on airbag modules available and in further development that help to overcome injuries caused by system behavior.

PRIMARY MEASURES - The P-folding of the airbag as well as the entire bag packaging can contribute to occupant load relief; this was demonstrated at the 39th Stapp Car Crash Conference [ADOM 95]. Venting techniques on the airbag can be foreseen, which could control the flow release of the airbag dependent on the deployment space in relation to the occupant. The airbag module can be retracted when ignited from the steering wheel (retractable module). In situations when the upper torso of the occupant is bent extremely forward, the airbag can be deployed behind the steering wheel rim. The cover design can be created so as to reduce possible injuries and diminish the punch-out effect. Constructive measures taken in module design can control the deployment trajectories of the airbag in defined ways. Furthermore, in OOP-situations with only very slight accident severity, the Real-Smart concept can prevent the crash of the occupant on the contact system of the vehicle due to the only slight forward displacement.

All these and further module design features in progress are to be realized in current application projects.

OOP-SENSING AND RESTRAINT ADAPTATION - A first step in the direction of OOP-sensing is, for example, the sensing of the vehicle speed. 70% of all accidents in which OOP-situations were determined at speeds below 32 kph. The firing of a mass flow reduced inflator stage hence seems to be load-reducing for speeds under 32 kph. However, this state can only be a compromise.

A non-compromising approach to OOP, however, is a "Real Smart" sensor system for the car interior in order to measure occupant positioning.

According to the car velocity, distance between occupant and airbag module as well as the size of the occupant, the reduced stage can either be ignited or the airbag can be entirely disabled, when an occupant-to-airbag distance is determined which would be more likely to induce additional injuries if the airbag is ignited.

OOP VERSUS DESIGN SHORTCOMINGS

These OOP analyses are somewhat misleading to the consumer community, law makers and even the world of experts.

Impressions are created and amplified that OOP situations are caused by misuse of the occupant alone.

It has to be considered, however, that the design performance of the airbag restraint as such can result in unfavorable, non-optimal restraint load application on airbag-interacting occupants.

The occupant´s size, geometry and driving attitude can differ fundamentally from laboratory test and system design conditions.

The airbag restraint system, however, is highly sensitive to these parameters, especially with non-belted occupants. Airbag size, shape and deployment trajectory, the folding concept and cover, mounting position of the module, bag dynamics during positioning and occupant interaction, and bag support in frontal car interior all play major roles for the design and philosophy of system performance.

As a result, the occupant kinematics can easily get out of control with a restraint system performance far off of the design targets. We suspect as a matter of interpretation, that many effects of airbags are therefore not only "related to occupant´s wrong usage", but are also related to poor (because non-robust) designs of the airbag restraint system.

Robust design, however, does not correlate with so-called "smart restraints", because of their complexity.

CONCLUSIONS

- A RS has a large adaptation requirement in the range of collision speed and crash type (crash pulse) as initial parameters. The largest adaptation requirement was determined for these loads (see Figure 4 and 5).
- The adaptation requirement of the RS is also dependent on the occupant parameters weight and position (longitudinal seat adjustments), but these parameters are significantly surpassed in importance by the crash parameters.
- Practically no adaptation requirement was demonstrated for the occupant size (with constant occupant weight).

Hence the comparison between a "smart restraint" represented by the multiple stage inflator and the "Real Smart" system (see Figures 19 to 21) can be summarized as follows:

- For the 30 kph-wall crash the "Real Smart" System displays a *significantly reduced forward displacement* with equal, in any case uncritical thorax load.
- Due to the in total more expressive characteristic of the restraining force acting on the thorax, the "Real Smart" system demonstrates a *significant reduction* of the maximal loads compared to the "smart restraint" or conventional RS for the 60 kph crash.
- For the ODB configuration the "Real Smart" system displays a significantly reduced forward displacement compared to the "smart restraint" system.
- Due to the only slight forward displacement of the occupant in the "Real Smart" system, in cases of only slight accident severity or with an offset crash, the risk of occupant contact with the steering wheel or dashboard is reduced.

For out-of-position occupant situations, both restraint system philosophies -"smart restraints" and "Real Smart" - demand the same characteristics of the system:

- Immediate implementation of the primary measures (folding concept and cover design, retractable module, bag dynamics and support) is necessary and possible.
- In the mid-term, a sensory development for the reliable determination of occupant behavior is necessary. These factors should deliver the initial data for a fire/non-fire decision.
- In addition, mass flow reducing inflator stages could be employed, in order to obtain residual benefits from the airbag.
- The entire system must be designed more "robustly" in order to attain full protection potential even with changing initial values of the accident occurrence and the occupants: the control of occupant kinematics during load applications by the restraint system must be reconsidered.

REFERENCES

[ADOM 93]
Dr.-Ing. H.-D. Adomeit. Konzepte von Rückhaltesystemen and Aspekte der Schnittstellen zum Fahrzeuginnenraum. Conference "Restraint Systems" in Haus der Technik e.V., Essen/Germany, 23./24. September 1993.

[ADOM 94]
Dr.-Ing. H.-D. Adomeit; Dipl.-Ing. D. Meißner. Airbag - Ergänzung im restraint system oder Basis für neue Konzepte des Insassenschutzes. Bag & Belt ´94, Third International Symposium for Car Occupant Restraint Systems; Cologne/Germany, 27.-29. April 1994.

[ADOM 95]
Dr.-Ing. H.-D. Adomeit; Dipl.-Ing. A. Malczyk. The Airbag Folding Pattern as a Means for Injury Reduction of Out-of-Position Occupants. 39th Stapp Car Crash Conference; San Diego; 8.-10. November, 1995.

BRAM 95]
)r.-Ing. L. Brambilla. 59. Jahrestagung der Deutschen Gesellschaft für Unfallchirugie e.V. (Annual Conference of German Society for Accident Surgery, ICC Berlin/German, 22.-25. November 1995.

BUND 94]
Statistisches Bundesamt. Verkehr, Fachserie 8, Reihe 7, Verkehrsunfälle (Traffic Accidents). Wiesbaden.

DECK 92]
I. Decker, L. Grösch, R. Justen, W. Schwede. Realitätsnahe frontale Crash-Versuche and daraus abgeleitete Schutzmaßnahmen in Mercedes-Benz Fahrzeugen. Automobiltechnische Zeitschrift (Professional Automobile Journal) 11/90.

HEYM 96]
A. Heym. Grundsatzuntersuchungen zum Einsatz von adaptiven Insassenschutzsystemen beim Frontalaufprall von Pkw. Diplomarbeit (Master's Thesis) 13/96, Technische Universität Berlin. Currently unpublished.

[PETR 96]
PETRI-internal investigation of biomechanical benefits of multiple stage inflators.

[SCHE 94]
D. Scheunert, W. Jahn, R. Zimmermann. Kleine Ursache, große Wirkung. Optimiertes Rückhaltesystem durch den Einsatz eines Gurtkraft-begrenzers. Bag & Belt ´96, Fourth International Symposium for Car Occupant Restraint Systems; Bonn/Germany, 24.-26. April 1996.

[SEIF 92]
Prof. Dr.-Ing. Ulrich Seifert. Fahrzeugsicherheit. Personenwagen (Driving Safety in Cars). VDI-Verlag, Wolfsburg/Germany, August 1992.

DEFINITIONS, ACRONYMS, ABBREVIATIONS

RS = Restraint System
ODB = Offset Deformable Barrier

		Basis	Variation No. 1	2	3	4	5	6	7	8	9	10	11	12	13	14	15	16	17	18	19	20	21	22	23
Crash Configuration	100%, 0°/ FMVSS 208	✓			✓	✓	✓	✓	✓	✓	✓	✓	✓							✓	✓	✓	✓	✓	✓
	100%, 30°		✓																						
	40% offset, deformable barrier			✓										✓	✓	✓	✓	✓	✓						
Collision Speed (ETS)	30 kph				✓															✓	✓	✓	✓	✓	✓
	40 kph					✓																			
	50 kph / FMVSS 208	✓	✓	✓				✓	✓	✓	✓	✓	✓	✓	✓	✓	✓	✓	✓						
	60 kph						✓																		
occupant´s mass	75 kg / FMVSS 208	✓	✓	✓	✓	✓	✓			✓	✓	✓	✓			✓	✓	✓	✓			✓	✓	✓	✓
	light (-30%)							✓						✓						✓					
	heavy (+30%)								✓						✓						✓				
occupant´s length	1.72 m / FMVSS 208	✓	✓	✓	✓	✓	✓	✓	✓			✓	✓	✓	✓			✓	✓	✓	✓			✓	✓
	small (-10%)									✓						✓						✓			
	large (+10%)										✓						✓						✓		
occupant´s positioning	acc. to FMVSS 208	✓	✓	✓	✓	✓	✓	✓	✓	✓	✓			✓	✓	✓	✓			✓	✓	✓	✓		
	forward (-20%)											✓						✓						✓	
	backward (+20%)												✓						✓						✓

Table 1: Overall view of the parameter variation related to FMVSS 208 [HEYM 96]

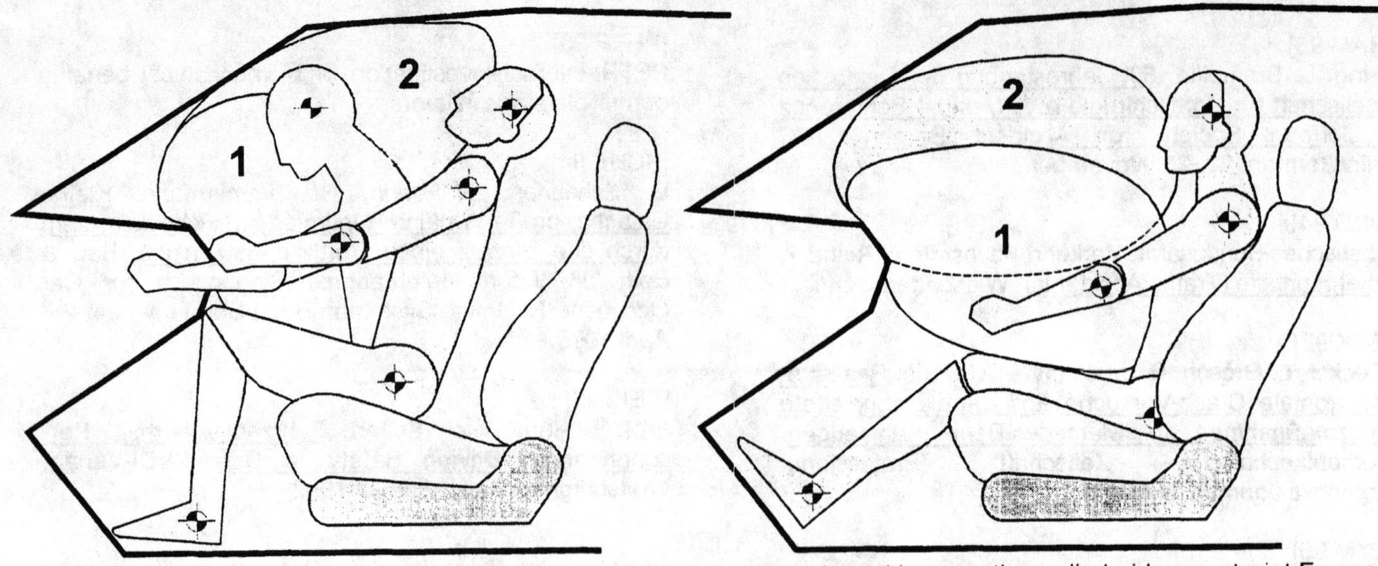

Figure 1: Correct (1) and incorrect (2) airbag deployment - correctly and incorrectly applied airbag restraint Forces

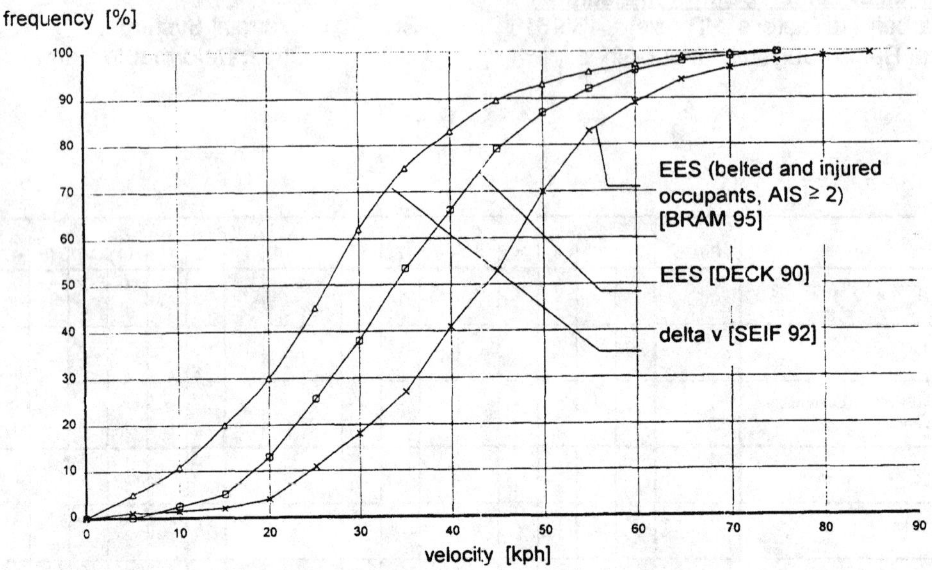

Figure 2: Cumulative frequency of Δv and EES in frontal collisions

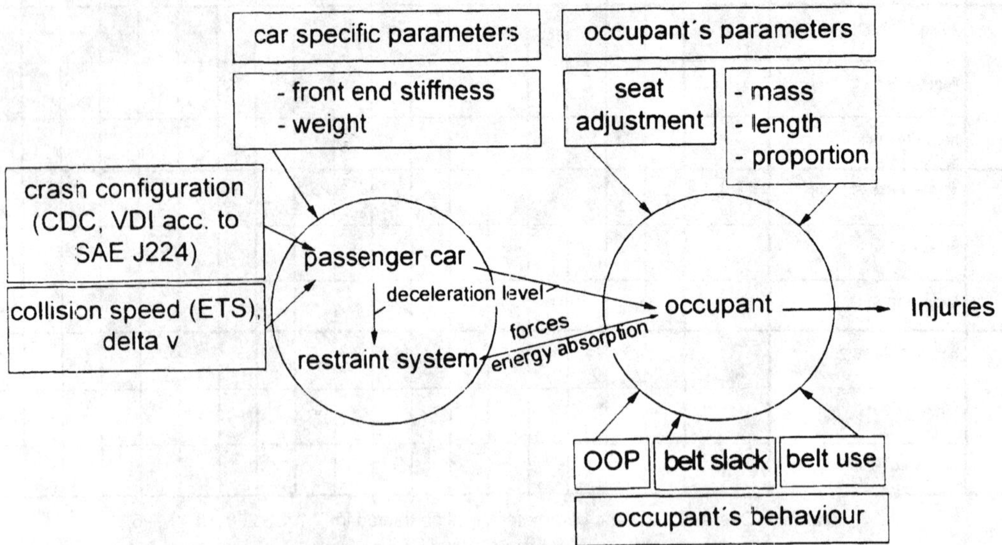

Figure 3: Crash input parameters and systems interactions in frontal crashs

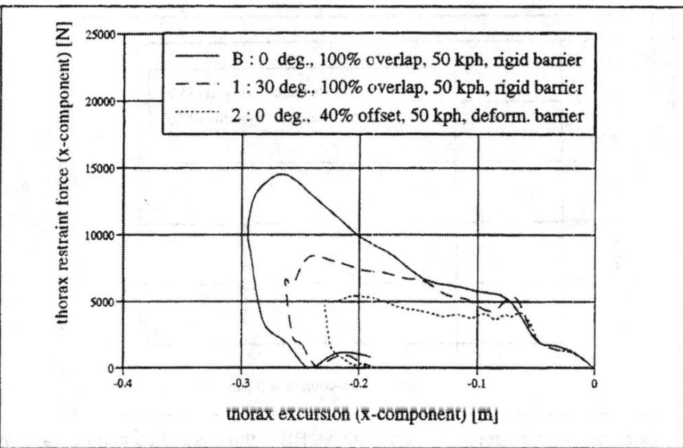

igure 4: Thorax load curves for different crash configurations [HEYM 96]

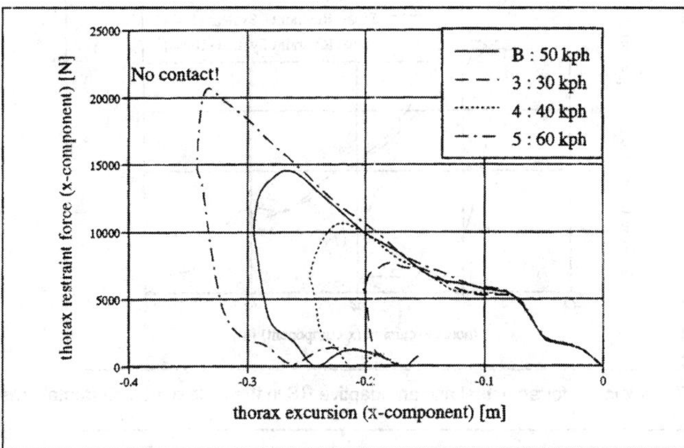

Figure 5: Thorax load curves for different collision speeds [HEYM 96]

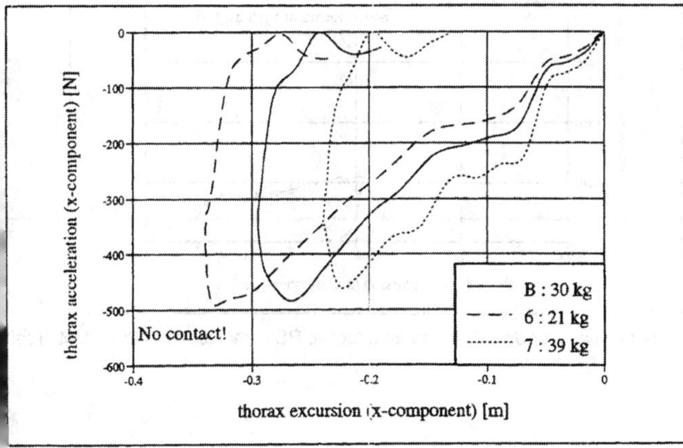

re 6: Thorax acceleration curves for different occupant masses [HEYM 96]

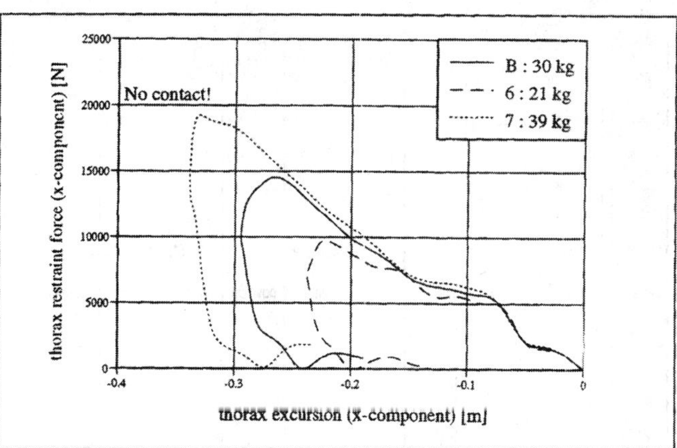

Figure 7: Thorax load curves for different occupant masses [HEYM 96]

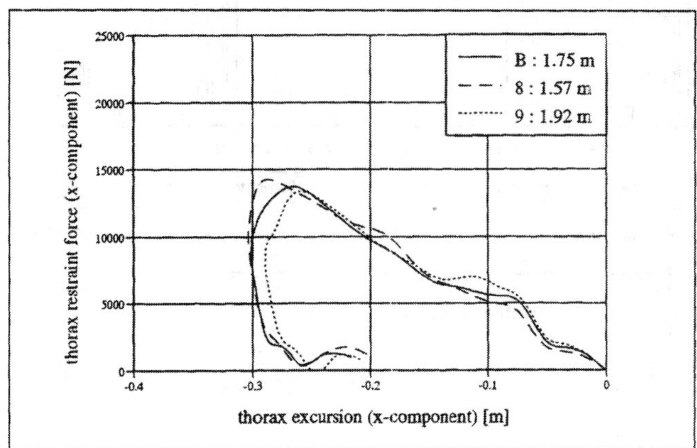

Figure 8: Thorax load curves for different occupant length [HEYM 96]

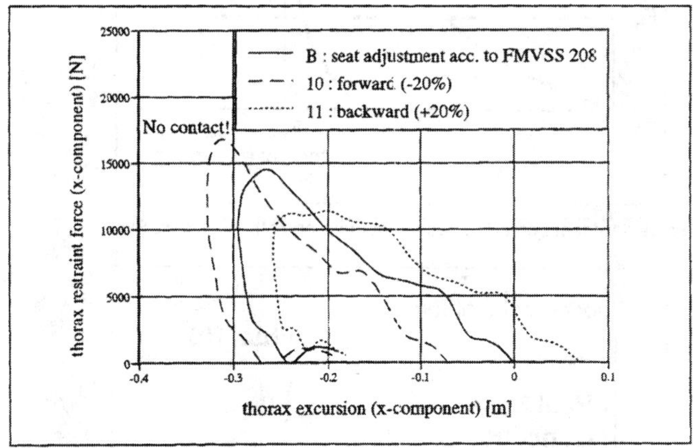

Figure 9: Thorax load curves for different seat adjustments [HEYM 96]

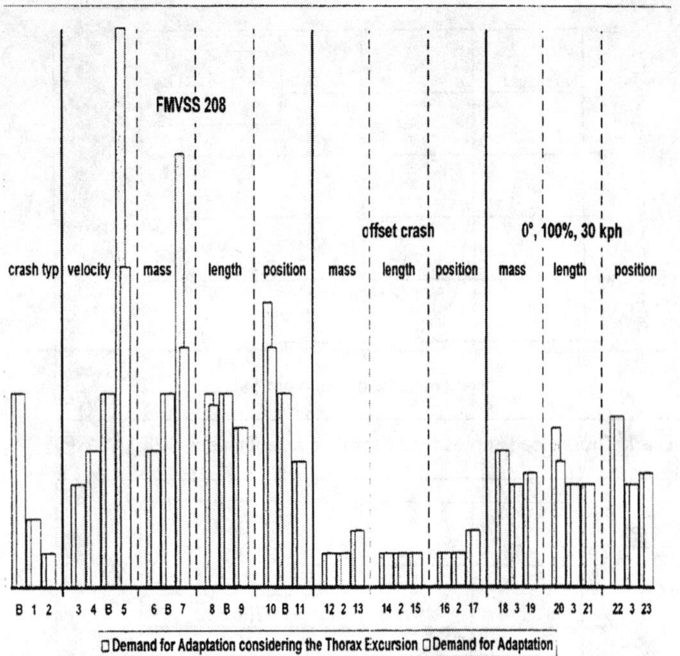

Figure 10: Relation between given force limit (a_{3ms}-Value) per regulation and real force values [HEYM 96]

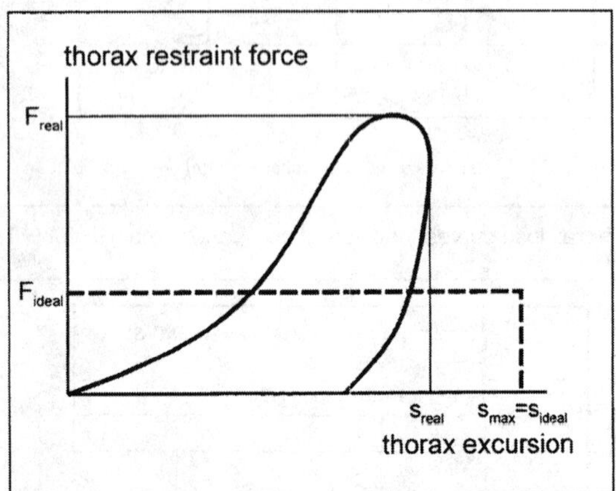

Figure 11: Ideal Force-Displacement Function of the Restraint System

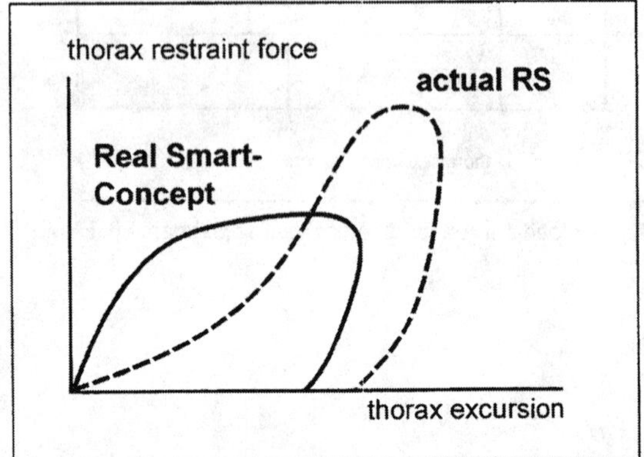

Figure 12: Actual and Real Smart force-fisplacement function of restraint systems

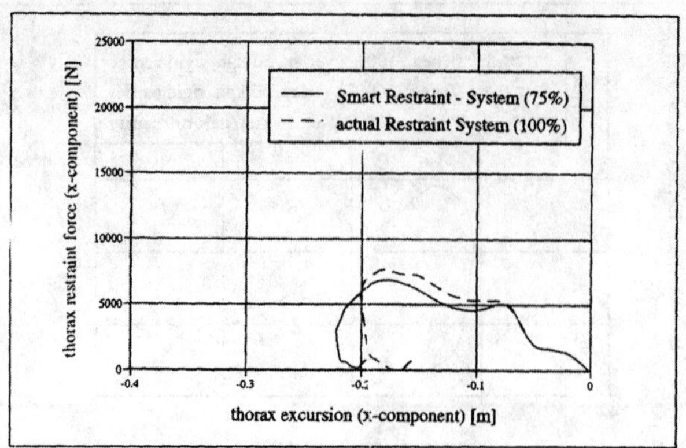

Figure 13: Thorax loads for an actual and an adaptive RS in the case of 30 kph frontal crash

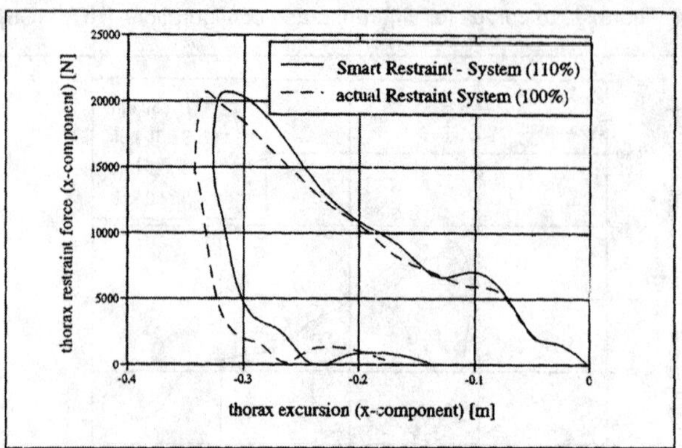

Figure 14: Thorax loads for an actual and an adaptive RS in the case of 60 kph frontal crash

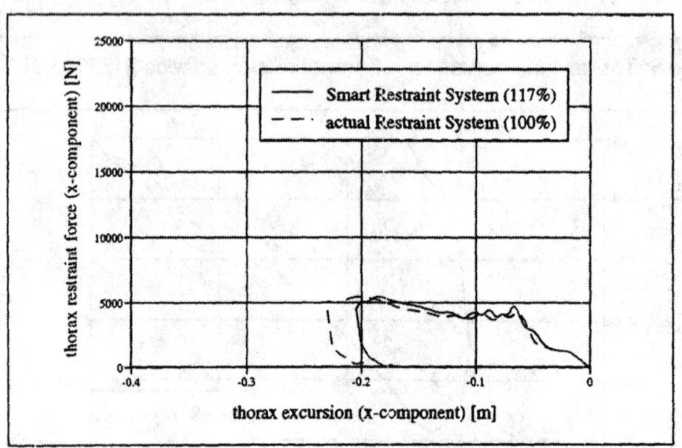

Figure 15: Thorax loads for an actual and an adaptive RS in the case of 40% offset crash

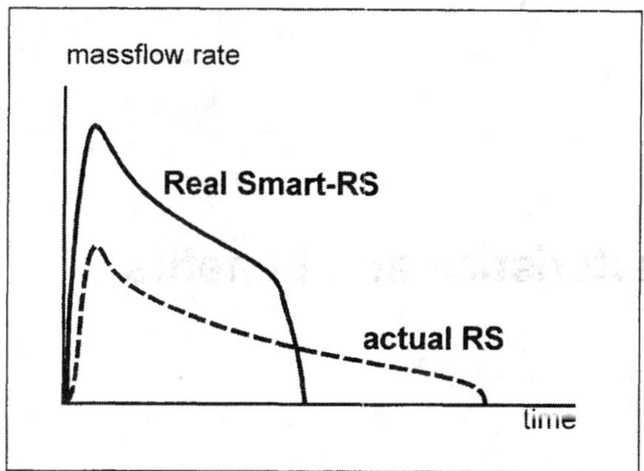

Figure 16: High efficiency massflow rate inflator

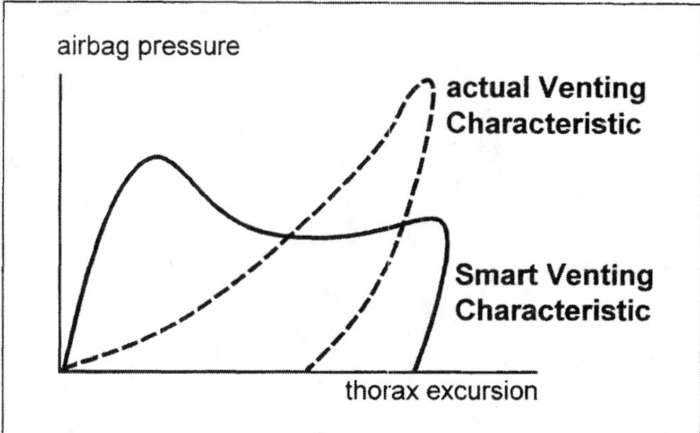

Figure 17: Smart Venting - Smart airbag pressure characteristic

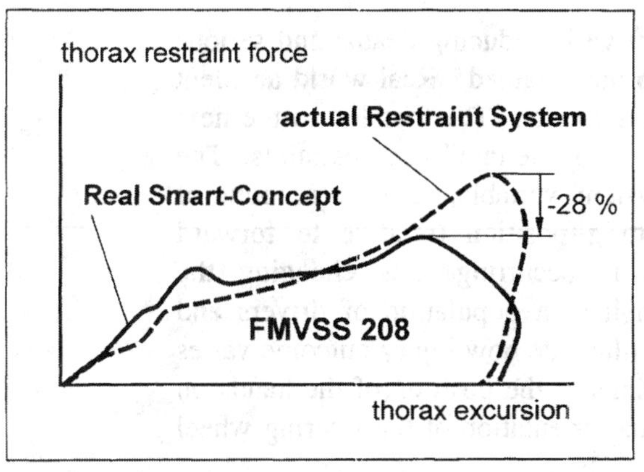

ure 18: Thorax loads in an actual and a "Real Smart" RS for a FMVSS 208 crash

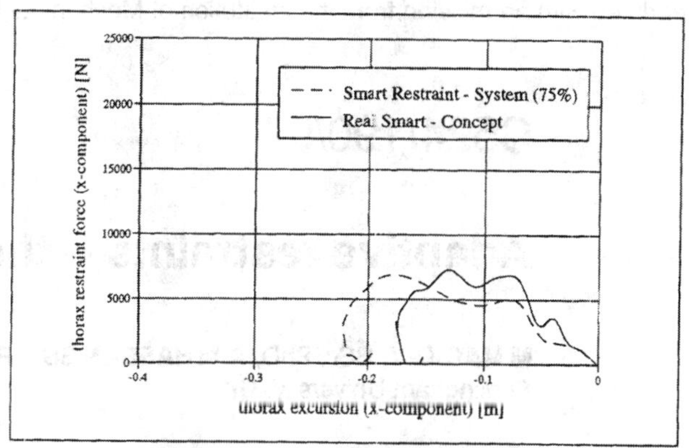

Figure 19: Thorax loads in an adaptive and a "Real Smart" RS for a 30 kph frontal crash

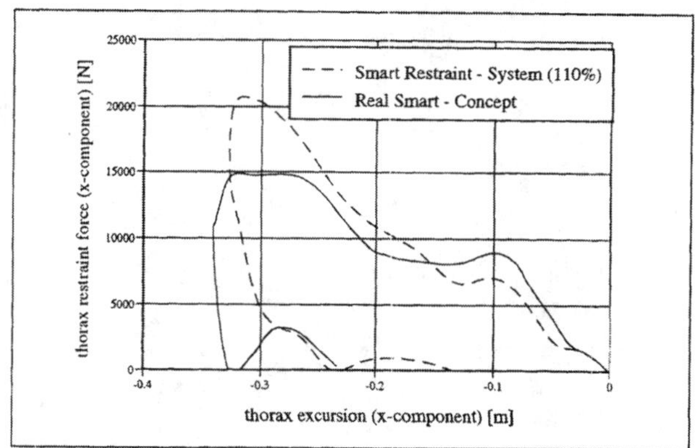

Figure 20: Thorax loads in an adaptive and a "Real Smart" RS for a 60 kph frontal crash

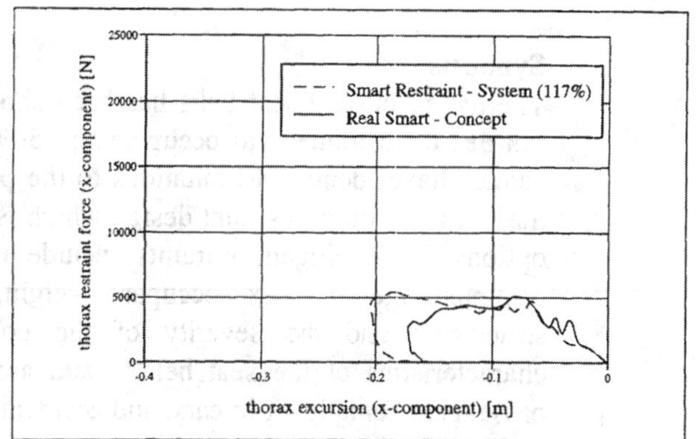

Figure 21: Thorax loads in an adaptive and a "Real Smart" RS for a 40% offset crash

C524/190/97

Adaptive restraints – their characteristics and benefits

M MACKAY, BSc, PhD, **A M HASSAN** BSc, PhD, and **J R HILL**
Birmingham University, UK

Synopsis
The use of current seat belts has been shown to be effective in reducing deaths and serious injuries to restrained car occupants by 50% compared to unrestrained. Real world accident studies have identified limitations to the performance of set belts. This has led to the next major evolution in restraint design which is the development of the intelligent restraints. The options for intelligent retraints include making the system variable, to take account of occupant age and sex, occupant weight, occupant sitting position (relative to forward structures) and the severity of the collision which is occurring thus changing the characteristics of the seat belt. Data are presented on how a population of drivers and passengers actually sit in cars, and accident analyses will illustrate how injury outcome varies with age and sex for restrained occupants. The implications of the position of the hands on the steering wheel during normal driving and the rotational orientation of the steering wheel during an impact for airbag design are also included.

2 INTRODUCTION

Current seat belts have been shown to be very effective in diminishing the frequency and severity of injuries to car occupants. So much so that high levels of seat belt use are a prime aim of all national transport safety policies in motorized countries. The limitations of the protective abilities of current seat belts have been well documented in many analyses of both field accident data and experimental studies (1).

Real world accident studies have identified five categories of limitations to the performance of current seat belts. These are:

1) Head and face contacts with the steering wheel by restrained drivers (2) - It is inherent in the kinematics of a restrained occupant that, in a severe collision at a velocity change of around 50 km/hr, the head will arc forwards and downwards, having a horizontal translation of some 60 to 70 cms Figure 1. If a normal steering wheel position is superimposed on such a trajectory, the head and face necessarily will strike the steering wheel. Such contacts usually produce AIS 1 to 3 injuries and are best addressed with the supplementary airbag systems becoming common throughout the new vehicle fleet.

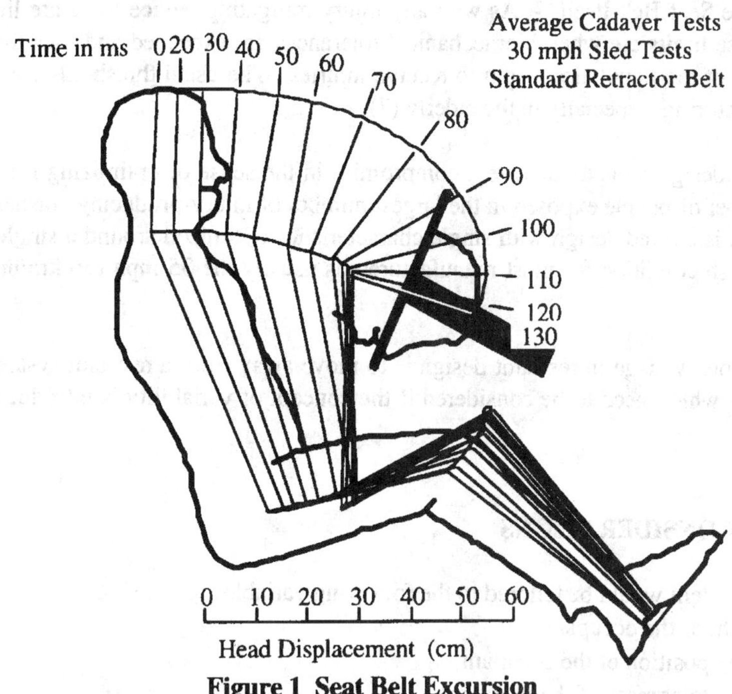

Figure 1 Seat Belt Excursion

2) Intrusion of Forward Structures - A seat belt requires a zone ahead of the occupant so that the occupant can be decelerated by the compliance of the restraint system. If intrusion compromises that space, then specific localized contacts can occur. The injury risk from such contacts may well be small if they are occurring with structures which have been engineered appropriately. Indeed, in the ultimate condition, it is better for the occupant to be decelerated not just by the seat belt alone but through a combination of belt loads and contact loads. Those contact loads are through the feet at the firewall, through the knees into the lower dash and through the airbag and belt at chest level. In severe collisions, however, major intrusions are destroying the passenger compartment so that exterior objects are actually striking the occupants. This is a feature of restrained fatalities in frontal impacts (3).

3) Rear Loading - Correctly restrained front seat occupants can receive injuries from unrestrained occupants, luggage or animals from the rear seats. Such events contribute to some 5% of restrained front seat fatalities (4).

4) Misuse of the Seat Belt - Seat belts must be positioned correctly on the human frame to work effectively. Dejeammes (5) in a survey of belt use in France found that some 1.6% of front seat occupants had the shoulder belt under the arm or behind the back whilst some 3.3% had

introduced slack because of the use of some clip or peg to relieve the retraction spring tension. A more important type of misuse relates to the positioning of the lap section. Many occupants, especially the overweight, place the lap section across the stomach instead of low across the pelvis. Indeed for the obese, it is often impossible to position the lap section so that it will engage on the iliac spines of the pelvis in a collision. These problems are reflected in abdominal injuries from the lap section of the seat belt (6).

5) Injuries from the Seat Belt Itself - As with any injury mitigating device there are limits to effectiveness. Those limits are when biomechanical tolerances are exceeded and thus the most vulnerable segment of the population begin to receive injuries. The usual thresholds are sternal and rib fractures occurring, especially in the elderly (7).

Current restraint design aims to achieve a compromise in the sense of optimizing protection for the largest number of people exposed in the largest number of injury-producing crashes. The end point, however, is a fixed design with single characteristics optimized around a single crash condition. That crash condition for most manufacturers is usually the 35 mph (56 km/hr) rigid barrier crash test.

The next evolutionary stage in restraint design is to move away from a restraint system with fixed characteristics which need to be considered if the concept of variability is introduced into restraint design.

2 POPULATION CONSIDERATIONS

The ideal restraint system would be tailored to the following variables:

- the specific weight of the occupant,
- the specific sitting position of the occupant,
- the biomechanical tolerances of the occupant,
- the severity of the specific crash which is occurring,
- the chances of specific passenger compartment intrusion occurring which might compromise restraint performance
- the specifics of the compartment geometry and crush properties of the car.

3 ANTHROPOMETRIC CONSIDERATIONS

Current dummies and modeling cover the 5th percentile female to 95th percentile male range. Assuming for simplicity that males and females are exposed equally and that there are few males smaller than the 5th percentile female or females larger than the 95th percentile male, these conventional limits put 2.5% (1 in 40) of the small population and 2.5% of the larger population beyond those limits; 5% or 1 in 20 overall.

Table 1 gives the 1% and 99% ranges for height, sitting height and weight. These data show what would be required if the design parameters were extended to cover this wider range, so that only 1 in 50 of car occupants would be outside the design parameters (8).

Table 1 Population Ranges for Height, Sitting Height and Weight

Adult	Height		Sitting Height		Weight	
	ins	cm	ins	cm	lbs	kg
1%ile female	57	145	28	72	82	37
5%ile female	59	150	29	75	90	41
95%ile male	73	185	37	93	225	102
99%ile male	75	190	38	96	236	107

More importantly, it is implicitly assumed in current designs that height (or sitting height) and hence sitting position are colinear with the weight of the occupant. In fact, there are data available to suggest that the relationship between height and weight are rather complex. For example, the body mass index (BMI) (i.e., the ratio of weight in kilograms to height in meters squared) varies to a greater degree in women than in men, and particularly at the 75th percentile and above, women have higher BMIs than men. In addition, the prevalence of overweight increases with age, more with females than males (9).

Therefore to optimize a restraint system it would appear appropriate that sitting position and body weight should be assessed independently if variability is to be introduced into restraint design.

4 POPULATION CHARACTERISTICS BY POSITION IN THE CAR

European data show that some 80% of drivers in injury-producing collisions are male, whilst some 65% of front seat passengers are female (10). Approximately one-third of rear passengers are children of 10 years of age or under (11). These simple frequencies suggest that restraint characteristics should not necessarily be the same for all sitting positions in the car.

4.1 Sitting Positions

Current design is predicated on the positions established for the three conventional dummies. Observational studies by Parkin et al (12) have demonstrated that there are substantial differences between those three positions and an actual population of drivers. Passive observations of drivers in the traffic stream have been made using video recording techniques, and drivers classified by sex and general age groups of young (35 years), middle (36-55 years) and elderly (56 years and older). Make and model of car were recorded and measurements made of the following distances:

- nasion to steering wheel upper rim and hub,
- top of head to side roof rail,
- back of head to head restraint, horizontally and vertically,
- shoulder in relation to 'B' pillar.

Such techniques allow thousands of observations to be made quickly and therefore population contours can be drawn. Figure 2 illustrates how particularly for the 5th percentile female population the actual sitting position is significantly closer than that of the 5th percentile dummy, by some 9.2cm. The 5th percentile, small female population sits some 38cm (15 inches) or closer to the hub of the steering wheel.

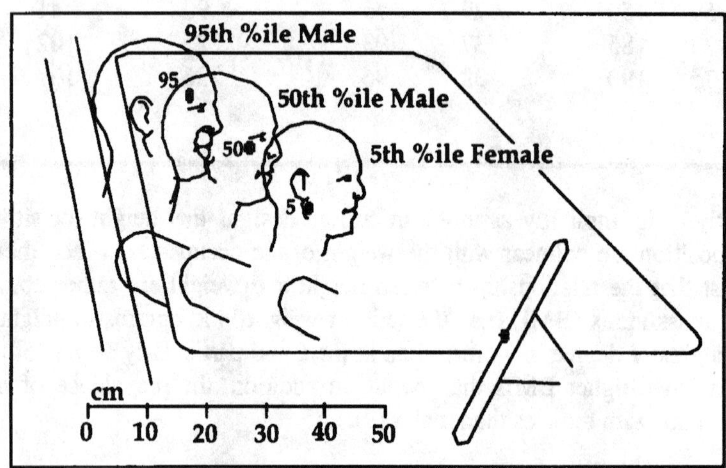

Figure 2 Drivers' Sitting Positions

5th 50th and 95th %ile naison positions are illustrated for "real drivers" (head outlines) and dummies (black spots).

5 BIOMECHANICAL VARIATION

An extensive literature exists concerning human response to impact forces, mostly conducted in an experimental context. A general conclusion from that body of knowledge is that for almost any parameter, there is a variation of at least a factor of 3 for the healthy population exposed to impact trauma in traffic collisions (13). That variation applies to variables which are relatively well researched such as the mechanical properties of bone strength, cartilage, ligamentous tissues and skin. It is likely to be even greater when applied to gross anatomical regions such as the thigh in compression, the thoracic cage, the neck or the brain.

How such variability is demonstrated in populations of collisions is less well understood. Data from a ten year period of the European Co-operative Crash Injury Study (CCIS) for restrained front seat occupants are given in Figures 3 and 4. The methodology of that work has been described elsewhere (14).

Figure 3 illustrates the effect of age on injury outcome in terms of the frequency of AIS 2 and greater injuries for three age groups. Data are presented for frontal impacts involving a principal direction of force (PDF) of 11 to 1 o'clock, controlling for crash severity by equivalent test speed (ETS). Injury severities were rated by Maximum Abbreviated Injury Scale (MAIS;(15).

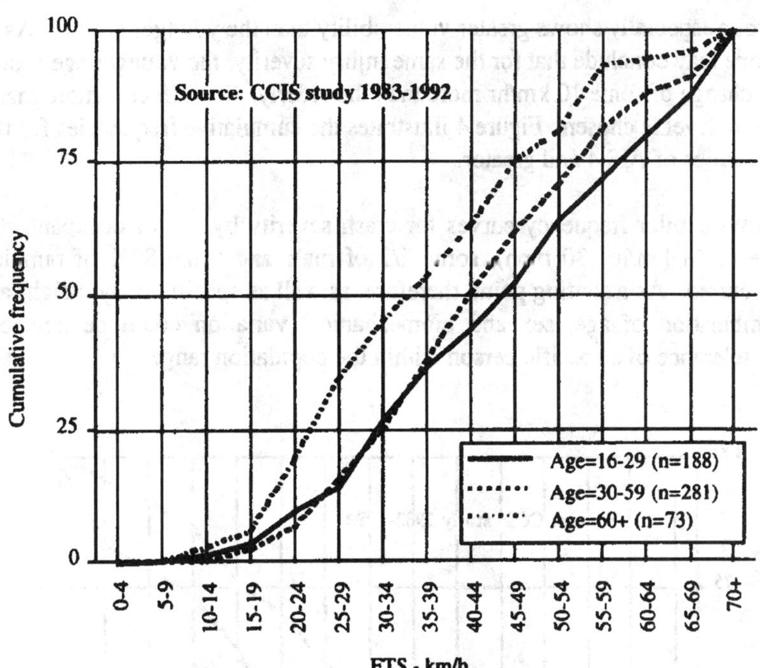

Figure 3 Crash speed distributions for frontal impacts (PDF of 11 to 1 o'clock) to drivers (by age groups) who experienced injuries with a MAIS > = 2

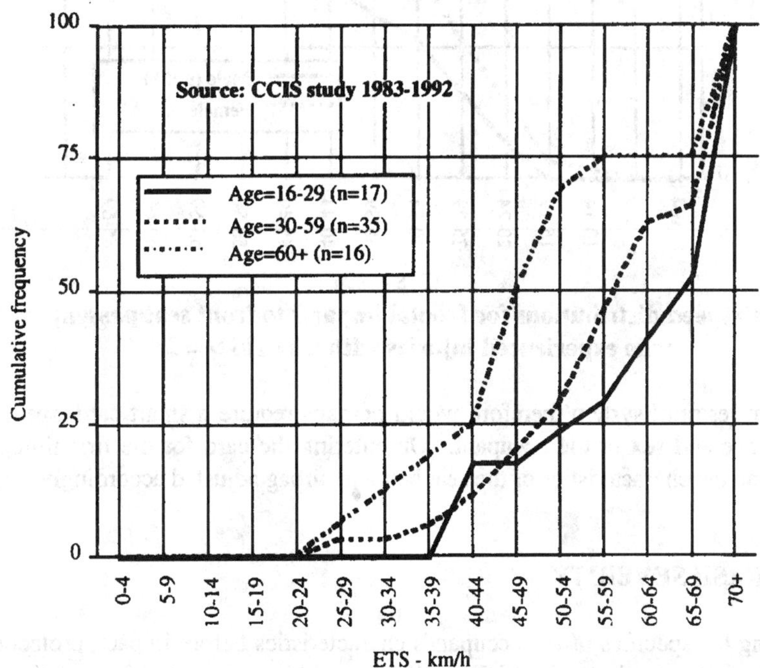

Figure 4 Crash speed distributions for frontal impacts (PDF of 11 to 1 o'clock) to front seat occupants (by age groups) who experienced injuries with a MAIS > = 4

The 60+ age group especially shows greater vulnerability than the younger groups. As a broad generalization one may conclude that for the same injury severity, the younger age groups must have a velocity change of some 10 km/hr more than the elderly. The effect is more marked if a more severe injury level is chosen. Figure 4 illustrates the cumulative frequencies for the three age groups for injuries of AIS 4 and greater.

Figure 5 shows similar frequency curves for crash severity by sex of occupant. Thus at a velocity change of 48 km/hr (30 mph), some 2/3 of male and some 80% of female AIS 2+ injuries have occurred. As a starting point, therefore, as well as specific body weight and sitting position, a combination of age, sex and biomechanical variation could be developed as a predictor of the tolerance of a specific person within the population range.

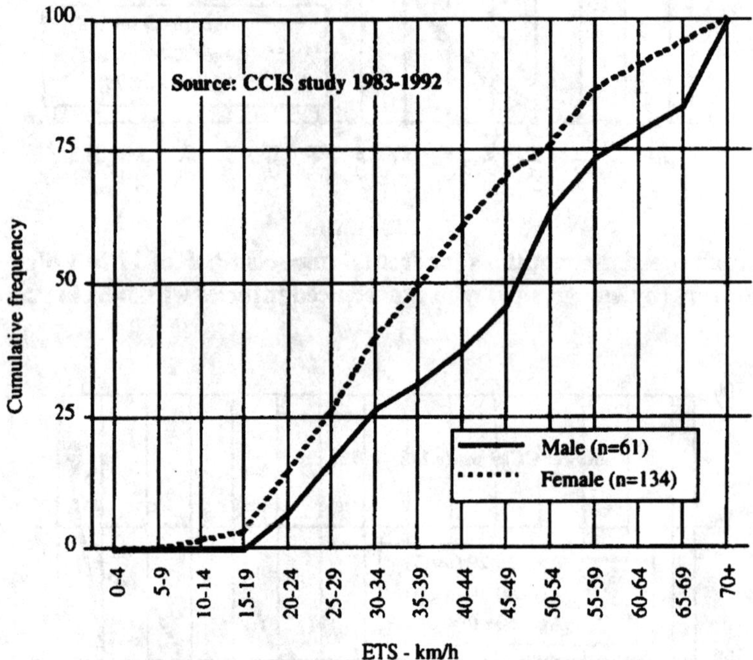

Figure 5 Crash speed distributions for frontal impacts to front seat passengers (by sex) who experienced injuries with a MAIS > = 2

An intelligent restraint system therefore would perhaps require a smart card, specifying the height, weight, age and sex of the occupant. On entering the card for the first time, the card would be read and the characteristics of the seat belt and airbag adjusted accordingly.

6 SENSING CRASH SEVERITY

Besides assessing the specifics of the occupant's characteristics before impact, protection could be enhanced if the nature and severity of the collision could be assessed early enough during the crash pulse so that the characteristics of the restraint system could be modified. That would require, for example, sensors to discriminate between distributed versus concentrated impacts, and between, for example, three levels of collision severity such as less than 30 km/hr, 30 to 50 km/hr, and greater than 50 km/hr. In addition, conceptually one might have an array of sensors which would detect the early development of compartment intrusion. Such electronic data could then instruct the restraint system to change its characteristics early enough during the crash phase to alter the characteristics of the restraint and thus the loads on and forward excursion of the occupant.

7 VARIABLE RESTRAINT CHARACTERISTICS

The advantages of a variable restraint system are illustrated by considering some examples. A front seat passenger, 70 years of age and female, weighing 45 kg sitting well back, in a 30 km/hr frontal collision with no intrusion, would be best protected by a relatively soft restraint system which would maximize the ride-down distance and minimize the seat belt loads. That would require a low pretensioning force, a long elongation belt characteristic provided by load limiters and a soft airbag.

Such a system is very different from what would be required by a 25 year old, 100 kg male, sitting close to the steering wheel in a 70 km/hr offset frontal collision. He would need a very stiff seat belt, an early deploying stiff airbag and a large amount of pretensioning load.

Consider thirdly a 9 year old girl, weighing 30 kg sitting in a rear seat in a 56 km/hr frontal impact. Maximizing her ride-down distance and minimizing the seat belt loads would require low pretensioning loads and a very soft belt system, but one which would still have a biomechanically satisfactory geometry at the forward limit of excursion. Possible techniques for introducing variability into restraint design are now discussed.

7.1 Variable Pretensioning Force

A retractor pretensioner might be devised which would have a variable stroking distance or perhaps two stages of pretensioning to address the population and crash severity requirements outlined above.

7.2 Combined Retractor Pretensioner & Buckle Pretensioner

Such a system of pretensioners might maintain good seat belt geometry especially for the small end of the population, such as the 9 year old girl in the rear seat, when soft restraint characteristics and hence large amounts of forward excursion are required.

7.3 Discretionary Web Locks

If the seat belt system needs to be stiffened for the heavy occupant with high biomechanical tolerance in a high speed crash, then the switching in of a web lock would be appropriate. Such a device would shorten the active amounts of webbing being loaded and diminish forward excursion at the expense of somewhat higher seat belt loads.

7.4 Discretionary Load Limiting Devices

One way of providing for biomechanical variability would be to have a load limiting mechanism which would be calibrated for the specifics of the occupant's age, sex and weight. Such a device could also be adjusted according to transient sitting position. Belt loads would be limited at the expense of increased forward excursion.

7.5 Variable Sitting Positions

Ultrasonic, infrared or other techniques of sensing might be used to monitor continuously the head position of each occupant. Such information could be used at a minimum to provide a warning that an occupant was sitting too far forward and in particular too close to the steering wheel. At a more advanced level it could be used to tune the seat belt and airbag characteristics to be optimized for that occupant in that specific position by adjusting the other restraint variables.

7.6 Variable Airbag Firing Threshold

The need for an airbag varies according to seated positions in the car and the characteristics and sitting position of the occupant. For most drivers in most sitting positions a supplementary steering wheel airbag becomes desirable in crash severities above 30 km/hr (2). For a front seat passenger however, particularly one who is towards the top end of the biomechanical tolerance spectrum and sitting well back, an airbag at 30 km/hr is unnecessary. For a child sitting a long way forward in such a crash, it might also be disadvantageous. Hence specific sensing techniques at a minimum could discriminate between the presence or absence of a passenger, and at the next level assess the need for the airbag to inflate or not.

7.7 Variable Airbag Characteristics

In response to the sensing data about the occupant's characteristics and transient sitting position, and the accelerometer data about the nature and severity of the collision which is occurring, the airbag properties could be varied. Specifically, gas volume and inflation rate could be changed. Compressed gas systems instead of chemical gas generators have the potential for providing those characteristics by having time-based adjustable inflation ports. This requires very advanced sensing and control systems but these aims could well be addressed through future research and development.

8 HAND POSITIONS AND STEERING WHEEL ORIENTATION

In addition to the seating position of the driver before impact, the position of the hands on the wheel and the orientation of the steering wheel at impact need to be considered. These factors may influence airbag characteristics.

8.1 Hand Positions on Steering Wheels

An observational study was carried out which looked at the position of the drivers hands on the steering wheel during normal driving condition on major roads with speed limits of 40 to 60mph (64 - 96km/h)in the UK and US, excluding motoways and freeways. Driving with only one hand on the steering wheel seems to be more common. Fifty eight percent of UK drivers used one hand only, while the proportion was much higher in the US at 70%.. Drivers were more inclined to hold the wheel in the upper semicircle, above the 3 and 9 o'clock positions. When two hand were used both tended to be at same height. The distribution of the positions of the hands on the steering wheel are shown in Figure 6 and Figure 7 where one and two handed positions have been counted together.

A quarter (26%) of the drivers were considered to be at risk of injuries because hands or arms were observed in very close proximity to the airbag module. The risk for US drivers was lower at 16%. Drivers were considered to be at risk of receiving injuries an airbag deployed while hands were (a) at the 1, 11 or 12 o'clock positions, or (b) at the 3 or 9 o'clock positions while resting inside the wheel rim on or near the airbag module. The risk of injury may be increased at junctions where 91% of the drivers in the UK and 98% of the drivers in the US were observed to cross arms while turning the wheel.

Therefore a significant group of the population may be at risk of injuries to the upper extremities if airbags deploy while the steering wheel is held near the top or while turning at a junction. The inclusion of sensors to assess arm position and tight steering manoeuvres at low speeds should be considered with smart restraints.

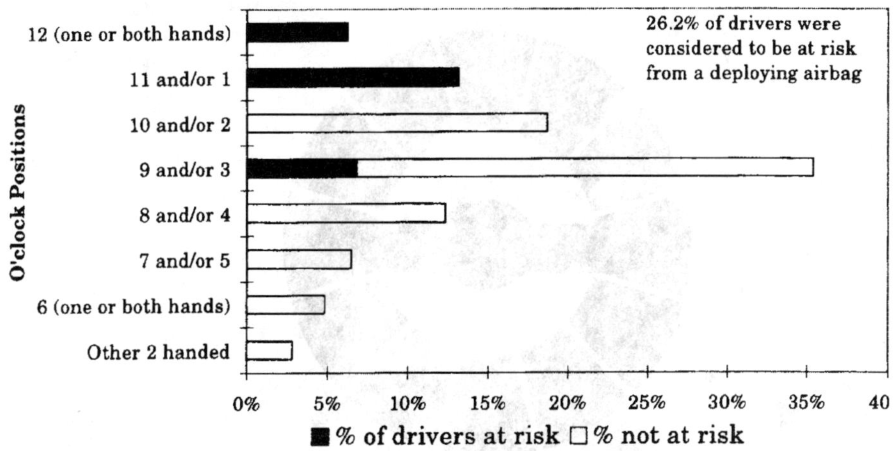

Figure 6 Hand positions on steering wheels observed for 850 U.K. drivers

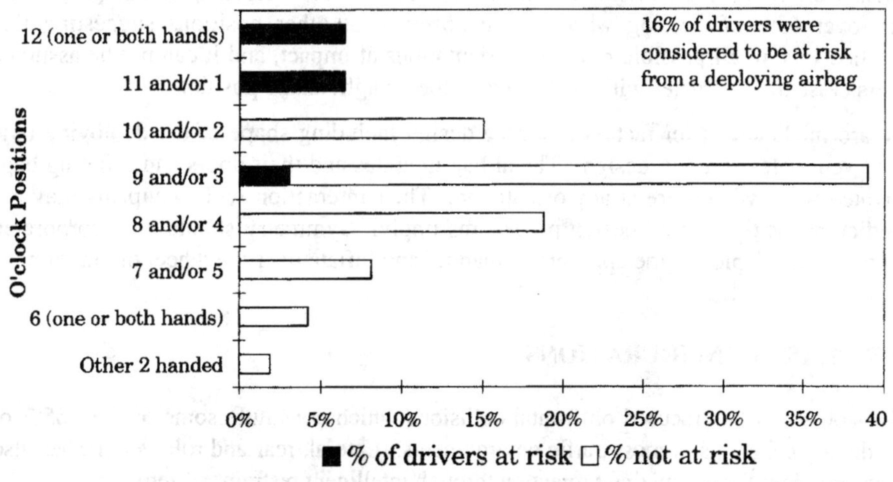

Figure 7 Hand positions on steering wheels observed for 850 US drivers

8.2 Steering Wheel Orientation

Accident records of cars, in the CCIS database, with steering wheels jammed by crushing at the time of impact were examined to determine the rotational orientation of the steering wheel. Only cars involved in single frontal impacts with a principle direction of force between 11 to 1 o'clock and with at least one front road wheel displaced rearwards (strutted) and firmly jammed by crush were included. Steering wheel orientation was assumed not to have changed post impact by considering factors such as degree of strutting, steering wheel damage, steering column damage and orientation of blood stains.

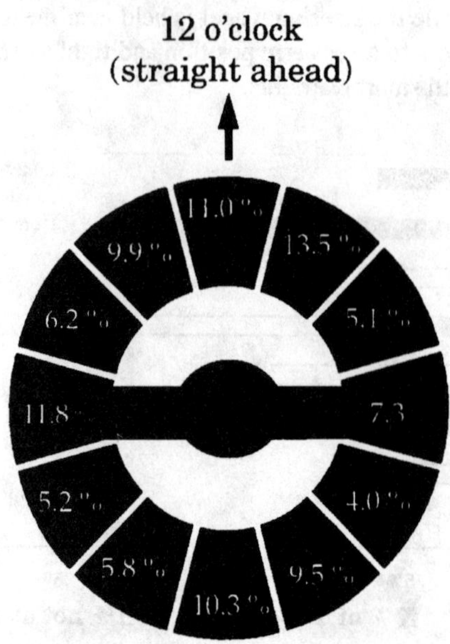

Figure 8 Observed steering wheel orientations grouped into 30° sectors (n = 272)

Wheels were more often orientated in the 11, 12 and 01 o'clock quadrant (Figure 8). However, over 65% of steering wheels were observed at other positions suggesting that steering wheels are in all possible rotational orientations at impact, and it can not be assumed that drivers crash their vehicles with the wheels in the straight ahead position.

There are implications for factors in airbag design including shape of the deploying bag, location of vent holes and port design. The airbag modules and their doors and inflating bags may activated when wheels are at any orientation. Their interaction with occupants may not be as predictable as the current design procedures imply. Symmetry should be incorporated so that components deploy in the appropriate manner and inflate over the wheel orientation.

9 OTHER CRASH CONFIGURATIONS

The discussion so far has focused on frontal collisions which constitute some 50% to 65% of injury producing collisions in most traffic environments. Lateral, rear and rollover crashes also suggest opportunities for optimizing protection through intelligent restraint systems.

9.1 Lateral Collisions

The technology is now developing for side impact airbags with two versions becoming available on 1995 model-year passenger cars. The observational data of Parkin et al (12) have illustrated the range of driver sitting positions which reflect the requirements of side impact airbag geometry to cover both the door and the B pillar. Because a significant part of the population, tall males, choose to sit as far rearward as possible, in a side impact in many four door vehicles the thorax would be loaded by the B pillar rather than the door.

A practical issue is the nature and position of the sensor for a side impact. Because of the extremely short time available for sensing, around 5 milliseconds, a simple switch system is appropriate (16). An analysis of a representative sample of AIS 3 plus lateral collisions has demonstrated that if a switch sensor is located in the lower rear quadrant of the front door then approximately 90% of all such side impacts would be sensed appropriately. A set of several sensors would be required to address the remaining few collisions, whilst rear seat occupant protection would also be addressed in large part by a sensor in the same position in the front door as is appropriate for front seat occupants (17).

9.2 Rear Impacts

Occupant protection in rear end collisions is addressed largely through the appropriate load deflection characteristics of seat backs and the provision of correctly positioned head restraints. The real world data of Parkin et al (12) demonstrates that head restraints are frequently positioned both too low and too far to the rear of the occupant's actual head position. The head position sensors discussed above could also be used for adjusting automatically both the vertical and horizontal position of the head restraint. Such a technology is relatively simple but the costs and reliability, as well as acceptability by the driving population, present serious practical problems.

9.3 Rollover Accidents

Actual mechanisms of injury in rollover accidents have been well researched by Bahling et al (18) for occupants in current seat belts. Conceptually one can suggest that a buckle pretensioner might have some benefits in rollover circumstances by diminishing the relative vertical motion of an occupant. However, in rollovers current dummies do not have the appropriate soft tissue or thoracic and lumbar spine response characteristics, in comparison to the human frame. The basic clearance of current bodyshell design and packaging limit intrinsically the ability of any restraint system to modify the nature of any roof contacts under the forces of actual rollover circumstances even with no roof deformation taking place. Raising current roof lines leads to many undesirable consequences. Nevertheless it would be of interest to explore occupant kinematics in rollovers using more realistic techniques with volunteer and cadaver subjects in the context of buckle pretensioners and the requirements of a sensor to detect incipient rollover.

10 CONCLUSIONS

This paper only attempts to outline in conceptual form some of the issues which need to be addressed in advancing from today's seat belts and airbags towards some form of intelligent restraint system. Of fundamental importance is to recognize the population issues of size, sitting position, biomechanical variation and changing crash exposures. Beyond these issues lies a larger amount of challenging research and development to actually produce the sensors and hardware to provide variability in a seat belt and airbag system. Proximity sensing has its advocates, and if radar techniques could actually discriminate an impending collision from a near miss or a passing object, then the provision of say 500 milliseconds warning would alter many of the restraint issues reviewed in this paper. However, the basic premise remains; the next generation of restraints must change from having single fixed characteristics towards variable ones which recognize the real world population variables of weight, sitting position, biomechanical tolerance and crash exposure.

11 ACKNOWLEDGEMENTS

This project was undertaken on behalf of the Co-operative Crash Injury Study Consortium. The study is funded by the Department of Transport, Ford Motor Company Limited, Nissan Motor Company, Rover Group, Toyota Motor Company and Honda Motor Company. The Project is managed by the Transport Research Laboratory. Grateful thanks are extended to everyone involved with the CCIS data collection process. The authors would also like to acknowledge the considerable effort made by Paul Kidman and Theo Gillam in the preparation of data used in this study.

12 REFERENCES

(1) BACON, D.G.C., The Effect of Restraint Design and Seat Position on the Crash Trajectory of the Hybrid III Dummy, 12th International Technical Conference on Experimental Safety Vehicles, 12:451-457, Göteborg, Sweden, 1989.

(2) ROGERS S., HILL J., and MACKAY M. Maxillofacial Injuries Following Steering Wheel Contacts by Drivers Using Seat Belts. Brit. J. Oral & Maxillofacial Surgery, 30:24-30, 1992.

(3) MACKAY G.M., CHENG L., SMITH M., and PARKIN S. Restrained Front Seat Car Occupant Fatalities. AAAM Proceedings 34:139-162, October 1990.

(4) GRIFFITHS D., HAYES M., GLOYNS P.F., RATTENBURY S., and MACKAY M., Car Occupant Fatalities and the Effects of Future Safety Legislation. Proceedings, 20th Stapp Car Crash Conference, 20:335-388, Society of Automotive Engineers, Warrendale, PA, 1976.

(5) DEJEAMMES M., ALAUZER A., TRAUCHESSEC R., Comfort of Passive Safety Devices in Cars: Methodology of a Long-Term Follow-Up Survey, SAE Paper No. 905199, Society of Automotive Engineers, Warrendale, PA, 1990.

(6) GALLUP B.M., St. LAURENT A.M. and NEWMAN J.A., Abdominal Injuries to Restrained Front Seat Occupants in Frontal Collisions, AAAM Proceedings, 26:131-145, October 1982.

(7) HILL J.R., MACKAY M., MORRIS A.P., SMITH M.T., and LITTLE S. Car Occupant Injury Patterns with Special Reference to Chest and Abdominal Injuries caused by Seat Belt Loading, Proc. IRCOBI Annual Conference, pp. 357-372, Verona, Italy, September 1992.

(8) Society of Actuaries Build and Blood Pressure Study, London, 1979.

(9) WILLIAMSON D.F., Descriptive Epidemiology of Body Weight and Weight Change in U.S. Adults. Ann Intern Med. Oct 1; 119(7 Pt 2):646-9, 1993.

(10) BULL J.B. and MACKAY G.M., Some Characteristics of Collisions, the Population of Car Occupant Casualties and Their Relevance to Performance Testing, Proceedings, IRCOBI 3rd Conference, pages 13-26, Lyon, France, September 1978.

(11) HUELKE D.F., The Rear Seat Occupant in Car Crashes, American Assn for Automotive Medicine Journal 9:21-24, 1987.

(12) PARKIN S., MACKAY M., and COOPER A. How Drivers Sit in Cars. Proceedings AAAM 37:375-388, November 1993.

(13) MCELHANEY J.H., ROBERTS V.L. and HILYARD J.F., Handbook of Human Tolerance, Japanese Automobile Research Institute, Tokyo, 1976.

(14) MACKAY M., ASHTON S., GALER M., and THOMAS P. Methodology of In-Depth Studies of Car Crashes in Britain. Proceedings, Accident Investigation Methodologies, SP159, pages 365-390, SAE Paper 850556, Society of Automotive Engineers, Warrendale, PA, 1985.

(15) AAAM, The Abbreviated Injury Scale. Ill, USA. 1985.

(16) HALAND Y. and PIPKORN B., The Protective Effect of Airbags and Padding in Side Impacts - Evaluation of a New Subsystem Test Method, 13th International Technical Conference on Experimental Safety Vehicles, 13:523-533, Paris, 1991.

(17) HASSAN A., MORRIS A., MACKAY M., and HALAND Y., The Best Place for a Side Impact Airbag Sensor. Proc. AAAM/IRCOBI Conference on Advances in Occupant Restraint Technologies, Lyon, France, September 1994.

(18) BAHLING, G.S., BUNDORF R.J., KASPZYK G.E., MOFFATT E.A., ORLOWSKI K.R., and STOCKE J.E. Rollover and Drop Tests. The Influence of Roof Strength on Injury Mechanisms Using Belted Dummies, Proceedings 34th Stapp Car Crash Conference, 34:101-112, Society of Automotive Engineers, Warrendale, PA, 1990.

CHAPTER 9:

STARTER INTERLOCK

OPTIMIZING SEAT BELT USAGE BY INTERLOCK SYSTEMS

Thomas Turbell
Swedish Road and Transport Research Institute (VTI)
Torbjörn Andersson
AUTOLIV Development AB
Anders Kullgren
FOLKSAM
Peter Larsson
Swedish National Road Administration
Björn Lundell
VOLVO Car Corp.
Per Lövsund
Chalmers University of Technology (CTH)
Christer Nilsson
SAAB Automobile AB
Claes Tingvall
Swedish National Road Administration
Sweden
Paper Number 96-S1-O-07

ABSTRACT

Seat belts are known to be very effective, reducing the risk of injury by approximately 50% when used. Such high effectiveness is, however, based on the fact that all car occupants use the available belts. In several studies it has been shown that, in severe accidents, the seat belt use was less than 50%.

In order to increase the wearing rate more drastic solutions than information, legislation etc. have to be used. A Swedish group, representing government, research, insurance companies, car and restraint systems industry has approached the problem by proposing a smart system that will force car occupants, that normally are unbelted, to use the seat belts by systems that will interfere with the normal use of the car. Different technical approaches, which not in any way will interfere with the normal belt user, will be put forward and evaluated. The problem will also be discussed from a cost-effectiveness point of view and the potential of saving lives in an international perspective will be analyzed. It is shown that more than 6.000 lives could be saved per year in the European Union if the existing seat belts were used.

INTRODUCTION

For a couple of years there has been a concern within the Swedish road safety community about the fact that the safety potential of the seat belts is not fully used.

One of the first alarms came in a report from 1992 where fatally injured car occupants in Stockholm were studied. (Kamrén, 1992). The belt use in this group was only 40% compared to 80% in the general population.

At the last ESV Conference preliminary thoughts on a seat belt interlock system were presented by the Folksam Research Group. (Kamrén, 1994)

Another Swedish study of the belt use and the injuries showed again that the belt usage rate among severely injured was 50% in rural and 33% in urban accidents. (Bylund 1995)

In a Finnish study (Rathmayer, 1994) a clear pattern of the behavior of non-users was observed. Seat belt users committed one traffic offense in every 13 km on highways while the non-users committed one offense every 5,5 km. In urban traffic the distance between offenses was 9 km for the belted and 2,5 km for the unbelted. Non-users also drove faster and had driving histories with longer violation records than the control drivers.

Seat belt usage rates are generally observed in daylight. It can be assumed the rates are lower in darkness and in other situations where the risk for being caught without a belt is lower.

There are also a number of international research results that confirms these findings regarding the belt use and the non-user.

Being aware of that further campaigns and enforcement could only have a limited effect and that technical solutions like automatic belts were not realistic the Swedish National Road Administration last year formed a group of people representing the administration, research, insurance and car industry in order to analyze the situation and propose solutions. This paper reflects the thoughts of that group so far.

THE US EXPERIENCE

Because of various delays in introducing mandatory automatic protection in the USA in the beginning of the seventies the starter interlock requirement was introduced for the period August 15, 1973 until August 15, 1975 for vehicles without automatic protection produced during that period. These systems were connected to both front seats in such a way that if any front seat belt in an occupied seat was not locked, the starter was disabled. If a buckle was opened later a buzzer-light system was activated. All 1974 model year cars sold in the United States came with this ignition interlock except a few thousand GM models that came with airbags that met the automatic protection requirement.

In March 1974 NHTSA described the public reaction to the ignition interlock as follows:" Public resistance to the belt-starter interlock system currently required has been substantial with current tallies of proper lap-shoulder belt usage at or below the 60% level. Even that figure is probably optimistic as a measure of results to be achieved, in light of the likelihood that as time passes the awareness that the forcing systems can be disabled, and the means for doing so will become more widely disseminated,....". There were also speeches on the floor of both houses of Congress expressing the public's anger at the interlock system. On October 27, 1974 President Ford signed into law a bill that prohibited any Federal Motor Vehicle Safety Standard from requiring or permitting the use of any seat belt interlock system. NHTSA then deleted the interlock option from October 31, 1994.

Thus the interlock systems were required in the USA for 14½ months instead of the 24 months that were originally intended. (Kratzke. 1995)

LESSONS TO BE LEARNED FROM THE US EXPERIENCE

The failure of the interlock systems in the USA 1974 can be explained by the following factors:

- Many people felt that it was an infringement of personal freedom. This is probably a typical US reaction that may not be valid for e.g. the European market.
- The voluntary seat belt use was very low.
- The belt systems that were used in the USA at that time were usually difficult to use and had a bad fit.
- The interlock system itself was too unsophisticated. It did not allow low speed maneuvers or sitting in the car with the engine idling.

NEW APPROACHES

Basic principles

Some basic principles for a new system have been established:

- The normal seat belt user shall not notice the system.
- It shall be more difficult and cumbersome to cheat on the system than using the belt.
- Permanent disconnection of the system shall be hard to make.
- The system must be very reliable and have a long lifetime.
- All seating positions in the car shall be covered by the system.
- The accident risk must not increase by any malfunctions in the system.
- Retrofit systems for old cars should be available.

Detection and processing

One input to any interaction system is the situation in the car. Which seats are occupied and are the belts properly used on these seats?

The basic sensor for an occupied seat is a contact that will detect a certain load on the seat. This concept can give false signals from e.g. luggage on the seats. Modern techniques with photocells, IR-detectors, inductance, pattern recognition and load measurements on the seatback can be used to overcome most of the problems.

To determine if the belts are properly used is maybe more complicated. The US experience showed that simple systems like a switch in the buckle or measuring the amount of webbing coming out from the retractor could easily be tampered with. There is more information available from the belt system that could be used e.g. angles and forces at anchor points. Also sophisticated systems like pattern recognition or transponders in the webbing could be used.

Information from the doors, the seats and from the belt system can be combined and analyzed in such a way that the proper conclusions can be made.

Other problems that must be considered are how child restraint systems will work in this new environment and how to handle the situation when a passenger disengages his belt during travel. In that case it is probably not possible to influence the behavior of the car other than gradually and after some proper warnings to the driver.

Interaction

Several ideas for interaction could be considered, of which some are presented here.

The starter-interlock as used in the USA is the most aggressive solution. As mentioned above there are several shortcomings with this system so it is not on the agenda for the new approaches.

External visual signals is a new concept that is worth considering. By flashing the headlights or the hazard warning flashers the surrounding traffic (and the police) will notice the vehicle with non-belt users. The social pressure and the risk of being caught ought to be a good incentive to use the seat belts.

Internal light and sound warnings are used already in cars today but they can be made more aggressive and more directed to the individual non-user.

Interactions with comfort and audio systems is another approach that is discussed. This is a "soft" countermeasure but by disabling the radio, the air-condition, opening the windows etc. some users may get the message.

Throttle pedal feedback can also be used so that the force on the pedal will increase at a certain speed. This will make it possible but very tiresome to exceed that speed. Another solution may be to introduce severe vibrations in the pedal at a certain speed level.

Maximum gear level makes it impossible to put in any gear than number 1 and reverse. This takes care of one of the main faults with the US starter interlock which made it impossible to garage the car without using the seat belt. It also makes it possible to remove a stuck car from e.g. a railway crossing or a burning garage.

Maximum speed is a similar solution to the maximum gear level. The limit that is discussed so far is 30 km/h.

The final solution may be a combination of these systems i.e. the sequence can start with a visual and audible warning and then increase in intensity and finally reduce the maximum speed.

POTENTIAL EFFECTS

Sweden has got one of the highest seat belt use rates in the world with a front seat use of about 88% in observational studies. Other countries in Europe have a marginally higher use with UK in top with 91%.

In Sweden, the 88% use is to be compared to the less than 50% use among fatalities. The following table on the number of fatalities can be derived from the present situation in Sweden.

Table 1.
Seat belt use among fatally injured car occupants in Sweden 1994, based on a sample of 32 cases, and estimated number of fatalities with 100% belt use

	1994	with 100% belt use
Seat belt used	155	272
Seat belt not used	234	-
Total	389	272

	1994	with 100% belt use
Saved lives in relation to current seat belt use in Sweden (50% effectiveness)		117
Saved lives in relation to 0% seat belt use in Sweden (50% effectiveness)	155	272

The potential number of savings is 272 fatalities per year, but we have only come to a level where we have used 57% of the potential savings. This also shows that we have a higher benefit per user from the last 15% than we have had from the 85% seat belt use that we have today. This is different from other areas where the major benefits comes from the first part of an investment and with a decreasing marginal benefit. The relation between the seat belt usage rate in the population and the potential effect based on the Swedish situation can be described by the curved line in the following curve. A low usage rate gives a very limited effect since these individuals drive very safe anyway. The last 10% probably represents the most accident prone group so this is where we find the largest benefits from the belt use. The straight line describes the common belief that there is a linear correlation between the usage rate and the effect.

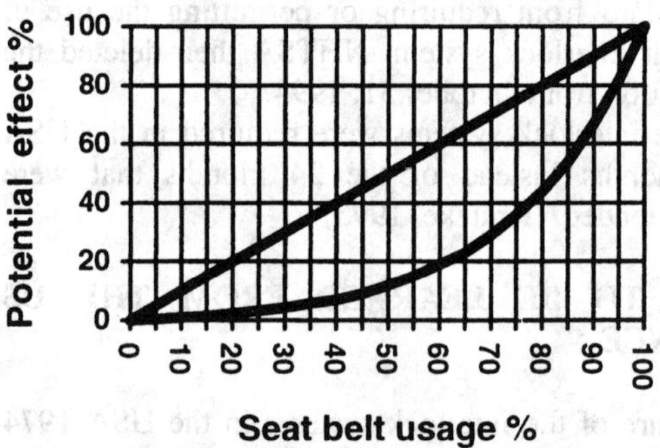

Figure 1. Correlation between seat belt usage and the potential effect on the fatalities.

If we use the Swedish figures and assume that the total European situation is not better, 15.200 unbelted occupants are killed every year in Europe. With a 100% seat belt use and a 50% injury reducing effectiveness, the total number of savings is around 7.600 per year. Given the Swedish situation, this is probably not an overestimation, although the potential savings may vary from country to country. It must be remembered that these figures apply only if the whole vehicle fleet is equipped with an interlock system.

A recent study from the European Transport Safety Council (ETSC) shows similar results with a potential reduction of 5.570 fatalities by a 95% belt usage rate.

Table 2.
ETSC estimations of seat belt use potential

Country	Killed car occupants 1993	Belt usage % Front seat 1991-95	Potential number of saved lives
Austria	747	70	175
Belgium	1050	55	277
Denmark	254	92	58
Finland	274	87	53
France	6168	85	1243
Germany	6128	92	1097
Greece	781	63	199
Ireland	187	53	51
Italy	3931	~55	998
Luxembourg	54	71	14
Netherlands	615	73	139
Portugal	1140	~63	234
Spain	3606	~75	834
Sweden	389	90	69
United Kingdom	1835	91	329
EU Total	27159	80	5570
USA	21987		

(IRTAD 1993) (ETSC 1996)

Only fatalities are discussed in this paper. An interlock system will of course also have a similar effect on the number of severely injured which is about 10 times larger than the number of fatalities.

ALTERNATIVE MEASURES

Preliminary calculations of the cost-effectiveness of an interlock system show that this is a very effective measure compared to some other ones.

By using the Swedish calculations of the willingness-to-pay for risk reductions, it is possible to calculate the possible economic benefits for an interlock system. It can be estimated that the savings from interlock in Sweden is in the region of more than 5 billion SEK/year (~700 million US$/year). With the medium age of cars that we have in Sweden for the moment , the cost that can be spent on each car for an interlock system is therefore approximately 20.000 SEK (~3.000 US$). With an anticipated cost of 200 SEK (~30 US$) per car for an interlock system, the ratio between benefit and cost is 100:1, which by margin is higher than for any other known safety measure. As an example, a 100% fitting of airbags from now on in Sweden would save approximately 50-60 lives annually, but for a cost that is ten times higher than for the interlock, still leaving us with a positive balance between cost and benefit, but serving as an indicator of the extreme benefits of interlock.

ATTITUDES

Preliminary results from a study made by the Swedish National Road Administration in 1995 based on interviews with 5914 persons aged 15-84 years show that there, in general, is a positive attitude for introducing interlock systems.

Table 3.
Swedish interviews

Do you agree or disagree that cars should not be able to run faster than 30 km/h if the driver is not using the seat belt?

%	Male	Female	Total
Strongly agree	24,1	36,6	30,2
Agree	17,7	19,8	18,7
Neither / or	14,1	15,1	14,6
Disagree	19,4	13,7	16,6
Strongly disagree	24,8	14,8	19,9
Total	100,0	100,0	100,0

Table 4.
Swedish interviews

Do you agree or disagree that cars should be equipped with buzzers and lights to warn that someone is not using the seat belt?

%	Male	Female	Total
Strongly agree	37,1	51,8	44,5
Agree	25,4	26,4	25,9
Neither / or	13,4	9,4	11,4
Disagree	12,5	7,2	9,8
Strongly disagree	11,6	5,3	8,4
Total	100,0	100,0	100,0

These two tables show that women are more positive than men and that the less aggressive buzzer-light system is preferred.

This investigation also shows that older persons are more positive to interlock systems than younger persons.

An alarming fact is that of those who state that they seldom or never buckle up in the front seat on rural roads we can find that 77% disagree or strongly disagree on a 30 km/h speed limiting interlock. 55% of this group are also against the buzzer and light warning system. This is actually our target group so we need to find out how to change their attitudes and how to prevent them from disconnecting the interlock system.

INCENTIVES FOR INSTALLATION

Since a legislation on a national level is difficult or impossible after Sweden has become a member of the European Union, other ways to have these systems installed in new and existing cars have been discussed.

- A majority of new cars in Sweden are bought as company cars. There is a possibility to lower the tax liability of the benefit in kind on cars if they are equipped with interlock systems.

• The general vehicle tax can also be moderated depending on the safety equipment of the car.

• Insurance companies are discussing to adopt the premiums along these lines.

• Another proposal is that drivers that are caught without using the belt will be obliged to install an interlock device in their cars.

FUTURE ACTIVITIES

Attitudes

During the spring of 1996 a survey of non-users will be made in Sweden. In cooperation with the police, non-users will be stopped and interviewed at 11 locations spread over Sweden. The interviews will concentrate on seat belts in general and reasons for non-wearing in particular.

System specification

A technical specification, probably in the format of a draft ECE-Regulation will be made during 1996. This draft will allow the car manufacturers to use different options but the goal will be a 99% belt use rate. Also retrofit systems for existing cars will be considered.

An interesting alternative to have detailed technical specifications is to measure the actual belt use rate in traffic for the different cars model years. The tax and other benefits could then be applied with about a one year delay from the introduction of a new car. This alternative will give the vehicle manufacturers free hands to do whatever they want to increase the use in their vehicles. If the coupling between the usage rate and the benefits for the car industry and the car user are strong enough this approach could lead also to a voluntary installation of interlock systems in the existing car fleet.

Implementation

Since this concept has been very positively accepted by the Swedish Ministry of Transport, discussions will go on in order to find the proper tax and other incentives to have some kind of system implemented on the Swedish market as soon as possible.

The car industry may object to this since they do not want to have different equipment on different markets. An international standard would of course be better for everybody but considering the time it will take - and the number of unbelted people killed during that time - our position is that we ought to do something wherever it is possible to get something done quickly in this field.

CONCLUSIONS

At least 6.000 lives can be saved in the European Union annually if the seat belts that are already in the cars are used by 100% of the occupants. The only way to reach this level is to have a technical solution that will make it impossible or very cumbersome to use the car without using the seat belts. There are several technical solutions available that could be implemented in a short time. The main obstacle to reach this goal is probably of a political nature.

REFERENCES

Bylund, Per-Olof; Björnstig, Ulf, Låg bältesanvändning bland allvarligt skadade bilister. Rapport nr 54, Olycksanalysgruppen Umeå, 1995

ETSC Seat belts and child restraints, Brussels, 1995

IRTAD, International Road Traffic and Accident Data, BASt, Cologne 1995.

Kamrén, Birgitta; Kullgren, Anders; Lie, Anders; Tingvall, Claes; The Construction of a Seat Belt System Increasing Seat Belt Use, ESV paper 94 S6 W 32, Munich 1994

Kamrén, Birgitta, Seat belt usage among fatally injured in the county of Stockholm 1991-1992, Folksam, Stockholm 1994

Kratzke, Stephen R, "Regulatory History of Automatic Crash Protection in FMVSS 208". NHTSA, SAE Technical Paper Series 950865. February 1995.

Rathmayer, Rita; Mäkinen, Tepani; Piipponen Seppo, Seat belt use and traffic offenses, Poliisiosaston julkaisuja 12/1994.

SECTION 3:

SEAT BELT RESTRAINT ISSUES

CHAPTER 10:

COMFORT AND FIT

890883

FMVSS 208 Belt Fit Evaluation–Possible Modification to Accommodate Larger People

Howard B. Pritz and Marian S. Ulman
National Highway Traffic Safety Admin.

ABSTRACT

This study was performed to examine the belt fit problem as it relates to various size vehicles, predict the additional seat belt length necessary to accommodate up to the 99th percentile person, and evaluate the safety aspects of additional belt length.

Ten vehicles that had been reported as having a belt fit problem were tested. Two were found to just meet the standard. Most had 6-7 inches of belt remaining even with the seat in the forwardmost position.

A series of vehicles were tested for belt length required as a function of occupant weight using ten subjects. It was determined that each additional inch of belting will increase the accommodated body weight by 7.5 pounds. Thus, to accommodate up to the 99th percentile person weighing 260 pounds or 45 pounds above the standard 95th percentile person requires an additional 6 inches of belt. It was also determined that the belt length gained by seat travel is 2-3 times the seat travel or about 15-17 inches of belt.

Six crash simulations were performed with the Hybrid III dummy at 30 mph in the passenger seat to evaluate the change in dynamics and excursions with additional belt lengths. Up to 10 inches of belt length above the minimum required were tested. It was found that the peak accelerations increased only nominally. The amount of belt spoolout increased from 1.6 to 3.0 inches. The total excursion of the head of the dummy increased 3 inches from about 18 to 21 inches.

THE LEGISLATION OF SEAT BELT use in many states has caused a large portion of the U. S. population to use seat belts for the first time. Some heavy people in these states are finding it difficult to comply with the law because their belts do not fit, or if they do just fit, are very restrictive and uncomfortable.

Seat belts are required by Federal Motor Vehicle Safety Standards 208 and 209 to adjust to fit, at least, occupants ranging in size from a 5th percentile adult female (weighing about 102 pounds) to a 95th percentile adult male (weighing about 215 pounds) with the seat in any position. Persons larger (in girth) than the 95th percentile male, particularly those short enough to require the seat to be positioned at the forward extreme of travel, are not within the range covered by the regulation, and therefore, are not accommodated in the belts of some motor vehicles. More people have difficulties in winter because winter clothing and heavy coats require even greater lengths of seatbelt webbing.

The difficulties encountered with safety belt length are not new, but in the past, if the belt did not fit, it usually was not worn. While seat belt wear was entirely voluntary in the U.S., only about 10% of the population wore them. For the few large people who did want to wear safety belts, extenders could be made. An extender is an additional length of belt with connectors on either end to fasten into a vehicle's existing seatbelt buckles. Extenders are custom-made for a specific person in a specific car. They are not transferrable to another car, and they must be removed for any other occupant of the seat.

The objectives of the study were to (1) identify the belt fit problem as it relates to vehicles of various sizes, (2) estimate the additional belt length necessary to accommodate up to the 99th percentile occupant, and (3) evaluate the safety effects of adding additional belt length.

RESULTS

OCCUPANT ANTHROPOMETRY - Data from the 1982 NASS, National Accident Severity Study, were used to examine height and weight percentiles for U. S. adults. After eliminating persons less than 18 years of age or older than 79 and those cases with missing or obviously erroneous anthropometric data, over 13,000 cases remained. The files yielded 8710 male data points and 4584 cases involving females. This data was handled first by sex and then as a set representative of the entire U.S. driving/riding public. Though the data set contained twice as much data for males as for females, the difference was not compensated in the calculations of averages and percentiles. The ratio of males to females is generally representative of the ratio of the sexes in the riding/driving public.

The overall distribution of the number of people in each height-weight category as indicated by the NASS data is shown in Table 1. The overall 99th percentile weight calculated from NASS data is about 260 pounds. This breaks down into a 99th percentile weight of 270 pounds for males and 230 pounds for females The data from NASS is comparable to figures obtained from the Health and Nutrition Examination Survey (HANES) [1]*, and the 95th percentile weight calculated from NASS agrees well with the weight incorporated into the current 95th percentile anthropormetric dummy. Belt fit problems are most common in persons who are short as well as fat, and require the seat in the full forward position. The number of short people in the various weight percentiles is shown in the table.

BELT FIT TESTS WITH 95TH PERCENTILE DUMMY - Belt fit testing was performed on nine vehicles. All of the automobiles were selected from or were representative of vehicles reported by one or more individuals as having belts that were too short. To evaluate the relationship between vehicle size and belt difficulties, the vehicles were from various size classifications.

Table 2 indicates the basic data for the vehicles tested using the 95th percentile dummy. All of the vehicles met the standard except the Vanagon which has a very large seat travel. The Accord and 4 door Chevette just fit the 95th dummy with the seat forward. Most other vehicles had 5 7 inches of belt remaining and the LTD had 13 inches remaining. It is interesting to note that there is a significant change in available seat belt length when the seat is moved to a different position. The change in the available length of belt is two to three times the distance of seat travel. Several inches of belt may be gained by moving the automobile seat rearward an inch or more. Similar results were obtained when shorter test subjects than the 95th dummy were used.

BELT FIT TESTS WITH HUMAN SUBJECTS - Ten test subjects were selected to represent a range of sizes so that a relationship between occupant weight and required belt length could be determined for each vehicle. Two of the ten are the 50th and 95th percentile dummies. Most of the tests were with the seat in the forwardmost position although each vehicle was examined for the change in available belt length with change in seat position. The measure of belt fit was taken as the length of webbing remaining on the retractor after a subject was secured into the seat. Where the belt was not long enough, the

*Numbers in brackets represent references at the end of this paper.

TABLE 1 -- Number of Persons in Each Height-Weight Category

	NASS 1982																			
HEIGHT / WEIGHT	49-50	51-52	53-54	55-56	57-58	59-60	61-62	63-64	65-66	67-68	69-70	71-72	73-74	75-76	77-78	79-80	81-82	83 +	PERCENT-AGE AT WEIGHT (%)	CUMULATIVE PERCENTAGE (%)
80-99	0	2	0	0	6	39	23	19	2	2	0	0	0	0	0	0	0	0	0.6995	0.6995
100-119	1	0	1	1	12	132	321	361	211	73	16	4	1	0	0	0	0	0	8.5289	9.2283
120-139	1	1	2	6	5	90	295	538	675	544	198	68	13	0	0	0	0	0	18.3213	27.5496
140-159	0	0	0	1	5	51	131	265	534	780	650	390	80	13	2	1	0	0	21.8336	49.3833
160-179	0	0	1	1	2	15	54	137	265	594	803	850	286	59	9	1	0	1	23.1498	72.5331
180-199	0	0	1	0	0	13	31	76	133	293	417	596	336	81	14	1	0	0	14.9819	87.5150
200-219	0	0	0	0	0	2	12	22	34	87	192	316	192	61	14	4	0	1	7.0472	94.5623
220-239	0	0	0	0	0	3	5	8	12	32	67	131	93	43	10	0	1	0	3.0460	97.6083
240-259	0	0	0	0	0	0	1	3	7	22	18	52	50	18	10	1	0	0	1.3688	98.9771
260-279	0	0	0	0	0	0	0	1	2	4	13	17	19	13	3	2	0	0	0.5566	99.5337
280-299	0	0	0	0	0	0	0	1	1	1	3	16	9	3	3	0	0	0	0.2783	99.8120
300-319	0	0	0	0	0	0	0	1	0	2	2	5	5	0	0	0	0	0	0.1128	99.9248
320-339	0	0	0	0	0	0	0	0	1	1	1	0	2	3	0	0	0	0	0.0526	99.9774
340-359	0	0	0	0	0	0	0	0	0	0	0	0	1	0	1	0	0	0	0.0150	99.9925
360-379	0	0	0	0	0	0	0	0	0	0	0	0	0	0	0	0	0	0	0.0000	99.9925
380-399	0	0	0	0	0	0	0	0	0	0	0	0	0	0	0	0	0	0	0.0000	99.9925
400-	0	0	0	0	0	0	0	0	0	0	0	0	1	0	0	0	0	0	0.0075	100.0000

TABLE 2 -- Belt Fit Test with 95th Percentile Dummy

Vehicle	Belt Remaining Seat Forward (in)	Belt Remaining Seat Rearward (in)	Belt Gained by Seat travel (in)
1980 Chevrolet Citation	6.3	23.5	17.2
1981 Ford LTD	13.0	26.4	13.4
1981 Chevrolet Chevette 4 dr	0.0	15.8	15.8
1981 Honda Accord	0.0	14.0	14.0
1985 Cutlass Ciera	7.0	23.3	16.3
1982 Chevrolet Chevette 2 dr	5.8	21.0	15.2
1979 Chevrolet Citation	4.8	24.8	20.0
1985 Chevrolet Celebrity	6.0	23.5	17.5
1983 Volkswagen Vanagon	-3.0		

belt attachment at the floor was disconnected and the distance between the belt end and the floor mounting point was measured.

Table 3 contains the results of the belt fit for the 5 vehicles tested and includes the height and weight for each test subject. Plots of the data for the remaining belt length for different subject weights are shown in Figures 1 through 5. Each plot shows a definite linear relationship especially if the two very light subjects are not included. Linear regression coeficients are shown in Table 3. The correlation coeficients of about .9 suggest a rather good straight line fit. For the vehicles tested, each additional inch of belt length accommodated from 6.4 to 8.3 additional pounds of body weight.

It is to be noted that in each of the five vehicles listed in Table 3 some seat belt webbing remained with the 95th percentile dummy in the forwardmost seat position as required by the current standard. The five vehicles had between 5 and 7 inches of belt webbing left and the LTD had 13 inches of belt webbing left. Yet these vehicles were reported to have had difficulties with people being able to use the belts. Of the five vehicles tested, the LTD and Ciera can accommodate people larger than the 99th percentile, while the Citation, Chevette, and the Celebrity need only 0.7 to 1.4 inches of additional belt webbing to accommodate the 99th percentile. One explanation may be the discomfort or inconvenience of using a belt without some extra belt length available. Each test subject indicated that a minimum of four to six inches of extra belt length is needed for comfort in using the seat belt.

The amount of seat belt length needed to accommodate winter clothing was also investigated briefly by putting a heavy winter coat on the 95th percentile dummy and placing him in the 1982 4dr Chevette. The chest circumference was increased from 44 inch to 46 inches and the waist circumference was increased from 42.5 inches to 46 inches. With the seat in the rearmost position in the 82 Chevette the amount

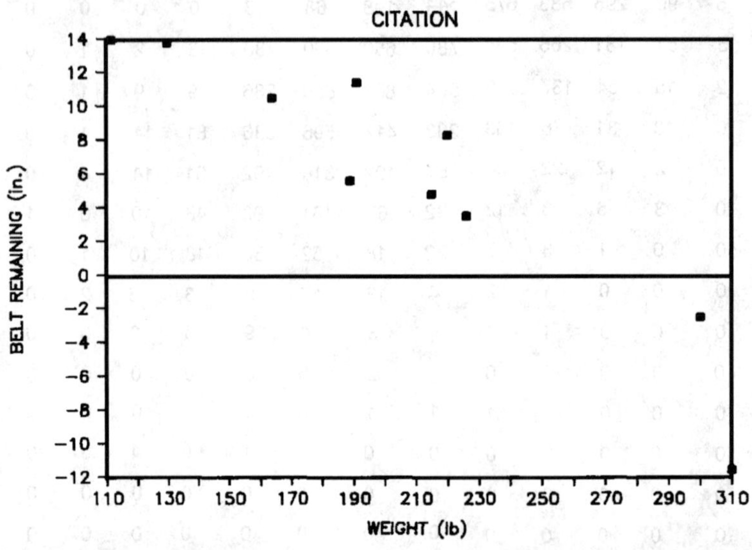

Fig. 1 - Available belt length -- 1979 Chevrolet Citation.

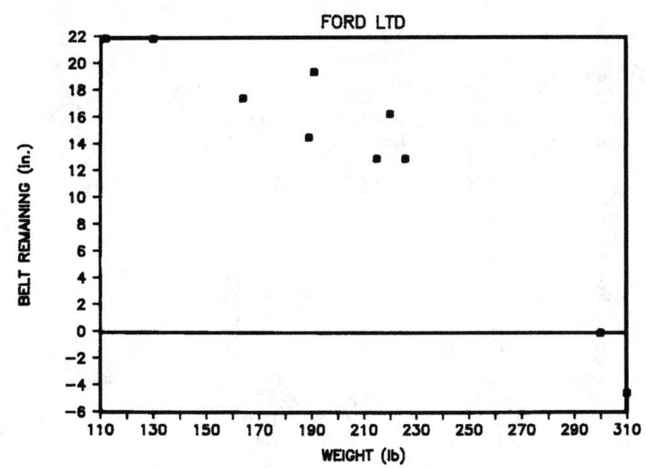

Fig. 2 - Available belt length -- 1981 Ford LTD

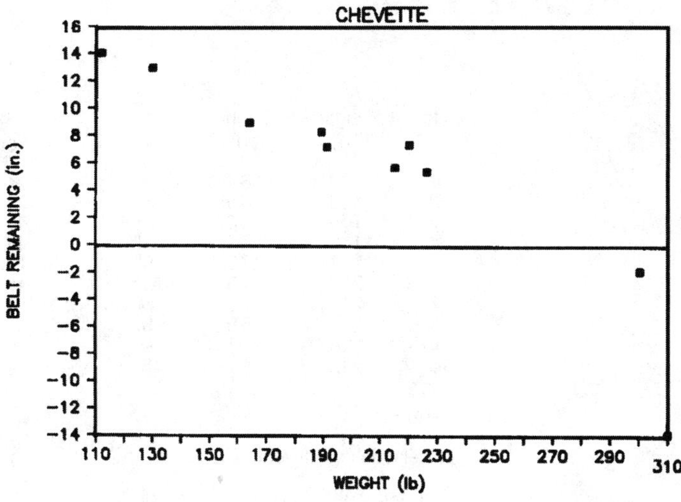

Fig. 3 - Available belt length -- 1980 Chevrolet Chevette

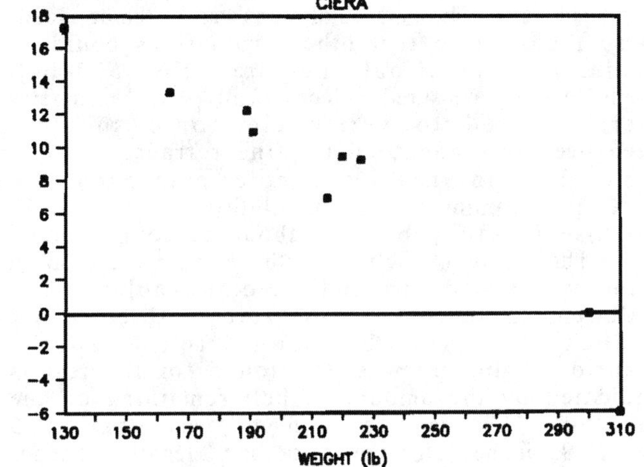

Fig. 4 - Available belt length -- 1985 Oldsmobile Ciera

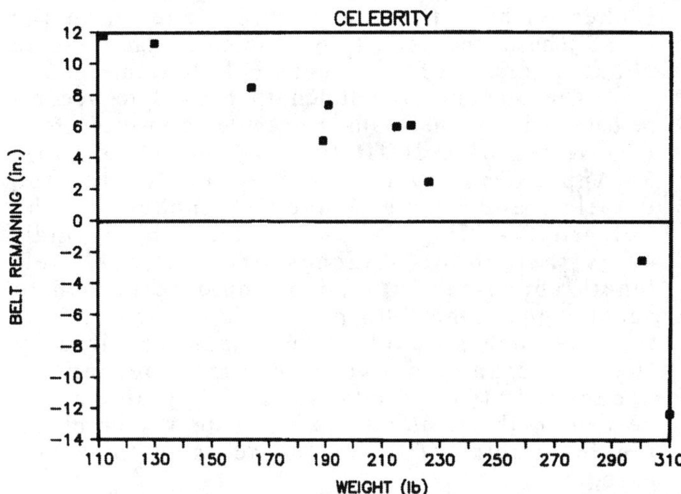

Fig 5 - Available belt length -- 1985 Chevrolet Celebrity

TABLE 3 -- Belt Fit Test Results

No	Weight (lbs)	Height (in)	Sex	Belt remaining (inch) 1981 Ford LTD	1979 Chevrolet Citation	1985 Chevrolet Celebrity	1982 Chevrolet Chevette	1985 Oldsmobile Ciera
1	112	64	F	21.9	14.0	11.8	14.1	NA
2	130	65	F	21.9	13.8	11.3	13.0	17.3
3	164	68	50th	17.5	10.5	8.5	9.0	13.4
4	189	72	M	14.6	5.6	5.1	8.4	12.3
5	191	73	M	19.5	11.4	7.4	7.3	11.0
6	215	73	95th	13.0	4.8	6.0	5.8	7.0
7	220	76	M	16.4	8.3	6.1	7.5	9.5
8	226	69	M	13.0	3.5	2.5	5.5	9.3
9	300	71	M	0.0	-2.5	-2.5	-1.8	0.0
10	310	67	F	-4.5	-11.5	-12.3	-13.8	-6.0
Linear regression results								
y=mx+b m=				-0.157	-0.133	-0.121	-0.133	-0.123
b =				46.8	33.9	30.1	33.7	35.0
correlation coefficient				0.948	0.916	0.913	0.898	0.962
Belt left with 95th				+13	+4.8	+6	+5.8	+7
Projected belt left for 99th (260 lbs)				+ 6	-0.7	-1.4	-0.9	+3
Difference in belt left (95th minus 99th)				- 7	-5.5	-7.4	-6.7	- 4

of belt webbing remaining changed from 15.8 inches without the coat to 9.8 inches with the coat. Thus, the effect of clothing can require about 6 inches of additional belt length.

The amount of belt length needed to accommodate up to the 99th percentile estimated from this series of belt fit tests is shown in Table 3. Approximately one inch is needed for the Citation and Chevette and 1.5 inches for the Celebrity. Adding 4-6 inches for comfort indicates that about 6 inches of additional belt length above current design would be needed to accommodate the 99th percentile occupant. This is 10-12 inches above the minimum required by the 208 standard using the 95th percentile occupant. Also, an additional 6 inches may be needed in the winter to accomodate winter clothing for a total of up to 18 inches in the worst case.

This method of establishing a relationship between subject weight and length of belt required is, of course, without a statistical base because of the small data sample and because of the effect on required belt length of height and shape of the seat occupants. Some measure of belt fit is required, however, to project the effect of adding belt length in accommodating a larger proportion of the population.

SLED TESTS - Six sled tests were conducted to examine the safety aspects of longer belts. Information obtained from the anthropometric study and from the belt fit measurements was used to define the belt lengths to be evaluated in sled tests. Tests were conducted in the front outboard passenger seat of a 1982 Celebrity test buck. Just the exterior plastic piece of the instrument panel was used primarily for orienting the dummy within the vehicle during the impact. The tests were conducted with the 50th percentile Hybrid III dummy with the seat in the mid fore and aft position. Tests were conducted using belt lengths that just meet to approximately 10 inches longer than required by the current 208 standard. All tests were run at 30 mph.

Dummy excursions were observed photographically. By comparing the excursions with the dimensions of the passenger compartments of other cars, the possibility of interior impacts could be evaluated. Standard data channels such as head and chest accelerations and femur loads were recorded so that other evaluations could be made. Belt spool-out and elongation and belt loads were measured electronically, and cameras were utilized to verify electronic spoolout measurements and to determine retractor lockup time. This information supplements excursion data in examining the possibility of any compromise in safety brought about by longer belts.

The actual belt lengths and the initial dummy positions are indicated in Table 4. The belt lengths increased from zero to three to 8.7 inches. The final belt condition when the Hybrid III dummy was positioned for the test is indicated by the amount of belt remaining on the retractor. This can be seen to vary from 18.5 to 28.9 inches for relative belt length changes of 0.0, 3.5, 4.3, 4.3, 9.0, and 10.4 inches.

The dummy was positioned using the procedure stated in the 208 standard. This procedure uses the device specified in SAE J826 to determine the H-point (OSCAR). Table 4 also indicates the H-point position as determined by OSCAR and the resulting H-point and head positions for the Hybrid III dummy.

All of the dynamic test results show only small increases with increased belt length since localized contact with likely impact points was intentionally avoided by using the passenger seat and having only the shell of the instrument panel in the vehicle.

The maximum HIC, peak accelerations, and belt loads measured during the sled tests are shown in Table 5. Peak head, chest, and pelvic accelerations show very little change with increasing belt length while HIC does show an increasing trend of about 12% (See Figure 6). All of the figures that follow are plotted using the relative additional belt webbing remaining on the retractor. The greatest change of about 17% occurs in the maximum neck flexion moment as shown in Figure 7. Figure 7 also shows the peak chest deflection with some scatter but with an increasing trend.

Femur loads have considerable scatter due to slight contact between the knees of the dummy

TABLE 4 -- Initial Sled Conditions

	T01	T02	T03	T04	T05	T06
Total belt length (in)	114.0	114.0	117.0	116.9	121.1	122.7
Relative difference	0.0	0.0	3.0	2.9	7.1	8.7
Remaining belt on retractor	18.5	22.0	22.8	22.8	27.5	28.9
Additional belt length, (in)	0.0	3.5	4.3	4.3	9.0	10.4
J826 H-point location - x	10.3	9.2	10.4	10.1	9.2	10.5
" " - z	8.6	8.8	8.8	9.2	9.4	9.0
Dummy position - H-point - x	10.0	9.2	9.9	10.0	9.2	10.3
" " " z	7.9	8.2	8.0	8.8	8.8	8.7
" " - head x	14.4	13.9	14.6	14.4	13.8	15.1
" " " z	34.8	34.1	34.2	34.5	34.9	34.8

TABLE 5 -- Dynamic Test Results - Maximum Values

	T01	T02	T03	T04	T05	T06
Resultant head accel. (g)	55	52	60	54	55	53
HIC	672	715	780	741	739	802
Resultant chest accel. (g)	39	40	39	41	39	43
Resultant chest defl. (in)	1.65	1.9	1.61	1.69	1.54	1.91
Neck moment flexion	103	110	123	114	118	121
Resultant pelvic accel.	41	43	42	42	41	47
Belt loads (lb)						
Outboard	1837	1993	2003	1958	2004	2105
Inboard	2962	2813	2652	3088	2768	3227
Femur - left (lb)	223	331	256	329	293	217

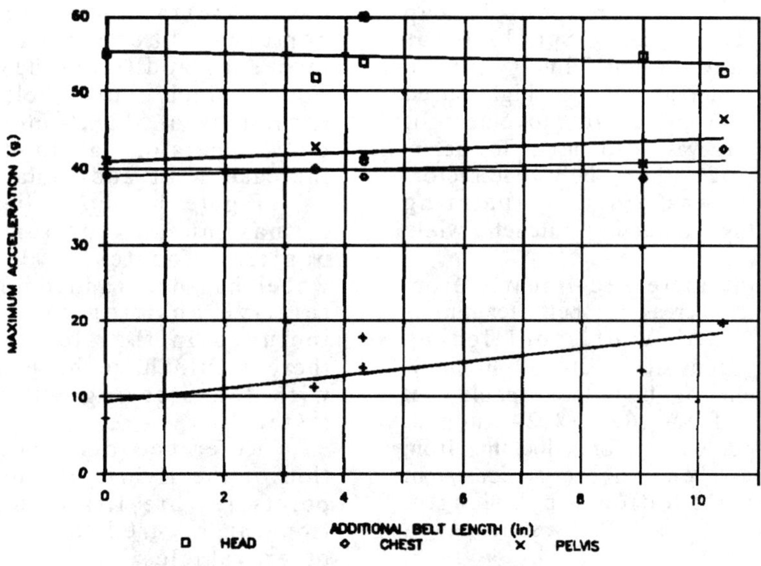

Fig. 6 - Maximum acceleration and HIC versus additional belt length

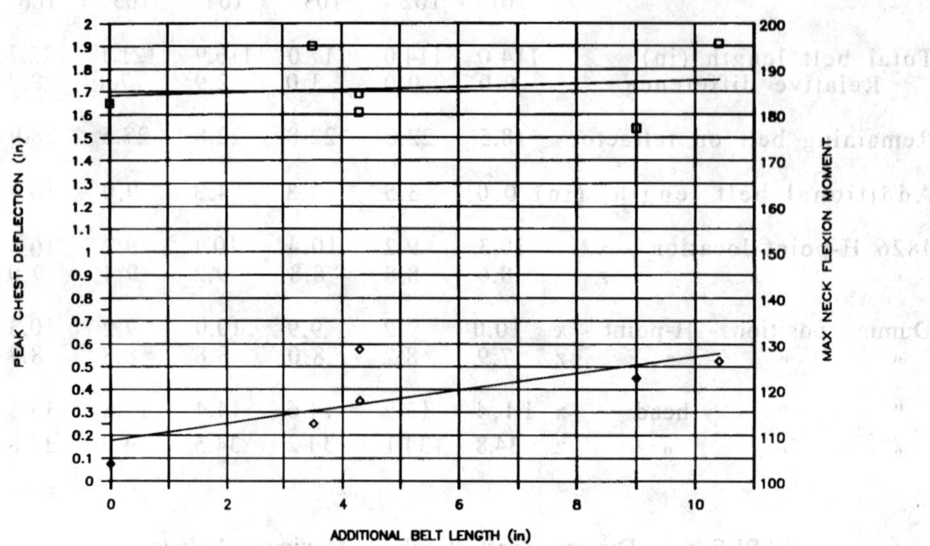

Fig. 7 - Peak chest deflection and maximum neck flexion moment versus additional belt lengt

and the dashboard. As mentioned above, only a partial dashboard without reinforcement was installed in the sled buck. Thus, this slight contact should not have much effect on the overall dynamics and trajectories, but it is likely that femur loads would have been higher if the full dashboard had been used.

Belt loads show an increasing trend as seen in Figure 8. The outboard belt load is the lap belt load. The inboard load cell was located between the attachment point and the buckle and, therefore, recorded combined lap and shoulder loads.

Six channel upper and lower neck load cells were included in the dummy instrumentation in these tests. A brief examination of the results show an increase in maximum neck flexion moment with increasing belt length.

Belt spoolout and retractor spool lockup times are listed in Table 6. Spoolout was established photographically using lines painted on the belt at 1 inch intervals and a high speed camera trained on the belt to record movement of the webbing. The camera was also used to determine lockup time of the belt retractor. Observation of lockup was aided by painting alternating teeth on the retractor ratchet white and red.

Total belt spoolout increased linearly from 1.6 to 3.0 inches with increasing belt length as shown in Figure 9. The time of spool lockup appears to be rather consistent at 30-40 milliseconds and the amount of belt that spools out prior to lockup varied from .62 - 1.24 inches. Subtracting the belt spoolout before lockup from the total belt spoolout shows a very consistent and linear relation with additional belt length. The correlation coeficient is .977 showing a definite linear relation.

Table 6 also indicates the belt strain measurements taken on an extensometer. The values were questionable for the first three tests. The final tests measured strain at about 7%.

The change in body excursion with increased belt length is significant. Measurement of excursions of the dummy was determined by film digitization. The results are presented in Table 7 for the x and z directions and the resultant distances for the head, T01 position at the base of the neck, and the H-point. The maximum excursions are plotted in Figure 10. The maximum resultant excursion of the head varies linearly from 17.9 inches to 20.9 inches for a 3 inch increase in travel. Maximum resultant excursion of T01 increased linearly also from 10.4 to 13.3 inches. T01 was examined as a guide to chest excursions because it was difficult to attach a target to the chest which would remain visible in the onboard side cameras. These results indicate about .25 inches of additional head travel for each additional inch of belt material. Maximum excursions of the H-point are smaller than upper body excursions at about 8.0 inches and do not increase with additional belt.

Figure 11 shows the position of the dummy at maximum excursion relative to the initial position for tests 01 and 06. The steering wheel has been added to the figure to show the relative position of the dummy but was <u>not</u> included in the sled tests. It does show that there would have been a substantial interaction with the steering wheel for all belt lengths tested.

Reference distances from the seated position of the Hybrid III dummy to probable contact points [2] are listed in Table 8. This information can be used to predict probable impacts in other vehicles.

For passenger occupants in the right front outboard seating position belts of any of the

TABLE 6 -- Belt Lockup and Spool Out Results

	T01	T02	T03	T04	T05	T06
Belt strain	NA	NA	NA	0.07	0.07	0.08
Total belt spoolout (in)	1.6	2.1	2.6	2.4	2.9	3.0
Time of belt spool lockup (msec)	37	35	39	41	35	33
Belt spool out before lockup (in)	0.62	1.0	1.24	1.03	0.97	0.84
Belt spool out after lockup (in)	0.98	1.1	1.36	1.37	1.93	2.16

TABLE 7 -- Dummy Excursion Results

	T01	T02	T03	T04	T05	T06
Head excursion x (in)	15.3	16.3	15.3	15.1	16.0	17.3
Head excursion z (in)	9.2	10.0	10.1	10.7	11.4	11.8
Head excursion resultant (in)	17.9	19.1	18.4	18.4	19.6	20.9
T01 excursion x (in)	10.1	11.2	10.5	10.6	11.1	12.9
T01 excursion z (in)	2.3	2.6	2.7	2.9	3.5	3.3
T01 excursion resultant (in)	10.4	11.5	10.9	11.0	11.6	13.3
H-point excursion x (in)	7.8	8.3	7.5	7.8	7.5	8.3
H-point excursion z (in)	1.4	1.5	1.8	2.5	1.7	1.3
H-point excursion resultant (in)	7.9	8.5	7.7	8.2	7.7	8.5
Additional belt length (in)	0.0	3.5	4.3	4.3	9.0	10.4

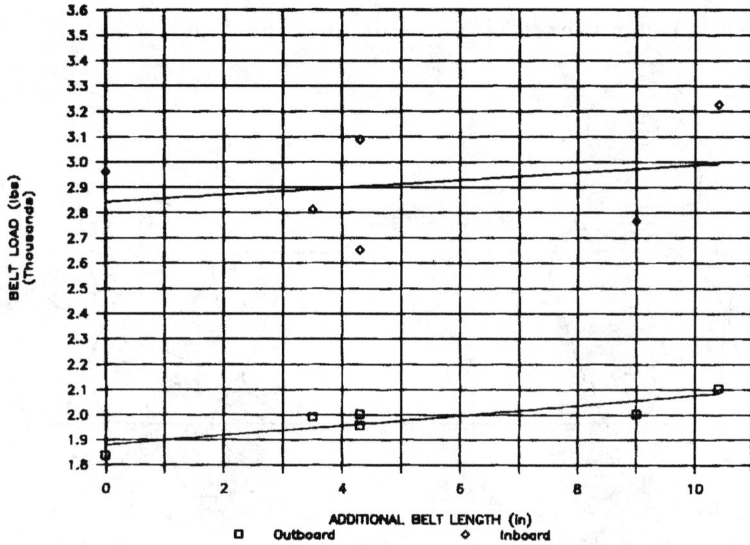

Fig. 8 - Maximum belt load versus additional belt length

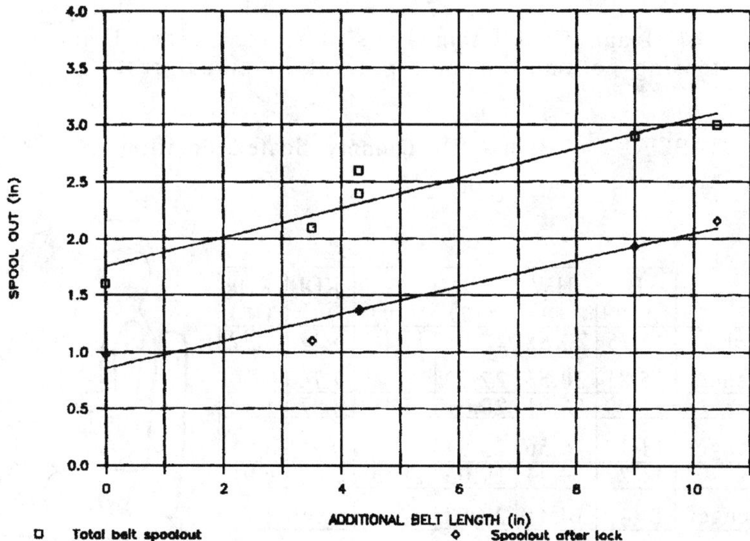

Fig. 9 - Belt spoolout versus additional belt length

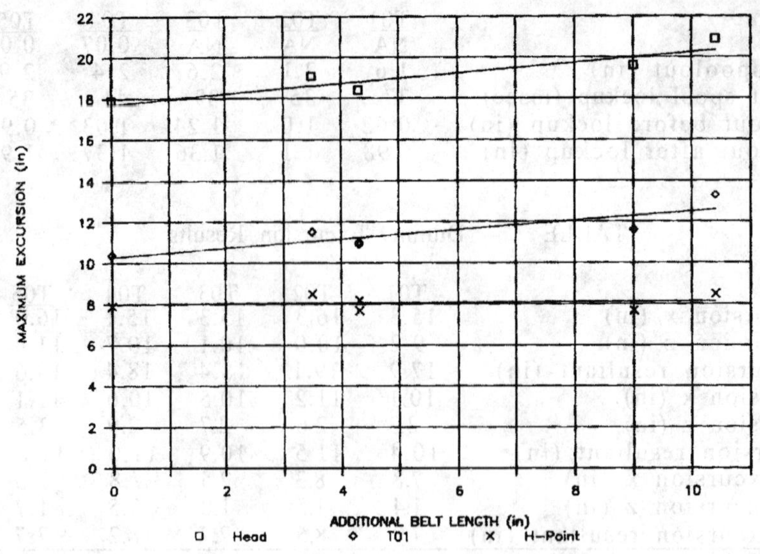

Fig. 10 - Dummy excursion versus additional belt length

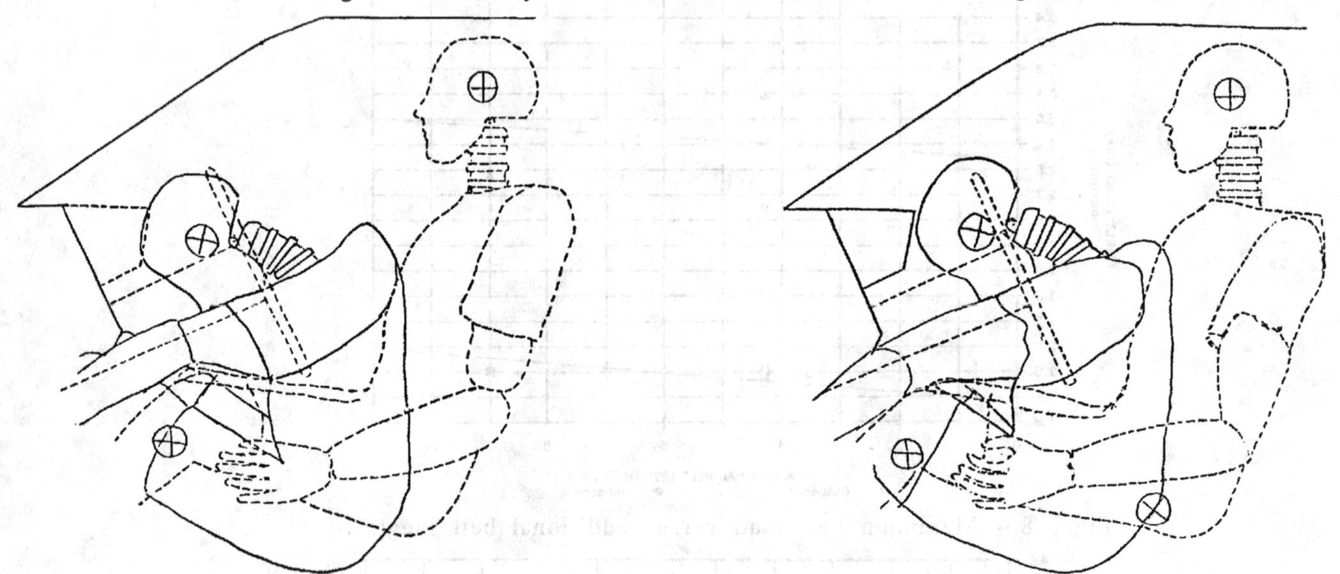

Fig. 11 - Initial and final dummy positions for Tests 01 and 06 -- Steering column was not present in actual tests

TABLE 8 -- Hybrid III Dummy Seated Position

VEHICLE		HH (in)	HW (in)	CD (in)	CS (in)	KDL (in)	KDR (in)
1984 Honda Accord	Driver	15.75	20.88	23.09	15.38	6.25	5.72
	Passenger	15.81	20.56	22.72	---	7.78	7.50
1984 Chev. Celebrity	Driver	11.00	16.91	20.09	13.41	4.97	5.16
	Passenger	11.13	16.56	25.91	---	4.91	4.91
1984 Dodge Omni	Driver	11.22	14.44	19.05	17.13	6.91	6.91
	Passenger	11.38	14.41	23.63	---	6.94	6.81

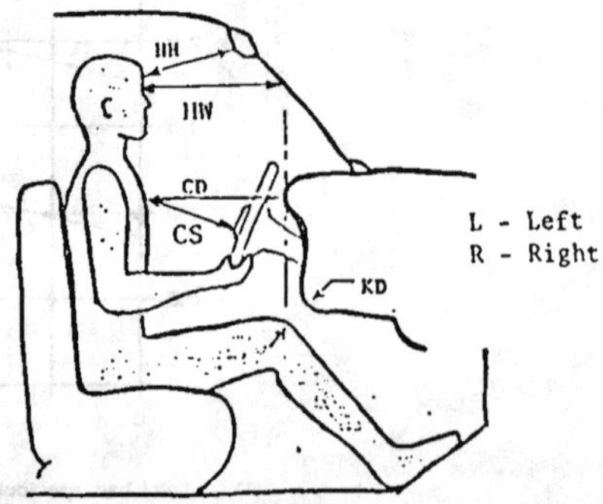

lengths tested would most likely protect the passenger from all but knee/dashboard impacts. Tests in this series used the right front outboard (passenger) position, and no upper body impact occurred in the Celebrity. It is also clear that all driver occupants would have significant interaction with the steering wheel.

DISCUSSION

DIFFICULTIES IN PROBLEM DEFINITION - To define potential modifications to the safety belt lengths to accommodate a larger percentage of the population, it is helpful to describe the sizes of persons who encounter difficulties with current belts. Because of the weight of a person that is predicted to fit in the vehicles tested and the large increase in available belt from moving the seat rearward, it is clear that the person having difficulty is likely to be fat and short in stature. Using the 95th percentile dummy, which represents a large and tall person, even with the seat forward does not represent the shape person having difficulty. Table 1 indicates the distribution of people for different heights and weights. A large number of people are of the same weight as the 95th (215 lbs) but shorter. It is likely that a person of the same weight as the 95th but shorter would not fit the belt because of increased waist and chest dimensions.

Besides defining seat belt length for sled tests, the belt fit tests produced other results as well. For cars tested in this study, seat belt fit problems did increase with the decreasing size of the vehicle. This agrees with the consensus of the 1980 Verve Corporation study which found a definite correlation between vehicle size and seat belt fit.[3] Reports from NAAFA (National Association for Aid to Fat Americans) indicate, however, that seat belt fit difficulty is not dependent entirely on vehicle size.[4] They mention problems with larger cars as well as with small ones. Difficulties are simply more common in small cars.

DISCUSSION OF THE SLED TEST RESULTS - The results of the sled tests indicate an increase in body excursion or travel within the vehicle with increasing belt length. Therefore, increasing the amount of belt webbing as a means of fitting a larger percentage of the population may cause a reduction in protection for others in situations where the additional excursion would allow or increase the severity of impacts with the vehicle interior. Assuming a linear relationship between head travel and additional belt length permits an estimate of 0.26 inches increase in head excursion for every inch of additional belt. Thus, to provide 6 inches of additional belt to include up to the 99th percentile will likely increase the head travel nominally 1.5 inches. The disbenefit of this additional belt would be high in the driver position because of the propensity for steering wheel impacts. Lesser disbenefits would result in passenger and rear seating positions with 3 point lap and shoulder belts since the overall propensity for head and thorax impacts against the vehicle interior is much lower.

Though it is desirable to fit 100% of the U.S. population in safety belts, this goal is not believed feasible by simply increasing the length of the seat belt. Safety requirements limit the options available to making seat belts fit more people. If belts are lengthened, spoolout and elongation will increase and increase the likelihood of impacts with the interior of the vehicle. In a small car, increases in elongation and spoolout could make the difference between minor and serious or fatal injury.

It is also likely that the overall effectiveness of the seat belt is reduced for fat automobile occupants. Submarining under the lap belt with resulting abdominal injuries is possible because of the general position of the lap belt relative to the location of the pelvic iliac crests.

EVALUATION OF SEATBELT EXTENDERS - Seatbelt extenders offer a satisfactory method for providing the extra belt needed for a fat person to wear the seat belt. The extender, however, is only good for the specific person and must not be used by anyone else. The importance of this is illustrated in the following example. The 95th percentile dummy was positioned in the LTD, the seatbelt was fastened, and photographs were taken. Then, an eight inch belt extender was inserted, and more photographs were taken. These photographs are shown in Figure 12. By inspection, it can be seen that the geometry of the belt is changed and that the buckle is closer to the center of the dummy when the extender is used.

Cautions on the belt extender specifically warn against using the extender when the lap strap will not adjust snugly on the hips and/or when the intersection of the lap and upper torso straps (measured along the lap strap) is less than 6 inches from an imaginary center line of the occupant's body. These conditions lead to an increased probability of abdominal injury due to submarining under the belt. Location of the intersection of the shoulder belt and the lap belt less than 6 inches from the centerline also increases the likelihood that the person will roll out of the belt and not be well protected in a crash.

RECOMMENDATIONS FOR FAT PEOPLE - There are several solutions to the difficulties of those persons whom seat belts do not fit. The Joint Committee of the State Department of Health of New York and the Governor's Traffic Safety Committee recommends that large persons who are primarily passengers in automobiles should ride in the rear seat when possible.[5] This appears to be a good recommendation.

The New York Committee also recommends belt extenders for fat persons, but the availability of these devices appears to be is low. They can

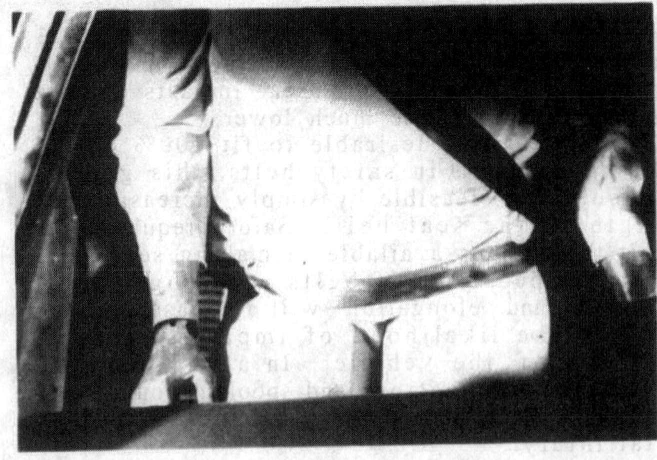

(a)

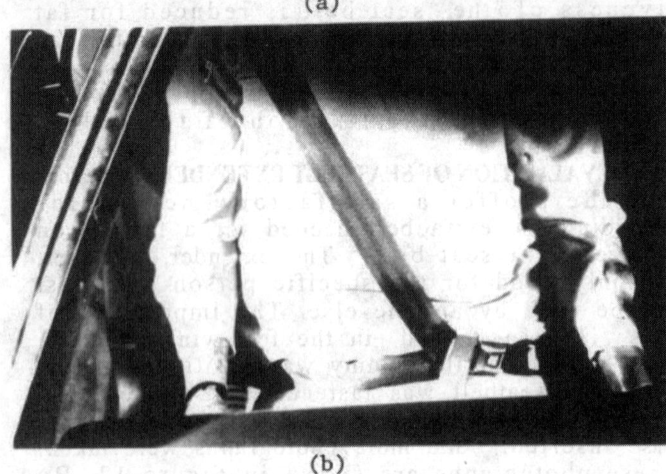

(b)

Fig. 12 - 95th percentile dummy in (a) standard LTD seatbelt and (b) LTD with eight inch seatbelt extender

be obtained through automobile dealers, but NAAFA reports that many dealers have not even heard of extenders.[4].

Belt extenders must be user specific. The general procedure is to have a persons' measurements taken at the dealer. The dealer sends the data to the manufacturer, and the manufacturer ships a properly sized extender to the customer. There appears to be a reluctance on the part of dealers to handle the extenders, however, because of possible product liability. For this project, an attempt was made to purchase a belt extender through a dealership. Many dealers claimed they knew nothing of belt extenders. One dealership was more helpful.

A simple solution to some belt fit problems is to move the seat back. Tests in this study show that for every inch the seat travels, the length of usable belt changes by 2 to 3 inches. Persons who position the seat as far rearward as possible take best advantage of the full length of the belt. People who normally position the seat in the forwardmost position may gain additional belt by moving the seat rearward a notch or two. In one test of an individual, it was found that vehicle foot pedals could be reached adequately if the seat was moved rearward two inches from the position at which she could reach the pedals most comfortably. There are also adjustments that may be made to enable a short person to reach the foot pedals of the vehicle by extending the pedals.

Another suggestion would be that fat drivers consider purchasing a vehicle with an air bag since optional driver air bag restraint systems are now available on many vehicles.

Also, because of the variations found in the seat belt length available in automobiles, fat persons should try on the seat belt in any car they are considering purchasing.

REFERENCES

1. "Weight and Height of Adults 18-74 Years of Age: United States, 1971-74," U. S. Department of Health Education and Welfare Publication No. (PHS) 79-1659, May 1979.

2. Hybrid III Positioning, Chamois Use, and Calibration Procedures. Status Report No. 2. Vehicle Research and Test Center, National Highway Traffic Safety Administration, U.S. Department of Transportation. March 1986.

3. Tom, J. C., Petersen, D. D.,Robbins, C. M., and Peters, R., "Evaluation of the Comfort and Convenience of Safety Belt Systems in 1980 and 1981 Model Vehicles," Verve Research Corporation, Department of Transportation Report DOT-HS-805 860, March 1981.

4. Fabrey, William J., Chairman of the Board and Founder, National Association to Aid Fat Americans. Letter to author. 25 March, 1985.

5. "Guidelines for Granting Medical Exemptions for Safety Belt Use," Governor's Traffic Safety Committee and the New York State Department of Health.

971137

Development of an Electronic Belt Fit Test Device

Y. I. Noy and V. Battista
Transport Canada

R. Carrier
Genicom Consultants Inc.

ABSTRACT

The purpose of this project was to develop and validate a computer-based version of the Belt Fit Test Device with a view towards exploring the potential of this technology to improve belt fitment for the general occupant population. The electronic BTD was initially developed and validated against two seats using the Transport Canada seat simulator. Preliminary validation indicated good correspondence between computed and measured BTD co-ordinates. The electronic BTD was then validated in ten vehicles. In total, 40 BTD scores were computed using the electronic BTD and compared with actual BTD values. In 30 of the 40 comparisons, the discrepancy between measured and computed values was less than one centimetre. In terms of test performance using the pass/fail criteria developed for the BTD, 37 of the 40 comparisons were in agreement. However, a number of refinements have been identified which could further improve the seat belt algorithm and the overall usefulness of the model.

INTRODUCTION

The Belt Fit Test Device (BTD) is a device used for he measurement and assessment of static seat belt geometry of automobile seat belts. The device was conceived and leveloped by Transport Canada to address abdominal injuries and upper body injuries that may result from a mismatch between belt geometry and the occupants' anthropometric characteristics. In effect, the BTD comprises an SAE 3-limensional H-point Machine with the additional attachment of special torso and lap forms that are designed to represent he 50th percentile Canadian adult. The surfaces of the lap and torso forms are marked with scales to permit quantifying belt position. When positioned on an automobile seat, the levice indicates whether the lap and shoulder belts fall within pecified bounds which have been established to minimise the isk of serious injuries to soft tissue and organs from belt ntrusion. Four criteria establish acceptable position limits with respect to the clavicle, sternum and lap scales. Thus, the BTD test would complement other occupant restraint requirements such as head injury criteria. Previous studies suggest that while the physical device is fixed in size, test compliance ensures adequate protection for a wide segment of the occupant population.

While the BTD criteria are considered important in as much as they provide a sensitive and reliable test of belt geometry and anchorage, issues concerning the extent to which the BTD criteria accommodate the full range of the occupant population and the usefulness of the device for design merit further development. Computer human modelling techniques offer an approach to addressing such issues.

The purpose of this project was to develop and validate a computer-based version of the BTD with a view towards exploring the potential of this technology to improve belt fitment for the general occupant population. In particular, the potential advantages of an electronic version of the BTD include:

- simplification of restraint testing
- extension of the BTD criteria for a wider range of occupants (correct positioning of lap and shoulder belts with respect to anatomical landmarks) and,
- provision for reverse engineering during the design stage to ensure compliance with criteria for safety and comfort.

The computer-based human model selected for this effort was the '*Safework*$_{TM}$' mannequin developed by Genicom Consultants. This paper describes the development and validation of the electronic[1] BTD within the *Safework*$_{TM}$ environment. Future phases in this series will include development of the BTD mannequin and criteria validation (for different segments of the population).

BELT FIT TEST DEVICE (BTD)

The development of the BTD began in the mid-seventies as an effort to minimise the incidence of lacerations and rupture of vital organs due to lap belt intrusion (Gibson et al. 1994). A review of the literature and analysis of collision

[1] The term "electronic" is used in this paper to denote a computer-based simulation.

data had identified geometric and anthropometric criteria for the correct positioning of the lap belt relative to the anterior superior iliac spines (ISIS). A need was identified for a reliable test of pelvic and thoracic belt fit in any vehicle. To ensure compatibility with existing automotive engineering practices, it was decided that the device would be based on the standard SAE H-point machine. Three dimensional lap and torso forms were constructed from anthropometric data obtained from a sample of the 50th percentile Canadian adult. Participants for this study were selected on the basis of their height and weight to reflect the 50th percentile values reported in a 1981 Fitness Canada survey. The height and weight screening criteria were 165 cm and 67 kg respectively. Details of the development of the lap and torso forms can be found in Gibson et al. (1994).

The proposed quality of fit requirements include four measurement criteria which establish belt position limits with respect to the clavicle, sternum, and inboard and outboard lap scales. The minimum acceptable scores are outlined in Table 1. For further information about the development and use of the BTD the reader is referred to Gibson et al. (1994) and Tylko et al. (1994).

Table 1 - BTD Criteria.

Measurement Criteria
1. Lap Form: $x > 1.5$ on inboard and outboard scales
2. Clavicle: $7 < x > 13$
3. Sternum: $12 < x > 22$
4. Belt contact at each of the clavicle and lap scales

DEVELOPMENT OF THE ELECTRONIC BTD

The software used in the development of the electronic BTD was *Safework*$_{TM}$, a program for human modelling applications. Three dimensional human mannequins representing diverse populations can be generated and integrated into a computer-based representation of the workplace

MODELLING THE H-POINT MACHINE - The first step in developing an electronic BTD was to develop an electronic model of the H-point machine. The two dimensional engineering drawings of the H-point machine were recreated in electronic format using the 3D CAD system, AUTOCAD. This 3D CAD drawing was then converted to a DXF file for its importation into *Safework*$_{TM}$.

The fully articulated electronic H-point machine permitted all the kinematics and anthropometric adjustments of the actual H-point machine.

CREATING THE ELECTRONIC BTD - The next step was to add the lap and torso forms to the H-point machine. These forms are the three-dimensional free-form surfaces representing the lap and torso shape of the 50th percentile Canadian adult. Drawings for the forms are available in the IGES format and an IGES file parser was developed to import them into *Safework*$_{TM}$.

After being imported into *Safework*$_{TM}$, the lap and torso forms were attached to the H-point machine according to the BTD configuration to form an electronic BTD.

PRELIMINARY VALIDATION OF ELECTRONIC BTD

The preliminary validation of the electronic BTD was accomplished by comparing actual and computed BTD scores on two different seats using the Transport Canada seat simulator[2]. The seats differed with respect to cushion compression characteristics, one seat being firm and the other soft.

While it is a relatively simple matter to simulate the seats within *Safework*$_{TM}$, there are no techniques yet available for 'seating' the electronic BTD in the seat. The resting position of an actual mannequin when placed on a seat depends upon numerous factors including seat cushion and seat back angles, the distribution of weight on regions of the buttocks and back, the deformation properties of the cushion, the shape of the cushion, the type of upholstery material used, belt contact points, etc. There are no algorithms available, at present, that can be used to determine the position of the H-point as a function of known seat and mannequin characteristics - this can only be determined empirically. Hence, for the purposes of this study, it was decided to position the electronic BTD in the seat by aligning its H-point with the corresponding digitised co-ordinate obtained using the actual BTD. While this procedure was necessitated by the lack of appropriate seating algorithms, it does not detract from the usefulness of the electronic BTD since the H-point[3] can be readily obtained from vehicle manufacturers.

MODELLING THE SEATS - Since the electronic BTD was to be placed with reference to the empirically determined H-point, seat characteristics would have no effect on the computation of the BTD scores. There was technically no need to model the seat and no attempt was made to model the actual seats tested on the seat simulator. However, a generic seat was modelled within *Safework*$_{TM}$ to enhance the visual appreciation of the seated electronic BTD. For this purpose, a typical bucket-type seat was selected and modelled from the manufacturer's design data.

MODELLING THE SEAT BELT - To simulate the physical effect of the seat belt, a computerised flexible seat belt model was created. The physical seat belt is, as a first approximation, constrained by forces at various anchor points as well as by the geometric shape of the lap and torso forms. The belt can be mathematically represented as a set of three spline curves lying on a surface with two tangent, directionally-constrained forces at each end. The shape of these spline curves is defined by the two forces at each end, the shape of the surface and by various anchor points. If there are no friction forces between the surface of the body and the belt, then the true definition of spline curves can be used. Also, the portion of the belt between the extreme contact points on the body and the anchor point will be a straight line

[2] The Transport Canada seat simulator is a platform that can accept a variety of automobile seats and contains adaptive hardware to install the associated occupant restraint system.

[3] The H-point is equivalent to the seating reference point (SRP) which is a standard design reference point universally employed within the automotive industry.

[4] The anchor point here is defined as the end of the flexible portion of the belt.

because the belt is made of soft fibre. These considerations define the shortest curve lying on the surface, i.e. the shortest curve among all the curves lying on the surface with the same starting and ending anchor points. The portion of the belt on the surface of the body is defined as the intersection between a plane containing the spline curve and the surface of the body. The lap and torso surfaces are B-spline surfaces. The intersection curve is obtained by using a numerical resolution algorithm implemented in C language. After being fully tested, the seat belt algorithm was integrated within *Safework*$_{TM}$.

METHOD OF DIGITISING BTD CO-ORDINATES - The initial validation of the electronic BTD was performed by comparing co-ordinate values of specific landmarks generated electronically against the actual BTD co-ordinates. The physical BTD was installed on two different seats (one having a soft cushion and the other a firm cushion) and seat belt fit measurements taken in accordance with the procedures outlined in the Operations Manual for the BTD.

The measurements on the physical device were performed at Biokinetics & Associates Ltd. in Ottawa using the seat simulator. The measurements of the belt location and reference landmarks on the BTD were obtained using a GP8-3D sonic digitiser.

COMPARISON OF ELECTRONIC AND PHYSICAL BTD RESULTS

Figure 1 is a computer generated image showing the actual measurement results as well as the computer model generated results for the clavicle scale on the hard seat. The points represent the actual co-ordinates superimposed over the computer model. This figure shows that computed co-ordinates were very close to measured co-ordinates. The small discrepancies were attributed mainly to measurement error of the GP8-3D device.

Figure 2 shows similar results for the sternum scale on the hard seat and Figure 3 shows the results for the left and right lap scale. Small errors were found between the computer model and the actual BTD but the correspondence was very good.

The results for the soft seat were as reliable as for the hard seat in that there was good correspondence between computed and measured co-ordinates.

Table 2 lists the principal anatomical landmarks and belt positions for which 3D co-ordinate values were obtained with the difference between the digitised and electronic data for both the hard and soft seats. The differences range between 0.36 mm and 12.87 mm. The full data set is found in the Appendix.

FIG. 1 - Hard seat clavicle scale

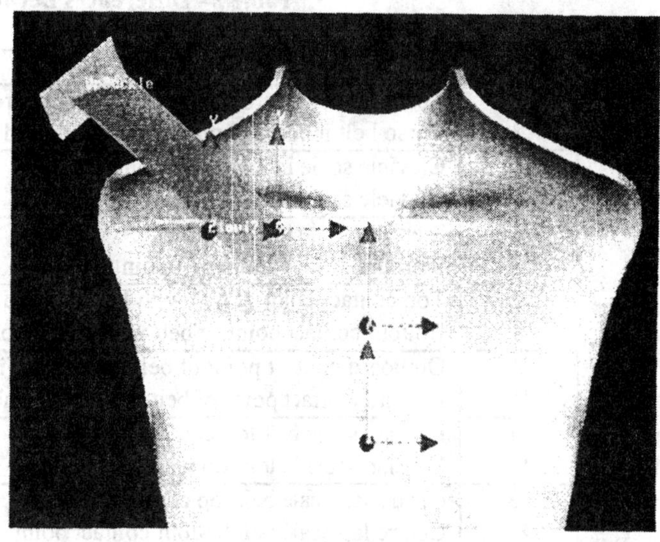

FIG. 2 - Hard seat sternum scale

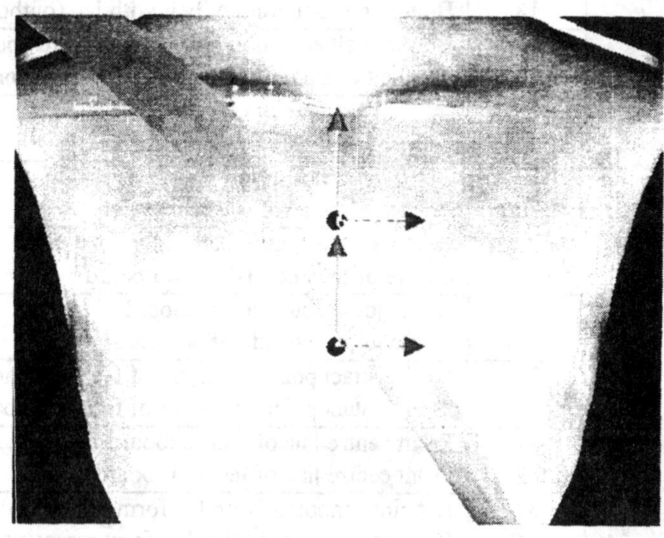

FIG. 3 - Hard seat lap scale

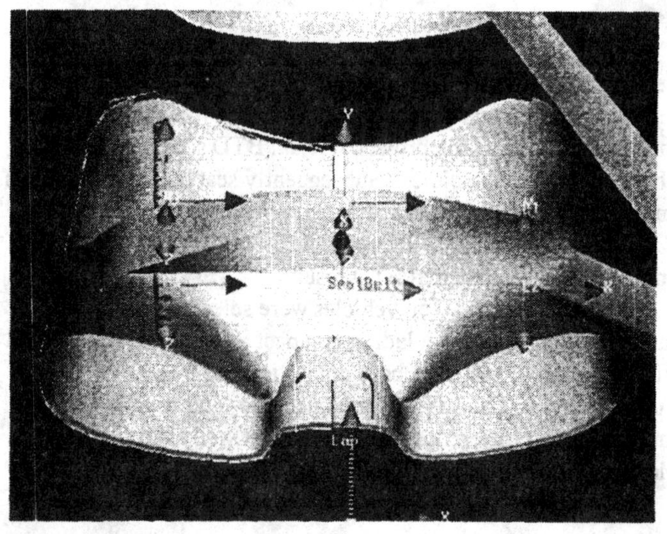

Table 2 - Differences Between Digitised and Electronic Data

POINT	DESCRIPTION	SOFT SEAT (mm.)	HARD SEAT (mm.)
1	Torso belt outboard contact point (outboard side)	3.80	10.59
2	Torso belt inboard contact point (outboard side)	0.36	1.00
3	Clavicle scale inboard belt point	5.86	9.94
4	Clavicle scale outboard belt point	11.67	4.06
5	Sternum scale top belt point	4.81	5.22
6	Sternum scale bottom belt point	12.49	6.00
7	Top contact point of belt with lap (inboard side)	12.08	5.01
8	Bottom contact point of belt with lap (inboard side)	3.83	5.30
9	Outboard contact point of belt with torso (inboard side)	0.66	3.40
10	Inboard contact point of belt with torso (inboard side)	2.91	5.17
11	Left lap scale belt top contact point	2.47	5.48
12	Left lap scale belt bottom contact point	3.73	8.46
13	Centre lap scale belt top contact point	11.37	5.61
14	Centre lap scale belt bottom contact point	7.80	4.42
15	Right lap scale belt top contact point	5.90	7.43
16	Right lap scale belt bottom contact point	7.75	8.14
17	Top contact point of belt with lap form (outboard side)	5.73	8.92
18	Bottom contact point of belt with lap (outboard side)	2.12	1.72
19	Bottom belt end point at lower outboard anchor	1.88	1.51
20	Top belt end point at lower outboard anchor	2.69	2.26
21	Inboard extreme of clavicle scale	4.55	2.49
22	Outboard extreme of clavicle scale	9.28	7.87
23	Top extreme of sternum scale	5.01	5.27
24	Bottom extreme of sternum scale	4.56	4.56
25	Centre of the knee T-bar - inboard	4.52	5.36
26	Centre of the knee T-bar - outboard	3.27	6.45
27	Ankle joints, outside of inboard	6.29	7.34
28	Ankle joints, outside of outboard	10.63	6.90
29	Heel contact points in centre of feet - inboard	10.79	10.35
30	Heel contact points in centre of feet - outboard	12.87	11.93
31	Front centre line of feet - inboard	0.86	9.51
32	Front centre line of feet - outboard	5.23	6.41
33	H-point - inboard (with lap form removed)	0.88	0.86
34	H-point - outboard (with lap form removed)	2.82	1.68

VALIDATION OF ELECTRONIC BTD

Since initial validation indicated good correspondence between computed and measured BTD co-ordinates of seat belt reference points, a more complete validation was undertaken in the second phase of the project. Actual BTD scores were obtained from ten vehicles. Subsequently seat belt anchor points and H-points were digitised and incorporated into the *Safework*$_{TM}$ algorithm. Electronic BTD scores were generated and compared with actual BTD results,

METHOD - Ten vehicles were selected to represent a range of vehicle sizes including compact and full size cars, small trucks and minivans from both domestic and foreign manufacturers. The vehicles included two- and four-door models having different seat belt types, as well as manual and electrical seat adjustment controls.

Measurements were taken using the physical BTD positioned in the passenger seat of each vehicle. The position of the anchor points of the belt system and the BTD were digitised for input into the electronic BTD model.

A number of digitisers were evaluated prior to selecting the Immersion Corp. MicroScribe 3D DL digitiser. Accuracy, installation cost, ability to reach distant and hidden points, sensitivity to metal interference, and ability to work within the confined space of a vehicle were all factors which played a role in the selection process. An Epson laptop 486 computer with version 1.8 of Mira Imaging Inc. Hyperspace Modeler for Windows was used to record, manipulate and visualise the digitised data directly on the screen.

The digitiser was verified for accuracy and consistency.

POINTS DIGITISED ON BTD - The H-points and seat belt anchor points were digitised twice for most of the vehicles tested. Because the H-points on the H-point machine were covered by the lap form, they were inaccessible for digitisation. It was not possible to remove the lap form without modifying the actual position of the H-point machine. It was therefore necessary to digitise other points on the BTD to derive the actual position of the H-points. Additional points were also digitised to permit further comparisons to be made. However, these comparisons are beyond the scope of this paper. The points that were digitised are shown in Fig. 4.

Fig. 4- Points Digitised on BTD

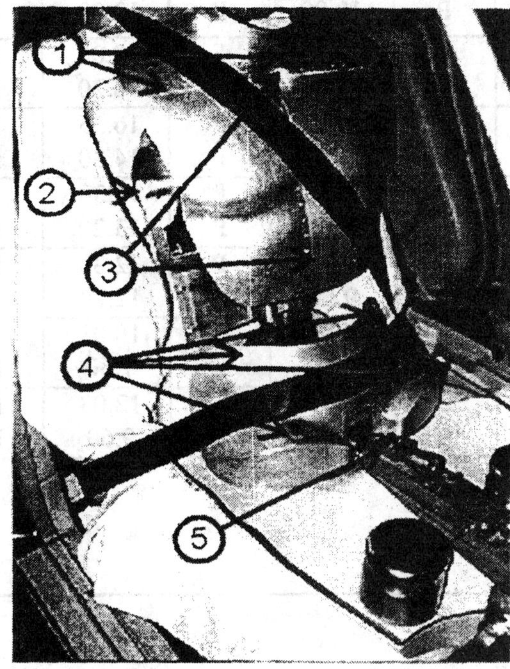

1: extremities of the clavicle scale - 2 points;
2: torso weight support - 2 points;
3: extremities of the sternum scale - 2 points;
4: extremities of the lap scale - 4 points;
5: centre screw of the plunger assembly - 1 point.

ESULTS

The results for the ten vehicles comparing the actual ıd the electronic BTD data are presented in Table 3. Graphical sults are presented in Fig. 5-8 for the inboard and outboard p, the clavicle and sternum scores respectively. The horizontal ference lines on these figures represent the test pass/fail iteria levels.

In total, 40 BTD scores were computed using the ectronic BTD and compared with actual BTD values. In 30 of e 40 comparisons, the discrepancy between measured and mputed values was less than one cm. The ten cases for which e discrepancy was greater than one centimetre are listed low:

- BMW sternum score
- Mercedes Benz clavicle score
- Mercedes Benz sternum score
- Plymouth Reliant clavicle score
- Plymouth Reliant sternum score
- Toyota Tercel clavicle score
- Toyota Tercel sternum score
- Mazda 626 sternum score
- Ford Windstar inboard lap score
- Chevrolet Jimmy outboard lap score

When BTD scores were expressed in terms of test performance using the pass/fail criteria indicated in Table 1, 37 of the 40 comparisons were in agreement. Two out of the three instances for which computed and measured performance differed, were associated with the torso form.

In terms of overall belt system performance, the electronic and actual BTD results were in agreement for eight out of the ten vehicles tested. One vehicle, the 1989 Toyota Tercel, failed the electronic BTD evaluation (both on clavicle and sternum scales) but passed with the actual device. It should be noted, however, that the clavicle and sternum scores were extremely close to criterion levels. The second vehicle, a 1985 Chevrolet Jimmy, failed the electronic BTD (due to outboard lap score) but passed the actual device. This is discussed further in the next section.

DISCUSSION

In the validation phase of the study there were 10 out of 40 instances in which BTD and electronic scores differed by more than one cm. Each case in which the discrepancies exceeded one cm was investigated to determine the possible reasons. The following observations provide at least a partial explanation for the discrepancies noted.

1. The seat belt algorithm was based on a restraint system comprising a single retractor system and a simple type of belt hardware. This configuration does not adequately represent the various belt systems in use (double retractor, multiple type of belt hardware, different seats, etc.). It is believed that expanding the algorithm to simulate various belt systems would improve the accuracy of the prediction.
2. In some vehicles, the seat squab was inclined significantly. It was found that seat inclination could have a significant effect on belt fit scores but this factor is not currently accounted for in the BTD model and seat belt algorithm.
3. The upper retractor in the Reliant was located further back relative to the H-point machine than the other vehicles tested. In the actual vehicle, the position of the retractor caused the seat belt to rest on top of the head restraint, rather than on top of the seat. Since the head restraint was not modelled, the seat belt algorithm determined an incorrect routing for the belt resulting in discrepant clavicle and sternum test scores.
4. It is believed that the discrepancies associated with the Tercel scores were due to technical problems that arose during the digitising process.

Table 3 - Comparison of Actual and Electronic BTD Test Scores

Vehicle Tested	Method	Lap Inboard	Pass/ Fail	Lap Outboard	Pass/ Fail	Clavicle	Pass/ Fail	Sternum	Pass/ Fail
BMW 319i (1985)	BTD	4.68	P	5.52	P	12.70	P	16.63	P
	Electronic	5.40	P	5.00	P	13.00	P	18.00	P
Chevrolet Jimmy (1985)	BTD	0.35	F	3.00	P	13.30	F	18.40	P
	Electronic	1.00	F	0.00	F	13.60	F	18.80	P
Chrysler Cirrus (1995)	BTD	4.80	P	4.70	P	12.65	P	15.73	P
	Electronic	4.80	P	3.80	P	12.60	P	16.00	P
Ford Windstar (1995)	BTD	4.28	P	4.50	P	10.95	P	13.00	P
	Electronic	3.20	P	4.00	P	10.90	P	13.00	P
Hyundai Excel (1987)	BTD	3.65	P	3.90	P	13.35	F	18.00	P
	Electronic	4.00	P	3.90	P	13.50	F	18.20	P
Mazda 626 LX (1989)	BTD	3.18	P	3.20	P	10.00	P	16.35	P
	Electronic	2.90	P	3.00	P	10.00	P	14.30	P
Mercedes Benz 190E (1984)	BTD	4.60	P	4.35	P	10.38	P	14.30	P
	Electronic	3.70	P	3.80	P	9.30	P	13.20	P
Nissan Sentra (1993)	BTD	3.17	P	3.83	P	9.83	P	12.33	P
	Electronic	3.60	P	3.80	P	9.60	P	12.20	P
Plymouth Reliant LE (1985)	BTD	2.63	P	2.45	P	6.83	F	10.03	F
	Electronic	2.60	P	2.40	P	4.60	F	7.60	F
Toyota Tercel (1989)	BTD	3.57	P	3.90	P	7.63	P	12.03	P
	Electronic	2.90	P	4.40	P	5.60	F	9.60	F
Pass criteria Lap score > 1.5 7 < Clavicle score < 13 12 < Sternum score < 22									

Fig. 5 - Lap Inboard

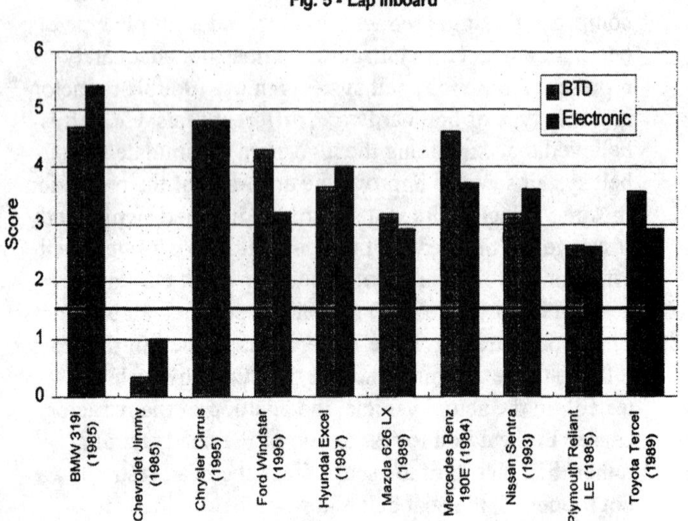

Fig. 6 - Lap Outboard

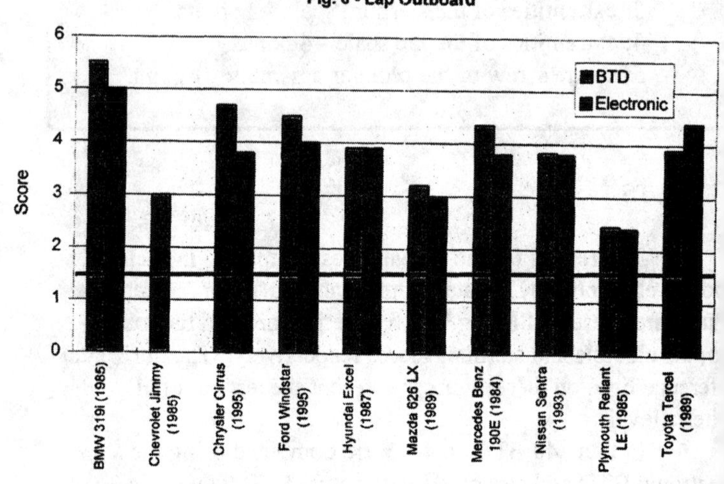

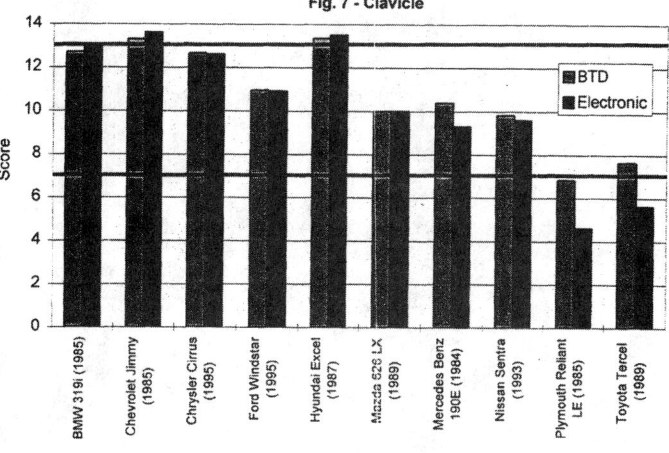

Fig. 7 - Clavicle

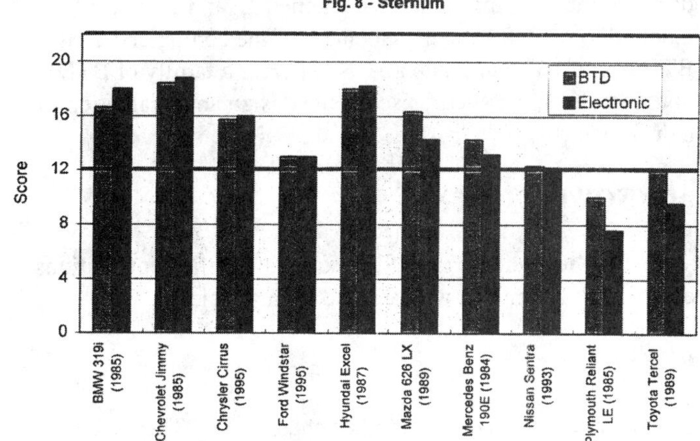

Fig. 8 - Sternum

5. The errors observed on the outboard side of the Jimmy lap scores may be due to the fact that the seat belt algorithm assumes that the belt forces on the torso and on the lap are equal in magnitude. In the case of vehicles with the dual retractor systems, such as the Jimmy, this assumption may not be tenable.

In summary, considering the limitations of the seat belt algorithm, the initial version of the electronic BTD produced remarkably good results. Certain limitations of the electronic BTD remain to be addressed. These are briefly discussed below.

LIMITATIONS (SEAT BELT ANCHORAGE) - At present, the seat belt algorithm calculates the shortest belt routing on the BTD surface for given starting and ending anchor points. These anchor points correspond to the end points of the flexible webbing (not including buckle, etc.) and do not necessarily correspond to the seat belt anchorage location[5]. Therefore, to implement the electronic BTD it was necessary to input the digitised values of the anchor points to generate the electronic seat belt.

Further development of the seat belt algorithm is required to permit the use of manufacturer provided anchorage co-ordinates as input to the model and eliminate the need to digitise belt starting and ending anchor points. Ideally, the input variables to the seat belt algorithm should include only characteristics that can be readily obtained from the manufacturers, such as seat belt type, specifications for associated hardware (belt buckle, plastic sleeves, etc.), co-ordinates of anchorage locations, H-point location, and seat position and geometry. The seat belt algorithm would use these characteristics to determine the natural routing of the belt, taking into account the buckle characteristics as well as seat contact points.

MODELLING SEATING - As discussed previously, there are no techniques to mathematically determine the position/posture of the BTD in a given seat. This was not considered problematic for the electronic BTD since it could be positioned accurately with reference to the manufacturer-provided H-point[6]. However, for mannequins representing other than 50th percentile adults, the design H-point, determined with the H-point machine, will not represent the position of the true hip point which depends critically on mannequin weight and other anthropometric characteristics.

Thus, in order to be able to extend the use of the model to evaluate belt fitment for a wider cross-section of the occupant population, it will be necessary to develop a general 'seating' model to, i.e., determine the hip point of any mannequin given relevant mannequin and seat characteristics.

CONCLUSIONS

A computer-based model of the BTD was developed and validated. The results of the validation study demonstrated good concordance with physical measurements obtained with the actual device. However, a number of refinements were identified which could further improve the seat belt algorithm and the overall utility of the model. In

---

5 The seat belt anchorage location is the location of the physical attachment of the harness to the vehicle structure or the seat, as the case may be.

6 Manufacturers normally provide H-point co-ordinates relative to a well defined vehicle structural landmark. The H-point is the hip point corresponding to the 50th percentile adult as represented by the H-point machine.

addition, further modelling is required for electronically deriving the H-point for a given mannequin and seat. It is planned to continue the development of the computer-based BTD model. The ultimate goal is to derive a family of BTD mannequins that could be used in the design and evaluation of belt fitment.

ACKNOWLEDGMENTS

The authors gratefully acknowledge the contributions of Jennifer Thompson who assisted in this project.

REFERENCES

1. Gibson, T., Tylko S., and Shewchenko, N., "The Belt-Fit Test Device: A Description of its Development and Function as an Evaluative Tool", Biokinetics & Associates, Document R93-05, June 1994.
2. Tylko, S., Gibson, T., Descôteaux and Fournier, E., " A Demonstration of the Capabilities of the Belt-Fit Test Device", Biokinetics & Associates, Document R93-06, June 1994.
3. Tylko, S. and Gibson, T.," Belt Fit Test Device Fleet Measurement", Biokinetics & Associates, Document R94-02, May 1994.
4. Transport Canada, "Operational Manual For The Belt Deployment Test Device", Road Safety, January 1993.
5. Y.I. Noy, R. Carrier, and V. Battista, "Development and Validation of the Electronic Belt Fit Test Device", Transport Canada Technical Memorandum TME 9601, September 1996.

CHAPTER 11:

PERFORMANCE ISSUES: RETRACTORS, AGING BELTS, ETC.

840396

Diagnosis of Seat Belt Usage in Accidents

Charles A. Moffatt
School of Public and Environmental Affairs
Indiana University
Bloomington, IN
Edward A. Moffatt
Biomech, Inc.
Orinda, CA
Ted R. Weiman
Bendix Safety Restraint Div.
Allied Corp.
Mt. Clemens, MI

ABSTRACT Determining whether restraint systems were used in an accident is an important task of the field accident investigator. Restraint systems protect people by causing the forces which must be applied to the person in a collision of a particular value of to be applied to stronger parts of the anatomy over a longer period of time, and by helping to prevent ejection. Restraint use may cause characteristic changes in the vehicle at the mounting points, in the retractor mechanism, the restraint system hardware, and the webbing. Occupant Injuries will reflect altered occupant kinematics and altered load application areas when restraints are used.

THE PURPOSE OF THIS PAPER is to bring together information that can lelp the investigator gather data in field accident studies to determine the use and performance of restraint systems in motor vehicle accidents.

Nearly the entire fleet of automobiles and light trucks in the United States was originally manufactured with some form of occupant restraint system. The value of seatbelts or other forms of occupant red straint in a crash is well recognizes and understood as a result of decade of engineering development and studies of actual accidents. Although seat belts potentially offer enormous savings in lives and injuries they are often unused. Only about 11% of the people in cars and an even lower percentage of the people in light trucks and utility vehicles use the restraint systems that are in their vehicles. Therefore, the rate of usage of restraint systems in that rate of usage are extremely significant figures. Since few people use their restraint systems, a relatively small error in determining the number of users would lead to a large error in the estimated rate of usage. This paper is intended to help the field accident investigator be more accurate and more conclusive in determining whether seat belts were used in an accident.

HOW RESTRAINT SYSTEMS PROTECT PEOPLE - The sudden change in speed of the human body which occurs during a collision can result in injuries because very large forces must be exerted on the body to change its speed quickly. The shorter the period of time in which the speed change occurs the larger the force required. Usually the amount of speed change is determined by the circumstances of the collision, but the suddeness with which the vehicle occupants change speed and the parts of their anatomy to which forces are applied can be affected by the restraint system.

A collision of vehicles with unrestrained occupants can be visualized by considering the first and second impacts. In the first impact the vehicle undergoes a speed change, or delta V. The speed change is determined by circumstances of the accident such as vehicle speeds and masses.

The first collision is usually

nearly complete by the time of the second collision in which unrestrained occupants impact upon the interior of the vehicle. In a crash where the deformation of the vehicle and struck object is measured in feet, the deformation of the occupant and the surfaces he strikes will be measured in inches or less. Thus, the occupant undergoes about the same speed change as the vehicle, but the occupant's speed changes in a much briefer time.

Restraint systems make the occupant participate in the first collision by coupling him closely with the vehicle: then the occupant changes speed while the vehicle changes speed. The occupant is much better off undergoing the speed change, which he must undergo in any case, during the relatively long first collision than during a brief second collision.

The second principal value of restraint systems is in helping to prevent ejection, or being thrown out of the vehicle. Inside a vehicle is statistically a much less hazardous place to be than outside the vehicle, or partly outside the vehicle, during an accident.

The third value of restraint systems is to direct the forces upon the occupant to parts of the anatomy which are more likely to withstand large forces without serious injury.

TERMINOLOGY

Many terms are used for describing restraint systems, because they are dealt with by people with different levels of technical expertise. It may be worthwhile, therefore, to review the terms used by engineers. Some of the main terms used in discussing restraint systems are defined below. More complete listings may be found in references (1)* and (2).

SEAT BELT SYSTEM - A restraint system consisting of straps or webbing, plus all the buckles, fasteners, attachment devices and other associated hardware is a seat belt system.

"Hardware" simply means the metal or rigid parts of a seat belt assembly.

"Buckle" refers to a quick- release connector which will release a seat belt system from restraining a person. This includes buckles in passive systems which are intended to be released only in extraordinary circumstances and buckles on child restraint systems which the child is not supposed to operate. The flat metal plate which is attached to one end of the webbing and extends inside the other part of the buckle is called the "latch plate" or the "tongue". The tongue usually has either a hole through it or notches in its sides. The hole or the notches are engaged in the buckle when it is latched. In this paper the term latch plate will be synonymous with tongue.

"Webbing" means a narrow fabric woven with continuous filling yarns and finished edges.

ATTACHMENT HARDWARE - The webbing or hardware is attached to the vehicle with "attachment hardware". The term includes anchors, bolts, guide rings, and other hardware fixed to the vehicle.

A "shoulder guide loop" is a ring through which the upper torso belt passes. It is usually mounted on the B pillar.

RETRACTOR - A device for storing part of the webbing by rolling it up is called a retractor.

A "non-locking retractor" is a retractor from which the webbing may be withdrawn at any time with little resistance.

An "automatic locking retractor" is a retractor that allows the webbing to be withdrawn then rewound but will not permit a second withdrawal until the webbing is almost completely rewound. These are commonly seen in lap belts in two-retractor systems and in rear seat lap belt systems. The occupant pulls the lap belt out to a length greater than he requires, and latches it. The automatic retractor rewinds the slack and then locks securely

An "emergency-locking retractor" is a retractor that allows the webbing to be withdrawn and rewound freely except when the retractor is caused to lock by vehicle acceleration, rapid webbing withdrawal, or some other automatic system.

"Vehicle-sensitive" retractors are emergency locking retractors which lock when the vehicle tilts or when it changes velocity sharply in any direction (figure 1). These retractors are required by Federal Motor Vehicle Safety Standard 209 (1) to lock with less than one inch of belt travel when the vehicle acceleration reaches 0.7g; typically they

*Numbers in parentheses designate references at end of paper.

lock at an acceleration less than that. These retractors are required not to lock when the vehicle tips 15 degrees or less in any direction. This amount of tipping corresponds to a horizontal acceleration of 0.26 g in any direction.

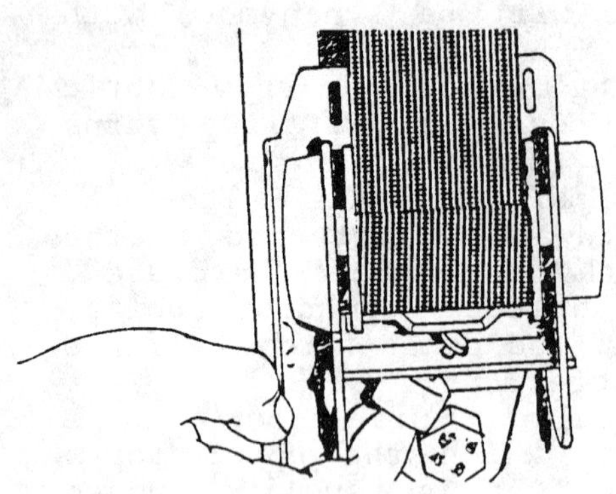

Figure 1 - The pendulum actuates this vehicle sensitive emergency locking retractor.

"Webbing-sensitive" retractors automatically lock when the webbing is suddenly withdrawn from the retractor, as in the early moments of a crash, but they do not lock when the webbing is withdrawn slowly in normal use. Webbing sensitive retractors are required by European regulation European vehicles, and those retractors also include the vehicle feaclude the vehicle sensitive feature as well.

A "tension release device" is a method for reducing the pressure of the upper torso retraint upon the occupant. This device can take the form of a clip which is positioned on the webbing to limit its retraction, as well as what is call window shade mechanism in the retractor.

Restraint systems are categorized by how many places they attach to the vehicle. A Type 1, or two point restraint system is a lap belt. A Type 2 or three point restraint system is a combination lap-shoulder belt. A Type 2A restraint is a separate shoulder belt which either anchors to the lap belt, forming a three point system or to the vehicle. A four point system is an entirely separate lap belt and shoulder belt. Three point restraint systems are often described by the manner in which the webbing moves for length adjustment and for accommodation of occupant position changes. The most common type today is "three point continuous loop." In this system one end of the lap belt is fixed to the vehicle, near the door sill, without a retractor. A continuous webbing extends across the occupant's lap, through a latch plate, then up across his shoulder to a guide assembly or to an emergency locking retractor. When the occupant places the restraint system around himself, he adjusts the lap belt length by sliding the webbing through the latchplate. Sometimes this involves tipping the latchplate, relative to the webbing, to allow it to slide, but in other designs it slides freely. The length of the upper torso restraint webbing is adjusted automatically by the retractor. Continuous loop restraint systems are also called "one retractor" restraint systems.

A type of three point system less commonly seen today is the "two retractor" restraint system. In this system the latch plate cannot slide along the webbing but has the pelvic and upper torso webbing permanently attached to it. The lap belt and upper torso belt are anchored at opposite ends to separate retractors. The lap belt retractor is usually automatic locking and the upper torso belt retractor is emergency locking.

The pair of terms "active" and "passive" and the pair of terms "automatic" and "manual" can be particularly confusing. Simply stated, an active restraint system requires the person using it to be "active" and put the restraint system around himself, "manually." A passive restraint system allows the person using it to be "passive" and the system will restrain him "automatically." Thus "active" is synonymous with "automatic" in restraint system.

VEHICLE FACTORS IN EVALUATING RESTRAINT USE

When a restraint system is worn by a person during a crash, tensile forces in the webbing are caused by the person's body. These forces can make changes at the anchor points, hardware, and the webbing which the investigator can detect. The signs of loading will be more pronounced in severe crashes and with heavier occupants: the absence of a particular finding, therefore, may not mean that the restraint system was not worn.

During the vehicle inspection the investigator should determine if the restraint system is in its normal position, or if it has been stowed in some way that precludes use. People stow belts away for such reasons as to make it easier to slide into the center position of a bench seat, to make it easier to get into the back seat of a two-door car, to get the belts off the seats, to prevent the buckles from rattling, or to defeat an interlock or buzzer system.

Some cars have warning buzzers and ignition interlocks which sense whether the belt has been pulled out from the retractor or whether it is latched. Cars with ignition interlocks are all 1974 models (but not all 1974 model cars have ignition interlocks). These warning systems are sometimes defeated by pulling the belt out and tying it in the car, or by buckling it. The buzzers are also defeated by disconnecting the electrical power lead to the buzzer or to the seat pad switch.

The investigator should determine if belt restraint systems which have a length adjustment are adjusted correctly for the person seated in each position in the car. Often, the investigator will not know the size of each person at the time he is inspecting the vehicle. He should try the restraint system on himself when possible, and note whether it fits, is larger or smaller, is much larger or much smaller than he would require. The adjustment position of the front seat should be noted, as well as whether the adjustment is jammed in place or may have been moved post crash. Before attempting to move the seat adjustment its original position should be marked and recorded. It is importatnt to be sure that the retractors allow the webbing to be withdrawn fully, and that the seat has not been displaced. The investigator should note whether he is wearing bulky clothing. Agreement between the length adjustments and the age, stature, and clothing of the occupant can be checked later.

Some should belts have a tension release device which works to reduce contact pressure of the belt on the occupant's shoulder. One form of tension release device, the comfort clip, can be placed at different positions along the upper torso belt, and limits retraction. This device can render the shoulder belt useless when it is positioned incorrectly.

The general condition of the restraint system should be noted; whether there is dust, glass particles, or blood on the webbing, and how these materials are distributed can be a useful clue. The investigator should determine if the material on the webbing was present before the accident, or was caused by the crash, towing, or storage of the vehicle. He should note whether the webbing has been moved since dust was deposited by looking at places where the webbing passes through guide rings, through the latch plate, where it is wrinkled, and where it passes into the retractors. Sometimes the part of the webbing inside the retractors is much cleaner than the rest; sometimes there is a silhouette where the belt lays on the seat or floor.

The total amount of use the webbing has had can often be gauged from wear of the edges of the belt where it touches the retractor cover, wear of the faces of the belt where its length is adjusted, and food stains, dust, and dirt on the belt. The edge of a lap belt that is nearer the occupant's torso tends to wear more than the opposite edge. The amount of belt use must, of course, be noted in comparison with the age of the vehicle. While the total amount of use does not necessarily relate to the use at the time of the accident, it is helpful in interpreting interview information about belt use habits.

INSPECTION OF THE TONGUE - The tongue goes into the body of the buckle and is engaged by positive locking parts when the restraint system is in use. Wear patterns on certain tongues can indicate the amount of use the restraint system has received. Most cars of American manufacture use a tongue with a single hole in it. The hole is either square or D-shaped. Latching is accomplished by a piece which fits into the hole and bears against the flat side of the hole when the restraint system is tensioned.

When the tongue is slipped into the buckle the latching mechanism rubs on it, in line with the D-shaped hole. after usage, an area of very fine scratches can be discerned on the bottom surface of the tongue. In extreme cases the eplating wears through, but usually the scratches can be detected only by holding the surface of the tongue so that light is reflected and noticing a dull area with lines in the direction of motion. Some cars use a tongue that has notches on the edges of the plate instead of a hole. As the tongue is pushed into the opening in the buckle the latching mechanism rubs its edges. The edges of the plate are not as smooth as the flat surfaces, so it is usually not possible to tell by looking at them how much the buckle has been used.

When a person uses a restraint system in a very severe crash, the force of his body on the belt can cause an impression on the tongue where it is engaged in the latching action. Impression or brinelling of the side of the hole in the tongue is a very insensitive indicator of belt use, and would almost certainly be accompanied by other indicators.

INSPECTION OF THE RETRACTORS - In an automatic locking retractor, there is typically one or two wheels with teeth resembling a coarse gear as shown in figure 2. When the retractor locks, a toothed wheel is engaged by a pawl, or locking bar, to prevent it from rotating and allowing the webbing to reel out. It is possible to inspect the engagement area of the teeth and locking bar to see if there is brinelling or local deformation from belt loading. The first place to look is on the leading edge of the locking bar. Then if marks are found on that, it can be possible to find corresponding marks on the engaging teeth, and thereby estimate the length of the webbing at the time of impact according to which tooth is marked. This is a relatively insensitive indicator of belt use, but in cars tested with dummies in a 30 mph rigid barrier impact it is possible to find marks on the retractor teeth, as a rule.

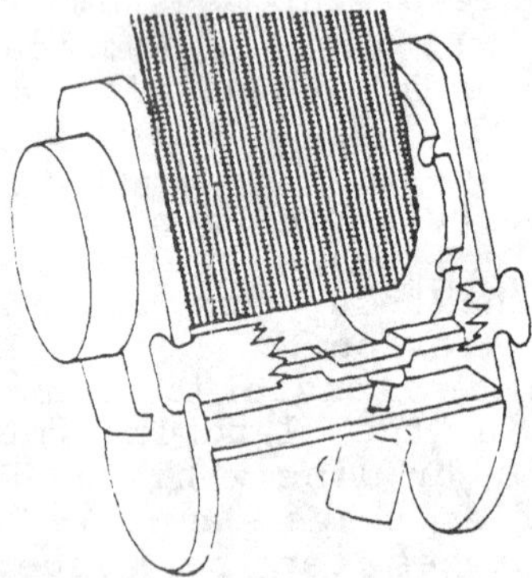

Figure 2 - Retractor latching mechanism.

Normal wear can mark the retractor mechanism, too. The locking bar wears more than the toothed wheel because the wear is distributed over several of the teeth of the wheel.

Most retractor designs use two toothed wheels and one locking bar. The toothed wheels are located at either end of the take-up spool. In some designs one of the toothed wheels is rotated slightly with respect to the other. This causes the locking bar to always engage the advanced wheel and not touch the other one as shown in figure 3. When the belt is worn in a crash, however, the large forces cause the locking bar to contact both toothed wheels. Wear marks will make the advanced wheel appear more polished than the unworn one. Marks from a severe crash, however, will show on both wheels.

Figure 3 - The locking bar will engage both toothed wheels only under heavy loading.

In a severe collision the locking bar can be jammed in the latched position, making it obvious that the belth was worn and exactly which tooth is involved.

Another indicator of belt use is deformation of the frame or the mounting plate of the retractor. Sometimes the plate of the retractor adjoining the mounting bolts is bent, the retractor sides are no longer square and parallel, or the extraction or retraction movement of the webbing is uneven. The investigator should consider the direction of force on the retractor, how it is likely to be deformed, and when necessary compare it to an undeformed exemplar retractor.

An investigator can demonstrate a vehicle sensitive retractor in a stationary car without disassembling anything by slapping the car body near the retractor with the palm of the hand, while reeling out the shoulder belt webbing. The slap can cause enough vibration to lock the retractor and stop the webbing withdrawal. This technique is convenient for vehicle inspections, but if the retractor does not lock it does not necessarily mean it does not work.

INSPECTION OF THE VEHICLE BODY

Evidence of belt use can also be found on the car body near the anchor points. The belt tension load would exceed 1000 lbs. in a 30 mile per hour barrier impact. This force can deform the sheet metal of a car body. A slight bulging occurs which is proportional to the load and perpendicular to the surface of the sheet metal, rather than in the direction the belt is pulling it. Outboard anchor points are usually located in the rocker panel or door sill area of the car. Inboard front seat anchor points are usually near the transmission hump. Rear inboard lap belt anchor points, however, are usually located in a flat area. The anchor point will be most susceptible to deformation if the metal is flat, less susceptible to deformation if the metal has a single curve, and least susceptible to deformation if the metal has a compound curve.

The portion of the vehicle body to which the shoulder guide loop is anchored, usually on the B pillar, can provide information about belt use. The force on the shoulder guide loop is downward, forward, and inward, and it tends to deform the mounting area in that direction. Usually the shoulder guide loop is mounted by a bolt which does not itself deform. The part of the body it threads into, however, gets twisted. An easy way to detect this is to remove the anchor bolt and thread a longer bolt into the hole. Misalignment of the hole can be observed by the angulation of the longer bolt.

INSPECTION OF PLASTIC PARTS - Metal parts of the restraint system often have a plastic coating or liner. If the webbing is loaded heavily in a crash it will imprint its fabric pattern upon this plastic as shown in figure 4. In severe crashes heat is generated where the belt slides. The temperature is high enought to melt the surface fibers of the belt. The plastic coating of parts contacting the belt will be ingrained with the webbing pattern. Places to look for this are guide rings, and the sliding tongue. The fixed anchor points, while plastic coated, are a less sensitive location because the belt does not slip through them. It is usually necessary to peel the belt back or cut it away in order to see the area where the highest pressure was exerted.

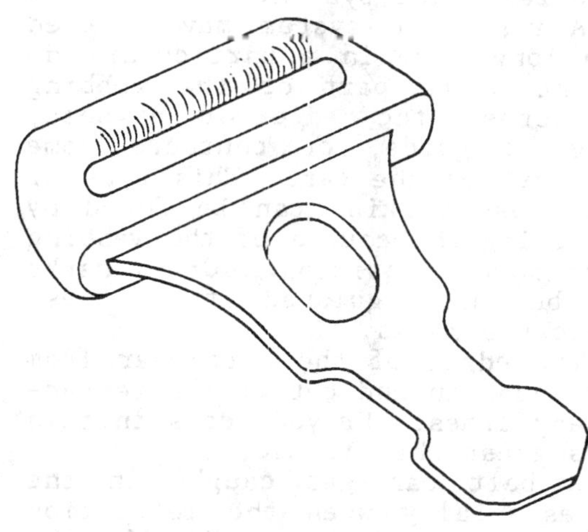

Figure 4. Fabric pattern of the webbing imprinted on a plastic part (figure drawn from a photograph)

INSPECTION OF THE WEBBING - Abrasion or rubbing on the webbing can be very useful in determining whether the restraint system was used during a collision. Both sides and both edges of the webbing should be examined over as much of the length as possible. The surface which the webbing rubbed on in forming any marks should be determined.

The webbing is sewn to itself at the end attachments. Very high belt loads, approaching the ultimate strength of the belt, will begin to break the stitches. If stitches are broken, other more sensitive indications of belt usse should be apparent as well. In many retractors the because there are several wraps of belt over them on the spool.

The cloth label on the belt can become puckered when the underlying belt is stretched. This assumes, however, that the label was drawn tight when it was originally sewn to the belt. Many belt labels are sewn on loosely enough that there is room to place a finger under them, so they would not become puckered under crash loading.

The restraint system webbing becomes harder, like starched laundry, when it is stretched, and it tends to curl. The belt regains its original softness, however, as it is moved and handled after stretching.

ABRASION MARKS DUE TO CHRONIC WEAR - In looking for objects that contacted the webbing, it is important to consider the dynamics of the crash as well as the normal position of the restraint system.

A restraint system may be used for a long time in a particular adjustment where part of the webbing lays across the edge of a seat, through a guide, or contacts some other part of the car. This type of chronic use marking can be found by considering the course of the webbing in its normal position. Crash marks must be distinguished from these chronic use marks.

The edges of the belt wear from being wound in and out of the retractor many times. Frayed edges in this area suggest chronic use.

A belt can get caught in the door especially when the retraction is slow or incomplete. Usually the door will not close on the same area of the restraint system every time so the marks will be scattered. Similarly, in a front engine, forward control van the webbing can get caught between the engine cover and the floor. Certain child safety seats clamp to the webbing and leave characteristic marks such as are shown in Figure 5.

Figure 5 - The lap belt webbing is marked by the clamp on this child safety seat.

Seat belt webbing is woven of threads (called 'ends') which are made of fibers which in turn are composed of finer filaments twisted together. Under long term wear the forces on the webbing are smaller than in a crash but they are repeated many times. This causes the webbing to lose its shininess and acquire a fuzziness. Under low power magnification the fuzziness may be seen to consist of uniform, short lengths of fine filaments of webbing, while the larger threads (ends) are still intact. The fine filaments do not show signs of heating and they are all about the same length. Examination under higher magnification discloses that the condition of free ends of the separated filaments are all similar.

WEBBING MARKS FROM CRASH LOADS- Severe forceful collision causes rubbing, different from the long term, gentle rubbing the seat belt is used for a long time in the same position.

Marks from a collision can occur at a point on the belt that does not come into contact with the surface upon which the belt rubbed unless the belt is heavily loaded. An example is where a floor anchored rear seat lap belt passes over the rear edge of the seat. Upon heavy loading of the belt by an occupant the seat edge can be pushed down, bringing the seat into contact with part of the webbing it would not normally touch, even with a person in the seat.

Scratch or abrasion marks caused by occupant loading in a collision will be formed in one, or a very few movements. Scratches formed by long-term use are more numerous, producing a uniform surface after a very long time. Even in a one-time loading there may be several parallel scratches formed. These need to be distinguished from chronic use marks, and the best approach is to identify a corresponding surface approximately equal in roughness to the scratches on the belt.

The most significant indicator of restraint system loading is heat. In a collision, the rubbing surfaces can melt plastic materials. This creates a streaky area on the webbing that is light in color, as shown in figure 6. Conditions that enhance the possibility of melting the fibers are: severity of the impact and weight of the occupant; curvature of the belt, for example, where it passes through a guide ring; and poor heat removing capacity of the rubbing surface. Webbing materials are poor conductors of heat, so a high temperature results when the webbing rubs against another poor conductor of heat.

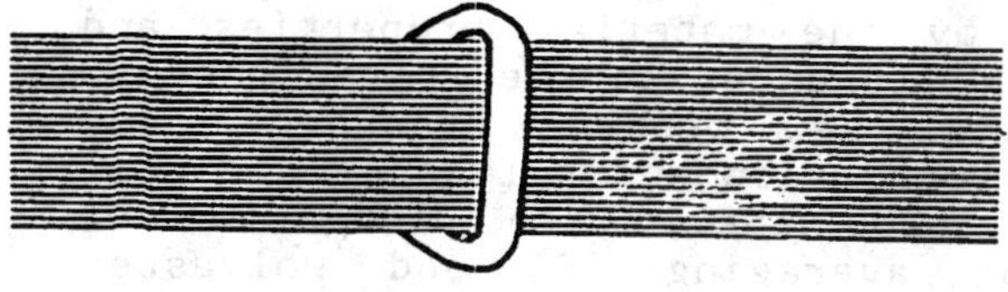

Figure 6 - Streaky area on the webbing is caused by rubbing (figure drawn from a photograph).

Plastics and clothing are generally much worse conductors of heater-than metals: heat marking on the webbing is most likely where it passes over plastic (coated) hardware. Where it passes over occupants clothing, fibers from the clothing can transfer to the webbing.

ANALYSIS OF CUT OR BROKEN WEBBING

The accident investigator may find that seat belt webbing has been separated. To complete the investigation he must determine the mechanism of the parting of the webbing. An initial distinction to be made is whether the seat belt webbing "broke" due to an overload, or if it was "cut" by some external force being directed against the webbing surface. If it was cut, then the question becomes whether it was intentionally cut either before or after the collision, or if it was accidentally cut during the collision by some surrounding structure.

When seat belt webbing is loaded to destruction and when it is cut, it obtains characteristics which are easily identifiable to the naked eye. Under peculiar circumstances such as photodegradation or when there is some question as to how the separation occurred, microscopic analysis can be helpful.

VISUAL ANALYSIS OF SEPARATED WEBBING

It is possible to use laboratory tests to show the characteristics of webbings which have been cut and webbings which have been broken in tension. Laboratory tensile tests of seat belt webbing and assemblies are routinely used by quality control to demonstrate compliance with FMVSS 209. The damage which results to the restraint assembly and webbing from tensile overload shows characteristic patterns which are useful in post crash restraint use investigation.

If seat belt webbing is loaded until some component breaks, while using its own hardware as its attachment points, the separation will either be from fractured hardware, broken stitching, or cut webbing where the webbing contacts the hardware. Webbing is never observed to break in tension away from the hardware during such tests. Figure 7 shows the typical appearance of webbing which has been cut by its attachment to anchor hardware in a laboratory tensile test.

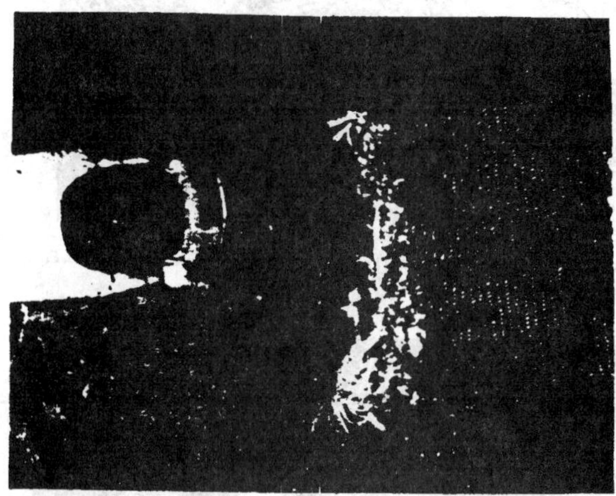

Figure 7 - Seat belt webbing cut at anchor attachment in a tension test.

A specially designed set of web jaws is used to test webbing strength. These jaws, as specified in FMVSS 209, are a split drum design with either a 2- or 4-inch diameter, as illustrated in figure 8. Using these jaws, webbing is pulled in tension until it breaks in mid span. The characteristic fracture pattern from the sudden release of stored energy is readily identifiable to the naked eye. Figures 9 and 10 show pieces of nylon and polyester webbing, respectively, which have been pulled apart in tension using webbing grips.

Figure 8 - Webbing jaws for tensile test.

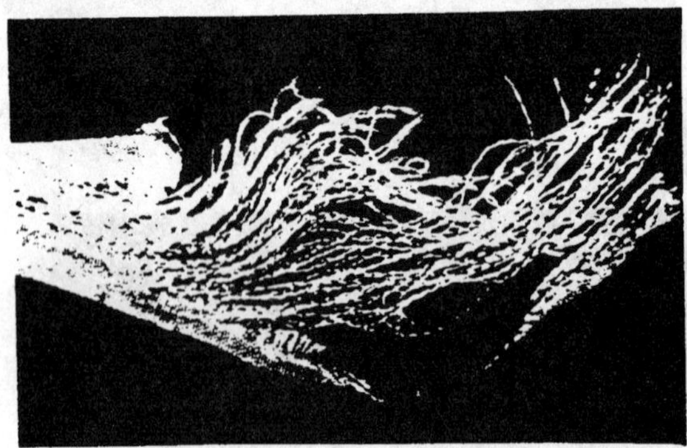

Figure 9 - Nylon webbing broken in a tension test.

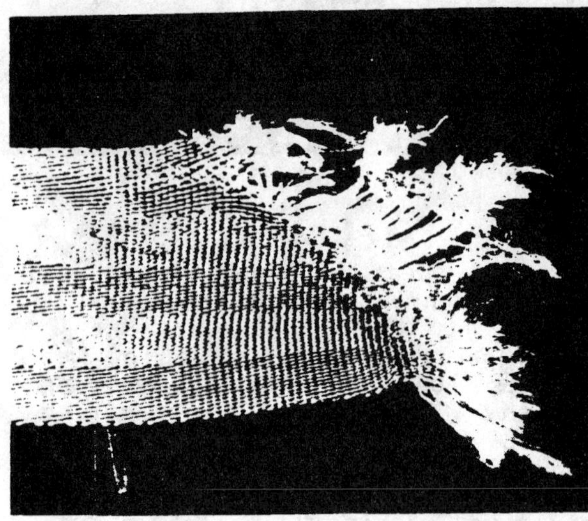

Figure 10 - Polyester webbing broken in a tension test.

A pattern is seen in these and other similarly loaded webbing overload fractures. The fibers separate in a random pattern, leading to a non-uniform disruption of the separated ends. In nylon, this has been described as "horsetailing." In polyester, the horsetailing is less pronounced. The differences between these two patterns is caused primarily by the material properties and the differences in the stored energy. Typical new webbing breakage loads for both nylon and polyester are well over 6000 lbs., with elongation of nylon averaging 17% and polyester elongation averaging 7%.

The separated fibers of the webbing are never all of even length across the width of the webbing when it is broken in pure tension. A typical piece of three-panel nylon webbing has 264 ends across its width, and comprises 15,000 to 18,000 filaments. Typical five-panel polyester webbing also has 264 ends, but is composed of 18,000 to 20,000 individual filaments. This many individual filaments simply cannot all break at the same length, unless a cutting surface is forced against them.

Webbing is often cut by vehicle owners, rescue operators, or even wrecker drivers who want to remove the belts. Intentionally cut webbing has a characteristic apperance. Cutting the webbing with a knife or shears results in uniform cuts with many of the fibers separated at exactly the same length. Sometimes a curious pattern cut end appears, as shown in figure 11, due to the belt being folded or roped at the time it is cut. Even in these cases sections of the fibers will still be separated in a uniform line. Figure 12 shows a piece of polyester webbing which was cut with a knife, and figure 13 shows a nylon webbing cut with scissors.

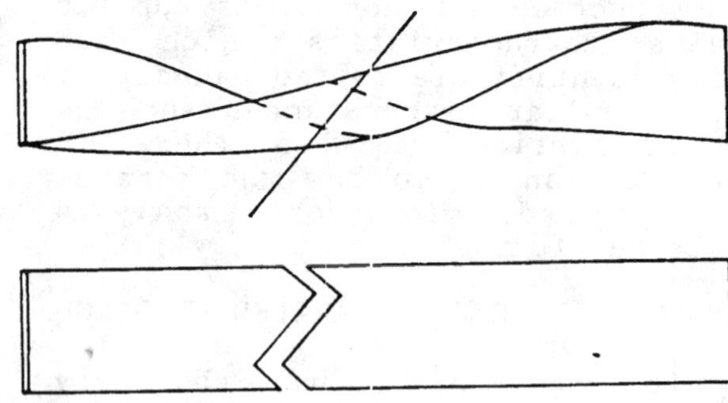

Figure 11 - Irregular cut end of a belt cut while folded over.

Figure 12 - Polyester webbing cut a knife.

Figure 13 - Nylon webbing cut by scissors.

Webbing may be cut accidentally during a collision. Sometimes this results from intruding side structures, such as a door in a side impact, which pinches the webbing at the seat or from a belt which is cut on an adjoining structure while being loaded by the occupant. There are two typical findings when this occurs. First, the location of the cut can be used to determine the cutting structure simply from its position in the vehicle. It is important, of course, to visualize the dynamic elastic deformation of the vehicle, and not simply to use the post-crash deformation. The cutting structure often has deformation, abrasions, or even pieces of belt fibers from its contact with the webbing. Second, unless the belt happens to be loaded across a sharp edge, the cut end will have an uneven appearance across its width without the characteristic even-length fibers as are seen from a knife cut. Figure 14 shows the separated end of nylon webbing totally severed due to loading across a piece of deformed metal during a collision.

Figure 14 - Nylon webbing cut during a collision.

MICROSCOPIC ANALYSIS OF SEPARATED WEBBING

Use of the microscope allows examination of individual filaments. Filaments which are fractured under known conditions have repeatable, characteristic shapes. A filament stretched to breaking has a considerable amount of stored energy which is suddenly released in breaking, causing the filament end to melt. The result of this energy dissipation is a "snapping back" which generally results in necking of the filament near the end. This necking in turn, frequently causes the filament to hook or sag due to its own weight during cooling, and forms an irregular mushroom-like ball on the end of the fiber. The "mushroom" or "shepherd's staff" appearance is shown in Figure 15. Slower loading rates as in laboratory evaluations of webbing and fibers will cause a different morphology to the separated filament end as shown in figure 16.

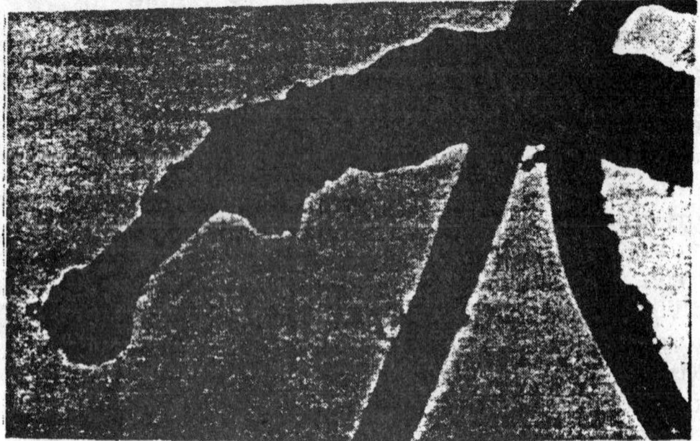

Figure 15 - Melted end of nylon fiber separated during a collision.

Figure 16 - Scanning electron micrograph of a nylon fiber fractured at a slow rate.

Fibers which have been cut with a knife generally have an end appearance that is different from fibers broken in tension, although individual fibers separated both ways may appear quite similar. Usually, knife cut fibers have a squared end with a slight upset on one side but without necking. Knife cuts often cause serrations across the fiber end, as shown in figures 17 and 18. The uniform length of fibers which so characteristically identifies a cut webbing to the naked eye can be clearly seen under the microscope, as shown in the electron microscope photograph, figure 18. Additional photomicrographs showing these effects have been published by Niederer (3).

Figure 17 - End of a cut fiber.

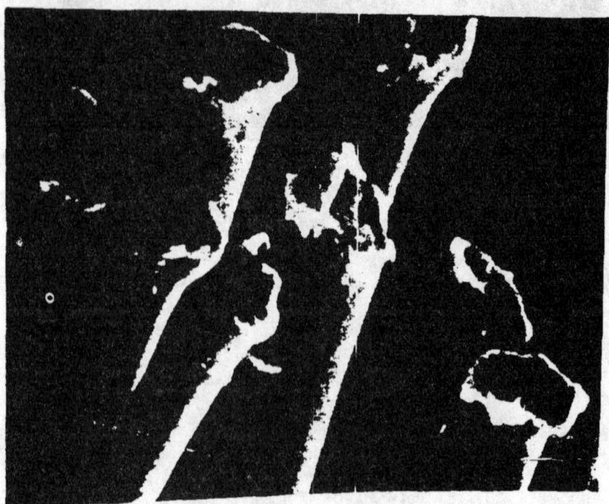

Figure 18 - Uniform length of a bunch of cut fibers.

Degradation of webbing by light is another condition which affects the microscopic appearance of the fiber. Due to material changes, webbing which has been exposed to sunlight is less pliable and has a lower ultimate strength. When fractured in tension it does not show the characteristic "mushroom" melt but, rather, a more squared end due to a more brittle fracture, as shown in figure 19.

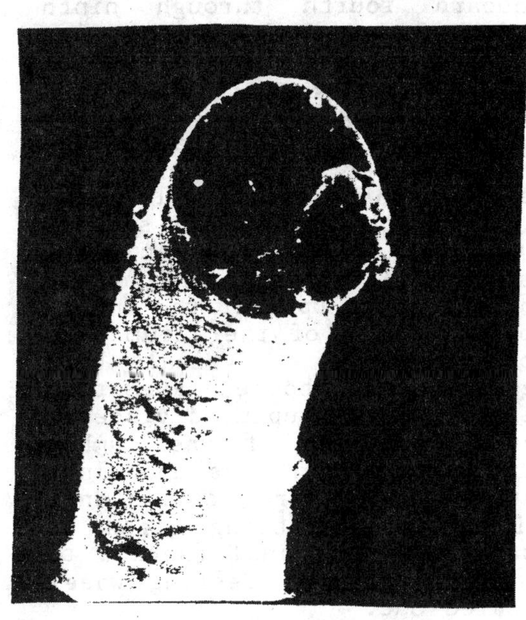

Figure 19 - Tensile separation of a photodegraded fiber.

Although there are characteristic shapes which filament ends assume when they are cut and broken, it can be extremely misleading to conclude that an entire piece of webbing was cut or broken solely upon the analysis of a small number of filaments in one area. Even when webbing is cut with a knife or scissors, some of the filaments will be broken in tension, as they bunch ahead of the cutting edge. Conversely, when a belt is pulled apart in tension one can always find a few fibers which have a cut or pinched appearance.

There are approximately 18,000 filaments in a seat belt webbing. With so many filaments the investigator can be misled by taking too small a sample. To conclude how a piece of webbing was separated on the basis of looking at a few filaments is unrealistic, particularly in light of the clear indicators of a cut or broken seat belt which are visible to the naked eye.

INJURY ANALYSIS TO DETERMINE IF A RESTRAINT SYSTEM WAS WORN

Restraint systems reduce a person's chance of being seriously injured or killed in a collision. Injuries to a restrained person result which would not have occurred if the person had been unrestrained. The restraint-caused injuries, however, are almost always less severe than the injuries that would have been suffered if restraints had not been used.

The accident investigator will find the pattern of injuries and altered kinematics of the occupant to be useful clues in determining if the seat belt system was used. First, however, he must have reconstructed the vehicular movement in the collision to estimate the principal direction of force and the change in velocity. Then he must analyze the occupant kinematics during the collision to determine where the person would have contacted the occupant compartment interior or other objects involved in the collision with and without the use of restraints. Finally, the investigator can use the actual contact points, along with his knowledge of human tolerance to impact to determine whether the mechanics of the injuries are consistent with restraint usage.

In many collisions this judgment is quite simple. For example, it is highly unlikely that a person will be totally ejected if he is correctly wearing a restraint system. Barring a cutting of the webbing, ejection is, therefore, a strong indication of not wearing the seat belt. Usually, however, the occupant is contained within the vehicle, and the restraint use determination is complicated by considerations such as occupant size, susceptibility to injury, seat position, pre-impact position, and restraint adjustment; vehicle size, pre-impact condition, and dynamic crush; and other vehicle or fixed object occupant contacts by the occupant. Obviously, a tall person is more apt to strike his head on the windshield header in a tiny car than a small person is on the windshield header of a large car. There are some observations, however, which generally hold, for frontal, side, rear and rollover collisions.

In frontal collisions the person moves forward relative to the occupant compartment. Typically, an unrestrained adult right front occupant will strike his knees on the lower instrument panel, chest on the upper instrument panel and head on the windshield or the windshield header. These impacts can leave marks on the car and cause injuries to the person. A restrained person will not move as far forward; often the knee, chest and head contact points and injuries do not occur. Their presence, of course depends on the severity of the crash and many other considerations. The typical forward movement of a restrained adult can be estimated from averaging laboratory tests upon cadavers with a barrier equivalent velocity of 50 km/h using a Volvo 244 three point restraint system which were reported by Kallieris (3). Kallieris found measured average displacements for the head of 56 cm., for the chest 40 cm., and the pelvis 37 cm. Typical belt loads measured in a 50 km/h barrier collision are 4000-6000 N, with the torso belt generally having a slightly higher load. When using these numbers to reconstruct a person's movement in an actual collision, it should be recalled that a 50 km/h barrier collision is a very severe impact, which results in occupant excursions which are larger than most real world collisions.

The driver in a frontal crash typically strikes his knees on the lower instrument panel, chest, abdomen and possibly face, on the steering wheel and head on the windshield. The steering wheel and energy absorbing steering column are good indicators of restraint usage. It is unlikely that any occupant induced compression of the energy absorbing device or any massive bending of the steering wheel rim will be caused by a restrained occupant, except in the most severe frontal collisions. With the excursions in the crash test noted above, however, it is possible to have some chest contact with the wheel and some face contact with the upper rim in severe frontal collisions, but with greatly reduced forces.

The injuries which a person incurs from contact with the restraint system depend upon the severity of the collision and the individual's tolerance to impact. Bruising, for example, can be a positive sign of belt usage. Bruising is most likely to occur at points on the body where the webbing has the high pressure against the body due to a change n angle of the belt. Bruising from the lap belt is most commonly seen at the anterior ilium. Bruising from the shoulder belt is most prevalent in the upper thorax or clavicle areas. The absence of bruising is not a definite sign of the absence of belt usage, however, due to the great variation in the resistance to bruising among different people.

In the more severe frontal collisions, bony injuries can result from the three point restraint system. Kallieris (4) found rib fractures to be the most prevalent injury resulting from shoulder belt usage. Laboratory tests showed these fractures to occur generally along the path of the shoulder belt. Patrick's (5) field studies showed fractures of the inboard fourth through ninth ribs, with the sixth being the most likely fractured. These fractures all occurred in the anterior rib on the side adjoining the belt. Clavicle fracture was not a prevalent injury from shoulder belt usage.

Lap belt injuries are well documented in the literature and summarized by Leung (6). Serious injuries can result from "submarining", where the lap belt slips up over the anterior superior spine of the ilium into the soft abdomen. Typical belt injuries to the abdomen are torn mesentery or bowel, ruptured viscera, distraction fracture of the lumbar spine and even torn rectus abdominus musculature in severe collisions. These injuries are strong signs of belt usage when abdominal contact to the lower rim of the steering wheel can be ruled out.

In side collisions effects of restraints on the kinematics and injuries are very dependent on whether the impact is to the near side or far side of the vehicle relative to the person. For a near side 90 degree side collision, the three point restraint has only a minor influence on the occupant kinematics due to interaction of the person with the door, so it is often difficult to ascertain from injuries alone whether the restraint was used. The far side occupant does benefit from the restraint, however, and has different kinematics and injuries than the unrestrained person. The pelvis is constrained, so rather than a pure translation laterally which is restricted only by leg and arm interferences, the torso will pivot bringing the head lateral and downward relative to the vehicle interior (7). This results in a lower head impact location in the car or no head impact at all. The lap belt contact injuries in this situation are limited to possible bruising of the pelvis, but with an unlikely bruising to the thorax due to the reduced contact force from the torso belt. Angled side impacts will have higher torso belt loadings than 90 degree side impacts with corresponding changes to the kinematics and injury patterns.

In rear collisions the motion of the person is rearward, away from the restraint system. Depending on collision severity and seat back stiffness, the lap belt can affect the person's kinematics by contacting the thighs as he begins to ramp up on the seat back. The authors of this paper have not seen field cases of thigh bruising in this manner, however.

Perhaps the best injury indication of restraint usage in rear collisions is the kinematics of the rebound, which can be considered as a low speed frontal collision.

In rollover collisions the most obvious indicator of belt usage is whether the person is ejected. For the contained occupant it is difficult to find bruising from the belts, because the decelerations are typically not severe, so the belt forces are not high. Occupant kinematics are changed by the belt system, but head contact into the roof directly adjacent to the seated position is still possible.

FORMING A CONCLUSION ABOUT RESTRAINT USE

It is well known from survey data that only about one in ten car occupants in the United States uses a restraint system. The investigator should not, however, have a presumption that unless he can find specific evidence that the belts were used that then they were not used. That would be contrary to the methods of a scientific inquiry. Every case must be considered all by itself.

Restraint systems help protect people from serious injury in a wide variety of crash types. When a severe accident results in only minor injuries the investigator should not simply rely upon this to conclude that the restraint system must have been in use. Injuries can be useful in addition to the other investigative techniques described in this paper to determine if restraints were being used. The descriptions given are guidelines, and exceptions will occur, so it is always necessary to consider all of the information available, in addition to the injury information.

Because the percentage of occupants using restraints is low, it is very important to make an accurate determination in every case. Significant research topics such as patterns of injuries to belted occupants, changes over time in the number of people wearing belts in crashes, and effectiveness of various restraint systems are affected by the quality of the investigator's work.

REFERENCES

(1) Federal Motor Vehicle Safety Standard 209, Code of Federal Regulations Title 49, Part 571.209.

(2) SAE Handbook, Society of Automotive Engineers, Inc., Chapter 33.

(3) Niederer, P., F. Walz, and V. Zollinger, "Adverse Effects of Seat Belts and Causes of Belt Failures in Sever Car Accidents in Switzerland During 1976," 21st Stapp Car Crash Conference, SAE, 1977.

(4) Kallieris, D., Mellander, H., Schmidt, G., Barz, J., Mattern, R., "Comparison Between Frontal Impact Tests with Cadavers and Dummies in a Simulated True Car Environment", 26th Stapp Car Crash Conference, SAE, 1982.

(5) Patrick, L., and Andersson, A., "Three- Point Harness Accident and Laboratory Data Comparison," 18th Stapp Car Crash Conference, SAE, 1974.

(6) Leung, Y., et al., "Submarining Injuries of Three-Point Belted Occupants in Frontal Collision - Description Mechanisms and Protection", 26th Stapp Car Crash Conference, SAE, 1982.

(7) Horsch, J., "Occupant Dynamics as a Function of Impact Angle and Belt Restraint", 24th Stapp Car Crash Conference, SAE.

ACKNOWLEDGEMENT

This paper was prepared partially from materials developed under funding from the U.S. Department of Transportation, National Highway Traffic Safety Administration, J. Vernon Roberts CTM. The opinions and recommendations expressed are those of the authors and not necessarily those of the National Highway Traffic Safety Administration or Allied Corporation, Bendix Safety Restraint Division.

2000-01-1317

Characteristics of Seat Belt Restraint System Markings

Jon E. Bready and Ronald P. Nordhagen
Collision Safety Engineering, L.C.

Richard W. Kent
University of Virginia Automobile Safety Laboratory

Mark W. Jakstis
Toyota Motor Sales

ABSTRACT

Markings or observable anomalies on seat belt webbing and hardware can be classified into two categories: (1) marks caused by collision forces, or "loading marks"; and (2) marks that are created by non-accident situations, or "noncollision marks". In a previous work, a survey of the driver's seat belt of 307 vehicles that had never experienced a collision was conducted, and several examples of marks created by normal, everyday usage, or "normal usage marks" were presented. It was found that some normal usage marks were visually similar to loading marks. This paper presents several examples comparing loading marks to visually similar normal usage marks and discusses the important similarities and differences.

INTRODUCTION

Accident investigators routinely visually inspect restraint system components for evidence of occupant loading in a collision. It is not unusual for the investigator to find several individual and distinct observable anomalies, or "marks," on restraint system and other proximate interior components during the inspection. The mere presence of marks does not necessarily indicate loading in a collision because marks from noncollision situations may also exist on the components. Consequently, the investigator must distinguish which of the marks (if any) occurred as a result of occupant loading in the collision. In order to correctly assess seat belt usage in a collision the investigator must have a knowledge of the characteristics of loading marks, and an understanding of reasonable alternative noncollision mechanisms that also could have produced marks.

There is a wide variety of marks resulting from noncollision situations, many of which will not easily be confused with loading marks. However, many noncollision markings can be misinterpreted because they are visually similar to loading marks. A knowledge of the characteristics and attributes of noncollision marks is therefore important to accurately assess occupant loading in a collision.

Normal usage marks can be classified further as: (1) marks created when the seat belt restraint is being used under normal conditions, or "usage marks"; and (2) marks occurring when the belt is not being used, or "stowage marks." In this paper, usage marks include those marks made during the process of donning and/or stowing the seat belt. Noncollision marks not made during normal usage include, for example, marks created when the webbing is used to tie damaged doors closed, or when loose parts are placed inside and rest on or against restraint system components.

This paper is the third in a series examining non-collision seat belt markings. In the previous two papers, noncollision seat belt restraint markings occurring under normal usage conditions were identified and classified [Bready, 1999 SAE], and the frequency of occurrence discussed [Bready, 1999 ISATA]. This paper presents several examples of loading marks and discusses their unique characteristics and attributes in comparison to visually similar normal usage marks.

The normal usage marks presented in this paper are from the previous survey of 307 vehicles that had never experienced a collision. The loading marks included in this paper represent a variety of vehicles and collision types, and a range of impact severities.

CHARACTERISTICS OF SEAT BELT MARKINGS

Typically, inspection of the seat belt is initially performed while the restraint is still installed in the vehicle, but occasionally seat belts are partially disassembled or completely removed for more elaborate laboratory examination. Most marks on restraint system components are located such that disassembly is not necessary. However, removal of the restraint facilitates microscopic examination of some marks, and affords examination of attachment and retractor hardware and lengths of webbing normally covered by interior panels. Occasionally, removal of the restraint is necessary in order to more clearly identify and document the characteristics of the marks. In addition, when examining marks on the seat belt components it can be instructive to use a surrogate occupant seated in the accident vehicle to locate the marks relative to guide loops, occupant, or other interior items in order to identify or clarify the potential sources of the marks. Subsequent use of an exemplar vehicle with the markings replicated on the seat belt and the seat correctly positioned can also be useful.

It is not always practical to remove the seat belt for close examination by microscope, nor is it always necessary because assessment of usage can often be accomplished with conventional visual methods (the naked eye, magnifying glass, or macro camera lens). However, microscopic examination of marks can be instructive. This paper includes photographs of marks magnified in order to more clearly illustrate the characteristics of selected marks.

D-RING/LATCH PLATE HARDWARE – In a collision, if the occupant's motion is not in a direction or of sufficient magnitude to load the belt heavily, occupant loading marks may be subtle or nonexistent. If a loading mark is subtle, it can potentially be misinterpreted as a normal usage mark or missed completely. To complicate analysis further, normal usage and subtle loading marks may be superimposed at the same location on the component.

Many accident investigations involve vehicles that, after the accident, are no longer usable due to severe damage. Often, however, the vehicle is repaired and returned to service. Among these vehicles, occasionally it is found that the original seat belt was not replaced during the repair process (perhaps due to apparent lack of stress or damage to system components, or simply neglect). Under such circumstances, evidence of loading on restraint system components may gradually deteriorate by further wear from normal usage or the observed marks could be mistakenly attributed to a subsequent rather than an earlier impact.

To help distinguish between subtle accident loading marks and marks occurring under normal usage on guide loop hardware, it is useful to discuss the mechanism by which the plastic material of a guide loop is being marked. During a collision, high forces on the restraint system can be generated by the deceleration of the occupant's mass. These forces are linked from occupant, to webbing, onto guide loop hardware and ultimately into the seat and/or vehicle body structures. The construction of the seat belt webbing provides a "ride down" effect for the occupant during the impact through stretching of the webbing fabric. Because of belt stretch, a short length of the webbing can be pulled through or over a guide loop even when the webbing is locked at the retractor. Friction resulting from contact pressure and relative motion between the webbing and guide loop can generate heat sufficient to soften or melt the plastic material to the point that it deforms or smears.

Figure 1a shows a latch plate that exhibits obvious signs of loading in a collision where the plastic has been melted and smeared at the interface with webbing contact. In some cases, when occupant loading is severe, a portion of the melted plastic of the guide loop has been observed to be smeared on and adhered to the belt webbing in the area of contact. This evidence of loading may, however, be fragile and can be lost by handling. The loaded D-ring of Figure 1b exhibits a rough abrasion without an obvious accumulation of melted plastic at one edge of the slot, rather, smearing of plastic is disbursed throughout the contact area. Figure 1c also shows a D-ring that experienced loading in a collision of relatively low severity. In this particular case, the webbing fabric pattern can be clearly seen imprinted in the surface of the plastic. Microscopic examination of this mark reveals the plastic material to have just begun to smear, and the individual webbing strands are also clearly visible in the melted plastic (see Figure 1d).

Under normal usage conditions, the level of force and heat required to cause imprints of the webbing fabric is not reached; the plastic material is gradually removed or polished away, rather than displaced, simply from cyclic motion of webbing against guide loop surfaces over a relatively long usage period. Figures 2a and 2b illustrate two typical examples of normal usage abrasions on guide loops. These abrasions are smooth, often having a polished appearance. Gradual removal of the guide loop material is due to direct contact and motion of the webbing and can be accelerated by dirt or dust particles, that normally accumulate on the webbing and can act as an abrasive agent. The abraded guide loop material is not typically seen accumulating on webbing, however, the webbing may exhibit discoloration or polishing.

The visual anomaly on the D-ring of Figure 3a appears at first to be melted plastic from loading but, upon closer examination, it is clear that this mark is actually comprised of several tiny cuts or lacerations in the surface of the plastic (see Figure 3b). When windows break in a collision, glass fragments can be thrown throughout the interior of the vehicle, particularly in rollover accidents. Tiny glass fragments being pulled along with the webbing through the guide loop slot created this particular mark. Cuts or lacerations such as these can occur at forces lower than those required to melt and smear plastic. However, because these marks require relative motion between the webbing and guide loop, the presence of these marks may indicate belt use during the collision.

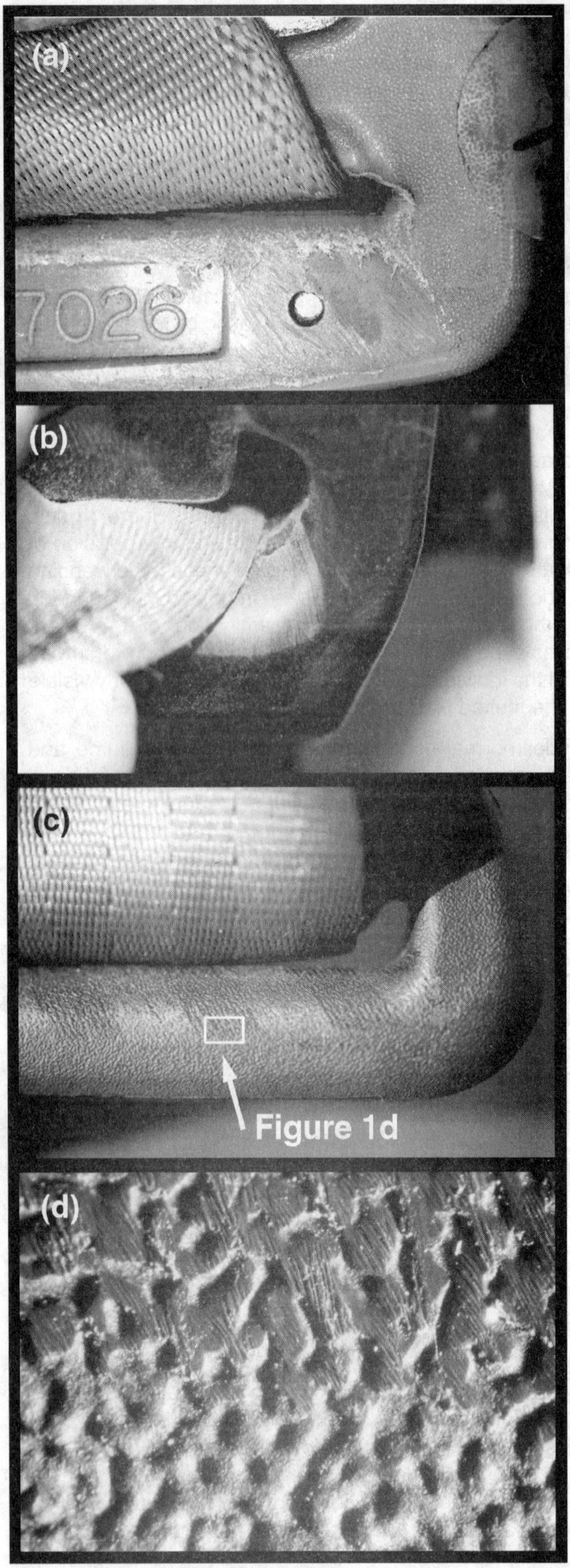

Figure 1. Collision Loading Marks: (a) Smeared Plastic, (b) Rough Abrasion, (c) Webbing Pattern Imprint, and (d) Strand Imprints.

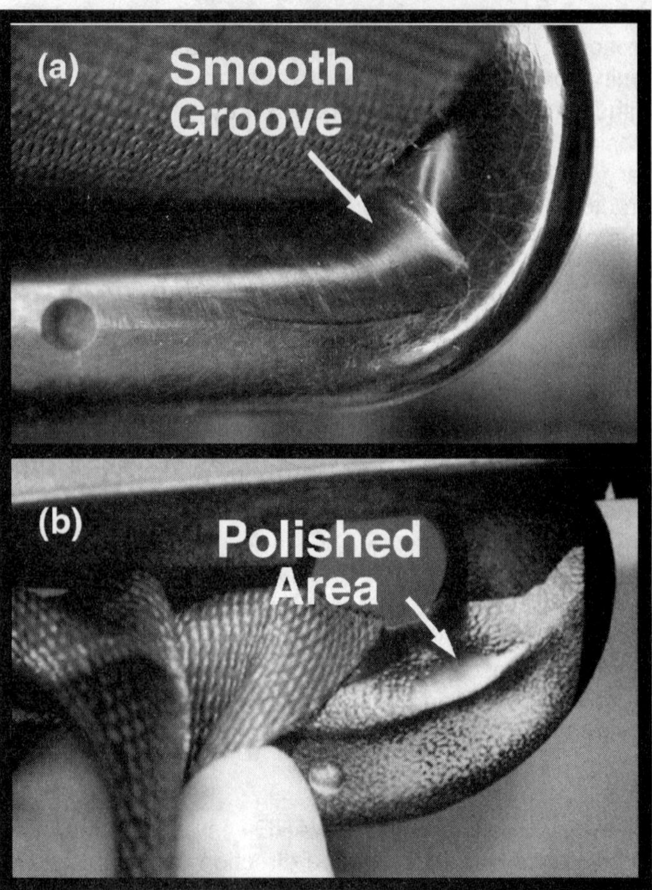

Figure 2. Normal Usage Marks: (a) Smooth Groove, and (b) Polished Areas.

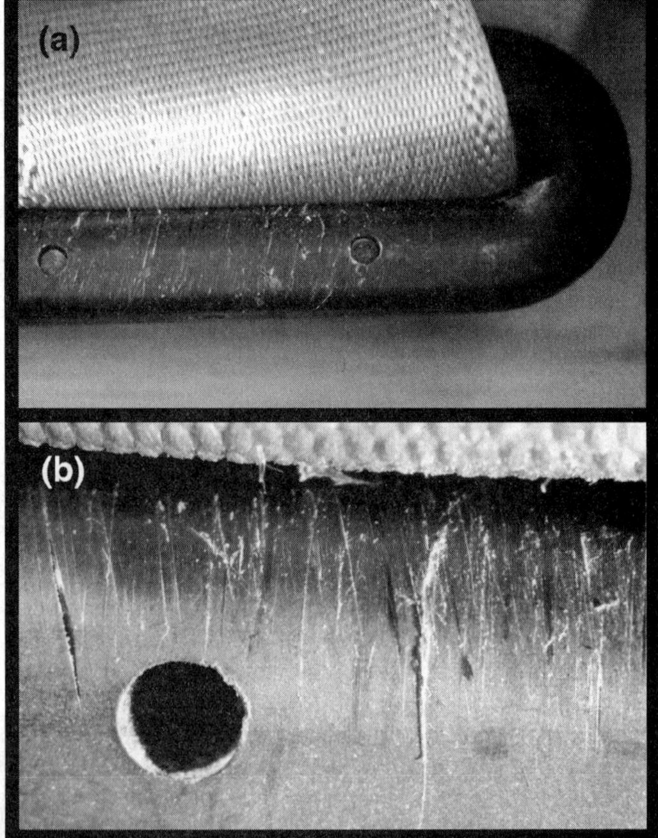

Figure 3. Lacerated Guide Loop from Broken Window Glass Fragments.

A build up of dirt/dust and other substances, referred to as "grime", commonly occurs in automobiles during normal usage. When the seat belt is used, some of the grime can rub off and accumulate on surfaces of contact. Figure 4a illustrates an accumulation of grime near the webbing slot of a latch plate, which is visually similar to the smeared plastic of a loaded latch plate because of its rough-looking texture. Grime on the belt webbing generally remains pliable and flexes with the webbing. In contrast, grime accumulating on a latch plate or other guide loop slot is typically more brittle. Grime can be easily scraped off using a fingernail, as illustrated in Figure 4b, revealing the original unabraded surface underneath.

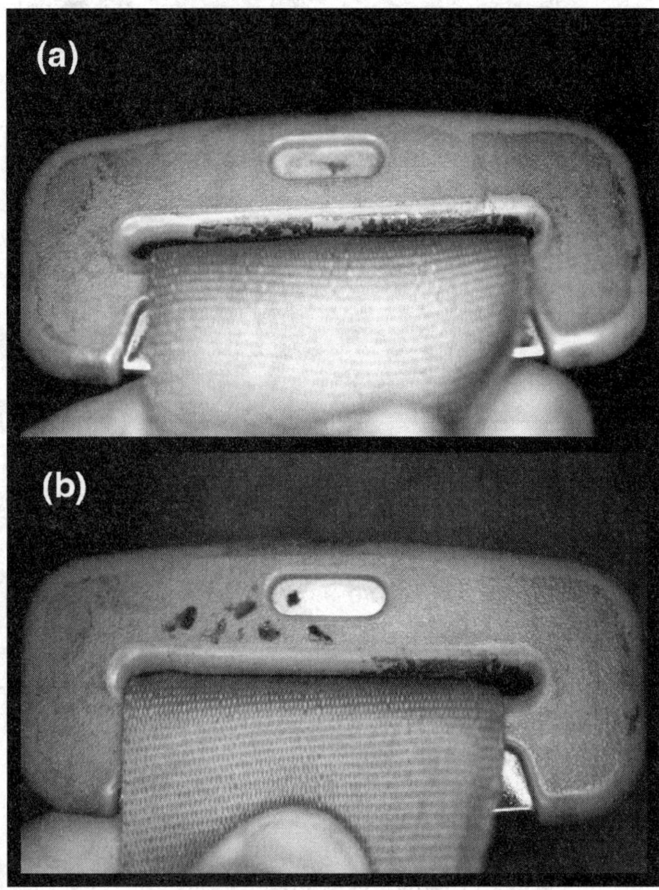

Figure 4. Grime on Latch Plate: (a) Normal Buildup, and (b) Scraped Off With Fingernail.

If grime is present on a latch plate or other guide loop during a collision, impact loading of the belt can break loose or rub off the accumulation, as illustrated in Figure 5a. In comparison, the grime on the unoccupied front passenger's latch plate of this same accident vehicle remains in place as seen in Figure 5b. Grime, broken loose in an impact, has also been observed to transfer onto the webbing as shown in Figure 5c. Guide loops have also been observed to exhibit both a build up of grime and abrasions from webbing motion under normal usage conditions, as illustrated in Figure 6.

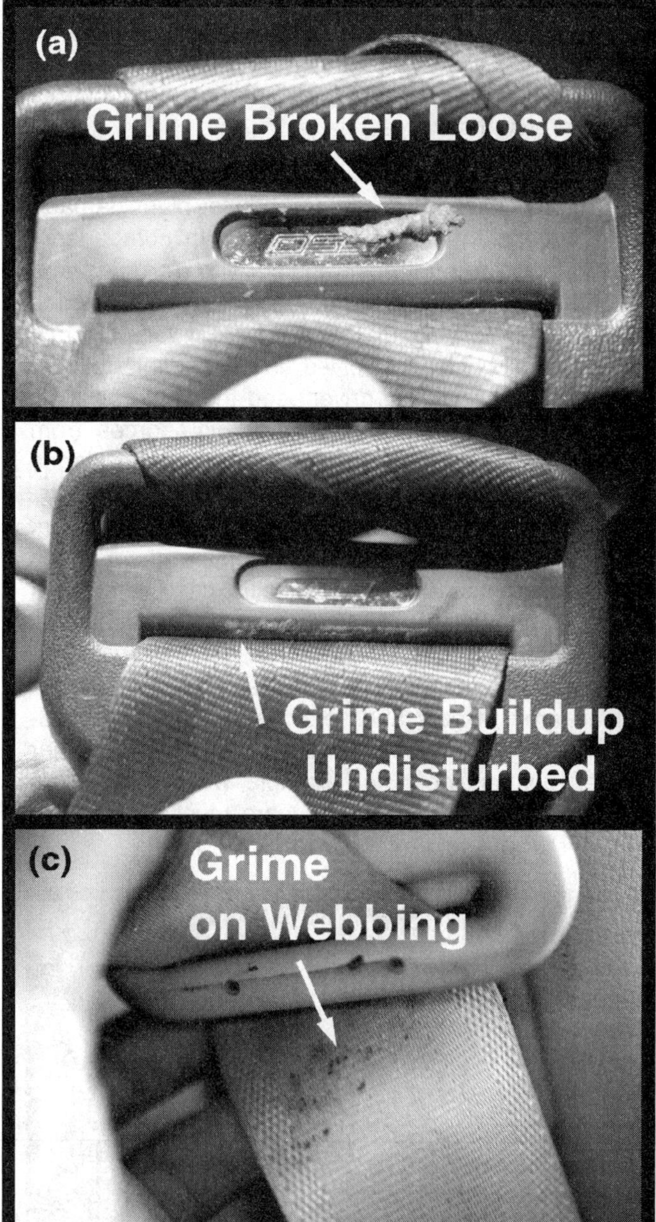

Figure 5. Impact Loading on Grime Buildup: (a) Broken Loose by Webbing, (b) Unloaded Latch Plate, and (c) Grime Transferred onto Webbing.

Figure 6. Normal Usage Grooves and Grime Buildup.

WEBBING – The seat belt webbing typically exhibits the widest variety of normal usage marks, a consequence of its routine interface with latch plate and guide loop hardware, occupant, and other interior components during usage and stowage.

Figure 7 shows a belt that has been closed in the door several times, resulting in grease stains, blunt damage and disruptions to the webbing fibers. In comparison, Figure 8 shows a length of webbing with disrupted and damaged fibers on the face, the appearance of which is similar to the webbing of Figure 7. However, this belt belongs to a vehicle involved in a multiple-rollover accident during which the driver used the seat belt. This particular mark is not the result of high forces from occupant loading, yet it is an example of evidence indicating the seat belt was used during the accident. Close examination of the webbing reveals that the mark is made up of several distinct, sharp cuts and scratches in the surface fibers. In this specific case, the mark occurred when a length of webbing, together with several window glass fragments, were pressed between the B-pillar cover and the driver's shoulder. There is also a matching area of cuts and scratches on the plastic B-pillar cover. Because seat belts are sometimes further damaged post-accident, for example when webbing is used to secure damaged doors, care must be taken to identify the true source of the cuts.

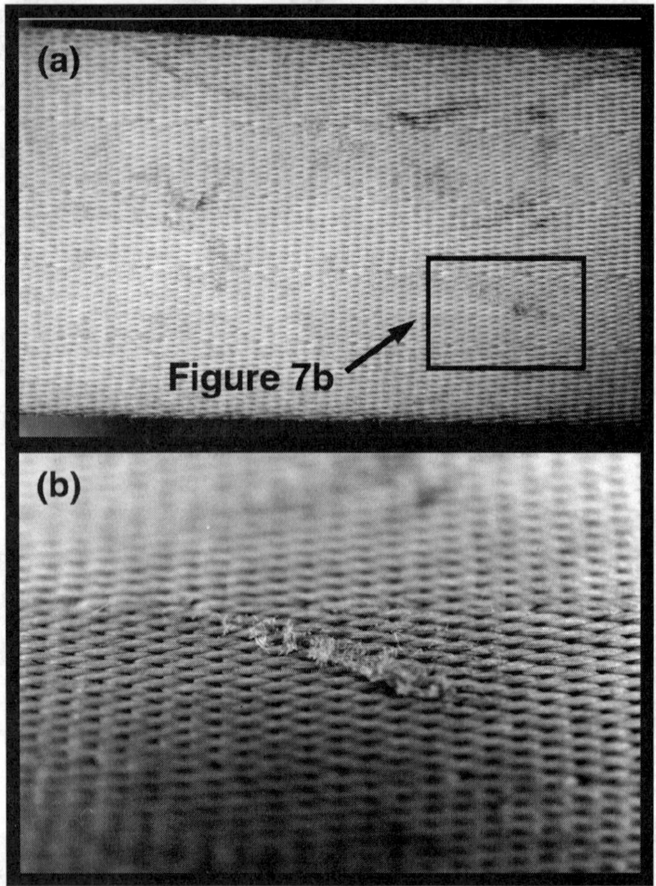

Figure 7. Damaged Webbing from Door Striker.

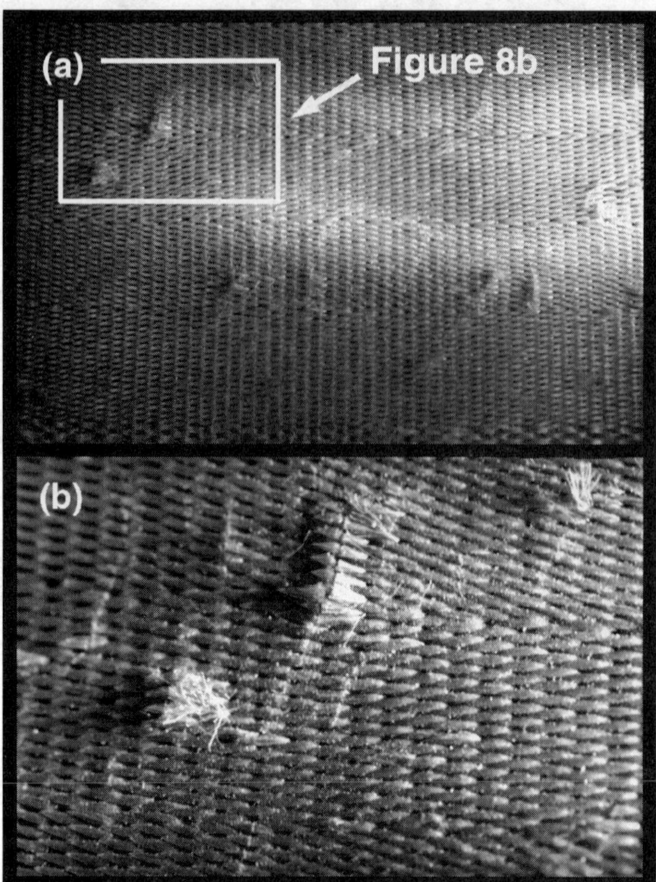

Figure 8. Damaged Webbing from Glass Cuts.

On certain restraints, when the retractor locks up and the belt is loaded, a series of one or more creases, imprints, and fiber disruptions can result from a web-grabber mechanism, shown in Figure 9a, where the longitudinal spacing of these creases will correspond to the geometry of the webbing lock "teeth". When examining a length of webbing near the retractor for evidence of occupant loading, familiarity with creases, imprints, and transfers occurring under normal usage conditions, referred to as "roll up" marks, is useful. Any discontinuity against which the webbing is tightly rolled, such as the manufacturer's label stitching, may cause a mark as shown in Figure 9b. Other noncollision marks occasionally seen on webbing near the spool include those made during manufacturing processes, for example, where the webbing has been secured by use of a clamp during quality control testing or by a clip used in shipping to prevent roll up.

Seat belt restraint systems are generally designed so that the webbing moves smoothly and evenly through various guide loops and access slots to minimize wear to the restraint system components, and to provide ease of use to the occupant. Occasionally, however, the webbing is pulled at such an angle or experiences a situation where it becomes permanently creased, twisted, folded, or frayed during normal usage or stowage. Several examples are shown in Figure 10.

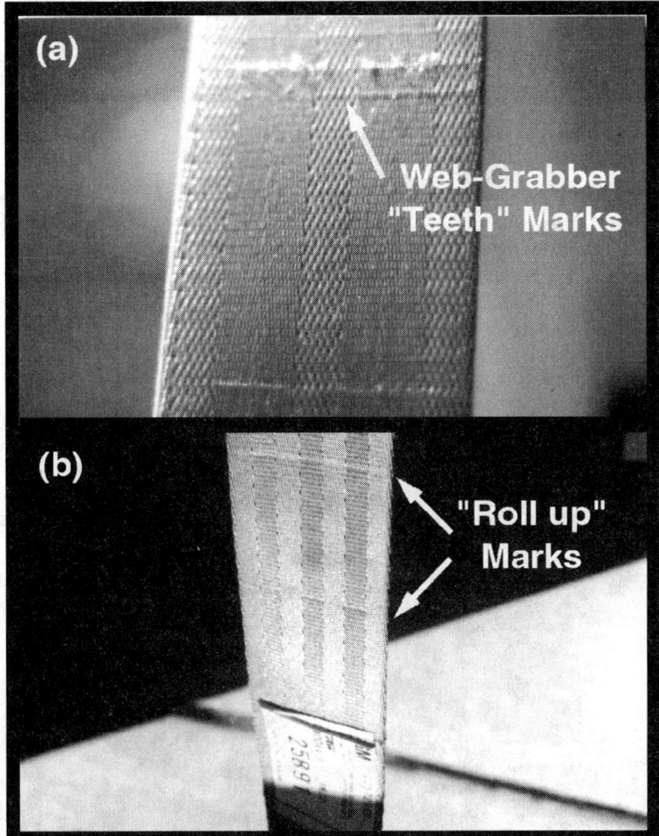

Figure 9. Markings Near the Retractor Spool: (a) Web-Grabber Marks, Which are Evidence of Occupant Loading, and (b) Normal Usage "Roll up" Marks.

During a collision, occupant loading may cause permanent deformation or damage to the webbing, commonly at points of load transfer. Signs of loading include creases or fiber disruptions oriented laterally to the webbing, for example, at a cinching latch plate or across a D-ring. Longitudinal loading creases may also result when, for example, the webbing is pulled to one end of the guide loop slot where it is creased or folded, and occasionally becomes jammed as shown in Figure 11.

There is a wide variety of loading marks, with many of these marks appearing remarkably similar to normal usage marks. It is not possible to describe all occupant loading marks or provide a clear method to positively assess each and every type of mark on loaded seat belt restraint systems. However, one insight gained from the previous work [Bready, 1999 SAE] regarding a way to positively assess occupant loading is to identify a specific characteristic of the marks that is consistent with relatively high forces produced in the collision. Some selected examples follow.

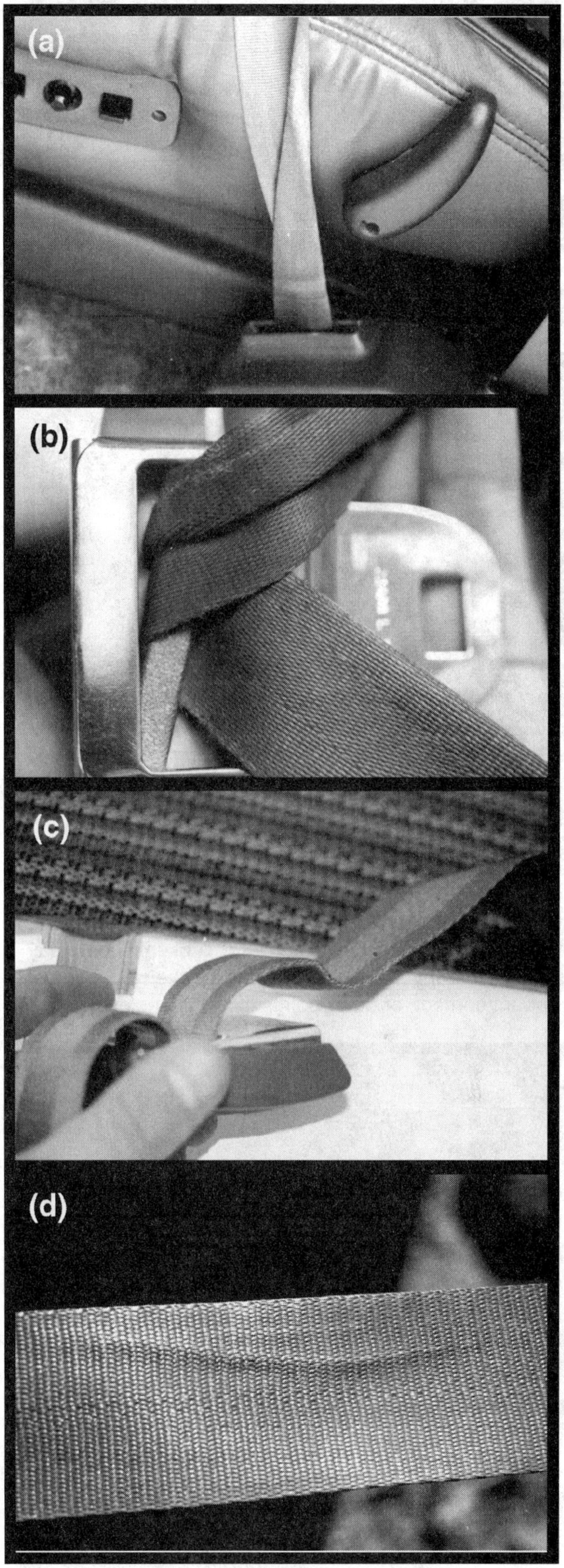

Figure 10. Normal Usage Marks on Webbing: Permanent Creases, Twists and Folds.

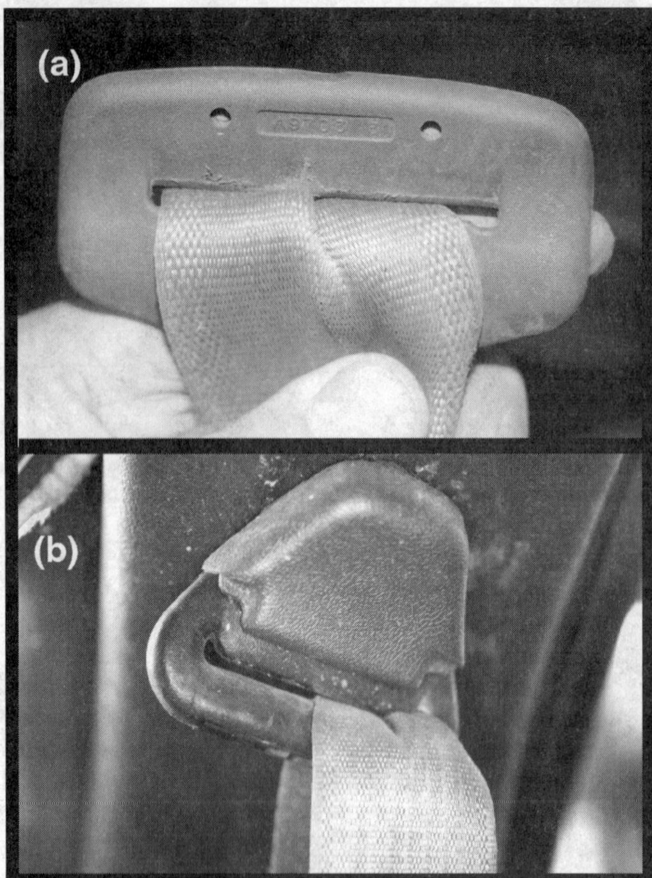

Figure 11. Jammed Webbing from Loading.

Figure 12 shows the lap portion of the webbing that has been permanently creased from normal usage by an habitual seat belt user; the crease is simply the result of the webbing lying across the occupant between the upper thigh and the abdomen. When occupant loading in a collision is severe, disrupted and broken fibers may also accompany such a crease at the location of occupant contact, or other points of load transfer.

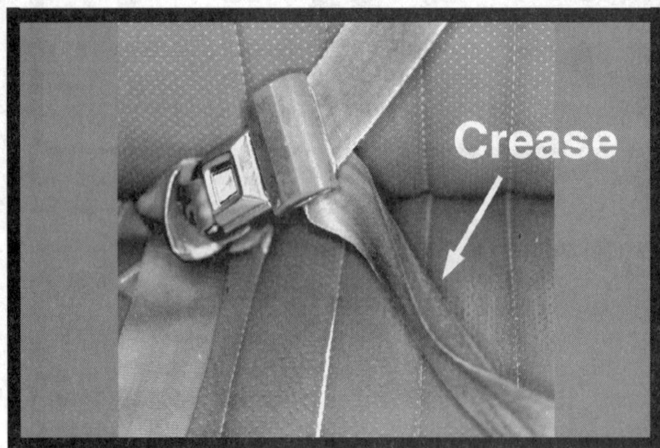

Figure 12. Crease from Normal Usage.

The scratches on the webbing of Figure 13a are not fiber disruptions; they are superficial scratches in a thin film of grime that can be easily produced, for example, by a fingernail, wrist watch or ring. A variety of visually similar loading marks has been observed to occur when occupant loading is severe. These include webbing fiber damage, smears of melted man-made clothing fibers (such as nylon fabric), and transfers from printed T-shirt graphics as shown in Figure 13b.

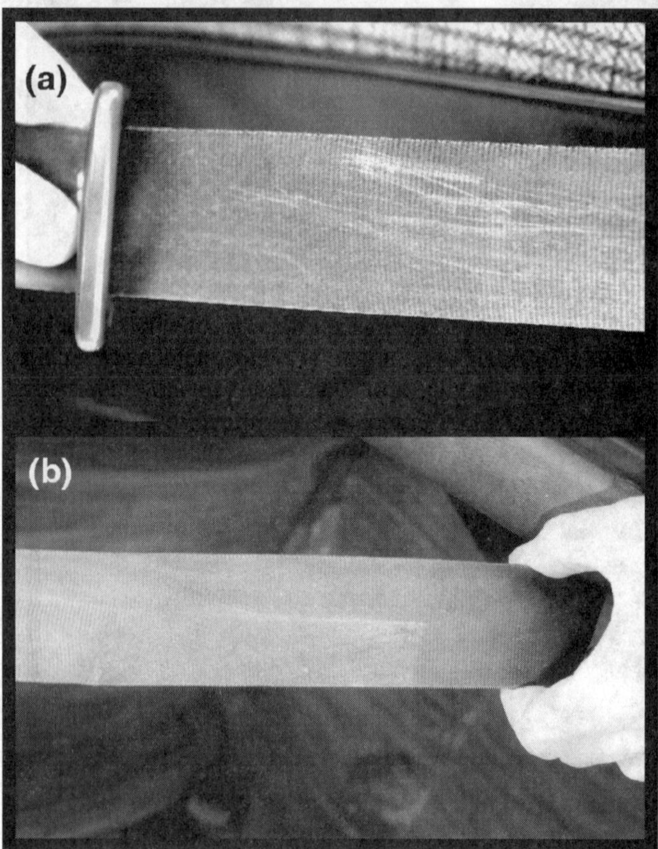

Figure 13. Marks on Webbing: (a) Scratches in Thin Film of Grime, and (b) Clothing Smears from Occupant Loading.

Under normal usage conditions, clothing fibers and hair are occasionally found loosely adhered to the webbing. When webbing is heavily loaded, fibers from clothing or seat upholstery in contact with the webbing have been observed to be abraded away and become smeared onto the webbing or even become caught within the expanded strands of the webbing and then held tightly when the stress is released.

BUCKLE – Some latch plates and buckles are provided with a plastic cover or housing. Bready et al. [Bready, 1999 SAE] found vehicles, which had never been in an accident, with broken covers (see Figures 14a and 14b). The broken covers observed were not found to have rendered the restraints inoperable, ineffective, or prevented them from being used on a regular basis. To recognize a broken cover which may have been damaged in an collision, it is useful to carefully examine all sides of the cover (including inside surfaces if accessible) for evidence of an applied force of sufficient magnitude to have caused the damage. Figure 15a shows a buckle cover which has been damaged from heavy occupant loading. Examination of the inside surface of the cover closest to the occupant's hip reveals contact prints of the housed buckle mechanism in the plastic as shown in Figure 15b.

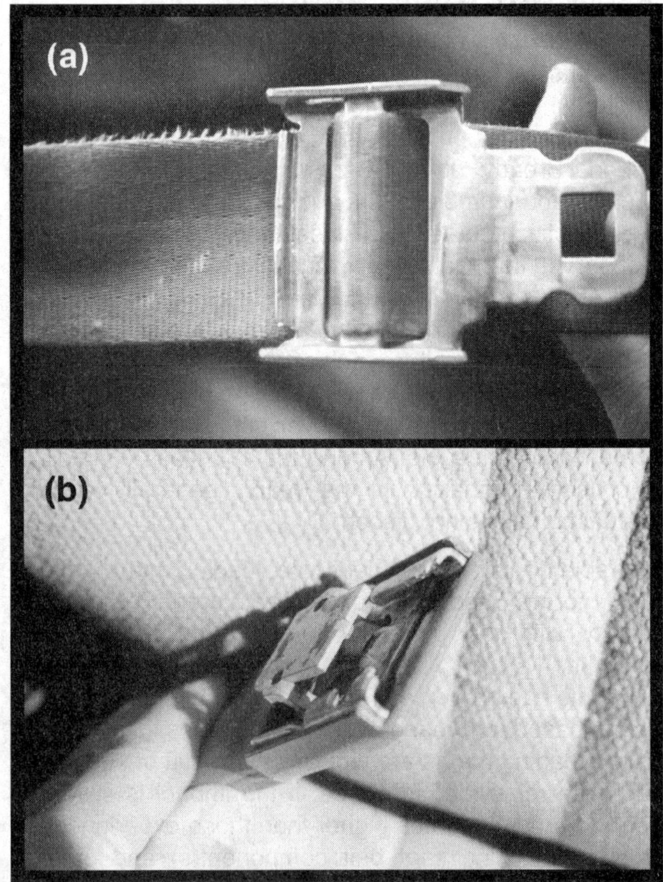

Figure 14. Broken Covers: (a) Latch Plate, and (b) Buckle.

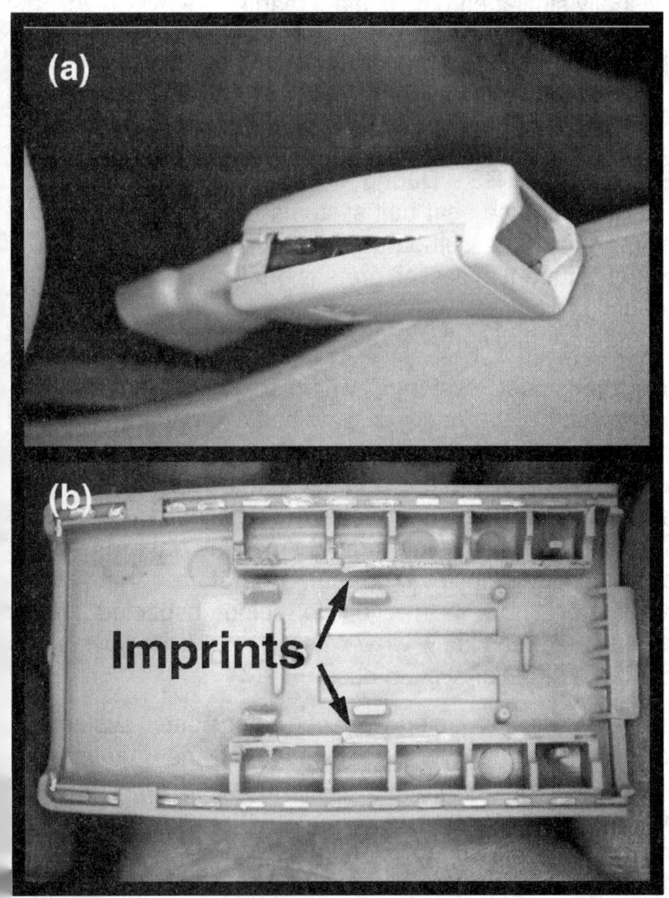

Figure 15. Imprints on Inside of Buckle Cover.

Contact marks such as smears, scratches, transfers, dents, or cracks in the outside surface may indicate a heavy impact or application of force to the cover by an external object. Conversely, the damage may be the result of tension in the belt webbing from occupant loading when the buckle is, for example, pulled and loaded against the seat bottom, seat recliner cover, or other interior surface/object. Of course, the consistency of this type of contact would depend on the direction and magnitude of the occupant's motion, and the location of interior components proximate to the buckle/latch plate.

If a seat belt has been pulled firmly against an adjacent object then there is a high probability that the opposing object will also exhibit some type of evidence of applied loading from the contact. Figures 16a and 16b illustrate a seat back recliner cover which has been loaded by the belt and loaded against the recliner mechanism causing cracks and areas of stressed plastic (light-colored areas). The investigator must also consider alternative mechanisms, such as an adjacent unrestrained occupant, that could potentially apply sufficient force to the recliner cover to cause this level of damage.

If damage to the latch plate is in question, the investigator must consider the possibility of the latch plate having been shut in the door during normal usage.

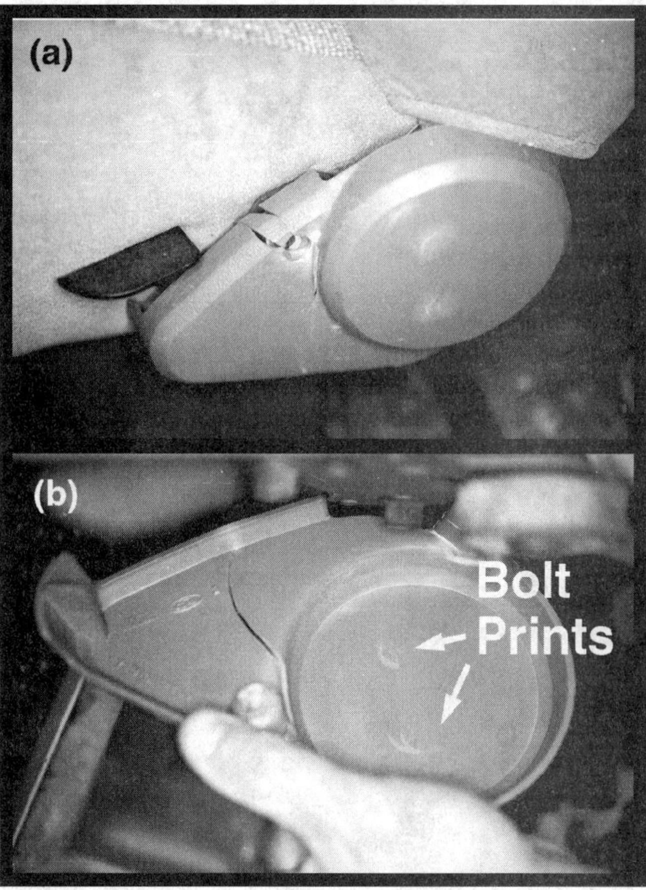

Figure 16. Damaged Seat Recliner Cover from Loaded Seat Belt: (a) Outside surface, and (b) Inside Surface.

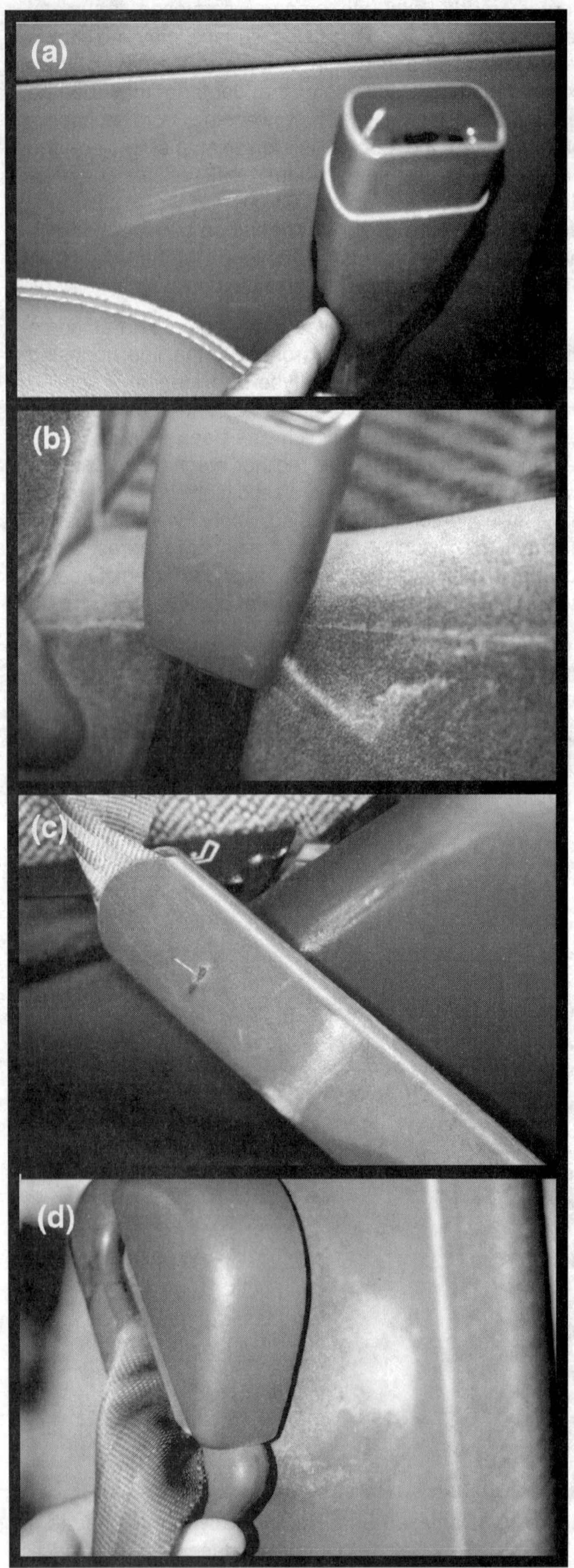

Figure 17. Chronic Wear Marks: (a) Center Console, (b) Seat Cushion, (c) Recliner Cover and Belt Sheath, and (d) B-Pillar Cover.

CHRONIC WEAR MARKS – Most seat belt components are routinely in contact with other non-seat belt components in both usage and stowed positions. For example, when the buckle is stowed it may rest or rub against the center console, seat bottom, or access slot surround. Bready et al. [Bready, 1999 ISATA] found scratches and abrasions on buckles, latch plates and their respective contacting surfaces to be quite common, and observed that the severity, location, appearance, and texture of these marks varied (see Figures 17a through 17d). The nature of normal usage marks such as these is that they typically result from chronic, relatively low-force contacts. Each individual contact adds to or enlarges the appearance of the mark, and, over time, the mark may become quite noticeable. Among the countless contacts that collectively make up the mark, generally each individual contact is not discernable.

Being able to positively identify a loading mark from among normal usage chronic wear marks requires familiarity with the general characteristics of chronic contact marks as described above so that correct judgements can be formed regarding seat belt usage in a collision based on reasonable comparisons. During a collision, loading marks are typically the result of a single impact with relatively high force. If the impact is severe enough to produce forces higher than those experienced between contacting seat belt component(s) and other interior surfaces under normal usage conditions, then the resulting loading mark (such as a scratch, indentation, crack, smear or transfer) may be unique when compared to visually similar chronic contact marks.

Figure 18 shows a seat base cover which exhibits both chronic wear marks as well as distinct loading marks from webbing contact during an impact.

PRETENSIONERS – During the past decade, pretensioner-style seat belt systems have become more common for front, outboard seating positions. There are several different types of pretensioner designs used in production vehicles with different triggering levels, force outputs and hardware.

All pretensioner systems, when triggered, have the potential to create marks on seat belt components, even with the seat belt in the stowed position. Since there exist various types of pretensioner systems, it is crucial to understand how the components are designed to operate to be able to correctly interpret the marking patterns. It is becoming evermore important, therefore, that the investigator know if the vehicle being inspected has pretensioners and know what types of marks are possible from that particular system design.

Some unique characteristics of pretensioner systems include the size, type and location of the markings. Examination of the inboard, outboard, and top interior surfaces of the D-ring loop can help reveal webbing location and movement during actuation. Figure 19a shows a D-ring with plastic smeared in the direction of webbing motion toward the retractor/pretensioner. In this figure, the mark is located too low on the inboard face for the belt to have been worn - this was an unoccupied seat belt when the pretensioner actuated. In addition, marks from pretensioner actuation can also be made on latch plate guide surfaces and stowage loop (if present) in either the stowed or usage positions. The exact location of pretensioner marks may provide information on belt geometry at impact and thus indicate whether or not the belt was being used.

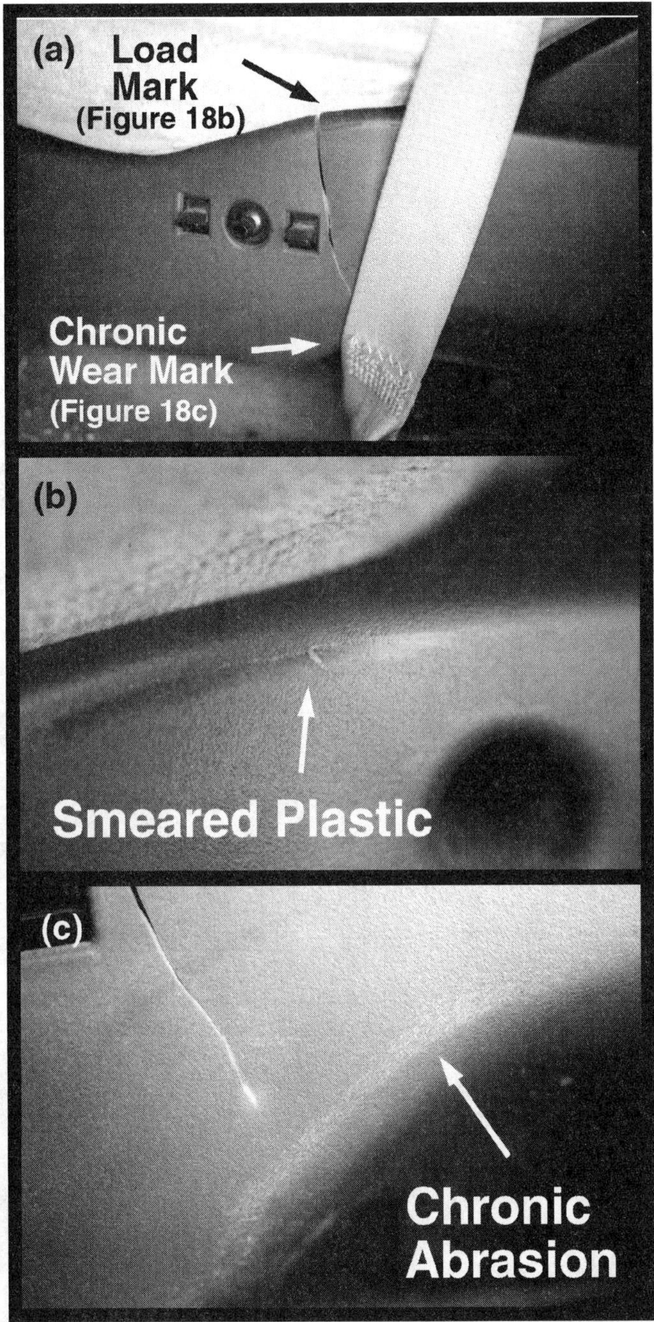

Figure 18. Loading and Chronic Wear Marks on Seat Base.

Figure 19b shows plastic transfers on the webbing made while in contact with the edge of the B-pillar cover access slot when the belt was pulled toward the retractor/pretensioner during actuation. The location of these marks indicate that this particular belt was stowed when the pretensioner was triggered.

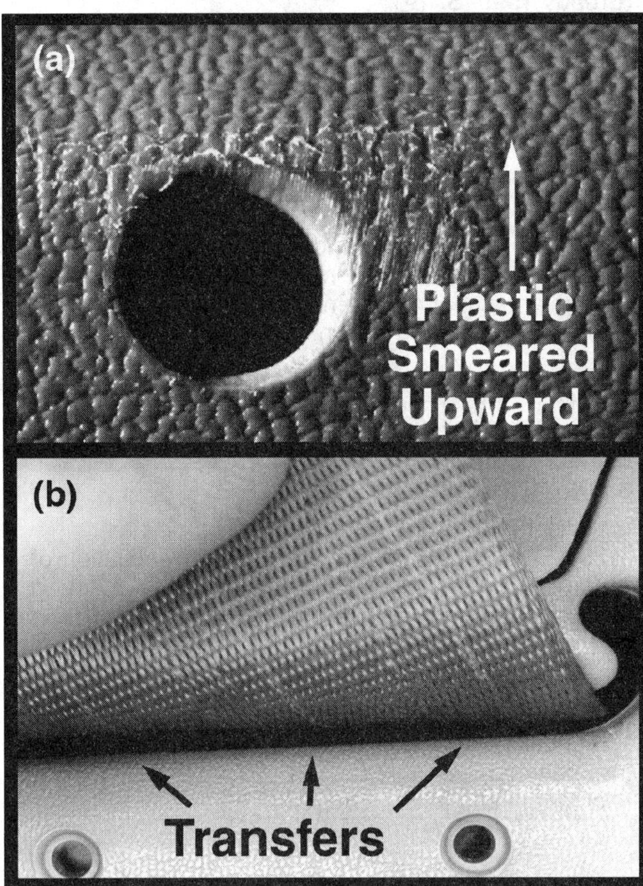

Figure 19. Marks from Seat Belt Pretensioner Actuation: (a) D-Ring Plastic Smeared Toward Retractor, and (b) Plastic Transfers from B-Pillar Cover Onto Webbing.

CONCLUSIONS

This paper presents several examples of marks on restraint system and other interior components, and discusses some of the unique characteristics of marks that may help the accident investigator distinguish loading marks from noncollision marks. The suggestions include: (a) consider a reasonable alternative noncollision source of the mark; (b) have a broad knowledge of noncollision markings so that correct judgements can be made based upon realistic comparisons; (c) consider how the direction and severity of the collision effect occupant motion; and (d) identify unique characteristics of the mark(s), such that those consistent with relatively high collision forces may be differentiated from low-force (normal usage) contacts.

REFERENCES

1. [Bready 1999 SAE]
Bready, J.E., Nordhagen, R.P., and Kent, R.W., "Seat Belt Survey: Identification and Assessment of Noncollision Markings," SAE 1999-01-0441
2. [Bready 1999 ISATA]
Bready, J.E., Nordhagen, R.P., and Kent, R.W., "Survey Results and Discussion of Noncollision Seat Belt Restraint System Markings," 31st ISATA 99SF033
3. [Cromack 1990]
Cromack, J.R., Schneider, D., and Blaisdell, D., "Occupant Kinematics & Belt Markings in Crash Tests with Unrestrained and Partially Restrained Test Dummies," 34th Annual Proceedings, Association for the Advancement of Automotive Medicine, Oct. 1990, Scottsdale, AZ
4. [Gorski 1990]
Gorski, Z.M., German, A., and Nowak, E.S., "Examination and Analysis of Seat Belt Loading Marks," Journal of Forensic Sciences, JFSCA, Vol. 35, No. 1, Jan. 1990, pp. 69-79
5. [Jakstis 1997]
Jakstis, M.W., and Brown, R., "Field Investigation of Airbag, Seat Belt, and Interior Markings", SAE Airbag Design and Performance TOPTEC, August, 1997, Costa Mesa, CA
6. [Moffatt 1984]
Moffatt, C.A., Moffatt, E.A., and Weiman, T.A., "Diagnosis of Seat Belt Usage in Accidents," International Congress & Exposition, Detriot Michigan, Feb., 1984, (SAE 840396)

930641

Inertial Seatbelt Release

Michael B. James, Douglas "L" Allsop, Thomas R. Perl, and Donald E. Struble
Collision Safety Engineering

ABSTRACT

Claims have been made that many production seatbelt buckles unlatch inertially during collisions. These claims are based on "parlor tricks" in which the back of the buckle is slapped, causing inertial release. Investigations of the potential for such inertial release in real-world collisions have consistently concluded that it is not a safety concern. Engineering analyses indicate that loadings experienced by buckles in real-world collisions are substantially different from the loadings which cause buckles to open in "parlor tricks."

INTRODUCTION

It has been claimed that many production seatbelt buckle designs can and do become unlatched during collisions. Specifically, the assertion is that buckle mechanisms with spring-loaded buttons can collide with an occupant's leg or hip during a crash, generating accelerations sufficient to cause enough relative movement between the button and the buckle to produce buckle release. To demonstrate this assertion, the back of the buckle is slapped against either a hard object, knee, or hip. Such demonstrations have been referred to as "parlor tricks." It is undisputed that many buckle systems can be opened by such "parlor tricks," but engineering analysis indicates that loadings experienced by the buckle during such tricks are substantially different, in two important aspects, from the loadings experienced by buckles in automobile collisions. First, accelerations generated by the tricks are often more severe, and second, the effects of belt tension are ignored.

Any mechanical system exposed to an environment different from that for which it was designed may appear to function or behave improperly. For example, an airplane cannot fly in a vacuum nor in a ten g environment. Even the strongest building might collapse if exposed to lateral forces substantially greater than hurricane force winds. On the planet Jupiter, where the acceleration of gravity is 2.64 times that of the earth, most people would not be able to support their own weight. These situations are not surprising; they are expected. They obviously do not indicate a faulty product nor faulty design. Neither do they shed any light on the usefulness of the product in its designated environment. Similarly, it is neither surprising nor enlightening to see that a seatbelt buckle may open when subjected to impacts more severe than those encountered in real-world accidents, or when these impacts are conducted in the absence of the counteracting forces created by seatbelt tension.

This paper presents a brief historical perspective on pushbutton seatbelt buckles, a discussion of theoretical and practical aspects of inertial buckle release, and empirical data of some production buckle systems.

HISTORICAL PERSPECTIVE

Spring-loaded buttons for the release mechanism of seatbelt buckles evolved over time. The first seatbelt buckles used other release mechanisms, such as lift covers, levers, clamps, etc. There were disadvantages associated with such designs, including ease of operation and the potential for accidental release. Eventually, almost all seatbelt buckle designs for automobiles evolved to recessed, spring loaded buttons. There are currently two general types of button designs: side-mounted buttons and end-mounted buttons. Both of these designs are capable of being opened by "parlor tricks," but the direction of the required acceleration for each design is different. Recent claims of potential problems have focused only on the side-mounted button design.

No federal standard or SAE recommended practice specifically addresses the issue of inertial seatbelt buckle release. The early SAE J4C Recommended Practice, which was used as a basis for Federal Motor Vehicle Safety Standard (FMVSS) 209, required that buckles be readily accessible and easily operable, with minimal possibility of accidental release. SAE J4C also included a recommendation that, with 150 pounds of tension in the webbing, the buckle release with no more than a 30 pound actuation force [SAE, 1965]. The current FMVSS 209 has essentially the same language. For pushbutton type buckles there is also a requirement for the size of the button [CFR, 1990]. The only federal standard which deals with inertial actuation of any vehicle component is FMVSS 206, which specifies that door latches not open when subjected to a longitudinal or lateral steady-state acceleration of 30 g's [CFR, 1990].

The notion of inertial seatbelt buckle release is as old as pushbutton buckles. In 1966 Ralph Nader testified before a Congressional committee that the new Ford and GM "deluxe" (i.e., pushbutton) seatbelt buckle designs were hazardous because they could be opened by an impact to the buckle. He suggested that such impacts can occur when "the buckle strikes the steering wheel rim, or one's arm or fist is shoved onto the buckle or a package is thrown against it" [Traffic Safety, 1966]. (Note that buckles at that time were located in front of the occupant.) Mr. Nader presented no data, field or laboratory, to substantiate these claims.

In 1971, the National Bureau of Standards (NBS) published its review of the performance of various seatbelt buckle designs. They evaluated a cross section of seatbelt assemblies sold in U.S. 1969 model year vehicles, and also several belt designs used in other countries [Castle, 1971]. The main issue addressed by the Bureau was whether the United States should mandate a single, standardized method of buckle release and, if so, which method should be mandated. They considered criteria associated with ease of use, but felt the potential for accidental release was the primary concern in their evaluation. Their conclusion was that a single type of release should be mandated, and that the pushbutton type design be adopted. The NBS noted that buckles with pushbutton designs had demonstrated a very high level of reliability since their introduction in late 1964. At the time of their evaluation, about 85% of all original equipment seatbelt buckle systems were pushbutton designs, and it was anticipated that it would rise to 97% for the following year.

As a result of their evaluation of belt performance, the NBS recommended several amendments to FMVSS 209, one of which was to require all buckles be pushbutton mechanisms. FMVSS 209 was amended that same year (1971), but the requirement that all buckles be pushbutton type was not adopted [FR, 1971]. The vast majority of buckles in the United States since that time have been pushbutton designs.

In 1977, a question arose at the National Highway Traffic Safety Administration (NHTSA) about the potential for the 1975 Chevrolet Monza buckle system to unlatch in an accident. A Multi-Disciplinary Accident Investigation reported that a sharp blow to the back of the buckle could cause it to open. The Office of Defects Investigation requested the Engineering Group of NHTSA to investigate [Bayer, 1978]. The buckle systems of three Monzas were tested in three modes: 1) striking the buckle with a rubber mallet, 2) pressing the buckle against the console, and 3) impacting the buckle with a spring-loaded plunger. It was determined that the Monza buckle could be inertially unlatched by a sharp blow, but not by pressing it against the console. A follow-up investigation was then undertaken by subjecting a large selection of buckle designs to the spring-loaded plunger. This device allowed the buckle to be tested either with no belt tension or with 30 lb of belt tension. The plunger, when actuated, exerted a force of 260 lb and was calibrated to repeatedly generate accelerations of 300 g's with no belt tension and 340 g's with 30 lb of belt tension. Of the 225 buckle systems tested, 50 opened. It was recognized that the plunger device loaded the buckles much more severely than they would be loaded in real-world collisions. It was suggested that additional work be done to develop "a more realistic impact" [Bayer2, 1978]. However, a third level of investigation examined actual field accident data and determined there were no claims which might indicate any problems in real-world accidents, and thus further experimental work was not required. The NHTSA concluded that while some buckles may be capable of inertial opening when high accelerations are applied, they are not susceptible to opening with the loadings actually experienced by the buckles during a crash [Chang-Memo, 1979].

In August of 1978, the NHTSA published its comprehensive In-House Survey of Safety Belt Problems [Bradford, 1978]. The study was designed to "search for defective safety belt components and locate possible safety hazards in conjunction with seatbelts." They analyzed all of the consumer complaint letters received by NHTSA from January 1970 to October 1977, and they also conducted a detailed survey of 453 NHTSA employees. Inertial buckle release was not mentioned in this survey as a potential problem.

More recently, the Institute for Injury Reduction petitioned NHTSA to initiate a new defect investigation, initiate rulemaking to amend FMVSS 209, and issue warnings to the public of the hazards associated with inertial belt buckle release [IIR, 1992]. The petition claims that buckle designs "with release buttons on the front face of the buckle" can be impacted by the occupant during a collision causing them to unlatch. The petition cites the results of the prior NHTSA testing where 50 of 225 belt buckle systems impacted by the spring-loaded plunger opened. It, however, does not refer to the final conclusion of that investigation. Significantly, the petition contains no evidence that inertial release occurs in real-world collisions, nor does it contain any scientific analysis which might refute the conclusions of the prior NHTSA Defect Investigation.

As a result of the petition, the NHTSA initiated an extensive evaluation of its crash test data, real-world field accident data and complaints filed with the "Auto Safety Hotline". The NHTSA also requested information from automobile manufacturers, safety belt manufacturers, and automobile safety agencies of foreign countries. They were unable to find any verified case of inertial buckle release. The Engineering Test Facility of the NHTSA conducted component buckle tests and full-scale crash tests with instrumented buckles. The agency's comprehensive review of all available information "did not provide any evidence that there is a safety problem associated with inertial unlatching of safety belts" [FR, 1992]. The petition to conduct a defect investigation was therefore denied.

INERTIAL BUCKLE UNLATCHING

An understanding of physics and a knowledge of the normal operation of pushbutton type buckles are required to understand the theory of inertial seatbelt release.

All pushbutton seatbelt buckles contain a spring-loaded "button" which is depressed by the occupant to open the

HYPOTHETICAL BUCKLE DIAGRAM

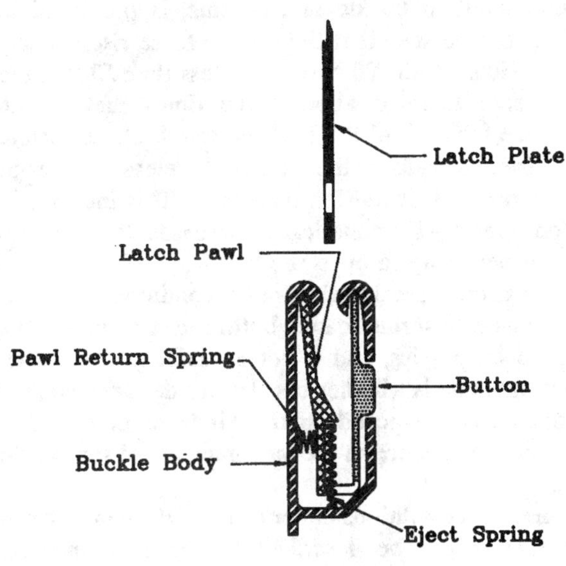

FIGURE 1

buckle. Figure 1 depicts a cross section of a hypothetical seatbelt buckle.

Figure 2 shows the motion of the various parts in normal operation. Before the latch plate is inserted into the buckle body, the pawl return spring (spring) is pushing the latch pawl against the button, and the button is supported against the buckle body, as shown in Figure 2A. As the latch plate is inserted, it contacts the latch pawl and moves it (to the left as depicted in the drawing), further compressing the spring (Figures 2B1 and 2B2). When the hole in the latch plate gets to the load-bearing surface of the latch pawl, the spring moves the latch pawl to the right and the load-bearing surface on the latch pawl blocks the forward surface of the latch plate hole, preventing the latch plate from coming out of the buckle body (Figure 2C).

NORMAL OPERATION (LATCHING)

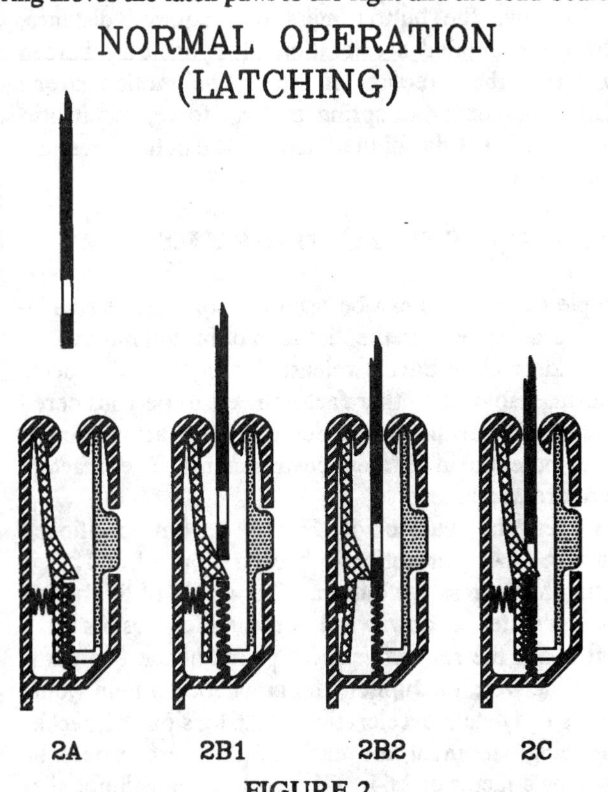

FIGURE 2

Figure 3 shows the sequence of events which unlatches the buckle in normal operation. The user depresses the button by applying a force to counteract the spring (Figure 3A). Both the button and latch pawl are displaced to the left. When the latch pawl has moved a sufficient distance (Figure 3B), the latch plate is no longer captured by the latch pawl and is pushed out of the buckle body by the eject spring. (Figure 3C).

NORMAL OPERATION (UNLATCHING)

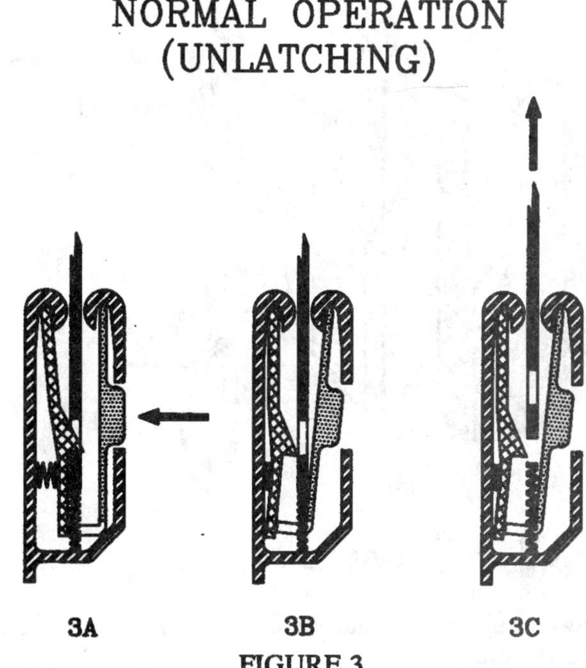

FIGURE 3

The sequence of events which leads to the inertial release of the hypothetical seatbelt buckle, as it is impacted on the back by a rubber mallet, is depicted in Figure 4. Figure 4A shows the initial impact of the mallet. The force of this impact causes the buckle body to accelerate and translate to the right. No direct contact is made between the mallet and either the latch pawl or the button. However, both of these elements are pivot-mounted to the buckle body at the top, and therefore, the upper end of these elements will accelerate and translate to the right. The spring preload force and spring stiffness, together with the inertia of the pawl and button, will determine whether, and to what amount, these elements will pivot relative to the buckle housing for a given acceleration. If acceleration is low, the pawl and button will accelerate and move with the buckle body, and the pawl will not pivot. If the acceleration of the buckle body is sufficiently high, the spring pre-load force will be overcome by the force required to accelerate these elements. The lower portions of the pawl and button will then move less than the upper portions, resulting in pivoting (Figure 4B). If the latch pawl pivots, the spring is further compressed resulting in a higher spring force than the initial spring pre-load. For continued latch pawl pivoting, and associated spring compression, the acceleration pulse to the buckle housing must be sufficient in both magnitude and duration to overcome this increasing spring force. If the

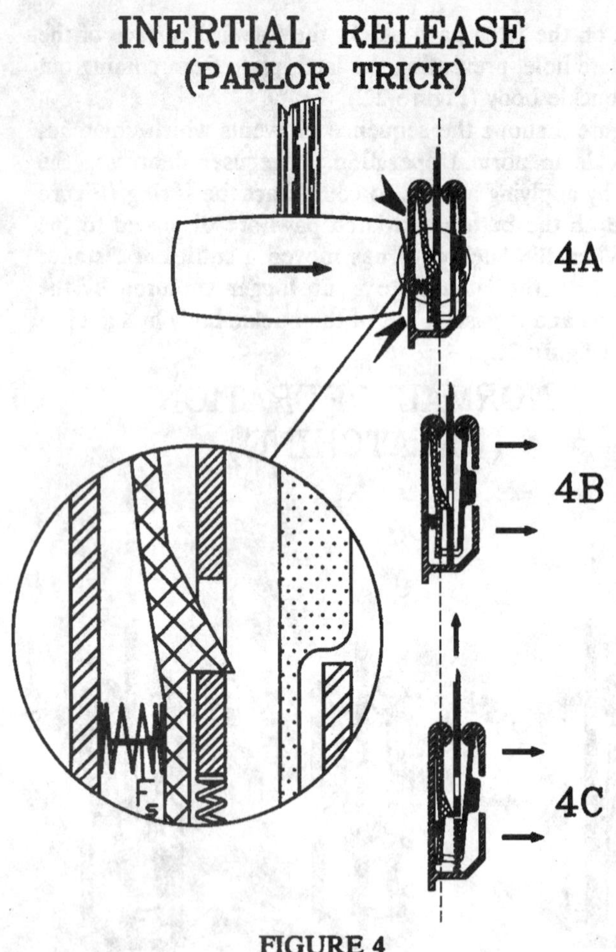

FIGURE 4

latch pawl displaces far enough the latch plate will be released. (Figure 4C).

In this example, a number of physical requirements must be met before the buckle unlatches:

1) The acceleration of the buckle body (due to the mallet impact) must be great enough to produce a reaction force on the spring/latch pawl interface sufficient to overcome the pre-load force of the spring.
2) The reaction force and inertia at the spring/latch pawl interface must be greater than the increasing spring force for the latch pawl to continue moving.
3) The reaction force at the spring/latch pawl interface must last long enough for the latch pawl to displace a sufficient distance for the latch plate to release.

WEBBING TENSION EFFECTS - In considering the potential for inertial buckle release in automobile collisions, there is an important factor not addressed in the above example. When the seatbelt is used by a vehicle occupant there is tension in the belt webbing. This tension increases rapidly as collision forces are applied to the vehicle and the occupant loads the belt. Belt tension results in both a normal force (F_N) and a lateral force (F_L) between the latch plate and the latch pawl (Figure 5). The magnitude of this lateral force depends on the friction between the surfaces and their orientation of contact. Under the scenario discussed above, for the buckle to inertially open, the acceleration to the buckle body must be sufficient to overcome not only the force of the spring, but also the lateral force between the latch pawl and latch plate. The effect of webbing tension on the button release force, measured by Collision Safety Engineering for 10 different side-mounted button buckle mechanisms, is presented in Figure 6. It can be seen that the release force rises rapidly with belt tension. With 200 N tension (less than 50 lb.), the seatbelt release force is about three times that of the unloaded case for each of the buckles tested. As expected, increased friction causes the required release force to continue to rise as belt tension increases. This increase in friction on the pawl dramatically increases the impulse requirement necessary to open the buckle.

Under extreme acceleration pulse conditions, seatbelt buckle systems with spring-loaded buttons can release. This fact is based on physics, and is not necessarily a problem. The important issue is whether buckles are designed so they do not open when exposed to the kinds of acceleration pulses buckles experience in the vast majority of real-world collisions.

The threshold conditions under which a given buckle will inertially release can be described by three parameters: acceleration magnitude, acceleration direction, and acceleration duration. The acceleration direction is important because only that component of an applied acceleration which is in the direction of motion between the buckle housing and the button will be effective in displacing the button. That is, only that component direction can be used when comparing the magnitude of the threshold acceleration and the magnitude of the applied acceleration.

For a buckle to inertially open, the button mass has to travel a prescribed distance against an increasing spring force. The product of this displacement and force determines the work or energy required to open the buckle. An applied acceleration, even if it is at a level above a steady state acceleration threshold, may not transfer sufficient energy to move the button mass the required distance. Therefore, the applied acceleration must not only have a component in the direction of the button motion strong enough to overcome the spring preload force, but it also must be of sufficient duration to depress the button mass the necessary distance.

CALCULATION OF BUCKLE PROPERTIES

Simple calculations can be used to estimate the steady-state acceleration level, in the direction of button movement, which could produce buckle release for any specific buckle mechanism. However, other factors need to be considered when using such calculations to determine the actual inertial unlatching potential in different environments. These factors will be addressed below.

Consider the buckle of Figure 3, but positioned horizontally. A gravitational acceleration of 1.0 g is attempting to depress the button. The weight of the button and pawl creates a downward force which results in a moment about the respective pivot points of the two parts. If this buckle were on Jupiter, the gravitational field would impose a steady-state acceleration of 2.64 g's on the buckle, and the hinge moment for each of the parts would be increased by a factor of 2.64. The acceleration will not alter

WEBBING TENSION EFFECTS

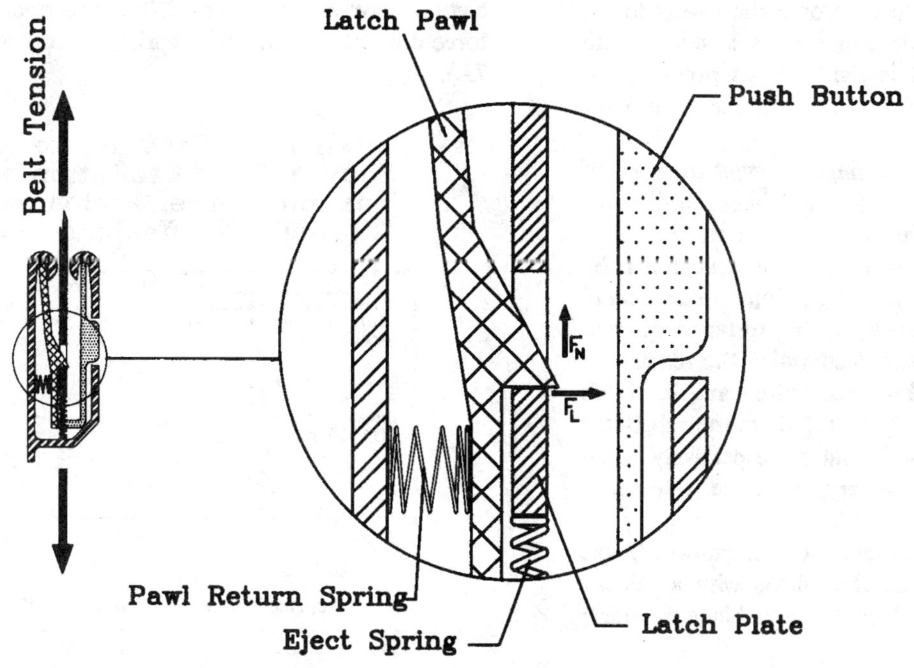

FIGURE 5

INFLUENCE OF WEBBING TENSION ON BUTTON RELEASE FORCE

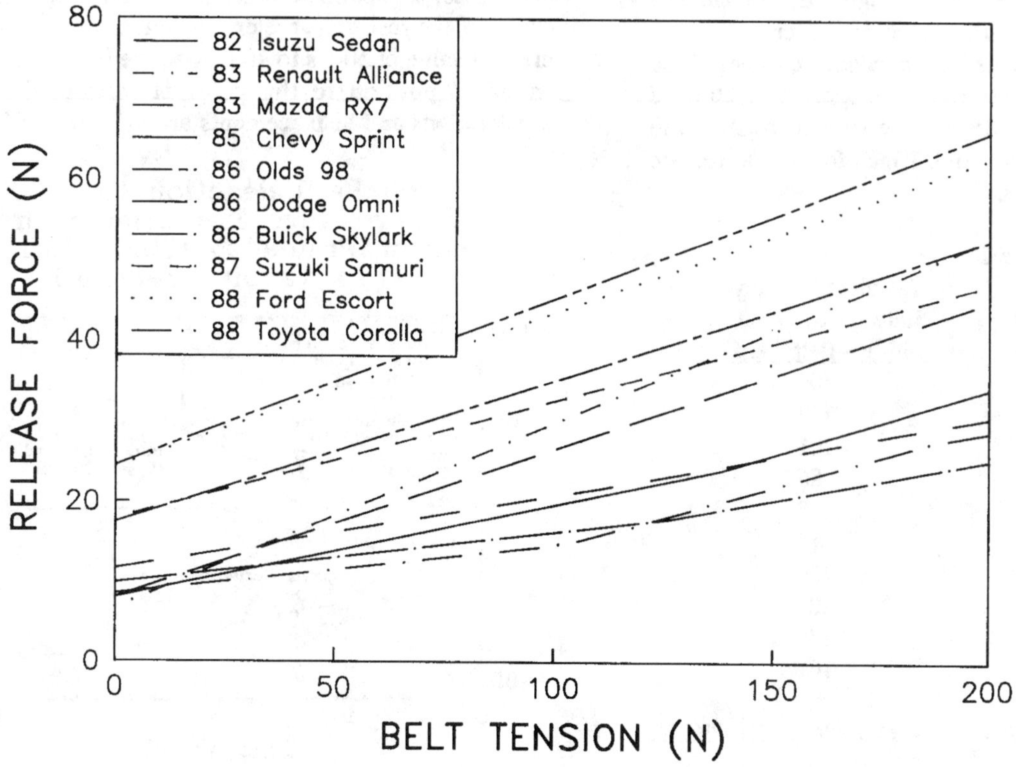

FIGURE 6

the spring force, and so even on Jupiter the button will not move due to its "weight." However, there is a level of steady-state acceleration, G, which would cause the button and pawl to "weigh" enough to overcome the spring force, and move. When the buckle experiences a steady-state acceleration of G, the effect is the same as pressing the button with force P_r at some distance x from the pivot point of the button.

In a one-g environment, the hinge moment for each of the moving parts is the weight of the part times the distance of its center of mass from its hinge point. For an irregularly shaped part, the hinge moment can be calculated by mathematically subdividing the part into many small elements, calculating the weight and the distance of each element from the hinge point, and summing the results. If the total hinge moment for both the button and the latch pawl at one-g acceleration is $M_p + M_b$ (where the subscripts p and b stand for the pawl and the button respectively), then the total hinge moment necessary to cause release is $G(M_p+M_b)$.

To say that the steady-state release acceleration G has the same effect as pressing on the button with a release force P_r is to say that they produce the same hinge moment; i.e.,

$$P_r x = G(M_p + M_b).$$

Solving for the steady-state release acceleration G yields:

$$G = P_r x / (M_p + M_b).$$

The value of G increases as the measured release force P_r increases (i.e. a stiffer spring). Since the hinge moments are reduced if the pawl and the button weigh less, the value of G also increases as these parts are made lighter.

A series of measurements and calculations, as described above, was made on the same 10 buckle systems which were evaluated for belt tension effects. The results are shown in Table 1. Pawl displacements of 0.1 inch for buckle release are typical of these designs.

Table 1

Vehicle Year, Make & Model	Calculated Steady-State g's to Open with no Belt Tension
82 Isuzu Sedan	100
83 Renault Alliance	113
83 Mazda RX7	101
85 Chevy Sprint	86
86 Olds 98	175
86 Dodge Omni	142
86 Buick Skylark	104
87 Suzuki Samurai	425
88 Ford Escort	186
88 Toyota Corolla	101

INERTIAL RELEASE OF A HYPOTHETICAL BUCKLE

Given a hypothetical buckle which has a steady state acceleration release threshold of 100 g's, a pawl that disengages the latch plate with 0.10 inch displacement, and a spring preload force that resists an acceleration of 50 g's, analysis shows that a 100 g square wave acceleration would have to have a duration of 2.28 ms to open it if the spring force did not increase with spring compression (see Figure 7A).

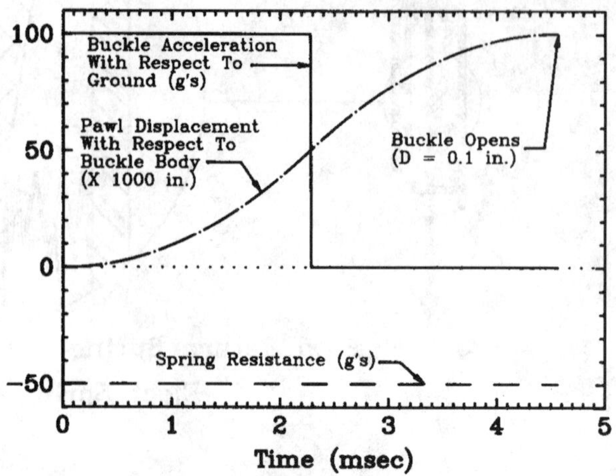

FIGURE 7A

However, as mentioned above, the resisting force of the spring does increase as the pawl displaces, so the resisting acceleration (i.e the spring force divided by the effective pawl mass) cannot be accurately depicted as constant. If, for example, the resisting acceleration increases linearly from its preload value of 50 g's to 100 g's as the pawl displaces from its initial position to the point of release, the resulting accelerations and displacements are as shown in Figure 7B.

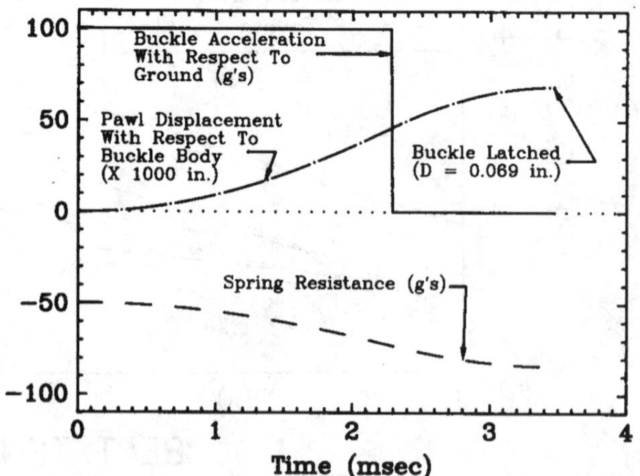

FIGURE 7B

The pawl does not displace far enough to open the buckle. Further analysis, including the effect of increasing spring force, indicates that the acceleration magnitude has to be increased to 118 g's (for a 2.28 ms pulse) in order to displace the pawl 0.10 inches. (See Figure 7C).

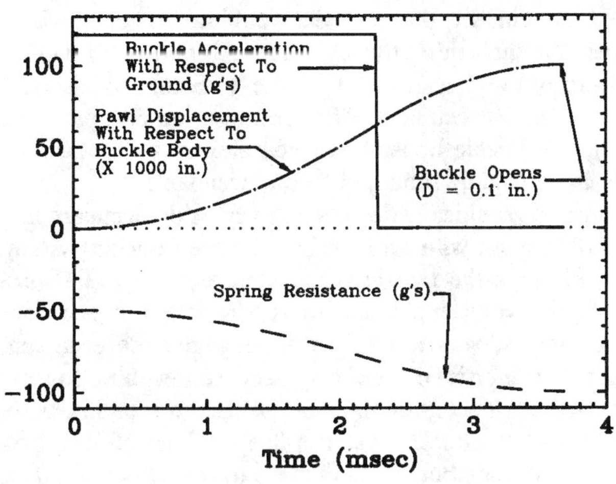

FIGURE 7C

The above examples have used a square wave to represent the acceleration of the buckle. This, however, does not represent accelerations experienced by buckles in real-world collisions. Impact accelerations are more closely approximated by triangular shaped waves. A more realistic plot of the buckle and pawl accelerations is shown in Figure 7D. This figure indicates that if the buckle is excited by a triangular wave form acceleration of 118 g's, the pawl displacement reaches a maximum of only 0.021 inches and thus does not come close to opening.

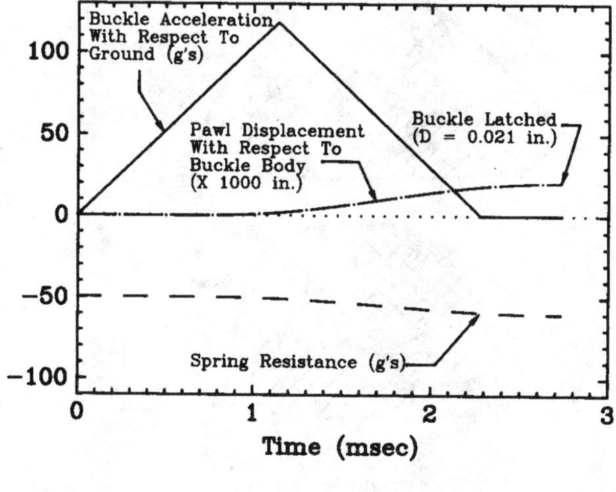

FIGURE 7D

To displace the pawl the required distance to open, a triangular wave with a peak acceleration of 226 g's is required, as shown in Figure 7E.

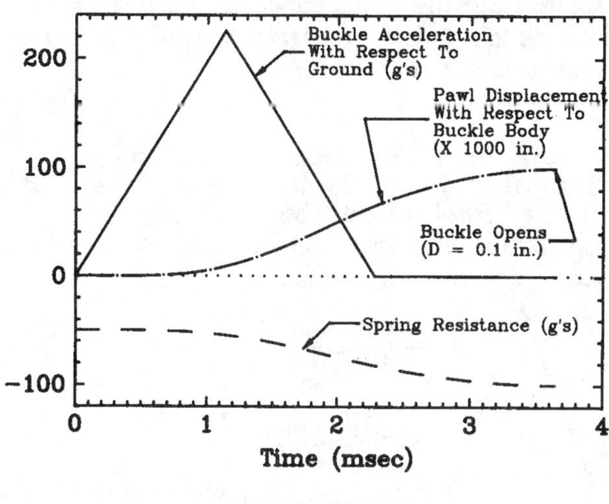

FIGURE 7E

In addition to pulse shape, webbing tension also has a significant effect on the acceleration required to open the buckle. Seatbelt tension creates resistive forces which have an influence similar to spring preload forces. As shown in Figure 6, a 50 lb webbing tension can increase the release force by a factor of 3. For this hypothetical buckle, the spring resistance is 100 g's at release. (It starts at 50 g's, the spring preload, and increases to 100 g's at the point of release.) With 50 lb of belt tension, the total resistance on the pawl at release will be 300 g's (100 g's due to the compressed spring and 200 g's due to friction on the pawl). For simplicity, assuming the friction resistance force to be

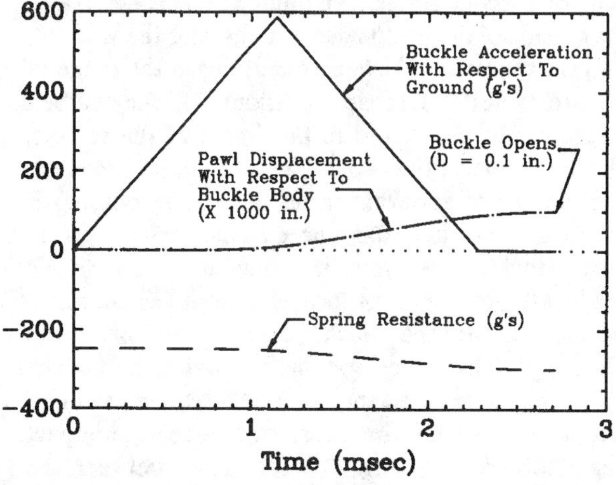

FIGURE 7F

constant, the initial resistance will be 250 g's (instead of the 50 g spring preload). Thus, if a 226 g triangular acceleration were applied, the pawl would not displace. In order to open the buckle with 50 lb of belt tension, a triangular acceleration with a peak of 587 g's, as shown in Figure 7F, would be required.

Another factor which affects the acceleration of the buckle is the acceleration pulse width. The effect of pulse width is demonstrated in Figure 7G. With the pulse width cut in half, from 2.28 to 1.14 ms, the buckle acceleration has to be increased to 915 g's to accelerate the pawl sufficiently to cause the latch plate to release.

Latch Pawl Reaction to a 915g Triangular Wave Excitation with a Linear, Preloaded (50g) "Spring" and 50 Pounds of Belt Load

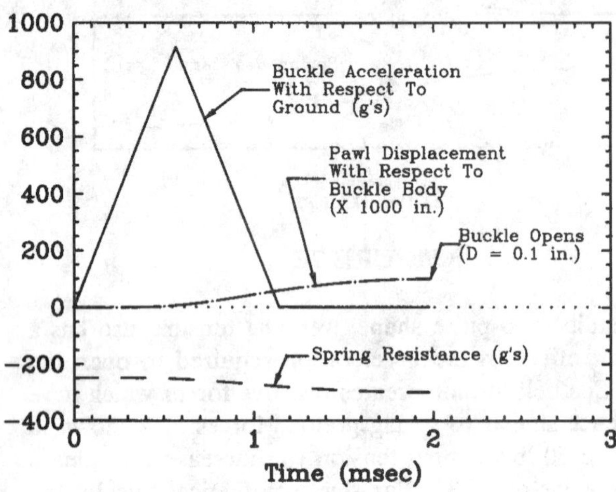

FIGURE 7G

Acceleration magnitude and pulse width are functions of occupant-buckle interaction. This interaction is complex and is affected by many factors. Some of these factors, and their associated effects, are described below.

VEHICLE COLLISION ENVIRONMENT

Vehicle compartment accelerations are typically measured by accelerometers mounted on solid frame members outside of the crush zone, usually near the vehicle's center of gravity. Peak vehicle accelerations in the range of 20-30 g's are typical for severe collisions. If the seatbelt buckle were rigidly connected to the frame of the vehicle, the vehicle acceleration would be of great interest. However, the acceleration acting on the buckle housing is not the same acceleration experienced by the vehicle's center of gravity. Buckles are typically mounted to the vehicle through a length of webbing or through a flexible stalk which cannot directly transfer the vehicle frame accelerations to the buckle housing. Thus, accelerations acting on the buckle, although they may be influenced by vehicle compartment accelerations, will be the direct result of a separate impact to the seatbelt buckle housing. It is the impact between the buckle and the occupant's hip which is most often claimed to cause inertial release. There has been very little investigation of this type of impact, yet understanding it is central to the determination of inertial buckle release potential.

KINEMATICS - In a vehicle collision, both the occupant and the buckle will undergo a change of speed as the vehicle accelerates or decelerates. Both are connected to the vehicle, although through different load paths. For example, in a far side impact the occupant is connected to the vehicle partially through the normal and frictional forces existing between him and the seat, and partially through forces exerted on him by the seatbelt webbing. The buckle is connected to the vehicle through the webbing and/or flexible stalk anchor, and also through the webbing which is in contact with the occupant. The motion of an occupant and the adjacent buckle housing is a complex event because the buckle is not rigidly mounted to the occupant.

Figure 8 graphically depicts a driver's side occupant in a right side impact with an inboard mounted buckle system. Prior to impact the buckle is oriented as shown in Figure 8A. As the vehicle accelerates to the left, the occupant compartment starts to move out from under the occupant, because the inertia of the occupant causes him to stay "stationary" until acted upon by an external force. Likewise, the inertia of the buckle tends to keep it from moving. As the vehicle and seatbelt anchors move to the left, the seatbelt is pulled across the slower moving occupant, increasing the tension in the belt and applying a force on the buckle which accelerates it to the left. There is now a relative velocity

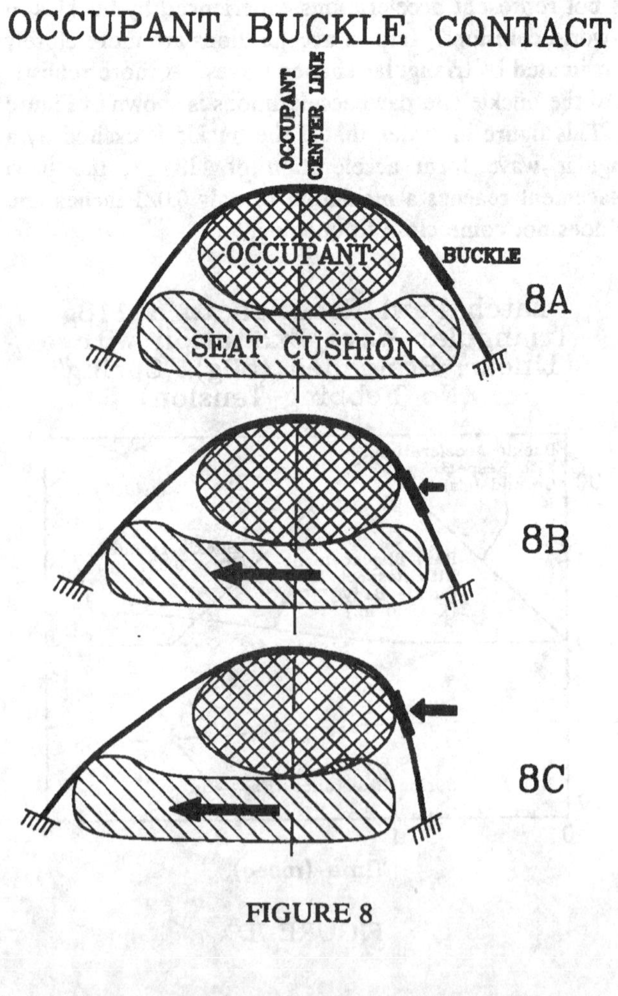

OCCUPANT BUCKLE CONTACT

OCCUPANT CENTER LINE

OCCUPANT

BUCKLE

SEAT CUSHION

8A

8B

8C

FIGURE 8

between the occupant and the buckle which decreases the distance between them, as shown in Figure 8B. Continued acceleration of the vehicle and belt anchors increases the belt tension and increases the closing velocity between the buckle and the occupant. The distance between the two continues to decrease and there is a potential contact (Figure 8C). If contact occurs, a force and associated acceleration will be applied to the buckle housing. Some of the factors which affect the acceleration during this contact are the relative velocity between the occupant and the buckle, and the stiffness of the occupant in the area of contact.

RELATIVE VELOCITY - The velocity between the occupant and the buckle, for a given vehicle acceleration, is a function of the initial distance between the two. A limited study was conducted of the range of initial distances between occupants and buckles in typical vehicles. Six occupants, approximating a 5th, 50th, and 95th percentile male and female, were seated in seven vehicles with side-mounted pushbutton type buckles. The occupants positioned the seat to their normal "comfortable" driving position. Results, presented in Table 2, indicated that in all cases the distance between the occupant and the buckle was very small. The maximum distance observed was 1.5 inches, and in over half of the cases some part of the buckle was in contact with the occupant. When the occupant is in full contact with the buckle prior to the collision, the only opportunity for any relative speed to develop between them will be during the period of body compression. In those cases where there was a gap between the occupant and the buckle, the distances were small enough to substantially limit the potential for a significant velocity difference to develop.

BUCKLE CONTACT LOCATIONS

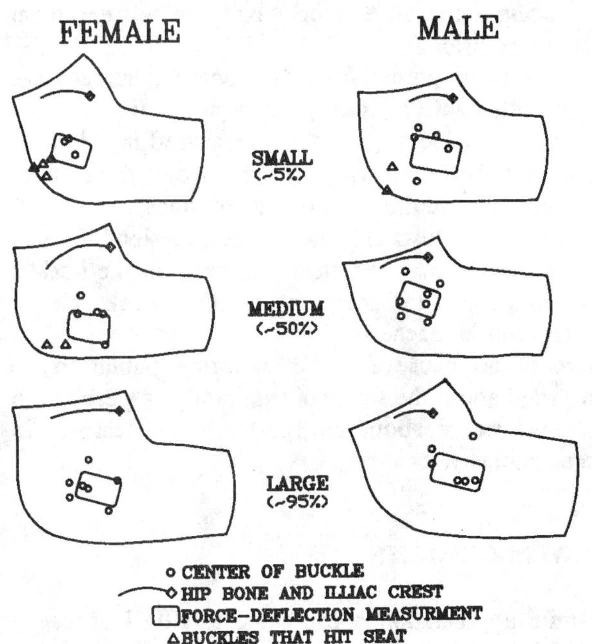

FIGURE 9

Table 2

DISTANCE FROM HIP TO BUCKLE

Occupant	Ave. Dist cm	(in)	Max. Dist cm	(in)	% Buckles in Contact
Small Female (5%)	1.8	(0.7)	3.8	(1.5)	40%
Med. Female (50%)	0.5	(0.2)	2.0	(0.8)	80%
Large Female(95%)	0.0	(0.0)	0.0	(0.0)	100%
Small Male (5%)	3.0	(1.2)	3.8	(1.5)	20%
Med. Male (50%)	0.3	(0.1)	2.0	(0.8)	71%
Large Male (95%)	0.5	(0.2)	3.8	(1.5)	86%

OCCUPANT STIFFNESS - In the above mentioned study, the contact or potential contact point between the buckle and the occupant was also recorded. These locations are presented in Figure 9, and indicate that contact tended to be on the lower portion of the buttocks, away from the relatively stiff area of the iliac crest. With the smaller occupants, the seatbelt buckle was in contact with the seat back in several instances, effectively eliminating the potential for contact with the occupant during a collision (see Figure 10).

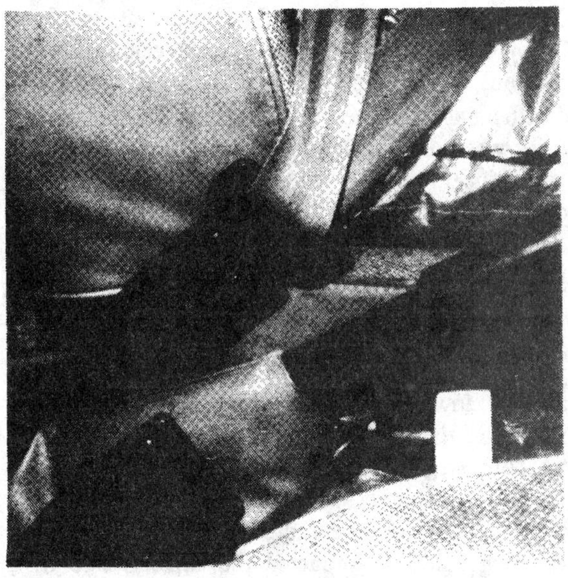

FIGURE 10

The force-deflection characteristic of each occupant was measured in an area approximately in the center of the potential contact points. A rigid deflector approximately the same size as a typical seatbelt buckle was used. After "snugging" the occupant's far side against a rigid surface, the deflector was slowly loaded against the occupant's hip. Applied load and deflection were recorded. A similar measurement was made on the Hybrid III dummy. Figure 11 indicates that under "static" loading, the dummy in this area, is two to three times stiffer than humans. Since the relationship between stiffness and acceleration is a square root function, this suggests that empirical buckle acceleration data gathered from crash tests with this dummy are roughly

1.5 to 1.75 times higher than that experienced in comparable real-world collisions.

FULL SCALE CRASH TESTS

A few full-scale crash tests with instrumented buckles have been recently reported in the literature. The NHTSA ran six such tests, two side impacts, one 20° frontal, and 3 straight frontal impacts [Howe, 1992]. No buckle releases were experienced in any of these tests. During these tests accelerations were monitored by mounting an accelerometer on the face of the buckle. Additionally, belt force was measured by mounting a webbing load cell on the shoulder belt. The data presented suggests that acceleration pulse shapes were nominally triangular, and that large seatbelt tension loads were generated before significant buckle accelerations occurred.

In a driver's side impact of a 1985 Chevy pickup, the driver's belt load was 40 lbs at the beginning of the acceleration pulse and 140 lb at the peak acceleration. Again, in a driver side impact of a 1989 Taurus, the driver's belt load began increasing prior to the main buckle acceleration pulse, and in this case, the belt load was 600 lb at the beginning of the pulse and 800 lb at the peak.

Failure Analysis Associates also published the results of instrumented buckles in 6 FMVSS 208-type rollover tests and fifteen sled tests [Thomas, 1992]. All of these tests involved a 1984 Chevrolet S10 Blazer. No buckle releases were evidenced in any of the tests. The buckle acceleration was measured with an accelerometer mounted on the buckle, and the belt force was measured using a specially designed belt tension transducer which was capable of accurate measurements from 0 to 100 lb. Peak acceleration, pulse duration and average belt load were reported for all buckle accelerations exceeding 20 g's. The instantaneous belt load at the time of peak g's was not given, but was likely greater than the average value. Plots of acceleration and belt load vs. time were not given, and no mention was made of the pulse shapes. However, the data presented support the trend observed in the NHTSA results, that significant belt loads accompany large seatbelt buckle accelerations.

Collision Safety Engineering conducted a very severe left side impact, with an instrumented buckle. A rigid moving barrier weighing 3,207 lb impacted the driver's side of a 3,185 lb, 1985 Buick Electra at 38 mph. The vehicle delta-V was approximately 19 mph. A Hybrid III dummy was positioned in the right front passenger seat and secured with the production three-point restraint system. Two accelerometers were mounted on the buckle, and strain gauges were mounted directly on both sides of the latch plate to measure tension experienced at the buckle. In the pre-impact seated position, the closest corner of the buckle was just touching the dummy, while the farthest corner was 1.8 inches away. This is similar to the fit of buckles in the previously mentioned occupant buckle position survey. The greatest buckle acceleration pulse was roughly triangular in shape, had a maximum amplitude of 68 g's, and was just under 4 msec in duration. The latch plate tension was 100 lb at the beginning of the pulse, increased to 185 lb by the end of the pulse, and continued to climb to over 1,250 lb. Despite the fact that impact damage and dummy kinematics were expectedly severe, the buckle remained latched. Thus, the buckle acceleration and latch plate tension corroborated the findings of significant belt loads during large buckle accelerations, observed in the NHTSA and Failure Analysis tests.

CONCLUSIONS

1. Several investigations of alleged inertial release behavior have been made, and it has been consistently concluded that there is no associated safety problem. There have been no verified cases of crash test dummies or occupants causing inertial buckle release since the introduction of pushbutton buckles in the 1960's.
2. Demonstrations of inertial release by means of "parlor tricks" are not indicators of inertial release potential in real-world collisions, because "parlor tricks" take the buckle out of the environment it was designed for, and expose the buckle to significantly different loadings.
3. Inertial release is a function of several factors, including the amplitude, duration, and pulse shape of the applied acceleration, and seatbelt tension.
4. Steady-state acceleration values calculated from static release force measurements are substantially lower than the impact acceleration required to open the buckle, even without belt tension.
5. In vehicle collisions, the occupant movement required to produce contact with the buckle creates tension in the belt, and belt tension significantly increases the release force and acceleration required to open a buckle.
6. Typical vehicle geometry results in very little, if any, distance between the occupant's hip and the buckle, minimizing the relative velocity build-up between them during a collision.
7. In the area of potential buckle contacts, humans are significantly softer than the Hybrid III dummy. Therefore, buckle accelerations measured in full scale crash tests are generally more severe than those experienced in similar real-world collisions.
8. Full scale crash tests indicate that belt tension begins to rise before the main acceleration pulse on the buckle, and continues to increase as the peak buckle acceleration is reached.
9. There is no cause for the motoring public to be concerned about the safety of side-mounted pushbutton type buckles, or about the "parlor tricks" featured in recent publicity.

ACKNOWLEDGEMENTS

Grateful appreciation is expressed to Ron Nordhagen Kevin Welch, and Brent Benson for assistance in hardware testing, to Andy May, Elaine Murray and Sue LeBaron for text and figure preparation, and to Charles Strother and Gregory Smith for their technical analysis.

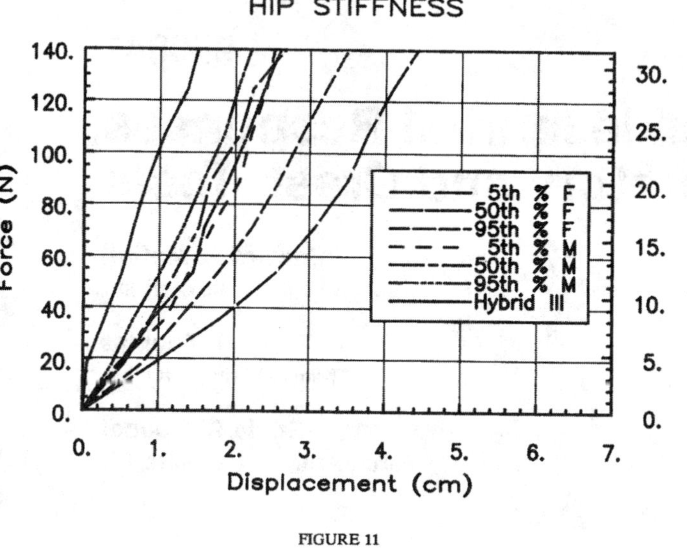

FIGURE 11

REFERENCES

[Bayer, 1978]
Anthony R. Bayer, Jr., Russell L. Kirkbridge, "Tests of 1975 Chevrolet Monza Seat Belt Latching Mechanisms," Engineering Test Facility. National Highway Traffic Safety Administration. Office of Defects Investigation. EA7-040. March, 1978.

[Bayer2, 1978]
Anthony R. Bayer, Jr., Russell L. Kirkbridge, "Survey of Seat Belt Latching Mechanisms Used on 1971-1978 Passenger Cars." Engineering Test Facility. NHTSA. Office of Defects Investigation. EA7-040. March, 1978.

Bradford, 1978]
Lynn L. Bradford, George D.C. Chang, Mark Meadows, "Safety Belt Problems, NHTSA In-House Survey," Office of Defects Investigation. National Highway Traffic Safety Administration. U.S. Dept. of Transportation. PB-286 949. August, 1978.

Castle, 1971]
A.B. Castle, Jr., "A Study of Seat Belt Buckle Release Methods," National Bureau of Standards Report. NBS Project 4080402. NBS Report 10387. U.S. Department of Commerce. January, 1971.

CFR, 1990]
Code of Federal Regulations. Transportation #49. Part 571. Revised as of October 1, 1990. U.S. Government Printing Office. Washington, D.C., 1990.

Chang-Memo, 1979]
Memorandum Subject: Alleged Failure of Seat Belt Buckle on 1975 Chevrolet Monza 2+2:EA7-040. U.S.DOT, NHTSA. October 24, 1979.

[FR, 1992]
57 FR 55298, Denial of Petition. November 24, 1992. (NHTSA Denies September 11, 1992, Institute for Injury Reduction Petition.)

[FR, 1971]
36 FR 4607, Amendment to FMVSS 209, Docket No. 69-23; No. 2. March 10, 1971.

[Howe, 1992]
Gavin Howe, Russ Kirkbride, Aloke Prasad, Mike Monk, "Tests Regarding Alleged Inertial Unlatching of Safety Belt Buckles" Vehicle Research and Test Center, NHTSA, November, 1992.

[IIR, 1992]
Petition for Defect, Rulemaking Action. Institute for Injury Reduction. Upper Marlboro, Md. September 11, 1992.

[SAE, 1965]
Motor Vehicle Seat Belt Assemblies - SAE J4C. 1966 SAE Handbook. Society of Automotive Engineers. July, 1965.

[Thomas, 1992]
Terry M. Thomas, Douglas A. Limbert, Robert C. Lange, "An Investigation of Seat Belt Buckle Dynamic Response to Inertial Loading Conditions," Failure Analysis Associates, FaAA-AZ-R-92-11-09, Prepared for General Motors Corp., November 12,1992.

[Traffic Safety, 1966]
Testimony of Ralph Nader. Hearings Before the Committee on Interstate and Foreign Commerce. House of Representatives. On H.R. 13228 and other bills relating to Traffic Safety. U.S. Government Printing Office. Washington, 1966.

950887

Safety Belt Buckle Inertial Responses in Laboratory and Crash Tests

Edward A. Moffatt
Biomech, Inc.

Terry M. Thomas
Thomas Engineering, Inc.

Eddie R. Cooper
Failure Analysis Associates, Inc.

ABSTRACT

Laboratory testing measured the response of a 1984 Chevrolet S-10 Blazer seatbelt buckle to impact on the back of the buckle. The peak acceleration, pulse duration and webbing tension were recorded to map the unique circumstances necessary to inertially unlatch the buckle. The conditions necessary to inertially unlatch the buckle in the laboratory were compared with the measured buckle responses in fifteen sled tests and six rollover crash tests using anthropomorphic dummies. All of the crash tested buckles remained latched and all had dynamic responses well below those required to produce inertial unlatching. Dummy hip areas were measured to be significantly stiffer than humans. Buckle accelerations measured in the "parlor trick" of intentionally striking the hip with a buckle are not representative of crash conditions.

BACKGROUND

Safety belt separations in crashes have been attributed to many causes such as cut webbing, fractured hardware, improper usage, inadvertent button activation, and inertial unlatching. This review focuses on the history of inertial unlatching.

Cichowski (1963) reported that General Motors had "observed inertial opening characteristics of buckles" when testing on their new sled facility (Ref 1). The buckles tested were the lift cover type, not push-button release.

The California Highway Patrol (CHP) developed a body block drop fixture in 1965, which resulted in push-button buckles inertially opening. Testing by seatbelt and vehicle manufacturers demonstrated that the CHP inertial release was caused by the buckle slapping into the body block during rebound with no load on the belt webbing. They concluded it was a condition resulting from the configuration of the test fixture and not representative of actual collisions. California officials subsequently agreed (Refs 2 and 4).

In 1966 the Committee on Interstate and Foreign Commerce of the House of Representatives conducted hearings on traffic safety. Ralph Nader testified that push-button buckles could be opened by entrapment between the abdomen and steering wheel during a crash. He also demonstrated inertial release by laying a buckle on the table and striking it. He labeled them "booby traps" (Ref 3). On behalf of automobile manufacturers, John Bugas of Ford testified that Ralph Nader's charges were unfounded. He referenced General Motors 1964 tests of the RCF-65 push-button buckles to determine if they "would unlatch under impact or inertial conditions." Body blocks had been accelerated into instrumented buckles on belts having various amounts of slack. None of the buckles opened despite buckle accelerations up to 105 g. He also cited Ford Motor Company tests at Wayne State University in 1965 to evaluate the effects of webbing slack on buckle inertial release. Cadavers and dummies had been tested at 25 mph impact velocities. They concluded, "In no case could the buckle be inertially unlatched because of belt slack." Mr. Bugas stated they had never seen one opened in any sled or crash test and had no accident reports of it happening. He labeled the inertial release a "parlor trick," claiming that it is unrelated to conditions in actual crashes (Refs 3 and 4).

The National Bureau of Standards studied buckles from 1969 model year vehicles to determine whether a single standardized buckle release method should be prescribed in FMVSS 209. In 1971 they recommended that push-button buckles be required, citing convenience and a high level of reliability (Ref 5).

Henderson (1973) reported an Australian rollover collision where a buckle "separated at impact." The type of buckle or cause of separation was not identified (Ref 6).

In 1973 the New South Wales Motor Transport Department reported a series of tests to inertially release various buckles. A push-button buckle opened at about 100 g from a 20 ms pulse. They stated that the probability of such a pulse occurring in an accident is unknown. They concluded that the Australian standard should be amended to include an inertial test on buckles with a proposed level of 40 g for 30 ms (Ref 7).

In 1978 the NHTSA investigated whether the buckle of a

1975 Chevrolet Monza could be inertially unlatched by striking with a rubber hammer or a spring-loaded plunger (Ref 8). They concluded that it could be, but that it was easier to open with zero lb. webbing tension than with 30 lb. (130 N) tension. Apparently no measurements of impact energy, velocity, or accelerations were undertaken. While noting that one buckle may have failed during a rollover crash, the NHTSA went on to survey 225 seatbelts from 1971 to 1978 model year cars for inertial release (Ref 9). They also impacted buckles with a spring-loaded plunger which was set to produce 300 g peak acceleration on belt assemblies with no web tension and 340 g on buckles with 30 lb. (130 N) of tension. Approximately one-third of the General Motors and Ford buckles opened and none of the Chrysler, AMC, or foreign buckles opened. The buckles were more likely to open if there was no web tension. They recommended that if further testing is conducted, a "more realistic force" simulating actual conditions be used. Finally, in 1979 the NHTSA in-house memorandum summarizing the Monza tests concluded that a buckle can be opened from a concentrated blow to its back side, but that it does not seem very probable it could receive such a blow during an accident. Further, "The absence of complaints about the malfunction would appear to support this point." They concluded, "The alleged defect does not appear to be a common problem, and no further investigation is warranted at the present time" (Ref 10).

The NHTSA Vehicle Research and Test Center (1992) reported on a test program similar to the study we are presenting here. They determined the dynamic conditions necessary to inertially unlatch some Nissan, Ford, and GM buckles. They compared those unlatching results with six crash tests where seatbelt buckle accelerations and concurrent webbing loads were measured. They concluded that conditions required to open a buckle are not achievable in real accidents because of the small distances that exist between occupants and properly worn safety belt buckles (Ref 11).

In November 1992, the NHTSA announced the results of their investigation regarding a petition from the Institute for Injury Reduction for a safety recall and rulemaking to deal with the alleged defect of inertial release. Following an extensive investigation, they found "absolutely no basis" for the allegation and stated it is not a phenomenon associated with real world crashes (Ref 12).

Collision Safety Engineering (1993) analyzed the mechanics of inertial release for 10 different seatbelt buckles. They found that the static button release force increases linearly with webbing tension. They noted that for inertial release to occur dynamically, there must be low webbing tension and sufficient acceleration pulse amplitude and duration to move the latch the required distance for releasing. Through static testing, they found the Hybrid III dummy to be significantly stiffer than human subjects in the hip area. They concluded that there is no cause for the motoring public to be concerned about the safety of side-mounted push-button-type buckles (Ref 13).

Arndt (1993) recognized two basic pieces of information were missing to determine the likelihood of inertial release. Needed first was a set of well-defined input conditions required to cause buckle release. Secondly, they noted the need for data reporting the acceleration histories of buckles in real-world crashes. Some laboratory conditions for release were provided through their drop fixture testing with various instrumented buckles (Ref 14).

INTRODUCTION

The possibility of opening some seatbelt buckles by intentionally striking the back of the buckle has been recognized for at least 30 years. The absence of this occurrence in thousands of sled and crash tests, however, has led to widespread belief that inertial release is either impossible or extremely unlikely to occur in actual crashes. Despite the historical lack of buckle inertial releases being observed in crash tests, new attention has been brought to inertial opening of seatbelt buckles. There is a need for a better understanding of the buckle impact conditions necessary for release to actually occur. Simply demonstrating that a buckle can be inertially opened by intentionally striking it does not demonstrate whether this can occur in actual vehicle crashes.

The research described here encompasses and expands that previously reported by Failure Analysis Associates, Inc. and Moffatt et al (Refs 15 and 16). The work defines the laboratory impact conditions necessary to cause a specific seatbelt buckle design to inertially unlatch due to impact on the back of the buckle. These conditions are compared with the accelerations measured in full-scale crash and sled tests. It is only through quantified comparisons of the buckle impacts measured in actual crashes with those impacts required for buckle release in the laboratory that one can judge the likelihood of inertial release.

The buckle used for all of this testing was the RCF-67, which is also known as the "Type 1" or "Minibuckle" manufactured by TRW (referred to as "Type 1" hereafter). It has a steel frame enclosing a pivoting latch held closed by a spring. When inserted into the buckle, the latch plate engages a pawl in the spring-loaded latch, preventing withdrawal. When the button is depressed 1.8 to 2.5 mm, it pivots the latch, causing release of the latch plate. Figure 1 illustrates the buckle.

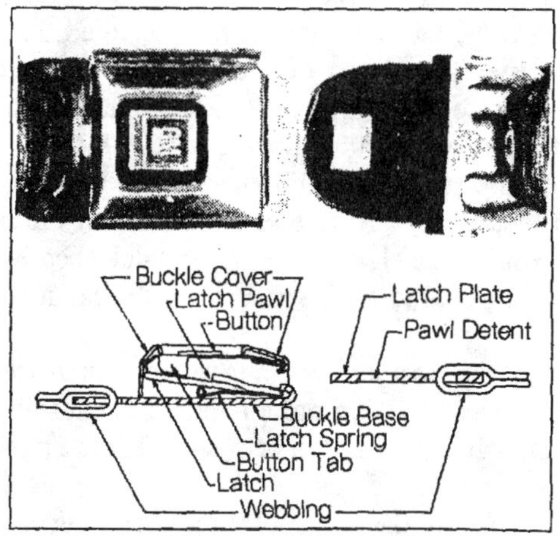

Figure 1: Type 1 TRW RCF-67 buckle

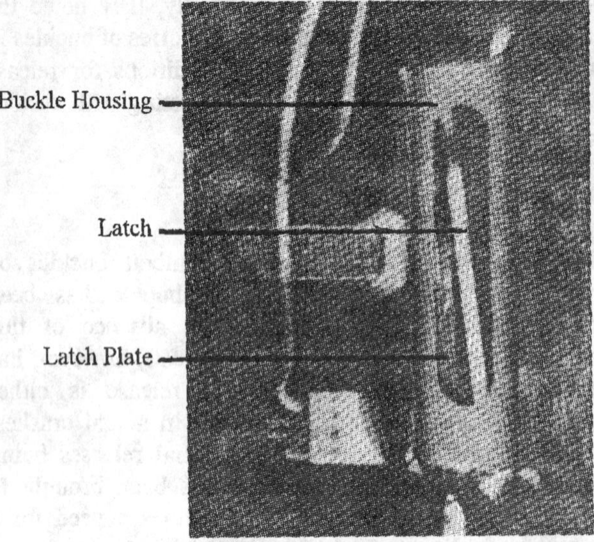

Fig. 2a: Buckle at rest prior to impact

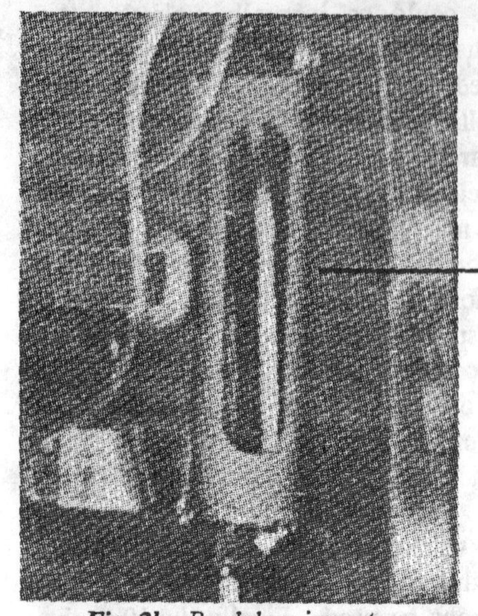

Fig. 2b: Pendulum impact -- latch moved relative to housing

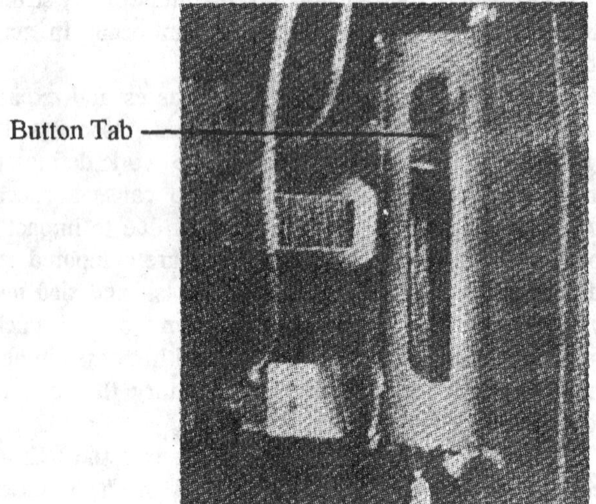

Fig. 2c: Button tab lags behind latch

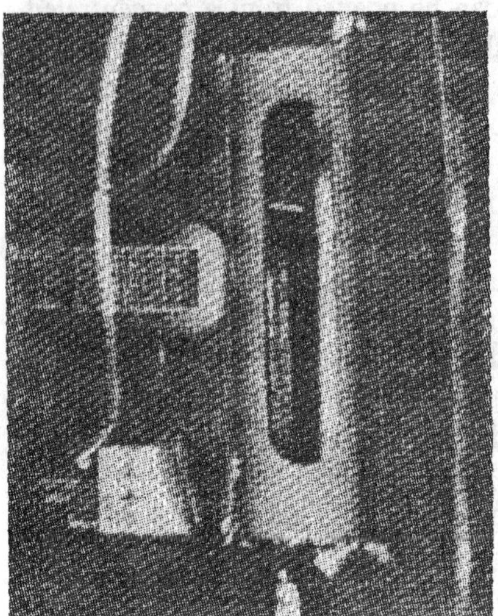

Fig. 2d: Latch fully open

Figure 2: High-speed film sequence of buckle impact; pendulum impact direction from right to left

The inertial release properties of this seatbelt buckle were analyzed in the Failure Analysis Associates, Inc. (FaAA) Test and Engineering Center (TEC) laboratory with high-speed filming of a pendulum impact to its back. With the side of the buckle cut away, and the internal parts painted for better visibility, the inertial release of the buckle was filmed at 7,000 frames per second. Figure 2 is a sequence from the high-speed film showing the motion of the buckle frame moving due to the pendulum impact while the light-colored latch remains stationary. In this test, the mass of the button was not a factor in inertially moving the latch, because the button tab lagged behind the movement of the latch (Figure 2c).

Tension in the seatbelt webbing greatly increases the magnitude of the acceleration pulse required to cause inertial release by increasing the friction between the latch pawl and the latch plate detent. For one TRW Type 1 buckle tested, with 13 N belt tension, the static force required to depress the button to the point of release was about 23 N. When the webbing tension was increased to 102 N, the button force required for release increased to approximately 53 N, and at 190 N tension, the button release load was approximately 67 N.

In order to inertially unlatch the buckle without direct contact to the button, the button release force must be produced by the resultant dynamic force that results from accelerating the moving parts (latch, spring, and button). This requires sharply accelerating the buckle frame toward the button with sufficient magnitude to actuate the latch mechanism. Even small webbing tensions essentially preclude inertial release because the accelerations required for release are too high.

FULL-SCALE ROLLOVER TESTS

A series of six full-scale FMVSS 208 dolly-type rollovers was conducted at FaAA TEC using a 1984 Chevrolet S-10 Blazer (Figure 3). The front bucket seat occupants had a three-point lap/shoulder belt system with two inertia-locking retractors, fixed latch plate, and a windowshade-style comfort feature on the torso belt. The front buckles were mounted at the inboard side of the front seats through a floor-mounted anchor connected to the buckle by webbing enclosed in a plastic sheath. The two rear occupant seating positions were fitted with lap belts only. The rear seatbelt buckles were mounted outboard of the seat positions. The rear retractors were center-mounted and could be operated in either the inertia-locking or automatic-locking mode. For these tests, the automatic-locking feature was defeated on the rear lap belts, so all tests were conducted using the inertia-locking mode.

Figure 3: 1984 Chevrolet S-10 Blazer on rollover dolly

Prior to testing, all glazing was removed from the vehicle, and internal rollcages were installed in the occupant compartment and under the front of the hood. Two onboard high-speed cameras filmed the motion of the seatbelt buckles for the front and rear occupants. Offboard photography included high-speed and real-time cameras. Vehicle accelerations were measured with triaxial accelerometers mounted on the floor near the base of each B-pillar.

The seatbelt instrumentation included FaAA-designed low-load webbing load cells stitched into the center front of the lap belt webbing (Figure 4). These load cells become overload protected at a limit of approximately 900 N and were calibrated in the range of 0-445 N. In the rollover tests, no measurements were taken of the front occupants' torso belt tensions. The seatbelt buckle accelerations were measured using damped 1,000-g Entran accelerometers, model EGAX-1000, mounted on an aluminum bridge affixed to the buckle over its center of gravity (Figure 5). The accelerometers were oriented to measure positive accelerations in the direction to cause inertial actuation of the latch mechanism. The accelerometer and belt tension data was acquired using an onboard transient data recorder with 2 kHz analog anti-aliasing filtering and a 10 kHz per channel data acquisition rate. Subsequent processing added SAE Class 600 filtering.

Each test included four anthropomorphic dummies. The left front and left rear seat positions were occupied by 50th-percentile Hybrid II male dummies, each with a seated pelvis. The right front seat held a 95th-percentile male Hybrid III dummy with a seated pelvis. In the right rear was a six-year-old child dummy with a seated/standing pelvis. There was no instrumentation in the dummies. Prior to each test, each dummy was seated in a normal position and photographed. The pretest lap belt tension for all dummies was determined solely by the retractor spring tension. There was no cinching of the lap belts. The front torso belt comfort feature was engaged to the first position, resulting in light chest contact.

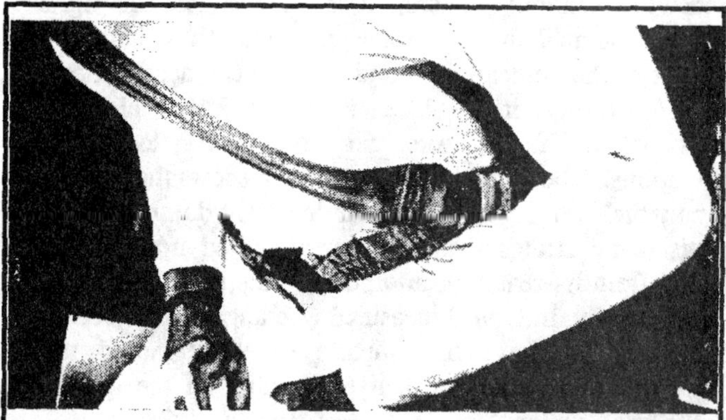

Figure 4: Load cell installation at center of lap belt

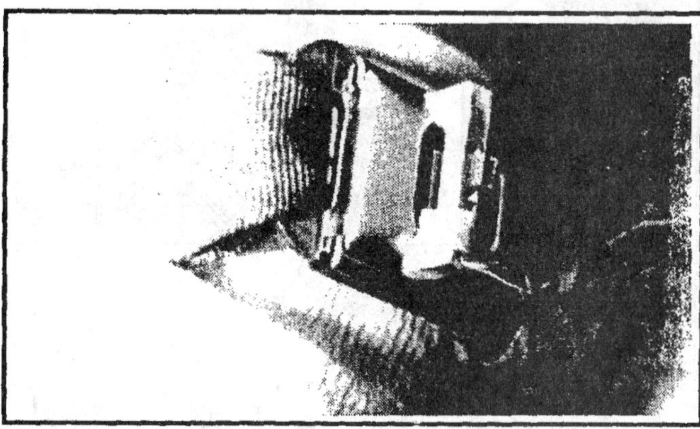

Figure 5: Accelerometer bridge on center of gravity of buckle

The same vehicle was used for all six rollover tests, with some minor body repair and axle replacement after some of the tests. The same seatbelts were used for all six tests, and all of the tests were conducted on flat asphalt. The rollover test conditions are shown in Table 1.

Table 1: Instrumented Rollover Test Conditions

Test ID	Vehicle Orientation	Test Speed
A	Passenger Side Leading	40.9 kph
B	Driver Side Leading	40.1 kph
C	Passenger Side Leading	57.1 kph
D	Driver Side Leading	57.0 kph
E	Rear Driver Side Leading	48.8 kph
F	Front Passenger Side Leading	48.8 kph

ROLLOVER TEST RESULTS

The six rollover tests resulted in a total of 18¼ vehicle rolls (where one roll is a 360 degree rotation), or about three rolls per test. With four dummies in each test, there were a total of 73 dummy-rolls. All of the seatbelts operated properly in all of the tests, and there were no releases of any buckles.

Figure 6 graphically presents the results of a typical test. The upper drawing illustrates the vehicle orientation at each ground impact. This rollover sequence was determined through high-speed film analysis which was correlated with the vehicle acceleration measurements. The upper data trace provides the resultant acceleration of the average of the right and left B-pillar/floor accelerometers. This trace indicates that most of the vehicle-to-ground impact peak accelerations were in the range of 5 - 15 g, interspersed with airborne periods where there were no significant floor pan accelerations. The second trace in Figure 6 shows the lap belt tension which was measured for the left front dummy. Film analysis demonstrates that the dummy moved upward and outboard from its seated position during the rolling airborne phases, resulting in typical measured webbing tensions of 200 - 650 N. Roof-to-ground impacts generally moved the dummy tighter into the lap belt (depending on the vehicle orientation), causing higher webbing tensions (such as at 0.55 sec and 1.6 sec). The wheel-to-ground impacts generally moved the dummy downward and laterally relative to the vehicle, and usually resulted in lower webbing tensions, depending upon the vehicle orientation and the resultant direction of movement of the dummy. Plotted beneath the webbing tension graph in Figure 6 is the acceleration trace for the left front buckle. Positive accelerations are in the direction of inertial depression of the button. This graph shows that the only significant buckle impact occurred as a result of a right wheels-to-ground impact. At about 2.3 sec, the right wheels (the "leading" side wheels) struck the ground, resulting in approximately a 10 g vehicle acceleration. The resultant movement of the left front dummy (which was seated on the side of the vehicle opposite the wheels striking the ground) caused a buckle peak acceleration of 62 g coincident with a 219 N webbing tension.

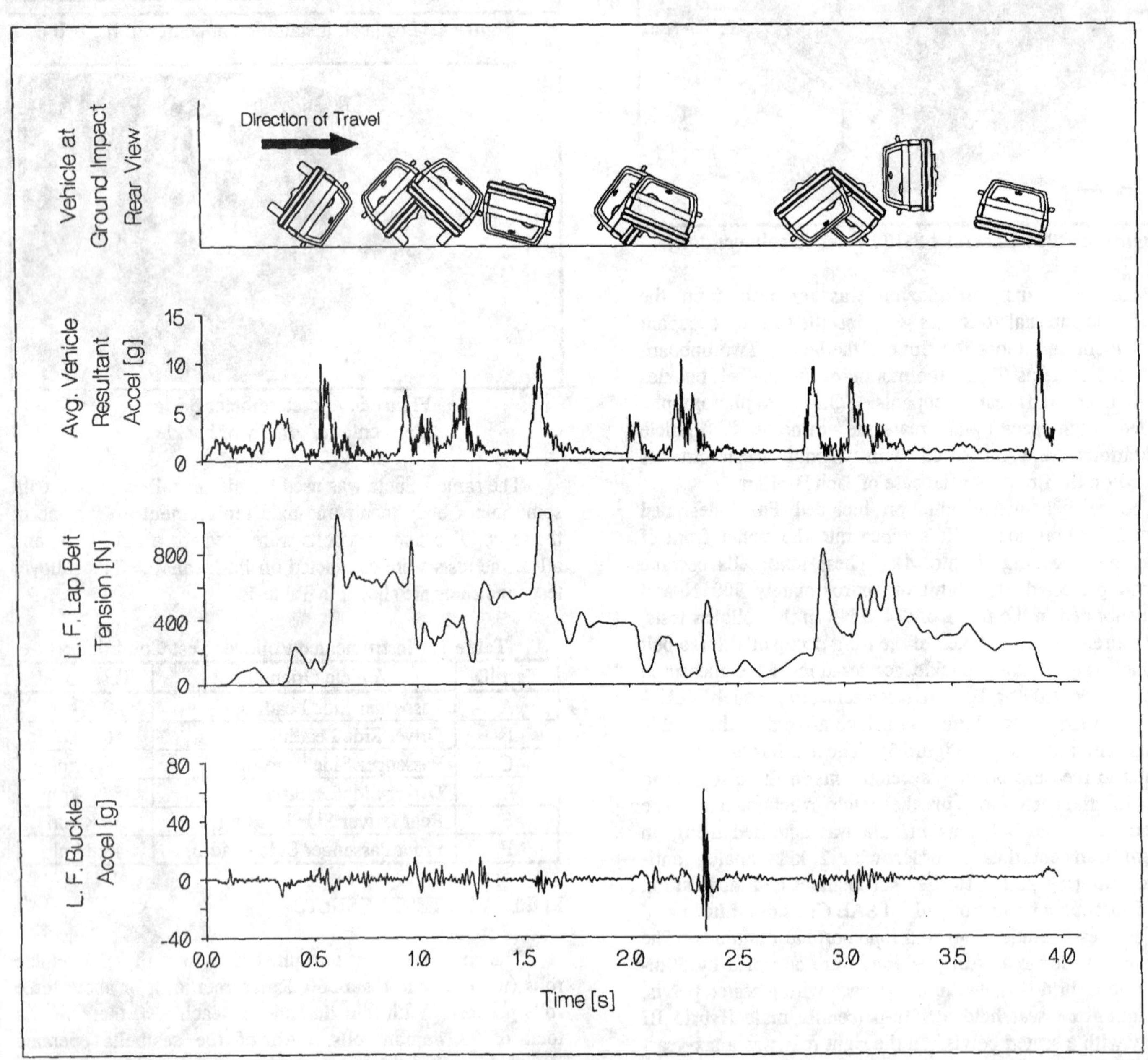

Figure 6: Example of rollover test results

A similar analysis was conducted for all impacts for all of the dummies in each of the six rollover tests. Every buckle acceleration exceeding 20 g was analyzed. The 20 g threshold for inclusion in analysis was selected to include most discrete pulses and yet eliminate insignificant vibrations. Table 2 is a summary of the most significant buckle impacts in the rollover tests. For those impacts where there were several buckle vibrations over 20 g associated with a single vehicle-to-ground impact, only the most severe impact is noted here.

As shown in Table 2, from the multitude of vehicle-to-ground impact orientations in these six rollover tests, essentially all of the significant buckle impacts occurred when the leading wheels struck the ground. This impact orientation caused dummy movement which maximized the combination of increased buckle acceleration and low webbing tension. The other vehicle-to-ground impact orientations (such as roof-to-ground) typically resulted in higher webbing tensions or lower buckle accelerations.

The most severe buckle impact in the six rollover tests was in Test E, when the right front dummy impacted the buckle, causing a 118 g peak acceleration with a concurrent 90 N webbing tension. Figure 7 illustrates this impact, where the Blazer had a substantial airborne leap before landing hard on its leading side rocker/frame/suspension. The resulting floor pan acceleration from this impact was 28.5 g, which was easily the most severe vehicle impact measured in any of the rollover tests. This ground impact was unusually severe because the left rear axle had been broken off in a previous roll during that same test, so there was no wheel and suspension to provide ride down. The onboard film analysis and the webbing load measurements show that there was low tension in the lap belt webbing prior to this impact as the right front dummy moved from the right to the left side of its bucket seat area within the confines of the seatbelt. As the dummy impacted the left side of the lap belt loop, the webbing tension increased from approximately 30 N to 300 N.

Table 2: Rollover Data Summary

Test (Description)	Dummy Loc.	Buckle Peak Accel. [g]	Pulse Duration Over 20 g [ms]	Event Min. Lap Belt Load [N]	Vehicle-to-Ground Impact Location (Time)
A (41 kph	LF	40.6	8.3	28	Leading Wheels (2533 ms)
Right Side	RF	[1]	--	--	
Leading)	LR	107.6	1.2	>445	Leading Roof Rail (574 ms)
# Rolls: 3	RR	[1]	--	--	
B (40 kph	LF	[1]	--	--	
Left Side	RF	60.0	6.6	[2]	Leading Wheels (2587 ms)
Leading)	LR	[1]	--	--	
# Rolls: 3	RR	61.9	6.2	386	Leading Roof Rail (550 ms)
		33.3	6.9	83	Trailing Roof Rail (707 ms)
C (57 kph	LF	61.6	7.3	219	Leading Wheels (2335 ms)
Right Side	RF	21.9	1.9	>445	Trailing Roof Rail (1611 ms)
Leading)		29.8	1.6	>445	Leading Roof Rail (2912 ms)
# Rolls: 3	LR	25.6	0.4	17	Leading Wheels (2283 ms)
	RR	27.1	1.2	6	Leading Wheels (3905 ms)
		26.9	4.2	35	Leading Wheels (3978 ms)
D (57 kph	LF	29.4	6.8	190	Leading Wheels/Rocker (2932 ms)
Left Side	RF	[1]	--	--	
Leading)	LR	[1]	--	--	
# Rolls: 4½	RR	241.9	0.8	>445	Leading Roof Rail (569 ms)
		20.6	0.2	>445	Leading Wheels/Quarter (2073 ms)
E (49 kph	LF	[1]	--	--	
Left Rear	RF	61.6	9.0	255	Leading Wheels (1792 ms)
Leading)		118.0	8.0	90	Leading Rocker/Susp. (3221 ms)
# Rolls: 2¾	LR	28.0	0.9	161	Leading Rocker/Susp. (3151 ms)
	RR	[1]	--	--	
F (49 kph	LF	26.4	3.1	434	Leading Wheels/Bumper (2075 ms)
Right Front	RF	30.1	8.8	247	Leading Wheels/Bumper (1930 ms)
Leading)	LR	47.9	4.5	322	Leading Roof Rail (948 ms)
# Rolls: 2		59.9	8.6	248	Leading Roof Rail (2421 ms)
	RR	[1]	--	--	

[1] This data contained no event exceeding the 20 g threshold.
[2] The belt load channel did not balance post-test; therefore, the load data may be unreliable.

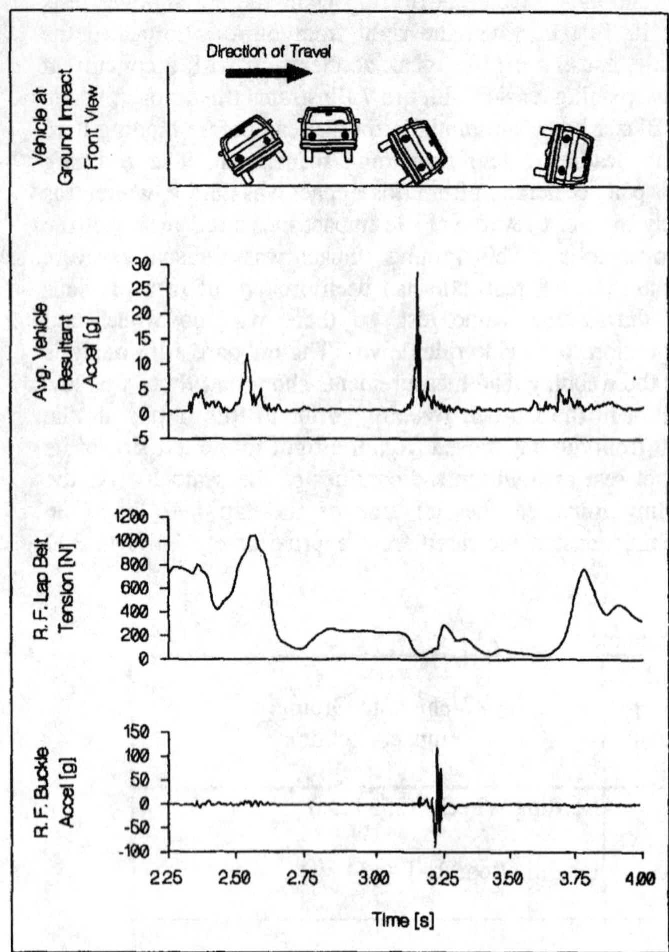

Figure 7: Test E, the most severe buckle impact

FULL-SCALE SLED TESTS

Fifteen sled tests were performed at FaAA TEC to measure the webbing loads and buckle accelerations in planar vehicle collisions which do not include rollovers. A 1984 Chevrolet S-10 Blazer body was mounted on the FaAA deceleration-type sled. This vehicle was similar to the rollover test vehicle, except that its original equipment included a center front console. The sled tests were run at 16, 32, and 48 kph nominal impact speeds at each of five different vehicle orientations: rear impact, left rear impact, left side impact, left front impact, and frontal impact. The sled deceleration pulses were designed to simulate the crash pulses in frontal, side, and rear impacts for the S-10 Blazer.

For the sled tests, the right front dummy was a 5th-percentile female, the right rear was the six-year-old child dummy, the left front was a 95th-percentile male Hybrid III, and the left rear was a 50th-percentile male Hybrid II. Each had a seated pelvis except the child dummy. The dummy and seatbelt pretest positioning was similar to the rollovers, where the lap belt tension was determined by the retractor tension and the front torso belt comfort feature was engaged at its first position, resulting in light chest contact. The same seatbelt instrumentation was used as in the rollover series, except that an additional load transducer was placed on the front torso belts slightly above the inboard hip. Onboard and offboard cameras documented each test. Some seats and seatbelts were replaced after certain tests due to damage. The center console was broken loose in Test 2C and not replaced. Figure 8 summarizes the vehicle orientations, test speeds, and dummy set-up for each of the 15 sled tests.

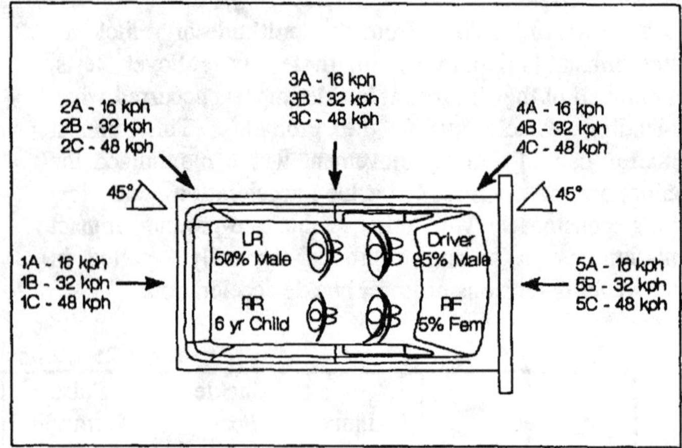

Figure 8: 1984 Chevrolet S-10 Blazer sled test conditions

SLED TEST RESULTS

Table 3 summarizes the sled test results. As in the rollover analysis, for each vehicle impact only the most significant buckle impact for each dummy is shown in the table. All of the seatbelts operated properly, and none of the buckles inertially unlatched in any of the 60 dummy-tests (four dummies times fifteen tests). The webbing tension for the front belts in Table 3 is the approximate vector sum of the lap and torso belt loads.

Several observations may be made from the results of the sled tests in Table 3. First, the lateral and lateral-rear impacts were more apt to have buckle accelerations over 20 g in combination with lower web tensions than other impact directions. In the frontal and frontal-oblique impacts the dummy movement put tension into the webbing early in the collision before any significant buckle accelerations occurred. Second, the higher impact speeds did not necessarily result in higher buckle accelerations. Also, the location of the buckle relative to the dummy (i.e., buckle-leading-dummy or buckle-trailing-dummy) did not have any significant effect. Finally, dummy size had no significant effect on the results as all the dummies were essentially of infinite mass relative to the buckle.

The two most severe buckle impacts in the sled test series occurred in the pure lateral impacts (Tests 3A and 3B). In both cases the left rear buckle (which is mounted on the outboard side of the dummy) was driven by the pelvis of the dummy into the upholstery-covered left rear wheel well.

Table 3: Sled Data Summary

Test (Description)	Dummy Loc.	Buckle Peak Accel. [g]	Pulse Duration Over 20 g [ms]	Event Min. Belt Load [3] [N]
1A (16 kph Nominal - Rear Leading)	LF	[1]	--	--
	RF	[1]	--	--
	LR	26.2	3.8	38
	RR	[1]	--	--
1B (32 kph Nominal - Rear Leading)	LF	50.5	5.7	28
	RF	35.4	0.6	39
	LR	53.8	9.4	145
	RR	40.0	4.2	31
1C (48 kph Nominal - Rear Leading)	LF	76.9	2.3	>445
	RF	30.9	3.1	16
	LR	180.4	3.3	[2]
	RR	39.5	3.6	136
2A (16 kph Nominal - Left Rear Leading)	LF	28.9	2.8	10
	RF	[1]	--	--
	LR	37.9	1.8	2
	RR	39.2	9.7	41
2B (32 kph Nominal - Left Rear Leading)	LF	35.3	7.5	12
	RF	92.0	2.6	>445
	LR	53.8	3.9	29
	RR	24.0	5.1	12
2C (48 kph Nominal - Left Rear Leading)	LF	[2]	--	--
	RF	49.8	2.6	299
	LR	[1]	--	--
	RR	280.3	4.4	>445
3A (16 kph Nominal - Left Side Leading)	LF	22.2	4.4	>445
	RF	25.5	3.0	238
	LR	68.5	6.4	11
	RR	65.2	8.8	303
3B (32 kph Nominal - Left Side Leading)	LF	105.2	4.5	>445
	RF	29.5	2.4	395
	LR	89.9	5.4	9
	RR	60.5	14.2	17
3C (48 kph Nominal - Left Side Leading)	LF	32.6	6.7	151
	RF	52.7	4.0	383
	LR	82.5	3.8	113
	RR	27.6	4.8	52
4A (16 kph Nominal - Left Front Leading)	LF	195.5	1.6	>445
	RF	[2]	--	--
	LR	[1]	--	--
	RR	37.2	7.9	83
4B (32 kph Nominal - Left Front Leading)	LF	[2]	--	--
	RF	54.2	3.8	247
	LR	451.2	1.4	>445
	RR	25.4	4.9	93
4C (48 kph Nominal - Left Front Leading)	LF	[2]	--	--
	RF	[1]	--	--
	LR	25.0	1.3	64
	RR	27.3	3.1	89

Table 3: Sled Data Summary (continued)

Test (Description)	Dummy Loc.	Buckle Peak Accel. [g]	Pulse Duration Over 20 g [ms]	Event Min. Belt Load [3] [N]
5A (16 kph Nominal - Front Leading)	LF	[1]	--	--
	RF	21.2	0.9	>445
	LR	129.1	1.3	>445
	RR	[1]	--	--
5B (32 kph Nominal - Front Leading)	LF	58.1	0.5	>445
	RF	71.7	3.5	>445
	LR	74.2	1.3	>445
	RR	135.7	2.9	>445
5C (48 kph Nominal - Front Leading)	LF	92.1	2.4	>445
	RF	435.7	3.8	>445
	LR	37.5	1.6	222
	RR	48.1	3.5	381

[1] This data contained no event exceeding the 20 g threshold.
[2] No data available because of data channel failure.
[3] For the front seats, the belt load indicated is the approximate vector sum of the lap and shoulder belt loads; the rear seats had lap belts only.

LABORATORY TESTS

The impact conditions required to inertially unlatch the test buckles were measured in the laboratory. Low-speed impacts were performed by striking the buckle with a pendulum, while higher-speed impacts utilized a shock table. The pendulum consisted of a 4-kg metal cylinder suspended by four wire ropes. Rubber pads were attached to the nose of the pendulum to produce acceleration pulses of different durations, and the drop height of the pendulum was adjusted to produce the variable peak buckle accelerations. Belt tension was applied by hanging weights vertically on the latch plate side of the buckle. A wire cable was attached to the rear of the pendulum to stop its forward motion after contact with the buckle back. Figure 9 shows the pendulum apparatus.

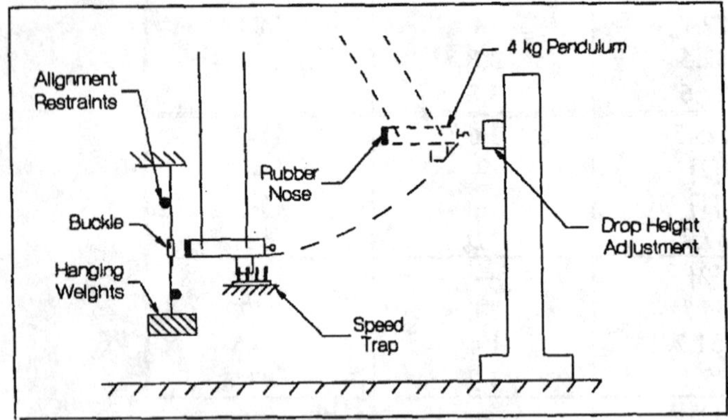

Figure 9: Pendulum fixture

Five Type 1 seatbelt buckles were characterized with short (nominally 1-2 ms) pulse durations, and five were characterized at medium (nominally 3-4 ms) pulse durations, each at three nominal belt tensions and at seven pendulum heights. Two additional Type 1 buckles were tested at medium pulse durations at the same belt tensions but at fewer pendulum heights, simply to obtain the lowest peak accelerations which would release them at those belt tensions. It was not possible to obtain high-amplitude acceleration pulses with the medium-duration rubber pad, so these tests were necessarily conducted at lower belt tensions and higher pendulum heights than the short-duration tests in order to obtain impacts sufficiently severe to result in buckle release. Acceleration pulses of 6 ms or longer were difficult to create with the pendulum because the compression of a softer foam rubber pad on the impacting surface of the pendulum became too large for these long-duration pulses.

A shock table, capable of producing 1 - 10 ms duration acceleration pulses, was designed and fabricated to supplement the pendulum test fixture for higher-speed impacts. The shock table consists of a buckle carrier mounted on a drop table guided by linear bearings. A schematic representation of the fixture and buckle carrier is presented in Figure 10. The primary velocity of the table is provided by four elastomeric bands, allowing an impact velocity of over 80 kph. The deceleration pulse is produced by stopping the table with a set of commercial shock table elastomers. This system is capable of producing 10 ms duration half-sine pulses of nearly 600 g peak amplitude. The buckle is supported in the carrier by the full contact area of the buckle frame back and held in place with two lateral clamps. The clamping force necessary to hold the buckle in contact with the back support is minimal and does not affect the function of the test buckle. Simulated belt tension is provided by a small elastomer attached to the latch plate and is reacted through a lip at the buckle latch plate inlet and not by the two clamps. The tension elastomer provides the energy necessary to extract the latch plate from the buckle when the latch pawl clears the detent. Comparison tests between the pendulum test fixture and the shock table were performed, and the same buckles tested on both fixtures produced similar results.

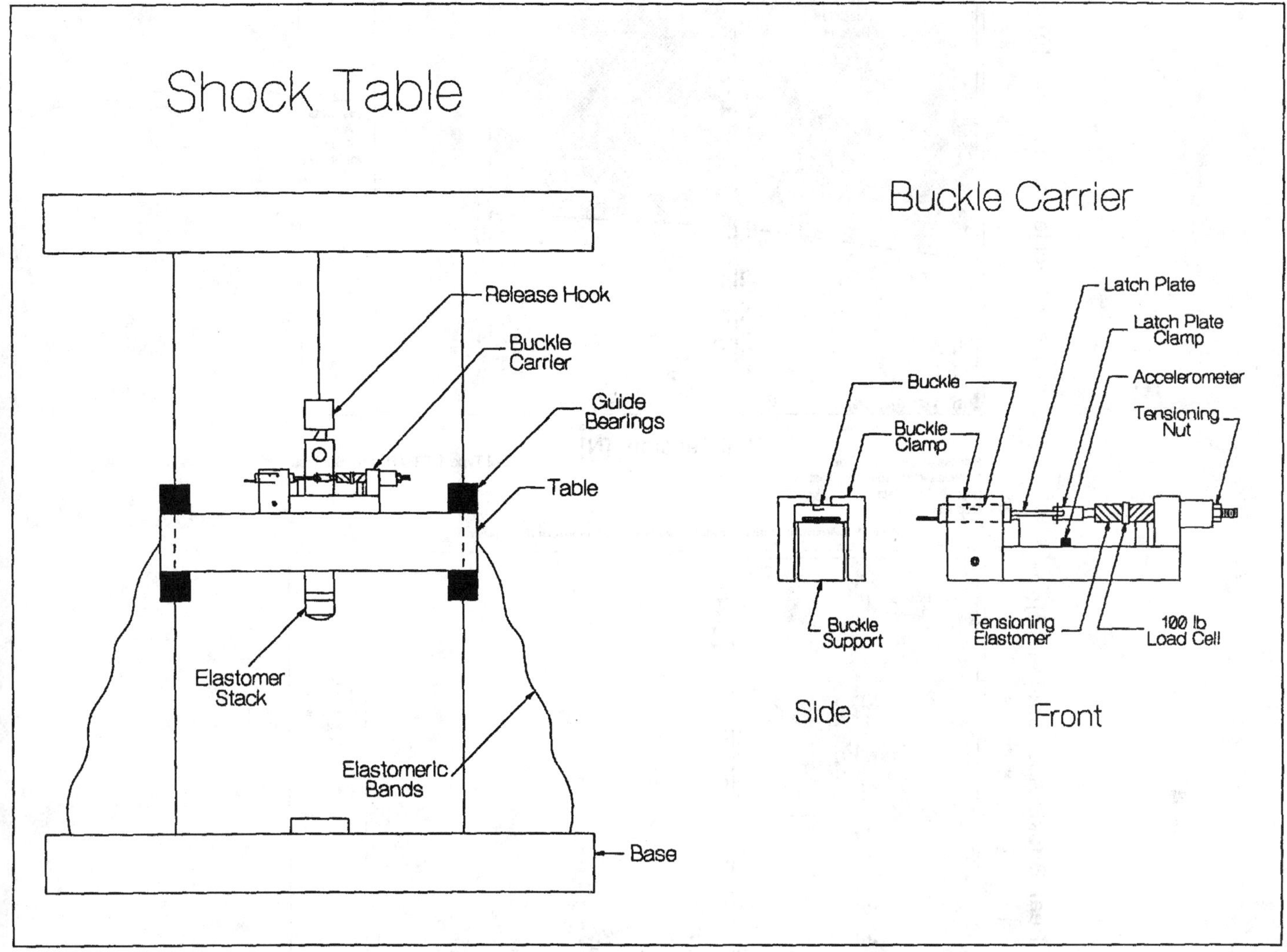

Figure 10: Shock table fixture

Six Type 1 seatbelt buckles were characterized on the hock table with long (nominally 8-10 ms) pulse durations, ach at three nominal belt tensions.

Similar data acquisition and processing techniques were ised for the crash testing and laboratory tests, except that the ninimum event belt tension is used for the crash test data, vhile the nominal belt tension is used for the laboratory test lata. The results of the laboratory tests are presented in the ext section.

ABORATORY AND CRASH TEST COMPARISON

igures 11, 12, and 13 show the laboratory "mapping" of the onditions necessary for inertial release along with the results f all of the sled and rollover tests, including the less ignificant events above the 20 g threshold. Each of the three raphs represents a different set of pulse durations. Figure 11 ives the results of the short-duration laboratory tests and rash tests for 1-2 ms pulse durations. The laboratory data is rouped in three columns at the three nominal webbing nsions (5 N, 183 N, and 361 N), indicated by the vertical nes; individual belt buckles are horizontally separated along the abscissa to avoid overlapping. Whether a buckle released at each impact level in a laboratory test is designated by a "Y" (for "Yes") or an "o" (for "No"). The location of the lowest "Y" on the figure shows the lowest peak acceleration for a given webbing tension and pulse duration where a specific buckle released in the laboratory. The "Possible Release" shaded area indicates the lowest release peak acceleration level recorded for the buckles tested in the data set. Figure 12 is a similar map of the 2-6 ms pulse duration crash test data with the medium-duration laboratory test data at 5 N, 94 N, and 183 N belt tensions, and Figure 13 shows the >6 ms pulse duration crash test data and long-duration laboratory data at 13 N, 93 N, and 182 N belt tensions. Note that Tables 2 and 3 include only the most severe crash test impacts, while Figures 11, 12, and 13 include all crash test events over 20 g for the specified durations. These graphs illustrate the magnitude of the peak buckle accelerations measured in the full-scale crash tests and the laboratory conditions that were found to be necessary for inertial release for three nominal pulse durations. All of the crash test buckle impact severities were well below the "possible release" line.

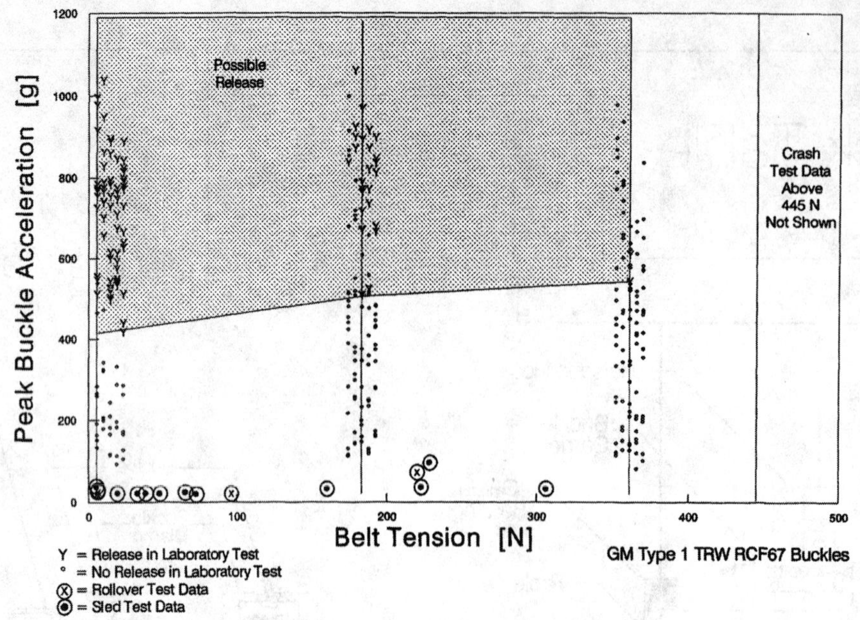

Figure 11: Laboratory and crash test results for pulse durations 1-2 ms and over 20 g amplitude

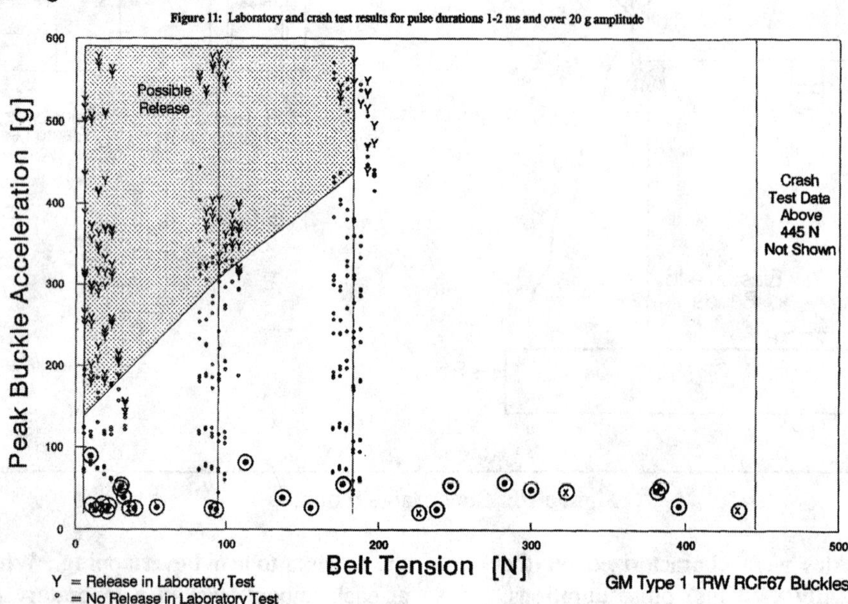

Figure 12: Laboratory and crash test results for pulse durations 2-6 ms and over 20 g amplitude

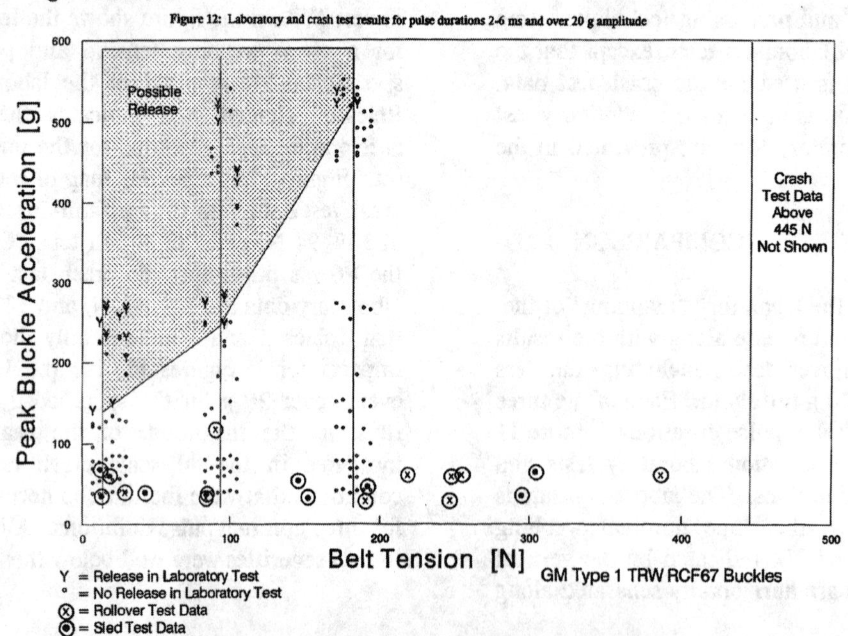

Figure 13: Laboratory and crash test results for pulse durations >6 ms and over 20 g amplitude

UMAN AND DUMMY HIP IMPACT TESTS

All of the sled and rollover tests utilized standard ıtomotive crash test dummies which are designed to ·plicate particular human characteristics. Although each ımmy has a human-like aluminum pelvis covered with soft .nyl foam, the dynamic force deflection biofidelity of the elvic area flesh in the seatbelt buckle area has not been erified to replicate human flesh.

James (Ref 13) addressed the relative stiffness of humans nd dummies by measuring the static force deflection haracteristics of the hip area of a Hybrid III dummy and umans. He found the dummy to be "two to three times iffer than humans."

Human soft tissue is viscoelastic in nature, so its ompliance is dependent upon the rate of loading. To easure the relative dynamic stiffness of the hip area of umans and the test dummies, FaAA developed a calibrated, oring-loaded hip impactor. This device, shown chematically in Figure 14, projects a properly ballasted TRW ype 1 buckle housing at known velocities and measures the uckle acceleration using the same instrumentation as used in e full-scale crash tests.

Twenty-seven human volunteers and four test dummies 95th-percentile male Hybrid III seated pelvis, 50th-percentile ale Hybrid II seated pelvis, 5th-percentile female Hybrid III eated pelvis, and six-year-old child seated/standing pelvis) ere the subjects used to compare dynamic response to npact in the hip area.

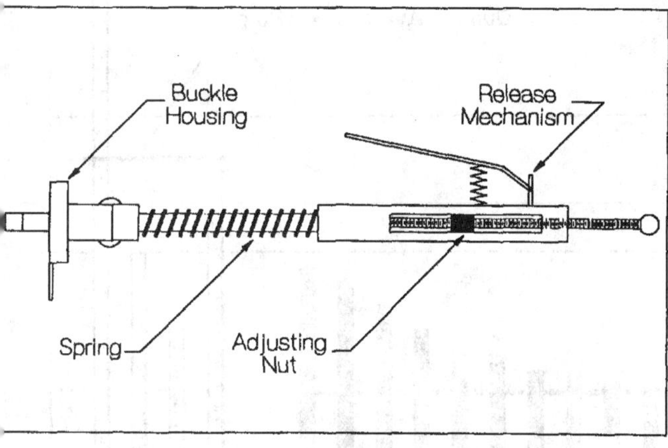

Figure 14: Hip impactor

Figure 15 illustrates the four impact locations. Point 1 as the anterior superior spine of the ilium, which is an ısily identifiable landmark on both humans and dummies. oint 2 was the greater trochanter, which is the lateral knob n the top of the femur. Dummies do not have trochanters, so is location was determined by measuring the same impact ea as on a human. It generally fell at the dummy H-point. oint 3 was located halfway between Points 1 and 2, which enerally corresponds with the fold between the wall of the odomen and the top of the thigh. Point 4 was a point on the teral buttock located by making an equilateral triangle with oints 1 and 2 with the subject seated.

The human subjects were all employees of FaAA TEC in Phoenix, Arizona, who were available during the two days of testing. Each was tested with whatever clothing they were wearing, which tended to be lightweight, since it was summertime in Arizona. No selection was made regarding body type, so the tests provide a diverse sample of human volunteers. Tables 4 and 5 summarize the subjects tested.

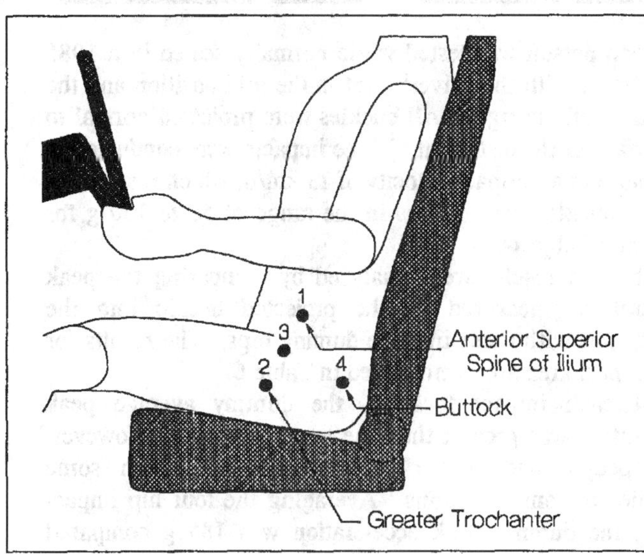

Figure 15: Hip area test locations on dummies and humans

Table 4: Hip Impact Human Subjects

Human Subject	Height [m]	Mass [kg]
#01	1.66	58.5
#02	1.83	122.5
#03	1.89	82.6
#04	1.85	116.6
#05	1.66	56.2
#06	1.73	73.9
#07	1.63	62.1
#08	1.60	54.0
#09	1.94	117.9
#10	1.57	57.2
#11	1.79	94.8
#12	1.69	88.9
#13	1.92	87.1
#14	1.83	75.7
#15	1.80	91.6
#16	1.63	84.4
#17	1.82	81.6
#18	1.55	45.4
#19	1.77	93.0
#20	1.75	79.8
#21	1.81	82.6
#22	1.82	85.3
#23	1.82	93.0
#24	1.77	79.8
#25	1.81	91.6
#26	1.87	79.4
#27	1.84	81.6

Table 5: Hip Impact Dummy Subjects

Dummy	Description
#1	5th-Percentile Female (H3X-5F) Dummy
#2	50th-Percentile Hybrid-II Male Dummy
#3	95th-Percentile Hybrid-III Male Dummy
#4	6-year-old Child Dummy

Each person was tested while normally seated in a 1985 S-10 Blazer with the driver's seat in the mid position and the seat back fully upright. All buckles were projected normal to the surface of the test point. Three impacts were conducted at each point at an impact velocity of 13 km/h, which resulted in average buckle accelerations in the range of 50 to 150 g for the human subjects.

The test results were analyzed by comparing the peak accelerations measured by the projected buckle into the human hips with those into the dummy hips. The results for each impact area are summarized in Table 6.

At each impact location, the dummy average peak acceleration was greater than the human average. However, some people had higher peak accelerations than some dummies in some locations. Averaging the four hip impact points, the dummy peak acceleration was 185 g compared with the human average of 98 g (Figure 16). The dummy hip area tends to be significantly stiffer than the average person under these impact conditions.

Table 6: Hip Impact Comparison

Point No.	Description	Human Avg.	Dummy Avg.
1	Anterior superior spine of ilium	109 g	355 g
2	Greater trochanter	90 g	141 g
3	Halfway between Points 1 and 2	91 g	114 g
4	Lateral buttocks	104 g	128 g

Additional tests were conducted to measure the effects of clothing on buckle acceleration. As expected, increased layers of clothing greatly softened the impact. For example, for Subject #01 at Point 1, the average peak acceleration was reduced approximately 50% by the addition of a cotton sweater.

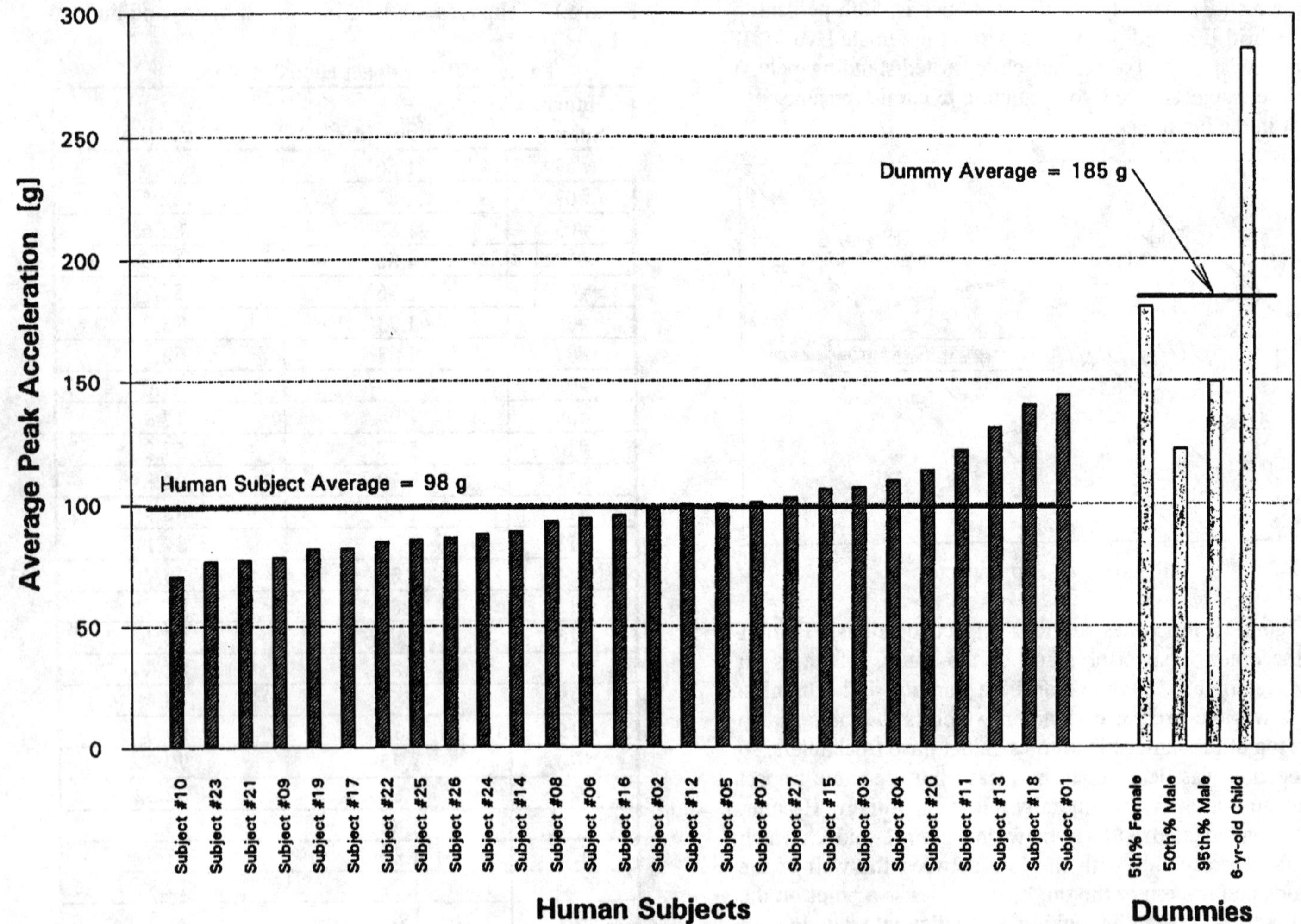

Figure 16: Hip impact test results, average of Points 1-4, three 13-kph impacts per point per subject

"PARLOR TRICK"

The term "parlor trick" was used in 1966 by John Bugas (Ref 3) and persists today as the common reference to intentionally opening a buckle by hitting its back side. The term is descriptive because the "parlor trick" is amazing to lay persons who first witness it. However, measurement of the forces in the typical "parlor trick" demonstrates that it is not representative of the forces in actual crashes.

For this study a series of "parlor trick" impacts was conducted by two subjects (E. A. Moffatt and E. R. Cooper) striking an instrumented buckle against themselves at the anterior superior spine area of the ilium. The buckle was also struck by a VHS video cassette case. As in the crash tests, the webbing tension adjacent to the latch plate and the buckle accelerations were measured. The buckle used in these "parlor tricks" had been previously tested on the laboratory shock table, so its release level was well-defined. As predicted by the laboratory testing, only those impacts with sufficiently high acceleration and low webbing tension opened the buckle.

Of particular interest, however, was the ease by which the seemingly mild "parlor trick" could far exceed any of the buckle impacts measured in any of the rollover or sled tests. Figure 17 is a bar graph showing the most significant buckle impacts measured in all of the full-scale crash tests and the peak accelerations from all of the "parlor trick" tests. All of the most significant crash test events with a measured belt tension below 445 N are shown. The numbers beneath the "parlor trick" results refer to the order in which they were performed. The severity of the buckle impacts which were sufficient to cause inertial unlatching in these "parlor tricks" is far more severe than any buckle impact in any of the crash tests. A videotape to demonstrate the relative severity of the crash tests and "parlor trick" tests was prepared to show all of the impacts in real time (Ref 17). Observations of the "parlor trick" tests in comparison with the crash results shown in Figure 17 demonstrate why intentionally hitting a buckle on the back side is a misrepresentation of typical real-world crash conditions.

DISCUSSION

The full-scale crash tests provide information regarding the forces on the 1984 Chevrolet S-10 Blazer seatbelts and buckles in a broad range of crash configurations. Other vehicles will not necessarily produce the same buckle forces due to differences in the seats and seatbelt installations, and due to human variations such as anatomical differences, clothing, seat position and how the belt is worn.

The "possible release" conditions measured by these laboratory buckle tests are specific for this TRW Type 1 buckle/latch plate assembly. Different design buckles and those from other manufacturers will not necessarily be similar.

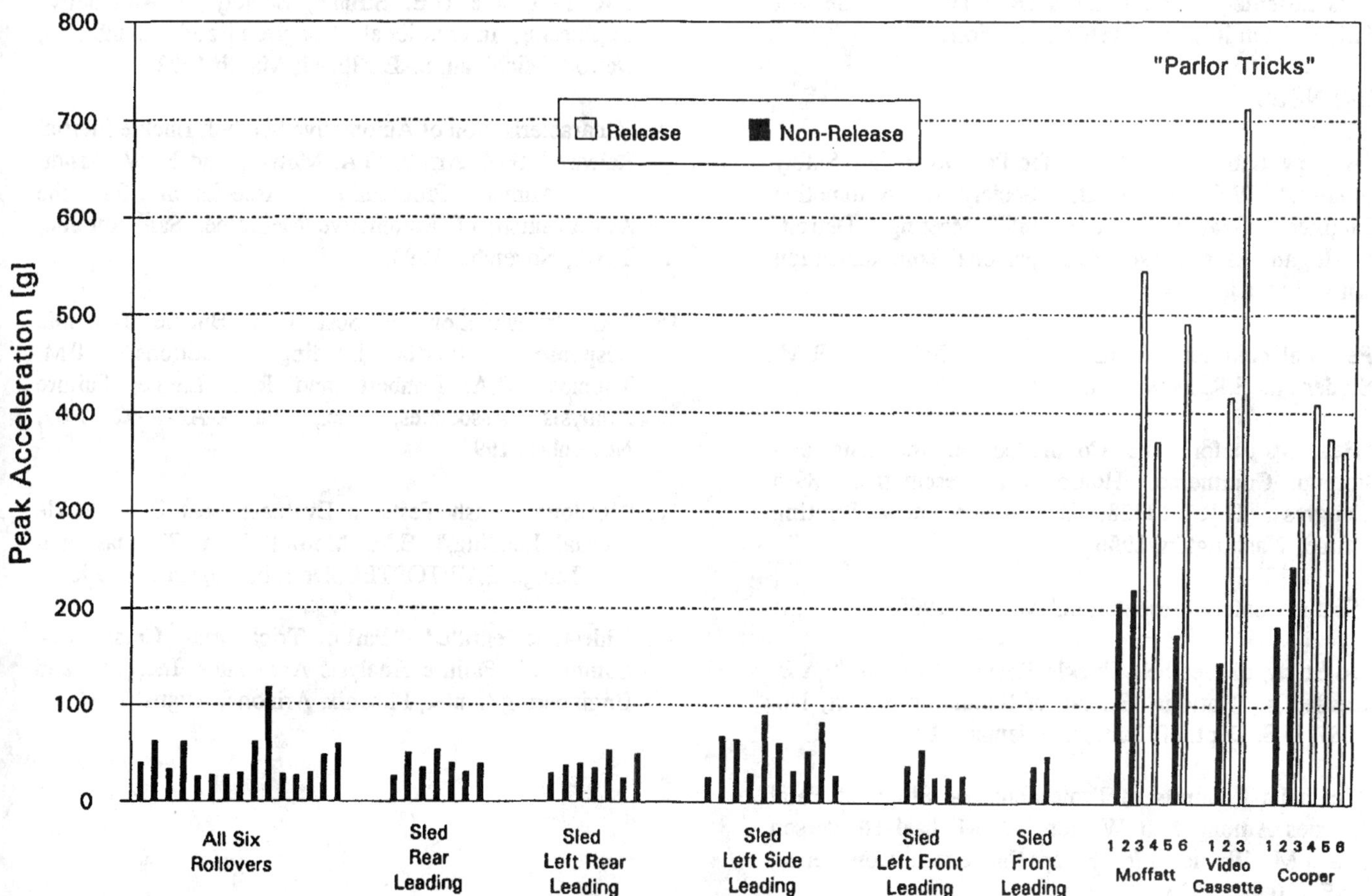

Figure 17: The most significant crash test impacts and all "parlor trick" results

The methods of data collection and processing used in this research were in accordance with SAE J211 specifications with the exception of the filter class on the belt load measurements. All belt load channels were filtered at SAE Class 600 for a more consistant comparison with the Class 600 acceleration channels. Additionally, FaAA found it necessary to use mechanically damped accelerometers to properly measure the accelerations experienced by the buckles. The relatively high natural frequency of the buckle assemblies tested in this research were found to be of sufficient levels to produce erroneous data if undamped accelerometers were employed. Caution should be used in making comparisons between the results presented here and the results obtained at other laboratories. Adherence to the SAE recommended practices and the use of properly mechanically damped accelerometers are necessary if comparisons are to be valid.

CONCLUSIONS

The necessary conditions for inertially unlatching the TRW Type 1 buckle were documented in the laboratory. The six rollover and fifteen sled tests provided a range of vehicle impact configurations and severities to evaluate the potential for these inertial release conditions to occur in 1984 Chevrolet S-10 Blazers. None of the crash tests came close to inertially unlatching any of the buckles. The significant differences between the conditions necessary for release compared with those measured in the crash tests indicate why it is extremely unlikely that a TRW Type 1 buckle will inertially unlatch in similar vehicle collisions.

REFERENCES

1. "A New Laboratory Device for Passenger Car Safety Studies," W.G. Cichowski, Society of Automotive Engineers National Automobile Meeting, Detroit, Michigan, March 1963 (and personal communication with W.G. Cichowski).

2. Personal communications with W.G. Cichowski, R.M. Studer and E.R. McKenna.

3. "Hearings before the Committee on Interstate and Foreign Commerce," House of Representatives, 89th Congress, H.R. 13228, U.S. Government Printing Office, Washington, 1966.

4. "Seatbelt Safety Presentation," May 17, 1966.

5. "A Study of Seatbelt Buckle Release Methods," A.B. Castle, Jr., National Bureau of Standards Report, No. 1087, U.S. Dept. of Commerce, January 1971.

6. "Seatbelts - Limits of Protection: A Study of Fatal Injuries Among Belt Wearers," J. Michael Henderson and J.M. Wyllie, 17th Stapp Car Crash Conference, SAE 730964, 1973.

7. "Dynamic Tests for Seatbelts," D. Herbert, et al., Traffic Accident Research Unit, Dept. of Motor Transport, New South Wales, March 1973.

8. "Tests of 1975 Chevrolet Monza Seatbelt Latching Mechanisms," A.R. Bayer, Jr. and R.L. Kirkbride, NHTSA Engineering Test Facility, March 1978.

9. "Survey of Seatbelt Latching Mechanisms Used on 1971-78 Passenger Cars," A.R. Bayer, Jr. and R.L. Kirkbride, NHTSA Engineering Test Facility, March 1978.

10. "Alleged Failure of Seatbelt Buckle on 1975 Chevrolet Monza 2Plus2: EA7-040," Letter from G. Chiang, Safety Defects Engineering to Acting Chief, Engineering Analysis Division, NHTSA, October 24, 1979.

11. "Tests Regarding Alleged Inertial Unlatching of Safety Belt Buckles," G. Howe, R. Kirkbride, A. Prasad, M. Monk, NHTSA Vehicle Research and Test Center, November 1992.

12. "Petition DP92-017 Inadvertent Release of Safety Belt Buckles," Office of Defects Investigation, NHTSA, November 18, 1992 (and press release regarding same).

13. "Inertial Seatbelt Release," M.B. James, D.L. Allsop, T.R. Perl, and D.E. Struble, Society of Automotive Engineers, International Congress and Exhibition, Detroit, Michigan, SAE 930641, March 1993.

14. "Characterization of Automotive Seatbelt Buckle Inertial Release," S.M. Arndt, G.A. Mowry, and M.W. Arndt, 37th Annual Proceedings, Association for the Advancement of Automotive Medicine, San Antonio, Texas, November 1993.

15. "An Investigation of Seat Belt Buckle Dynamic Response to Inertial Loading Conditions," T.M. Thomas., D.A. Limbert, and R.C. Lange, Failure Analysis Associates, Inc., FaAA-AZ-R-92-11-09, November, 1992.

16. "Rollover Crash Tests to Evaluate Seat Belt Buckle Inertial Loading," E.M. Moffatt, T.M. Thomas, and R.C. Lange, SAE/TOPTEC, Detroit, August 25, 1993.

17. Videotape entitled "Parlor Trick and Crash Test Summary," Failure Analysis Associates, Inc. Test and Engineering Center, Phoenix, Arizona, August 1994.

CHAPTER 12:

SIMULATION

930635

Evaluation of Belt Modelling Techniques

E. Fraterman and H. A. Lupker
TNO Crash-Safety Research Centre

ABSTRACT

An important assumption of conventional belt models is that the belt can slide over points that are fixed to bodies or the inertial space. Using finite elements sliding of the belt in all directions can be modelled. A study comparing both belt modelling techniques using the MADYMO program is described.

A mid-severity frontal impact sled test using a 50th percentile Hybrid III dummy is simulated. In addition, a performance study for oblique impact situations is presented to determine the effect of multi-directional belt slip of the shoulder belt. For the situations studied, it was found that both belt models can be used if belt roll-out at the shoulder does not occur. However CPU times are significantly higher for the simulations with finite element belts. An efficient approach for those cases where multi-directional belt slip of the shoulder belt occurs is to apply the finite element belt model only after a parameter study using the conventional belt model.

INTRODUCTION

The introduction of seat belts in passenger cars decreased the injury risk of car occupants in frontal crashes significantly. The first regulations concerning seat belts were proposed by the US government in 1963. In the same year that the first US belt regulations were drawn up, McHenry presented a 2D numerical model to describe the motion of a vehicle occupant in a collision event [1]*. In 1970, the first 3D mathematical model of the occupant was published by Robbins [2]. Since then, increasingly advanced models of a seatbelted occupant in an automotive crash environment have been introduced [3].

MADYMO has been designed especially for the study of the complex dynamical response of humans or human surrogates under extreme loading conditions [4, 5, 6 and 7]. The program has also been applied successfully for the simulation of other dynamic events, like vehicle riding and handling [8]. MADYMO combines multibody and finite element techniques with several force interaction models in one program (Figure 1). The multibody module uses a relative description for the kinematics of systems of rigid bodies. Several standard joints as well as user-defined joints can be applied. The finite element module uses a displacement based formulation with a Lagrangian material description allowing large translations and rotations. Constant strain triangular membrane elements with material models suitable for fabrics have been implemented. The method of

* Numbers in parentheses designate references at the end of the paper

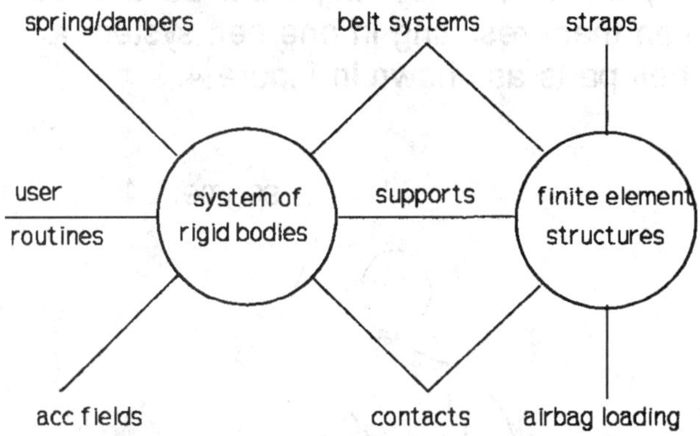

Fig. 1 MADYMO structure

isentropic compressible flow is used for the simulation of airbag gas dynamics [9], giving a good agreement between measured and calculated results [10, 11 and 12] with CPU times that are a fraction of the CPU times of finite element fluid models. Ellipsoids and planes can be connected to the rigid bodies for visualisation and contact interaction with other bodies or the environment. Additional force models which work on bodies are a belt model, acceleration fields, several types of spring and damper elements and user-defined force models. Finite element structures can be connected to inertial space, rigid bodies or belt segments. Contact interactions can be defined between FEM structures and ellipsoids and planes.

In this paper two belt modelling techniques will be compared. First both models will be described followed by a comparison when applied to a frontal impact sled test. In addition a performance study for oblique impact situations is presented. A discussion on the major findings of this study concludes this paper.

BELT MODELLING TECHNIQUES

A simple belt system can be modelled using spring-damper elements. For more complicated belt systems, MADYMO offers a belt model which allows slip of belt material from one segment to another. Further, this conventional belt model accounts for initial belt slack or pretension and rupture of belt segments. Elastic characteristics can be specified individually for each belt segment. Sliprings, retractors and pretensioners can also be modelled. An important assumption for this model is that the belt can slide over points that are fixed to bodies or to the inertial space, i.e., sliding over the body in the direction perpendicular to the belt segments is prevented. Multi-directional sliding can be realized by connecting the belt segments to a separate small ellipsoid which contacts the occupant surface. Such an approach has been applied to model submarining of the lap belt using a detailed model of the pelvis [13 and 14].

Multi-directional sliding can also be realized by using membrane elements to model those belt parts that are in contact with the occupant, and connecting these parts to belt segments of the conventional belt model. All options of the conventional belt model can still be used.

Another approach in seat belt modelling can be found in the advanced harness-belt system described by Butler and Fleck [15]. It features compliance of surfaces, multi-directional friction, with two different friction coefficients for two perpendicular sliding directions, and tie-points, i.e. connections of several belts together. A similar approach is described by Deng [16] to model submarining of the lap belt. Recently, prototype software for an Advanced Belt Model, ABM, has been realized in MADYMO which is based on the same theory as the advanced harness-belt system.

MADYMO CONVENTIONAL BELT MODEL

In the MADYMO belt model seat belts are modelled by a number of belt segments. Each belt segment is positioned between a rigid body and inertial space or between two rigid bodies by means of user-defined belt attachment points. The attachment points are fixed to the bodies or inertial space. This model therefore does not account for belt slip in the direction perpendicular to the belt segments. A belt segment is modelled as a massless spring with continuously changing untensioned length. The actual length of each segment is determined by the location of the two belt attachment points which represent its begin and end point (Figure 2).

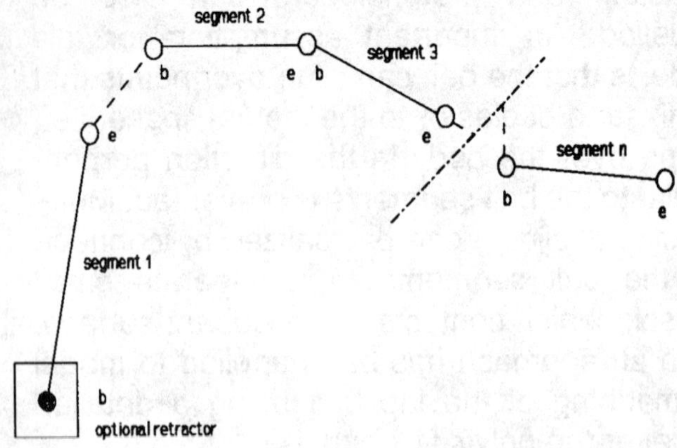

Fig. 2 Set-up of a belt system

Belt attachment points on the dummy are generally chosen to coincide with the location of the outer contact points between belt and occupant at the time of peak belt loads. Belt material between attachment points on the surface of the dummy is included by adding additional length to the adjacent belt segments. Exchange of belt material is possible between two adjacent belt segments of the same belt system. The belt material exchange between two belt segments is controlled by the belt forces, the specified friction coefficient and the angle between the two belt segments. The slip calculation is based on a pulley model (Figure 3).

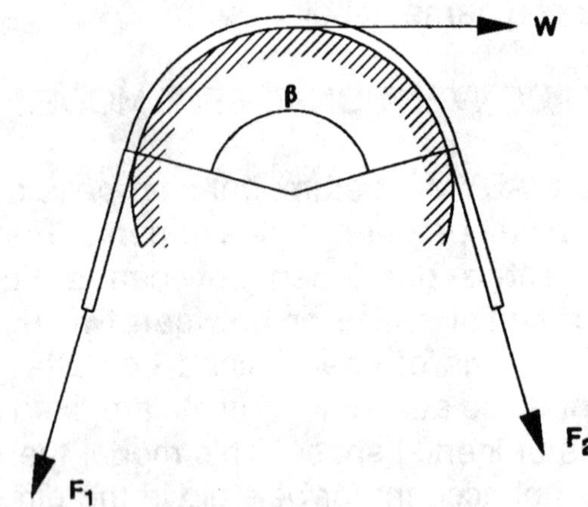

Fig. 3 Slip between the rim of a pulley and a belt

By defining the end point of a belt segment and the begin point of the adjacent belt segment at the same location a slipring is modelled. To model belt slip between shoulder and lap part of a 3-point belt a slip ring must be defined between them resulting in one belt system for both belt parts as shown in Figure 4.

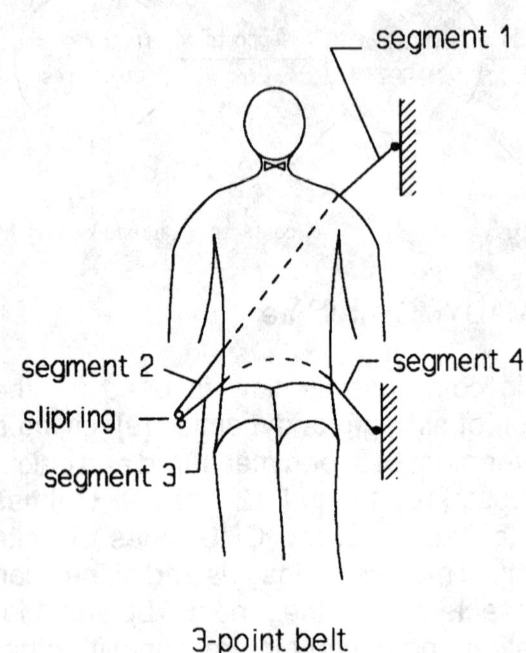

Fig. 4 Model of a 3-point belt

The belt stiffness is defined by means of a non-linear force-relative elongation characteristic which may also include hysteresis. At the start of the simulation, initial slack or pretension in a belt segment can be defined by means of a pretension factor.

A retractor with an optional pretensioner and a webbing grabber can be applied at the beginning of the first belt segment. A retractor can lock on a specified time or if an acceleration level is exceeded during a fixed time interval. Several conditions can be used for activating the pretensioner.

A buckle pretensioner can be simulated by modelling a body between the stalk and the vehicle connection point. For the connection to the vehicle an initially locked translational joint is used. For this translational joint a pretensioned spring and conditions for unlocking must be defined.

The belt model was applied successfully in several comprehensive validation studies. In [17], the experimental results of Hybrid III sled tests were compared with results of simulations using the 15 segment 50th percentile Hybrid III dummy database of MADYMO. A validation study for the 22 segment Advanced 50th percentile Hybrid III dummy using the belt model is described in [18].

MADYMO FEM BELT

As mentioned above the conventional model does not account for belt slip in the direction perpendicular to the belt segments. The belt webbing can be modelled using membrane elements because it possesses almost no bending stiffness. By modelling those belt parts that are in contact with the occupant by finite elements and connecting these parts to belt segments of the conventional belt model, multi-directional sliding can be realized and all options of the conventional belt model can be applied. The membrane element portion of the belt must be defined such that, during the simulation, it is the only part of the belt contacting the occupant surface. Linear isotropic and linear orthotropic materials can be specified for the membrane elements. It is important to note that hysteresis can not be modelled with these material models and must therefore be integrated in the characteristic of the parts of the belt that are modelled with the conventional belt model. With the conventional belt model, a lap belt would be modelled by two belt segments with identical force-relative elongation characteristics, i.e. identical material behaviour. If membrane elements are included, the properties of the different belt parts must be determined based on the force-elongation characteristic of the lap belt. All membrane elements are given the same elastic constants and thickness. The resulting stiffness of the FEM part must be larger than the hysteresis slope of the lap belt to obtain the correct characteristic also during the unloading phase. Both belt segments are given identical force-relative elongation characteristics, i.e. different force-elongation characteristics if they have different untensioned lengths. The characteristic for the belt segments can be calculated by considering the lap belt as a system of three springs in series which has the same resulting force-elongation characteristic as the lap belt. The contact interaction between the membrane elements and the rigid bodies is based on transfer of momentum between the nodes, the vertices of the triangular elements, and the ellipsoids and planes representing the outer surface of the bodies. The contact forces parallel to the contact surfaces are based on the Coulomb friction model, i.e. the tangential force equals the product of the normal force and the friction coefficient. Thus, for sliding in the belt direction and perpendicular to the belt direction, the same friction coefficient is used. Due to compliance of the occupant surface it can be argued that different friction coefficients should be used for both directions as is the case in the Advanced Belt Model.

The combination of membrane elements and belt segments is very effective if sliprings are involved. If just finite elements are used, a very fine mesh is necessary to model the sliding of the belt through the sliprings. To account for the contact interaction with the occupant surface a much coarser mesh is normally sufficient.

UNI-DIRECTIONAL BELT SLIP CONDITIONS

To compare both belt modelling options first a ECE-R 16 mid-severity frontal impact test configuration (20g and 50 km/h) has been simulated. The belt system consist of separate shoulder and lap belts with 8% webbing (8% elongation at 10.8 kN static load), with fixed anchor points and each having their own buckle. Descriptions of the sled environment can be found in [18].

Most input data are obtained from a simulation which was used to validate the MADYMO Advanced 50th percentile Hybrid III dummy database [18, 19]. In the simulation, the sled is connected to the inertial space and the measured sled deceleration is applied as an acceleration field on the dummy. The rigid seat on the sled is modelled by contact planes with very stiff force-deflection characteristics. The dummy is represented by the Advanced 50th percentile Hybrid III male dummy database released in 1992 [19]. This database is equipped with a separate sternum body, which is connected to the upper torso by means of point-restraints, to account for chest deflection.

For the simulation using only the conventional belt model, the belts are modelled with two belt systems. The belt system representing the shoulder belt consists of three belt segments, i.e., between the upper anchor point and the shoulder, between the shoulder and the flexible sternum, and between the sternum and the first buckle. The belt system representing the lap belt consists of two belt segments, i.e., between the second buckle and the pelvis and between the pelvis and the lower anchor point. The compliance of the pelvis is accounted for by adapting the lap belt characteristic. The measured initial force of 90 N in the shoulder belt has been modelled by applying a pretension factor for the belt segments.

In the second simulation the belt parts that contact the dummy surface are represented by membrane elements with an isotropic material behaviour. A pre-simulation was performed to calculate the initial nodal positions. Four belt systems, each consisting of one belt segment were applied: between the upper anchor point and the FEM shoulder belt, the FEM shoulder belt and the first buckle, the second buckle and the FEM lap belt and between the FEM lap belt and the lower anchor point. The overall stiffness of the FEM belt parts is proportional to the product of the cross sectional area and the Young's modulus. By decreasing the Young's modulus and increasing the thickness without changing the overall stiffness the maximum stable integration time step increases. Increasing the thickness does however also increase the mass of the finite elements. To keep simulation results unaffected, the thickness was only increased to a value that results in belt masses still much smaller than the masses of the bodies contacting the belts. After adjusting the thickness of the belts the elasticity constants for the FEM parts and the force-relative elongation characteristics for the belt segments were calculated by considering the lap belt as well as the shoulder belt as a system of three springs in series. The resulting overall belt force-elongation characteristics are equal to those used in the model applying only the conventional belt model. Pretension was not accounted for. An extra ellipsoid was attached to the spine to prevent the membrane elements from sliding behind the abdomen when the chest is deflecting.

RESULTS - The kinematics of the simulation using the conventional belt model are shown in Figure 5. The kinematics of the simulation using the FEM belt are shown in Figure 6. The resultant linear acceleration of the pelvis, chest, and head and the chest deflection are presented in Figure 7. The belt forces in different belt parts are presented in Figure 8: in the belt parts between upper anchor and shoulder, between sternum and buckle, between buckle and pelvis and between pelvis and lower anchor.

A fairly good overall agreement can be found between experimental results and results of both simulations. The linear acceleration of the lower torso is well predicted by both simulations. The first peak in the resultant acceleration of the upper torso is predicted satisfactorily. The second peak, however, is visible only in the results of the conventional belt model. Both simulation models slightly underestimate the first peak in the resultant head acceleration. The second peak in the head acceleration comes out too low for both models. The maximum chest deflection is better predicted by the FEM belt model. Both models overestimate the peak in the belt force of the buckle-pelvis part, and slightly overestimate the peak in the belt part between pelvis and lower anchor. For both simulation models a time-shift is visible for the peak in the belt force of the sternum-buckle part. The belt forces between upper anchor and shoulder are underestimated by both models.

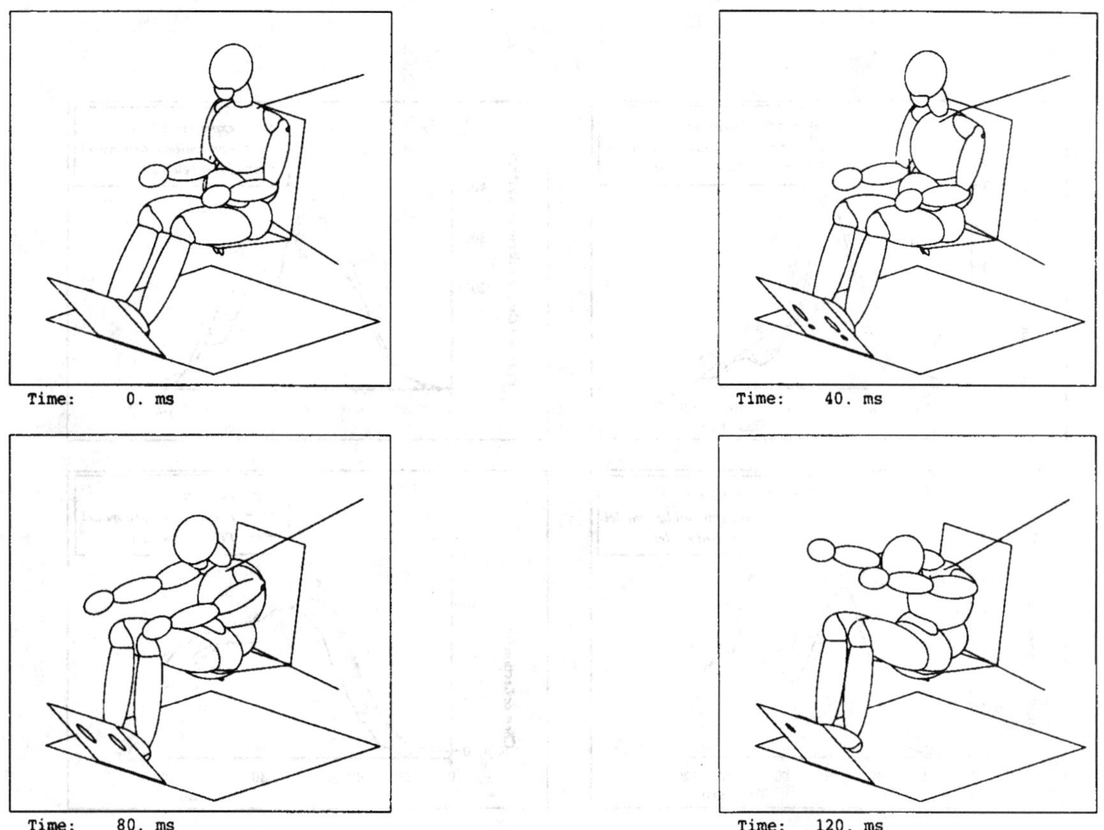

Fig. 5 Conventional belt model kinematics

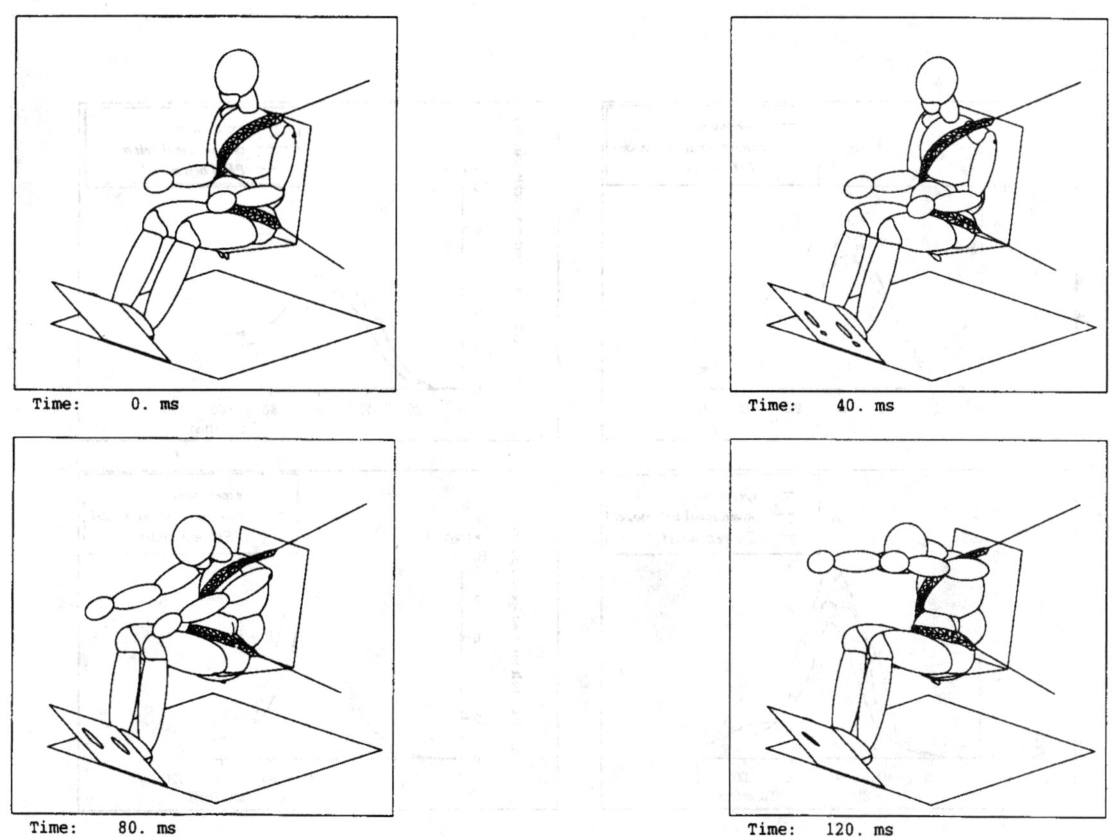

Fig. 6 Kinematics for partly FEM belt

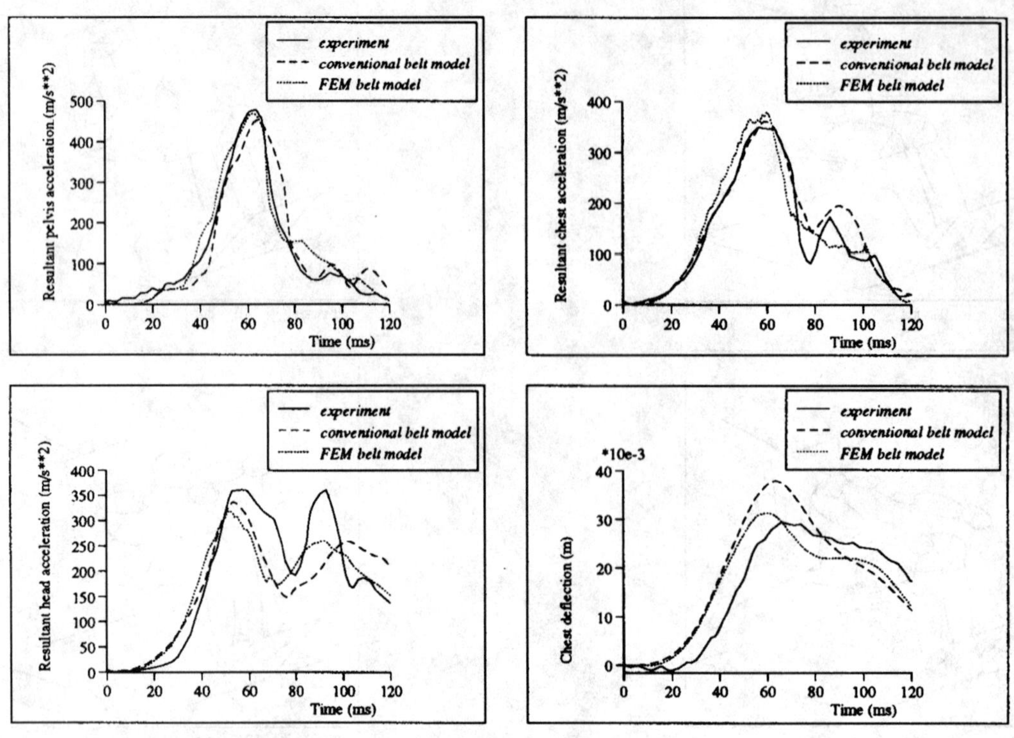

Fig. 7 Resultant accelerations and chest deflection

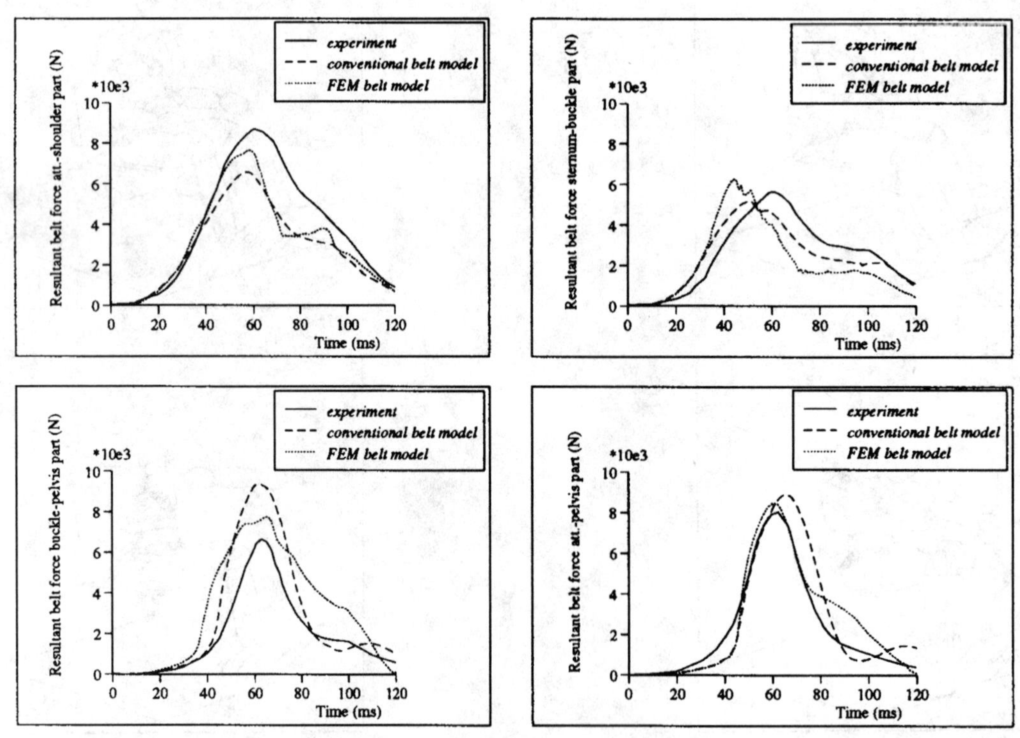

Fig. 8 Resultant belt forces

MULTI-DIRECTIONAL BELT SLIP CONDITIONS

To compare both models in a multi-directional belt sliding situation, the impact direction was varied to obtain belt sliding over the dummy shoulder. This was realized by changing the orientation of the acceleration field in the horizontal plane from 0 degrees, i.e. frontal impact, to 30 degrees by steps of 10 degrees. The position of the attachment points was taken equal for all conventional belt model simulations. There was no slack or pretension in these simulations.

RESULTS - The kinematics of the two belt models and different impact directions are shown at 80 ms for orientations of 0 and 10 degrees in Figure 9 and for orientations of 20 and 30 degrees in Figure 10. Resultant belt forces in the belt segment between the upper anchor point and the shoulder and chest deflections are shown for both simulation models at different orientations of the acceleration field in Figure 11. In the oblique FEM belt simulations with an orientation angle larger than 10 degrees, the belt slides over the shoulder and is caught between upper arm and shoulder resulting in a pronounced second peak in the belt force of the segment between upper anchor point and shoulder. Flexion and torsion angles of the upper neck joints are shown in Figure 12. It can be observed that the torsion angle increases more rapidly in the model with FEM elements than in the model with only the conventional belt segments. Flexion and torsion angles of the lower neck joint are shown in Figure 13. No large differences can be found between the results of both modelling techniques. Table 1 shows the resulting HIC values for the two modelling techniques.

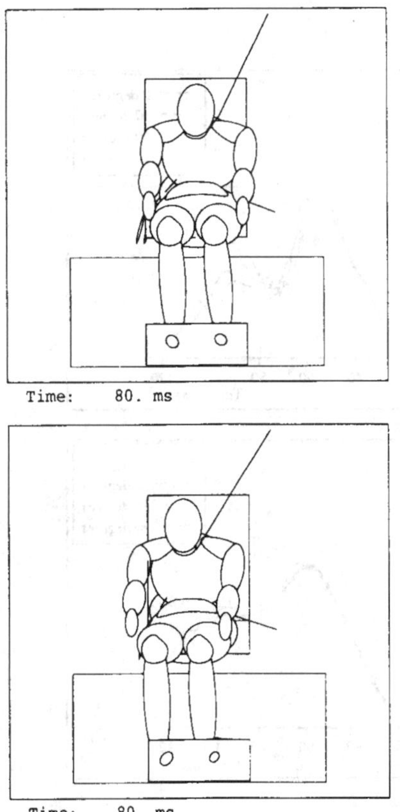

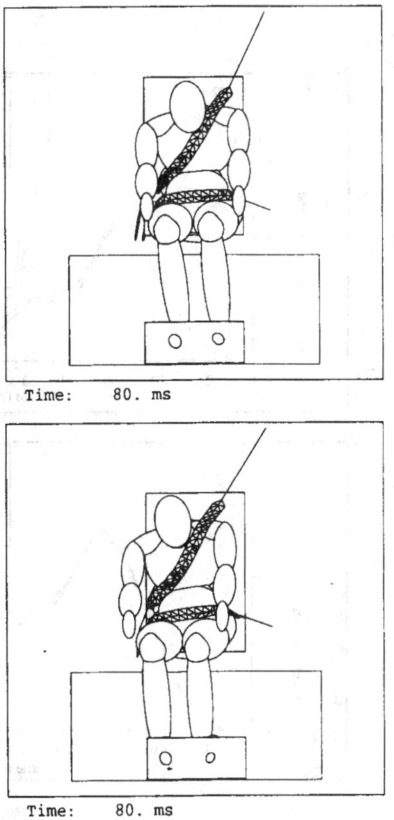

Fig. 9 Kinematics at 80 ms for orientations of 0 and 10 degrees

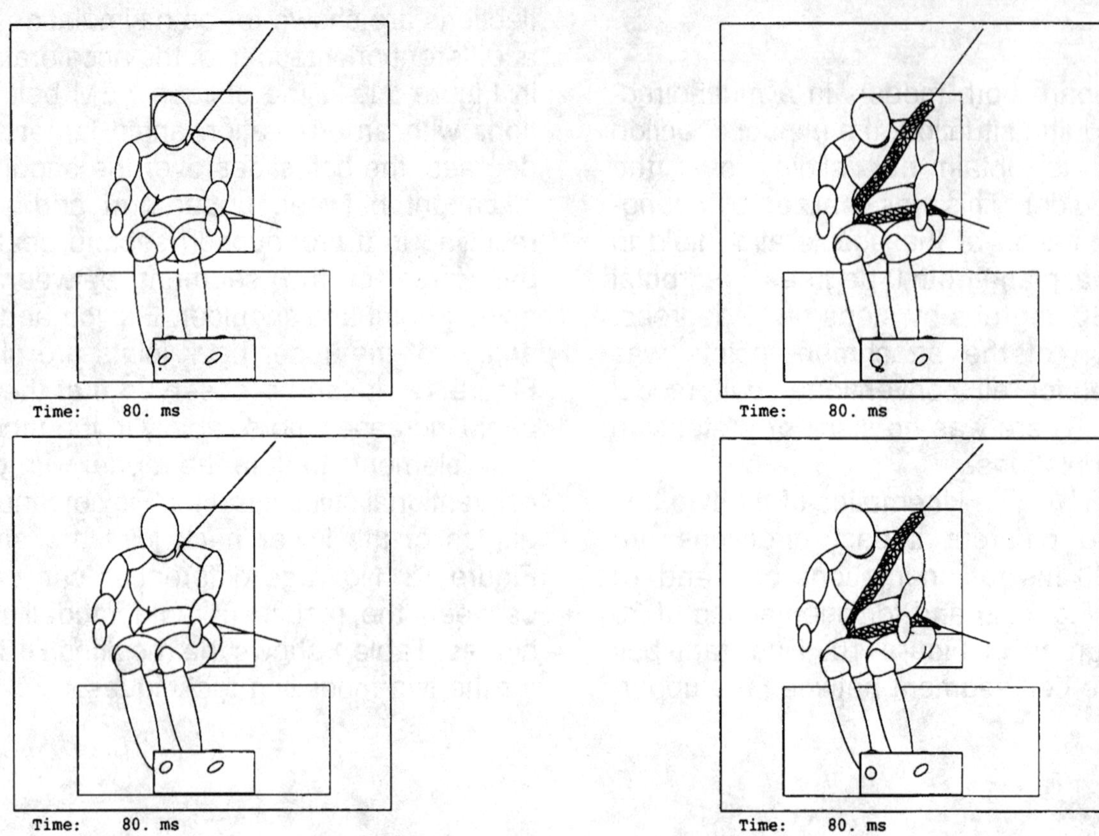

Fig. 10 Kinematics at 80 ms for orientations of 20 and 30 degrees

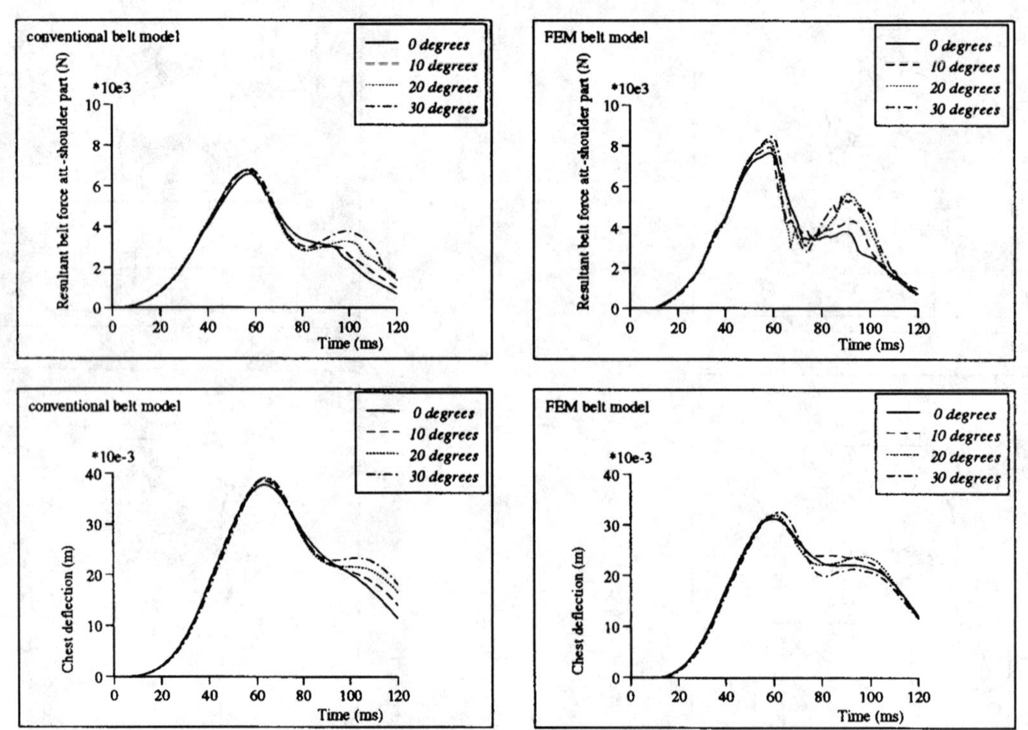

Fig. 11 Resultant belt forces in the belt segment between upper anchor point and shoulder and chest deflections for orientations of 0, 10, 20 and 30 degrees

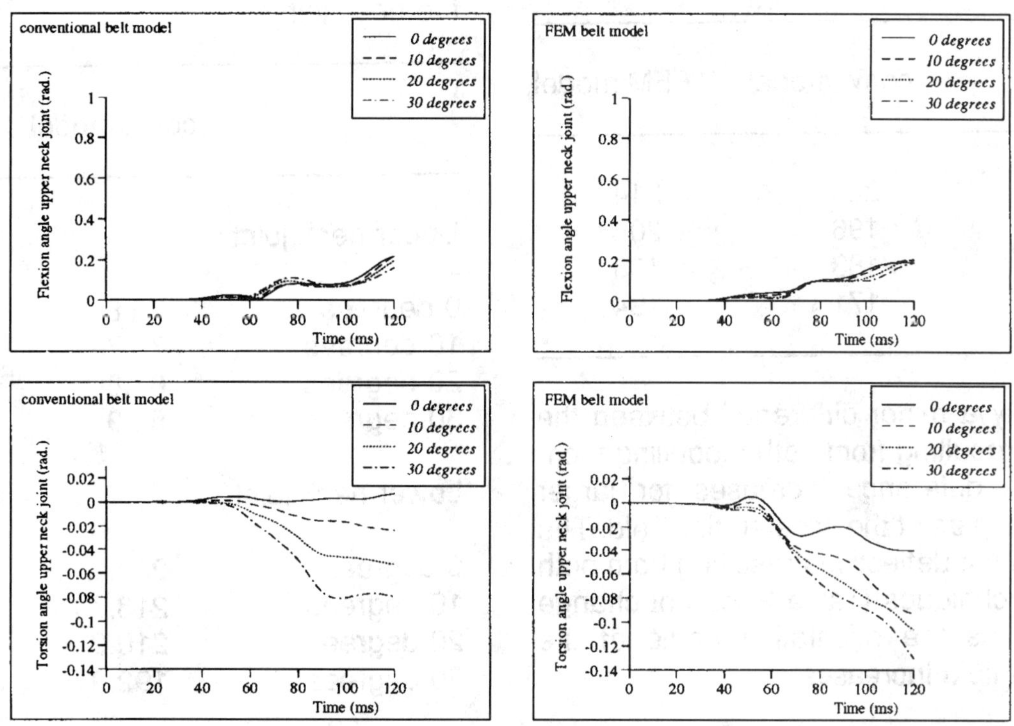

Fig. 12 Flexion and torsion angles in the upper neck joint for orientations of 0, 10, 20 and 30 degrees

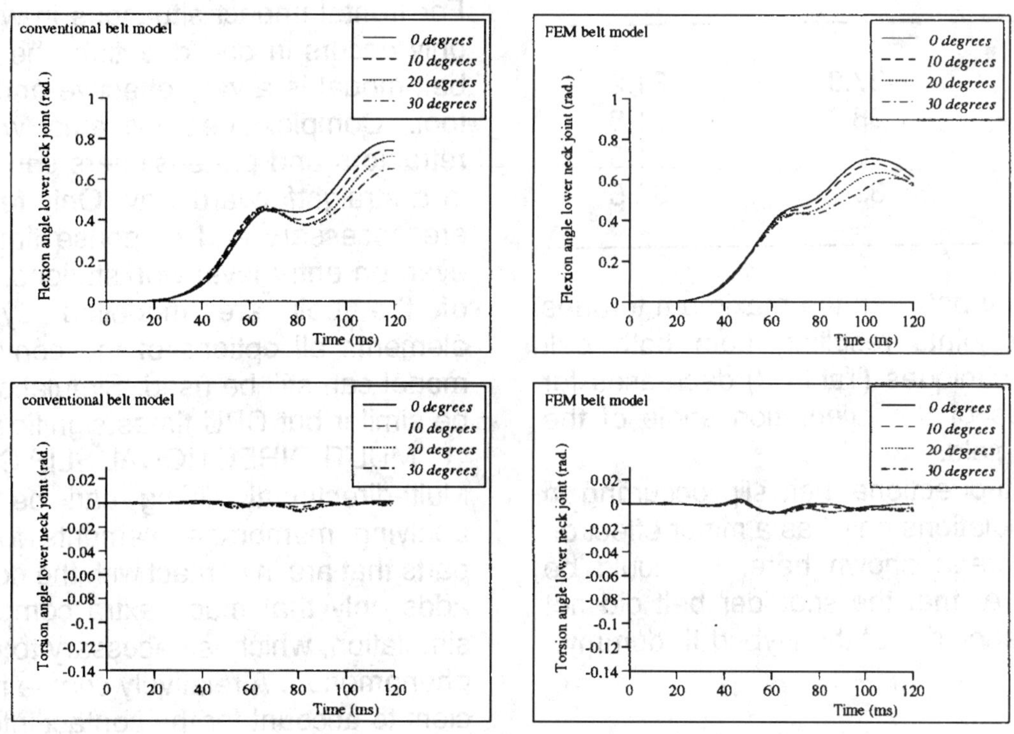

Fig. 13 Flexion and torsion angles in the lower neck joint for orientations of 0, 10, 20 and 30 degrees

Table 1 HIC values

	conv. model	FEM model
0 degrees	220	214
10 degrees	196	204
20 degrees	183	198
30 degrees	171	194

There is only a minor difference between the HIC-values resulting from both modelling techniques. The difference increases for larger orientation angles of the acceleration field. The maximum chest deflections resulting from both modelling techniques (Table 2) do not change significantly as the orientation angle of the acceleration field increases.

Table 2 Maximum chest deflections (mm)

	conv. model	FEM model
0 degrees	37.8	31.2
10 degrees	38.6	31.8
20 degrees	39.1	31.9
30 degrees	39.1	32.5

The difference between the maximum torques in the neck joints resulting from both belt modelling techniques (Table 3) decreases for higher values of the orientation angle of the acceleration field.

The multi-directional belt slip occurring in the FEM simulations only has a minor effect on the injury criteria shown here. It should be noted however that the shoulder belt did not slide of the shoulder of the Hybrid III dummy.

Table 3 Maximum resultant torques (Nm) in the neck joints

	conv. model	FEM model
Upper neck joint		
0 degrees	70.8	54.1
10 degrees	73.7	53.3
20 degrees	67.8	50.7
30 degrees	56.9	53.7
Lower neck joint		
0 degrees	230.1	207.2
10 degrees	218.7	198.3
20 degrees	210.2	184.7
30 degrees	192.9	177.3

DISCUSSION AND CONCLUSIONS

UNI-DIRECTIONAL SLIP CONDITIONS - For frontal impact situations in which belt slip only occurs in one direction, the conventional belt model is a very effective and userfriendly tool. Complex belt systems with sliprings, retractors and pretensioners can be modelled in a straightforward way. Only few input data are necessary and response times are good even on entry level workstations. Also if parts of the belt are modelled by membrane elements all options of the conventional belt model can still be used. Simulation results will be similar but CPU times significantly higher.

MULTI-DIRECTIONAL SLIP CONDITIONS Multi-directional sliding can be realized by applying membrane elements for those belt parts that are in contact with the occupant. This adds only that much extra complexity to the simulation, which is necessary to describe this phenomenon. A relatively coarse mesh is sufficient to account for the contact interaction with the occupant surface. If however the whole belt is modelled by finite elements, a very fine mesh is necessary to model the sliding of the belt through the slipring which would increase the CPU times significantly. Connecting the FEM belt part to belt segments of the conventional belt model is crucial to model hysteresis. When belt roll-out does not occur, no major influence of multi-directional belt slip at the shoulder on injury criteria is observed. The FEM belt model is capable of simulating belt roll-out. In Figure 14, the kinematics at 100 ms of two simulations are shown, one using the conventional belt model and the other using the FEM belt model. In these simulations the upper anchor point location was moved closer to the dummy shoulder, slack was introduced in the shoulder belt and an orientation angle of 30 degrees for the acceleration field was used. Belt roll-out at the shoulder occurred in the simulation using the FEM-belt model. A large difference between the injury criteria resulting from both modelling techniques was observed. Therefore, considering the CPU times, an effective way to work with both options would be to use the conventional belt model for parameter variations and the FEM belt model only to check the most appropriate set-up resulting from this parameter study.

SUBMARINING - Application of the FEM belt approach to model submarining of the lap belt is expected to be an interesting subject for future studies. MADYMO contact surfaces can be used to model the abdomen and the iliac crest of the pelvis similar as discussed in [13 and 14]. The contact surfaces are assumed rigid for the contact evaluations between finite elements and contact surfaces. Pelvis compliance could be taken into account by a separate body with connected contact surfaces attached by point-restraints to the lower torso body.

ABM - Prototype software for an Advanced Belt Model, ABM, in MADYMO has recently been realized. It features compliance of surfaces as well as multi-directional friction, with two different friction coefficients for two perpendicular sliding directions, which both can not be modelled in a direct way with the conventional and FEM belt model. The ABM assumes second degree ellipsoids thus arbitrary contours can not be accounted for. Moreover in a first evaluation it was found that the current model is not very robust. Due to these important limitations this model will not be released in its current stage.

FUTURE DEVELOPMENTS - Two future developments in MADYMO which will be very beneficial for its application in the field of seat belt modelling are non-linear material behaviour for the FEM elements and arbitrary contact contours. The non-linear material description allows the user to model the actual material behaviour also of those belt parts represented by finite elements. Connecting arbitrary contact contours to the rigid bodies, will make it possible to define a more detailed representation of the dummy outer surface.

Time: 100. ms

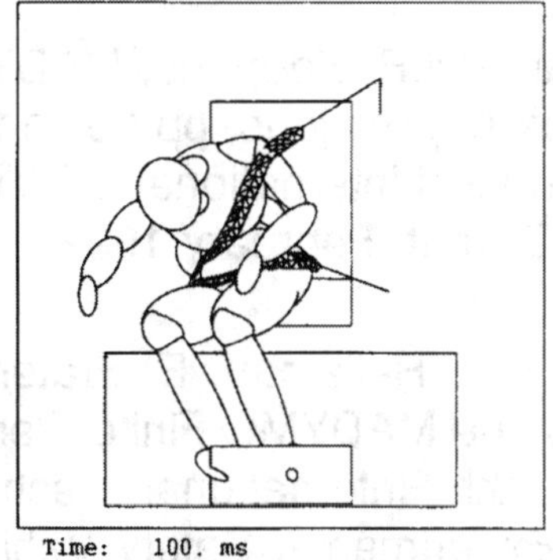

Fig. 14 Kinematics at t = 100 ms for a new upper anchor point location, slack and an orientation angle of 30 degrees for the acceleration field

REFERENCES

1. R.R. McHenry; Analysis of the dynamics of automobile passenger-restraint systems; 7th Stapp Car Crash Conference, 1963.

2. D.H. Robbins; Three-dimensional simulation of advanced automotive restraint systems; International Automotive Safety Conference Compendium, paper no. 700421, 1970.

3. P. Prasad and C.C. Chou; A Review of Mathematical Occupant Simulation Models, pp 95-112 Crashworthiness and Occupant Protection in Transportation Systems AMD-Vol. 106, The winter annual meeting of the American society of mechanical engineers, San Francisco, December 1989.

4. H.A. Lupker, P.J.A. de Coo, J.J. Nieboer and J. Wismans; Advances in MADYMO Crash Simulations, SAE 910879; International Congress and Exposition, Detroit, February 1991.

5. P.J.A. de Coo, E.G. Janssen, A.P. Goudswaard and J. Wismans; Simulation Model for Vehicle Performance Improvement in Lateral Collisions, 13th International Technical Conference on Experimental Safety Vehicles, paper no. 91-S9-O-23, Paris, November 1991.

6. J.J. Nieboer, A.P. Goudswaard and J. Wismans; Computer Simulation of Motorcycle Airbag Systems, 13th International Technical Conference on Experimental Safety Vehicles, paper no. 91-S3-O-02, Paris, November 1991.

7. J.J. Nieboer, J. Wismans and R. Verschut; Occupant Simulation as an Aspect of Flight Safety Research, AGARD 73rd Aerospace Medical Panel Symposium on Aircraft Accidents: Trends in Aerospace Medical Investigation Techniques, Cesme, April 1992.

8. H.A. Lupker and W.P. Koppens; MADYMO Vehicle Dynamics Applications, pp 101-113, Proceedings of the 3rd International MADYMO Users' Meeting, Detroit, February 1992.

9. H.A. Lupker, H.B. Helleman, E. Fraterman and J. Wismans; The MADYMO Finite Element Airbag Model, 13th International Technical Conference on Experimental Safety Vehicles, paper no. 91-S9-O-23, Paris, November 1991.

10. W.E.M. Bruijs, P.J.A. de Coo, R.J. Ashmore and A.R. Giles; Airbag Simulations with the MADYMO Fem Module, SAE 920121, International Congress and Exposition, Detroit, February 1992.

11. A. Hirth, H.J. Petit und W. Bacher; Validierung eines FE-Airbagmodells für Schlittenversuchs-Simulationen, pp 557-578, VDI Congress Berechnung im Automobilbau, Würzburg, September 1992.

12. C.S. O'Connor and M.K. Rao; Development of a model of a Three-Year-Old Child Dummy used in Air Bag Applications, 36th Stapp Car Crash Conference, paper no. 922517, Seattle, 1992.

13. A.C. Bosio; Simulation of submarining with MADYMO, Proceedings of the Second International MADYMO Users' Meeting, Noordwijk 1990.

14. Y. Håland and G. Nilson; Seat Belt Pretensioners to Avoid the Risk of Submarining -A Study of Lap-Belt Slippage Factors, 13th International Technical Conference on Experimental Safety Vehicles, paper no. 91-S9-O-10, Paris, November 1991.

15. F.E. Butler, J.T. Fleck; Advanced Restraint System Modeling, Air Force Aerospace Medical Research Laboratory Report, Number AFAMRL-TR-8014 Ohio, 1980.

16. Deng Y.C.; Development of a Submarining Model in the CAL3D Program, 36th Stapp Car Crash Conference, paper no. 922530, Seattle, 1992.

17. J. Wismans and J.H.A. Hermans; MADYMO 3D Simulations of Hybrid III Sled Tests, SAE 880645, International Congress and Exposition, Detroit, 1988.

18. M. Heinz, B. Pletschen, H. Wester and T. Scharnhorst; An Advanced 50th Percentile Hybrid-III Dummy Database Validation, SAE 910658, International Congress and Exposition, Detroit, 1991.

19. MADYMO Databases Version 5.0, MADYMO Applications Version 5.0, MADYMO User's Manual 3D, Version 5.0, TNO Road-Vehicles Research Institute, Delft, July 1992.

900549

An Improved Belt Model in CAL3D and Its Application

Yih-Charng Deng
Engineering Mechanics Dept.
GM Research Laboratories

ABSTRACT

The 'Harness Model' in the CAL3D occupant simulation program is improved to incorporate a reference point generation scheme and a new belt slip algorithm. The reference point generation scheme results in more accurate belt geometry and alleviates the need to manually specify the reference points coordinates on the occupant body. The new belt slip algorithm balances the belt force and the friction force at each reference point in a successive fashion. This approach gives satisfactory results and does not have the convergence problem found in the old slip algorithm. These new enhancements were used to developed a model to simulate the Hybrid III dummy response in a barrier test. Good correlation was obtained between the model response and the test results. Parametric studies indicated that the shoulder belt stiffness has a significant effect on the head motion, the abdomen deformation, and the peak shoulder belt force. On the other hand, the lap belt stiffness affects the abdomen deformation, the femur loads, and the peak lap belt force. The knee bolster stiffness affects the pelvis acceleration and the femur loads. The guide loop location affects the head motion and the peak shoulder belt force.

IN THE DEVELOPMENT OF ADVANCED AUTOMOBILE SAFETY DEVICES such as airbags and automatic belt restraint systems, a mathematical model is a very useful tool for assessing different design concepts. In the past few years, considerable effort has been made at the GM Research Laboratories to enhance the modeling capabilities of the CAL3D[1]* simulation program to facilitate this purpose. Some of these efforts can be found in Ref. 2 and 3. In Ref. 2 a belted occupant model was developed and correlated with a frontal impact sled test. The belt was simulated using the 'Harness Algorithm'[4] in the CAL3D program. In this algorithm the belt is modeled by a number of user-specified reference points. The distance between the reference points during the occupant motion is used to compute the belt strain and the belt force. Although the harness algorithm is effective in simulating the belt restraint force on the occupant, a number of shortcomings were also noticed in that study. First, it is difficult and cumbersome to specify a number of appropriate reference points on the occupant torso. Considerable adjustment is necessary to achieve a smooth configuration with the reference points. Second, it was found that the belt slip feature in this algorithm resulted in convergence problems.

To resolve the first difficulty an algorithm was developed to automatically generate the reference points on the occupant body. This new feature only requires the user to specify the desired number of points on each segment. The program produces equally-spaced reference points and selects the appropriate points for the belt configuration.

To resolve the second problem in the harness algorithm, i.e., the belt slip, a new algorithm was developed in this study. This feature is important due to the following considerations. First, if the belt does not slip the webbing length between any two reference points on the same rigid segment remains unchanged. This implies that the belt force is only present at the two most outer reference points for each segment instead of distributed at each of the reference points along the belt configuration. Second, without a slip feature to transport the belt material at the reference point it is impossible to model certain components in the belt restraint system such as the guide loop. In addition, the investigation of problems associated with occupant submarining clearly requires the capability to model belt slip. There is a major distinction between slip at the guide loop and that which occurs in submarining. The former involves slip

* Numbers in brackets designate references at the end of the paper

along the beltline direction and the latter involves slip along a direction perpendicular to the belt. This study will concentrate on the beltline direction slip.

DESCRIPTION OF THE HARNESS ALGORITHM IMPROVEMENTS

REFERENCE POINT GENERATION SCHEME - Two schemes were developed for generating reference points on the occupant torso. The first scheme is designed to model belts that are contained in a plane. The belt plane is defined by three points - two end points (anchor points), and one selected point on the occupant body. It is not necessary to place the third point precisely on the ellipsoid surface in the input data. The program will reposition the third point on the surface. To generate reference points, one can specify all the possible contact ellipsoids for this belt. As an example, a belt is specified by two anchor points, one point on the torso; and three contact ellipsoids as shown in Fig. 1. The program first generates a plane from the three given points and the intersection between this plane and the contact ellipsoids is identified. The program then connects lines between the end points and the centers of these ellipses to determine the sectors on the ellipses for placing the reference points. The user needs to specify the desired number of reference points on each ellipse sector and the program will generate these points with equal angle intervals. For the ellipse that contains the specified point, the left- and the right-portion of the ellipse sector with respect to the specified point is considered separately and the specified number of points are generated on both sides of the sectors.

The second scheme is designed to model belts that are not contained in a plane. In such cases, the user needs to specify one point on each contact ellipsoid. As an example, a belt is defined by two anchor points, and three points on three different ellipsoids A, B, and C, as shown in Fig. 2. The program uses the left anchor point, the specified points on ellipsoids A and B to generate

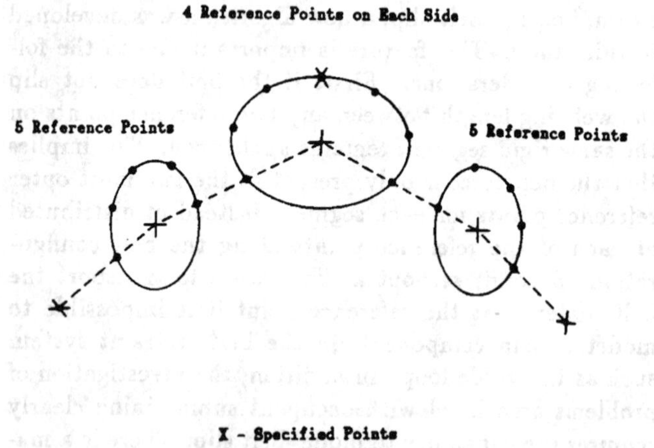

4 Reference Points on Each Side

5 Reference Points

5 Reference Points

X - Specified Points

Fig. 1. Reference points generation for planer belt.

the first plane. The intersection between this plane and ellipsoids A and B will be identified. As in the previous scheme, lines between the anchor points and the centers of these two ellipses are used to determine the sector on the ellipse A for placing the reference points. Then, the specified number of reference points are generated on both sides of the specified point. Next, the program will take the specified points on ellipsoids A, B, and C to define the second plane and generate points on ellipsoid B. Finally, points are generated on ellipsoid C on a plane containing the specified points on ellipsoids B, C, and the right anchor.

After the reference points are generated, the program then goes through another algorithm to select the appropriate points to form the belt configuration. Three criteria are used: (1). all the user-specified points are included; (2). points inside another ellipse are rejected; (3). points that fail the convexity test are rejected. The third criterion requires the point to have a convex belt configuration with adjacent reference points as shown in Fig. 3. This implies that the angle between the belt segments and the normal should be greater than 90 degrees. To illustrate a typical point selection procedure, let us consider a possible lap belt plane in which reference points are generated on the pelvis and the left and the right

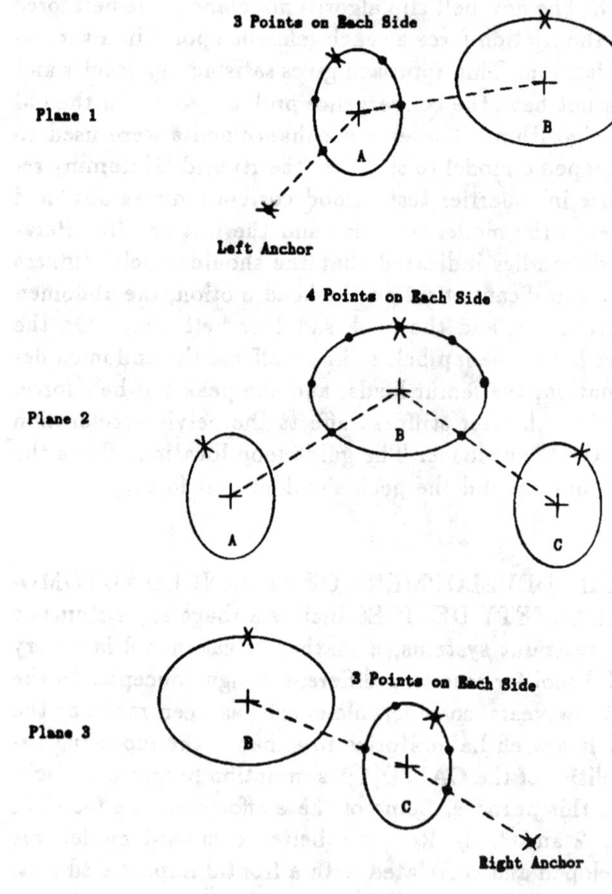

Fig. 2. Reference points generation for non-planer belt.

upper legs as shown in Fig. 4. The program will evaluate each reference point in a sequential fashion. Since reference point 1 is a user-specified point (left anchor), it will be included in the belt configuration. To evaluate point 2, the program connects points (1,2) and (2,3). As depicted in Fig. 4, line (1,2) fails the convexity criterion and, thus, point 2 is rejected. Once point 2 is rejected, point 3 will be evaluated based on the triplet of (1,3,4). If point 2 were accepted, point 3 will be evaluated using the triplet of (2,3,4). This procedure continues until all the points have been considered.

One shortcoming arose when this scheme was tested. A non-symmetric belt configuration can occur even though all the inputs are symmetric with respect to the sagittal plane. This is due to the fact that the reference point evaluation is conducted in a sequential fashion; which implies that points on the left-hand side of the torso and points on the right-hand side of the torso are evaluated in opposite orders. On the left-hand side, the reference points are evaluated in a sequence going from the left anchor towards the center of the pelvis. However, on the right-hand side, the reference points are evaluated in a sequence going from the center of the pelvis towards the right anchor. To remedy this shortcoming, a supplementary set of points was generated during the evaluation process. This set contains all the qualified points from the selection process. In addition, points that fail the convexity test without involving a user-specified point in that particular belt segment are also retained in this set for further evaluation. For example, as illustrated in Fig. 4, points 2, 3, 4, 5, and 7 were rejected because of the convexity test failure. The first four points were not retained in the supplementary array because their evaluation was based on belt segments (1,2), (1,3), (1,4), (1,5) which involved point 1 - a specified point. Since point 1 is always included in the belt configuration points 2, 3, 4, 5 will always fail the convexity test. On the other hand, point 7 will be retained in the supplementary array since this point is evaluated based on belt segments (6,7) and (7,14) which do not involve any specified point. This implies that the rejection of point 7 will depend on whether points 6 and 14 are accepted or not. In this example, after the first round of evaluation the points selected are (1,6,14,15,24,29) and the points in the supplementary array are (1,6,7,14,15,16,23,24,29). Thus, the first round of evaluation did not result in a symmetric point configuration. This is because of the following difference in evaluation sequence: On the left-hand side, (6,7,14) rejects 7 since segment (7,14) fails the convexity test, but (6,14,15) accepts 14. On the right-hand side, (15,16,23) rejects 16 since segment (16,23) fails the convexity test, (15,23,24) rejects 23 since segment (15,23) fails the convexity test. No consideration was given to the combination of (15,16,24) which is the corresponding set for (6,14,15). To overcome this difficiency, the program will compare these two arrays. Once differences are detected in the comparison, the program will re-evaluate all the extra points in the supplementary array with all the possible combinations. Such a scheme will reveal that point 16 is also an acceptable point and it will be added to the final point array and the final lap belt configuration is depicted in Fig. 5. This two-step point selection scheme will not dismiss any potential points and symmetry should be preserved if so arranged in the input data.

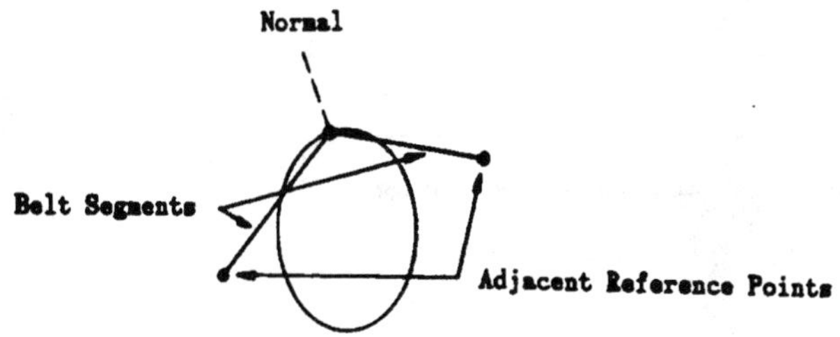

Fig. 3. Convexity considerations.

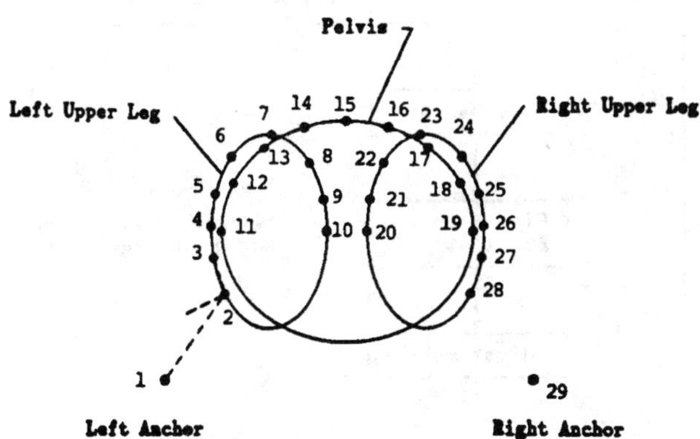

Fig. 4. Reference points for lap belt.

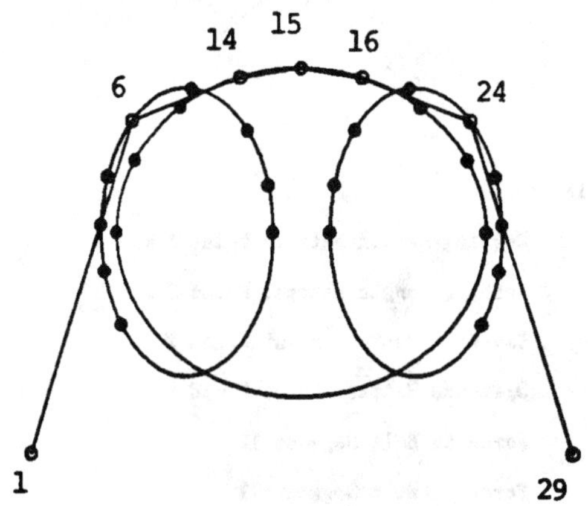

Fig. 5. The final lap belt configuration.

NEW BELT SLIP ALGORITHM - Consider a 3 point belt configuration as shown in Fig. 6 in which the belt can slip at point 2. Using the nomenclature defined in Fig. 1 the belt tension in each segment can be computed by

$$t_{12} = F(\frac{l_{12} - L_{12}}{L_{12}})$$

$$t_{23} = F(\frac{l_{23} - L_{23}}{L_{23}})$$

in which, t_{12} and t_{23} are the scaler values of the force vectors $\bar{t}_{12}$ and $\bar{t}_{23}$, and F( ) is the force-strain function of the belt. The belt force at point 2 can be decomposed into the normal and the tangential components

$$\bar{t}_{12} = \bar{t}^n_{12} + \bar{t}^t_{12}$$

$$\bar{t}_{23} = \bar{t}^n_{23} + \bar{t}^t_{23}$$

The friction force at point 2 can be computed by

$$f_2 = (t^n_{12} + t^n_{23}) \cdot \sigma_2$$

The tangential forces in the two belt segments are in the opposite directions at point 2. If the net tangential force is less than the friction force there will be no belt slip, i.e.,

$$\Delta t_2 = 0 \,, \qquad if \;\; |t^t_{12} - t^t_{23}| < f_2$$

in which Δt_2 represents the force unbalance at point 2. If the net tangential force at point 2 is greater than the friction force, slip must take place. In this case, the force unbalance can be computed as

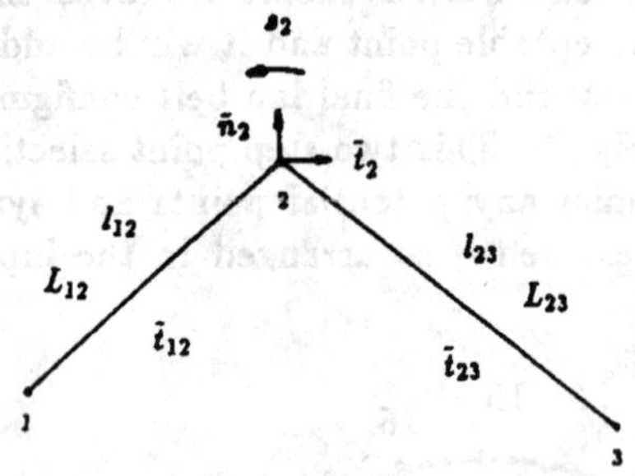

Nomenclature :

L_{12}	-	Webbing Length Between Point 1 and 2
L_{23}	-	Webbing Length Between Point 2 and 3
l_{12}	-	Distance Between Point 1 and 2
l_{23}	-	Distance Between Point 2 and 3
$\bar{t}_{12}$	-	Force in Belt Segment 12
$\bar{t}_{23}$	-	Force in Belt Segment 23
$\bar{t}_2$	-	Unit Vector Along Beltline Direction at Point 2
$\bar{n}_2$	-	Unit Vector Along Normal Direction at Point 2
s_2	-	Belt Slip at Point 2
σ_2	-	Friction Coefficient at Point 2
f_2	-	Friction Force at Point 2

Fig. 6. Belt slip at a reference point.

$$\Delta t_2 = |t^t_{12} - t^t_{23}| - f_2 \,, \qquad if \;\; |t^t_{12} - t^t_{23}| > f_2$$

This force unbalance must be 'relaxed' by an appropriate amount of belt slip at point 2. Since in general we will have a set of nonlinear equations, iteration methods have to be used to solve for the unknown belt slip.

For a multiple-point configuration shown in Fig. 7 the force unbalance Δt_i can be computed for each reference point. We first check to see if all the force unbalances are below a specified tolerance level (Δt_{max}). If this level is not exceeded no slip will occur. On the other hand, if some points have a force unbalance greater than this tolerance level, belt slip will take place. To compute the amount of belt slip we will consider one point at a time starting with the point having the maximum force unbalance. The algorithm derived in the previous section can be used to compute the belt slip such that the force unbalance will be reduced below a specified level (Δt_0). Standard numerical methods such as Newton's method and modified regula falsi[5] are used to solve for the slip in this step. The latter is used only if Newton's method over-predicts in the iteration process such that Δt_i reverses sign. After the slip is computed, the webbing lengths of the two belt segments are adjusted. This would change the force unbalances at the adjacent refer-

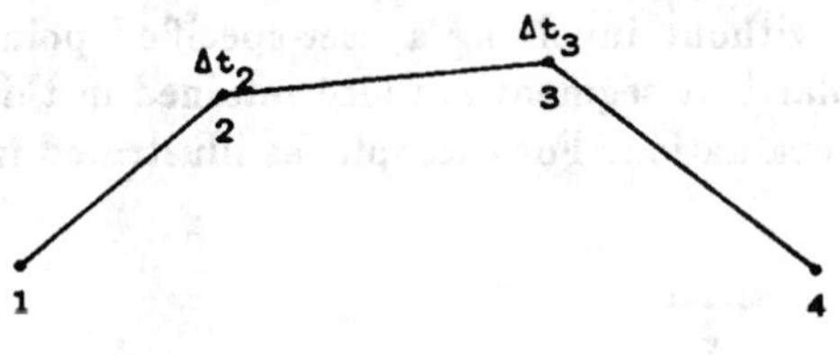

Successive Relaxation

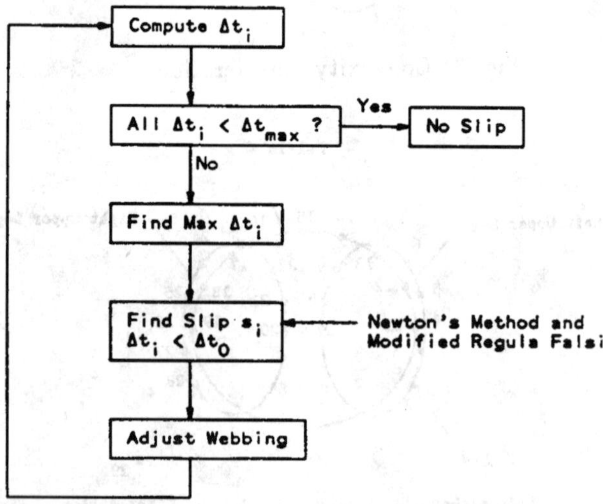

Fig. 7. Solving multiple-point slip by successive relaxation.

ence points. Thus, we have to go back to the first step to recompute the Δt_i for the reference points affected by the belt slip. This procedure is repeated until all the force unbalances in the system are below the specified level Δt_{max}.

This algorithm relaxes the force unbalances in the belt reference points in a successive fashion. There are two control parameters in this process, i.e., Δt_{max} and Δt_0. The former is the level of force unbalance to trigger the relaxation process and the latter is the level of force unbalance after the relaxation. Slip can be computed with high accuracy by choosing: (i). a very small Δt_{max} and an even smaller Δt_0, or, (ii). a finite Δt_{max} and a Δt_0 only slightly smaller than Δt_{max}. In scheme (i), the relaxation process is carried out once a small amount of force unbalance is detected. In scheme (ii), the force unbalance is allowed to build up to a certain level but the relaxation is carried out in an infinitesimal manner from point to point. Both schemes result in a great amount of computation time. To achieve a compromise between accuracy and computation efficiency these two parameters can be selected by trying various values and comparing the results. For simulating the belt slip in a barrier test it was found that a value of 10 kg (98 N) for Δt_{max} and a value of 1 kg (9.8 N) for Δt_0 were reasonable choices.

The slip algorithm was first tested using a simple belt configuration to evaluate the results. The test cases are depicted in Fig. 8 in which a belt goes through 5 points with given friction coefficients for the 3 intermediate points. These points are fixed in space and are 5 cm apart from the adjacent point. In test case 1, the initial webbing lengths of the 4 belt segments were: {3, 5, 5, 5}. This implies a 66.6 percent strain in the first segment and 0 strain in the other 3 segments. The friction coefficient is 0, which would induce belt slip from the 3 unstretched segments to the first stretched segment. As can be seen from Fig. 3, after executing the slip algorithm the webbing lengths in these belt segments became identical.

Test case 2 is similar to test case 1 except that the friction coefficient at the reference points is 0.5. The results indicated belt material has slipped to the first segment. There was more slip at the point closer to the first segment. A reversed configuration gave symmetric results for test case 2. In test case 3 only 3 belt segments are considered with webbing lengths {5, 3, 5} and friction coefficient 0.5. In this case, belt material slips from both segment 1 and 3 to segment 2 and the final configuration shows identical webbing length in segment 1 and 3, and less webbing in segment 2. This is also a reasonable result. To illustrate the effect of a different friction coefficient, test case 3 was also executed with friction coefficient equal to 0.25. As depicted in Fig. 3, a smaller friction coefficient allows more webbing material to slip from segments 1 and 3 to segment 2.

Test Cases

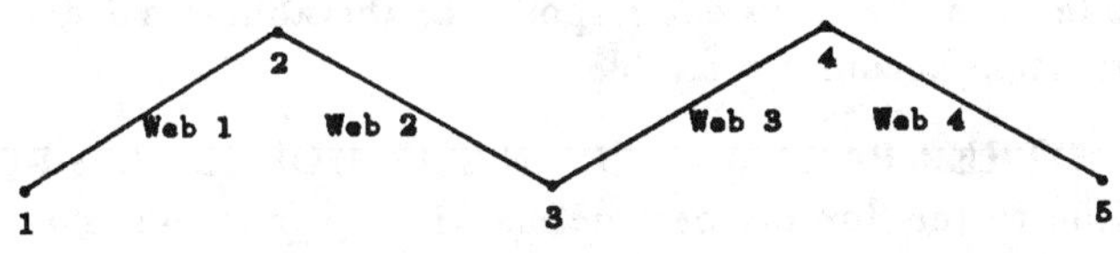

	Webbing Length				Friction
Ex 1 :					
Before Slip	3.0	5.0	5.0	5.0	0.
After Slip	4.5	4.5	4.5	4.5	
Ex 2 :					
Before Slip	3.0	5.0	5.0	5.0	0.5
After Slip	3.9	4.4	4.7	4.8	
Before Slip	5.0	5.0	5.0	3.0	0.5
After Slip	4.8	4.7	4.4	3.9	
Ex 3 :					
Before Slip	5.0	3.0	5.0		0.5
After Slip	4.50	4.01	4.50		
Before Slip	5.0	3.0	5.0		0.25
After Slip	4.41	4.18	4.41		

Fig. 8. Test cases for the slip algorithm.

BARRIER TEST SIMULATION

After the new algorithms were implemented into the CAL3D program, the motion of a belt-restrained occupant in a barrier test was simulated for further validation. Figure 9 illustrates the model configuration. The occupant is a Hybrid III dummy[6] sitting in the right front passenger position of a compact-size car. The belt is an automatic 3-point system with the lap and the shoulder belt separated at the latch plate. A web lock was used in the shoulder belt retractor to reduce the spool-out. We

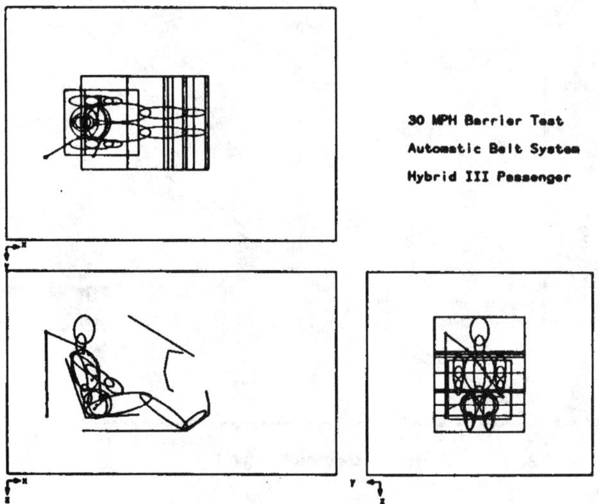

30 MPH Barrier Test
Automatic Belt System
Hybrid III Passenger

Fig. 9. Model configuration for simulating barrier test.

will first discuss the modeling considerations for the Hybrid III dummy and the belt restraint system. Then, the simulation results will be discussed.

THE HYBRID III DUMMY MODEL - A Hybrid III dummy model has been established in a previous study based on the characteristics of each body region of the dummy obtained from static tests[2,7]. The model has been validated against the dummy response in a sled test. For the present study this model was further improved in the thorax and abdomen regions.

The thorax model in Ref. 2 consists of two segments representing the rib cage and the thoracic spine. These two segments were connected by a linear spring and a linear damper to simulate the chest compliance. Since the static test results indicated some stiffening behavior, using a nonlinear spring element may produce better simulation results, especially for large deformation. Thus, the previous thorax model was modified to include a quadratic spring element. The spring constants were determined by matching the static test results as shown in Fig. 10. The damping coefficient was determined by correlating the model response to the chest deformation time-history recorded in the barrier test. However, it was found that if the damping coefficient was adjusted to match the maximum chest compression the rebound of the chest model was too slow compared to the test. On the other hand, if a damping coefficient was chosen such that the rebound behavior was correlated, the maximum compression in the model became too high in the model. This observation seemed to suggest that the Hybrid III dummy thorax structures have different damping characteristic during loading and unloading. In order to correlate both the maximum deformation and the rebound behavior the model was given a damping coefficient of 10 $kg \cdot s/cm$ for the loading phase, and 2.5 $kg \cdot s/cm$ for the unloading phase.

The second improvement in the Hybrid III model was in the abdomen. Due to the recent concerns of occupant submarining and abdominal injury[8], the lower torso of the dummy was modeled by two segments representing the pelvis and the abdomen. In order to obtain the abdomen characteristics under lap belt load, a Hybrid III pelvis with the 'abdomen insert' was tested on an MTS machine. The test was conducted in a similar fashion as the thorax test described in Ref. 2. It should be pointed out that in the test the force is recorded against the piston displacement of the MTS machine. The actual abdominal deformation is less than the piston displacement due to belt stretch. Since there was no transducer in the abdomen to measure its deformation directly, a ruler was used in the test to record the abdomen deformation when the piston reached the top position. The ratio between the maximum abdomen deformation and the maximum piston displacement was then used to scale the test results into the abdomen deformation curve as shown in Fig. 11.

Like the thorax model, the abdomen model also consists of two segments, one representing the pelvis and the other representing the deformable abdomen. These two segments are connected by a quadratic spring and a linear damper. The spring constants are determined by matching the static test results as shown in Fig. 11. Since no information for the abdominal deformation was recorded in the barrier test the damping coefficient can only be estimated by correlating the pelvis response and some judgement on the possible peak abdomen deformation. Further improvement of the damping coefficient can be made once the transient response of the abdominal deformation becomes available.

OTHER PARAMETERS IN THE MODEL - In addition to the dummy parameters, the belt restraint system also needs to be described by appropriate parameters

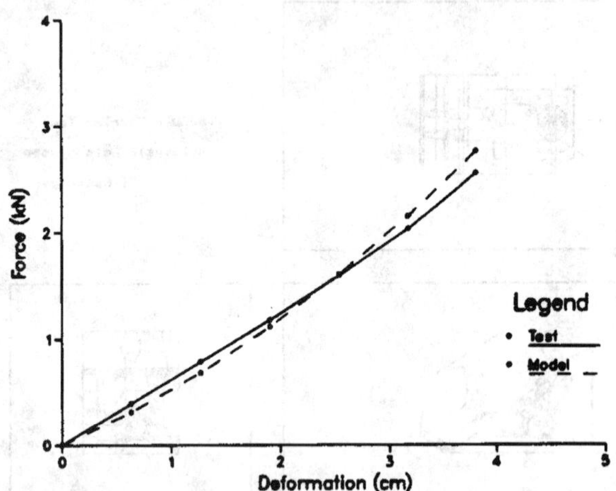

Fig. 10. Hybrid III dummy thorax deformation due to quasi-static belt load.

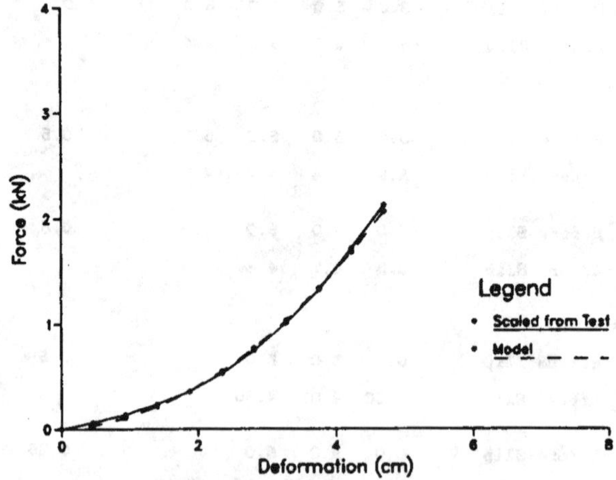

Fig. 11. Hybrid III dummy abdomen deformation due to quasi-static belt load.

in the model. These parameters include shoulder and lap belt loading/unloading characterstics, friction coefficients between the belt and the occupant torso, friction coefficient of the guide loop, and the knee/IP impact loading/unloading characteristics. Although the force-deflection relationship of the webbing material is available in Ref. 2 there are other deformable components in the restraint system that affect the belt force. These deformable components include the anchors, the retractors, the buckle, and the guide loop. In this study, the effects of these components will be lumped into a single effective force-deflection function for each belt.

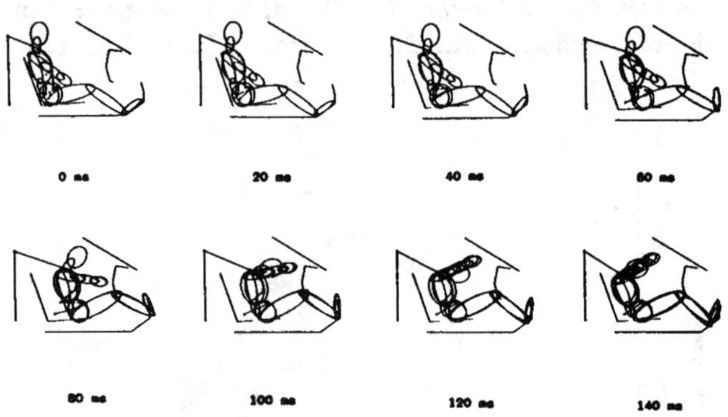

Fig. 12. Simulated occupant motion in barrier impact.

One possible approach to determine these parameters is to start with some assumed values in the model and make adjustment until the model responses conform to the test results. Since each parameter can affect the occupant response in a number of body regions it is necessary to examine the model response against the test data for all the body regions simultaneously. Once good correlations are achieved between the test and the simulation for both the occupant and the belt response these unknown system parameters should also have been identified with reasonable accuracy.

SIMULATION RESULTS - Figure 12 illustrates the occupant motion at 20 ms intervals. The sequence bears reasonable resemblance to the occupant motion observed from the high speed films. The comparison between the test and the simulation for the head resultant acceleration, the spine resultant acceleration and the pelvis resultant acceleration is shown in Fig. 13. The correlation appears to be favorable for both the the waveform shapes and the timing. Figure 14 shows the head displacement in the longitudinal and vertical directions. The test data are digitized from the high speed films. The simulation re-

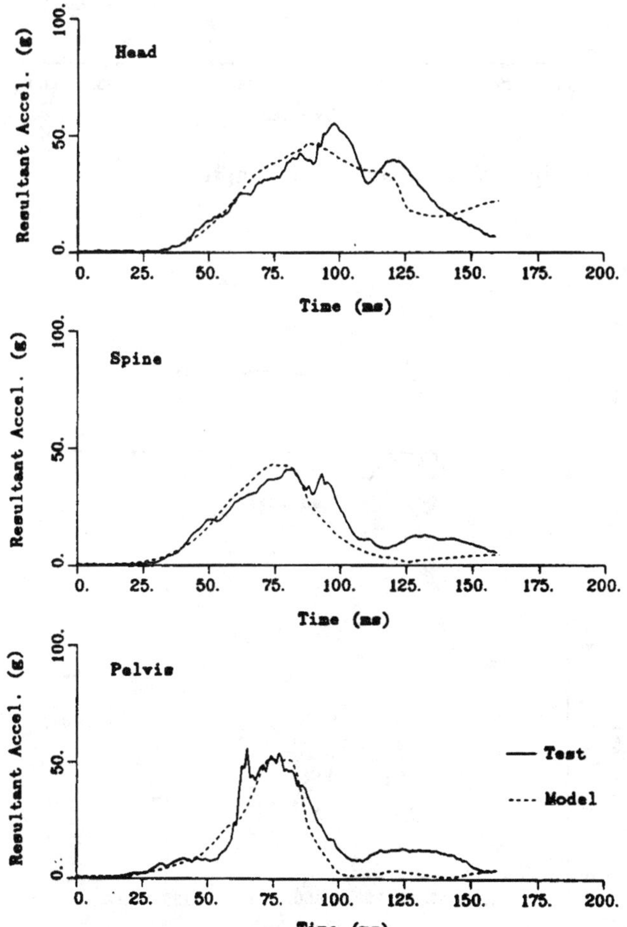

Fig. 13. Occupant acceleration response.

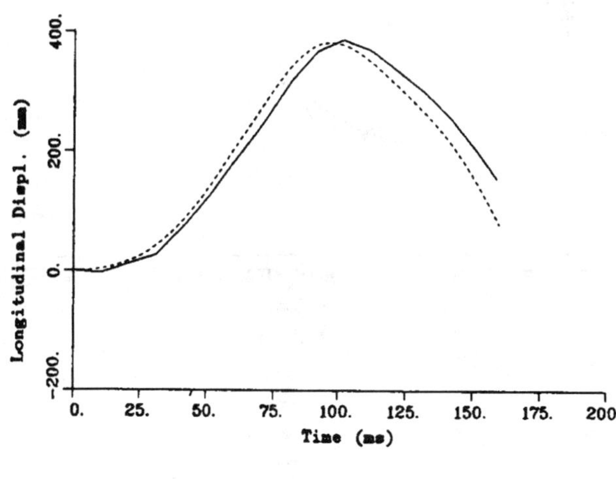

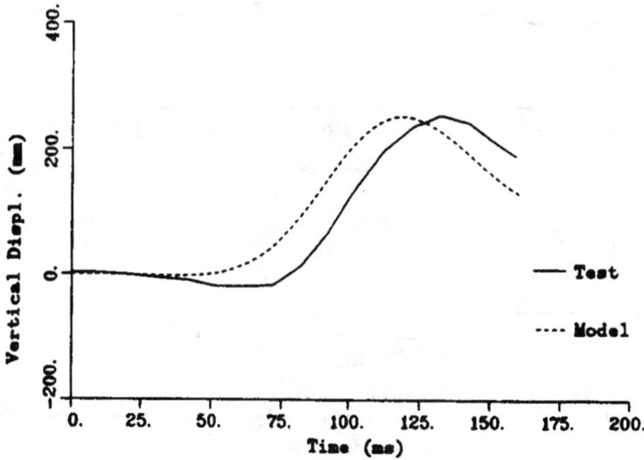

Fig. 14. Occupant head displacement.

sults are very similar to the test data except for some timing difference in the vertical component. The head/neck joint response of the model as illustrated in Fig. 15 also shows good correlation with the test. Figure 16 depicts the chest compression from the test and the simulation. The improved chest model shows good correlation with the test results except for some timing difference. Since injury assessment usually is based on the peak values, the timing difference appeared in Fig. 16 is not considered to be critical. Figure 17 illustrates the left and right femur load from the test and the simulation. Finally, Fig. 18 shows the lap and shoulder belt force. In the test, the lap belt force was measured at the outboard anchor and the shoulder belt force was measured between the guide loop and the occupant. From the test film it was observed that there was about 7.6 to 10.2 cm (3 to 4 inches) of belt slip at the guide loop. The model indicated there was 8.9 cm slip at the guide loop, which also compared favorably to the test.

The simulation results shown above were obtained from using a set of parameters listed in Table 1. The good correlation between the test and the simulation suggests this set of parameters is a reasonable representation of the physical system. A parametric study based on these baseline values was conducted to illustrate possible model applications.

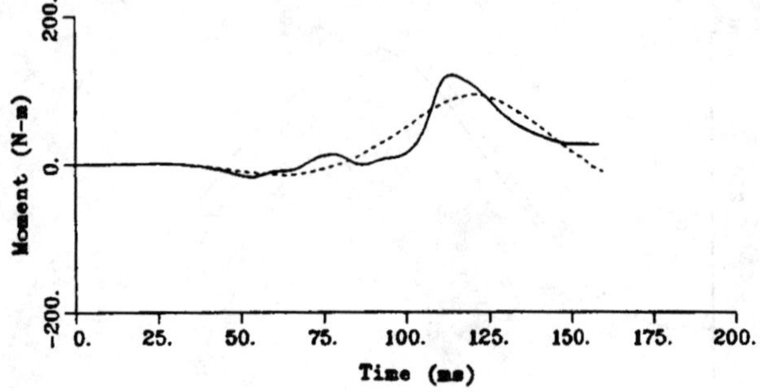

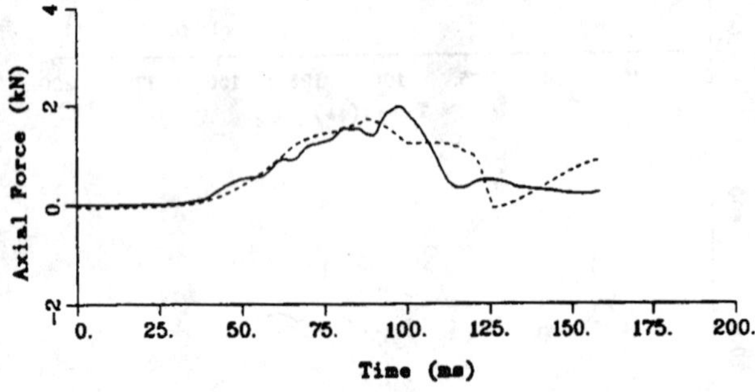

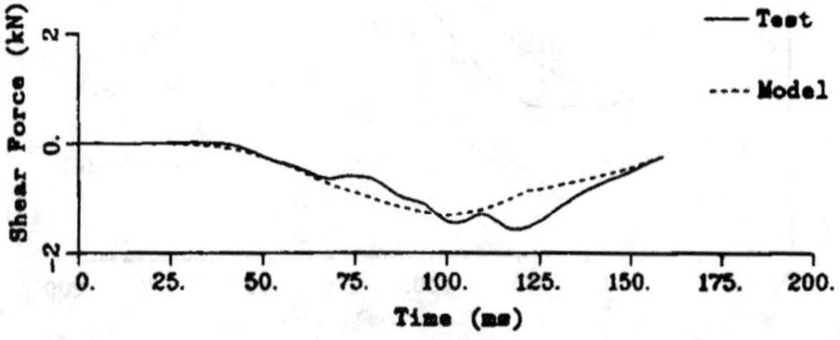

Fig. 15. Occupant neck response.

PARAMETRIC STUDY

In this Section, a number of model parameters were altered from the baseline values to investigate their effects on the occupant kinematics. These model parameters include the effective shoulder belt stiffness, the effective lap belt stiffness, the knee bolster stiffness, the guide loop location, and the guide loop friction coefficient. Specifically, each of the stiffness functions were multiplied by a factor of 0.5 and 2, the guide loop location was moved forward by 20 cm, and the guide loop friction coefficient was changed from 0.02 to 0.1. Table

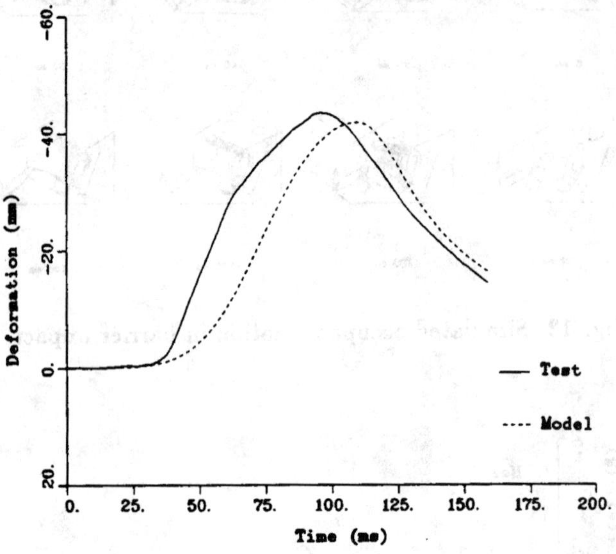

Fig. 16. Occupant chest compression

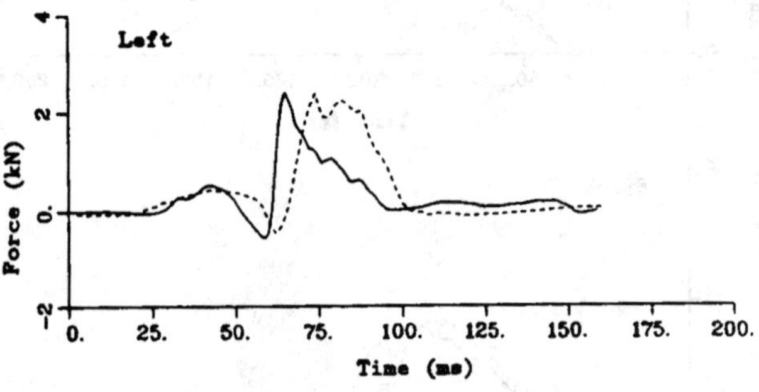

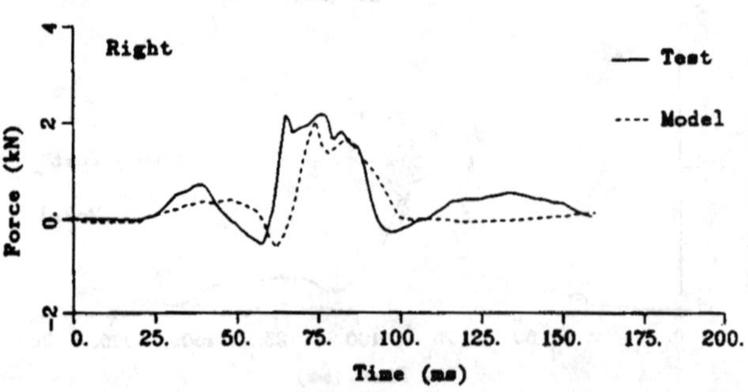

Fig. 17. Occupant femur load.

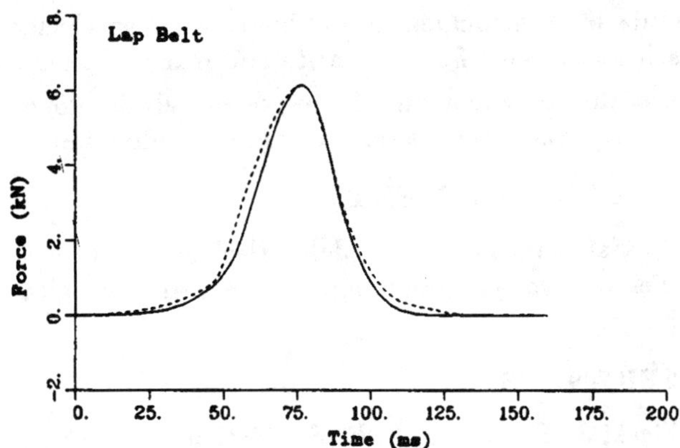

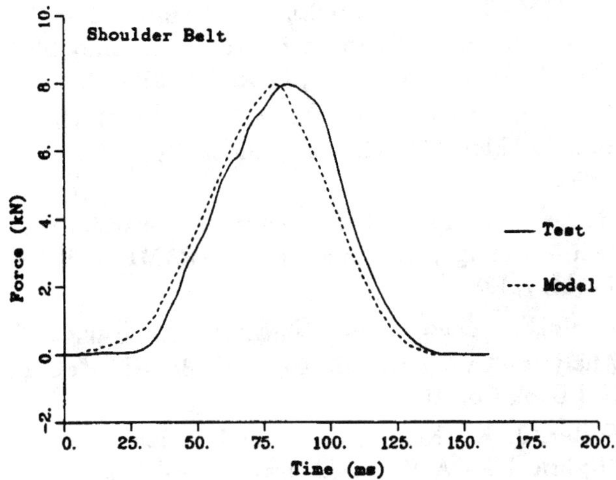

Fig. 18. belt force response.

2 contains the occupant response changes including HIC, peak head resultant acceleration, maximum head longitudinal displacement, peak spine resultant acceleration, maximum chest deformation, peak pelvis resultant acceleration, maximum abdominal deformation, and peak left and right femur loads. In addition, peak shoulder belt force and peak lap belt force are also examined. The numbers listed in Table 2 are the percentage change relative to the baseline values.

It can be observed from Table 2 that the shoulder belt stiffness significantly affects the occupant head response including the HIC, the resultant acceleration, and the longitudinal displacement. A stiffer shoulder belt would reduce the peak values and a softer shoulder belt would

Effective Lap Belt Stiffness =

49 N (5 Kg) / 1% strain, if 0. < strain < 20%

196 N (20 Kg) / 1% strain, if 20% < strain

Effective Shoulder Belt Stiffness =

392 N (40 Kg) / 1% strain, if 0. < strain < 2%

637 N (65 Kg) / 1% strain, if 2% < strain

Knee/IP Contact Fuction = 245 N (25 Kg) / cm

Belt/Torso Friction Coeff. = 0.1

Belt/Guide Loop Friction Coeff. = 0.02

Table 1. Derived system parameters.

Response Sensitivity

(Change in Percentage)

	HIC	Head Accel.	Head Displ.	Spine Accel.	Chest Def.	Pelvis Accel.	Abdomen Def.	L. Femr Load	R. Femr Load	S Belt Force	L Belt Force
Test	430	54	38.8	41.1	4.34	56.9	-	290	260	800	620
Model	404.6	47.1	38.3	43.1	4.19	52.9	3.71	245.2	206.7	814.1	626.8
S Blt X 0.5	+44.8	+21.9	+29.2	- 0.2	- 0.2	+ 4.3	+16.2	+ 1.3	+ 4.5	-18.2	+ 6.8
Stiff X 2.0	-26.7	-17.2	-23.0	- 7.2	- 4.5	- 4.0	-13.5	- 0.3	- 2.5	+16.7	- 6.4
Lp Blt X 0.5	+ 6.4	+ 1.7	- 1.6	+ 2.3	+ 7.2	- 1.7	-21.3	+24.9	+39.9	+ 5.3	-31.1
Stiff X 2.0	+ 1.4	- 1.5	+ 1.8	- 3.9	- 4.8	- 4.7	+14.8	-29.7	-34.2	- 5.0	+27.6
Kn B X 0.5	+ 2.9	- 0.8	- 0.5	- 2.3	+ 1.7	- 6.8	+ 3.8	-32.6	-25.9	+ 0.4	+ 0.8
Stiff X 2.0	- 0.4	- 2.3	+ 0.8	+ 5.3	- 0.2	+13.2	- 4.6	+68.9	+52.4	- 0.2	- 0.5
GL Dx + 20.	+35.8	+12.7	+11.2	+ 6.0	+ 5.5	+ 2.3	+ 4.8	- 9.0	+ 0.6	+11.7	+ 1.5
GL Frc X 5.0	- 7.9	- 4.0	- 2.9	- 0.5	- 2.4	- 0.8	- 1.1	+ 1.1	- 0.5	+ 2.2	- 0.5

Table 2. Parametric study results.

increase the peak values. Less sensitivity was found in the spine acceleration, the pelvis accelerations, and the chest compression. However, it appears that a stiffer shoulder belt would reduce the abdominal deformation by a substantial margin. No significant changes were found in the femur loads. Finally, a stiffer shoulder belt would result in higher peak force in the shoulder belt and lower peak force in the lap belt.

On the other hand, the lap belt stiffness does not have a significant effect on the head kinematics, the spine and pelvis accelerations, and chest deformation. However, substantial changes were found in the abdominal deformation and the femur loads. A stiffer lap belt would have a much higher peak force in the lap belt and somewhat lower force in the shoulder belt. The knee bolster stiffness change does not have much effect on the head or upper torso response. A softer knee bolster would have slightly lower pelvis acceleration and a very distinct reduction on the femur loads.

The guide loop location variation indicates a more forward guide loop would result in higher values in most of the occupant responses especially in the head. Considerable increase was also found in the shoulder belt force. The only reduction in occupant response was the left femur load. This is conceivably due to the fact that a more forward guide loop would generate less contact force on the right shoulder, which results in less torso rotation and thus lower left femur load. Finally, a greater guide loop friction coefficient would lower the HIC value and the head acceleration. Other occupant response numbers are not significantly affected by this change. A slight increase in the shoulder belt load is observed as a result of greater restraint to the occupant upper body.

The trends observed in the above parameteric study are in general consistent with past experience. However, it is possible that certain trends shown in this model are pertinent to the particular belt system used in this study. Other belt systems may exhibit different behavior.

CLOSING REMARKS

The reference point generation scheme developed in this work provides a convenient and more accurate way to define the belt geometry. The new slip feature offers a number of improvements over the model in Ref. 2. First, this feature enables the modeling of belt system components (such as the guide loop and the buckle) in which the webbing can slip which allows the specific contributions of these components to be assessed. Belt slip can have a significant effect on the belt force, which is an important design consideration. The belt slip also results in a more realistic belt force distribution on the occupant body.

Currently, we are working towards a more detailed belt restraint system model in which each component will be modeled individually instead of using an effective stiffness function to simulate the whole system characteristics. Such a model would provide insights to the importance of each component and how the design change of each component affects the belt performance. The slip feature, the guide loop and the buckle models developed in this study are importantant steps in that direction.

ACKNOWLEDGEMENT

Special thanks go to Dr. John Fleck of J & J Technologies Inc. for programming assistance and consultation.

REFERENCES

1. Fleck, J. T., Butler, F. E., Validation of the Crash Victim Simulator, DOT HS-806 279-282, 1981.
2. Deng, Y.-C., "Analytical Study of the Interaction Between the Seat Belt and a Hybrid III Dummy Sled Tests.", SAE Transactions, Paper No. 880648.
3. Wang, J. T., and Nefske, D., "A New CAL3D Airbag Inflation Model", SAE Transactions, Paper No. 880654.
4. Butler, F. E., Fleck, J. T., "Advanced Restraint System Modeling", Air Force Report AFAMRL-TR-80-14, May, 1980.
5. Conte, S. D., and de Boor, C., Elementary Numerical Analysis - An Algorithmic Approach, 3rd ed., McGraw-Hill Book Co., 1980.
6. Foster, J. K., Kortge, J. O., and Wolanin, M. J., "Hybrid III - A Biomechanically-Based Crash Test Dummy", Proceedings of the 21st Stapp Car Crash Conference, SAE Paper No. 770938.
7. Deng, Y.-C., "Anthropomorphic Dummy Neck Modeling and Injury Considerations", Accident Analysis & Prevention, Vol. 21, No. 1, pp. 85-100, 1989.
8. Rouhana, S. W., Viano, D. C., Jedrzejczak, E. A., McCleary, J. D., "Assessing Submarining and Abdominal Injury Risk in the Hybrid III Family of Dummies", Proceedings of the 33rd Stapp Car Crash Conference, Washington, D. C., 1989.

933108

Finite Element Simulation of the Occupant/Belt Interaction: Chest and Pelvis Deformation, Belt Sliding and Submarining

D. Song
Peugeot/Renault

P. Mack and C. Tarriere
Renault DSE

F. Brun-Cassan and J. Y. Le Coz
Peugeot/Renault

F. Lavaste
ENSAM Paris

ABSTRACT

In frontal impact, the occupant/belt interaction is essential to obtain a good simulation of the occupant dynamic behaviour. Nevertheless, current mathematical models do not allow a realistic representation of this interaction to be obtained. Especially they are not adapted to simulate two important phenomena : the chest and pelvis deformation under the belt loading, and the belt sliding on the occupant.

This paper deals with a tridimensional finite element model which allows an improved simulation of this interaction. The Hybrid III dummy, restrained by a 3-point retractor belt, was aimed, with a finite element program (RADIOSS). The model consisted of two parts : a deformable part representing, by means of springs and shell elements, the belt system, the thorax and the- pelvis ; a rigid part representing, with rigid shell elements, the other components of the system. The belt was simulated by shell elements with a elasto-plastic material law. For the pelvis/lap belt modelling, special attention was given to the iliac spine function and the abdomen deformation effect as well as the belt sliding over the pelvis, essential for the submarining investigation. For the chest/shoulder belt interaction, the whole chest surface was considered as being deformable, in order to take account of its influence on belt sliding. A series of subsystem tests was firstly used to validate the chest and pelvis model : a belt loaded by an impactor, the pelvis loaded by a lap belt, the thorax loaded by an impactor and by a shoulder belt. Then the complete model was evaluated by comparing with sled tests, special attention being paid to the submarining. The correlation of analytical predictions with test data is presented.

1. INTRODUCTION

In frontal impact, the dynamic response of a belted occupant is essentially conditioned by the interaction of his torso with the belt system. Conditions of occupant impacts with others protection devices, such as air bag for example, are also closely linked with this interaction. For frontal impact analysis with mathematical models, the quality of the occupant/belt interaction approach is therefore primordial.

The principal method used in the past for simulating this interaction is based on a spring and rigid body approach. The belt is divided into different segments and each segment is modelled by a spring, while the occupant torso is represented by a system composed of rigid bodies. Springs simulating the segments which restrain directly the occupant are fixed to these rigid bodies in order to simulate the occupant/belt interaction. This is the technique used by most of the rigid body CVS softwares, such as MVMA2D, CAL-3D and MADYMO [1,2,3].

Compared to the reality of the occupant/belt interaction, this method has several shortcomings, in particular :

1) It only allows simulation of the belt slip along its own length. Slip in others directions is impossible to account for, such as for example the upward slip of the lap belt over the iliac spines, which can sometimes lead to the submarining. Recourses to artificial astuteness, such as attaching the lap belt to small ellipsoids [4], seem too far from the physical reality.

2) It does not allow a realistic repartition of belt loads over the occupant torso. In fact, with this method, belt loads on the occupant are simplified as point forces and the application points are fixed during impact. So effects of belt bearing point change over occupant can not be accounted for.

3) With this method, it is delicate to define characteristics of the springs simulating belt segments. They can not be deduced directly from the webbing characteristics, because the torso compliance has to be integrated in their definition. In order to do that, subsystem tests are necessary and these have to be repeated when webbing characteristics or belt path over the torso are modified.

In order to reduce these inconveniences, some rigid body CVS software packages were enhanced by introducing more advanced features. In CAL-3D a "harness belt system" option allows a description of belt slip by using a collection of reference points and by introducing slip conditions which control the movement of these points on ellipsoids representing the occupant torso [5,6]. In MADYMO a finite element feature was added which allows belt description with linear triangular membrane elements [7]. Nevertheless applications of these features for belt slip modelling are limited by the fact that the occupant torso has to be modelled by ellipsoid rigid bodies.

The finite element modelling of both occupant torso part and belt part appears to be the most promising option for improving the simulation of the occupant/belt interaction. Nevertheless, until now, investigations into this subject remain insufficient. The few approaches presented in the literature seem still too simplified to allow a satisfactory simulation of this interaction [8,9].

This paper describes the development of an FEM approach which allows simulation of the principal aspects of the occupant/belt interaction in a realistic way. The emphasis was in particular to simulate the belt slip over the chest and pelvis and the deformation of these under belt loading. Special attention was also directed towards the simplicity of approach in order to make it a frontal impact analysis tool, more powerful and realistic than the spring/rigid body approach, and less CPU-time consuming than a detailed finite element representation.

This study was carried out with an available finite element package RADIOSS [10]. The target occupant was the 50th percentile frontal impact dummy HYBRID-III [11].

2. SUBSYSTEM MODELLING

Three key elements of the occupant/belt interaction - belt, thorax and pelvis - were firstly modelled and validated with respect to a series of subsystem tests.

2.1 BELT MODELLING

The quality of the occupant/belt interaction simulation is conditioned by that of the belt representation.

It was decided to investigate firstly the belt modelling in the following configuration as it represented the simplest problem with a minimum of variables. It involved the simulation of the impact of a cylindrical mass upon a belt strap which was fixed at its two extremities to a rigid support. Figure 1 shows the experimental set up for the test configuration. The choice of such an impact loading of the belt was to observe its behaviour under dynamic contact. The impactor mass is 8.25 kg ; two impact velocities used were 2.5 m/s and 3.7m/s. The displacement and acceleration of impactor were measured, and from this the belt tension is deduced.

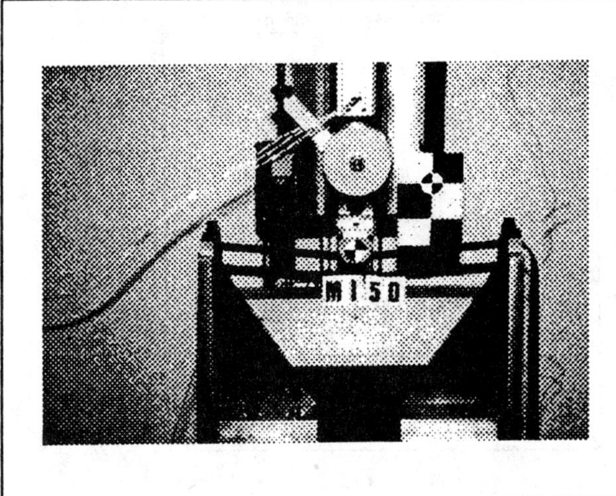

Figure 1. Belt test set up

Figure 2 shows the model simulating these tests. Shell elements were used to model the belt. The "One point integration in thickness" option was used to obtain the membrane behaviour of shell elements. The size of each element is identical and corresponds to a third of belt breadth. The impactor was treated as a rigid body. Its contact surface is represented by rigid shell elements.

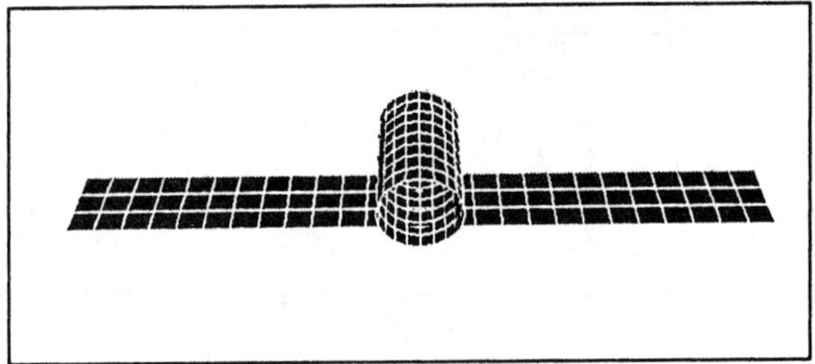

Figure 2. Belt and impactor model set up

The elasto-plastic law is used to characterize these elements.

$$\sigma = \begin{cases} E\varepsilon & (\varepsilon \leq \varepsilon_0) \\ (A + B\varepsilon_p^n) & (\varepsilon \geq \varepsilon_0) \end{cases}$$

Where E is Young's modulus, A elastic limit, B hardening coefficient and ε_p plastic deformation

Figure 3 shows the definition of the law according to experimental data from belt static tests. Poisson's ratio was assumed 0.2 and the thickness of elements is that of the belt.

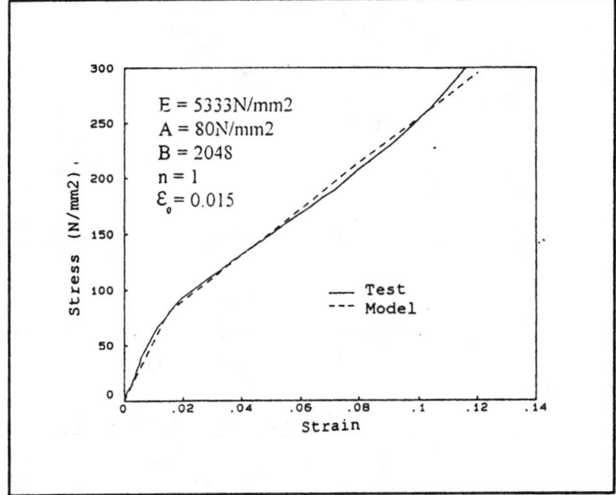

Figure 3. Approach of quasi-static belt elongation test data by a elesto-plastic law

Figure 4 compares the time histories of the displacement and acceleration of the impactor, and the belt tension for model and experiments. The belt model prediction matches well with experimental results.

Based on this model, a parametric study was carried out for belt thickness and Poisson's ratio. The thickness does not influence the model behaviour if the force-deflection relationship is conserved. By varying the Poisson's ratio between 0.1 and 0.4, model responses are almost identical.

2.2 THORAX MODELLING

Figure 5 shows the thorax model developed. It is composed of a rigid body carrying its inertia and an envelope in shell elements representing the real thorax contact surface. This envelope is attached to the "thorax" rigid body by general springs (springs which allow control of the 6-DOF movement between two nodes by defining the force-deflection and damping velocity relationships for both translation and rotation). Such a spring is created behind each node of the frontal surface of the thorax.

The thorax surface is divided into two parts : the frontal surface discretized in deformable shell elements and the thorax back in rigid shell

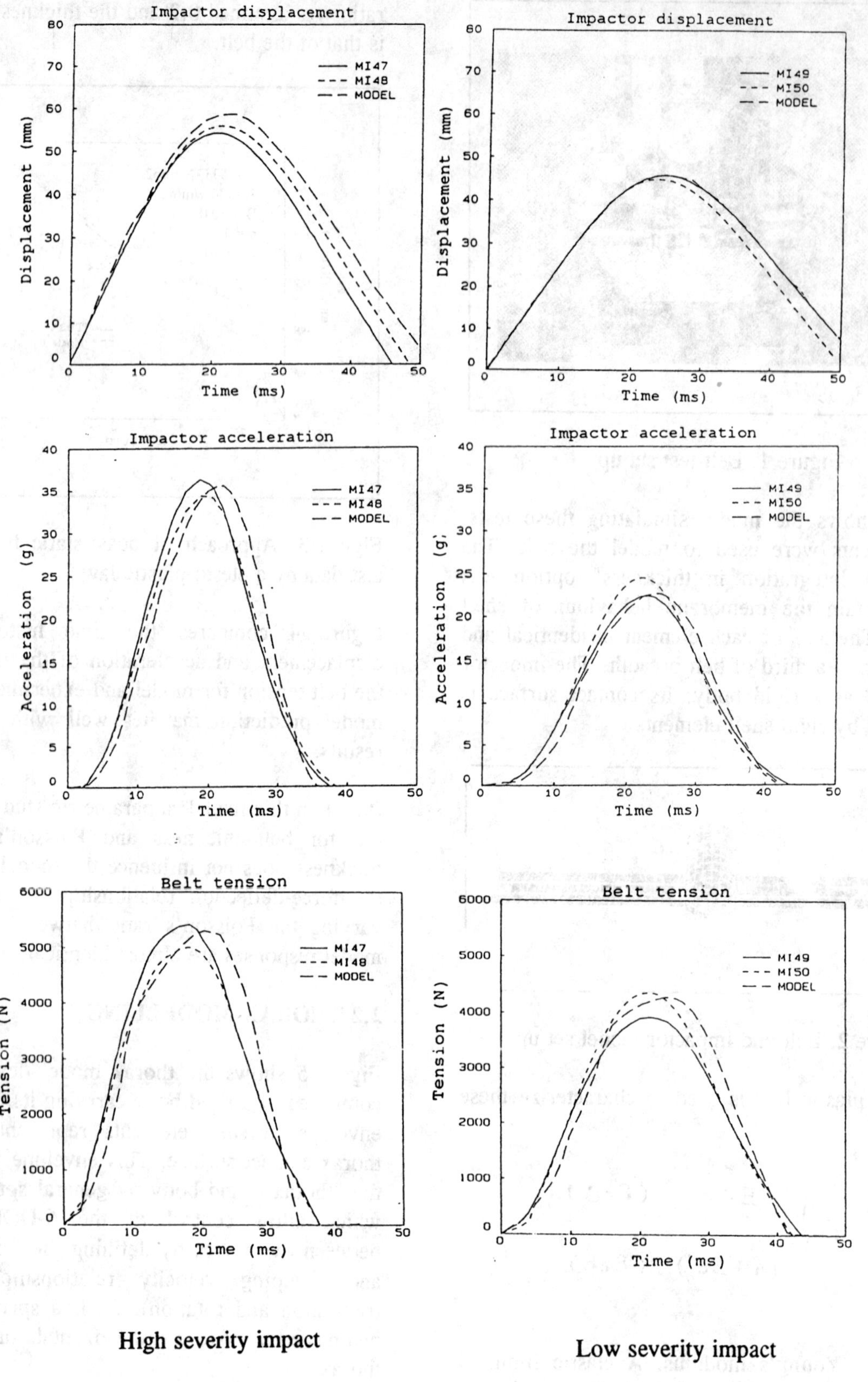

Figure 4. Belt model response compared to experimental results

elements. These latter ones were inserted into the "thorax" rigid body. General springs were classed into three groups according to their stiffness :

Group 1 : "Sternum" springs
Group 2 : "lower thorax" springs.
Group 3 : "shoulder" springs

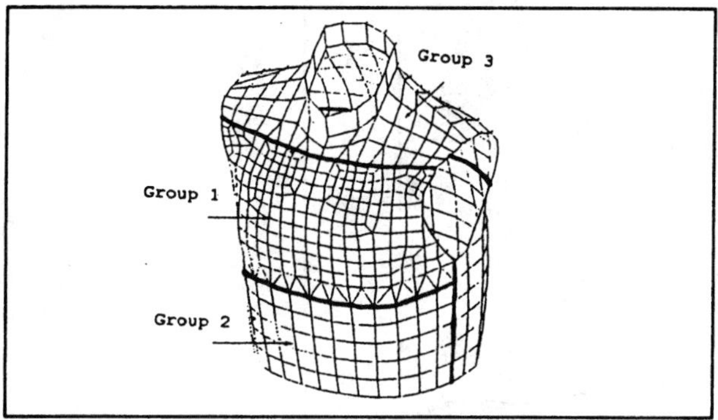

Figure 5. Thorax model

Characteristics of shell elements and general springs used are :

Shell elements

Young's Modulus = 1.5 N/mm2

Poisson's Ratio = 0.2

Thickness = 3 mm

Springs

Group 1 : K = 0.155 N/mm, C = 0.55 N.s/m

Group 2 : K = 0.6 N/mm, C = 0.55 N.s/m

Group 3 : K = 15.5 N/mm, C = 0.55 N.s/m

In all, 997 nodes, 558 shell elements (156 rigid elements) and 419 springs were used to model the thoracic portion. The number of nodes and elements were significantly reduced compared to a full finite element thorax description, such as the model developed by Yang et al [12] where, for the description of the ribs and spine box alone, 1996 nodes and 1582 elements had been used.

Two types of test were used to evaluate this model.

Evaluation with impactor tests : This was based on the HYBRID-III thorax calibration test results. Two standard impact velocities - 4.3 m/s and 6.7 m/s respectively - were simulated.

Taking into account the decoupling movement between the thorax assembly and the others segments of the occupant in the first 20 milliseconds, i.e. the loading phase of impact, an effective mass of 24.9 kg was used in the model. This mass corresponds to the sum of the thorax and uppers extremities masses. The impactor was treated as a rigid body with its contact surface discretized in shell elements.

Figure 6 illustrates the thorax shape at 0 and 20 ms for the 6.7 m/s test. A comparison of the impact force and chest deflection between the model and the test is presented in figure 8. The agreement is satisfactory for the loading phase. Similar results can be obtained for the 4.3 m/s test.

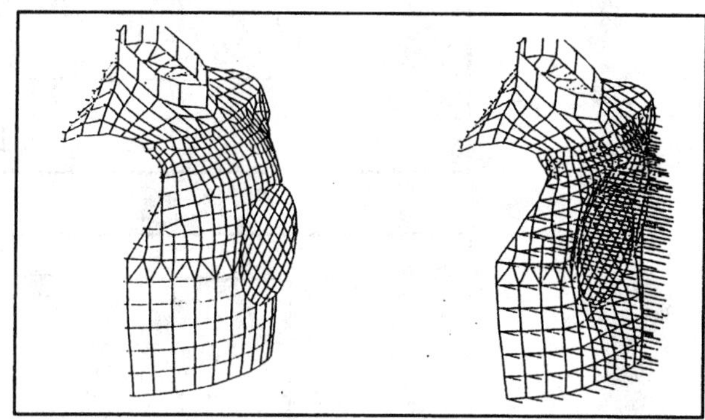

Figure 6. Impacted thorax at 0 and 20 ms

Evaluation with belt tests : A series of tests was conducted by INRETS to study the HYBRID-III thorax behaviour under belt loading [13]. As shown in figure 7, the dummy is laying on a rigid flat surface. A seat belt strap is placed across the thorax and compresses this one when the belt is loaded by an impactor.

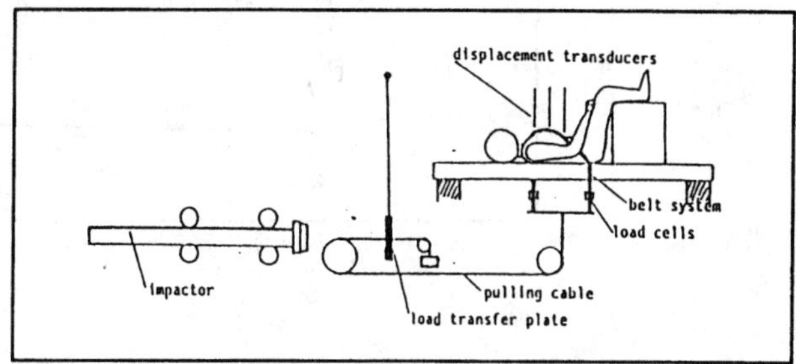

Figure 7. Thorax belt test set up

Two tests were chosen to evaluate the thorax model behaviour for its interaction with a belt :

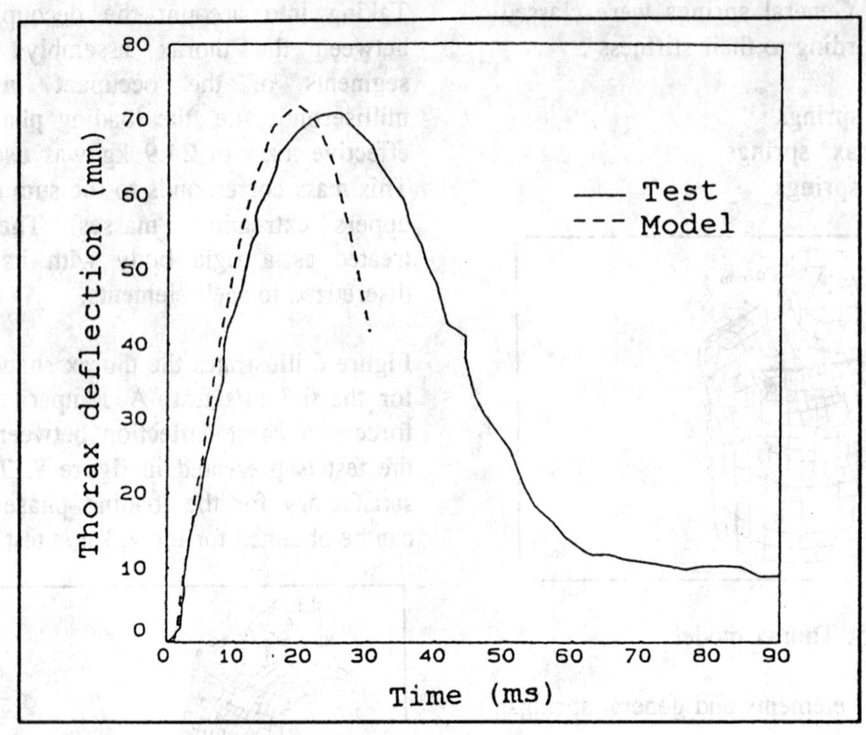

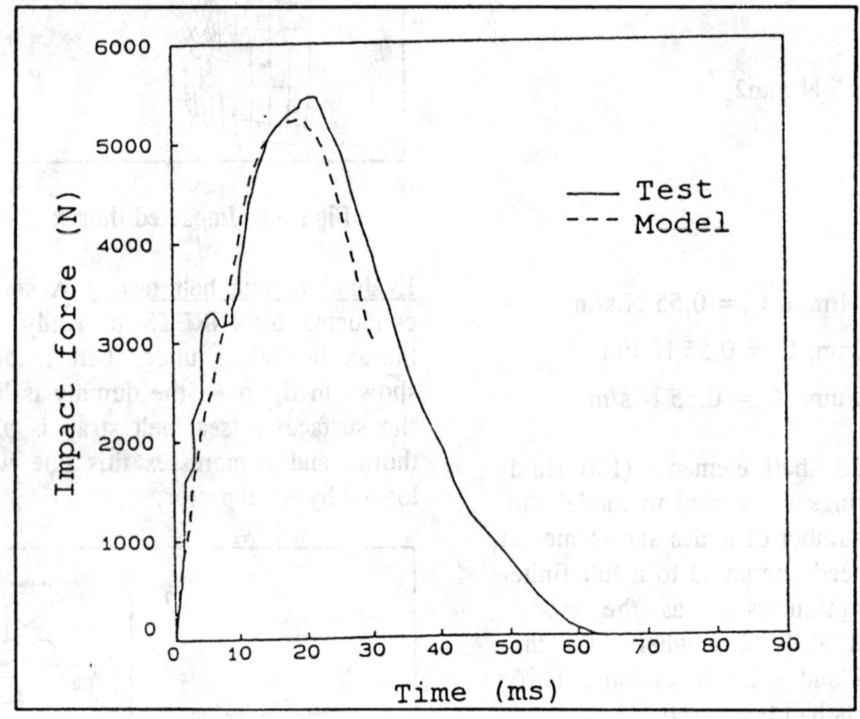

Figure 8. Comparison of contact force and thorax deflection between model and experiment for 6.7 m/s pendulum impact

THC89 :
Impactor mass = 22.4 kg
Impact velocity = 7.3 m/s
THC83 :
Impactor mass = 76.1 kg
Impact velocity = 3.4 m/s

Figure 9 shows the model simulating these tests. The cylindrical surfaces in shell elements in the model were to simulate pulleys used in the tests which guide the belt during impact. The chest/belt interaction is modelled by using a master surface/slave surface contact algorithm. The interfaces were defined between belt and shoulder, sternum and lower thorax surface.

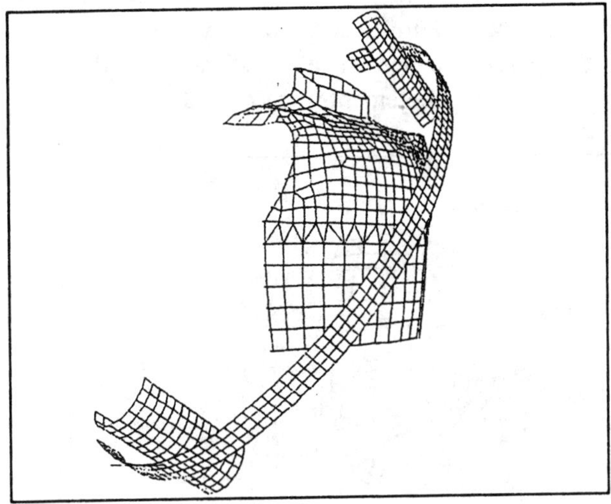

Figure 9. Model for thorax belt test

The belt was loaded in two ways. The first one consists in distributing the impactor mass to the boundary belt nodes and assigning the initial impactor velocity to the nodes.

A comparison of calculated and measured values of thorax deflection and belt tension is presented in figure 10 for test THC 83. The principal difference lies in the time for reaching peak values. This can be explained partly by existing plays in the experimental set up and the energy loss in the cable path.

Another way for loading the belt is to assign directly the measured belt tension to the boundary belt nodes. In this case the model behaviour is evaluated by comparing the chest deflection. Figure 11 shows the agreement between model and experiment for test THC83. Figure 12 shows the model configuration at 35 ms for test THC89. Figure 13 illustrates the calculated thorax deformation form and the experimentally reconstructed one, for test THC83, at the instant where the maximum thoracic deformation was reached. The thorax compliance under belt loading is quite satisfactory.

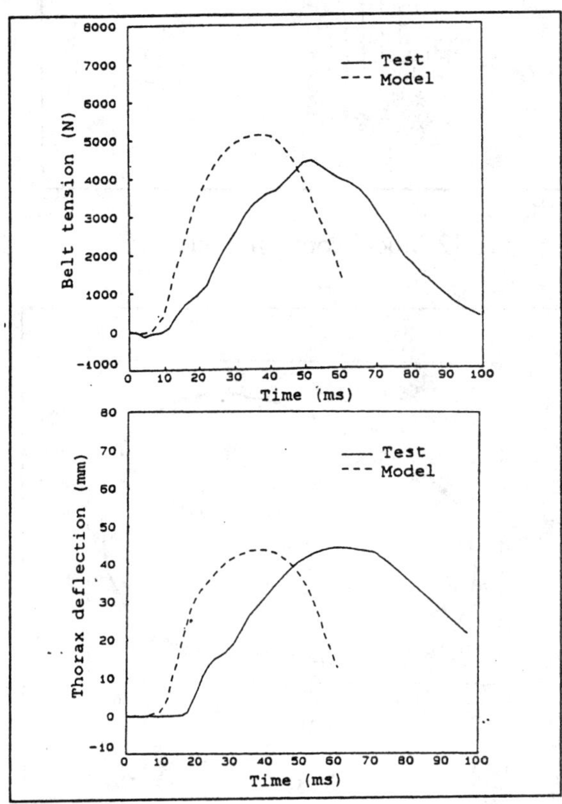

Figure 10. Comparison of belt tension and thorax deflection between model and experiment for test THC 83.

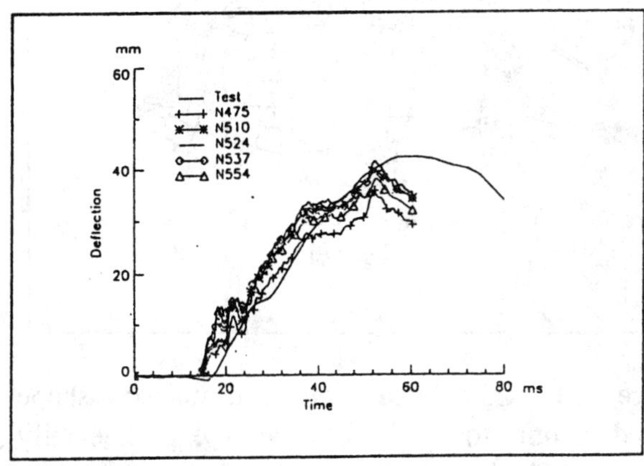

Figure 11. Comparison of thorax deflection for test THC 83

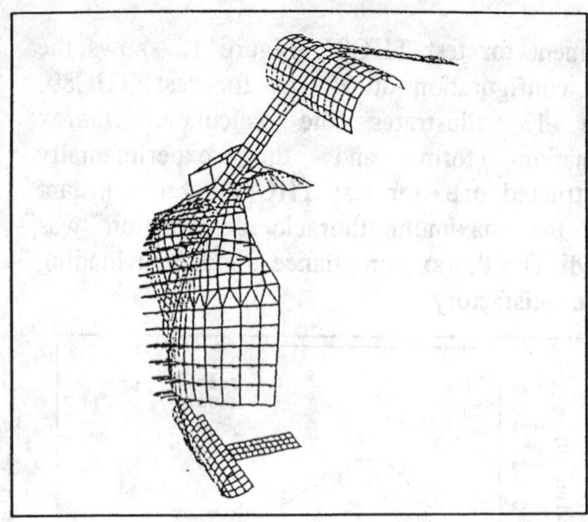

Figure 12 Model configuration at 35 ms

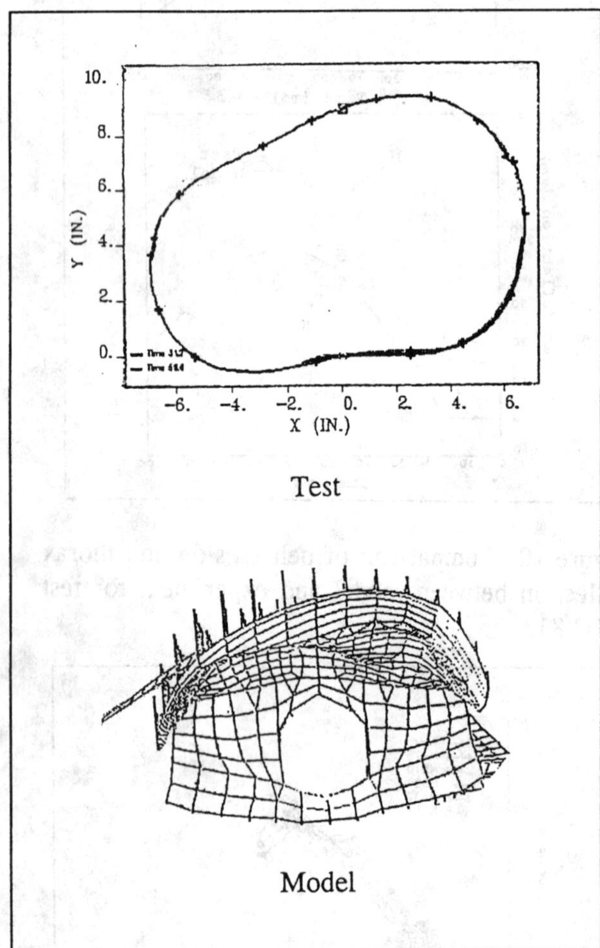

Figure 13. Calculated deformed thorax shape viewed from toe to head and experimentally reconstructed thorax deformation contour at the moment where the maximum thoracic deformation was reached

2.3 PELVIS MODELLING

A pelvis model should account for the essential role of the iliac spine for the pelvis/lap belt interaction, in particular for the submarining risk. To do this, it is necessary to model the iliac foam and skin compression under belt loading, and also the iliac spine shape effects.

Figure 14 shows the pelvis model developed. Four shell element groups were used to model the exterior pelvis surface as well as the front iliac spine shape. Groups 1 and 2 represent the pelvis surface part which can enter into contact with the lap belt. The group 3, representing the rest of the surface, was used for the pelvis/seat cushion modelling. The group 4 describes the iliac spine front shape which influences directly the belt slip over the pelvis surface. In all, 995 nodes and 919 elements (639 rigid elements) were used.

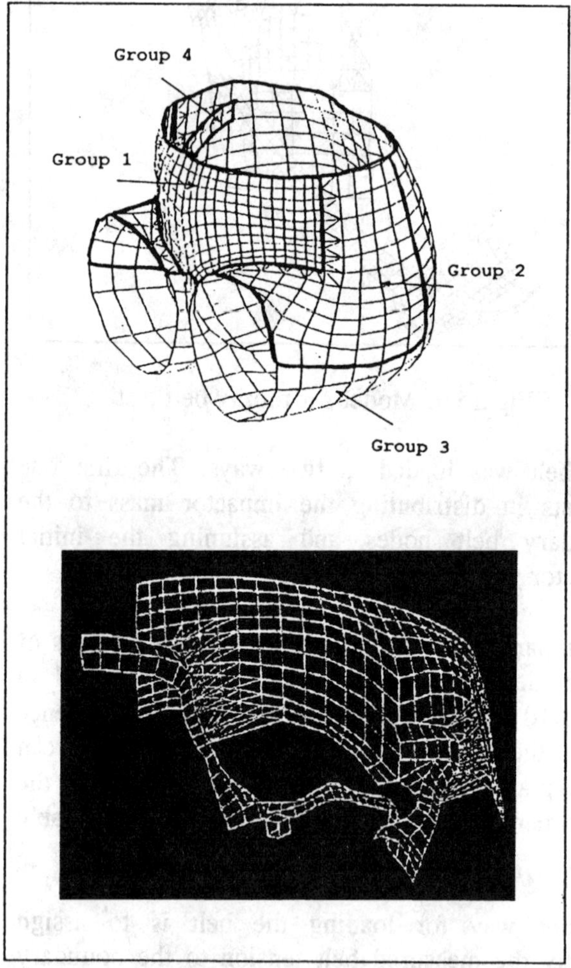

Figure 14. Pelvis model

The pelvis inertia is carried by a "pelvis" rigid body. Shell elements of groups 3 and 4 were inserted in this rigid body and consequently are rigid. Elements of group 1 are deformable and those of group 2 rigid. These elements were disconnected from group 3 and were attached to iliac spine elements with 64 general springs. These springs, with the aim of simulating the foam in front of the iliac spine, were guided along the A-P direction of the pelvis. Their characteristics were calculated according to the foam strain/stress relationship and their length. In total, four stiffness were distinguished (Figure 15).

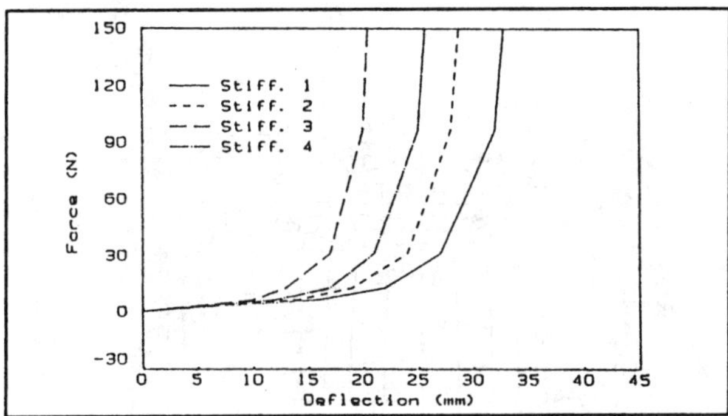

Figure 15. Iliac springs stiffness

Quasi-static tests were carried out to evaluate this model in a relatively simple configuration. Figure 16 shows the experimental set up. The pelvis is fixed rigidly and was compressed by loading the belt with two cranks. The belt path over the pelvis can be adjusted by moving two pulleys guiding the belt during test. A piece of tissue was placed between belt and pelvis to account for the clothing influence on friction. Two configurations were tested, one leading to submarining and the other not.

Figure 17 shows the model simulating the tests. A constant velocity of 0.5 m/s was assigned to the boundary belt nodes for loading the belt.

Simulation of configuration without submarining : In this case, as illustrated in figure 19 (a), shell elements and general springs were compressed under belt loading and the belt was hooked to the iliac spine. Figure 18 shows a comparison of the belt tension between model and experiment.

Figure 16. Pelvis belt test set up

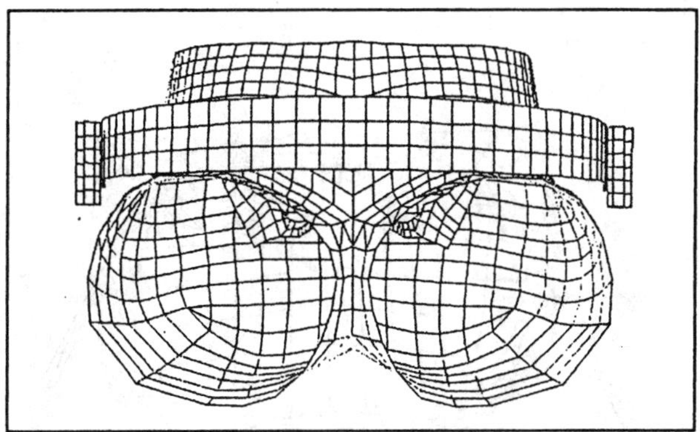

Figure 17. Model for pelvis belt test

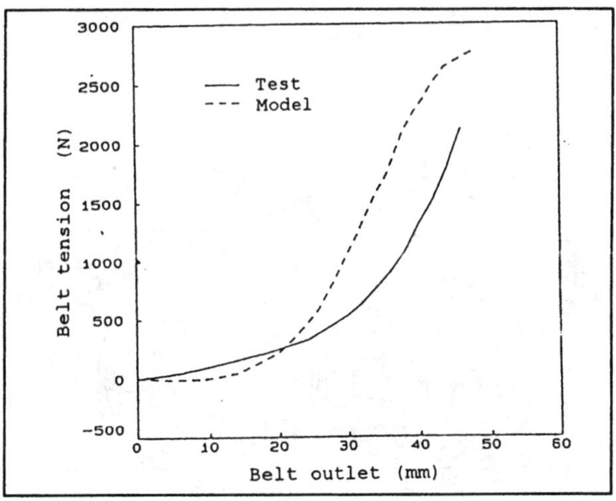

Figure 18 Comparison of predicted belt tension with quasi-static test results

Simulation of configuration with submarining : We can observe in this case an upward belt slip which ends up passing over the iliac crest ; such movement corresponding to the occurence of submarining. Figure 19 (b) shows the model shape at the moment where the belt reached its slip limit.

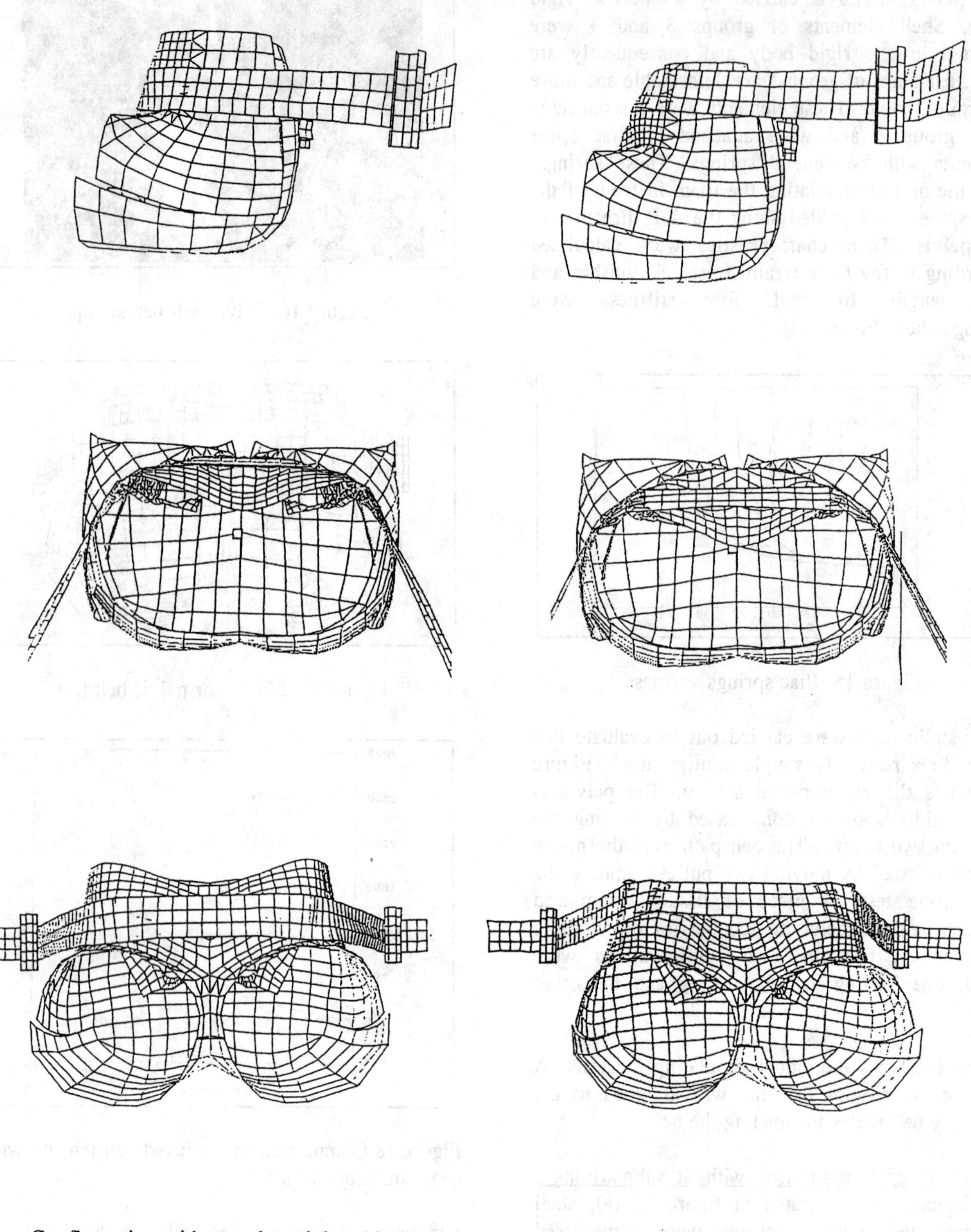

Configuration without submarining (a)

Configuration with submarining (b)

Figure 19. Comparison of belt position and deformed pelvis for configurations with and without submarining

3. OCCUPANT/BELT INTERACTION MODELLING IN SLED TESTS

Based on the subsystem models of the belt, thorax and pelvis outlined above, the occupant/belt interaction in a vehicle crash environment will be examined in this section. The target system was composed of the HYBRID-III dummy restrained by a 3 point retractor belt, a seat and a toeboard.

3.1 MODEL DESCRIPTION

Occupant

As shown in figure 20, the full dummy model is a combination of preceding thorax and pelvis models with a RADIOSS rigid body H-III model. The latter was constructed using data in the literature [14,15] and consisted of 15 rigid bodies representing respectively the head, neck, thorax, lumbar spine, pelvis, upper legs, lower legs, feet, upper arms and lower arms/hands assemblies. Connections between these segments were modelled with general springs. The relative rotation between adjacent segments was governed by a torque-angle relation, a dry friction constant and a damping constant for each rotation axis. The surface of each segment was approached by ellipsoid dicretized in shell elements.

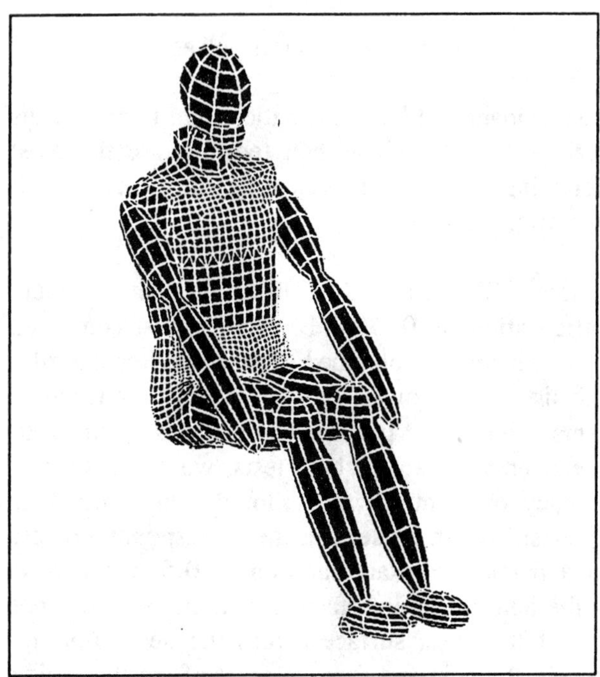

Figure 20. Full dummy model

Restraint System

The idea of modelling the total belt by shell elements was thrown out. On the one hand, the surface shape of the belt material as a whole is very complex, in particular for the D-ring part and stalk buckle part where belt surface warping is difficult to represent ; on the other, it is not more advantageous to model with shell elements the belt segments which can not enter into contact with the occupant surface than with springs.

A combination of springs, rigid bodies and shell elements was used to model the belt system (Figure 21).

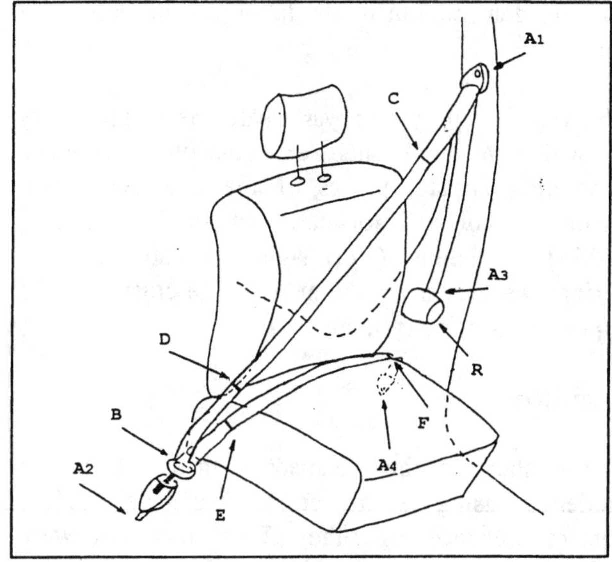

Figure 21. Belt system layout

1) <u>Belt parts in contact with occupant :</u> Segments **CD** and **EF**, in contact with the occupant surface, were modelled with shell elements. At the ends of two segments triangular rigid shell elements were used in order to obtain a good link between belt shell elements and springs modelling the others parts of the belt.

2) <u>Belt parts without contact with occupant :</u> Segments **A_3A_1C**, **DBE** and **FA_4** can not enter into contact with occupant and were modelled by spring elements -a simple spring for segment **FA_4** and two 3-node springs respectively for segments **A_3A_1C** and **DBE** in order to account for the belt material transfer at the D-ring and the stalk buckle.

3) Friction : Tension in the upper shoulder belt is greater than that in the retractor part owing to the friction between D-ring and belt material. It is the same for the lower shoulder belt and inboard lap belt. The friction effect was accounted for by installing two simple springs (between point A_1 and point **C** for D-ring friction and between point **B** and point **E** for stalk buckle friction).

4) Film spooling effect : This was simulated by a 1D spring which links the extremity point A_3 of the spring A_3A_1C to the anchorage **R**.

5) Stalk : A rigid body was used to represent the stalk. It is attached to the sled by a general spring with which the stalk rotation about anchorage A_2 and the deformation of the latter can be accounted for.

The whole sled system was treated as a rigid body to which 4 rigid shell elements were attached simulating contact planes of the seat back, seat cushion, floor and toeboard. In all 3439 nodes, 2655 shell elemnts (1751 rigid elements) and 577 springs were used for the model. The crash pulse is applied as a velocity input.

Interfaces

The contact of the occupant with the belt was modelled using a master surface/slave surface contact interface algorithm. Two interfaces were defined between shoulder belt and thorax (shoulder, sternum and lower thorax) and between lap belt and pelvis . The master surfaces are those of the thorax and the pelvis.

Contacts between occupant and sled interior surface are between rigid surfaces. A rigid surface contact interface algorithm was used for modelling these contacts. Each contact is defined by a force-penetration function and a dry friction constant. The following interfaces were defined :

Pelvis/seat cushion
Right foot/toeboard
Left foot/toeboard

3.2 MODEL EVALUATION WITH RESPECT TO SLED TESTS

The model evaluation proceeded in two steps. In the first step, the system model was evaluated for configurations without submarining. In the second, several important parameters that affect significantly the submarining were varied in order to evaluate the model behaviour for the submarining simulation.

Simulations without submarining

To serve as experimental references for model evaluation, two sled tests were conducted with a change in velocity of 35-65 km/h, and a peak acceleration of 15-35 g (Figure 22). The HYBRID-III dummy was restrained by a 3-point retractor belt, a rigid seat and a toeboard.

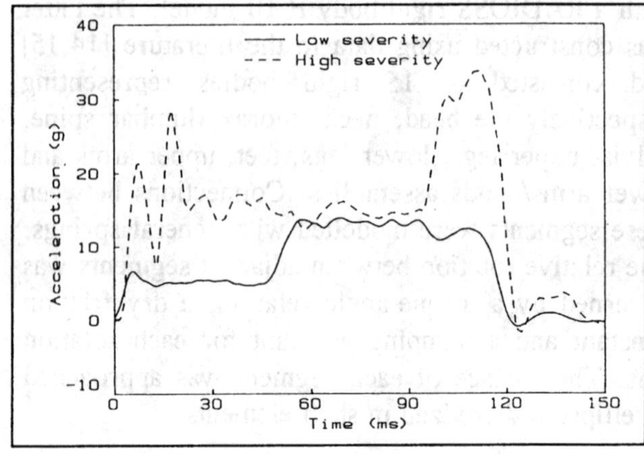

Figure 22. Impact pulses

The occupant and belt kinematics, thorax and pelvis accelerations as well as belt tensions are the most interesting criteria for evaluating the occupant/belt interaction simulation.

Figures 23 and 24 illustrate the model configurations at 0, 70, 110 and 150 ms, compared to the kinematics obtained from experimental film analysis for the low severity test. The sequence of events given by the model matches well with experimental tests. For both tests, we can observe a tendency of right-upwards shoulder belt slip over the chest, as can be seen in the corresponding tests. But a friction constant superior to 0.5 was needed for the belt/thorax interface so that the belt does not slip off the chest surface across the neck. For the lap belt, the skin and foam in front of the iliac spine were collapsed under belt loading; the belt was hooked to the iliac spine with a friction constant of 0.3 between pelvis skin and belt material. Figure 25

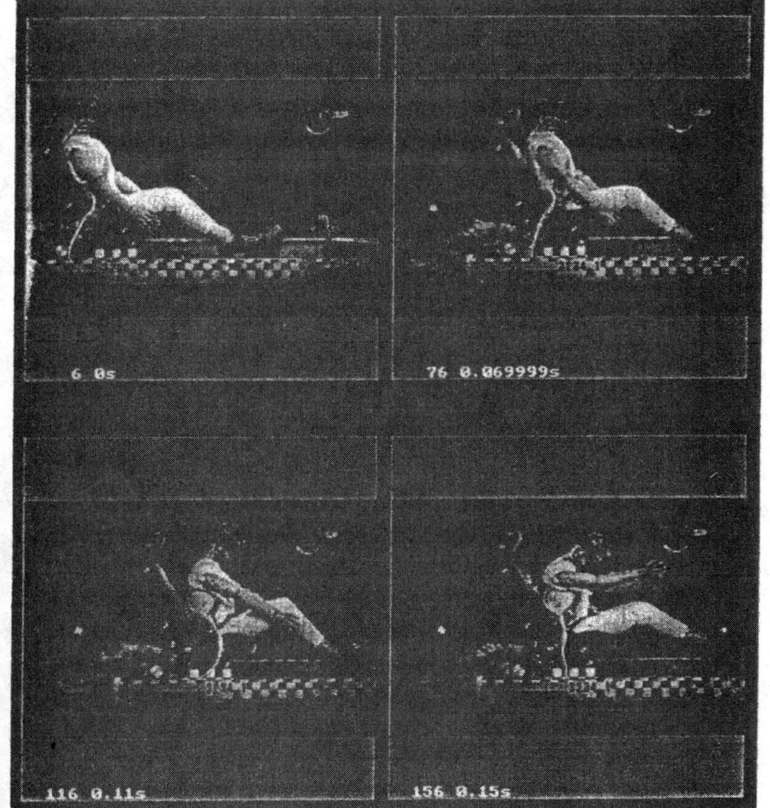

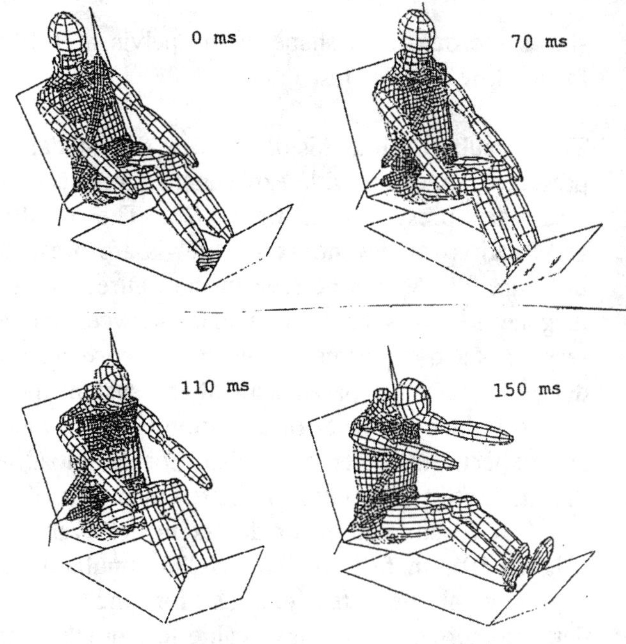

Figure 24. Model kinematics in perspective view

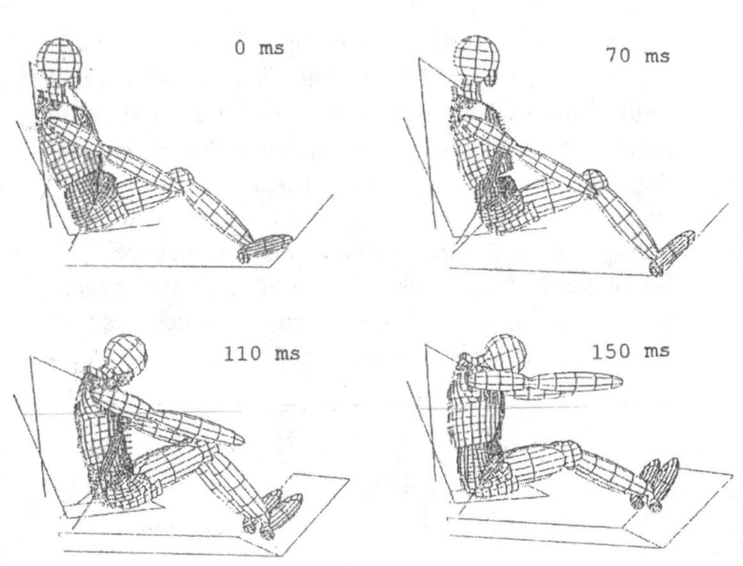

Figure 23. Model kinematics compared to experiment

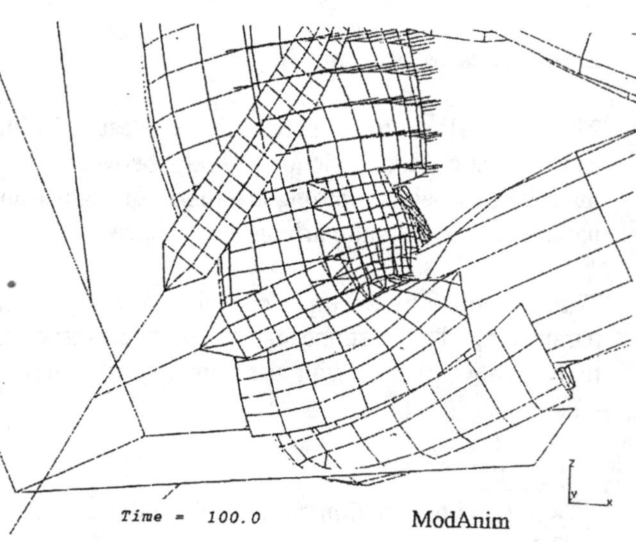

Figure 25. Lap belt position relative to the deformed pelvis for configuration without submarining

shows the deformed shape of the pelvis at 100 ms for the low severity test.

The calculated time histories with the model are presented together with experimental results. Belt loads are presented in figure 26. The resultant acceleration for the thorax and pelvis are presented in figure 27. As can be seen in the figures, there is in general a reasonable agreement between models responses and experimental results. A more accurate definition of some parameters of the system model should reduce the level of deviation between model and experiment. It is noted that, for belt loading, the model/test correlation is better than that of the test THC83 simulation for thorax model evaluation (Figure 10). In fact, in the THC83 simulation the experimental belt tension, as for the thoracic deflection, reached his peak value less quickly than the model prediction because of existing plays in the experimental set up and the energy loss in the cable path. These are difficult to simulate in the model representation of the experimental set up.

Simulations with submarining

The lap belt - to - pelvis angle, seat cushion characteristic and friction level between belt material and pelvis clothing are the most significant parameters on the submarining occurrence.

Three simulations were performed by varying these parameters. The baseline dataset used corresponds to the above model with the low severity impact pulse.

Run 1 :
Reducing the friction constant from 0.3 to 0.1
Run 2 :
Replacing the rigid cushion stiffness by a softer one (Figure 28)
Run 3 :
Moving lap belt anchorages back 100 mm.

For run 1, the model kinematics show that the lap belt was always maintained in front of the iliac spine during impact and there was not submarining. For run 3, similar belt kinematics can be observed.

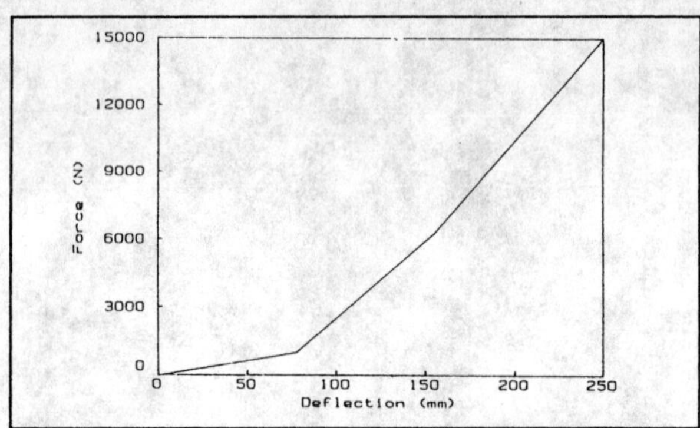

Figure 28. Stiffness of soft seat cushion

For run 2, the seat cushion stiffness was insufficient to resist the downwards motion and rear rotation of the pelvis. These motions changed progressively the lap belt - to -pelvis angle. When this angle reached a critical angle, the lap belt slipped over the iliac crest, that corresponding to the occurrence of submarining. Figure 29 shows the deformed pelvis and the belt position relative to the pelvis about this moment. After its passage over iliac crests, the lap belt was totally unhooked from the pelvis, as illustrated in figure 30

Figure 31 presents calculated lap belt loads for run 2 along with the baseline results. One can observe a sudden drop of lap belt load for run 2, a typical indication of submarining occurrence. Nevertheless, this model can not yet calculate the force transmitted by lap belt to the abdomen after its passage above the iliac spine. As a consequence, the calculated belt load does not present a second peak, as in experiments with submarining, which results from the belt interaction with the lumbar spine.

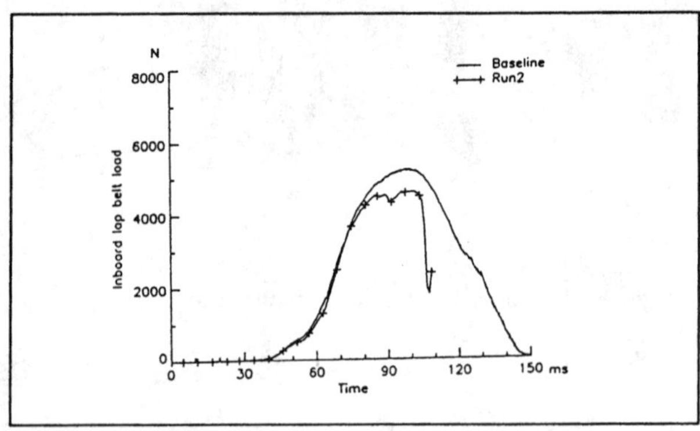

Figure 31. Comparison of lap belt load for run 2 (submarining) and baseline run (no submarining)

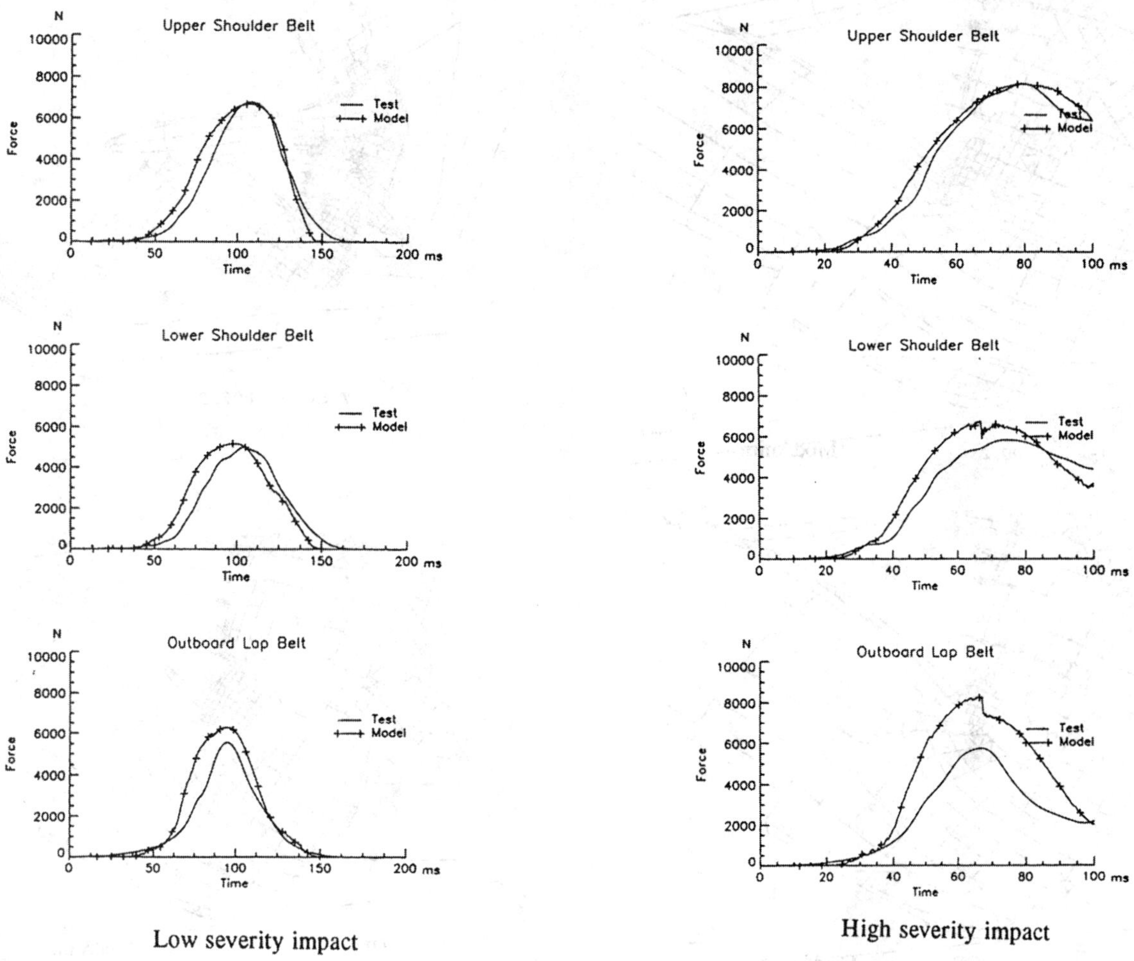

Figure 26. Comparison of belt loads

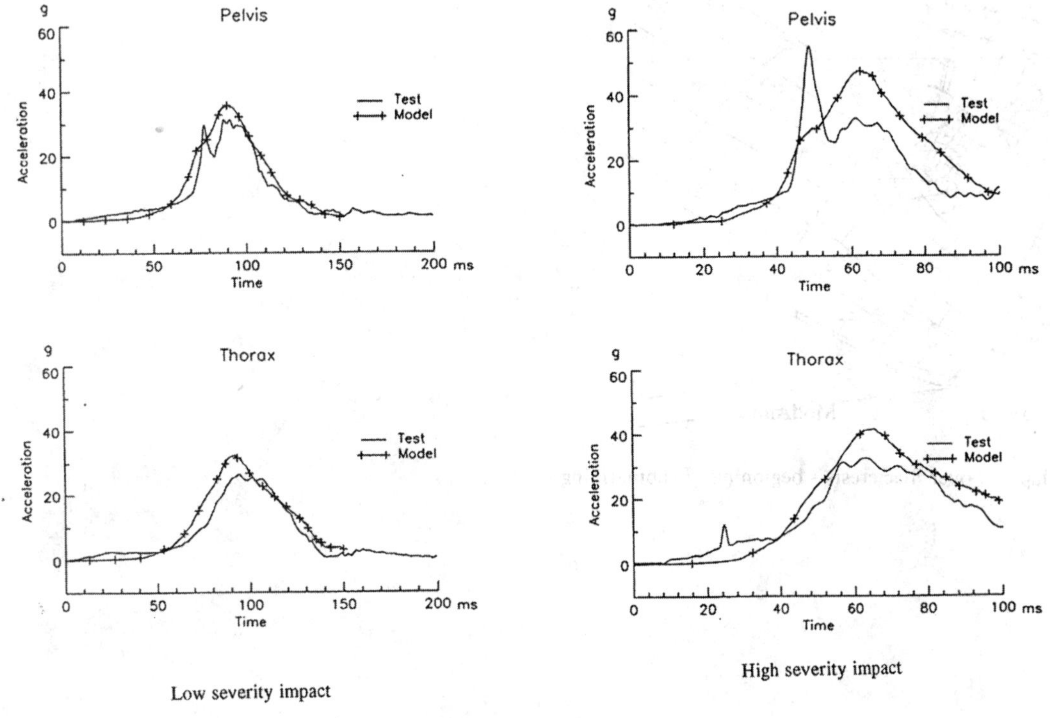

Figure 27. Comparison of accelerations

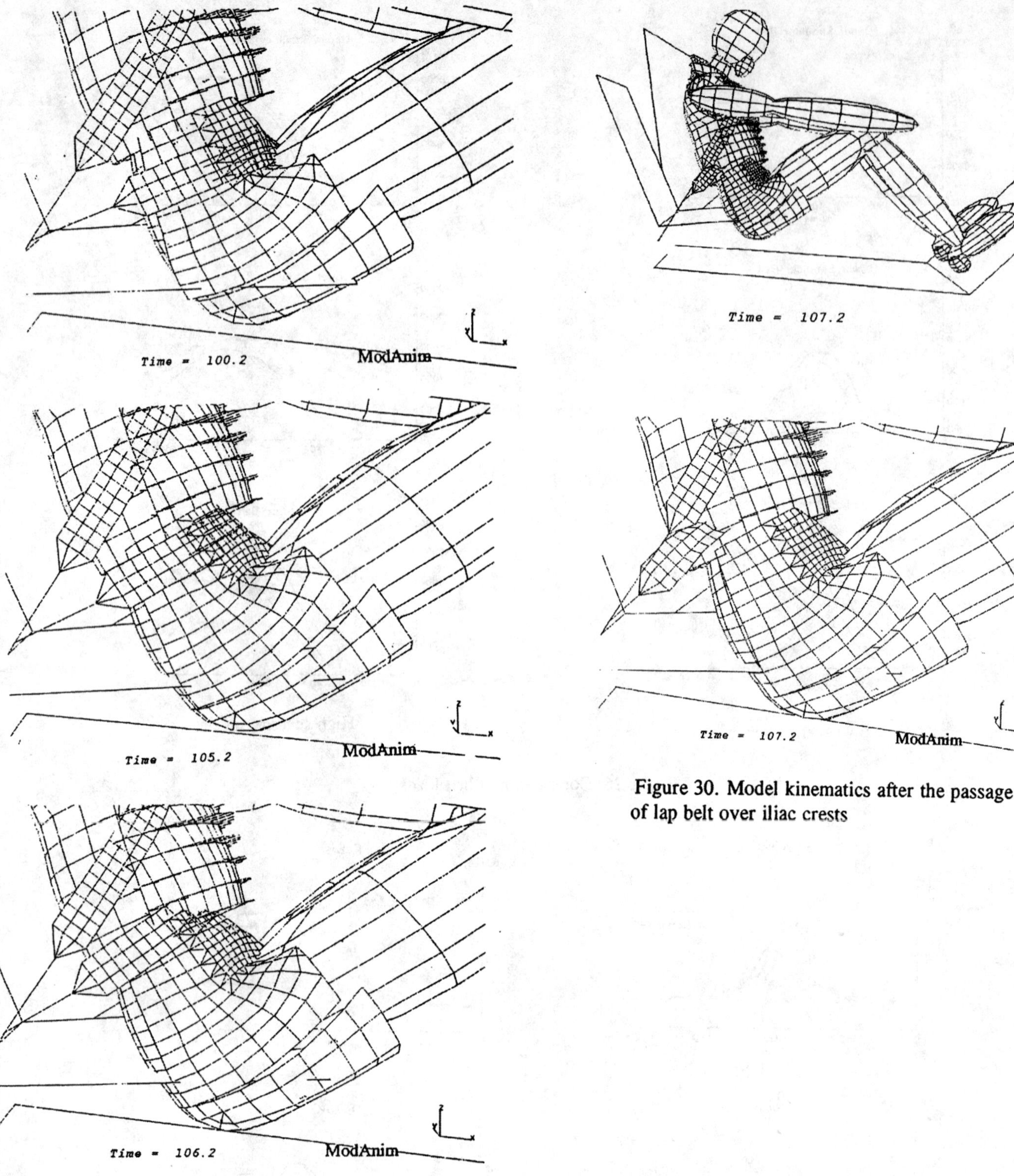

Figure 29. Passage of lap belt over iliac crests - beginning of submarining

Figure 30. Model kinematics after the passage of lap belt over iliac crests

4. SUMMARY AND CONCLUSION

A finite element approach to the occupant/belt interaction in frontal impact was developed. The target system was the 50th percentile HYBRID III dummy restrained by a 3-point retractor belt. Three key elements of the occupant/belt interaction - belt, thorax and pelvis - were firstly modelled and validated with respect to four subsystem configurations : a belt loaded by a cylindrical surface impactor, the dummy pelvis compressed by a lap belt, the dummy thorax loaded by an impactor and by a shoulder belt. Based on these subsystem models, the occupant/belt interaction in a vehicle crash environment was modelled. The model was evaluated with respect to two sled tests of different severities for configurations without submarining. The model behaviour for submarining simulation was examined by varying some parameters that affect significantly such type of occupant kinematics.

This approach allows simulation, in a realistic way, of the principal aspects of the occupant/belt interaction, and in particular the belt slip over the thorax and pelvis and the deformation of these under belt loading. In general, a good agreement between model and experiment can be observed. For submarining simulation, the model allows the detection of submarining occurrence through lap belt kinematics and its tension time history.

This model is an approach more powerful and realistic than spring/rigid body approach and less CPU-consuming than a detailed finite element representation. With enhancements of some aspects of the model, such as a more realistic description of the seat structure and a direct simulation of friction in the D-ring and stalk buckle, this approach should allow a satisfactory simulation of a belted occupant response. The modelling methodology could be also extended easily for simulating the interaction of others human surrogates with a belt system.

ACKNOWLEDGEMENTS

The authors wish to thank J. Uriot, M. Moutreuil, S. Mairesse and F. Bendjellal for their assistance in many aspects of this project.

REFERENCES

1. Bowman, B.M., Bennett, R.O., and Robbins, D.H., "MVMA Two-Dimensional Crash Victim Simulation", Version 4, Final Report UM-HSRI-79-5-1, Hight Traffic Safety Research Institute, June 1979.

2. Fleck, J.T., Butler, F. and Deleys N., "Validation of the Crash Victim Simulator", Final Report DOT-HS-806280, Volume I-IV, August 1982

3. MADYMO Users Manuals, Version 4.3, April 90, TNO Road-Vehicles Research Institute, Delft, the Netherlands

4. Bosio, A.C., "Simulation of Submarining with MADYMO", Proceedings of the Second International MADYMO Users Meeting, 1990.

5. Butler, F.E., Fleck, J.T., "Advanced Restraint System Modeling", Air Force Report AFAMRL-TR-80-14, May, 1980.

6. Deng, Y.C., "Developement of a Submarining Model in the CAL3D Program", SAE Paper No. 922530, 36th Stapp Car Crash Conference, 1992.

7. Fraterman, E. and Lupker, H.A., "Evaluation of Belt Modelling Techniques", SAE Paper No. 930635, SAE International Congress & Exposition, Detroit, Michigan, 1993.

8. Sturt, R., Walker, B.D, MILES, J.C. and Giles, A. "Modelling the Occupant in a Vehicle Context-an Intergrated Approch", 13th International Conference on Experimental Safety Vehicles, 1991.

9. Midoun, D.E., Rao, M.K., and Kalidindi, B., "Dummy Models for Crash Simulation in Finite Element Programs", SAE Paper No. 912912, 35th Stapp Car Crash Conference, 1991.

10. RADIOSS Users Manuals, Version 2.1, July 1992, MECALOG, les Ulis, France

11. Faster, J.K., Kortge J.O., and Wolanin M.J., "Hybrid III - A Biomecanically-Based Crash Test Dummy", SAE Paper No. 770938, 21th Stapp Car Crash Conference, 1977.

12. Yang, K.H., Pan, H., Lasry, D. and Hoffmann, R., "Mathematical Modelling of the Hybid III Dummy Chest with Chest Foam", SAE Paper No. 912892, 35th Stapp Car Crash Conference, 1991.

13. Cesari, D., and Bouquet, R; "Behaviour of Human Surrogates Thorax under Belt Loading", SAE Paper No. 902310, 34th Stapp Car Crash Conference, 1990.

14. Wismans, J., and Hermans, J.H.H, "MADYMO 3D Simulation of HYBRID III Dummy Sled Tests", SAE Paper No. 880645, SAE International Congress & Exposition, Detroit, Michigan, 1988.

15. Kaleps, I., and Whitestone J., "HYBRID III Geometrical and Inertial Properties", SAE Paper No. 880638, SAE International Congress & Exposition, Detroit, Michigan, 1988.

SECTION 4:

ALTERNATIVES TO MANUAL, BODY MOUNTED 3-POINT BELTS

CHAPTER 13:

AUTOMATIC BELTS / PASSIVE BELTS

870221

Historical Review of Automatic Seat Belt Restraint Systems

H. George Johannessen
OmniSafe, Inc.

ABSTRACT

Following early rulemaking by the National Highway Traffic Safety Administration requiring passive occupant crash protection for occupants in vehicles covered by Motor Vehicle Safety Standard 208, intensive development began on passive seatbelts as promising alternatives to the inflatable systems being considered. The general types of passive seatbelt systems are identified. Passive systems that have been offered to date as optional installations in production car lines are described, as well as the production models being introduced in Model Year 1987 car lines in response to recent rulemaking. Design considerations are discussed. Recent usage rates of installed passive systems and manual belt systems are compared.

THE U.S. CONGRESS ENACTED the Highway Traffic Safety Act of 1966 (1)* to address the problem of traffic fatalities and injuries on American roadways. The first director of the agency established to carry out the provisions of the new law - namely the National Highway Safety Bureau (NHSB) in the Department of Commerce, which was later superseded by the National Highway Traffic Safety Administration (NHTSA) in the Department of Transportation- was Dr. William Haddon, Jr., a medical doctor with a long and distinguished role in public health and traffic safety in the State of New York. He viewed traffic fatalities and injuries as a public health problem of epidemic proportions that could best be overcome by passive means -that is, by means requiring no overt action on the part of the parties affected. He cited past examples of successful passive solutions to major public health problems, such as inoculation of entire populations to overcome dread diseases, pasteurization of milk, and use of automatic sprinkler systems to control fires. He strongly advocated passive solutions to the problem of highway fatalities and injuries, including passive means for occupant crash protection.

*Numbers in parentheses designate references at end of paper.

PASSIVE SYSTEMS FOR PASSENGER CARS

The early patent literature relating to passive restraint systems goes back to 1949 when a collision mat was described (2). The first patent relating to a passive belt system (3) was issued in 1958, in which a system was described having a safety belt that moved into position by closing the door and then automatically released upon opening one or both doors. The patent described a single belt used to restrain all the passengers in the front seat, which could be one, two or three passengers. Such a conceptual flaw was symptomatic of the inadequacies of many of the early concepts for passive restraints that did not embody performance characteristics consistent with requirements dictated by the realities of crash dynamics, occupant kinematics, and limits of human tolerance to injury. Meaningful development of passive occupant protection systems required the concurrent advance in the knowledge and technology relating to the real-world car crash event.

In November 1970, the U.S. Federal Motor Vehicle Safety Standard 208 (MVSS 208) (4) which addresses vehicle occupant crash protection, was amended (5) to incorporate the original requirements for passive occupant protection in the front seating positions of all passenger cars manufactured on or after the first day of July 1973 for sale in the U.S. car market. The original requirements and original effective date embodied in the amendment were challenged and further considered in a series of subsequent actions from 1970 to the present. The most recent action by NHTSA (6) calls for passive occupant protection in front outboard seating positions on a phased-in schedule with installations in a minimum of 10 percent of each manufacturer's car build after September 1, 1986, 25 percent after September 1, 1987, 40 percent after September 1, 1988, and 100 percent after September 1, 1989. The passive requirement will be rescinded if states representing two-thirds of the nation's population enact mandatory seatbelt usage laws before April 1, 1989 that will be in effect and enforced by September 1, 1989.

The passive system for vehicle occupant protection that received substantial attention following the issuance of the original rulemaking in 1970 was the inflatable restraint (or "airbag" or "air cushion"), which was already under development. Alternative noninflatable means, however, were considered as well. Many categories of such noninflatable passive candidate systems were identified and explored (Table 1). These included transparent shields, deployable nets and blankets, cushions, restraining arms and barriers, integrated seat systems, and seatbelt systems.

DEVELOPMENTAL PASSIVE SEATBELTS

Of these various categories, passive seatbelts showed the greatest promise by far. Many conceptual designs of passive seatbelts were investigated. Several distinctive classes may be identified (Table 2). They may be 2-point, 3-point, or 4-point systems based on the number of load-bearing connections to vehicle structure. They may be all-mechanical or may be all- or partially-motorized. They may have retractors located in either inboard or outboard locations and fixed anchors located either inboard or outboard. Some may require knee bolsters to control lower body movement. All must have some provision to permit disengaging the seatbelt after an accident. Prototypes of all these classes of systems have been developed and installed in exemplar production cars.

INSTALLATIONS IN PRODUCTION CAR LINES

In spite of the very substantial amount of effort directed to the automatic belted restraint systems and the large number of prototype systems that were developed and presented to vehicle manufacturers and to NHSTA, the first production-ready automatic seatbelt system was not installed in a production car line until the 1975 Model Year (Table 3).

The first production model to appear was a 2-point system, installed as an extra-cost option in a Volkswagen Rabbit in Model Year 1975 and subsequent years. Figure 1 shows the conceptual design of the VW Rabbit installation. This system includes a 2-point upper torso belt with the retractor inboard and the fixed anchor attached outboard on the door. The door connection incorporates a buckle for emergency disconnection. The buckle has an electrical interlock with the starter system to insure that the buckle is connected in normal driving use. The lower body movement is controlled by a knee bolster and a specially designed seat pan.

The first production automatic seatbelt installation in a car produced in the United States appeared in the Chevette manufactured by the Chevrolet Division of General Motors in Model Year 1978 1/2. It was quite similar to the VW installation in the Rabbit, except that it used a seat of more conventional design and included a manual lap belt. Figure 2 shows the conceptual design, with an automatic 2-point shoulder belt, a knee bolster, and a manual lap belt. Figure 3 shows the system as installed in the Chevette.

The next production installation appeared in the Chevette in Model Year 1980. This was a 3-point system with an inboard retractor and two fixed anchor points on the door. Figure 4 shows the conceptual design of this system. A buckle on the lower anchor on the door was provided for emergency release. This buckle also incorporated a switch to provide an electrical connection with a buzzer to insure that the buckle was engaged during normal use. Figure 5 shows the system as installed.

A more recent automatic system was installed in the Toyota Cressida in Model Year 1981 and subsequent years. It is a motorized 2-point system with an inboard emergency-locking retractor and an outboard anchor location on the B-pillar. Figure 6 shows the conceptual design of the Toyota system. A manual lap belt is included to control lower body movement. The lower portion of the instrument panel provides a knee bolster as well. The motorized outboard anchor point is driven along the upper portion of the B-pillar, along the roof rail, and the upper portion of the A-pillar. This action moves the shoulder belt forward to the A-pillar for easy occupant ingress into the vehicle, and then returns it to the B-pillar to don the shoulder belt on the occupant after he is seated and the door is closed. The traverse of the belt has been carefully timed to move the belt quickly, yet not so quickly as to startle the occupant.

CURRENT INSTALLATIONS OF AUTOMATIC SEATBELTS

In addition to the automatic seatbelt systems installed in cars prior to Model Year 1987, more systems will appear in Model Year 1987 in selected car models from all manufacturers producing cars for sale in the U.S.A. as a result of the federal requirement in MVSS 208 (6). The seatbelt installations include motorized and non-motorized systems (Table 4).

The systems installed by the Ford Motor Company in their Ford Escort and Mercury Lynx car models closely follow the Toyota system concept. Figure 7 shows the installed system.

The systems adopted by General Motors for installation in their Oldsmobile Calais and Delta 88 models are three-point systems having separate emergency-locking retractors for lap belt and shoulder belt mounted within the door and a fixed inboard buckle end, as shown in Figure 8. Figure 9 shows the system as installed in an Oldsmobile Delta 88 Sedan.

DESIGN CONSIDERATIONS FOR AUTOMATIC SEATBELT SYSTEMS

The general design considerations for all seatbelts are shown in Table 5. The list includes performance in accidents and evasive maneuvers, simplicity in design and operation, reliability, durability, vulnerability to damage during normal use or in accidents that could render them inoperable, comfort during use, convenience in use, packaging size and shape for accommodation in the vehicle, esthetics, and cost. Automatic seatbelts have the added requirement for automatic donning and doffing.

In any given system design, tradeoffs inevitably are encountered such that optimization of one variable is accompanied by some sacrifice in optimization of another. Some obvious examples are (1) the possible degradation of system performance in emergencies for the sake of increased occupant comfort in normal usage by the incorporation of a tension-relieving device on the shoulder belt, and (2) the possible decrease in simplicity and reliability for the sake of increased convenience and improved stowage through the use of retractors.

USAGE AND PERFORMANCE

The usage rates and performance of automatic seatbelts already in the field have been quite satisfactory. Figure 10 shows the usage rates for automatic systems as compared with the usage rates for manual belts, representing data obtained in the ongoing program sponsored by NHTSA in which usage rates are monitored in 19 cities across the United States. Average driver usage of manual belts in these 19 cities increased from 13.3 percent in 1983 to 14.4 percent in 1984 and 21.4 percent in 1985, with 23.3 percent in the last half of 1985. The upward trend was due mainly to the intensive efforts of NHTSA and others in the safety community to increase seatbelt usage in the United States and the increasing number of mandatory use laws adopted by states. Usage rates varied appreciably from one city to another in the 19-city study, with a low rate in 1985 of 11.1 percent and the highest rate of 46.3 percent.

The usage rate for the VW system has persisted in the range of 70 to 80 percent through the years since 1975, when it was introduced, with observed usage rates of 75 percent in 1983 and 76 percent in 1984 and 71 percent in 1985. The Toyota Cressida system has done even better, with observed usage rates of 95 percent in 1983, 97 percent in 1984, and 92 percent in 1985.

Performance data for these systems in accidents are limited because of the relatively short time they have been exposed on the roadways and the relatively small number involved in accidents. The data acquired to the present through the National Accident Sampling System (NASS) operated in the United States by NHTSA indicate the performance of these automatic systems in the real world is generally comparable to the performance of manual systems.

THE FUTURE OF AUTOMATIC SEATBELTS

The key to the need for automatic belts is the usage rate of seatbelts. In the many countries and jurisdictions around the world having mandatory seatbelt use laws, starting with Australia in 1970, usage rates of 75 to 95 percent are achieved with manual seatbelt systems when continuing enforcement is practiced and fines are imposed. It is reasonable to assume that comparable usage rates can be achieved in the states in the U.S. that have mandatory usage laws, provided that appropriate enforcement is practiced and fines are imposed.

Automatic seatbelts are not inherently better performers than manual belts, but they do appear to have induced higher usage rates when installed as extra-cost options. The public acceptance and usage of automatic seatbelts when installed as standard equipment remains to be seen. Automatic seat belts can be defeated, and their presence in the car does not guarantee their being used by car occupants. The degree of success in enacting and enforcing mandatory seatbelt usage laws in the United States will determine whether automatic seatbelt systems will be preferred in the long term.

ACKNOWLEDGEMENT

The author acknowledges with thanks the support provided by the American Seat Belt Council in the preparation of this report, and the cooperation of vehicle manufacturers and ASBC member companies in providing information and illustrations.

REFERENCES

1. 15 USC 1391 et seq., National Traffic and Motor Vehicle Safety Act

2. U.S. Patent 2,477,933 (August 2, 1949) - Collision Mat for Vehicles

3. U.S. Patent 2,858,144 (October 28, 1958) - Safety Belt for Vehicles

4. 49 CFR 571.208 Standard No. 208; Occupant Crash Protection

5. 35 FR 16927 - Docket 69-07; Notice 7 - Occupant Crash Protection

6. 51 FR 9800 - Docket 74-14; Notice 43 - Occupant Crash Protection

INFLATABLE

NON-INFLATABLE

- TRANSPARENT SHIELDS
- DEPLOYABLE NETS
- DEPLOYABLE BLANKETS
- SEAT-INTEGRATED SYSTEMS
- RESTRAINING ARMS
- CUSHIONS AND BOLSTERS
- SEATBELT SYSTEMS

Table 1 - Automatic restraint systems

- 2-POINT* — AUTOMATIC SHOULDER BELT ONLY
- 3-POINT* — AUTOMATIC LAP AND SHOULDER BELTS
- 4-POINT* — SEPARATE AUTOMATIC LAP AND SHOULDER BELTS

* MECHANICAL AND MOTORIZED MODELS IN ALL TYPES

Table 2 - Automatic seatbelt system types

CAR LINE	MODEL YEAR INTRODUCED	TYPE
VW RABBIT	1975	2-POINT
GM CHEVETTE	1978 1/2	2-POINT
GM CHEVETTE	1980	3-POINT
TOYOTA	1981	2-POINT -MOTORIZED

Table 3 - Pre-1987 automatic seatbelt systems

MANUFACTURER	CAR LINE
MOTORIZED SEATBELTS	
ALFA ROMEO	SPYDER–QUADRIFOGLIO
CHRYSLER–PLYMOUTH	CONQUEST
FORD	FORD ESCORT MERCURY LYNX
MAZDA	626 4-DOOR SEDAN
MITSUBISHI	STARION
NISSAN	MAXIMA
SAAB	900 S 3-D
SUBARU	XT COUPE
TOYOTA	CRESSIDA CAMRY
SEATBELTS (NOT MOTORIZED)	
AMERICAN MOTORS	ALLIANCE L AND DL
CHRYSLER –PLYMOUTH	LE BARON COUPE DODGE DAYTONA
GENERAL MOTORS	BUICK SOMERSET BUICK SKYLARK BUICK LE SABRE PONTIAC GRAND AM PONTIAC BONNEVILLE OLDSMOBLE CALAIS OLDSMOBLE DELTA 88
HONDA	ACCORD HB
VOLKSWAGON	GOLF JETTA

Table 4 - Model Year 1987 automatic seatbelt Systems

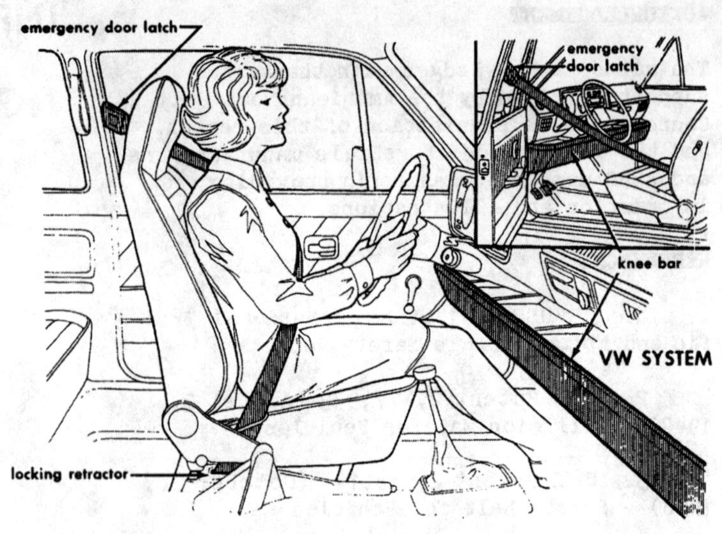

Fig. 1 - First automatic seatbelt system in production car line

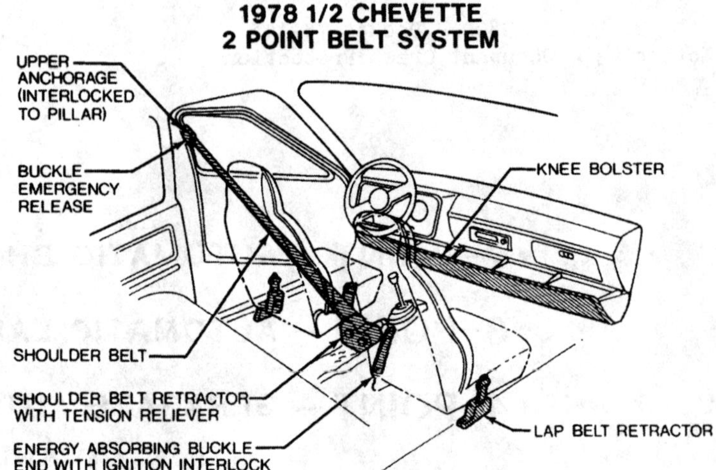

Fig. 2 - First automatic seatbelt system in U. S. domestic production car line

Fig. 3 - 1978 1/2 Chevette automatic seatbelt installation

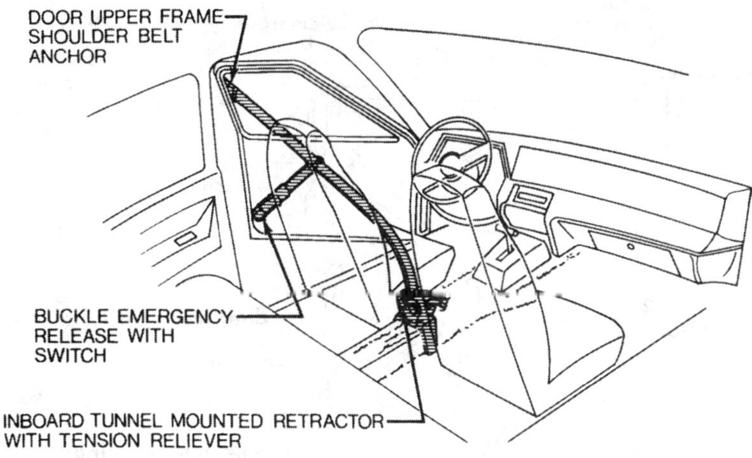

Fig. 4 - Second generation automatic seatbelt system in U. S. domestic production car line

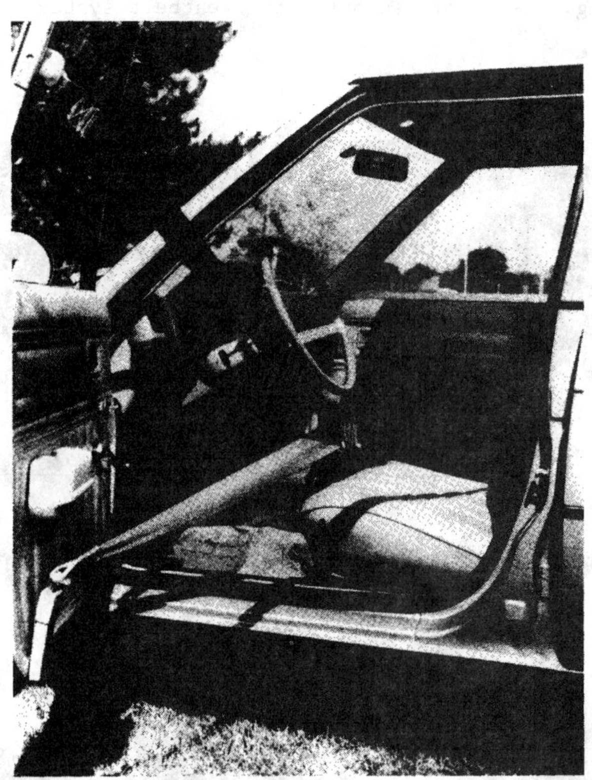

Fig. 5 - 1980 Chevette automatic seatbelt installation

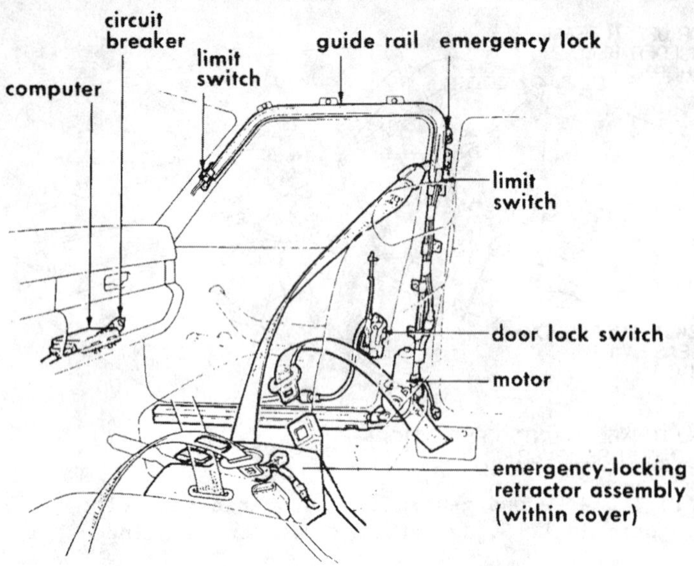

Fig. 6 - Motorized automatic seatbelt system

1987 OLDSMOBILE CALAIS PASSIVE RESTRAINT

THREE POINT BELT SYSTEM

UPPER GUIDE LOOP

INTERLOCK TO PILLAR OR QUARTER

SHOULDER BELT RETRACTOR

BELT SNORKLE

EMERGENCY RELEASE BUCKLE

LAP BELT RETRACTOR

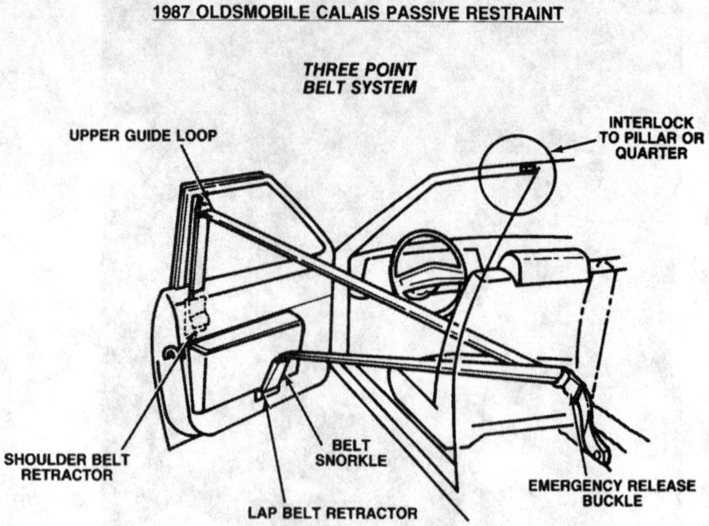

Fig. 8 - 1987 Door-mounted 3-point automatic seatbelt system

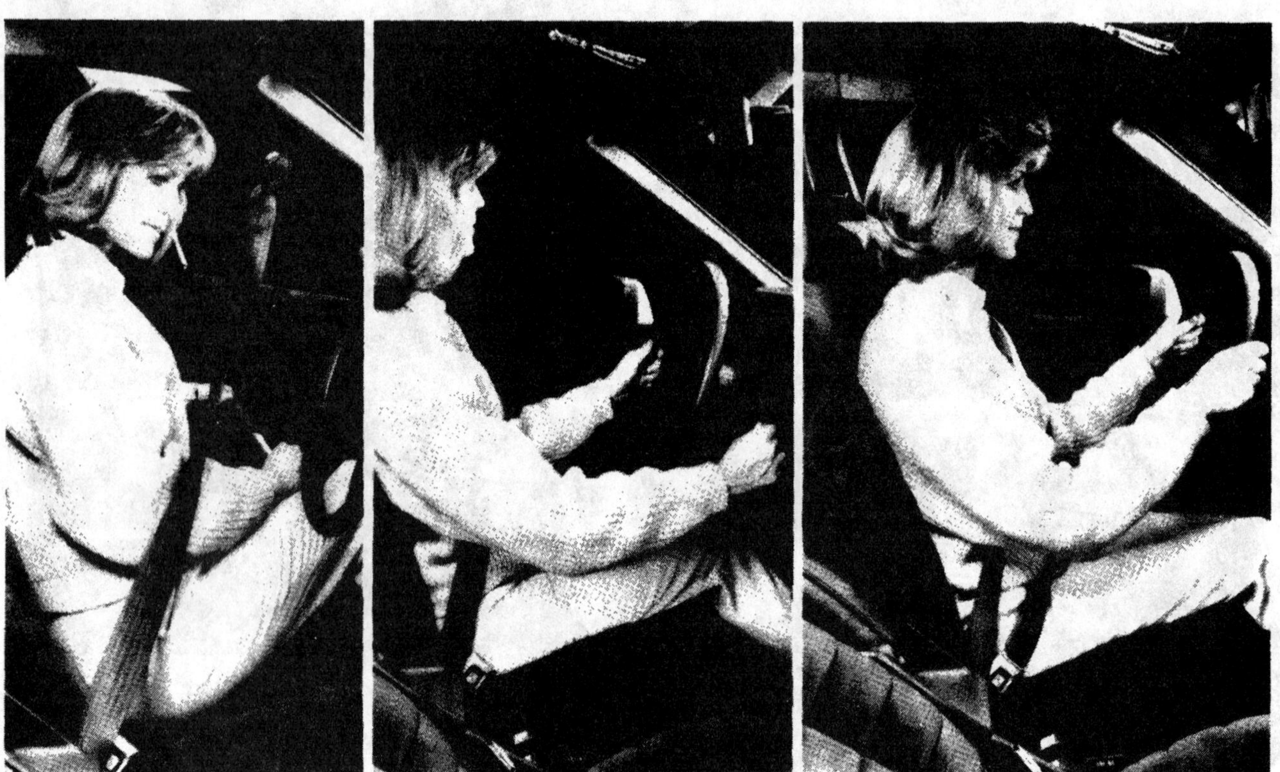

Fig. 7 - Ford automatic seatbelt installation

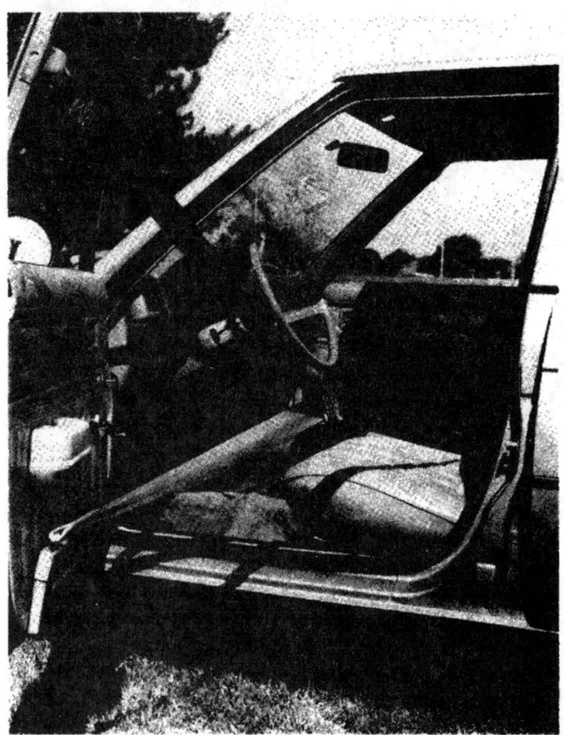

Fig. 9 - General Motors automatic seatbelt installation

MANUAL SEATBELT SYSTEMS

- PERFORMANCE IN EMERGENCY EVENTS
- RELIABILITY
- DURABILITY
- VULNERABILITY TO DAMAGE
- VULNERABILITY TO DELIBERATE OR INADVERTENT TAMPERING
- SIMPLICITY IN OPERATION
- COMFORT IN NORMAL OPERATION
- CONVENIENCE IN NORMAL OPERATION
- ACCEPTABLE PACKAGING IN CAR
- ESTHETIC CONSIDERATIONS
- COST

AUTOMATIC SEATBELT SYSTEMS

- ALL OF THE CONSIDERATIONS LISTED ABOVE
- AUTOMATIC DONNING AND DOFFING

Table 5 - Seatbelt design considerations

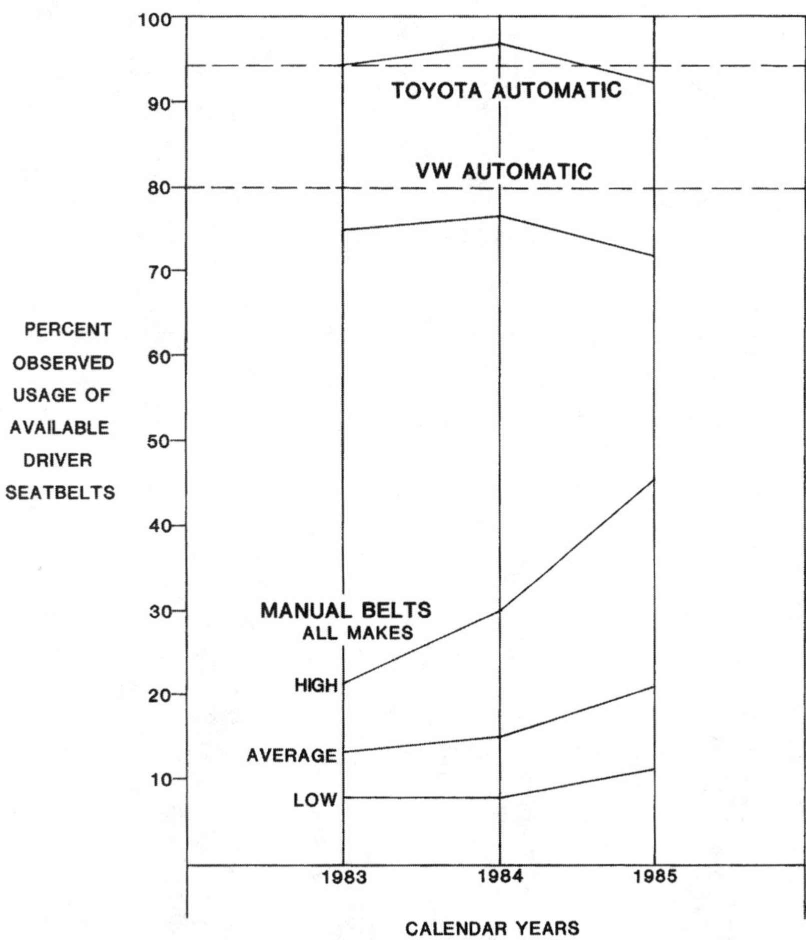

RECENT DRIVER SEATBELT USAGE RATES

Fig. 10 - Recent driver seatbelt usage rates

CHAPTER 14:

ABTS (ALL BELTS TO SEAT)

860053

Seat-Integrated Safety Belt

Heinz P. Cremer
Keiper Recaro GmbH & Co.

ABSTRACT

In the case of fronttal or rear collision, the vehicle seat and safety belt act as a retaining system that is supposed to protect the person and prevent injury as much as possible. A double shoulder belt (harness belt) was integrated into the seat as a 4-point belt in order to examine the possibilities of improving personal protection. Due to the transfer of the test requirements for the safety belt to the seat with integrated safety belt, the load level in the direction of motion increases considerably. The normal test conditions were increased by using a 95% dummy rather than a 50% dummy. According to these requirements, a seat back with scat back adjuster was designed, built, and tested.

The test showed that by proper deformation of the seat back, the load on the person is reduced and that the seat with integrated 4-point belt provides an improvement of the retaining system for the frontal as well as the rear crash.

Seat with Double Shoulder Belt (Harness Belt) - Various examinations showed that a belt system with a double shoulder belt, i.e. a 4-point belt, provides better protection than the currently used diagonal belt.

The goal of this development was to examine:

- whether there are further advantages when a 4-point belt is completely integrated into the vehicle seat.
- whether an acceptable seat back structure with seat back adjuster can be built, that can take the increased load.
- whether the seat back structure can be designed in such a way that the energy will be changed by proper shaping to achieve a lower load level than in other retaining systems with comparable testing conditions.

Not included in this report is the highly resistant seat track and swivel mechanism necessary for belt integration.

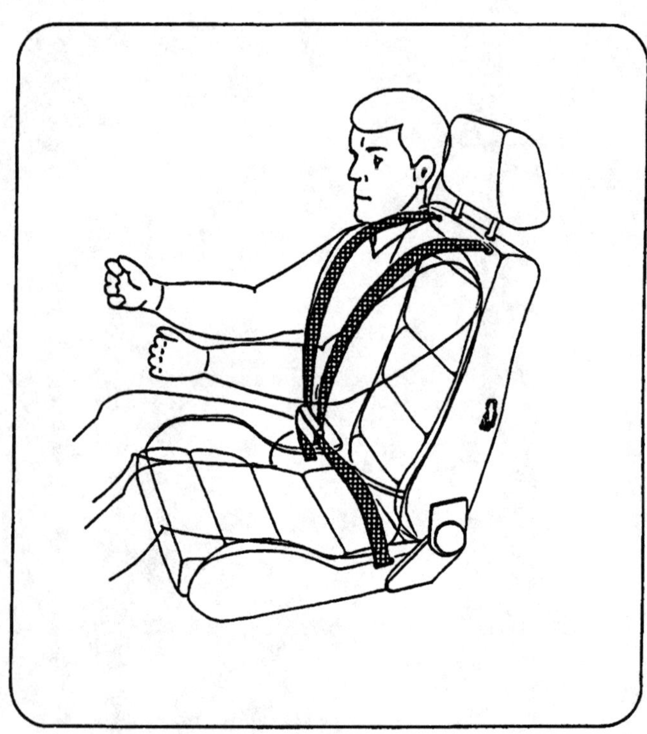

Seat integreated safety belt

Since the maximum protection of the person is at hand, it was accepted that:

- up to this day the 4-point belt is only used in motor racing sport vehicles, i.e. not common in assembly line production.
- the request for a one-hand buckling operation of the belt has not been met satisfactorily yet.

Generally, it can be stated that the 4-point belt has proven itself under extreme conditions in the motor sport for years.

The issue was not to determine which 4-point belt system is the most appropriate, since this was already done elsewhere.

Driving and Sitting Comfort of Seat with Belt - In order to provide drivers of various heights with good vision from the vehicle, the vehicle seats are equipped with a height adjuster. The currently used seat height adjusters primarily adjust the entire seat in height.

For the seat-integrated safety belt system, however, it is of advantage to adjust the height of the seat cushion next to the stationary back.

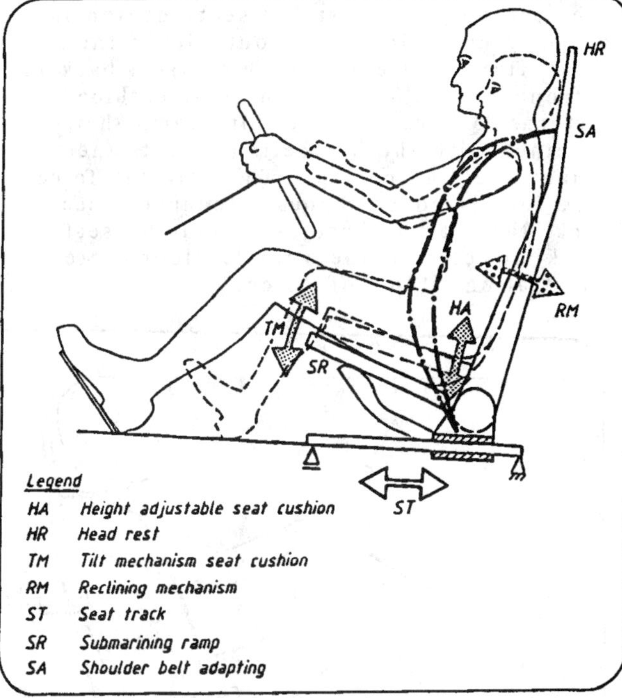

Height adjustable seat cushion

The height-adjustable seat cushion achieves:

- the actual purpose of adjusting the visual point of people of various heights.
- an adjustment of the shoulders of people of various heights to the level where the seat belt emerges from the seat back.
- an improved positioning of the head in relationship to the headrest.
- an adjusted location of the seat belt lock, since it is appropriately attached to the upper part of the seat track.

The seat-integrated 4-point safety belt achieves:

- improved wearing comfort provided by better belt location for people of various heights.
- improved wearing comfort for men and women provided by symmetrical belt location.
- decreased and more even pull-out and retraction force of the belt due to smaller change in direction and shorter belt.
- improved comfort when putting on safety belt, since the parts of the belt are always in the same position and always easily reachable, independent from the position of the seat.
- symmetrical and, therefore, natural movement of putting the belt on, however, with both hands.
- improved ease of getting in and out of the vehicle, especially for the rear passengers in 2-door automobiles.
- reduction in the number of anchor points for seat and safety belt, less robot mounting.
- new design of the floor pan for the different force introduction only by the seat.
- more freedom of design in the area of the B-column.
- no problem when used in vehicles without B-column; convertible, sliding doors, etc.

Better Crash Record is Provided by Seat with Belt - Seat and safety belt are the systems that support and hold the person in an accident. Even though the position of the person in a vehicle is determined by the position of the seat, the safety belt in most cases is still attached to the car body. Increasingly, the safety belt lock is attached to the seat or moved back and forth with the seat movement in order to reach a better position of the person, seat and safety belt at this point.

At an accident, the "slack", i.e. the distance a body can move freely before he is caught by the restraining system, seat or safety belt, is of great importance. The greater the slack, the greater the impact on the body. See in this aspect the "Crash Dynamic" illustration on the following page.

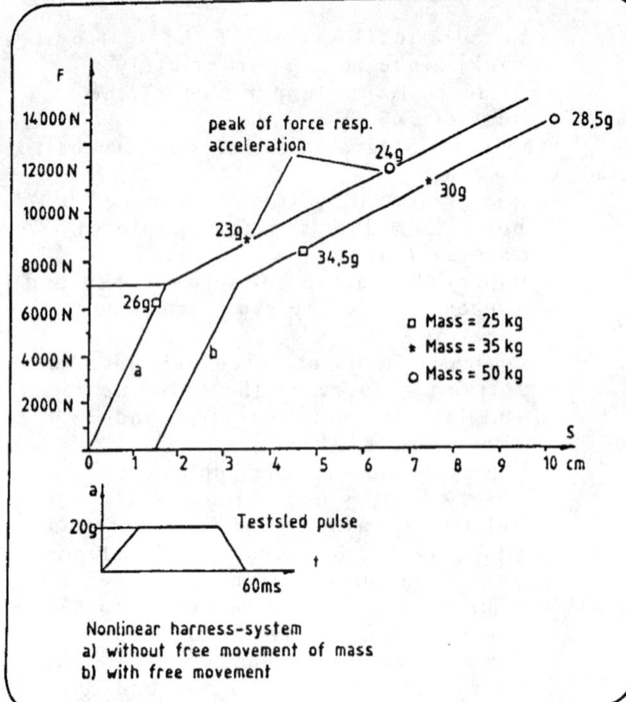

Crash Dynamic

Today, floor pans are designed in a way that they include:

- anchoring points for the seat and
- anchoring points for the safety belt.

For the seat with the integrated safety belt the anchoring points can be reduced to the anchor points for the seat. This, however, due to the greater force, requires a new concept of the floor pan in the area of the front seats.

The following features are characteristics for a seat with a 4-point integrated safety belt:

FRONTAL CRASH

- symmetrical force on the body, therefore, no torsion.
- good position of the shoulder belts for people of different heights.
- good positioning of the hip belt, since the belt lock as well as the belt anchoring point is attached to the upper part of the seat track.
- less surface pressure on the body due to two shoulder belts.
- shorter people sit higher due to the height-adjustable seat cushion, so that the lesser weight acts on a larger lever
- the slack is less due to the closed position of the body, seat and safety belt and can be minimized even more with a belt tightener.
- less elongation of the belt strap due to stretching, since the total length of the belt is shorter due to the integration of the belt roller in the seat back.
- less elongation of the belt strap due to film spool effect, since a lesser belt length has to be on the spool.
- energy change by the seat back is greater than by the B-column.
- earlier retention of the body by the forward-shaped back during rebound.

REAR CRASH

- shoulder belts resist unwanted upward movement of the body.
- timely retention of the body during rebound.

DIAGONAL OR SIDE IMPACT

- the body is well retained by the closed retaining system seat - safety belt.

<u>The Highly Load-Resistant Seat Back</u> - A larger force acts upon the seat structure when the shoulder belts are attached to the upper edge of the seat back as compared to the force created only by the rear crash (when the belt is attached to the vehicle body). For this reason the construction of the seat structure is to be designed in such a way that as few structural elements as possible are located in the high force flux. This, however, should not impair the functions that belong to a modern seat. Here too, the height-adjustable seat cushion is advantageous, since it is outside of the force flux, that comes from the seat back in a front or rear crash. The seat cushion just has to have a submarining ramp that, together with the hip belt, prevents the plunging of the person. This way the force flux in this concept goes through the seat back, the seat back adjuster and the seat track directly in the vehicle floor. See illustration "Flux of force."

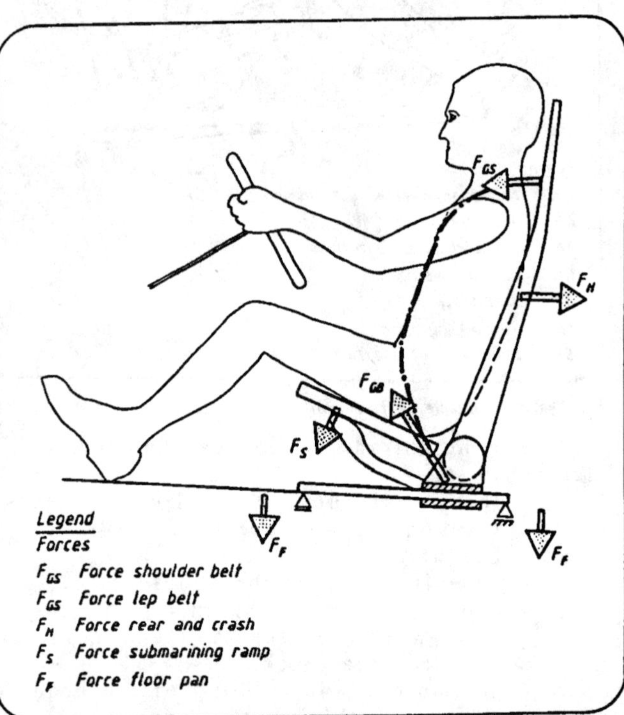

Flux of force

Seat Back Adjuster - The seat back also includes a seat back adjuster that is suitable for the task and the high force. Principally, an adjuster based on a gear (gear type reclining mechanism) is suitable for the task, since the flux of force is not interrupted, even during adjustment. This requirement of being locked constantly seems to be inevitable for the belt that is attached to the seat back and is fulfilled by the TAUMEL seat back adjuster (TAUMEL reclining mechanism).

Calculation of the Seat Back Structure - In the seat back structure with integrated safety belt, the force passes through the belt or the back and exits through the seat back adjuster into the seat track. The shorter the distance of the force, the better the problem solution. In order to find an optimal structure for the seat back with integrated safety belt, calculation procedures for stability calculations, FEM and a crash simulation were used. The symmetrical force on the structure due to the double shoulder belt is advantageous. However, the non-symmetrical force during a diagonal impact and the side stability during the side impact were not neglected.

For series production, the structure of the seat has to be adapted to the deforming characteristics and the delay impulse of the floor pan as well as the available area of the respective vehicle; this achieves a minimum force on the body, while taking advantage of the forward movement of the person in the vehicle.

Vehicle, seat, and person are a dynamic system that can lead to optimal results only when all components are taken into consideration.

For the crash simulation an 8-mass dummy model was used. Random impulses can be entered, as a normal signal or acceleration impulses of floor pans of certain vehicles, as well as the retaining outlines of the seat back and safety belt. See illustration in the next column.

According to the entered force, the following can be calculated:

- forces of contact to the seat (max. 6)
  - - force on the submarining ramp
- the belt forces
- the acceleration of head, chest, and pelvis
- base forces at the floor pan

The simulation can be shown on the screen in motion so that the deformation of the structural parts as well as the movement of the dummy can be observed. See Crash Simulation illustration in the next column.

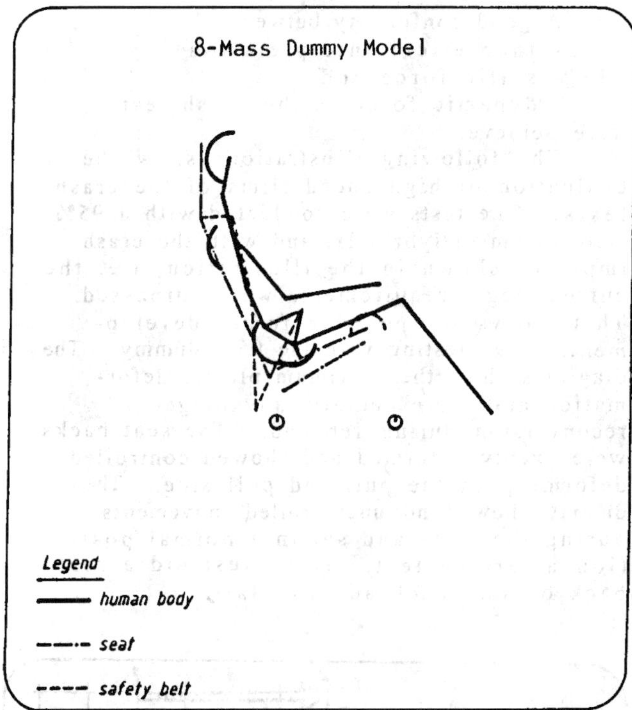

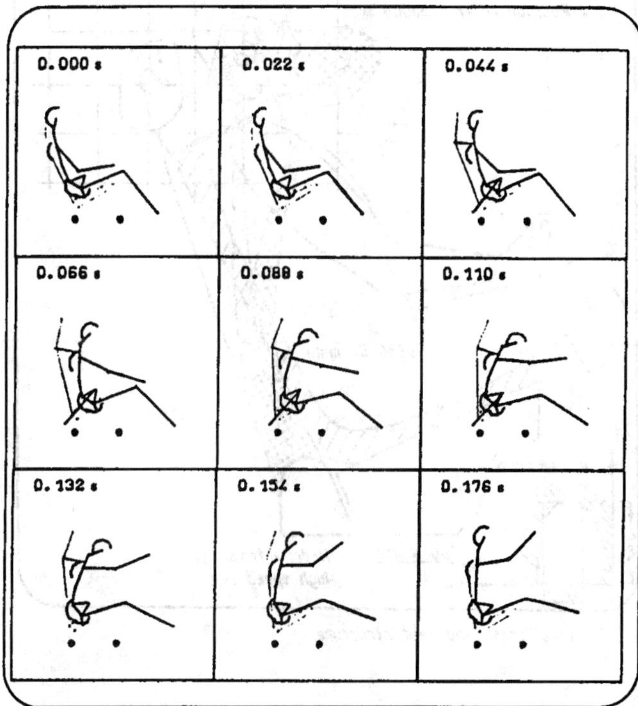

Crash Simulation

Test Results - According to the previous mentioned criteria and requirements, prototypes of the seat back structures with seat back adjusters of the TAUMEL type and seat-integrated safety belt were built and tested. The seat structure was mounted firmly onto the test fixture.

A good conformity between

- theoretical interpretation
- static force and
- dynamic force in the crash test

were achieved.

The following illustrations show the evaluation of high speed films of the crash tests. The tests were conducted with a 95% male dummy (Hybrid II) and with the crash impulses shown in the illustration, i.e. the current legal requirements were surpassed. This allows for possible future developments, e.g. testing with the 95% dummy. The diagrams show the maximum plastic deformation and the especially advantageous reconversion during rebound. The seat backs were evenly deformed and showed controlled deforming on the pull and pull side. The dummy showed no uncontrolled movements during the tests and sat in a normal position after the test. In no test did a seat back or seat back adjuster fail.

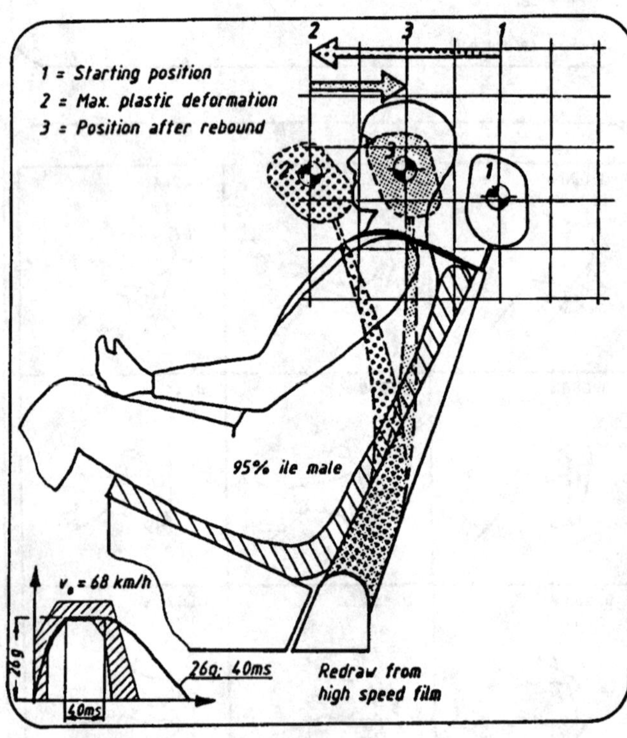

Frontal crash and seat structure

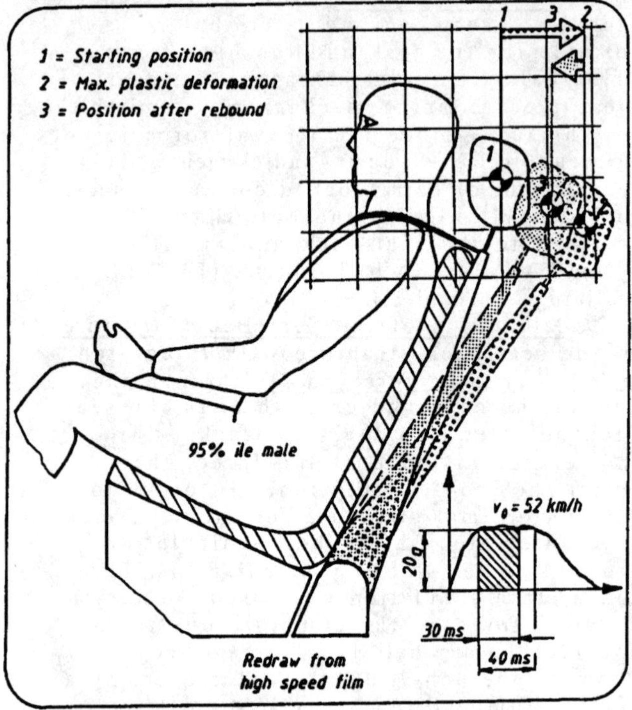

Rearend crash and seat structure

The improved protection of the person becomes clear when observing the process of the deformation of the seat back. From the original position 1) the seat back is deformed forward or backward by the belts or the upper body. At the end of this force the body is moved in the other direction, so that the force is taken off the seat back. The elastic deformation disappears and the seat back takes the position of the maximum plastic deformation 2). In the following instant, the seat back is being deformed into the final position 3) by the remaining energy from the rebound. This timely retention of the body during rebound is only possible with the seat-integrated 4-point safety belt.

The weight of the seat back with seat back adjuster is 25% to 50% higher than a conventional seat back, whereby, however, a 300% to 500% higher load capacitance is achieved. It is to be expected that an acceptable increase in the total weight will be necessary for a uniform layout of seat, safety belt, floor pan and B-column, whereby, however, simultaneously a considerably better protection of the person is achieved by the seat-integrated safety belt. Should only the currently required test conditions be used, the above mentioned weight can be reduced.

Comparison Values - For comparison, the following were tested:

- seat with integrated 3-point belt
- 3-point belt with shoulder belt attached to B-column

The illustration "Different seat belt systems" shows the measured results. A distinct advantage for the seat-integrated 4-point safety belt can be recognized.

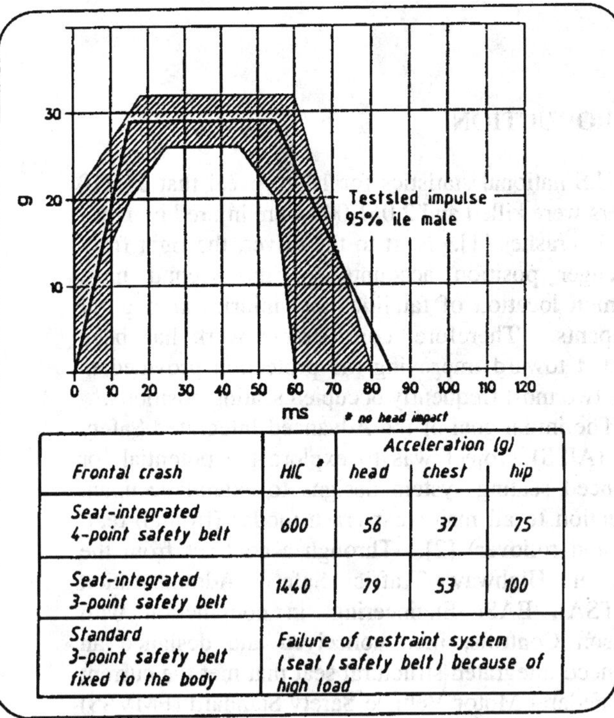

Frontal crash	HIC *	Acceleration (g) head	chest	hip
Seat-integrated 4-point safety belt	600	56	37	75
Seat-integrated 3-point safety belt	1440	79	53	100
Standard 3-point safety belt fixed to the body	Failure of restraint system (seat / safety belt) because of high load			

Different seat belt systems

ANALYTICAL EVALUATION OF AN ADVANCED INTEGRATED SAFETY SEAT DESIGN IN FRONTAL, REAR, SIDE, AND ROLLOVER CRASHES

Mostafa Rashidy, Balachandra Deshpande, Gunasekar T.J., Russel Morris
EASi Engineering
Robert A. Munson, Jeffrey A. Lindberg
Johnson Controls, Inc.
Lori Summers
National Highway Traffic Safety Administration
United States of America
Paper No. 305

ABSTRACT

Analytical computer simulations were used to optimize and fabricate an Advanced Integrated Safety Seat (AISS) for frontal, rear, side, and rollover crash protection. The AISS restraint features included: dual linear recliners, pyrotechnic lap belt pretensioner, 4 kN load-limiter, extended head restraint system, rear impact energy absorber, seat-integrated belt system, and side impact air bag system. The evaluation and optimization of the AISS design was achieved through analytical simulations using MADYMO multi-body analysis software, LS-DYNA3D finite element software, and through LS-DYNA3D/MADYMO coupling. Frontal and rear impact sled tests were also conducted with physical AISS prototypes and baseline integrated seats to verify performance.

Both the analytical modeling and the experimental sled testing demonstrated safety improvements over the baseline integrated seat. The AISS pyrotechnic lap belt pretensioner and 4 kN load-limiter contributed to a 26 percent reduction in occupant chest acceleration in the frontal impact mode. In the rear impact mode, the AISS dual linear recliners, rear impact energy absorber and extended head restraint system contributed to reducing the occupant upper neck injury parameters. Full vehicle finite element models were used in both the side impact and rollover simulations to evaluate occupant restraint performance. Two generic AISS restraint features were modeled for side impact protection: an inflatable tubular cushion air bag system and a combination head/thorax side impact air bag system. The combination head/thorax side impact air bag system model was found to provide improved occupant protection due to its ability to cover both head and thorax regions and provided a softer reaction surface for the occupant. Upper and lower rib Thoracic Trauma Index (TTI) were reduced 22.1 percent and 14.8 percent, respectively in the side impact simulations. The AISS extended head restraint, pyrotechnic lap belt pretensioner, and seat-integrated belt system also provided benefit in restraining the occupant and minimizing roof crush in the rollover simulations.

INTRODUCTION

U.S national statistics for 1998 reveal that 25,210 drivers were killed and 2,061,000 were injured in motor vehicle crashes [1]. Next to the driver, the right front passenger position accounts for the second most common location of fatalities and injuries among car occupants. Therefore, considerable work has been directed toward improving the protection provided in these two most frequently occupied seating positions.

The initial goal of the Advanced Integrated Safety Seat (AISS) project was to explore the potential for advanced seating system designs to extend occupant protection to all major car crash modes (frontal, rear, side and rollover) [2]. Through a contract from the National Highway Traffic Safety Administration (NHTSA), EASi Engineering, in conjunction with Johnson Controls, Inc., conceived and designed an advanced integrated structural seat that met the current U.S. Federal Motor Vehicle Safety Standard (FMVSS) requirements and improved occupant protection for frontal, rear, side, and rollover crashes and contributed to passenger compartment intrusion resistance.

By focusing on the seat structure modifications, it was thought that the resulting design would be simple and cost efficient. Previous studies have shown that it is possible to develop an integrated structural seat to provide the improved occupant protection and comfort of an integrated seat belt system without the associated cost and weight increases of more complex systems [3]. Therefore, an integrated structural seat was chosen as the baseline seat in the program. Integrated seats have the belt anchorages on the seat itself as opposed to conventional seats where the shoulder belt upper anchorage is located on the car upper body structure. Therefore, the belt fit is considerably improved regardless of the seating position, and the assembly of the seat in the car becomes much easier with this design as the belts are part of the seat. The AISS seat system is also designed to function with the body structure to resist passenger compartment intrusion in side and rollover crashes.

This paper reports on the results of analytical modeling conducted in the frontal, rear, side, and rollover crash modes, and frontal and rear impact sled tests conducted with AISS prototypes.

RESULTS AND DISCUSSION

Frontal Impact

In the U.S., frontal crashes are the most significant cause of motor vehicle fatalities [1]. Research has found that lap/shoulder belts, when used properly, reduce the risk of fatal injury to front seat passenger car occupants by 45 percent and the risk of moderate-to-critical injury by 50 percent [4]. For light truck occupants, seat belts reduce the risk of fatal injury by 60 percent and moderate-to-critical injury by 65 percent [4]. NHTSA also estimates that drivers protected by air bags experience a reduced fatality risk of 19 percent in all frontal crashes [5]. However, even after full implementation of driver and passenger air bags by FMVSS No. 208, it has also been estimated that frontal impacts will still account for over 8,500 fatalities and 120,000 moderate-to-critical injuries each year [6]. Therefore, the AISS project explored potential seat enhancements that could improve occupant protection beyond the current FMVSS.

AISS Features for Frontal Impact: The AISS design has the following countermeasures for improved occupant crash protection in frontal impacts: pyrotechnic lap belt pretensioner, 4 kN load limiter, integrated seat belt system, and extended head restraint system.

In a frontal impact crash, the pyrotechnic lap belt pretensioner takes up slack in the seat belt and induces energy absorption during the early forward travel of the occupant. When the device is fired, the pretensioner pulls down on a cable that is attached to the belt buckle. This effectively removes the extra slack, and cinches the occupant to the seat. Some studies have reported that this also prevents submarining by narrowing the opening for the pelvis to slide through [7].

A 4kN load limiter was included in the AISS design to reduce the belt loads on the chest while maintaining enough restraint to keep the occupant within the compartment during side and rollover crashes. The upper anchorage of the torso belt on the seat back structure of current integrated seats is the greatest source of seat back bending moment and shear load on the seat structure. By limiting the torso belt loads on the integrated seat, it allows a weight reduction of the seat back structure and reduced floor pan shear while also reducing occupant injuries in frontal crashes. The load limiting retractor of the AISS is designed to let the belt spool out of the retractor at a prescribed load of 4 kN. The 4 kN load limiting designation was based on biomechanical analyses conducted by Kallieris, et. al. [8]. In the AISS design, the retractor is fitted with a torsion bar to provide the load limiting feature.

Analytical Modeling: A coupled LS-DYNA3D/MADYMO model of the AISS was developed and used for design optimization and countermeasure evaluation. The AISS model consisted of the following major structural assemblies: seat pan, seat track, retractor housing, seat belt, recliner, and energy absorber. Figure 1 provides four illustrations of the analytical AISS model. Details of the model have been reported in [3, 9, 10].

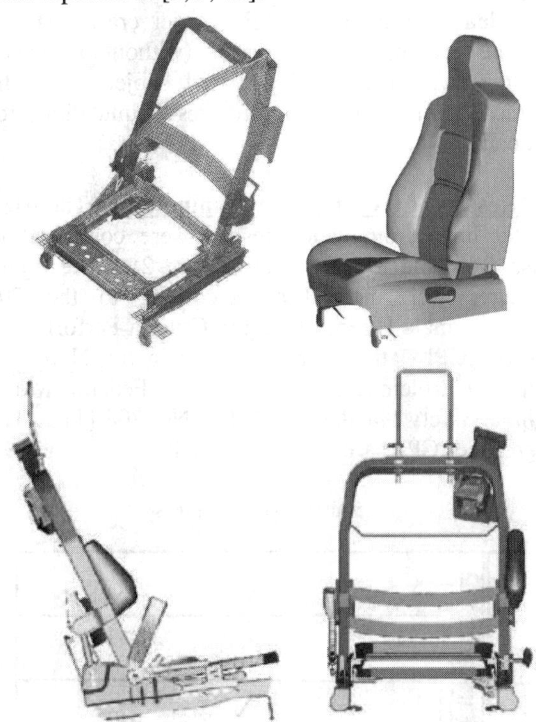

Figure 1: Analytical model of the Advanced Integrated Safety Seat (AISS).

For the frontal impact simulation, the MADYMO Hybrid III 50th percentile male dummy model was coupled with the AISS model. The seat belt was modeled and fitted around the dummy using EASi-CRASH software and the belt was anchored at the appropriate locations. The slip ring feature in MADYMO was used to simulate the seat belt sliding at the buckle location. The seat cushion, seat back foam, and knee bolster were modeled as solid elements. The floor pan and toe board assemblies were modeled as shell elements. The finite element geometry of the seat model was based on IGES data provided by Johnson Controls, Inc.

The material and section properties were also supplied by Johnson Controls, Inc. [9]. The welded structural components of the AISS were modeled using nodal rigid bodies. Bar elements were used to simulate the bolts/pins between components. Sliding contact surfaces were defined between other finite element parts. Geometric contact entities were used to simulate the contact interaction between the seat structure and the dummy. A 0.5 coefficient of friction was applied. The seat back frame, the dual linear recliners, and the rear impact energy absorbers were optimized for frontal and rearward loading. For frontal impact, the seat back was designed not to buckle at a force level of 4 kN, plus an additional 10% percent [9].

The AISS analytical model with a belted MADYMO Hybrid III 50th percentile male dummy was subjected to a frontal impact sled test simulation equivalent to a 56 kmph rigid barrier crash test. A baseline integrated seat model (without advanced features) was also developed and subjected to the identical frontal impact sled test simulation for comparison.

Results of the Analytical Simulations: The AISS frontal impact simulation results were compared to those from the baseline seat (Figure 2). The injury measures were normalized according to the 50th percentile male Hybrid III Injury Criteria Performance Levels (ICPLs) that were proposed in the Notice of Proposed Rulemaking (NPRM) for Federal Motor Vehicle Safety Standard (FMVSS) No. 208 [11]. The referenced ICPLs are listed in Table 1.

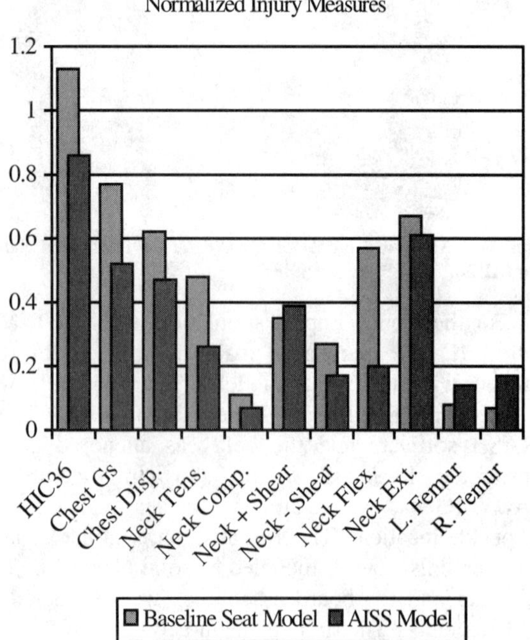

Figure 2: Frontal impact analytical simulation results.

The analytical simulations predicted that almost all of the 50th percentile male injury measures were below the ICPLs for both the AISS and the baseline integrated seat in a simulated 56 kmph rigid barrier crash. The results further showed that the AISS would improve most occupant injury measures for the 50th percentile male dummy over a baseline integrated seat. The greatest improvements in the AISS simulation were the chest acceleration results which decreased 32%, and chest deflection and HIC results which both decreased 24%. These improvements were primarily attributed to the pretensioner and force limiting features of the seat belt.

Table 1. FMVSS No. 208 NPRM Injury Criteria Performance Limits (ICPLs)	**Mid-Sized Male**
Head Criteria: HIC (36 ms)	1,000
Neck Criteria:	
Peak Tension (N)	3,300
Peak Compression (N)	4,000
Fore-and-Aft Shear (N)	3,100
Flexion Bending Moment (Nm)	190
Extension Bending Moment (Nm)	57
Thoracic Criteria	
1. Chest Acceleration (G)	60
2. Chest Deflection (mm)	63
Lower Ext. Criteria:	
Femur Load (N)	10,008

AISS Prototypes: Four AISS prototypes were fabricated for testing in the research program. Two seats were evaluated in frontal impact sled tests and two were evaluated in rear impact sled tests.

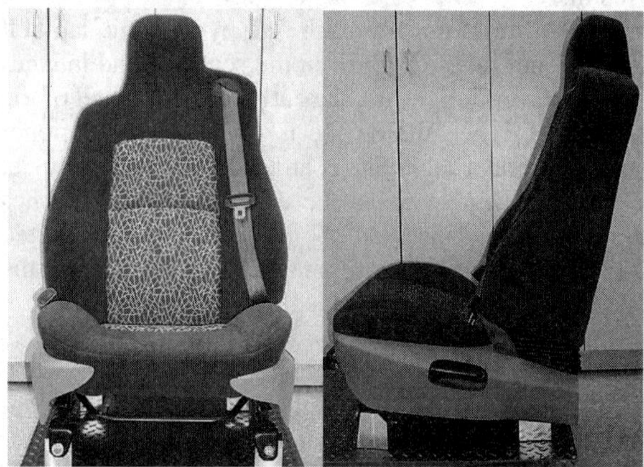

Figure 3: AISS prototype (front and side view).

Quasar Industries conducted the soft tooling and was the part source for all structural metal for the AISS prototypes. Breed Technologies, Inc. provided the off-the-shelf retractors, pretensioners, and side air bag parts. Johnson Controls Inc. assembled the seat structure and also fabricated and assembled the foam and trim. The photos in Figure 3 are of one of the final prototypes.

Sled Testing: The frontal impact sled test setup is shown in Figure 4. An instrumented 50th percentile male Hybrid III dummy was seated on the AISS and restrained by the integrated seat belt system. As in the analytical modeling, the sled test simulated a 56 kmph full frontal rigid barrier vehicle crash. The peak deceleration was approximately 34.5 G and the pulse duration was 85 msec. The crash pulse is illustrated in Figure 5. A production driver side air bag was fired at 20 msec into the simulated crash event.

For comparison, a baseline seat was also tested under the same sled test conditions. The baseline seat was an integrated seat with a single recliner and adjustable head restraint.

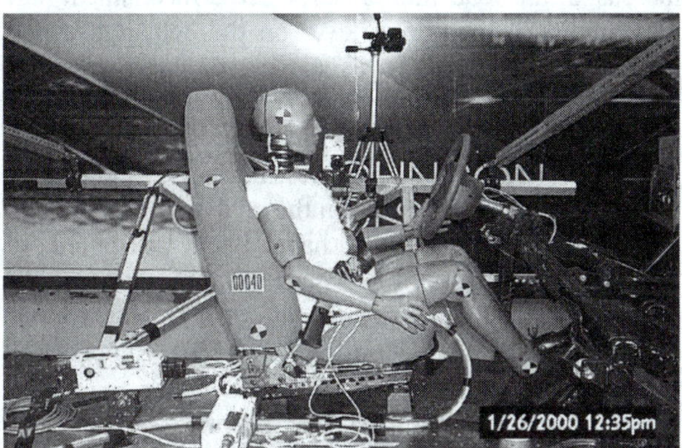

Figure 4: Frontal impact sled test set-up.

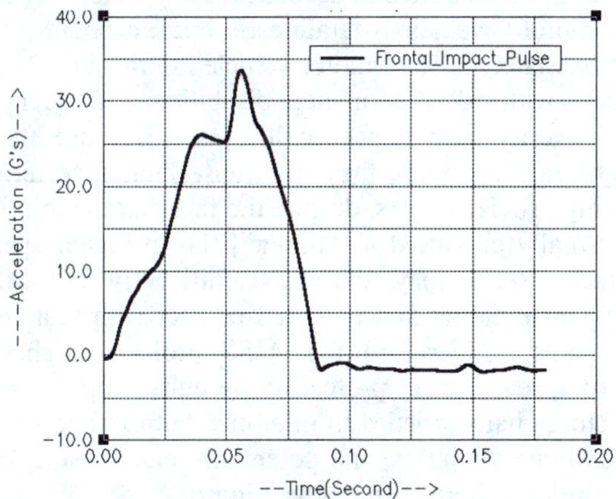

Figure 5: Frontal impact sled test pulse.

Two sled tests were conducted in the frontal impact mode with two of the AISS prototypes (each tested once). In the first AISS sled test, the pretensioner did not function as intended. A metal tab designed to constrain rotation of the unit was bent and slipped away from the slot in the casing. An attempt was made to correct this in the second test. Therefore the normalized injury measures from the second AISS sled test were compared against the results from the baseline seat sled test in the present analysis (Figure 6). As in the frontal impact analytical simulations, most of the 50th percentile male dummy injury measures from the AISS sled test were lower than those from the baseline sled test. Additionally, all of the injury measures were below the ICPLs for the AISS sled test.

Unlike the analytical simulations, the head accelerations were relatively comparable between the baseline seat and the AISS. The peak head acceleration in the baseline seat was 79 Gs and in the AISS it was 80 Gs. The HIC results were also relatively similar; the baseline seat has a HIC result of 923 and the AISS had a HIC result of 911.

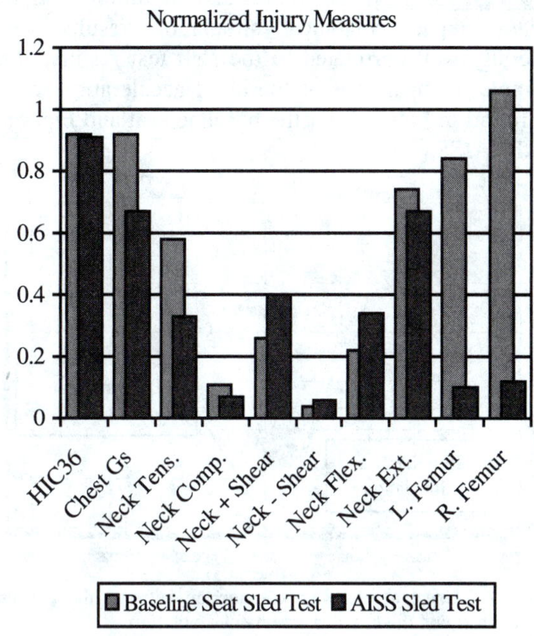

Figure 6: Normalized results from the frontal impact sled tests.

The AISS sled test demonstrated large improvements in peak chest acceleration responses when compared to the baseline seat. (This was also predicted by the analytical simulations.) The dummy in the baseline seat sled test had a chest acceleration result of 55 Gs; whereas the dummy in the AISS sled test had a chest acceleration result of only 40 Gs. This was an improvement of approximately 26%. In the AISS sled test, the load limiter was effective in keeping the chest acceleration levels relatively constant (approximately 40 Gs for 20 msec.). The addition of the pretensioner in the AISS was also able to begin restraining the chest earlier in the event. However post-test inspection of the hardware indicated that the pretensioner had a slight internal malfunction and pulled in only 40 mm of the intended 80 mm of belt. (Thus, further chest acceleration improvements may be expected with full-functionality of the pretensioner.)

Axial neck forces and neck extension moments were also slightly lower in the AISS sled test; whereas neck shear and neck flexion moments resulted in a slight increase. However, all neck injury measures were well below the ICPLs. This is consistent with other frontal impact testing using the 50th percentile male dummy [12]. Low femur loads resulted in the AISS sled test and were potentially due to anti-submarining effects of the buckle pretensioner. Femur loads in the baseline sled test may have been artificially high due to contact with the steel knee bolster support bracket.

Correlation of the Analytical Simulations to the Sled Test Results: The trends and the timing from the frontal impact analytical simulation results were generally well correlated to the sled test results. For example, comparisons of the chest acceleration traces are found in Figure 7 for the baseline seat and Figure 8 for the AISS.

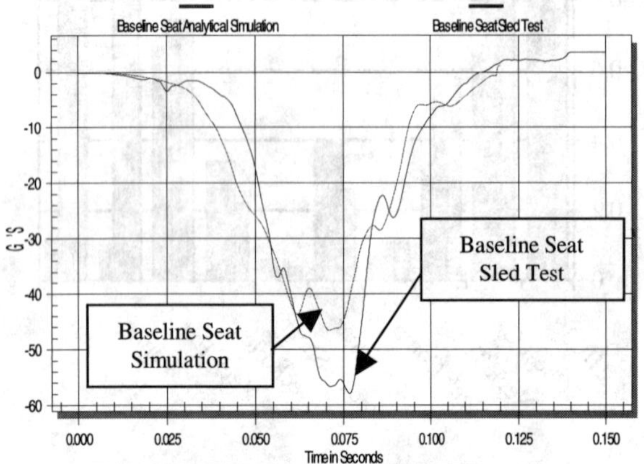

Figure 7: Chest acceleration comparison between the baseline seat sled test and the baseline analytical simulation.

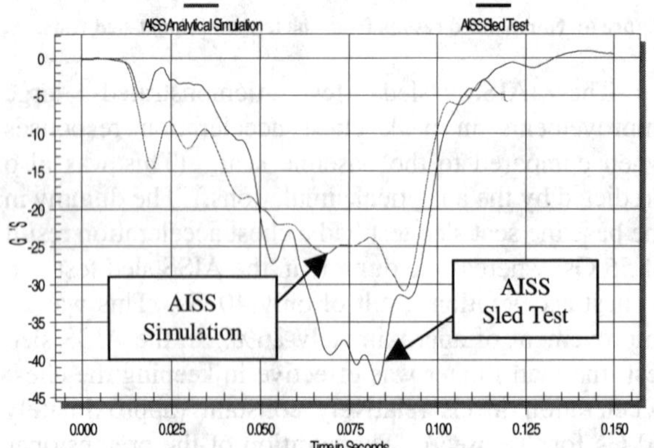

Figure 8: Chest acceleration comparison between the AISS sled test and the AISS analytical simulation.

A few notable differences between the sled test and the simulations resulted in the differences in the peak values. As previously mentioned, the pretensioner in the AISS sled test did not function to its full potential. This is illustrated by the small initial chest acceleration peak in the AISS sled test at 20 msec (Figure 8). However, in the analytical simulation of the AISS, the pretensioner had pulled in the full 80 mm of belt, and the restraining forces were applied to the dummy's chest at a much higher level (during the same 20 msec time frame in Figure 8). Therefore, this resulted in lower overall peak chest acceleration for the AISS analytical simulation.

The mid portion of the chest acceleration signal for the AISS sled test also exhibited less force limiting effects than the AISS simulation (Figure 8). The AISS sled test maintained a chest acceleration level of 40 Gs for approximately 20 msec; whereas in the AISS simulation, the chest acceleration was maintained at a lower level of 25 Gs for approximately 30 msec. This may be attributed to the physical properties of the mechanical load limiter device. As the belt load increased, the load limiter deformed a mechanical part to maintain a 4 kN load level on the belt, and in the physical device, there was a limited length that the loads could be maintained. However the AISS simulation did not have this limitation and spooled out additional belt material, as necessary.

Other variability between the analytical simulations and the sled tests may have been attributed to assumptions about the material properties and differences in the modeling set-up. This is further discussed in the AISS Modification 2 Final Report [10].

Rear Impact Testing And Simulation

Rear impact crashes account for approximately 5% of motor vehicle fatalities, but account for approximately 30% of motor vehicle injuries [7]. Neck injuries with risk of permanent disability are frequent in low severity rear impact collisions. Because these crashes are common, they can cause human suffering and high societal costs, despite the fact that the injuries are usually classified as "minor"[7]. In higher speed impacts, seats may deform significantly, allowing occupant ramping and potential contact with rear seat occupants. Therefore the AISS project sought to improve rear impact protection by enhancing the seat structure characteristics to minimize both frequent low speed impact injuries, and potentially more severe, but less frequent, high speed impact injuries.

AISS Features for Rear Impact: The AISS design has the following countermeasures for improved occupant crash protection in rear impacts: an integrated seat belt system, dual linear recliners, energy absorbers, and an extended head restraint system. In the AISS design, the integrated seat belt system helped reduce seat back ramping and potential contact with rear seat occupants. The dual linear recliner resulted in uniform loading of the seat back and provided torsional resistance from seat back twist. The energy absorbers also minimized occupant rebound and ramping by adding structural elements that deform plastically in a controlled manner. The mechanical energy absorbing elements were added in series to both recliners on either side of the seat. Finally, the AISS extended head restraint was fixed in position (i.e., not adjustable). This eliminated the possibility of the head restraint not being positioned correctly for the reduction of whiplash injuries in rear impacts.

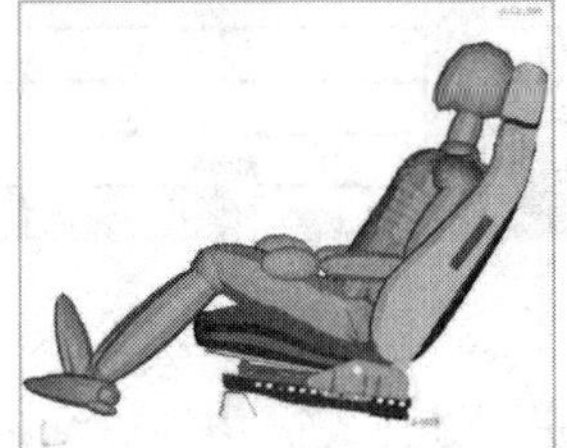

Figure 9: Rear impact simulation of the AISS.

Analytical Modeling: The AISS model (discussed in the frontal section) was also used for conducting rear impact simulations and for designing rear impact countermeasures. As previously mentioned, the seat back frame, the dual linear recliners, and the energy absorbers were optimized for frontal and rearward loading. For rear impacts, the seat back was designed not to yield at 1800 Nm about the h-point [9].

Models of the AISS recliners and energy absorbers are shown in Figure 10. The AISS recliners are bracketed to the seat back frame and are also secured to the seat track supporting plate by a bolt that passes through both the recliner rod end piece and the supporting plate. The energy absorber is positioned underneath the recliner rods and is constructed of two concentric bracket shells. When loaded by the recliner rods, the energy absorbers are designed to deform plastically in a controlled manner (i.e., as the recliner is forced downwards the metal in the energy absorber plates gets rolled inward).

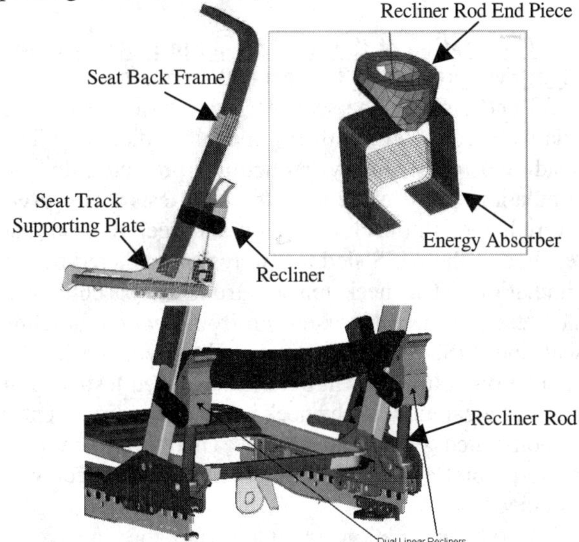

Figure 10: Analytical modeling of the dual linear recliners and rear impact energy absorbers.

The energy absorber was designed to withstand occupant loads to the seat back during normal use, but was designed to yield (a maximum of 30 degrees) when the seat back begins to fail in a rear impact crash [9]. The packaging of the energy absorber bracket in the seat track of the AISS was verified by Johnson Controls, Inc.

Finally, the complete AISS analytical model was coupled with the MADYMO Hybrid III 50th percentile male dummy model and was subjected to a simulated rear impact crash pulse (approximately 18 G peak, 90 msec pulse duration). A baseline integrated seat model (without the energy absorber and dual linear recliner) was also developed and subjected to the identical rear impact simulation for comparison.

Results of the Analytical Simulations: The AISS rear impact simulation results were compared to those from the baseline seat. Figure 11 compares the normalized injury measures. The analytical simulation predicted that all the 50th percentile male dummy injury measures were well below the ICPLs for both the AISS and the baseline integrated seat. The results further showed that the AISS would improve most occupant injury measures for the 50th percentile male dummy in a rear impact crash over a baseline integrated seat. However, the simulations only predicted marginal improvements.

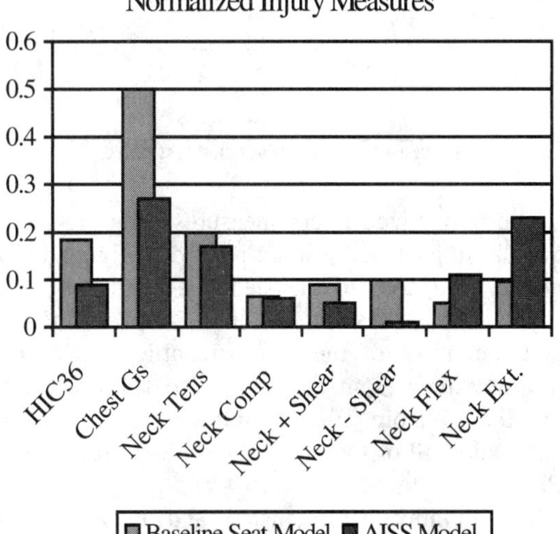

Figure 11: Rear impact analytical simulation results.

Sled Testing: Two of the AISS prototypes were used in rear impact sled tests. The sled test setup is shown in Figure 12. An instrumented Hybrid III dummy was seated on the AISS prototype and restrained by the integrated belt system. The sled test simulated a rear impact pulse with a 18 G peak, 90 msec duration. The sled pulse is illustrated in Figure

13. A baseline seat was also tested under the same sled conditions as the AISS for comparison. The baseline seat was an integrated seat with a single recliner and an adjustable head restraint. The structural energy absorbing elements were also not included.

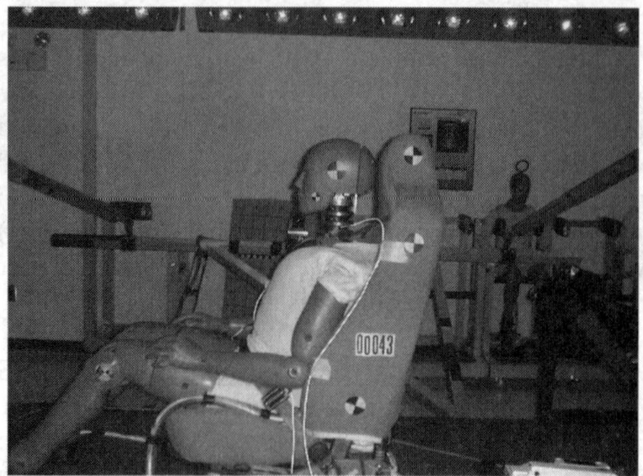

Figure 12: Rear impact sled test set-up.

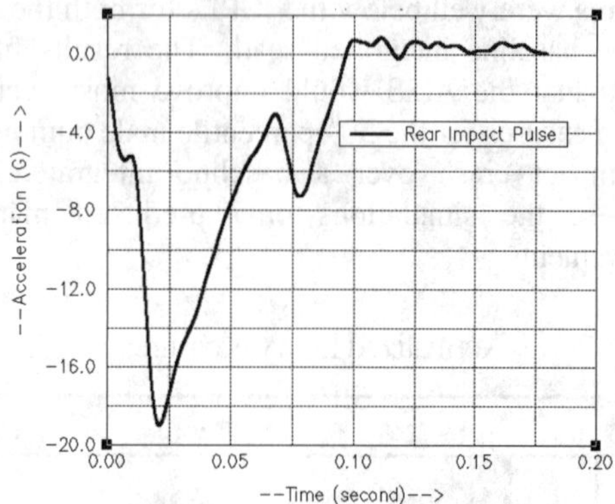

Figure 13: Rear impact sled test pulse.

The normalized injury measures from the second[1] AISS rear impact sled test are plotted in Figure 14, and they are compared to the sled test results using the baseline seat. As in the rear impact analytical simulation, most of the 50th percentile male dummy injury measures from the AISS sled test were lower than those from the baseline seat sled test. Additionally, all of the injury measures were below the ICPLs.

The dummy neck tension and neck extension results were greatly reduced in the AISS sled test. Peak neck tension in the AISS sled test was 38% lower than in the baseline seat sled test. Similarly the dummy's peak upper neck extension moment in the AISS sled test was 29% lower than the baseline seat sled test. It is believed that the extended head restraint and the energy absorbing elements contributed to the lower neck injury measures.

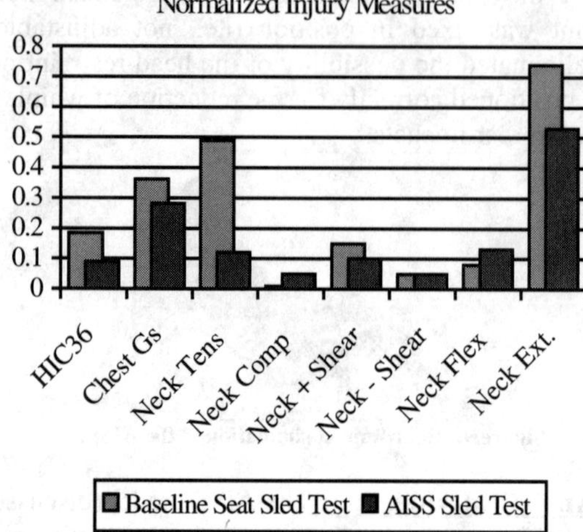

Figure 14: Normalized results from the rear impact sled tests.

The head and chest responses were slightly lower in the AISS rear impact sled test. Initially the dummy started sinking into the seat back until it reached full penetration (at approximately 50 msec). The energy absorber then started to crush and allowed the seat back to yield in a controlled manner (from approximately 60 msec to the end of seat back rotation). This resulted in a reduction of the dummy head and chest accelerations. However, the baseline seat, without the energy absorbing elements, resulted in higher injury measures in the rear impact sled tests. The peak head acceleration in the baseline seat sled test was 39 Gs and the HIC was 184. In the AISS sled test, the peak head acceleration was 31 Gs and the HIC was 90. Chest acceleration results were also reduced from 21 Gs in the baseline sled test to 16 Gs in the AISS sled test.

Correlation of the Analytical Simulations to the Sled Test Results: The general trends between the AISS and the baseline seat rear impact sled tests were relatively consistent with the model predictions. The head and chest injury reductions predicted in the simulations were evident in the sled tests. However, the reductions in neck extension and neck tension that resulted in the AISS sled test were not predicted by the simulation. The neck tension from the baseline seat sled test and the neck extension from both the baseline seat and AISS sled tests were under-predicted by the simulation. However, in both the two sled tests and in the analytical models, the neck responses independently demonstrated good repeatability. Therefore, there may be some biofidelic shortcomings in how the neck was modeled.

Good correlation resulted in the analytical simulation of the AISS energy absorbing elements.

[1] The second AISS test had improved dummy positioning and was used for comparison.

Figure 15 illustrates the comparison of the analytical model results with the actual post-test hardware from the AISS rear impact sled test.

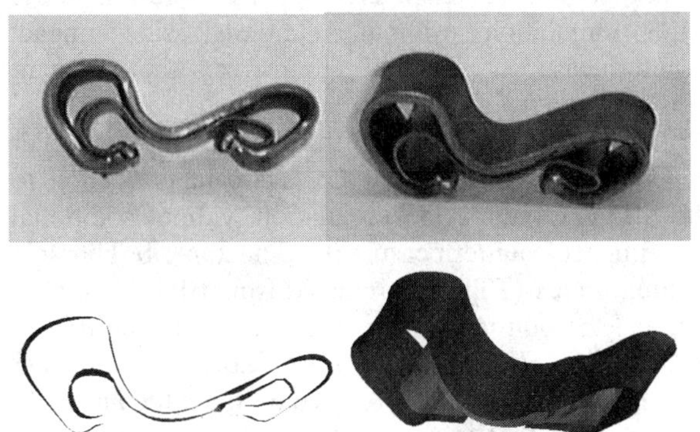

Figure 15: Comparison of analytical modeling of the energy absorbing elements vs. the post-test hardware from the rear impact sled test.

Side Impact Simulation

Side impact crashes of light vehicles, (i.e. passenger cars, light trucks and vans), result in approximately 9,800 fatalities and over 1,020,000 injuries each year. This corresponds to about 30% of vehicles involved in tow away crashes [13]. Therefore, considerable work has been done in the development of effective side impact countermeasures. Some have been incorporated into the vehicle structure while others have been mounted on the seat. The AISS project focused on evaluating the potential of those that could be mounted on the seat itself.

Side Impact Countermeasures: Two side impact countermeasures proposed for the AISS design were independently evaluated through analytical simulations. The countermeasures included: an Inflatable Tubular Cushion (ITC), and a combination head/thorax side impact air bag system.

The ITC system is a seat-mounted tubular bladder that is designed to reduce lateral loads on the occupant and prevent occupant ejection during a side impact collision (Figure 17). The ITC is an inflated tube which is anchored on the outboard side of the seat at two end points (toward the front of the seat pan and the top of the seat back). When deployed, the ITC expands and shortens at the same time. Due to the unique properties of the structural materials and the anchored end points, tension is introduced into the system as the bag inflates. The advantage of the ITC is that it eliminates the need for a reaction surface and consequentially creates a self-supporting inflated unit.

The combination head/thorax side impact air bag system is mounted on the outboard side of the seat frame (Figure 18). The seat trim has a special seam at the air bag location that splits open from the inflation pressure and allows the air bag to deploy into the vehicle interior. The combination head/thorax side impact air bag has two inflation chambers separated by a vented partition wall. The head/thorax side impact air bag system is designed to protect the occupant by filling the airspace between the intruding door and the occupant as early as possible in the crash event. Upon activation, the thorax chamber is filled directly by the inflator within the first 20 msec. The head chamber is subsequently filled as gas is passed through the vent holes in the chamber partition wall.

Analytical Modeling Description: In the evaluation of the side impact countermeasures, coupled LS-DYNA3D/MADYMO simulations were conducted using the U.S. FMVSS No. 214 side impact crash mode configuration (Figure 16). The AISS model (previously discussed) was incorporated into a full vehicle finite element model of a 1991 Ford Taurus [14]. The MADYMO Side Impact Dummy (SID) model was seated in the AISS model and positioned in the driver location of the full vehicle model. The vehicle was assumed to be stationary in the setup. A moving deformable barrier was modeled and given an initial velocity of 53.4 kmph, crabbed at an angle of 27 degrees. Simulations were conducted to compare the occupant protection provided to the SID in the following three configurations: the AISS with no side impact countermeasure, the AISS with an integrated ITC system, and the AISS with an integrated combination head/thorax side impact air bag system.

Figure 16: Side impact simulation set-up.

In modeling the two side impact countermeasures for this study, a *representative* model was developed for each system. Neither represented an exact model of a specific system. The primary goals were to have a model of the ITC with approximately the correct inflated size and shape, and a realistic system tension and internal pressure at the time of contact with the occupant. Using this approach, it was not necessary to model the exact details of the woven ITC material, the inflator properties, the packaging or the inflation process. Instead, the ITC was modeled as an initially flat tube with two end straps for connecting the ITC to the seat structure (Figure 17). Simula Technologies provided a "generic" ITC model and system data to approximate the inflated geometry and internal pressure [9].

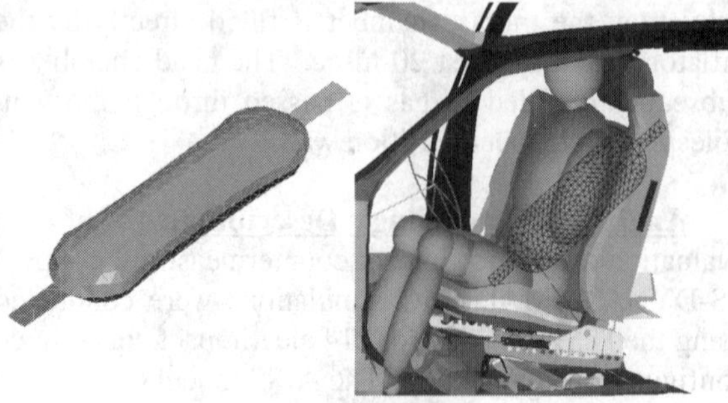

Figure 17: Inflatable Tubular Cushion (ITC) side impact countermeasure and simulation set-up.

Similarly, the goal in modeling the combination head/thorax side impact air bag system was to develop a *representative* model in the side impact simulation (Figure 18). Breed Technologies provided a "generic" combination head/thorax side impact air bag model [10].

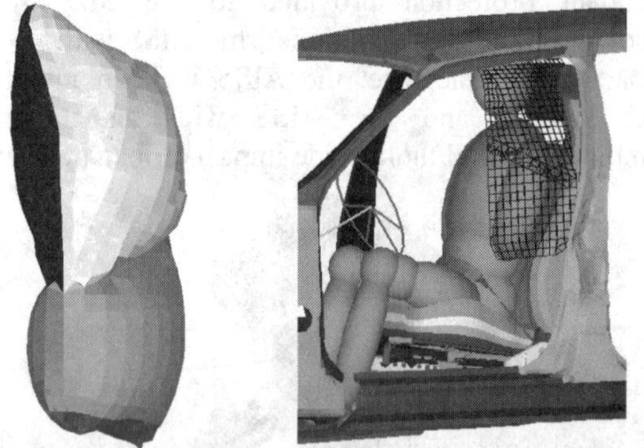

Figure 18: Combination head/thorax side impact air bag system countermeasure and simulation set-up.

Results of the Analytical Simulations: The side impact simulation results of the AISS with the ITC and the AISS with the combination head/thorax side impact air bag system were compared against the results of the AISS simulation without additional side impact countermeasures. Table 2 summarizes the occupant injury results from each simulation.

First, the combination head/thorax side impact air bag model provided improved occupant protection to the SID (over the AISS simulation without additional side impact countermeasures). The top rib Thoracic Trauma Index (TTI) was reduced from 61.1 Gs to 47.6 Gs, and the bottom rib TTI was reduced from 60.2 Gs to 51.3 Gs. The combination head/thorax side impact air bag model filled the air-space between the intruding door and the SID very quickly. This resulted in early restraint being applied to the SID ribs (around 15 msec with the combination head/thorax side impact air bag vs. 28 msec without the air bag). This early restraining force, combined with the energy absorption of the thorax chamber (as gas was vented into the head chamber) resulted in lower peak rib accelerations.

Conversely, the side impact simulation using the AISS and ITC system model resulted in increasing the peak rib accelerations of the SID. Unlike the combination head/thorax side impact air bag, the ITC model was not vented and became very stiff under tension. Even though the ITC model quickly filled the air-space between the intruding door and the SID, the ITC model acted like a rigid spacer, transferring load directly from the door to the SID.

Neither the AISS with the combination head/thorax side impact air bag model nor the AISS with the ITC model predicted improved occupant protection for the pelvis region. Since the combination head/thorax side impact air bag model did not provide coverage of the pelvic region, it had little effect on the pelvic response. However, the AISS with the ITC model actually increased the peak pelvis acceleration. This was attributed to the fact that the ITC model was stiffer in the deployed state and interacted, to some extent, with

Table 2. Comparison of side impact simulation injury numbers.

	Top Rib		Bottom Rib		Pelvis		T12	
	TTI (G's)	Δ TTI (%)	TTI (G's)	Δ TTI (%)	Peak Acc. (G's)	Δ Peak (%)	Peak Acc. (G's)	Δ Peak (%)
AISS	61.1		60.2		113.6		66.5	
AISS + ITC	66.0	+8.0	73.5	+22.1	127.0	+11.8	77.0	+15.8
AISS +Combination Head/Thorax Bag	47.6	-22.1	51.3	-14.8	110.8	-2.4	64.5	-3.0

the pelvis and left femur of the SID. This resulted in higher levels of pelvic acceleration when trapped between the dummy and the intruding door structure. A similar trend resulted in the T12 response of the SID in the AISS with ITC model; whereas the combination head/thorax side impact air bag model resulted in an earlier interaction and reduced peaks in the T12 acceleration. Therefore, of the two countermeasure models, the AISS with the combination head/thorax side impact air bag predicted improved occupant protection for the SID.

Rollover Simulation

According to the 1999 Fatality Analysis Reporting System (FARS), 10,142 people were killed as occupants in light vehicle rollovers, including 8,345 killed in single-vehicle rollovers [15]. Using data from the 1995-1999 National Automotive Sampling System (NASS), an estimated 253,000 light vehicles were towed from a rollover crash each year (on average), and 27,000 occupants of these vehicles were seriously injured [15]. Therefore, the data presented above demonstrates that rollover crashes create a serious safety problem to motor vehicle occupants . The AISS project therefore explored potential seat countermeasures that could improve rollover crash protection and also contribute to passenger compartment intrusion resistance.

Rollover Countermeasures: Seat belts are currently the most significant restraint for occupants in rollover crashes because they help prevent occupant ejection from the vehicle. The AISS incorporates a seat integrated belt system which improves occupant belt fit and couples the occupant to the seat. The torso belt was also designed to limit belt system loads to 4 kN (to optimize frontal impact protection) while sustaining loads capable of retaining the occupant within the compartment during rollover crashes. A buckle pretensioner was incorporated in the AISS to take-up slack in the belt system and reduce occupant excursion toward the roof in a rollover event. The AISS was also designed with an 815 mm tall head restraint system to reduce structural intrusion from roof crush. The head restraint design provides a minimal backset between the occupant's head and the head restraint.

Analytical Modeling Description: The AISS analytical model (discussed previously) was incorporated into the driver position of a full vehicle finite element model of a 1991 Ford Taurus [14]. A MADYMO Hybrid III 50th percentile male dummy model was positioned in the AISS. The integrated seat belt was attached to eliminate occupant ejection, and 80 mm of belt was pulled into the retractor to simulate the pretensioner activation at time=0 (during the initial phase of rollover).

Coupled LS-DYNA3D/MADYMO simulations of a vehicle rollover test were conducted. The particular test set-up involved a vehicle mounted on a rollover cart, which was inclined at an angle of 60 degrees with respect to a horizontal datum, at a height of 1.22 meters. The vehicle was placed on the cart such that the driver's side of the vehicle was closest to the ground. The cart was then accelerated to a velocity of 48 kmph and suddenly decelerated to a 0 kmph stop. The sudden change in the rollover cart velocity sent the vehicle into flight at 48 kmph. The vehicle then rolled in the air and hit the ground with the leading edge of the roof (on the driver's side of the vehicle).

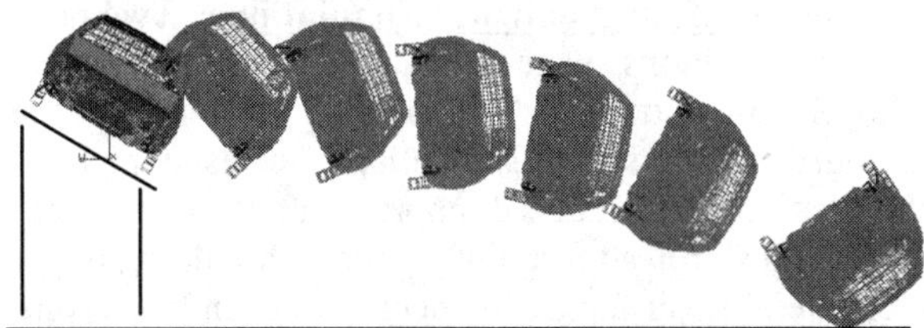

Figure 19: The rigid free-flight phase of the rollover simulation.

The analytical simulation of the rollover test was carried out in two phases. The first was a rigid free-flight phase (Figure 19) and second was a deformable vehicle-to-ground impact phase (Figure 20).

The first phase of simulation began with the free-flight of the vehicle into the air where no ground contact or deformation to the vehicle occurred. To take advantage of this fact, the entire vehicle was considered rigid to improve the required simulation time. Once the vehicle hit the ground, at the end of this phase, the velocities were recorded for use in the second phase.

The vehicle contacted the ground plane with a 48 kmph velocity in the y-direction and a 20.5 kmph velocity in the z-direction at approximately 580 msec into the simulation. The vehicle model also rotated approximately 105 degrees about the x-axis from the initial position on the cart. These final velocities and rotation from the rigid phase were used as initial conditions for the second (deformable) phase.

The second phase of the simulation started when the vehicle hit the ground (Figure 20). In this phase, the yaw and pitch of the vehicle were not considered, as they were not significant compared to the roll of the vehicle, and a constant rotation of 180 degrees/sec was assumed. The vehicle model was deformable in this phase, and the occupant was free to move (as opposed to the first phase in which the occupant was restrained in position). The contact friction between the ground and the vehicle was assumed to be 0.3.

The rollover simulations were performed with two different head restraint positions of the AISS. The top surface of the head restraint at highest position was flush with the top of the crown of a seated 95% percentile male occupant head. The lower head restraint was positioned such that the top of the head restraint was flush with top of the head of a seated 50% percentile male occupant.

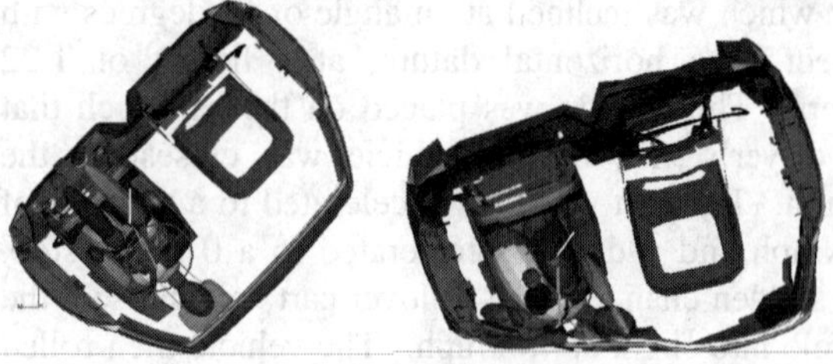

Figure 20: The deformable phase of the rollover simulation.

Results of the Analytical Simulations: Two head restraint positions were evaluated in the analytical rollover simulation, as discussed above. In the extended head restraint simulation, the resultant head acceleration was reduced approximately 14 percent from the lower head restraint position. Similarly, upper neck lateral shear forces and neck extension loads were also reduced 38 percent and 32 percent, respectively. However, only small reductions in upper neck fore/aft shear and axial loading resulted in the extended head restraint position.

While the extended head restraint prevented larger intrusion into the occupant compartment during roof crush, it only yielded marginal improvements in occupant protection over the lower head restraint position. This resulted for a number of reasons. The rollover test simulated in the model was particularly demanding on the structural integrity of the roof. Since the model was not fully-validated in the rollover crash mode, significant structural deformation of the vehicle roof resulted (Figure 20). The inner structural details of the model can affect occupant injury, as they determine the specific hit location of the occupant inside the compartment. Also, there was not a significant difference between the height of the extended head restraint and the lower head restraint position (27 mm). Therefore, only marginal gains in reducing the injury parameters resulted. Future work could involve improving the vehicle finite element model validation in the rollover crash mode and conducting additional simulations to better demonstrate the AISS benefits of the integrated belt system, the pretensioner activation, and the extended head restraint system.

CONCLUSION

An Advanced Integrated Safety Seat (AISS) was evaluated in the front, rear, side, and rollover crash modes using coupled LS-DYNA3D/MADYMO analytical simulations. Frontal and rear impact crash modes were also experimentally evaluated using sled tests with AISS prototypes. Comparative tests with a baseline integrated seat were also conducted.

In the frontal crash mode, the AISS pretensioner and 4 kN force limiter contributed to reducing the chest injury measures on the Hybrid III dummy. This was reflected in both the analytical simulations and the prototype sled test results. The analytical model also demonstrated relatively good trend correlation with the experimental sled testing.

In the rear impact mode, the dual linear recliner, energy absorber, and extended head restraint system also helped to reduce head, neck, and chest injury numbers on the Hybrid III dummy. The analytical model was also able to predict the energy absorber bracket deformation that resulted in the sled test.

A full vehicle finite element model of the 1991 Ford Taurus was used in the side impact simulations of the FMVSS No. 214 crash test configuration. Analytical models of a "generic" Inflatable Tubular Cushion and a combination head/thorax side impact air bag were each incorporated into the AISS model for comparison. The combination head/thorax side impact air bag model predicted improved occupant protection for the SID, overall, when placed between the intruding door structure and the occupant. The early restraining force, combined with the energy absorption of the thorax chamber (as gas was vented into the head chamber) resulted in lower injury measures. Future AISS improvements could be made in protecting the pelvis of the SID dummy in side impact crashes.

Rollover simulations were also conducted with a full vehicle finite element model of the 1991 Ford Taurus. The simulations predicted that the AISS extended head restraint prevented larger intrusion into the occupant compartment during roof crush and reduced occupant injury. However, future work in refining the full vehicle finite element model in the rollover crash mode could better demonstrate the rollover benefits of the AISS integrated belt system, buckle pretensioner, and extended head restraint system.

ACKNOWLEDGMENTS

The authors wish to thank Simula Technologies for the Inflatable Tubular Cushion data, and Breed Technologies for the head-thorax combination side impact air bag data.

REFERENCES

[1] National Highway Traffic Safety Administration, *Traffic Safety Facts 1999: A Compilation of Motor Vehicle Crash Data from the Fatality Analysis Reporting System and the General Estimates System*, DOT HS 809 100, December 2000, www.nhtsa.dot.gov.

[2] Gupta, V., Menon, R., Gupta, S., Mani, A., Shanmugavelu, I., *Advanced Integrated Structural Seat Final Report,* February 1997, Docket No. NHTSA-1998-4064.

[3] Cole, J.H., *Developing a Cost Effective Integrated Structural Seat,* SAE Paper No. 930109, Society of Automotive Engineers, Warrendale, PA, 1993.

[4] Interim Final Rule for FMVSS No. 208, National Highway Traffic Safety Administration, Federal Register, Volume 65, No. 93, page 30680, May 12, 2000, NHTSA Docket No. NHTSA-2000-7013.

[5] *Effectiveness of Occupant Protection Systems and Their Use,* U.S. Department of Transportation, National Highway Traffic Safety Administration, National Center for Statistics and Analysis, May, 1999.

[6] Hollowell, W.T., Gabler, H.C., Stucki, S.L., Summers, S., Hackney, J.R., *Updated Review of Potential Test Procedures for FMVSS No. 208*, October 1999, NHTSA Docket No. NHTSA-2000-7013.

[7] Gupta, V., Menon, R., Gupta, S., Mani, A., Shanmugavelu, I., Kossar, J., *Improved Occupant Protection through Advanced Seat Design,* Proceedings from the Experimental Safety Vehicle Conference, Melbourne, Australia, 1996, Paper 96-S1-O-08.

[8] Kallieris, D., Rizzetti, A., Mattern, R., Morgan, R., Eppinger, R., Keenan, L., *On the Synergism of the Driver Air Bag and the 3-Point Belt in Frontal Collisions*, 39th Stapp Car Crash Conference Proceedings, Society of Automotive Engineers, Warrendale, PA, 1995.

[9] Rashidy, M., Gunasekar, T.J., Gupta, S., Griffioen, J., and Howard, J., *Advanced Integrated Safety Seat, DTNH22-97-C-07003, Task Order #01, Final Report,* July 1999.

[10] Rashidy, M., Gunasekar, T.J., Morris, R., Munson, B., and Lindberg, J.A., *Advanced Integrated Safety Seat, DTNH22-97-C-07003, Task Order #01,Modification 2, Final Report,* August 2000.

[11] Notice of Proposed Rulemaking for FMVSS No. 208, National Highway Traffic Safety Administration, Federal Register, Volume 63, No. 181, September 18, 1998, page 49987, NHTSA Docket No. NHTSA-1998-4405.

[12] NHTSA Docket Number NHTSA-2000-7013-16. www.dms.gov, June 20, 2000.

[13] Samaha, R.R., *B.02.02.03 Upgrade Side Crash Protection*, NHTSA Research & Development Project Summaries, www.nhtsa.dot.gov, June 1998.

[14] Varadappa, S., Shyo, S., Mani, A., *Development of a Passenger Vehicle Finite Element Model*, DOT HS 808 145, Contract DTNH22-92-D-07323, November 1993.

[15] Notice of Final Decision for Consumer Information Regulations on Rollover Resistance, National Highway Traffic Safety Administration, Federal Register, Volume 66, No. 9, January 12, 2001, page 3388.

[16] Bardini, R., and Hiller, M., *Contribution of Occupant and Vehicle Dynamics Simulation to Testing Occupant Safety in Passenger Cars During Rollover.* SAE 1999-01-0431, Society of Automotive Engineers, Warrendale, PA, 1999.

[17] Chace, M.A., and Wielenga, T.J. *A test and Simulation Process to Improve Rollover Resistance*, SAE 1999-01-0125, Society of Automotive Engineers, Warrendale, PA, 1999.

[18] Gupta, V., Gunasekar, T.J., Rao, A., Kamarajan, J., *Reverse Engineering method for developing full vehicle finite element models*, SAE 1999-01-0083, Society of Automotive Engineers, Warrendale, PA, 1999.

CHAPTER 15:

BELT INNOVATIONS: INFLATABELTS, 4 POINTS, ETC.

912905

Investigation of Inflatable Belt Restraints

John D. Horsch, Gerald Horn, and Joseph D. McCleary
Biomedical Science Dept.
General Motors Research Labs.

ABSTRACT

Studies conducted in the 1970's suggested that inflatable belt restraints might provide a high level of occupant protection based on experiments with dummies, cadavers and volunteers. Although inflating the belt was one factor which contributed to achieving these experimental results, much of the reported performance was associated with other features in the restraint system. Exploratory experiments with the Hybrid III dummy indicated similar trends to previous studies, belt inflation reducing dummy response amplitudes by pretensioning and energy absorption while reducing displacement. The potential advantage of an increased loaded area by an inflatable belt could not be objectively demonstrated from previous studies or from dummy responses.

Clearly, belt inflation can be one component of a belt restraint system which tends to reduce test response amplitudes. However, other belt system configurations have demonstrated similar test response amplitudes. Additionally, packaging and comfort issues for inflatable belts have not been completely resolved. Thus inflatable belts do not appear competitive with other approaches to achieve belt restraint system performance.

IN THE EARLY 1970's, Allied Chemical developed an inflatable belt restraint to achieve FMVSS 208 performance [1]. They suggested that inflatable belts might be a superior restraint system compared with belts or air bags, by providing lower dummy response amplitudes than conventional belts or air bags in frontal tests. In lateral and oblique crashes, and potential ejection situations the inflatable belt could perform better than air bags and at least as good as belts. They suggested that inflatable belts might cost less than an air bag system and have lower replacement costs, less stringent triggering requirements, and fewer inflations based on a higher inflation threshold because of the protection provided by the uninflated restraint. However the belt would need to be "used" to protect an occupant.

Allied Chemical conducted various general and specific development efforts. These efforts included volunteer tests sponsored by the NHTSA and conducted at Southwest Research Institute in which no significant injury was noted in sled tests to 32.5 mph. Sled tests with dummies achieved HIC and chest acceleration well below "208" limits at 32.5 mph. Other efforts chose inflatable belt restraint systems for front or rear seating positions in Research Safety Vehicles with good test results well above 30 mph barrier crash severities [1].

In spite of these favorable demonstration efforts, the system never achieved sufficient development to be placed in a fleet of experimental vehicles. Unresolved design and packaging issues, and the use of other technologies to achieve protection may be among the reasons that inflatable belt design never reached a fully developed product.

Based on the previous test results, the trend of increased belt usage, and the increase of elderly car occupants who generally have lower tolerance to belt loading, a study was undertaken at the General Motors Research Laboratory in the mid-1980's to evaluate inflatable belts. The study's objective was to review the state-of-art for inflatable belts in terms of performance and design by reviewing the literature and conducting exploratory sled tests with current dummies and advanced injury assessment techniques.

TEST RESPONSES

PREVIOUS STUDIES - Studies of inflatable belt restraints had been summarized by Allied Chemical [1]. Digges and Morris provide an overview of previous efforts [2]. All of the studies achieved low dummy response amplitudes for HIC and chest acceleration when compared to "208" requirements and test severity, Table 1. HIC amplitudes of <300 and chest accelerations of <25 g's were observed in 32 mph sled tests conducted by Southwest Research Institute in conjunction with their volunteer program [1]. Factors to be considered in extrapolating these 1970's sled test results to the 1990's include the use of a Sierra 1050 dummy, minimal injury assessment instrumentation, simulation of "older" vehicles, no steering assembly, no belt retractors, and little velocity change at the time of inflation (early inflation). This was one reason for us to conduct exploratory sled tests with inflatable belts.

Development efforts generally demonstrated large amplitude reductions in measured responses when comparing the original conventional belt system with the inflatable belt system, Table 2. Analysis of the data provided by Reference [1] indicates that good system test responses and large amplitude reductions from the base conventional belt system were the result of not only belt inflation, but also due to other system modifications and/or enhancements. Estimates of the influence of belt inflation independent of other factors are provided in Figure 1. The estimated ratio of response amplitudes comparing inflated belts to conventional webbing ranges from 0.41 to 0.77 for HIC and from 0.67 to 1.02 for chest acceleration.

The rear seat ESV and the Minicars study appear to provide the least subjective estimate of the influence of belt inflation for HIC and chest acceleration. In the ESV rear seat development conducted at 47 mph [1], the initial conventional belt system tests resulted in HIC's averaging about 2400 while the final inflatable belt system averaged about 1050. Importantly, this improvement was due to many changes and additions including the seat, head rest, anchor locations, allowing "partial submarining", and removing the front seat structure. The belt failed to inflate in four development tests, providing the most direct comparison as to the influence of inflation. On the basis of tests with every thing similar except belt inflation indicates that inflation of the belt provided 27% of the total HIC reduction from 2400 to 1050.

Minicars [1] reported a reduction of HIC from 1720 with conventional belts to 426 with an inflatable belt system in a 54 mph sled test [1]. A major feature of the inflating belt system was a force limiter at all three anchor locations. The force limiters alone resulted in a reduction of HIC from 1720 to 693, 79% of the improvement. Clearly belt inflation provided an important contribution to system HIC performance, but force limiting appears to be a more important factor in these tests.

Inflatable belts were studied as the restraint system for front occupants in two vehicle applications. Minicars applied an inflatable belt system with force limiting anchors to a modified Pinto. A HIC of 302 and chest peak acceleration of 36 g's were reported for a 42 mph frontal barrier crash. These results are for the right-front location, and thus did not have a steering assembly.

A Calspan/Chrysler Research Safety Vehicle used a modified Simca 1308 with a target weight of 2700 pounds. A webbing lock to reduce spooling from the retractors and a force limiting mechanism were part of the inflatable belt system. HIC's of 1422 and 2161, and chest acceleration's of 60 and 50 g's for Hybrid II dummies in the left front and right front respectively, were reported for a 45.8 mph frontal barrier crash. In a frontal head-on crash having a 26.3 mph change of velocity for the RSV, HIC's of 716 (340 not considering B-pillar contact) and 537 and chest accelerations of 37 and 46 for Hybrid II dummies in the left front and right front were reported.

Dummy responses reported in the literature for inflatable belt systems are in most cases representative of an overall system effort to reduce HIC and chest acceleration. Clearly belt inflation was an important feature in obtaining these results. However many other features such as force limiting anchors also played important roles in achieving the reported response magnitudes. As with any restraint system, the performance is that of the system and cannot be based only on single features of that system. Thus adding an inflatable belt to an existing belt restraint will not assure achievement of the reported results.

EXPERIMENTAL STUDY - Exploratory Hyge sled tests were conducted in the mid-1980's at the GM Research Laboratories with inflatable belt restraints to help separate belt inflation from other system features, to evaluate response amplitudes using the Hybrid III dummy and additional injury assessment techniques, and to develop an experimental model for possible future studies. A simplistic fixture containing a bucket seat, "rigid" belt anchors, and a floor/toe-pan surface was used for these tests. Anchor locations and the amount of webbing on the retractors approximated door-mounted lap-shoulder belt systems parameters. Importantly, the overall system did not fully model these restraint systems, differences include the rigid fixture, lack of an instrument panel or steering assembly, and a "severe" generic sled pulse, chosen to provide HIC and chest acceleration near the 208 limits as a "sensitive" region to evaluate the influence of an inflated belt restraint. This differs from previous studies which made system changes to provide low dummy responses. The fixture with a conventional belt webbing restraint are shown in Figure 2. Sled pulses are shown in Figure 3.

Instrumentation included standard head and thorax triaxial accelerometers, axial chest compression, and six-axis top-of-the-neck force and moment transducer. An accelerometer array was used to determine head dynamics and head contact forces [3].

The inflatable belt hardware was obtained from Morton International for the exploratory sled tests, having a gas generator located inside of the inflatable cushion, Figure 4. The inflatable cushion, constructed from a coated material, was attached to conventional webbing by either of two methods. For both configurations conventional webbing carried the tensile loads. For one configuration, the cushion was located between the dummy and the conventional webbing, Figure 5a. The other configuration placed the cushion between two conventional belts, similar to that used by Minicars [1], Figure 5b. A third configuration developed by Allied Chemical [1] was not tested. In this configuration the cushion was the tension member, attached to conventional webbing at each end, Figure 5c. Systems were obtained in four groups of three restraints to allow parameter adjustments based on test results.

TEST RESULTS - The first group of three restraints had a cover over the inflating portion of the belt. One restraint had a cushion on both the lap and shoulder portion, the other two had a cushion only on the shoulder belt portion. The cushions were located between the belt and the dummy as shown in Figure 5a. The first and second tests had "late" cushion inflation, 56 and 33 ms instead of the 10 ms desired. In all three tests, the gas was forced into the free length over the shoulder, compounded by the covers not fully opening, Figure 6. The responses were similar to tests with conventional webbing.

Table 1

Summary of Sled Test Results for Systems Having Inflatable Belts (1)
Sierra 1050 Dummy, Frontal Impact

Test Series	Sled Velocity (mph)	Number Observations	Mean Response: HIC	Mean Response: Chest Accel. (3 ms-g)
Allied Initial Demonstration	33	2	425	--
Rear Seat ESV	27	3	375	33
	34	3	740	50
	47	6	1049	55
Allied Development				
Pinto Fixture	32	4	467	42
1957 Ford Fixture	32	4	560	44
Southwest Research Institute	32	2	235	21
Minicars (Hybrid II Dummy)	54	2	426	47

Table 2

Sled Test Results Comparing Initial Conventional Belt System with the Final Inflatable Belt System (1)
Sierra 1050 Dummy, Frontal Impact

Test Series	HIC: Initial System	HIC: Final System	Chest Acceleration (3 ms-g): Initial System	Chest Acceleration (3 ms-g): Final System
Allied Initial Demonstration	676	425	--	--
Rear Seat ESV	2431	1049	67	50
Allied Development				
Pinto Fixture	730	467	63	42
1957 Ford Fixture	1379	560	53	44
Minicars (Hybrid II Dummy)	1720	426	68	47

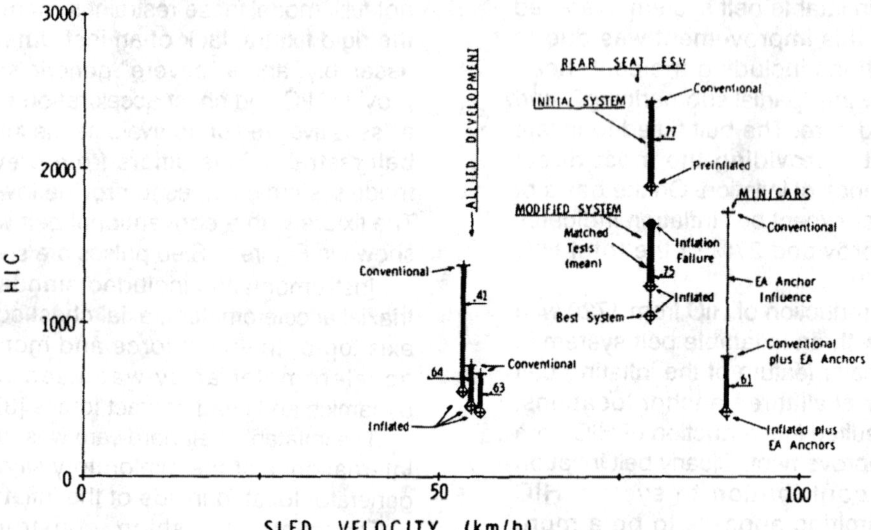

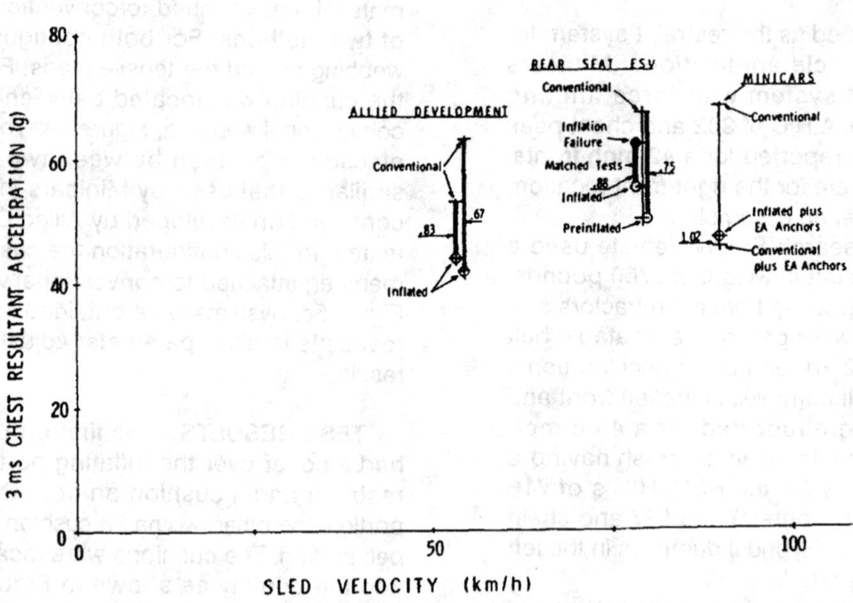

Figure 1. Estimates of the influence of belt inflation from data reported in the literature [1], indicated by heavy bars. Ratio of inflated belt system response to the corresponding conventional belt system response is given for each estimate.

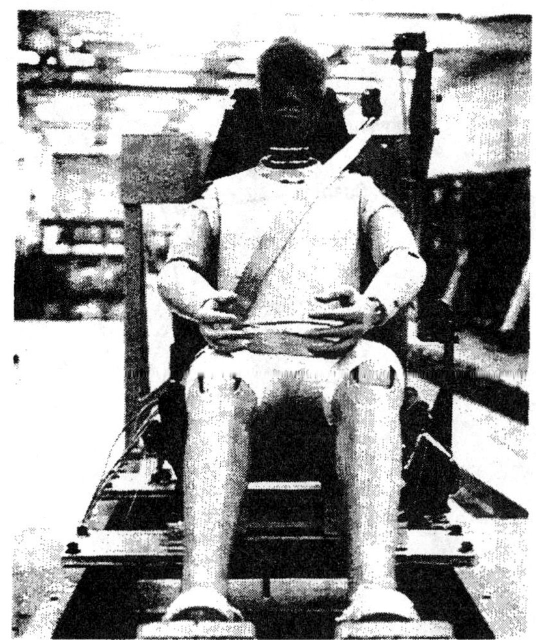

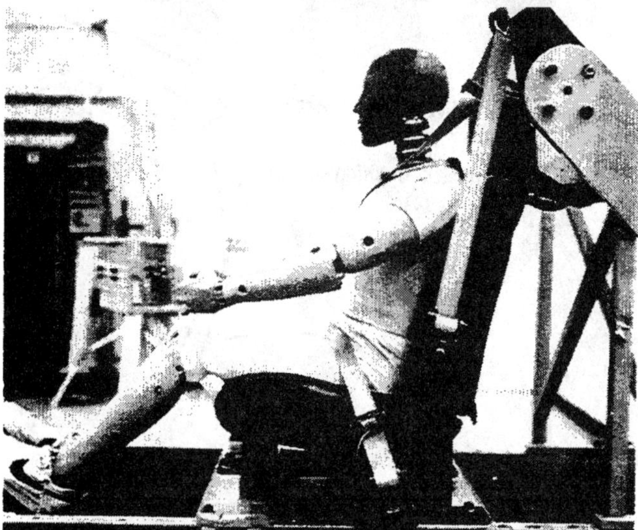

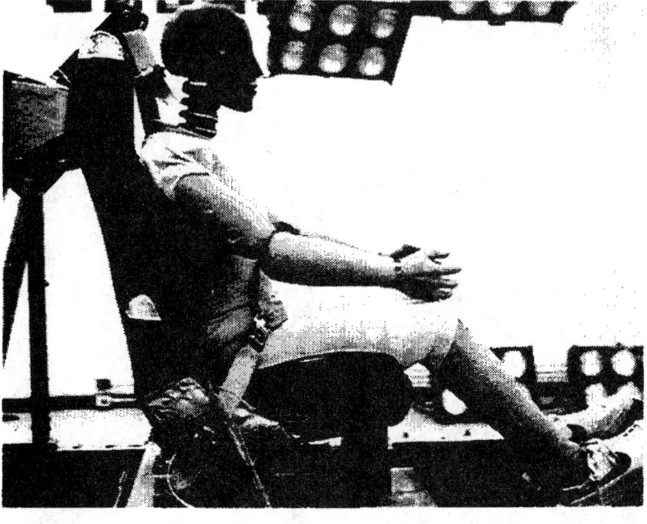

Figure 2. Sled test fixture shown with the conventional webbing restraint.

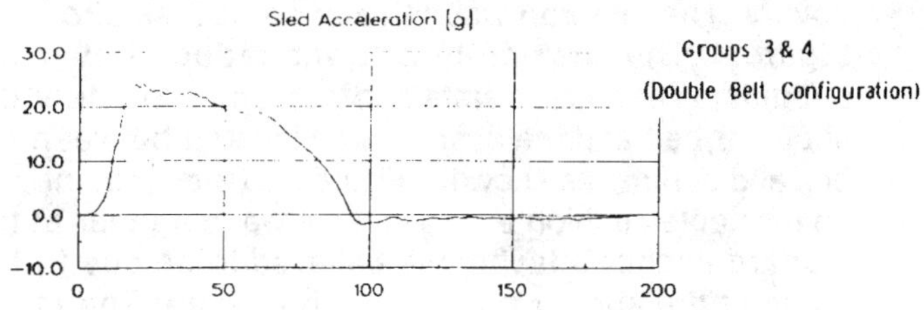

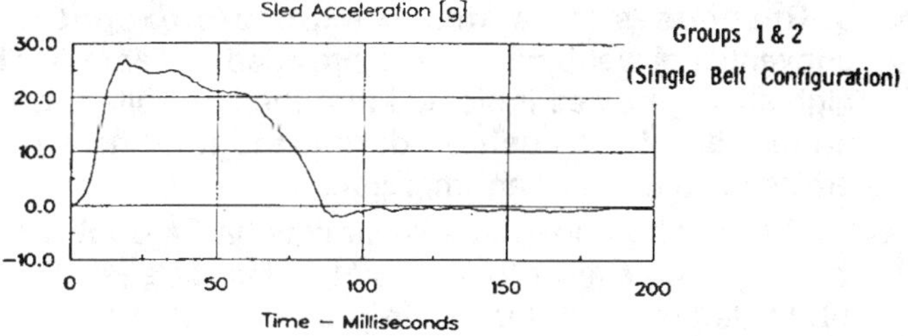

Figure 3. Sled acceleration pulses used for the exploratory tests.

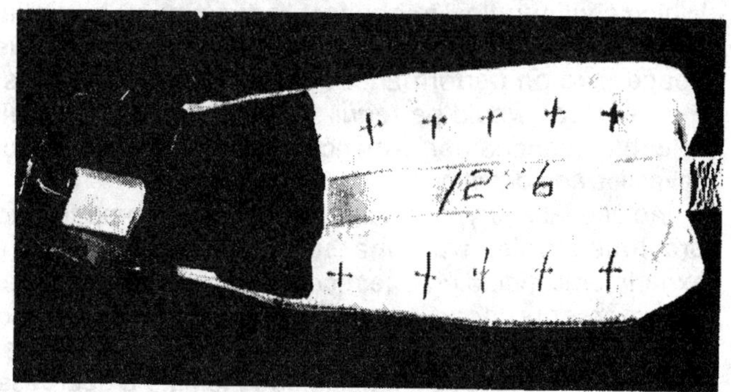

Figure 4. Inflatable belt cushion and inflator as tested for the single belt configuration.

The second group of three restraints did not have covers. The cushion was initially spread as shown in Figure 4. The cushion length was reduced over the shoulder. These restraints had a cushion on only the shoulder belt and the cushion was located between the belt and dummy as shown in Figure 5a. One restraint with an inflatable cushion experienced a webbing tear at the inboard anchor attachment, believed to be due to the reuse of the anchor attachment hardware. The mean responses from the other two tests are provided in Table 3. Responses for a similar exposure except with conventional webbing are also provided in Table 3. The high speed movies indicated that the belt rolled to the underside of the cushion during the loading which provided an EA mechanism, Figure 7.

The third group of three restraints had a double belt configuration similar to that used by Minicars [1], Figure 5b to determine if this could prevent "rolling". These restraints maintained the same cushion size and inflation as the previous group. The test severity was reduced to eliminate the potential for webbing tearing with the experimental hardware, the sled acceleration shown in Figure 3. The mean responses from these tests are provided in Table 3. The high speed movies indicated reduced rolling of the cushion with the double belt configuration, Figure 8. Responses for a similar exposure except with conventional webbing are also provided in Table 3.

The final group of three restraints used the double belt configuration, Figure 5b, but with larger circumference cushions of 480 and 430 mm compared with 330 mm in previous restraints. One cushion had a 16 mm diameter vent hole. Responses were similar to those of the previous group of restraints. Cushion diameter and venting are likely important performance parameters. These exploratory experiments might not have adjusted system parameters such as inflator output to best match these parameters.

Due to the difference in test severity between inflatable belt configurations, responses for the "single" and "double" belt configurations should be related by corresponding conventional webbing tests. On this basis, the single belt configuration (Figure 5a) reduced dummy responses more than the double belt configuration (Figure 5b), but allowed a greater displacement of the dummy. Thus the rolling action of the single belt configuration appears to provide an energy absorbing displacement. Which configuration might best reduce injury potential likely depends on the application, such as the available space, and on performance in other crash situations. Further study would be required to determine how belt rolling influences performance over a range of impact severities and directions.

Additional injury assessment responses compared to previous studies were measured in the exploratory experiments including chest compression, neck forces and moments, head angular acceleration and head contact force. These responses provide additional information on occupant loading and indicate a decrease of loading magnitude associated with belt inflation except for the head contact force due to the cushioning of the chin by the inflated belt, Figure 9.

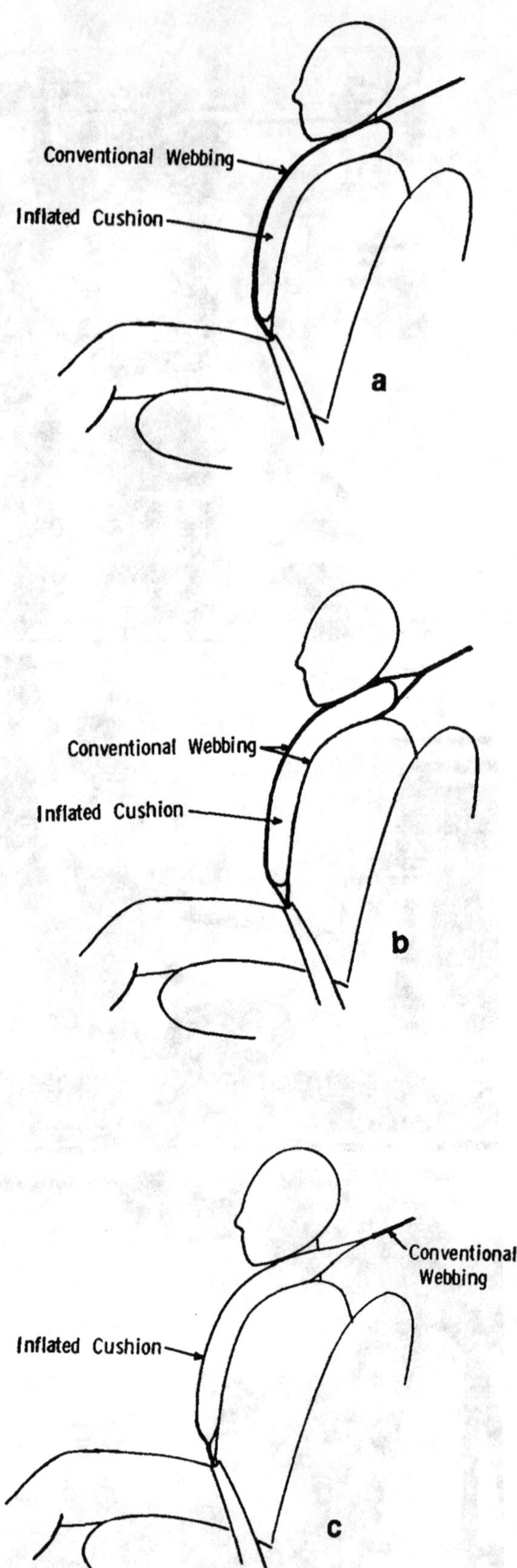

Figure 5. Inflatable belt configurations. a) The inflatable cushion located between the subject and the conventional webbing. b) The inflatable cushion located between conventional webbing. c) The inflatable cushion attached to conventional webbing at each end.

Figure 6. Initial test in which the cushion was inflated late and was filled only behind the shoulder. The black "webbing" seen in the side view is a cover used in the initial tests.

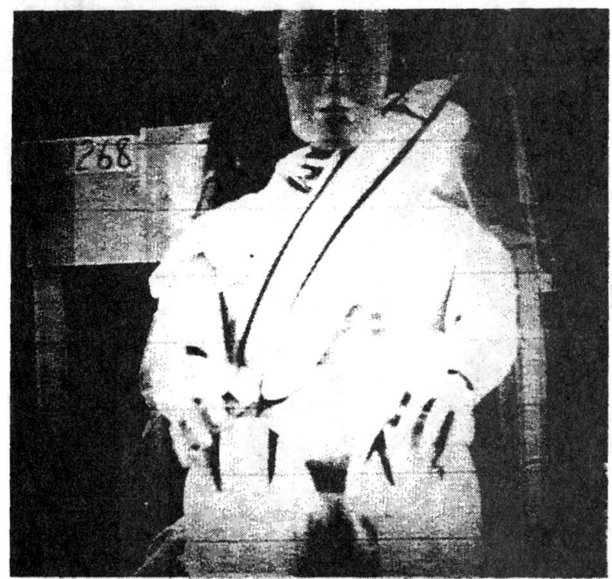

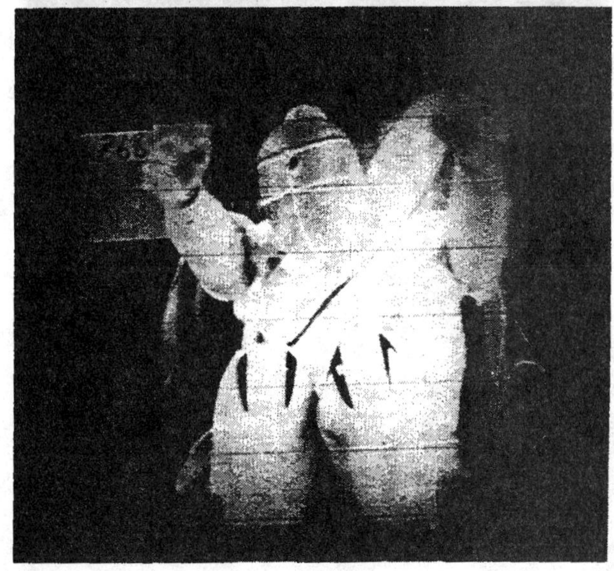

Figure 7. High-speed movie showing "rolling" of the single belt configuration.

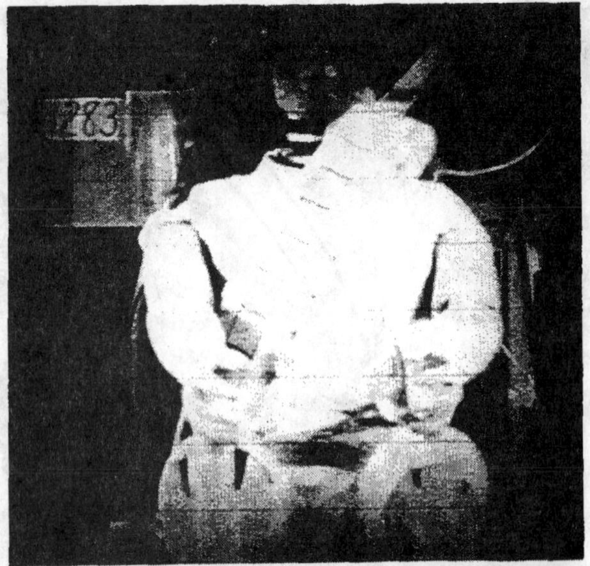

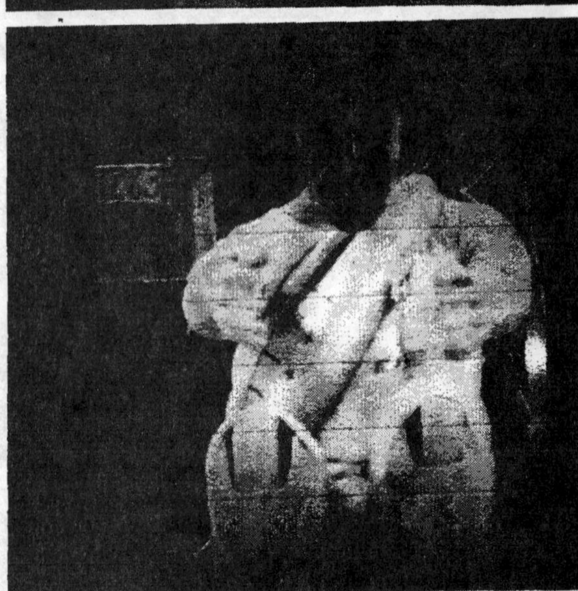

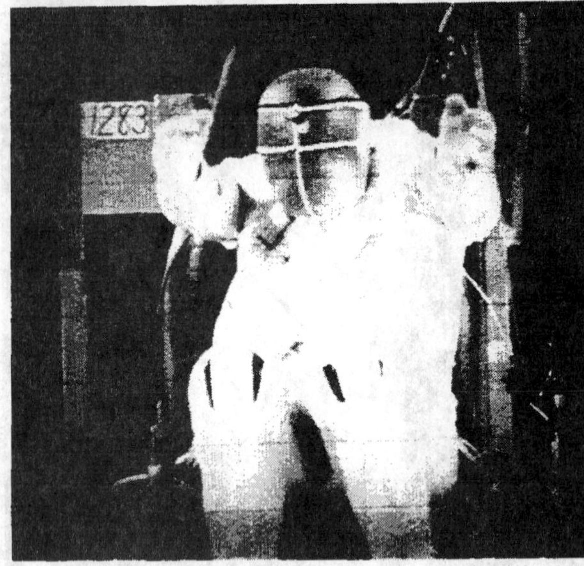

Figure 8. High-speed movies showing reduced "rolling" of the double belt configuration compared with the single belt configuration.

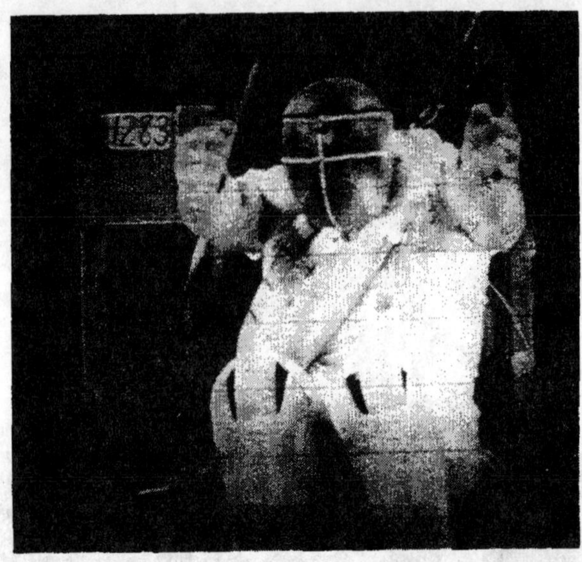

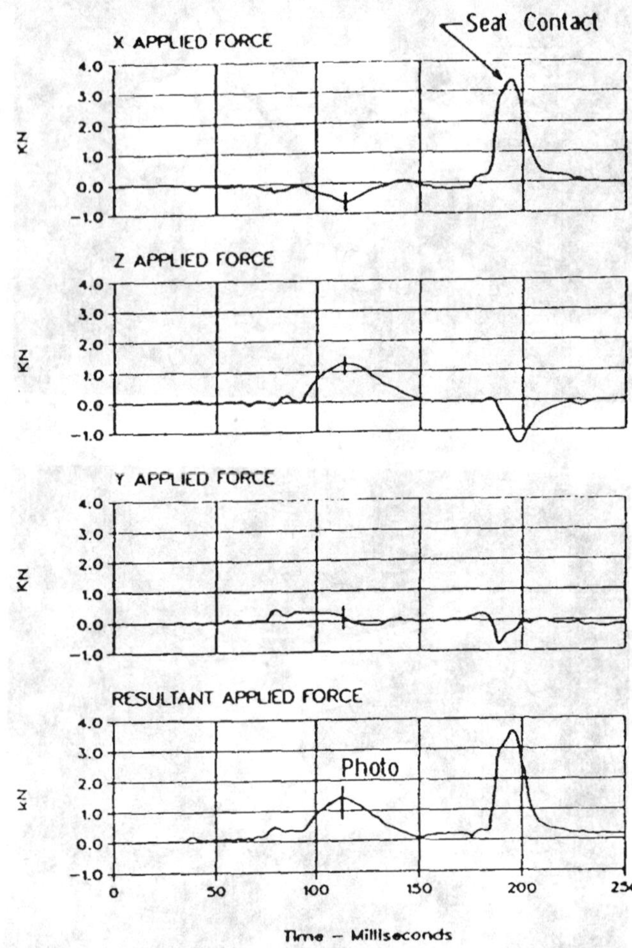

Figure 9. Computed external interaction force acting on the head from the head dynamics instrumentation, illustrating potential for the inflated cushion to restrain the chin.

Table 3

Exploratory Sled Test Responses. Hybrid III Dummy.

Response*	Double Belt Configuration (Fig. 5b) Inflated	Conventional	Ratio	Single Belt Configuration (Fig. 5a) Inflated	Conventional	Ratio
Head						
HIC	796	995	0.80	968	1289	0.75
Max Ang Accel (rad/sec^2)	3350	4050	0.83	3450	4300	0.80
Max Ang Vel (rad/sec)	51	58	0.88	48	61	0.79
Max Chin Force (kN)	1.40	0.67	-	1.29	0.98	-
Max Horizontal Disp (mm)	441	528	0.84	499	554	0.90
Neck						
Max X Force (kN)	1.50	1.78	0.84	1.69	2.09	0.81
Max Z Force (kN)	2.96	3.42	0.87	2.85	3.47	0.82
Max Y Moment (N;m)	71.7	141	0.51	124	158	0.79
Chest						
Max Comp (mm)	56	58	0.97	35	58	0.60
Max Comp VC (m/s)	0.34	0.37	0.92	0.19	0.31	0.61
3 ms Res Accel (g)	37.5	38.4	0.98	36.0	40.7	0.88
Max Shldr Belt Tension (kN)	10.3	11.0	0.94	9.9	11.9	0.83

* Inflated belt data is the mean of two tests. Conventional is a single test.

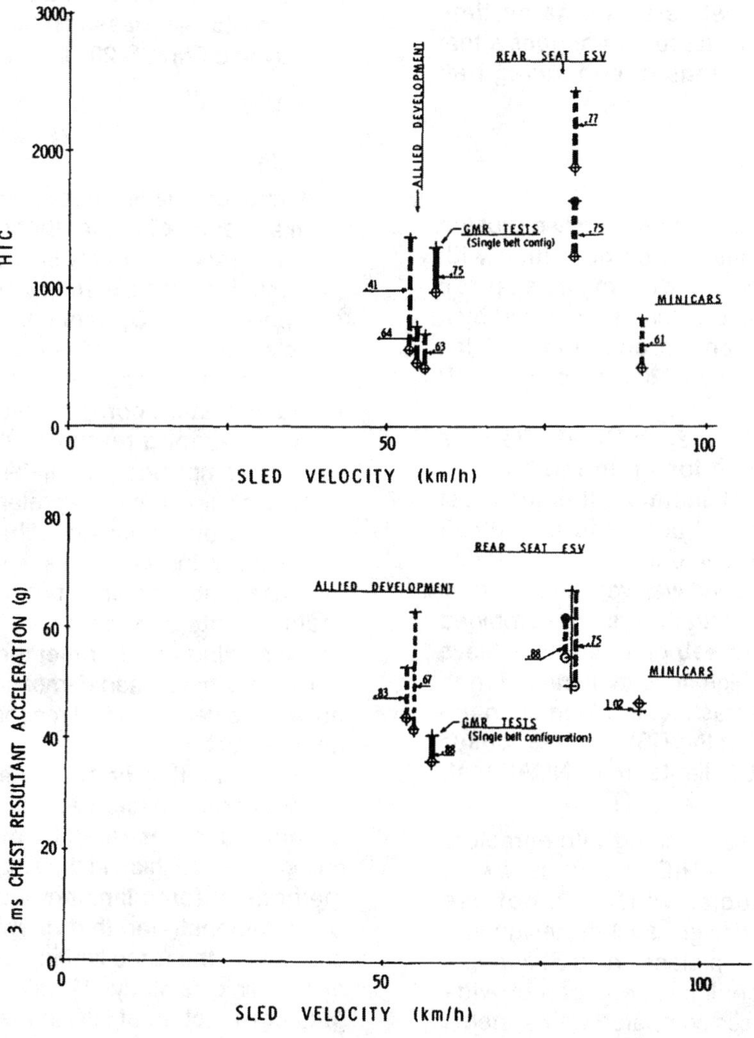

Figure 10. Comparison of the influence of belt inflation observed in the exploratory sled tests for the single belt configuration (Fig. 5a) with estimates made from data provided in the literature (from Fig. 1).

COMPARISON OF EXPERIMENTAL RESULTS - The single belt configuration (Figure 5a) provided the greatest reduction of response amplitudes and is used as the basis for comparison with previous studies for the improvement of HIC and chest acceleration. When only the influence of inflation is considered in the previous studies, the results of the exploratory experiments fall within the range of previous studies. The estimated changes of HIC and chest acceleration due to belt inflation are shown in Figure 10. Clearly such a comparison is at best suggestive of relative performance since different test environments, different dummies, and different restraints were used among the studies.

It is unlikely that optimum performance for an inflatable belt was achieved in the exploratory experiments. On this basis, the results may be conservative for predicting the potential for inflatable belts as part of a restraint system. Additionally, performance cannot be fully demonstrated by testing at a single severity [4], or based on only dummy responses of HIC and chest acceleration [5]. What might be perceived as best likely depends on the application and on the criteria used to judge performance. In the exploratory tests, all responses indicated a less severe dummy interaction associated with belt inflation compared with conventional webbing, while at the same time reducing dummy displacement. This further suggests that belt inflation might be one means of enhancing belt system test performance.

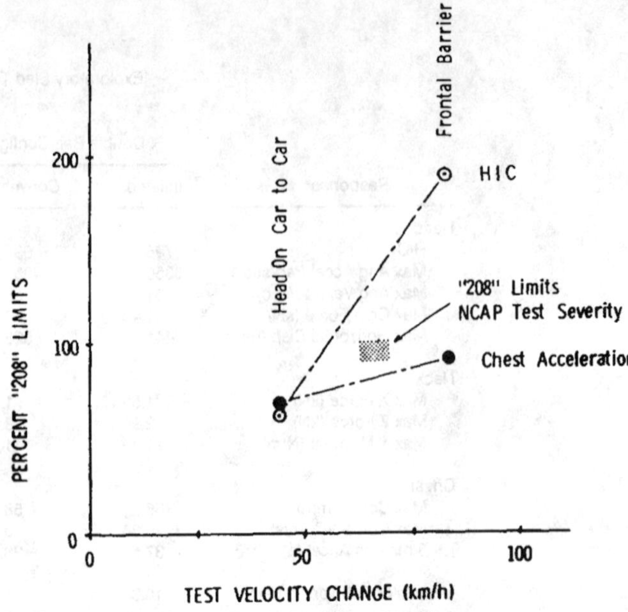

Figure 11. Comparison of test responses obtained for the Calspan/Chrysler Research Safety Vehicle with NCAP test severity and FMVSS 208 limits.

DISCUSSION

Inflatable belt restraints have been previously evaluated as a restraint system component in combination with other components to reduce dummy response amplitudes. Low dummy responses can be achieved by a variety of restraint systems. Some production vehicles have response amplitudes less than "208" limits in NCAP 35 mph barrier crashes.

The Calspan/Chrysler Research Safety Vehicle provides the best comparison for an inflatable belt restraint relative to the current situation. It is the most recent inflatable belt development and used an FMVSS dummy, the Hybrid II. This experimental vehicle had belt retractors, a steering assembly, and was vehicle based as opposed to a sled fixture. The restraint system combined belt inflation, force limiters, and web grabbers to achieve "low" response amplitudes in high severity tests (45 mph frontal barrier crash). Frontal crash tests which span the NCAP test severity suggest that the RSV test responses would have been close to "208" limits in an NCAP test, Figure 11.

A study [6] using conventional webbing with retractors in 50 kmph sled tests achieved a HIC of 280, as low as that from inflatable belt studies which did not use retractors, by the use of seat design, anchor design and location, webbing grabbers, and pretensioning. It appears probable that other system configurations can provide equivalent dummy test responses compared with systems that used inflatable belts. What is perceived as best is dependent to some extent on the test environment, the overall system, the available space for dummy displacement, what criteria are used to define performance, the test goals, and non-safety related issues such as complexity, availability, cost, comfort, etc.

Test response amplitudes are dependent on the interaction of vehicle and restraint system components and parameters. System constraints such as space and vehicle crash pulse influence system configuration and restraint system requirements. Thus a variety of effective restraint system components and configurations have evolved. Among features which tend to reduce test response amplitudes are pretensioning and force limiting.

Belt inflation provides pretensioning by shortening the belt and/or by expansion of the cushion between the belt and dummy. Increased cushion inflated thickness should increase pretensioning, but will result in increased gas requirements and increased bulk. Other less complex pretensioning or other methods to reduce test response amplitudes have been demonstrated. Pretensioning might not always decrease test response amplitudes dependent on other factors.

Force limiting or energy absorption is provided by cushion compression and enhanced by venting of the cushion (demonstrated by previous testing [1]) or by rolling of the cushion in the single belt configuration. Other methods of force limiting and energy absorption have been demonstrated that directly limit belt tension. One example is the force limiting belt anchors demonstrated by the Minicars study [1] which appears to provide much greater reduction of HIC in the specific test environment than for any of the reported inflatable belt results in their respective environments. Force limiting results in increased displacement which should be consistent with available space.

LOAD DISTRIBUTION - An inflatable belt appears to provide increased load area on the dummy and on this basis probably a more uniform load distribution on an occupant and a lower potential for lateral loading of the thorax than other belt systems or enhancements due to the inflated belt being able to "roll", Figure 12. The Allied Chemical configuration in which the inflated cushion is also the tension member, Figure 5c, might achieve a better load distribution than would either of the configurations which have conventional webbing as the tension member. The double belt configuration, Figure 5b, would locate conventional webbing against the occupant. The single belt configuration, Figure 5a, could place the conventional webbing against the occupant with sufficient roll.

The hypothesis that improved load distribution would reduce injury potential seems correct. However tests with dummies do not appear to directly demonstrate this possible important feature of inflatable belt restraints. Thus inflatable belts might not appear as competitive as other methods to achieve restraint system responses if the influence of load distribution cannot be objectively demonstrated.

<u>Volunteer Tests</u> - Human volunteer tests conducted at Southwest Research Institute [1] provide a more relevant injury model and no significant injury was found with test velocities to 32.5 mph. In evaluating the significance of these tests relating to load distribution, it would be helpful if results were available for conventional webbing tests having similar restraining forces.

All volunteers were in their 20's, were thoroughly examined to assure that they were not likely to be injury prone, and all had previous sled experience. "Pretest briefing emphasized the importance of coordinated body bracing." The volunteers had peak shoulder belt loads less than 65% of those of the test dummy in spite of the volunteers being heavier, likely due to "coordinated body bracing". Thus injury risk was influenced by the relatively low restraint load due to bracing and good system performance and a probable high tolerance of the volunteers.

The peak shoulder belt tension as a function of the volunteer's age is compared with car occupant injury as a function of shoulder belt tension reported by Foret-Bruno [7], Figure 13. On this basis the belt tension would need to double or the volunteer's age increased to about 60 years to have experienced thoracic injury if a conventional webbing had been used with the same level of restraint loads. The volunteer tests were conducted at too low of a severity to establish the potential benefit of the greater loaded area associated with an inflatable belt.

<u>Cadaver Tests</u> - Sled tests with human cadaver subjects were reported by Calspan [8]. Less severe injury was associated with tests having an inflatable belt restraint system compared with a belt restraint system having conventional webbing, even though the inflatable belt tests were conducted at a greater severity. However there were important age and skeletal differences among the cadaver subjects and important differences among the test situations than inflatable vs conventional webbing.

The conventional webbing tests used a Citroen fixture, the inflatable belts a Pinto fixture. Two of the three tests with the conventional webbing had a steering assembly which was contacted, the third test with conventional webbing and the two tests with the inflatable belt did not have a steering assembly. The tests with the inflatable belt restraint also had force limiting anchors, the conventional webbing tests did not. The mean age for the subjects was 59 years with the conventional webbing and 51 years with the inflatable restraint (65 vs 51 years for tests without a steering assembly). The mean subject weight was 60.9 kg for tests with conventional webbing compared with 73.2 kg for tests with inflatable belts (56.8 kg vs 73.2 kg with out a steering assembly).

Shoulder belt tension was chosen from the reported test responses as the best measure of subject loading severity. This estimates the influence of the inflatable belt compared with conventional webbing separated from the other system differences. The mean belt tension for three tests with conventional webbing was 8.3 kN (9.1 kN for the test without steering assembly). This compares with 6.0 kN for the force limited inflatable restraint. The second test with the force limited inflatable restraint did not have shoulder belt tension reported.

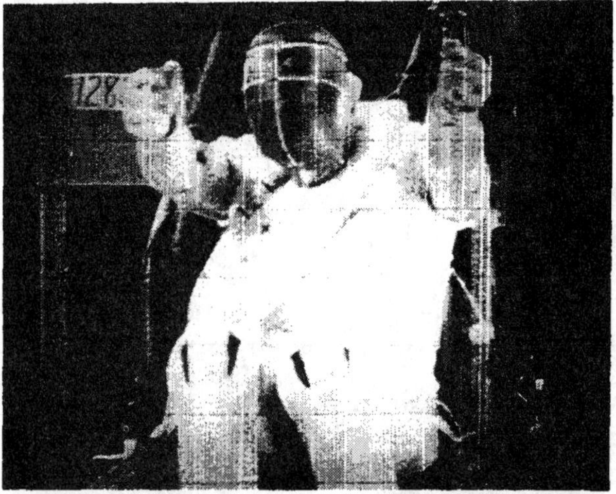

Figure 12. Visual comparison of load distribution for inflated cushion and conventional webbing.

Eppinger [9] developed an empirical relationship between the number of rib fractures and the loading severity in terms of shoulder belt tension for belt restrained cadavers. His analysis included the age and weight of the subject. Applying his relationship to the tests without the steering assembly predicted 31 rib fractures compared with 32 observed for the conventional webbing test. The autopsy stated, "generalized osteoporosis of the skeletal structures were noted, this could represent disuse atrophy, however, specific pathologic changes other than the age alone may be involved, such as --".

Eppinger's relationship predicted 13 rib fractures compared with 8 observed for the in the first test with the inflatable belt restraint. Shoulder belt tension was not reported for the second test having force limiting. If it is assumed that shoulder belt tension in this second test was the same as in the first test due to force limiting, Eppinger's relationship predicts 14 rib fractures as compared with only one observed. Although this data could be interpreted to support the hypothesis that inflatable belts provide a beneficial increase in load area, the results are too few and the restraints, environments and subjects have too many differences to consider this a proof of the hypothesis.

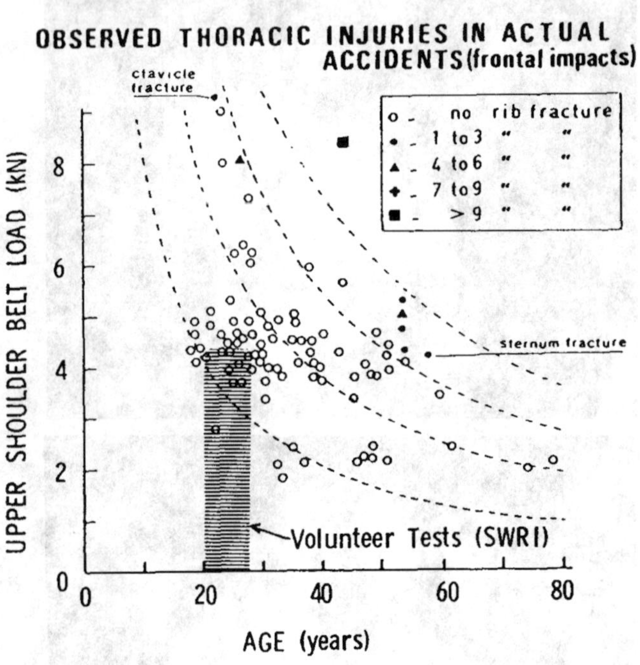

Figure 13. Peak shoulder belt load (tension) vs occupant age with thoracic skeletal fractures noted, derived from car crash investigations by Foret-Bruno [7]. Superimposed (shaded area) from the Southwest Research Institute volunteer tests with inflatable belt restraints is the range of peak shoulder belt tensions and age.

DESIGN ISSUES - A variety of system configurations were considered by previous inflatable belt studies including both manual and automatic placing of the belt on the occupant. Both front and rear seat restraint systems were studied. Selection of restraint system features will likely be influenced by dummy responses and other issues such as comfort, and component availability. Inflatable belts will continue at a disadvantage if reduction of injury risk due to increased load area can not be objectively demonstrated and if design issues are not fully resolved.

Several fundamental design issues have not been fully resolved, the most important being the location of the gas generator and possible discomfort from the "bulky" nature of the inflatable portion of the belt. Gas for inflation of the belt was provided either from stored high-pressure gas or generated by pyrotechnics in previous studies. The hardware used in the exploratory experiments had a pyrotechnic inflator (64 mm by 48 mm by 22 mm) located inside the cushion, Figure 4. Location of the inflator inside of the cushion adds to the "bulk" of the inflatable belt with potential comfort implications. Location of the gas source external to the belt requires "piping" from the gas source to the cushion with packaging issues related to buckles and/or retractors. Allied Chemical studied both internal and external location of the inflator. Inflation of the belt was successful for either location; however practical problems were not fully resolved for either location.

The inflatable portion of an air belt restraint would include that portion over the sternum, shoulder, and beside the neck. The extra "bulk" might represent a comfort issue, with decreased use rates if not properly addressed. The attachment of a cushion to conventional webbing as shown in Figures 5a and 5b will likely have greater "bulk" than the use of the cushion as the tension member as shown in Figure 5c. In this configuration, the cushion must be strong enough for both the pressure generated tensions and the restraint tension. Allied Chemical developed cushions which successfully carried restraint loads in severe environments with most design problems relating to the cushion-to-belt attachment. With advances in materials and processes, further advances in inflatable belt design might be possible.

Our study did not evaluate inflation of the lap-belt. This should provide pretensioning, force limiting, and increased load area for lower body restraint. How an inflated belt would influence the belt riding over the pelvic structure and directly load the abdominal region might be an issue.

SUMMARY

Our study clearly indicates that an inflatable belt can be one component of a belt restraint system which tends to reduce test response amplitudes. However, other belt restraint system configurations have demonstrated equivalent test response amplitudes to those having an inflatable belt. Inflatable belt restraints on this basis appear to have a disadvantage because comfort and packaging of inflatable belts have not been fully resolved in spite of significant efforts in the past.

Inflatable belt restraints were developed by Allied Chemical in the 1970's. These and later studies demonstrated "low" dummy response amplitudes in crash severities up to 50 mph and no significant injuries to volunteers up to 32.5 mph. Belt inflation contributed to these improvements. However other restraint system features were also important to obtain the reported system test results.

Exploratory sled tests conducted at the GM Research Laboratories in the mid-1980's produced results with an inflatable belt restraint and Hybrid III dummy that were consistent with previous studies. These tests demonstrated that not only were reductions of HIC and chest acceleration associated with belt inflation, but also neck forces and moments, chest compression, head angular acceleration, and upper body displacement.

Inflatable belt restraints can provide pretensioning, energy absorption, and increased load area. Pretensioning and energy absorption can be achieved by less complex means, the desirability of these features depends on system characteristics and constraints. The potential benefit of increased load area could not be determined from current dummy responses. Because of "system" approaches and insufficient severity, previous volunteer and cadaver test results do not objectively demonstrate the benefit of increased load of an inflatable belt restraint compared with conventional webbing.

Advancement of the state-of-art for inflatable belt restraints might be best served by an objective demonstration of the influence of the increased loaded area to reduce belt related injury risks and by addressing design issues such as packaging and comfort than by another study which achieves "low" dummy HIC and chest acceleration.

ACKNOWLEDGMENT

The inflatable cushions, inflators and attachment to the conventional webbing were obtained from Morton International. David Dahle of Morton International coordinated the adaptation of the test hardware which had been developed for a helicopter crew restraint system. Richard Frantom of the Bendix Safety Restraints Group of Allied Signal Inc. provided helpful discussion and a technical summary of previous inflatable belt studies. David Viano, John Melvin, Steve Rouhana, David Dahle, and Richard Frantom provided technical review of the paper.

REFERENCES

1. "Technical Summary of Inflataband Development," A summary of inflatable belt development efforts made available by the Bendix Safety Restraints Division of Allied Automotive.

2. K.H. Digges and J.B. Morris, "Opportunities for Frontal Crash Protection at Speeds Greater than 35 MPH", SAE 910807, SAE Congress, 1991.

3. D.C. Viano, J.W. Melvin, R. Madeira, J.D. McCleary, R. Shee, and J.D. Horsch, "Measurement of Head Dynamics and Facial Contact Forces in the Hybrid III Dummy," SAE 861891, 30th Stapp Car Crash Conference, 1986.

4. J.D. Horsch, " Evaluation of Occupant Protection from Responses Measured in Laboratory Tests," SAE No 870222, SP-690, SAE Congress, February, 1987.

5. J.D. Horsch, J.W. Melvin, D.C. Viano, and H. Mertz, "Thoracic Injury Assessment for Belt Loading" In preparation for 1991 Stapp

6. L.G. Svensson, "Means for Effective Improvement of the Three-Point Seat Belt in Frontal Crashes," SAE 780898, 22ond Stapp Car Crash Conference, 1978.

7. J.Y. Foret-Bruno, F. Hartemann, C. Thomas, A. Fayon, C. Tarrier, C. Got, and A. Patel, "Correlation Between Thoracic Lesions and Force Values Measured at the Shoulder Belt of 92 Belted Occupants Involved in Real Accidents," SAE 780892, 22ond Stapp Car Crash Conference, 1987.

8. M.J. Walsh, "Sled Tests of Three-Point Systems Including Air Belt Restraints," Final Report, Contract No. DOT-HS-5-01017, August 1976.

9. R.H. Eppinger, "Prediction of Thoracic Injury Using Measurable Experimental Parameters". Report 6th International Technical Conference on Experimental Safety Vehicles, pp 770-779. National Highway Traffic Safety Adminstration, Washington, D.C.

1999-01-0085

Injury Mitigating Benefits of an Inflatable Shoulder Belt for Seat Integrated Application

Shyam S. Karigiri and Robert V. McClenathan
Universal Propulsion Company, Inc.

James A. Kargol and Ilango Shanmugavelu
Johnson Controls, Inc.

ABSTRACT

This study examines the added safety improvement effects of the Integrated (Belted) Structural Seat (ISS^{TM}) combined with an inflatable shoulder belt that inflates to form a cylindrical airbag during an automobile frontal crash. Although inflatable shoulder belts have been studied for many years with proven results, occupant impact response data and sled testing has been very limited to date when combining these two relatively new types of automotive safety systems.

Accident studies show that frontal collisions, are by far the crashes with the most serious consequences. In frontal collisions, the main lesions placed upon the occupants are thoracic, mainly high chest deceleration, head injuries which result from the high deceleration levels due to occupant's contact with the vehicle interiors, and rib fractures resulting from the high seat belt contact forces. In order to improve this situation, it becomes important to ensure that the means used to restrain the automotive occupant(s) work as efficiently as possible early in the crash event to reduce the motion of the torso and head.

With this perspective, it becomes important to have an occupant restraint system which has the overall injury mitigating benefits of the conventional restraint systems, but at the same time eliminates certain other potential injury mechanisms. The seat integrated inflatable shoulder belt is one such self contained restraint system which has improved occupant protection capabilities.

HyGe^{TM} sled tests were conducted to evaluate the injury mitigating benefits of the seat integrated inflatable shoulder belt restraint system. Sled tests were conducted for the three different sized Hybrid III dummies and using a 35 mph crash pulse. Detailed results of the sled tests are presented later in this paper, and the results are also compared with the published NCAP results for a mid-size production vehicle that has a similar vehicle crash pulse and a conventional 3-point seat belt and airbag system.

INTRODUCTION

Occupant protection in vehicular environments during a crash relies on decelerating the occupant over the longest time possible while preventing striking to any interior objects. Significant improvements in occupant protection can be provided by an inflatable shoulder belt which inflates very early in the crash event. It inflates in approximately 10 milliseconds after the crash sensor signal and simultaneously tightens (pretensions) the belt. Inflation reduces the belt length by approximately 12 % and applies a pre-load of about 2 kN on the occupant's upper torso. This causes belt slack to be eliminated so that it initiates occupant deceleration with the smallest possible delay. This type of restraint system also has the advantage that it inflates away from the occupant instead of deploying towards the occupant at high speeds. The seat integrated inflatable belt when inflated, provides a cushioned control area of approximately 0.16 m^2. This causes distribution of the restraint load on the occupant and is thereby effective in reducing the well recognized occurrence of belt induced trauma. Moreover, the occupant's head and neck motion are significantly reduced as compared to conventional restraint systems.

Historically, inflatable shoulder belts have shown immense promise as an effective automobile occupant restraint system. However, the problems associated with such a system back in the 1970s were those of stowage, bulkiness of the inflatable section, large and heavy gas generators, large and unreliable sensors, etc. Technological developments and improvements in the area of inflators, high tenacity lightweight fabric, etc., have addressed and solved many of the older concerns such as inflator output gas temperature, packaging, styling and comfort.

Many agencies and companies have evaluated the inflatable seat belt system in the past. Extensive research and development work has been carried out in the area of inflatable seat belt systems. Most of the prior published data are from crash tests that were carried out under conditions involving lower impact speeds and non-production intent hardware. The integrated inflatable belted seat sled data presented in this paper, uses NCAP level test conditions and production intent hardware.

DYNAMIC SCIENCES, INC. – Dynamic Sciences Inc., prepared a report (Report No. 8300-77-146) for NHTSA (Contract DOT-HS-6-01307). The objective of this basic research program was to establish the occupant performance limits of four advanced restraint systems . One of these was a force-limited air belt. These restraint systems were integrated into a production compact automobile which had shown acceptable crashworthiness performance. The Volvo 244 and the Ford Torino were the cars chosen. The crash test conditions were selected to determine the performance limits of the combined advanced restraints and production structural systems.

The test configuration involved different crash modes including car to car Head-on, Oblique, Offset and 90° Barrier impacts. The closing speeds involved in the car-to-car tests exceeded 80 mph. The speeds involved in the barrier tests exceeded 45 mph. HIC values as low as 205 and Chest 3 ms Criteria values of 30 g were observed on some of these tests.

It was concluded that the inflatable seat belt concept restraint provided adequate protection for frontal impact at speeds up to 48 mph and excellent protection for the 40-45 mph speed range.

IMPACT TESTING OF ALLIED CHEMICAL "INFLATABAND" – A series of 69 dynamic sled tests were conducted in 1975 to evaluate the effectiveness of the "Inflataband" restraint system as a viable method of occupant protection in simulated head-on automotive crashes. Of these, 30 tests involved dummies and the other 39 involved human volunteers. Kinematic performance of anthropomorphic dummies and human volunteers under simulated impact conditions was evaluated.

The report was prepared for NHTSA (Contract No. DOT-HS-4-00933 October 1975). The deceleration pulse utilized for the 32.5 mph total velocity change tests, was a representative 30 mph barrier crash pulse for a 1972 Pinto with a peak deceleration of less than 20 g.

In no case did the observed severity indicators (HIC, Chest Acceleration 3 ms clip) approach or exceed the existing human tolerance levels for these indicators. HIC values as low as 174 and Chest 3 ms Criteria values of 18 g were observed on some of these tests. Injuries to the human subjects were minimal consisting primarily of mild erythema to the face and neck. At the higher impact severities, some residual neck soreness which existed for a short period of time was documented. Impact forces on the upper torso and abdomen were effectively distributed without major discomfort.

AIR BELT RESTRAINTS – A experimental test program was conducted in 1976. This program included six sled tests using five unembalmed cadavers and one anthropomorphic test dummy to evaluate a standard three point belt system and an air belt which inflated during impact. Two of these cadaver tests were conducted to evaluate the airbelt restraint system. The other tests were for the standard three-point seat belt system. The report was submitted to NHTSA (Contract No. DOT-5-01017). The tests were conducted by simulating 30 mph frontal collisions for the three-point seat belts and 47 mph frontal collisions for the air belt systems.

Cadaver instrumentation recorded levels of acceleration at the head below those accepted as fatal limits. The chest peak resultant acceleration measured was in the range of 48 g and 50 g. This coupled with a Chest Severity Index (CSI) in the range of 510 and 600, must lead to the interpretation that these 47 mph crashes would have been painful but not life endangering since :

Head and NeckAIS 0
ChestAIS 2/3
AbdominalAIS 0

MATERIALS AND METHODS

Universal Propulsion Co., Inc. has since revisited and developed the Inflataband™ inflatable seat belt restraint system. This system utilizes production intent hardware. For the purposes of evaluating the occupant protection performance of the seat integrated inflatable shoulder belt system, a series of ten sled tests were conducted.

Seat integrated belt restraint systems have gained interest and popularity because of their ability to provide good occupant protection, improved seat belt packaging and comfort. The integrated structural seat system is an automobile seat which has the belt retractor assemblies mounted onto the seat. The integrated structural seat enhances occupant protection in all of the basic crash modes. The belt positioning on the occupant is consistent and improved regardless of the seating position. The overall length of the seat belt webbing is reduced, leading to less webbing elongation and hence less occupant excursion. Additional safety features such as side airbags can be integrated to the seat to further enhance occupant protection.

The seat selected for this study was a standard production JCI Integrated Structural Seat (ISS™) which has conventional seat belt restraint system on it. The seat structure was modified to replace the existing seat belt with the Inflataband™ system. The upper torso retractor assembly was relocated. A guide tube assembly was added in the seat back to house and route the inflatable shoulder belt system attached to the inflator. The appropriate buckle and latch plate hardware was selected and mounted on the production seat. The rest of the seat structure was left unchanged. Figure 2 is a schematic of the dummy positioned on the inflatable belt integrated seat system. The belt and retractor assembly mounting locations are clearly depicted. Figure 1 shows a cut away section of the seat back frame of the production ISS™ modified to accommodate the inflatable seat belt system and guide tube assembly.

The Inflataband™ is essentially a 6 to 7 inches diameter fabric tube when inflated. In the stowed configuration, it forms a seatbelt as shown in Figure 3. The inflatable section is enclosed by a fabric cover which tears open in a controlled manner during inflation. The assembly is attached at it's bottom end to a seatbelt buckle latch plate using a short length of conventional lap belt webbing. The lap belt section of the restraint system is similar to the conventional seat belt system. The upper end of the inflatable shoulder belt through which inflation occurs is connected to a Universal Propulsion Company's proprietary Direct-Thermal™ Inflator (DTI). This inflator type produces the cool gas that is required for this restraint system as it deploys in close proximity to the occupant. Deployment tests have demonstrated a temperature increase of 10° C on the inflatable belt fabric surface. This restraint system can be adapted to either a vehicle integrated or a seat integrated environment.

SLED SETUP – All tests were performed on HyGe™ sleds. In all the tests, the inflatable shoulder belt integrated seat was mounted on a rigid flat plate which in turn was mounted on the sled. Automobile interior environment structures like the steering wheel, instrument panel, knee bolster etc. were not used in any of these 10 tests. This condition basically represented and tested the performance of the seat integrated inflatable seat belt restraint system as a stand alone system.

The accelerometers mounted on the sled recorded the sled acceleration. The dummies were fitted with tri-axial accelerometers in the head, chest and pelvis to record the acceleration response. Load cells mounted at the dummy upper neck recorded the upper neck shear and axial forces and neck bending moments. The chest pot on the dummy recorded the sternum deflection. A pressure transducer measured the internal pressure of the inflatable shoulder belt. A thermocouple mounted on the inflatable seat belt measured the surface temperature. Load cells were mounted on the upper torso loop and the lap belt loop webbing to record the restraint system loads. Output signal conditioning was achieved using the standard recommended SAE filters. Table 1 summarizes the various channels of data collected and the corresponding filters used.

Five 13 mm wide angle lens high speed cameras were appropriately positioned to record the dummy kinematics and restraint system hardware performance during the crash event. These cameras captured the side, front and top views.

All the sled tests were conducted using a typical 35 mph mid-sized vehicle NCAP crash pulse depicted in Figure 4. Tests 1 through 6 were conducted using the 50th percentile Male Hybrid III dummy. Sled tests 7 and 8 evaluated the restraint system for a 5th percentile Hybrid III dummy. Sled Tests 9 and 10 evaluated the restraint system for a 95th percentile Male Hybrid III dummy.

RESULTS

Results of the ten sled tests for the three different sized occupants restrained by the inflatable shoulder belt system are presented. The data is presented under three different sections, one each for the three different size dummies. The data presented includes the plots of the resultant acceleration of the dummy head, chest and pelvis; the dummy sternum deflection plot and the dummy upper neck force and bending moment plots.

50TH PERCENTILE DUMMIES – Figure 5 illustrates the kinematics of the 50th %ile Hybrid III male dummy during the crash event, as recorded by the driver side off-board high speed camera. Frame (a) illustrates the dummy positioned in the seat and restrained by the inflatable seat belt at time T_{zero}. Frame (b) shows the fully deployed inflatable shoulder belt at about 20 ms. Frames (c) through (f) shows the sequence of the dummy kinematics during the crash event.

The dummy head resultant acceleration curves for Tests 1 through 6 are as shown in Figure 6. The rise in the head g between 10 and 20 ms is due to the deployment of the inflatable seat belt and it's interaction with the dummy's head. The head resultant peak acceleration vary between 45 g and 55 g. The corresponding HIC values obtained from these curves are compared in Figure 7 and are well within the FMVSS 208 allowable limit of 1000.

The pretensioning effect of the inflatable seat belt is clearly visible from the dummy chest resultant acceleration curves shown in Figure 8. The rise in chest g at about 10 ms indicates the deployment and subsequent tightening of the inflatable seat belt around the dummy's upper torso. This causes the dummy chest excursion to be minimal. The chest resultant acceleration curves closely follow the sled acceleration trace. This velocity coupling is the primary design intent of the inflatable seat belt. The 3 ms clip of the peak chest resultant acceleration is compared in Figure 9 and is well within the FMVSS 208 allowable limit of 60 g.

The loading on the dummy chest by the inflatable seat belt has a direct bearing on the dummy chest deflection. The chest deflection traces as obtained from the dummy chest pot are indicated in Figure 10. Out of the six, two tests indicated faulty dummy chest pot instrumentation. The peak chest deflection values are well below the FMVSS 208 allowable limit of 76 mm.

The dummy pelvis resultant acceleration plots are illustrated in Figure 11. Out of the six tests, one test indicated faulty dummy pelvic accelerometer. The peaks in the pelvis resultant acceleration at about 50 ms is a clear indication of the dummy's pelvis contacting the anti-submarining ramp of the seat structure.

Table 1. Sled Test Data Channels and Filters

Channel Description	Filter Class
Sled Acceleration	60.0
Dummy Head Acceleration	1000.0
Dummy Chest Acceleration	180.0
Dummy Chest Deflection	180.0
Dummy Pelvic Acceleration	1000.0
Dummy Upper Neck Force	1000.0
Dummy Upper Neck Moment	600.0
Inflatable Seat Belt Pressure	1000.0
Inflatable Seat Belt Thermocouple	1000.0
Inflatable Seat Belt Lebow	60.0
Lap Belt Lebow	60.0

The inflatable seat belt is in close proximity to the dummy's neck. Hence, it becomes necessary to closely monitor the dummy upper neck forces and bending moment. The dummy upper neck fore-aft shear force curves are depicted in Figure 12. The fore shear force reaches it's maximum value around the time when the dummy's head is beginning to rebound. The peak upper neck fore-aft shear force is well within the allowable maximum of 3100 N. The aft shear force is low and well within the allowable maximum of 3100 N. The upper neck peak fore-aft shear forces are compared with the allowable maximum in Figures 13 and 14.

The dummy upper neck lateral shear force as indicated in Figure 15 shows some peaks early in the crash event. This is due to the deployment of the inflatable seat belt and subsequent lateral loading on the dummy upper neck. Later in the crash event, the force pattern reverses direction and reaches it's maximum value around the time when the dummy's head is beginning to rebound. The peak upper neck lateral shear force is compared in Figures 16 and 17 and is well within the allowable limit of 3100 N.

The dummy upper neck axial force plots are depicted in Figure 18. The plot shows initial peaks between 10 and 20 ms. This is due to the deployment of the inflatable seat belt which causes the dummy to be pushed down into the seat. The inflatable seat belt supports the head and neck during the crash event which causes the neck axial loads to be low when compared to the allowable limits.. Figures 19 and 20 compare the neck axial load with the allowable limits of 3300 N in tension and 4000 N in compression.

The dummy upper neck flexion-extension bending moment plots are illustrated in Figure 21. The peak values of the flexion and extension bending moments are compared in Figures 22 and 23. It can be noted that the peak upper neck flexion and extension bending moments are well within the allowable limits of 190 N-m and 57 N-m respectively.

5TH PERCENTILE DUMMIES – The dummy head resultant curves for Tests 7 and 8 are as shown in Figure 24. The rise in resultant head acceleration g as a result of the interaction of the inflatable seat belt with the dummy's head is clearly visible between 10 and 20 ms in the crash event. In Test 8, due to faulty instrumentation, the Z channel data of the dummy head acceleration was lost. The resultant acceleration has been calculated ignoring this component. The corresponding HIC values as obtained from these curves are compared in Figure 40. and are well within the allowable limit of 1113.

The exposed length of the inflatable section of the restraint system, outside the seat trim is the least for the 5th percentile dummy when compared to the 50th and the 95th percentile dummies. This results in higher operating pressures inside the inflatable seat belt. As a result, higher pre-tensioning of the dummy occurs and this is obvious from the chest resultant acceleration curves depicted in Figure 25. The initial peak in the chest resultant curves at about 10 to 20 ms due to the early restraining effect of the inflatable seat belt causes the dummy excursion to be minimal. The 3 ms clip values as obtained from these chest resultant acceleration plots are compared in Figure 41 and are below the allowable limit of 73 g.

The dummy pelvis resultant accelerations curves are as shown in Figure 26. This gives details about the dummy excursion. The 5th percentile dummy pelvis shows minimal interaction with the anti-submarining ramp of the seat structure. The absence of spikes in the pelvis resultant acceleration traces support this statement.

On Test 8, the dummy had a faulty chest pot which recorded erroneous chest deflection values. The chest deflection trace for dummy in Test 7 is as shown in Figure 27. The peak chest deflection values exceeded the allowable limit of 46 mm by a small percentage. As only one data point is available, further testing is required to raise the statistical confidence in the results. Also, the restraint system can be suitably modified, for example by providing a vent in the inflatable section, to rectify this situation.

The inflatable seat belt rides closest to the neck on the 5th percentile dummy compared to the other sized dummies. It is thus essential to evaluate the loading pattern on the dummy upper neck by the inflatable seat belt. The upper neck fore-aft shear force plot is depicted in Figure 28. The aft shear force exerted on the dummy upper neck is very low and is compared in Figure 43 with the allowable limit of 2068 N . The fore-aft shear force exerted on the dummy is compared in Figure 42 and Figure 43 and is well within the allowable limit of 2068 N.

The lateral shear force on the dummy upper neck shows initial peaks due to the deployment of the inflatable seat belt as shown in Figure 29. The peak upper neck lateral force values are compared in Figure 44 and Figure 45 and are well below the allowable limit of 2068 N.

The dummy upper neck axial force plots are depicted in Figure 30. The plot shows initial peaks between 10 and 20 ms. This is due to the deployment of the inflatable seat belt which causes the dummy to be pushed down into the seat. The inflatable seat belt supports the head and neck during the crash event which causes the neck axial loads to be low when compared to the allowable limits.. Figures 46 and 47 compare the neck axial load with the allowable limits of 2201 N in tension and 2668 N in compression respectively.

In test 7, the dummy upper neck extension bending moment exceeded the allowable limits by a small percentage. This situation can be easily corrected by slight modifications to the restraint system like venting of the inflatable section during the crash event. The flexion-extension bending moment curves are as shown in Figure 31. The peak values obtained from these curves are low and are compared in Figures 48 and 49 with the allowable limit of 104 N-m in flexion and 31 N-m in extension respectively.

95TH PERCENTILE DUMMIES – The dummy head resultant acceleration curves for Tests 9 and 10 are shown in Figure 32. The rise in the head resultant acceleration between 10 and 20 ms is due to the deployment of the inflatable seat belt and it's interaction with the dummy's head. The peak head resultant acceleration values vary between 45 g and 55 g. The corresponding HIC values as obtained from these curves are compared in Figure 40 and are within the allowable limit of 957.

The pretensioning effect of the inflatable seat belt is clearly visible from the dummy chest resultant acceleration curves in Figure 33. The rise in chest resultant acceleration at about 15 ms indicates the deployment and subsequent tightening of the inflatable seat belt around the dummy's upper torso. This causes the dummy chest excursion to be minimal. The chest resultant acceleration reaches the maximum value at about 85 ms due to the loading of the restraint system by the heavier 95th percentile dummy. The 3 ms clip of the peak chest resultant g's is compared in Figure 41 and is well within the allowable limit of 54 g.

The loading on the dummy chest by the inflatable seat belt has a direct bearing on the dummy chest deflection. The chest deflection traces as obtained from the dummy chest pot are indicated in Figure 35. Out of the two tests with the 95th percentile dummy, one had faulty dummy chest pot instrumentation. The peak chest deflection values are well below the allowable limit of 55 mm.

The dummy pelvis resultant acceleration plots are illustrated in Figure 34. The peak values range between 47 g and 52 g. The interaction between the dummy pelvis and the anti-submarining ramp of the seat structure is the highest with the 95th percentile dummy.

The dummy upper neck fore-aft shear force curves are depicted in Figure 36. The fore shear force reaches it's maximum value around the time when the dummy's head is beginning to rebound. The peak upper neck fore-aft shear force is compared in Figures 42 and 43 and is well within the allowable limit of 3807 N.

The dummy upper neck lateral shear force as indicated in Figure 37 shows some peaks early in the crash event. This is due to the deployment of the inflatable seat belt and subsequent lateral loading on the dummy upper neck. The lateral shear force reaches it's maximum value around the time when the dummy's head is beginning to rebound. The peak upper neck lateral shear force is well within the allowable limit of 3807 N and is compared in Figures 44 and 45.

The dummy upper neck axial force plots are depicted in Figure 38. The plot shows initial peaks between 10 and 20 ms. This is due to the deployment of the inflatable seat belt which causes the dummy to be pushed down into the seat. The inflatable seat belt supports the head and neck during the crash event which causes the neck axial loads to be low when compared to the allowable limits.. Figures 46 and 47 compare the neck axial load with the allowable limits of 4052 N in tension and 4912 N in compression respectively.

The dummy upper neck flexion extension bending moment plots are as illustrated in Figure 39. The peak values of the flexion and extension bending moments are compared in Figures 48 and 49. It can be noted that the peak upper neck bending moments are well below the allowable limits of 258 N-m in flexion and 78 N-m in extension respectively.

DISCUSSION

From a detailed review of the sled test results, it can be noted that all the dummy injury values like the HIC; dummy chest resultant acceleration 3 ms criteria; dummy chest deflection; dummy upper neck fore-aft, lateral and axial force; and dummy upper neck flexion-extension bending moment, (except one neck extension moment for the 5th percentile dummy) for the three different sized occupants are well below the recommended maximums. This clearly indicates that the seat integrated inflatable shoulder belt is very effective as a automobile occupant restraint system. Table 2 summarizes the various injury criteria as obtained from the ten sled tests and compares them with the allowable limits.

COMPARISON TO NCAP VALUES AND FMVSS LIMITS – Table 6 shows a direct comparison of the injury values from the sled tests with the 50th percentile dummy and those from the passenger dummy of the NCAP test for a mid-size production vehicle that has a conventional 3-point seat belt and airbag system, and a similar vehicle crash pulse.

It can be observed that, except for three of the severity indicators from the inflatable seat belt sled tests, the rest are lower than the values recorded on the NCAP test. Although three of the severity indicators were higher than the NCAP test results, they were well below the injury assesment reference values as seen from Table 3.

LIMITATIONS OF THE FINDINGS – The seat integrated inflatable seat belt system so far has been tested on a rigid plate sled, not in an actual vehicle environment. Although this represents a more stringent condition, the restraint system has to be tested in a vehicle to identify the enhancement of the occupant protection in that environment. Although the crash pulse chosen represents a fairly severe one, the restraint system has to be validated for other test conditions which may include more severe crash pulses. The seat and inflatable belt design parameters need to be optimized further, which, is hoped, would only increase the protection benefits further.

PLANS FOR FUTURE DEVELOPMENTS – A MADYMO computer simulation model for a frontal impact will be jointly developed with TNO-MADYMO and will be validated with sled tests of the inflatable belt integrated seat. The model will be used to assess the occupant protection of the inflatable belt integrated seat for additional, including more severe, conditions and crash pulses. The inflatable belt integrated seat will be sled tested with child dummies, and with forward and rear facing infant seats. Protection performance will be evaluated for out of position occupants. Misuse and abuse due care tests will be completed. The inflatable belt integrated seat will be evaluated for occupant protection in side impact. The belt tensioning and load limiting performance of the inflatable belt integrated seat will be compared with performance of non-inflatable belt system which has belt pre-tensioner and load limiting belt retractors. The crash protection performance of the seat integrated inflatable shoulder belt will be compared against the performance of the conventional 3-point seat belt system and a 3-point seat belt system with an airbag. Additional tests will be conducted to assess the performance in cases which involve interaction of the inflatable belt with frontal driver and passenger airbags. The performance of the inflatable belt integrated seat without the frontal airbags needs to be assessed for application in markets that do not have a passive restraint regulation.

Table 2. Sled Test Results Summary and Comparison with the Allowable Limits

	5th Percentile Female Dummy		**50th Percentile Male Dummy**		**95th Percentile Male Dummy**	
Injury Criteria	**Sled Test (Avg. of 2)**	**Allowable Maximum**	**Sled Test (Avg. of 6)**	**Allowable Maximum**	**Sled Test (Avg. of 2)**	**Allowable Maximum**
HIC	506	1113	482	1000	705	957
Peak Chest 'g' (3 ms Clip)	54	73	48	60	44	54
Chest Deflection (mm)	53.2	46	32.3	76	24.3	61
Upper Neck Fore-Aft Shear Force (N)	+ 10 / - 907	+/- 2068	+ 209 / -133	+/- 3100	+ 12 / - 1288	+/- 3807
Upper Neck Lateral Shear Force (N)	+ 39 / - 584	+/- 2068	+ 53 / - 160	+/- 3100	+ 91 / - 587	+/- 3807
Upper Neck Axial Force (N)	+ 1922 / - 56	+ 2201 / - 2668	+ 1987 / - 375	+ 3300 / - 4000	+ 2014 / - 62	+ 4052 / - 4912
Upper Neck Flexion Extension Bending Moment (N-m)	+ 51 / - 33	+ 104 / - 31	+ 68 / - 39	+ 190 / - 57	+ 83 / - 19	+258 / - 78

Table 3. Comparison of Sled Test Results vs. Published NCAP Results

Injury Criteria	Recommended Maximum	NCAP (Passenger)	Inflatable Belt Integrated Seat (avg. of 6)	Change (%)
Test conditions	For 50th Percentile Male ATD	Rigid barrier crash test, 50th percentile dummy, belted, airbag, full vehicle	Sled test, rigid floor, 50th percentile dummy, belted, no airbag, no knee bolster, no steering column	
Sled 'g', max (g @ msec)		33 @ 43	33	
Velocity change (mph)		35	33	
HIC	1000	698	482	-31
Head 'g', max (g@msec)		97.25@67	52	-46
Chest 'g' (3 m. sec clip)	60	54	48	-11
Chest deflection (inches)	3	1.14	1.27	+11*
Neck moment, max (Flexion / Extension - (Nm@msec)	+190 / -57	+36@49 / -61@91	+68 / -39	+88*/-36
Neck axial load, max (Tensile/ Compression - (N@msec)	+3300 / -4000	+2495@73 / -526@196	+1987 / -375	-20/-28
Neck shear load, max (Fore-Aft - (N@msec)	+3100 / -3100	+584@49 / -545@198	+209 / -133	-64/-75
Neck shear load, max (Lateral - (N@msec)	+3100 / -3100	+351@139 / -401@69	+53 / -160	-85/-60
Shoulder belt load, max (N)	6700	4534	4862	+7*
Lap belt load, max (N)	11160	5081	10929**	

* The measured values are well below the tolerance values

** The increase in the lap belt load is primarily due to the absence of the knee bolster in the sled test

CONCLUSION

The inflatable shoulder belt was combined with the integrated structural seat to create the inflatable belt integrated seat system that demonstrates the enhanced safety features of the two automotive safety systems. Sled tests were conducted for the inflatable belt integrated seat system with 5th, 50th, and 95th percentile Hybrid III dummies to evaluate the frontal impact protection capabilities of this type of restraint system.

The sled tests showed acceptable occupant protection indices for the seat integrated inflatable seat belt restraint system in terms of low HIC, low Chest 3 ms Criteria, low chest deflection, low neck force and bending moment values as compared to the published NCAP test results and the established Injury Assessment Reference Values (IARV).

ACKNOWLEDGMENTS

The authors gratefully acknowledge Mr. Alex Devonport (UPCO) for his help in the sled testing activities and Mr. Marc LaClair (UPCO) for his editorial assistance.

REFERENCES

1. Vehicle Integration and Evaluation of Advanced Restraint Systems. Test Reports - Phase A and Phase B. Report No. 8300-77-146. Contract DOT-HS-6-01307
2. Charles Strother, Michael U. Fitzpatrick, and Timothy P. Egbert. Development of Advanced Restraint Systems for Minicars RSV. SAE 766043
3. Mertz, H. J. Injury Assessment Values Used to Evaluate Hybrid III Response Measurements, General Motors Corporation NHTSA Docket Submission VSG 2284 Part III, Attachment I, Enclosure 2, 1984

4. Patrick, L. M. and Chou, C. Response of the Human Neck in flexion, extension, and lateral flexion. Vehicle Research Institute Report No. VRI-7-3. Society of Automotive Engineers, Warrendale, Pennsylvania, 1976

5. Michael J. Walsh, Sled Tests of Three-Point Systems Including Air Belt Restraints, Prepared for US DOT NHTSA, January 1976 (DOT-HS-5-01017)

6. James M. Burkes, J. Robert Cromack and Haskell Ziperman, Impact Testing of Allied Chemical "Inflataband" with Dummies and Human Volunteers, Volume II, Prepared for US DOT NHTSA, October 1975 (DOT-HS-4-00933)

7. Melvin, John W. and Nahum, Alan M., Accidental Injury : Biomechanics and Prevention

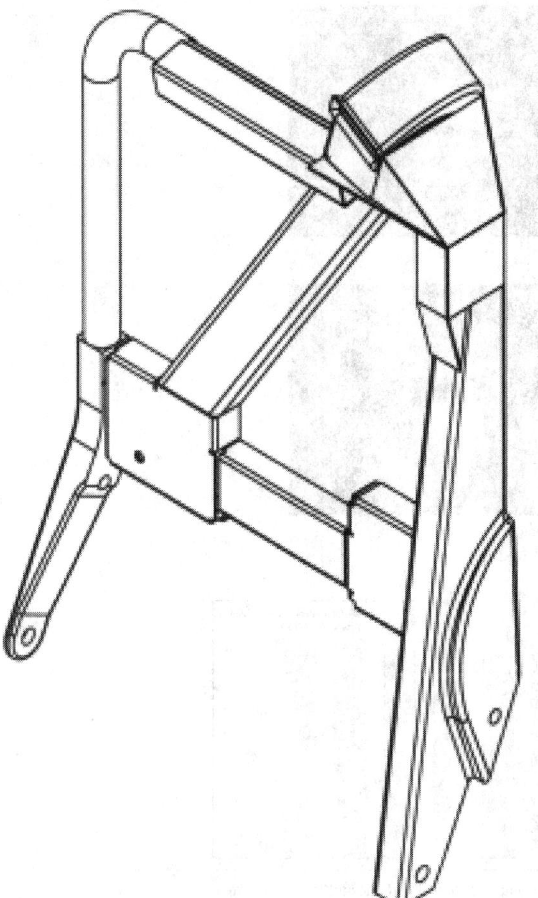

Figure 1.

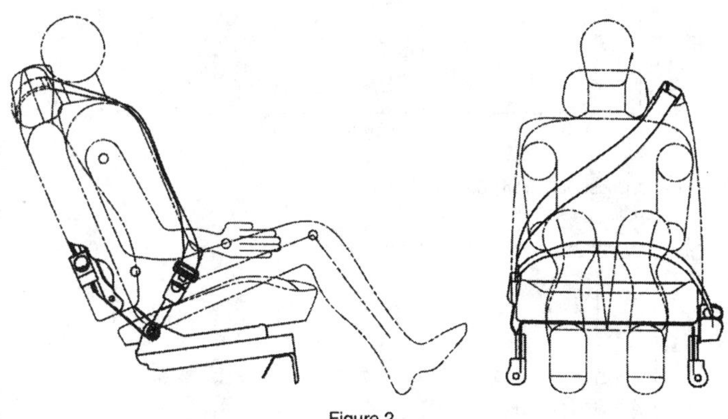

Figure 2.

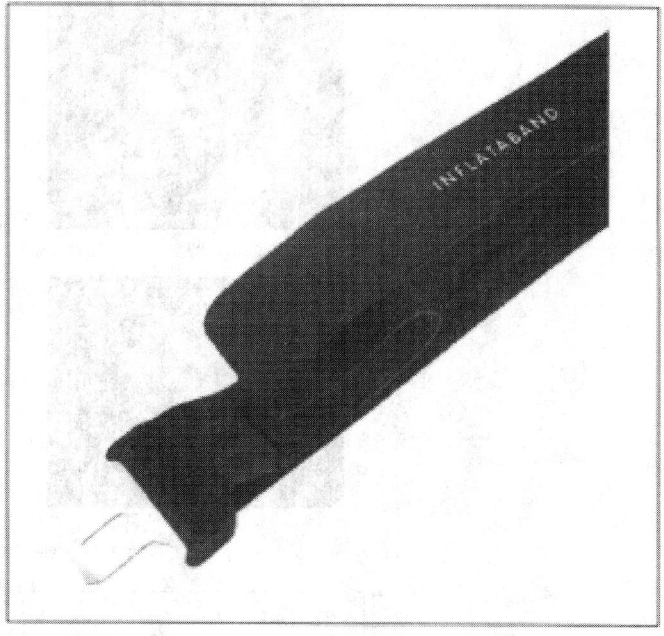

Figure 3. The Inflatable Seat Belt

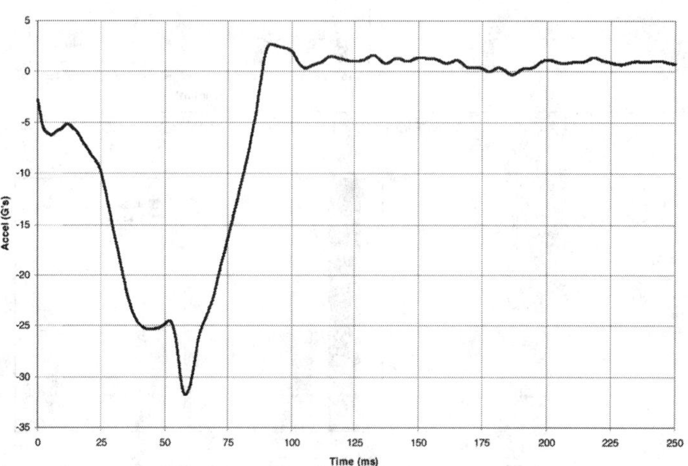

Figure 4. JCI Generic 35 mph Crash Pulse

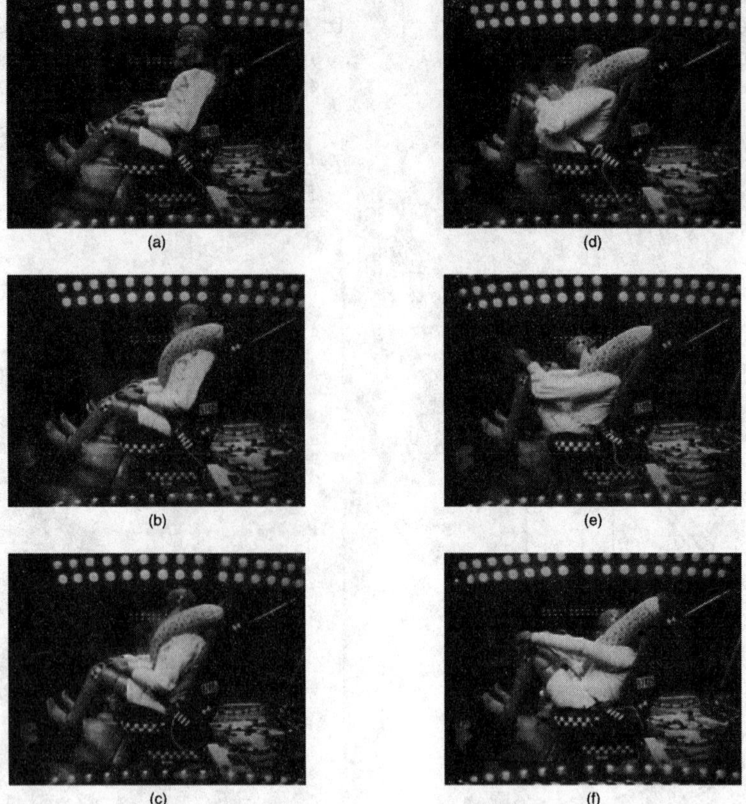

Figure 5. 50th Percentile Dummy Kinematics - Sled Test

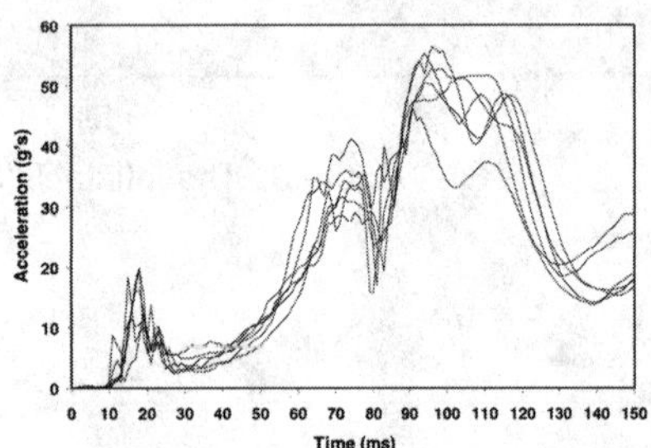

Figure 6. Head Resultant Acceleration - 50%ile

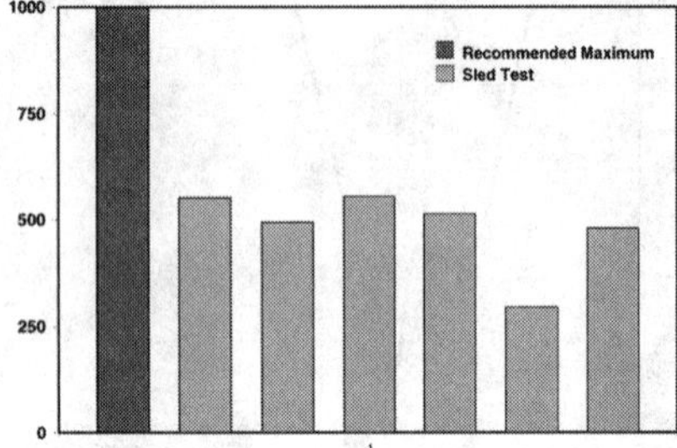

Figure 7. Comparison of Head Injury Criteria (HIC) - 50%ile

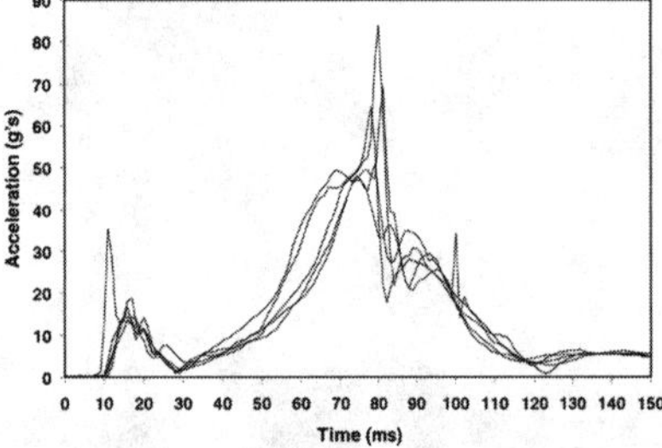

Figure 8. Chest Resultant Acceleration - 50%ile

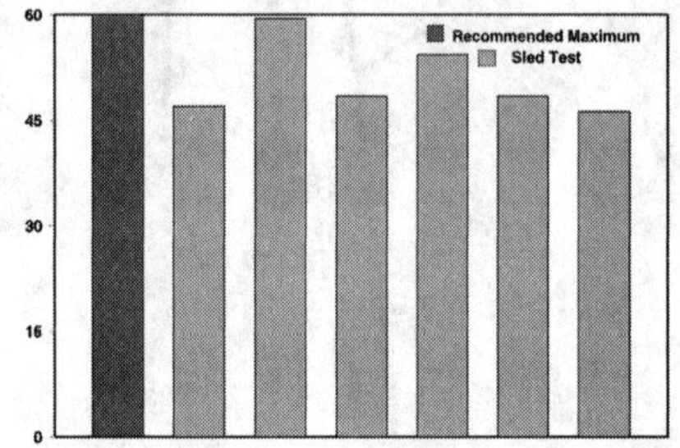

Figure 9. Comparison of 3ms Criteria (3ms) - 50%ile

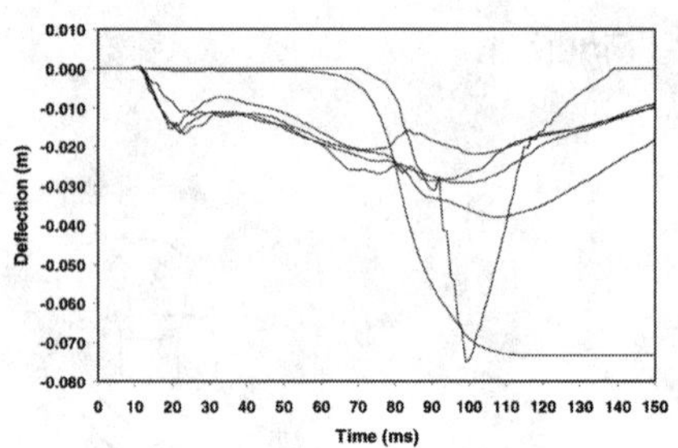

Figure 10. Chest Deflection - 50%ile

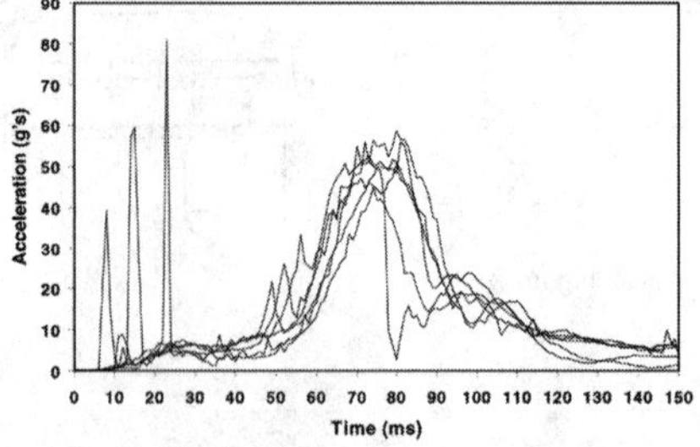

Figure 11. Pelvis Resultant Acceleration - 50%ile

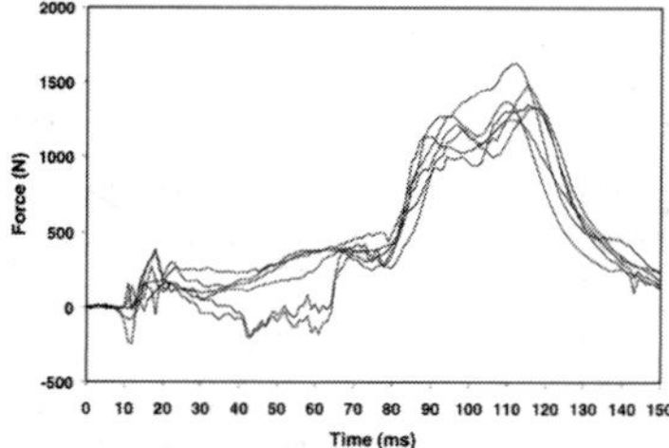

Figure 12. Neck Shear Force (Fx) - 50%ile

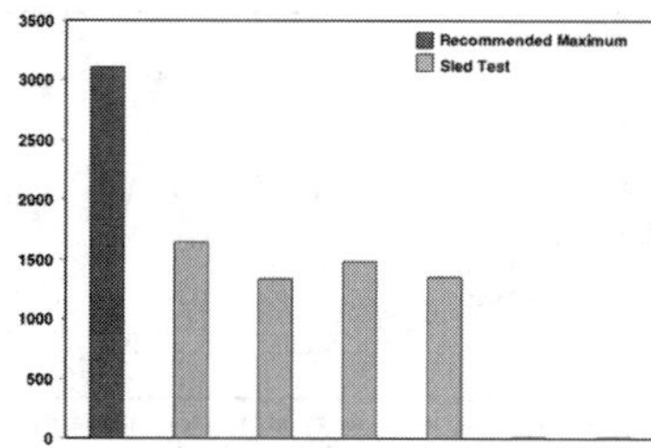

Figure 13. Comparison of Peak Neck Shear Force (+Fx) – 50%ile

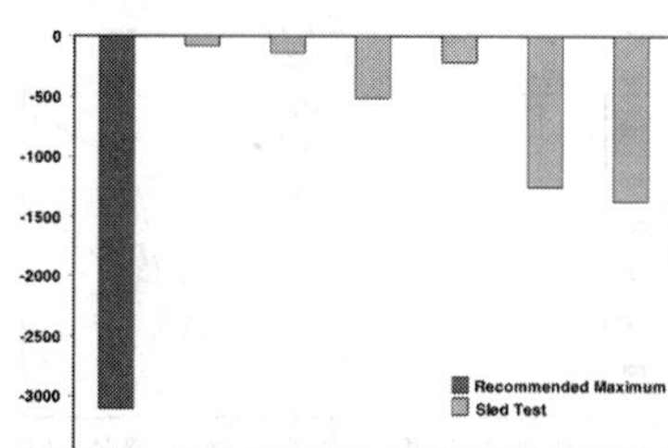

Figure 14. Comparison of Peak Neck Shear Force (-Fx) - 50%ile

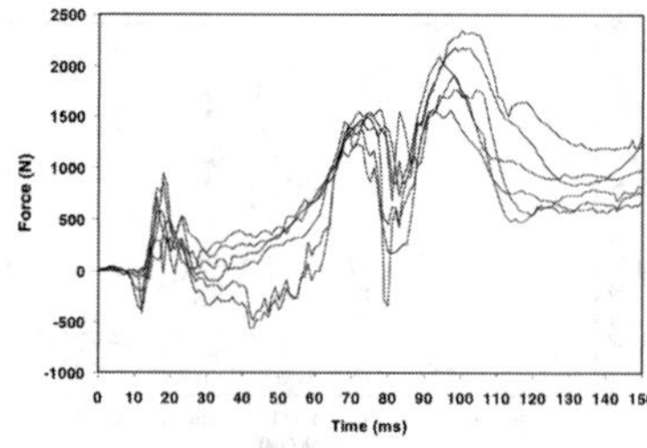

Figure 18. Neck Axial Force (Fz) - 50%ile

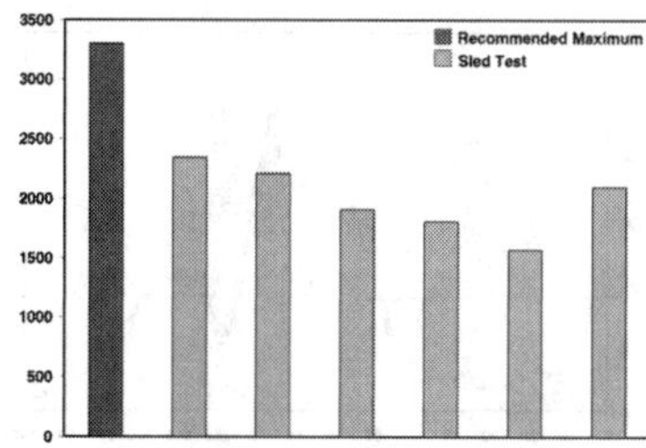

Figure 19. Comparison of Peak Tensile Axial Neck Force (+Fz)- 50%ile

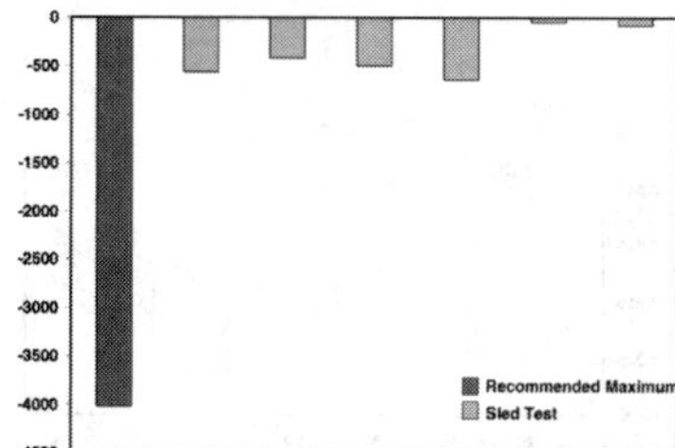

Figure 20. Comparison of Peak Compressive Axial Neck Force (-Fz) - 50%ile

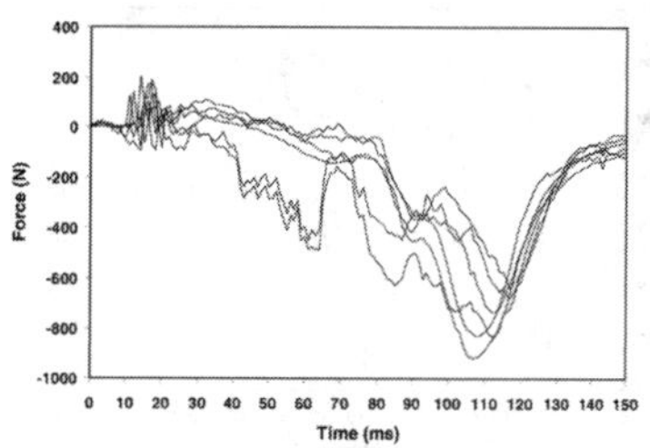

Figure 15. Neck Shear Force (Fy) - 50%ile

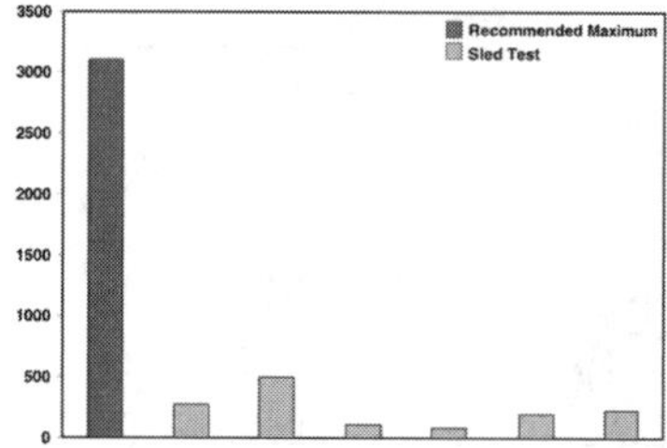

Figure 16. Comparison of Peak Neck Shear Force (+Fy) - 50%ile

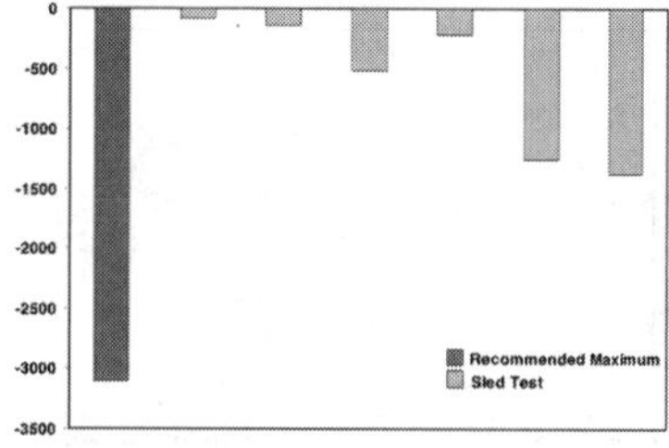

Figure 17. Comparison of Peak Shear Force (-Fy) - 50%ile

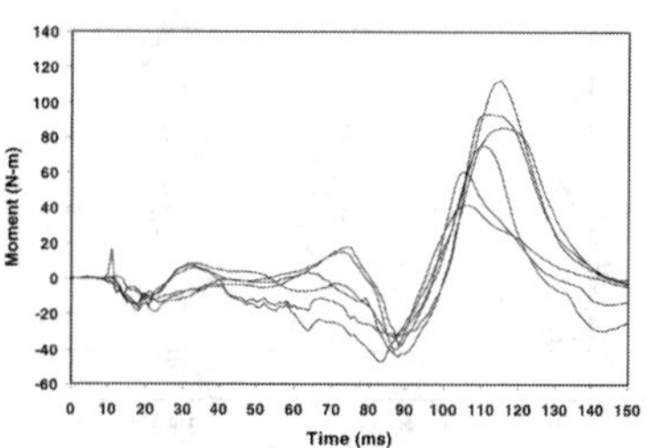

Figure 21. Neck Flexion/Extension Bending Moment (My) 50%ile

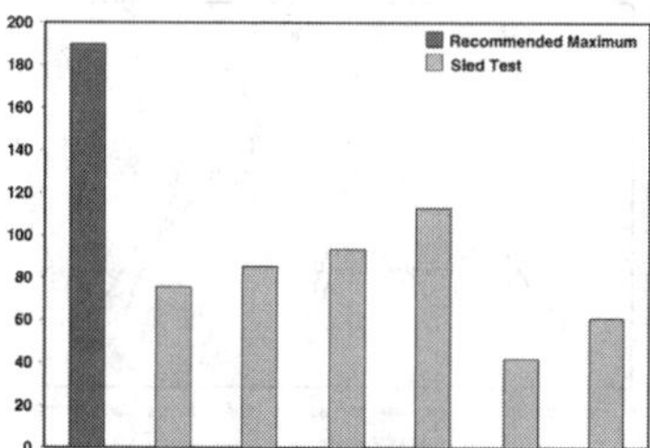

Figure 22. Comparison of Peak Neck Flexion Moment (+My) - 50%ile

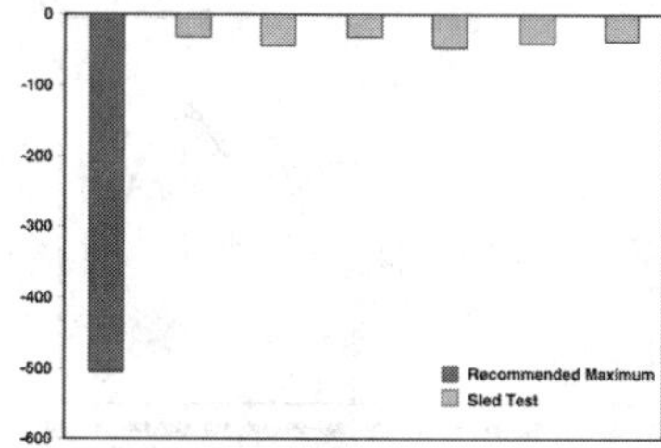

Figure 23. Comparison of Peak Neck Extension Moment (-My) - 50%ile

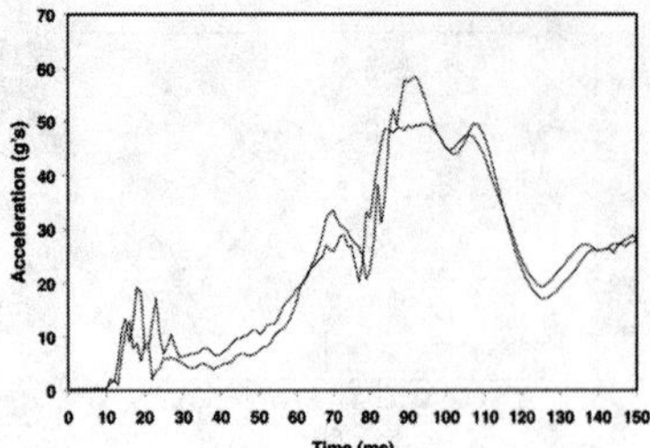

Figure 24. Head Resultant Acceleration - 5%ile

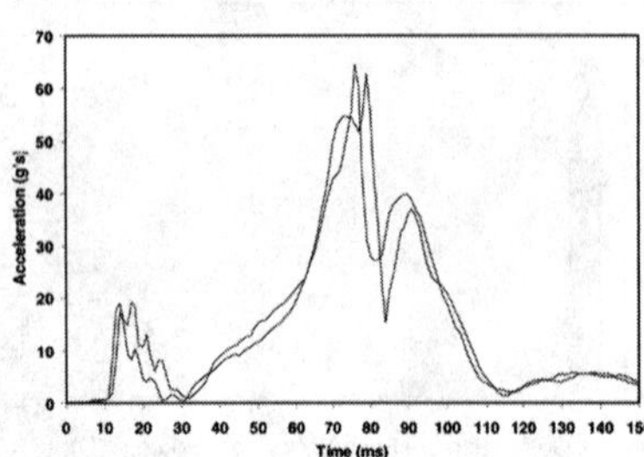

Figure 25. Chest Resultant Acceleration - 5%ile

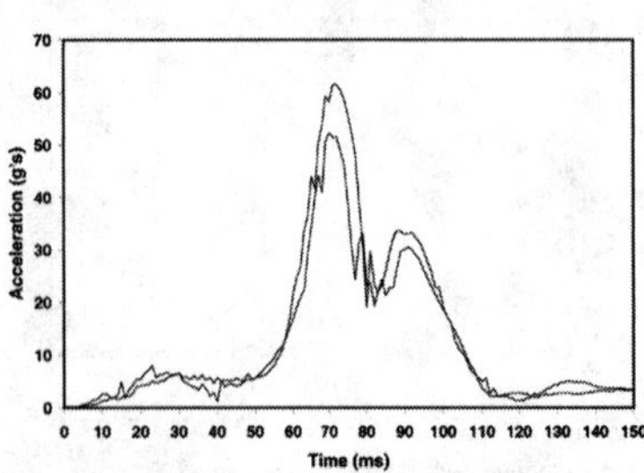

Figure 26. Pelvis Resultant Acceleration – 5%ile

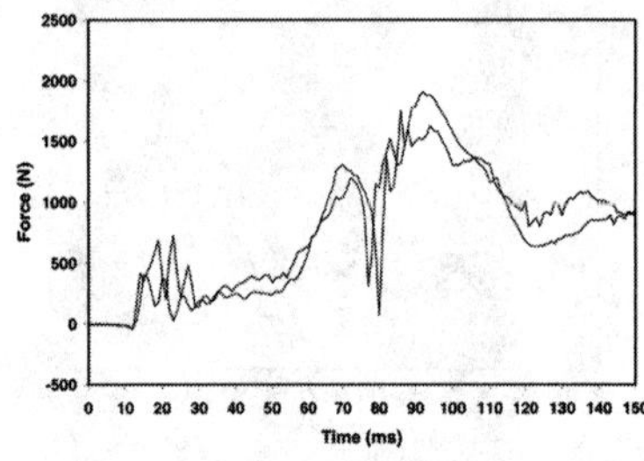

Figure 30. Neck Axial Force (Fz) - 5%ile

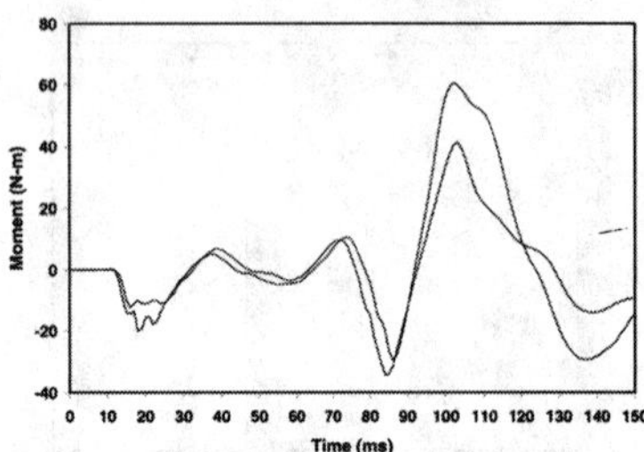

Figure 31. Neck Flexion/Extension Bending Moment (My) – 5%ile

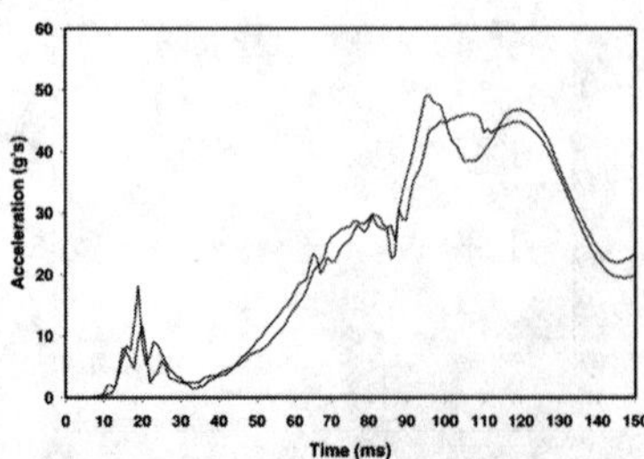

Figure 32. Head Resultant Acceleration – 95%ile

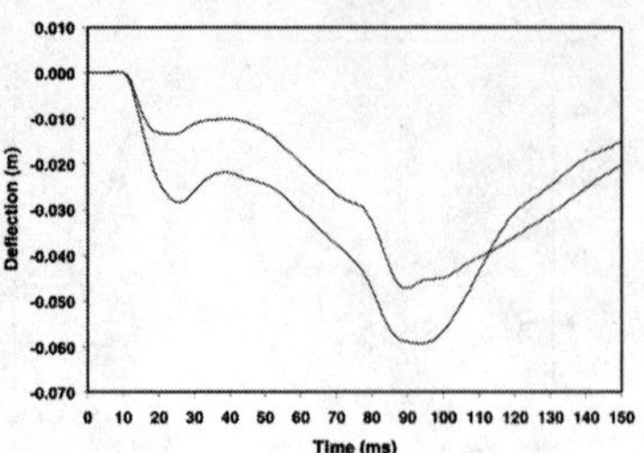

Figure 27. Chest Deflection - 5%ile

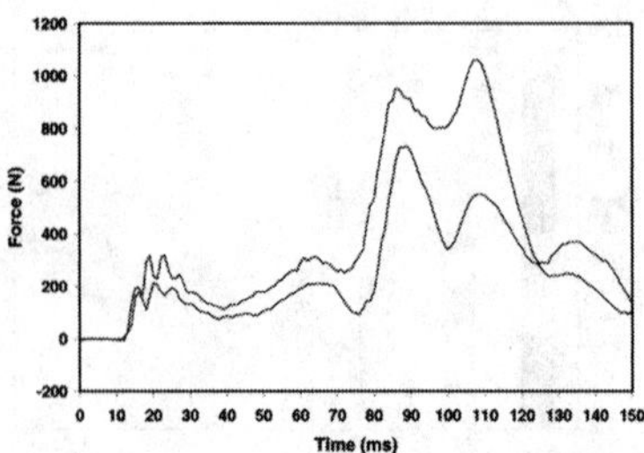

Figure 28. Neck Shear Force (Fx) - 5%ile

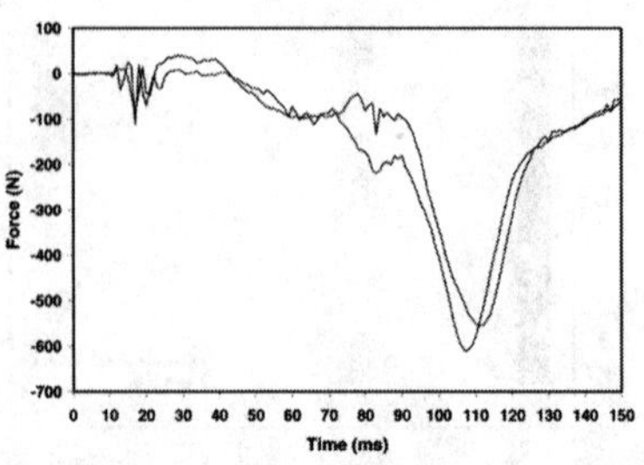

Figure 29. Neck Shear Force (Fy) - 5%ile

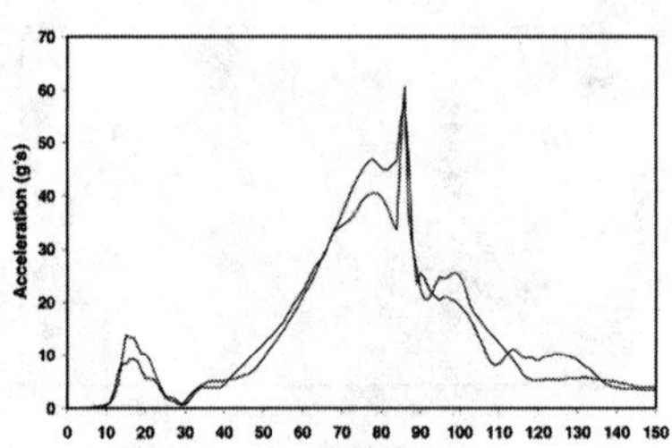

Figure 33. Chest Resultant Acceleration – 95%ile

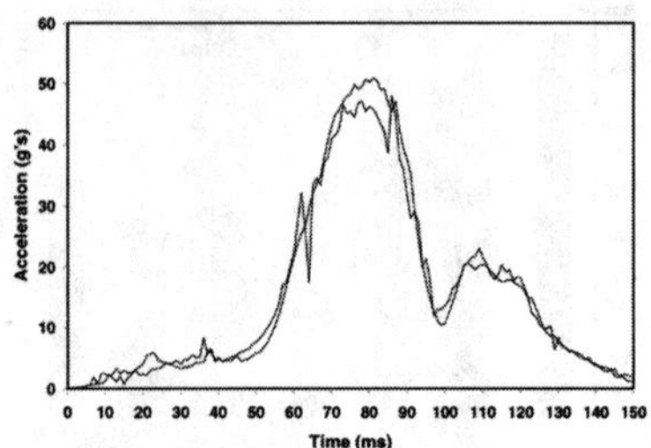

Figure 34. Pelvis Resultant Acceleration - 95%ile

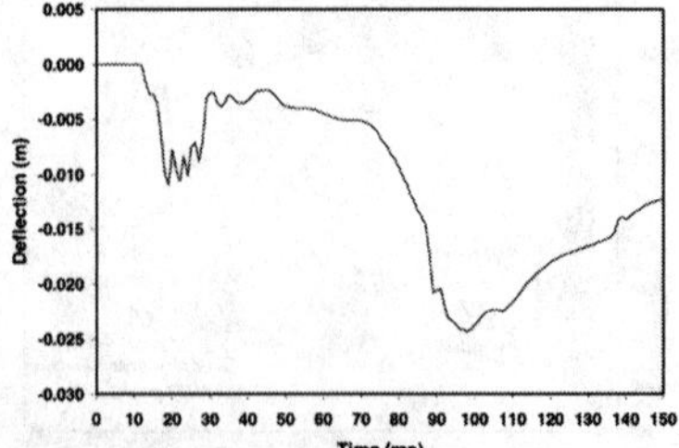

Figure 35. Chest Deflection - 95%ile

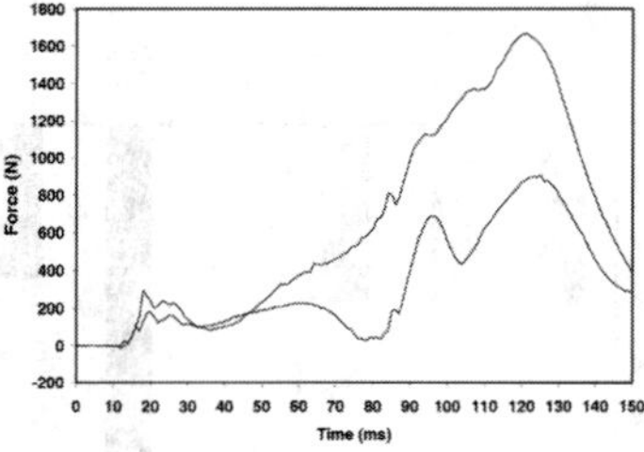

Figure 36. Neck Shear Force (Fx) - 95%ile

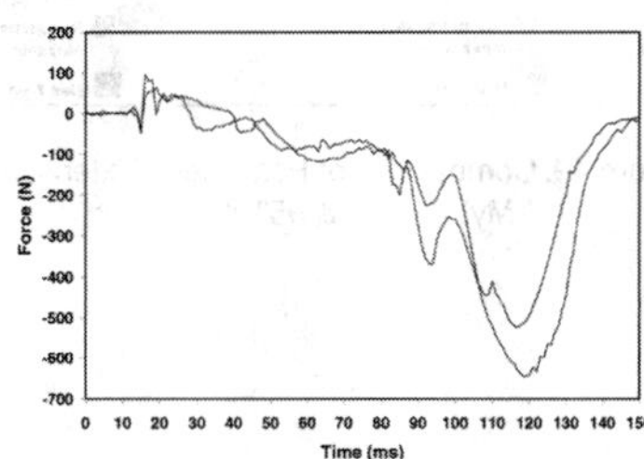

Figure 37. Neck Shear Force (Fy) - 95%ile

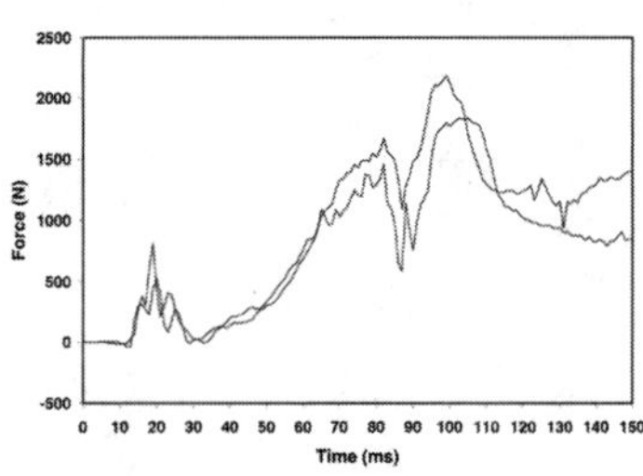

Figure 38. Neck Axial Force (Fz) - 95%ile

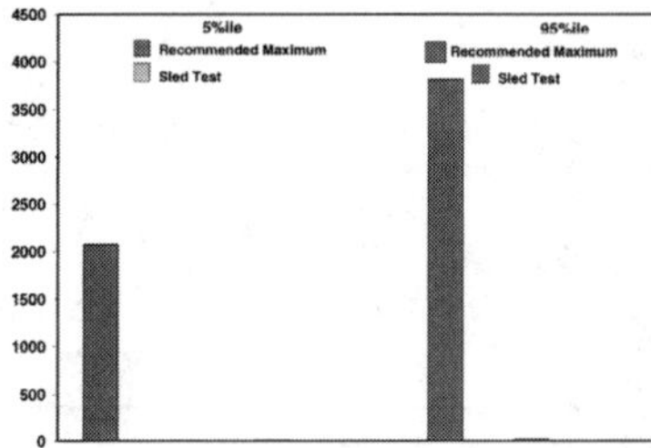

Figure 42. Comparison of Peak Neck Shear Force (+Fx) - 5%ile & 95%ile

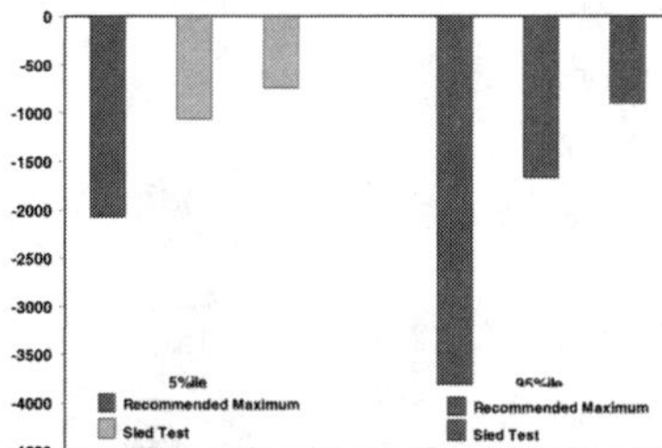

Figure 43. Comparison of Peak Neck Shear Force (-Fx) - 5%ile & 95%ile

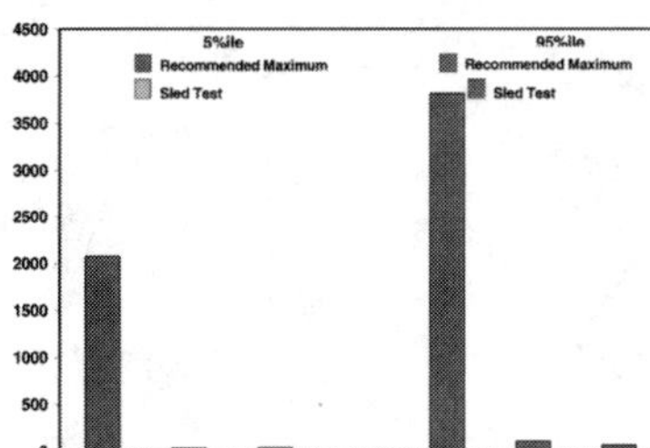

Figure 44. Comparison of Peak Neck Shear Force (+Fy) - 5%ile & 95%ile

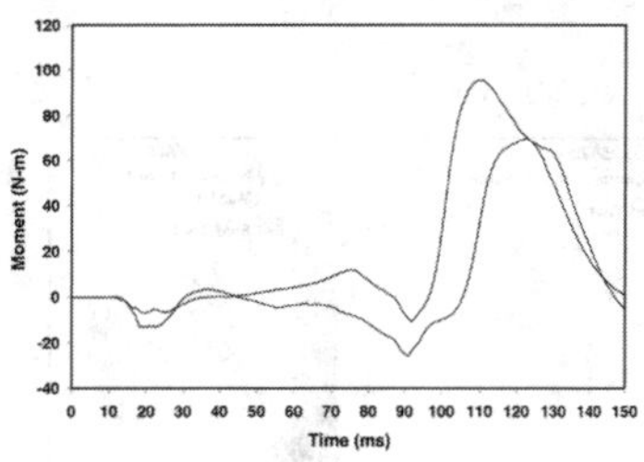

Figure 39. Neck Flexion/Extension Bending Moment (My) – 95%ile

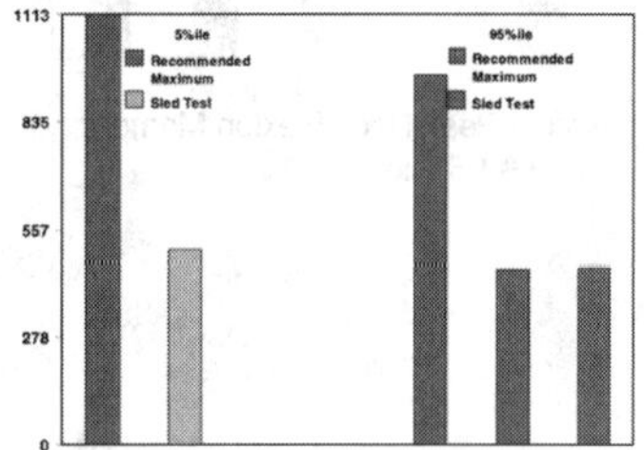

Figure 40. Comparison of Head Injury Criteria (HIC) - 5%ile & 95%ile

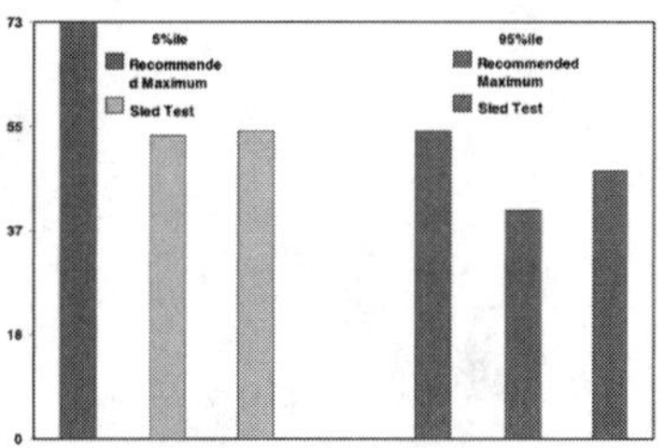

Figure 41. Comparison of 3ms Criteria (3ms) - 5%ile & 95%ile

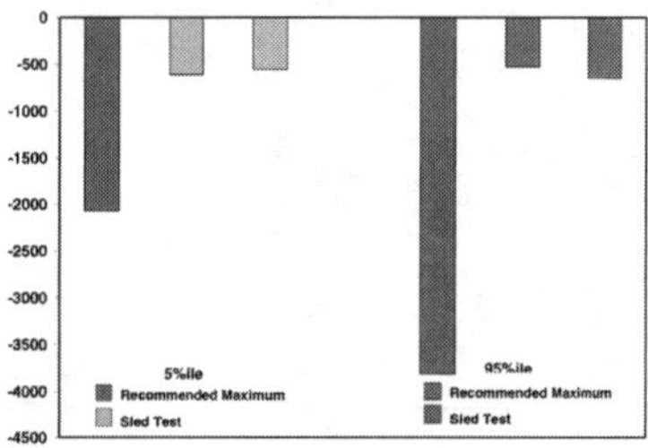

Figure 45. Comparison of Peak Shear Force (-Fy) - 5%ile & 95%ile

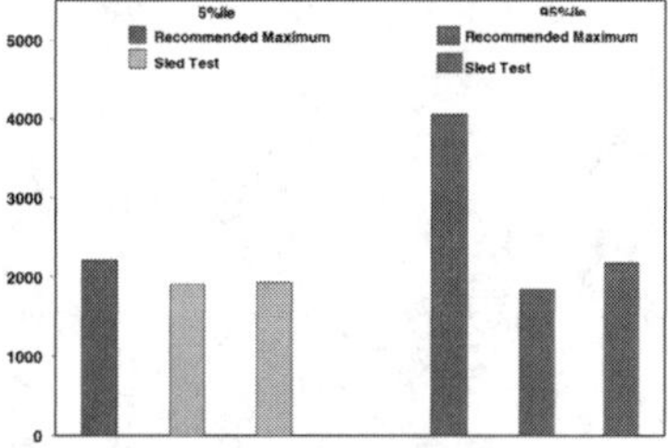

Figure 46. Comparison of Peak Tensile Axial Neck Force (+Fz)- 5%ile & 95%ile

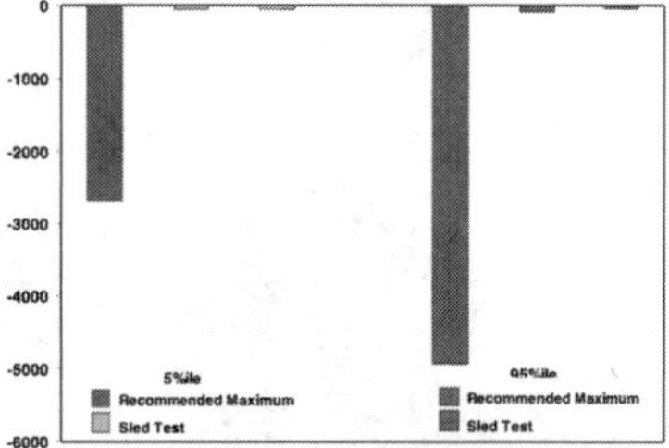

Figure 47. Comparison of Peak Compressive Axial Neck Force (-Fz) - 5%ile & 95%ile

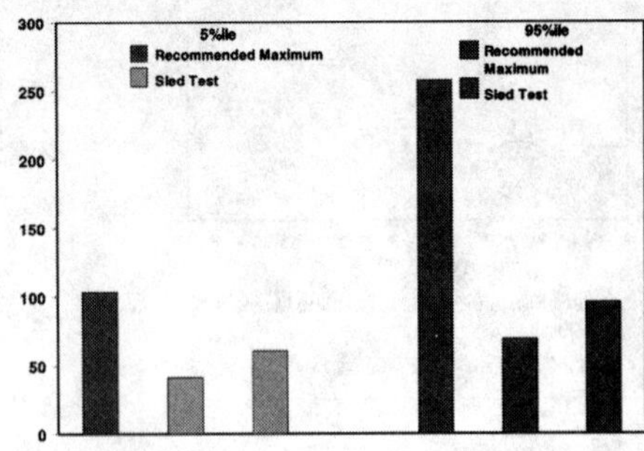

Figure 48. Comparison of Peak Neck Flexion Moment (+My) - 5%ile & 95%ile

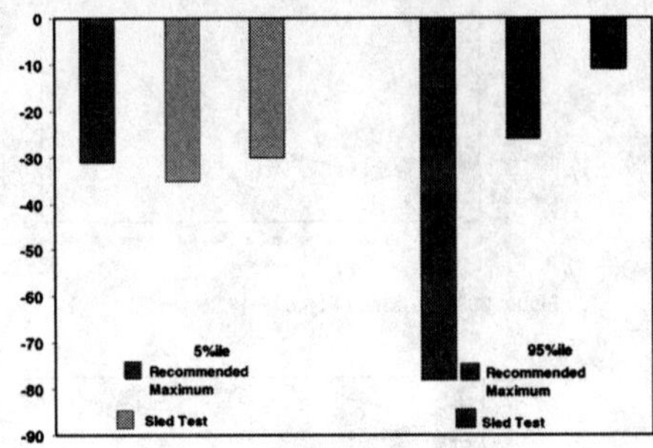

Figure 49. Comparison of Peak Neck Extension Moment (-My) - 5%ile & 95%ile

SECTION 5:

REAR SEAT OCCUPANTS AND CHILDREN

CHAPTER 16:

REAR SEAT OCCUPANTS

THE INJURY EXPERIENCE OF ADULT REAR SEAT CAR PASSENGERS

Richard W. Cuerden, Andrew W. Scott, Ahamedali M. Hassan and Murray Mackay

Accident Research Centre
The University of Birmingham
Edgbaston, Birmingham, UK.

ABSTRACT

The effectiveness of rear lap and diagonal seat belts at limiting injuries during car collisions is estimated to be 40%. The nature, severity and frequency of injuries to belted and unbelted outboard rear seat passengers is described. The mechanisms and sources of injury are prioritised and discussed. It is shown that population differences, varying kinematics and restraint characteristics all combine to produce somewhat different levels of use and effectiveness for rear occupants in comparison to those in the front.

REAR SEAT CAR PASSENGERS accounted for 14.4% (252) of the UK's car occupant fatalities in 1995, and 74.6% (188) were adults (Road Accidents Great Britain, 1995). Cars sold within the UK since April 1987 have been required to have rear seat belts fitted. However, legislation didn't require rear adult passengers to use available seat belts until 1993. Surveys investigating car occupants seat belt wearing rates, have found that between 43% and 48% of rear passengers 14 years and older use their belts in the UK (Restraint Use By Car Occupants, 1996). This is in contrast to front belt usage where observed rates are around 90%.

Surprisingly little research has been devoted to the question of the use and effectiveness of rear seat belts in Europe. It is often assumed that their acceptability and performance in crashes is the same as for the front seats, and yet there are good and obvious reasons why that is not the case. The rear seat occupant population is very different. There are many children in the age range 8 to 15 years who are around the approximate anthropometry of the 5th percentile female. There are proportionally more elderly and more female people in the rear compared to the front seat. Such population considerations have an influence on injury tolerance and hence seat belt effectiveness and usage.

There are interior design differences which affect belt performance. Most rear seats are of the bench variety rather than bucket seats. This can provide a different seat belt geometry in some cases, with less favourable lap section location relative to the iliac spines. In the front there is always an instrument panel which is contacted by the knees in most instances where the velocity change exceeds 30 km/hr. This means that in higher energy front crashes a substantial proportion of an occupant's energy is transferred through these knee contacts, reducing seat belt loads. It also has the effect of limiting submarining, the rotation of the pelvis out from under the lap section of the belt. The kinematics of the restrained rear seat occupant are different as there are no equivalent limiting knee contacts. The backs of the front seats are much more compliant and deformable, hence the rear restraint systems have to manage proportionally more of the crash energy. Therefore, it is a more challenging condition from the point of view of rear restraint design. Limiting submarining in the rear seats for instance becomes a particular concern.

There is a low incidence of intrusion into the rear section of the passenger compartment. This means that in very high energy crashes the restraint system will be stressed severely in comparison to the front seat zone, where intrusion from forward and side structures is controlling the occupant's energy.

Hence field accident analysis can provide a useful indication as to the relative importance of some of these issues. Because rear seat adult occupancy is only around 10%, the data sources for adequate statistical analysis are limited. This paper is an attempt to address some of these issues.

METHODOLOGY

Six hundred and twenty rear seat car passengers who were 15 years or older, and had been involved in collisions documented by the UK's Co-operative Crash Injury Study (CCIS) were analysed. The CCIS is an ongoing project which has collected real world in-depth data since 1983 (Mackay et al., 1985; Hassan et al., 1995). Vehicle examinations are undertaken at recovery garages several days after the collision. Car occupant injury information is collected from hospital records, coroners reports and questionnaires sent to survivors. Injury severities are rated according to the Abbreviated Injury Scale (AIS; AAAM 1990 Revision). Accidents are investigated according to a stratified sampling procedure which favours cars containing fatal or seriously injured occupants as defined by the British Government definitions of fatal, serious and slight. This study analysed car and occupant records for accidents that occurred between April 1992 and December 1995. The data collected before this period is currently not compatible with later cases.

All adult rear seat occupants involved in all impact types were initially selected. Table 1 details the injury rates of front and rear car occupants aged

15 years and older. The percentage of occupants with a Maximum Abbreviated Injury Scale (MAIS) greater than 1 is shown, and MAIS>2 in the parenthesis. The sources of injury for outboard Rear Seat Passengers (RSPs) are investigated. The analysis specifically considers only adult outboard occupants whose cars experienced a front, a non-struck side or a rollover collision. Lap and diagonal seat belts are generally considered to be effective at reducing the risk of injury for these collision types. In parallel, Front Seat Passengers (FSPs) were selected who also experienced a front, a non-struck side or a rollover collision, and their injury rates compared with those of the RSPs. Occupants with and without seat belts were compared. Finally, only front impacts were selected for RSPs and their injury rates recorded against their crash severity. The crash severity was assessed from vehicle damage dimensions using the CRASH3 programme (NHTSA, 1982). The selection required the car's collision to have been described by an Estimated Test Speed (ETS); in most cases this is broadly equivalent to change in velocity.

Table 1: Occupants Injury Severity according to the British Government Definitions, and percent MAIS > 1.

Injury Severity	Front Car Occupants Only (Drivers & Front Passengers)			Rear Passengers Only		
	N	%	% MAIS>1	N	%	% MAIS>1
Fatal	272	4.9	97 (96)	18	2.9	100 (100)
Serious	1584	28.5	66 (24)	152	24.5	68 (30)
Slight	2383	42.8	10 (1)	297	47.9	10 (0.7)
No injury	1243	22.3	0.5 (0)	137	22.1	0
Not Known	81	1.5	17 (4)	16	2.5	44 (13)
Total	5563	100	-	620	100	-

Values in parenthesis are the % MAIS $\geq$ 3 per severity measure.

There were 79 adult occupants who were known to have been sitting in the centre rear position. Only 5 of these were known to have used their seat belts (59 were known to have been unbelted). No cars in our study to date have been fitted with a centre lap and diagonal seat belt. The only restraints fitted to the centre rear position in our sample were static lap belts. The analysis of these occupants will form a separate study.

ANALYSIS

The collision types for front and rear outboard occupants (over 14 years old) are presented in Figure 1. Only three point lap and diagonal automatic retractor seat belts are included in the following analysis. Occupants were grouped by their MAIS (MAIS=0, MAIS=1 and MAIS>1). There are considerably more front seat occupants than rear ones and some caution must be applied to these simple comparisons. However, some basic trends are apparent. Front impacts for belted occupants can be seen to account for approximately twice the percentage of MAIS>1 injuries for front compared with

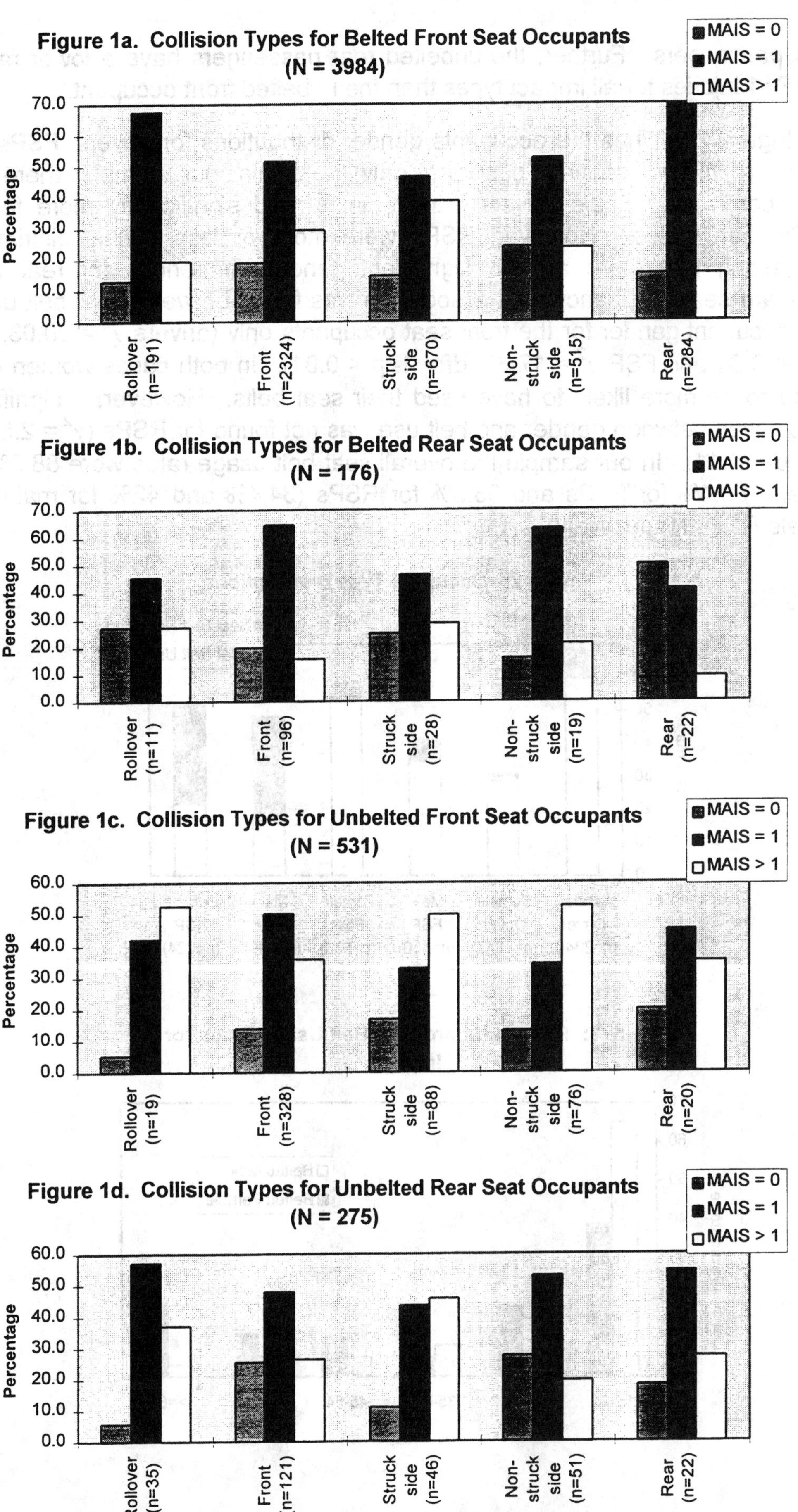
Figure 1a. Collision Types for Belted Front Seat Occupants
(N = 3984)
MAIS = 0
MAIS = 1
MAIS > 1
Percentage
Rollover (n=191)
Front (n=2324)
Struck side (n=670)
Non-struck side (n=515)
Rear (n=284)
Figure 1b. Collision Types for Belted Rear Seat Occupants
(N = 176)
MAIS = 0
MAIS = 1
MAIS > 1
Percentage
Rollover (n=11)
Front (n=96)
Struck side (n=28)
Non-struck side (n=19)
Rear (n=22)
Figure 1c. Collision Types for Unbelted Front Seat Occupants
(N = 531)
MAIS = 0
MAIS = 1
MAIS > 1
Percentage
Rollover (n=19)
Front (n=328)
Struck side (n=88)
Non-struck side (n=76)
Rear (n=20)
Figure 1d. Collision Types for Unbelted Rear Seat Occupants
(N = 275)
MAIS = 0
MAIS = 1
MAIS > 1
Percentage
Rollover (n=35)
Front (n=121)
Struck side (n=46)
Non-struck side (n=51)
Rear (n=22)

rear passengers. Further, the unbelted rear passengers have a lower rate of MAIS>1 injuries for all impact types than the unbelted front occupant.

Figure 2 outlines the occupants gender distributions for drivers, FSPs and RSPs (outboard seating positions only). Within our sample there are significantly more male drivers than females, and significantly more female FSPs than males. However, RSPs were more or less evenly distributed, suggesting that there are no significant gender differences for rear adult outboard seat occupancy. An association was found between seat belt usage and occupant gender for the front seat occupants only (drivers $\chi^2 = 30.03$, df = 1, $p < 0.01$ and FSP $\chi^2 = 17.33$, df = 1, $p < 0.01$). In both cases women were found to be more likely to have used their seat belts. However, a significant association between gender and belt use was not found for RSPs ($\chi^2 = 2.58$, df = 1, p = 0.11). In our sample the overall seat belt usage rates were 88.5% for drivers, 88.4% for FSPs and 38.3% for RSPs (34.4% and 42% for male and female RSPs respectively)

Figure 2: Occupant Type Distributions

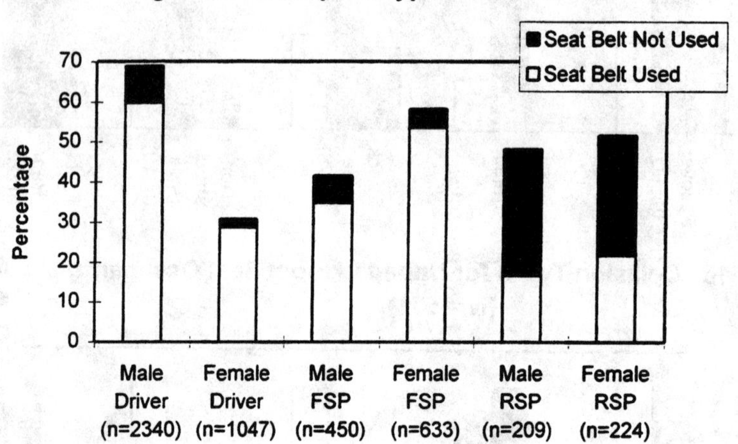

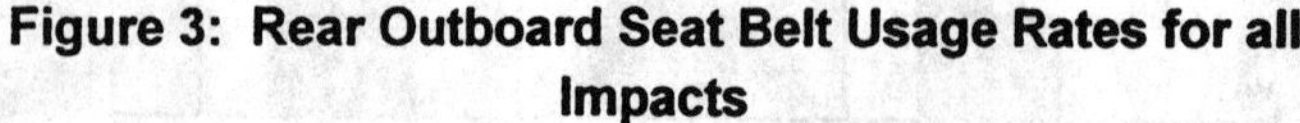

Figure 3: Rear Outboard Seat Belt Usage Rates for all Impacts

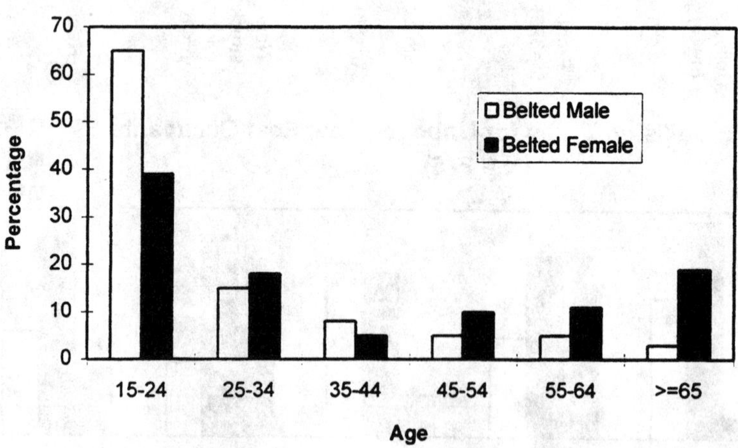

The male and female RSPs restraint usage rates are shown against age in figure 3. When seat belts were worn a significant difference was found between males and females ages (median ages were:- male = 21.5 years and female = 29.5 years; $\chi^2 = 8.96$, df = 1, $p < 0.01$). Further, restrained RSPs were significantly older than unrestrained ones (median ages were :- belted = 26 years and unbelted = 22 years; $\chi^2 = 5.39$, df = 1, $p < 0.05$).

REAR LAP AND DIAGONAL SEAT BELT EFFECTIVENESS

Tables 2a and 2b show that outboard rear lap and diagonal seat belts are significantly associated with a reduction in the risk of MAIS > 1 injury for all impact types ($\chi^2 = 8.42$, df = 2, $p < 0.05$) and frontal collisions only ($\chi^2 = 6.37$, df = 2, $p < 0.05$).

Table 2a: All impacts - MAIS by belt use

MAIS	Seat Belt Used	Seat Belt Not used	Total
0	43 (38.4)	56 (60.6)	99
1	102 (93.2)	138 (146.8)	240
2 to 6	34 (47.4)	88 (74.6)	122
Total	179	282	461

Table 2b: Front impacts - MAIS by belt use

MAIS	Seat Belt Used	Seat Belt Not used	Total
0	19 (22.1)	31 (27.9)	50
1	62 (53.1)	58 (66.9)	120
2 to 6	15 (20.8)	32 (26.2)	47
Total	96	121	217

(Values in parenthesis are expected values for the χ^2 statistics.)

The effectiveness of lap and diagonal seat belts is defined as the percent reduction in the chance of an occupant sustaining injury (MAIS > 1) compared to the unrestrained condition, and is calculated as below:-

$$\text{effectiveness} = \frac{(\text{unbelted MAIS} > 1 \text{ rate}) - (\text{belted MAIS} > 1 \text{ rate})}{\text{unbelted MAIS} > 1 \text{ rate}} \%$$

The effectiveness of rear seat belts in preventing MAIS > 1 for our sample is 39% for all crash types (Table 2a) and 41% for frontal impacts (Table 2b).

Table 2c: Ejection against Seat Belt use

	Belt Used	No Belt Used
No Ejection	178 (168.3)	248 (257.7)
Ejection	1 (10.7)	26 (16.3)
Total	179	274

Table 2d: MAIS by Ejection for all impacts and belt use

	MAIS 0	MAIS 1	MAIS 2	MAIS 3 to 6
No Ejection	127 (118.6)	258 (251.3)	69 (72.9)	42 (53.2)
Ejection	0 (8.4)	11 (17.7)	9 (5.1)	15 (3.8)
Total	127	269	78	57

(Values in parenthesis are expected values for the χ^2 statistics.)

Within our sample there were 16 RSP fatalities. Six of these were involved in lateral collisions and seated on the struck side. Only one belted passenger died. He was 77 years old and sustained an AIS 5 chest injury due to loading from the diagonal portion of the belt. The impact was unclassifiable and no ETS could be calculated. Eight of the fatalities were either fully or partially ejected from their cars. Belt usage was found to influence the incidence of ejection. There were 27 RSPs who were ejected from their vehicles, and only one was wearing a lap and diagonal seat belt (Table 2c, $\chi^2 = 15.41$, df = 1, $p < 0.01$). The use of rear belts was found to significantly limit ejection and thus reduce the risk of serious injury (Table 2d, $\chi^2 = 50.82$, df = 3, $p < 0.01$). Previous studies have quantified that rear seat passengers experience almost twice as many rollover impacts as other passengers and have a seven times greater chance of being ejected (Bodiwala, 1989; Tunbridge, 1988).

Table 3 describes the injury sources for the major body regions for belted and unbelted outboard RSPs for all impact types. The highest injury severity per body region (MAIS) is recorded for each passenger and correlated to a source . The lower extremities are detailed for only the maximum leg injury. If both legs sustained injuries of an equal AIS severity, these are summed, and the total given in the parenthesis. The injuries to the upper extremities are not included in this analysis, as there were few known contact sources. There were 368 and 564 injuries correlated for belted and unbelted RSPs respectively. Of these injuries 12.2% (45) for belted and 22% (124) for unbelted were AIS > 1. Therefore, in our sample seat belts reduced the risk of an AIS > 1 injury for all impact types by approximately 45%.

The most common cause of belted RSP's injury at AIS > 1, was chest injury due to seat belt loads. Nine individuals (5%) sustained such an injury from the diagonal portion of the belt. Three (1.7%) sustained thoracic injuries greater than AIS 2 (two received AIS 4, and one AIS 5). Unbelted occupants sustained an AIS > 1 most frequently (13.5%) to the head (cranium). Notable contact sources were the seat or head restraint in front of them, external objects and the car's roof.

Some 15.6% of belted RSPs sustained minor neck strains (AIS = 1) without head or neck impacts, whereas only 4.6% of unbelted occupants received this injury. Further, belted RSPs are documented as having head or face contacts with the head restraint or seat in front of them. There was no clear evidence of serious injury (AIS > 1) due to these contacts. However, femur and leg fractures are attributed to seat contacts for belted passengers.

No attempt has been made to investigate the incidence of restrained front occupants injuries due to restrained or unrestrained RSPs. Several authors including Huelke(1974), Rattenbury(1979) and Griffiths(1976), have found an increased risk of front occupant injury when loaded by an unrestrained RSP. Figure 3h and Figure 5 describe belted RSPs sustaining leg fractures due to contacts with the front seats. Therefore, an increased risk of front occupant injury from interaction with RSPs could exist when all the car's occupants are belted.

Table 3: Injury sources for 179 belted and 282 unbelted RSPs.

Table 3a. Injury Source for Cranium

Injury Source	Belted				Non-Belted			
	MAIS = 1		MAIS ≥ 2		MAIS = 1		MAIS ≥ 2	
	N	%	N	%	N	%	N	%
B or other pillar	1	4.2	3	33.3	-	-	-	-
Non contact	-	-	-	-	-	-	2	5.3
External object	-	-	-	-	-	-	4	10.5
Flying glass	-	-	-	-	1	2.4	-	-
Head Restraint	3	12.5	1	11.1	2	4.8	3	7.9
Seat	3	12.5	-	-	2	4.8	3	7.9
Side roof rail	-	-	-	-	1	2.4	1	2.6
Side glass	2	8.3	2	22.2	3	7.1	1	2.6
Roof	1	4.2	1	11.1	1	2.4	4	10.5
Windscreen	-	-	-	-	1	2.4	1	2.4
Other	1	4.2	-	-	2	4.8	3	7.9
Not known	13	54.2	2	22.2	29	69.0	15	39.5
Total	24	100.0	9	100.0	42	100.0	38	100.0

Table 3b. Injury Source for Face

Injury Source	Belted				Non-Belted			
	MAIS = 1		MAIS ≥ 2		MAIS = 1		MAIS ≥ 2	
	N	%	N	%	N	%	N	%
B or other pillar	1	3.3	-	-	1	1.0	-	-
External Object	-	-	-	-	1	1.0	3	37.5
Flying glass	1	3.3	-	-	1	1.0	-	-
Head Restraint	5	16.7	-	-	12	12.5	-	-
Seat	6	20.0	-	-	19	19.8	-	-
Side roof rail	-	-	-	-	2	2.1	2	25.0
Side glass	3	3	-	-	10	10.4	-	-
Windscreen	-	-	-	-	3	3.1	-	-
Other	1	3.3	-	-	10	10.4	-	-
Not known	13	43.3	-	-	37	38.5	3	37.5
Total	30	100.0	-	-	96	100.0	8	100.0

Table 3c. Injury Source for Neck

Injury Source	Belted				Non-Belted			
	MAIS = 1		MAIS ≥ 2		MAIS = 1		MAIS ≥ 2	
	N	%	N	%	N	%	N	%
Neck impact	3	6.5	-	-	3	5.4	1	20.0
Head impact	13	28.3	-	-	23	41.1	2	40.0
No impact	28	60.9	3	100.0	13	23.2	1	20.0
Not known	2	4.3	-	-	17	30.4	1	20.0
Total	46	100.0	3	100.0	56	100.0	5	100.0

Table 3d. Injury Source for Thorax

Injury Source	Belted				Non-Belted			
	MAIS = 1		MAIS ≥ 2		MAIS = 1		MAIS ≥ 2	
	N	%	N	%	N	%	N	%
B or other pillar	-	-	-	-	1	3.0	-	-
Door	3	6.5	-	-	3	9.1	7	35.0
Seat	-	-	-	-	10	30.3	1	5.0
Seat belt	36	78.3	8	88.9	-	-	-	-
Side, other	-	-	-	-	-	-	2	10.0
Other	1	2.2	-	-	3	9.1	7	35.0
Not known	6	13.0	1	11.1	16	48.5	3	15.0
Total	46	100.0	9	100.0	33	100.0	20	100.0

Table 3e. Injury Source for Abdomen

Injury Source	Belted				Non-Belted			
	MAIS = 1		MAIS ≥ 2		MAIS = 1		MAIS ≥ 2	
	N	%	N	%	N	%	N	%
Door	1	2.5	-	-	3	9.1	5	55.6
Seat	-	-	-	-	10	33.3	1	11.1
Seat belt	34	85.0	3	60.0	-	-	-	-
Side, other	-	-	-	-	1	3.0	-	-
Other	1	2.5	-	-	5	15.2	2	22.2
Not known	4	10.0	2	40.0	14	42.4	1	11.1
Total	40	100.0	5	100.0	33	100.0	9	100.0

Table 3f. Injury Source for Thoracic and Lumbar Spine

Injury Source	Belted				Non-Belted			
	MAIS = 1		MAIS ≥ 2		MAIS = 1		MAIS ≥ 2	
	N	%	N	%	N	%	N	%
Spine impact	2	18.2	-	-	3	25.0	2	18.2
Indirect loading	7	63.6	-	-	3	25.0	3	27.3
Not known	2	18.2	1	100.0	6	50.0	6	54.5
Total	11	100.0	1	100.0	12	100.0	11	100.0

Table 3g. Injury Source for Pelvis

Injury Source	Belted				Non-Belted			
	MAIS = 1		MAIS ≥ 2		MAIS = 1		MAIS ≥ 2	
	N	%	N	%	N	%	N	%
Direct front load	-	-	1	10.0	-	-	2	20.0
Direct Side Load	-	-	2	20.0	-	-	4	40.0
Other	-	-	-	-	-	-	2	20.0
Not known	45	100.0	7	70.0	-	-	2	20.0
Total	45	100.0	10	100.0	-	-	10	100.0

Table 3h. Injury Source for Right and Left Thighs (highest AIS per passenger)
Values in parenthesis are the total number of injuries to both thighs per MAIS group.

Injury Source	Belted				Non-Belted			
	MAIS = 1		MAIS ≥ 2		MAIS = 1		MAIS ≥ 2	
	N	%	N	%	N	%	N	%
Compartment Side	0	-	1	16.7	4	18.2	2	22.2
Seat	2	40.0	3	50.0	7	31.8	3	33.3
Other	-	-	-	-	-	-	1	11.1
Not known	3	60.0	2	33.3	11	50.0	3	33.3
Total	5 (7)	100.0	6 (7)	100.0	22 (28)	100.0	9 (11)	100.0

Table 3i. Injury Source for Right and Left Knees (highest AIS per passenger)
Values in parenthesis are the total number of injuries to both knees per MAIS group.

Injury Source	Belted				Non-Belted			
	MAIS = 1		MAIS ≥ 2		MAIS = 1		MAIS ≥ 2	
	N	%	N	%	N	%	N	%
Compartment Side	-	-	-	-	2	6.7	-	-
Seat	3	23.1	1	50.0	14	46.7	1	33.3
Other	-	-	-	-	2	6.7	1	33.3
Not known	10	76.9	1	50.0	12	40.0	1	33.3
Total	13 (17)	100.0	2 (2)	100.0	30 (37)	100.0	3 (3)	100.0

Table 3j. Injury Source for Right and Left lower legs (highest AIS per passenger)
Values in parenthesis are the total number of injuries to both legs per MAIS group.

Injury Source	Belted				Non-Belted			
	MAIS = 1		MAIS ≥ 2		MAIS = 1		MAIS ≥ 2	
	N	%	N	%	N	%	N	%
Compartment Side	-	-	-	-	-	-	2	28.6
Facia	-	-	-	-	2	4.1	-	-
Seat	16	47.1	-	-	28	57.1	1	14.3
Other	1	2.9	-	-	2	4.1	-	-
Not known	17	50.0	-	-	17	34.7	4	57.1
Total	34 (41)	100.0	-	-	49 (72)	100.0	7 (8)	100.0

Table 3k. Injury Source for both Ankles and Feet (highest AIS per passenger)
Values in parenthesis are the total number of injuries to both ankles and feet per MAIS group.

Injury Source	Belted				Non-Belted			
	MAIS = 1		MAIS ≥ 2		MAIS = 1		MAIS ≥ 2	
	N	%	N	%	N	%	N	%
Under Seat	6	50.0	-	-	11	45.8	3	75.0
Other	2	16.7	-	-	1	4.2	1	25.0
Not known	4	33.3	-	-	12	50.0	-	-
Total	12 (15)	100.0	-	-	24 (31)	100.0	4 (4)	100.0

FRONT, NON-STRUCK SIDE AND ROLLOVER COLLISIONS

The injury rates per body region have been compared in Table 4 for belted and unbelted front and rear passengers. Only front, non-struck side and rollover collisions were considered as the lap and diagonal restraint is generally accepted to be most effective during these collisions. There were only 24 RSPs in the sample who wore lap and diagonal belts and sustained an overall MAIS > 1 for this subset. Therefore, potential analysis is limited.

FRONT IMPACTS ONLY (WITH KNOWN IMPACT SEVERITY)

Figure 4 shows belted and unbelted RSPs injury outcomes following front impacts. MAIS values are grouped, 0 and 1; and 2 to 6 (MAIS > 1). No significant difference was found between the median ETS for belted and unbelted MAIS > 1 cumulative percent curves (median ETS = 36 km/hr, χ^2 = 0.015, df = 1, p = 0.9). However, approximately 70% (8) of the belted passengers were rated MAIS = 2 only. Whereas, 48% (11) of the unbelted passengers were rated MAIS > 2. Therefore, there was a tendency for the unrestrained to be more severely injured. However, the small sample size should be borne in mind when these results are considered. There was no significant association found between gender or age and belt use at MAIS > 1. Further, no significant difference was found for the distribution of ETS at MAIS>1 between belted FSPs and RSPs (median ETS = 36 km/hr, χ^2 = 0.0007, df = 1, p = 0.98). There is some suggestion that restrained occupants are somewhat more vulnerable to minor injury in the rear than unrestrained occupants.

Table 4: MAIS > 1 and MAIS > 2 Injury rates for the major boby regions.

	FSPs Belted (N=755)		FSPs Unbelted (N=114)		RSPs Belted (N=126)		RSPs Unbelted (N=207)	
	% MAIS >1	% MAIS >2	% MAIS >1	% MAIS >2	% MAIS >1	% MAIS >2	% MAIS >1	% MAIS >2
Head (Cranium)	7.2	2.5	12.3	6.1	3.2	0.8	10.1	3.4
Face	0.8	-	0.9	-	-	-	3.4	0.5
Neck (Incl. Cervical Spine)	2.4	0.9	1.8	-	0.8	-	1.0	0.5
Upper Extremities	8.3	1.7	13.2	3.5	6.3	-	8.7	1.0
Thorax	10.9	3.8	6.1	5.3	6.3	1.6	5.3	4.3
Abdomen	2.1	1.2	2.6	1.8	3.2	1.6	1.9	0.5
Spine (Thoracic & Lumbar)	2.8	0.5	-	-	0.8	-	3.9	1.4
Lower Extremities	7.0	2.8	7.9	2.6	3.2	2.4	7.7	4.8

Clearly there are different mechanisms responsible for belted and unbelted RSPs injuries. The belted RSPs who sustained a MAIS > 2 following a frontal impact are detailed in table 5. Their injuries can be summarised as to the chest, abdomen and femur fractures.

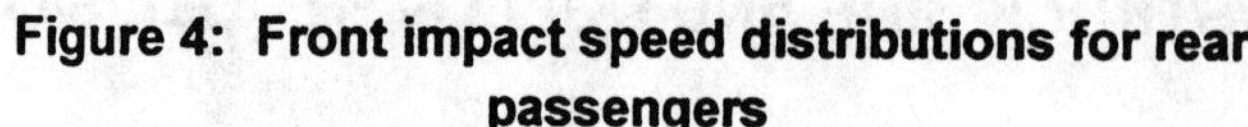

Figure 4: Front impact speed distributions for rear passengers

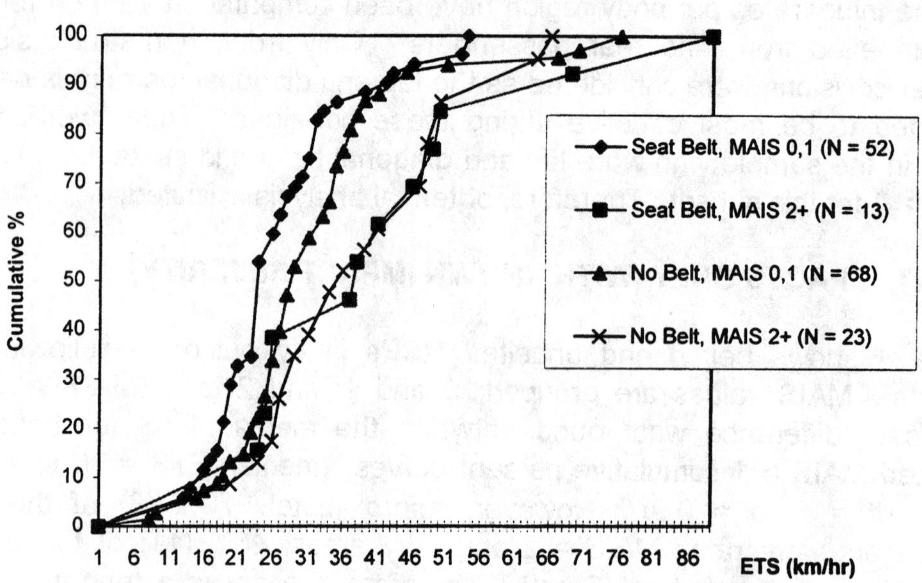

Table 5: Belted Adult RSPs Involved In Frontal Impacts Who Sustained MAIS > 2.

Occ No.	Injury Descriptions (AIS in parenthesis)	MAIS	ETS km/hr	Sex	Age	Height (m)	Weight (kg)
1	Multiple rib fractures with pneumo-thorax (4) & spleen contusion (3).	4	36	male	47	n/k	n/k
2	Femur fracture (3).	3	48	male	23	n/k	n/k
3	Small bowel perforation (3).	3	68	male	20	1.8	57
4	Flail chest (4); liver & kidney lacerations (2) & Cervical spine disc fracture (2).	4	40	female	36	1.6	54
5	Rib fractures (2) & two femur fractures (3).	3	45	female	n/k	n/k	n/k

CONCLUSIONS

This study has begun to discriminate the injury outcomes for restrained and unrestrained rear seat occupants and to compare them to those in the front seats. Population differences have a clear influence on effectiveness with elderly restrained rear occupants appearing as a factor which diminishes the overall effectiveness of restraints. By implication, some of the serious chest and abdominal injuries described above relate to rear belt geometry and the specific way in which the seat belt was being worn.

Forward contacts by restrained rear occupants wearing lap and diagonal belts striking their heads on front seat backs give indications of the trajectories of their heads in collisions. Belted RSPs head and leg contacts with front seats must be expected, and the possible interaction between front and rear occupants could increase the risk of injury to all car occupants.

With increasing numbers in our data base we will be able to examine these issues more precisely in the future.

ACKNOWLEDGEMENTS

The Co-operative Crash Injury Study is funded by the Department of Transport, Ford Motor Company Limited, Nissan Motor Company, Rover Group, Toyota Motor Company and Honda Motor Company. The project is managed by the Transport Research Laboratory. Grateful thanks are extended to everyone involved with the CCIS data collection process.

REFERENCES

American Association for Automotive Medicine. The Abbreviated Injury Scale 1990 Revision. AAAM, Arlington Heights, Illinois, USA; 1990.

Bodiwala et. al. Protective Effect of Rear Seat Restraints During Car Collisions. The Lancet, February 18th 1989.

Griffiths D. K. et. al. Car Occupant Fatality and the Effect of Future Safety Legislation. 20th Stapp Car Crash Conference. SAE Warrendale. USA. 1976
Haberl, J., Eichinger, S., Wintershoff, W. New Rear Safety Belt Geometry - A Contribution to Increase Belt Usage and Restraint Effectiveness. S.A.E. 870488

Hassan, A. M., Hill, J.R., Parkin, S., Mackay, G.M. Secondary Safety Developments: Some Applications of Field Data. Autotech 1995; IMechE; 1995.

Huelke D. F. et. al. The Hazard of the Unrestrained Occupant. 18th American Association for Automotive Medicine. 1974.

Kraft, M., Nygren, C., Tingvall C. Rear seat occupant protection. A study of children and adults in the rear seat of cars in relation to restraint use and car characteristics. Journal of Traffic Medicine 1990; 18:51-60.

Lundell, B., Mellander, H., Carlsson, I. Safety Performance of a Rear Seat Belt System with Optimized Seat Cushion Design. Passenger Car Meeting, Dearborn, Michigan. June 8-12, 1981. S.A.E.; 1981:810796

Mackay, G. M., Galer, M. D., Ashton, S. J., Thomas, P. D. The Methodology of In-depth Studies of Car Crashes in Britain. International Congress and Exposition; 1985; Detroit, Michigan, USA.: S.A.E.; 1985:365-390.

NHTSA. CRASH3 User's Guide. USA. Department of Transportation, National Highway Traffic Safety Administration; 1982.

NHTSA. Rear Seat Lap Shoulder Belts in Passenger Cars. Office of Regulatory Analysis Plans and Policy. Final Regulatory Evaluation. April 1989

Padmanaban, J., Ray, R. Safety Performance of Rear Seat Occupant Restraint Systems. S.A.E. 922524. 1992.

Rattenbury S. J. The Biomechanical Limits of Seat belt Protection. American Association for Automotive Medicine Conference. 1979.

Restraint Use by Car Occupants, 1994-96. LF 2074. Transport Research Laboratory. UK. 1996.

Road Accidents Great Britain: 1995. The Casualty Report. The Department of Transport. HMSO, London. 1995.

Tunbridge et. al. An In-Depth Study of Road Accident Casualties and their Injury Patterns. RR 136. Transport Research Laboratory. UK. 1988.

Residual Injuries to Restrained Car Occupants in Front- and Rear-Seat Positions

Dietmar Otte,
Norbert Südkamp,
Hermann Appel,
Accident Research Unit Hannover,
Federal Republic of Germany

Abstract

The highly protective quality of the safety belt remains undisputed. Nevertheless, it is evident, that almost the same injuries may occur whether or not a safety belt is used. Cognitions can be derived from this study in which accident and collision situations injuries are inflicted to belt-protected drivers, co-drivers and rear-seat passengers. On the basis of documented traffic accidents, investigated by a scientific research team, a total of 1,865 single injuries incurred by 605 belt-protected front-seat and 35 rear-seat car passengers were analyzed. It became evident that in frontal collisions and protection by safety belt, a high injury risk still remained for car passengers, for instance for the knee region by the dashboard, for thorax and head region by the steering wheel. In lateral collisions, injuries to the whole body were caused by the side structures of the doors. This situation must be regarded as a challenge for modification by car constructors.

Introduction

Ever since it was introduced, there have been those in favour of the safety belt, pointing out the high protective quality for injury avoidance and mitigation. But there were also those who opposed belt usage, accentuating the negative effects of the belt or belt strap. Through the arguments many car occupants were in doubt if they would not be safer without belt in accidents like overturning or car fires. Especially in view of the fact that serious abdominal injuries could be attributed to the belt. Consequently, there was quite a difference of opinion about the acceptance of this system. In 1983, the rate of belt-wearing in Germany amounted to just 45 percent for drivers in town districts—on motorways, however, to 81 percent. This rate increased to over 90 percent for both road systems, only after legislation of a fine. It was presumed that rear-seat passengers faced relatively little risk to get injured, due to their seating position. Only as late as 1979, safety belts were installed in rear seats in Germany. Since 1986, non-observance of belt-usage laws is fined in the German Federal Republic. The rate of belt usage for rear-seat passengers amounts to approximately 39 percent inside towns, and to approximately 66 percent on federal motorways.

This development was due to a multitude of effectivity analyses. Especially in frontal collisions of a vehicle, the highly protective effect of the safety belt was clearly demonstrated. The passengers were restrained by the elastic material, and an impact with the interior structures was prevented by the belt. In lateral collisions, however, the protective effect is limited. The passenger on the impact side is mainly injured, due to intrusion of the compartment, the belt, however, reduces the risk of being flung out. For passengers sitting in the opposite direction, an increased protective effect is apparent.

From all accident analyses it was evident that basically injuries can not be completely eliminated by the belt, and injuries can still occur despite belt usage. Publications (2,3) often deal with the problem of intra-abdominal injuries caused by the belt. The downward-directed movement of the passenger led to the conception of the so-called 'submarining movement', a motion causing intra-abdominal injuries, attributed to the belt strap. However, detailed accident analyses proved this to be a wrong conclusion, as basically these injuries can be caused in accident situations with intense intrusion of the passenger compartment(4).

On the one hand, the great protective effect of the safety belt is recognized; on the other hand, a multitude of injuries may still occur with belt usage. It is the objective of this study to demonstrate and analyse the injuries which still occur, despite belt usage, in certain accident situations.

In order to indicate injuries and accident mechanisms, a detailed accident analysis is required, which can not be obtained from police reports or isolated medical reports. The basis for such special detailed accident analyses is provided by special documentations of the 'investigations at the scene of accident'.

Description of the Investigation Material

In the greater vicinity of Hannover (town and country district of Hannover), an interdisciplinary research team of doctors and engineers drive to the place of accident immediately after the accident event. The team is informed of the accident by the Rescue Headquarters in Hannover. It drives to the place of the accident in vehicles equipped with blue lights and sirens. Accident traces like vehicle deformations, impact points on people are investigated and recorded immediately. Photogrammetric true-to-scale drawings are produced by a stereo-camera. These drawings enable a reconstruction and analysis of the accident as well as the collision phases. The injuries are documented divided into types, localisation and severity degree (AIS), and the injury mechanisms assigned within the framework of a technical/medical accident reconstruction.

608 front car passengers with automatic belts, in 445 cars built in 1976 and later, were analyzed, after collisions with other cars and objects like trees, walls, leading planks etc. Table 1 illustrates the distribution of cars, according to collision type, year of construction, weight category and impact deceleration, established by reconstruction and indicated by $\triangle V$. Frontal collisions of the accident vehicles predominate in the accident scene with 51.4 percent. Multiple collisions, i.e. when a vehicle collides in the accident incident with several sides, are very frequent with 21.6 percent.

Injury Situation of Front-seat Passengers

Injury severity. The highest injury severity degrees MAIS were established for frontal impacts with impact direction, almost in axial length of the vehicle ($\pm / -30°$) and for the rectangular impact ($\pm / -30°$) with impact in the compartment region, and also for the multiple colliding car occupants as shown in table 1: More than 50 percent of the occupants remained still unhurt in impacts outside the compartment region. There were no injured occupants with severity degrees 5 and 6 in these collision constellations.

Impact decelerations to passengers occurred with $\triangle$ values up to 30 km/h. In this group, 65.6 percent of the persons remained uninjured at this speed level, while a distinct increase in the maximum injury severity degree MAIS can be observed with a higher $\triangle V$. In vehicles of heavier weight categories, more passengers with lesser injury severity degrees were found than in vehicles of lighter weight.

Injury causes. In frontal collisions the moment of inertia causes the car occupants to continue in their original direction of movement, and their exposed body regions impact the vehicle parts in their way, after deceleration of the vehicle. The belt reduces a luxation of the body trunk. The relatively free moving body parts like the head, arms and legs, however, are

Table 1. Frequencies of impact directions in traffic accidents in relation to injury severity grades; Vehicle registration year, vehicle weight and $\triangle V$.

		cars		occupants		Maximum Injury Severity-grade (<-- 100 % -->)			
		n	%	n	%	uninjured	MAIS 1/2	MAIS 3/4	MAIS 5/6
total		445	100.0	608	100.0	44.9	43.2	8.6	4.1
impact direction									
frontal	frontal + 30	130	29.2	174	28.6	39.1	43.7	12.6	4.6
frontal	oblique/ rectangular	97	21.8	130	21.4	64.6	32.3	2.3	0 8
lateral	inside compartment rectangular + 30	25	5.6	37	6.1	18.9	56.8	10.8	13.5
lateral	oblique	31	7.0	38	6.3	44.7	39.5	7.9	7.9
lateral	outside compartment rectangular + 30	9	2.0	13	2.1	76.9	23.1	-	-
lateral	oblique	12	2.7	17	2.8	58.8	35.3	5.9	-
rear collision		23	5.2	34	5.6	79.1	20.9	-	-
multiple collision		118	26.5	165	27.1	27.3	56.4	11.5	4.8
vehicle-registration-year									
1976 - 1979		245	55.1	334	54.9	46.4	41.6	9.6	2.4
1980 - 1983		139	31.2	190	31.3	45.2	40.0	7.4	7.4
1984 -		22	4.9	30	4.9	50.0	50.0	-	-
unknown		39	8.8	54	8.9	22.2	61.1	11.1	5 6
vehicle weight									
- 900 kp		73	16.4	84	13.8	36.9	44.1	11.9	7 1
900 kp - 1200 kp		218	49.0	299	49.2	41.1	46.2	8.0	4.7
1200 kp -		139	31.2	205	33.7	52.7	36.1	8.8	2.4
unknown		15	3.4	20	3.3	30.0	70.0	-	-
delta-v									
- 30 km/h		210	47.2	288	47.4	65.6	32.3	1.8	0.3
31 km/h - 50 km/h		112	25.2	154	25.3	29.9	54.6	12.3	3.2
51 km/h - 70 km/h		25	5.6	33	5.4	9.1	39.4	27.3	24.2
71 km/h - 90 km/h		3	0.7	3	0.5	-	-	-	100.0
91 km/h -		4	0.9	4	0.7	-	-	50.0	50.0
unknown		91	20.4	126	20.7	23.8	57.9	13.5	4.8

6/87-1

still mobile. A possible luxation of the body trunk is influenced by the elastic quality of the belt, the response of the belt strap and its slackness in at-ease position, and further by the thickness of the clothes and the body build, i.e. whether not-so-slim or obese. Depending on the seating position and body size, an impact of the extended body parts is likely.

It was established that, even with automatic safety belts, 30.4 percent of the drivers suffered a head impact, as shown in table 2. This figure was distinctly lower for co-drivers with 18.4 percent. Abdominal injuries occurred with 8.3 percent almost twice as often for drivers as for co-drivers with 4.6 percent.

Table 2. Kinds of impact by driver and co-driver in relation to injured body regions (100% = all persons each kind of impact separately for drivers and co-drivers).

injured body region driver/ co-driver	kind of impact: frontal %	lateral, occupant on impact side %	lateral, occupant on opponent side %	multiple collision %
driver				
total (n)	217	38	39	109
head	30.4	47.4	35.9	43.1
neck	8.8	21.1	2.6	12.8
thorax	28.6	31.6	25.6	39.4
arm	18.9	28.9	17.9	30.3
abdomen	8.3	15.8	7.7	14.7
pelvis	6.0	15.8	5.1	11.9
leg	28.1	42.1	23.1	27.5
co-driver				
total (n)	87	17	11	56
head	18.4	23.5	27.3	41.1
neck	9.2	-	-	16.1
thorax	33.3	11.8	27.3	46.4
arm	19.5	23.5	36.4	35.7
abdomen	4.6	5.9	-	12.5
pelvis	5.7	11.8	9.1	16.1
leg	24.1	29.4	9.1	44.6

6/87-2

There were hardly any differences between injuries for drivers and co-drivers to all other body parts. Beside the head which is for the driver the most frequently injured body part, the thorax (28.6% for drivers) and the lower extremities (28.1%) must be mentioned as especially endangered body parts, as can be seen in table 2. 23 percent of the drivers incurred soft-part injuries of the face, 10.6 percent suffered a skull-brain trauma. The proportion of facial fractures is with 4.6 percent for co-drivers relatively high, in view of the fact that on the whole they quite rarely suffer head impacts (see table 3). The head impacts here often, i.e. with 12.4%, the steering wheel, to 6% the upper A-posts and the windscreen region to 8.8%, as shown in table 4 for drivers.

Throat injuries especially occur after an impact with the front structures, within an indirect trauma. This happens to approximately 8.8% of the drivers and 9.2% of co-drivers. They are as a rule distortions, so-called whiplash injuries (4.6%). Fractures of the cervical vertebra were not recorded with belt-protected co-drivers, after frontal collisions.

The thorax is another body region exposed to great injury risk. Approximately one third of all drivers and co-drivers incur such injuries (table 2), especially soft-part lesions. These injuries can primarily be attributed to the belt, as 19.4% of the injured in frontal collisions incurred thorax injuries by the belt.

Table 3. Localization and kind of injuries for drivers and co-drivers in frontal collisions and for front seat occupants in lateral and multiple collisions (100% = all persons each column).

localisation and kind of injuries	kind of impact				
	driver frontal %	co-driver frontal %	lateral, occupant on impact side %	lateral, occupant on opposite side %	multiple collision %
total (n)	217	87	55	50	165
skull soft part	7.4	4.6	21.8	20.0	24.8
face soft part	23.0	10.3	23.6	22.0	23.0
skull fracture	4.6	1.1	3.6	4.0	3.0
facial fracture	5.1	4.6	1.8	-	3.6
cerebral injury	10.6	8.0	10.9	10.0	17.6
eyes	6.5	4.6	-	2.0	8.5
neck soft part	3.7	4.6	9.1	2.0	7.3
cervical spine					
-distorsion	4.6	4.6	5.5	-	4.2
-fracture	0.5	-	1.8	-	2.4
thorax soft part	25.3	27.6	16.4	22.0	33.3
rib fracture	6.0	3.4	10.9	4.0	9.7
sternum fracture	1.8	3.4	-	2.0	1.8
lung	2.8	-	3.6	2.0	4.2
diaphragm	0.9	1.1	-	2.0	0.6
heart, vessels	1.8	-	3.6	-	1.2
thoracic spine	0.5	-	-	-	-
clavicle fracture	1.8	1.1	-	2.0	3.6
scapula fracture	-	-	-	-	0.6
abdomen soft part	5.1	3.4	7.3	4.0	9.7
vessels	0.9	-	3.6	-	0.6
mesentery	0.9	-	-	-	1.2
liver, gall	0.9	-	-	2.0	1.8
spleen	-	-	5.5	-	2.4
kidney	0.5	-	-	2.0	1.2
stomach, intestine	0.5	-	-	-	1.2
lumbar spine	1.8	1.1	-	-	1.8
pelvis soft part	4.6	4.6	7.3	4.0	9.7
pelvis fracture	1.8	1.1	7.3	4.0	4.2
symphysis pubis	-	-	1.8	-	-
vessels	-	-	1.8	-	0.6
organs	0.5	-	1.8	-	31.5
legs soft part	26.3	21.8	36.4	20.0	31.5
thigh fracture	8.3	4.6	9.1	2.0	2.4
knee fracture/lux.	2.3	1.1	-	-	1.2
lower leg fracture	1.8	2.3	5.5	2.0	1.8
foot fracture/lux.	2.3	5.7	-	-	1.8
arms soft part	13.8	18.4	25.5	20.0	29.7
upper arm fracture	1.8	2.3	-	-	2.4
elbow fracture/lux.	0.5	-	-	-	-
forearm fracture	5.1	1.1	-	-	-
hand fracture/lux.	2.3	1.1	1.8	-	1.2

6/87-3

9.7% of the thorax injuries were caused by the steering wheel. With co-drivers, no injuries were found to the lung, the heart and the dorsal vertebra. They did, however, more frequently incur sternum fractures (3.4% of the co-drivers and 1.8% of drivers).

Injuries to the abdominal organs for drivers can with less than 1% be considered as quite rare. For co-drivers they were not registered at all in this collision situation. In frontal collisions, injuries to the spleen were not registered for either drivers or co-drivers. All injuries to the abdomen were exclusively soft-part injuries. They were found to 3.4% for co-drivers, and to 5.1% for drivers. For belt users in frontal collisions, bony pelvis injuries are almost exclusively indirect traumas which would lead to hip luxations and luxation fractures, due to a knee impact and power transversion, via the thigh and hip region. This applies to 1% of drivers and 1.1% of co-drivers. Thigh fractures were established for 8.3% and knee fractures or knee-joint luxations for 2.3% of the drivers; fractures of the ankle-joint and foot region

Table 4. Injured body regions and impacted vehicle parts of drivers by kinds of impact (100% = all persons each column).

injured body region and impacted vehicle parts	kind of impact			
	frontal %	lateral, occupant on impact side %	lateral, occupant on opposite side %	multiple collision %
total (n)	217	38	39	109
head				
windscreen region	8.8	5.3	10.3	13.8
dashboard	2.8	-	-	-
steering wheel	12.4	2.6	5.1	8.3
upper A-pillar	6.0	5.3	5.1	5.5
door	1.8	26.3	10.3	6.4
upper B-pillar	-	7.9	2.6	4.6
rear compartment	-	-	5.1	0.9
roof	-	-	2.6	3.7
seat belt	-	-	-	0.9
neck				
dashboard	0.9	-	-	-
steering wheel	-	-	-	0.9
upper A-pillar	0.5	-	-	0.9
door	-	5.3	-	0.9
B-pillar	-	-	-	0.9
seat belt	2.3	5.3	2.6	5.5
indirect	4.6	10.5	-	3.7
thorax				
windscreen region	0.9	-	-	0.9
dashboard	1.8	-	-	-
steering wheel	9.7	-	2.6	7.3
door	0.5	18.4	5.1	11.0
B-pillar	-	2.6	-	1.8
roof	-	-	-	0.9
seat belt	19.4	10.5	20.5	20.2
upper extremities				
windscreen region	5.1	2.6	5.1	7.3
dashboard	4.1	5.3	2.6	10.1
steering wheel	5.5	7.9	7.7	7.3
upper A-pillar	1.4	-	-	2.8
door	5.1	13.2	7.7	6.4
B-pillar	-	-	-	1.8
seat belt	0.5	-	-	-
abdomen				
steering wheel	3.2	-	2.6	4.6
door	-	10.5	2.6	0.9
seat belt	3.7	2.6	2.6	8.3
pelvis				
dashboard	-	-	-	1.8
door	-	10.5	2.6	4.6
B-pillar	-	-	2.6	-
seat belt	4.6	2.6	2.6	5.5
lower extremities				
dashboard	21.2	13.2	15.4	18.3
steering wheelent	0.9	-	-	4.6
front leg room	12.9	21.1	10.3	13.8
door	0.9	13.2	2.6	4.6
seat belt	1.4	2.6	-	0.9

6/87-4

for co-drivers are frequent with 5%. Approximately 25% of all front passengers incur soft-part injuries to the legs. The knee is especially involved, due to an impact to the dashboard, as is shown in table 4.

In lateral collisions, passengers experience an additional relative movement component in lateral direction, beside a frontal-directed deceleration. In this case, the passenger on the impact side experiences an extreme encumbrance by the deformed vehicle side structures. For the passenger sitting impact directed, almost all body parts can be considered as injury regions. For the passenger sitting impact averted, only injuries to the exposed body parts like head, thorax, lower and upper extremities are frequent, as can also be seen in table 2. Impact averted as well as impact directed sitting passengers incurred to 42% and 45.4% respectively soft-part injuries to the head (table 3). These included, up to approximately 20% injuries to the vault of the cranium. Passengers sitting impact directed very often receive soft-part injuries to the throat region (9.1%). Distortions and fractures of the cervical vertebra are established exclusively for this passenger group. Rib fractures are also extremely frequent (10%) for those sitting impact directed. They also incurred in 5.5% ruptures of the spleen. These ruptures were exclusively established in this group with belt usage. To 7.3% bony injuries in the pelvis region were also found in this group; this is considered as bursting of symphysis and ileosacral with these persons. Another frequent injury cause for impact directed sitting passengers are door structures (table 4). They were responsible to 26.3% for head injuries, to 18.4% for thorax injuries, to 10.5% for injuries to abdomen and pelvis, and to 13.2% for injuries to the lower extremities. The B post is also a frequent cause for injuries. It was responsible for head injuries to 7.9% of car occupants, who were sitting impact directed. However, not all injuries in lateral collisions are inflicted by parts of the compartment side. In oblique impacts and subsequent oblique relative movement of the occupant, injuries are also caused by front parts, for instance by the dashboard, especially to those sitting impact directed. In oblique impacts, a motion of the upper body trunk and the head region may occur, as far as the front interior parts like A posts, dashboard and steering wheel. The occupant sitting opposite directed, more often incurs minor injuries (90.6% severity degree AIS 1 and 2). A higher injury frequency by the belt can be registered for this person, due to a more frequent restraint of the body (20.5% impact averted, 10.5% impact directed).

Multiple collisions, in which the vehicle is submitted to several successive collisions during the whole accident phase, are of special significance, as far as the injury severity is concerned. It appears that in this case injuries may be caused by almost all parts of the

Table 5. Proportion of all injuries with AIS 1/2 for front-seat passengers in frontal and lateral collisions in relation to injured body regions and injury causing parts (100% = all injuries each causing parts and body region).

injured body region and injury causing parts	portion of all injuries with AIS 1/2			
	driver frontal	co-driver frontal	lateral, occupant on impact side	lateral, occupant on opposite side
	%	%	%	%
head				
windscreen region	88.1	96.0	100.0	100.0
dashboard	100.0	87.5	-	-
steering wheel	98.0	-	100.0	100.0
upper A-pillar	83.9	-	85.7	100.0
door	100.0	-	92.0	90.0
upper B-pillar	-	-	42.9	100.0
rear compartment	-	-	-	66.7
roof	-	-	-	100.0
neck				
dashboard	100.0	-	-	-
upper A-pillar	100.0	-	-	-
door	-	-	50.0	-
seat belt	100.0	100.0	100.0	100.0
indirect	100.0	100.0	80.0	-
thorax				
windscreen region	100.0	-	-	-
dashboard	100.0	77.8	-	-
steering wheel	56.5	-	-	100.0
door	100.0	100.0	42.9	-
upper B-pillar	-	-	100.0	-
seat belt	100.0	85.7	100.0	92.3
upper extremities				
windscreen region	100.0	100.0	100.0	100.0
dashboard	100.0	93.3	100.0	100.0
steering wheel	68.2	-	100.0	100.0
upper A-pillar	57.1	-	-	-
door	88.9	100.0	100.0	100.0
seat belt	100.0	-	-	-
abdomen				
steering wheel	44.4	-	-	-
door	-	-	33.3	-
seat belt	100.0	100.0	100.0	100.0
pelvis				
front compartment	60.0	-	-	-
door	-	-	62.5	75.0
B-pillar	-	-	-	100.0
seat belt	100.0	100.0	100.0	100.0
lower extremities				
dashboard	94.2	83.8	85.7	92.3
front leg room	88.9	94.4	85.0	100.0
door	66.7	100.0	62.5	100.0
seat belt	100.0	-	100.0	-

6/87-5

interior, due to the actual impact situation and the consequent relative motion of the occupants.

For this collision type it also became evident that the steering wheel, the dashboard and the lateral door structures, inclusive A and B posts, are the most frequent cause for injuries. Here 47.8% of the occupants involved in multiple collisions incured soft-part injuries to the head (table 3), and 17.6% a skull-brain trauma. Eye injuries are also exceptionally frequent, with 8.5%. Fractures of the cervical vertebra result from a shoving-forward process (6). For this reason, such injuries occur more frequently (to 2.4%) in multiple collisions. In multiple collisions, all types of injuries to almost all body regions are exceptionally frequent.

Injury severity. As only 56.5% of thorax injuries caused by the steering wheel are of severity degree AIS 1 and 2, as shown in table 5, and 68.2% arm and 44.4% abdominal injuries, the steering wheel proved to be an especially injury-causing part for drivers in frontal collisions. Head injuries caused by the steering wheel are to 98% minor ones (AIS 1/2).

Serious injuries are quite rare for belt-wearing co-drivers in frontal collisions. The most frequent injury cause for them is the dashboard. They receive mostly injuries to the head (12.5% AIS >2) and the lower extremities (16.2% AIS >2). The high proportion of serious injuries inflicted by the dashboard (22.2% AIS > 2) must be attributed to an extensive intrusion of the passenger compartment.

The impact-directed sitting occupant often incurs serious injuries, especially to the head, in lateral impacts, for instance by the lateral posts (57.1%—AIS > 2, the A-post 14.3%—AIS > 2). The thorax, with 57.1%, abdomen with 66.7%, pelvis with 37.5% are seriously injured by the door structures.

Injury Situation of Rear-Seat Passengers

With a total of n = 35 persons, only a small number of belt-protected rear-seat passengers were at our disposal for a detailed analysis. Due to this fact, a comprehensive illustration of the injury mechanisms was not possible. Instead, we prepared an abbreviated form about the injury situation of rear-seat passengers. 18 persons out of 35 incurred slighter injuries of severity degree 1 and 2, only one person incurred injuries of severity degree MAIS 4. The remaining 16 persons (45.5%) were uninjured. Approximately 50% of the rear-seat passengers were involved in frontal collisions, only three were seated in the middle and were protected by static belt. 20 persons were protected by automatic belt, three wore a static belt, and 12 used childrens' restraint systems. The primarily injured body regions were the head and lower and upper extremities. Exclusively soft-part injuries were established for thorax and pelvis which were attributed to the belt. It became evident that the relative movement of the belt-wearing rear-seat passenger procures an impact of the rather free moving extremities. In this process, the head impacts mainly head-rest and back-rest of the front-seat. The legs, and especially the tibia and knee hit the frame of the back-rest, the arms hit the lateral interior structures. As far as childrens' restraint systems are concerned, only soft-part injuries of the head and upper extremities were established for the 12 investigated cases.

Influence of Accident Severity

The definition of the vehicle deformation is an example for the accident severity. The stress caused to the occupants is mainly due to deceleration during the collision. This must be seen in connection with the inflicted injuries. The occurring accident stress is expressed by $\triangle V$. For the above study, the injuries to passengers were analyzed, divided into body regions and injury cause, in dependance of $\triangle V$ values (figures 1 to 4). In frontal collisions, soft-part head injuries to belt-wearing front passengers occurred only above $\triangle V$ values of 20 km/h. With lower $\triangle V$ values, head injuries were only caused by the steering wheel. Especially in the driver's position, the distance to the steering wheel, which is often too short, could be the injury cause. 3.6 percent of involved drivers suffered soft-part head injuries by impacts to the windscreen region. Fractures were observed only with $\triangle V$ above 30 km/h. While head injuries by the dashboard only occurred with $\triangle V$ above 30 km/h, soft-part injuries caused by the upper A-post were already observed with $\triangle V$ above 10 km/h. It is evident that with a higher $\triangle V$ the injury frequency increases.

A soft-part injury to the thorax and pelvis region caused by the belt, the so-called belt mark, can be observed with $\triangle V$ values exceeding 10 km/h.

In conclusion, a significant risk for soft-part injuries with $\triangle V$ above 20 km/h is apparent, with the exception of injuries to the lower extremities and the belt-strap hematoma, which may already occur at 10 km/h. The injury risk for fractures is apparent above 30 km/h. The dashboard is to be regarded as especially dangerous, in view of the very great increase in frequency for leg injuries, with increasing $\triangle V$.

In lateral collisions, soft-part injuries occur as a rule already with lesser $\triangle V$ values. Especially soft-part injuries to the relatively free moving head occur below 10 km/h, to the occupant sitting in impact direction. In contrast, soft-part injuries to the thorax by seat belt and side doors were observed above 10 km/h. The legs incurred injuries by the dashboard and the front foot room at speeds above 20 km/h. Door structures represent a great risk for fractures to the impact-directed occupant. Here 11.1 percent of the occupants in collisions of $\triangle V$ up to 10 km/h suffered pelvis fractures.

A low injury frequency for the whole $\triangle V$ region is evident for the impact-averted sitting passengers. For these persons fractures were observed only from $\triangle V$ of 40 km/h. This is due to the highly protective effect of the safety belt, which controls the relative movement of the impact-averted sitting occupant. This explains the fact that soft-part injuries by the belt may occur already with low $\triangle V$ values. Rib fractures may even occur in $\triangle V$ of 20 km/h.

Discussion

This study of real accidents proves that almost all injuries may occur despite belt usage, even if they are distinctly reduced in number by the highly protective function of the belt, and occur mainly in more serious accidents. Injuries which could be attributed to the belt are as a rule defined as hematoma and are caused by pressure of the belt strap to the soft parts. With belt usage, intra-abdominal injuries occur only due to incorrect belt fitting in the pelvis region, with soft and yielding seat bolsters and intense deformation of the department, superimposing the forward movement. A submarining movement causing intra-abdominal inju-

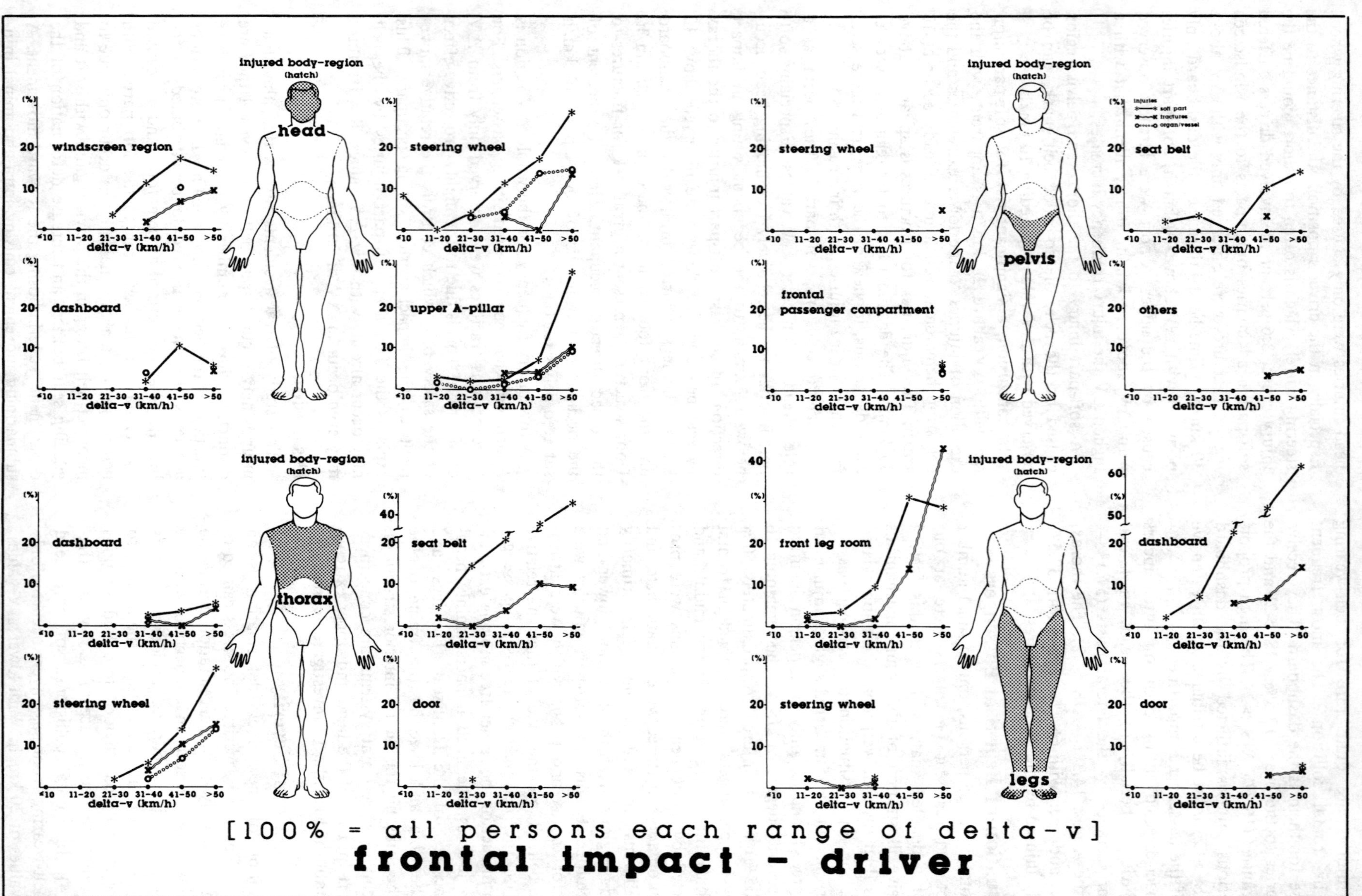

Figure 1. Relevant injury causing parts of selected body regions in relation to ΔV

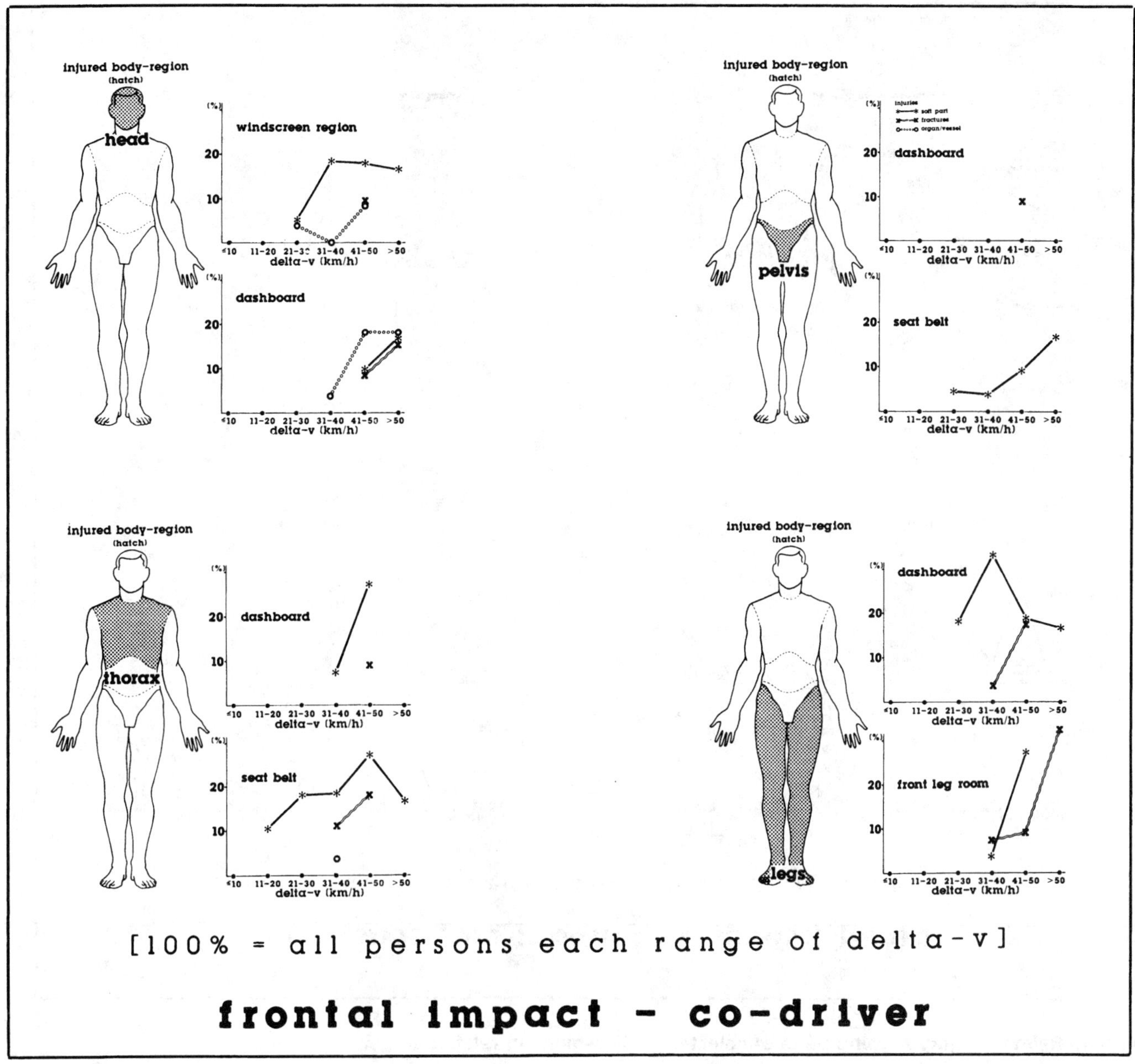

Figure 2. Relevant injury causing parts of selected body regions in relation to △V

ries could be established in a detailed analysis of 1.6 percent of all belt-using front passengers. It appears to be an essential cognition that intra-abdominal injuries must not necessarily be caused by a submarining movement, but that in fact other accident conditions like high intrusion or insufficient distance from the steering wheel could be responsible for the injury patterns. In cases of submarining, established beyond doubt, belt contusion marks were found in the abdominal and thorax region with characteristic patterns. These are

- marks of the belt strap on or just above both iliac crestlateral, while the belt impact mark in the front medial region of the stomach was clearly higher situated;
- mark on the thorax appears to have a distinct lower border, while the upper edge continuous blurred. This results in a broader belt mark than the original width of the belt.

Apart from these facts, the study demonstrates that, even with belt usage, a great injury risk still exists in frontal collisions, for the knee region by the dashboard, for the thorax and head region by the steering wheel, and in lateral impacts for the whole body by

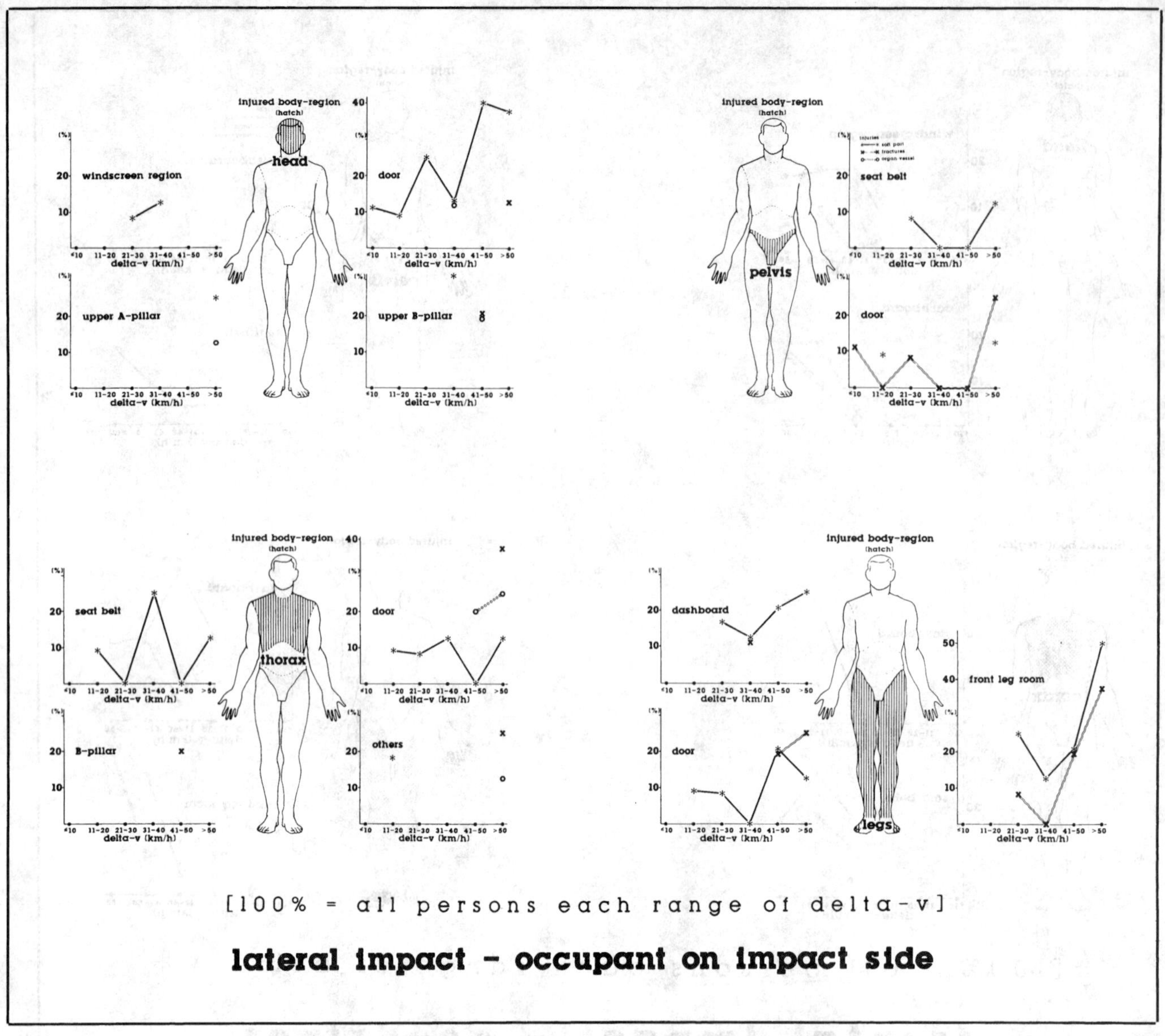

Figure 3. Relevant injury causing parts of selected body regions in relation to △V

the side structures of the doors. Demands on the car constructors are, therefore,

- to extend the moving space for occupants,
- to bolster the front, but especially the lateral interior on impact exposed points.

In this connection, the efforts of some car manufacturers must be acknowledged, who are considering to reduce the injury risk by the steering wheel through modifications like airbag and Pro-Con Ten⁺. Also for the rear-seat region, the demand for the widest possible moving space and bolstering of the back rest and lateral vehicle structures has to remain.

Finally, the protective quality of the safety belt which was proved to be of benefit to front as well as rear-seat passengers, should be accentuated. This applies in lateral collisions especially for passengers sitting impact averted. In lateral and multiple collisions, it prevents the dangerous flinging out. It is a special attribution of the safety belt that the number of persons killed annually in road traffic, as well as the injury severity degree and also the total number of single injuries to an accident casualty has been distinctly reduced.

⁺Programmed contraction and tension by AUDI

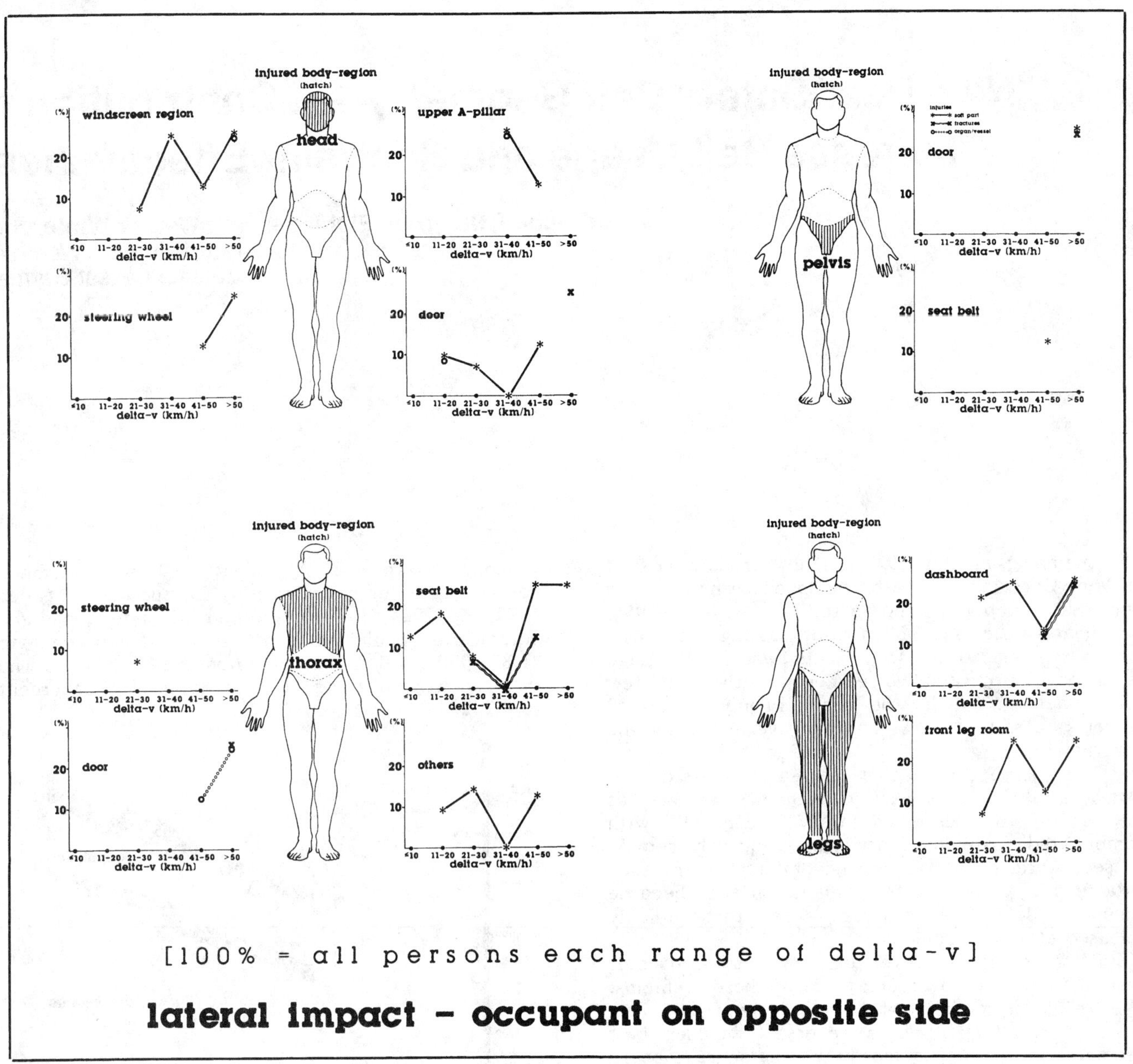

Figure 4. Relevant injury causing parts of selected body regions in relation to ΔV

References:

1. Marburger, E.A.; Meyer, L.; Radscheit, H. Gurtanlegequoten in Pkw Zeitschr. f. Verk.sich. 33, 40-41, 1987
2. Leung, Y.C., Tarmere, C.; Lestrelin, D. Submarining Injuries of 3Pt. Belted Occupants in Frontal Collisions—Description, Mechanisms and Protection SAE-Paper 821158, 26th Stapp Car Crash Conf. 173-205, Ann Arbor 1982
3. Sabey, B.E.; Grant, B.E.; Hobbs, C.A. Alleviation of Injuries by Use of Seat-Belts 6th Int. Conf. IAATM, 480-484, Melbourne 1977
4. Otte, D.; Südkamp, N.; Appel, H. Variations of Injury Patterns of Seat-Belt Users SAE 870226, 61-72, Int. Congr. and Exposition, Detroit 1987
5. American Association for Automotive Medicine The Abbreviated Injury Scale—Revision 85 Morton Grove, Illinois 1985
6. Otte, D. Risk and Mechanisms of Injuries to The Cervical Spine in Traffic Accidents IRCOBI Conf., 17-32, Göteborg 1985

870488

New Rear Safety Belt Geometry—A Contribution to Increase Belt Usage and Restraint Effectiveness

Josef Haberl, Siegfried Eichinger, and Werner Wintershoff
Bayerische Motoren Werke AG
München, West Germany

ABSTRACT

Beginning in May 1979, all new production cars in West Germany had to be equipped with front and rear safety belts, and effective August 1984, a mandatory use law applied to all seating positions.

In contrast to the front seats, where belt usage rates have exceeded 90 percent in the last few years, belt usage in the rear is only about 50 percent, in spite of laws and punitive action by the government.

The poor acceptance of rear safety belts can be attributed to comfort and convenience factors, as well as the low perceived safety value. Belts with complicated fastening procedures frequently remain unused, while belts that are uncomfortable are soon unfastened. Some belts remain unused because occupants doubt the ability of a belt to be quickly released after an accident.

BMW decided to develop a new rear safety belt system that would overcome most of these problems. The result is a belt whose most significant new design characteristic is a reversed shoulder belt geometry - the upper mounting points are inboard and the diagonal shoulder belt angle across the torso is in a direction opposite to what is customary. This enables placement of the female half of the buckle on the outboard side, integrated into the seat and well forward for easy access. Buckling these belts is a one-hand operation that is as quick and simple as putting on a front safety belt. After opening the rear door, people are 'invited' to put the belts on.

The forward location of the outboard buckle also makes pelvic restraint more effective in head-on collisions, reducing the likelihood of submarining and abdominal injury.

BELT USAGE IN GERMANY

It was not until the early nineteen-seventies that Federal German statistics for fatal accidents to automobile occupants in relation to total traffic mileage and vehicle population began to take a most welcome downturn. Although the total distance covered by automobiles rose by 55.9 percent between 1970 and 1985, to 313.4 billion kilometers, and the automobile population also rose by 85.3 percent to 25,800,000, the number of automobile occupants who suffered a fatal accident went down by 46.5 percent to 4,182.(1)*

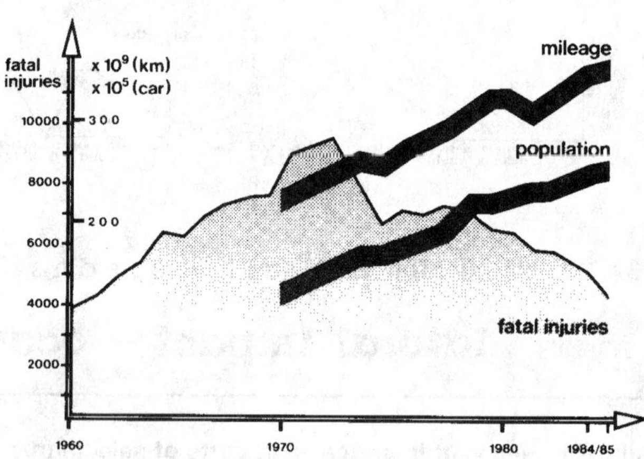

Figure 1: Graph of total automobile mileage, automobile population and fatal injuries to automobile occupants in the Federal Republic of Germany from 1960 to 1985 (1)

This positive trend, which is repeated in the figures for injuries to automobile occupants, was brought about to a significant extent by an increase in both the provision and actual use of occupant restraint systems.

Despite this welcome result, it is clearly essential to reduce the number of accident victims still further, for both humanitarian and macro-economic reasons. We must therefore ask ourselves what potential still remains for improvements in various areas.

*Numbers in parentheses designate references at end of paper

The maximum safety benefit is undoubtedly obtained if the safety equipment already installed in the automobile in particular the safety belts is indeed used by its occupants.

The chart shows the safety belt wearing behavior of automobile occupants in West Germany and the temporal relationship between this and various statutory measures aimed at increasing the proportion of belt users.

If we commence by studying the belt wearing habits of front seat occupants, we find that administrative measures incurring fines and a reduction in insurance claim entitlements have proven more effective than promotional or educational campaigns.

The 98 percent front safety belt usage rate recorded for Federal German highways by the Federal German Road Transport Office (Bundesanstalt für Straßenwesen) in September 1986 in all probability represents the pratical upper limit. (2) More stringent controls, however, could effect a further improvement in the figures of 95.5 and 93 percent recorded for ordinary main roads and urban traffic respectively.

Although the same legal requirements have applied to rear seat passengers since July 1, 1986, the usage rates for cars equipped with rear safety belts have so far not exceeded 66, 60 and 38 percent, again for highways, ordinary main roads and urban traffic respectively. (2)

According to an estimate made by the German Accident and Motor Insurers (HUK), between 450 and 700 rear seat occupants are killed and up to 8,000 seriously injured every year on the roads of the Federal Republic of Germany. (3) By increasing belt usage to values similar to those now achieved in the case of front seat occupants, it should be possible to effect a considerable reduction in these figures. Furthermore, rear seat occupants not wearing their safety belts constitute a considerable hazard to the front seat occupants in the event of an accident.

Why are those occupying the rear seats of automobiles unwilling to make use of existing safety belt systems as a means of reducing the risk of injury to which they are exposed, despite the fact that the reduction in the severity of injury has been demonstrated to reach 50 to 70 percent? (4)

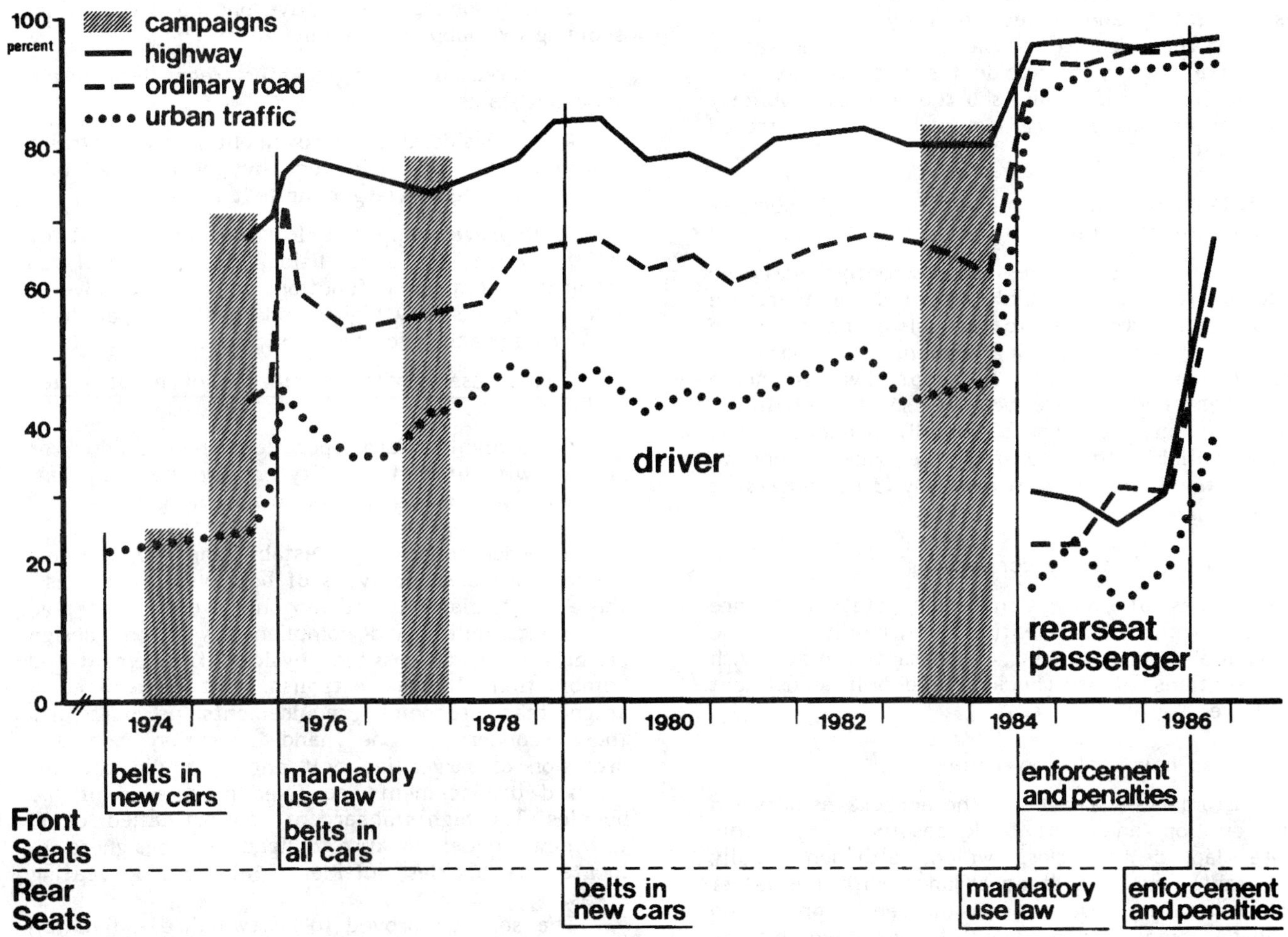

Figure 2: Safety belt usage rate versus time of introduction of usage laws and enforcement (2)

REAR SAFETY BELT PROBLEM AREAS

In order to provide an answer to this question, BMW conducted a survey of rear seat passenger safety, and established that the following problem areas exist which in some cases are bound to have a decisive influence on the introduction of new and more advanced safety belt concepts:

1. Lack of rear safety belt systems

The rate of rear belt installation in automobiles has risen continuously in recent years and approximately 75 percent of in-use vehicles are now equipped with rear belts. Assuming an average car life of 11 years and in view of the fact that the installation of rear safety belts has been compulsory in Germany since May 1, 1979, an availability rate of almost 100 percent will not be achieved until 1990 or later.

2. Underestimation by occupant of the injury risk

Since the structural elements of the car and the seats appear to be immensely strong, and the occupants are surrounded by padded interior, those at the rear in particular are deluded into a sense of absolute safety and protection. They fail to identify the very real risk to occupants not wearing safety belts in a variety of accident situations. Nor is it possible to rectify this shortcoming by publicity alone, as various past campaigns have demonstrated only too well.

3. Difficulty in fastening belts and inadequate comfort in use

The safety belt buckles of currently marketed systems are usually flexibly mounted and therefore compel the wearers to use two hands when attaching and releasing them, a basically non-ergonomic procedure. It is not always clear to which seat each buckle belongs, so that confusion may arise on this account as well. In addition, the areas available for operating the buckles tend to overlap when the full seat capacity (3 passengers) is in use.

4. Reduced seating comfort

The buckles of conventional rear safety belts are located in the actual seating area, resulting in the occupants' bodies making unpleasant contact with hard sections of the buckles and belt adjustment hardware.

5. Lap belt angle too acute

The customary belt path to the anchorages between seat cushion and seat back results in relatively acute lap belt angles, which, although legally acceptable, are not in accordance with the latest technical safety research. The degree of protection against submarining provided, for example, by the front safety belts cannot be attained for this reason, and there is a risk of injury as a result of the lap belt slipping up as far as the abdominal region, particularly in the event of severe frontal collisions.

6. Insufficient body support from seat

Rear seats designed with the emphasis on comfort, with relatively large spaces beneath, do not provide sufficient support for the body from beneath in the event of a frontal collision.This can increase the risk of abdominal injury.

DEVELOPMENT OF THE NEW SYSTEM

In order to eliminate this shortcoming, a restraint system was specified for the new BMW 7 Series from a very early stage of development onwards which would be the equivalent of the front seat systems in terms of occupant protection and comfort.

Here are the main details of the design criteria for such a system:

1. An increased level of acceptance and therefore a higher usage rate, to be achieved by a more attractive and practical design with clearly identifiable component functions.

2. Ergonomically effective one hand operation, avoiding overlapping fastening zones.

3. Increased wearing comfort regardless of the occupant's size.

4. A considerable improvement in the restraint function, to be achieved by optimizing belt geometry, including larger lap belt angles.

5. Improved support below the occupant's body in the event of a collision, by means of an integrated seat ramp function, but without adverse effects on the very high standard of occupant ride comfort demanded for the new model.

6. Increased protection in the event of a side collision.

7. Straightforward operation when children or persons with limited mobility occupy the rear seat, and also for occupant rescue in an emergency.

In order to assist in establishing the optimum concept, all possible types of belt systems (harness, three-point, diagonal and lap only) were first tested for practicability and compliance with the design criteria in every possible physical arrangement and combination. It was established at a very early stage that ergonomic requirements, in particular those concerning one hand operation and the provision of larger lap belt angles, would call for forward displacement and fixed positioning of the buckles. The high standard of comfort called for on the new model in turn obliged the designers to locate the buckles outside of all possible seating areas.

The solution proved to be two three-point (lap and diagonal) belts for the outer seated positions, with their geometry reversed by comparsion with conventional systems, together with a retractor lap belt for the center rear seat occupant.

Figure 3: The ergonomic rear safety belt system in the new BMW 7 Series model, with two outer lap and diagonal retractor belts and one central retractor lap-only belt

The location of the retractors necessitated by this solution and the resulting introduction of restraint forces into a relatively 'soft' area of the body made it necessary to perform tests on seat belt anchorages in accordance with a wide variety of national regulations before a decision could be made in favor of the new system.

After only a few optimizing measures, it proved possible to achieve the required structural strength for maintenance of the correct belt geometry, including the additional safety margin required for such systems. This paved the way for what has become the "BMW Ergonomic Safety Belt System".

DETAILED DESCRIPTION OF THE SYSTEM

The safety belt buckles are bolted firmly to the body by means of anchorage fittings and are integrated into the seat. They are easily visible in an ergonomically satisfactory position on the outer edge of the seat.

Figure 4: Buckle positioned outside the seating area for maximum comfort, and firmly located to facilitate one-hand operation

This position permits the large lap belt angle so essential for occupant restraint to be achieved without adversely affecting wearing comfort. Fixed buckle location is essential for one hand operation. At the belt end anchorage points the seat bench is cut away as far as the visible edge between seat cushion and seat back so that the belt can run from the anchorage point, which is also brought forward, to the occupant lap belt area without excessive deflection from its path.

In addition, all center seat lap belt elements located in this area are recessed.

The shoulder belt retractors with integrated belt guide devices are mounted on the shelf below the rear window and offset by 175 mm from the center line of the vehicle.

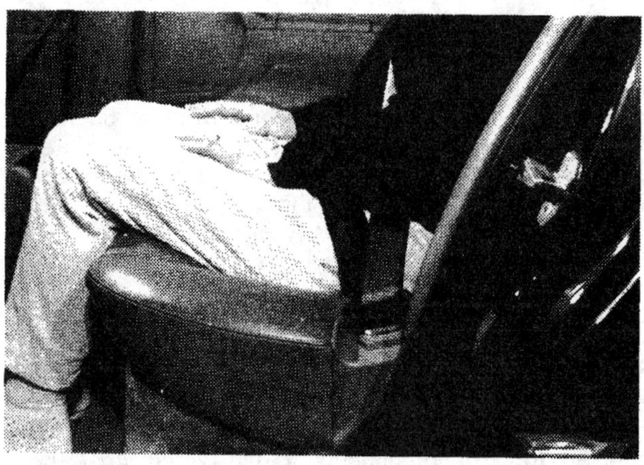

Figure 5: Moving the seat belt buckle forward increases the angle of the lap belt and therefore improves the degree of occupant restraint

Figure 6: At the belt end anchorage points, the seat is cut away locally and the anchorage also moved forward to ensure a greater lap belt angle

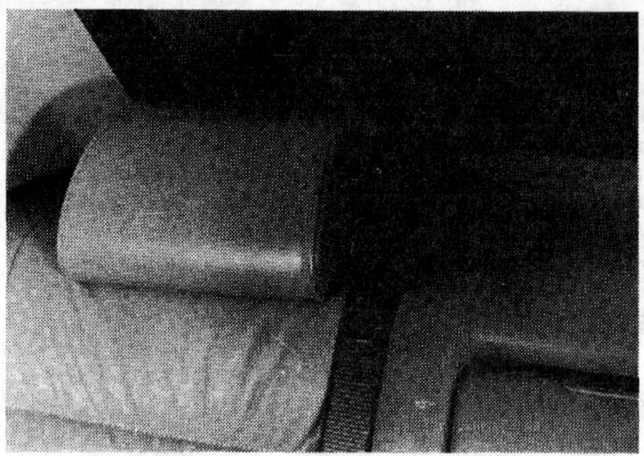

Figure 7: Retractor with belt guide devices are integrated into the shelf below the rear window

The belts in retracted position (matched in design and color to the car's trim) leading vertically from the floor anchorages to the retractors in conjunction with the arrangement of the buckles yield a visually attractive result and one which encourages the passenger to take hold of the safety belt and put it on when entering the car.

In addition, the various elements of the belt system create no obstruction as people enter the car.

The safety belt latch tongues, which have to be located by someone wishing to put on the belt, are convenient to reach in the seat back area and clearly identifiable as belonging to one or the other outside rear seat. They can be grasped, pulled out and down towards the buckle and inserted with the outboard hand only.

The above arrangement ideally interprets the desired ergonomic requirements.

Figure 8: The high ergonomic standard of the new belt system enables the belt to be fastened with one-hand only

Step 1: Grasping the latch tongue

Step 2: Pulling out the belt

Step 3: Inserting tongue into the buckle

When the belts are being worn, occupants of all sizes enjoy the benefit of a belt geometry which is ideal in both technical and comfort terms, without any additional height adjustment being necessary. Since the retractors are well-placed, belt return forces are extremely low, giving no cause for complaint even during long time journeys.

Figure 9: Location of the various components of the system ensures a safe, comfortable belt geometry for person of both small and large stature

Figure 10: Belt geometry for larger occupants

The belt can also be released with one hand, and returns to its 'parked' position without requiring any corrective action on the part of the occupants.

Since the buckles are so easy to reach, it becomes easier to carry children in safety on the rear seats and also persons whose movements are limited.

Figure 11: Positioning the buckles on the outside makes it much easier to secure children with the rear safety belts

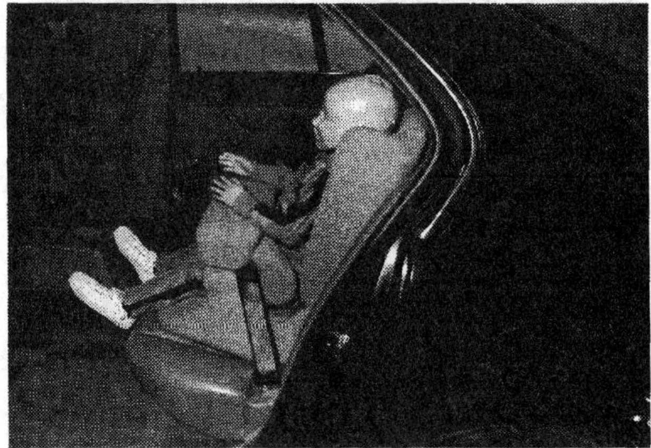

Figure 12: In an emergency, the easily reached buckles permit persons outside the car to release and rescue the occupants of the rear seat more rapidly

Operation of the safety belts is altogether more straightforward and the occupants can be more easily released in an emergency.

The rear seat itself represents a further element in the overall system. A tray integrated into the all-foam seat cushion is shaped to act as a ramp, so that the body is supported from below. In conjunction with the optimized belt geometry, the risk of the pelvic region slipping out under the belt (described as 'submarining') is effectively avoided.

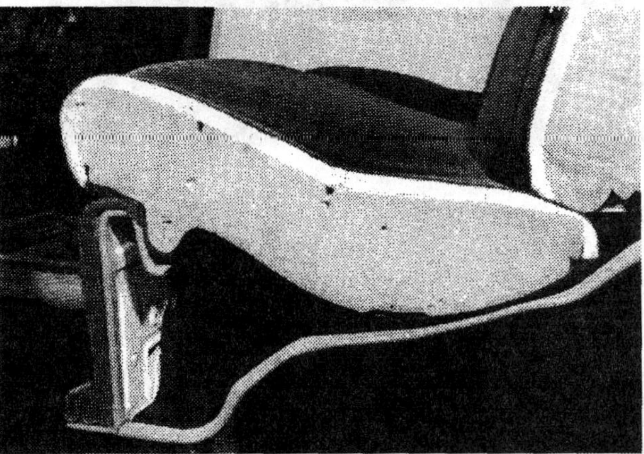

Figure 13: The seat cushion, which contains a tray shaped to act as a ramp, prevents the occupants from 'submarining' in the event of a collision

The superior protective capability of the new BMW ergonomic rear safety belt system compared with conventional safety belts is illustrated by the sequences of pictures taken during 30 mph frontal and simulated side impacts.

The 0-degree frontal test shown here as representative of the many other sled and crash tests conducted in every conceivable set of conditions is impressive evidence of the new belt system's efficiency. Both the lap and shoulder belts remain within the optimum occupant restraint zones throughout the entire test. Occupant energy absorption takes place with the minimum possible risk, and the effect is fully equivalent to that achieved by good front safety belts. (see figure 14)
If severe lateral acceleration occurs, as in a side impact or roll-over situation, the diagonal belt straps in conjunction with the marked shell-shaped configuration of the rear seat back impose a limit on displacement of the occupants' upper bodies towards the centerline of the vehicle, so that the risk of the rear seat occupants striking each other is considerably reduced. (see figure 15)

In conclusion, it can be stated that the new ergonomic safety belt system introduced on the BMW 7 Series models succeeds for the first time in achieving an ideal combination of potentially contradictory requirements with respect to safety, comfort and ergonomics.
We are convinced that this system will be accepted by our customers and approved by safety experts, and will therefore fulfil a pioneering function in increasing occupant safety in the automobile.

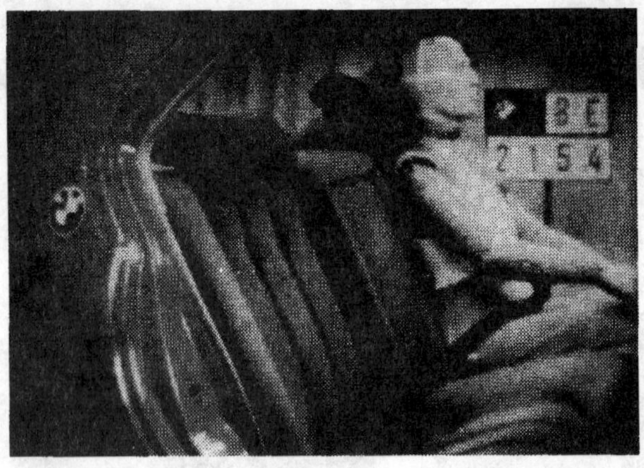

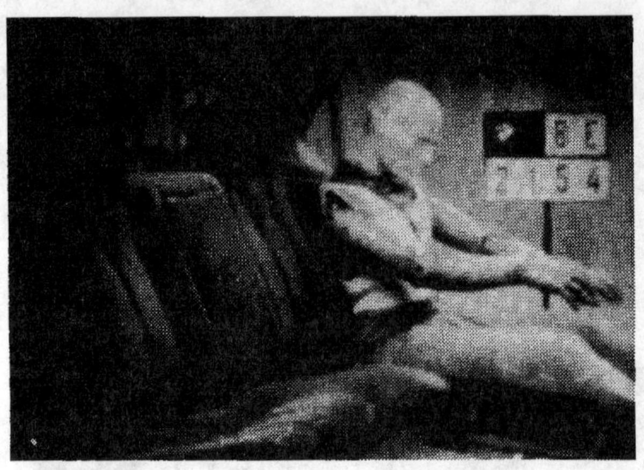

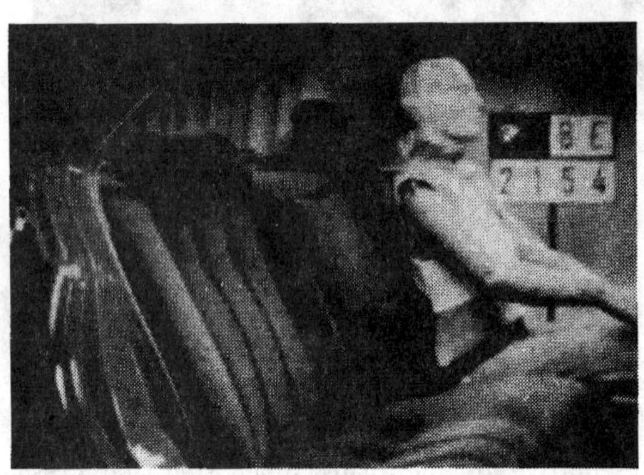

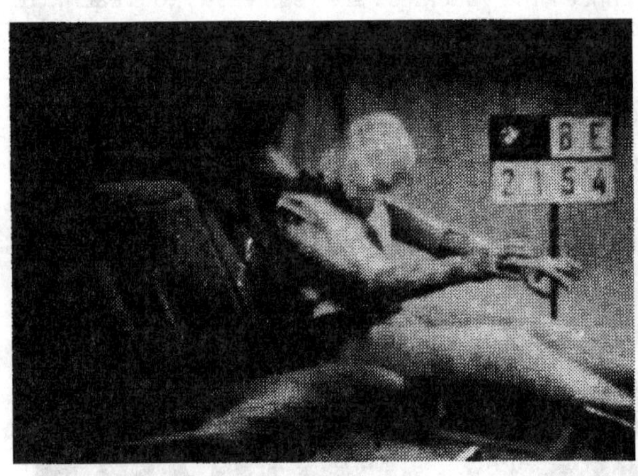

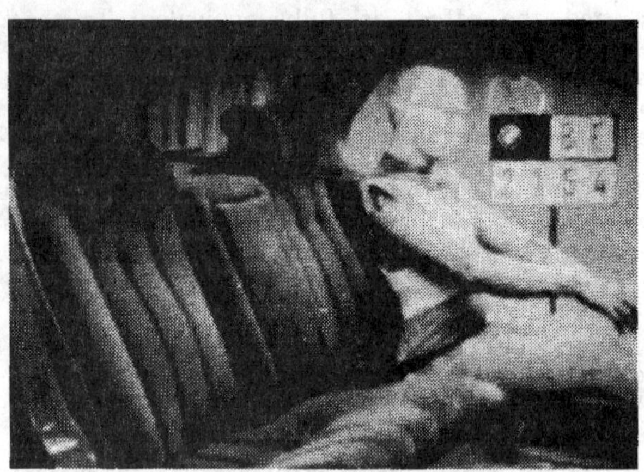

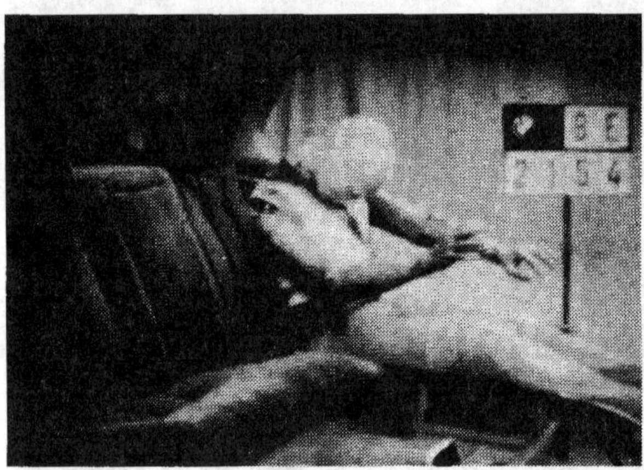

Figure 14: Simulated 30 mph frontal collision test of the ergonomic rear safety belt system with a 50-percentile Hybrid II dummy

Figure 15: Simulated side impact test of the new safety belt system with two 50-percentile Hybrid II dummies

REFERENCES

1. Statistisches Bundesamt Wiesbaden: "Straßenverkehrsunfälle"

2. Bundesanstalt für Straßenwesen: "Die Sicherheitsgurtanlegequoten von Pkw-Insassen", Ergebnisse der Erhebung vom September 1986

3. K. Langwieder: "Das Verletzungsrisiko am Rücksitz bei Pkw-Unfällen - Mögliche Beeinflussung durch Angurten", Euro-Seatbelt 84, ENKA AG, Wuppertal, März 1984

4. K. Langwieder, T. Hummel: "Der heutige Stand des Pkw-Insassenschutzes und Möglichkeiten für weitere Verbesserungen in der Zukunft aus der Sicht der Unfallforschung

CHAPTER 17:

CHILDREN, BELT RESTRAINTS AND SEATS

933098

The Seat Belt Syndrome in Children

J. C. Lane
Monash Univ.

ABSTRACT

Lap belts, fitted to the centre seats of Australian cars for the past twenty-two years, have come under criticism as being injurious to children. The weight of evidence is that lap belts provide substantial protection, though less than three-point belts. A specific injury, the seat belt syndrome (SBS), to abdominal viscera and/or lumbar spine has been particularly associated with lap belts, an association confirmed by a hospital-based study in Melbourne. Roadside observations of belt use and Transport Accident Commission claims permitted the calculation of the incidence of SBS and the relative risks of SBS by seated position. The centre rear seat (lap belt) carried about twice the risk of SBS as outboard rear seats (three-point belts) which in turn have 2.7 times the risk of the outboard front seat. The number of SBS cases in Victoria has increased with penetration of the car fleet by 1971 and later cars. Suggestions are made for improvements in the restraint system.

FOR THE PAST TWENTY-TWO YEARS, Australian Design Rule 5A has required cars and station wagons to be fitted with three-point seat belts (lap and shoulder belts) in all seat except for the front and rear centre seats, which have lap belts. Lap belts have come under criticism on the ground that, in rears seats, they provide little effective protection and because of an association with a particular injury, the so-called seat belt syndrome (SBS).

In a review of the literature, the great weight of evidence indicates that lap belts provide substantial protection, ranging from 18% to 50% reduction in injuries, though less protection than three-point belts in comparable crashes. Reports from Sweden and the United States show that lap belts provide protection for, in particular, children and the elderly (Kraft et al.,1989; Morris, 1983; Orsay et al., 1989; Partyka, 1988).

The occurrence of specific serious injury presumptively associated with the belt itself was first reported by Kulowski and Rost (1956). An attempt to estimate the incidence of the visceral injury was made in 1962 by Garrett and Braunstein, who coined the term "seat belt syndrome" (SBS). They analysed data from the Cornell Crash Injury Research files of highway accidents in which at least one occupant was wearing a belt - lap belts at that time. Of 3325 belt wearers, 944 were injured and, of these, seven had "reported or possible" abdominal injuries, seven had pelvic injuries and twelve had lumbar spine injuries.

Analysis of two large samples of hospital admissions in the U.K., for the year before and the year after the introduction of a law requiring belts to be worn in the front seats of cars, presumably with mostly three-point belts, demonstrated an increase in injuries to abdominal and pelvic organs in the after period, except for kidney injuries, for which there was a decrease. Changes in the occurrence of lumbar spine injuries were inconsistent (Rutherford et al, 1985; Tunbridge, 1989).

Review of records of car occupant casualties at a trauma centre, over a five year period (1984-88) during which seat belt usage increased substantially, showed that lumbar spine fractures and injuries to the small and large intestine also increased (Anderson et al., 1991). There was high belt usage, of both three-point and lap belts, by the casualties.

A comparison of injuries usually associated with belts, in a large sample of hospital admissions from motor vehicle crashes, showed that gastro-intestinal injuries had a higher incidence in the belted than the unbelted group, but lumbar spine injuries were not significantly different in incidence (Rutledge et al., 1991).

Thus kidney injuries should be excluded from the list of injuries associated with belts. Injuries of the spleen, when the only visceral injury, should probably also be excluded because of the propensity of this organ to be damaged in any blunt impact. But, because of practical difficulties of reclassification, the seat belt syndrome (SBS) is defined as injury to the abdominal viscera and/or injury to the lumbar spine.

REPORTED CHILD SBS CASUALTIES

There are many case reports of seat belt injuries in general in the medical literature. As regards children, a search of the English language literature through 1991 yielded single cases or small series in which individual cases are detailed, to a total of 69 children, 19 of whom had visceral injuries only, 11 spinal injuries only, and 39 both visceral and spinal (sources and details are given by Lane, 1992). These cases cannot be regarded a a sample, but it can be noted that the modal age was ten years and 81% were reported to have been wearing lap belts - three quarters of these in the rear seat. It seems likely that selection for reporting was biased towards severity.

A series of papers by Agran et al. (1985 - 1990) refers to paediatric patients from car accidents at emergency rooms of hospitals serving a community of 1.9 million people. Two hundred and twenty-nine belt wearers yielded one case of small bowel laceration and one of ruptured spleen (in a three-point belt wearer). There were three cases of bladder contusion but no spine fractures, dislocations or cord injury.

None of the 160 paediatric patients reviewed by Orsay et al. (1989) suffered a visceral or lumbar spine injury. There were no SBS cases in 46 restrained children from 231 crashes in which at least one child less than eight years had been transported by ambulance (Corben and Herbert, 1981). Eight hundred and twenty responses to a newspaper questionnaire referred to 288 unrestrained and 865 restrained children. There were four SBS cases (0.5%) in the restrained and four (1.38%) in the unrestrained children: none had lumbar spine injuries (Langweider and Hummel, 1989). Clearly the incidence of SBS in children is not high.

Paediatric casualties from motor vehicle accidents admitted to the Royal Children's Hospital in Melbourne for the period 1984 to 1989 have been reviewed by Hoy and Cole (1992). (This hospital admits children up to age 16 directly or on transfer from the whole State of Victoria.) Of 541 casualties, 29 had belt injuries of the abdomen and of these seven had Chance fractures of the spine. One had a cord injury without radiological abnormality. Four of the SBS cases were in the first half of the period and 25 in the second, a nearly four-fold increase. Lap belts were used by 19 of the 28 cases for which the type of restraint was ascertained. The crashes generating the cases were severe: 21 came from vehicles in which at least one occupant was killed.

For rear seated children (a child in a safety seat excluded) 19 were in the centre (lap belt) seat and 5 in outboard seats (with three-point belts). There are more child occupants in outboard than centre seats, so there was a pronounced tendency for SBS injuries to be associated with lap belts.

MASS DATA ANALYSIS

In the State of Victoria, no fault injury compensation of casualties from motor vehicle accidents is the function of the Transport Accident Commission. The TAC has amassed a great volume of data on casualties, to which access was provided to the Monash University Accident Research Centre for mass data analysis.

For estimating the incidence of SBS, a file was used containing casualty information on claims from July 1978, to June 1988 derived from crashes involving 1975 and later vehicles. Total claimants were approximately 77,000, of whom 26,863 were passengers, including 3369 aged 0-14 years.

SBS cases were defined as those car occupants with lumbar spine injuries (ICD 9 codes 805.4, 805.5, 806.4, 806.5 and 952.2) and/or abdominal visceral injuries (ICD 9 codes 863.0 through 866.9 and 868.0 through 869.9) (World Health Organisation, 1975). These rubrics embrace the wider definition of SBS referred to above. The nature of the restraint available could be inferred from the seating position, but information on belt use could not be obtained from this file. This was supplied from the surveys. The case frequencies are shown in Table 1.

Table 1 - SBS in Car Occupants in Crashes, July 1978 - June 1988, of Post 1975 Cars

Age	L Front	C Front	O/B Rear	C Rear	Total
0-4	1	0	5	2	8
5-9	2	1	11	5	19
10-14	6	1	7	5	19
All children	9	2	23	12	46
>14	312	3	138	30	483
Total	321	5	161	42	529

(Australian road traffic is on the left; the driver's seat is the right front)

The number of SBS cases among children (14 years or less) for whom a claim was made to TAC in the ten year period was 46. This compares with 32 (estimate derived from 29 in nine years) at the Royal Children's Hospital. The time periods overlap: TAC 1978-1988; RCH 1980-1989. The RCH casualties may be regarded as a subset containing the more severe SBS injuries in Victoria as a whole.

Although those SBS cases in children reported in the literature represent only an unknown fraction of those which occur, the rather small number of cases listed in 23 years suggests either that the risk of occurrence is low and/or that the exposure has been low. Most of the case reports are from North America where the use of restraints in the rear seat has been low until recent years. In addition, the size of the population from which the cases are derived is generally unknown, with some exception in the series of Agran et al.

The TAC data permit an estimate in relation to the populations of children and of vehicles in the State of Victoria, as shown in Table 2.

Table 2 - Incidence of Child SBS Casualties in Victoria

Cases (1978-1988)	46
Vehicles (1975 and later)	1.5×10^{6}*
Cases/10,000 Vehicles p.a.	0.058*
Population 0-14, Persons	929×10^{3}*
Cases/100,000 p.a.	1.10*
All Child Occupant Casualties #	3369
SBS as % of Child Occupant Casualties #	1.37%

*for the year 1987. # 1975 and later cars.

The number of child SBS cases 1978 - 1988, 46, is unsuitable for an estimate of incidence, because children's exposure to risk (by being restrained) increased substantially over the ten year period (1978 and 1988 were half years). Australian Design Rules 4 and 5A required belts in rear seats of cars and station wagons (hereafter referred to as "cars") manufactured after January 1971. The Victorian belt wearing law of 1976 required children less than eight to be restrained if riding in a front seat. This is likely to have had the effect of moving children to the back seat, where they were not required to be restrained and where, in any case, there were often no belts available.

The belt-wearing law was changed in December 1981 to require all children, wherever seated, to be restrained (if a restraint was available). In the back seat, this change in the law effectively applied to 1971 and later cars. The number of these cars increased almost linearly with calendar year, from about 940,000 in 1978 to 1,850,000 in 1988. This, together with probable increase in compliance, greatly increased the number of child passengers exposed to risk. In effect, the 1981 law required all children to be restrained and the 1971 design rule provided the means of restraint in the back seat. Consequently the number of SBS cases in children also increased.

Because of this increase, instead of the mean annual number of cases, it is appropriate to consider the expected frequency from the regression of case frequency on calendar year. For 1987 the expected number is 8.7, relating to 1975 and later cars, and must be factored up to account for 1971 to 1974 cars. The factor, from the 1988 vehicle census, is 1.17, and the expected number of child SBS cases in Victoria in 1987 is thus 10.2, rounded to 10.

EXPOSURE - Information on seating position occupancy by age and restraint use was, derived from surveys with matched observation sites, on Melbourne arterial roads in 1985, 1986 and 1988, and for restraint type usage in 1990.

The survey observations can be summarised as follows. Centre front seats: low occupancy for all age groups; belt use moderately high (60%) for 0-7, medium (40%) for 8-13, belt use low for adults.

For the centre rear seat: low occupancy but wearing rate improving from 1985 to 1988 for all age groups. The number of belt wearers in the centre front seat has tended to increase in the period 1985-88, in the centre rear seat it has remained constant.

Child occupants of the centre front seat constitute 2.9% of all child passengers; child occupants of the centre rear seat constitute 18% of all child passengers. Overall, child occupants of the centre seats constitute 4% of all car passengers.

RELATIVE RISK OF CENTRE SEATS - The TAC data (Table 1) provide the cases recorded in the 10 year period mid-1978 to mid-1988. Measures of exposure are provided by the survey data. To make use of he survey data it is necessary to assume that the arterial road samples are reasonably representative of Victoria both at the times of observation and for some years earlier.

The second assumption is, from the viewpoint of risk calculations, conservative, since the restraint use rate is likely to have been lower in the years 1978 to 1984. As only "wearers" are used in the estimation of exposure, the estimate of wearers is likely to be inflated and the estimate of relative rates of SBS correspondingly reduced. The observed frequencies in the various seating positions, factored by the survey wearing rates, are used as measures of relative exposure. In addition, in the youngest age groups, some "wearers" (though perhaps not as many as the 35% shown in the 1990 survey) will have been using the restraints such as child seats, generally regarded as safer than lap belts. They have been counted as lap belt wearers, so the estimate of relative risk is conservative on this count also. The child group 0-13 years in the survey has been adjusted to conform with 0-14 in the TAC data. The risk calculations are shown in Table 3. The frequencies in the centre front seats are too small for significance calculations.

Table 3 - Relative Risks of SBS in Various Seating Positions with Present Restraints

A. CHILDREN - CENTRE REAR VS OUTBOARD REAR

	Distribution of exposure to risk of SBS*	SBS Cases fo	SBS Cases fe	Relative Rate of SBS (%)
OBR	79.7%	23	27.9	0.77
CR	20.3%	12	7.1	1.57
	100%	35	35	

* based on sample of 3764 wearers; chi-square = 4.24, 0.02<p<0.05.

B. CHILDREN - LEFT FRONT VS OUTBOARD REAR

	Distribution of exposure to risk of SBS*	SBS Cases fo	SBS Cases fe	Relative Rate of SBS (%)
LF	30.1%	9	9.64	0.696
OBR	69.9%	23	22.36	0.767
	100%	32	32	

* based on sample of 19232 wearers; The relative rates are not significantly different.

C. ADULTS - CENTRE REAR VS OUTBOARD REAR

	Distribution of exposure to risk of SBS*	SBS Cases fo	SBS Cases fe	Relative Rate of SBS (%)
OBR	93%	138	156.17	4.38
CR	7%	30	11.83	12.56
	100%	168	168	

* based on sample of 22278 wearers;chi-square = 30.02, p<0.001.

D. ADULTS - LEFT FRONT VS OUTBOARD REAR

	Distribution of exposure to risk of SBS*	SBS Cases fo	SBS Cases fe	Relative Rate of SBS (%)
LF	85.8%	312	386.29	1.63
OBR	14.2%	138	63.71	4.38
	100%	450	450	

* based on a sample of 22278 wearers;chi-square = 100.92, p<<0.001.

E. CHILDREN VS ADULTS (SUMMARY)

	relative rate of SBS (%) Child	Adult	chi-square	p
CR	1.57	12.56	52.5	<.001
OBR	0.77	4.38	76.5	<.001
LF	0.696	1.63	6.85	<.01

In rear seats, the centre seat is seen to confer a significant increased risk of SBS. The increase is by a factor of two (1.57/0.77) for children and by a factor of almost three for adults (12.56/4.38). The assumptions about exposure referred to above make these estimates conservative, especially for children. Children appear to be less at risk of SBS than adults in the same seating positions.

Adults in outboard rear seats are at greater risk of SBS, by a factor of 2.7 (Table 3D, 4.38/1.63), than occupants of the left front seat. The increase in risk for children is small and non-significant (.767/.696). Since front seats are well known to be less safe than rear seats (for example, Evans and Frick, 1988), this is an unexpected result suggesting that there is a deficiency in the rear seat belt installations.

DISCUSSION

It appears that lap belts were initially accepted uncritically by those concerned with crash protection: more recently they have been perhaps unreasonably condemned. The limitations of the lap belt are, first, it provides insufficient protection, by failing to prevent the head and upper body from contact with unyielding surfaces. Second, it may cause injuries to the abdomen and lumbar spine by direct loading combined with the body motion, "submarining", that the belt induces under impact. The standard Emergency Locking Retractor (ELR) three point combination shares these shortcomings but to a much smaller degree.

Despite much case description, it has been difficult to estimate the numerical size of the SBS problem, through the collections of Agran and associates and casualty series based on accidents suggest that the incidence has been low. The increased child case frequencies noted in recent years in Australia can be related to increased restraint use in child passengers as post 1974, and by inference, post 1970, cars have penetrated the car fleet (Figure 1).

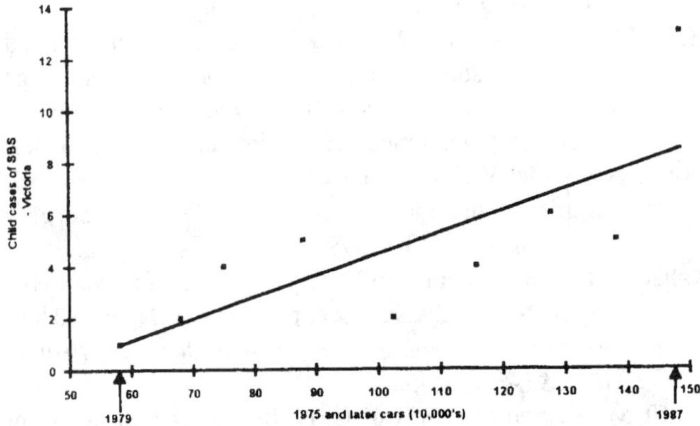

Figure 1 - Relation of child SBS to post-1974 cars

The fatality rate in casualties with SBS derived from published papers is 14.5%, but, as noted earlier, this collection seems selectively biased towards severity and also unduly weighted by the five fatalities from the NTSB series of 26 severe accidents (1986). If these are excluded, the rate is 8.3%. The casualties may have had other serious injuries, as they did in the Royal Children's Hospital series but in which there were no deaths.

SBS is a serious condition, usually requiring emergency surgery and carrying the risk of missed early diagnosis. The characteristic visceral injury is to the gastro-intestinal tract, especially to the small and large intestines. When there is a lumbar spine fracture, there is risk of spinal cord damage and paraplegia. There was 3.5% paraplegia in the Royal Children's Hospital series. A particular type of lumbar spine fracture, the Chance fracture, is especially associated with lap belts.

The substantial case literature indicates a preponderance of lap belt restraint, though the association is confounded with rear seat position in many reports. This association of lap belts with SBS is confirmed in the Royal Children's Hospital series.

From the Victorian mass data the relative risk of incurring SBS from a lap belt is now estimated, for adults, as three times that from a three point belt in the rear seat. For children the relative risk is twice that of a rear-seat three point belt. In addition, rear seat three point installations have, themselves, nearly three times the risk of front seat passenger three point belts.

A deficiency in the published information is any clear estimate of the incidence of SBS. The mass data analysed above indicate a case rate of about ten child cases per annum (for 1987) in the State of Victoria (less confidently, about 14 for 1991). Not all these children were using lap belts: in fact, twice as many children sustained SBS when using adult three point belts, because, despite the greater risk in lap belts, more children are seated in outboard than in centre seats.

Although the focus of this investigation was on children, there are many times more SBS passenger casualties in adults than children. In addition there were about 485 SBS cases in drivers in the years surveyed. Some of the drivers' visceral injuries may come from steering wheel contacts, but the lumbar spine injuries must be ascribed to the belt. Adult car occupants with SBS amount to 159 (the expected number for 1987 in 1975 and later cars), eighteen times the number of child cases. This approximates to 186 adult cases for all cars for 1987. Overall there is a case frequency for all car occupants, in Victoria, of about 196 per annum for the year 1987.

It is to be noted that these totals may contain an uncertain number of occupants who were not wearing belts. They cannot be eliminated from the data because the TAC file does not contain information on belt wearing and because it has been necessary, to use a broad definition of the injuries contributing to the Seat Belt Syndrome.

COUNTERMEASURES

A substantial benefit may be gained by providing upper body restraint in the centre seat positions. More than four fifths of centre-seated occupants are in the rear (87% of children, 77% of adults). Most recent cars do not provide a centre front seat.

The rear centre seat can be provided with a tethered harness for children of appropriate body size. For adults and larger children a three point belt is needed. For new cars, this could become standard practice - a few sedan car models already have three point belts in the centre rear seat. Replacing the lap belt with a lap-sash belt could be expected to eliminate about two thirds of the SBS cases in occupants of the centre rear seat.

Reducing SBS in all seats already required to be fitted with three point belts requires attention to the geometry of installations. In brief, the needed improvements are in making the lap belt angle steeper, appropriate compliance and profile of seat cushions, better access o the buckle in outboard rear seats and vertically adjustable D rings (Fildes et al, 1991). Belt tensioners would be very useful additions, to minimise belt slack (Haland and Nilson, 1991).

For smaller children, for whom adult belts with or without a booster are unsuitable, there is the available child seat, or, best for children up to 9 kg, a backward facing seat or capsule (Turbell, 1990; Lutter, Kramer and Appel, 1991).

Choice, by parents, of restraint type for young children, positioning of belts and the installation of child restraint seats all need to be greatly improved.

CONCLUSIONS

1. Lap belts provide substantial protection to occupants, both adult and child, of both front and rear seats. Lap belts should always be used if no better restraint is available.

2. Both lap belts and three point belts have a disbenefit, being liable to cause a particular type of injury, the seat belt syndrome (SBS) consisting of abdominal visceral injury and/or lumbar spine injury. Lap belts appear to have two to three times the propensity to cause this injury as three point lap sash belts. Children appear to be less at risk than adults.

3. Of adults restrained by three point belts, rear outboard occupants have a greater liability to SBS than left front passengers.

4. Lap and three point belts, as a group, cause about ten cases of SBS in children and 186 cases in adults in Victoria per annum (1987 expected totals). Annual child frequencies are increasing in parallel with increased numbers of post-1970 cars.

5. Two- thirds of the SBS cases in lap belt wearers in the centre rear seat could be expected to be eliminated by replacing the lap belts with three point belts, or, for children, adding a tethered harness. For children of appropriate sizes, a tethered booster with adult lap-sash belt, a tethered booster with lap belt and harness, or child seat are the preferred restraints. For infants and small children backward facing devices are the restraints of choice.

6. Reduction of SBS in general (three quarters of all passenger cases occur in three point wearers) requires improved seat and belt design.

ACKNOWLEDGMENT

The author is indebted to Mr Max Cameron, Dr Brian Fildes, Dr Peter Vulcan, Ms Jennie Oxley, Mr Foong Chee Wai and Mr Raphael Saldana of the Accident Research Centre.

Information on roadside surveys and on the relevant laws was kindly provided by Mrs Pat Rogerson, Ms Sandra Torpey and Ms Pam Francis of Vic Roads. The Transport Accident Commission made available a file of claims arising from crashes to post 1974 cars. Professor W G Cole and Dr Gregory Hoy, of the Royal Children's Hospital generously granted the use of material from a paper in publication.

The project was supported by the Federal Office of Road Safety of the Department of Transport and Communications.

REFERENCES

Agran P F, Dunkle D E, Winn D G, (1985). Motor vehicle Accident trauma and restraint usage patterns in children less than 4 years of age. *Pediatrics, 76*, 382-386.

Agran P F, Dunkle D E, Winn D G, (1987). Injuries to a sample of seat-belted children evaluated and treated in a hospital emergency room. *The Journal of Trauma, 27*, 58-64.

Agran P F, Winn D G, (1987). Traumatic injuries among children using lap belts and lap/shoulder belts in motor vehicle collisions. *31st Proceedings, American Association for Automotive Medicine*, 283-296.

Agran P F, Winn D G, Dunkle D E, (1989). Injuries among 4 to 9 year old restrained motor vehicle occupants by seat location and crash impact site. *American Journal Diseases of Children, 143*, 1317-1321.

Agran P F, Castillo D, Winn D G, (1990). Childhood motor vehicle occupant injuries. *American Journal Diseases of Children, 144*, 653-662.

Anderson P A, Rivara F P, Maier R V, Drake C, (1991). The epidemiology of seat belt-associated injuries. *The Journal of Trauma, 31(1)*, 60-67.

Corben C W, Herbert D C, (1981). Children wearing approved restraints and adult's belts in crashes. Traffic Accident Research Unit Report 1/81. Department of Motor Transport, New South Wales, Australia.

Evans L & Frick M C, (1988). Seating position in cars and fatality risk. *American Journal of Public Health, 78(11)*, 1456-1458.

Fildes B N, Lane J C, Lenard J, Vulcan A P, (1991). Passenger Cars and Occupant Injury. Report No. CR 95, Federal Office of Road Safety, Canberra.

Garrett J W & Braunstein P W, (1962). The seat belt syndrome. *The Journal of Trauma*, 2, 230-238.

Haland Y and Nilson G (1991). Seat belt pretensioners to avoid the risk of submarining - a study of lap-belt slippage factors. Paper 91-S9-0-10. Thirteeenth International Technical Conference on Experimental Safety Vehicles.

Hoy G A & Cole W G, (1992). Concurrent paediatric seat belt injuries of the abdomen and spine. *Pediatric Surgery International*, 7, 376-379.

Kahane C J, (1987). Fatality and injury reducing effectiveness of lap belts for back seat occupants. SAE Paper 870486, *Restraint Technologies: Rear Seat Occupant Protection, SP-691*, 45-52.

Krafft M, Nygren C, Tingvall C, (1989). Rear seat occupant protection. A study of children and adults in the rear seat of cars in relation to restraint use and car characteristics. *12th International Conference on Experimental Safety Vehicles, 2*, 1145-1149. Also in *Journal of Traffic Medicine (1990)*, 18(2), 51-53.

Kulowski J & Rost W B, (1956). Intra-abdominal injury from safety belt in auto accidents. *Archives of Surgery, 73*, 970-971.

Lane, J C, (1992). The child in the center seat. Report CR 107, Federal Office of Road Safety, Canberra.

Langwieder K, Hummel T, (1989). Children in cars - their injury risks and the influence of child protection systems. *12th International Conference on Experimental Safety Vehicles.* NHTSA, US Department of Transportation, Section 3, 39-49.

Lutter G, Kramer F, Appel H, (1991). Evaluation of child safety systems on the basis of suitable assessment criteria. *Proceedings IRCOBI Conference*, 157-169.

Morris J B, (1983). Protection for 5-12 year old children. SAE Paper 831654, *SAE Child injury and restraint Conference Proceedings, P-135*, 89-100.

National Transportation Safety Board, (1986). Performance of lap belts in 26 frontal crashes. Report no. NTSB/SS-86/03. United States Government, Washington D C.

Orsay E M, Turnbull T L, Dunne M, Barrett J A, Langenberg P, Orsay C P, (1989). The effect of occupant restraints on children and the elderly in motor vehicle crashes. *12th International Technical Conference on Experimental Safety Vehicles, 2*, 1213-1215.

Partyka S, (1988). Lives saved by child restraints from 1982 through 1987. Technical report HS 807 371, NHTSA, Department of Transportation.

Rutherford W H, Greenfield T, Hayes H R, Nelson J K, (1985). The medical effects of seat belt legislation in the United Kingdom. Research Report No. 13. Department of Health & Social Security, Her Majesty's Stationery Office.

Rutledge R, Thomason M, Oller D, Meredith W, Moylan J, Clancy T, Cunningham P, Baker C, (1991). The spectrum of abdominal injuries associated with the use of seat belts. *The Journal of Trauma, 31(6)*, 820-826.

Tunbridge R J, (1989). The long term effect of seat belt legislation on road user injury patterns. Research Report 239, Transport & Road Research Laboratory, Department of Transport.

Turbell T, (1990). Safety of children in cars - Use of child restraint systems in Sweden. Swedish Road & Traffic Research Institute, VTI.

Crash restraint of children by adult seat belts and booster cushions

D G C Bacon, BSc, PhD
Vehicle Safety Division, Motor Industry Research Association, Watling Street, Nuneaton, Warwickshire

Child dummies in the age range three to ten years old, restrained by adult seat belts alone and in combination with booster cushions, were tested in simulated perpendicular impacts of cars at 50 km/h. The project was divided into two phases in which various aspects of the restraint of child dummies were examined. In Phase 1 basic data was obtained on the way child dummies behave when restrained by adult seat belts. Proprietary booster cushions were tested for Phase 2 and they were found to be of benefit to the child. Overall the best combination was a firm tethered booster cushion with a three-point automatic seat belt.

1 INTRODUCTION

A group project was organized at the Motor Industry Research Association (MIRA) to examine aspects of the restraint of children in vehicles during a crash. In Europe, ECE regulation 44 had recently been introduced and manufacturers were exploring the problems of compliance with the specified tests. In the USA there was a similar federal standard and Australia had already a design rule on child restraints. Perhaps a greater influence to the fitting of child restraints was the UK legislation for the mandatory wearing of seat belts. This requirement extended to children over the age of one year sitting in the front passenger seat of cars. Whilst special belts are available for children, these are only installed by a small percentage of people in the rear of cars and they are most unlikely to be fitted in the front of cars. There is no doubt that a child restrained by almost any belt is likely to be safer than one without a seat belt but there was a need to provide more background information on their performance in an accident.

Booster cushions, which raise the child to improve the fit of the seat belt around him and also improve the child's visibility from the car, were put forward as solving the problem of a child restraint for the front seats of cars.

The project was arranged to examine the performance of adult seat belts in restraining children during a crash. A range of booster cushions was tested and an assessment made of their effectiveness. The tests were carried out in car bodyshells using current seat belt systems and a wide range of child dummy sizes were employed to investigate the effect of size on restraint performance.

2 TEST PROGRAMME

2.1 Phase 1

The thirty-three tests in Phase 1 tests were designed, mainly, to provide baseline data on the way child dummies behave when restrained only by adult belts and to show the effect of employing simple non-proprietary booster cushions of very hard and very soft consistency in a tethered and untethered condition, on the dummy movement. A secondary aim was to compare the performance of two ten year old types of dummies with the intention of selecting one for the remainder of the project.

Test conditions were chosen to be a simulation of a 50 km/h perpendicular barrier impact of the vehicle concerned and child dummies in the range three to ten year old were used. In addition, a fifth percentile female was used to represent a teenage child.

2.2 Phase 2

The thirty-four tests in Phase 2 involved testing of proprietary booster cushions. Four cushions were selected on the basis of providing an example of the main types available on the market internationally. These cushions were tested in conjunction with the most commonly found seat belt systems in cars. In a similar manner to Phase 1, the crash environment was that of a 50 km/h perpendicular impact and a similar range of dummies was tested.

3 TEST PROCEDURE

3.1 General technique

There was strong emphasis in obtaining test information in as representative a manner as could be achieved with current test methods available. The most realistic method would have been to crash a vehicle equipped with child dummies and restraint systems but obviously this would have been prohibitively expensive considering the range of parameters under investigation. Therefore it was decided to use sled testing methods which would be essentially non-destructive to the test vehicle.

The tests were performed using the vehicle bodyshell with sufficient trim detail to provide a representative environment for the occupant in a crash situation. For instance it was important to have at least the squabs of the front seats in position when testing rear seat installations to cater for possible head impact with the seat. Similarly, trim detail would be required on doors in the eventuality of contact with this area. The preferred choice for this work would have been a fully trimmed bodyshell.

The MS was received on 12 July 1983 and was accepted for publication on 12 April 1984.

Again, in the interests of realism, it was necessary to represent the correct crash environment for the test vehicle rather than a standardized or legislated set of conditions. It was decided that the series of sled tests would be performed on the MIRA Hyge reverse acceleration facility since this system allows accurate reproduction of vehicle decelerations during a crash with good repeatability.

The bodyshell of the test vehicle was secured to the Hyge sled through its suspension mounting points by a purpose built rigid frame. Particular attention was paid to making certain that the bodyshell was well secured in view of the large number of dynamic tests to which it would be subjected.

High-speed cameras, running at 1000 pictures per second, were mounted on outriggers to film the movements of the child occupants. These cameras travelled with the sled during the tests. In some tests, an overhead camera was employed but this did not move with the sled.

The child was represented by dummies having the morphology of various weights or ages of the human population. These devices were installed in the vehicles using the standard range of seat belt configurations and booster cushions.

The Hyge works in reverse when compared with conventional trolley rigs (where the trolley is arrested from speed by a suitable retarder), in that the sled is motionless just prior to the simulated impact. This has certain advantages in that the test engineer can be sure that the initial conditions of dummy installation and inertia mechanisms (seat belt reels) are correct and have not been disturbed in a run-up phase. During the test, the Hyge pneumatic ram accelerates the sled violently rearwards to the velocity of the simulated crash. The acceleration pulse can be shaped by various equipment parameters to match that of the crash data supplied.

3.2 Test measurements

Pre-test measurements were made of dummy location within the vehicle and the position of the cameras in relation to the dummy and reference points.

Abdomen penetration on the dummy was calculated from the pre- and post-test dimensions of a plasticene indicator (see Section 5.1.1).

For all the tests, the acceleration and velocity of the sled was recorded.

3.3 Test assessment

A standard chart was prepared for summarizing the test data and making an assessment of the child restraint system. The assessment was divided into three areas with an overall summary at the end.

The first area dealt with the dummy performance, with aspects of its movement evaluated from the high-speed film. Submarining was assessed subjectively from film by following the position of the lap belt on the dummy abdomen during the test. The deformation of the plasticene insert was also recorded as a measure of the loading on the abdomen. Head and chest impacts, if any, were evaluated for severity from the film by examining locations of contact and velocity and angle of impact. The forward movement of the head was measured for comparison to the limits set by ECE 44 of 550 mm horizontally and 800 mm vertically. There was a complication in making this comparison because the limits of movement are related to a point on the ECE 44 tests seat (Cr-point). However, by seating each dummy in turn on the ECE 44 seat, the relationship of its H-point to the Cr-point was ascertained and knowing the forward movement of the dummy in the test vehicle relative to H-point then the ECE 44 value could be calculated.

The second area of the chart was concerned with the seat belt. Various aspects of the fit of the belt were evaluated before the test and the buckle was checked for its operation and position. During the test the dynamic configuration was examined from the high-speed film. After the test, the belt was examined for failures and webbing damage. The operation of the buckle was also rechecked for release. Belt loads were added to the chart, when measured.

The third area of the chart was devoted to an assessment of the performance of the booster cushion. Retention on the seat was important for the cushion and this was evaluated from high-speed film. The effectiveness of the cushion was examined from various standpoints, the 'belt lie', the severity of submarining and to a certain extent the performance of the restraint without the cushion. Combined car seat and booster cushion vertical deflections were recorded as well.

The last section of the chart was a considered judgement of the overall system effectiveness.

4 TEST MATERIAL

4.1 Vehicle bodyshells

4.1.1 Vehicle selection and seat belt anchorage locations

The two vehicles selected to provide the crash environment for the child dummy occupants were a Ford Cortina and a British Leyland Metro. It was considered that these two vehicles were typical of the size and design of many cars on the roads in the UK and the results of tests could perhaps be more meaningful in a wider context. Another important factor in the selection of the vehicles, was the requirement for testing a wide range of seat belt anchorage locations. The effective anchorage locations for the Cortina and Metro were measured and some of the results presented in Fig. 1 show a satisfactorily wide variation.

4.1.2 Vehicle preparation

In the preparation of these vehicles for the test programme, the emphasis was on providing a realistic passenger compartment from the point of view of impact by an occupant. It was important to have at least the squabs of the front seats in position for testing of the rear installations and similarly the facia panel assembly should be in position for front seat occupants. In the event, the Cortina was delivered in the fully trimmed condition and the Metro was equipped with all necessary trim in impact areas. The Cortina was a left hand drive version and in order to rationalize the installation of both vehicles on the Hyge sled, the Metro was ordered in the same hand of drive. Attachment to the

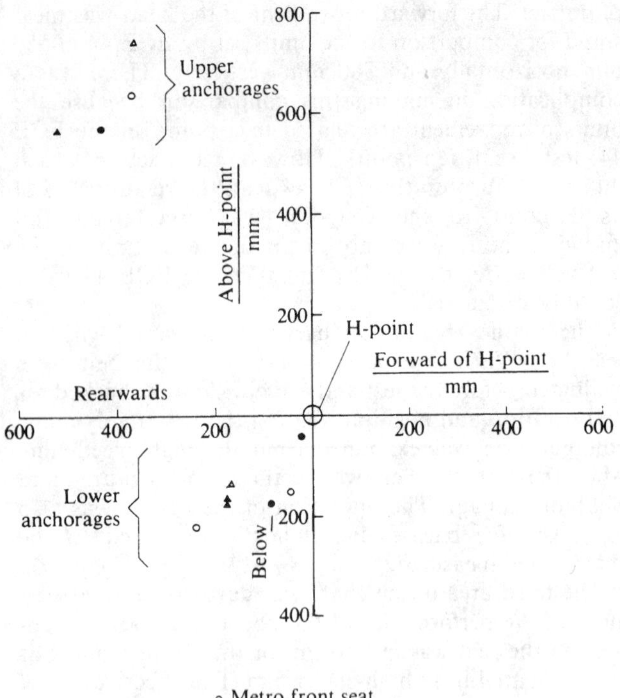

Fig. 1 Side elevation showing the relationship of Metro and Cortina effective anchorages to the H-point of the SAE mannikin

sled was made through the suspension mounting points. Figure 2 shows a typical installation on the Hyge for a test in this project.

The vehicle manufacturers also supplied several sets of seats to replace those damaged during testing. The life of a seat base was in the range of five to ten dynamic tests.

4.2 Test dummies

Anthropomorphic data for children are rather sparse when it comes to specifying a dynamic performance in a crash situation. The available dummies are normally constructed to the dimensions and segment masses of the fiftieth percentile US and European child of a given weight. Other more complex parameters such as centres of gravity for segments, ranges of motion and deflection characteristics of the chest tend to be either a function of the construction methods adopted or based on limited data. In some cases construction methods are scaled from adult dummies where there is more extensive literature available.

The degree of sophistication in the joint arrangements can affect the articulation of segments and obviously affect the kinematics under dynamic loading.

The skeletal development for a given weight of child can vary according to the age of the child. Thus in this work dummies are loosely described as being three, six, or ten years old but more strictly they should be referred to by their weights. Child dummies are in the early stages of development and no doubt with growing need for work in this area more data will be forthcoming on the impact performance and a new generation of child dummies will be produced.

4.2.1 Child dummies specified by legislation and other types available

The Australian standard for child restraints (AS 1754-1975) gives a range of child dummies from which dummies of the appropriate weight are selected for a given restraint system. Three of the dummies are from the US Sierra range, 'Toddler', 'Sammy', and 'Suzie', (15.5, 21.5 and 37 kg mass respectively) and the other two, representing babies, are Taru 'Teresa' and Taru 'Simon' (4 and 8 kg mass respectively) designed by the Traffic Accident Research Unit of the Department of

Fig. 2 Typical child restraint test on Hyge: Cortina bodyshell, three and six year old dummies in front and rear respectively

Motor Transport, New South Wales. (Sierra have recently stopped manufacture of dummies.)

Humanoid Systems, California, manufacture six year old, three year old and six month old dummies, having masses of 22, 15 and 8 kg. These conform to the Part 572 specification. A similar range is produced by Alderson Research Laboratories, Connecticut. The three year old and six month old dummies are specified by American legislation, FMVSS 213 for the assessment of child restraint systems for children who weigh not more than 23 kg.

In the UK, David Ogle Limited, Letchworth have been producing dummies for about ten years and include child dummies in their range, namely a six year old and a ten year old of weights 22 and 35 kg.

A new generation of TNO dummies has been developed recently to meet the requirements as laid down in the ECE 44 Regulation, 'Uniform provisions concerning the approval of restraining devices for child occupants of power-driven vehicles'. Their four dummies designated the $P\frac{3}{4}$, P3, P6 and P10 have a mass of respectively 9, 15, 22 and 32 kg and are now widely used in Europe for homologation tests. It was decided that the TNO range of dummies was most appropriate to this project but only the P3, P6 and P10 would be employed.

For comparison purposes it was desirable to obtain at least one more dummy from another manufacturer, in particular a second ten year old was thought important to evaluate the submarining response. After some search, an Ogle ten year old dummy was located in the UK and loaned to MIRA for the duration of the project.

To investigate the kinematics of children at the top end of the size range, a standard fifth percentile female dummy was included to represent, say, a teenager.

4.2.2 Dummy specification

The three TNO dummies are well described in their literature and ECE 44 contains comprehensive details of their construction. The dummies contain an abdominal insert made of soft open cell foam. Between the abdominal insert and the lumbar vertebrae a space was available to take some kind of transducer to determine abdominal penetration. Whilst TNO say that a final method has not yet been established, ECE 44 requires the use of a block of plasticene in that area and measurement of the decrease in thickness to ascertain whether abdominal loading has taken place, this technique was employed for this project.

The Ogle ten year old dummy is shown in Fig. 3. It was generally of more sophisticated skeletal construction than the TNO. The head was constructed of polyurethane-skinned aluminium mounted on a moulded rubber neck. The rib cage was manufactured from sheet plastic connected to a welded steel thoracic spine. By virtue of sliding joints it was possible for the shoulders to 'shrug' and a full range of articulation was possible in the arms. The pelvis was a plastic dipped casting of a representative shape to that of a child. Ball joints were used for the hip and the knee had a hinge joint. The 'flesh' consisted of moulded polyurethane sections. Clothing was in the form of a one-piece garment covering the torso, held together by a zip.

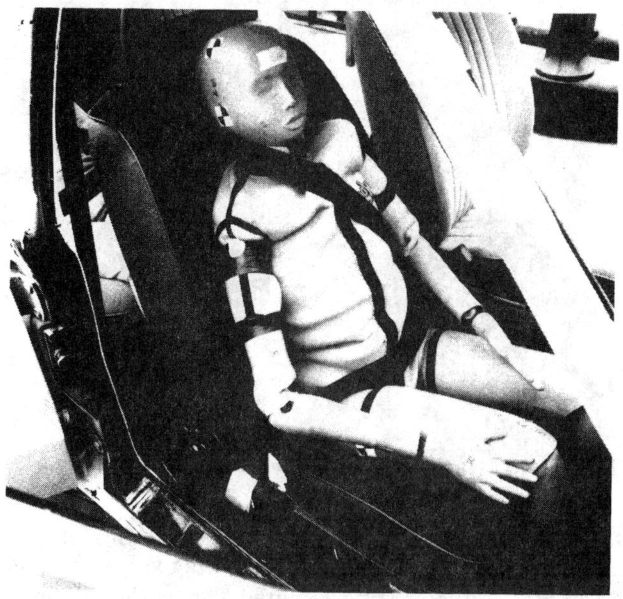

Fig. 3 Ogle ten year old child dummy

The fifth percentile female dummy, employed to represent an older child, was made by Alderson to comply with American specifications. This dummy was a standard device meeting the joint mobility requirements established in SAE J963. In the joint load adjustability and instrumentation provisions, it conformed to FMVSS 208.

4.3 Booster cushions

Booster cushion designs are quite varied but they have certain features in common, namely:

(a) a thickness of about 100 mm to raise the child;
(b) a plan area of about 300–400 mm square to spread the dynamic loads on the car seat;
(c) a means of retention either to the seat belt or the car seat to prevent the cushion moving forward in an accident.

Other features are added to the cushion by the manufacturer such as padding to make the device more comfortable or an attractive cover to make it more acceptable.

A variety of cushions was required for this project so that the range of designs could be evaluated for their performance.

Only two cushions were on the market in the UK during the project and the remainder had to be found from Europe and Australia.

As Australia has had child restraint legislation for some years, many booster cushions were offered by manufacturers in that country. Contact was made with these firms and samples despatched to MIRA for assessment. One of these Australian made cushions was selected for the test programme. Another was made under licence in Germany and this was also chosen. The compliment of four cushions (A–D) for the main test programme was completed by the two sold in the UK. Figure 4 shows the cushions selected for test and Fig. 5 shows some of the other types of cushion available. Cushion E was tested in a small programme because it

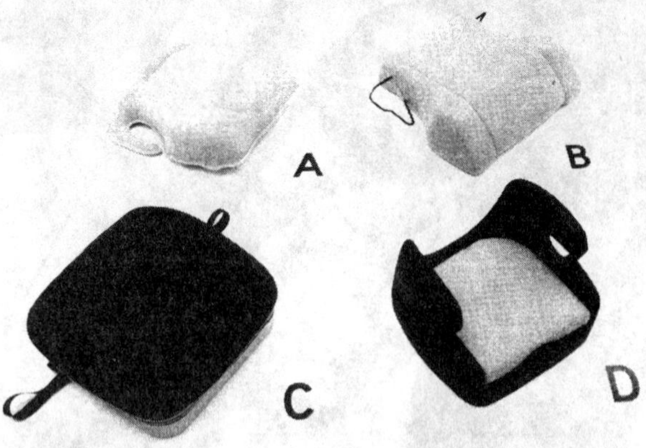

Fig. 4 Booster cushions selected for main test programme

was quite different from the other types in relying on an integral squab for retention and also being very soft.

The Phase 1 of testing required two cushions, one very hard and the other very soft. These cushions were made at MIRA from slabs of foam of two different hardness values and covered in a strong denim fabric.

Loops of webbing were sewn at each side for attachment to the seat belt system of the car. The assembled standard cushion can be seen in Fig. 6. The foam hardness values for the cushions were measured to the specification of BS 4443 and it was found that the hard cushion was thirty times stiffer than the soft version.

4.3.1 Description of selected booster cushions

Cushion A This cushion had a cover over a foam core. The horizontal ears were reinforced by an ABS plastic moulding. The whole assembly was fabricated in a single process. A separate belt was supplied for the retention of the cushion when used on a front seat.

Cushion B This cushion had a fabric stretch cover over an expanded polystyrene base. The retention was by wire ears that also acted as a reinforcement to the structure.

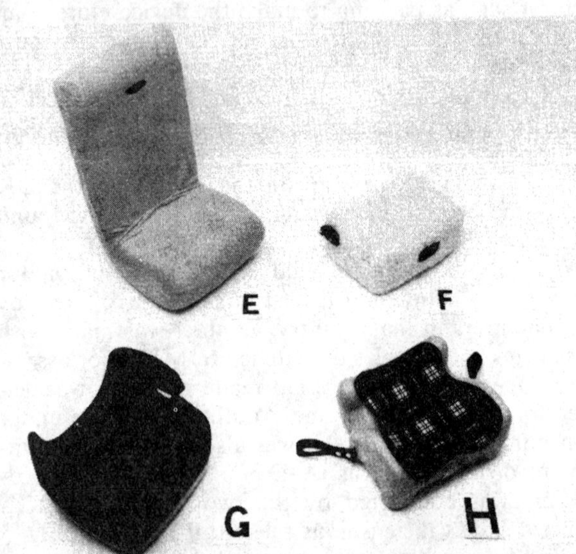

Fig. 5 Some other types of booster cushion

Fig. 6 Standard cushion

Cushion C This cushion was of more complex construction than the others having a plastic, steel reinforced base and a tartan fabric foam top. Two anchorage devices were incorporated, either a webbing loop or tubes with webbing clamps.

Cushion D This cushion had sides and a back and was designed more like a seat. Construction was in black-skinned polyurethane and vertical ears provided retention to the seat belt. A separate small soft foam cushion was tied to the seat for comfort.

Cushion E This cushion consisted of an integral base and squab made from soft foam covered in fabric. No separate anchoring system was provided as the cushion was retained by the squab portion behind the child.

4.4 Seat belts

Three main types of seat belt were used, namely:

(a) three-point, lap and diagonal, inertia reel automatic belt (three-point auto);
(b) three-point, lap and diagonal, static belt (three-point static);
(c) two-point, lap only, static belt (two-point static)

These were selected as being the installations most commonly found. In UK cars, the front seating positions are required by national type approval to be fitted with the inertia type of belt and for this reason 'e' marked three-point auto belts were used exclusively in the front of the Metro and Cortina. In the rear of the UK cars the situation is more varied in that there is no legal requirement to fit belts. Those belts which are installed as an option can be any of the three types described. For this project the majority of rear belts were three-point auto with about 10 per cent of rear installations having three-point static belts and 20 per cent two-point static belts. The latter two belt variants were only readily available for the Cortina rear and hence the full range was tested only in the Cortina. All the tests in the Metro rear were equipped with three-point auto belts.

Three different types of inboard stalk on which the receptacle or buckle was mounted were employed in the four seating positions. The front seat of the Metro had a wire stalk attached to the top of the transmission tunnel whereas the Cortina front had a seat mounted buckle attached by a steel plate. Both rear installations were equipped with buckles mounted on webbing stalks for all the seat belt variants tested.

5 RESULTS AND ASSESSMENT OF TESTS

5.1 Phase 1: base-line tests

This phase had the aim of providing base-line data on the performance of adult belts on child dummies and also investigated the benefit, or otherwise, of a very soft and very hard simple non-proprietary cushion. Dummy kinematics were compared for the two ten year old dummies.

The detail test programme is shown in Table 1. Thirty-three tests were performed, of which thirteen did not employ a cushion. The tests with cushions were almost equally split between the hard and soft version. The majority of tests were carried out with three-point auto seat belts and of the remainder, five had two-point static belts and one had a three-point static belt. Test assessment forms were completed for each test which provided a comprehensive record of the results but only summarized results are presented here. The acceleration pulse for the simulated impact of the Metro and Cortina bodyshells is described in the Appendix.

5.1.1 Submarining: Phase 1

'Submarining' is a term generally used to describe the slipping of the lap belt off the pelvis bone structure onto the soft parts of the abdomen of a vehicle occupant during a crash.

Table 1 Phase 1: tests performed

	Cortina rear	Metro rear	Cortina front	Metro front
*No cushion tests**				
3 year old		×	×	
6 year old	×	×	×	×
10 year old (TNO)	×			×
5th % female		×	×	
10 year old (Ogle)	×			×
*Hard cushion tests**				
3 year old			×	
6 year old	×	×	× -ut	×
10 year old (TNO)	×			
5th % female		×		
10 year old (Ogle)				
*Soft cushion tests**				
3 year old			×	
6 year old		×	× -ut	×
10 year old (TNO)	×			×
5th % female				
10 year old (Ogle)	×			×
Two-point static seat belt tests				
3 year old	× soft			
6 year old	× none			
10 year old (TNO)	× soft			
5th % female	× hard			
10 year old (Ogle)	× soft			
Three-point static seat belt tests				
6 year old	× hard			

* Three-point auto seatbelts
ut untethered

The submarining of the dummies was assessed in two ways, by measuring the decrease in thickness of the plasticene block in the case of the TNO dummies and by film analysis for all the dummies. The results of this exercise for the Phase 1 tests are given in Table 2. Some comments can be made at this stage on the correlation of the plasticene deformation with observed submarining severity. In many cases where the lap belt was well located on the pelvis, there was still a small reduction in plasticene thickness, for example, the three year old in the rear of the Cortina with a soft cushion and the six year old in the front of the Metro with a hard cushion. This phenomenon was attributed to distortion of the plasticene block as it conformed to the shape of the lumbar spine during flexing of this area in the test. Another anomaly could occur in the case of very severe submarining when the lap belt rode high up on the abdomen and tightened under the rib cage. In this case the plasticene deformation would be small, for example, the six year old with hard cushion and three-point static belt. With these provisos in mind there appeared to be some correlation of observed submarining with plasticine deformation. However, it was not considered advisable to use penetration measurements without the information from film. To clarify the descriptions of observed submarining, 'slight' meant usually a riding up of the lap belt from the pelvis into the lower abdomen late in the test sequence, 'moderate' meant lap belt penetration of the 'abdomen throughout the test but reasonable dummy kinematics otherwise, and 'severe' meant high ride up of the lap belt into the dummy abdomen and the pelvis almost horizontal. These are general descriptions and obviously, due to the variety of dummy movements, there were 'grey' areas between categories which were left to the judgement of the test engineer for assessment.

Returning to the results of the Phase 1 tests, submarining of various degrees of severity was experienced by all the child dummies when only restrained by an adult seat belt. The most severe submarining was evident in the case of the ten year old dummies in the rear of the Cortina. This was mainly because the dummies had to be seated initially in a slouched position with the pelvis away from the squab to enable the legs to fold at the knees over the edge of the cushion and miss the front seat (see Fig. 7a). The smaller dummies could be seated with their backs against the seat squab and legs outstretched (see Fig. 7b and 7c). The use of a hard cushion tethered to the seat belt reduced the severity of submarining in all cases with three-point auto belts and in some cases reduced it to none. The cushion improved the sitting position of the ten year old (see Fig. 7d), and the kinematics of the dummy during the test were more well controlled, though submarining still occurred. The soft cushion had little beneficial effect and in the case of the six year old in the front of the Cortina with an untethered version, it induced more severe submarining than the seat belt alone.

As would be expected, a two-point static belt was not so inducive of submarining because the dummy torso folded forward. However, a three-point static belt in the rear of the Cortina caused very severe submarining even

Table 2 Assessment of submarining for Phase 1 tests with reference to abdomen penetration and film analysis

Dummy	Cushion	Cortina front ABD	Cortina front FA	Cortina rear ABD	Cortina rear FA	Metro front ABD	Metro front FA	Metro rear ABD	Metro rear FA
3 year old	None	nm	Mod	—		—		5	Mod
	Soft	5	Mod	2	None (2-pt)	—		—	
	Hard	2	None	—		—		—	
6 year old	None	nm	Slight	5	Mod (2-pt)	10	Severe	12	Mod
				nm	Severe (3-pt auto)				
	Soft	10	Severe (ut)	—		5	None	2	None
	Hard	2	None (ut)	5	Severe (3-pt)	5	None	nm	None
				5	Slight (3-pt auto)				
10 year old, TNO	None	—		nm	Severe	10	Mod	—	
	Soft	—		5	Slight (2-pt)	5	None	—	
				8	Severe (3-pt auto)				
	Hard	—		5	Mod	—		—	
10 year old, Ogle	None	—			Severe		Mod	—	
	Soft	—			Slight (2-pt)		Mod	—	
					Mod (3-pt auto)	—		—	
	Hard	—		—		—		—	
5th % female	None		Slight	—		—			None
	Hard	—			None (2-pt)	—			None

Notes
ut = untethered cushion
— = no test performed
ABD = abdomen penetration in mm
FA = film analysis assessment
nm = not measured
Mod = moderate submarining
Unless otherwise stated, seatbelts were three-point auto
Penetration measurement not possible on fifth percentile female or ten year old Ogle
2-pt = two-point static seatbelt, no diagonal
3-pt = three-point static seatbelt

Fig. 7a Ten year old in rear of Cortina

Fig. 7b Three year old in rear of Cortina

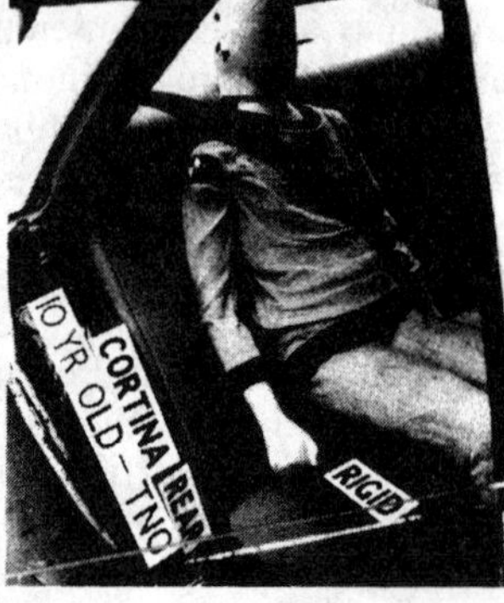

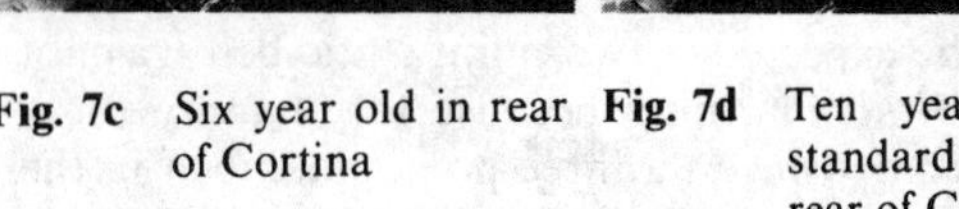
Fig. 7c Six year old in rear of Cortina

Fig. 7d Ten year old on standard cushion in rear of Cortina

in conjunction with a hard cushion. This type of belt system is very stiff and the torso was restrained early causing the dummy eventually to adopt an almost horizontal position with the lap belt under the rib cage and the diagonal belt high under the arm and chin. The untethered hard cushion test with the six year old in the front of the Cortina gave a satisfactory result with the cushion staying in place. It was the opinion of the Steering Group, that this result should not be interpreted to mean that firm cushions need not be tethered. The retention would be very much a function of coefficients of friction, and, as these booster cushions are supposed to be universal, tests would need to be performed in conjunction with a variety of seat trim materials. To try to evaluate all combinations of trim, seat design and dummy size was a daunting task and the recommendation was that booster cushions should always be tethered by some means.

When restrained by a three-point auto belt, the kinematics of the fifth percentile female were very well controlled, with only the slightest hint of submarining on one test. As most modern car installations of seat belts are designed to cover the range of adult size from fifth percentile female to ninety-fifth percentile male, this was not an unexpected result.

5.1.2 Diagonal belt performance and head forward movement: Phase 1

The dummy torso movement was considered to be somewhat artificial in that it was predominantly a function of the chest construction of the dummies. The TNO range were thought to have an unrealistically stiff chest when compared to a human child of the same size.

Table 3 Assessment of diagonal belt performance and measured head forward movement for Phase 1 tests

Dummy	Cushion	Cortina front		Cortina rear		Metro front		Metro rear	
		HFM	DBP	HFM	DBP	HFM	DBP	HFM	DBP
3 year old	None	430	Neck					480	Neck
	Soft	440	Neck	970*	(2-pt)				
	Hard	400	Mid						
6 year old	None	490	Neck	410	Neck	370	Neck	500	Mid
				870*	(2-pt)				
	Soft	510	Mid (ut)			610	Out	690	Out
	Hard	460	Neck (ut)	540	Mid	590	Out	550	Out
				400	Neck (3-pt)				
10 year old, TNO	None			360	Neck	480	Neck		
	Soft			560	Out	620	Out		
				970*	(2-pt)				
	Hard			710	Out				
10 year old, Ogle	None			200	Neck	420	Neck		
	Soft			570	Out	510	Neck		
				1090*	(2-pt)				
	Hard								
5th % female	None	—	Mid					—	Mid
	Hard							—	Out

Notes
HFM = head forward movement relative to ECE 44 Cr-point, in mm
DBP = diagonal belt performance
Neck = high position of diagonal, close or touching neck
Out = tendency for torso to roll out of belt, diagonal low
Mid = balanced restraint of torso
2-pt = two-point static seatbelt, no diagonal
3-pt = three-point static seatbelt
ut = untethered
* = head impact
Unless otherwise stated, seat belts were three-point auto

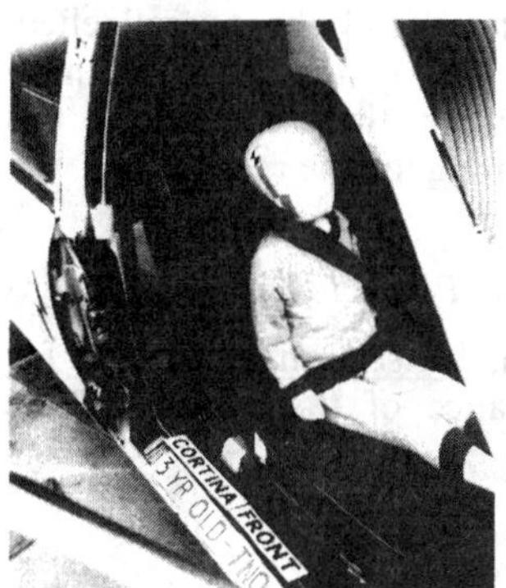

Fig. 8a Three year old in front of Cortina: no cushion

Fig. 8b Three year old in front of Cortina: with standard cushion

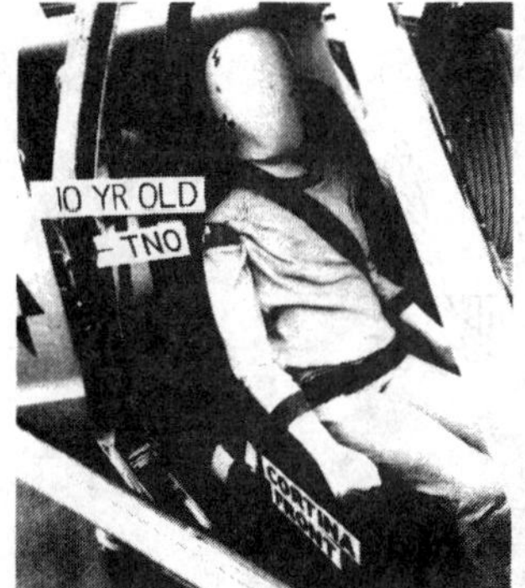

Fig. 8c Ten year old in front of Cortina: no cushion

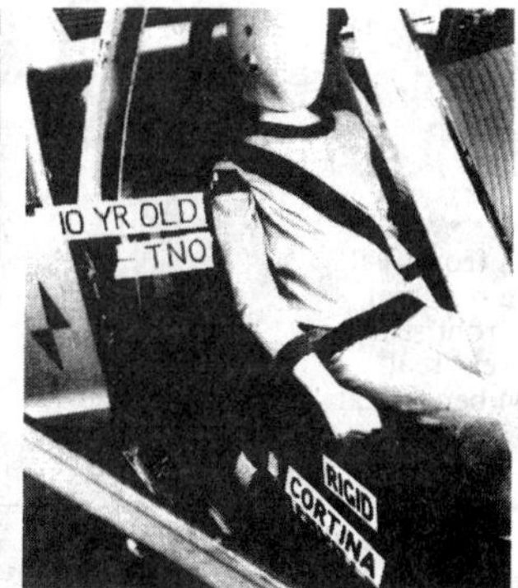

Fig. 8d Ten year old in front of Cortina: with standard cushion

There was a feeling that the restraint of the dummy chest was an 'unstable equilibrium' condition in that if the belt was slightly high it would ride up and if it was slightly low the dummy would tend to roll out of the belt. There was evidence from adult cadaver tests in the USA and Europe that the position of the diagonal belt was not so critical because the real chest and shoulder are more compliant. This was considered the most likely situation for the child as well.

However, the assessment of diagonal belt performance, Table 3, is a reasonable indicator for the location of the diagonal on the chest of the various dummy sizes. There was obviously a benefit in using a hard cushion for the three year old in the front of the Cortina (see Figs. 8a and 8b), which was borne out by the dynamic result. With the larger dummies it was evident that whilst a cushion improved the submarining performance there was a tendency for the torso to roll out of the belt because the diagonal was low (see Figs. 8c and 8d).

Where the diagonal belt rode up into the neck area, the movement of the head was reduced in some cases by a large amount, for example, ten year old Ogle in the rear of the Cortina with no cushion. The well-balanced torso restraint tests allowed head forward movement normally just within the ECE 44 limit of 550 mm. Where the torso of the dummy rolled out of the belt this limit was exceeded. However, it was considered that this was unlikely to happen with real children in an accident.

The tests with two-point static belts, of course, did not have the benefit of a diagonal belt and the dummy in each case would 'jack-knife' forward, and hit the seat in front and perhaps also follow through to hit his own legs or the seat base. The head impacts tended to be glancing blows or contact with soft trim and it was estimated that these were not of a serious nature. The child was obviously better off with the two-point belt than with none at all.

5.1.3 *Ten year old dummy comparison: Phase 1*

The kinematics of the two ten year old dummies were very similar in most respects. Certainly, there were no significant differences in submarining, which was the main aim of the comparison. On this basis, either dummy could have been selected for continuing the project work and initial preferences were for the TNO dummy as it is widely used in Europe. However, more detailed film analysis of performance in the chest and shoulder areas showed that the Ogle dummy was better able to maintain location with the diagonal belt because of its more sophisticated chest and 'shrugable' shoulders. For this reason it was recommended that testing continue with the Ogle dummy. In the event, because of test breakages in the Ogle dummy, both were used in the rest of the project to maintain the testing schedule, but where a choice of dummy was available, the Ogle was used.

5.1.4 *The effect of car seat stiffness: Phase 1*

Just as the hardness of the booster cushion was seen to have a significant effect on the restraint system, it was realized that the hardness of the car seat cushion must be equally important. A small programme of work was initiated to investigate this aspect.

The stiffness or force/deflection relationship for each of the four seating positions, front and rear of Metro and Cortina, was measured statically by loading up the hip pan of the SAE manikin and recording the deflection. The same operation was carried out on an ECE 44 bench seat for direct comparison. The results are summarized in Fig. 9, where it can be seen that the Cortina front seat is the stiffest and the Cortina rear the softest car installation. The Metro front and rear stiffness were closer together and in between the Cortina values. What was surprising was the softness of the ECE 44 seat, which was nearly three times less stiff than the Cortina front seat and about half the stiffness of the Cortina rear seat for a high applied load. At a low load the ECE 44 seat was in the middle of the stiffness range, due to its non-linear characteristics. Whilst with just a child sitting on the seat the deflection would be comparable to the car installations considered here, under dynamic loading which can exceed the static values used for Fig. 9 (800 N), the deflections would be excessive. This may well have repercussions on the submarining performance of the child dummies which are measured by abdomen penetration in ECE 44.

The Phase 1 test programme was examined for possible comparative tests of the front and rear of the Cortina. Two comparisons were found, namely, the six year old, with no cushion and the hard cushion. The detailed test results for these four tests have been given in Table 4. These tests gave an opportunity to assess the effect of seat stiffness on dummy kinematics, with the proviso that these seating positions had different seat belt anchorage locations (see Fig. 1). The lap belt effective anchorages were very close in side elevation. The largest difference was in the position of the upper anchorage, where the Cortina rear anchorage was 157 mm rearward of and 173 mm below the Cortina front anchorage. The test results for the six year old, without a cushion and with the hard cushion, are worse for the softer seat. Submarining was more severe in the rear of the Cortina without a cushion and the overall effectiveness was rated 'bad'. Using the hard cushion reduced the difference between the installations but there was still more submarining for the rear.

5.1.5 *Overall assessment of the Phase 1 tests*

The test assessment chart had a section at the end for rating the overall performance of the system. These

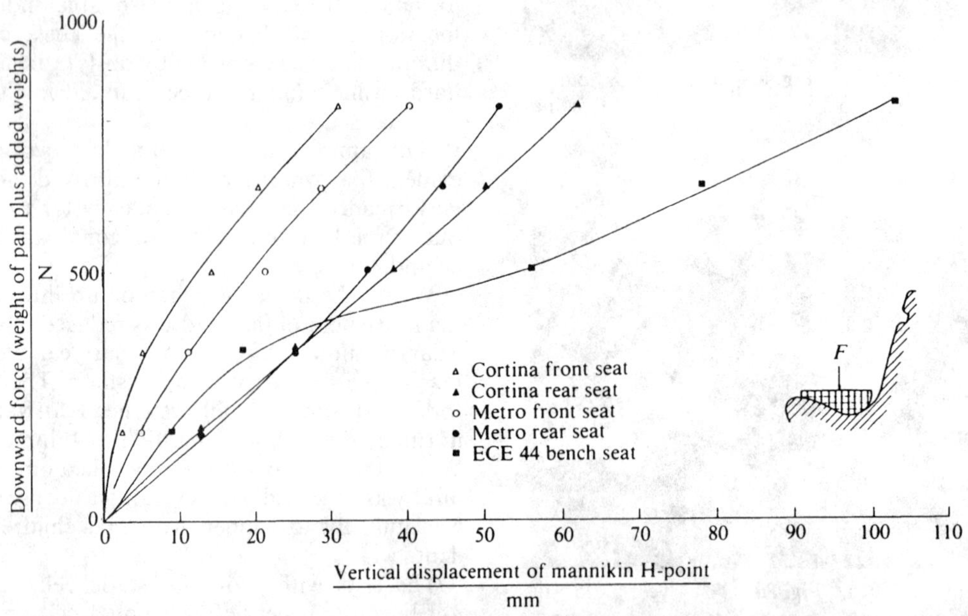

Fig. 9 Seat stiffness comparison for the Cortina and Metro front and rear seats and the ECE 44 bench seat

Table 4 Effect of car seat stiffness on dummy kinematics
Stiff seat: Cortina front
Soft seat: Cortina rear
Note also that the seats have different anchorage locations

	Front	Rear
6 year old: no cushion		
Submarining	Slight	Severe
Abdomen penetration	nm	nm
Forward movement	490 mm	410 mm
Dynamic configuration	Acceptable	Bad
Hip vertical movement	Down 5 mm	Up 30 mm
Overall effectiveness	Acceptable	Bad
6 year old: rigid cushion		
Submarining	None	Slight
Abdomen penetration	2 mm	5 mm
Forward movement	460	540
Dynamic configuration	Good	Good
Hip vertical movement	Down 70 mm	Down 138 mm
Overall effectiveness	Good	Acceptable

Note: nm = not measured

Table 5 Overall system assessment of Phase 1 tests

		Cushion		
		None	Soft	Hard
3 year old				
Metro	front			
	rear	**		
Cortina	front	**	***	****
	rear		** (2-pt)	
6 year old				
Metro	front	**	***	***
	rear	**	***	****
Cortina	front	***	** (ut)	**** (ut)
	rear	*		***
		* (2-pt)		* (3-pt)
10 year old, TNO				
Metro	front	**	****	
	rear			
Cortina	front			
	rear	*	**	**
			** (2-pt)	
10 year old, Ogle				
Metro	front	**	**	
	rear			
Cortina	front			
	rear	*	**	
			** (2-pt)	
5th % female				
Metro	front			
	rear	****		****
Cortina	front	****		
	rear			* (2-pt)

Notes
* bad
** poor
*** acceptable
**** good
***** very good
ut = untethered
2-pt = two-point static seat belt
3-pt = 3-point static seat belt
Tests employed three-point auto seat belts unless otherwise stated

ratings have been listed for all the Phase 1 tests in Table 5. A general improvement in system effectiveness is seen, going from no cushion, through a soft cushion to the hard cushion, where data is available. The valid exception to this trend was the test involving a three-point static belt, where severe submarining was induced by the stiffer belt system.

5.2 Phase 2: booster cushion tests

In this phase, proprietary booster cushions were evaluated in simulations of 50 km/h perpendicular barrier impacts. Cushions from four different manufacturers were selected for the majority of the work and another cushion was selected for a small programme. These cushions are described in Section 4.3.

In preparing the test programme for this phase, it was decided not to continue testing of the fifth percentile female, used to represent a teenage child, because it was outside the size range specified for booster cushions. The smaller child was considered to give most problems for restraint design and for this reason more emphasis was placed in testing the three year old at the expense of some ten year old tests. The finalized programme is shown in Table 6.

There was debate on whether cushions should be tested in some conditions of misuse, for example, not hooking the seat belt through the loops. However, it was decided to test them as designed and, in this context, cushion A required an additional belt to attach the cushion securely to the front seats of cars.

Phase 2 contained thirty-four dynamic tests with eight tests performed on each of the four main cushions and two on the cushion E. Most of the tests employed three-point auto belts with about 10 per cent each of two-point and three-point static belts. As the Cortina seats were of the extremes of stiffness and also because the static belt installations were only installed in the

Table 6 Phase 2 tests performed

	Cortina		Metro	
Cushion	Front	Rear	Front	Rear
Booster cushion A				
3 year old	×	× (2-pt)	×	
6 year old	×	×	×	×
10 year old		× (3-pt)		
Booster cushion B				
3 year old	×			×
6 year old	×	× (2-pt)	×	×
10 year old		× (3-pt)		×
Booster cushion C				
3 year old	×	× (3-pt)	×	
6 year old	×	× (2-pt)	×	×
10 year old			×	
Booster cushion D				
3 year old	×	×	×	
6 year old	×	×	×	×
10 year old		× (2-pt)		
Booster cushion E				
3 year old	×			
6 year old	×			

Notes
2-pt = two-point static seatbelt
3-pt = three-point static seatbelt
Remaining tests use three-point automatic seatbelts

rear of the Cortina, a greater proportion of tests were carried out using this vehicle than the Metro.

The crash pulse used was that of the vehicle impacting a perpendicular barrier at 50 km/h; this is more fully described in the Appendix.

5.2.1 Booster cushion installation and convenience of use: Phase 2

The ease of fitting the booster cushions to the seat belt system of the car and their convenience of use were considered as important factors in the encouragement or discouragement of a high rate of use. In the course of performing the many tests on proprietary booster cushions, this aspect was noted, for the various sizes of dummies and different car seating positions. In addition, observations of the way real children sit on the cushions supplemented the assessment. (This information came from a small number of users.)

Typical installations of the four selected booster cushions and the six year old dummy are shown in Fig. 10 for the front seat of the Metro. These photographs are worthy of close examination in conjunction with Fig. 4 as they not only demonstrate the different methods of retention of the cushion but also show the different path of the lap belt and the significant effect this has on the lie of both the lap and diagonal belts.

Cushion A employed a separate cushion attachment strap, hooked over the ears of the cushion and tied behind the seat (only for front seat applications where the seat is wide). This strap was easy to use but was obviously an additional operation for the system. It was then a simple matter to put the seat belt on the child as

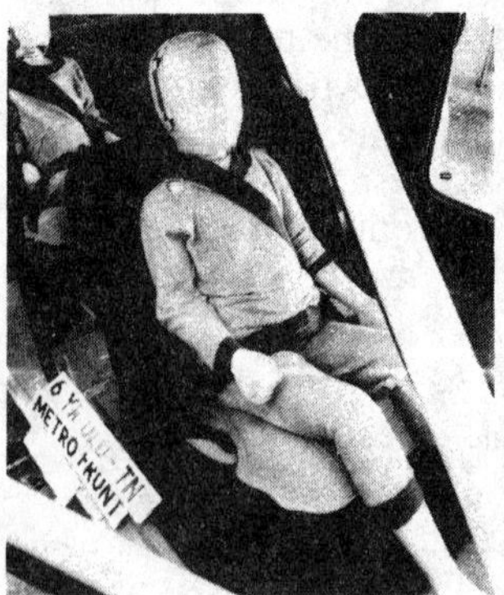

Fig. 10a Six year old in front of Metro: with cushion A

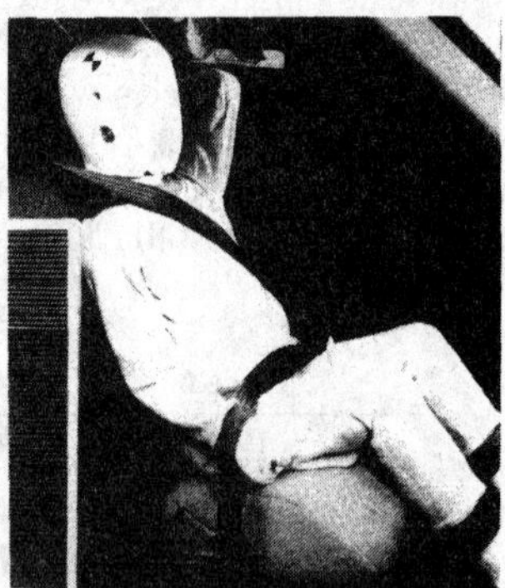

Fig. 10b Six year old in front of Metro: with cushion B

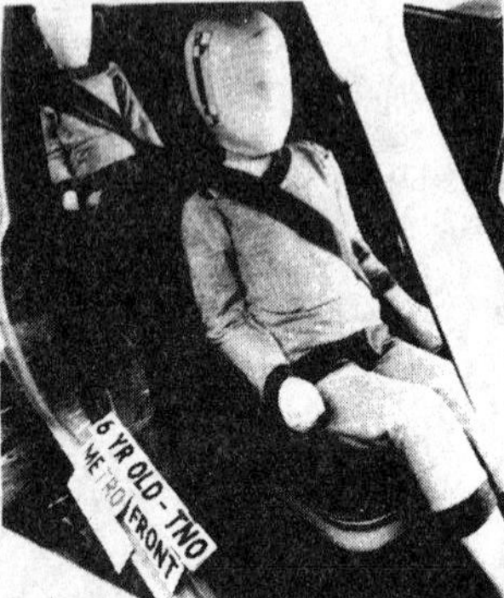

Fig. 10c Six year old in front of Metro: with cushion C

Fig. 10d Six year old in front of Metro: with cushion D

the lap belt did not have to be attached to the cushion. The lap and diagonal belt could now take an uninterrupted line over the torso of the child but this tended to place the diagonal high on the chest and near the neck for the Metro front seat.

Cushion B, of similar concept to A, had rigid wire loops forming ears which could be used in two ways to retain the cushion; either the buckle and belt could be passed through the loop or hooked round the ear. Both inner and outer portions of the lap belt were hooked onto the ears of the cushion for the Metro (see Fig. 10b). This tended to pull the buckle towards the occupant and lowered the inboard end of the diagonal. The result of this manoeuvre was to change the 'effective' anchorage locations of the seat belt by large amounts and radically alter the lie of the lap and diagonal belts around the child. It was fairly difficult to install the child and it was most unlikely that a child could attach the belt properly by itself.

Cushion C employed two tubular rails which were attached to each side of the cushion and then curved round to the rear. These were designed for belts to be hooked onto and brought around to the side where plastic eccentric clips were used to clamp the webbing in place. Short webbing loops were also provided on each side to fit over a buckle where there was little or insufficient webbing. In front seating positions, the outboard portion of the lap belt would be fitted to the relevant tube and on the inboard side the webbing loop would be put over the buckle and stalk. In the case of the Cortina installation the loop was placed just above the latch plate. This cushion was also quite difficult to fit. The diagonal belt tended to be high on the chest.

Cushion D had moulded vertical ears and was relatively easy to install. The outboard portion of the lap belt was hooked through one ear and both belts hooked through thc inboard ear on the Metro front installation. On some seating positions and with the larger child dummies the diagonal belt lie was better with only the inboard lap portion through the cushion ear. The diagonal belt adopted a reasonable position on the chest for the Metro front seat.

Rear installations for the four cushions are shown in Fig. 11. Fitting techniques were similar to that for the front seat with a few exceptions. The cushion A was installed by hooking onto the lap belt with its ears. If there was a stalk made from webbing of sufficient length this would be inserted through the ear loop for cushion B and hooked onto the inboard tube and clamped for cushion C.

For rear or front seats the easiest installation was with cushion D and the most difficult with cushion C, especially where both lap portions were fitted to the tube system. A and B were fairly difficult to install. If a child was given these cushions to install, only D stood a high probability of being dong correctly. Actually, once the cushions had been fitted to the belt system, C was perhaps even more foolproof in that the portions of the belt were clipped in place and the cushion could not be dislodged when not in use. Generally, though, it was felt that booster cushions and the seat belt should be fitted under adult supervision to ensure correct installation.

Other aspects of the cushion design, relating to the comfort of the child, were also considered. It was obviously difficult for the manufacturer to satisfy the needs of the wide range of child sizes that are catered for. Cushion length and child thigh length were not always compatible and there was a decision to be made on whether the legs should be straight or bent over the edge. The test engineer used his subjective judgement to set up what appeared to be a reasonably comfortable installation, which meant in some cases straight legs and in others bent legs (see Figs. 11c and 10c). It was apparent that there was little thigh support from some cushions, for example, Fig. 11d. Evidence from the field for child seated positions was inconclusive, which was not surprising as children do not sit still in one position for more than a few minutes. A young child had been seen at one time sitting cross-legged on cushion C and at another time sitting in a slumped position with legs bent over the front. To reduce the number of test variables in this programme of work, a rather stereotyped seating position was used so that some comparison could be made between cushions.

The other question relating to cushion design was for what age range they could be used. The literature with the Australian cushions recommended that they were suitable for children aged two and a half to eight years, or more precisely, body weights of 14–30 kg. The TNO three year old dummy, at 15 kg, was slightly large for the lower end of the scale and at the top end there was a difference: TNO ten year old, 32 kg; Ogle ten year old, 35 kg. It was apparent that the ten year old could be seated quite satisfactorily in some installations without a cushion (see Fig. 8c) and that with a cushion he was perhaps too high (see Fig. 12), in particular the rear of the Metro. This last statement refers to the lie of the

Fig. 12a Ten year old in rear of Cortina: with cushion C

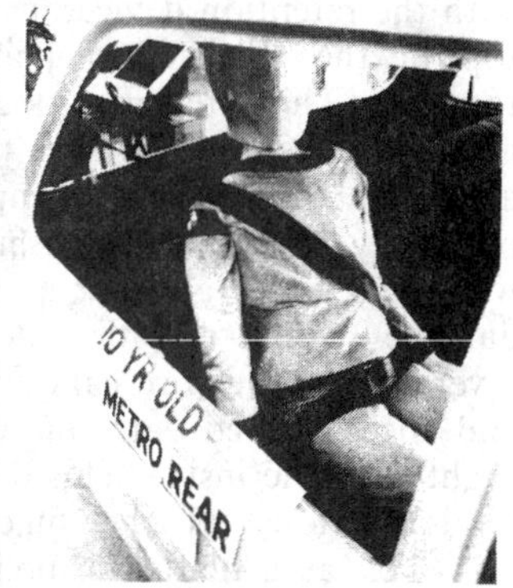

Fig. 12b Ten year old in rear of Metro: with cushion C

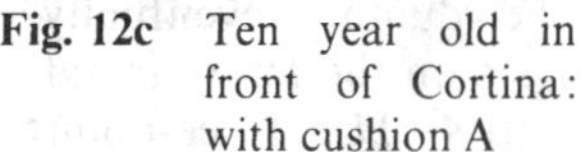

Fig. 12c Ten year old in front of Cortina: with cushion A

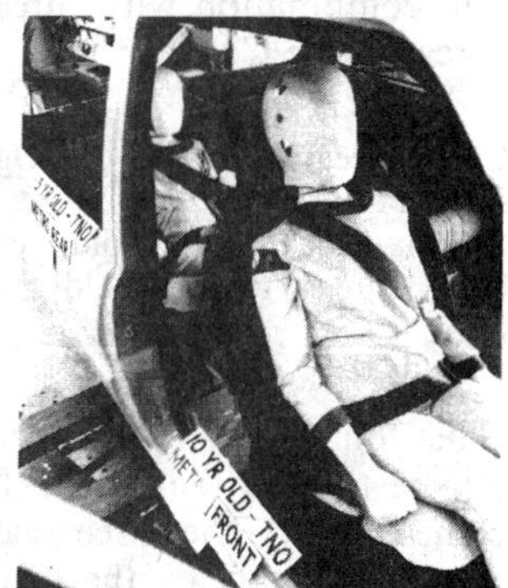

Fig. 12d Ten year old in front of Metro: with cushion A

belt over the child, mainly the diagonal, and the position of the head relative to the back of the seat; Section 6.1.1 has shown that considerable reduction in submarining can result from the use of a booster cushion with the ten year old and those results for this phase, described in the next section, show that submarining can be completely eliminated with proprietary cushions. Because there was a tendency for the diagonal belt to be low on the chest, dummy roll-out of the belt was expected to be a problem. (This is more fully described in Section 5.2.3.) The three year old benefited considerably from the use of a booster cushion by the improved lie of the seat belt (for example, Fig. 11c) and the better visibility of his surroundings. The final assessment of a suitable age range for booster cushions depends on not only the appraisal of the static installation, but also the other factors investigated in this project and these are brought together in Section 5.2.5.

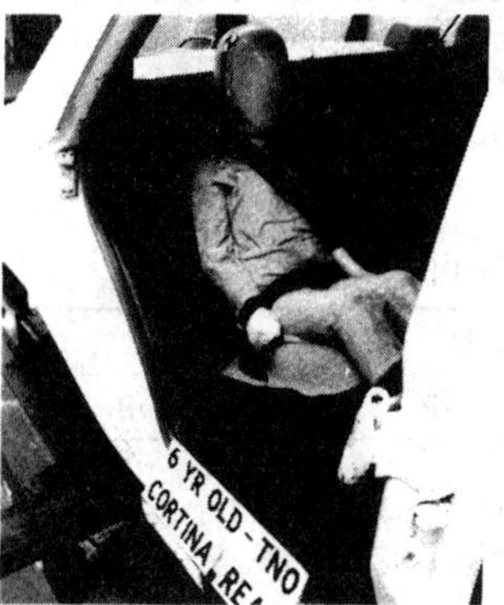

Fig. 11a Six year old in rear of Cortina: with cushion A

Fig. 11b Ten year old in rear of Cortina: with cushion B

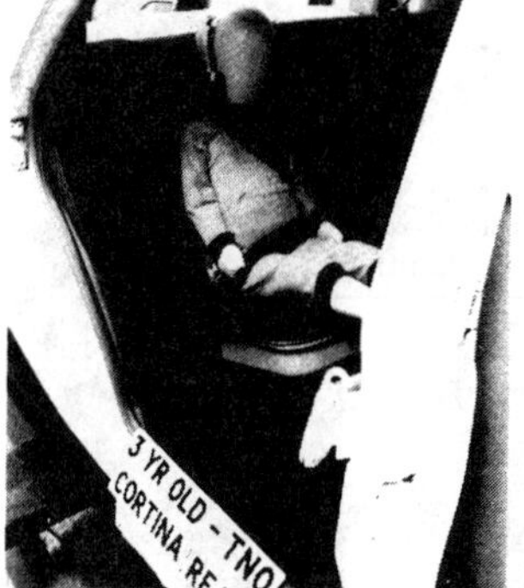

Fig. 11c Three year old in rear of Cortina: with cushion C

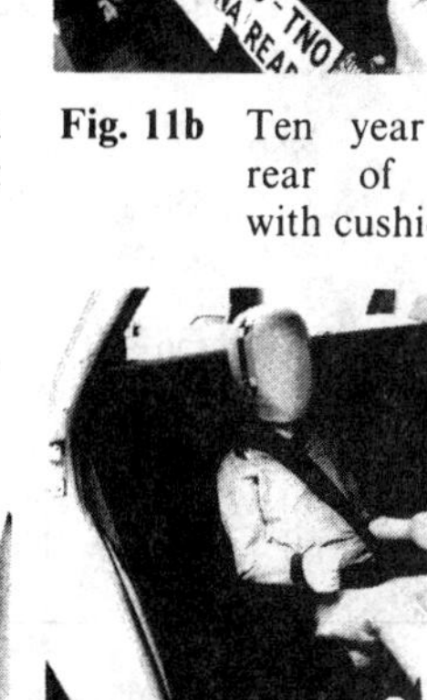

Fig. 11d Six year old in rear of Cortina: with cushion D

5.2.2 Assessment of submarining: Phase 2

During the thirty-four dynamic tests employing booster cushions, the position and movement of the lap belt around the child dummies was examined on high-speed film. At the end of the test the position of the lap belt was noted and any other relevant information was included. The proprietary booster cushions had a marked effect on submarining by completely eliminating it for all cushions in combination with three-point auto and two-point static seat belts. The near vertical run of the lap belt on some of the installations in conjunction with the retention devices, caused the lap to tighten on top of the thighs and pull the cushion and child together. This prevented the belt riding up into the abdomen, even on the three year old dummy which has a vestigial pelvis. This clamping force to the cushion obviously also kept the cushion in place and retention was rated either good or acceptable for all tests. Some damage to the retention device was experienced on several tests, the wire ears on cushion B would distort and the moulded ears on A would occasionally tear slightly on the inside. This damage was considered not to be serious as the function was not impaired. However, as a matter of policy, new booster cushions were used for each test so that there was no question of the integrity of the cushion.

Submarining did occur in some tests, but these were all in combination with three-point static seat belts in the rear of the Cortina:

Cortina rear, ten year old Ogle, cushion B, three-point static
Cortina rear, three year old, cushion C, three-point static
Cortina rear, ten year old Ogle, cushion A, three-point static

This follows on the experience from the three-point static test from the Phase 1 programme which submarined badly. The three year old with cushion C was pulled up sharply by the diagonal, the lap belt moved up onto the abdomen and the dummy eventually became almost horizontal. The tests on the ten year old resulted in very similar kinematics. The three-point static belt had low elongation and restrained the chest early, whereas a three-point auto belt had more webbing to stretch and also webbing came off the retractor as it tightened on the spool. For good dummy movement, it was apparent that the right balance of diagonal to lap belt stiffness was required and a three-point static belt was too stiff in the diagonal. This effect is also found in the restraint of adults where a well-designed seat belt will allow controlled forward movement of the torso.

5.2.3 Diagonal belt performance and head forward movement: Phase 2

Similar caution was exercised to that described in Section 5.1.2 when evaluating the movement of the torso of the dummies under the restraint of the diagonal belt. This movement was considered to be largely a function of the stiff chest construction of the TNO dummies, though the ten year old Ogle was more compliant.

The dynamic tests were examined for diagonal belt performance and the results have been summarized in Table 7.

The torso movement of the 6 year old in the front of the Metro with the four cushions follows what would be expected from the initial installation (see Fig. 10), except for the test with cushion B which, due to an omission, had not been installed with the inboard portion of the lap belt hooked to the cushion. Cushion B, fitted in this manner, caused roll-out, whereas the other cushions either gave a well-balanced torso restraint in the case of D or the diagonal moved to touch the neck.

Head forward movements were generally smaller for the diagonal belt close to the neck and largest where there was torso roll-out. In comparison with the ECE 44 limit of 550 mm, the balanced restraint tests (mid) were very near the limit and some exceeded it. The ten year old, TNO dummy in the front seat of the Metro, sitting on cushion C gave a particularly low position for

Table 7 Assessment of diagonal belt performance and measured head forward movement for Phase 2 tests

	Cushion A		Cushion B		Cushion C		Cushion D		Cushion E	
Installation	HFM	DBP	HFM	DBP	HFM	DBP	HFM	DBP	HFM	DBP
Metro front										
3 year old	320	Neck			670	Out	440	Mid		
6 year old	420	Neck	760	Out[3]	350	Neck	450	Mid		
10 year old, TNO					805[2]	Out				
Metro rear										
3 year old			670	Out						
6 year old	590[1]	Mid	600	Out	530	Mid	550	Mid		
10 year old, Ogle			690	Mid						
Cortina front										
3 year old	405	Neck	480	Neck	405	Neck	520	Neck	530	Neck
6 year old	480	Neck	360	Neck	450	Neck	505	Neck	640	Mid
Cortina rear										
3 year old	980[2]	(2-pt)			435	Neck (3-pt)	475	Neck		
6 year old	530	Neck	930[2]	(2-pt)	910[2]	(2-pt)	490	Neck		
10 year old, Ogle	525	Mid (3-pt)	445	Neck (3-pt)			1000[2]	(2-pt)		

Notes
HFM = head forward movement relative to ECE 44 Cr-point, in mm
DBP = diagonal belt performance
Neck = high position of diagonal, close or touching neck
Out = tendency for torso to roll out of belt, diagonal low
Mid = balanced restraint of torso
2-pt = two-point static seatbelt, no diagonal
3-pt = three-point static seatbelt
1 = late reel lock-up
2 = head impact
3 = inboard lap belt not hooked to cushion
Unless otherwise stated, seat belts were three-point auto

the diagonal. This enabled the dummy to roll out of the belt early in the impact sequence, the diagonal sank into the abdomen and the head struck the facia. It was of interest to note that the plasticene block had been flattened by 13 mm as a result of the diagonal belt penetration.

The vertical ears of cushion D appeared to have a beneficial effect on the location of the diagonal for the three year old by placing it in a central position.

The Cortina front seat belt was close to the neck on most tests and for that reason, head displacements were reasonable. It could be argued that it was probably better to have a belt lie close to the neck if there was the possibility of the child slipping out when it was lower down.

In a similar conclusion to the discussion of this aspect for Phase 1 tests, it was considered that this 'delicate balance' of the torso on the diagonal belt was an artificial result of dummy construction and that real children were retained in most cases.

5.2.4 Assessment of cushion E: Phase 2

This cushion was selected in addition to the main four, because of its different construction. It was in the form of a soft foam chaise, with integral seat and back and had no means of retention other than its back portion behind the child (see Figs. 13a and 13b). Two tests were performed in the front of the Cortina with the three and six year old dummies. Because of the softness of the cushion, the forward movement of the dummies was accompanied by large vertical displacements of 125 and 140 mm for the three and six year old respectively. Vertical movements with other cushions were of the order of 60 mm. The belt lie was good dynamically even though it was close to the neck for the three year old. Forward movements of the head relative to initial dummy H-point location were just acceptable, but ECE 44 forward limits of motion are related to survival space and therefore are established relative to the seat.

Because this cushion has a back portion, the dummy had already used up some of the survival space before the test and in the case of the six year old the forward movement relative to Cr point was 640 mm (see Table 7). The post-test positions of the dummies were almost as the pre-test condition which was an unusual characteristic of this installation.

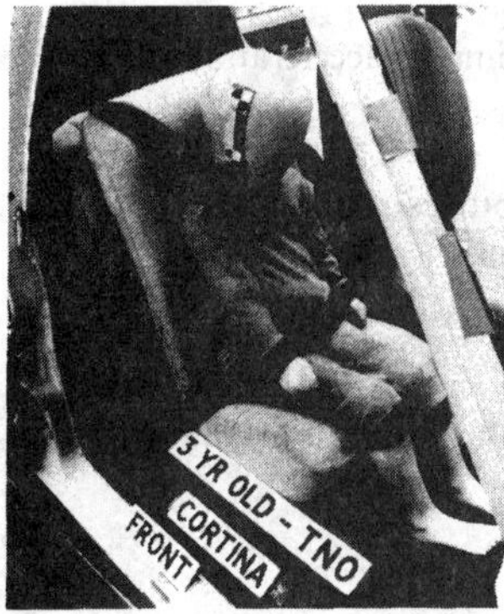

Fig. 13a Three year old in front of Cortina: with cushion E

Fig. 13b Six year old in front of Cortina: with cushion E

This type of cushion has an advantage in the ease of installation and appears to offer a comfortable seating position for a child.

5.2.5 Overall assessment of the Phase 2 tests

The 'overall assessments' have been listed for all the Phase 2 tests in Table 8. Where booster cushions were used in combination with three-point auto seat belts, submarining was eliminated and the main reason for any inferior performance was excessive forward movement of the head and torso because the chest rolled out of the diagonal. If it can be accepted that this movement is a function of dummy construction rather than the retention of a child in a real accident then all such seat belt and booster cushion combinations are acceptable. The problems occur with tests involving two-point and three-point static belts where either the forward movement is excessive or submarining is induced. The two-point installation is not improved by the use of a booster cushion because submarining is not the problem, whereas forward movement of the head is for two-point belts. The booster cushion was seen to reduce the severity of submarining for three-point static seat belts and could, therefore, be considered of benefit.

The suitable age range for booster cushions depends on various factors. From the test results in this phase there would be no hesitation in recommending a booster cushion for a three year old or child weighing

Table 8 Overall system assessment for Phase 2 tests

	3 year old		6 year old		10 year old	
Metro front						
Cushion A	***		****			
B			**[3]			
C	**[1]		****		*[1]	
D	*****		****			
None			**		**	
Metro rear						
Cushion A			***[2]			
B	**[1]		**[1]		***	
C			****			
D			****			
None	**		**			
Cortina front						
Cushion A	*****		****			
B	***		****			
C	***		****			
D	****		****			
E	****		****			
None	**		***			
Cortina rear						
Cushion A	**	(2-pt)	*****		**	(3-pt)
B			**	(2-pt)	**	(3-pt)
C	*	(3-pt)	*	(2-pt)		
D	****		****		*	(2-pt)
None			*		*	

Notes
* bad
** poor
*** acceptable
**** good
***** very good
1 diagonal low 2 late reel lock-up
3 inboard lap belt not hooked to cushion
2-pt = two-point static seatbelt
3-pt = three-point static seatbelt
Test results with no cushion from Phase 1
Tests employed three-point auto seat belts unless otherwise stated

15 kg, where there was considerable improvement in the movement of dummies and submarining was eliminated. The difficulty is to draw conclusions on the largest suitable child size. A cushion stopped the submarining of the ten year old but there were indications that the diagonal belt lie was low on the chest. The size of the car and the height of the seat and its upper anchorage all have an effect on this aspect. The ten year old dummies appeared to be at about the top limit for size for use of booster cushions, i.e., children in the weight range 32–35 kg. The fifth percentile female, representing a teenager, was obviously too high when using the hard cushion in the Phase 1 series. Australian booster cushion manufacturers recommend an eight year old as the upper limit which is probably about right. Stature or sitting height are the critical parameters in every case and if it were not for the risk of submarining, the ten year old dummies could be restrained with the adult seat belt alone. The new British Standard BS AU 185 1983 specifies that these cushions are intended for use by persons of small stature, of mass 15 kg or more, for the purposes of improving the fit of the seat belt. However, the preceding observations indicate the limitations of booster cushion usage.

6 CONCLUSIONS

This project investigated the performance of adult seat belts alone and in combination with a range of booster cushions when employed to restrain child dummies of various sizes in simulated perpendicular impacts of cars.

Sixty-seven dynamic tests were performed in the course of the project using two vehicle bodyshells, four types of child dummy, three variants of seat belt and five booster cushions of a range of designs.

Submarining of various degrees of severity was experienced by the child dummies when restrained only by the adult seat belt in simulated 50 km/h perpendicular impacts.

A soft cushion was of little benefit and could induce submarining if untethered.

Proprietary booster cushions eliminated submarining in almost every case except where three-point static seat belts were used. In the latter case however a reduction was observed.

The best combination was a firm tethered booster cushion with a three-point automatic seat belt and this combination was acceptable for the restraint of child dummies in the age range three to ten years.

Differences in restraint performances between the various proprietary designs of booster cushions were small and the choice for a particular installation may therefore be influenced largely by ease of fitting and convenience of daily use.

ACKNOWLEDGEMENTS

MIRA is grateful for the assistance in terms of test material, information and guidance, from the participating organizations in this project. In particular, MIRA wishes to acknowledge the support of the Department of Industry, Requirements Board, for their partial funding of the project test work.

APPENDIX

Crash pulses

The Hyge pneumatic system employs a graduated pin that is drawn through an orifice to control gas flow and ultimately the acceleration of the sled and occupants. Various operating parameters can also be adjusted to modify the duration and energy of the pulse. An existing 'two-hump' pin was selected to provide the simulated accelerations for the Cortina and Metro bodyshells. By referring to past crash tests of the two vehicles, the manufacturers were able to supply acceleration–time histories in 50 km/h perpendicular impacts with barriers. The Hyge operating parameters were then developed with a few trial runs to provide a reasonable approximation to the crash pulse. The ISO draft standard for 'In car structure frontal test for occupant protection systems' was used as a guide for the accuracy of the 'equivalent pulse'. The velocity change included the rebound component.

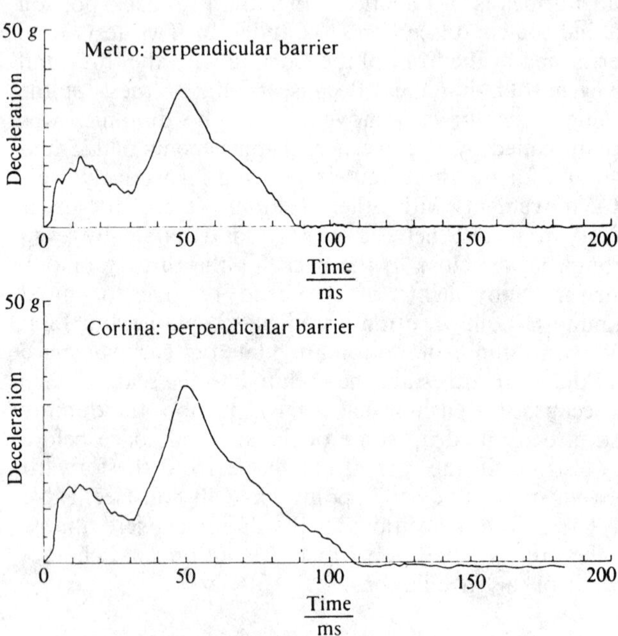

Fig. 14 Simulated vehicle impact accelerations

The two simulated vehicle pulses are shown in Fig. 14. For reasons of confidentiality the actual vehicle crash pulses have been omitted.

942222

Survey of Older Children in Automotive Restraints

Kathleen DeSantis Klinich
Transportation Research Center, Inc.

Howard B. Pritz, Michael S. Beebe, and Kenneth E. Welty
Vehicle Research and Test Center

ABSTRACT

This paper describes results from a survey of older children with respect to vehicle and booster restraints. The work first consisted of a rudimentary anthropometry study of 155 volunteers aged between 7 and 12 years. The data were compared to an extensive child anthropometry study conducted by the University of Michigan in 1975. Height and sitting height data matched well, while children in the current study appeared heavier. In the restraint fit survey, each child sat in the rear seat alone and in three belt-positioning booster seats (Volvo, Kangaroo, Century CR-3) in three vehicles (Ford Taurus, Pontiac Sunbird, Dodge Caravan). Booster seats greatly improved belt fit over the rear seat alone. The majority of children in this study had better belt fit with the boosters than with the rear seat alone, regardless of size. However, children who could fit well in the boosters and had good or fair belt fits were generally 36 kg or less. The sitting heights of these children ranged from 58 to 76 cm, and their standing heights varied from 117 to 152 cm. In general, the minimum size child for using three-point belts alone is a sitting height of 74 cm, standing height of 148 cm, and a weight of 37 kg. A possible cause of poor belt fit that is specialized to this group of children in the "slouch factor". Children of these ages will scoot forward in a seat to allow comfortable leg positions rather than sitting up straight and putting pressure on the backs of their lower legs. Booster seats seem to prevent slouching by allowing a comfortable leg position while sitting upright.

CHILD RESTRAINT RESEARCH has naturally focused on younger children (aged 0 to 5 years), because their distinctly smaller sizes most obviously would not be accommodated by vehicle restraints designed to protect adults. However, both vehicle and child restraint researchers have recently devoted more efforts to the needs of older children (aged 6 to 12 years). In particular, child seat manufacturers have developed belt-positioning booster seats intended for children with maximum weights between 30 and 45 kg. The National Highway Traffic Safety Administration (NHTSA) initiated this program to study the older child population with respect to the automotive environment[1]. The survey discussed in this paper investigates how well current vehicle restraints and belt-positioning booster seats meet the needs of children aged 7 to 12 years. The restraint fit study recorded how seat belts fit 155 older children seated in three different vehicles when they were used with and without belt-positioning booster seats. One specific goal was to determine the largest size child who benefitted from booster seats, and the smallest sized child that could effectively use three-point belts alone. As part of the restraint fit study, a simple anthropometric survey of the child volunteers was also conducted. The anthropometry study showed the height, weights, and sitting heights of the volunteers, and compared them to the last detailed anthropometric study conducted in 1975.

ANTHROPOMETRY STUDY

VOLUNTEER SELECTION - Because this program required child volunteers, it was conducted in compliance with NHTSA Order 700-1, "Protection of the Rights and Welfare of Human Subjects in NHTSA-sponsored Experiments". The principal at Edgewood Elementary School in Marysville, Ohio agreed to let students in grades one through five participate in the survey. The school was chosen because its size would provide enough volunteers (approximately thirty per grade) and because of its proximity to NHTSA's Vehicle Research and Test Center (VRTC). The school population does not necessarily match the national distribution by race or economic status. However, comparisons between the size distributions of the VRTC volunteers and those from more extensive measurement studies would help to ensure that the sample population was reasonable.

To implement the program, VRTC first sent letters to parents explaining the program. At that time, parents could refuse to have their children be considered for participation in the program. After the students returned their initial permission forms, they heard presentations on crash test dummies and the measurement program during their physical education classes. The students were then asked to volunteer. Most of the students in each classroom were interested, so all the names were collected and eight names were selected from each of four classrooms per grade. Names of students whose parents previously refused permission were not included. Students who did not volunteer were usually overweight, or boys who did not want to miss part of their physical education class to participate. The volunteers received another, more formal permission slip, which was signed by their parents before they were able to participate. A total of 155 volunteers received permission. The distribution of volunteers by gender and age appears in Table 1.

TABLE 1 --Distribution of Volunteers by Gender and Age

	7	8	9	10	11	12
Boys	14	13	17	16	14	2
Girls	13	16	17	15	17	1
Total	27	29	34	31	31	3

Fig. 1 – Front view of a volunteer seated in reference chair

ANTHROPOMETRIC MEASUREMENTS - A simple anthropometry study of the volunteers constituted the first phase of the program. Each child's weight, height, sitting height, age, and birthdate were recorded. The age used was an exact calculation as of March 1, 1993. The volunteers were also photographed from the front and side while sitting in a reference chair. These photos would allow comparison between volunteers and approximations of pertinent anatomical measurements. The children also wore a target "necklace" which would be helpful when reviewing the reference chair pictures and in-vehicle videotapes. Figures 1 and 2 show samples of the photos. Each child was also assigned an identification number. The distribution of volunteers according to height and weight is found in Table 2.

ANTHROPOMETRIC ANALYSIS - To check the validity of the VRTC data, the weight, height and sitting height data were compared to data from other studies. The most extensive anthropometric study of older children was conducted by the University of Michigan (UM) in the mid 1970's[2]. The study measured the anthropometric characteristics of 4127 children aged 0 to 18 years. The children were selected to represent the United States child population with regard to race, demographic, and socio-economic factors. The large sample size and careful sample selection process make the UM study an excellent benchmark

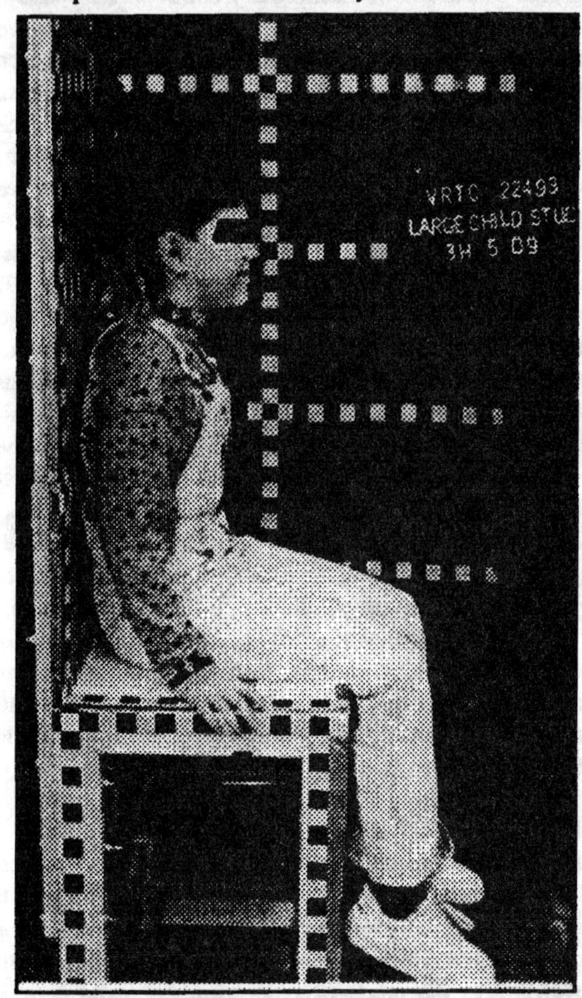

Fig. 2 – Side view of a volunteer seated in reference chair

TABLE 2 --Distribution of Volunteers by Weight and Height

Weight (kg)	Height (cm) 112 - 117	118 - 124	125 - 132	133 - 140	141 - 147	148 - 155	156 - 168
< 23	5	13	3				
24-27		6	18	6			
28-32	1	1	9	17	11		
33-36			2	7	8	2	
37-41			2	3	4	7	
42-50				2	7	7	1
51 +					2	5	6

for comparison. The average, minimum, and maximum weights, heights, and sitting heights by age for the VRTC and UM studies appear in Tables 3 and 4.

The sitting height measurements from the VRTC study match the UM data the best. This indicates that the VRTC sample of children may be generally considered as well-distributed as the UM study, even though the number of participants was much smaller. The height measurements from the current study are slightly higher than those of the UM study. However, the VRTC volunteers wore shoes, while the UM participants were barefoot. In a test using five of the VRTC volunteers, the difference in height caused by shoes averaged 2.2 cm, accounting for some of the difference.

The average weights between the two studies are the most different. However, some of the differences may be caused by differences in clothing. The average weight for each age group in the current study is higher than in the UM study. The differences become larger with increasing age. To check this finding, two other studies were also compared. The first is a 1979 study by the National Center for Health Statistics (NCHS)[2]. The second is a VRTC analysis of child occupants in the 1988-1991 National Accident Sampling System (NASS) database. The average heights and weights by

TABLE 3 -- VRTC Large Child Measurement Survey
Weight and Heights by Age

Age	N	Weight (kg)			Height (cm)			Sitting Height (cm)		
		Avg	Lo	Hi	Avg	Lo	Hi	Avg	Lo	Hi
7	27	24.3	15.0	39.9	123	112	137	64	58	70
8	29	28.1	20.4	45.4	130	119	146	67	60	75
9	34	32.0	20.9	59.9	137	119	150	71	61	79
10	31	36.5	23.6	56.2	142	126	156	73	64	79
11	31	43.0	22.2	81.2	148	130	166	75	65	88
12	3	45.7	38.1	52.2	150	147	152	77	76	78

TABLE 4 -- UM Child Measurement Study
Weight and Heights by Age

Age	N	Weight (kg)			Height (cm)			Sitting Height (cm)		
		Avg	Lo	Hi	Avg	Lo	Hi	Avg	Lo	Hi
7	227	23.7	15.4	43.6	121.2	106.3	133.8	65.8	56.6	73.5
8	198	26.6	17.6	54.2	126.9	111.5	140.6	68.2	59.7	75.6
9	257	29.7	19.6	60.0	133.0	117.8	150.3	70.3	61.2	78.5
10	258	33.1	22.2	69.0	137.7	120.1	159.0	72.1	64.5	81.6
11	282	37.2	20.7	80.0	143.3	122.0	151.1	74.5	65.2	84.6
12	287	40.3	26.1	77.8	148.8	132.8	172.4	76.7	67.1	87.7

TABLE 5 -- Average Heights and Weights by Age								
Age	Average Height (cm)				Average Weight (kg)			
	UM (1975)	NCHS (1979)	NASS ('88-'91)	VRTC (1993)	UM (1975)	NCHS (1979)	NASS ('88-'91)	VRTC (1993)
7	121.2	121	118	123	23.7	23	26	24
8	126.9	127	127	130	26.6	25	30	28
9	133.0	132	132	137	29.7	28	33	32
10	137.7	137	137	142	33.1	32	37	37
11	143.3	143	145	148	37.2	36	41	43
12	148.8	150	151	150	40.3	40	47	46

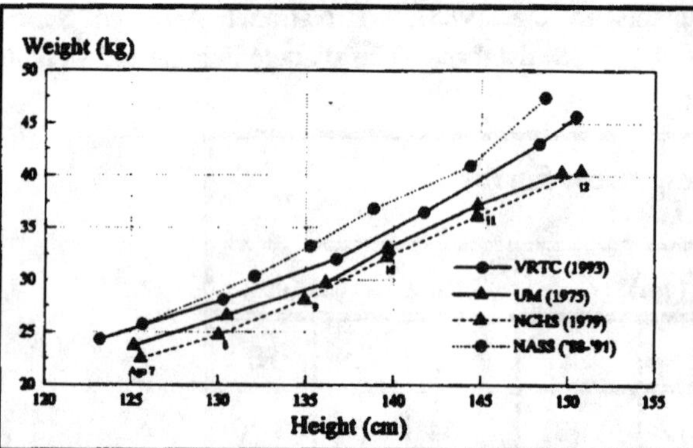

Fig. 3 -- Average heights and weights by age in four studies

age are found in Table 5. Figure 3 plots average height vs. weight by age for all four studies. Comparison between the older and newer studies shows that while average heights have not changed, average weights seem to have increased in older children.

BOOSTER AND RESTRAINT FIT SURVEY

SURVEY DESIGN - During the survey's initial planning stages, four belt-positioning booster seats were commercially available. The recommended height and weight ranges for each seat appear in Table 6. The instructions also cautioned against using the seats if the center of the child's head rose above the vehicle seat back. The Century CR-3 and Gerry Double Guard boosters have similar designs, so only one of these two was tested. The Century was selected because of its higher sales volume and higher upper weight limit. Figure 4 shows the Century, Volvo, and Kangaroo booster seats.

Vehicle selection resulted from compromise among several factors. An intermediate car, compact car, and minivan (Ford Taurus, Pontiac Sunbird, and Dodge Caravan) were chosen. These vehicles represented the prominent manufacturers and highest-selling makes and types of vehicles. While small pick-ups were more frequently purchased than minivans, a minivan was selected for this program because more families with children would be expected to prefer minivans. The test vehicles have bucket, bench, and captain's chair seats. The Taurus is a four-door, while the Sunbird is a two-door, which would vary interior compartment space. All vehicles have three-point belts in the back seats.

TABLE 6 --Recommended Height and Weight Ranges for Belt-Positioning Booster Seats

Seat	Min Height (cm)	Max Height (cm)	Min Weight (kg)	Max Weight (kg)
Volvo Booster	117	137	23	36
Kangaroo Booster			23	36
Century CR-3			11	30
Gerry Double Guard	84	130	14	27

Fig. 4 -- Century, Volvo, and Kangaroo booster seats

SURVEY PROCEDURE - Once the anthropometric measurements were taken, the vehicle restraint survey was begun. The volunteers were excused from their physical education classes one at a time to participate. Since the vehicles were outside, the volunteers wore coats as required by the weather. In the first session with each child, the volunteer tried out the booster seats to see if he or she could fit. Unless the child was uncomfortable, the seat was used regardless of the recommended height/weight requirements. If the student could not fit into any of the booster seats, he or she sat in the front seat instead, with the seat adjusted to the foremost and rearmost positions. Barring a few make-up days, all of the volunteers were surveyed in one vehicle before proceeding to the next one. Each session took approximately ten minutes.

The volunteers first sat in the vehicle's rear seat on the passenger side. If needed, the project technician assisted the students in buckling the three-point belt. The students were asked to try to sit up as straight as they could, which they did to varying degrees. Each volunteer also wore a target "necklace", which could be matched to the photographs taken earlier in the reference chair. The engineer then videotaped the volunteer from three to five angles, depending on the vehicle. After the rear seat position was filmed, the process was repeated with the three booster seats. However, the Kangaroo was unable to fit in the contoured captain's chair seat of the minivan, as illustrated in Figure 5. The volunteers were asked to remember which seat/position they liked best; the favorites were recorded at the end of each session.

Fig. 5 – Contoured seat of Caravan would not allow proper positioning of Kangaroo booster

EVALUATION PROCEDURES - On completion of the survey, an evaluation scheme was developed to assess each vehicle/seat combination, hereafter referred to as an observation. Several researchers tested the proposed rating scheme on a sample of 10 volunteers, to arrive at "consensus" definitions for each category. The chief evaluator, who also conducted the survey, then rated each observation in the categories found below using the group definitions. Photos that illustrate different degrees of lap belt fit, shoulder belt fit, slouch, and leg angle appear in Appendix A.

Shoulder Belt Fit - Position of the shoulder belt over the child. Ratings include in face, against neck, on shoulder but too close to neck, centered on shoulder, on shoulder but too close to arm, and over arm.

Lap Belt Fit - Position of the lap belt over the child. Ratings include flat over legs, low over pelvis, over pelvis, over abdomen, and approaching ribcage.

Booster Fit - Fit of child in booster seat; each seat required a different rating system. For the Century, ratings include could not fit, snug fit, and comfortable fit. For the Kangaroo, ratings were primarily based on the position of the child's shoulders relative to targets located on the seat back sides. Ratings include could not fit, tight fit (shoulders covered entire target), medium fit (shoulders covered half of targets), snug fit (shoulders covered part of targets), and comfortable fit (shoulders within seat back sides). For the Volvo, ratings include could not fit, tight fit (child could fit in seat, but belt could not be routed properly), snug fit, and comfortable fit.

Slouch - Child's posture in the seat, usually judged from pelvis angle and space between the seat back and child's buttocks. (The contoured seat of the Caravan made it difficult to judge slouch in that vehicle). Ratings include sitting up straight, almost straight, slightly slouched, and extremely slouched.

Seat Position - Child's position in the seat. Ratings include extremely inboard, slightly inboard, centered in seat, slightly outboard, and extremely outboard.

Leg Angle - Angle of child's thighs relative to horizontal landmarks in the vehicle. Ratings include significantly above horizontal, slightly above horizontal, horizontal, slightly below horizontal, and significantly below horizontal.

Feet Position - Location of child's feet. Ratings include on floor, almost on floor, hanging over seat, parallel to seat, and on seat.

Sitting Height - Location of child's head relative to rear seat back. Ratings include entire head below seat, ears below seat, ears along seat, ears above seat, chin above seat, and head almost touching roof.

Clothing Effect - Effect of clothing on lap and shoulder belt fit. Ratings include no effect on shoulder belt fit, minor effect on shoulder belt fit, major effect on shoulder belt fit, no effect on lap belt fit, minor effect on lap belt fit, and major effect on lap belt fit.

Chest Mark - Length of belt needed to reach center of child's chest, estimated from markings on the belt.

Belt Mark - Length of belt needed to reach center of child's pelvis, estimated from markings on the belt.

Target Location - Position of a target "necklace" worn by the volunteers, relative to the shoulder belt.

After all of the films were evaluated, the data were entered into a spreadsheet for analysis. A volunteer usually had eleven entries: rear seat and three boosters for the Taurus and Sunbird, plus rear seat and two boosters for the Caravan. A total of 1658 observations was obtained in this study. Each observation contains all of the rated data for the particular vehicle/seat/child combination, plus anthropometric data for the child. Several derived variables were also included. The data were converted for input into SAS, which has convenient cross-tabulation and statistical capabilities.

VEHICLE COMPARISONS: ALL OBSERVATIONS - The vehicle comparison section uses all observations completed for each vehicle, and includes rear seat, front seat, and booster seat combinations. The differences in shoulder belt fit appear in Table 7. The compact car (the Sunbird) had the highest rate of good fit, which is considered centered on the shoulder. Over half (53.2%) were good with the Sunbird, while 38% were good with the Taurus and only 19% were good with the Caravan. Fair fits (slightly off center of the shoulder, both inboard and outboard), were achieved approximately one-third of the time in all three vehicles. Outboard fair and poor fits (too close to or on the arm) happened with only a fraction of the observations in the cars, but not the minivan. Inboard poor fit of the shoulder belt (too close to the neck or face) occurred in 14% of observations for the compact car, 28% of observations for the intermediate car, and 47% of the time with the minivan. Greater z-distance to the shoulder anchor point, plus a steep angle in the x-z plane seem to make shoulder belt fit worse. Because these observations include the vehicle seats alone and with the boosters, this also means that the shoulder belt fit may be less correctable with increasing distance to the shoulder anchor.

Contrastingly, distribution of lap belt fit did not vary with vehicle type. Overall, lap belt scores were 65% good, 21% fair, and 14% poor. Good fits were horizontal over the legs or low on the pelvis, fair were over the pelvis but slightly too high, and poor were over the child's abdomen or approaching the rib cage.

A combined belt fit rating was also developed. If both lap and shoulder belt had good ratings, belt fit was also good. If either one was fair, and the other good, or if both were fair, the observation's belt fit was deemed fair. If either one was poor, the overall fit was regarded as poor. The combined belt fit rankings by vehicle appear in Figure 6. About one-third of the observations for each vehicle had fair fit. The Sunbird had the highest percentage of good and the lowest percentage of poor fit, while the opposite was true for the Caravan. Almost half of the belt fits were poor in the minivan. Belt fit in the Taurus was almost equally distributed among the three ratings.

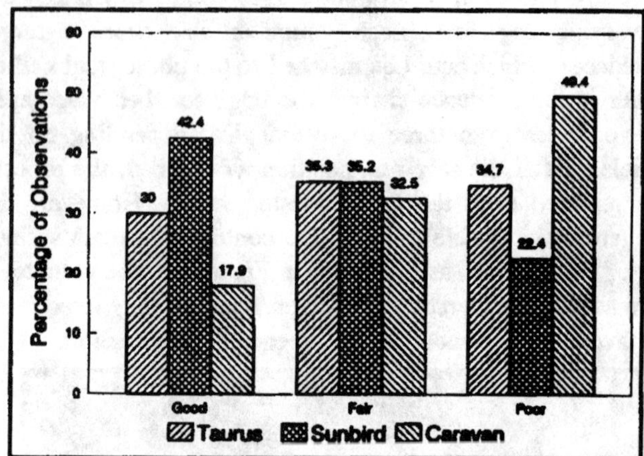

Fig. 6 – Combined belt fit by vehicle

TABLE 7 – Shoulder Belt Fit by Vehicle

Rating	Score	Observations			Percentages		
		T	S	C	T	S	C
In Face	Poor	42	8	47	7.2	1.4	12.1
Against Neck	Poor	119	70	137	20.5	12.6	35.1
On Shoulder; Too Close to Neck	Fair	198	170	131	34.1	30.7	33.6
Centered on Shoulder	Good	219	295	75	37.8	53.2	19.2
Too Close to Arm	Fair	2	10	0	0.3	1.8	0.0
On Arm	Poor	0	1	0	0.0	0.2	0.0
Total		580	554	390	100.0	100.0	100.0

T = Taurus S = Sunbird C = Caravan

When studying fit within a vehicle, belt fit alone does not necessarily describe the environment. Belt fit does partially depend on posture, but posture should be considered on its own as well. Three main categories were examined in evaluating posture for the volunteers: sitting height, leg angle, and slouch. For sitting height, if the volunteers' ears were along or below the top of the seat back, it was judged acceptable. Any heights higher than that were bad, because the head would not be supported in a rear collision. The basis for including vehicle sitting height as a posture issue is the specification in Federal Motor Vehicle Safety Standard No. 213 that a dummy placed in a booster seat must not have its head center of gravity rise above the seat back. The Sunbird had the highest rate of unacceptable sitting heights, with approximately 25% of the seats placing the volunteers too high relative to the seat back. Only 15% were too high with the Taurus. All of the volunteers had good sitting heights in the Caravan. This corresponds with observations of the seat back designs in the three vehicles. The top of the compact car's rear seat back was horizontal, while the Taurus rear seat back was shaped to form a partial headrest. The captain's chair seat back in the minivan was tall enough for all volunteers to have an acceptable sitting height.

The other two posture considerations must be reviewed with potential for submarining[4]. If the thigh is angled below horizontal, the occupant may slip under the lap belt regardless of how low it rests over the pelvis. A leg angle significantly above horizontal induces the pelvis to tilt under the lap belt, which also may cause submarining. The best leg angle is horizontal or slightly above horizontal. The degree of slouch also affects the potential for submarining. The more the child slouches down in the seat, the greater potential for the lap belt to slide over the pelvis and load the abdomen.

A combined posture scale was derived to rate these three factors together. If the volunteer sat up straight or almost straight, had an acceptable sitting height, and a horizontal or slightly above horizontal leg angle, he or she received a good posture rating. If the sitting height was unacceptable, the leg angle significantly below horizontal, or the child slouched significantly, a poor rating was given. The remaining combinations received fair ratings. The combined posture results are found in Table 8. The vehicle with the best belt fits (Sunbird) had the worst postures, while the vehicle with the worst belt fits (Caravan) had the best postures. Combining belt fit and posture scores to look at overall fit evened out the ratings between the vehicles. Of the vehicle/seat combinations, 17% were good, 23% were fair, 32% were mediocre, and 29% were poor. Since only 40% of the observations in the survey were good or acceptable, it shows that restraint design with respect to older children could be improved.

SEAT COMPARISONS: ALL OBSERVATIONS - The following section reviews the differences in belt fit and postures by seat, regardless of vehicle. The first step is to study the volunteer fit in each booster seat. All of the observations were counted, instead of just one fit per volunteer, because the clothing sometimes affected how the

TABLE 8 --Combined Posture by Vehicle

	Observations			Percentages		
	T	S	C	T	S	C
Good	299	205	232	52.0	37.0	59.5
Fair	118	133	118	20.5	24.0	30.3
Poor	158	216	40	27.5	39.0	10.3
Total	575	554	390	100.0	100.0	100.0

booster seat fit. The differences in seat designs called for the different gradings of fit for each booster. The results on booster seat fit appear in Figure 7. As expected, the Century had the highest rate of nonfit, because it had the lowest recommended maximum weight. Of the remaining observations with the Century, approximately one-third were a little too snug while the rest appeared to fit comfortably in the seat. About half of the observations with the Volvo had comfortable fit, while 21% were snug. The 23% with tight fit were able to use the Volvo, but the lap belt was not able to be routed as intended. For the Kangaroo, all of the children had shoulders too broad to fit all of the way back into the seat. Their shoulders extended over the side pieces of the seat with varying degrees, as shown in the table. Figures in Appendix B illustrate the different ratings for each booster seat.

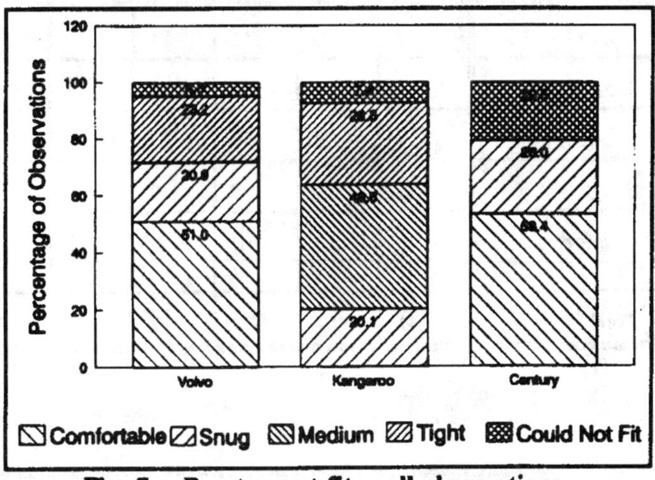

Fig. 7 -- Booster seat fit -- all observations

Another look at booster seat fit shows fit rating by weight class for each booster seat, as presented in Tables 9, 10, and 11. The fit for children 33-36 kg (highest recommended weight range for Volvo and Kangaroo seats) appears in Figure 8. In the highest recommended weight class for the Volvo, 18% had comfortable fit, 62% had snug fit, and 20% had tight fit; all volunteers at or below that weight class were able to sit in the booster. For the Kangaroo in the highest recommended weight class, 3% could not fit at all, 35% had tight fit, 62% had medium fit, and none had snug or comfortable fits. The Kangaroo does not seem to fit the majority of cases in its highest, or even next to highest, recommended weight classes. The Century, although its recommended maximum weight is 6 kg lower than that for the

TABLE 9 – Booster Seat Fit by Weight Class for Volvo														
	Observations							Percentages						
Weight (kg)	< 23	24-27	28-32	33-36	37-41	42-50	51 +	< 23	24-27	28-32	33-36	37-41	42-50	51 +
Could Not Fit	0	0	0	0	0	2	19	0.0	0.0	0.0	0.0	0.0	4.4	54.3
Tight Fit	0	0	6	11	27	41	16	0.0	0.0	5.4	20.0	58.7	91.1	45.7
Snug Fit	0	8	29	34	18	2	0	0.0	9.5	25.9	61.8	39.1	4.4	0.0
Comfortable Fit	58	76	77	10	1	0	0	100.0	90.5	68.8	18.2	2.2	0.0	0.0
Total	58	84	112	55	46	45	35	100.0	100.0	100.0	100.0	100.0	100.0	100.0

TABLE 10 – Booster Seat Fit by Weight Class for Kangaroo														
	Observations							Percentages						
Weight (kg)	< 23	24-27	28-32	33-36	37-41	42-50	51 +	< 23	24-27	28-32	33-36	37-41	42-50	51 +
Could Not Fit	0	0	1	1	1	5	14	0.0	0.0	1.3	2.7	3.3	16.1	60.9
Tight Fit	1	4	11	13	23	25	9	2.4	6.7	14.1	35.1	76.7	80.6	39.1
Medium Fit	9	33	59	23	6	1	0	22.0	55.9	75.6	62.2	20.0	3.2	0.0
Snug Fit	31	22	7	0	0	0	0	75.6	37.3	9.0	0.0	0.0	0.0	0.0
Comfortable Fit	0	0	0	0	0	0	0	0.0	0.0	0.0	0.0	0.0	0.0	0.0
Total	41	59	78	37	30	31	23	100.0	100.0	100.0	100.0	100.0	100.0	100.0

TABLE 11 – Booster Seat Fit by Weight Class for Century														
	Observations							Percentages						
Weight (kg)	< 23	24-27	28-32	33-36	37-41	42-50	51 +	< 23	24-27	28-32	33-36	37-41	42-50	51 +
Could Not Fit	0	0	2	1	16	39	32	0	0	1.7	1.8	34.8	86.7	91.4
Snug Fit	1	5	28	42	29	6	3	1.7	6.0	24.3	76.4	63.0	13.3	8.6
Comfortable Fit	57	79	85	12	1	0	0	98.3	94.0	73.9	21.8	2.2	0.0	0.0
Total	58	84	115	55	46	45	35	100.0	100.0	100.0	100.0	100.0	100.0	100.0

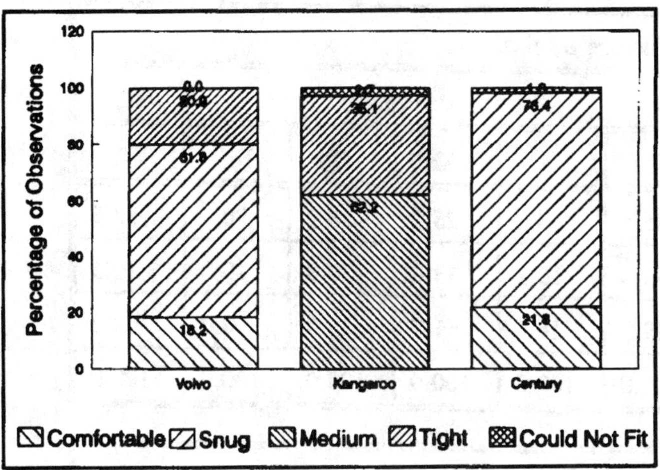

Fig. 8 -- Booster seat fit of children weighing 33-36 kg

previous two boosters, appears to provide the same degree of fit as the Volvo in the higher weight class. Only 2% could not fit, and 76% had snug fit while 22% had comfortable fit. Children heavier than recommended may be able to comfortably use the Century.

The following discussion compares belt fit and overall fit between seats, regardless of how well the boosters fit. Figure 9 contains shoulder belt fit by seat. For the rear seat alone, only 15.6% of the observations had good fit. In over half the cases (50.2%), the belt rubbed against the neck or face. All of the booster seats drastically cut the occurrence of this shoulder belt fit, down to an average near 17% for all three boosters. Almost half the shoulder belt fits were centered on the shoulder with each booster; about one-third were on the shoulder but too close to the neck. Only a small percentage of observations with the booster seats had the shoulder belt too close to the arm.

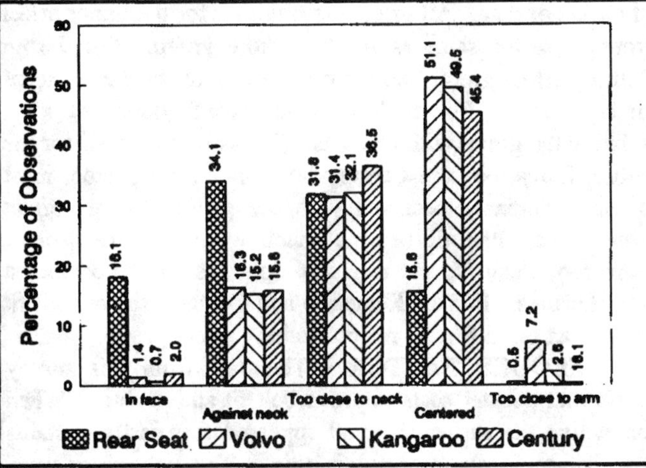

Fig. 9 -- Shoulder belt fit for rear seat alone and booster seats

Lap belt fit also drastically improves with the use of booster seats, as shown in Figure 10. Almost 40% of the observations in the rear seat alone had lap belt fits high on the abdomen. The Volvo positioned the lap belt low over the legs or pelvis 93% of the time. The other two boosters placed the lap belt low just over 70% of the time.

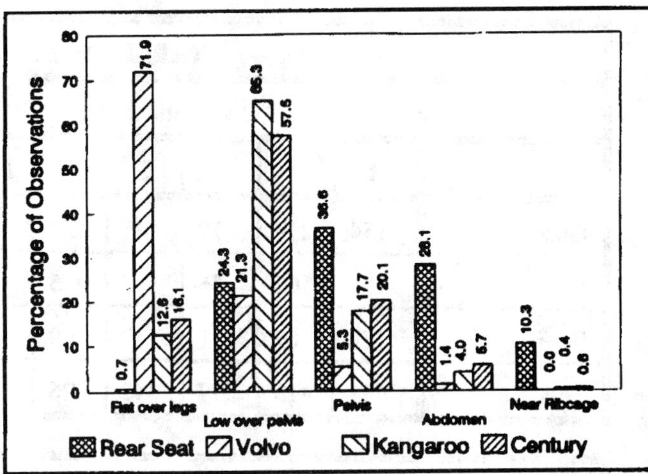

Fig. 10 -- Lap belt fit for rear seat and booster seats

Looking at the combined belt fit in Figure 11, the rear seat alone has good or fair fit only 34.1% of the time. The percentage of good/fair fit increases to around 80% for all three booster seats. The combined posture ratings by seat appear in Table 12. The rear seat postures were almost evenly split between the good, fair, and poor classes. The Volvo had the best posture, allowing only 17% to be considered poor. The Century also did well, with only 14% poor posture, although the mix of good vs. fair was not quite as favorable as with the Volvo. Contrary to the other boosters, the Kangaroo did not generally improve posture. Its high cushion depth raised more volunteers above an acceptable sitting height. The soft foam of the seat did not the support the legs as well, which caused more thigh angles to fall below horizontal.

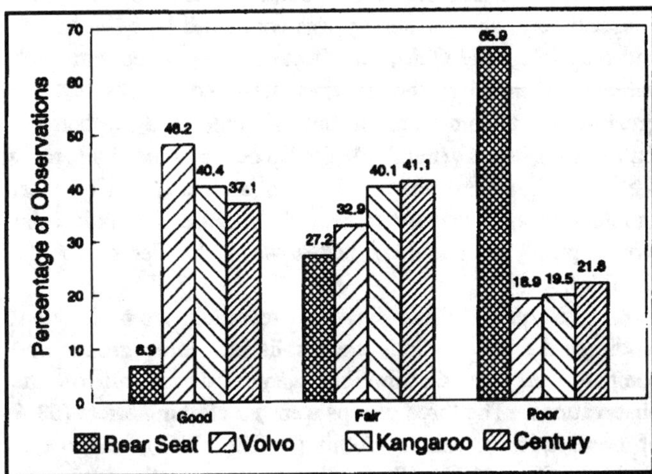

Fig. 11 -- Combined belt fit for rear seats and booster seats

The final step in the seat comparison analysis is to look at the overall fit, shown in Table 13. The rear seat allowed reasonable fit only 19% of the time. The Volvo and Century offered significant increases in protection by having good or fair fit in 61% and 54% of the observations. Some of the fair and poor fits for these two boosters resulted from the volunteer being too large to sit properly in the booster. The Kangaroo improved the fit slightly over the rear seat, but neglected the needs of most of the volunteers.

TABLE 12 -- Combined Posture by Seat*												
	Observations						Percentages					
Fit	R	V	K	C	FF	FR	R	V	K	C	FF	FR
Good	154	279	70	195	21	17	35.4	67.2	25.2	56.2	80.8	89.5
Fair	154	67	36	105	5	2	35.4	16.1	13.0	30.3	19.3	10.5
Poor	127	69	171	47	0	0	29.2	16.6	61.7	13.5	0.0	0.0
Total	435	415	277	347	26	19	100.0	100.0	100.0	100.0	100.0	100.0

TABLE 13 -- Overall Fit by Seat*												
	Observations						Percentages					
	R	V	K	C	FF	FR	R	V	K	C	FF	FR
Good	20	128	21	82	1	4	4.6	30.8	7.6	23.6	3.8	21.1
Fair	63	125	34	106	12	6	14.5	30.1	12.3	30.5	46.2	31.6
Mediocre	124	119	120	101	10	8	28.5	28.7	43.3	29.1	38.5	42.1
Poor	228	43	102	58	3	1	52.4	10.4	36.8	16.7	11.5	5.3
Total	435	415	277	347	26	19	100.0	100.0	100.0	100.0	100.0	100.0

*R = Rear seat alone V = Volvo booster FF = Front seat, forward position
C = Century booster K = Kangaroo booster FR = Front seat, rear position

CLOTHING EFFECT - Since the vehicles were tested consecutively, and the survey was conducted between March and May in central Ohio, the effect of coats varied with each vehicle. Generally, the intermediate-sized Taurus had the greatest effect from coats on both shoulder and lap belt fit. Bulky coats would generally make lap belt fit worse by forcing the belt higher up the volunteer's abdomen. However, shoulder belt fit usually improved, because a bulky collar would push the shoulder belt toward the center of the shoulder.

To quantify the changes resulting from different clothing, the observations were split into three groups and compared to the distributions generated for all of the observations. The three groups were no clothing effect (63% of observations), minor clothing effect (28%), and major clothing effect (9%). Discarding the observations wearing coats was suggested, because coats add a complicating factor to the fit issue. However, some coats had an effect, while others did not, depending on how bulky they were and how they happened to be placed within the seat belt. In addition, some of the clothing effect resulted from just a bulky sweatshirt or sweater. Another suggestion was to delete the observations with any clothing effect at all. However, this would reduce the number of observations by over one-third. It also seemed to contradict the original goal of studying fit on actual children, because sometimes children wear coats or bulky clothing, and should therefore be included.

To check how clothing affected results, the distributions of five fit variables (shoulder fit, lap fit, combined belt fit, posture, and overall fit) were reviewed for the three groups, and compared to the distributions for the whole set of data. All of the distributions for the minor effect group were the same as for the whole group. For the no clothing effect group, the distributions were the same except for lap belt fit. For the Taurus, lap belt fit improved when isolating the group without coats. The same happened for the Volvo, Kangaroo, and Century boosters. As expected, most of the differences occurred with the major clothing effect group. Lap belt fit became much worse for the Volvo, Kangaroo, Century, and rear seat alone, as did the combined belt fit scores. For the Kangaroo and Century, the overall fit was not as good for the major clothing effect group.

SLOUCH FACTOR - The data from this survey shows that booster seats improve belt fit and posture. When conducting the survey, a trend appeared that partly explains how booster seats accomplish this. The volunteers were always asked to sit up as straight as possible. They generally did some of the time. After surveying a number of children, the reason why some of them were not sitting up straight became apparent. Figure 12 illustrates the phenomena termed the "slouch factor". When older children sit up straight, the widest part of their calves rests on the edge of the seat, as shown in Figure 12a. This puts pressure on their legs and causes discomfort. Instead of sitting like this, they will scoot

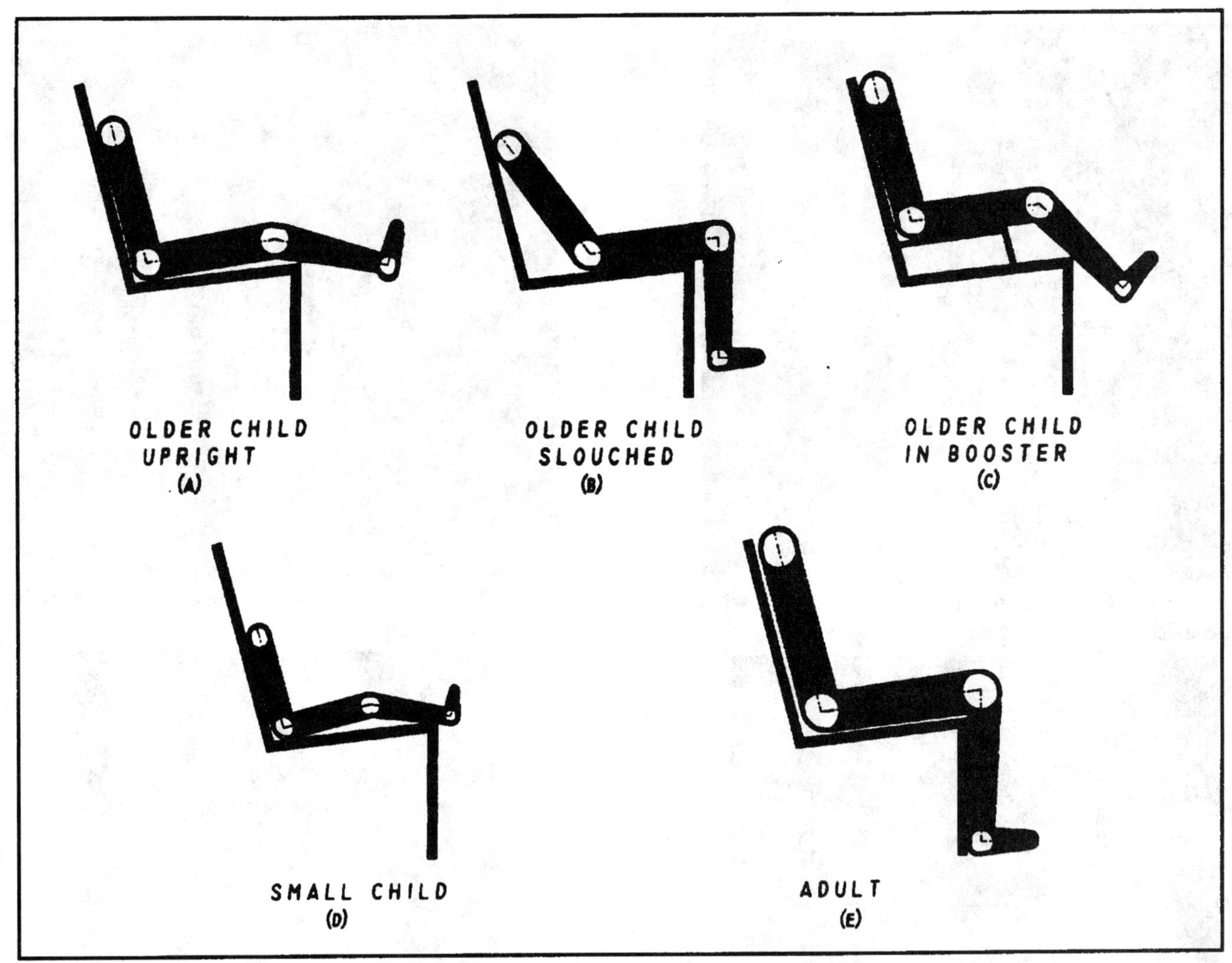

Fig. 12 -- Illustration of "slouch factor"

forward to let their legs hang more comfortably, as shown in 12b. This makes them slouch, often causing the lap belt to slide up onto their abdomens and their faces to become closer to the shoulder belt. One of the reasons why booster seats work is that they discourage this slouching. As Figure 12c illustrates, the booster lifts them up so their legs rest more comfortably on the edge of the seat. Very small children do not have this problem (Figure 12d) because they can comfortably rest their entire legs on the seat; neither do adults (Figure 12e) because their legs are long enough to reach the floor comfortably. Figures 13 and 14 show a nine-year-old child sitting in a generic automotive seat in upright and slouched positions. The child was uncomfortable when asked to sit upright, because it put pressure on the backs of her lower legs. This posture, which is normal for younger child dummies, is not one that a real older child would be able to sit in.

CHILD SIZE VS. FIT - The preceding analysis of older children showed that in general, booster seats improved fit. However, some children had good fit in the rear seat alone, and with the booster. Contrary to these children, some had poor fit in both the rear seat alone and in the boosters. Others had different combinations for both rear seats and boosters. The following section tries to define the anthropometric characteristics of the children that indicate the minimum size necessary for good fit with a three-point belt, and the maximum size for using or benefitting from booster seats.

Booster seat manufacturers primarily define the sizes of children that could use a particular seat by setting weight ranges, and sometimes height ranges. The two children pictured in Figures 15 and 16 show that using weight alone is an oversimplification. Both of these volunteers weigh 36 kg, but have a height difference of 15 cm, and a sitting height difference of 7.6 cm. The shorter child had 9 out of 11 poor combined belt fits, while the taller girl had only 1 out of 10 poor belt fits. In addition, the shorter girl had more uncomfortable fits in the boosters than the taller one. Weight, height, and sitting height must all be considered when looking at relationships between size and belt fit.

A first step in this analysis was to look at the relationship between sitting height and belt fit. Figure 17 shows the percentage of good and poor lap belt fits for the rear seats and booster seats separately. The boosters generally

Fig. 13 -- Child sitting upright with uncomfortable leg position

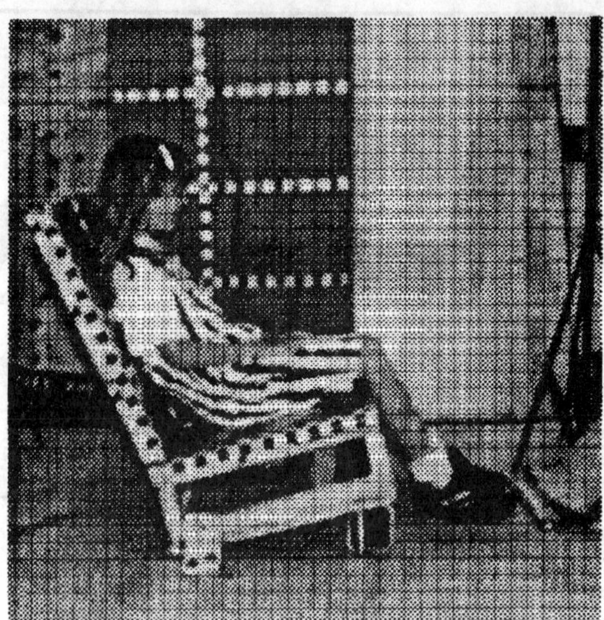

Fig. 14 -- Child seated slouched, with legs in comfortable position

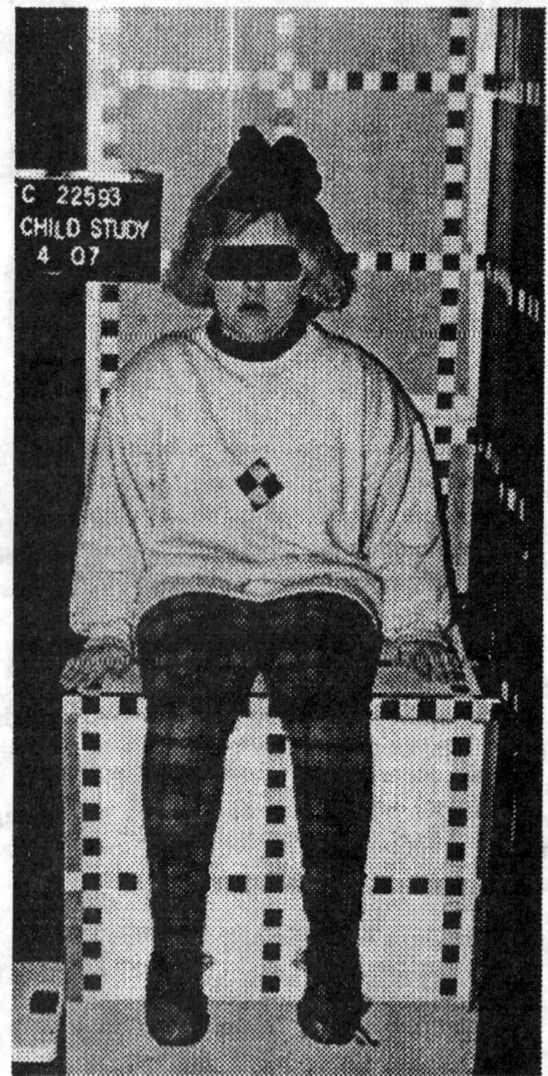

Fig. 15 -- Child who weighs 36 kg, stands 128 cm tall, and had 9/11 poor belt fits

Fig. 16 -- Child who weighs 36 kg, stands 145 cm tall, and has 2/9 poor belt fits

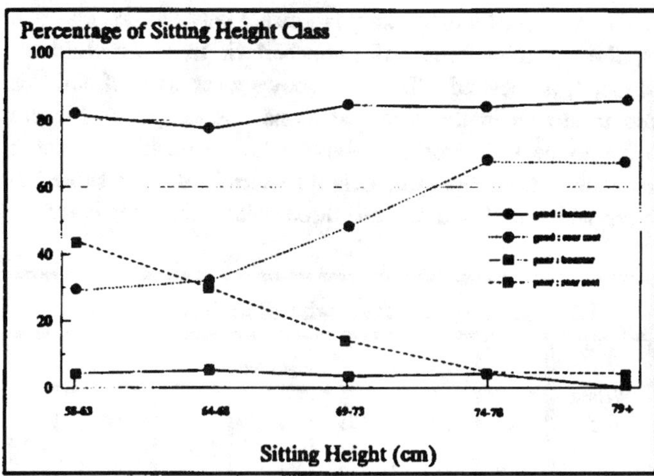

Fig. 17 -- Good and poor lap belt fits for boosters and rear seats by sitting height

have the same rates of good and poor lap belt fit regardless of sitting height. Over 80% are good, and fewer than 10% are poor for every sitting height. This indicates that most of the children seated in boosters had good lap belt fit regardless of sitting height.

The rear seat observations are much different. The poor lap belt fits do not fall below 10% until a sitting height of 74-78 cm is reached; the taller sitting height classes have about 70% good lap belt fit. For the 69-73 cm sitting height class, the poor lap belt fits are about 15%, but the good lap belt fits are only about 50%, indicating that children with these sitting heights can probably not use the belt alone and get a good lap belt fit.

For the shoulder belt fit shown in Figure 18, the differences between rear seat and booster seat for shoulder belt are not as large. For almost every sitting height class, the percentage good with the booster is higher than the percentage good for the rear seat, and the percentage poor with the booster is less than the percentage poor with the rear seat. Until the 74-78 cm sitting height class is reached, the poor percentages are much higher than the good percentages. At that sitting height and above, the levels of good and poor are comparable.

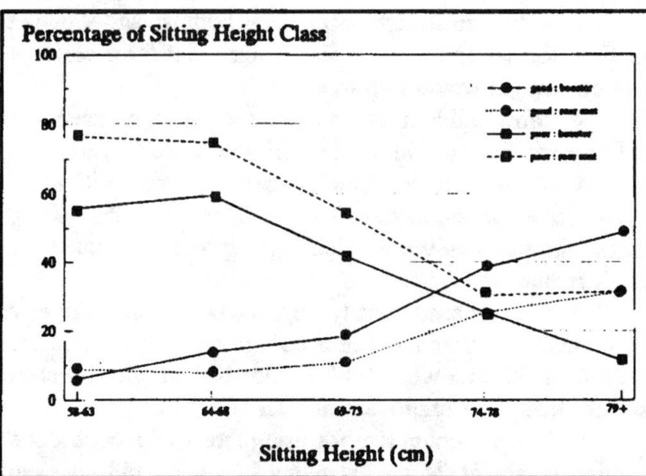

Fig. 18 -- Good and poor shoulder belt fits for boosters and rear seats by sitting height

When considering both lap and shoulder belt fit, children with sitting heights 73 cm and below cannot use three-point belts effectively, while those above 73 cm can do so some of the time. It appears that all sitting heights can benefit from boosters. However, the observations here include children who benefitted from the booster, even though the booster fit too tightly or improperly. Children who benefit from boosters are not the same as those who fit in boosters.

To look at size vs. fit a different way, the children were reviewed according to height/weight categories and combined belt fit. To try to determine the smallest-sized child that can effectively use a three-point belt, the three rear seat entries for each volunteer were examined. If they had poor combined belt fit two or three times in the rear seat, they were considered poor. If they had only one poor score, they were considered fair. If they had no poor scores they were considered good. Then, the height/weight class chart was used to map out how many children in each particular height/weight class had overall poor combined belt fit. The results appear in Table 14. Each cell contains the number of children with overall poor belt fit, and the number of children in that particular height/weight class. The boxes are shaded where the majority of the height/weight class was considered poor. A height of 148 cm seems to be the threshold above which poor belt fit is not a problem. A weight of 36 kg might be considered a threshold if the chubbiest children (the shortest for a given weight) are ignored. The table does show that the poor belt fits occur most often with the chubbiest children.

Table 15 contains the number of children whose overall combined belt fit was better with booster seats than with belts alone. This table therefore includes those who benefit from boosters, though they may not necessarily fit in the boosters very well. In a manner similar to that used with the three rear seat entries, if 6 or more out of 8 belt fits with the booster were good, the overall booster score was good; if fewer than 3 out of 8 belt fits with the booster were good, the overall booster score was poor; the remaining combinations were considered fair. The shaded squares indicate that a majority of children in that particular height/weight class had improved fit with a booster. Children who had the same fit in belts and boosters (good/good, fair/fair, and poor/poor) are not included. The large number of shaded squares indicates that boosters improve belt fit for almost all height/weight combinations. The smallest children did not benefit from belt-positioning boosters; the results were poor with and without the boosters because of their small sizes.

As mentioned previously, children who benefit from boosters and those who fit in boosters are not necessarily the same children. Table 16 tries to illustrate the difference. This table includes children who had overall good or fair combined belt fits with the boosters, and had comfortable or snug fits in the boosters. Few children over 36 kg were able to fit well in the boosters, so a threshold is evident. However, since the previous table indicates that children over 36 kg could benefit from boosters, it may be advantageous for booster seat manufacturers to build wider seats that could accommodate and help these heavier children.

TABLE 14 -- Children with Overall Poor Combined Belt Fit in Rear Seat Alone

Weight (kg)	Height (cm)						
	112 - 117	118 - 124	125 - 132	133 - 140	141 - 147	148 - 155	156 - 168
< 23	5/5	12/13	2/3	\\\\\	\\\\\	\\\\\	\\\\\
24-27	\\\\\	5/6	14/18	4/6	\\\\\	\\\\\	\\\\\
28-32	1/1	1/1	9/9	11/17	6/11	\\\\\	\\\\\
33-36	\\\\\	\\\\\	2/2	7/7	3/8	1/2	\\\\\
37-41	\\\\\	\\\\\	2/2	1/3	0/4	2/7	\\\\\
42-50	\\\\\	\\\\\	\\\\\	1/2	4/7	3/7	0/1
51 +	\\\\\	\\\\\	\\\\\	\\\\\	2/2	2/5	0/6

TABLE 15 -- Combined Belt Fit with Boosters Better than Combined Belt Fit in Rear Seat

Weight (kg)	Height (cm)						
	112 - 117	118 - 124	125 - 132	133 - 140	141 - 147	148 - 155	156 - 168
< 23	1/5	11/13	3/3	\\\\\	\\\\\	\\\\\	\\\\\
24-27	\\\\\	4/6	15/18	5/6	\\\\\	\\\\\	\\\\\
28-32	0/1	1/1	9/9	13/17	9/11	\\\\\	\\\\\
33-36	\\\\\	\\\\\	1/2	6/7	6/8	1/2	\\\\\
37-41	\\\\\	\\\\\	2/2	3/3	1/4	4/7	\\\\\
42-50	\\\\\	\\\\\	\\\\\	1/2	5/7	6/7	1/1
51 +	\\\\\	\\\\\	\\\\\	\\\\\	1/2	3/5	0/6

TABLE 16 -- Boosters Fit and Combined Belt Fit Good or Fair

Weight (kg)	Height (cm)						
	112 - 117	118 - 124	125 - 132	133 - 140	141 - 147	148 - 155	156 - 168
< 23	1/5	11/13	3/3	\\\\\	\\\\\	\\\\\	\\\\\
24-27	\\\\\	4/6	15/18	2/6	\\\\\	\\\\\	\\\\\
28-32	0/1	1/1	9/9	17/17	11/11	\\\\\	\\\\\
33-36	\\\\\	\\\\\	1/2	5/7	7/8	2/2	\\\\\
37-41	\\\\\	\\\\\	1/2	2/3	2/4	2/7	\\\\\
42-50	\\\\\	\\\\\	\\\\\	1/2	0/7	0/7	0/1
51 +	\\\\\	\\\\\	\\\\\	\\\\\	0/2	0/5	0/6

As a final look at the relationship between size and belt fit, the apparent cause of poor belt fit in older children, "slouch", is recalled. Table 17 shows what sizes of children tend to slouch in the rear seat alone. A height of 147 cm seems to be the threshold above which slouching is not a problem. This is the same height (determined from Table 14) above which children can use three-point belts effectively.

TABLE 17 -- Children who Slouch in Rear Seat

Weight (kg)	Height (cm)						
	112 - 117	118 - 124	125 - 132	133 - 140	141 - 147	148 - 155	156 - 168
< 23	4/5	11/13	3/3	\\\\\	\\\\\	\\\\\	\\\\\
24-27	\\\\\	5/6	16/18	6/6	\\\\\	\\\\\	\\\\\
28-32	0/1	0/1	8/9	15/17	7/11	\\\\\	\\\\\
33-36	\\\\\	\\\\\	2/2	6/7	7/8	1/2	\\\\\
37-41	\\\\\	\\\\\	1/2	3/3	2/4	1/7	\\\\\
42-50	\\\\\	\\\\\	\\\\\	1/2	3/7	1/6	0/1
51 +	\\\\\	\\\\\	\\\\\	\\\\\	1/2	3/5	0/6

CHILD SURVEY DISCUSSION - The anthropometry and belt fit studies of 155 older children have illustrated a great deal about the automotive environment for older children. The following list summarizes the major findings:

- For a given height, children today seem heavier than they were twenty years ago. The difference may partly result from different clothing, but a more extensive anthropometry study of older children may be appropriate to confirm this.
- The Caravan had the worst shoulder belt fits in the rear seat alone and with the boosters, mostly resulting from the large distance to the shoulder belt anchor in the x-z plane. The Sunbird had the best belt fits in the rear seat and boosters.
- Sitting up straighter improves both lap and shoulder belt fit. Booster and vehicle seat designs that force children to sit up straight would improve belt fit.
- Most children who met the weight requirements of the Kangaroo booster did not fit well within the booster.
- The "slouch factor", where children will scoot forward to allow comfortable leg positions rather than sit up straight and put pressure on their legs, appears to cause poor belt fit frequently.
- Booster seats greatly improve belt fit over the rear seat alone, going from two-thirds poor to one-fifth. However, the 20% of children who still had poor belt fit with boosters indicates that more improvements can be made.
- The minimum size for using three-point belts alone is sitting height of 74 cm, standing height of 148 cm, and weight of 37 kg. A child with the height requirements but a smaller weight would also probably have an acceptable belt fit,

while a child with the weight requirement but not the height requirements may not have an acceptable belt fit with the belt alone.

- The majority of children in this study had better belt fit with the boosters than with the rear seat alone, regardless of size.
- Children who could fit well in boosters and had good/fair belt fits were generally 36 kg or less. Weight seems the most important factor, since their sitting heights varied from 58 to 76 cm, and their standing heights ranged from 117 to 152 cm.
- Children who weigh more than the recommended weight can still have improved belt fit with belt-positioning booster seats. This conclusion is based only on static belt fit tests and may not be true in dynamic testing.

ACKNOWLEDGEMENTS

The authors would like to thank the faculty and students of Edgewood Elementary School for participating in this survey. We also wish to recognize Gerda England, Herman Jooss, and Susie Weiser for their efforts in assembling this paper.

REFERENCES

1. Klinich, K. D., Pritz, H. B., Beebe, M. S., Welty, K., Burton, R. W., NASS Injury Analysis and Restraint/Booster Fit Survey of Older Children, National Highway Traffic Safety Administration, publication pending.

2. Snyder, R.G., Schneider, L.W., Owings, C.L., Reynolds, H.M., Golomb, D.H., Schork, M.A., Anthropometry of Infants, Children, and Youths to Age 18 for Product Safety Design, Highway Safety Research Institute, The University of Michigan, UM-HSRI-77-17, May 1977.

3. Hamill, P.V.V., Drizd, T.A., Johnson, D.L., Reed R.B., Roche, A.F., and Moore, W.M., "Physical growth: National Center for Health Statistics Percentiles," AMJ CLIN NUTR 32:607-627, 1979.

4. MacLaughlin, T.F., Sullivan, L.K., and O'Connor, C.S., "Rear Seat Submarining Investigation," National Highway Traffic Safety Administration, DOT HS-807-347, May 1988.

ADULT SEAT BELTS: HOW SAFE ARE THEY FOR CHILDREN?

Michael Henderson
Michael Henderson Research

Julie Brown
Michael Griffiths
Roads and Traffic Authority
Australia

Paper Number 96-S7-O-04

ABSTRACT

Investigation of crashes involving 121 children aged up to 14 in adult three-point lap/shoulder (lap/sash) belts showed that irrespective of age they were generally well protected even in severe frontal crashes, and none sustained belt-induced inertial neck injury. The prime cause of injury among these children was contact with the interior surfaces of the car, predominantly in side impacts. Lap-belted children sustained a higher proportion of belt-induced abdominal injuries and a similar proportion of head injuries despite mostly being seated in centre positions away from the side of the car. Sled tests with 18-month, three-year-old and six-year-old dummies produced data consistent with the conclusion that adding torso restraint slightly increases the risk of minor (AIS 1 or 2) neck injury, but has the major benefit of reducing the risk of serious head and abdominal injuries. The conclusion of this work is that adult lap/shoulder belts do not present a significant risk of injury to young children.

INTRODUCTION

A point of interest for many years has been the extent to which children may be placed at risk by using adult belts, because of the incompatibility of the size and shape of the typical child with the geometry of the typical seat-belt installation.

In an early study of restrained children by the Traffic Accident Research Unit in New South Wales (Vazey, 1977), a selection of reasonably severe crashes involving 65 case occupants were examined to provide evidence that would address the question posed by Snyder and O'Neill (1975): *"Are 1974/1975 automotive belt systems hazardous to children?"*. At that time in the United States, and to a considerable extent even now in that country, few children were restrained in adult belts that had shoulder belts incorporated. In Australia, however, lap/shoulder belts had been required in outboard positions both front and rear since 1971, quite independently of mandatory-wearing requirements.

Snyder and O'Neill had noted that in Australia the original legislation requiring the use of seat belts did not apply to children under eight years, but they wrongly concluded that the reason was concern for the safety of children using adult belts. In fact, at least in NSW it was for administrative reasons associated with the age of legal responsibility, and soon all passengers over the age of one year had to be restrained in a seat belt, including an adult lap/shoulder belt for children if that was the only restraint available.

The conclusion of the 1977 NSW study and those that succeeded it (Corben and Herbert, 1981) was that in practice children appeared to be afforded good protection by adult three-point lap/shoulder (lap/sash) belts, even down to two years of age, as long as the restraint was properly adjusted. At that time few seat belts in the rear positions had automatically adjusting and locking retractors, whereas modern cars are now so equipped. When firmly restrained in well-adjusted belts, the children were found to withstand crash forces as well or better when wearing adult restraints than adults in the same car, even in crashes of 50 km/h change of velocity (delta-V).

The importance of this question has diminished over recent years with the ever-increasing availability and use of dedicated child restraints (including booster seats to be used with adult belts) that are much more appropriate for different ages and sizes of children. Nevertheless, many children still use adult belts alone, and clearly in many cases seat-belt restraint is not optimal for small occupants.

To review issues of child occupant protection, a field crash investigation study undertaken through 1993 was aimed at studying crashes involving vehicle passengers aged 14 years or under.*

Table 1.
Summary of Results: NSW Study, All Restraints

	Child restraint		Lap/ shoulder belt		Lap-only belt		No restraint	
MAIS	N	%	N	%	N	%	N	%
0	28	39.4	11	9.1	5	14.3	0	0.0
1	30	42.3	83	68.6	21	60.0	6	31.6
2	7	9.9	13	10.7	3	8.6	3	15.8
3	1	1.4	7	5.8	2	5.7	5	26.3
4/5	2	2.8	1	0.8	1	2.9	0	0.0
6	3	4.2	6	5.0	3	8.6	5	26.3
Total	71	100.0	121	100.0	35	100.0	19	100.0
MAIS 2+	13	18.3	27	22.3	9	25.7	13	68.4

Among the main research questions were these:

- What injuries, especially to the neck, were occurring despite the use of restraints?
- How was restraint effectiveness related to the age of the children?
- How might effectiveness be improved?
- What were the effects of misuse, if any?
- In very severe crashes, which children were surviving, and why? Did their survival give any guidance on the tolerance to injury of children?

The study aim was to include for analysis the following cases:

- all fatally injured child occupants from throughout the State;
- any child occupant involved in a crash in which another occupant was killed anywhere in the State;
- and all child occupants presenting (but not necessarily admitted) to hospitals accepting trauma patients in the greater metropolitan Sydney area.

A summary of the results of this study is shown in Table 1.

It can be seen that throughout the entire sample of children, in all kinds of crashes, dedicated child restraints generally performed the best, followed by adult lap/shoulder belts and then by lap-only belts. As would be expected, children without restraints fared badly. Although this was not a random sample of crashes, the difference in injury risk between restraint and no restraint is of the same order of magnitude as shown in studies based on comprehensive statistical data.

Some of the results of the New South Wales field study in relation to dedicated child restraints have been reported previously (Henderson *et al*, 1994). The present paper focuses on the use of adult belts by children in the study. It describes the case histories of the most severely-injured children and also those who were not injured in severe crashes. The paper further adds data from a comparative series of laboratory sled tests using new-model child anthropomorphic dummies in adult lap/shoulder and lap-only seat belts. The dummies were made by First Technology Safety Systems. They represented children aged 18 months (CRABI dummy), three years and six years of age (Hybrid III).

THE FIELD STUDY: FACTORS RELATED TO INJURY SEVERITY

Belt Types in Use

It will be seen from Table 1 that in the sample of children aged 14 years or under entering into the Australian study, easily the commonest single type of restraint used was the lap/shoulder seat belt, without any kind of booster seat or cushion. This is not surprising, because the age range covered by this kind of restraint is much wider than that for child seats and infant capsules. There were 121 children in the study wearing lap/shoulder belts.

Thirty-five children were using lap-only belts. Since January 1971 all outboard seating positions in passenger cars sold in Australia have been required to be equipped with lap/shoulder seat belts. Thus, the only lap-only belts in the present sample were worn by children in centre-rear (23) and centre-front (4) positions, plus eight children lap-belted in third-row seats in multi-passenger vehicles.

* The study was performed by Michael Henderson Research and Roads and Traffic Authority personnel, in association with the Child Accident Prevention Foundation of Australia (CAPFA).

Table 2 summarises the maximum AIS (MAIS) score for children in the sample. It shows a lower proportion of MAIS 3-6 injuries among children using lap/shoulder belts rather than lap-only belts, but the difference is not statistically significant. The average MAIS for children in lap/shoulder belts was 1.405 (SD 1.262) and in lap belts 1.571 (SD 1.614).

Table 2.
Maximum AIS by type of seat belt

MAIS	0-2		3-6		Total
	N	%	N	%	N
Lap/shoulder belt	107	88.4	14	11.6	121
Lap belt	29	82.9	6	17.1	35
Total	136		20		156

Age and Injury Severity

Table 3 shows that six (5.0%) of the children using lap/shoulder belts were killed (MAIS 6), 21 (17.4%) suffered injuries with a maximum AIS of 2-4, and the majority (94, 77.7%) had injuries of AIS 1 or were uninjured.

Table 3.
Lap/shoulder Belts, Maximum AIS by Age of Child

Age (years)	MAIS 0	1	2	3	4/5	6	Total
< 3	1	5	0	0	0	1	7
4 - 6	1	15	2	1	0	1	20
7 - 9	6	21	4	4	0	1	36
10-14	3	42	7	2	1	3	58
Total	11	83	13	7	1	6	121

The ages of the 121 children using adult lap/shoulder belts ranged from one year to 14 (the sample maximum), with a mean age of nine. Table 3 indicates no relationship between age and injury as represented by maximum AIS. This is confirmed by correlation analysis (Pearson correlation r=0.003). This lack of correlation remains if the analysis is restricted to generally-frontal crashes with a principal direction of force between 10:00 and 2:00, and also if only crashes with a delta-V of over 45 km/h are analysed.

For 45 frontal crashes with known delta-V, there was also no correlation between age and the delta-V (r=-0.079). In other words, there was no suggestion that the younger the child, the lower the impact speed at which injury occurred.

Of the 121 children in lap/shoulder belts, 27 (22.3%) were aged six years or less, and would probably have been better served by a child restraint or a booster seat in combination with the adult belt. With such appropriate restraint, they might well have received no injuries at all. However, for this sample at least, they appear to have received as much benefit from the adult belts as older children.

The ages of the 35 children wearing lap-only belts ranged from two years to 14, with a mean of between eight and nine (see Table 4). Thirteen were aged two to six years, indicating that they were very small to have been restrained in any kind of adult belt. Relating maximum AIS to age, two of the 11 children (18%) aged five or under received MAIS 2+ injuries, and seven of the 24 (29%) aged six and over. The cell sizes are too small to read any significance into these figures.

Table 4.
Lap Belts, Maximum AIS by Age of Child

Age (years)	MAIS 0	1	2	3	4/5	6	Total
< 3	1	2	1	0	0	0	4
4 - 6	2	5	1	0	0	1	9
7 - 9	1	4	0	0	0	1	6
10-14	1	10	1	2	1	1	16
Total	5	21	3	2	1	3	35

Injury Severity by Body Region

Table 5 shows injuries by severity and body region for lap/shoulder belts, and Table 6 the same for lap belts. Many children sustained more than one injury, but only the two most severe injuries for each case have been included. For each kind of seat belt, the percentage of all children who sustained an injury to each body region is shown.

Thus, Table 5 shows that 16.5% of children wearing lap/shoulder belts sustained a head injury and Table 6 that 22.9% of those using lap-only belts did so, but that the majority of injuries were minor. There was a higher incidence of AIS 1 thoracic and neck injuries among lap/shoulder belt users, almost all being bruising from belt loads. However, shoulder belt loads did not result in any more than trivial injuries.

Table 5.
Two Most Severe Injuries by Body Region, Lap/shoulder Belts

AIS	1	2	3	4/5	6	Total	% in cases
Head	13	2	3		2	20	16.5
Face	23	3				26	21.5
Neck	12					13	10.7
Thorax	11		1	1	1	14	11.6
Abdomen	9	3	1			13	10.7
Spine	1	1			1	3	2.5
Extremities	19	12	1			32	26.4
External	31	1	1		3	36	29.7
Total injuries						156	
Total cases						121	

Table 6.
Two Most Severe Injuries by Body Region, Lap Belts

AIS	1	2	3	4/5	6	Total	% in cases
Head	5	1	1		1	8	22.9
Face	6		1			7	20.0
Thorax	1				1	2	5.7
Abdomen	7	2				9	25.7
Spine	1		1		1	3	8.6
Extremities	5					5	14.3
External	5	1		1	1	8	22.9
Total injuries						42	
Total cases						35	

There was a higher incidence of spinal and abdominal injuries among wearers of lap belts, although the numbers are too small to draw firm conclusions.

Crash Severity

In the case of lap/shoulder-belted children, for generally-frontal crashes with a principal direction of force between 10:00 and 2:00, the delta-V* ranged from 14 to 89 km/h, mean 44.5 km/h, SD 19.8 km/h. For these crashes, the delta-V was positively and significantly related to the MAIS for children in lap/shoulder belts ($r=0.438$, $p=0.003$).

For frontal crashes involving lap-belted children, the delta-V ranged from 11 to 77 km/h, with a mean of 44.8 km/h, SD 21.1 km/h. There was also a relationship between delta-V and MAIS, but it was weaker ($r=0.207$) and non-significant.

Seating Position

Those wearing lap/shoulder belts were evenly distributed through front passenger and the two rear outboard seating positions. For the whole sample, there was no significant relationship between MAIS and whether the child was seated in the front left position or in either of the two rear outboard positions.

Four children of the 35 wearing lap-only belts were in centre front seats, 23 in the centre rear, and eight children were in the third row of multi-passenger vehicles (six of these in outboard positions).

As would be expected, the risk of injury in side impacts is strongly related to the seating position. There were 14 children in lap/shoulder belts in side impacts who sustained injuries of MAIS 2 or more. Eleven of these children were sitting on the side of the impact, and only three on the other side. This difference is statistically significant ($p<0.05$).

Contact Points

Contact points (where identified) for the two principal injuries among children restrained by lap/shoulder belts are shown in Table 7 and for lap belts

---

* The delta-V was calculated following measurements of vehicle crush and included scene data when available. Calculations were performed with the aid of the EDVAP vehicle analysis package (Engineering Dynamics Corporation, 1989).

in Table 8. Again, because of the rather small numbers of lap belted children in the sample, comparisons should be made with caution.

Table 7.
Contact Points for Two Most Severe Injuries, Lap/shoulder Belts

Body region	Point of contact						
	Seat belt	Wind-screen or pillar	Dash	Door or win-dow	Car seat	Glass	External to car
Head		3	2	4	2	1	1
Face		2	5	9	1	3	
Neck	5	1		1			
Thorax	10		1		1		
Abdomen	10			2			
Spine		1					
Extrem's	2		2	9	8	3	2
Totals	27	7	10	25	12	7	

Table 8.
Contact Points for Two Most Severe Injuries, Lap Belts

Body region	Point of contact						
	Seat belt	Con-sole	Dash	Door or win-dow	Car seat	Roof	External to car
Head		1			4	1	1
Face		1		1	3		
Neck							
Thorax	1			1			
Abdomen	9						
Spine		1			1		1
Extrem's		2	1	1			
Totals	10	5	1	3	8	1	2

However, 20% of the lap-belted children (seven out of 35) sustained head contacts as opposed to about 11% (13 out of 121) of the lap/shoulder-belted children. Also, there was a higher proportion of belt-induced abdominal injuries among those using lap belts (nine out of 35, 26%) rather than lap/shoulder belts (12 out of 121, 10%).

CASE HISTORIES

Fatalities in Lap/shoulder Belts

The youngest child to be killed when wearing a lap/shoulder seat belt was riding in the front seat of a 1982 Toyota Cressida that ran heavily into the side of an oncoming out-of-control Chrysler Sigma on a country road, and after the crash ended up on its side. The Sigma caught fire as a result of the crash, and the fire spread in part to the Cressida. The child was a boy aged three years and eleven months. The delta-V was some 65 to 70 km/h. The child suffered AIS 6 internal thoracic and abdominal injuries, plus a fractured cervical spine at C1/C2. The female driver of the car was also fatally injured, with a fractured skull and internal abdominal injuries.

There was considerable distortion rearwards in the region of the A pillar on the passenger side. The child's head probably hit the windscreen, which had a bloodstained contact point. There were also contact marks on the lid of the glovebox. The seat belt showed evidence of heavy loading, unusual with a light child, and it is likely that the internal injuries were inflicted by the belt, possibly as he was partly ejected out of it. The crash was survivable in the absence of contact with internal surfaces, as shown by the fact that a five-year-old girl restrained by a lap/shoulder belt in the rear left seat suffered only minor concussion and belt bruising, and a nine-year-old girl similarly restrained in the right rear seat received only belt bruises.

This crash was the only one in which a child was killed in a frontal crash while wearing an adult three-point belt.

Four of the six crashes in which lap/shoulder-belted children were killed were side impacts associated with intrusion on the side the child was sitting, and the other was a rollover. In two of the side-impact crashes, the intrusion was very extensive and the crash probably unsurvivable in the child's seating position.

In the first of these two, an eleven-year-old girl was in the left rear seat of a 1981 Toyota Landcruiser. The car drifted off the left side of a country highway, came abruptly back on to the road surface and spun into the path of an oncoming van which hit at about 60 km/h. A

70-year-old female sitting next to her in the centre rear position was also killed, as was the adult male in the front passenger seat. A nine-year-old girl in the right rear seat survived with internal injuries. In the other case involving gross intrusion, an out-of-control utility truck impacted the left side of an oncoming Toyota Corolla at over 60 km/h. In the left rear seat was a 13-year-old boy who was killed immediately as a result of multiple skull fractures and brain tissue disruption.

Another death resulting from intrusion, following a side-swipe collision with a tree, was that of a 13-year-old boy who was sitting in the right rear seat of a 1989 Holden Commodore. Although the intrusion was not severe, the collision was at high speed and the impact sufficient to cause fatal chest injuries. Rather similar damage was caused to a 1971 Volkswagen 1500 Fastback sedan in a side-swipe collision with an oncoming panel van. A girl aged seven was seated in the right rear position, adjacent to the maximum point of intrusion, and she died from very severe head injuries. In this case there was conflicting evidence about belt wearing, as the car was carrying six people plus the driver, and the age of the vehicle made any loading marks on the dirt-impregnated seat belts impossible to detect. In any event, the configuration of the crash made that seating position probably unsurvivable, whether a seat belt was worn or not.

The only other death among children restrained by lap/shoulder belts occurred to a five-year-old boy who was ejected from the lap/shoulder belt in the right rear seat of a 1991 Toyota Landcruiser. Sadly, the child had shortly beforehand been using a booster seat in association with the restraint, but the booster was taken away to allow him to lie down on the rear seat, with effectively only the lap part of the restraint holding him in place. This was insufficient when the Landcruiser rolled after hitting a roadside rock face on a country highway, and he was ejected through the rear window of the vehicle. Two adults in the car, plus an eighteen-month-old girl in a child seat, all survived. The roof of the vehicle was distorted to some extent, but survival space was generally sufficient for all occupants.

Children Aged One to Four Years in High-speed Impacts

Although they would have been more appropriately seated in dedicated child restraints, some very small children using lap/shoulder belts in high-speed crashes nevertheless received only minor injuries.

The youngest was a one-year-old boy in the left rear position of a 1981 Ford Cortina that rolled in a single-vehicle crash on an open country road. There was extensive external damage to the car, and some distortion of the roof. However, both the one-year-old boy and his three-year-old sister, who was restrained in the right rear seat also with a lap/shoulder belt, escaped with minor bruising. A female adult sitting between them, restrained by a lap-only belt, died as a result of fracture dislocation of the cervical spine and disruption of the spinal cord, most probably as result of contact with the roof of the rolling car.

A girl aged two and a half years was in the left rear seat of a 1990 Daihatsu Charade that crashed head-on with a Nissan Patrol four-wheel drive. The delta-V was about 55 km/h. She suffered soft tissue neck injury that required admission for exclusion of cord injury, plus belt bruising on the left shoulder and both hips, but no more serious injuries. The restrained female driver sustained lacerations to the head and left knee, and a small child in a forward-facing child restraint was unhurt.

A slightly older child, a girl aged four survived a high-speed head-on crash in a Mitsubishi Sigma, with a delta-V of about 65 km/h, while seated in the front passenger seat. She received belt bruising only. The female adult driver suffered fractured facial and leg bones.

Children Aged Four to 14 Years in High-speed Impacts

Some older children also survived very destructive crashes in lap/shoulder belts. For example, there were two 10-year-old girls riding in a Mitsubishi Colt hatchback that came in to head-on collision on a country road with an oncoming Mazda RX7.

The delta-V was 65 to 70 km/h, and the driver of the Colt died from unsurvivable (AIS 6) chest and brain injuries. The child in the front passenger seat suffered no more than belt bruises, leaving loading marks on the webbing. The child in the rear seat did suffer abdominal injuries, but was discharged from hospital within three weeks.

A Toyota Tarago (Previa) multi-passenger vehicle was involved in a head-on collision with a truck at a delta-V of at least 60 km/h together with very extensive damage. In the third row of seats were riding, in the outboard positions, an eight-year-old girl and a ten-year-old girl. Both were wearing the available lap/shoulder restraints. Also in the vehicle were three adults and a three-year-old boy in a child seat with harness. All three adults were killed, and all three children survived.

An older-style 1985 model Toyota Tarago drifted across to the wrong side of a country highway and hit a tree in the dead centre of the front of the vehicle. The delta-V was in the order of 60 km/h. The male driver suffered head injuries and fractured limbs, and the female front-seat passenger also received lower leg fractures. The vehicle was filled with eight children, in addition to the two adults in the front seats. Some of these children were unrestrained and lying on the floor, but four were wearing lap/shoulder seat belts. All suffered bruising from belt loading, and loading marks were apparent on the webbing. In addition, three of the children sitting in the outboard positions in the second and third row of seats - a male aged 13, another male aged 13, and a female aged 13 - received fractures to the limbs adjacent to the interior.

In addition, a female aged 13 sitting in the left outboard seat in the third row suffered a traumatic amputation of her left arm. The mechanism of this injury was not clear. The occupants in this crash survived the high deceleration loads but were injured by contact with the generally unyielding interior of this particular vehicle.

In yet another high-speed crash involving a carload full of children, a 1989 Toyota Landcruiser carrying two adults and five children slid off a dirt road and hit a tree at an angle that drove it into the left side of the car from the front left towards the vehicle's centre. The male adult in the front passenger seat was fatally injured, receiving chest and other critical injuries from the intruding tree. The female driver was injured by contact with the steering wheel and surrounding components. A 12-year-old female wearing a lap/shoulder belt was seated behind the front-seat passenger, and her pelvis was fractured from contact with the vehicle interior. A 16-year-old wearing a lap/shoulder belt in the right seat behind the driver was not significantly injured, with abrasions only. The remaining children were wearing lap-only belts, and will be described in the next section of this paper.

The only lap/shoulder belt-induced injury more than minor (AIS 1) was an AIS 2 haematoma of the liver received by an eleven-year-old male in the left rear seat of a 1972 Holden Torana. This ageing compact car did not have a locking retractor reel for the seat belt, and it is likely that the heavy belt loadings the boy received were as a result of wearing the belt rather loosely. The impact was dead centre front into a telegraph pole, with a delta-V in the order of 45 km/h.

As already noted, significant injury in the sample as a whole was more likely to be associated with side impacts than other configurations of crash. However, children wearing lap/shoulder belts survived some serious side impacts when seated away from the impacted side. Having gone through a red light, a 1990 Ford Falcon was hit heavily on the driver's side (right) door by a bus at a crossroad, and the driver was killed. The delta-V for the Falcon was around 50 km/h. A 14-year-old boy and a 10-year-old girl were in the front and rear left seats respectively, and suffered minor injuries only.

On the other hand, sitting adjacent to the impacted side in a pure side-impact collision was far more likely to result in injury. There were 14 seat-belted children injured (maximum AIS 2 or more, including fatals) in side impact crashes; eleven of the 14 were seated on the side of the main impact.

Lap-only Seat Belts

There were three fatalities among lap-belted children.

One of the children killed was a girl aged five and a half restrained in the centre rear seating position of a 1986 Toyota Cressida. Coming off a suburban medium-speed highway, the car hit a telegraph pole in the centre of the front of the car, with a resulting delta-V of about 45 km/h. Vehicle examination showed the lap belt to have been adjusted so that it would have been very loose for this child. Examination also showed the console between the two front seats to have been extensively damaged. In the frontal impact, the child received a pattern of injuries that is typical of lap-belt restraint in a frontal impact at more than a slow speed. She suffered minor facial injuries and some external leg injuries, probably from contact with the console. On the front and right and left sides of her hip were bruises and abrasions typical of lap-belt loading, and she had some associated internal haemorrhage between the bladder and the pubic symphysis. She died almost immediately from a fracture-dislocation of the second and third cervical vertebrae with associated cord damage, and the detail of the injury showed that this injury was caused by distraction of the spinal column.

In brief, this child was allowed to move forwards to an excessive extent by the loose lap belt, thus allowing head and face contact with the console in front of her.

She flexed violently over the lap belt, causing the bruising in the hip region and intra-abdominal injury. In the flexed position, her head was stretching the spinal cord as a result of the forces of deceleration, and this fact - in association, probably, with the relatively insignificant head contact - caused the distraction fracture-dislocation of the neck that was the fatal injury.

This was a survivable crash, giving rise only to moderate deceleration from the change of velocity. The driver suffered some lacerations to his face as he hit the windscreen, and a 13-year-old girl in the left outboard seat, in a lap-sash belt, suffered cracked ribs and bruising from belt loads.

Another death was an eight-year-old female who was seated in the outboard left seat of the third row of a light passenger-carrying van (a 1983 Holden Shuttle). The vehicle left a country road and rolled down a six-metre embankment. The damage to the vehicle indicated that during the roll the vehicle hit heavily on its left side towards the rear, causing some intrusion directly adjacent to the deceased child's seat. The girl died from severe pulmonary contusion affecting all parts of both lungs, probably in association with the vehicle damage in the region. Another child in the same vehicle, an 11-year old male in the outboard right seat in the third row, also wearing a lap belt, also suffered chest injuries that were not fatal.

In a crash that was essentially unsurvivable in the child's seating position, a 1975 Toyota Corona left an outer suburban road at high speed and half-rolled into a telegraph pole so that the pole deeply intruded into the roof of the car. A 13-year-old boy in the centre rear position died of multiple skull and facial fractures with associated brain damage.

Among those non-fatally injured in frontal crashes, in a very severe crash a girl aged 12 years was seated in the centre rear position of a 1983 Holden Statesman. The car drifted over to the wrong side of a country highway and collided head-on with an oncoming semi-trailer. The delta-V was in the order of 75 km/h, but with relatively low deceleration as the car under-rode the truck. Both adult front seat passengers were killed. The girl was wearing a lap-only belt, and received several facial fractures from contact with the console structure in front of her, and internal abdominal injuries from the lap belt. Her 18-year-old brother, in a lap/shoulder belt in the left rear position beside her, sustained minor chest and abdominal injuries from belt loading.

In the 1989 Landcruiser that hit a tree, to which reference has already been made in the section above, there were three children restrained by lap-only belts. A male aged 14 was in the centre seat of the bench immediately behind the front seats. He suffered moderate head and spinal injuries. In the third row of seats, another bench, were seated a 12-year-old boy on the left and a 13-year-old boy on the right. The first, seated adjacent to the intruding tree, suffered fractures to the skull and cervical spine. The other, seated away from the intrusion, received only left-side abrasions.

A 1989 Nissan Skyline collided with a cliff face at the side of the road. Three children were on the back seat, all restrained in available belts. The driver and the two children seated in the outboard rear positions (one a one-year-old in a child seat) received minor belt bruising only; the six-year-old boy in the centre rear position, however, was admitted to hospital with injuries to the bowel. The belt was apparently correctly adjusted.

A 14-year-old male was seated in the front centre passenger seat of a Toyota Landcruiser that was struck on its left side by an oncoming van, causing severe intrusion damage. He suffered internal abdominal injuries. Two adults (left front and centre rear) and a child (left rear) died in this crash.

SLED TESTING

In order to more clearly identify any differences between three-point lap/shoulder (lap/sash) and lap-only belts in their capacity to protect children in frontal impacts, a series of comparative sled tests was undertaken. These were all conducted on an MTS Monterey "Impac" rebound sled at a nominal delta-V of 48 to 49 km/h (30 miles/hr).* The configuration of this sled gives rise to a short-duration, near-sinusoidal pulse, with a rapid rise of acceleration. For the given delta-V, therefore, these tests represent a violent and rather "stiff" crash. The peak sled acceleration for all runs was within the range 26.8 g to 27.5 g, typically at 40 msec.

Three anthropomorphic dummies were employed, representing for the desired age ranges the most biofidelic examples currently available. All were manufactured by First Technology Safety Systems Inc, of Plymouth, Michigan. They were as follows.

---

* The original protocol required runs at 56 km/h, but calibration tests revealed the probability of damage to the dummies in the lap-belted configuration at this delta-V.

- CRABI ("Child Restraint Airbag Interaction") Eighteen-month Old Infant Dummy (Version 1).

- Hybrid III Three-year-old Dummy (prototype status, in verification testing stage, especially made available for this research by First Technology Safety Systems).

- Hybrid III Six-year-old Dummy (Model 127-0000).

Their basic dimensions are shown in Table 9.

Table 9.
Basic Dimensions, Child Dummies

Dummy	Weight		Erect sitting height	
	kg	pounds	mm	inches
CRABI 18-month Old	11.2	24.7	505	19.9
Hybrid III Three-year-old	14.5	32.0	546	21.5
Hybrid III Six-year-old	22.8	50.2	640	25.2

Each of the three dummies was tested with a lap/shoulder belt and a lap-only belt, with two sled runs for each configuration. The belt was replaced by a new one after each run. The acceleration/time characteristics of each run were measured by accelerometers mounted on the sled.

The lap/shoulder belt in each case was of running-loop configuration, with a dual inertia-locking and webbing-sensitive emergency-locking retractor (as required in Australian cars) mounted at the upper end of the shoulder belt. The positioning of the belt anchor points was in accordance with the requirements of Australian Standard 3629.1-1991, *Methods of testing child restraints; Part 1: Dynamic testing* (Standards Australia, 1991). This positioning is consistent with the Australian Design Rules for motor vehicle safety.

The seat used for the tests was a stylised generic rear passenger-vehicle seat, also in accordance with the requirements of Australian Standard 3629.1-1991. The base of this seat is a polyurethane slab, density 28-29 kg/m^3, 156 mm thick, on a rectangular frame. The seat back is 70 mm thick.

All three dummies were instrumented as follows:

- head acceleration: 3-axis accelerometers;
- upper neck forces and moments: 6-axis transducers;
- chest acceleration: 3-axis accelerometer;
- pelvis acceleration: 3-axis accelerometer.

In addition, the lumbar region of the 18-month CRABI carried a 6-axis transducer for forces and moments. Belt force transducers were mounted on the webbing straps and buckle mounts.

Sign conventions, head acceleration coordinates and data filter classes were as specified in SAE J211 (Society of Automotive Engineers, 1988). The condition of the dummies was monitored after each test by visual inspection and instrument checks. Faces were painted to detect contact points.

All runs were filmed by a stationary high-speed camera positioned to the side of the sled. The cameras were operated at 1000 frames per second.

Results of sled testing

The data on accelerations and loads are summarised in Table 10. Figures given there relate to those values suggested as being of primary importance to injury risk, and show each of the two runs for each configuration. The purpose at this stage of the research was simply to provide some base data comparing performance in lap/shoulder and lap-only belts.*

The *CRABI Eighteen-month-old* dummy was restrained by the lap/shoulder adult belt surprisingly well, although as might be expected the kinematics were far from satisfactory. The dummy still had considerable forwards rotation despite the upper torso remaining constrained by the shoulder belt.

When restrained by a lap belt, the dummy rotated much further forwards, so that the belt moved down on to the upper surface of the thighs. This allowed considerable excursion, and the dummy's head impacted the front of the seat, including the wooden frame supporting the 156 mm deep cushion. The result was a

* Fully detailed data from this series of sled runs will be published in a separate comprehensive technical report.

high resultant head acceleration and HIC value for both the lap-belt runs.

Table 10.
Summary of Principal Results, Lap/shoulder (L/S) and Lap-only Belts: Two Runs for Each Case (Figures Rounded to Nearest Whole Number)

	18-month-old CRABI		Three-year-old Hybrid III		Six-year-old Hybrid III	
	L/S	Lap	L/S	Lap	L/S	Lap
Head						
GRes (g)	87 88	135 391	72 79	184 189	140 84	339 474
HIC	1004 1056	865 2567	822 868	2196 2159	1488 1163	2753 3605
Neck						
Forwards shear +Fx (N)	53 87	1310 1310	44 85	906 1139	66 46	1001 1201
Rearwards shear -Fx (N)	-931 -951	-763 -790	-678 -790	-418 -548	-1430 -1290	-560 -685
Axial tension +Fz (N)	2168 2287	1816 2154	1665 1757	2128 2255	3090 2622	2496 3912
Forwards moment +My (Nm)	26 22	67 93	53 56	23 41	75 60	20 33
Chest						
-Gx (g)	-53 -51	-55 -76	-50 -55	-64 -63	-59 -53	-37 -54
Belts						
Lap (N)	1038 1069	1945 1910	2109 2025	3890 4017	3780 3117	6220 5918
Buckle (N)	3370 3333	2067 2048	4350 4349	3215 3261	710 1885	5176 5222
Shoulder (N)	2676 2754		2571 2760		4481 4366	

Lap belt loads were much higher in the lap-only configuration, as would be expected. Adding a shoulder belt about halved the loads on the lap belt.

Peak neck shear loads (+Fx and -Fx) were +1310 N for the lap-belted configuration and -951 N for the lap/shoulder belt. Axial tensile loads (+Fz) were slightly higher with the lap/shoulder belt, but forward neck bending moment (+My) lower with the lap/shoulder belt.

The *Hybrid III Three-year-old* was much better restrained by the adult belt than the smaller dummy. There was submarining to the extent that the lap portion of the belt rode over the rudimentary pelvic structure, but otherwise the dummy's motion was driven by its high centre of gravity consistent with a child of an equivalent age. The top mounting of the shoulder belt was high (as it is in passenger cars) in relation to the sitting height of the dummy, and the shoulder belt allowed considerable downwards motion of the dummy's torso as it moved forwards within it. However, the net excursion of the dummy head was within acceptable levels, and did not extend beyond the front of the seat base (410 mm from its angle with the seat back).

The Hybrid III three-year-old flexed sharply over the lap-only belt. In the first of the two runs in this configuration, the lap belt ruptured the dummy's vinyl "skin" on the abdominal region, between the pelvic and rib structures. In both the lap-belt runs the head of the dummy impacted the forward side of the seat and its base.

Lap belt loads were much higher with the lap-only belt than with the shoulder belt added.

Just as for the 18-month-old CRABI, head accelerations and +Fx neck shear loads were lower with the lap/shoulder belt, but -Fx neck loads were higher. The peak neck shear load was +1139 N for the lap-belt configuration.

Axial tensile (+Fz) neck loading was higher with the lap belt alone than with the lap/shoulder belt. The two lap/shoulder runs gave figures of 1665 N and 1757 N, and with the lap-only belt, the figures were 2128 N and 2254 N. The +My moment, however, was higher (at 53 and 56 Nm) with the lap/shoulder belt than the lap belt (23 and 41 Nm).

The *Hybrid III Six-year-old* was well restrained by the adult lap/shoulder belt, with acceptable excursion and kinematics generally. The rotation forward seen in the smaller dummies, with their relatively high centres of gravity, was not apparent.

With the lap-only belt, as with the other dummies,

there was sharp flexion and excessive head excursion. The dummy head struck the forward face of the rigid base of the seat with high resultant head accelerations and HIC values.

As with the other dummies, loads in the lap belt were nearly twice as high without a shoulder belt. In one run the lap belt became wedged between the ribs and the abdominal insert.

Head accelerations and HIC were higher with the lap belt, because of head strike. Neck +Fx shear was higher with the lap belt, and -Fx shear higher with the lap/shoulder belt. Axial loads were slightly higher with the lap/shoulder belt, contrary to the values for the three-year-old dummy.

In summary, head accelerations and HIC were higher with the lap belt, as were lap belt loads. Neck +Fx shear forces were all higher with the lap belt, but -Fx shear forces all lower. Results for +Fz axial tensile forces were mixed, none varying drastically from others. Results for forwards bending moment were also mixed, being higher with the lap/shoulder belt for the two larger dummies and lower with the smaller CRABI dummy.

DISCUSSION

Field and Statistical Studies

It is now well accepted that any child riding in a passenger vehicle should be restrained in dedicated restraint equipment of a type appropriate to the child's size and age. Surveys indicate that until the child weighs more than 36 kg, or has a sitting height of about 760 mm (roughly equivalent to an age of 11 or 12 years), the seat belt will not fit in an ideal manner (Klinich *et al*, 1994). However, the fact is that countless children worldwide, much smaller than this, commonly do ride in motor vehicles while restrained only by adult seat belts. It is a reasonable expectation that from time to time vehicles with children thus restrained will crash. It would be a matter of great concern if this mismatching led to a commensurate increase in risk of injury to the restrained child. As it happens, available epidemiological data do not point to restrained children of at least ten years or so being at especial risk (Evans, 1988). However, studies of the effects of adult belts on child injury reduction and injury patterns are rare, and predictions of injury risk (especially for smaller children) are based on a narrow knowledge base. This paper reports field and laboratory data that are intended to build on existing knowledge.

In the field study reported in part in this paper, the sample of children using adult lap/shoulder belts totalled 121. There were also 35 children in the study who were restrained by lap-only seat belts.

The prime cause of injury among children restrained by adult lap/shoulder belts was contact with the interior surfaces of the car, often in association with side impacts and related intrusion. The incidence of injury to the head and face was much the same among lap-belted and lap-sash belted children, but children using lap/shoulder belts in outboard positions received most of their head injuries by contact with the adjacent doors and window structures. Those wearing lap belts were mostly using centre seats, and sustained head and face injuries from the consoles and seat backs in front of them.

The incidence of injury to the heads and faces of lap-belted and lap/shoulder-belted children was comparable to that found by Khaewpong *et al* (1995) in a very similar field study in Washington D.C. Tingvall (1987), in Sweden, also found no difference in injury rate between children restrained by lap belts in the "safer" centre seat than in outboard seats with three-point belts. Further, using police-reported data in Canada, Chipman and Hu (1995) found that injury risk was similar in front seats (where shoulder belts are more common) and rear seats (where lap belts persist in that country).

However, several studies, including a recent one by Huelke and Compton (1994), have shown that rear seats represent a safer environment than front seats. Thus, if the injury-protective effect of lap/shoulder and lap belts was similar, rear-seat occupants should consistently be at lower risk of injury. This does not appear to be the case. In both the Australian and Swedish studies, because those wearing lap belts were using centre seats, many of the observed head injuries should have been preventable because upper torso restraint would have minimised the forward excursion that allowed contact with structures in front, such as consoles and front seats. Simply put, the net effect of the lap belt in terms of head-injury prevention appears to be that it nullifies the benefit of riding in the rear of a passenger vehicle.

The field study reported here, earlier Australian studies (Henderson *et al*, 1976), and recent studies of fatalities in the United Kingdom (Rattenbury and Gloyns, 1993) have all confirmed the overwhelming importance of head injury in determining the outcome for a restrained child in a crash. While protection of the neck of the child is important, it is more important to limit the excursion of the head and upper torso. Where there is conflict between ways to bring about these aims, the protection of the head should take priority.

For an adult belt to be effective, of course, the child must be properly held within it. The field study did not find significant performance degradation from suboptimal repositioning of the seat belts by restless children. This is in contrast to suggestions by, for example, Agran and Winn (1988) and Meissner *et al* (1994). To the extent that repositioning occurs, it did not appear to affect protection to a significant extent; indeed, it should not, because an adequate restraint should be tolerant of minor "misuse".

The field study did disclose, however, deliberate degradation of the system by some adults. One fatality resulted from a child being removed by his adult carers from a booster seat and then being placed in only the lap portion of a lap/shoulder belt, from which he was ejected. Further, while as already noted the number of cars with manually-adjustable seat belts is these days very small, we did find one case of belt-induced abdominal injury in a child restrained in a manually-adjustable lap/shoulder belt that was being worn too loosely. A substantial benefit of the emergency-locking retractor reel is that it keeps the webbing in reasonable proximity even to the restless child.

Although the use of lap-only belts by children in our sample prevented many of them from more serious injury, the evidence of this study is that the lap belt is an incomplete restraint, to be used only when no better system is available. We found a significantly greater incidence of belt-induced abdominal injury among lap-belt wearers than lap/shoulder belt users, which supports the conclusion of Lane (1992) that the child in the centre seat (with a lap belt) is at significantly greater risk of seat-belt induced injury.

In addition, there is the additional factor that in Australian cars (unlike the typical American car, where lap belts are much more common and lap belts have retractors) the lap belt is almost always manually adjustable only. This compounds the problem of misuse by too easily allowing the belt to be worn loosely and thus increasing the risk not only of abdominal injury but also head injury through excessive excursion of the torso.

There has been a recent movement of some manufacturers, including major Australian ones in the mass market for family cars, away from the use of centre-seat lap belts and towards lap/shoulder belts. This provides more effective restraint for the very positions that children are most likely to use them, and thus maximises the benefit to any child of riding in the rear.

The Risk of Cervical Spine Injury

The alternative fitment of a lap/shoulder seat belt in the centre rear seat would provide better protection overall. However, some critics have suggested that to restrain the upper torso, especially that of a child, places the neck at greater risk than if the torso is allowed to swing unrestrained. Anatomical considerations (Burdi *et al*, 1969; Huelke *et al*, 1992), coupled with case reports of cervical spine injury to forward-facing children (Fuchs *et al*, 1989; Langwieder and Hummel, 1989; Huelke *et al*, 1992) have caused considerable international attention to be drawn to the issue of cervical and high thoracic spinal cord injury to infants and young children in forward-facing restraint systems.

However, data searches in Australia have failed to show that the lap/shoulder seat belt poses a significant threat to a child's spine. In any event, serious spinal injury is rare. In the United States, after reviewing about 60,000 crashes for 1980 to 1989 in the National Accident Survey Study (NASS) files, Huelke *et al* (1992) found only nine children aged 10 years or less who had a cervical spine injury of AIS 3 or greater. None were in a child restraint, three were wearing lap belts in the rear seat, and the others were unrestrained.

On the other hand, over the years Australian case histories have included a high proportion of well documented crashes, at much higher changes of velocity than the 48 km/h barrier equivalent, that did not result in more than minor cervical spine injury to children restrained facing forwards in adult belts or child restraints.

In the USA, Kelleher-Walsh *et al* (1993) also found no injuries to the cervical spine in their retrospective case review of 198 children injured in forward-facing child restraints. Other studies have indicated that although the use of some kinds of restraint can increase the overall risk of neck injury, such injuries are generally minor while there is a decreased risk of injury overall for both children (Agran and Winn, 1987; Norin *et al*, 1984) and adults (Bourbeau *et al* 1993). In particular, torso restraint of any kind appears to increase the risk of minor (AIS 1) injuries to the cervical spine as a trade-off for improved protection from more severe injury (Yoganandan *et al* 1989).

Reporting a series of 66 deaths among children in the UK using adult lap/shoulder belts, Rattenbury and Gloyns (1993) found (while conceding the small number of cases) "little evidence of a major risk of life-threatening injuries being caused by the diagonal section of the adult belt, except perhaps for very young children . . . The authors' view is that direct belt induced neck injury for children in adult belts (with or without booster cushions) is not as great a problem as some people have feared".

The field study reported in the present paper has confirmed earlier findings from Australia and elsewhere that children - even very small ones - do well in severe crashes when using lap/shoulder seat belts, and that concerns about vulnerability based on purely anatomical considerations may be misplaced. This was also the case for child restraints (Henderson *et al* 1994).

In the present study, neck injury in children using adult lap/shoulder belts was not found to exceed very minor degrees of severity even when belt loadings had caused significant bruising of the soft tissues of the thorax and nearby neck. Although the field study did not on its own establish an upper limit of tolerance for cervical spine injury in restrained children facing forwards, it indicated that the limit may be higher than might be deduced from clinical studies of injured children. The present field study included children who were not significantly injured despite the severity of the crash, and who would not therefore have been included in the typical trauma system databases. To study only those children who are injured can obscure the beneficial effect of safety equipment and give a false impression of vulnerability.

There appears to be a difference between the incidence of cervical spine injury (without head contact) among children in child restraints in the United States and Europe on the one hand, and Australia (and perhaps the United Kingdom) on the other. A key difference is that most Australian child restraints have a high top tether. Without such a tether, as in most US and European child seats, the seat and child together can rotate forwards and subject the cervical spine to the axial tensile +Fz forces that result in the type of distraction injuries reported in most of their cases by Langwieder and Hummel (1989) and Huelke *et al* (1992). Brown *et al* (1995) recently described a series of tests with a six-month-old CRABI dummy, with and without high and low mounted top tether straps for a child restraint. They found that axial Fz forces were reduced 30-40% (as well as resultant neck forces) in the presence of a top tether that held the restraint and the child upright.

These findings initially appear to be contrary to those of Weber *et al* (1993) and Janssen *et al* (1993) that changing the parameters for child restraint mounting do not much affect values for neck forces. However, Australian child restraints have top tethers as an integral requirement, and the tethers for most restraints have (by good fortune) become high mounted. The seats cannot perform properly without them. In contrast, the comparisons of top tethers conducted by Weber and others used seats with *low mounted* top tethers. These seats were also capable of being restrained *without* top tethers.

The point of the findings of Brown *et al* (1995) is that more vertical restraint of a child's upper torso seems to result in reduced axial neck forces. However, the rigidity in fixation of a child restraint is probably an order of magnitude greater than an adult seat belt.

After analysing a selection of cases of real-world spinal cord injuries in children, Stalnaker (1993) concluded that as long as the injuries are not caused by external forces applied to the head, spinal column tension is by far the most important parameter for limiting distraction injuries for children of the age group he analysed, up to five years. Trosseille and Tarriere (1993) correctly complicate the issue by pointing out that different forces in different crashes involving children of different ages produce different injuries and thus lead to the definition of different tolerances. Nevertheless, although the relative importance of shear, compression and tensile forces in bringing about injuries to children's necks is yet to be fully elucidated, much contemporary work stresses the importance of axial tensile forces.

Unfortunately, the neck of the Hybrid III dummy - having been designed very much with flexion and extension as priorities (Deng, 1989) - is poorly biofidelic in regard to axial forces (Pintar *et al*, 1990). Essentially, it is too stiff. That could be one explanation for the rather similar and non-discriminatory values for +Fx tensile forces for all the sled tests for all three dummies reported in the present paper.

Sled test data for three dummies

To support and build upon the field data reported in this paper, the first objective of our related sled study was to assess the effects of using three-point lap/shoulder seat belts for the restraint of a selection of child anthropomorphic test dummies, in comparison with the effects under the same test conditions but using lap-only seat belts. (Later analysis in a technical report will discuss further the matter of injury tolerance.) Particular attention was paid to neck forces and seat-belt loads. This appears to be the first time that such direct comparisons have been undertaken in a systematic manner.

Until recently there have not been available sufficiently biofidelic child dummies to attempt crash simulation studies. However, an 18-month-old CRABI dummy is now available, as is an early model of the new Hybrid III six-year-old dummy. Further, the prototype Hybrid III three-year-old dummy became available in Australia for a limited period in early 1996 thanks to First Technology Safety Systems Inc, and was used for this research.

The sled test data for the three dummies showed mixed results for neck shear, axial tension and forward bending moment. Head accelerations and lap belt loads were consistently higher with the lap belt alone, with the shoulder belt sharing loads when used. In summary, accepting some inconsistencies in the results from dummy to dummy, the results are in accord with the field data: broadly, that in return for a greatly reduced risk of head and abdominal injury, a lap/shoulder belt may present a slightly higher risk of minor inertial neck injury, equivalent to AIS 1 or 2. However, there is nothing in this set of sled test results to indicate that adding a shoulder belt to a lap belt places a child at a higher risk of serious neck injury.

There have of course been several previous studies of neck loads on impact, intended at least in the early stages to develop neck tolerances for adults. The work with adult volunteers and cadavers by Mertz and Patrick (1971) indicated a risk of injury with a bending moment in flexion of 189 Nm, with a possibility of muscular injury at lower levels. For tensile loading these authors suggested a tolerance of 1160 N during postero-anterior acceleration of the torso, in rough accordance with the conclusions of Sances *et al* (1982). Shea *et al* (1991) reported a tensile load to failure of about 500 N in the absence of muscle tone. Mertz has summarised tolerance levels for several neck values in order to evaluate the responses of the Hybrid III (adult) dummy (Mertz, 1984).

Turning to children in child restraints (as opposed to adult belts), Weber *et al* (1993) used a six-month-old CRABI dummy (7.8 kg) in reproducing a crash of 50 km/h delta-V in which a six-month-old child had sustained a spinal cord contusion at T2. The child seat had been used without an effective top tether. They recorded a resultant force in the upper neck of 1260 N and in the lower neck 1159 N. The resultant moments were -6 Nm in the upper neck and 45 Nm in the lower neck.

Planath *et al* (1992) reported data following reconstruction on sled runs of crashes involving two children sustaining fatal head/neck injuries in forwards-facing child seats. The forwards-facing sled tests were performed with a Type P572C (Hybrid II) three-year-old dummy with a replacement neck that could be instrumented at the craniocervical junction (upper neck). Runs were at 40 km/h and 50 km/h, but sled g levels were not reported. The 50 km/h runs reproduced a crash with a 15-month-old child in a forwards-facing child seat. The child sustained fatal brain contusion without skull fracture, and no neck injury. The average figures for the tests were for HIC 809, shear (Fx) 280 N, tension (Fz) 2570 N, and flexion (My) 33 Nm.

Planath *et al* also brought together data from sled tests with rearwards-facing seats, plus data from scaling down data for adults. In addition, they noted the work on child/airbag interactions of Prasad and Daniel (1984) and Mertz and Weber (1982) with matched sets of tests with a three-year-old "airbag dummy" and piglet child surrogates.

The synthesis of all these data led them to conclude that the following values could be used as guidelines for neck protection criteria for assessment of the risk of neck injury for a child of about four years of age: tensile axial force, 1000 N; shear force, 300 N; forward bending moment, 30 Nm.

Janssen *et al* (1993) used similar reconstruction and scaling techniques, and employed a TNO 3/4 (9-month-old, 9 kg) dummy for their series of sled runs. The neck of the standard TNO dummy cannot be instrumented, and it was modified for this research. The restraint system was a four-point child harness in a child seat in all cases. They proposed maximum shear and tension forces for guidelines for protection criteria for children through all age ranges. For a three-year-old, the suggested maxima for neck tension and shear would be about 1000 N, and for bending moment about 30 Nm.

Trosseille and Tarriere (1993), again using crash reconstruction techniques (four crashes, including one also used by Weber in her work), found for six-month-old children no injury under Fx 950 N and My 41 Nm, but injury over Fz 1200 N. They note the importance of obtaining data from *uninjured* children, which we also stress. They agree with Planath's (1992) suggestion of a limit of Fx of 300 N for three-year-old children, but note the substantial and rapidly-changing influence of age: there were children of 4.5 years who sustained no injury with Fx of 750 N. They also found no injury at Fz of 2500 N for this age group, and suggest further work to explain this.

Planath *et al* (1992) cautiously suggest that their figures might be unduly conservative. Further, Janssen *et al* point to the fact that measures taken to lower neck forces might increase excursion of the restraint and the child and thus the risk of head injury.

All the research groups noted above stress the dangers in comparing results from different dummies, and variations in design could explain some of the differences between dummies that we found. There is also the matter of time dependency. There is general agreement in the literature that there will be a higher tolerance to forces of very short duration, usually published as "peak" forces, whereas forces applied over 30 msec or more would have a better association with injury tolerance. We support all the above cautions.

Relating Field and Laboratory Data

Generally, the values for force and bending moment in our sled test series were high in comparison with other research. However, in our field data there were 19 children aged two years to 14 who were restrained in lap/shoulder belts in generally frontal crashes at a calculated delta-V of 45 km/h or over (see Table 11). More than two-thirds of these children (13) were in frontal crashes of 65 km/h or over, which gives some allowance for errors in delta-V calculations. It is probable, therefore, that all these children were exposed to forces of the same order of magnitude that we found in our series of sled runs, generally above the tolerance criteria suggested for guidance by Planath *et al* (1992) and other workers.

Yet, among these children there was only one neck injury of AIS 3 or more. This AIS 6 (fatal) injury in a three-year-old directly resulted from a heavy head contact with the windshield, and was described in this paper among the case histories. The other neck injuries were all AIS 1 or 2, being soft tissue injuries commonly associated with bruising and abrasions from belt loading.

In their crash reconstructions, both Planath *et al* (1992) and Trosseille and Tarriere (1993) recorded neck tensile (Fz) forces of over 2.5 kN when no neck injuries had been sustained in the real crashes. Although we did not make direct comparisons of individual crashes with individual sled runs for the purpose of this paper, these relationships are in accordance with our observations.

Table 11.
Lap/shoulder belts, frontal crashes with delta-V of 45 km/h or more

	Maximum AIS		
Age (years)	0 - 2	3 - 6	Total
< 3	1	1	2
4 - 6	2	0	2
7 - 9	4	1	5
10 - 14	8	2	10
Total	15	4	19

This is not, of course, to suggest that children's necks are immune from inertial injury in high-speed frontal impacts. This is manifestly not the case. Much larger studies, including *uninjured* children, are required properly to assess the degree of risk. But it may well be the case that some of the crash reconstruction studies in the literature, being based on children whose spines were known to have been severely injured, are consequently based on outlier cases involving crash-related or child-related factors not typically representative.

The spine and the head together make up an exceedingly complex system, and spinal injury mechanisms are sensitive to countless variations in the way that potentially injurious loads are applied. There is a very great deal of work yet to be done before tolerance levels for the cervical spine can be firmly established, and in respect to children this work is at a very early stage. Children, by definition, are growing up quickly and tolerances may be expected to change year by year for each child, yet vary from child to child at a given age. The problems are compounded by the difficulties in performing cadaver experimentation with children, and there have only recently been improvements in the biofidelity of test dummies. Animal models are generally inappropriate, and now rarely used. Thus, field and epidemiological research has a particularly important part to play.

There are many more head than neck injuries in the data from field studies. Lap-belt-induced injury of the abdominal organs and lumbar spine are also far more common than inertial injuries to the cervical spine. In the development of design or performance criteria, for the minimisation of the risk of cervical spine injury it is important not to unreasonably raise the risk of other serious injuries.

SUMMARY

To obtain maximum protection, children should be restrained in dedicated child seats, or adult belts supplemented by booster seats, until they are of a size appropriate to the use of adult belts.

However, field data from investigating crashes involving 121 children aged one to 14 years in adult lap/shoulder belts show that they were generally well protected even in severe frontal crashes, and none sustained belt-induced inertial neck injury. Change of velocity was related to injury risk, but age was not.

Lap-belted children sustained a higher proportion of abdominal injuries and a similar proportion of head injuries despite almost all being seated in centre positions away from the side of the car.

A series of sled runs with 18-month, three-year-old and six-year-old new-generation dummies showed generally slightly higher values for neck shear forces and moments among those using lap/shoulder belts, but much lower values for abdominal belt loads and head accelerations. These data were consistent with the conclusion that adding torso restraint slightly increases the risk of minor (AIS 1 or 2) neck injury, but has the major benefit of reducing the risk of serious head and abdominal injuries.

The analysis of field data based on uninjured as well as injured children should lead to the derivation of realistic tolerance criteria, whereas the use only of injured children can lead to over-conservative estimates.

Present data indicate that adult lap/shoulder belts do not present a significant risk of severe injury to young children.

ACKNOWLEDGMENTS

The authors and the Roads and Traffic Authority of NSW wish to thank the Motor Accidents Authority of New South Wales for providing financial assistance for the field study, the Federal Office of Road Safety for sharing the financial cost of the sled studies, and First Technology Safety Systems for providing the prototype Three-year-old Hybrid III dummy on special loan to the RTA.

REFERENCES

Agran P F and Winn D G , Injuries Among 4 to 9 Year Old Restrained Motor Vehicle Occupants by Seat Location and Crash Impact Site, in *Proceedings, 32nd Annual Conference, Association for the Advancement of Automotive Medicine*, Seattle, AAAM, 1988.

Agran P F , and Winn D , Traumatic Injuries Among Children using Lap Belts and Lap/shoulder Belts in Motor Vehicle Collisions, in *Proceedings, 31st Annual Conference, American Association for Automotive Medicine*, New Orleans, AAAM, 1987.

Bourbeau R, Desjardins D, Magg U and Laberge-Nadeau C, Neck injuries among belted and unbelted occupants of the front seat of cars, *The Journal of Trauma*, 35(5):794-799, 1993.

Brown J, Kelly P, Griffiths M, Tong S, Pak R and Gibson T, The Performance of Tethered and Untethered Forward Facing Child Restraints, in *Proceedings, 1995 International IRCOBI Conference on the Biomechanics of Impact*, IRCOBI, September 1995.

Burdi A R, Huelke D F, Snyder R G and Lowrey G H, Infants and children in the adult world of automobile safety design: paediatric and anatomical considerations for design of child restraints, *Biomechanics*, 2: 267-280, 1969.

Chipman M L and Lu X, The Effectiveness of Safety Belts in Preventing Fatalities and Major Injuries Among School-aged Children, in *39th Annual Procceedings, Association for the Advancement of Automotive Medicine*, AAAM 1995.

Corben C W and Herbert D C, *Children Wearing*

Approved Restraints and Adults' Belts in Crashes, Traffic Accident Research Unit, Report 1/81, Department of Motor Transport, NSW, 1981.

Deng Y-C, Anthropomorphic dummy neck modelling and injury considerations, *Accident Analysis and Prevention*, 21(1): 85-100, 1989.

Evans L, Risk of fatality from physical trauma versus sex and age, *Journal of Trauma*, 28: 368-378, 1988.

Fuchs S, Cervical spine fractures sustained by young children in forward-facing car seats, *Paediatrics*, 84(2), 1989.

Henderson J M, Herbert D C, Vazey B A and Stott J D, Performance of Child Restraints in Crashes and in Laboratory Tests, in *Proceedings, Seat Belt Seminar*, Melbourne 1976, Commonwealth Department of Transport, Canberra, 1976.

Henderson M, Brown J and Paine M, Injuries to Restrained Children, in *38th Annual Proceedings, Association for the Advancement of Automotive Medicine*, Lyon, AAAM 1994.

Huelke D F, Mackay G M, Morris A, and Bradford M, Car Crashes and Non-Head Impact Cervical Spine Injuries in Infants and Children, SAE 920562, in *Proceedings, SAE International Congress*, Detroit, Society of Automotive Engineers, Warrendale PA, 1992.

Huelke D F and Compton C P, The Effects of Seat Belts on Injury Severity of Front and Rear Seat Occupants in the Same Frontal Crash, in *38th Annual Proceedings, Association for the Advancement of Automotive Medicine*, Lyon, AAAM 1994.

Janssen E G, Huijskens C G, Verschut R and Twisk D, Cervical Spine Loads Induced in Restrained Child Dummies II, SAE paper 933102, in *Child Occupant Protection*, SAE SP-986, Society of Automotive Engineers, Warrendale PA, 1993.

Kelleher-Walsh B, Walsh M J, States J D and Duffy L C, Trauma to Children in Forward-Facing Car Seats, SAE paper 933095, in *Child Occupant Protection*, SAE SP-986, Society of Automotive Engineers, Warrendale PA, 1993.

Klinich K DeS, Pritz H B, Beebe M S and Welty K E, Survey of Older Children in Automotive Restraints, SAE 942222, in *Proceedings, 38th Stapp Car Crash Conference*, Society of Automotive Engineers, Warrendale PA, 1994.

Khaewpong N, Nguyen T T, Bents F D, Eichelberger M R, Gotschall C S and Morrissey R, Injury Severity in Restrained Children in Motor Vehicle Crashes, SAE 952711, in *Proceedings, 39th Stapp Car Crash Conference*, Society of Automotive Engineers, Warrendale PA, 1995.

Lane J C, *The Child in the Centre Seat*, Monash University Accident Research Centre, for the Federal Office of Road Safety, Report CR 107, Department of Transport and Communications, Canberra, 1992.

Langwieder K, and Hummel T, Neck Injuries to Restrained Children, in *Proceedings, Annual Conference of the International Research Committee for the Biokinetics of Impact: Workshop on the Future of Child Restraints*, 1989.

Meissner U, Stephens G and Alfredson L, Children in Restraints, in *38th Annual Proceedings, Association for the Advancement of Automotive Medicine*, Lyon, AAAM 1994.

Mertz H J and Patrick L M, Strength and Response of the Human Neck, SAE 710855, in *Proceedings, 15th Stapp Car Crash Conference*, Society of Automotive Engineers, Warrendale PA, 1971.

Mertz H J, *Injury Assessment Values Used to Evaluate Hybrid III Response Measurements*, General Motors Corporation, 1984.

Mertz H J and Weber D A, Interpretations of the Impact Responses of a 3-year-ol Child Dummy Relative to Child Injury Potential, *Proceedings, 9th International Technical Conference on Experimental Safety Vehicles*, Kyoto, 1982.

Norin H, Carlsson G and Korner J, Seat Belt Usage in Sweden and its Injury Reducing Effect, in *Advances in Belt Restraint Systems Design Performance and Usage*, P-141, Society of Automotive Engineers, Warrendale PA, 1984.

Pintar F A, Sances A, Yoganandan N et al, Biodynamics of the Total Human Cadaveric Cervical Spine, SAE 902309, *Proceedings, 34th Stapp Car Crash Conference*, Society of Automotive Engineers, Warrendale PA, 1990.

Planath I, Synthesis of Data Towards Neck Protection Criteria for Children, in *Proceedings, 1992 International IRCOBI Conference on the Biomechanics of Impact*, IRCOBI, September 1992.

Prasad P and Daniel R P, A Biomechanical Analysis of Head, Neck and Torso Injuries to Child Surrogates Due to Sudden Torso Acceleration, SAE 841656, *Proceedings, 28th Stapp Car Crash Conference*, Society of Automotive Engineers, Warrendale PA, 1984.

Rattenbury S J and Gloyns P F, A Population Study of UK Car Accidents in Which Restrained Children Were Killed, in *Child Occupant Protection*, SP-986, Society of Automotive Engineers, Warrendale PA, 1993.

Sances A Jr *et al*, *Head and Spine Injuries*, AGARD Conference on Injury Mechanism, Prevention and Cost, Koln Germany, 1982; cited in McElhaney J H and Myers B S, *Biomechanical Aspects of Cervical Trauma*, in *Accidental Injury: Biomechanics and Prevention*, Nahum and Melvin (Editors), Springer-Verlag, 1993.

Shea M, Edwards W T, White A A and Hayes WC, Variations of stiffness and strength along the human cervical spine, *Journal of Biomechanics*, 24(2):95-107, 1991.

Snyder R G and O'Neill B, Are 1974/1975 automotive belt systems hazardous to children?, *American Journal of Diseases of Children*, 129: 946., 1975.

Society of Automotive Engineers, *SAE J211, Instrumentation for Impact Tests*, SAE 1988

Stalnaker R L, Spinal Cord Injuries to Children in Real World Accidents, SAE paper 933100, in *Child Occupant Protection*, SP-986, Society of Automotive Engineers, Warrendale PA, 1993.

Standards Australia, *Methods of testing child restraints; Part 1: Dynamic testing, Australian Standard 3629.1-1991*, Standards Australia and Standards Association of New Zealand, North Sydney, NSW, Australia, and Wellington NZ, 1991.

Tingvall C, Children in Cars, Some Aspects of the Safety of Children as Car Passengers in Road Traffic Accidents, *Acta Paediatr. Scand., Suppl.*, 339: 1-35, 1987.

Trossielle X and Tarriere C, Neck Injury Criteria for Children from Real Crash Reconstructions, SAE paper 933103, in *Child Occupant Protection*, SAE SP-986, Society of Automotive Engineers, Warrendale PA, 1993.

Vazey B, *Child Restraint Field Study*, Traffic Accident Research Unit Report 7/77, NSW Department of Motor Transport, 1977.

Weber K, Dalmotas D and Hendrick B, Investigation of Dummy Response and Restraint Configuration Factors Associated with Upper Spinal Cord Injury in a Forward-Facing Child Restraint, SAE paper 933101, in *Child Occupant Protection*, SP-986, Society of Automotive Engineers, Warrendale PA, 1993.

Yoganandan N, Haffner M, Maiman D J et al, Epidemiology and Injury Biomechanics of Motor Vehicle Related Trauma to the Human Spine, in *Proceedings, 33rd Stapp Car Crash Conference*, Society of Automotive Engineers, 1989.

973314

The Effects of Belt Pretensioners on Various Child Restraint Designs in Frontal Impacts

Waldemar Czernakowski
Britax Römer Kindersicherheit GmbH

Robert Bell
Britax Child Safety, Inc

I. ABSTRACT

The performance advantage of seat belt pretensioners is well documented. But what is the effect for child restraints? We tested four categories of CRS with and without seat belt pretensioners on adult seat belts, using the ECE44.03 dynamic test. We compared outcomes in terms of head excursion, head and chest deceleration and HIC ("Head Injury Criterion") and used a common type of pyrotechnic pretensioner. For infant restraints, pretensioned conditions lead to a reduction in head excursion and HIC. Two forward facing toddler seats showed lower HIC, and one a large reduction in excursion. Pretensioning led to reductions in HIC for forward facing impact shield boosters and for belt positioning boosters. Three CRS were also tested in a pre-impact braking mode, with and without pretensioner. Here the differences were less pronounced.

On the whole, pretensioners did improve the outcome for CRS, in particular for designs which did less well in the "normal" mode. The tests revealed no major ill effects for these models of CRS when used in adult seat belts with pretensioners.

II. INTRODUCTION

Belt pretensioners have been used in motor vehicles in Europe for more than 15 years as a means to improve the performance of adult seat belts. Their main purpose is to eliminate or reduce potential belt slack resulting in less head displacement and lower deceleration loads onto the adult occupant. Within the literature the positive effect of belt pretensioners on adult occupants is well documented. This paper poses the question: How do pretensioners affect child restraints?

In 1977 and 1978 Rüter [1], [2] reported on the reduction of head displacement of 10 cm and an improved effectiveness of automatic seat belt systems of 50%.

1984 Mitzgus [3] described in detail design and performance of pyrotechnic belt pretensioners and emphasised the need for a very early ignition. Ignitions later than 30 milliseconds after crash had no effect on reducing belt slack.

1989 Zuppichini [4] investigated mechanical pretensioners and expected advantages in the prevention of lesions caused by loose seat belts in spite of the risk of a more violent whiplash motion.
Deng [5] described the function of buckle pretensioners and Muser [6] required pretensioners in designing low-mass vehicles. In conjunction with smart seat belt systems a number of authors recently considered pretensioners as a necessary design feature such as Johannessen [7], Miller [8] and Bernat [9].
In the latter two papers it has been stated that "pretensioners are designed for taking up webbing slack by capturing the occupant earlier in the crash event thus riding down the crash as the vehicle structure begins to deform" respectively "the effect of pretensioners to have a rapid onset of restraint force thus minimising the loads on the occupants".

The described advantages of belt pretensioners refer to adult occupants who are directly restrained by 3-point lap/shoulder belts equipped with such pretensioners. However, do these advantages also apply to restrained children? It is remarkable that to our knowledge the effects of belt pretensioners on restrained children have not been published. Due to the different methods used to restrain children versus adults the advantages valid to the latter may not necessarily apply to children.

In most cases children are restrained within children's car seats and therefore - as opposed to adults - are not making direct use of adult seat belts. This is particularly true for rearward facing infants as well as rearward and forward facing toddlers. The belts are used to retain the seat shell and not the child. The different routes used to guide the adult seat belt through a seat shell are manifold. How do such different belt geometries affect the performance of belt pretensioners? Could the rapid onset of restraint force do harm to children, particularly infants?

This paper attempts to deal with some of these issues in so far as dynamic sled tests may illuminate the above questions.

III. TEST PROGRAM

1. CHILD RESTRAINT SYSTEMS (CRS)
At the outset the authors decided to examine a total of 8 different CRS with distinct styles and design features and to include and compare at least two different CRS for each mass/age group:
Two rear facing infant seats with 3-point harness (up to 9 kg / 10 months)
Two toddler seats with 5-point harness (9-18 kg / 10 months to 4 years)
Two toddler seats with impact table (9-18 kg / 10 months to 4 years)
Two booster seats with backrest (9/18 kg to 32 kg / 10 months respectively 4 years to 10 years)

Within each style, the pairs of CRS were selected for different shoulder belt geometry described in the following chapters.

2. PRETENSIONER
For this study, the same pretensioner type was used in all tests, namely the most common shoulder belt pyrotechnic pretensioner installed on an automatic reel (type Pystra, manufacturer TRW Repa). It is activated by a decelerometer at 7-10g sled deceleration at which the belt retraction is initiated. The retraction lasts over 10-12 milliseconds and is completed around 25 milliseconds after impact. With no load the pretensioner takes up 200mm of webbing. The schematic design of such a pretensioner is shown in illustration 1.

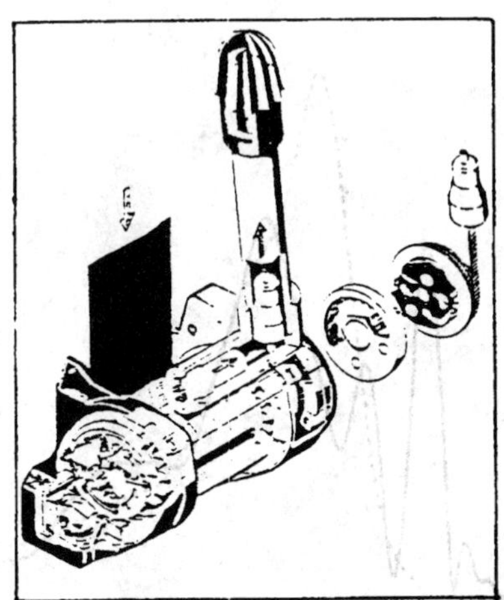

Illustration 1: Schematic Automatic Reel with Pyrotechnic Belt Pretensioner

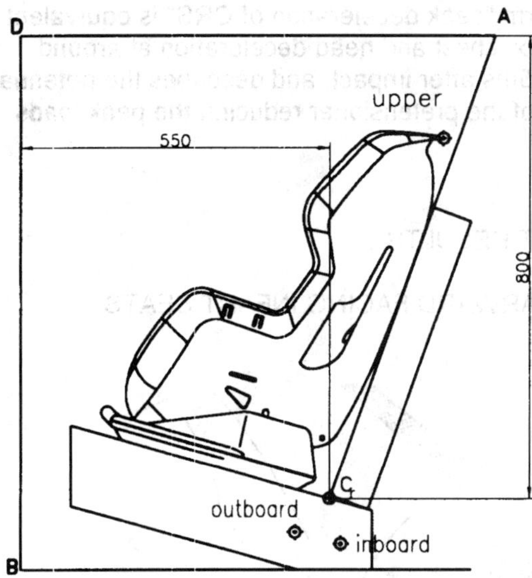

Illustration 2: Location of Anchorages and max. horizontal and vertical head excursion as per ECE44/03, CRS model B1 shown

3. TEST CONDITIONS/MEASUREMENTS
The dynamic tests were done to ECE44/03 conditions using asymmetrical lower belt anchorages. (See illustration 2)
The automatic reel was equipped with the selected pretensioner.
The impact velocity was 50 - 2 km/h.
Chest deceleration, head excursion and - in addition to ECE44-requirements - head deceleration were measured and HIC ("Head Injury Criterion") at 36 ms max. was calculated.
The average sled deceleration was 22,3 g.
The type of TNO dummy used is specified in the following tables. Unfortunately, the type of dummy used did not allow measuring neck loads. Therefore the effect of belt pretensioning on children's neck loads was not considered.

4. ASSESSMENT
In order to assess the effect of the pretensioner on each CRS a dynamic test without pretensioner was performed first as baseline data. The results of these baseline tests are shown in brackets in the tables respectively in dotted lines in the diagrams of the following chapters, whereas the tests with pretensioner are shown in full lines. Within the tables the results of all tests with pretensioner are given as positive or negative difference to the baseline test without pretensioner.

Within the following tables the term "initial deceleration caused by the pretensioner" is used. This deceleration occurs upon tightening of the belt slack at around t=25ms after impact.

The term "peak deceleration of CRS" is equivalent to the max. chest and head deceleration at around t=45-55ms after impact, and describes the potential effect of the pretensioner reducing the peak loads.

The time at which both terms occur is illustrated in diagram A1a.

5.TEST RESULTS

A) REARWARD FACING INFANT SEATS

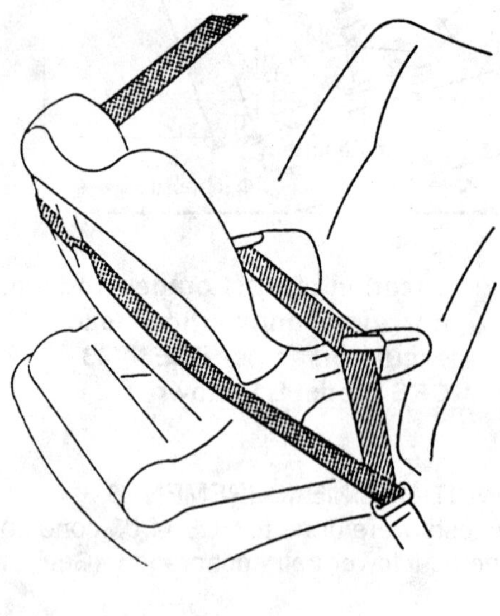

Illustration A1

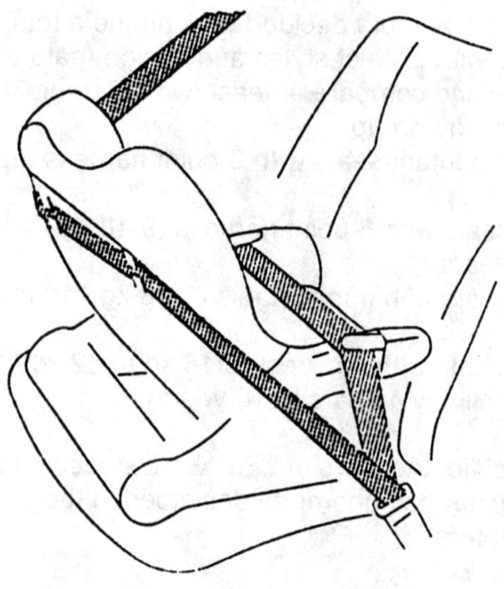

Illustration A 2

Except for the attachment of the shoulder belt to the CRS shell both seats were of similar design. However, the shell of model A2 was much stiffer than of A1. Model A1 had one central belt guide with clamping device on the rear side of the seat back, whereas model A2 had two symmetrical horizontally placed guides. On both models the adult shoulder belt is guided around the seat back and the lap belt across the seat shell.

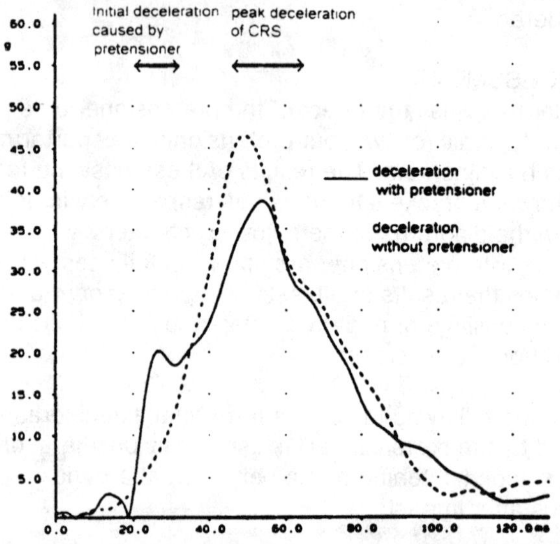

Diagram A1a: Resultant Chest Deceleration

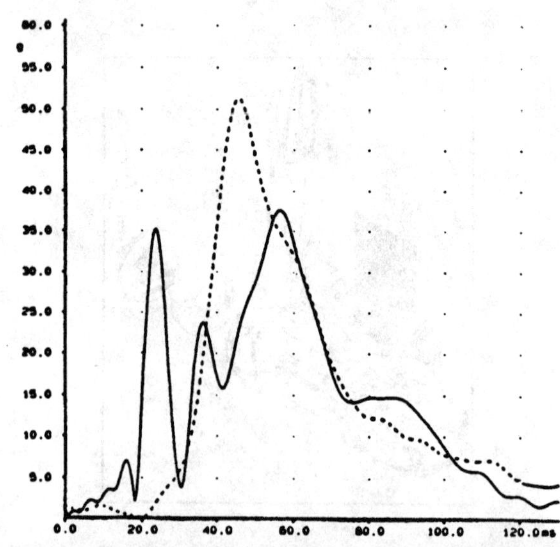

Diagram A1b: Resultant Head Deceleration

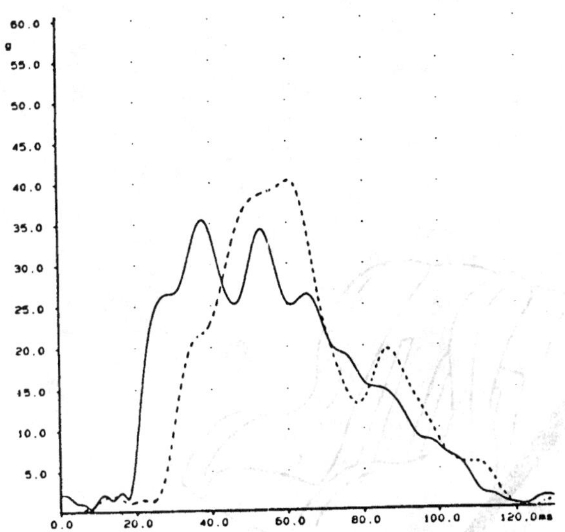

Diagram A2a: Resultant Chest Deceleration

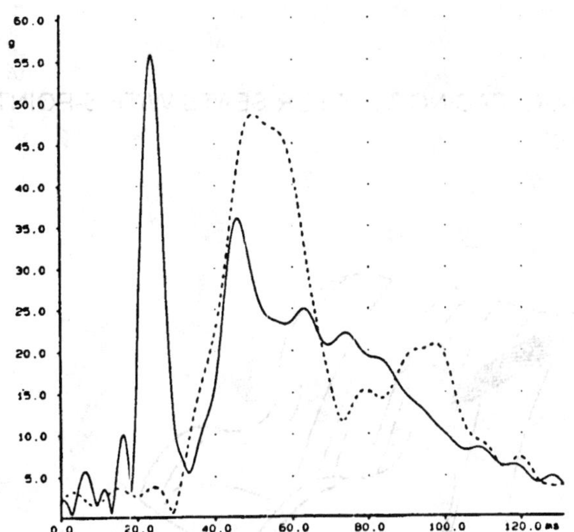

Diagram A2b:Resultant Head Deceleration

Table A

CRS model	Dummy used	Belt retraction (mm)	Reduction of		Initial deceleration caused by pretensioner		Peak deceleration of CRS	
			Head excursion (mm)	HIC	Chest (g)	Head (g)	Chest (g)	Head (g)
A1	P ¾	140	(540) -70	(258) -131	(10) +10	(10) +24	(46) -8	(52) -4
A2	P ¾	130	(530) -65	(267) -151	(21) +4	(16) +40	(40) -4	49 +7

Comments on test results

- The belt retraction between 130-140 mm on both CRS models was comparatively high, resulting in a reduction of head excursion of 50% of the belt take-up. However there was very little belt retraction through the shoulder belt guides on both models.

- On model A1 belt pretensioning resulted mainly in a lateral deformation and some lateral and vertical displacement of the shell. The stiffer shell of model A2 did only allow for little shell deformation, and therefore led to a more violent displacement of the shell.

- The reduction of HIC on both CRS was in excess of 50%.

- There was little initial effect on the chest deceleration on both CRS. However, the initial effect on head deceleration is higher in comparison to forward facing CRS. In fact, the model A2 showed a significant increase of 40g as compared to 24g in model A1and was mainly caused by the violent displacement of the stiff shell. The authors are not in a position to comment on potential injuries due to an initial head deceleration of 56g for infants, particularly newborns. However, since this effect seems to be design-induced, it can be avoided.

- Chest and head decelerations are only little affected by the pretensioner during the usual peak loads around 45-55 ms, as rear facing restraints benefit from the longer ride down in the non-pretensioned condition.

B) FORWARD FACING TODDLER SEATS WITH 5-POINT HARNESS

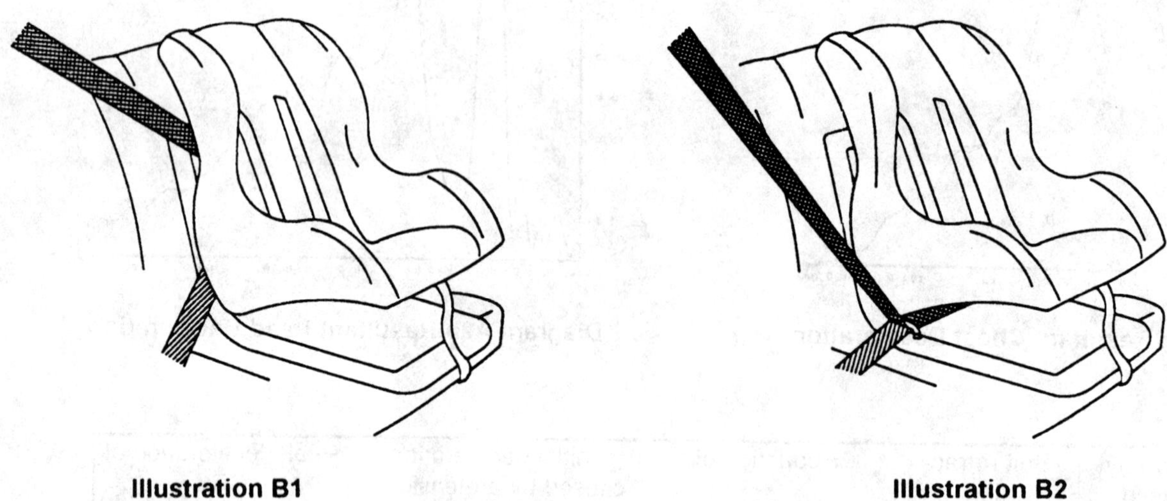

Illustration B1 **Illustration B2**

Both forward facing toddler type CRS models (illustrations B1 and B2) use a 5-point harness. The shoulder strap of the adult belt on model B1 is guided through a pair of beams behind the shell at about center of gravity level and is kept tight by a clamping device resulting in a stable fit on the seat bench. On model B2 both shoulder and lap straps are located on top of each other underneath the CRS shell and do not allow for a tight and stable position of the CRS on the car seat.

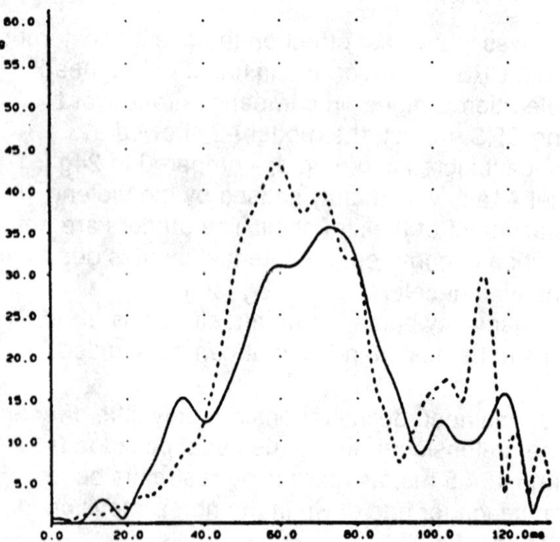

Diagram B1a: Resultant Chest Deceleration

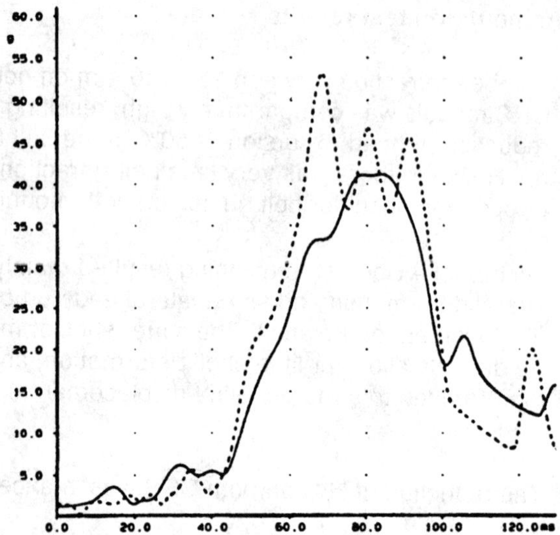

Diagram B1b: Resultant Head Deceleration

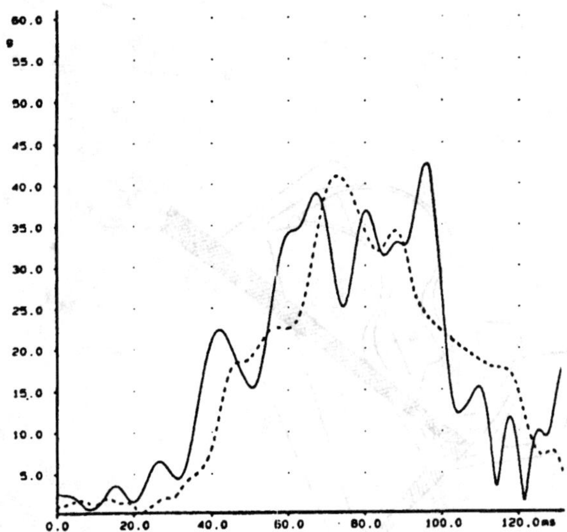

Diagram B2a: Resultant Chest Deceleration

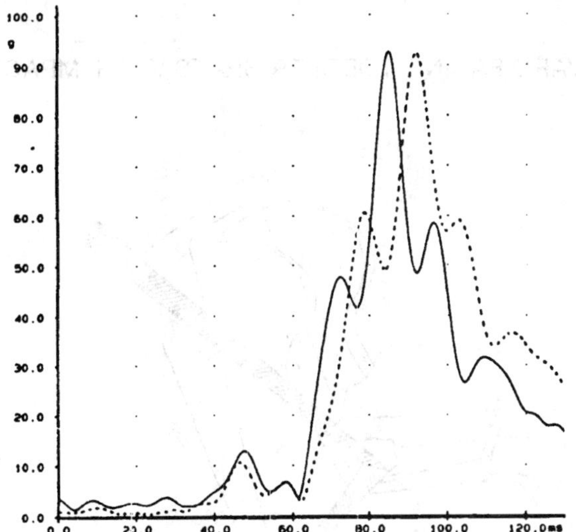

Diagram B2b: Resultant Head Deceleration

Table B

CRS model	Dummy used	Belt retraction (mm)	Reduction of		Initial deceleration caused by pretensioner		Peak deceleration of CRS	
			Head excursion (mm)	HIC	Chest (g)	Head (g)	Chest (g)	Head (g)
B1	P3	80	(490) -25	(412) -131	(5) +10	(25) -	(43) -7	(59) -17
B2	P3	130	(675) -125	(1031) -220	(25) -	(25) -	(41) -3	(91) +2

Comments on test results:

- On model B1 the pretensioner achieved a belt retraction of only 80mm due to its initial tight fit with little belt slack. On model B2 the belt hook diverting the shoulder strap underneath the shell broke upon pretensioning causing excessive belt slack which was taken up by the pretensioner upon retracting the shoulder belt by 130 mm.

- On model B1 the initial low head excursion was slightly reduced. B2 had an initial head excursion well above ECE44 requirement, with the pretensioner the head excursion was brought down just to the upper limit of ECE44.

- Good results were achieved on reduction of HIC for both models. The pretensioner managed to reduce the initially high HIC of model B2 from above 1000 by a margin of about 20 %.

- On both models there was basically little initial effect on chest decelerations due to the pretensioner. Its effect is restricted to the car seat belt and is on the whole not transferred to the CRS internal harness system restraining the dummy.

- In principle this also applies to the peak head deceleration on both models. In particular the initially high head deceleration on model B2 stayed high, whereas the low head deceleration on model B1 was somewhat reduced by the pretensioner.

C) FORWARD FACING TODDLER SEATS WITH IMPACT TABLE

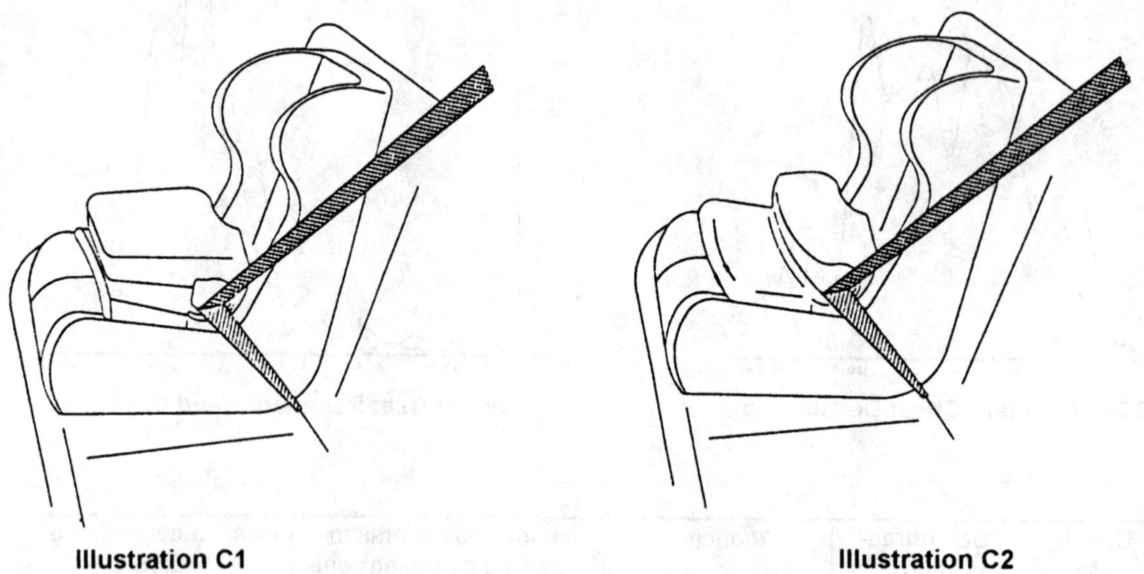

Illustration C1 **Illustration C2**

The two toddler type CRS C1 and C2 have identical seat shells but different impact tables. On C2 the head of the dummy does not make contact with the L-shaped impact table upon max. head excursion, whereas this occurs on model C1. On both systems the lap/shoulder straps are guided through the impact table, and both have the same effective height of the upper torso retention area.

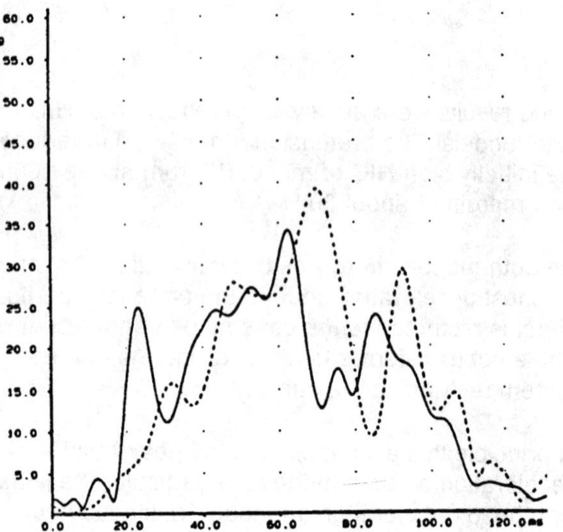

Diagram C1a: Resultant Chest Deceleration

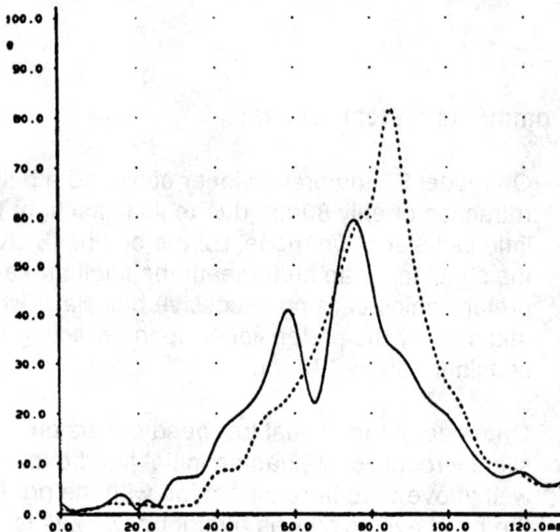

Diagram C1b: Resultant Head Deceleration

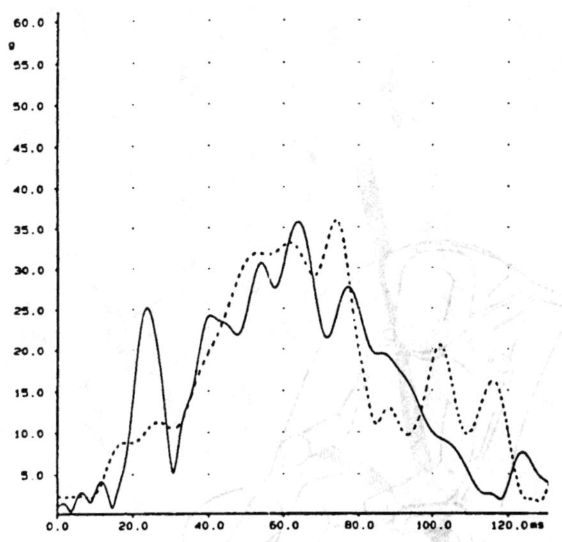

Diagram C2a: Resultant Chest Deceleration

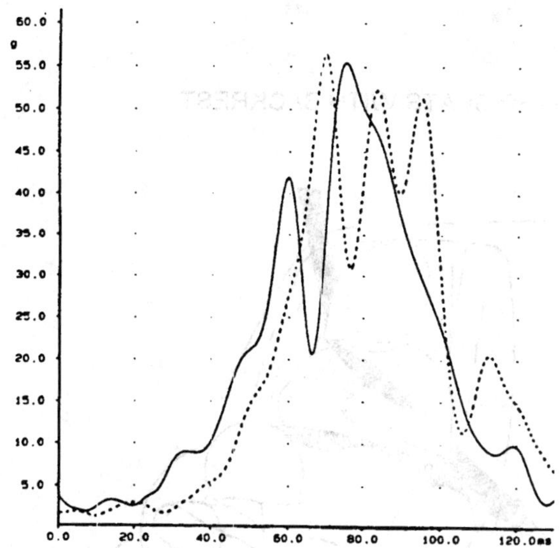

Diagram C2b: Resultant Head Deceleration

Table C

CRS model	Dummy used	Belt retraction (mm)	Reduction of		Initial deceleration caused by pretensioner		Peak deceleration of CRS	
			Head excursion (mm)	HIC	Chest (g)	Head (g)	Chest (g)	Head (g)
C1	P3	105	(490) -85	(648) -299	(16) +8	< 3 -	(39) -5	81 -22
C2	P3	105	(455) -55	(465) -157	(12) +13	< 3 -	(36) -	(56) -

Comments on test results

- Not surprisingly, the belt retraction of 105 mm was identical on both models.
  The impact tables were slightly lifted by the shoulder belt. The pretensioner pulled both the impact table closer towards the dummy as well as the CRS shell into the seat back due to the relatively plain shoulder belt angle between impact table and pillar loop.

- Without pretensioner the head excursion on the baseline tests showed a small difference - perhaps due to different prior belt tension -, with pretensioner both models resulted in similar max. head excursion.

- Coming from a higher HIC on model C1 the pretensioner reduced the HIC to a value equivalent to model C2.

- The initial deceleration effect was similar on both models with chest loads slightly rising due to pretensioning, whereas the head loads remained the same.

- The peak chest decelerations were not or little affected by the pretensioner. This also applies to the head load on C2. On C1, however, the head loads were reduced to a level close to the C2-results, comparable to the outcomes in terms of HIC described above.

D) BOOSTER SEATS WITH BACKREST

Illustration D1

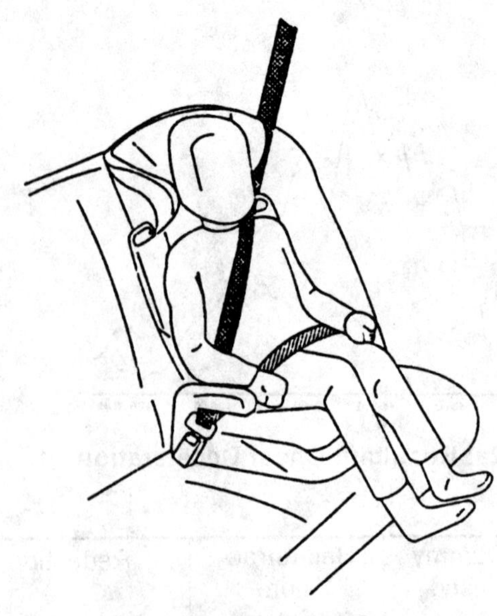

Illustration D2

Both CRS models D1 and D2 are booster type CRS and use the adult 3-point lap/shoulder belt as direct restraint for the child. Model D1 has a polystyrene shell with an integrated backrest, whereas D2 is used with an add-on backrest. On both models the shoulder belt is guided through the backrest. The small P ¾ dummy was used on model D1 in order to examine the effect on a potentially unfavourable belt diversion.

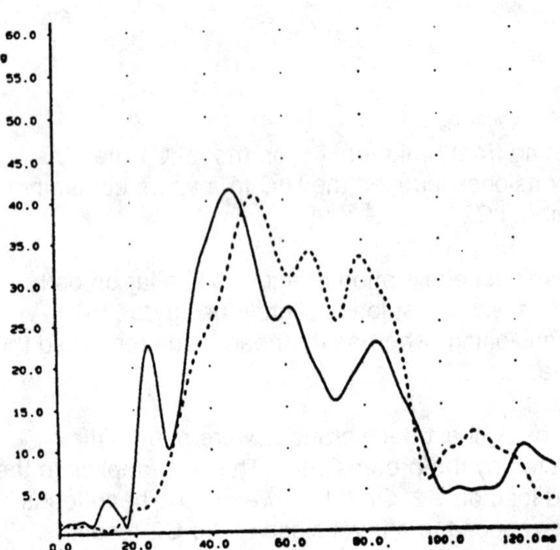

Diagram D1a: Resultant Chest Deceleration

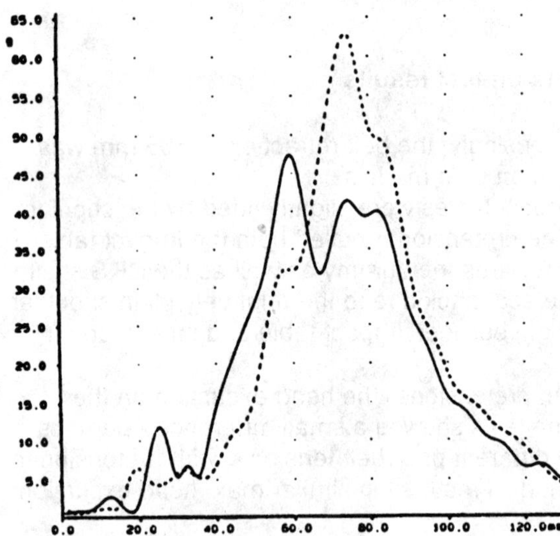

Diagram D1b: Resultant Head Deceleration

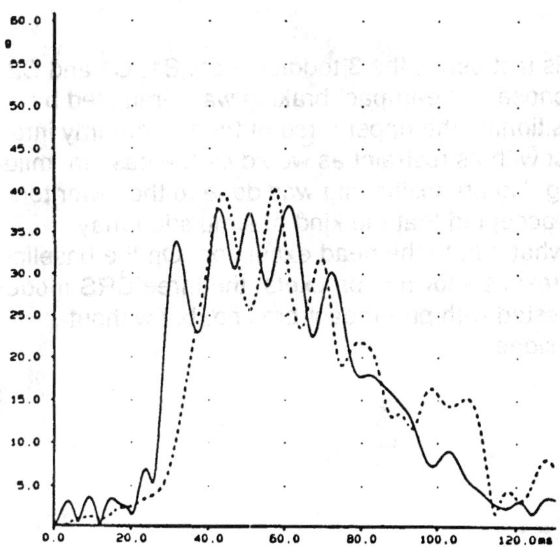

Diagram D2a: Resultant Chest Deceleration

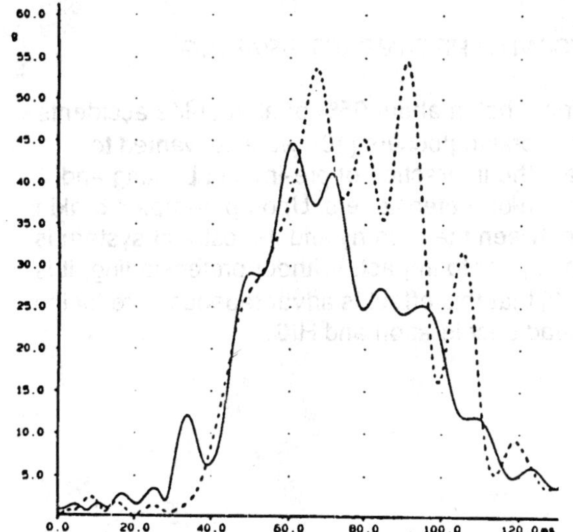

Diagram D2b: Resultant Head Deceleration

Table D

CRS model	Dummy used	Belt retraction (mm)	Reduction of		Initial deceleration caused by pretensioner		Peak deceleration of CRS	
			Head excursion (mm)	HIC	Chest (g)	Head (g)	Chest (g)	Head (g)
D1	P ¾	90	(415) -95	(559) -222	(14) +8	(5) +7	(40) +1	(65) -18
D2	P 3	55	(385) +15	(460) -171	(12) +16	< 5 + 3	(40) - 5	(54) - 4

Comments on test results

- Due to the initial short belt take-up on both models, the belt retraction was comparatively low. On model D1 the shoulder belt diversion caused partial breakage of the upper part of the polystyrene backrest until the belt reached a straight line adding to the belt take-up.

- The head excursion on D1 used with the small dummy was substantially reduced.

- On model D2 belt pretensioning promoted twisting of the dummy torso out of the shoulder belt resulting in a slightly higher head excursion. This effect is not typical and is due to the pre-impact position of the dummy and - as a consequence - of the belt on the dummy's shoulder. In a further test the head excursion was in fact reduced by 75 mm.

- On both models the HIC data was substantially reduced.

- There was a slight increase of decelerations in chest and head due to the initial effect of the pretensioner.

- The head peak deceleration on the small dummy in D1 was substantially reduced, on the bigger dummy in D2 slightly.

6. TESTS WITH PRE-IMPACT BRAKING

Considering that in about 65% of all real-life accidents pre-impact braking occurs [10], we also wanted to investigate the interaction of pre-impact braking and the performance of pretensioners. Upon pre-impact braking, contact between the dummy and its restraint system is established prior to impact. Without pretensioning, it is known [10] that this effect is advantageous in reducing mainly head deceleration and HIC.

For this test series the 3 toddler seats B1, C1 and C2 were chosen. Pre-impact braking was simulated by prepositioning the upper torso of the P 3 dummy into contact with its restraint as would be the case in mild braking. No prepositioning was done to the lower torso. It was accepted that the kind of simulation may somewhat add to the head excursion. On the baseline tests (results shown in brackets) the three CRS models were tested with pre-impact braking, but without pretensioner.

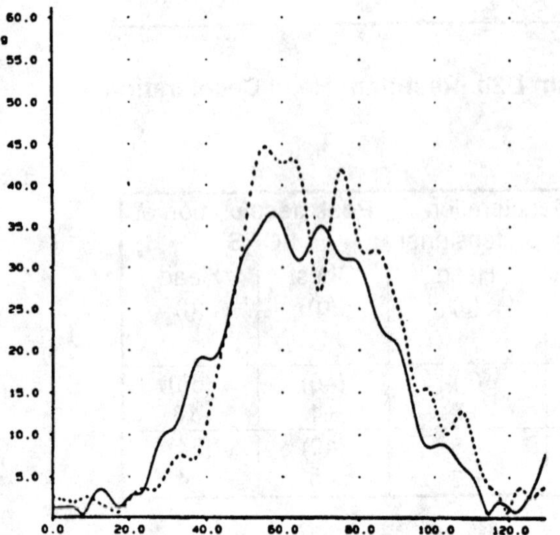

Diagram B1a': Resultant Chest Deceleration

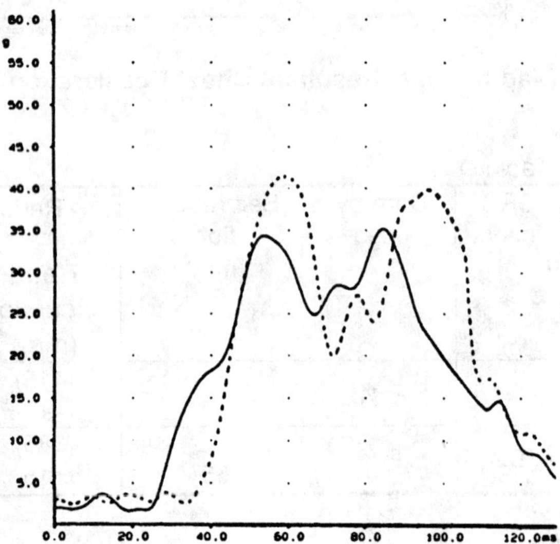

Diagram B1b': Resultant Head Deceleration

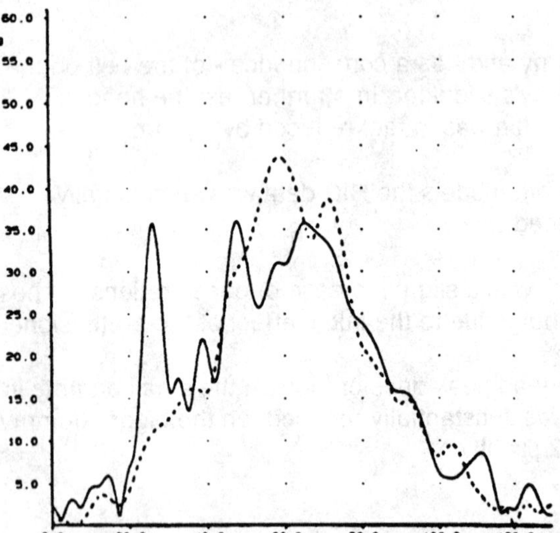

Diagram C1a': Resultant Chest Deceleration

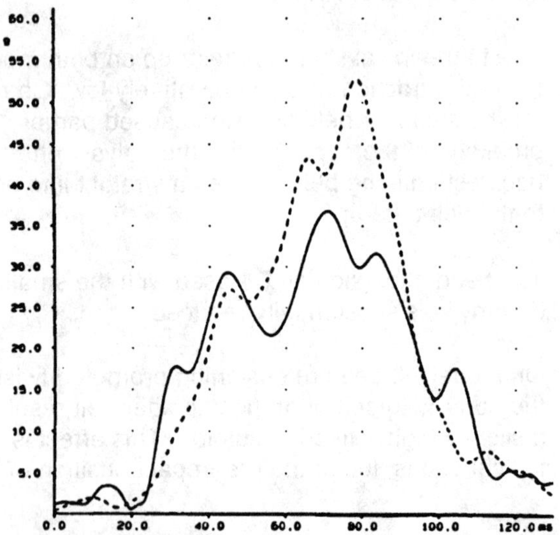

Diagram C1b': Resultant Head Deceleration

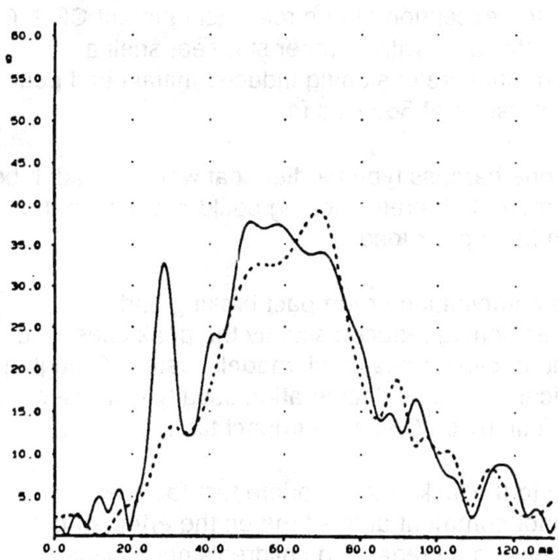

Diagram C2a': Resultant Chest Deceleration

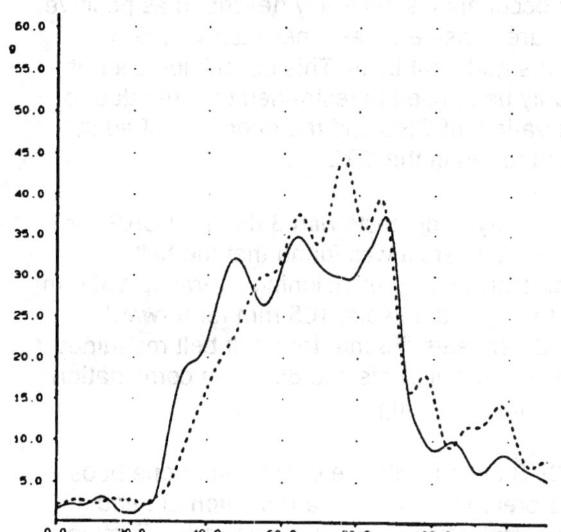

Diagram C2b': Resultant Head Deceleration

Table E

CRS model	Dummy used	Belt retraction (mm)	Reduction of		Initial deceleration caused by pretensioner		Peak deceleration of CRS	
			Head excursion (mm)	HIC	Chest (g)	Head (g)	Chest (g)	Head (g)
B1	P 3	50	(550) -25	(197) -13	(< 5) +6	(< 5) +6	(44) -7	(42) -7
C1	P 3	70	(520) -55	(390) -211	(12) +23	(< 3) + 8	(43) - 7	(52) -15
C2	P 3	75	(520) -50	(278) -71	(10) +22	(< 3) -	(39) - 1	(44) - 6

Comments on test results

- Due to the early tight load onto the shoulder belt the belt retraction was less on all three CRS models in conditions that simulated pre-impact braking.
- In spite of the low belt retraction some reduction on head excursion was observed.
- The low HIC data due to pre-impact braking were further reduced by the pretensioner on all three CRS models.
- The initial pretensioner effect on CRS model B1 with its indirect use of the lap/shoulder belt was again minor. The two impact table seats directly applying the car seat belt showed an initial increase of more than 20g in chest deceleration, whereas the head deceleration stayed almost unchanged.
- On all 3 models the low peak in chest and head deceleration achieved in the baseline tests was slightly reduced by pretensioning.

V. SUMMARY AND CONCLUSIONS

- Within the literature, the effect of belt pretensioners on adult occupants is generally described as positive, and they are considered as a necessary design feature of smart seat belts. This conclusion cannot necessarily be applied to restrained children due to the vast variety of CRS and the geometry of adult belts used to retain the CRS.

- Based upon dynamic tests with 8 different CRS for infants and toddlers it was found that the belt retraction caused by pretensioning averaged 135mm for rear facing infant seats, 105 mm for forward facing toddler seats, 73 mm for adult belt restrained children on booster seats and 65mm in combination with pre-impact braking.

- For all CRS tested, with the exception of one booster seat, the pretensioner led to a reduction of head excursion. Belt retraction was particularly effective on a toddler seat with poor adult belt geometry.

- Pretensioning resulted in substantial reductions of HIC: averaging 54% for dummies in infant seats, 27% in harness type toddler seats, 40% in CRS with impact table and 39% on boosters. This general positive result occured in spite of a comparatively little reduction of head peak decelerations on most CRS tested. The pretensioner-induced early capturing of the child is beneficial for children as much as for adults.

- The initial deceleration load of the pretensioner occurring at about 25 milliseconds after impact stayed below 20g for the chest and 10g for the head with the exception of one rear facing infant CRS. On the latter CRS with a rather stiff seat shell a substantial pretensioning induced initial head peak deceleration of 56g was found.

- On one harness type toddler seat with poor adult belt geometry belt pretensioning could not reduce the high head peak loads.

- The combination pre-impact braking and pretensioning reduced slightly the peak chest and head decelerations on all models tested. Though not critical, the initial deceleration usually increased particularly on CRS with impact table.

- Due to the lack of appropriate test facilities we cannot comment at this time on the effect of belt pretensioning regarding children's neck loads.

- In spite of the generally positive results on the CRS tested, the authors urge manufacturers to consider the potential effect of belt pretensioners upon designing CRS particularly regarding adult belt routing.

V. REFERENCES

[1] G. Rüter et al. "Untersuchungen von Einzelelementen zur Erhöhung der Wirksamkeit von Sicherheitsgurten"
Batelle-Institut, Frankfurt, 1977

[2] G. Rüler "Schutzwirkung von Sicherheitsgurten" Band 2, Literaturanalyse
Batelle-Institut, Frankfurt, 1978

[3] J. E. Mitzgus, H. Eyrainer "Three-Point Belt Improvements for Increased Occupant Protection", "Advances in Belt Restraint Systems: design, performance and usage", SAE Conference Detroit 1984, SAE Paper 840395

[4] F. Zuppichini "Effectiveness of a Mechanical Pretensioner on the Performance of Seat Belts", International IRCOBI Conference Proceedings, Stockholm, 1989

[5] Yih-Charng Deng "How Air Bags and Seat Belts Work Together in Frontal Crashes", 39th STAPP Car Crash Conference Proceedings, San Diego 1995, SAE Paper 952702

[6] M. H. Muser et al "Optimised Restraint Systems for Low Mass Vehicles", 40th STAPP Car Crash Conference Proceedings, Albuquerque 1996, SAE Paper 962435

[7] H. G. Johannessen, M. Mackay "Why 'Intelligent' Automotive Occupant Restraint Systems?", 39th Annual Proceedings of AAAM Conference, Chicago 1995

[8] H. J. Miller "Injury Reduction with Smart Restraint Systems", 39th Annual Proceedings of AAAM Conference, Chicago 1995

[9] A. R. Bernat "Smart Safety Belts for Injury Reduction", 39th Annual Proceedings of AAAM Conference, Chicago 1995

[10] W. Czernakowski, D. Otte "The Effect of Pre-Impact Braking on the Performance of Child Restraint Systems in Real-Life Accidents and under Varying Test Conditions", Child Occupant Protection Symposium, San Antonio, Texas November 1993, SAE Paper 933097

About the Editor

Dr. Viano is a specialist in injury biomechanics and impact protection in automotive crashes, sport impacts, and defense/law-enforcement actions. He is Adjunct Professor of Traffic Medicine at Chalmers University of Technology and Biomedical Engineering at Wayne State University, where he serves as the Director of the Sport Biomechanics Laboratory. He has advised and graduated five doctoral students. He is a Fellow of SAE, ASME, AIMBE and AAAM, and serves on the Board of Directors of several scientific, medical and charitable organizations. Dr. Viano is Editor-in-Chief of *Traffic Injury Prevention.* He has published over 200 papers on injury biomechanics and impact protection; and he recently published the book, *Role of the Seat in Rear Crash Safety* with SAE. He has more than a dozen patents, most notably the first commercial active head restraint system to prevent whiplash and the high retention seat for rear impact safety. He is recently retired as Principal Scientist from General Motors Corporation, and now serves as a consultant to the National Football League in their efforts to understand and prevent concussion. He works through his company ProBiomechanics LLC. Dr. Viano served on the National Academy of Science committee that wrote *Injury in America*, and received the Award of Engineering Excellence from NHTSA/DOT as well as numerous SAE awards. Dr. Viano received his PhD in Applied Mechanics from the California Institute of Technology, Pasadena, California and Dr. med. from the Karolinska Institute and Medical University, Stockholm, Sweden.